Textbook of
MEDICAL PHYSIOLOGY

ARTHUR C. GUYTON, M.D.

Professor and Chairman of the
Department of Physiology and Biophysics,
University of Mississippi School of Medicine

FOURTH EDITION – ILLUSTRATED

W. B. SAUNDERS COMPANY • PHILADELPHIA • LONDON • TORONTO

W. B. Saunders Company: West Washington Square
Philadelphia, Pa. 19105

12 Dyott Street
London WC1A 1DB

1835 Yonge Street
Toronto 7, Ontario

Listed here is the latest translated edition of this book together with the language of the translation and the publisher.

Spanish (*4th Edition*) — Nueva Editorial Interamericana, S.A., de C.V., Mexico

Textbook of Medical Physiology   ISBN 0-7216-4392-2

Print Number: 9   8   7   6   5   4   3

Dedicated to

MY FATHER

for the uncompromising principles that have guided his life

MY MOTHER

for leading her children into intellectual pursuits

MY WIFE

for her magnificent devotion to her family

MY CHILDREN

for making everything worthwhile

PREFACE

When I wrote the first edition of this textbook, I had the naïve belief that once the book was completed, subsequent revisions would require only simple changes. However, with each new edition I find that continuing strides in the field of physiology belie this belief, so much so that large proportions of the text must be completely recast and rewritten, and even the emphasis of the subject matter must often be changed as our knowledge becomes more penetrating. Therefore, the reader will find hardly a chapter in this edition that has not been significantly altered. Also, many of the figures have been either changed in at least some feature or completely replaced, and over 50 new figures have been added.

If I should characterize the major change in both this and the last edition, it would be a striking trend toward more fundamental physiology. The reason for this is mainly the greater success of the research physiologist in probing more basic mechanisms of function than was true a few years ago. Yet I have still attempted to present physiology as an integrated study of the body's functional systems, attempting to utilize the new fundamental knowledge to build a better understanding of the mechanisms upon which life depends.

I have also made a serious attempt to devise and achieve techniques of expression that will help bring the physical and chemical principles of the body's complexities into the medical student's realm of understanding. In pursuit of this goal, I have kept records of those types of material with which the student has difficulty; I have quizzed students in detail to determine their levels of comprehension; and I have attempted to note inconsistencies in logic that might appear in student discussions. All these data have been used to help choose the material and methods of presentation.

Thus, it has been my desire to make this book a "teaching" text as well as one that covers essentially all the basic physiology required of a student of medicine.

A special emphasis of the book is a more detailed attention than is given in other textbooks of physiology to the body's many regulatory systems. It is these that provide what physiologists call *homeostasis*. The reason for this special emphasis is that most disease conditions of the body are made manifest by abnormal function of one or more of the regulatory systems. Therefore, the student's comprehension of "medical" physiology depends perhaps more on a knowledge of these systems than on any other facet of physiology.

Another goal in the preparation of this text has been to make the text as accurate as possible. To help attain this, suggestions and critiques from many physiologists, students, and clinicians throughout the United States and other parts of the world have been received and utilized in checking the factual accuracy of the text. Yet, even so, because of the likelihood of error in sorting through many thousands of bits of information, I wish to issue an invitation — in fact, much more than merely an invitation, actually a request — to all readers to send along criticisms of error or inaccuracy. Indeed, physiologists understand how important feedback is to proper function of the human body; so, too, is feedback equally important for progressive development of a textbook of medical physiology. I hope also that those many persons who have already helped will accept my sincerest appreciation for their efforts.

A word of explanation is needed about two features of the text — first, the references, and, second, the two print sizes. The references have been chosen primarily for their up-todateness and for the quality of their own bibliographies. Use of these references as well as of cross-references from them can give the student almost complete coverage of the entire field of physiology.

The print is set in two sizes. The material presented in small print is of several different kinds: first, anatomical, chemical, and other information that is needed for the immediate discussion but that most students will learn in more detail in other courses; second, information that is of special importance to certain clinical fields of medicine but that is not necessary to fundamental understanding of the body's basic physiologic mechanisms; and, third, information that will be of value to those students who wish to pursue a subject more deeply than does the average medical student. On the other hand, the material in large print, which represents about two-thirds of the text,

constitutes the major bulk of physiological information that the student will require in his medical studies and that he will not obtain in other courses. For those teachers who would like to present a more limited course of physiology, the student's study can be restricted primarily to the large type.

Again, I wish to express my deepest appreciation to many others who have helped in the preparation of this book. I am particularly grateful to Mrs. Billie Howard and Mrs. Judy Bass for their secretarial services, to Mrs. Carolyn Hull for the new illustrations, and to the staff of the W. B. Saunders Company for its continued editorial and production excellence.

ARTHUR C. GUYTON
Jackson, Mississippi

CONTENTS

PART II

BLOOD CELLS, IMMUNITY, AND BLOOD CLOTTING

Chapter 8

Chapter 9

Chapter 10

Chapter 11

PART III

THE HEART

PART IV

THE CIRCULATION

Chapter 18

Chapter 26

Chapter 27

Chapter 28

PART V

THE BODY FLUIDS AND KIDNEYS

PART VI

RESPIRATION

PART VII

AVIATION, SPACE, AND DEEP SEA DIVING PHYSIOLOGY

PART VIII

THE NERVOUS SYSTEM AND SPECIAL SENSES

PART IX

THE GASTROINTESTINAL TRACT

PART X

METABOLISM AND TEMPERATURE REGULATION

PART XI

ENDOCRINOLOGY AND REPRODUCTION

PART I

INTRODUCTION TO PHYSIOLOGY: THE CELL AND GENERAL PHYSIOLOGY

CHAPTER 1

Functional Organization of the Human Body and Control of the "Internal Environment"

Physiology is the study of function in living matter; it attempts to explain the physical and chemical factors that are responsible for the origin, development, and progression of life. Each type of life, from the monomolecular virus up to the largest tree or to the complicated human being, has its own functional characteristics. Therefore, the vast field of physiology can be divided into *viral physiology, bacterial physiology, cellular physiology, plant physiology, human physiology*, and many more subdivisions.

Human Physiology. In human physiology we attempt to explain the chemical reactions that occur in the cells, the transmission of nerve impulses from one part of the body to another, contraction of the muscles, reproduction, and even the minute details of transformation of light energy into chemical energy to excite the eyes, thus allowing us to see the world. The very fact that we are alive is almost beyond our own control, for hunger makes us seek food and fear makes us seek refuge. Sensations of cold make us provide warmth, and other forces cause us to seek fellowship and reproduce. Thus, the human being is actually an automaton, and the fact that we are sensing, feeling, and knowledgeable beings is part of this automatic sequence of life; these special attributes allow us to exist under widely varying conditions, which otherwise would make life impossible.

CELLS AS THE LIVING UNITS OF THE BODY

The basic living unit of the body is the cell, and each organ is actually an aggregate of many different cells held together by intercellular supporting structures. Each type of cell is specially adapted to perform one particular function. For instance, the red blood cells, twenty-five trillion in all, transport oxygen from the lungs to the tissues. Though this type of cell is perhaps the most abundant of any in the whole body, there are approximately another 75 trillion cells. The entire body, then, contains about 100 trillion cells.

However much the many cells of the body differ from each other, all of them have certain basic characteristics that are alike. For instance, each cell requires nutrition for maintenance of life, and all cells utilize almost identically the same types of nutrients. All cells use oxygen as one of the major substances from which energy is derived; the oxygen combines with carbohydrate, fat, or protein to release the energy required for cell function. The general mechanisms for changing nutrients into energy are basically the same in all cells, and all cells also deliver end-products of their chemical reactions into the surrounding fluids.

Almost all cells also have the ability to reproduce, and whenever cells of a particular type are destroyed for one cause or another, the remaining cells of this type usually divide again and again until the appropriate number is replenished.

THE EXTRACELLULAR FLUID—THE INTERNAL ENVIRONMENT

About 56 per cent of the adult human body is fluid. Some of this fluid is inside the cells and is called, collectively, the *intracellular fluid*. The fluid in the spaces outside the cells is called the *extracellular fluid*. Among the dissolved constituents of the extracellular fluids are the ions and the nutrients needed by the cells for maintenance of life. The extracellular fluid is in constant motion throughout the body and is constantly mixed by the blood circulation and by diffusion between the blood and the tissue spaces. Therefore, all cells live in essentially the same environment, for which reason the extracellular fluid is often called the *internal environment* of the body, or the *milieu intérieur*, the French equivalent first used a hundred years ago by the great 19th century French physiologist, Claude Bernard.

Cells are automatons that are capable of living, growing, and providing their special functions so long as the proper concentrations of oxygen, glucose, the different electrolytes, amino acids, and fatty substances are available in the internal environment.

Differences Between Extracellular and Intracellular Fluids. The extracellular fluid contains large amounts of sodium, chloride, and bicarbonate ions, plus nutrients for the cells, such as oxygen, glucose, fatty acids, and amino acids. It also contains carbon dioxide, which is being transported from the cells to the lungs, and other cellular excretory products, which are being transported to the kidneys.

The intracellular fluid is much the same from one cell to another, but it differs significantly from the extracellular fluid; particularly, the intracellular fluid contains large amounts of potassium, magnesium, and phosphate ions instead of the sodium and chloride ions found in the extracellular fluid. Special mechanisms for transporting ions through the cell membranes maintain these differences. These mechanisms will be discussed in detail in Chapter 4.

"HOMEOSTATIC" MECHANISMS OF THE MAJOR FUNCTIONAL SYSTEMS

HOMEOSTASIS

The term *homeostasis* is used by physiologists to mean *maintenance of static, or constant, conditions in the internal environment.* Essentially all the organs and tissues of the body perform functions that help to maintain these constant conditions. For instance, the lungs provide new oxygen as it is required by the cells, the kidneys maintain constant electrolyte concentrations, and the gut provides nutrients. A large segment of this text is concerned with the manner in which each organ or tissue contributes to homeostasis. To begin this discussion, the different functional systems of the body and their homeostatic mechanisms will be outlined briefly; then the basic theory of the control systems that cause the functional systems to operate in harmony with each other will be discussed.

THE FLUID TRANSPORT SYSTEM

Extracellular fluid is transported to all parts of the body in two different stages. The first stage entails movement of blood around and around the circulatory system, and the second, movement of fluid between the blood capillaries and the cells. Figure 1–1 illustrates the overall circulation of blood, showing that the heart is actually two separate pumps, one of which propels blood through the lungs and the other through the systemic circulation. All the blood in the circulation traverses the entire circuit of the circulation an average of once each minute when a person is at rest and as many as five times each minute when he becomes extremely active.

Figure 1–1 General organization of the circulatory system.

Figure 1–2 Diffusion of fluids through the capillary walls and through the interstitial spaces.

As blood passes through the capillaries, continual exchange occurs between the plasma portion of the blood and the interstitial fluid in the spaces surrounding the capillaries. This process is illustrated in Figure 1–2. Note that the capillaries are porous so that large amounts of fluid can *diffuse* back and forth between the blood and the tissue spaces, as illustrated by the arrows. This process of diffusion is caused by kinetic motion of the molecules in both the plasma and the extracellular fluid. That is, all fluid and dissolved molecules are continually moving and bouncing in all directions, through the pores, through the tissue spaces, and so forth. Almost no cell is located more than 25 to 50 microns from a capillary, which allows movement of any substance from the capillary to the cell within a few seconds. Thus, the extracellular fluid throughout the body is continually mixed and thereby maintains almost complete homogeneity.

ORIGIN OF NUTRIENTS IN THE EXTRACELLULAR FLUID

The Respiratory System. Figure 1–1 shows that each time the blood passes through the body it also flows through the lungs. The blood picks up oxygen in the alveoli, thus acquiring the oxygen needed by the cells. The membrane between the alveoli and the lumen of the pulmonary capillaries is only 1 to 4 microns in thickness, and oxygen diffuses through this membrane into the blood in exactly the same manner that water, nutrients, and excreta diffuse through the tissue capillaries.

The Gastrointestinal Tract. Figure 1–1 also

shows that a large portion of the blood pumped by the heart passes through the walls of the gastrointestinal organs. Here, different dissolved nutrients, including carbohydrates, fatty acids, amino acids, and others, are absorbed into the extracellular fluid.

The Liver and Other Organs that Perform Primarily Metabolic Functions. Not all substances absorbed from the gastrointestinal tract can be used in their absorbed form by the cells. The liver alters the chemical compositions of many of these to more usable forms, and other tissues of the body—such as the fat cells, the gastrointestinal mucosa, the kidneys, and the endocrine glands—help to modify the absorbed substances or store them until they are needed at a later time.

Musculoskeletal System. Sometimes the question is asked: How does the musculoskeletal system fit into the homeostatic functions of the body? Were it not for this system, the body could not move to the appropriate place at the appropriate time to obtain the foods required for nutrition. The musculoskeletal system also provides motility for protection against adverse surroundings, without which the entire body, and along with it all the homeostatic mechanisms, could be destroyed instantaneously.

REMOVAL OF METABOLIC END-PRODUCTS

Removal of Carbon Dioxide by the Lungs. At the same time that blood picks up oxygen in the lungs, carbon dioxide is released from the blood into the alveoli, and the respiratory movement of air into and out of the alveoli carries the carbon dioxide to the atmosphere. Carbon dioxide is the most abundant of all the end-products of metabolism.

The Kidneys. Passage of the blood through the kidneys removes most substances from the plasma that are not needed by the cells. These substances include especially different end-products of metabolism and excesses of electrolytes or water that might have accumulated in the extracellular fluids. The kidneys perform their function by, first, filtering large quantities of plasma through the glomeruli into the tubules and then reabsorbing into the blood those substances needed by the body, such as glucose, amino acids, large amounts of water, and many of the electrolytes. However, substances not needed by the body generally are not reabsorbed but, instead, pass on through the renal tubules into the urine.

REGULATION OF BODY FUNCTIONS

The Nervous System. The nervous system is composed of three major parts: the *sensory portion*, the *central nervous system*, or *integrative portion*, and the *motor portion*. Sensory nerves detect the state of the body or the state of the surroundings. For instance, nerves present everywhere in the skin apprise one every time an object touches him at any point. The eyes are sensory organs that give one a visual image of the surrounding area. The ears also are sensory organs. The central nervous system is comprised of the brain and spinal cord. The brain can store information, generate thoughts, create ambition, and determine reactions that the body should perform in response to the sensations. Appropriate signals are then transmitted through the motor portion of the nervous system to carry out the person's desires.

A large segment of the nervous system is called the *autonomic system*. It operates at a subconscious level and controls many functions of the internal organs, including the action of the heart, the movements of the gastrointestinal tract, and the secretion by different glands.

The Hormonal System of Regulation. Located in the body are eight major endocrine glands that secrete chemical substances called *hormones*. Hormones are transported in the extracellular fluids to all parts of the body to help regulate function. For instance, thyroid hormone increases the rates of almost all chemical reactions in all cells. In this way thyroid hormone helps to set the tempo of bodily activity. Likewise, insulin controls glucose metabolism; adrenocortical hormones, electrolyte and protein metabolism; and parathormone, bone metabolism. Thus, the hormones represent a system of regulation that complements that of the nervous system. The nervous system, in general, regulates rapid muscular and secretory activities of the body, whereas the hormonal system regulates mainly the slowly reacting metabolic functions.

REPRODUCTION

Reproduction is another function of the body that sometimes is not considered to be a homeostatic function. But it does help to maintain a static situation by generating new beings to take the place of ones that are dying. This perhaps sounds like a farfetched usage of the term homeostasis, but it does illustrate that, in the final analysis, essentially all structures of the body are so organized that they help to maintain continuity of life.

THE CONTROL SYSTEMS OF THE BODY

The human body has literally thousands of control systems in it. Some of these operate within the cell to control intracellular function, a subject that will be discussed in detail in Chapter 3. Others operate within the organs to control functions of the individual parts of the organs, while others operate throughout the entire body to control the interrelationships between the different organs. For instance, the respiratory system regulates the concentration of carbon dioxide in the extracellular fluids. The liver and the pancreas regulate the concentration of glucose in the extracellular fluids. And the kidneys regulate the concentrations of hydrogen, sodium, potassium, phosphate, and other ions in the extracellular fluids.

EXAMPLES OF CONTROL MECHANISMS

Regulation of Oxygen and Carbon Dioxide Concentrations in the Extracellular Fluid. Since oxygen is one of the major substances required for chemical reactions in the cells, the rates of the chemical reactions are, to a great extent, dependent on the concentration of oxygen in the extracellular fluid. For this reason a special control mechanism is available for maintenance of an almost exact and constant oxygen concentration in the extracellular fluids. This mechanism depends principally on the chemical characteristics of *hemoglobin*, which is present in all the red blood cells. Hemoglobin combines with oxygen as blood passes through the lungs. In the tissue capillaries it will not release oxygen into the interstitial fluid if too much oxygen is already there but, if the oxygen concentration is too little, sufficient oxygen will be released to reestablish an adequate interstitial fluid concentration. Thus, the regulation of oxygen concentration in the tissues is vested principally in the chemical characteristics of hemoglobin itself. This regulation is called the *oxygen-buffering function of hemoglobin.*

Carbon dioxide is one of the major end-products of the oxidative reactions in cells. If all the carbon dioxide formed in the cells should continue to accumulate in the tissue fluids, the *mass action* of the carbon dioxide itself would soon halt all the energy-giving reactions of the cells. Fortunately, the following mechanism maintains a constant and reasonable concentration of carbon dioxide in the extracellular fluids: A very high carbon dioxide concentration *excites the respiratory center*, causing the person to breathe rapidly and deeply. This increases the rate of removal of carbon dioxide from the blood and extracellular fluid, and the process continues until the concentration returns to normal.

Regulation of Sodium Ion Concentration. Figure 1–3 illustrates the control of extracellular sodium ion concentration under two different conditions. First, the sodium ion concentration is at the normal level of 142 mEq./liter. Then the person ingests a large amount of sodium in some of his food, which increases his sodium concentration to a much higher level. Within the next few hours, the rate of aldosterone secretion by the adrenal cortex decreases to one-half to one-third normal. Because aldosterone is required by the kidneys to promote normal reabsorption of sodium ions in the tubules, this decrease in aldosterone causes the kidneys to excrete very large quantities of sodium in the urine. Therefore, within several more hours the excess extracellular sodium ion concentration falls back almost to normal. Simultaneously, the rate of aldosterone secretion rises toward normal. Thus, the control system automatically readjusts the extracellular sodium ion concentration back toward its normal mean level.

The next day the same person exercises very strenuously, during the course of which he sweats an extreme amount. As a result, the sodium ion concentration falls considerably below normal. Within the next half to one hour the rate of aldosterone secretion increases to a high level, causing the kidneys to retain sodium tenaciously, thereby preventing loss of significant amounts of sodium in the urine. Simultaneously, the person's appetite for salt increases greatly because the appetite center of his brain is stimulated by low concentration of extracellular sodium. Then, because of increased intake of salt and decreased excretion of sodium, the sodium concentration rises back toward normal, and simultaneously the rate of aldosterone secretion also falls back toward the normal level.

Note especially that the sodium concentration in both of these conditions did not change more than 3 per cent from the normal value. This illustrates how accurately some of the control systems can maintain constant conditions in the extracellular fluids.

Regulation of Arterial Pressure. Several different systems contribute to the regulation of arterial pressure. One of these, the *baroreceptor system*, is very simple and an excellent example of a control mechanism. In the walls of most of the great arteries of the upper body, especially the bifurcation region of the carotids and the arch of the aorta, are many nerve receptors, called *baroreceptors*, which are stimulated by stretch of the arterial wall. When the arterial pressure becomes great, these baroreceptors are stimulated excessively, and impulses are transmitted to the medulla of the brain. Here the impulses inhibit the *vasomotor center*, which in turn decreases the number of impulses transmitted through the sympathetic nervous system to the heart and blood vessels. Lack of these impulses causes diminished pumping activity by the heart and increased ease of blood flow through the peripheral vessels, both of which lower the arterial pressure back toward normal. Conversely, a fall in arterial pressure relaxes the stretch receptors, allowing the vasomotor center to become more active than usual and thereby causing the arterial pressure to rise back toward normal.

CHARACTERISTICS OF CONTROL SYSTEMS

The above examples of homeostatic control mechanisms are only a few of the many hundreds in the body, all of which have certain characteristics in common. These are explained in the following pages.

Negative Feedback Nature of Control Systems. The control systems of the body act by a process of *negative feedback*, which can be explained best by reviewing some of the homeostatic control systems mentioned above. In the regulation of carbon dioxide concentration, a high level of carbon dioxide in the extracellular fluid causes increased pulmonary ventilation, and this in turn causes decreased carbon dioxide concentration. In other words, the response is *negative* to the initiating stimulus. Conversely, if the carbon dioxide concentration falls too low, this causes feedback through

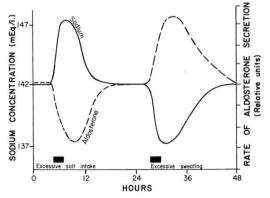

Figure 1–3 Control of sodium concentration in the body fluids by the rate of aldosterone secretion under two conditions of stress: excessive salt intake and excessive sweating.

the control system to raise the carbon dioxide concentration. This response also is negative to the initiating stimulus.

In the aldosterone-sodium control mechanism a high concentration of sodium ions causes decreased aldosterone secretion, which in turn diminishes sodium concentration. Thus, the response is *negative* to the elevated sodium level that initiates the response. Conversely, low sodium concentration promotes increased aldosterone secretion and increased appetite for salt, both of which cause the sodium concentration to rise back toward normal. Thus, a low sodium level causes a higher level to ensue, or, in other words, a response that is *negative* to the initiating stimulus.

Finally, in the arterial pressure-regulating mechanism, a high pressure causes a series of reactions that promote a lowered pressure, or a low pressure causes a series of reactions that promote an elevated pressure. In both instances these effects are negative with respect to the initiating stimulus.

Therefore, in general, if some factor becomes excessive or too little, a control system initiates *negative feedback*, which consists of a series of changes that returns the factor toward a certain mean value, thus maintaining homeostasis.

AMPLIFICATION, OR GAIN, OF A CONTROL SYSTEM. The degree of effectiveness with which a control system maintains constant conditions is called the *amplification*, or *gain*, of the system. For instance, let us assume that a person has been in a room for 24 hours, the temperature of which has been 60° F. Then suddenly the room temperature is increased to 110° F., a total increase of 50°. However, the person's body temperature rises only from 98.0° to 99.0° F., because the body automatically controls its own temperature within very narrow limits rather than following the change in atmospheric temperature. In this case, the change in atmospheric temperature is 49° more than the change in body temperature, which changes only 1°. Therefore, we say that the *gain* of the control system is 49. In other words, for each degree change in body temperature that does occur, there would be 49 times that much additional change were it not for the control system.

The gain of the baroreceptor system for control of arterial pressure, as measured in dogs, is approximately 2. That is, an extraneous factor that tends to increase or decrease the arterial pressure does so only one-third as much as would occur if this control system were not present. Therefore, one can see that the temperature control system is much more effective than the baroreceptor system.

Positive Feedback as a Cause of Death — Vicious Cycles. One might ask the question: Why do essentially all control systems of the body operate by negative feedback rather than positive feedback? However, if he will consider the nature of positive feedback, he will immediately see that positive feedback does not lead to stability but to instability and often to death.

Figure 1–4 illustrates an instance in which death can ensue from positive feedback. This figure depicts the pumping effectiveness of the heart, showing that the heart of the normal human being pumps about 5 liters of blood per minute. However, if the person is suddenly bled 2 liters, the amount of blood in his body is decreased to such a low level that not enough is available for his heart to pump effectively. As a result, the arterial pressure falls, and the flow of blood to the heart muscle through the coronary vessels diminishes. This results in weakening of the heart, further diminished pumping, further decrease in coronary blood flow, and still more weakness of the heart; the cycle repeats itself again and again until death. Note that each cycle in the feedback results in further weakness of the heart. In other words, the initiating stimulus causes more of the same, which is *positive feedback*.

Positive feedback is better known as a "vicious cycle," but actually a mild degree of positive feedback can be overcome by the negative feedback control mechanisms of the body, and a vicious cycle will fail to develop. For instance, if the person in the above example were bled only 1 liter instead of 2 liters, the normal negative feedback mechanisms for controlling cardiac output and arterial pressure would over-

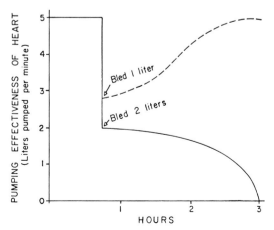

Figure 1–4   Death caused by positive feedback when 2 liters of blood is removed from the circulation.

balance the positive feedback, and the person would recover, as shown by the dashed curve of Figure 1–4.

AUTOMATICITY OF THE BODY

The purpose of this chapter has been to point out, first, the overall organization of the body and, second, the means by which the different parts of the body operate in harmony. To summarize, the body is actually an *aggregate of about 100 trillion cells* organized into different functional structures, some of which are called *organs*. Each functional structure provides its share in the maintenance of homeostatic conditions in the extracellular fluid, which is often called the *internal environment*. As long as normal conditions are maintained in the internal environment, the cells of the body will continue to live and function properly. Thus, each cell benefits from homeostasis, and in turn each cell contributes its share toward the maintenance of this state. This reciprocal interplay provides continuous automaticity of the body until one or more functional systems lose their ability to contribute their share of function. When this happens, all the cells of the body suffer. Extreme dysfunction of any integrated system leads to death, while moderate dysfunction leads to sickness.

APPENDIX

BASIC PHYSICAL PRINCIPLES OF CONTROL SYSTEMS

Until recently the principles of control systems as applied to the human body have been taught only qualitatively rather than quantitatively. For instance, in the case of baroreceptor regulation of arterial pressure, physiologists have simply stated that an increase in pressure causes a reflex decrease in pressure back toward normal. However, this means very little unless we know *how much* effect occurs. But within the past 15 years, application of the much more quantitative physical principles of control systems as worked out by engineers has begun to make physiology a much more exact science than it has been in the past.

Basic Symbols Used in Control System Analysis. Figure 1–5 illustrates the most common basic symbols used in control system analysis. These are the following:

THE ADDITION-SUBTRACTION SYMBOL. This symbol is shown in Figure 1–5A. For instance, let x represent the rate of intake of salt in the solid food eaten each day, y the intake of salt in liquids drunk each day, and z the rate of loss of salt in the urine each day. Then the net rate of change of salt

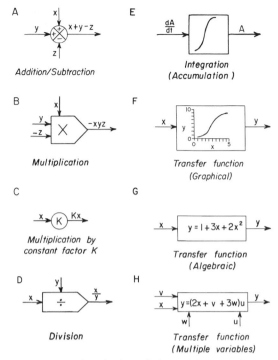

Figure 1–5   Standard symbols used in control system diagrams.

in the body will be $x + y - z$, as indicated by the arrow.

THE MULTIPLICATION SYMBOL. Figure 1–5B illustrates the multiplication of three quantities, x, y, and $-z$, to give $-xyz$. For instance, let us assume that three separate factors are affecting arterial pressure and that these factors multiply each other. Thus, the baroreceptor system might be causing x effect to elevate arterial pressure; a hormone secreted by one of the endocrine glands might be causing y effect to elevate arterial pressure; and hemorrhage might be causing $-z$ effect to decrease arterial pressure. The net effect would be $-xyz$.

MULTIPLICATION BY A CONSTANT FACTOR. Figure 1–5C illustrates multiplication by a constant factor. Let us assume that x is the *concentration* of sodium in the extracellular fluids and K is the volume of extracellular fluid; the total *quantity* of sodium in extracellular fluid would then be Kx.

THE DIVISION SYMBOL. Figure 1–5D shows the value x divided by the value y to give x/y. For example, if x is the total *quantity* of sodium in the body and y is the extracellular fluid *volume*, then the output of this block, x/y, is the concentration of sodium in the extracellular fluid.

THE INTEGRATION OR ACCUMULATION SYMBOL. Many functions of the body depend upon slow accumulation of some factor. Thus, in Figure 1–5E the *rate of change* of aldosterone in the body is represented by the differential term, dA/dt. The output of the block is the *quantity* of aldosterone that has accumulated in the body at any given instant; the symbol in this block is called the *integration symbol*, or, in other words, the rate dA/dt is integrated to give A. If dA/dt is positive, the quantity of aldosterone in the body will be increasing, whereas if dA/dt is negative, the quantity of aldosterone in the body will be decreasing.

TRANSFER FUNCTIONS. Figures 1–5F, G, and H illustrate transfer functions that show a quantity x entering a block and a quantity y leaving the block. Each block means that y is related to x in accordance with the function that is placed inside the box. In Figure 1–5F this function is represented graphically. In Figure 1–5G it is represented by an algebraic equation. Figure 1–5H illustrates four inputs and one output, showing that three of the inputs add to each other and the other multiplies the first three.

As an example, Figure 1–5F might represent the relationship between glucose concentration x in the extracellular fluids and the rate of insulin secretion, y, by the pancreas. This transfer function shows that at low glucose concentrations essentially no insulin is secreted, but at high concentrations very large quantities of insulin are secreted.

General Analysis of a Control System. Figure 1–6A illustrates a general analysis that can be applied to essentially any control system in the body. Note the line that is labeled *Controlled quantity*. This is the concentration, the rate, or the quantity that is to be controlled by the control system. The exact value of the controlled quantity is equal to the normal value of that quantity ± any *disturbance* that might try to change this value from the normal and ± the *compensation* of the control system.

Figure 1–6B illustrates an application of this general diagram. In this case the controlled quantity is the body temperature (BT). A disturbance has occurred because the air temperature of the surroundings has risen to 50° F. above normal. If there were no compensation in the body, i.e., no sweating, no dilation of the blood vessels, and no increased evaporation of sweat from the skin, the increase in air temperature would be expected to increase the body temperature the same amount, 50°. However, a slight increase in body temperature sets off a series of control reactions which give

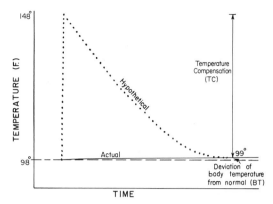

Figure 1–7 Hypothetical and actual effects on body temperature caused by suddenly increasing the air temperature 50° F.

a temperature compensation (TC) of 49°. The transfer function of the feedback control system is that shown in the block of Figure 1–6B. And the gain of this feedback is equal to 49. This means simply that the amount of temperature compensation is 49 times as great as the final deviation of body temperature from the normal.

Figure 1–7 illustrates graphically the function of the body temperature control system illustrated in Figure 1–6B. Let us assume that the normal temperature is 98° F. and that the person is suddenly exposed to an air temperature of 50° above normal. If no control system should exist, the hypothetical increase in body temperature would be up to 148°. Then, if the feedback control system were allowed to act, it would bring the body temperature back down to 99°. In actuality, the hypothetical increase in temperature to 148° never occurs because the feedback takes place *at the same time that the body temperature is rising*. Therefore, the body temperature simply rises and stabilizes at a level of 99°. To the right in the figure is shown the degree of temperature compensation and the deviation of body temperature from normal. The negative feedback gain of the control system is equal to the temperature compensation divided by the deviation of body temperature from normal, or a total gain of 49.

More Complex Analysis of a Control System. Figure 1–8 illustrates an analysis of the control system for regulation of sodium concentration in the extracellular fluids. In this case, a person suddenly begins to drink concentrated salt solution at a continuous rate. Eventually the sodium concentration

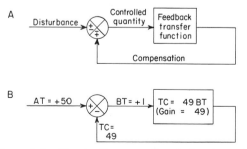

Figure 1–6 Simple diagrams of control systems: (A) general diagram that can be applied to almost any control system of the body, and (B) effect on body temperature (*BT*) and subsequent temperature compensation (*TC*) caused by increasing the air temperature 50° F.

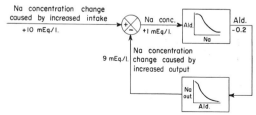

Figure 1–8 Steady state analysis of the effect of increased sodium intake on body sodium concentration.

in his extracellular fluids would increase 10 mEq./ liter if no compensation occurred for the increased rate of sodium intake. However, the increasing concentration of sodium in the extracellular fluids causes decreased aldosterone concentration, which in turn causes increased output of sodium by the kidneys. This effect is sufficient to cause a decrease in sodium concentration of 9 mEq./liter. Therefore, the large intake of sodium solution causes the sodium concentration in the body fluids to rise only 1 mEq./liter, which is a very slight amount.

Note that the diagram of Figure 1–8 shows two transfer functions, one showing the effect of sodium concentration on the aldosterone concentration, and the other showing the effect of aldosterone concentration on the sodium concentration. Some control systems have 5, 10, or more steps in them, and each of these steps can be represented by a transfer function.

Steady State versus Transient Analysis of Control Systems. The diagrams in Figures 1–6 and 1–8 represent steady state analyses of two control systems. These analyses do not tell anything about the time courses of the changes; instead they simply analyze what the conditions will be after a complete state of equilibrium has developed. In some instances, these steady state analyses are adequate to display the concepts of the different control systems of the body, but more often it is also important to know the time courses of changes in concentrations of different substances or changes in rates of activity. An analysis of a control system that can give the time courses of the changes is called a *transient analysis*.

Figure 1–9 illustrates a transient analysis of the same control system analyzed by the steady state method in Figure 1–8. Here again, it is still the sodium concentration in the extracellular fluids that is being regulated, and the analysis shows what will happen to the sodium concentration when the rate of sodium intake is altered. The following is a description of this analysis. Block 1: The *net rate of sodium intake or output* from the body is equal to the *rate of sodium intake* minus the *rate of sodium output* through the kidneys. Block 2: The *quantity of sodium* in the body at any given time is the integral of the net rate of sodium intake or output. Block 3: The *concentration of sodium* in the body is equal to the quantity of sodium in the body divided by the *volume of fluid* (K_1) in which the sodium is distributed. Block 4: The *rate of aldosterone secretion* is a function of the sodium concentration. Block 5: The *rate of aldosterone destruction* is proportional to the aldosterone concentration in the body fluids. Block 6: The net *rate of change of aldosterone* in the body is equal to the rate of aldosterone secretion minus its rate of destruction. Block 7: The total *quantity of aldosterone* in the body at any given time is the integral of the net rate of aldosterone change in the body. Block 8: The *aldosterone concentration* is equal to the quantity of aldosterone multiplied by the reciprocal of the *volume of fluid* (K_2) in which it is distributed. Block 9: The *rate of output of sodium* by the kidney is a function of the aldosterone concentration. Thus, we have developed a *closed loop* system in which a series of complex reactions cause a feedback that prevents the sodium concentration in the

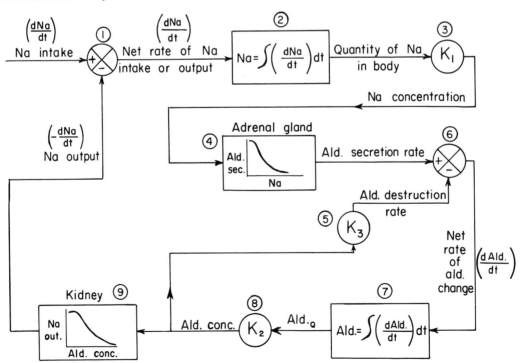

Figure 1–9 Transient control system diagram for analyzing the effect of sodium intake on sodium concentration in the extracellular fluids.

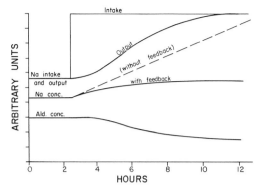

Figure 1-10 Transient changes in sodium output, sodium concentration, and aldosterone concentration when sodium intake is suddenly increased.

body fluids from changing significantly even though the sodium intake increases or decreases drastically.

Figure 1-10 illustrates the function of the closed loop system in Figure 1-9. Let us assume that the sodium intake has been constant for 2 days and that the sodium and aldosterone concentrations have reached a state of equilibrium. Then, suddenly, the person begins to ingest greatly increased quantities of sodium. Immediately, the sodium concentration in the body fluids begins to rise, and, if there were no feedback to make the person excrete increased quantities of sodium, the sodium concentration would rise indefinitely, as indicated by the dashed curve, until it eventually reached a value that would kill the person. However, with feedback regulation the rate of output eventually rises to equal the intake once again so that the concentration of sodium in the body reaches a new steady state. Likewise, the aldosterone concentration also reaches a new steady state, the sodium concentration above normal and the aldosterone concentration below normal. And these values will remain essentially constant so long as the excessive salt intake lasts but will change to new values when the rate of intake changes again.

OSCILLATION OF CONTROL SYSTEMS

Unfortunately, even negative feedback control systems sometimes become unstable and oscillate. Figure 1-11 illustrates blood pressure waves caused by oscillation of the baroreceptor system described earlier in the chapter. The cause of the oscillation is the following: Some extraneous factor causes the arterial pressure to become too high, and this activates the baroreceptor reflex. The pressure begins to fall and eventually falls even below normal. Then the very low arterial pressure activates the baroreceptor reflex in the opposite direction, causing the pressure to increase to a level above normal once again. This initiates a new cycle of oscillation, which can continue indefinitely.

Damping of Control Systems. Fortunately, oscillation of the arterial pressure, as shown in Figure 1-11, does not occur very often but only under certain peculiar conditions such as decreased blood volume or compression of the brain. The reason for lack of oscillation is that this control system, as is also true of almost all other control systems of the body, is highly *damped*, which results from the basic organization of the control system itself. For instance, the sodium control system of Figures 1-9 and 1-10 does not show oscillation, primarily because the volume of fluid in the body is so large that even a greatly changed rate of sodium accumulation requires many hours to alter the sodium concentration significantly. In the meantime, the other parts of the control system can adapt themselves almost completely to the changing sodium concentration, thereby preventing any overresponse of the control system.

Figure 1-11 shows *waxing oscillation, driving oscillation*, and *damped oscillation*. In waxing oscillation, the system is said to have a *damping ratio* of less than zero, which means that even the slightest disturbance can cause mild oscillation at first which grows and grows to greater and greater oscillation. If the damping ratio is exactly zero, then the oscillations, once they start, will continue indefinitely, neither increasing nor decreasing, as shown in the middle panel of Figure 1-11. On the other hand, if the damping ratio is greater than zero but less than 1, any oscillation that does occur will gradually fade away. Finally, if the damping ratio is 1 (*critically damped*) or greater than 1 there will be no oscillation at all; instead, sudden changes, such as the sudden increase in sodium intake in the analysis of Figure 1-10, will simply cause slow changes in the other elements of the system, with the system gradually approaching a new state of equilibrium without any overresponse at all, and, therefore, without any oscillation.

A few control systems continue to oscillate indefinitely. One of these is the system that causes respiration, which results from oscillation, or "reverberation," of impulses within the respiratory center of the brain. Another continuous oscillation is responsible for the monthly sexual cycle of the

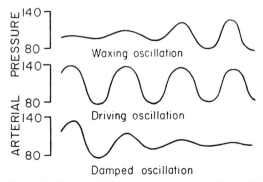

Figure 1-11 Arterial pressure waves caused by oscillation of the baroreceptor system.

female, in which hormones secreted by the anterior pituitary gland stimulate the ovaries to produce ovarian hormones, which in turn act on the pituitary to inhibit its secretion. Lack of the pituitary hormones then causes the ovaries to stop secreting their hormones. This in turn allows the pituitary to begin secreting once again. As a result, the cycle repeats itself again and again from puberty to the menopause.

REFERENCES

Adolph, E. F.: Early concepts of physiological regulations. *Physiol. Rev., 41*:737, 1961.

Adolph, E. F.: Origins of Physiological Regulations. New York, Academic Press, Inc., 1968.

Adolph, E. F.: Perspectives of adaptation: some general properties. *In* Handbook of Physiology. Baltimore, The Williams & Wilkins Co., 1964, Sec. 4, p. 27.

Adolph, E. F.: The Development of Homeostasis. New York, Academic Press, Inc., 1962.

Aschoff, J.: Comparative physiology: diurnal rhythms. *Ann. Rev. Physiol., 25*:581, 1963.

Bellman, R.: Control theory. *Sci. Amer., 211*:186(3), 1964.

Cannon, W. B.: Organization for physiological homeostasis. *Physiol. Rev., 9*:399, 1930.

Cannon, W. B.: The Wisdom of the Body. New York, W. W. Norton & Company, Inc., 1932.

Engelberg, J.: Physiological regulation: the steady state. *Physiologist, 9(2)*:69, 1966.

Guyton, A. C., and Coleman, T. G.: Quantitative analysis of the pathophysiology of hypertension. *Circ. Res., 14*:I-1, 1969.

Milhorn, H. T.: The Application of Control Theory to Physiological Systems. Philadelphia, W. B. Saunders Company, 1966.

Physiology Training Committee: A view of systems physiology. *Physiologist, 11(2)*:115, 1968.

Reeve, E. B., and Guyton, A. C.: Physical Bases of Circulatory Transport: Regulation and Exchange. Philadelphia, W. B. Saunders Company, 1967.

Sherrington, C. S.: Man on His Nature. New York, The Macmillan Company, 1941.

Söderberg, U.: Neurophysiological aspects of homeostasis. *Ann. Rev. Physiol., 26*:271, 1964.

Warner, H. R.: Use of analogue computers in the study of control mechanisms in the circulation. *Fed. Proc., 21*:87, 1962.

Wolf, W.: Rhythmic functions in the living system. *Ann. N.Y. Acad. Sci., 98*:753, 1962.

CHAPTER 2

The Cell and Its Function

Each of the 100 trillion cells in the human being is a living structure that can survive indefinitely and in most instances can even reproduce itself provided its surrounding fluids simply remain constant. To understand the function of organs and other structures of the body, it is essential that we first understand the basic organization of the cell and the functions of its component parts.

ORGANIZATION OF THE CELL— PROTOPLASM

A typical cell, as seen by the light microscope, is illustrated in Figure 2–1. Its two major parts are the *nucleus*, which is filled with *nucleoplasm*, and the *cytoplasm*. The nucleus is separated from the cytoplasm by a *nuclear membrane*, and the cytoplasm is separated from the surrounding fluids by a *cell membrane*.

The different substances that make up the cell are collectively called *protoplasm*. Protoplasm is composed mainly of five basic substances: water, electrolytes, proteins, lipids, and carbohydrates.

Water. The fluid medium of all protoplasm is water, which is present in a concentration between 70 and 85 per cent. Many cellular chemicals are dissolved in the water, while others are suspended in small particulate form. Chemical reactions take place between the dissolved chemicals or at the surface boundaries between the suspended particles and the water. The fluid nature of water allows both the dissolved and suspended substances to diffuse to different parts of the cell, thereby providing transport of the substances from one part of the cell to another.

Electrolytes. The most important electrolytes in the cell are *potassium, magnesium, phosphate, sulfate, bicarbonate*, and small quantities of *sodium* and *chloride*. These will be discussed in much greater detail in Chapter 4, which will consider the interrelationships between the different fluids of the body.

The electrolytes are dissolved in the water of protoplasm, and they provide inorganic chemicals for cellular reactions. Also, they are necessary for operation of some of the cellular control mechanisms. For instance, electrolytes acting at the cell membrane allow transmission of electrochemical impulses in nerve and muscle fibers, and the intracellular electrolytes determine the activity of different enzymatically catalyzed reactions that are necessary for cellular metabolism.

Proteins. Next to water, the most abundant substance in most cells is proteins, which normally constitute 10 to 20 per cent of the cell mass. These can be divided into two different types, *structural proteins* and *enzymes*.

To get an idea of what is meant by *structural proteins*, one needs only to note that leather is composed principally of structural proteins, and that hair is almost entirely a structural protein. Proteins of this type are present in the cell membrane, in the nuclear membrane, and in membranes surrounding special intracellular structures, such as the endoplasmic reticulum and the mitochondria. Thus, structural proteins hold the structures of the cell together. Most structural proteins are *fibrillar*; that is, the individual protein molecules are polymer-

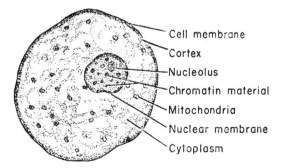

Figure 2–1 Structure of the cell as seen with the light microscope.

Labels:
Cell membrane
Cortex
Nucleolus
Chromatin material
Mitochondria
Nuclear membrane
Cytoplasm

ized into long fibrous threads. The threads in turn provide tensile strength for the cellular structures.

Enzymes, on the other hand, are an entirely different type of protein, composed usually of individual protein molecules or at most of aggregates of a few molecules in a *globular* form rather than a fibrillar form. These proteins, in contrast to the fibrillar proteins, are often soluble in the fluid of the cell or are adsorbed to the surface of membranous structures inside the cell. The enzymes come into direct contact with other substances inside the cell and catalyze chemical reactions. For instance, the chemical reactions that split glucose into its component parts and then combine these with oxygen to form carbon dioxide and water are catalyzed by a series of protein enzymes. Thus, enzyme proteins control the metabolic functions of the cell.

Special types of proteins are present in different parts of the cell. Of particular importance are the *nucleoproteins,* present both in the nucleus and in the cytoplasm. The nucleoproteins of the nucleus are mainly the *genes,* which control both the overall function of the cell and the transmission of hereditary characteristics from cell to cell. These substances are so important that they will be considered in detail in Chapter 3. In addition, the chemical nature of proteins will be considered in Chapter 69, and the different structural and enzymatic functions of proteins will be subjects of discussion at numerous points throughout this text.

Lipids. Lipids are several different types of substances that are grouped together because of their common property of being soluble in fat solvents. For instance, the usual fat of animal tissues is a lipid called *neutral fat.* However, in addition to neutral fat two other types of lipids, *phospholipids* and *cholesterol,* are very common throughout cells.

The usual cell contains 2 to 3 per cent lipids which are dispersed throughout the cell but are present in especially high concentrations in the cell membrane, the nuclear membrane, and the membranes lining intracytoplasmic organelles, such as the endoplasmic reticulum and the mitochondria.

The special importance of lipids in the cell is that they are either insoluble or only partially soluble in water. They combine with structural proteins to form the different membranes that separate the different water compartments of the cell from each other. The lipids of each membrane form a boundary between the solutions on the two sides of the membrane, making it impervious to many dissolved substances.

The chemical natures of the different types of fats and their functions in the body will be discussed in Chapter 68.

Carbohydrates. In general, carbohydrates have very little structural function in the cell, but they play a major role in nutrition of the cell. Most human cells do not maintain large stores of carbohydrates, usually averaging about 1 per cent of their total mass. However, carbohydrate, in the form of *glucose,* is always present in the surrounding extracellular fluid so that it is readily available to the cell. The small amount of carbohydrates stored in the cells is almost entirely in the form of *glycogen,* which is an insoluble polymer of glucose.

PHYSICAL STRUCTURE OF THE CELL

The cell is not merely a bag of fluid, enzymes, and chemicals; it also contains highly organized physical structures called *organelles,* which are equally as important to the function of the cell as the cell's chemical constituents. For instance, without one of the organelles, the *mitochondria,* the energy supply of the cell would cease almost entirely. Some principal organelles of the cell, are the *cell membrane, nuclear membrane, endoplasmic reticulum, mitochondria,* and *lysosomes,* as illustrated in Figure 2–2. Others not shown in the figure are the *Golgi complex, centrioles, cilia,* and microtubules.

MEMBRANES OF THE CELL

Essentially all physical structures of the cell are lined by a membrane composed primarily

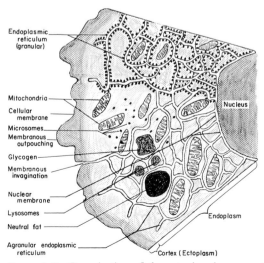

Figure 2–2 Organization of the cytoplasmic compartment of the cell.

of lipids and proteins. All of the different membranes are similar in structure, and this common type of structure is called the "unit membrane," which will be described below in relation to the cell membrane. The different membranes include the *cell membrane*, the *nuclear membrane*, the *membrane of the endoplasmic reticulum*, and the *membranes of the mitochondria, lysosomes, Golgi complexes*, and so forth.

The Cell Membrane. The cell membrane is thin (approximately 75 to 100 Ångstroms) and elastic. It is composed almost entirely of proteins and lipids, with a percentage composition approximately as follows: proteins, 55 per cent; lipids, 40 per cent; polysaccharides, perhaps 5 per cent. The proteins in the cell are mainly a type of protein called *stromatin*, an insoluble structural protein having elastic properties. The lipids are approximately 65 per cent phospholipids, 25 per cent cholesterol, and 10 per cent other lipids.

The precise molecular organization of the cell membrane is unknown, but many experiments point to the structure illustrated in Figure 2–3A. This shows a central layer of lipids covered by protein layers and a thin mucopolysaccharide layer on the outside surface. The presence of protein and mucopolysaccharide on the surfaces supposedly makes the membrane *hydrophilic*, which means that water adheres easily to the membrane. The lipid center of the membrane makes the membrane mainly impervious to lipid-insoluble substances. The small knobbed structures lying at the bases of the protein molecules in Figure 2–3A are phospholipid molecules; the fat portion of the phospholipid molecule is attracted to the central lipid phase of the cell membrane, and the polar (ionized) portion of the molecule protrudes toward either surface, where it is bound electrochemically with the inner and outer layers of proteins. The thin layer of mucopolysaccharide on the outside of the cell membrane makes the outside different from the inside, thus polarizing the membrane so that the chemical reactivities of the cell's inner surface are different from those of the outer surface.

PORES IN THE MEMBRANE. The membrane is believed to have many minute pores that pass from one side to the other as shown in Figure 2–3B. These pores have never been demonstrated even with an electron microscope, but functional experiments to study the movement of molecules of different sizes between the extra- and intracellular fluids have demonstrated free diffusion of molecules up to a size of approximately 8 Ångstroms. It is through these pores that lipid-insoluble substances of very small sizes, such as water and urea molecules, are believed to pass with relative ease between the interior and exterior of the cell.

The Nuclear Membrane. The nuclear membrane is actually a double membrane having a wide space between two "unit" membranes. Each unit membrane is almost identical to the cell membrane, having lipids in its center and protein on its two surfaces, but having no mucopolysaccharide layer. The nuclear membrane probably has characteristics similar to those of the cell membrane, except that very large holes, or "pores," of several hundred Ångstroms diameter are present in the membrane so that almost all dissolved substances can move with ease between the fluids of the nucleus and those of the cytoplasm.

The Endoplasmic Reticulum. Figure 2–3 illustrates in the cytoplasm a network of tubular and vesicular structures, constructed of a system of unit membranes, called the *endoplasmic reticulum*. The detailed structure of this organelle is illustrated in Figure 2–4. The space

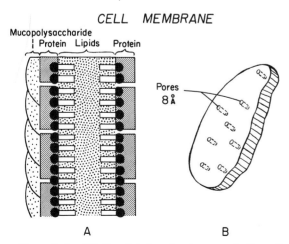

CELL MEMBRANE

Mucopolysaccharide
| Protein Lipids Protein

Pores
8 Å

A   B

Figure 2–3 (A) Postulated molecular organization of the cell membrane. (B) Pores in the cell membrane.

RNP particles

Membranes

Matrix

Figure 2–4 Structure of the membranes of the endoplasmic reticulum. (Redrawn from De Robertis, Nowinski, and Saez: Cell Biology. 4th Ed.)

inside the tubules and vesicles is filled with *endoplasmic matrix*, a fluid medium that is different from the fluid outside the endoplasmic reticulum.

Electron micrographs show that the space inside the endoplasmic reticulum is connected with the space between the two membranes of the double nuclear membrane. Also, in some instances the endoplasmic reticulum connects directly through small openings with the exterior of the cell. Substances formed in different parts of the cell are believed to enter the space between the two nuclear membranes and then to be conducted to other parts of the cell through the endoplasmic reticular tubules.

RIBOSOMES. Attached to the outer surfaces of many parts of the endoplasmic reticulum are large numbers of small granular particles called ribosomes. Where these are present, the reticulum is frequently called the *granular endoplasmic reticulum*, or the *ergastoplasm*. The ribosomes are composed mainly of ribonucleic acid, which functions in the synthesis of protein in the cells, as is discussed in the following chapter.

Part of the endoplasmic reticulum has no attached ribosomes. This part is called the *agranular*, or *smooth, endoplasmic reticulum*. It is believed that the agranular reticulum helps in the synthesis of lipid substances and probably also acts as a medium for transporting secretory substances to the exterior of the cell, as is discussed later in the chapter.

Golgi Complex. The Golgi complex, illustrated in Figure 2–5, is probably a specialized portion of the endoplasmic reticulum. It has membranes similar to those of the agranular endoplasmic reticulum and is composed of four or more layers of thin vesicles. Electron micrographs show direct connections between the endoplasmic reticulum and parts of the Golgi complex.

The Golgi complex is very prominent in secretory cells; in these, it is located on the side of the cell from which substances will be secreted. Its function is believed to be temporary storage of secretory substances and preparation of these substances for final secretion.

PHYSICAL NATURE OF THE CYTOPLASM

The cytoplasm is filled with both minute and large dispersed particles ranging in size from a few Ångstroms to 1 micron in diameter. The clear fluid portion of the cytoplasm in which the particles are dispersed is called *hyaloplasm*; this contains mainly dissolved proteins, electrolytes, glucose, and small quantities of phospholipids, cholesterol, and esterified fatty acids.

The portion of the cytoplasm immediately beneath the cell membrane is frequently gelled into a semi-solid called the *cortex*, or *ectoplasm*. The cytoplasm between the cortex and the nuclear membrane is liquefied and is called the *endoplasm*.

Among the large dispersed particles in the cytoplasm are neutral fat globules, glycogen granules, ribosomes, secretory granules, and two important organelles—the *mitochondria* and *lysosomes*—which are discussed below.

Colloidal Nature of the Cytoplasm. All the particles dispersed in the cytoplasm, whether the large lysosomes and mitochondria or the small granules, are hydrophilic—that is, attracted to water—because of electrical charges on their surfaces. The lining membranes of the mitochondria and lysosomes have charges on their surfaces, because of the manner in which the protein molecules on the membrane surfaces are oriented. The particles remain dispersed in the cytoplasm because of mutual repulsion of the charges on the different particles. Thus, the cytoplasm is actually a colloidal solution. Likewise, the nucleoplasm is a colloidal solution.

Mitochondria. Mitochondria are present in the cytoplasm of all cells, as illustrated in Figure 2–2, but the number per cell varies from a few hundred to many thousand, depending on the amount of energy required by each cell to perform its functions. Mitochondria are also very variable in size and shape; some are only a few hundred millimicrons in diameter and

Figure 2–5   A typical Golgi complex.

Figure 2-6   Structure of a mitochondrion. (Redrawn from De Robertis, Nowinski, and Saez: Cell Biology, 4th Ed.)

globular in shape while others are as large as 1 micron in diameter, as long as 7 microns, and filamentous in shape.

The basic structure of a mitochondrion is illustrated in Figure 2-6, which shows it to be surrounded by a double-unit membrane that is similar in structure to the nuclear membrane. Many infoldings of the inner unit membrane form shelves on which almost all the oxidative enzymes of the cell are believed to be adsorbed. When nutrients and oxygen come in contact with these enzymes in the mitochondrion, they combine to form carbon dioxide and water, and the liberated energy is used to synthesize a substance called *adenosine triphosphate (ATP)*. The ATP then diffuses throughout the cell and releases its stored energy wherever it is needed for performing cellular functions. The function of ATP is so important to the cell that it is discussed in detail later in the chapter.

Mitochondria are probably self-replicative, which means that one mitochondrion can probably form a second one, a third one, and so on, whenever there is need in the cell for increased amounts of ATP. Indeed, the mitochondria contain a special type of deoxyribonucleic acid similar to, but different from, that found in the nucleus. In the following chapter we shall see that deoxyribonucleic acid is the basic substance that controls replication of the entire cell; it is reasonable, therefore, to suspect that this substance might play a similar role in the mitochondrion.

Lysosomes.   Another structure recently discovered in cells is the lysosome. The lysosome is 250 to 750 millimicrons in diameter, and is surrounded by a lipoprotein unit membrane. It is filled with large numbers of small granules 55 to 80 Ångstroms in diameter which are protein aggregates of hydrolytic (digestive) enzymes. A hydrolytic enzyme is capable of splitting an organic compound into two or more parts by combining hydrogen from a water molecule with part of the compound and by combining the hydroxyl portion of the water molecule with the other part of the compound. For instance, protein is hydrolyzed to form amino acids, and glycogen is hydrolyzed to form glucose. More than a dozen different *acid hydrolases* have been found in lysosomes, and the principal substances that they digest are proteins, nucleic acids, mucopolysaccharides, and glycogen.

Ordinarily, the membrane surrounding the lysosome prevents the enclosed hydrolytic enzymes from coming in contact with other substances in the cell. However, many different conditions of the cell will break the membranes of some of the lysosomes, allowing release of the enzymes. These enzymes then split the organic substances with which they come in contact into small, highly diffusible substances, such as amino acids and glucose. Some of the more specific functions of lysosomes are discussed later in the chapter.

Microtubules.   Located in many cells are fine tubular structures that are approximately 250 Å in diameter and that have lengths of 1 to many microns. These structures, called *microtubules*, are very thin in relation to their length, but they are usually arranged in bundles, which gives them, en masse, considerable structural strength. Furthermore, microtubules are usually stiff structures that break if bent too severely. Figure 2-7 illustrates some typical microtubules that have been teased from the flagellum of a sperm. Another example of microtubules is the tubular filaments that give cilia their structural strength, radiating upward from within the cell cytoplasm to the tip of the cilium.

The primary function of microtubules appears to be to act as a *cytoskeleton*, providing rigid physical structures for certain parts of cells such as the cilia just mentioned. However, the tubular nature of their structure also suggests that substances might be transported through the tubules. Indeed, cytoplasmic streaming has been observed in the vicinity of microtubules, indicating that these tubular structures might play a role in causing movement of cytoplasm.

Secretory Granules.   One of the most important functions of many cells is secretion of special substances. The secretory substances are usually formed inside the cell and are held there until an appropriate time for release to the exterior. The storage depots within the cells are called secretory granules; these are

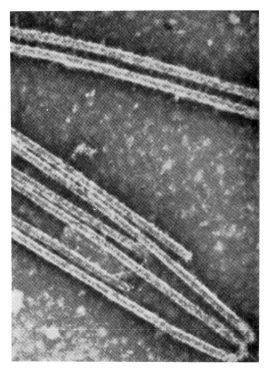

Figure 2–7   Microtubules teased from the flagellum of a sperm. (From Porter, K. R.: Ciba Foundation Symposium: Principles of Biomolecular Organization. Little, Brown and Co., 1966.)

Figure 2–8   Secretory granules in acinar cells of the pancreas.

illustrated by the dark spots in Figure 2–8, which shows pancreatic acinar cells that have formed and stored enzymes that will be secreted later into the intestinal tract.

Many secretory granules lie inside the tubules and vesicles of the endoplasmic reticulum and Golgi complex, while others are free in the cytoplasm.

Other Organelles of the Cytoplasm.   The cytoplasm of each cell contains two *centrioles*, which are small cylindrical structures that play a major role in cell division, as will be discussed in Chapter 3. Also, most cells contain small *lipid droplets* and *glycogen granules* that play important roles in energy metabolism of the cell. And certain cells contain highly specialized structures such as the *cilia* of ciliated cells which are actually outgrowths from the cytoplasm, and the *myofibrils* of muscle cells. All of these are discussed in detail at different points in this text.

THE NUCLEUS

The nucleus is the control center of the cell. It controls both the chemical reactions that occur in the cell and the reproduction of the cell. Briefly, the nucleus contains large quantities of

deoxyribonucleic acid, which we have called *genes* for many years. The genes control the characteristics of the protein enzymes of the cytoplasm, and in this way control cytoplasmic activities. To control reproduction, the genes reproduce themselves, and after this is accomplished the cell splits by a special process called *mitosis* to form two daughter cells, each of which receives one of the two sets of genes. These activities of the nucleus are considered in detail in the following chapter.

The appearance of the nucleus under the microscope does not give much of a clue to the mechanisms by which it performs its control activities. Figure 2–9 illustrates the interphase nucleus (period between mitoses), showing darkly staining, granular *chromatin material* throughout the *nuclear sap*. During mitosis, the chromatin material becomes readily identifiable as part of the highly structured *chromosomes*, which can be seen easily with a light microscope. Even during the interphase of cel-

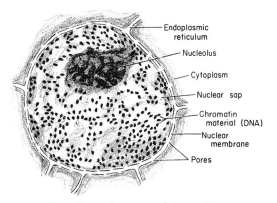

Figure 2–9   Structure of the nucleus.

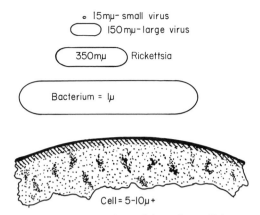

Figure 2–10 Comparison of sizes of precellular organisms with that of the average cell in the human body.

lular activity the granular chromatin material is still organized into chromosomal structures, but this is impossible to see except in a few types of cells.

Nucleoli. The nuclei of many cells contain one or more lightly staining structures called nucleoli. The nucleolus, unlike most of the organelles that we have discussed, does not have a limiting membrane. Instead, it is simply an aggregate of loosely bound granules composed mainly of *ribonucleic acid*. It usually becomes considerably enlarged when a cell is actively synthesizing proteins. The genes of the chromosomes are believed to synthesize the ribonucleic acid and then to store it in the nucleolus; this ribonucleic acid later disperses from the nucleolus into the cytoplasm where it controls cytoplasmic function. The details of these mechanisms are discussed in the following chapter.

COMPARISON OF THE ANIMAL CELL WITH PRECELLULAR FORMS OF LIFE

Many of us think of the cell as the lowest level of animal life. However, the cell is a very complicated organism, which probably required several billion years to develop after the earliest form of life, some organism similar to the present-day *virus*, first appeared on Earth. Figure 2–10 illustrates the relative sizes of the smallest known virus, a large virus, a *rickettsia*, a *bacterium*, and a cell, showing that the cell has a diameter about 1000 times that of the smallest virus, and, therefore, a volume about 1 billion times that of the smallest virus. Correspondingly, the functions and anatomical organization of the cell are also far more complex than those of the virus.

The principal constituent of the very small virus is a *nucleic acid* embedded in a coat of protein. This nucleic acid is similar to that of the cell, and it is capable of reproducing itself if appropriate nutrients are available. Thus, the virus is capable of propagating its lineage from generation to generation, and, therefore, is a living structure in the same way that the cell and the human being are living structures.

As life evolved, other chemicals besides nucleic acid and simple proteins became integral parts of the organism, and specialized functions began to develop in different parts of the virus. A membrane formed around the virus, and inside the membrane a fluid matrix appeared. Specialized chemicals developed inside the matrix to perform special functions; many protein enzymes appeared which were capable of catalyzing chemical reactions, and, therefore, of controlling the organism's activities.

In still later stages, particularly in the rickettsial and bacterial stages, *organelles* developed inside the organism, these representing aggregates of chemical compounds that perform functions in a more efficient manner than can be achieved by dispersed chemicals throughout the fluid matrix. And, finally, in the cell, still more complex organelles developed, the most important of which is the *nucleus*. The nucleus distinguishes the cell from all lower forms of life; this structure provides a control center for all cellular activities, and it also provides for very exact reproduction of new cells generation after generation, each new cell having essentially the same structure as its progenitor.

FUNCTIONAL SYSTEMS OF THE CELL

In the remainder of this chapter some of the more important functional systems of the cell are discussed. However, one of the most important functions of all, control of protein synthesis and of other cellular functions by the genes of the nucleus, is so important that the following entire chapter is devoted to it.

FUNCTIONS OF THE ENDOPLASMIC RETICULUM

The endoplasmic reticulum exists in many different forms in different cells, sometimes highly granular with a large number of ribosomes on its surface, sometimes agranular, sometimes tubular, sometimes vesicular with large shelf-like surfaces, and so forth. Therefore, simply from anatomical considerations alone, it is certain that the endoplasmic reticulum performs a very large share of the cell's functions. Yet, understanding of its functions has been extremely slow to develop because of the difficulty of studying function with the electron microscope.

On the vast surfaces of the endoplasmic reticulum are adsorbed many of the protein enzymes of the cell, and perhaps still other enzymes are actually integral parts of the reticulum itself. Many of these enzymes synthesize substances in the cells, and others undoubtedly act to transport substances through the membrane of the endoplasmic reticulum from the hyaloplasm of the cytoplasm into the matrix of the reticulum or in the opposite direction. Some of the proved functions of the endoplasmic reticulum are the following:

Secretion of Proteins by Secretory Cells. Many cells, particularly cells in the various glands of the body, form special proteins that are secreted to the outside of the cells. The mechanism for this involves the endoplasmic reticulum and Golgi complex as illustrated in Figure 2–11. The ribosomes on the surface of the endoplasmic reticulum synthesize the protein that is to be secreted. This protein is either discharged directly into the tubules of the endoplasmic reticulum by the ribosomes or is immediately transported into the tubules to form small protein granules. These granules then move slowly through the tubules toward the Golgi complex, arriving there a few minutes to an hour or more later. In the Golgi complex the granules are condensed into coalesced granules that then evaginate outward through the membrane of the Golgi complex into the cytoplasm of the cell to form *secretory granules.* Each of these granules carries with it part of the membrane of the Golgi apparatus which provides a membrane around the secretory granule and prevents it from dispersing in the cytoplasm. Gradually, the secretory granules move toward the surface of the cell where their membranes become miscible with the membrane of the cell itself, and in some way not completely understood they expel their substances to the exterior. It is in this way that protein enzymes, for instance, are secreted by the exocrine glands of the gastrointestinal tract.

Lipid Secretion. Almost exactly the same mechanism applies to lipid secretion. The one major difference is that lipids are synthesized by the agranular portion of the endoplasmic reticulum. The Golgi complex also provides much the same function in lipid secretion as in protein secretion, for here lipids are stored for long periods of time before finally being extruded into the cytoplasm as lipid droplets and thence to the exterior of the cells.

Release of Glucose from Glycogen Stores of Cells. Electron micrographs show that glycogen is stored in cells as minute granules lying in close apposition to agranular tubules of the endoplasmic reticulum. The actual chemical reactions that cause polymerization of glucose to form glycogen granules probably occur in the hyaloplasm and not in the wall of the endoplasmic reticulum, but the reticulum seems to play a transport role in bringing glucose to the site of glycogen formation and in carrying glucose away from the site when the glycogen is later broken down.

Other Possible Functions of the Endoplasmic Reticulum. The functions presented thus far for the endoplasmic reticulum have involved, first, synthesis of substances and, second, transport of these substances to the exterior of the cell. It is probable that many other substances are synthesized by the endoplasmic reticulum and are then secreted to the interior of the cell rather than to the exterior. It is likewise probable that the Golgi complex plays a role in these internal secretory processes, for large Golgi complexes are found in some cells, such as nerve cells, that do not perform any external secretory function.

Because of the multitude of different anatomical forms of the endoplasmic reticulum, it is certain that we will discover many more functional roles that it plays besides those few that have been discussed here.

INGESTION BY THE CELL—PINOCYTOSIS

If a cell is to live and grow, it must obtain nutrients and other substances from the surrounding fluids. Substances can pass through a cell membrane in three separate ways: (1) by *diffusion* through the pores in the membrane or through the membrane matrix itself; (2) by *active transport* through the membrane, a mechanism in which enzyme systems and special carrier substances "carry" the substances through the membrane; and (3) by *pinocytosis*, a mechanism by which the membrane actually engulfs some of the extracellular fluid and its contents. Transport of substances through the mem-

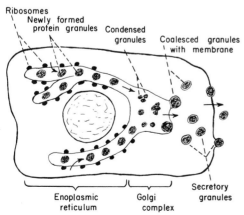

Figure 2–11 Function of the endoplasmic reticulum and Golgi complex in secreting proteins.

Figure 2–12   Mechanism of pinocytosis.

brane is such an important subject that it will be considered in detail in Chapter 4, but one of these mechanisms of transport, pinocytosis, is such a specialized cellular function that it deserves mention here as well.

Figure 2–12 illustrates the mechanism of pinocytosis. Figure 2–12A shows three molecules of protein in the extracellular fluid approaching the surface of the cell. In Figure 2–12B these molecules have become attached to the surface, presumably by the simple process of adsorption. The presence of these proteins causes the surface tension properties of the cell surface to change in such a way that the membrane invaginates as shown in Figure 2–12C. Immediately thereafter, the invaginated portion of the membrane breaks away from the surface of the cell, forming a *pinocytic vesicle* which then penetrates deep into the cytoplasm away from the cell membrane.

Pinocytosis occurs only in response to certain types of substances that contact the cell membrane, the two most important of which are *proteins* and *strong solutions of electrolytes*. It is especially significant that proteins cause pinocytosis, because pinocytosis is the only means by which proteins can pass through the cell membrane.

Phagocytosis. Phagocytosis means the ingestion of large particulate matter by a cell, such as the ingestion of (a) a bacterium, (b) some other cell, or (c) particles of degenerating tissue. Pinocytosis was not discovered until the advent of the electron microscope because pinocytic vesicles are smaller than the resolution of the light microscope. However, phagocytosis has been known to occur from the earliest studies using the light microscope.

The mechanism of phagocytosis is almost identical with that of pinocytosis. The particle to be phagocytized must have a surface that can become adsorbed to the cell membrane. Many bacteria have membranes that, on contact with the cell membrane, actually become

miscible with the cell membrane so that the cell membrane simply spreads around the bacterium and invaginates to form a *phagocytic vesicle* containing the bacterium, a vesicle that is essentially the same as the pinocytic vesicle shown in Figure 2–12 but much larger. Phagocytosis will be discussed at further length in Chapter 9 in relation to function of the white blood cells.

THE DIGESTIVE ORGAN OF THE CELL— THE LYSOSOMES

Almost immediately after a pinocytic or phagocytic vesicle appears inside a cell, one or more lysosomes become attached to the vesicle and empty their hydrolases into the vesicle, as illustrated in Figure 2–13. Thus, a *digestive vesicle* is formed in which the hydrolases begin hydrolyzing the proteins, glycogen, nucleic acids, mucopolysaccharides, and other substances in the vesicle. The products of digestion are small molecules of amino acids, glucose, phosphates, and so forth that can then diffuse through the membrane of the vesicle into the cytoplasm. What is left of the digestive vesicle, called the *residual body,* is then either excreted or undergoes dissolution inside the cytoplasm. Thus, the lysosomes may be called the *digestive organ* of the cells.

Regression of Tissues and Autolysis of Cells. Often, tissues of the body regress to a much smaller size than previously. For instance, this occurs in the uterus following pregnancy, in muscles during long periods of inactivity, and in mammary glands at the end of the period of lactation. Lysosomes are probably responsible for much if not most of this regression, for one can show that the lysosomes become very active during these periods of regression. However, the mechanism by which the lack of activity in a tissue causes the lysosomes to

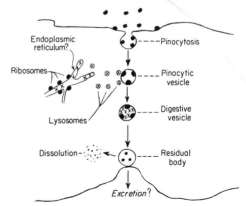

Figure 2–13   Digestion of substances in pinocytic vesicles by enzymes derived from lysosomes. (Modified from C. De Duve, in Lysosomes, Ed. by Reuck and Cameron, Little, Brown and Co., 1963.)

increase their activity is completely unknown.

Another very special role of the lysosomes is the removal of damaged cells or damaged portions of cells from tissues—cells damaged by heat, cold, trauma, chemicals, or any other factor. Damage to the cell causes lysosomes to rupture, and the released hydrolases begin immediately to digest the surrounding organic substances. If the damage is slight, only a portion of the cell will be removed, followed by repair of the cell. However, if the damage is severe the entire cell will be digested, a process called *autolysis*. In this way, the cell is completely removed and a new cell of the same type ordinarily is formed by mitotic reproduction of an adjacent cell to take the place of the old one.

EXTRACTION OF ENERGY FROM NUTRIENTS

The principal nutrients from which cells extract energy are oxygen and one or more of the foodstuffs—carbohydrates, fats, and proteins. In the human body essentially all carbohydrates are converted into glucose before they reach the cell, the proteins are converted into amino acids, and the fats are converted into fatty acids. Figure 2–14 shows oxygen and the foodstuffs—glucose, fatty acids, and amino acids—all entering the cell. Inside the cell, the foodstuffs react chemically with the oxygen under the influence of various enzymes that control their rates of reactions and channel the energy that is released in the proper direction.

Formation of Adenosine Triphosphate (ATP). The energy released from the nutrients is used to form adenosine triphosphate, generally called ATP, the formula for which is:

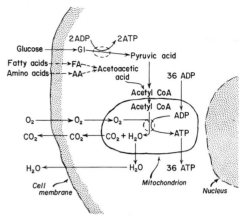

Figure 2–14   Formation of adenosine triphosphate in the cell, showing that most of the ATP is formed in the mitochondria.

Note that ATP is a nucleotide composed of the nitrogenous base *adenine*, the pentose sugar *ribose*, and three *phosphate radicals*. The last two phosphate radicals are connected with the remainder of the molecule by so-called *high energy phosphate bonds*, which are represented by the symbol ~. Each of these bonds contains about 8000 calories of energy per mole of ATP under the physical conditions of the body (7000 calories under standard conditions), which is many times the energy stored in the average chemical bond, thus giving rise to the term "high energy" bond. Furthermore, the high energy phosphate bond is very labile so that it can be split instantly on demand whenever energy is required to promote other cellular reactions.

When ATP releases its energy, a phosphoric

acid radical is split away, and *adenosine diphosphate (ADP)* is formed. Then, once again, energy derived from the cellular nutrients causes the ADP and phosphoric acid to recombine to form new ATP, the entire process continuing over and over again. For these reasons, ATP has been called the *energy currency* of the cell, for it can be spent and remade again and again.

CHEMICAL PROCESSES IN THE FORMATION OF ATP. On entry into the cells, glucose, fatty acids, and amino acids are subjected to enzymes in the cytoplasm or the nucleoplasm that convert glucose into *pyruvic acid* (a process called *glycolysis*), and convert fatty acids and most amino acids into *acetoacetic acid.* The chemical reactions of these conversions will be presented in Chapters 67 through 69. A small amount of ADP is changed into ATP by energy released during the conversion of glucose to pyruvic acid, but this amount is so slight that it plays only a minor role in the overall energy metabolism of the cell.

By far the major portion of the ATP formed in the cell is formed in the mitochondria. The pyruvic and acetoacetic acids are both converted into the compound *acetyl co-A* in the cytoplasm, and this is transported along with oxygen through the mitochondrial membrane into the matrix of the mitochondria. Here this substance is acted upon by a series of enzymes and undergoes dissolution in a sequence of chemical reactions called the *tricarboxylic acid cycle*, or *Krebs cycle*. These chemical reactions will be explained in detail in Chapter 67.

In the tricarboxylic acid cycle, acetyl co-A is split into its component parts, hydrogen atoms and carbon dioxide. The carbon dioxide in turn diffuses out of the mitochondria and eventually out of the cell. The hydrogen atoms combine with carrier substances and are carried to the surfaces of the shelves that protrude into the mitochondria, as shown in Figure 2–6. Attached to these shelves are so-called *oxidative enzymes* which, by a series of additional reactions, cause the hydrogen atoms to combine with oxygen. The enzymes are arranged on the surfaces of the shelves in such a way that the products of one chemical reaction are immediately relayed to the next enzyme, then to the next, and so on until the complete sequence of reactions has taken place. During the course of these reactions, the energy released from the combination of hydrogen with oxygen is used to manufacture tremendous quantities of ATP from ADP. The ATP then diffuses out of the mitochondria into all parts of the cytoplasm and nucleoplasm where its energy is used to energize the functions of the cell.

ANAEROBIC FORMATION OF ATP DURING GLYCOLYSIS. Note in Figure 2–14 that no oxygen is required for the splitting of glucose to form pyruvic acid, but even so a small amount of ATP is formed during this process. Therefore, this mechanism (glycolysis) for providing energy to the cell is called *anaerobic energy metabolism.*

OXIDATIVE ENERGY METABOLISM IN THE MITOCHONDRIA. Note, on the other hand, that tremendous quantities of ATP are formed by the chemical reactions in the mitochondria. Because these reactions require oxygen, this mechanism for providing energy is called *oxidative energy metabolism.* Approximately 90 per cent of all ATP formed in the cell is formed in the mitochondria, in comparison with only 10 per cent outside the mitochondria: 5 per cent by anaerobic metabolism, as described previously, and 5 per cent by another method of oxidative metabolism that will be discussed in Chapter 67. Because of the tremendous quantity of ATP formed in the mitochondria and because of the need for ATP to supply energy for all cellular functions, the mitochondria are called the *power-houses* of the cell.

Uses of ATP for Cellular Function. ATP is used to promote three major categories of cellular functions: (1) *membrane transport*, (2) *synthesis of chemical compounds* throughout the cell, and (3) *mechanical work.* These three different uses of ATP are illustrated in Figure 2–15: (a) to supply energy for the transport of glucose through the membrane, (b) to promote protein synthesis by the ribosomes, and (c) to supply the energy needed during muscle contraction.

Figure 2–15 Use of adenosine triphosphate to provide energy for three major cellular functions: membrane transport, protein synthesis, and muscle contraction.

In addition to membrane transport of glucose, energy from ATP is required for transport of sodium ions, potassium ions, and, in certain cells, calcium ions, phosphate ions, chloride ions, uric acid, hydrogen ions, and still many other special substances. Membrane transport is so important to cellular function that some cells utilize as much as 30 per cent of the ATP formed in the cells for this purpose alone.

In addition to synthesizing proteins, cells also synthesize phospholipids, cholesterol, purines, pyrimidines, and a great host of other substances. Synthesis of almost any chemical compound requires energy. For instance, a single protein molecule might be composed of as many as several thousand amino acids attached to each other by peptide linkages, and the formation of each of these linkages requires the breakdown of an ATP molecule to ADP; thus several thousand ATP molecules must release their energy as each protein molecule is formed. Indeed, cells often utilize as much as 75 per cent of all the ATP formed in the cell simply to synthesize new chemical compounds; this is particularly true during the growth phase of cells.

The final major use of ATP is to supply energy for special cells to perform mechanical work. We shall see in Chapter 7 that each contraction of a muscle fibril requires expenditure of tremendous quantities of ATP. This is true whether the fibril is in skeletal muscle, smooth muscle, or cardiac muscle. Other cells perform mechanical work in two additional ways, by *ciliary* or *ameboid motion*, both of which will be described in the following section. The source of energy for all these types of mechanical work is ATP.

In summary, therefore, ATP is always available to release its energy rapidly and almost explosively wherever in the cell it is needed. To replace the ATP used by the cell, other but much slower chemical reactions break down carbohydrates, fats, and proteins and use the energy derived from these to form new ATP.

CELL MOVEMENT

By far the most important type of cell movement that occurs in the body is that of the specialized muscle cells in skeletal, cardiac, and smooth muscle, which comprise almost 50 per cent of the entire body mass. The specialized functions of these cells will be discussed in Chapter 7. However, two other types of movement occur in other cells, *ameboid movement and ciliary movement.*

Ameboid Motion. Ameboid motion means movement of an entire cell in relation to its surroundings, such as the movement of white blood cells through tissues. Typically ameboid motion begins with protrusion of a pseudopodium from one end of the cell. The pseudopodium projects far out, away from the cell body; and then the remainder of the cell moves toward the pseudopodium. Formerly it was believed that the protruding pseudopodium attached itself far away from the cell and then pulled the remainder of the cell toward it. However in recent years, new studies have changed this idea to a "streaming" concept, as follows: Figure 2–16 illustrates diagrammatically an elongated cell, the right-hand end of which is a protruding pseudopodium. The membrane of this end of the cell is continually expanding forward while the membrane at the left-hand end of the cell is continually contracting and following along as the cell moves. It has been postulated that this movement is caused by "streaming" of the endoplasmic portion of the cytoplasm through the center of the cell from left to right. On reaching the right-hand end, the endoplasm turns toward the outside wall of the cell, becomes gelled, and therefore becomes ectoplasm that is fixed in position. Simultaneously, at the back end of the cell, the gelated ectoplasm becomes dissoluted and streams forward as endoplasm to make the pseudopodium move forward still farther. The continuous repetition of this process presumably makes the cell move in the direction in which the pseudopodium projects. One can readily see that this streaming movement inside the cell is analogous to the revolving movement of the track of a Caterpillar tractor.

TYPES OF CELLS THAT EXHIBIT AMEBOID MOTION. The most common cells to exhibit ameboid motion in the human body are the *white blood cells* moving out of the blood into the tissues in the form of *tissue macrophages* or *microphages.* However, many other types of cells can

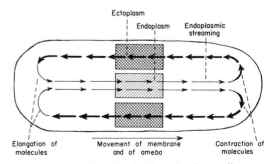

Ectoplasm

Endoplasm   Endoplasmic streaming

Elongation of molecules   Movement of membrane and of ameba   Contraction of molecules

Figure 2–16   Ameboid motion by a cell.

move by ameboid motion under certain circumstances. For instance, fibroblasts will move into any damaged area to help repair the damage, and even some of the germinal cells of the skin, though ordinarily completely sessile cells, will move by ameboid motion toward a cut area to repair the rent. Finally, ameboid motion is especially important in the development of the fetus, for embryonic cells often migrate long distances from the primordial site of origin to new areas during the development of special structures.

CONTROL OF AMEBOID MOTION—"CHEMOTAXIS." At the onset of ameboid motion, an electrochemical change occurs at the cell membrane and causes an *action potential* similar to that which occurs in nerves for the transmission of nerve impulses, as explained in Chapter 5. It is presumably this action potential that initiates the forward projection of the pseudopodium.

The most important factor that usually initiates ameboid motion, presumably by causing action potentials in the cell membrane, is the appearance of certain chemical substances in the tissues. This phenomenon is called *chemotaxis*, and the chemical substance causing it to occur is called a *chemotaxic substance*. Most ameboid cells move toward the source of the chemotaxic substance—that is, from an area of lower concentration toward an area of higher concentration—which is called *positive chemotaxis*. However, some cells move away from the source, which is called *negative chemotaxis*.

Movement of Cilia. A second type of cellular motion, *ciliary movement*, occurs along the surface of cells in the respiratory tract and in some portions of the reproductive tract. As illustrated in Figure 2–17, a cilium looks like a minute, sharp-pointed hair that projects 3 to 4 microns from the surface of the cell. Many cilia can project from a single cell.

The cilium is covered by an outcropping of the cell membrane, and it is supported by 11 microtubular filaments, 9 double tubular filaments located around the periphery of the cilium and 2 single tubular filaments down the center, as shown in the cross-section illustrated in Figure 2–17. Each cilium is an outgrowth of a structure that lies, immediately beneath the cell membrane called the *basal body* of the cilium.

In the inset of Figure 2–17 movement of the cilium is illustrated. The cilium moves forward with a sudden rapid stroke, bending sharply where it projects from the surface of the cell. Then it moves backward very slowly in a whip-like manner. The rapid forward movement pushes the fluid lying adjacent to the cell in the direction that the cilium moves, then the slow whiplike movement in the other direction has almost no effect on the fluid. As a result, fluid is continually propelled in the direction of the forward stroke. Since most ciliated cells have large numbers of cilia on their surfaces, and since ciliated cells on a surface are oriented in the same direction, this is a very satisfactory means for moving fluids from one part of the surface to another, for instance, for moving mucus out of the lungs or for moving the ovum along the fallopian tube.

MECHANISM OF CILIARY MOVEMENT. The mechanism of ciliary movement is unknown, though we do know some facts about it. First, the movement requires energy supplied by ATP. Second, an electrochemical impulse similar to that transmitted over nerves is transmitted over the membrane of the cilium to produce the movement. Third, the two filaments in the center of the cilium are necessary for movement to occur.

The filaments of the cilium are believed to be nothing more than elastic cytoskeletal mi-

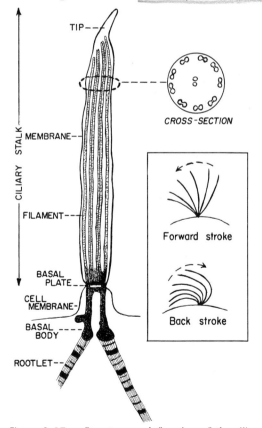

Figure 2–17 Structure and function of the cilium. (Modified from Satir, *Sci. Amer., 204*[2]:108, 1961.)

crotubules that give the cilium its physical rigidity. These filaments are not known to have the capability of contracting, even though various theories have postulated that contraction occurs. It is more likely that some cytoplasmic element surrounding the filaments causes the forward movement of the cilium, while the filaments, by virtue of their elastic rigidity, cause the slower whiplike return movement of the cilium. It is possible that the tubular structure of the filaments plays a role in transporting necessary substances into the cilium for the contractile process to take place.

REPRODUCTION OF CILIA. Cilia have the peculiar ability to reproduce themselves. This is achieved by the basal body, which is almost identical to the centriole, an important structure in the reproduction of whole cells, as we shall see in the next chapter. The basal body, as does the centriole, has the ability to reproduce itself by means not yet understood. After it reproduces itself, the new basal body then grows an additional cilium from the surface of the cell.

REFERENCES

Bourne, G. H.: Cytology and Cell Physiology. 3rd ed., New York, Academic Press, Inc., 1964.

Brachet, J., and Mirsky, A. E.: The Cell. 6 volumes. New York, Academic Press, Inc., 1961.

Brieger, E. M.: Structure and Ultrastructure of Microorganisms. Academic Press, Inc., 1963.

De Duve, C., and Baudhuin, P.: Peroxisomes (microbodies and related particles). Physiol. Rev., 46:323, 1966.

De Duve, C., and Wattiaux, R.: Functions of lysosomes. Ann. Rev. Physiol., 28:435, 1966.

De Robertis, E. D. P.; Nowinski, W. W.; and Saez, F. A.: Cell Biology. 4th ed., Philadelphia, W. B. Saunders Company, 1965.

Fawcett, D. W.: The Cell. Philadelphia, W. B. Saunders Company, 1966.

Giese, A. C.: Cell Physiology. 3rd ed., Philadelphia, W. B. Saunders Company, 1968.

Goodwin, B. C.: Temporal Organization in Cells. New York, Academic Press, Inc., 1963.

Green, D. E.: The mitochondrion. Sci. Amer., 210:63(1), 1964.

Grundmann, E.: General Cytology. Baltimore, The Williams & Wilkins Co., 1966.

Hayashi, T., and Szent-György: Molecular Architecture in Cell Physiology. Englewood Cliffs, New Jersey, Prentice-Hall, Inc., 1966.

Jahn, T. L., and Bovee, E. C.: Protoplasmic movements within cells. Physiol. Rev., 49:793, 1969.

Kinosita, H., and Murakami, A.: Control of ciliary motion. Physiol. Rev., 47:53, 1967.

Packer, L.: Cell Physiology. New York, Academic Press, Inc., 1967.

Paul, J.: Cell Biology. Stanford, California, Stanford University Press, 1964.

Porter, K. R., and Bonneville, M. A.: An Introduction to the Fine Structure of Cells and Tissues. London, Henry Kimpton, 1963.

Porter, K. R., and Franzini-Armstrong, C.: The sarcoplasmic reticulum. Sci. Amer., 212:72(3), 1965.

Racker, E.: The membrane of the mitochondrion. Sci. Amer., 218:32(2), 1968.

Rhodin, J. A. G.: Atlas of Ultrastructure. Philadelphia, W. B. Saunders Company, 1963.

Roodyn, D.: Enzyme Cytology. London, Academic Press, Inc., 1967.

Threadgold, L. T.: The Ultrastructure of the Animal Cell. Oxford, England, Pergamon Press, Inc., 1967.

Wolstenholme, G. E. W., and O'Connor, M.: Ciba Symposium on Principles of Biomolecular Organization. Boston, Little, Brown and Company, 1966.

CHAPTER 3

Genetic Control of Protein Synthesis, Cell Function, and Cell Reproduction

Almost everyone knows that the genes control heredity from parents to children, but most persons do not realize that the same genes control the reproduction and also the day-by-day function of all cells. The genes control function of the cell by determining what substances will be synthesized within the cell — what structures, what enzymes, what chemicals.

Figure 3–1 illustrates the general schema by which the genes control cellular function. The gene, which is a nucleic acid called *deoxyribonucleic acid (DNA)*, automatically controls the formation of another nucleic acid, *ribonucleic acid (RNA)*, which spreads throughout the cell and controls the formation of the different proteins. Some of these proteins are *structural proteins* which, in association with various lipids, form the structures of the various organelles that were discussed in the preceding chapter. But by far the majority of the proteins are *enzymes* that promote the different chemical reactions in the cells. For instance, enzymes promote all the oxidative reactions that supply energy to the cell, and they promote the synthesis of various chemicals such as lipids, glycogen, adenosine triphosphate, etc.

Though the number of genes in a nucleus of the human cell is unknown, it is known that the genes control the formation of at least several thousand separate types of protein essential to function of the different cells.

THE GENE

The gene is a long, double-stranded, helical molecule of *deoxyribonucleic acid (DNA)* having a molecular weight usually measured in the millions. A very short segment of such a molecule is illustrated in Figure 3–2. This molecule is composed of several simple chemical compounds arranged in a regular pattern explained in the following few paragraphs.

The Basic Building Blocks of DNA. Figure 3–3 illustrates the basic chemical compounds involved in the formation of DNA. These include *phosphoric acid,* a sugar called *deoxyribose,* and four nitrogenous bases (two purines, *adenine* and *guanine,* and two pyrimidines, *thymine* and *cytosine*). The phosphoric acid and deoxyribose form the two helical strands of DNA, and the bases lie between the strands and connect them.

Gene (DNA)

↓

RNA formation

↓

Protein formation

Cell structure   Cell enzymes

Cell function

Figure 3–1   General schema by which the genes control cell function.

Figure 3–2   The helical, double-stranded structure of the gene. The outside strands are composed of phosphoric acid and the sugar deoxyribose. The internal molecules connecting the two strands of the helix are purine and pyrimidine bases; these determine the "code" of the gene.

The Nucleotides. The first stage in the formation of DNA is the combination of one molecule of phosphoric acid, one molecule of deoxyribose, and one of the four bases to form a nucleotide. Four separate nucleotides are thus formed: *adenylic, thymidylic, guanylic,* and *cytidylic acids.* Figure 3–4A illustrates the chemical structure of adenylic acid, and Figure 3–4B illustrates simple symbols for all the four basic nucleotides that form DNA.

Note also in Figure 3–4B that the nucleotides are separated into two pairs. Adenylic acid and thymidylic acid form one pair. Guanylic acid and cytidylic acid form the other pair. The pairs of nucleotides provide the means by which the two strands of the DNA helix are bound together: one nucleotide of a pair is on one strand of DNA, the other nucleotide of that pair is in a corresponding position on the other strand, and these are bound together by loose and reversible bonds.

Organization of the Nucleotides to Form DNA. Figure 3–5 illustrates the manner in which multiple numbers of nucleotides are bound together to form DNA. Note that these are combined in such a way that phosphoric acid and

A. NUCLEOTIDE

Adenylic acid

B. THE FOUR BASIC NUCLEOTIDES OF DNA

PAIR #1

A
—P—D—
(Adenylic acid)

T
—P—D—
(Thymidylic acid)

PAIR #2

G
—P—D—
(Guanylic acid)

C
—P—D—
(Cytidylic acid)

Figure 3–4 Combinations of the basic building blocks of DNA to form nucleotides. (*A*, adenine; *C*, cytosine; *D*, deoxyribose; *G*, guanine; *P*, phosphoric acid; *T*, thymine.)

deoxyribose alternate with each other in the two separate strands, and these strands are held together by the respective pairs of bases. Thus, in Figure 3–5 the respective pairs of bases are CG, CG, GC, TA, CG, TA, GC, AT, and AT. However, the base pairs are bound together by very loose chemical bonds, represented in the figure by dashed lines. Because of the looseness of these bonds, the two strands can pull apart with ease, and they do so many times during the course of their function in the cell.

Now, to put the DNA of Figure 3–5 into its proper physical perspective, one needs merely to pick up the two ends and twist them into a helix. Ten pairs of nucleotides are present in

PHOSPHORIC ACID:

DEOXYRIBOSE:

BASES:

Adenine

Thymine

Guanine

Cytosine

PURINES

PYRIMIDINES

Figure 3–3 The basic building blocks of DNA.

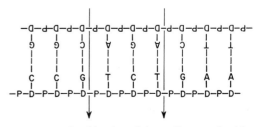

Figure 3–5 Combination of deoxyribose nucleotides to form DNA.

each full turn of the helix in the DNA molecule as illustrated in Figure 3–2.

THE GENETIC CODE

The importance of the gene lies in its ability to control the formation of other substances in the cell. It does this by means of the so-called genetic code. When the two strands of a DNA molecule are split apart, this exposes a succession of purine and pyrimidine bases projecting to the side of each strand. It is these projecting bases that form the code.

Research studies in the past few years have demonstrated that the so-called *code words* consist of "triplets" of bases—that is, each three successive bases are a code word. And the successive code words control the sequence of amino acids in a protein molecule to be synthesized in the cell. Note in Figure 3–5 that each of the two strands of the DNA molecule carries its own genetic code. For instance, the top strand, reading from left to right, has the genetic code GGC, AGA, CTT, the code words being separated from each other by the arrows. As we follow this genetic code through Figures 3–6 and 3–7, we shall see that the code words are responsible for placement of the three amino acids, *proline, serine,* and *glutamic acid,* in a molecule of protein. Furthermore, these three amino acids will be lined up in the protein molecule in exactly the same way that the genetic code is lined up in this strand of DNA.

It is also important that some code words do not cause amino acids to incorporate into proteins but, instead, perform other functions in the synthesis of protein molecules, such as initiating and stopping the formation of a protein molecule. For instance, some DNA molecules of viruses have molecular weights of 40,000,000 or more and cause the formation of many more than a single protein molecule—usually 8, 10, or 20 molecules. It is the code words that do not cause incorporation of amino acids into a protein molecule that signal when to stop the formation of one protein and to start the formation of another.

RIBONUCLEIC ACID (RNA)

Since almost all DNA is located in the nucleus of the cell and yet most of the functions of the cell are carried out in the cytoplasm, some means must be available for the genes of the nucleus to control the chemical reactions of the cytoplasm. This is done through the intermediary of another type of nucleic acid, ribonucleic acid (RNA), the formation of which is controlled by the DNA of the nucleus. The RNA is then transported into the cytoplasmic cavity where it controls protein synthesis.

At least three separate types of RNA are important to protein synthesis: *messenger RNA, transfer RNA,* and *ribosomal RNA.* Before we describe the functions of messenger and transfer RNA's in the synthesis of proteins, let us see how DNA controls the formation of RNA.

Synthesis of RNA. One strand of the DNA molecule, which constitutes the gene, acts as a template for synthesis of each type of RNA molecule. (The other strand of the DNA has no genetic function but does function for replication of the gene itself, which will be discussed later in the chapter.) The code words in DNA cause the formation of *complementary* code words (or *codons*) in RNA. The stages of RNA synthesis are as follows:

THE BASIC BUILDING BLOCKS OF RNA. The basic building blocks of RNA are almost the same as those of DNA except for two differences. First, the sugar deoxyribose is not used in the formation of RNA. In its place is another sugar of very slightly different composition, *ribose.* Referring back to Figure 3–3, the deoxyribose becomes ribose if the hydrogen radical to the right is replaced by a hydroxyl radical.

The second difference in the basic building blocks is that thymine is replaced by another pyrimidine, *uracil.* If in Figure 3–3 the methane radical of the thymine is replaced by a single hydrogen atom, the structure then becomes uracil rather than thymine.

Formation of RNA Nucleotides. The basic building blocks of RNA first form nucleotides exactly as described above for the synthesis of DNA. Here again, four separate nucleotides are used in the formation of RNA. These nucleotides contain the bases *adenine, guanine, cytosine,* and *uracil,* respectively, the uracil replacing the thymine found in the four nucleotides that make up DNA. Also, uracil takes the place of thymine in pairing with the purine adenine, as we shall see in the following paragraphs.

Activation of the Nucleotides. The next step in the synthesis of RNA is activation of the nucleotides. This occurs by addition to each nucleotide of two phosphate radicals to form triphosphates of the nucleotides. These last two phosphates are combined with the nucleotide by *high energy phosphate bonds* derived from adenosine triphosphate in the cell.

The result of this activation process is that large quantities of energy are made available to each of the nucleotides, and it is this energy that is used in promoting the chemical reactions that eventuate in the formation of the RNA chain.

Combination of the Activated Nucleotides with the DNA Strand. The next stage in the formation of RNA is the splitting apart of the two strands of the DNA molecule. Then, activated nucleotides become attached to the bases on the DNA strand that is the gene, as illustrated by the top panel in Figure 3–6. Note that a ribose nucleotide base always combines with the deoxyribose bases in the following combinations:

DNA base	*RNA base*
guanine	cytosine
cytosine	guanine
adenine	uracil
thymine	adenine

Polymerization of the RNA Chain. Once the ribose nucleotides have combined with the DNA strand as shown in the top panel of Figure 3–6, they are then lined up in proper sequence to form code words (codons) that are complementary to those in the DNA molecule. At this time an enzyme called *RNA polymerase* causes the two extra phosphates on each nucleotide to split away and, at the same time, to liberate enough energy to cause bonds to form between the successive ribose and phosphoric acid radicals. As this happens, the RNA strand automatically separates from the DNA

Figure 3–6 Combination of ribose nucleotides with a strand of DNA to form a molecule of ribonucleic acid (RNA) that carries the DNA code from the gene to the cytoplasm.

strand and becomes a free molecule of RNA. This bonding of the ribose and phosphoric acid radicals and the simultaneous splitting of the RNA from the DNA is illustrated in the lower part of Figure 3–6.

Once the RNA molecules are formed, they are transported, by means yet unknown, into all parts of the cytoplasm where they perform further functions.

MESSENGER RNA—THE PROCESS OF TRANSCRIPTION

The type of RNA that carries the genetic code to the cytoplasm for formation of proteins is called messenger RNA. Molecules of messenger RNA are usually composed of several hundred to several thousand nucleotides in a single strand containing codons that are exactly complementary to the code words of the genes. Figure 3–7 illustrates a small segment of a molecule of messenger RNA. Its codons are CCG, UCU, and GAA. These are the codons for proline, serine, and glutamic acid. If we now refer back to Figure 3–6, we see that the events recorded in this figure represent transfer of this particular genetic code from the DNA strand to the RNA strand. This process of transferring the genetic code from DNA to messenger RNA is called *transcription.*

Messenger RNA molecules are long straight strands that are suspended in the cytoplasm. These molecules migrate to the ribosomes where protein molecules are manufactured, which is explained below.

RNA Codons. Table 3–1 gives the RNA codons for the 20 common amino acids found in protein molecules. Note that several of the amino acids are represented by more than one codon; and, as was pointed out before, some codons represent such signals as "start manufacturing a protein molecule" or "stop manufacturing a protein molecule." In Table 3–1, these two codons are designated CI for "chain-initiating" and CT for "chain-terminating."

TRANSFER RNA

Another type of RNA that plays a prominent role in protein synthesis is called transfer RNA (also called soluble RNA) because it transfers amino acid molecules to protein molecules as the protein is synthesized. There is a separate type of transfer RNA for each of the 20 amino acids that are incorporated into proteins. Furthermore, each type of transfer RNA usually combines with only one type of amino

TABLE 3-1.   RNA Codons for the Different Amino Acids and for Start and Stop

Amino Acid	RNA Codons				
Alanine	GCU	GCC	GCA	GCG	
Arginine	CGU	CGC	CGA	CGG	AGA	AGG
Asparagine	AAU	AAC			
Aspartic acid	GAU	GAC			
Cysteine	UGU	UGC			
Glutamic acid	GAA	GAG			
Glutamine	CAA	CAG			
Glycine	GGU	GGC	GGA	GGG	
Histidine	CAU	CAC			
Isoleucine	AUU	AUC	AUA		
Leucine	CUU	CUC	CUA	CUG	
Lysine	AAA	AAG			
Methionine	AUG				
Phenylalanine	UUU	UUC			
Proline	CCU	CCC	CCA	CCG	
Serine	UCU	UCC	UCA	UCG	
Threonine	ACU	ACC	ACA	ACG	
Tryptophan	UGG				
Tyrosine			UAU	UAC	
Valine	GUU	GUC	GUA	GUG	
Start (CI)	AUG	GUG			
Stop (CT)	UAA	UAG	UGA		

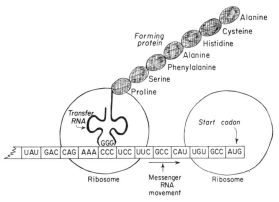

Figure 3-8 Postulated mechanism by which a protein molecule is formed in ribosomes in association with messenger RNA and soluble RNA.

acid—one and no more. Transfer RNA then acts as a *carrier* to transport its specific type of amino acid to the ribosomes where protein molecules are formed. In the ribosomes, each specific type of transfer RNA recognizes a particular code word on the messenger RNA, as is described below, and thereby delivers the appropriate amino acid to the appropriate place in the chain of the newly forming protein molecule.

Transfer RNA, containing only about 75 nucleotides, is a relatively small molecule in comparison with messenger RNA. It is a folded chain of nucleotides believed to have a cloverleaf appearance similar to that illustrated in Figure 3-8. At one end of the molecule is always an adenylic acid; it is to this that the transported amino acid attaches to a hydroxyl group of the ribose in the adenylic acid. A specific enzyme causes this attachment for each specific type of transfer RNA; this enzyme also determines the type of amino acid that will attach to the respective type of transfer RNA.

Figure 3-7   Portion of a ribonucleic acid molecule, showing three "code" words, CCG, UCU, and GAA, which represent the three amino acids *proline, serine, and glutamic acid.*

Since the function of transfer RNA is to cause attachment of a specific amino acid to a forming protein chain, it is essential that each type of transfer RNA also has specificity for a particular codon in messenger RNA. The specific prosthetic group in the transfer RNA that allows it to recognize a specific codon is called an *anticodon,* and this is believed to be located approximately in the middle of the transfer RNA molecule (at the bottom of the cloverleaf configuration illustrated in Figure 3-8). During formation of a protein molecule, the anticodon bases combine loosely by hydrogen bonding with the codon bases of the messenger RNA. In this way the respective amino acids are lined up one after another along the messenger RNA chain, thus establishing the appropriate sequence of amino acids in the protein molecule.

RIBOSOMAL RNA

The third type of RNA in the cell is that found in the ribosomes; it constitutes between 40 and 50 per cent of the ribosome. The remainder of the ribosome is protein.

The function of ribosomal RNA is mainly unknown, but it is known that it exists in particles of several different sizes. The transfer RNA, along with its attached amino acid, first binds with one of the ribosomal particles. Then as the messenger RNA passes across or through the ribosome, the amino acid is released to the forming protein while the transfer RNA is released back into the cytoplasm to combine again with another molecule of amino acid of the same type. Thus the ribosome acts as a physical structure upon which the protein is formed, and its RNA is in some way necessary for this process to take place.

FORMATION OF PROTEINS ON THE RIBOSOMES — THE PROCESS OF TRANSLATION

When a molecule of messenger RNA comes in contact with a ribosome, it travels through or across the surface of the ribosome beginning at a predetermined end specified by one of the "start" (or "chain-initiating") codons. Then, as illustrated in Figure 3–8, while the messenger RNA travels past the ribosome, a protein molecule is formed. The exact events in the formation of this protein molecule are still a mystery, though it is presumed that the ribosome reads the code of the messenger RNA in much the same way that a tape is "read" as it passes through the playback head of a tape recorder, a process called *translation*. As each codon slips past the ribosome, the specific transfer RNA corresponding to each specific codon supposedly attaches to the codon of the messenger RNA, and the ribosome in some unexplained way causes the successive amino acids to combine with each other by the process of peptide linkage. At the same time, the transfer RNA is freed from the amino acid, and returns to the cytoplasm to pick up another amino acid of its specific type.

Thus, as the messenger RNA passes through the ribosome, a protein molecule is formed as shown in Figure 3–8. Then, when a "stop" (or "chain-terminating) codon slips past the ribosome, the end of a protein molecule is signaled, and the entire molecule is freed into the cytoplasm.

Polyribosomes. A single messenger RNA molecule can form protein molecules on several different ribosomes at the same time, the molecule attaching to a successive ribosome as it leaves the first, as shown in Figure 3–8. The protein molecules obviously are in different stages of development. As a result, clusters of ribosomes frequently occur, four or five or even more ribosomes being attached together by a single messenger RNA at the same time. These clusters are called polyribosomes.

It is especially important that a messenger RNA can cause the formation of a protein molecule in any ribosome, and that there is no specificity of ribosomes for given types of protein. The ribosome seems to be simply the physical structure in which or on which the chemical reactions take place.

Chemical Steps in Protein Synthesis. Some of the chemical events that occur in synthesis of a protein molecule are illustrated in Figure 3–9. This figure shows representative reactions for three separate amino acids, AA_1, AA_2, and AA_{20}. The stages of the reactions are the following : (1) Each amino acid is *activated* by a chemical process in which ATP combines with the amino acid to form an *adenosine monophosphate complex with the amino acid.* (2) The activated amino acid, having an excess of energy, then *combines with its specific transfer RNA to form an amino acid – tRNA complex* and, at the same time, releases the adenosine monophosphate. (3) The transfer RNA carrying the amino acid complex then comes in contact with the messenger RNA molecule in the ribosome where the anticodon of the transfer RNA attaches temporarily to its specific codon of the messenger RNA thus lining up the amino acids in appropriate sequence to form a protein molecule. The energy from *guanine triphosphate,* another high energy phosphate substance almost identical with ATP, is utilized to form the chemical bonds between the successive amino acids, thus creating the protein.

PEPTIDE LINKAGE. The successive amino acids in the protein chain combine with each other according to the following typical reaction:

$$\underset{\substack{|\\R-C-C-\ OH}}{\overset{\substack{NH_2\ \ O\\|\qquad\|}}{}} + H\ \underset{\substack{|\ \ |\\-N-C-COOH}}{\overset{\substack{H\ \ R\\|\ \ |}}{}} \rightarrow$$

$$\underset{\substack{|\qquad|\ \ |\ \ |\\R-C-C-N-C-COOH}}{\overset{\substack{NH_2\ \ O\ \ H\ \ R\\|\qquad\|\ \ |\ \ |}}{}} + H_2O$$

In this chemical reaction, a hydroxyl radical is removed from the COOH portion of one amino acid while a hydrogen of the NH_2 portion of the other amino acid is removed. These combine to form water, and the two reactive sites left on the two successive amino acids combine, resulting in a single molecule. This process is called *peptide linkage.*

SYNTHESIS OF OTHER SUBSTANCES IN THE CELL

Many hundreds or perhaps a thousand or more protein enzymes formed in the manner just described control essentially all the other chemical reactions that take place in cells. These enzymes promote synthesis of lipids, glycogen, purines, pyrimidines, and hundreds of other substances. We will discuss many of these synthetic processes in relation to carbohydrate, lipid, and protein metabolism in Chapters 67 through 69. It is by means of all these different substances that the many functions of the cells are performed.

CONTROL OF GENETIC FUNCTION AND BIOCHEMICAL ACTIVITY IN CELLS

There are basically two different methods by which the biochemical activities in the cell are controlled. One of these can be called

Figure 3–9   Chemical events in the formation of a protein molecule.

genetic regulation, in which the activities of the genes themselves are controlled, and the other can be called *enzyme regulation,* in which the rates of activities of the enzymes within the cell are controlled.

GENETIC REGULATION

Figure 3–10 illustrates the general plan by which function of the genes is regulated. At the top of the figure is a *regulatory gene* that has the ability to regulate the degree of activity of other genes. It does this by controlling the formation of a *repressor substance,* a compound of small molecular weight, which in turn represses the activities of other genes. The repressor substance does not act directly on these other genes but instead on a small part of the DNA helix called an *operator,* which lies adjacent to the genes that are to be controlled. The genetic operator excites the genes, but when the operator is repressed by the repressor substance, the genes become inactive.

The genes controlled by the operator are called *structural genes;* it is they that cause the eventual formation of enzymes in the cell. The group of structural genes under the influence

of each genetic operator is known as an *operon.* In Figure 3–10, the structural genes initiate the formation of enzymes A, B, and C, which in turn control biochemical reactions in the cell leading to the formation of some specific chemical product.

As noted in Figure 3–10, the repressor substance from the regulatory gene can be either *inhibited* or *activated.* If inhibited, the genetic operator is no longer repressed and therefore

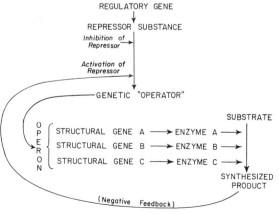

Figure 3–10   Control system that regulates function of genes. See text for detailed explanation.

becomes very active. As a result the bio-chemical synthesis proceeds unabated. On the other hand, if the repressor substance is acti-vated, it immediately represses the genetic operator, and the events leading to the synthe-sized product cease within minutes or hours.

Control of Cellular Constituents by Feedback Repression of the Genes. The importance of the genetic regulatory system of Figure 3–10 is that it allows the concentrations of different cellular substances to be controlled. The mechanism of this is the following: The syn-thesized product often can activate the repres-sor substance, and the repressor substance then inactivates the genetic operator, leading to decreased or cessation of production of the enzymes required for forming the synthesized product. Thus, negative feedback occurs in which the concentration of the synthesized product controls the product's rate of synthe-sis. When its concentration becomes too great, the rate of synthesis falls; when its concentra-tion falls too low, its rate of synthesis increases.

Genetic control systems of this type are espe-cially important in controlling the intracellular concentrations of amino acids, amino acid derivatives, and some of the intermediary substances of carbohydrate, lipid, and protein metabolism.

Induction of Enzyme Formation. Note in Figure 3–10 that the repressor substance can be inhibited as well as activated. When it is inhibited, the repressor no longer holds the operon in abeyance, and the structural genes operate to their full capacity in forming en-zymes. Therefore, a simple way to "induce" the formation of a particular group of enzymes is to inhibit the repressor. This is probably one of the means by which hormones act to in-crease or decrease specific cellular activities. For instance, the steroid hormone, aldoster-one, increases the quantities of enzymes required for transport of sodium through membranes, and the steroid sex hormones in-duce the formation of enzymes required for protein synthesis in the sex organs. These effects will be discussed later at greater length in relation to the functions of these hormones.

Induction also occurs in response to sub-stances inside the cell itself. For instance, cells grown in tissue culture in the absence of lac-tose do not have the ability to utilize lactose when first exposed to it. However, within a few hours after exposure, because of inhibition of the appropriate repressor substance, the en-zyme required for splitting lactose into its monosaccharides appears in the cell. Thus, lactose induces a specific enzyme synthesis and, in so doing, makes it possible for the cell to utilize the lactose.

CONTROL OF ENZYME ACTIVITY

In the same way that inhibitors and acti-vators can affect the genetic regulatory system, so also can the enzymes themselves be directly controlled by other inhibitors or activators. This, then, represents a second mechanism by which cellular biochemical functions can be controlled.

Enzyme Inhibition. A great many of the chemical substances formed in the cell have a direct feedback effect on the respective enzyme systems that synthesize them. Thus, in Figure 3–10, the synthesized product is shown to act directly back on enzyme A to inactivate it. If enzyme A is inactivated, then none of the sub-strate will begin to be converted into the synthesized product. Almost always in enzyme inhibition, the synthesized product acts on the first enzyme in a sequence, rather than on the subsequent enzymes. One can readily recog-nize the importance of inactivating this first enzyme: it prevents buildup of intermediary products that will not be utilized.

This process of enzyme inhibition is another example of negative feedback control; it is responsible for controlling the intracellular concentrations of some of the amino acids not controlled by the genetic mechanism, of the purines, the pyrimidines, vitamins, and other substances.

Enzyme Activation. Enzymes that are either normally inactive or that have been inacti-vated by some inhibitor substance can often be activated. An example of this is the action of cyclic adenosine monophosphate (AMP) in causing glycogen to split so that the released glucose molecules can be used to form high energy ATP, as discussed in the previous chapter. When most of the adenosine triphos-phate has been depleted in the cell, a consider-able amount of cyclic AMP begins to be formed as a breakdown product of the ATP; the pres-ence of cyclic AMP in the cell indicates that the reserves of ATP have approached a low ebb. The cyclic AMP immediately activates the glycogen-splitting enzyme phosphorylase, lib-erating glucose molecules that are rapidly used for replenishment of the ATP stores. Thus, in this case the cyclic AMP acts as an enzyme activator and thereby helps to control intra-cellular ATP concentration.

Another interesting instance of both enzyme

inhibition and enzyme activation occurs in the formation of the purines and the pyrimidines. These substances are needed by the cell in approximately equal quantities for formation of DNA and RNA. When purines are formed, they inhibit the enzymes that are required for formation of additional purines, but they activate the enzymes for formation of the pyrimidines. Conversely, the pyrimidines inhibit their own enzymes but activate the purine enzymes. In this way there is continual cross-feed between the synthesizing systems for these two substances, resulting in almost exactly equal amounts of the two substances in the cells at all times.

In summary, there are two different methods by which the cells control proper proportions and proper quantities of different cellular constituents: the mechanism of genetic regulation and the mechanism of enzyme regulation. The genes can be either activated or inhibited, and, likewise, the enzymes can be either activated or inhibited. Furthermore, it is principally the substances formed in the cells that cause activation or inhibition; but on occasion, substances from without the cell (especially some of the hormones which will be discussed later in this text) also control the intracellular biochemical reactions.

CELL REPRODUCTION

Most cells are continually growing and reproducing. The new cells take the place of the old ones that die, thus maintaining a complete complement of cells in the body.

Also, one can remove most types of cells from the human body and grow them in tissue culture where they will continue to grow and reproduce so long as appropriate nutrients are supplied and so long as the end-products of the cells' metabolism are not allowed to accumulate in the nutrient medium. Thus, the life lineage of cells is indefinite.

As is true of almost all other events in the cell, reproduction also begins in the nucleus itself. The first step is *replication (duplication) of all genes and of all chromosomes*. The next step is division of the two sets of genes between two separate nuclei. And the final step is splitting of the cell itself to form two new daughter cells, a process called *mitosis*.

The complete life cycle of a cell that is not inhibited in some way is about 10 to 30 hours from reproduction to reproduction, and the period of mitosis lasts for approximately one-half hour. The period between mitoses is called *interphase*. However, in the body there are almost always inhibitory controls that slow or stop the uninhibited life cycle of the cell and give cells life cycle periods that vary from as little as 10 hours for stimulated bone marrow cells to an entire lifetime of the human body for nerve cells.

REPLICATION OF THE GENES

The genes are reproduced several hours before mitosis takes place, and the duration of gene replication is only about one hour. When replication begins, it occurs for all genes at the same time and not for only part of them. Furthermore, the genes are duplicated only once. The net result is two exact duplicates of all genes, which respectively become the genes in the two new daughter cells that will be formed at mitosis. Following replication of the genes, the nucleus continues to function normally for several hours before mitosis begins abruptly.

Chemical and Physical Events. The genes are duplicated in almost exactly the same way that RNA is formed from DNA. First, the two strands of the DNA helix of the gene pull apart. Second, each of these strands combines with deoxyribose nucleotides of the four types described early in the chapter as the basic building blocks of DNA. Each of the bases on each strand of DNA in the chain attracts a nucleotide containing the appropriate *complementary* base. In this way, the appropriate nucleotides are lined up side by side. Third, appropriate enzyme mechanisms then provide energy and cause polymerization of the nucleotides to form a new DNA strand. The only difference between this formation of the new strand of DNA and the formation of an RNA strand is that the new strand of DNA remains attached to the old strand that has formed it, thus forming a new double-stranded DNA helix. At the same time the other strand of the original helix forms its complementary DNA strand and thereby also forms a new double-stranded helix.

THE CHROMOSOMES AND THEIR REPLICATION

Unfortunately, we know much less about replication of the chromosomes than about replication of the genes themselves. The primary reason for this is that the structure of the chromosomes itself is almost unknown.

The chromosomes consist of two major parts: the genes, consisting of DNA, and protein. The DNA is bound loosely with the protein and sometimes during the life cycle of the cell seems to become separated from the protein. The combination of the two is known as a *nucleoprotein*.

Recent experiments indicate that all the DNA of a particular chromosome is arranged in one long double helix and that the genes are attached end-on-end with each other. Such a molecule, if spread out linearly, would be approximately 2 cm. long, or several hundred times as long as the diameter of the nucleus itself; but the experiments also indicate that this long double helix is folded like a spring and is held in this folded position by its linkages to protein molecules. The protein has nothing to do with the genetic potency of the chromosome, and the protein molecules can be replaced by new protein molecules without any alteration in the functions of the genes.

Replication of the chromosomes follows as a natural result of replication of the DNA strand. When the new double helix separates from the original double helix, it presumably carries some of the old protein with it or combines with new protein, the DNA acting as the backbone of the newly replicated chromosome and the protein acting only as an accessory to the chromosomal structure.

Number of Chromosomes in the Human Cell. Each human cell contains 46 chromosomes arranged in 23 pairs. In general, the genes in the two chromosomes of each pair are almost identical with each other, so that it is usually stated that the different genes exist in pairs, though occasionally this is not the case. Also, most genes have several duplicates even on the same chromosome, so that loss of any one usually will not completely remove a particular characteristic from a cell.

MITOSIS

The actual process by which the cell splits into two new cells is called mitosis. Once the genes have been duplicated and each chromosome has split to form two new chromosomes, each of which is now called a *chromatid*, mitosis follows automatically, almost without fail, within a few hours.

Replication of the Centrioles. The first event of mitosis takes place in the cytoplasm, occurring during the latter part of interphase in the small structures called *centrioles*. As illustrated in Figure 3–11, two pairs of centrioles lie close to each other near one pole of the nucleus. Each centriole is a small cylindrical body about 0.4 micron long and about 0.15 micron in diameter, consisting mainly of nine parallel, tubular-like structures arranged around the inner wall of the cylinder. The two centrioles of each pair lie at right angles to each other.

In so far as is known, the two pairs of centrioles remain dormant during interphase until shortly before mitosis is to take place. At that time, the two pairs begin to move apart from each other. This is caused by protein microtubules growing between the respective pairs and actually pushing them apart. At the same time, microtubules grow radially away from each of the pairs. Some of these penetrate the nucleus. The set of microtubules connecting the two centriole pairs is called the *spindle*, and the entire set of microtubules plus the two centrioles is called the *mitotic apparatus*.

While the spindle is forming, the *chromatin material* of the nucleus (the DNA) becomes condensed into well-defined but disoriented chromosomes, as shown in Figures 3–11A and B.

Prophase. The first stage of mitosis, called prophase, is shown in Figures 3–11C and D. The nuclear envelope dissolutes, and some of the microtubules from the forming mitotic apparatus become

Figure 3–11 Stages in the reproduction of the cell. A and B, late interphase; C and D, prophase; E, metaphase; F, anaphase; G and H, telophase. (Redrawn from Mazia: *Sci. Amer.*, 205[3]:102, 1961.)

attached to the chromosomes. This attachment always occurs at the same point on each chromosome, at a small condensed portion called the *centromere*.

Metaphase. During metaphase (Fig. 3–11E) the centriole pairs are pushed far apart by the growing spindle, and the chromosomes are thereby pulled tightly by the attached microtubules to the very center of the cell, lining up in the equatorial plane of the mitotic spindle.

Anaphase. With still further growth of the spindle, each pair of chromosomes is now broken apart, a stage of mitosis called anaphase (Fig. 3–11F). A microtubule connecting with one pair of centrioles pulls one chromatid and a microtubule connecting with the other centriole pair pulls the opposite chromatid. Thus, all 46 pairs of chromatids are separated, forming 46 daughter chromosomes that are pulled toward one mitotic spindle and another 46 *complementary* chromosomes that are pulled toward the other mitotic spindle.

Telophase. In telophase (Figs. 3–11G and H) the mitotic spindle grows still longer, pulling the two sets of daughter chromosomes completely apart. Then a new nuclear membrane develops around each set of chromosomes, this membrane perhaps being formed from portions of the endoplasmic reticulum that are already present in the cytoplasm. Simultaneously, the mitotic apparatus undergoes dissolution, and the cell pinches in two midway between the two nuclei, for reasons totally unexplained at present.

Note, also, that each of the two pairs of centrioles

are replicated during telophase, the mechanism of which is not understood. These new pairs of centrioles remain dormant through the next interphase until a mitotic apparatus is required for the next cell division.

CONTROL OF CELL GROWTH AND REPRODUCTION

Cell growth and reproduction usually go together, growth occurring simultaneously with replication of the genes of the nucleus, followed a few hours later by mitosis.

In the normal human body, regulation of cell growth and reproduction is mainly a mystery. We know that certain cells grow and reproduce all the time, such as the blood-forming cells of the bone marrow, the germinal layers of the skin, and the epithelium of the gut. However, many other cells, such as some muscle cells, do not reproduce for many years. And a few cells, such as the neurons, do not reproduce during the entire life of the person.

Thus, most cells of the human body do have the ability to reproduce continually, though the rate of reproduction usually remains greatly suppressed. Yet if there is an insufficiency of a given type of cell in the body, this type of cell will grow and reproduce very rapidly until appropriate numbers of them are again available. For instance, seven-eighths of the liver can be removed surgically, and the cells of the remaining one-eighth will grow and divide until the liver mass returns almost to normal. The same effect occurs for almost all glandular cells, for cells of the bone marrow, the subcutaneous tissue, the intestinal epithelium, and almost any other tissue except highly differentiated cells, such as nerve cells.

We know very little about the mechanisms that maintain proper numbers of the different types of cells in the body. It is assumed that control substances are secreted by the different cells and cause feedback effects to stop or slow their growth when too many of them have been formed, though only a few such substances have been found. We know that cells of any type removed from the body and grown in tissue culture can grow and reproduce rapidly and indefinitely if the medium in which they grow is continually replenished. Yet they will stop growing when even small amounts of their own secretions are allowed to collect in the medium, which supports the idea that control substances limit cellular growth.

Regulation of Cell Size. Cell size is determined almost entirely by the amount of DNA in the nucleus. If replication of the genes does not occur, the cell grows to a certain size and thereafter remains at that size. On the other hand, it is possible, by use of the chemical *colchicine*, to prevent mitosis even though replication of the genes continues. In this event, the nucleus then contains far greater quantities of DNA than normally, and the cell grows proportionately larger. It is assumed that this results simply from increased production of RNA and cell proteins, which in turn cause the cell to grow larger.

CANCER

Cancer can occur in any tissue of the body. It results from a change in certain cells that allows them to disrespect normal growth limits, no longer obeying the feedback controls that normally stop cellular growth and reproduction after a given number of such cells have developed. As pointed out above, even normal cells when removed from the body and grown in tissue culture can grow and proliferate indefinitely if the growth medium is continually changed. Therefore, in tissue culture, normal tissue cells behave exactly as cancer cells, but in the body, normal tissue cells behave differently, for they are subject to limits, whereas cancer cells are not.

What is the difference between the cancer cell and the normal tissue cell that allows the cancer cell to grow and reproduce unabated? The answer to this question is not known, but researchers have found the genetic make-up of some cancer cells to be different from those of normal cells. This has led to the idea that cancer usually results from mutation of part of the genetic system in the nucleus (or from a change of this genetic system induced by the presence of a virus in the cell). The mutated or otherwise changed "genome" eliminates the feedback mechanisms that normally limit growth and reproduction of the cell. Once even a single such cell is formed, it obviously will grow and proliferate indefinitely, its number increasing exponentially.

Cancerous tissue competes with normal tissues for nutrients, and because cancer cells continue to proliferate indefinitely, their number multiplying day-by-day, one can readily understand that the cancer cells will soon demand essentially all the nutrition available to the body. As a result, the normal tissues gradually suffer nutritive death.

CELL DIFFERENTIATION

A special characteristic of cell growth and cell division is *cell differentiation*, which means changes in physical and functional properties of cells as they proliferate in the embryo to form the different bodily structures. However, our problem is not to describe the stages of differentiation but simply to discuss the theories concerning the means by which the cells change their characteristics to form all the different organs and tissues of the body.

The earliest and simplest theory for explaining differentiation was that the genetic composition of the nucleus undergoes changes during successive generations of cells in such a way that one daughter cell develops one set of genetic characteristics while another daughter cell develops entirely different genetic characteristics.

However, this theory has now been almost completely disproved by the following simple experiment. The nucleus from an intestinal cell of a frog, when surgically implanted into a frog ovum from which the original nucleus has been removed, will often cause the formation of a completely normal frog. This demonstrates that even the intestinal cell, which is a reasonably well differentiated cell, still carries all the necessary genetic information for development of all structures required in the frog's body.

Therefore, the present idea is that instead of loss of genetic material during the process of differentiation, there occurs selective repression of different genetic operons. This presumably results from the buildup of different repressor substances in the cytoplasm, the repressor substances in one cell acting to repress one genetic characteristic and the repressor substances in another cell acting on a different set of genetic characteristics.

Embryological experiments show also that certain cells in an embryo control the differentiation of adjacent cells. For instance, the *primordial chordamesoderm* is called the *primary organizer* of the embryo because it forms a focus around which the rest of the embryo develops. It differentiates into a *mesodermal axis* containing segmentally arranged *somites* and, as a result of complex *inductions* in the surrounding tissues, causes formation of essentially all the organs of the body.

Another instance of induction occurs when the developing eye vesicles come in contact with the ectoderm of the head and cause it to thicken into a lens plate that folds inward to form the lens of the eye. It is possible that the entire embryo develops as a result of complex inductions, one part of the body affecting another part, and this part affecting still other parts.

Thus, our understanding of cell differentiation is still hazy. We know many different control mechanisms by which differentiations *could* occur. Yet the overall basic controlling factors in cell differentiation are yet to be discovered.

REFERENCES

Allen, J. M.: The Molecular Control of Cellular Activity. New York, McGraw-Hill Book Company, 1962.

Allen, R. D.: Amoeboid motion. *Sci. Amer., 236*(2):112, 1962.

Brenner, S.: Theories of gene regulation. *Brit. Med. Bull., 21*:244, 1965.

Ciba Foundation Symposium: Cell Differentiation. Boston, Little, Brown and Company, 1968.

Clark, B. F. C., and Marcker, K. A.: How proteins start. *Sci. Amer., 218*:36(1), 1968.

Crick, F. H. C.: The genetic code. *Sci. Amer., 215*:55(4), 1966.

Eckhart, W.: Transformation of animal cells by oncogenic DNA viruses. *Physiol. Rev., 48*:513, 1968.

Fitzgerald, P. H.: Chromosomal abnormalities in man. *In* Bittar, E. Edward, and Bittar, Neville (eds.): The Biological Basis of Medicine. New York, Academic Press, Inc., 1969, Vol. 4, p. 133.

Gaze, R. M.: Growth and differentiation. *Ann. Rev. Physiol., 29*:59, 1967.

Gross, P. R.: Control of protein synthesis. *Proc. Int. Union of Physiol. Sciences, 6*:158, 1968.

Gurdon, J. B.: Transplanted nuclei and cell differentiation. *Sci. Amer., 219*:24(6), 1968.

Harris, R. J. C. (ed.): Cell Movement and Cell Contact. *Exp. Cell Res.*, Suppl. 8, 1961.

Harris, R. J. C. (ed.): Symposia of the International Society for Cell Biology: Cell Growth and Cell Division. New York, Academic Press, Inc., 1963, Vol. 2.

Hayflick, L.: Human cells and aging. *Sci. Amer., 218*:32(3), 1968.

Holwill, M. E. J.: Physical aspects of flagellar movement. *Physiol. Rev., 46*:696, 1966.

Levine, L.: The Cell in Mitosis. New York, Academic Press, Inc., 1963.

Marifield, R. B.: The automatic synthesis of proteins. *Sci. Amer., 218*:56(3), 1968.

Michelson, A. M.: The Chemistry of Nucleosides and Nucleotides. New York, Academic Press, Inc., 1963.

Moyed, H. S., and Umbarger, H. E.: Regulation of biosynthetic pathways. *Physiol. Rev., 42*:444, 1962.

New York Heart Association: Differentiation and Development. Little, Brown and Company, 1964.

Post, J., and Hoffman, J.: Cell renewal patterns. *New Eng. J. Med., 279*:248, 1968.

Siddiqi, O.: The mechanism of gene action and regulation. *In* Bittar, E. Edward, and Bittar, Neville (eds.): The Biological Basis of Medicine. New York, Academic Press, Inc., 1969, Vol. 4, p. 43.

Sirlin, J. L.: The cell nucleus as the centre of RNA synthesis. *In* Bittar, E. Edward, and Bittar, Neville (eds.): The Biological Basis of Medicine. New York, Academic Press, Inc., 1968, Vol. 1, p. 283.

Skoglund, S.: Growth and differentiation, with special emphasis on the central nervous system. *Ann. Rev. Physiol., 31*:19, 1969.

Smellie, R. M. S.: The biosynthesis and function of nucleic acids. *In* Bittar, E. Edward, and Bittar, Neville (eds.): The Biological Basis of Medicine. New York, Academic Press, Inc., Vol. 1, p. 243.

Starr, J. L., and Sells, B. H.: Methylated ribonucleic acids. *Physiol. Rev., 49*:623, 1969.

Synthesis and Structure of Macromolecules. *Cold Spr. Harb. Symp. Quant. Biol., XXVIII*, 1964.

Taylor, J. H.: Molecular Genetics, Part I. New York, Academic Press, Inc., 1963.

Watson, J. D.: Molecular Biology of the Gene. New York, Benjamin Company, Inc., 1965.

CHAPTER 4

Transport Through the Cell Membrane

The fluid inside the cells of the body, called *intracellular fluid*, is very different from that outside the cells, called *extracellular fluid*. The extracellular fluid circulates in the spaces between the cells and also mixes freely with the fluid of the blood through the capillary walls. Thus, it is the extracellular fluid that supplies the cells with nutrients and other substances needed for cellular function. But before the cell can utilize these substances, they must be transported through the cell membrane.

Note in Figure 4–1 that the extracellular fluid contains large quantities of *sodium* but only small quantities of *potassium*. Exactly the opposite is true of the intracellular fluid. Also, the extracellular fluid contains large quantities of chloride, while the intracellular fluid contains very little. On the other hand, the concentration of phosphates in the intracellular fluid is considerably greater than that in the extracellular fluid. All these differences between the components of the intracellular and extracellular fluids are extremely important to the life of the cell. It is the purpose of this chapter to explain how these differences are brought about by the transport mechanisms in the cell membrane.

Substances are transported through the cell membrane by two major processes, *diffusion* and *active transport*. Though there are many different variations of these two basic mechanisms, as we shall see later in this chapter, basically, diffusion means movement of substances in a random fashion caused by the normal kinetic motion of matter, whereas active transport means movement of substances as a result of chemical processes that impart energy to cause the movement.

DIFFUSION

All molecules and ions in the body fluids, including both water molecules and dissolved substances, are in constant motion, each particle moving its own separate way. Motion of these particles is what physicists call heat—the greater the motion, the higher is the temperature—and motion never ceases under any conditions except absolute zero temperature. When a moving particle, A, bounces against a stationary particle, B, its electrostatic forces repel particle B, momentarily adding some of its energy of motion to particle B. Consequently, particle B gains kinetic energy of motion while particle A slows down, losing some of its kinetic energy. Thus, as shown in Figure 4–2, a single molecule in solution bounces among the other molecules first in one direction, then another, then another, and so forth, bouncing hundreds to millions of times each second. At times it travels a far distance before striking the next molecule, but at other times only a short distance.

This continual movement of molecules among each other in liquids, or in gases, is

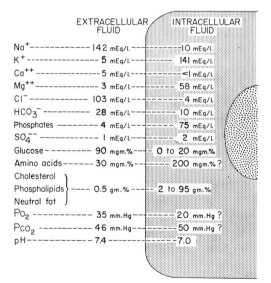

	EXTRACELLULAR FLUID	INTRACELLULAR FLUID
Na^+	142 mEq/l.	10 mEq/l.
K^+	5 mEq/l.	141 mEq/l.
Ca^{++}	5 mEq/l.	<1 mEq/l
Mg^{++}	3 mEq/l	58 mEq/l.
Cl^-	103 mEq/l.	4 mEq/l.
HCO_3^-	28 mEq/l.	10 mEq/l.
Phosphates	4 mEq/l.	75 mEq/l.
SO_4^{--}	1 mEq/l.	2 mEq/l
Glucose	90 mgm.%	0 to 20 mgm.%
Amino acids	30 mgm.%	200 mgm.% ?
Cholesterol Phospholipids Neutral fat	0.5 gm.%	2 to 95 gm.%
Po_2	35 mm.Hg	20 mm.Hg ?
Pco_2	46 mm.Hg	50 mm.Hg ?
pH	7.4	7.0

Figure 4–1 Chemical compositions of extracellular and intracellular fluids.

called *diffusion*. Ions diffuse in exactly the same manner as whole molecules, and even suspended colloid particles diffuse in a similar manner, except that because of their very large sizes they diffuse far less rapidly than molecular substances.

KINETICS OF DIFFUSION—THE CONCENTRATION DIFFERENCE

When a large amount of dissolved substance is placed in a solvent at one end of a chamber, it immediately begins to diffuse toward the opposite end of the chamber. If the same amount of substance is placed in the opposite end of the chamber, it begins to diffuse toward the first end, the same amount diffusing in each direction. As a result, the *net rate of diffusion* from one end to the other is zero. If, however, the concentration of the substances is greater at one end of the chamber than at the other end, the net rate of diffusion from the area of high concentration to low concentration is directly proportional to the larger concentration minus the lower concentration. This concentration change along the axis of the chamber is called a *concentration difference* or a *diffusion gradient*.

The rapidity with which a molecule diffuses from one point to another is less the greater the molecular size, because large particles are not so easily impelled by collisions with other molecules. In gases, the rate of diffusion is approximately inversely proportional to the square root of the molecular weight, but in fluids it is more nearly inversely proportional to the molecular weight itself.

If we consider all the different factors that affect the rate of diffusion of a substance from one area to another, they are the following: (1) The greater the concentration difference, the greater is the rate of diffusion. (2) The less the molecular weight, the greater is the rate of diffusion. (3) The shorter the distance, the greater is the rate. (4) The greater the cross-section of the chamber in which diffusion is taking place, the greater is the rate of diffusion. (5) The greater the temperature, the greater is the molecular motion and also the greater is the rate of diffusion. All these can be placed in an approximate formula, found at the bottom of this page, for diffusion in solutions.

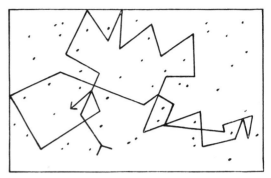

Figure 4–2   Diffusion of a fluid molecule during a fraction of a second.

DIFFUSION THROUGH THE CELL MEMBRANE

The cell membrane is essentially a sheet of lipid material, called the *lipid matrix*, covered on each surface by a layer of protein, the detailed structure of which was discussed in Chapter 2 and shown in the diagram of Figure 2–3. The fluids on each side of the membrane are believed to penetrate the protein portion of the membrane with ease so that any dissolved substances can diffuse into this portion of the cell membrane without any impedient. However, the lipid portion of the membrane is an entirely different type of fluid medium, acting as a limiting membrane between the extracellular and intracellular fluids.

Two different methods by which the substances can diffuse through the membrane are: (a) by becoming dissolved in the lipid and diffusing through it in the same way that diffusion occurs in water, and (b) by diffusing through minute pores that pass directly through the membrane at wide intervals over its surface.

Diffusion in the Dissolved State Through the Lipid Portion of the Membrane.   A few substances are soluble in the lipid of the cell membrane as well as in water. These include oxygen, carbon dioxide, alcohol, fatty acids, and a few other less important substances. When one of these comes in contact with the membrane it immediately becomes dissolved in the lipid and continues to diffuse in exactly the same manner that it diffuses within the watery medium on either side of the membrane. In other words, the molecule continues its random motion within the substance of the membrane in

$$\text{Diffusion rate} \propto \frac{\text{Concentration difference} \times \text{Cross-sectional area} \times \text{Temperature}}{\text{Molecular weight} \times \text{Distance}}$$

exactly the same way that it undergoes random motion in the surrounding fluids.

EFFECT OF LIPID SOLUBILITY ON DIFFUSION THROUGH THE LIPID MATRIX. The primary factor that determines how rapidly a substance can diffuse through the lipid matrix of a cell membrane is its solubility in lipids. If it is very soluble, it becomes dissolved in the membrane very easily and therefore passes on through. On the other hand, a substance that dissolves very poorly in lipids will be greatly retarded. For instance, oxygen, which is very soluble, passes through the lipid matrix with ease, as shown in Figure 4–3, whereas water, which is almost completely insoluble in lipids, passes through the lipid matrix almost not at all.

FACILITATED DIFFUSION. Some substances are very insoluble in lipids and yet can still pass through the lipid matrix by a process called facilitated diffusion. This is the means by which different sugars in particular cross the membrane. The most important of these sugars is glucose, the membrane transport of which is illustrated in the lower part of Figure 4–3. This shows that glucose Gl combines with a *carrier* substance C at point 1 to form the compound CG1. This combination is soluble in the lipid so that it can diffuse (or simply move by rotation of the large carrier molecule) to the other side of the membrane, where the glucose breaks away from the carrier (point 2) and passes to the inside of the cell, while the carrier moves back to the outside surface of the membrane to pick up still more glucose and transport it also to the inside. Thus, the effect of the carrier is to make the glucose soluble in the membrane; without it, glucose cannot pass through the membrane.

The rate at which a substance passes through a membrane by facilitated diffusion depends on the difference in concentration of the substance on the two sides of the membrane, the amount of carrier available, and the rapidity with which the chemical reactions can take place. In the case of glucose transport, the overall rate is greatly increased by insulin, which is the primary hormone secreted by the pancreas. Large quantities of insulin can increase the rate of glucose transport about seven- to ten-fold, though it is not known whether this is caused by an effect of insulin to increase the quantity of carrier in the membrane or to increase the rate at which the chemical reactions take place between glucose and the carrier.

Very little is known about the carrier substance for transport of glucose through the membrane. It binds specifically with certain types of monosaccharides, but other monosaccharides are not transported at all or only slightly. A protein having a molecular weight of approximately 45,000 and also having the property of combining specifically and reversibly with the same types of monosaccharides that are transported has been discovered in the cell membrane. It is likely that this protein is the carrier substance for transport of glucose.

Many of the dynamics of facilitated diffusion are similar to those of active transport, which is discussed in much greater detail later in the chapter. The primary difference between the two is that active transport can move substances through a membrane from a low concentration on one side to a high concentration on the other side, whereas in facilitated diffusion, the substance can move only from a high concentration toward a low concentration.

Diffusion Through the Membrane Pores. Some substances, such as water and many of the dissolved ions, seem to go through holes in the cell membrane called *membrane pores*. The nature of these pores is unknown. They might be round holes, small slits in the cell membrane, or possibly some large hydrophilic molecule penetrating all the way through the cell membrane and providing a pathway for movement of water-soluble substances along the axis of the molecule. At any rate, these so-called "pores" in the cell membrane behave as if they were minute round holes approximately 8 Å in diameter and as if the total area of the pores should equal approximately 1/1600 of the total surface area of the cell. Despite this very minute total area of the pores, molecules and ions diffuse so rapidly that the entire volume of fluid in a cell can easily pass through the pores within a few hundredths of a second.

Figure 4–4 illustrates a postulated structure of a pore, indicating that its surface is probably

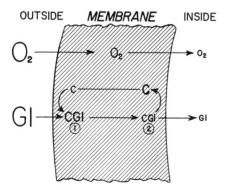

Figure 4–3 Diffusion of substances through the lipid matrix of the membrane. The upper part of the figure shows *free diffusion* of oxygen through the membrane and the lower part shows *facilitated diffusion* of glucose.

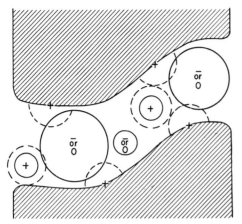

Figure 4–4   Postulated structure of the pore in the mammalian red cell membrane, showing the sphere of influence exerted by charges along the surface of the pore. (Modified from Solomon: *Sci. Amer., 203*[6]:146, 1960.)

TABLE 4–1.   Relationship of Effective Diameters of Different Substances to Pore Diameter*

Substance	Diameter Å	Ratio to Pore Diameter	Approximate Relative Diffusion Rate
Water molecule	3	0.38	50,000,000
Urea molecule	3.6	0.45	40,000,000
Hydrated chloride ion			
(red cell)	3.86	0.48	36,000,000
(nerve membrane)	—	—	200
Hydrated potassium ion	3.96	0.49	100
Hydrated sodium ion	5.12	0.64	1
Lactate ion	5.2	0.65	?
Glycerol molecule	6.2	0.77	?
Ribose molecule	7.4	0.93	?
Pore size	8 (Ave.)	1.00	—
Galactose	8.4	1.03	?
Glucose	8.6	1.04	0.4
Mannitol	8.6	1.04	?
Sucrose	10.4	1.30	?
Lactose	10.8	1.35	?

* These data have been gathered from different sources but relate primarily to the red cell membrane. Other cell membranes have different characteristics.

lined with positively charged prosthetic groups. This figure shows several small particles passing through the pore and also shows that the maximum size of the particle that can pass through is approximately equal to the size of the pore itself.

EFFECT OF PORE SIZE ON DIFFUSION THROUGH THE PORE.   Table 4–1 gives the effective diameters of various substances in comparison with the diameter of the pore. Note that some substances, such as the water molecule, urea molecule, and chloride ion, are considerably smaller in size than the pore. All these pass through the pore with great ease. For instance, the rate per second of diffusion of water in each direction through the pores of a cell is about one hundred times as great as the volume of the cell itself. It is fortunate that an identical amount of water normally diffuses in each direction, which keeps the cell from either swelling or shrinking despite the rapid rate of diffusion. The rates of diffusion of urea and chloride ions through the membrane are somewhat less than that of water, which is in keeping with the fact that their effective diameters are slightly greater than that of water.

Table 4–1 also shows that most of the sugars, including glucose, have effective diameters that are slightly greater than that of the pores. Obviously, not even a single molecule as large as these could go through a pore that is smaller than its size. For this reason essentially none of the sugars can pass through the pores; instead most of these pass through the lipid matrix by the process of facilitated diffusion.

EFFECTS OF ELECTRICAL CHARGE ON TRANSPORT OF IONS THROUGH THE MEMBRANE.   Positively charged ions, such as sodium and potassium ions, pass through the cell membrane with extreme difficulty, as shown in Table 4–1. The reason for this is believed to be the presence of positive charges of proteins or adsorbed positive ions, such as calcium ions, lining the pores. Figure 4–4 illustrates that each positive charge causes a sphere of electrostatic space charge to protrude into the lumen of the pore. A positive ion attempting to pass through a pore also exerts a sphere of positive electrostatic charge so that the two positive charges repel each other. This repulsion, therefore, blocks or greatly impedes movement of the positive ion through the pore.

In contrast to the positive ions, negatively charged ions pass through mammalian cell membrane pores much more easily. Thus, for the nerve membrane, chloride ions permeate the membrane about two times as easily as potassium ions and 100 to 200 times as easily as sodium ions. Moreover, for the red cell membrane, chloride ions permeate the membrane still another half million times as easily. It is believed that these differences are caused by lack of, or few numbers of, negative charges lining the pores in contrast to the large numbers of positive charges.

In addition to electrical charge, the *effective diameter* of an ion also determines how easily it can pass through a membrane. For instance, the hydrated diameter of the sodium ion is approximately 30 per cent greater than that of the potassium ion, and because of this slight difference the sodium ion diffuses through the usual pore only $1/100$ as easily as the potassium ion.

EFFECT OF DIFFERENT FACTORS ON PORE PERMEABILITY. Pore permeability does not always remain exactly the same under different conditions. For instance, excess *calcium* in the extracellular fluid causes the permeability to decrease, and diminished calcium causes considerably increased permeability. This is extremely important in the function of nerves, for the enlarged pores that occur in extracellular fluid calcium deficiency cause excessive diffusion of ions, which results in spurious nerve discharge throughout the body.

Another factor that has an important effect on pore permeability of many cells is *antidiuretic hormone,* which is secreted in the hypothalamus. This hormone has an especially important effect on the membranes of the cells lining the kidney tubules. Increased quantities of the hormone increase the pore diameter, which allows water and other substances to diffuse out of the tubules and back into the blood with ease.

NET DIFFUSION THROUGH THE CELL MEMBRANE AND FACTORS THAT AFFECT IT

From the preceding discussion, it is evident that many different substances can diffuse either through the lipid matrix of the cell membrane or through the pores. It should be noted, however, that substances that diffuse in one direction can also diffuse in the opposite direction. It is not the total quantity of substances diffusing in both directions through the cell membrane that is important to the cell but instead the *net quantity* diffusing either into or out of the cell.

In addition to the permeability of the membrane, which has already been discussed, three other factors determine the rate of net diffusion of a substance: the concentration difference of the substance across the membrane, the electrical potential difference across the membrane, and the pressure difference across the membrane.

Effect of a Concentration Difference. Figure 4–5A illustrates a membrane with a substance in high concentration on the outside and low concentration on the inside. The rate at which the substance diffuses *inward* is proportional to the concentration of molecules on the outside, for this concentration determines how many of the molecules strike the pore each second. On the other hand, the rate at which the molecules

A. CONCENTRATION DIFFERENCE

B. ELECTRICAL POTENTIAL DIFFERENCE

C. PRESSURE DIFFERENCE

Figure 4–5 Effect of (A) concentration difference, (B) electrical difference, and (C) pressure difference on diffusion of molecules and ions through a cell membrane.

diffuse *outward* is proportional to their concentration inside the membrane. Obviously, therefore, the rate of net diffusion is proportional to the concentration on the outside *minus* the concentration on the inside, or

$$\text{Net diffusion} \propto P(C_1 - C_2)$$

in which C_1, is the concentration on the outside, C_2 is the concentration on the inside, and P is the permeability of the membrane for the substance.

Effect of an Electrical Potential Difference. If an electrical potential is applied across the membrane as shown in Figure 4–5B, ions, because of their electrical charges, will move through the membrane even though no concentration difference exists to cause their movement. Thus, to the left in Figure 4–5B, the concentrations of negative ions are exactly the same on both sides of the membrane, but a positive charge has been applied to the right side and a negative charge to the left, creating an electrical gradient across the membrane. The positive charge attracts the negative ions while the negative charge repels them. Therefore, net diffusion occurs from left to right. After a while large quantities of negative ions will have moved to the right, creating the condition illustrated on the right in Figure 4–5B, in which a concentration difference of the same ions has developed in the direction opposite to the electrical potential difference. Obviously, the concentration difference is tending to move the ions to the left, while the electrical difference is tending to move them to the right. When the concentration difference rises high enough, the two effects exactly balance each other. At normal body temperature (38° C.), the electrical difference that will exactly balance a given concentration difference can be determined from the following formula:

$$\text{EMF (in millivolts)} = \pm 61 \log \frac{C_1}{C_2}$$

in which EMF is the electromotive force (voltage) between side 1 and side 2 of the membrane, C_1 is the concentration on side 1, and C_2 is the concentration on side 2. The required polarity is + for negative ions and − for positive ions. This relationship is extremely important in understanding the transmission of nerve impulses, for which reason it is discussed in even greater detail in the following chapter.

Effect of a Pressure Difference. At times considerable pressure difference develops between the two sides of a membrane. This occurs, for instance, at the capillary membrane, which has a pressure approximately 25 mm. Hg greater inside the capillary than outside. Pressure actually means the sum of all the forces of the different molecules striking a unit surface area at a given instant. Therefore, when the pressure is increased on one side of a membrane, this means that the sum of all the molecular forces striking the pores on that side of the membrane is greater than on the other side. This can result either from greater numbers of molecules striking the membrane per second or from greater kinetic energy of the average molecule striking the membrane. In either event, increased amounts of energy are available to cause net movement of molecules from the high pressure side toward the low pressure side. This effect is illustrated in Figure 4–5C, which shows a piston developing high pressure on one side of a cell membrane, thereby causing net diffusion through the membrane to the other side.

For the usual red cell membrane, 1 mm. Hg pressure difference causes approximately 10^{-4} cubic microns net diffusion of water through each square micron of membrane each second. This appears to be only a very minute rate of fluid movement through the membrane, but in relation to the normal cell size and the very large potential diffusion pressures that exist at the cell membrane, as is discussed later in the chapter, this rate represents tremendous transport of fluid in only a few seconds.

NET DIFFUSION OF WATER ACROSS CELL MEMBRANES — OSMOSIS

By far the most abundant substance to diffuse through the cell membrane is water, but it should be recalled again that enough water ordinarily diffuses in each direction through the red cell membrane per second to equal about *100 times the volume of the cell itself.* Yet, *normally,* the amount that diffuses in the two directions is so precisely balanced that not even the slightest *net* diffusion of water occurs. Therefore, the volume of the cell remains constant. However, under certain conditions, a *concentration difference for water* can develop across a membrane, just as concentration differences for other substances can also occur. When this happens, net diffusion of water does occur across the cell membrane, causing the cell either to swell or to shrink, depending on the direction of the net diffusion. This process of net diffusion of water caused by a concentration difference is called *osmosis.*

To give an example of osmosis, let us assume that we have the conditions shown in Figure 4–6, with pure water on one side of the cell membrane and a solution of sodium chloride on the other side. Referring back to Table 4–1, we see that water molecules pass through the cell membrane with extreme ease while sodium ions pass through only with extreme difficulty. And chloride ions cannot pass through the membrane because the positive charge of the sodium ions holds the negatively charged chloride ions back. Therefore, sodium chloride solution is actually a mixture of diffusible water molecules and nondiffusible sodium ions and chloride ions. Yet, the presence of the sodium and chloride has reduced the concentration of water molecules below that of pure water. As a result, in the example of Figure 4–6, more water molecules strike the pores on the left side where there is pure water than on the right side where the water concentration has been reduced. Thus, osmosis occurs from left to right.

Osmotic Pressure. If in Figure 4–6 pressure were applied to the sodium chloride solution, osmosis of water into this solution could be slowed or even stopped. The amount of pressure required to stop osmosis completely is called the osmotic pressure of the sodium chloride solution.

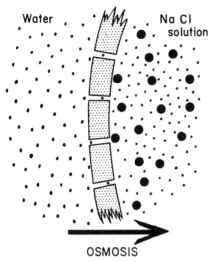

Figure 4-6   Osmosis at a cell membrane when a sodium chloride solution is placed on one side of the membrane and water on the other side.

The principle of a pressure difference opposing osmosis is illustrated in Figure 4-7, which shows a semipermeable membrane separating two separate columns of fluid, one containing water and the other containing a solution of water and some solute that will not penetrate the membrane. Osmosis of water from chamber B into chamber A causes the levels of the fluid columns to become farther and farther apart, until eventually a pressure difference is developed that is great enough to oppose the osmotic effect. The pressure difference across the membrane at this time is the

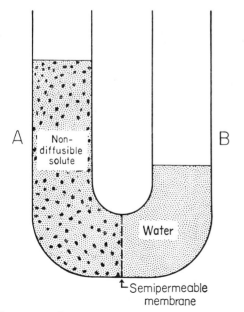

Figure 4-7   Demonstration of osmotic pressure on the two sides of a semipermeable membrane.

osmotic pressure of the solution containing the nondiffusible solute.

KINETICS OF OSMOTIC PRESSURE.   Figure 4-8 gives an analysis that can be very useful in understanding the basic cause of osmosis and osmotic pressure. The actual pressure at the surface of a solution, such as a solution of sodium chloride, will be equal to atmospheric pressure. Thus, at sea level the pressure at the surface of a solution is 760 mm. Hg. Yet, strangely enough, pressure differences of many thousand mm. Hg frequently occur between the two sides of a semipermeable membrane when the solutions on the two sides of the membrane are different from each other. This ability of the solutions to cause such pressure can be expressed in terms of *potential pressure*. The potential pressure of each solution is usually expressed as *chemical activity* of the different molecules and ions present in the solution, and the amount of chemical activity (or potential pressure) of each type of molecule or ion is directly proportional to its fractional concentration in the solution.

Figure 4-8 illustrates some representative potential pressures as they occur at cell membranes. The first potential pressure, 5500 mm. Hg, represents the total potential pressure of the nondiffusible particles in extracellular fluid — such particles as sodium, potassium, magnesium, calcium, chloride, bicarbonate, and protein ions, as well as non-ionic substances such as glucose.

In the middle of the figure are shown the potential pressures of water on both sides of the membrane. On the lower side there exists pure water, and 1,073,000 mm. Hg is the theoretical total potential pressure of pure water if its entire chemical activity were converted into actual pressure. That is, if pure water were on one side of a semipermeable membrane and pure electrolyte were on the opposite side, water theoretically could move by osmosis through the semipermeable membrane with sufficient pressure to create an actual pressure across the membrane of 1,073,000 mm. Hg. On the top of the membrane, the potential pressure of water molecules is shown to be 5500 mm. Hg less than that on the lower side. The reason for this is that the water on top has been diluted by the nondiffusible particles, making the fractional concentration of water above the membrane slightly less than that below.

Because of the greater potential pressure of water molecules below the membrane in Figure 4-8 than above, more water will diffuse from below upward than in the opposite

(pressures are in mm.Hg)

Figure 4–8   Kinetics of osmosis. Explained in the text.

direction, causing osmosis; and the amount of hydrostatic pressure that would be required on top of the membrane to prevent osmosis is 5500 mm. Hg. Note that this value is equal to the potential pressure of the nondiffusible particles in the upper solution.

Thus, the potential pressure of nondiffusible particles (caused by the chemical activity of these particles) is what is called osmotic pressure of the solution. It causes osmosis of water in the direction opposite to the osmotic pressure. Therefore, the term "osmotic pressure" is actually a misnomer because it is not an actual pressure but only a potential pressure.

LACK OF EFFECT OF MOLECULAR AND IONIC MASS ON OSMOTIC PRESSURE. The osmotic pressure exerted by nondiffusible particles in a solution, whether they be molecules or ions, is determined by the *numbers* of particles per unit volume of fluid and not the mass of the particles. The reason for this is that each particle in a solution, regardless of its mass, exerts, on the average, the same amount of pressure against the membrane. That is, all particles are bouncing among each other with, on the average, equal energy. If some particles have greater kinetic energy of movement than others, their impact with the low energy particles will impart part of their energy to these, thus decreasing the energy level of the high energy particles while increasing the energy level of the others. The large particles, which have greater mass (m) than the small particles, move at low velocity (v), while the small particles move at a high velocity in such a way that their average energies, determined by the equation $k = \dfrac{mv^2}{2}$, are equal to each other. Therefore, on the average, the energy that a particle has when it strikes a membrane is approximately equal for all size particles, so that the factor that determines the osmotic pressure of a solution is the concentration of the solution in terms of numbers of particles and not in terms of mass of the solute.

EFFECT OF INTERACTION BETWEEN DISSOLVED PARTICLES ON OSMOTIC PRESSURE. If particles interact with each other by either chemical or physical bonds, the osmotic effect becomes changed. Thus, sodium and chloride ions in solution have tremendous attractive power for each other so that neither of these ions can move totally unrestrained; therefore, their chemical activities are normally only about three quarters of what they would be were it not for this attraction. Because of interionic attraction and less powerful types of attraction between nonionized particles, the amount of osmotic pressure in the body fluids is actually only 0.7 to 0.8 times the maximum amount that would be calculated on the basis of individual numbers of particles. However, since all the fluids of the body are affected approximately equally in this way, for practical purposes this interaction between particles is usually ignored in studying osmotic effects in the body.

OSMOLALITY. Since the amount of osmotic pressure exerted by a solute is proportional to the concentration of the solute in numbers of molecules or ions, expressing the solute concentration in terms of mass is of no value in determining osmotic pressure. To express the concentration in terms of numbers of particles, the unit called the *osmol* is used in place of grams.

One osmol is the number of particles in 1 gram molecular weight of undissociated solute. Thus, 180 grams of glucose is equal to 1 osmol of glucose, because glucose does not dissociate. On the other hand, if the solute dissociates into two ions, 1 gram molecular weight of the solute equals 2 osmols, because the number of osmotically active particles is now twice as great as is the case in the undissociated solute. Therefore, 1 gram molecular weight of sodium chloride, 58.5 gm., is equal to 2 osmols.

A solution that has 1 osmol of solute dissolved in each liter of water is said to have a concentration of 1 osmol per liter, and a solution that has $1/1000$ osmol dissolved per liter has a concentration of 1 milliosmol per liter. The normal concentration of the extracellular and intracellular fluids is about 300 milliosmols per liter.

RELATIONSHIP OF OSMOLALITY TO OSMOTIC PRESSURE. At normal body temperature, a concentration of 1 osmol per liter will cause 19,300 mm. Hg osmotic pressure in the solution. Likewise, 1 milliosmol per liter concentration is equivalent to 19.3 mm. Hg osmotic pressure.

ACTIVE TRANSPORT

Often only a minute concentration of a substance is present in the extracellular fluid, and yet a large concentration of the substance is required in the intracellular fluid. For instance, this is true of potassium ions. Conversely, other substances frequently enter cells and must be removed even though the concentration inside is far less than that outside. This is true of sodium ions.

From the discussion thus far it is evident that *no substances can diffuse against a concentration difference,* or, as is often said, "uphill." To cause movement of substances uphill, energy must be imparted to the substance. This is analogous to the compression of air by a pump. After compression, the concentration of the air molecules is far greater than before compression, but to create this greater concentration, energy must be expended by the piston of the pump as it compresses the air molecules. Likewise, as molecules are transported through a cell membrane from a dilute solution to a concentrated solution, energy must be expended. The process of moving molecules uphill against a concentration difference (or uphill against an electrical or pressure difference) is called *active transport.* Among the different substances that are actively transported through the cell membrane in at least some parts of the body are sodium ions, potassium ions, calcium ions, iron ions, hydrogen ions, chloride ions, iodide ions, ureate ions, several different sugars, and the amino acids.

BASIC MECHANISM OF ACTIVE TRANSPORT

The mechanism of active transport is believed to be similar for all substances and to be dependent on transport by *carriers.* Figure 4–9 illustrates the basic mechanism, showing a substance S entering the outside surface of the membrane where it combines with carrier C. At the inside surface of the membrane, S separates from the carrier and is released to the inside of the cell. C then moves back to the outside to pick up more S.

One will immediately recognize the similarity between this mechanism of active transport and that of facilitated diffusion discussed earlier in the chapter. The difference, however, is that energy is imparted to the system in the course of transport, so that transport can occur *against a concentration difference* (or against an electrical or pressure difference).

Unfortunately, little is known about the mechanisms by which energy is utilized in the transport mechanism. The energy could cause combination of the carrier with the substance, or it could cause splitting of the substance away from the carrier. Regardless of which of these is true, the energy is probably supplied to the transport system at the inside surface of the cell membrane, because it is inside the cell that large quantities of the energy-giving substance ATP are available for promoting chemical reactions.

It is also evident from Figure 4–9 that the chemical reactions are believed to be promoted by specific enzymes which catalyze the chemical reactions. Indeed, the carrier might itself be an enzyme as well as a carrier.

Chemical Nature of the Carrier. Carrier substances are believed to be either proteins or lipoproteins, the protein moiety providing a specific site for attachment of the substance to be transported and the lipid moiety providing solubility in the lipid phase of the cell membrane.

It has been suggested that the carrier might transport substances by a simple process of

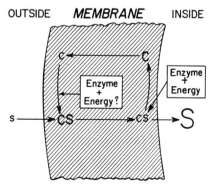

Figure 4–9   Basic mechanism of active transport.

thermal motion, such as by rotating to expose its carrier site first to the outside surface of the cell membrane and then to the inside, or by sliding a substance from one reactive site of the carrier to another until the substance enters the cell. Whether or not a "shuttling" or a "fixed" type of carrier exists, nevertheless, essentially the same principles of transport through the membrane would still apply.

Specificity of Carrier Systems. Several different carrier systems exist in cell membranes, each of which transports only certain specific substances. One carrier system, for instance, transports sodium to the outside of the membrane and probably transports potassium to the inside at the same time. Another system actively transports sugars through the membranes of certain cells, whereas still other specific carrier systems transport different ones of the amino acids.

The specificity of active transport systems for substances is determined either by the chemical nature of the carrier, which allows it to combine only with certain substances, or by the nature of the enzymes that catalyze the specific chemical reactions. Unfortunately, very little is known about the precise function of these two components of the carrier systems.

Energetics of Active Transport. The amount of energy required to transport a substance actively through a membrane (aside from energy lost as heat in the chemical reactions) is determined by the degree that the substance is concentrated during transport. Compared to the energy required to concentrate a substance 10-fold, to concentrate it 100-fold requires twice as much energy, and to concentrate it 1000-fold requires three times as much. In other words, the energy required is proportional to the logarithm of the degree that the substance is concentrated, as expressed by the following formula:

$$\text{Energy (in calories per osmol)} = 1400 \log \frac{C_1}{C_2}$$

Thus, in terms of calories, the amount of energy required to concentrate 1 osmol of substance 10-fold is about 1400 calories. One can see that the energy expenditure for concentrating substances in cells or for removing substances from cells against a concentration gradient can be tremendous. Some cells, such as those lining the renal tubules, expend as much as 50 per cent of their energy for this purpose alone.

TRANSPORT THROUGH INTRACELLULAR MEMBRANES

Though much less is known about the transport processes of the membranes within the cell than those of the cell membrane, we do know that at least two of these membranes, that of the mitochondrion and that of the endoplasmic reticulum, obey the same laws of both diffusion and active transport as does the cell membrane. However, their permeabilities and specificities for transport are often different. Indeed, the mitochondrion has a double membrane, with an outer membrane that is much more permeable than the inner membrane. Undoubtedly, transport through these membranes plays important roles in function of the intracellular organelles, though most of these roles are still a mystery.

ACTIVE TRANSPORT THROUGH CELLULAR SHEETS

In many places in the body substances must be transported through an entire *cellular layer* instead of simply through the cell membrane itself. Transport of this type occurs through the intestinal epithelium, the epithelium of the renal tubules, the epithelium of all exocrine glands, the membrane of the choroid plexus of the brain, and many other membranes.

The general mechanism of active transport through a sheet of cells is illustrated in Figure 4–10. This figure shows two adjacent cells in a typical epithelial membrane as found in the intestine, in the gallbladder, and in certain other areas of the body. On the luminal surface of the cells is a brush border that is highly permeable to both water and solutes, allowing both of these to diffuse readily from the lumen of the intestine to the interior of the cell. Once inside the cell, some of the solutes are actively transported into the space between the cells. This space is closed at the brush border of the epithelium but is wide open at the base of the cells where they rest on the basement membrane. Furthermore, the basement membrane is extremely permeable. Therefore, substances transported into this channel between the cells flow toward the connective tissue.

One of the most important substances actively transported in this manner is sodium; and when

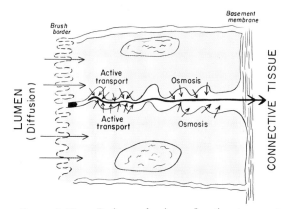

Figure 4–10   Basic mechanism of active transport through a layer of cells.

sodium ions are transported into the space between the cells, their positive electrostatic charges pull negatively charged chloride ions through the membrane as well. Then, when the concentration of sodium and chloride ions increases, this in turn causes osmosis of water out of the cell and into the intercellular space. In consequence, the sodium and chloride flow along with the water into the connective tissue behind the basement membrane where the water and ions diffuse into the blood capillaries. These same principles of transport apply generally wherever such transport occurs through cellular sheets, whether in the intestine, gallbladder, kidneys, or elsewhere.

ACTIVE TRANSPORT OF SODIUM, POTASSIUM, AND OTHER ELECTROLYTES

Referring back to Figure 4–1, one sees that the sodium concentration outside the cell is very high in comparison with its concentration inside, and the converse is true of potassium. Also, Table 4–1 shows that minute quantities of sodium and potassium can diffuse through the pores of the cell. If such diffusion should take place over a long period of time, the concentrations of the two ions would eventually become equal inside and outside the cell unless there were some means to remove the sodium from the inside and to transport potassium back in.

Fortunately, a system for active transport of sodium and potassium is present probably in all cells of the body. The mechanism has been postulated to be that illustrated in Figure 4–11, which shows sodium (Na) inside the cell com-

bining with carrier Y at the membrane surface to form large quantities of the combination NaY. This then moves to the outer surface where sodium is released and the carrier Y changes its chemical composition slightly to become carrier X. This carrier then combines with potassium K to form KX, which moves to the inner surface of the membrane where energy is provided to split K from X under the influence of the enzyme ATPase, the energy being derived from MgATP.

The transport mechanism in Figure 4–11 is believed to be more effective in transporting sodium than in transporting potassium, usually transporting about three sodium ions for every two potassium ions. The carrier is probably a lipoprotein, and it is likely that this same lipoprotein molecule acts as the enzyme ATPase to release the energy required for transport.

The sodium transport mechanism is so important to many different functioning systems of the body—such as to nerve and muscle fibers for transmission of impulses, various glands for the secretion of different substances, and all cells of the body to prevent cellular swelling—that it is frequently called the *sodium pump*. We will discuss the sodium pump at many places in this text.

Importance of the Sodium Pump in Controlling Cell Size. One of the most important functions of the sodium pump is to prevent continual swelling of the cells. This may be explained as follows: All cells form many intracellular substances that are nondiffusible through the cell membrane, such as protein molecules, phosphocreatine, and adenosine triphosphate. These substances tend to cause osmosis of water to the interior of the cell all the time. Also, electrolytes tend to leak along with the water to the inside of the cell. If some factor should not overcome the continual tendency for water and electrolytes to enter the cell, the cell would eventually swell until it burst. However, the sodium transport mechanism opposes this tendency of the cell to swell by continually transporting sodium to the exterior, which initiates an opposite osmotic tendency to move water out of the cell. Whenever the metabolism of the cell ceases so that energy from adenosine triphosphate is not available to keep the sodium pump operating, the cell begins to swell immediately.

Other Electrolytes. Calcium and magnesium are probably transported by all cell membranes in much the same manner that sodium and potassium are transported, and certain cells of the body have the ability to transport still other ions. For instance, the glandular cell membranes of the thyroid gland can transport large quantities of iodide ion; the epithelial cells of the intestine can transport sodium, chloride, calcium, iron, bicarbonate, and probably many other ions; and the epithelial

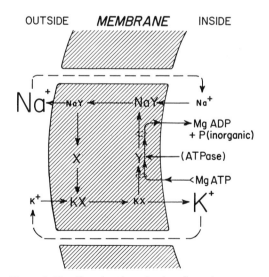

Figure 4–11 Postulated mechanism for active transport of sodium and potassium through the cell membrane, showing coupling of the two transport mechanisms and delivery of energy to the system at the inner surface of the membrane.

cells of the renal tubules can transport hydrogen, calcium, sodium, potassium, and a number of other ions.

ACTIVE TRANSPORT OF SUGARS

Facilitated diffusion of glucose and certain other sugars occurs in essentially all cells of the body, but active transport of sugars occurs in only a few places in the body. For instance, in the intestine and renal tubules, glucose and several other monosaccharides are continually transported into the blood even though their concentrations be minute in the lumens. Therefore, essentially none of these sugars is lost either in the intestinal excreta or the urine.

Though not all sugars are actively transported, almost all monosaccharides that are important to the body *are* actively transported, including *glucose, galactose, fructose, mannose, xylose, arabinose*, and *sorbose*. On the other hand, the disaccharides such as sucrose, lactose, and maltose are not actively transported at all.

As is true of essentially all other active transport mechanisms, the precise carrier system and chemical reactions responsible for transport of the monosaccharides are yet unknown. The one common denominator in transport of sugars seems to be the necessity for an intact —OH group on the C_2 carbon of the monosaccharide molecule. It is presumed that the monosaccharide attaches to the carrier at this point.

Transport of the monosaccharides will not occur when active transport of sodium is blocked. Because of this, it is believed that active transport of sodium ions, by means of coupled reactions with the glucose transport mechanism, provides the energy required to move the monosaccharides through the membrane. Therefore, it is stated that monosaccharide transport is *secondary active transport*, in contrast to *primary active transport* represented by the transport of sodium through the membrane.

ACTIVE TRANSPORT OF AMINO ACIDS

Amino acids and the closely related substances, the amines, are actively transported through the membranes of all cells that have been studied. Active transport through the epithelium of the intestine, the renal tubules, and many glands is especially important. Probably at least four different carrier systems exist for transporting different groups of amino acids. However, the carrier systems for transport of amino acids are equally as poorly understood as the systems for transporting other substances. One of the few known features of the amino acid carrier system is that transport of at least some amino acids depends on pyridoxine (vitamin B_6). Therefore, deficiency of this vitamin causes protein deficiency.

Hormonal Regulation of Amino Acid Transport. At least four different hormones are important in controlling amino acid transport: (1) *Growth hormone*, secreted by the adenohypophysis, increases amino acid transport into essentially all cells. (2) *Insulin* and (3) *glucocorticoids* increase amino acid transport at least into liver cells and possibly into other cells as well, though much less is known about these. (4) *Estradiol*, the most important of the female sex hormones, causes rapid transport of amino acids into the musculature of the uterus, thereby promoting development of this organ.

Thus, several of the hormones exert much, if not most, of their effects in the body by controlling active transport of amino acids into all or certain cells.

PINOCYTOSIS AND PHAGOCYTOSIS

It was pointed out in Chapter 2 that the cell membrane has the ability to imbibe small amounts of substances from the extracellular fluid by the process called *pinocytosis*. *Phagocytosis* occurs by essentially the same mechanism as pinocytosis, but phagocytosis means ingestion of large particulate matter, such as bacteria or cell fragments, that is free in the extracellular fluid.

The real importance of pinocytosis to the body is that this is the only known means by which very large molecules, such as those of protein, can be transported to the interior of cells. The importance of phagocytosis is that it is used by special cells, such as the white blood cells, to rid the body of bacteria and unwanted debris in the tissues.

REFERENCES

Adolph, E. F.: Physiology of Man in the Desert. New York, Interscience Publishers, 1947.

Biological Membranes (Symposium). *Ann. N.Y. Acad. Sci.*, 137: 403, 1966.

Bourgoignie, J.; Klahr, S.; Yates, J.; Guerra, L.; and Bricker, N. S.: Characteristics of ATPase system of dogfish erythrocyte membranes. *Amer. J. Physiol.*, 218:314, 1970.

Charnock, J. S., and Opit, L. J.: Membrane metabolism and ion transport. *In* Bittar, E. Edward, and Bittar, Neville (eds.): The Biological Basis of Medicine. New York, Academic Press, Inc., 1968, Vol. 1, p. 69.

Csaky, T. Z.: Transport through biological membranes. *Ann. Rev. Physiol.*, 27:415, 1965.

Curran, P. F., and MacIntosh, J. R.: A model system for biological water transport. *Nature*, 193:347, 1962.

Diamond, J. M., and Tormey, J. M.: Studies on the structural basis of water transport across epithelial membranes. *Fed. Proc.*, 25:1458, 1968.

Diamond, J. M., and Wright, E. M.: Biological membranes: the physical bases of ion and non-electrolyte selectivity. *Ann. Rev. Physiol., 31*:581, 1969.

Edelman, I. S.: Transport through biological membranes. *Ann. Rev. Physiol., 23*:37, 1961.

Finn, A. L.: Effects of potassium and amphotericin B on ion transport in the toad bladder. *Amer. J. Physiol., 218*:463, 1970.

Finn, A. L.; J. S. Handler; and J. Orloff: Active chloride transport in the isolated toad bladder. *Amer. J. Physiol., 213*:179, 1967.

Gamble, J. L.: Chemical Anatomy, Physiology and Pathology of Extracellular Fluid: A Lecture Syllabus. 6th ed., Cambridge, Massachusetts, Harvard University Press, 1954.

Harris, E. J.: Transport and Accumulation in Biological Systems. 2nd ed., New York, Academic Press, Inc., 1960.

Heinz, E.: Transport through biological membranes. *Ann. Rev. Physiol., 29*:21, 1967.

Hokin, L., and Hokin, M.: The chemistry of cell membranes. *Sci. Amer., 213*:78(4), 1965.

Lehninger, A. L.: Water uptake and extrusion by mitochondria in relation to oxidative phosphorylation. *Physiol. Rev., 42*:467, 1962.

Levin, M. L.; Rector, F. C., Jr.; and Seldin, D. W.: Effects of potassium and ouabain on sodium transport in human red cells. *Amer. J. Physiol., 214*:1328, 1968.

Lovtrup, S.: On the rate of water exchange across the surface of animal cells. *J. Theor. Biol., 5*:341, 1963.

Malhotra, S. K., and Van Harreveld, A.: Molecular organization of membranes of cells and cellular organelles. *In* Bittar, E. Edward, and Bittar, Neville (eds.): The Biological Basis of Medicine. New York, Academic Press, Inc., Vol. 1, p. 3.

Nims, L. F.: Steady state material transfer through biological barriers. *Amer. J. Physiol., 201*:987, 1961.

Post, R. L., and Jolly, P. C.: The linkage of sodium, potassium, and ammonium active transport across the human erythrocyte membrane. *Biochim. biophys. Acta, 25*:118, 1957.

Post, R. L.; Merritt, C. R.; Kinsolving, C. R.; and Albright, C. D.: Membrane adenosine triphosphatase as a participant in the active transport of sodium and potassium in the human erythrocyte. *J. Biol. Chem., 235*:1796, 1960.

Renkin, E. M.: Exchangeability of tissue potassium in skeletal muscle. *Amer. J. Physiol., 197*:1211, 1959.

Rothstein, A.: Membrane phenomena. *Ann. Rev. Physiol., 30*:15, 1968.

Rustad, R. C.: Pinocytosis. *Sci. Amer., 204*(4):120, 1961.

Sen, A. K., and Post, R. L.: Stoichiometry and localization of adenosine triphosphate-dependent sodium and potassium transport in the erythrocyte. *J. Biol. Chem., 239*:345, 1964.

Skou, J. C.: Enzymatic basis for active transport of Na^+ and K^+ across cell membrane. *Physiol. Rev., 45*:596, 1965.

Solomon, A. K.: Ion transport in single cell populations. *Biophys. J., 2*(2)Pt. 2:79, 1962.

Structure and Function of Membranes (Symposium). *Brit. Med. Bull., 24*:99, 1968.

Tosteson, D. C.: Membrane transport of Na and K. *Physiologist, 9*:89, 1966.

Villegas, R.: On the mechanisms of transport across cell membranes with special reference to the axolemma. *Proc. Int. Union of Physiol. Sciences, 6*:125, 1968.

Whittam, R., and Wheeler, K. P.: Transport across cell membranes. *Ann. Rev. Physiol., 32*:21, 1970.

CHAPTER 5

Membrane Potentials, Action Potentials, Excitation, and Rhythmicity

Electrical potentials exist across the membranes of essentially all cells of the body, and some cells, such as nerve and muscle cells, are "excitable"—that is, capable of transmitting electrochemical impulses along their membranes. And in still other types of cells, such as glandular cells, macrophages, and ciliated cells, changes in membrane potentials probably play significant roles in controlling many of the cell's functions. However, the present discussion is concerned with membrane potentials generated both at rest and during action by nerve and muscle cells.

BASIC PHYSICS OF MEMBRANE POTENTIALS

Before beginning this discussion, let us first recall that the fluids both inside and outside the cells are electrolytic solutions containing approximately 155 mEq./liter of anions and the same concentration of cations. Generally, an excess number of negative ions (anions) accumulates immediately inside the cell membrane along its inner surface, and an equal number of positive ions (cations) accumulates immediately outside the membrane. The result of this is the development of a *membrane potential*.

The two basic means by which membrane potentials can develop are: (1) active transport of ions through the membrane, thus creating an imbalance of negative and positive charges on the two sides of the membrane, and (2) diffusion of ions through the membrane as a result of a concentration difference between the two sides of the membrane, also creating an imbalance of charges.

Membrane Potentials Caused by Active Transport. Figure 5–1A illustrates how active transport can create a membrane potential. This

shows the sodium "pump," which was explained in Chapter 4, transporting *positively charged* sodium ions to the exterior.

THE CELL MEMBRANE AS A CAPACITOR. Note in Figure 5–1A that the negative and positive ionic charges are shown lined up against the membrane. This occurs because the negative charges inside the membrane and the positive charges on the outside pull each other together at the membrane barrier, thus creating an abrupt change in electrical potential across the cell membrane.

This alignment of electrical charges on the two sides of the membrane is exactly the same process that takes place when an electrical capacitor becomes charged with electricity. In the cell membrane, the lipid matrix of the cell is the *dielectric*, much as mica is frequently used as a dielectric in

Figure 5–1 (A) Establishment of a membrane potential as a result of active transport of sodium ions. (B) Establishment of a membrane potential of opposite polarity caused by rapid diffusion of sodium ions from an area of high concentration to an area of low concentration.

an electrical capacitor. It will be recalled that a capacitor's capacity for holding electrical charges is inversely proportional to the thickness of the membrane. Because of the extreme thinness of the cell membrane (70 to 100 Ångstroms), the capacitance of the cell membrane is tremendous for its area.

Membrane Potentials Caused by Diffusion. Figure 5–1B illustrates the nerve fiber under another condition in which the permeability of the membrane to sodium has increased so much that diffusion of sodium through the membrane is now great in comparison with the transport of sodium by active transport. Since the sodium concentration is great outside the cells and slight inside, the positively charged sodium ions now move rapidly to the inside of the membrane, obviously causing the inside to become positively charged in relation to the outside.

RELATIONSHIP OF THE DIFFUSION POTENTIAL TO THE CONCENTRATION DIFFERENCE. When a concentration difference of ions across a membrane causes diffusion of ions through the membrane, thus creating a membrane potential, the magnitude of the potential is determined by the difference in tendency of the ions to diffuse in one direction versus the other direction, which is determined by the following formula (at body temperature, 38°C.):

$$EMF \text{ (in millivolts)} = 61 \log \frac{\text{Conc. inside}}{\text{Conc. outside}} \quad (1)$$

Thus, when the concentration of ions on one side of a membrane is 10 times that on the other, the log of 10 is 1, and the potential difference calculates to be 61 millivolts.

However, two conditions are necessary for a membrane potential to develop as a result of diffusion: (1) The membrane must be semipermeable, allowing ions of one charge to diffuse through the pores more readily than ions of the opposite charge. (2) The concentration of the diffusible ions must be greater on one side of the membrane than on the other side.

CALCULATION OF THE MEMBRANE POTENTIAL WHEN THE MEMBRANE IS PERMEABLE TO SEVERAL DIFFERENT IONS. When a membrane is permeable to several different ions, the diffusion potential that will develop depends on three factors: (1) the polarity of the electrical charge of each ion, (2) the permeability of the membrane (P) to each ion, and (3) the concentration of the respective ions on the two sides of the membrane. Thus, the following formula, called the *Nernst equation*, gives the calculated membrane potential when two cations (C) and two anions (A) are involved.

$$EMF \text{ (in millivolts)} =$$
$$-61 \log \frac{C_{1_i}P_1 + C_{2_i}P_2 + A_{3_o}P_3 + A_{4_o}P_4}{C_{1_o}P_1 + C_{2_o}P_2 + A_{3_i}P_3 + A_{4_i}P_4} \quad (2)$$

Note that a cation gradient from the inside (i) to the outside (o) of a membrane causes electronegativity inside the membrane, while an anion gradient in *exactly the opposite direction* also causes electronegativity on the inside.

ORIGIN OF THE CELL MEMBRANE POTENTIAL

Role of Sodium and Potassium Pumps and of Nondiffusible Anions in the Origin of the Cell Membrane Potential. Before attempting to explain the origin of the cell membrane potential, several basic facts need to be understood as follows:

1. The nerve membrane is endowed with a sodium and a potassium pump, sodium being pumped to the exterior and potassium to the interior. These pumps were discussed in the previous chapter.

2. The resting nerve membrane is normally 50 to 100 times as permeable to potassium as to sodium. Therefore, potassium diffuses with relative ease through the resting membrane, whereas sodium diffuses only with difficulty.

3. Inside the nerve fiber are large numbers of anions that cannot diffuse through the nerve membrane at all or that diffuse very poorly. These anions include especially organic phosphate ions, organic sulfate ions, and protein ions.

Now let us put the above facts together to see how the resting nerve membrane potential comes about. First, the sodium pump decreases the concentration of sodium ions inside the nerve fiber to a very low level, only 10 mEq./liter in contrast to the very high level in the extracellular fluids of 142 mEq./liter. Second, the potassium pump causes a high concentration of potassium ions inside the nerve fiber, 140 mEq./liter in contrast to only 5 mEq./liter in the extracellular fluids. These concentration differences are illustrated in Figure 5–2. Third, also illustrated in Figure 5–2 is a very high concentration of nondiffusible anions, 150 mEq./liter, inside the nerve fiber in contrast to a very low concentration of only 5 mEq./liter outside the fiber.

With the above distributions of sodium, potassium, and nondiffusible anions on the two sides of the membrane, a membrane potential develops in accordance with the Nernst equation discussed previously. Potassium ions dif-

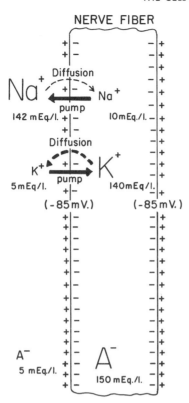

Figure 5–2 Establishment of a membrane potential of −85 millivolts in the normal resting nerve fiber, and development of concentration differences of sodium, potassium, and chloride ions between the two sides of the membrane. The dashed arrows represent diffusion and the solid arrows represent active transport (pumps).

fuse through the membrane relatively easily while the sodium ions and nondiffusible anions diffuse very poorly. Therefore, the main tendency for diffusion of electrical charges will be caused by diffusion of the positive potassium ions from inside the nerve membrane to the outside. This effect obviously removes positive charges from inside the membrane while building up positive charges outside. As a result, charges line up on the two sides of the membrane as shown in Figure 5–2, giving negativity on the inside and positivity on the outside.

IMPORTANCE OF POTASSIUM IN DETERMINING THE MAGNITUDE OF THE MEMBRANE POTENTIAL. Since potassium can diffuse through the membrane very easily in comparison with either sodium or the nondiffusible anions, the Nernst equation predicts that it is almost entirely the concentration difference of potassium across the membrane that determines the magnitude of the membrane potential. Using Equation 1 discussed earlier in the chapter, one can calculate that the diffusion potential created by the concentration difference of potassium shown

in Figure 5–2 will be approximately 90 millivolts. This is very near to the usually measured potential across the resting nerve membrane, which averages about 85 millivolts. However, the calculated value for the potassium diffusion potential is normally slightly greater than the actual measured value of the resting membrane potential because sodium ions have a slight tendency to diffuse to the interior of the nerve fiber, and some of the so-called nondiffusible anions have a very slight tendency to diffuse to the outside. Obviously, either of these effects would reduce the potential to slightly below that calculated for potassium diffusion alone.

Role of Chloride and Other Ions. Chloride ions are not pumped through the nerve membrane in either direction. However, they diffuse through the membrane about twice as easily as potassium ions. Because there is no pump to force buildup of a concentration difference of chloride ions across the membrane, the distribution of chloride ions on the two sides of the membrane is determined entirely by electrical potential. That is, the negativity that develops inside the membrane repels chloride ions from inside the fiber and causes the chloride concentration to fall to a very low value, about 4 mEq./liter in comparison with an extracellular fluid concentration of about 101 mEq./liter. Thus, chloride ions play only a passive role. Later in the chapter we will discuss instantaneous changes in membrane potential caused by the so-called action potential. Under these conditions, chloride ions move rapidly through the membrane and affect the duration and the magnitude of the action potential, but again their role is a passive one.

Essentially the same principles as those for sodium, potassium, and chloride distribution across the membrane also apply to other ions. Magnesium ions are affected in much the same way as potassium, and calcium ions in much the same manner as sodium. However, the concentrations of magnesium and calcium ions are small enough that the total number of electrical charges actually involved is also small. Yet both of these ions affect membrane potentials in another way by altering the permeability of the membrane to other ions, a subject that is discussed at greater length later in the chapter.

Membrane Potentials Measured in Nerve and Muscle Fibers. Figure 5–3 illustrates a method that has been used for measuring the resting membrane potential. A micropipet is made from a minute capillary glass tube so that the tip of the pipet has a diameter of only 0.5 to 2 microns. Inside the pipet is a strong solution of potassium chloride, which acts as an electrical conductor. The fiber whose membrane is to be measured is pierced with the pipet and electrical connections are made from the pipet to an appropriate meter, as illustrated in the fig-

Figure 5–3   Measurement of the membrane potential of the nerve fiber using a microelectrode.

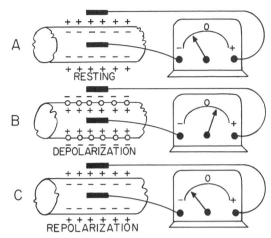

Figure 5–4   Sequential events during the action potential, showing: (A) the normal resting potential, (B) development of a reversal potential during depolarization, and (C) re-establishment of the normal resting potential during repolarization.

ure. The resting membrane potential measured in many different nerve and muscle fibers of mammals has usually ranged between −75 and −95 millivolts, with −85 millivolts as a reasonable average of the different measurements.

THE ACTION POTENTIAL

So long as the membrane of the nerve fiber remains completely undisturbed, the membrane potential remains approximately −85 millivolts, which is called the *resting potential*. However, any factor that suddenly increases the permeability of the membrane to sodium is likely to elicit a *sequence of rapid changes* in membrane potential lasting a minute fraction of a second, followed immediately by return of the membrane potential to its resting value. This sequence of potential changes is called the *action potential*.

Some of the factors that can elicit an action potential are *electrical stimulation* of the membrane, *application of chemicals* to the membrane to cause increased permeability to sodium, *mechanical damage* to the membrane, *heat, cold,* or almost any other factor that momentarily disturbs the normal resting state of the membrane.

Depolarization and Repolarization. The action potential occurs in two separate stages called depolarization and repolarization, which may be explained by referring to Figure 5–4. Figure 5–4A illustrates the resting state of the membrane, with negativity inside and positivity outside. When the permeability of the membrane to sodium ions suddenly increases, many of the sodium ions rush to the inside of the fiber, carrying enough positive charges to the inside to cause complete disappearance of the normal resting potential and usually enough charges actually to develop a positive state inside the fiber instead of the normal negative

state, as illustrated in Figure 5–4B. This positive state inside the fiber is called the *reversal potential.*

Almost immediately after depolarization takes place, the pores of the membrane again become almost totally impermeable to sodium ions. Because of this, the reversal potential inside the fiber disappears, and the normal resting membrane potential returns. This is called *repolarization* (Fig. 5–4C).

Now, let us explain in more detail *how* this sequence of events occurs.

CONDUCTANCE CHANGES IN THE MEMBRANE DURING THE ACTION POTENTIAL

Sudden Increase in Sodium Conductance at the Onset of the Action Potential—"Activation" of the Membrane. The first event in the action potential is a sudden increase in membrane permeability to sodium. The exact cause of this is still a mystery, though one of the theories to explain this effect is the following: First, it is known that an excess of calcium ions in the extracellular fluid decreases the permeability of the nerve membrane to sodium. Also, calcium ions have a high binding power with proteins. Therefore, it is postulated that under normal circumstances the sodium "channels"—whatever they might be: actual pores, protein molecules extending through the membrane, or whatever—are lined with calcium ions bound to the walls of the channels. Because of their positive charges, the calcium ions repel sodium and other positive ions and resist their passage through the channels. Yet if some calcium ions can be dislodged from their binding sites, a few sodium ions can begin to move inward. The inward movement of the inrushing sodium ions then theoretically can displace more and more calcium ions from their binding sites; as a result, still more sodium ions rush inward; and the process accelerates until essentially no calcium ions remain to block the movement of sodium through the channels.

The change in sodium conductance is illustrated by the lowermost curve of Figure 5–5, which shows that at the onset of the action potential the conductance of the membrane for sodium ions increases several hundred- to several thousand-fold in only a small fraction of a millisecond. This tremendous increase in sodium conductance is called *activation of the membrane.*

Subsequent Decrease in Sodium Conductance. As sodium ions diffuse to the inside of the membrane, they carry positive charges also to the inside, causing a *reversal potential* with the development of positivity inside the membrane and negativity outside. The positivity inside the membrane now repulses further inflow of the positively charged sodium ions. This obviously stops the movement of the sodium ions in the sodium channels and theoretically allows calcium ions to begin rebinding with the binding sites along the channel walls. Now, another vicious cycle develops, operating in the opposite direction: as the calcium ions bind, sodium conductance decreases, which allows still more calcium ions to bind, which causes sodium conductance to decrease still more, and so forth. This cycle continues until the membrane becomes almost totally impermeable to sodium once again. (It must be remembered that this is only a theory at present, not fact.)

Thus, the lowermost curve in Figure 5–5 shows that immediately after the sodium conductance increases several hundred- to several thousand-fold at the onset of the action potential, it decreases back to its original resting level in another small fraction of a millisecond.

Changes in Potassium Conductance During the Action Potential. Referring once again to Figure 5–5, we see that the normal resting membrane conductance for potassium is about 100 times that for sodium. Also, this figure shows that *potassium conductance does not change significantly during the first half of the action potential.* Yet toward the end of the action potential the potassium conductance increases about 30- to 40-fold. This increase is minute compared with the tremendous increase in sodium conductance, but it nevertheless plays a significant role in the membrane potential changes that occur during the course of the action potential.

The mechanism of the increased potassium conductance during the stages of the action potential, like that for changes in sodium conductance, is also almost completely hypothetical. However, the loss of negativity inside the membrane allows the highly concentrated potassium ions to begin moving *outward* through so-called *potassium "channels."* These channels are believed to be separate from the sodium channels, but this might not be true. It is assumed that this movement of potassium through the channels opens them to potassium in the same manner that sodium movement opens the sodium channels. Regardless of the mechanism that causes the increased potassium conductance, the potassium curve in Figure 5–5 shows that potassium conductance remains far above normal for almost 1 millisecond in large nerve fibers.

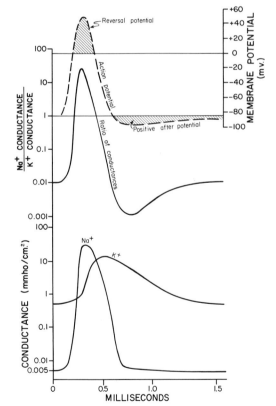

Figure 5–5 Changes in sodium and potassium conductances during the course of the action potential. Note that sodium conductance increases several thousand-fold during the early stages of the action potential, while potassium conductance increases only about 30-fold during the latter stages of the action potential and for a short period thereafter. (Curves constructed from data in Hodgkin and Huxley papers but transposed from squid axon to large mammalian nerve fibers.)

MEANS BY WHICH CONDUCTANCE CHANGES CAUSE THE ACTION POTENTIAL

Development of the Reversal Potential — Depolarization.   The second curve from the top in Figure 5–5 depicts the *ratio of sodium conductance to potassium conductance*, showing this to be 0.01 before the action potential begins. However, immediately after the membrane begins to become activated, the sodium conductance increases tremendously while the potassium conductance hardly changes. As a result, the ratio becomes reversed so that now sodium conductance is 30 times as great as potassium conductance. In other words, the permeability to sodium ions now becomes much greater than that for potassium ions. Therefore, the membrane potential is now determined almost entirely by diffusion of sodium ions to the interior of the nerve fiber, carrying positivity to the inside and creating negativity on the outside. Referring back to Equation 1 discussed at the beginning of this chapter, we can calculate the amount of diffusion potential one would expect. The concentration of sodium ions outside the membrane is 142 mEq./liter and inside 10 mEq./liter, a ratio of 14 to 1. For this ratio, the calculated diffusion potential would be approximately +70 mv. However, the action potential shown at the top of Figure 5–5 shows a recorded reversal potential of +50 mv. instead of the calculated +70 mv. If it is recalled that the membrane is still somewhat permeable to potassium ions, which causes a diffusion potential of opposite polarity, it is easy to understand why the recorded reversal potential does not become quite as great as the calculated value based on sodium diffusion alone.

Return of the Membrane Potential to the Resting Level — Repolarization.   The reversal potential lasts for only a fraction of a millisecond in large nerve fibers, only for that short period of time during which sodium conductance is greater than potassium conductance; this can be seen by comparing the top two curves of Figure 5–5. As the membrane becomes "inactivated" — that is, as sodium conductance returns toward its normal state — the membrane potential also returns toward its resting level.

Potassium Diffusion as the Cause of Return of the Resting Potential.   On first thought, one might suspect that return of the resting membrane potential to its original resting level would be caused by movement of sodium ions out of the fiber by the sodium pump. However, the sodium pump plays almost no role in this process because it operates much too slowly to recharge the membrane in a fraction of a millisecond. Instead, the return of the resting membrane potential is caused almost entirely by diffusion of potassium ions outward through the membrane. The potassium conductance curve of Figure 5–5 shows that at the same time that the sodium conductance returns toward its resting level, the potassium conductance increases to approximately 30 times its normal level. Thus, potassium ions now diffuse outward through the membrane extremely rapidly, while essentially no sodium ions diffuse through the membrane. Thus, we now have a situation in which a diffusion potential develops across the membrane caused almost entirely by potassium ions and almost none at all by the other ions. With an intracellular potassium concentration of 140 mEq./liter and an extracellular concentration of 5 mEq./liter, this is a 28-fold concentration ratio. Referring again to Equation 1, we see that a 28-fold ratio in potassium concentration calculates to give a membrane potential of −88 millivolts. Thus, one can readily understand that the rapid diffusion of potassium to the outside is more than capable of returning the membrane potential back to its resting level of −85 mv.

PROPAGATION OF THE ACTION POTENTIAL

In the preceding paragraphs we have discussed the action potential as if it occurs all at one spot on the membrane. However, an action potential elicited at any one point on an excitable membrane usually excites adjacent portions of the membrane, resulting in propagation of the action potential. The mechanism of this is illustrated in Figure 5–6. Figure 5–6A shows a normal resting nerve fiber, and Figure

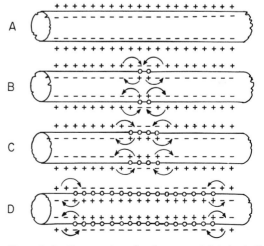

Figure 5–6   Propagation of action potentials in both directions along a conductive fiber.

5-6B shows a nerve fiber that has been excited in its midportion—that is, the midportion has suddenly developed increased permeability to sodium. The arrows illustrate a *local circuit* of current flow between the depolarized and the resting membrane areas; current flows inward through the depolarized membrane and outward through the resting membrane, thus completing a circuit. In some way not understood, *the current flow through the* resting *membrane now increases the membrane's permeability to sodium*, which immediately allows sodium ions to diffuse inward through the membrane, thus setting up the vicious cycle of membrane activation discussed earlier in the chapter. As a result, depolarization occurs at this area of the membrane as well. Therefore, as illustrated in Figure 5-6C and D, successive portions of the membrane become depolarized. And these newly depolarized areas cause local circuits of current flow still farther along the membrane, causing progressively more and more depolarization. Thus, the depolarization process travels in both directions along the entire extent of the fiber. The transmission of the depolarization process along a nerve or muscle fiber is called a *nerve* or *muscle impulse.*

Direction of Propagation. It is now obvious that an excitable membrane has no single direction of propagation, but that the impulse can travel in all directions away from the stimulus—even along all branches of a nerve fiber—until the entire membrane has become depolarized.

The All-or-Nothing Law. It is equally obvious that, once an action potential has been elicited at any point on the membrane of a normal fiber, the depolarization process will travel over the entire membrane. This is called the all-or-nothing law, and it applies to all normal excitable tissues. Occasionally, though, when the fiber is in an abnormal state the impulse will reach a point on the membrane at which the action potential does not generate sufficient voltage to stimulate the adjacent area of the membrane. When this occurs the spread of depolarization will stop. Therefore, for normal propagation of an impulse to occur, the ratio of action potential to threshold for excitation, called the *safety factor*, must at all times be greater than unity.

Propagation of Repolarization. The action potential normally lasts almost exactly the same length of time at each point along a fiber. Therefore, repolarization occurs at the point of stimulus first and then spreads progressively along the membrane, moving in the same direction that depolarization had previously spread.

REPOLARIZATION

Figure 5–7   Propagation of repolarization in both directions along a conductive fiber.

Figure 5-7 illustrates the same nerve fiber as that in Figure 5-6, showing that the repolarization process is propagated in the same direction as the depolarization process but a few ten-thousandths of a second later.

Importance of Energy Metabolism for "Recharging" the Fiber Membrane. Since transmission of large numbers of impulses along the nerve fiber reduces the ionic concentration gradients, these gradients must be re-established. This is achieved by the action of the sodium pump in exactly the same way as that described in the first part of the chapter for establishment of the original resting potential. That is, the sodium ions that have diffused during the action potentials to the interior of the cell and the potassium ions that have diffused to the exterior are returned to their original state by the sodium and potassium pumps. Since both the sodium and potassium pumps require energy for operation, this process of "recharging" the nerve fiber is an active metabolic one, utilizing energy derived from the adenosine triphosphate energy "currency" system of the cell.

Figure 5-8 illustrates the relationship of heat production in a nerve to the number of impulses transmitted by the nerve each second. The rate of heat production is roughly a measure of the rate of metabolism in the nerve, because heat is always liberated as a by-product of the chemical reactions of energy metabo-

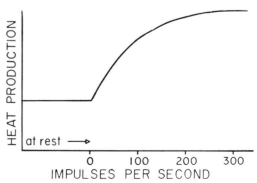

Figure 5–8   Heat production in a nerve at rest and at progressively increasing rates of stimulation.

lism. Note that the heat production increases markedly as the number of impulses per second increases. It is this increased energy that causes the "recharging" process.

Unimportance of Energy Metabolism During Actual Transmission of the Action Potential. The role of the sodium and potassium pumps to recharge the nerve fiber is simply to maintain adequate concentration differences of sodium and potassium across the nerve membrane. Even if metabolism in the nerve fiber suddenly ceased and the sodium and potassium ions were not returned to their original sides of the membrane, the nerve fiber would still be able to transmit approximately half a million nerve impulses before it would be necessary to recharge the fiber again. The reason for this is that the quantities of sodium and potassium ions that diffuse through the nerve membrane with each action potential are extremely minute, allowing tremendous numbers of action potentials to occur before the recharging process is necessary.

Thus, the action potential itself is a passive process in contrast to the recharging process, which is an active metabolic process requiring utilization of ATP as an energy source.

THE SPIKE POTENTIAL AND THE AFTER-POTENTIALS

Figure 5–9 illustrates an action potential recorded with a much slower time scale than that illustrated in Figure 5–5; many milliseconds of recording are shown in comparison with only the first 1.5 milliseconds of the action potential in Figure 5–5.

The Spike Potential. The initial very large change in membrane potential shown in Figure 5–9 is called the spike potential. In large type A nerve fibers it lasts for about 0.4 millisecond. The spike potential is analogous to the action potential that has been discussed in the preceding paragraphs, and it is the spike potential that is called the *nerve impulse.*

The Negative After-Potential. At the termination

of the spike potential, the membrane potential fails to return all the way to its resting level for another few milliseconds, as shown in Figure 5–9. This is called the *negative after-potential.* It is believed to result from a buildup of potassium ions immediately outside the membrane; this causes the concentration ratio of potassium across the membrane to be temporarily less than normal and therefore prevents full return of the normal resting membrane for a few additional milliseconds.

The Positive After-Potential. Once the membrane potential has returned to its resting value, it becomes a little more negative than its normal resting value; this excess negativity is called the *positive after-potential.* It is a fraction of a millivolt to, at most, a few millivolts more negative than the normal resting membrane potential, but it lasts from 50 milliseconds to as long as many seconds.

This positive after-potential is caused principally by the pumping of sodium outward through the nerve fiber membrane, which is the recharging process that was discussed previously. If the active transport processes are poisoned, the positive after-potential is lost, though both the action potential and the negative after-potential continue to occur.

(The student might wonder why greater negativity in the resting membrane potential is called positive rather than negative, and likewise, why the so-called negative after-potential is not named positive. The reason for this is that these potentials were first measured *outside* the nerve fibers rather than inside, and all potential changes on the outside are of exactly opposite polarity, whereas modern terminology expresses membrane potentials in terms of the inside potential rather than the outside potential.)

PLATEAU IN THE ACTION POTENTIAL

In some instances the excitable membrane does not repolarize immediately after depolarization, but, instead, the potential remains on a plateau near the peak of the spike sometimes for many milliseconds before repolarization begins. Such a plateau is illustrated in Figure 5–10, from which one can readily see that the plateau greatly prolongs the period of depolarization. It is this type of action potential that occurs in the heart, where the plateau lasts for as long as two- to three-tenths second and causes contraction of the heart muscle during this entire period of time.

The cause of the action potential plateau in some membranes is not known, but it has been postulated that the efflux of potassium at the end of the spike potential remains too low in these membranes to initiate early membrane inactivation. Yet when the process does begin, the inactivation process proceeds unabated, and the membrane becomes repolarized rapidly, as illustrated in Figure 5–10 by the rapid

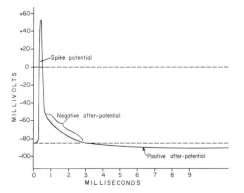

Figure 5–9 An idealized action potential, showing: the initial spike followed by a negative after-potential and a positive after-potential.

Figure 5–10   An action potential from cardiac muscle, showing a "plateau."

Figure 5–12   A myelinated nerve fiber. (Redrawn from Ramon y Cajal: Histology, William Wood and Co.)

decline of the potential at the end of the plateau.

SPECIAL ASPECTS OF IMPULSE TRANSMISSION IN NERVES

"Myelinated" and Unmyelinated Nerve Fibers.   Figure 5–11 illustrates a cross-section of a typical small nerve trunk, showing a few very large nerve fibers that comprise most of the cross-sectional area and many more small fibers lying between the large ones. The large fibers are *myelinated* and the small ones are *unmyelinated.* The average nerve trunk contains about twice as many unmyelinated fibers as myelinated fibers.

Figure 5–12 illustrates a typical myelinated fiber. The central core of the fiber is the *axon,* and the membrane of the axon is the actual *conductive membrane.* The axon is filled in its center with *axoplasm,* which is a viscid intracellular fluid. Surrounding the axon is a *myelin sheath* that is approximately as thick as the axon itself, and about once every milli-

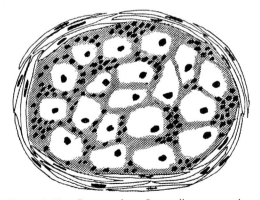

Figure 5–11   Cross-section of a small nerve trunk containing myelinated and unmyelinated fibers.

meter along the extent of the axon the myelin sheath is interrupted by a *node of Ranvier.*

The myelin sheath is deposited around the axon by Schwann cells in the following manner: The membrane of a Schwann cell first envelopes the axon. Then it rotates around the axon several times, laying down multiple layers of cellular membrane containing the lipid substance *sphingomyelin.* This substance is an excellent insulator that prevents almost all flow of ions. However, at the juncture between each two successive Schwann cells along the axon, a small uninsulated area remains where ions can flow with ease between the extracellular fluid and the axon. This is the node of Ranvier.

Saltatory Conduction in Myelinated Fibers. Even though ions cannot flow to a significant extent through the thick myelin sheaths of myelinated nerves, they can flow with considerable ease through the nodes of Ranvier. Indeed, the membrane at this point is 500 times as permeable as the membranes of some unmyelinated fibers. Impulses are conducted from node to node by the myelinated nerve rather than, as occurs in unmyelinated fiber, continuously along the entire fiber. This process, illustrated in Figure 5–13, is called *saltatory conduction.* That is, electrical current flows through the surrounding extracellular fluids and through the axoplasm from node to node, exciting successive nodes one after another. Thus, the impulse jumps down the fiber, which is the origin of the term "saltatory."

Saltatory conduction is believed to be of value for two reasons: First, by causing the depolarization process to jump long intervals along the axis of the nerve fiber, it is reasonable to believe that this mechanism helps to explain the high velocity of nerve transmission in myelinated fibers. Second, saltatory conduction conserves energy for the axon, for only

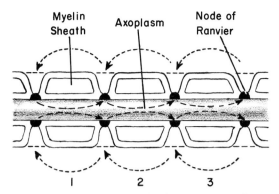

Figure 5–13   Saltatory conduction along a myelinated axon.

the nodes depolarize, allowing very little loss of ions and requiring little extra metabolism for retransporting the ions across the membrane.

VELOCITY OF CONDUCTION IN NERVE FIBERS

The velocity of conduction in nerve fibers varies from as little as 0.5 meter per second in very small fibers up to as high as 130 meters per second (the length of a football field) in very large fibers. Thus the velocity increases with the fiber diameter.

EXCITATION—THE PROCESS OF ELICITING THE ACTION POTENTIAL

Chemical Stimulation.   Basically, any factor that causes sodium ions to begin to diffuse inward through the membrane in sufficient numbers will set off the automatic "activation" mechanism, noted earlier in the chapter, that eventuates in the action potential. Thus, certain chemicals can stimulate a nerve fiber by increasing the membrane permeability. Such chemicals include acids, bases, almost any salt solution of very strong concentration, and, most importantly, the substance *acetylcholine*. In the body many nerve fibers, when stimulated, secrete acetylcholine at their endings where they synapse with other neurons or where they end on muscle fibers. The acetylcholine in turn stimulates the successive neuron or muscle fiber. This is discussed in much greater detail in the following chapter. This is probably the most important means in the body by which nerve and muscle fibers are stimulated. Under some conditions *norepinephrine* secreted by sympathetic nerve endings can stimulate smooth or cardiac muscle fibers, and it is possible that still other hormonal substances that can stimulate successive neurons are secreted in the central nervous system.

Mechanical Stimulation.   Crushing, pinching, or pricking a nerve fiber can cause a sudden surge of sodium influx and, for obvious reasons, can elicit an action potential. Even slight pressure on some specialized nerve endings can stimulate these; this will be discussed in Chapter 48 in relation to sensory perception.

Electrical Stimulation.   Electrical stimulation also can initiate an action potential. An electrical charge artificially induced across the membrane can cause excess flow of ions through the membrane; this in turn initiates an action potential. However, not all methods of applying electrical stimuli result in excitation, and, since this is the usual means by which nerve fibers are excited when they are studied in the laboratory, the process of electrical excitation deserves more detailed comment.

CATHODAL VERSUS ANODAL CURRENTS.   Figure 5–14 illustrates a battery connected to two electrodes on the surface of a nerve fiber. At the cathode, or negative electrode, the potential outside the membrane is negative with respect to that on the inside, and the current that flows through the membrane at this point is called *cathodal current*. At the anode, the electrode is positive with respect to the potential immediately inside the membrane, and the current flow at this point is called *anodal current*.

A cathodal current excites the fiber whereas an anodal current actually makes the fiber more resistant to excitation than normal. Though the cause of this difference between the two types of current cannot be explained completely, it is known that the normal impermeability of the membrane to sodium results partially from the high resting membrane potential across the membrane, and any condition that lessens this potential causes the membrane to become progressively more permeable to sodium. Obviously, at the cathode the applied voltage is opposite to the resting potential of the membrane, and this reduces the net potential. As a result, the membrane becomes far more permeable than usual to sodium followed by subsequent development of an action potential.

On the other hand, at the anode, the applied potential actually enhances the membrane potential. This makes the membrane less permeable to

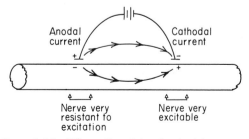

Figure 5–14   Effects of anodal and cathodal currents on excitability of the nerve membrane.

sodium than ever, resulting in increased resistance of the membrane to stimulation by other means.

THRESHOLD FOR EXCITATION AND "ACUTE SUB-THRESHOLD POTENTIAL." A very weak cathodal potential cannot excite the fiber. But, when this potential is progressively increased, there comes a point at which excitation takes place. Figure 5-15 illustrates the effects of successively applied cathodal stimuli of progressing strength. A very weak stimulus at point A causes the membrane potential to change from −85 to −80 millivolts, but this is not a sufficient change for the automatic processes of the action potential to develop. At point B the stimulus is greater, but, here again, the intensity is not enough to set off the automatic action potential. Nevertheless, the membrane voltage is disturbed for as long as a millisecond or more after both of the weak stimuli; the potential changes during these small intervals of time are called *acute subthreshold potentials,* as illustrated in the figure.

At point C in Figure 5-15 the stimulus elicits an acute membrane potential that is not subthreshold but slightly more than the threshold value, and, after a short "latent period," it initiates an action potential. At point D the stimulus is still stronger, and the acute membrane potential initiates the action potential even sooner. Thus, this figure shows that even a very weak stimulus always causes a local potential change at the membrane, but that the intensity of the *local potential* must rise to a *threshold value* before the automatic action potential will be set off.

EXCITABILITY CURVE OF NERVE FIBERS. A so-called "excitability curve" of a nerve fiber is shown in Figure 5-16. To obtain this curve a high voltage stimulus (4 volts in this instance) is first applied to the fiber, and the minimum duration of stimulus required to excite the fiber is found. The voltage and minimal time are plotted as point A. Then a stimulus voltage of 3 volts is applied, and the minimal time required is again determined; the results are plotted as point B. The same is repeated at 2 volts, 1 volt, 0.5 volt, and so forth, until the least voltage possible at which the membrane is stimulated has been reached. On connecting these

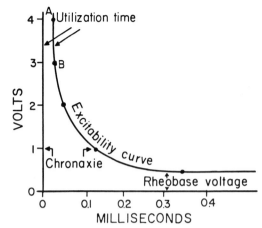

Figure 5-16 Excitability of a large myelinated nerve fiber.

points, the so-called excitability curve of the excitable tissue is determined.

The excitability curve of Figure 5-16 is that of a large myelinated nerve fiber. The least possible time required for stimulation even with very high voltage of stimulus is called the *utilization time.* On the other hand, the least possible voltage at which it will fire, however long the stimulus, is called the *rheobase.* Then, if the voltage is increased to twice the rheobase voltage, the time required to stimulate the fiber is called the *chronaxie,* and this is often used as a means of expressing relative excitabilities of different excitable tissues. For instance, the chronaxie of a large type A fiber is about 0.0001 to 0.0002 second; of smaller myelinated nerve fibers, approximately 0.0003 second; of unmyelinated fibers, 0.0005 second; of skeletal muscle fibers, 0.00025 to 0.001 second; and of heart muscle, 0.001 to 0.003 second.

"ACCOMMODATION" TO STIMULI. When a cathodal potential applied to a nerve fiber is made to increase very slowly, rather than rapidly, the threshold voltage required to cause firing is considerably increased. This phenomenon is called *accommodation;* in other words, the excitable membrane is said to "accommodate" itself to slowly increasing potentials rather than firing. It is probable that a slowly increasing stimulatory current allows time for ions to build up (or become depleted) in the areas immediately adjacent to the fibers, these ions partially opposing the stimulus.

The Refractory Period. A second action potential cannot occur in an excitable fiber as long as the membrane is still depolarized from the preceding action potential. Therefore, even an electrical stimulus of maximum strength applied before the first spike potential is almost over will not elicit a second one. This interval of inexcitability is called the *absolute refractory period.* The absolute refractory period of large myelinated nerve fibers is about $1/2500$ second. Therefore, one can readily calculate that such a fiber can carry a maximum

Figure 5-15 Effect of stimuli on the potential of the excitable membrane, showing the development of "acute subthreshold potentials" when the stimuli are below the threshold value required for eliciting an action potential.

of about 2500 impulses per second. Following the absolute refractory period is a *relative refractory period* lasting about one quarter as long. During this period, stronger than normal stimuli are required to excite the fiber. In some types of fibers, a short period of supernormal excitability follows the relative refractory period.

FACTORS THAT INCREASE MEMBRANE EXCITABILITY

Any condition that increases the natural permeability of the membrane usually causes it to become more excitable than usual. For instance, the drug *veratrine* has a direct action on the membrane to increase its permeability to sodium, and, as a consequence, the strength of stimulus needed to elicit an impulse is greatly reduced; or, on occasion, the fiber becomes so excitable that spontaneous impulses are generated without any extraneous excitation.

Low Calcium Tetany. An extremely important potentiator of excitability is low concentration of calcium ions in the extracellular fluids. Earlier in the chapter it was pointed out that calcium ions normally decrease the permeability of the membrane to sodium. If sufficient calcium ions are not available, however, the permeability becomes increased, and, as a result, the membrane excitability greatly increases – sometimes so greatly that many spontaneous impulses result and cause muscular spasm. This condition is known as low calcium tetany. It often occurs in patients with rickets or in patients who have lost their parathyroid glands and who therefore cannot maintain normal calcium ion concentrations. These conditions will be discussed in Chapter 79.

INHIBITION OF EXCITABILITY – "STABILIZERS" AND LOCAL ANESTHETICS

Contrary to the factors that increase excitability, still others called *membrane stabilizing factors* can decrease excitability. For instance, a *high calcium ion concentration* decreases the membrane permeability and simultaneously reduces its excitability. Therefore, calcium ions are said to be a "stabilizer." Also, *low potassium concentration* in the extracellular fluids, because it increases the resting membrane potential, likewise acts as a stabilizer and reduces membrane excitability. Indeed, in a hereditary disease known as *familial periodic paralysis,* the extracellular potassium concentration is often so greatly reduced that the person actually becomes paralyzed, but he reverts to normalcy instantly after intravenous administration of potassium.

Local Anesthetics and the "Safety Factor." Among the most important stabilizers are the many substances used clinically as local anesthetics, including *cocaine, procaine, tetracaine,* and many other drugs. These act directly on the membrane, decreasing its permeability to sodium and, therefore, also reducing membrane excitability. When the excitability has been reduced so low that the ratio of *action potential strength to excitability threshold* (called the "safety factor") is reduced below unity, a nerve impulse fails to pass through the anesthetized area.

RHYTHMICITY OF CERTAIN EXCITABLE TISSUES – REPETITIVE DISCHARGE

All excitable tissues can discharge repetitively if the threshold for stimulation is reduced low enough. For instance, even nerve fibers and skeletal muscle fibers, which normally are highly stable, discharge repetitively when they are veratrinized or when the calcium ion concentration falls below a critical value. Repetitive discharges, or rhythmicity, occur normally in the heart, in most smooth muscle, and probably also in some of the neurons of the central nervous system.

The Re-excitation Process Necessary for Rhythmicity. For rhythmicity to occur, the membrane, even in its natural state, is already permeable enough to allow automatic membrane activation and consequent depolarization. That is, (a) sodium ions flow inward, (b) this further increases the membrane permeability, (c) still more sodium ions flow inward, (d) the permeability increases more, and so forth, until the spike potential is generated. Then the spike declines until the resting membrane potential is re-established. Shortly thereafter a new action potential occurs spontaneously.

Yet why does the membrane not remain depolarized all the time instead of re-establishing a membrane potential, only to become depolarized again shortly thereafter? The answer to this can be found by referring back to Figure 5–5, which shows that toward the end of the action potential, and continuing for a short period thereafter, the membrane becomes excessively permeable to potassium. The excessive outflow of potassium carries tremendous numbers of positive charges to the outside of the membrane, creating inside the fiber considerably more negativity than would otherwise occur. This is a state called *hyperpolarization.* As long as this state exists, reexcitation will not occur; but gradually the excess potassium conductance (and the state of hyperpolarization) disappears, thereby allowing the onset of a new action potential.

Figure 5–17 illustrates this relationship between repetitive action potentials and potassium conductance. The state of hyperpolari-

Figure 5–17 Rhythmic action potentials, and their relationship to potassium conductance and to the state of hyperpolarization.

zation is established immediately after each preceding action potential; but it gradually recedes, and the membrane potential correspondingly increases until it reaches the threshold for excitation; then suddenly a new action potential results, the process occurring again and again.

RECORDING MEMBRANE POTENTIALS AND ACTION POTENTIALS

The Cathode Ray Oscilloscope. In this chapter we have noted that the membrane potential changes very rapidly throughout the course of an action potential. Indeed, most of the action potential complex of large nerve fibers takes place in less than $1/1000$ second. In some figures of this chapter a meter has been shown recording these potential changes. However, it must be understood that any meter capable of recording them must be capable of responding extremely rapidly. For practical purposes the only type of meter that is capable of responding accurately to the very rapid membrane potential changes of most excitable fibers is the cathode ray oscilloscope.

Figure 5–18 illustrates the basic components of a cathode ray oscilloscope. The cathode ray tube itself is composed basically of an *electron gun* and a *fluorescent surface* against which electrons are fired. Where the electrons hit the surface, the fluorescent material glows. If the electron beam is

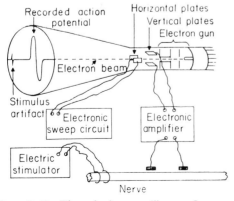

Figure 5–18 The cathode ray oscilloscope for recording transient action potentials.

moved across the surface, the spot of glowing light also moves.

In addition to the electron gun and fluorescent surface, the cathode ray tube is provided with two sets of plates: one set, called the horizontal deflection plates, positioned on either side of the electron beam and the other set, called the vertical deflection plates, positioned above and below the beam. If a negative charge is applied to the left-hand plate and a positive charge to the right-hand plate, the electron beam will be repelled away from the left plate and attracted toward the right plate, thus bending the beam toward the right, and this will cause the spot of light on the fluorescent surface of the cathode ray screen to move to the right. Likewise, positive and negative charges can be applied to the vertical deflection plates to move the beam up or down.

Since electrons travel at extremely rapid velocity and since the plates of the cathode ray tube can be alternately charged positively or negatively within less than a millionth of a second, it is obvious that the spot of light on the face of the tube can also be moved to almost any position in less than a millionth of a second. For this reason, the cathode ray tube oscilloscope can be considered to be an inertialess meter capable of recording with extreme fidelity almost any change in membrane potential.

To use the cathode ray tube for recording action potentials, two electrical circuits must be employed. These are (1) an *electronic sweep circuit* that controls the voltages on the horizontal deflection plates and (2) an *electronic amplifier* that controls the voltages on the vertical deflection plates. The sweep circuit automatically causes the spot of light to begin at the left-hand side and move slowly toward the right. When the spot reaches the right side it jumps back immediately to the left-hand side and starts a new trace.

The electronic amplifier amplifies signals that come from the nerve. If a change in membrane potential occurs while the spot of light is moving across the screen, this change in potential will be amplified and will cause the spot to rise above or fall below the mean level of the trace, as illustrated in the figure. In other words, the sweep circuit provides the lateral movement of the electron beam while the amplifier provides the vertical movement in direct proportion to the changes in membrane potentials picked up by appropriate electrodes.

Figure 5–18 also shows an electric stimulator used to stimulate the nerve. When the nerve is stimulated, a small *stimulus artifact* usually appears on the oscilloscope screen prior to the action potential.

Recording the Monophasic Action Potential. Throughout this chapter monophasic action potentials have been shown in the different diagrams. To record these, an electrode such as that illustrated earlier in the chapter in Figure 5–7 must be inserted into the interior of the fiber. Then, as the

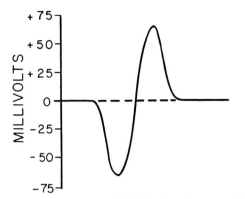

Figure 5–19 Recording of a biphasic action potential.

action potential spreads down the fiber, the changes in the potential inside the fiber are recorded as illustrated earlier in the chapter in Figures 5–5, 5–9, and 5–10.

Recording a Biphasic Action Potential. When one wishes to record impulses from a whole nerve trunk, it is not feasible to place electrodes inside the nerve fibers. Therefore, the usual method of recording is to place two electrodes on the outside of fibers. However, the record that is obtained is then biphasic for the following reasons: When an action potential moving down the nerve fiber reaches the first electrode, it becomes charged negatively while the second electrode is still unaffected. This causes the oscilloscope to record in the negative direction. Then as the action potential passes beyond the first electrode and reaches the second electrode, the membrane beneath the first electrode becomes repolarized while the second electrode is still negative, and the oscilloscope records in the opposite direction. When these changes are recorded continuously by the oscilloscope, a graphic record such as that illustrated in Figure 5–19 is recorded, showing a potential change first in one direction and then in the opposite direction.

REFERENCES

Baker, P. F.; Hodgkin, A. L.; and Shaw, T. I.: The effects of changes in internal ionic concentrations on the electrical properties of perfused giant axons. *J. Physiol. (Lond.), 164*:355, 1962.

Baker, P. F.; Hodgkin, A. L.; and Shaw, T. I.: Replacement of the axoplasm of giant nerve fibers with artificial solutions. *J. Physiol. (Lond.), 164*:330, 1962.

Caldwell, P. C.: Factors governing movement and distribution of inorganic ions in nerve and muscle. *Physiol. Rev., 48*:1, 1968.

Cole, K. S.: An analysis of the membrane potential along a clamped squid axon. *Biophys. J., 1*:401, 1961.

Cole, K. S.: Electrodiffusion models for the membrane of squid giant axon. *Physiol. Rev., 45*:340, 1965.

Effects of Anesthesia on Metabolism and Cellular Function (Symposium). *Pharmacol. Rev., 17*:183, 1965.

Frankenhaeuser, B.: Nerve membrane. *Proc. Int. Union of Physiol. Sciences, 6*:105, 1968.

Gonzalez, C. F.; Shamoo, Y. E.; Wyssbrod, H. R.; Solinger, R. E.; and Brodsky, W. A.: Electrical nature of sodium transport across the isolated turtle bladder. *Amer. J. Physiol., 213*:333, 1967.

Graham, G. D.; Bennett, R. B.; and Ware, F.: Potassium effects on transmembrane potentials in frog ventricle. *Amer. J. Physiol., 216*:1360, 1969.

Hodgkin, A. L.: The Conduction of the Nervous Impulse. Springfield, Ill., Charles C Thomas, Publisher, 1963.

Hodgkin, A. L., and Horowicz, P.: The effect of sudden changes in ionic concentrations on the membrane potential of single muscle fibres. *J. Physiol. (Lond.), 153*:370, 1960.

Hodgkin, A. L., and Huxley, A. F.: Movement of sodium and potassium ions during nervous activity. *Cold Spr. Harb. Symp. Quant. Biol., 17*:43, 1952.

Hodgkin, A. L., and Huxley, A. F.: Quantitative description of membrane current and its application to conduction and excitation in nerve. *J. Physiol. (Lond.), 117*:500, 1952.

Houck, D. J.: Effect of alcohols on potentials of lobster axons. *Amer. J. Physiol., 216*:364, 1969.

Huxley, A. F.: Ion movements during nerve activity. *Ann. N.Y. Acad. Sci., 81*:221, 1959.

Illanes, A.; Carpenter, R.; and Reuss, L.: Effects of temperature on membrane potentials of myocardium from toad and guinea pig. *Amer. J. Physiol., 218*:536, 1970.

Katz, B.: Nerve, Muscle, and Synapse. New York, McGraw-Hill Book Company, 1968.

Keynes, R. D.: Thermal and optical changes during the nerve impulse. *Proc. Int. Union of Physiol. Sciences, 6*:74, 1968.

Machne, X., and Orozco, R.: Effects of temperature and calcium variation on membrane potentials of crayfish axons. *Amer. J. Physiol., 214*:672, 1968.

Martin, A. R., and Veale, J. L.: The nervous system at the cellular level. *Ann. Rev. Physiol., 29*:401, 1967.

Maruhashi, J., and Wright, E. B.: Effect of oxygen lack on the single isolated mammalian (rat) nerve fiber. *J. Neurophysiol., 30*:434, 1967.

Moore, E. N., and Bloom, M.: A method for intracellular stimulation and recording using a single microelectrode. *J. Appl. Physiol., 27*:734, 1969.

Moore, J. W., and Adelman, W. J., Jr.: Electronic measurement of the intracellular concentration and net flux of sodium in the squid axon. *J. Gen. Physiol., 45*:77, 1961.

Mueller, P., and Rudin, D. O.: Resting and action potentials in experimental bimolecular lipid membranes. *J. Theor. Biol., 18*:222, 1968.

Nathan, P. W., and Sears, T. A.: Some factors concerned in differential nerve block by local anaesthetics. *J. Physiol. (Lond.), 157*:565, 1961.

Noble, D.: Applications of Hodgkin-Huxley equations to excitable tissues. *Physiol. Rev., 46*:1, 1966.

Perkins, M. S., and Wright, E. B.: The crustacean axon. I. Metabolic properties: ATPase activity, calcium, binding, and bioelectric correlations. *J. Neurophysiol., 32*:930, 1969.

Sachs, J. R., and Conrad, M. E.: Effect of tetraethylammonium on the active cation transport system of the red blood cell. *Amer. J. Physiol., 215*:795, 1968.

Schoepfle, G. M.: Effects of cyanide at different concentrations on single nerve fibers. *Amer. J. Physiol., 204*:77, 1963.

Schoepfle, G. M.; Johns, G. C.; and Molnar, C. E.: Stimulated responses of depressed and hyperpolarized medullated nerve fibers. *Amer. J. Physiol., 216*:932, 1969.

Segal, J. R.: Effect of metabolism on the excitability of the squid giant axon. *Amer. J. Physiol., 215*:467, 1968.

Seyama, I.: Effect of high calcium concentration on the action potential of the skate heart. *Amer. J. Physiol., 216*:687, 1969.

Shanes, A. M.: Electrochemical aspects of physiological and pharmacological action in excitable cells. *Pharmacol. Rev., 10*:59, 1958; *10*:165, 1958.

Strumwasser, F.: Nervous function at the cellular level. *Ann. Rev. Physiol., 27*:451, 1965.

Tasaki, I.: A new measurement of action currents developed by single nodes of Ranvier. *J. Neurophysiol., 27*:1199, 1964.

Tasaki, I.; Lerman, L.; and Watanabe, A.: Analysis of excitation process in squid giant axons under bi-ionic conditions. *Amer. J. Physiol., 216*:130, 1969.

Teorell, T.: Excitability phenomena in artificial membranes. *Biophys. J., 2*(2)Pt. 2:27, 1962.

Ugolev, A. M.: Membrane (contact) digestion. *Physiol. Rev., 45*: 555, 1965.

Wright, E. B., and Tomita, T.: Separation of sodium and potassium ion carrier systems in crustacean motor axon. *Amer. J. Physiol., 202*:856, 1962.

CHAPTER 6

Function of Neuronal Synapses and Neuromuscular Junctions

Every medical student is aware that information is transmitted in the central nervous system through a succession of neurons, one after another. However, it is not immediately apparent that the impulse may be (a) blocked in its transmission from one neuron to the next, (b) changed from a single impulse into repetitive impulses, or (c) integrated with impulses from other neurons to cause highly intricate patterns of impulses in successive neurons. All these functions can be classified as *synaptic functions of neurons.*

PHYSIOLOGIC ANATOMY OF THE SYNAPSE

The juncture between one neuron and the next is called a *synapse.* Figure 6–1 illustrates a typical *motoneuron* in the anterior horn of the spinal cord. It is comprised of three major parts, the *soma,* which is the main body of the neuron, the *axon,* which extends from the soma into the peripheral nerve, and the *dendrites,* which are relatively short projections of the soma into the surrounding areas of the cord.

It should be noted, also, that literally hundreds of small *presynaptic terminals* lie on the surfaces of the dendrites and soma. These terminals are the ends of nerve fibrils that originate in many other neurons, and, usually, not more than a few of the terminals are derived from any single previous neuron. Later it will become evident that many of these presynaptic terminals are *excitatory* and still others are *inhibitory.*

Neurons in other parts of the cord and brain differ markedly from the motoneuron in (1) the size of the cell body, (2) the length, size, and number of dendrites, ranging in length from almost none at all up to as long as one meter (the peripheral sensory nerve fiber), (3) the length and size of the axon, and (4) the number of presynaptic terminals, which may range from 100 to several thousand. It is these differences that make neurons in different parts of the nervous system react differently to incoming signals and therefore to perform different functions, as will be explained in Chapters 46 through 62.

Electron microscope studies of the presynaptic terminals show that they have varied anatomical forms, but most resemble small knobs and therefore are frequently called *terminal knobs or boutons.*

EXCITATORY FUNCTIONS OF THE SYNAPSES

MECHANISM OF EXCITATION AND THE EXCITATORY TRANSMITTER SUBSTANCE

The mechanism by which the presynaptic terminals excite the postsynaptic neuron is believed to be by secretion of an excitatory substance, called the *excitatory transmitter.* Figure 6–2 illustrates the basic structure of the presynaptic terminal. It is separated from the neuronal soma by a *synaptic cleft* having an average width of approximately 200 Ångstroms. The terminal has two internal structures im-

Figure 6–1 A typical motoneuron, showing synapses with presynaptic terminals that originate from other neurons.

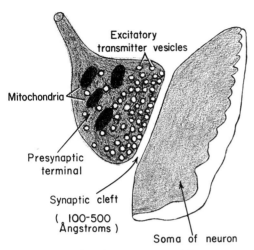

Figure 6-2   Physiologic anatomy of the synapse.

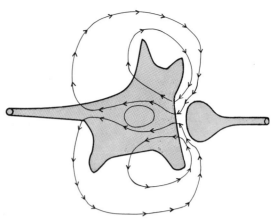

Figure 6-3   Flow of currents in the synaptic cleft and around the neuron, showing that a synaptic discharge causes decreased negativity inside the neuron not merely in the near vicinity of the synapse.

portant to the transmission of impulses through the synapse, the *synaptic vesicles* and the *mitochondria*. The synaptic vesicles contain the *excitatory transmitter,* which, when released into the synaptic cleft, excites the membrane of the neuronal soma. The ATP required for synthesis of new excitatory transmitter is provided by the mitochondria, and it is possible that the mitochondria even synthesize the excitatory transmitter itself. This transmitter must be synthesized extremely rapidly because the amount stored in the vesicles is sufficient to last for only a few minutes of normal function.

When an action potential spreads over a presynaptic terminal, the membrane depolarization causes automatic emptying of a small number of vesicles into the cleft; and the released excitatory transmitter in turn causes an immediate increase in permeability of the subsynaptic somal membrane, which allows sodium ions to flow freely from the extracellular fluids to the inside of the neuron, causing decreased electronegativity on the inside. This decreased electronegativity spreads almost equally throughout the inside of the neuron cell body because of the high degree of conductivity of the intracellular fluid. Therefore, as illustrated in Figure 6-3, the effect does not occur only in the vicinity of the synapse. If sufficient numbers of sodium ions diffuse inward, the neuron will discharge. This is discussed in detail in the following paragraphs.

Mechanism by Which the Action Potential Causes Release of Excitatory Transmitter Vesicles.   Unfortunately, we can only guess at the mechanism by which an action potential on reaching the terminal knob causes the vesicles to release transmitter substance into the synaptic cleft. However, the number of vesicles released with each action potential is greatly reduced (a) when the quantity of calcium ions in the extracellular fluid is diminished, (b) when the quantity of sodium ions in the extracellular fluid is diminished, (c) when the quantity of magnesium ions in the extracellular fluids is increased, or (d) when the membrane of the presynaptic terminal has already been partially depolarized prior to transmission of the action potential so that the action potential is weaker than usual. On the basis of these characteristics, it has been suggested that the spread of the action potential over the membrane of the terminal knob causes electrical attraction of the transmitter vesicles to the membrane and simultaneously greatly increases the porosity of the membrane to the substances in the vesicles, thus allowing these to spill into the synaptic cleft. Influx of calcium ions through the membrane of the terminal knob at the time of the action potential perhaps is the stimulus that ruptures the vesicles.

At any rate, it is known that each time an action potential travels over the surface of the terminal knob, whole vesicles of transmitter substance are emptied all at once into the synaptic cleft. It may be a few vesicles or many vesicles, depending upon the factors mentioned above.

One transmitter substance that occurs in certain parts of the nervous system is acetylcholine, as is discussed later. It has been calculated that about 3000 molecules of acetylcholine are present in each vesicle, and enough vesicles are present in each terminal knob to transmit about 10,000 impulses. As the vesicles are used up, new vesicles are synthesized to take their places.

Chemical Nature of the Excitatory Transmitter. The precise chemical nature of the excitatory transmitter secreted in the central nervous system synapses is doubtful, but it is almost certain that the excitatory transmitter at the synapses in the autonomic ganglia is *acetylcholine.* The major reasons for believing this are: (a) acetylcholine-like substances will stimulate the postganglionic neurons in the autonomic ganglia, (b) substances that prevent the destruction of acetylcholine by cholinesterase at the synapses will potentiate the transmission through the autonomic ganglia, and (c) measurable quantities of acetylcholine can be recovered from fluids perfusing the autonomic ganglia after strong stimulation of the preganglionic neurons.

In the central nervous system, injected acetylcholine will stimulate some of the neurons. Therefore, it is believed that at least some of the central nervous system excitatory presynaptic terminals secrete acetylcholine, and perhaps this is true of most of them.

However, it is probable that more than one type of excitatory transmitter substance is secreted in the central nervous system, because some excitatory nerve endings do not seem to have the necessary enzymes to synthesize acetylcholine. On the other hand, some do have the requisite enzymes to synthesize *norepinephrine,* which is the transmitter substance of the peripheral sympathetic nerve endings, as will be discussed later in the chapter. Also, this transmitter can excite at least some of the neurons in the central nervous system. Therefore it, too, is probably an excitatory transmitter at some synapses, particularly those on the intermediolateral horn cells of the spinal cord and perhaps some in the hypothalamus and other basal regions of the brain.

Other substances that have been suggested as possible excitatory transmitters include serotonin and 1–glutamic acid. However, far too little is yet known about the function of these substances to be certain of their action as excitatory transmitters.

The Excitatory Postsynaptic Potential and Its Relationship to the Threshold for Excitation of the Neuron. When an excitatory presynaptic terminal releases its transmitter, the neuronal membrane becomes highly permeable to both sodium and potassium ions for about 1 millisecond. During this time sufficient sodium diffuses to the interior of the cell to neutralize part of the normal resting potential caused primarily by potassium ion diffusion. That is,

the potential inside the cell becomes less negative. This change in resting membrane potential is called the *excitatory postsynaptic potential* (EPSP), and it persists for as long as 15 milliseconds. In the large motoneuron, the excitatory postsynaptic potential disappears with a half-time of approximately 4 milliseconds. That is, each 4 milliseconds it decreases by one-half.

Figure 6–4 illustrates several typical excitatory postsynaptic potentials. This figure also shows that the normal resting membrane potential of the soma of the neuron is approximately −70 mv. in contradistinction to some −85 mv. in the peripheral axons. The bottom excitatory postsynaptic potential in Figure 6–4 was *theoretically* caused by stimulation of only four excitatory synapses; then stimulation of two times as many synapses caused the next higher potential; and finally stimulation of four times as many caused a still much higher excitatory postsynaptic potential, this time causing an action potential to be generated in the neuron.

Thus, the basic effect of exciting the presynaptic terminal is to create an excitatory postsynaptic potential in the neuron. If this potential is very slight, the neuron will not discharge at all, but, if the potential rises above a certain level, called the *threshold for excitation* of the neuron, action potentials begin to appear in the neuron. The average threshold for excitation of the anterior motoneuron is 11 mv. above the resting membrane potential— that is, when the potential rises from its normal value of −70 to −59 mv., an action potential occurs.

Summation of Excitatory Postsynaptic Potentials. SPATIAL SUMMATION AND THE LOW THRESHOLD FOR EXCITATION OF THE "INITIAL

Figure 6–4 Excitatory postsynaptic potentials, showing that simultaneous firing of only a few synapses will not cause sufficient summated potential to elicit an action potential, but that simultaneous firing of many synapses will raise the summated potential to the threshold for excitation and cause a superimposed action potential. (The exact number of synapses firing is hypothetical.)

SEGMENT" OF THE AXON. Figure 6–4 shows that simultaneous excitation of successively greater numbers of excitatory presynaptic terminals causes progressive increase in the postsynaptic potential. This is called *spatial summation.*

It should be noted particularly that spatial summation does not occur only in the immediate vicinity of the excited terminals but, instead, *occurs over the entire membrane of the soma and initial segment of the axon, all at the same time.* Also, postsynaptic potentials generated in the dendrites will summate partially with those of the soma despite the wide spatial separation. The reason for all this summation is that the intracellular electrical resistance is very low, which allows a change in potential anywhere in the cell to spread to all areas of the inner surface of the cell membrane almost instantaneously.

Electrical recordings made directly from large neurons show that the first part of the neuron to initiate an action potential regardless of the positions of the excited presynaptic terminals on the neuron, is the *initial segment of the axon,* that is, the axon hillock and first 50 to 100 microns of the axon, which is non-myelinated. This part of the neuron discharges first because its threshold for excitation (in the anterior motoneuron) is only 11 mv. in contrast to 25 mv. for the soma. After the initial segment discharges, the action potential then travels forward in the axon and backwards over the soma.

This ability of the initial segment of the axon to respond to discharges from widely separated presynaptic terminals means that all the presynaptic discharges can summate together and not simply the discharges of presynaptic terminals that lie near each other on the surface of the cell.

TEMPORAL SUMMATION. Not only can discharges from separate presynaptic terminals summate with each other, but rapidly successive discharges *from the same presynaptic terminal* can also summate. Each excitatory postsynaptic potential lasts as long as 15 milliseconds, and, if the terminal discharges a second time before the 15 milliseconds is over, the new postsynaptic potential simply adds to what is left of the old. This effect, called *temporal summation,* is illustrated in Figure 6–5, which shows at the lower left two postsynaptic potentials separated by 15 milliseconds; the stimulation is then increased to once every 3 milliseconds, which causes a moderate amount of temporal summation. Finally, the rate is increased to once every millisecond, causing the summation

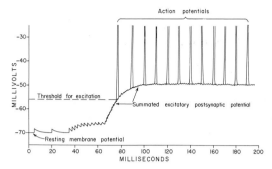

Figure 6–5 Temporal summation of excitatory postsynaptic potentials, showing that potentials widely spaced in time do not summate, potentials occurring once every 3 milliseconds summate moderately, and potentials occurring once every millisecond summate markedly.

to become very great and resulting in a very considerable rise in the summated postsynaptic potential. This time the potential becomes greater than the threshold for excitation of the neuron, and action potentials now appear repetitively because the presynaptic potential has risen above the *threshold* for excitation; these action potentials continue as long as the threshold is exceeded by this potential.

Facilitation of the Neuron; Excitation versus Facilitation. The postsynaptic potential caused by discharge of a single presynaptic terminal is almost always very slight—almost never sufficient to cause *excitation* of the neuron. Instead, either spatial summation or temporal summation or both must occur for the potential to become great enough to excite the neuron.

However, even subthreshold levels of postsynaptic potential cause a neuron to become *facilitated.* This means that the neuron can then be excited much more easily by additional presynaptic discharges than would normally be true. The degree of facilitation may be *almost* great enough to cause excitation, in which case even a single additional discharge of an excitatory terminal might cause an action potential in the neuron. On the other hand, the neuron may be facilitated only a very minute amount, in which case a large number of other presynaptic discharges will still be necessary to excite the neuron.

In summary, if the summated postsynaptic potential is greater than the threshold for excitation of the neuron, an action potential occurs, but if it is less than the threshold, then the neuron becomes facilitated but not excited.

Relationship of Summated Postsynaptic Potential to Rate of Discharge of the Neuron. When the summated postsynaptic potential rises only barely above the threshold for excitation, an action potential occurs in the neuron. If the summated postsynaptic potential remains at this level, a second action potential will appear after another 50 to 200 milliseconds: that is, when the neuron has recovered completely from the previous action potential. On the other hand, if the summated postsynaptic potential is far greater than the threshold level for excitation, the neuron can now discharge every 1 to 5 milliseconds because recovery now needs to be only partial before a new discharge can take place. In other words, a neuron is completely *refractory* to stimulation a second time while an action potential is still being transmitted. However, immediately after the action potential is over (within less than 1 millisecond in a large neuron), the neuron can be excited again but only with a summated postsynaptic potential far above threshold value. That is, the neuron is in a *relative refractory state*. This period lasts for 50 to 200 milliseconds, during which time the neuron progressively recovers and gradually becomes as excitable as ever. Thus, a barely threshold excitatory potential causes the neuron to fire approximately once every 50 to 200 milliseconds, or at a rate of 5 to 20 times per second. Then as the postsynaptic potential rises progressively above threshold, the repetitive discharges appear closer and closer together, rising to as many as 1000 or more impulses per second in some neurons.

Figure 6–6 demonstrates these relationships between level of summated postsynaptic potential and rate of firing of the neuron. Note that no action potential occurs until the postsynaptic potential rises to threshold level. As it rises somewhat above threshold, discharges then occur every 25 milliseconds. Shortly thereafter the postsynaptic potential falls back exactly to threshold level, and 90 milliseconds then lapse between successive action poten-

tials. Finally, toward the end of the record the postsynaptic potential rises to a very high level, and action potentials then occur about 8 milliseconds apart, giving a repetitive firing rate of 125 discharges per second.

Response Characteristics of Different Neurons to Increasing Levels of Excitatory Postsynaptic Potential. Histological study of the nervous system immediately convinces one of the widely varying types of neurons in different parts of the nervous system. And, physiologically, the different types of neurons perform different functions. Therefore, as would be expected, the ability to respond to presynaptic discharges varies from one type of neuron to another.

Figure 6–7 illustrates theoretical responses of three different types of neurons to varying levels of summated postsynaptic potential. Note that neuron number 1 will not discharge at all until the postsynaptic potential rises to 5 millivolts. Then, as the summated potential rises progressively to 35 mv. the frequency of discharge rises to slightly over 500 per second.

Neuron number 2 is quite a different type, having a threshold for excitation of 8 mv. and a maximum rate of discharge of only 190 per second. Finally, neuron number 3 has a high threshold for excitation, about 18 mv., but its discharge rate rises rapidly to over 600 per second as the postsynaptic potential rises only slightly above threshold.

Thus, different neurons respond differently, have different thresholds for excitation, and have widely differing maximal frequencies of discharge. With a little imagination one can readily understand the importance of having neurons with many different types of response characteristics to perform the widely varying functions of the nervous system.

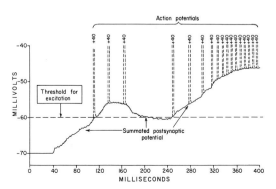

Figure 6–6   Relationship between the level of summated excitatory postsynaptic potential and the rate of discharge of the neuron.

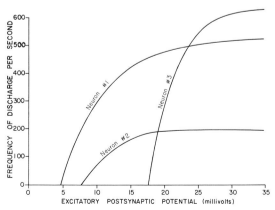

Figure 6–7   Response characteristics of different types of neurons to progressively increasing levels of excitatory postsynaptic potential.

INHIBITORY FUNCTION OF SYNAPSES

The Inhibitory Postsynaptic Potential and the Mechanism of "Postsynaptic" Inhibition. Two different types of inhibition occur at the synapse—*postsynaptic inhibition* and *presynaptic inhibition*. Postsynaptic inhibition results from secretion by certain presynaptic terminals of an *inhibitory transmitter* instead of an excitatory transmitter. These terminals derive from special types of neurons that secrete only inhibitory transmitter, rather than excitatory transmitter, at all of their endings, as will be discussed below.

The mechanism of postsynaptic inhibition is not certain, but it has been postulated that the inhibitory transmitter increases the permeability of the neuronal membrane to potassium and chloride ions but not to sodium ions. Since hydrated sodium ions have a diameter only about 1 Å greater than that of hydrated potassium and chloride ions, it is difficult to understand how the pore size could suddenly be increased to a size barely large enough to allow potassium and chloride ions to pass through freely without at least some of the pores becoming large enough for passage of sodium ions. Therefore, as a temporary expedient, it has been suggested that there are really two different types of pores in the neuronal membrane: one that has a maximum size slightly less than 5 Å in diameter, which is large enough to pass potassium and chloride but not sodium, and another that has a pore size somewhat larger than 5 Å, which is large enough to allow passage of sodium as well. It has also been assumed that most of these pores are normally blocked by the presence of electrical charges in the channels. Furthermore, it is believed that the inhibitory transmitter opens the smaller pores, perhaps by removing electrical charges, thereby allowing greatly increased permeability of the membrane to potassium and chloride ions. On the other hand, it is assumed that the excitatory transmitter opens the larger pores, thereby allowing passage of essentially all ions; and since there is a far greater electrochemical gradient for movement of sodium ions through the pores than there is for either of the other two ions, the excitatory transmitter causes primarily sodium ion movement in contrast to the potassium and chloride ion movement caused by the inhibitory transmitter.

ELECTRICAL EVENTS AT THE MEMBRANE DURING INHIBITION. When the membrane becomes highly permeable to both potassium and chloride ions, potassium ions move through the membrane rapidly because there is an excess of potassium inside the cell in relation to the electrical potential that is attempting to hold them inside. On the other hand, chloride ions move either not at all or hardly at all because the chloride ions are distributed across the membrane in exact equilibrium with the electrochemical potential for chloride, as was discussed in Chapter 4. Movement of potassium ions out of the neuron causes positive charges to move from inside the membrane to the outside, leaving a greater degree of negativity in the neuron. In the spinal motoneuron, this negativity can increase from the normal value of approximately −70 mv. to a value as great as −80 mv., which is 10 mv. more negative than the resting somal membrane potential. This state, called *hyperpolarization,* is illustrated in Figure 6–8A, which shows an initial resting membrane potential and then a hyperpolarizing *inhibitory postsynaptic potential* (IPSP) superimposed upon the resting membrane potential. The inhibitory postsynaptic potential, like the excitatory postsynaptic potential, lasts for 10 to 15 milliseconds. This long persistence is also shown in Figure 6–8A.

Figure 6–8B illustrates the manner in which the inhibitory postsynaptic potential inhibits excitation. This shows a positive increase in the summated postsynaptic potential when only excitatory presynaptic terminals are discharging. Then, while these same excitatory terminals continue to discharge, inhibitory terminals also discharge and decrease the summated potential back to a more negative value. Note that as long as only excitatory terminals are discharging, the postsynaptic potential is above the threshold for excitation, and action potentials appear; but after inhibition begins, the action potentials cease. Inhibitory signals can also inhibit neurons that fire repetitively because of their own intrinsic excitability.

In some neurons an inhibitory postsynaptic potential does not occur when the inhibitory transmitter is released at the presynaptic terminals, but even so the neuron still be-

Figure 6–8 A mechanism of postsynaptic inhibition, showing: (A) a single inhibitory postsynaptic potential and (B) decreased summated postsynaptic potential caused by superimposition of an inhibitory postsynaptic potential on an excitatory postsynaptic potential.

comes inhibited. The reason for this is that free movement of potassium and chloride ions through the somal membrane acts to "clamp" the membrane potential at its resting level. That is, this free movement of both chloride and potassium ions through the membrane makes it necessary for far greater than usual numbers of sodium ions to move through the membrane to cause excitation. This effect can be understood by referring to the Nernst equation given in Chapter 4. The effectiveness of ions in determining the membrane potential is directly proportional to the permeability of the membrane to each of the ions. Therefore, opening the pores of the membrane to potassium and chloride will make these two ions very effective in maintaining a very negative membrane potential even in the face of increased permeability of the membrane to sodium.

THE INHIBITORY TRANSMITTER. The precise chemical nature of the inhibitory transmitter, like that of the excitatory transmitter, is unknown. A substance called *gamma aminobutyric acid* (GABA), which is present in the cerebral cortex, is known to inhibit the excitatory action of excitatory presynaptic terminals in some lower animals. Therefore, this substance has been postulated to be an inhibitory transmitter. Another substance that will inhibit neurons and that might possibly act as an inhibitory transmitter in some areas of the nervous system is the amino acid glycine.

Inhibitory Neurons. The inhibitory presynaptic terminals all derive from special types of neurons called inhibitory neurons, and all terminal fibers branching from these neurons secrete only inhibitory transmitter and never excitatory transmitter. (Likewise, excitatory neurons secrete only excitatory transmitter.) When we study the organization of the nervous system in Chapters 46 through 62, it will become apparent that the mechanism by which most inhibition occurs in the nervous system is the passage of a signal first through one of these inhibitory neurons, which in turn secretes the inhibitory transmitter at the next synapse, resulting in inhibition.

"Presynaptic" Inhibition. The second type of inhibition, called presynaptic inhibition, is believed to be caused by excitatory terminals lying on the surfaces of excitatory presynaptic nerve fibrils. When these terminals release the excitatory transmitter, the presynaptic fibrils and their terminals are believed to become partially depolarized, and this prevents their releasing normal quantities of excitatory transmitter. Though the exact mechanism of presynaptic inhibition is still not clear, it is known that the amount of excitatory transmitter released is decreased.

Function of Inhibition in the Central Nervous System. Many nervous pathways are routed through inhibitory neurons, thereby causing inhibition of successive neurons rather than excitation. This inhibition prevents nervous activity that otherwise would occur and thereby helps to determine the overall function of the nervous system. For instance, all parts of the central nervous system are continually barraged by large numbers of incoming nerve impulses from peripheral nerves. If it were not for adequate inhibition of most of these input signals, the brain would remain in a state of such continual excitation that it would be almost useless. Thus, inhibition plays a role in *selecting* those signals that are important while blocking those that are unimportant. Such effects will be discussed much more fully in Chapters 46 through 62 in relation to central nervous system organization.

SOME SPECIAL CHARACTERISTICS OF SYNAPTIC TRANSMISSION

One-Way Conduction through Synapses. From the above discussion, it should be evident by now that impulses are conducted through synapses only from the presynaptic terminals to the successive neurons and never in the reverse direction. This is called the *Bell-Magendie law.*

Synaptic Delay. In transmission of an impulse from a presynaptic terminal to a postsynaptic neuron, a certain amount of time is consumed in the process of (a) discharge of the transmitter substance by the presynaptic terminal, (b) diffusion of the transmitter to the subsynaptic neuronal membrane, (c) action of the transmitter on the membrane, and (d) inward diffusion of sodium to raise the excitatory postsynaptic potential to a high enough value to elicit an action potential. The minimal period of time required for all these events to take place, even when large numbers of presynaptic terminals are stimulated simultaneously, is approximately 0.5 millisecond. This is called the *synaptic delay.* It is important for the following reason: Neurophysiologists can measure the minimal delay time between an input volley of impulses and an output volley and from this can estimate the number of series neurons in the circuit.

Fatigue of Synaptic Transmission. When the presynaptic terminals are repetitively stimulated at a rapid rate, the number of discharges by the postsynaptic neuron is at first very great but becomes progressively less in succeeding milliseconds or seconds. This is called fatigue of synaptic transmission. Fatigue is an exceedingly important characteristic of synaptic function, for when areas of the nervous system become overexcited, fatigue causes them to lose this excess excitability after a while. For example, fatigue is probably the means by which the excess excitability of the brain during an epileptic fit is finally subdued so that the fit ceases. Thus, the development of fatigue is a protective mechanism against excess neuronal activity. This will be discussed further in the description of reverberating neuronal circuits in Chapter 47.

The mechanism of fatigue has been presumed to be simply exhaustion of the stores of transmitter substance in the presynaptic terminals, particularly since it has been calculated that the excitatory presynaptic terminals can store enough excitatory transmitter for only 10,000 normal synaptic transmissions, an amount that can be exhausted in only a few seconds.

Post-Tetanic Facilitation. When a rapidly repetitive (tetanizing) stimulus is applied to the presynaptic terminals for a short period of time (less than the time for fatigue to develop), the neuron will usually become even more responsive to incoming impulses than normally. This is called post-tetanic facilitation.

One of the most likely explanations of post-tetanic facilitation is that the repetitive stimulation alters the membrane of the presynaptic terminals to cause more rapid emptying of transmitter vesicles.

The physiological significance of post-tetanic facilitation is yet very doubtful, and it may have no real significance at all. However, since post-tetanic facilitation can last from a fraction of a second in some neurons to many minutes in others, it is immediately apparent that neurons could possibly store information by this mechanism. Therefore post-tetanic facilitation might well be a mechanism of "short-term" memory in the central nervous system. This possibility will be discussed at further length in Chapter 61 in relation to the memory function of the cerebral cortex.

Increase in Numbers of Presynaptic Terminals Following Prolonged Repetitive Stimulation. Prolonged repetitive stimulation of synapses over a period of many days has been reported by some anatomists to cause the number of presynaptic terminals at each synapse to increase and also to cause the shapes and sizes of the terminals to change. *It is suggested that these physical changes might be the basis of memory.*

That is, incoming sensory signals could theoretically cause structural changes in the synapses in such a way that the synapses could be stimulated at a later time much more easily than the first time, so easily sometimes that even a stray impulse from anywhere in the nervous system could elicit the same signals as those elicited by the original sensory experience. These relationships of possible synaptic changes to memory will be discussed in much greater detail in Chapter 61.

Effect of Acidosis and Alkalosis on Synaptic Transmission. The neurons are highly responsive to changes in pH of the surrounding interstitial fluids. *Alkalosis greatly increases neuronal excitability.* For instance, a rise in arterial pH from the normal of 7.4 to about 7.8 often causes cerebral convulsions because of increased excitability of the neurons. This can be demonstrated especially well by having a person who is normally predisposed to epileptic fits overbreathe. The overbreathing elevates the pH of the blood only momentarily, but even this short interval can often precipitate an epileptic convulsive attack.

On the other hand, *acidosis greatly depresses neuronal activity;* a fall in pH from 7.4 to about 7.0 usually causes a comatose state. For instance, in very severe diabetic or uremic acidosis, coma always develops.

Effect of Hypoxia on Synaptic Transmission. Neuronal excitability is also highly dependent on an adequate supply of oxygen. Cessation of oxygen supply for only a few seconds can cause complete inexcitability of the neurons. This is often seen when the cerebral circulation is temporarily interrupted, for within less than 3 to 5 seconds the person becomes unconscious.

Effect of Drugs on Synaptic Transmission. Many different drugs are known to increase the excitability of neurons, and others are known to decrease the excitability. For instance, caffeine, theophylline, and theobromine, which are found in coffee, tea, and cocoa, respectively, all increase neuronal excitability, presumably by reducing the threshold for excitation of the neurons. However, strychnine, which is one of the best known of all the agents that increase the excitability of neurons, does not reduce the threshold for excitation of the neurons at all but, instead, *inhibits the action of the inhibitory transmitter* on the neurons. In consequence, the effects of the excitatory transmitter from the excitatory terminals become overwhelming, and the neurons become so excited that they go into rapidly repetitive discharge, resulting in severe convulsions.

The hypnotics and anesthetics have long been believed to increase the threshold for excitation of the neurons and thereby to decrease neuronal activity throughout the body. This is based principally on the fact that most of the volatile anesthetics are chemically inert compounds but are lipid soluble. Therefore, it has been reasoned that these substances might change the physical characteristics of the neuronal membranes, making them less excitable to excitatory agents. It still remains a possibility, however, that at least some hypnotics and anesthetics cause their anesthetic actions by (a) reducing the quantity of excitatory transmitter released or (b) enhancing the inhibitory effects of the inhibitory transmitter.

TRANSMISSION OF IMPULSES FROM NERVES TO SKELETAL MUSCLE FIBERS: THE NEUROMUSCULAR JUNCTION

Physiologic Anatomy of the Neuromuscular Junction. Figure 6–9A illustrates the neuromuscular junction between a large myelinated nerve fiber and a skeletal muscle fiber. The nerve fiber branches at its end to form the complicated structure called the *end-plate,* which invaginates into the muscle fiber but lies entirely outside the muscle fiber membrane. At the tips of the many nerve branches in the end-plate are *sole feet.*

Figure 6–9B shows an electron micrographic sketch of the juncture between a sole foot and the muscle fiber membrane. The invagination of the membrane is called the *synaptic gutter,* and the space between the sole foot and the fiber membrane is called the *synaptic cleft.* The synaptic cleft is filled with a gelatinous

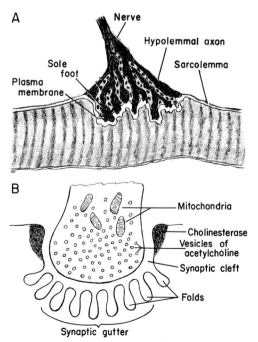

Figure 6–9 (A) The neuromuscular junction. (B) Invagination of a sole foot into the membrane of the muscle fiber.

"ground" substance through which diffuses extracellular fluid. At the bottom of the gutter are numerous *folds* of the muscle membrane which greatly increase the surface area at which the synaptic transmitter can act. In the sole foot are many mitochondria that supposedly synthesize the excitatory transmitter *acetylcholine* or that at least supply the energy for this synthesis. The acetylcholine is stored in the sole foot in large numbers of small *vesicles.* Around the rim of the synaptic gutter are large aggregates of the enzyme *cholinesterase,* which is capable of destroying acetylcholine, as is explained in further detail below.

Secretion of Acetylcholine by the Sole Feet. When a nerve impulse reaches the neuromuscular junction, some 50 to 100 vesicles of acetylcholine are released by the sole feet into the synaptic clefts between the sole feet and the muscle fiber membrane. It has been postulated that the nerve impulse causes calcium ions to move from the extracellular fluid into the membranes of the sole feet, and that it is these ions that in turn cause the vesicles of acetylcholine to rupture through the membrane. In the absence of calcium or in the presence of excess magnesium, the release of acetylcholine is greatly depressed.

DIFFUSION OF THE RELEASED ACETYLCHOLINE AWAY FROM THE MUSCLE FIBER MEMBRANE, AND ITS DESTRUCTION BY CHOLINESTERASE. Within approximately 2 milliseconds after acetylcholine is released by the sole feet, it diffuses out of the synaptic gutter and no longer acts on the muscle fiber membrane. The cholinesterase accumulated around the borders of the gutters begins immediately to destroy the acetylcholine, and this destruction is probably complete within another few milliseconds. The very short period of time that the acetylcholine remains in contact with the muscle fiber membrane, about 2 milliseconds, is almost always sufficient to excite the muscle fiber, and yet the rapid removal of the acetylcholine prevents re-excitation after the muscle fiber has recovered from the first action potential.

The "End-Plate Potential" and Excitation of the Skeletal Muscle Fiber. Even though the acetylcholine released into the space between the end-plate and the muscle membrane lasts for only a minute fraction of a second, nevertheless, even during this period of time it can affect the muscle membrane sufficiently to make it very permeable to sodium ions, allowing rapid influx of sodium into the muscle fiber. As a result, the membrane potential rises a few millivolts, creating a potential called the end-plate potential, which is entirely analogous to the excitatory postsynaptic potential that excites a neuron, as explained earlier in the chapter.

A typical end-plate potential is illustrated at point A in Figure 6–10. This potential was recorded after the muscle had been poisoned with curare, which made the end-plate potential too low to excite an action potential in the muscle fiber. However, in the normal muscle fiber, the effect shown at point B occurs: The end-plate potential begins, but before it can complete its course, the more intense electrical charges created by this potential stimulate the adjacent portions of the muscle fiber and initiate an action potential. At point C, only a small quantity of acetylcholine is released at the end-plate, causing the end-plate potential again to be less than the threshold value required to cause an action potential. In this case the end-plate potential is analogous to the subthreshold excitatory postsynaptic potential that follows discharge of too few presynaptic terminals to cause discharge of a neuron.

"Safety Factor" for Transmission at the Neuromuscular Junction; Fatigue of the Junction. Ordinarily, each impulse that arrives at the neuromuscular junction creates an end-plate potential about three to four times that required to stimulate the muscle fiber. Therefore, the normal neuromuscular junction is said to have a very high *safety factor.* Stimulation of the nerve fiber at rates greater than 150 times per second for many minutes often diminishes the quantity of acetylcholine released with each impulse so that impulses then often fail to pass into the muscle fiber. This is called *fatigue* of the neuromuscular junction, and it is analogous to fatigue of the synapse. However, under normal functioning conditions, fatigue of the neuromuscular junc-

Figure 6–10 End-plate potentials: *A*, normal end-plate potential recorded in a curarized muscle so that the muscle fiber itself could not generate an action potential; *B*, end-plate potential eliciting a muscle action potential; and *C*, end-plate potential too weak to elicit a muscle action potential.

tion almost never occurs, because it is rare that more than 150 impulses per second reach even the most active neuromuscular junctions.

Drugs that Affect Transmission at the Neuromuscular Junction. Drugs that Stimulate the Muscle Fiber by Acetylcholine-Like Action. Many different compounds, including *methacholine, carbachol,* and *nicotine,* have the same effect on the muscle fiber as does acetylcholine. The difference between these drugs and acetylcholine is that they are not destroyed by cholinesterase or are destroyed very slowly, so that when once applied to the muscle fiber the action persists for many minutes to several hours. Moderate quantities of the above three drugs applied to a muscle fiber cause localized areas of depolarization, and every time the muscle fiber becomes repolarized elsewhere, these depolarized areas, by virtue of their leaking ions, cause new action potentials, thereby causing a state of spasm. On the other hand, when extreme dosages of these drugs are used, so much of the membranes becomes depolarized that the fibers can no longer pass impulses at all, and a state of flaccid paralysis exists instead of the spasm that occurs with moderate dosages.

Drugs that Block Transmission at the Neuromuscular Junction. One group of drugs, known as the *curariform drugs,* can prevent passage of impulses from the end-plate into the muscle. Curare itself affects the membrane in such a way that acetylcholine cannot increase the permeability of the membrane sufficiently to initiate a depolarization wave. That is, the end-plate potential does not rise to threshold level. Curare apparently inhibits an *acetylcholine receptor substance* at the end-plate, which substance is necessary for the action of acetylcholine on the membrane.

A second group of drugs prevents transmission of impulses into muscle fibers by a different mechanism. These, of which *decamethonium* is a principal example, act in the same manner as *large doses* of nicotine, methacholine, and carbachol to depolarize the muscle fibers and to keep them depolarized. As a result, no impulses can be transmitted over the muscle fiber membranes even though the end-plate of the motor nerve fiber functions entirely normally.

Drugs that Stimulate the Neuromuscular Junction by Inactivating Cholinesterase. Three particularly well-known drugs, *neostigmine, physostigmine,* and *diisopropyl fluorophosphate,* inactivate cholinesterase so that the cholinesterase normally in the muscle fibers will not hydrolyze acetylcholine released at the end-plate. As a result, acetylcholine increases in quantity with successive nerve impulses so that extreme amounts of acetylcholine can accumulate and repetitively stimulate the muscle fiber. This causes *muscular spasm* when even a few impulses reach the muscle; this can cause death that is due to laryngeal spasm, which smothers the person.

Neostigmine and physostigmine combine with cholinesterase to inactivate it for several hours, after which they are displaced from the cholinesterase so that it once again becomes active. On the other hand, diisopropyl fluorophosphate, which has military potential as a very powerful "nerve" gas, actually inactivates cholinesterase for several weeks, which makes this a particularly lethal drug.

Figure 6–11 Junction of terminal nerve fibrils with smooth muscle fibers, showing a rare discrete terminal but many points where naked axons touch smooth muscle fibers, which points are also believed to act as transmission junctions.

MYASTHENIA GRAVIS

A disease known as myasthenia gravis occurs in human beings possibly or probably as a result of the inability of the end-plate to secrete adequate acetylcholine. (Other possible causes are the presence of excessive quantities of cholinesterase at the nerve endings or nonresponsiveness of the muscle fiber membrane to acetylcholine.) As a result, the end-plate potentials developed at the nerve endings are too weak to stimulate the muscle fibers adequately. If the disease is intense enough, the patient dies of paralysis—in particular, of paralysis of the respiratory muscles. However, the disease can usually be ameliorated with several different drugs, as follows:

Treatment with Drugs. When a patient with myasthenia gravis is treated with a drug, such as neostigmine, that is capable of inactivating or destroying cholinesterase, the acetylcholine secreted by the end-plate is not destroyed immediately. If a sequence of nerve impulses arrives at the end-plate, the quantity of acetylcholine present at the membrane increases progressively until finally the end-plate potential caused by the acetylcholine rises above threshold value for stimulating the muscle fiber. Thus, it is possible by diminishing the quantity of cholinesterase in the muscles of a patient with myasthenia gravis to allow even the inadequate quantities of acetylcholine secreted at the end-plates to effect almost normal muscular activity.

EXCITATION OF SMOOTH MUSCLE

Physiologic Anatomy of Smooth Muscle Neuromuscular Junctions. Figure 6–11 illustrates a nerve fiber arborizing into a fine reticulum of terminal fibrils that spread among smooth muscle fibers. These terminal fibrils are insulated by sheaths of Schwann cells, but where the fibers come in contact with muscle fibers the sheath is often interrupted, allowing the naked axons to touch the muscle fibers. In rare instances, a minute nerve terminal invaginates into the membrane of a smooth muscle fiber and is thus similar to the end-plate of the skeletal neuromuscular junction. However, it is believed that most transmission occurs at the points of contact between the naked fibrils and the smooth muscle fibers. Also, some terminal nerve fibers probably release transmitter substance diffusely among the smooth muscle fibers, and the transmitter substance then diffuses in all directions to excite the smooth muscle.

Excitatory and Inhibitory Transmitter Substances at the Smooth Muscle Neuromuscular Junction. Two different transmitter substances known to be secreted by the autonomic nerves innervating smooth muscle are *acetylcholine* and *norepinephrine*. Acetylcholine is an excitatory substance for smooth muscle fibers in some organs but an inhibitory substance for smooth muscle in other organs. And when acetylcholine excites a muscle fiber, norepinephrine ordinarily inhibits it, or when acetylcholine inhibits a fiber norepinephrine excites it.

It is believed that *receptor substances* in the membranes of the different smooth muscle fibers determine which will excite them, acetylcholine or norepinephrine. These receptor substances will be discussed in more detail in Chapter 59 in relation to function of the autonomic nervous system.

Transmission of Action Potentials to Smooth Muscle Fibers—The Junctional Potential. Transmission of impulses from terminal nerve fibers to smooth muscle fibers occurs in very much the same manner as transmission at the neuromuscular junction of skeletal muscle fibers except for temporal differences. That is, depolarization of the smooth muscle fiber is slow to develop, and once it does develop it lasts for a very long period of time. To be more exact, when an action potential reaches the terminal of an excitatory nerve fibril, there is a latent period of about 50 milliseconds before any change in the membrane potential of the smooth muscle fiber can be detected. Then the potential rises to a maximal level in approximately 100 milliseconds. If an action potential does not occur, this potential gradually disappears at a rate of approximately one-half every 200 to 500 milliseconds. This complete sequence of potential changes is called the *junctional potential*; it is analogous to both the excitatory postsynaptic potential of a neuron and the end-plate potential of the skeletal muscle fibers except that its duration is 50 to 200 times as long.

If the junctional potential rises to the threshold level for discharge of the smooth muscle membrane, an action potential will occur in the smooth muscle fiber in exactly the same way that an action potential occurs in a skeletal muscle fiber. A typical smooth muscle fiber has a normal resting membrane potential of −50 to −60 mv., and the threshold potential at which the action potential occurs is about −30 to −40 mv.

INHIBITION AT THE SMOOTH MUSCLE NEUROMUSCULAR JUNCTION.   When an inhibitory transmitter substance is released instead of an excitatory transmitter, the membrane potential of the muscle fiber becomes more negative than ever; that is, it becomes *hyperpolarized* and therefore becomes much more difficult to excite than is usually the case. Thus, inhibition occurs at the smooth muscle fiber in identically the same way that inhibition occurs at the neuronal synapse.

TRANSMISSION TO HEART MUSCLE

Autonomic nerve fibers that enter the heart arborize and synapse with cardiac muscle fibers in much the same manner as with smooth muscle fibers, and transmission to the cardiac fibers probably also occurs in much the same manner. Norepinephrine acts as an excitatory transmitter, and acetylcholine acts as an inhibitory transmitter.

Heart muscle, as is also true of some types of smooth muscle, has an intrinsic self-excitatory mechanism that causes continual rhythmic cardiac contraction. (This intrinsic rhythmicity of excitable membranes was explained in Chapter 5.) When norepinephrine is released among heart muscle fibers, the membrane potential becomes less negative than usual, which makes the membrane more excitable and therefore increases the heart rate. When acetylcholine is released, on the other hand, the membrane potential becomes more negative, developing a state of *hyperpolarization*. The muscle then becomes less excitable than normal and the heart rate slows, or in some instances the rhythmic self-excitation stops entirely.

Norepinephrine is believed to depolarize the heart muscle membrane by increasing its permeability to all ions, which allows leakage of positively charged sodium ions to the inside of the fiber. And acetylcholine is believed to hyperpolarize the membrane by selectively increasing its permeability to potassium ions, which allows excess positive charges to leak to the outside of the fiber, thereby increasing the negativity of the membrane potential. Thus, the depolarizing and hyperpolarizing effects of excitatory and inhibitory transmitter substances function in the same manner at the heart muscle as at the central nervous system neuronal synapse and at the smooth muscle junction.

REFERENCES

Bennett, M. R., and Burnstock, G.: Electrophysiology of the innervation of intestinal smooth muscle. *In* Handbook of Physiology. Baltimore, The Williams & Wilkins Co., 1968, Vol. IV, Sec. 6, p. 1709.

Bradley, P. B.: Pharmacology of the central nervous system. *Brit. Med. Bull., 21:*1, 1965.

Burgen, A. S. V.: The actions of drugs on post-junctional receptors. *Proc. Int. Union of Physiol. Sciences, 6:*283, 1968.

Drachman, D. B., and Houk, J.: Effect of botulinum toxin on speed of skeletal muscle contraction. *Amer. J. Physiol., 216:* 1453, 1969.

Eccles, J.: The Physiology of Synapses. New York, Academic Press, Inc., 1964.

Eccles, J.: The synapse. *Sci. Amer., 212:*56(1), 1965.

Elmqvist, D.: Neuromuscular transmission with special reference to myasthenia gravis. *Acta Physiol. Scand., 64,* Suppl. 249, 1965.

Guyton, A. C., and MacDonald, M. A.: Physiology of botulinus toxin. *Arch. Neurol. & Psychiat.,* 57:578, 1947.

Guyton, A. C., and Reeder, R. C.: The dynamics of curarization. *J. Pharmacol. Exp. Ther.,* 97:322, 1949.

Hebb, C.: CNS at the cellular level: identity of transmitter agents. *Ann. Rev. Physiol.,* 32:165, 1970.

Hebb, C.: Synthesis and storage of acetylcholine considered as a special example of a synaptic transmitter. *Proc. Int. Union of Physiol. Sciences, 6:*277, 1968.

Hofmann, W. W.: Caffeine effects on transmitter depletion and mobilization at motor nerve terminals. *Amer. J. Physiol., 216:*621, 1969.

Kandel, E. R., and Kupfermann, I.: The functional organization of invertebrate ganglia. *Ann. Rev. Physiol.,* 32:193, 1970.

Libet, B., and Tosaka, T.: Slow inhibitory and excitatory postsynaptic responses in single cells of mammalian sympathetic ganglia. *J. Neurophysiol.,* 32:43, 1969.

Maeno, T.: Analysis of mobilization and demobilization processes in neuromuscular transmission in the frog. *J. Neurophysiol., 32:*793, 1969.

Maeno, T., and Edwards, C.: Neuromuscular facilitation with low-frequency stimulation and effects of some drugs. *J. Neurophysiol.,* 32:785, 1969.

Martin, A. R.: Mechanisms of transmitter release. *Proc. Int. Union of Physiol. Sciences, 6:*279, 1968.

Martin, A. R.: Quantal nature of synaptic transmission. *Physiol. Rev., 46(1):*51, 1966.

McLennan, H. D.: Synaptic Transmission. Philadelphia, W. B. Saunders Company, 1963.

Molbech, S., and Johansen, S. H.: Endurance time in static work during partial curarization. *J. Appl. Physiol.,* 27:44, 1969.

Murphy, J. T., and Renaud, L. P.: Mechanisms of inhibition in the ventromedial nucleus of the hypothalamus. *J. Neurophysiol., 32:*85, 1969.

Orlov, R. S.: Nervous activation of smooth muscle cells of intact animals. *Proc. Int. Union of Physiol. Sciences, 6*:146, 1968.

Parsons, S. L.: Mechanism of neuromuscular blockade by tetraethylammonium. *Amer. J. Physiol., 216*:925, 1969.

Sair, R. A.; Lister, D.; Moody, W. G.; Cassens, R. G.; Hoekstra, W. G.; and Briskey, E. J.: Action of curare and magnesium on striated muscle of stress-susceptible pigs. *Amer. J. Physiol., 218*:108, 1970.

Shapovalov, A. I.: Post-tetanic potentiation of monosynaptic and disynaptic actions from supraspinal structures on lumbar motoneurons. *J. Neurophysiol., 32*:948, 1969.

Simpson, J. A.: The defect in myasthenia gravis. *In* Bittar, E. Edward, and Bittar, Neville (eds.): The Biological Basis of Medicine. New York, Academic Press, Inc., 1969, Vol. 3, p. 345.

Stefani, E., and Gerschenfeld, H. M.: Comparative study of acetylcholine and 5-hydroxytryptamine receptors on single snail neurons. *J. Neurophysiol., 32*:64, 1969.

Takeuchi, A.: Action of transmitters on synaptic membranes. *Proc. Int. Union of Physiol. Sciences, 6*:281, 1968.

Tauc, L.: Transmission in invertebrate and vertebrate ganglia. *Physiol. Rev., 47(3)*:521, 1967.

CHAPTER 7

Contraction of Muscle

Approximately one half of the body is skeletal, smooth, or cardiac muscle. Many of the same principles of contraction apply to all these different types of muscle, but in the present chapter the function of skeletal and smooth muscle is considered mainly, whereas the specialized functions of cardiac muscle will be discussed in Chapter 14.

PHYSIOLOGIC ANATOMY OF SKELETAL MUSCLE

The Skeletal Muscle Fiber. All skeletal muscles of the body are made up of numerous muscle fibers ranging between 10 and 100 microns in diameter. In most muscles the fibers extend the entire length of the muscle, and except for about 2 per cent of the fibers, each of these is innervated by only one neuromuscular junction located almost exactly in the middle of the fiber.

Each muscle fiber contains several hundred to several thousand myofibrils, which are illustrated by the small dots in the cross-sectional view of Figure 7–1. Each myofibril in turn has, lying side-by-side, about 1500 *myosin filaments* and two times this many *actin filaments,* which are large polymerized protein molecules

Figure 7–1 Side and cross-sectional views of a skeletal muscle fiber.

that are responsible for muscle contraction. These can be seen in the electron micrograph of Figure 7–2, and are represented diagrammatically in Figure 7–3. The thick filaments are *myosin* and the thin filaments are *actin.* Note that the myosin and actin filaments interdigitate and thus the myofibrils have alternate light and dark bands. The light bands, which contain only actin filaments, are called *I bands* because they are *isotropic* to polarized light. The dark bands, which contain the myosin filaments as well as the ends of the actin filaments where they overlap the myosin, are called *A bands* because they are *anisotropic* to polarized light. The combination of an A and an I band is called a sarcomere, the total length of which in the resting state is about 2 microns.

Note in Figure 7–3 that the actin filaments are attached to each other at the so-called *Z line* or *Z membrane.* The Z membrane passes from one myofibril to the next, attaching them to each other all the way across the muscle fiber and causing the respective sarcomeres of adjacent myofibrils to lie side by side.

When a muscle fiber is stretched beyond its natural length, the ends of the actin filaments pull apart, leaving a light area in the center of the A band. This light area, called the *H zone,* is illustrated in Figure 7–3. Such an H zone does not occur in the normal functioning muscle; nevertheless, it demonstrates that the actin filaments can slide inward or outward in the spaces between the myosin filaments to allow muscle contraction or relaxation.

The actin filaments are longer than the myosin filaments, 2.05 microns in comparison with 1.60 microns. Therefore, as the actin filaments are pulled toward the center of the myosin filaments, their ends overlap. Indeed, the normal muscle functions with the ends of the actin filaments usually slightly overlapped, as will be discussed later.

THE SARCOPLASM.   The myofibrils are suspended in a matrix called *sarcoplasm,* which is

Figure 7–2 Electron micrograph of muscle myofibrils, showing the detailed organization of actin and myosin filaments in the fibril. Note the mitochondria lying between the myofibrils. (From Fawcett: The Cell.)

composed of usual intracellular constituents. The fluid of the sarcoplasm contains large quantities of potassium, magnesium, phosphate, and protein enzymes. Also present are large numbers of mitochondria that lie mainly in close apposition to the actin filaments of the I bands, suggesting that the actin filaments play a major role in utilizing the ATP formed by the mitochondria.

THE SARCOPLASMIC RETICULUM. Also in the sarcoplasm is an extensive endoplasmic reticulum, which in the muscle fiber is called the *sarcoplasmic reticulum.* This reticulum has a special organization that is extremely important in the control of muscle contraction, which will be discussed later in the chapter. Figure 7–4 illustrates the arrangement of this sarcoplasmic reticulum and shows how extensive it can be. The more rapidly contracting types of muscle have especially extensive sarcoplasmic reticulums, indicating that this structure is important in causing rapid muscle contraction, as will also be discussed later.

The sarcoplasmic reticulum is composed of two distinct and separate types of tubules called the *transverse,* or *T, tubules* and the *longitudinal tubules.* Figure 7–4 illustrates primarily the longitudinal tubules, which lie parallel to the myofibrils. However, it will be noted that each end of each longitudinal tubule terminates in a bulbous cistern abutting against a very small T tubule cut in cross-section. This area of contact between the longitudinal system and transverse system is called the *triad* because it is composed of a central small tubule and, on its sides, two bulbous cisterns of the longitudinal tubules. One triad

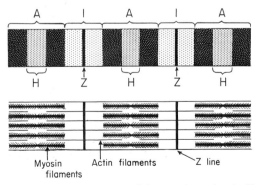

Figure 7–3 Arrangement of the myosin and actin filaments in the sarcomeres.

occurs at each point where the actin and myosin filaments overlap, thus making two triads to each sarcomere.

The T tubules collectively make up the *T system*, which provides a means of communication from the outside of the muscle fiber to its innermost portions. The T tubules pass all the way through the fiber from one side of the membrane to the other; they open to the exterior rather than to the interior of the cell and contain extracellular fluid rather than intracellular fluid. When an action potential spreads over the muscle fiber membrane, electrical currents are transmitted to the interior of the muscle fiber by way of the T tubule system.

CONDUCTION OF IMPULSES BY SKELETAL MUSCLE FIBERS

Almost everything said in Chapter 5 regarding excitation and conduction of impulses in nerve fibers applies equally as well, except for quantitative differences, to skeletal muscle fibers. However, several specific qualities of skeletal muscle excitation and impulse transmission must be mentioned.

The Resting Potential and Action Potential in Skeletal Muscle. The normal resting potential of skeletal muscle fibers is essentially the same as that of nerve fibers, about 80 to 90 millivolts. Likewise, the spike potential generated by excitation of a skeletal muscle fiber has essentially the same voltage as that in large nerve fibers, but it lasts 5 to 10 milliseconds instead of 0.5 milliseconds — that is, about 15 times as long.

A muscle fiber can be stimulated much more easily when the stimulus is applied at the neuromuscular junction than at other points on the fiber. This fact is used clinically to determine the presence or absence of muscular innervation. An electrode is moved back and forth on the skin overlying the muscle in question, and, if a particular point can be found at which the muscle is much more strongly stimulated than at other points, one can reasonably expect that this is the locus of viable neuromuscular junctions. These points of excessive excitability are called *motor points*.

Velocity of Conduction in Skeletal Muscle Fibers. The average velocity of conduction in skeletal muscle fibers is approximately 5 meters per second, about the same as that in very small myelinated nerve fibers.

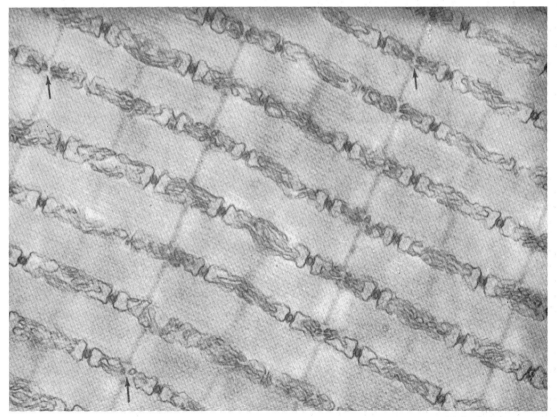

Figure 7–4 Sarcoplasmic reticulum surrounding the myofibril, showing mainly the longitudinal tubule system paralleling the myofibrils but also showing in cross-section the small T tubules that lead to the exterior of the fiber membrane and that contain extracellular fluid (arrows). (From Fawcett: The Cell.)

SKELETAL MUSCLE CONTRACTION

INITIATION OF CONTRACTION BY THE ACTION POTENTIAL

When an action potential spreads along the muscle fiber, the fiber begins to contract after an initial latent period of about 3 milliseconds. To cause contraction of the myofibrils deep in the muscle fiber, the action potential causes electrical current to flow deep into the interior of the muscle fiber by way of the T tubules. It is believed that this current crosses the T tubule membrane at the triad and causes current also to flow in the longitudinal tubules as well. To complete the electrical circuit, the current then traverses the walls of the longitudinal tubules into the sarcoplasm and thence back outward through the cell membrane. Thus, a local circuit of ionic current flow occurs throughout the entire muscle fiber with each action potential.

An electrical potential applied directly to the opening of a T tubule at the surface of a muscle fiber will cause contraction of that half of the sarcomere supplied by the tubule. Therefore, the spread of electrical current inward through the T tubules when an action potential passes obviously has direct electrical effect on the interior of the muscle fiber to cause attraction between the actin and myosin filaments.

CALCIUM IONS IN THE INITIATION OF CONTRACTION. The presence of very low concentrations of calcium ions in the myofibrils, only 2×10^{-4} molar, will cause maximum muscle contraction, but the normal concentration of calcium ions is only 10^{-7} molar, too little to elicit contraction. Yet, the flow of electrical current through the walls of the longitudinal tubules causes them to release calcium ions into the surrounding sarcoplasm. Within a few ten-thousandths of a second, these ions diffuse to the interior of the myofibrils and there initiate the contractile process. It is believed that this is the mechanism by which the action potential causes muscle contraction.

However, the calcium ions do not remain in the myofibrils for more than a few milliseconds, because once the electrical current caused by the action potential is over the longitudinal tubules almost immediately reabsorb the calcium ions out of the sarcoplasm. Therefore, in effect, the action potential causes a short pulse of calcium ions in the myofibril, and it is during this time that the contractile process is activated. At the termination of this pulse of calcium ions, the muscle immediately relaxes.

The exact mechanism by which the longitudinal tubules release and then rebind calcium ions is yet to be determined. However, the resting longitudinal tubules have high concentrations of calcium ions in the cisterns where they abut upon the transverse tubules. Therefore, it is suggested that electrical current crossing from the transverse tubule into the cisterns at the time of the action potential is responsible for release of the calcium ions. Between action potentials, the membranes of the longitudinal tubules supposedly actively reabsorb the calcium ions back to the interior of the tubules. This active absorption requires energy to remove the calcium ions from the sarcoplasm; it receives this energy from ATP.

THE CONTRACTILE PROCESS

Figure 7–5 illustrates the most likely method of skeletal muscle contraction, showing in the upper drawing a relaxed state of the fiber and in the lower drawing a contracted state. It will be noted from this figure that neither the myosin nor the actin filaments decrease in length during contraction; instead the actin filaments simply slide like pistons inward among the myosin filaments. In the resting, elongated state of the muscle, the ends of the actin filaments are usually barely overlapping. Then during contraction, these ends overlap considerably while the two Z membranes approach the ends of the myosin filaments.

Basic Structures of Myosin and Actin Filaments. Electron micrographs show *crossbridges* projecting outward from the myosin filaments toward the actin filaments every 435

Figure 7–5 The relaxed and contracted states of a myofibril, showing sliding of the actin filaments into the channels between the myosin filaments.

Å along the axis of the myosin. Furthermore, since there are six actin filaments arranged hexagonally around the myosin filament, the cross-bridges are repeated six times around the circumference of the myosin filament, arranged helically along the axis of the myosin.

On the other hand, the actin filament does not have actual projections in the form of cross-bridges, but it is composed of two long fibrillar actin molecules wound spirally around each other with a complete turn each 700 Å. Also, it has a reactive site along its axis occurring once every 405 Å. It is believed that these reactive sites in some way interact with the ends of the cross-bridges to provide the force required for pulling the actin filaments inward among the myosin filaments.

Chemical Considerations in the Contractile Process. Unfortunately, the precise chemical and physical events that take place during contraction to cause forceful movement of the actin filaments inward among the myosin filaments are yet unknown. However, a number of chemical properties of actin, myosin, and other constituents of the myofibrils have been discovered that allow one to speculate about the chemical basis for muscle contraction. Some of the important discoveries are the following:

1. Actin filaments appear to be negatively charged at their reactive sites, presumably because of negative ionic sites on the polymerized actin protein molecules that make up the actin filaments.

2. The cross-bridges of the myosin filaments are believed to attach ATP, which also has a strong electronegative charge. Since both the actin and myosin filaments are normally negatively charged, these filaments remain separated from each other under resting conditions. However, this state of affairs can exist only in the absence of calcium ions.

3. It is postulated that presence of a very low concentration of calcium ions causes the calcium ions to combine with the negative electrical charges of both the actin and myosin filaments; this binds these filaments to each other. In some way not understood, this binding also causes inward movement of the actin filaments among the myosin filaments. Possible mechanisms for causing this will be discussed subsequently.

4. The presence of calcium ions also causes the myosin molecules of the myosin filament to develop an ATPase capability; that is, the myosin now has the capability of splitting ATP and liberating energy from this molecule. It is probably the ATP bound to the cross-bridges that is split.

5. Approximately 40 per cent of the actin filament is composed, in addition to actin, of another protein called tropomyosin. In the absence of this tropomyosin, the actin molecules take on entirely different characteristics and bind under abnormal conditions with myosin molecules. Therefore, tropomyosin, in some way not yet understood, is a part of the actin filament that is essential for normal function.

6. Still other types of proteins are also present in the myofibrils, some of which have the ability to inhibit or suppress the ATPase activity of myosin and have the ability to modify the binding characteristics of myosin with actin. Though the functions of these are not understood, they seem to play important roles in the control of muscle contraction and muscle relaxation.

With the above bits of chemical information in mind, we can now attempt to discuss theories concerning the means by which contraction takes place.

THE RATCHET THEORY FOR MUSCLE CONTRACTION. It was only 15 years ago that the sliding mechanism of the filaments was discovered, and it was suggested then by Huxley, one of the discoverers of this mechanism, that the cross-bridges provide a possible "ratchet" mechanism for causing forceful movement of the actin filaments along the myosin filaments. The theory has been expanded with time and is now essentially the following: First, it is believed that, in the resting state, the negative charges of the ATP bound with the cross-bridges of the myosin filaments and the negative charges of the actin filaments cause these two to remain in the uncombined state with no attractive forces between them. Second, on the appearance of calcium ions the following events occur: (a) The calcium ions bind with negative reactive sites of the ATP on the myosin cross-bridges and, at the same time, with the negative reactive sites on the actin filaments, pulling these filaments together. (b) It is assumed that the cross-bridges, having strong electronegativity in the resting state because of the presence of ATP, normally project straight outward from the myosin filaments because the shank of the filament also is negatively charged. That is, repulsion of the two charges makes the cross-bridges project outward. (c) When calcium ions bind with the ATP on the cross-bridges, the negativity of the cross-bridges becomes neutralized. Therefore, the bridges now bend inward toward the axis of the myosin filament. This also pulls the actin filament, thus shortening the muscle. (d) When the cross-bridges fold in against the shank of the myosin filament, the ATPase activity of the myosin filament causes the ATP to split immediately to ADP. This breaks the calcium-linked connections between the myosin cross-bridges and the actin, but in the meantime the actin filament has been pulled along the axis of the myosin filament. (e) Subsequent similar reactions occur at other cross-bridges, and the actin filament is pulled another step. (f) Energy from other sources, such as from high energy creatine phosphate, causes almost immediate reconstitution of the ADP to ATP. Therefore, the cross-bridges that had folded inward now bend outward again and bind with other calcium ions to pull the actin filament another step.

Thus, by a series of "ratchets" — that is, bonding and pulling, bonding and pulling — the actin filaments theoretically are pulled inward among the myosin filaments. However, one can readily see that this is strictly a theory, indeed, one with so many possible flaws that it will be very surprising if many of its details turn out in the end to be correct. It merely shows how movement of the actin filaments along the myosin filaments could possibly occur.

Function of ATP in Contraction of the Myofibril. One can see from the theory of myofibrillar contraction just explained that ATP is postulated to perform two principal functions: First, to keep the myosin and actin filaments from bonding during the resting state, and second, to pull the actin filaments forward when the cross-bridges fold inward before the ATP bound with these cross-bridges is degraded into ADP. It is this second function, the breakdown of ATP, that provides the energy to cause movement of actin filaments inward among the myosin filaments.

THE ELECTROSTATIC SOLENOID THEORY OF MUSCLE CONTRACTION. Though the ratchet theory previously discussed has received widespread recognition, there is one well-known fact that may make it untenable: As the myofibrils contract they also swell in diameter. As a result the filaments spread farther apart, increasing their interdistances as much as 50 Å during the contractile process. Because of the small sizes of the calcium ions supposedly linking the actin and myosin filaments, it is doubtful that these could act through such long distances. Therefore, it seems more likely that the attractive forces between the actin and myosin filaments are perhaps electrostatic in nature — and able to act through long distances — rather than actual chemical bonds. In this context, several different theories have been offered to explain the mechanism of muscle contraction. One of these is the simple theory of the so-called electrostatic solenoid, which is as follows:

Let us assume that normally both the actin and myosin filaments are negatively charged so that there are no attractive forces between the two types of filaments. Then suppose that calcium ions bind with the myosin molecules to make these electropositive while the actin filaments remain electronegative. Such a condition would cause the actin and the myosin filaments to attract each

other electrostatically. Let us assume further that the filaments can slide among each other but cannot touch each other. This will cause forces between the myosin and actin filaments as shown by the diagonal lines in Figure 7–6. Note that at the ends of the filaments the forces have vectors that extend linearly along their axes, which will cause the actin filaments to pull inward among the myosin filaments. When the ends of the actin filaments overlap each other, the linear forces of attraction between the filaments will then occur mainly at the ends of the myosin filaments and not at the ends of the actin filaments. Therefore, the strength of contraction theoretically should fall off; this, indeed, is known to occur.

The electrostatic solenoid mechanism is simply another possible means by which muscle contraction could occur. One can demonstrate with metal plates, interdigitated in the manner as shown for the myosin and actin filaments in Figure 7–6, that a difference in electrical charges on the plates will cause inward movement. Furthermore, calculation of the electrical potential difference between the myosin and actin filaments that would be required to cause full strength skeletal muscle contraction (2 to 3 kg. per square centimeter) shows that an electrical potential difference of only 70 mv. would be sufficient. This is less than the action potential itself, illustrating that the potential required for function of the electrostatic solenoid mechanism is easily attainable.

Relationship Between Actin and Myosin Filament Overlap and Tension Developed by the Contracting Muscle. Figure 7–7 illustrates the relationship between the length of sarcomere and the tension developed by the contracting muscle fiber. To the right are illustrated the degrees of overlap of the myosin and actin filaments at different stages of contraction. At point D on the diagram, the actin filament has pulled all the way out to the end of the myosin filament with no overlap at all. At this point, the tension developed by the activated muscle is zero. Then as the sarcomere shortens and the actin filament overlaps the myosin filament progressively more and more, the

Figure 7–6 Schema to illustrate forces (the oblique lines) that theoretically develop between filaments charged differently from each other, illustrating the so-called "electrostatic solenoid" concept of muscle contraction.

Figure 7–7 Length-tension diagram for a single sarcomere, illustrating maximum strength of contraction when the sarcomere is 2.0 to 2.2 microns in length. At the upper right are shown the relative positions of the actin and myosin filaments at different sarcomere lengths from point A to point D. (Modified from Gordon, Huxley, and Julian: *J. Physiol., 171*:28P, 1964.)

tension increases progressively until the sarcomere length decreases to about 2.2 microns. At this point the actin filament has already overlapped all the cross-bridges of the myosin filament but has not yet reached the center of the myosin filament. Upon further shortening, the sarcomere maintains full tension until point B at a sarcomere length of approximately 2.0 microns. It is at this point that the ends of the two actin filaments begin to overlap. As the sarcomere length falls from 2 microns down to about 1.65 microns at point A, the strength of contraction decreases.

This diagram illustrates that maximum contraction occurs when there is maximum overlap between the actin filaments and the cross-bridges of the myosin filaments, and it supports the idea that it is an attractive force between the cross-bridges and the actin filaments that causes the contraction. The decrease in contractile strength when the actin filaments overlap each other has been postulated to be caused by backward pull of the myosin cross-bridges on the overlapped portions of the actin filaments, either by the ratchet mechanism or electrostatically.

Relation of Force of Contraction of the Intact Muscle to Muscle Length. Figure 7–8 illustrates a diagram similar to that in Figure 7–7, but this time for the intact whole muscle rather than for the isolated muscle fiber. The whole muscle has a large amount of connective tissue in it, and the sarcomeres in different parts of the muscle do not necessarily contract exactly in unison. Therefore, the curve has somewhat different dimensions from those illustrated

for the individual muscle fiber, but it nevertheless exhibits the same form.

Note in Figure 7–8 that when the muscle is at its normal resting *stretched* length and is then activated, it contracts with maximum force of contraction. If the muscle is stretched to greater than normal length prior to contraction, a large amount of *resting tension* develops in the muscle even before contraction takes place. That is, the two ends of the muscle are pulled toward each other by the elastic forces of the connective tissue, of the sarcolemma, the blood vessels, the nerves, and so forth. However, the *increase* in tension during contraction, called *contractile tension*, decreases as the muscle is stretched beyond its normal length.

Note also in Figure 7–8 that when the resting muscle is shortened to less than its normal fully stretched length, the maximum tension of contraction decreases progressively and reaches zero when the muscle has shortened to approximately 60 per cent of its maximum resting length.

Relation of Velocity of Contraction to Load. A muscle contracts extremely rapidly when it contracts against no load—to a state of full contraction in approximately 1/20 second for the average muscle. However, when loads are applied, the velocity of contraction becomes progressively less as the load increases, as illustrated in Figure 7–9. When the load increases to equal the maximum force that the muscle can exert, then the velocity of contraction becomes zero, and no contraction at all results, despite activation of the muscle fiber.

This decreasing velocity seems to be associated mainly with the rate at which reactive sites between the filaments can be activated: that is, the greater the load, the greater the

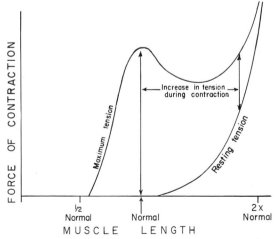

Figure 7–8 Relation of muscle length to force of contraction.

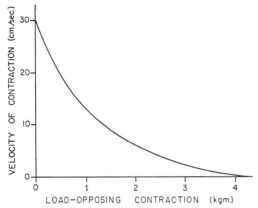

Figure 7–9 Relation of load to velocity of contraction in a skeletal muscle 8 cm. long.

number of sites required to lift the load itself, and it is only the remaining sites that cause the filaments to move.

Relation of Muscle Energy Expenditure to Work Performed—the "Fenn" Effect. The shortening process of muscles can lift objects or move objects against force and thereby perform *work*. The amounts of oxygen and other nutrients consumed by the muscle increase greatly when the muscle performs work rather than simply contracting without causing work. This is called the "Fenn" effect. Though this seems to be an obvious effect that one would expect, nevertheless, the chemical basis for it has not been discovered. In some way the contraction of a muscle against a load causes the rate of breakdown of ATP to ADP to increase. This possibly results from the fact that increased numbers of reactive sites must be activated to overcome the load.

EFFICIENCY OF MUSCLE CONTRACTION

The "efficiency" of an engine or a motor is calculated as the percentage of energy input that is converted into work instead of heat. The percentage of the input energy to a muscle (the chemical energy in the nutrients) that can be converted into work is less than 20 to 25 per cent, the remainder becoming heat. Maximum efficiency can be realized only when the muscle contracts at a moderate velocity. If the muscle contracts very slowly, large amounts of maintenance heat are released during the process of contraction, thereby decreasing the efficiency. On the other hand, if contraction is too rapid, large proportions of the energy are used to overcome the viscous friction within the muscle itself, and this, too, reduces the efficiency of contraction. Ordinarily, maximum efficiency is developed when the velocity of contraction is about 30 per cent of maximum.

CHARACTERISTICS OF A SINGLE MUSCLE TWITCH

Many features of muscle contraction can be especially well demonstrated by eliciting single *muscle twitches.* This can be accomplished by instantaneously exciting the nerve to a muscle or by passing a short electrical stimulus through the muscle itself, giving rise to a single, sudden contraction lasting for a fraction of a second.

Isometric versus Isotonic Contraction. Muscle contraction is said to be *isometric* when the muscle does not shorten during contraction and *isotonic* when it shortens but the tension on the muscle remains constant. Systems for recording the two types of muscle contraction are illustrated in Figure 7–10.

To the right is the isometric system in which the muscle is suspended between a solid rod and a lever of an electronic force transducer. This transducer records force with almost zero movement of the lever; therefore, in effect, the muscle is bound between fixed points so that it cannot contract significantly. To the left is shown an isotonic recording system: the muscle simply lifts a pan of weights so that the force against which the muscle contracts remains constant, though the length of the muscle changes considerably.

There are several basic differences between isometric and isotonic contractions. First, isometric contraction does not require sliding of myofibrils among each other. Second, in isotonic contraction a load is moved, which involves the phenomenon of inertia. That is, the weight or other type of object being moved must first be accelerated, and once a velocity has been attained the load has momentum that causes it to continue moving even after the contraction is over. Therefore an isotonic contraction is likely to last considerably longer than an isometric contraction of the same muscle. Third, isotonic contraction entails the performance of external work. Therefore, in accordance with the Fenn effect discussed

Figure 7–10 Isotonic and isometric recording systems.

previously, a greater amount of energy is used by the muscle.

In comparing the rapidity of contraction of different types of muscles, isometric recordings such as those illustrated in Figure 7–11 are usually used instead of isotonic recordings because the duration of an isotonic recording is almost as dependent on the inertia of the recording system as upon the contraction itself, and this makes it difficult to compare time relationships of contractions from one muscle to another.

Muscles can contract both isometrically and isotonically in the body, but most contractions are actually a mixture of the two. When a person stands, he tenses his quadriceps muscles to tighten the knee joints and to keep the legs stiff. This is isometric contraction. On the other hand, when a person lifts a weight using his biceps, this is mainly an isotonic contraction. Finally, contractions of leg muscles during running are a mixture of isometric and isotonic contractions—isometric mainly to keep the limbs stiff when the legs hit the ground and isotonic mainly to move the limbs.

THE SERIES ELASTIC COMPONENT OF MUSCLE CONTRACTION. When muscle fibers contract against a load, those portions of the muscle that do not contract—the tendons, the sarcolemmal ends of the muscle fibers where they attach to the tendons, and perhaps even the membranes of the myofibrils—will stretch slightly as the tension increases. Consequently, the muscle must shorten an extra 3 to 5 per cent to make up for the stretch of these elements. The elements of the muscle that stretch during contraction are called the *series elastic component* of the muscle.

Characteristics of Isometric Twitches Recorded from Different Muscles. The body has many different sizes of skeletal muscles—from the very small stapedius muscle of only a few millimeters length and a millimeter or so in diameter up to the very large quadriceps muscle. Furthermore, the fibers may be as small as 10 microns in diameter or as large as 100 microns. And, finally, the energetics of muscle contraction vary considerably from one muscle to another. These different physical and chemical characteristics often manifest themselves in the form of different characteristics of contraction, some muscles contracting rapidly while others contract slowly.

Figure 7–11 illustrates isometric contractions of three different types of skeletal muscles: an ocular muscle, which has a duration of contraction of less than $1/100$ second; the gastrocnemius muscle, which has a duration of contraction of about $1/30$ second; and the soleus muscle, which has a duration of contraction of about $1/10$ second. It is interesting that these durations of contraction are adapted to the function of each of the respective muscles, for ocular movements must be extremely rapid to maintain fixation of the eyes upon specific objects, the gastrocnemius muscle must contract moderately rapidly to provide sufficient velocity of limb movement for running and jumping, while the soleus muscle is concerned principally with slow reactions for continual support of the body against gravity.

All-or-Nothing Effect. In Chapter 5 it was pointed out that, following stimulation of a nerve fiber, the action potential either travels over the entire neuron or fails to travel at all. This is called the all-or-nothing effect. The same principle applies in muscle—that is, a stimulus to a muscle fiber causes an action potential to travel over the entire fiber or fails to stimulate it at all. Therefore, either all of the fiber contracts or none of it contracts. Yet the force of contraction that results when the muscle does become excited can vary, depending on the contractile state of the fiber when it is stimulated. That is, if appropriate nutrients are not available, the contraction might well be weak, or the same will be true if the fiber is already fatigued. On the other hand, the fiber might be in a highly contractile state, such as occurs when the muscle is warmed, and will contract even more strongly than normally.

CONTRACTION OF SKELETAL MUSCLE IN THE BODY

THE MOTOR UNIT

Each motor neuron that leaves the spinal cord usually innervates many different mus-

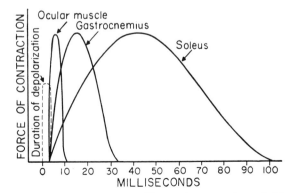

Figure 7–11 Duration of isometric contractions of different types of mammalian muscles, showing also a latent period between the action potential and muscle contraction.

cle fibers, the number depending on the type of muscle. All the muscle fibers innervated by a single motor nerve fiber are called a *motor unit.* In general, muscles that react rapidly and whose control is exact have few muscle fibers (as few as 10 to 15 in ocular muscles) in each motor unit and have a large number of nerve fibers going to each muscle. On the other hand, the slowly acting postural muscles, which do not require a very fine degree of control, may have as many as 300 to 800 muscle fibers in each motor unit. An average figure for all the muscles of the body can be considered to be about 180 muscle fibers to the motor unit.

Usually muscle fibers of adjacent motor units overlap, with small bundles of 10 to 15 fibers from one motor unit lying among similar bundles of the second motor unit. This interdigitation allows the separate motor units to contract in support of each other rather than entirely as individual segments.

Macromotor Units. Loss of some of the nerve fibers to a muscle belly causes the remaining nerve fibers to sprout forth and innervate many of the paralyzed muscle fibers. When this occurs, such as following poliomyelitis, one occasionally develops *macromotor units,* which can contain as many as 10 times the normal number of muscle fibers. This obviously decreases the degree of control that one has over his muscles, but, nevertheless, allows the muscle to regain function.

SUMMATION OF MUSCLE CONTRACTION

Summation means the adding together of individual muscle twitches to make strong and concerted muscle movements. In general, summation occurs in two different ways: (1) by increasing the number of motor units contracting simultaneously and (2) by increasing the rapidity of contraction of individual motor units. These are called, respectively, *multiple motor unit summation* and *wave summation,* or spatial summation and temporal summation.

Multiple Motor Unit Summation. Figure 7–12 illustrates multiple motor unit summation, showing that the force of contraction increases progressively as the number of contracting motor units increases from 1 to 8. That is, as more and more motor units are stimulated simultaneously, the greater becomes the strength of contraction. Even within a single muscle, the numbers of muscle fibers and their sizes vary tremendously, so that one motor unit may be as much as 50 times as strong as another. The smaller motor units are

Figure 7–12 Multiple motor unit summation.

far more easily excited than are the larger ones because they are innervated by smaller nerve fibers whose cell bodies in the spinal cord have a naturally high level of excitability. This effect causes the gradations of muscle strength during weak muscle contraction to occur in very small steps, while the steps become progressively greater as the intensity of contraction increases.

Wave Summation. Figure 7–13 illustrates the principles of wave summation, showing in the lower left-hand corner a single muscle twitch followed by successive muscle twitches at various frequencies. When the frequency of twitches is 10 per second, the first muscle twitch is not completely over by the time the second one begins. Therefore, since the muscle is already in a partially contracted state when the second twitch begins, the degree of muscle shortening this time is slightly greater than that which occurs with the single muscle twitch. The third, fourth, and additional twitches add still more shortening.

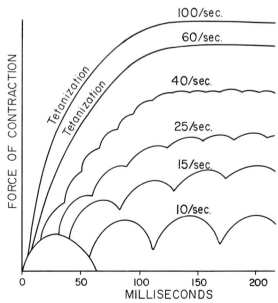

Figure 7–13 Wave summation and tetanization.

At more rapid rates of contraction, the degree of summation of successive contractions becomes greater and greater, because the successive contractions appear at earlier times following the preceding contraction.

TETANIZATION. When a muscle is stimulated at progressively greater rates, a frequency of stimulation is finally reached at which the successive contractions fuse together and cannot be distinguished one from the other. This state is called *tetanization,* and the lowest frequency at which it occurs is called the *critical frequency.*

Tetanization results partly from the viscous properties of the muscle and partly from the nature of the contractile process itself. The muscle fibers are filled with sarcoplasm, which is a viscous fluid, and the fibers are encased in fasciae and muscle sheaths that have a viscous resistance to change in length. Therefore, undoubtedly these viscous factors play a role in causing the successive contractions to fuse with each other.

But in addition to the viscous property of muscle, the activation process itself lasts for a definite period of time, and successive pulsatile states of activation of the muscle fiber can occur so rapidly that they fuse into a long continual state of activation. That is, free calcium ions persist continuously in the myofibrils and provide an uninterrupted stimulus for maintenance of contraction. Once the critical frequency for tetanization is reached, further increase in rate of stimulation increases the force of contraction only a few more per cent, as shown in Figure 7–13.

Asynchronous Summation of Motor Units. Actually it is rare for either multiple motor unit summation or wave summation to occur separately from each other in normal muscle function. Instead, special neurogenic mechanisms in the spinal cord normally increase both the impulse rate and the number of motor units firing at the same time. If a motor unit fires at all, it usually fires at least five times per second, but this can increase to as high as several hundred per second, to frequencies sufficient to cause complete tetanization.

Yet, even when tetanization of individual motor units of a muscle is not occurring, the tension exerted by the whole muscle is still continuous and nonjerky because *the different motor units fire asynchronously.* That is, while one is contracting another is relaxing; then another fires, followed by still another, and so forth. Consequently, even when motor units fire as infrequently as five times per second, the muscle contraction, though weak, is nevertheless very smooth.

Maximum Strength of Contraction. The maximum strength of tetanic contraction of a muscle operating at a normal muscle length is about 3 kilograms per square centimeter of muscle, or 42 pounds per square inch. Since a quadriceps muscle can at times have as much as 16 square inches of muscle belly, as much as 600 to 700 pounds of tension may at times be applied to the patellar tendon. One can readily understand, therefore, how it is possible for muscles sometimes to pull their tendons out of the insertions in bones. This often occurs where the patellar tendon inserts in the tibia, and even more frequently where the Achilles tendon of the gastrocnemius muscle inserts at the heel.

Changes in Muscle Strength at the Onset of Contraction — The Staircase Effect (Treppe). When a muscle begins to contract after a long period of rest, its initial strength of contraction may be as little as one-half its strength 30 to 50 muscle twitches later. That is, the strength of contraction increases to a plateau, a phenomenon called the *staircase effect* or *treppe.* This phenomenon has interested physiologists greatly because it gives possible insights into the mechanism of muscle contraction.

Though all the possible causes of the staircase effect are not yet known, it is believed to be caused primarily by electrolyte changes that occur when a series of contractions begins. For instance, there is a net increase in calcium ions inside the muscle fiber because of movement of calcium ions inward through the membrane with each action potential. There is probably also further increase of calcium ions in the sarcoplasm; this increase is caused by release of these ions from the longitudinal tubules of the sarcoplasmic reticulum and by the failure to recapture the ions immediately. In addition, there is decreased potassium inside the cell as well as increased sodium; it has been suggested that the changes in these two ions increase the rate of liberation of calcium ions from the sarcoplasmic reticulum. Recalling the earlier discussion of the relationship of calcium ions to the contractile process, one can readily understand that progressive increase in calcium ion concentration in the sarcoplasm, caused either directly or as a consequence of sodium and potassium movement, could progressively increase the strength of muscle contraction, giving rise to the staircase effect.

SKELETAL MUSCLE TONE

Even when muscles are at rest, a certain amount of tautness usually remains. This residual degree of contraction in skeletal muscle is called *muscle tone.* Since skeletal muscle fibers do not contract without an actual action potential to stimulate the fiber except in certain pathological conditions, it is believed that skeletal muscle tone results entirely from nerve impulses coming from the spinal cord. These in turn are controlled partly by impulses transmitted from the brain to the appropriate anterior motor neurons and partly by impulses that originate in *muscle spindles* located in the muscle itself.

Muscle spindles are sensory receptors that exist throughout essentially all skeletal muscles to detect the degree of muscle contraction. These will be discussed in detail in Chapter 56, but, briefly, they transmit impulses almost continually through the posterior roots into the spinal cord, where they excite the anterior motor neurons, which in turn provide the necessary nerve stimuli for muscle tone. Simply cutting the posterior roots, thereby blocking the muscle spindle impulses, usually reduces muscle tone to such a low level that the muscle becomes almost completely flaccid.

Many other neurogenic factors, originating especially in the brain, enter into the control of muscle tone. These will be discussed in detail in relation to muscle spindle and spinal cord function in Chapter 56.

MUSCLE FATIGUE

Prolonged and strong contraction of a muscle leads to the well-known state of muscle fatigue. This results simply from inability of the contractile and metabolic processes of the muscle fibers to continue supplying the same work output. The nerve continues to function properly, the nerve impulses pass normally through the neuromuscular junction into the muscle fiber, and even normal action potentials spread over the muscle fibers, but the contraction becomes weaker and weaker because of depletion of energy supplies in the muscle fibers themselves.

Interruption of blood flow through a contracting muscle leads to almost complete muscle fatigue in a minute or more because of the obvious loss of nutrient supply.

Physiological Contracture. If a muscle becomes fatigued to an extreme extent, it is likely to become continually contracted, remaining contracted and rigid for many minutes even without an action potential as a stimulus. This is called physiological contracture of the muscle. It is believed to result from a peculiar feature of the contractile process itself: As pointed out earlier, ATP is required to make actin and myosin separate during the process of muscle relaxation. Otherwise the myosin and actin filaments will continue to be attracted to each other. In extreme degrees of muscle fatigue, essentially all the ATP has been depleted, thus accounting for the rigid binding of actin and myosin filaments.

THE LEVER SYSTEMS OF THE BODY

Muscles obviously operate by applying tension to their points of insertion into bones, and the bones in turn form various types of lever systems. Figure 7–14 illustrates the lever system activated by the biceps muscle to lift the forearm. If we assume that a large biceps muscle has a cross-sectional area of 6 square inches, then the maximum force of contraction would be about 250

Figure 7–14 The lever system activated by the biceps muscle.

pounds. When the forearm is exactly at right angles with the upper arm, the tendon attachment of the biceps is about 2 inches anterior to the fulcrum at the elbow, and the total length of the forearm lever is about 14 inches. Therefore, the amount of lifting power that the biceps would have at the hand would be only one seventh of the 250 pounds force, or about 36 pounds. Yet when the arm is in the fully extended position the attachment of the biceps is much less than 2 inches anterior to the fulcrum, and the force with which the forearm can be brought forward is less than 36 pounds.

In short, an analysis of the lever systems of the body depends on (a) a discrete knowledge of the points of muscular insertion and (b) their distances from the fulcrum of the lever, as well as (c) the length of the lever arm and (d) the position of the lever. Obviously, many different types of movement are required in the body, some of which need great strength and others large distances of movement. For this reason there are all varieties of muscles; some are long and contract a long distance and some are short but have large cross-sectional areas and therefore can provide extreme strengths of contraction over short distances. The study of different types of muscles, lever systems, and their movements is called *kinesiology* and is a very important phase of human physioanatomy.

Accommodation of Muscle Length to the Length of the Lever System. If a bone is broken and then heals in a shortened state, the force of contraction of the muscles lying along this broken bone would obviously become decreased because of the shortened lengths of muscles. However, muscles shortened in this manner undergo *physical shortening* during the next few weeks. That is, the muscle fibers actually shorten and re-establish new muscle lengths approximately equal to the maximum length of the lever system itself, thus re-establishing optimum force of contraction by the muscles.

The same shortening process also occurs in muscles of limbs immobilized for several weeks in casts if the muscles during this time are in a shortened position. When the cast is removed, the muscles must often be restretched over a period of weeks before full mobility is restored.

SPECIAL FEATURES AND ABNORMALITIES OF SKELETAL MUSCLE FUNCTION

MUSCULAR HYPERTROPHY

Forceful muscular activity causes the muscle size to increase, a phenomenon called hypertrophy. The diameters of the individual muscle fibers increase, and the fibers gain in total numbers of myofibrils as well as in various nutrient and intermediary metabolic substances, such as adenosine triphosphate, phosphocreatine, glycogen, and so forth. Briefly, muscular hypertrophy increases both the motive power of the muscle and the nutrient mechanisms for maintaining increased motive power.

Weak muscular activity, even when sustained over long periods of time, does not result in significant hypertrophy. Instead, hypertrophy results mainly from *very* forceful muscle activity, though it might occur for only a few minutes each day. For this reason, strength can be developed in muscles much more rapidly when "resistive" or "isometric" exercise is used rather than simply prolonged mild exercise. Indeed, essentially no new myofibrils develop unless the muscle contracts to at least 75 per cent of its maximum tension.

MUSCULAR ATROPHY

Muscular atrophy is the reverse of muscular hypertrophy; it results any time a muscle is not used or even when a muscle is used only for very weak contractions. Atrophy is particularly likely to occur when limbs are placed in casts, thereby preventing muscular contraction. As little as one to two months of disuse can sometimes decrease the muscular size to one-half normal.

Reaction of Muscle to Denervation. When a muscle is denervated it immediately begins to atrophy, and the muscle continues to decrease in size for several years. If the muscle becomes re-innervated during the first three to four months, full function of the muscle usually returns, but after four months of denervation some of the muscle fibers usually will have degenerated. Re-innervation after two years rarely results in return of any function at all. Pathological studies show that the muscle fibers have by that time been replaced by fat and fibrous tissue.

PREVENTION OF MUSCULAR ATROPHY BY ELECTRICAL STIMULATION. Strong electrical stimulation of denervated muscles, particularly when the resulting contractions occur against loads, will delay and in some instances prevent muscular atrophy despite denervation. This procedure is used to keep muscles alive until re-innervation can take place.

PHYSICAL CONTRACTURE OF MUSCLE FOLLOWING DENERVATION. When a muscle is denervated, its fibers tend to shorten if the muscle is kept in a shortened position, and even the associated nerves and fasciae shorten. All this is a natural characteristic of protein fibers called "creep." That is, unless continual movement keeps stretching the muscle and other structures, they will creep toward a shortened length. This is one of the most difficult problems in the treatment of patients with denervated muscles, such as occur in poliomyelitis or nerve trauma. Unless passive stretching is applied daily to the muscles, they may become so shortened that even when re-innervated they will be of little value. But, more important, the shortening can often result in extremely contorted positions of different parts of the body.

RIGOR MORTIS

Several hours after death all the muscles of the body go into a state of contracture called rigor mortis, which is similar to physiological contracture. That is, the muscle contracts and becomes rigid even without action potentials. It is believed that this rigidity, like physiological contracture, is caused by loss of all the ATP, which is required to cause separation of the actin and myosin filaments during the relaxation process. The muscles remain in rigor until the muscle proteins are destroyed, which usually occurs as a result of bacterial putrefaction some 15 to 25 hours later.

FAMILIAL PERIODIC PARALYSIS

Occasionally, a hereditary disease called familial periodic paralysis occurs. In persons so afflicted, the extracellular fluid potassium concentration periodically falls to very low levels, causing various degrees of paralysis. It is believed that the paralysis is caused in the following manner: A great decrease in extracellular fluid potassium increases the muscle fiber membrane potential to a very high value. This results in strong hyperpolarization of the membrane, making the fiber almost totally inexcitable; that is, the membrane potential is so high that the normal end-plate potential at the neuromuscular junction is incapable of reducing it to the threshold level for excitation.

MYOTONIA CONGENITA

This is also hereditary, but in this condition after the muscle has contracted it will not relax immediately after an action potential because the action potential is followed by a series of repetitive weak depolarizations of the muscle membrane, a process that stops slowly over a period of seconds. Administration of calcium ions to these patients will temporarily improve the condition, and since calcium normally decreases the permeability of conductive membranes, it is assumed that myotonia congenita results from excessively permeable muscle membranes that are difficult to polarize enough to prevent spontaneous discharge.

THE ELECTROMYOGRAM

Each time an action potential passes along a muscle fiber a small portion of the electrical current spreads away from the muscle as far as the skin. If many muscle fibers contract simultaneously, the summated electrical potentials at the skin may be very great. By placing two electrodes on the skin or inserting electrodes into the muscle, an electrical recording called the electromyogram can be made when the muscle is stimulated. Figure 7–15 illustrates a typical electromyographic recording from the gastrocnemius muscle during a moderate contraction. Electromyograms are frequently used clinically to discern abnormalities of muscle excitation. Two such abnormalities are muscle *fasciculation* and *fibrillation*.

Muscle Fasciculation. When an abnormal impulse occurs in a motor nerve fiber its whole motor unit contracts. This often causes sufficient contraction in the muscle that one can see a slight ripple in the skin over the muscle. This process is called fasciculation.

Fasciculation occurs especially following destruction of anterior motor neurons in poliomyelitis or following traumatic interruption of a nerve. As the peripheral nerve fibers die, spontaneous impulses are generated during the first few days, and fasciculatory muscle movements result in the muscle. Typical electromyographic records of weak periodic potentials can be obtained from the skin overlying the muscle.

Muscle Fibrillation. After all nerves to a muscle have been destroyed and the nerve fibers themselves have become nonfunctional, which requires three to five days, spontaneous impulses begin to appear in the denervated muscle fibers. At first, these occur at a rate of once every few seconds, but, after a few more days or a few weeks, the impulses become as rapid as 3 to 10 times per second. Thus, skeletal muscle fibers, when released from innervation, develop an intrinsic rhythmicity. After several more weeks the muscle fibers atrophy to such an extent that the fibrillatory impulses finally cease. To record an electromyogram of fibrillation, minute bipolar electrodes must be inserted into the muscle belly itself because adjacent muscle fibers do not fire simultaneously and, therefore, do not summate. As a result, the potentials are not strong enough to record at the surface of the skin.

Figure 7–15 Electromyogram recorded during contraction of the gastrocnemius muscle.

CONTRACTION OF CARDIAC MUSCLE

The basic mechanism of cardiac muscle contraction is very nearly the same as that in skeletal muscle except that the manner of excitation is partially different. Cardiac muscle is striated like skeletal muscle, and each sarcomere contains the same arrangement of myosin and actin filaments and a sarcoplasmic reticulum that is not greatly different from that in skeletal muscle.

However, cardiac muscle fibers, in contrast to skeletal muscle fibers, are formed of a series of distinct cells that are connected together at junctures called *intercalated disks.* These intercalated disks have slight electrical resistance which affects conduction of the action potential, slowing the conduction down to velocities as low as 0.5 meter per second. In contrast, conduction in skeletal muscle is 10 times as great. Also, the cardiac muscle fibers of the atria and those of the ventricles are arranged respectively in two large latticeworks, called *functional syncytiums,* so that impulses transmitted over the membrane of any single fiber spread to all the fibers either of the atria or of the ventricles. This will be discussed in Chapter 14 in relation to conduction of the cardiac impulse.

The character of the action potential and duration of depolarization is also very different in cardiac muscle from that in skeletal muscle. The action potential in skeletal muscle lasts less than 0.01 second, whereas the action potential in cardiac muscle has a plateau which makes it have a duration of about 0.15 second in the atria and 0.3 second in the ventricles. During this entire period, the chemical process for eliciting contraction continues, which means that cardiac muscle remains contracted for 0.15 to 0.3 second, rather than for only a very short twitch as is true of skeletal muscle.

Finally, cardiac muscle is intrinsically self-excitable. Its membrane is considerably more permeable to sodium than is the skeletal muscle fiber membrane, which makes the heart muscle membrane potential discharge periodically, thus giving the rhythmic heartbeat. The heart is such an important organ and the details of its conductive and contractile processes are so important that these will be presented fully in Chapter 14.

CONTRACTION OF SMOOTH MUSCLE

Though the smooth muscle in each organ is usually different from that in other organs,

Figure 7–16 Visceral and multiunit smooth muscle fibers.

smooth muscle can be divided into two major types as illustrated in Figure 7–16: *multiunit smooth muscle* and *visceral smooth muscle*.

Multiunit Smooth Muscle. This type of smooth muscle is composed of discrete smooth muscle fibers, usually having lengths of 100 to 200 microns and diameters averaging about 10 microns. This is the type of smooth muscle found in the blood vessels, in the iris of the eye, and in the nictitating membrane that covers the eyes of some lower animals. In general, the multiunit type of smooth muscle contracts only when stimulated by a nerve.

Visceral Smooth Muscle. Visceral smooth muscle fibers are very similar to multiunit fibers except that they are crowded together and lie in such close contact with each other that it is sometimes difficult to see the limits between the adjacent cells. For this reason, this type of smooth muscle is also known as *unitary smooth muscle*. This type of muscle is found in most of the organs of the body, especially in the walls of the gut, the bile ducts, the ureters, the uterus, and so forth.

When even one fiber in a visceral muscle tissue is stimulated, the impulse ordinarily is conducted to the surrounding fibers by *ephaptic conduction*. This means that the action potential generated in one of the fibers is strong enough electrically to excite the adjacent fibers without secretion of any excitatory substance. Visceral smooth muscle fibers can transmit impulses one to another in this manner because the fiber membranes where the fibers contact each other develop enhanced permeability so great that the electrical resistance between the inside of one fiber and the next is only a fraction of the normal membrane resistance. This obviously allows easy flow of current from the interior of one cell to the next and therefore allows ease of transmission of impulses over the surface membranes from one fiber to the next.

Membrane Potentials in Smooth Muscle. The membrane potential of smooth muscle is variable from one type of smooth muscle to another and also depends upon the momentary condition of the muscle. However, in the normal resting condition, the membrane potential is usually about −50 millivolts. This is greatly increased by inhibitory transmitters, which cause *hyperpolarization*, resulting in membrane potentials of −70 or more millivolts and making the fibers very inexcitable. The potential can be decreased to lower values by excitatory transmitters, by stretching the smooth muscle fiber, by hypoxia, by various types of hormones and many other factors. When decreased to about −40 millivolts, action potentials usually occur.

Characteristics of Action Potentials in Smooth Muscle. Because of the extreme variability of smooth muscle in different parts of the body, one would expect many variations in the characteristics of the action potentials. These appear in three different forms as follows:

SPIKE POTENTIALS. Typical spike action potentials, such as those seen in skeletal muscle, occasionally occur in smooth muscle fibers, such as in uterine muscle fibers that have been exposed to large quantities of estrogen. However, this is not a very common form of smooth muscle action potential.

ACTION POTENTIALS WITH PLATEAUS. One of the most common types of action potentials is that illustrated in Figure 7–17, which shows a typical plateau type of action potential lasting in this case for 0.3 seconds. Such action potentials may last from 0.05 second to as long as 0.5 second, depending on the type of smooth muscle. The muscle remains contracted as long as this action potential persists.

SMALL SPIKES AND RIPPLES. The resting membrane potential also frequently shows small spikes of only a few millivolts instead of the 70 or more millivolt spikes one sees in the usual action potentials. And many small ripples of 1 to 3 millivolts frequently appear in the resting membrane potential. The small

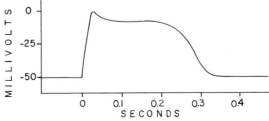

Figure 7–17 Monophasic action potential from a smooth muscle fiber of the rat uterus.

spikes and ripples probably result from local discharges along the surface of the fiber, discharges that do not develop into a full action potential.

Rhythmicity of Visceral Muscle. Most types of visceral muscle exhibit rhythmicity. This results from the presence of at least some smooth muscle fibers within the muscle mass that are self-excitable—that is, their membranes are excessively permeable to sodium, which allows spontaneous generation of action potentials as was explained in Chapter 5. After recovery from each action potential, self-excitation occurs again a few seconds later. As an example, the muscle in the pelvis of the kidney is self-excitable and transmits impulses along the ureter, causing a peristaltic wave to travel along this tubular structure, thus forcing urine into the bladder. Similar effects occur in the bile ducts, in other gland ducts, in the vas deferens, in the gut, and in many other tubular structures of the body.

THE CONTRACTILE PROCESS IN SMOOTH MUSCLE

The Chemical Basis for Contraction. Smooth muscle contains exactly the same chemical substances as skeletal muscle, including both actin and myosin. However, the actin and myosin filaments do not have the same periodic spatial arrangement as that seen in the sarcomeres of skeletal muscle. Instead the filaments sometimes lie at odd angles to each other, though mostly oriented in the longitudinal direction of the muscle fibers. Typical actin filaments can be seen in electron micrographs; the tips of some of these are apparently attached to the cell membrane, thereby providing a method for mechanically coupling the contractile process inside the fiber with the tissues outside the fiber. On the other hand, typical myosin filaments as seen in striated muscle are rarely seen in smooth muscle, so that there is actually some doubt that myosin filaments of the type known to occur in skeletal muscle are necessary for smooth muscle contraction. Yet, chemically there is approximately the same ratio of myosin molecules to actin molecules in smooth muscle as in skeletal muscle. It is possible that some of the smaller filaments that have tentatively been identified as actin filaments are instead myosin filaments, or it could be that unpolymerized myosin molecules too small to be seen with the electron microscope could provide binding forces between the actin filaments during contraction.

Nevertheless, the contractile process appears in many other ways to be the same as in skeletal muscle: that is, it is activated by calcium ions; ATP is converted to ADP, just as occurs in skeletal muscle; and contraction occurs during depolarization of the membrane and stops following repolarization.

One significant difference between smooth muscle contraction and skeletal muscle contraction is the timing of the process; contraction develops only one-fourth to one-twentieth as rapidly in smooth muscle as in skeletal muscle. Also, relaxation at the end of the action potential is equally slow. This difference probably results from slowness of the activation and inactivation processes in smooth muscle, and this presumably is caused by the structural differences between the two types of muscle fibers. Especially important, smooth muscle does not have a T tubule system, and it does not have a well developed internal sarcoplasmic reticulum system for rapid release or reabsorption of calcium ions. Therefore, the calcium ions required to activate smooth muscle contraction probably come through the cell membrane itself from the interstitial fluid or are released inside the smooth muscle cell from structures that are directly susceptible to the action potential at the cell membrane surface.

It has also been suggested that the contractile elements in smooth muscle tend to stick to each other so that they "drag" against each other during contraction and relaxation. This could explain the ability of smooth muscle to maintain large amounts of tension over long periods of time without utilizing much ATP, an effect that is quite different from that in skeletal muscle.

Gross Characteristics of Smooth Muscle Contractions. From the foregoing discussion of the different types of action potentials that can occur in smooth muscle, and from the different characteristics of transmission of action potentials, one can readily understand that smooth muscle in different parts of the body can have many different characteristics of contraction. For instance, smooth muscle of the blood vessels contracts mainly in response to nerve impulses, but in many types of visceral smooth muscle, such as in the ureter, the bile ducts, other glandular ducts, and so forth, the self-excitatory process causes continuous rhythmic contraction, promoting peristalsis along the tubular structures.

TONUS OF SMOOTH MUSCLE. Smooth muscle can also maintain tonic contraction over long periods of time, and the degree of tonic contraction may be entirely distinct from

Figure 7–18 Record of rhythmic and tonic smooth muscle contraction.

superimposed rhythmic contractions. This effect is illustrated in Figure 7–18, which shows at first rhythmic contractions and then a strong tonic contraction that is sustained. During the tonic contraction, the rhythmicity may disappear completely, it may change its frequency, or it may become less intense or more intense. In other words, the two processes are not necessarily related to each other.

Tonic contraction probably results in most instances from summation of repetitive action potentials in the same way that summation occurs in skeletal muscle. If the repetitive action potentials are infrequent, a rhythmic contraction results, but if the repetitive action potentials occur in rapid succession, one can obtain typical tetanization.

Tonus also frequently results from continuous depolarization of the muscle cell membrane caused by excitatory transmitter substances, hormones, heat, stretch, or many other factors.

Distance of Contraction of Smooth Muscle. A special characteristic of smooth muscle—one that is different from skeletal muscle—is its ability to contract over far greater distances than can skeletal muscle. Skeletal muscle has a useful distance of contraction equal to only 25 to 35 per cent of its length, whereas smooth muscle can contract quite effectively from a length two times its normal length to as short as one-half its normal length, giving as much as a four-fold distance of contraction. This allows smooth muscle to perform important functions in the hollow viscera—for instance, allowing the gut, the bladder, blood vessels, and other internal structures of the body to change their lumen diameters from almost zero up to very large values.

Stress-Relaxation of Smooth Muscle. A very important characteristic of smooth muscle is its ability to change length greatly without marked changes in tension. This results from a phenomenon called stress-relaxation, which may be explained as follows:

If a segment of smooth muscle 1 inch long is suddenly stretched to 2 inches, the tension between the two ends increases tremendously at first, but the extra tension also begins to disappear immediately, and within a few minutes it has returned almost to its level prior to the stretch, even though the muscle is now twice as long. This probably results from the fact that the actin and myosin filaments are arranged randomly in smooth muscle. Over a period of time, the filaments of the stretched muscle presumably rearrange their bonds and gradually allow the sliding process to take place, thus allowing the tension to return almost to its original amount.

Exactly the converse effect occurs when smooth muscle is shortened. Thus, if the 2 inch segment of smooth muscle is shortened back to 1 inch, essentially all tension will be lost from the muscle immediately. Gradually, over a period of 1 minute or more, the tension returns, this again presumably resulting from slow sliding of the filaments. This is called *reverse stress-relaxation*.

CREEP AND PLASTICITY OF SMOOTH MUSCLE. A phenomenon similar to stress-relaxation, called *creep*, occurs when a constant tension is applied to a segment of smooth muscle. Gradually, the muscle fibers creep to a longer length, sometimes continuing to creep for many hours or even days. Then, if the tension is reduced, the muscle creeps to a shorter length, which is called *reverse creep*. These phenomena of creep and reverse creep are also frequently called *plasticity* of smooth muscle.

If we apply these principles to hollow organs, such as the gut, one can readily understand that only a slight increase in filling pressure will cause the organ to increase its volume gradually to accommodate whatever amount of material enters—that is, up to the limits of stretch of the connective tissue surrounding the smooth muscle. Thus, a hollow viscus can accommodate either small or large amounts of material without greatly altering its internal pressure.

REFERENCES

Astrand, P., and Rodahl, K.: Textbook of Work Physiology. New York, McGraw-Hill Book Company, 1968.
Barr, L.: Smooth muscle as an electrical syncytium. *In* Fishman, A. P., and Hecht, H. H. (eds.): The Pulmonary Circulation and Interstitial Space. Chicago, University of Chicago Press, 1969, p. 161.

Bohr, David F., and Uchida, Eiichi: Activation of vascular smooth muscle. *In* Fishman, A. P., and Hecht, H. H. (eds.): The Pulmonary Circulation and Interstitial Space. Chicago, University of Chicago Press, 1969, p. 133.

Bozler, E.: Role of calcium in initiation of activity of smooth muscle. *Amer. J. Physiol., 216*:671, 1969.

Ekelund, L.: Exercise, including weightlessness. *Ann. Rev. Physiol., 31*:85, 1969.

Fales, J. T.. and Zierler, K. L.: Relation between length, tension, and heat: frog sartorius muscle, brief tetani. *Amer. J. Physiol., 216*:70, 1969.

Falls, Harold B.: Exercise Physiology. New York, Academic Press, Inc., 1968.

Guyton, A. C., and Reeder, R. C.: Pain and contracture in poliomyelitis. *Arch. Neurol. & Psychiat., 63*:954, 1950.

Hajdu, S.: Mechanism of the Woodworth staircase phenomenon in heart and skeletal muscle. *Amer. J. Physiol., 216*:206, 1969.

Hill, A. L.: Physiology of voluntary muscle. *Brit. Med. Bull., 12*(Sept.): 1956.

Holman, M. E.: An introduction to electrophysiology of visceral smooth muscle. *In* Handbook of Physiology. Baltimore, The Williams & Wilkins Co., 1968, Vol. IV, Sec. 6, p. 1665.

Hotta, K., and Bowen, W. J.: Contraction of ATPase activity of glycerinated muscle fibers and myofibrillar fragments. *Amer. J. Physiol., 218*:332, 1970.

Hoyle, G.: Comparative aspects of muscle. *Ann. Rev. Physiol., 31*:43, 1969.

Hurwitz, L., and Joiner, P. D.: Mobilization of cellular calcium for contraction in intestinal smooth muscle. *Amer. J. Physiol., 218*:12, 1970.

Hurwitz, L.; Joiner, P. D.; Von Hagen, S.; and Davenport, G. R.: Calcium accumulation and mechanical response of ileal muscle from rat and guinea pig. *Amer. J. Physiol., 216*:215, 1969.

Huxley, A. F., and Gordon, A. M.: Striation patterns in active and passive shortening of muscle. *Nature (Lond.), 193*:280, 1962.

Huxley, A. F.: Muscle. *Ann. Rev. Physiol., 26*:131, 1964.

Huxley, H. E.: The mechanism of muscular contraction. *Sci. Amer., 213*(6):18, 1965.

Lesch, M.; Parmley, M. W.; Hamosh, M.; Kaufman, S.; and Sonnenblick, E. H.: Effects of acute hypertrophy on the contractile properties of skeletal muscle. *Amer. J. Physiol., 214*: 685, 1968.

Mommaerts, W. F. H. M.: Energetics of muscular contraction. *Physiol. Rev., 49*:427, 1969.

Mommaerts, W. F. H. M.: Muscle energetics: biochemical differences between muscles as determined by the innervation. *Proc. Int. Union of Physiol. Sciences, 6*:116, 1968.

New York Heart Association: The Contractile Process. Boston, Little, Brown and Company, 1967.

Niedergerke, R.: The role of calcium ions as coupling agents. *Proc. Int. Union of Physiol. Sciences, 6*:143, 1968.

Page, S.: Structure of the sarcoplasmic reticulum in vertebrate muscle. *Brit. Med. Bull., 24*:170, 1968.

Peachey, L. D.: Muscle. *Ann. Rev. Physiol., 30*:401, 1968.

Podolsky, R. J.: Contraction of striated muscle fibers. *Proc. Int. Union of Physiol. Sciences, 6*:89, 1968.

Sandow, A.: Excitation-contraction coupling in skeletal muscle. *Pharmacol. Rev., 17*:265, 1965.

Sandow, A.: Skeletal muscle. *Ann. Rev. Physiol., 32*:87, 1970.

Wilson, D. M.; Smith, D. O.; and Dempster, P.: Length and tension hysteresis during sinusoidal and step function stimulation of arthropod muscle. *Amer. J. Physiol., 218*:916, 1970.

Wright, D. L., and Sonnenschein, R. R.: Muscle force and electromyogram with alteration in flow and composition of blood. *Amer. J. Physiol., 216*:1075, 1969.

PART II

BLOOD CELLS, IMMUNITY, AND BLOOD CLOTTING

CHAPTER 8

Red Blood Cells, Anemia, and Polycythemia

With this chapter we begin a discussion of blood, which is the vehicle for transport of nutrients, cellular excreta, hormones, electrolytes, and other substances from one part of the body to another. First, we will describe the cells of the blood, and in later chapters we will discuss blood plasma and the relationships of blood to the extravascular fluids.

THE RED BLOOD CELLS

The major function of red blood cells is to transport hemoglobin, which in turn carries oxygen from the lungs to the tissues. In some lower animals hemoglobin circulates as free protein in the plasma, not enclosed in red blood cells. However, when it is free in the plasma of the human being, approximately 3 per cent of it leaks through the capillary membrane into the tissue spaces or through the glomerular membrane of the kidney into Bowman's capsule each time the blood passes through the capillaries. Therefore, for hemoglobin to remain in the blood stream, it must exist inside red blood cells.

The red blood cells have other functions besides simply transport of hemoglobin. For instance, they contain a large quantity of carbonic anhydrase, which catalyzes the reaction between carbon dioxide and water, increasing the rate of this reaction about 250 times. The rapidity of this reaction makes it possible for blood to react with large quantities of carbon dioxide and thereby transport it from the tissues to the lungs. Finally, the hemoglobin in the cells is an excellent acid-base buffer (as is true of all proteins), so that the red blood cells are responsible for approximately 70 per cent of all the buffering power of whole blood. These specific functions are discussed elsewhere in the text and, consequently, are merely mentioned at the present time.

The Shape and Size of Red Blood Cells. Normal red blood cells are biconcave disks having a mean diameter of approximately 8 microns and a thickness at the thickest point of 2 microns and in the center of 1 micron or less. The average volume of the red blood cell is 87 (±5) cubic microns.

The shapes of red blood cells can change remarkably as the cells pass through capillaries. Actually, the red blood cell is a "bag" that can be deformed into almost any shape. Furthermore, because the normal cell has a great excess of cell membrane for the quantity of material inside, deformation does not stretch the membrane, and consequently does not rupture the cell as would be the case with many other cells.

Concentration of Red Blood Cells in the Blood. In the normal man the average number of red blood cells per cubic millimeter is 5,400,000 (±600,000) and in the normal woman 4,600,000 (±500,000). The number of red blood cells varies in the different sexes and at different ages, as shown in Figure 8–1. Also, the altitude at which the person lives and his degree of exercise affect the number of red blood cells; these are discussed later.

Quantity of Hemoglobin in the Cells. Red blood cells have the ability to concentrate hemoglobin in the cell fluid up to approximately 34 grams per 100 ml. of cells. Almost never does the concentration rise above this value, for this is a metabolic limit of the cell's hemoglobin-forming mechanism.

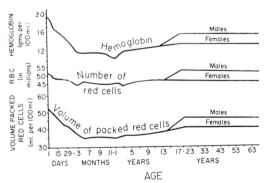

Figure 8–1 Relationship of age and sex to the hemoglobin content, red blood cell count, and hematocrit of the blood. (Redrawn from Wintrobe: Clinical Hematology. Lea & Febiger.)

Furthermore, in normal persons the percentage of hemoglobin is almost always near the maximum in each cell. However, when hemoglobin formation is deficient in the bone marrow, the percentage of hemoglobin in the cells may fall to as low as 15 grams per cent or less.

When the hematocrit (the percentage of the blood that is cells—normally about 40 per cent) and the quantity of hemoglobin in each respective cell are normal, the whole blood of men contains an average of 16 grams of hemoglobin per 100 ml. and of women an average of 14 grams per 100 ml. As will be discussed in connection with the transport of oxygen in Chapter 41, each gram of hemoglobin is capable of combining with approximately 1.33 ml. of oxygen. Therefore, in normal man, 21 ml. of oxygen can be carried in combination with hemoglobin in each 100 ml. of blood, and in normal woman 18 ml. of oxygen can be carried.

PRODUCTION OF RED BLOOD CELLS

Areas of the Body that Produce Red Blood Cells. In the early few weeks of embryonic life, primitive red blood cells are produced in the yolk sac. During the middle trimester of gestation the liver is the main organ for production of red blood cells, and at the same time a reasonable quantity of red blood cells is also produced by the spleen and lymph nodes. Then, during the latter part of gestation and after birth, red blood cells are produced almost exclusively by the bone marrow.

As illustrated by Figure 8–2, the bone marrow of essentially all bones produces red blood cells until adolescence, but the marrow of the long bones

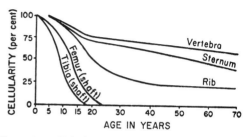

Figure 8–2 Relative rates of red blood cell production in the different bones at different ages. (Redrawn from Wintrobe: Clinical Hematology. Lea & Febiger.)

becomes quite fatty and produces no more red blood cells after approximately the age of 20. Beyond this time only the marrow of the membranous bones, such as the vertebrae, the sternum, and the ribs, produces cells. Even in these bones the marrow becomes less productive as age increases; as a result, progressive, mild anemia usually develops in old age.

Sometimes when various factors stimulate the bone marrow to produce tremendous quantities of red blood cells, much of the marrow that has already stopped producing red blood cells can once again become productive, and marrow that is still producing red blood cells becomes greatly hyperplastic and produces far greater than normal quantities. Indeed, even the spleen and occasionally the liver as well may re-establish their hemopoietic functions long after birth when extreme stimuli persist for prolonged periods of time.

Genesis of the Red Blood Cell. The blood cells are derived from a cell known as the *hemocytoblast*, which is illustrated in Figure 8–3, and new hemo-

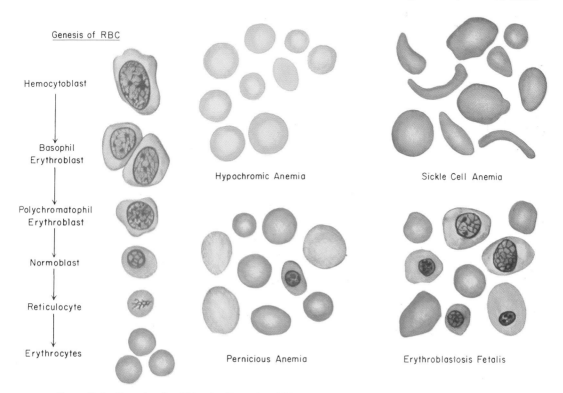

Genesis of RBC

Hemocytoblast

Basophil Erythroblast

Polychromatophil Erythroblast

Normoblast

Reticulocyte

Erythrocytes

Hypochromic Anemia

Sickle Cell Anemia

Pernicious Anemia

Erythroblastosis Fetalis

Figure 8–3 Genesis of red blood cells, and red blood cells in different types of anemias.

cytoblasts are continually being formed from primordial *stem* cells located throughout the bone marrow.

As illustrated in Figure 8–3, the hemocytoblast first forms the *basophil erythroblast* which begins the synthesis of hemoglobin. The erythroblast then becomes a *polychromatophil erythroblast*, so-called because of a mixture of basophilic material and the red hemoglobin. Following this, the nucleus of the cell shrinks while still greater quantities of hemoglobin are formed, and the cell becomes a *normoblast*. During all these stages the different cells continue to divide so that greater and greater numbers of cells are formed. Finally, after the cytoplasm of the normoblast has become filled with hemoglobin to a concentration of approximately 34 per cent, the nucleus is autolyzed and absorbed. The cell that is finally formed, the *erythrocyte*, usually contains no nuclear material when it passes by the process of diapedesis (squeezing through the pores of the membrane) into the blood capillaries.

Some, if not most, of the erythrocytes entering the blood contain small amounts of basophilic reticulum interspersed among the hemoglobin in the cytoplasm. This reticulum is chiefly the remains of the endoplasmic reticulum that produces hemoglobin in the early cell, and hemoglobin continues to be produced as long as the reticulum persists, a period of up to two to three days. In this stage the cells are known as *reticulocytes*. Ordinarily, the total proportion of reticulocytes in the blood is less than 0.5 per cent.

Regulation of Red Blood Cell Production

The total mass of red blood cells in the circulatory system is regulated within very narrow limits so that an adequate number of red cells is always available to provide sufficient tissue oxygenation and, yet, so that the cells are not so concentrated that they impede blood flow. The little we know about this control mechanism is the following:

Tissue Oxygenation as the Basic Regulator of Red Blood Cell Production. Any condition that causes the quantity of oxygen transported to the tissues to decrease ordinarily increases the rate of red blood cell production. Thus, when a person becomes extremely *anemic* as a result of hemorrhage or any other condition, the bone marrow immediately begins to produce large quantities of red blood cells. Also, destruction of portions of the bone marrow by any means, especially x-ray therapy, causes hyperplasia of the remaining bone marrow, thereby supplying the demand for red blood cells in the body.

At very *high altitudes,* where the quantity of oxygen in the air is greatly decreased, insufficient oxygen is transported to the tissues, and red cells are produced so rapidly that their number in the blood is considerably increased. Therefore, it is obvious that it is not the concentration of red blood cells in the blood that controls the rate of red cell production, but instead it is the functional ability of the cells to transport oxygen to the tissues.

Various diseases of the circulation that cause decreased blood flow through the peripheral vessels or that cause failure of oxygen absorption by the blood as it passes through the lungs also increase the rate of red cell production. This is especially apparent in prolonged *cardiac failure,* for the hypoxia resulting from sluggish flow of blood through the tissues in this condition increases the rate of red cell production, with resultant increase in the hematocrit and usually some increase in the total blood volume.

Finally the degree of physical activity of a person determines to a great extent the rate at which he will produce red blood cells. A very active athlete or laborer may produce red blood cells twice as rapidly as a person lying in bed, and, correspondingly, the athlete will have a red blood cell count as high as 6.5 million per cubic millimeter whereas the asthenic person will have a count as low as 3 million per cubic millimeter. This fact that exercise can increase the rate of red blood cell production indicates that it is *hypoxia of the tissues* themselves that causes red cell production, because the tissues become depleted of oxygen in exercise.

Erythropoietin, Its Response to Hypoxia, and Its Function in Regulating Red Blood Cell Production. Erythropoietin, also called *erythropoietic stimulating factor* or *hemopoietin*, is a glycoprotein that has a molecular weight between 25,000 and 45,000 and that is formed in the tissues of the body in response to hypoxia. Erythropoietin in turn acts on the bone marrow to increase the rate of red blood cell production. Especially large quantities of erythropoietin are formed in the kidneys during hypoxia, but lesser amounts seem also to be formed in the liver and possibly even in other tissues.

There is no direct response of the bone marrow to hypoxia. Instead, hypoxia stimulates red blood cell production only through the mechanism of erythropoietin.

Though erythropoietin begins to be formed by the kidneys almost immediately upon placing an animal or person in an atmosphere of low oxygen, very few new red blood cells appear in the circulating blood within the first two days; and it is only after three to four days that the maximum rate of new red cell production is reached. Thereafter, cells continue to be produced as long as the person remains in the low oxygen state or until he has produced enough red blood cells to carry adequate amounts of oxygen to his tissues despite the low oxygen.

Upon removal of a person from a state of low oxygen, his rate of oxygen transport to the tissues rises above normal, which causes his rate of erythropoietin formation to decrease to zero almost instantaneously and his rate of red blood cell production to fall essentially to zero within 24 to 36 hours. Red cell production remains at this extremely low level until enough cells have lived out their life spans and degenerated, so that the tissues again receive only their normal complement of oxygen.

The site of action of erythropoietin seems to serve to increase both maturation rate and numbers of mitotic cells, this effect occurring at all stages in the genesis of the red cell but mainly *at the stem cell* itself to increase the rate of conversion of stem cells into hemocytoblasts.

In the complete absence of erythropoietin, very few red blood cells are formed by the bone marrow. At the other extreme, when extreme quantities of erythropoietin are formed, the rate of red blood cell production can rise to as high as 25 to 50 times normal. Therefore, the erythropoietin control mechanism for red blood cell production is an extremely powerful one.

SYSTEM DIAGRAM FOR CONTROL OF RED CELL NUMBERS. Figure 8–4 illustrates a control diagram for the feedback mechanism that regulates red blood cell concentration, drawn using the denotations described in Chapter 1. This shows that hypoxia causes production of erythropoietin. Erythropoietin then causes red blood cell production, and red blood cells finally cause a negative effect on the hypoxia — that is, they transport oxygen to the tissues, and this alleviates the hypoxia. Because of the

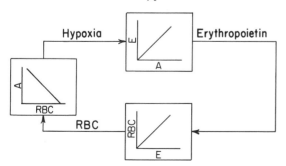

Figure 8–4   Control diagram, showing the negative feedback mechanism by which hypoxia regulates red blood cell concentration.

extremely powerful control of erythropoietin over red cell production, this negative feedback mechanism can adjust the red blood cell concentration to almost precisely that amount required for adequate oxygenation of the tissues, and yet not great enough to make the blood overly viscous because of cells.

Effect of Rate of Red Blood Cell Formation on the Type of Cell Released into the Blood

When the bone marrow produces red blood cells at a very rapid rate, many of the cells are released into the blood before they are mature erythrocytes. Thus, very rapid red cell production can cause the percentage of reticulocytes in the circulating blood to rise as high as 30 to 50 per cent of the total number of red blood cells, and, if the rate of production is even greater, a large number of normoblasts may appear in the circulating blood. In some severe anemias, such as erythroblastosis fetalis, which is discussed later in the chapter, the number of normoblasts may occasionally rise to as high as 5 to 20 per cent of all the circulating red blood cells, and a small number of erythroblasts and possibly even an occasional hemocytoblast may appear in the circulating blood.

Vitamins Needed for Formation of Red Blood Cells

The Maturation Factor — Vitamin B_{12} (Cyanocobalamin).   Vitamin B_{12} is an essential nutrient for all cells of the body, and growth of tissues in general is greatly depressed when this vitamin is lacking. This results from the fact that vitamin B_{12} is required for conversion of ribose nucleotides into deoxyribose nucleotides, one of the essential steps in DNA formation. Therefore, lack of this vitamin causes failure of nuclear maturation and division. Since tissues that produce red blood cells are among the most rapidly growing and proliferating of all the tissues, lack of vitamin B_{12} inhibits the rate of red blood cell production especially. Furthermore, the erythroblastic cells of the bone marrow, in addition to failing to proliferate rapidly, become larger than normal, developing into so-called *megaloblasts,* and the adult erythrocyte, called a *macrocyte,* has a flimsy membrane and is often irregular, large, and oval instead of the usual biconcave disk. These poorly formed macrocytes, after entering the circulating blood, are quite capable of carrying oxygen, but their fragility causes them to have a short life, measured in weeks instead of months as is true of normal cells. Therefore, it is said that vitamin B_{12} deficiency causes *maturation failure* in the process of erythropoiesis.

The occurrence of maturation failure does not prevent normal formation of hemoglobin. Indeed, the hemoglobin *concentration* in the cell is approximately 34 grams per cent, the same as normal, and, moreover, the average *quantity* of hemoglobin in the macrocyte is considerably greater than normal because the average volume of each cell is greater than normal.

MATURATION FAILURE CAUSED BY POOR ABSORPTION OF VITAMIN B_{12}—PERNICIOUS ANEMIA. The usual cause of maturation failure is not a lack of vitamin B_{12} in the diet but instead failure to absorb vitamin B_{12} from the gastrointestinal tract. This most commonly occurs in the disease called *pernicious anemia*, in which the basic abnormality is an *atrophic gastric mucosa* that fails to secrete normal gastric secretions. In the mucus secreted by the fundus and the body of the stomach is a mucopolysaccharide or mucopolypeptide (molecular weight about 50,000) called *intrinsic factor*, which combines with vitamin B_{12} of the food and makes the B_{12} available for absorption by the gut. It does this in the following way: First, the intrinsic factor binds tightly with the vitamin B_{12}. In this bound state the B_{12} is protected from digestion by the gastrointestinal enzymes. Second, still in the bound state, the B_{12} and intrinsic factor become adsorbed to the membranes of the gastrointestinal mucosal cells. Third, the combination is then transported into the cells in pinocytic vesicles, and about four hours later free vitamin B_{12} is released into the blood.

Lack of intrinsic factor, therefore, causes loss of much of the vitamin because of enzyme action in the gut, and also results in failure of absorption.

Once vitamin B_{12} has been absorbed from the gastrointestinal tract it is stored in large quantities in the liver and then released slowly as needed to the bone marrow and other tissues of the body. The total amount of vitamin B_{12} required each day to maintain normal red cell maturation is less than 1 microgram, and the normal store in the liver is about 1000 times this amount. Therefore, many months of defective B_{12} absorption is required to cause maturation failure anemia.

RELATIONSHIP OF FOLIC ACID (PTEROYL-GLUTAMIC ACID) TO VITAMIN B_{12}. Occasionally a patient with maturation failure anemia responds equally as well to folic acid as to vitamin B_{12}, so that it has become apparent that this vitamin is also concerned with the maturation of red blood cells. Folic acid, like B_{12}, is required for formation of DNA but in a different way. It promotes the methylation of deoxyuridylate to form deoxythymidylate, one of the nucleotides required for DNA synthesis.

FORMATION OF HEMOGLOBIN

Synthesis of hemoglobin begins in the erythroblasts and continues through the normoblastic stage. Even when young red blood cells leave the bone marrow and pass into the blood stream, they continue to form hemoglobin for a few days.

Figure 8–5 gives the basic chemical steps in the formation of hemoglobin. From tracer studies with isotopes it is known that hemoglobin is synthesized mainly from acetic acid and glycine. It is believed that the acetic acid is changed in the Krebs cycle, which will be explained in Chapter 67, into α-ketoglutaric acid, and then two molecules of this combine with one molecule of glycine to form a pyrrole compound. In turn, four pyrrole compounds combine to form a protoporphyrin compound. One of the protoporphyrin compounds, known as protoporphyrin III, then combines with iron to form the heme mole-

Figure 8–5   Formation of hemoglobin.

cules. Finally, four heme molecules combine with one molecule of globin, a globulin, to form hemoglobin, the formula for which is shown in Figure 8–5 E. Hemoglobin has a molecular weight of 68,000.

Combination of Hemoglobin with Oxygen. The most important feature of the hemoglobin molecule is its ability to combine loosely and reversibly with oxygen. This ability will be discussed in detail in Chapter 41 in relation to respiration, for the primary function of hemoglobin in the body depends upon its ability to combine with oxygen in the lungs and then to release this oxygen readily in the tissue capillaries where the gaseous tension of oxygen is much lower than in the lungs.

Oxygen *does not* combine with the two positive valences of the ferrous iron in the hemoglobin molecule. Instead, it binds loosely with two of the six "coordination" valences of the iron atom. This is an extremely loose bond so that the combination is easily reversible. Furthermore, the oxygen does not become ionic oxygen but is carried as molecular oxygen to the tissues where, because of the loose, readily reversible combination, it is released into the tissue fluids in the form of dissolved molecular oxygen rather than ionic oxygen.

It should be recalled, also, that each molecule of hemoglobin contains four molecules of heme. Therefore, one molecule of hemoglobin contains four iron atoms and can carry four molecules of oxygen.

Accessory Substances Needed for Formation of Hemoglobin. In addition to amino acids and iron, which are needed directly for formation of the hemoglobin molecule, a number of other substances act as catalysts or enzymes during different stages of hemoglobin formation. For instance, an average human adult requires approximately 2 mg. of copper each day in his diet if normal hemoglobin formation is to take place, and addition of small quan-

tities of copper to the diet of patients who have hypochromic anemia occasionally accelerates the rate of hemoglobin formation. This occurs even though copper is not one of the substances needed as a building stone in the formation of hemoglobin. Fortunately, the quantity of copper in the normal daily diet is sufficient that copper deficiency anemia is almost unknown in the human being.

Lack of *pyridoxine* in the diet of some animals not only decreases the rate of red blood cell formation but also depresses the rate of hemoglobin formation to an even greater extent. Also, *cobalt* deficiency can greatly depress the formation of hemoglobin in some animals, whereas, strangely enough, a great excess of cobalt can cause formation of greater than normal numbers of red blood cells that contain normal quantities of hemoglobin. Finally, *nickel* has been found to take the place of cobalt to a moderate extent in aiding the synthesis of hemoglobin in the bone marrow.

Though the function of these different substances in the formation of hemoglobin is not known, the above listing of them serves mainly to emphasize the fact that hemoglobin formation results from a series of synthesis reactions, each of which depends upon appropriate building materials and also upon appropriate controlling catalysts and enzymes.

IRON METABOLISM

Because iron is important for formation of hemoglobin, myoglobin, and other substances, such as the cytochromes, cytochrome oxidase, peroxidase, and catalase, it is essential to understand the means by which iron is utilized in the body.

The total quantity of iron in the body averages about 4 grams, approximately 65 per cent of which is present in the form of hemoglobin. About 4 per cent is present in the form of myoglobin, 1 per cent in the form of the various heme compounds that control intracellular oxidation, 0.1 per cent in the form of trans-

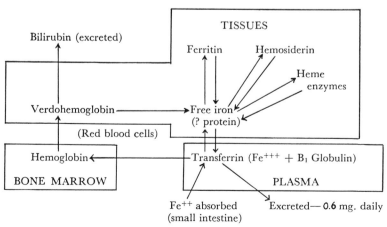

Figure 8–6 Iron transport and metabolism.

ferrin in the blood plasma, and 15 to 30 per cent stored mainly in the form of ferritin.

Transport and Storage of Iron. Transport, storage, and metabolism of iron in the body is illustrated in Figure 8–6, which may be explained as follows: When iron is absorbed from the small intestine, it immediately combines with a beta globulin to form the compound *transferrin*, in which form it is transported in the blood plasma. The iron in this compound is very loosely combined with the globulin molecule and, consequently, can be released to any of the tissue cells at any point in the body. Excess iron in the blood is deposited in all cells of the body *but especially in the liver cells,* where more than 60 per cent of the excess is stored. There it combines with the protein *apoferritin* to form *ferritin.* Apoferritin has a molecular weight of approximately 460,000, and varying quantities of iron can combine in clusters of iron radicals with this large molecule; therefore, ferritin may contain only a small amount of iron or a relatively large amount. This iron stored in ferritin is called *storage iron.*

When the total quantity of iron in the body is more than the apoferritin storage pool can accommodate, some of it is stored in an extremely insoluble form called *hemosiderin.* Hemosiderin forms large clusters in the cells and consequently can be stained and observed in tissue slices by usual histologic techniques, whereas ferritin cannot because it is distributed diffusely.

When the quantity of iron in the plasma falls very low, iron is absorbed from ferritin quite easily but less easily from hemosiderin. The absorbed iron is then transported to the portions of the body where it is needed.

When red blood cells have lived their life span and are destroyed, the hemoglobin released from the cells is ingested by the reticuloendothelial cells. There free iron is liberated, and it can then either be stored in the ferritin pool or be reused for formation of hemoglobin.

Daily Loss of Iron. About 0.6 mg. of iron is excreted each day by the male, mainly into the feces. Additional quantities of iron are lost whenever bleeding occurs. Thus, in the female, the menstrual loss of blood brings the average iron loss to a value of approximately 2.1 mg. per day.

Obviously, the quantity of iron derived from the diet each day must at least equal that lost from the body.

Absorption of Iron from the Gastrointestinal Tract. Iron is absorbed almost entirely in the upper part of the small intestine, mainly in the duodenum. It is absorbed by an active absorptive process, though the precise mechanism of this active absorption is unknown.

Iron is absorbed mainly in the ferrous rather than ferric form, for, in general, the greater the degree of positive charge of any ion the more difficult is it to be absorbed from the intestinal tract. This has practical significance because it means that ferrous iron compounds are more effective in treating iron deficiency than are ferric compounds.

The rate of iron absorption is extremely slow, with a maximum rate of only a fraction of a gram per day. This means that when tremendous quantities of iron are present in the food, only small proportions of this will be absorbed. On the other hand, if only minute quantities are present, essentially all will be absorbed.

Regulation of Total Body Iron by Alteration of Rate of Absorption. When the body has become saturated with iron so that essentially all of the apoferritin in the iron storage areas is already combined with iron, the rate of absorption of iron from the intestinal tract becomes greatly decreased. On the other hand, when the iron stores have been depleted of iron, the rate of absorption becomes greatly accelerated, 20 or more times as great as when the iron stores are saturated. Thus, the total body iron is regulated primarily by alterations in rate of absorption.

Alterations in rate of excretion can also help to regulate total body iron, but to a much less extent. The rate of excretion of iron falls from about 0.6 mg. per day in the normal male to as low as 0.1 mg. per day, a six-fold change, in a person with severe iron deficiency.

FEEDBACK MECHANISMS BY WHICH IRON ABSORPTION AND IRON EXCRETION ARE REGULATED. When essentially all the apoferritin in the body has become saturated with iron, it becomes difficult for transferrin to release iron to the tissues. As a consequence, the transferrin, which is normally only one-third saturated with iron, now becomes almost fully bound with iron so that the transferrin accepts almost no new iron from the mucosal cells. Then, as a final stage of this process, the buildup of excess iron in the mucosal cells themselves depresses active absorption of iron from the intestinal lumen and at the same time enhances the rate of excretion of iron from the mucosa.

Though the details of the method by which excess iron depresses active absorption of iron by the mucosa are yet mostly unknown, some are beginning to emerge. When the epithelial cells of the mucosa are first formed in the crypts of the intestinal mucosa, the degree of their ability to absorb iron seems to be pre-ordained. If there is excess iron already in the blood and tissues, small vesicles filled with ferritin, called *ferritin bodies,* are present in these newly formed epithelial cells. For reasons that are not understood, the presence of such bodies in the cells prevents or markedly reduces the active absorption of iron by the cells. The epithelial cells migrate in three to four days from the crypts outward over the villi of the intestines, at the end of which time they are desquamated into the intestinal lumen. During this entire period, their ability to absorb iron is greatly depressed.

On the other hand, if the body is deficient in iron, ferritin bodies do not appear in the newly formed epithelial cells, so that the rate of iron absorption through the intestinal mucosa becomes maximal. However, if the animal then suddenly comes into a great supply of iron, some of this combines slowly with apoferritin in the epithelial cells to form ferritin bodies, and these cells also become poor absorbers of iron. Thus, the presence of excess iron in the blood and tissues causes a *mucosal block* for the absorption of further iron.

The rate of iron excretion also increases when mucosal block occurs. The reason for this is that a considerable quantity of iron becomes stored in the epithelial cells, and these are later desquamated from the tips of the villi into the intestine and carry with them this stored iron. In addition, phagocytic cells phagocytize iron in the mucosal wall and are discharged along with the desquamating epithelial cells, thereby carrying still additional iron into the feces. Thus, the process of iron excretion also helps to regulate the concentration of iron in the body.

DESTRUCTION OF RED BLOOD CELLS

When red blood cells are delivered from the bone marrow into the circulatory system they normally circulate 100 to 120 days before being destroyed. Even though red cells do not have a nucleus, they do still have cytoplasmic enzymes for metabolizing glucose and other substances and for the utilization of oxygen, but these metabolic systems become progressively less active with time. As the cells become older they become progressively more fragile, presumably because their life processes simply wear out.

Once the red cell membrane becomes very fragile, it ruptures during passage through some tight spot of the circulation. Many of the red cells fragment in the spleen, perhaps where the cells squeeze through the red pulp of the spleen. When the spleen is removed, the number of abnormal cells and old cells circulating in the blood increases considerably.

Destruction of Hemoglobin. The hemoglobin released from the cells when they burst is phagocytized and digested almost immediately by reticuloepithelial cells, releasing iron back into the blood to be carried by transferrin either to the bone marrow for production of new red blood cells or to the liver and other tissues for storage in the form of ferritin. The heme portion of the hemoglobin molecule is converted through a series of stages into the bile pigment *bilirubin,* which is secreted by the liver into the bile; this will be discussed in relation to liver function in Chapter 74.

THE ANEMIAS

Anemia means a deficiency of red blood cells, which can be caused either by too rapid loss or too slow production of red blood cells. Some types of anemia and their physiological causes are the following:

Blood Loss Anemia. After rapid hemorrhage the body replaces the plasma within one to three days, but this leaves a low concentration of red blood cells. If a second hemorrhage does not occur, the red blood cell concentration returns to normal within three to four weeks.

In chronic blood loss, the person frequently cannot absorb enough iron from the intestines to form hemoglobin as rapidly as it is lost. Therefore, red cells are frequently produced with too little hemoglobin inside them, giving rise to so-called *hypochromic anemia.*

Aplastic Anemia. *Bone marrow aplasia* means lack of a functioning bone marrow. For instance, the person exposed to gamma ray radiation from a nuclear bomb blast is likely to sustain complete destruction of his bone marrow, followed in a few weeks by lethal anemia. Likewise, excessive x-ray treatment, certain industrial chemicals, and even drugs to which the person might be sensitive can cause the same effect.

Maturation Failure Anemia. From the earlier discussion in this chapter of vitamin B_{12}, folic acid, and intrinsic factor from the stomach mucosa, one can readily understand that loss of any one of these factors can lead to failure of maturation of the red blood cells. Thus, atrophy of the stomach mucosa, as occurs in *pernicious anemia,* or loss of the entire stomach as the result of total gastrectomy can lead to maturation failure. Also, patients who have intestinal sprue, in which folic acid, B_{12}, and other vitamin B compounds are poorly absorbed, often develop maturation failure. Because the bone marrow cannot proliferate rapidly enough to form normal numbers of red blood cells in maturation failure, the cells that are formed are oversized, of bizarre shapes, and have fragile membranes. Therefore, these cells rupture easily, leaving the person in dire need of an adequate number of red cells.

Hemolytic Anemia. Many different abnormalities of the red blood cells, most of which are hereditarily acquired, make the cells very fragile, so that they rupture easily as they go through the capillaries. Therefore, even though the number of red blood cells formed is completely normal, the red cell life span is so short that serious anemia results. Some of these types of anemia are the following:

In *familial hemolytic anemia* the red cells are very small in size, and they are *spherical* in shape rather than being biconcave discs. These cells cannot be compressed because they do not have the normal loose bag-like cell membrane structure of the biconcave discs. Therefore, on passing through small capillaries they are easily ruptured by even the slightest compression.

In *sickle cell anemia,* which is present in approximately 7 per cent of West African and American Negroes, the cells contain an abnormal type of hemoglobin called *hemoglobin S.* When this hemoglobin is exposed to low concentrations of oxygen, it precipitates into long crystals inside the red blood cell. These crystals elongate the cell and give it the appearance of being a sickle rather than a biconcave disc. The precipitated hemoglobin also damages the cell membrane so that the cells become highly fragile, leading to serious anemia. Such patients frequently get into a vicious cycle in which low oxygen tension in the tissues causes sickling, which causes impediment of blood flow through the tissue, causing still further decrease in oxygen tension. Thus, once the process starts, it progresses rapidly, leading to a serious decrease in red blood cell mass within a few hours and, often, to death.

Thalassemia, which is also known as *Cooley's anemia* or *Mediterranean anemia,* is another hereditary type of hemolytic anemia in which the cells are small and have fragile membranes. Here again, the cells are easily ruptured upon passing through the tissues.

In *erythroblastosis fetalis,* Rh positive red blood cells in the fetus are attacked by antibodies from an Rh negative mother. These antibodies make the cells fragile and cause the child to be born with serious anemia. This will be discussed in detail in Chapter 11 in relation to the Rh factor of blood.

Hemolysis also occasionally results from transfusion reactions, from malaria, and from reactions to certain drugs.

EFFECTS OF ANEMIA ON THE CIRCULATORY SYSTEM

The viscosity of the blood, which will be discussed in detail in Chapter 18, is dependent almost entirely on the concentration of red blood cells. In severe anemia the blood viscosity may fall to as low as one and one-half times that of water rather than the normal value of approximately three times the viscosity of water. The greatly decreased viscosity decreases the resistance to blood flow in the peripheral vessels so that far greater than normal quantities of blood return to the heart. As a consequence, the cardiac output increases to as much as two or more times normal as a result of decreased viscosity.

Moreover, hypoxia due to diminished transport of oxygen by the blood causes the tissue vessels to dilate, allowing further increased return of blood to the heart, increasing the cardiac output to a higher level than ever. Thus, one of the major effects of anemia is greatly *increased work load on the heart.*

The increased cardiac output in anemia offsets many of the symptoms of anemia, for, even though each unit quantity of blood carries only small quantities of oxygen, the rate of blood flow may be increased to such an extent that almost normal quantities of oxygen are delivered to the tissues.

As long as an anemic person's rate of activity is low, he can live without fatal hypoxia of the tissues, even though his concentration of red blood cells may be reduced to one-fourth normal. However, when he begins to exercise, his heart is not capable of pumping much greater quantities of blood than it is already pumping. Consequently, during exercise, which greatly increases the demands for oxygen, extreme tissue hypoxia results, and acute cardiac failure often ensues.

POLYCYTHEMIA

Physiologic Polycythemia. When a person remains at high altitudes for weeks his blood-forming organs automatically produce large quantities of red blood cells. This becomes so extreme at altitudes of 14,000 to 15,000 feet that all persons residing permanently at these levels have continued polycythemia, with blood cell counts of 6 to 8 million per cubic millimeter, which are far above the normal 4.5 to 5.5 million.

Polycythemia Vera (Erythremia). In addition to those persons who have physiologic polycythemia, others have a condition known as *polycythemia vera* in which the red blood cell count may be as high as 11 million and the hematocrit as high as 70 to 80 per cent. Polycythemia vera is a tumorous condition of the organs that produce blood cells. It causes excess production of red blood cells in the same manner that a tumor of a breast causes excess production of a specific type of breast cell. It also causes excess production of white blood cells and platelets, but these are masked by the great overabundance of red cells.

In polycythemia vera not only does the hematocrit increase but the total blood volume occasionally also increases to as much as twice normal. As a result, the entire vascular system becomes intensely engorged. In addition, many of the capillaries become plugged by the very viscous blood, for the viscosity of the blood in polycythemia vera sometimes increases from the normal of 3 times the viscosity of water to 15 times that of water.

EFFECT OF POLYCYTHEMIA ON THE CIRCULATORY SYSTEM

Because of the greatly increased viscosity of the blood in polycythemia, the flow of blood through the vessels is extremely sluggish. In accordance with the factors that regulate the return of blood to the heart as discussed in Chapter 27, it is obvious that increasing the viscosity tends to decrease the rate of venous return to the heart. On the other hand, the blood volume is greatly increased in polycythemia, which tends to increase the venous return. Actually, the cardiac output in polycythemia is not far from normal because these two factors more or less neutralize each other.

Because the total blood volume in polycythemia is sometimes twice normal, the circulation time through the body may also be increased to twice normal. In other words, the mean circulation time occasionally is as great as 120 seconds instead of the normal of approximately 60 seconds. Thus, the velocity of blood flow in any given vessel is considerably decreased in polycythemia.

The arterial pressure is normal in most persons with polycythemia, though in approximately one third of them the pressure is elevated. This means that the blood pressure regulating mechanisms can usually offset the tendency for increased blood viscosity to increase peripheral resistance and thereby increase arterial pressure. Yet, beyond certain limits, these regulations fail.

The color of the skin is dependent to a great extent on the quantity of blood in the subpapillary venous plexus. In polycythemia vera the quantity of blood in this plexus is greatly increased. Furthermore, because the blood passes sluggishly through the skin capillaries before entering the venous plexus, a larger than normal proportion of the hemoglobin is reduced before the blood enters the plexus. The blue color of this reduced hemoglobin is prepotent over the red color of oxygenated hemoglobin. Therefore, a person with polycythemia vera ordinarily has a cyanotic (blue) skin.

REFERENCES

Atherton, R. W., and Timiras, P. S.: Erythropoietic and somatic development in chick embryos at high altitude. Amer. J. Physiol., 218:75, 1970.

Boggs, D. R.: Homeostatic regulatory mechanisms of hematopoiesis. Ann. Rev. Physiol., 28:39, 1966.

Callender, S. T.: The interstitial mucosa and iron absorption. Brit. Med. Bull., 23:263, 1967.

Card, R. T.; McGrath, M. J.; Paulson, E. J.; and Valberg, L. S.: Life-span and autohemolysis of macrocytic erythrocytes produced in response to hemorrhage. Amer. J. Physiol., 216:974, 1969.

Castle, W. B.: Gastric intrinsic factor and vitamin B_{12} absorption. In Handbook of Physiology. Baltimore, The Williams & Wilkins Co., 1968, Vol. III, Sec. 6, p. 1529.

Clarke, C. A.: The prevention of "rhesus" babies. Sci. Amer., 219:46, 1968.

Cooper, B. A.: Complex of intrinsic factor and B_{12} in human ileum during vitamin B_{12} absorption. Amer. J. Physiol., 214:832, 1968.

Crosby, W. H.: Iron absorption. In Handbook of Physiology. Baltimore, The Williams & Wilkins Co., 1968, Vol. III, Sec. 6, p. 1553.

Firkin, B. G.: The fine structure and chemistry of haemopoietic tissue. In Bittar, E. Edward, and Bittar, Neville (eds.): The Biological Basis of Medicine. New York, Academic Press, Inc., 1968, Vol. 3, p. 3.

Fisher, J. W.: The structure and physiology of erythropoietin. In Bittar, E. Edward, and Bittar, Neville (eds.): The Biological Basis of Medicine. New York, Academic Press, Inc., 1969, Vol. 3, p. 41.

Hamstra, R. D., and Block, M. H.: Erythropoiesis in response to blood loss in man. J. Appl. Physiol., 27:503, 1969.

Ingram, V. M.: On the biosynthesis of hemoglobin. Harvey Lectures, 1965–66, p. 43.

Linman, J. W.: Physiologic and pathophysiologic effects of anemia. New Eng. J. Med., 279:819, 1968.

Murray, J. F.; Escobar, E.; and Rapaport, E.: Effects of blood viscosity on hemodynamic responses in acute normovolemic anemia. Amer. J. Physiol., 216:638, 1969.

Necas, E., and Neuwirt, J.: Response of erythropoiesis to blood loss in hyperbaric air. Amer. J. Physiol., 216:800, 1969.

Rosenthal, H. L.; Cutler, L.; and Sobieszczanska: Uptake and transport of vitamin B_{12} in subcellular fractions of intestinal mucosa. Amer. J. Physiol., 218:358, 1970.

Sorbie, J., and Valberg, L. S.: Splenic sequestration of stress erythrocytes in the rabbit. Amer. J. Physiol., 218:647, 1970.

Stokstad, E. L. R., and Koch, J.: Folic acid metabolism. Physiol. Rev., 47(1):83, 1967.

Weissbach, H., and Dickerman, H.: Biochemical role of vitamin B_{12}. Physiol. Rev., 45:80, 1965.

Wintrobe, M. M.: Clinical Hematology. 6th ed., Philadelphia, Lea & Febiger, 1967.

CHAPTER 9

Resistance of the Body to Infection—The Reticuloendothelial System, Leukocytes, and Inflammation

The body is constantly exposed to bacteria, these occurring especially in the mouth, the respiratory passageways, the colon, the mucous membranes of the eyes, and even the urinary tract. Many of these bacteria are capable of causing disease if they invade the deeper tissues. In addition, a person is intermittently exposed to highly virulent bacteria and viruses from outside the body which can cause specific diseases, such as pneumonia, streptococcal infections, and typhoid fever.

On the other hand, a group of tissues including the *reticuloendothelial system* and the leukocytes constantly combat any infectious agent that tries to invade the body. These tissues function in two different ways to prevent disease: (1) by actually destroying invading agents by the process of phagocytosis and (2) by forming antibodies against the invading agent, the antibodies in turn destroying the invader. The present chapter is concerned with the first of these methods, while the following chapter is concerned with the second.

THE RETICULOENDOTHELIAL SYSTEM

The reticuloendothelial system is comprised of those cells that line many of the vascular and lymph channels and are capable of phagocytizing bacteria, viruses, or other foreign agents or are capable of forming immune bodies against these. They include the phagocytic cells of the bone marrow, spleen, liver, and lymph nodes. All these cells are closely related to each other, and are called *reticulum cells.*

The Reticulum Cell.   The reticulum cell is a primordial cell not greatly different from the

original germinal epithelial cells. It is capable of differentiating into many other types of cells. For instance, reticulum cells in the bone marrow are continually forming (a) *hemocytoblasts,* which then form red blood cells, and (b) *myeloblasts,* which form white blood cells. But equally as important is their capability to develop into *lymphocytes* in the lymph nodes and into *tissue histiocytes* which wander through the tissues to perform the function of *phagocytosis,* that is, the ingestion of foreign organisms or foreign particulate matter of any type.

Reticulum cells are also found in great numbers in the spleen, liver, and lymph nodes. At all these different points they exhibit the property of phagocytosis.

Finally, reticulum cells in almost any part of the body can form *plasma cells* which, as is discussed in detail in the following chapter, are principally responsible for the formation of immune bodies. Therefore, the reticulum cell is the basis of a system that provides the body with much of its resistance to infection.

The Lymph Nodes.   The lymphatic system is discussed in Chapter 21 where it is pointed out that essentially no particulate matter or even large molecular compounds can be absorbed directly through the capillary membranes into the blood stream. Instead, any invading organisms or large molecules that enter the tissue spaces must pass by way of the lymphatic system into the blood. Lymph nodes are located intermittently along the course of the lymphatics. Figure 9–1 illustrates the general organization of a lymph node, showing lymph entering by way of the *afferent lymphatics,* flowing through the outer capsule of the lymph node into *subcapsular sinuses,* then gradually through the *medullary sinuses,* and finally out

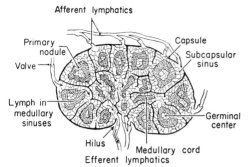

Figure 9-1 Functional diagram of a lymph node. (Redrawn from Ham: Histology. J. B. Lippincott Co.)

the *hilus* into the *efferent lymphatics.* Large numbers of *reticulum cells* line the sinuses, and if any particles or proteins foreign to the person's own body are in the lymph, these cells phagocytize them and prevent general dissemination throughout the body.

THE RETICULAR STRUCTURES OF THE CIRCULATORY SYSTEM

If invading agents succeed in getting past the lymph nodes or if they should by chance enter the circulation directly without going through the lymph nodes, there are still other lines of defense for their removal even after they enter the circulation. In addition to the granulocytes, circulating in the blood, which can phagocytize the invaders, three special structures of the circulation contain large numbers of reticulum cells—the spleen, the liver, and the bone marrow.

Figure 9-2 illustrates the general structure of the *spleen,* with an artery entering the splenic capsule, then passing into the *splenic pulp,* and probably connecting through small capillaries with the large *venous sinuses.* The capillaries seem to be highly porous, allowing large quan-

tities of whole blood cells to pass out of the capillaries into the *red pulp* and gradually to *squeeze* through the pulp and then back through the sinus walls into the venous system. The red pulp is loaded with reticulum cells, and, in addition, the venous sinuses are also lined with reticulum cells. This peculiar passage of cells through the red pulp provides an exceptionally effective means for phagocytosis of unwanted debris in the blood, especially old and abnormal red blood cells.

The *sinuses of the liver* are also lined with reticulum cells, called *Kupffer cells,* which are illustrated in Figure 9-3. Though in any small portion of the liver the Kupffer cells may not be so numerous as the reticulum cells of the spleen, the total size of the liver is so much greater than that of the spleen that the liver ranks along with the spleen as an exceptionally important blood purifier. The reticulum system of the liver is especially valuable because all portal blood must pass through the liver before entering the general circulation. Large numbers of bacteria constantly pass by diapedesis through the gastrointestinal mucosa into the portal blood, but the liver filtration system is so effective that essentially none of these bacteria succeeds in passing from the portal blood into the general systemic circulation.

The reticulum cells of the *bone marrow* are

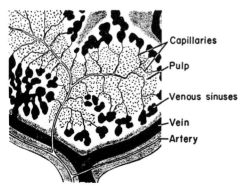

Figure 9-2 Functional structures of the spleen. (Modified from Maximow and Bloom: Textbook of Histology.)

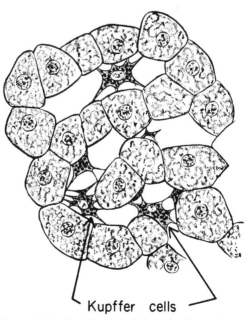

Figure 9-3 Kupffer cells lining the liver sinusoids, showing phagocytosis of India ink particles. (Redrawn from Copenhaver and Johnson: Bailey's Textbook of Histology. The Williams and Wilkins Co.)

of major importance as precursors of the red and white blood cells, but they are also capable of phagocytizing foreign substances that have entered the circulation. It is interesting that the reticulum cells of the bone marrow are even more capable than those of the spleen and liver for removing very fine particles, such as protein toxins.

In summary, the different structures of the circulatory system that contain large numbers of reticulum cells operate continuously as a filtration system for removing unwanted materials from the circulation. Indeed, even old blood cells that have ended their useful life span and have finally ruptured are removed by the reticulum cells—especially by those of the spleen.

The Tissue Histiocyte and Defense of the Tissues Against Infection. In virtually all tissues of the body are many cells called *histiocytes* or *clasmatocytes* that have properties similar to those of reticulum cells. These are always available to remove foreign substances, and they can even phagocytize necrotic tissue cells and any other debris that might enter the tissues.

In addition to their phagocytic ability, histiocytes in damaged tissues are capable of changing into fibroblasts which then lay down collagen fibers to repair the damage. Furthermore, histiocytes can swell up and move by ameboid motion through the tissues, whereupon they are then called *macrophages*. These cells seek out areas of inflammation and help to devour the infectious agents or necrotic tissues. Two of the white blood cells, the monocytes and lymphocytes, upon migrating into the tissues can also become macrophages and therefore are closely allied to histiocytes. Histiocyte function is discussed further in connection with the functions of the leukocytes.

THE LEUKOCYTES (WHITE BLOOD CELLS)

The leukocytes are the *mobile units* of the body's protective system. They are formed partially in the bone marrow (the *granulocytes*) and partially in the lymph nodes (*lymphocytes* and *monocytes*), but after formation they are transported in the blood to the different parts of the body where they are to be used. The real value of the white blood cells is that they are specifically transported to areas of serious inflammation, thereby providing a rapid and potent defense against any infectious agent that might be present.

GENERAL CHARACTERISTICS OF LEUKOCYTES

The Types of White Blood Cells. Five different types of white blood cells are normally found in the blood. These are *polymorphonuclear neutrophils,* *polymorphonuclear eosinophils,* *polymorphonuclear basophils, monocytes,* and *lymphocytes.* In addition, there are large numbers of *platelets,* which are fragments of a sixth type of white blood cell, the *megakaryocyte.* The three types of polymorphonuclear cells have a granular appearance, as illustrated in Figure 9–4, for which reason they are called *granulocytes,* or in clinical terminology they are often called simply "polys."

The granulocytes and the monocytes protect the body against invading organisms by ingesting them—that is, by the process of *phagocytosis.* One of the functions of the lymph glands and lymphocytes is formation of monocytes which in turn destroy invading organisms. Finally, the function of platelets is to activate the blood clotting mechanism. All these functions are protective mechanisms of one type or another.

Concentrations of the Different White Blood Cells in the Blood. The adult human being has approximately 7000 white blood cells per cubic millimeter of blood. The normal percentages of the different types of white blood cells are approximately the following:

Polymorphonuclear neutrophils	62.0%
Polymorphonuclear eosinophils	2.3%
Polymorphonuclear basophils	0.4%
Monocytes	5.3%
Lymphocytes	30.0%

The number of platelets in each cubic millimeter of blood is normally about 300,000.

GENESIS OF THE LEUKOCYTES

Figure 9–4 illustrates the stages in the development of white blood cells. The polymorphonuclear cells are normally formed only in the bone marrow. On the other hand, lymphocytes and monocytes are produced in the various lymphogenous organs, including the lymph glands, the spleen, the thymus, the tonsils, and various lymphoid rests in the gut and elsewhere.

Some of the white blood cells formed in the bone marrow are stored within the marrow until they are needed in the circulatory system. Then when the need arises, various factors that are discussed later cause them to be released.

As illustrated in Figure 9–4, megakaryocytes are also formed in the bone marrow and are part of the myelogenous group of white blood cells. It is generally taught that these megakaryocytes fragment in the bone marrow, the small fragments known as *platelets* passing then into the blood, but

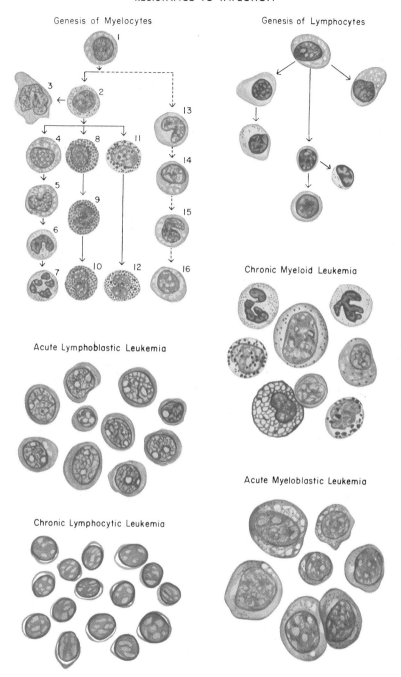

Figure 9–4 Genesis of the white blood cells, and the blood picture in different types of leukemia. The different cells of the myelogenous series are: *1*, myeloblast; *2*, promyelocyte; *3*, megakaryocyte; *4*, neutrophil myelocyte; *5*, young neutrophil metamyelocyte; *6*, "band" neutrophil metamyelocyte; *7*, polymorphonuclear neutrophil; *8*, eosinophil myelocyte; *9*, eosinophil metamyelocyte; *10*, polymorphonuclear eosinophil; *11*, basophil myelocyte; *12*, polymorphonuclear basophil; *13–16*, stages of monocyte formation. (Redrawn in part from Piney: A Clinical Atlas of Blood Diseases. The Blakiston Co.)

it is possible that the megakaryocytes, which are very large cells, are released directly into the blood and fragment later as they attempt to traverse the capillaries.

Materials Needed for Formation of White Blood Cells. In general, the white blood cells need essentially the same vitamins and amino acids as most of the other cells of the body for their forma-

tion. Especially does lack of folic acid, a compound of the vitamin B complex, block the formation of white blood cells as well as prevent maturation of red blood cells, which was pointed out in Chapter 8. Also, in extreme debilitation, production of white blood cells may be greatly reduced, despite the fact that these cells are needed more during such a state than usually.

LIFE SPAN OF THE WHITE BLOOD CELLS

The main reason white blood cells are present in the blood is simply to be transported from the bone marrow or lymphoid tissue to the areas of the body where they are needed. Therefore, it is to be expected that the life of the white blood cells in the blood would be short.

The life span of the granulocytes in the blood averages about 12 hours, though this may be as short as only two to three hours during times of serious tissue infection or as great as several days when the granulocytes are not needed in the tissues.

The life span of the monocytes in the blood is a complete mystery, for monocytes apparently wander back and forth between the tissues and the blood. The monocytes probably live for considerably longer periods of time, especially in the tissues, than do the polymorphonuclear cells, for they persist in infectious areas much longer than the polys.

Lymphocytes enter the circulatory system continually along with the drainage of fluid from the lymph nodes. The total number entering the blood from the thoracic duct in each 24 hours is usually several times the total number of lymphocytes present in the blood stream at any given time. Therefore, the span of time that the lymphocytes remain in the blood must be only a few hours. However, studies using radioactive lymphocytes have shown that almost all of these pass by diapedesis into the tissues and then re-enter the lymph and return to the lymph nodes. And many of these cells have life spans of 100 to 300 days, but this also depends on the tissue's need for these cells.

The platelets in the blood are totally replaced approximately once every four days; in other words, about 75,000 platelets are formed each day for each cubic millimeter of blood.

PROPERTIES OF WHITE BLOOD CELLS

Diapedesis. White blood cells can squeeze through the pores of the blood vessels by the process of diapedesis. That is, even though a pore is much smaller than the size of the cell, a small portion of the cell slides through the pore at a time, the portion sliding through being momentarily constricted to the size of the pore, as illustrated in Figure 9–5.

Ameboid Motion. Once the cells have entered the tissue spaces, the polymorphonuclear leukocytes especially, and the large lymphocytes and monocytes to a lesser degree, move through the tissues by ameboid motion, which was described in Chapter 2. Some of the cells can move through the tissues at a rate as great as 40 microns per minute—that is, they can move at least three times their own length each minute.

Chemotaxis. A number of different chemi-

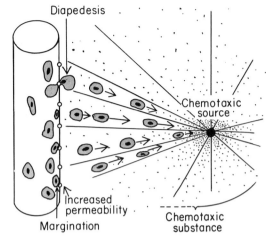

Figure 9–5 Movement of neutrophils by the process of *chemotaxis* toward an area of tissue damage.

cal substances in the tissues cause the leukocytes to move either toward or away from the source of the chemical. This phenomenon is known as chemotaxis. Degenerative products of inflamed tissues, especially tissue polysaccharides, can *cause neutrophils to move toward the area of inflammation*. Also, a number of bacterial toxins can cause chemotaxis of leukocytes. Some of these toxins cause *positive chemotaxis,* which means that they attract the cells toward the source of the toxin, whereas others cause *negative chemotaxis,* which means that they repel the leukocytes from the source of the toxin. It is interesting that a source of positively charged ion exchange resin causes positive chemotaxis, whereas a source of negatively charged resin causes negative chemotaxis.

As illustrated in Figure 9–5, chemotaxis depends on the existence of a concentration gradient of the chemotaxic substance. The concentration is greatest near the source, and as the substance spreads by diffusion away from the source its concentration decreases approximately in proportion to the square of the distance. Therefore, on the side of the cell away from the source of the chemotaxic substance the concentration is less than on the side facing the source. In positive chemotaxis, the greater concentration on one side of the cell causes pseudopodia to project toward the source of the substance, and in negative chemotaxis the opposite effect occurs.

Phagocytosis. The most important function of the neutrophils and monocytes is phagocytosis.

Obviously, the phagocytes must be selective in the material that is phagocytized, or otherwise some of the structures of the body itself would be ingested. Whether or not phagocytosis will occur depends especially upon three selective procedures. First, if the surface of a particle is rough, the likelihood of phagocytosis is increased. Second, most natural substances of the body have electronegative surface charges and therefore repel the phagocytes, which also carry electronegative surface charges. On the other hand, dead tissues and foreign particles are frequently electropositive and are therefore subject to phagocytosis. Third, the body has a means for promoting phagocytosis of specific foreign materials by first combining them with globulin molecules called *opsonins*. After the opsonin has combined with the particle, it allows the phagocyte to adhere to the surface of the particle, which promotes phagocytosis. The special relationship of opsonization to immunity is discussed in the following chapter.

The *macrophages*, which are derived mainly from the monocytes of the blood, are much more powerful phagocytes than the neutrophils. They have the ability to engulf much larger particles and often five or more times as many particles as the neutrophils. And they can even phagocytize whole red blood cells or malarial parasites, whereas neutrophils are not capable of phagocytizing particles much larger than bacteria. Also, macrophages have much greater ability than neutrophils to phagocytize necrotic tissue, which is a very important function performed by these cells in chronic infection.

If the inflammation becomes chronic and lasts for several weeks or months, such as in tuberculosis and chronic salpingitis (an infection of the fallopian tubes), the concentration of monocytes in the blood may rise from 5 per cent of the total leukocytes to as high as 30 to 50 per cent. The reason for this effect is not known, though it is presumed to result from the fact that tremendous numbers of lymphocytes, on entering inflamed tissues, swell over a period of hours to become macrophages which later wander into the blood.

ENZYMATIC DIGESTION IN THE PHAGOCYTES. Once a foreign particle has been phagocytized, the cell immediately begins digesting the particle. Neutrophils and macrophages both have an abundance of lysosomes filled with *proteolytic enzymes* especially geared for digesting bacteria and other foreign protein matter. The lysosomes of macrophages also contain large amounts of *lipases,* which digest the thick lipid membranes possessed by tubercle bacteria, leprosy bacteria, and others.

In addition to the enzymes that actually digest the ingested particles, phagocytic cells also contain bactericidal agents, which kill bacteria before they can multiply and destroy the phagocyte itself. The neutrophil, for instance, contains in its cytoplasm considerable quantities of *lysozyme* and *phagocytin,* especially important bactericidal agents.

DEATH OF THE PHAGOCYTES AS A RESULT OF PHAGOCYTOSIS. Phagocytes continue to ingest and digest foreign particles until sufficient quantities of breakdown substances from the foreign particles accumulate within the cytoplasm of the phagocyte to kill the phagocyte itself. Thus, a polymorphonuclear neutrophil is usually capable of phagocytizing about 5 to 25 bacteria before death occurs, but a macrophage sometimes engulfs as many as 100 bacteria before death.

INFLAMMATION AND THE FUNCTION OF LEUKOCYTES

THE PROCESS OF INFLAMMATION

Inflammation comprises the changes in the tissues in response to injury. When tissue injury occurs, whether it be caused by bacteria, trauma, chemicals, heat, or any other phenomenon, the substance *histamine,* along with other humoral substances, is liberated by the damaged tissue into the surrounding fluids. This increases the local blood flow and also increases the permeability of the capillaries, allowing large quantities of fluid and protein, including fibrinogen, to leak into the tissues. Local extracellular edema results, and the extracellular fluid and lymphatic fluid both clot because of the coagulating effect of tissue exudates on the leaking fibrinogen. Thus, *brawny edema* develops in the spaces surrounding the injured cells.

The "Walling Off" Effect of Inflammation. It can be seen from the preceding discussion that the net result of inflammation is to "wall off" the area of injury from the remaining portions of the body. The tissue spaces and the lymphatics are blocked by fibrinogen clots so that fluid flows through the spaces very slowly. Therefore, walling off the area of injury delays the spread of bacteria or toxic products.

The intensity of the inflammatory process is proportional to the degree of tissue injury. For instance, staphylococci invading the tissues liberate extremely lethal cellular toxins. As a result, the process of inflammation develops rapidly—indeed, much more rapidly than the staphylococci themselves can multiply and spread. Therefore, staphylococcal infection is characteristically "walled off" rapidly. On the other hand, streptococci do not cause such intense local destruction of the cells. Therefore the walling off process develops slowly while the streptococci reproduce and migrate. As a result, streptococci have far greater tendency to spread through the body and cause death than do staphylococci, even though staphylococci are far more destructive to the tissues.

Attraction of Neutrophils to the Area of Inflammation. When tissues are damaged, several effects occur to cause movement of neutrophils into the damaged area. First, the neutrophils stick to the walls of the damaged capillary, resulting in a process known as "margination," which is shown in Figure 9–5. Gradually the cells pass by diapedesis into the tissue spaces.

The second effect is chemotaxis of the neutrophils toward the damaged area; this is caused by bacterial or cellular products that attract the neutrophils. Thus, within a few hours after tissue damage begins, the area becomes well supplied with neutrophils.

Neutrophilia During Inflammation—Leukocytosis Promoting Factor. The term *neutrophilia* means an increase above normal in the number of neutrophils in the circulatory system and the term *leukocytosis* means an excess total number of white blood cells.

A globulin substance known as *leukocytosis promoting factor* is liberated by most inflamed tissues. This factor diffuses into the blood and finally to the bone marrow where it has two actions: First, it causes large numbers of granulocytes, especially neutrophils, to be released within a few minutes to a few hours into the blood from the storage areas of the bone marrow, thus increasing the total number of neutrophils per cubic millimeter of blood sometimes to as high as 20,000 to 30,000. Second, it increases the rate of granulocyte production by the bone marrow. Within a day or two after onset of the inflammation, the bone marrow becomes hyperplastic and then continues to produce large numbers of granulocytes as long as leukocyte promoting factor is formed in the inflamed tissues. Yet it is only the stored white blood cells that provide the granulocytes for the first day or so of inflammation until the bone marrow has had time to become hyperplastic. The quantity of stored neutrophils in the bone marrow fortunately is 30 to 40 times the quantity of these cells normally circulating in the blood; this obviously means that there is a large reserve available that can be called into service within hours.

THE MACROPHAGE RESPONSE IN CHRONIC INFLAMMATION

The monocytic cells—including the tissue histiocytes, the blood monocytes, and the lymphocytes—also play a major role in protecting the body against infection. First, the tissue histiocytes change into macrophages within minutes, develop ameboid motion, and migrate chemotactically toward the area of inflammation. These cells provide the first line of defense against the infection within the first hour or so, but their numbers are not very great. Within the next few hours, the neutrophils become the primary defense, reaching their maximum effectiveness in about 6 to 12 hours. By that time large numbers of monocytes have begun to enter the tissues from the blood. They change their characteristics drastically during the first few hours; they start to swell, to exhibit increased ameboid motion, and to move chemotactically toward the damaged tissues.

Finally, the largest source of all for developing macrophages is the lymphocytes of the blood, for tremendous numbers of lymphocytes invade the damaged tissue by about the 10th to 12th hour. Then within the next hour or so, these undergo progressive changes first into monocytes and then into macrophages.

Therefore, in summary, very large numbers of macrophages are derived from three different sources: the tissue histiocytes, the monocytes of the blood, and the lymphocytes of the blood.

Another reason the macrophagic response is of importance in chronic infection is that the affected area usually becomes acidic, and neutrophils cannot live in an acid environment, whereas macrophages live well under acid conditions, and their proteolytic enzymes actually become more active than ever. Consequently, after the initial stages of inflammation are over, the neutrophils are no longer nearly as effective phagocytes as the macrophages, and the shift from neutrophils to macrophages appears to be a purposeful reaction.

Formation of Pus. When the neutrophils and macrophages engulf large amounts of bacteria and necrotic tissue they themselves eventually die. After several days, a cavity is often excavated in the inflamed tissues containing varying portions of necrotic tissue, dead neutrophils, and dead macrophages. Such a mixture is commonly known as *pus*.

Ordinarily, pus formation continues until all infection is suppressed. Sometimes the pus cavity eats its way to the surface of the body or into an internal cavity and in this way empties itself. At other times the pus cavity remains enclosed even after tissue destruction has ceased. When this happens the dead cells and necrotic tissue in the pus gradually autolyze over a period of days, and the end-products of autolysis are absorbed into the surrounding tissues until most of the evidence of tissue damage is gone.

NEUTROPHILIA CAUSED BY OTHER CONDITIONS BESIDES INFLAMMATION

Almost any factor that causes some degree of tissue destruction will cause neutrophilia. For instance, persons debilitated by cancer exhibit an increase in neutrophils from the normal of 4500 per cubic millimeter sometimes up to 15,000 or more, and even extreme fatigue can cause neutrophilia. Acute hemorrhage, poisoning, operative procedures, very slight hemorrhage into the peritoneal cavity, and injection of foreign protein into the body all cause considerable increase in the number of neutrophils in the circulatory system. In summary, neutrophilia results from almost any tissue-damaging process in the body, whether or not this process is associated with inflammation.

For instance, neutrophilia is one of the special diagnostic features of coronary thrombosis. Presumably, when the coronary vessel becomes blocked the ischemic musculature of the heart begins to necrose, and degenerative substances liberated into the blood promote the release of neutrophils from the bone marrow.

Physiological Neutrophilia. The number of neutrophils in the circulatory system can increase to as much as 3 to 4 times normal after a single minute of extremely hard exercise or after injection of norepinephrine. This can be explained as follows: When blood flow is sluggish through the tissues, large numbers of white blood cells, especially neutrophils, adhere to the walls of the capillaries and, therefore, are sequestered from the usual circulation. Hard exercise or stimulation of the circulation by norepinephrine, with rapid flow of blood through essentially all capillaries, can mobilize the leukocytes.

Approximately one hour after physiological neutrophilia has resulted from exercise or any other stimulus, the number of leukocytes in the blood is usually back to normal because most of the leukocytes will again be sequestered in the capillaries.

THE EOSINOPHILS

The function of the eosinophils is almost totally unknown even though they normally comprise 1 to 3 per cent of all the leukocytes. Eosinophils are weak phagocytes, and they exhibit chemotaxis, but in comparison with the neutrophils, it is doubtful that the eosinophils are of significant importance in protection against usual types of infection.

Eosinophils enter the blood in large numbers after foreign protein injection. Furthermore, many eosinophils are present in the mucosa of the intestinal tract and in the tissues of the lungs, where foreign proteins normally enter the body. It has been suggested that the function of these is to detoxify the proteins before they can cause damage to the body.

Eosinophils also migrate into blood clots where they probably release the substance *profibrinolysin.* This substance then becomes activated to form *fibrinolysin,* which is an enzyme that digests fibrin, a subject that will be discussed in Chapter 12. Therefore, eosinophils are possibly important for dissolution of old clots.

The total number of eosinophils increases greatly in the circulating blood during allergic reactions, and eosinophils collect at sites of antigen-antibody reactions in the tissues. Here again there is no reasonable theory to explain these effects. It is possible that the allergic reactions release toxic products from the tissues and that these in turn cause the number of eosinophils in the tissues and in the circulating blood to increase.

Probably the most common cause of extremely large numbers of eosinophils in the blood is infection with parasites. In the condition known as *trichinosis,* which results from invasion of the muscles by the *Trichinella* parasite ("pork worm") after eating uncooked pork, the percentage of eosinophils in the circulating blood may rise to as high as 25 to 50 per cent of all the leukocytes. The means by which parasitic infections cause an increase in eosinophils also is not known, though if the function of eosinophils is to detoxify proteins, it is possible that parasites continually elaborate protein substances that must be detoxified.

THE BASOPHILS

The basophils in the circulating blood are very similar to the large *mast* cells located immediately outside many of the capillaries in the body. These cells liberate *heparin* into the blood, a substance that can prevent blood coagulation and that can also speed the removal of fat particles from the blood after a fatty meal. Therefore, it is probable that the basophils in the circulating blood perform similar functions within the blood stream, or it is even possible that the blood simply transports basophils to tissues where they then become mast cells and perform the function of heparin liberation.

The total number of basophils in the blood is ordinarily very small, ranging about four such cells for every 1000 leukocytes. This number increases during the healing phase of inflammation, and it also increases a small amount during prolonged chronic inflammation. It is a well-known fact that prolonged inflammation causes a tendency for red blood cells to clump. Therefore, it is possible that this increase in basophilic elements in the blood and in the tissues might be a means for combating agglutinative or coagulative processes.

The granules of mast cells (and presumably also of the basophils) contain, in addition to tremendous quantities of heparin, moderate to large quantities of histamine, hyaluronic acid, and serotonin. Whether these play any role in the function of basophils is not known. All of them are very important to the body, histamine and serotonin as vasoactive agents, and hyaluronic acid as "ground substance" in the interstitial spaces.

THE LYMPHOCYTES

Until recently it was believed that the lymphocytes represent a homogeneous group of cells, with a possible distinction between the so-called small lymphocytes and large lymphocytes. However, it is now known that the cell known as the lymphocyte comprises a number of different types of cells, all of which have essentially the same staining characteristics. A few of these cells, probably produced primarily in the bone marrow, seem to be multipotential cells that are similar, if not identical, to the stem cell from which almost any other type of cell can be formed. These multipotential cells can be changed into erythroblasts, myeloblasts, monocytes, plasma cells, other types of small lymphocytes, fibroblasts, and so forth.

Many of the lymphocytes play special roles in the process of immunity. These will be discussed in detail in Chapter 10.

AGRANULOCYTOSIS

A clinical condition known as "agranulocytosis" occasionally occurs, in which the bone marrow stops producing white blood cells, leaving the body unprotected against bacteria and other agents that might invade the tissues. The failure to produce platelets may cause internal hemorrhage.

Actually, the human body lives in symbiosis with many bacteria, for all the mucous membranes of the body are constantly exposed to large numbers of bacteria. The mouth almost always contains various spirochetal, fusiform, pneumococcal, and streptococcal bacteria, and these same bacteria are present to a lesser extent in the entire respiratory tract. The gastrointestinal tract is especially loaded with colon bacilli. Furthermore, one can almost always find bacteria in the eyes, the urethra, and the vagina. Therefore, any decrease

in the number of neutrophils and monocytes immediately allows invasion of the tissues by the bacteria that are already present in the body. Within two days after the bone marrow stops producing white blood cells, ulcers appear in the mouth and colon, or the person develops some form of severe respiratory infection. Bacteria from the ulcers then rapidly invade the surrounding tissues and the blood. Without treatment, death usually ensues three to six days after agranulocytosis begins.

Irradiation of the body by gamma rays caused by a nuclear explosion or drugs and chemicals containing the benzene or anthracene nuclei are quite likely to cause aplasia of the bone marrow. Indeed, some of the common drugs, such as the sulfonamides, chloramphenicol, thiouracil (used to treat thyrotoxicosis), and even the various barbiturate hypnotics, on occasion cause agranulocytosis, thus setting off the entire infective sequence of this malady.

After irradiation injury to the bone marrow, a large number of stem cells and hemocytoblasts usually remain undestroyed and are capable of regenerating the bone marrow, provided sufficient time is available. Therefore, the patient properly treated with antibiotics and other drugs to ward off infection will usually develop enough new bone marrow within several weeks to several months that his blood leukocyte concentration can return to normal.

THE LEUKEMIAS

Uncontrolled production of white blood cells is caused by cancerous mutation of a myelogenous or a lymphogenous cell. It usually causes leukemia, which means greatly increased numbers of white blood cells in the circulating blood.

Types of Leukemia. Ordinarily, leukemias are divided into two general types: the *lymphogenous leukemias* and the *myelogenous leukemias*. The lymphogenous leukemias are caused by cancerous production of lymphoid cells, beginning first in a lymph node or other lymphogenous tissue and then spreading to other areas of the body. The second type of leukemia, myelogenous leukemia, begins by cancerous production of young myelogenous cells in the bone marrow and then spreads throughout the body so that white blood cells are produced in many extramedullary organs.

In myelogenous leukemia, the cancerous process occasionally produces reasonably differentiated cells, resulting in *neutrophilic leukemia, eosinophilic leukemia, basophilic leukemia*, or *megakaryocytic leukemia*. More frequently, however, the leukemia cells are bizarre, undifferentiated, and not identical with any of the granulocytes.

Leukemic cells are usually nonfunctional, so that they cannot provide the usual protection associated with white blood cells.

EFFECTS OF LEUKEMIA ON THE BODY

The first effect of leukemia is metastatic growth of leukemic cells in abnormal areas of the body. The leukemic cells of the bone marrow may reproduce so greatly that they invade the surrounding bone, causing a tendency to easy fracture. Almost all leukemias spread to the spleen, the lymph nodes, the liver, and other especially vascular regions, regardless of whether the origin of the leukemia is in the bone marrow or in the lymph nodes. In each of these areas the rapidly growing cells invade the surrounding tissues, utilizing the metabolic elements of these tissues and consequently causing tissue destruction.

A very common effect in leukemia is the development of severe anemia. This results mainly from displacement of the normal bone marrow by the leukemic process.

Finally, perhaps the most important effect of leukemia on the body is the excessive use of metabolic substrates by the growing cancerous cells. The leukemic tissues reproduce new cells so rapidly that tremendous demands are made on the body fluids for foodstuffs, especially the amino acids and vitamins. Consequently, the energy of the patient is greatly depleted, and the excessive utilization of amino acids causes rapid deterioration of the normal protein tissues of the body. Thus, while the leukemic tissues grow, the other tissues are debilitated. Obviously, after metabolic starvation has continued long enough, this alone is sufficient to cause death.

REFERENCES

Archer, R. K.: The Eosinophil Leucocyte. Philadelphia, F. A. Davis Co., 1963.

Athens, J. W.: Blood: leukocytes. Ann. Rev. Physiol., 25:195, 1963.

Biology and function of the lymphocyte (Symposium). Fed. Proc., p. 1711, 1966.

Burnet, M.: The thymus gland. Sci. Amer., 207(5):50, 1962.

Cline, M. J.: Metabolism of the circulating leukocyte. Physiol. Rev., 45:674, 1965.

Frei, E., 3rd, and Freireich: Leukemia. Sci. Amer., 210:88(5), 1964.

Heller, J. H.: Host defence and the reticulo-endothelial system. In Bittar, E. Edward, and Bittar, Neville (eds.): The Biological Basis of Medicine. New York, Academic Press, Inc., 1969, Vol. 4, p. 181.

Huennekens, F. M.: The biochemistry of normal and leukemic leukocytes. In Bittar, E. Edward, and Bittar, Neville (eds.): The Biological Basis of Medicine. New York, Academic Press, Inc., 1969, Vol. 3, p. 129.

Jandl, J. H.: The spleen and reticuloendothelial system. In Sodeman, William A., and Sodeman, William A., Jr. (eds.): Pathologic Physiology. 4th ed., Philadelphia, W. B. Saunders Company, 1967, p. 897.

Karnovsky, M. L.: Metabolic basis of phagocytic activity. Physiol. Rev., 42:143, 1962.

Reed, P. W., and Tepperman, J.: Phagocytosis-associated metabolism and enzymes in the rat polymorphonuclear leukocyte. Amer. J. Physiol., 216:223, 1969.

Saba, T. M., and DiLuzio, N. R.: Effect of x-irradiation on reticuloendothelial phagocytic function and serum opsonic activity. Amer. J. Physiol., 216:910, 1969.

Saba, T. M., and DiLuzio, N. R.: Reticuloendothelial blockade and recovery as a function of opsonic activity. Amer. J. Physiol., 216:197, 1969.

Speirs, R. S.: How cells attack antigens. Sci. Amer., 210:58(2), 1964.

Zweifach, B.: The Inflammatory Process. New York, Academic Press, Inc., 1965.

CHAPTER 10

Immunity and Allergy

In addition to the ability of the phagocytes of the body to provide considerable natural resistance to infection and toxicity in the body, the blood contains still other substances that can react directly with invading organisms or toxins to destroy them. Some of these are: (1) *lysozyme*, a mucolytic polysaccharide that attacks bacteria and causes them to dissolute; (2) *basic polypeptides*, which react with and inactivate certain types of gram-positive bacteria; (3) *properdin*, a very large protein that can react directly with gram-negative bacteria and destroy them, thereby protecting the body against diseases such as dysentery; and (4) *interferon*, another large protein that inactivates many viruses.

Thus, the body is said to have a large amount of *natural immunity*, which makes it resistant to many bacteria, viruses, and toxins even without any specific immune process. For instance, the human body is entirely resistant to distemper virus that results in death of 25 to 30 per cent of all dogs that are afflicted with it. Likewise, the human body has native resistance against herpes virus, which is almost always lethal in the rabbit but which rarely causes more than fever blisters in the human being.

However, in addition to its natural immunity, the human body also has the ability to develop immunity against specific invading agents such as lethal bacteria, viruses, toxins, and even foreign tissues from other animals. It is with this immune mechanism and with the very closely allied sequela of the immune mechanism, allergy, that this chapter will be concerned.

THE IMMUNE PROCESS

The immune process is initiated by exposure to an invading agent. Once exposed, the body begins developing in its lymphoid tissues (1) specific globulin molecules or (2) sensitized lymphocytes that are capable of reacting with and destroying the invading agent. However, it takes a week to several months to develop immunity against the invading agent. Therefore, the person is likely to receive considerable harm from the first exposure before immunity has developed, but upon subsequent exposure the invading agent is often totally blocked before any damage can be done. Indeed, exposure to some types of invading agents, such as tetanus toxin, in very minute doses (not enough to cause death) can initiate such intense immunity against the toxin that the person can subsequently withstand 100,000 times the normally lethal dose, thus illustrating the extreme capability of the immune process to protect the body against foreign agents.

Two basic types of immunity can develop in the body. In one of these the body develops circulating *antibodies*, which are globulin molecules that are capable of attacking the invading agent. This type of immunity is called *humoral immunity*. The second type of immunity is achieved through the lymphocyte system in such a way that lymphocytes become specifically sensitized against the foreign agent and later can attach themselves to the agent and destroy it. This type of immunity is frequently called *cellular immunity* or *lymphocytic immunity*.

Antigens. Any invading agent that can elicit an immune response is called an *antigen*. The usual antigen to which the human body is exposed is a large protein, a large polysaccharide, or a combination of these. For instance, essentially all toxins elaborated by bacteria are proteins, large polysaccharides, or mucopolysaccharides. Also, bacteria and viruses themselves contain mainly these same substances.

For a substance to be antigenic it usually must have a high molecular weight, 10,000 or greater. Furthermore, the process of antigenicity probably depends upon regularly recurring prosthetic radicals on the surface of the large molecule, which perhaps explains why proteins and polysaccharides are almost always antigenic, for they do have this type of stereochemical characteristic.

HAPTENS. Though substances with molecular weights less than 10,000 can almost never act as antigens, immunity can nevertheless be developed against substances of low molecular weight in a very special way, as follows: If the low molecular weight compound, which is called a *hapten,* first combines with a substance that *is* antigenic, such as a protein, then the combination will elicit an immune response. The antibodies that develop against the combination can then react either against the protein or against the hapten. Therefore, on second exposure to the hapten, the antibodies react with it before it can spread through the body and cause damage.

The haptens that elicit immune responses of this type are usually drugs, chemical constituents in dust, breakdown products of dandruff from animals, degenerative products of scaling skin, various industrial chemicals, and so forth.

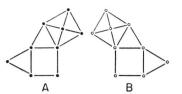

Figure 10-1 Polar forces: (A) on an antigen and (B) on an antibody.

HUMORAL IMMUNITY — ANTIBODIES

Humoral immunity depends upon the development of *antibodies* in the circulating body fluid. These antibodies are formed primarily by plasma cells in the lymph nodes; this mechanism will be discussed later. At least five different types of antibodies have been discovered, ranging in molecular weight from less than 100,000 up to approximately 800,000. However, the vast majority of the antibodies are gamma globulin molecules, having a mean molecular weight of approximately 160,000.

Specific Chemical Reactivity of Antibodies with Specific Antigens. Gamma globulins formed in response to a particular antigen can react specifically only with that one type of antigen and rarely with any other type. The bonds between the antigen and antibody, however, are relatively weak, probably being comprised mainly of hydrogen bonds.

The exact reason for the specificity of antibodies for specific antigens is not known, but a logical theory to explain this is based on spatial orientation of electrostatic (or hydrogen bond) forces at the surface of the molecule, as illustrated in Figure 10-1. Figure 10-1A illustrates the spatial orientation of "polar forces" on the surface of an antigen molecule. By some means yet obscure, the antibodies supposedly develop polar forces of opposite electrostatic charge and arranged as a mirror image to the forces on the antigen molecule, as illustrated in Figure 10-1B.

Each antibody has either one or two reactive sites — that is, areas on its molecular surface that have the appropriate image polar forces to react with an antigen — but never more than two reactive sites appear on the same antibody. Therefore, antibodies are said to be either *univalent or bivalent.* Antigens, in contrast to antibodies, can have large numbers of reactive sites: three, five, ten, or more.

Mechanism of Antibody Formation — Function of Plasma Cells

Almost all antibodies are formed in *plasma cells,* which are found in the lymphoid tissue of lymph nodes, the spleen, the gastrointestinal tract, and, to a slight extent, other tissues.

The plasma cells, before exposure to an antigen, remain essentially in a dormant state and are called *plasmablasts.* However, upon entry of a foreign antigen into the lymphoid tissue, some of the plasmablasts begin to divide at a rate of approximately once every 10 hours for about nine divisions, giving in four days a total population of about 500 cells for each originally stimulated plasmablast. During this division process, the maturing plasma cells develop an extensive granular endoplasmic reticulum that produces gamma globulin antibodies at an extremely rapid rate, a rate of at least 100 molecules per second. Indeed, radioactive amino acids injected intravenously can become embodied in an active antibody within 15 to 20 minutes, illustrating the rapidity with which antibodies are formed by plasma cells.

Mechanisms for Producing Specific Antibodies. Even though an antigen causes the plasma cells to divide and to form antibodies, the mature plasma cells do not contain any of the antigen. Therefore, the antigen must cause the plasma cells to produce their specific antibodies in some indirect way. Though we do not know the whole mechanism of this, it is known that the antigen is first entrapped by the *fixed reticulum cells (macrophages)* of the lymphoid tissue and probably remains there until it is destroyed. The macrophages have long tentacles that extend outward into the surrounding lymphoid tissue, forming a *reticulum* throughout the tissue's meshwork. This reticulum surrounds the plasmablasts and supposedly initiates their development into mature plasma cells.

There are two basic theories that attempt to explain how plasma cells produce antibodies against the specific antigen entrapped in the macrophages. These are (1) the template theory and (2) the clonal theory.

The *template theory* assumes that the antigen causes development of some secondary material, called a *superantigen,* within the macrophage. This superantigen presumably is transferred from the macrophage into the plasma cells, carrying appropriate information from the original antigen to cause production of antibodies having a specific stereochemical organization of reactive sites. Once the plasmablast has been induced to begin its division process, it thereafter produces only that one type of antibody.

The second theory, the *clonal theory,* has recently become attractive because it explains some peculiar aspects of immunity, such as failure of antibodies to attack the human being's own proteins, a subject that will be discussed later in the chapter. This theory proposes that there are literally thousands of different types of plasmablasts, each type capable of being stimulated by only one specific type of antigen. It is assumed that the stimulated plasmablast, upon exposure to this specific antigen, then begins to produce antibodies against the antigen. Thus, the function of the antigen is simply to select the type of plasmablast that will be activated.

Regardless of which of the above theories is correct, or indeed whether either one of them is correct, a single plasma cell almost always produces only a single type of antibody. If the lymphoid tissue is producing several different types of antibodies simultaneously, it has several different types of plasma cells functioning simultaneously.

The Primary Response to an Antigen and the Secondary Response. When a person is first exposed to an antigen, only a very few of the plasmablasts in the lymphoid tissue are activated. Then antibodies appear in the circulating blood a week to ten days later, and they continue to be produced for four to six weeks. This is called the *primary response,* the course of which is illustrated in Figure 10–2. If the person is exposed to the same antigen a few weeks or a few years later, a so-called *secondary response,* also illustrated in Figure 10–2, is likely to occur. This time the outpouring of antibodies begins within two to three days, and the quantity of antibodies released is far greater and extends over a much longer period of time than occurs in the primary response. The reason for this difference is believed to be the following: After the primary response is over, a large number of the sensitized plasma cells remain in the plasmablast stage rather than proceeding to mature plasma cells. Therefore, upon second exposure to the same antigen, this antigen now finds a large number of plasmablasts immediately susceptible to its stimulation in contrast to only a few plasmablasts that were susceptible during the primary response. Obviously, therefore, the response occurs much more rapidly and more forcefully.

Mechanism by which Antibodies Inactivate Antigens — The Antigen-Antibody Reaction

Reaction of antibodies with antigens can either destroy or inactivate the antigens. If the antigen is part of a bacterial body, antibodies can either

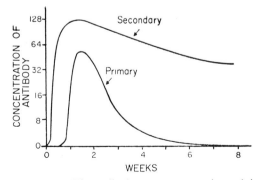

Figure 10–2 The antibody response to a primary injection of antigen and to a secondary injection of antigen several weeks later.

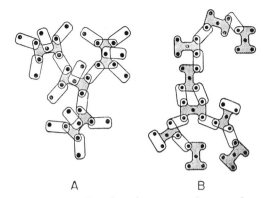

Figure 10–3 Reactions between antigens and antibodies. The antibodies are represented by rectangular structures, and the antigens are represented by stippled structures. (A) Reaction when excess antibody is present. (B) Reaction when excess antigen is present. (Redrawn from Raffel: Immunity. Appleton-Century-Crofts.)

inactivate the bacterium or completely destroy it. There are several different mechanisms by which antibodies perform these functions:

Neutralization of Antigens by Antibodies. When an antigen is toxic to the body, combining this toxic antigen with antibodies usually *neutralizes* the toxin. This is believed to result simply by the antibodies covering up the toxic prosthetic groups on the toxin. Usually, after the antibodies have reacted with the toxin, the combined antibody and antigen is engulfed by macrophages of the reticuloendothelial system, whereupon both the antigen and antibody are simultaneously digested and thereby destroyed.

Viruses as well as toxins can be neutralized in this same manner, the antibody making the virus noninvasive and also exposing it to phagocytosis.

Antigen-Antibody Precipitation. When appropriate proportions of antigen and antibody molecules come in contact with each other they will usually precipitate. Precipitating combinations of antigen and antibody are illustrated in Figure 10–3. The small white bars with two reactive sites represent bivalent antibodies, and the dumbbell structures with five reactive sites represent antigens. Note how the antibody combines successive antigens one to the other. When enough of these become combined into one large solid mass, the agglomerate becomes insoluble, and the insoluble particles usually are engulfed by macrophages and digested.

Agglutination of Bacteria. In the same manner that antibodies can attach antigen molecules together, they can also attach one bacterium to the next, thereby binding the bacteria together in clumps. This process is called *agglutination.* The agglutination process prevents tissue invasion by the bacteria and also exposes the bacteria to phagocytosis.

Lysis of Cellular Structures as a Result of the Antigen-Antibody Reaction. Antibodies can also cause destruction of bacteria, viruses, and foreign animal cells. It does this in the following way: The antibody first attaches to the cell membrane, and then a substance called *complement* combines with the antigen. Complement is a complex lipoprotein that has the ability to penetrate the cell membrane and cause its rupture. Complement is present normally in the circulating body fluids, but it will not attack the foreign cells until an antibody first attaches to the cell and then to complement. However, when this process occurs, the cell membrane often breaks open immediately, and the organism spills its contents.

Opsonization. Sometimes when antibodies and complement are attached to bacteria or other foreign cells they still do not have the power to destroy the cells. Yet, the presence of complement on the surface of the cell makes it far more susceptible than normally to phagocytosis. This phenomenon is known as *opsonization,* and the antibodies that cause this response are called *opsonins.* The presence of specific opsonizing antibodies and complement may increase the rate of phagocytosis as much as 100-fold.

The mechanism of opsonization seems to be based on the physical nature of complement; complement is miscible with the membrane of the phagocyte, and allows the phagocyte to engulf the opsonized bacterium or other type of foreign cell.

FUNCTION OF LYMPHOCYTES IN IMMUNITY (CELLULAR OR LYMPHOCYTE IMMUNITY)

Often immunity can be developed against a foreign agent even without formation of humoral antibodies. In these instances, large numbers of small lymphocytes take on the characteristics of whole cell "antibodies"; that is, specific lymphocytes become sensitized against specific antigens in the same way that specific antibodies are formed against specific antigens in humoral immunity.

Characteristics of the Lymphocyte Type of Immunity. The lymphocyte type of immunity is different from the humoral type in several important ways. First, once lymphocytes have been sensitized to a particular antigen, they continue to circulate in the blood and body tissues, sometimes for years. Moreover, these lymphocytes can even divide and form new lymphocytes having the same specific capability. Therefore, the lymphocyte type of immunity usually lasts far longer than does the humoral type of immunity.

Second, the lymphocyte type of immunity will frequently develop upon exposure to extremely minute amounts of antigens, even those amounts too minute to activate the plasmablasts. Therefore, immunity can be developed against some antigens by this mechanism when it cannot be developed by the humoral mechanism.

Formation of Sensitized Lymphocytes. Much less is known about the manner in which specifically sensitized lymphocytes are created than about the mechanism for formation of humoral antibodies. However, since lymphocytes are generated in the germinal centers of lymph nodes in close association with the generation of plasma cells, it is presumed that the many small specifically sensitized lymphocytes that develop during the course of the immune response have an origin similar to that of the plasma cells. However, unlike the plasma cells, the sensitized lymphocytes fail to secrete antibodies. Instead they are released from the lymph nodes to be circulated throughout the body. To state this another way, the normal resting lymphocytes of the lymph nodes are actually dormant *lymphoblasts.* Upon exposure to a foreign antigen, these lymphoblasts become sensitized to the antigens and proliferate in the same manner as plasmablasts. Once the lymphocytes are sensitized in this manner they are called *committed lymphocytes* because thereafter they cannot be changed into lymphocytes sensitized against other types of foreign antigens.

The mechanism by which lymphocytes become sensitized to foreign antigens is probably very similar to the mechanism for formation of antibodies by the plasma cells. Presumably, the small lymphocytes develop antibodies within their cytoplasm or on their cell surfaces, and it is these antibodies that allow the lymphocyte thereafter to attack foreign antigens. In support of this concept is the fact that released products from the lymphocytes do indeed react with specific foreign antigens in a manner similar to the reaction of gamma globulin antibodies.

Mechanism by which Sensitized Lymphocytes Attack Antigens. When an organism to which sensitized lymphocytes have been formed invades a tissue, the sensitized lymphocytes (1) swell in size, (2) take on characteristics similar to those of monocytes, and (3) soon thereafter attach side by side to the invading organism. The membrane of the invading organism then gradually undergoes autolysis, the organism itself swelling and eventually rupturing. Thus, the lymphocytic method for destroying foreign invaders appears to be very much the same as the lysis mechanism utilized by humoral antibodies to cause dissolution of invading cells.

Types of Invading Agents Attacked by Sensitized Lymphocytes. In general, the lymphocytic type of immunity can be developed against almost any type of antigen, either an invading organism or an invading toxic substance. However, certain types of antigens seem especially to promote this type of immunity rather than the humoral type. For instance, some bacteria, such as the tubercle bacillus, cause little of the humoral type of immunity but elicit important immune responses of the lymphocytic type. Thus, invading tubercle bacilli become surrounded by swollen lymphocytes, giving rise to a walling off process that prevents further invasion by the tubercle bacillus. Fibrous tissue grows into this walled off area, thereby developing the so-called *tubercle.*

Another type of antigen that is specifically attacked by the lymphocytic type of immunity is foreign animal cells such as those in a grafted heart or other tissue. Beginning approximately a week after the graft, sensitized lymphocytes against the graft begin to appear in the blood, and these rapidly invade the graft itself. Within another four days to a few weeks the lymphocytes gradually cause lysis of the cells in the graft and, subsequently, complete destruction.

It is probable that some types of cancer cells that happen to be foreign to the body are also attacked by lymphocytes in this same manner. Indeed, it is possible that this mechanism prevents the development of more cancers than actually do occur.

FUNCTION OF THE THYMUS GLAND IN IMMUNITY

In the fetus and newborn animal, the thymus gland is a large organ lying in the upper mediastinum beneath the sternum. In the very young fetus it is composed primarily of epithelioid cells, but it has a reticular structure that entraps stem cells circulating in the blood. For reasons not understood, these trapped stem cells are then stimulated to proliferate extremely rapidly, and the cells that are formed are the first lymphocytes to appear in the fetus. These lymphocytes migrate into the blood and thence throughout the body, seeding the lymph nodes and other lymphoid tissue. The seeded cells then become either plasmablasts, the progenitors of the plasma cells, or lymphoblasts, the progenitors of the small lymphocytes formed in the lymphoid tissue. Thus, the thymus plays an important role in the original genesis of the entire lymphoid system of the body.

Failure of the Immune System to Develop in the Absence of the Thymus Gland. If the thymus gland is removed early in fetal life, the lymphoid tissue of the entire body fails to become seeded with lymphocytes, and consequently, essentially no immune system develops. Even if the thymus is removed shortly before birth of the baby or within the first few days after birth, the immune system is still greatly impaired. Furthermore, the lymphocytic type of immunity seems to be much more impaired than the humoral type, indicating that plasmablast seeding of the lymphoid tissue occurs earlier than lymphoblast seeding.

However, if the thymus gland is removed from an animal a few weeks to a few months after birth, very little harm is done to the immune system, because by that time enough seeding of the lymphoid tissue has occurred for the immune system to function throughout life. One exception to this occurs as follows: If the person is subsequently exposed to some effect, such as a nuclear bomb blast, that destroys most of his immune system, he will have difficulty reseeding his lymphoid tissue without the availability of a thymus gland. Here again, lymphocytic immunity is affected much more severely than is humoral immunity.

Possibility of a Thymic Hormone. Even though the thymus gland is absolutely necessary for the early seeding of lymphoid tissue with lymphoblasts and plasmablasts, once this seeding process has occurred one might expect that the lymphoblasts and plasmablasts could proliferate within the lymphoid tissue without further intervention from the thymus gland. Yet, even so, removal of the thymus gland still stops or greatly impedes further growth of the lymphoid tissue. If the thymus gland is then re-implanted in the animal, but inside a porous capsule so that no lymphocytes can get out of the thymus gland, the lymphoid tissue throughout the body immediately resumes its growth. Therefore, it is believed that some hormonal substance released by the thymus gland diffuses throughout the body to activate the growth of the lymphoid tissue. This substance is called simply the *thymic hormone.*

Thus far all attempts to extract from the thymus gland a specific substance that will perform all the functions of the so-called thymic hormone have failed. Nevertheless, there is still reason to believe that such a hormone might exist and that eventually it will be extracted.

IMMUNOLOGIC TOLERANCE OF ONE'S OWN PROTEINS

Obviously, if one should develop immunity against his own body proteins, he would soon wreak severe damage within his body or actually cause his own death. Fortunately, during development of the immune system, the system learns to recognize the body's own proteins and not to form antibodies against them, while nevertheless retaining the ability to form antibodies against any foreign invader. This specific ability of the immune system to recognize the body's own proteins is called *immunologic tolerance.* Its mechanism has always been extremely perplexing, and even the theories that attempt to explain it are still vague and questionable. Yet, one of these theories deserves comment, as follows:

The Clonal Theory for Immunologic Tolerance. When the thymus forms the first early lymphocytes, it does so at an extremely rapid rate, and it has been postulated that this rapid rate of formation causes tremendous numbers of genetic mutations so that literally thousands of different types of lymphoblasts and plasmablasts seed the lymphoid tissue, each type of lymphoblast and plasmablast having the capability of forming immunity against one particular type of antigen—one and no more. Presumably, some of these early lymphoblasts and plasmablasts can form immunity against the body's own proteins. However, it is postulated that those that are capable of acting against the body's own proteins are quickly bound with the body's proteins and are either inactivated or destroyed before they can seed the lymphoid tissue. This theoretically explains why the immune system thereafter will not react with the body's own tissues but nevertheless will react with foreign antigens that enter the body.

Creation of Artificial Immunologic Tolerance. Immunologic tolerance can on occasion be developed against a foreign antigen in one of two ways: (1) By overwhelming the immune system with large quantities of the foreign antigen or (2) by injecting very minute amounts of the foreign antigen slowly over a period of many weeks. In the first instance, the extreme amount of foreign antigen supposedly neutralizes all of the specific lymphoblasts and plasmablasts capable of forming immunity and destroys them before they can begin to proliferate. In the second case, when only minute amounts of antigen are injected, this supposedly is not enough to cause proliferation but is enough to cause slow destruction of the specific lymphoblasts and plasmablasts. The second mechanism rarely occurs, however.

Tolerance to foreign antigens can be developed in fetal life much more readily than in later life, presumably because of the relatively few lymphocytic cells present in the fetus.

AUTOIMMUNITY

Though in the normal person the immune system fails to develop immunity against the body's own tissue proteins, on occasion such immunity does occur, and the result is called *autoimmunity*. Autoimmunity is believed to be caused by one of several untoward events, as follows:

1. First, certain types of antigens in the body seem to be sequestered so that they are not exposed to the immune system in the early life of the person. If ever such antigen is released from its sequestered position, it can elicit an immune response and thereby cause autoimmunity. For instance, the protein in the lens of the eye is normally sequestered, but once this protein is released from one of the lenses, autoimmunity can occur and thereafter cause cloudiness of the lens of the other eye. It is believed also that the disease myasthenia gravis, which was discussed in Chapter 6, might be an autoimmunity disease in which the body has become immunized against some of its own muscle proteins. It is believed that the antibodies then attack the muscle cells and make them less responsive to acetylcholine.

2. Autoimmunity also occasionally develops because of similarity of the body's own proteins to foreign antigens that enter the body. For instance, rheumatic fever might occur in this manner. A certain strain of streptococci releases a toxin that has an antigen structure perhaps similar to that of heart muscle, heart valves, and synovial membranes. This antigen causes the immune system to develop autoimmunity against the specific heart and joint structures. In turn, the immunity causes destructive lesions in the heart and joints, often leaving the heart valves scarred for life, as will be discussed in Chapter 30 in relation to valvular heart disease.

3. Autoimmunity is also believed to occur following exposure of the body to certain haptens. These haptens presumably combine with some of the body's natural proteins, and in turn, the body forms immunity against both the hapten and its conjugated protein. Thereafter, the immune system attacks the body's own proteins.

Thus, in several different ways the body can develop immunity against some of its own tissues. Perhaps the most common cause is the release of abnormal protein products into the body during infectious diseases, the protein products having antigenicities similar to those of the body's own tissues. At any rate, a large number of the nonspecific diseases that occur in middle and late life are now believed to be autoimmune diseases.

ALLERGY

Allergy is one of the side effects of immunity. Some allergic responses result from typical immune reactions taking place under odd conditions so that damage occurs in some of the body's tissues. On the other hand, other allergies are caused by abnormal immune reactions occurring only in the so-called "allergic" person who has inherited an abnormal immune system.

Allergies are generally divided into three different types: (1) the *precipitin allergies,* (2) the *delayed reaction allergies,* and (3) the *atopic allergies.* The first two of these can occur in anyone, whereas the last, atopic allergy, occurs only in the hereditarily predisposed allergic individual.

PRECIPITIN ALLERGIES

Precipitin allergies occur only when there is a very high titer of specific antibodies in the circulating body fluids, a titer high enough to cause actual precipitation reactions between the antigen and antibody. When such an antigen-antibody reaction occurs in direct contact with the body tissues, it causes severe tissue damage. Some types of precipitin allergies are the following:

Anaphylaxis. Anaphylaxis is an acute allergic reaction that occurs generally throughout the body. It develops when an antigen is injected directly into the blood of a person whose circulating blood has a high titer of specific antibodies. The antigen-antibody reaction occurs in close association with the white cells in the circulating blood and also in close association with the tissues immediately surrounding the blood vessels, thus causing local cellular damage. These cells then release pharmacologic agents that can be very detrimental to the body. Particularly important is *histamine,* which is supposedly released by the mast cells that surround the capillaries. In addition, serotonin, acetylcholine, various polypeptides, and other substances are released, though it is the histamine that seemingly causes most of the damage. In the human being, histamine causes a number of different reactions all simultaneously, including (1) severe peripheral vasodilatation, (2) loss of fluid through the capillary membranes, and (3) sometimes bronchiolar spasm. The most usual result is severe circulatory collapse with perhaps some respiratory difficulty, and the person is likely to die within the next few minutes if the reaction is severe enough.

Serum Sickness. Serum sickness is similar to anaphylaxis except that it develops slowly. When serum from a foreign animal is injected into a human being, he develops antibodies against the serum during the next 7 to 12 days. These antibodies then begin to attack the serum proteins, causing the same types of reactions as those that occur in anaphylaxis, such as loss of fluid into the tissues with the occurrence of edema, some degree of circulatory collapse, and fever. However, because the process occurs slowly, the body is capable of compensating for the disturbance, and the person gets over the reaction within another week or so instead of dying as frequently occurs in anaphylaxis.

The Arthus Reaction. When an antigen is injected into a local tissue area of a person whose body fluids contain large quantities of specific precipitin antibodies, a serious local destructive reaction occurs called the *Arthus reaction*. The local blood vessels become spastic, the endothelium degenerates, and granulocytes plug the vessels and cause ischemic necrosis of the tissues.

THE DELAYED REACTION ALLERGIES

The delayed reaction allergies can also occur in normal persons who do not have hereditary predisposition to allergy. The delayed reaction allergies are not caused by antibodies in the circulating body fluids but instead by sensitized lymphocytes. A typical delayed reaction allergy is that which occurs in response to poison ivy. Upon first exposure to the poison ivy, nothing at all occurs except that the body develops lymphocytes sensitized to the poison ivy toxin. Then, upon second exposure, a delayed reaction occurs, taking place a day or so after the exposure. After the toxin penetrates the skin, the sensitized lymphocytes slowly infiltrate the area and gradually initiate their allergic effect on the local tissue, thus resulting in the typical dermatitis of poison ivy.

Delayed reaction allergies also occur in response to different chemicals or drugs; and delayed reaction allergies develop in certain types of infection such as tuberculosis, brucellosis, and histoplasmosis. Once the person has become sensitized to these invading agents, lymphocytic reactions around the invaders cause local tissue destruction and ingrowth of fibroblasts, thus replacing the infected area of the tissue with scar tissue. This is perhaps an important reaction in preventing the spread of the disease.

THE ATOPIC ALLERGIES (REAGINS)

The atopic allergies are those that occur in the hereditarily predisposed person, and they result from abnormal immunity against everyday occur-

ring antigens such as pollens from plants, scalings from the skin of animals, and organic material in dust. These so-called "allergic" individuals form a special types of antibodies, called *reagins* or *sensitizing antibodies,* that are not found in the normal person. They are very weak antibodies and therefore are relatively useless for giving immunologic protection, but on the other hand, they have an extreme propensity for causing local tissue damage when the antibody-antigen reaction takes place.

Among the different types of atopic allergies caused by reagins are urticaria, hay fever, asthma, eczema, and gastrointestinal disturbances.

Urticaria seems to be almost the same type of reaction as that which occurs in anaphylaxis, except that it takes place in many localized areas of the body, presumably at points where large quantities of reagins are attached to the tissues or are accumulated for other reasons. At these places, the blood vessels become greatly dilated and the capillaries become very leaky, resulting in fluid exudation into the tissues, and the appearance of large welts all over the surface of the body and small edematous areas throughout the internal organs. Antihistaminic drugs are very effective against urticaria, indicating that much of this reaction might be caused by histamine released during the allergic reaction.

Hay fever results from antigen-antibody reactions in the nose. Much of the effect is caused by released histamine, which causes local blood vessel dilatation and leaky capillaries with resultant tissue edema. Antihistaminic drugs and sympathomimetic drugs can counteract edema formation in the nose in hay fever, but they do not prevent the local nasal irritation, indicating that the allergic reaction probably releases other damaging substances in the tissues as well as histamine.

Asthma results from exposure to airborne antigens to which the person is sensitized, such as ragweed, pollen, or house dust. These result in localized antigen-antibody reactions in the lungs that cause bronchiolar constriction. Though it is known that histamine can cause bronchiolar constriction, antihistaminic drugs provide almost no benefit, indicating that the bronchiolar constriction probably results from some factor, other than histamine that is released by the allergic reaction.

Desensitization as a Means for Treating Allergy. A person with different types of atopic allergies can frequently be treated by desensitization to the allergen. To do this, one simply collects some of the allergen, sterilizes it, and injects it into the allergic person in small but progressively larger quantities over a period of weeks or years. At first, the injections cause mild allergic reactions, but the reactions usually become weaker and weaker until the person has become desensitized.

The mechanism of desensitization seems to be two-fold: First, the person forms normal or "complete" antibodies against the allergen, and these are so powerful that they are capable of destroying the allergen before it can cause the allergic reaction with the reagins. Second, in the later stages of the desensitization process, the quantity of reagins, for inexplicable reasons, seems to decrease.

REFERENCES

Burnet, F. M.: Autoimmune disease. *In* Sodeman, William A., and Sodeman, William A., Jr. (eds.): Pathologic Physiology. 4th ed., Philadelphia, W. B. Saunders Company, 1967, p. 15.

Ciba Foundation: Antilymphocytic Serum. Boston, Little, Brown and Company, 1967.

Ciba Foundation: The Thymus. Boston, Little, Brown and Company, 1966.

Gatti, R. A.; Stutman, O.; and Good, R. A.: The lymphoid system. *Ann. Rev. Physiol., 32*:529, 1970.

Gowland, G.: The mechanism of immunological tolerance and sensitivity. *In* Bittar, E. Edward, and Bittar, Neville (eds.): The Biological Basis of Medicine. New York, Academic Press, Inc., 1969, Vol. 4, p. 305.

Harboe, M.: Structure and metabolism of immunoglobulins. *In* Bittar, E. Edward, and Bittar, Neville (eds.): The Biological Basis of Medicine. New York, Academic Press, Inc., 1969, Vol. 4, p. 197.

Haurowitz, F.: Antibody formation. *Physiol. Rev., 45*:1, 1965.

Kahlson, G., and Rosengren, E.: New approaches to the physiology of histamine. *Physiol. Rev., 48*:155, 1968.

Miescher, P. A., and Muller-Eberhard, H. J. (eds.): Textbook of Immunopathology. New York, Grune & Stratton, Inc., 1968, Vols. I and II.

Miller, J. F. A. P.: The thymus in immunity. *In* Bittar, E. Edward, and Bittar, Neville (eds.): The Biological Basis of Medicine. New York, Academic Press, Inc., 1969, Vol. 4, p. 225.

Miller, J. F. A. P., and Osoba, D.: Current concepts of the immunological function of the thymus. *Physiol. Rev., 47*:437, 1967.

Mongar, J. L., and Schild, H. O.: Cellular mechanisms in anaphylaxis. *Physiol. Rev., 42*:226, 1962.

Najjar, V. A.: Some aspects of antibody-antigen reactions and theoretical considerations of the immunologic response. *Physiol. Rev., 43*:243, 1963.

Physiologic functions of histamine (Symposium). *Fed. Proc., 24*:1293, 1965.

Taylor, K. B.: Autoimmunity and disease. *In* Bittar, E. Edward, and Bittar, Neville (eds.): The Biological Basis of Medicine. New York, Academic Press, Inc., 1969, Vol. 4, p. 239.

Turk, J. L.: Delayed hypersensitivity: specific cell-mediated immunity. *Brit. Med. Bull., 23*:1, 1967.

Tuttle, R. S., and McCleary, M.: Centrally mediated and direct effects of anaphylaxis on cat vascular smooth muscle. *Amer. J. Physiol., 216(6)*:1306, 1969.

Uhr, J. W.: Delayed hypersensitivity. *Physiol. Rev., 46*:359, 1966.

Zmijewski, C. M.: Immunohematology. New York, Appleton-Century-Crofts, 1968.

CHAPTER 11

Blood Groups; Transfusion; Tissue and Organ Transplantation

ANTIGENICITY AND IMMUNE REACTIONS OF BLOOD

When blood transfusions from one person to another were first attempted the transfusions were successful in some instances, but, in many more, immediate or delayed agglutination and hemolysis of the red blood cells occurred. Soon it was discovered that the bloods of different persons usually have different antigenic and immune properties so that antibodies in the plasma of one blood react with antigens in the cells of another. Furthermore, the antigens and the antibodies are almost never precisely the same in one person as in another. For this reason, it is easy for blood from a donor to be mismatched with that of a recipient. Fortunately, if proper precautions are taken, one can determine ahead of time whether or not appropriate antibodies and antigens are present in the donor and recipient bloods to cause a reaction, but, on the other hand, lack of proper precautions often results in varying degrees of red cell agglutination and hemolysis, resulting in a typical transfusion reaction that can lead to death.

Multiplicity of Antigens in the Blood Cells. As pointed out in Chapter 10, a person does not normally form antibodies against the antigens in his own cells, but if cells from one person are transfused into another person, antibodies will be developed against all the antigens not in the recipient's own blood. Fortunately, many of the antigens are common from one person to another, and,bloods are grouped and typed on the basis of the major types of antigens appearing in the cells.

At least 30 commonly occurring antigens, each of which can at times cause antigen-antibody reactions, have been found in human blood cells, especially in the cell membranes. In addition to these, more than 100 others of less potency or that occur in individual families rather than having widespread occurrence are known to exist. Among the 30 or more common antigens, certain ones are highly antigenic and regularly cause transfusion reactions if proper precautions are not taken, whereas others are of importance principally for studying the inheritance of genes and therefore for establishing parentage, race, and so forth. However, even though these less important antigens do not normally cause serious transfusion reactions, they do cause reactions in transplanted tissues and organs, as is discussed later in the chapter.

Two particular groups of antigens are more likely than the others to cause blood transfusion reactions. These are the so-called *O-A-B* system of antigens and the *Rh-Hr* system. Bloods are divided into different *groups* and *types* in accordance with the types of antigens present in the cells.

THE O-A-B BLOOD GROUPS

THE A AND B ANTIGENS — THE AGGLUTINOGENS

Two different but related antigens—type A and type B—occur in the cells of different persons. Because of the way these antigens are inherited, a person may have neither of them in his cells, or he may have one or both simultaneously.

As will be discussed below, transfused blood often contains strong antibodies that react specifically with either the type A or type B antigens in the cells, usually causing agglutination. Therefore, because the type A and type B antigens in the cells make the cells susceptible to agglutination, these antigens are called *agglutinogens*, and it is on the basis of presence or absence of agglutinogens in the blood that different types of blood for transfusion are grouped. The agglutinogens are also frequently called *group specific substances* because they are used to specify the O-A-B blood groups.

The Four Major O-A-B Blood Groups. In transfusing blood from one person to another, the bloods of donors and recipients are normally classified into four major O-A-B groups, as illustrated in Table 11–1, depending on the

TABLE 11–1. The Blood Groups, with Their Genotypes and Their Constituent Agglutinogens and Agglutinins

Genotypes	Blood Groups	Agglutinogens	Agglutinins
OO	O	–	Anti-A and Anti-B
OA or AA	A	A	Anti-B
OB or BB	B	B	Anti-A
AB	AB	A and B	–

presence or absence of the two agglutinogens. When neither A nor B agglutinogen is present, the blood group is *group O*. When only type A agglutinogen is present, the blood is *group A*. When only type B agglutinogen is present, the blood is *group B*. And when both A and B agglutinogens are present, the blood is *group AB*.

RELATIVE FREQUENCY OF THE DIFFERENT BLOOD TYPES. The prevalence of the different blood types among white persons is approximately as follows:

Type	Per cent
O	47
A	41
B	9
AB	3

It is obvious from these percentages that the O and A genes occur frequently but the B gene is infrequent.

Inheritance of the Agglutinogens. Figure 11–1A illustrates two genes on two adjacent chromosomes, one gene on each chromosome, that determine the blood groups. These are allelomorphic genes that can be any of three different types: type O, type A, or type B. There is no dominance among the three different allelomorphs. However, the type O gene is either functionless or almost functionless, so that it causes either no type O agglutinogen in the cells or such a weak agglutinogen that it is normally insignificant. On the other hand, the type A and type B genes do cause strong agglutinogens in the cells. Therefore, if either of the genes on the two respective chromosomes is type A, the red blood cells will contain type A agglutinogen, and likewise, if either of the two genes is type B, the red blood cells will contain type B agglutinogen.

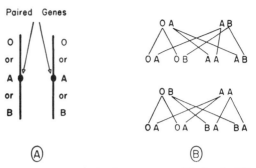

Figure 11–1 (A) Paired genes responsible for the different blood group genotypes. (B) Two examples of father and mother genotypes and the genotypes of their offspring.

The six possible combinations of genes, as shown in Table 11–1, are OO, OA, OB, AA, BB, and AB. These different combinations of genes are known as the *genotypes,* and each person is one of the six different genotypes. Figure 11–1B shows how mating fathers and mothers of different genotypes can cause inheritance of the different group specific substances in the offspring. Only two different types of matings are shown in Figure 11–1B, but 21 different matings of genotypes are possible between the male and the female.

One can observe from Table 11–1 that a person with genotype OO produces no agglutinogens at all, and, therefore, his blood group is O. A person with either genotype OA or AA produces type A agglutinogens and, therefore, has blood group A. Genotypes OB and BB give group B blood, and genotype AB gives group AB blood.

THE AGGLUTININS

When type A agglutinogen *is not present* in a person's red blood cells, antibodies known as "anti-A" agglutinins develop in his plasma. Also, when type B agglutinogen *is not present* in the red blood cells, antibodies known as "anti-B" agglutinins develop in the plasma.

Thus referring once again to Table 11–1, it will be observed that group O blood, though containing no agglutinogens, does contain both *anti-A* and *anti-B agglutinins,* whereas group A blood contains type A agglutinogens and *anti-B agglutinins,* and group B blood contains type B agglutinogens and *anti-A agglutinins.* Finally, group AB blood contains both A and B agglutinogens but no agglutinins at all.

Titer of the Agglutinins at Different Ages. Immediately after birth the quantity of agglutinins in the plasma is almost zero. Two to eight months after birth, the infant begins to produce agglutinins—anti-A agglutinins when type A agglutinogens are not present in the cells and anti-B agglutinins when type B ag-

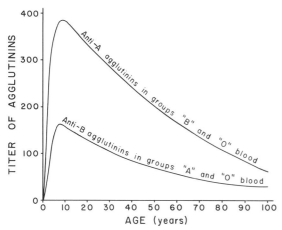

Figure 11–2 Average titers of anti-A and anti-B agglutinins in the blood of persons in group B and group A at different ages.

glutinogens are not in the cells. Figure 11–2 illustrates the changing titer of alpha and beta agglutinins at different ages. A maximum titer is usually reached at 8 to 10 years of age, and this gradually declines throughout the remaining years of life.

Origin of the Agglutinins in the Plasma. The agglutinins are gamma globulins, as are other antibodies, and are produced by the same cells that produce antibodies to infectious diseases.

It is difficult to understand how agglutinins are produced in individuals who do not have the respective antigenic substances in their red blood cells. However, small amounts of group A and B antigens enter the body in the food, in bacteria, or by other means, and these substances, it is believed, initiate the development of anti-A or anti-B agglutinins. One of the reasons for believing this is that injection of group A or group B antigen into a recipient of another blood type causes a typical immune response with formation of greater quantities of agglutinins than ever. Also, the newborn baby has few if any agglutinins, showing that the agglutinin formation occurs after birth.

THE AGGLUTINATION PROCESS IN TRANSFUSION REACTIONS

When bloods are mismatched so that anti-A or anti-B agglutinins are mixed with red blood cells containing A or B agglutinogens respectively, the red cells agglutinate by the following process: The agglutinins attach themselves to the red blood cells. Because the agglutinins are bivalent, a single agglutinin can attach to two different red blood cells at the same time, thereby causing the cells to adhere to each

other. This causes the cells to clump and then plug up small blood vessels throughout the circulatory system. During the ensuing few hours to few days, the reticuloendothelial system destroys the cells, releasing hemoglobin into the blood.

Hemolysis in Transfusion Reactions. Sometimes *anti-A* or *anti-B hemolysins,* as well as agglutinins, develop in the plasma. Usually hemolysins are present when the titer of agglutinins is very high, but the hemolysins are actually different antibodies from the agglutinins. The hemolysins, instead of causing the blood cells to clump, cause lysis of the red blood cells by reaction with the hemolysins and complement as described in Chapter 10. Because hemolysins are less commonly present in plasma than are agglutinins, the immune reaction that occurs between bloods of different groups is usually agglutination rather then hemolysis. However, even agglutination eventually causes hemolysis of the cells because the reticuloendothelial system destroys the agglutinated cells within a few hours after agglutination has taken place.

BLOOD TYPING

Prior to giving a transfusion, it is necessary to determine the blood group of the recipient and the group of the donor blood so that the bloods will be appropriately matched. This "typing" of blood is performed as follows:

Occasionally, a type A person has extremely strong anti-B agglutinins in his plasma, and, also occasionally, a type B person has extremely strong anti-A agglutinins in his plasma. When such persons are found, their plasmas, after removal of the fibrinogen which causes clotting, are used as laboratory "typing sera." In other words, two sera are prepared, the first containing a very strong titer of anti-A agglutinins and the second a very strong titer of anti-B agglutinins.

The usual method of blood typing is the slide technique. In using this technique a drop or more of blood is removed from the person to be typed. This is then diluted approximately 10 times with saline so that clotting will not occur. This leaves essentially a suspension of red blood cells in saline. Two separate drops of this suspension are placed on a microscope slide, and a drop of anti-A agglutinin serum is mixed with one of the drops of cell suspension while a drop of anti-B agglutinin serum is mixed with the second drop of cell suspension. After allowing several minutes for the agglutination process to take place, the slide is observed under a microscope to determine whether or not the cells have clumped. If they have clumped, one knows that an immune reaction has resulted between the serum and the cells.

Table 11–2 illustrates the reactions that occur with each of the four different types of blood. Group O red blood cells have no agglutinogens and, therefore, do not react with either the anti-A or the anti-B serum. Group A blood has A agglutinogens and therefore agglutinates with anti-A agglutinins. Group B blood has B agglutinogens and agglutinates with the anti-B serum. Group AB blood has both A and B agglutinogens and agglutinates with both types of serum.

TABLE 11–2. Blood Typing—Showing Agglutination of Cells of the Different Blood Groups with Anti-A and Anti-B Agglutinins

Red Blood Cells	Sera	
	Anti-A	Anti-B
O	−	−
A	+	−
B	−	+
AB	+	+

CROSS-MATCHING

If a person's life depends on immediate transfusion, it is not absolutely essential to type the bloods of the donor and the recipient, for one can simply test the bloods against each other to determine whether or not agglutination will occur. To do this, one prepares, first, a suspension of red cells from the donor, and, second, a small quantity of defibrinated serum from the recipient. Then he mixes the serum from the recipient with the cells from the donor to determine whether or not agglutination occurs. In a second test, the cells of the recipient are "cross-matched" against the serum of the donor. If no agglutination of either the donor's or the recipient's cells occurs, it can be assumed that the two bloods are probably compatible enough to proceed with a transfusion even though the actual blood type is unknown.

THE Rh-Hr BLOOD TYPES

In addition to the O-A-B blood group system, several other systems are sometimes important in the transfusion of blood, the most important of which is the Rh-Hr system. The one major difference between the O-A-B system and the Rh-Hr system is as follows: In the O-A-B system, the agglutinins responsible for causing transfusion reactions develop spontaneously, whereas in the Rh-Hr system spontaneous agglutinins almost never occur. Instead, the person must first be massively exposed to some antigen of the system, usually by transfusion of blood into him, before he will develop enough agglutinins to cause significant transfusion reaction.

The Rh Agglutinogens. There are at least eight different types of Rh agglutinogens, each of which is called an *Rh factor*. The symbols for each of these are given in Table 11–3. The table also shows the symbols for the genes responsible for inheritance of the agglutinogens and, too, the frequencies of the different Rh factors in Caucasoids. Note that three Rh factors, designated rh, Rh_1, and Rh_2, are common among the population, whereas the other five Rh factors are uncommon.

TABLE 11–3.   The Rh Series of Agglutinogens*

	Agglutin-ogens	Frequencies among NYC Caucasoids (per cent)	Corre-sponding Genes	Gene Frequencies (per cent)	Reactions with Rh Antisera			Reactions with Hr Antisera		
					Anti-Rh_0	Anti-rh'	Anti-rh''	Anti-hr'	Anti-hr''	Anti-hr
Rh Negative	rh	61.6	r	38.0	−	−	−	+	+	+
	rh'	2.8	r'	1.4	−	+	−	−	+	−
	rh''	1.0	r''	0.5	−	−	+	+	−	−
	rh_y	0.02	r^y	0.01	−	+	+	−	−	−
Rh Positive	Rh_0	6.3	R^0	3.2	+	−	−	+	+	+
	Rh_1	64.5	R^1	40.4	+	+	−	−	+	−
	Rh_2	29.3	R^2	16.4	+	−	+	+	−	−
	Rh_z	0.2	R^z	0.1	+	+	+	−	−	−

* Modified from A. S. Wiener and I. B. Wexler: An Rh-Hr Syllabus. Grune & Stratton, 1963.

INHERITANCE OF THE Rh FACTORS. Though there exists some confusion about the exact means of inheritance of the Rh factors, it is now believed that a single pair of genes is responsible for this inheritance. The eight genes shown in Table 11–3 are "allelomorphic," which means that any one of these eight genes can occur at the appropriate gene site on each respective chromosome. The Rh factors are mendelian dominant, so that if either chromosome contains a particular gene, the corresponding Rh factor is present in the blood. Thus, if the gene r appears on one chromosome and the gene r' on the other chromosome, the red cells of that particular person would have both rh and rh' agglutinogens. This person would be *heterozygous* for rh factors. If he should have an r gene on each of the chromosomes, he would have only one agglutinogen, rh, in his red blood cells; and he would be *homozygous*.

The Hr Factors. A factor similar to the Rh factor, but having far less tendency to cause transfusion reactions, is the Hr factor. It has been suggested that for each type of Rh factor there is a corresponding Hr factor and that these are reciprocally related to the Rh factors. That is, when the Rh factor is missing, the Hr factor supposedly is present. However, only a few of the theoretically possible Hr factors have actually been demonstrated, the three most important of which are hr, hr', and hr".

Typing Bloods for Rh Factors. Table 11–3 illustrates the reactions of the eight different types of Rh agglutinogens with six different types of Rh and Hr antisera. Note from this chart that there is a great amount of cross-reactivity of the antisera for different ones of the Rh factors rather than reactivity for only specific factors. To determine the Rh type, the six different types of antisera listed in the table are each mixed with a saline suspension of red blood cells and a small amount of plasma protein. From the agglutination reactions that occur, one can use Table 11–3 to determine the different types of Rh factor present in the blood.

Rh POSITIVE AND Rh NEGATIVE PERSONS

Certain ones of the Rh agglutinogens are very likely to cause transfusion reactions. These are capitalized in Table 11–3: Rh_0, Rh_1, Rh_2, and Rh_z. Note also in Table 11–3 that all these agglutinogens react with anti-Rh_0 antiserum. Therefore, any time a person's blood reacts with anti-Rh_0 antiserum, his blood contains an Rh factor that can cause a serious transfusion reaction. For this reason he is said to be Rh positive.

When a person's red blood cells will not react with anti-Rh_0 antiserum, he is not likely to have an Rh factor that can cause a significant transfusion reaction. Therefore, he is said to be Rh negative; however, his blood might contain some less antigenic Rh factor or Hr factor that on occasion can cause a mild reaction.

Approximately 85 per cent of all Caucasoids are Rh positive and approximately 15 per cent are Rh negative.

THE Rh IMMUNE RESPONSE

Formation of Anti-Rh Agglutinins. When red blood cells containing Rh factor, or even protein breakdown products of such cells, are injected into an Rh negative person, anti-Rh agglutinins develop very slowly, the maximum concentration of agglutinins occurring approximately two to four months later. This immune response occurs to a much greater extent in some people than in others. On multiple exposure to the Rh factor, the Rh negative person eventually becomes strongly "sensitized" to the Rh factor—that is, he develops a much higher titer of anti-Rh agglutinins.

Characteristics of Rh Agglutination. The anti-Rh agglutinins are similar to the anti-A and anti-B agglutinins discussed previously; they attach to the Rh positive red blood cells and cause them to agglutinate. However, univalent anti-Rh antibodies also frequently develop. These will attach to the cells but cannot form bridges from cell to cell to cause agglutination. This attachment of these "incomplete" antibodies greatly reduces the available reactive sites for the agglutinins and especially interferes with serologic tests for detecting the presence of the Rh agglutinogens. Indeed, the Rh agglutination reaction can be detected by the usual methods of blood typing only when extremely high titers of agglutinins have formed. More often, agglutination will take place only when the red blood cells are suspended in a solution of plasma protein which causes the reaction to take place much more rapidly than it takes place in the absence of the protein.

Anti-Rh antibodies do not cause hemolysis directly, but whenever agglutination of the cells occurs, the agglutinated cells plug the capillaries in the peripheral circulatory system, and the cells are gradually destroyed by phagocytes during the next few hours to few days so that the final effect of the agglutination reaction is hemolysis.

Erythroblastosis Fetalis. Erythroblastosis fetalis is a disease of the newborn infant characterized by progressive agglutination and subsequent phagocytosis of the red blood cells. In most instances of erythroblastosis fetalis the mother is Rh negative, and the father is Rh positive; the baby has inherited the Rh positive characteristic from the father, and the mother has developed anti-Rh agglutinins that have diffused into the fetus to cause red blood cell agglutination.

PREVALENCE OF THE DISEASE. An Rh negative mother having her first Rh positive child usually does not develop sufficient anti-Rh agglutinins to cause any harm. However, an Rh negative mother having her second Rh positive child often will have become "sensitized" by the first child and therefore will often develop anti-Rh agglutinins rapidly upon becoming pregnant with the second child. Approximately 3 per cent of these second babies exhibit some signs of erythroblastosis fetalis; approximately 10 per cent of the third babies exhibit the disease; and the incidence rises progressively with subsequent pregnancies.

The Rh negative mother develops anti-Rh agglutinins only when the fetus is Rh positive. Many of the Rh positive fathers are heterozygous (about 55 per cent), causing about one fourth of the offspring to be Rh negative. Therefore, after an erythroblastotic child has been born, it is not certain that future children will also be erythroblastotic, for some of them could be Rh negative and not susceptible. Also, it is probable that the anti-Rh immune response in the mother becomes less and less intense the longer the interval between pregnancies; as a result, long intervals between pregnancies might be considerably safer than short intervals when the mother is Rh negative and the father is positive.

PREVENTION OF ERYTHROBLASTOSIS IN BABIES WITH RH NEGATIVE MOTHERS. The Rh negative mother usually becomes sensitized to the Rh positive factor in her child during the period from a few days before birth of the baby to the first few days following birth. At this time large quantities of fetal tissues, especially degenerating products of the placenta, release their antigens into the blood of the mother, giving her a healthy dose of the baby's Rh positive antigen. If this Rh positive antigen can be destroyed at this time before it can initiate the antibody response, the mother will not become presensitized against the Rh factor for subsequent pregnancies. This result can be achieved by passively immunizing the mother against the Rh positive factor. This is done by injecting into the mother serum from another Rh negative person who has already formed anti-Rh agglutinins. These agglutinins circulate in the mother's blood for three to eight weeks and destroy all the Rh positive factor from the baby, thus preventing development of the mother's own anti-Rh agglutinins. With this new treatment it should be possible almost to eliminate future occurrences of erythroblastosis fetalis.

EFFECT OF THE MOTHER'S ANTIBODIES ON THE FETUS. After anti-Rh antibodies have formed in the mother, they diffuse very slowly through the placental membrane into the fetus' blood. There they cause slow agglutination of the fetus' blood, and clumps of blood cells occlude small blood vessels. The red blood cells of these clumps gradually hemolyze, releasing hemoglobin into the blood. The reticuloendothelial system then converts the hemoglobin into bilirubin, which causes yellowness (jaundice) of the skin. The antibodies probably also attack and damage many of the other cells of the body.

CLINICAL PICTURE OF ERYTHROBLASTOSIS. The newborn, jaundiced, erythroblastotic baby is usually anemic at birth, and the anti-Rh agglutinins from the mother usually circulate in the baby's blood for one to two months after birth, destroying more and more red blood cells. Therefore, the hemoglobin level of the erythroblastotic baby often falls for approximately the first 45 days after birth, and, if the level falls below 6 to 8 grams per cent, the baby usually dies.

The hemopoietic tissues of the baby attempt to replace the hemolyzing red blood cells. The liver and the spleen become greatly enlarged and produce red blood cells in the same manner that they normally produce red blood cells during the middle of gestation. Because of the very rapid production of cells, many early forms, including many erythroblasts, are emptied into the circulatory system, and it is because of the presence of these in the blood that the disease has been called "erythroblastosis fetalis."

Though the severe anemia of erythroblastosis fetalis is usually the cause of death, it is not the only cause, for severely erythroblastotic children are very likely to have many other degenerative abnormalities throughout the body, including especially damage to the brain. Many children who barely survive from the anemia exhibit permanent mental impairment or damage to motor areas of the brain. Much of this damage is caused by precipitation of bilirubin in the neuronal cells, causing their destruction, a condition called *kernicterus*.

TREATMENT OF THE ERYTHROBLASTOTIC BABY. The usual treatment for erythroblastosis fetalis is to replace the newborn infant's blood with Rh negative blood. Approximately 400 ml. of Rh negative blood is infused over a period of one and a half or more hours while the baby's own Rh positive blood is being removed. Once the Rh negative blood is in the baby, little further destruction of red blood cells occurs, and second transfusions are usually unnecessary. By the time the Rh negative cells are replaced with the baby's own cells, a process that requires six or more weeks, the anti-Rh agglutinins will have been destroyed.

ERYTHROBLASTOSIS FETALIS OCCURRING IN BABIES OF RH POSITIVE MOTHERS. Approximately 7 per cent of the babies who have erythroblastosis fetalis are born of Rh positive mothers rather than Rh negative mothers. A very few cases are caused by anti-A or anti-B agglutinins that diffuse from the mother into the fetus and agglutinate the fetus' red blood cells. Some of the other instances result from the fact that the so-called "Rh positive" mother is sometimes Rh positive for only one or two Rh factors and yet Rh negative for the other Rh factors. Finally, similar reactions have also been noted on rare occasions as a result of the Hr factor or other less well-known antigenic factors in the fetus' blood, some of which are discussed in the following section.

OTHER BLOOD FACTORS

Many antigenic proteins besides the O, A, B, Rh, and Hr factors are present in red blood cells of different persons, but these other factors very rarely cause transfusion reactions and, therefore, are mainly of academic and legal importance. Some of these different blood factors are the M, N, S, s, P, Kell, Lewis, Duffy, Kidd, Diego, and Lutheran factors. Occasionally, transfusion of red cells containing one of these factors causes specific agglutinins to develop in the recipient, and immune reactions can then occur when the same type of blood is infused a second time. However, these different antigenic factors do not exhibit extreme degrees of antigenicity and, therefore, usually cause very weak transfusion reactions or no reactions at all.

Inheritance of most of the different blood factors follows almost the same principles as those observed for inheritance of the O-A-B agglutinogens and for the Rh and Hr factors. The genes for the different factors are almost all dominant, and whenever any one of the genes is inherited from a parent the factor is usually present in the child's blood.

Method for Studying Obscure Blood Factors. One of the means by which different blood factors, including the Rh factor, have been studied in the human being has been to immunize lower animals, such as rabbits, with human red blood cells. Specific antibodies develop in the animal against the different antigens of the human cells, and specific immune sera prepared from the plasma of these animals can then be used for determining the presence or absence of the same antigens in red blood cells of other persons.

Blood Typing in Forensic Medicine. In the past three decades the use of blood typing has become an important legal procedure in cases of disputed parentage. There are three different genes in the O-A-B system of blood groups and eight major ones in the Rh-Hr system. Including all the other blood types as well, more than 30 different blood group genes can be determined for each person by blood-typing procedures. After the mother's and child's genes have been determined, many of the father's genes, and his corresponding blood factors, are then known immediately, because any gene present in a child but not present in the mother must be present in the father. If a suspected man is missing any one of the necessary blood factors, he is not the father. A falsely charged man can be cleared in about 75 per cent of the cases using the usually available antisera for blood typing, and in a much higher percentage of cases using all possible types of antisera.

TRANSFUSION

Indications for Transfusion. The most common reason for transfusion is decreased blood volume, which will be discussed in detail in Chapter 28 in relation to shock. Also, transfusions are often used for treating anemia or to supply the recipient with some other constituent of whole blood besides red blood cells, such as to supply a thrombocytopenic patient with new platelets. Also, hemophilic patients can be rendered temporarily nonhemophilic by plasma transfusion, and, occasionally, the quantity of "complement" in a recipient must be supplemented by fresh plasma infusions before certain antigen-antibody reactions can take place.

TRANSFUSION REACTIONS RESULTING FROM MISMATCHED BLOOD GROUPS

If blood of one group is transfused to a recipient of another group, a transfusion reaction is likely to occur in which the red blood cells of the donor blood are agglutinated. However, even though the bloods might be mismatched, it is very rare that the transfused blood ever causes agglutination of the *recipient's cells* for the following reason: The plasma portion of the donor blood immediately becomes diluted by all the plasma of the recipient, thereby decreasing the titer of the infused agglutinins to a level too low to cause agglutination. On the other hand, the infused blood does not dilute the agglutinins in the recipient's plasma to a major extent. Therefore, they can still agglutinate the donor cells.

Universal Donor Blood. Because the cells of group O blood contain neither of the two agglutinogens, small quantities of this blood can be transfused into almost any recipient without immediate agglutination occurring. For this reason group O blood is sometimes called universal donor blood. However, transfusion of especially large amounts of group O blood into a mismatched recipient can cause either immediate or delayed agglutination of the recipient's own cells because the infused agglutinins then are not diluted sufficiently to prevent the reaction.

Nevertheless, for use in extreme emergencies, when sufficient time is not available for blood typing or cross-matching, some blood banks keep small quantities of group O blood specially available.

The Universal Recipient. Group AB persons are sometimes called universal recipients because their plasma contains neither alpha nor beta agglutinins. Small quantities of all other blood groups can be infused without causing a transfusion reaction. However, here again, if large quantities of these mismatched bloods should be administered, the agglutinins in the donor blood might accumulate in sufficient quantities to agglutinate the recipient's type AB cells.

Hemolysis of Red Cells Following Transfusion Reactions. All transfusion reactions resulting from mismatched blood groups eventually cause hemolysis of the red blood cells. Occasionally, specific *hemolysins* are present in the plasma of the recipient along with the agglutinins and therefore cause immediate hemolysis, but more frequently the cells agglutinate first and then are entrapped in the peripheral vessels. Over a period of hours to days the entrapped cells degenerate and hemolyze, liberating hemoglobin into the circulatory system.

When the rate of hemolysis is rapid, the concentration of hemoglobin in the blood can rise to extremely high values. A small quantity of hemoglobin, up to 100 mg./100 ml. of plasma, becomes attached to one of the plasma proteins, *haptoglobin,* and continues to circulate in the blood without causing any harm. However, above the threshold value of 100 mg./100 ml. of plasma, the excess hemoglobin remains in the free form and diffuses out of the circulation into the tissue spaces or through the renal glomeruli into the kidney tubules, as is discussed below. The hemoglobin remaining in the circulation or passing into the tissue spaces is gradually ingested by reticuloendothelial cells and converted into bilirubin, which will be discussed in Chapter 74. The concentration of bilirubin in the body fluids sometimes rises high enough to cause *jaundice* — that is, the person's tissues become tinged with yellow pigment. But jaundice usually does not appear unless 300 to 500 ml. of blood is hemolyzed in less than a day.

Acute Kidney Shutdown Following Transfusion Reactions. One of the most lethal effects of transfusion reactions is acute kidney shutdown, which can begin within a few minutes to a few hours and continue until the person dies of renal failure.

The kidney shutdown seems to result from two different causes: First, the antigen-antibody reaction of the transfusion reaction releases toxic substances from the hemolyzing blood that cause powerful renal vasoconstriction. Second, if the total amount of free hemoglobin in the circulating blood is greater than that quantity which can bind with haptoglobin, much of the excess leaks through the glomerular membranes into the kidney tubules. If this amount is still slight, it can be reabsorbed through the tubular epithelium into the blood and will cause no harm, but, if it is great, then only a small percentage is reabsorbed. Yet water continues to be reabsorbed, causing the tubular hemoglobin concentration to rise so high that it precipitates and blocks many of the tubules. Thus, renal vasoconstriction and tubular blockage add together to cause acute renal shutdown. If the shutdown is complete, the patient dies within a week to 12 days, as explained in Chapter 38.

Physiological Principles of Treatment Following Transfusion Reactions. A transfusion reaction can cause considerable degrees of fever and certain amounts of general toxicity, but these effects, in the absence of renal shutdown, are almost never lethal. Therefore, treatment is usually directed toward preventing renal damage, especially preventing hemoglobin precipitation in the kidney tubules, as follows: (1) rapid infusion of dilute intravenous fluids to cause water diuresis, (2) administration of diuretics to prevent reabsorption of water by the tubules and also to overcome the renal vasoconstriction, and (3) alkalinization of the body fluids because alkaline tubular fluid can dissolve more hemoglobin than can acid tubular fluid.

TRANSFUSION REACTIONS OCCURRING WHEN Rh FACTORS ARE MISMATCHED

If the recipient has previously received a blood transfusion containing a strongly antigenic Rh factor not present in his own blood, then he may have built up appropriate anti-Rh antibodies that can cause a transfusion reaction on subsequent exposure to the same type of Rh blood. However, since anti-Rh agglutination is usually very weak, the intensity of the reaction is usually much less severe than when there is an anti-A or anti-B reaction.

OTHER TYPES OF TRANSFUSION REACTIONS

Pyrogenic Reactions. Pyrogenic reactions occur more frequently than either agglutination or hemolysis. These reactions cause fever in the recipient but do not destroy red blood cells. Most pyrogenic reactions probably result from the presence in the donor plasma of proteins to which the recipient is allergic, or they result from the presence of breakdown products in old, deteriorating donor blood. If the donor has eaten certain foods to which the recipient is allergic, allergens from these foods can sometimes be transmitted to the recipient, thereby causing allergic reactions. The usual pyrogenic reaction causes severe chills, and occasionally the blood pressure falls, but ordinarily no dire results follow the reaction.

Transfusion Reactions Resulting from Anticoagulants. The usual anticoagulant used for transfusion is a citrate salt. As discussed in the following chapter, citrate operates as an anticoagulant by combining with the calcium ions of the plasma so that these become nonionizable. Without the presence of ionizable calcium the coagulation process cannot take place. Normal nerve, muscle, and heart function also cannot occur in the absence of calcium ions. Therefore, if blood containing large quantities of citrate anticoagulant is administered rapidly, the recipient is likely to experience typical tetany due to low calcium, which is discussed in detail in Chapter 79. Such tetany can kill the patient within a few minutes because of respiratory muscle spasm.

Ordinarily the liver can remove citrate from the blood within a few minutes and polymerize it into glycogen or utilize it directly for energy. Therefore, when blood transfusions are given at slow rates, the citrate type of reaction is not likely to occur. On the other hand, if the patient has liver damage, the rate of transfusion must be decreased even more than usually to prevent a low calcium ion reaction.

TRANSPLANTATION OF TISSUES AND ORGANS

Relation of Genotypes to Transplantation. In this modern age of surgery, many attempts are being made to transplant tissues and organs from one person to another, or, occasionally, from lower animals to the human being. All the different antigenic proteins of red blood cells that cause transfusion reactions, and still many more, are present in the other cells of the body. Consequently, any foreign cells transplanted into a recipient can cause immune responses and immune reactions. In other words, most recipients are just as able to resist invasion by foreign cells as to resist invasion by foreign bacteria.

Autologous, Homologous, and Heterologous Transplants. A transplant of tissue or an organ from one part of a person's body to some other part of his body is called an *autologous transplant*. Such transplants can be skin grafts, bone grafts, cartilage grafts, and so forth. A transplant from one human being to another or from any other animal to another member of the same species is called a *homologous transplant*. Finally, a transplant from a lower animal to a human being or from an animal of one species to one of another species is called a *heterologous transplant*.

Cellular Transplants. In the case of autologous transplants, cells in the transplant will almost always live indefinitely if an adequate blood supply is provided, but, in the case of homologous and heterologous transplants, immune reactions almost always occur, causing death of all the cells in the transplant 3 to 10 weeks after transplantation unless some specific therapy is used to prevent the immune reaction. The cells of homologous transplants usually persist slightly longer than those of heterologous transplants because the antigenic structure of the homologous transplant is more nearly the same as that of the recipient's tissues than is true of the heterologous transplant. And when the tissues are properly "typed" prior to transplant, completely successful homologous transplants occasionally result. The greater the difference in antigenic structure, the more rapid and the more severe are the immune reactions to the graft.

Transplants from one identical twin to another are one example in which homologous transplants of cellular tissues are usually successful. The reason for this is that the antigenic proteins of both twins are determined by identical genes derived originally from the single fertilized ovum.

Some of the different cellular tissues and organs that have been transplanted either experimentally or for temporary benefit from one person to another are skin, kidney, heart, liver, glandular tissue, bone marrow, and lung. A few kidney homologous transplants have been successful for as long as five years, a rare liver and heart transplant for one to two years, and lung transplants for one month.

Transplantation of Noncellular Tissues. Certain tissues that have no cells or in which the cells are unimportant to the purpose of the graft, such as the cornea, tendon, fascia, and bone, can usually be grafted from one person to another with considerable success. In these instances the grafts act merely as a supporting latticework into which or around which the surrounding living tissues of the recipient grow. Indeed, some such grafts – bone and fascia – are occasionally successful even when they come from a lower animal rather than from another human being.

A special problem arises in the case of a corneal transplant from one human eye to another, for, not only must it maintain structural continuity, but it must also remain transparent. About 50 to 90 per cent of the corneas remain transparent, the exact percentage depending principally on the preliminary condition of the recipient's eye. The remainder suddenly become cloudy at the end of two weeks to two years. This cloudiness is the result of an antigen-antibody reaction. It is believed that corneas that do not become cloudy fail to do so for any of three different reasons: First, their antigenicity might be low enough that they fail to cause the production of sufficient antibodies to elicit the response. Second, the vascularity of the cornea is so slight that the antigens of the cornea may never infuse the host in sufficient quantities to elicit an immune response. Or, third, it is possible that, if the reaction is delayed long enough, the corneal proteins adapt to the immunologic structure of the recipient so that a reaction can no longer occur against the corneal tissue.

ATTEMPTS TO OVERCOME THE ANTIGEN-ANTIBODY REACTIONS IN TRANSPLANTED TISSUE

Because of the extreme potential importance of transplanting certain tissues and organs, such as skin, kidneys, and lungs, serious attempts have been made to prevent the antigen-antibody reactions associated with transplants. The following specific procedures have met with certain degrees of success.

Use of Anti-Lymphocyte Serum. It was pointed out in the previous chapter that grafted tissues are usually destroyed by lymphocytes that have become sensitized against the graft. These cells invade the graft and presumably secrete substances that gradually destroy the grafted cells. Upon contact with the lymphocytes, the cells of the graft begin to swell, their membranes become very permeable, and finally they rupture. Within a few days to a few weeks after this process begins, the tissue is completely destroyed, even though the graft had been completely viable and functioning normally for the few days to few weeks after its original transplantation.

Therefore, one of the most effective procedures in preventing rejection of grafted tissues has been to inoculate the recipient with *anti-lymphocyte serum.* This serum is made in horses by injecting human lymphocytes into them; the antibodies that develop in these animals will then attack and destroy the human lymphocytes. When serum from one of these animals is injected into the transplanted recipient, the number of circulating small lymphocytes can be decreased to as little as 5 to 10 per cent of the normal number, and there is corresponding decrease in the intensity of the rejection reaction. Unfortunately, the procedure does not continue to work very satisfactorily after the first few injections of the antiserum because the recipient soon begins to build up antibodies against the antiserum itself.

Glucocorticoid Therapy (Cortisone, Hydrocortisone, and ACTH). The glucocorticoid hormones from the adrenal gland greatly suppress the formation of antibodies and immunologically competent lymphocytes. Therefore, administration of large quantities of these, or of ACTH, which causes the adrenal gland to produce glucocorticoids, often allows transplants to persist much longer than usual in the recipient. However, this is only a delaying process rather than a preventive one.

Suppression of Antibody Formation. Occasionally, a human being has naturally suppressed antibody formation resulting from (a) congenital aglobulinemia, in which case gamma globulins are not produced, and (b) destructive diseases of the reticuloendothelial system such as result occasionally in multiple myeloma. Transplants of homologous tissues into such individuals are occasionally successful, or at least their destruction is de-layed. Also, irradiative destruction of most of the lymphoid tissue by either x-rays or gamma rays renders a person much more susceptible than usual to a homologous transplant. And treatment with certain drugs, such as azathioprine (Imuran) and thioguanine—both of which suppress antibody formation—also increases the likelihood of success with homologous transplants. Unfortunately, all of these procedures leave the person unprotected from disease.

REMOVAL OF THE THYMUS GLAND TO PREVENT ANTIBODY FORMATION. It was pointed out in Chapter 10 that the thymus gland during fetal and early postnatal life forms thymic cells that migrate throughout the body and become the precursors of the immunologic lymphocytes and the plasma cells. Also, for the first few weeks or months after birth, the thymus seems to secrete a hormone that greatly enhances the proliferation of lymphoid tissue throughout the body. Therefore, if the thymus gland is removed during fetal life or immediately after birth, the immune system is greatly impaired throughout the remainder of the animal's life. Homologous transplants are frequently successful in such animals, but the loss of immunity to disease makes it difficult for the animal to live.

To summarize, transplantation of living tissues in human beings up to the present has been hardly more than an experiment. But when someone succeeds in blocking the immune response of the recipient to a donor organ without at the same time destroying the recipient's specific immunity for disease, this story will undoubtedly change overnight.

REFERENCES

Billingham, R. E.: The Biology of Graft-versus-Host Reactions. Harvey Lectures, 1966–67, p. 21.

Brent, L.: Transplantation of tissues and organs. *Brit. Med. Bull.,* 21:97, 1965.

Calne, R. Y.: Renal transplantation. Baltimore, The Williams & Wilkins Co., 1963.

Ciba Foundation: Anti-Lymphocytic Serum. Boston, Little, Brown and Company, 1967.

Coombs, R. R. A., and Lachmann, P. S.: Immunological reactions at the cell surface. *Brit. Med. Bull.,* 24:113, 1968.

Gamma globulins [Series of papers]. *New Eng. J. Med.,* 275:480, 536, 591, 652, 709, 769, 826; 1966.

Merrill, J. P.: Transplantation of normal tissues. *Physiol. Rev.,* 39:860, 1959.

Miller, J. F. A. P.: Immunology of tissue transplantation. *In* Bittar, E. Edward, and Bittar, Neville (eds.): The Biological Basis of Medicine. New York, Academic Press, Inc., 1969, Vol. 4, p. 265.

Mollison, P. L.: Blood Transfusion in Clinical Medicine. 4th ed., Philadelphia, F. A. Davis Co., 1967.

Race, R. R., and Sanger, R.: Blood Groups in Man. 4th ed., Philadelphia, F. A. Davis Co., 1962.

Wiener, A. S.: Elements of blood group nomenclature with special reference to the Rh-Hr blood types. *J.A.M.A., 199*:985, 1967.

Wiener, A. S.: Modern blood group nomenclature. *J. Amer. Med. Tech.,* 30:174, 1968.

Wiener, A. S., and Wexler, I. B.: An Rh-Hr Syllabus: The Types and Their Applications. 2nd ed., New York, Grune & Stratton, Inc., 1963.

CHAPTER 12

Hemostasis and Blood Coagulation

EVENTS IN HEMOSTASIS

The term hemostasis means prevention of blood loss. Whenever a vessel is severed or ruptured, hemostasis is achieved by several different mechanisms including (1) vascular spasm, (2) formation of a platelet plug, (3) blood coagulation, and (4) growth of fibrous tissue into the blood clot to close the hole in the vessel permanently.

VASCULAR SPASM

Immediately after a blood vessel is cut or ruptured, the wall of the vessel contracts; this instantaneously reduces the flow of blood from the vessel rupture. The contraction results from both nervous reflexes and local myogenic spasm. The nervous reflexes presumably are initiated by pain impulses originating from the traumatized vessel or from nearby tissues. The reflex signals travel first to the spinal cord and then back through the sympathetic nerves to cause spasm of the vessel for several centimeters in either direction from the rupture.

However, this neurogenic spasm lasts for only a fraction of a minute to a few minutes. In its place a local myogenic vascular spasm ensues. This is initiated by direct damage to the vascular wall, which presumably causes transmission of action potentials along the vessel wall for several centimeters and results in constriction of the vessel. The more of the vessel that is traumatized, the greater is the degree of spasm; this means that the sharply cut blood vessel usually bleeds much more than does the vessel ruptured by crushing. This local vascular spasm lasts for as long as 20 minutes to half an hour, during which time the ensuing processes of platelet plugging and blood coagulation can take place.

The value of vascular spasm as a means of hemostasis is illustrated by the fact that persons whose legs have been severed by crushing types of trauma sometimes have such intense spasm in vessels as large as the anterior tibial artery that there is not serious loss of blood.

FORMATION OF THE PLATELET PLUG

The second event in hemostasis is an attempt by the platelets to plug the rent in the vessel. To understand this it is important that we first understand the nature of platelets themselves.

Platelets are minute round or oval discs about 2 to 5 microns in diameter. On coming in contact with wettable surfaces, they change their characteristics drastically, assuming bizarre and irregular forms with numerous irradiating processes protruding from their surfaces. Platelets are breakdown products of *megakaryocytes,* which are extremely large cells of the granulocytic series formed in the bone marrow. The megakaryocytes disintegrate into platelets while still in the bone marrow and release the platelets into the blood. The normal concentration of platelets in the blood is between 150,000 and 300,000 per cubic mm.

Mechanism of the Platelet Plug. The mechanism of the platelet repair of vascular openings seems to be the following: At the site of any rent in a vessel, the endothelium loses its normal smoothness and also loses its normal nonwettability. As a result, platelets immediately begin to adhere to the endothelium at this point. On adhering, the platelets themselves then take on new characteristics; most important of all, they become sticky to other platelets. Therefore, a second layer of platelets adheres to the first layer, and third and fourth layers adhere until a small platelet plug actually fills the rent in the vessel.

Importance of the Platelet Method for Closing Vascular Holes. If the rent in a vessel is small, the platelet plug by itself can stop blood loss completely, but if there is a large hole, a blood clot in addition to the platelet plug is required to stop the bleeding.

The platelet plugging mechanism is extremely important to close the minute ruptures in very small blood vessels that occur hundreds of times daily. A person who has very few platelets develops literally hundreds of small hemorrhagic areas under his skin and throughout his internal tissues, but this does not occur in the normal person. The platelet plugging mechanism does not occlude the vessel itself but merely plugs the hole, so that the vessel continues to function normally. This usually is not true when a blood clot occurs.

1. Severed vessel

2. Platelets agglutinate

3. Fibrin appears

4. Fibrin clot forms

5. Clot retraction occurs

Figure 12–1   The clotting process in the traumatized blood vessel. (Redrawn from Seegers: Hemostatic Agents. Charles C Thomas.)

CLOTTING IN THE RUPTURED VESSEL

The third mechanism for hemostasis is formation of the blood clot. The clot begins to develop in 15 to 20 seconds if the trauma of the vascular wall has been severe and in one to two minutes if the trauma has been minor. Activator substances both from the traumatized vascular wall and from platelets adhering to the traumatized vascular wall initiate the clotting process. The physical events of this process are illustrated in Figure 12–1, and the chemical events will be discussed in detail later in the chapter.

Within three to six minutes after rupture of a vessel, the entire end of the vessel is filled with clot. After 30 minutes to an hour, the clot reacts; this closes the vessel still further. Platelets play an important role in this clot retraction, as will also be discussed later in the chapter.

FIBROUS ORGANIZATION OF THE BLOOD CLOT

Once a blood clot has formed, it can follow two separate courses: it can become invaded by fibroblasts, which subsequently form connective tissue all through the clot; or it can dissolute. The usual course for a clot that forms in a small hole of a vessel wall is invasion by fibroblasts, beginning within a few hours after the clot is formed, and complete organization of the clot into fibrous tissue within approximately 7 to 10 days. On the other hand, when a large mass of blood clots, such as blood that has leaked into tissues, special substances within the tissue usually activate a mechanism for dissoluting most of the clot, which will be discussed later in the chapter.

MECHANISM OF BLOOD COAGULATION

Basic Theory.   Though tremendous amounts of effort have been expended studying the mechanisms of blood clotting, the precise means by which this occurs is still unknown. Over 30 different substances that affect blood coagulation have been found in the blood and related tissues, some promoting coagulation, called *procoagulants*, and others inhibiting coagulation, called *anticoagulants*. Whether or not the blood will coagulate depends on a balance between these two groups of substances. Normally the anticoagulants are in the ascendancy, and the blood does not coagulate, but when a vessel is ruptured the activity of the procoagulants in the area of damage becomes much greater than that of the anticoagulants, and then a clot does develop.

General Mechanism.   Almost all research workers in the field of blood coagulation agree that clotting takes place in three essential steps:

First, a substance called *prothrombin activator* is formed in response to rupture of the vessel or damage to the blood itself.

Second, the prothrombin activator catalyzes the conversion of prothrombin into *thrombin*.

Third, the thrombin acts as an enzyme to convert fibrinogen into *fibrin threads* that enmesh red blood cells and plasma to form the clot itself.

Unfortunately, the detailed mechanisms by which prothrombin activator is formed are still extremely doubtful, so much so that almost no two research workers in the blood clotting field support the same scheme of events. On the other hand, the mechanisms by which prothrombin is converted into thrombin and by which thrombin then acts to cause the formation of fibrin threads are much better

Figure 12-2   Schema for conversion of prothrombin to thrombin, and polymerization of fibrinogen to form fibrin threads.

understood. Therefore, let us first discuss the basic mechanism by which the blood clot is formed, beginning with the activation of prothrombin; then we will come back to the initiating stages in the clotting process by which prothrombin activator is formed.

CONVERSION OF PROTHROMBIN TO THROMBIN

After prothrombin activator has been formed as a result of rupture of the blood vessel or as a result of damage to the platelets in the blood, it can cause conversion of prothrombin to thrombin, which in turn causes polymerization of fibrinogen molecules into fibrin threads within 10 to 15 seconds. Thus the rate limiting factor in causing blood coagulation is usually the formation of prothrombin activator and not the subsequent reactions beyond that point.

Prothrombin.   Prothrombin is a plasma protein, an alpha$_2$ globulin, having a molecular weight of 68,700. It is present in normal plasma in a concentration of about 15 mg./100 ml. It is an unstable protein that can split easily into smaller compounds, one of which is thrombin, which has a molecular weight of 33,700, almost exactly half that of prothrombin.

Prothrombin is formed continually by the liver, and it is continually being used throughout the body for blood clotting. If the liver fails to produce prothrombin, its concentration in the plasma falls too low within 24 hours to provide normal blood coagulation. Vitamin K is required by the liver for normal formation of prothombin; therefore, either lack of vitamin K or the presence of liver disease that prevents normal prothrombin formation can often decrease the prothrombin level so low that a bleeding tendency results.

Effect of Prothrombin Activator on Prothrombin.   Figure 12-2 illustrates the conversion of prothrombin to thrombin under the influence of prothrombin activator and calcium ions. The rate of formation of thrombin from pro-

thrombin is almost directly proportional to the quantity of prothrombin activator available, which in turn is approximately proportional to the degree of trauma to the vessel wall or to the blood. In turn, the rapidity of the clotting process is proportional to the quantity of thrombin formed.

Recent studies by Seegers on the basic chemistry of prothombin indicate that prothrombin is first split to form two intermediate compounds before formation of thrombin, *prethrombin* and *autoprothrombin C*. The autoprothrombin C is an enzyme that immediately acts on the prethrombin to convert it into thrombin. These intermediate stages in the formation of thrombin are important because some of the blood factors that activate the clotting process enhance formation of autoprothrombin C from prothrombin, whereas others enhance the enzymatic action of autoprothrombin C on prethrombin. Though the precise mechanisms involved in these processes are yet to be completely elucidated, the discovery of these new steps in the formation of thrombin is helping to characterize the events in the prothrombin activation process.

CONVERSION OF FIBRINOGEN TO FIBRIN — FORMATION OF THE CLOT

Fibrinogen.   Fibrinogen is a high molecular weight protein (340,000) occurring in the plasma in quantities of 100 to 700 mg./100 ml. Most if not all of the fibrinogen in the circulating blood is formed in the liver, and liver disease occasionally decreases the quantity of circulating fibrinogen, as it does the quantities of prothrombin, which was pointed out previously.

Because of its large molecular size, fibrinogen normally leaks into the interstitial fluids to only a slight extent, and, since it is one of the essential factors in the coagulation process, interstitial fluids ordinarily coagulate poorly if at all. Yet, when the permeability of the capillaries becomes pathologically increased, fibrinogen does then appear in the tissue fluids and in lymph in sufficient quantities to allow clotting of these fluids in much the same way that plasma and whole blood clot.

Action of Thrombin on Fibrinogen to Form Fibrin.   Thrombin is a protein enzyme with proteolytic capabilities. It acts on fibrinogen to remove two low molecular weight peptides from each molecule of fibrinogen, forming a molecule of *activated fibrin*, also called *fibrin monomer*. Many fibrin monomer molecules polymerize rapidly into *long fibrin threads*, which form the *reticulum* of the clot. During this polymerization, *calcium ions* and another factor called *protein stabilizing factor* enhance the bonding between the fibrin monomer molecules, as well as between the polymerized chains themselves, and thereby add stability and strength to the fibrin threads.

The Blood Clot. The clot is composed of a meshwork of fibrin threads running in all directions and entrapping blood cells, platelets, and plasma. The fibrin threads adhere to damaged surfaces of blood vessels; therefore, the blood clot becomes adherent to any vascular opening and thereby prevents blood loss.

Clot Retraction—Serum. Within a few minutes after a clot is formed, it begins to contract and usually expresses most of the plasma from the clot within 30 to 60 minutes. The plasma expressed from the clot is called *serum,* for all of its fibrinogen and most of the other clotting factors have been removed. Serum obviously cannot clot because of lack of these factors.

For reasons not completely understood, large numbers of platelets are necessary for clot retraction to occur. Electron micrographs of platelets in blood clots show that they become attached to the fibrin threads in such a way that they actually bond different threads together. Furthermore, platelets contain tremendous quantities of the high energy compound adenosine triphosphate. It has been suggested that the energy in this compound plays a role in bonding the successive molecules in the fibrin threads or that it causes increased folding of the threads, thereby decreasing the lengths of the threads and thus expressing the serum from the clot.

As the clot retracts, the edges of the broken blood vessel are pulled together, thus contributing to the ultimate state of hemostasis.

THE VICIOUS CYCLE OF CLOT FORMATION

Once a blood clot has started to develop, it normally extends within minutes into the surrounding blood. That is, the clot itself initiates a vicious cycle to promote more clotting. One of the most important causes of this is the fact that the proteolytic action of thrombin allows it to act on many of the other blood clotting factors in addition to fibrinogen. For instance, thrombin has a direct proteolytic effect on prothrombin itself, tending to split this into still more thrombin, and it acts on some of the blood clotting factors responsible for the formation of prothrombin activator. Thus, once a critical amount of thrombin is formed, a vicious cycle develops that causes still more thrombin to be formed, and the blood clot continues to grow until something stops its growth.

Blockage of Clot Growth in Flowing Blood. Fortunately, when a clot develops inside a blood vessel, the vicious cycle of clot formation occurs only where the blood is not moving, because flowing blood carries the thrombin and other procoagulants away so rapidly that their concentrations cannot rise high enough to promote further clotting. Thus, extension of a clot almost always stops when it comes in contact with blood that is flowing faster than a certain velocity.

INITIATION OF COAGULATION: FORMATION OF PROTHROMBIN ACTIVATOR

Though we know most of the chemical events for the formation of the clot itself, the basic control system that turns the clotting process on—that is, the basic process for the formation of prothrombin activator—is still mainly shrouded in mystery. In the following few paragraphs we will attempt to relate what is known about this, even though the chemical interrelationships are still poorly understood.

The Extrinsic and Intrinsic Mechanisms for Initiating Clotting

Blood clotting can be initiated in two separate ways: (1) by the *extrinsic mechanism,* in which an extract from the damaged tissues is mixed with the blood, or (2) by the *intrinsic mechanism,* in which the blood itself is traumatized.

It is primarily the extrinsic mechanism that causes clotting when a blood vessel is damaged, since the broken wall of the blood vessel exudes a tissue extract that initiates the clotting process.

It is primarily the intrinsic mechanism that initiates clotting of blood that has been removed from the body and held in a container. If the walls of this container are completely nonwettable, such as a siliconized and very clean vessel, then clotting of the blood may be delayed as long as an hour or more. On the other hand, if the surface of the vessel is easily wettable, such as the usual glass surface, clotting usually begins within a few minutes. Contact between certain elements within the blood and the vessel wall initiates this clotting process.

The extrinsic clotting process usually occurs

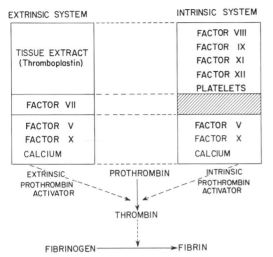

Figure 12–3 The extrinsic and intrinsic systems for blood clotting, illustrating the different factors that are necessary for each of these two systems. (Modified from Hougie: Fundamentals of Blood Coagulation in Clinical Medicine. McGraw-Hill Book Co.)

much more rapidly than the intrinsic one, illustrating that the clotting factors derived from tissue extracts are very powerful activators of the clotting mechanism.

Figure 12–3 depicts the basic essentials of the extrinsic and intrinsic systems for initiating clotting, illustrating that many different factors, most of which are special types of proteins in the blood plasma, supposedly enter into the reactions that initiate clotting. Note that the final products for activation of clotting are called, respectively, *extrinsic* and *intrinsic prothrombin activators*. It is possible that these two activators are the same chemical substance, but it is more likely that each of them is really a combination of several of the individual factors in the system. Furthermore, these separate factors perhaps activate separate parts of the chemical process for conversion of prothrombin into thrombin; for instance, it was just pointed out in relation to the formation of thrombin that some blood factors seem to accelerate the formation of autoprothrombin C and others accelerate the effect of autoprothrombin C on prethrombin to convert this into thrombin. Thus, the term prothrombin activator is actually a collective term meaning the sum of all the factors within a particular system that causes the breakdown of prothrombin into thrombin.

Characteristics of Different Clotting Factors. The clotting factors illustrated in the extrinsic and intrinsic schemata of Figure 12–3 are of three different types: calcium ions, phospholipids, and proteins. The functions of these are as follows.

CALCIUM IONS. In the absence of calcium ions neither the extrinsic nor the intrinsic system will operate. However, in normal blood, calcium ions are almost always present in sufficient excess that this is almost never a factor in determining the kinetics of the clotting reactions. Note that calcium does not actually enter into any of the reactions other than to act as a cofactor to cause the reactions to take place.

PHOSPHOLIPIDS. In both the extrinsic and intrinsic systems, a phospholipid plays a major role in the formation of the final prothrombin activator. In the extrinsic system, this phospholipid is in thromboplastin. In the intrinsic system, it is in platelet factor 3, which is a product of the platelets. In both instances, the phospholipid responsible for the reactions is either a cephalin or a cephalin-like substance.

PROTEIN. All the blood clotting factors from factor V through factor XII are plasma proteins; most of these are beta globulins, but in a few instances are alpha or gamma globulins. Certain of these deserve special mention.

Factor V is a labile substance easily destroyed by slight heat and therefore is often called *labile factor*.

Factors VII, IX, and X have characteristics similar to those of prothrombin and are formed in the liver along with prothrombin. Indeed, some research workers believe that these three factors actually represent variant forms of prothrombin. All these three factors as well as prothrombin itself require vitamin K for their formation, and in the absence of vitamin K, their concentrations in the plasma diminish rapidly.

Absence of or malfunction of any one of the blood clotting factors from the plasma can cause a bleeding tendency, as is discussed later in the chapter. Factor VIII is most frequently implicated; when it fails to function the person has the disease called *hemophilia*. For this reason, factor VIII is frequently called *antihemophilic factor*.

Almost every investigator working in the field of blood coagulation develops new terms for different blood clotting factors. For this reason, there is tremendous confusion regarding nomenclature. The nomenclature using Roman numerals used in this chapter was adopted by an international convention on blood clotting, but synonyms in common use for the different clotting factors are given in Table 12–1.

Thromboplastin, the Initiating Factor in the Extrinsic System for Clotting. The clotting substance (or substances) in a tissue extract

TABLE 12-1. Clotting Factors in the Blood and their Synonyms

Clotting Factor	Synonym
Fibrinogen	Factor I
Platelets	Platelet factor 3
Prothrombin	Factor II
Factor V	Proaccelerin; labile factor; Ac-globulin; Ac-G
Factor VII	Serum prothrombin conversion accelerator; SPCA; proconvertin; autoprothrombin I
Factor VIII	Antihemophilic factor; AHF: antihemophilic globulin; AHG
Factor IX	Plasma thromboplastin component; PTC; Christmas factor; CF; autoprothrombin II
Factor X	Stuart factor; Stuart-Prower factor; Prower factor; autoprothrombin I_c
Factor XI	Plasma thromboplastin antecedent; PTA
Factor XII	Hageman factor
Prothrombin activator	Thrombokinase; complete thromboplastin

that initiates the extrinsic mechanism for clotting is called *thromboplastin*. This is composed primarily of lipoproteins containing one or more phospholipids. The phospholipid *cephalin* is especially active in promoting clotting.

However, thromboplastin by itself is not capable of splitting prothrombin into thrombin but instead must interact with several clotting factors of the blood itself before it becomes prothrombin activator. The following plasma factors are required in this process: *factor V, factor VII, factor X, and calcium ions*.

Blood Factors that Initiate the Intrinsic System for Blood Coagulation. In the intrinsic system, blood platelets play an integral role in the clotting mechanism. When platelets adhere to a broken surface of the blood vessel, many of them disintegrate and release into the blood *platelet factor 3*, which contains a phospholipid that is similar in many ways to the thromboplastin of tissue extract. However, platelet factor 3 lacks the intensity of tissue thromboplastin in initiating the clotting process, and it too must interact with several other blood factors before clotting will occur. Two of these factors are the same as those utilized for the extrinsic system, *factor V* and *factor X*; four additional factors are *factors VIII, IX, XI, and XII*.

In addition to the ability of platelets to disintegrate and thereby help activate the intrinsic mechanism for blood clotting, two other factors are supposedly involved in activating this system under some conditions. These are *factors XI and XII*. When they come in contact with a roughened or wettable surface, such as the glass wall of the test tube, they react together to form a *contact activation product*; this then acts as an enzyme to activate the other factors in the intrinsic system. When a person does not have one of these two factors, the intrinsic system for blood clotting becomes defective.

Summary

Even though the detailed mechanisms for initiating blood clotting are unknown, the physical factors that initiate clotting are well understood: namely, trauma to tissues, trauma to the endothelial surface of the vascular wall, or trauma to some of the intrinsic clotting factors of the blood itself.

Most of what is known about the different blood clotting factors has been derived from the discovery of persons who have deficiencies of one or more of these factors. When the factors are absent, specific peculiarities of the clotting process appear.

PREVENTION OF BLOOD CLOTTING IN THE NORMAL VASCULAR SYSTEM—THE INTRAVASCULAR ANTICOAGULANTS

Endothelial Surface Factors. Probably the two most important factors for preventing clotting in the normal vascular system are, first, the smoothness of the endothelium which prevents contact activation of the intrinsic clotting system and, second, a monomolecular layer of negatively charged protein adsorbed to the inner surface of the endothelium that repels the clotting factors and platelets, thereby preventing activation of clotting. When the endothelial wall is damaged, its smoothness and its negative electrical charge are both lost, which is believed to initiate the reaction between factors XI and XII, forming a contact activation product that initiates the intrinsic system of blood clotting.

Antithrombin and the Antithrombin Action of Fibrin. Among the most important anticoagulants in the blood itself are those that remove thrombin from the blood. The two most powerful of these are the *fibrin threads* that are formed during the process of clotting and an alpha globulin called *antithrombin*.

While a clot is forming, approximately 85 to 90 per cent of the thrombin formed from the prothrombin becomes adsorbed to the fibrin threads as they develop. This obviously prevents the spread of thrombin into the remaining blood and therefore prevents excessive spread of the clot.

The thrombin that does not adsorb to the fibrin threads combines with antithrombin and thereby becomes inactivated during the next 12 to 20 minutes.

Heparin. Heparin, a powerful anticoagulant, is a conjugated polysaccharide found in the cytoplasm of many types of cells, including even the cytoplasm of unicellular animals. Therefore, heparin probably is produced by many different cells of the human body, though especially large quantities are formed by the basophilic *mast cells* located in the pericapillary connective tissue throughout the body. It is believed that these mast cells continually secrete small quantities of heparin and that the heparin then diffuses into the circulatory system. The *basophil cells* of the blood, which seem to be functionally almost identical with the mast cells, possibly also release minute quantities of heparin into the plasma, but the number of basophils is so slight that their importance for heparin secretion is probably negligible.

Mast cells are extremely abundant in the tissue surrounding the capillaries of the lungs and to a less extent the capillaries of the liver. It is easy to understand why large quantities of heparin might be needed in these areas, for the capillaries of the lungs and liver receive many embolic clots formed in the slowly flowing venous blood; sufficient formation of heparin might prevent further growth of the clots.

The concentration of heparin in normal blood, on the basis of yet inaccurate methods of measurement, has been estimated to be as much as 0.01 mg./100 ml. of blood. Though this concentration is 10 to 100 times less than that often used clinically to prevent blood clotting, it is probably sufficient to aid in preventing blood coagulation in the normal circulatory system, for only minute quantities of procoagulants are normally formed, and only minute amounts of heparin are therefore needed to prevent clotting.

MECHANISM OF HEPARIN ACTION. We still do not know all the different mechanisms by which heparin prevents blood coagulation. However, it is known to have at least the following four effects:

First, in low concentrations, heparin prevents the formation of intrinsic prothrombin activator, though the mechanism of this is unknown.

Second, heparin, in association with an albumin cofactor, inhibits the action of thrombin on fibrinogen and thereby prevents the conversion of fibrinogen into fibrin threads.

Third, heparin increases the rapidity with which thrombin interacts with antithrombin and therefore helps in the deactivation of thrombin.

And, fourth, heparin increases the amount of thrombin adsorbed by fibrin.

LYSIS OF BLOOD CLOTS — PLASMIN

The plasma proteins contain a euglobulin called *plasminogen* or *profibrinolysin* which, when activated, becomes a substance called *plasmin* or *fibrinolysin*. Plasmin is a proteolytic enzyme that resembles trypsin, the most important digestive enzyme of pancreatic secretion. It digests the fibrin threads and also digests other substances in the surrounding blood, such as fibrinogen, factor V, factor VIII, prothrombin, and factor XII. Therefore, whenever plasmin is formed in the blood, it can cause lysis of a clot and also destruction of the clotting factors, thereby causing hypocoagulability of the blood.

Activation of Plasmin and Lysis of Clots. When a clot is formed a large amount of plasminogen is incorporated in the clot along with other plasma proteins. However, this will not become plasmin and will not cause lysis of the clot until it is activated. Fortunately, the tissues contain substances that can activate plasminogen to plasmin, though the exact natures of these are not known. Nevertheless, within a day or two after blood has leaked into a tissue and clotted, these activators cause the formation of plasmin which in turn dissolutes the clot.

Clots that occur inside blood vessels can also be dissoluted, though this occurs less often than does dissolution of tissue clots. This illustrates that activator systems also occur within the blood itself.

An activator called *urokinase* is found in the urine and is believed to be important in the lysis of clots that develop in the urinary system. It is possible that urokinase also plays a role as an intravascular activator before it is excreted into the urine by the kidneys.

Certain bacteria also release activator enzymes; for instance, streptococci release a substance called *streptokinase,* which acts on plasminogen to form plasmin. When streptococcal infection occurs in the tissues, the plasmin generated in response to the streptokinase dissolves clotted lymph and clotted tissue fluids and allows the streptococci to spread extensively through the tissues instead of being blocked by the body's "walling off" process described in Chapter 9.

Significance of the Fibrinolysin System. The lysis of blood clots allows rapid clearing of extraneous blood from the tissues and in a few instances allows reopening of clotted vessels. Unfortunately, reopening of large vessels occurs only rarely. Perhaps an important function of the fibrinolysin system is to remove very minute clots from the millions of tiny peripheral vessels that eventually would all become occluded were there no way to cleanse them.

CONDITIONS THAT CAUSE EXCESSIVE BLEEDING IN HUMAN BEINGS

Excessive bleeding can result from deficiency of any one of the many different blood clotting factors. Three particular types of bleeding tendencies that have been studied to the greatest extent will be discussed: (1) bleeding caused by vitamin K deficiency, (2) hemophilia, and (3) thrombocytopenia (platelet deficiency).

DECREASED PROTHROMBIN, FACTOR VII, FACTOR IX, AND FACTOR X—VITAMIN K DEFICIENCY

Hepatitis, cirrhosis, acute yellow atrophy, and other diseases of the liver, can all depress the formation of prothrombin and factors VII, IX, and X so greatly that the patient develops a severe tendency to bleed.

Another cause of depressed levels of these substances is vitamin K deficiency. Vitamin K is necessary for some of the intermediate stages in the formation of all of them. Some research workers believe that prothrombin is the primary product formed and that the other factors are variants of prothrombin. Fortunately, vitamin K is continually synthesized in the gastrointestinal tract by bacteria so that vitamin K deficiency rarely if ever occurs because of its absence from the diet. However, vitamin K deficiency does often occur as a result of poor absorption of fats from the gastrointestinal tract because vitamin K is fat soluble and ordinarily is absorbed into the blood along with the fats.

One of the most prevalent causes of vitamin K deficiency is failure of the liver to secrete bile into the gastrointestinal tract (which occurs either as a result of obstruction of the bile ducts or as a result of liver disease), for lack of bile prevents adequate fat digestion and absorption. Therefore, liver disease often causes decreased production of prothrombin and the other factors both because of poor vitamin K absorption and because of dysfunctional liver cells. Because of this, vitamin K is injected into all patients with liver disease or obstructed bile ducts prior to performing any surgical procedure. Ordinarily, if vitamin K is given to a deficient patient four to eight hours prior to operation and the liver parenchymal cells are at least one-half normal, sufficient clotting factors will be produced to prevent excessive bleeding during the operation.

HEMOPHILIA

The term hemophilia is loosely applied to several different hereditary deficiencies of coagulation, all of which cause bleeding tendencies hardly distinguishable one from the other. The three most common causes of the hemophilic syndrome are deficiency of (1) factor VIII (classical hemophilia)—about 75 per cent of the total, (2) factor IX—about 15 per cent, and (3) factor XI—about 5 to 10 per cent.

Many persons with hemophilia die in early life, though still many others with less severe bleeding have a normal life span. Very commonly, the person's joints become severely damaged because of repeated joint hemorrhage following exercise.

Regardless of the precise cause of the different types of hemophilia, transfusion of normal plasma into the hemophilic person temporarily relieves his bleeding tendency for a few days.

THROMBOCYTOPENIA

Thrombocytopenia means the presence of a very low quantity of platelets in the circulatory system. Persons with thrombocytopenia have a tendency to bleed as do hemophiliacs, except that the bleeding is usually from many small capillaries rather than from large vessels as in hemophilia. As a result, small punctate hemorrhages occur throughout all the body tissues. The skin of such a person displays many small, purplish blotches, giving the disease the name *thrombocytopenic purpura.* It will be remembered that platelets are especially important for repair of minute breaks in capillaries and other small vessels. Indeed, platelets can agglutinate to fill such ruptures without actually causing clots.

Ordinarily, excessive bleeding does not occur until the number of platelets in the blood falls below a value of approximately 70,000 per cubic millimeter rather than the normal of 150,000 to 350,000.

Even without making specific platelet counts on the blood, one can sometimes demonstrate the existence of thrombocytopenia by simply noting whether or not a clot of the person's blood retracts, for, as pointed out above, clot retraction is dependent upon the presence of large numbers of platelets entrapped in the fibrin mesh of the clot.

Most persons with thrombocytopenia have the disease known as *idiopathic thrombocytopenia,* which means simply "thrombocytopenia of unknown cause." However, in the past few years it has been discovered that in most of these persons specific antibodies are destroying the platelets. Occasionally these have developed because of transfusions from other persons, but usually they result from development of autoimmunity to the person's own platelets, the cause of which, however, is not known.

In addition to idiopathic thrombocytopenia, the number of thrombocytes (platelets) in the blood may be greatly depressed by any abnormality that causes aplasia of the bone marrow. For instance, *irradiation injury* to the bone marrow, aplasia of the bone marrow resulting from *drug sensitivity,* and even *pernicious anemia,* can all cause sufficient decrease in the total number of platelets that thrombocytopenic bleeding results.

Relief from bleeding for one to four days can often be effected in the thrombocytopenic patient by giving him *whole blood transfusions.* To do this, the blood must be removed from the donor into a siliconized chamber and then rapidly placed in the recipient so that the platelets are damaged as little as possible. *Cortisone,* which suppresses immune reactions, is often beneficial in the idiopathic type of thrombocytopenia. And, *splenectomy* often helps because the spleen removes large numbers of platelets, particularly damaged ones, from the blood.

THROMBOEMBOLIC CONDITIONS IN THE HUMAN BEING

Thrombi and Emboli. An abnormal clot that develops in a blood vessel is called a *thrombus.* Once a clot has developed, continued flow of blood past the clot is likely to break it away from its attachment, and such freely flowing clots are known as *emboli.* Emboli generally do not stop flowing until they come to a narrow point in the circulatory system. Thus, emboli originating in large arteries or in the left side of the heart eventually plug either smaller arteries or arterioles. On the other hand, emboli originating in the venous system or in the right side of the heart flow into the vessels of the lung to cause pulmonary embolism.

CAUSES OF THROMBOEMBOLIC CONDITIONS. The causes of thromboembolic conditions in the human being are usually two-fold: First, any *roughened endothelial surface of a vessel* — as may be caused by arteriosclerosis, infection, or trauma — is likely to initiate the clotting process. Second, blood often clots *when it flows very slowly* through blood vessels, for small quantities of thrombin and other procoagulants are always being formed. If the blood is flowing too slowly, the concentrations of the procoagulants in local areas often rise high enough to initiate clotting, but when the blood flows rapidly these are rapidly mixed with large quantities of blood and never reach sufficient concentrations to cause clotting.

FEMORAL THROMBOSIS AND MASSIVE PULMONARY EMBOLISM

Because clotting almost always occurs when blood flow is blocked in any vessel of the body, the immobility of bed patients plus the practice of propping the knees up with pillows often causes stasis of blood in one or more of the leg veins for as much as an hour at a time, and this stasis initiates the clotting process. Then the clot grows in all directions, especially in the direction of the slowly moving blood, sometimes growing the entire length of the leg veins and occasionally even into the common iliac vein and inferior vena cava. Then, about 1 time out of every 10, the clot disengages from its attachments to the vessel wall and flows freely with the venous blood into the right side of the heart and thence into the pulmonary arteries to cause *massive pulmonary embolism.* If the clot is large enough to occlude both the pulmonary arteries, immediate death ensues. If only one pulmonary artery or a smaller branch is blocked, death may not occur, or the embolism may lead to death a few hours to several days later because of further growth of the clot within the pulmonary vessels.

ANTICOAGULANTS FOR CLINICAL USE

In some thromboembolic conditions it is desirable to delay the coagulation process to a certain degree. Therefore, various anticoagulants have been developed for treatment of these conditions. The ones most useful clinically are heparin and Dicumarol.

HEPARIN AS AN INTRAVENOUS ANTICOAGULANT

Commercial heparin is extracted from animal tissues, especially from the lungs, and is prepared in almost pure form. Injection of relatively small quantities, approximately 0.5 to 1 mg. per kilogram of body weight, causes the blood clotting time to increase from a normal of approximately 6 minutes to 30 or more minutes. Furthermore, this change in clotting time occurs instantaneously, thereby immediately preventing further development of the thromboembolic condition.

The action of heparin lasts approximately three to four hours. It is believed that the injected heparin is destroyed by an enzyme in the blood known as *heparinase.* Also, much of the injected heparin diffuses into the interstitial fluids and therefore becomes unavailable as a blood anticoagulant.

In the treatment of a patient with heparin, too much heparin is sometimes given, and *serious* bleeding crises occur. In these instances, *protamine* acts specifically as an antiheparin, and the clotting mechanism can be reverted to normal by administering this substance. This substance probably combines with heparin and inactivates it because it carries positive electrical charges, whereas heparin carries negative charges.

DICUMAROL AS AN ANTICOAGULANT

When dicumarol is given to a patient, the plasma levels of prothrombin and factors VII, IX, and X, all formed by the liver, begin to fall, indicating that Dicumarol has a potent depressant effect on liver formation of all these compounds.

Dicumarol causes this effect by competing with vitamin K for reactive sites in the intermediate processes for formation of prothrombin and the other three clotting factors, thereby blocking the action of vitamin K.

After administration of an effective dose of Dicumarol, the coagulant activity of the blood decreases to approximately 50 per cent of normal by the end of 12 hours and to approximately 20 per cent of normal by the end of 24 hours. In other words, the coagulation process is not blocked immediately, but must await the consumption of the prothrombin and other factors already present in the plasma. Normal coagulation returns one to three days after discontinuing therapy.

PREVENTION OF BLOOD COAGULATION OUTSIDE THE BODY

Though blood removed from the body normally clots in three to six minutes, blood collected in *siliconized containers* often does not clot for as long as an hour or more. The reason for this delay in clotting is that preparing the surfaces of the containers with these substances prevents contact activation of factors XI and XII, which initiates the intrinsic clotting mechanism. On the other hand, untreated glass containers allow contact activation and rapid development of clots.

Heparin can be used for preventing coagulation of blood outside the body as well as in the body, and heparin is occasionally used as an anticoagulant when blood is removed from a donor to be transfused later into a recipient. Also, heparin is used in almost all animal experiments in which the blood is passed through external tubes and then back into the animal. In this instance, the quantity of heparin that must be administered to the animal is 3 to 5 mg. per kilogram of body weight.

Various substances that *decrease the concentration of calcium ions* in the blood can be used for preventing blood coagulation outside the body. For instance, soluble *oxalate* compounds mixed in very small quantity with a sample of blood cause precipitation of calcium oxalate from the plasma and thereby decrease the ionic calcium level so much that blood coagulation is blocked.

A second calcium deionizing agent used for preventing coagulation is *sodium, ammonium,* or *potassium citrate.* The citrate ion combines with calcium in the blood to cause an unionized calcium compound, and the lack of ionic calcium prevents coagulation. Citrate anticoagulants have a very important advantage over the oxalate anticoagulants, for oxalate is toxic to the body, whereas small quantities of citrate can be injected intravenously. After injection, the citrate ion is removed from the blood within a few minutes by the liver and is polymerized into glucose, and then metabolized in the usual manner. Consequently, 500 ml. of blood that has been rendered incoagulable by sodium citrate can ordinarily be injected into a recipient within a few minutes without any dire consequences. If the liver is damaged or if large quantities of citrated blood or plasma are given too rapidly, the citrate ion may not be removed quickly enough, and the citrate can then greatly depress the level of calcium ion in the blood, which results in tetany and convulsive death.

BLOOD COAGULATION TESTS

BLEEDING TIME

When a sharp knife is used to pierce the tip of the finger or lobe of the ear, bleeding ordinarily lasts three to six minutes. However, the time depends largely on the depth of the wound and on the degree of hyperemia in the finger at the time of the test. Consequently, "bleeding times" are not very significant in studying the ability of the blood to coagulate.

CLOTTING TIME

Many methods have been devised for determining clotting times. One most widely used is to collect blood in a chemically clean glass test tube and then to tip the tube back and forth approximately every 30 seconds until the blood has clotted. By this method, the normal clotting time ranges between five and eight minutes.

Procedures using multiple test tubes have been devised for determining clotting time more accurately. However, none of the clotting times is accurate because they are dependent to a high degree on the condition of the glass itself and even on the size of the tube.

PROTHROMBIN TIME

The prothrombin time gives an indication of the total quantity of prothrombin in the blood. Figure 12–4 shows the relationship of prothrombin concentration to prothrombin time. The means for determining prothrombin time is the following:

Blood removed from the patient is immediately oxalated so that none of the prothrombin can change into thrombin. At any time later, a large excess of calcium ion and tissue extract (thromboplastin) is suddenly mixed with the oxalated blood. The calcium nullifies the effect of the oxalate, and the thromboplastin activates the prothrombin-to-thrombin reaction. The time required for coagulation to take place is known as the "prothrombin time." The normal prothrombin time is approximately 12 seconds, though this depends to a certain extent on the exact procedure employed. In each laboratory a curve relating concentration to prothrombin time, such as that shown in Figure 12–4, is ordinarily drawn for the method used so that the significance of the prothrombin time can be evaluated.

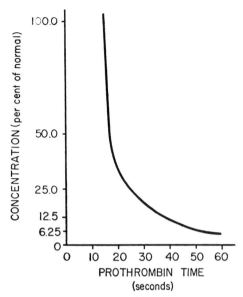

Figure 12-4 Relationship of prothrombin concentration in the blood to the prothrombin time. (Redrawn from Marple: Thromboembolic Conditions and Their Treatment with Anticoagulants. Charles C Thomas.)

OTHER TESTS RELATING TO COAGULATION

The normal *platelet count* in the blood as measured by most platelet-counting techniques ranges between approximately 150,000 and 300,000 per cubic mm. Obviously, if a decreased platelet count is found, thrombocytopenia exists.

Another factor of importance in so far as platelets are concerned is *clot retraction*. Once a clot has formed in the bottom of a test tube, clot retraction normally occurs in 30 to 60 minutes. If clot retraction does not occur, the number of platelets in the blood may be greatly reduced.

Tests similar to that for prothrombin time have been devised to determine the relative quantities of other clotting factors in the blood. In performing the tests excesses of all the factors besides the one being tested are added to oxalated blood all at once, and then the time of coagulation is determined in the same manner as the usual prothrombin time. If the factor is deficient, the time will be considerably prolonged. In determining the usual prothrombin time, a deficiency of some of these other factors can also prolong the measured time. Therefore, an increased prothrombin time as performed in the usual hospital laboratory does not always mean decreased quantity of prothrombin but may mean decreased quantity of some other factor.

REFERENCES

Atencio, A. C.; Chao, P. Y.; Chen, A. Y.; and Reeve, E. B.: Fibrinogen response to corticotropin preparations in rabbits. *Amer. J. Physiol.*, *216*:773, 1969.

Atencio, A. C.; Joiner, K.; and Reeve, E. B.: Experimental and control systems studies of plasma fibringen regulation in rabbits. *Amer. J. Physiol.*, *216*:764, 1969.

Barnhart, M. I.: Protein synthesis: an example of hepatic function. *J. Histochem. Cytochem.*, *13*:740, 1965.

Biggs, Rosemary: The nature of blood coagulation and its defects. *In* Bittar, E. Edward, and Bittar, Neville (eds.): The Biological Basis of Medicine. New York, Academic Press, Inc., 1969, Vol. 3, p. 183.

Biggs, R., and Macfarlane, R. G.: Treatment of Haemophilia and Other Coagulation Disorders. Philadelphia, F. A. Davis Co., 1965.

Crowell, J. W., and Houston, B.: Effect of acidity on blood coagulation. *Amer. J. Physiol.*, *201*:379, 1961.

Dietrich, C. P.: Enzymatic degradation of heparin. *Biochem. J.*, *111*:91, 1969.

Dodds, W. J.; Packham, M. A.; Rowsell, H. C.; and Mustard, J. F.: Factor VII survival and turnover in dogs. *Amer. J. Physiol.*, *213*:36, 1967.

Fibrinogen and fibrin (Symposium). *Fed. Proc.*, *24*:783, 1965.

Hoak, J. C.; Warner, E. D.; and Connor, W. E.: Platelets, fatty acids, and thrombosis. *Circ. Res.*, *20*:11, 1967.

Hurt, J. P., and Krigman, M. R.: Selected procoagulants in the guinea pig. *Amer. J. Physiol.*, *218*:832, 1970.

Jaques, L. B.: Effects of hormones and drugs on hemostasis. *In* Drugs in Relation to Blood Coagulation, Haemostasis, and Thrombosis. New York, Pergamon Press, Inc., 1968.

Kazmier, F. J.; Spittel, J. A., Jr.; Bowie, E. J. W.; Thompson, J. H., Jr.; and Owen, C. A., Jr.: Release of vitamin-K dependent coagulation factors by isolated perfused rat liver. *Amer. J. Physiol.*, *214*:919, 1968.

Kline, D. L.: Blood coagulation: reactions leading to prothrombin. *Ann. Rev. Physiol.*, *27*:285, 1965.

McDonagh, J.; Kiesselbach, T. H.; and Wagner, R. H.: Factor XIII and antiplasmin activity in human platelets. *Amer. J. Physiol.*, *216*:508, 1969.

McMillan, C. W., and Mason, R. G.: Plasma factor VIII activity in newborn dogs. *Amer. J. Physiol.*, *214*:395, 1968.

Mills, D. C. B.: Platelet aggregation. *In* Bittar, E. Edward, and Bittar, Neville (eds.): The Biological Basis of Medicine. New York, Academic Press, Inc., 1969, Vol. 3, p. 163.

Morrison, J. T.; Colvin, H. W., Jr.; Skaife, D. D.; and Marsh, R. E.: Effect of bishydroxycoumarin and diet on blood coagulation in ground squirrels. *Amer. J. Physiol.*, *216*:429, 1969.

Moschos, C. B.; Lehan, P. H.; Koroxenidis, G. T.; Weisse, A. B.; Oldewurtel, H. A.; and Regan, T. J.: Regional and systemic clotting and fibrinolytic activity after myocardial injury. *Amer. J. Physiol.*, *216*:308, 1969.

Nossel, H. L.: The Contact Phase of Blood Coagulation. Philadelphia, F. A. Davis Co., 1964.

Quick, A. J.: Hemorrhagic Diseases and Thrombosis. 2nd ed., Philadelphia, Lea & Febiger, 1966.

Seegers, W. H. (ed.): Blood Clotting Enzymology. New York, Academic Press, Inc., 1967.

Seegers, W. H.: Blood clotting mechanisms: three basic reactions. *Ann. Rev. Physiol.*, *31*:269, 1969.

Seegers, W. H.: Prothrombin. Cambridge, Massachusetts, Harvard University Press, 1962.

Siddique, I. H., and Cooper, G. W., Jr.: Hemodynamic effect of listerial hemolysin in dogs. *Amer. J. Physiol.*, *216*:1399, 1969.

Silver, M. J.: Platelet aggregation and plug formation: a model test system. *Amer. J. Physiol.*, *218*:384, 1970.

Stohlman, F., Jr.: Conference on platelet transfusions. Introductory remarks. *Transfusion*, *6*:1, 1966.

Vitamin K (Symposium): *Fed. Proc.*, *20*:989, 1961.

PART III

THE HEART

The Heart as a Pump, the Cardiac Cycle, Cardiac Contractility, and Stroke Volume Output

With this chapter we begin discussion of the heart and circulatory system. The heart is a pulsatile, four-chamber pump composed of two atria and two ventricles. The atria function principally as entryways to the ventricles, but they also pump weakly to help move the blood on through the atria into the ventricles. The ventricles supply the main force that propels blood through the lungs and through the peripheral circulatory system.

In previous chapters the basic principles of membrane potentials, action potentials, and muscle contraction have been discussed. Special mechanisms in the heart maintain cardiac rhythmicity and transmit action potentials throughout the cardiac musculature to initiate its contraction. These mechanisms will be explained in detail in the following chapter. In the present chapter we wish to explain how the heart operates as a pump: that is, to explain the function of the muscle, of the valves, and of the various chambers of the heart. Therefore, we will discuss first the basic physiology of cardiac muscle itself, especially how it differs from skeletal muscle, which was discussed in Chapter 7.

PHYSIOLOGY OF CARDIAC MUSCLE

The heart is composed of three major types of cardiac muscle: atrial muscle, ventricular muscle, and specialized excitatory and conductive muscle fibers. The atrial and ventricular types of muscle contract in much the same manner as skeletal muscle fibers. On the other hand, the specialized excitatory and conductive fibers contract only feebly because they contain few contractile fibrils; instead, they provide an excitatory system for the heart and a transmission system for rapid conduction of impulses throughout the heart.

PHYSIOLOGIC ANATOMY OF CARDIAC MUSCLE

Figure 13–1 illustrates a typical histologic section of cardiac muscle, showing arrangement of the cardiac muscle fibers in a latticework, the fibers dividing, then recombining, and spreading in all directions. One notes immediately from this figure that cardiac muscle is *striated* in the same manner as typical skeletal muscle. Furthermore, cardiac muscle has typical myofibrils that contain *actin* and *myosin filaments* almost identical to those found in skeletal muscle, and these filaments interdigitate and slide among each other during the process of contraction in the same manner as occurs in skeletal muscle. (See Chap. 7.)

Cardiac Muscle as a "Functional Syncytium." The angulated dark areas crossing the cardiac muscle fibers in Figure 13–1 are called *intercalated discs;* however, they are actually cell membranes that separate individual cardiac muscle cells from each other. Thus, the cardiac muscle fibers are in reality a series of cardiac muscle cells connected in series with each other. Yet electrical resistance through the intercalated disc is 400 times less than the re-

Intercalated discs

Figure 13–1 The "syncytial" nature of cardiac muscle.

sistance through the outside membrane of the cardiac muscle fiber. Therefore, from a functional point of view, ions flow with relative ease along the axes of the cardiac muscle fibers so that action potentials travel from one cardiac muscle cell to another, past the intercalated discs, without significant hindrance. Therefore, cardiac muscle is a *functional syncytium,* in which the cardiac muscle cells are so tightly bound that when one of these cells becomes excited, the action potential spreads to all of them, spreading from cell to cell and spreading laterally through the latticework interconnections.

The heart is composed of two separate functional syncytiums, the *atrial syncytium* and the *ventricular syncytium.* These are normally connected with each other only by way of the specialized conductive system, the *A-V bundle,* which transmits impulses from the atrial muscle into the ventricular muscle. Other than for this connection, the atrial and ventricular muscles are separated from each other by the fibrous tissue surrounding the valvular rings.

All-or-Nothing Principle as Applied to the Heart. Because of the syncytial nature of cardiac muscle, stimulation of any single atrial muscle fiber causes the action potential to travel over the entire atrial muscle mass, and, similarly, stimulation of any single ventricular fiber causes excitation of the entire ventricular muscle mass. If the A-V bundle is intact, the action potential passes also from atria to ventricles. This is called the all-or-nothing principle; and it is precisely the same as that discussed in Chapter 7 for skeletal muscle, except that the cardiac muscle fibers interconnect with each other, which means that the all-or-nothing principle applies to the entire functional syncytium of the heart rather than to single muscle fibers as is true in the case of skeletal muscle fibers.

ACTION POTENTIALS IN CARDIAC MUSCLE

The *resting membrane potential* of normal cardiac muscle is approximately 80 to 85 mv., and approximately 90 to 100 mv. in the specialized conductive fibers, the Purkinje fibers, which are discussed later.

The total *action potential* recorded in cardiac muscle, shown in Figure 13-2, is about 105 mv., which means that there is a *reversal potential* or "overshoot" beyond the resting potential of about 20 mv. The basic principles of the genesis of resting, action, and reversal potentials were discussed in Chapter 5.

Cardiac muscle has a peculiar type of action

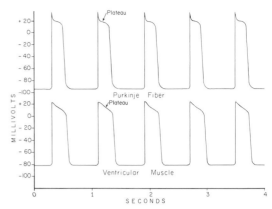

Figure 13-2 Rhythmic action potentials from a Purkinje fiber and from a ventricular muscle fiber recorded by means of microelectrodes.

potential. After the initial *spike* the membrane remains depolarized for 0.15 to 0.3 second, exhibiting a *plateau* as illustrated in Figure 13-2, followed at the end of the plateau by abrupt repolarization. The presence of this plateau in the action potential causes the action potential to last 20 to 50 times as long in cardiac muscle as in skeletal muscle and causes a correspondingly increased period of contraction.

The cause of the plateau in the action potential is yet unknown. However, conductance of the membrane for potassium is less than normal during the period of the plateau, which is exactly opposite to the potassium conductance in large nerve fibers and in skeletal muscle fibers immediately following the spike. The decreased potassium conductance impedes leakage of potassium out of the fiber and therefore prevents immediate recovery from the depolarization process. A speculative theory that has been offered to explain this effect is that the depolarization process in cardiac muscle allows such large quantities of calcium ions to diffuse inward through the membrane that the calcium ion concentration outside the membrane becomes depleted. This depletion in turn supposedly decreases the membrane permeability to potassium. Toward the end of the plateau, enough calcium presumably diffuses from the surrounding extracellular fluids into the depleted area to restore normal membrane permeability, and the normal repolarization process then begins.

Velocity of Conduction in Cardiac Muscle. The velocity of conduction of the action potential in both atrial and ventricular muscle fibers is about 0.3 to 0.4 meter per second, or about 1/300 the velocity in very large nerve fibers and about 1/10 the velocity in skeletal muscle fibers. The velocity of conduction in the specialized conductive system varies from only a small fraction of a meter to several meters per second in different parts of the system, as is explained in the following chapter.

Refractory Period of Cardiac Muscle. Cardiac muscle, like all excitable tissue, is refractory to restimulation *during the spike* of the action potential. An extremely strong electrical stimulus can sometimes initiate a new spike while the plateau of an action potential still exists. However, the normal cardiac impulse does not have the extreme stimulatory power of a high voltage electrical stimulus and cannot re-excite the muscle until the end of the plateau. Therefore, the refractory period of the heart is usually stated in terms of the *functional refractory period,* which is the interval of time during which an action potential from another part of the heart fails to re-excite an already excited area of cardiac muscle. The normal functional refractory period of the ventricle is approximately 0.25 second, which is illustrated to the left in Figure 13–3. There is an additional *relative refractory period* of about 0.05 second during which the muscle is more difficult than normal to excite but nevertheless can be excited.

The functional refractory period of atrial muscle is about 0.15 second, and the relative refractory period another 0.03 second.

CONTRACTION OF CARDIAC MUSCLE

Excitation-Contraction Coupling — Function of Calcium Ions. It is not the action potential itself that causes the myofibrils of cardiac muscle to contract. Instead, it is movement of calcium ions into the myofibrils. This probably occurs in the following manner: When the action potential travels over the muscle, it also causes flow of electrical current to the interior

of the fibers by way of the T tubules that were described in relation to skeletal muscle contraction in Chapter 7. The electrical current, in turn, is believed to cause release of calcium ions from the longitudinal sarcoplasmic reticulum that lies linearly along the myofibrils. These calcium ions diffuse rapidly into the myofibrils and there catalyze the chemical reactions that promote sliding of the actin and myosin filaments along each other; this in turn produces the muscle contraction. Immediately after the action potential is over, the calcium ions begin to bind once again with the longitudinal reticulum, so that within a few milliseconds the density of calcium ions falls below that needed to maintain contraction; as a result, the muscle relaxes. This contractile process and its coupling with the action potential have been discussed in much greater detail in Chapter 7 in relation to skeletal muscle fibers. The basic principles are the same for both skeletal and cardiac muscle contraction.

Relation of Action Potential Strength to Strength of Contraction. The greater the voltage of the action potential and the longer its duration, the greater also is the work output caused by the cardiac muscle contraction. The reason for this seems to be that either increased voltage or increased duration of action potential increases the quantity of calcium ions released to the interior of the muscle fibers, which in turn enhances the chemical reactions of the contractile process itself.

Duration of Contraction. Cardiac muscle begins to contract a few milliseconds after the action potential begins and continues to contract for a few milliseconds after the action potential ends. Therefore, the duration of contraction of cardiac muscle is mainly a function of the duration of the action potential — about 0.15 second in atrial muscle and 0.3 second in ventricular muscle.

Effect of Heart Rate on Duration of Contraction. When the heart rate increases, the duration of each cycle of the heart obviously decreases. The duration of the action potential and the period of contraction also decrease but not as much percentagewise as does the entire cycle period. Therefore, the ratio of contraction time to relaxation time increases as the heart rate increases. At a normal heart rate of 72 beats per minute, the period of contraction is about 0.4 of the entire cycle. At three times normal heart rate, this period is about 0.65 of the entire cycle, which means that the heart under some conditions does not remain relaxed long enough to allow complete filling of the cardiac chambers prior to the next contraction.

Effect of Premature Contraction on Contraction Strength. When the heart muscle is stimulated a second time soon after a previous contraction is over, the strength of the second contraction is usually depressed below that of the first, an effect that is illustrated in the second

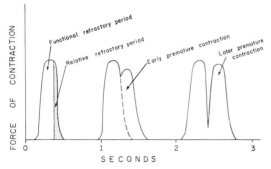

Figure 13–3 Contraction of the heart, showing: the durations of the functional refractory period and the relative refractory period, the effect of an early premature contraction, and the effect of a later premature contraction. Note that the premature contractions do not cause wave summation as occurs in skeletal muscle.

record of Figure 13–3. The presumptive reason for this is that when a second contraction occurs rapidly after the first some of the chemical substances necessary for contraction are slightly depleted, which results in diminished force of contraction. As the interval between two successive contractions increases, as illustrated in the third record of Figure 13–3, the strength of the second contraction rises to approach that of the first and sometimes actually exceeds it.

THE CARDIAC CYCLE

The period from the end of one heart contraction to the end of the next contraction is called the *cardiac cycle*. Each cycle is initiated by spontaneous generation of an action potential in the S-A node, as will be explained in detail in the following chapter. This node is located in the posterior wall of the right atrium near the opening of the superior vena cava, and the action potential travels rapidly from the S-A node through both atria and thence through the A- V node and A- V bundle into the ventricles. However, because of a special arrangement of the conducting system from the atria into the ventricles, there is a delay of more than 1/10 second between passage of the cardiac impulse through the atria and then through the ventricles. This allows the atria to contract ahead of the ventricles, thereby pumping blood into the ventricles prior to the very strong ventricular contraction. Thus, the atria act as primer pumps for the ventricles, and the ventricles then provide the major source of power for moving blood through the vascular system.

SYSTOLE AND DIASTOLE

The cardiac cycle consists of a period of relaxation called *diastole* followed by a period of contraction called *systole*.

Figure 13–4 illustrates the different events during the cardiac cycle. The top three curves show the pressure changes in the aorta, the left ventricle, and the left atrium, respectively. The fourth curve depicts the changes in ventricular volume, the fifth the electrocardiogram, and the sixth a recording of the sounds produced by the heart as it pumps. It is especially important that the student study in detail the diagram of this figure and understand the causes of all the events illustrated. These are explained as follows:

RELATIONSHIP OF THE ELECTROCARDIOGRAM TO THE CARDIAC CYCLE

The electrocardiogram in Figure 13–4 shows the *P, Q, R, S,* and *T* waves, which will be discussed in the following chapters. On look-

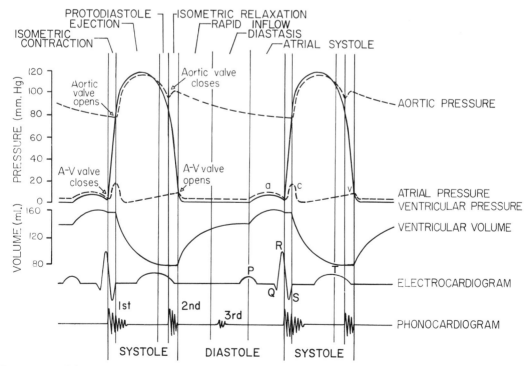

Figure 13–4   The events of the cardiac cycle, showing changes in left atrial pressure, left ventricular pressure, aortic pressure, ventricular volume, the electrocardiogram, and the phonocardiogram.

ing at the atrial pressure curve, it is evident that the P wave begins immediately prior to a rise in atrial pressure; that is, the *P wave* is caused by the *spread of depolarization* through the atria. The state of depolarization, which lasts for about 0.15 second in the atria, causes the atria to contract and raise the atrial pressure. Approximately 0.16 second after the P wave, the *QRS complex* appears as a result of depolarization of the ventricles. This initiates contraction of the ventricles and causes the ventricular pressure to begin rising, as illustrated in the figure. Therefore, the QRS complex begins slightly before the onset of ventricular systole.

Finally, one observes the *ventricular T wave* in the electrocardiogram. This represents the stage of repolarization of the ventricles at which time the ventricular muscle fibers begin to relax. Therefore, the T wave occurs slightly prior to the end of ventricular contraction.

FUNCTION OF THE ATRIA AS PUMPS

Blood normally flows continually from the great veins into the atria, and ordinarily approximately 70 per cent of this flows directly into the ventricles even before the atria contract. Then atrial contraction causes an additional 30 per cent filling. Therefore, the primer function of the atria increases the effectiveness of the ventricles as pumps approximately 40 to 50 per cent. Yet, the heart can continue to operate quite satisfactorily under normal resting conditions even without this extra 40 to 50 per cent effectiveness of pumping. The reason for this is that the heart normally has the capability of pumping 300 to 400 per cent more blood than is required by the body, so that the decrease in pumping effectiveness caused by atrial inactivity hardly affects the ability of the heart to supply the resting needs of the body. Indeed, the person is likely not to notice the difference unless he exercises.

Pressure Changes in the Atria—The a, c, and v waves. If one observes the atrial pressure curve of Figure 13–4, he will note three major pressure elevations called the *a, c,* and *v atrial pressure waves.*

The *a wave* is caused by atrial contraction. Ordinarily, the *right* atrial pressure rises 4 to 6 mm. Hg during atrial contraction, while the *left* atrial pressure rises about 7 to 8 mm. Hg.

The *c wave* occurs when the ventricles begin to contract, and it is caused by three factors: (1) reflux of blood out of the ventricles into the atria during ventricular contraction, (2)

bulging of the A-V valves toward the atria because of increasing pressure in the ventricles, and (3) pulling on the atrial muscle where it is attached to the ventricular muscle by the contracting ventricles.

The *v wave* occurs toward the end of ventricular contraction, and it results from slow buildup of blood in the atria because the A-V valves are closed during this time. Then, when ventricular contraction is over, the A-V valves open, allowing blood to flow rapidly into the ventricles and causing the v wave to disappear.

FUNCTION OF THE VENTRICLES AS PUMPS

Filling of the Ventricles. By the end of systole large amounts of blood have accumulated in the atria; as a consequence the atrial pressures are considerably elevated. Therefore, just as soon as systole is over and the ventricular pressures fall again to their low diastolic values, the high pressures in the atria immediately push open the A-V valves and allow blood to flow rapidly into the ventricles, as shown by the ventricular volume curve in Figure 13–4. This is called the *period of rapid filling of the ventricles.* The atrial pressures fall to within a fraction of a millimeter of the ventricular pressures, because the normal A-V valve openings are so large that they offer almost no resistance to blood flow.

The period of rapid filling lasts approximately the first third of diastole. During the middle third of diastole only a small amount of blood normally flows into the ventricles; this is blood that continues to empty into the atria from the veins and passes on through the atria directly into the ventricles. This middle third of diastole, when the inflow of blood into the ventricles is almost at a standstill, is called *diastasis.*

During the latter third of diastole, the atria contract and give an additional thrust to the inflow of blood into the ventricles; this accounts for approximately 20 to 30 per cent of the filling of the ventricles during each heart cycle.

Emptying of the Ventricles During Systole. PERIOD OF ISOMETRIC CONTRACTION. Immediately after ventricular contraction begins, the ventricular pressure abruptly rises, as shown in Figure 13–4, causing the A-V valves to close. Then an additional 0.02 to 0.03 second is required for the ventricle to build up sufficient pressure to push the semilunar (aortic and pulmonary) valves open against the pressures in the aorta and pulmonary artery. Therefore, during this period of time, contraction is occurring in the ventricles but there is no emptying. This period is called the period of isometric contraction, meaning by this term that tension is increasing in the muscle but no shortening of the muscle fibers is occurring.

PERIOD OF EJECTION. When the left ventricular pressure rises slightly above 80 mm. Hg and the right ventricular arterial pressure slightly above 8 mm. Hg, the ventricular pressures now push the semilunar valves open. Immediately, blood begins to pour out of the ventricles, about half of the emptying occurring during the first quarter of systole and most of the remaining half during the next two quarters. These three-quarters of systole are called the period of ejection.

PROTODIASTOLE. During the last one-fifth to one-fourth of ventricular systole, almost no blood flows from the ventricles into the large arteries; yet the ventricular musculature remains contracted. This period is called protodiastole. The arterial pressure falls during this period, because almost no blood is entering the arteries even though large quantities of blood are flowing from the arteries through the peripheral vessels.

The ventricular pressure actually falls to a value *below* that in the aorta during protodiastole despite the fact that there is a small quantity of blood still remaining in the ventricles. The reason for this is that the blood flowing out of the ventricles has built up momentum. As this momentum is lost during the latter part of systole, the kinetic energy of the momentum is converted into pressure in the large arteries, which makes the arterial pressure slightly greater than the pressure inside the ventricles.

PERIOD OF ISOMETRIC RELAXATION. At the end of systole, ventricular relaxation begins suddenly, allowing the intraventricular pressures to fall rapidly. The elevated pressures in the large arteries immediately push blood back toward the ventricles, which snaps the aortic and pulmonary valves closed. For another 0.03 to 0.06 second, the ventricular muscle continues to relax, and the intraventricular pressures fall rapidly back toward their very low diastolic levels. As the ventricular pressures fall below the atrial pressures, the A-V valves open to begin a new cycle of ventricular pumping.

FUNCTION OF THE VALVES

The *A-V valves* obviously prevent backflow of blood from the ventricles to the atria during systole, and the *semilunar valves* prevent backflow from the aorta and pulmonary arteries into the ventricles during diastole. All these valves, illustrated in Figure 13–5, close and open *passively.* That is, they close when a backward pressure gradient pushes blood backward, and they open when a forward pressure gradient forces blood in the forward direction. For obvious anatomical reasons, the thin, filmy A-V valves require almost no backflow to cause closure while the much heavier semilunar valves require rather rapid backflow.

Function of the Papillary Muscles. Figure 13–5 also illustrates the papillary muscles attached to the vanes of the A-V valves by the *chordae tendineae.* The papillary muscles contract when the ventricular walls contract, and, contrary to what might be expected, they *do not* help to close the valves. Instead, they pull the vanes of the valves inward toward the ventricles to prevent their bulging too far backward toward the atria during ventricular contraction. If a chorda tendinea becomes ruptured or if one of the papillary muscles becomes paralyzed as a result of ischemia or other effects, the valve bulges far backward, sometimes so far that it leaks severely and results in severe or even lethal cardiac incapacity.

The Aortic and Pulmonary Valves. There are some differences between the operation of the aortic and pulmonary valves and that of the A-V valves. First, the high pressure in the arteries at the end of systole causes the semilunar valves to snap to the closed position in

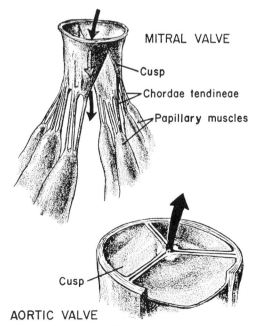

Figure 13–5 Mitral and aortic valves.

comparison with a much softer closure of the A-V valves. Second, the velocity of blood ejection through the aortic and pulmonary valves is far greater than that through the much larger A-V valves. Because of the rapid closure and rapid ejection, the edges of the semilunar valves are subjected to much greater mechanical abrasion than are the A-V valves, which also are supported by the chordae tendineae. It is obvious from the anatomy of the aortic and pulmonary valves, as illustrated in Figure 13–5, that they are well adapted to withstand this extra physical trauma.

THE AORTIC PRESSURE CURVE

The pressure that develops in the aorta depends on many other factors in addition to contraction of the heart. The aortic pressure curve will be discussed in detail in Chapter 19, but it is desirable now simply to point out those features of the curve that relate especially to the cardiac cycle.

When the left ventricle contracts, the ventricular pressure rises rapidly until the aortic valve opens. Then the pressure in the ventricle rises only slightly thereafter, as illustrated in Figure 13–4, because blood immediately flows out of the ventricle into the aorta.

The entry of blood into the arteries causes the walls of these arteries to stretch and the pressure in the arterial system to rise. Then, at the end of systole, even after the left ventricle stops ejecting blood, the extra filling of the arteries maintains high pressure in the arteries while the ventricular pressure falls. This creates a pressure differential backward across the aortic valve, and the aortic valve closes.

A so-called *incisura* occurs in the aortic pressure curve when the aortic valve closes. This is caused by a short period of backward flow of blood immediately prior to closure of the valve.

After the aortic valve has closed, pressure in the aorta falls slowly throughout diastole because blood stored in the distended elastic arteries flows continually through the peripheral vessels back to the veins. By the time the ventricle contracts again, the aortic pressure usually has fallen to approximately 80 mm. Hg (diastolic pressure), which is two-thirds the maximal pressure of 120 mm. Hg (systolic pressure) occurring in the aorta during ventricular contraction.

Obviously, the pressure curve in the pulmonary artery is similar to that in the aorta, except that the pressures are much less, as will be discussed in Chapter 23.

THE VENTRICULAR VOLUME CURVE

Also shown in Figure 13–4 is a curve depicting the relative changes in ventricular volume from instant to instant. Such a curve can be recorded by placing a cylinder, called an "oncometer," around the ventricle. The cylinder is closed at one end and has a snug rubber gasket between its other end and the atrioventricular ring of the heart. Every time the ventricles increase in size, they displace air from the cylinder. The cylinder is connected by means of a rubber tube to a tambour recording system, and the increase and decrease in ventricular volume is recorded.

End-Diastolic and End-Systolic Volumes of the Ventricles.   During diastole, filling of the ventricles normally increases the volume of each ventricle to about 120 to 130 ml. This volume is known as the *end-diastolic volume*. Then, as the ventricles empty during systole, the volume decreases about 70 ml., which is called the *stroke volume*. The remaining volume in each ventricle, about 50 to 60 ml. is called the *end-systolic volume.*

When the heart contracts strongly, the end-systolic volume can fall to as little as 10 to 30 ml. On the other hand, when large amounts of blood flow into the ventricles during diastole, their end-diastolic volumes can become as great as 200 to 250 ml. in the normal heart. Thus, by increasing the end-diastolic volume and decreasing the end-systolic volume, the stroke volume output can be increased to more than double normal.

RELATIONSHIP OF THE HEART SOUNDS TO HEART PUMPING

When one listens to the heart with a stethoscope, he does not hear the opening of the valves, for this is a relatively slowly developing process that makes no noise. However, when the valves close, the vanes of the valves and the surrounding fluids vibrate under the influence of the sudden pressure differentials which develop, giving off sound that travels in all directions through the chest. When the ventricles first contract, one hears a sound that is caused by closure of the A-V valves. The vibration is low in pitch and relatively long continued and is known as the *first heart sound.* When the aortic and pulmonary valves close, one hears a relatively rapid snap, for these valves close extremely rapidly, and the surroundings vibrate for only a short period of time. This sound is known as the *second heart sound.* The precise causes of these sounds will be discussed in Chapter 30 in relation to auscultation.

Occasionally, one can hear an *atrial sound* when the atria beat, presumably because of vibrations associated with the flow of blood into the ventricles. Also, a *third heart sound* sometimes occurs at the end of the first third of diastole or in the middle of diastole. This is said to be caused by blood flowing with a rumbling motion into the almost-filled ventricle. The atrial sound and the third heart sound can usually be recorded with special recording instruments but can be heard with the stethoscope only with great difficulty.

The Terms "Systole" and "Diastole" in Clinical Usage. Strictly speaking, systole means "contraction." Therefore, physiologically it is probably best that systole be considered to begin approximately with the closure of the A-V valves and to end approximately with the opening of the A-V valves.

Clinically, it is not possible to determine when the A-V valves open, but it is easy to determine when the A-V valves and the aortic and pulmonary valves close by listening to the heart sounds. Because only a short interval elapses between the closure of the aortic and pulmonary valves and the opening of the A-V valves, the clinician measures systole as the time interval between the first heart sound and the second heart sound or, in other words, the time interval between closure of the A-V valves and closure of the semilunar valves. Diastole is considered to be the interval between closure of the aortic and pulmonary valves and closure of the A-V valves.

STROKE VOLUME OUTPUT OF THE HEART

The stroke volume output of the heart is the quantity of blood pumped from each ventricle with each heartbeat. Normally this is about 70 ml., but under conditions compatible with life it can decrease to as little as a few ml. per beat and can increase to about 140 ml. per beat in the normal heart and to over 200 ml. per beat in persons with very large hearts, such as in some athletes.

WORK OUTPUT OF THE HEART

The work output of the heart is the amount of energy that the heart transfers to the blood while pumping it into the arteries. Energy is transferred to the blood in two forms: First, by far the major proportion is used to move the blood from the low pressure veins to the high pressure arteries. This is *potential energy of pressure.* Second, a minor proportion of the energy is used to accelerate the blood to its velocity of ejection through the aortic and pulmonary valves. This is *kinetic energy of blood flow.*

Energy Expended to Create Potential Energy of Pressure. The work performed by the left ventricle to raise the pressure of the blood during each heartbeat is equal to *stroke volume output × (left ventricular mean ejection pressure* minus *left atrial pressure).* Likewise, the work performed by the right ventricle to raise the pressure of the blood is equal to *stroke volume output × (right ventricular*

mean ejection pressure minus *right atrial pressure).* When pressure is expressed in *dynes per square centimeter* and stroke volume output in *milliliters,* the work output is in *ergs.* Right ventricular work output is usually about one-seventh the work output of the left ventricle because of the difference in systolic pressure against which the two ventricles must pump.

The Kinetic Energy of Blood Flow. The work output of each ventricle required to create kinetic energy of blood flow is proportional to the mass of the blood ejected times the square of the velocity of ejection. That is,

$$\text{Kinetic energy} = \frac{mv^2}{2}$$

When the mass is expressed in *grams* of blood ejected and the velocity in *centimeters per second,* the work output is in *ergs.*

Ordinarily, the work output of the left ventricle required to create kinetic energy of blood flow is 2 to 4 per cent of the total work output of the ventricle. Most of this energy is required to cause the rapid acceleration of the blood during the first quarter of systole. In certain abnormal conditions, such as aortic stenosis, in which the blood flows with great velocity through the stenosed valve, as much as 50 per cent of the total work output may be required to create kinetic energy of blood flow.

ENERGY FOR CARDIAC CONTRACTION

Heart muscle, like skeletal muscle, utilizes chemical energy to provide the work of contraction. This energy is derived mainly from metabolism of glucose and fatty acids with oxygen and, to a lesser extent, from metabolism of other nutrients with oxygen. The different reactions that liberate this energy will be discussed in detail in Chapters 67 and 68.

The amount of energy expended by the heart is related to its work load in the following manner: The energy expended is *approximately proportional to the degree of tension generated by the heart musculature during contraction × the amount of time that the tension is maintained.*

Efficiency of Cardiac Contraction. During muscular contraction most of the chemical energy is converted into heat and a small portion into work output. The ratio of work output to chemical energy expenditure is called the efficiency of cardiac contraction, or simply *efficiency of the heart.* The efficiency of the normal heart, beating under normal load, is usually low, some 5 to 10 per cent. However, during maximum work output it rises as high as 15 to 20 per cent in the normal heart and perhaps 25 per cent in the heart of the well-trained athlete.

REGULATION OF CARDIAC FUNCTION

When a person is at rest, the heart must pump only 4 to 6 liters of blood each minute.

However, during severe exercise it may be required to pump as much as five times this amount. The present section discusses the means by which the heart can adapt itself to such extreme increases in cardiac output.

The two basic means by which the pumping action of the heart is regulated are (1) intrinsic autoregulation in response to changes in volume of blood flowing into the heart and (2) reflex control of the heart by the autonomic nervous system.

INTRINSIC AUTOREGULATION OF CARDIAC PUMPING—THE FRANK-STARLING LAW OF THE HEART

In Chapter 27 we shall see that one of the major factors determining the amount of blood pumped by the heart each minute is the rate of blood flow into the heart from the veins, which is called *venous return*. That is, each peripheral tissue of the body controls its own blood flow, and whatever amount of blood flows through the peripheral tissues returns by way of the veins to the right atrium. The heart in turn automatically pumps this incoming blood on into the systemic arteries so that it can flow around the circuit again. Thus, the heart must adapt itself from moment to moment or even second to second to widely varying inputs of blood, sometimes falling as low as 2 to 3 liters per minute and at other times rising as high as 25 or more liters per minute.

This intrinsic ability of the heart to adapt itself to changing loads of inflowing blood is called the Frank-Starling law of the heart, in honor of Frank and Starling, two of the great physiologists of half a century ago. Basically, the Frank-Starling law states that the greater the heart is filled during diastole, the greater will be the quantity of blood pumped into the aorta. Or another way to express this law is: *within physiologic limits, the heart pumps all the blood that comes to it without allowing excessive damming of blood in the veins.* In other words, the heart can pump either a small amount of blood or a large amount, depending on the amount that flows into it from the veins; and it automatically adapts to whatever this load might be as long as the total quantity does not rise above a physiologic limit that the heart can pump.

The primary mechanism by which the heart adapts to changing input volumes of blood is the following: When the cardiac muscle becomes stretched, as it does when extra amounts of blood enter the heart chambers, the muscle contracts with a greatly increased force, thereby automatically pumping the extra blood into the arteries. The ability of stretched muscle to contract with increased force is characteristic of all striated muscle and not simply of cardiac muscle. Referring back to Figure 7–13 in Chapter 7, one will see that stretching a skeletal muscle, within its physiological limit, also increases its force of contraction. As was also pointed out in Chapter 7, the increased force of contraction is probably caused by the fact that the actin and myosin filaments are brought to an optimal degree of interdigitation for achieving contraction. This ability of the heart to contract with increased force as its chambers are stretched is sometimes called *heterometric autoregulation of the heart.*

In addition to the important effect of stretching the heart muscle, at least two other less important factors increase heart pumping effectiveness when its volume is increased. First, stretch of the right atrial walls increases the heart rate by as much as 10 to 15 per cent;

Figure 13–6 Left and right ventricular function curves in a dog, depicting ventricular stroke work output as a function of left and right mean atrial pressures. (Curves reconstructed from data in Sarnoff: *Physiol. Rev., 35:*107, 1955.)

this in itself can increase the amount of blood pumped each minute. Second, changes in heart metabolism that occur when the heart is stretched cause an additional increase in contraction strength. It takes approximately 30 seconds for this effect to develop fully, and it is called *homeometric autoregulation* because there is no further stretch of the muscle fibers. However, these two additional factors, increased heart rate and homeometric autoregulation, are of far less importance in adaptation of the heart to increasing volume input than is heterometric autoregulation.

Ventricular Function Curves. One of the best ways to express the functional ability of the ventricles is by ventricular function curves, Figures 13–6 and 13–7. Figure 13–6 illustrates a type of ventricular function curve called the *stroke work output curve*. Note that as the atrial pressures increase, the stroke work output also increases until it reaches the limit of the heart.

Figure 13–7 illustrates another type of ventricular function curve called the *minute ventricular output curve*. These two curves represent function of the two ventricles of the human heart based on data extrapolated from lower animals. As each atrial pressure rises, the respective ventricular volume output per minute also increases.

Thus, ventricular function curves are another way of expressing the Frank-Starling law of the heart. That is, as the ventricles fill to higher atrial pressures, the strength of cardiac contraction increases, causing the heart to pump increased quantities of blood into the arteries. In later chapters we shall see that ventricular function curves are exceedingly important in analyzing overall function of the circulation, for it is by such means that one can express the quantitative capabilities of the heart as a pump.

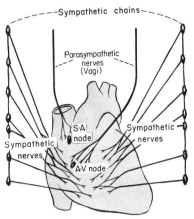

Figure 13–8 The cardiac nerves.

CONTROL OF THE HEART BY NERVES

The heart is well supplied with both sympathetic and parasympathetic nerves, as illustrated in Figure 13–8. These nerves affect cardiac pumping in two ways: (1) by changing the heart rate and (2) by changing the strength of contraction of the heart. The effect of nerve stimulation on heart rate and rhythm will be discussed in detail in the following chapter. For the present, suffice it to say that parasympathetic stimulation decreases heart rate and sympathetic stimulation increases heart rate. The maximum range of control is from as little as zero heart beats per minute up to as high as 250 or, rarely, 300 heart beats per minute.

Effect of Heart Rate on Function of the Heart as a Pump. In general, the more times the heart beats per minute, the more blood it can pump, but there are important limitations to this effect. For instance, once the heart rate rises above a critical level, the heart strength itself decreases, presumably because of overutilization of metabolic substrates in the cardiac muscle. In addition, the period of diastole between the contractions becomes so reduced that blood does not have time to flow adequately from the atria into the ventricles. For these reasons, when the heart rate is increased artificially by electrical stimulation, the heart has its peak ability to pump large quantities of blood at a heart rate of about 150 beats per minute. On the other hand, when its rate is increased by sympathetic stimulation, it reaches its peak ability to pump blood at a heart rate of about 200 beats per minute. The reason for this difference is that sympathetic stimulation not only increases the heart rate but also increases heart strength as well.

Nervous Regulation of Contractile Strength of the Heart. The two atria are supplied with large numbers of both sympathetic and para-

Figure 13–7 Approximate normal right and left ventricular output curves for the human heart as extrapolated from data obtained in dogs.

sympathetic nerves, but the ventricles are supplied almost exclusively by sympathetic nerves
with perhaps a few straggling parasympathetic fibers. Therefore, for practical purposes,
the overall effect of the autonomic nerves on
heart strength is mediated primarily through
the sympathetic nerves. In general, sympathetic stimulation increases the strength of
heart muscle contraction, whereas parasympathetic stimulation decreases the strength of
contraction.

Under normal conditions the sympathetic
nerve fibers to the heart continually discharge
at a slow rate that maintains a strength of
ventricular contraction about 20 per cent
above its strength with no sympathetic stimulation at all. Therefore, one method by which
the nervous system can decrease the strength
of ventricular contraction is simply to slow or
stop the transmission of sympathetic impulses
to the heart. On the other hand, maximal
sympathetic stimulation can increase the
strength of ventricular contraction to approximately double normal.

Maximal parasympathetic stimulation of the
heart decreases ventricular contractile strength
perhaps 10 to 30 per cent. Thus, the parasympathetic effect is, by contrast with the
sympathetic effect, relatively unimportant.

**Effect of Autonomic Control of Cardiac Func-
tion as Depicted by Cardiac Function Curves.**
Figure 13–9 illustrates four separate cardiac
function curves. These are much the same as
the ventricular function curves of Figures
13–6 and 13–7, except that they represent

function of the entire heart rather than of a
single ventricle; they show the relationship
between the right atrial pressure at the input
of the heart and cardiac output from the left
ventricle.

The curves of Figure 13–9 demonstrate that
at any given right atrial pressure, the cardiac
output increases with increasing sympathetic
stimulation and decreases with increasing
parasympathetic stimulation. Note in particular that the changes in the function curves
of Figure 13–9 are brought about by both
changes in heart rate and *changes in contractile
strength of the heart,* for both of them affect
cardiac output.

EFFECT OF HEART DEBILITY ON CARDIAC FUNCTION – THE HYPOEFFECTIVE HEART

Any factor that damages the heart, whether
it be damage to the myocardium, to the valves,
to the conducting system, or otherwise, is
likely to make the heart a poorer pump, and
the heart under these conditions is called a
hypoeffective heart. Figure 13–10 illustrates
by the very dark curve the normal cardiac
function curve and by the three curves below
this the effect of different degrees of hypo-
effectiveness on cardiac function. Obviously,
the more serious the damage, the less will be
the cardiac output at any given right atrial
pressure.

Different factors that can cause a hypoeffective heart include:

> Myocardial infarction
> Valvular heart disease
> Vagal stimulation of the heart
> Inhibition of the sympathetics to the heart
> Congenital heart disease
> Myocarditis
> Cardiac anoxia
> Diphtheritic or other types of myocardial
>   damage

EFFECT OF EXERCISE ON THE HEART – THE HYPEREFFECTIVE HEART

Chronic heavy exercise over a period of
many weeks or months leads to hypertrophy
of the cardiac muscle and also to enlargement
of the ventricular chambers. As a result, the
overall strength of the heart becomes greatly
enhanced, and the effectiveness of the heart
as a pump obviously increases. The upper

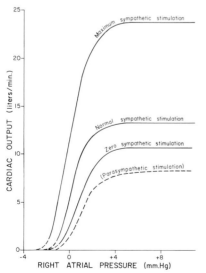

Figure 13–9   Effect on the cardiac output curve of different degrees of sympathetic and parasympathetic stimulation.

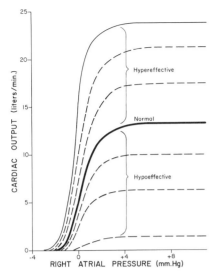

Figure 13-10 Cardiac output curves for various degrees of hypo- and hypereffective hearts. (From Guyton: Cardiac Output and Its Regulation.)

three function curves of Figure 13–10 illustrate the effect of different degrees of hypereffectiveness on the cardiac function curve, showing that maximal degrees of hypereffectiveness can increase pumping by the heart perhaps 70 to 100 per cent.

Other factors besides hypertrophy that can cause a hypereffective heart are:

Sympathetic stimulation
Parasympathetic inhibition
Decreased arterial pressure against which the heart must pump

EFFECT OF VARIOUS IONS ON HEART FUNCTION

In the discussion of membrane potentials in Chapter 5, it was pointed out that three particular cations—potassium, sodium, and calcium—all have marked effects on action potential transmission, and in Chapter 7 it was noted that calcium ions play an especially important role in catalyzing some reactions in the muscle contractile process. Therefore, it is to be expected that the concentrations of these three ions in the extracellular fluids will have marked effects on cardiac function.

The action of individual ions on the heart can be studied by perfusing the isolated heart, as illustrated in Figure 13–11, using the so-called *Langendorf preparation*. In making this preparation, the heart is removed rapidly from the animal's body, and the aorta is cannulated immediately. The cannula is connected to a perfusion bottle located approximately three feet higher than the heart itself so that the fluid in the aorta will be under pressure. The pressure causes the aortic valve to close and the fluid to flow through the coronary vessels. When the perfusion fluid contains appropriate nutrients (glucose and oxygen) and ions and the heart is kept appropriately warmed, the perfused mammalian heart will beat for many hours.

Effect of Potassium Ions. Excess potassium in the extracellular fluids causes the heart to become extremely dilated and flaccid and slows the heart rate. Very large quantities can also block conduction of the cardiac impulse from the atria to the ventricles through the A-V bundle. Elevation of potassium concentration to only 8 to 15 mEq./liter—two to three times the normal value—will usually cause such weakness of the heart that it will cause death.

All these effects of potassium excess are believed to be caused by decreased resting membrane potentials which result from the high potassium concentration in the extracellular fluids. As the membrane potential decreases, the intensity of the action potential also decreases, which makes the contraction of the heart become progressively weaker and weaker, for, as discussed earlier in the chapter, the strength of the action potential determines to a great extent the strength of contraction.

Effect of Calcium Ions. An excess of calcium ions causes effects almost exactly opposite to those of potassium ions, causing the heart to

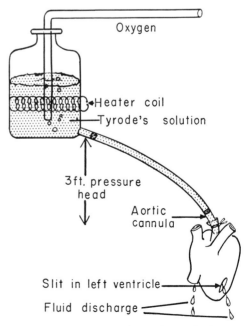

Figure 13-11 System for perfusing the heart.

go into spastic contraction. This is probably caused by the direct effect of calcium ions in exciting the cardiac contractile process, as explained earlier in the chapter. Conversely, a deficiency of calcium ions causes cardiac flaccidity, similar to the effect of potassium.

However, the calcium ion concentration rarely changes sufficiently during life to alter cardiac function greatly, for greatly diminished calcium ion concentration will kill a person because of tetany before it can significantly affect the heart, and elevation of the calcium ion concentration to a level that will significantly affect the heart almost never occurs because calcium ions are precipitated in the body's tissues as insoluble calcium salts before such a level can be reached.

Effect of Sodium Ions. An excess of sodium ions depresses cardiac function, an effect similar to that of potassium ions but for an entirely different reason. Sodium ions compete with calcium ions at some yet unexplained point in the contractile process of muscle in such a way that the greater the sodium ion concentration in the extracellular fluids the less the effectiveness of the calcium ions in causing contraction when an action potential occurs.

However, again, from a practical point of view, the sodium ion concentration in the extracellular fluids probably never becomes abnormal enough even in serious pathological conditions to cause significant change in cardiac function.

EFFECT OF TEMPERATURE ON THE HEART

The effect of temperature on the heart can also be studied using the perfusion system of Figure 13–11. Increased temperature causes greatly increased heart rate, and decreased temperature causes greatly decreased rate.

These effects of temperature presumably result from increased permeability of the muscle membrane to the different ions, resulting in acceleration of the self-excitation process.

THE HEART-LUNG PREPARATION

A method frequently used to demonstrate function of the heart is the heart-lung preparation illustrated in Figure 13–12. In this preparation, blood leaving the heart by way of the aorta is diverted into an external system of tubes and then back again into the right atrium. The only parts of the animal's body that remain alive are the heart and lungs. The blood is oxygenated by artificial respiration of the lungs, and a glucose solution is added periodically to the blood to supply nutrition for cardiac contraction.

As blood flows through the external circuit, right atrial pressure, aortic pressure, and blood flow are

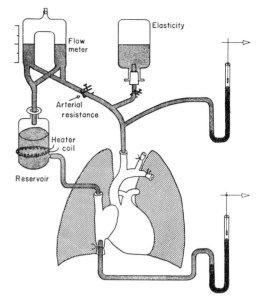

Figure 13–12   The heart-lung preparation.

measured. Also, an adjustable venous reservoir allows the pressure in the right atrium to be increased or decreased, and a heater coil is wound around the reservoir to maintain normal blood temperature. A screw clamp is provided on the main outflow tube from the heart so that the arterial resistance can be increased or decreased. Finally, a special reservoir containing air in its upper part and blood in its lower part is used as an "elastic buffer" so that the blood pressure will not rise and fall excessively with each contraction of the heart. Some of the principles of cardiac function that can be demonstrated with the heart-lung preparation are the following:

First, the *relationship of end-diastolic volume to stroke volume output* can be demonstrated by raising and lowering the venous reservoir. When the reservoir is raised, blood flows into the heart rapidly, distending the cardiac chambers and causing increased stroke volume output, thus demonstrating the law of the heart.

Second, one can show that *increasing the arterial resistance does not greatly decrease the stroke volume output as long as the right atrial pressure remains constant.* This illustrates that the aortic pressure, within physiological limits, has rather little effect on the output, though very high pressures can overload the heart so much that the stroke volume output does then become reduced.

Third, the *effect of different factors on heart rate* can be studied, such as the slight increase in heart rate that occurs when the load of inflowing blood increases, the increase in heart rate with increasing temperature, the increase in heart rate when norepinephrine is injected into blood perfusing the heart, the decrease in heart rate when acetylcholine is injected, and changes in heart rate when the nerves to the heart are stimulated.

REFERENCES

Armour, J. A.; Pace, J. B.; and Randall, W. C.: Interrelationship of architecture and function of the right ventricle. *Amer. J. Physiol., 218*:174, 1970.

Bing, R. J.: Cardiac metabolism. *Physiol. Rev., 45*:171, 1965.

Bishop, V. S.; Horwitz, L. D.; Stone, H. L.; Stegall, H. F.; and Engelken, E. J.: Left ventricular internal diameter and cardiac function in conscious dogs. *J. Appl. Physiol., 27*:619, 1969.

Brady, Allan J.: Active state in cardiac muscle. *Physiol. Rev., 48*:570, 1968.

Braunwald, E.; Ross, J.; and Sonnenblick, E.: Mechanism of Contractility of the Normal and Failing Heart. Boston, Little, Brown and Company, 1968.

Brecher, G. A., and Galletti, P. M.: Functional anatomy of cardiac pumping. *In* Handbook of Physiology. Baltimore, The Williams & Wilkins Co., 1963, Vol. II, Sec. II, p. 759.

Chandler, B. M.; Sonnenblick, E. H.; and Pool, Peter E.: Mechanochemistry of cardiac muscle: III. Effects of norepinephrine on the utilization of high-energy phosphates. *Circ. Res., 22*:729, 1968.

Clancy, R. L.; Graham, T. P., Jr.; Ross, J., Jr.; Sonnenblick, E. H.; and Braunwald, E.: Influence of aortic pressure-induced homeometric autoregulation on myocardial performance. *Amer. J. Physiol., 214*:1186, 1968.

Coleman, H. N.; Dempsey, P. J.; and Cooper, T.: Myocardial oxygen consumption following chronic cardiac denervation. *Amer. J. Physiol., 218*:475, 1970.

Cronin, R.; Armour, J. A.; and Randall, W. C.: Function of the in-situ papillary muscle in the canine left ventricle. *Circ. Res., 25*:67, 1969.

De Ceretti, E. R. P.; Webb, J. L.; and Schanne, O. F.: Interatrial conduction in the rat. *Amer. J. Physiol., 215*:671, 1968.

Entman, M. L.; Levey, G. S.; and Epstein, S. E.: Mechanism of action of epinephrine and glucagon on the canine heart: evidence for increase in sarcotubular calcium stores mediated by cyclic 3',5'-AMP. *Circ. Res., 25*:429, 1969.

Falsetti, H. L.; Mates, R. E.; Grant, C.; Greene, D. G.; and Bunnell, I. L.: Left ventricular wall stress calculated from one-plane cineangiography: an approach to force-velocity analysis in man. *Circ. Res., 26*:71, 1970.

Grimm, A. F., and Whitehorn, W. V.: Myocardial length-tension sarcomere relationships. *Amer. J. Physiol., 214*:1378, 1968.

Guyton, A. C.: Determination of cardiac output by equating venous return curves with cardiac response curves. *Physiol. Rev., 35*:123, 1955.

Hamlin, R. L., and Smith, C. R.: Effects of vagal stimulation on S-A and A-V nodes. *Amer. J. Physiol., 215*:560, 1968.

Hawthorne, E. W.: Heart. *Ann. Rev. Physiol., 27*:351, 1965.

Katz, L. N.; Wise, W.; and Jochim, K.: Dynamics of isolated heart and heart-lung preparation of dog. *Amer. J. Physiol., 143*:463, 1945.

Levy, M. N., and Berne, R. M.: Heart. *Ann. Rev. Physiol., 32*:373, 1970.

Levy, M. N., and Zieske, H.: Comparison of the cardiac effects of vagus nerve stimulation and of acetylcholine infusions. *Amer. J. Physiol., 216*:890, 1969.

Monroe, R. G.; Gamble, W. J.; LaFarge, C. G.; Kumar, A. E.; and Manasek, F. J.: Left ventricular performance at high end-diastolic pressures in isolated, perfused dog hearts. *Circ. Res., 26*:85, 1970.

Moulopoulos, S. D.; Anthopoulos, L. P.; and Stamatelopoulos, S. F.: Myocardial response changes after short loading periods. *J. Appl. Physiol., 28*:28, 1970.

Napolitano, L. M.; Willman, V. L.; Hanlon, C. R.; and Cooper, T.: Intrinsic innervation of the heart. *Amer. J. Physiol., 208*:455, 1965.

Pollack, G. H.: Maximum velocity as an index of contractility in cardiac muscle: a critical evaluation. *Circ. Res., 26*:111, 1970.

Randall, W.: Nervous Control of the Heart. Baltimore, The Williams & Wilkins Co., 1965.

Randall, W. C.; Wechster, J. S.; Pace, J. B.; and Szentivanyi, M.: Alterations in myocardial contractility during stimulation of the cardiac nerves. *Amer. J. Physiol., 214*:1205, 1968.

Sabatini-Smith, S., and Holland, W. C.: Influence of manganese and ouabain on the rate of action of calcium on atrial contractions. *Amer. J. Physiol., 216*:244, 1969.

Sarnoff, S. J.: Myocardial contractility as described by ventricular function curves. *Physiol. Rev., 35*:107, 1955.

Sarnoff, S. J., and Mitchell, J. H.: The control of the function of the heart. *In* Handbook of Physiology. Baltimore, The Williams & Wilkins Co., 1962, Vol. 1, Sec. II, p. 489.

Schaper, W.: Heart. *Ann. Rev. Physiol., 29*:259, 1967.

Sonnenblick, E. H.: Force-velocity relations in mammalian heart muscle. *Amer. J. Physiol., 202*:931, 1962.

Sonnenblick, E. H., and Stam, A. C., Jr.: Cardiac muscle: activation and contraction. *Ann. Rev. Physiol., 31*:647, 1969.

Starling, E. H.: The Linacre Lecture on the Law of the Heart. London, Longmans Green & Co., 1918.

Stone, H. L.; Bishop, V. S.; and Guyton, A. C.: Cardiac function after embolization of coronaries with microspheres. *Amer. J. Physiol., 204*:16, 1963.

Tanz, Ralph D.: Factors Influencing Myocardial Contractility. New York, Academic Press, Inc., 1967.

Taylor, R. R.; Covell, J. W.; and Ross, J., Jr.: Volume-tension diagrams of ejecting and isovolumic contractions in left ventricle. *Amer. J. Physiol., 216*:1097, 1969.

Wildenthal, K.; Mierzwiak, D. S.; and Mitchell, J. H.: Effect of sudden changes in aortic pressure on left ventricular dp/dt. *Amer. J. Physiol., 216*:185, 1969.

Wildenthal, K.; Mierzwiak, D. S.; Wyatt, H. L.; and Mitchell, J. H.: Influence of efferent vagal stimulation on left ventricular function in dogs. *Amer. J. Physiol., 216*:577, 1969.

Wildenthal, K., and Mitchell, J. H.: Dimensional analysis of the left ventricle in unanesthetized dogs. *J. Appl. Physiol., 27*:115, 1969.

CHAPTER 14

Rhythmic Excitation of the Heart

The heart is endowed with a special system (a) for generating rhythmical impulses to cause rhythmical contraction of the heart muscle and (b) for conducting these impulses throughout the heart. A major share of the ills of the heart is based on abnormalities of this special excitatory and conductive system.

THE SPECIAL EXCITATORY AND CONDUCTIVE SYSTEM OF THE HEART

The adult human heart normally contracts at a rhythmic rate of about 72 beats per minute. Figure 14–1 illustrates the special excitatory and conductive system of the heart that controls these cardiac contractions. The figure shows: (A) the *S-A node* in which the normal rhythmic self-excitatory impulse is generated, (B) the *A-V node* in which the impulse from the atria is delayed before passing into the ventricles, (C) the *A-V bundle,* which conducts the impulse from the atria into the ventricles, and (D) the *left* and *right bundles of Purkinje fibers*, which conduct the cardiac impulse to all parts of the ventricles.

THE SINO-ATRIAL NODE

The sino-atrial (S-A) node is a small strip of specialized muscle approximately 3 mm. wide and 1 cm. long; it is located in the posterior wall of the right atrium immediately beneath and medial to the opening of the superior vena cava. The fibers of this node are only 3 to 5 microns in diameter in contrast to a diameter of 15 to 20 microns for the surrounding atrial muscle fibers. However, the S-A fibers are continuous with the atrial fibers so that any action potential that begins in the S-A node spreads immediately into the atria.

Rhythmicity of the Sino-Atrial Fibers. The resting membrane potential of the S-A fibers is quite low, only 55 to 60 mv. in comparison with 80 to 90 mv. in other cardiac muscle fibers. The cause of this low resting potential is presumably a very high membrane conductance of these fibers for sodium, which allows rapid leakage of sodium through the membrane. And it is this leakage of sodium that also causes self-excitation of the S-A fibers. Figure 14–2 illustrates the self-excitation process; each time the resting membrane potential is re-established, it gradually decays (returns toward zero) until it reaches the threshold for self-excitation. At this point an action potential suddenly appears and lasts about 0.15 second. Immediately after the action potential is over, the conductance of the membrane for potassium becomes very high, and rapid outward diffusion of potassium carries positive charges to the outside of the fiber, increasing the negativity inside the fiber to a level much greater than normal, a condition called *hyperpolarization.* As long as this increased potassium conductance persists, the membrane potential remains more negative than the threshold for self-excitation; but, gradually, the conductivity for potassium decreases, allowing the high

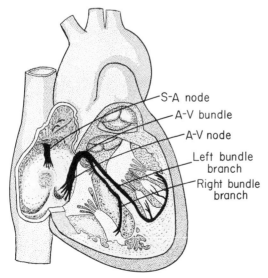

Figure 14–1 The S-A node and the Purkinje system of the heart.

Figure 14-2 Rhythmic discharge of an S-A nodal fiber.

sodium influx to excite the fiber again. Thus, the process repeats itself again and again, producing the rhythmic excitation of the heart.

TRANSMISSION OF THE CARDIAC IMPULSE THROUGH THE ATRIA.   Each time a rhythmic impulse is generated in any single fiber of the S-A node, it spreads immediately into the surrounding atrial muscle and is conducted in all directions at a velocity of about 0.3 meter per second. Though all atrial muscle fibers probably participate in conducting the cardiac impulse, a few specialized fibers seem to conduct the impulse with extra speed directly from the S-A node to the A-V node. These fibers are part of the rapid conducting system of the heart and are comparable to the very important Purkinje fibers of the ventricles, which will be discussed subsequently.

As the cardiac impulse travels through the atria, the atrial muscle contracts, forcing blood through the atrioventricular valves into the ventricles.

THE ATRIOVENTRICULAR (A-V) NODE AND THE PURKINJE SYSTEM

Delay in Transmission at the A-V Node.   Fortunately, the conductive system is organized so that the cardiac impulse will not travel from the atria into the ventricles too rapidly, this allows time for the atria to empty their contents into the ventricles before ventricular contraction begins. It is primarily the A-V node and its associated conductive fibers that delays this transmission of the cardiac impulse from the atria into the ventricles.

Figure 14-3 shows diagrammatically the different parts of the A-V node and its connections with the atrial muscle and the A-V bundle. The figure also shows the approximate intervals of time in fractions of a second between the genesis of the cardiac impulse in the S-A node and its appearance at different points in its conductive pathway. Note that the impulse, after traveling through the atrial muscle, reaches the A-V node approximately 0.04 second after its origin in the S-A node.

However, between this time and the time that the impulse emerges in the A-V bundle, another 0.11 second elapses. About one half of this time lapse occurs in the *junctional fibers*, which are very small fibers that connect the normal atrial fibers with the fibers of the node itself (illustrated in Figures 14-3 and 14-4). The velocity of conduction in these fibers is 0.05 meter per second or less (about one-tenth that in normal cardiac muscle), which greatly delays entrance of the impulse into the A-V node. After entering the node, the velocity of conduction in the *nodal fibers* is still quite low, only 0.1 meter per second, about one-third to one-fourth the conduction velocity in normal cardiac muscle. Therefore, a further delay in transmission occurs as the impulse travels through the A-V node into the *transitional fibers* and finally into the *A-V bundle*.

Figure 14-4 illustrates a functional diagram of the A-V nodal region showing the atrial fibers leading into the minute junctional fibers, then progressive enlargement of the fibers again as they spread through the node, through the transitional region, and into the A-V bundle. The characteristics of the action potential in the different parts of the node are illustrated at the bottom.

TRANSMISSION IN THE PURKINJE SYSTEM

The *Purkinje fibers* that lead from the A-V node through the A-V bundle and into the ventricles have functional characteristics quite the opposite to those of the A-V nodal fibers; they are very large fibers, even larger than the normal ventricular muscle fibers, and they transmit impulses at a velocity of 1.5 to 2.5 meters per second, a velocity about six times that in the usual cardiac muscle and 40 times that in the junctional fibers. This allows al-

Figure 14-3 Organization of the A-V node. The numbers represent the interval of time from the origin of the impulse in the S-A node. The values have been extrapolated to the human being. (This figure is based on studies in lower animals discussed and illustrated in Hoffman and Cranefield: Electrophysiology of the Heart. McGraw-Hill Book Co.)

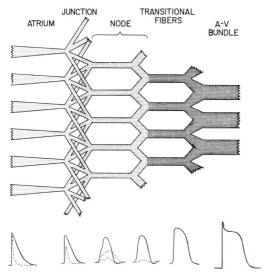

Figure 14–4 Functional diagram of the anatomical relationships in the region of the atrioventricular node. The action potentials at the bottom were recorded under normal conditions (solid lines) and under the influence of acetylcholine (dashed lines). (From Hoffman and Cranefield: Electrophysiology of the Heart. McGraw-Hill Book Co.)

most immediate transmission of the cardiac impulse throughout the entire ventricular system.

DISTRIBUTION OF THE PURKINJE FIBERS IN THE VENTRICLES. The Purkinje fibers, after originating in the A-V node, form the A-V bundle, which then threads between the valves of the heart and thence into the ventricular septum as shown in Figure 14–1. The A-V bundle divides almost immediately into the *left* and *right bundle branches* that lie beneath the endocardium of the respective sides of the septum. Each of these branches spreads downward toward the apex of the respective ventricle, then curves around the tip of the ventricular chamber and finally back toward the base of the heart along the lateral wall. About one-third the way up the lateral walls of the ventricles, the two bundle branches break up into many small branches of Purkinje fibers that spread in all directions beneath the endocardium of the ventricles. The terminal Purkinje fibers enter the ventricular muscle from the endocardial surface and terminate on the muscle fibers. At this termination, the cardiac impulse can pass readily from the Purkinje fiber into the cardiac muscle fiber; but, for reasons not yet understood, the impulse probably cannot pass backward from the muscle into the Purkinje fiber.

From the time that the cardiac impulse first enters the A-V bundle until it reaches the terminations of the Purkinje fibers, the total time that lapses is only 0.03 second; therefore, once a cardiac impulse enters the Purkinje system, it spreads almost immediately to the entire endocardium of the ventricles.

TRANSMISSION OF THE CARDIAC IMPULSE IN THE VENTRICULAR MUSCLE

Once the cardiac impulse has reached the ends of the Purkinje fibers, it is then transmitted through the ventricular muscle mass by the ventricular muscle fibers themselves. The velocity of transmission is now only 0.3 to 0.4 meter per second, one-sixth that in the Purkinje fibers.

The cardiac muscle is arranged in whorls with fibrous septa between the whorls; therefore, the cardiac impulse does not necessarily travel directly outward toward the surface of the heart but instead angulates toward the surface along the directions of the whorls. Because of this, transmission from the endocardial surface to the epicardial surface of the ventricle requires as much as another 0.03 second, approximately equal to the time required for transmission through the entire Purkinje system. Thus, the total time for transmission of the cardiac impulse from the origin of the Purkinje system to the last of the ventricular muscle fibers is about 0.06 second.

One can see from this description of conduction in the ventricles that the first part of the ventricular muscle mass to be excited is the septum; this is followed rapidly by the endocardial surfaces of the apices and lateral walls of the ventricles and finally by the epicardial surfaces of the ventricles. The precise time intervals of spread of the impulse to the different anterior and posterior epicardial surfaces of the ventricles are illustrated in Figure 14–5, which shows actual measurements from human hearts.

SUMMARY OF THE SPREAD OF THE CARDIAC IMPULSE THROUGH THE HEART

Figure 14–6 illustrates in summary form the transmission of the cardiac impulse through the human heart. The numbers on the figure represent the intervals of time in fractions of a second that lapse between the origin of the cardiac impulse in the S-A node and its appearance at each respective point in the heart. Note that the impulse spreads at moderate velocity through the atria but is delayed at least one-tenth second in the A-V nodal region before appearing in the A-V bundle. Once it has entered the bundle, it spreads rapidly

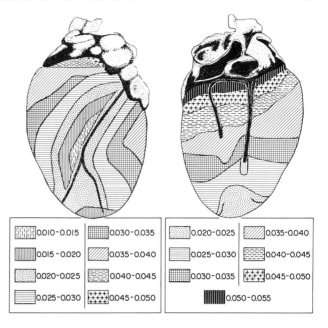

Figure 14–5 Sequence of activation of the anterior and posterior surfaces of the heart during the normal cardiac impulse. The numbers represent fractions of a second after first appearance of the impulse in the ventricular septum. (From Sodi-Pallares and Calder: New Bases of Electrocardiography. The C. V. Mosby Co.)

0.010-0.015   0.030-0.035   0.020-0.025   0.035-0.040

0.015-0.020   0.035-0.040   0.025-0.030   0.040-0.045

0.020-0.025   0.040-0.045   0.030-0.035   0.045-0.050

0.025-0.030   0.045-0.050   0.050-0.055

through the Purkinje fibers to the entire endocardial surfaces of the ventricles. Then the impulse spreads slowly through the ventricular muscle to the epicardial surfaces.

It is extremely important that the student learn in detail the course of the cardiac impulse through the heart and the times of its appearance in each separate part of the heart, for a quantitative knowledge of this process is essential to the understanding of electrocardiography, which is discussed in the following three chapters.

CONTROL OF EXCITATION AND CONDUCTION IN THE HEART

THE S-A NODE AS THE PACEMAKER OF THE HEART

In the above discussion of the genesis and transmission of the cardiac impulse through the heart, it was stated that the impulse normally arises in the S-A node. However, this need not be the case under abnormal conditions, for other parts of the heart can exhibit rhythmic contraction in the same way that the fibers of the S-A node can; this is particularly true of the A-V nodal and Purkinje fibers.

The A-V nodal fibers, when not stimulated from some outside source, discharge at an intrinsic rhythmic rate of 40 to 60 times per minute, and the Purkinje fibers discharge at a rate of somewhere between 15 to 40 times per minute. These rates are in contrast to the

normal rate of the S-A node of 70 to 80 times per minute.

Therefore, the question that we must ask is: Why does the S-A node control the heart's rhythmicity rather than the A-V node or the Purkinje fibers? The answer to this is simply that the rate of the S-A node is considerably greater than that of either the A-V node or the Purkinje fibers. Each time the S-A node discharges, its impulse is conducted both into the A-V node and into the Purkinje fibers, discharging their excitable membranes. Then all these tissues begin their cycle of recovery. But the S-A node recovers much more rapidly than can either of the other two and emits still another impulse before either one of them can reach its own threshold for self-excitation. The new impulse again discharges both the A-V node and Purkinje fibers. This process continues on and on, the S-A node always exciting

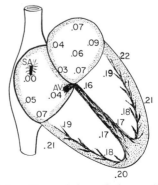

Figure 14–6 Transmission of the cardiac impulse through the heart, showing the time of appearance (in fractions of a second) of the impulse in different parts of the heart.

these other potentially self-excitatory tissues before self-excitation can actually occur.

Thus, the S-A node controls the beat of the heart because its rate of rhythmic discharge is greater than that of any other part of the heart. Therefore, it is said that the S-A node is the normal *pacemaker* of the heart.

Abnormal Pacemakers—The Ectopic Pacemaker. Occasionally some other part of the heart develops a rhythmic discharge rate that is more rapid than that of the S-A node. For instance, this frequently occurs in the A-V node or in the Purkinje fibers. In either of these cases, the pacemaker of the heart shifts from the S-A node to the A-V node or to the excitable Purkinje fibers. Under rare conditions a point in the atrial or ventricular muscle develops excessive excitability and becomes the pacemaker.

A pacemaker elsewhere than the S-A node is called an *ectopic pacemaker.* Obviously, an ectopic pacemaker causes an abnormal sequence of contraction of the different parts of the heart.

Ectopic Beats. Isolated parts of the conductive system or cardiac muscle occasionally emit single spurious impulses that appear at odd times. If such a spurious impulse occurs during the nonrefractory period of the cardiac muscle, it will travel throughout the heart and cause an abnormal beat. This is called an ectopic beat.

CONTROL OF HEART RHYTHMICITY AND CONDUCTION BY THE AUTONOMIC NERVES

The heart is supplied with both sympathetic and parasympathetic nerves, as illustrated in Figure 13–5 of the previous chapter. The parasympathetic nerves are distributed mainly to the S-A and A-V nodes; to a lesser extent to the muscle of the two atria; but very sparsely, if at all, to the ventricular muscle. The sympathetic nerves are distributed to these same areas but with a strong representation to the ventricular muscle.

Effect of Parasympathetic (Vagal) Stimulation on Cardiac Function—Ventricular Escape. Stimulation of the parasympathetic nerves to the heart (the vagi) causes the hormone acetylcholine to be released at the vagal endings. This hormone has two major effects on the heart. First, it decreases the rate of rhythm of the S-A node, and, second, it decreases the excitability of the A-V junctional fibers between the atrial musculature and the A-V node, thereby slowing transmission of the cardiac impulse into the ventricles. Very strong stimulation of the vagi can completely stop the rhythmic contraction of the S-A node or completely block transmission of the cardiac impulse through the A-V junction. In either case, rhythmic impulses are no longer transmitted into the ventricles. The ventricles stop beating for 4 to 10 seconds, but then some point in the Purkinje fibers, usually in the A-V bundle, develops a rhythm of its own and causes ventricular contraction at a rate of 15 to 40 beats per minute. This phenomenon is called *ventricular escape.*

MECHANISM OF THE VAGAL EFFECTS. The acetylcholine released at the vagal nerve endings greatly increases the permeability of the fiber membranes to potassium, which allows rapid leakage of potassium to the exterior. This causes increased negativity inside the fibers, an effect called *hyperpolarization,* which makes excitable tissue much less excitable, as was explained in Chapter 5.

In the A-V node, the state of hyperpolarization makes it difficult for the minute junctional fibers, which can generate only small quantities of current during the action potential, to excite the nodal fibers. Therefore, the *safety factor* for transmission of the cardiac impulse through the junctional fibers and into the nodal fibers decreases. A moderate decrease in this simply delays conduction of the impulse, but a decrease in safety factor below unity (which means so low that the action potential of one fiber cannot cause an action potential in the successive fiber) completely blocks conduction.

Effect of Sympathetic Stimulation on Cardiac Function. Sympathetic stimulation causes essentially the opposite effects on the heart to those caused by vagal stimulation as follows: First, it increases the rate of S-A nodal discharge. Second, it increases the excitability of all portions of the heart. Third, it increases greatly the force of contraction of all the cardiac musculature, both atrial and ventricular, as discussed in the previous chapter.

In short, sympathetic stimulation increases the overall activity of the heart. Maximal stimulation can almost triple the rate of heartbeat and can increase the strength of heart contraction as much as two- to three-fold.

MECHANISM OF THE SYMPATHETIC EFFECT. Stimulation of the sympathetic nerves releases the hormone norepinephrine at the sympathetic nerve endings. The precise mechanism by which this hormone acts on cardiac muscle fibers is still somewhat doubtful, but the present belief is that it increases the permeability of the fiber membrane to sodium. In the S-A node, an increase of sodium permeability would cause increased tendency for the resting membrane potential to decay to the threshold level for self-excitation, which obviously would accelerate the onset of self-excitation after each successive heartbeat and therefore increase the rate of the heart.

In the A-V node, increased sodium permeability would make it easier for each fiber to excite the succeeding fiber. Therefore, the minute junctional fibers of the A-V nodal region could then easily excite the nodal fibers, thereby decreasing the conduction time from the atria to the ventricles.

ABNORMAL RHYTHMS OF THE HEART

Abnormal cardiac rhythms can be caused by (1) abnormal rhythmicity of the pacemaker itself, (2) shift of the pacemaker from the S-A node to other parts of the heart, (3) blocks at different points in the transmission of the impulse through the heart, (4) abnormal pathways of impulse transmission through the heart, and (5) spontaneous generation of abnormal impulses in almost any part of the heart. Some of these will be discussed in Chapter 17 in relation to electrocardiographic analysis of cardiac arrhythmias, but the major disturbances and their causes are presented here to illustrate some aberrations that can occur in the rhythmicity and conducting system of the heart.

PREMATURE CONTRACTIONS — ECTOPIC FOCI

Often, a small area of the heart becomes much more excitable than normal and causes an occasional abnormal impulse to be generated in between the normal impulses. A depolarization wave spreads outward from the irritable area and initiates a *premature contraction* of the heart. The focus at which the abnormal impulse is generated is called an *ectopic focus*.

The usual cause of an ectopic focus is an irritable area of cardiac muscle resulting from overuse of stimulants, such as caffeine or nicotine, or

resulting from lack of sleep, anxiety, or other debilitating states.

Shift of the Pacemaker to an Ectopic Focus. Sometimes an ectopic focus becomes so irritable that it establishes a rhythmic contraction of its own at a more rapid rate than that of the S-A node. When this occurs the ectopic focus becomes the pacemaker of the heart. The most common point for development of an ectopic pacemaker is the A-V node itself or the A-V bundle, as will be discussed in more detail in Chapter 16.

HEART BLOCK

Occasionally, transmission of the impulse through the heart is blocked at critical points in the conductive system. One of the most common of these points is between the atria and the ventricles; this condition is called *atrioventricular block*. Another common point is in one of the *bundle branches* of the Purkinje system. Rarely, a block also develops between the S-A node and the atrial musculature.

Atrioventricular Block. In the human being, block from the atria to the ventricles can result from localized damage or depression of the *A-V junctional* fibers or of the *A-V bundle*. The causes include different types of infectious processes, excessive stimulation by the vagus nerves (which depresses conductivity of the junctional fibers), localized destruction of the A-V bundle as a result of a coronary infarct, pressure on the A-V bundle by arteriosclerotic plaques, or depression caused by various drugs.

Figure 14–7 illustrates a typical record of atrial and ventricular contraction while the A-V bundle was being progressively compressed to cause successive stages of block. During the first three contractions of the record, ventricular contraction followed in orderly sequence approximately 0.16 second after atrial contraction. Then, A-V bundle compression was begun, and the interval of time between the beginning of atrial contraction and the beginning of ventricular contraction increased steadily during the next five heartbeats from 0.16 to 0.32 second. Beyond this point further compression completely blocked impulse transmission. Thereafter, the atria continued to beat at their normal rate of rhythm, while the ventricles failed to contract at all for approximately 7 seconds. Then,

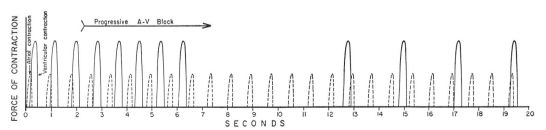

Figure 14–7 Contraction of the atria and ventricles of a heart, showing the effect of progressive A-V block. Note the progressive increase in the interval between the onset of atrial contraction and the onset of ventricular contraction. Note also the final complete asynchronism between atrial and ventricular contraction.

"ventricular escape" occurred, and a rhythmic focus in the ventricles suddenly began to act as a ventricular pacemaker, causing ventricular contractions at a rate of approximately 30 per minute; these were completely dissociated from the atrial contraction.

FLUTTER AND FIBRILLATION

Frequently, either the atria or the ventricles begin to contract extremely rapidly and often incoordinately. The lower frequency and more coordinate contractions up to 200 to 300 beats per minute are generally called *flutter* and the very high frequency and incoordinate contractions *fibrillation.*

Two basic theories of flutter and fibrillation have been proposed; these are (1) a *single* or *multiple ectopic foci* emitting many impulses one after another in rapid succession and (2) a *circus movement,* in which the impulse travels around and around through the heart muscle, never stopping.

Ectopic Foci. The ectopic focus theory is easy to understand. An area of the heart simply becomes so irritable that it keeps sending rapid impulses in all directions, resulting in rapid rates of contraction. There is reason to believe that at least some instances of atrial flutter may result from this cause.

The Circus Movement. There is also reason to believe that many instances of atrial flutter result from circus movements around the atria, and since only the circus movement theory has explained adequately the course of events in most instances of atrial and ventricular fibrillation, we will discuss it in detail.

Figure 14–8 illustrates several small cardiac muscle strips cut in the form of circles. If such a strip is stimulated at the 12 o'clock position *so that the impulse travels in only one direction,* the impulse spreads progressively around the circle until it returns to the 12 o'clock position. If the originally stimulated muscle fibers are still in a refractory state, the impulse then dies out, for refractory muscle cannot transmit a second impulse. However, there are three different conditions that could cause this impulse to continue to travel around the circle:

First, if the *length of the pathway around the circle is long,* by the time the impulse returns to the 12 o'clock position the originally stimulated muscle will no longer be refractory, and the impulse will continue around the circle again and again.

Second, if the length of the pathway remains constant but the *velocity of conduction becomes decreased* enough, an increased interval of time will elapse before the impulse returns to the 12 o'clock position. By this time the originally stimulated muscle might be out of the refractory state, and the impulse can continue around the circle again and again.

Third, *the refractory period of the muscle might have become greatly shortened.* In this case, the impulse could also continue around and around the circle.

All three of these conditions occur in different pathological states of the human heart as follows: (1) A long pathway frequently occurs in dilated hearts. (2) Decreased rate of conduction frequently results from blockage of the Purkinje system. (3) A shortened refractory period frequently occurs in response to various drugs, such as epinephrine, or following intense vagal stimulation. Thus, in many different cardiac disturbances circus movements can cause abnormal cardiac rhythmicity that completely ignores the pacesetting effects of the S-A node.

Atrial Flutter Resulting from a Circus Pathway. Figure 14–9 illustrates a circus pathway around and around the atria from top to bottom, the pathway lying to the left of the superior and inferior venae cavae. Such circus pathways have been initiated experimentally in the atria of dogs' hearts, and electrocardiographic records, which will be discussed in Chapter 17, indicate that this type of circus pathway also develops in the human heart when the atria become greatly dilated as a result of valvular heart disease. The rate of flutter is usually 200 to 350 times per minute.

Partial Block at the A-V Node During Atrial Flutter. The functional refractory period of the Purkinje fibers and ventricular muscle is approximately 1/200 minute so that not over 200 impulses per minute can be transmitted into the ventricles. Therefore, when the atrium contracts as rapidly as 300 times per minute, only one out of every two impulses passes into the ventricles, thus causing the atria to beat at a rate two times that of the ventricles. The heart is then said to have a 2:1 rhythm. Occasionally, a 3:1 and, rarely, a 4:1 rhythm of the heart develops in the same manner.

The "Chain Reaction" Mechanism of Fibrillation. Fibrillation, whether it occurs in the atria or in the ventricles, is a very different condition from flutter. One can see many separate contractile waves spreading in different directions over the cardiac muscle at the same time in either atrial or ven-

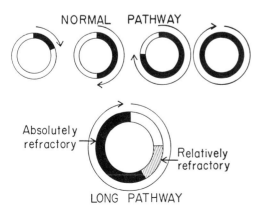

Figure 14–8 The circus movement, showing annihilation of the impulse in the short pathway and continued propagation of the impulse in the long pathway.

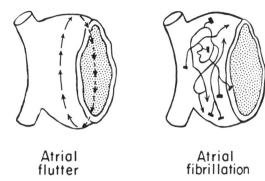

Atrial flutter

Atrial fibrillation

Figure 14-9 Pathways of impulses in atrial flutter and atrial fibrillation.

tricular fibrillation. Obviously, then, the circus movement in fibrillation is entirely different from that in flutter. One of the best ways to explain the mechanism of fibrillation is to describe the initiation of fibrillation by stimulation with 60 cycle alternating electrical current.

FIBRILLATION CAUSED BY 60 CYCLE ALTERNATING CURRENT. At a central point in the ventricles of heart A in Figure 14-10 a 60 cycle electrical stimulus is applied through a stimulating electrode. The first cycle of the electrical stimulus causes a depolarization wave to spread in all directions, leaving all the muscle beneath the electrode in a refractory state. After about 0.25 second, this muscle begins to come out of the refractory state, some portions of the muscle coming out of refractoriness prior to other portions. This state of events is depicted in heart A by many light patches, which represent excitable cardiac muscle, and dark patches, which represent refractory muscle. Stimuli from the electrode can now cause impulses to travel in certain directions through the heart but not in all directions. It will be observed in heart A that certain impulses travel for short distances until they reach refractory areas of the heart and then are blocked. Other impulses, however, pass between the refractory areas and continue to travel in the excitable patches of muscle. Now, several events transpire in rapid succession, all occurring simultaneously and eventuating in a state of fibrillation. These are:

First, block of the impulses in some directions but successful transmission in other directions creates one of the necessary conditions for a circus movement to develop—that is, *transmission of at least some of the depolarization waves around the heart in only one direction.* As a result, these waves do not annihilate themselves on the opposite side of the heart but can continue around and around the ventricles.

Second, the rapid stimulation of the heart causes two changes in the cardiac muscle itself, both of which predispose to circus movement: (1) The *velocity of conduction through the heart becomes decreased,* which allows a longer time interval for the impulses to travel around the heart. (2) The *refractory period of the muscle becomes*

shortened, allowing re-entry of the impulse into previously excited heart muscle within a much shorter period of time than normally.

Third, one of the most important features of fibrillation is the *division of impulses,* as illustrated in heart A. When a depolarization wave reaches a refractory area in the heart, it travels to both sides around the area. Thus, a single impulse becomes two impulses. Then when each of these reaches another refractory area it, too, divides to form still two more impulses. In this way many different impulses are continually being formed in the heart by a progressive *chain reaction* until, finally, there are many impulses traveling in many different directions at the same time. Furthermore, this irregular pattern of impulse travel causes a *circuitous route for the impulses to travel, greatly lengthening the conductive pathway, which is one of the conditions leading to fibrillation.* It also results in a continual irregular pattern of patchy refractory areas in the heart. One can readily see that a vicious cycle has been initiated: more and more impulses are formed; these cause more and more patches of refractory muscle; and the refractory patches cause more and more division of the impulses. Therefore, any time a single area of cardiac muscle comes out of refractoriness, an impulse is always close at hand to re-excite the area.

Heart B in Figure 14-10 illustrates the final state that develops in fibrillation. Here one can see many impulses traveling in all directions, some dividing and increasing the number of impulses while others are blocked entirely by refractory areas. In the final state of fibrillation, the number of new impulses being formed exactly equals the number of impulses that are being blocked by refractory areas. Thus, a steady state has developed with a certain average number of impulses traveling all the time in all directions through the cardiac syncytium.

As is the case with many other cardiac arrhythmias, fibrillation is usually confined to either the atria or the ventricles alone and not to both syncytial masses of muscle at the same time.

DEMONSTRATION OF THE CHAIN REACTION MECHANISM OF VENTRICULAR FIBRILLATION IN THE "IRON

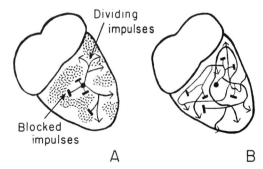

A   **B**

Figure 14-10 (A) Initiation of fibrillation in a heart when patches of refractory musculature are present. (B) Continued propagation of fibrillatory impulses in the fibrillating ventricle.

HEART." In the human heart the chain reaction mechanism is difficult to demonstrate for two reasons: (1) It is impossible to follow exactly the wave fronts of the electrical impulses traveling through the heart muscle and (2) there are so many minute waves of contraction spreading at the same time that the eyes cannot follow these continually. An *iron heart model* has been developed in our laboratory that does show easily the chain reaction mechanism. This is a large iron bob suspended in nitric acid. Under appropriate conditions an oxide film develops on the surface of the iron. Then, a single electrical stimulus to the surface will cause a single excitation wave to travel over the entire surface comparable to normal stimulation of the heart. But multiple stimuli either all at once or in rapid succession will cause the typical chain reaction as just described, with resultant fibrillation. One can easily see the chain reaction on the surface of the iron bob, and it can be recorded electrically or photographically. Also, typical fibrillatory electrocardiographic patterns are recorded.

Atrial Fibrillation. Atrial fibrillation is completely different from atrial flutter because the circus movement does not travel in a regular pathway. Instead, many different excitation waves can be seen to travel over the surface of the atria at the same time. Atrial fibrillation occurs very frequently when the atria become greatly overdilated — in fact, many times as frequently as flutter. When flutter does occur, it usually becomes fibrillation after a few days or weeks. To the right in Figure 14–9 are illustrated the pathways of fibrillatory impulses traveling through the atria. Obviously, atrial fibrillation results in complete incoordination of atrial contraction so that atrial pumping ceases entirely.

EFFECT OF ATRIAL FIBRILLATION ON THE OVERALL PUMPING EFFECTIVENESS OF THE HEART. The normal function of the atria is only to help fill the ventricles. However, the atria are probably responsible for not more than 25 to 30 per cent of the normal ventricular filling, which was explained in Chapter 13. Therefore, even when the atria fail to act as primer pumps, the ventricles can still fill enough that the effectiveness of the heart as a pump is reduced only 25 to 30 per cent, which is well within the "cardiac reserve" of all but severely weakened hearts. For this reason, atrial fibrillation can continue for many years without profound cardiac debility.

IRREGULARITY OF VENTRICULAR RATE DURING ATRIAL FIBRILLATION. When the atria are fibrillating, impulses arrive at the A-V node rapidly but also irregularly. Since the A-V bundle will not pass a second impulse for approximately 0.3 second after a previous one, at least 0.3 second must elapse between one ventricular contraction and the next, and an additional interval of 0 to 0.6 second usually occurs before one of the irregular fibrillatory impulses happens to arrive at the A-V node. Thus, the interval between successive ventricular contractions varies from 0.3 second to about 0.9 second, causing a very irregular heart beat. In fact, this irregularity is one of the clinical findings used to diagnose the condition.

Ventricular Fibrillation. Ventricular fibrillation is extremely important because at least one quarter of all persons die in ventricular fibrillation. For instance, the hearts of most patients with coronary infarcts fibrillate shortly before death. In only a few instances on record have fibrillating human ventricles been known to return of their own accord to a rhythmic beat.

The likelihood of circus movements in the ventricles and, consequently, of ventricular fibrillation is greatly increased when the ventricles are dilated or when the rapidly conducting *Purkinje system is blocked* so that impulses cannot be transmitted rapidly. Also, *electrical stimuli,* as noted above, or *ectopic foci,* which are discussed below, are common initiating causes of ventricular fibrillation.

INABILITY OF THE HEART TO PUMP BLOOD DURING VENTRICULAR FIBRILLATION. When the ventricles begin to fibrillate, the different parts of the ventricles no longer contract simultaneously. For the first few seconds the ventricular muscle undergoes rather coarse contractions which may pump a few milliliters of blood with each contraction. However, the impulses in the ventricles rapidly become divided into many much smaller impulses, and the contractions become fine rather than coarse, pumping no blood whatsoever. The ventricles dilate because of failure to pump the blood that is flowing into them, and within 60 to 90 seconds the ventricular muscle becomes too weak, because of lack of coronary blood supply, to contract strongly, even if coordinate contraction should return. Therefore, death is immediate when ventricular fibrillation begins.

IRRITABLE FOCI AS THE USUAL CAUSE OF VENTRICULAR FIBRILLATION. The cause of ventricular fibrillation is most often an irritable focus in the ventricular muscle caused by coronary insufficiency or by compression from an arteriosclerotic plaque. Most impulses originating in an irritable focus travel around the heart in all directions, meet on the opposite side of the heart, and annihilate themselves. However, when impulses from a focus become frequent the refractory period of the stimulated muscle becomes shortened, impulse conduction becomes slowed, and patchy areas of refractoriness develop in the ventricles. These are the essential conditions for initiating the chain reaction mechanism, and fibrillation begins in the same manner as that described above for stimulation with a 60 cycle electrical current. Thus, any ectopic focus that emits rapid impulses is likely to initiate fibrillation.

ELECTRICAL DEFIBRILLATION OF THE VENTRICLES. Though a weak alternating current almost invariably throws the ventricles into fibrillation, a very strong electrical current passed through the ventricles for a short interval of time can stop fibrillation by throwing all the ventricular muscle into refractoriness simultaneously. This is accomplished by passing intense current through electrodes placed on two sides of the heart. The current penetrates most of the fibers of the ventricles, thus stimulating essentially all parts of the ventricles simultaneously and causing them to become refractory. All impulses stop, and the heart then remains quiescent for 3 to 5 seconds, after which it begins to beat again, with the S-A node or, often, some other part of the heart becoming the pacemaker. Occasionally, however, the same irritable focus that had originally thrown the ventricles into fibrillation is still present, and fibrillation begins again immediately.

When electrodes are applied directly to the two sides of the heart, fibrillation can usually be stopped with 70 to 100 volts of 60 cycle alternating current applied for $^1\!/_{10}$ second or 1000 volts direct current applied for a few thousandths of a second. When applied through the chest wall, as illustrated in Figure 14–11, approximately 440 volts of 60 cycle alternating current or several thousand volts direct current is required. In our laboratory the heart of a single anesthetized dog was defibrillated 130 times through the chest wall, and the animal remained in perfectly normal condition.

Hand Pumping of the Heart ("Cardiac Massage") as an Aid to Defibrillation. Unless defibrillated within one minute after fibrillation begins, the heart is usually too weak to be revived. However, it is still possible to revive the heart by preliminarily pumping it by hand. In this way small quantities of blood are delivered into the aorta, and a renewed coronary blood supply develops. After a few minutes, electrical defibrillation often becomes possible. Indeed, fibrillating hearts have been pumped by hand as long as 90 minutes before defibrillation. In recent years, a technique of pumping the heart without opening the chest has been developed; this technique consists of intermittent, very powerful thrusts of pressure on the chest wall.

Lack of blood flow to the brain for more than 5 to 10 minutes usually results in permanent mental impairment or even total destruction of the brain. Even though the heart should be revived, the person might die from the effects of brain damage or live with permanent mental impairment.

CARDIAC ARREST

When cardiac metabolism becomes greatly disturbed as a result of any one of many possible conditions, the rhythmic contractions of the heart occasionally stop. One of the most common causes of cardiac arrest is hypoxia of the heart, for severe hypoxia prevents the muscle fibers from maintaining normal ionic differentials across their membranes. Therefore, polarization of the membranes becomes reduced, and the excitability may be so affected that the automatic rhythmicity disappears.

Occasionally, patients with severe myocardial disease develop cardiac arrest, which obviously can lead to death. In a few cases, however, rhyth-

Figure 14–11 Application of alternating current to the chest to stop ventricular fibrillation.

mic electrical impulses from an implanted electronic cardiac "pacemaker" have been successfully used in keeping patients alive for several years.

REFERENCES

Bigger, J. T., Jr.; Weinberg, D. I.; Kovalik, A. T. W.; Harris, P. D.; Cranefield, P. C.; and Hoffman, B. F.: Effects of diphenylhydantoin on excitability and automaticity in the canine heart. Circ. Res., 26:1, 1970.

Brady, A. J.: Excitation and excitation-contraction coupling in cardiac muscle. Ann. Rev. Physiol., 26:341, 1964.

Eick, R. E. T., and Hoffman, B. F.: Chronotropic effect of cardiac glycosides in cats, dogs, and rabbits. Circ. Res., 25:365, 1969.

Govier, W. C., and Holland, W. C.: Effects of ouabain on tissue calcium exchange in pacemaker of turtle heart. Amer. J. Physiol., 207:195, 1964.

Guyton, A. C., and Satterfield, J.: Factors concerned in electrical defibrillation of the heart, particularly through the unopened chest. Amer. J. Physiol., 167:81, 1951.

Hasimoto, K.; Suzuki, Y.; and Chiba, S.: Effect of potassium excess on pacemaker activity of canine sinoatrial node in vivo. Amer. J. Physiol., 218:83, 1970.

Hayden, W. G.; Hurley, E. J.; and Rytand, D. A.: The mechanism of canine atrial flutter. Circ. Res., 20:496, 1967.

Hoffman, B. F.: Physiology of atrioventricular transmission. Circulation, 24:506, 1961.

Hoffman, B. F., and Cranefield, P. F.: Electrophysiology of the Heart. New York, Blakiston Div., McGraw-Hill Book Company, 1960.

Janse, M. J.: Influence of the direction of the atrial wave front on A-V nodal transmission in isolated hearts of rabbits. Circ. Res., 25:439, 1969.

Jose, A. D., and Stitt, F.: Effects of hypoxia and metabolic inhibitors on the intrinsic heart rate and myocardial contractility in dogs. Circ. Res., 25:53, 1969.

Levy, M. N.; Martin, P. J.; Iano, T.; and Zieske, H.: Paradoxical effect of vagus nerve stimulation on heart rate in dogs. Circ. Res., 25:303, 1969.

Levy, M. N., and Zieske, H.: Autonomic control of cardiac pacemaker activity and arterioventricular transmission. J. Appl. Physiol., 27:465, 1969.

Mendez, C.; Mueller, W. J.; and Urguiaga, X.: Propagation of impulses across the Purkinje fiber-muscle junctions in the dog heart. Circ. Res., 26:135, 1970.

Sano, T., and Sawanobori, T.: Mechanism initiating ventricular fibrillation demonstrated in cultured ventricular muscle tissue. Circ. Res., 26:201, 1970.

Scher, A. M.: Excitation of the heart. In Handbook of Physiology. Baltimore, The Williams & Wilkins Co., 1962, Vol. I, Sec. II, 287.

Scher, A. M.; Rodriguez, M. I.; Liikane, J.; and Young, A. C.: The mechanism of atrioventricular conduction. Circ. Res., 7:54, 1959.

Scherf, D., and Cohen, J.: The Atrioventricular Node and Selected Cardiac Arrhythmias. New York, Grune & Stratton, Inc., 1964.

Tacker, W. A., Jr.; Geddes, L. A.; McFarlane, J.; Milnor, W.; Gullett, J.; Havens, W.; Green, E.; and Moore, J.: Optimum current duration for capacitor-discharge defibrillation of canine ventricles. *J. Appl. Physiol., 27*:480, 1969.

Torres, J. C., and Angelakos, E. T.: Effect of hypothermia on excitation and propagation in the isolated atrium. *Amer. J. Physiol., 207*:199, 1964.

Tsagaris, T. J.; Sutton, R. B.; and Kuida, H.: Hemodynamic effects of varying pacemaker sites. *Amer. J. Physiol., 218*:246, 1970.

Tyberg, J. V.; Parmley, W. W.; and Sonnenblick, E. H.: In-vitro studies of myocardial asynchrony and regional hypoxia. *Circ. Res., 25*:569, 1969.

Vick, R. L.: Effects of increased transmural pressures upon atrial and ventricular rhythms in the dog heart-lung preparation. *Circ. Res., 13*:39, 1963.

Watanabe, Y., and Dreifus, L. S.: Sites of impulse formation within the atrioventricular junction of the rabbit. *Circ. Res., 22*:717, 1968.

Yamada, K.; Okajima, M.; Hori, K.; Fujino, T.; Muraki, H.; Hishida, H.; and Kobayashi, T.: On the genesis of the absolute ventricular arrhythmia associated with atrial fibrillation. *Circ. Res., 22*:707, 1968.

CHAPTER 15

The Normal Electrocardiogram

Transmission of the impulse through the heart has been discussed in detail in Chapter 14. As the impulse passes through the heart, electrical currents spread into the tissues surrounding the heart, and a small proportion of these spreads all the way to the surface of the body. If electrodes are placed on the body on opposite sides of the heart, the electrical potentials generated by the heart can be recorded; the recording is known as an *electrocardiogram.* A normal electrocardiogram for two beats of the heart is illustrated in Figure 15–1.

CHARACTERISTICS OF THE NORMAL ELECTROCARDIOGRAM

The normal electrocardiogram is composed of a P wave, a "QRS complex," and a T wave. The QRS complex is actually three separate waves, the Q wave, the R wave, and the S wave, all of which are caused by passage of the cardiac impulse through the ventricles. In the normal electrocardiogram, the Q and S waves are often much less prominent than the R wave and sometimes actually absent, but, even so, the wave is still known as the QRS complex, or simply QRS wave.

The P wave is caused by electrical currents generated as the atria depolarize prior to contraction, and the QRS complex is caused by currents generated when the ventricles depolarize prior to contraction. Therefore, both the P wave and the components of the QRS complex are *depolarization waves.* The T wave is caused by currents generated as the ventricles recover from the state of depolarization. This process occurs in ventricular muscle

about 0.25 second after depolarization, and this wave is known as a *repolarization wave.* Thus, the electrocardiogram is composed of both depolarization and repolarization waves. The principles of depolarization and repolarization were discussed in Chapter 5. However, the distinction between depolarization waves and repolarization waves is so important in electrocardiography that further clarification is needed, as follows:

DEPOLARIZATION WAVES VERSUS REPOLARIZATION WAVES

Figure 15–2 illustrates a muscle fiber in four different stages of depolarization and repolarization. During the process of "depolarization" the normal negative potential inside the fiber is lost and the membrane potential actually reverses; that is, it becomes slightly positive inside and negative outside.

In Figure 15–2A the process of depolarization, illustrated by the positivity inside and negativity outside, is traveling from left to right, and the first half of the fiber is already depolarized while the remaining half is still polarized. Therefore, the left electrode on the fiber is in an area of negativity where it touches the outside of the fiber, while the right electrode is in an area of positivity; this causes the meter to record positively. To the right of the muscle fiber is illustrated a record of the potential between the electrodes as recorded by a high-speed recording meter at this particular stage of depolarization. Note that when depolarization has reached this halfway mark, the record has risen to a maximum positive value.

In Figure 15–2B depolarization has extended over the entire muscle fiber, and the recording to the right has returned to the zero base line because both electrodes are now in areas of equal negativity. The completed wave is a *depolarization wave* because it results from spread of depolarization along the extent of the muscle fiber.

Figure 15–2C illustrates the repolarization process in the muscle fiber, which has proceeded halfway along the extent of the fiber from left to right. At this point, the left electrode is in an area of positivity, while the right electrode is in an area of negativity. This is opposite to the polarities in Figure 15–2A. Consequently, the recording, as illustrated to the right, becomes negative.

Figure 15–1 The normal electrocardiogram.

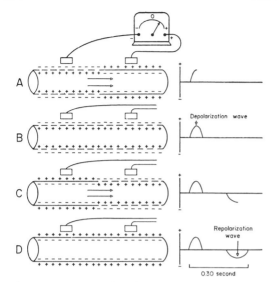

Figure 15–2 Recording the *depolarization wave* and the *repolarization* wave from a cardiac muscle fiber.

Finally, in Figure 15–2D the muscle fiber has completely repolarized, and both electrodes are in areas of positivity so that no potential is recorded between them. Thus, in the recording to the right, the potential returns once more to the zero level. This completed negative wave is a *repolarization wave* because it results from spread of the repolarization process over the muscle fiber.

Relationship of the Monophasic Action Potential of Cardiac Muscle to the QRS and T Waves. The monophasic action potential of ventricular muscle, which was discussed in the preceding chapter, normally lasts between 0.25 and 0.30 second. The top part of Figure 15–3 illustrates a monophasic action potential recorded from a microelectrode inserted into the inside of a single ventricular muscle fiber. The upsweep of this action potential is caused by *depolarization,* and the return of the potential to the base line is caused by *repolarization.* Note below the simultaneous recording of the electrocardiogram from this same ventricle, which shows the QRS wave appearing at the beginning of the monophasic action potential and the T wave appearing at the end. Note especially that *no potential at all is recorded in the electrocardiogram when the ventricular muscle is either completely polarized or completely depolarized.* It is only when the muscle is partly polarized and partly depolarized that current flows from one part of the ventricles to another part and therefore also flows to the surface of the body to cause the electrocardiogram.

RELATIONSHIP OF ATRIAL AND VENTRICULAR CONTRACTION TO THE WAVES OF THE ELECTROCARDIOGRAM

Before contraction of muscle can occur, a depolarization wave must spread through the muscle to initiate the chemical processes of contraction.

The P wave results from spread of the depolarization wave through the atria, and the QRS wave from spread of the depolarization wave through the ventricles. Therefore, the P wave occurs at the *beginning of contraction of the atria,* and the QRS wave occurs at the *beginning of contraction of the ventricles.*

The atria repolarize approximately 0.10 to 0.20 second after the depolarization wave. However, this is just at the moment that the QRS wave is being recorded in the electrocardiogram. Therefore, the atrial repolarization wave, known as the *atrial T wave,* is usually totally obscured by the much larger QRS wave. For this reason, an atrial T wave is rarely observed in the electrocardiogram.

On the other hand, the ventricular repolarization wave is the T wave of the normal electrocardiogram. Ordinarily, ventricular muscle begins to repolarize in some fibers approximately 0.15 second after the beginning of the depolarization wave and completes its repolarization in all fibers approximately 0.30 second after onset of depolarization. Thus, the process of repolarization extends over a fairly long period of time, about 0.15 second. For this reason the T wave in the normal electrocardiogram is a fairly prolonged wave, but the voltage of the T wave is considerably less than the voltage of the QRS complex, partly because of its prolonged length.

VOLTAGE AND TIME CALIBRATION OF THE ELECTROCARDIOGRAM

All recordings of electrocardiograms are made with appropriate calibration lines on the recording paper. Either these calibration lines are already ruled on the paper, as is the case when the pen recorder is used, or they are recorded on the paper at the same time that the electrocardiogram is recorded, which is the case with the optical types of electrocardiograph.

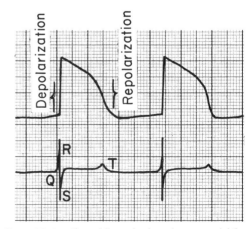

Figure 15–3 *Above:* Monophasic action potential from a ventricular muscle fiber during normal cardiac function, showing rapid depolarization and then repolarization occurring slowly during the plateau stage but very rapidly toward the end. *Below:* Electrocardiogram recorded simultaneously.

As illustrated in Figure 15–1, the horizontal calibration lines are arranged so that 10 small divisions (1 cm.) in the vertical direction in the standard electrocardiogram represents 1 millivolt.

The vertical lines on the electrocardiogram are time calibration lines. Each inch of the standard electrocardiogram is 1 second. Each inch in turn is usually broken into five segments by dark vertical lines, the distance between which represent 0.20 second. The intervals between the dark vertical lines are broken into five smaller intervals by thin lines, and the distance between each two of the smaller lines represents 0.04 second.

NORMAL VOLTAGES IN THE ELECTROCARDIOGRAM. The voltages of the waves in the normal electrocardiogram depend on the manner in which the electrodes are applied to the surface of the body. When one electrode is placed directly over the heart and the second electrode is placed elsewhere on the body, the voltage of the QRS complex may be as much as 3 to 4 mv. Even this voltage is very small in comparison with the monophasic action potential of 135 mv. when recorded directly from heart muscle. When electrocardiograms are recorded from electrodes on the two arms or on one arm and one leg, the voltage of the QRS complex usually is approximately 1 mv. from the top of the R wave to the bottom of the S wave, the voltage of the P wave between 0.1 and 0.3 mv., and that of the T wave between 0.2 and 0.3 mv.

The P-Q Interval. The duration of time between the beginning of the P wave and the beginning of the QRS wave is the interval between the beginning of contraction of the atrium and the beginning of contraction of the ventricle. This period of time is called the P-Q interval. The normal P-Q interval is approximately 0.16 second. This interval is sometimes also called the P-R interval because the Q wave frequently is absent.

The Q-T Interval. Contraction of the ventricle lasts essentially between the Q wave and the end of the T wave. This interval of time is called the Q-T interval and ordinarily is approximately 0.30 second.

The Rate of the Heart as Determined from Electrocardiograms. The rate of heartbeat can be determined easily from electrocardiograms, because the time interval between two successive beats is the reciprocal of the heart rate. If the interval between two beats as determined from the time calibration lines is 1 second, the heart rate is 60 beats per minute. The normal interval between two successive QRS complexes is approximately 0.8 second. This is a heart rate of 60/0.8 times per minute, or 75 beats per minute.

METHODS FOR RECORDING ELECTROCARDIOGRAMS

The electrical currents generated by the cardiac muscle during each beat of the heart sometimes change potentials and polarity in less than 0.03 second. Therefore, it is essential that any apparatus for recording electrocardiograms be capable of responding rapidly to these changes in electrical potentials. In general, two different types of recording apparatuses are used for this purpose, as follows:

THE PEN RECORDER

Most older types of electrocardiographic apparatus used optical methods for recording the electrocardiogram. In these a beam of light was focused on a mirror mounted on a galvanometer coil located in a powerful magnetic field. When current flowed through the coil, the coil rotated in the field, and the mirror swept the reflected beam of light across a moving photographic paper. However, because the records required photographic development before the recording could be viewed, direct pen writing recorders have become the vogue in recent years. These write the electrocardiogram with a pen directly on a moving sheet of paper. The pen is often a thin tube connected at one end to an inkwell, and its recording end is connected to a powerful electromagnet system that is capable of moving the pen back and forth at high speed. As the paper moves forward, the pen records the electrocardiogram. The movement of the pen in turn is controlled by means of appropriate amplifiers connected to electrocardiographic electrodes on the subject.

Other pen recording systems use special paper that does not require ink in the recording stylus. One such paper turns black when it is exposed to heat; the stylus itself is made very hot by electrical current flowing through its tip. Another type turns black when electrical current flows from the tip of the stylus through the paper to an electrode at its back. This leaves a black line at every point on the paper that the stylus touches.

RECORDING ELECTROCARDIOGRAMS WITH THE OSCILLOSCOPE

Electrocardiograms can also be recorded on the screen of an oscilloscope by the method discussed for nerve potentials in Chapter 5, or they can be photographed from the oscilloscopic screen. However, because of the cost of the oscilloscope, and because extremely high frequency electrical potentials do not need to be recorded, the less complicated and less expensive pen recorder just described is ordinarily used in clinical electrocardiography.

FLOW OF CURRENT AROUND THE HEART DURING THE CARDIAC CYCLE

RECORDING ELECTRICAL POTENTIALS FROM A PARTIALLY DEPOLARIZED MASS OF SYNCYTIAL CARDIAC MUSCLE

Figure 15–4 illustrates a syncytial mass of cardiac muscle that has been stimulated at its

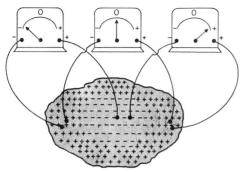

Figure 15–4 Instantaneous potentials developed on the surface of a cardiac muscle mass that has been depolarized in its center.

centralmost point. Prior to stimulation, all of the exteriors of the muscle cells had been positive and the interiors negative. However, for reasons presented in Chapter 5 in the discussion of membrane potentials, as soon as an area of the cardiac syncytium becomes depolarized, negative charges leak to the outside of the depolarized area, making this particular surface area electronegative, as represented by the negative signs, with respect to the remaining surface of the heart which is still polarized in the normal manner, as represented by the positive signs. Therefore, a meter connected with its negative terminal on the area of depolarization and its positive terminal on one of the still-polarized areas, as illustrated to the right in the figure, will record positively.

Two other possible electrode placements and meter readings are also illustrated in Figure 15–4. These should be studied carefully. Obviously, since the process of depolarization spreads in all directions through the heart, the potential differences shown in the figure last for only a few milliseconds, and the actual voltage measurements can be accomplished only with a high-speed recording apparatus.

Figure 15–5 Flow of current in the chest around a partially depolarized heart.

FLOW OF ELECTRICAL CURRENTS AROUND THE HEART IN THE CHEST

Figure 15–5 illustrates the ventricular muscle mass lying within the chest. Even the lungs, though filled with air, conduct electricity to a surprising extent, and fluids of the other tissues surrounding the heart conduct electricity even more easily. Therefore, the heart is actually suspended in a conductive medium. When one portion of the ventricles becomes electronegative with respect to the remainder, electrical current flows from the depolarized area to the polarized area in large circuitous routes, as noted in the figure.

It will be recalled from the discussion of the Purkinje system in Chapter 14 that the cardiac impulse first arrives in the ventricles in the walls of the septum and almost immediately thereafter on the endocardial surfaces of the remainder of the ventricles, as shown by the shaded areas and the negative signs in Figure 15–5. This provides electronegativity on the insides of the ventricles and electropositivity on the outer walls of the ventricles. If one algebraically averages all the lines of current flow (the elliptical lines in Figure 15–5), he finds that the average current flow is from the base of the heart toward the apex. During most of the remainder of the cycle of depolarization, the current continues to flow in this direction as the impulse spreads from the endocardial surface outward through the ventricular muscle. However, immediately before the depolarization wave has completed its course through the ventricles, the direction of current flow reverses for about 1/100 second, flowing then from the apex toward the base because the very last part of the heart to become depolarized is the outer walls of the ventricles near their base.

Thus, in the normal heart it may be considered that current flows primarily in the direction from the base toward the apex during almost the entire cycle of depolarization except at the very end. Therefore, if a meter is connected to the surface of the body as shown in Figure 15–5, the electrode nearer the base will be negative with respect to the electrode nearer the apex, so that the meter shows a slight potential between the two electrodes. In making electrocardiographic recordings, various standard positions for placement of electrodes are used, and whether the polarity of the recording during each cardiac cycle is positive or negative is determined by the orientation of electrodes with respect to the current flow in the heart. Some of the conventional electrode systems, commonly called *electrocardiographic leads,* are discussed below.

ELECTROCARDIOGRAPHIC LEADS

THE THREE STANDARD LIMB LEADS

Figure 15–6 illustrates electrical connections between the limbs and the electrocardiograph for recording electrocardiograms from the so-called "standard" limb leads. The electrocardiograph in each instance is illustrated by mechanical meters in the diagram, though the actual electrocardiograph is a high-speed recording meter.

Lead I. In recording limb lead I, the negative terminal of the electrocardiograph is connected to the right arm and the positive terminal to the left arm. Therefore, when the point on the chest where the right arm connects to the chest is electronegative with respect to the point where the left arm connects, the electrocardiograph records positively — that is, above the zero voltage line in the electrocardiogram. When the opposite is true, the electrocardiograph records below the line.

Lead II. In recording limb lead II, the negative terminal of the electrocardiograph is connected to the right arm and the positive terminal to the left leg. Thus, when the right arm is negative with respect to the left leg, the electrocardiograph records positively.

Lead III. In recording limb lead III, the negative terminal of the electrocardiograph is connected to the left arm and the positive terminal to the left leg. This means that the electrocardiograph records positively when the left arm is negative with respect to the left leg.

Einthoven's Triangle. In Figure 15–6, a triangle, called *Einthoven's triangle,* is drawn around the area of the heart. This is a diagrammatic means of illustrating that the two arms and the left leg form apices of a triangle surrounding the heart. The two apices at the upper part of the triangle represent the points at which the two arms connect electrically with the fluids around the heart and the lower apex is the point at which the left leg connects with the fluids.

EINTHOVEN'S LAW. Einthoven's law states simply that if the electrical potentials of any two of the three standard electrocardiographic leads are recorded, the third one can be determined mathematically from the first two by simply summing the first two. For instance, let us assume that momentarily, as noted in Figure 15–6, the right arm is 0.2 mv. negative with respect to the average potential in the body, the left arm is 0.3 mv. positive, and the left leg is 1.0 mv. positive. Observing the meters in the figure, it will be seen that lead I records a positive potential of 0.5 mv., lead III records a positive potential of 0.7 mv., and lead II records a positive potential of 1.2 mv., because these are the instantaneous potential differences between the respective pairs of limbs.

Note that *the sum of the voltages in leads I and III equals the voltage in lead II.* That is, 0.5 plus 0.7 equals 1.2. Mathematically, this principle, called Einthoven's law, holds true at any given instant while the electrocardiogram is being recorded.

Normal Electrocardiograms Recorded by the Three Standard Leads. Figure 15–7 illustrates simultaneous recordings of the electrocardiogram in leads I, II, and III. It is obvious from this figure

Figure 15–6 Conventional arrangement of electrodes for recording the standard electrocardiographic leads. Einthoven's triangle is superimposed on the chest.

that the electrocardiograms in these three standard leads are very similar to each other, for they all record positive P waves and positive T waves, and the major proportion of the QRS complex is positive in each electrocardiogram.

On analysis of the three electrocardiograms, it can be shown with careful measurements that at

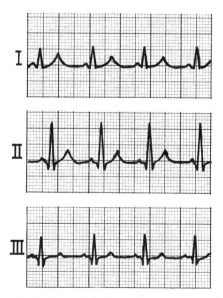

Figure 15–7 Normal electrocardiograms recorded from the three standard electrocardiographic leads.

any given instant the sum of the potentials in leads I and III equals the potential in lead II, thus illustrating the validity of Einthoven's law.

Because all normal electrocardiograms are very similar to each other, it does not matter greatly which electrocardiographic lead is recorded when one wishes to diagnose the different cardiac arrhythmias, for diagnosis of arrhythmias depends mainly on the time relationships between the different waves of the cardiac cycle. On the other hand, when one wishes to determine the extent and type of damage in the ventricles or in the atria, it does matter greatly which leads are recorded, for abnormalities of the cardiac muscle change the patterns of the electrocardiograms markedly in some leads and yet may not affect other leads.

Electrocardiographic interpretation of these two types of conditions—cardiac myopathies and cardiac arrhythmias—are discussed separately in the following two chapters.

CHEST LEADS (PRECORDIAL LEADS)

Often electrocardiograms are recorded with one electrode placed on the anterior aspect of the chest over the heart, as illustrated by the six separate points in Figure 15–8. This electrode is connected to the positive terminal of the electrocardiograph, and the negative electrode, called the *indifferent electrode,* is normally connected simultaneously through electrical resistances to the right arm, left arm, and left leg, as also shown in the figure. Usually six different standard chest leads are recorded from the anterior chest wall, the chest electrode being placed respectively at the six points illustrated in the diagram. The different leads recorded by the method illustrated in Figure 15–8 are known as leads V_1, V_2, V_3, V_4, V_5, and V_6.

Figure 15–9 illustrates the electrocardiograms of the normal heart as recorded in the six standard chest leads. Because the heart surfaces are close

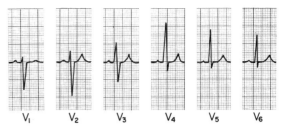

Figure 15–9 Normal electrocardiograms recorded from the six standard chest leads.

to the chest wall, each chest lead records mainly the electrical potential of the cardiac musculature immediately beneath the electrode. Therefore, relatively minute abnormalities in the ventricles, particularly in the anterior ventricular wall, frequently cause marked changes in the electrocardiograms recorded from chest leads.

In leads V_1 and V_2, the QRS recordings of the normal heart are mainly negative because, as illustrated in Figure 15–8, the chest electrode in these leads is nearer the base of the heart than the apex, which is the direction of electronegativity during most of the ventricular depolarization process. On the other hand, the QRS complexes in leads V_4, V_5, and V_6 are mainly positive because the chest electrode in these leads is near the apex, which is the direction of electropositivity during depolarization.

AUGMENTED UNIPOLAR LIMB LEADS

Another system of leads in wide use is the "augmented unipolar limb lead." In this type of recording, two of the limbs are connected through electrical resistances to the negative terminal of the electrocardiograph while the third limb is connected to the positive terminal. When the positive terminal is on the right arm, the lead is known as the aV_R lead; when on the left arm, as the aV_L lead; and when on the left leg, as the aV_F lead.

Normal recordings of the augmented unipolar limb leads are shown in Figure 15–10. These are all similar to the standard limb lead recordings except that the aV_R lead is inverted. The reason for this inversion is that the polarity of the electrocardiograph in this instance is connected back-

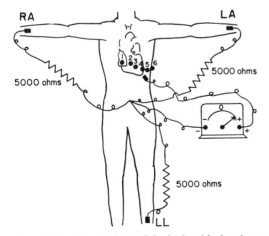

Figure 15–8 Connections of the body with the electrocardiograph for recording chest leads. (Modified from Burch and Winsor: A Primer of Electrocardiography. Lea & Febiger.)

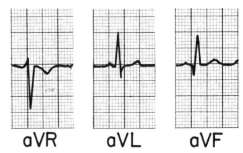

Figure 15–10 Normal electrocardiograms recorded from the three augmented unipolar limb leads.

ward to the major direction of current flow in the heart during the cardiac cycle.

Each augmented unipolar limb lead records the potential of the heart on the side nearest to the respective limb. Thus, when the recording in the aV_R lead is negative, this means that the side of the heart nearest to the right arm is negative in relation to the remainder of the heart; when the recording in the aV_F lead is positive, this means that the bottom of the heart is positive with respect to the remainder of the heart.

REFERENCES

See bibliography at end of Chapter 17.

CHAPTER 16

Electrocardiographic Interpretation in Cardiac Myopathies — Vectorial Analysis

From the discussion in Chapter 14 of impulse transmission through the heart, it is obvious that any change in the pattern of this transmission can cause abnormal electrical currents around the heart and, consequently, can alter the shapes of the waves in the electrocardiogram. For this reason, almost all serious abnormalities of the heart muscle can be detected by analyzing the contours of the different waves in the different electrocardiographic leads. The purpose of the present chapter is to discuss the various alterations in the electrocardiograms when the muscle of the heart, especially of the ventricles, functions abnormally.

PRINCIPLES OF VECTORIAL ANALYSIS OF ELECTROCARDIOGRAMS

USE OF VECTORS TO REPRESENT ELECTRICAL POTENTIALS

Before it is possible for one to understand how cardiac abnormalities affect the contours of waves in the electrocardiogram, he must first become familiar with the concept of vectors and vectorial analysis as applied to electrical currents flowing in and around the heart.

Several times in the preceding chapter it was pointed out that heart currents flow in a particular direction at any given instant in the cardiac cycle. A vector is an arrow that points in the direction of current flow *with the arrowpoint in the positive direction*. Also, by convention, the length of the arrow is drawn *proportional to the voltage generated by the current flow*.

In Figure 16–1, several different vectors are shown between areas of negativity and positivity in a syncytial mass of cardiac muscle. Note, first, that each vector points from negative to positive and, second, that the length of each vector is proportional to the amount of charges causing the current flow.

The Summated Vector in the Heart at Any Given

Instant. Figure 16–2 shows, via the shaded area and the negative signs, depolarization of the ventricular septum and parts of the lateral endocardial walls of the two ventricles. Electrical currents flow from the septum and from the lateral endocardial walls to the outside of the heart as indicated by the elliptical arrows. Currents also flow inside the heart chambers directly from the depolarized areas toward the polarized areas. Even though a small amount of current flows upward inside the heart, a considerably greater quantity flows downward on the outside of the ventricles toward the apex. Therefore, the summated vector of currents at this particular instant is drawn through the center of the ventricles in a direction from the base of the heart toward the apex. Furthermore, because these currents are considerable in quantity, the vector is relatively long.

Transposition of Vectors in Space. Two other vectors besides the one through the center of the heart are shown in Figure 16–2. These vectors extend in exactly the same direction as that through the center of the heart, and they are exactly the same length as the vector through the heart. They also

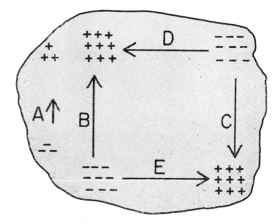

Figure 16–1 Vectors depicting current flow within a mass of cardiac muscle. Discussed in the text.

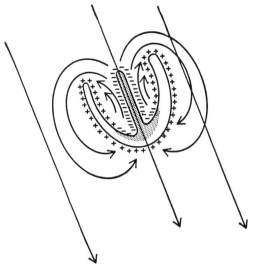

Figure 16-2 A summated vector through the partially depolarized heart. Two additional vectors are shown, one to each side of the heart. These have the same significance as the vector through the center of the heart (discussed in the text).

have exactly the same meaning as the single vector drawn through the center of the heart, because a vector indicates only (1) the direction of current flow and (2) the potential caused by the current. Therefore, the vector can be drawn to pass directly through the center of the heart, or it can be drawn to the side of or at one end of the heart. As long as the direction and the length of the vector are appropriate, its meaning remains exactly the same regardless of its position with respect to the heart.

DENOTING THE DIRECTION OF A VECTOR IN TERMS OF DEGREES

When a vector is horizontal and directed toward the subject's left side, it is said that the vec-

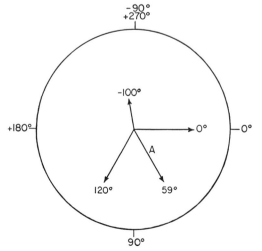

Figure 16-3 Vectors drawn to represent direction of current flow and potentials for several different hearts.

tor extends in the direction of 0 degrees, as is illustrated in Figure 16-3. From this zero reference point, the scale of vectors rotates clockwise: when the vector extends from above downward, it has a direction of 90 degrees; when it extends from the subject's left to his right, it has a direction of 180 degrees; and, when it extends upward, it has a direction of −90 or +270 degrees.

In a normal heart the average direction of the vector of the heart during spread of the depolarization wave is approximately 59 degrees, which is illustrated by vector A drawn from the center of Figure 16-3 in the 59 degree direction. This means that during most of the depolarization wave, the apex of the heart remains positive with respect to the base of the heart, as is discussed later in the chapter.

"AXIS" OF EACH OF THE STANDARD AND UNIPOLAR LEADS

In the preceding chapter the three standard leads and the three unipolar leads were described. Each lead is actually a pair of electrodes connected to the body on opposite sides of the heart, and the direction from the negative to the positive electrode is called the *axis of the lead*. A horizontal axis from the right side of the subject to his left is said to be 0 degrees, and the degrees are then recorded in the clockwise direction, as was just explained for vectors. Lead I is recorded from two electrodes placed respectively on the two arms. Since the electrodes lie in the horizontal direction with the positive electrode to the left, the axis of lead I is 0 degrees.

In recording lead II, electrodes are placed on the right arm and left leg. The right arm connects to the torso in the upper right-hand corner and the left leg to the lower left-hand corner. Therefore, the direction of this lead is approximately 60 degrees.

By a similar analysis it can be seen that lead III has an axis of approximately 120 degrees, lead aV_R 210 degrees, aV_F 90 degrees, and aV_L −30 degrees. The directions of the axes of all these different leads are shown in Figure 16-4. Also, the polarities of the electrodes are illustrated by the plus and minus signs. *The student must learn these axes and their polarities, particularly for the standard leads I, II, and III, if he wishes to understand the remainder of this chapter.*

VECTORIAL ANALYSIS OF POTENTIALS RECORDED IN DIFFERENT LEADS

Now that we have discussed, first, the conventions for representing current flow and potentials across the heart by means of vectors and, second, the axes of the leads, it becomes possible to put these two together to determine the potential that will be recorded in each lead for a given vector in the heart.

Figure 16-5 illustrates a partially depolarized heart; vector A represents the direction of current

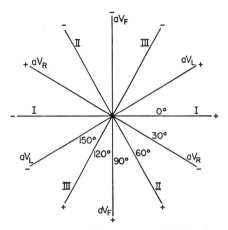

Figure 16–4 Axes of the three standard leads and of the three unipolar leads.

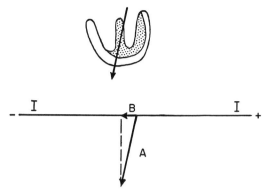

Figure 16–6 Determination of a resultant vector *B* along the axis of lead I when vector *A* represents the current flow in the ventricles.

flow in the heart and its potential. In this instance the direction of current flow is 55 degrees, and the potential will be assumed to be 2 mv. This vector *A* is now moved to a convenient point below the heart, but its direction and length remain exactly the same. Through its base is drawn the axis of lead I in the 0 degree direction. From the tip of vector *A* a perpendicular is dropped to the lead I axis, and a so-called *resultant vector* (*B*) is drawn along this axis. The head of this vector is in the positive direction, which means that the record momentarily being recorded in the electrocardiogram of lead I will be positive. The voltage recorded will be equal to length of *B* divided by length of *A* times 2 mv., or approximately 1 mv.

In Figure 16–6, vector *A* represents the current flow at an instantaneous point during ventricular depolarization in another heart in which the left side of the heart becomes depolarized somewhat more rapidly than the right. In this instance the vector has a direction of 100 degrees, and the voltage is again 2 mv. To determine the potential in lead I, we drop a perpendicular to the lead I axis and find the resultant vector *B*. Vector *B* is very short and this time in the negative direction, indicating that at this particular instant the recording in lead I will be negative (below the zero line), and the voltage recorded will be slight. This

figure illustrates that *when the vector in the heart is in a direction almost perpendicular to the axis of the lead, the voltage recorded in the lead is very low.* On the other hand, *when the vector has almost exactly the same axis as the lead, essentially the entire voltage of the vector will be recorded.*

Vectorial Analysis of Potentials in the Three Standard Leads. In Figure 16–7, vector *A* depicts the instantaneous direction and potential of current in a partially depolarized ventricle. To determine the potential recorded at this instant in each one of the three standard leads, perpendiculars are dropped to all the different leads as illustrated in the figure. The resultant vector *B* depicts the potential recorded in lead I, vector *C* depicts the potential in lead II, and vector *D* in lead III. In each of these the record is positive because the resultant vectors lie in the positive directions along the axes of the leads. The potential in lead I is approximately one-half that of the vector through the heart; in lead II it is almost exactly equal to that in the heart; and in lead III it is about one-third that in the heart.

An identical analysis can be used to determine potentials recorded in augmented limb leads, except that the respective axes of these leads are used in place of the standard lead axes used in Figure 16–7.

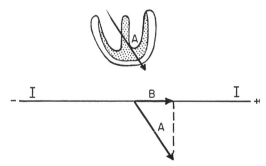

Figure 16–5 Determination of a resultant vector *B* along the axis of lead I when vector *A* represents the current flow in the ventricles.

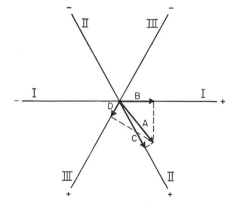

Figure 16–7 Determination of resultant vectors in leads I, II, and III when vector *A* represents the current flow in the ventricles.

VECTORIAL ANALYSIS OF THE NORMAL ELECTROCARDIOGRAM

VECTORS OCCURRING DURING DEPOLARIZATION OF THE VENTRICLES —THE QRS COMPLEX

When the cardiac impulse enters the ventricles through the A-V bundle, the first part of the ventricles to become depolarized is the left endocardial surface of the septum. This depolarization spreads rapidly to involve both endocardial surfaces of the septum, as illustrated by the shaded portion of the ventricle in Figure 16–8A. Then the depolarization spreads along the endocardial surfaces of the two ventricles, as shown in Figures 16–8B and C. Finally, it spreads through the ventricular muscle to the outside of the heart, as shown progressively in Figures 16–8C, D, and E.

At each stage of depolarization of the ventricles in Figures 16–8A to E, the instantaneous, summated current flow is represented by a vector superimposed on the ventricle in each figure. Each of these vectors is analyzed by the method described in the preceding section to determine the voltages that will be recorded at each instant in each of the three standard electrocardiographic leads. To the right is shown the progressive development of the QRS complex.

Before proceeding with any further consideration of vectorial analysis, it is essential that this analysis of the successive normal vectors presented in Figure 16–8 be understood. Each of these analyses should be studied in detail by the procedure given above. A short summary of this sequence is the following:

In Figure 16–8A the vector is short because only a small portion of the ventricles – the septum – is depolarized. Therefore, all electrocardiographic voltages are low at this instant. The voltage in lead II is greater than the voltages in leads I and III because the vector extends mainly in the same direction as the axis of lead II.

In Figure 16–8B the vector is long because much of the ventricles has now become depolarized. Therefore, the voltages in all electrocardiographic leads have increased.

In Figure 16–8C the vector is becoming shorter and the recorded EKG voltages less because the outside of the apex of the heart is now electronegative, neutralizing much of the negativity on

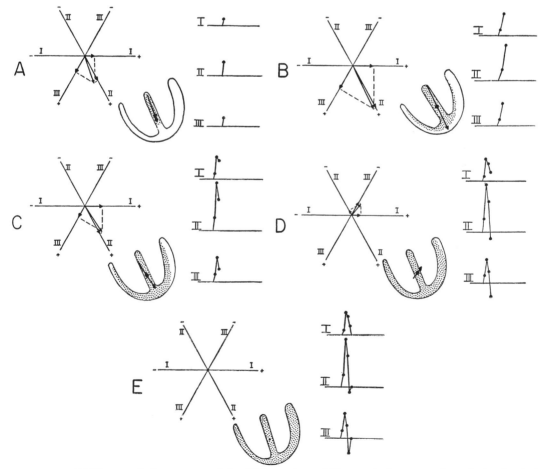

Figure 16–8   (A) The ventricular vectors and the QRS complexes 0.01 second after onset of ventricular depolarization, (B) 0.03 second after onset of depolarization, (C) 0.05 second after onset of depolarization, (D) 0.06 second after onset of depolarization, and (E) after depolarization of the ventricles is complete.

the endocardial surfaces of the heart. Also, the axis of the vector is shifting toward the left side of the chest because the left ventricle is slightly slower to depolarize than the right. The ratio of the voltage in lead I to that in lead III is increasing.

In Figure 16–8D the vector points toward the base of the left ventricle, and it is short because only a minute portion of the ventricular muscle is still polarized. Because of the direction of the vector at this time, the voltages recorded in leads II and III are both negative.

In Figure 16–8E the entire ventricular muscle mass is depolarized so that no current flows around the heart at all. The vector becomes zero, and the voltages in all leads become zero.

Thus, the QRS complexes are recorded in the three standard electrocardiographic leads. The QRS complex sometimes has a slight negative depression at its beginning; this is the Q wave. When it occurs, it is caused by initial depolarization of the septum nearer to the apex of the septum than to the base, which occasionally creates a weak vector in the apex-to-base direction for a minute fraction of a second before the usual base-to-apex vector occurs. The major positive deflection is the R wave, and the final negative deflection is the S wave.

THE ELECTROCARDIOGRAM DURING REPOLARIZATION—THE T WAVE

Once ventricular muscle becomes depolarized, approximately 0.15 second elapses before sufficient repolarization begins for it to be observed in the electrocardiogram, and repolarization is complete in about 0.30 second. It is this process of repolarization that causes the T wave in the electrocardiogram. Because the septum depolarizes first, it seems logical that it should repolarize first, but this is not the usual case because the septum has a longer period of contraction and depolarization than do other areas of the heart. Actually, many sections of the ventricles begin to repolarize almost simultaneously. Yet *the greatest portion of ventricular muscle to repolarize first is that located exteriorly near the apex of the heart.* And the endocardial surfaces, on the average, repolarize last. The reason for this abnormal sequence of repolarization is reputed to be that high pressure in the ventricles during contraction greatly reduces coronary blood flow to the endocardium, thereby slowing the repolarization process on the endocardial surfaces. The successive stages of repolarization are shown by the white areas in the ventricles of Figure 16–9.

Thus, *the predominant direction of the vector through the heart during* repolarization *of the ventricles is from base to apex, which is also the predominant direction of the vector during* depolarization. *As a result, the T wave in the normal electrocardiogram is positive, which is also the polarity of most of the normal QRS complex.*

In Figure 16–9 five stages of repolarization of

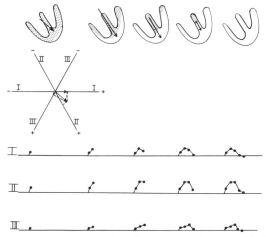

Figure 16–9   Generation of the T wave during repolarization of the ventricles, showing a vectorial analysis of the first stage.

the ventricles are noted. In each of these, the vector extends from the base toward the apex. At first the vector is relatively small because the area of repolarization is small. Later the vector becomes stronger and stronger because of greater degrees of repolarization. Finally, the vector becomes weaker again because the areas of depolarization still persisting become so slight that the total quantity of current flow begins to decrease. These changes illustrate that the vector is greatest when approximately half the heart is in the polarized state and approximately half is depolarized.

The changes in the electrocardiograms of the three standard leads during the process of repolarization are noted under each of the ventricles, depicting the progressive stages of repolarization. Over approximately 0.15 second, the period required for the whole process to take place, the T wave of the electrocardiogram is generated.

DEPOLARIZATION OF THE ATRIA—THE P WAVE

Depolarization of the atria begins in the S-A node and spreads in all directions over the atria. Therefore, the point of original electronegativity in the atria is approximately at the base of the superior vena cava, and the direction of current flow in the atrium at the beginning of depolarization is in the direction noted in Figure 16–10. Furthermore, the vector remains generally in this direction throughout the process of atrial depolarization.

Thus, the vector of current flow during depolarization in the atria points in almost the same direction as that in the ventricles. And, because this direction is in the direction of the axes of the standard leads I, II, and III, the electrocardiogram recorded from the atria during the process of depolarization usually is positive in all three standard leads, as illustrated in Figure 16–10. The record of atrial depolarization is known as the P wave.

Figure 16–10 Depolarization of the atria and generation of the P wave, showing the vector through the atria and the resultant vectors in the three standard leads. To the right are shown the atrial P and T waves.

Repolarization of the Atria — The Atrial T Wave. Spread of the depolarization wave through the atrial muscle is *extremely slow.* Therefore, the musculature around the S-A node becomes depolarized a long time before the musculature in distal parts of the atria. Because of this, *the area in the atria that becomes repolarized first is the area that becomes depolarized first,* which is an entirely different situation from that in the ventricles. Thus, repolarization begins in the region surrounding the S-A node. At this time the region around the S-A node becomes positive with respect to the rest of the atria, and this same general direction of the vector is maintained throughout the repolarization process — that is, *backward to the vector of depolarization.* Note again that this is opposite to the effect that occurs in the ventricles. Therefore, as noted to the right in Figure 16–10, the so-called atrial T wave follows about 0.15 second after the atrial P wave, this T wave being on the opposite side of the zero reference line from the P wave — that is, it is normally negative rather than positive. In the normal electrocardiogram this T wave appears at approximately the same time that the QRS complex of the ventricle appears. Therefore, it is almost always totally obscured by the larger QRS complex, though in some abnormal states it does play a role in the recorded electrocardiogram.

THE VECTORCARDIOGRAM

It was noted in the preceding discussions that the vector of current flow through the heart changes rapidly as the impulse spreads through the myocardium. It changes in two aspects: First, the vector increases and decreases in length because of the increasing and decreasing potential of the vector. Second, the vector changes its direction because of changes in the average direction of current flow through the heart. The so-called vectorcardiogram depicts these changes in the vectors at the different times during the cardiac cycle, as illustrated in Figure 16–11.

In the vectorcardiogram of Figure 16–11, point 5 is the zero reference point, and this point is the negative end of all the vectors. While the heart is quiescent, the positive end of the vector also remains at the zero point because there is no current flow. However, just as soon as current begins to flow through the heart, the positive end of the vector leaves the zero reference point.

When the septum first becomes depolarized, the vector extends downward toward the apex of the heart, but it is relatively weak, thus generating the first portion of the vectorcardiogram, as illustrated by the positive end of vector 1. As more of the heart becomes depolarized, the vector becomes stronger and stronger, usually swinging slightly to one side. Thus vector 2 of Figure 16–11 represents the state of depolarization of the heart about 0.02 second after vector 1. After another 0.02 second, vector 3 represents the current flow in the heart, and vector 4 occurs in still another 0.01 second. Finally, the heart becomes totally depolarized, and the vector becomes zero once again, as shown at point 5.

The elliptical figure generated by the positive ends of the vectors is called the *QRS vectorcardiogram.*

Vectorcardiograms can be recorded instantaneously on an oscilloscope by connecting electrodes from above and below the heart to the vertical plates of the oscilloscope and connecting electrodes from each side of the heart to the horizontal plates. When the vector changes, the spot of light on the oscilloscope follows the course of the positive end of the changing vector, thus inscribing the vectorcardiogram on the oscilloscopic screen.

Changing vectors in the heart do not occur only during the depolarization process, for vectors depicting current flow around the ventricles reappear during repolarization. Therefore, a second, smaller vectorcardiogram — the *T vectorcardiogram* — is inscribed during repolarization of the muscular mass; this is depicted to the right in Figure 16–10. Also, a still smaller *P vectorcardiogram* is inscribed during atrial depolarization.

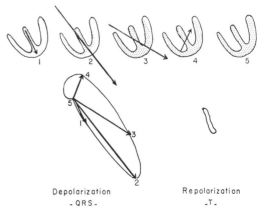

Figure 16–11 The QRS and T vectorcardiograms.

THE MEAN ELECTRICAL AXIS OF THE VENTRICLE

The vectorcardiogram of the ventricular de-polarization wave shown in Figure 16–11 is that of a normal heart. Note from this vectorcardio-gram that the preponderant direction of the vec-tors of the ventricles is normally toward the apex of the heart—that is, during most of the cycle of ventricular depolarization, current flows from the base of the ventricle toward the apex. This pre-ponderant direction of current flow during depolar-ization is called the mean electrical axis of the ventricles. The mean electrical axis of the normal ventricles is 59 degrees. However, in certain path-ological conditions of the heart this direction of current flow is changed markedly—sometimes even to opposite poles of the heart.

DETERMINING THE ELECTRICAL AXIS FROM STANDARD LEAD ELECTROCARDIOGRAMS

Clinically, the electrical axis of the heart is usually determined from the standard lead elec-trocardiograms. Figure 16–12 illustrates a method for doing this. After recording the various stand-ard leads, one determines the potential and polar-ity of the recording in two of the leads. In lead I of the figure the recording is positive and in lead III the recording is mainly positive, but negative during part of the cycle. If any part of the record-ing is negative, *this negative potential is sub-tracted from the positive potential.* After subtract-ing the negative portion of the QRS wave in lead III from the positive portion, each net vector is plotted on the axes of the respective leads, with the base of the vector at the point of intersection of the axes, as illustrated in Figure 16–12.

If the vector of lead I is positive, it is plotted in a positive direction along the line depicting lead I. On the other hand, if this vector is nega-tive, it is plotted in a negative direction. Also, for lead III, the vector is placed with its base at the point of intersection, and, if positive, it is plotted in the positive direction along the line depicting

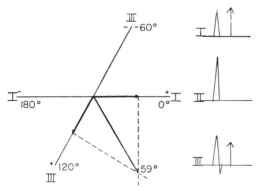

Figure 16–12 Plotting the mean electrical axis of the heart from two electrocardiographic leads.

lead III. If it is negative it is plotted in the nega-tive direction.

In order to determine the actual vector of cur-rent flow in the heart, one draws perpendicular lines from the apices of the two vectors of leads I and III, respectively. The point of intersection of these two perpendicular lines represents, by vec-torial analysis, the apex of the actual vector in the heart, and the point of intersection of the two lead axes represents the negative end of the actual vec-tor. The average potential generated in the direc-tion of current flow is represented by the length of the actual vector, and the mean electrical axis is represented by the direction of the vector. Thus, the vector of the mean electrical axis of the heart as determined in Figure 16–12 is 59 degrees.

The vector of current flow through the ven-tricles determined by this method does not de-termine the vector at any given instant in the depolarization process of the ventricles but, in-stead, determines approximately how much and in what direction the average current flows during the entire depolarization period.

VENTRICULAR CONDITIONS THAT CAUSE AXIS DEVIATION

Though the mean electrical axis of the ven-tricles averages approximately 59 degrees, this average can swing to the left even in the normal heart to approximately 20 degrees or to the right to approximately 100 degrees. The causes of the normal variations are simply anatomical differ-ences in the Purkinje fiber distribution or in the musculature itself of different hearts. Yet, a num-ber of conditions can cause axis deviation even beyond these normal limits, as follows:

Change in the Position of the Heart. Obviously, if the heart itself is angulated to the left, the mean electrical axis of the heart will also shift to the left. Such shift occurs (a) during expiration, (b) when a person lies down, because the abdominal contents press upward against the diaphragm, and (c) quite frequently in stocky, fat persons whose dia-phragms normally press upward against the heart.

Likewise, shift of the heart to the right causes the mean electrical axis of the ventricles to shift to the right. This condition occurs (a) during in-spiration, (b) when a person stands up, and (c) normally in tall, lanky persons whose hearts hang downward.

Hypertrophy of One Ventricle. When one ven-tricle greatly hypertrophies, the axis of the heart shifts toward the hypertrophied ventricle for two reasons: First, far greater quantity of muscle exists on the hypertrophied side of the heart than on the other side, and this allows excess generation of electrical currents on that side. Second, more time is required for the depolarization wave to travel through the hypertrophied ventricle than through the normal ventricle. Consequently, the normal ventricle becomes depolarized considerably in advance of the hypertrophied ventricle, and this

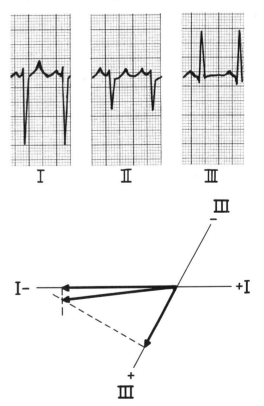

Figure 16–13   Left axis deviation in hypertensive heart disease. Note the slightly prolonged QRS complex.

causes a strong vector from the normal side of the heart toward the hypertrophied side. Thus the axis deviates toward the hypertrophied ventricle.

LEFT AXIS DEVIATION RESULTING FROM HYPERTROPHY OF THE LEFT VENTRICLE.   Figure 16–13 illustrates the three standard leads of an electrocardiogram in which an analysis of the axis direction shows left axis deviation with the mean electrical axis point in the −15 degree direction. This is a typical electrocardiogram resulting from increased muscular mass of the left ventricle. In this instance the axis deviation was caused by *hypertension,* which caused the left ventricle to hypertrophy in order to pump blood against the elevated systemic arterial pressure. However, a similar picture of left axis deviation occurs when the left ventricle hypertrophies as a result of aortic valvular stenosis, aortic valvular regurgitation, or any of a number of congenital heart conditions in which the left ventricle enlarges while the right side of the heart remains relatively normal in size.

RIGHT AXIS DEVIATION RESULTING FROM HYPERTROPHY OF THE RIGHT VENTRICLE.   The electrocardiogram of Figure 16–14 illustrates intense right axis deviation with an electrical axis of approximately 170 degrees, which is 111 degrees to the right of the normal mean electrical axis of the ventricles. The right axis deviation illustrated in this figure was caused by hypertrophy of the right ventricle as a result of pulmonary stenosis. However, right axis deviation may also occur in other congenital heart conditions, such as tetralogy of Fallot or interventricular septal defect. Also, hypertrophy of the right ventricle as a result of increased pulmonary vascular resistance, and even *massive pulmonary embolism* with excessive di-

lation of the right side of the heart can cause right axis deviation.

Bundle Branch Block.   Ordinarily, the two lateral walls of the ventricles depolarize at almost the same time, because both the left and right bundle branches transmit the cardiac impulse to the endocardial surfaces of the two ventricular walls at almost the same instant. As a result, the currents flowing from the walls of the two ventricles almost exactly neutralize each other. However, if one of the major bundle branches is blocked, depolarization of the two ventricles does not occur even nearly simultaneously, and the depolarization currents do not neutralize each other. As a result, axis deviation occurs as follows:

LEFT BUNDLE BRANCH BLOCK.   When the left bundle branch is blocked, cardiac depolarization spreads through the right ventricle approximately three times as rapidly as through the left ventricle. Consequently, much of the left ventricle remains polarized for a long time after the right ventricle has become totally depolarized. Thus, the right ventricle is electronegative with respect to the left ventricle during most of the depolarization process, and a very strong vector projects from the right ventricle toward the left ventricle. In other words, there is intense left axis deviation because the positive end of the vector points toward the left ventricle. This is illustrated in Figure 16–15, which shows typical left axis deviation re-

Figure 16–14   High-voltage electrocardiogram due to pulmonary stenosis with right ventricular hypertrophy. Also, there is intense right axis deviation.

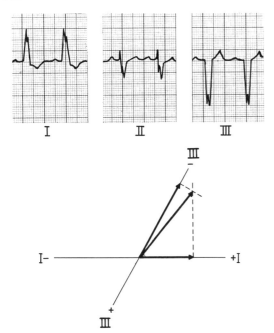

Figure 16–15 Left axis deviation due to left bundle branch block. Note the greatly prolonged QRS complex.

sulting from left bundle branch block. Note that the axis is approximately −50 degrees.

Because of slowness of impulse conduction when the Purkinje system is blocked, axis deviation resulting from bundle branch block greatly increases the duration of the QRS complex, as discussed in greater detail later in the chapter. It is the prolonged QRS complex that differentiates this condition from axis deviation occurring with hypertrophy.

RIGHT BUNDLE BRANCH BLOCK. When the right bundle branch is blocked, the left ventricle depolarizes far more rapidly than the right ventricle so that the left becomes electronegative while the right remains electropositive. A very strong vector develops with its negative end toward the left ventricle and its positive end toward the right ventricle. In other words, intense right axis deviation occurs.

Right axis deviation caused by right bundle branch block is illustrated in Figure 16–16, which shows an axis of approximately 105 degrees and a prolonged QRS complex because of blocked conduction.

Axis Deviation Caused by Muscular Destruction. Following heart attacks that result in destruction of part of the cardiac muscle and replacement of this muscle with fibrous tissue, very abnormal deviations of the axis of the ventricles can occur. The major causes of the axis deviation are twofold: First, part of the muscular mass itself is destroyed and is replaced by scar tissue so that a smaller quantity of muscle may then be available on one side of the heart than on the other side to generate electrical currents. Second, and probably more important, is blockage of the conduction

of depolarization at one or more local points in the smaller branches of the Purkinje system. When blockage occurs, the impulse must then be conducted by the muscle itself. This conduction occurs at a slow velocity, and it often has to detour around the scar tissue, allowing the muscular tissue beyond the block to remain electropositive for long periods of time after other portions of the heart have already become totally depolarized. Thus, even local blocks in the ventricles can cause considerable shift of the axis of the heart.

Muscular destruction causes the axis to shift in the direction opposite to the side of destruction, but conduction block causes the axis to shift toward the side of destruction. Thus, the two effects may or may not neutralize each other. Consequently, old infarctions in the heart do not produce consistent changes in axis deviation. Therefore, when other signs of old infarction are present in the electrocardiogram, such as low voltage (as shown in Figure 16–17), increased duration of the QRS complex, and bizarre spiking patterns of the QRS complex, the diagnosis is usually definite, and it is generally unimportant to determine the axis of the heart.

CONDITIONS THAT CAUSE ABNORMAL VOLTAGES OF THE QRS COMPLEX

INCREASED VOLTAGE IN THE STANDARD LEADS

Normally, the voltages in the three standard electrocardiographic leads, as measured from the peak of the R wave to the bottom of the S wave, vary between 0.5 and 2.0 mv., with lead III usually recording the lowest voltage and lead II the highest voltage. However, these relationships are not invariably true even in the normal heart. In general, when the sum of the voltages of all the QRS complexes of the three standard leads is greater

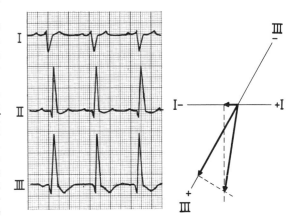

Figure 16–16 Right axis deviation due to right bundle branch block. Note the greatly prolonged QRS complex.

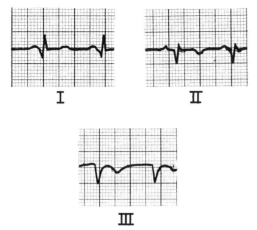

Figure 16–17 Low-voltage electrocardiogram, with evidences of local conduction blocks throughout the ventricles caused by old myocardial infarction.

than 4 mv., one considers that the patient has a high-voltage electrocardiogram.

The cause of high-voltage QRS complexes is increased muscular mass of the heart, which ordinarily results from hypertrophy of the muscle in response to excessive load on one part of the heart or the other. For instance, the right ventricle hypertrophies when it must pump blood through a stenotic pulmonary valve, and the left ventricle hypertrophies in systemic hypertension. The increased quantity of muscle allows generation of increased quantities of current around the heart. As a result, the electrical potentials recorded in the electrocardiographic leads are considerably greater than normal, as shown in Figures 16–13 and 16–14.

DECREASED VOLTAGE IN THE STANDARD LEADS

There are three major causes of decreased voltage in the electrocardiograms of the three standard electrocardiographic leads. These are, first, abnormalities of the cardiac muscle itself that prevent generation of large quantities of currents; second, abnormal conditions around the heart so that currents cannot be conducted from the heart to the surface of the body with ease; and, third, rotation of the apex of the heart anteriorly so that the electrical currents of the heart flow mainly anteroposteriorly in the chest rather than in the frontal plane of the body.

Decreased Voltage Caused by Cardiac Myopathies. One of the most usual causes of decreased voltage of the QRS complex is a series of *local intraventricular blocks* to the conduction of the cardiac impulse associated with *diminished muscle mass*. This causes electrical current to flow through the ventricles slowly and prevents major portions of the heart from becoming massively depolarized all at once. Consequently, this condition causes moderate prolongation of the QRS complex along with decreased voltage. Figure 16–17 illustrates

the typical low-voltage electrocardiogram with prolongation of the QRS complex, which one often finds after multiple small infarctions of the heart that have resulted in local blocks and loss of muscle mass throughout the ventricles.

Decreased Voltage Caused by Conditions Surrounding the Heart. One of the most important causes of decreased voltage in the electrocardiographic leads is *fluid in the pericardium*. Because extracellular fluid conducts electrical currents with great ease, a large portion of currents flowing out of the heart are conducted from one part of the heart to another through the pericardial effusion, so that the quantity of current reaching the surface of the body is greatly diminished. That is, the pericardial fluid "short circuits" the potentials generated by the heart. *Pleural effusion* to a lesser extent can also "short" the currents around the heart so that the voltages at the surface of the body and in the electrocardiograms are decreased.

Pulmonary emphysema can decrease the electrocardiographic potentials but by a different method from that of pericardial effusion. In pulmonary emphysema, conduction of electrical current through the lungs is considerably depressed because of the excessive quantity of air in the lungs. Also, the chest cavity enlarges, and the lungs tend to envelop the heart to a greater extent than normally. Therefore, the lungs act as an insulator to prevent spread of currents from the heart to the surface of the body, and this in general results in decreased electrocardiographic potentials in the various leads.

PROLONGED AND BIZARRE PATTERNS OF THE QRS COMPLEX

PROLONGED QRS COMPLEX AS A RESULT OF CARDIAC HYPERTROPHY OR DILATATION

The QRS complex lasts as long as the process of depolarization continues to spread through the ventricles—that is, as long as part of the ventricles is depolarized and part is still polarized. Therefore, the cause of a prolonged QRS complex is always *delayed conduction* of the impulse through the ventricles. Such delay often occurs when one or both ventricles are hypertrophied or dilated, owing to the longer pathway that the impulse must then travel. The normal QRS complex lasts about 0.06 second, whereas in hypertrophy or dilatation of the left or right ventricle the QRS complex may be prolonged to 0.09 second or occasionally 0.10 second.

PROLONGED QRS COMPLEX RESULTING FROM PURKINJE SYSTEM BLOCKS

Block of the Purkinje fibers requires that the cardiac impulse be conducted by the ventricular muscle instead, thereby decreasing the rate of impulse conduction to approximately one-third to one-fourth normal. Therefore, if complete block of one of the bundle branches occurs, the duration of the QRS complex is increased to approximately 0.16 to 0.18 second.

In general a QRS complex is considered to be abnormally long when it lasts more than 0.08 second, and when it lasts more than 0.12 second the prolongation is almost certain to be caused by pathologic block of the conduction system somewhere in the ventricles, as illustrated in Figures 16–15 and 16–16.

CONDITIONS CAUSING BIZARRE QRS COMPLEXES

Bizarre patterns of the QRS complex are most frequently caused by two conditions: first, destruction of cardiac muscle in various areas throughout the ventricular system with replacement of this muscle by scar tissue, and second, local blocks in the conduction of impulses by the Purkinje system.

Sometimes local blocks occur in both the right and left ventricles. If the blocks are such that the impulse reaches the blocked area in the right ventricle much later than it reaches the blocked area in the left ventricle, but the total quantity of blocked muscle in the left ventricle is greater than the total quantity in the right ventricle, then a situation occurs in which the axis of the heart first shifts strongly to the left and then—when the left ventricle has become totally depolarized but a small portion of the right ventricle still remains polarized—rapidly shifts diametrically oppositely toward the right ventricle. This effect causes double or even triple peaks in some of the standard electrocardiographic leads, such as those illustrated in Figure 16–15.

CURRENT OF INJURY

Many different cardiac abnormalities, especially those that damage the heart muscle itself, often cause part of the heart to remain *depolarized all the time.* When this occurs, current flows between the pathologically depolarized and the normally polarized areas. This is called the *current of injury.*

Some of the abnormalities that can cause current of injury are: (1) *mechanical trauma,* which makes the membranes remain so permeable that repolarization cannot take place, (2) *infectious processes* that damage the muscle membranes, and (3) *ischemia of the muscle caused by coronary occlusion,* which is by far the most common cause of current of injury in the heart; during ischemia enough energy simply is not available to maintain normal integrity of the membranes.

EFFECT OF CURRENT OF INJURY ON THE QRS COMPLEX

In Figure 16–18 a shaded area in the base of the left ventricle is newly infarcted. Therefore, during the T-P interval—that is, when the normal ventricular muscle is polarized—current flows from the base of the left ventricle toward the rest of the ventricles. The vector of this "current of injury" is in a direction of approximately 125 degrees. As illustrated in the lower portions of the figure, even before the QRS complex begins, this current flow causes an initial record in lead I below the zero potential line because the resultant vector of the current of injury points toward the negative end of the lead I axis. In lead II the record is above the line because the resultant vector points toward the positive terminal of lead II. In lead III the vector of the current flow is in the same direction as the polarity of lead III so that the record is positive. Furthermore, because the vector of the current of injury lies almost exactly along the axis of lead III, the potential of the current of injury in lead III is much higher than in either of the other two records.

As the heart then proceeds through its normal process of depolarization, part of the septum first becomes depolarized, and the depolarization spreads down to the apex and back toward the bases of the ventricles. The last portion of the ventricles to become totally depolarized is the base of the right ventricle, because the base of the left ventricle is already totally and permanently depolarized. By vectorial analysis, as illustrated in the figure, the electrocardiogram generated by the depolarization wave traveling through the ventricles may be constructed graphically.

When the heart becomes totally depolarized at the end of the depolarization process, as noted by the next to last stage in Figure 16–18, all of the ventricular muscle is in a negative state. Therefore, at this instant in the electrocardiogram,

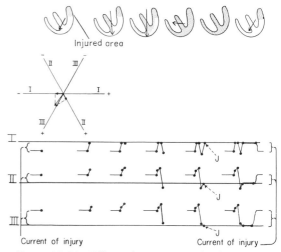

Figure 16–18 Effect of a current of injury on the electrocardiogram.

absolutely no current flows around the musculature of the ventricles, and the current of injury is abolished.

As repolarization then takes place, all of the heart finally repolarizes except the area of permanent depolarization in the injured base of the left ventricle. Thus, the repolarization causes a return of the current of injury in each lead, as noted at the far right in Figure 16–18.

THE J POINT—THE ZERO REFERENCE POTENTIAL OF THE ELECTROCARDIOGRAM

One would think that the electrocardiograph machines for recording electrocardiograms could determine when no currents are flowing around the heart. However, many stray currents exist in the body, such as currents resulting from "skin potentials" and from differences in ionic concentrations in different parts of the body. Therefore, when two electrodes are connected between the arms or between an arm and a leg, these stray currents make it impossible for one to predetermine the exact zero reference level in the electrocardiogram. To determine a zero potential line in the record of the electrocardiogram one notes *the exact point at which the wave of depolarization just completes its passage through the heart,* which obviously occurs at the very end of the QRS complex. This point is known as the "J point" in the electrocardiogram, as illustrated in Figures 16–18 and 16–19.

For further analysis of the electrical axis of the

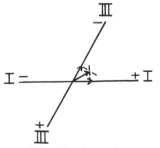

Figure 16–19   The "J" point as the zero reference voltage of the electrocardiogram. Also, method for plotting the axis of a current of injury.

current of injury, a horizontal line is drawn through the electrocardiogram at the level of the J point, and this horizontal line is the zero potential line in the electrocardiogram.

Use of the J Point in Plotting the Axis of a Current of Injury.   Figure 16–19 illustrates electrocardiograms recorded from leads I and III, both of which show currents of injury. In other words, the J point and the S-T segment of each of these two electrocardiograms are not on the same line as the T-P segment. A horizontal line has been drawn through the J point to represent the zero potential level in each of the two recordings. The potential of the current of injury in each lead is the difference between the level of the T-P segment of the electrocardiogram and the zero potential line, as illustrated by the arrows. In lead I the recorded current of injury is above the zero potential line and is, therefore, positive. On the other hand, in lead III the T-P segment is below the zero potential line; therefore, the current of injury in lead III is negative.

At the bottom in Figure 16–19, the potentials of the current of injury in leads I and III are plotted on the coordinates of these leads, and the vector of the current of injury is determined by the method already described. In this instance, the vector of the current of injury extends from the right side of the body toward the left and slightly upward, with an axis of approximately −30 degrees.

If one places the vector of the current of injury directly over the ventricles, *the negative end of the vector points toward the depolarized area of the ventricles.* In the instance illustrated in Figure 16–19, the damaged area would be in the lateral wall of the right ventricle.

CORONARY ISCHEMIA AS A CAUSE OF CURRENT OF INJURY

Insufficient blood flow to the cardiac muscle depresses metabolism of the muscle for three different reasons: oxygen lack, excess carbon dioxide, and lack of sufficient nutrients. Consequently, polarization of the membranes cannot occur in areas of severe myocardial ischemia. Often, the heart muscle does not die, because the blood flow is sufficient to maintain life of the muscle even though it is not sufficient to cause repolarization of the membranes. As long as this state exists, a current of injury continues to flow.

Obviously, extreme ischemia of the cardiac muscle occurs following coronary occlusion, and strong currents of injury flow from the infarcted area of the ventricles during the period between heartbeats, as is illustrated in Figures 16–20 and 16–21. Therefore, one of the most important diagnostic features of electrocardiograms recorded following acute coronary thrombosis is the current of injury.

Acute Anterior Wall Infarction.   Figure 16–20 illustrates the electrocardiogram in the three standard leads and in one chest lead recorded from a patient with acute anterior wall cardiac

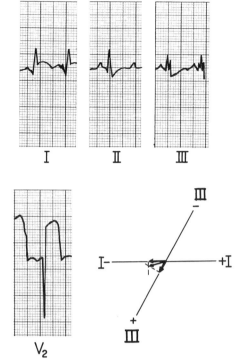

Figure 16–20 Current of injury in acute anterior wall infarction. Note the intense current of injury in lead V_2.

infarction. The most important diagnostic feature of this electrocardiogram is the current of injury in the chest lead. If one draws a zero potential line through the J point of this lead, he finds a strong negative current of injury during diastole. In other words, the negative end of the current of injury vector is against the chest wall. This means that the current of injury is emanating from the anterior wall of the ventricles, which is the main reason for diagnosing this condition as anterior wall infarction.

If one analyzes the currents of injury in leads I and III, he finds a negative vector for the current of injury in lead I and a positive vector for the current of injury in lead III. This means that the resultant vector of the current of injury in the heart is approximately +150 degrees, with the negative end of the vector pointing toward the left ventricle and the positive end pointing toward the right ventricle. Thus, in this particular electrocardiogram the current of injury appears to be coming mainly from the left ventricle and from the anterior wall of the heart. Therefore, one would suspect that this anterior wall infarction is probably caused by thrombosis of the descending limb of the left coronary artery.

Posterior Wall Infarction. Figure 16–21 illustrates the three standard leads and one chest lead from a patient with posterior wall infarction. The major diagnostic feature of this electrocardiogram is also in the chest lead. If a zero potential reference line is drawn through the J joint of this lead, it is readily apparent that during the T-P

interval the vector of the current of injury is positive. This means that the positive end of the vector is at the chest wall and the negative end is away from the chest wall. In other words, the current of injury is coming from the opposite side of the heart from that portion adjacent to the chest wall, which is the reason why this type of electrocardiogram is the basis for diagnosing posterior wall infarction.

If one analyzes the currents of injury in leads II and III of Figure 16–21, it is readily apparent that the current of injury is negative in both leads. By vectorial analysis as shown in the figure, one finds that the vector of the current of injury is approximately −95 degrees with the negative end of the vector pointing downward and the positive end upward. Thus, because the infarct, as indicated by the chest lead, is on the posterior wall of the heart and, as indicated by the currents of injury in leads II and III, is in the apical portion of the heart, one would suspect that this infarct is close to the apex on the posterior wall of the left ventricle, possibly spreading over the septum into the wall of the right ventricle to a slight extent.

Infarction in Other Parts of the Heart. By the same procedures as those illustrated in the preceding two discussions of anterior and posterior wall infarctions, it is possible to determine the locus of an infarcted area emitting a current of injury, regardless of which part of the heart is involved. In making such vectorial analyses, it must always be remembered that *the positive end of the vector points toward the normal cardiac muscle and the negative end points toward the abnormal portion of the heart that is emitting the current of injury.*

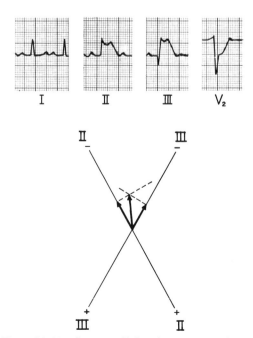

Figure 16–21 Current of injury in acute posterior wall, apical infarction.

Recovery from Coronary Thrombosis. Figure 16–22 illustrates the chest lead from a patient with posterior infarction, showing the change in this chest lead from the day of the attack to one week later, then three weeks later, and finally one year later. From this electrocardiogram it can be seen that the current of injury is strong immediately after the acute attack, but after approximately one week the current of injury has diminished considerably and after three weeks it is completely gone. After that, the electrocardiogram does not change greatly during the following year. This is the usual recovery pattern following cardiac infarction of moderate degree when the collateral coronary blood flow is sufficient to reestablish appropriate nutrition to the infarcted area.

On the other hand, when all the coronary vessels throughout the heart are fairly well sclerosed, it may not be possible for the adjacent coronary vessels to supply sufficient blood to the infarcted area for recovery. Therefore, in some patients with coronary infarction, the infarcted area never redevelops an adequate coronary blood supply, and relative coronary insufficiency persists in this area of the heart indefinitely. If the muscle does not die and become replaced by scar tissue, it continually emits a current of injury as long as the relative ischemia exists.

Old Recovered Myocardial Infarction. Figure 16–23 illustrates leads I and III in anterior infarction and posterior infarction as these leads appear approximately a year after the acute episode. These are what might be called the "ideal" configurations of the QRS complex in these types of recovered myocardial infarction. It is illustrated in these figures that usually a Q wave develops at the beginning of the QRS complex in lead I in anterior infarction, whereas in posterior infarction a Q wave develops at the beginning of the QRS complex in lead III.

These configurations are certainly not those found in all cases of old anterior and posterior cardiac infarction. As pointed out above, local loss of muscle and local areas of conduction block can cause the following abnormalities of the QRS complex: many bizarre patterns, decreased voltage, and prolongation.

Anterior   **Posterior**

I   III   I   III

Figure 16–23 Electrocardiograms of old anterior and posterior wall infarctions, illustrating the Q wave in lead I in old anterior wall infarction and the Q wave in lead III in old posterior wall infarction.

Current of Injury in Angina Pectoris. "Angina pectoris" means simply pain in the pectoral regions of the upper chest, this pain usually radiating down the left arm and into the neck. The pain is caused by *relative* ischemia of the heart, for no pain may be felt as long as the person is perfectly quiet, but, just as soon as he overworks his heart, the pain appears.

A current of injury often occurs during an attack of severe angina pectoris, for the relative coronary insufficiency occasionally becomes great enough to prevent adequate repolarization of the membranes in some areas of the heart during diastole.

ABNORMALITIES IN THE T WAVE

Earlier in the chapter it was pointed out that the T wave is normally positive in all the standard leads and that this is caused by repolarization of the apex of the heart ahead of the endocardial surfaces of the ventricles. This is backward to the direction in which depolarization takes place. (If the basic principles of the upright T wave in the standard leads have not been understood by now, become familiar with the earlier, more detailed discussion of this before proceeding to the following few sections.)

The T wave becomes abnormal when the normal sequence of repolarization does not occur. Several factors can change this sequence of repolarization, as follows:

EFFECT OF SLOW CONDUCTION OF THE DEPOLARIZATION WAVE ON THE T WAVE

Referring back to Figure 16–15, note that the QRS complex is considerably prolonged. The reason for this prolongation is delayed conduction in the left ventricle as a result of left bundle branch block. The left ventricle becomes depolarized approximately 0.10 second after depolarization of the right ventricle. The refractory periods of the right and left ventricular muscle masses are not greatly different from each other. Therefore, the right ventricle begins to repolarize long before the left ventricle; this causes positivity in the right ventricle and negativity in the left ventricle. In

Same day   I wk.   3 wks.   I yr.

Figure 16–22 Recovery of the myocardium following moderate posterior wall infarction, illustrating disappearance of the current of injury (lead V_3).

other words, the mean axis of the T wave is from left to right, which is opposite to the mean electrical axis of the QRS complex in this same electrocardiogram. Thus, when conduction of the impulse through the ventricles is greatly delayed, the T wave is almost always of opposite polarity to that of the QRS complex.

In Figure 16–16 and in several figures in the following chapter, conduction also does not occur through the Purkinje system. As a result, the rate of conduction is greatly slowed, and in each instance the T wave is of opposite polarity to that of the QRS complex, whether the condition causing this delayed conduction happens to be left bundle branch block, right bundle branch block, ventricular extrasystole, or otherwise.

PROLONGED PERIODS OF DEPOLARIZATION IN PORTIONS OF THE VENTRICULAR MUSCLE AS A CAUSE OF ABNORMALITIES IN THE T WAVE

If the apex of the ventricles should have an abnormally long period of depolarization, then repolarization of the ventricles would not begin at the apex as it normally does. Instead, the base of the ventricles would repolarize ahead of the apex, and the vector of repolarization would point from the apex toward the base of the heart. This is opposite to the usual vector of repolarization, and, consequently, the T wave in all three standard leads would be negative rather than the usual positive. Thus, the simple fact that the apical muscle of the heart has a prolonged period of depolarization is sufficient to cause marked changes in the T wave, even to the extent of changing the entire polarity, as is illustrated in Figure 16–24.

Ischemia is by far the most common cause of increased period of depolarization of cardiac muscle, and, when the ischemia occurs in only one area of the heart, the depolarization period of this area increases out of proportion to that in other portions. As a result, definite changes in the T wave may take place. The ischemia may result from chronic, progressive coronary occlusion, acute coronary occlusion, or relative coronary insufficiency occurring during exercise.

One means for detecting mild coronary insufficiency is to have the patient exercise and then record the electrocardiogram immediately thereafter, noting whether or not changes occur in the T waves. The changes in the T waves need not be

Figure 16–25 Biphasic T wave caused by digitalis toxicity.

specific, for any change in the T wave in any lead – inversion, for instance, or a biphasic wave – is evidence enough that some portion of the ventricular muscle has increased its period of depolarization out of proportion to the rest of the heart, and this is probably caused by relative coronary insufficiency.

All the other conditions that can cause currents of injury, including pericarditis, myocarditis, and mechanical trauma of the heart, can also cause changes in the T wave. A current of injury occurs when the period of depolarization of some muscle is so long that the muscle fails to repolarize before the next cardiac cycle begins. Therefore, a current of injury is actually an exacerbated form of abnormal T wave, for both of these result from an increased depolarization period of one or several portions of cardiac muscle, and the difference is only a matter of degree.

Effect of Digitalis on the T Wave. As discussed in Chapter 29, digitalis is a drug that can be used during relative coronary insufficiency to increase the strength of cardiac muscular contraction. However, digitalis also increases the period of depolarization of cardiac muscle. It usually increases this period by approximately the same proportion in all or most of the ventricular muscle, but when overdosages of digitalis are given, the period of one part of the heart may be increased out of proportion to that of other parts. As a result, nonspecific changes, such as T wave inversion or biphasic T waves, may occur in one or more of the electrocardiographic leads. A bizarre T wave caused by excessive administration of digitalis is illustrated in Figure 16–25. There is a slight amount of current of injury, too. This probably results from continuous depolarization of part of the ventricular muscle.

Changes in the T wave during digitalis administration are the earliest signs of digitalis toxicity. If still more digitalis is given to the patient, strong currents of injury may develop, and, as noted in the following chapter, digitalis can also block conduction of the cardiac impulse to various portions of the heart, so that various arrhythmias result. It is desirable clinically to keep the effects of digitalis from going beyond the stage of mild T wave abnormalities. Therefore, the electrocardiograph is used routinely in following digitalized patients.

REFERENCES

Figure 16–24 Inverted T wave resulting from mild ischemia of the apex of the ventricles.

See bibliography at end of Chapter 17

Electrocardiographic Interpretation of Cardiac Arrhythmias

The rhythmicity of the heart and some abnormalities of rhythmicity were discussed in Chapter 14. The major purpose of the present chapter is to discuss still other abnormalities of rhythm and to describe the electrocardiograms recorded in the conditions known clinically as "cardiac arrhythmias."

ABNORMAL SINUS RHYTHMS

TACHYCARDIA

The term "tachycardia" means a fast heart rate; an electrocardiogram recorded from a patient with tachycardia is illustrated in Figure 17–1. This electrocardiogram is normal except that the rate of heartbeat, as determined from the time intervals between QRS complexes, is approximately 150 per minute instead of the normal 72 per minute.

The three general causes of tachycardia are *increased body temperature, stimulation of the heart by the autonomic nerves,* and *toxic conditions of the heart.*

The rate of the heart increases approximately 10 beats per minute for each degree Fahrenheit increase in body temperature up to a body temperature of about 105°; beyond this the heart rate may actually decrease, owing to progressive weakening of the heart muscle as a result of the fever. Fever causes tachycardia because increased temperature increases the rate of metabolism of the cardiac muscle, which in turn directly increases the excitability and the rate of rhythm of the heart.

The various factors that can cause the autonomics to excite the heart will be discussed in relation to the reflexes of the circulatory system in Chapter 25. For instance, when a patient loses blood and passes into a state of shock or semishock, reflex stimulation of the heart increases its rate to as much as 150 to 180 beats per minute. Also, simple weakening of the myocardium usually increases the heart rate because the weakened heart does not pump blood into the arterial tree to a normal extent, and this elicits sufficient reflexes to increase the rate of the heart.

BRADYCARDIA

The term "bradycardia" means, simply, a slowed heart rate; bradycardia is illustrated by the electrocardiogram in Figure 17–2.

Bradycardia in Athletes. The athlete's heart is considerably stronger than that of a normal person, a fact that allows his heart to pump a greater stroke volume output per beat. The large quantities of blood pumped into the arterial tree with each beat presumably initiate sufficient circulatory reflexes to cause bradycardia.

Vagal Stimulation as a Cause of Bradycardia. Obviously, any circulatory reflex that stimulates the vagus nerve can cause the heart rate to decrease considerably. Perhaps the most striking example of this occurs in patients with the *carotid sinus syndrome.* In these patients an arteriosclerotic process in the carotid sinus region of the carotid artery causes excessive sensitivity of the pressure receptors located in the arterial wall; as a result, mild pressure on the neck elicits a strong pressoreceptor reflex, causing intense vagal stimulation of the heart and extreme bradycardia. Indeed, such pressure may actually stop the heart.

SINUS ARRHYTHMIA

Figure 17–3 illustrates a *cardiotachometer* recording of the heart rate during normal and deep respiration. A cardiotachometer is an instrument that records by the height of successive spikes the

Figure 17–1   Sinus tachycardia (lead I).

Figure 17–2   Sinus bradycardia (lead III).

Figure 17-3 Sinus arrhythmia as detected by a cardio-tachometer. To the left is the recording taken when the subject was breathing normally; to the right, when breathing deeply.

Figure 17-4 S-A nodal block with A-V nodal rhythm (lead III).

duration of the interval between each two QRS complexes in the electrocardiogram. Note from this record that the heart rate increases and decreases approximately 5 per cent during the various phases of the normal respiratory cycle. However, during deep respiration, as is shown to the right in Figure 17-3, the heart rate increases and decreases by as much as 30 per cent.

Sinus arrhythmia is a result of several circulatory reflexes that will be described in detail in Chapter 25. First, when the blood pressure rises and falls during each cycle of respiration, the *baroreceptors are alternately stimulated* and depressed, causing reflex slowing and speeding up of the heart. Second, during each respiratory cycle, the negative intrapleural pressure increases and decreases, which increases and decreases the effective pressure in the veins of the chest. This elicits a *waxing and waning Bainbridge reflex* that also increases and decreases the cardiac rate. Third, when the respiratory center of the medulla is excited during each respiratory cycle, some of the *impulses "spill over" from the respiratory center into the vasomotor center,* causing alternate increase and decrease in the number of impulses transmitted to the heart through the sympathetics and vagi.

ABNORMAL RHYTHMS RESULTING FROM BLOCK IN IMPULSE CONDUCTION

SINO-ATRIAL BLOCK

In rare instances the impulse from the S-A node is blocked before it enters the atrial muscle. This phenomenon is illustrated in Figure 17-4, which shows the sudden cessation of P waves due to standstill of the atrium. However, the ventricle picks up a new rhythm, the impulse usually originating in the A-V node or in the A-V bundle so that the ventricular QRS-T complex is not altered.

ATRIOVENTRICULAR BLOCK

The only means by which impulses can ordinarily pass from the atria into the ventricles is through the *A-V bundle,* which is also known as the *bundle of His.* The different conditions that can

either decrease the rate of conduction of the impulse through this bundle or can totally block the impulse are:

1. *Ischemia of the A-V junctional fibers* often delays or blocks conduction from the atria to the ventricles. Coronary insufficiency can cause ischemia of the A-V junction in the same manner that it can cause ischemia of the myocardium.

2. *Compression of the A-V bundle* by scar tissue or by calcified portions of the heart can depress or block conduction from the atria to the ventricles.

3. *Inflammation of the A-V bundle or fibers of the A-V junction* can depress conductivity between the atria and the ventricles. Inflammation results frequently from different types of myocarditis such as occur in diphtheria and rheumatic fever.

4. *Extreme stimulation of the heart by the vagi* in rare instances blocks impulse conduction through the A-V junctional fibers. Such vagal excitation occasionally results from strong stimulation of the baroreceptors in persons with the *carotid sinus syndrome,* which was just discussed in relation to bradycardia.

Incomplete Heart Block. PROLONGED P-R (OR P-Q) INTERVAL—"FIRST DEGREE BLOCK." The normal lapse of time between the *beginning* of the P wave and the *beginning* of the QRS complex is approximately 0.16 second when the heart is beating at a normal rate. This P-R interval decreases in length with faster heartbeat and increases with slower heartbeat. In general, when the P-R interval increases above a value of approximately 0.20 second, the P-R interval is said to be prolonged, and the patient is said to have *first degree incomplete heart block.* Figure 17-5 illustrates an electrocardiogram with a prolonged P-R interval, the interval in this instance being approximately 0.30 second.

The P-R interval rarely increases above 0.35 to 0.50 second, for by the time the rate of conduction through the A-V bundle is depressed to this extent, the intensity of the action potentials in the junctional fibers is usually so depressed that conduction stops entirely. Thus, when a patient's P-R interval is approaching these limits, additional slight increase in the severity of the condition will completely block impulse conduction rather than simply delaying conduction further. Therefore,

Figure 17-5 Prolonged P-R interval (lead II).

Figure 17-6 Partial atrioventricular block (lead V₃).

one of the means for determining the severity of some heart diseases—rheumatic fever, for instance—is to measure the P-R interval.

SECOND DEGREE BLOCK. When conduction through the A-V junction is slowed until the P-R interval is 0.25 to 0.50 second, the action potentials traveling through the A-V junctional fibers are sometimes strong enough to pass on into the A-V node and at other times are not strong enough. Often the impulse passes into the ventricles on one heartbeat and fails to pass on the next one or two beats, thus alternating between conduction and nonconduction. In this instance, the atria beat at a considerably faster rate than the ventricles, and it is said that there are "dropped beats" of the ventricles. This condition is called *second degree incomplete heart block.*

Figure 17-6 illustrates P-R intervals as long as 0.30 second, and it also illustrates one dropped beat as a result of failure of conduction from the atria to the ventricles.

At times every other beat of the ventricles is dropped so that a "2:1 rhythm" develops in the heart, with the atria beating twice for every single beat of the ventricles. Sometimes other rhythms, such as 3:2 or 3:1, also develop.

Complete Atrioventricular Block. When the condition causing poor conduction in the A-V bundle becomes extremely severe, complete block of the impulse from the atria into the ventricles occurs. In this instance the P waves become completely dissociated from the QRS-T complexes, as illustrated in Figure 17-7. Note that the rate of rhythm of the atria in this electrocardiogram is approximately 100 beats per minute, while the rate of ventricular beat is less than 40 per minute. Furthermore, there is no relationship whatsoever between the rhythm of the atria and that of the ventricles, for the ventricles have "escaped" from control by the atria, and they are beating at their own natural rate.

STOKES-ADAMS SYNDROME—VENTRICULAR ESCAPE. In some patients with atrioventricular block, total block comes and goes—that is, all impulses are conducted from the atria into the ventricles for a period of time, and then suddenly no impulses at all are transmitted. The duration of total block may be a few seconds, a few minutes, a few hours, or even weeks or more before conduction returns. Particularly does this condition occur in hearts with borderline ischemia.

Immediately after A-V conduction is first blocked, the ventricles stop contracting entirely for about 5 to 10 seconds. Then some part of the Purkinje system beyond the block, usually in the A-V bundle itself, begins discharging rhythmically at a rate of 15 to 40 times per minute and acting as the pacemaker of the ventricles. This is called *ventricular escape.* Because the brain cannot remain active for more than 3 to 5 seconds without blood supply, patients usually faint between block of conduction and "escape" of the ventricles. These periodic fainting spells are known as the Stokes-Adams syndrome.

Occasionally the interval of ventricular standstill is so long that it either becomes detrimental to the health of the patient or causes death. Consequently, many of these patients would be better off with total heart block than constantly shifting back and forth between total heart block and atrial control of the ventricles. Such patients are frequently treated with drugs that depress conduction in the A-V bundle so that total heart block persists all the time. Commonly, digitalis is the drug used for this purpose, but quinidine or barium ion can also convert a Stokes-Adams syndrome into total heart block.

Also, many of these patients are now provided with an *artificial pacemaker,* which is a small battery-operated electrical stimulator planted beneath the skin, the electrodes from which are placed on the heart. This pacemaker provides continued rhythmic impulses that take over control of the heart. The batteries are replaced about once a year.

INCOMPLETE INTRAVENTRICULAR BLOCK— ELECTRICAL ALTERNANS

The same factors that can cause A-V block, except vagal stimulation, can also block impulse conduction in peripheral portions of the Purkinje system. At times, *incomplete* block occurs so that the impulse is sometimes transmitted and sometimes not transmitted, causing slowing or block of the impulse during some heart cycles and not during others. The QRS complex may be considerably abnormal during those cycles in which the impulse is slowed or blocked. Figure 17-8 illustrates the condition known as *electrical alternans,* which results from intraventricular block every other heartbeat. This electrocardiogram also illustrates rather extreme tachycardia, which is probably the reason the block has occurred, for when the rate of the heart is very rapid, it may be impossible for portions of the Purkinje system to recover from the refractory period rapidly enough to respond during each succeeding heartbeat. Also, many conditions

Figure 17-7 Complete atrioventricular block (lead II).

Figure 17–8 Partial intraventricular block—electrical alternans (lead III).

that depress the heart, such as ischemia, myocarditis, and digitalis toxicity, can cause incomplete intraventricular block and electrical alternans.

PREMATURE BEATS

A premature beat is a contraction of the heart prior to the time when normal contraction would have been expected. This condition is also frequently called *extrasystole*.

Causes of Premature Beats. Most premature beats result from *ectopic foci* in the heart, which emit abnormal impulses at odd times during the cardiac rhythm. The possible causes of ectopic foci are (1) local areas of ischemia, (2) small calcified plaques at different points in the heart, which press against the adjacent cardiac muscle so that some of the fibers are irritated, and (3) toxic irritation of the A-V node, Purkinje system, or myocardium caused by drugs, nicotine, caffeine, etc. Mechanical initiation of ectopic foci is also frequent during cardiac catheterization, large numbers of premature beats occurring when the catheter enters the right ventricle and presses against the endocardium.

ATRIAL PREMATURE BEATS

Figure 17–9 illustrates an electrocardiogram showing a single atrial premature beat. The P wave of this beat is relatively normal and the QRS complex is also normal, but the interval between the preceding beat and the premature beat is shortened. Also, the interval between the premature beat and the next succeeding beat is slightly prolonged. The reason for this is that the premature beat originated in the atrium some distance from the S-A node, and the impulse of the premature beat had to travel through a considerable amount of atrial muscle before it discharged the S-A node. Consequently, the S-A node discharged very late in the cycle and made the succeeding heartbeat also late in appearing.

Atrial premature beats occur frequently in healthy persons and, indeed, are often found in athletes or others whose hearts are certain to be in healthy condition. Yet mild toxic conditions resulting from such factors as excess smoking, lack of sleep, too much coffee, alcoholism, and use of various drugs can also initiate such beats.

Pulse Deficit. When the heart contracts ahead of schedule, the ventricles will not have filled normally, and the stroke volume output during that contraction is depressed or sometimes almost none at all. Therefore, the pulse wave passing to the periphery following a premature beat may be so weak that the pulse cannot be felt at all in the radial artery, whereas the succeeding pulse may be extra strong because of compensatory overfilling. Thus, a deficit in the number of pulses felt in the radial pulse occurs in relation to the number of beats of the heart.

Bigeminal Pulse. At times every other beat of the heart may be a premature beat. This causes the patient to have a bigeminal pulse—that is, a strong pulse alternating with a weak pulse.

A-V NODAL OR A-V BUNDLE PREMATURE BEATS

Figure 17–10 illustrates a premature beat originating either in the A-V node or in the A-V bundle. The P wave is missing from the record of the premature beat. Instead, the P wave is superimposed on the QRS-T complex of the premature beat; this distorts the complex, but the P wave itself cannot be seen.

In general, A-V nodal premature beats have the same significance and causes as atrial premature beats.

VENTRICULAR PREMATURE BEATS

The electrocardiogram of Figure 17–11 illustrates a series of ventricular premature beats alternating with normal beats. Ventricular premature beats cause several special effects in the electrocardiogram, as follows: First, the QRS complex is usually considerably prolonged. The reason for this is that the impulse is conducted mainly through the muscle of the ventricle rather than through the Purkinje system.

Second, the QRS complex has a very high voltage, for the following reason: When the normal impulse passes through the heart, it passes through both ventricles approximately simul-

Figure 17–9 Atrial premature beat (lead I).

Figure 17–10 A-V nodal premature beat (lead III).

taneously; consequently, the depolarization waves of the two sides of the heart partially neutralize each other. However, when a ventricular premature beat occurs, the impulse travels in only one direction so that there is no such neutralization effect, and one entire side of the heart is depolarized while the other entire side is still polarized; this causes intense electrical potentials.

Third, following almost all ventricular premature beats the T wave has a potential opposite to that of the QRS complex because the *slow conduction of the impulse* through the cardiac muscle causes the area first depolarized also to repolarize first. As a result, the direction of current flow in the heart during repolarization is opposite to that during depolarization. This is not true of the normal T wave, as was explained in Chapter 16.

Some ventricular premature beats are benign in their origin and result from simple factors such as cigarettes, coffee, lack of sleep, various mild toxic states, and even emotional irritability. On the other hand, a large share of ventricular premature beats results from actual pathology of the heart. For instance, many ventricular premature beats occur following coronary thrombosis because of stray impulses originating around the borders of the infarcted area of the heart.

Vector Analysis of the Origin of an Ectopic Ventricular Premature Beat. In Chapter 16 the principles of vectorial analysis were explained. Applying these principles one can determine from the electrocardiogram in Figure 17–11 the point of origin of the ventricular premature beat as follows: Note that the vectors of the premature beats in leads II and III are both strongly positive.

Plotting these vectors on the axes of leads II and III and solving by vectorial analysis for the actual vector in the heart, one finds that the vector of the ectopic beat has its negative end at the base of the heart and its positive end toward the apex. Thus, the first portion of the heart to become depolarized during the premature beat lies near the base of the heart, which therefore is the locus of the ectopic focus.

PAROXYSMAL TACHYCARDIA

Abnormalities in any portion of the heart, including the atria, the Purkinje system, and the ventricles, can sometimes cause rapid rhythmic discharge of impulses which spread in all directions throughout the heart. Because of the rapid rhythm in the irritable focus, this focus becomes the pacemaker of the heart.

The term "paroxysmal" means that the heart rate usually becomes very rapid in paroxysms, the paroxysms beginning suddenly and lasting for a few seconds, a few minutes, a few hours, or sometimes much longer. Then the paroxysms usually end as suddenly as they had begun, the pacemaker of the heart shifting back to the S-A node.

Paroxysmal tachycardia can often be stopped by eliciting a vagal reflex. A strange type of vagal reflex sometimes used for this purpose is one that occurs when painful pressure is applied to the eyes. Also, pressure on the carotid sinuses can sometimes elicit enough vagal reflex to stop the tachycardia. Various drugs may also be used to stop the tachycardia. One frequently used is quinidine, which depresses cardiac muscle and often blocks the rhythmic discharge of the irritable focus that is causing the paroxysmal attack.

ATRIAL PAROXYSMAL TACHYCARDIA

Figure 17–12 illustrates a sudden increase in rate of heartbeat from approximately 95 beats per minute to approximately 150 beats per minute. On close observation of the electrocardiogram it will be seen that an inverted P wave occurs before each of the QRS-T complexes during the paroxysm of rapid heartbeat, though this P wave is partially superimposed on the normal T wave of the preceding beat. This indicates that the origin of this paroxysmal tachycardia is in the atrium, but, because the P wave is abnormal, the origin is not near the S-A node.

A-V Nodal Paroxysmal Tachycardia. Paroxysmal tachycardia sometimes results from an irritable

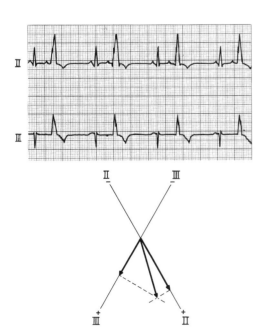

Figure 17–11 Ventricular premature beats (leads II and III). Axis of the premature beats is plotted in accord with the principles of vectorial analysis explained in Chapter 16.

Figure 17–12 Atrial paroxysmal tachycardia—onset in middle of record (lead I).

focus in either the A-V node or the A-V bundle. This initiates normal impulses in the ventricles, thereby causing normal QRS-T complexes. However, P waves do not appear, because the atrial impulse travels backward from the A-V node through the atrium at the same time that the ventricular impulse travels through the ventricles. Therefore, the P wave is obscured by the QRS complex.

Atrial or A-V nodal paroxysmal tachycardia usually occurs in young, otherwise healthy persons for reasons totally unknown. Yet, factors that irritate the heart, such as coffee or lack of sleep, can predispose to the condition. In general, atrial paroxysmal tachycardia frightens the individual tremendously and may cause him to be weak during the paroxysms, but ordinarily no permanent harm comes from the attacks.

VENTRICULAR PAROXYSMAL TACHYCARDIA

Figure 17–13 illustrates a typical short paroxysm of ventricular tachycardia. The electrocardiogram of ventricular paroxysmal tachycardia has the appearance of a series of ventricular premature beats occurring one after another without any normal beats interspersed.

Ventricular paroxysmal tachycardia is usually a serious condition for two reasons. First, this type of tachycardia usually does not occur unless considerable damage is present in the ventricles. Second, ventricular tachycardia predisposes to ventricular fibrillation, which is almost invariably fatal. The reasons why ventricular tachycardia predisposes to ventricular fibrillation are: First, rapid stimulation decreases the refractory period of the heart and slows conduction, both of which predispose to circus movements. Second, one of the rapid ventricular impulses frequently occurs while part of the ventricular muscle is in a refractory state and part in a nonrefractory state; this causes the impulse to take a devious route through the heart, avoiding the refractory areas and initiating a long circus pathway, as was discussed in Chapter 14. The long circus pathway allows the impulse to continue around and around the ventricles, resulting in ventricular fibrillation.

Digitalis intoxication sometimes causes ventricular tachycardia. On the other hand, quinidine, which increases the refractory period of cardiac muscle, may be used to block irritable foci causing ventricular tachycardia. Another drug, procaine amide, decreases the irritability of cardiac muscle and can also frequently block an irritable focus.

ABNORMAL RHYTHMS RESULTING FROM CIRCUS MOVEMENTS

The circus movement phenomenon was discussed in detail in Chapter 14, and it was pointed out that these movements can cause atrial flutter, atrial fibrillation, and ventricular fibrillation.

ATRIAL FLUTTER

Figure 17–14 illustrates lead II of the electrocardiogram in atrial flutter. The rate of atrial contraction (P waves) is approximately 300 times per minute, while the rate of ventricular contraction (QRS-T waves) is only 125 times per minute. From the record it will be seen that sometimes a 2:1 rhythm occurs and sometimes a 3:1 rhythm. In other words, the atria beat two or three times for every one impulse that is conducted through the A-V bundle into the ventricles.

An important distinguishing feature of atrial flutter is the relatively high voltage, 0.2 to 0.3 mv., of the P waves in lead II and sometimes in lead III. Leads II and III record higher voltage than lead I because the atrial flutter pathway passes around and around the atria from top to bottom. This means that the current flow in the atria extends up and down the body more or less parallel to the electrical connections of leads II and III, which causes strong voltages in these leads but weak voltages in lead I.

ATRIAL FIBRILLATION

Figure 17–15 illustrates the electrocardiogram during atrial fibrillation. As discussed in Chapter 14, numerous impulses spread in all directions through the atria during atrial fibrillation. The intervals between impulses arriving at the A-V node are extremely variable. Therefore, an impulse may arrive at the A-V node immediately after the node itself is out of its refractory period from its previous discharge, or it may not arrive there for several tenths of a second. Consequently, the rhythm of the ventricles is very irregular, many of the ventricular beats falling quite close together and many far apart, the overall ventricular rate being 125 to 150 beats per minute in most instances.

The fibrillatory process in the atria creates only minute voltages in the electrocardiogram because the impulses are traveling in all directions, and no major mass of muscle depolarizes at any one

Figure 17–13 Ventricular paroxysmal tachycardia (lead III).

Figure 17–14 Atrial flutter — 2:1 and 3:1 rhythm (lead II).

Figure 17–15   Atrial fibrillation (lead I).

Figure 17–16   Ventricular fibrillation (lead II).

time. The potentials from the small fibrillatory impulses in the atria almost totally neutralize each other so that only a fine wavy record, if that much, is recorded in the electrocardiogram between ventricular beats, as is shown in Figure 17–15. On the other hand, the QRS-T complexes are entirely normal unless there is some simultaneous pathology of the ventricles.

The pumping efficiency of the heart in atrial fibrillation is considerably depressed because the ventricles often do not have sufficient time to fill between beats. Therefore, one of the principles of treating patients with atrial fibrillation is to slow down the rate of the ventricles. This is ordinarily done by administering digitalis, which causes reduced conduction through the A-V junctional tissue and A-V bundle, thereby decreasing the number of impulses conducted from the atria into the ventricles and also preventing any two ventricular beats from occurring close together. Heart function may be greatly improved when the rate is thus depressed to less than 100 beats per minute.

Pulse Deficit in Atrial Fibrillation. In the same manner that premature beats can cause a pulse deficit in the radial pulse, so can ventricular beats occurring close together in atrial fibrillation also cause a pulse deficit. Therefore, before treatment of atrial fibrillation one frequently finds a pulse deficit, whereas, after slowing the ventricular rate, the pulse deficit will have disappeared.

VENTRICULAR FIBRILLATION

In ventricular fibrillation the electrocardiogram is extremely bizarre, as shown in Figure 17–16, and ordinarily shows no tendency toward a rhythm of any type. In the early phases of ventricular fibrillation relatively large masses of muscle contract simultaneously, and this causes strong though irregular voltages in the electrocardiogram. However, after only a few seconds, the coarse contractions of the ventricles disappear, and the electrocardiogram changes into a new pattern of low-voltage, very irregular waves. Thus, no characteristic electrocardiographic pattern can be ascribed to ventricular fibrillation except that the electrical potentials constantly and spasmodically change, because the currents in the heart flow first in one direction, then in another, and rarely repeat any specific cycle.

The voltages of the waves in the electrocardiogram in ventricular fibrillation are usually about 1 mv. when ventricular fibrillation first begins, but these decay rapidly so that after 20 to 30 seconds they may be only 0.2 to 0.3 mv. Minute voltages of 0.1 mv. or less may be recorded for 10 minutes or longer after ventricular fibrillation begins.

REFERENCES

Bellet, S.: Clinical Disorders of the Heart Beat. 2nd ed., Philadelphia, Lea & Febiger, 1963.

Burch, G. E., and DePasquale, N. P.: Electrocardiography in the Diagnosis of Congenital Heart Disease. Philadelphia, Lea & Febiger, 1967.

Burch, G. E., and Winsor, T.: A Primer of Electrocardiography. 5th ed., Philadelphia, Lea & Febiger, 1966.

Chou, T., and Helm, R. A.: Clinical Vectorcardiography. New York, Grune & Stratton, Inc., 1967.

Dimond, E. G.: Electrocardiography and Vectorcardiography. 4th ed., Boston, Little, Brown and Company, 1967.

Dreifus, L. S., and Likoff, W.: Mechanisms and Therapy of Cardiac Arrhythmias. New York, Grune & Stratton, Inc., 1966.

Guyton, A. C., and Crowell, J. W.: A stereovectorcardiograph. *J. Lab. Clin. Med.*, 40:726, 1952.

Hoffman, I.: Vectorcardiography. Philadelphia, J. B. Lippincott Co., 1966.

Hurst, J. W., and Myerburg, R. J.: Introduction to Electrocardiography. New York, McGraw-Hill Book Company, 1968.

Lamb, L. E.: Electrocardiography and Vectorcardiography. Philadelphia, W. B. Saunders Company, 1965.

Marriott, H. J. L.: Practical Electrocardiography. 4th ed., Baltimore, Williams & Wilkins Co., 1968.

Riseman, J. E. F.: P-Q-R-S-T: A Guide to Electrocardiogram Interpretation. 5th ed., New York, The Macmillan Company, 1968.

Schaefer, H., and Haas, H. G.: Electrocardiography. *In* Handbook of Physiology. Baltimore, The Williams & Wilkins Co., 1962, Vol. 1, Sec. II, p. 323.

Schamroth, L.: An Introduction to Electrocardiography. 3rd ed., Philadelphia, F. A. Davis Co., 1966.

Scher, A. M., and Young, A. C.: Frequency analysis of the electrocardiogram. *Circ. Res.*, 8:344, 1960.

Sodi-Pallares, D., and Calder, R. M.: New Bases of Electrocardiography. St. Louis, The C. V. Mosby Co., 1956.

Stephenson, H. E.: Cardiac Arrest and Resuscitation. 3rd ed., St. Louis, The C. V. Mosby Co., 1969.

Tyberg, J. V.; Parmley, W. W.; and Sonnenblick, E. H.: In-vitro studies of myocardial asynchrony and regional hypoxia. *Circ. Res.*, 25:569, 1969.

PART IV

THE CIRCULATION

CHAPTER 18

Physics of Blood, Blood Flow, and Pressure: Hemodynamics

THE CIRCULATORY SYSTEM AS A "CIRCUIT"

The most important feature of the circulation that must always be kept in mind is that it is a continuous circuit. That is, if a given amount of blood is pumped by the heart, this same amount must also flow through each respective subdivision of the circulation. Furthermore, if blood is displaced from one segment of the circulation, some other segment of the circulation must expand unless the blood is lost from the circulation.

Figure 18–1 illustrates the general plan of the circulation, showing the two major subdivisions, the *systemic circulation* and the *pulmonary circulation.* In the figure the arteries of each subdivision are represented by a single distensible chamber and all the veins by another even larger distensible chamber, and the arterioles and capillaries represent very small connections between the arteries and veins. Blood flows with almost no resistance in all the larger vessels of the circulation, but this is not the case in the arterioles and capillaries, where considerable resistance does occur. To cause blood to flow through these small "resistance" vessels, the heart pumps blood into the arteries under high pressure—normally at a systolic pressure of about 120 mm. Hg in the systemic circulation and about 22 mm. Hg in the pulmonary system.

As a first step toward explaining the overall function of the circulation, this chapter will discuss the physical characteristics of blood itself and then the physical principles of blood flow through the vessels, including especially the interrelationships between pressure, flow, and resistance. The study of these interrelationships and other basic physical principles of blood circulation is called *hemodynamics.*

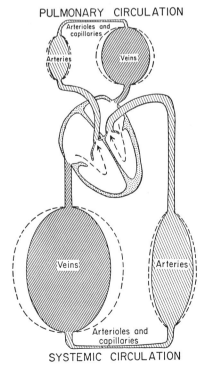

Figure 18–1 Schematic representation of the circulation, showing both the distensible and the resistive portions of the systemic and pulmonary circulations.

THE PHYSICAL CHARACTERISTICS OF BLOOD

Blood is a viscous fluid composed of *cells* and *plasma.* More than 99 per cent of the cells are red blood cells; this means that for practical purposes the white blood cells play almost no role in determining the physical characteristics of the blood. The fluid of the plasma is part of the extracellular fluid, and the fluid inside the cells is intracellular fluid.

THE HEMATOCRIT

The hematocrit of blood is the per cent of the blood that is cells. Thus, if one says that a person has a hematocrit of 40, he means that 40 per cent of the blood volume is cells and the remainder is plasma. The hematocrit of normal man averages about 42, whereas that of normal woman averages about 38. These values vary tremendously, depending upon whether or not the person has anemia, upon his degree of bodily activity, and on the altitude at which he resides. These effects were discussed in relation to the red blood cells and their function in Chapter 8.

Blood hematocrit is determined by centrifuging blood in a calibrated tube such as that shown in Figure 18–2. The calibration allows direct reading of the per cent of cells.

Effect of Hematocrit on Blood Viscosity. The greater the percentage of cells in the blood—that is, the greater the hematocrit—the more friction there is between successive layers of blood, and it is this friction that determines viscosity. Therefore, the viscosity of blood increases drastically as the hematocrit increases, as illustrated in Figure 18–3. If we arbitrarily consider the viscosity of water to be 1, then the viscosity of whole blood is about 3 or 4; this means that 3 to 4 times as much pressure is required to force whole blood through a given tube than to force water through the same tube. Note that when the hematocrit rises to 60 or 70, which it often does in polycythemia as discussed in Chapter 8, the blood viscosity

Figure 18–3 Effect of hematocrit on viscosity.

can become as great as 10 times that of water, and its flow through blood vessels becomes extremely difficult.

Another factor that affects blood viscosity is the concentration and types of proteins in the plasma, but these effects are so much less important than the effect of hematocrit that they are not significant considerations in most hemodynamic studies. The viscosity of blood plasma is approximately 1.5 times that of water.

Since most resistance in the circulatory system occurs in the very small blood vessels, it is especially important to know how blood viscosity affects blood flow in these minute vessels. At least three additional factors besides hematocrit and plasma proteins affect blood viscosity in these vessels:

1. Blood flow in very minute vessels exhibits far less viscous effect than it does in large vessels. This effect, called the *Fahraeus-Lindqvist effect,* begins to appear when the vessel diameter falls below approximately 1.5 mm. In vessels as small as capillaries, this effect is so prominent that the viscosity of whole blood theoretically could be as little as one-half that in large vessels. The Fahraeus-Lindqvist effect is probably caused by alignment of the red cells as they pass through the vessels. That is, the red cells, instead of moving randomly, line up and move through the vessels as a single plug, thus eliminating the viscous resistance that occurs internally in the blood itself. The Fahraeus-Lindqvist effect, however, is mainly offset by the following two effects.

2. The viscosity of blood increases drastically as its velocity of flow decreases. Since the velocity of blood flow in the small vessels is extremely minute, often less than 1 mm. per second, blood viscosity can increase as much as 10-fold from this factor alone. This effect is presumably caused by adherence of the red cells to each other and to the vessel walls.

3. Cells also often become stuck in small blood vessels and this happens especially in capillaries where the nuclei of endothelial cells protrude into

NORMAL ANEMIA POLYCYTHEMIA

Figure 18–2 Hematocrits in the normal person and in anemia and polycythemia.

the capillary. When this occurs, blood flow can become totally blocked for a fraction of a second, for several seconds, or for much longer periods of time, thus giving an apparent effect of greatly increased viscosity.

Because of these special effects that occur in the minute vessels of the circulatory system, it has been impossible to determine the exact manner in which hematocrit affects viscosity in the minute vessels, which is the place in the circulatory system where viscosity plays its most important role. Nevertheless, because some of these effects tend to decrease viscosity and others tend to increase viscosity, it is perhaps best at present simply to assume that the viscous effects in the small vessels are approximately equivalent to those that occur in the large vessels.

PLASMA

Plasma is part of the extracellular fluid of the body. It is almost identical to the interstitial fluid found between the tissue cells except for one major difference: plasma contains about 7 per cent proteins, whereas interstitial fluid contains an average of only 2 per cent proteins. The reason for this difference is that plasma proteins filter through the capillary pores into the interstitial spaces very poorly. As a result, most of the plasma proteins are held in the circulatory system, giving it a protein concentration about three and one-half times that of the fluid outside the capillaries.

The Proteins of Plasma. The plasma proteins are divided into three major types, as follows:

	Grams Per Cent
Albumin	4.5
Globulins	2.5
Fibrinogen	0.3

The primary function of the *albumin* (and of the other proteins to a lesser extent) is to cause osmotic pressure at the capillary membrane. This pressure, called *colloid osmotic pressure,* prevents the fluid of the plasma from leaking out of the capillaries into the interstitial spaces. This function is so important that it will be discussed in detail in Chapter 20.

The globulins are divided into three major types: alpha, beta, and gamma globulins. The *alpha* and *beta globulins* perform diverse functions in the circulation, such as transporting other substances by combining with them, acting as substrates that react with other substances, and, transporting protein itself from one part of the body to another. The *gamma globulins,* and to a lesser extent the *beta globulins,* play a special role in protecting the body against infection, for it is these globulins that are mainly the *antibodies* that resist infection

and toxicity, thus providing the body with what we call *immunity.* The function of immunity was discussed in detail in Chapter 10.

The *fibrinogen* of plasma is of basic importance in blood clotting and was discussed in Chapter 12.

INTERRELATIONSHIPS BETWEEN PRESSURE, FLOW, AND RESISTANCE

Flow through a blood vessel is determined entirely by two factors: (1) the *pressure difference* tending to push blood through the vessel and (2) the impediment to blood flow through the vessel, which is called vascular *resistance.* Figure 18–4 illustrates these relationships quantitatively, showing a blood vessel segment located anywhere in the circulatory system.

P_1 represents the pressure at the origin of the vessel; at the other end the pressure is P_2. Since the vessel is relatively small, the blood is experiencing difficulty in flowing. This is the resistance. Stated another way, a *pressure difference* between the two ends of the vessel causes *blood* to *flow* from the high pressure area to the low pressure area while *resistance* impedes the flow. This can be expressed mathematically as follows:

$$Q = \frac{\Delta P}{R} \tag{1}$$

in which Q is blood flow, ΔP is the pressure difference between the two ends of the vessel, and R is the resistance.

It should be noted especially that it is the *difference* in pressure between the two ends of the vessel that determines the rate of flow and not the absolute pressure in the vessel. For instance, if the pressure at both ends of the segment were 100 mm. Hg and yet no difference existed between the two ends, there would be no flow despite the presence of the 100 mm. Hg pressure.

The above formula expresses the most important of all the relationships that the student needs to understand to comprehend the hemodynamics of the circulation. Because of the extreme importance of this formula the

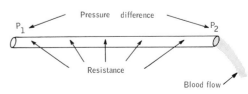

Figure 18–4 Relationship between pressure, resistance, and blood flow.

student should also become familiar with its other two algebraic forms:

$$\Delta P = Q \times R \qquad (2)$$

$$R = \frac{\Delta P}{Q} \qquad (3)$$

BLOOD FLOW

Blood flow means simply the quantity of blood that passes a given point in the circulation in a given period of time. Ordinarily, blood flow is expressed in milliliters or liters per minute, but it can be expressed in milliliters per second or in any other unit of flow.

The overall blood flow in the circulation of an adult person at rest is about 5000 ml. per minute. This is called the *cardiac output* because it is the amount of blood pumped by each ventricle of the heart in a unit period of time. It is also obvious that this same amount of blood must pass through both the systemic and pulmonary circulations.

Methods for Measuring Blood Flow. Many different mechanical or mechanoelectrical devices can be inserted in series with a blood vessel or in some instances applied to the outside of the vessel to measure flow. These are called simply *flowmeters.*

THE ROTAMETER. Figure 18–5 illustrates a *rotameter.* In using this the blood vessel must be cut and the blood passed through the flowmeter. Also, an anticoagulant, such as heparin, must be used to prevent blood clotting.

Figure 18–5 A flowmeter of the rotameter type.

Blood enters the rotameter from the bottom and flows upward through a cone-shaped cavity. In this cavity is a small float. When no blood at all is flowing, the float falls to the bottom of the cavity, but as blood flows faster and faster, the float rises higher and higher. In using the rotameter, one determines the blood flow by measuring the height to which the float rises, or, if he wishes to record the blood flow continuously, an arrangement such as that shown in the figure can be used. A small steel shaft projects upward from the float into a coil of wire. The inductance of the coil increases as the steel shaft moves farther up into the coil, and by using appropriate electronic apparatus the blood flow can be recorded continuously.

THE ELECTROMAGNETIC FLOWMETER. In recent years several new devices have been developed which can be used to measure blood flow in a vessel without opening it. One of the most important of these is the electromagnetic flowmeter, the principles of which are illustrated in Figure 18–6. Figure 18–6A shows the generation of electromotive force in a wire that is moved rapidly through a magnetic field. This is the well-known principle for production of electricity by the electric generator. Figure 18–6B shows that exactly the same principle applies for generation of electromotive force in blood when it moves through a magnetic field. In this case, a blood vessel is placed between the poles of a strong magnet, and electrodes are placed on the two sides of the vessel perpendicular to the magnetic lines of force. When blood flows through the vessel, electrical voltage is generated between the two electrodes, and this is recorded using an appropriate meter or electronic apparatus. Figure 18–6C illustrates an actual "probe" that is placed on a large blood vessel to record its blood flow. This probe contains both the strong magnet and the electrodes.

An additional advantage of the electromagnetic flowmeter is that it can record changes in flow that occur in less than 0.01 second, allowing accurate recording of pulsatile changes in flow as well as steady state changes.

THE ULTRASONIC FLOWMETER. Another type of flowmeter that can be applied to the outside of the vessel and that has the same advantages as the electromagnetic flowmeter is the ultrasonic flowmeter illustrated in Figure 18–7. A minute piezoelectric crystal is mounted in the walls of each half of the device; one of these crystals transmits sound diagonally along the vessel and the other receives the sound. An electronic apparatus alternates the direction of sound transmission several hundred times per second, transmitting first downstream, then upstream. Sound waves travel downstream with greater velocity than upstream. An appropriate electronic apparatus measures the difference between these two velocities, which is a measure of blood flow. The sound frequency can be anywhere between 100,000 and 4,000,000 cycles per second.

THE PLETHYSMOGRAPH. Figure 18–8 illustrates an apparatus known as a "plethysmograph,"

Figure 18-6 A flowmeter of the electromagnetic type, showing (A) generation of an electromotive force in a wire as it passes through an electromagnetic field, (B) generation of an electromotive force in electrodes on a blood vessel when the vessel is placed in a strong magnetic field and blood flows through the vessel, and (C) a modern electromagnetic flowmeter "probe" for chronic implantation around blood vessels.

which is used for estimating the quantity of blood flow through the forearm. This apparatus operates as follows: The forearm is placed in a chamber, and an airtight seal is made between the proximal portion of the chamber and the forearm near the elbow. The chamber is connected to a tambour recorder so that an increase in volume of the arm inside the airtight chamber causes the membrane of the tambour to rise. To determine the quantity of blood flowing through the forearm, a blood pressure cuff on the arm above the plethysmograph is suddenly inflated to a pressure above venous blood pressure but below arterial blood pressure — 40 mm. Hg, for instance. The pressure in the cuff occludes the veins so that blood cannot return from the forearm into the general circulatory system, but blood does continue to flow through the unoccluded arteries into the forearm. Therefore, when the blood pressure cuff is first inflated, the forearm swells, and the quantity of blood flowing into the forearm is recorded on the kymograph as illustrated in Figure 18-8A and B. The plethysmo-

graph is then calibrated by injecting into the chamber standard volumes of air or fluid from the syringe shown also in the diagram.

In making measurements of blood flow from the plethysmograph, a tangent is drawn to the plethysmographic recording at the very onset of the rise, as illustrated in Figure 18-8B. The reason for this is that, after a few seconds of blood flow into the arm, all the blood vessels of the forearm become distended and exert a considerable amount of back pressure, which opposes further inflow of blood. Therefore, it is only immediately after the cuff pressure is elevated that the recording actually represents the normal rate of blood flow into the forearm. In the recording of the figure, measurements from the tangent show that the initial rate of blood flow is 30 ml. in 20 seconds, or 90 ml. per minute.

Plethysmographs can be used for recording blood flow in a leg, a toe, a finger, or even in a number of internal organs of the body, for plethysmographs known as "oncometers" can be placed around the spleen, the kidney, the liver, various glands, and even the heart. Unfortunately, technical difficulties in the use of plethysmographs are likely to make their recordings inaccurate — sometimes 20 per cent or more.

Laminar Flow of Blood in Vessels. When blood flows at a continuous rate through a long, smooth vessel, the velocity of flow in the center of the vessel is far greater than that toward the outer edges. This is illustrated by the experiment shown in Figure 18-9. In

Figure 18-7 Basic construction of the ultrasonic flowmeter probe.

Figure 18–8 (A) The plethysmograph. (B) A plethysmographic record.

vessel A are two different fluids, the one to the left colored by a dye and the one to the right a clear fluid, but there is no flow in the vessel. Then the fluids are made to flow; and a parabolic interface develops between the two fluids, as shown 1 second later in vessel B, illustrating that the portion of fluid adjacent to the walls has hardly moved at all, the portion slightly away from the wall has moved a small distance, and the portion in the center of the vessel has moved a long distance. This effect is called *laminar flow* or *streamline flow*.

The cause of laminar flow is the following: The fluid molecules touching the wall move essentially none at all because of adherence to the vessel wall. The next layer of molecules slips over these, the third layer over the second, the fourth layer over the third, and so forth. Therefore, the fluid in the middle of the vessel can move rapidly because many layers of molecules exist between the middle of the vessel and the vessel wall, all of these capable of slipping over each other, while those portions of fluid near the wall do not have this advantage.

Turbulent Flow of Blood Under Some Conditions. When the rate of blood flow becomes too great, when it passes by an obstruction in a vessel, when it makes a sharp turn, or when it passes over a rough surface, the flow may then become

Figure 18–9 An experiment illustrating laminar blood flow, showing (A) two separate fluids before flow begins and (B) the same fluids 1 second after flow begins.

turbulent rather than streamline. Turbulent flow means that the blood flows crosswise in the vessel as well as along the vessel, usually forming whorls in the blood called *eddy currents*. These are similar to the whirlpools that one frequently sees in a rapidly flowing river at a point of obstruction.

When eddy currents are present, blood flows with much greater resistance than when the flow is streamline because the eddies add tremendously to the overall friction of flow in the vessel.

The tendency for turbulent flow increases in direct proportion to the velocity of blood flow, in direct proportion to the radius of the blood vessel, and inversely proportional to the viscosity of the blood divided by its density in accordance with the following equation:

$$\text{Re} = \frac{v \cdot r}{\dfrac{\eta}{\rho}} \qquad (4)$$

in which Re is *Reynolds' number* and is the measure of the tendency for turbulence to occur, v is the velocity of blood flow (in centimeters per second), r is the radius of the vessel (in centimeters), η is the viscosity (in poises), and ρ is density. The viscosity of blood is normally about $1/30$ poise, and the density is only slightly greater than 1. When Reynolds' number rises above 200, turbulent flow will occur at the branches of the arterial tree but will die out along the smooth portions of the vessels. However, when Reynolds' number rises above approximately 1000, turbulence can occur even in a straight, smooth vessel. Reynolds' number for flow in the vascular system rises above 200 in the large arteries; as a result there can be considerable turbulence of flow at the branches of these vessels. In the proximal aorta and pulmonary artery, Reynolds' number can rise to several thousand during the rapid phase of ejection of the ventricles, and this causes marked turbulence in the aorta and pulmonary arteries immediately distal to the aortic and pulmonary valves.

BLOOD PRESSURE

The Standard Units of Pressure. Blood pressure is almost always measured in *millimeters of mercury (mm. Hg)* because the mercury manometer (shown in Figure 18–10) has been used as the standard reference for measuring blood pressure for many years. Actually, blood pressure means the *force exerted by the blood against any unit area of the vessel wall.* When one says that the pressure in a vessel is 50 mm. Hg, this means that the force exerted would be sufficient to push a column of mercury up to a level of 50 mm. If the pressure were 100 mm. Hg, it would push the column of mercury up to 100 mm.

Occasionally, pressure is measured in *centi-*

Figure 18–10 Recording arterial pressure with a mercury manometer.

meters of water. A pressure of 10 cm. of water means a pressure sufficient to raise a column of water to a height of 10 cm. *One millimeter of mercury equals 1.36 cm. of water* because the density of mercury is 13.6 times that of water, and 1 cm. is 10 times as great as 1 mm. Dividing 13.6 by 10, we derive the factor 1.36.

Measurement of Blood Pressure Using the Mercury Manometer. Figure 18–10 illustrates a standard mercury manometer for measuring blood pressure. A cannula is inserted into an artery, a vein, or even into the heart, and the pressure from the cannula is transmitted to the left-hand side of the manometer where it pushes the mercury down while raising the right-hand mercury column. The difference between the two levels of mercury is approximately equal to the pressure in the circulation in terms of millimeters of mercury. (To be more exact, it is equal to 104 per cent of the true pressure because of the weight of the water on the left-hand column of mercury.)

High-Fidelity Methods for Measuring Blood Pressure. Unfortunately, the mercury in the mercury manometer has so much *inertia* that it cannot rise and fall rapidly. For this reason the mercury manometer, though excellent for recording steady pressures, cannot respond to pressure changes that occur more rapidly than approximately one cycle every 2 to 3 seconds. Whenever it is desired to record rapidly changing pressures, some other type of pressure recorder is needed. Figure 18–11 demonstrates the basic principles of three electronic pressure *transducers* commonly used for converting pressure into electrical signals and then recording the pressure on a high-speed electrical recorder. Each of these transducers em-

ploys a very thin and highly stretched metal membrane which forms one wall of the fluid chamber. The fluid chamber in turn is connected through a needle with the vessel in which the pressure is to be measured. Pressure variations in the vessel cause changes of pressure in the chamber beneath the membrane. When the pressure is high, the membrane bulges outward slightly and when low it returns toward its resting position. In Figure 18–11A a simple metal plate is placed a few thousandths of an inch above the membrane. When the membrane bulges outward, the *capacitance* between the plate and membrane increases, and this change in capacitance can be recorded by an appropriate electronic system. In Figure 18–11B a small iron slug rests on the membrane, and this can be displaced upward into a coil. Movement of the iron changes the *inductance* of the coil, and this, too, can be recorded electronically. Finally, in Figure 18–11C a very thin, stretched resistance wire is connected to the membrane. When this wire is greatly stretched its resistance increases, and when less stretched the resistance decreases. These changes also can be recorded by means of an electronic system.

It is possible also to connect a mirror to the edge of the membrane so that it angulates as the membrane bulges. A beam of light reflected from the mirror reflects a moving spot of light that can record pressure changes on a moving photographic paper.

Using some of these high fidelity types of recording systems, pressure cycles up to 500 cycles per second have been recorded accurately, and in common use are recorders capable of registering pressure changes as rapidly as 20 to 100 cycles per second.

Figure 18–11 Principles of three different types of electronic transducers for recording rapidly changing blood pressures.

RESISTANCE TO BLOOD FLOW

Units of Resistance. Resistance is the impediment to blood flow in a vessel, but it cannot be measured by any direct means. Instead, resistance must be calculated from measurements of blood flow and pressure difference in the vessel. If the pressure difference between two points in a vessel is 1 mm. Hg and the flow is 1 ml./sec., then the resistance is said to be 1 *peripheral resistance unit,* usually abbreviated *PRU.*

EXPRESSION OF RESISTANCE IN CGS UNITS. Occasionally, a basic physical unit called the CGS unit is used to express resistance. This unit is *dyne seconds/centimeters*[5]. Resistance in these units can be calculated by the following formula:

$$R \left(in \frac{\text{dynes sec.}}{\text{cm.}^5} \right) = \frac{1333 \times \text{mm. Hg}}{\text{ml./sec.}} \quad (5)$$

Total Peripheral Resistance and Total Pulmonary Resistance. The rate of blood flow through the circulatory system when a person is at rest is close to 100 ml./sec., and the pressure gradient from the systemic arteries to the systemic veins is about 100 mm. Hg. Therefore, in round figures the resistance of the entire systemic circulation, called the *total peripheral resistance,* is approximately 100/100 or 1 PRU. In some conditions in which the blood vessels throughout the body become strongly constricted, the total peripheral resistance rises to as high as 4 PRU, and when the vessels become greatly dilated it can fall to as little as 0.25 PRU.

In the pulmonary system the mean arterial pressure averages 13 mm. Hg and the mean left atrial pressure averages 4 mm. Hg, giving a net pressure gradient of 9 mm. Therefore, in round figures the *total pulmonary resistance* at rest calculates to be about 0.09 PRU. This can increase in disease conditions to as high as 1 PRU and can fall in certain physiological states, such as exercise, to as low as 0.03 PRU.

"Conductance" of Blood in a Vessel and Its Relationship to Resistance. Conductance is a measure of the amount of blood flow that can pass through a vessel in a given time for a given pressure gradient. This is generally expressed in terms of ml./sec./mm. Hg pressure, but it can also be expressed in terms of liters/sec./mm. Hg or in any other units of blood flow and pressure.

It is immediately evident that conductance is the reciprocal of resistance in accord with the following equation:

$$\text{Conductance} = \frac{1}{\text{Resistance}} \quad (6)$$

EFFECT OF VASCULAR DIAMETER ON CONDUCTANCE — POISEUILLE'S LAW. Slight changes in diameter of a vessel cause tremendous changes in its ability to conduct blood. This is illustrated forcefully by the experiment in Figure 18–12A, which shows three separate vessels with relative diameters of 1, 2, and 4 but with the same pressure difference of 100 mm. Hg between the two ends of the vessels. Though the diameters of these vessels increase only four-fold, the respective flows are 1, 16, and 256 ml./min., which is a 256-fold increase in flow. Thus, the conductance of the vessel increases in proportion to the *fourth power of the diameter,* in accord with the following formula:

$$\text{Conductance} \propto \text{Diameter}^4 \quad (7)$$

The cause of this great increase in conductance with an increase in diameter can be explained by referring to Figure 18–12B. This illustrates cross-sections of a large and a small vessel. The concentric rings inside these vessels indicate that the velocity of flow in each ring is different from that in the other rings because of *laminar* flow, which was discussed earlier in the chapter. That is, the blood in the ring touching the wall of the vessel is flowing hardly at all because of its adherence to the vascular endothelium. The next ring of blood slips past the first ring and, therefore, flows at a more rapid velocity. The third, fourth, fifth, and sixth rings likewise flow at progressively increasing velocities. Thus, the blood that is very near the wall of the vessel flows extremely slowly, while that in the middle of the vessel flows extremely rapidly.

In the small vessel essentially all of the blood is very near the wall so that the extremely rapidly

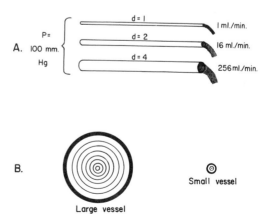

Figure 18–12 (A) Demonstration of the effect of vessel diameter on blood flow. (B) Concentric rings of blood flowing at different velocities; the farther away from the vessel wall, the faster the flow.

flowing central stream of blood simply does not exist.

By integrating the velocities of all the concentric rings of flowing blood one can derive the following formula relating *mean velocity* of blood flow to vascular diameter:

$$v = \frac{\Delta P \cdot r^2}{8\eta l} \tag{8}$$

in which v is velocity (in centimeters per second), ΔP is the pressure gradient (in dynes per square centimeter), r is the radius of the vessel (in centimeters), η is the viscosity (in poises), and l is the length of the vessel (in centimeters).

The quantity of blood that will flow through a vessel in a given period of time is equal to the velocity of flow times the cross-sectional area according to the following equation:

$$Q = v\pi r^2 \tag{9}$$

in which Q is the rate of blood flow (in milliliters per second) and πr^2 is the cross-sectional area (in square centimeters).

Now let us substitute the value for velocity of blood flow from Equation 8 into Equation 9. This gives the following equation, which is known as Poiseuille's law:

$$Q = \frac{\pi \Delta P r^4}{8\eta l} \tag{10}$$

Note particularly in this equation that the rate of blood flow is directly proportional to the *fourth power of the radius* of the vessel, which illustrates once again that the diameter of a blood vessel plays by far the greatest role of all factors in determining the rate of blood flow through the vessel.

SUMMARY OF THE DIFFERENT FACTORS THAT AFFECT CONDUCTANCE AND RESISTANCE. In the equation representing Poiseuille's law, Q represents flow, and ΔP represents the pressure difference. The remainder of the equation represents the conductance in accordance with the following equation:

$$C = \frac{\pi r^4}{8\eta l} \tag{11}$$

And, since conductance is the reciprocal of resistance, the following equation shows the factors that affect resistance:

$$R = \frac{8\eta l}{\pi r^4} \tag{12}$$

Thus, note that the resistance of a vessel is directly proportional to the *blood viscosity* and *length* of the vessel but inversely proportional to the *fourth power of the radius.*

Resistance to Blood Flow Through Series Vessels. In Figure 18–13A are two vessels connected in series, having resistances of R_1 and R_2. It is im-

Figure 18–13 Vascular resistances: (A) in series and (B) in parallel.

mediately evident that the total resistance is equal to the sum of the two, or

$$R_{(total)} = R_1 + R_2 \tag{13}$$

Furthermore, it is equally evident that any number of resistances in series with each other must be added together. For instance, the total peripheral resistance is equal to the resistance of the arteries plus that of the arterioles plus that of the capillaries plus that of the veins.

Resistance of Vessels in Parallel. Shown in Figure 18–13B are four vessels, *connected in parallel,* with respective resistances of R_1, R_2, R_3, and R_4. However, the diameters of these vessels are not exactly the same. It is obvious that for a given pressure difference, far greater amounts of blood will flow through this system than through any one of the vessels alone. Therefore, the total resistance is far less than the resistance of any single vessel.

To calculate the total resistance in Figure 18–13B one first determines the *conductance* of each of the vessels, which is equal to the reciprocal of the resistance, or $\frac{1}{R_1}$, $\frac{1}{R_2}$, $\frac{1}{R_3}$, and $\frac{1}{R_4}$. The total conductance of all the vessels, $\frac{1}{R_{(total)}}$, is equal to the sum of the individual conductances:

$$\frac{1}{R_{(total)}} = \frac{1}{R_1} + \frac{1}{R_2} + \frac{1}{R_3} + \frac{1}{R_4} \tag{14}$$

And resistance through the parallel circuit is the reciprocal of the total conductance, or

$$R_{(total)} = \frac{1}{\dfrac{1}{R_1} + \dfrac{1}{R_2} + \dfrac{1}{R_3} + \dfrac{1}{R_4}} \tag{15}$$

Effect of Pressure on Vascular Resistance — Critical Closing Pressure

Since all blood vessels are distensible, increasing the pressure inside the vessels causes the vascular diameters also to increase. This in turn reduces the resistance of the vessel. Conversely, reduction in vascular pressures increases the resistance.

The middle curve of Figure 18–14 illustrates the effect on blood flow through a small tissue artery caused by changing the arterial pressure. As the arterial pressure falls from 130 mm. Hg, the flow decreases rapidly at first because of two factors: (1) the decreasing pressure and (2) the decreasing diameter of the vessel. However, as the pressure continues to fall, the rate of decrease in flow becomes progressively less. Then, at 20 mm. Hg blood flow ceases entirely. This point at which the blood stops flowing is called the *critical closing pressure*, because at this point the small vessels, the arterioles in particular, close so completely that all flow through the tissue ceases. The mechanism of this can be explained as follows:

Mechanism of Critical Closing Pressure. The vasomotor tone of the arterioles is always attempting to constrict these vessels to smaller diameters, while on the other hand the pressure inside the arterioles is tending to dilate them. As the pressure falls progressively lower and lower, it finally reaches a point at which the pressure inside the vessel is no longer capable of keeping the vessel open; this pressure is called the critical closing pressure.

A physical law, the *law of Laplace,* helps to explain the closure of vessels at an exact critical pressure level. This law states that the distending force tending to stretch the muscle fibers in the vascular wall is proportional to the *diameter of the vessel* times the *pressure,* or F ∝ D · P. Therefore, as the pressure in the vessel falls, thus allowing the vascular diameter also to decrease, the force tending to keep the vascular wall stretched decreases extremely rapidly, much more rapidly than would be accounted for by the decrease in pressure alone.

Another factor that contributes to the cessation of blood flow when the arterial pressure falls low is the size of the blood cells themselves, for, when the arteriolar diameter falls below a certain criti-

cal diameter, the red cells cannot pass through, and they will actually block the flow of plasma as well. The average critical closing pressure is about 20 mm. Hg when whole blood is flowing through the vessels and 5 to 10 mm. Hg when only plasma is flowing through.

Effect of Sympathetic Inhibition and Stimulation on Vascular Flow. Essentially all blood vessels of the body are normally excited by sympathetic impulses even under resting conditions. Different circulatory reflexes, which will be discussed in Chapter 25, can cause these impulses either to disappear, called *sympathetic inhibition,* or to increase to many times their normal rate, called *sympathetic stimulation.* Sympathetic impulses in most parts of the body cause an increase in tone of the vascular smooth muscle. Therefore, as illustrated by the dashed curves of Figure 18–14, sympathetic inhibition allows far more blood to flow through a tissue for a given pressure gradient than is normally true, whereas sympathetic stimulation vastly decreases the rate of blood flow.

Note also that sympathetic inhibition decreases the critical closing pressure, sometimes to as low as 5 mm. Hg. On the other hand, sympathetic stimulation can increase the critical closing pressure to as high as 100 or more mm. Hg, which means that strong sympathetic stimulation can actually stop blood flow through some tissues.

VASCULAR DISTENSIBILITY— PRESSURE-VOLUME CURVES

The diameter of blood vessels, unlike that of metal pipes and glass tubes, increases as the internal pressure increases, because blood vessels are *distensible.* However, the vascular distensibilities differ greatly in different segments of the circulation, and, as we shall see, this affects significantly the operation of the circulatory system under many changing physiological conditions.

Units of Vascular Distensibility. Vascular distensibility is normally expressed as the fractional increase in volume for each millimeter mercury rise in pressure in accordance with the following formula:

$$\text{Vascular distensibility} = \frac{\text{Increase in volume}}{\text{Increase in pressure} \times \text{Original volume}} \quad (16)$$

That is, if 1 mm. Hg causes a vessel originally containing 10 ml. of blood to increase its volume by 1 ml., then the distensibility would be 0.1 per mm. Hg, or 10 per cent per mm. Hg.

DIFFERENCE IN DISTENSIBILITY OF THE ARTERIES AND THE VEINS. Anatomically, the walls of arteries are far stronger than those of veins.

Figure 18–14 Effect of arterial pressure on blood flow through a blood vessel at different degrees of vascular tone, showing also the effect of vascular tone on critical closing pressure.

Consequently, the veins, on the average, are about 6 to 10 times as distensible as the arteries. That is, a given rise in pressure will cause about 6 to 10 times as much extra blood to fill a vein as an artery of comparable size.

In the pulmonary circulation the veins are very similar to those of the systemic veins. However, the pulmonary artery, which normally operates under pressures about one-fifth those in the systemic arterial system, has a distensibility only about one-half that of veins, rather than one-eighth, as is true of the systemic arteries.

VASCULAR COMPLIANCE

Usually in hemodynamic studies it is much more important to know the *total quantity of blood* that can be stored in a given portion of the circulation for each mm. Hg pressure rise than to know the distensibility of the individual vessels. This value is sometimes called the *overall distensibility* or *total distensibility,* or it can be expressed still more precisely by either of the terms *compliance* or *capacitance,* which are physical terms meaning the increase in volume caused by a given increase in pressure as follows:

Vascular compliance =
$$\frac{\text{Increase in volume}}{\text{Increase in pressure}} \quad (17)$$

Compliance and distensibility are quite different. A highly distensible vessel which has a very slight volume may have far less compliance than a much less distensible vessel which has a very large volume, for *compliance is equal to distensibility × volume.*

The compliance of a vein is about 24 times that of its corresponding artery because it is about 8 times as distensible and it has a volume about 3 times as great (8 × 3 = 24).

PRESSURE-VOLUME CURVES OF THE ARTERIAL AND VENOUS CIRCULATIONS

A convenient method for expressing the relationship of pressure to volume in a vessel or in a large portion of the circulation is the so-called *pressure-volume curve.* The two solid curves of Figure 18–15 represent respectively the pressure-volume curves of the normal arterial and venous systems, showing that the arterial system, including the larger arteries, small arteries, and arterioles, contains approximately 750 ml. of blood when the mean ar-

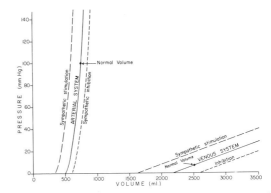

Figure 18–15 Pressure-volume curves of the systemic arterial and venous systems, showing also the effects of sympathetic stimulation and sympathetic inhibition.

terial pressure is 100 mm. Hg but only 500 ml. when the pressure has fallen to zero.

The volume of blood normally in the entire venous tree is about 2500 ml., but only a slight change in venous pressure changes this volume tremendously.

Difference in Compliance of the Arterial and Venous Systems. Referring once again to Figure 18–15, one can see that a change of 1 mm. Hg increases the venous volume a very large amount but increases the arterial volume only slightly. That is, the *compliance of the venous system is far greater than the compliance of the arteries—about 24 times as great.*

This difference in compliance is particularly important because it means that tremendous amounts of blood can be stored in the veins with only slight changes in pressure. Therefore, the veins are frequently called the *storage areas* of the circulation.

Effect of Sympathetic Stimulation or Sympathetic Inhibition on the Pressure-Volume Relationships of the Arterial and Venous Systems. Also shown in Figure 18–15 are the pressure-volume curves of the arterial and venous systems during moderate sympathetic stimulation and also during sympathetic inhibition. It is evident that sympathetic stimulation, with its concomitant increase in smooth muscle tone in the vascular walls, increases the pressure at each volume of the arteries or veins while, on the other hand, sympathetic inhibition decreases the pressure at each respective volume. Obviously, control of the vessels in this manner by the sympathetics can be valuable for diminishing or enhancing the dimensions of one segment of the circulation, thus transferring blood to or from other segments. For instance, an increase in vascular tone throughout the systemic circulation without a similar increase in the pulmonary circulation often causes large volumes of blood to shift out of the systemic vessels into the lungs.

Sympathetic control of vascular capacity is also especially important during hemorrhage. Enhancement of the sympathetic tone of the vessels, especially of the veins, simply reduces the dimensions of the circulatory system, and the circulation continues to operate almost normally even when as much as 25 per cent of the total blood volume has been lost.

"MEAN CIRCULATORY PRESSURE" AND PRESSURE-VOLUME CURVES OF THE ENTIRE CIRCULATORY SYSTEM

THE MEAN CIRCULATORY PRESSURE

The mean circulatory pressure is a measure of the degree of filling of the circulatory system. That is, it is the pressure that would be measured in the circulation if one could instantaneously stop all blood flow and bring all the pressures in the circulation immediately to equilibrium. The mean circulatory pressure has been measured reasonably accurately in dogs within 2 to 3 seconds after the heart has been stopped. To do this the heart is thrown into fibrillation by an electrical stimulus, and blood is pumped rapidly from the systemic arteries to the veins to cause equilibrium between the two major chambers of the circulation.

The mean circulatory pressure measured in the above manner is almost exactly 7 mm. Hg and almost never varies more than 1 mm. from this value in the normal resting animal. However, many different factors can change it, including especially change in the blood volume and increased or decreased sympathetic stimulation.

The mean circulatory pressure is *one of the major factors that determines the rate at which blood flows from the vascular tree into the right atrium of the heart and, therefore, that controls the cardiac output itself.* This is so important that it will be explained in detail in Chapter 27; it will also be discussed in relation to blood volume regulation in Chapter 36.

PRESSURE-VOLUME CURVES OF THE ENTIRE CIRCULATION

Figure 18–16 illustrates the changes in mean circulatory pressure as the total blood volume increases (1) under normal conditions, (2) during strong sympathetic stimulation, and (3)

during complete sympathetic inhibition. The point marked by the arrow is the operating point of the normal circulation: a mean circulatory pressure of 7 mm. Hg and a blood volume of 5000 ml. However, *if blood is lost from the circulatory system,* the mean circulatory pressure falls to a lower value. If increased amounts of blood are added, the mean circulatory pressure rises accordingly.

The compliance of the entire circulatory system in the human being, as estimated from experiments in dogs, is approximately 100 ml. for each 1 mm. rise in mean circulatory pressure.

Sympathetic stimulation and *inhibition* affect the pressure-volume curves of the entire circulatory system in the same way that they affect the pressure-volume curves of the individual parts of the circulation, as illustrated by the two dashed curves of Figure 18–16. That is, for any given blood volume, the mean circulatory pressure rises two- to fourfold with strong sympathetic stimulation and falls markedly when the sympathetics are inhibited. This is an extremely important factor in the regulation of blood flow into the heart and thereby for regulating the cardiac output. For instance, during exercise, sympathetic activity increases the mean circulatory pressure and correspondingly helps to increase the cardiac output.

RELATIONSHIP BETWEEN MEAN CIRCULATORY PRESSURE, MEAN SYSTEMIC PRESSURE, AND MEAN PULMONARY PRESSURE

The term *mean circulatory pressure* refers to the pressure that would be measured in the entire circulatory system if all pressures were instantane-

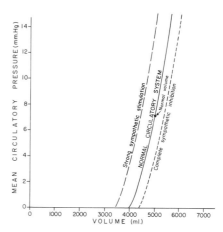

Figure 18–16 Pressure-volume curves of the entire circulation, illustrating the effect of strong sympathetic stimulation and complete sympathetic inhibition.

ously brought to equilibrium. *Mean systemic pressure* is the pressure that would be measured in the systemic vessels if the root of the aorta and the great veins entering the heart were suddenly clamped and all pressures in the systemic system were brought instantaneously to equilibrium. Finally, *mean pulmonary pressure* is the pressure that would be measured in the pulmonary system if the pulmonary artery and large pulmonary veins were suddenly clamped and all pulmonary pressures brought instantaneously to equilibrium.

In measurements made in dogs, the normal mean systemic pressure has averaged about 7 mm. Hg, which is equal to the mean circulatory pressure. The mean pulmonary pressure has measured about 10 mm. Hg. However, these values change markedly in certain pathological conditions for the following reason:

If the right heart is much weaker than the left heart, blood becomes dammed in the systemic circulation, while it is pumped out of the pulmonary system. As a consequence, the mean systemic pressure rises while the mean pulmonary pressure falls. Because of small compliance of the pulmonary system compared with that of the systemic system (about one seventh as great), the decrease in mean pulmonary pressure is about seven times as great as the rise in mean systemic pressure. Conversely, damage to the left heart causes exactly the opposite effect.

DELAYED COMPLIANCE (STRESS-RELAXATION) OF VESSELS

The term "delayed compliance" means that a vessel whose pressure is increased by increased volume gradually loses much of this pressure over a period of many minutes, or a vessel exposed to increased pressure becomes progressively enlarged. Figure 18–17 illustrates diagrammatically one of the effects of delayed compliance. In this figure, the pressure is being recorded in a small segment of a vein that is occluded at both ends. Then, an extra volume of blood is suddenly injected into the segment until the pressure rises from 5 to 12 mm. Hg. Even though none of the blood is removed after it is injected, the pressure nevertheless begins to fall immediately and approaches approximately 9 mm. Hg after several minutes. In other words, the volume of blood injected caused immediate *elastic* distention of the vein, but then the smooth muscle fibers of the vein began to adjust themselves to new tensions as a result of the stretch imposed on them. This is a characteristic of all smooth muscle tissue called *stress-relaxation*, which was explained in Chapter 7.

After the delayed increase in compliance had taken place in the experiment illustrated in Figure 18–17, the extra blood volume was suddenly removed, and the pressure immediately fell to a very low value. Subsequently, the smooth muscle fibers began to readjust their tensions back to their

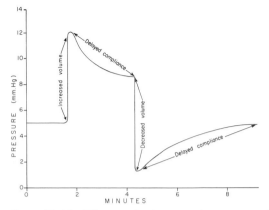

Figure 18–17 Effect on the intravascular pressure of injecting a small volume of blood into a venous segment, illustrating the principles of stress relaxation.

initial values, and after a number of minutes the normal vascular pressure of 5 mm. Hg had returned.

Delayed compliance occurs only slightly in the arteries but to a much greater extent in the veins. As a result, prolonged elevation of venous pressure can often double the blood volume in the venous tree. This is a valuable mechanism by which the circulation can accommodate much extra blood when necessary, such as following too large a transfusion. Also, delayed compliance in the reverse direction is a valuable means by which the circulation automatically adjusts itself to diminished blood volume after serious hemorrhage.

REFERENCES

Benis, A. M., and Lacoste, J.: Study of erythrocyte aggregation by blood viscometry at low shear rates using a balance method. *Circ. Res.,* 22:29, 1968.

Brooks, D. E.; Goodwin, J. W.; and Seaman, G. V. F.: Interactions among erythrocytes under shear. *J. Appl. Physiol.,* 28:172, 1970.

Burton, A. C.: The importance of mechanics and geometry of the red cell. *Proc. Int. Union of Physiol. Sciences,* 6:9, 1968.

Canham, P. G., and Burton, A. C.: Distribution of size and shape in populations of normal human red cells. *Circ. Res.,* 22:405, 1968.

Dahn, I.; Johnson, B.; and Nilsen, R.: Plethysmographic in vivo determinations of elastic properties of arteries in man. *J. Appl. Physiol.,* 28:328, 1970.

Green, H. D.: Circulation: physical principles. *In* Glasser, O. (ed.): Medical Physics. Chicago, Year Book Medical Publishers, 1944.

Green, H. D.; Rapela, C. E.; and Conrad, M. C.: Resistance (conductance) and capacitance phenomena in terminal vascular beds. *In* Handbook of Physiology. Baltimore, The Williams & Wilkins Co., 1963, Vol. II, Sec. II, p. 935.

Guyton, A. C.: Peripheral circulation. *Ann. Rev. Physiol.,* 21:239, 1959.

Guyton, A. C.; Armstrong, G. G.; and Chipley, P. L.: Pressure-volume curves of the entire arterial and venous systems in the living animal. *Amer. J. Physiol.,* 184:253, 1956.

Guyton, A. C., and Greganti, F. P.: A physiologic reference point for measuring circulatory pressures in the dog – particularly venous pressure. *Amer. J. Physiol.,* 185:137, 1956.

Harlan, J. C.; Smith, E. E.; and Richardson, T. Q.: Pressure-volume curves of systemic and pulmonary circuit. *Amer. J. Physiol.,* 213:1499, 1967.

Hochberger, A. I., and Zweifach, B. W.: Analysis of critical closing pressure in the perfused rabbit ear. *Amer. J. Physiol., 214*:962, 1968.

Johnson, P. C.: Hemodynamics. *Ann. Rev. Physiol., 31*:331, 1969.

Mason, S. G.: The flow behavior of particulate suspensions. *In* Wolstenholme, G. E. W., and Knight, J. (eds.): CIBA Foundation Symposium on Circulatory and Respiratory Mass Transport. London, Churchill Ltd., 1969, p. 105.

McDonald, D. A.: Hemodynamics. *Ann. Rev. Physiol., 30*:525, 1968.

Merrill, E. W.: Rheology of blood. *Physiol. Rev., 49*:863, 1969.

Patel, Dali J.: Mechanical properties of large blood vessels. *In* Fishman, A. P., and Hecht, H. H. (eds.): The Pulmonary Circulation and Interstitial Space. Chicago, University of Chicago Press, 1969, p. 355.

Porciuncula, C. I.; Armstrong, G. G.; Guyton, A. C.; and Stone, H. L.: Delayed compliance in external jugular vein of the dog. *Amer. J. Physiol., 207*:728, 1964.

Rand, P. W.; Barker, N.; and Lacombe, E.: Effects of plasma viscosity and aggregation on whole-blood viscosity. *Amer. J. Physiol., 218*:681, 1970.

Remington, J. W.: The physiology of the aorta and major arteries. *In* Handbook of Physiology. Baltimore, The Williams & Wilkins Co., 1963, Vol. II, Sec. II, p. 799.

Richardson, T. Q.; Stallings, J. O.; and Guyton, A. C.: Pressure-volume curves in live, intact dogs. *Amer. J. Physiol., 201*:471, 1961.

Schmid-Schonbein, H.; Wells, R.; and Goldstone, J.: Influence of deformability of human red cells upon blood viscosity. *Circ. Res., 25*:131, 1969.

Schrier, R. W.; McDonald, K. M.; Wells, R. E.; and Lauler, D. P.: Influence of hematocrit and colloid on whole blood viscosity during volume expansion. *Amer. J. Physiol., 218*:346, 1970.

Stone, H. O.; Thompson, H. K., Jr.; and Schmidt-Nielsen, K.: Influence of erythrocytes on blood viscosity. *Amer. J. Physiol., 214*:913, 1968.

Taylor, M. G.: The elastic properties of arteries in relation to the physiological functions of the arterial system. *Gastroenterology, 52*:358, 1967.

Tickner, E. G., and Sacks, A. H.: Engineering simulation of the viscous behavior of whole blood using suspensions of flexible particles. *Circ. Res., 25*:389, 1969.

Wayland, H.: Rheology and the microcirculation. *Gastroenterology, 52*:342, 1967.

Wells, R., and Schmid-Schonbein, H.: Red cell deformation and fluidity of concentrated cell suspensions. *J. Appl. Physiol., 27*:213, 1969.

CHAPTER 19

The Systemic Circulation

The circulation is divided into the *systemic circulation* and the *pulmonary circulation*. Since the systemic circulation supplies all the tissues of the body except the lungs, with blood flow, it is also frequently called the *greater circulation* or *peripheral circulation*.

Though the blood vessels in each part of the body have their own special characteristics, some general principles of vascular function nevertheless apply in all parts of the systemic circulation. It is the purpose of the present chapter to discuss these general principles.

The Functional Parts of the Systemic Circulation. Before attempting to discuss the details of function in the systemic circulation, it is important to understand the overall role of each of its parts, as follows:

The function of the *arteries* is to transport blood under high pressure to the tissues. For this reason the arteries have strong vascular walls, and blood flows rapidly in the arteries to the tissues.

The *arterioles* are the last small branches of the arterial system, and they act as control valves through which blood is released into the capillaries. The arteriole has a strong muscular wall that is capable of closing the arteriole completely or of being dilated several fold, thereby vastly altering blood flow to the capillaries.

The function of the *capillaries* is to exchange fluid and nutrients between the blood and the interstitial spaces. For this role, the capillary walls are very thin and permeable to small molecular substances.

The *venules* collect blood from the capillaries; they gradually coalesce into progressively larger veins.

The *veins* function as conduits for transport of blood from the tissues back to the heart. Since the pressure in the venous system is very low, the venous walls are thin. Even so, they are muscular, and this allows them to contract or expand and thereby to store either a small or large amount of blood, depending upon the needs of the body.

PHYSICAL CHARACTERISTICS OF THE SYSTEMIC CIRCULATION

Quantities of Blood in the Different Parts of the Circulation. By far the greatest amount of the blood in the circulation is contained in the systemic veins. Thus, Figure 19-1 shows that approximately 79 per cent of the entire blood volume of the body is in the systemic circulation, with 59 per cent in the veins, 15 per cent in the arteries, and 5 per cent in the capillaries. The heart contains 9 per cent of the blood, and the pulmonary vessels contain 12 per cent. Most surprising is the very low blood volume in the capillaries of the systemic circulation, only about 5 per cent of the total, for it is here that the most important function of the systemic circulation occurs; namely, diffusion of substances back and forth between the blood and the interstitial fluid. This function is so important that it will be discussed in detail in Chapter 20.

Cross-Sectional Areas and Velocities of Blood Flow. If all vessels of each type were put side by side, their total cross-sectional areas would be:

	cm^2
Aorta	2.5
Small arteries	20
Arterioles	40
Capillaries	2500
Venules	250
Small veins	80
Venae cavae	8

Note particularly the large cross-sectional areas of the veins, averaging about four times those of the corresponding arteries. This explains the very large storage of blood in the venous system in comparison with the arterial system.

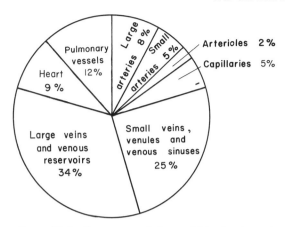

Figure 19–1 Percentage of the total blood volume in each portion of the circulatory system.

The velocity of blood flow in each segment of the circulation is inversely proportional to its cross-sectional area. Thus, under resting conditions, the velocity averages 33 cm. per second in the aorta, but 1/1000 this in the capillaries, or about 0.3 mm per second. However, since the capillaries have a typical length of only 0.3 mm. to 1 mm., each unit of blood remains in the capillaries for only 1 to 3 seconds, a very surprising fact, since all diffusion that is going to take place through the capillary walls must occur in this exceedingly short time.

Pressures and Resistances in the Various Portions of the Systemic Circulation. Since the heart pumps blood continually into the aorta, the pressure in the aorta is obviously high, averaging approximately 100 mm. Hg. And, since pumping by the heart is intermittent, the arterial pressure fluctuates between a *systolic level* of 120 mm. Hg and a *diastolic level* of 80 mm. Hg, as illustrated in Figure 19–2. As the blood flows through the systemic circulation, its pressure falls progressively to approximately 0 mm. Hg by the time its reaches the right atrium.

The decrease in arterial pressure in each part of the systemic circulation is directly proportional to the vascular resistance. Thus, in the aorta the resistance is almost zero; therefore, the mean arterial pressure at the end of the aorta is still amost 100 mm. Hg. Likewise, the resistance in the large arteries is very slight so that the mean arterial pressure in arteries as small as 3 mm. in diameter is still 95 to 97 mm. Hg. Then the resistance begins to increase rapidly in the very small arteries, causing the pressure to drop to approximately 85 mm. Hg at the beginning of the arterioles.

The resistance of the arterioles is the greatest of any part of the systemic circulation, accounting for about half the resistance in the entire systemic circulation. Thus, the pressure decreases about 55 mm. Hg in the arterioles so that the pressure of the blood as it leaves the arterioles to enter the capillaries is only about 30 mm. Hg. Arteriolar resistance is so important to the regulation of blood flow in different tissues of the body that it is discussed in detail later in the chapter and also in Chapter 24 and 25, which consider the regulation of the systemic circulation.

The pressure at the arterial ends of the capillaries is normally about 30 mm. Hg and at the venous ends about 10 mm. Hg. Thus, the pressure decrease in the capillaries is only 20 mm. Hg, which illustrates that the capillary resistance is about two-fifths that of the arterioles.

The pressure at the beginning of the veins is about 10 mm. Hg, and this decreases to about 0 mm. Hg at the right atrium. This large decrease in pressure in the veins indicates that the veins have far more resistance than one would expect for vessels of their large sizes. This resistance is caused by compression of the veins from the outside, which keeps many of them, especially the venae cavae, collapsed a large share of the time. This effect is discussed in detail later in the chapter.

PRESSURE PULSES IN THE ARTERIES

Since the heart is a pulsatile pump, blood enters the arteries intermittently, causing *pressure pulses* in the arterial system. In the normal young adult the pressure at the height of a pulse, the *systolic pressure*, is about 120 mm. Hg and at its lowest point, the *diastolic pressure,* is about 80 mm. Hg. The difference between these two pressures, about 40 mm. Hg, is called the *pulse pressure.*

Figure 19–3 illustrates a typical *pressure*

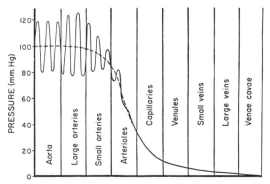

Figure 19–2 Blood pressures in the different portions of the systemic circulatory system.

pulse curve recorded in the ascending aorta of a human being, showing a very rapid rise in arterial pressure during ventricular systole, followed by a maintained high level of pressure for 0.2 to 0.3 second. This is terminated by a sharp *incisura* at the end of systole, followed by a slow decline of pressure back to the diastolic level. The incisura occurs at the same time that the aortic valve closes, and is caused as follows: During systole the pressure in the arteries rises to a high value. When the ventricle relaxes the intraventricular pressure begins to fall rapidly, and backflow of blood from the aorta into the ventricle allows the aortic pressure also to begin falling. However, the backflow suddenly snaps the aortic valve closed. The momentum that has built up in the backflowing blood brings still more blood into the root of the aorta, raising the pressure again and thus giving the incisura in the record.

After systole is over, the pressure in the central aorta decreases rapidly at first but progressively more and more slowly as diastole proceeds. The reason for this is that blood flows through the peripheral vessels much more rapidly when the pressure is high than when it is low.

FACTORS THAT AFFECT THE PULSE PRESSURE

Effect of Stroke Volume and Arterial Compliance. There are two major factors that affect the pulse pressure: (1) the *stroke volume output* of the heart and (2) the *compliance (total distensibility)* of the arterial tree; a third, less important, factor is the character of ejection from the heart during systole.

In general, the greater the stroke volume output, the greater is the amount of blood that must be accommodated in the arterial tree with each heartbeat and, therefore, the greater is the pressure rise during systole and the pressure fall during diastole, thus causing a greater pulse pressure.

On the other hand, the greater the compliance of the arterial system the less will be the rise in pressure for a given stroke volume of blood pumped into the arteries. In effect, then, the pulse pressure is determined approximately by the *ratio of stroke volume output to compliance* of the arterial tree. Therefore, any condition of the circulation that affects either of these two factors will also affect the pulse pressure.

Factors that Affect the Pulse Pressure by Changing the Stroke Volume Output. So many different circulatory conditions change the stroke volume output that only a few of these can be mentioned:

An *increase in heart rate* while the cardiac output remains constant causes the stroke volume output to decrease in inverse proportion to the increased rate, and the pulse pressure decreases accordingly.

A *decrease in total peripheral resistance* allows rapid flow of blood from the arteries to the veins. This increases the venous return to the heart and increases the stroke volume output. Therefore, the pulse pressure is also greatly increased.

An *increase in mean circulatory pressure,* if all other circulatory factors remain constant, increases the rate of venous return to the heart and consequently increases the stroke volume output. Here again the pulse pressure is accordingly increased.

The *character of ejection from the heart* affects the pulse pressure in two ways: First, if the duration of ejection is long, a large portion of the stroke volume output runs off through the systemic circulation while it is being ejected into the aorta; therefore, the magnitude of the stroke volume output effect on pulse pressure is decreased. Second, sudden ejection of blood from the heart causes the pressure in the initial portion of the aorta to rise very high before the blood can run off to the more distal portions of the aorta. Therefore, sudden ejection causes a greater pulse pressure than does more prolonged ejection.

Factors that Affect the Pulse Pressure by Altering the Arterial Compliance. In contrast to the many different conditions that can change the stroke volume output, there are only two major factors that often alter the arterial compliance. These are (1) *change in mean arterial pressure* and (2) *pathological changes that affect the distensibility of the arterial walls.*

In the preceding chapter it was pointed out that the compliance of a vascular segment is the volume of blood that can be accommodated for a given rise in pressure. As the arterial pressure rises from low to high values, the compliance decreases slightly. Therefore, in the low pressure range considerably more blood can be accommodated for a given rise in pressure than in the high pressure range. For this reason, *in a person who has high arterial pressure but a normal stroke volume output, the pulse pressure is considerably increased.*

In old age, the arterial walls lose much of their elastic and muscular tissues, and these are replaced by fibrous tissue and sometimes even calci-

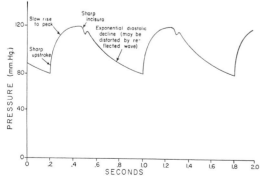

Figure 19–3 A normal pressure pulse contour recorded from the ascending aorta. (From Opdyke: *Fed. Proc., 11:*734, 1952.)

fied plaques that cannot stretch a significant amount. These changes greatly decrease the compliance of the arterial system, which in turn causes the arterial pressure to rise very high during systole and to fall greatly during diastole as blood runs off from the arteries to the veins.

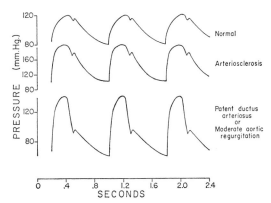

ABNORMAL PRESSURE PULSE CONTOURS

Some conditions of the circulation cause abnormal contours to the pressure pulse wave in addition to altering the pulse pressure. Especially distinctive among these are arteriosclerosis, patent ductus arteriosus, and aortic regurgitation.

Arteriosclerosis. In arteriosclerosis the arteries become fibrous and sometimes calcified, thereby resulting in greatly reduced arterial compliance and also resulting in markedly increased pulse pressure. Usually a mild degree of hypertension also accompanies arteriosclerosis so that not only does the pulse pressure rise but the mean pressure also. The middle curve of Figure 19–4 illustrates a characteristic aortic pressure pulse contour in arteriosclerosis, showing a markedly elevated systolic pressure, a slightly elevated diastolic pressure, and also a great increase in pulse pressure. The curve is more peaked than that of the normal curve, and the incisura lies at a lower point on the downslope of the curve than in the normal.

Patent Ductus Arteriosus. In patent ductus arteriosus blood flows from the aorta through the open ductus into the pulmonary artery, allowing very rapid runoff of blood from the arterial tree after each heartbeat. However, this is compensated by a greater than normal stroke volume output, giving the pressure pulse contour shown by the lowest curve in Figure 19–4. Here one finds an elevated systolic pressure, a greatly depressed diastolic pressure, a greatly increased pulse pressure, and the incisura occurring again at a very low point on the downslope because an excessively large amount of blood runs off from the arteries even before systole is over.

Aortic Regurgitation. In aortic regurgitation much of the blood that is pumped into the aorta during systole flows back into the left ventricle during diastole. However, this backflow is compensated by a much greater than normal stroke volume output during systole. Thus, the condition is very similar to that of patent ductus arteriosus but not always identical, for in aortic regurgitation the valve sometimes fails entirely to close. When this is true the incisura illustrated in the lower curve of Figure 19–4 is entirely absent.

TRANSMISSION OF THE PRESSURE PULSE
TO THE PERIPHERY

When the heart ejects blood into the aorta during systole, only the proximal portion of the aorta becomes distended at first, and it is only in this

Figure 19–4 Pressure pulse contours in arteriosclerosis, patent ductus arteriosus, and moderate aortic regurgitation.

portion of the arterial tree that the pressure rises immediately. The cause of this is the inertia of the blood in the aorta, which prevents its sudden movement away from the central arteries into the peripheral arteries.

However, the rising pressure in the central aorta gradually overcomes the inertia of the blood, causing the pressure to rise progressively farther and farther out in the arterial tree. Figure 19–5 illustrates this *transmission of the pressure pulse* down the aorta, the aorta becoming distended as the pressure wave moves forward.

The velocity of transmission of the pressure pulse along the normal aorta is 3 to 5 meters per second, along the large arterial branches 7 to 10 meters per second, and in the smaller arteries 15 to 35 meters per second. In general, the less the compliance of each vascular segment, the faster is the velocity of transmission, which explains the slowness of transmission in the aorta versus

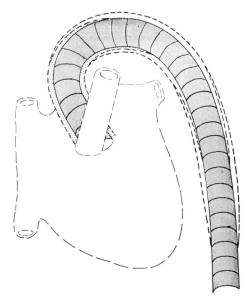

Figure 19–5 Progressive stages in the transmission of the pressure pulse along the aorta.

the rapidity of transmission in the far less compliant small, distal arteries.

The *velocity of transmission of the pressure pulse is much greater than the velocity of blood flow,* which was discussed earlier in the chapter. In transmission of the pressure pulse, only a small amount of blood entering the proximal aorta pushes the more distal blood forward sufficiently to elevate the pressure in the very distal arteries. Therefore, the blood ejected by the heart may have traveled only a few centimeters by the time the pressure wave has already reached the distal ends of the arteries. In the aorta, the velocity of the pressure pulse is approximately 15 times that of blood flow, while in the more distal arteries the velocity of the pressure pulse may be as great as 100 times the velocity of blood flow.

Augmentation of the Pulse Pressure in the Peripheral Arteries. An interesting phenomenon that often occurs in transmission of pressure pulses to the periphery is an increase in pulse pressure. This effect is illustrated in Figure 19–6, which shows considerable enhancement of the pulse pressure more and more peripherally.

At least the following three factors are known to play significant roles in this pulse pressure augmentation effect:

1. Part of the augmentation undoubtedly results from progressive decrease in compliance of the more distal portions of the large arteries. Transmission of the pressure pulse is characterized by a high level of momentum of the blood in the advancing edge of the pressure wave. When this wave suddenly reaches the less compliant areas of the arterial tree, much of the kinetic energy of this momentum is changed into pressure, resulting in augmentation of the pressure.

2. Pressure waves are reflected to some extent by the peripheral arteries. That is, when a pressure pulse enters the peripheral arteries and distends them, the pressure in these peripheral arteries causes the pulse wave to begin traveling backward. This is analogous to a wave traveling

in a bowl of water until it hits the edge. On striking the edge, the wave turns around and travels back onto the surface of the water. If the returning wave strikes an oncoming wave, the two "summate," causing a much higher wave than would otherwise occur. Such is the case in the arterial tree. The first portion of the pressure wave is reflected before the latter portion of the same wave ever reaches the peripheral arteries. Therefore, the first portion summates with the latter portion, causing higher pressures than would otherwise be recorded.

3. The pulse pressure is also augmented by transmission of certain parts of the pulse wave at more rapid rates than other parts. For instance, the high pressure portion is transmitted more rapidly than the low pressure portion because the arterial distensibility is less at high pressure than at low pressure. This causes crowding together of certain portions of the wave and, therefore, "peaking" of the pressure pulse as illustrated for the dorsalis pedis in Figure 19–6.

Though the relative significance of the different causes of augmentation of the pulse pressure is disputed by different physiologists, this phenomenon must be remembered whenever pressure measurements are made in peripheral arteries, for systolic pressure is sometimes as much as 20 to 30 per cent above that in the central aorta, and diastolic pressure is often reduced as much as 10 to 15 per cent.

Damping of the Pressure Pulse in the Small Arteries and Arterioles. The pressure pulse becomes less and less intense as it passes through the small arteries and arterioles, until it becomes almost absent in the capillaries. This was illustrated in Figure 19–2 and is called *damping* of the pressure pulse.

Damping of the pressure pulse is caused mainly by a combined effect of vascular distensibility and vascular resistance. That is, for a pressure wave to travel from one area of an artery to another area, a small amount of blood must flow between the two areas. The resistance in the small arteries and arterioles is great enough that this flow of blood, and consequently the transmission of pressure, is greatly impeded. At the same time, the distensibility of the small vessels is very great, so that the small amount of blood that is caused to flow during a pressure pulse produces progressively less and less pressure rise in the more distal vessels. Therefore, even though the pressure rises and falls quite significantly in the large arteries, the combined resistance-distensibility effects cause the pressure pulse to become progressively smaller and smaller as it proceeds farther and farther into the small vessels.

Figure 19–7 illustrates the extreme effect of resistance on damping. The curve to the left is a normal pressure pulse contour recorded from the dorsalis pedis artery in a dog. To the right is the same pressure pulse contour recorded during contraction of the femoral artery. The damping effect of the increased resistance is readily apparent.

Figure 19–6 Pressure pulse contours in different segments of the arterial tree in man, showing (1) the delay in the pressure pulse as it spreads more and more peripherally and (2) augmentation of the pulse pressure especially in the dorsalis pedis artery. (Modified from Remington and Wood: *J. Appl. Physiol.,* 9:440, 1956.)

Figure 19–7 Pressure pulse contour in the dorsalis pedis artery of a dog recorded, first, under normal conditions and, second, during stimulation of the sympathetic nerves supplying the femoral artery. (Modified from Alexander and Kantowitz: *Surgery, 33:42*, 1953.)

CAPILLARY PULSATION. Occasionally pressure pulses are not completely damped by the time they reach the capillaries, and abnormal capillary pulsations result; this can be readily demonstrated by pressing on the anterior portion of the fingernail so that the blood is pressed out of the anterior capillaries of the nail bed. This causes blanching anteriorly while the posterior half of the nail remains red because of blood still in the capillaries. If significant pulsation is occurring in the capillaries, the border between the red and white areas will shift forward as more capillaries become filled during the high pressure phase and backward during the low pressure phase.

Two major factors can cause capillary pulsation: (1) It occurs when the central pressure pulse is greatly exacerbated, as occurs when the *heart rate is very slow* or when the stroke volume output is greatly increased as a result of *aortic regurgitation, patent ductus arteriosus, or even extreme increase in venous* return. (2) Another cause of capillary pulsation is *extreme dilatation of the small arteries and arterioles,* which reduces the resistance of these vessels and thereby reduces the damping.

THE RADIAL PULSE

Clinically, it has been the habit for many years for a physician to feel the radial pulse of each patient. This is done to determine the rate of the heartbeat or, frequently, because of the psychic contact that it gives the doctor with his patient. Under certain circumstances, however, the character of the pulse can also be of value in the diagnosis of circulatory diseases.

Weak Pulse. A weak pulse at the radial artery usually indicates either (1) greatly decreased central pulse pressure, such as occurs when the stroke volume output is low, or (2) increased damping of the pulse wave caused by vascular spasm; the latter occurs when the sympathetic nervous system becomes overly active, for instance, following blood loss or when a person is having a chill.

Pulsus Paradoxus. Occasionally the strength of the pulse becomes strong, then weak, then strong, continuing in synchrony with the phases of respiration. This is caused by alternate increase and decrease in cardiac output with each respiration. During inspiration, all the blood vessels of the lungs increase in size because of increased negative pressure in the thorax. Therefore, blood collects in the lungs, and the stroke volume output and pulse strength decrease. During expiration, opposite effects occur. This is a normal phenomenon in all persons, but it becomes extremely distinct in some conditions such as cardiac tamponade and very deep breathing.

Pulse Deficit. The rhythm of the heart is very irregular in atrial fibrillation or in the case of ectopic pacemakers. In these arrhythmias, which were discussed in Chapter 17, two beats of the heart often come so close together that the second beat pumps no blood or very little blood because the left ventricle has too little time to fill between the two beats. In this circumstance, one can hear the second beat of the heart with a stethoscope applied directly over the heart but cannot feel a pulsation in the radial artery, an effect called a *pulse deficit.* The greater the rate of pulse deficit each minute, the more serious, ordinarily, is the arrhythmia.

Pulsus Alternans. In a few conditions the heart beats strongly with one beat and then weakly with the next, causing similar alternation in the strength of the pulse, a condition called *pulsus alternans.* This alternation is frequently caused by intoxication with digitalis, partial block of the Purkinje system, or enlarged ventricles.

THE ARTERIOLES AND CAPILLARIES

Blood flow in each tissue is controlled almost entirely by the degree of contraction or dilatation of the arterioles, and it is in the capillaries where the important process of exchange between blood and the interstitial fluid occurs. These two segments of the circulation are so important that they will receive special discussion in the chapters to follow: capillary exchange phenomena in Chapters 20, 21, and 22 and arteriolar regulation of blood flow in Chapters 24 and 25.

Upon leaving the small arteries, blood courses through the arterioles, which are only a few millimeters in length and have diameters from 8 to 50 microns. Each arteriole branches many times to supply perhaps 10 to 100 capillaries.

There are approximately 10 billion capillaries in the peripheral tissues, having more than 100 square meters of surface area. The thickness of the capillary wall is usually less than 1 micron, and there are small pores in the wall (as will be explained in Chapter 20) through which substances can diffuse.

Though the capillary walls are very thin and, therefore, very weak, their diameters are also very small. It will be recalled from the discussion of vascular wall tension and vascular pressure in the previous chapter that the wall tension is directly proportional to the pressure times the diameter (law of Laplace). Therefore, since the diameter is extremely minute, the tension developed in the wall is also extremely minute, which explains why the very thin-walled capillaries can withstand the pressure therein.

THE VEINS AND THEIR FUNCTIONS

For years the veins have been considered to be nothing more than passageways for flow of blood into the heart, but it is rapidly becoming apparent that the veins are capable of constricting and enlarging, of storing large quantities of blood and making this blood available when it is required by the remainder of the circulation, of actually propelling blood forward by means of the so-called venous pump that is described later, and even helping to regulate cardiac output, a function so important that it will be described in detail in Chapter 27.

RIGHT ATRIAL PRESSURE AND ITS REGULATION

To understand the various functions of the veins, it is first necessary to know something about the pressures in the veins and how they are regulated. Blood from all the systemic veins flows into the right atrium; therefore, the pressure in the right atrium is frequently called the *central venous pressure*. The pressures in the peripheral veins are to a great extent dependent on the level of this pressure, so that anything that affects this pressure usually affects venous pressure everywhere in the body.

Right atrial pressure is regulated by a balance between, first, the *ability of the heart to pump blood* and, second, the *tendency for blood to flow from the peripheral vessels back to the heart.* Therefore, if the heart is pumping strongly, the right atrial pressure tends to decrease, whereas, on the other hand, weakness of the heart tends to elevate the right atrial pressure. Likewise, any effect that causes rapid inflow of blood into the right atrium from the veins tends to elevate the right atrial pressure. Some of the factors that increase this tendency for venous return are (1) increased blood volume, (2) increase in vascular tone throughout the body with resultant increased peripheral venous pressures, and (3) dilatation of the systemic small vessels, which decreases the peripheral resistance.

The same factors that regulate right atrial pressure also enter into the regulation of cardiac output, for the amount of blood pumped by the heart is dependent both on the ability of the heart to pump and the tendency for blood to flow into the heart from the peripheral vessels. Therefore, we will discuss the regulation of right atrial pressure with much more precision in Chapter 27 in connection with the regulation of cardiac output.

The *normal right atrial pressure* is approximately 0 mm. Hg, which is about equal to the atmospheric pressure around the body. However, it can rise to as high as 20 to 30 mm. Hg under very abnormal conditions, such as (a) in serious heart failure or (b) following massive transfusion when excessive quantities of blood are attempting to flow into the heart from the peripheral vessels.

The lower limit to the right atrial pressure is about −4 to −5 mm. Hg, which is the pressure in the pericardial and intrapleural spaces that surround the heart. The right atrial pressure approaches these very low values when the heart pumps with exceptional vigor or when the flow of blood into the heart from the peripheral vessels is greatly depressed, such as following severe hemorrhage.

VENOUS RESISTANCE AND PERIPHERAL VENOUS PRESSURE

Large veins have almost no resistance *when they are distended*. However, as illustrated in Figure 19–8, most of the large veins entering the thorax are compressed at many points so that blood flow is impeded. For instance, the veins from the arms are compressed by their

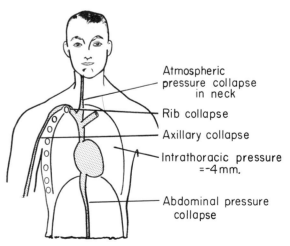

Figure 19–8 Factors tending to collapse the veins entering the thorax.

sharp angulation over the first rib. Second, the pressure in the neck veins often falls so low that the atmospheric pressure on the outside of the neck causes them to collapse. Finally, veins coursing through the abdomen are compressed by different organs and by the intraabdominal pressure so that often they are almost totally collapsed. For these reasons the *large veins do usually offer considerable resistance to blood flow*, and because of this the pressure in the peripheral veins is usually 6 to 9 mm. Hg greater than the right atrial pressure.

Note, however, that the veins inside the thorax *are not collapsed* because the *negative pressure* inside the chest distends these veins.

Effect of High Right Atrial Pressure on Peripheral Venous Pressure. When the right atrial pressure rises above its normal value of 0 mm. Hg, it begins to back blood up in the large veins and to open them up. The pressures in the peripheral veins do not rise until all the collapsed points between the peripheral veins and the large veins have opened up. This usually occurs when the right atrial pressure rises to about 4 to 6 mm. Hg. If the right atrial pressure then rises still further, the additional increase is reflected by a corresponding rise in peripheral venous pressure. Since the heart must be greatly weakened to cause a rise in right atrial pressure to as high as 4 to 6 mm. Hg, one often finds that the peripheral venous pressure is not elevated in the early phases of cardiac failure.

Effect of Abdominal Pressure on Venous Pressures of the Leg. The normal pressure in the peritoneal cavity averages about 2 mm. Hg, but at times it can rise to as high as 15 to 20 mm. Hg as a result of pregnancy, large tumors, or excessive ascites in the peritoneal cavity as explained in Chapter 22. When this happens, the pressure in the veins of the legs must rise *above* the abdominal pressure before the abdominal veins will open. Thus, if the intra-abdominal pressure is 20 mm. Hg, the lowest possible pressure in the femoral veins is 20 mm. Hg.

EFFECT OF HYDROSTATIC PRESSURE ON VENOUS PRESSURE

In a large chamber of water, the pressure at the surface of the water is equal to atmospheric pressure, but the pressure rises 1 mm. Hg for each 13.6 mm. distance below the surface of the water. This pressure results from the weight of the water and therefore is called *hydrostatic pressure.*

Hydrostatic pressure also occurs in the vascular system of the human being, as is illustrated in Figure 19–9. When a person is standing, the pressure in the right atrium remains approximately 0 mm. Hg because the heart pumps into the arteries any excess blood that attempts to accumulate at this point. However, in an adult *who is standing absolutely still* the pressure in the veins of the feet is approximately +90 mm. Hg simply because of the distance from the feet to the heart. The venous pressures at other levels of the body lie proportionately between 0 and 90 mm. Hg.

In the arm veins, the pressure at the level of the top rib is usually about +6 mm. Hg because of compression of the subclavian vein as it passes over this rib. The hydrostatic pressures down the length of the arm are then measured below the level of this rib. Thus, if the hydrostatic difference between the level of the rib and the hand is 29 mm. Hg, this hydrostatic pressure must be added to the pressure caused by collapse of the vein as it crosses the rib, making a total of 35 mm. Hg pressure in the veins of the hand.

The neck veins collapse almost completely all the way to the skull owing to atmospheric pressure on the outside of the neck. This collapse causes the pressure in these veins to remain zero along their entire extent. The reason for this is that any tendency for the pressure to rise above this level opens the veins

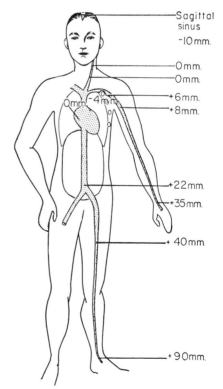

Figure 19–9 Effect of hydrostatic pressure on the venous pressures throughout the body.

and allows the pressure to fall back to zero, and any tendency for the pressure to fall below this level collapses the veins still more, which increases their resistance and again returns the pressure back to zero.

The veins inside the skull, however, are in a noncollapsible chamber, and they will not collapse. Consequently, *negative pressure can exist in the dural sinuses of the head;* in the standing position the venous pressure in the sagittal sinus is calculated to be approximately −10 mm. Hg because of the hydrostatic "suction" between the top of the skull and the base of the skull. Thus, if the sagittal sinus is entered during surgery, air can be sucked immediately into this vein; and the air may even pass downward to cause air embolism in the heart so that the heart valves do not function satisfactorily, and death ensues.

Effect of the Hydrostatic Factor on Other Pressures. The hydrostatic factor also affects the peripheral pressures in the arteries and capillaries as well as in the veins. For instance, a standing person who has an arterial pressure of 100 mm. Hg at the level of his heart has an arterial pressure in his feet of about 190 mm. Hg. Therefore, any time one states that the arterial pressure is 100 mm. Hg, he generally means that this is the pressure at the hydrostatic level of the heart.

VENOUS VALVES AND THE "VENOUS PUMP"

Because of hydrostatic pressure, the venous pressure in the feet would always remain about +90 mm. Hg in a standing adult were it not for the valves in the veins, which are shown in Figure 19–10. Every time one moves his legs he compresses the muscles against the fascia and the fascia against the skin so that the veins of the legs are compressed, and blood tends to flow away from the areas of compression. The valves in the veins are arranged so that the direction of blood flow can be only toward the heart, as is evident in the figure. Consequently, every time a person moves his legs or even tenses his muscles, a certain amount of blood is propelled toward the heart, and the pressure in the dependent veins of the body is lowered. This pumping system is known as the "venous pump" or "muscle pump" and it is efficient enough that under ordinary circumstances the venous pressure in the feet of a walking adult remains less than 25 mm. Hg.

If the human being stands perfectly still, the venous pump does not work, and the venous pressures in the lower part of the leg can rise to

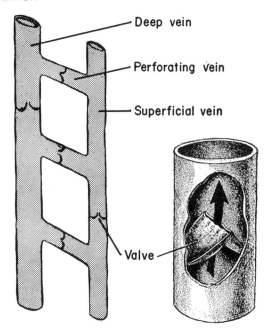

Figure 19–10 The venous valves of the leg.

the full hydrostatic value of 90 mm. Hg in about 30 seconds. Under such circumstances the pressures within the capillaries also increase greatly, and fluid leaks from the circulatory system into the tissue spaces. As a result, the legs swell, and the blood volume diminishes. Indeed, as much as 15 to 20 per cent of the blood volume is frequently lost from the circulatory system within the first 15 minutes of quiet standing.

Varicose Veins. Pathologically, the valves of the venous system are frequently destroyed. This destruction occurs particularly when the veins have been overstretched by an excess of venous pressure for a prolonged period of time, as occurs in pregnancy or when one stands on his feet most of the time. Stretching the veins, obviously, increases their cross-sectional areas, but the valves do not concurrently increase in size. Therefore, the valves will no longer block reverse blood flow in the enlarged veins. When such a situation develops, the pressure in the veins of the legs increases owing to failure of the venous pump; this further increases the size of the veins and finally destroys the function of the valves entirely. Thus, the person develops "varicose veins," which are characterized by large bulbous protrusions of the veins beneath the skin of the entire leg and particularly of the lower leg. The venous and capillary pressures become very high because of the incompetent venous pump, and this causes constant edema in the legs of these persons whenever they stand for more than a few minutes. The edema in turn prevents adequate diffusion of nutritional materials from the capillaries to the muscle and skin cells so that muscles become painful and weak, and the skin frequently atrophies and finally ulcerates. Obviously, the best treatment for such a condition is continual elevation of the legs to a level at least as high as the heart, but tight binders on the legs are also of considerable aid.

REFERENCE POINT FOR MEASURING VENOUS AND OTHER CIRCULATORY PRESSURES

In previous discussions we have often spoken of right atrial pressure as being 0 mm. Hg and arterial pressure as being 100 mm. Hg, but we have not stated the hydrostatic level in the circulatory system to which this pressure is referred. There is one point in the circulatory system at which hydrostatic pressure factors do not affect the pressure measurement by more than 1 mm. Hg. This is at the level of the tricuspid valve, as shown by the crossed axes in Figure 19–11. Therefore, all pressure measurements discussed in this text are referred to the level of the tricuspid valve, which is called the *reference point for pressure measurement.*

The reason for lack of hydrostatic effects at the tricuspid valve is that the heart automatically prevents significant hydrostatic changes in pressure at this point in the following way:

If the pressure at the tricuspid valve rises slightly above normal, then the right ventricle fills to a greater extent than usual, causing the heart to pump blood more rapidly than usual and thereby decreasing the pressure at the tricuspid valve back toward the normal mean value. On the other hand, if the pressure at this point falls, the right ventricle fails to fill adequately, its pumping decreases, and blood dams up in the venous system until the tricuspid pressure again rises to a normal value. In other words, *the heart acts as a feedback regulator of pressure* at the tricuspid valve.

The axes of the reference point, as extrapolated from animals to the human being, are (1) *anterior to the back approximately 61 per cent of the thickness of the chest,* (2) almost *exactly in the midline,* and (3) approximately *one quarter of the distance above the lower end of the sternum.* A person can be in the standing or lying or even head-down position, and his central venous pressure referred to this reference point will remain almost exactly constant, regardless of the position of the body.

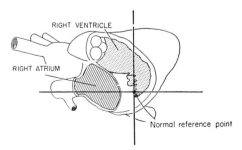

Figure 19–11 Location of the reference point for pressure measurement at the tricuspid valve.

This does not mean that the pressure at this point is always zero. It may be as low as −4 mm. Hg or as high as +20 mm. Hg, but, whatever its value, changing the position of the body does not alter it more than 1 mm. Hg.

In making arterial pressure measurements, the precise hydrostatic point in the chest to which pressures are referred matters little because percentagewise a hydrostatic error of as much as 10 or more centimeters (equivalent to 7.4 mm. Hg error) still does not affect the arterial pressure reading to a great extent. However, in venous pressure measurements the reference level must be exact if the measurements are to be significant, for abnormalities of venous pressure as small as 1 mm. Hg can result in changes in cardiac output as great as 50 to 100 per cent.

PRESSURE PULSES IN THE VEINS

The pressure pulses from the arteries are almost always damped out completely before they pass through the capillaries into the systemic veins. However, right atrial pulsations are sometimes transmitted backward along the veins to cause pressure pulses of an entirely different character. This usually does not occur to a significant extent in the normal circulation because most of the veins, where they lead into the thoracic cavity, are compressed by surrounding tissues; this causes sufficient resistance to damp out the pulsations before they can be transmitted backward into the peripheral veins. However, whenever the mean right atrial pressure is high, especially in heart failure, the veins are then well filled with blood and can easily transmit the pulsations.

The atrial pulsations were discussed in Chapter 13. Briefly, they are (1) the *a wave,* caused by contraction of the atrium, (2) the *c wave,* caused by contraction of the ventricle, and (3) the *v wave,* caused by continued inflow of blood into the atrium when the A-V valves are closed during ventricular systole.

In severe cardiac failure the venous pressure waves usually become prominent enough that one can see the veins of the patient's neck pulsating. In earlier days these pulsations were recorded for diagnostic purposes. The *a-c interval* in the recording is approximately equal to the P-Q interval of the electrocardiogram, and, before the days of electrocardiography, this was often used as a measure of the delay between atrial excitation and ventricular excitation. These waves, therefore, were then of considerable value in diagnosing different degrees of heart block.

Measurement of Venous Pressure. CLINICAL ESTIMATION OF VENOUS PRESSURE. The venous pressure can be estimated by simply observing the distention of the peripheral veins—especially the neck veins. For instance, in the sitting position, the neck veins are never distended in the normal person. However, when the right atrial pressure becomes increased to as much as 10 mm. Hg, the lower veins of the neck begin to protrude even when one is sitting; when the right atrial pressure rises to as high as 15 mm. Hg, essentially all the veins in the neck become greatly distended, and the venous pulse becomes prominent in these veins.

Rough estimates of the venous pressure can also be made by raising or lowering an arm while observing the degree of distention of the antecubital or hand veins. As the arm is progressively raised, the veins suddenly collapse, and the level at which they collapse, when referred to the level of the heart, is a rough measure of the peripheral venous pressure.

DIRECT MEASUREMENT OF VENOUS PRESSURE AND RIGHT ATRIAL PRESSURE. Venous pressure can be measured with ease by inserting a syringe needle connected to a water manometer directly into a vein. The venous pressure is expressed in relation to the level of the tricuspid valve, i.e., the height in centimeters of water above the level of the tricuspid valve; this can be converted to mm. Hg by dividing by a factor of 1.36.

The only means by which *right atrial pressure* can be measured accurately is by inserting a catheter through the veins into the right atrium. This catheter can then be connected to a water manometer and the pressure measured as noted previously.

BLOOD RESERVOIR FUNCTION OF THE VEINS

In discussing the general characteristics of the systemic circulation earlier in the chapter, it was pointed out that over 50 per cent of all the blood in the circulatory system is in the systemic veins. For this reason it is frequently said that the systemic veins act as a *blood reservoir* for the circulation. Also, relatively large quantities of blood are present in the veins of the lungs so that these, too, are considered to be blood reservoirs.

When blood is lost from the body to the extent that the arterial pressure begins to fall, pressure reflexes are elicited from the carotid sinuses and other pressure sensitive areas of the circulation, as will be detailed in Chapter 25; these in turn cause sympathetic constriction of the veins, which automatically takes up much of the slack in the circulatory system caused by the lost blood. Indeed, even after as much as 20 to 25 per cent of the total blood volume has been lost, the circulatory system often functions almost normally because of this variable reservoir system of the veins.

Specific Blood Reservoirs. Certain portions of the circulatory system are so extensive that they are specifically called "blood reservoirs." These include (1) the *spleen,* which can sometimes decrease in size sufficiently to release as much as 150 ml. of blood into other areas of the circulation, (2) the *liver,* the sinuses of which can release several hundred milliliters of blood into the remainder of the circulation, (3) the *large abdominal veins,* which can contribute as much as 300 ml., and (4) the *venous plexus beneath the skin,* which can probably contribute several hundred milliliters. The *heart* itself and the *lungs,* though not parts of the systemic venous reservoir system, must also be considered to be blood reservoirs. The heart, for instance, becomes reduced in size during sympathetic stimulation and in this way can contribute about 100 ml. of blood, and the lungs can contribute another 100 to 200 ml. when the pulmonary pressures fall to low values.

ASSESSMENT OF VENOUS FUNCTION BY MEASUREMENT OF MEAN CIRCULATORY PRESSURE

It was pointed out in the previous chapter that the *mean circulatory pressure* can be measured by stopping the heart and pumping blood rapidly from the systemic arteries to the veins, bringing the pressures in the two major vascular reservoirs to equilibrium within a few seconds; this equilibrium pressure is the mean circulatory pressure. This pressure is actually a measure of how tightly the vascular system is filled with blood. Therefore, the greater the degree of filling (especially of the veins), the higher is the mean circulatory pressure. The normal value is approximately 7 mm. Hg, but following massive transfusion this value can rise to as high as 30 to 40 mm. Hg; even in congestive heart failure, in which the circulatory blood volume is greatly increased, it can rise to as high as 20 to 25 mm. Hg.

Measurements of mean circulatory pressure are also important in assessing venous contraction for the following reasons: Since the systemic venous system has a very high compliance in comparison with that of the rest of the circulation, the mean circulatory pressure is about four-fifths determined by venous tone and venous filling with blood in contrast with only about one-fifth by all the remainder of the circulation. Therefore, even slight changes in degree of venous contraction will cause marked changes in mean circulatory pressure, whereas even large changes in degree of contraction of other segments of the circulation, such as of the arterioles, the pulmonary vessels, and so forth, have very little effect. Therefore, one of the most sensitive barometers of the degree of contraction of the venous system is the mean circulatory pressure.

REFERENCES

Alexander, R. S.: The peripheral venous system. Handbook of Physiology. Baltimore, The Williams & Wilkins Co., 1963, Vol. II, Sec. II, p. 1075.

Attinger, E. (ed.): Pulsatile Blood Flow. New York, McGraw-Hill Book Company, 1964.

Baez, S.: Simultaneous measurements of radii and wall thickness of microvessels in the anesthetized rat. *Circ. Res.,* 25:315, 1969.

D'Agrosa, L. S.: Patterns of venous vasomotion in the bat wing. *Amer. J. Physiol.,* 218:530, 1970.

Davis, D. L.: Sympathetic stimulation and small artery constriction. *Amer. J. Physiol.,* 206:262, 1964.

Emerson, T. E., Jr.; Hinshaw, L. B.; and Brake, C. M.: Vascular effects of angiotensin and norepinephrine in the dog, cat and monkey. *Amer. J. Physiol.,* 208:260, 1965.

Folkow, B.: Role of the nervous system in the control of vascular tone. *Circulation,* 21:706, 1960.

Friedman, S. M., and Friedman, C. L.: Effects of ions on vascular smooth muscle. *In* Handbook of Physiology. Baltimore, The Williams & Wilkins Co., 1963, Vol. II, Sec. II, p. 1135.

Frohlich, E. D.; Scott, J. B.; and Haddy, F. J.: Effect of cations on resistance and responsiveness of renal and forelimb vascular beds. *Amer. J. Physiol.,* 203:583, 1962.

Green, H. D., and Kepchar, J. H.: Control of peripheral resistance in major systemic vascular beds. *Physiol. Rev.,* 39:617, 1959.

Greenfield, A. D. M.: The peripheral circulation. *Ann. Rev. Physiol.,* 27:323, 1965.

Grigor'eva, A.: The Innervation of Blood Vessels. New York, The Macmillan Company, 1963.

Guyton, A. C.: The venous system and its role in the circulation. *Mod. Conc. Cardiov. Dis.,* 27:483, 1958.

Guyton, A. C.: Peripheral circulation. *Ann. Rev. Physiol.,* 21:239, 1959.

Guyton, A. C.; Ross, J. M.; Carrier, O., Jr.; and Walker, J. R.: Evidence for tissue oxygen demand as the major factor causing autoregulation. *Circ. Res.,* 14:60, 1964.

Haddy, F. J.: Local effects of sodium, calcium and magnesium upon small and large blood vessels of the dog forelimb. *Circ. Res.,* 8:57, 1960.

Herd, J. A.: Overall regulation of the circulation. *Ann. Rev. Physiol.,* 32:289, 1970.

Korner, P. I.: Circulatory adaptations in hypoxia. *Physiol. Rev.,* 39:687, 1959.

Mauskopf, J. M.; Gray, S. D.; and Renkin, E. M.: Transient and persistent components of sympathetic cholinergic vasodilatation. *Amer. J. Physiol.,* 216:92, 1969.

McDonald, D. A.: Regional pulse-wave velocity in the arterial tree. *J. Appl. Physiol.,* 24:73, 1968.

Mellander, S.: Systemic circulation. *Ann. Rev. Physiol.,* 32:313, 1970.

O'Rourke, M. F.: Pressure and flow waves in systemic arteries and the anatomical design of the arterial system. *J. Appl. Physiol.,* 23:139, 1967.

O'Rourke, M. F., and Taylor, M. G.: Input impedance of the systemic circulation. *Circ. Res.,* 20:365, 1967.

Patel, D. J.; Janicki, J. S.; and Carew, T. E.: Static anisotropic elastic properties of the aorta in living dogs. *Circ. Res.,* 25:765, 1969.

Remington, J. W.: Contour changes of the aortic pulse during propagation. *Amer. J. Physiol.,* 199:331, 1960.

Remington, J. W., and O'Brien, L. J.: Construction of aortic flow pulse from pressure pulse. *Amer. J. Physiol.,* 218:437, 1970.

Rovick, A. A., and Randall, W. C.: Systemic circulation. *Ann. Rev. Physiol.,* 29:225, 1967.

Sapirstein, L. A.: The indicator fractionation technique for the study of regional blood flow. *Gastroenterology,* 52:365, 1967.

Schmidt-Nielsen, K., and Pennycuik, P.: Capillary density in mammals in relation to body size and oxygen consumption. *Amer. J. Physiol.,* 200:746, 1961.

Somlyo, A. V.; Woo, C. Y.; and Somlyo, A. P.: Responses of nerve-free vessels to vasoactive amines and polypeptides. *Amer. J. Physiol.,* 208:748, 1965.

Sonnenschein, R. R., and White, F. N.: Systemic circulation. *Ann. Rev. Physiol.,* 30:147, 1968.

Spencer, M. P.: Systemic circulation. *Ann. Rev. Physiol.,* 28:311, 1966.

Stainsby, W. N.: Autoregulation of blood flow in skeletal muscle during increased metabolic activity. *Amer. J. Physiol.,* 202:273, 1962.

Stainsby, W. N., and Renkin, E. M.: Autoregulation of blood flow in resting skeletal muscle. *Amer. J. Physiol.,* 201:117, 1961.

Uchida, E., and Bohr, D. F.: Myogenic tone in isolated perfused resistance vessels from rats. *Amer. J. Physiol.,* 216:1343, 1969.

Uchida, E., and Bohr, D. F.: Myogenic tone in isolated perfused vessels: occurrence among vascular beds and along vascular trees. *Circ. Res.,* 25:549, 1969.

von Euler, U. S.: Noradrenaline. Springfield, Ill., Charles C Thomas, Publisher, 1956.

Wiedeman, Mary P.: Dimensions of blood vessels from distributing artery to collecting vein. *Circ. Res.,* 12:375, 1963.

Winkelmann, R. K.; Sams, W. M., Jr.; and Bohr, D. F.: Effect of nicotinate ester, acetylcholine, and other vasodilating agents on cutaneous and mesenteric vascular smooth muscle. *Circ. Res.,* 25:687, 1969.

Wood, E.: The Veins. Boston, Little, Brown and Company, 1965.

CHAPTER 20

Capillary Dynamics, and Exchange of Fluid Between the Blood and Interstitial Fluid

It is in the capillaries that the most purposeful function of the circulation occurs, namely, interchange of nutrients and cellular excreta between the tissues and the circulating blood. About 10 billion capillaries, having a total surface area probably greater than 100 square meters, provide this function. Indeed, it is rare that any single functional cell of the body is more than 20 to 30 microns away from a capillary.

It is the purpose of this chapter to discuss the transfer of substances between the blood and interstitial fluid and especially to discuss the factors that affect the transfer of fluid volume itself between the circulating blood and the interstitial fluids.

STRUCTURE OF THE CAPILLARY SYSTEM

Figure 20–1 illustrates the structure of a "unit" capillary bed, illustrating that blood enters the capillaries through an *arteriole* and leaves by way of a *venule*. Blood from the arteriole passes into a series of *meta-arterioles,* which have a structure midway between that of arterioles and capillaries. After leaving the meta-arteriole, the blood enters the *capillaries,* some of which are large and are called the *preferential channels* and others of which are small and are the *true capillaries.* After passing through the capillaries the blood enters the venule and returns to the general circulation.

Figure 20–1   Overall structure of a capillary bed. (From Zweifach: Factors Regulating Blood Pressure. Josiah Macy, Jr., Foundation, 1950.)

The arterioles are highly muscular, and their diameters can change many fold, as was discussed in Chapter 18. The meta-arterioles do not have a continuous muscular coat, but smooth muscle fibers encircle the vessel at intermediate points, as illustrated in Figure 20–1 by the large black dots to the sides of the meta-arteriole.

At the origin of the true capillaries from the meta-arterioles a smooth muscle fiber usually encircles the capillary. This is called the *precapillary sphincter.*

The venules are considerably larger than the arterioles and have a much weaker muscular coat. Yet, it must be remembered that the pressure in the venules is much less than that in the arterioles so that the venules can perhaps contract as much as can the arterioles.

This typical arrangement of the capillary bed is not found in all parts of the body; however, some similar arrangement is found, beginning with arterioles, passing through meta-arterioles and capillaries, and returning through venules.

Structure of the Capillary Wall.   Figure 20–2 illustrates an electron micrograph of the capillary wall. Note that the wall is primarily composed of a thin, unicellular layer of endothelial cells and is surrounded by a thin basement membrane on the outside. The total thickness of the membrane is about 0.5 micron. Within the endothelial cells are found the usual intracellular structures, such as mitochondria.

The diameter of the capillary is 7 to 9 microns, barely large enough for red blood cells and other blood cells to squeeze through.

PORES IN THE CAPILLARY MEMBRANE.   On studying Figure 20–2 again, one can see a minute passageway connecting the interior of the capillary with the exterior. However, such passageways are located far apart from one another, and they represent no more than 0.001 of the total surface area of the capillary. These passageways are the *pores* of the capillary membrane through which water and

Figure 20–2 An electron micrograph showing a long slitlike pore that exists at the junction (*J*) between adjacent capillary endothelial cells (*E*$_1$ and *E*$_2$). (Approximate magnification 30,000×.) *L*, capillary lumen; *B*, basement membrane; *P*, pericyte; *V*, pinocytic vesicle. (Courtesy of Drs. Guido Majno and Ramzi Cotran.)

many dissolved substances pass between the lumen of the capillary and interstitial spaces, as is discussed later in the chapter. These pores have a diameter between 80 and 90 Ångstroms.

FLOW OF BLOOD IN THE CAPILLARIES — VASOMOTION

Blood does not flow at a continuous rate through the capillaries. Instead, it flows in intermittent spurts. The cause of this intermittency is the phenomenon called *vasomotion,* which means intermittent contraction of the meta-arterioles and precapillary sphincters. These constrict and relax in an alternating cycle 6 to 12 times per minute.

Regulation of Vasomotion. The most important factor found thus far to affect the degree of opening and closing of the meta-arterioles and precapillary sphincters is the concentration of *oxygen* in the tissues. When the oxygen concentration is very low, the spurts of blood flow occur more often, and the duration of each period of flow lasts for a more prolonged interval, thereby allowing the blood to carry increased quantities of oxygen to the tissues. Therefore, one of the theories of regulation of vasomotion, and therefore also for regulation of rate of blood flow through the nutrient capillaries, is that the primary regulatory factor is the need of the tissue for oxygen. The lesser the amount of oxygen in the blood, the greater will be the amount of blood allowed to flow through the tissue. Also, the greater the use of oxygen by the tissue, the greater will be the amount of blood that flows. Thus, by this intermittent opening and closing of the precapillary sphincters and meta-arterioles, the blood flow to the tissue is theoretically *auto-regulated,* as will be discussed in Chapter 24.

Blood Flow Through Non-nutrient Vessels. In addition to blood flow through the "nutrient capillaries," there exist in most tissues direct vascular connections between the arterioles and venules that do not participate in transport of nutrients to the tissues. For instance, some of the *thoroughfare channels* seem not to exchange large quantities of substances between the blood and interstitial fluids. In addition, some large direct vascular connections called *arteriovenous anastomoses* have strong muscular coats so that the blood flow through them can be altered from very slow to tremendous rates. In some parts of the skin these A-V anastomoses provide a means for allowing rapid inflow of warm arterial blood into the large venous plexuses of the skin, thereby warming the skin.

Therefore, the blood flow through the tissues is divided into (1) *nutritive blood flow* and (2) *non-nutritive blood flow.* The nutritive blood flow is apparently regulated in response to the need of the tissues for nutrients (mainly oxygen); and the non-nutritive blood flow is regulated by other factors, such as nervous stimuli.

AVERAGE FUNCTION OF THE CAPILLARY SYSTEM

Despite the fact that blood flow through each capillary is intermittent, so many capillaries are present in the tissues that their overall function becomes averaged. That is, there is an *average rate of blood flow* through each tissue capillary bed, an *average capillary pressure* within the capillaries, and an *average rate of transfer of substances* between the blood of the capillaries and the surrounding interstitial fluid. In the remainder of this chapter, we will be concerned with these average functions of the capillaries, though one must remember that the average functions are in reality the functions of literally billions of individual capillaries each operating intermittently.

EXCHANGE OF NUTRIENTS AND OTHER SUBSTANCES BETWEEN THE BLOOD AND INTERSTITIAL FLUID

DIFFUSION THROUGH THE CAPILLARY MEMBRANE

By far the most important means by which substances are transferred between the plasma and interstitial fluids is by diffusion. Figure 20–3 illustrates this process, showing that as the blood traverses the capillary, tremendous numbers of water molecules and dissolved particles diffuse back and forth through the

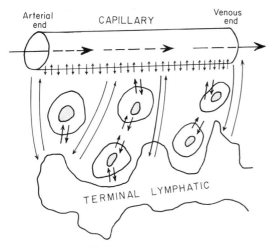

Figure 20–3 Diffusion of fluid and dissolved substances between the capillary and interstitial fluid spaces.

capillary wall, providing continual mixing between the interstitial fluids and the plasma as was explained in Chapter 4. Diffusion results from thermal motion of the water molecules and dissolved substances in the fluid, the different particles moving first in one direction, then another, moving randomly in every direction.

Diffusion of Water-Soluble, Lipid-Insoluble Substances through the Capillary Membrane. Many substances needed by the tissues are soluble in water but are completely insoluble in the lipid membranes of the capillaries; such substances include sodium ions, chloride ions, glucose, and so forth. These substances diffuse between the plasma and interstitial fluids only through the pores, which themselves are filled with water.

Despite the fact that not over $1/1000$ of the surface area of the capillaries is represented by pores, the rate of thermal motion in fluids is so great that even this small area of pores is sufficient to allow tremendous diffusion of water-soluble substances across the membrane. To give one an idea of the extreme rapidity with which substances diffuse, *the rate at which water molecules diffuse through the capillary membrane is approximately 40 times as great as the rate at which plasma itself flows linearly along the capillary.* That is, the water of the plasma is exchanged with the water of the interstitial fluids 40 times before the plasma can go the entire distance through the capillary. Yet, despite this rapid rate of diffusion of water, the rates of diffusion *into* the capillary and *out* of the capillary are so nearly equal that the volume of plasma in the capillary remains almost exactly constant.

EFFECT OF MOLECULAR SIZE ON PORE PER-

MEABILITY. The diameter of the capillary pores, 80 to 90 Ångstroms, is about 25 times the diameter of the water molecule, which is the smallest molecule that normally passes through the capillary pores. However, the diameters of plasma protein molecules are slightly greater than that of the pores, and other substances, such as glucose and urea, have intermittent diameters. Therefore, it is obvious that the permeability of the capillary pores for different substances varies according to their molecular diameters.

TABLE 20–1. Relative Permeability of the Capillary Pores to Different Size Molecules

Substance	Molecular Weight	Permeability
Water	18	1.00
NaCl	58.5	0.96
Urea	60	0.8
Glucose	180	0.6
Sucrose	342	0.4
Inulin	5,000	0.2
Myoglobin	17,600	0.03
Hemoglobin	68,000	0.01
Albumin	69,000	<0.0001

Modified from Pappenheimer.

Table 20–1 gives the relative capillary permeabilities for substances commonly encountered by the capillary membrane, illustrating for instance that the permeability for glucose molecules is 0.6 times as great as that for water molecules, whereas the permeability for albumin molecules is less than $1/10,000$ that for water molecules. Thus, the membrane is almost impermeable to albumin, which causes a significant concentration difference to develop between the albumin of the plasma and that of the interstitial fluid, as will become evident later in the chapter.

Diffusion of Lipid-Soluble Substances Through the Capillary Membrane. If a substance is lipid-soluble, it can diffuse directly through the cell membranes of the capillary without having to go through the pores. Lipid-soluble substances normally transported in the blood include especially oxygen and carbon dioxide. Since these substances can permeate all areas of the capillary membrane, not merely the pores, their rates of transport through the membrane are many times greater than the rates for water and other lipid-insoluble substances.

Other lipid-soluble substances that are rapidly transported through the capillary membrane include various anesthetic gases and alcohol.

Effect of Concentration Difference on Net Rate of Diffusion Through the Capillary Membrane. As explained in Chapter 4, the "net" exchange of substances through the capillary membrane is proportional to the *concentration difference* between the two sides of the membrane. That is, the greater the difference between the concentrations of any given substance on the two sides of the capillary membrane, the greater will be the net movement of the substance through the membrane. Thus, the concentration of oxygen in the blood is normally greater than that in the interstitial fluids. Therefore, large quantities of oxygen normally move from the blood toward the tissues. Conversely, the concentration of carbon dioxide is greater in the tissues than in the blood, which obviously causes carbon dioxide to move into the blood and to be carried away from the tissues.

Fortunately, the rates of diffusion through the capillary membranes of most nutritionally important substances are so great that only slight concentration differences suffice to cause more than adequate transport between the plasma and interstitial fluid. For instance, the concentration of oxygen in the interstitial fluid immediately outside the capillary is normally no more than 1 per cent less than the concentration in the blood, and yet this 1 per cent difference causes enough oxygen to move from the blood into the interstitial spaces to provide all the oxygen required for tissue metabolism.

TRANSPORT THROUGH THE CAPILLARY MEMBRANE BY PINOCYTOSIS

Another means by which minute quantities of substances can be transported through the capillary membrane is by pinocytosis, a phenomenon that was explained in Chapter 4. Basically, this process means that the endothelial cells of the capillary wall ingest small amounts of plasma or interstitial fluid, and small vesicles of the ingested fluid then migrate from one surface of the endothelial cell to the other where the fluid is released.

Anatomists, who have had difficulty demonstrating pores in electron micrographs of the capillary membrane but who can always see pinocytic vesicles in the endothelial cells, have suggested that pinocytosis is probably the major means by which substances are transported through the capillary membrane. Yet dynamic studies have shown that low molecular weight substances, such as water, sodium ions, chloride ions, glucose and urea, are transported through the membrane at rates more than *10 million times* as great as could

possibly be accounted for by pinocytosis. Therefore, it is inconceivable that pinocytosis is the major means for transporting substances through the capillary membrane.

On the other hand, transport of high molecular weight substances—for instance, plasma proteins, glycoproteins and large polysaccharides such as dextran—might well occur to a significant extent by pinocytosis.

TRANSFER OF FLUID VOLUME BETWEEN THE PLASMA AND INTERSTITIAL FLUIDS

One of the most important considerations of capillary dynamics is the means by which the plasma is held in the circulation and not allowed to transude continuously through the capillary membrane into the interstitial fluid. Figure 20–4 illustrates the four primary factors that determine whether fluid will move out of the blood into the interstitial fluid or in the opposite direction; these are:

1. The *capillary pressure* (Pc), which tends to move fluid outward through the capillary membrane.

2. The *interstitial fluid pressure* (Pif), which tends to move fluid inward through the capillary membrane.

3. The *plasma colloid osmotic pressure* (πp), which tends to cause osmosis of fluid inward through the membrane.

4. The *interstitial fluid colloid osmotic pressure* (πif), which tends to cause osmosis of fluid outward through the membrane.

The regulation of fluid volumes in the blood and interstitial fluid is so important that each of these factors is discussed in turn in the following sections.

CAPILLARY PRESSURE

Unfortunately, the exact capillary pressure is not known because it has been impossible to measure capillary pressure under absolutely normal conditions. Yet two different methods have been used to estimate the capillary pressure: (1) *direct cannulation of the capillaries,*

Figure 20–4 Forces operative at the capillary membrane tending to move fluid either outward or inward through the membrane.

which has given an average mean capillary pressure of about 25 mm. Hg, and (2) *indirect functional measurement of the capillary pressure*, which has given a capillary pressure averaging about 17 mm. Hg. These methods are the following:

Cannulation Method for Measuring Capillary Pressure. To measure pressure in a capillary by cannulation, a microscopic glass pipet is thrust directly into the capillary, and the pressure is measured by an appropriate manometer system. Using this method, capillary pressures have been measured in capillaries of the exposed mesentery of lower animals and in large capillary loops of the eponychium at the base of the fingernail in human beings. These measurements have given pressures of 30 to 40 mm. Hg in the arterial ends of the capillaries, 10 to 15 mm. Hg in the venous ends, and about 25 mm. Hg in the middle.

Isogravimetric and Isovolumetric Methods for Measuring Mean Capillary Pressure. Figure 20–5 illustrates an *isogravimetric* method for estimating capillary pressure. This figure shows a section of gut held by one arm of a gravimetric balance. Blood is perfused through the gut. When the arterial pressure is decreased, the resulting decrease in capillary pressure causes osmotic absorption of fluid out of the gut wall and makes the weight of the gut decrease. This immediately causes movement of the balance arm. However, to prevent this weight decrease, the venous pressure is raised a sufficient amount to overcome the effect of decreasing the arterial pressure.

In the lower part of the figure, the changes

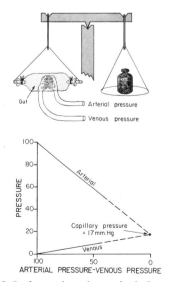

Figure 20–5 Isogravimetric method for measuring capillary pressure. Explained in the text.

in arterial and venous pressures that exactly balance each other in their effect on the weight of the gut are illustrated. The arterial and venous curves cross each other at a value of 17 mm. Hg. Therefore, the capillary pressure must have remained at this same level of 17 mm. Hg throughout these maneuvers. Thus, in a roundabout way, the "functional" capillary pressure is measured to be about 17 mm. Hg.

The *isovolumetric* method for measuring capillary pressure is essentially the same as the isogravimetric method, except that the *volume* of the tissue is recorded rather than the weight. The arterial and venous pressures are gradually brought toward each other and are continually adjusted so that the tissue neither gains nor loses volume. On extrapolating the arterial and venous pressures, one again finds that the capillary pressure estimated by this means is considerably lower than that measured by direct cannulation, again averaging about 17 mm. Hg.

Functional Capillary Pressure. Since the cannulation method does not give the same pressure measurement as the isogravimetric and isovolumetric methods, one must decide which of these measurements is probably the true functional capillary pressure of the tissues. The conditions of the isovolumetric and isogravimetric measurements are probably much more nearly normal than are those of the micropipet measurements, because insertion of a micropipet into a capillary often blocks its blood flow or otherwise alters its dynamics, which causes abnormal pressure. Therefore, it is likely that the lower pressure measurement, about 17 mm. Hg as estimated by the isogravimetric and isovolumetric methods, is more nearly the true functional capillary pressure than is the mean capillary pressure of 25 mm. Hg as estimated by the micropipet method.

A much more important reason for believing that the lower pressure is the true *functional* capillary pressure is that the meta-arterioles and precapillary sphincters of the capillary system are normally closed during a greater part of the vasomotion cycle than they are open. When they are closed the pressure in the entire capillary system beyond the closures should be almost exactly equal to the pressure at the venous ends of the capillaries, about 10 mm. Hg. Therefore, if one considers a *weighted* average of the pressures in all capillaries, one would expect the *functional* mean capillary pressure to be much nearer to the pressure in the venous ends of the capillaries than to the pressure in the arterial ends.

Thus, there are many reasons for believing *that the normal functional mean capillary pressure is about 17 mm. Hg.*

It is also reasonable to believe that the average pressure in the venous ends of the capillaries is very near to that in the venules, or about 8 to 10 mm. Hg.

Yet what is the *average* functional pressure in the arterial ends of the capillaries? This we do not know, but it must be considerably below the 30 to 40 mm. Hg measured with the micropipet in *open* arterial capillaries, because even the arterial ends of the capillaries are closed off from the meta-arterioles most of the time. We might estimate this pressure to be about 25 mm. Hg, which is somewhat below the pipet measurements and somewhat above the functional mean capillary pressure.

INTERSTITIAL FLUID PRESSURE

The interstitial fluid pressure, like capillary pressure, has been difficult to measure, primarily because the maximum width of interstitial spaces is about 1 micron. Thus far, it has been impossible to cannulate these spaces directly to make measurements of the pressure. Therefore, several indirect methods have been used to estimate the interstitial fluid pressure. These are the following:

Measurement of Interstitial Fluid Pressure in Implanted Perforated Spheres. Figure 20–6 illustrates an indirect method for measuring interstitial fluid pressure which may be explained as follows: A small hollow plastic sphere perforated by several hundred small holes is implanted in the tissues, and the surgical wound is allowed to heal for approximately one month. At the end of that time, tissue will have grown inward through the holes to line the inner surface of the sphere. Furthermore, the cavity is filled with fluid that

Figure 20–7   The equilibrium method for measuring interstitial fluid pressure.

flows freely through the perforations back and forth between the fluid of the interstitial fluid spaces and the fluid of the cavity. Therefore, the pressure in the cavity should equal the pressure in the interstitial fluid spaces. A needle is inserted through the skin and through one of the perforations to the interior of the cavity, and the pressure is measured by use of an appropriate manometer.

Interstitial fluid pressure measured by this method in normal tissues averages about −7 mm. Hg. That is, the pressure is actually *less than atmospheric pressure* or, in other words, is a semi-vacuum. The significance of the negativity of interstitial fluid pressure and the mechanism of its development is discussed in the following chapter in relation to interstitial fluid dynamics.

Measurement of Interstitial Fluid Pressure in Fluid Spaces Developed Underneath the Skin. Figure 20–7 illustrates another method that has been used to measure the interstitial fluid pressure. A small vacuum cup is placed over the skin, and the skin is sucked up into the cup. Gradually, over about 24 hours, free fluid collects beneath the mound of skin, and its pressure comes to equilibrium. A needle is then inserted into the tissue so that its tip is within this free fluid, and the pressure is recorded. Since the fluid is in free communication with the fluid in the surrounding tissues, the pressure measured here should equal the pressure in the other interstitial spaces. The pressure measured by this method is also negative, averaging again about −7 mm. Hg.

Measurement of Interstitial Fluid Pressure by Means of a Cotton Wick.   A new method used recently is to insert into a tissue a small Teflon tube with about eight cotton fibers protruding from its end. Each cotton fiber is itself a very minute tube containing large numbers of small side openings. The fluid inside the cotton fibers, therefore, makes excellent contact with the tissue fluids and transmits interstitial fluid pressure into the Teflon tube; the pressure can then be measured from the tube by usual manometric means. Pressures measured by this technique have also been negative, similar to those measured by the two techniques previously discussed.

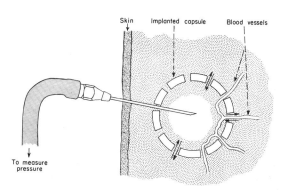

Figure 20–6   The perforated capsule method for interstitial fluid pressure.

PLASMA COLLOID OSMOTIC PRESSURE

Colloid Osmotic Pressure Caused by Proteins. The proteins are the only dissolved substances of the plasma that do not diffuse readily into the interstitial fluid. Furthermore, when small quantities of protein do diffuse into the interstitial fluid, these are soon removed from the interstitial spaces by way of the lymph vessels. Therefore, the concentration of protein in the plasma averages about four times as much as that in the interstitial fluid, 7.3 gm./100 ml. in the plasma versus 1.8 gm./100 ml. in the interstitial fluid.

In the discussion of osmotic pressure in Chapter 4, it was pointed out that only those substances that fail to pass through the pores of a semipermeable membrane exert osmotic pressure. Therefore, the dissolved proteins of the plasma and interstitial fluids are responsible for the osmotic pressure that develops at the capillary membrane. To distinguish this osmotic pressure from that which occurs at the cell membrane, it is called *colloid osmotic pressure* or *oncotic pressure*. The term "colloid" osmotic pressure is derived from the fact that a protein solution resembles a colloidal solution despite the fact that it is actually a true solution. The osmotic pressure that results at the cell membrane is often called *total osmotic pressure* because essentially all dissolved substances of the body fluids exert osmotic pressure at the cell membrane.

EFFECT OF THE DONNAN EQUILIBRIUM ON THE COLLOID OSMOTIC PRESSURE. A special effect, called the *Donnan equilibrium,* causes the colloid osmotic pressure to be about 50 per cent greater than that caused by the proteins alone. This results from the fact that the proteins are negative ions, and to balance these negative ions a large number of positively charged ions (cations), mainly sodium ions, must remain on the same side of the membrane as the proteins. These extra cations increase the number of osmotically active substances and increase the total colloid osmotic pressure. Even more important, however, (for mathematical reasons that cannot be explained here), the Donnan equilibrium effect becomes progressively more significant the higher the concentration of proteins. This means, as illustrated in Figure 20–8, that the initial few grams of protein in each 100 ml. of plasma or interstitial fluid has much less colloid osmotic effect than do the next few grams.

Normal Values for Plasma Colloid Osmotic Pressure. The colloid osmotic pressure of normal human plasma averages approximately 28 mm. Hg; 19 mm. of this is caused by

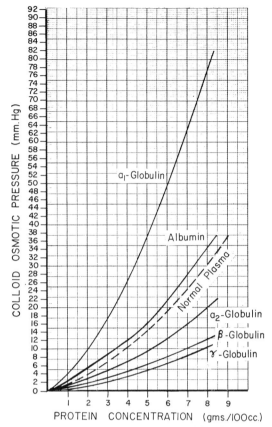

Figure 20–8 The osmotic pressure of five fractions of the plasma proteins at different concentrations. Also, the dashed line shows the osmotic pressure of normal plasma proteins, which are a mixture of the others. (Modified from Ott: *Klin. Wschr., 34:*1079, 1956.)

the dissolved protein, and 9 mm. by the cations held in the plasma by proteins as was just discussed.

Note particularly that the 28 mm. Hg colloid osmotic pressure that can develop at the capillary membrane is only about ¹/₂₀₀ the total osmotic pressure that would develop at a *cell* membrane if normal interstitial fluid were on one side of the membrane and pure water on the other side. Thus, the colloid osmotic pressure of the plasma is actually a weak osmotic force; but, even so, it plays an exceedingly important role in the maintenance of normal blood and interstitial fluid volumes, as will be evident in the remainder of this chapter and in the following chapters.

Effect of the Different Plasma Proteins on Colloid Osmotic Pressure. The plasma proteins are a mixture of proteins that contain albumin, with an average molecular weight of 69,000; globulins, 140,000; and fibrinogen, 400,000. Thus, 1 gram of globulin contains only half as many molecules as 1 gram of albumin, and 1 gram of fibrinogen contains only one-sixth as many molecules as 1 gram of albumin. Furthermore, the average relative concentrations of these different types of proteins in the plasma are:

	gm. %
Albumin	4.5
Globulins	2.5
Fibrinogen	0.3
TOTAL	7.3

Because each gram of albumin exerts twice the osmotic pressure of a gram of globulins and because there is almost twice as much albumin in the plasma as globulins, about 70 per cent of the total colloid osmotic pressure of the plasma results from the albumin fraction and only about 30 per cent from the globulins and fibrinogens. Therefore, from the point of view of capillary dynamics, it is mainly albumin that is important.

Figure 20–8 illustrates graphically the colloid osmotic pressure exerted by different concentrations of albumin and four different fractions of globulins. Note especially that each gram of dissolved protein exerts considerably less colloid osmotic pressure at low concentrations than at higher concentrations; this is caused by the Donnan equilibrium factor discussed previously.

INTERSTITIAL FLUID COLLOID OSMOTIC PRESSURE

Though the size of the usual capillary pore is smaller than the molecular sizes of the plasma proteins, this is not true of all the pores. Therefore, small amounts of plasma proteins do leak into the interstitial spaces. Albumin molecules, because they are smaller than most globulin molecules, normally leak 1.6 times as readily as the globulins, causing the proteins in the interstitial fluids to have a disproportionately high albumin to globulin ratio.

The total quantity of protein in the entire 12 liters of interstitial fluid of the body is almost exactly equal to the total quantity of protein in the plasma itself, but since this volume is four times the volume of plasma the average protein *concentration* of the interstitial fluid is only one-fourth that in plasma, or approximately 1.8 grams per cent. Referring to the diagram of Figure 20–8, one finds that the average colloid osmotic pressure of this concentration of proteins in the interstitial fluids is approximately 4.5 mm. Hg.

EXCHANGE OF FLUID VOLUME THROUGH THE CAPILLARY MEMBRANE

Now that the different factors affecting capillary membrane dynamics have been discussed, it is possible to put all these together to see how normal capillaries function.

The capillary pressure at the arterial ends of the capillaries is 15 to 20 mm. Hg greater than at the venous ends. Because of this difference, fluid "filters" out of the capillaries at their arterial ends and then is reabsorbed into the capillaries at their venous ends. Thus, a small amount of fluid actually "flows" through the tissues from the arterial ends of the capillaries to the venous ends. The dynamics of this flow are the following:

Analysis of the Forces Causing Filtration at the Arterial End of the Capillary. The forces operative at the arterial end of the capillary to cause movement through the capillary membrane are:

Forces tending to move fluid outward:

	mm. Hg
Capillary pressure	25
Negative interstitial fluid pressure	7
Interstitial fluid colloid osmotic pressure	4.5
TOTAL OUTWARD FORCE	36.5

Force tending to move fluid inward:

	mm. Hg
Plasma colloid osmotic pressure	28
TOTAL INWARD FORCE	28

Summation of forces:

	mm. Hg
Outward	36.5
Inward	28
NET OUTWARD FORCE	8.5

Thus, the summation of forces at the arterial end of the capillary shows a net *filtration pressure* of about 8.5 mm. Hg tending to move fluid out of the arterial ends of the capillaries into the interstitial spaces.

This 8.5 mm. Hg filtration pressure causes about 0.5 per cent of the plasma entering the capillaries to filter out of the arterial ends of the capillaries into the interstitial spaces. (This percentage varies tremendously from very, very low in the brain to very high in the liver.)

Analysis of Reabsorption at the Venous End of the Capillary. The low pressure at the venous end of the capillary changes the balance of forces in favor of absorption as follows:

Force tending to move fluid inward:

	mm. Hg
Plasma colloid osmotic pressure	28
TOTAL INWARD FORCE	28

Forces tending to move fluid outward:

	mm. Hg
Capillary pressure	9
Negative interstitial fluid pressure	7
Interstitial fluid colloid osmotic pressure	4.5
TOTAL OUTWARD FORCE	20.5

Summation of forces:

	mm. Hg
Inward	28
Outward	20.5
NET INWARD FORCE	7.5

Thus, the force that causes fluid to move into the capillary, 28 mm. Hg, is greater than that opposing reabsorption, 20.5 mm. Hg. The difference, 7.5 mm. Hg, is the *reabsorption pressure*.

The reabsorption pressure causes about nine-tenths of the fluid that had filtered out the arterial ends of the capillaries to be reabsorbed at the various ends. The other one-tenth flows into the lymph vessels, as is discussed in the following chapter.

Flow of Fluid Through the Interstitial Spaces. The 0.5 per cent of the plasma that filters out of the arterial ends of the capillaries *flows* through the tissue spaces to the venous ends of the capillaries where all but about $1/10$ of it is reabsorbed. (Much more than this is reabsorbed in the legs and arms, and much less is reabsorbed in the liver.)

It is especially important to distinguish the difference between *filtration* and *diffusion* through the capillary membrane. Diffusion occurs in both directions, while filtration is the *net* movement of fluid out of the capillary at the arterial ends. The rate of diffusion of water through all the capillary membranes of the entire body is about 120,000 ml./minute, while the normal rate of filtration at the arterial ends of all the capillaries is only 16 ml./minute, a difference of about 8000-fold. Thus, the quantitative rate of diffusion of water and nutrients back and forth between the capillaries and the interstitial spaces is tremendous in comparison with the minute rate of "flow" of fluid through the tissues.

THE STARLING EQUILIBRIUM OF CAPILLARY EXCHANGE

E. H. Starling pointed out three-quarters of a century ago that under normal conditions a state of equilibrium exists at the capillary membrane whereby the amount of fluid leaving the circulation through the capillaries exactly equals that quantity of fluid that is returned to the circulation by reabsorption at the venous ends of the capillaries and by flow through the lymphatics. This equilibrium is caused by equilibration of the *mean* forces tending to move fluid through the capillary membranes. If we assume that the mean capillary pressure is 17 mm. Hg, the normal mean equilibrium dynamics of the capillary are the following:

Mean forces tending to move fluid outward:

	mm. Hg
Mean capillary pressure	17
Negative interstitial fluid pressure	7
Interstitial fluid colloid osmotic pressure	4.5
TOTAL OUTWARD FORCE	28.5

Mean force tending to move fluid inward:

	mm. Hg
Plasma colloid osmotic pressure	28
TOTAL INWARD FORCE	28

Summation of mean forces:

	mm. Hg
Outward	28.5
Inward	28.0
NET OUTWARD FORCE	0.5

Thus, we find a slight imbalance of forces at the capillary membranes that causes slightly more filtration of fluid into the interstitial spaces than reabsorption. This slight excess of filtration is called the *net filtration*, and it is balanced by fluid return to the circulation through the lymphatics. The normal rate of net filtration in the entire body is about 1.7 ml. per minute. This figure also represents the rate of fluid flow into the lymphatics each minute.

Effect of Imbalance of Forces at the Capillary Membrane. If the mean capillary pressure rises above 17 mm. Hg the net force tending to cause filtration of fluid into the tissue spaces obviously rises. Thus a 20 mm. Hg rise in mean capillary pressure causes an increase in the net filtration pressure from 0.5 mm. Hg to 20.5 mm. Hg, which results in 41 times as much net filtration of fluid into the interstitial spaces as normally occurs, and this is also 41 times the normal flow of fluid into the lymphatic system, an amount that is usually too much for the lymphatics to carry away. As a result, fluid begins to accumulate in the interstitial spaces, and edema results.

Conversely, if the capillary pressure falls very low, net reabsorption of fluid into the capillaries occurs instead of net filtration, and the blood volume increases at the expense of the interstitial fluid volume. The effects of these imbalances at the capillary membrane are discussed in the following chapter in relation to the formation of edema.

FLUID EXCHANGE AT THE PULMONARY CAPILLARY

The dynamics of fluid exchange at the pulmonary capillary membranes is almost identical to the dynamics in other parts of the circulation, except that the pulmonary capillary pressure is far lower than the capillary pressure elsewhere in the body, only about 7 mm. Hg in contrast to 17 mm. This low pressure will be discussed in Chapter 23. Figure 20–9 illustrates the forces operative at the pulmonary capillary membrane:

Forces tending to move fluid outward:

	mm. Hg
Pulmonary capillary pressure	7
Interstitial colloid osmotic pressure	4.5
Negative interstitial fluid pressure in the lungs calculates to be about	17
TOTAL OUTWARD FORCE	28.5

Force tending to move fluid inward:

	mm. Hg
Plasma colloid osmotic pressure	28
TOTAL INWARD FORCE	28

Summation of forces:

	mm. Hg
Outward	28.5
Inward	28.0
NET OUTWARD FORCE	0.5

Thus, there is a slight net "filtration" force causing a very slight excess of fluid to filter out of the pulmonary capillaries over that reabsorbed. This excess returns to the circulation by way of the lymph vessels, as it does in the systemic circulation.

Pulmonary Interstitial Fluid Pressure. The interstitial fluid pressure in the lungs has been measured by the capsule technique (described previously) to be negative, as elsewhere in the body, ranging as low as −16 mm. Hg. However the measurements are not accurate because of infection and tissue changes around the capsules. The other three forces, besides interstitial pressure, acting at the membrane have been estimated with a degree of accuracy within 2 to 3 mm. Hg. Therefore, applying the principles of the Starling equilibrium of the capillaries, as just given in the calculation, it is readily apparent that the interstitial fluid pressure of the lungs must be very negative, about −17 mm. Hg.

Mechanism for Keeping the Alveoli Dry. Figure 20–9 also illustrates the alveolar epithelial lining, as well as the capillary membrane. This *alveolar membrane* is continuous and does not have large pores similar to those in the capillary membrane. Instead, the only substances that diffuse readily through the alveolar membrane are those substances that can diffuse through the alveolar epithelial cells themselves. For instance, lipid-soluble substances, such as oxygen and carbon dioxide, diffuse through this membrane extremely rapidly. Likewise, water diffuses rapidly because the cell pores are larger than the water molecules. On the other hand, substances such as sodium and glucose require many minutes to hours to diffuse through the membrane. Thus, the alveolar membrane is quite different from the pulmonary capillary membrane.

Yet the negative pressure in the interstitial fluid spaces, −17 mm. Hg, creates enough imbalance across the alveolar membrane to cause absorption of fluid from the alveolus into the interstitial fluid spaces. Thus, the negative interstitial fluid pressure provides a mechanism for keeping the alveoli dry.

REFERENCES

Blood flow and exchange at microcirculatory level (Symposium). *Fed. Proc., 24*:1074, 1965.

Chinard, F. P.: Starling's hypothesis in the formation of edema. *Bull. N. Y. Acad. Med., 38*:375, 1962.

Duling, B. R., and Berne, R. M.: Propagated vasodilation in the microcirculation of the hamster cheek pouch. *Circ. Res., 26*:163, 1970.

Folkow, B.; Lundgren, O.; and Wallentin, I.: Studies on the relationship between flow resistance, capillary filtration coefficient and regional blood volume in the intestine of the cat. *Acta Physiol. Scand., 57*:270, 1963.

Friedman, J. J.: Single-passage extraction of 86Rb from the circulation of skeletal muscle. *Amer. J. Physiol., 216*:460, 1969.

Guyton, A. C.: Concept of negative interstitial pressure based on pressures in implanted perforated capsules. *Circ. Res., 12*:399, 1963.

Guyton, A. C.: Interstitial fluid pressure: II. Pressure-volume curves of interstitial space. *Circ. Res., 16*:452, 1965.

Guyton, A. C.: Interstitial fluid pressure-volume relationships and their regulation. *In* Wolstenholme, G. E. W., and Knight, J. (eds.): Ciba Foundation Symposium on Circulatory and Respiratory Mass Transport. London, J. & A. Churchill Ltd., 1969, p. 4.

Guyton, A. C.: Regulation of interstitial fluid volume and pressure. *Proc. Int. Union of Physiol. Sciences, 6*:71, 1968.

Guyton, A. C.; Prather, J.; Scheel, K.; and McGehee, J.: Interstitial fluid pressure: IV. Its effect on fluid movement through the capillary wall. *Cir. Res., 19*:1022, 1966.

Guyton, A. C.; Scheel, K.; and Murphree, D.: Interstitial fluid pressure: III. Its effect on resistance to tissue fluid mobility. *Circ. Res., 19*:412, 1966.

Hyman, C., and Wong, W. H.: Capillary filtration coefficient in the extremities of man in high environmental temperatures. *Circ. Res., 22*:251, 1968.

Figure 20–9 Capillary dynamics at the pulmonary capillary membrane. The negative interstitial fluid pressure causes continual transudation of fluid out of the alveoli, thus keeping them dry. *Pc,* capillary pressure; *Pif,* interstitial fluid pressure; *πif,* interstitial fluid colloid osmotic pressure; *πp,* plasma colloid osmotic pressure.

Landis, E. M.: Capillary pressure and capillary permeability. *Physiol. Rev., 14*:404, 1934.

Landis, E. M., and Pappenheimer, J. R.: Exchange of substances through the capillary walls. *In* Handbook of Physiology. Baltimore, The Williams & Wilkins Co., 1963, Sec. II, Vol. II, p. 961.

Mayerson, H. S.: The physiologic importance of lymph. *In* Handbook of Physiology. Baltimore, The Williams & Wilkins Co., 1963, Sec. II, Vol. II, p. 1035.

Microcirculation (Symposium). *Fed. Proc., 25*:1742, 1966.

Nicoll, P. A., and Webb, R. L.: Vascular patterns and active vasomotion as determiners of flow through minute vessels. *Angiology, 6*:291, 1955.

Pappenheimer, J. R.: Passage of molecules through capillary walls. *Physiol. Rev., 33*:387, 1953.

Prather, John W.; Gaar, Kermit A., Jr.; and Guyton, A. C.: Direct continuous recording of plasma colloid osmotic pressure of whole blood. *J. Appl. Physiol., 24*:602, 1968.

Renkin, E. M.: Capillary blood flow and transcapillary exchange. *Physiologist, 9*:361, 1966.

Renkin, E. M.: Exchange of substances through capillary walls. *In* Wolstenholme, G. E. W., and Knight, J. (eds.): CIBA Foundation Symposium on Circulatory and Respiratory Mass Transport. London, J. & A. Churchill Ltd., 1969, p. 50.

Shirley, H. H., Jr.; Wolfram, C. G.; Wasserman, K.; and Mayerson, H. S.: Capillary permeability to macromolecules: stretched pore phenomenon. *Amer. J. Physiol., 189*:190, 1957.

Sonnenschein, R. R.; Wright, D. L.; and Mellander, S.: Effects of vasodilators on filtration coefficient and distal arterial pressure in muscle. *Amer. J. Physiol., 213*:706, 1967.

Stromme, S. B.; Maggert, J. E.; and Scholander, P. F.: Interstitial fluid pressure in terrestrial and semiterrestrial animals. *J. Appl. Physiol., 27*:123, 1969.

Wolf, A. V.; Fuller, J. B.; Goldman, E. J.; and Mahony, T. D.: New refractometric methods for the determination of total proteins in serum and in urine. *Clin. Chem., 8*:158, 1962.

Yudilevich, D. L., and Alvarez, O. A.: Water, sodium, and thiourea transcapillary diffusion in the dog heart. *Amer. J. Physiol., 213*:308, 1967.

Zweifach, B. W.: Functional Behavior of the Microcirculation. Springfield, Illinois, Charles C Thomas, Publisher, 1961.

CHAPTER 21

The Lymphatic System, Interstitial Fluid Dynamics, and Edema

THE LYMPHATIC SYSTEM

The lymphatic system represents an accessory route by which fluids can flow from the interstitial spaces into the blood. And, most important of all, the lymphatics can carry proteins and even large particulate matter away from the tissue spaces, neither of which can be removed by absorption directly into the blood capillary. We shall see that this removal of proteins from the interstitial spaces is an absolutely essential function, without which we would die within about 24 hours.

THE LYMPH CHANNELS OF THE BODY

All tissues of the body, with the exception of a very few, have lymphatic channels that drain excess fluid directly from the interstitial spaces. The exceptions include the superficial portions of the skin, the central nervous system, deeper portions of peripheral nerves, the endomysium of muscles, and the bones. However, even these tissues have minute interstitial channels through which interstitial fluid can flow, and eventually this fluid flows into lymphatic vessels along the periphery of the tissue or, in the case of the brain, flows into the cerebrospinal fluid and thence directly back into the blood.

Essentially all the lymph from the lower part of the body—even some of that from the legs—flows up the *thoracic duct* and empties into the venous system at the juncture of the left internal jugular vein and subclavian vein, as illustrated in Figure 21–1. However, small amounts of lymph from the lower part of the body can enter the veins in the inguinal region and perhaps also at various points in the abdomen.

Lymph from the left side of the head, the left arm, and left chest region also enters the thoracic duct before it joins the venous system. Lymph from the right side of the neck and head, from the right arm, and from parts of the right thorax enters the *right lymph duct,* which then empties into the venous system at the juncture of the right subclavian vein and right internal jugular vein.

The Terminal Lymphatic Capillaries and their Permeability. Most of the fluid filtering from the arterial capillaries flows among the cells and is finally reabsorbed back into the venous capillaries; but on the average, about one-tenth of the fluid enters the terminal lymphatic capillaries and returns to the blood through the lymphatic system rather than directly through the venous capillaries.

The minute quantity of fluid that returns to the circulation by way of the lymphatics is extremely important because substances of high molecular weight, such as proteins, cannot pass with ease through the pores of the venous capillaries, but can enter the lymphatic capillaries almost completely unimpeded. The reason for this is a special structure

CERVICAL NODES

SENTINEL NODE

SUBCLAVIAN VEIN
R. LYMPH DUCT
THORACIC DUCT
AXILLARY NODES

CISTERNA CHYLI
ABDOMINAL NODES

INGUINAL NODES

Figure 21–1   The lymphatic system.

of the terminal lymphatic capillaries, illustrated in Figure 21–2. This figure shows the endothelial cells of the terminal lymphatic attached by *anchoring filaments* to the connective tissue between the surrounding tissue cells. However, at the junctions between the endothelial cells there are no connections between the cells. Instead, the edge of one endothelial cell overlaps the edge of the adjacent one in such a way that the overlapping edge is free to flap. Therefore, the structure forms a minute valve that opens to the interior of the lymphatic. Interstitial fluid, along with its suspended particles, can push the valve open and flow directly into the terminal lymphatic. But this fluid cannot leave the lymphatic once it has entered because any backflow will close the flap valve. Thus, the lymphatics have valves at the very tips of the terminal lymphatics as well as valves along their entire extent up to the point where they empty into the blood vessels.

FORMATION OF LYMPH

Lymph is interstitial fluid that flows into the lymphatics. Therefore, lymph has almost the identical composition as the tissue fluid in the part of the body from which the lymph flows.

The protein concentration in the interstitial fluid averages about 1.8 gms. per cent, and this is also the protein concentration of lymph flowing from most of the peripheral tissues. On the other hand, lymph formed in the liver has a protein concentration as high as 6 gms. per cent, and lymph formed in the intestines has a protein concentration as high as 3 to 4 gms. per cent. Since approximately half of the lymph is derived from the liver and intestines, the thoracic lymph, which is a mixture of lymph from all areas of the body, usually has a protein concentration of 3 to 5 gms. per cent.

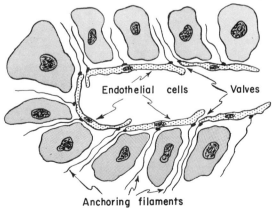

Figure 21–2 Special structure of the terminal lymphatic capillaries that permits passage of substances of high molecular weight back into the circulation.

The lymphatic system is also one of the major channels for absorption of nutrients from the gastrointestinal tract, being responsible principally for the absorption of fats, as will be discussed in Chapter 65. Therefore, after a fatty meal, thoracic duct lymph sometimes contains as much as 1 to 2 per cent of fat.

Finally, even large particles, such as bacteria, can push their way between the endothelial cells of the terminal lymphatics and in this way enter the lymph. As discussed in Chapter 9, these particles are then removed by the lymph nodes, where they are destroyed.

The Mechanism by which Proteins Become Concentrated in Interstitial Fluids. Fluid filtering from the arterial ends of the capillaries in peripheral tissues such as the subcutaneous tissue usually has a protein concentration of less than 0.15 per cent, though the average concentration of protein in the interstitial fluids is some 10 times this value. The reason for this difference is that almost none of the protein that leaks into the tissue spaces is reabsorbed at the venous ends of the capillaries, whereas most of the water is reabsorbed. In this way, the proteins become concentrated in the interstitial fluid before it flows into the lymphatics.

TOTAL RATE OF FLOW THROUGH THE LYMPHATIC CHANNELS

Approximately 100 ml. of lymph flow through the thoracic duct of a resting man per hour, and perhaps another 20 ml. of lymph flow into the circulation each hour through other channels, making a total estimated lymph flow of perhaps 120 ml. per hour. This is about 1/100,000 of the calculated rate of fluid diffusion back and forth through the capillary membranes, and it is also only one-tenth the rate of fluid filtration from the arterial ends of the capillaries into the tissue spaces in the entire body. These facts illustrate that the flow of lymph is relatively small in comparison with the total exchange of fluid between the plasma and the interstitial fluid.

Factors That Determine the Rate of Lymph Flow

Interstitial Fluid Pressure. Elevation of interstitial fluid pressure above its normal level of −7 mm. Hg increases the flow of interstitial fluid into the terminal lymphatics and consequently also increases the rate of lymph flow. This increase in flow is approximately linear until the interstitial fluid pressure reaches 0, at which point the flow rate is 10 to 50 times normal. When the pressure rises above 0, the flow continues at its very rapid rate but does not increase further because the interstitial fluid then begins to compress the outsides of the lymphatics and therefore begins to increase their resistance to flow. Therefore, any factor, besides obstruction of the lymphatic system itself, that tends to increase interstitial pressure increases the rate of lymph flow. Such factors include:

Elevated capillary pressure
Decreased plasma colloid osmotic pressure
Increased interstitial fluid protein
Increased permeability of the capillaries

The Lymphatic Pump. Valves exist in all lymph channels, even down to the tips of the terminal lymphatics, as shown in Figure 21–2. In the larger lymphatics, valves exist every few millimeters, and in the smaller lymphatics the valves are much closer together than this, which illustrates the widespread existence of the valves. Every time the lymph vessel is compressed by pressure from any source, lymph tends to be squeezed in both directions, but because the valves open only in the central direction, the fluid moves unidirectionally. In order of their importance, the factors that often compress the lymphatics are:

Contraction of muscles
Passive movement of the parts of the body
Arterial pulsations
Compression of the tissues from the outside

Obviously, the lymphatic pump becomes very active during exercise but sluggish under resting conditions. During exercise the rate of lymph flow can increase to as high as 3 to 14 times normal because of the increased activity.

In summary, then, *the rate of lymph flow is determined principally by the product of tissue pressure × the activity of the lymphatic pump.*

Flow of Lymph into the Lymphatic Vessels Despite Negative Pressure in the Interstitial Spaces. It has been difficult to understand how fluid can flow out of the interstitial spaces into the lymphatics in the face of the fact that the pressure in the interstitial spaces is normally −7 mm. Hg, as was discussed in the previous chapter. The answer to this is that the −7 mm. Hg pressure is a *mean* pressure, and the actual interstitial fluid pressure rises and falls tremendously during intermittent motion of the tissues. When the interstitial pressure rises above atmospheric pressure in any minute

portion of a tissue, some of the interstitial fluid of this area immediately flows into a terminal lymphatic; then, because of the valves in the lymphatics, the fluid will not flow backward into the interstitial spaces when the interstitial pressure falls again to a more negative value.

CONTROL OF INTERSTITIAL FLUID PROTEIN CONCENTRATION AND INTERSTITIAL FLUID PRESSURE

The fact that interstitial fluid pressure is negative (that is, subatmospheric) was discovered only a few years ago, though it has now been confirmed by a number of different independent methods that are described in the previous chapter. Even so, it has been difficult for many students and even professional physiologists to understand the basic mechanism by which the negative pressure is generated. To explain this, it is first necessary to discuss the regulation of interstitial fluid protein concentration because the problem of interstitial fluid pressure is inextricably bound with that of interstitial fluid protein concentration, as we shall see in the following paragraphs.

REGULATION OF INTERSTITIAL FLUID PROTEIN BY THE LYMPHATICS

Since protein continually leaks from the capillaries into the interstitial fluid spaces, it must also be removed continually, or otherwise the tissue colloid osmotic pressure will become so high that normal capillary dynamics can no longer continue. Unfortunately, only a small proportion of the protein that leaks into the tissue spaces can diffuse back into the capillaries because there is already four times as much protein in the plasma as in the interstitial fluid. Therefore, by far the most important of all the lymphatic functions is the maintenance of low protein concentration in the interstitial fluid. The mechanism of this is the following:

As fluid leaks from the arterial ends of the capillaries into the interstitial spaces, only small quantities of protein accompany it, but then, as the fluid is reabsorbed at the venous ends of the capillaries, most of the protein is left behind. Therefore, *protein progressively accumulates in the interstitial fluid* and this in turn *increases the tissue colloid osmotic pressure.* The osmotic pressure decreases reabsorption of fluid by the capillaries, thereby *promoting increased tissue fluid volume* and *increased tissue pressure*. The increased pressure then forces interstitial fluid into the lymphatic channels, and the fluid carries with it the excess protein that has accumulated. As a result, normal capillary dynamics ensue once again.

To summarize, an increase in tissue fluid protein increases the rate of lymph flow, and this washes the proteins out of the tissue spaces, automatically returning the protein concentration to its normal low level.

The importance of this function of the lymphatics cannot be stressed too strongly, for *there is no other route besides the lymphatics through which proteins can return to the circulatory system*. If it were not for this continual removal of proteins, the dynamics of the capillaries would become so abnormal within only a few hours that life could no longer continue. There is certainly no other function of the lymphatics that can even approach this in importance.

MECHANISM OF NEGATIVE INTERSTITIAL FLUID PRESSURE

Until recent measurements of the interstitial fluid pressure demonstrated that the interstitial fluid pressure is negative rather than positive, as explained in the preceding chapter, it had been taught that the normal interstitial pressure ranges between +1 and +4 mm. Hg, and it has still been difficult to understand how negative pressures can develop in the interstitial fluid spaces. However, we can explain this negative interstitial fluid pressure by the following considerations:

First, it was pointed out above that fluid can flow into lymphatic vessels from interstitial spaces even when the mean interstitial fluid pressure is negative, because this pressure rises and falls every time a tissue moves. Each time the pressure rises to a value slightly above atmospheric pressure, fluid flows out of the interstitial spaces into the lymphatics and carries with it proteins that have accumulated in the interstitial fluid. This continual movement of interstitial fluid into the lymphatics keeps the protein concentration of the interstitial fluid at a low value and thereby keeps the interstitial fluid colloid osmotic pressure also at a low value, usually at about 4 mm. Hg.

Second, the negativity of the interstitial fluid pressure can then be explained on the basis of the Starling equilibrium of the capillaries. If we add all the other forces besides the interstitial fluid pressure that cause movement of

fluid across the capillary membrane, we find the following:

Outward forces:

	mm. Hg
Capillary pressure	17
Interstitial fluid colloid osmotic pressure	4
TOTAL	21

Inward forces:

	mm. Hg
Colloid osmotic pressure	28
DIFFERENCE (Interstitial fluid pressure)	−7

Thus, we see that the interstitial fluid pressure required to balance the other forces across the capillary membrane, in accordance with Starling's equilibrium of the capillaries, is −7 mm. Hg.

Role of the Lymphatic System in Regulation of Interstitial Fluid Pressure. In the above explanation of the mechanism of negative interstitial fluid pressure, one immediately recognizes that the lymphatics play a dominant role, for it is the lymphatics that maintain the low protein concentration in the interstitial fluid. Therefore, in effect, the lymphatics are also one of the primary regulators of the interstitial fluid pressure.

Effect of Tissue Motion on the Interstitial Fluid Pressure. One of the primary factors controlling the rate of lymph flow is the degree of motion of the tissues. The greater the motion, the greater the lymph flow and the lower becomes the protein concentration in the interstitial fluid. The low protein concentration in turn allows increased reabsorption of fluid into the capillaries, which causes increased negativity of the interstitial fluid pressure. Conversely, complete lack of motion makes it difficult for lymph to flow along the lymphatic channels because then the lymphatic pump is activated only by the arterial pulsations. As a result, protein accumulates in the interstitial fluid spaces so that reabsorption of fluid into the capillaries becomes reduced. Consequently, the interstitial fluid pressure rises from its normal negative value of about −7 mm. Hg up to about −2 mm. Hg or, occasionally, even to a slightly positive level.

In short, the greater the degree of motion in a tissue the more negative is the interstitial fluid pressure, whereas in motionless tissues the pressure rises to approach zero, and in some cases even rises above zero.

Significance of Negative Interstitial Fluid Pressure as a Means for Holding the Body Tissues Together. In the past it has been assumed that the different tissues of the body are held together entirely by connective tissue fibers. However, at many places in the body, connective tissue fibers are actually absent. This occurs particularly at points where tissues slide over each other. Yet, even at these places, the tissues are held together by the negative interstitial fluid pressure, which is actually a partial vacuum. When the tissues lose their negative pressure, fluid accumulates in the spaces, and the condition known as *edema* occurs, which is discussed later in the chapter.

Significance of the Normally "Dry" State of the Interstitial Spaces. The normal tendency for the capillaries to absorb fluid from the interstitial spaces and thereby to create a partial vacuum causes all the minute structures of the interstitial spaces to be *compacted* together. Figure 21–3 illustrates a physical model of the tissues constructed to illustrate this effect. To the left, positive pressure is present and excessive quantities of fluid are present in the "interstitial spaces." To the right, negative pressure has been applied through the perforated tube, which simulates a capillary, and the tissue elements are pulled tightly together. This represents a "dry" state; that is, no *excess* fluid is present besides that required simply to fill the crevices between the tissue elements.

The "dry" state of the tissues is particularly important for optimal nutrition of the tissues, because nutrients pass from the blood to the cells by diffusion; and the rate of diffusion between two points is inversely proportional to the distance between the cells and the capillaries. Therefore, it is essential that the distances be maintained at a minimum, or otherwise nutritive damage to the cells can result.

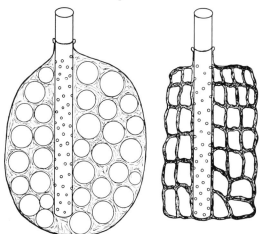

Figure 21–3 Physical model of the tissues constructed of a rubber bag, a perforated tube to simulate a capillary, balloons filled with water to simulate cells, and cotton between the balloons to simulate intercellular elements. *Left:* edematous state. *Right:* nonedematous state.

EDEMA

Edema means the presence of excess interstitial fluid in the tissues. Thus, referring once again to Figure 21–3, we see that the left-hand portion of the figure represents the edematous state, while the right-hand portion represents the nonedematous state.

Obviously, any factor that increases the interstitial fluid pressure high enough can cause development of excess interstitial fluid volume and thereby cause edema to develop. However, to explain the conditions under which edema develops, we must first characterize the *pressure-volume curve* of the interstitial fluid spaces.

PRESSURE-VOLUME CURVE OF THE INTERSTITIAL FLUID SPACES

Figure 21–4 illustrates the average relationship between pressure and volume in the interstitial fluid spaces in the human body as extrapolated from measurements in the isolated hind leg of a dog. The shape of the curve was determined in the following manner: The dog's leg was removed from his body and then perfused with concentrated dextran solution having a colloid osmotic pressure about twice that of normal plasma. This high colloid osmotic force inside the capillaries caused absorption of fluid from the interstitial spaces and caused the weight of the leg to decrease. Measurement of this change in weight provided a means for measuring the decrease in interstitial fluid volume. Simultaneously, the pressure in the interstitial fluid spaces was measured using the capsule method described in the preceding chapter. Later in the experiment, fluid having no colloid osmotic pressure was perfused through the limb, and this caused tremendous quantities of fluid to leak out of the capillaries into the interstitial spaces, thereby increasing the interstitial fluid volume. The curve of Figure 21–4 represents the results obtained in experiments of this type.

The Slight Interstitial Fluid Volume Change in the Negative Pressure Range. One of the most significant features of the curve in Figure 21–4 is that as long as the interstitial fluid pressure remains in the negative range there is little change in interstitial fluid volume despite marked changes in pressure. Therefore, edema will not occur so long as the interstitial fluid pressure remains negative. Indeed, in several hundred measurements of interstitial fluid pressure made in experimental animals, no edema has ever been recorded in the presence of negative interstitial pressure.

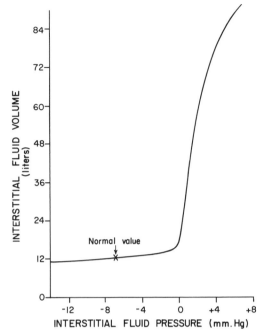

Figure 21–4 Pressure-volume curve of the interstitial spaces. (Extrapolated to the human being from data obtained in dogs.)

Tremendous Increase in Interstitial Fluid Volume When the Interstitial Fluid Pressure Becomes Positive. Note in Figure 21–4 that just as soon as the interstitial fluid pressure rises to equal atmospheric pressure (zero pressure), the slope of the pressure-volume curve suddenly changes and the volume increases precipitously. An additional increase in interstitial fluid pressure of only 1 to 3 mm. Hg now causes the interstitial fluid volume to increase several hundred per cent. Finally, at the very top of the figure, the skin begins to be stretched so that the volume now increases much less rapidly.

Similarity of the Tissue Spaces to a Bag. If one will think for a moment, he will realize that a pressure-volume curve similar to that illustrated in Figure 21–4 can also be recorded from almost any collapsible bag, such as a rubber balloon. When negative pressure is applied to the balloon, it collapses completely; and, however negative the pressure becomes, the volume of the bag still remains almost nothing. But when the pressure is increased above atmospheric pressure, the balloon suddenly begins to fill. Almost no additional pressure is then required to fill the balloon until its walls begin to stretch. Then, further filling is impeded by the stretching wall. These are precisely the characteristics illustrated in Figure 21–4 for the interstitial fluid spaces of the body. They are in effect a collapsed bag which

expands greatly when the interstitial fluid pressure rises above the surrounding atmospheric pressure.

Tissue Space Compliance in Different Pressure Ranges. Another way to express the pressure-volume characteristics of the interstitial fluid spaces is in terms of compliance, which is defined as the change in volume for a given change in pressure. In the negative pressure range, the compliance of the interstitial spaces is slight, about 400 ml./mm. Hg for the entire human body (as extrapolated from measurements in dogs). But, just as soon as interstitial fluid pressure rises into the positive range, this compliance increases tremendously, rising to about 10,000 ml./mm. Hg. Thus, the compliance increases approximately 25-fold between the negative pressure range and the positive pressure range.

POSITIVE INTERSTITIAL FLUID PRESSURE AS THE PHYSICAL BASIS FOR EDEMA

After studying the pressure-volume curve of Figure 21–4, one can readily see that just as soon as the interstitial fluid pressure rises above the surrounding atmospheric pressure, the tissue spaces begin to swell. Therefore, *the physical cause of edema is positive pressure in the interstitial fluid spaces.*

Degree of Edema in Relation to the Degree of Positive Pressure. The solid curve of Figure 21–5 is the same pressure-volume curve as that shown in Figure 21–4, but an edema scale has

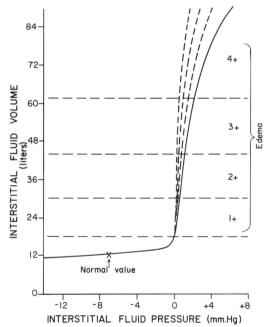

Figure 21–5 Relationship of edema to the pressure-volume curve of the interstitial spaces. The dashed curves show the effect of prolonged edema on the pressure-volume curve.

been added to the figure. 1+ edema means edema that is barely detectable, and 4+ edema means edema in which the limbs are swollen to diameters $1\frac{1}{2}$ to 2 times normal.

Note in Figure 21-5 that edema usually is not detectable in tissues until the interstitial fluid volume has risen to about 30 per cent above normal. And note also that the interstitial fluid volume increases to several hundred per cent above normal in seriously edematous tissues.

Stretch of the Tissue Spaces in Chronic Edema. If edema persists even for a few hours, and especially if it persists for weeks, months, or years, the tissue spaces gradually become stretched. As a result, the pressure-volume curve changes from the solid curve in Figure 21-5 to the progressively higher dashed curves. In other words, in chronic edema the tissue "bag" expands, which increases the ease with which the tissues can develop severe edema. Even a 1 to 2 mm. Hg pressure rise above atmospheric pressure can cause 4+ edema once the tissue spaces have been stretched for many days. This phenomenon of stretch of the tissues is called *delayed compliance.*

The Phenomenon of "Pitting" Edema. If one presses with his finger on the skin over an edematous area and then suddenly removes the finger, a small depression called a "pit" remains. Gradually, within 5 to 30 seconds, the pit disappears. The cause of "pitting" is that edema fluid has been translocated away from the area beneath the pressure point. The fluid simply flows to other tissue areas. Then, when the finger is removed, 5 to 30 seconds is required for the fluid to flow back into the area from which it had been compressed.

Nonpitting Edema. Occasionally, the fluid in seriously edematous tissues cannot be mobilized to other areas by pressure. The usual cause of this is coagulation of the fluid in the tissues. For instance, in an infected or traumatized area, large quantities of fluid can collect, but coagulation of the fluid prevents the fluid from being expulsed by pressure. This type of edema is frequently called *brawny* edema.

THE CONCEPT OF A "SAFETY FACTOR" BEFORE EDEMA DEVELOPS

Safety Factor Caused by the Normal Negative Interstitial Fluid Pressure. One can readily see from Figures 21-4 and 21-5 that the interstitial fluid pressure must rise from the normal value of -7 to above zero mm. Hg before edema begins to appear. Thus, there is a safety factor of 7 mm. Hg caused by the normal negative interstitial fluid pressure before edema appears.

Safety Factor Caused by Flow of Lymph from the Tissues. Another safety factor that helps to prevent edema is increased lymph flow.

When the interstitial fluid volume rises only slightly above the normal value of -7 mm. Hg, lymph flow increases greatly, which removes a portion of the extra fluid entering the interstitial spaces. And this obviously helps to prevent development of edema.

One can estimate that maximally increased lymph flow gives approximately a 6 mm. Hg safety factor, for the extra lymph flow can carry away from the tissues approximately the extra amount of fluid that is formed by a 6 mm. Hg excess in mean capillary pressure.

Safety Factor Caused by Washout of Protein from the Interstitial Spaces. In addition to removal of fluid volume from the interstitial fluid spaces, increased lymph flow also washes out most of the proteins from the interstitial fluid spaces, decreasing the colloid osmotic pressure of the interstitial fluid from the normal value of 4.5 mm. Hg down to nearly zero mm. Hg. This provides another 4 mm. Hg safety factor.

Total Safety Factor and Its Significance. Now, let us add all the above safety factors:

	mm. Hg
Negative interstitial fluid pressure	7
Lymphatic flow	6
Lymphatic washout of proteins	4
TOTAL	17

Thus we find that a total safety factor of about 17 mm. Hg is present to prevent edema. This means that the capillary pressure must rise about 17 mm. Hg above its normal value — that is, above 34 mm. Hg — before edema begins to appear. Or the plasma colloid osmotic pressure must fall from the normal level of 28 mm. Hg to below 11 mm. Hg before edema begins to appear. This explains why the normal human being does not become edematous until severe abnormalities occur in his circulatory system.

EDEMA RESULTING FROM ABNORMAL CAPILLARY DYNAMICS

From the discussions of capillary and interstitial fluid dynamics in the preceding and present chapters, it is already evident that several different abnormalities in these dynamics can increase the tissue pressure and in turn cause extracellular fluid edema. The different causes of extracellular fluid edema are:

Increased Capillary Pressure as a Cause of Edema. Figure 21-6A shows the effect of increased mean capillary pressure on the dynamics of fluid exchange at the capillary membrane. When the mean capillary pressure first becomes abnormally high, more fluid flows out of the capillary than re-

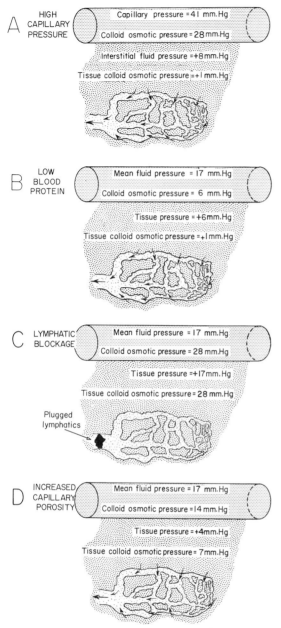

Figure 21-6 Various causes of edema.

struction or arteriolar dilatation. Large venous blood clots frequently result in local areas of venous obstruction, which blocks the return of blood to the heart and promotes edema in the tissues normally drained by the obstructed veins.

More frequently, capillary pressure is increased by obstruction of venous return due to cardiac failure, for, when the heart no longer pumps blood out of the veins with ease, blood dams up in the venous system. The capillary pressure rises, and serious "cardiac edema" occurs. The dynamics of this type of edema are complicated, however, and will be discussed in detail in Chapter 29.

When arteriolar dilatation occurs in localized areas of the body, blood flows rapidly through the locally dilated arterioles and the capillary pressure increases tremendously. Therefore, local edema results. Such local edema occurs commonly in allergic conditions and in the condition known as "angioneurotic edema." Allergic reactions (discussed in Chapter 10) cause the release of histamine into the tissues; histamine relaxes the smooth muscle of arterioles and, when present in large amounts, constricts the venules. The localized edematous areas that result are called "hives" or *urticaria*.

Angioneurotic edema apparently is caused by localized decrease in arteriolar tone due to abnormal vascular control by the autonomic nervous system. When a person is emotionally upset, such edema frequently occurs in the larynx and causes hoarseness.

Decreased Plasma Proteins as a Cause of Edema. Figure 21-6B illustrates the abnormal dynamics that occur at the capillary membrane when the quantity of plasma protein falls to abnormally low values. The major effect is a markedly lowered colloid osmotic pressure of the plasma. Consequently, the capillary pressure far overbalances the colloid osmotic pressure, increasing the tendency for fluid to leave the capillaries and enter the tissue spaces. As a result, fluid collects in the tissue spaces, and the interstitial fluid pressure rises. As long as the pressure remains excessively elevated, the tissue spaces continually enlarge, with the edema becoming progressively worse.

As is also the case with changes in capillary pressure, the decrease in plasma colloid osmotic pressure must be extreme before edema begins to develop. Since the normal safety factor is about 17 mm. Hg, one can calculate that edema begins to appear when the plasma colloid osmotic pressure falls below approximately 11 mm. Hg.

CONDITIONS THAT DECREASE THE PLASMA PROTEIN CONCENTRATION. Albumin is often lost from the plasma in large quantities when the skin is extensively burned. Therefore, one of the complications of severe burns is not only severe edema in the tissues surrounding the burned area but also edema throughout the body because of lowered colloid osmotic pressure.

Often large quantities of protein, especially albumin, are lost through the kidneys into the urine in the disease known as *nephrosis*. Sometimes as much as 20 to 30 grams of albumin are

turns into the capillary and, therefore, it collects in the tissue spaces until the interstitial fluid pressure rises high enough to balance the excessive level of pressure in the capillaries. In Figure 21-6A the mean capillary pressure is 41 mm. Hg instead of the usual normal 17 mm. Hg. Consequently, in this instance enough fluid flows into the tissue spaces to raise the interstitial fluid pressure to +8 mm. Hg. This is far above the atmospheric pressure, and therefore causes progressive enlargement of the tissue spaces with tremendous expansion of the extracellular fluid volume.

CAUSES OF INCREASED CAPILLARY PRESSURE. Increased capillary pressure can result from any clinical condition that causes either venous ob-lost each day, and the colloid osmotic pressure of the plasma may fall to one-half normal or even less. This results in severe edema, and the edema itself is likely to contribute to death by means that are discussed later in the chapter.

Finally, persons who do not have sufficient protein in their diets are unable to form adequate quantities of plasma protein and, therefore, are likely to develop protein deficiency edema, which is called *nutritional edema*. This occurs frequently in war-torn areas.

Lymphatic Obstruction as a Cause of Edema. A small amount of protein filters continually from the capillaries into the tissue spaces, but this protein cannot be reabsorbed into the circulatory system through the capillary membrane. The only route by which the protein can be returned to the circulatory system is through the lymphatics. If the lymphatic drainage from any area of the body becomes blocked, more and more protein collects in the local tissue spaces until finally the concentration of this protein may approach the concentration of protein in the plasma. As shown in Figure 21–6C, the colloid osmotic pressure of the tissue fluids may theoretically rise to as high as 28 mm. Hg, and, to balance this, fluid collects in the tissues until the interstitial fluid pressure rises to a value equal to the capillary pressure, about 17 mm. Hg. Such elevated tissue pressure rapidly expands the tissue spaces, with resultant edema of the severest kind.

CAUSES OF LYMPHATIC OBSTRUCTION. One of the most common causes of lymphatic obstruction is *filariasis* — that is, infection by nematodes of the superfamily Filarioidea. The disease is widespread in the tropics, where larvae (microfilariae) are transmitted to human hosts by mosquitoes. The larvae pass out of the capillaries into the interstitial fluid and then by way of the lymph into the lymph nodes. Subsequent inflammatory reactions progressively obstruct the lymphatic channels of these nodes with scar tissue. After several years, the lymphatic drainage from one of the peripheral parts of the body may become almost totally occluded. Thus, a leg can swell to such a size that it might weigh as much as all the remainder of the body. Because of this extreme degree of edema, the swollen condition is frequently called *elephantiasis*. A very interesting type of elephantiasis is that which occasionally occurs in the scrotum, which has been known to enlarge so much that the person must carry it in a wheelbarrow in order to move about.

Lymphatic obstruction also occurs following operations for removal of cancerous tissue. Because the lymph nodes draining a cancerous area of the body must be removed in order to prevent possible spread of the cancer, the return of lymph to the circulatory system from that area will be blocked. Occasionally a radical mastectomy for removal of a cancerous breast causes the corresponding arm to swell to as much as twice its normal size, but usually the swelling regresses during the following two to three months as new lymph channels develop.

Increased Permeability of the Capillaries as a Cause of Edema. Figure 21–6D illustrates a capillary whose membrane has become so permeable that even protein molecules pass from the plasma into the interstitial spaces with ease. The protein content of the plasma decreases while that of the interstitial spaces increases. In the example of the figure the tissue pressure rises to +7 mm. Hg in order to balance the changes in plasma and tissue colloid osmotic pressure occasioned by the leakage of protein. The elevated interstitial fluid pressure in turn causes progressive edema.

CAUSE OF INCREASED CAPILLARY PERMEABILITY. Capillaries become excessively permeable when any factor destroys the integrity of the capillary endothelium. Burns are a frequent cause of increased permeability of the capillaries because overheated capillaries become friable, and their pores enlarge. Allergic reactions also frequently cause the release of histamine or various polypeptides that damage the capillary membranes and cause increased permeability.

A bacterial toxin produced by *Clostridium oedematiens* can often cause such extreme increase in capillary permeability that plasma loss into the tissues kills the patient within a few hours.

Recently it has been shown that overincreasing the pressure in the capillaries can stretch the capillary walls so greatly that the pores themselves enlarge. This is called the *stretched pore phenomenon*. Even a few minutes of overstretching can cause the condition, and at least 4 to 10 hours are required for the pores to return to normal condition.

EDEMA CAUSED BY KIDNEY RETENTION OF FLUID

When the kidney fails to excrete adequate quantities of urine, and the person continues to drink normal amounts of water and eat normal amounts of electrolytes, the total amount of extracellular fluid in the body progressively increases. This fluid is absorbed from the gut into the blood and elevates the capillary pressure a slight amount. This in turn causes most of the fluid to pass into the interstitial fluid spaces, elevating the interstitial fluid pressure there as well. As more and more fluid is retained, both the capillary pressure and the tissue pressure rise concurrently, causing an increase especially in tissue fluid but also some increase in plasma volume. Therefore, simple retention of fluids by the kidneys can also result in extensive edema. Furthermore, if the retained fluids are mainly water, which contains few electrolytes, intracellular edema also results, as will be discussed in Chapter 33.

PULMONARY EDEMA

Almost exactly the same principles apply to the development of pulmonary edema as to the development of edema in the peripheral tissues. However, there is a significant quantitative difference because the normal pulmonary capillary pressure is about 7 mm. Hg in contrast to 17 mm. Hg capillary pressure in the peripheral tissues. Therefore, the lungs have considerably more of a safety factor against edema than do the other tissues of the body, as will be discussed in Chapter 23 in relation to the pulmonary circulation.

TOTAL TISSUE PRESSURE AND SOLID TISSUE PRESSURE

Throughout the present chapter and the preceding chapter, the term *interstitial fluid pressure* has been used to designate the pressure of the *fluid* in the minute tissue spaces. This pressure has also been called simply *tissue pressure* by many physiologists. However, another pressure has also been called tissue pressure, which we will call here "*total tissue pressure.*"

Total tissue pressure is the summated pressure of all the elements of the tissue. And it can be greatly different from the interstitial fluid pressure.

One can envisage the difference between total tissue pressure and interstitial fluid pressure by imagining a suitcase filled tightly with clothes and yet with a vacuum applied to the air within the suitcase. The pressure of the air between the clothes would be negative while the pressure of the solid elements (the clothes themselves) against each other would be positive. Thus, the pressure of the solid elements of the tissue (the cells, fibers, and so forth) against each other can be called *solid tissue pressure*. And the sum of the "solid" and "fluid" tissue pressures is the total tissue pressure.

Significance of Total Tissue Pressure. The particular importance of total tissue pressure is that it is this pressure that presses against the blood vessels and that tends to collapse these vessels. Thus, there is normally approximately +2 mm. Hg pressure of the solid elements plus the fluid of the tissue against the outside of the vessels tending to collapse them. When a blood pressure cuff is inflated around the arm, pressure in the cuff is transmitted to the blood vessels by way of both the fluid and the solid elements and not only by way of the fluid between these solid elements. Under these conditions the total tissue pressure pressing on the outsides of the blood vessels may be as great as several hundred mm. Hg.

Condition in Which Interstitial Fluid Pressure Equals Total Tissue Pressure. Once the tissue becomes edematous and the solid elements are spread apart from each other, the interstitial fluid pressure then becomes equal to the total tissue pressure. That is, the solid elements are no longer in contact with each other so that there is no longer any solid tissue pressure. However, as fluid is removed from the tissues, the solid elements begin to crowd together, and the total pressure then becomes some value greater than the interstitial fluid pressure.

Measurement of Total Tissue Pressure—The Balloon and Needle Methods. Total tissue pressure has been measured in several different ways and is +1 to +3 mm. Hg. This is in contrast to the −7 mm. Hg for interstitial fluid pressure.

The most easily understandable method for measuring total tissue pressure is to implant a minute flaccid balloon into a tissue and then measure the pressure that is required barely to begin filling the balloon. Since both fluid and solid elements compress the outside of this balloon, the pressure required to push the walls of the balloon apart is equal to the total tissue pressure, and the value measured in this way is +1 to +3 mm. Hg.

However, the method that has been used most frequently in the past to measure total tissue pressure has been to insert a small needle into the tissue and then apply increasing pressure to the fluid in the needle until it barely begins to move into the tissue. Here again, this pressure measures to be about +1 to +3 mm. Hg. If flow of fluid through the needle is stopped for more than 10 to 15 seconds, the pressure measured through the needle is likely to become progressively negative or to become very erratic because the tip of the needle becomes blocked. Therefore, it is essential to make this pressure measurement within a few seconds after a small amount of fluid has moved into the tissue.

Calculation of Solid Tissue Pressure. It has not been possible to measure solid tissue pressure, but one can calculate this by subtracting interstitial fluid pressure from total tissue pressure. Assuming that the average total tissue pressure is +2 mm. Hg and that the average interstitial fluid pressure is −7 mm. Hg, one then calculates the average solid tissue pressure to be +9 mm. Hg. That is, the solid elements of the tissues are compacted against each other with a force of +9 mm. Hg. Obviously, this solid tissue pressure is the summation of large numbers of minute forces exerted at the contact points between the solid elements of the tissues. (If these solid contact points cover a significant proportion of the surface area of the tissue spaces, the equation for the interrelationship between interstitial fluid pressure, solid tissue pressure, and total tissue pressure has to be modified slightly because the interstitial fluid pressure then acts only on part of interstitial space surface rather than on all of it.)

Significance of Solid Tissue Pressure. Solid tissue pressure is the force that tends to deform the solid elements of the tissues. For instance, it is this pressure that causes cells to become contorted so that the cells will fill the minute interstices. In edematous tissues, where there is no solid tissue pressure, the cells tend to be rounded, an effect well known to the pathologist. On the other hand, in normal tissues the cells tend to have multi-faceted shapes to allow better fitting into the tissue spaces.

THE PRESENCE AND IMPORTANCE OF GEL IN THE INTERSTITIAL SPACES

Up to this point we have talked about interstitial fluid as if it were entirely in a mobile state. But, both biochemical and physical studies indicate that under normal conditions almost all of the interstitial fluid is held in a gel that fills the spaces between the cells. This gel contains large quantities of mucopolysaccharides, the most abundant of which is hyaluronic acid.

Most of the principles of interstitial fluid hemodynamics still hold true even though most of the fluid is in a gel state. For instance, diffusion of small molecules can occur through the gel perhaps 95 per cent as effectively as in free fluid. Therefore, nutrients can diffuse from the capillaries to the cells almost equally as well through the gel as through free fluid.

On the other hand, the fluid in the gel cannot be readily moved from one part of the tissue to another part. That is, the gel structure holds the interstitial fluid in place, except for minute rivulets of free interstitial fluid that move along the surfaces of the gel between the gel and the cells.

Relation of Edema Fluid to the Gel. When edema occurs, a small portion of the edema fluid is probably imbibed by the gel, but once the extracellular volume has increased beyond approximately 20 to 30 per cent, large quantities of freely moving fluid appear in the tissues. When this occurs, pressure on one tissue area will cause fluid to move freely to another area, which is the phenomenon called *pitting.* Also, the fluid flows downward in the tissues because of the pull of gravity, giving rise to the phenomenon called *dependent edema;* that is, edema in the lower parts of the body.

Theoretical Importance of Having a Gel in the Interstitial Spaces. One can readily understand that if all cells abutted directly onto each other without any interstitial spaces, those cells not directly in contact with capillaries could receive nutrition only through other cells; any many substances, such as glucose, amino acids, and so forth, would not be able to reach the outlying cells in adequate quantities. At the opposite extreme, the presence of tremendous quantities of extracellular fluid, with consequent very large tissue spaces, would increase the distances from the capillaries to the outlying cells so much that again the diffusion of nutrients to outlying cells would be diminished. Obviously, therefore, there is a certain interstitial fluid space size for optimum transport of substances from the capillaries to the cells. It has been postulated that the function of the gel in the interstitial spaces is to create interstitial fluid spaces of approproate size for optimum cell nutrition to occur.

Relationship of Interstitial Fluid Gel to the Regulation of Interstitial Fluid Volume. Since approximately 17 per cent of the average tissue is composed of interstitial fluid but almost none of this is in the mobile state, one can derive the following theory for the regulation of interstitial fluid volume. The mechanism previously described for creating a negative pressure in the tissue spaces is actually a "drying" mechanism that attempts all the time to remove any excess fluid that appears in the tissues. Thus, essentially all mobile fluid would be removed. The question remains then, why does this drying mechanism not also remove fluid from the gel? The answer to this is that the gel has an osmotic pressure caused by the Donnan equilibrium effect: that is, the gel lattice has electrostatic charges that hold free ions within the gel. These ions in turn cause osmosis of water into the gel. The quantity of mucopolysaccharides in tissue gel is sufficient to give an osmotic pressure in the gel of about 9 mm. Hg. Therefore, the gel can resist dehydration caused by the -7 mm. Hg pressure in the free fluid in the tissue.

REFERENCES

Abramson, D. I. (ed.): Blood Vessels and Lymphatics. New York, Academic Press, Inc., 1962.

Allen, L.: Lymphatics and lymphoid tissues. *Ann. Rev. Physiol.,* 29:197, 1967.

Biophysics and physical chemistry of connective tissues (Symposium). *Fed. Proc.,* 25:951 and 1799, 1966.

Drinker, C. K.: The Lymphatic System. Stanford, Stanford University Press, 1942.

Gauer, O. H.; Henry, J. P.; and Behn, C.: The regulation of extracellular fluid volume. *Ann. Rev. Physiol., 32*:547, 1970.

Guyton, A. C.: Interstitial fluid pressure: II. Pressure-volume curves of interstitial space. *Circ. Res., 16*:452, 1965.

Guyton, A. C.: Interstitial fluid pressure-volume relationships and their regulation. *In* Wolstenholme, G. E. W., and Knight, J. (eds.): CIBA Foundation Symposium on Circulatory and Respiratory Mass Transport. London, J. & A. Churchill, Ltd., 1969, p. 4.

Guyton, A. C.: Introduction to Part I: Pulmonary alveolar-capillary interface and interstitium. *In* Fishman, A. P., and Hecht, H. H. (eds.): The Pulmonary Circulation and Interstitial Space. Chicago, University of Chicago Press, 1969, p. 3.

Guyton, A. C.: Pressure-volume relationships in the interstitial spaces. *Invest. Ophthal., 4*:1075, 1965.

Guyton, A. C., and Coleman, T. G.: Regulation of interstitial fluid volume and pressure. *Ann. N.Y. Acad. Sci.,150*:537, 1968.

Guyton, A. C., and Lindsey, A. W.: Effect of elevated left atrial pressure and decreased plasma protein concentration on the development of pulmonary edema. *Circ. Res., 7*:649, 1959.

Haagensen, C. D., et al.: The Lymphatics in Cancer. Philadelphia, W. B. Saunders Company, Vol. I. In press.

Leak, L. V., and Burke, J. F.: Ultrastructural studies on the lymphatic anchoring of filaments. *J. Cell Biol., 36*:129, 1968.

Mayerson, H. S.: The physiologic importance of lymph. *In* Handbook of Physiology. Baltimore, The Williams & Wilkins Co., 1963, Sec. II, Vol. II, p. 1035.

McMaster, P. D.: The lymphatic system. *Ann. Rev. Physiol., 5*:207, 1943.

Rosenthal, S. M.: Production and measurement of traumatic swelling. *Amer. J. Physiol., 216*:630, 1969.

Rusznyak, I.; Foldi, M.; and Szabo, G.: Lymphatics and Lymph Circulation: Physiology and Pathology. New York, The Macmillan Company, 1960.

Starling, E. H.: The influence of mechanical factors on lymph production. *J. Physiol. (Lond.), 16*:224, 1894.

Uhley, H. N.; Leeds, S. E.; Sampson, J. J.; and Friedman, M.: Right duct lymph flow in experimental heart failure following acute elevation of left atrial pressure. *Circ. Res., 20*:306, 1967.

Visscher, M. B.; Haddy, F. J.; and Stephens, G.: The physiology and pharmacology of lung edema. *Pharm. Rev., 8*:389, 1956.

Yoffey, J. M., and Courtice, F. C.: Lymphatics, Lymph, and Lymphoid Tissue. Baltimore, The Williams & Wilkins Co., 1967.

CHAPTER 22

The Special Fluid Systems of the Body—Cerebrospinal, Ocular, Pleural, Pericardial, Peritoneal, and Synovial

Several special fluid systems exist in the body, each performing functions peculiar to itself. For instance, the cerebrospinal fluid supports the brain in the cranial vault, the intraocular fluid maintains distention of the eyeballs so that the optical dimensions of the eye remain constant, and the potential spaces, such as the pleural and pericardial spaces, provide lubricated chambers in which the internal organs can move. All these fluid systems have characteristics that are similar to each other and that are also similar to those of the interstitial fluid system. However, they are also sufficiently different that they require special consideration.

THE CEREBROSPINAL FLUID SYSTEM

The entire cavity enclosing the brain and spinal cord has a volume of approximately 1650 ml., and about 140 ml. of this volume is occupied by cerebrospinal fluid. This fluid, as shown in Figure 22–1, is found in the ventricles of the brain, in the cisterns around the brain, and in the subarachnoid space around both the brain and the spinal cord. All these chambers are connected with each other, and the pressure of the fluid is regulated at a constant level.

CUSHIONING FUNCTION OF THE CEREBROSPINAL FLUID

A major function of the cerebrospinal fluid is to cushion the brain within its solid vault. Were it not for this fluid, any blow to the head would cause the brain to be juggled around and severely damaged. However, the brain and the cerebrospinal fluid have approximately the same specific gravity, so that the brain simply floats in the fluid. Therefore, blows on the head move the entire brain simultaneously, causing no one portion of the brain to be momentarily contorted by the blow.

Contrecoup. When a blow to the head is extremely severe, it usually does not damage the brain on the side of the head where the blow is struck, but, instead, on the opposite side. This phenomenon is known as "contrecoup," and the reason for this effect is the following: When the blow is struck, the fluid on the struck side is so incompressible that, as the skull moves, the fluid pushes the brain at the same time. However, on the opposite side, the sudden movement of the skull causes it to pull away from the brain momentarily, creating for a short instant a vacuum in the cranial vault at this point. Then, when the skull is no longer being accelerated by the blow, the vacuum suddenly collapses and the brain strikes the inner surface of the skull. For instance, the damage to the brain of a boxer usually does not occur in the frontal regions but, instead, in the occipital regions.

Cerebral Damage from Rotational Blows. Even though the cerebrospinal fluid can cushion the brain against linear blows, it is not capable of preventing damage from *rotational acceleration* of the

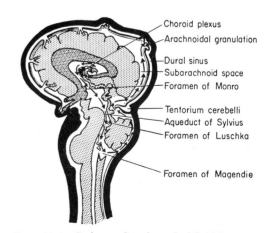

Figure 22–1 Pathway of cerebrospinal fluid flow. (Modified from Ranson and Clark: Anatomy of the Nervous System.)

Choroid plexus
Arachnoidal granulation
Dural sinus
Subarachnoid space
Foramen of Monro
Tentorium cerebelli
Aqueduct of Sylvius
Foramen of Luschka
Foramen of Magendie

skull. The brain, like all objects that have mass, has inertia. Therefore, when the skull is suddenly rotated, the brain tends to remain stationary inside the cranial vault. This causes a shearing effect where the brain is attached to the skull, sometimes resulting in tearing of the arachnoidal tissues or even rupture of the epiploic blood vessels passing from the substance of the brain through the meninges into the skull. This ability of rotational movement to damage the surface structures of the brain probably explains why sudden rotational movements of the head when a person has a headache are likely to aggravate the headache, and it possibly also explains why a rotational blow to the head during a boxing match, such as an uppercut or a sidewise blow to the chin, is likely to knock out the opponent even though a linear blow fails.

FORMATION OF CEREBROSPINAL FLUID

Cerebrospinal fluid is formed in several different ways; the major portion is formed by the choroid plexuses in the ventricles, smaller portions by the blood vessels of the meningeal and ependymal linings of the cerebrospinal fluid chambers, and still smaller portions by the blood vessels of the brain and spinal cord.

Formation of Cerebrospinal Fluid by the Choroid Plexus. The choroid plexus, which is illustrated in Figure 22-2, is a cauliflower growth of blood vessels covered by a thin coat of epithelial cells. This plexus projects into (a) the temporal horns of the lateral ventricles, (b) the posterior portions of the third ventricle, and (c) the roof of the fourth ventricle.

Cerebrospinal fluid continually exudes from the surface of the choroid plexus. This fluid is not

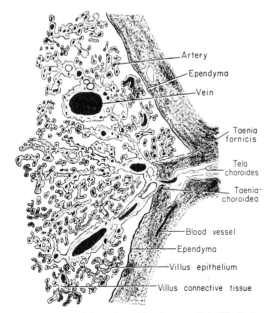

Figure 22-2 The choroid plexus. (Modified from Clara: Das Nervensystem des Menschen. Barth.)

exactly like the other extracellular fluids. Instead, its concentration of sodium is 7 per cent greater than the concentration in extracellular fluid; its glucose concentration is 30 per cent less and potassium 40 per cent less. Obviously, then, the fluid arising from the choroid plexus is not a simple filtrate from the capillaries, but a *choroid secretion*.

The probable mechanism by which the choroid plexus secretes fluid is the following: The cuboidal epithelial cells of the choroid plexus actively secrete sodium ions, which develop a positive charge in the cerebrospinal fluid. This in turn pulls negatively charged ions, particularly chloride ions, also into the cerebrospinal fluid. Thus an excess of ions develops in the fluids of the ventricles. As a result, the osmotic pressure of the ventricular fluid is elevated to approximately 160 mm. Hg greater than that of the plasma, and this osmotic force causes large quantities of water and other dissolved substances to move through the choroidal membrane into the cerebrospinal fluid. Since glucose is not as diffusible as water, its concentration remains somewhat low. The low concentration of potassium is probably caused by potassium transport in the opposite direction through the epithelial cells. The major basis for this osmotic theory is that the concentration of osmotically active substances in the cerebrospinal fluid is 9 milliosmols greater than that of plasma; this difference would exert the 160 mm. Hg osmotic force noted above.

The rate of choroidal secretion is estimated to be about 750 ml. each day, which is about five times as much as the total volume of fluid in the entire cerebrospinal cavity.

Diffusion into the Cerebrospinal Fluid Through the Meningeal and Ependymal Surfaces. The surfaces of the ventricles are lined with a thin cuboidal epithelium called the *ependyma,* and the cerebrospinal fluid is in contact with this surface at all points. In addition, cerebrospinal fluid fills the *subarachnoid space* between the *pia mater* that covers the brain and the *arachnoidal membrane.* The cerebrospinal fluid, therefore, is in contact with larger surface areas of ependyma and meninges, and diffusion occurs continually between the cerebrospinal fluid and the brain substance beneath the ependyma and also between the cerebrospinal fluid and the blood vessels of the meninges, especially the subarachnoidal vessels. Some substances diffuse through these membranes extremely rapidly, but others diffuse very poorly. For instance, lipid-soluble substances, such as alcohol, diffuse so rapidly that equilibrium occurs between alcohol in the cerebrospinal fluid and alcohol in the substance of the brain within less than an hour; other substances such as proteins, sucrose, and so forth, may never reach equilibrium between their concentrations in the cerebrospinal fluid and in the interstitial fluid of the brain substance.

The Perivascular Spaces and Cerebrospinal Fluid. The blood vessels to the substance of the brain pass first along the surface of the brain and then

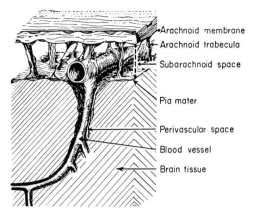

Arachnoid membrane
Arachnoid trabecula
Subarachnoid space
Pia mater
Perivascular space
Blood vessel
Brain tissue

Figure 22–3 Drainage of the perivascular spaces into the subarachnoid space. (From Ranson and Clark: Anatomy of the Nervous System.)

penetrate inward, carrying a layer of *pia mater* with them, as shown in Figure 22–3. The pia is only loosely adherent to the vessels, so that a space, the *perivascular space,* exists between it and each vessel. Perivascular spaces follow both the arteries and the veins into the brain as far as the arterioles and venules but not to the capillaries.

THE LYMPHATIC FUNCTION OF THE PERIVASCULAR SPACES. As is true elsewhere in the body, a small amount of protein leaks out of the parenchymal capillaries into the interstitial spaces of the brain, and since no true lymphatics are present in brain tissue, this protein leaves the tissues mainly through the perivascular spaces but partly also by direct diffusion through the pia mater into the subarachnoid spaces. On reaching the subarachnoid spaces, the protein flows along with the cerebrospinal fluid to be absorbed through the *arachnoidal villi* into the cerebral veins, as is discussed subsequently. Therefore, the perivascular spaces, in effect, are a modified lymphatic system for the brain.

In addition to transporting fluid and proteins, the perivascular spaces also transport extraneous particulate matter from the brain into the subarachnoid space. For instance, whenever infection occurs in the brain, dead white blood cells are carried away through the perivascular spaces.

ABSORPTION OF CEREBROSPINAL FLUID— THE ARACHNOIDAL VILLI

Almost all the cerebrospinal fluid formed each day is reabsorbed into the blood through special structures called *arachnoidal villi* or *granulations,* which project from the subarachnoid spaces into the venous sinuses of the brain and, rarely, also into the veins of the spinal canal. The arachnoidal villi are actually arachnoidal trabeculae that protrude through the venous walls, resulting in extremely permeable areas that allow relatively free flow of cerebrospinal fluid as well as protein molecules and even small particles (less than 1 micron in size) into the blood.

FLOW OF FLUID IN THE CEREBROSPINAL FLUID SYSTEM

The main channel of fluid flow from the choroid plexuses of the ventricles to the arachnoidal villi is illustrated in Figure 22–1. Fluid formed in the lateral ventricles passes into the third ventricle through the *foramina of Monro,* combines with that secreted in the third ventricle, and passes along the *aqueduct of Sylvius* into the fourth ventricle, where still more fluid is formed. It then passes into the *cisterna magna* through two lateral *foramina of Luschka* and a midline *foramen of Magendie.* From here it flows through the *subarachnoid spaces* upward toward the cerebrum, where almost all of the arachnoidal villi are located; but first the fluid must pass through the subarachnoid spaces of the small *tentorial opening* around the mesencephalon. Finally, the fluid reaches the arachnoidal villi and empties into the venous sinuses.

CEREBROSPINAL FLUID PRESSURE

The normal pressure in the cerebrospinal fluid system when one is lying in a horizontal position averages 130 mm. water (10 mm. Hg), though this may be as low as 70 mm. water or as high as 180 mm. water even in the normal person. These values are considerably greater than the −7 mm. Hg pressure in the interstitial spaces elsewhere in the body.

The cerebrospinal fluid pressure is regulated by the product of, first, the *rate of fluid formation* and, second, the *resistance to absorption through the arachnoidal villi.* When either of these is increased, the pressure rises; and when either is decreased, the pressure falls.

Cerebrospinal Fluid Pressure in Pathologic Conditions of the Brain. Often a large *brain tumor* elevates the cerebrospinal fluid pressure by decreasing the rate of absorption of fluid. For instance, if the tumor is above the tentorium and becomes so large that it compresses the brain downward, the upward flow of fluid through the tentorial opening may become blocked and the absorption of fluid greatly curtailed. As a result, the cerebrospinal fluid pressure can rise to as high as 500 mm. water (37 mm. Hg) or more. Also, if the tumor is near the surface of the brain, it can cause inflamed meninges, which then exude large quantities of fluid and protein into the cerebrospinal fluid, thereby raising the pressure in this way as well.

The pressure also rises considerably when *hemorrhage* or *infection* occurs in the cranial vault. In both of these conditions, large numbers of cells suddenly appear in the cerebrospinal fluid, and these can almost totally block absorption through the arachnoidal villi. This sometimes elevates the cerebrospinal fluid pressure to as high as 400 to 600 mm. water.

Many babies are born with very high cerebrospinal fluid pressure. Most frequently it is caused by excess formation of fluid by the choroid plexuses, but abnormal absorption is also an occasional cause. This is discussed later in connection with *hydrocephalus*.

Measurement of Cerebrospinal Fluid Pressure. The usual procedure for measuring cerebrospinal fluid pressure is the following: First, the subject lies on his side so that the spinal fluid pressure is equal to the pressure in the cranial vault. A spinal needle is then inserted into the lumbar spinal canal below the lower end of the cord and is connected to a glass tube. The spinal fluid is allowed to rise in the tube as high as it will. If it rises to a level 100 mm. above the level of the needle, the pressure is said to be 100 mm. water pressure or, dividing this by 13.6, which is the specific gravity of mercury, about 7.5 mm. Hg pressure.

Effect of High Cerebrospinal Fluid Pressure on the Optic Disc — Papilledema. Anatomically the dura of the brain extends as a sheath around the optic nerve and then connects with the sclera of the eye. When the pressure rises in the cerebrospinal fluid system, it also rises in the optic nerve sheath. The retinal artery and vein pierce this sheath a few millimeters behind the eye and then pass with the optic nerve into the eye itself. The high pressure in the optic sheath impedes the flow of blood in the retinal veins, thereby also increasing the retinal capillary pressure throughout the eye, which results in retinal edema. The tissues of the optic disc are much more distensible than those of the remainder of the retina, so that the disc becomes far more edematous than the remainder of the retina and swells into the cavity of the eye. The swelling of the disc, which can be observed with an ophthalmoscope, is called *papilledema*, and neurologists can estimate the cerebrospinal fluid pressure by assessing the extent to which the optic disc protrudes into the eyeball.

OBSTRUCTION TO THE FLOW OF CEREBROSPINAL FLUID

Hydrocephalus. "Hydrocephalus" means excess water in the cranial vault. This condition is frequently divided into *communicating hydrocephalus* and *noncommunicating hydrocephalus*. In communicating hydrocephalus fluid flows readily from the ventricular system into the subarachnoid space, whereas in noncommunicating hydrocephalus fluid flow out of one or more of the ventricles is blocked.

Usually the noncommunicating type of hydrocephalus is caused by a block in the aqueduct of Sylvius so that as fluid is formed by the choroid plexuses of the two lateral and the third ventricles, the volumes of these three ventricles increase greatly. This flattens the brain into a thin shell against the skull. In newborn babies the increased pressure also causes the whole head to swell. This type of hydrocephalus can occasionally be treated satisfactorily by making an artificial opening between the ventricular system and the subarachnoid space, such as between the third ventricle and the interpeduncular cistern, or through a plastic tube from one of the ventricles into the cisterna magna.

The communicating type of hydrocephalus is usually caused by overdevelopment of the choroid plexuses in the newborn infant. As a result, far more fluid is formed than can re-enter the venous system through the arachnoidal villi. Fluid therefore collects both inside the ventricles and on the outside of the brain, causing the head to swell tremendously and often damaging the brain severely. To correct this condition the choroid plexuses in both lateral ventricles and sometimes in the third ventricle are destroyed. Thereafter, far less fluid is formed, the cerebrospinal fluid pressure falls, and the hydrocephalus does not become any worse.

Blockage of Fluid Flow in the Spinal Canal. Only small quantities of cerebrospinal fluid can be absorbed into the lymphatics and capillaries of the spinal canal. Instead, most of the fluid formed in the spinal regions flows all the way to the arachnoidal granulations in the cranial vault to be absorbed. Consequently, when the spinal canal is blocked by a tumor, a blood clot, or cord transection, the spinal fluid pressure below the block usually rises from the normal value of 130 mm. water to as high as 300 to 500 mm. water, even though the pressure above the block, measured by passing a spinal needle beneath the skull into the cisterna magna, remains normal.

Another important aid for diagnosing spinal canal block is *Queckenstedt's test*. To perform this, the observer presses tightly on both sides of the neck to compress the internal jugular veins. This temporarily impedes the return of blood from the venous system of the brain and causes the venous sinuses in the cranial vault to swell. The swelling sinuses in turn press against the cerebrospinal fluid to increase the cerebrospinal fluid pressure. If the spinal canal is not blocked, the pressure in the lower canal, as measured by a spinal fluid manometer, immediately rises to double or triple normal. On the other hand, if the canal is blocked, the pressure does not change.

THE BLOOD-CEREBROSPINAL FLUID AND BLOOD-BRAIN BARRIERS

It has already been pointed out that the constituents of the cerebrospinal fluid are not exactly the same as those of the extracellular fluid elsewhere in the body. In addition, many large molecular substances hardly pass at all from the blood into the cerebrospinal fluid or into the interstitial fluids of the brain even though these same substances pass readily into the usual interstitial fluids of the body. Therefore, it is said that barriers, called the *blood-cerebrospinal* and *blood-brain barriers*, exist between the blood and the cerebrospinal fluid and brain fluids, respectively. These barriers exist in the choroid plexus and in essentially all areas of the brain parenchyma except the hypothalamic region. It is perhaps significant that the hypothalamus is peculiarly sensitive to alterations in extracellular fluid constituents.

In general, the blood-cerebrospinal fluid and blood-brain barriers are highly permeable to water, carbon dioxide, and oxygen, slightly permeable to the electrolytes, such as sodium, chloride, and potassium, and almost totally impermeable to substances such as arsenic, sulfur and gold. Though these latter substances are not of physiological importance, they are important occasionally for certain types of drug therapy, the blood-cerebrospinal fluid and blood-brain barriers often making it impossible to achieve effective concentrations of some such drugs in the cerebrospinal fluid or parenchyma of the brain.

The cause of the low permeability of the blood-cerebrospinal barrier is the almost total impermeability of the secretory cells of the choroid plexus. All substances entering the cerebrospinal fluid must be transported through the secretory cells in addition to passing through the capillary membranes of the choroid plexus.

The cause of the low permeability of the blood-brain barrier is yet unknown. Some physiologists believe that many small "feet" from glial cells envelop the capillaries in the brain and prevent easy passage of ions and other substances into the brain parenchyma. However, other physiologists believe that the capillary membranes themselves are less permeable in the brain than elsewhere in the body.

Interstitial Fluid of the Brain. Research studies have suggested that the interstitial fluid in brain tissue is about 12 per cent of the tissue weight, in contrast to 17 per cent elsewhere in the body. The cause of this is lack of collagen fibers between the cells.

THE INTRAOCULAR FLUID

The eye is filled with intraocular fluid which maintains sufficient pressure in the eyeball to keep it distended. Figure 22–4 illustrates that this fluid can be divided into two portions, the *aqueous humor,* which lies in front and to the sides of the

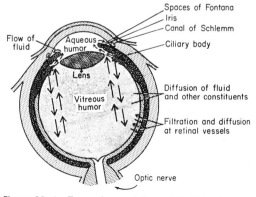

Figure 22–4   Formation and flow of fluid in the eye.

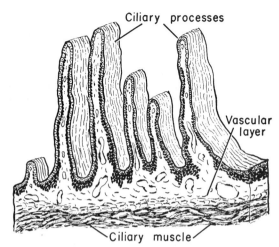

Figure 22–5   Anatomy of the ciliary processes.

lens, and the *vitreous humor,* which lies between the lens and the retina. The aqueous humor is a freely flowing clear fluid, whereas the vitreous humor, sometimes called the *vitreous body,* is a gelatinous mass held together by a fine fibrillar network. Substances can *diffuse* slowly in the vitreous humor, but there is little *flow* of fluid.

Aqueous humor is continually being formed and reabsorbed. The balance between formation and reabsorption of aqueous humor regulates the total volume and pressure of the intraocular fluid.

FORMATION OF AQUEOUS HUMOR BY THE CILIARY BODY

Aqueous humor is formed in the human eye *at an average rate of about 12 cubic millimeters each minute.* Essentially all of this is secreted by the *ciliary processes,* which are linear folds projecting from the *ciliary body* into a space called the *posterior chamber* that is located all around the eye immediately behind the lens ligaments. A cross-section of these ciliary processes is illustrated in Figure 22–5, and their relationship to the fluid chambers of the eye can be seen in Figure 22–4. Because of their folding architecture, the total surface area of the ciliary processes is approximately 6 sq. cm. in each eye, a large area, considering the small size of the ciliary body. The surfaces of these processes are covered by a single layer of epithelial cells, and immediately beneath these is a highly vascular area.

Aqueous humor is formed by the ciliary processes in much the same manner that cerebrospinal fluid is formed by the choroid plexus. The theoretical mechanism is the following: It is believed that the ciliary epithelium actively secretes sodium ions into the aqueous humor, and chloride and bicarbonate ions are probably pulled along with the sodium by the positive charge of the sodium. These substances increase the osmolar activity of the aqueous humor, which in turn causes osmosis of large amounts of water through the ciliary epithelium to the inside of the eye. The reasons for postulating this mechanism are: First, the osmolar activity of the aqueous humor is 3 to 6 milliosmols greater than that of plasma, which could exert an osmotic pressure at the ciliary membrane of 50 to 100 mm. Hg, far more than required to promote adequate transfer of water through the membrane. Second, direct studies of the epithelial membrane of the ciliary processes have demonstrated active transport of sodium from the plasma into the aqueous humor, thus providing the necessary osmotic activity to cause osmosis of water. A few other substances are also actively transported, including especially the amino acids.

DIFFUSION INTO INTRAOCULAR FLUID FROM OTHER STRUCTURES OF THE EYE

In addition to the capillaries of the ciliary body, there are three other capillary systems in the eye: that of the choroid, that of the retina, and that of the iris. In so far as is known, the capillaries of the choroid and retina function purely for nutrition of the retina, similar to capillary function elsewhere in the body. However, small amounts of the fluid from these sources diffuse into the aqueous humor.

The capillary system of the iris also is involved principally with nutrition of the iridic structures, but because of the location of the iris in the very middle of the aqueous humor, the rates of diffusion of substances both into and out of the aqueous humor from the iris are rapid, accounting for as much as 20 to 70 per cent of some of the ions that enter the aqueous humor. In addition, the iris is covered with an epithelium that is capable of phagocytizing proteins and small particulate matter from the aqueous humor, thereby helping to maintain a perfectly clear fluid.

OUTFLOW OF AQUEOUS HUMOR FROM THE EYE

After aqueous humor is formed by the ciliary processes, it flows, as shown in Figure 22–4, *between the ligaments of the lens*, then *through the pupil*, and finally *into the anterior chamber of the eye*. Here, the fluid flows into the *angle between the cornea and the iris* and thence through a latticework of *trabeculae*, finally into the *canal of Schlemm*. Figure 22–6 illustrates the anatomical structures at the irido-corneal angle, showing that the spaces between the trabeculae extend all the way from the anterior chamber to the canal of Schlemm. The canal of Schlemm in turn is a thin-walled vein that extends circumferentially all the way around the eye. Its endothelial membrane is so porous that even large protein molecules, as well as small particulate matter, can pass from the anterior chamber into the canal of Schlemm. Even though the canal of Schlemm is actually a venous blood vessel, so much aqueous humor normally flows into it that it is filled only with aqueous humor rather than with blood. Also, the small

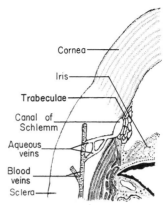

Figure 22–6 Anatomy of the irido-corneal angle, showing the system for outflow of aqueous humor into the conjunctival veins.

veins that lead from the canal of Schlemm to the larger veins of the eye usually contain only aqueous humor, and these are called *aqueous veins.*

Removal of Protein from the Aqueous Humor. It has previously been pointed out that all extravascular fluid chambers of the body must have a means for removing protein; otherwise the colloid osmotic pressure will rise high enough to cause abnormal dynamics. In most areas of the body this function is subserved by the lymphatics, but in the eye the extreme permeability of the canal of Schlemm allows the proteins to flow directly into the venous system, which, obviously, is different from the method of protein return in any other structure of the body.

INTRAOCULAR PRESSURE

The average normal intraocular pressure, when measured through a small needle placed directly in the anterior chamber, is approximately 20 mm. Hg, with a range from 15 to 25 mm. Hg.

Tonometry. Because it is impractical to pass a needle into a patient's eye for measurement of intraocular pressure, this pressure is measured clinically by means of a tonometer, the principle of which is illustrated in Figure 22–7. The cornea of the eye is anesthetized with a local anesthetic, and the footplate of the tonometer is placed on the cornea. A small force is then applied to a central plunger, causing the central part of the cornea to be displaced inward. The amount of displacement is recorded on the scale of the tonometer, and this in turn is calibrated in terms of intraocular pressure.

Regulation of Intraocular Pressure. The intraocular pressure of the normal eye remains almost exactly constant throughout life, illustrating that the pressure regulating mechanism is very effective. But the precise operation of this mechanism is not clear. It is believed that the pressure is regulated mainly by the outflow resistance from the anterior chamber into the canal of Schlemm in the following way: A rise in pressure above normal

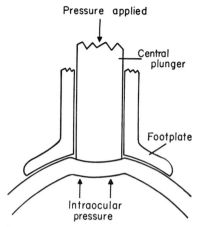

Figure 22–7 Principles of the tonometer.

supposedly distends the spaces between the trabeculae leading to the canal of Schlemm, thus allowing greatly increased outflow and consequently a return of the pressure to normal. On the other hand, a decrease in pressure below normal causes the spaces to narrow, causing retention of fluid until the pressure rises back to normal. One of the reasons for believing that intraocular pressure is regulated by some mechanism at the iridocorneal angle is that pathologic changes at this point almost always elevate the intraocular pressure.

Other factors that might also play important roles in intraocular pressure regulation are: (1) variations in the rate of formation of aqueous humor by the ciliary processes, (2) changes in resistance to flow of aqueous humor through the aqueous veins into the blood veins, and (3) changes in pressure in the blood veins where the aqueous veins empty.

Glaucoma.   Glaucoma is a disease of the eye in which the intraocular pressure becomes pathologically high, sometimes rising to as high as 80 to 90 mm. Hg. Pressures rising above as little as 30 to 40 mm. Hg can cause severe pain and even blindness. As the pressure rises, the retinal artery, which enters the eyeball at the optic disc, is compressed, thus reducing the nutrition to the retina. This often results in permanent atrophy of the retina and optic nerve with consequent blindness.

Glaucoma is one of the most common causes of blindness. Very high pressures lasting only a few days can at times cause total and permanent blindness, but, in cases with only mildly elevated pressures, the blindness may develop progressively over a period of many years.

In essentially all cases of glaucoma the abnormally high pressure results from increased resistance to fluid outflow at the irido-corneal junction. In most patients, the cause of this is unknown, but in others it results from infection or trauma to the eye. In these persons large quantities of blood, white blood cells, proteins, and other matter collect in the aqueous humor and then flow into the trabecular spaces of the irido-corneal angle where

they block the outflow of fluid, thereby greatly increasing the intraocular pressure.

Administration of *Diamox* is usually effective in reducing the pressure in glaucoma. This drug acts by reducing the rate of formation of aqueous humor, possibly because of a yet unexplained effect that reduces sodium secretion.

Sometimes, when all other therapeutic procedures fail to reinstitute adequate outflow of aqueous humor, a *trephine* operation is performed – that is, a small hole is made through the corneoscleral junction into the anterior chamber, a portion of the iris is removed, and the conjunctiva is sewed over the hole. Fluid thereafter leaks from the anterior chamber into the areolar tissue beneath the conjunctiva and is absorbed into the lymphatic channels.

FLUID CIRCULATION IN THE POTENTIAL SPACES OF THE BODY

Among the potential spaces of the body are the *pleural cavity,* the *pericardial cavity,* the *peritoneal cavity,* the cavity of the *tunica vaginalis* surrounding the testis, and the *synovial cavities* including both the joint cavities and the bursae. Normally, these are all empty except for a few milliliters of fluid that lubricates movement of the surfaces against each other. But in certain abnormal conditions the spaces can swell to contain tremendous quantities of fluid, which is the reason they are called "potential" spaces.

FLUID EXCHANGE BETWEEN THE CAPILLARIES AND THE POTENTIAL SPACES

The membrane of a potential space does not offer significant resistance to the passage of fluid and electrolytes back and forth between the space and the surrounding interstitial fluid. Consequently, fluid leaving a capillary adjacent to the membrane, as shown in Figure 22–8, not only diffuses into the interstitial fluid but also into the potential space. Likewise, fluid can diffuse back out of the space into the interstitial fluid and thence into the capillary.

Lymphatic Drainage of the Potential Spaces.   In the same manner that protein collects in the interstitial spaces because of leakage out of the capillaries, so does it tend to collect in the potential spaces, and it must be removed through lymphatics or some other channel. Fortunately, each potential space is either directly or indirectly connected with lymphatic drainage systems, which are discussed in connection with each different type of potential space.

Function of the Lymphatic Pump in the Drainage of the Potential Spaces.   Almost all the potential spaces are located where there is continual movement: the joints are moved, the bursae are compressed, the heart moves in the pericardial cavity, with respiration, and the peritoneal spaces are squeezed also with each respiration or with other body movement. Each movement compresses fluid into the lymphatics surrounding the potential spaces, and, because of the valves in the lymphatics, the fluid flows continually away from the space. Therefore, anytime that excess protein collects in the potential space and osmotically pulls any excess fluid whatever into the space, this fluid automatically washes the protein immediately into the lymphatics and back into the blood stream. Loss of the protein reduces the osmotic pressure, and the space remains in a collapsed state.

Altered Capillary Dynamics as a Cause of Increased Pressure and Fluid in the Potential Spaces. Any abnormal changes that can occur in capillary dynamics to cause extracellular edema of the tissues, as described in the preceding chapter, can also cause increased pressure and fluid in the potential spaces. Thus, *increased capillary permeability, increased capillary pressure, decreased plasma colloid osmotic pressure,* and *blockage of the lymphatics* from a potential space can all cause swelling of the space. The fluid that collects is usually called a *transudate* if it is not infected and an *exudate* if it is infected. Excessive fluid in the peritoneal space, one of the spaces most prone to develop extra fluid, is called *ascites.*

One of the most common causes of swelling in a potential space is *infection.* White blood cells and other debris caused by the infection block the lymphatics, resulting in (1) buildup of protein in the space, (2) increased colloid osmotic pressure, and (3) consequent failure of fluid reabsorption.

THE PLEURAL CAVITY

Figure 22–8 shows specifically the diffusion of fluid into and out of the pleural cavity at the parietal and the visceral pleural surfaces. This occurs in precisely the same manner as in the usual tissue spaces, except that a very porous mesenchymal *serous membrane,* the *pleura,* is interspersed between the capillaries and the pleural cavity.

Large numbers of lymphatics drain from the

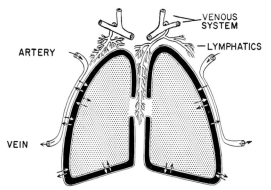

Figure 22–8 Dynamics of fluid exchange in the intrapleural spaces.

mediastinal and lateral surfaces of the parietal pleura, and with each expiration the intrapleural pressure rises, forcing small amounts of fluid into the lymphatics; also the respiratory movements alternately compress the lymphatic vessels, promoting continuous flow along the lymphatic channels.

Maintenance of a Negative Pressure in the Pleural Cavity. The visceral pleura of the lungs continually absorbs fluid with considerable "absorptive force." This is caused by the low capillary pressure —about 5 to 10 mm. Hg—in the pulmonary system. In contrast to this low pressure, the plasma proteins exert about 28 mm. Hg colloid osmotic pressure, causing an absorption pressure at the visceral pleura of approximately 20 mm. Hg at all times. As a result, the pressure of the *fluid* in the intrapleural space remains negative at all times, averaging −10 to −12 mm. Hg. This negative pressure is much greater than the elastic force of the lungs (−4 mm. Hg) that tends to collapse the lungs. Therefore, it keeps the lungs expanded.

THE PERICARDIAL CAVITY

The space around the heart operates with essentially the same dynamics as those of the pleural cavity. The pressure of the fluid in the pericardial cavity, like that in the pleural cavity, is negative. Here again, during expiration as well as during excessive filling of the heart, the pericardial pressure often rises intermittently to a positive value, forcing excess fluid into lymphatic channels of the mediastinum.

THE PERITONEAL CAVITY

The peritoneal cavity of the abdomen is subject to the same fluid dynamics as are the other potential spaces; fluid is filtered into the peritoneal space through the serous membrane called the *visceral peritoneum* covering the viscera and the *parietal peritoneum* lining the outer walls of the abdominal cavity. Fluid is also absorbed through the peritoneum.

The peritoneal cavity is more susceptible to the development of excess fluids than are most of the other cavities, for two reasons: (1) Any time the pressure in the liver sinusoids rises more than 5 to 10 mm. Hg, fluid containing large amounts of protein begins to transude through the liver surface into the abdominal cavity, and (2) the capillary pressure in the visceral peritoneum is higher than elsewhere in the body; this higher pressure is caused by the resistance to portal blood flow through the liver. High venous pressure caused by heart failure or extra resistance in the liver in pathologic states, such as *cirrhosis, carcinoma,* or *portal vein obstruction,* frequently results in marked collection of fluid in the abdomen, a condition called *ascites.*

Numerous large lymphatic channels lead from

the peritoneal cavity, especially from the lower surface of the diaphragm. With each diaphragmatic movement relatively large quantities of lymph flow out of the peritoneal cavity into the thoracic duct. This can be shown effectively by injecting radioactive red blood cells into the abdomen. A large proportion of these cells, still in the whole form, is found in the blood within 10 to 20 minutes. Occasionally, however, cancer spreads so widely throughout the abdomen that it blocks the lymphatics, thereby preventing return of protein to the blood. As a result, the colloid osmotic pressure also rises and ascites ensues.

THE SYNOVIAL CAVITIES

The joint cavities and the bursae are known as *synovial cavities*. The synovial membrane has been said not to be a true membrane at all but only a collection of fibrous tissue that lines the surface between the interstitial spaces and the cavities. For this reason these cavities might be considered to be nothing more than enlarged tissue spaces. However, the synovial cavities do contain large amounts of mucopolysaccharides, much more than normally present in the interstitial fluids. The origin of this is not known, though presumably it is secreted by the surrounding connective tissue cells.

In the synovial cavities, as in the other potential spaces, excess proteins are likely to collect, and these must be returned to the circulatory system through the lymphatics; otherwise the space swells. Since the synovial membrane offers little or no barrier to the transfer of fluid into the surrounding tissues, the protein likewise can flow into the lymphatics of the area.

The pressure in joint cavities often measures as low as −8 to −10 mm. Hg. This negative pressure presumably results from the same factors that cause negative pressures throughout most of the interstitial spaces of the body; these factors were discussed in detail in the preceding chapter.

REFERENCES

Allen, L., and Weatherford, T.: Role of fenestrated basement membrane in lymphatic absorption from peritoneal cavity. *Amer. J. Physiol., 197*:551, 1959.

Atkinson, A. J., Jr., and Weiss, M. F.: Kinetics of blood-cerebrospinal fluid glucose transfer in the normal dog. *Amer. J. Physiol., 216*:1120, 1969.

Bowsher, D.: Cerebrospinal Fluid Dynamics in Health and Disease. Springfield, Illinois, Charles C Thomas, Publisher, 1960.

Brooks, G. F.; Koch, A.; and Wong, J.: Kinetics of sodium transfer from blood to brain of the dog. *Amer. J. Physiol., 218*:693, 1970.

Collins, C. C.; Bach-y-Rita, P.; and Loeb, D. R.: Intraocular pressure variation with oculorotary muscle tension. *Amer. J. Physiol., 213*:1039, 1967.

Davson, H.: Intracranial and intraocular fluids. *In* Handbook of Physiology. Baltimore, The Williams & Wilkins Co., 1960, Sec. I, Vol. III, p. 1761.

Davson, H.: The Physiology of the Cerebrospinal Fluid. Boston, Little, Brown and Company, 1967.

Davson, H., and Spaziani, E.: The fate of substances injected into the anterior chamber of the eye. *J. Physiol. (Lond.), 151*:202, 1960.

Hamerman, D.; Barland, P.; and Janis, R.: The structure and chemistry of the synovial membrane in health and disease. *In* Bittar, E. Edward, and Bittar, Neville (eds.): The Biological Basis of Medicine. New York, Academic Press, Inc., 1969, Vol. 3, p. 269.

Heisey, S. R.; Held, D.; and Pappenheimer, J. R.: Bulk flow and diffusion in the cerebrospinal fluid system of the goat. *Amer. J. Physiol., 203*:775, 1962.

Langham, M. E.: Aqueous humor and control of intra-ocular pressure. *Physiol. Rev., 38*:215, 1958.

Millen, J. W., and Woollam, D. H. M.: The Anatomy of the Cerebrospinal Fluid. New York, Oxford University Press, 1962.

Nicholls, J. G., and Kuffler, S. W.: Extracellular space as a pathway for exchange between blood and neurons in the central nervous system of the leech: ionic composition of glial cells and neurons. *J. Neurophysiol., 27*:645, 1964.

Nicholls, J. G., and Wolfe, D. E.: Distribution of 14C-labeled sucrose, inulin, and dextran in extracellular spaces and in cells of the leech central nervous system. *J. Neurophysiol., 30*:1574, 1967.

Pappenheimer, J. R.; Heisey, S. R.; Jordan, E. F.; and Downer, J. D.: Perfusion of the cerebral ventricular system in unanesthetized goats. *Amer. J. Physiol., 203*:763, 1962.

Shulman, K., and Verdier, G. R.: Cerebral vascular resistance changes in response to cerebrospinal fluid pressure. *Amer. J. Physiol., 213*:1084, 1967.

Welch, K.: Secretion of cerebrospinal fluid by choroid plexus of the rabbit. *Amer. J. Physiol., 205*:617, 1963.

CHAPTER 23

The Pulmonary Circulation

The quantity of blood flowing through the lungs is essentially equal to that flowing through the systemic circulation. However, there are problems related to distribution of blood flow and other hemodynamics that are special to the pulmonary circulation. Therefore, the present discussion is concerned specifically with the peculiarities of blood flow in the pulmonary circuit and the function of the right side of the heart in maintaining this flow.

PHYSIOLOGIC ANATOMY OF THE PULMONARY CIRCULATORY SYSTEM

The Right Side of the Heart. As illustrated in Figure 23–1, the right ventricle is wrapped halfway around the left ventricle. The cause of this is the difference in pressures developed by the two ventricles during systole. Because the left ventricle contracts with extreme force in comparison with the right ventricle, the left ventricle assumes a globular shape, and the septum protrudes into the right heart. Yet each side of the heart pumps essentially the same quantity of blood; therefore, the external wall of the right ventricle bulges far outward and extends around a large portion of the left ventricle, in this way accommodating about the same quantity of blood as the left ventricle. The semilunar shape of the right ventricle allows it to pump with little shortening of its muscle fibers.

The muscle of the right ventricle is little more than one third as thick as that of the left ventricle; this also results from the difference in pressures between the two sides of the heart. Indeed, the wall of the right ventricle is only about three times as thick as the atrial walls, whereas the left ventricular muscle is about eight times as thick.

The Pulmonary Vessels. The pulmonary artery extends only 4 centimeters beyond the apex of the right ventricle and then divides into the right and left main branches, which supply blood to the two respective lungs. The pulmonary artery is also a thin structure with a wall thickness approximately twice that of the venae cavae and one-third the thickness of the aorta. The pulmonary arterial branches are all very short. However, all the pulmonary arteries, even the small arteries and ar-

terioles, have much larger diameters than their counterpart systemic arteries. This, combined with the fact that the vessels are very thin and distensible, gives the pulmonary arterial tree a compliance perhaps as great as that of the systemic arterial tree, thereby allowing the pulmonary arteries to accommodate the stroke volume output of the right ventricle.

The pulmonary veins, like the pulmonary arteries, are also short, but their distensibility characteristics are similar to those of the systemic circulation.

The Bronchial Vessels. A minor accessory arterial blood supply to the lungs exists directly from the aorta through usually one bronchial artery to the right lung and two bronchial arteries to the left lung. The blood flowing in the bronchial arteries is *oxygenated* blood, in contrast to the partially deoxygenated blood in the pulmonary arteries. It supplies the supporting tissues of the lungs, including the connective tissue, the septa, and the large and small bronchi. After this bronchial arterial blood has passed through the sup-

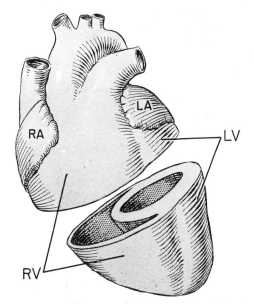

Figure 23–1 The anatomical relationship of the right ventricle to the left ventricle, showing the globular shape of the left ventricle and the half-moon shape of the right ventricle as it drapes around the left ventricle.

porting tissues, it empties into the pulmonary veins and *enters the left atrium* rather than passing back to the right atrium. An average of 1 to 2 per cent of the total cardiac output takes this route, thus making the left ventricular output slightly greater than the right ventricular output.

The Lymphatics. Lymphatics extend from all the supportive tissues of the lung to the hilum of the lung and thence into the thoracic duct. Particulate matter entering the alveoli is usually removed rapidly via these channels, and protein is also removed from the lung tissues, thereby preventing edema.

PRESSURES IN THE PULMONARY SYSTEM

THE PRESSURE PULSE CURVE IN THE RIGHT VENTRICLE

The pressure pulse curves of the right ventricle and pulmonary artery are illustrated in the lower portion of Figure 23-2. These are contrasted with the much higher aortic pressure curve shown above. Approximately 0.16 second prior to ventricular systole, the atrium contracts, pumping a small quantity of blood into the right ventricle, and thereby causing about 4 mm. Hg initial rise in the right ventricular diastolic pressure even before the ventricle contracts. Immediately following this priming by the right atrium, the right ventricle contracts, and the right ventricular pressure rises rapidly until it equals the pressure in the pulmonary artery. The pulmonary valve opens, and for approximately 0.3 second blood flows from the right ventricle into the pulmonary artery. When the right ventricle relaxes, the pulmonary valve closes, and the right ventricular pressure falls to its diastolic level of about zero.

Figure 23-2 Pressure pulse contours in the right ventricle, pulmonary artery, and aorta. (Redrawn from Cournand: *Circulation*, 2:641, 1950.)

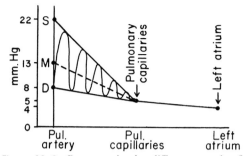

Figure 23-3 Pressures in the different vessels of the lungs. (Redrawn from Cournand: *Circulation*, 2:641, 1950.)

The systolic pressure in the right ventricle of the normal human being averages approximately 22 mm. Hg, and the diastolic pressure averages about 0 to 1 mm. Hg.

PRESSURES IN THE PULMONARY ARTERY

During systole, the pressure in the pulmonary artery is essentially equal to the pressure in the right ventricle, as shown in Figure 23-2. However, after the pulmonary valve closes at the end of systole, the ventricular pressure falls, while the pulmonary arterial pressure remains elevated and then falls gradually as blood flows through the capillaries of the lungs.

As shown in Figure 23-3, the systolic pulmonary arterial pressure averages approximately 22 mm. Hg in the normal human being, the diastolic pulmonary arterial pressure approximately 8 mm. Hg, and the mean pulmonary arterial pressure 13 mm. Hg.

Pulmonary Arterial Pulse Pressure. The pulse pressure in the pulmonary arteries averages 14 mm. Hg, which is almost two thirds as much as the systolic pressure. In previous discussions of pulse pressure, it has been pointed out that the less the compliance of an elastic reservoir that receives pulsatile injections of blood, the greater will be the pulse pressure. If it were not for the large compliance of the thin pulmonary arteries, the pulmonary arterial pulse pressure would be even greater than it is.

LEFT ATRIAL AND PULMONARY VENOUS PRESSURE

The mean pressure in the left atrium and in the major pulmonary veins averages approximately 4 mm. Hg in the human being. However, the pressure in the left atrium varies, even among normal individuals, from as low as 1 mm. Hg to as high as 6 mm. Hg.

The pulse pressures in the left atrium are

several times as great as those in the right atrium mainly because the left ventricle contracts with far greater force than the right ventricle but also because the combined compliance of the pulmonary veins and left atrium is considerably less than the combined compliance of the right atrium and the large systemic veins.

EFFECT OF PULMONARY ARTERIAL PRESSURE ON PULMONARY RESISTANCE

Figure 23–4 illustrates the relationship between pulmonary arterial pressure and pulmonary resistance, showing that at low pulmonary arterial pressures the pulmonary resistance is very high because of collapsed blood vessels; but at high pressures, during which the pulmonary vessels distend markedly, the resistance falls to very low values.

The cause of the great increase in pulmonary resistance at low pressures is compression of the vessels by extravascular structures. Once the pressure in the vessels is high enough to overcome this compression, the resistance becomes reasonably constant, as shown at pulmonary arterial pressure levels above 20 mm. Hg in Figure 23–4.

BACKWARD TRANSMISSION OF PRESSURE FROM THE LEFT ATRIUM TO THE PULMONARY ARTERY

As illustrated in Figure 23–5, a rise in the left atrial pressure from 0 to about 7 mm. Hg has little effect on the pulmonary arterial pressure. Yet, at atrial pressures greater than 7 mm. Hg, the pulmonary arterial pressure increases almost proportionately to further

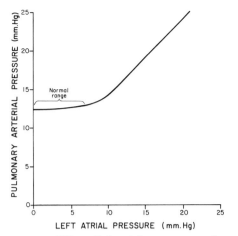

Figure 23–5 Effect of left atrial pressure on pulmonary arterial pressure.

increases in left atrial pressure. The cause of this effect is two-fold: first, collapse of many of the small pulmonary vessels at pressures below about 7 mm. Hg due to compression of the vessels from the outside, and, second, collapse of vessels in the upper parts of the lungs because left atrial pressure is not great enough to overcome the negative hydrostatic pressure difference between heart level and the upper lung level. However, when the left atrial pressure has risen to 7 mm. Hg, the small vessels have by then almost all opened up. With further rise of left atrial pressure above 7 mm. Hg, the pulmonary system acts as if it were a system of almost rigid tubes so that the pulmonary arterial pressure then rises approximately an equal amount.

Significance of These Relationships. During normal operation of the left ventricle, marked increases in systemic resistance can overload the ventricle enough to make the left atrial pressure rise from its normal level of about 4 mm. Hg to as much as 6 to 7 mm. Hg but not more than this unless the left ventricle fails. This increase is still not above 7 mm. Hg, so that it is transmitted backward into the pulmonary arterial tree hardly at all. Consequently, the right heart continues to operate against a normal pulmonary arterial pressure load despite the increased load on the left ventricle. This means, also, that the rate of right ventricular output is not measurably affected by changes in systemic resistance unless the left ventricle fails.

Right Ventricular Compensation and Failure as the Pulmonary Arterial Pressure Rises. When the left heart does fail, the left atrial pressure can rise to values as high as 15 to 40 mm. Hg. The mean pulmonary arterial pressure then also rises to high levels, placing a consider-

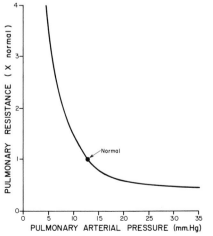

Figure 23–4 Effect of pulmonary arterial pressure on pulmonary resistance.

ably greater than normal pressure load on the right ventricle. Up to 30 to 40 mm. Hg mean pulmonary arterial pressure, the right ventricle continues to pump essentially normal quantities of blood with only a slight rise in right atrial pressure. However, above this pressure the right ventricle begins to fail, so that further increases in pulmonary arterial pressure cause inordinate increases in right atrial pressure; and it is only now that the cardiac output decreases significantly.

Figure 23–6 Effect on the pulmonary arterial pressure of increasing the cardiac output.

EFFECT OF INCREASED CARDIAC OUTPUT ON PULMONARY ARTERIAL PRESSURE

The cardiac output can increase to approximately four times normal before pulmonary arterial pressure becomes seriously elevated; this effect is illustrated in Figure 23–6. As the blood flow into the lungs increases, an initial slight rise in pressure causes the pulmonary arterioles and capillaries to expand and allows the increased blood flow to pass on through the capillary system without excessive increase in pulmonary arterial pressure. However, after cardiac output has reached three to four times normal, the expansion of the previously collapsed vessels is essentially complete, and further increase in cardiac output then causes almost proportionate increase in pulmonary arterial pressure.

This ability of the lungs to accommodate greatly increased blood flow with little increase in pulmonary arterial pressure is important because it conserves the energy of the heart. Actually, the only reason for having blood flow through the lungs is to pick up oxygen and to release carbon dioxide. If the pulmonary arterial pressure should increase as a result of increased blood flow, this would cause a further unnecessary increase in load on the right ventricle.

BLOOD FLOW THROUGH THE LUNGS AND ITS DISTRIBUTION

The rate of blood flow through the lungs is equal to the cardiac output (except for the 1 to 2 per cent that goes through the bronchial circulation), and this is controlled almost entirely by factors elsewhere in the circulation rather than the lungs themselves. Under most conditions, the pulmonary vessels act as passive, distensible tubes that enlarge with increasing pressure and narrow with decreasing pressure. But, for the blood to be aerated adequately, it is important for it to be distributed as evenly as possible to all the different segments of the lung. Some of the problems related to this are the following:

Maldistribution of Blood Flow Between the Top and Bottom of the Lung. Hydrostatic factors affect pressures in the lungs in the same way they do in the systemic circulation. Therefore, when the person is in the upright position, the pulmonary vascular pressures at the top of the lungs are about 10 mm. Hg less than they are in the middle of the lung at the level of the heart; and the pressures at the bottom of the lung are about 10 mm. Hg greater. This effect causes serious differences in blood flow between the upper and lower lung because the mean pulmonary arterial pressure is 13 mm. Hg at the level of the heart, about 3 mm. Hg at the apex of the lungs, and about 23 mm. Hg at the bottom. Consequently, blood flow and blood volume in the bottom of the lung are both much greater than at heart level, whereas at the apices blood flows only during systole. These effects are illustrated by the lowermost curve of Figure 23–7.

Fortunately, there is also greater ventilation of the lowermost alveoli of the lungs in comparison with those at the lung apices, but not quite enough to keep step with the increased blood flow.

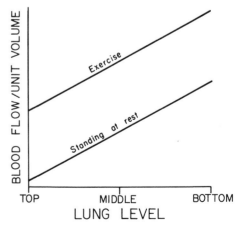

Figure 23–7 Blood flow at different levels in the lung of an upright person, both at rest and during exercise. Note that when the person is at rest, the blood flow is almost zero at the top of the lungs and most of the flow is through the lower lung.

In the lying position, there is a different pattern of maldistribution of blood volume and blood flow, with extra blood volume and blood flow in the posterior parts of the lungs with respect to the anterior, but the effect is not nearly so severe because the differences in hydrostatic levels are much less.

Exercise partially overcomes this maldistribution, as illustrated by the upper curve of Figure 23–7, because the pulmonary arterial pressure then rises high enough to cause reasonable flow even in the lung apices. Indeed, a very good balance develops between pulmonary blood flow and ventilation because there is then almost exactly the same amount of extra flow in the lower lungs as there is extra ventilation.

Effect of Low Alveolar Oxygen Pressure on Pulmonary Vascular Resistance — Autoregulation of Local Pulmonary Blood Distribution. When alveolar oxygen concentration becomes very low, the adjacent blood vessels slowly constrict during the ensuing 5 to 10 minutes, the vascular resistance increasing to as much as double normal. It should be noted specifically that this effect of oxygen lack is *opposite to the effect* normally observed in systemic vessels, and its cause is yet unexplained.

This effect of low oxygen on pulmonary vascular resistance has an important purpose: to distribute blood flow where it is most effective. That is, when some of the alveoli are poorly ventilated and the oxygen concentration in them is low, the local vessels constrict. This in turn causes most of the blood to flow through other areas of the lungs that are better aerated, thus providing an automatic control system for distributing blood flow through different pulmonary areas in proportion to their degrees of ventilation.

It is still questionable how important this autoregulation of blood flow distribution might be. Some research workers have even failed to demonstrate its existence, though most do agree that it occurs at least to a moderate extent.

Paucity of Autonomic Nervous Control of Blood Flow in the Lungs. Though autonomic nerves, both parasympathetic and sympathetic, innervate the larger pulmonary vessels, it is doubtful that these have any major function in normal control of pulmonary blood flow. Normally, stimulation of the vagal fibers to the lungs causes a slight decrease in pulmonary resistance, and stimulation of the sympathetics causes a slight increase in resistance. Likewise, injection of acetylcholine normally causes a slight decrease in resistance, while injection of norepinephrine or epinephrine causes a slight increase. But these effects are all so slight that they are difficult to demonstrate either in experimental animals or in the human being.

Often research workers describe reflexes in the pulmonary vascular system that might be of clinical importance. For instance, it has been claimed that small emboli occluding the small pulmonary arteries cause a reflex that elicits sympathetic vasoconstriction throughout the lungs, this vasoconstriction then leading to an intense increase in pulmonary arterial pressure. It has also been claimed that obstruction of a pulmonary vein causes a sympathetic reflex to constrict the arteries that supply the segment of the lungs feeding into the obstructed vein. Thus far, neither of these reflexes has been proved beyond doubt.

PATHOLOGIC CONDITIONS THAT OBSTRUCT BLOOD FLOW THROUGH THE LUNGS

Removal of Lung Tissue. When one *entire lung* is removed, the flow of blood through the remaining lung usually is well within the limits of compensation as long as the patient remains inactive. However, the patient thereafter has far less *pulmonary circulatory reserve* than does the normal person, for if the cardiac output increases more than 100 per cent above normal, his pulmonary arterial pressure begins to rise rapidly. This compares with the normal individual whose pulmonary circulatory reserve is about 300 per cent before the pressure rises seriously.

Massive Pulmonary Embolism. One of the most severe postoperative calamities in surgical practice is massive pulmonary embolism. Patients lying immobile in bed tend to develop extensive clots especially in the veins of the legs because of sluggish blood flow. Also, women, after delivery of babies, frequently develop massive clots in the hypogastric veins. Such clots often break away from the initial sites of formation, particularly when the patient first walks after a long period of immobilization. The clots then flow to the right side of the heart and into the pulmonary artery. Such a free-moving clot is called an "embolus."

Total blockage of only one of the major branches of the pulmonary artery usually is not immediately fatal because the opposite lung can accommodate all the blood flow. However, blood clots, as was discussed in Chapter 12, have a tendency to grow. Consequently, the embolus becomes larger and larger, and, as it extends into the other major branch of the pulmonary artery, the few remaining vessels that do not become plugged are taxed beyond their limit, and death ensues because of right-sided heart failure.

If the blood is rendered less coagulable by administration of anticoagulants so that the clot cannot grow, the life of the patient who has suffered massive pulmonary embolism can often be saved.

Diffuse Pulmonary Embolism. Sometimes many small blood clot emboli or small emboli of other substances besides blood clots enter the venous blood and eventually clog the vessels of the lungs. One such instance is *fat embolism*. This occurs when a large volume of fatty tissue, such as that in the breast, becomes traumatized or infected so that the fatty material liquefies and enters the venous blood. It passes to the lungs, and because of the high surface tension of fat, the pulmonary arterial pressure often cannot force the fat globules through the small capillary vessels. Consequently, many of the small vessels of the lungs become blocked.

Another instance is *air embolism*. The air, like the fat, creates a large surface tension at the interface between the blood and air so that the globules of air cannot be deformed enough to be pushed through the capillaries.

The physical effects of diffuse pulmonary embolism on the pulmonary circulatory system are similar to those of massive pulmonary embolism — that is, increased pulmonary circulatory resistance with resulting failure of the right side of the heart. However, with diffuse embolism the pulmonary vessels appear to develop considerable vasospasm which adds additional resistance to flow besides that caused by the emboli themselves. It is believed that this vasospasm is caused by a sympathetic reflex, and, clinically, the stellate ganglion is often anesthetized to block the reflex.

Patients with this type of embolism, as well as those with massive embolism, exhibit a rapid respiratory rate because of local irritation by the emboli in the lungs and also because of resultant ischemia throughout the body.

Emphysema. Pulmonary emphysema literally means too much air in the lungs, and it is usually characterized by destruction of many of the alveolar walls. This causes the adjacent alveoli to become confluent, thereby forming large *emphysematous cavities* rather than the usual small alveoli. Obviously, loss of the alveolar septa greatly decreases the total alveolar surface area of the lungs and hinders gas exchange between the alveoli and the blood. This effect will be discussed in connection with the gas exchange functions of the lungs in Chapter 43.

Emphysema also has an important effect on the pulmonary vasculature, for each time an alveolar wall is destroyed, some of the small blood vessels of the pulmonary system are also destroyed, thus progressively increasing the pulmonary resistance and elevating the pulmonary arterial pressure.

In addition, a *physiologic increase in vascular resistance* also occurs in emphysema in the following manner: Some of the emphysematous alveoli have poor exchange of air with the atmosphere. This poor ventilation results in alveolar anoxia so that the adjacent blood vessels come continually under the influence of the hypoxic stimulus discussed above, resulting in chronic vasoconstriction, which compounds the difficulties already caused by the pathologic destruction of pulmonary vessels. Furthermore, the generalized hypoxia throughout the body causes increased cardiac output, and this, combined with the increased pulmonary resistance, sometimes results in serious pulmonary hypertension and right heart failure.

The physiologic increase in resistance can often be treated effectively by simple oxygen therapy. Therefore, emphysematous patients in right heart failure often experience rapid reversal of the failure along with decrease of the dyspneic symptoms after only a few hours of oxygen breathing.

Unfortunately, the prevalence of emphysema is increasing rapidly because of smoking. Therefore, the physiologic principles of its treatment are becoming progressively more important.

Diffuse Sclerosis of the Lungs. A number of pathologic conditions cause excessive fibrosis in the supportive tissues in the lungs, and the fibrous tissue in turn contracts around the vessels. Some of these conditions are silicosis, tuberculosis, syphilis, and, to a lesser extent, anthracosis.

In early stages of diffuse sclerosis, the pulmonary arterial pressure is normal as long as the person is not exercising, but just as soon as he performs even mild exercise the pulmonary arterial pressure often rises inordinately because the vessels do not have the ability to expand. In late stages of diffuse sclerosis the pulmonary arterial pressure remains elevated constantly; as a result, the right ventricle hypertrophies, and it may fail. The alveolar membranes also thicken, or their surface area decreases so much that the rate of gaseous exchange between the alveoli and the blood is diminished.

Atelectasis. Atelectasis is the clinical term for collapse of a lung or part of a lung. This occurs often when the bronchi become plugged because the blood in the capillaries rapidly absorbs the air in the entrapped alveoli, which causes the alveoli to collapse.

Atelectasis also occurs when the chest cavity is opened to atmospheric pressure, for, when air is allowed to enter the pleural space, the elastic nature of the lungs causes them to collapse immediately.

When the elastic lungs contract during atelectasis, they constrict not only the alveoli but also the blood vessels. This automatically decreases blood flow in the atelectatic portions of the lungs to about one-fourth normal, shifting the remaining three-fourths to the aerated portions. This is an important safety mechanism, for it prevents flow of major quantities of blood through pulmonary areas that are not aerated.

THE BLOOD VOLUME OF THE LUNGS

The blood volume of the lungs, as measured by indirect methods, is approximately 12 per cent of the total blood volume of the circulatory system. In other words, in the average human being the two lungs probably contain approximately 600 ml. of blood. About 70 ml. of this is in the capillaries, and the remainder is divided about equally between the arteries and veins.

The Lungs as a Blood Reservoir. Under different physiologic and pathologic conditions, the quantity of blood in the lungs can vary from as little as 50 per cent of normal up to as high as 200 per cent of normal. For instance, when a person blows air out so hard that he builds up high pressure in his lungs— such as when blowing a trumpet—as much as 300 ml. of blood can be expelled from the pulmonary circulatory system into the systemic circulation. Also, loss of blood from the systemic circulation by hemorrhage can be partly compensated by automatic shift of blood from the lungs into the systemic vessels.

SHIFT OF BLOOD BETWEEN THE PULMONARY AND SYSTEMIC CIRCULATORY SYSTEMS AS A RESULT OF CARDIAC PATHOLOGY.   Failure of the left side of the heart or increased resistance to blood flow through the mitral valve as a result of mitral stenosis or mitral regurgitation causes blood to dam up in the pulmonary circulation, thus greatly increasing the pulmonary blood volume while decreasing the systemic volume. Concurrently, the pressures in the lung increase while the systemic pressures decrease.

On the other hand, exactly the opposite effects take place when the right side of the heart fails.

Because the volume of the systemic circulation is about seven times that of the pulmonary system, a shift of blood from one system to the other affects the pulmonary system greatly but usually has only mild systemic effects.

CAPILLARY DYNAMICS IN THE LUNGS

PULMONARY CAPILLARY PRESSURE

Unfortunately, no direct measurements of pulmonary capillary pressure have been made. However, isogravimetric measurement of pulmonary capillary pressure, using the technique described in Chapter 20, has given a value of 7 mm. Hg. This is probably very nearly correct because the mean left atrial pressure is about 4 mm. Hg and the mean pulmonary arterial pressure is only 13 mm. Hg, so that the mean pulmonary capillary pressure must be between these two values. Since venous resistance is less than arterial resistance, the pulmonary capillary pressure should be closer to the venous value of 4 mm. Hg than to the arterial value of 13 mm. Hg.

Pulmonary Wedge Pressure.   By inserting a catheter through the pulmonary artery into one of the small branch arteries and then pushing the catheter until it wedges tightly in the artery, a pressure measurement called the "wedge pressure" can be made. The measured wedge pressure averages approximately 5 mm. Hg, and some physiologists have suggested that this represents a measurement of capillary pressure because the tip of the catheter is nearly at the capillaries. However, since the arterial blood flow is blocked by the catheter but blood can still flow from the tip of the catheter to the pulmonary veins through the capillaries, it is probable that wedge pressure is more nearly a measure of left atrial pressure than of capillary pressure.

LENGTH OF TIME BLOOD STAYS IN THE CAPILLARIES

From histologic study of the total cross-sectional area of all the pulmonary capillaries, it can be calculated that when the cardiac output is normal, blood passes through the pulmonary capillaries in about 1 second. Increasing the cardiac output shortens this time sometimes to less than 0.4 second, but the shortening would not be as great as it is were it not for the fact that additional capillaries, which normally remain collapsed, open up to accommodate the increased blood flow. Thus, in less than 1 second, blood passing through the capillaries becomes oxygenated and loses its excess carbon dioxide.

CAPILLARY MEMBRANE DYNAMICS IN THE LUNGS

Negative Interstitial Fluid Pressure in the Lungs and Its Significance.   Fluid exchange through the capillary membrane was discussed in Chapter 20, where it was also pointed out that the only significant difference between pulmonary capillary dynamics and capillary dynamics elsewhere in the body is the very low capillary pressure in the lungs, about 7 mm. Hg, in comparison with a considerably higher capillary pressure elsewhere in the body, probably about 17 mm. Hg. Because of the very low pulmonary capillary pressure, the hydrostatic force tending to push fluid out the capillary pores into the interstitial spaces is also very slight. Yet, the colloid osmotic pressure of the plasma, about 28 mm. Hg, is a large force tending to pull fluid into the capillaries. Therefore, there is continual osmotic tendency to dehydrate the interstitial spaces of the lungs. Referring back to Chapter 20, it was calculated that the normal interstitial fluid pressure of the human lungs is probably about −17 mm. Hg. That is, there is approximately 17 mm. Hg tending to pull the alveolar epithelial membrane toward the capillary membrane, thus squeezing the pulmonary interstitial space down to almost nothing. Electron micrographic studies have demonstrated exactly this fact, the interstitial space at times being so narrow that the alveolar epithelial cells appear to be directly adherent to the pulmonary capillary endothelial cells. As a result, the distance between the air in the alveoli and the blood in the capillaries is minimal, averaging less than 0.4 micron in distance; and this obviously allows very rapid diffusion of oxygen and carbon dioxide.

MECHANISM BY WHICH THE ALVEOLI REMAIN DRY. Another consequence of the negative pressure in the interstitial spaces is that it pulls fluid from the alveoli through the alveolar membrane and into the interstitial spaces, thereby tending to cause dryness in the alveoli.

Pulmonary Edema. Pulmonary edema means excessive quantities of fluid either in the pulmonary interstitial spaces or in the alveoli. The normal pulmonary interstitial fluid volume is approximately 20 per cent of the lung mass, but this can increase several-fold in serious pulmonary edema. In addition, tremendous volumes of fluid can enter the alveoli and cause *intra-alveolar edema;* intra-alveolar fluid sometimes reaches 1000 per cent of the normal interstitial fluid volume.

SAFETY FACTOR AGAINST PULMONARY EDEMA. The most common cause of pulmonary edema is greatly elevated capillary pressure resulting from failure of the left heart and consequent damming of blood in the lungs. However, the pulmonary capillary pressure usually must rise to very high values before serious pulmonary edema will develop. The reason for this is the very high dehydrating force of the colloid osmotic pressure of the blood in the lungs. This effect is illustrated in Figure 23–8, which shows development of lung edema in dogs subjected to progressively increasing left atrial pressure. In this experiment no edema fluid developed in the lungs until left atrial pressure rose above 23 mm. Hg, which was approximately 3 mm. Hg greater than the colloid

osmotic pressure of dog blood, about 20 mm. Hg. This experiment demonstrates that the hydrostatic pressure in the pulmonary capillary usually must rise a few mm. Hg above the colloid osmotic pressure before serious pulmonary edema can ensue. In the human being the colloid osmotic pressure is about 28 mm. Hg, so that pulmonary edema will rarely develop below 30 mm. Hg pulmonary capillary pressure. Thus, if the capillary pressure in the lungs is normally 7 mm. Hg and this pressure must usually rise above 30 mm. Hg before edema will occur, the lungs have a *safety factor against edema* of approximately 23 mm. Hg.

SAFETY FACTOR AGAINST EDEMA IN CHRONIC ELEVATION OF CAPILLARY PRESSURE. Patients with chronic elevation of pulmonary capillary pressure occasionally will not develop pulmonary edema even with pulmonary capillary pressures as high as 45 mm. Hg. The reason for this is probably extremely rapid run-off of fluid from the pulmonary interstitial spaces through the lymphatics. If the pulmonary capillary pressure remains elevated for more than approximately two weeks, the pulmonary lymphatics enlarge as much as 6 to 10-fold, and lymphatic flow can increase perhaps 20 to 50-fold above the normal resting level. This extra lymphatic flow gives a safety factor perhaps 15 mm. Hg above that which one normally has against edema.

PULMONARY EDEMA AS A RESULT OF CAPILLARY DAMAGE. Pulmonary edema can also result from local capillary damage in the lungs. This effect is often caused by bacterial infection, such as occurs in pneumonia, or by irritant gases such as chlorine, phosgene, lewisite, or mustard gas. All of these directly damage the alveolar epithelium and the endothelium of the capillaries, allowing rapid transudation of both fluid and protein into the alveoli and interstitial spaces.

OCCURRENCE OF ALVEOLAR FLUID IN PULMONARY EDEMA. Though mild degrees of pulmonary edema can be limited to an increase only of the pulmonary interstitial fluid, serious edema almost always causes fluid transudation into the alveoli themselves. It seems that the alveolar epithelium does not have enough strength to resist any significant degree of positive pressure in the interstitial spaces. Therefore, minute or large ruptures occur in the epithelium, and fluid flows readily out of the interstitial spaces into the alveoli.

REFERENCES

Aviado, D. M.: The Lung Circulation. 2 volumes. New York, Pergamon Press, Inc., 1965.

Cournand, A.: Some aspects of the pulmonary circulation in normal man and in chronic cardiopulmonary diseases. *Circulation,* 2:641, 1950.

Fishman, A. P.: Dynamics of the pulmonary circulation. *In* Handbook of Physiology. Baltimore, The Williams & Wilkins Co., 1963, Sec. II, Vol. II, p. 1667.

Gaar, K. A., Jr.; Taylor, A. E.; and Guyton, A. C.: Effect of lung edema on pulmonary capillary pressure. *Amer. J. Physiol.,* 216:1370, 1969.

Gaar, K. A., Jr.; Taylor, A. E.; Owens, L. J.; and Guyton, A. C.:

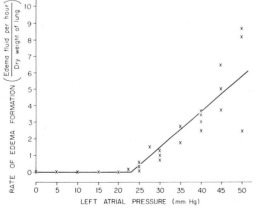

Figure 23–8 Rate of fluid loss into the lung tissues when the left atrial pressure (and also pulmonary capillary pressure) is increased.

Effect of capillary pressure and plasma protein on development of pulmonary edema. *Amer. J. Physiol., 213*:79, 1967.

Gaar, K. A., Jr.; Taylor, A. E.; Owens, L. J.; and Guyton, A. C.: Pulmonary capillary pressure and filtration coefficient in the isolated perfused lung. *Amer. J. Physiol., 213*:910, 1967.

Gorsky, B. H., and Lloyd, T. C., Jr.: Effects of perfusate composition on hypoxic vasoconstriction in isolated lung lobes. *J. Appl. Physiol., 23*:683, 1967.

Grand, M. G., and Downing, S. E.: Metabolic and reflex influences on pulmonary vasomotion. *Amer. J. Physiol., 218*:654, 1970.

Guyton, A. C.: Introduction to Part I: Pulmonary alveolar-capillary interface and interstitium. *In* Fishman, A. P., and Hecht, H. H. (eds.): The Pulmonary Circulation and Interstitial Space. Chicago, University of Chicago Press, 1969, p. 3.

Harlan, J. C.; Smith, E. E.; and Richardson, T. Q.: Pressure-volume curves of systemic and pulmonary circuits. *Amer. J. Physiol., 213*:1499, 1967.

Hebb, C.: Motor innervation of the pulmonary blood vessels of mammals. *In* Fishman, A. P., and Hecht, H. H. (eds.): The Pulmonary Circulation and Interstitial Space. Chicago, University of Chicago Press, 1969, p. 195.

Hyman, A. L.: Effects of large increases in pulmonary blood flow on pulmonary venous pressure. *J. Appl. Physiol., 27*:179, 1969.

Karatzas, N. B., and Lee, G. de J.: Propagation of blood flow pulse in the normal human pulmonary arterial system: analysis of the pulsatile capillary flow. *Circ. Res., 25*:11, 1969.

Karp, R. B.; Graf, P. D.; and Nadel, J. A.: Regulation of pulmonary capillary blood volume by pulmonary arterial and left atrial pressures. *Circ. Res., 22*:1, 1968.

Kira, S., and Hukushima, Y.: Effect of negative-pressure inflation on pulmonary vascular flow. *J. Appl. Physiol., 25*:42, 1968.

Kirk, B. W.: Effect of alterations in pulmonary blood flow on lung-exchangeable water in the dog. *J. Appl. Physiol., 27*:607, 1969.

Knopp, T. J., and Bassingthwaighte, J. B.: Effect of flow on transpulmonary circulatory transport function. *J. Appl. Physiol., 27*:36, 1969.

Levine, O. R.; Mellins, R. B.; and Senior, R. M.: Extravascular lung water and distribution of pulmonary blood flow in the dog. *J. Appl. Physiol., 28*:166, 1970.

Mead, J., and Whittenberger, J. L.: Lung inflation and hemodynamics. *In* Handbook of Physiology. Baltimore, The Williams & Wilkins Co., 1964, Sec. III, Vol. I, p. 477.

Meyer, B. J.; Meyer, A.; and Guyton, A. C.: Interstitial fluid pressure V: Negative pressure in the lungs. *Circ. Res., 22*:263, 1968.

Murray, J. F.; Karp, R. B.; and Nadel, J. A.: Viscosity effects on pressure-flow relations and vascular resistance in dogs' lungs. *J. Appl. Physiol., 27*:336, 1969.

Permutt, S.; Howell, J. B.; Proctor, D. F.; and Riley, R. L.: Effect of lung inflation on static pressure-volume characteristics of pulmonary vessels. *J. Appl. Physiol., 16*:64, 1961.

Reed, J. H., Jr., and Wood, E. H.: Effect of body position on vertical distribution of pulmonary blood flow. *J. Appl. Physiol., 28*:303, 1970.

Szidon, J. P., and Fishman, A. P.: Autonomic control of the pulmonary circulation. *In* Fishman, A. P., and Hecht, H. H. (eds.): The Pulmonary Circulation and Interstitial Space. Chicago, University of Chicago Press, 1969, p. 239.

Taylor, A. E.; Gaar, K.; and Guyton, A. C.: Na^{24} space, D_2O space, and blood volume in isolated dog lung. *Amer. J. Physiol., 211*:66, 1966.

Taylor, A. E.; Guyton, A. C.; and Bishop, V. S.: Permeability of the alveolar membrane to solutes. *Circ. Res., 16*:353, 1965.

West, J. B.: Pulmonary microcirculation. *Proc. Int. Union of Physiol. Sciences, 6*:169, 1968.

West, J. B.: Ventilation/Blood Flow and Gas Exchange. Oxford, Blackwell Scientific Publications Ltd., 1965.

West, J. B.; Glazier, J. B.; Hughes, J. M. B.; and Maloney, J. E.: Pulmonary capillary flow, diffusion ventilation and gas exchange. *In* Wolstenholme, G. E. W., and Knight, J. (eds.): CIBA Foundation Symposium on Circulatory and Respiratory Mass Transport. London, J. & A. Churchill Ltd., 1969, p. 256.

Williams, M. H.: Clinical Applications of Cardiopulmonary Physiology. New York, Hoeber Medical Division, Harper & Row, Publishers, 1960.

CHAPTER 24

Intrinsic Regulations of the Circulation

Many students are surprised to learn that the circulatory system does not require continual control by the nervous system. Indeed, the circulation of a dog whose head has been guillotined and whose spinal cord has been destroyed by injection of alcohol still has intrinsic controls for controlling arterial pressure, cardiac output, blood volume, and local blood flow. It is these controls that regulate the circulation day in and day out and that are mainly responsible for adjusting the blood flow in each individual tissue to accord with the tissue's needs. The purpose of the present chapter is to discuss these intrinsic controls. Then, in the following chapter we will consider the mechanisms by which the nervous system alters these intrinsic controls and adds its additional advantages to circulatory regulation.

Three basic intrinsic controls of the circulation are responsible for day in, day out control of the circulation: (1) The ability of the heart to respond automatically to increased input of blood; this effect is called the Frank-Starling mechanism. (2) The ability of local tissue blood vessels to adjust their blood flow in response to the needs of the surrounding tissues; this phenomenon is called *local blood flow regulation.* (3) The ability of the circulation, operating in conjunction with the kidneys and interstitial fluid system, to regulate extracellular fluid volume and blood volume, which in turn play major roles in controlling blood input to the heart and, therefore, in controlling cardiac output and arterial pressure.

The first of these three intrinsic controls, the Frank-Starling mechanism of the heart, was discussed in detail in Chapter 13; the other two mechanisms are discussed here.

ACUTE LOCAL BLOOD FLOW REGULATION IN RESPONSE TO TISSUE NEED FOR FLOW

In the majority of the tissues, the mechanism of local blood flow regulation maintains flow through each tissue almost exactly at that level required to supply the tissue's needs—no more, no less. To explain this mechanism more fully, let us recall that blood supplying each local tissue area flows first through very small *arterioles* and *meta-arterioles* before passing into the capillaries. The arterioles have a strong muscular coat, the meta-arterioles are surrounded by sparse but highly active smooth muscle fibers, and, finally, at each point where a capillary leaves a meta-arteriole a small muscular *precapillary sphincter* surrounds the origin of the capillary. All of these structures were illustrated in Figure 20–1 of Chapter 20. The smooth muscle of the meta-arterioles, of the precapillary sphincters, and, to a less extent, of the smaller arterioles ordinarily has a high degree of intrinsic tone—that is, even without nervous impulses or humoral stimulation, the smooth muscle remains contracted to a considerable degree. Various local factors in the tissues can cause this smooth muscle to relax or to contract still further, thereby increasing or decreasing the blood flow through the local area. It is by means of these local factors that each local tissue controls its own blood flow, which is the phenomenon of local blood flow regulation.

Table 24–1 gives the relative blood flows through different organs of the body. Note the tremendous flows through the brain, liver, and kidneys despite the fact that these organs represent only a small fraction of the total body mass. Yet, in the case of the brain and liver, the metabolic rates are extremely high. Thus, the blood flow through these two organs is proportioned approximately according to their metabolic needs. In the case of the kidneys, the need is not a metabolic one, but instead it is a need to supply substances required for excretion in the urine, which is discussed later in the chapter.

The skeletal muscle of the body represents 35 to 40 per cent of the total body mass, and yet, in the inactive state, the blood flow

TABLE 24–1. Blood Flow to Different Organs and Tissues Under Basal Conditions*

	Per cent	*ml./min.*
Brain	14	700
Heart	3	150
Bronchial	3	150
Kidneys	22	1100
Liver	27	1350
Portal	(21)	(1050)
Arterial	(6)	(300)
Muscle (inactive state)	15	750
Bone	5	250
Skin (cool weather)	6	300
Thyroid gland	1	50
Adrenal glands	0.5	25
Other tissues	3.5	175
Total	100.0	5000

* Based mainly on data compiled by Dr. L. A. Sapirstein.

Figure 24–2 Effect of arterial oxygen saturation on blood flow through an isolated dog leg.

through all the skeletal muscle is only 15 to 20 per cent of the total cardiac output. This accords with the fact that inactive muscle has a very low metabolic rate. Yet, when the muscles become active, their metabolic rate sometimes increases as much as 50-fold, and the blood flow in individual muscles can increase as much as 20-fold, illustrating a marked increase in blood flow in response to the increased need of the muscle for nutrients.

Figure 24–1 illustrates the approximate quantitative effect on blood flow of increasing the rate of metabolism in a local tissue such as muscle. Note that an increase of metabolism up to eight times normal increases the blood flow about four-fold. The increase in flow at first is less than the increase in metabolism. However, once the metabolism rises high enough to remove most of the nutrients from the blood, further increase in metabolism can

occur only with a concomitant increase in blood flow to supply the required nutrients.

Local Blood Flow Regulation When Oxygen Availability Changes. One of the most necessary of the nutrients is oxygen. Whenever the availability of oxygen to the tissues decreases, such as at high altitude, in pneumonia, in carbon monoxide poisoning, or in cyanide poisoning, the blood flow through the tissues increases markedly. Thus, Figure 24–2 shows that as the arterial oxygen saturation falls to about 25 per cent of normal, the blood flow through an isolated leg increases about three-fold; that is, the blood flow increases almost enough, but not quite, to make up for the decreased amount of oxygen in the blood, thus automatically maintaining an almost constant supply of oxygen to the tissues. Cyanide poisoning of local tissue areas can cause a local blood flow increase as much as seven-fold, thus illustrating the extreme effect of oxygen deficiency in increasing blood flow.

The Oxygen Demand Theory of Local Blood Flow Regulation. The discussion thus far has simply demonstrated the ability of each local tissue to regulate its own blood flow in proportion to its own individual needs but has not explained the mechanism by which this regulation occurs. Unfortunately, the precise cause of local blood flow regulation is not known, but many experiments in recent years point mainly to oxygen concentration in the tissues as the regulator of local blood flow in most (but not all) organs—the less the oxygen concentration, the greater the flow. Two of the reasons for believing this are:

1. Minute arteries isolated from a number of different tissues respond to changes in oxygen in exactly the same way that local tissues do. Increased oxygen in the blood causes constriction; decreased oxygen causes dilatation. Furthermore the smaller the artery, the greater the intensity of this effect, indicating that the minute arterioles, meta-arterioles, and precapillary sphincters could be extremely sensi-

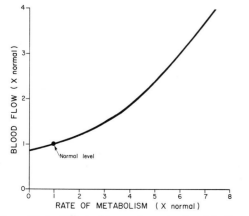

Figure 24–1 Effect of increasing rate of metabolism on tissue blood flow.

tive to slight alterations in local oxygen concentration.

2. Many different experiments have demonstrated that any factor that alters the availability of oxygen to the tissues, as illustrated in Figure 24–2, causes immediate and marked change in local blood flow.

MECHANISM BY WHICH OXYGEN CONCENTRATION COULD REGULATE BLOOD FLOW. Figure 24–3 illustrates what might be called a tissue unit, showing blood flowing into a tissue through a minute vessel and then entering the capillary; at the origin of the vessel is a precapillary sphincter. When one observes a local tissue under the microscope, he sees that the precapillary sphincters are normally either completely open or completely closed. And the number of precapillary sphincters that are open at any given time is approximately proportional to the requirements of the tissue for nutrition. In some tissues the precapillary sphincters open and close cyclically several times per minute, the duration of the open phases being approximately proportional to the metabolic needs of the tissues. This cyclic opening and closure of the sphincter is called *vasomotion*.

Now, let us explain how oxygen concentration in the local tissue *could* regulate blood flow through the area. It is well known that smooth muscle requires oxygen to remain contracted. Therefore, we might assume that the strength of contraction of the precapillary sphincter would increase with an increase in oxygen concentration. Conse-

quently, when the oxygen concentration in the tissue rose above a certain level, the precapillary sphincter presumably would close and remain closed until the tissue cells consumed the excess oxygen. Then, when the oxygen concentration fell low enough, the sphincter would open once more to begin the cycle again. There is some experimental evidence that this mechanism actually does exist: When a micro-"oxygen" electrode, which measures oxygen concentration, is inserted into a tissue adjacent to a precapillary sphincter, the recorded oxygen concentration goes up and down as the sphincter opens and closes, and the phase relationships are such that accord with the theory just explained.

When many capillaries are present in the same local tissue area, this alternate opening and closure of precapillary sphincters does not occur. Instead, a certain proportion of the sphincters remains open while the others remain closed. Furthermore, the proportion that remains open is that number required to keep the oxygen concentration approximately midway between that necessary to cause closure of the sphincters and that required to cause opening.

Thus, the precapillary sphincter mechanism illustrated in Figure 24–3 could easily regulate local blood flow in such a way that the tissue oxygen concentration would remain almost exactly constant, whatever the requirements of the tissue for oxygen.

The Vasodilator Theory for Local Blood Flow Regulation. Another theory to explain regulation of local blood flow is the vasodilator theory. According to this theory, the greater the rate of metabolism or the less the availability of nutrients to a tissue, the greater becomes the rate of formation of a *vasodilator substance*. The vasodilator substance then supposedly diffuses back to the precapillary sphincters, meta-arterioles, and arterioles to cause dilatation. Some of the different vasodilator substances that have been suggested are carbon dioxide, lactic acid, adenosine, adenosine phosphate compounds, histamine, potassium ions, and hydrogen ions. Thus far, none of these vasodilator substances has been proved to be formed in large enough quantities to cause local regulation, though many physiologists place greatest faith in *adenosine* as a possible dilator substance.

Some of the vasodilator theories assume that the vasodilator substance is released from the tissue in response to oxygen lack. For instance, it has been demonstrated that oxygen lack can cause both lactic acid and adenosine to be released from the tissues, and these are substances that can cause vasodilatation and therefore could be responsible, or partially responsible, for the local blood flow regulation mechanism.

Figure 24–3 Diagram of a tissue unit area for explanation of local feedback control of blood flow.

SPECIAL EXAMPLES OF ACUTE LOCAL BLOOD FLOW REGULATION

"Autoregulation" of Blood Flow When the Arterial Pressure Changes. If the arterial pressure should decrease, one might expect the decrease in pressure to decrease the blood flow at least as much as the pressure decreases. However, in metabolically active tissues, the decrease in flow is not very great, because local factors in the tissues automatically dilate the arterioles and precapillary sphincters. Likewise, if the arterial pressure rises above normal, the blood flow increases very little because excess blood flow provides too much nutrients, which causes automatic vasoconstriction and return of the blood flow almost to its normal level. These effects are illustrated by the solid curve in Figure 24–4, which shows that a change in arterial pressure between 80 mm. Hg and 175 mm. Hg does not cause marked change in blood flow through muscle but instead only a slight change that is far less than would have occurred were it not for the local blood flow regulation phenomenon. This ability of the tissues to maintain essentially normal local blood flow despite changes in arterial pressure is called *autoregulation of blood flow.* In its broad sense, this term is also sometimes used synonymously with the term local blood flow regulation.

At pressures below 80 mm. Hg or above 175 mm. Hg, the elastic changes caused in the vessel diameter by the pressure itself are too great to be overcome by the acute autoregulation mechanism. "Long-term" autoregulation is much more powerful than acute autoregulation and therefore can function much more effectively and over a much wider pressure range, as illustrated by the dashed curve of Figure 24–4. This long-term local blood flow regulation will be discussed later in the chapter.

Reactive Hyperemia. Another example of local regulation is the change in blood flow after temporary occlusion of a tissue's blood supply. If the blood supply to a leg is occluded for three minutes and then the occlusion is relieved, flow through the leg rises within 10 seconds to about four times normal and then gradually returns to normal during the next minute or so. This phenomenon is called reactive hyperemia. For short periods of occlusion, the extra blood flow lasts long enough to repay almost exactly the tissue oxygen deficit that has accrued during the period of occlusion.

SPECIAL TYPES OF LOCAL BLOOD FLOW REGULATION

In certain tissues, blood flow is regulated in proportion to other factors besides the requirement for nutrients. This is true particularly in the kidney and in the brain, as follows:

"Autoregulation" in the Kidneys. The rate of blood flow through the kidneys remains constant, even more constant than in most other tissues, despite marked changes in arterial pressure. Yet the availability of oxygen to the kidneys can be changed tremendously without changing renal blood flow significantly. Therefore, renal local blood flow regulation ("autoregulation" of flow) is undoubtedly different from the usual type of local blood flow regulation.

The two factors that can change renal blood flow to the greatest extent are sodium concentration in the blood and the concentration of end-products of protein metabolism; an increase in either of these increases renal blood flow. Recent experiments suggest that it is the concentrations of some of these substances in the renal tubules that control renal blood flow. It has been postulated that feedback control of afferent arteriolar constriction occurs at the juxtaglomerular apparatus where the distal tubule comes in contact with the afferent arteriole. The degree of constriction of the afferent arteriole of each nephron supposedly changes in proportion to the concentrations of certain substances in the distal tubule. In Chapter 35 this mechanism will be discussed separately in relation to the physiology of the kidney.

Local Regulation of Cerebral Blood Flow in Response to Tissue P_{CO_2}. The oxygen type of local blood flow regulation can occur in the brain in the same way that it occurs in most of the rest of the body. However, in addition to this type of regulation, a much more powerful regulation based on carbon dioxide concentration also occurs in the brain. When the brain P_{CO_2} increases, the blood vessels dilate, which allows more rapid blood flow to remove much of the carbon dioxide from the tissues, thereby reducing the P_{CO_2} back toward a normal mean value. Conversely, a decrease in P_{CO_2} causes vasoconstriction, which allows carbon dioxide to accumulate in the tissues until it rises back toward the normal mean

Figure 24–4 Effect on blood flow through a muscle of increasing arterial pressure. The solid curve shows the effect if the arterial pressure is raised over a period of a few minutes. The dashed curve shows the effect if the arterial pressure is raised extremely slowly over a period of many weeks.

value. Thus, the carbon dioxide regulation mechanism in the brain regulates the P_{CO_2} in the brain tissues. This is an important regulation because changes in P_{CO_2} drastically change the degree of activity of all neurons.

LONG-TERM LOCAL BLOOD FLOW REGULATION

The local blood flow regulation that has been discussed thus far occurs acutely, within a minute or more after local tissue conditions have changed. For instance, if the arterial pressure is suddenly increased from 100 mm. Hg to 150 mm. Hg, the blood flow increases almost instantaneously about 100 per cent, partly because the increase in pressure directly increases the flow and partly because the high pressure dilates the peripheral vessels. However, within one to two minutes the blood flow decreases back to only 25 per cent above the original control value. This illustrates the rapidity of the acute type of local regulation, but the regulation is still very incomplete because there remains a 25 per cent increase in blood flow.

However, over a period of hours, days, and weeks a long-term type of local blood flow regulation develops in addition to the acute regulation, and this long-term regulation gives far more complete regulation than does the acute mechanism. For instance, in the above example, even if the arterial pressure should remain at 150 mm. Hg indefinitely, within a few weeks the blood flow through the tissues would gradually reapproach almost exactly the normal value.

Long-term regulation also occurs when the metabolic demands of a tissue change. Thus, if a tissue becomes chronically overactive and therefore requires chronically increased quantities of nutrients, the blood supply gradually increases to match the needs of the tissue.

Figure 24–4 illustrates the far greater effectiveness of the long-term regulation mechanism. This figure shows that acute local blood flow regulation is partially capable of maintaining constant blood flow through a muscle when the arterial pressure changes between about 80 mm. Hg and 170 mm. Hg, but even in this range there is still considerable increase in flow with pressure. On the other hand, the long-term type of regulation will eventually return the blood flow almost completely to normal over an even wider range of pressure. Thus, one can see that the long-term mechanism is extremely valuable in adjusting the blood flow to each tissue to the needs of that tissue.

Long-Term Local Flow Regulation in Coarctation of the Aorta. Coarctation of the aorta means occlusion of the aorta somewhere along its course. When such an occlusion occurs in the descending thoracic aorta, blood must flow to the lower part of the body through collateral blood vessels. Therefore, the arterial pressure in the upper part of the body· becomes very high, sometimes two times normal, while the arterial pressure in the lower body falls slightly below normal. Yet, despite this tremendous difference in pressure between the upper and lower body, in several weeks the blood flow per unit mass of tissue is essentially the same above and below the coarctation. It has been such studies of coarctation in animals that have given us most of our quantitative information on the powerful nature of the long-term local flow regulatory mechanism. Indeed, this long-term mechanism seems to be several times as important in circulatory dynamics as is the mechanism of acute local flow regulation.

Change in Tissue Vascularity as the Mechanism of Long-Term Regulation. The mechanism of long-term regulation is almost certainly a change in the degree of vascularity of the tissues. That is, if the arterial pressure falls to 60 mm. Hg and remains at this level for many weeks, the number and size of vessels in the tissue increase; if the pressure then rises to a very high level, the number and size of vessels decrease. Likewise, if the metabolism in a given tissue becomes elevated for a prolonged period of time, vascularity increases; or if the metabolism is decreased, vascularity decreases.

Thus, there is continual day-by-day reconstruction of the tissue vasculature to meet the needs of the tissues. This reconstruction occurs very rapidly in extremely young animals. It also occurs rapidly in new growth of tissue, such as scar tissue or new growth of glandular tissue; but on the other hand, it occurs very slowly in old, well established tissues. Therefore, the time required for long-term regulation to take place may be only a few days in the newborn or as long as months or even years in the elderly person. Furthermore, the final degree of response is much greater in younger tissues than in older, so that in the newborn the vascularity will adjust to match almost exactly the needs of the tissue for blood flow; whereas in older tissues, vascularity frequently lags far behind the needs of the tissues.

ROLE OF OXYGEN IN LONG-TERM REGULATION. The probable stimulus for increased or decreased vascularity in many if not most instances is need of the tissue for oxygen. The reason for believing this is that hypoxia causes increased vascularity and hyperoxia causes decreased vascularity. This effect is demonstrated in newborn infants who are put into an oxygen tent for therapeutic purposes. The excess oxygen causes almost immediate cessation of new vascular growth in the retina of the eye and even degeneration of some of the capillaries that have already formed. Then when the baby is taken out of the oxygen, there is an explosive overgrowth of new vessels to make up for the sudden decrease in available oxygen; indeed, there is so much overgrowth that the vessels grow into the vitreous humor and eventually cause blindness.

The mechanisms by which changes in oxygen availability also cause changes in vascularity are not yet clear. Lack of oxygen causes enlargement of the vessels already in the tissue and, under appropriate conditions, growth of new vessels as well. It is possible that the change in vascularity results secondarily to the acute local blood flow regulation mechanism in the following way: In the absence of oxygen, the blood vessels dilate. Then, because the tension in the vascular wall increases directly in proportion to the diameter of the vessel, the vascular wall elements become subjected to far greater than normal stretch. This stretch then theoretically could cause permanent enlargement of the vessel or outgrowth of new vessels. In support of this concept is the fact that venous occlusion causes more growth of new small vessels than does arterial occlusion, both types of occlusion reducing tissue oxygenation but only venous occlusion increasing the intraluminal pressure.

COLLATERAL CIRCULATION AS A PHENOMENON OF LONG-TERM LOCAL BLOOD FLOW REGULATION

The development of collateral circulation when blood flow to a tissue is blocked is a type of long-term local blood flow regulation. Thus, if the femoral artery becomes occluded, the small vessels that by-pass the femoral occlusion become greatly enlarged, and the leg usually will develop an adequate blood supply. Within seconds after the occlusion, blood flow through these collateral vessels is only about one-eighth the normal blood flow to the leg. But acute dilatation of the collateral vessels occurs within the first 1 to 2 minutes, so that blood flow to the leg returns to approximately one-half normal. Then, over a period of a week or more

the blood flow returns almost all the way to normal, indicating progressive opening of the collateral vessels.

Thus, the opening of collateral vessels follows almost identically the same pattern as that observed in other types of acute and long-term local blood flow regulation. Furthermore, the acute opening of collateral vessels seems to be metabolically initiated because reduction of the animal's arterial pressure to 50 mm. Hg causes the collateral vessels to open just as much as does occlusion of an artery; that is, insufficiency of blood flow to the tissues opens the collaterals, even though the main artery is not closed.

SIGNIFICANCE OF LONG-TERM LOCAL REGULATION — THE METABOLIC MASS TO TISSUE VASCULARITY PROPORTIONALITY

From these discussions it should already be apparent to the student that there is a built-in mechanism in most tissues to keep the degree of vascularity of the tissue almost exactly that required to supply the metabolic needs of the tissue. Thus, one can state as a general rule that the vascularity of most tissues of the body is directly proportional to the local metabolism times the tissue mass. If ever this proportionality constant becomes abnormal, the long-term local regulatory mechanism automatically readjusts the degree of vascularity over a period of weeks or months. In young persons this degree of readjustment is usually very exact; in old persons it is only partial.

ROLES OF LOCAL BLOOD FLOW REGULATION AND OF THE HEART IN INTRINSIC CONTROL OF CARDIAC OUTPUT

It was pointed out in Chapter 13 that the heart is capable of pumping blood indefinitely without any external control. That is, the rhythmical beating process of the heart will continue indefinitely, and because of the Frank-Starling mechanism of the heart, the heart will pump whatever amount of blood flows into its atria up to its physiologic limit. In other words, the heart is an automatic pump, adjusting itself to its input load. As we shall see in the following chapter, this intrinsic ability of the heart can be enhanced by nervous or humoral controls; but in most resting conditions of bodily function these controls are not necessary, since the ability of the heart to adapt to increasing input load is far more than enough to supply the needs of the circulation. On the other hand, during conditions of severe circulatory stress, such as during heavy exercise, the extrinsic controls are also needed.

One can readily see that it is not the heart that intrinsically controls cardiac output under normal conditions. Instead, it is the flow of blood into the heart that is the controller; and this input flow to the heart is in turn controlled by the flows through all the tissues of the body. Therefore, one can make the simple statement that *cardiac output is determined by the sum of the blood flows through all the individual tissues.*

Following this logic still further, one finds that each tissue has its own local intrinsic ability to regulate its blood flow. Therefore, in essence, cardiac output is controlled primarily by the sum of all the local regulatory processes in all the peripheral vasculature of the body. To state this in still another way, cardiac output is controlled by the tissues themselves in proportion to their needs for flow.

The more one studies the circulation, the more he recognizes the beauty of this mechanism, because it adjusts the blood flow through the tissues to that amount required for proper servicing of the tissues—no more, no less; at the same time it also adjusts the cardiac output to that amount required by the whole body, no more, no less. Were it not for this ability of the tissues to control cardiac output, it would be almost impossible for the cardiac output to be adjusted so beautifully in accord with the needs of the body, and either the cardiac output would be so high that the heart would soon wear out or the blood flow through individual tissues would be so low that the tissues would die.

INTRINSIC EFFECTS OF BLOOD AND EXTRACELLULAR FLUID VOLUMES ON CIRCULATORY CONTROL

Though the intrinsic mechanism just discussed for control of cardiac output based on local tissue regulation of blood flow and intrinsic adaptation of the heart to increasing input loads is a very beautiful one, unfortunately it will not function properly without two other types of controls: (1) control of the blood volume so that there is always blood to be pumped and (2) control of the arterial pressure so that there is a pressure head to force blood through the peripheral tissues when the local regulatory mechanisms open or close the vessels. These two controls are provided primarily by the following basic negative feedback mechanism involving the circulatory system and the kidneys:

1. A person ingests daily both water and salt, and these tend continually to increase his extracellular fluid volume and blood volume.

2. The increase in blood volume increases the availability of blood to be pumped by the heart, thus increasing the cardiac output and arterial pressure.

3. The increase in arterial pressure causes the kidneys to excrete water and salt, thus returning the fluid volumes and arterial pressures back toward their normal levels.

When the arterial pressure has become high enough to make the kidneys excrete water and salt at the same rate that water and salt is ingested each day, extracellular fluid volume, blood volume, and arterial pressure all become stabilized. The stabilized level of blood volume is that amount of blood needed to keep the circulatory system properly filled, and the pressure level set in this manner is high enough to cause adequate blood flow through the peripheral tissues.

SYSTEM ANALYSIS OF THE FLUID VOLUME FEEDBACK MECHANISM

Though it is very easy to understand the basic principles of the interplay between the kidneys, the blood volume, and the circulation, there are many minute features of this feedback mechanism that are exceedingly important in understanding both normal and abnormal circulation. Furthermore, it is not adequate to consider this mechanism simply from a qualitative point of view because some of the basic quantitative factors in the feedback make tremendous differences in one's conceptual thinking about circulatory function. Therefore, in the next few paragraphs, we will utilize the principles of systems analysis discussed in Chapter 1 to formulate a simple analysis of the fluid volume feedback mechanism.

Effect of Arterial Pressure on Urinary Output of Extracellular Fluid. One of the most important steps in the fluid volume feedback mechanism is the effect of arterial pressure on urinary output. This effect is illustrated in Figure 24–5, which shows that at normal arterial pressure of 100 mm. Hg, the two kidneys of the normal human being excrete approximately 1 ml. of extracellular fluid each minute. This fluid contains both water and salt, the principal constituents of extracellular fluid. Figure 24–5 also illustrates that increasing the arterial pressure up to 200 mm. Hg

Figure 24–5   Effect of arterial pressure (AP) on urinary output of extracellular fluid [dE/dt$_{(o)}$].

increases the excretion of extracellular fluid (both water and salt) to about six times normal; whereas decreasing the arterial pressure below its normal value greatly diminishes urinary output, urinary output becoming zero when the arterial pressure falls to about 60 mm. Hg. The details of these effects of arterial pressure on urinary output will be discussed in Chapter 35.

Thus, the rate at which the body loses extracellular fluid is highly dependent upon the arterial pressure, the rate of loss increasing with pressure; therefore, one can already begin to understand how the kidneys can regulate both the arterial pressure and the fluid volumes.

Balance Between Intake and Output of Extracellular Fluid.   The balance between intake and output of extracellular fluid can be expressed by the simple system diagram in Figure 24–6. In this diagram (dE/dt)$_i$ represents the intake of extracellular fluid volume — that is, the intake of water and salt. The symbol (dE/dt)$_o$ represents the output of extracellular fluid by the kidneys. The symbol Σ in the circle stands for summation. Note that the input of

extracellular fluid has a plus sign by it and the output has a minus sign. These two sum, and the output from the summing point is dE/dt, which represents the *rate of change* of the extracellular fluid volume. The extracellular fluid volume will be increasing if the intake is greater than the output, or it will be decreasing if the output is greater than the intake, or it will remain absolutely constant if the intake and output are equal to each other.

Accumulation of Extracellular Fluid Volume. If the rate of change of extracellular fluid volume, dE/dt, is zero, then the extracellular fluid volume will be neither increasing nor decreasing; but, on the other hand, if dE/dt is positive, then the extracellular fluid volume will be increasing. Thus, continual accumulation of extracellular fluid will cause the extracellular fluid volume to increase, or, conversely, continual loss of extracellular fluid will cause the volume to decrease. This effect is illustrated by the block in Figure 24–7, which shows an integral sign in the square. The integral sign simply means accumulation, and this block means that over a period of time the extracellular fluid volume will accumulate either positively or negatively in response to positive or negative rates of change of extracellular fluid volume. Therefore, the output of this block gives the extracellular fluid volume at any given instant.

Effect of Extracellular Fluid Volume on Blood Volume.   Figure 24–8 illustrates the approximate normal relationship between extracellular fluid volume and blood volume. When the extracellular fluid volume is 15 liters, the blood volume is approximately 5 liters. On the other hand, when the extracellular fluid volume increases, approximately one-third of the extra fluid remains in the blood and therefore increases the blood volume as well. This effect continues in the human being until the extracellular fluid volume reaches about 22 liters. At this point, further accumulation of extracellular fluid causes essentially no additional rise in blood volume, because the extra fluid then passes into the tissues to cause edema, for reasons explained in Chapter 21. This effect

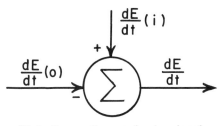

Figure 24–6   System diagram showing that the rate of change of extracellular fluid volume (dE/dt) is equal to the intake of extracellular fluid in the form of water and salt [(dE/dt)$_{(i)}$] minus the output of extracellular fluid [(dE/dt)$_{(o)}$] through the kidneys.

Figure 24–7   Integration of the rate of change of extracellular fluid volume (dE/dt) to give extracellular fluid volume (ECFV).

Figure 24–8 Relationship between extracellular fluid volume (ECFV) and blood volume (BV).

actually acts as a safety valve against the blood volume's rising too high.

Yet, it can be seen from Figure 24–8 that there is a direct correspondence between extracellular fluid volume and blood volume in the normal operating range of the circulation.

Effect of Blood Volume on Cardiac Output. Now we need to see how blood volume in turn plays a role in regulating cardiac output and arterial pressure; this will complete the feedback loop. To do this it is important to understand how the blood fills the circulatory system and especially how it fills the systemic portion of the circulation. It was pointed out in Chapter 18 that a measurement that reflects the degree of filling of the systemic circulation is the *mean systemic pressure.* That is, a high mean systemic pressure means that the circulatory system is partly filled with blood, whereas a low mean systemic pressure means that it is poorly filled with blood.

EFFECT OF BLOOD VOLUME ON MEAN SYSTEMIC PRESSURE. Figure 24–9 illustrates the approximate relationship between blood volume and mean systemic pressure, showing that at a normal blood volume of 5 liters the normal mean systemic pressure is 7 mm. Hg. If the blood volume is increased to about 6.7 liters, the mean systemic pressure approximately doubles. On the other hand, a decrease in blood volume to approximately 2.8 liters decreases the mean systemic pressure to zero. The values given in this graph are only approximate because this relationship varies tremendously with the length of time that a given blood volume is in the circulation, as was discussed in Chapter 18. However, it is very obvious that the greater the blood volume, the more tightly filled becomes the circulatory system.

EFFECT OF MEAN SYSTEMIC PRESSURE ON VENOUS RETURN AND CARDIAC OUTPUT. One can intuitively understand that if there were no filling pressure whatsoever in the circulation, it would be impossible for blood to flow from any one part of the circulation to another part and therefore venous return and cardiac output would become zero. Likewise, one can intuitively understand that should this system become progressively filled to greater mean systemic pressures, the pressure gradients for flow through the system could increase in proportion to filling pressure. Though it is too difficult to go into this at the present time, it will be discussed in detail in Chapter 27. One can show mathematically and experimentally that the tendency for blood to flow from the peripheral vessels toward the heart is directly affected by the mean systemic pressure and negatively affected by right atrial pressure, the venous return being proportional to *mean systemic pressure* minus *right atrial pressure,* which is called the *pressure gradient for venous return.* This relationship is shown in system analysis terminology in Figure 24–10.

EFFECT OF "PRESSURE GRADIENT FOR VENOUS RETURN" AND "RESISTANCE TO VENOUS RETURN" ON VENOUS RETURN AND CARDIAC OUTPUT. In addition to the pressure gradient for venous return (mean systemic pressure *minus* right atrial pressure), still another factor affects return of blood to the heart from the periphery. This factor is the resistance of blood flow through the peripheral vessels, especially through the veins; and this effect is called *resistance to venous return.* Flow of blood from the peripheral vessels toward the heart

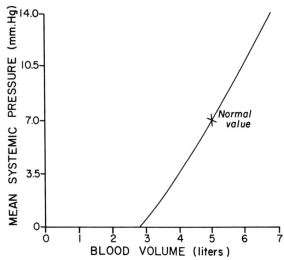

Figure 24–9 Relationship between blood volume (BV) and mean systemic pressure (MSP).

Figure 24-10 Effect of mean systemic pressure (MSP) and right atrial pressure (RAP) on pressure gradient for venous return (MSP-RAP).

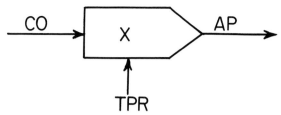

Figure 24-12 Effect of cardiac output (CO) and total peripheral resistance (TPR) on arterial pressure (AP).

is inversely proportional to this resistance. Here again, it is too difficult to discuss at this point why the veins play the major role in this resistance to venous return, but some aspects of this will be discussed in Chapter 27 in relation to venous return and cardiac output.

The interrelationships between pressure gradient for venous return, resistance to venous return, and venous return are illustrated in Figure 24-11, which shows that pressure gradient for venous return divided by resistance to venous return equals venous return. And since venous return equals cardiac output, this relationship also holds for cardiac output.

Effect of Cardiac Output and Total Peripheral Resistance on Arterial Pressure. As a last step in completing the fluid volume feedback system, we now need to know how cardiac output affects arterial pressure. It does so by the simple relationship that *arterial pressure equals cardiac output times total peripheral resistance,* which was discussed in Chapter 18 in relation to basic hemodynamics. This relationship is expressed in systems analysis terminology in Figure 24-12.

Now we are back to where we started, because our starting point in this mechanism was the effect of arterial pressure on kidney function.

The Composite Fluid Control System. All the effects that have been discussed in the last few paragraphs and illustrated in Figures 24-5 through 24-12 are collected into a complete system in Figure 24-13. This system constitutes a "closed loop feedback system," since each one of the function blocks leads directly to some other block and the whole system goes

Figure 24-11 Effect of pressure gradient for venous return (MSP-RAP) and resistance to venous return (RVR) on venous return (VR) or cardiac output (CO).

through a sequence of steps until it comes back to the starting point.

Let us take several examples to see how this closed loop operates:

1. First, let us assume that the arterial pressure becomes abnormally high. Block 1 in the diagram shows that there will be very rapid loss of extracellular fluid, a decrease in extracellular fluid and blood volumes, a decrease in systemic pressure, a decrease in venous return and cardiac output, and, therefore, a decrease in arterial pressure back toward its normal mean value. The arterial pressure will reach its equilibrium value when the daily output of extracellular fluid through the kidneys equals the daily intake of extracellular fluid. This event will occur when each one of the values in Blocks 1, 4, and 5 have reached the points marked by the crosses.

2. Let us now assume that the person has been overinfused with fluid and that his blood volume is 7 liters instead of the normal 5 liters. This will increase the mean systemic pressure to more than double normal; and it will more than double the cardiac output, increase the arterial pressure very greatly, and therefore, increase the rate of loss of extracellular fluid from the body. Consequently, the extracellular fluid volume falls, and the blood volume returns back to its normal value of 5 liters. Thus, once again the entire system readjusts itself back to its normal equilibrium state.

In essence, therefore, the system illustrated in Figure 24-13 is a self-equilibrating system that always readjusts to an equilibrium state whenever some abnormality throws it out of equilibrium. In subsequent chapters we shall see that abnormalities of the circulation, such as abnormalities of the kidneys or of the heart, can at times change the parameter levels at which the system equilibrates. Under such abnormal conditions, the equilibrium value for arterial pressure might be 175 mm. Hg, which is a state of hypertension; or extracellular fluid volume might be 40 liters instead of 15, which is a state of severe edema; or the cardiac output might be 3 liters per minute instead of 5

Figure 24-13   Total system analysis for fluid feedback control of circulatory function, combining the blocks from Figures 24-5 through 24-12.

liters per minute, which is a state of hypodynamic action of the heart. To understand these abnormalities, other factors must be added to the basic control system. These other factors modify the system tremendously, as we shall see in the following sections and the following chapters.

EFFECT OF LOCAL BLOOD FLOW REGULATION ON THE BASIC FLUID PRESSURE CONTROL SYSTEM

There are times when the mechanisms for local regulation of blood flow through isolated tissues conflict with the basic fluid pressure control system depicted in Figure 24-13. For example, if any factor causes an increase in blood volume and therefore also an increase in cardiac output, too much blood will then flow through the tissues to supply the nutrition required. According to the local flow regulation concept, the local tissue vessels should then constrict progressively over a period of days, weeks, or months until finally the blood flow through the tissues would return near to normal. Therefore, eventually, despite the elevated blood volume, the cardiac output would be returned almost to normal. But, in the meantime, the total peripheral resistance would increase markedly, and this in turn would also elevate the arterial pressure markedly.

Now let us put these intuitive thoughts into a quantitative form in the next few paragraphs.

Effect of Cardiac Output on Total Peripheral Resistance.   Figure 24-14 illustrates the approximate relationship between cardiac output and resting total peripheral resistance (when the metabolic rate of the body is neither elevated nor depressed). At the input side of this diagram, it is shown that the total peripheral resistance does not change instantaneously when the cardiac output changes. Instead, the approach to the new equilibrium level for total peripheral resistance occurs with a half-time of approximately two days. Thus, during the first day the cardiac output can be greatly elevated or greatly depressed without any effect on total peripheral resistance; but then gradually, as the local blood flow regulatory mechanism comes into play through the next few days or weeks, the total peripheral resistance changes markedly.

Figure 24-14   Effect of cardiac output (CO) on total peripheral resistance (TPR) under resting conditions. The half-time for the changes to occur is about 2 days.

Figure 24-15 Effect of metabolic dilatation on total peripheral resistance (TPR).

Effect of Metabolic Dilatation on Total Peripheral Resistance. An increase in metabolism resulting from exercise or other factors causes a different type of local blood flow regulation, resulting in almost instantaneous decrease in the total peripheral resistance. Figure 24–15 illustrates this relationship, showing that the normal resting total peripheral resistance becomes markedly decreased as metabolic dilatation takes place.

COMPOSITE ANALYSIS SHOWING THE EFFECT OF LOCAL BLOOD FLOW REGULATION ON THE FLUID FEEDBACK CONTROL SYSTEM

Figure 24–16 illustrates the fluid feedback control system with the system blocks of Figures 24–14 and 24–15 added to the diagram. Also, the proportionality constant between total peripheral resistance and resistance to venous return, 0.07, is shown in Block 9 of this composite figure. This means simply that the resistance to venous return is normally 0.07 times the total peripheral resistance.

The addition of Blocks 9, 10, and 11 to the analysis creates several new complexities, because it gives two additional loops that were not present in the original system. One of these loops is from cardiac output to local peripheral resistance, and then back to arterial pressure. The other loop is from cardiac output, to total peripheral resistance, to resistance to venous return, and then back to cardiac output. These two loops are especially important in dissociating cardiac output regulation from arterial pressure regulation, as follows:

Importance of the Local Blood Flow Regulation Mechanism in Dissociating Cardiac Output Regulation from Arterial Pressure Regulation. The best way to illustrate how the local blood flow regulation mechanism allows dissociation between cardiac output regulation and arterial pressure regulation is to take an example:

Let us assume that an animal ingests tremendous quantities of water and salt, which means an increase in $(dE/dt)_i$ at Block 2 in the figure. This increases the extracellular fluid volume, the blood volume, the mean systemic pressure, cardiac output, and arterial pressure. Thus, immediately, both cardiac output and arterial pressure will rise to a high value and will continue to rise until the output of the kidney from Block 1 equals the intake of water and salt at Block 2. All of this takes place during the first day or so before the local flow regulation mechanism can act significantly. Then the long-term local blood flow regulation mechanism begins to enter the picture, and the excess cardiac output above that

Figure 24-16 Composite system analysis showing interaction between the local blood flow regulation mechanism and the fluid feedback control system.

amount required to supply the needs of the tissues causes the total peripheral resistance to begin rising. As the total peripheral resistance rises, the resistance to venous return also rises (Block 7), and this reduces the cardiac output back toward normal. Yet, the effect of the rising total peripheral resistance at Block 8 is to increase the arterial pressure. Therefore, as the cardiac output falls back toward normal, the arterial pressure does not fall but still remains elevated because the percentage increase in total peripheral resistance is almost exactly equal to the percentage decrease in cardiac output.

Thus, in this example, despite continued excess fluid volumes and continued elevation of mean systemic pressure, the local blood flow regulation mechanism causes the cardiac output to return almost all the way back to normal, while the arterial pressure is adjusted to that level needed by the kidneys to make it put out an appropriate amount of water and salts.

To state this another way, it is the local blood flow regulation mechanism that in the long run sets the cardiac output; and it is the balance between intake and output of fluid, operating in conjunction with the kidneys, that sets the arterial pressure. Furthermore, the local blood flow regulation mechanism allows eventual dissociation between cardiac output regulation and arterial pressure regulation, the cardiac output being adjusted to that value required to supply the needs of the tissues and the arterial pressure being adjusted to that value required to give adequate function of the kidneys as well as to maintain an adequate pressure head in the arteries to keep blood flowing through the system.

SUMMARY

The purpose of this chapter has been to point out that most of the really important features of circulatory control are vested in intrinsic functions of the circulatory elements themselves and that the circulatory system can function quite well without superimposed nervous or humoral controls. We do not mean that these other superimposed controls are not important, for they are; it is merely that the intrinsic controls are capable of performing most of the basic day in and day out controls of the circulation.

The intrinsic controls of the circulation that have been presented in this chapter are the following:

1. The phenomenon of local blood flow regulation controls blood flow in individual tissues in accordance with the needs of the tissues for flow. Acute local regulation gives a small degree of control that takes place in a matter of minutes. Long-term regulation gives an intense degree of control, but it is slow to develop fully, requiring days, weeks, or months.

2. The intrinsic controls of the heart, coupled with the local blood flow regulation mechanism, are basic to the control of cardiac output. That is, the heart automatically pumps the blood that comes to it from the tissues, and the amount of blood that comes from the tissues is automatically set by the sum of the local regulatory mechanisms in the tissues. Therefore, except when the heart becomes overloaded, long-term regulation of cardiac output is controlled primarily by the local blood flow regulatory mechanism of the tissues.

3. The fluid volume control system allows control of the extracellular fluid volume, blood volume, and arterial pressure. Control of these factors is essential to the operation of the local blood flow regulatory mechanism, for without blood volume no blood at all can circulate, and without pressure blood will not flow through the tissues when local tissue vasodilatation occurs.

4. The local blood flow regulation mechanism allows eventual dissociation between arterial pressure regulation and cardiac output regulation because the long-term local regulation mechanism, once it has had time to take effect, sets the level of cardiac output; while on the other hand, the fluid feedback system and the kidneys set the arterial pressure at that level required for adequate urinary output of water and salt.

5. The fluid feedback control system and the long-term local blood flow regulation mechanism are both slow to come to equilibrium, requiring days in the case of the fluid mechanism and days, weeks, or months in the case of the long-term local flow regulation mechanism. It is because of the slowness of action of these two mechanisms that the circulatory system needs additional rapidly acting controls—the nervous and humoral controls—to handle acute emergencies. The next chapter will deal with these rapidly acting controls.

6. The three factors that play the dominant role in the overall intrinsic regulation of the circulation, therefore, are: (a) the intrinsic ability of the heart to pump blood automatically into the arteries whenever blood appears at the input side of the heart (the Frank-Starling mechanism), (b) the long-term local

blood flow regulation mechanism, and (c) the effect of arterial pressure on output of extracellular fluid from the kidneys into the urine.

If the student is not by now thoroughly familiar with the above concepts, he should restudy each of them in detail before going to subsequent chapters, because the future discussions are all highly dependent upon the basic principles presented in this chapter.

REFERENCES

Carrier, O., Jr.; Walker, J. R.; and Guyton, A. C.: Role of oxygen in autoregulation of blood flow in isolated vessels. *Amer. J. Physiol.,* 206:951, 1964.

Coe, J.; Detar, R.; and Bohr, D. F.: Substrates and vascular smooth muscle contraction. *Amer. J. Physiol.,* 214:245, 1968.

Costin, J. C., and Skinner, N. S., Jr.: Effects of systemic hypoxemia on vascular resistance in dog skeletal muscle. *Amer. J. Physiol.,* 218:886, 1970.

Crawford, D. G.; Fairchild, H. M.; and Guyton, A. C.: Oxygen lack as a possible cause of reactive hyperemia. *Amer. J. Physiol.,* 197:613, 1959.

Daugherty, R. M., Jr.; Scott, J. B.; Dabney, J. M.; and Haddy, F. J.: Local effects of O_2 and CO_2 on limb, renal, and coronary vascular resistances. *Amer. J. Physiol.,* 213:1102, 1967.

Davis, D. L., and Hammond, M. C.: Effects of constant-inflow and constant-pressure perfusion on vascular responses. *Amer. J. Physiol.,* 216:1292, 1969.

Detar, R., and Bohr, D. F.: Oxygen and vascular smooth muscle contraction. *Amer. J. Physiol.,* 214:241, 1968.

Donald, D. E.; Rowlands, D. J.; and Ferguson, D. A.: Similarity of blood flow in the normal and the sympathectomized dog hind limb during graded exercise. *Circ. Res.,* 26:185, 1970.

Duling, B. R., and Berne, R. M.: Propagated vasodilation in the microcirculation of the hamster cheek pouch. *Circ. Res.,* 26:163, 1970.

Fairchild, H. M.; Ross, J.; and Guyton, A. C.: Failure of recovery from reactive hyperemia in the absence of oxygen. *Amer. J. Physiol.,* 210:490, 1966.

Granger, H. J., and Guyton, A. C.: Autoregulation of the total systemic circulation following destruction of the central nervous system in the dog. *Circ. Res.,* 25:379, 1969.

Guyton, A. C., and Coleman, T. G.: Long-term regulation of the circulation interrelationships with body fluid volumes. *In* Reeve, B., and Guyton, A. C. (eds.): Physical Bases of Circulatory Transport: Regulation and Exchange. Philadelphia, W. B. Saunders Company, 1967, p. 179.

Guyton, A. C.; Farish, C. A.; and Williams, J. W.: An improved arteriovenous oxygen difference recorder. *J. Appl. Physiol.,* 14:145, 1959.

Guyton, A. C.; Ross, J. M.; Carrier, O., Jr.; and Walker, J. R.: Evidence for tissue oxygen demand as the major factor causing autoregulation. *Circ. Res.,* 14:60, 1964.

Hall, V. E.: The interrelations between blood flow and metabolic rate: a graphic representation. *Physiologist,* 11:207, 1968.

Harrison, R. G.: Functional aspects of the vascularization of tissues. *Scient. Basis Med. Ann. Rev.,* 76, 1963.

Herd, J. A.: Overall regulation of the circulation. *Ann. Rev. Physiol.,* 32:289, 1970.

Herndon, C. W., and Sagawa, K.: Combined effects of aortic and right atrial pressures on aortic flow. *Amer. J. Physiol.,* 217:65, 1969.

Hyman, C.; Paldino, R. L.; and Zimmermann, E.: Local regulation of effective blood flow in muscles. *Circ. Res.,* 12:176, 1963.

Johnson, P. C.: Autoregulation of intestinal blood flow. *Amer. J. Physiol.,* 199:311, 1960.

Johnson, P. C.: Autoregulatory responses of cat mesenteric arterioles measured in vivo. *Circ. Res.,* 22:199, 1968.

Kjellmer, I.: Studies on exercise hyperemia. *Acta Physiol. Scand.,* 64(Suppl. 244):1, 1965.

Kontos, H. A.; Levasseur, J. E.; Richardson, D. W.; Mauck, H. P., Jr.; and Patterson, J. L., Jr.: Comparative circulatory responses to systemic hypoxia in man and in unanesthetized dogs. *J. Appl. Physiol.,* 23:381, 1967.

Lamport, H., and Baez, S.: Physical properties of small arterial vessels. *Physiol. Rev.,* 42 (Suppl. 5):328, 1962.

Nagle, F. J.; Scott, J. B.; Swindall, B. T.; and Haddy, F. J.: Venous resistances in skeletal muscle and skin during local blood flow regulation. *Amer. J. Physiol.,* 214:885, 1968.

Pinakatt, T., and Richardson, A. W.: Distribution of cardiac output in dogs. *Amer. J. Physiol.,* 213:905, 1967.

Reeve, E. B., and Guyton, A. C.: Physical Bases of Circulatory Transport: Regulation and Exchange. Philadelphia, W. B. Saunders Company, 1967.

Reich, T.; Tuckman, J.; Naftchi, N. E.; and Jacobson, J. H.: Effect of normo- and hyperbaric oxygenation on resting and postexercise calf blood flow. *J. Appl. Physiol.,* 28:275, 1970.

Rosenthal, S. L., and Guyton, A. C.: Hemodynamics of collateral vasodilatation following femoral artery occlusion in anesthetized dogs. *Circ. Res.,* 23:239, 1968.

Sagawa, K.: Overall circulatory regulation. *Ann. Rev. Physiol.,* 31:295, 1969.

Stainsby, W. N.: Autoregulation of blood flow in skeletal muscle during increased metabolic activity. *Amer. J. Physiol.,* 202:273, 1962.

Stainsby, W. N., and Renkin, E. M.: Autoregulation of blood flow in resting skeletal muscle. *Amer. J. Physiol.,* 201:117, 1961.

Walker, J. R., and Guyton, A. C.: Influence of blood oxygen saturation on pressure-flow curve of dog hindleg. *Amer. J. Physiol.,* 212:506, 1967.

Wiedman, M. P.: Blood flow through terminal arterial vessels after denervation of the bat wing. *Circ. Res.,* 22:83, 1968.

CHAPTER 25

Nervous and Humoral Regulation of the Circulation—the Cardiovascular Reflexes

Superimposed onto the intrinsic regulations of the circulation discussed in the previous chapter are two additional types of regulation: (1) *nervous* and (2) *humoral.* These regulations are not necessary for basic function of the circulation, but they do provide greatly increased effectiveness of control. The nervous system provides rapid control that can function until the more slowly acting intrinsic controls can take hold. Some of the humoral controls also act rapidly through release of epinephrine and norepinephrine from the adrenal medullae, and these substances support the nervous reflexes; other factors, such as the secretion of aldosterone by the adrenal cortex, aid the intrinsic fluid control system (discussed in the previous chapter) by providing more effective control of water and salt excretion by the kidneys.

NERVOUS REGULATION OF THE CIRCULATION

There are two very important features of nervous regulation of the circulation: First, nervous regulation can function extremely rapidly, some of the nervous reflexes beginning to take effect within two to three seconds and reaching full development within 20 to 30 seconds. Second, the nervous system provides a means for controlling large parts of the circulation simultaneously, irrespective of the needs of the individual tissues. For instance, when it is important to raise the arterial pressure temporarily, the nervous system can arbitrarily cut off, or at least greatly decrease, blood flow to major segments of the circulation despite the fact that the local blood flow regulatory mechanisms oppose this.

On the other hand, nervous control of the circulation usually cannot last for long periods of time because of "adaptation" of parts of the nervous control system and also because the local blood flow regulatory mechanisms eventually become powerfully activated and either override or partially override the nervous controls. Nevertheless, for minutes and sometimes even for hours the nervous system can dominate circulatory control; but usually within a few hours to one to two days, the nervous effects will have been taken over by the basic intrinsic controls.

AUTONOMIC REGULATION OF THE CIRCULATION

The autonomic nervous system will be discussed in detail in Chapter 59. However, it is so important to the regulation of the circulation that its specific anatomical and functional characteristics relating to the circulation deserve special attention.

By far the most important part of the autonomic nervous system for regulation of circulation is the *sympathetic nervous system.* The parasympathetic nervous system is important only in its regulation of heart function, as we shall see later in the chapter.

The Sympathetic Nervous System. Figure 25–1 illustrates the anatomy of sympathetic nervous control of the circulation. Sympathetic vasomotor nerve fibers leave the spinal cord through all the thoracic and the first one to two lumbar spinal nerves. These pass into the sympathetic chain and thence by two routes to the blood vessels throughout the body: (1) through the *peripheral sympathetic nerves* and (2) through the *spinal nerves.* The precise connections of these fibers in the spinal cord and in the sympathetic chains will be discussed in Chapter 59. Suffice to say here that with rare exceptions,

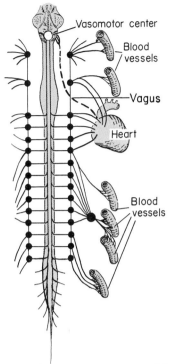

Figure 25-1   The vasomotor center and its control of the circulatory system through the sympathetic and vagus nerves.

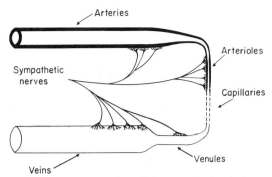

Figure 25-2   Innervation of the systemic circulation.

all vessels of the body are supplied with sympathetic nerve fibers.

DISTRIBUTION OF SYMPATHETIC NERVE FIBERS TO THE PERIPHERAL VASCULATURE.   Figure 25-2 illustrates the distribution of sympathetic nerve fibers to the peripheral blood vessels, and shows that all the vessels except the capillaries are innervated.

The innervation of the arterioles, small arteries, venules, and small veins allows sympathetic stimulation to alter the *resistance* of the vessels and thereby to change the rate of blood flow through the tissues. The innervation of the large vessels, particularly of the veins, makes it possible for sympathetic stimulation to change the volume of these vessels and thereby to alter the volume of the total circulatory system, which plays a major role in the regulation of cardiovascular function, as we shall see later in this chapter and especially in Chapter 27.

SYMPATHETIC NERVE FIBERS TO THE HEART. In addition to fibers supplying the blood vessels, sympathetic fibers spread also to the heart from the upper four thoracic sympathetic ganglia, from the stellate ganglion, and from the cervical ganglia. This innervation was discussed in Chapter 13; and it will be recalled that sympathetic stimulation markedly increases the activity of the heart, increasing the

heart rate and enhancing its strength of pumping.

Parasympathetic Control of the Circulation—Control of the Heart.   Though the parasympathetic nervous system is exceedingly important in many other autonomic functions of the body, it plays only a minor role in regulation of the circulation. Its only important effect is its control of heart rate, and it probably has a slight but relatively unimportant influence on the control of cardiac contractility; however, this latter effect is far overshadowed by that of the sympathetic nervous system on the heart. Parasympathetic nerves pass to the heart in the vagus nerve, as illustrated in Figure 25-1. The effects of parasympathetic stimulation on heart function were discussed in detail in Chapter 13. Principally, parasympathetic stimulation causes a marked decrease in heart rate, and inhibition of the parasympathetics causes an increase in heart rate.

The Sympathetic Vasoconstrictor System and Its Control by the Central Nervous System

The sympathetic nerves carry both vasoconstrictor and vasodilator fibers, but by far the most important of these are the *sympathetic vasoconstrictor* fibers. Sympathetic vasoconstrictor fibers are distributed to essentially all segments of the circulation. However, this distribution is greater in some tissues than in others. It is rather poor in both skeletal and cardiac muscle and in the brain, while it is powerful in the kidneys, the gut, the spleen, and the skin.

The Vasomotor Center and Its Control of the Vasoconstrictor System.   Located bilaterally in the reticular substance of the lower third of the pons and upper two thirds of the medulla, as illustrated in Figure 25-3, is an area called the *vasomotor center*. This center transmits impulses downward through the cord and thence through the vasoconstrictor fibers to all the blood vessels of the body.

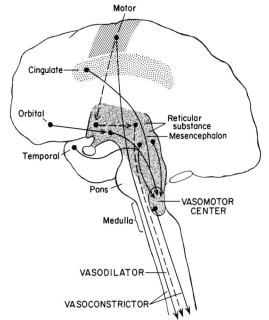

Figure 25–3 Areas of the brain that play important roles in the nervous regulation of the circulation.

The lateral portions of the vasomotor center are *tonically active*. That is, they have an inherent tendency to transmit nerve impulses all the time, thereby maintaining even normally a slow rate of firing in essentially all vasoconstrictor nerve fibers of the body at a rate of about one-half to two impulses per second. This continual firing is called *sympathetic vasoconstrictor tone*. These impulses maintain a partial state of contraction in the blood vessels, a state called *vasomotor tone*.

The brain stem can be severed above the lower third of the pons without significantly changing the normal activity of the vasomotor center. This center remains tonically active and continues to transmit approximately normal numbers of impulses to the sympathetic vasoconstrictor fibers throughout the body.

Figure 25–4 demonstrates the significance of vasoconstrictor tone. In the experiment of this figure, total spinal anesthesia was administered to an animal, which completely blocked all transmission of nerve impulses from the central nervous system to the periphery. As a result, the arterial pressure fell from 100 to 50 mm. Hg, illustrating the effect of loss of vasoconstrictor tone throughout the body. A few minutes later a small amount of the hormone norepinephrine was injected—norepinephrine is the substance secreted at the endings of sympathetic nerve fibers throughout the body. As this hormone was transported in the blood to all the blood vessels the vessels once again became constricted, and the arterial pressure

rose to a level even greater than normal for a minute or two until the norepinephrine was destroyed.

THE INHIBITORY AREA OF THE VASOMOTOR CENTER. The medial portion of the vasomotor center does not participate in excitation of the vasoconstrictor fibers. Instead, stimulation of this medial area transmits *inhibitory* impulses into the lateral parts of the vasomotor center, thereby *decreasing the degree of sympathetic vasoconstrictor tone* and consequently allowing dilatation of the blood vessels. Thus, the vasomotor center is composed of two parts, a bilateral *excitatory part* that can excite the vasoconstrictor fibers and cause vascular constriction and a medial *inhibitory part* that can inhibit vasoconstriction, thus allowing vasodilatation.

CONTROL OF HEART ACTIVITY BY THE VASOMOTOR CENTER. At the same time that the vasomotor center is controlling the degree of vascular constriction, it regulates heart activity. The lateral portions of the vasomotor center transmit excitatory impulses through the sympathetic nerve fibers to the heart, while the medial portion of the vasomotor center, which lies in immediate apposition to the *dorsal motor nucleus of the vagus nerve*, transmits impulses through the parasympathetic fibers of the vagus nerve to the heart to decrease heart rate. Therefore, the vasomotor center can either increase or decrease heart activity, the activity ordinarily increasing at the same time that vasoconstriction occurs throughout the body and ordinarily decreasing at the same time that vasoconstriction is inhibited. However, these interrelationships are not invariable, because some nerve impulses passing down the vagus nerves to the heart bypass the vasomotor center.

Figure 25–4 Effect of total spinal anesthesia on the arterial pressure, showing a marked fall in pressure resulting from loss of vasomotor tone.

CONTROL OF THE VASOMOTOR CENTER BY HIGHER NERVOUS CENTERS. Large numbers of areas throughout the *reticular substance* of the *pons, mesencephalon,* and *diencephalon* can either excite or inhibit the vasomotor center. This reticular substance is illustrated in Figure 25–3 by the diffuse shaded area. In general, the more lateral portions of the reticular substance cause excitation while the more medial portions cause inhibition.

The *hypothalamus* plays a special role in the control of the vasoconstrictor system, for it can exert powerful excitatory or inhibitory effects on the vasomotor center. The *posterolateral portions* of the hypothalamus cause mainly excitation, while the *anterior part* can cause either excitation or inhibition, depending on the precise part of the anterior hypothalamus stimulated.

Many different parts of the *cerebral cortex* can also excite or inhibit the vasomotor center. Stimulation of the *motor cortex,* for instance, excites the vasomotor center because of impulses transmitted downward into the hypothalamus and thence to the vasomotor center. Also, stimulation of the *anterior temporal lobe,* the *orbital areas of the frontal cortex,* or the *anterior part of the cingulate gyrus* can either excite or inhibit the vasomotor center, depending on the precise portion of these areas that is stimulated and upon the intensity of the stimulus.

Thus, widespread areas of the brain can have profound effects on the vasomotor center and in turn on the sympathetic vasoconstrictor system of the body, either further enhancing the degree of vasoconstriction or inhibiting the vasoconstrictor tone and thereby causing vasodilatation.

Direct Vasoconstrictor Pathway That By-passes the Vasomotor Center. In addition to the vasoconstrictor pathway that operates through the vasomotor center, there is a vasoconstrictor pathway that completely by-passes both the hypothalamus and the vasomotor center. This pathway begins in the *motor cortex* and courses downward with the fibers of the corticospinal tract, passing directly to the lateral horn cells of the spinal cord and there stimulating the preganglionic neurons of the sympathetic nerves. This pathway, in contrast to that which operates through the vasomotor center, is not tonically active but becomes activated when the motor portion of the brain initiates muscular activity anywhere in the body. The vasoconstrictor impulses cause vasoconstriction throughout the body, thereby raising the arterial pressure, which makes increased quantities of blood available to the muscles.

Norepinephrine — The Sympathetic Vasoconstrictor Transmitter Substance. The substance secreted at the endings of the vasoconstrictor nerves is norepinephrine. Norepinephrine acts directly on the smooth muscle of the vessels to cause vasoconstriction, as will be discussed in Chapter 59.

The Adrenal Medullae and Their Relationship to the Sympathetic Vasoconstrictor System. Sympathetic vasoconstrictor impulses are transmitted to the adrenal medullae at the same time that they are transmitted to all the blood vessels. These impulses cause the medullae to secrete both norepinephrine and epinephrine into the circulating blood, as will be described in Chapter 59. These two hormones are carried in the bloodstream to all parts of the body, where they act directly on the blood vessels usually to cause vasoconstriction, but sometimes the epinephrine causes vasodilatation, as will be discussed later in the chapter.

The Sympathetic Vasodilator System and Its Control by the Central Nervous System

The sympathetic nerves to skeletal muscles carry sympathetic vasodilator fibers as well as constrictor fibers. However, these fibers release *acetylcholine* at their endings, which acts on the smooth muscle of the blood vessels to cause vasodilatation in contrast to the vasoconstrictor effect of norepinephrine.

The pathways for central nervous system control of the vasodilator system are illustrated by the dashed lines in Figure 25–3. The principal area of the brain controlling this system is the *anterior hypothalamus,* which transmits impulses to a relay station in the *subcollicular region* of the mesencephalon. From here impulses are transmitted down the cord to the *sympathetic preganglionic neurons* in the lateral horns of the cord. These then transmit vasodilator impulses to the blood vessels of the muscles.

Note also in Figure 25–3 that stimulation of the *motor cortex excites* the *sympathetic vasodilator system* by transmitting impulses downward into the hypothalamus.

Importance of the Sympathetic Vasodilator System. It is doubtful that the sympathetic vasodilator system plays a very important role in the control of the circulation, because complete block of the sympathetic nerves to the muscles hardly affects the ability of the muscles to control their own blood flow in response to their needs. Yet, it is possible, if not probable, that at the onset of exercise the sympathetic vasodilator system causes initial vasodilation in the skeletal muscles to allow an *anticipatory increase in blood flow* even before the muscles require increased nutrients.

"PATTERNS" OF CIRCULATORY RESPONSES ELICITED BY DIFFERENT CENTRAL NERVOUS SYSTEM CENTERS

Stimulation of the Vasomotor Center — The Mass Action Effect. Stimulation of the lateral portions of the vasomotor center causes widespread activation of the vasoconstrictor fibers throughout the body, while stimulation of the medial portions of the vasomotor center causes widespread *inhibition of vasoconstriction.* In many conditions the entire vasomotor center acts as a unit, stimulating all vasoconstrictors throughout the body, stimulating the heart, and stimulating the adrenal medullae to secrete norepinephrine and epinephrine that circulate in the blood to excite the circulation still further. The results of this "mass action" are three-fold: First, the peripheral resistance increases in all parts of the circulation, thereby elevating the arterial pressure. Second, the capacity vessels, particularly the veins, are excited at the same time, greatly decreasing their capacity; this forces increased quantities of blood into the heart, thereby increasing the cardiac output. Third, the heart is simultaneously stimulated so that it can handle the increased cardiac output.

Stimulation of the Hypothalamus — The "Alarm" Pattern. Under normal conditions, the hypothalamus probably does not transmit large numbers of impulses into the sympathetic vascular control system. However, on occasion the hypothalamus becomes strongly stimulated and can then activate both the vasoconstrictor and vasodilator systems or either of them separately.

Diffuse stimulation of the hypothalamus activates the vasodilator system to the muscles, thereby increasing the blood flow through the muscles; at the same time it causes intense vasoconstriction throughout the remainder of the body and an intense increase in heart activity. The arterial pressure rises, the cardiac output increases, the heart rate increases, and the circulation is ready to supply nutrients to the muscles if there be need. Also, impulses are transmitted simultaneously throughout the central nervous system to cause a state of generalized excitement and attentiveness, these often increasing to such a pitch that the overall pattern of the reaction is that of *alarm.* This pattern seems to have the purposeful effect of preparing the animal or person to perform on a second's notice whatever activity is required.

The "Motor" Pattern of Circulatory Stimulation. When the motor cortex transmits signals to the skeletal muscle to cause motor activities, it also sends signals to the circulatory system to cause the following effects: First, it excites the alarm system just described, and this generally activates the heart and increases arterial pressure, making the circulatory system ready to supply increased blood flow to the muscles.

This effect also causes vasoconstriction in most nonmuscular parts of the body, such as the kidneys, the skin, and the gut, thereby forcing a larger share of the blood flow through the active muscles. Second, impulses pass directly from the motor cortex to the sympathetic neurons of the spinal cord. These enhance the vasoconstriction in the nonmuscular parts of the body, thereby raising the arterial pressure still more and helping to increase muscle blood flow.

The motor pattern of circulatory responses can be elicited by stimulation of several areas of the reticular substance in the brain stem. Stimulation of one of these areas, the *fields of Forel,* elicits a response almost precisely the same as that caused by direct activation of the motor cortex. Therefore, it is possible, if not likely, that this area of the reticular substance plays a major role in control of the circulatory system in exercise.

Emotional Patterns of Circulatory Responses. Most emotional reactions of the nervous system elicit some type of vasomotor activity. For instance, increasing levels of excitement are usually associated with increasing degrees of vasoconstriction. Likewise, states of depression often cause the opposite effect. These changes in circulatory responses are what one would expect when he realizes that many of the emotional patterns are initiated in the same areas of the brain as those that exert strong influences on the circulation. These areas include (a) the hypothalamus and associated areas of the diencephalon, (b) the anterior part of the cingulate gyrus of the cerebral cortex, (c) the orbitofrontal portions of the cortex, and (d) the anterior temporal lobe.

EMOTIONAL FAINTING. A particularly interesting circulatory reaction occurs in persons who faint because of intense emotional experiences. In this condition, the muscle cholinergic vasodilator system becomes powerfully activated so that blood flow through the muscles increases several fold. This decreases the arterial pressure, which in turn reduces the blood flow to the brain and causes the person to lose consciousness. It is probable, therefore, that emotional fainting results from powerful stimulation of the anterior hypothalamic vasodilator center.

EFFECT OF CARBON DIOXIDE ON THE ACTIVITY OF THE VASOMOTOR CENTER

Carbon dioxide concentration in the blood is one of the most powerful of all stimuli that affect the vasoconstrictor activity of the vasomotor center — a very high carbon dioxide concentration can increase the mean arterial pressure from a normal value of 100 mm. Hg to as high as 270 mm. Hg.

The exact mechanism by which carbon dioxide affects the vasomotor center is not known. However, it is presumed that carbon dioxide has a direct stimulatory effect on the neuronal cells. Some of the theories for this effect will be discussed in Chapter 42 in connection with the regulation of respiration, for the respiratory center reacts to carbon dioxide in much the same manner as the vasomotor center.

Importance of Carbon Dioxide as a Regulator of Arterial Pressure. The amount of carbon dioxide produced by the tissues of the body increases in direct proportion to their rate of metabolism. And, when the metabolism increases, the amount of blood flow required by the tissues to meet the metabolic needs is correspondingly increased. It is perhaps valuable, then, that the elevated carbon dioxide concentration increases the arterial pressure and that this in turn helps to force increased quantities of blood through the vascular system.

EFFECT OF ISCHEMIA ON THE ACTIVITY OF THE VASOMOTOR CENTER — THE CNS ISCHEMIC RESPONSE

Ischemia means lack of sufficient blood flow to maintain normal metabolic function of a tissue. When ischemia of the vasomotor center occurs as a result of occluding the arteries to the brain, systemic arterial pressure rises markedly. It is believed that this is caused mainly by failure of the slowly flowing blood to carry carbon dioxide away from the vasomotor center; the local concentration of carbon dioxide supposedly increases greatly and elevates the arterial pressure. Or perhaps other factors, such as the buildup of lactic acid in the vasomotor center, also contribute to the marked elevation in pressure. This arterial pressure elevation in response to cerebral ischemia is known as the *central nervous system ischemic response* or simply CNS ischemic response.

The magnitude of the ischemic effect on vasomotor activity is almost identical to that caused by excess carbon dioxide; both elevate the mean arterial pressure sometimes to as high as 270 mm. Hg. *The degree of sympathetic vasoconstriction caused by intense cerebral ischemia is often so great that some of the peripheral vessels become totally or almost totally occluded.* The kidneys, for instance, will entirely cease their production of urine because of arteriolar constriction in response to the sympathetic discharge. Therefore, *the CNS ischemic response is one of the most powerful of all the activators of the sympathetic vasoconstrictor system.*

Importance of the CNS Ischemic Response as a Regulator of Arterial Pressure. Despite the extremely powerful nature of the CNS ischemic response, it does not become very active until the arterial pressure falls far below normal, down to levels of 20 to 40 mm. Hg. Therefore, it is not one of the mechanisms for regulating normal arterial pressure. Instead it operates principally as an *emergency arterial pressure control system that acts rapidly and extremely powerfully to prevent further decrease in arterial pressure whenever blood flow to the brain decreases dangerously close to the lethal level.*

The Cushing Reaction. The so-called Cushing reaction is a special type of CNS ischemic response that results from increased pressure in the cerebrospinal fluid system. When the cerebrospinal fluid pressure rises to equal the arterial pressure, it compresses the arteries in the cranial vault and cuts off the blood supply to the brain. Obviously, this initiates a CNS ischemic response, which causes the arterial pressure to rise. When the arterial pressure has risen to a level higher than the cerebrospinal fluid pressure, blood flows once again into the vessels of the brain to relieve the ischemia. Ordinarily, the blood pressure comes to a new equilibrium level slightly higher than the cerebrospinal fluid pressure, thus allowing blood to continue flowing to the brain at all times. A typical Cushing reaction is illustrated in Figure 25–5.

Obviously, the Cushing reaction helps to protect the vital centers of the brain from loss of nutrition if ever the cerebrospinal fluid pressure rises high enough to compress the cerebral arteries.

Figure 25–5 The Cushing reaction, showing a rise in arterial pressure resulting from increased cerebrospinal fluid pressure.

Depressant Effect of Extreme Ischemia on the Vasomotor Center. If cerebral ischemia becomes so severe that maximum rise in mean arterial pressure still cannot relieve the ischemia, the neuronal cells begin to suffer metabolically, and within 3 to 10 minutes they become totally inactive. The arterial pressure then falls to about 40 to 50 mm. Hg, which is the level to which the pressure falls when all tonic vasoconstrictor activity is lost. Therefore, it is fortunate that the ischemic response is extremely powerful so that arterial pressure can usually rise high enough to correct the ischemia before it causes nutritional depression and death of the neuronal cells.

THE ARTERIAL BARORECEPTOR CONTROL SYSTEM — BARORECEPTOR REFLEXES

The nervous system also controls many circulatory functions by means of circulatory reflexes. By far the most important and best known of the circulatory reflexes is the *baroreceptor reflex.* Basically, this reflex is initiated by pressure receptors, called either *baroreceptors* or *pressoreceptors,* located in the walls of the large systemic arteries. A rise in pressure causes the baroreceptors to transmit signals into the central nervous system, and other signals are in turn sent to the circulation to reduce arterial pressure back toward the normal level.

Physiologic Anatomy of the Baroreceptors. Baroreceptors are spray-type nerve endings that are stimulated when stretched. A few baroreceptors are located in the wall of almost every large artery of the thoracic and neck regions; but, as illustrated in Figure 25–6, baroreceptors are extremely abundant, in (1) the walls of the internal carotid arteries slightly above the carotid bifurcations, areas known as the *carotid sinuses,* and (2) the walls of the aortic arch.

Figure 25–6 also shows that impulses are transmitted from each carotid sinus through the very small *Hering's nerve* to the glossopharyngeal nerve and thence to the medulla. Impulses from the arch of the aorta are transmitted through the vagus nerves to the medulla. Hering's nerve is especially important in physiologic experiments because baroreceptor impulses can be recorded from it with ease.

Response of the Baroreceptors to Pressure. Figure 25–7 illustrates the effect of different arterial pressures on the rate of impulse transmission in a Hering's nerve. Note that the baroreceptors are not stimulated at all by pressures between 0 and 50 mm. Hg, but above 50 mm. Hg they respond progressively more and more rapidly and reach a maximum at about 200 mm. Hg.

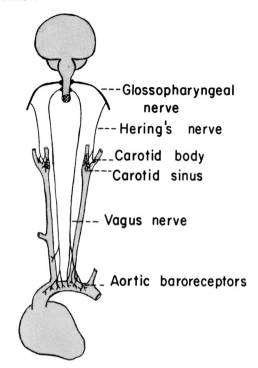

Glossopharyngeal nerve

Hering's nerve

Carotid body

Carotid sinus

Vagus nerve

Aortic baroreceptors

Figure 25–6 The baroreceptor system.

It is especially important that the increase in number of impulses for each unit change in arterial pressure, expressed as $\Delta I/\Delta P$ in the figure, is greatest at about the level of the normal mean arterial pressure. This means that the baroreceptors respond most markedly to changes in arterial pressure just at the level where the response needs to be most marked; that is, in the normal operating range of arterial pressure even a slight change in pressure at this level causes strong sympathetic reflexes

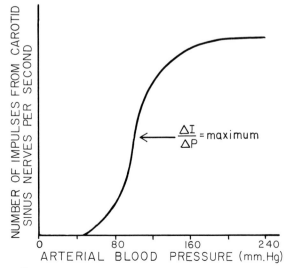

$$\frac{\Delta I}{\Delta P} = maximum$$

Figure 25–7 Response of the baroreceptors at different levels of arterial pressure.

to readjust the arterial pressure when it becomes even slightly abnormal.

The baroreceptors respond extremely rapidly to changes in arterial pressure; the number of impulses even increases greatly during systole and decreases again during diastole. Furthermore, the baroreceptors *respond much more to a rising pressure* than to a stationary pressure and, also, respond less to a falling pressure. That is, if the mean arterial pressure is 150 mm. Hg but at that moment is rising rapidly, the rate of impulse transmission may be as much as twice that when the pressure is stationary at 150 mm. Hg. On the other hand, if the pressure is falling, the rate might be as little as one-quarter that for a stationary pressure. This effect allows the baroreceptors to inform the vasomotor center not only of the actual mean arterial pressure but also whether or not the pressure is instantaneously rising or falling.

The Reflex Initiated by the Baroreceptors. The baroreceptor impulses *inhibit the vasomotor center* of the medulla and *excite the vagal center.* The net effects are (1) *vasodilatation* throughout the peripheral circulatory system and (2) *decreased cardiac rate* and *strength of contraction.* Therefore, excitation of the baroreceptors by pressure in the arteries reflexly *causes the arterial pressure to decrease.* Conversely, low pressure has opposite effects, reflexly causing the pressure to rise back toward normal.

Figure 25-8 illustrates a typical reflex change in arterial pressure caused by clamping the common carotids. This procedure reduces the carotid sinus pressure; as a result, the

Figure 25-9 illustrates a typical reflex change in arterial pressure caused by clamping the common carotids. This procedure reduces the carotid sinus pressure; as a result, the

Figure 25-9 Arterial pressure changes caused by alternate injection and removal of 50 ml. of blood into and from a dog before and after denervation of the (1) left carotid sinus, (2) right carotid sinus, (3) left vagus nerve, and (4) right vagus nerve.

baroreceptors become inactive and lose their inhibitory effect on the vasomotor center. The vasomotor center then becomes much more active than usual, causing the arterial pressure to rise and to remain elevated as long as the carotids are clamped. Removal of the clamps allows the pressure to fall immediately to slightly below normal as a momentary overcompensation and then to return to normal in another minute or so.

THE "BUFFER" FUNCTION OF THE BARORECEPTOR CONTROL SYSTEM. Because the baroreceptor system opposes increases and decreases in arterial pressure, it is often called a *buffer system,* and the nerves from the pressoreceptors are called *buffer nerves.*

Figure 25-9 illustrates the importance of this buffer function of the baroreceptors. In this experiment 50 ml. of blood was transfused into and then removed from the circulatory system of a 10 kg. dog once every minute, and the arterial pressure rose and fell. In the normal dog the inflow and outflow of blood caused a 20 mm. Hg rise and fall in pressure. After one carotid sinus had been denervated, the pressure rose and fell 24 mm. Hg with each cycle, and after denervation of all the pressoreceptors, the pressure rose and fell 60 mm. Hg. This illustrates that *the baroreceptor system reduces changes in arterial pressure to about one-third what they would be were the baroreceptor system not present.*

FUNCTION OF THE BARORECEPTORS DURING CHANGES IN BODY POSTURE. The ability of the baroreceptors to maintain relatively constant arterial pressure is extremely important when a person sits or stands after having been lying down. Immediately upon standing, the

Figure 25-8 Typical carotid sinus reflex effect on arterial pressure caused by clamping both common carotids.

arterial pressure in the head and upper part of the body obviously tends to fall, and marked reduction of this pressure can cause loss of consciousness. Fortunately, however, the falling pressure at the baroreceptors elicits an immediate reflex, resulting in strong sympathetic discharge throughout the body, and this minimizes the decrease in pressure in the head and upper body.

Unimportance of the Baroreceptor System for Long-Term Regulation of Arterial Pressure — Adaptation of the Baroreceptors. The baroreceptor control system is probably of no importance whatsoever in long-term regulation of arterial pressure for a very simple reason: The baroreceptors themselves eventually adapt to whatever pressure level they are exposed to. That is, if the pressure rises from the normal value of 100 mm. Hg up to 200 mm. Hg, extreme numbers of baroreceptor impulses are at first transmitted. During the next few seconds, the rate of firing diminishes considerably; then it diminishes very slowly during the next two to four days, at the end of which time the rate will have returned essentially to the normal level despite the fact that the arterial pressure remains 200 mm. Hg. Conversely, when the arterial pressure falls to a very low value, the baroreceptors at first transmit no impulses at all, but gradually over a period of several days the rate of baroreceptor firing returns again to the original control level.

This adaptation of the baroreceptors obviously prevents the baroreceptor reflex from continuing to function as a control system for longer than a few days at a time. Therefore, prolonged regulation of arterial pressure requires other control systems, principally the intrinsic control systems discussed in the previous chapter.

The Carotid Sinus Syndrome. Strong pressure on the neck over the bifurcations of the carotids in the human being can excite the baroreceptors of the carotid sinuses, causing the arterial pressure to fall as much as 20 mm. Hg in the normal person. In some older persons, particularly after calcified arteriosclerotic plaques have developed in the carotid arteries, pressure on the carotid sinuses often causes such strong responses that the heart stops completely, or at least the arterial pressure falls drastically. Even tight collars can cause the arterial pressure to fall low enough to cause fainting in these persons, an effect called *carotid sinus syncope.* Fortunately, when the reflex causes the heart to stop, the ventricles usually "escape" from the vagal inhibition in approximately 7 to 10 seconds and begin to beat with their own intrinsic rhythm. However, occasionally, the ventricles fail to escape, and the patient dies of cardiac arrest. Treatment of the conditions consists of surgically stripping the nerves from the carotid arteries above and below the bifurcations.

LESS IMPORTANT CIRCULATORY REFLEXES

Many other circulatory reflexes have been described, though none of them seems to be nearly as powerful or as important as the baroreceptor reflexes. Some of these reflexes are the following:

Chemoreceptor Reflexes — Effect of Oxygen Lack on Arterial Pressure. Located in the bifurcations of the carotid arteries and also along the arch of the aorta are several small bodies, 1 to 2 mm. in size, called respectively *carotid* and *aortic bodies.* These contain specialized sensory *receptors sensitive to oxygen lack,* called *chemoreceptors.* They stimulate nerve fibers that pass respectively along Hering's nerves and the vagus nerves into the vasomotor center. Each carotid or aortic body is supplied with an abundant blood flow through a small nutrient artery so that the chemoreceptors are always in close contact with arterial blood.

Whenever the oxygen concentration in the arterial blood falls too low, the chemoreceptors become excited, and impulses are transmitted into the vasomotor center to *excite* the vasomotor center, thus reflexly elevating the arterial pressure. Obviously, this reflex helps to increase the quantity of oxygen carried to the tissues whenever the arterial blood becomes deficient in oxygen. The oxygen lack reflex is not powerful at all in the normal arterial pressure range, but it does exert reasonable feedback effects on arterial pressure when the pressure is in the range of 40 to 80 mm. Hg.

The chemoreceptors will be discussed in much more detail in Chapter 42 in relation to respiratory control.

Atrial Reflexes. Increased pressure in the atria can cause reflex changes in circulatory function, but research workers have disagreed about the exact nature of the effects and also about whether the origin of the reflexes is in the left atrium, right atrium, or both. It seems that an increase in either atrial pressure probably causes increased heart rate, a slight increase in arterial pressure, and an increase in urinary output. The afferent pathway for the reflexes is through the vagus nerves, whereas the efferent pathways are the usual vasomotor pathways. The effect of elevated right atrial pressure to cause increased heart rate is called the *Bainbridge reflex,* but it is still questionable how important this reflex is, since an increase in right atrial pressure also has a direct effect — because of stretch of the S-A node — to increase the heart rate as much as 10 to 15 per cent, as was discussed in Chapter 14.

The increase in urinary output in response to increased atrial pressure has been called a *volume reflex,* and the atrial receptors responsible for this are frequently called *volume receptors* because this effect helps to control the blood volume.

These atrial reflexes, like the baroreceptor reflex, are known to be transient in nature, lasting no longer than a few days at most. Therefore, sustained elevation of atrial pressures, as occurs in heart failure, does not cause long-term continuation of the reflex effects.

Spinal Cord Reflexes. Ordinarily, essentially all of the circulatory reflexes operate through the lower areas of the brain, and the spinal cord is primarily a way station for passage of appropriate sympathetic signals. However, after the spinal cord has been transected in the neck, several spinal cord vasomotor reflexes become apparent. For instance, changes in local temperature in different parts of the body then cause changes in the degree of constriction of the local blood vessels, this effect being mediated through the spinal cord and local autonomic nerves. Warmth of the skin increases blood flow, and cold decreases flow. Secondly, pain anywhere in the body below the level of the spinal cord transection can set off powerful vasomotor signals through the entire spinal cord, sometimes increasing the arterial pressure 50 per cent or more. This effect is usually associated with the so-called "mass reflex" that involves other motor activities of the spinal cord as well, as will be described in Chapter 56.

PARTICIPATION OF THE VEINS IN NERVOUS REGULATION OF THE CIRCULATION

Thus far, we have discussed primarily the ability of the nervous system to regulate arterial resistance and thereby to control arterial pressure. However, much of the regulatory effect of the nervous system is carried out through sympathetic vasoconstrictor fibers to the veins. Indeed, the veins constrict in response to even weaker sympathetic stimuli than do the arterioles and arteries.

Sympathetic Alterations in Venous Capacity and Cardiac Output. The veins offer relatively little resistance to blood flow in comparison with the arterioles and arteries. Therefore, sympathetic constriction of the veins does not change significantly the overall total peripheral resistance. Instead, the important effect of sympathetic stimulation of the veins is a *decrease in their capacity*. This means that the veins then hold less blood at any given venous pressure, which increases the translocation of blood out of the systemic veins into the heart, lungs, and systemic arteries. The distention of the heart in turn causes the heart to pump with increasing effectiveness, in accordance with the Frank-Starling law of the heart, as discussed in Chapter 13. Therefore, the net effect of sympathetic stimulation of the veins is to increase the cardiac output.

ROLE OF THE SKELETAL NERVES AND SKELETAL MUSCLES IN CIRCULATORY CONTROL

Though most nervous control of the circulation is effected through the autonomic nervous system, at least two conditions in which the skeletal nerves and muscles also play major roles in circulatory responses are the following:

The Abdominal Compression Reflex. When a baroreceptor or chemoreceptor reflex is elicited, or whenever almost any other factor stimulates the sympathetic vasoconstrictor system, the vasomotor center and other areas of the reticular substance of the brain stem transmit impulses simultaneously through skeletal nerves to skeletal muscles of the body, particularly to the abdominal muscles. This obviously increases the basal tone of these muscles, and their contraction compresses all the venous reservoirs of the abdomen, helping to translocate blood out of these reservoirs toward the heart. As a result, increased quantities of blood are made available to the heart to be pumped. This overall response is called the *abdominal compression reflex*. The resulting effect on the circulation is exactly the same as that caused by sympathetic vasoconstrictor impulses when they constrict the veins. The two effects operate together to increase the cardiac output.

The abdominal compression reflex is probably much more important than has been realized in the past, for it is well known that persons whose skeletal muscles have been paralyzed are considerably more prone to hypotensive episodes than are normal persons.

Increased Mean Systemic Pressure Caused by Skeletal Muscle Contraction During Exercise. When the skeletal muscles contract during exercise, they compress blood vessels throughout the body. And even anticipation of exercise tightens the muscles, thereby compressing the vessels. The resulting effect is to increase the *mean systemic pressure* from its normal value of about 7 mm. Hg to as high as 20 to 30 mm. Hg. We shall see in Chapter 27 that this high mean systemic pressure acts to translocate blood from the peripheral vessels into the heart and lungs and therefore to increase the cardiac output. This is probably an essential effect to help cause the 5- to 6-fold increase in cardiac output that sometimes occurs in severe exercise.

VASOMOTOR WAVES — OSCILLATION OF THE PRESSURE REFLEX CONTROL SYSTEMS

Often when one is recording arterial pressure in an animal, he notes that the blood pressure rises 10 to 20 mm. Hg, then falls, then rises again, continuing cyclically on and on. The usual duration of each cycle is from 15 to 40 seconds, averaging 26 seconds in a dog. These waves are called *vasomotor waves* and they are also frequently called "Mayer waves" or "Traube-Hering waves." Such records are illustrated in Figure 25–10, showing the cyclical rise and fall in arterial pressure.

The cause of vasomotor waves is usually oscillation of one or more nervous pressure control mechanisms, some of which are the following:

Oscillation of the Baroreceptor and Chemoreceptor Reflexes. The vasomotor waves of Figure 25–10B are the common vasomotor waves that are seen almost daily in experimental pressure recordings. They are caused mainly by oscillation of the *baroreceptor reflex.* That is, a high pressure excites the baroreceptors; this then inhibits the sympathetic nervous system and lowers the pressure. The decreased pressure reduces the baroreceptor stimulation and allows the vasomotor center to become active once again, elevating the pressure to a high value. This high pressure then initiates another cycle, and the oscillation continues on and on.

The *chemoreceptor reflex* can also oscillate to give the same type of waves. Usually, this reflex oscillates simultaneously with the baroreceptor reflex. In fact, it probably plays the major role in causing vasomotor waves when the arterial pressure is in the range of 40 to 80 mm. Hg, because in this range chemoreceptor control of the circulation becomes powerful while baroreceptor control becomes very weak.

Oscillation of the CNS Ischemic Response. The record in Figure 25–10A resulted from oscillation of the medullary ischemic pressure control mechanism. In this experiment the cerebrospinal fluid pressure was raised to 160 mm. Hg, which compressed the cerebral vessels and initiated a CNS ischemic response. When the arterial pressure rose above 160 mm. Hg, the ischemia was relieved, and the sympathetic nervous system became inactive. As a result, the arterial pressure fell rapidly back to a lower value, causing medullary ischemia once again. The ischemia then initiated another rise in pressure. Then, again the ischemia was relieved, and the pressure fell. This repeated itself cyclically

as long as the cerebrospinal fluid pressure remained elevated.

Thus, any pressure control mechanism can oscillate if the intensity of "feedback" is strong enough. The vasomotor waves are of considerable theoretical importance because they show that the nervous reflexes that control arterial pressure obey identically the same principles as those applicable to mechanical and electrical control systems. For instance, if the feedback is too great in the guiding mechanism of an automatic pilot for an airplane, the plane oscillates from side to side instead of following a straight course.

RESPIRATORY WAVES IN THE ARTERIAL PRESSURE

With each cycle of respiration, the arterial pressure rises and falls because of several different effects, some of which are reflex in nature, as follows:

First, many impulses arising in the respiratory center of the medulla "spill over" into the vasomotor center.

Second, every time a person inspires, the pressure in the thoracic cavity becomes more negative than usual, causing the blood vessels in the chest to expand. This reduces the quantity of blood returning to the left side of the heart and thereby momentarily decreases the cardiac output and arterial pressure.

Third, the pressure changes in thoracic vessels caused by respiration can excite baroreceptors.

Though it is difficult to analyze the exact relationship of all these factors in causing the so-called *respiratory pressure waves,* the net result during normal respiration is usually an increase in arterial pressure during the very late part of inspiration and early part of expiration and a fall in pressure during the remainder of the respiratory cycle. During deep respiration the blood pressure can rise and fall as much as 20 mm. Hg with each respiratory cycle.

HUMORAL REGULATION OF THE CIRCULATION

The term humoral regulation means regulation by substances — such as hormones, ions, and so forth — in the body fluids. Among the most important of these factors are the following:

EFFECT OF ALDOSTERONE IN CIRCULATORY REGULATION

The most important of all humoral factors affecting circulatory regulation is the hormone *aldosterone* secreted by the adrenal cortex. This

Figure 25–10 (A) Vasomotor waves caused by oscillation of the CNS ischemic response. (B) Vasomotor waves caused by pressoreceptor reflex oscillation.

hormone helps to regulate salt and water in the extracellular fluid, thereby also helping to regulate blood volume. Since this hormone will be discussed in detail in Chapter 36 in relation to renal function and in Chapter 76 in relation to the hormones, only its function in overall circulatory regulation will be presented briefly here.

Whenever the extracellular fluid volume, the blood volume, the sodium concentration in the body fluids, or the cardiac output decreases, the rate of secretion of aldosterone automatically increases. The precise cause or causes of the increase in aldosterone secretion in all these instances has not yet been determined. Yet, the fact remains that almost any type of circulatory debility causes increased output of aldosterone.

The aldosterone in turn acts on the kidney to cause increased reabsorption of sodium, and secondary results of this effect are increased reabsorption of chloride ions and water, thereby increasing the extracellular fluid volume and blood volume. Obviously, these effects are additive to the effects of the intrinsic fluid control system that was discussed in the previous chapter, making the intrinsic system perhaps two or three times as effective as it would be without the aldosterone feedback mechanism. Therefore, one can readily understand the extreme importance of this humoral factor in the control of the circulation.

OTHER HUMORAL FACTORS

Norepinephrine and Epinephrine. Early in the chapter it was pointed out that the adrenal medullae secrete both norepinephrine and epinephrine. When the sympathetic nervous system throughout the body is stimulated to cause direct effects on the blood vessels, it also causes the adrenal medullae to secrete these two hormones, which then circulate everywhere in the body fluids and act on all vasculature. Norepinephrine has vasoconstrictor effects in almost all vascular beds of the body, and epinephrine has similar effects in some, but not all, beds. For instance, epinephrine causes vasodilatation in both skeletal and cardiac muscle. These two hormones will be discussed in more detail in Chapter 59 in the discussion of the autonomic nervous system.

The Kinins. Several substances called *kinins* that can cause vasodilatation have been isolated from blood and tissue fluids. One of these substances is *bradykinin.*

The kinins are small polypeptides that are split away from an alpha$_2$ globulin in the plasma or tissue fluids. Different types of proteolytic enzymes can split the kinins from the globulin, but an enzyme of particular importance is *kallikrein,* which is present in the blood and tissue fluids, in an inactive form. Kallikrein can be activated in several different ways, such as by maceration of the blood, dilution of the blood, contact of the blood with glass, and other similar chemical and physical effects on the blood. As kallikrein becomes activated, it acts immediately on the alpha$_2$ globulin to release *kallidin,* which is a mixture of bradykinin and another similar kinin. Once formed, the kinins persist for only a few minutes, because they are digested by the enzyme *carboxypeptidase.* The activated kallikrein enzyme is destroyed by a kallikrein inhibitor also present in the body fluids.

The kinins cause very powerful vasodilatation. For instance, injection of 1 *microgram* of bradykinin into the brachial artery of a man increases the blood flow through the arm as much as sixfold.

Though unfortunately we know little about the function of the kinins in the control of the circulation, this extremely powerful vasodilator effect, coupled with the fact that kinins can develop anywhere in the circulatory system with ease, indicates that they must play important roles in circulatory regulation. There is some reason to believe that the kinins play roles in regulating skin blood flow and blood flow in some of the glands.

Angiotensin. *Angiotensin* is a hormone formed in the blood in response to *renin,* a substance found in the juxtaglomerular cells of the kidney. Ordinarily, almost no renin leaves the kidney, but under special conditions, such as when the kidneys are acutely exposed to low arterial pressure or to low concentration of sodium in the blood, large quantities of renin do spill into the general circulation and cause the formation of angiotensin.

Angiotensin causes powerful vasoconstriction of the arterioles but has little effect on the veins. These effects will be discussed at greater length in connection with other functions of angiotensin when we consider hypertension in Chapter 26.

Vasopressin. *Vasopressin* is a hormone formed in the hypothalamus but secreted through the posterior pituitary gland. It has a powerful effect on the arterioles that is similar to that of angiotensin, but it has almost no effect on the veins. Yet it is doubtful that vasopressin plays much of a role in regulation of peripheral resistance, because the quantities secreted into the circulation are extremely minute. The primary function of vasopressin is to control reabsorption of water from the renal tubules (see Chapter 36), which it does in concentrations in the order of one-hundredth that required to cause significant vasoconstriction. This effect, obviously, is important to circulatory function, though, from a quantitative point of view, it is only a fraction as important as the aldosterone mechanism just discussed.

Serotonin. Serotonin (5-hydroxytryptamine) is present in large concentrations in the chromaffin tissue of the intestine and other abdominal structures. Also, it is present in high concentration in the platelets. Serotonin can have either a vasodilator or vasoconstrictor effect, depending on the condition or the area of the circulation. And, even though these effects can sometimes be powerful, the functions of serotonin in regulation of the circulation are almost entirely unknown. Occasionally, tumors composed of chromaffin tissue develop, called *carcinoid tumors*. These secrete tremendous quantities of serotonin and cause mottled areas of vasodilatation in the skin; but the very fact that these tremendous quantities of serotonin do not drastically disturb the circulation makes it doubtful that serotonin plays a widespread general role in the regulation of circulatory function.

Histamine. Histamine is released by essentially every tissue of the body whenever it becomes damaged. Most of the histamine is probably derived from eosinophils and mast cells in the damaged tissues.

Histamine has a powerful vasodilator effect on the arterioles but at the same time constricts the veins. Though the role of histamine in normal regulation of the circulation is unknown, in many pathological conditions the intense arteriolar dilatation and venous constriction caused by histamine increases the capillary pressure so greatly that tremendous quantities of fluid leak out of the circulation into the tissues to cause edema. The effects of histamine were discussed in Chapter 10 in relation to allergic reactions.

EFFECTS OF CHEMICAL FACTORS ON VASCULAR CONSTRICTION

Many different chemical factors can either dilate or constrict local blood vessels. Though the roles of these substances in the overall *regulation* of the circulation generally are not known, their specific effects can be listed as follows:

An increase in *calcium ion* concentration causes vasoconstriction. This results from the general effect of calcium to stimulate smooth muscle contraction, as discussed in Chapter 7.

An increase in *potassium ion* concentration causes vasodilatation. This results from the general effect of potassium ions to inhibit smooth muscle contraction.

An increase in *magnesium ion* concentration causes powerful vasodilatation, for magnesium ions, like potassium ions, inhibit smooth muscle generally.

Increased *sodium ion* concentration causes arteriolar dilatation. This results from an increase in osmolality of the fluids rather than from a specific effect of sodium ion itself. *Increased osmolality* of the blood caused by increased quantities of *glucose* or other nonvasoactive substances also causes arteriolar dilatation. Decreased osmolality causes arteriolar constriction.

The only anions to have significant effects on blood vessels are *acetate* and *citrate*, both of which cause mild degrees of vasodilatation.

An *increase in hydrogen ion* concentration (decrease in pH) causes dilatation of the arterioles. A slight *decrease in hydrogen ion* concentration causes arteriolar constriction, but an intense decrease causes dilatation, which is the same effect as that which occurs with increased hydrogen ion concentration.

An increase in carbon dioxide concentration causes slight vasodilatation in most tissues but marked vasodilatation in the brain. However, carbon dioxide, acting on the vasomotor center, indirectly has an extremely powerful vasoconstrictor effect, as described earlier in the chapter.

The adenosine compounds, including adenosine itself, AMP, ADP, and ATP, all cause vasodilatation.

REFERENCES

Agarwal, J. B.; Baile, E. M.; and Palmer, W. H.: Reflex systemic hypotension due to hypertonic solutions in pulmonary circulation. *J. Appl. Physiol.*, 27:251, 1969.

Allison, J. L.; Sagawa, K.; and Kumada, M.: An open-loop analysis of the aortic arch barostatic reflex. *Amer. J. Physiol.*, 217:1576, 1969.

Bond, R. F., and Green, H. D.: Cardiac output redistribution during common carotid occlusion. *Amer. J. Physiol.*, 216:393, 1969.

Brody, M. J., and Shaffer, R. A.: Distribution of vasodilator nerves in the canine hindlimb. *Amer. J. Physiol.*, 218:470, 1970.

Cushing, H.: Concerning a definite regulatory mechanism of the vasomotor center which controls blood pressure during cerebral compression. *Bull. Hopkins Hosp.*, 12:290, 1901.

Diana, J. N., and Kaiser, R. S.: Pre- and postcapillary resistance during histamine infusion in isolated dog hindlimb. *Amer. J. Physiol.*, 218:132, 1970.

Donald, D. E., and Ferguson, D. A.: Study of the sympathetic vasoconstrictor nerves to the vessels of the dog hind limb. *Circ. Res.*, 26:171, 1970.

Emerson, T. E., Jr.; Scott, J. B.; and Haddy, F. J.: Effects of acute multiple changes in plasma electrolyte levels of dog blood pressure. *Amer. J. Physiol.*, 218:234, 1970.

Eyzaguirre, C., and Lewin, J.: Chemoreceptor activity of the carotid body of the cat. *J. Physiol. (Lond.)*, 159:222, 1961.

Gleser, M. A., and Grupp, G.: Mathematical model of response to carotid sinus nerve stimulation. *Amer. J. Physiol.*, 216:263, 1969.

Gupta, P. D.; Henry, J. P.; Sinclair, R.; and von Baumgarten, R.: Responses of atrial and aortic baroreceptors to nonhypotensive hemorrhage and to transfusion. *Amer. J. Physiol.*, 211:1429, 1966.

Guyton, A. C.: Acute hypertension in dogs with cerebral ischemia. *Amer. J. Physiol.*, 154:45, 1948.

Guyton, A. C.; Batson, H. M.; Smith, C. M.; and Armstrong, G. G.: Method for studying competence of the body's blood pressure regulatory mechanisms and effect of pressoreceptor denervation. *Amer. J. Physiol.*, 164:360, 1951.

Guyton, A. C., and Harris, J. W.: Pressoreceptor-autonomic oscillation: a probable cause of vasomotor waves. *Amer. J. Physiol.*, 165:158, 1951.

Guyton, A. C., and Satterfield, J. H.: Vasomotor waves possibly resulting from CNS ischemic reflex oscillation. *Amer. J. Physiol.*, 170:601, 1952.

Haddy, F. J.: Local effects of sodium, calcium and magnesium upon small and large blood vessels of the dog forelimb. *Circ. Res.*, 8:57, 1960.

Hainsworth, R.; Ledsome, J. R.; and Carswell, F.: Reflex responses from aortic baroreceptors. *Amer. J. Physiol.*, 218:423, 1970.

Hammond, M. C.; Davis, D. L.; and Dow, P.: Rate of development of constrictor responses of the dog paw vasculature. *Amer. J. Physiol.*, 216:414, 1969.

Hilton, S. M.: Hypothalamic regulation of the cardiovascular system. *Brit. Med. Bull.*, 22:243, 1966.

Horwitz, L. D.; Bishop, V. S.; Stone, H. L.; and Stegall, H. F.:

Cardiovascular effects of low-oxygen atmospheres in conscious and anesthetized dogs. *J. Appl. Physiol.,* 27:370, 1969.

Iriuchijima, J., and Koike, H.: Carotid flow, intrasinusal pressure, and collateral flow during carotid occlusion. *Amer. J. Physiol.,* 218:876, 1970.

Korner, P. I.: Circulatory adaptations in hypoxia. *Physiol. Rev.,* 39:687, 1959.

Koushanpour, E., and McGee, J. P.: Effect of mean pressure on carotid sinus baroreceptor response to pulsatile pressure. *Amer. J. Physiol.,* 216:599, 1969.

Krasney, J. A.: Effect of sino-aortic denervation on regional circulatory responses to cyanide. *Amer. J. Physiol.,* 218:56, 1970.

Krieger, E. M.: Time course of baroreceptor resetting in acute hypertension. *Amer. J. Physiol.,* 218:486, 1970.

Mellander, S.: Systemic circulation. *Ann. Rev. Physiol.,* 32:313, 1970.

Mellander, S., and Johansson, B.: Control of resistance, exchange, and capacitance functions in the peripheral circulation. *Pharmcol. Rev.,* 20:117, 1968.

Myers, H. A., and Honig, C. R.: Influence of initial resistance on magnitude of response to vasomotor stimuli. *Amer. J. Physiol.,* 216:1429, 1969.

Ninomiya, I., and Irisawa, H.: Summation of baroceptor reflex effects on sympathetic nerve activities. *Amer. J. Physiol.,* 216:1330, 1969.

Sagawa, K.; Carrier, O.; and Guyton, A. C.: Elicitation of theoretically predicted feedback oscillation in arterial pressure. *Amer. J. Physiol.,* 203:141, 1962.

Sagawa, K.; Ross, J. M.; and Guyton, A. C.: Quantitation of cerebral ischemic pressor response in dogs. *Amer. J. Physiol.,* 200:1164, 1961.

Sagawa, K.; Taylor, A. E.; and Guyton, A. C.: Dynamic performance and stability of cerebral ischemic pressor response. *Amer. J. Physiol.,* 201:1164, 1961.

Sagawa, K., and Watanabe, K.: Summation of bilateral carotid sinus signals in the barostatic reflex. *Amer. J. Physiol.,* 209:1278, 1965.

Schachter, M.: Kallikreins and kinins. *Physiol. Rev.,* 49:509, 1969.

Scher, A. M., and Young, A. C.: Reflex control of heart rate in the unanesthetized dog. *Amer. J. Physiol.,* 218:780, 1970.

Scott, J. B.; Rudko, M.; Radawski, D.; and Haddy, F. J.: Role of osmolarity, K^+, H^+, Mg^{++}, and O_2 in local blood flow regulation. *Amer. J. Physiol.,* 218:338, 1970.

Smith, E. E., and Guyton, A. C.: Center of arterial pressure regulation during rotation of normal and abnormal dogs. *Amer. J. Physiol.,* 204:979, 1963.

Song, S. H.; Lee, W. K.; Chung, Y. A.; and Hong, S. K.: Mechanism of apneic bradycardia in man. *J. Appl. Physiol.,* 27:323, 1969.

Thames, M. D., and Kontos, H. A.: Mechanisms of baroreceptor-induced changes in heart rate. *Amer. J. Physiol.,* 218:251, 1970.

Tsakiris, A. G.; Donald, D. E.; Rutishauser, W. J.; Banchero, N.; and Wood, E. H.: Cardiovascular responses to hypertension and hypotension in dogs with denervated hearts. *J. Appl. Physiol.,* 27:817, 1969.

Warner, H. R., and Cox, A.: A mathematical model of heart rate control by sympathetic and vagus efferent information. *J. Appl. Physiol.,* 17:349, 1962.

Youmans, W. B.; Murphy, Q. R.; Turner, J. K.; Davis, L. D.; Briggs, D. I.; and Hoye, A. S.: The Abdominal Compression Reaction. Baltimore, The Williams & Wilkins Co., 1963.

Zapata, P.; Hess, A.; and Eyzaguirre, C.: Reinnervation of carotid body and sinus with superior laryngeal nerve fibers. *J. Neurophysiol.,* 32:215, 1969.

CHAPTER 26

Regulation of Mean Arterial Pressure; and Hypertension

Most of the basic principles of arterial pressure regulation have been presented in the previous two chapters, Chapter 24 dealing with the intrinsic mechanisms of pressure regulation and Chapter 25 dealing with the rapid reflex controls of pressure. It will be the purpose of the present chapter to see how these two types of regulatory mechanisms work together to maintain normal mean arterial pressure. We will also discuss the clinical problem of *hypertension,* or "high blood pressure," which is caused by abnormalities of arterial pressure regulation. But first let us discuss some of the normal values of arterial pressure in the human being and the clinical method for measuring these pressures.

NORMAL ARTERIAL PRESSURES

Arterial Pressures at Different Ages. Figure 26–1 illustrates the typical diastolic, systolic, and mean arterial pressures from birth to 80 years of age. From this figure it can be seen that the systolic pressure of a normal young adult averages about 120 mm. Hg and his diastolic pressure about 80 mm. Hg—that is, his arterial pressure is said to be 120/80. The

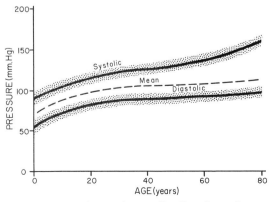

Figure 26–1 Changes in systolic, diastolic, and mean arterial pressures with age. The shaded areas show the normal range.

shaded areas on either side of the curves depict the normal ranges of systolic and diastolic pressures, showing considerable variation from person to person.

The increase in arterial pressures at older ages is usually associated with developing arteriosclerosis. The systolic pressure increases especially; in approximately one-tenth of all old people it rises above 200 mm. Hg.

THE MEAN ARTERIAL PRESSURE

The mean arterial pressure is the average pressure throughout the pressure pulse cycle. Off hand one might suspect that it would be equal to the average of systolic and diastolic pressures, but this is not true; the arterial pressure usually remains nearer to diastolic level than systolic level during a greater portion of the pulse cycle. Therefore, the mean arterial pressure is usually slightly less than the average of systolic and diastolic pressures, as is evident in Figure 26–1.

Mean Arterial Pressure in the Human Being. The mean arterial pressure of the normal young adult averages about 96 mm. Hg, which is slightly less than the average of his systolic and diastolic pressures, 120 and 80 mm. Hg, respectively. However, for purposes of discussion, the mean arterial pressure is usually considered to be 100 mm. Hg because this value is easy to remember.

The mean arterial pressure, like the systolic and diastolic pressures, is lowest immediately after birth. As illustrated in Figure 26–1, it measures about 70 mm. Hg at birth and reaches an average of about 110 mm. Hg in the normal old person or as high as 140 mm. Hg in the person with arteriosclerosis. From adolescence to middle age the mean arterial pressure does not vary greatly from the normal value of 100 mm. Hg.

Significance of Mean Arterial Pressure. Mean arterial pressure is the average pressure tending to push blood through the systemic circulatory system. Therefore, *from the point of view of tissue blood flow, it is the mean arterial pressure that is important.*

CLINICAL METHODS FOR MEASURING SYSTOLIC AND DIASTOLIC PRESSURES

Obviously, it is impossible to use the various rapid response recorders, utilizing needle insertion into an artery as described in Chapter 18, for making routine pressure measurements in human patients, although they are used on occasion when special studies are necessary. Instead, the clinician determines systolic and diastolic pressures by indirect means, most usually by the auscultatory method.

The Auscultatory Method. Figure 26–2 illustrates the auscultatory method for determining systolic and diastolic arterial pressures. A stethoscope is placed over the antecubital artery while a blood pressure cuff is inflated around the upper arm. As long as the cuff presses against the arm with so little pressure that the artery remains distended with blood, no sounds whatsoever are heard by the stethoscope despite the fact that the blood pressure within the artery is pulsating. But when the cuff pressure is great enough to close the artery during part of the arterial pressure cycle, a sound is heard in the stethoscope with each pulsation. These sounds are called *Korotkow sounds.*

The exact cause of Korotkow sounds is still questionable, but they are believed to be caused by blood jetting through the compressed vessel when it opens just enough to allow flow. The jet causes turbulence that sets up the vibrations heard through the stethoscope. Regardless of what causes the sounds to occur, it is an experimental fact that they do occur in the artery when the vessel opens during each pressure cycle.

In determining blood pressure by the auscultatory method, the pressure in the cuff is first elevated well above arterial systolic pressure. Then the cuff pressure is gradually reduced. As long as this pressure is higher than systolic pressure, the brachial artery remains collapsed and no blood whatsoever flows into the lower artery during any part of the pressure cycle. Therefore, no Korotkow sounds are heard in the lower artery. Just as soon as the pressure in the cuff falls below systolic pressure, blood slips through the artery beneath the cuff during the peak of systolic pressure, and one begins to hear *tapping*

Figure 26–2 The auscultatory method for measuring systolic and diastolic pressures.

sounds in the antecubital artery in synchrony with the heart beat. As soon as these sounds are heard, the level is approximately equal to the systolic pressure.

As the pressure in the cuff is lowered still more, the Korotkow sounds continue their tapping quality until the pressure in the cuff falls to equal diastolic pressure. Below this pressure level the artery no longer closes completely during diastole, which means that the basic factor causing the sound is no longer present. Therefore, the tapping quality of the sound suddenly changes to a muffled, thumping quality, and the sounds usually disappear entirely after another 5 to 10 mm. drop in cuff pressure. One notes the manometer pressure when the Karotkow sound quality changes from tapping to thumping, and this pressure is approximately equal to the diastolic pressure.

The auscultatory method for determining systolic and diastolic pressures is not entirely accurate, but it usually gives values within 10 per cent of those determined by direct measurements from the arteries.

Radial Pulse and Oscillometric Methods for Estimating Arterial Pressure. Arterial blood pressure can also be estimated by feeling the radial pulse or by recording the pulsation in the lower arm with an oscillometer while a cuff is inflated over the upper arm. An oscillometer consists of a recording apparatus that can register pulsations in a lightly inflated blood pressure cuff around the forearm. A blood pressure recorder of the aneroid type makes an excellent oscillometer for estimating arterial pressure.

In applying these methods, the pressure in the cuff on the upper arm is raised well above the systolic level and then it is progressively reduced. No pulsation will be felt in the radial artery or registered by the oscillometer until the cuff pressure falls below the systolic pressure level. But, just as soon as this point is reached, a distinct radial pulse can be felt, or a distinct oscillation will be recorded by the oscillometer. Further decrease in the upper arm cuff pressure results, at first, in progressive increase in the intensity of radial or oscillometer pulsation. Then there is a slight decrease in these pulsations at approximately the diastolic level of pressure, but this change is so indistinct that estimation of diastolic pressure is likely to be in severe error.

The oscillometric method is valuable for measuring pressures when distinct sounds cannot be heard from the forearm arteries. This often results in (1) very young children or (2) adults when the arteries are in a state of spasm, such as when a patient is in *shock*. These instances are often among the most important in which it is essential to have at least some estimation of arterial pressure, thus illustrating the value of knowing this accessory method for measuring systolic arterial pressure in addition to the auscultatory method.

REGULATION OF ARTERIAL PRESSURE — THE OVERALL SYSTEM: NERVOUS, INTRINSIC, HUMORAL

Before discussing the overall regulation of cardiac output, it is good to remember the basic relationship between arterial pressure, cardiac output, and total peripheral resistance, which was discussed in detail in Chapter 18, as follows:

Pressure = Cardiac Output
× Total Peripheral Resistance

It is obvious from this formula that any factor that increases either the cardiac output or total peripheral resistance will cause an increase in mean arterial pressure. We have seen in the previous two chapters that both of these factors are manipulated in the control of arterial pressure, but except under temporary conditions, it is usually total peripheral resistance that is manipulated. We shall have more to say about this later in the chapter.

It is good also to remember that approximately 50 per cent of the total resistance to blood flow in the systemic circulation occurs in the arterioles. Furthermore, the arterioles can be constricted or dilated tremendously by nervous, humoral, or local tissue factors. Therefore, the arterioles play the major role in the control of arterial pressure.

RAPIDLY ACTING ARTERIAL PRESSURE REGULATION — NERVOUS REGULATION

It is primarily the circulatory reflexes that keep the arterial pressure from rising extremely high or falling extremely low from minute to minute. For instance, when a person first stands up there is a hydrostatic tendency

for arterial pressure in his head to fall very low, but the nervous reflexes take over immediately and cause vasoconstriction throughout the body, thus maintaining an almost normal pressure in the head. Likewise, when a person bleeds severely and his pressure tends to fall, the nervous reflexes play an immediate role in bringing the arterial pressure back toward normal.

There are three circulatory reflexes that play major roles in arterial pressure regulation: (1) the baroreceptor reflex, (2) the chemoreceptor reflex, and (3) the central nervous system (CNS) ischemic response. However, each of these reflexes operates in a different pressure range, the baroreceptor mechanism operating from 60 to 200 mm. Hg, the chemoreceptor mechanism from 40 to 100 mm. Hg, and the CNS ischemic mechanism from 15 to 50 mm. Hg. Some of the special features of each of these mechanisms are the following:

The Baroreceptor Reflex. When the arterial pressure begins to fall, the baroreceptors are immediately stimulated, and signals are transmitted to the vasomotor center to elicit a reflex increase in peripheral vasoconstriction, thus "moderating" the fall in pressure. The gain of this baroreceptor reflex is approximately 2; this means that two-thirds of the fall in pressure will be corrected by the baroreceptor mechanism. Thus, if the pressure tends to fall from 100 to 40 mm. Hg, it will instead fall only to 80 mm. Hg.

Chemoreceptor Reflex. When the arterial pressure falls below 100 mm. Hg and especially when it falls below 80 mm. Hg, blood flow to the aortic and carotid bodies is depressed. Therefore, delivery of oxygen to these bodies is deficient, and they in turn transmit signals to the vasomotor center to return the pressure back toward normal. Though the gain of this mechanism has not been measured exactly, it probably has approximately the same gain as that of the baroreceptor mechanism and can correct about two-thirds of the further fall in pressure.

The CNS Ischemic Response — The Last-Ditch Stand. When the arterial pressure falls below 50 mm. Hg and especially when it reaches 30 to 20 mm. Hg, the CNS ischemic response becomes extremely active and elicits throughout the body a sympathetic discharge that is about as powerful as can ever be had. Indeed, the gain of the system has been measured to be as high as 11, which means that eleven-twelfths of a further fall in pressure can be corrected. Therefore, when the arterial pressure falls to 20 to 30 mm. Hg, this tremendously powerful sympathetic discharge acts as a "last-ditch stand" to prevent final demise of the animal.

Temporary Nature of the Nervous Regulatory Mechanisms

Unfortunately, these rapidly acting nervous regulatory mechanisms can act maximally for only a few hours or, at most, a few days, for several reasons. First, the baroreceptors adapt within a few days, as illustrated in Figure 26–3, so that they no longer send appropriate signals. Second, the circulatory system adapts partially within a few hours to the sympathetic stimuli. Third, the local blood flow regulatory mechanisms overcome much of the sympathetic response. Therefore, it is doubtful that the rapidly acting nervous arterial pressure regulatory mechanisms are significant in long-term regulation of arterial pressure.

INTERMEDIATELY ACTING ARTERIAL PRESSURE REGULATIONS

In contrast to the nervous regulating mechanisms, which can begin to act within seconds, most intrinsic regulations of the circulation usually require much longer periods of time. However, two of these begin to act within the first few minutes and reach full function within approximately an hour. These are: (1) the capillary fluid shift mechanism and (2) the vascular stress-relaxation mechanism.

Capillary Fluid Shift. When the arterial pressure changes, this is usually also associated with a similar change in capillary pressure. For reasons discussed in Chapter 20, one can readily understand that this will cause fluid to begin moving across the capillary membrane between the blood and the interstitial fluid compartment. Within a few minutes to an hour a new state of equilibrium usually will be achieved, but in the meantime this shift of fluid has played a very beneficial role in the control of arterial pressure. For instance, if the arterial pressure rises too high, loss of fluid through the capillaries into the interstitial spaces causes the blood volume to fall and thereby causes return of arterial pressure back toward normal. The gain of this system is

about 3 to 4, which means that it has about twice as great an effect as the baroreceptor mechanism in returning the arterial pressure back toward its normal mean value; but it takes effect much more slowly than does the baroreceptor mechanism.

Stress-Relaxation. When the arterial pressure falls, the pressure usually also falls in most of the blood storage areas such as in the veins, the liver, the spleen, the lungs, and so forth. Conversely, a rise in arterial pressure is often associated with a rise in pressure in the same storage areas. As pointed out in Chapter 18, a pressure change causes vessels gradually to adapt to a new size, thereby accommodating the amount of blood that is available. This phenomenon is called *stress-relaxation* or *reverse stress-relaxation*. Thus, following massive transfusion, the arterial pressure rises markedly at first, but because of continued relaxation of the circulation during the next 10 minutes to an hour, the arterial pressure returns essentially to normal even though the blood volume may be as great as 30 per cent above normal. Conversely, following severe bleeding, the reverse stress-relaxation mechanism can cause the blood vessels gradually to tighten around the amount of blood that is left, thereby again re-establishing almost normal circulatory dynamics.

Unfortunately, the stress-relaxation mechanism has very definite limits, so that acute changes in blood volume of more than approximately plus 30 per cent or minus 15 per cent cannot be corrected by this mechanism.

LONG-TERM ARTERIAL PRESSURE REGULATION—ROLE OF THE KIDNEY

Now we come to the truly important mechanism for regulating arterial pressure, the long-term regulation of pressure day in and day out and month in and month out, without which we could not live. This regulation seems to be vested mainly, if not almost entirely, in the kidneys, and the kidneys seem to function mainly by controlling the fluid volumes of the body, as was discussed in relation to the intrinsic fluid volume mechanism for overall circulatory regulation in Chapter 24. This basic mechanism is so important that it will be described briefly once more at this point, but it would also be wise to study its details in Chapter 24 if these are not already well understood.

The Renal–Fluid Volume Mechanism for Arterial Pressure Regulation. The output of the kidneys is highly responsive to the arterial

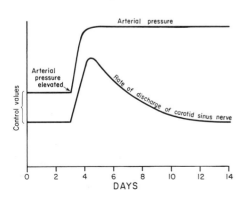

Figure 26–3 Adaptation of the carotid sinus baroreceptors following initial strong stimulation by increased arterial pressure.

pressure level. For instance, an increase in arterial pressure from 100 to 200 mm. Hg increases the urinary output of both water and salt approximately six-fold. Conversely, a decrease in arterial pressure causes fluid and salt retention. Thus, the kidney is geared to act as a feedback regulator of body fluid volumes and, at the same time, arterial pressure. That is, if the arterial pressure falls too low, the extracellular fluid volume continues to rise until the arterial pressure returns to normal. The overall schema for this effect is illustrated in Figure 26-4. Conversely, if the arterial pressure rises too high, excess quantities of fluid are lost, causing decreased extracellular fluid volume, blood volume, mean systolic pressure, venous return, and arterial pressure. Thus, the arterial pressure once again returns to normal. The phenomenon of long-term local tissue blood flow regulation (autoregulation) plays an especially important role in keeping the cardiac output relatively normal even though the renal-fluid volume mechanism alters the arterial pressure markedly. The details of this dissociation of arterial pressure and cardiac output regulation were presented in Chapter 24.

GRAPHICAL REPRESENTATION OF THE RENAL-FLUID VOLUME–ARTERIAL PRESSURE REGULATION MECHANISM. The renal–fluid volume mechanism for regulating arterial pressure can be represented graphically in a very simple manner, as illustrated in Figure 26-5. This figure contains four separate curves representing two levels of fluid intake and two different functional states of the kidney. Look

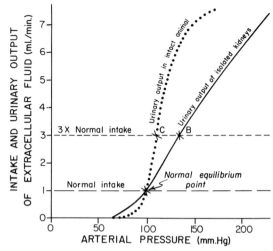

Figure 26-5 Equilibration of extracellular fluid intake and output. This figure shows two levels of fluid intake and, also, two different function curves for the kidneys: (1) for isolated kidneys and (2) for kidneys in the intact person in whom the aldosterone feedback mechanism is operative.

first at the curve representing normal intake of fluid and also at the curve representing urinary output of the isolated kidneys. If the arterial pressure were 150 mm. Hg, the urinary output as shown on the kidney function curve would be approximately four times the normal intake. Therefore, the body would continue to lose extracellular fluid until the arterial pressure would fall back to 100 mm. Hg, which is the point where the output of urine from the two isolated kidneys would equal the normal intake of fluid.

As another example, let us assume that the intake of extracellular fluid, as shown by the upper horizontal curve, increases to three times normal but that the arterial pressure is still 100 mg. Hg. Now, the output of extracellular fluid is one-third the intake, and therefore the extracellular fluid volume and arterial pressure continue to rise until the pressure reaches 133 mm. Hg, at which point (point B in Figure 26-5) the urinary output once again equals the intake.

Thus, there is continual readjustment of balance between intake and output of extracellular fluid, and in the process of these readjustments the arterial pressure is controlled to that level which will give exact balance between intake and output.

Importance of Aldosterone in the Renal-Fluid Volume–Arterial Pressure Regulating System. In Chapter 25 it was pointed out that aldosterone greatly helps to control extracellular fluid salt and water and, also, extracellular fluid volume. We can now see how aldosterone

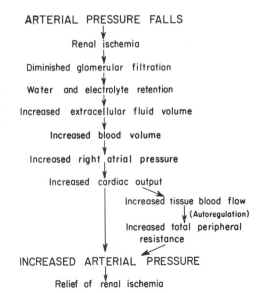

Figure 26-4 The extracellular fluid volume mechanism of arterial pressure regulation.

plays an important role in the overall renal-fluid volume–arterial pressure regulating system.

When arterial pressure falls too low, so that the tissues become poorly perfused with blood, the adrenal cortices secrete increased quantities of aldosterone. The aldosterone in turn acts on the kidneys to prevent loss of salt and water. Conversely, whenever there is more than adequate perfusion of the tissues, the output of aldosterone decreases. Unfortunately, we still do not know the precise factor that regulates the output of aldosterone; we know only that the rate of aldosterone secretion is in some way inversely related to the adequacy of circulation throughout the body. The theories relating to this mechanism will be discussed in Chapters 36 and 76.

Yet, even though we do not know the precise factors that increase or decrease aldosterone secretion, we can use the fact that aldosterone secretion tends to decrease at high arterial pressures and to increase at low pressures to show how this mechanism interacts with the renal–arterial pressure regulating mechanism, as follows:

When arterial pressure increases, there is a direct effect of the increased pressure acting on the kidneys to increase urinary output. Also, there is an indirect effect operating through the aldosterone mechanism to increase urinary output; that is, the increased pressure decreases the aldosterone output, and the decrease in aldosterone allows excessive loss of salt and water from the kidneys. Therefore, at each given increase in arterial pressure, the output of water and salt is much greater from the kidneys if the aldosterone feedback mechanism is operative than is true if it is not operative.

Conversely, at low pressures the decrease in arterial pressure decreases urinary output (a) directly and (b) indirectly through the aldosterone mechanism; that is, decreased pressure causes excessive production of aldosterone that in turn causes the kidneys to retain salt and water. Therefore, for any given decrease in arterial pressure, there is much more decrease in urinary output when the aldosterone mechanism is operative than when it is not operative.

The net effect, therefore, of the aldosterone mechanism is to increase greatly the steepness of the pressure–renal output curve, and this can be seen by comparing the dotted renal function curve of Figure 26–5 with the solid curve. This increased steepness makes the renal–fluid volume mechanism for regulating arterial pressure far more effective than it would be were it not for the aldosterone mechanism.

Now, let us consider again a three-fold increase in extracellular fluid intake. The equilibrium point between the dotted curve and the elevated intake curve in Figure 26–5 is at point C. This is in contrast to point B, when the aldosterone mechanism was not functioning. The pressure at point C is 9 mm. Hg above normal, whereas the pressure at point B was 33 mm. Hg above normal, almost a four-fold difference. Therefore, one would predict that the aldosterone mechanism should increase the effectiveness of the renal–fluid volume mechanism for controlling arterial pressure by perhaps three to four-fold. And, as would be expected, arterial pressure is considerably less stable in the absence of the adrenal cortices than it is when the cortices are intact and functioning normally. (Unfortunately, only a few points on the dotted curve of Figure 26–5 have thus far been measured experimentally. Consequently, the overall quantitative importance of the aldosterone mechanism is still questionable. The dotted curve, therefore, is more an estimate to demonstrate the principle than it is an exact quantitative truth.)

Time Required for the Renal–Fluid Volume Mechanism to Act. In acute situations, such as following severe loss of blood, the fluid volume mechanism begins to return arterial pressure back toward normal within hours. But often, almost a day will lapse before significant response can be seen; and in general, it takes 3 to 10 days for the mechanism to come to complete equilibrium. This delay in response is caused both by the time required to increase or decrease the fluid volume in the body and the time required for the aldosterone feedback mechanism to act. Therefore, one can understand the extreme importance of the rapidly acting and intermediately acting pressure control mechanisms discussed earlier, for these tide the circulation over until the long-term regulatory mechanism can take effect.

Nervous Effect on the Long-Term Renal Mechanism for Regulating Arterial Pressure. The kidneys, like most other tissues of the body, are strongly supplied by sympathetic nerves, and the degree of sympathetic stimulation can alter renal function tremendously. Therefore, nervous signals can alter the long-term regulation of arterial pressure. For instance, experiments have shown that when the sympathetic nerves to the kidneys are stimulated continuously for several weeks, renal retention of fluid occurs and causes chronically elevated arterial pressure as long as the sympathetic stimulation continues. Therefore, it is possible for nervous stimulation of the kidneys to cause chronic elevation of arterial pressure.

Role of Renin and Angiotensin in the Control of Arterial Pressure. Later in the chapter we shall see that the kidneys are capable of secreting an enzyme called *renin,* and the renin in turn acts on the plasma proteins to cause release of a vasoconstrictor substance called *angiotensin.* The angiotensin in turn increases total peripheral resistance and also increases aldosterone secretion by the adrenal cortex. Therefore, it is tempting to believe that renin and angiotensin might also play roles in the normal renal mechanism for control of arterial pressure. At present there is no proof that the renin–angiotensin system does play a significant role in normal regulation of arterial pressure, but there is reason to believe that this system does cause hypertension in certain abnormal states. Therefore, it will be discussed in detail later in the chapter in relation to hypertension and is mentioned here only to recognize the fact that it might also be an accessory factor in regulating even normal arterial pressure.

HYPERTENSION

Hypertension means, simply, high arterial pressure. This subject is extremely important because about 12 per cent of all persons die as a direct result of hypertension, and about 20 per cent of all people can expect to have high blood pressure at some time during their lives.

Relative Importance of Total Peripheral Resistance and Cardiac Output in Causing Hypertension. Since arterial pressure is the product of cardiac output times total peripheral resistance, it is obvious that an increase in either of these two factors can cause hypertension. However, in most types of hypertension, one finds that the total peripheral resistance is greatly increased while the cardiac output is very near to normal. This has usually been interpreted as meaning that hypertension is caused by some primary factor that increases the resistance. As a result, many research workers have searched for vasoconstrictor substances circulating in the blood, hoping to prove that one of these constricts the peripheral vessels and thereby causes hypertension. To date, there is no proof that such a vasoconstrictor substance is the usual cause of hypertension, though several substances have been postulated, as will be discussed later in the chapter.

CAN A PRIMARY INCREASE IN CARDIAC OUTPUT CAUSE INCREASED TOTAL PERIPHERAL RESISTANCE? If there should be no vasoconstrictor substance to increase the total peripheral resistance in hyper-

tension, how could the measured increase in resistance come about? One answer to this would be a primary increase in cardiac output followed by activation of the local blood flow regulation mechanism discussed in Chapter 24. That is, if the blood volume increases and in turn causes an increase in cardiac output, there would be too much blood flow through the tissues; in response, the local flow regulating mechanism in the tissues would gradually cause the total peripheral resistance to increase. It was pointed out in Chapter 24 that the long-term local flow regulatory mechanism is probably very powerful. It takes days to weeks to react fully, but when it does do so, a primary increase in cardiac output is followed by an increase in total peripheral resistance but yet a return of the cardiac output essentially to its normal value.

Therefore, the concept that most hypertension is caused by a primary increase in total peripheral resistance could easily be a false assumption, for it is obviously possible for a primary increase in cardiac output to cause a secondary rise in total peripheral resistance.

Under What Conditions Can an Increase in Total Peripheral Resistance Cause Hypertension? Lessons to be Learned from the A-V Fistula. A recent computer analysis of arterial pressure regulation has suggested that an increase in total peripheral resistance *per se* without an increase in the resistance of blood flow through the kidneys will cause only temporary hypertension. The reason for this is that elevation of arterial pressure will immediately cause excessive loss of extracellular fluid through the kidneys, thereby setting into play the fluid volume mechanism for regulation of arterial pressure and returning the arterial pressure back to that level at which renal output of fluid equals the daily intake of fluid; that pressure level is the normal arterial pressure. That this computer analysis is correct is proved by the fact that opening or closing an A-V fistula, which can decrease total peripheral resistance as much as 60 per cent when the fistula is opened or can increase total peripheral resistance as much as 150 per cent when it is closed, has only a temporary effect on arterial pressure. For instance, the 150 per cent increase in total peripheral resistance when a fistula is closed might elevate the arterial pressure temporarily as much as 60 mm. Hg. However, within hours and, certainly, within days the arterial pressure is already back to normal. This decrease in pressure back to normal is accompanied by a decrease in extracellular fluid volume and blood volume. Therefore, one can see that primary changes in total peripheral resistance are not enough by themselves to regulate arterial pressure.

Another example of this effect occurs in a human being who has lost all four limbs. His total peripheral resistance is approximately 60 per cent above normal, and yet his arterial pressure is exactly normal.

However, in each of the above examples, the A-V fistula and the person who has lost all four limbs, there is no change in resistance to blood flow through the kidneys. On the other hand, many conditions increase resistance to renal blood flow at the same time that they increase resistance elsewhere in the body. When this occurs, the kidneys retain fluid, so that the secondary ability of the kidneys to reduce the arterial pressure back to normal does not result.

Therefore, one can readily see that increased total peripheral resistance can cause chronic hypertension only when the renal resistance participates in this increase. Further, it is mainly increased afferent arteriolar resistance in the kidneys that is important, because it is this resistance that mainly causes fluid retention. This overriding importance of renal resistance in relation to hypertension has recently been demonstrated by injecting microspheres into the circulation. When these are injected at the root of the aorta so that they are dispersed everywhere in the circulation, thereby blocking many small vessels and increasing peripheral resistance everywhere in the body, only temporary hypertension results. When they are injected into the renal artery to increase renal arteriolar resistance, sustained hypertension develops.

Keeping these basic interrelationships between cardiac output and total peripheral resistance in mind, and especially keeping in mind the overall importance of the kidneys and renal resistance in the development of hypertension, we can now discuss some of the basic causes of hypertension.

HYPERTENSION INVOLVING THE KIDNEY

In general, anything that decreases the overall function of the kidney causes extracellular fluid retention and hypertension. Consequently, hypertension can be caused by many different conditions involving the kidneys. These include the following:

Removal of Kidney Tissue. Removing one kidney and then removing more and more of the second kidney causes progressive elevation of the arterial pressure, even though the remaining portion of the kidney continues to function normally. Usually, more than one half of the second kidney must be removed before significant hypertension results.

Entire Removal of Both Kidneys — "Renoprival Hypertension." With the advent of artificial kidneys, more and more human beings are being kept alive with no functional kidney mass at all. When the person without kidneys is properly dialyzed, his arterial pressure can almost always be kept in a normal range. But, on the other hand, if the amount of water and salt in the patient's body is allowed to build up, so also does his arterial pressure increase. The increase in pressure does not occur instantaneously with the increase in extracellular fluid volume; instead, it occurs a few days later, preceded by a transient increase in cardiac output. This is what would be predicted from the renal-fluid volume–arterial pressure regulatory mechanism discussed in Chapter 24 in relation to intrinsic circulatory control. That is, increase in fluid volume causes a transient increase in cardiac output, and the excess blood flow through the tissues during the transient increase in cardiac output causes progressive increase in total peripheral resistance because of the ability of the tissues to readjust their flows to normal by increasing the resistance.

If the diseased kidneys are left in the patient, it is sometimes very difficult to decrease his arterial pressure to a normal level by dialysis. The cause of this effect is believed to be secretion of vasoconstrictor substances by the diseased kidneys, and this will be discussed in more detail later.

Hypertension Caused by Renal Insufficiency Plus Water and Salt Loading. Renal hypertension occurs very commonly in patients with severely damaged kidneys that cannot excrete adequate quantities of salt. Retention of salt in the body causes secondary retention of water, as will be explained in Chapter 36, thereby building up the extracellular fluid volume. Concomitantly, the patient develops hypertension. It is surprising how little the extracellular fluid volume has to be increased above normal, only 5 to 10 per cent, to cause this effect, but the smallness of this increase is predicted by the fluid volume mechanism for arterial pressure regulation that was discussed in Chapter 24.

This type of hypertension can be created experimentally in animals, a typical example of which is illustrated in Figure 26–6. This figure illustrates average results from four dogs in which 70 per cent of the renal tissue was removed during the first few weeks of the experiment. The arterial pressure remained normal until the animals were required to drink 0.9 per cent saline solution instead of the usual drinking water. Within a few days the arterial pressure had increased almost 50 per cent. Two weeks later the animals were given pure water to drink once again, and the arterial pressure fell back to normal within 24 hours. A few weeks later the animals were again required to drink 0.9 per cent saline solution, and hypertension reappeared immediately. Measurements of cardiac output and total peripheral resistance in these experiments showed that the cardiac output rises about 35 per cent immediately after the animals begin drinking the saline solution, but the total peripheral resistance does not begin to rise until two to three days later. Then, the total peripheral resistance rises approximately 40 per cent while the cardiac output returns almost all the way back to normal. It has been experiments such as these that have demonstrated the importance of fluid volume retention and the local tissue blood flow regulatory mechanism in the causation of many types of renal hypertension.

Goldblatt Hypertension. Almost 40 years ago Goldblatt demonstrated that hypertension develops regularly in animals whose renal arteries are constricted. Furthermore, the degree of hypertension can be altered by the degree of renal artery constriction. Therefore, this preparation has proved to be extremely valuable in studying renal

Figure 26–6 Effect on arterial pressure of requiring four dogs that had had 70 per cent of their renal tissue removed to drink 0.9 per cent saline solution instead of water. (Courtesy Dr. Jimmy B. Langston.)

hypertension, for which reason the hypertension caused in this manner is called "Goldblatt" hypertension.

Figure 26–7 illustrates a typical arterial pressure record in a dog whose renal arteries are constricted. Note that immediately after constriction the renal arterial pressure falls to a very low value, but gradually over a period of one to two weeks the systemic arterial pressure rises to a very high level while the renal arterial pressure returns either to normal or almost to normal. Thus, once again, the principle that the kidney regulates its own arterial pressure to a normal mean value is demonstrated.

Goldblatt hypertension is caused at least partially by an increase in extracellular fluid volume that is caused by failure of the kidneys to excrete fluid when the clamps are first applied. The volume increases about 35 per cent and remains increased until the arterial pressure rises to a high value, at which time the volume decreases back almost to the normal level.

However, there is also reason to believe that Goldblatt kidneys secrete renin, which in turn causes vascular constriction throughout the body, as subsequently described. Therefore, part of the increase in pressure in Goldblatt hypertension might also be ascribed to this vasoconstrictor mechanism.

The Renin-Angiotensin Mechanism. We have several times suggested that some types of hypertension might be caused or partially caused by a vasoconstrictor substance secreted by the kidneys. The substance *renin* is secreted into the bloodstream from the kidneys under some conditions, and it can indirectly cause vasoconstriction throughout the body. Therefore, it deserves special comment as follows:

When blood flow through the kidneys is so poor that it causes renal ischemia, the *juxtaglomerular cells* (cells located in the walls of the afferent arterioles immediately adjacent to the glomeruli) are believed to secrete renin, which, it is postulated, can elevate the arterial pressure and thereby correct the poor renal blood flow. Figure 26–8 illustrates the schema of events that occurs in this mechanism, as follows:

Renin itself is an enzyme that catalyzes the conversion of one of the plasma proteins, a glycoprotein called *angiotensinogen,* into a substance called *angiotensin I.* This reaction occurs within seconds to minutes after the renin has been secreted into the blood. Within another minute or so, angiotensin I is converted into *angiotensin II* by an enzyme already present in the lung tissue and called *converting enzyme.* Angiotensin II persists in the blood for 30 seconds to 2 minutes, but it is gradually inactivated by a number of different blood and tissue enzymes collectively called *angiotensinase.*

During its persistence in the blood, angiotensin II has two effects which can elevate arterial pressure: (1) It causes *vasoconstriction of the systemic arterioles* throughout the body; the arteriolar constriction in turn increases the total peripheral resistance and thereby increases the arterial pressure. (2) It causes *increased production of aldosterone* by the adrenal cortex, and the aldosterone in turn acts on the kidneys to cause salt and water retention, thereby elevating the arterial pressure as a result of increased extracellular fluid volume.

Quantitative Importance of the Renin-Angiotensin Mechanism in Renal Hypertension. In most types of hypertension, chemical measurements of the concentrations of renin or angiotensin in the circulating blood have shown that the quantities of these are probably much too little to be the cause of the hypertension. Therefore, most renal hypertension is probably caused by water and salt retention or other factors besides the renin-angiotensin mech-

Figure 26-7 Changes in systemic arterial pressure and renal arterial pressure distal to the clamp following renal artery constriction.

anism. However, there are at least two conditions in which the renin-angiotensin mechanism does seem to play a role: (1) in hypertension caused by unilateral kidney disease and (2) in hypertension caused by diseased kidneys in renoprival patients.

"GOLDBLATT" HYPERTENSION CAUSED BY UNI-LATERAL KIDNEY DISEASE. Occasionally a person develops severe hypertension when one of his kidneys is diseased even though the second kidney is entirely normal. It has been difficult to understand how hypertension could develop under these conditions because the normal kidney theoretically would excrete large volumes of extracellular fluid in response to the elevated arterial pressure, thereby returning the arterial pressure essentially

Figure 26-8 The renin-angiotensin theory of arterial pressure regulation.

to normal. At present, the only logical explanation for this type of hypertension is based on the renin-angiotensin mechanism in the following way: When angiotensin is formed in response to the diseased kidney, this theoretically could affect the opposite kidney to decrease its loss of extracellular fluid. It could do this in two separate ways: The angiotensin could cause direct vasoconstriction of the afferent arterioles of the good kidney, or the angiotensin could cause secretion of aldosterone which in turn would inhibit loss of salt and water by the good kidney. In either event, the good kidney would retain water and salt and thereby would not return the arterial pressure back to normal.

HYPERTENSION CAUSED BY DISEASED KIDNEYS IN THE RENOPRIVAL STATE. Occasionally a person's kidneys have been damaged so severely that he has essentially zero renal excretory function, and his life must be maintained by dialysis with the artificial kidney. Yet, as long as the two diseased kidneys are still in his body, it is often very difficult to return this arterial pressure to normal. Upon removal of the two diseased kidneys, the hypertension can usually then be easily corrected by dialysis. Therefore, here again it is believed that the renin-angiotensin mechanism might be responsible for a peripheral vasoconstrictor state that makes it difficult for the dialysis procedure to control the hypertension.

HYPERTENSION IN COARCTATION OF THE AORTA. One of the most interesting examples of the Goldblatt type of hypertension occurs in persons who have *coarctation of the aorta;* in this condition, the aorta is greatly constricted or totally occluded, usually in its thoracic portion. Therefore, blood cannot flow through the aorta to the lower part of the body but instead must flow through collateral vessels, such as the internal mammary arteries and arteries along the lateral chest wall and ribs. The resistance to blood flow from the upper part of the body to the lower part through these accessory channels is great. As a result, the arterial pressure in the upper part of the body is usually 50 to 100 mm. Hg greater than that in the lower part. Yet the pressure below the aortic constriction is essentially normal because the kidneys maintain their pressure at a normal value, while pressure in the upper aorta is always greatly elevated.

It is even more interesting that the blood flow to the tissues is approximately normal in both the upper and lower body. That is, the vascular resistance is very high in the upper body but normal in the lower body, and the normal resistance in the lower body demonstrates that there probably is little if any generalized vasoconstrictor substance acting on the circulation.

Summary of the Mechanisms of Renal Hypertension

From the foregoing discussions it is clear that there are probably two ways in which renal insufficiency and renal disease can elevate the arterial pressure. The first of these is simply retention of water and salt, causing a buildup of fluid volumes and subsequent rise in arterial pressure as described in Chapter 24 in relation to the fluid volume mechanism for arterial pressure regulation. The second factor that might at times be at least partially responsible for renal hypertension is secretion of renin, which in turn causes formation of angiotensin, the angiotensin then causing vasoconstriction or increased secretion of aldosterone. Though a tremendous amount has been written about this renin mechanism, its significance has been proved only under special conditions rather than in the vast majority of cases.

NEUROGENIC TYPES OF HYPERTENSION

Strong stimulation of the sympathetic nervous system, as might occur in severe emotional crises or in severe states of anxiety, can cause very severe acute elevation of arterial pressure, sometimes to mean pressure levels as high as 200 mm. Hg and systolic pressures above 350 mm. Hg. However, these hypertensive episodes usually last only for a few minutes or at most a few hours. On the other hand, a few neurogenic conditions can cause chronic elevation of the arterial pressure as follows:

Brain Ischemia. Prolonged ischemia of the brain caused experimentally by progressively ligating almost all arteries to the brain will cause chronic hypertension, presumably by initiating the CNS ischemic response, which in turn causes chronic stimulation of the sympathetic nervous system. This type of hypertension can also be caused by direct compression of the vasomotor center by a brain tumor, but this occurs only rarely.

Baroreceptor Denervation. Denervation of all the baroreceptors in an experimental animal sometimes causes chronic hypertension, though at other times, for peculiar reasons, it causes only a transient hypertension lasting for a few days. The cause of this hypertension is believed to be release of the vasomotor center from the normal inhibitory tone caused by baroreceptor impulses. The vasomotor center thereafter supposedly transmits excess numbers of sympathetic impulses throughout the body.

Pheochromocytoma. Hypertension is occasionally caused by a small tumor, called a pheochromocytoma, that develops in one of the adrenal medullae or, rarely, in other chromaffin tissue of the sympatho-adrenal system. This tumor is composed of neuron-like cells similar to the normal secreting cells of the adrenal medulla. It secretes epinephrine and norepinephrine. However, it usually secretes these only in response to sympathetic stimuli in the same manner that the normal adrenal medullae secrete these two hormones.

The pheochromocytoma secretes tremendously more epinephrine and norepinephrine than do the normal adrenal medullae. Therefore, when a person who has a pheochromocytoma becomes excited or for any other reason increases his level of sym-

pathetic activity, these two hormones pour into his circulation, and his arterial pressure goes sky-high. The person can frequently be treated by giving a drug that blocks the activity of the sympathetic nervous system, but obviously the more effective procedure is to remove the tumor.

Sometimes a pheochromocytoma secretes norepinephrine and epinephrine continually. Occasionally as much as 600 to 2700 micrograms of these hormones appear in the urine each day in contrast to the normal rate of 15 to 40 micrograms. This continual secretion of these vasoconstrictor hormones obviously can cause continued elevation of arterial pressure. The hormones cause constriction of the afferent arterioles of the kidney as well as constriction of blood vessels elsewhere in the body; therefore, the usual ability of the kidneys to return the arterial pressure back to the normal level is blocked by the direct action of the hormone on the kidneys themselves.

"ESSENTIAL" HYPERTENSION

The term essential hypertension means hypertension without known cause. This hypertension is also frequently called *idiopathic hypertension,* a term that also means hypertension without known cause. Any person having an arterial pressure greater than 140/90 and who also has no obvious cause for the hypertension is considered to have essential hypertension. About 90 per cent of all persons who have hypertension have this essential variety.

Inheritance of Essential Hypertension. Almost all essential hypertensive patients are the descendants of one or more parents or grandparents who have also had essential hypertension. Thus, the condition appears to be inherited.

Role of the Kidneys in Essential Hypertension. Because of the dominant role of the kidneys in so many known types of hypertension, most research workers have the feeling that the kidneys are in some way involved in essential hypertension. Indeed, one important functional abnormality of the kidneys in this disease is known to occur as follows: The kidneys require a far higher than normal arterial pressure to make them excrete a normal amount of urine. For instance, if the arterial pressure of an essential hypertensive person is dropped from his usual level of 150 mm. Hg down to 100 mm. Hg, which is the proper level for a normal person, his kidneys excrete only one-quarter to one-sixth the normal amount of urine. Thus, the person with essential hypertension definitely requires a far higher than normal arterial pressure to make his kidneys operate normally.

Therefore, though one cannot show definite disease in the kidneys of persons who have essential hypertension (except for thickened arteriolar walls), the kidneys are nevertheless functionally different from those of the normal person. Some research workers believe that this functional difference is an effect of the hypertension rather than the cause, but there is equally as much reason to believe that it is this functional abnormality that actually causes the hypertension.

Treatment of Essential Hypertension by Decreasing Extracellular Fluid Volume. The kidneys of the essentially hypertensive person can be made to excrete extra quantities of extracellular fluid by giving a drug called a *natriuretic,* which promotes the loss of salt into the urine; and the salt carries with it an appropriate volume of water as well. Therefore, the extracellular fluid volume becomes decreased, and the pressure of many essential hypertensive patients can in this way be lowered to normal. Indeed, it is amazing how little extracellular fluid volume must be removed from the patient to reduce his arterial pressure, an amount usually less than 1 liter.

Likewise, limiting the amount of salt in the food can help to decrease the extracellular fluid volume and thereby help to control the arterial pressure.

The fact that essential hypertension can usually be effectively treated by reducing extracellular fluid volume is further reason to believe that essential hypertension is caused basically by functional retention of water and salt by the kidneys.

HYPERTENSION RESULTING FROM ADRENOCORTICAL HYPERSECRETION — PRIMARY ALDOSTERONISM

Rarely a small tumor develops in this *zona glomerulosa* of the *adrenal cortex* which secretes large quantities of aldosterone, causing the condition called *primary aldosteronism,* which will be discussed in detail in Chapter 76. The aldosterone usually increases the sodium concentration and extracellular fluid volume only slightly, though in severe degrees of primary aldosteronism the person is likely to become edematous, illustrating that basically the condition can cause extreme increases in extracellular fluid volume. And it causes hypertension in almost all patients.

The hypertension in this condition is probably another example of the increased extracellular fluid volume type of hypertension. However, some research workers have suggested that aldosterone has a direct effect on the arterioles to cause arteriolar constriction or to sensitize the arterioles to the constrictor influences of the sympathetic nerve impulses.

EFFECTS OF HYPERTENSION ON THE BODY

Hypertension can be very damaging because of two primary effects: (1) increased work load on the heart and (2) damage to the arteries themselves by the excessive pressure.

Effects of Increased Work Load on the Heart. Cardiac muscle, like skeletal muscle, hypertrophies when its work load increases. In hypertension, the very high pressure against which the left ventricle must beat causes it to increase in weight to as great as 300 to 400 gm. instead of the usual weight of about 150 gm. This increase is not accompanied

by quite as much increase in coronary blood supply as there is increase in muscle tissue itself. Therefore, *relative ischemia* of the left ventricle develops as the hypertension becomes more and more severe. In the late stages of hypertension, this can become serious enough that the person develops angina pectoris. Also, the very high pressure in the coronary arteries causes rapid development of coronary arteriosclerosis so that hypertensive patients tend to die of coronary occlusion at much earlier ages than do normal persons.

Effects of the High Pressure in the Arteries. High pressure in the arteries not only causes coronary sclerosis but also sclerosis of blood vessels throughout all the remainder of the body. As explained in Chapter 31, the arteriosclerosis process causes blood clots to develop in the vessels and also causes the blood vessels to become weakened. Therefore these vessels frequently thrombose, or they rupture and bleed severely. In either case, marked damage can occur in organs throughout the body. The two most important types of damage which occur in hypertension are:

1. Cerebral hemorrhage, which means bleeding of a cerebral vessel with resultant destruction of local areas of brain tissue.

2. Hemorrhage of the renal vessels, which destroys large areas of the kidneys and therefore causes progressive deterioration of the kidneys and further exacerbation of the hypertension.

To illustrate the damaging effects of hypertension on the small vessels of the microcirculation, Figure 26-9 shows the retina of a person with malignant hypertension. This figure shows edema of the optic disc, spasm of the arterioles, dark areas in the retina that are the result of hemorrhages from the small arteries, and very light areas in the retina that represent protein that has exuded out of the blood vessels into the surrounding tissues.

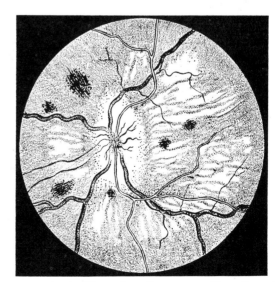

Figure 26-9 The retina of a person with malignant hypertension, illustrating edema of the optic disc, spasm of the arterioles, dark areas of retinal hemorrhages, and light areas of protein exudate in the retina.

REFERENCES

Abbrecht, P. H.; Vander, A. J.; and Turcotte, J. G.: Effects of saline loading on the renin, erythropoietin, and blood pressure responses to canine renal allotransplantation. *Circ. Res., 25*:99, 1969.

Alexander, N., and DeCuir, M.: Sinoaortic baroreflex system and early pressure rise in renal hypertensive rabbits. *Amer. J. Physiol., 213*:701, 1967.

Alexander, N.; Goldfarb, T.; and Drury, D. R.: Cardiac performance of hypertensive aorta-constricted rabbits. *Circ. Res., 10*:11, 1962.

Blalock, A.: Experimental hypertension. *Physiol. Rev., 20*:159, 1940.

Brubacher, E. S., and Vander, A. J.: Sodium deprivation and renin secretion in unanesthetized dogs. *Amer. J. Physiol., 214*: 15, 1968.

Carretero, O., and Gross, F.: Renin substrate in plasma under various experimental conditions in the rat. *Amer. J. Physiol., 213*:695, 1967.

Champlain, J. de; Mueller, R. A.; and Axelrod, J.: Turnover and synthesis of norepinephrine in experimental hypertension in rats. *Circ. Res., 25*:285, 1969.

Coleman, T. G., and Guyton, A. C.: Hypertension caused by salt loading in the dog. III. Onset transients of cardiac output and other circulatory variables. *Circ. Res., 25*:153, 1969.

Conway, J.: Changes in sodium balance and hemodynamics during development of experimental renal hypertension in dogs. *Circ. Res., 22*:763, 1968.

Crawford, M. P.; Richardson, T. Q.; and Guyton, A. C.: Renal servocontrol of arterial blood pressure. *J. Appl. Physiol., 22*:139, 1967.

Dahl, L. K.; Knudsen, K. D.; Heine, M. A.; and Leitl, G. J.: Effects of chronic excess salt ingestion: modification of experimental hypertension in the rat by variations in the diet. *Circ. Res., 22*:11, 1968.

Douglas, B. H.; Guyton, A. C.; Langston, J. B.; and Bishop, V. S.: Hypertension caused by salt loading. II: Fluid volume and tissue pressure changes. *Amer. J. Physiol., 207*:669, 1964.

Ebihara, A., and Grollman, A.: Pressor activity of renal venous effluent following constriction of renal artery in dogs. *Amer. J. Physiol., 214*:1, 1968.

Fregly, M. J.: Production of hypertension in adrenalectomized rats given hypertonic salt solution to drink. *Endocrinology, 66*:240, 1960.

Freis, E. D.: Hemodynamics of hypertension. *Physiol. Rev., 40*:27, 1960.

Friedman, S. M., and Friedman, C. L.: Electrolytes in the regulation of vascular reactivity. *In* Fishman, A. P., and Hecht, H. H. (eds.): The Pulmonary Circulation and Interstitial Space. Chicago, University of Chicago Press, 1969, p. 173.

Granger, H. J., and Guyton, A. C.: Autoregulation of the total systemic circulation following destruction of the central nervous system in the dog. *Circ. Res., 25*:379, 1969.

Guyton, A. C.: Acute hypertension in dogs with cerebral ischemia. *Amer. J. Physiol., 154*:45, 1948.

Guyton, A. C.: The ratio of arterial to venous resistance as a determinant of arterial pressure. *Amer. J. Physiol., 183*:623, 1955.

Guyton, A. C., and Coleman, T. G.: Quantitative analysis of the pathophysiology of hypertension. *Circ. Res., 24*:I-1, 1969.

Guyton, A. C.; Coleman, T. G.; Fourcade, J.; and Navar, L. G.: Physiological control of arterial pressure. *Bull. N. Y. Acad. Med., 45*:811, 1969.

Haas, E., and Goldblatt, H.: Kinetic constants of the human renin and human angiotensinogen reaction. *Circ. Res., 20*:45, 1967.

Horton, E. W.: Hypotheses on physiological roles of prostaglandins. *Physiol. Rev., 49*:112, 1969.

Johnston, C. I.; Davis, J. O.; Robb, C. A.; and Mackenzie, J. W.: Plasma renin in chronic experimental heart failure and during renal sodium "escape" from mineralocorticoids. *Circ. Res., 22*:113, 1968.

Kezdi, P.: Baroreceptors and Hypertension. New York, Pergamon Press, Inc., 1968.

Kezdi, P., and Geller, E.: Baroreceptor control of postganglionic sympathetic nerve discharge. *Amer. J. Physiol., 214*:427, 1968.

Langston, J. B.; Guyton, A. C.; Douglas, B. H.; and Dorsett, P. E.: Effect of changes in salt intake on arterial pressure and renal function in partially nephrectomized dogs. *Circ. Res., 12*:508, 1963.

Louis, W. J., and Jerums, G.: Tachyphylaxis to alpha- and beta-angiotensin in dogs perfused with Ringer's solution or blood. *Circ. Res., 22*:75, 1968.

McCubbin, J., and Page, I.: Renal Hypertension. Chicago, Year Book Medical Publishers, Inc., 1968.

Meyer, P.; Menard, J.; Papanicolaou, N.; Alexandre, J. M.; Devaus, C.; and Milliez, P.: Mechanism of renin release following furosemide diuresis in rabbit. *Amer. J. Physiol., 215*: 908, 1968.

Miller, R. E.; Vander, A. J.; Kowalczyk, R. S.; and Geelhoed, G. W.: Aldosterone secretion and plasma renin during renin infusion and acute salt depletion. *Amer. J. Physiol., 214*:228, 1968.

Mogil, R. A.; Itskovitz, H. D.; Russell, J. H.; and Murphy, J. J.: Renal innervation and renin activity in salt metabolism and hypertension. *Amer. J. Physiol., 216*:693, 1969.

Montague, D.: Kinetics of renin-angiotensinogen reaction in plasma of normal and nephrectomized rats. *Amer. J. Physiol., 215*:78, 1968.

Nishimura, H.; Oguri, M.; Ogawa, M.; Sokabe, H.; and Imai, M.: Absence of renin in kidneys of elasmobranchs and cyclostomes. *Amer. J. Physiol., 218*:911, 1970.

Oken, D. E., and Biber, T. U. L.: Biologically effective immunization against angiotensin. *Amer. J. Physiol., 214*:791, 1968.

Peart, W. S.: The renin-angiotensin system. *Pharm. Rev., 17*:143, 1965.

Rapp, J. P.: Adrenal steroidogenesis and serum renin in rats bred for juxtaglomerular granularity. *Amer. J. Physiol., 216*:860, 1969.

Richardson, T. Q.; Fermoso, J. D.; and Guyton, A. C.: Increase in mean circulatory pressure in Goldblatt hypertension. *Amer. J. Physiol., 207*:751, 1964.

Schroeder, H. A.; Nason, A. P; Prior, R. E.; Reed, J. B.; and Haessler, W. T.: Influence of cadmium on renal ischemic hypertension in rats. *Amer. J. Physiol., 214*:469, 1968.

Sen, S.; Smeby, R. R.; and Bumpus, F. M.: Plasma renin activity in hypertensive rats after treatment with renin preinhibitor. *Amer. J. Physiol., 216*:499, 1969.

Skeggs, L. T.; Lentz, K. E.; Kahn, J. R.; Dorer, F. E.; and Levine, M.: Pseudorenin: a new angiotensin-forming enzyme. *Circ. Res., 25*:451, 1969.

Tarazi, R. C.; Melsher, H. J.; Dustan, H. P.; and Frohlich, E. D.: Plasma volume with upright tilt: studies in hypertension and in syncope. *J. Appl. Physiol., 28*:121, 1968.

Tobian, L.; Olson, R.; and Chesley, G.: Water content of arteriolar wall in renovascular hypertension. *Amer. J. Physiol., 216*:22, 1969.

Vander, A. J.: Control of renin release. *Physiol. Rev., 47(3)*:359, 1967.

Yagi, S.; Kramsch, D. M.; Madoff, I. M.; and Hollander, W.: Plasma renin activity in hypertension associated with coarctation of the aorta. *Amer. J. Physiol., 215*:605, 1968.

CHAPTER 27

Cardiac Output, Venous Return, and Their Regulation

Cardiac output is perhaps the single most important factor that we have to consider in relation to the circulation, for it is cardiac output that is responsible for transport of blood with its contained nutrients.

Cardiac output is the quantity of blood pumped by the left ventricle into the aorta each minute, and *venous return* is the quantity of blood flowing from the veins into the right atrium each minute. Obviously, over any prolonged period of time, venous return must equal cardiac output. However, for short intervals, venous return and cardiac output need not be the same, since blood can be temporarily stored in or lost from the central circulation.

NORMAL VALUES FOR CARDIAC OUTPUT

The normal cardiac output for the young healthy male adult averages approximately 5.6 liters per minute. However, if we consider all adults, including older people and females, the average cardiac output is very close to 5 liters per minute. In general, the cardiac output of females is about 10 per cent less than that of males of the same body size.

Cardiac Index. The cardiac output changes markedly with body size. Therefore, it has been important to find some means by which the cardiac outputs of different sized persons can be compared with each other. Experiments have shown that the cardiac output increases approximately in proportion to the surface area of the body. Therefore, the cardiac output is frequently stated in terms of the cardiac index, which is the *cardiac output per square meter of body surface area.* The normal human being weighing 70 kg. has a body surface area of approximately 1.7 square meters, which means that the normal average cardiac index for adults of all ages, both males and females, is approximately 3.0 liters per minute. The body surface area of persons of different

Figure 27–1 Cardiac index at different ages. (From Guyton: Cardiac Output and Its Regulation.)

heights and weights can be determined from the chart in Figure 70–7, Chapter 70.

Effect of Age. Figure 27–1 illustrates the change in cardiac index with age. Rising rapidly to a level greater than 4 liters per minute per square meter at 10 years of age, the cardiac index declines to only slightly above 2 liters per minute at the age of 80.

Effect of Posture. When a person rises from the reclining to the standing position, the cardiac output falls about 20 per cent if he stands quietly. However, if his muscles become taut, as occurs when one prepares for exercise, the cardiac output rises 1 to 2 liters per minute. The causes of these changes are explained later in the chapter.

Effect of Metabolism and Exercise. The cardiac output usually remains almost proportional to the overall metabolism of the body. That is, the greater the degree of activity of the muscles and other organs, the greater also will be the cardiac output. This relationship is illustrated in Figure 27–2, which shows that as the work output during exercise increases, the cardiac output also increases in almost linear proportion. Note that in very intense

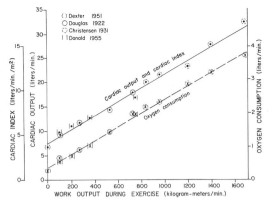

Figure 27–2 Relationship between cardiac output and work output (solid curve) and between oxygen consumption and work output (dashed curve) during different levels of exercise. (From Guyton: Cardiac Output and Its Regulation.)

exercise the cardiac output can rise to as high as 30 to 35 liters per minute in the young, well-trained athlete, which is about five to six times the normal control value.

Figure 27–2 also demonstrates that oxygen consumption increases in almost direct proportion to work output during exercise. We shall see later in the chapter that the increase in cardiac output probably results primarily from increased oxygen consumption.

REGULATION OF CARDIAC OUTPUT

PRIMARY ROLE OF PERIPHERAL CIRCULATION AND PERMISSIVE ROLE OF HEART IN CARDIAC OUTPUT REGULATION

The peripheral circulatory system is primarily responsible for controlling cardiac output in the normal person. This seems strange, because at first one would think that the heart would be the controller of cardiac output. Instead, the heart plays what may be described as a *permissive role* in cardiac output regulation. That is, it is capable of pumping a certain amount of blood each minute and, therefore, will *permit* the cardiac output to be regulated at any value below this given permitted level. Thus, the normal human heart, under resting conditions, permits a maximum heart pumping of about 13 to 15 liters per minute, but the actual cardiac output under resting conditions is only approximately 5 liters per minute. It is the peripheral circulatory system, not the heart, that sets this level of 5 liters per minute.

The concept that the heart plays a permissive role in cardiac output regulation is so important that it needs to be explained in still another way, as follows: The normal heart,

beating at a normal heart rate, and having a normal strength of contraction, neither excessively stimulated by the autonomic nervous system nor suppressed, will pump whatever amount of blood flows into the right atrium, up to about 13 to 15 liters per minute. If any more than this tries to flow into the right atrium, the heart will not pump it without cardiac stimulation. Under normal resting conditions the amount of blood that normally flows into the right atrium from the peripheral circulation is about 5 liters; and since this 5 liters is within the permissive range of heart pumping, it is pumped on into the aorta. Therefore, the peripheral circulatory system controls the cardiac output whenever the permissive level of heart pumping is greater than the required cardiac output.

Increase in Permissive Level for Heart Pumping in Hypertrophied Hearts. Heavy athletic training causes the heart to enlarge sometimes as much as 50 per cent. Coincident with this enlargement is an increase in the permissive level to which the heart can pump. Thus, even under resting conditions, the permissive level for a well-trained athlete might be greater than 20 liters per minute, rather than the normal value of about 13 to 15 liters per minute.

Increase in Permissive Level of Heart Pumping by Autonomic Stimulation of the Heart. There are times when the cardiac output must rise temporarily to levels greater than the normal permissive level of the heart. For instance, in heavy exercise by well-trained athletes, cardiac outputs as high as 35 liters per minute have been measured. Obviously, the resting heart would not be able to pump this amount of blood. On the other hand, stimulation of the heart by the sympathetic nervous system increases the permissive level of heart pumping to approximately double normal. This effect comes about by autonomic enhancement of both heart rate and strength of heart contraction. Furthermore, the increase in permissive level of heart pumping occurs within a few seconds after exercise begins, long before the cardiac output itself increases.

Reduction in Permissive Level for Heart Pumping in Heart Disease. Though the normal permissive level for heart pumping is usually much higher than the cardiac output, this is not necessarily true when the heart is diseased. Such conditions as myocardial infarction, valvular heart disease, myocarditis, congenital heart abnormalities, and so forth, can reduce the pumping effectiveness of the heart. In these instances, the permissive level for heart pumping may fall as low as 5 liters per minute or, for a few hours, even as low as 2 to 3 liters per minute. When this happens, the heart becomes unable to cope with the amount of blood that is attempting to flow into the right atrium from the peripheral circulation. Therefore, the heart is said to *fail*, meaning simply that it fails to pump the amount of blood that is demanded of it.

Peripheral Circulatory Factors that Play Significant Roles in Cardiac Output Control — Local Tissue Blood Flow Regulation. We have discussed in previous chapters, especially in Chapter 24, the concept of local control of blood flow in the tissues themselves. In general, blood flow through each individual tissue is locally adjusted to the changing needs of the tissues for oxygen or other nutrients; or in some tissues, blood flow is controlled by other factors such as the blood carbon dioxide concentration in the brain or blood electrolyte concentrations in the kidneys. Therefore, control of blood flow into the heart, which is called *venous return*, is controlled mainly by the tissues themselves.

However, local blood flow regulation is not the only peripheral circulatory factor that plays a significant role in cardiac output regulation. For instance, skin blood flow is controlled almost entirely by nerve signals from the brain, which are activated by the heat control centers of the body, as discussed in Chapter 71. That is, when a person is very hot, skin blood flow may be as much as 2 liters per minute or more; when one is very cold, skin blood flow may be as little as 100 milliliters per minute. Likewise, pathologic conditions frequently cause blood to flow directly from arteries to veins without having any relation to the nutrition of the tissues. This occurs in A-V fistulae, in dilated arteriovenous anastomoses, and through some very large capillaries that are concerned more with direct transmission of blood from arteries to veins than with nutrition. Such direct flow of blood from arteries to veins also represents part of the venous return to the heart.

Thus, any peripheral factor that increases the flow of blood from the arteries to the veins increases venous return and thereby increases the cardiac output.

CARDIAC OUTPUT REGULATION AS THE SUM OF LOCAL TISSUE BLOOD FLOW REGULATIONS. It is obvious from this discussion that under normal conditions the amount of blood that flows into the heart is determined by the sum of the blood flows through all the local tissues throughout the body. Therefore, it follows also that normal cardiac output regulation is achieved by the sum of the local peripheral blood flow regulations. This does not hold true when the pumping ability of the heart is so greatly curtailed that the heart is unable to pump all the blood required by the tissues, as we shall see subsequently.

Effect of Mean Systemic Pressure on Cardiac Output Regulation. Up to this point we have discussed changes in peripheral resistance to blood flow as the primary factor regulating cardiac output. However, another circulatory factor that can have temporary and extremely important effects on cardiac output regulation is the mean systemic pressure, which was discussed in Chapter 18. Basically, the mean systemic pressure is a measure of the degree of filling of the systemic circulation with blood. It is measured by stopping the circulation and pumping blood from the arteries to the veins within a few seconds so that all pressures in the systemic circulation come to an equilibrium level. This equilibrium level of pressure is the mean systemic pressure, and its normal value is about 7 mm. Hg.

Obviously, if the systemic circulation contains a small amount of blood, despite a normal heart and despite regulatory mechanisms for changing resistance in the tissues, there still will be essentially zero cardiac output. Conversely, if the circulatory system is greatly overloaded with blood, cardiac output can be increased to levels as great as three to four times normal. Therefore, blood volume, because of its effect on the mean systemic pressure, is another extremely important factor in the regulation of cardiac output.

In addition to the effect of blood volume on mean systemic pressure, the mean systemic pressure can also be altered by sympathetic stimulation or by tightening the skeletal muscles around the peripheral blood vessels. For instance, maximal sympathetic stimulation increases the mean systolic pressure as much as two and one-half–fold, and skeletal muscle contraction can increase it as much as fourfold. In each instance the blood in the peripheral vessels is squeezed toward the heart, thereby increasing cardiac output.

REGULATION OF CARDIAC OUTPUT IN IMPORTANT EXPERIMENTAL CONDITIONS

Effect of Arteriovenous Fistulae on Cardiac Output. The effect of an A-V fistula on cardiac output is discussed here because of the extremely important lessons that can be learned from this condition. An A-V fistula is a direct opening from an artery into a vein and allows rapid flow of blood directly from the artery to the vein. In experimental animals, such a fistula can be opened and closed artificially, and one can study the changes in fistula flow, cardiac output, and arterial pressure. A typical experiment of this type in the dog is illustrated in Figure 27–3, showing an initial cardiac output of 1300 ml. per minute. After 15 seconds of recording normal conditions, a fistula is opened through which 1200 ml. of blood flows per minute. Note that the cardiac output increases instantaneously about 1000 ml. per minute, or about 84 per cent as much increase in cardiac output as there is fistula flow. Then, when the fistula is closed, the cardiac output returns instantaneously back to its original level.

Now for the lesson to be learned from this study: When the fistula is opened, the pumping ability of the heart is not changed at all; there is only a change in the peripheral circulation—opening of the A-V fistula. Nevertheless, within seconds after the fistula is opened, the cardiac output increases almost as much as there is increase in blood flow through the fistula. In other words, the fistula flow is added to the flow through the remainder of the body, and the two of these summate to determine the cardiac output. This type of experiment demonstrates the two cardinal features of cardiac output regulation: (1) almost all normal cardiac output regulation is determined by peripheral circulatory factors, and (2) the normal heart has a permissive level of pumping that is normally considerably above the actual cardiac output, so that the heart does not have to be stimulated for the increase in cardiac output to come about.

Regulation of Cardiac Output in Heavy Exercise. Heavy exercise is one of the most stressful situations to which the body is ever subjected. The tissues can require as much as 20 times the normal amounts of oxygen and other nutrients, so that simply to transport enough oxygen from the lungs to the tissues sometimes demands a minimal cardiac output increase of five to six-fold. To insure this result, almost every factor that is known to increase cardiac output is pulled into play. These may be listed as follows:

1. Even before exercise begins, the thought of exercise stimulates the autonomic nervous system, which increases the heart rate and strength of the heart contraction, thus immediately increasing the permissive level of heart pumping from the normal level of 13 to 15 liters per minute up to as high as 25 to 30 liters per minute. Simultaneously, the sympathetic stimulation constricts the veins throughout the body, and this can increase the mean circulatory pressure to as high as two and one-half times normal, thus pushing extra quantities of blood from the peripheral circulation toward the heart. Together, these effects increase the cardiac output instantaneously, even before exercise begins, as much as 50 per cent.

2. At the very onset of exercise, the motor cortex transmits signals through the sympathetic cholinergic nerve fibers (described in Chapter 25) to cause vasodilatation of the muscle blood vessels. This causes a further instantaneous increase in cardiac output, perhaps to as high as double normal.

3. Also, at the onset of exercise, the motor cortex transmits signals directly into the sympathetic nervous system to intensify further the degree of sympathetic activity, further enhancing heart activity, mean systemic pressure, and arterial pressure.

4. Still another effect at the very onset of exercise is tightening of the skeletal muscles and of the abdominal wall around the peripheral blood vessels. This additionally increases the mean systemic pressure, now to as high as four times normal. However, compression of blood vessels in the muscles during muscle contraction has a mechanical effect to increase their resistance, and this partially offsets the value of the additional increase in mean systemic pressure.

5. Finally, the most important effect of all is the direct effect of increased metabolism in the muscle itself. This increased metabolism causes a tremendously increased usage of oxygen and other nutrients, as well as formation of vasodilator substances, all of which cause marked local vasodilatation, thereby greatly increasing local blood flow. This local vasodila-

Figure 27–3 Effect of suddenly opening and suddenly closing an A-V fistula, showing changes in fistula flow and cardiac output.

tation requires 5 to 20 seconds to reach full development after a person begins to exercise strongly; but, once it does reach full development, it by all means is the most important of all the factors that increase cardiac output during exercise.

In summary, there is tremendous setting of the background conditions of the circulatory system so that it can supply the required blood flow to the muscles during heavy exercise. These conditions include increased mean systemic pressure, increased arterial pressure, and greatly increased activity of the heart. But it is the local vasodilatation in the muscles themselves, occurring as a direct consequence of the muscle activity, that finally sets the level to which the cardiac output rises.

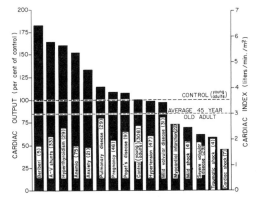

Figure 27–4 Cardiac output in different pathological conditions. (From Guyton: Cardiac Output and Its Regulation.)

CARDIAC OUTPUT IN ABNORMAL CONDITIONS

In the discussions of cardiac output regulation thus far, it has been evident that cardiac output is normally regulated by peripheral factors and that the heart mainly acts as a *sump pump,* automatically pumping whatever amount of blood that flows into it from the periphery. On the other hand, in many abnormal conditions the heart becomes unable to pump the amount of blood that attempts to flow into it. When this happens, the heart then becomes the limiting and determining factor in cardiac output regulation. Therefore, there are two entirely different sets of abnormal factors which can affect cardiac output: (1) abnormal factors affecting peripheral control of venous return and (2) abnormal factors affecting cardiac pumping ability.

Abnormalities Affecting Peripheral Control of Venous Return. Figure 27–4 illustrates the cardiac output in different pathological conditions, showing excess cardiac output in some conditions and decreased cardiac output in others.

INCREASED CARDIAC OUTPUT CAUSED BY DECREASED TOTAL PERIPHERAL RESISTANCE. In almost all conditions in which there is excess cardiac output, one finds decreased total peripheral resistance. For instance, in *beriberi* the total peripheral resistance is sometimes reduced to one-half normal. The same is true in *A-V fistulae, anoxia, pulmonary disease, pregnancy,* and *Paget's disease.* In beriberi the decreased total peripheral resistance results from dilatation of the peripheral vessels that is caused by thiamine deficiency. In A-V fistula, decreased total peripheral resistance is caused

by direct flow of blood from the arteries to the veins. In hyperthyroidism it is caused by increased peripheral metabolism, which results in local vasodilatation. In hypoxia (anoxia) it is caused by local vasodilatation resulting from lack of oxygen. In pregnancy it is caused by excess blood flow through the placenta. And in Paget's disease it is caused by large numbers of minute A-V anastomoses in the bones. Obviously, all of these conditions can increase return of blood to the heart.

DECREASE IN CARDIAC OUTPUT CAUSED BY PERIPHERAL FACTORS—DECREASED BLOOD VOLUME. The only peripheral factor that usually decreases cardiac output is decreased mean systemic pressure, most often caused by decreased blood volume. This obviously reduces the peripheral pressures necessary to return blood to the heart. Ordinarily, rapid bleeding of a person of approximately 25 to 30 per cent of his blood volume will reduce the cardiac output to zero: and slow bleeding, over a period of an hour of 40 to 50 per cent, will do the same thing. This is such an important problem in the condition known as "circulatory shock" that it will be discussed in detail in Chapter 28.

Decrease in Cardiac Output Caused by Cardiac Factors. Most students will be surprised to learn that an increase in cardiac activity caused by direct electrical stimulation of the heart or caused by sympathetic stimulation of the heart (without simultaneous sympathetic stimulation elsewhere in the body) will hardly affect cardiac output. What autonomic stimulation does is simply to increase the permissive level of the heart pumping rather than to increase the actual level.

On the other hand, whenever the heart becomes severely damaged, its permissive level of pumping falls below that needed for adequate blood flow to tissues. Therefore, the heart can be the cause of *decreased* cardiac output, even though it ordinarily cannot by itself be the cause of increased cardiac output. Thus, note in Figure 27–4 that cardiac output is decreased in myocardial infarction, in severe valvular heart disease, and in cardiac shock; in all these conditions the heart simply is not capable of pumping an adequate cardiac output. These conditions will be discussed more fully in Chapter 29 in relation to cardiac failure.

GRAPHICAL ANALYSIS OF CARDIAC OUTPUT REGULATION

These discussions have emphasized the basic principles of cardiac output regulation. However, it is often important to obtain a much more quantitative understanding of cardiac output regulation. To do this, it is necessary to analyze, first, the pumping ability of the heart and, second, the peripheral factors that affect flow of blood into the heart. Then, these two factors, the cardiac and the peripheral, must be put together in a quantitative fashion to see how they interact with each other to determine cardiac output. This process has been frequently accomplished by using electronic computers, but it can also be done more simply by using a type of graphical analysis that one can carry in his mind and that can be used daily in solving some of the quantitative problems of cardiac output regulation. In the following sections we will discuss this type of analysis. To begin, we first characterize the pumping ability of the heart by use of *cardiac function curves*, and then we characterize the peripheral factors that affect venous return by use of *systemic function curves*.

CARDIAC FUNCTION CURVES

Use of cardiac function curves to depict the ability of the heart to pump blood was discussed in Chapter 13, and the effects of several different factors on these function curves were demonstrated in Figures 13–9 and 13–10. For the purpose of understanding cardiac output regulation, the type of cardiac function curve required is that illustrated in Figure 27–5 and in Figure 27–6, which depicts the *cardiac output at different right atrial pressures*. The normal cardiac function curve is labeled in each figure, while the other curves represent cardiac function in special conditions.

The basic factors that determine the precise characteristics of the cardiac function curve are (1) the *effectiveness of the heart as a pump* and (2) the *extracardiac pressure*.

Effectiveness of the Heart as a Pump. A heart that is capable of pumping greater quantities of blood than can the normal heart, such as a heart strongly stimulated by sympathetic impulses or a hypertrophied heart, is called a *hypereffective heart*. A weakened heart is called a *hypoeffective heart*. Figure 27–5 illustrates the effect of hyper-

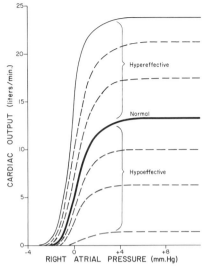

Figure 27–5 Cardiac function curves for the normal heart and for hypo- and hypereffective hearts. (From Guyton: Cardiac Output and Its Regulation.)

effectivity or hypoeffectivity on the cardiac function curves, showing that the cardiac function curve rises greatly in the case of the hypereffective heart and falls equally as much in the case of the hypoeffective heart. Obviously, there are all degrees of hypereffectivity and hypoeffectivity.

FACTORS THAT CAN CAUSE A HYPEREFFECTIVE HEART. Factors that can make the heart a better pump than normal include:

1. *Sympathetic stimulation,* which can increase the cardiac function curve to approximately $1\frac{2}{3}$ times normal.

2. *Hypertrophy of the heart,* which can increase the cardiac function curve to perhaps as much as two times normal.

3. *Decreased systemic arterial pressure,* which allows the normal heart to pump perhaps as much as 10 to 20 per cent extra amounts of blood.

4. *Inhibition of the parasympathetics to the heart,* which removes the parasympathetic tone, allowing the heart rate to increase, and, therefore, increases the pumping effectiveness of the heart.

FACTORS THAT CAN CAUSE A HYPOEFFECTIVE HEART. Obviously, any factor that decreases the

Figure 27–6 Cardiac function curves caused by changes in intrapleural pressure and by cardiac tamponade. (From Guyton: Cardiac Output and Its Regulation.)

ability of the heart to pump blood can cause hypo-effectivity. Some of these numerous factors are:

Myocardial infarction
Valvular heart disease
Parasympathetic stimulation of the heart
Inhibition of the sympathetics to the heart
Congenital heart disease
Myocarditis
Cardiac anoxia
Diphtheritic or other types of myocardial damage or toxicity

Effect of Extracardiac Pressure on Cardiac Function Curves. Figure 27–6 illustrates the effect of changes in extracardiac pressure on the cardiac output curve. The normal extracardiac pressure, which is also equal to the normal intrapleural pressure, is −4 mm. Hg. Note in the figure that a rise in intrapleural pressure to −2 mm. Hg shifts the entire curve to the right by this same amount. This shift occurs because an extra 2 mm. Hg right atrial pressure is now required to overcome the increased pressure on the outside of the heart. Likewise, an increase in intrapleural pressure to +2 mm. Hg requires a 6 mm. Hg increase in right atrial pressure, which obviously shifts the entire curve 6 mm. Hg to the right. And, in the same manner, a more negative intrapleural pressure shifts the curve to the left.

Some of the factors that alter the intrapleural pressure and thereby shift the cardiac function curve include:

1. *Cyclic changes during normal respiration.*

2. *Breathing against a negative pressure,* which shifts the curve to a more negative right atrial pressure.

3. *Positive pressure breathing,* which shifts the curve to the right.

4. *Opening the thoracic cage,* which increases the intrapleural pressure and shifts the curve to the right.

5. *Cardiac tamponade,* which means an accumulation of large quantities of fluid in the pericardial cavity around the heart with resultant increase in extracardiac pressure and shifting of the curve to the right. Note in Figure 27–6 that cardiac tamponade shifts the upper parts of the curves farther to the right than the lower parts because the extracardiac pressure rises to higher values as the chambers of the heart fill to increased volumes during high cardiac output.

Combinations of Different Patterns of Cardiac Function Curves. Figure 27–7 illustrates that the cardiac output curve can change as a result of simultaneous changes in extracardiac pressure and effectiveness of the heart as a pump. Thus, if one knows what is happening to the extracardiac pressure and also the approximate capability of the heart as a pump, he can express the momentary ability of the heart to pump blood by a single cardiac function curve.

SYSTEMIC FUNCTION CURVES

There still remains the entire systemic circulation that must be considered before a total analysis of circulatory function can be made. To analyze the

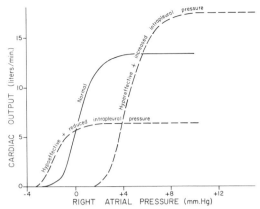

Figure 27–7 Combinations of two major patterns of cardiac output curves, showing the effect of alterations in both extracardiac pressure and effectiveness of the heart as a pump. (From Guyton: Cardiac Output and Its Regulation.)

function of the systemic circulation, we remove the heart and lungs from the circulation of an animal and replace these by a pump and artificial oxygenator system. Then, different factors, such as changes in blood volume and changes in right atrial pressure, are altered to determine how the systemic circulation operates in different circulatory states. In these studies, one finds two principal factors that affect the function of the systemic circulation in relation to cardiac output regulation. These are: (1) *the degree of filling of the systemic circulation,* which is measured by the *mean systemic pressure,* which was discussed in Chapter 18, and (2) the *resistance to blood flow* in the different segments of the systemic circulation.

The Normal Systemic Function Curve. In the same way that the cardiac function curve relates cardiac output to right atrial pressure, the *systemic function curve relates venous return to right atrial pressure.* For this reason, the systemic function curve has also been called *a venous return curve.*

The very dark curve in Figure 27–8 is the normal systemic function curve. This curve shows that as the right atrial pressure increases, it causes back pressure on the systemic circulation and thereby decreases venous return of blood to the heart. *If all circulatory reflexes are prevented from acting,* venous return decreases to zero when the

Figure 27–8 Systemic function curves, showing the normal curve when the mean systemic pressure (*Pms*) is 7 mm. Hg, and showing the effect of altering the mean systemic pressure. (From Guyton: Cardiac Output and Its Regulation.)

right atrial pressure rises to equal the normal mean systemic pressure of about 7 mm. Hg. Such a slight rise in right atrial pressure causes a drastic decrease in venous return because the systemic circulation is actually a very distensible bag so that any increase in back pressure causes blood to dam up in this bag instead of returning to the heart. Lack of blood flow into the heart decreases the cardiac output, thereby decreasing the arterial pressure. Thus, at the same time that the right atrial pressure is rising, the arterial pressure is falling, and these two pressures come to equilibrium when all flow in the circulation ceases at a pressure of 7 mm. Hg, which, by definition, is the mean systemic pressure.

PLATEAU IN THE SYSTEMIC FUNCTION CURVE CAUSED BY COLLAPSE OF THE VEINS. When the right atrial pressure falls *below* zero—that is, below atmospheric pressure—the venous return does not increase significantly. And, by the time the right atrial pressure has fallen to about −2 mm. Hg, the venous return will have reached a plateau; then it remains at this plateau level even though the right atrial pressure falls as low as −20 to −50 mm. Hg. This plateau is caused by collapse of the veins entering the chest. Low right atrial pressure sucks the walls of the veins together, which prevents the negative pressure from sucking blood through the veins. Instead, the pressure in the veins immediately outside the chest remains almost exactly equal to atmospheric pressure (zero pressure). Therefore, for all practical purposes, the central venous pressure never falls below a value of 0 mm. Hg despite the fact that the right atrial pressure may fall to very negative values.

Effect on the Systemic Function Curve of Changes in Mean Systemic Pressure. Figure 27–8 illustrates the effect of increasing or decreasing the mean systemic pressure. By definition, the mean systemic pressure is the pressure in all parts of the systemic circulation when there is no flow in the circulation. Therefore, at the point where the systemic function curve reaches the zero venous return level, the right atrial pressure becomes equal to the mean systemic pressure. Note in Figure 27–8 that the normal mean systemic pressure (*Pms*) is 7 mm. Hg. For the uppermost curve in the figure, the mean systemic pressure is 14 mm. Hg, and for the lowermost curve it is 3.5 mm. Hg. These curves demonstrate that the greater the mean systemic pressure, which also means the greater the "tightness" with which the circulatory system is filled with blood, the more the systemic function curve shifts upward and to the right. Conversely, the lower the mean systemic pressure, the more the curve shifts downward and to the left. To express this another way, the greater the degree of tightness with which the system is filled, the easier is it for blood to flow into the heart. And the less the tightness, the more difficult is it for blood to flow into the heart.

FACTORS THAT CAN ALTER THE MEAN SYSTEMIC PRESSURE. The factors that can alter the mean systemic pressure were discussed in Chapter 18. The three primary factors are: (1) The mean systemic pressure increases rapidly with an *increase in blood volume,* an acute increase in blood volume of 15 per cent increasing the mean systemic pressure to approximately double normal. (2) Maximal *sympathetic stimulation* can increase the mean systemic pressure from 7 mm. Hg to about 17 mm. Hg, approximately a 2½ fold increase. (3) An *increase in contraction* of all the skeletal muscles throughout the body can increase the mean systemic pressure by compressing the blood vessels from the outside to about 25 mm. Hg.

Factors that decrease the mean systemic pressure include principally (1) *loss of sympathetic tone,* which decreases the mean systemic pressure from 7 mm. Hg to about 4 mm. Hg, or (2) *loss of blood volume,* which can decrease the mean systemic pressure to as little as zero.

PRESSURE GRADIENT FOR VENOUS RETURN. When the right atrial pressure rises to equal the mean systemic pressure, all other pressures in the systemic circulation also approach this mean systemic pressure. Therefore, there becomes no pressure gradient for flow of blood toward the heart. However, as the right atrial pressure falls progressively lower than the mean systemic pressure, the flow to the heart increases proportionately, as one can see by studying any of the systemic function curves in Figure 27–8. That is, *the greater the difference between the mean systemic pressure and the right atrial pressure the greater becomes the venous return.* Therefore, the difference between these two pressures is called the *pressure gradient for venous return.*

Effect of Systemic Resistance on the Systemic Function Curve. Figure 27–9 illustrates the effect of different systemic resistances on the systemic function curve, showing that a decrease in resistance to one-half normal *rotates the curve upward to the right,* while an increase in resistance *rotates the curve downward to the left.* The venous return is still equal to zero when the right atrial pressure rises to equal the mean systemic pressure, because when there is no pressure gradient to cause flow of blood, it makes no difference what the resistance is in the circulation—the flow is still zero.

Obviously, any factor that dilates the peripheral

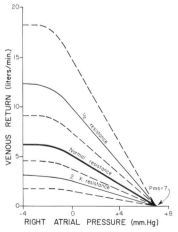

Figure 27–9 Systemic function curves, depicting the effect of altering the "resistance to venous return." (From Guyton: Cardiac Output and Its Regulation.)

vessels decreases the resistance and rotates the function curve upward to the right, while any factor that constricts the peripheral vessels rotates the curve downward.

Combinations of Systemic Function Curve Patterns. Figure 27–10 illustrates the effects on the systemic function curve caused by simultaneous changes in mean systemic pressure and resistance to venous return, illustrating that both of these factors can operate simultaneously.

ANALYSIS OF CARDIAC OUTPUT AND RIGHT ATRIAL PRESSURE USING CARDIAC AND SYSTEMIC FUNCTION CURVES

Up to this point we have discussed, first, the individual factors that affect the pumping ability of the heart as depicted by *cardiac* function curves and, second, the individual factors that affect the ability of blood to flow through the systemic circulation as depicted by *systemic* function curves. In the complete circulation, however, the heart and the systemic circulation must operate together. This means: (1) The venous return from the systemic circulation must equal the cardiac output from the heart. (2) The right atrial pressure is the same for both the heart and for the systemic circulation.

Therefore, one can predict the cardiac output and right atrial pressure in the following way: (1) determine the momentary pumping ability of the heart and depict this in the form of a cardiac function curve, (2) determine the momentary state of the systemic circulation and depict this in the form of a systemic function curve, and (3) "equate" these two curves against each other as shown in Figure 27–11.

The two solid curves of Figure 27–11 depict both the *normal cardiac function curve* and the *normal systemic function curve*. Obviously, there is only one point on the graph, point A, at which the venous return equals the cardiac output and at which the right atrial pressure is the same in rela-

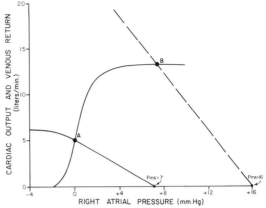

Figure 27–11 The two solid curves demonstrate an analysis of cardiac output and right atrial pressure when the cardiac and systemic function curves are known. Transfusion of blood equal to 20 per cent of the blood volume causes the systemic function curve to become the dashed curve; as a result the cardiac output and right atrial pressure shift from point A to point B. *Pms,* mean systemic pressure.

tion to both the cardiac function curve and the systemic function curve. Therefore, in the normal circulation the right atrial pressure, cardiac output, and venous return are all depicted by point A, called the *equilibrium point.*

Effect of Increased Blood Volume on Cardiac Output

A sudden increase in blood volume of about 20 per cent increases the cardiac output to about 2½ to 3 times normal. An analysis of this effect is illustrated by the dashed curve of Figure 27–11. Immediately upon infusing the large quantity of extra blood, the increased filling of the system causes the mean systemic pressure to rise to 16 mm. Hg, which shifts the systemic function curve upward and to the right. At the same time, the increased blood volume distends the blood vessels and thereby rotates the curve upward to the right. As a result of these two effects, the systemic function curve of Figure 27–11 changes from the solid curve to the dashed curve. This new curve equates with the cardiac function curve at point B, showing that the cardiac output increases 2½ to 3 times and that the right atrial pressure rises to about +8 mm. Hg.

Compensatory Effects Initiated by the Increased Blood Volume. The increased cardiac output caused by the increased blood volume lasts only a few minutes, because several different compensatory effects immediately begin to occur: (1) The increased cardiac output increases the capillary pressure so that fluid begins to transude out of the capillaries into the tissues, thereby returning the blood volume toward normal. (2) The increased pressure in the veins causes the veins to distend gradually, especially causing the venous blood reservoirs such as the liver and spleen to distend. These factors cause the mean systemic pressure to return back toward normal and also allow the re-

Figure 27–10 Combinations of the major patterns of systemic function curves, illustrating the effects of simultaneous changes in mean systemic pressure (*Pms*) and in "resistance to venous return." (From Guyton: Cardiac Output and Its Regulation.)

sistance vessels of the systemic circulation to constrict once again. Therefore, gradually over a period of about 40 minutes, the cardiac output returns to normal.

Effect of Sympathetic Stimulation on Cardiac Output

Sympathetic stimulation affects both the heart and the systemic circulation: (1) It makes the heart a stronger pump. (2) In the systemic circulation, it increases the mean systemic pressure and also increases the resistance to venous return. Figure 27–12 illustrates an analysis of the effects of moderate and maximal sympathetic stimulation on the cardiac output and right atrial pressure. The normal cardiac and systemic function curves are depicted by the very dark lines; moderate sympathetic stimulation is depicted by the long-dashed curves, and maximal sympathetic stimulation by the dot-dash curves.

Note that maximal sympathetic stimulation increases the mean systemic pressure (depicted by the point at which the systemic function curve reaches the zero venous return level) to 17 mm. Hg, and it also increases the pumping effectiveness of the heart by about 70 per cent. As a result, the cardiac output rises from the normal value at equilibrium point A to approximately double normal at equilibrium point D. Thus, different degrees of sympathetic stimulation can increase the cardiac output progressively to about two times normal.

Effect of Sympathetic Inhibition

The sympathetic nervous system can be blocked completely by inducing total spinal anesthesia or by using some drug, such as hexamethonium, that blocks transmission of nerve impulses through the autonomic ganglia. The two lowermost curves in Figure 27–12, the short-dashed curves, show the effect of total sympathetic inhibition caused by spinal anesthesia, illustrating that (1) the mean systemic pressure falls to about 4 mm. Hg, and (2) the effectiveness of the heart as a pump decreases to about 80 per cent of normal. The cardiac output falls from point A to point B, which is a decrease to about 60 per cent of normal.

Increase in Cardiac Output in Exercise

Several different factors operate in exercise to increase the cardiac output. These include: (1) intense sympathetic stimulation, which strengthens the heart and also increases the mean systemic pressure, (2) contraction of the muscles around the blood vessels, which further increases the mean systemic pressure, and (3) intense dilatation of the resistance vessels in the muscles, thereby decreasing the resistance to venous return. Figure 27–13 depicts the effects of these factors on the circulation, showing that (1) compression of the blood vessels by the muscles plus sympathetic stimulation increases the mean systemic pressure from 7 mm. Hg to 20 mm. Hg, (2) sympathetic stimulation increases the strength of the heart as a pump by about 70 per cent, and (3) the local vasodilatation in the muscles rotates the systemic function curve upward to the right. As a result, the cardiac and systemic curves in exercise (the dashed curves) equate at point B, which is a cardiac output of 22 liters per minute, or 4½ times normal.

RIGHT VENTRICULAR OUTPUT VERSUS LEFT VENTRICULAR OUTPUT — BALANCE BETWEEN THE VENTRICLES

Obviously, the output of one ventricle must remain almost exactly the same as that of the other. Therefore, an intrinsic mechanism must be available for automatically adjusting the outputs of the two ventricles to each other. This operates as follows:

Let us assume that the strength of the left ventricle decreases suddenly so that left ventricular output falls below right ventricular output. This immediately causes more blood to be pumped into the lungs than is pumped into the systemic circulation. Consequently, the mean pulmonary pressure and the left atrial pressure rise while the mean systemic pressure and the right atrial pressure fall. If we apply the principles of analysis presented previously to each of the ventricles separately, we will see that the increased *mean pulmonary pressure* increases left heart venous return and left ventricular output while the decrease in *mean systemic pressure* decreases right heart venous

Figure 27–12   Analysis of the effect on cardiac output of (1) moderate sympathetic stimulation, (2) maximal sympathetic stimulation, and (3) sympathetic inhibition caused by total spinal anesthesia. (From Guyton: Cardiac Output and Its Regulation.)

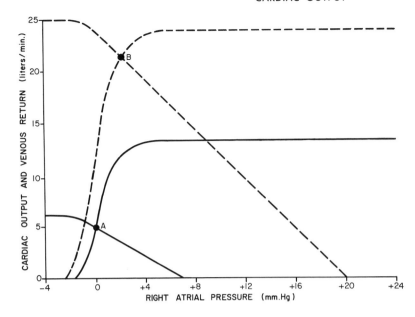

Figure 27–13 Graphical analysis of the changes in cardiac output and right atrial pressure with the onset of strenuous exercise.

return and right ventricular output. This process continues until the output of the left ventricle rises to equal the falling output of the right ventricle. Thus, the outputs of the two ventricles become rebalanced with each other within a few beats of the heart. The same effects occur in the opposite direction when the strength of the right heart diminishes.

These problems of balance between the two ventricles are especially important when myocardial failure occurs in one ventricle or when valvular lesions cause poor pumping by one side of the heart. These conditions will be discussed in more detail in Chapter 29 in relation to cardiac failure.

METHODS FOR MEASURING CARDIAC OUTPUT

In the lower animal, the aorta, the pulmonary artery or the great veins entering the heart can be cannulated and the flow measured by any type of flowmeter. Also, an electromagnetic or ultrasonic flowmeter can be placed on the aorta or pulmonary artery to measure cardiac output. However, in the human being, cardiac output is measured by indirect methods that do not require surgery. Two of these indirect methods commonly used are the *oxygen Fick method* and the *indicator dilution method*.

PULSATILE OUTPUT OF THE HEART AS MEASURED BY AN ELECTROMAGNETIC OR ULTRASONIC FLOWMETER

Figure 27–14 illustrates a recording of blood flow in the root of the aorta made using an electro-

magnetic flowmeter. It demonstrates that the blood flow rises rapidly to a peak during systole and, then, at the end of systole, actually reverses for a small fraction of a second. It is this reverse flow that causes the aortic valve to close. And a minute amount of reverse flow continues throughout diastole to supply blood to the coronary vessels.

MEASUREMENT OF CARDIAC OUTPUT BY THE OXYGEN FICK METHOD

The Fick procedure is best explained by Figure 27–15, which shows the absorption of 200 ml. of oxygen from the lungs into the pulmonary blood each minute and also illustrates that the blood entering the right side of the heart has an oxygen concentration of approximately 16 volumes per cent (16 ml. of oxygen per 100 ml. of blood) while that leaving the left side has an oxygen concentration of approximately 20 volumes per cent. From these data we see that each 100 ml. of blood passing through the lungs picks up 4 ml. of oxygen. And, since the total quantity of oxygen absorbed into the blood from the lungs each minute is 200 ml., a total of 50 100-ml. portions of blood must pass through the pulmonary circulation each minute to absorb this amount of oxygen. Therefore, the quantity of blood flowing through the lungs each minute is 5000 ml., which is also a measure of

Figure 27–14 Pulsatile blood flow in the root of the aorta recorded by an electromagnetic flowmeter.

Figure 27–15 The Fick principle for determining cardiac output.

the cardiac output. Thus, the cardiac output can be calculated by the following formula:

Cardiac output (ml./min.) =

$$\frac{\text{Oxygen absorbed per minute by the lungs (ml./min.)}}{\text{Arteriovenous oxygen difference (ml./ml. of blood)}}$$

In applying the Fick procedure, it is the mean concentrations of oxygen in the venous and arterial bloods that must be used in the calculations. To obtain an accurate determination of venous oxygen concentration it is necessary to take the blood sample directly from the right ventricle or preferably the pulmonary artery, for blood in any one vein of the body usually has a concentration of oxygen different from that in other veins, and even blood in the right atrium usually has not mixed satisfactorily. To obtain such a sample of mixed venous blood a catheter is usually inserted up the brachial vein of the forearm, through the subclavian vein, down to the right atrium, and finally into the right ventricle or pulmonary artery.

Blood used for determining the oxygen saturation in arterial blood can be obtained from any artery in the body, because all arterial blood is thoroughly mixed before it leaves the heart and therefore has the same oxygen concentration.

The rate of oxygen absorption by the lungs is usually measured by a "respirometer," which will be described in Chapter 70. In essence, this device is a floating chamber that sinks in the water as oxygen is removed from it, thus measuring the use of oxygen by the rate at which it falls.

THE INDICATOR DILUTION METHOD

In measuring the cardiac output by the indicator dilution method a small amount of indicator, such as a dye, is injected into a large vein or preferably into the right side of the heart itself. This then passes rapidly through the right heart, the lungs, the left heart, and finally into the arterial system. If one records the concentration of the dye as it passes through one of the peripheral arteries, a curve such as one of those illustrated in Figure 27–16 will be obtained. In each of these instances

5 mg. of T-1824 dye was injected at zero time. In the top recording none of the dye passed into the arterial tree until approximately 3 seconds after the injection, but then the arterial concentration of the dye rose rapidly to a maximum in approximately 6 to 7 seconds. After that, the concentration fell rapidly. However, before the concentration reached the zero point, some of the dye had already circulated all the way through some of the peripheral vessels and returned through the heart for a second time. Consequently, the dye concentration in the artery began to rise again. For the purpose of calculation, however, it is necessary to extrapolate the early downslope of the curve to the zero point, as shown by the dashed portion of the curve. In this way, the *time-concentration curve* of the dye in an artery can be measured in its first portion and estimated reasonably accurately in its latter portion.

Once the time-concentration curve has been determined, one can then calculate the mean concentration of dye in the arterial blood for the duration of the curve. In Figure 27–16, this was done by measuring the area under the entire curve, and then averaging the concentration of dye for the duration of the curve; one can see from the shaded rectangle straddling the upper curve of the figure that the average concentration of dye was approximately 0.25 mg./100 ml. blood and that the duration of the curve was 12 seconds. However, a total of 5 mg. of dye was injected at the beginning of the experiment. In order for blood carrying only 0.25 mg. of dye in each 100 ml. to carry the entire 5 mg. of dye through the heart and lungs in 12 seconds, it would be necessary for a total of 20 100-ml. portions of blood to pass through the heart during this time, which would be the same as a cardiac output of 2 liters per 12 seconds, or 10 liters per minute.

In the bottom curve of Figure 27–16 the blood flow through the heart was considerably slower, and the dye did not appear in the arterial system until approximately 6 seconds after it had been

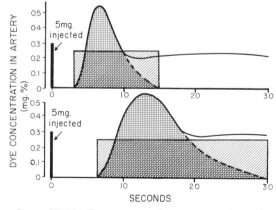

Figure 27–16 Dye concentration curves used to calculate the cardiac output by the dilution method. (The rectangular areas are the calculated average concentrations of dye in the arterial blood for the durations of the respective curves.)

injected. It reached a maximum height in 12 to 13 seconds and was extrapolated to 0 at approximately 30 seconds. Averaging the dye concentrations over the 24-second duration of the curve one finds again an average concentration of 0.25 mg. of dye in each 100 ml. of blood, but this time for a 24-second time interval instead of 12 seconds. To transport the total 5 mg. of dye, 20 100-ml. portions of blood would have had to pass through the heart during the 24-second time interval. Therefore, the cardiac output was 2 liters per 24 seconds, or 5 liters per minute.

To summarize, the cardiac output can be determined from the following formula:

$$\text{Cardiac output (ml./min.)} =$$

$$\frac{\text{Milligrams of dye injected} \times 60}{\left(\begin{array}{c}\text{Average concentration of dye} \\ \text{in each milliliter of blood} \\ \text{for the duration of the curve}\end{array}\right) \times \left(\begin{array}{c}\text{Duration of} \\ \text{the curve} \\ \text{in seconds}\end{array}\right)}$$

Substances That Can Be Injected for Determining Cardiac Output by the Indicator Dilution Method. Almost any substance that can be analyzed satisfactorily in the arterial blood can be injected when making use of the indicator dilution method for determining cardiac output. However, for optimum accuracy it is necessary that the injected substance not be lost into the tissues of the lungs during its passage to the sampling site. The most widely used substance is *Cardio-Green,* a dye that combines with the plasma proteins and, therefore, is not lost from the blood.

ESTIMATION OF CARDIAC OUTPUT BY THE PULSE PRESSURE METHOD

A fairly reliable estimation of cardiac output can also be made from recordings of pressure pulse contours from the aorta. The basic theory of this method is the following: During diastole no blood flows into the aorta from the heart, but it does continue to flow from the central arteries into the peripheral blood vessels, thus allowing the pressure in the central arteries to decline. Obviously, the greater the rate of blood flow, the greater also is the rate of blood pressure decline during diastole. By using empirical formulas, the cardiac output can be calculated from this downward slope of the pressure during diastole, and other characteristics of the pressure pulse curve can be used to make the calculation even more valid.

The beauty of this method for estimating cardiac output is that a measurement of cardiac output can be made each beat of the heart, which allows one to observe rapid changes in cardiac output. Unfortunately, the characteristics of the pressure pulse curve depend on the distensibility of the arteries as well as on the rate of run-off of blood through the peripheral vessels, thus making the method sometimes seriously in error.

REFERENCES

Bishop, V. S., and Stone, H. L.: Quantitative description of ventricular output curves in conscious dogs. *Circ. Res., 20:*581, 1967.

Bishop, V. S.; Stone, H. L.; and Guyton, A. C.: Cardiac function curves in conscious dogs. *Amer. J. Physiol., 207:*677, 1964.

Brecher, G. A.: Venous Return. New York, Grune & Stratton, Inc., 1956.

Butler, J.: Measurement of cardiac output using soluble gases. *In* Handbook of Physiology. Baltimore, The Williams & Wilkins Co., 1965, Sec. III, Vol. II, p. 1489.

Clausen, J. P.; Larsen, O. A.; and Trap-Jensen, J.: Cardiac output in middle-aged patients determined with CO_2 rebreathing method. *J. Appl. Physiol., 28:*337, 1970.

Coleman, T. G., and Criddle, F. J., Jr.: Computerized analysis of indicator-dilution curves. *J. Appl. Physiol., 28:*358, 1970.

Donald, D. E., and Shepherd, J. T.: Response to exercise in dogs with cardiac denervation. *Amer. J. Physiol., 205:*393, 1963.

Dow, P.: Estimations of cardiac output and central blood volume by dye dilution. *Physiol. Rev., 36:*77, 1956.

Ferguson, R. J.; Faulkner, J. A.; Julius, S.; and Conway, J.: Comparison of cardiac output determined by CO_2 rebreathing and dye-dilution method. *J. Appl. Physiol., 25:*450, 1968.

Fermoso, J. D.; Richardson, T. Q.; and Guyton, A. C.: Mechanism of decrease in cardiac output caused by opening the chest. *Amer. J. Physiol., 207:*1112, 1964.

Grodins, F. S.: Integrative cardiovascular physiology: a mathematical synthesis of cardiac and blood vessel hemodynamics. *Quart. Rev. Biol., 34:*93, 1959.

Guyton, A. C.: Circulatory Physiology: Cardiac Output and Its Regulation. Philadelphia, W. B. Saunders Company, 1963.

Guyton, A. C.: Determination of cardiac output by equating venous return curves with cardiac response curves. *Physiol. Rev., 35:*123, 1955.

Guyton, A. C.: Venous return. *In* Handbook of Physiology. Baltimore, The Williams & Wilkins Co., 1963, Sec. II, Vol. II, p. 1099.

Guyton, A. C.; Abernathy, J. B.; Langston, J. B.; Kaufmann, B. N.; and Fairchild, H. M.: Relative importance of venous and arterial resistance in controlling venous return and cardiac output. *Amer. J. Physiol., 196:*1008, 1959.

Guyton, A. C., and Adkins, L.: Quantitative aspects of the collapse factor in relation to venous return. *Amer. J. Physiol., 177:*523, 1954.

Guyton, A. C., and Sagawa, K.: Compensations of cardiac output and other circulatory functions in areflex dogs with large A-V fistulas. *Amer. J. Physiol., 200:*1157, 1961.

Guyton, A. C.; Douglas, B. H.; Langston, J. B.; and Richardson, T. Q.: Instantaneous increase in mean circulatory pressure and cardiac output at onset of muscular activity. *Circ. Res., 11:*431, 1962.

Guyton, A. C.; Farish, C. A.; and Abernathy, J. B.: A continuous cardiac output recorder employing the Fick principle. *Circ. Res., 7:*661, 1959.

Guyton, A. C.; Lindsey, A. W.; Abernathy, J. B.; and Richardson, T. Q.: Venous return at various right atrial pressures and the normal venous return curve. *Amer. J. Physiol., 189:*609, 1957.

Guyton, A. C.; Lindsey, A. W.; and Kaufmann, B. N.: Effect of mean circulatory filling pressure and other peripheral circulatory factors on cardiac output. *Amer. J. Physiol., 180:*463, 1955.

Hamilton, W. F.: Measurement of the cardiac output. *In* Handbook of Physiology. Baltimore, The Williams & Wilkins Co., 1962, Sec. II, Vol. I, p. 551.

Jones, W. B.; Finchum, R. N.; Russell, R. O., Jr.; and Reeves, T. J.: Transient cardiac output response to multiple levels of supine exercise. *J. Appl. Physiol., 28:*183, 1970.

Keroes, J.; Ecker, R. R.; and Rapaport, E.: Ventricular function curves in the exercising dog: effects of rapid intravenous infusions and of propranolol. *Circ. Res., 25:*557, 1969.

Khalil, H. H.; Richardson, T. Q.; and Guyton, A. C.: Measurement of cardiac output by thermal-dilution and direct Fick methods in dogs. *J. Appl. Physiol., 21:*1131, 1966.

Muiesan, G.; Sorbini, C. A.; Solinas, E.; Grassi, V.; Casucci, G.; and Petz, E.: Comparison of CO_2-rebreathing and direct Fick methods for determining cardiac output. *J. Appl. Physiol., 24:*424, 1968.

Neutze, J. F.; Wyler, F.; and Rudolph, A. M.: Use of radioactive microspheres to assess distribution of cardiac output in rabbits. *Amer. J. Physiol., 215*:486, 1968.

Prather, J. W.; Taylor, A. E.; and Guyton, A. C.: Effect of blood volume, mean circulatory pressure, and stress relaxation on cardiac output. *Amer. J. Physiol., 216*:467, 1969.

Sarnoff, S., and Mitchell, J. H.: The regulation of the performance of the heart. *Amer. J. Med., 30*:747, 1961.

Smith, E. E., and Crowell, J. W.: Influence of hypoxia on mean circulatory pressure and cardiac output. *Amer. J. Physiol., 212*:1067, 1967.

Stone, H. L.; Bishop, V. S.; and Dong, E., Jr.: Ventricular function in cardiac-denervated and cardiac-sympathectomized conscious dogs. *Circ. Res., 20*:587, 1967.

Sugimoto, Tsuneaki; Sagawa, Kiichi; and Guyton, A. C.: Effect of tachycardia on cardiac output during normal and increased venous return. *Amer. J. Physiol., 211*:288, 1966.

Wade, O. L., and Bishop, J. M.: Cardiac Output and Regional Blood Flow. Philadelphia, F. A. Davis Co., 1962.

Whitty, A. J., and Shepard, R. S.: Role of the vagus in control of cardiac output in the unanesthetized dog. *Amer. J. Physiol., 213*:1520, 1967.

CHAPTER 28

Circulatory Shock and Physiology of Its Treatment

Cardiac Output in Shock. *Circulatory shock* means reduced cardiac output to the extent that the tissues of the body are damaged for lack of adequate tissue blood flow. Even the cardiovascular system itself—the heart musculature, the walls of the blood vessels, the vasomotor center, and other parts—begins to weaken so that the shock itself often becomes progressively worse and worse. Therefore, in this discussion shock is divided into two major stages, the *nonprogressive* and *progressive stages.* The nonprogressive stage means cardiac output that is deficient, but not deficient enough to cause a vicious cycle of cardiovascular deterioration. The progressive stage means shock beyond the point at which the circulatory system begins to deteriorate, thus leading to death unless treatment is instituted.

Arterial Pressure in Shock. Shock is the result of diminished cardiac output and not diminished arterial pressure. There are times when a person can be in severe shock and still have a normal arterial pressure because of nervous reflexes that keep the pressure from falling. At other times the arterial pressure can fall to as low as one-half normal but the person can still have a normal cardiac output and therefore not be in shock.

However, in many types of shock, especially that caused by severe blood loss, the arterial pressure does usually fall at the same time that the cardiac output decreases. Therefore, measurements of arterial pressure are often of value in assessing the degree of shock.

PHYSIOLOGICAL CAUSES OF SHOCK

Since shock results from inadequate cardiac output, any factor that can reduce cardiac output can also cause shock. The different factors that can do this were discussed in Chapter 27 and were grouped into two categories:

1. Those that decrease the ability of the heart to pump blood.

2. Those that tend to decrease the venous return.

Thus, serious myocardial infarction or any other factor that damages the heart so severely that it cannot pump adequate quantities of blood can cause a type of shock called *cardiac shock,* which will be discussed in the following chapter. Also, all the factors that reduce venous return, including (a) diminished blood volume, (b) decreased vasomotor tone, or (c) greatly increased resistance to blood flow, can result in shock. The present chapter will deal primarily with these factors.

SHOCK CAUSED BY HYPOVOLEMIA —HEMORRHAGIC SHOCK

Hypovolemia means diminished blood volume, and hemorrhage is perhaps the most common cause of hypovolemic shock. Therefore, a discussion of hemorrhagic shock will serve to explain many of the basic principles of the shock problem.

Hemorrhage *decreases the mean systemic pressure* and as a consequence decreases venous return. As a result, the cardiac output falls below normal, and shock ensues. Obviously, all degrees of shock can result from hemorrhage, from the mildest diminishment of cardiac output to almost cessation of output.

RELATIONSHIP OF BLEEDING VOLUME TO CARDIAC OUTPUT AND ARTERIAL PRESSURE

Figure 28–1 illustrates the effect on both cardiac output and arterial pressure of removing blood from the circulatory system over a period of about half an hour. Approximately 10 per cent of the total blood volume can be removed with no significant effect on arterial pressure or cardiac output, but greater blood loss usu-

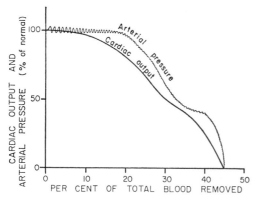

Figure 28–1 Effect of hemorrhage on cardiac output and arterial pressure.

ally diminishes the cardiac output first and later the pressure, both of these falling to zero when about 35 to 45 per cent of the total blood volume has been removed.

Sympathetic Reflex Compensation in Shock. Fortunately, the decrease in arterial pressure caused by blood loss initiates powerful sympathetic reflexes that stimulate the sympathetic vasoconstrictor system throughout the body, resulting in three important effects: (1) The arterioles constrict in most parts of the body, thereby greatly increasing the total peripheral resistance. (2) The veins and venous reservoirs constrict, thereby helping to maintain adequate venous return despite diminished blood volume. And (3) heart activity increases markedly, sometimes increasing the heart rate from the normal value of 72 beats per minute to as much as 200 beats per minute.

VALUE OF THE REFLEXES. In the absence of the sympathetic reflexes, only 15 to 20 per cent of the blood volume can be removed over a period of half an hour before a person will die; this is in contrast to 30 to 35 per cent when the reflexes are intact. Therefore, the reflexes extend the amount of blood loss that can occur without causing death to about two times that which would be possible in their absence.

GREATER EFFECT OF THE REFLEXES IN MAINTAINING ARTERIAL PRESSURE THAN IN MAINTAINING OUTPUT. Referring again to Figure 28–1, it is evident that the arterial pressure is maintained at or near normal levels in the hemorrhaging person longer than is the cardiac output. The reason for this is that the sympathetic reflexes are geared more for maintenance of arterial pressure than for maintenance of output. They increase the arterial pressure primarily by increasing the total peripheral resistance, which has no beneficial effect on cardiac output; instead, cardiac output is regulated more by local factors in the circulation than by the nervous system as was discussed in Chapter 27.

Especially interesting are the two plateaus in the arterial pressure curve of Figure 28–1. The failure of the arterial pressure to decrease when one first begins to remove blood from the circulation is caused mainly by strong stimulation of the *baroreceptor reflexes.* Then, as the arterial pressure falls to about 30 mm. Hg, a second plateau results from activation of the *CNS ischemic response,* which was discussed in Chapter 25. This effect of the CNS ischemic response can be called the "last-ditch stand" of the sympathetic reflexes in their attempt to keep the arterial pressure from falling too low.

PROTECTION OF CORONARY AND CEREBRAL BLOOD FLOW BY THE REFLEXES. A special value of the maintenance of normal arterial pressure even in the face of decreasing cardiac output is protection of blood flow through the coronary and cerebral circulatory systems. Sympathetic stimulation does not cause significant constriction of either the cerebral or cardiac vessels—if anything, it causes slight vasodilatation of the coronary vessels. In addition, in both these vascular beds local autoregulation is excellent, which prevents moderate changes in arterial pressure from significantly affecting their blood flows. Therefore, blood flow through the heart and brain is maintained essentially at normal levels as long as the arterial pressure does not fall below about 70 mm. Hg, despite the fact that blood flow in many other areas of the body might be decreased almost to zero because of vasospasm.

NONPROGRESSIVE AND PROGRESSIVE HEMORRHAGIC SHOCK

Figure 28–2 illustrates an experiment in dogs that demonstrates the effects of different degrees of hemorrhage on the subsequent course of arterial pressure. The dogs were bled rapidly until their arterial pressures fell to different levels. Those dogs whose pressures fell no lower than 45 mm. Hg all eventually

Figure 28–2 Course of arterial pressure in dogs after different degrees of acute hemorrhage. Each curve represents the average results from six dogs.

recovered; the recovery occurred rapidly if the pressure fell only slightly and occurred slowly if it fell almost to the 45 mm. Hg level. On the other hand, when the arterial pressure fell below 45 mm. Hg, all the dogs died, though many of them hovered between life and death for many hours before the circulatory system began to deteriorate.

This experiment demonstrates that the circulatory system can recover as long as the degree of hemorrhage is no greater than a certain critical amount. However, crossing this critical amount by even a few milliliters of blood loss makes the eventual difference between life and death. Thus, hemorrhage beyond a certain critical level causes shock to become *progressive.* That is, *the shock itself causes still more shock,* the condition becoming a vicious cycle that leads eventually to complete deterioration of the circulation and to death.

Nonprogressive Shock—Compensated Shock. If shock is not severe enough to cause its own progression, the person eventually recovers. Therefore, shock of this lesser degree can be called *nonprogressive shock.* It is also frequently called *compensated shock,* meaning that the sympathetic reflexes and other factors have compensated enough to prevent deterioration of the circulation.

The factors that cause a person to recover from moderate degrees of shock are the negative feedback control mechanisms of the circulation that attempt to return cardiac output and arterial pressure to normal levels. These include (1) the *baroreceptor reflexes,* which elicit powerful sympathetic stimulation of the circulation; (2) the *central nervous system ischemic response,* which elicits even more powerful sympathetic stimulation throughout the body but is not activated until the arterial pressure falls below 50 mm. Hg; (3) *reverse stress relaxation of the circulatory system,* which causes the blood vessels to contract down around the diminished blood volume so that the blood volume that is available will more adequately fill the circulation; and (4) *compensatory mechanisms that return the blood volume back toward normal,* including absorption of large quantities of fluid from the intestinal tract, absorption of fluid from the interstitial spaces of the body, and increased thirst and increased appetite for salt which make the person drink water and eat salty foods if he is able.

The sympathetic reflexes provide immediate help in bringing about recovery, for they become maximally activated within 30 seconds after hemorrhage. The reverse stress relaxation that causes contraction of the blood ves-

sels and venous reservoirs around the blood requires some 10 minutes to an hour to occur completely, but, nevertheless, this aids greatly in increasing the mean systemic pressure and thereby increasing the return of blood to the heart. Finally, the readjustment of blood volume by absorption of fluid from the interstitial spaces and from the intestinal tract, as well as the ingestion and absorption of additional quantities of fluid and salt, may require from 1 to 48 hours, but eventually recovery takes place provided the shock does not become severe enough to enter the progressive stage.

Progressive Shock—The Vicious Cycle of Cardiovascular Deterioration. Once shock has become severe enough, the structures of the circulatory system themselves begin to deteriorate, and various types of positive feedback develop that can cause a vicious cycle of progressively decreasing cardiac output. Figure 28–3 illustrates some of these different types of positive feedback that further depress the cardiac output in shock. These are the following:

CARDIAC DEPRESSION. When the arterial pressure falls low enough, coronary blood flow decreases below that required for adequate nutrition of the myocardium itself. This obviously weakens the heart and thereby decreases the cardiac output still more. As a consequence, the arterial pressure falls still further, and the coronary blood flow decreases some more, making the heart still weaker. Thus, a positive feedback cycle has developed whereby the shock becomes more and more severe.

Figure 28–4 illustrates cardiac function curves from experiments in dogs, showing progressive deterioration of the heart at different times following the onset of shock. The dog was bled until the arterial pressure fell to 30 mm. Hg, and the pressure was held at this level by further bleeding or retransfusion of blood as required. Note that there was little deterioration of the heart during the first two hours, but by four hours the heart had deteriorated about 40 per cent; and, finally, during the last hour of the experiment the heart deteriorated almost completely.

Thus, one of the important features of progressive shock, whether it be hemorrhagic shock or any other type of shock, is eventual progressive deterioration of the heart. In the early stages of shock, this plays very little role in the condition of the person, partly because deterioration of the heart itself is not very severe during the first hour or so of shock but mainly because the heart has a tremendous reserve that makes it normally capable of pumping 300 to 400 per cent more blood than

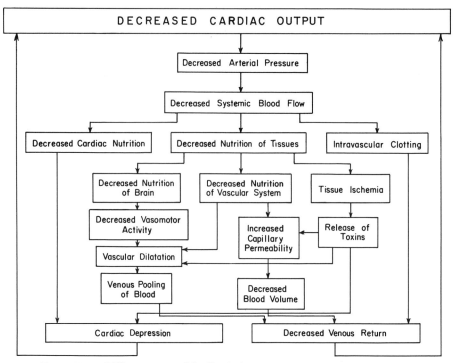

Figure 28–3 Different types of feedback that can lead to progression of shock.

is required by the body for adequate nutrition. However, in the late stages of shock, deterioration of the heart is probably by far the most important factor in the further progression of the shock.

VASOMOTOR FAILURE. In the early stages of shock, various circulatory reflexes cause intense activity of the sympathetic nervous system. This, as discussed above, helps to delay depression of the cardiac output and especially helps to prevent decreased arterial pressure. However, there comes a point at which diminished blood flow to the vasomotor center itself so depresses the center that it becomes progressively less active and finally totally inactive. For instance, complete circulatory arrest to the brain causes, during the first four to eight minutes, the most intense of all sympathetic discharges, resulting in a typical CNS ischemic response, but by the end of 10 to 15 minutes the vasomotor center becomes so depressed that no evidence of sympathetic discharge at all can be demonstrated. Fortunately, though, the vasomotor center does not usually fail in the early stages of shock—only in the late stages.

VASCULAR FAILURE. In addition to vascular dilatation caused by vasomotor failure, dilatation can also result from direct diminishment of nutrients to the vessels during shock. Arteriolar dilatation can reduce the arterial pressure

and thereby reduce the blood flow to the heart and brain, in this way causing further progression of the shock. And venous dilatation can cause blood to "pool" in the veins, thereby depressing cardiac output.

THROMBOSIS OF THE MINUTE VESSELS— SLUDGED BLOOD. Recently, several research workers have shown that thrombosis of many of the minute vessels in the circulatory system can be one of the causes of progression of

Figure 28–4 Function curves of the heart at different times after hemorrhagic shock begins. (These curves are extrapolated to the human heart from data obtained in dog experiments by Dr. J. W. Crowell.)

shock. That is, blood flow through many tissues becomes extremely sluggish, but tissue metabolism continues so that large amounts of acid, either carbonic acid or lactic acid, continue to empty into the blood. This acid, plus other deterioration products from the ischemic tissues, causes blood agglutination or actual blood clots, thus leading to minute plugs in the small vessels. Even if the vessels do not become plugged, the agglutination of cells makes it more difficult for blood to flow through the microvasculature, giving rise to the term *sludged blood.*

INCREASED CAPILLARY PERMEABILITY. After many hours of capillary hypoxia the permeability of the capillaries gradually increases and large quantities of fluid begin to transude into the tissues. This decreases the blood volume, with resultant further decrease in cardiac output, thereby making the shock still more severe.

Fortunately, capillary hypoxia does not cause increased capillary permeability until the very late stages of extremely prolonged shock. Therefore, this factor plays a significant role in only few instances of shock.

RELEASE OF TOXINS BY ISCHEMIC TISSUES. Many different research workers in the field of shock have suggested time and again that shock causes tissues to release toxic substances and that these toxic substances in turn cause further deterioration of the circulatory system. For instance, one of the suggested toxins is histamine; the ischemic tissues supposedly release histamine, and the histamine in turn causes (a) a weakened heart, (b) arteriolar paralysis, and (c) increased permeability of the capillaries. However, critical studies have failed to demonstrate release of sufficient quantities of histamine in most cases of shock for this to be one of the primary causes of progression of shock.

Another toxic factor that probably plays a role is *endotoxin,* which is a toxin released from the bodies of dead gram-negative bacteria in the intestines. Diminished blood flow to the intestines causes enhanced absorption of this toxic substance from the intestines, and it then causes extensive vascular dilatation and cardiac depression. Though this toxin can play a major role in some types of shock, as is discussed later in the chapter, it is still unknown how much of the endotoxin is released during hemorrhagic shock and whether or not it is an important factor in the progression of this type of shock.

Relationship of Positive Feedback to Vicious Cycles in Shock. All the factors just discussed that can lead to further progression of shock are types of *positive feedback.* That is, in each instance, a decrease in cardiac output causes a further decrease in cardiac output.

However, positive feedback does not always necessarily lead to a vicious cycle. Whether or not a vicious cycle develops depends on the intensity of the positive feedback. In mild degrees of shock, the negative feedback mechanisms — the sympathetic reflexes, the plasticity of the circulation, and the recovery of blood volume — can easily overcome the positive feedback influences and can therefore cause recovery. In severe degrees of shock, however, the positive feedback mechanisms become more and more powerful, thus leading to such rapid deterioration of the circulation that the negative feedback systems cannot return the cardiac output to normal.

Thus, if one will consider once again the principles of positive feedback as discussed in Chapter 1, he will see that a vicious cycle of deterioration occurs in shock only when the positive feedback effect to decrease cardiac output further is greater than the initial decrease in output. This explains why there is a critical cardiac output level above which a person in shock recovers and below which he goes into a vicious cycle of circulatory deterioration. Above the critical level the negative feedback factors attempting to cause recovery can overcome the positive feedback factors. Below the critical level, the positive feedback response becomes sufficiently great that even a slight further decrease in cardiac output can lead to a vicious cycle that eventuates in death.

IRREVERSIBLE SHOCK

After shock has progressed to a certain stage, transfusion or any other type of therapy becomes incapable of saving the life of the person. Therefore, the person is then said to be in the *irreversible stage of shock.* Paradoxically, therapy often returns the arterial pressure and even the cardiac output to normal, but the circulatory system continues to deteriorate and death ensues in another few minutes to hours.

Figure 28–5 illustrates this effect, showing that transfusion during the irreversible stage can cause the cardiac output (as well as the arterial pressure) to return to normal. However, the cardiac output soon begins to fall again, and subsequent transfusions cause less and less effect. Thus, something changes in the overall function of the circulatory system during shock that does not affect the *immediate* ability of the heart to pump blood but over a

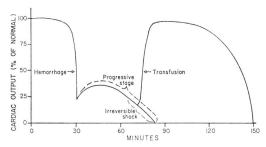

Figure 28–5 Failure of transfusion to prevent death in irreversible shock.

long period of time does depress this ability and eventuates in death. Now the question remains: What factor or factors lead to the eventual deterioration of the circulation?

Relationship of Oxygen Deficit to the Irreversible Stage of Shock. The diminished cardiac output during shock obviously causes diminished transport of oxygen to the tissues. Therefore, the tissues have a *deficit* of oxygen utilization during shock. Recently, Crowell has demonstrated that shock becomes irreversible in an animal when his oxygen deficit has reached an average of 120 ml. of oxygen per kg. This effect is illustrated by the *solid curve* in Figure 28–6, which shows the survival rates of dogs following different degrees of oxygen deficit occurring in hemorrhagic shock. Up to an oxygen deficit of 100 ml./kg., none of the dogs died. Beyond 150 ml./kg., all of them died. And half of them died at 120 ml./kg.

In the experiment represented by Figure 28–6, it did not matter how long or under what conditions the dogs accumulated the oxygen deficit. If the shock was severe, the dogs accumulated the deficit in as little as 30 minutes, or if the shock was slight, many hours were required. But once the dogs had accumulated the critical oxygen deficit, regardless of how long a period of time it required, the shock

Figure 28–6 Survival rates of dogs following various degrees of oxygen deficit developed during the course of shock. The solid curve represents survival without treatment with ouabain and the dashed curve with ouabain treatment, showing considerable benefit from the cardiotonic drug. (The numbers represent numbers of dogs.) (Courtesy of Dr. J. W. Crowell.)

became irreversible to normal types of therapy, and the animals died.

This experiment demonstrates the importance of oxygen in preventing death from hemorrhagic shock and probably also in preventing death from other types of shock. Indeed, animals administered oxygen in a pressure tank in which oxygen pressure can be increased to extremely high levels can survive longer periods of hypotension than can animals exposed only to air. Therefore, it is probable that the most detrimental factor in shock is diminished transport of oxygen to the tissues and that it is mainly lack of oxygen that leads to the severe deterioration of the heart and other parts of the circulatory system, finally eventuating in irreversible shock and death.

Deterioration of the Heart in the Irreversible Stage of Shock. Earlier in the chapter, it was pointed out that many different parts of the circulatory system can deteriorate in shock. Yet it is almost certain that deterioration of the heart itself is the primary factor that causes shock to become irreversible. The reason for believing this is that even in this stage of shock one can provide more than adequate venous return by simply administering transfusions to the person. However, despite more than adequate venous return, the heart cannot pump the extra blood flowing into the right atrium.

The dashed curve of Figure 28–6 illustrates that when the hearts of dogs were strengthened by the use of the cardiotonic drug ouabain, the dogs were able to survive far greater oxygen deficit during shock than was possible without this drug. Thus, this curve demonstrates that if cardiac deterioration can be prevented, an animal will not go into irreversible shock nearly so easily, thus showing again that it is primarily the heart itself that becomes damaged beyond repair in irreversible shock.

OTHER CAUSES OF HYPOVOLEMIC SHOCK

Plasma Loss Shock. Loss of plasma from the circulatory system can sometimes be severe enough to reduce the total blood volume markedly, in this way causing typical hypovolemic shock similar in almost all details to that caused by hemorrhage. Severe plasma loss occurs in the following conditions:

1. *Intestinal obstruction* is often a cause of reduced plasma volume. The distention of the intestine in this condition causes fluid to leak from the intestinal capillaries into the intestinal walls and intestinal lumen. This increased loss of fluid might result from elevated capillary pressure caused by increased resistance in the stretched veins over the surface of the intestine, or it might

be caused by direct damage to the capillaries themselves. Nevertheless, the lost fluid has a very high content of protein, thereby reducing the total plasma protein as well as the plasma volume.

2. Often, in patients who have _severe burns_ or other denuding conditions of the skin, so much plasma is lost through the exposed areas that the plasma volume becomes markedly reduced.

The hypovolemic shock that results from plasma loss has almost the same characteristics as the shock caused by hemorrhage, except for one additional complicating factor – the blood viscosity increases greatly as a result of plasma loss, and this further exacerbates the sluggishness of blood flow.

3. Loss of fluid from all fluid compartments of the body is called _dehydration;_ this, too, can reduce the blood volume and cause hypovolemic shock very similar to that resulting from hemorrhage. Some of the causes of this type of shock are: (a) excessive sweating; (b) fluid loss in severe diarrhea or vomiting; (c) excess loss of fluid by nephrotic kidneys; (d) inadequate intake of fluid and electrolytes; (e) destruction of the adrenal cortices, with consequent failure of the kidneys to reabsorb sodium, chloride, and water; and (f) loss of the secretion of antidiuretic hormone by the supra-optico-hypophyseal system.

NEUROGENIC SHOCK – INCREASED VASCULAR CAPACITY

Occasionally shock results without any loss of blood volume whatsoever. Instead, the _vascular capacity_ increases so much that even the normal amount of blood becomes incapable of adequately filling the circulatory system. One of the major causes of this is _loss of vasomotor tone_ throughout the body, and the resulting condition is then known as _neurogenic shock._

The relationship of vascular capacity to blood volume was discussed in Chapter 18, where it was pointed out that either an increase in vascular capacity or a decrease in blood volume _reduces the mean systemic pressure,_ which in turn reduces the venous return to the heart. This effect is often called "venous pooling" of blood.

Importance of Body Position in Neurogenic Shock. Ordinarily, sudden loss of vasomotor tone throughout the body will not cause shock if the person is lying in a slightly head-down position (Trendelenburg position). If he is lying in a completely horizontal position, only mild to moderate degrees of reduced cardiac output result. But, if the person is in the upright position, the vessels of the lower part of the body become so distended that the blood "pools" and fails to flow uphill in sufficient quantities to maintain cardiac output. As a result, extremely severe shock results.

Causes of Neurogenic Shock. Some of the different factors that can cause sudden loss of vasomotor tone include:

1. _Deep general anesthesia_ often depresses the vasomotor center enough to cause vasomotor collapse, with resulting neurogenic shock.

2. _Spinal anesthesia,_ especially when this extends all the way up the spinal cord, blocks the sympathetic outflow from the nervous system and is a common cause of neurogenic shock.

3. _Brain damage_ is often a cause of vasomotor collapse. Many patients who have had brain concussion or contusion of the basal regions of the brain develop profound neurogenic shock. Also, even though short periods of medullary ischemia cause extreme vasomotor activity, prolonged ischemia can result in inactivation of the vasomotor neurons and can cause development of severe neurogenic shock.

4. In _fainting,_ the peripheral blood vessels become greatly dilated. As a result, blood pools, and the cardiac output falls drastically. If such a person is held in an upright position, he will go into the progressive stage of shock and can die as a result. Fortunately, on fainting, a person usually falls to a horizontal position so that almost normal cardiac output ordinarily re-ensues almost immediately.

VASOVAGAL SYNCOPE – EMOTIONAL FAINTING. The circulatory collapse that results from "emotional" fainting usually is not caused by vasomotor failure but instead by strong emotional excitation of the parasympathetic nerves to the heart and of the vasodilator nerves to the skeletal muscles, thereby slowing the heart and reducing the arterial pressure. Therefore, the fainting that results from an emotional disturbance is called _vasovagal syncope_ to differentiate it from the other types of fainting which result from diminished sympathetic activity throughout the body or from other causes of reduced cardiac output.

OTHER TYPES OF SHOCK THAT RESULT FROM INCREASED VASCULAR CAPACITY

Occasionally, the vascular capacity is increased in other ways besides reduction in vasomotor tone. For instance, a person who has had severe abdominal ascites often develops shock when the ascitic fluid is removed rapidly. This results from loss of the pressure that the ascites had been exerting on all the blood vessels of the abdominal cavity and on the blood reservoirs – the liver and spleen. Suddenly, therefore, an increased amount of blood is needed to fill the circulatory system. Since this is not available, shock often ensues.

Likewise, shock often occurs after a large abdominal tumor has been removed, and a _tendency_ to shock even occurs when a mother gives birth to a baby because of suddenly increased vascular capacity, illustrating the importance of the extra blood volume that a mother accumulates prior to parturition (Chapter 82).

ANAPHYLACTIC SHOCK

"Anaphylaxis" is an allergic condition in which the cardiac output and arterial pressure often fall

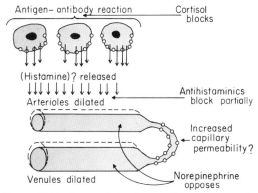

Figure 28–7 Some of the factors that cause anaphylactic shock.

drastically. This was discussed in Chapter 10. It results primarily from an antigen-antibody reaction that takes place all through the body immediately after an antigen to which the person is sensitive has entered the circulatory system. Such a reaction is detrimental to the circulatory system in several important ways, as illustrated in Figure 28–7. First, if the antigen-antibody reaction takes place in direct contact with the vascular walls or cardiac musculature, damage to these tissues presumably can result directly. Second, cells damaged anywhere in the body by the antigen-antibody reaction release several highly toxic substances into the blood. Among these is *histamine or a histamine-like substance* that has a strong vasodilator effect. The histamine in turn causes (1) an increase in vascular capacity because of venous dilatation and (2) dilatation of the arterioles with resultant greatly reduced arterial pressure. Unfortunately, all the precise relationships of the above factors in anaphylaxis have not been determined, but the sum total is a great reduction in venous return and often such serious shock that the person dies within minutes.

Intravenous injection of large amounts of histamine causes "histamine shock," which has characteristics almost identical with those of anaphylactic shock, though usually less severe.

SEPTIC SHOCK

The condition that was formerly known by the popular name of "blood poisoning" is now called *septic shock* by most clinicians. This means simply widely disseminated infection in many areas of the body, the infection often being borne through the blood from one tissue to another and causing extensive damage. Actually, there are many different varieties of septic shock because of the many different types of bacterial infection that can cause it and also because infection in one part of the body will produce different effects from those of infection elsewhere in the body.

Septic shock is extremely important to the clinician because it is this type of shock that, more frequently than any other kind of shock besides cardiac shock, causes death in the modern hospital. Some of the typical causes of septic shock include:

1. Peritonitis caused by spread of infection from the uterus and fallopian tubes, frequently resulting from instrumental abortion

2. Peritonitis resulting from rupture of the gut, sometimes caused by intestinal disease and sometimes by wounds

3. Generalized infection resulting from spread of a simple skin infection such as streptococcal or staphylococcal infection

4. Generalized gangrenous infection resulting specifically from gas gangrene bacilli, spreading first through the tissues themselves and finally by way of the blood to the internal organs, especially to the liver.

Special Features of Septic Shock. Because of the multiple types of septic shock, it is difficult to give categorically an exact picture of this condition. However, some features often seen in septic shock are the following:

1. High fever

2. Marked vasodilatation throughout the body, especially in the infected tissues

3. High cardiac output probably caused by the vasodilatation in the infected tissues and also by high metabolic rate elsewhere in the body resulting from the high body temperature

4. Sludging of the blood, presumably caused by red cell agglutination in response to bacterial toxins or in response to degenerating tissues.

In the early stages of septic shock the patient usually does not have signs of circulatory collapse but, instead, simply signs of the bacterial infection itself. However, as the infection becomes more severe, the circulatory system usually becomes involved either directly or as a secondary result of toxins from the bacteria, and there finally comes a point at which deterioration of the circulation becomes progressive in the same way that progression occurs in all other types of shock. Therefore, the end stages of septic shock are not greatly different from the end stages of hemorrhagic shock, even though the initiating factors are markedly different in the two conditions.

Endotoxin Shock. A special type of septic shock is known as endotoxin shock. It frequently occurs when a large segment of the gut becomes strangulated and loses most of its blood supply. The gut rapidly becomes gangrenous, and the bacteria in the gut multiply rapidly. Most of these bacteria are so-called "gram-negative" bacteria, mainly colon bacilli, that contain a toxin called *endotoxin.*

On entering the circulation, endotoxin causes an effect very similar to that of anaphylaxis, sometimes resulting in severe shock. Indeed, the cause of this shock is probably an anaphylactic reaction to the endotoxin itself, the mechanism of which is the following:

All persons are exposed throughout life to minute quantities of endotoxin released from degenerating colon bacilli in the gastrointestinal tract. This small amount of toxin causes the body to develop moderate amounts of antibodies against the endotoxin. Then, when a large dose of endotoxin enters the circulatory system during some gastrointestinal crisis, an anaphylactic reaction occurs, resulting in the severe shock often seen in these conditions.

Severe shock also occurs during the course of diseases caused by other gram-negative bacterial infections, such as dysentery, tularemia, brucellosis, and typhoid fever. It is likely that this shock results at least partially from endotoxin released from the gram negative bacteria causing these conditions.

TRAUMATIC SHOCK

One of the most common causes of circulatory shock is trauma to the body. Often the shock results simply from hemorrhage caused by the trauma, but it can also occur even without hemorrhage, for contusion of the body can often damage the capillaries sufficiently to allow excessive loss of plasma into the tissues. This results in greatly reduced plasma volume with resultant hypovolemic shock. Thus, whether or not hemorrhage occurs when a person is severely traumatized, the blood volume can still be markedly reduced.

The *pain* associated with serious trauma can be an additional aggravating factor in traumatic shock, for pain sometimes strongly inhibits the vasomotor center, thereby increasing the vascular capacitance and reducing the venous return. Various attempts have also been made to implicate toxic factors released by the traumatized tissues as one of the causes of shock following trauma. However, cross-transfusion experiments have failed to show any such toxic element.

In summary, traumatic shock seems to result mainly from hypovolemia, though there may also be a moderate degree of concomitant neurogenic shock caused by the pain.

EFFECTS OF SHOCK ON THE BODY

Decreased Metabolism. The decreased cardiac output in shock reduces the amount of oxygen and other nutrients available to the different tissues, and this in turn reduces the level of metabolism that can be maintained by the different cells of the body. Usually a person can continue to live for only a few hours if his cardiac output falls to as low as one-half normal.

Muscular Weakness. One of the earliest symptoms of shock is severe muscular weakness which is also associated with profound and rapid fatigue whenever one attempts to use his muscles. This obviously results from the diminished supply of nutrients—especially oxygen—to the muscles.

Body Temperature. Because of the depressed metabolism in shock, the amount of heat liberated in the body is reduced in shock. As a result, the body temperature tends to decrease if the body is exposed to even the slightest cold.

Mental Function. In the early stages of shock the person is usually conscious, though he may exhibit signs of mental haziness. As the shock progresses, he falls into a state of stupor, and in the last stages of shock even his subconscious mental functions, including vasomotor control and respiration, fail.

A person who recovers from shock usually exhibits no permanent impairment of mental functions. However, following complete circulatory arrest, in which the blood flow is completely cut off from the brain for many minutes, the brain does often suffer severe permanent impairment, the cause of which is discussed in a separate section on circulatory arrest later in the chapter.

Reduced Renal Function. The very low blood flow during shock greatly diminishes urine output because glomerular pressure falls below the critical value required for filtration of fluid into Bowman's capsule, as explained in Chapter 34. Also, the kidney has such a high rate of metabolism and requires such large amounts of nutrients that the reduced blood flow often causes *tubular necrosis,* which means death of the tubular epithelial cells, with subsequent sloughing and blockage of the tubules, causing total loss of function of the respective nephrons. This is often a serious aftereffect of shock that occurs during major surgical operations; the patient sometimes survives the shock associated with the surgical procedure and then dies a week or so later of uremia.

PHYSIOLOGY OF TREATMENT IN SHOCK

Basically, shock can be treated by any type of therapy that increases the cardiac output. In general, the best therapy is to correct the condition that is causing the shock. Aside from the treatment of cardiac shock, which will be discussed in Chapter 29, the two most common types of treatment of the different types of shock are (1) replacement therapy with whole blood or other fluids and (2) use of sympathomimetic or sympatholytic drugs.

REPLACEMENT THERAPY

Blood and Plasma Transfusion. If a person is in shock caused by hemorrhage, the best possible therapy is usually transfusion of whole blood. If the shock is caused by plasma loss, the best therapy is administration of plasma; when dehydration is the cause, administration of the appropriate electrolytic solution can correct the shock.

The basic principles of blood typing and the use of whole blood and plasma for transfusion were discussed in Chapter 11. Unfortunately, properly matched whole blood is not always available, such as under battlefield conditions. However, plasma can usually substitute adequately for whole blood because it increases the blood volume and restores normal hemodynamics. Plasma cannot restore a normal hematocrit, but the human being can usually stand a decrease in hematocrit to about one-third normal before serious consequences result. Therefore, in acute conditions it is reasonable to use plasma in place of whole blood for treatment of hemorrhagic and most other types of hypovolemic shock.

Sometimes plasma also is unavailable. For these instances, various *plasma substitutes* have been developed that perform almost exactly the same hemodynamic functions as plasma. One of these is the following:

DEXTRAN SOLUTION AS A PLASMA SUBSTITUTE. The principal requirement of a truly effective plasma substitute is that it remain in the circulatory system—that is, not filter through the capillary pores into the tissue spaces. But, in addition, the solution must be nontoxic and must contain appropriate electrolytes to prevent derangement of the extracellular fluid electrolytes on administration. To remain in the circulation the plasma substitute must contain some substance that has a large enough molecular size to exert colloid osmotic pressure.

One of the most satisfactory substances that has been developed thus far for this purpose is dextran, a large polysaccharide polymer of glucose. Certain bacteria secrete dextran as a by-product of their growth, and commercial dextran is manufactured by a bacterial culture procedure. By varying the growth conditions of the bacteria, the molecular weight of dextran can be controlled to any desired value. Dextrans of appropriate molecular size do not pass through the capillary pores and, therefore, can replace plasma proteins as colloid osmotic agents.

Fortunately, few toxic reactions have been observed when using dextran to provide colloid osmotic pressure; therefore, solutions of this substance have proved to be a satisfactory substitute for plasma in fluid replacement therapy.

TREATMENT OF SHOCK WITH SYMPATHO-MIMETIC AND SYMPATHOLYTIC DRUGS OR OTHER THERAPY

A *sympathomimetic drug* mimics sympathetic stimulation. These drugs include norepinephrine, epinephrine, and a large number of long-acting drugs that have the same effects as epinephrine and norepinephrine. A *sympatholytic drug,* on the other hand, blocks the action of normal sympathetic stimuli on the circulatory system.

Both sympathomimetic and sympatholytic drugs have been used in treating different types of shock, including hemorrhagic shock. However, there is no proof that either plays any very beneficial role in this type of shock. The sympathomimetic drugs usually have little benefit because the circulatory reflexes are usually already maximally activated in severe hypovolemic shock. On the other hand, the proponents of the use of sympatholytic drugs have claimed that these drugs increase blood flow through the tissues and thereby prevent nutritional damage. Unfortunately, these drugs also usually cause paralysis of the venous vasculature, resulting in increased pooling of blood in the venous system, which reduces the cardiac output and negates much if not all of the direct effect of the drugs in increasing local blood flow. Yet a combination of *transfusion plus sympatholytic drugs* probably is beneficial in the treatment of many patients with shock; the transfusion fills up the expanded vascular bed and prevents the pooling, while the sympatholytic drug allows rapid blood flow to the tissues.

Treatment of Neurogenic Shock. In neurogenic shock, sympathomimetic drugs actually reverse the basic cause of the shock itself by increasing the vasomotor tone throughout the body, but even in this condition these drugs may not be required, for as long as the patient is in the horizontal or head-down position, adequate venous return to the heart usually still occurs and maintains a sufficient cardiac output to prevent progressive shock.

Treatment of Anaphylactic Shock. Unfortunately, anaphylactic shock occurs so rapidly that one often cannot institute any therapy before death ensues, but if therapy can be instituted the condition can often be ameliorated or almost completely reverted by rapid administration of norepinephrine or some other sympathomimetic drug. This does not correct the basic cause of the anaphylaxis, but it does cause vasoconstriction, which opposes the vasodilatation caused by histamine in anaphylaxis. Also, if one suspects that anaphylaxis might occur in a patient, its severity can usually be reduced by preliminary administration of *cortisol,* which attenuates the allergic reaction responsible for the anaphylaxis, or preliminary administration of *antihistaminics,* which reduce the effects of the histamine released during anaphylaxis. However, in both these instances the therapy must be given *before* anaphylaxis takes place, for which reason they are usually of little value in therapy of anaphylaxis.

Pain-Relieving Drugs in Traumatic Shock. The use of pain-relieving drugs, such as morphine, has been found clinically to reduce the severity of most traumatic shock, supposedly by removing the neurogenic element of the shock.

CIRCULATORY ARREST

A condition closely allied to circulatory shock is circulatory arrest, in which all blood flow completely stops. This occurs frequently on the surgical operating table as a result of *cardiac arrest* or of *ventricular fibrillation.*

Ventricular fibrillation can usually be stopped by the following procedure: The heart is pumped by hand for 5 to 10 minutes to restore adequate coronary flow to the ventricular muscle. Then a strong electrical current is passed through the heart from two electrode plates pressing against the sides of the heart. This throws essentially all the heart into a state of refractoriness, and usually a normal cardiac rhythm is restored thereafter. The basic principles of this procedure were described in Chapter 14.

Cardiac arrest usually results from too little oxygen in the anesthetic gaseous mixture or from a depressant effect of the anesthesia itself. A normal cardiac rhythm can usually be restored by removing the anesthetic and then pumping the heart by hand for a few minutes while supplying the patient's lungs with adequate quantities of ventilatory oxygen.

EFFECT OF CIRCULATORY ARREST ON THE BRAIN

The real problem in circulatory arrest is usually not to restore cardiac function but instead to prevent detrimental effects in the brain as a result of the circulatory arrest. In general, four to five minutes of circulatory arrest causes permanent brain damage in over half the patients, and circulatory arrest for as long as 10 minutes almost universally destroys most, if not all, of the mental powers.

For many years it has been taught that these detrimental effects on the brain were caused by the cerebral hypoxia that occurs during circulatory arrest. However, recent studies by Crowell and others have shown that dogs can almost universally stand 30 minutes of circulatory arrest without brain damage if the blood is removed from the circulation prior to the arrest. On the basis of these studies, it is postulated that the circulatory arrest causes vascular *clots* to develop throughout the brain and that these cause permanent or semipermanent ischemia of brain areas. This accords well with the results in human beings who have undergone long periods of circulatory arrest, for complete destruction of large areas in one side of the brain often occurs while corresponding areas in the opposite side of the brain, which should also be affected if hypoxia were the cause of the damage, are not affected even in the slightest.

REFERENCES

Chick, W. L.; Weiner, R.; Cascarno, J.; and Zweifach, B. W.: Influence of Krebs-cycle intermediates on survival in hemorrhagic shock. *Amer. J. Physiol.,* 215:1107, 1968.

Chien, S.: Role of the sympathetic nervous system in hemorrhage. *Physiol. Rev.,* 47:214, 1967.

Crowell, J. W.: Cardiac deterioration as the cause of irreversibility in shock. *In* Mills, Lewis, J., and Moyer, John H. (eds.): Shock and Hypotension: Pathogenesis and Treatment. Grune & Stratton, Inc., 1965, p. 605.

Crowell, J. W., and Guyton, A. C.: Evidence favoring a cardiac mechanism in irreversible hemorrhagic shock. *Amer. J. Physiol.,* 201:893, 1961.

Crowell. J. W., and Read, W. L.: *In vivo* coagulation—a probable cause of irreversible shock. *Amer. J. Physiol.,* 183:565, 1955.

Crowell, J. W., and Smith, E. E.: Oxygen deficit and irreversible hemorrhagic shock. *Amer. J. Physiol.,* 206:313, 1964.

Crowell, J. W.; Bounds, S.; and Johnson, W. W.: The effect of varying the hematocrit ratio on the susceptibility to hemorrhagic shock. *Amer. J. Physiol.,* 192:171, 1958.

Crowell, J. W.; Jones, C. E.; and Smith, E. E.: Effect of allopurinol on hemorrhagic shock. *Amer. J. Physiol.,* 216:744, 1969.

Crowell, J. W.; Sharpe, G. P.; Lambright, R. L.; and Read, W. L.: The mechanism of death after resuscitation following acute circulatory failure. *Surgery,* 38:696, 1955.

Elsberry, D. D.; Rhoda, D. A.; and Beisel, W. R.: Hemodynamics of staphylococcal B enterotoxemia and other types of shock in monkeys. *J. Appl. Physiol.,* 27:164, 1969.

Granata, L.; Huvos, A.; Pasque, A.; and Gregg, D. E.: Left coronary hemodynamics during hemorrhagic hypotension and shock. *Amer. J. Physiol.,* 216:1583, 1969.

Guyton, A. C., and Crowell, J. W.: Dynamics of the heart in shock. *Fed. Proc.,* 20:51, 1961.

Halmagyi, D. F. J.; Goodman, A. H.; and Neering, I. R.: Hindlimb blood flow and oxygen usage in hemorrhagic shock. *J. Appl. Physiol.,* 27:508, 1969.

Harris, P. D.; Greenwald, E. K.; and Nicoll, P. A.: Neural mechanisms in small-vessel response to hemorrhage in the unanesthetized bat. *Amer. J. Physiol.,* 218:560, 1970.

Hirshfeld, J. W., Jr., and Fell, C.: Changes in regional blood content during hemorrhagic shock in dogs. *Amer. J. Physiol.,* 216:380, 1969.

Jones, C. E.; Crowell, J. W.; and Smith, E. E.: Significance of increased blood uric acid following extensive hemorrhage. *Amer. J. Physiol.,* 214:1374, 1968.

Jones, C. E.; Crowell, J. W.; and Smith, E. E.: A cause-effect relationship between oxygen deficit and irreversible hemorrhagic shock. *Surgery,* 127:93, 1968.

Kaihara, S.; Rutherford, R. B.; Schwentker, E. P.; and Wagner, H. N., Jr.: Distribution of cardiac output in experimental hemorrhagic shock in dogs. *Amer. J. Physiol.,* 27:218, 1969.

Kim, S. I.; Desai, J. M.; and Shoemaker, W. C.: Sequence of cardiorespiratory alterations after gradual prolonged hemorrhagic shock in conscious dogs. *Amer. J. Physiol.,* 216:1044, 1969.

Lefer, A. M., and Martin, J.: Mechanism of the protective effect of corticosteroids in hemorrhagic shock. *Amer. J. Physiol.,* 216:314, 1969.

Lefer, A. M., and Martin, J.: Relationship of plasma peptides to the myocardial depressant factor in hemorrhagic shock in cats. *Circ. Res.,* 26:59, 1970.

Moyer, C. A., and Butcher, H. R.: Burns, Shock, and Plasma Volume Regulation. St. Louis, The C. V. Mosby Co., 1967.

Rothe, C. F.; Love, J. R.; and Selkurt, E. E.: Control of total vascular resistance in hemorrhagic shock in the dog. *Circ. Res.,* 12:667, 1963.

Siegel, H. W., and Downing, S. E.: Reduction of left ventricular contractility during acute hemorrhagic shock. *Amer. J. Physiol.,* 218:772, 1970.

Smith, E. E., and Crowell, J. W.: Effect of hemorrhagic hypo-
    tension on oxygen consumption of dogs. *Amer. J. Physiol.,*
    207:647, 1964.
Smith, E. E.; Crowell, J. W.; Moran, C. J.; and Smith, R. A.:
Intestinal fluid loss in dogs during irrevesible hemorrhagic
    shock. *Surgery, 125*:45, 1967.
Weil, M. H., and Shubin, H.: Shock. Baltimore, The Williams &
    Wilkins Co., 1967.

CHAPTER 29

Cardiac Failure

Perhaps the most important ailment that must be treated by the physician is cardiac failure, which can result from any heart condition that reduces the ability of the heart to pump blood. Usually the cause is decreased contractility of the myocardium caused by diminished coronary blood flow, but failure to pump adequate quantities of blood can also be caused by damage to the heart valves, external pressure around the heart, vitamin deficiency, or any other abnormality that makes the heart a hypoeffective pump. The subjects of heart valvular disease and coronary disease are so important that they will be discussed in detail in Chapters 30 and 31, respectively.

Definition of Cardiac Failure. The term cardiac failure means simply *failure of the heart to pump blood adequately.* This does not mean that the cardiac output in all instances of failure is less than normal, for the output can be normal or sometimes above normal provided the tendency for venous return is high enough to offset the diminished strength of the heart. Therefore, cardiac failure may be manifest in either of two ways: (1) by a decrease in cardiac output or (2) by an increase in either left or right atrial pressure even though the cardiac output is normal or above normal.

Unilateral versus Bilateral Cardiac Failure. Since the left and right sides of the heart are two separate pumping systems, it is possible for one of these to fail independently of the other. For instance, unilateral failure can result from coronary thrombosis in one or the other of the ventricles. Because debilitating thrombosis occurs approximately 30 times as often in the left ventricle as in the right ventricle, there is a tendency among clinicians to view failure following myocardial infarction as almost always primarily left-sided. Occasionally, however, right-sided failure does occur with no left-sided failure at all; this happens most frequently in persons with pulmonary stenosis or some other congenital disease affecting primarily the right heart.

Usually, though, when one side of the heart becomes weakened, this causes a sequence of events that makes the opposite side of the heart also fail. For instance, in left-sided failure the left atrial pressure increases greatly, resulting in considerable back pressure in the pulmonary system and a rise in pulmonary arterial pressure sometimes to two to three times normal. This loads the right ventricle, causing combined failure of both ventricles, even though the initiating cause was in the left side of the heart only. On the other hand, primary damage to the right ventricle sometimes reduces the overall cardiac output, resulting in diminished coronary blood flow and at least a mild degree of left heart weakening.

In the first part of this chapter on cardiac failure we will consider the whole heart failing as a unit and then return to the specific features of unilateral left- and right-sided failure.

DYNAMICS OF THE CIRCULATION IN CARDIAC FAILURE

In the following discussion we will consider the progressive changes in the circulation following acute cardiac failure, first, when the failure is moderately severe and, second, when it is almost lethal. However, *before attempting to understand the analyses in this chapter, one must become familiar with the graphical procedure presented in Chapter 27 for equating cardiac function curves with systemic function curves.*

DYNAMICS OF MODERATE CARDIAC FAILURE

Acute Effects. If a heart suddenly becomes severely damaged in any way, such as by myocardial infarction, the pumping ability of the heart is immediately depressed. As a result, two essential effects occur: (a) reduced cardiac output and (b) increased systemic venous pressure. These effects are shown graphically in Figure 29–1, which illustrates, first, a normal cardiac function curve and a normal systemic function curve, depicting the state of the circulation prior to the cardiac damage. The point at which these two curves cross, point A, represents the normal equilibrium state of the

Figure 29–1 Immediate effect of acute myocardial damage on cardiac output and right atrial pressure.

circulation, showing that the normal cardiac output under resting conditions is 5 liters per minute and the right atrial pressure essentially 0 mm. Hg.

Immediately after the heart becomes damaged, the cardiac function curve becomes greatly reduced, falling to the lower, dashed curve. However, at first, nothing at all happens to the systemic function curve, for the systemic circulation has not yet been altered. Therefore, within a few seconds after the acute heart attack, a new equilibrium is established at point B rather than point A, showing that the cardiac output has fallen to 2 liters per minute, or about two-fifths normal, while the right atrial pressure has risen to 4 mm. Hg. This low cardiac output is still sufficient to sustain life, but it is not sufficient to allow a person to stand or even to sit up without fainting. Fortunately, this acute stage lasts for only a few seconds because autonomic reflexes occur immediately that can compensate to a great extent for the damaged heart as follows:

Compensation for Acute Cardiac Failure by Sympathetic Reflexes. When the cardiac output falls precariously low, many of the different circulatory reflexes discussed in Chapter 25 immediately become active. The best known of these reflexes is the baroreceptor reflex, which is activated by diminished arterial pressure. It is probable that the central nervous system ischemic response and possibly even reflexes originating in the damaged heart itself also become activated. Whatever all the reflexes might be, the sympathetics become strongly stimulated within a few seconds, and the parasympathetics become reciprocally inhibited at the same time.

Strong sympathetic stimulation has two major effects on the circulation, first, on the heart itself, and, second, on the peripheral vasculature. Even a damaged myocardium usually responds with increased force of contraction following sympathetic stimulation. If all the musculature is diffusely damaged, sympathetic stimulation usually strengthens this damaged musculature. Likewise, if part of the muscle is totally nonfunctional while part of it is still normal, the normal muscle is strongly stimulated by sympathetic stimulation, in this way compensating for the nonfunctional muscle. Thus, *the heart one way or another becomes a stronger pump, often as much as 100 per cent stronger, under the influence of the sympathetic impulses.*

Sympathetic stimulation also increases the tendency for venous return, for it increases the tone of most of the blood vessels of the circulation, *raising the mean systemic pressure* to 12 to 14 mm. Hg, almost 100 per cent above normal. As will be recalled from the discussion in Chapter 27, this greatly increases the tendency for blood to flow back to the heart. Therefore, the damaged heart becomes primed with more inflowing blood than usual, and, at least under certain circumstances, this increased priming helps the heart to pump larger quantities of blood.

The sympathetic reflexes become maximally developed in about 30 seconds, and the resulting increased ability of the heart and increased tendency for venous return add together to cause the cardiac output to become once again almost normal. As a result, the person who has a sudden moderate heart attack might experience nothing more than a few seconds of fainting. Shortly thereafter, with the aid of the autonomic compensation, his circulation may be completely adequate as long as he remains quiet.

GRAPHICAL REPRESENTATION OF THE EFFECTS OF THE SYMPATHETIC REFLEXES. Two of the curves shown in Figure 29–1 are repeated in Figure 29–2, the normal systemic function curve and the cardiac curve of the acutely damaged heart, these two curves crossing at equilibrium point B. Within 30 seconds both the cardiac and systemic curves change to the two dashed curves. These two new curves equilibrate at point C, showing that the cardiac output within a minute has risen back to approximately 4 liters per minute, which is four-fifths normal, while the right atrial pressure has risen to 5 mm. Hg, a rise of approximately another millimeter. Note especially in this analysis that the *changes in the heart and in the systemic circulation must both be considered simultaneously;* this graphical procedure provides a means for doing this.

The Chronic Stage of Failure—Compensation by Renal Retention of Fluid. A low cardiac output has a profound effect on renal function, sometimes causing anuria when the cardiac output falls to as low as one-half to two-thirds normal. In general, the urinary output is less than normal as long as the cardiac output is less than normal, and it usually does not return to normal after an acute heart attack until the cardiac output rises either all the way back to normal or almost to normal. This re-

Figure 29–2 Effect of autonomic reflexes in helping to compensate for depressed cardiac output in acute failure.

lationship of renal function to cardiac output is one of the most important of all the factors affecting the dynamics of the circulation in chronic cardiac failure.

CAUSES OF RENAL RETENTION OF FLUID IN CARDIAC FAILURE. There are two known causes of the reduced renal output during cardiac failure, both of which are perhaps equally important.

First, a decrease in cardiac output has a tendency to reduce the glomerular pressure in the kidneys because of (a) *reduced arterial pressure* and (b) *intense sympathetic constriction of the afferent arterioles of the kidney.* As a consequence, the glomerular filtration rate becomes less than normal. When the cardiac output falls to about one-half normal, this factor alone can result in almost complete anuria.

The second factor that causes reduced renal output in cardiac failure is an *increase in aldosterone secretion* by the adrenal cortices caused by the reduced cardiac output, as will be discussed in Chapter 76. The aldosterone increases the rate of reabsorption of sodium by the renal tubules, and this in turn leads to a secondary increase in water reabsorption for two reasons: First, as the sodium is reabsorbed, it reduces the osmotic pressure in the tubules while increasing the osmotic pressure in the renal interstitial fluids; these changes promote osmosis of water into the blood. Second, the absorbed sodium increases the osmotic concentration of the extracellular fluid and elicits *antidiuretic hormone* secretion by the supraopticohypophyseal system, which is discussed in Chapter 36. The antidiuretic hormone then promotes increased tubular reabsorption of water.

EFFECT OF FLUID RETENTION ON CARDIOVASCULAR DYNAMICS. Fluid retention does not have any significant effect on the pumping ability of the heart; but it does have an extreme effect on the tendency for venous return. The fluid retained by the kidneys causes a progressive increase in extracellular fluid volume, and part of this increased fluid usually remains in the blood, causing an increase in blood volume as well. For two different reasons this increases the tendency for venous return to the heart: First, both the increased extracellular fluid and blood volume increase the mean systemic pressure, which *increases the pressure gradient for flow of blood toward the heart.* Second, *reduced venous resistance* caused by distention of the veins allows increased flow of blood toward the heart.

Detrimental Effects of Fluid Retention in the Late Stages of Cardiac Failure. In moderate cardiac failure, retention of fluid helps to compensate for the failure by allowing more blood to flow into the heart and thereby priming the heart with increased quantities of blood. In severe cardiac failure, however, the amount of blood priming the heart often becomes so great that the cardiac muscle fibers are stretched beyond their normal physiological lengths, and this sometimes decreases the effectiveness of the heart as a pump. Thus, in the late stages, as will be discussed in

subsequent sections of the chapter, further fluid retention can be very detrimental.

RECOVERY OF THE MYOCARDIUM FOLLOWING MYOCARDIAL INFARCTION. After a heart becomes suddenly damaged as a result of coronary thrombosis, the natural reparative processes of the body begin immediately to help restore normal cardiac function. Thus, a new collateral blood supply begins to penetrate the peripheral portions of the infarcted area, often completely restoring the muscle function. Also, the undamaged musculature hypertrophies, in this way offsetting much of the cardiac damage.

Obviously, the degree of recovery depends on the type of cardiac damage, and it varies from no recovery at all to almost complete recovery. Ordinarily, after myocardial infarction the heart will have reached its final state of recovery within two to six months.

GRAPHICAL ANALYSIS OF THE CHANGES IN CIRCULATORY DYNAMICS DURING CHRONIC FAILURE. The two dashed curves of Figure 29–3 are the same as the dashed curves of Figure 29–2, illustrating the dynamics of the circulation about 30 seconds after an acute heart attack—that is, after the sympathetic reflexes have become active and have partially compensated for the failure. During the ensuing few days to few weeks the retention of fluid causes the systemic curve to change gradually to the upper curve illustrated by the dots and dashes. Partial recovery of the heart likewise increases the cardiac curve to the upper one shown in Figure 29–3. Thus, after these two compensations have taken place, the new point of equilibrium between cardiac output and venous return rises from point C to point D, which is now at a cardiac output of 5 liters per minute, though the right atrial pressure has increased to 6 mm. Hg.

Since the cardiac output has returned to normal, renal output also will have returned to normal and no further fluid retention will occur. Therefore, except for the high right atrial pressure represented by point D in the figure, the person now has essentially normal cardiovascular dynamics as long as he remains at rest.

If the heart itself recovers to a significant extent and if adequate fluid retention occurs, the sympathetic stimulation gradually abates toward normal for the following reasons: The partial recovery of the heart can do the same thing for the cardiac curve as sympathetic stimulation, and fluid retention in the circulatory system can do the same

Figure 29–3 Effect of (a) fluid retention by the kidneys and (b) partial recovery of the heart to cause return of cardiac output toward normal in chronic cardiac failure.

thing for venous return as sympathetic stimulation. Thus, as these two factors develop, the fast pulse rate, cold skin, sweating, and pallor resulting from sympathetic stimulation in the acute stage of cardiac failure gradually disappear.

SUMMARY OF THE CHANGES THAT OCCUR FOLLOWING ACUTE CARDIAC FAILURE — "COMPENSATED HEART FAILURE"

To summarize the events discussed in the past few sections describing the dynamics of circulatory changes following an acute, moderate heart attack, we may divide the stages into (1) the instantaneous effect of the cardiac damage, (2) compensation by the sympathetic nervous system, and (3) chronic compensations resulting from partial cardiac recovery and renal retention of fluid. All these changes are shown graphically in Figure 29-4, which is a composite of the preceding three figures. This figure shows the normal state of the circulation (point A), the state a few seconds after the heart attack but before sympathetic reflexes have occurred (point B), the rise in cardiac output toward normal caused by sympathetic stimulation (point C), and final return of the cardiac output to normal following several days to several weeks of cardiac recovery and fluid retention (point D). This final state is called compensated heart failure.

Compensated Heart Failure. Note especially in Figure 29-4 that the pumping ability of the heart, as depicted by the cardiac function curve, is still depressed to less than one-half normal. This illustrates that factors that increase the tendency for venous return can return the cardiac output to a normal level despite continued weakness of the heart itself. However, one of the results of chronic cardiac weakness is increased right atrial pressure; in Figure 29-4 it is shown to be 6 mm. Hg. There are many persons, especially in old age, who have completely normal resting cardiac outputs but mildly to moderately elevated right atrial pressures because of compensated heart failure. These persons may not know that they have cardiac damage because the damage more often than not has occurred a little at a time, and the com-

pensation has occurred concurrently with the progressive stages of damage.

The slight elevation of right atrial pressure has almost no harmful effects on the circulatory system. However, the person with compensated heart failure certainly does not have a normal circulatory system, for should he try to exercise strongly or should any other stress be placed on his circulatory system, such as might occur in some disease condition, his heart would be unable to respond normally. The reason for this is that the compensatory mechanisms normally used to increase the cardiac output are already partially in use simply to provide a normal output. Therefore, the remaining compensation that can be invoked has been greatly reduced. To state this another way, the normal person has far greater *cardiac reserve* than the person with compensated heart disease. Cardiac reserve is discussed at greater length later in the chapter.

DYNAMICS OF SEVERE CARDIAC FAILURE — DECOMPENSATED HEART FAILURE

If the heart becomes severely damaged, then no amount of compensation, either by sympathetic nervous reflexes or by fluid retention, can make this weakened heart pump a normal cardiac output. As a consequence, the cardiac output cannot rise to a high enough value to bring about return of normal renal function. Fluid continues to be retained, the person develops progressively more and more edema, and this state of events, eventually leads to his death. This is called decompensated heart failure. The basis of decompensated heart failure is *failure of the heart to pump sufficient blood to make the kidneys function adequately.*

Graphical Analysis of Decompensated Heart Disease. Figure 29–5 illustrates by the two heavy lines (1) the normal systemic function curve (normal venous return curve) and (2) a cardiac output curve depicting the function of a heart that has become extremely weakened and that cannot be strengthened. The equilibrium point of these two curves, point A, does not last more than a few seconds, because sympathetic compensation increases the systemic function curve to the second curve, labeled "autonomic compensation." If we assume that the heart is so weakened that it still remains very weak even after sympathetic compensation, the new point of equilibrium is point B, illustrating that the cardiac output has risen a slight amount as a result of the compensation, but at the same time the right atrial pressure also has risen to approximately 5 mm. Hg. Since the cardiac output is still 1 liter per minute less than the critical cardiac output level required for normal renal function, the output of the kidneys is still greatly depressed, and fluid continues to be retained.

After a day of fluid retention the systemic curve shifts still farther to the right, as shown by the curve labeled "2nd day," and the cardiac output becomes that depicted by point C. Here again, be-

Figure 29–4   Diagram showing the progressive changes in cardiac output and right atrial pressure during different stages of cardiac failure. This figure is a composite of the three preceding figures.

Figure 29-5 Graphical analysis of decompensated heart disease, showing progressive shift of the venous return curve to the right as a result of fluid retention.

cause the heart is still incapable of pumping the 5 liters cardiac output per minute required for normal renal function, fluid continues to be retained, so that the cardiac output and right atrial pressure progress to points D and E. By this time, the heart has become so *overstretched* that the overstretching alone begins to reduce its capabilities as a pump, or it will have begun to become *edematous*. Further retention of fluid in the body can now only *reduce cardiac output*. By the eighth day, additional fluid retention shifts the systemic curve still farther to the right, and depresses cardiac output tremendously, down to point F. This represents the final stage of decompensation that occurs shortly before death, for retention of only small quantities of fluid from this point on causes rapid deterioration of cardiac pumping.

Thus, one can see from this graphical analysis that failure of the cardiac output ever to rise to the critical level required for normal renal function results in (a) progressive retention of fluid, which causes (b) progressive elevation of the mean systemic pressure, then (c) progressive shift of the systemic function curve to the right, and (d) progressive elevation of the right atrial pressure until finally the heart is so overstretched or so edematous that it becomes unable to pump even moderate quantities of blood, and, therefore, fails completely. Clinically, one detects this serious condition of decompensation principally by the progressive edema. All clinicians know that failure to institute appropriate therapy when this state of events occurs leads to rapid death.

Treatment of Decompensation. The two ways in which the decompensation process can often be stopped are: (1) by strengthening the heart in any one of several ways so that it can pump adequate quantities of blood to make the kidneys function normally again or (2) by administering diuretics and reducing water and salt intake, which brings about a balance between fluid intake and output despite the low cardiac output.

Both methods stop the decompensation process by re-establishing normal fluid balance so that at least as much fluid leaves the body as enters it.

TREATMENT BY STRENGTHENING THE HEART. A weakened heart can frequently be strengthened by administering digitalis or a similar cardiotonic drug. Recent experimental evidence indicates that

these drugs strengthen the heart by increasing the rate of diffusion of calcium through the membrane to the interior of the muscle fibers; the calcium then promotes increased force of contraction by the actomyosin complex, as explained in Chapter 7. The chronically weakened heart can often be strengthened by as much as 50 per cent by these drugs.

Another means by which the heart often becomes strengthened is simply prolonged bed rest, for this protects the heart against excessive overloading and allows maximum recovery from any myocardial disease.

Figure 29-6 illustrates graphically the effect of strengthening the heart on the dynamics of the circulation. The two light curves of the figure represent (1) the cardiac output curve of Figure 29-5 and (2) the systemic curve of Figure 29-5 on the sixth day. Point E is the equilibrium point with a cardiac output of 4 liters per minute and a right atrial pressure of 11 mm. Hg. At this point the patient is in a state of serious decompensation and within a few more days would be dead. Within a few hours after administering digitalis, the strength of the heart might well be increased to the dark cardiac curve of the figure. The systemic curve does not change significantly at first because all the fluid that had been retained in the body is still present. Therefore, a few hours after digitalization a new equilibrium is established at point G with a cardiac output now of almost 6 liters per minute and a right atrial pressure of 9 mm. Hg. Thus, simply strengthening the heart reduces the right atrial pressure a small amount and elevates the cardiac output. Note especially that the cardiac output is now greater than the critical level for normal fluid balance, and the kidneys now begin to put out more urine than normally, causing a net loss of fluid from the body. Therefore, during the subsequent few days, with progressively more and more fluid loss, the systemic curve shifts to the left until it becomes the dark one of the figure; the new equilibrium point is now point H with a cardiac output of 5 liters per minute and a right atrial pressure of 4.5 mm. Hg. The circulatory system is now in a stable state.

An especially interesting point is the extra flow of urine, or *diuresis,* that occurs after the adminis-

Figure 29-6 Treatment of decompensated heart disease, showing the effect of digitalis in elevating the cardiac output curve, this in turn causing progressive shift of the venous return curve to the left.

tration of digitalis. Because of this, digitalis is frequently said to be a "diuretic." Actually, its diuretic properties probably result not from a direct action on the kidney but from the strengthening of the heart; this allows the heart to pump increased quantities of blood into the kidneys.

The *circulatory dynamics become stable only when the systemic and cardiac curves cross each other at the cardiac output level that provides normal renal function.* If these two curves cross above the 5-liter level, the renal output is too great, causing fluid loss and shift of the systemic curve to the left until the cardiac output falls back to the 5-liter level. On the other hand, if the equilibrium point is below the 5-liter mark, fluid is retained until the cardiac output rises back to the 5-liter level.

TREATMENT OF DECOMPENSATION BY DIETARY RESTRICTION OF WATER AND SALT AND ADMINISTRATION OF DIURETICS. The second means by which the process of decompensation can be stopped is to administer a diuretic that makes the kidneys put out far greater quantities of urine than normally or to restrict the person's intake of salt and water. By these two measures the output of fluid can be increased to a greater value than the intake, thereby preventing further retention of fluid even though the cardiac output is considerably less than normal.

Often, it is possible to stop the process of decompensation simply by excluding salt from the diet, for, if this is done, the extracellular fluid tends to become slightly less concentrated than usual; this inhibits the osmoreceptor system, thereby allowing extra quantities of water to be lost into the urine. Thus, reduction of dietary salt alone in the earlier stages of cardiac failure is often sufficient to prevent or even to stop the process of decompensation, though, in severe cases, water retention may also be required.

EDEMA IN CARDIAC FAILURE

Much of the preceding discussion has emphasized the elevated right atrial pressure in cardiac failure equally as much as the reduced cardiac output. Referring to Chapter 21, it will be recalled that any time the right atrial pressure rises the capillary pressure likewise is likely to rise; this causes transfer of fluid into the interstitial spaces. Thus, one of the most important problems in cardiac failure is the development of severe *cardiac edema.*

Dynamics of Edema Formation in Cardiac Failure. On first thought, it seems that cardiac edema is a simple problem—that increased right atrial pressure causes increased capillary pressure, which causes the edema. Unfortunately, this is not true, for reasons that can be explained best by Figure 29-7. When the heart progressively fails as a pump so that the cardiac output falls progressively from normal to zero, the right atrial pressure rises

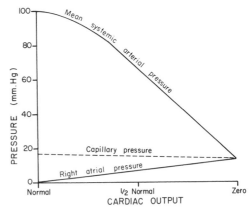

Figure 29-7 Progressive changes in mean systemic arterial pressure and right atrial pressure as the cardiac output falls from normal to zero. This figure also shows the probable effect of decreased cardiac output on capillary pressure.

while the systemic arterial pressure falls. These approach each other at an equilibrium value of about 13 mm. Hg. It is obvious that capillary pressure must also fall from its normal value of 17 mm. Hg to the equilibrium pressure of 13 mm. Hg. Thus, it is reasonable to believe that severe *acute cardiac failure would actually cause a fall in capillary pressure rather than a rise.* And animal experiments as well as experience in human beings show that acute cardiac failure does not cause immediate development of peripheral edema. In fact, fluid is actually absorbed into the capillaries from the interstitial spaces.

The edema, on the other hand, develops during the succeeding days *because of fluid retention by the kidneys.* The retention of fluid increases the mean systemic pressure, resulting in increased tendency for blood to return to the heart. This now elevates the right atrial pressure to a still higher value, and at the same time it keeps the cardiac output and arterial pressure from decreasing very much despite the failing myocardium. Therefore, the rising right atrial pressure now *causes the capillary pressure to rise markedly,* thus causing loss of fluid into the tissues and development of severe edema.

Edema of cardiac failure can occur either in the systemic circulation or in the lungs, or in both, depending on which side or sides of the heart might be failing. Pulmonary edema is further discussed below in relation to unilateral left heart failure.

CARDIAC SHOCK

In the description of acute heart failure, it was pointed out that the cardiac output can fall very low immediately after heart damage occurs. This obviously leads to greatly diminished blood flow throughout the body and can lead to a typical picture of circulatory shock as described in the previous chapter. This type of shock is called *cardiac shock.*

Cardiac shock is extremely important to the clinician because approximately one-tenth of all patients who have severe acute myocardial infarction will die of an ensuing shock syndrome before the physiologic compensatory measures can come into play to save life. Once cardiac shock has become well established after myocardial infarction, all the typical events that occur in the late stages of other types of circulatory shock, as described in the previous chapter, also take place in this condition. That is, the shock itself causes further progression of the shock, and eventually a serious, vicious cycle of circulatory deterioration sets in. Indeed, this problem of a vicious cycle is even more severe in cardiac shock than in other types of shock for the following reasons:

Vicious Cycle of Cardiac Deterioration in Cardiac Shock. The discussion of circulatory shock in the previous chapter emphasized the tendency for the heart itself to become progressively damaged when its coronary blood supply is reduced during the course of shock. That is, shock reduces the coronary supply, which makes the heart still weaker, which makes the shock still worse. In cardiac shock caused by myocardial infarction, this problem is further compounded by the already existing coronary thrombosis. For instance, in a normal heart, the arterial pressure usually must be reduced below about 45 mm. Hg before cardiac deterioration sets in. However, in a heart that already has a major coronary vessel blocked, a similar degree of deterioration will set in when the arterial pressure falls as low as 80 to 90 mm. Hg. In other words, even the minutest amount of fall in arterial pressure can then set off a vicious cycle of cardiac deterioration, though this cycle will occur in a normal heart only in the very late stages of shock. For this reason, in treating myocardial infarction it is extremely important to prevent even short periods of hypotension.

Physiology of Treatment. Often a patient dies of cardiac shock before the various compensatory processes can return his cardiac output to a life-sustaining level. Therefore, treatment of this condition is one of the most important problems in the management of acute heart attacks. Immediate digitalization of the heart is often employed for strengthening the heart, but more frequently intravenous infusion of whole blood, plasma, or a blood pressure raising drug is used to sustain the arterial pressure. If this is done, the coronary blood flow can often be elevated to a high enough value to prevent the vicious cycle of deterioration until appropriate compensatory mechanisms in the body have corrected the shock.

UNILATERAL CARDIAC FAILURE

In all the discussions thus far in this chapter, we have considered failure of the heart as a whole. The principles discussed apply to almost all cases of cardiac failure because only rarely does one side of the heart fail without some degree of failure of the other side. Yet in a large number of patients with early cardiac failure, left-sided failure predominates over right-sided failure, and in rare instances, especially in congenital heart disease, the right side may fail without significant failure of the left side. Therefore, we now need to discuss the special features of unilateral cardiac failure.

UNILATERAL LEFT HEART FAILURE

When the left side of the heart fails without concomitant failure of the right side, blood continues to be pumped into the lungs with usual right heart vigor while it is not pumped adequately out of the lungs into the systemic circulation. As a result, the *mean pulmonary pressure* rises while the *mean systemic pressure* falls because of shift of large volumes of blood from the systemic circulation into the pulmonary circulation.

As the volume of blood in the lungs increases, the pulmonary vessels enlarge, and, if the pulmonary capillary pressure rises above 28 mm. Hg, that is, above the colloid osmotic pressure of the plasma, fluid begins to filter out of the capillaries into the interstitial spaces and alveoli, resulting in pulmonary edema.

Thus, among the most important problems of left heart failure are *pulmonary vascular congestion* and *pulmonary edema,* which are discussed in detail in Chapters 23 and 21 in relation to the pulmonary circulation and capillary dynamics. As long as the pulmonary capillary pressure remains less than the normal colloid osmotic pressure of the blood, about 28 mm. Hg, the lungs remain "dry." But even a millimeter rise in capillary pressure above this critical level causes progressive transudation of fluid into the interstitial spaces and alveoli, leading rapidly to death. Pulmonary edema can occur so rapidly that it can cause death after only 20 to 30 minutes of severe acute left heart failure.

Figure 29–8 The overall effects, acute and chronic, of left-sided heart failure.

Course of Events in Chronic Left Heart Failure. In chronic left heart failure, one additional feature must be added to the acute picture. This is retention of fluid resulting from reduced renal function. In moderate acute left heart failure the pulmonary capillary pressure sometimes rises only to 15 to 20 mm. Hg, not enough to cause pulmonary edema. Yet following retention of fluid for the next few days, the blood volume increases, and more blood is pumped into the lungs by the right ventricle. Then, the pulmonary capillary pressure often rises above the colloid osmotic pressure, resulting in severe pulmonary edema, as shown in Figure 29–8. Indeed, this is a common occurrence, the patient suddenly developing severe pulmonary edema a week or more after the acute attack and dying a respiratory death — not a death resulting from diminished cardiac output.

Graphical Analysis of Acute Left Heart Failure. Exactly the same principles that have been applied in the earlier parts of this chapter for the analysis of the entire circulatory system can be applied in analyzing the pumping dynamics of the left heart alone. This is illustrated by Figure 29–9. The normal left ventricular function curve is very similar to that for the entire heart except that it is shifted to considerably higher atrial pressure

levels. This time the *left* atrial pressure instead of *right* atrial pressure is plotted against cardiac output. Likewise, the function curve of the pulmonary circulation is similar to that of the systemic circulation except that none of the veins of the pulmonary system is exposed to atmospheric pressure, and, therefore, a plateau does not occur in the negative pressure range. The two dark curves of the figure represent the normal left ventricular curve and the normal pulmonary curve. Point A is the equilibrium state, with a cardiac output of 5 liters per minute and a left atrial pressure of about 4 mm. Hg. When the left ventricle suddenly fails, the ventricular curve falls to the lower curve, and this immediately dams large quantities of blood in the lungs, shifting the pulmonary function curve far to the right. The new equilibrium point, point B, illustrates that the cardiac output falls a slight amount, to about 4.5 liters per minute, while the left atrial pressure rises to 13 mm. Hg.

This figure shows the effects only of *acute* left heart failure. In chronic failure, progressive increase in pulmonary blood volume gradually shifts the pulmonary curve still farther to the right, raising the left atrial pressure higher and higher and finally causing pulmonary edema if the failure is severe enough.

UNILATERAL RIGHT HEART FAILURE

In unilateral right heart failure, blood is pumped normally by the left ventricle from the lungs into the systemic circulation, but it is not pumped adequately from the systemic circulation into the lungs. Therefore, blood shifts from the lungs into the systemic circulation, causing systemic congestion. However, in acute right heart failure this congestion is hardly noticeable for the following reasons: The total amount of blood in the lungs is only about one-seventh that in the systemic circulation. Therefore, even in severe right heart failure, the systemic blood volume increases only a few per cent, and this is not sufficient to cause significant systemic congestion.

Low Cardiac Output in Acute Right Heart Failure. On the other hand, acute right heart failure causes far greater depression of the cardiac output than acute left heart failure of the same degree. This again stems from the far greater compliance of the systemic circulation than of the pulmonary circulation. Not enough blood can transfer from the lungs into the systemic vessels to raise the systemic pressures to a very high level, and these pressures are not enough to make the weakened right ventricle pump adequate quantities of blood. Therefore, the cardiac output falls greatly in acute right heart failure, often resulting in cardiac shock.

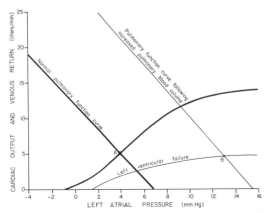

Figure 29–9 Graphical analysis of left ventricular dynamics following acute left ventricular failure.

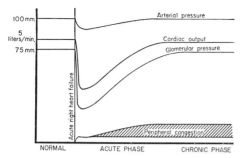

Figure 29–10 The overall effects, acute and chronic, of right-sided heart failure.

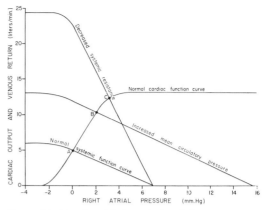

Figure 29–11 Graphical analysis of the two types of factors that can cause high cardiac output failure.

The Chronic Stage of Unilateral Right Heart Failure. The chronic stage of unilateral right heart failure is much the same as that discussed earlier for the entire heart. The depressed cardiac output results in progressively more and more retention of fluid by the kidneys until the cardiac output either rises back nearly to normal or until the person goes into decompensation and dies. Figure 29–10 illustrates the progressive changes that occur in chronic right-sided heart failure, showing a gradual return of cardiac output to normal or near normal, a return along with increased urinary output, and, finally, progressive development of peripheral congestion and edema.

"HIGH CARDIAC OUTPUT FAILURE" — OVERLOADING OF THE HEART

A condition called "high cardiac output failure" frequently occurs in persons who have cardiac outputs much higher than normal but who appear to be in cardiac failure because the left or right or both atrial pressures are very high. Sometimes there is also accumulation of edema. The true problem is often not failure of the pumping ability of the heart at all but *overloading of the heart with too much venous return.* The basic causes of increased cardiac output are (1) decreased systemic resistance or (2) increased mean systemic pressure. This *may* or *may not* be associated with a weak heart, depending on the condition causing the high output failure.

Decreased Systemic Resistance. Diseases that cause high cardiac output failure by decreasing the systemic resistance include (a) *arteriovenous fistulae,* in which the blood is shunted directly from the arteries to the veins, (b) *beriberi heart disease,* in which avitaminosis of the B vitamins results in profound systemic vasodilatation, and (c) *thyrotoxicosis,* which, because of its stimulatory effect on metabolism, causes generalized systemic vasodilatation. Obviously, the decreased resistance results in rapid venous return, often causing more blood to flow into the heart than can be properly handled and resulting in greatly increased right

and left atrial pressures but at the same time resulting in a cardiac output sometimes as much as two to three times normal.

Increased Mean Systemic Pressure. The conditions that cause high cardiac output failure by increasing the mean systemic pressure include (a) too rapid and too much *transfusion* and (b) fluid retention resulting from excess secretion of different *steroid hormones*—especially aldosterone. These *increase the pressure gradient for venous return,* thereby forcing large excesses of blood into the heart and overloading it.

Graphical Analysis of High Cardiac Output Failure. Figure 29–11 gives a graphical analysis of cardiac dynamics in high cardiac output failure. This shows the normal cardiac and systemic function curves with normal equilibrium at point A. The equilibrium point shifts from point A to point B when the mean systemic pressure rises. It shifts to point C when the systemic resistance greatly decreases. In both these conditions, the cardiac output is far above normal while the right atrial pressure is considerably elevated.

Thus, it is evident that nothing need be wrong with the pumping ability of the heart in high cardiac output failure. However, in some diseases, such as thyrotoxicosis and beriberi heart disease, weakness of the heart might be superimposed onto the peripheral factors, making the signs of congestion even more apparent than would be evident from the peripheral factors alone.

PHYSIOLOGICAL CLASSIFICATION OF CARDIAC FAILURE

From the above discussions, it is apparent that the symptoms of cardiac failure fall into the following three physiological classifications:

> Low cardiac output
> Pulmonary congestion
> Systemic congestion.

Low cardiac output usually occurs immediately

after a heart attack. If the attack is mainly right-sided, a slight amount of systemic congestion may occur along with the low cardiac output, but in the early stages of acute attacks of any type the degree of systemic congestion is slight until fluids are retained by the kidneys. If the acute heart attack is mainly left-sided, concurrent pulmonary congestion usually occurs along with the low cardiac output, but here again the pulmonary congestion may be mild until after considerable fluid has been retained by the kidneys. Thus, low cardiac output may be the only significant clinical effect observed in many persons who have sudden heart attacks. This results in the following symptoms:

1. Generalized weakness
2. Fainting
3. Symptoms of increased sympathetic activity such as high heart rate, thready pulse, cold skin, sweating, and so forth.

Pulmonary congestion may be the only effect in patients with pure left-sided heart failure, for, after the fluid that shifts from the systemic circulation into the lungs during acute left-sided heart failure has been replenished by renal retention of fluid, the heart sometimes pumps normal quantities of blood, and the right side of the heart may not fail at all. Therefore, pulmonary congestive symptoms alone can occur with essentially no systemic congestion nor low cardiac output.

Systemic congestion alone can occur in pure right-sided heart failure. In this condition there is no pulmonary congestion, and, if sufficient fluids have been retained in the blood to prime the heart sufficiently, the heart may pump a normal cardiac output.

Obviously, all the above classes of heart failure can occur together or in any combination.

CARDIAC RESERVE

Fortunately, the normal heart can increase its output to four to five times normal under conditions of stress. The maximum percentage that the cardiac output can increase above normal is called the cardiac reserve. In the normal average adult the cardiac reserve is slightly more than 300 per cent. In the athletically trained person it is occasionally as high as 500 per cent, while in the asthenic person it may be as low as 200 per cent. As an example, during severe exercise, the cardiac output of the normal adult can rise to about four times normal; this is an increase above normal of

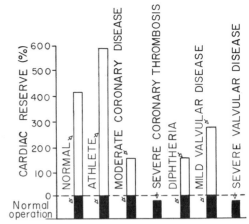

Figure 29–12 Cardiac reserve in different conditions.

300 per cent—that is, the normal adult has a cardiac reserve of about 300 per cent.

Any factor that prevents the heart from pumping blood satisfactorily decreases the cardiac reserve. Such a decrease can result from coronary thrombosis, diphtheritic damage to the myocardium, valvular heart disease, and many other factors, some of which are illustrated in Figure 29–12.

If the cardiac abnormality is moderate, the person will not know that he has heart disease so long as he does not attempt strenuous activity, but, if he attempts to exercise, his heart soon uses up all its cardiac reserve despite the fact that the cardiac output still has not risen nearly so high as occurs in the normal person. Immediate and sometimes extreme shortness of breath (dyspnea) results, for when the heart does not pump sufficient blood, the tissues throughout the body become ischemic, creating a sensation of air hunger. Also, the skeletal muscles fatigue readily because of poor blood supply, thus limiting one's ability to continue with the exercise.

REFERENCES

Barger, A. C.: The kidney in congestive heart failure. *Circulation,* 21:124, 1960.

Barger, A. C.; Yates, F. E.; and Rudolph, A. M.: Renal hemodynamics and sodium excretion in dogs with graded valvular damage, and in congestive failure. *Amer. J. Physiol., 200*:601, 1961.

Bishop, S. P., and Altschuld, R. A.: Increased glycolytic metabolism in cardiac hypertrophy and congestive failure. *Amer. J. Physiol., 218*:153, 1970.

Braunwald, E.; Ross, J.; and Sonnenblick, E.: Mechanism of Contractility of the Normal and Failing Heart. Boston, Little, Brown and Company, 1968.

Brender, D.; Vanhoutte, P. M.; and Shepherd, J. T.: Potentiation of adrenergic venomotor responses in dogs by cardiac glycosides. *Circ. Res., 25*:597, 1969.

Driscol, T. E.; Maultsby, M. M.; Farrell, G. L.; and Berne, R. M.: Aldosterone secretion in experimental congestive heart failure. *Amer. J. Physiol., 191*:140, 1957.

Friedberg, C. K.: Diseases of the Heart. 3rd ed., Philadelphia, W. B. Saunders Company, 1966.

Gersmeyer, G., and Holland, W. C.: Influence of ouabain on contractile force, resting tension, Ca^{45} entry and tissue Ca content in rat atria. *Circ. Res., 12*:620, 1963.

Guyton, A. C.: Cardiac output and venous return in heart failure. *In* Luisada, A. A. (ed.): Cardiology. New York, McGraw-Hill Book Co., 1959, Vol. IV, p. 18–8.

Guyton, A. C.: The systemic venous system in cardiac failure. *J. Chronic Dis., 9*:465, 1959.

Harigaya, S., and Schwartz, A.: Rate of calcium binding and uptake in normal animal and failing human cardiac muscle: membrane vesicles (relaxing system) and mitochondria. *Circ. Res., 25*:781, 1969.

Harrison, T. R., and Reeves, T. J.: Principles and Problems of Ischemic Heart Disease. Chicago, Year Book Medical Publishers, Inc., 1968.

Hurst, W.: The Heart. New York, McGraw-Hill Book Company, 1966.

Metcalfe, J.; Dhindsa, D. S.; Edwards, M. J.; and Mourdjinis, A.: Decreased affinity of blood for oxygen in patients with low-output heart failure. *Circ. Res., 25*:47, 1969.

Myerson, R. M., and Pastor, B. H.: Congestive Heart Failure. St. Louis, The C. V. Mosby Co., 1967.

Stone, H. L.; Bishop, V. S.; and Guyton, A. C.: Progressive changes in cardiovascular function after unilateral heart irradiation. *Amer. J. Physiol., 206*:289, 1964.

CHAPTER 30

Heart Sounds; Dynamics of Valvular and Congenital Heart Defects

THE HEART SOUNDS

The function of the heart valves was discussed in Chapter 13, and it was pointed out that closure of the valves is associated with an audible sound, though no sound whatsoever occurs when the valves open. The purpose of the present section is to discuss the factors that cause the sounds in the heart, under both normal and abnormal conditions.

NORMAL HEART SOUNDS

When one listens with a stethoscope to a normal heart, he hears a sound usually described as "lub, dub, lub, dub ----." The "lub" is associated with closure of the A-V valves at the beginning of systole and the "dub" with closure of the semilunar valves at the beginning of diastole. The "lub" sound is called the *first heart sound* and the "dub" the *second heart sound* because the normal cycle of the heart is considered to start with the beginning of systole and to end with the end of diastole.

Causes of the First and Second Heart Sounds. Closure of the valves in any pump system usually causes a certain amount of noise because the valves close solidly and suddenly over some opening, setting up vibrations in the fluid or walls of the pump. In the heart, the valves are cushioned by blood, so that it is difficult to understand why these valves create as much sound as they do.

The earliest suggestion for the cause of the heart sounds was that slapping together of the vanes of the valves sets up vibrations, but this has now been shown to cause little if any of the sound because of the cushioning effect of the blood. Instead, the cause is *vibration of the walls of the heart and major vessels around the heart*. For instance, contraction of the ventricles causes

sudden backflow of blood against the A-V valves, causing the valves to close and to bulge toward the atria. The elasticity of the valves then causes the back-surging blood to bounce backward into each respective ventricle. This effect sets the walls of the ventricles into vibration, and the vibrations travel away from the valves. When the vibrations reach the chest wall where it contacts the heart, sound waves are created and can be heard by the stethoscope. This is the cause of the major part of the first heart sound.

A possible cause of a small portion of the first heart sound is contraction of the heart muscle itself, for intrinsic rearrangement of muscular elements during contraction in any muscle causes sounds that can be heard through an overlying stethoscope. However, this contribution to heart sounds is extremely slight because the first heart sound disappears almost entirely when the valves are removed.

The second heart sound results from vibration of the walls of the pulmonary artery, the aorta, and, to a much less extent, the ventricles. When the semilunar valves close, they bulge backward toward the ventricles, and then the elastic stretch recoils the blood in the bulge back into the arteries, which causes a short period of reverberation of blood back and forth between the walls of the arteries and the valves. The vibrations set up in the arterial walls are then transmitted along the arteries at the velocity of the pulse wave. These vibrations are not transmitted as compression waves characteristic of sound but instead as lateral vibrations of the arterial walls. However, when the vibrations of the walls come into contact with a "sounding board," such as the chest wall, they create sound that can be heard.

Duration and Frequencies of the First and Second Heart Sounds. The duration of each of the heart sounds is slightly more than 0.1 second; the first sound lasts about 0.14 second and the second about 0.11 second. And both of them are described as very low-pitch sounds, the first lower than the second.

Sound consists of vibrations of different frequencies. Figure 30–1 illustrates by the shaded area the amplitudes of the different frequencies in the heart sounds and murmurs, illustrating that these are composed of frequencies ranging all the way from a few cycles per second to more than 1000 cycles per second, with the maximum amplitude of vibration occurring at a frequency of about 24 cycles per second.

Also shown in Figure 30–1 is another curve called the "threshold of audibility," which depicts the capability of the ear to hear sounds of different amplitudes. Note that in the very low frequency range the heart vibrations have a high degree of amplitude, but the threshold of audibility is so high that ordinarily the heart vibrations below approximately 30 to 50 cycles per second are not heard by the ears. Then above about 500 cycles per second, the heart sounds are so weak that despite a low threshold of audibility no frequencies in this range are heard. For practical purposes, then, we can consider that all the *audible* heart sounds lie in the range of approximately 40 to 500 cycles per second despite the fact that the maximum amplitude of vibration occurs at the very low frequency of 24 cycles per second.

Both the first and second heart sounds have a mixture of frequencies in the entire audible range of the heart sounds, though the first sound has slightly more low frequency sound than does the second heart sound.

The reason the frequency of the first heart sound is lower than that of the second sound is probably two-fold: First, the *elastic modulus* of the A-V valves and of the walls of the ventricles is far less than the elastic modulus of the semilunar valves and the arterial walls. It is well known that any mechanical vibrating system having a low elastic modulus oscillates at a lower frequency than a system having a greater modulus. Second, the *mass of blood* in the great vessels is less than that in the ventricles, which means that the inertia of the vibrating mass is less. This also would cause the second heart sound to have a higher frequency than the first.

The reason the second heart sound is shorter in duration than the first is probably that the second sound is "damped" out by the vascular walls much more rapidly than is the first heart sound by the ventricular walls.

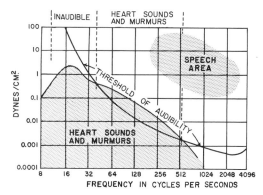

Figure 30–1   Amplitude of different frequency vibrations in the heart sounds and heart murmurs in relation to the threshold of audibility, showing that the range of sounds that can be heard is between about 40 and 500 cycles per second. (Modified from Butterworth, Chassin, and McGrath: Cardiac Auscultation. Grune & Stratton.)

Loudness of the First and Second Heart Sounds.   The loudness of the first and second heart sounds is almost directly proportional to the *rate of change* of the pressure differences across the A-V and semilunar valves, respectively. For instance, when the onset of systole is very rapid, the intraventricular pressure rises very rapidly during the isometric period of ventricular contraction; yet the pressure on the opposite side of the A-V valves remains only a few mm. Hg. Thus, the rate of change of the pressure difference between the two sides of the valves is very great, and as a result the first heart sound is loud. Also, when the heart is very active, such as during and immediately following exercise, the force of contraction of the ventricle is greatly enhanced, so that the first heart sound is in this instance also greatly accentuated. Conversely, in a weakened heart in which the onset of contraction is sluggish, the loudness of the first sound is greatly diminished.

In the case of the second heart sound, it is the rate of decrease in ventricular pressure at the end of systole that determines the loudness. The magnitude of this rate of decrease is determined mainly by the level of ventricular systolic pressure at the time the valve closes. In a person who has hypertension, the pressure at the time the aortic valve closes may be as great as 200 mm. Hg, so that the intraventricular pressure falls twice as rapidly as normally, all the way to zero in a few hundredths of a second. Therefore, the aortic sound is markedly accentuated. Likewise, in pulmonary hypertension the pulmonic sound is greatly accentuated. On the other hand, when the arterial pressure is low, such as in shock or in the terminal stages of cardiac failure, the second heart sound is diminished to a very low intensity.

The Third Heart Sound. Occasionally a third heart sound is heard at the beginning of the middle third of diastole. The most logical explanation of this sound is oscillation of blood back and forth between the walls of the ventricles initiated by inrushing blood from the atria. This is analogous to running water from a faucet into a sack, the inrushing water reverberating back and forth between the walls of the sack to cause vibrations in the walls.

The third heart sound is an extremely weak rumble of such low frequency that it usually cannot be heard with a stethoscope, but it can be recorded frequently in the phonocardiogram. The very low frequency of this sound presumably results from the flaccid, inelastic condition of the heart during diastole. Also, the reason the third heart sound does not occur until after the first third of diastole is over is presumably that in the early part of diastole the heart is not filled sufficiently to create even the small amount of elastic tension in the ventricles necessary for reverberation. The reason the third heart sound does not continue into the latter part of diastole is presumably that little blood flows into the ventricles during the latter part of diastole, so that no initiating stimulus then exists for causing reverberation.

The Atrial Heart Sound (Fourth Heart Sound). An atrial heart sound can be recorded in many persons in the phonocardiogram, but it can almost never be heard with a stethoscope because of its low frequency—usually 20 cycles per second or less. This sound occurs when the atria contract, and presumably it is caused by an inrush of blood into the ventricles, which initiates vibrations similar to those of the third heart sound.

AREAS FOR AUSCULTATION OF NORMAL HEART SOUNDS

Listening to the sounds of the body, usually with the aid of a stethoscope, is called *auscultation.* Figure 30–2 illustrates the areas of the chest wall from which the heart sounds from

Figure 30–2 Chest areas from which each valve sound is best heard. (Redrawn from Adams: Physical Diagnosis. The Williams & Wilkins Co.)

each valve are usually best heard. With the stethoscope placed in any one of the special valvular areas, the sounds from all the other valves can still be heard, though the sound from the special valve is as loud, *relative to the other sounds,* as it ever will be. The cardiologist distinguishes the sounds from the different valves by a process of elimination; that is, he moves the stethoscope from one area to another, noting the loudness of the sounds in different areas and gradually picking out the sound components from each valve.

The areas for listening to the different heart sounds are not directly over the valves themselves. The aortic area is upward along the aorta, the pulmonic area is upward along the pulmonary artery, the tricuspid area is over the right ventricle, and the mitral area is over the apex of the heart, the portion of the left ventricle nearest to the surface of the chest. In other words, the sounds caused by the A-V valves are transmitted to the chest wall through each respective ventricle, and the sounds from the semilunar valves are transmitted especially along the great vessels leading from the heart. This transmission of sounds is in keeping with the concept that vibrations in the walls of the ventricles or large arteries are the cause of the heart sounds.

THE PHONOCARDIOGRAM

If a microphone specially designed to detect low-frequency sound waves is placed on the chest, the heart sounds can be amplified and recorded by a special high-speed recording apparatus, such as an oscilloscope or a high-speed pen recorder, which are described in Chapters 5 and 15. The recording is called a *phonocardiogram,* and the heart sounds appear as waves in the record, as illustrated in Figure 30–3. Record A illustrates a recording of normal heart sounds, showing the vibrations of the first, second, and third heart sounds and even the atrial sound. Note specifically that the third and atrial heart sounds are each a very low rumble. The third heart sound can be recorded in only one third to one half of all persons, and the atrial heart sound can be recorded in perhaps one fourth of all persons.

ABNORMAL HEART SOUNDS

Rheumatic Valvular Lesions. By far the greatest number of valvular lesions result from rheumatic fever. Rheumatic fever is an autoimmune disease in which the heart valves are likely to be damaged or destroyed. It is initiated by streptococcal toxin in the following manner:

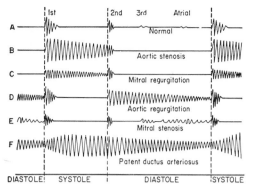

Figure 30-3 Phonocardiograms from normal and abnormal hearts.

The entire sequence of events almost always begins with a preliminary streptococcal infection caused by a special group A strain of streptococci, such as a sore throat, scarlet fever, or middle ear infection. The streptococci release several different proteins against which antibodies are formed. Perhaps some of the streptococcal products also combine with the body's own proteins, and this combination, or the bacterial antigens alone, initiate an immune response so that specific antibodies are released into the blood. The antibodies then react with many different tissues of the body, and these reactions continue to take place as long as the antibodies persist in the blood — six months or more. As a result, rheumatic fever causes damage in many parts of the body but especially in certain very susceptible areas such as the heart valves. The principles of immunity relating to this type of reaction were discussed in Chapter 10, and it is also noted in Chapter 38 that acute glomerular nephritis has a similar basis.

In rheumatic fever, large hemorrhagic, bulbous lesions grow along the inflamed edges of the heart valves. Because the mitral valve receives more trauma during valvular action than do any of the other valves, this valve is the one most often affected, and the aortic valve is second most frequently involved. The tricuspid and pulmonary valves are rarely involved because the stresses acting on these valves are slight compared to those in the left ventricle.

SCARRING OF THE VALVES. The bulbous, hemorrhagic lesions of acute rheumatic fever frequently adhere to adjacent valve leaflets simultaneously so that the edges of the leaflets become stuck together. Then, in the late stages of rheumatic fever, the lesions become scar tissue, permanently fusing portions of the leaflets. Also, the free edges of the leaflets, which are normally filmy and free-flapping, become solid, scarred masses.

A valve in which the leaflets adhere to each other so extensively that blood cannot pass satisfactorily is said to be *stenosed*. On the other hand, when the valve edges are so destroyed by scar tissue that they cannot close together, *regurgitation*, or backflow, of blood occurs when the valve should be closed. Stenosis usually does not occur without

coexistence of at least some degree of regurgitation, and vice versa. Therefore, when a person is said to have stenosis or regurgitation, it is usually meant that one merely predominates over the other.

Other Causes of Valvular Lesions. Until the last few years, a prominent cause of aortic valvular destruction was vascular syphilis. However, with the advent of modern treatment, syphilis now rarely proceeds to such a severe state. In syphilitic lesions, aortic regurgitation almost always occurs without any stenosis of the valve, differing in this respect from regurgitation caused by rheumatic fever.

Occasionally, stenosis or lack of one or more leaflets of a valve occurs as a congenital defect. Complete lack of leaflets is rare, though stenosis is common, as is discussed later in this chapter.

Abnormal Heart Sounds Caused by Valvular Lesions. As illustrated by the phonocardiograms of Figure 30-3, many abnormal heart sounds, known as "murmurs," occur in abnormalities of the valves, as follows:

THE MURMUR OF AORTIC STENOSIS. In aortic stenosis, blood is ejected from the left ventricle through only a small opening at the aortic valve. Because of the resistance to ejection, the pressure in the left ventricle rises sometimes to as high as 450 mm. Hg while the pressure in the aorta is still normal. Thus, a nozzle effect is created *during systole,* with blood jetting at tremendous velocity through the small opening of the valve. This causes severe *turbulence* in the root of the aorta. The turbulent blood impinging against the aortic walls causes intense vibration, and a loud murmur is transmitted throughout the upper aorta and even into the larger vessels of the neck. This sound is harsh and frequently so loud that it can be heard several feet away from the patient. Also, the sound vibrations can often be felt with the hand on the upper chest and lower neck, a phenomenon known as a "thrill."

THE MURMUR OF AORTIC REGURGITATION. In aortic regurgitation no sound is heard during systole, but *during diastole* blood flows backward from the aorta into the left ventricle, causing a "blowing" murmur of relatively high pitch heard maximally over the left ventricle. This murmur does not have the low pitch of blood reverberating between the walls of the ventricle. Consequently, the blowing quality probably results from the *turbulence* of blood jetting backward into the blood already in the left ventricle.

The sound of aortic regurgitation is not nearly so loud as that of aortic stenosis, mainly because the pressure differential between the aorta and left ventricle in regurgitation is not nearly so great as it can sometimes be in stenosis.

If the aortic valve is so badly destroyed that essentially all the return of blood from the arterial tree into the heart takes place during the first portion of diastole, the sound of aortic regurgitation may not be heard at all during the latter portion of diastole. Therefore, loud aortic regurgitation murmurs lasting throughout diastole sometimes mean less severely damaged valves than weaker murmurs heard only during the early part.

THE MURMUR OF MITRAL REGURGITATION. In mitral regurgitation blood flows backward through the mitral valve *during systole*. This also causes a high frequency "blowing" sound, which is transmitted most strongly into the left atrium, but the left atrium is so deep within the chest that it is difficult to hear this sound directly over the atrium. As a result, the sound of mitral regurgitation is transmitted to the chest wall mainly through the left ventricle, and it is usually heard best at the apex of the heart.

The blowing quality of the mitral regurgitation murmur, like that of aortic regurgitation, is presumably caused by the turbulence of blood ejected backward through the mitral valve into the blood of the left atrium or against the atrial wall. The quality of the mitral regurgitation murmur is almost exactly the same as that of aortic regurgitation, but it occurs during systole rather than diastole.

THE MURMUR OF MITRAL STENOSIS. In mitral stenosis, blood passes with difficulty from the left atrium into the left ventricle and, because the pressure in the left atrium never rises to a very high value, a great pressure differential forcing blood from the left atrium into the left ventricle never develops. Consequently, the abnormal sounds heard in mitral stenosis are usually extremely weak.

During the early part of diastole, the ventricle has so little blood in it and its walls are so flabby that blood does not reverberate back and forth between the walls of the ventricle. For this reason, even in severe mitral stenosis, no murmur at all might be heard during the first third of diastole. However, after the first third the ventricle is stretched enough for blood to reverberate, and a murmur then often begins. This murmur is of such low pitch that it is difficult to hear, but, with the aid of a proper stethoscope (the "bell" type), one can usually discern very low frequency sounds of 30 to 50 cycles per second. On the other hand, one can often feel intense low frequency vibrations, called a "thrill," over the apex of the heart despite the fact that the sound itself may be weak. The reason for this is that the frequency of vibration in the ventricle is often so low that it cannot be heard but yet can be felt.

Often in early stages of mitral stenosis only a *presystolic murmur* may be heard. It is probable that this presystolic murmur is caused by the momentarily increased left atrial pressure resulting from atrial contraction.

PHONOCARDIOGRAMS IN VALVULAR MURMURS. Phonocardiograms B, C, D, and E of Figure 30–3 illustrate, respectively, idealized records obtained from patients with aortic stenosis, mitral regurgitation, aortic regurgitation, and mitral stenosis. It is obvious from these phonocardiograms that the aortic stenotic lesion causes the loudest of all these murmurs, and the mitral stenotic lesion causes the weakest, a murmur of very low frequency and rumbling quality. The phonocardiograms show how the intensity of the murmurs varies during different portions of systole and diastole, and the relative timing of each murmur is also evident. Note especially that the murmurs of aortic stenosis and mitral regurgitation occur only during systole, whereas the murmurs of aortic regurgitation and mitral stenosis occur only during diastole—if the student does not understand this timing, he should pause a moment until he does understand it.

ABNORMAL CIRCULATORY DYNAMICS IN VALVULAR HEART DISEASE

DYNAMICS OF THE CIRCULATION IN AORTIC STENOSIS AND AORTIC REGURGITATION

Aortic stenosis means a constricted aortic valve with a tubular opening too small for easy ejection of blood from the left ventricle. *Aortic regurgitation* means failure of the aortic valve to close completely and, therefore, failure to prevent backflow of blood from the aorta into the left ventricle during diastole. In aortic stenosis the left ventricle fails to empty adequately, whereas in aortic regurgitation blood returns to the ventricle after it has been emptied. Therefore, in either case, the net stroke volume output of the heart is reduced, and this in turn tends to reduce the cardiac output, resulting eventually in typical circulatory failure.

However, several important compensations take place that can ameliorate the severity of the circulatory defects. Some of these are the following:

Hypertrophy of the Left Ventricle. In both aortic stenosis and regurgitation, the left ventricle enlarges and its musculature hypertrophies. Sometimes the left ventricle muscle mass increases as much as four- to five-fold, creating a tremendously large left heart. When the aortic valve is seriously stenosed, this hypertrophied muscle allows the left ventricle sometimes to develop as much as 500 mm. Hg intraventricular pressure; and in aortic regurgitation, the hypertrophied muscle allows the left ventricle sometimes to pump a stroke volume output as high as 300 ml., though most of this blood returns to the ventricle during diastole.

Increase in Blood Volume. Another effect that helps to compensate for the diminished net pumping by the left ventricle is increased blood volume. This results from a slight fall in arterial pressure plus peripheral circulatory reflexes that this decrease induces, both of which diminish renal output of urine. Also, red cell mass increases because of a slight degree of tissue hypoxia.

The increase in blood volume tends to increase venous return to the heart. This in turn increases the ventricular end-diastolic volume, causing the left ventricle to pump its very high pressure in aortic stenosis or its very high stroke volume output in aortic regurgitation.

Eventual Failure of the Left Ventricle, and Development of Pulmonary Edema

In the early stages of aortic stenosis or aortic regurgitation, the intrinsic ability of the left ventricle to adapt to increasing loads prevents severe abnormalities in circulatory function other than increased work output required of the left ventricle. Therefore, marked degrees of aortic stenosis or aortic regurgitation often occur before the person knows that he has serious heart disease, such as left ventricular systolic pressures as high as 250 mm. Hg in aortic stenosis or left ventricular stroke volume outputs as high as double normal in aortic regurgitation.

However, beyond the critical degrees of these two aortic lesions, the left ventricle finally cannot keep up with the work demand, and as a consequence the cardiac output begins to fall while blood simultaneously begins to dam up in the left atrium and lungs, behind the left ventricle. The left atrial pressure rises progressively, and at pressures above 30 to 45 mm. Hg, edema appears in the lungs, as was discussed in detail in Chapter 23.

The progressive changes in circulatory dynamics are shown diagrammatically in Figures 30–4 and 30–5 for aortic stenosis and regurgitation, respectively.

DYNAMICS OF MITRAL STENOSIS AND MITRAL REGURGITATION

In mitral stenosis blood flow from the left atrium into the left ventricle is impeded, and in mitral regurgitation much of the blood that has flowed into the left ventricle is pumped back into the left atrium during systole rather than being pumped

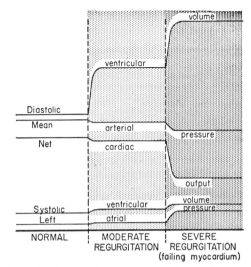

Figure 30–5 Circulatory dynamics in aortic regurgitation.

into the aorta. Therefore, the effect is reduced net movement of blood from the left atrium into the left ventricle.

Pulmonary Edema in Mitral Valvular Disease. Obviously, the buildup of blood in the left atrium causes progressive increase in left atrial pressure, and this can result eventually in the development of serious pulmonary edema. Ordinarily, lethal edema will not occur until the left atrial pressure rises at least above 30 mm. Hg; and more often it must rise to as high as 45 mm. Hg because the lymphatic vasculature enlarges many-fold and can carry fluid away from the lung tissues extremely rapidly.

Enlarged Left Atrium and Atrial Fibrillation. The high left atrial pressure also causes progressive enlargement of the left atrium, which increases the distance that the cardiac impulse must travel through the atrial walls. Eventually, this pathway becomes so long that it predisposes to the development of circus movements. Therefore, in late stages of mitral valvular disease, atrial fibrillation is an almost invariable result. This state further reduces the pumping effectiveness of the heart and, therefore, causes still further cardiac debility.

Compensations in Mitral Valvular Disease. As also occurs in aortic valvular disease and in many types of congenital heart disease, the blood volume increases in mitral valvular disease. This tends to increase venous return to the heart, thereby helping to overcome the cardiac debility. Therefore, cardiac output does not fall greatly until the late stages of mitral valvular disease.

As the left atrial and pulmonary capillary pressures rise, blood also begins to dam up in the pulmonary artery, increasing pulmonary arterial pressure sometimes to more than double normal. This, in turn, causes enlargement of the right heart, which partially compensates for its increased work load.

Figure 30–6 illustrates the progressive changes

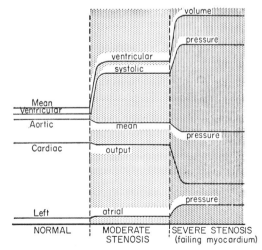

Figure 30–4 Circulatory dynamics in aortic stenosis.

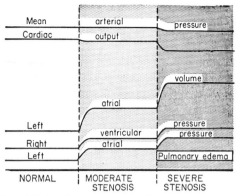

Figure 30–6   Circulatory dynamics in mitral stenosis.

in circulatory dynamics in mitral stenosis. Essentially the same effects occur in mitral regurgitation. Indeed, these two conditions more often than not occur together in varying degrees.

ABNORMAL CIRCULATORY DYNAMICS IN CONGENITAL HEART DEFECTS

Occasionally, the heart or its associated blood vessels are malformed during fetal life; the defect is called a *congenital anomaly*. Basically, there are three major types of congenital anomalies of the heart and its associated vessels: (1) *stenosis* of the channel of blood flow at some point in the heart or in a closely allied major vessel, (2) an abnormality that allows blood to flow directly from the left heart or aorta to the right heart or pulmonary artery, thus by-passing the systemic circulation — this is called a *left-to-right shunt* — and (3) an abnormality that allows blood to flow from the right heart or pulmonary artery directly into the left heart or aorta, thus by-passing the lungs — this is called a *right-to-left shunt.*

The three most common types of stenotic lesions are *coarctation of the aorta, congenital pulmonary stenosis,* and *congenital aortic stenosis.* The effect of coarctation of the aorta on the circulation in relation to the regulation of arterial pressure was discussed in Chapter 26, and congenital aortic stenosis causes almost exactly the same alterations in circulatory dynamics as those noted above for acquired aortic stenosis. Therefore, in this present section only the dynamics of pulmonary stenosis are discussed.

In all left-to-right shunts, blood is shunted directly from the left heart back to the right heart without going through the systemic circulation, and compensatory effects occur in the circulatory system to increase the total cardiac output, thereby making up for most of the extra blood flow through the shunt. For instance, if 50 per cent of all the blood pumped by the left heart is shunted directly back to the right heart, then the compensatory mechanisms normally increase the cardiac

output to almost double normal so that the amount of blood flowing through the systemic vessels is still almost normal despite the shunt. However, this causes an extra load on the heart itself and therefore usually causes the heart to fail at an early age. Also, it reduces the cardiac reserve of the person throughout his life. Some of the types of left-to-right shunts are *patent ductus arteriosus, interatrial septal defect, interventricular septal defect,* and a *direct communication between the aorta and pulmonary artery at their bases.*

In all right-to-left shunts, much of the blood by-passes the lungs and therefore fails to become oxygenated. Consequently, large quantities of venous blood enter the arterial system directly, and the patient is usually cyanotic (blue) all the time. The most common type of right-to-left shunt is the *tetralogy of Fallot,* which is described later; several other right-to-left shunts are some variation of this abnormality.

Causes of Congenital Anomalies.   One of the most common causes of congenital heart defects is a virus infection of the mother during the first trimester of pregnancy when the fetal heart is being formed. Defects are particularly prone to develop when the mother contracts German measles at this time — so often indeed that obstetricians advise termination of pregnancy if German measles occurs in the first trimester. However, some congenital defects of the heart are believed to be hereditary because the same defect has been known to occur in identical twins and also in succeeding generations. Congenital defects of the heart are frequently also associated with other congenital defects of the body.

PATENT DUCTUS ARTERIOSUS — A LEFT-TO-RIGHT SHUNT

During fetal life the lungs are collapsed, and the elastic factors that keep the alveoli collapsed also keep the blood vessels collapsed. Therefore, in the collapsed lung the resistance to blood flow is 5 to 10 times as great as it is in the inflated lung. For this reason, and because of the very low systemic resistance through the large vessels of the placenta, the pressure in the pulmonary artery is higher than in the aorta, causing almost all the pulmonary arterial blood to flow through the ductus arteriosus into the aorta rather than through the lungs. This allows immediate recirculation of the blood through the systemic arteries of the fetus. Obviously, this lack of blood flow through the lungs is of no detriment to the fetus because the blood is oxygenated by the placenta of the mother.

Closure of the Ductus.   As soon as the baby is born, his lungs inflate; and not only do the alveoli fill, but the resistance to blood flow through the pulmonary vascular tree decreases tremendously, allowing pulmonary arterial pressure to fall. Simultaneously, the aortic pressure rises because of sudden cessation of blood flow through the placenta. Thus, the pressure in the pulmonary artery falls, while that in the aorta rises. As a result, forward blood flow through the ductus ceases suddenly at birth, and blood even flows backward from the aorta to the pulmonary artery. This new state of blood flow causes the ductus arteriosus to become occluded within a few hours to a few days in most babies so that blood flow through the ductus does not persist. The possible causes of ductus closure will be discussed in Chapter 83. In many instances it takes several months for the ductus to close completely, and in about 1 out of every 5500 babies the ductus never closes, causing the condition known as *patent ductus arteriosus,* which is illustrated in Figure 30-7.

Dynamics of Persistent Patent Ductus. During the early months of an infant's life a patent ductus usually does not cause severely abnormal dynamics because the blood pressure of the aorta then is not much higher than the pressure in the pulmonary artery, and only a small amount of blood flows backward into the pulmonary system. However, as the child grows older, the differential between the pressure in the aorta and that in the pulmonary artery progressively increases with corresponding increase in the backward flow of blood from the aorta to the pulmonary artery. Also, the diameter of the partially closed ductus often increases with time, making the condition worse.

RECIRCULATION THROUGH THE LUNGS. In the older child with a patent ductus, as much as half to three fourths of the aortic blood flows into the pulmonary artery, then through the lungs, into the left atrium, and finally back into the left ventricle, passing through this circuit two or more times for every one time that it passes through the systemic circulation.

These persons do not show cyanosis until the heart fails or until the lungs become congested. Indeed, the arterial blood is often better oxygenated than normally because of the extra passage of blood through the lungs. Furthermore, in the early stages of patent ductus arteriosus the quantity of blood flowing into the systemic aorta remains essentially normal because the quantity of blood returning to the heart from the peripheral circu-

latory system is normal. Yet, because of the tremendous accessory flow of blood around and around through the lungs and left side of the heart, the output of the left ventricle in patent ductus arteriosus is usually two to four times normal.

DIMINISHED CARDIAC AND RESPIRATORY RESERVE. The major effect of patent ductus arteriosus on the patient is low cardiac and respiratory reserve. The left ventricle is already pumping approximately two or more times the normal cardiac output, and the maximum that it can possibly pump is about four to six times normal. Therefore, during exercise the cardiac output can be increased much less than usual. Under basal conditions, the patient usually appears normal except for possible heaving of his chest with each beat of the heart, but just as soon as he tries to perform even moderately strenuous exercise he is likely to become weak and occasionally even faint from momentary heart failure. Also, the high pressures in the pulmonary vessels soon lead to pulmonary congestion.

The entire heart usually hypertrophies greatly in patent ductus arteriosus. The left ventricle hypertrophies because of the excessive work load that it must perform in pumping a far greater than normal cardiac output, while the right ventricle hypertrophies because of increased pulmonary arterial pressure resulting from, first, increased flow of blood through the lungs caused by the extra blood from the patent ductus and, second, increased resistance to blood flow through the lungs caused by progressive sclerosing of the vessels as they are exposed year in and year out to excessive pulmonary blood flow.

As a result of the increased load on the heart and because of the pulmonary congestion, most patients with patent ductus die between the ages of 20 and 40 unless the defect is corrected by surgery.

The Machinery Murmur. In the infant with patent ductus arteriosus, occasionally no abnormal heart sounds are heard because the quantity of reversed blood flow may be insignificant. As the baby grows older, reaching the age of one to three years, a harsh, blowing murmur begins to be heard in the pulmonic area of the chest. This sound is much more intense during systole when the aortic pressure is high and much less intense during diastole, so that it waxes and wanes with each beat of the heart, creating the so-called "machinery murmur." The idealized phonocardiogram of this murmur is shown in Figure 30-3F.

Surgical Treatment. Surgical treatment of patent ductus arteriosus is extremely simple, for all one needs to do is to ligate the patent ductus.

INTERVENTRICULAR SEPTAL DEFECT — A LEFT-TO-RIGHT SHUNT

Because the systolic pressure in the left ventricle is normally about six times that in the right ventricle, a large amount of blood flows from the left to the right ventricle whenever an interventricular defect is present. As a result of elevated pressure in the right ventricle, the right ventricle hyper-

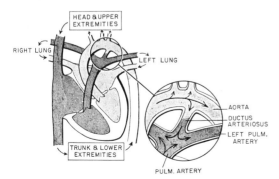

Figure 30-7 Patent ductus arteriosus, illustrating the degree of blood oxygenation in the different parts of the circulation.

trophies, sometimes to such an extent that its muscular wall approximately equals that of the left ventricle.

Diagnosis of an interventricular septal defect is based on (1) the presence of a systolic blowing murmur heard over the anterior projection of the heart, (2) high right ventricular systolic pressure recorded from a catheter, and (3) the presence of oxygenated blood in a blood sample removed through the catheter from the right ventricle, this blood having leaked backward from the left ventricle.

Blood flowing from the left ventricle into the right ventricle passes one or more times through the lungs and then back to the left ventricle again before finally entering the peripheral circulatory system. This condition, therefore, is analogous to patent ductus arteriosus except that the flow of blood from the systemic circulation to the pulmonary circulation occurs only during systole rather than during both systole and diastole as in patent ductus.

INTERATRIAL SEPTAL DEFECTS — A LEFT-TO-RIGHT SHUNT

Closure of the Foramen Ovale. During fetal life much of the blood entering the right atrium fails to pass into the right ventricle but instead courses directly through the foramen ovale into the left atrium and thence out into the systemic circulation. This mechanism aids the ductus arteriosus in shunting blood around the lungs, thereby relieving the fetal heart from unnecessary load. Immediately after birth of the child, the pressure in the pulmonary artery decreases while that in the aorta increases, as discussed above. These changes decrease and increase respectively the loads on the right and left ventricles so that the right atrial pressure decreases and the left atrial pressure increases. As a result, blood then attempts to flow from the left atrium back into the right atrium. However, the foramen ovale is covered by a small valvelike vane, which closes over the foramen. In two thirds of all persons the foramen later becomes totally occluded by fibrous tissue, but in one third the foramen never becomes totally occluded. Yet this nonoccluded opening does not cause any physiological abnormality in the heart because the left atrial pressure remains 3 to 4 mm. Hg higher than the right atrial pressure, keeping the valve permanently closed.

Dynamics of Interatrial Defects. Occasionally, the valve does not cover the foramen ovale, and a hole persists permanently between the left atrium and the right atrium. If this hole is small, only a small amount of blood passes from the left atrium back to the right atrium, but occasionally almost the entire wall between the two atria is missing. In this case a tremendous quantity of blood flows from the left atrium into the right atrium and circulates repeatedly through the right ventricle and lungs before finally passing into the systemic circulation. Therefore, even though a small opening between the atria does not make a great deal of difference in the function of the heart, large openings greatly reduce the cardiac reserve, and early death results from right ventricular failure and pulmonary congestion.

The diagnosis of an interatrial septal defect is difficult to make, for no murmurs can be heard, the patient's arterial blood is well oxygenated, and the usual x-ray studies are relatively nonspecific. The diagnosis can be made accurately by (1) angiocardiograms — that is, x-rays made after injecting a radiopaque dye into the heart — or (2) finding oxygenated blood in a catheter specimen from the right atrium in the absence of a detectable murmur.

TETRALOGY OF FALLOT — A RIGHT-TO-LEFT SHUNT

Tetralogy of Fallot is illustrated in Figure 30-8, from which it will be noted that four different abnormalities of the heart occur simultaneously.

First, the aorta originates from the right ventricle rather than the left, or it overrides the septum as shown in the figure.

Second, the pulmonary artery is stenosed so that a much less than normal amount of blood passes from the right side of the heart into the lungs; instead the blood passes into the aorta.

Third, blood from the left ventricle flows through a ventricular septal defect into the right ventricle and then into the aorta or directly into the overriding aorta.

Fourth, because the right side of the heart must pump large quantities of blood against the high pressure in the aorta, its musculature is highly developed, causing an enlarged right ventricle.

Abnormal Dynamics. It is readily apparent that the major physiological difficulty caused by tetralogy of Fallot is shunt of blood past the lungs without its becoming oxygenated. As much as 75 per cent of the venous blood returning to the heart may pass directly from the right ventricle into the aorta without becoming oxygenated. Tetralogy of Fallot is the major cause of cyanosis in babies ("blue babies").

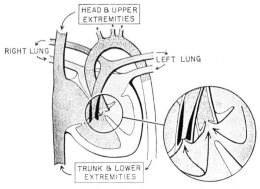

Figure 30-8 Tetralogy of Fallot, illustrating the degree of blood oxygenation in the different parts of the circulation.

A diagnosis of tetralogy of Fallot is usually based on (1) the fact that the baby is blue, (2) records of high systolic pressure in the right ventricle recorded through a catheter, (3) characteristic changes in the x-ray silhouette of the heart, and (4) angiograms showing abnormal blood flow through the interventricular septal defect and the pulmonary stenotic area.

Surgical Treatment. In recent years tetralogy of Fallot has been treated relatively successfully by surgery. One type of treatment is to create an artificial ductus arteriosus by making a small opening between the aorta and pulmonary artery, thereby correcting the cyanosis and increasing the life expectancy of the patient with tetralogy of Fallot from approximately 1 to 10 years to an age of perhaps 50 or more years. Another operation is to open the pulmonary stenosis and close the septal defect in those cases in which this is possible.

PULMONARY STENOSIS

Often a child is born with pulmonary stenosis but without other congenital abnormalities. Such severe pulmonary stenosis occasionally occurs that the right side of the heart is likely to fail at an early age because blood flow from the right ventricle into the lungs is greatly impeded, the right ventricular systolic pressure rising to 75 to 100 mm. Hg instead of the normal 22 mm. Hg. The right side of the heart dilates, and the muscle becomes greatly hypertrophied in order to withstand the load. Also, a loud stenotic murmur is heard over the pulmonary valve area. In many cases of pulmonary stenosis, the stenotic area can be enlarged by surgical maneuvers so that the heart resumes normal function.

USE OF EXTRACORPOREAL CIRCULATION IN CARDIAC SURGERY

It is almost impossible to repair intracardiac defects while the heart is still pumping. Therefore, many different types of artificial *heart-lung machines* have been developed to take the place of the heart and lungs during the course of operation. Such a system is called an *extracorporeal circulation.* The system consists principally of (1) a pump and (2) an oxygenating device. Almost any type of pump that does not cause hemolysis of the blood seems to be suitable.

The different principles that have been used for oxygenating blood are (1) bubbling oxygen through the blood and then removing the bubbles from the blood before passing it back into the patient, (2) dripping the blood downward over the surfaces of large areas of screen wire, (3) passing the blood over the surfaces of rotating discs, and (4) passing the blood between thin membranes that are porous to oxygen and carbon dioxide.

The different oxygenators have been fraught with many difficulties, including hemolysis of the blood, development of small clots in the blood, likelihood of small bubbles of oxygen or small emboli of antifoaming agent passing into the arteries of the patient, necessity for large quantities of blood to prime the entire system, failure to exchange adequate quantities of oxygen, and the necessity to use heparin in the system to prevent blood coagulation, the heparin also preventing adequate hemostasis during the surgical procedure. Yet, despite these difficulties, in the hands of experts patients can be kept on artificial heart-lung machines for several hours while operations are performed on the inside of the heart.

HYPERTROPHY OF THE HEART IN VALVULAR AND CONGENITAL HEART DISEASE

Hypertrophy of cardiac muscle is one of the most important mechanisms by which the heart adapts to increased work loads, whether these loads be caused by increased pressure against which the heart muscle must contract or by increased volume that must be pumped. Some investigators believe that it is the increased work load itself that causes the hypertrophy; others believe the increased metabolic rate of the muscle to be the primary stimulus and the work load simply to be the cause of the increase in the metabolic rate. Regardless of which of these is correct, one can calculate approximately how much hypertrophy will occur in each chamber of the heart by multiplying ventricular output times the pressure against which the ventricle must work. Thus, hypertrophy occurs in most types of valvular and congenital disease, as follows:

In *aortic stenosis* and *aortic regurgitation,* the left ventricular musculature hypertrophies tremendously, sometimes to as much as four to five times normal, so that the weight of the heart on occasion may be as great as 1000 grams instead of the normal 300 grams.

In mitral disease, the left atrium hypertrophies and dilates, but no left ventricular hypertrophy occurs in *mitral stenosis,* and only mild to moderate hypertrophy of the left ventricle occurs in *mitral regurgitation.* Some hypertrophy of the right ventricle may develop owing to back pressure effects through the lungs causing elevation of the pulmonary arterial pressure.

In *patent ductus arteriosus,* the work load of both ventricles is increased. The left ventricle pumps an average of twice the quantity of blood that it normally pumps; therefore, it would be expected that the left ventricle would hypertrophy to approximately twice normal size. On the other hand, the right ventricle must pump its blood against a much higher than normal pulmonary arterial pressure because of the large quantity of blood refluxing from the aorta into the pulmonary artery. Consequently, the right ventricle also hypertrophies.

In *tetralogy of Fallot,* one of the cardinal signs is the marked right ventricular hypertrophy, for the right ventricle must pump against the pressure in the aorta. On the other hand, the work load of the left ventricle is actually less than normal because of the reduced blood flow from the lungs into the left heart. Therefore, the right ventricle in tetralogy of Fallot is often larger than the normal left ventricle, while the left ventricle may be relatively small.

Detrimental Effects of the Late Stages of Hypertrophy. Though physiological hypertrophy of heart muscle is usually very beneficial to cardiac function, extreme degrees of hypertrophy sometimes lead to failure. One of the reasons for this is that coronary blood flow does not always increase to the same extent as the mass of muscle. A second reason is that fibrosis often develops in the muscle, the fibrous tissue replacing degenerating muscle fibers. This means that relative ischemia develops as the muscle hypertrophies, and coronary insufficiency easily ensues. Therefore, anginal pain is a frequent accompaniment of many valvular and congenital heart diseases.

REFERENCES

Bellhouse, B., and Bellhouse, F.: Fluid mechanics of model normal and stenosed aortic valves. *Circ. Res., 25*:693, 1969.

Buccino, R. A.; Harris, E.; Spann, J. F., Jr.; and Sonnenblick, E.: Response of myocardial connective tissue to development of experimental hypertrophy. *Amer. J. Physiol., 216*:425, 1969.

Butterworth, J. S.; Chassin, M. R.; McGrath, R.; and Reppert, E. H.: Cardiac Auscultation—Including Audio-Visual Principles. 2nd ed., New York, Grune & Stratton, Inc., 1960.

Cooley, D. A., and Hallman, G. L.: Surgical Treatment of Congenital Heart Disease. Philadelphia, Lea & Febiger, 1966.

Dart, C. H., Jr., and Holloszy, J. O.: Hypertrophied non-failing rat heart: partial biochemical characterization. *Circ. Res., 25*:245, 1969.

Eitzman, D. V.; Hessler, J. R.; and Cassin, S.: Changes in blood flow and resistance in neonatal goats during hypoxia. *Amer. J. Physiol., 216*:1577, 1969.

Faber, J. J., and Burton, A. C.: Spread of heart sounds over chest wall. *Circ. Res., 11*:96, 1962.

Galletti, P. M., and Brecher, G. A.: Heart-Lung Bypass—Principles and Techniques of Extracorporeal Circulation. New York, Grune & Stratton, Inc., 1962.

Grimm, A. F.; Torre, L. de la; and Porta, M. la, Jr.: Ventricular nuclei-DNA relationships with myocardial growth and hypertrophy in the rat. *Circ. Res., 26*:45, 1970.

Grove, D.; Nair, K. G.; and Zak, R.: Biochemical correlates of cardiac hypertrophy: III. Changes in DNA content; the relative contributions of polyploidy and mitotic activity. *Circ. Res., 25*:463, 1969.

Grove, D.; Zak, R.; Nair, K. G.; and Aschenbrenner, V.: Biochemical correlates of cardiac hypertrophy: IV. Observations on the cellular organization of growth during myocardial hypertrophy in the rat. *Circ. Res., 25*:473, 1969.

Hudson, R. E. B.: Cardiovascular Pathology. 2 volumes. Baltimore, The Williams & Wilkins Co., 1965.

Lipana, J. G., and Fanburg, B. L.: Heart growth in response to aortic constriction in the hypophysectomized rat. *Amer. J. Physiol., 218*:641, 1970.

Ravin, A.: Auscultation of the Heart. Baltimore, The Williams & Wilkins Co., 1967.

Stept, M. E.; Heid, C. E.; Shaver, J. A.; Leon, D. F.; and Leonard, J. J.: Effect of altering P-R interval on the amplitude of the first heart sound in the anesthetized dog. *Circ. Res., 25*:255, 1969.

Taussig, H.: Congenital Malformations of the Heart. Vol. 1: General Considerations, 2nd ed., Cambridge, Mass., Harvard University Press, 1960.

Taussig, H.: Congenital Malformations of the Heart. Vol. 2: Specific Malformations, 2nd ed., Cambridge, Mass., Harvard University Press, 1961.

Tavel, M. E.: Clinical Phonocardiography and External Pulse Recording. Chicago, Year Book Medical Publishers, Inc., 1967.

Ur, A., and Gordon, M.: Origin of Korotkoff sounds. *Amer. J. Physiol., 218*:524, 1970.

Wannemacher, R. W., Jr., and McCoy, J. R.: Regulation of protein synthesis in the ventricular myocardium of hypertrophic hearts. *Amer. J. Physiol., 216*:781, 1969.

Wood, E. H.: Diagnostic applications of indicator-dilution technics in congenital heart disease. *Circ. Res., 10*:531, 1962.

CHAPTER 31

The Coronary Circulation

Approximately one third of all deaths result from coronary artery disease, and almost all elderly persons have at least some impairment of coronary artery circulation. For this reason, the normal and pathologic physiology of the coronary circulation is one of the most important subjects in the entire field of medicine. The purpose of this chapter is to present in detail the subject of coronary circulation, emphasizing especially the physiology of recovery from coronary thrombosis.

NORMAL CORONARY BLOOD FLOW AND ITS VARIATIONS

PHYSIOLOGIC ANATOMY OF THE CORONARY BLOOD SUPPLY

The *left coronary artery* supplies mainly the left ventricle, and the *right coronary artery* supplies mainly the right ventricle but usually also part of the left ventricle as well. In about one-half of all human beings, more blood flows through the right coronary artery than through the left, whereas the left artery predominates in only 20 per cent.

Most of the venous blood flow from the left ventricle leaves by way of the *coronary sinus*— about 70 per cent of the total coronary blood flow—and most of the venous blood from the right ventricle flows through the small *anterior cardiac veins*, which empty directly into the right atrium and are not connected with the coronary sinus. A small amount of coronary blood flows back into the heart through *thebesian veins*, which empty directly into all chambers of the heart.

NORMAL CORONARY BLOOD FLOW

The resting coronary blood flow in the human being averages approximately 225 ml. per minute, which is about 0.8 ml. per gram of heart muscle, or 4 to 5 per cent of the total cardiac output.

In strenuous exercise the heart increases its cardiac output as much as four- to six-fold, and it pumps this blood against a higher than normal arterial pressure. Consequently, the work output of the heart under severe conditions may increase as much as 8- to 10-fold. The coronary blood flow increases four- to five-fold to supply the extra nutrients needed by the heart. Obviously, this increase is not as much as the increase in work load, which means that the ratio of coronary blood flow to energy expenditure by the heart decreases. However, the "efficiency" of cardiac contraction increases to make up for this relative deficiency of blood supply.

Phasic Changes in Coronary Blood Flow. Figure 31–1 illustrates the blood flow *through the small nutrient vessels* of the coronary system in milliliters per minute in the human heart during systole and diastole as *calculated* from experiments in lower animals. Note from this diagram that the blood flow through the capillaries of the left ventricle falls to a very low value during *systole*. The reason for this is the strong compression of the cardiac muscle around the intramuscular vessels during systole.

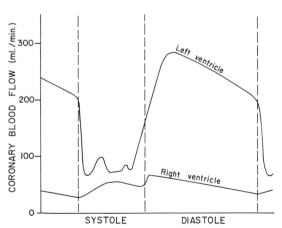

Figure 31–1 Phasic flow of blood through the coronary capillaries of the left and right ventricles.

During *diastole*, the cardiac muscle relaxes completely and no longer obstructs the blood flow through the left ventricular capillaries, so that blood now flows rapidly during all of diastole.

Blood flow through the coronary capillaries of the right ventricle undergoes phasic changes similar to those in the coronary capillaries of the left ventricle during the cardiac cycle, but, because the force of contraction of the right ventricle is far less than that of the left ventricle, these phasic changes are relatively mild compared to those in the left ventricle, as is shown in Figure 30–1.

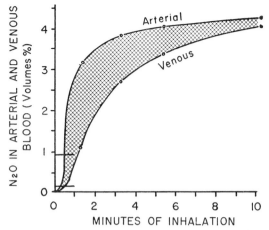

Figure 31–2 Method for measuring coronary blood flow in man as explained in the text. (Modified from Kety: Methods in Medical Research, 1948.)

MEASUREMENT OF CORONARY BLOOD FLOW IN MAN

Obviously, it would be almost impossible to insert a flowmeter into the coronary circulatory system of man. However, a satisfactory indirect method for measuring coronary flow has been achieved, the principles of which are the following:

The blood flow through any organ of the body can be determined by means of the so-called Fick principle. That is, if a constituent of the blood is removed as it flows through an organ, then the blood flow can be determined from two measurements: first, the amount removed from each milliliter of blood as it goes through the organ and, second, the total quantity of the substance that is removed by the organ in a given period of time. The total blood flow during the period of measurement can be calculated by dividing the amount of substance removed from each milliliter of blood into the total quantity removed.

In measuring blood flow through the heart the subject suddenly begins to breathe nitrous oxide of a given concentration. Samples of arterial blood are removed from any artery in the body and samples of venous blood are obtained from the coronary sinus through a catheter that has been passed into the sinus by way of the venous system. The concentrations of the nitrous oxide in both of these bloods are then plotted for 10 minutes, as shown in Figure 31–2. The quantity of nitrous oxide lost to the heart by each milliliter of blood as it passes through the heart at each instant of the experiment can be determined from the difference in height of the two curves. This is called the *A-V difference*. The A-V difference is determined each minute for a period of 10 minutes and then averaged.

Now, to determine the amount of nitrous oxide that is removed from the blood by the cardiac muscle, one proceeds as follows: It will be noted from the curves of Figure 31–2 that the concentrations of nitrous oxide begin to reach an equilibrium level toward the end of the 10 minute experiment. It is also known that the total amount of nitrous oxide absorbed by the cardiac muscle is directly proportional to its concentration in the arterial blood at the end of the 10 minute run. A *proportionality factor* has been determined in experiments on isolated cardiac muscle so that at the end of 10 minutes the total amount of nitrous oxide absorbed by each 100 grams of cardiac muscle can be determined by multiplying the arterial concentration of nitrous oxide times the proportionality factor.

Once the average *A-V difference* and the *amount of nitrous oxide absorbed* by each 100 grams of heart muscle have been determined, the coronary blood flow per 100 grams of muscle is then calculated by dividing the A-V difference into the quantity absorbed per minute. As an example, if 8 ml. of nitrous oxide is absorbed by an 100 grams of heart in 10 minutes and the average A-V difference during this period of time is 0.01 ml. for each ml. of blood, a total of 800 ml. of blood would have passed through the 100 grams of heart during the 10 minute interval. In other words, the blood flow would be 80 ml. per 100 grams of heart per minute. Assuming a heart mass of 280 grams, the blood flow would be 224 ml. per minute.

CONTROL OF CORONARY BLOOD FLOW

LOCAL BLOOD FLOW REGULATION AS THE PRIMARY CONTROLLER OF CORONARY FLOW

Blood flow through the coronary system is regulated almost entirely by local blood flow regulation in the heart in response to the needs of the cardiac musculature for nutrition. This mechanism works equally well when the nerves to the heart are intact or are removed. Whenever the vigor of contraction is increased, regardless of cause, the rate of coronary blood flow simultaneously increases, and, conversely, decreased activity is accompanied by decreased coronary flow. It is immediately obvious that this local autoregulation of blood flow is almost identical with that which occurs in many other tissues, especially in the skeletal muscles of all the body.

Oxygen Demand as the Mechanism of Local Blood Flow Regulation. Blood flow in the coronaries is regulated almost exactly in proportion to the need of the cardiac musculature for oxygen. Even in the normal resting state, about 65 per cent of the oxygen in the arterial blood is removed as the blood passes through the heart; and, because not much oxygen is left in the blood, little additional oxygen can be removed from the blood unless the flow increases. Fortunately, the blood flow does increase, and almost directly in proportion to the need of the heart for oxygen. Therefore, it is believed that *oxygen lack dilates the coronary arterioles.*

Yet the means by which oxygen lack causes coronary dilatation has not been determined. The two principal possibilities that have been suggested are: (1) Decreased oxygen tension in the cardiac tissues reduces the amount of oxygen available to the coronary vessels themselves and this causes the coronaries to become weakened and, therefore, to dilate automatically. (2) Oxygen lack causes vasodilator substances, such as adenosine compounds, to be released by the tissues. (Small amounts of adenosine are released during oxygen lack, but it has not been shown that it is released in sufficient quantity to cause the extreme degree of vasodilatation that results.) Therefore, for the present we can simply say that oxygen lack is followed by coronary vasodilatation and in this way regulates blood flow to the cardiac musculature in proportion to the metabolic need for oxygen by the muscle fibers.

The Determinants of Oxygen Consumption. Since the rate of oxygen consumption is the major factor that determines coronary blood flow, it is important to know the different factors that can alter myocardial oxygen consumption.

In general, the rate of cardiac oxygen consumption is closely related to work performed by the heart—the greater the work, the greater the oxygen consumption, and consequently the greater the coronary flow. However, this is not invariably true for the following reasons: Work output of the heart is determined by the arterial pressure against which the heart is pumping × the cardiac output pumped per minute. When the pressure increases, the oxygen consumption increases almost directly in proportion to the increase in pressure. On the other hand, when the cardiac output increases without an increase in pressure, the oxygen consumption increases only a slight to moderate amount rather than in direct proportion to the increase in cardiac output.

TENSION × TIME AS THE PRIMARY DETERMINANT OF OXYGEN CONSUMPTION. Perhaps the best relationship that has yet been found between cardiac function and oxygen consumption is: *oxygen consumption is proportional to myocardial muscle tension × time of contraction.* Thus, when the arterial pressure rises, the muscle tension increases and oxygen consumption also increases. Likewise, when the heart dilates, which makes it necessary (because of the law of Laplace) for the muscle to generate increased tension to pump against even a normal arterial pressure, oxygen consumption also increases even though the work output of the heart does not increase.

OTHER FACTORS THAT INCREASE OXYGEN CONSUMPTION. Other factors that increase cardiac oxygen consumption are stimulation of the heart with epinephrine or norepinephrine, stimulation with thyroxine, and increased temperature of the heart. All these factors increase the metabolic activity of the cardiac muscle fibers themselves, which in turn increases the rate of oxygen usage even though they might not increase the work output of the heart.

Reactive Hyperemia in the Coronary System. If the coronary flow to the heart is completely occluded for a few seconds to a few minutes and then suddenly unoccluded, the blood flow increases to as high as three to six times normal, as shown in Figure 31–3. It remains high for a few seconds to several minutes, depending on the period of occlu-

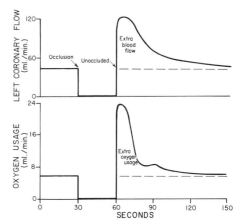

Figure 31–3 Reactive hyperemia in the coronary system caused by a 30 second period of coronary occlusion. Note the extra blood flow called "reactive hyperemia" and the extra oxygen usage after the period of occlusion was over. (Reconstructed from data in Gregg and Fisher: Handbook of Physiology, Section II, Vol. I. The Williams & Wilkins Co.)

sion. This extra flow of blood is called reactive hyperemia, which was explained in Chapter 24. During the period of excess flow, the heart removes a large amount of extra oxygen from the blood, as shown by the bottom curve of the figure, to make up for the deficiency of oxygen during the period of occlusion. Reactive hyperemia is simply another manifestation of the ability of the coronary system to adjust its flow to the need of the heart for oxygen.

NERVOUS CONTROL OF CORONARY BLOOD FLOW

Stimulation of the autonomic nerves to the heart can affect coronary blood flow in two ways—directly and indirectly. The direct effects result from direct action of the nervous transmitter substances, acetylcholine and norepinephrine, on the coronary vessels themselves. The indirect effects result from secondary changes in coronary blood flow caused by increased or decreased activity of the heart.

The indirect effects probably play the more important role in control of coronary blood flow. Thus, sympathetic stimulation increases both heart rate and heart contractility as well as its rate of metabolism. In turn, the increased activity of the heart sets off local blood flow regulatory mechanisms for increasing coronary blood flow, the blood flow increasing approximately in proportion to the metabolic needs of the heart muscle.

The distribution of parasympathetic (vagal) nerve fibers to the ventricular coronary system is so slight that parasympathetic stimulation has an almost negligible effect. On the other hand, there is extensive sympathetic innervation of the coronary vessels, and the norepinephrine released from the sympathetic endings generally has a strong vasodilator effect on the coronary vessels. Thus, when sympathetic stimulation increases the overall activity of the heart, it also has a direct effect on the coronary vessels to increase coronary flow simultaneously.

Effects of Norepinephrine, Epinephrine, or Acetylcholine in the Blood. Infusion of epinephrine or norepinephrine into the coronary blood usually causes vasodilatation, exactly as occurs following sympathetic stimulation. Likewise, infusion of acetylcholine causes considerable vasodilatation, and this is in contrast to the almost negligible effect of vagal stimulation. This difference is presumably caused by the fact that very few vagal nerve fibers end on the coronary vessels.

Occasional Vasoconstrictor Effect Caused by Sympathetic Stimulation or Norepinephrine or Epinephrine Infusion. In Chapter 59 it will be pointed out that the sympathetic transmitter substances, norepinephrine and epinephrine, can have either dilator or constrictor effects, depending on the presence or absence of specific receptors in the blood vessels. The constrictor receptors are called *alpha* receptors and the dilator receptors *beta* receptors. In the normal coronary system, beta receptors predominate, and this explains why sympathetic stimulation or norepinephrine or epinephrine infusion normally causes coronary vasodilatation. However, even in the normal heart some alpha receptors are present; and in the occasional animal these receptors predominate, in which case vasoconstriction rather than vasodilatation occurs. Though this effect has not been proved to occur in the human being, it is possible that sympathetic stimulation can, under some conditions, cause coronary vasoconstriction. For this reason, it has been postulated that some persons develop anginal pain of the heart as a result of sympathetic vasoconstriction, though this too has not been proved.

In general, the nervous influences on coronary blood flow are unimportant in comparison with the local ability of the blood vessels to adjust blood flow according to the metabolic needs of the cardiac musculature. But it is possible that sympathetic stimulation of the coronaries does on some occasions moderately enhance or diminish blood flow to the heart muscle.

EFFECT OF DRUGS ON CORONARY BLOOD FLOW

In addition to the vasodilating effect of norepinephrine, epinephrine, and acetylcholine, which have just been discussed, other drugs that have important vasodilating effects on the coronary circulation include the *nitrites, papaverine,* and *aminophylline*. These drugs, along with a few newer ones that have not yet been studied extensively, are of possible value clinically to increase coronary blood flow in conditions in which the cardiac muscle needs additional flow, such as in *angina pectoris*.

CORONARY OCCLUSION

The coronary blood vessels sometimes become occluded rapidly—within a few minutes or a few hours—as a result of blood clots or other abnormalities that can plug the lumina of the vessels. At other times the vessels are slowly and progressively occluded over a period of years, in which case a collateral blood supply can usually develop to take over the function of the primary coronary blood supply. Unfortunately, the collateral blood supply usually does not become well developed until after the coronaries are occluded. Therefore, whether or not the heart is greatly damaged by coronary occlusion depends mainly on the rapidity with which the occlusion occurs.

Collateral Circulation in the Heart. Relatively few communications exist between the larger coronary arteries, but there are many anastomoses between the small arteries approximately 20 to 350 microns in diameter, as shown in Figure 31–4.

When a sudden occlusion occurs in one of the larger coronary vessels, the sizes of the minute anastomoses increase to their maximum physical diameters within a few seconds. The blood flow through these minute collaterals is less than one-half that needed to keep alive the cardiac muscle that they supply; and unfortunately the diameters of the collateral vessels do not enlarge further for the next 8 to 24 hours. But then collateral flow begins to increase, doubling in the next one to two days, often reaching normal or almost normal coronary supply within about one month; and it is capable even of increasing further with increased metabolic loads. It is because of these developing collateral channels that a patient recovers from the various types of coronary occlusion.

The coronary arteries also anastomose through minute channels with the arteries of the parietal pericardium, these anastomoses traveling along the pericardial reflections

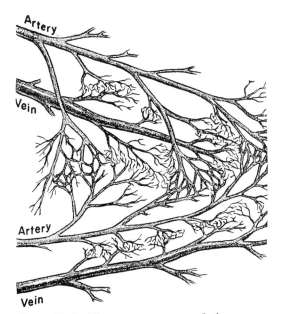

Figure 31–4 Minute anastomoses of the coronary arterial system.

around the major vessels of the heart. It will be recalled that the heart floats within the pericardial cavity, and the only possible communication between the coronary vessels and other vessels outside the pericardium would be by way of these reflections. A few patients at autopsy have been found to have both major coronary arteries totally occluded, and yet the small anastomotic vessels in the pericardial reflections have become large enough to supply the entire heart with blood.

Atherosclerosis and Coronary Thrombosis. The most frequent cause of coronary occlusion is *thrombosis* resulting from atherosclerosis. The atherosclerotic process will be discussed in connection with lipid metabolism in Chapter 68, but, to summarize this process, in various diseases affecting lipid metabolism and also in old age, lipids containing mainly cholesterol and cholesterol salts are deposited beneath the intima of the major arteries. These deposits become calcified over a period of years, and considerable fibrous tissue also invades the walls of the degenerating arteries. Furthermore, "atherosclerotic plaques" occasionally break through the intima of the blood vessel and protrude into the lumen. The presence of such a rough surface inside a vessel initiates the clotting process. When a small clot has developed, platelets become entrapped and cause more clot to develop until the vessel is plugged; or the clot breaks away and plugs a smaller vessel farther downstream. This mechanism causes most coronary occlusions.

If the blood flow to cardiac muscle is diminished beyond a critical level, the muscle not only becomes nonfunctional but actually begins to die. Death of the muscle fibers begins within about one hour in total ischemia and is complete in four to five hours.

VALUE OF REST IN TREATING CORONARY OCCLUSION. The degree of cellular death is determined by the *ratio of the degree of ischemia to the degree of metabolism* of the heart muscle. When the metabolism of the heart muscle is greatly increased, such as during exercise, in severe emotional strain, or as a result of fatigue, the heart musculature needs increased oxygen and other nutrients for sustaining its life. Furthermore, anastomotic blood vessels that supply blood to ischemic areas of the heart must still supply the areas of the heart that they normally supply. When the heart becomes excessively active, the vessels of the normal musculature become greatly dilated, and this allows most of the blood flowing into the coronary vessels to flow through the normal muscle tissue, thus leaving little blood to flow through the small anastomotic channels into the ischemic areas. Consequently, one of the most important factors in the treatment of a patient with acute coronary ischemia is observance of absolute rest. The greater the degree of rest during the first few weeks following sudden coronary occlusion, the less the degree of cellular death.

CAUSES OF DEATH IN SEVERE CORONARY ISCHEMIA

The four major causes of death following coronary occlusion are decreased cardiac output; damming of blood in the pulmonary or systemic veins with death resulting from edema, especially pulmonary edema; fibrillation of the heart; and, occasionally, rupture of the heart.

Decreased Cardiac Output — Systolic Stretch. When some of the cardiac muscular fibers are not functioning at all and others are too weak to contract with great force, the overall pumping ability of the affected ventricle is proportionately depressed. Indeed, the overall pumping strength of the heart is often decreased more than one might expect because of the phenomenon of *systolic stretch,* which is illustrated in Figure 31–5. When the normal portions of the ventricular muscle contract, the ischemic muscle, whether this be dead or simply nonfunctional, instead of contracting is actually forced outward by the pressure that develops inside the ventricle. If this area did not stretch, the pumping force of the ventricle would be directly proportional to the strength of contraction of the remainder of the ventricle. However, owing to stretch of the ischemic area, much of the pumping force of the ventricle is dissipated by bulging of the area of nonfunctional cardiac muscle.

When the heart becomes incapable of contract-

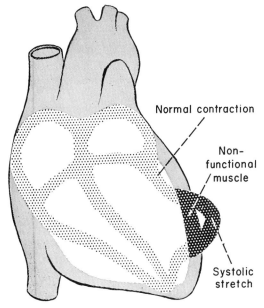

Figure 31–5 Systolic stretch in an area of ischemic cardiac muscle.

ing with sufficient force to pump enough blood into the arterial tree, cardiac failure and death of the peripheral tissues ensue as a result of peripheral ischemia. The dynamics of this was discussed in Chapter 29.

Damming of Blood in the Venous System. When the heart is not pumping blood forward, it must be damming blood in the venous systems of the lungs or the systemic circulation. In the early stages of coronary thrombosis this is not usually as serious as it is later, for, as was discussed in Chapter 29 in relation to cardiac failure, during the first few days after a coronary occlusion progressively more and more fluid collects in the body because of renal shutdown and adds progressively to the venous congestive symptoms. When the congestion becomes intense, death frequently results from pulmonary edema or, rarely, from systemic congestive symptoms.

Rupture of the Ischemic Area. During the first day of an acute coronary infarct there is little danger of rupture of the ischemic portion of the heart, but a few days after a large infarct occurs, the dead muscle fibers begin to degenerate, and the dead heart musculature is likely to become very thin. If this happens, the degree of systolic stretch becomes greater and greater until finally the heart ruptures. In fact, one of the means used in assessing the progress of a severe myocardial infarction is to record by x-ray whether the degree of systolic stretch is getting worse or better.

When a ventricle does rupture, the loss of blood into the pericardial space causes rapid development of *cardiac tamponade* — that is, compression of the heart from the outside by blood collecting in the pericardial cavity. Because the heart is compressed, blood cannot flow into the right atrium with ease, and the patient dies of decreased cardiac output.

Fibrillation of the Ventricles Following Coronary Occlusion. Many persons who die of coronary occlusion die because of ventricular fibrillation. The tendency to develop fibrillation is especially great following a large occlusion, but fibrillation occasionally occurs even following a small occlusion.

There are two especially dangerous periods during which fibrillation is likely to occur. The first of these is during the first 10 minutes after the occlusion occurs. Then there is a period of relative safety, followed by a secondary period of cardiac irritability beginning three to five hours after the occlusion and lasting for another five to six hours.

At least three different factors enter into the tendency for the heart to fibrillate:

First, acute loss of blood supply to the cardiac muscle causes rapid depletion of potassium from the ischemic musculature. This increases the potassium concentration in the extracellular fluids surrounding the cardiac muscle fibers. Experiments in which potassium has been injected into the coronary system have demonstrated that an elevated extracellular potassium concentration increases the irritability of the cardiac musculature.

Normal contraction

Non-functional muscle

Systolic stretch

Second, ischemia of the muscle causes an "injury current," which was described in Chapter 16 in relation to electrocardiograms in patients with acute coronary occlusion. The ischemic musculature cannot repolarize its membranes so that this muscle remains negative with respect to the normal polarized cardiac muscle membrane. Therefore, electrical current flows from this ischemic area of the heart to the normal area and can elicit abnormal impulses which can cause fibrillation.

Third, powerful sympathetic reflexes develop following massive coronary occlusion, principally because the heart does not pump an adequate volume of blood into the arterial tree. The sympathetic stimulation also increases the irritability of the cardiac muscle and thereby predisposes to fibrillation.

THE STAGES OF RECOVERY FROM ACUTE CORONARY THROMBOSIS

The upper part of Figure 31–6 illustrates the effects of acute coronary thrombosis, on the left, in a patient with a small area of muscle ischemia and, on the right, in a patient with a large area of ischemia. When the area of ischemia is small, little or no death of the muscle cells may occur, but part of the muscle often does become temporarily nonfunctional because of inadequate nutrition to support muscle contraction.

When the area of ischemia is large, some of the muscle fibers in the very center of the area die rapidly, within a few hours in an area of total cessation of coronary blood supply. Immediately around the dead area is an area that is nonfunctional because of failure of conduction or contraction. Then, extending circumferentially around the nonfunctional area, is an area that is still contracting, but weakly so because of mild ischemia. In this mildly ischemic area the small arterial collaterals are sufficient to supply the cardiac musculature, provided that the patient is kept at rest, but even this area may become nonfunctional when the coronary blood flow is diverted to normal musculature during exercise. This again illustrates the necessity for rest in patients following coronary heart attacks.

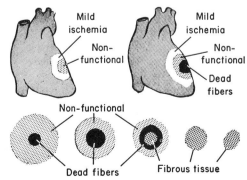

Figure 31–6 *Top:* small and large areas of coronary ischemia. *Bottom:* stages of recovery from myocardial infarction.

Replacement of Dead Muscle by Scar Tissue. In the lower part of Figure 31–6, the various stages of recovery following a coronary occlusion are illustrated. Shortly after the occlusion, the muscle fibers die only in the very center of the ischemic area, but during the ensuing days this area of dead fibers grows because many of the marginal fibers finally succumb to the prolonged ischemia. At the same time, owing to the enlargement of the collateral arterial channels, the nonfunctional area of the heart muscle becomes smaller and smaller. After approximately two to three weeks most of the nonfunctional area of muscle becomes functional again. In the meantime, fibrous tissue begins developing among the dead fibers, for ischemia stimulates growth of fibroblasts and promotes development of greater than normal quantities of fibrous tissue. Therefore, the dead muscular tissue is gradually replaced by fibrous tissue. Then, because it is a general property of fibrous tissue to undergo progressive elastomeric contraction and dissolution, the size of the fibrous scar becomes smaller and smaller over a period of several months to a year.

During progressive recovery of the infarcted area of the heart, the development of a strong, fibrous scar which becomes smaller and smaller finally stops the original systolic stretch, and the functional musculature once again becomes capable of exerting its entire force for pumping blood rather than for stretching the dead area of the heart. Furthermore, the normal areas of the heart gradually hypertrophy to compensate at least partially for the lost cardiac musculature. By these means the heart recovers.

FUNCTION OF THE HEART FOLLOWING RECOVERY FROM CORONARY OCCLUSION

Occasionally, a heart that has recovered from a large coronary occlusion returns to full functional capability, but more frequently its pumping capability is decreased below that of a normal heart. This does not mean that the person is necessarily a cardiac invalid or that his cardiac output is depressed below normal, because the normal person's heart is capable of pumping 400 per cent more blood per minute than his body requires – that is, he has a "cardiac reserve" of 400 per cent. Even when the cardiac reserve is reduced to as little as 100 per cent, the person can still perform normal activity of a quiet, restful type, but he cannot perform strenuous exercise that would rapidly overload his heart.

The effect of coronary occlusion and other myocardial debilities on cardiac output was discussed in Chapter 29 in relation to the regulation of cardiac output in cardiac failure.

PAIN IN CORONARY DISEASE – ANGINA PECTORIS

Normally, a person cannot "feel" his heart, but ischemic cardiac muscle does exhibit pain sensation. Exactly what causes this pain is not known, but it is believed that ischemia causes the muscle to release acidic or other pain-promoting products, such as histamine or kinins, that are not removed rapidly enough by the slowly moving blood. The high concentrations of these abnormal products then stimulate the pain endings in the cardiac muscles, and pain impulses are conducted through the sympathetic afferent nerve fibers into the central nervous system. The pain is usually referred to surface areas of the body, mainly the left arm and left shoulder; this will be discussed in detail in Chapter 50.

Ischemic pain that occurs intermittently from the heart is known as *angina pectoris,* and it occurs whenever the load on the heart becomes too great in relation to the blood flow through the coronary vessels. In general, most persons who have chronic angina pectoris feel the pain only when they exercise or when they experience emotions that increase the metabolism of the heart. However, some patients have such severe and lasting cardiac ischemia that pain is present all the time.

SURGICAL TREATMENT OF CORONARY DISEASE

Treatment of the Pain of Angina Pectoris. For treating intractable angina pectoris the sympathetic chain is occasionally removed from T-2 through T-5 to block the pathway of cardiac pain fibers into the spinal cord. Such an operation occasionally is successful when performed only on the left side, but often it is necessary to remove the sympathetic chain on both sides of the vertebral column.

Treatment of Ischemia. Various attempts have been made to create additional collateral channels of blood flow into the heart. One of the most promising procedures has been to implant the internal mammary artery, freely bleeding at its end, into a tunnel in the myocardium. Collaterals develop to the coronaries and can supply as much as 30 per cent of the coronary blood flow. About half of the patients treated in this way receive significant benefit from the operation.

REFERENCES

Afonso, S.: Coronary vasodilator responses to hypoxia and induced tachycardia before and after lidoflazine. *Amer. J. Physiol., 216:*297, 1969.

Beneken, J. E.; Guyton, A. C.; and Sagawa, K.: Coronary perfusion pressure and left ventricular function. *Pfluegar. Arch. Ges. Physiol., 305:*76, 1969.

Berne, R. M.: Regulation of coronary blood flow. *Physiol. Rev., 44:*1, 1964.

Blumgart, H.: Symposium on Coronary Heart Disease. 2nd ed., American Heart Association, 1967.

Brandfonbrener, M.: Left ventricular blood supply from sources other than left coronary artery. *J. Appl. Physiol., 27:*313, 1969.

Coffman, J. D., and Gregg, D. E.: Oxygen metabolism and oxygen debt repayment after myocardial ischemia. *Amer. J. Physiol., 201:*881, 1961.

Domenech, R. J.; Hoffman, J. I. E.; Noble, M. I. M.; Saunders, K. B.; Henson, J. R.; and Subijanto, S.: Total and regional coronary blood flow measured by radioactive microspheres in conscious and anesthetized dogs. *Circ. Res., 25:*581, 1969.

Driscol, T. E., and Eckstein, R. W.: Coronary inflow and outflow responses to coronary artery occlusion. *Circ. Res., 20:*485, 1967.

Elliot, E. C.; Jones, E. L.; Bloor, C. M.; Leon, A. S.; and Gregg, D. E.: Day-to-day changes in coronary hemodynamics secondary to constriction of circumflex branch of left coronary artery in conscious dogs. *Circ. Res., 22:*237, 1968.

Elzinga, W. E.: Ameroid constrictor: uniform closure rates and a calibration procedure. *J. Appl. Physiol., 27:*419, 1969.

Fam, W. M., and McGregor, M.: Pressure-flow relationships in the coronary circulation. *Circ. Res., 25:*293, 1969.

Feigl, E. O.: Parasympathetic control of coronary blood flow in dogs. *Circ. Res., 25:*509, 1969.

Feigl, E. O.: Sympathetic control of coronary circulation. *Circ. Res., 20:*262, 1967.

Gerlings, E. D.; Miller, D. T.; and Gilmore, J. P.: Oxygen availability: a determinant of myocardial potassium balance. *Amer. J. Physiol., 216:*559, 1969.

Gregg, D. E.: Coronary Circulation in Health and Disease. Philadelphia, Lea & Febiger, 1950.

Harris, A. S., and Bisteni, A.: Effects of sympathetic blockade drugs on ventricular tachycardia resulting from myocardial infarction. *Amer. J. Physiol., 181:*559, 1955.

Howe, B. B., and Winbury, M. M.: Effect of acute coronary occlusion on myocardial nutritional circulation in swine. *Amer. J. Physiol., 216:*1069, 1969.

Joyce, E. E., and Gregg, D. E.: Coronary artery occlusion in the intact unanesthetized dog: intercoronary reflexes. *Amer. J. Physiol., 213:*64, 1967.

Kline, H. J.; Marano, A. J.; Johnson, C. D.; Goodman, P.; Jacobson, J. H., II; and Kuhn, L. A.: Hemodynamic and metabolic effects of hyperbaric oxygenation in myocardial infarction. *J. Appl. Physiol., 28:*256, 1970.

Levy, M. N., and Chansky, M.: Collateral circulation after coronary artery constriction. *Amer. J. Physiol., 208:*144, 1965.

Olsson, R. A., and Gregg, D. E.: Metabolic responses during myocardial reactive hyperemia in the unanesthetized dog. *Amer. J. Physiol., 208:*231, 1965.

Pitt, B.; Sugishita, Y.; and Gregg, D. E.: Coronary hemodynamic effects of calcium in the unanesthetized dog. *Amer. J. Physiol., 216:*1456, 1969.

Rees, J. R., and Redding, V. J.: Experimental myocardial infarction in the dog: comparison of myocardial blood flow within, near, and distant from infarct. *Circ. Res., 25:*161, 1969.

Rubio, R., and Berne, R. M.: Release of adenosine by the normal myocardium in dogs and its relationship to the regulation of coronary resistance. *Circ. Res., 25:*407, 1969.

Rubio, R.; Berne, R. M.; and Katori, M.: Release of adenosine in reactive hyperemia of the dog heart. *Amer. J. Physiol., 216:*56, 1969.

Sarnoff, S. J.; Braunwald, E.; Welch, G. H., Jr.; Case, R. B.; Stainsby, W. N.; and Macruz, R.: Hemodynamic determinants of oxygen consumption of the heart with special reference to the tension-time index. *Amer. J. Physiol., 192:*148, 1958.

Stone, H. L.; Bishop, V. S.; and Guyton, A. C.: Cardiac function after embolization of coronaries with microspheres. *Amer. J. Physiol., 204:*16, 1963.

Stone, H. L.; Bishop, V. S.; and Guyton, A. C.: Progressive changes in cardiovascular function after unilateral heart irradiation. *Amer. J. Physiol., 206:*289, 1964.

Sugimoto, T.; Sagawa, K.; and Guyton, A. C.: Quantitative effect of low coronary pressure on left ventricular performance. *Jap. Heart J., 9:*46, 1968.

Weglicki, W. B.; Rubenstein, C. J.; Entman, M. L.; Thompson, H. K., Jr.; and McIntosh, H. D.: Effects of hyperbaric oxygenation on myocardial blood flow and myocardial metabolism in the dog. *Amer. J. Physiol., 216:*1219, 1969.

Weisfeldt, M. L., and Shock, N. W.: Effect of perfusion pressure on coronary flow and oxygen usage of nonworking heart. *Amer. J. Physiol., 218:*95, 1970.

CHAPTER 32

Blood Flow Through Special Areas of the Body

The blood flow in many special areas of the body, such as the lungs and the heart, has already been discussed in previous chapters, and uterine and fetal blood flow during reproduction will be discussed later in the text. In the present chapter the characteristics of blood flow in some of the other important circulatory systems, such as the brain, the splanchnic system, the muscles, and the skin, are presented.

CEREBRAL CIRCULATION

PHYSIOLOGIC ANATOMY OF BLOOD FLOW THROUGH THE BRAIN

Blood flows into the brain through the two *carotid arteries* and, to a less extent, the two *vertebral arteries.* The vertebral arteries coalesce to form the *basilar artery* of the hind-brain, and the basilar artery connects with the two internal carotid arteries, which also connect with each other, by way of the *circle of Willis.* Therefore, an adequate collateral circulation exists among the large vessels to the brain. Because of these large collateral connections, both carotid arteries can usually be occluded in young persons without any detrimental effects to the function of the cerebrum. In older persons, however, the large arteries at the base of the brain are often so sclerotic that blocking both common carotids results in considerably diminished blood supply, thus causing temporary or permanent impairment of cerebral function.

NORMAL RATE OF CEREBRAL BLOOD FLOW

The normal blood flow through brain tissue averages 50 to 55 ml. per 100 grams of brain per minute. For the entire brain of the average adult, this is approximately 750 ml. per minute, or 15 per cent of the total resting cardiac output. This blood flow through the brain, even under extreme conditions, usually does not vary greatly from the normal value be-

cause the control systems are especially geared to maintain constant cerebral blood flow. Some exceptions to this occur when the brain is subjected to excess carbon dioxide or extreme lack of oxygen, as is discussed subsequently.

REGULATION OF CEREBRAL CIRCULATION

Regulation in Response to Excess Carbon Dioxide. The rate of blood flow through the brain is regulated principally by the concentration of carbon dioxide in the cerebral tissues. Excess carbon dioxide causes marked vasodilation. Therefore, any time the cerebral blood flow becomes sluggish, causing a buildup of carbon dioxide in the cerebral blood, the blood flow automatically increases, thereby correcting this condition. Here again, one finds an important local regulatory mechanism for maintenance of blood flow to the brain in proportion to its metabolism.

Value of Carbon Dioxide Control of the Cerebral Blood Flow. Figure 32–1 illustrates the relationship between P_{CO_2} in the arterial blood and the cerebral blood flow, showing that doubling the normal P_{CO_2}, such as by breathing 7 per cent CO_2, approximately doubles the

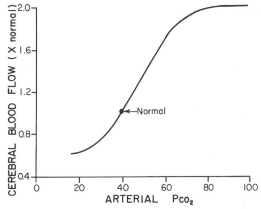

Figure 32–1 Relationship between arterial P_{CO_2} and cerebral blood flow.

blood flow. This is the most powerful single stimulus known to affect cerebral blood flow.

The excitability of neurons is greatly influenced by carbon dioxide concentration in the cerebral fluids; increased carbon dioxide concentration depresses neuronal activity, and diminished carbon dioxide concentration increases neuronal activity. Therefore, it is fortunate that an increase in carbon dioxide concentration increases the blood flow, which in turn carries carbon dioxide away from the brain tissues at an increased rate. Conversely, decreased carbon dioxide reduces the blood flow and allows the carbon dioxide concentration to rise toward normal.

Oxygen Deficiency as a Regulator of Cerebral Blood Flow. The rate of metabolism and also the rate of oxygen utilization by the brain are extremely constant under many different conditions, varying little during intense mental activity, during muscular exercise, or even during sleep. Indeed, the rate of oxygen utilization by brain tissue remains within a few per cent of 3.5 ml. of oxygen per 100 gm. of brain tissue per minute. If ever the blood flow to the brain becomes insufficient to supply this amount of oxygen, the oxygen deficiency mechanism for vasodilatation, present in almost all tissues of the body, immediately causes vasodilatation, allowing the transport of oxygen to the cerebral tissues to return toward normal. Thus, this autoregulatory mechanism is much the same as that which exists in the coronary and muscle circulations and in many other circulatory areas of the body. The need for constant excitability of the neurons—never depressed and never overexcited—makes this mechanism of special importance in the cerebral circulation.

A decrease in the oxygen saturation of the arterial blood to about 75 per cent increases the cerebral blood flow approximately 40 per cent, thus showing that oxygen deficiency within the physiologic range is perhaps one half as potent as carbon dioxide excess in increasing cerebral blood flow. At very low oxygen saturations, oxygen deficiency probably becomes an even more potent vasodilator than carbon dioxide excess.

Effect of Cerebral Activity on Blood Flow. Only rarely does neuronal activity increase sufficiently to increase the overall rate of metabolism in the brain. And, in those few instances in which this does occur, the cerebral blood flow increases only a moderate amount. For instance, a convulsive attack in an animal, which causes extreme activity throughout the entire brain, can result in an overall increase in cerebral blood flow of as much as 50 per cent. On the other hand, administration of anesthetics sometimes reduces brain metabolism and also cerebral blood flow as much as 30 to 40 per cent.

Figure 32–2 Increase in blood flow to the occipital regions of the brain when a light is flashed in the eyes of an animal.

Blood flow in localized portions of the brain can increase as much as 40 to 50 per cent as a result of intense localized neuronal activity even though the total cerebral flow is hardly affected. This effect is illustrated in Figure 32-2, which shows the effect of shining an intense light into one eye of a cat; this results in increased blood flow in the occipital cortex that becomes excited.

Effect of Arterial Pressure on Cerebral Blood Flow. Figure 32–3 shows cerebral blood flow measured in human beings having different blood pressures as a result of different clinical abnormalities. This figure shows especially the extreme constancy of cerebral blood flow in different persons who had widely varying arterial pressures between the limits of 60 and 180 mm. Hg mean arterial pressure. Cerebral blood flow suffers only when the arterial pressure falls below approximately the 60 mm. Hg mark.

Even when the arterial pressure is changed acutely from one pressure to another, the flow changes only slightly with changes in pressure in the physiologic range. This illustrates how well the autoregulatory mechanism functions in the cerebral circulation.

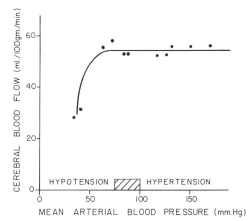

Figure 32–3 Relationship of mean arterial pressure to cerebral blood flow in normotensive, hypotensive, and hypertensive persons. (Modified from Lassen: *Physiol. Rev., 39:*183, 1959.)

Minor Importance of Autonomic Nerves in Regulating Cerebral Blood Flow. Sympathetic nerves from the cervical sympathetic chain pass upward along the cerebral arteries to supply at least some of the cerebral vasculature. Also, parasympathetic fibers from the great superficial petrosal and facial nerves supply some of the vessels. However, transection of either the sympathetic or parasympathetic nerves causes no measurable effect on cerebral blood flow. On the other hand, stimulation does affect blood flow slightly, the sympathetics causing mild vasoconstriction of the large vessels but not of the small vessels, and the parasympathetics causing mild vasodilatation. It remains to be proved that either of these has any functional importance in regulation of cerebral blood flow.

MEASUREMENT OF CEREBRAL BLOOD FLOW IN THE HUMAN BEING

Cerebral blood flow in the human being is measured in the same way that coronary flow is measured, as was explained in the preceding chapter. The person breathes a certain concentration of nitrous oxide for a period of 10 minutes. Simultaneous blood samples are removed approximately each minute from (a) a systemic artery and (b) the jugular bulb where the venous blood flows out of the brain. The A-V difference of nitrous oxide is calculated by the procedure discussed in the previous chapter and illustrated in Figure 31-2. The uptake of nitrous oxide in the brain is calculated from the known partition coefficient for nitrous oxide in the brain. From these values, the blood flow to the brain can be calculated by applying the Fick principle – that is, by dividing the A-V nitrous oxide difference into the nitrous oxide uptake by the brain. The Fick principle was described in detail in Chapter 27 in relation to measurement of cardiac output.

THE SPLANCHNIC CIRCULATION

A large share of the cardiac output flows through the vessels of the intestines and through the spleen, finally coursing into the portal venous system and then through the liver, as illustrated in Figure 32-4. This is called the portal circulatory system, and it, plus the arterial blood flow into the liver, is called the splanchnic circulation.

BLOOD FLOW THROUGH THE LIVER

About 1100 ml. of portal blood enter the liver each minute. This flows through the *hepatic sinuses* in close contact with the cords of liver parenchymal cells. Then it enters the *central veins* of the liver and from there flows into the vena cava.

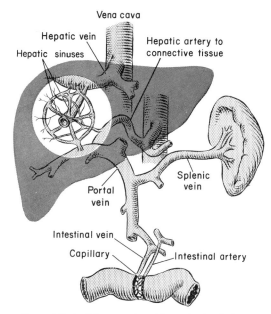

Figure 32-4 The portal and hepatic circulations.

In addition to the portal blood flow, approximately 350 ml. of blood flows into the liver each minute through the hepatic artery, making a total hepatic flow of almost 1500 ml. per minute, or an average of 29 per cent of the total cardiac output. This arterial blood flow maintains nutrition of the connective tissue and especially of the walls of the bile ducts. Therefore, loss of hepatic arterial blood flow can be lethal because this often causes necrosis of the basic liver structures. The blood from the hepatic artery, after it supplies the structural elements of the liver, empties into the hepatic sinuses to mix with the portal blood.

Reservoir Function of the Liver. Because the liver is an expandable and compressible organ, large quantities of blood can be stored in its blood vessels. Its normal blood volume, including both that in the hepatic veins and hepatic sinuses, is about 650 ml., or 13 per cent of the total blood volume. However, when high pressure in the right atrium causes back pressure in the liver, the liver expands, and as much as 1 liter of extra blood occasionally is thereby stored in the hepatic veins and sinuses. This occurs especially in cardiac failure with peripheral congestion, which was discussed in Chapter 29.

Thus, in effect, the liver is a large expandable venous organ capable of acting as a valuable blood reservoir in time of excess blood volume and capable of supplying extra blood in times of diminished blood volume.

Permeability of the Hepatic Sinuses. The hepatic sinuses are lined with an endothelium similar to that of the capillaries, but its permeability is extreme in comparison with that of usual capillaries — so much so that even the proteins of the blood diffuse into the extravascular spaces of the liver almost as easily as fluids. Fortunately, the pressure in the sinuses is only 6 to 8 mm. Hg so that most of the proteins that diffuse out of the sinuses also diffuse back in. Yet, the remainder passes into the liver lymphatics, so that liver lymph normally contains almost as great a protein concentration as that of the plasma itself. Also, liver lymph accounts for one-half to two-thirds of the total lymph flow.

This extreme permeability of the liver sinuses brings the fluids of the hepatic blood into extremely close contact with the liver parenchymal cells, thus facilitating rapid exchange of nutrient materials between the blood and the liver cells.

Measurement of Blood Flow Through the Liver in Man. Several indirect methods for measuring blood flow through the liver similar to those used for measuring blood flow through the heart and brain have been developed. The most important of these is the following: The substance bromosulfophthalein, when injected into the blood, is excreted into the gut by the liver, and almost none of it is lost from the blood in any other way. This material is injected continuously into the circulation and blood samples are removed from two points in the circulation: (1) from a catheter inserted into a peripheral vein and (2) from another catheter inserted through a peripheral vein all the way into the hepatic vein. The difference between the concentrations in a usual peripheral vein and in the hepatic vein represents the quantity of bromosulfophthalein removed from each unit of blood passing through the liver. To determine the quantity of bromosulfophthalein removed by the liver each minute, the substance is injected into the circulation continuously and its rate of injection is adjusted so that its concentration in the blood is neither increasing nor decreasing. When this is achieved the amount being injected is equal to the rate of liver excretion. On dividing the A-V difference into the rate being secreted, one determines the total liver blood flow per minute.

The Blood Cleansing Function of the Liver. Blood flowing through the intestinal capillaries picks up many bacteria from the intestines. Indeed, a sample of blood from the portal system almost always grows colon bacilli when cultured, whereas growth of colon bacilli from blood in the systemic circulation is extremely rare. Special high-speed motion pictures of the action of Kupffer cells, the large phagocytic cells that line the hepatic sinuses, have demonstrated that these cells can cleanse blood extremely efficiently as it passes through the sinuses; when a bacterium comes into momentary contact with a Kupffer cell, in less than 0.01 second the bacterium passes inward through the wall of the Kupffer cell to become permanently lodged therein until it is digested. Probably not over 1 per cent of the bacteria entering the portal blood from the intestines succeeds in passing through the liver into the systemic circulation.

The Metabolic "Buffering" Function of the Liver. The function of blood flow through the intestines is to absorb into the body fluids the products of digestion. These products, after entering the portal system, pass through the venous sinuses of the liver in close proximity to the liver parenchymal cells. Before the blood leaves the sinuses, the liver cells remove most of the glucose absorbed from the intestinal tract, most of the amino acids, and probably most of the fat that occasionally is absorbed in special forms into the portal blood. Removal of these substances from the portal blood prevents excessive increase in their concentrations in the systemic blood immediately after a meal. Then, during the following 8 to 24 hours or more, the removed products of digestion are released by the liver into the systemic circulatory system so that they flow continually at well-controlled blood concentrations to the remainder of the body. Also, the liver changes some of these products of digestion into new forms so that they can be used to better advantage by the remainder of the body, as will be discussed later in the chapters on metabolism.

BLOOD FLOW THROUGH THE INTESTINAL VESSELS

About four-fifths of the portal blood flow originates in the intestines and stomach (about 850 ml. per minute), and the remaining one-fifth originates in the spleen and pancreas.

Control of Gastrointestinal Blood Flow. Blood flow in the gastrointestinal tract seems to be controlled in almost exactly the same way as in most other areas of the body: that is, mainly by local regulatory mechanisms. Furthermore, blood flow to the mucosa and submucosa, where the glands are located and where absorption occurs, is controlled separately from blood flow to the musculature. When glandular secretion increases, so does mucosal and submucosal blood flow. Likewise, when motor activity of the gut increases, blood flow in the muscle layers increases.

However, the precise mechanisms by which alterations in gastrointestinal activity alter the blood flow are not completely understood. It is known that decreased availability of oxygen to the gut increases local blood flow in the same way that this occurs elsewhere in the body, so that local regulation of blood flow in the gut might occur entirely secondarily to changes in metabolic rate. On the other hand, it has been claimed that stimulation of most gastrointestinal glands causes the glands to release an enzyme that acts on a plasma protein to form the substance bradykinin. The bradykinin then supposedly causes local vasodilatation, thereby increasing the glandular blood flow. Though this mechanism has received much publicity, especially in relation to salivary glands, recent experiments in which the bradykinin mechanism was blocked show that, in spite of the block, salivary blood flow increases just as much as normally during glandular secretion. Therefore, it is doubtful that this mechanism is a correct one, despite its wide acceptance at the present time.

NERVOUS CONTROL OF GASTROINTESTINAL BLOOD FLOW. Stimulation of the parasympathetic nerves to the stomach and lower colon increases local blood flow at the same time that it increases glandular secretion. This increased flow probably results from nervous enhancement of glandular activity. On the other hand, parasympathetic stimulation has little effect on blood flow in the small intestine, where glandular secretion is controlled primarily by local mechanisms rather than by mere nerve stimulation.

Sympathetic stimulation, on the other hand, has a direct effect on essentially all blood vessels of the gastrointestinal tract to cause intense vasoconstriction, sometimes reducing intestinal blood flow almost to zero. A major value of sympathetic vasoconstriction in the gut is that it allows shutting off of blood flow when increased flow is needed by the muscles and heart during extreme exercise.

Effect of Gastrointestinal Distention on Intestinal Blood Flow. Blood vessels enter the gut from the mesenteric margin and then pass circumferentially in both directions around the gut. Therefore, when the gut becomes greatly distended, the blood vessels are stretched, which tends especially to occlude the veins. With extreme distention, the increased resistance in the veins elevates the capillary pressure to a very high level, resulting in rapid loss of fluid and protein into the gut as well as severe edema in the gut wall. Thus, a large amount of plasma can be lost from the circulation in a short period of time. Consequently, severe and widespread distention of the gut is a frequent cause of plasma loss shock.

PORTAL VENOUS PRESSURE

The liver offers a moderate amount of resistance to the blood flow from the portal system to the vena cava. As a result, the pressure in the portal vein averages 10 mm. Hg, which is considerably higher than the almost zero pressure in the vena cava. Because of this high portal venous pressure, the pressures in the portal venules and capillaries have a much greater tendency to become abnormally high than is true elsewhere in the body.

Blockage of the Portal System. Frequently, extreme amounts of fibrous tissue develop within the liver structure, destroying many of the parenchymal cells and eventually contracting around the blood vessels, thereby greatly impeding the flow of portal blood through the liver. This disease process is known as *cirrhosis of the liver,* and it may occur secondary to alcoholism, ingestion of poisons such as carbon tetrachloride, virus diseases such as infectious hepatitis, or infectious processes in the bile ducts.

The portal system is also occasionally blocked by a large clot developing in the portal vein or in its major branches.

When the portal system is suddenly blocked, the return of blood from the intestines and spleen to the systemic circulation is tremendously impeded, the capillary pressure rising as much as 20 to 30 mm. Hg, and the patient usually dies within a few hours because of excessive loss of fluid from the capillaries into the lumina and walls of the intestines.

COLLATERAL CIRCULATION FROM THE PORTAL VEINS TO THE SYSTEMIC VEINS. When portal blood flow is blocked slowly rather than suddenly, such as occurs in slowly developing cirrhosis of the liver, large collateral vessels develop between the portal veins and the systemic veins. The most important of these are collaterals from the splenic to the esophageal veins and these collaterals frequently become so large that they protrude deeply into the lumen of the esophagus and are then called *esophageal varicosities.* The esophageal mucosa overlying these varicosities eventually becomes eroded; and in many if not most of these patients, the erosion finally penetrates all the way into the varicosity and causes the person to bleed severely, often to bleed to death.

ASCITES AS A RESULT OF PORTAL OBSTRUCTION. Ascites is free fluid in the peritoneal cavity. It results from exudation of fluid either from the surface of the liver or from the surfaces of the gut and its mesentery. Ascites usually will develop only in case outflow of blood from the liver into the inferior vena cava is blocked. This causes extremely high pressure in the liver sinusoids, which in turn causes fluid to weep from the surfaces of the liver. The weeping fluid is almost pure plasma, containing tremendous quantities of protein. The protein, because it causes a high colloid osmotic pressure in the abdominal fluid, then pulls (by osmosis) additional fluid from the surfaces of the gut and mesentery.

On the other hand, obstruction of the portal vein, without directly involving the liver, rarely causes ascites. If obstruction occurs acutely, the person is likely to die of shock within hours because of plasma loss into the gut; and if it occurs slowly, collateral vessels can usually develop rapidly enough to prevent ascites.

THE SPLENIC CIRCULATION

The Spleen as a Reservoir. The capsule of the spleen in many lower animals contains large amounts of smooth muscle; and sympathetic stimulation causes intense contraction of this capsule, whereas sympathetic inhibition results in considerable splenic expansion with consequent storage of blood.

In man the splenic capsule is nonmuscular, but even so, constriction or dilatation of vessels within the spleen can still cause the spleen to store several hundred milliliters of blood at times, and then, under the influence of sympathetic stimulation, to express most of this blood into the general circulation. Unfortunately, these effects in man are poorly understood.

As illustrated in Figure 32–5, two areas exist in the spleen for the storage of blood; these are the venous sinuses and the pulp. Small vessels flow directly into the venous sinuses, and when the spleen distends, the venous sinuses swell, storing large quantities of blood.

In the splenic pulp, the capillaries are very permeable, so that much of the blood passes first into the pulp and then oozes through this before entering the venous sinuses. As the spleen enlarges, many cells become stored in the pulp. Therefore, the net quantity of red blood cells in the general circulation decreases slightly when the spleen enlarges. The spleen can store enough cells that splenic contraction can cause the hematocrit of the systemic blood to increase as much as 3 to 4 per cent.

This ability of the spleen to store and release

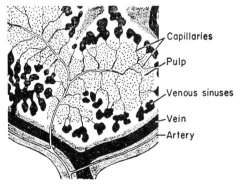

Figure 32–5 The functional structures of the spleen. (Modified from Bloom and Fawcett: Textbook of Histology.)

blood is important, at least in lower animals, in times of stress such as during strenuous exercise, for the circulatory system then needs an increased volume of blood so that rapid return of blood to the heart can be maintained. At the same time, the blood that flows through the muscles needs to carry larger quantities of oxygen than normally. Therefore, the increased hematocrit of the circulating blood resulting from splenic contraction is also an aid to the body during these periods of stress.

The Blood Cleansing Function of the Spleen— Removal of Old Cells. Blood passing through the splenic pulp before it enters the sinuses undergoes thorough squeezing. Indeed, it is supposed that the red blood cells re-enter the venous sinuses from the pulp by "diapedesis"; that is, by squeezing through pores much smaller than the size of the cells themselves. Under these circumstances it is to be expected that fragile red blood cells would not withstand the trauma. For this reason, many of the red blood cells destroyed in the body have their final demise in the spleen. After the cells rupture, the released hemoglobin and the cell stroma are ingested by the reticuloendothelial cells of the spleen.

RETICULOENDOTHELIAL CELLS OF THE SPLEEN. The pulp of the spleen contains many large phagocytic reticuloendothelial cells, and the venous sinuses are lined with similar cells. These cells act as a cleansing system for the blood, similar to that in the venous sinuses of the liver. When the blood is invaded by infectious agents, the reticuloendothelial cells of the spleen rapidly remove debris, bacteria, parasites, etc. Also, in many infectious processes the spleen enlarges in the same manner that lymph glands enlarge and performs its cleansing function even more adequately.

Much of the spleen is filled with *white pulp,* which is in reality a large quantity of lymphocytes and plasma cells. These function in exactly the same way in the spleen as in the lymph glands to cause either humoral or lymphocytic immunity against toxins, bacteria, and so forth, as described in Chapter 10.

The Spleen as a Hemopoietic Organ. During fetal life, the splenic pulp produces blood cells in exactly the same manner that the red bone marrow in the adult produces cells. As the normal fetus approaches parturition, the spleen normally loses this ability to produce cells, but, in some diseases, the spleen continues to produce cells even after birth. For instance, in the disease *erythroblastosis fetalis,* which results from excessive destruction of red blood cells by abnormal antibodies in the plasma, as discussed in Chapter 8, the fetus must produce 10 or more times as many red blood cells as normally. As a result, the hemopoietic function of the spleen persists into babyhood.

BLOOD FLOW THROUGH THE SKELETAL MUSCLES

RATE OF BLOOD FLOW THROUGH THE MUSCLES

During rest, blood flow through skeletal muscle averages 4 to 7 ml. per 100 grams of muscle. However, during extreme exercise this rate can increase as much as 15- to 20-fold, rising to more than 100 ml. per 100 grams of muscle.

Intermittent Flow During Muscle Contraction. Figure 32–6 illustrates a study of blood flow changes in the calf muscles of the human leg during strong rhythmic contraction. Note that the flow increases and decreases with each muscle contraction, decreasing during the contraction phase and increasing between contractions. At the end of the rhythmic contractions, the blood flow remains very high for a minute or so and then gradually fades toward normal.

The cause of the decreased flow during sustained muscle contraction is compression of the blood vessels by the contracted muscle. During strong *tetanic* contraction, blood flow can be almost totally stopped.

Opening of Muscle Capillaries During Exercise. During rest, only about 10 per cent of the muscle capillaries are open. But during strenuous exercise all the capillaries open up, which can be demonstrated by studying histologic specimens removed from active muscles. It is this opening up of dormant capillaries that allows most of the increased blood flow. It also diminishes the distance that oxygen and other nutrients must diffuse from the capillaries to the muscle fibers.

Figure 32–6 Effect of muscle exercise on blood flow in the calf of a leg during strong rhythmic contraction. The blood flow was much less during contraction than between contractions. (From Barcroft and Dornhorst: *J. Physiol.,* *109:*402, 1949.)

CONTROL OF BLOOD FLOW THROUGH THE SKELETAL MUSCLES

Local Regulation. The tremendous increase in blood flow that occurs during skeletal muscle contraction is caused primarily by local effects in the muscles acting directly on the arterioles to cause vasodilatation. These effects occur even after all sympathetic nerves to the muscle have been cut. Furthermore, the degree of dilatation is equally as great (except at the very beginning of muscle activity, as will be noted later) without these nerves as it is when they are present.

The local increase in blood flow during muscle contraction is probably caused by several different factors all operating at the same time. One of the most important of these is reduction of dissolved oxygen in the muscle tissues. That is, during muscle activity the muscle utilizes oxygen very rapidly, thereby decreasing its concentration in the tissue fluids. This in turn causes vasodilatation either because the vessel walls cannot maintain contraction in the absence of oxygen or because oxygen deficiency causes release of some vasodilator material. The vasodilator material that has been suggested most widely in recent years has been adenosine.

Other vasodilator substances released during muscle contraction include potassium ions, acetylcholine, adenosine triphosphate, lactic acid, and carbon dioxide. Unfortunately, we still do not know quantitatively how much role each of these plays in increasing muscle blood flow during muscle activity.

Finally, an additional factor that increases blood flow during muscle contraction is elevation of arterial pressure, an effect that usually accompanies muscle exercise. Because even slight elevation in pressure stretches the arterioles considerably, as little as a 20 to 30 mm. Hg rise in pressure can often double the blood flow.

Nervous Control of Muscular Blood Flow. In addition to the basic autoregulatory mechanism, the skeletal muscles are also provided with both sympathetic vasoconstrictor fibers and sympathetic vasodilator fibers.

SYMPATHETIC VASOCONSTRICTOR FIBERS. The sympathetic vasoconstrictor fibers secrete norepinephrine and when maximally stimulated can decrease blood flow through the muscles to about one-fourth normal. This represents rather poor vasoconstriction in comparison with that caused by sympathetic nerves in some other areas of the body in which blood flow can be almost completely blocked. Yet under some conditions even this degree of vasoconstriction may be of physiological importance, such as during shock and other periods of stress when it would be desirable to reduce blood flow through the many muscles of the body.

In addition to the norepinephrine secreted at the sympathetic vasoconstrictor nerve endings, the adrenal medullae secrete large amounts of additional norepinephrine and epinephrine into the circulating blood during strenuous exercise. The circulating norepinephrine acts on the muscle vessels to cause a vasoconstrictor effect similar to that caused by direct sympathetic nerve stimulation. The epinephrine, on the other hand, often has a vasodilator effect. It is believed that this results from the fact that epinephrine excites the *beta* receptors of the vessels, which are vasodilator receptors, in contrast to the *alpha* constrictors excited by the norepinephrine. These receptors are discussed in Chapter 59.

SYMPATHETIC VASODILATOR FIBERS. The sympathetic vasodilator fibers to the skeletal muscles secrete acetylcholine and on maximal stimulation can increase blood flow by 400 per cent. These vasodilator fibers are activated by a special nervous pathway beginning in the cerebral cortex in close association with the motor areas for control of muscular activity and passing downward through the hypothalamus and brain stem into the spinal cord, as was explained in Chapter 25. When the motor cortex initiates muscle activity it simultaneously excites the vasodilator fibers to the active muscles, and vasodilatation occurs immediately, several seconds before the local regulatory vasodilatation can take place. Thus, it seems that this vasodilator system has the important *function of initiating extra blood flow through the muscles at the onset of muscular activity.*

The vasodilator fibers are probably of little or no importance in maintaining increased blood flow during prolonged muscular activity, for both in lower animals and in human beings the final degree of vasodilatation which occurs during muscular exercise is not discernibly different when the muscle is normally innervated with sympathetics or denervated. Therefore, it is evident that the local regulatory mechanism by itself is capable of providing essentially all the blood flow that is needed.

EFFECT OF MUSCULAR BLOOD FLOW ON CARDIAC OUTPUT

When muscles become active in the body, the cardiac output rises simultaneously to supply the increased flow of blood through the muscles. The cause of this increased cardiac output is three-fold: First, the sympathetic centers of the medulla become strongly stimulated during muscular exercise. Sympathetic impulses in turn constrict many blood reservoirs, thereby elevating the mean circulatory pressure, which increases venous return.

Second, contraction of the muscles compresses the intramuscular vessels, which also increases

the venous return to the heart. As much as 8 per cent of the total blood volume can be squeezed out of the muscles, particularly from the veins, into the heart and lungs, which in turn makes the heart pump with greatly increased force of contraction.

Third, and most important of all, the vasodilatation of the muscle vessels themselves causes blood to flow rapidly into the veins, thereby increasing right atrial pressure and also increasing cardiac output. The increase in cardiac output in exercise was discussed in detail in Chapter 27.

CIRCULATION IN THE SKIN

PHYSIOLOGIC ANATOMY OF THE CUTANEOUS CIRCULATION

Circulation through the skin subserves two major functions: first, *nutrition of the skin tissues* and, second, *conduction of heat* from the internal structures of the body to the skin so that the heat can be removed from the body. To perform these two functions the circulatory apparatus of the skin is characterized by two major types of vessels, illustrated diagrammatically in Figure 32–7: (1) the usual nutritive arteries, capillaries, and veins and (2) vascular structures concerned with heating the skin, consisting principally of (a) an extensive *subcutaneous venous plexus,* which holds large quantities of blood that can heat the surface of the skin, and (b) in some skin areas, *arteriovenous anastomoses,* which are large vascular communications directly between the arteries and the venous plexuses. The walls of these anastomoses have strong muscular coats innervated by sympathetic vasoconstrictor nerve fibers that secrete norepinephrine. When constricted, they reduce the flow of blood into the venous plexuses to almost nothing; or when maximally dilated, they allow extremely rapid flow of warm blood into the plexuses. The arteriovenous anastomoses are found principally in the volar surfaces of the hands and

Figure 32–7 Diagrammatic representation of the skin circulation.

feet, the lips, the nose, and the ears, which are areas of the body most often exposed to maximal cooling.

Rate of Blood Flow Through the Skin. The rate of blood flow through the skin is among the most variable of those in any part of the body, because the flow required to regulate body temperature changes markedly in response to, first, the rate of metabolic activity of the body, and, second, the temperature of the surroundings. This will be discussed in detail in Chapter 71. The blood flow required for nutrition is slight, so that this plays almost no role in controlling normal skin blood flow. At ordinary skin temperatures, the amount of blood flowing through the skin vessels to subserve heat regulation is about 10 times as much as that needed to supply the nutritive needs of the tissues. But, when the skin is exposed to extreme cold, the blood flow may become so slight that nutrition begins to suffer—even to the extent, for instance, that the fingernails grow considerably more slowly in arctic climates than in temperature climates.

Under ordinary cool conditions the blood flow to the skin is about 0.25 liter/sq. meter of body surface area, or a total of about 400 ml. per minute, in the average adult. On the other hand, when the skin is heated until maximal vasodilatation has resulted, the blood flow can be as much as 7 times this value, or a total of about 2.8 liters per minute, thus illustrating both the extreme variability of skin blood flow and the great drain on cardiac output that can occur under hot conditions. Indeed, many persons with borderline cardiac failure become decompensated in hot weather because of the extra load on the heart and recompensated in cool weather.

REGULATION OF BLOOD FLOW IN THE SKIN

Nervous Control of Cutaneous Blood Flow. Since the principal function of blood flow through the skin is to control body temperature, and since this function in turn is regulated by the nervous system, the blood flow through the skin is principally regulated by nervous mechanisms rather than by local regulation, which is opposite to the regulation in most parts of the body.

TEMPERATURE CONTROL CENTER OF THE HYPOTHALAMUS. Located in the preoptic region of the anterior hypothalamus is a small center that is capable of controlling body temperature. Heating this area causes vasodilation of essentially all the skin vessels of the body and also causes sweating. Cooling of the center causes vasoconstriction and cessation of sweating. The detailed function of this center will be discussed in Chapter 71 in relation to body temperature. The important point in the present discussion is that the hypothalamus controls blood flow through the skin in response to changes in body temperature by two mechanisms: (1) a sympathetic vasoconstrictor mechanism and (2) a sympathetic vasodilator mechanism.

THE SYMPATHETIC VASOCONSTRICTOR MECHANISM. The skin throughout the body is supplied with sympathetic vasoconstrictor fibers that secrete norepinephrine at their endings. This constrictor system is extremely powerful in the feet, hands, lips, nose, and ears, which are the areas most frequently exposed to severe cold and which are also the areas where large numbers of arteriovenous anastomoses are found. At normal body temperature the sympathetic vasoconstrictor nerves keep these anastomoses almost totally closed, but when the body becomes overly heated the number of sympathetic impulses is greatly reduced so that the anastomoses dilate and allow large quantities of warm blood to flow into the venous plexuses, thereby promoting loss of heat from the body.

In the remainder of the body—that is, over the surfaces of the arms, legs, and body trunk—almost no arteriovenous anastomoses are present, but nevertheless vasoconstrictor control of the nutritive vessels can still effect major changes in blood flow. When the body becomes overheated, the sympathetic vasoconstrictor impulses cease, and the blood flow to the skin vessels increases about two-fold.

THE SYMPATHETIC VASODILATOR MECHANISM. When the body temperature becomes excessive and sweating begins to occur, the blood flow through the skin of the forearms and trunk increases an additional three-fold, which occurs as a result of so-called "active" vasodilatation of the vessels. However, the basic mechanism by which this active dilation occurs is not completely known. Some physiologists believe that sympathetic vasodilator fibers that secrete acetylcholine at their endings exist in the skin of human beings. Yet this additional increase in blood flow does not occur except in the presence of sweating, and it does not occur in lower animals that do not have sweat glands. Therefore, another theory is that the sympathetic fibers that secrete acetylcholine in the sweat glands cause a secondary vasodilatation as follows: The increased activity of the sweat glands is postulated to cause these glands to release an enzyme called *kallikrein,* which in turn splits a polypeptide called *bradykinin* from globulin in the interstitial fluids. Bradykinin in turn is a powerful vasodilator substance that could account for the greatly increased blood flow when sweating begins to occur. In opposition to this theory, however, is the fact that inhibition of the bradykinin mechanism does not block the increased blood flow associated with sweating.

Effect of Cold on the Skin Circulation. When cold is applied directly to the skin, the skin vessels constrict more and more down to a temperature of about 15° C., at which point they reach their maximum degree of constriction. This constriction results primarily from increased sensitivity of the vessels to nerve stimulation, but it probably also results from a reflex that passes to the cord and then back to the vessels. At temperatures below 15° C., the vessels begin to dilate. This dilation is caused by a direct local effect of the cold on the vessels themselves — probably paralysis of the contractile mechanism of the vessel wall or block of the nerve impulses coming to the vessels. In any event, at temperatures approaching 0° C. the vessels frequently reach maximum vasodilation. This intense vasodilatation in severe degrees of cold plays a purposeful role in preventing freezing of the exposed portions of the body, particularly the hands and ears.

The Triple Response. When the skin is stroked firmly with a pointed object, three local reactions frequently occur affecting the local skin circulation called the *triple response.* These reactions are development of a red line, a flare, and a linear wheal.

The *red line* is caused by direct local damage to the tissues and probably results from the release of histamine or some other vasodilator substance from the damaged tissues.

The *red flare* develops 20 to 40 seconds after the stroke and results from the *axon reflex.* The axon reflex is caused by stimulation of pain nerve fibers from the skin. The impulses travel up these fibers and then back down the other branches of the same fibers. At the ends of these branches some vasodilator substance is released, which causes vasodilatation for a distance 2 to 3 cm. on each side of the line of stroke. The nature of the vasodilator substance is unknown.

In certain people who have appropriately sensitive skin, a raised *wheal* develops for as much as 1 cm. on each side of the line of stroke, beginning approximately one minute after the scratch and reaching full development within five minutes. This wheal is probably caused by excessive sensitivity of the tissues to histamine or to similar substances released from the damaged tissues. The histamine causes such intense arteriolar vasodilatation and venular constriction that the capillary pressure rises enough to promote transudation of fluid into the tissues.

Local Regulation of Blood Flow in the Skin. Though local regulation of blood flow usually plays almost no role in skin blood flow, it does occur in those few instances in which skin blood flow becomes greatly decreased. For instance, if one sits on one portion of his skin for 30 minutes or more and then removes the pressure, he will note intense reddening of the skin. That is, the blood flow to the skin area has now increased to a marked extent, which is typical *reactive hyperemia* resulting from diminished availability of nutrients to the tissues during the period of compression. Thus, the local regulatory mechanism present in essentially all other tissues of the body is also present in the skin and can be called into play when needed to prevent nutritional damage to the tissues.

Shift of Blood from the Skin to the Remainder of the Circulation in Times of Circulatory Stress. The skin venous plexuses of the entire body, including those of the hands, feet, arms, legs, and trunk, are all strongly supplied with sympathetic vasoconstrictor innervation. In times of circulatory stress, such as during exercise, following severe hemorrhage, or even in states of anxiety, sympathetic stimulation of these venous plexuses can force large quantities of blood, estimated to be as much as 5 to 10 per cent of the blood volume, into the internal vessels. Thus, the subcutaneous veins of the skin act as an important blood reservoir, often providing blood to subserve other circulatory functions when needed.

COLOR OF THE SKIN IN RELATION TO SKIN TEMPERATURE

Much of the skin color is due to the color of the blood in the skin capillaries and veins. Therefore, when the skin is hot and arterial blood is flowing rapidly into these vessels, the skin is usually red. On the other hand, when the skin is cold and the blood is flowing extremely slowly, most of the oxygen is removed for nutritive purposes before the blood can leave the capillaries. Consequently, the capillaries and veins then contain large amounts of dark, deoxygenated blood that gives the skin a bluish hue.

Another cutaneous vascular effect that often affects skin color is severe constriction of the cutaneous vessels, which expresses most of the blood out of the skin into other parts of the circulation. When this occurs the skin takes on the color of the subcutaneous connective tissue, which is composed principally of collagen fibers and has a whitish hue. Thus, in conditions in which the cutaneous vessels become greatly constricted, the skin has an ashen white pallor.

PHYSIOLOGY OF VASCULAR DISEASES OF THE LIMBS

RAYNAUD'S DISEASE

The local vasoconstrictor reflex that occurs normally during exposure to cold becomes especially marked in the limbs of some persons, particularly in many women who live in northern climates. This effect, called *Raynaud's disease,* occurs most frequently in the hands, though occasionally also in the feet. When the hands become even slightly cooled, they are likely to exhibit extreme vascular spasm—so much so that the hands do not receive sufficient flow of blood even for maintaining adequate metabolism. The deep tissues, especially the bones, then develop *ischemic pain* in exactly the same manner that the heart develops pain in angina pectoris and as other tissues develop pain during ischemia. After the ischemia persists for several minutes, *hyperemia* supervenes, causing dilatation of all the vessels of the hands so that the final stage is sometimes *intense reddening* of the hands. Strangely enough, the pain becomes even more severe at this moment, owing either to mobilization of the abnormal metabolic products that cause pain or, perhaps, in part to the greatly increased blood pressure within the capillaries upon dilatation of all the arterioles.

Though Raynaud's disease in its early stages is merely a physiologic entity that passes through the phases of blanching of the skin, pain, and reddening of the skin, after it progresses for years the vessels may become continuously constricted. In severe cases the blood flow can become decreased so much that the fingers become gangrenous.

Sympathectomy for Raynaud's Disease. To treat Raynaud's disease the sympathetic reflex must be blocked. This is done best by removing the sympathetic chain at T-2 and T-3, which interrupts the sympathetic nerve impulses from the spinal cord to the hand. The postganglionic fibers to the hand originate mainly in the stellate ganglion, but the stellate ganglion is not removed, for removal of postganglionic sympathetic fibers allows the blood vessels to become excessively sensitized to circulating norepinephrine and epinephrine. If this happens the hands then exhibit Raynaud's syndrome every time the adrenal glands are excited. This phenomenon of sensitization will be discussed in Chapter 59 in connection with the various reactions of the sympathetic nervous system.

BUERGER'S DISEASE

Buerger's disease, also known as *thromboangiitis obliterans,* is caused by an inflammatory reaction in the triangular sheaths containing the nerve, the artery, and the vein in peripheral areas. This condition occurs most frequently in the legs but occasionally in the arms, and it is much more common among Jewish people than among other peoples. Even though the disease is caused by an inflammatory process that constricts the vessels, it is made much worse by the effects of nicotine incident to the use of tobacco.

The patient with Buerger's disease suffers a symptom known as *intermittent claudication*—that is, when he walks, his calf becomes extremely painful because of ischemic products collecting in his muscles and not removed by adequate blood flow. The ischemic pain is similar to that of Raynaud's disease, and even to the ischemic cardiac pain of angina pectoris.

Sympathectomy for Buerger's Disease. Buerger's disease can be benefited greatly by sectioning the sympathetic nerves to the area afflicted. This is done for a leg by removing the L-1, L-2, and L-3 ganglia of the sympathetic chain. Even though the pathologic changes of Buerger's disease are not caused by excessive sympathetic activity, removal of the sympathetic nerves blocks normal vasomotor tone, which allows vasodilatation in the affected areas. Approximately 90 per cent of the patients who otherwise might lose toes or even legs from Buerger's disease can save these by sympathectomy.

PERIPHERAL ARTERIOSCLEROSIS

Arteriosclerosis, which causes intra-arterial plaques to develop, has already been discussed in connection with the coronary arteries. This same process occurs throughout the arterial tree and not simply in the coronaries. Therefore, in old age it is likely that many of the peripheral arteries will have their intimal surfaces almost completely covered with lipid deposits or calcified plaques, and the intima will be destroyed.

Progressive obliteration of the peripheral arterial tree by arteriosclerosis results in all the symptoms that one would expect. It causes pain in the tissues because of ischemia, and it is likely to cause gangrene and ulceration of the skin because of poor blood supply.

A special difference between arteriosclerotic peripheral vascular disease and either Buerger's disease or Raynaud's disease is that sympathectomy is not nearly so effective for treatment as it is in the other two diseases. This is true probably because this vascular disease is caused by obstruction of the large arteries rather than the small arteries, and the small arteries, which are affected most by the sympathetics, are probably already vasodilated even before the sympathetics are removed.

REFERENCES

Alpert, J. S., and Coffman, J. D.: Effect of intravenous epinephrine on skeletal muscle, skin, and subcutaneous blood flow. *Amer. J. Physiol., 216*:156, 1969.

Barcroft, H.: Circulation in skeletal muscle. *In* Handbook of Physiology. Baltimore, The Williams & Wilkins Co., 1963, Sec. II, Vol. II, p. 1353.

Bevegard, B. S., and Shepherd, J. T.: Regulation of the circulation during exercise in man. *Physiol. Rev., 47*:178, 1967.

Bradley, S. E.: The hepatic circulation. *In* Handbook of Physiology. Baltimore, The Williams & Wilkins Co., 1963, Sec. II, Vol. II, p. 1387.

Cerebral blood flow (Symposium). *Acta Physiol. Scand., 66,* Suppl. 258, 1966.

Chapman, C. B.: The physiology of exercise. *Sci. Amer., 212*:88, 1965.

Cohen, P. J.; Alexander, S. C.; Smith, T. C.; Reivich, M.; and Wollman, H.: Effects of hypoxia and normocarbia on cerebral blood flow and metabolism in conscious man. *J. Appl. Physiol.,* 23:183, 1967.

Dobbing, J.: The blood-brain barrier. *Physiol. Rev., 41*:130, 1961.

Ekelund, L.: Circulatory and respiratory adaptation during prolonged exercise. *Acta Physiol. Scand., 69,* Suppl. 292, 1967.

Folkow, B.: Regional adjustments of intestinal blood flow. *Gastroenterology, 52*:423, 1967.

Gray, S. D., and Staub, N. S.: Resistance to blood flow in leg muscles of dog during tetanic isometric contraction. *Amer. J. Physiol., 213*:677, 1967.

Grayson, J., and Mendel, D.: Physiology of the Splanchnic Circulation. Baltimore, The Williams & Wilkins Co., 1965.

Greenfield, A. D. M.: The circulation through the skin. *In* Handbook of Physiology. Baltimore, The Williams & Wilkins Co., 1963, Sec. II, Vol. II, p. 1325.

Green, H. D., and Kepchar, J. H.: Control of peripheral resistance in major systemic vascular beds. *Physiol. Rev., 39*:617, 1959.

Guntheroth, W. G.; McGough, G. A.; and Mullins, G. L.: Continuous recording of splenic diameter, vein flow, and hematocrit in intact dogs. *Amer. J. Physiol., 213*:690, 1967.

Haddy, F. J.; Molnar, J. I.; Borden, C. W.; and Texter, E. C., Jr.: Comparison of direct effects of angiotensin and other vasoactive agents on small and large blood vessels in several vascular beds. *Circulation, 25*:239, 1962.

Jacobson, E. D.: Recent advances in the gastrointestinal circulation and related areas: comments on a symposium. *Gastroenterology, 52*:332, 1967.

Jacobson, E. D.: Secretion and blood flow in the gastrointestinal tract. *In* Handbook of Physiology. Baltimore, The Williams & Wilkins Co., 1967, Sec. VI, Vol. II, p. 1043.

Jacobson, E. D.: The gastrointestinal circulation. *Ann. Rev. Physiol., 30*:133, 1968.

James, I. M.; Millar, R. A.; and Purves, M. J.: Observations on the extrinsic neural control of cerebral blood flow in the baboon. *Circ. Res., 25*:77, 1969.

Jones, R. D., and Berne, R. M.: Vasodilation in skeletal muscle. *Amer. J. Physiol., 204*:461, 1963.

Kaplan, H. A., and Ford, D. H.: The Brain Vascular System. New York, American Elsevier Publishing Co., Inc., 1966.

Lassen, N. A.: Cerebral circulation. *Proc. Int. Union of Physiol. Sciences,* 6:173, 1968.

Lopez-Majano, V.; Rhodes, B. A.; and Wagner, H. N., Jr.: Arteriovenous shunting in extremities. *J. Appl. Physiol., 27*:782, 1969.

Mellander, S.: Circulation in skeletal muscle. *Proc. Int. Union of Physiol. Sciences,* 6:171, 1968.

Montagna, W., and Lobitz, W. C.: The Epidermis. New York, Academic Press, Inc., 1964.

Myers, S. J., and Sullivan, W. P.: Effect of circulatory occlusion on time to muscular fatigue. *J. Appl. Physiol., 24*:54, 1968.

Potter, B.: The physiology of the skin. *Ann. Rev. Physiol., 28*:159, 1966.

Price, W. E.; Shehadeh, E. Z.; Thompson, G. H.; Underwood, L. D.; and Jacobson, E. D.: Effects of acetylcholine on intestinal blood flow and motility. *Amer. J. Physiol., 216*:343, 1969.

Ross, G., and Kurrasch, M.: Adrenergic responses of the hepatic circulation. *Amer. J. Physiol., 216*:1380, 1969.

Sedgewick, C. E., and Poulantzas, J. K.: Portal Hypertension. Boston, Little, Brown & Company, 1967.

Sejrsen, P.: Blood flow in cutaneous tissue in man studied by washout of radioactive xenon. *Circ. Res., 25*:215, 1969.

Shehadeh, Z.; Price, W. E.; and Jacobson, E. D.: Effects of vasoactive agents on intestinal blood flow and motility in the dog. *Amer. J. Physiol., 216*:386, 1969.

Shepherd, J. T.: Physiology of Circulation in the Limbs. Philadelphia, W. B. Saunders Company, 1963.

Sokoloff, L., and Kety, S. S.: Regulation of cerebral circulation. *Physiol. Rev., 40*(Suppl. 4):38, 1960.

Takeuchi, J.; Kubo, T.; Tone, T.; Takada, T.; Kitagawa, T.; and Yoshida, H.: Autoregulation and interaction between two vascular systems in dog liver. *J. Appl. Physiol., 27*:77, 1969.

Wahren, J.: Quantitative aspects of blood flow and oxygen uptake in the human forearm during rhythmic exercise. *Acta Physiol. Scand., 67,* Suppl. 269, 1966.

Wallentin, I.: Studies on intestinal circulation. *Acta Physiol. Scand., 69,* Suppl. 279, 1966.

Webb-Peploe, M. M.: The isovolumetric spleen: index of reflex changes in splanchnic vascular capacity. *Amer. J. Physiol., 216*:407, 1969.

PART V

THE BODY FLUIDS AND KIDNEYS

The Body Fluids: Osmotic Equilibria Between Extracellular and Intracellular Fluids

The body fluids are so important to the basic physiology of bodily function that we discuss them at several points in this text but each time in different contexts. In Chapter 4 it was pointed out that the body fluids can be divided mainly into extracellular and intracellular fluids, and the basic differences between these were discussed, as well as the manner in which these differences come about as the result of cell membrane transport. In Chapter 24 the relationship of blood volume to the overall regulation of the circulation was presented, and in Chapters 20 through 22 the interrelationships between capillary dynamics and interstitial fluids were discussed.

In the present section on body fluids and kidneys, we will be concerned with the body fluids from the total body point of view, including regulation of body fluid volume, regulation of the constituents in the extracellular fluid, regulation of acid-base balance, and factors that govern gross interchange of fluid between the extracellular and the intracellular compartments—especially the osmotic relationships between these compartments. The present chapter will discuss the general distribution of body fluids and their osmotic interrelationships.

TOTAL BODY WATER

The total amount of water in a man of average weight (70 kg.) is approximately 40 liters (see Fig. 33–1), averaging 57 per cent of his total body weight. In a newborn infant this may be as high as 75 per cent of the body weight, but it progressively decreases from birth to old age, most of the decrease occurring in the first 10 years of life. Also, obesity decreases the percentage of water in the body, sometimes down to as low as 45 per cent.

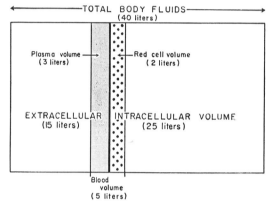

Figure 33–1 Diagrammatic representation of the body fluids, showing the extracellular fluid volume, intracellular fluid volume, blood volume, and total body fluids.

BODY FLUID COMPARTMENTS

THE INTRACELLULAR COMPARTMENT

About 25 of the 40 liters of fluid in the body are inside the 100 trillion cells of the body and are collectively called the *intracellular fluid.* The fluid of each cell contains its own individual mixture of different constituents, but the concentrations of these constituents are reasonably similar from one cell to another. For this reason the intracellular fluid of all the different cells is considered to be one large fluid compartment, though in reality it is an aggregate of trillions of minute compartments.

THE EXTRACELLULAR FLUID COMPARTMENT

All the fluids outside the cells are called *extracellular fluid,* and these fluids are constantly mixing, as was explained in Chapter 1. The total amount of fluid in the extracellular compartment averages 15 liters in a 70 kg. adult.

The extracellular fluid can be divided into *interstitial fluid, plasma, cerebrospinal fluid, intraocular fluid, fluids of the gastrointestinal tract,* and *fluids of the potential spaces.*

Interstitial Fluid. The interstitial fluid lies in the spaces between the cells. A small portion of it is free in the form of actual flowing fluid, while the major portion is held tightly by hydrated substances in the interstitial spaces. For instance, large quantities of water are adsorbed by collagen fibers, and still larger quantities are held in a gel matrix composed mainly of polymerized hyaluronic acid. Yet, despite this binding of water, dissolved substances can still move through all this water by the process of diffusion, which was explained in detail in Chapter 4.

Plasma. The plasma is the noncellular portion of blood. It is part of the extracellular fluid and communicates continually with the interstitial fluid through pores in the capillaries. Loss of plasma from the circulatory system through the capillary pores is minimized by the colloid osmotic pressure exerted by the plasma proteins, which was explained in Chapter 20. Yet the capillaries are porous enough for most dissolved substances and water molecules to *diffuse* through them freely, allowing constant mixing between the plasma and interstitial fluid of almost all substances except the protein.

The plasma volume averages 3 liters in the normal adult.

Fluid in Other Extracellular Spaces. The *cerebrospinal fluid* comprises all the fluid in the ventricles of the brain and in the subarachnoid spaces surrounding both the brain and spinal cord. This fluid has slightly different constituents from the interstitial fluid and plasma because of somewhat restricted diffusion back and forth between it and plasma and because of active secretion of a few substances by the choroid plexus. Nevertheless, from the point of view of the present discussion, cerebrospinal fluid is so nearly like interstitial fluid that it is considered to be actually a part of it. The special dynamics of the cerebrospinal fluid were presented in detail in Chapter 22.

The *intraocular fluid,* the fluid in the eyes, has properties similar to those of the cerebrospinal fluid, and here again the fluid is a product of both diffusion and secretion. These fluids, too, were discussed in detail in Chapter 22, and, for the purpose of the present discussion, are considered to be part of the interstitial fluid.

Many spaces exist in the body that normally contain little fluid but under special conditions can become filled with large amounts. These are called *potential spaces.* An example of a potential space is the space between the visceral and parietal pleurae of the lungs. Normally only 10 to 15 ml. of very viscid fluid is present in this space, but under abnormal circumstances the amount can become as great as several liters. Other potential spaces are the *peritoneal cavity,* the *pericardial cavity,* all the *joint spaces,* and the *bursae.* The fluids of these potential spaces communicate freely with the surrounding interstitial fluids and therefore are considered to be part of the interstitial fluid.

Finally, moderate amounts of extracellular fluid are normally inside the *gastrointestinal tract.* The quantity of these fluids varies greatly at different times of the day in relation to the intake and digestion of food. As much as a liter of digestive juices is sometimes in the gastrointestinal tract at once, and under certain pathologic conditions, such as gastrointestinal obstruction, this can become as much as 10 liters. The gastrointestinal fluids, except for a few of the glandular secretions, have electrolyte compositions similar to those of the interstitial fluid, and these fluids are usually considered to be part of the extracellular fluid.

BLOOD VOLUME

Blood contains both extracellular fluid (the fluid of the plasma) and intracellular fluid (the fluid in the red blood cells). However, since blood is contained in a closed chamber all its own — the circulatory system — its volume and its special dynamics are exceedingly important.

The average blood volume of a normal adult is almost exactly 5000 ml. On the average, approximately 3000 ml. of this is plasma, and the remainder, 2000 ml., is red blood cells. However, these values vary greatly in different individuals; also, sex, weight, and many other factors affect the blood volume.

Effect of Weight and Sex on Blood Volume. In persons who have a minimum of adipose tissue the blood volume varies almost directly in proportion to the body weight, normally averaging about 79 ml./kg. ±10% for both *lean* males and *lean* females. However, the greater the obesity, the less the blood volume per unit weight, because fat tissue has little vascular volume. Figure 33–2 illustrates this effect of obesity, showing that the heavier the person becomes, the less on the aver-

Figure 33–2 Relationship of blood volume to sex and weight (Redrawn from Gibson and Evans: *J. Clin. Invest.,* 16:317, 1937.)

age becomes the blood volume in relation to weight. Furthermore, in very heavy females, who are especially prone to have high ratios of fat tissue to lean tissue, the blood volume per kilogram decreases much more than for males. The average female, because of a far greater fat-to-lean tissue ratio than that of the average male, has a blood volume about 20 per cent below that of the average male.

The Hematocrit. The hematocrit is the percentage of red blood cells in the blood as determined by centrifuging blood in "hematocrit tubes" until the cells become packed tightly in the bottoms of the tubes. The percentage of red blood cells in the blood can be determined roughly from the levels of the packed cells. Such centrifuged blood in hematocrit tubes was illustrated in Figure 18–2 of Chapter 18. Unfortunately, it is impossible for the red blood cells to be packed completely together; about 3 to 8 per cent plasma remains entrapped among the cells. Therefore, the true cell percentage (H) averages about 96 per cent of the measured hematocrit (Hct); that is:

$$H = 0.96 \text{ Hct}$$

The normal hematocrit (H) is approximately 40 for a man and 36 for a woman.

In severe anemia, the hematocrit may fall to as low as 10, but this small quantity of red blood cells is barely sufficient to sustain life. On the other hand, a few conditions cause excessive production of red blood cells, resulting in polycythemia. In these instances, the hematocrit often rises to 65, and occasionally to 80. Obviously, there is an upper limit to the level of the hematocrit in polycythemic blood because excessive hematocrit causes the blood to become so viscous that death ensues as a result of multiple plugging of the peripheral vascular tree.

THE BODY HEMATOCRIT. The hematocrit of the blood in the capillaries, arterioles, and other very small vessels of the body is considerably less than that in the large veins and arteries. The cause of this is *axial streaming* of blood cells in blood vessels, which was explained in Chapter 18. In general, red cells cannot flow near the walls of the vessels nearly so easily as can plasma. Therefore, the cells tend to migrate to the center of the vessels, while a large portion of the plasma remains near the walls. In the very large vessels the ratio of wall surface to total volume is slight, so that the accumulation of plasma near the walls does not affect the hematocrit significantly. However, in the small vessels this ratio

of wall surface to volume is great, causing the ratio of plasma to cells to be far greater than in the large vessels.

If one averages the hematocrit in both the large and small vessels, he determines a value called *body hematocrit.* For normal man the body hematocrit (H_0) averages 91 per cent of the large vessel hematocrit; that is:

$$H_0 = 0.91 \text{ H} = 0.87 \text{ Hct}$$

MEASUREMENT OF BODY FLUID VOLUMES

THE DILUTION PRINCIPLE FOR MEASURING FLUID VOLUMES

The volume of any fluid compartment of the body can be measured by placing a substance in the compartment, allowing it to disperse evenly throughout the fluid, and then measuring the extent to which the substance becomes diluted. Figure 33–3 illustrates this "dilution" principle for measuring the volume of any fluid compartment of the body. A small quantity of dye or other foreign substance is placed in fluid chamber A, and the substance is allowed to disperse throughout the chamber until it becomes mixed in equal concentrations in all areas, as shown in Chamber B. Then a sample of the dispersed fluid is removed and the concentration of the substance is analyzed chemically, photoelectrically, or by any other means. The volume of the chamber can then be determined from the following formula:

Volume in ml.

$$= \frac{\text{Quantity of test substance instilled}}{\text{Concentration per ml. of dispersed fluid}}$$

Figure 33–3 Principles of the dilution method for measuring fluid volumes (explained in the text).

Note that all one needs to know is (1) the *total quantity of the test substance* put into the chamber and (2) the *concentration in the fluid after dispersement.*

DETERMINATION OF BLOOD VOLUME

Substances Used in Determining Blood Volume. A substance used for measuring blood volume must be capable of dispersing throughout the blood with ease, and it must remain in the circulatory system long enough for measurements to be made. The two major groups of substances that satisfy these conditions for measurement of blood volume are substances that combine with the red blood cells or substances that combine with the plasma proteins, for both the red blood cells and the plasma proteins remain reasonably well in the circulatory system, and any foreign substance that combines with either of them likewise remains in the blood stream.

Substances that combine with red blood cells and that are used for determining blood volume are *radioactive iron, radioactive chromium,* and *radioactive phosphate.* Substances that combine with plasma proteins are the *vital dyes* and *radioactive iodine.*

RADIOACTIVE RED BLOOD CELLS. The method most often used to make red blood cells radioactive is to tag the red blood cells with radioactive chromium (Cr^{51}). A small quantity of Cr^{51} is mixed with a few milliliters of blood removed from the person, and this is incubated at 36° C. for half an hour or more. After this time, most of the Cr^{51} will have entered the red blood cells, but to remove the extra chromium from the mixture, the red blood cells are washed in saline. Their total content of Cr^{51} is then determined with a Geiger or scintillation counter (apparatuses for measuring the total number of radioactive disintegrations occurring in the sample per minute). Then the radioactive cells are reinjected into the person. After mixing in the circulatory system has continued for approximately 10 minutes, blood is removed from the circulatory system, and the radioactivity in this blood is determined. Using the above dilution formula, the total blood volume is calculated.

To be accurate, this calculated blood volume must now be corrected to determine the true value, because the blood sample is removed from the veins where the hematocrit is not equal to the body hematocrit, as discussed earlier in the chapter. This correction is made as follows:

Actual blood volume
$$= 1.1 \times \text{Measured blood volume}$$

VITAL DYES FOR MEASUREMENT OF PLASMA VOLUME. A number of dyes, generally known as "vital dyes," have the ability to combine with proteins. When such a dye is injected into the blood, it immediately forms a slowly dissociable union with the plasma proteins. Thereafter, the dye travels where the proteins travel.

The dye almost universally used for measuring plasma volume is *T-1824,* also called *Evans blue.* In making determinations of plasma volume, a known quantity of the dye is injected, and it immediately combines with the proteins and disperses throughout the circulatory system within approximately 10 minutes. A sample of the blood is then taken, and the red blood cells are removed from the plasma by centrifugation. Then, by spectrophotometric analysis of the plasma, one can determine the exact quantity of dye in the sample of plasma. From the determined quantity of dye in each milliliter of plasma and the known quantity of dye injected, the *plasma volume* is calculated using the dilution formula noted above.

To be even more exact in measuring the plasma volume, the rate of loss of dye from the circulatory system during the interval of mixing must also be considered. On the average, T-1824 is lost from the circulatory system at a rate of about 5 per cent per hour, part of this being excreted into the urine and part being carried into the interstitial spaces by the leakage of plasma proteins through the capillary walls. Under abnormal conditions the rate of loss can be as great as 20 to 50 per cent per hour, which can cause considerable error in the measurement of plasma volume. To offset this error, three different samples of blood are usually removed from the circulation at 10, 20, and 30 minutes. Then the plasma volumes are calculated from these samples and plotted on semilog graph paper, as illustrated in Figure 33-4. A straight line is drawn through the measured points and extrapolated back to zero time. Since the percentage loss is approximately the same during the first 10 minutes as during each of the two succeeding 10-minute intervals, the extrapolated point indicates approximately the true plasma volume. In Figure 33-4 this value is 2500 ml.

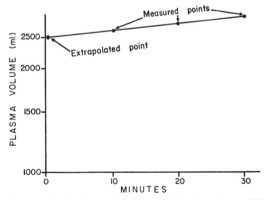

Figure 33-4 Determination of blood volume. The measured points represent plasma volumes calculated from blood samples removed 10, 20, and 30 minutes after injection of T-1824. The extrapolated point represents the calculated blood volume at the instant of dye injection.

Note that neither T-1824 nor any other vital dye enters the red blood cells. Therefore, this method *does not measure the total blood volume*. However, the blood volume can be calculated from the plasma volume, provided that the hematocrit is determined by using the following formula:

Blood volume

$$= \text{Plasma volume} \times \frac{100}{100 - 0.87 \text{ Hematocrit}}$$

RADIOACTIVE PROTEIN. If a sample of plasma is allowed to incubate with radioactive iodine (I^{131}) for 30 minutes or more, some of the protein combines with the iodine, and the iodinated protein can be separated from the remaining iodine by dialysis. The radioactive protein is then injected into the subject, and plasma and blood volumes are determined in the same manner as that discussed for the vital dyes.

MEASUREMENT OF THE EXTRACELLULAR FLUID VOLUME

To use the dilution principle for measuring the volume of the extracellular fluid one injects into the blood stream a substance that can diffuse readily throughout the entire extracellular fluid chamber, passing easily through the capillary membranes but as little as possible through the cell membranes into the cells. After half an hour or more of mixing, a sample of extracellular fluid is obtained by removing blood and separating the plasma from the cells by centrifugation. The plasma, which is actually a part of the extracellular fluid, is then analyzed for the injected substance.

Substances Used in Measuring Extracellular Fluid Volume — The Concept of "Fluid Space." Substances that have been used for measuring extracellular fluid volume are *radioactive sodium, radioactive chloride, radioactive bromide, thiosulfate ion, thiocyanate ion, inulin,* and *sucrose.* Some of these, sucrose and inulin especially, do not diffuse readily into all the out-of-the-way places of the extracellular fluid compartment. Therefore, the volume of extracellular fluid measured with these is likely to be lower than the actual volume of the compartment. On the other hand, others of these substances — radioactive chloride, radioactive bromide, radioactive sodium, and thiocyanate ion, for instance — are likely to penetrate into the cells to a slight extent and, therefore, are likely to measure a space somewhat in excess of the extracellular fluid volume.

Because there is no single substance that measures the exact extracellular fluid volume, one usually speaks of the *sodium space,* the *thiocyanate space,* the *inulin space,* and so forth, rather than the extracellular volume. At present it is impossible to say exactly which "space" most nearly approximates the true extracellular volume. Measurements for the normal 70 kg. adult, when different ones of the above substances have been used, have ranged from 8 liters to 22 liters; but the average measurement has been about 15 liters.

Correction for Loss of the Test Substance. In measuring the extracellular fluid volume of a normal person, at least 30 minutes is required for almost complete dispersion of the test substance throughout the extracellular compartment, and, if major amounts of the substance are lost from the compartment during this time, an error will be made in the measurement. For this reason, measurements employing each of the different test substances require different systems of correction but they generally employ an extrapolation back to zero time as explained above for plasma volume determination.

MEASUREMENT OF TOTAL BODY WATER

The total body water can be measured in exactly the same way as the extracellular fluid volume except that a substance must be used that will diffuse into the cells as well as throughout the extracellular compartment. The substance that gives best results is *heavy water,* which can be analyzed quantitatively either by accurate specific gravity measurements of water samples or by infrared spectrophotometry. After administration of the heavy water, several hours are required for complete mixing with all the water of the body, and appropriate corrections must be made for any fluid that is lost either into the urine or otherwise during this period of mixing. The concentration of the heavy water in the total body water can, at the end of the period of mixing, be determined by simply measuring the heavy water concentration in the plasma.

Another substance that has proved satisfactory for measuring total body water is *antipyrine,* which diffuses almost uniformly into all cells of the body and which can be analyzed readily by chemical means.

Measurements of total body water in the 70 kg. adult have ranged from as low as 30 liters up to as high as 50 liters, with a reasonable average of approximately 40 liters, or 57 per cent of the total body mass.

CALCULATION OF INTERSTITIAL FLUID VOLUME

Since any substance that passes into the interstitial fluid also passes into almost all other portions of the extracellular fluid, there is no direct method for measuring interstitial fluid volume separately from the entire extracellular fluid volume. However, if the extracellular fluid volume and plasma volume have both been measured, the interstitial fluid volume can be approximated by *subtracting the plasma volume from the total extracellular fluid volume.* This calculation gives a normal interstitial fluid volume of 12 liters in the 70 kg. adult.

CONSTITUENTS OF EXTRACELLULAR AND INTRACELLULAR FLUIDS

Figure 33–5 illustrates diagramatically the major constituents of the extracellular and intracellular fluids. (The actual values for these constituents were given in tabular form in Fig. 4–1 of Chap. 4.) The quantities of the different substances are represented in Figure 33–5 in *milliequivalents* or *millimols per liter.* However, the protein molecules and some of the nonelectrolyte molecules are extremely large in comparison with the more numerous small ions. Therefore, *in terms of mass,* the proteins and nonelectrolytes actually comprise about 90 per cent of the dissolved constituents

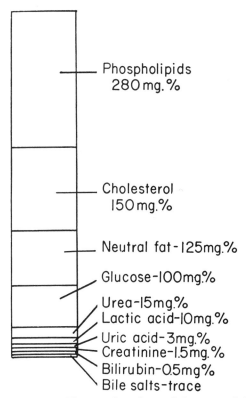

Figure 33–6 The nonelectrolytes of the extracellular fluid.

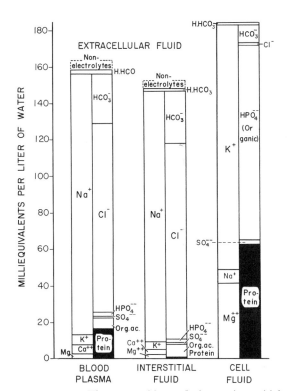

Figure 33–5 The compositions of plasma, interstitial fluid, and intracellular fluid. (Modified from Gamble: Extracellular Fluid. Harvard University Press.)

in the plasma, about 60 per cent of those in the interstitial fluid, and about 97 per cent of those in the intracellular fluid.

Figure 33–6 illustrates the distribution of the nonelectrolytes in the plasma, and most of these same substances are also present in almost equal concentrations in the interstitial fluid, except that some of the fatty compounds exist in the blood in large suspended particles, the *lipoproteins,* and, therefore, do not pass to any significant extent into the interstitial spaces.

From the point of view of most chemical and physical reactions of the fluids, it is usually the *concentration of molecules or ions* of a particular substance that is important. Therefore, in this text all concentrations are generally expressed as presented in Figure 33–5; that is, either in millimols of nonelectrolytes or milliequivalents of electrolytes.

Constituents of the Extracellular Fluid. Referring to Figure 33–5, it is evident that extracellular fluid, both that of the blood plasma and of the interstitial fluid, contains large quantities of *sodium* and *chloride ions,* reasonably large quantities of *bicarbonate ion,* but only small quantities of potassium, calcium, magnesium, phosphate, sulfate, and organic acid ions. In addition, plasma contains a large amount of protein while interstitial fluid contains much less. (The proteins in these fluids and their significance are discussed in detail in Chap. 20.)

In Chapter 1 it was pointed out that the extracellular fluid is called the *internal environment* of the body and that its constituents are accurately regulated so that the cells remain bathed continually in a fluid containing the proper electrolytes and nutrients for continued cellular function. The regulation of most of these constituents will be presented in Chapter 36.

Constituents of the Intracellular Fluid. From Figure 33–5 it is also readily apparent that the intracellular fluid contains only small quantities of sodium and chloride ions and almost no calcium ions; but it does contain large quantities of *potassium* and *phosphate* and moderate quantities of *magnesium* and *sulfate ions,* all of which are present in only small concentrations in the extracellular fluid: In addition, cells contain large amounts of protein, approximately four times as much as the plasma.

OSMOTIC EQUILIBRIA AND FLUID SHIFTS BETWEEN THE EXTRACELLULAR AND INTRACELLULAR FLUIDS

One of the most troublesome of all problems in clinical medicine is maintenance of adequate body fluids and proper balance between the extracellular and intracellular fluid volumes in seriously ill patients. The purpose of the following discussion, therefore, is to explain the interrelationships between extracellular and intracellular fluid volumes and the osmotic factors that cause shifts of fluid between the extracellular and intracellular compartments.

BASIC PRINCIPLES OF OSMOSIS AND OSMOTIC PRESSURE

The basic principles of osmosis and osmotic pressure were presented in Chapter 4. However, these principles are so important to the following discussion that they are reviewed briefly here.

Whenever a membrane between two fluid compartments is permeable to water but not to some of the dissolved solutes (this is called a *semipermeable membrane*) and the concentration of nondiffusible substances is greater on one side of the membrane than on the other, water passes through the membrane toward the side with the greater concentration of nondiffusible substances. This phenomenon is called *osmosis.*

Osmosis results from the kinetic motion of the molecules in the solutions on the two sides of the membrane and can be explained in the following manner: The individual molecules on both sides of the membrane are equally active because the temperature, which is a measure of the kinetic activity of the molecules, is the same on both sides. However, the nondiffusible solute on one side of the membrane displaces some of the water molecules, thereby reducing the concentration of water molecules. As a result, the so-called *total chemical activity* of water molecules on this side is less than on the other side, so that fewer water molecules strike each pore of the membrane each second on the solute side of the pore than on the pure water side, resulting in net diffusion of water molecules from the water side to the solute side. This net rate of diffusion is the *rate of osmosis.*

Osmosis of water molecules can be opposed by applying a pressure across the semipermeable membrane in the direction opposite to that of the osmosis. The amount of pressure required exactly to oppose the osmosis is called the *osmotic pressure.*

Relationship of the Molecular Concentration of a Solution to Its Osmotic Pressure. Each nondiffusible molecule dissolved in water dilutes the "activity" of the water molecules by a given amount. Consequently, the tendency for the water in the solution to diffuse through a membrane is reduced in direct proportion to the numbers of nondiffusible molecules. And, as a corollary, the osmotic pressure of the solution is also proportional to the numbers of nondiffusible molecules in the solution. This relationship holds true for all nondiffusible molecules almost regardless of their molecular weight. For instance, one molecule of albumin with a molecular weight of 70,000 has the same osmotic effect as a molecule of glucose with a molecular weight of 180.

Osmotic Effect of Ions. Nondiffusible ions cause osmosis and osmotic pressure in exactly the same manner as do nondiffusible molecules. Furthermore, when a molecule dissociates into two or more ions, each of the ions then exerts osmotic pressure individually. Therefore, to determine the osmotic effect, all the nondiffusible ions must be added to all the nondiffusible molecules; but note that a bivalent ion, such as calcium, exerts no more osmotic pressure than does a univalent ion, such as sodium.

Osmols. The ability of solutes to cause osmosis and osmotic pressure is measured in terms of "osmols," and the osmol is a measure of solute concentration in terms of the number of dissolved particles. *One gram mol of dissolved nondiffusible and nonionizable substance is equal to 1 osmol.* On the other hand, if a substance ionizes into two ions (sodium chloride into sodium and chloride ions, for instance), then 0.5 gram mol of the substance would equal 1 osmol. The obvious reason for using the osmol is that osmotic pressure is proportional to the concentration in number of particles instead of the concentration in terms of dissolved mass.

In general, the osmol is too large a unit for satisfactory use in expressing osmotic activity of solutions in the body. Therefore, the term *milliosmol, which equals* $1/1000$ *osmol*, is commonly used. The osmotic pressure of a solution *at body temperature* can be approximately determined from the following formula:

Osmotic pressure (mm. Hg)
$$= 19.3 \times \text{Concentration in milliosmols}$$

OSMOLALITY OF THE BODY FLUIDS

Table 33–1 lists the osmotically active substances in plasma, interstitial fluid, and intracellular fluid. The milliosmols of each of these

per liter of water is given. Note especially that approximately four-fifths of the total osmolality of the interstitial fluid and plasma is caused by sodium and chloride ions, while approximately half of the intracellular osmolality is caused by potassium ions, the remainder being divided among the many other intracellular substances.

As noted at the bottom of Table 33–1, the total osmolality of each of the three compartments is almost exactly 300 milliosmols, with that of the plasma 1.3 milliosmols greater than that of the interstitial and intracellular fluids. This slight difference between plasma and interstitial fluid is caused by the osmotic effect of the plasma proteins, which maintains about 24 mm. Hg greater pressure in the capillaries than in the surrounding interstitial fluid spaces, as was explained in Chapter 20.

Corrected Osmolar Activity of the Body Fluids. At the bottom of Table 33–1 is shown a corrected osmolar activity of plasma, interstitial fluid, and intracellular fluid. The reason for this correction is the following: All molecules and ions in solution exert either *intermolecular attraction* or *intermolecular repulsion,* and these two effects can cause, respectively, a decrease or an increase in the osmotic "activity" of the dissolved substance. In general, there is more intermolecular attraction than repulsion, so that the overall osmotic activity of the substances is only about 93 per cent of that which one would calculate from the number of milliosmols present. For this reason, the actual

TABLE 33–1. Osmolar Substances in Extracellular and Intracellular Fluids

	Plasma (mOsmols./L. of H_2O)	Interstitial (mOsmols./L. of H_2O)	Intracellular (mOsmols./L. of H_2O)
Na$^+$	144	137	10
K$^+$	5	4.7	141
Ca^{++}	2.5	2.4	0
Mg^{++}	1.5	1.4	31
Cl$^-$	107	112.7	4
HCO$_3^-$	27	28.3	10
HPO$_4^{--}$ H$_2$PO$_4^-$	2	2	11
SO$_4$	0.5	0.5	1
Phosphocreatine			45
Carnosine			14
Amino acids	2	2	8
Creatine	0.2	0.2	9
Lactate	1.2	1.2	1.5
Adenosine triphosphate			5
Hexose monophosphate			3.7
Glucose	5.6	5.6	
Protein	1.2	0.2	4
Urea	4	4	4
TOTAL mOsmols.	303.7	302.2	302.2
Corrected osmolar activity (mOsmols.)	282.6	281.3	281.3
Total osmotic pressure at 37° C. (mm. Hg)	5454	5430	5430

osmotic pressure of the body fluids is proportional to the corrected osmolar activity, which amounts to approximately 280 milliosmols/liter.

However, since osmotic effects in the body are usually determined by the relative, rather than the absolute, osmolar concentrations, the correction factor is often ignored.

Total Osmotic Pressure Exerted by the Body Fluids. At the bottom of Table 33–1 is shown the total osmotic pressure in mm. Hg that would be exerted by each of the different fluids if it were placed on one side of a cell membrane with pure water on the other side. Note that this total pressure averages about 5450 mm. Hg and also that the osmotic pressure of plasma is 24 mm. Hg greater than that of the interstitial fluids, this difference equaling the approximate hydrostatic pressure difference between the pressure of the blood inside the capillaries and the pressure in the interstitial fluid outside the capillaries.

MAINTENANCE OF OSMOTIC EQUILIBRIUM BETWEEN EXTRACELLULAR AND INTRACELLULAR FLUIDS

The tremendous osmotic pressure that can develop across the cell membrane when one side is exposed to pure water—about 5400 mm. Hg—illustrates how much force could become available to push water molecules through the membrane should the solutions on the two sides of the membrane not be in osmotic equilibrium. For instance, in Figure 33–7A, a cell is placed in a solution that has an osmolality far less than that of the intracellular fluid. As a result, osmosis of water begins immediately from the extracellular fluid to the intracellular fluid, causing the cell to swell and diluting the intracellular fluid, while concentrating the extracellular fluid. When the fluid inside the cell becomes diluted sufficiently to equal the concentration of the fluid on the outside, further osmosis then ceases. This condition is shown in Figure 33–7B. In Figure 33–7C, a cell is placed in a solution having a much higher concentration outside the cell than inside. This time, water passes by osmosis to the exterior, diluting the extracellular fluid while concentrating the intracellular fluid. In this process the cell shrinks until the two concentrations become equal, as shown in Figure 33–7D.

Rapidity of Attaining Extracellular and Intracellular Osmotic Equilibrium. The transfer of water through the cell membrane by osmosis occurs so rapidly that any lack of osmotic equilibrium between the two fluid compartments is usually corrected within a few seconds

and at most within a minute. Because of this rapidity of equilibration, it is generally considered that a state of osmotic equilibrium between the two compartments is maintained constantly.

Isotonicity, Hypotonicity, and Hypertonicity. A fluid into which normal body cells can be placed without causing either swelling or shrinkage of the cells is said to be *isotonic* with the cells. A 0.9 per cent solution of sodium chloride or a 5 per cent glucose solution is approximately isotonic.

A solution that will cause the cells to swell is said to be *hypotonic;* any solution of sodium chloride with less than 0.9 per cent concentration is hypotonic.

A solution that will cause the cells to shrink is said to be *hypertonic;* sodium chloride solutions of greater than 0.9 per cent concentration are all hypertonic.

CHANGES IN THE VOLUMES AND OSMOLALITIES OF THE EXTRACELLULAR AND INTRACELLULAR FLUID COMPARTMENTS IN ABNORMAL STATES

CALCULATION OF FLUID SHIFTS BETWEEN THE EXTRACELLULAR AND INTRACELLULAR FLUID COMPARTMENTS

Among the different factors that can cause extracellular or intracellular volumes to change markedly are ingestion of water, dehydration,

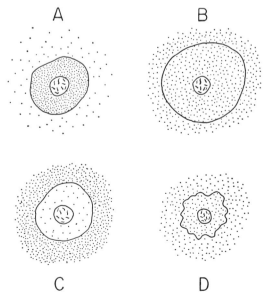

Figure 33–7 Establishment of osmotic equilibrium when cells are placed in a hypo- or hypertonic solution.

intravenous infusion of different types of solutions, loss of large quantities of fluids from the gastro-intestinal tract, or loss of abnormal quantities of fluid by sweating or through the kidneys.

The changes in both extracellular and intra-cellular fluid volumes can be calculated easily if one will keep the following two basic principles in mind:

1. The osmolalities of the extracellular and intracellular fluids remain exactly the same except during the first few seconds to few minutes after a change in one of the fluids occurs.

2. The number of osmols of osmotically active substance in each compartment, in the extracellular fluid or in the intracellular fluid, remains constant unless one of the osmotically active substances moves through the cell membranes to the other compartment or is lost from or added to one of the two compartments in some other way.

Using these two basic principles, we can now analyze the effects of different abnormal fluid conditions on extracellular and intracellular fluid volumes and osmolalities.

EFFECT OF ADDING WATER TO THE EXTRACELLULAR FLUID

Water can be added to the extracellular fluid by injection into the bloodstream, injection beneath the skin, or by ingesting water and this followed by absorption from the gastrointestinal tract into the blood. The water dilutes the extracellular fluid, causing it to become hypotonic with respect to the intracellular fluids. Osmosis begins immediately at the cell membranes, with large amounts of water passing to the interior of the cells. Within a few minutes the water becomes distributed almost evenly among all the extracellular and intra-cellular fluid compartments.

Calculation of the Changes in Extracellular and Intracellular Volumes and Osmolalities. Table 33–2 illustrates the progressive changes that would result from injecting 10 liters of water into the extracellular fluid. This table gives the volumes and milliosmols per liter of the extracellular fluid, the intracellular fluid, and the total body water.

Initially the extracellular fluid volume is 15 liters, the intracellular fluid volume 25 liters, and the total body water 40 liters; the milliosmols per liter is 300 in each of these. The solution that is added, 10 liters of water, has no osmolality, and when first injected it is added both to the extracellular fluid and to the total body water, but not to the intracellular fluid.

The third line of the table shows the instantaneous effect on the volumes and osmolalities caused by addition of this fluid (before any osmosis occurs). The extracellular fluid volume rises to 25 liters, and its concentration becomes diluted immediately to 180 milliosmols/liter. During this instantaneous interval of time nothing has happened to the intracellular fluids. But within a few seconds to a few minutes, the dilute extracellular fluid diffuses throughout the interstitial spaces of the body and comes in contact with the cells, allowing large portions of the water to pass by osmosis into the intracellular fluid. Osmosis will not stop until the milliosmolalities of the two fluids become equal. As a result, the milliosmolality of the extracellular fluid rises from 180 while that of the intracellular fluid falls from 300 until the two equal each other. After a few moments the conditions in the last line of the table obtain, with the extracellular fluid volume 18.75 liters, the intracellular volume 31.25 liters, and a milliosmolality in both compartments of 240.

To calculate the changes occurring in the above

TABLE 33–2. Effect of Administering 10 Liters of Water Intravenously

	Extracellular			Intracellular			Total Body Water		
	Volume (Liters)	Concentration (mOsmols./L.)	Total mOsmols.	Volume (Liters)	Concentration (mOsmols./L.)	Total mOsmols.	Volume (Liters)	Concentration (mOsmols./L.)	Total mOsmols.
Initial	15	300	4500	25	300	7500	40	300	12000
Solution added	10	0	0	0	0	0	10	0	0
Instantaneous effect	25	180	4500	25	300	7500	50	No equilibrium	12000
After osmotic equilibrium	18.75	240	4500	31.25	240	7500	50	240	12000

example, one needs only to keep accurate accounting of the total number of milliosmols in each fluid compartment and also in the total body water in the following manner: After the 10 liters of water are added, the total body water becomes 50 liters instead of 40, but the total milliosmols in the whole body remain the same, 12,000. Dividing 50 into 12,000, the average milliosmolar concentration in each liter of the body fluids under the new conditions is found to be 240. One can readily understand that, after osmotic equilibrium has occurred throughout the body, this calculated value of 240 is the milliosmolality in both the extracellular and intracellular fluid compartments. Then, dividing 240 milliosmols into the total milliosmols in the extracellular fluid compartment, 4500, one finds that the new volume of the extracellular fluid is 18.75 liters. Dividing 240 into the total intracellular milliosmols, 7500, gives a new intracellular fluid volume of 31.25 liters.

EFFECT OF DEHYDRATION

Water can be removed from the body by evaporation from the skin, evaporation from the lungs, or excretion of a very dilute urine. In all these conditions the water leaves the extracellular fluid compartment, but on doing so some of the intracellular water passes immediately into the extracellular compartment by osmosis, thus keeping the osmolalities of the extracellular and intracellular fluids equal to each other. The overall effect is called *dehydration*.

EFFECT OF ADDING SALINE SOLUTIONS TO THE EXTRACELLULAR FLUID

If an *isotonic* saline solution is added to the extracellular fluid compartment, the osmolality of the extracellular fluid does not change, and no osmosis results. The only effect is an increase in extracellular fluid volume.

However, if a *hypertonic* solution is added to the extracellular fluid, the osmolality increases and

causes osmosis of water out of the cells into the extracellular compartment.

Finally, if a *hypotonic* solution is added, the osmolality of the extracellular fluid decreases, and some of the extracellular fluids pass into the cells.

Table 33–3 gives the calculations of the effects which would occur if 2 liters of 4.4 per cent (five times isotonic concentration) sodium chloride solution were added to the extracellular fluid compartment. The added solution contains a large total number of milliosmols which are added to the extracellular fluid compartment and also to the total body fluids, and the volume of the total body fluids rises from 40 to 42. Dividing 42 into the new total milliosmols, 15,000, one finds that the new milliosmolar concentration is 357. Now, dividing 357 into the total milliosmols in the extracellular fluid, 7500, one finds that the new extracellular fluid volume is 21 liters. And, dividing 357 into the total intracellular milliosmols, 7500, the intracellular fluid is also found to be 21 liters.

The student should become completely familiar with this method of calculation, for an understanding of the mathematical aspects of osmotic equilibria between the two compartments is essential to the understanding of any fluid problem of the body.

EFFECT OF INFUSING HYPERTONIC GLUCOSE, MANNITOL, OR SUCROSE SOLUTIONS

Very concentrated glucose, mannitol, or sucrose solutions are often injected into patients to cause immediate decrease in intracellular fluid volume. For instance, often in severe cerebral edema the patient dies because of too much pressure in the cranial vault, which obstructs the flow of blood to the brain. The condition can be relieved, however, in a few minutes by injecting a hypertonic solution of a substance that will not enter the intracellular compartment. The dynamics of the resulting changes are shown in Figure 33–8, illustrating that the intracellular fluid volume can be decreased by several liters in a few minutes.

TABLE 33–3.   Effect of Adding 2 Liters of 4.4 Per Cent Sodium Chloride Solution

	Extracellular			Intracellular			Total Body Water		
	Volume (Liters)	Concentration (mOsmols./L.)	Total mOsmols.	Volume (Liters)	Concentration (mOsmols./L.)	Total mOsmols.	Volume (Liters)	Concentration (mOsmols./L.)	Total mOsmols.
Initial	15	300	4500	25	300	7500	40	300	12000
Solution added	2	1500	3000	0	0	0	2	1500	3000
Instantaneous effect	17	441	7500	25	300	7500	42	No equilibrium	15000
After osmotic equilibrium	21	357	7500	21	357	7500	42	357	15000

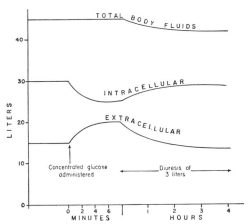

Figure 33-8 Time course of the body fluid changes following infusion of very concentrated glucose into the extracellular fluids.

But glucose, mannitol, and sucrose are all excreted rapidly by the kidneys, and glucose is also metabolized by the cells for energy. Therefore, within two to four hours the osmotic effects of these substances are lost so that large quantities of water can then rediffuse into the intracellular compartment. Thus, this procedure is temporarily beneficial, lasting only a few hours. Nevertheless, from the standpoint of saving a life of a patient it is often very valuable.

GLUCOSE AND OTHER SOLUTIONS ADMINISTERED FOR NUTRITIVE PURPOSES

Many different types of solutions are often administered intravenously to provide nutrition to patients who cannot otherwise take adequate amounts of food. Especially used are glucose solutions, to a lesser extent amino acid solutions, and, rarely, homogenized fat solutions. In administering all these, their concentrations are adjusted nearly to isotonicity, or they are given slowly enough that they do not upset the osmotic equilibria of the body fluids. However, after the glucose or other nutrient is metabolized, an excess of water often remains. Ordinarily, the kidneys excrete this in the form of a very dilute urine. Thus, the net result is only addition of the nutrient to the body.

But occasionally the kidneys are functioning poorly, such as often occurs after a surgical operation, and the body becomes greatly overhydrated, resulting sometimes in "water intoxication," which is characterized by mental irritability and even convulsions.

SOLUTIONS USED FOR PHYSIOLOGIC AND CLINICAL REPLACEMENT PURPOSES

Below are listed the compositions of several solutions commonly used for replacing fluids lost from the body or for nutrition:

Isotonic saline (normal saline) solution for extracellular fluid replacement:

Na^+ 155 mEq./l. Cl^- 155 mEq./l.

Glucose solutions for nutrition:

  I. Glucose   2.5, 5, or 10%
 II. Glucose   2.5, 5, or 10%
    Na^+ 155 mEq./l. Cl^- 155 mEq./l.

Tyrode's solution for extracellular fluid replacement:

Na^+	149.4 mEq./l.	Cl^-	145.1 mEq./l.
K^+	2.7	HCO_3^-	12.0
Ca^{++}	3.6	HPO_4^{--}	0.7
Mg^{++}	2.1		
Glucose 0.1%			

Ringer's solution for frog and turtle experiments:

Na^+	115 mEq./l.	Cl^-	106 mEq./l.
K^+	1	HCO_3^-	12
Ca^{++}	2		

Ringer's solution for extracellular fluid replacement:

Na^+	146 mEq./l.	Cl^-	155.4 mEq./l.
K^+	4		
Ca^{++}	5.4		

Modified Hartmann's solution for extracellular fluid replacement:

Na^+	128 mEq./l.	Cl^-	110.6 mEq./l.
K^+	4	Lactate$^-$	25
Ca^{++}	3.6		

Darrow's solution for potassium and extracellular fluid replacement:

Na^+	122 mEq./l.	Cl^-	104 mEq./l.
K^+	35	Lactate$^-$	52

Sodium lactate solution (1/6 molar) for correcting acidosis:

Na^+ 167 mEq./l.   Lactate$^-$ 167 mEq./l.

Ammonium chloride solution (1/6 molar) for correcting alkalosis:

NH_4^+ 167 mEq./l. Cl^- 167 mEq./l.

Amino acid solution for nutrition:

Amino acid   5 or 10%
Glucose   5%

Plasma for nutrition or replacement:

Protein   7%
  (for other constituents, see Table 8)

Fat emulsion for nutrition:

Fat   15%

REFERENCES

Bajusz, E.: Electrolytes and Cardiovascular Disease. Baltimore, The Williams & Wilkins Co., 1966, Vol. I.

Black, D.: Essentials of fluid balance. 4th ed., Philadelphia, F. A. Davis Co., 1968.

Bland, J. H. (ed.): Clinical Metabolism of Body Water and Electrolytes. Philadelphia, W. B. Saunders Company, 1963.

Cardus, D.; McTaggart, W. G.; and Young, C. L.: Effect of exercise on determination of total body water by tritium oxide. *J. Appl. Physiol., 27*:1, 1969.

Chiaraviglio, E. del C., and Wolf, A. V.: Diagnosis of drowning. *Arch. Path., 75*:337, 1963.

Maxwell, M. H., and Kleeman, C. R.: Clinical Disorders of Fluid and Electrolyte Metabolism. New York, McGraw-Hill Book Company, 1962.

Mendez, J.; Prokop, E.; Picon-Reategui, E.; Akers, R.; and Buskirk, E. R.: Total body water by D_2O dilution using saliva samples and gas chromatography. *J. Appl. Physiol., 28*:354, 1970.

Moore, F. D.; Olesen, K. H.; McMurrey, J. D.; Parker, H. V.; Ball, M. R.; and Boyden, C. M.: The Body Cell Mass and Its Supporting Environment. Philadelphia, W. B. Saunders Company, 1963.

Potts, W. T. W.: Osmotic and ionic regulation. *Ann. Rev. Physiol., 30*:73, 1968.

Ray, C. T.: Water and electrolyte balance. *In* Sodeman, W. A., and Sodeman, W. A., Jr. (eds.): Pathologic Physiology. 4th ed., Philadelphia, W. B. Saunders Company, 1967, p. 109.

Sandor, G.: Serum Proteins in Health and Disease. Baltimore, The Williams & Wilkins Co., 1967.

Statland, H.: Fluid and Electrolytes in Practice. 3rd ed., Philadelphia, J. B. Lippincott Co., 1963.

Sunderman, F. W., and Sunderman, F. W., Jr.: The Serum Proteins and the Dysproteinemias. Philadelphia, J. B. Lippincott Co., 1964.

Tosteson, D. C., and Hoffman, J. F.: Regulation of cell volume by active cation transport in high and low potassium sheep red cells. *J. Gen. Physiol., 44*:169, 1960.

Welt, L. G.: Water balance in health and disease. *In* Duncan, Garfield G. (ed.): Diseases of Metabolism. 5th ed., Philadelphia, W. B. Saunders Company, 1964, p. 449.

Wolf, A. V., and Crowder, N. A.: Introduction to Body Fluid Metabolism. Baltimore, The Williams & Wilkins Co., 1964.

Wolf, A. V.; Prentiss, P. G.; Douglas, L. G.; and Swett, R. J.: Potability of sea water with special reference to the cat. *Amer. J. Physiol., 196*:633, 1959.

CHAPTER 34

Formation of Urine by the Kidney: I. Glomerular Filtration and Tubular Function

The kidneys perform two major functions: first, they excrete most of the end-products of bodily metabolism, and, second, they control the concentrations of most of the constituents of the body fluids. The purpose of the present chapter is to discuss the principles of urine formation and especially the mechanisms by which the kidneys excrete the end-products of metabolism.

PHYSIOLOGIC ANATOMY OF THE KIDNEY

Each kidney contains about 1,000,000 nephrons, each of which is capable of forming urine by itself. Therefore, in most instances, it is not necessary to discuss the entire kidney but merely the activities in the single nephron in order to explain the function of the kidney.

The nephron is composed basically of (1) a *glomerulus* through which fluid is filtered out of the blood and (2) a long *tubule* in which the filtered fluid is converted into urine on its way to the *pelvis* of the kidney. Figure 34–1 shows the general organizational plan of the kidney, illustrating especially the distinction between the *cortex* of the kidney and the *medulla*. Figure 34–2 illustrates the basic anatomy of the nephron, which may be described as follows: Blood enters the glomerulus through the *afferent arteriole* and then leaves through the *efferent arteriole*. The glomerulus is a network of up to 50 parallel capillaries encased in *Bowman's capsule*. Pressure of the blood in the glomerulus causes fluid to filter into Bowman's capsule, from which it flows first into the *proximal tubule* that lies in the *cortex* of the kidney along with the glomerulus. From there the fluid passes into the *loop of Henle*. Those nephrons that have glomeruli lying very close to the renal *medulla* are called the *juxtamedullary nephrons*, and they have long extended loops of Henle that dip deep into the medulla; the

lower portion of the loop has a very thin wall and therefore is called the *thin segment* of the loop of Henle. From the loop of Henle the fluid flows next into the *distal tubule*, which again is in the renal cortex. Finally, the fluid flows into the *collecting tubule*, also called the *collecting duct*, which collects fluid from several nephrons. The collecting tubule passes from the cortex back downward through the medulla, paralleling the loops of Henle. Then it empties into the pelvis of the kidney.

As the glomerular filtrate flows through the tubules, most of its water and varying amounts of its solutes are reabsorbed into the tubular capillaries. The water and solutes that are not reabsorbed become urine.

Nephrons that have glomeruli lying close to

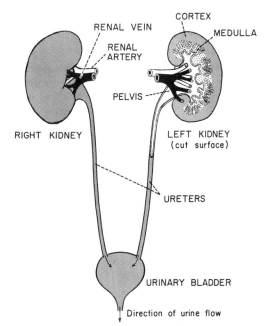

Figure 34–1 The general organizational plan of the kidney.

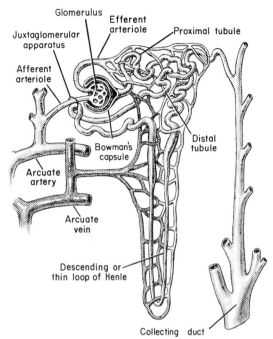

Figure 34-2 The nephron. (Redrawn from Smith: The Kidney. Oxford University Press.)

the surface of the kidney do not have long thin segments of their loops of Henle, and their loops of Henle fail to penetrate all the way into the medulla. Other than this difference, these cortical nephrons seem to function in much the same manner as the juxtamedullary nephrons.

After blood passes into the efferent arteriole from the glomerulus, most of it flows through the *peritubular capillary network* that surrounds the tubules. From this network, straight capillary loops, called *vasa recta*, extend downward into the medulla to envelop the lower parts of the thin segments before looping back upward to empty into the cortical veins.

Functional Diagram of the Nephron. Figure 34–3 illustrates a simplified diagram of the "physiologic nephron." This diagram contains most of the nephron's functional structures, and it is used in the present discussion to explain many aspects of renal function.

BASIC THEORY OF NEPHRON FUNCTION

The basic function of the nephron is to clean, or "clear," the blood plasma of unwanted substances as it passes through the nephron. The substances that must be cleared include particularly the end-products of metabolism, such as urea, creatinine, uric acid, sulfates and phenols. In addition, many nonmetabolic substances, such as sodium ions,

potassium ions, and chloride ions, frequently accumulate in the body in excess quantities; it is the function of the nephron also to clear the plasma of these excesses.

The mechanism by which the nephron clears the plasma of unwanted substances is: (1) It filters a large proportion of the plasma, usually about one-fifth of it, through the glomerular membrane into the tubules of the nephron. (2) Then, as this filtered fluid flows through the tubules, the unwanted substances fail to be reabsorbed while the wanted substances, especially the water and many of the electrolytes, are reabsorbed back into the plasma of the peritubular capillaries. In other words, the tubules separate the unwanted portions of the tubular fluid from the wanted portions, and the wanted portions are returned to the blood, while the unwanted portions pass into the urine.

RENAL BLOOD FLOW AND PRESSURES

BLOOD FLOW THROUGH THE KIDNEYS

The rate of blood flow through both kidneys of a 70 kg. man is about 1200 ml./minute, but even under normal conditions this varies from one person to another as much as ±500 ml./minute.

A small proportion of the renal blood flow, 2 to 3 per cent, is believed to pass through

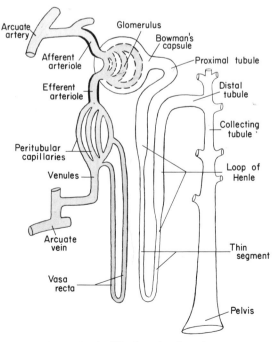

Figure 34-3 The functional nephron.

arteriovenous shunts called "Trueta shunts" that by-pass the glomeruli and peritubular capillary plexuses. However, for practical purposes, one can consider that all the renal blood passes through the nephrons prior to entering the veins.

The Renal Fraction. The portion of the total cardiac output that passes through the kidneys is called the *renal fraction*. Since the normal cardiac output of a 70 kg. adult male is about 5600 ml./minute, and the blood flow through both kidneys is about 1200 ml./minute, one can calculate that the normal renal fraction is about 21 per cent. This can vary from as little as 12 per cent to as high as 30 per cent even in the normal resting person.

Special Aspects of Blood Flow Through the Nephron. Note in Figure 34–3 that there are two capillary beds supplying the nephron: (1) the *glomerulus* and (2) the *peritubular capillaries.* These two capillary beds are separated from each other by the *efferent arteriole,* which offers considerable resistance to blood flow. As a result, the glomerular capillary bed is a *high pressure bed* while the peritubular capillary bed is a *low pressure bed.* Because of the high pressure in the glomerulus, it functions in much the same way as the usual arterial ends of the tissue capillaries, with fluid filtering continually out of the glomerulus into Bowman's capsule. On the other hand, the low pressure in the peritubular capillary system causes it to function in much the same way as the venous ends of the tissue capillaries, with fluid being absorbed continually into the capillaries.

THE VASA RECTA. A special portion of the peritubular capillary system is the vasa recta, which are a network of capillaries that descend around the lower portions of the loops of Henle. These capillaries form loops into the medulla of the kidney and then return to the cortex before emptying into the veins. The vasa recta play a special role in the formation of concentrated urine, as is discussed later in the chapter.

Only a small proportion of the total renal blood flow, about 1 to 2 per cent, flows through the vasa recta. In other words, blood flow through the medulla of the kidney is sluggish in contrast to the rapid blood flow in the cortex.

At the outer rim of the medulla where the medulla comes in contact with the cortex there is an extensive peritubular capillary plexus through which approximately 11 per cent of the total renal blood flow passes. Unfortunately, the significance of this plexus is yet unknown, but the large blood flow in it is in sharp contrast to the poor blood flow (only

1 to 2 per cent of the total) to the inner medulla through the vasa recta.

PRESSURES IN THE RENAL CIRCULATION

Figure 34–4 gives the pressures in the different parts of the renal circulation, showing an initial pressure of approximately 100 mm. Hg in the large arcuate arteries and about 8 mm. Hg in the arcuate veins into which the blood finally drains. The two major areas of resistance to blood flow through the nephron are (1) the *afferent arteriole* and (2) the *efferent arteriole.* In the afferent arteriole the pressure falls from 100 mm. Hg at its arterial end to an estimated mean pressure of about 70 mm. Hg in the glomerulus. (This pressure is still in serious doubt, having been "calculated" in two different recent experimental studies to be 90 and 55 mm. Hg, respectively. Therefore, 70 is a reasonable average estimate.) As the blood flows through the efferent arterioles from the glomerulus to the peritubular capillary system, the pressure falls another 57 mm. Hg to a mean peritubular capillary pressure of 13 mm. Hg. Thus, the high pressure capillary bed in the glomerulus operates at a mean pressure of

Figure 34–4 Pressures at different points in the vessels and tubules of the functional nephron and in the interstitial fluid.

about 70 mm. Hg while the low pressure capillary bed in the peritubular capillary system operates at a mean capillary pressure of about 13 mm. Hg.

INTRARENAL PRESSURE AND RENAL INTERSTITIAL FLUID PRESSURE

The kidney is encased in a tight *fibrous capsule*. When a needle is inserted into the kidney and the pressure in the needle is gradually raised until fluid flows into the kidney tissue, the pressure at which flow begins is between 8 and 15 mm. Hg, averaging perhaps 10 mm. Hg. This "needle" pressure is called the *intrarenal pressure*. It was pointed out in Chapter 20 that needle pressures do not measure the interstitial fluid pressure but instead measure the *total tissue pressure* (the pressure that tends to collapse blood vessels and tubules). That the intrarenal pressure is indeed total tissue pressure rather than interstitial fluid pressure is demonstrated by the fact that the pressure inside the proximal tubules is almost exactly equal to the intrarenal pressure.

Recent attempts to measure the interstitial fluid pressure of the kidney utilizing implanted perforated capsules as described in Chapter 20 have given values sometimes slightly negative with respect to atmospheric pressure and sometimes slightly positive. The average mean value has been approximately equal to atmospheric pressure, which, at present, is the best estimate of interstitial fluid pressure of the kidney.

FUNCTION OF THE PERITUBULAR CAPILLARIES

Tremendous quantities of fluid, about 180 liters each day, are reabsorbed from the tubules into the renal interstitial spaces and thence into the peritubular capillaries. This represents about four times as much fluid as that reabsorbed at the venous ends of all the other capillaries of the entire body. Therefore, one can readily see that reabsorption of fluid into the peritubular capillaries presents a special problem. However, the peritubular capillaries are extremely porous in comparison with those in other body tissues, so that extremely rapid osmosis of fluid resulting from the colloid osmotic pressure of the plasma proteins probably can account for the rapid absorption that is required.

GLOMERULAR FILTRATION AND THE GLOMERULAR FILTRATE

The Glomerular Membrane. The fluid that filters through the glomerular membrane into Bowman's capsule is called *glomerular filtrate,* and the membrane of the glomerular capillaries is called the *glomerular membrane.* Though, in general, this membrane is similar to that of other capillaries throughout the body, it has several differences. First, it has three major layers: (1) the endothelial layer of the capillary itself, (2) a basement membrane, and (3) a layer of epithelial cells that line the surfaces of Bowman's capsule. Yet, despite the number of layers, the permeability of the glomerular membrane is about 25 times as great as that of the usual capillary.

The endothelial layer of the glomerular membrane has very large pores between its cells, for which reason many electron microscopists doubt that these pores could represent the filtration barrier of the membrane. On the other hand, the basement membrane is continuous, giving rise to the suggestion that the lattice structure of the basement membrane acts as the molecular sieve of the glomerular membrane. But there are also narrow slits between the epithelial cells on the Bowman's capsule surface of the glomerular membrane, and these slits, too, could represent the molecular sieve.

Thus, at present, we still do not know the exact structural basis for the ability of the glomerular membrane to act as a filter, allowing some substances to pass through the membrane while holding others back. Nevertheless, the glomerular membrane is extremely permeable to almost all substances having molecular sizes less than those of plasma proteins (molecular weight less than 70,000), while almost no plasma proteins or other substances of larger molecular size are able to filter into Bowman's capsule.

Composition of the Glomerular Filtrate. The glomerular filtrate has almost exactly the same composition as the fluid that filters from the arterial ends of the capillaries into the interstitial fluids. It contains no red blood cells and less than 0.03 per cent protein, or less than $1/200$ the protein in the plasma.

The electrolyte and crystalloid composition of glomerular filtrate is also similar to that of the interstitial fluid. Because of the paucity of the negatively charged protein ions in the filtrate, the concentration of the other negative ions, including chloride and bicarbonate ions, is about 5 per cent higher in both interstitial fluid and glomerular filtrate than in plasma; and the concentration of positive ions is about 5 per cent lower. Also, the concentrations of the nonionized crystalloidal substances, such as urea, creatinine, and glucose, are increased about 4 per cent because of the almost total lack of proteins.

To summarize: for all practical purposes, glomerular filtrate is the same as plasma except that it has no significant amount of proteins.

THE GLOMERULAR FILTRATION RATE

The quantity of glomerular filtrate formed each minute in all nephrons of both kidneys is called the *glomerular filtration rate*. In the normal person this averages approximately 125 ml./min.; however, in different normal functional states of the kidneys, it can vary from a few milliliters to 200 ml./min. To express this differently, the total quantity of glomerular filtrate formed each day averages about 180 liters, or more than two times the total weight of the body. Over 99 per cent of the filtrate is usually reabsorbed in the tubules, the remainder passing into the urine, which is explained later in the chapter.

The Filtration Fraction. The filtration fraction is the fraction of the renal plasma flow that becomes glomerular filtrate. Since the normal plasma flow through both kidneys is 650 ml./min. and the normal glomerular filtration rate in both kidneys is 125 ml., *the average filtration fraction is approximately* $125/650$, *or 19 per cent*. Here again, this value can vary tremendously, both physiologically and pathologically.

DYNAMICS OF GLOMERULAR FILTRATION

Glomerular filtration occurs in almost exactly the same manner that fluid filters out of any high pressure capillary in the body. That is, *pressure inside the glomerular capillaries* causes filtration of fluid through the capillary membrane into Bowman's capsule. On the other hand, *colloid osmotic pressure in the blood and pressure in Bowman's capsule* oppose the filtration. Ordinarily, the amount of protein in Bowman's capsule is too slight to be of any significance, but if this ever becomes increased to a significant amount, its colloid osmotic pressure will obviously also be active at the membrane, promoting increased filtration of fluid through the membrane.

Glomerular Pressure. The glomerular pressure is the average pressure in the glomerular capillaries. This unfortunately has never been measured by either direct or indirect means. Yet, on the basis of physiological experiments, estimates have been made — as low as 55 mm.

Hg and as high as 90 mm. Hg. A *reasonable average value can be considered to be 70 mm. Hg,* though, as noted below, this can increase or decrease considerably under varying conditions.

Pressure in Bowman's Capsule. In lower animals, pressure measurements have actually been made in Bowman's capsule and at different points along the renal tubules by inserting micropipets into the lumen. On the basis of these studies, *capsular pressure in the human being is estimated to be 14 mm. Hg.*

Colloid Osmotic Pressure in the Glomerular Capillaries. Because approximately one fifth of the plasma in the capillaries filters into the capsule, the protein concentration increases about 20 per cent as the blood passes from the arterial to the venous ends of the glomerular capillaries. If the normal colloid osmotic pressure of blood entering the capillaries is 28 mm. Hg, it rises to approximately 36 mm. Hg by the time the blood reaches the venous ends of the capillaries, and the average colloid osmotic pressure is about 32 mm. Hg. (See Fig. 20-8 in Chap. 20, which shows the relationship between protein concentration and colloid osmotic pressure.)

Filtration Pressure. The filtration pressure is the net pressure forcing fluid through the glomerular membrane, and this is *equal to the glomerular pressure minus the sum of glomerular colloid osmotic pressure and capsular pressure.* In Figure 34-5A, *the normal filtration pressure is shown to be about 24 mm. Hg.*

REGULATION OF GLOMERULAR FILTRATION RATE

The rate at which glomerular filtrate is formed is directly proportional to the filtration pressure. Therefore, any factor that changes the filtration pressure also changes the filtration rate.

Effect of Afferent Arteriolar Constriction. Afferent arteriolar constriction decreases the rate of blood flow into the glomerulus and thereby decreases the glomerular pressure and the filtration rate. This effect is illustrated in Figure 34-5B. Conversely, dilatation of the afferent arteriole increases the glomerular pressure, with a corresponding increase in glomerular filtration rate.

Effect of Efferent Arteriolar Constriction. Constriction of the efferent arteriole increases the resistance to outflow from the glomeruli. This obviously increases the glomerular pressure and usually increases the glomerular

Figure 34–5 (A) Normal pressures at different points in the nephron, and the normal filtration pressure. (B) Effect of afferent arteriolar constriction on pressures in the nephron and on filtration pressure. (C) Effect of efferent arteriolar constriction on pressures in the nephron and on filtration pressure.

filtration rate as well, which is illustrated in Figure 34–5C. However, if the degree of efferent arteriolar constriction is severe, the blood flow through the glomerulus may become sluggish. Then, because the plasma remains for a long period of time in the glomerulus and extra large portions of plasma are lost into the capsule, the plasma colloid osmotic pressure rises to excessive levels, causing glomerular filtration to fall paradoxically to a low value despite the elevated glomerular pressure.

Effect of Sympathetic Stimulation. In mild sympathetic stimulation of the kidneys, the afferent and efferent arterioles constrict approximately proportionately to each other so that the glomerular filtration rate neither rises nor falls. This allows blood flow to be shunted from the kidneys to other parts of the body during emergency states without significantly altering renal function.

With strong sympathetic stimulation, however, the effects are different, because all the arterioles then become so greatly constricted that glomerular blood flow is reduced almost to zero, thereby greatly reducing the filtration rate. Thus, strong sympathetic stimulation reduces renal output almost to zero.

Effect of Arterial Pressure. One would expect an increase in arterial pressure to cause a proportionate increase in all pressures in the nephron and therefore to increase the glomerular filtration rate greatly. In actual fact, this is not the case because of the phenomenon called *autoregulation,* which is explained later in the chapter. Briefly, when the arterial pressure rises, afferent arteriolar constriction occurs automatically in the nephron; this prevents a significant rise in glomerular pressure despite the rise in arterial pressure. Therefore, the glomerular filtration rate increases only 5 to 15 per cent, even though the arterial pres-

sure rises from its normal value of 100 mm. Hg to as high as 200 mm. Hg.

Nevertheless, we shall also see later in the chapter that even a slight percentage increase in glomerular filtration rate can cause a many-fold increase in urinary output. Therefore, an increase in arterial pressure can greatly increase urinary output even though it affects the glomerular filtration rate only slightly.

Effect of Changes in Plasma Colloid Osmotic Pressure. Any effect that increases or decreases plasma protein concentration can modify glomerular filtration rate by changing the colloid osmotic pressure—the higher the colloid osmotic pressure, the less the filtration rate; and the lower the colloid osmotic pressure, the greater the filtration rate. For instance, after large quantities of saline solution are ingested intravenously, the consequent dilution of the plasma can at times change the plasma colloid osmotic pressure as much as 5 mm. Hg; this in turn can increase the glomerular filtration rate as much as 15 to 20 per cent.

REABSORPTION AND SECRETION IN THE TUBULES

The glomerular filtrate entering the tubules of the nephron flows (1) through the *proximal tubule,* (2) through the *loop of Henle,* (3) through the *distal tubule,* and (4) through the *collecting duct* into the pelvis of the kidney. Along this course, substances are selectively reabsorbed or secreted by the tubular epithelium, and the resultant fluid entering the pelvis is *urine.* Reabsorption plays a much greater role than does secretion in this formation of urine, but secretion is especially important in determining the amounts of potassium ions, hydrogen ions, and a few other substances in the urine, as is discussed later.

Ordinarily, more than 99 per cent of the water in the glomerular filtrate is reabsorbed as it passes through the tubules. Therefore, if some dissolved constituent of the glomerular filtrate is not reabsorbed at all along the entire course of the tubules, this reabsorption of water obviously concentrates the substance more than 99-fold. On the other hand, some constituents, such as glucose and amino acids, are reabsorbed almost entirely so that their concentrations decrease almost to zero before the fluid becomes urine. In this way the tubules separate substances that are to be conserved in the body from those that are to be eliminated in the urine.

BASIC MECHANISMS OF ABSORPTION AND SECRETION IN THE TUBULES

The basic mechanisms for transport through the tubular membrane are identical with those for transport through essentially any other membrane of the body. These can be divided into *active transport* and *passive transport*. The basic essentials of these mechanisms are described here, but for details refer to Chapter 4.

Active Transport. Figure 34–6 illustrates the mechanism for active transport of sodium from the lumen of the proximal tubule into the peritubular fluid. Active transport of sodium occurs through only one side of the epithelial cell membrane, the side adjacent to the peritubular fluid; on the other hand, the "brush" border of the cell is reasonably permeable to the diffusion of sodium. The active transport of sodium out of the cell and into the peritubular fluid diminishes the concentration of sodium inside the cell. Then, because of the low concentration inside the cell, sodium ions diffuse from the lumen of the tubule to the interior of the cell; once there, the active transport process carries the sodium the rest of the way into the peritubular fluid.

Figure 34–6   Mechanism for active transport of sodium from the tubular lumen into the peritubular fluid, illustrating active transport at the base of the epithelial cell and diffusion through the luminal border of the cell.

Note also in Figure 34–6 the differences in electrical potential at different points in the fluids. The active transport of sodium out of the tubular cell creates an electrical potential of -70 mv. inside the cell with respect to the peritubular fluids. And the diffusion of sodium from the lumen of the tubule into the interior of the epithelial cell decreases the electrical potential of the tubular lumen to approximately -20 mv. Thus, there is an electrical difference of 50 mv. between the lumen of the tubule and the interior of the epithelial cell, with negativity inside the cell. This electrical difference acts as a force to pull sodium ions into the epithelial cell from the tubular lumen. Obviously, this is a second factor in addition to the concentration differences that causes diffusion of sodium from the tubular lumen to the interior of the cell. Therefore, both a *concentration difference* (also called a "chemical gradient") and an *electrical difference* move sodium through the brush border of the cell. This combined effect is called an *electrochemical gradient*.

The basic mechanism for active transport of sodium through the cell membrane itself at the cell's posterior border was discussed in Chapter 4. In brief, the sodium combines with a *carrier* in the substance of the membrane and in this form diffuses to the opposite wall of the membrane where it is released. Metabolic processes supply the energy for the chemical reactions that cause sodium to be transported through the membrane.

Other substances beside sodium that are actively absorbed through the tubular epithelial cells include *glucose, amino acids, calcium ions, potassium ions, phosphate ions,* and *urate ions.* The basic principles of their absorption are also discussed in Chapter 4.

In addition, some substances, especially *hydrogen ions,* are *actively secreted* into certain portions of the tubules. Active secretion occurs in the same way as active absorption except that the cell membrane transports the secreted substance in the opposite direction.

Diffusion Through the Tubular Membrane. Diffusion can occur through the tubular membrane all the way from the lumen of the tubule to the peritubular fluid or in the opposite direction as the result of either a concentration difference or an electrical difference.

DIFFUSION AS A RESULT OF A CONCENTRATION DIFFERENCE — OSMOSIS OF WATER.   To give an example of diffusion as a result of a concentration difference, let us first consider the transport of water through the tubular membrane.

When sodium and other solutes are transported from the lumen of the tubule, this decreases the concentration of solutes in the tubular fluid and thereby *increases* the tubular concentration of water. Conversely, the increase in solute concentration in the peritubular fluid *decreases* the concentration of water there. As a result, a concentration difference of water develops between the tubular lumen and the peritubular fluid and causes net diffusion of water from the lumen into the peritubular fluid. This process is called *osmosis*. The basic principles of diffusion and osmosis were also described in detail in Chapter 4.

DIFFUSION OF UREA AND OTHER NONACTIVELY TRANSPORTED SOLUTES. When water is reabsorbed by osmosis, the concentration of urea in the tubular fluid rises, which obviously establishes a concentration difference for urea between the tubular and peritubular fluids. This in turn causes the urea to diffuse from the tubular fluid into the peritubular fluid. These effects also occur for other tubular solutes that are not actively reabsorbed.

The rate of reabsorption of a nonactively reabsorbed solute is determined by (1) the amount of water that is reabsorbed, because this determines the tubular concentration of the solute, and (2) the permeability of the tubular membrane for the solute. The permeability of the membrane for urea is far less than that for water, which means that less urea is reabsorbed than water. The permeability of the tubular membrane for reabsorption of creatinine, inulin (a large polysaccharide), mannitol (a disaccharide), and sucrose is zero, which means that once these substances have filtered into the glomerular filtrate their total quantity appears in the urine.

DIFFUSION CAUSED BY ELECTRICAL DIFFERENCES ACROSS THE TUBULAR MEMBRANE. The active absorption of sodium from the tubule creates negativity inside the tubule with respect to the peritubular fluid, as was illustrated in Figure 34–6. In the proximal tubule, this electrical difference is approximately 20 mv. In the distal tubule, it averages 50 to 60 mv., but can rise to as high as 120 mv. when sodium reabsorption occurs rapidly.

Note particularly that the peritubular fluid is *positive* with respect to the tubular fluid. This causes *negative ions,* such as *chloride, phosphate,* and *bicarbonate ions,* all to be attracted from the tubular fluid toward the peritubular fluid. The factor that determines how rapidly these ions will diffuse along this electrical gradient is the degree of permeability of the epithelial cells to the different ions. The permeability to chloride ions is especially great, so that chloride ions diffuse in this manner to a great extent.

PASSIVE SECRETION OF POTASSIUM BY DIFFUSION. Though the process of diffusion is mainly utilized for tubular absorption, potassium is secreted by the process of diffusion in the distal tubules and collecting tubules. The mechanism for this is as follows: Sodium reabsorption from the tubules causes a very negative electrical potential in their lumens, sometimes as great as −120 millivolts; this potential in turn pulls potassium passively through the epithelial cells into the lumen. This is a very important mechanism for control of potassium concentration in the body fluids and will be discussed further in Chapter 36.

ABSORPTIVE CAPABILITIES OF DIFFERENT TUBULE SEGMENTS

In subsequent sections of this chapter the absorption and secretion of specific substances in different segments of the tubular system will be discussed. However, it is first important to point out basic differences between the absorptive and secretory capabilities of the different types of tubules.

Figure 34–7 illustrates the cellular characteristics of the tubular membrane in (1) the proximal tubule, (2) the thin segment of the loop of Henle, and (3) the distal tubule. The proximal tubular cells have the appearance of being highly metabolic cells, having large numbers of mitochondria to support extremely rapid active transport processes; and, true enough, one finds that about 80 per cent of all reabsorptive and secretory processes that occur in the tubular system take place in the proximal tubules. That is, only 20 per cent of the glomerular filtrate normally passes the whole distance through the proximal tubules, the remaining 80 per cent being reabsorbed before reaching the loops of Henle.

The thin segments of the loops of Henle neither reabsorb nor secrete large quantities of fluid and solutes, but they play a specific role in concentrating sodium and chloride in the

Proximal tubule   Thin segment   Distal tubule

Figure 34–7 Characteristics of the epithelial cells in different tubular segments.

peritubular fluids of the medulla, as will be discussed in the following chapter in relation to the formation of concentrated urine.

The distal tubules and collecting ducts normally reabsorb about 12 to 15 per cent of the total volume of the glomerular filtrate. Therefore, the epithelium of these tubules does not require the tremendous metabolic activity that is required in the proximal tubules. On the other hand, the distal tubules and collecting ducts are composed of much less permeable cells than those of the proximal tubules, which allows them to reabsorb substances against far greater electrochemical gradients without the interfering problem of rapid back-diffusion through the membrane. Therefore, these tubules play specific roles in the final conditioning of the fluid that is to be passed into the urine.

With this brief introduction to the capabilities of the different types of tubules, we will now discuss reabsorption and secretion of specific substances at different points in the tubular system.

REABSORPTION AND SECRETION OF INDIVIDUAL SUBSTANCES IN DIFFERENT SEGMENTS OF THE TUBULES

Transport of Water and Flow of Fluid at Different Points in the Tubular System. Water transport occurs entirely by osmotic diffusion. That is, whenever some solute in the glomerular filtrate is absorbed either by active absorption or by diffusion caused by an electrochemical gradient, the resulting decreased concentration of solute in the tubular fluid and increased concentration in the peritubular fluid causes osmosis of water out of the tubules.

Figure 34–8 depicts aggregate rate of water flow axially along the lumens of each segment of the tubular system. In both kidneys of man, the total volumes of water flowing into each segment of the tubular system each minute (under normal resting conditions) are the following:

	ml./min.
Glomerular filtrate	125
Flowing into the loops of Henle	25
Flowing into the distal tubules	18
Flowing into the collecting ducts	6
Flowing into the urine	1

From this chart one can also deduce the per cent of the glomerular filtrate water that is reabsorbed in each segment of the tubules, as follows:

Figure 34–8 Volume flow of fluid in each segment of the tubular system per minute. Note that the flow is plotted on a semi-logarithmic scale, illustrating the tremendous difference in flow between the earlier and later segments of the tubules.

	Per Cent
Proximal tubules	80
Loop of Henle	6
Distal tubules	9
Collecting ducts	4
Passing into the urine	1

We shall see later in the chapter that these values vary greatly under different operational conditions of the kidney, particularly when the kidney is forming very dilute or very concentrated urine.

PRESSURES AT DIFFERENT POINTS IN THE TUBULAR SYSTEM. Figure 34–8 also illustrates the pressures in different segments of the tubular system. These pressures have been measured in small mammals, in the rat for instance, by inserting a micropipet into different parts of the tubular system. At the beginning of the proximal tubules, the pressure is approximately 14 mm. Hg and in the distal tubules about 7 mm. Hg, thus giving a pressure difference of 7 mm. Hg to move the fluid this distance along the tubules. The pressure at the tips of the collecting ducts where they empty into the pelves of the kidneys is about zero, giving another 7 mm. Hg pressure difference for causing flow along the distal tubules and collecting ducts. The distal tubules and collecting ducts have a certain amount of rigidity which keeps them from collapsing, for the intrarenal pressure on the outsides of the tubules is about 10 mm. Hg.

Transport of Substances of Nutritional Value to the Body—Glucose, Proteins, Amino Acids, Acetoacetate Ions, and Vitamins. Five different substances in the glomerular filtrate of particular importance to bodily nutrition are glucose, proteins, amino acids, acetoacetate ions, and the vitamins. Normally all of these are completely or almost completely reab-

Figure 34–9 Reabsorption of the nutritionally important substances in the proximal tubules, and poor reabsorption of the metabolic end-products in all segments of the tubules. Note the total absence of reabsorption of inulin and the secretion of para-aminohippuric acid into the proximal tubules.

sorbed by active processes in the *proximal tubules* of the kidney, as shown in Figure 34–9 for glucose, protein, and amino acids. Therefore, almost none of these substances remain in the tubular fluid entering the loop of Henle.

SPECIAL MECHANISM FOR ABSORPTION OF PROTEIN. As much as 30 grams of protein filters into the glomerular filtrate each day. This would be a great metabolic drain on the body if the protein were not returned to the body fluids. Because the protein molecule is much too large to be transported by the usual transport processes, protein is absorbed through the brush border of the proximal tubular epithelium by pinocytosis, which means simply that the protein attaches itself to the membrane and this portion of the membrane then invaginates to the interior of the cell. Once inside the cell the protein is probably digested into its constituent amino acids which are then actively absorbed through the base of the cell into the peritubular fluids. Details of the pinocytosis mechanism were discussed in Chapter 4.

Poor Transport of the Metabolic End-Products Urea, Creatinine, Urates, and Others. Figure 34–9 illustrates the rates of flow of three major metabolic end-products in the different segments of the tubular system — urea, creatinine, and urate ions. Note, especially, that only small quantities of *urea* — about 40 per cent of the total — are reabsorbed during the entire course through the tubular system.

Creatinine is not reabsorbed in the tubules

at all; and, indeed, small quantities of creatinine are actually secreted into the tubules by the proximal tubules so that the total quantity of creatinine increases about 20 per cent.

The *urate ion* is absorbed much more than urea — about 86 per cent reabsorption. But even so, large quantities of urate remain in the fluid that finally issues into the urine.

Several other end-products, such as *sulfates, phosphates,* and *nitrates,* are transported in much the same way as urate ions. All of these are normally reabsorbed to a far less extent than is water so that their concentrations become greatly increased as they flow along the tubules. Yet, each is actively reabsorbed under some conditions, which keeps their concentrations in the extracellular fluid from ever falling too low.

Transport of Inulin and Para-aminohippuric Acid by the Tubules. Note also in Figure 34–9 that the rate of flow of *inulin,* a large polysaccharide, remains exactly the same throughout the entire tubular system. The cause of this is simply that inulin is neither reabsorbed nor secreted in any segment of the tubules.

Finally, Figure 34–9 shows that the rate of flow of *para-aminohippuric acid* (PAH) increases about five-fold as the tubular fluid passes through the proximal tubules; then its rate of flow remains constant in the other tubules. This is because large quantities of PAH are *secreted* into the tubular fluid by the proximal tubular epithelial cells, and it is not reabsorbed in any segment of the tubular system.

These two substances play an important role in experimental studies of tubular function, as will be discussed later in the chapter.

Transport of Different Ions by the Tubular Epithelium — Sodium, Chloride, Bicarbonate, and Potassium. Figure 34–10 illustrates the rates of flow of different important ions — sodium, chloride, bicarbonate, and potassium — in different segments of the tubular system. Note that all these decrease markedly as the tubular fluid progresses from glomerular filtrate to urine.

Positive ions are generally transported through the tubular epithelium by active transport processes (except for potassium secretion, as noted subsequently), whereas negative ions are usually transported passively as a result of electrical differences developed across the membrane when the positive ions are transported. For instance, when sodium ions are transported out of the tubular fluid, the resulting electronegativity that develops in the tubular fluid causes chloride ions to follow in the wake of the sodium ions. But, despite the general rule that negative ions are usually passively reabsorbed, under some conditions, such as in extreme metabolic alkalosis, chloride ions probably can be actively reabsorbed in the distal tubules.

The increased flow of all ions in the first half of the loop of Henle is caused by *diffusion* of ions into the tubules from the peritubular fluids of the medulla, which is explained later in the chapter in relation to the mechanism for concentrating urine.

SECRETION OF POTASSIUM AND HYDROGEN IONS IN THE DISTAL TUBULES AND COLLECTING DUCTS. Note also in Figure 34–10 that the flow of potassium ions normally increases four or more fold as the tubular fluid passes through the distal tubules and collecting ducts. This results from *passive secretion of potassium.* That is, the strong electronegativity that develops in the distal tubules and collecting ducts actually pulls potassium into the tubules. This will be discussed at greater length in the following chapter in relation to the regulation of potassium concentration in the extracellular fluids.

Hydrogen ions are actively secreted in the proximal tubules, distal tubules, and collecting ducts. This secretion is controlled by the hydrogen ion concentration of the extracellular fluid, which will be discussed in Chapter 37.

SPECIAL ASPECTS OF BICARBONATE ION

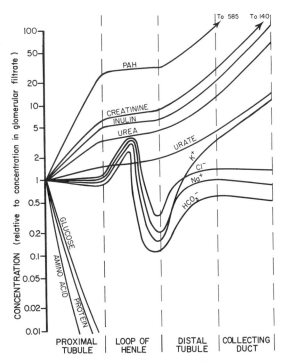

Figure 34–11 Composite figure, showing average changes in concentrations of different substances at different points in the tubular system.

TRANSPORT. Bicarbonate ion is primarily reabsorbed in the form of carbon dioxide rather than in the form of bicarbonate ion itself. This occurs as follows: The bicarbonate ions first combine with hydrogen ions that are secreted into the tubular fluid by the epithelial cells. The reaction forms carbonic acid which then dissociates into water and carbon dioxide. The carbon dioxide, being highly lipid-soluble, diffuses rapidly through the tubular membrane into the peritubular fluids. When more bicarbonate ions are present than there are available hydrogen ions, almost all of the excess bicarbonate ions flow on into the urine because the tubules are only slightly permeable to them.

TRANSPORT OF OTHER IONS. Though we know much less about transport of other ions besides the four illustrated in Figure 34–11, in general essentially all of them can be reabsorbed either by active transport or as a result of electrical differences across the membrane. Thus, calcium, magnesium, and other positive ions are actively reabsorbed, and many of the negative ions are reabsorbed as a result of electrical differences that develop when the positive ions are reabsorbed. In addition, certain negative ions—urate, phosphates, sulfates, and nitrates—can probably be reabsorbed by active transport, this occurring primarily in the proximal tubules.

Figure 34–10 Reabsorption of different ions in the tubules. Note, also, the secretion of potassium in the distal tubule and collecting duct.

CONCENTRATION OF DIFFERENT SUBSTANCES AT DIFFERENT POINTS IN THE TUBULES

Whether or not a substance becomes concentrated in the tubular fluid as it moves along the tubules is determined by the *relative reabsorption of the substance versus the reabsorption of water*. If a greater percentage of water is reabsorbed, the substance becomes progressively more and more concentrated. Conversely, if a greater percentage of the substance is reabsorbed, it becomes more and more dilute. Therefore, we can combine the curves of Figures 34–8, 34–9, and 34–10 to determine which substances become concentrated in the tubular system and which become diluted. Figure 34–11 illustrates this combination, showing three different classes of substances as follows:

First, the nutritionally important substances —glucose, protein, and amino acids—are reabsorbed so much more rapidly than water that their concentrations fall extremely rapidly in the proximal tubules and remain essentially zero throughout the remainder of the tubular system as well as in the urine.

Second, the concentration of the metabolic end-products as well as of inulin and PAH become progressively greater throughout the tubular system, because all of these substances are reabsorbed to a far less extent than is water. Note that potassium ions also fall into this category because a far greater proportion of potassium ions is normally removed each day from the extracellular fluid than is water.

Third, many other ions are normally excreted into the urine in concentrations not greatly different from those in the glomerular filtrate and extracellular fluid. That is, water, sodium, chloride, and bicarbonate ions, on the average, are normally reabsorbed from the tubules in not too dissimilar proportions.

TABLE 34–1.   Relative Concentrations of Substances in the Glomerular Filtrate and in the Urine

	Glomerular Filtrate (125 ml./min.)		Urine (1 ml./min.)		Conc. Urine/ Conc. Plasma (Plasma Clearance per Minute)
	Quantity/min.	Concentration	Quantity/min.	Concentration	
Na+	17.7 mEq.	142 mEq./l.	0.128 mEq.	128 mEq./l.	0.9
K+	0.63	5	0.06	60	12
Ca++	0.5	4	0.0048	4.8	1.2
Mg++	0.38	3	0.015	15	5.0
Cl-	12.9	103	0.134	134	1.3
HCO₃-	3.5	28	0.014	14	0.5
$H_2PO_4^-$ HPO_4^{--}	0.25	2	0.05	50	25
SO₄--	0.09	0.7	0.033	33	47
Glucose	125 mg.	100 mg.%	0 mg.	0 mg.%	0.0
Urea	33	26	18.2	1820	70
Uric acid	3.8	3	0.42	42	14
Creatinine	1.4	1.1	1.96	196	140
Inulin	....	....	....	....	125
Diodrast	....	....	....	....	560
PAH	....	....	....	....	585

Table 34–1 summarizes the concentrating ability of the tubular system for different substances. It also gives the actual quantities of the different substances normally handled by the tubules each minute.

Variability of Concentrating Power. A word of caution must be given at this point to emphasize that the quantitative values for reabsorption of different substances, as given in Figures 34–8, 34–9, and 34–10, and for concentrating power of the tubules, as given in Figure 34–11 and Table 34–1, are *average values.* These values change greatly in different physiological states of the kidney. For instance, when essentially no antidiuretic hormone is secreted by the neurohypophyseal mechanism, reabsorption of water becomes greatly reduced; the presence of the extra water in the urine dilutes essentially all the substances listed in Table 34–1 by as much as four- to five-fold. Conversely, secretion of large quantities of antidiuretic hormone causes such excessive reabsorption of water that all the substances can be concentrated to as much as four to five times the values given in Table 34–1.

REFERENCES

See bibliography at end of Chapter 35.

Formation of Urine by the Kidney: II. Regulation of Urine Concentration; Autoregulation of Renal Blood Flow; and the Concept of Clearance

In the previous chapter we discussed primarily the mechanisms by which glomerular filtrate is formed and how it is then processed in the tubules to form urine. However, the composition of urine can change from moment to moment to reflect the needs of the body for excreting different substances. For instance, the body at times has great excesses of water and, therefore, must excrete a dilute urine; at other times the body has a deficit of water and, therefore, must excrete minimum water in the urine while still excreting the end-products of metabolism and other substances that must be removed from the body fluids.

The purpose of the present chapter, therefore, is to begin discussing the mechanisms by which the kidneys adapt to the changing needs of the body for excretion of different substances. We will discuss in the present chapter some of the major principles of renal regulation and, in the following chapter, the regulation of specific solutes in the extracellular fluids.

CONCENTRATING AND DILUTING MECHANISM OF THE KIDNEY — THE COUNTER-CURRENT MECHANISM

Since water is reabsorbed from the tubules entirely passively while many solutes in the tubular fluid are absorbed by active absorption, it is easy to understand how a dilute urine could be formed: This could be achieved by simply absorbing the solutes from the glomerular filtrate and leaving the water behind to be excreted in the urine. This is basically the means by which a dilute urine is formed, as we shall see in the following paragraphs.

However, the process for concentrating urine is not nearly so simple, and there are times when it is exceedingly important to concentrate the urine as much as possible to conserve water in the body — for instance, when one is exposed to desert conditions with an inadequate supply of water. Fortunately, the kidneys have developed a special mechanism for concentrating the urine, called the *counter-current mechanism.*

The counter-current mechanism depends on a special anatomical arrangement of the loops of Henle and vasa recta. The *loops of Henle* of approximately one fifth of the nephrons, called the *juxtamedullary nephrons,* dip deep into the medulla and then return to the cortex, some dipping all the way to the tips of the papillae that project into the renal pelvis. Also, the straight peritubular capillaries called the *vasa recta* loop down into the medulla and then back out of the medulla to the cortex. These arrangements of the different parts of the juxtamedullary nephron and of the vasa recta are illustrated diagrammatically in Figure 35–1.

Hyperosmolality of the Medullary Fluid. The normal osmolality of the glomerular filtrate as it enters the proximal tubules is about 300 milliosmols/liter. However, as shown by the numbers in Figure 35–1, the osmolality of the interstitial fluid in the medulla of the kidney becomes progressively greater the more deeply one goes into the medulla, increasing from 300 milliosmols/liter in the cortex to 1200 milliosmols/liter at the pelvic tip of the medulla.

| VASA RECTA | INT. FLUID | LOOP OF HENLE | INT. FLUID | COLLECTING DUCT |

Figure 35–1   The "counter-current" mechanism for concentrating the urine.

The cause of this greatly increased osmolality of the medullary interstitial fluid is active transport of sodium out of the loop of Henle into the medullary interstitial fluid. The very dark arrows through the wall of the ascending limb of the loop of Henle illustrate this active absorption of sodium. In addition, a slight amount of sodium is actively absorbed through the walls of the collecting duct into the interstitial fluid, and it is likely that sodium is actively absorbed from the descending limb of the loop of Henle. As the sodium is actively absorbed, an electrical difference develops across the walls of the tubules which pulls chloride ions and some other negative ions into the interstitial fluid as well. Thus, the concentrations of these ions increase markedly in the interstitial fluid of the medulla.

The question that must now be asked is: Why does the osmolar concentration of solutes rise so high in the medulla but not in the fluid of the cortex where equally large quantities of sodium are reabsorbed from the proximal and distal tubules? The answer to this is two-fold: First, the blood flow through the vasa recta is extremely sluggish. Only about 1 to 2 per cent of the total renal blood flow passes through these vessels in comparison with 90 per cent through the cortical peritubular capillaries. Obviously, the sluggish flow in the medulla would help to prevent removal of the large quantities of sodium chloride being continually absorbed into the medullary interstitial fluid.

The second cause of the high concentration of osmotically active substances in the medulla is the presence of counter-current mechanisms operating in both the loop of Henle and the vasa recta. These may be explained as follows:

Counter-Current "Multiplier" of the Loop of Henle. A counter-current fluid mechanism is one in which fluid flows through a long U tube, the fluid flowing in one arm of the U and then out the opposite arm. When the fluids in the two parallel streams of flow interact appropriately with each other, tremendous concentrations of solute can be built up at the tip of the loop. As an example, let us first observe the loop of Henle. The descending limb of this loop is highly permeable to sodium chloride, while the ascending limb has a strong transport mechanism to remove sodium chloride from the tubule into the interstitial fluid. Therefore, each time sodium chloride is transported out of the ascending limb, it almost immediately diffuses into the descending limb, thereby increasing the concentration of sodium chloride in the tubular fluid that flows downward toward the tip of the loop. As illustrated by the arrows in the diagram, this sodium chloride flows on around the loop and is then actively transported again out of the ascending limb. In addition, new sodium chloride is entering the tubular system in the glomerular filtrate and is also transported out of the ascending limb. Thus, by this retransport again and again of sodium chloride, plus constant addition of more and more sodium chloride, one can readily see why the concentration of sodium chloride in the medullary fluid rises very high. This mechanism of the loop of Henle is called a counter-current multiplier.

Counter-Current Mechanism in the Vasa Recta. Another factor that helps explain the very high concentration of sodium chloride in the medulla is the *counter-current flow of blood in the vasa recta.* This is illustrated to the left in Figure 35–1. As blood flows down the descending limbs of the vasa recta, sodium chloride diffuses into the blood from the interstitial fluid, causing the osmolar concentration to rise progressively higher, to a maximum concentration of 1200 milliosmols/liter at the tips of the vasa recta. Then as the blood flows back up the loop, the extreme diffusibility of sodium chloride through the capillary membrane allows essentially all of the extra sodium chloride to be lost once again back into the interstitial fluid. Therefore, by the time the blood finally leaves the medulla, its osmolar concentration is only slightly greater than that of the blood that had initially entered the vasa recta. As a result, blood flowing through the vasa recta carries only a minute amount of sodium chloride away from the medulla.

Dilution of the Tubular Fluid in the Ascending Limb of the Loop of Henle. The ascending limb of the loop of Henle is highly impermeable to water, and yet tremendous quantities of sodium chloride are transported out of the ascending limb into the interstitial fluid. Therefore, as illustrated in Figure 35–1, the fluid in the ascending limb of the loop of Henle becomes progressively more dilute as it ascends toward the cortex, decreasing to an osmolality of about 100 milliosmols/liter before leaving the loop of Henle.

Mechanism for Excreting a Dilute Urine. Now that some of the basic aspects of the counter-current mechanism of the kidney have been described, we can explain easily how the kidney secretes a dilute urine. Since the tubular fluid has a concentration of only 100 milliosmols/liter as it leaves the ascending limb of the loop of Henle, the kidney can excrete a dilute urine by simply allowing this fluid to empty directly into the pelvis of the kidney. And the kidney has a mechanism for doing this. In the absence of *antidiuretic hormone,* which is secreted by a hypothalamic-neurohypophyseal mechanism, the pores of the epithelial cells in the distal tubules and collecting ducts become very small, too small for absorption of water or almost any other substance that is normally absorbed by diffusion. As a result, in the absence of this hormone almost no water is absorbed from the distal tubules and collecting ducts,

and the dilute tubular fluid flushes rapidly on through these tubules into the urine. However, a slight amount of sodium and a few other substances are actively absorbed, which reduces the osmolality sometimes to as low as 70 milliosmols/liter by the time the fluid enters the urine; this is about one-fourth the osmolality of the plasma and glomerular filtrate.

Mechanism for Excreting a Concentrated Urine. The mechanism for excreting a concentrated urine is exactly opposite to that for excreting a dilute urine. In this case, large quantities of *antidiuretic hormone* are secreted into the body fluids, and this in turn increases the sizes of the pores in the distal tubules and collecting ducts until water molecules can be absorbed with ease. The very dilute fluid leaving the ascending limb of the loop of Henle rapidly loses water in the distal tubule by osmosis, and its osmolality rises to equilibrium with that in the cortical interstitial fluids, rising to 300 milliosmols/liter as illustrated in Figure 35–1. Now, the tubular fluid passes back through the medulla a second time, this time downward through the collecting duct. Here it is exposed to the hyperosmolality of the medullary interstitial fluid, for the collecting duct passes directly through the medulla on its way to the pelvis of the kidney. Therefore, as illustrated in Figure 35–1, large quantities of water are absorbed by osmosis into the medullary interstitial fluid from the collecting duct. Therefore, the concentration of the fluid in the duct rises to approach the osmolality of the fluid in the interstitial fluid, about 1200 milliosmols per liter, so that the fluid leaving the collecting duct to enter the pelvis of the kidney is very concentrated urine. Thus, the maximum concentration of the urine that can usually be achieved is about four times the osmolality of the plasma and glomerular filtrate.

Summary of the Osmolar Concentration Changes in the Different Segments of the Tubules. Figure 35–2 illustrates the changes in osmolality of the fluid as it passes through the different segments of the tubules. In the proximal tubules, the tubular membrane is so highly permeable to water that any time a solute is transported across the membrane almost exactly a proportionate amount of water crosses the membrane by osmosis at the same time; therefore, the osmolality of the fluid remains almost exactly equal to that of the glomerular filtrate, 300 milliosmols/liter, throughout the entire extent of the proximal tubule.

In the descending limb of the loop of Henle the osmolality rises rapidly because of (a) dif-

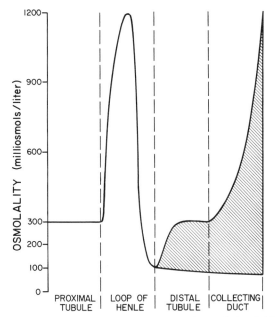

Figure 35–2 Changes in osmolality of the tubular fluid as it passes through the tubular system.

fusion of sodium and chloride ions into the tubular fluid as it descends in the loop and (b) osmosis of water out of the loop into the interstitial fluid at the same time.

In the ascending limb of the loop of Henle the transport of large quantities of sodium and chloride out of the fluid, and yet almost total impermeability of the membrane to water, causes the osmolality to fall drastically, to almost exactly 100 milliosmols/liter at the point where the fluid leaves the ascending limb of the loop of Henle.

Beyond the loop of Henle the osmolality of the tubular fluid depends on the presence or absence of antidiuretic hormone. In the absence of antidiuretic hormone the fluid actually becomes a little more dilute, as shown by the lower border of the darkened area in Figure 35–2, because of active transport of sodium and other substances through the membranes of the distal tubule and collecting duct, falling to a minimum of about 70 milliosmols/liter. In the presence of excess antidiuretic hormone, water diffuses through the distal tubular membrane so rapidly that the distal tubular fluid comes almost to complete equilibrium with the cortical interstitial fluid; that is, to an osmolality of 300 milliosmols/liter. Then, as the fluid passes down the collecting duct its osmolality comes almost to complete equilibrium with the osmolality of the interstitial fluids of the renal medulla, rising to an osmolality of 1200 milliosmols/liter as it issues into the urine.

All degrees of concentration of the urine can occur between the two extremes of 70 and 1200 milliosmols/liter. This, of course, depends mainly on the quantity of antidiuretic hormone present in the body fluid at any given time. The importance of the antidiuretic hormone and its relationship to control of osmolality of the body fluids is discussed in the following chapter.

REGULATION OF RENAL BLOOD FLOW AND GLOMERULAR FILTRATION RATE

AUTOREGULATION

In the earlier sections of this chapter, we noted that blood flow through the kidneys and the glomerular filtration rate both remain nearly constant despite widely varying arterial pressure. For instance, the arterial pressure can change from as low as 70 mm. Hg to as high as 200 or more mm. Hg, and yet the renal blood flow and glomerular filtration rate change only a few per cent. This effect is illustrated in Figure 35–3. It is called *autoregulation of renal blood flow and autoregulation of glomerular filtration rate.* The exact cause of this phenomenon of autoregulation has not been discovered. However, an exciting concept for which there is at least some experimental support is that autoregulation in the kidney results from a feedback mechanism operating from the distal tubule to control blood flow into the glomerulus, as follows:

The Juxtaglomerular Apparatus and Its Possible Feedback Role in the Nephron. Figure 35–4 illustrates a peculiar characteristic of each of the two million nephrons in the two kidneys.

Figure 35–3 Autoregulation of renal blood flow (*RBF*) and glomerular filtration rate (*GFR*), and lack of autoregulation of urine flow.

Immediately beyond the point where the ascending limb of the loop of Henle empties into the distal tubule, the distal tubule returns to the glomerulus and passes between the afferent and efferent arterioles. The epithelial cells of the distal tubule where it lies in apposition to the afferent arteriole become condensed into small, tightly packed cells which are collectively called the *macula densa*. The basement membrane outside the macula densa disappears so that the contact between the macula densa and afferent arteriole is intimate. The smooth muscle cells of the afferent arteriole at this point of contact are swollen and are filled with granules of the substance *renin,* which was discussed in relation to hypertension in Chapter 26.

It is almost certain that the intimate relationship between the distal tubule and the afferent arteriole is the basis for some type of feedback mechanism whereby conditions in the distal tubule control blood flow through the afferent arteriole.

TYPES OF FEEDBACK THAT MIGHT OCCUR. Two types of feedback have received special attention. One of these is *osmotic feedback* and the other is *feedback based on the concentration of sodium ions* in the distal tubule.

In the *osmotic theory,* it is assumed that too rapid formation of glomerular filtrate causes the osmolality of the fluid at the macula densa to decrease below normal. The dilute fluid at the macula densa then causes the afferent arteriole to constrict, which decreases glomerular blood flow and formation of glomerular filtrate back to a normal level. Too little formation of glomerular filtrate would cause excess osmolality of the fluid at the macula densa, which would cause dilatation of the afferent arteriole with resultant increase in glomerular filtration back toward normal. Thus, this feedback would maintain essentially constant glomerular filtration rate and constant renal blood flow—that is, it would cause renal autoregulation.

The *sodium theory* assumes that too rapid flow of glomerular filtrate through the tubules allows the sodium concentration to rise in the distal tubule. This in turn causes the afferent arteriole to constrict, decreasing the glomerular filtration rate back toward normal. Converse effects would occur if the flow of glomerular filtrate were insufficient. Thus, this mechanism also could cause autoregulation.

AUTOREGULATION IN EACH SINGLE NEPHRON. Note especially that if a feedback type of autoregulation occurs at the juxtaglomerular apparatus, it would autoregulate the glomerular filtration rate in each nephron. This would

allow exactly the amount of glomerular filtrate to enter the tubules that can be processed each minute, no more, no less.

RELATIONSHIP OF GLOMERULAR FILTRATION RATE TO URINE OUTPUT—THE OVERFLOW PHENOMENON

Note in Figure 35–3 that, even though the glomerular filtration rate *is* autoregulated, *the rate of urine output is not autoregulated.* Instead, the urinary output rises greatly as the arterial pressure increases. The cause of this difference is the *overflow phenomenon,* which can be explained as follows:

Later in the chapter we will see that most substances actively absorbed by the tubules have a maximum rate of absorption called a "transport maximum." If the glomerular filtration rate is low, the tubular load of these substances becomes so little that essentially all of them are reabsorbed. Since this absorption of solute causes almost proportionate osmosis of water out of the tubules, most of the water is also reabsorbed. On the other hand, as the glomerular filtration rate rises, the amount of substances that enters the tubules each minute eventually exceeds the amount that can be reabsorbed. Beyond this glomerular filtration rate, the substances begin to spill into the urine, and proportionate quantities of water fail to be reabsorbed by the tubules, thus increasing the quantity of water also in the urine. In other words, when the absorptive ability of the tubules for each substance is exceeded, the phenomenon of *overflow* takes place. Thus, as the glomerular filtration rate increases higher and higher, more and more substances and more and more water overflow, increasing the urine flow tremendously.

Now, let us return to Figure 35–3 and observe the relationship between glomerular filtration rate and urine flow. Between the limits of 100 and 200 mm. Hg arterial pressure, the glomerular filtration rate increases only about 15 per cent. Yet, because of the overflow phenomenon, the urine flow increases 450 per cent, which is 30 times as great an increase as the glomerular filtration rate increase. Thus, one can readily see that extremely slight changes in glomerular filtration rate can cause tremendous changes in urine output.

Obligatory Urine Output. A few substances in the tubule, such as creatinine and urea, can be reabsorbed only slightly or not at all. Therefore, as long as there is any glomerular filtration rate at all, at least some of these substances remain in the tubules and prevent a small amount of water from being reabsorbed. This

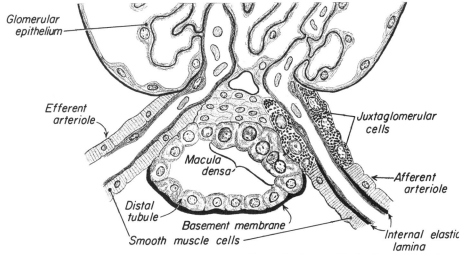

Figure 35–4 Structure of the juxtaglomerular apparatus, illustrating its possible feedback role in the control of nephron function. (Modified from Ham: Histology. J. B. Lippincott Co.)

fluid flows slowly through the tubules into the urine and is called the *obligatory urine output.* This minimum amount of urine output required to rid the body of the normal metabolic end-products each day is about 400 ml.

THE CONCEPT OF "PLASMA CLEARANCE"

The term "plasma clearance" is used to express the ability of the kidneys to remove various substances from the plasma. If the plasma passing through the kidneys contains 0.1 gram of a substance in each 100 ml., and 0.1 gram of this substance passes into the urine each minute, then 100 ml. of the plasma are cleaned or "cleared" of the substance per minute.

Referring to Table 34–1 in the previous chapter, note that the normal concentration of urea in each 100 ml. of plasma and glomerular filtrate is 26 mg., and the quantity of urea that passes into the urine each minute is approximately 18.2 mg. Therefore, the quantity of plasma that completely loses its entire content of urea each minute may be calculated by dividing the quantity of urea entering the urine each minute by the quantity of urea in each milliliter of plasma. Thus, $18.2 \div 0.26 = 70$; that is, 70 ml. of plasma, by filtering through the glomerulus and then being reabsorbed by the tubules, *is cleaned* or *cleared* of urea each minute. This is known as the *plasma clearance* of urea.

Plasma clearance for any substance can be calculated by the following formula:

The concept of plasma clearance is important because it is an excellent measure of kidney function, and the clearance of different substances can be determined by simply analyzing the concentrations of the substances simultaneously in the plasma and in the urine.

The plasma clearances of the usual constituents of urine are shown in the last column of Table 34–1.

INULIN CLEARANCE AS A MEASURE OF GLOMERULAR FILTRATION RATE

Inulin is a polysaccharide that has the specific attributes of not being reabsorbed to any extent by the tubules of the nephron and yet being of small enough molecular weight (about 5200) that it passes through the glomerular membrane as freely as the crystalloids and water of the plasma. Also, inulin is not actively secreted even in the minutest amount by the tubules. Consequently, glomerular filtrate contains the same concentration of inulin as does plasma, and as the filtrate flows down the tubules, all the inulin continues on into the urine. Thus, *all the glomerular filtrate formed is cleared of inulin, and this is equal to the amount of plasma that is simultaneously "cleared."* Therefore, the plasma clearance per minute of inulin is also equal to the glomerular filtration rate.

As an example, let us assume that it is found by chemical analysis that the inulin concentration in the plasma is 0.1 gram in each 100 ml., and that 0.125 gram of inulin passes into the urine per minute. Therefore, by dividing 0.1 into 0.125, one finds that 1.25 *100-milliliter portions* of glomerular filtrate must be formed each minute in order to deliver to the urine the analyzed quantity of inulin. In other words, 125 ml. of glomerular fil-

$$\text{Plasma clearance (ml./min.)} = \frac{\text{Quantity of urine (ml./min.)} \times \text{Concentration in urine}}{\text{Concentration in plasma}}$$

trate is formed per minute, and this is also the plasma clearance of inulin.

Inulin is not the only compound that can be used for determining the quantity of glomerular filtrate formed each minute, for the plasma clearance of any other substance that is totally diffusible through the glomerular membrane but is neither absorbed nor secreted by the tubular walls is equal to the glomerular filtration rate. *Mannitol* is another polysaccharide that is frequently used instead of inulin for such measurements.

PARA-AMINOHIPPURIC ACID (PAH) CLEARANCE AS A MEASURE OF PLASMA FLOW AND BLOOD FLOW THROUGH THE KIDNEYS

PAH, like inulin, passes through the glomerular membrane with perfect ease along with the remainder of the glomerular filtrate. However, it is different from inulin in that almost all the PAH remaining in the plasma after the glomerular filtrate is formed is secreted into the tubules by the tubular epithelium (if the plasma load of the PAH is considerably less than the Tm for PAH). Indeed, less than one tenth of the original PAH remains in the plasma when the blood leaves the kidneys.

One can use the clearance of PAH for estimating the *flow of plasma* through the kidneys. As an example, let us assume that 1 mg. of PAH is present in each 100 ml. of plasma, and that 5.85 mg. of PAH passes into the urine per minute. Consequently, 585 ml. of plasma is cleared of PAH each minute. Obviously, if this much plasma is cleared of PAH, at least this much plasma must have passed through the kidneys in this same period of time. And, since it is known that almost all the PAH is cleared from the blood as it passes through the kidneys, 585 ml. would be a reasonable first approximation of the actual plasma flow per minute.

Yet, to be still more accurate, one can correct for the average amount of PAH that is still in the blood when it leaves the kidney. In many different experiments the PAH clearance has averaged 91 per cent of the plasma load entering the kidneys. Therefore, the 585 ml. of plasma calculated above would be only 91 per cent of the total amount of plasma flowing through the kidneys. Dividing 585 by 0.91 gives a total plasma flow per minute of approximately 650 ml.

The 91 per cent removal of PAH as it passes through the kidneys is called the *extraction ratio*.

One can calculate the *total blood flow* through the kidneys each minute from the plasma flow and the hematocrit (the percentage of red blood cells in the blood). If the hematocrit is 45 per cent and the plasma flow 650 ml./min., the total blood flow through both kidneys is $650 \times (100/55)$, or 1182 ml./min.

Clearance of *Diodrast* can also be used for determining plasma flow or blood flow through the two kidneys. However, the extraction ratio of Diodrast averages about 0.85, slightly less than that of PAH.

CALCULATING THE FILTRATION FRACTION FROM PLASMA CLEARANCES

To calculate the filtration fraction—that is, the fraction of the plasma that filters through the glomerular membrane—one must determine (1) the plasma flow through the two kidneys and (2) the glomerular filtration rate per minute. Using 650 ml. plasma flow and 125 ml. glomerular filtration rate as values, we find that the calculated filtration fraction is 125/650, or, to express this as a percentage, 19 per cent.

EFFECT OF "TUBULAR LOAD" AND "TRANSPORT MAXIMUM" ON URINE CONSTITUENTS

Plasma Load and Tubular Load. The *plasma load* of a substance means the total amount of the substance in the plasma that passes through the kidney each minute. For instance, if the concentration of glucose in the plasma is 100 mg./100 ml., and 600 ml. of plasma passes through both kidneys each minute, the plasma *load of glucose* is 600 mg./min.

Normally a fraction of the plasma load filters into the glomeruli, and this portion is called the *tubular load*. In the example just given, if 125 ml. of glomerular filtrate is formed each minute with a glucose concentration of 100 mg. per cent, the tubular load of glucose is 100 mg. × 1.25, or 125 mg. of glucose per minute. Similarly, the load of sodium that enters the tubules each minute is approximately 18 mEq./min., the load of chloride ion is about 13 mEq./min., and the load of urea is approximately 33 mg./min., etc.

Maximum Rate of Transport of Actively Reabsorbed or Secreted Substances—The "Transport Maximum" (Tm). Since each substance that is actively reabsorbed requires a specific transport system in the tubular epithelial cells, the maximum amount that can be reabsorbed often depends on the maximum rate at which the transport system itself can operate, and this in turn depends on the total amounts of carrier and specific enzymes available. Consequently, for almost every actively reabsorbed substance, there is a maximum rate at which it can be reabsorbed; this is called the *transport maximum* for the substance, and the abbreviation is *Tm*. For instance, the Tm for glucose averages 320 mg./min. for the adult human being, and if the tubular load becomes greater than 320 mg./min., the excess above this amount is not reabsorbed but instead passes on into the urine. Ordinarily, though, the tubular load of glucose is only 125 mg./min., so that for practical purposes, all of it is reabsorbed.

Figure 35–5 demonstrates the relationship between tubular load of glucose, transport maximum for glucose, and rate of glucose loss into the urine. Note that when the tubular load is at its normal level of 125 mg./min. there is no detectable loss of glucose into the urine. However, when the tubular load rises above about 220 mg./min., significant quantities of glucose begin to appear in the urine. And, once the load has risen above approximately 400 mg./min., the loss into the urine thereafter remains *320 mg./min. less than the tubular load.* Thus, at 400 mg./min. tubular load, the loss is 80 mg./min., and at 800 mg./min. tubular load it is 480 mg./min. In other words, 320 mg./min. of the tubular load, which represents the transport maximum for glucose, is reabsorbed and all the remainder is lost into the urine.

THRESHOLD FOR SUBSTANCES THAT HAVE A TRANSPORT MAXIMUM. Every substance that has a transport maximum also has a "threshold" concentration in the plasma below which none of it appears in the urine and above which progressively larger quantities appear in the urine.

Thus, Figure 35–5 shows that glucose begins to spill into the urine when its tubular load exceeds 220 mg./min. The threshold concentration of glucose in plasma that gives this tubular load is 180 mg. per cent when the kidneys are operating at their normal glomerular filtration rate of 125 ml./min.

TRANSPORT MAXIMUMS OF IMPORTANT SUB-STANCES ABSORBED FROM THE TUBULES. Some of the important transport maximums for substances absorbed from the tubules are the following:

Glucose	320	mg./min.
Phosphate	0.1	mM./min.
Sulfate	0.06	mM./min.
Amino acids	1.5	mM./min.
Vitamin C	1.77	mg./min.
Urate	15	mg./min.
Plasma protein	30	mg./min.
Hemoglobin	1	mg./min.
Lactate	75	mg./min.
Acetoacetate	variable (about 30 mg./min.)	

It is particularly significant that sodium ions *do not* exhibit a transport maximum.

TRANSPORT MAXIMUMS FOR SECRETION. Many substances secreted by the tubules also exhibit transport maximums as follows:

	mg./min.
Creatinine	16
PAH	80
Diodrast	57 (of iodine)
Phenol red	56

DETERMINATION OF THE TRANSPORT MAXIMUM

The transport maximum for any substance can be determined by loading the tubular cells with far more of the substance than can be transported and then analyzing the amount that actually is transported.

The secretory transport maximum of either Diodrast (TmD) or para-aminohippuric acid (TmPAH) is often measured to estimate the amount of active tubular tissue in both kidneys. The normal TmD is 57 mg./min. (measured as iodine content), while the normal TmPAH is 80 mg./min. in the human adult. Whenever the transport maximum for either of these substances is less than normal, this is a sign of diminished total amount of tubular cells or tubular enzyme carrier systems for transport of the substance being measured. Therefore, transport maximum measurements can often give one an excellent assessment of tubular function in the kidneys.

A typical calculation of transport maximum for Diodrast is the following: When the concentration of Diodrast in the plasma rises above approximately 5 mg. per cent (measured in terms of its iodine content), the tubular walls become unable to excrete all of it from the plasma because the plasma load will have exceeded the transport maximum. To determine the transport maximum, inulin clearance must first be determined to measure the quantity of glomerular filtrate formed per minute. Let us assume that 125 ml. of glomerular

Figure 35–5 Relationship of tubular load of glucose to loss of glucose into the urine.

filtrate is formed each minute. Also, analysis of the plasma Diodrast iodine shows a concentration of 40 mg./100 ml., and analysis of the urine shows 107 mg. of Diodrast iodine entering the urine each minute. From these values it can be calculated that the amount of Diodrast iodine entering the urine by way of the glomerular filtrate is 40 mg. × 1.25, or 50 mg./min. Subtracting this quantity from the total 107 mg., one finds that the Diodrast transport maximum (TmD) is 57 mg./minute.

The transport maximum for glucose or for other substances that are actively reabsorbed can be determined by loading the plasma so greatly that the tubular load becomes greater than that which can be reabsorbed. For instance, if the tubular load of glucose is 400 mg./minute and the amount that passes into the urine is 80 mg./minute, then the transport maximum for glucose would be 320 mg./minute.

REFERENCES

Baines, A. D.; Gottschalk, C. W.; and Lassite, W. E.: Microinjection study of p-aminohippurate excretion by rat kidneys. *Amer. J. Physiol.,* 214:703, 1968.

Bennett, C. M.; Clapp, J. R.; and Berliner, R. W.: Micropuncture study of the proximal and distal tubule in the dog. *Amer. J. Physiol.,* 213:1254, 1967.

Bergeron, M., and Morel, F.: Amino acid transport in rat renal tubules. *Amer. J. Physiol.,* 216:1139, 1969.

Bowman, F. J., and Foulkes, E. C.: Antidiuretic hormone and urea permeability of collecting ducts. *Amer. J. Physiol.,* 218:231, 1970.

Brunette, M.; Wen, S. F.; Evanson, R. L.; and Dirks, J. H.: Micropuncture study of magnesium reabsorption in the proximal tubule of the dog. *Amer. J. Physiol.,* 216:1510, 1969.

De Wardener, H. E.: The Kidney. 3rd ed., Boston, Little, Brown and Company, 1968.

Dies, F.; Ramos, G.; Avelar, E.; and Matos, M.: Relationship between renal substrate uptake and tubular sodium reabsorption in the dog. *Amer. J. Physiol.,* 218:411, 1970.

Fanelli, G. M., Jr.; Bohn, D.; Stafford, S.: Functional characteristics of renal urate transport in the Cebus monkey. *Amer. J. Physiol.,* 218:627, 1970.

Forster, R. P.: Kidney, water, and electrolytes. *Ann. Rev. Physiol.,* 27:183, 1965.

Gagnon, J. A.; Mailloux, L. U.; Doolittle, J. E.; and Teschan, P. E.: An isotopic method for instantaneous measurements of effective renal blood flow. *Amer. J. Physiol.,* 218:180, 1970.

Giebisch, G.: Kidney, water and electrolyte metabolism. *Ann. Rev. Physiol.,* 24:357, 1962.

Giebisch, G., and Malnic, G.: Lessons from micropuncture. *Proc. Int. Union of Physiol. Sciences,* 6:67, 1968.

Giebisch, G., and Windhager, E. E.: Characterization of renal tubular transport of sodium chloride and water as studied in single nephrons, *Amer. J. Med.,* 34:1, 1963.

Giebisch, G., and Windhager, E. E.: Measurement of chloride movement across single proximal tubules of Necturus kidney. *Amer. J. Physiol.,* 204:387, 1963.

Glabman, S.; Klose, R. M.; and Giebisch, G.: Micropuncture study of ammonia excretion in the rat. *Amer. J. Physiol.,* 205:127, 1963.

Gottschalk, C. W.; Lassiter, W. E.; and Mylle, M.: Localization on urine acidification in the mammalian kidney. *Amer. J. Physiol.,* 198:581, 1960.

Gottschalk, C. W.; Lassiter, W. E.; Mylle, M.; Ulrich, K. J.; Schmidt-Nielsen, B.; O'Dell, R.; and Pehling, G.: Micropuncture study of composition of loop of Henle fluid in desert rodents. *Amer. J. Physiol.,* 204:532, 1963.

Hayslett, J. P.; Domoto, D. T.; Kashgarian, M.; and Epstein, F. H.: Role of physical factors in the natriuresis induced by acetylcholine. *Amer. J. Physiol.,* 218:880, 1970.

Hollenberg, N. K.; Epstein, M.; Guttmann, R. D.; Conroy, M.; Basch, R. I.; and Merrill, J. P.: Effect of sodium balance on intrarenal distribution of blood flow in normal man. *J. Appl. Physiol.,* 28:312, 1970.

Jamison, R. L.: Micropuncture study of superficial and juxtamedullary nephrons in the rat. *Amer. J. Physiol.,* 218:46, 1970.

Knox, F. G.: Effect of increased proximal delivery on furosemide natriuresis. *Amer. J. Physiol.,* 218:819, 1970.

Landwehr, D. M.; Schnermann, J.; Klose, R. M.; and Giebisch, G.: Effect of reduction in filtration rate on renal tubular sodium and water reabsorption. *Amer. J. Physiol.,* 215:687, 1968.

Lassiter, W. E.; Gottschalk, C. W.; and Mylle, M.: Micropuncture study of net transtubular movement of water and urea in nondiuretic mammalian kidney. *Amer. J. Physiol.,* 200:1139, 1961.

Lewy, J. E., and Windhager, E. E.: Peritubular control of proximal tubular fluid reabsorption in the rat kidney. *Amer. J. Physiol.,* 214:943, 1968.

Maack, T., and Kinter, W. B.: Transport of protein by flounder kidney tubules during long-term incubation. *Amer. J. Physiol.,* 216:1034, 1969.

Malnic, G., and Aires, M. M.: Microperfusion study of anion transfer in proximal tubules of rat kidney. *Amer. J. Physiol.,* 218:27, 1970.

Malnic, G.; Klose, R. M.; and Giebisch, G.: Micropuncture study of renal potassium excretion in the rat. *Amer. J. Physiol.,* 206:674, 1964.

Malvin, R. L.; Wilde, W. S.; and Sullivan, L. P.: Localization of nephron transport by stop flow analysis. *Amer. J. Physiol.,* 194:135, 1958.

Marsh, D. J.: Solute and water flows in thin limbs of Henle's loop in the hamster kidney. *Amer. J. Physiol.,* 218:824, 1970.

Maude, D. L.: Effect of K and ouabain on fluid transport and cell Na in proximal tubule in vitro. *Amer. J. Physiol.,* 216:1199, 1969.

Morgan, T.; Tadokoro, M.; Martin, D.; and Berliner, R. W.: Effect of furosemide on Na+ and K+ transport studied by microperfusion of the rat nephron. *Amer. J. Physiol.,* 218:292, 1970.

Mudge, G. H.; Berndt, W. O.; Lockhart, J.; and Saunders, A.: Renal tubular secretion-reabsorption of thiosulfate in the dog. *Amer. J. Physiol.,* 216:843, 1969.

Navar, L. G.; Guyton, Arthur C.; and Langston, J. B.: Effect of alterations in plasma osmolality on renal blood flow autoregulation. *Amer. J. Physiol.,* 211:1387, 1966.

Pitts, R.: Physiology of the Kidney and Body Fluids. 2nd ed., Chicago, Year Book Medical Publishers, Inc., 1968.

Shalhoub, R.; Webber, W.; Glabman, S.; Canessa-Fischer, M.; Klein, J.; deHaas, J.; and Pitts, R. F.: Extraction of amino acids from and their addition to renal blood plasma. *Amer. J. Physiol.,* 204:181, 1963.

Silverman, M.; Aganon, M. A.; and Chinard, F. P.: Specificity of monosaccharide transport in dog kidney. *Amer. J. Physiol.,* 218:743, 1970.

Slotkoff, L. M.; Eisner, G. M.; and Lilienfield, L. S.: Functional separation of renal cortical-medullary circulation: significance of Diodrast extraction. *Amer. J. Physiol.,* 214:935, 1968.

Smith, H. W.: The Kidney: Structure and Function in Health and Disease. New York, Oxford University Press, Inc., 1951.

Thurau, K.; Valtin, H.; and Schnermann, J.: Kidney. *Ann. Rev. Physiol.,* 30:441, 1968.

Waugh, W. H., and Shanks, R. G.: Cause of genuine autoregulation of the renal circulation. *Circ. Res.,* 8:871, 1960.

Wedeen, R. P., and Jernow, H. I.: Autoradiographic study of cellular transport of Hippuran-125I in the rat nephron. *Amer. J. Physiol.,* 214:776, 1968.

Wesson, L. G., Jr.: Physiology of the Human Kidney. New York, Grune & Stratton, Inc., 1969.

Wiederholt, M.; Hierholzer, K.; Windhager, E. E.; and Giebisch, G.: Microperfusion study of fluid reabsorption in proximal tubules of rat kidneys. *Amer. J. Physiol.,* 213:809, 1967.

Wilson, D. M.; Apter, J. T.; and Schwartz, F. D.: A model for measuring renal blood flow from plasma disappearance of iodopyracet. *J. Appl. Physiol.,* 28:79, 1970.

Wilson, O. H., and Scriver, C. R.: Specificity of transport of neutral and basic amino acids in rat kidney. *Amer. J. Physiol.,* 213:185, 1967.

Windhager, E. E.: Kidney, water, and electrolytes. *Ann. Rev. Physiol.,* 31:117, 1969.

Wolf, A. V.: Urinary concentrative properties. *Amer. J. Med.,* 32:329, 1962.

CHAPTER 36

Regulation of Extracellular Fluid Composition and Volume, and Regulation of Blood Volume by the Kidneys and by the Thirst Mechanism

In the discussion of homeostasis in Chapter 1 it was pointed out that the cells of the body can continue to live and function normally as long as the composition of the extracellular fluid remains within reasonably normal limits. Though most organs of the body are at least partially concerned with maintenance of constancy in the extracellular fluids, the kidneys are more directly responsible than any other organ for regulation of most of the constituents.

The different characteristics of the body fluids that the kidneys help to regulate are (1) the specific concentrations of the different ions, (2) the osmolality of the fluids—that is, the ratio of water to dissolved substances, (3) the volume of extracellular fluids, (4) the volume of blood, and (5) the hydrogen ion concentration of the extracellular fluids. The first four of these are discussed in the present chapter, and the last, regulation of hydrogen ion concentration, is discussed in Chapter 37.

REGULATION OF ION CONCENTRATIONS IN THE EXTRACELLULAR FLUIDS

The most important positive ions of the extracellular fluid are sodium, potassium, calcium, and magnesium ions; the most important negative ions are chloride, bicarbonate, and phosphate ions; two less important negative ions are sulfate and nitrate ions.

The positive ions especially need to be ex-actly regulated, because variations in these can cause drastic changes in bodily function. For instance, an abnormally low extracellular fluid potassium concentration increases the resting membrane potential of nerve and muscle fibers, thereby causing paralysis. Also, a sufficient decrease in sodium concentration can decrease the voltage of the action potentials, which can cause weakness of muscle contraction. And the concentration of calcium ions regulates the permeability of cell membranes. When the concentration is high, the permeability becomes decreased, whereas a low concentration increases the permeability. The precise function of magnesium is not known; however, high levels depress the nervous system and also depress muscle contraction.

There is not the extreme need for regulating the specific concentrations of the negative ions as is the case with the positive ions because many of these negative ions can be substituted for each other without causing severe effects. However, the *ratio* of chloride to bicarbonate ions plays an important role in the regulation of acid-base balance, as is discussed in detail in the following chapter. And phosphate ion concentration must be regulated within reasonable limits for two reasons: First, it is principally this ion that combines with calcium to provide the bone salts, and a low level prevents proper calcification of bone. Second, phosphate is needed constantly for maintenance of intracellular chemical reactions, and too low a level in the extracellular fluids can result in impaired cellular function.

REGULATION OF SODIUM ION CONCENTRATION – THE ALDOSTERONE FEEDBACK MECHANISM

Sodium ions represent approximately 90 per cent of all the extracellular cations. Therefore, sodium is the single most important ion that needs to be regulated. A specific mechanism for regulating its concentration is vested in the kidneys and the adrenal cortex. In the preceding chapter it was noted that sodium is actively reabsorbed by the tubules and that this active reabsorption of sodium seems to occur in all segments of the tubules. The sodium may be almost completely reabsorbed, with essentially none of it passing into the urine, or, on the other hand, large portions of the sodium may fail to be reabsorbed and are lost in the urine. In this way, the rate of sodium loss from the extracellular fluid can be altered from as little as a fraction of a gram a day to as great as 30 or more grams per day. The precise amount of sodium reabsorbed is regulated mainly by the concentration of *aldosterone*, a hormone secreted by the adrenal cortices, in the body fluids. The mechanism of this regulation is explained in the following paragraphs.

Aldosterone and Its Effect on Sodium Reabsorption in the Tubules. Aldosterone is one of several steroid hormones secreted by the adrenal cortex; these will be discussed in detail in Chapter 76 in relation to the overall function of the adrenal gland. There are basically three different types of adrenocortical hormones: (1) the mineralocorticoids, (2) the glucocorticoids, and (3) the androgenic hormones. The mineralocorticoids act on the renal tubules to increase the rate of sodium reabsorption. By far the most important of the mineralocorticoids is aldosterone, which accounts for about 95 per cent of all the sodium reabsorption caused by adrenal steroids and which has at least 30 times as much potency for promoting tubular reabsorption of sodium as does any other adrenal steroid.

Aldosterone increases the rate of sodium reabsorption in all segments of the tubular system; but from a practical point of view, its function in the ascending limbs of the loops of Henle, the distal tubules, and the collecting ducts is more important than in the proximal tubules and thin segments of the loops of Henle. In the absence of aldosterone, almost no sodium is reabsorbed by the distal segments of the tubular system. On the other hand, in the presence of excess aldosterone almost all sodium that reaches these segments is reabsorbed. Therefore, the presence or absence of aldosterone determines whether or not significant quantities of sodium will be lost in the urine.

In the absence of aldosterone the two kidneys will usually lose 15 to 30 grams of sodium chloride in the urine each day, and this amounts to a loss of one-sixth to one-tenth of all the sodium chloride in the body. Therefore, in the absence of aldosterone secretion by the adrenal cortices, a person can become seriously depleted of sodium and chloride, the main ionic constituents of the extracellular fluid, within a few days.

Regulation of Aldosterone Secretion. The rate of secretion of aldosterone can be increased greatly by any one or a combination of the following stimuli: (1) reduced sodium concentration in the extracellular fluids, (2) high extracellular concentration of potassium, (3) reduced blood volume or cardiac output, and (4) physical stress caused by trauma, burns, and so forth.

Effect of Low Sodium Concentration. One of the most important of the stimuli that increase the rate of aldosterone secretion is low sodium concentration in the extracellular fluid. A 5 per cent decrease in sodium concentration continued over a period of several days approximately doubles the rate of aldosterone secretion. Conversely, a similar increase in sodium concentration reduces aldosterone secretion to levels far below normal.

Obviously, this effect of sodium concentration on aldosterone secretion provides a feedback system for maintaining a constant concentration of sodium ions in the extracellular fluid. When their concentration falls below normal, increased aldosterone secretion causes increased reabsorption of sodium by the renal tubules, thus conserving sodium in the body while the daily intake of sodium returns the sodium concentration to normal. Conversely, elevated sodium concentration reduces aldosterone secretion, which allows rapid loss of sodium into the urine with consequent return of the sodium concentration to normal.

Effect of High Potassium Concentration. Elevation of the extracellular fluid potassium concentration by approximately 10 per cent also doubles the rate of aldosterone secretion. Yet, if we remember that a 10 per cent change in potassium concentration is a change of only 0.4 mEq./liter, we see that aldosterone secretion is very sensitive to only slight quantitative changes in extracellular potassium concentration.

Increased Aldosterone Secretion Caused by Reduced Blood Volume or Cardiac Output. For reasons not completely understood, decreased blood volume or cardiac output, with consequent decreased transport of adequate nutrients to the tissues, increases the rate of aldosterone secretion.

This in turn increases the sodium concentration in the body fluids. Later in the chapter we shall see that an increase in sodium concentration causes increased extracellular fluid and blood volumes, which in turn raise the cardiac output back toward normal. Thus, the rate of aldosterone secretion and its consequent effects on sodium concentration in the body fluids is closely allied to the regulation of blood volume and cardiac output and to the supply of adequate blood flow to the tissues.

INCREASED ALDOSTERONE SECRETION FOLLOWING PHYSICAL STRESS. We shall see in Chapter 76 that different types of physical stress, such as a broken bone, burns of the skin, extreme exercise, or any other stress that can damage or tax the abilities of the body tissues, cause the adenohypophysis to secrete increased quantities of the hormone *corticotropin;* this in turn passes through the blood to the adrenal cortex to increase the output of glucocorticoids, another group of adrenocortical hormones with actions quite different from those of aldosterone. However, corticotropin also has a *mild* stimulatory effect on secretion of aldosterone as well.

Mechanisms by which the Adrenal Cortex Is Stimulated to Secrete Aldosterone. Even though we know that each of the four stimuli just discussed can increase the rate of aldosterone secretion, the precise mechanisms by which the different stimuli act on the adrenal cortex are unknown except in the case of the action of corticotropin, which causes only a mild increase in aldosterone secretion. Some of the theories for explaining the tremendous increases in aldosterone secretion caused by the other stimuli are the following:

DIRECT EFFECT OF DECREASED SODIUM OR INCREASED POTASSIUM CONCENTRATION ON THE ADRENAL CORTEX. The adrenal gland can be transplanted from its normal position above the kidney into the subcutaneous tissue where different concentrations of sodium and potassium can be infused directly into the adrenal artery, and venous blood can be removed to measure the rate of secretion of aldosterone. The rate of aldosterone secretion measured in this way increases to about double normal when the sodium concentration is decreased 10 per cent below normal or when potassium concentration is increased 10 per cent above normal.

Thus, it is possible that changes in plasma sodium or potassium concentration control aldosterone secretion mainly if not entirely by their direct effects on the adrenal glands themselves.

THE RENIN-ANGIOTENSIN THEORY FOR REGULATION OF ALDOSTERONE SECRETION. *Renin* is a hormone secreted by the juxtaglomerular cells of the kidney. On occasion, excessively large quantities of this hormone are secreted in the blood and circulate throughout the body. It reacts with the blood plasma to cause the formation of *angiotensin;* the mechanism of this reaction was discussed in Chapter 26 in relation to the possible role of angiotensin in hypertension. Angiotensin in turn acts directly on the adrenal cortex to increase the rate of secretion of aldosterone.

Furthermore, either reduced blood flow through the kidneys or reduced sodium concentration in the extracellular fluid increases the rate of renin formation by the kidneys. Therefore, the following theory has been postulated for the regulation of sodium ion concentration in the extracellular fluid:

When sodium concentration falls to a low value, renin is formed by the kidneys, which in turn increases the concentration of angiotensin in the blood. The angiotensin increases the secretion of aldosterone, which causes increased reabsorption of sodium by the renal tubules, thus conserving sodium in the extracellular fluid until its concentration can rise back to normal.

Conversely, a high level of sodium diminishes renin secretion and the opposite sequence of events leads to decreased extracellular sodium.

This mechanism could also explain the increased sodium retention by the kidneys when the blood volume or cardiac output falls below normal, for diminished blood flow through the kidneys increases the formation of renin, which would cause (a) sodium retention, (b) increased extracellular and blood volumes, and (c) increase of the cardiac output back toward normal.

Though all these steps in the renin-angiotensin-aldosterone theory for regulation of sodium ion concentration appear to be logical, some of the different steps have not yet been proved to be quantitatively important enough to explain the extremely critical degree of sodium ion concentration regulation that is known to exist.

THE "ADRENOGLOMERULOTROPIN" THEORY FOR REGULATION OF ALDOSTERONE SECRETION AND SODIUM ION CONCENTRATION. A large series of experiments in which diencephalic areas of the brain have been selectively destroyed has indicated that in the absence of these areas decreased blood volume and cardiac output do not cause the usual marked increase in aldosterone secretion. It has been postulated that a hormone called *adrenoglomerulotropin* is secreted in response to diminished cardiac output and that the adrenoglomerulotropin stimulates the *zona glomerulosa* of the adrenal cortex to increase its secretion of aldosterone. Thus, decreased cardiac output could set off a sequence of events that would increase the sodium concentration, increase the extracellular fluid volume, and return the cardiac output back to normal. Unfortunately, not all the steps in this mechanism have yet been proved. But, until the precise mechanism for control of aldosterone secretion is found, this mechanism must be remembered along with the others.

Summary of the Regulation of Aldosterone Secretion and of Extracellular Fluid Sodium Composition. Figure 36–1 gives a summary of the three different postulated mechanisms for regulation of both the rate of aldosterone secretion and extracellular fluid sodium concentration. Any of the three mech-

Figure 36–1 Three postulated mechanisms of feedback control of sodium concentration and cardiac output. (An arrow pointing downward means a decrease, and an arrow pointing upward means an increase.)

anisms could theoretically provide a complete feedback loop for regulation of sodium concentration and, secondarily, for helping to regulate cardiac output as well. The renin-angiotensin mechanism has recently received the most attention by physiologists. But the actual *quantitative* data recently available perhaps favor the *direct effect* of sodium and potassium on the adrenal cortex to be a more important means by which extracellular sodium and potassium concentrations control aldosterone secretion.

Effect of Arterial Pressure on Sodium Conservation in the Body. It was pointed out in the preceding chapter that an increase in arterial pressure greatly increases the rate of urine formation and loss of sodium into the urine. To emphasize what a marked effect arterial pressure has on sodium output by the kidneys, an acute increase in arterial pressure from 100 mm. Hg to 200 mm. Hg increases the rate of sodium loss 6- to 8-fold, illustrating the extreme importance of arterial pressure in regulating the output of sodium by the kidneys. This effect and its importance also in the regulation of arterial pressure was discussed in detail in Chapter 24.

REGULATION OF OTHER POSITIVE IONS

Even though the concentrations of the other positive ions in the extracellular fluids are also regulated exactly, the mechanisms for their regulation are less well known than for sodium. Nevertheless, we know that the kidneys play key roles in the regulation of at least potassium, calcium, and magnesium.

Regulation of Potassium Ion Concentration. Potassium ion concentration is regulated in at least two different ways: (1) by an aldosterone feedback method that functions exactly oppositely to that for sodium regulation and (2) by direct secretion of potassium into the distal tubules and collecting ducts when the extracellular fluid potassium concentration rises too high.

ALDOSTERONE FEEDBACK REGULATION OF POTASSIUM. The aldosterone feedback mechanism for control of potassium ion concentration is the following: When aldosterone causes sodium to be reabsorbed in the distal tubules and collecting ducts, this causes large quantities of positive ions to transfer from the tubules into the peritubular fluids. As a result, strong electronegativity develops in the tubules, −40 to −120 millivolts, and pulls positive potassium ions from the tubular cells into the tubules, thereby causing passive secretion of potassium into the urine. Thus, in effect, exchange of potassium ions for sodium ions occurs in the distal tubules, this exchange increasing many-fold under the influence of aldosterone.

To complete the feedback loop, high potassium concentration in the extracellular fluid increases the rate of aldosterone secretion, as was discussed earlier. The aldosterone then simultaneously increases sodium reabsorption and potassium excretion, decreasing the potassium back to a normal level. Thus, in effect, the aldosterone feedback mechanism controls the *sodium to potassium ratio* in the extracellular fluids.

EXCRETION OF POTASSIUM AS A DIRECT RESPONSE TO INCREASED EXTRACELLULAR FLUID POTASSIUM. In addition to the aldosterone mechanism for control of extracellular fluid potassium ion concentration, an increase in this concentration also has a direct effect on the distal tubules and collecting ducts to cause potassium secretion into the tubular fluid. This has been postulated to result from the following mechanism: Increased potassium concentration in the extracellular fluids causes increased potassium ion concentration in the epithelial cells of the distal tubules and collecting ducts. This in turn provides increased quantities of potassium to diffuse into the tubules. Under the influence of the electronegativity in the tubules, the potassium is passively secreted.

The combined effects of the aldosterone feedback mechanism and the direct mechanism for secretion of potassium cause the potassium ion concentration in the extracellular fluids to remain almost exactly 4 to 5 milliequivalents per liter, rarely rising or falling more than ±10 per cent.

Regulation of Calcium Ion Concentration. A low level of calcium ions in the extracellular fluids causes the heart to dilate and eventually to stop in a flaccid state. And, even more important, it causes extreme numbers of spontaneous impulses to originate in the peripheral nerves, resulting in tetanic contraction of muscles throughout the body and in death due to spastic respiratory paralysis.

The calcium ion concentration is controlled principally by the parathyroid glands in the following manner: A low level of calcium in the extracellular fluids has a direct effect on the parathyroid glands to promote increased secretion of parathyroid hormone, which in turn acts directly on the bones to increase the rate of reabsorption of bone salts. This releases large amounts of calcium and phosphate ions into the extracellular fluids and thereby elevates the calcium level to normal.

However, calcium is also regulated to some extent by the kidney tubules as well, for large amounts of calcium are known to be lost into the urine when its extracellular concentration is high, while very little is lost when the concentration is low. Unfortunately, the total mechanism for this regulation is unknown, though at least part of it results from a direct effect of parathyroid hormone on the renal tubules to increase calcium reabsorption (see Chap. 79).

Regulation of Magnesium Concentration. Almost nothing is understood about the regulation of magnesium concentration other than that decreased magnesium concentration in the extracellular fluids promotes increased reabsorption of magnesium from the tubules while increased concentration promotes decreased reabsorption. The few available experiments indicate that the kidneys regulate extracellular magnesium concentration in much the same way that they regulate potassium concentration. This is logical since both these ions function principally as intracellular rather than as extracellular ions.

REGULATION OF NEGATIVE ION CONCENTRATIONS IN THE EXTRACELLULAR FLUIDS

Regulation of Total Negative Ion Concentration. In general, the total negative ion concentration is regulated secondarily to the regulation of the positive ions for the following reasons: Each time a positive ion is absorbed from the tubules, a state of electronegativity is created in the lumens of the tubules. Such a potential across the walls of the tubules immediately causes negative ions, primarily chloride ions, to diffuse through the membrane to provide electrical neutrality.

Therefore, the same regulatory mechanisms that increase the reabsorption of positive ions also promote reabsorption of negative ions at the same time. For instance, secretion of aldosterone by the adrenal glands directly increases the reabsorption of sodium and therefore indirectly promotes absorption of negative ions at the same time; and since about three-fourths of the negative ions in the glomerular filtrate are chloride ions, it frequently is said that aldosterone promotes chloride reabsorption.

Regulation of Chloride to Bicarbonate Ratio—Relation to Acid-Base Balance. Much more important than the absolute concentrations of chloride and bicarbonate in the extracellular fluid is the *ratio of chloride to bicarbonate.* When the extracellular fluids become exceedingly acidic, the renal tubules reabsorb large quantities of bicarbonate ions, and chloride ion reabsorption becomes greatly diminished. The excessive reabsorption of bicarbonate ions shifts the pH of the buffer systems in the extracellular fluid toward a normal pH. Conversely, when the extracellular fluids are very alkalotic, excess quantities of chloride ions are reabsorbed, while bicarbonate ions are lost into the urine. The reabsorption of chloride ions displaces bicarbonate ions from the extracellular fluids, which causes the pH again to return toward normal. These relationships between chloride and bicarbonate ions are discussed in detail in the following chapter.

Regulation of Phosphate Concentration. Phosphate concentration is regulated primarily by an *overflow mechanism,* which can be explained as follows: The renal tubules have a normal transport maximum of 0.1 millimol of phosphate per minute. When less than this "load" of phosphate is present in the glomerular filtrate, all of it is reabsorbed. When more than this amount is present, the excess is excreted. Therefore, normally, phosphate ion spills into the urine when its concentration in the plasma is above the threshold value of approximately 0.8 millimol/liter. Any time the concentration falls below this value, all the phosphate is conserved in the plasma, and the daily ingested phosphate accumulates in the extracellular fluid until its concentration rises above the threshold. On the other hand, whenever the phosphate concentration rises above this level, the excess is excreted into the urine. Since most people ingest large quantities of phosphate day in and day out, either in milk or in meat, the concentration of phosphate is usually maintained at a level of about 1.0 millimol/liter, a level at which there is continual overflow of excess phosphate into the urine.

ROLE OF PARATHYROID HORMONE IN PHOSPHATE ION REGULATION. Parathyroid hormone, which plays a major role in regulation of calcium ion concentration as explained above, also affects phosphate ion concentration in two different ways. First, parathyroid hormone promotes bone reabsorption, thereby dumping large quantities of phosphate ions into the extracellular fluid. Second, this hormone decreases the transport maximum for phosphate by the renal tubules so that a greater proportion of the tubular phosphate is lost in the urine. The combination of these two factors causes marked loss of phosphates in the urine. Yet one of these factors tends to increase the extracellular concentration of phosphate while the other tends to decrease it. Therefore, whenever excess para-thyroid hormone is secreted, the phosphate concentration hardly changes; if anything it decreases slightly. These interrelationships between calcium and phosphate ion regulation will be discussed in detail in Chapter 79.

Regulation of Other Negative Ions. Other less important negative ions in the body fluid include sulfates, nitrates, urates, lactates, and the amino acids. Essentially all of these, like phosphate, have definite transport maximums. When the concentration of each is below its respective threshold, it is conserved in the extracellular fluid, but when above this threshold the excess begins to spill into the urine. Thus, the concentration of most of these negative ions is regulated by the overflow mechanism in the same way that phosphate ion concentration is regulated.

REGULATION OF BODY WATER

INTAKE VERSUS OUTPUT OF WATER

Daily Intake of Water. Most of our daily intake of water enters by the oral route. Approximately two-thirds is in the form of pure water or some other beverage, and the remainder is in the food that is eaten. A small amount is also synthesized in the body as the result of oxidation of hydrogen in the food; this quantity ranges between 150 and 250 ml. per day, depending on the rate of metabolism. The normal intake of fluid, including that synthesized in the body, averages about 2400 ml. per day.

Daily Loss of Body Water. Table 36–1 shows the routes by which water is lost from the body under different conditions. Normally, at an atmospheric temperature of about 68 degrees, approximately 1400 ml. of the 2400 ml. of water intake is lost in the *urine,* 100 ml. is lost in the *sweat,* and 200 ml. in the *feces.* The remaining 700 ml. is lost by *evaporation through the lungs* or by *diffusion through the skin.*

INSENSIBLE WATER LOSS. Loss of water by diffusion through the skin and by evaporation from the lungs is known as *insensible water loss* because the person does not know that he is actually losing water at the time that it is leaving the body.

The average loss of water by diffusion through the skin is approximately 300 to 400 ml. per day; this amount is lost even in a person who is born with congenital lack of sweat glands. In other words, the water molecules actually diffuse through the cells of the skin themselves. Fortunately, the cholesterol-filled, cornified layer of the skin acts as a protector against still much greater loss of water by diffusion. However, when the cornified layer becomes denuded, such as after extensive burns, the rate of evaporation can increase to as much as 3 to 5 liters each day.

All air that enters the lungs becomes totally saturated with moisture, to a vapor pressure of approximately 47 mm. Hg, before it is expelled. Since the vapor pressure of the inspired atmospheric air is usually far less than 47 mm. Hg, the average water loss through the lungs is about 300 to 400 ml. per day. The atmospheric vapor pressure normally decreases with decreasing temperature so that the loss is greatest in very cold and least in very warm weather. This explains the dry feeling in the respiratory passages in cold weather.

LOSS OF WATER IN HOT WEATHER AND DURING EXERCISE. In very hot weather, water loss in the sweat is occasionally increased to as much as 3.5 liters an hour, which obviously can rapidly deplete the body fluids. Sweating will be discussed in Chapter 71.

Exercise increases the loss of water in two ways: First, it increases the rate of respiration, which promotes increased water loss through the lungs in proportion to the increased ventilatory rate. Second, and much more important, exercise increases the body heat and consequently is likely to result in excessive sweating.

REGULATION OF WATER LOSS FROM THE BODY — THE OSMORECEPTOR-ANTIDIURETIC HORMONE SYSTEM

In addition to regulation of the different ions in the extracellular fluids, regulation of the total water is equally as important, for this controls the osmotic concentration in the extracellular fluid. This regulation is vested in the so-called *osmoreceptor-antidiuretic hormone system* involving the *hypothalamus,* the *posterior pituitary gland, antidiuretic hormone,* and the *renal tubules.* This system, when activated, promotes water conservation in a complex manner, as described in the following paragraphs.

Antidiuretic Hormone and Its Effect on the Kidney Tubules. *Antidiuretic hormone* (ADH) is secreted into the blood by the supraoptico-hypophyseal axis of the hypothalamus and posterior pituitary gland. This hormone pro-

TABLE 36–1.   Daily Loss of Water (in Milliliters)

	Normal Temperature	Hot Weather	Prolonged Heavy Exercise
Insensible Loss:			
Skin	350	350	350
Lungs	350	250	650
Urine	1400	1200	500
Sweat	100	1400	5000
Feces	200	200	200
Total	2400	3400	6700

motes increased water reabsorption from both the distal tubules and the collecting ducts of the kidneys. When no ADH is secreted, the amount of water passing into the urine each day is 5 to 15 times normal. And, because of this, the extracellular fluids become more concentrated. On the other hand, when large quantities of ADH are secreted, water is reabsorbed to an extreme degree so that the volume of urine formed each day may be as little as 400 to 500 ml., or one-third the normal volume. As a result, water is conserved, and the extracellular fluids become more dilute.

The mechanisms by which ADH acts on the kidneys to control the degree of dilution or concentration of the urine were discussed in the preceding chapter, and its overall function as a hormone will be discussed in Chapter 75.

Function of the Osmoreceptors in Regulating Osmolality. Antidiuretic hormone would not be of any value to the body unless there existed a concomitant system for regulating the secretion of antidiuretic hormone in proportion to its need. To provide this, special neurons of the anterior hypothalamus, located, as shown in Figure 36–2, mainly in the supraoptic nuclei and called *osmoreceptors,* respond specifically to changes in extracellular fluid concentration. The osmoreceptors are specialized neuronal cells that respond to changes in osmolality of the extracellular fluid. When the osmolality becomes very low, osmosis of water into the osmoreceptors causes them to swell, which decreases their rate of impulse discharge. Conversely, increased osmolality in the extracellular fluid pulls water out of the osmoreceptors, causing them to shrink, and thereby increasing their rate of discharge.

The impulses from the osmoreceptors are transmitted from the supraoptic nucleus through the pituitary stalk into the posterior pituitary gland where they promote the release of antidiuretic hormone (ADH), the details of which will be discussed in Chapter 75 in relation to the endocrinology of the pituitary gland.

Thus, ADH is secreted in relation to the osmolality of the extracellular fluids — the greater the osmolality, the greater the rate of ADH secretion; and the less the osmolality, the less the rate of ADH secretion.

Summary of Feedback Control of Extracellular Fluid Osmolality. Now to synthesize the whole mechanism: An increase in osmolality of the extracellular fluids excites the osmoreceptors; this promotes ADH secretion and causes marked reabsorption of water by the renal tubules while solutes continue to be lost into the urine. As a consequence, the extracellular fluids become diluted and their concentration returns toward normal. On the other hand, low osmolality of the extracellular fluids decreases the activity of the osmoreceptors, thus decreasing ADH secretion, and large amounts of water are lost into the urine until the extracellular fluid osmolality returns to normal.

The osmoreceptor system is sufficiently active that overconcentration of the extracellular fluid constituents of only 1 to 2 per cent causes marked retention of water, or a similar decrease causes rapid loss of water.

Water Diuresis. When a person drinks a large amount of water, a phenomenon called *water diuresis* ensues, a typical record of which is shown in Figure 36–3. In this example a person drank 1 liter of water, and approximately 30 minutes later his urine output increased to eight times normal, where it remained for two hours — that is, until the osmolality of the extracellular fluid had returned essentially to normal. The delay in onset of water diuresis is caused partly by a delay in absorption of the water from the gastrointestinal tract but

Hypothalamus

Supraoptic nuclei

Excited by concentrated extracellular fluid

Antidiuretic hormone released by posterior pituitary gland

Excess reabsorption of water corrects excess concentration of extracellular fluid

Urine decreased but concentrated

Figure 36–2 Control of extracellular fluid osmolality by the osmoreceptor system.

Figure 36–3 Water diuresis in a human being following ingestion of 1000 ml. of water. (Redrawn from Smith: The Kidney. Oxford University Press.)

mainly by the time required for destruction of the antidiuretic hormone that has already been released by the pituitary gland prior to drinking the water.

Diabetes Insipidus. Destruction of the supraoptic nuclei or high level destruction of the nerve tract from the supraoptic nuclei to the posterior pituitary gland causes antidiuretic hormone production to cease or to become greatly reduced. When this happens the person thereafter secretes nothing but a dilute urine, and his daily urine volume may be anywhere from 5 to 15 liters per day. This causes, first, slightly reduced body fluid volumes and, second, increase in the osmolar concentration of the fluids. However, the rapid loss of fluids ordinarily creates a strong sensation of thirst, and the person drinks water rapidly enough to make up for the deficit. As long as he does so his body fluids remain almost normal, though any condition that depresses his sense of thirst, such as unconsciousness, might easily result in a state of dehydration and hyperosmolality of the body fluids.

The different lesions in the hypothalamus or pituitary stalk that can cause diabetes insipidus are discussed in Chapter 75.

THIRST

The phenomenon of thirst is equally as important for regulating body water as are the renal mechanisms discussed above, for the amount of water in the body is determined by the balance between both *intake* and *output* of water each day. Thirst is the primary regulator of the intake. A definition of thirst is the *conscious desire for water.*

Causes of Thirst

The four most important causes of thirst are *extracellular dehydration, low cardiac output, intracellular dehydration,* and *dryness of the mouth.*

Extracellular Dehydration. In any instance in which the extracellular fluid volume is reduced,

an animal greatly increases its rate of drinking, or the human being develops intense thirst and therefore seeks water.

An especially interesting type of extracellular dehydration that causes thirst occurs when an animal is maintained for long periods of time on a low salt intake. In such an instance the osmolar concentration of the extracellular fluids falls very low, and the extracellular fluid volume also decreases, sometimes to as little as two-thirds normal. In such instances the animal begins to drink large volumes of water, but unless sodium chloride is available in the water, the normal extracellular fluid volume is not restored, and his excessive drinking continues on and on. Yet it disappears immediately when an adequate amount of sodium chloride is placed in his water to return the extracellular fluid volume to normal. This illustrates that low extracellular fluid volume and not lack of water is important for initiating thirst.

Hemorrhage and Low Cardiac Output. A small amount of hemorrhage ordinarily does not cause thirst, but this also almost never reduces the cardiac output and does not in any other way impair the circulatory system. However, when hemorrhage is sufficient to cause reduced cardiac output, experimental animals and human beings experience intense desire for water. This same effect also occurs in *acute cardiac failure,* in which the circulatory system is compromised in much the same way as following hemorrhage.

Intracellular Dehydration. At times the extracellular fluid volume is completely normal, but the intracellular fluid volume is markedly decreased. This, too, causes thirst. For instance, injection of hypertonic sodium chloride or hypertonic sucrose solution into the blood, which causes osmotic removal of water from the intracellular fluid compartment, induces immediate and intense thirst within seconds. The subject normally drinks a sufficient amount of water to reduce the extracellular fluid osmolality to normal, which allows normal amounts of water to return to the cells.

Dryness of the Mouth. Almost everyone is aware of the fact that a dry mouth is often associated with thirst. A partial explanation of this is that a dry mouth also is often associated with extracellular and intracellular dehydration. Yet when a drug, such as atropine, is given to a person to prevent salivation, his extracellular and intracellular fluid volumes may remain entirely normal, but his mouth becomes dry. Under these conditions the dryness of the mouth can promote sensations in the mouth that cause the person to drink water. However, these sensations are not the same as thirst, as everyone can attest, though they do cause increased drinking.

Temporary Relief of Thirst Caused by the Act of Drinking

A thirsty person receives relief from his thirst immediately after drinking water even before the water has been absorbed from the gastrointestinal tract. In fact, in persons who have esophageal fistulae (a condition in which the water never goes into the gastrointestinal tract) relief of thirst still occurs following the act of drinking, but this relief is only temporary and thirst returns after 15 minutes or more. If the water does enter the gut, distention of the upper gastrointestinal tract, particularly the stomach, provides still further temporary relief of thirst. For instance, simple inflation of a balloon in the stomach can relieve thirst for 5 to 30 minutes.

One might wonder what the value of this temporary relief of thirst could be, but there is a good reason for its occurrence as follows: After a person has drunk water, as long as a half to one hour may be required for the water to be absorbed into the extracellular fluids. Were his thirst sensations not temporarily relieved following the drinking of water, he would continue to drink more and more. When all this water should finally become absorbed, his body water would be far greater than normal, and he would have created a condition opposite to that which he was attempting to correct. It is well known that a thirsty animal almost never drinks more than the amount of water needed to relieve his state of dehydration but ordinarily drinks almost exactly the right amount.

Neural Integration of Thirst — The "Thirst" or "Drinking" Center

An area called the *drinking center* has been localized in each side of the hypothalamus, lateral and caudal to the supraoptic nuclei. This center, illustrated in Figure 36–4, overlaps with the area of the hypothalamus that causes the release of ADH, thus explaining why factors that cause increased intake of water often cause increased conservation of water by the kidneys at the same time.

Electrical stimulation of the thirst center by implanted electrodes causes an animal to begin drinking within seconds and to continue drinking until the electrical stimulus is stopped. Also, injection of hypertonic salt solutions into the area, which causes osmosis of water out of the neuronal cells and results in intracellular dehydration, also causes drinking.

Basic Stimulus for Exciting the Thirst Center — Intracellular Dehydration. In the above discussion of the many factors affecting thirst it was pointed out that *extracellular dehydration, intracellular dehydration,* and *circulatory failure* can all promote thirst. Furthermore, the *act of drinking* and *fullness of the gastrointestinal tract* can inhibit thirst. It is almost certain that the act of drinking and fullness of the gastrointestinal tract excite peripheral sensory receptors that transmit impulses into the thirst

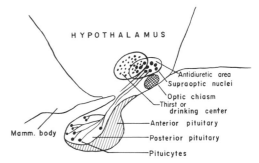

Figure 36–4 Location of the thirst center in the hypothalamus in relation to the supraoptico-pituitary antidiuretic system.

center to inhibit it. However, the other three stimuli seem to affect the neurons of the thirst center directly, and it is possible to explain all their effects on the basis of one single stimulus; namely, *intracellular dehydration.* For instance, extracellular dehydration is known to promote loss of fluid from inside cells. Thus, extracellular dehydration could well cause sufficient dehydration of the neurons in the thirst center to promote thirst. In the case of circulatory failure it is also known that an inadequate supply of nutrients to the cells depresses the active transport mechanisms that are responsible for maintaining normal intracellular electrolyte compositions, and this, too, under some conditions can promote loss of intracellular electrolytes with resultant intracellular dehydration.

However, this attempt to unify the different stimuli that excite thirst remains at present mainly speculation.

APPETITE AND CRAVING FOR SALT

Maintenance of normal extracellular fluid composition requires not only a balance between the intake and output of water but also a balance between the intake and output of the different electrolytes. The only electrolyte intakes that have received adequate attention are those of sodium and chloride (salt), and even here almost nothing is known of the basic mechanisms involved. Yet it is known that salt-depleted persons develop a desire for salt; this occurs especially in persons who have *Addison's disease,* a condition in which the adrenal cortices no longer secrete aldosterone, so that the salt stores of the body become very depleted. The salt-depleted person craves and eats naturally salty foods. Likewise, it is well known that animals living in areas far removed from the seashore actively search out "salt licks." This craving for salt is analogous to thirst, and it is also analogous to appetite for other types of foods, which is still another homeostatic mechanism that will be discussed in Chapter 72.

REGULATION OF EXTRACELLULAR FLUID VOLUME AND BLOOD VOLUME

Many aspects of extracellular fluid volume and blood volume have already been discussed — especially the regulation of interstitial fluid volume, in the discussion of capillary dynamics in Chapter 21, and the regulation of blood volume, in the discussion of overall regulation of the circulation in Chapter 24. Therefore, the purpose of the present section will be to put together in a concise form the overall picture of these volume regulations.

Basically, there are three different mechanisms for extracellular fluid volume and blood volume regulation: (1) an intrinsic mechanical system, which includes the dynamics of fluid exchange at the capillary membranes and the effects of hemodynamics on fluid excretion by the kidneys; (2) hormonal factors that affect fluid volume regulation, especially those that affect the ability of the kidneys to excrete extracellular fluid; and (3) nervous reflexes that also affect primarily the rate of excretion of fluid by the kidneys.

THE INTRINSIC MECHANICAL SYSTEM FOR EXTRACELLULAR FLUID VOLUME AND BLOOD VOLUME REGULATION

Regulation of Extracellular Fluid Volume. Extracellular fluid is primarily the plasma and the interstitial fluid. Furthermore, as discussed in Chapter 21, the interstitial fluid is composed of two separate components: (1) mobile fluid that is capable of moving freely through the interstitial fluid spaces and (2) immobile fluid that is bound in the ground substance of the interstitial spaces. It was also pointed out in Chapter 21 that the quantity of mobile fluid in the interstitial spaces is probably almost zero because two basic mechanisms cause mobile fluid to be absorbed almost as fast as it is formed. These mechanisms are (a) pumping of fluid from the spaces by the lymphatic system and (b) osmotic absorption of fluid directly into the capillaries. It is these two absorptive mechanisms that keep the tissue spaces normally "dry" of almost all *mobile* fluid, leaving only minute amounts of mobile fluid to fill the last small interstices in the tissue spaces.

Therefore, the normal interstitial fluid is almost entirely bound fluid, and its volume is controlled by the amount of mucopolysaccharides in the interstitial spaces. For instance,

in young children the quantity of mucopolysaccharides is far greater than in older people, and this causes a much higher percentage of interstitial fluid in the tissues of a baby than in those of an adult. Likewise, in the "myxedematous" state of hypothyroidism, large quantities of mucopolysaccharides are formed, giving rise again to a high volume of interstitial fluid in the tissues.

Therefore, under *normal* conditions interstitial fluid volume is controlled mainly by the tissues themselves and not, to a great extent, by the circulatory system. On the other hand, abnormal retention of fluid in the body, such as occurs in kidney disease or as a result of abnormal capillary dynamics, can cause excess quantities of *mobile* fluid in the tissues, giving rise to the condition called *edema,* which was discussed in detail in Chapter 21. When this occurs, the interstitial fluid volume becomes mainly a reflection of the function of the circulatory system and of the kidneys.

Regulation of Blood Volume. An overall schema for the regulation of extracellular fluid volume, blood volume, cardiac output, arterial pressure, and total peripheral resistance was presented in Chapter 24 to show the interrelationships between all these separate factors. It is important at the present time that the student review this schema to understand the intrinsic mechanical system for regulation of blood volume.

Basically, this system is the following: When the extracellular fluid volume becomes too high, about one-third of the excess extracellular fluid remains in the blood plasma and, therefore, increases the blood volume. The increase in blood volume in turn increases venous return to the heart, and this increases cardiac output. The increase in cardiac output then has two effects to increase arterial pressure: First, increased blood flow through the tissue blood vessels has a direct effect to increase arterial pressure; second, this increased blood flow causes a progressive increase in the total peripheral resistance, which secondarily increases arterial pressure an additional amount. Finally, the increase in arterial pressure causes the kidneys to excrete the excess extracellular fluid volume; this reduces the extracellular fluid volume back toward normal and thereby also returns the plasma volume and blood volume back toward normal.

This description of the intrinsic mechanical feedback mechanism for regulation of extracellular fluid volume and blood volume is only a skeleton of the factors involved, but if the student will return to Chapter 24 and study all of these factors in detail he will understand the beautiful integration of the circulatory and fluid volume regulatory systems.

Importance of the Mechanical Fluid Volume Regulatory Systems

The intrinsic systems for regulating fluid volumes are ever present, always helping to return the fluid volumes back toward their normal values. Even in an animal whose head has been removed and the spinal cord destroyed, this mechanical system still operates, indicating that it is capable of functioning even in the absence of nervous reflexes. Likewise, when the ADH system for regulating renal output of water is destroyed or when the aldosterone mechanism for controlling most of the extracellular fluid ions has been abrogated, the intrinsic mechanical system is still operative. Therefore, the other factors that help to regulate fluid volumes of the body are mainly superimposed onto the basic intrinsic system that is always present.

HORMONAL FACTORS THAT AFFECT VOLUME REGULATION

Two hormonal systems play major roles in fluid volume regulation. These are (1) the aldosterone system for regulating many of the ions in the extracellular fluids, and (2) the ADH system for regulating water excretion. These have already been discussed earlier in the chapter, but it is important also to understand how they affect the overall regulation of extracellular fluid volume and blood volume.

Effect of the Aldosterone System. As discussed earlier, aldosterone has a direct effect on the kidneys to increase the retention of both sodium chloride and water, thereby increasing the extracellular fluid volume.

Yet, despite this significant effect of aldosterone in causing retention of extracellular fluid by the kidneys, after the blood volume has increased to 5 to 10 per cent above normal and the arterial pressure has increased to 15 to 20 mm. Hg above normal, the intrinsic mechanical system discussed previously causes enough increase in fluid excretion by the kidneys to balance the aldosterone's retaining effect. Therefore, even large quantities of aldosterone will usually increase the extracellular fluid and blood volumes only a few per cent and can only rarely cause enough retention of extracellular fluid to cause edema.

In the absence of aldosterone secretion, the kidneys lose tremendous quantities of sodium and chloride in the urine; consequently, large amounts of water are also lost, thereby reducing extracellular fluid volume and blood volume below normal. If this loss continues for more than 4 to 5 days the effect can actually cause death because of development of circulatory shock. This condition is called *Addison's disease,* which is discussed in more detail in Chapter 76.

The ADH System. The ADH system can alter by a few per cent the total quantity of water in the body fluids, excess ADH increasing body water as much as 3 to 5 per cent and lack of ADH decreasing body water perhaps 3 to 5 per cent. Likewise, abnormalities of the thirst mechanism can cause a person to build up excess quantities of water or to diminish the water in his body.

The ADH system, like the aldosterone system, is to a great extent opposed by the intrinsic mechanical system for regulation of extracellular fluid volume and blood volume. Therefore, the ADH system is not nearly so important for regulation of fluid volumes as it is for regulation of body fluid osmolality.

Possibility of a Natriuretic Hormone that Helps to Regulate Extracellular Fluid Volume. A number of research workers in recent years have claimed that an excess of sodium in the extracellular fluid causes formation somewhere in the body of a hormone that acts on the renal tubules to cause rapid loss of sodium into the urine. This hormone has been called the *third factor* or the *natriuretic hormone.* Some experiments have indicated that such a hormone is formed somewhere in the head region of the body, and other experiments have failed to confirm the presence of such a hormone at all. If the hormone does exist, it supposedly reduces the reabsorption of sodium by the tubules. This hormone is mentioned here not because of its proved importance but because it has been invoked by many physicians to explain some of the perplexing problems of renal excretion, especially in cardiac failure.

NERVOUS REFLEXES THAT AFFECT VOLUME REGULATION

Almost all the circulatory reflexes play at least some role in the regulation of blood volume. For instance, the *baroreceptor reflex,* when elicited by low arterial pressure, causes afferent arteriolar constriction in the kidney and thereby causes retention of extracellular fluid. The buildup in extracellular fluid eventually helps to increase the arterial pressure back toward normal. Likewise, the *CNS ischemic response* has a tremendous effect in decreasing urinary output, sometimes reducing urinary output to absolute zero.

However, a few circulatory reflexes seem to play more specific roles in helping to control blood volume; these are mainly reflexes that have sensors that are excited by excess filling of the atria and great veins. Located in the walls of these structures, particularly in the walls of the left atrium, are specific stretch receptors called *volume receptors*, which elicit strong renal reflexes to cause increased excretion of urine. Also, signals are transmitted to the hypothalamus to reduce the output of ADH, which also causes the kidneys to lose extra quantities of water. Thus, whenever the atria and great veins become overfilled with blood, these volume receptors set into play almost instantaneous effects to increase the loss of fluid from the body, thereby helping to readjust both extracellular fluid volume and blood volume back toward normal.

REFERENCES

Adolph, E. F.: Regulation of water intake in relation to body water content. *In* Handbook of Physiology. Baltimore, The Williams & Wilkins Co., 1967, Sec. VI, Vol. I, p. 163.

Andersson, B.; Dallman, M. F.; and Olsson, K.: Studies of hypothalamic regulation of water and electrolyte balance in the goat. *Proc. Int. Union of Physiol. Sciences,* 6:200, 1968.

Berkowitz, H. D.; Miller, L. D.; and Itskovitz, H. D.: Renal function and the renin-angiotensin system in the isolated perfused kidney. *Amer. J. Physiol., 213*:928, 1967.

Berliner, R. W.: The formation of concentrated urine. *Proc. Int. Union of Physiol. Sciences,* 6:53, 1968.

Bossert, W. H., and Schwartz, W. B.: Relation of pressure and flow to control of sodium reabsorption in the proximal tubule. *Amer. J. Physiol., 213*:793, 1967.

Cortney, M. A.: Renal tubular transfer of water and electrolytes in adrenalectomized rats. *Amer. J. Physiol., 216*:589, 1969.

Denton, D. A.: Salt appetite. *Proc. Int. Union of Physiol. Sciences,* 6:61, 1968.

Fitzsimons, J. T.: The hypothalamus and drinking. *Brit. Med. Bull.,* 22:232, 1966.

Frazier, H. S.: Renal regulation of sodium balance. *New Eng. J. Med.,* 279:868, 1968.

Gann, D. S.; Delea, C. S.; Gill, J. R., Jr.; Thomas, J. P.; and Bartter, F. C.: Control of aldosterone secretion by change of body potassium in normal man. *Amer. J. Physiol., 207*:104, 1964.

Gilmore, J. P., and Michaelis, L. L.: Influence of anesthesia on renal response of the foxhound to intravascular volume expansion. *Amer. J. Physiol., 216*:1367, 1969.

Gottschalk, C. W., and Lassiter, W. E.: Intrarenal control mechanisms for salt and water reabsorption. *Proc. Int. Union of Physiol. Sciences,* 6:109, 1968.

Gottschalk, C. W., and Mylle, M.: Micropuncture study of the mammalian urinary concentrating mechanism: evidence for the countercurrent hypothesis. *Amer. J. Physiol., 196*:927, 1959.

Guyton, A. C.; Langston, J. B.; and Navar, G.: Theory for renal autoregulation by feedback at the juxtaglomerular apparatus. *Circ. Res., 14*:187, 1964.

Holmes, J. H.: Thirst and fluid intake as clinical problems. *In* Handbook of Physiology. Baltimore, The Williams & Wilkins Co., 1967, Sec. VI, Vol. I, p. 147.

Johnston, C. I.; Davis, J. O.; Howards, S. S.; and Wright, F. S.: Cross-circulation experiments on the mechanism of the natriuresis during saline loading in the dog. *Circ. Res., 20*:1, 1967.

Know, F. G.; Davis, B. B.; and Berliner, R. W.: Effect of chronic cardiac denervation on renal response to saline infusion. *Amer. J. Physiol., 213*:174, 1967.

Lauson, H. D.: Antidiuretic hormone. *Fed. Proc., 24*:731, 1965.

Lassiter, W. E.; Gottschalk, C. W.; and Mylle, M.: Micropuncture study of net transtubular movement of water and urea in nondiuretic mammalian kidney. *Amer. J. Physiol., 200*:1139, 1961.

Leyssac, P. P.: The regulation of proximal tubular reabsorption in the mammalian kidney. *Acta Physiol. Scand., 70*, Suppl. 291, 1967.

Malvin, R. L., and Vander, A. V.: Effects of angiotensin infusion on renal function in the unanesthetized rat. *Amer. J. Physiol., 213*:1205, 1967.

Massry, S. G.; Coburn, J. W.; and Kleeman, C. R.: Renal handling of magnesium in the dog. *Amer. J. Physiol., 216*:1460, 1969.

McDonald, K. M.; Rosenthal, A.; Schrier, R. W.; Galicich, J.; and Lauler, D. P.: Effect of interruption of neural pathways on renal response to volume expansion. *Amer. J. Physiol., 218*:510, 1970.

McKenna, O. C., and Angelakos, E. T.: Adrenergic innervation of the canine kidney. *Circ. Res., 22*:345, 1968.

Pomerznz, B. H.; Birtch, A. G.; and Barger, A. C.: Neural control of intrarenal blood flow. *Amer. J. Physiol., 215*:1067, 1968.

Santos-Martinez, J., and Selkurt, E. E.: Renal lymph and its relationship to the countercurrent multiplier system of the kidney. *Amer. J. Physiol., 216*:1548, 1969.

Schmid, H. E., Jr., and Spencer, M. P.: Characteristics of pressure-flow regulation by the kidney. *J. Appl. Physiol., 17*:201, 1962.

Schmidt-Nielsen, B.: Organ systems in adaptation: the excretory system. *In* Handbook of Physiology. Baltimore, The Williams & Wilkins Co., 1964, Sec. IV, p. 215.

Schmidt-Nielsen, B., and Laws, D. F.: Invertebrate mechanisms for diluting and concentrating the urine. *Ann. Rev. Physiol., 25*:631, 1963.

Share, L.: Effects of carotid occlusion and left atrial distention on plasma vasopressin titer. *Amer. J. Physiol., 208*:219, 1965.

Stevenson, J. A. F.: Central mechanisms controlling water intake. *In* Handbook of Physiology. Baltimore, The Williams & Wilkins Co., 1967, Sec. VI, Vol. I, p. 173.

Towbin, E. J.: Gustatory response to water in thirst regulation. *In* Handbook of Physiology. Baltimore, The Williams & Wilkins Co., 1967, Sec. VI, Vol. I, p. 191.

Verney, E. B.: Absorption and excretion of water; antidiuretic hormone. *Lancet,* 2:739, 1946.

Wathen, R. L., and Selkurt, E. E.: Intrarenal regulatory factors of salt excretion during renal venous pressure elevation. *Amer. J. Physiol., 216*:1517, 1969.

Wilde, W. S., and Vorburger, C.: Albumin multiplier in kidney vasa recta analyzed by microspectrophotometry of T-1824. *Amer. J. Physiol., 213*:1233, 1967.

Windhager, E. E.: Glomerulo-tubular balance of salt and water. *Physiologist, 11*:103, 1968.

Wolf, A. V.: Thirst: Physiology of the Urge to Drink and Problems of Water Lack. Springfield, Ill., Charles C Thomas, Publishers, 1958.

Wolf, G.: Hypothalamic regulation of sodium intake: relations to preoptic and tegmental function. *Amer. J. Physiol., 213*:1433, 1967.

CHAPTER 37

Regulation of Acid-Base Balance

When one speaks of the regulation of acid-base balance he actually means regulation of hydrogen ion concentration in the body fluids. The hydrogen ion concentration in different solutions can vary from less than 10^{-14} equivalents per liter to higher than 10^1, which means a total variation of more than a quadrillion fold. On a logarithmic basis, the hydrogen ion concentration in the human body is approximately midway between these two extremes.

Only slight changes in hydrogen ion concentration from the normal value can cause marked alterations in the rates of chemical reactions in the cells, some being depressed and others accelerated. For this reason the regulation of hydrogen ion concentration is one of the most important aspects of homeostasis. Later in the chapter the overall effects of high hydrogen ion concentration (acidosis) and low hydrogen ion concentration (alkalosis) are discussed. In general, when a person becomes acidotic he is likely to die in coma, and when he becomes alkalotic he may die of tetany or convulsions.

Relationship of Hydrogen Ions to Hydronium Ions. Essentially all hydrogen ions in the body fluids combine with water molecules in accordance with the following equation:

$$H^+ + H_2O \rightleftarrows H_3O^+$$

The resulting substance formed, H_3O^+, is called the hydronium ion. However, this ion can split immediately to form a hydrogen ion once again. Therefore, from a practical point of view, one can speak of hydrogen ion concentration to include both the free hydrogen ions and the great mass of hydronium ions. This is the way in which we will speak of hydrogen ions in this chapter, but always realizing that almost all of the hydrogen ions are normally bound with water molecules. This bonding with water molecules decreases the mobility of the hydrogen ions and limits their penetration into cells.

Another problem in speaking of acid-base balance is the two different terminologies that are used to designate a base. In most chemical laboratories, the term "base" is usually used to mean a compound that will react with an acid to neutralize it. On the other hand, an entirely different type of terminology considers a base to be any negative ion which, when bound with hydrogen ions, forms an acid. To prevent confusion of the student, we have used here the former, more common terminology for a base.

Normal Hydrogen Ion Concentration and Normal pH of the Body Fluids—Acidosis and Alkalosis. The hydrogen ion concentration in the extracellular fluid is normally regulated at a constant value of approximately 4×10^{-8} Eq./liter; this value can vary from as low as 1.6×10^{-8} to as high as 1.2×10^{-7}.

From these values, it is already apparent that expressing hydrogen ion concentration in terms of actual concentrations is a cumbersome procedure. Therefore, the symbol *pH* has come into usage for expressing the concentration, and pH is related to the actual hydrogen ion concentration by the following formula (when H^+ conc. is expressed in equivalents per liter):

$$pH = \log \frac{1}{H^+ \text{ conc.}} = -\log H^+ \text{ conc.} \quad (1)$$

Note from this formula that a low pH corresponds to a high hydrogen ion concentration, which is called *acidosis;* and, conversely, a high pH corresponds to a low hydrogen ion concentration, which is called *alkalosis.*

The normal pH of arterial blood is 7.4, while the pH of venous blood and of interstitial fluids is about 7.35 because of extra quantities of carbon dioxide which form carbonic acid in these fluids.

Since the normal pH of the arterial blood is 7.4, a person is considered to have acidosis whenever the pH is below this value and to have alkalosis when it rises above 7.4. The lower limit at which a person can live more than a few minutes is about 7.0, and the upper limit approximately 7.8.

Intracellular pH. The pH of the fluids inside the cells has never been measured, but on the basis of indirect measurements, it is believed that the intracellular pH usually ranges between 4.5 and 8.0 in different cells, perhaps averaging about 7.0. A rapid rate of metabolism in the cells increases the rate of carbon dioxide formation and consequently decreases the pH. Also, poor blood flow to any tissue causes carbon dioxide accumulation and a decrease in pH.

DEFENSE AGAINST CHANGES IN HYDROGEN ION CONCENTRATION

To prevent acidosis or alkalosis, several special control systems are available: (1) All the body fluids are supplied with acid-base *buffer systems* that immediately combine with any acid or alkali and thereby prevent excessive changes in hydrogen ion concentration. (2) If the hydrogen ion concentration does change measurably, the *respiratory center is immediately stimulated* to alter the rate of pulmonary ventilation. As a result, the rate of carbon dioxide removal from the body fluids is automatically changed, and, for reasons that will be presented later, this causes the hydrogen ion concentration to return toward normal. (3) When the hydrogen ion concentration changes from normal, *the kidneys excrete either an acid or alkaline urine,* thereby also helping to readjust the hydrogen ion concentration of the body fluids toward normal.

The buffer systems can act within a fraction of a second to prevent excessive changes in hydrogen ion concentration. On the other hand, it takes one to three minutes for the respiratory system to readjust the hydrogen ion concentration after a sudden change has occurred. Finally, the kidneys, though providing the most powerful of all the acid-base regulatory systems, require several hours to a day or more to readjust the hydrogen ion concentration.

FUNCTION OF ACID-BASE BUFFERS

An acid-base buffer is a solution of two or more chemical compounds that prevents marked changes in hydrogen ion concentration when either an acid or a base is added to the solution. As an example, if only a few drops of concentrated hydrochloric acid are added to a beaker of pure water, the pH of the water might immediately fall to as low as 1.0. However, if a satisfactory buffer system is present, the hydrochloric acid combines instantaneously with the buffer, and the pH falls only slightly. Perhaps the best way to explain the action of an acid-base buffer is to consider an actual simple buffer system, such as the bicarbonate buffer, which is extremely important in regulation of acid-base balance in the body.

THE BICARBONATE BUFFER SYSTEM

A typical bicarbonate buffer system consists of a mixture of carbonic acid (H_2CO_3) and sodium bicarbonate ($NaHCO_3$) in the same solution. It must first be noted that carbonic acid is a very weak acid for two reasons: First, its degree of dissociation into hydrogen ions and bicarbonate ions is poor in comparison with that of many other acids. Second, about 999 parts out of 1000 of any carbonic acid in a solution almost immediately dissociates into carbon dioxide and water, the net result being a high concentration of dissolved carbon dioxide but only a weak concentration of acid.

When an acid such as hydrochloric acid is added to a solution containing bicarbonate buffer, the following reaction takes place:

$$HCl + NaHCO_3 \rightarrow H_2CO_3 + NaCl \quad (2)$$

From this equation it can be seen that the strong hydrochloric acid is converted into the very weak carbonic acid. Therefore, the HCl lowers the pH of the solution only slightly.

On the other hand, if a strong base, such as sodium hydroxide, is added to the solution, the following reaction takes place:

$$NaOH + H_2CO_3 \rightarrow NaHCO_3 + H_2O \quad (3)$$

This equation shows that the hydroxyl ion of the sodium hydroxide combines with the hydrogen ion from the carbonic acid to form water and that the other product formed is sodium bicarbonate. The net result is loss of carbonic acid from the solution. However, carbonic acid is only a weak acid and, consequently, its loss does not cause the hydrogen ion concentration to rise greatly.

Though this bicarbonate buffer system has been illustrated in the above reactions as a mixture of carbonic acid and sodium bicarbonate, any other salt of bicarbonate besides sodium bicarbonate has identically the same function. Therefore, the small quantities of potassium bicarbonate, calcium bicarbonate, and magnesium bicarbonate in the extracellular fluids are also effective in the bicarbonate buffer system. In the intracellular fluid, where little sodium bicarbonate is present, the bicarbonate ion is provided mainly by potassium and magnesium bicarbonate.

Quantitative Dynamics of Buffer Systems

Dissociation of Carbonic Acid. All acids are ionized to a certain extent, and the percentage of ionization is called the *degree of dissociation.* Equation 4 illustrates the reversible relationship between undissociated carbonic acid and the two ions that it forms, H^+ and HCO_3^-.

$$H_2CO_3 \rightleftarrows H^+ + HCO_3^- \qquad (4)$$

A physicochemical law that has been found to apply to the dissociation of all acids is expressed by the following formula:

$$\frac{H^+ \times HCO_3^-}{H_2CO_3} = K' \qquad (5)$$

This formula states that in any given solution the concentration of hydrogen ions times the concentration of bicarbonate ions divided by the concentration of undissociated carbonic acid is equal to a constant, K'.

However, it is almost impossible to measure the concentration of undissociated carbonic acid in a solution because it is also continually in reversible equilibrium with dissolved carbon dioxide in the solution. Ordinarily, the amount of dissolved carbon dioxide is approximately 1000 times the concentration of the undissociated acid. On the other hand, it is possible to measure the total amount of dissolved carbon dioxide, and, since the amount of undissociated carbonic acid is proportional to the amount of dissolved carbon dioxide, Formula 5 above can also be expressed as follows:

$$\frac{H^+ \times HCO_3^-}{CO_2} = K \qquad (6)$$

The only real difference between the above two formulas is that the constant K' is approximately 1000 times the constant K.

Formula 6 can be changed into the following form:

$$H^+ = K \cdot \frac{CO_2}{HCO_3^-} \qquad (7)$$

If we take the logarithm of each of the two sides of Formula 7, it becomes the following:

$$\log H^+ = \log K + \log \frac{CO_2}{HCO_3^-} \qquad (8)$$

Now, the signs of the log H^+ and of log K are changed from positive to negative and the carbon dioxide and bicarbonate are inverted in the last term, which is the same as changing its sign also, giving the following formula:

$$-\log H^+ = -\log K + \log \frac{HCO_3^-}{CO_2} \qquad (9)$$

It will be recalled from earlier in the chapter that $-\log H^+$ is equal to the pH of the solution. Likewise, *−log K is called the pK of a buffer.* Therefore, this formula can be changed still further to the following:

$$pH = pK + \log \frac{HCO_3^-}{CO_2} \qquad (10)$$

The Henderson-Hasselbalch Equation. For the bicarbonate buffer system the pK is 6.1, and Formula 10 may be expressed as follows:

$$pH = 6.1 + \log \frac{HCO_3^-}{CO_2} \qquad (11)$$

This is called the Henderson-Hasselbalch equation, and by using it one can calculate the pH of a solution with reasonable accuracy after measuring the concentrations of bicarbonate ion and dissolved carbon dioxide, respectively. If the bicarbonate concentration is equal to the dissolved carbon dioxide concentration, the second member of the right-hand portion of the equation becomes log of 1, which is equal to zero. Therefore, under these conditions the pH of the solution is equal to the pK.

From the Henderson-Hasselbalch equation one can readily see that an increase in bicarbonate ion concentration causes the pH to rise, or, in other words, shifts the acid-base balance toward the alkaline side. On the other hand, an increase in the concentration of dissolved carbon dioxide decreases the pH or shifts the acid-base balance toward the acid side. It will be apparent later in this chapter that one can change the concentration of dissolved CO_2 in his body fluids by increasing or decreasing his rate of respiration. In this way the respiratory system can control to a certain extent the pH of the body fluids. On the other hand, the kidneys can increase or decrease the concentration of bicarbonate ion in the body fluids, in this way increasing or decreasing the pH. Thus, these two major hydrogen ion regulatory systems operate principally by altering one or the other of the two elements of the bicarbonate buffer system.

The Reaction Curve of the Bicarbonate Buffer System. Figure 37–1 shows the changes in pH of the body fluids when the ratio of bicarbonate ion to carbon dioxide changes. Note that when the concentrations of the two elements of the buffer are equal, the pH of the solution is 6.1, which is equal to the pK of the bicarbonate buffer system. When base is added to the buffer a large proportion of the

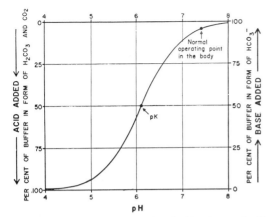

Figure 37–1   Reaction curve for the bicarbonate buffer system.

carbon dioxide is converted into bicarbonate ions, and the ratio is altered. As a result, the pH rises, as indicated by the upward slope of the curve. On the other hand, when acid is added, a large proportion of the bicarbonate ion is converted first into carbonic acid and then into dissolved carbon dioxide so that the ratio of the two elements of the system changes in favor of the acidic side, and the pH falls, as illustrated by the downslope of the curve.

BUFFERING POWER OF THE BICARBONATE BUFFER SYSTEM. Referring once again to Figure 37–1, note that at the central point of the curve, addition of a slight amount of acid or base causes minimal change in pH. However, toward each end of the curve addition of a slight amount of acid or base causes the pH to change greatly. Thus, the so-called *buffering power* of the buffer system *is greatest when the pH is equal to the pK*. The buffering power is still reasonably effective until the ratio of one element of the buffer system to the other reaches as much as 8:1 or 1:8, but beyond this point the buffering power diminishes rapidly. And, when all the carbon dioxide has been converted into bicarbonate ion or when all the bicarbonate ion has been converted into carbon dioxide, the system has no more buffering power at all.

A second factor that determines the buffering power is the concentrations of the two elements in the buffer solution. Obviously, if the concentrations are slight, only a small amount of acid or base need be added to the solution to change the pH considerably. Thus, *the buffering power of a buffer is directly proportional to the concentrations of the buffer substances*.

THE BUFFER SYSTEMS OF THE BODY FLUIDS

The three major buffer systems of the body fluids are the *bicarbonate buffer*, which has been described above, the *phosphate buffer*, and the *protein buffer*. Each of these performs major buffering functions under different conditions.

The Bicarbonate Buffer System. The bicarbonate system is not an exceedingly powerful buffer for two reasons. First, the pH in the extracellular fluids is about 7.4, while the pK of the bicarbonate buffer system is 6.1. This means that approximately 20 times as much of the bicarbonate buffer is in the form of bicarbonate ion as in the form of dissolved carbon dioxide. For this reason, the system is operating on a portion of its buffering curve where the buffering power is poor. Second, the concentrations of the bicarbonate buffer system are not exceedingly great.

Yet despite the fact that the bicarbonate buffer system is not especially powerful, it is probably equally as important as all the others in the body *because the concentrations of each of the two elements of the bicarbonate system can be regulated*, carbon dioxide by the respiratory system and bicarbonate ion by the kidneys. As a result, the pH of the blood can be shifted up or down by the respiratory and renal regulatory systems.

The Phosphate Buffer System. The phosphate buffer system acts in almost identically the same manner as the bicarbonate buffer system, but it is composed of the following two elements: NaH_2PO_4 and Na_2HPO_4. When a strong acid, such as hydrochloric acid, is added to a mixture of these two substances, the following reaction occurs:

$$HCl + Na_2HPO_4 \rightarrow NaH_2PO_4 + NaCl \qquad (12)$$

The net result of this reaction is that the hydrochloric acid is removed, and in its place an additional quantity of NaH_2PO_4 is formed. Since NaH_2PO_4 is only weakly acidic, a strong acid has been traded for a very weak acid, and the pH changes relatively slightly. Conversely, if a strong base, such as sodium hydroxide, is added to the buffer system, the following reaction takes place:

$$NaOH + NaH_2PO_4 \rightarrow Na_2HPO_4 + H_2O \qquad (13)$$

Here sodium hydroxide is decomposed to form water and Na_2HOP_4. That is, a strong base is traded for the very weak base Na_2HPO_4, allowing only a slight shift in pH toward the alkaline side.

The phosphate buffer system has a pK of 6.8, which is not far from the normal pH of 7.4 in the body fluids; this allows the phosphate system to operate near its maximum buffering power. However, despite the fact that this buffer system operates in a reasonably good portion of the buffer curve, its concentration in the extracellular fluids is only one-sixth that of the bicarbonate buffer. Therefore, its total buffering power is even less than that of the bicarbonate system.

The phosphate buffer is especially important in the tubular fluids of the kidneys for two reasons: First, phosphate usually becomes greatly concentrated in the tubules, thereby increasing the buffering power of the phosphate system. Second, the tubular fluid also usually becomes more acidic than the extracellular fluids, bringing the operating range even closer to the pK of the system.

The phosphate buffer is also extremely important in the intracellular fluids because the concentration of phosphate in these fluids is many times that in the extracellular fluids, and also because the pH of the intracellular fluids is usually closer to the pK of the phosphate buffer system than is the pH of the extracellular fluid.

The Protein Buffer System. By far the most plentiful buffer of the body is the proteins of the cells and plasma. There is a slight amount of diffusion of hydrogen ions through the cell membrane, and even more important, carbon dioxide can diffuse readily through cell membranes and bicarbonate ions can diffuse to some extent. The diffusion of the two elements of the bicarbonate buffer system causes the pH in the intracellular fluids to change approximately in proportion to the changes in pH in the extracellular fluids. Thus, all the buffer systems inside the cells help to buffer the extracellular fluids as well. These include the extremely large amounts of proteins inside the cells. Indeed, experimental studies have shown that at least three quarters of all the *chemical* buffering power of the body fluids is inside the cells, and most of this results from the intracellular proteins.

The method by which the protein buffer system operates is precisely the same as that of the bicarbonate buffer system. It will be recalled that a protein is composed of amino acids bound together by peptide linkages, but some of the different amino acids have free acidic radicals in the form of $-COOH$, and these can dissociate into $-COO^-$ and H^+. Also, some have free basic radicals in the form of $-NH_3OH$, which can dissociate into $-NH_3^+$ and OH^-. The OH^- in turn can react with hydrogen ions to form water, in this way reducing the hydrogen ion concentration. Thus, proteins can operate in both *acidic* and *basic* buffering systems, the basic systems operating oppositely but similarly to the acidic systems. Furthermore, the pK's of the different weak acidic and basic buffering systems are not far from 7.4. This, too, helps to make the protein buffering systems by far the most powerful of the body.

THE ISOHYDRIC PRINCIPLE

Each of the above buffer systems has been discussed as if it could operate individually in the body fluids. However, they all actually work together, for the hydrogen ion is common to the chemical reactions of all the systems. Therefore, whenever any condition causes the hydrogen ion concentration to change, it causes the balance of all the buffer systems to change at the same time. This phenomenon, called the *isohydric principle,* is represented by the following formula:

$$H^+ = \frac{K_1 \times HA_1}{A_1^-} = \frac{K_2 \times HA_2}{A_2^-} = \frac{K_3 \times HA_3}{A_3^-} \quad (14)$$

in which K_1, K_2, and K_3 are the dissociation constants of three respective acids, HA_1, HA_2, and HA_3, and A_1^-, A_2^-, and A_3^- are the concentrations of the free negative ions of the acids.

The important feature of this principle is that any condition that changes the balance of any one of the buffer systems also changes the balance of all the others, for *the buffer systems actually buffer each other.*

RESPIRATORY REGULATION OF ACID-BASE BALANCE

In the discussion of the Henderson-Hasselbalch equation, it was noted that an increase in carbon dioxide concentration in the body fluids lowers the pH toward the acidic side, whereas a decrease in carbon dioxide raises the pH toward the alkaline side. It is on the basis of this effect that the respiratory system is capable of altering the pH either up or down.

Balance between Metabolic Formation of Carbon Dioxide and Pulmonary Expiration of Carbon Dioxide. Carbon dioxide is continually being formed in the body by the different intracellular metabolic processes, the carbon in the foods being oxidized by oxygen to form carbon dioxide. This in turn diffuses into the interstitial fluids and blood, and is transported to the lungs where it diffuses into the alveoli and is transferred to the atmosphere by pul-

Figure 37-2   Relationship between metabolic formation of carbon dioxide and rate of carbon dioxide expiration by the lungs.

monary ventilation. However, several minutes are required for this passage of carbon dioxide from the cells to the atmosphere. Since the carbon dioxide is not removed instantaneously, an average of 1.2 millimols/liter of dissolved carbon dioxide is normally in the extracellular fluids.

If the rate of metabolic formation of carbon dioxide becomes increased, its concentration in the extracellular fluids is likewise increased. Conversely, decreased metabolism decreases the carbon dioxide concentration.

On the other hand, if the rate of pulmonary ventilation is increased, the rate of expiration of carbon dioxide becomes increased, and this decreases the amount of accumulated carbon dioxide in the extracellular fluids.

Figure 37-2 illustrates the relationship between metabolic formation of carbon dioxide and its expiration by the lungs. The horizontal solid line shows that under resting conditions, the normal metabolic formation of carbon dioxide is approximately constant at about 200 ml./minute. The diagonal solid curve illustrates that, with normal respiration, the rate of carbon dioxide expiration is directly proportional to the carbon dioxide concentration in the plasma. The point at which these two curves cross, point A, is the actual operating conditions in the body, because at this point the rate of expiration of carbon dioxide equals the rate of metabolic formation. Thus, as shown by the figure, the normal carbon dioxide level in the plasma is 1.2 millimols/liter.

The dashed lines of Figure 37-2 illustrate what happens when the rate of carbon dioxide formation and the rate of expiration change. In this instance the rate of formation has risen to 300 ml./minute, and the pulmo-

nary ventilation has increased approximately 25 per cent. At point B where these two curves equilibrate with each other, the CO_2 concentration is approximately 1.4 millimols/liter. Thus, it is evident that the carbon dioxide concentration in the extracellular fluids is controlled by a balance between the rate of metabolism and the rate of pulmonary excretion.

Effect of Increasing or Decreasing the Alveolar Ventilation on pH of the Extracellular Fluids. If we assume that the rate of metabolic formation of carbon dioxide remains constant, then the only factor that affects the carbon dioxide concentration in the body fluids is the rate of alveolar ventilation as expressed by the following formula:

$$CO_2 \propto \frac{1}{\text{Alveolar ventilation}} \qquad (15)$$

And, since an increase in carbon dioxide decreases the pH, changes in alveolar ventilation also change the hydrogen ion concentration.

Figure 37-3 illustrates the approximate change in pH in the blood that can be effected by increasing or decreasing the rate of alveolar ventilation. Note that an increase in alveolar ventilation to two times normal raises the pH of the extracellular fluids by about 0.23 pH unit. This means that if the pH of the body fluids has been 7.4 with normal alveolar ventilation, doubling the ventilation raises the pH to 7.63. Conversely, a decrease in alveolar ventilation to one-quarter normal reduces the pH 0.4 pH unit. That is, if at normal alveolar ventilation the pH had been 7.4, reducing the ventilation to one-quarter normal reduces the pH 0.4 unit to 7.0. Since alveolar ventilation can be reduced to zero ventilation or increased to about 15 times normal, one can readily understand how much the pH of the body

Figure 37-3   Approximate change in body fluid pH caused by increased or decreased rate of alveolar ventilation.

fluids can be changed by alterations in the activity of the respiratory system.

EFFECT OF HYDROGEN ION CONCENTRATION ON ALVEOLAR VENTILATION

Not only does the rate of alveolar ventilation affect the hydrogen ion concentration of the body fluids, but, in turn, the hydrogen ion concentration can affect the rate of alveolar ventilation. This results from a *direct action of hydrogen ions on the respiratory center in the medulla oblongata* that controls breathing.

Figure 37–4 illustrates the changes in alveolar ventilation caused by changing the pH of arterial blood from 7.0 to 7.6. From this graph it is evident that a decrease in pH from the normal value of 7.4 to the strongly acidic range can increase the rate of alveolar ventilation to as much as four to five times normal, while an increase in pH into the alkaline range can decrease the rate of alveolar ventilation to as little as 50 to 75 per cent of normal.

Feedback Regulation of Hydrogen Ion Concentration by the Respiratory System. Because of the ability of the respiratory center to respond to hydrogen ion concentration and because changes in alveolar ventilation in turn alter the hydrogen ion concentration in the body fluids, the respiratory system acts as a typical feedback regulatory system for controlling hydrogen ion concentration. That is, any time the hydrogen ion concentration becomes high, the respiratory system becomes more active, and alveolar ventilation increases. As a result, the carbon dioxide concentration in the extracellular fluids decreases, thus reducing the hydrogen ion concentration back toward a normal value. Conversely, if the hydrogen ion concentration falls too low, the respiratory center becomes depressed, alveolar ventilation also decreases, and the hydrogen ion concentration rises back toward normal.

EFFICIENCY OF RESPIRATORY REGULATION OF HYDROGEN ION CONCENTRATION. Unfortunately, respiratory control cannot return the hydrogen ion concentration all the way to the normal value of 7.4 when some abnormality outside the respiratory system has altered the pH from the normal. The reason for this is that, as the pH returns toward normal, the stimulus that has been causing either increased or decreased respiration will itself begin to be lost. Ordinarily, the respiratory mechanism for regulation of hydrogen ion concentration has a control effectiveness of between 50 and 75 per cent. That is, if the hydrogen ion concentration should suddenly be decreased from 7.4 to 7.0 by some extraneous factor, the respiratory system, in a minute or more, returns the pH to a value of about 7.2 to 7.3.

BUFFERING POWER OF THE RESPIRATORY SYSTEM. In effect, respiratory regulation of acid-base balance is a *physiological type of buffer* system having almost identically the same importance as the chemical buffering systems of the body discussed earlier in the chapter. The overall "buffering power" of the respiratory system is one to two times as great as that of all the chemical buffers combined. That is, one to two times as much acid or base can normally be buffered by this mechanism as by the chemical buffers.

RENAL REGULATION OF HYDROGEN ION CONCENTRATION

In the earlier discussion of the Henderson-Hasselbalch equation it was pointed out that the kidneys regulate hydrogen ion concentration principally by increasing or decreasing the bicarbonate ion concentration in the body fluid. To do this, a complex series of reactions occurs in the renal tubules, including reactions for hydrogen ion secretion, sodium ion reabsorption, bicarbonate ion excretion into the urine, and ammonia secretion into the tubules. The following sections describe the interplay of these different tubular mechanisms and the manner in which they help to regulate the hydrogen ion concentration of the body fluids.

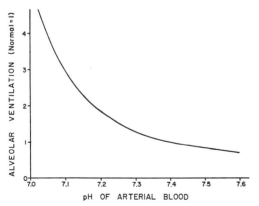

Figure 37–4   Effect of blood pH on the rate of alveolar ventilation. (Constructed from data obtained by Gray: Pulmonary Ventilation and Its Regulation. Charles C Thomas.)

TUBULAR SECRETION OF HYDROGEN IONS

The epithelial cells of the proximal tubules, distal tubules, and collecting ducts all secrete hydrogen ions into the tubular fluid. The mechanism by which this occurs is illustrated in Figure 37–5, which shows that hydrogen ion excretion *begins with carbon dioxide* in the tubular epithelial cells. The carbon dioxide, under the influence of an enzyme, *carbonic anhydrase,* combines with water to form *carbonic acid.* This then dissociates into *bicarbonate ion* and *hydrogen ion,* and the hydrogen ion is secreted through the cell membrane into the tubule.

The precise mechanism by which hydrogen ions are finally secreted through the cell membrane is not known. However, in the collecting ducts this process can continue until the concentration of hydrogen ions in the tubules becomes as much as 900 times that in the extracellular fluid or, in other words, until the pH of the tubular fluids falls to about 4.5. This represents a limit to the ability of the tubular epithelium to secrete hydrogen ions.

About 84 per cent of all the hydrogen ions secreted by the tubules are secreted in the proximal tubules, but the maximum concentration gradient that can be achieved here is only about three-fold instead of the 900-fold that can be achieved in the collecting ducts. That is, the pH can be decreased about 0.4 to 0.5 pH unit below 7.4, the pH of the glomerular filtrate. The ability of the distal tubules to secrete hydrogen ions is between the abilities of the proximal tubules and collecting ducts.

Reabsorption of Sodium Ions in Exchange for Hydrogen Ions. Figure 37–5 also illustrates that hydrogen ion secretion is closely coupled to sodium ion reabsorption. That is, each time a hydrogen ion is secreted a sodium ion is reabsorbed.

Regulation of Hydrogen Ion Secretion by the Carbon Dioxide Concentration in the Extracellular Fluid. Since the chemical reactions for secretion of hydrogen ions begin with carbon dioxide, the greater the carbon dioxide concentration in the extracellular fluid, the more rapidly the reactions proceed, and the greater becomes the rate of hydrogen ion secretion. Therefore, any factor that increases the carbon dioxide concentration in the extracellular fluids, such as decreased respiration or increased metabolic rate, also increases the rate of hydrogen ion secretion. Conversely, any factor that decreases the carbon dioxide, such as excess pulmonary ventilation or decreased metabolic rate, decreases the rate of hydrogen ion secretion.

At normal carbon dioxide concentrations, the rate of hydrogen ion secretion is about 3.5 millimols per minute, but this rises or falls directly in proportion to any changes in extracellular carbon dioxide.

INTERACTION OF BICARBONATE IONS WITH HYDROGEN IONS IN THE TUBULES — "REABSORPTION" OF BICARBONATE IONS

Bicarbonate ion is continually being filtered through the glomerular membrane into the glomerular filtrate, and this usually is in combination with sodium ions, or, to a lesser extent, with other positive ions of the extracellular fluid. When the sodium and bicarbonate ions enter the tubules, many of the sodium ions are absorbed in interchange for secreted hydrogen ions. Simultaneously, the hydrogen ions combine with the bicarbonate ions to form carbonic acid, as illustrated in Figure 37–5. This carbonic acid in turn dissociates into carbon dioxide and water; essentially all the carbon dioxide then diffuses through the epithelial cells into the peritubular fluids while the water passes into the urine. Thus, *if plenty of hydrogen ions are available, the bicarbonate ions are almost completely removed from the tubules,* so that for practical purposes *none* remains to pass into the urine.

If we now note in Figure 37–5 the chemical reactions that are responsible for formation of hydrogen ions in the epithelial cells, we will see that a bicarbonate ion is formed inside these cells each time a hydrogen ion is formed by the dissociation of H_2CO_3. The bicarbonate ion then diffuses into the peritubular fluid in combination with the sodium ion that has been absorbed from the tubule.

The net effect of all these reactions is a mechanism for "reabsorption" of bicarbonate ions from the tubules, though the bicarbonate ions that enter the peritubular fluid are not the same bicarbonate ions that had been in the tubular fluid.

Figure 37–5 Chemical reactions for (1) hydrogen ion secretion, (2) sodium ion absorption in exchange for a hydrogen ion, and (3) combination of hydrogen ions with bicarbonate ions in the tubules.

Normal Rates of Bicarbonate Ion Filtration and Hydrogen Ion Secretion into the Tubules. Under normal conditions, the rate of hydrogen ion secretion is about 3.50 millimols/minute, and the rate of filtration of bicarbonate ions is about 3.49 millimols/minute. Therefore, essentially all the bicarbonate ions are normally "reabsorbed" from the tubules by the process just described, while a slight excess of hydrogen ions remains in the tubules to react with other substances and to be excreted into the urine.

RENAL CORRECTION OF ALKALOSIS— DECREASE IN BICARBONATE IONS IN THE EXTRACELLULAR FLUID

Now that we have described the mechanisms by which the renal tubules secrete hydrogen ions and reabsorb bicarbonate ions, we can explain the manner in which the kidneys readjust the pH of the extracellular fluids when it becomes abnormal.

The initial step in this explanation is to understand what happens to the concentrations of carbon dioxide and bicarbonate ions in the extracellular fluids in alkalosis and acidosis. First, let us consider *alkalosis*. Referring again to Equation 11, the Henderson-Hasselbalch equation, we see that the *ratio* of bicarbonate ions to dissolved carbon dioxide molecules increases when the pH rises into the alkalosis range above 7.4. The effect of this in the tubules is to increase the *ratio* of bicarbonate ions filtered into the tubules to hydrogen ions secreted. Therefore, the fine balance that normally exists between these two no longer occurs. Instead, far greater quantities of bicarbonate ions now enter the tubules than hydrogen ions. Since no bicarbonate ions can be reabsorbed without first reacting with hydrogen ions, all the excess bicarbonate ions pass into the urine and carry with them sodium ions or other positive ions. Thus, in effect, sodium bicarbonate is removed from the extracellular fluid.

Loss of sodium bicarbonate from the extracellular fluid decreases the bicarbonate ion portion of the bicarbonate buffer system, and, in accordance with the Henderson-Hasselbalch equation, this shifts the pH of the body fluids back in the acid direction. Thus, the alkalosis is corrected.

RENAL CORRECTION OF ACIDOSIS— INCREASE IN BICARBONATE IONS IN THE EXTRACELLULAR FLUID

In acidosis, the *ratio* of carbon dioxide to bicarbonate ions in the extracellular fluid increases, which is exactly opposite to the effect in alkalosis. Therefore, in acidosis, the *rate of hydrogen ion secretion* rises to a level far greater than the *rate of bicarbonate ion filtration* into the tubules. As a result, excess hydrogen ions are secreted into the tubules and have no bicarbonate ions to react with. These combine with the buffers in the tubular fluid, as is explained in following paragraphs, and are excreted into the urine.

Figure 37–5 shows that each time a hydrogen ion is secreted into the tubules two other effects occur simultaneously: first, a bicarbonate ion is formed in the tubular epithelial cell and, second, a sodium ion is absorbed from the tubule into the epithelial cell. The sodium ion and bicarbonate ion then diffuse together from the epithelial cell into the peritubular fluid. Thus, *the net effect of secreting excess hydrogen ions into the tubules is to increase the quantity of sodium bicarbonate in the extracellular fluid.* This increases the bicarbonate salt portion of the bicarbonate buffer system, which, in accordance with the Henderson-Hasselbalch equation, increases the pH and thereby corrects the acidosis.

TRANSPORT OF EXCESS HYDROGEN IONS FROM THE TUBULES INTO THE URINE

When excess hydrogen ions are secreted into the tubules, they combine with other substances in the tubules and are transported into the urine. Two distinct mechanisms exist by which the excess hydrogen ions are transported from the tubules: first, by combination with the buffers of the tubular fluid and, second, by combination with ammonia secreted by the tubular epithelium.

Combination of Excess Hydrogen Ions with Tubular Buffers. The principal buffer in the tubular fluid (besides the bicarbonate buffer, which has already been discussed above) is the phosphate buffer. The quantity of Na_2HPO_4 in the glomerular filtrate is normally about four times as great as that of NaH_2PO_4. Excess hydrogen ions entering the tubules combine with the Na_2HPO_4, as illustrated in Figure 37–6, forming NaH_2PO_4, which passes on into the urine. Thus, sodium is liberated and is absorbed from the tubules in place of the hydrogen ion involved in the reaction. The sodium that is absorbed combines with the *bicarbonate ion* formed in the process of secreting hydrogen ions. Thus, the net effect of this reaction is to increase the amount of sodium bicarbonate in the extracellular fluids, which is the kidney's

Figure 37–6 Chemical reactions in the tubules involving hydrogen ions, sodium ions, and the phosphate buffer system.

way of reducing the degree of acidosis in the body fluids.

Ammonia Secretion by the Tubular Epithelium. Another mechanism for transporting hydrogen ions from the tubules to the exterior is to combine them with ammonia. The epithelial cells of the proximal tubules, distal tubules, and collecting ducts all continually synthesize ammonia and this diffuses into the tubules. The ammonia then reacts with hydrogen ions, as illustrated in Figure 37–7, forming ammonium ions that are excreted into the urine in combination with chloride or other tubular anions. Note in the figure that the net effect of these reactions is again *to increase the sodium bicarbonate concentration* in the extracellular fluid.

This mechanism for removal of hydrogen ions from the tubules is especially important because most of the negative ions of the tubular fluid are chloride. Only a few hydrogen ions could be transported into the urine in direct combination with chloride, because hydrochloric acid is a very strong acid and the tubular pH would fall rapidly to the critical value of 4.5, below which further hydrogen ion secretion would cease. However, when hydrogen ions combine with ammonia and the

Figure 37–7 Secretion of ammonia by the tubular epithelial cells, and reaction of the ammonia with hydrogen ions in the tubules.

resulting ammonium ions then combine with chloride, the pH does not fall significantly because ammonium chloride is a neutral salt instead of an acid.

Sixty per cent of the ammonia secreted by the tubular epithelium is derived from *glutamine,* and the remaining 40 per cent from asparagine or different ones of the amino acids.

If the tubular fluids remain highly acidic for long periods of time, the formation of ammonia steadily increases during the first two to three days, rising as much as 10-fold. For instance, immediately after acidosis begins, as little as 30 millimols of ammonia might be secreted each day, but after two days as much as 200 to 300 millimols can be secreted, illustrating that the ammonia-secreting mechanism can adapt readily to handle greatly increased loads of acid elimination.

RAPIDITY OF ACID-BASE REGULATION BY THE KIDNEYS

Figure 37–8 illustrates the effect of extracellular fluid pH on the rate at which bicarbonate ions are lost from or gained by the body fluids each minute, illustrating that at a pH of 7.0 approximately 2.3 millimols of bicarbonate ions is gained each minute, but as the pH returns toward normal the rate of gain falls off markedly. Then, when the pH becomes significantly greater than normal, bicarbonate ions are lost by the extracellular fluids. For instance, at a pH of 7.6 about 1.5 millimols of bicarbonate ions are lost each minute.

The total amount of buffers in the entire body is approximately 1000 millimols. If all these should be suddenly shifted to the alkaline or acidic side by injecting an alkali or an acid, the kidneys would be able to return the pH of the body fluids back almost to normal in 10 to 20 hours. Though this mechanism is slow to act, it continues acting until the pH returns almost exactly to normal rather than a certain percentage of the way. Therefore, the real value of the renal mechanism for regulating hydrogen ion concentration is not rapidity of action but instead its ability in the end to neutralize completely any excess acid or alkali that enters the body fluids.

Ordinarily, the kidneys can remove up to about 500 millimols of acid or alkali each day. If greater quantities of acid or alkali than this enter the body fluids, the kidneys are unable to cope with the extra load, and severe acidosis or alkalosis ensues.

Figure 37–8 Effect of extracellular fluid pH on the rate of loss or gain of bicarbonate ions in the body fluids each minute.

Range of Urinary pH. In the process of adjusting the hydrogen ion concentration of the extracellular fluids, the kidneys often excrete urine at pH's as low as 4.5 or as high as 8.0. When acid is being excreted the pH falls, and when alkali is being excreted the pH rises. Even when the pH of the extracellular fluids is at the normal value of 7.4, a fraction of a millimol of acid is still lost each minute. The reason for this is that about 50 to 100 millimols more acid than alkali are formed in the body each day, and this acid must be removed continually. Because of the presence of this excess acid in the urine, the normal urine pH is about 6.0 instead of 7.4, the pH of the blood.

SPECIAL FACTORS THAT CAN ALTER ACID-BASE EQUILIBRIUM

At least three special factors can affect acid-base equilibrium even under otherwise normal conditions. These are (1) the plasma concentration of chloride ions, (2) the plasma concentration of potassium ions, and (3) the rate of aldosterone secretion by the adrenal glands.

Alkalosis Caused by Low Plasma Chloride and Acidosis Caused by High Plasma Chloride. When the plasma level of chloride ions falls, the quantity of chloride ions filtering into the glomerular filtrate also decreases. This makes less chloride ions available to be reabsorbed with sodium ions. As a consequence, greater than normal quantities of hydrogen ions are secreted in exchange for sodium ions. For each extra hydrogen ion secreted, a bicarbonate ion appears in the extracellular fluid as explained above. Thus, low chloride concentration in the plasma leads to an increase in bicarbonate ions, which shifts the pH of the body fluids toward alkalosis.

A high level of chloride ions causes exactly the opposite effect, decreased hydrogen ion secretion into the tubules and therefore decreased reabsorption of bicarbonate ions from the tubules. This leads to acidosis.

Acidosis Caused by High Plasma Potassium. Potassium ions, like hydrogen ions, are also secreted by the tubular epithelium, particularly in the distal tubules and collecting ducts, as explained in Chapter 34. And potassium ions are exchanged for sodium ions in exactly the same way that hydrogen ions are. When the plasma concentration of potassium ions is high, the tubular epithelium secretes large quantities of these into the tubules, and these ions compete with hydrogen ions for exchange with sodium. Thus, secretion of potassium ions *decreases* secretion of hydrogen ions. And a decrease in hydrogen ions allows greater spillage of bicarbonate ions into the urine, causing loss of bicarbonate ions from the extracellular fluid, which leads to *acidosis*.

Alkalosis Caused by Excess Aldosterone; Acidosis Caused by Deficiency of Aldosterone. When excess quantities of aldosterone are secreted by the adrenal glands, the extracellular fluid becomes slightly alkalotic. This is believed to be caused in the following way: the aldosterone promotes extensive reabsorption of sodium ions, making more of these available to exchange with secreted hydrogen ions. Therefore the rate of hydrogen ion secretion becomes greater than normal; and, for each hydrogen ion secreted, a molecule of sodium bicarbonate enters the extracellular fluid as was explained above. This excess bicarbonate salt shifts the pH toward the alkaline side.

Deficiency of aldosterone secretion causes exactly the opposite effects: too little sodium ion reabsorption, too little hydrogen ion secretion, spillage of bicarbonate into the urine with loss of this from the extracellular fluid, and consequent development of a mild degree of acidosis.

CLINICAL ABNORMALITIES OF ACID-BASE BALANCE

RESPIRATORY ACIDOSIS AND ALKALOSIS

From the discussions earlier in the chapter it is obvious that any factor that decreases the rate of pulmonary ventilation increases the concentration of dissolved carbon dioxide in the extracellular fluid, which in turn leads to increased carbonic acid and hydrogen ions, thus resulting in acidosis. Because this type of acidosis is caused by an abnormality of respiration, it is called *respiratory acidosis.*

On the other hand, excessive pulmonary ventilation reverses the process and decreases the hydrogen ion concentration, thus resulting in alkalosis; this condition is called *respiratory alkalosis.*

A person can cause respiratory acidosis in himself by simply holding his breath, which he can do until the pH of the body fluids falls to as low as perhaps 7.1. On the other hand, he can voluntarily overbreathe and cause alkalosis to a pH of about 7.8.

Respiratory acidosis frequently results from pathologic conditions. For instance, damage to the respiratory center in the medulla oblongata, causing reduced breathing, obstruction of the passageways in the respiratory tract, pneumonia, decreased pulmonary surface area, and any other factor that interferes with the exchange of gases between the blood and alveolar air, results in respiratory acidosis.

On the other hand, only rarely do pathologic conditions cause *respiratory alkalosis.* However, an occasional psychoneurosis causes a person to overbreathe to such an extent that he becomes alkalotic. A physiological type of respiratory alkalosis occurs when a person ascends to a high altitude. The low oxygen content of the air stimulates respiration, which causes excess loss of carbon dioxide and development of mild respiratory alkalosis.

METABOLIC ACIDOSIS AND ALKALOSIS

The terms "metabolic acidosis" and "metabolic alkalosis" refer to all other abnormalities of acid-base balance besides those caused by excess or insufficient carbon dioxide in the body fluids. Use of the word "metabolic" in this instance is unfortunate, because carbon dioxide is also a metabolic product. Yet, by convention, carbonic acid resulting from dissolved carbon dioxide is called a *respiratory acid* while any other acid in the body, whether it be formed by metabolism or simply ingested by the person, is called a *metabolic acid.*

Causes of Metabolic Acidosis. Metabolic acidosis can result from (1) the formation of metabolic acids in the body, (2) intravenous administration of metabolic acids, or (3) addition of metabolic acids by way of the gastrointestinal tract. Metabolic acidosis can result also from (4) loss of alkali from the body fluids. Some of the specific conditions that cause metabolic acidosis are the following:

Severe diarrhea is one of the most frequent causes of metabolic acidosis for the following reasons: The gastrointestinal secretions normally contain large amounts of sodium bicarbonate. Therefore, excessive loss of these secretions during a bout of diarrhea is exactly the same as excretion of large amounts of sodium bicarbonate into the urine. In accordance with the Henderson-Hasselbalch equation, this results in a shift of the bicarbonate buffer system toward the acid side and results in metabolic acidosis. In fact, acidosis resulting from severe diarrhea can be so serious that it is one of the most common causes of death in young children.

A second cause of metabolic acidosis is *vomiting.* Vomiting of gastric contents alone, which occurs rarely, causes a loss of acid and leads to alkalosis, but vomiting of contents from deeper in the gastrointestinal tract, which often occurs, causes loss of alkali and results in metabolic acidosis.

A third common type of acidosis is *uremic acidosis,* which occurs in severe renal disease. The cause of this is failure of the kidneys to rid the body of even the normal amounts of acids formed each day by the metabolic processes of the body.

A fourth and extremely important cause of metabolic acidosis is *diabetes mellitus.* In this condition, lack of insulin secretion by the pancreas prevents normal use of glucose for metabolism. Instead, the stored fats are split into acetoacetic acid, and this in turn is metabolized by the tissues for energy in place of glucose. Simultaneously, the concentration of acetoacetic acid in the extracellular fluids often rises very high, and large quantities of it are excreted in the urine, sometimes as much as 500 to 1000 millimols per day. This condition unfortunately causes metabolic acidosis not only because of the presence of acetoacetic acid in the bloodstream but also because the acetoacetic acid combines with sodium bicarbonate in the tubules and thereby prevents reabsorption of the bicarbonate into the extracellular fluids. Therefore, in diabetes mellitus the acidosis results, first, from the presence of acetoacetic acid itself in the extracellular fluids, and second, from the loss of large amounts of sodium bicarbonate, the alkaline half of the bicarbonate buffer system, from the extracellular fluids.

Causes of Metabolic Alkalosis. Metabolic alkalosis does not occur nearly so often as metabolic acidosis. It most frequently follows *excessive ingestion of alkaline drugs,* such as sodium bicarbonate, for the treatment of gastritis or peptic ulcer. However, metabolic alkalosis occasionally results from *excessive vomiting of gastric contents* without vomiting of lower gastrointestinal contents, which causes excessive loss of the hydrochloric acid secreted by the stomach mucosa. The net result is loss of acid from the extracellular fluids and development of metabolic alkalosis.

EFFECTS OF ACIDOSIS AND ALKALOSIS ON THE BODY

Acidosis. The major effect of acidosis is depression of the *central nervous system.* When the pH of the blood falls below 7.0, the nervous system becomes so depressed that the person first becomes disoriented and, later, comatose. Therefore, patients dying of diabetic acidosis, uremic acidosis, and other types of acidosis usually die in a state of coma.

In metabolic acidosis the high hydrogen ion concentration causes increased rate and depth of respiration. Therefore, one of the diagnostic signs of *metabolic* acidosis is increased pulmonary ventilation. On the other hand, *in respiratory acidosis, respiration is usually depressed,* which is opposite to the effect in metabolic acidosis.

Alkalosis. The major effect of alkalosis on the body is *overexcitability of the nervous system.* This effect occurs both in the central nervous system and in the peripheral nerves, but usually the peripheral nerves are affected before the central nervous system. The nerves become so excitable that they automatically and repetitively fire even when they are not stimulated by normal stimuli. As a result, the muscles go into a state of *tetany,* which means a state of tonic spasm. This tetany usually appears first in the muscles of the forearm, then spreads rapidly to the muscles of the face, and finally all over the body. Alkalotic patients may die from tetany of the respiratory muscles.

Only occasionally does an alkalotic person develop severe symptoms of central nervous system overexcitability. The symptoms may manifest themselves as extreme nervousness or, in susceptible persons, as convulsions. For instance, in persons who are predisposed to epileptic fits, simply overbreathing often results in an attack. Indeed, this is one of the clinical methods for assessing one's degree of epileptic predisposition.

RESPIRATORY COMPENSATION OF METABOLIC ACIDOSIS OR ALKALOSIS

It was pointed out above that the high hydrogen ion concentration of metabolic acidosis causes increased pulmonary ventilation, which in turn results in rapid removal of carbon dioxide from the body fluids and reduces the hydrogen ion concentration. Thus, this respiratory effect helps to compensate for the metabolic acidosis. However, this compensation occurs only partway. Ordinarily, the respiratory system is capable of compensating between 50 and 75 per cent. That is, if the metabolic factor makes the pH of the blood fall to 7.0 at normal pulmonary ventilation, the rate of pulmonary ventilation normally increases sufficiently to return the pH of the blood to 7.2 to 7.3, as was pointed out earlier in the chapter.

The opposite effect occurs in metabolic alkalosis. That is, alkalosis diminishes the pulmonary ventilation, which in turn causes increased hydrogen ion concentration. Here again the compensation can take place to about 50 to 75 per cent.

RENAL COMPENSATION OF RESPIRATORY ACIDOSIS OR ALKALOSIS

If a person develops respiratory acidosis that continues for a prolonged period of time, the kidneys will secrete an excess of hydrogen ions, resulting in an increase in sodium bicarbonate in the extracellular fluids. After half a day or more, the pH of the body fluids will have returned essentially to normal even though the person continues to breathe poorly. Thus, renal compensation of respiratory acidosis can be almost complete rather than only partial.

Exactly the converse effect occurs in respiratory alkalosis. Large amounts of sodium bicarbonate are lost into the urine, thereby decreasing the pH to normal. Here again the compensation goes almost to completion.

PHYSIOLOGY OF TREATMENT IN ACIDOSIS OR ALKALOSIS

Obviously, the best treatment for acidosis or alkalosis is to remove the condition causing the abnormality, but, if this cannot be effected, different drugs can be used to neutralize the excess acid or alkali.

To neutralize excess acid, large amounts of sodium bicarbonate can be ingested by mouth. This is absorbed into the blood stream and increases the bicarbonate ion portion of the bicarbonate buffer, thereby shifting the pH to the alkaline side. Sodium bicarbonate is occasionally used also for intravenous therapy, but this has such strong and often dangerous physiological effects that other substances are more often used, such as sodium lactate, sodium gluconate, or other organic compounds of sodium. The lactate and gluconate portions of the molecules are metabolized in the body, leaving the sodium in the extracellular fluids in the form of sodium bicarbonate, and thereby shifting the pH of the fluids in the alkaline direction.

For treatment of alkalosis, ammonium chloride is usually administered by mouth. When this is absorbed into the blood, the ammonia portion of the ammonium chloride is converted by the liver into urea; this reaction liberates hydrochloric acid, which immediately reacts with the buffers of the body fluids to shift the hydrogen ion concentration in the acid direction. Occasionally, ammonium chloride is infused intravenously, but the ammonium ion is highly toxic and this procedure can be dangerous.

INTERRELATIONSHIPS OF pH, DISSOLVED CO_2, AND HCO_3^- CONCENTRATION

pH Measurements. In studying a patient with acidosis or alkalosis one needs preferably to know the pH of his body fluids. This can be determined easily by measuring the pH of the plasma with a glass electrode pH meter. However, extreme care must be exercised in removing the plasma and in making the measurement, because even the slightest diffusion of carbon dioxide out of the plasma into the air shifts the bicarbonate buffer system in the alkaline direction, resulting in a much elevated pH measurement.

The pH-Bicarbonate Diagram. The so-called pH-bicarbonate diagram, which is illustrated in Figure 37–9, can be used to determine the *type* of acidosis or alkalosis a person has and its *severity.* Its use may be explained as follows:

The more vertical curves of the diagram depict different carbon dioxide concentrations. The normal carbon dioxide concentration of 1.2 mM./liter is denoted by the heavy line. The points along this line represent the possible combinations of bicarbonate concentration and pH that can exist in the body fluids when the carbon dioxide concentration is normal.

The more horizontal lines represent the concen-

Figure 37–9 The pH-bicarbonate diagram for determining relative degrees of metabolic and respiratory acidosis or alkalosis in a patient. (Modified from Davenport: The ABC of Acid-Base Chemistry. University of Chicago Press.)

trations of metabolic acids and bases in the body fluids. The heavy line, denoted by the numeral zero, indicates a balance between both of these. That is, the points on this line represent the possible combinations of bicarbonate concentration and pH that can occur as long as the metabolic acids and bases in the body fluids are normal. The upper two horizontal lines indicate respectively additions of 5 or 10 mM./liter of extra metabolic base to the body fluids, and the two lower horizontal lines indicate additions of 5 to 10 extra mM./liter of metabolic acid.

To use this diagram we simply find the pH of the blood and the bicarbonate concentration, then plot the appropriate point on the diagram. For instance, if the pH is found to be the normal value of 7.4 and bicarbonate concentration the normal value of 26 mM./liter, we plot point A, which represents the normal condition.

Now using data obtained from another patient, we plot a new point with a pH of 7.63 and a bicarbonate concentration of 28 mM./liter. This is point B on the diagram, which represents a carbon dioxide concentration of 0.8 mM./liter and 7 mM./liter extra metabolic base. Thus, this person has *metabolic alkalosis* because he has considerable extra metabolic base in his body fluids, but he also has *respiratory alkalosis* because he is overventilating to an extent that his carbon dioxide concentration is considerably less than normal.

Similarly, from other patients, we plot points C, D, and E. Point C represents 6 mM./liter of *metabolic acidosis* and sufficient *respiratory alkalosis* to reduce the carbon dioxide concentration to 0.7 mM./liter. A person with such a finding as this could have respiratory alkalosis that has been partially compensated by metabolic acidosis produced by the kidneys. Or, conversely, it is possible that the metabolic acidosis was primary, and the person has overcompensated with respiratory alkalosis.

Point D represents *mild metabolic acidosis,* 2 mM./liter, combined with *severe respiratory*

acidosis. A person could get into this state with severe primary respiratory acidosis and mild metabolic acidosis resulting from some other cause.

Point E represents *mild respiratory acidosis and severe metabolic alkalosis.* Presumably, in this case the metabolic alkalosis was primary, and respiratory compensation caused mild respiratory acidosis in an attempt to compensate for the metabolic alkalosis.

In summary, by using the pH-bicarbonate diagram, one can determine both the degree of metabolic acidosis or alkalosis and the degree of respiratory acidosis or alkalosis in a patient at the same time.

The Carbon Dioxide Combining Power of the Blood. Unfortunately, it is almost impossible to determine the plasma carbon dioxide concentration in a clinical laboratory, and it is somewhat difficult to determine pH and bicarbonate concentration because the blood must be removed from the patient and not be exposed to air before the appropriate measurements are made. Therefore, the usual clinical laboratory measures the so-called *carbon dioxide combining power of the plasma,* which is also called the *alkali reserve.* This measurement is made as follows: Blood is removed from the patient, it is rendered incoagulable with potassium oxalate, and the plasma is separated from the cells. No special precautions are taken about exposure to air. Instead, in the laboratory the plasma is re-exposed to 7 per cent carbon dioxide until it equilibrates. Then the total carbon dioxide in the plasma is determined chemically, including all the dissolved carbon dioxide as well as that which can be liberated from the bicarbonate in the plasma by the addition of acid. Since about $^{19}/_{20}$ is in the form of bicarbonate, this is principally a measure of plasma bicarbonate, and it is called the *carbon dioxide combining power of the plasma.*

The normal carbon dioxide combining power is 28 mM./liter, but in severe metabolic acidosis it may be reduced to as low as 5 to 10 mM./liter. Conversely, in severe metabolic alkalosis, it may rise to as high as 40 to 50 mM./liter. Ordinarily, these concentrations are expressed in terms of the *volume* of carbon dioxide that can be released from each 100 ml. of plasma rather than in millimols. The normal amount is 58 volumes per cent (58 ml./100 ml. of plasma). In severe metabolic alkalosis this occasionally rises to as high as 80 to 90 volumes per cent, and in severe metabolic acidosis it can fall to as low as 10 to 15 volumes per cent.

Note that the carbon dioxide combining power tells one nothing at all about *respiratory* acidosis or alkalosis, for the plasma is not equilibrated with the alveolar carbon dioxide of the patient prior to the test but with 7 per cent carbon dioxide provided in the laboratory.

REFERENCES

Bernstein, B. A., and Clapp, J. R.: Micropuncture study of bicarbonate reabsorption by the dog nephron. *Amer. J. Physiol.,* 214:251, 1968.

Brown, E. B., Jr., and Goott, B.: Intracellular hydrogen ion changes and potassium movement. *Amer. J. Physiol.*, *204*:765, 1963.

Christensen, H. N.: Body Fluids and the Acid-Base Balance. 2nd ed., Philadelphia, W. B. Saunders Company, 1964.

Churchill, P. C., and Malvin, R. L.: Relation of renal gluconeogenesis to ammonia production in the dog. *Amer. J. Physiol.*, *218*:241, 1970.

Cohen, J. J.: Selective Cl retention in repair of metabolic alkalosis without increasing filtered load. *Amer. J. Physiol.*, *218*: 165, 1970.

Engel, K.; Dell, R. B.; Rahill, W. J.; Denning, C. R.; and Winters, R. W.: Quantitative displacement of acid-base equilibrium in chronic respiratory acidosis. *J. Appl. Physiol.*, *24*:288, 1968.

Glabman, S.; Klose, R. M.; and Giebisch, G.: Micropuncture study of ammonia excretion in the rat. *Amer. J. Physiol.*, *205*:127, 1963.

Goldberger, E.: A Primer of Water, Electrolyte, and Acid-Base Syndromes. 3rd ed., Philadelphia, Lea & Febiger, 1965.

Gottschalk, C. W.; Lassiter, W. E.; and Mylle, M.: Localization of urine acidification in the mammalian kidney. *Amer. J. Physiol.*, *198*:581, 1960.

Davenport, H. W.: The ABC of Acid-Base Chemistry. 4th ed., Chicago, University of Chicago Press, 1958.

Howell, B. J.; Baumgardner, F. W.; Bondi, K.; and Rahn, H.: Acid-base balance in cold-blooded vertebrates as a function of body temperature. *Amer. J. Physiol.*, *218*:600, 1970.

Malnic, G.; Aires, M. M.; and Vieira, F. L.: Chloride excretion in nephrons of rat kidney during alterations of acid-base equilibrium. *Amer. J. Physiol.*, *218*:20, 1970.

Pitts, R. F.: The renal metabolism of ammonia. *Physiologist*, *9*:97, 1966.

Robinson, J. R.: Fundamentals of Acid-Base Regulation. 3rd ed., Philadelphia, F. A. Davis Co., 1967.

Roth, D. G., and Gamble, J. L., Jr.: Deoxycorticosterone-induced alkalosis in dogs. *Amer. J. Physiol.*, *208*:90, 1965.

Stern, L. I., and Simmons, D. H.: Estimation of nonrespiratory acid-base abnormalities. *J. Appl. Physiol.*, *27*:21, 1969.

Vieira, F. L., and Malnic, G.: Hydrogen ion secretion by rat renal cortical tubules as studied by an antimony microelectrode. *Amer. J. Physiol.*, *214*:710, 1968.

Waddell, W. J., and Bates, R. G.: Intracellular pH. *Physiol. Rev.*, *49*:285, 1969.

Weinstein, S. W.: Micropuncture studies of the effects of acetazolamide on nephron function in the rat. *Amer. J. Physiol.*, *214*:222, 1968.

CHAPTER 38

Micturition, Renal Disease, and Diuresis

MICTURITION

Micturition is the process by which the urinary bladder empties itself when it becomes filled. Basically the bladder (1) progressively fills until the tension in its walls rises above a threshold value, at which time (2) a nervous reflex called the "micturition reflex" is elicited that (3) greatly exacerbates the pressure in the bladder and simultaneously causes a conscious desire to urinate. The micturition reflex also initiates appropriate signals from the nervous system to relax the external sphincter of the bladder, thereby allowing micturition.

PHYSIOLOGIC ANATOMY OF THE BLADDER AND ITS NERVOUS CONNECTIONS

The urinary bladder, which is illustrated in Figure 38–1, is mainly a smooth muscle vesicle composed of two principal parts: (a) the *body,* which is comprised mainly of the *detrusor muscle,* and (b) the *trigone,* a small triangular area near the mouth of the bladder through which both the *ureters* and the *urethra* pass. During bladder expansion the body of the bladder stretches, and during the micturition reflex the detrusor muscle contracts to empty the bladder.

The trigonal muscle is interlaced around the

Figure 38–1   The urinary bladder and its innervation.

opening of the urethra and maintains tonic closure of the urethral opening until the pressure in the bladder rises high enough to overcome the tone of the trigonal muscle. The trigonal muscle, therefore, is called the *internal sphincter of the bladder.* About 2 cm. beyond the bladder, the urethra passes through the *urogenital diaphragm,* the muscle of which constitutes the *external sphincter* of the bladder. This muscle is a voluntary skeletal muscle, in contrast to the other muscle of the bladder, which is entirely smooth muscle. Normally, this muscle remains tonically contracted, which prevents constant dribbling of urine, but it can be reflexly or voluntarily relaxed at the time of micturition.

Figure 38–1 also illustrates the basic nervous connections with the spinal cord for bladder control. Parasympathetic excitation causes contraction of the detrusor muscle and perhaps some opening of the internal sphincter. Sensory nerve fibers leave the bladder mainly in association with the parasympathetic nerves and enter the spinal cord through the pelvic nerves and sacral plexus. The external sphincter of the bladder, which is a voluntary skeletal muscle, is controlled by the *pudic nerve,* which has its origin in the first two sacral segments of the spinal cord. Figure 38–1 also shows sympathetic nerve fibers passing from the lumbar regions of the spinal cord through the hypogastric plexus to the bladder. Stimulation of these nerves causes the bladder to relax, which is opposite to the effect of parasympathetic stimulation. However, the sympathetic nerves do not normally enter into micturition control.

TRANSPORT OF URINE THROUGH THE URETERS

The ureters are small smooth muscle tubes that originate in the pelves of the two kidneys and pass downward to enter the bladder. Each ureter is innervated by both sympathetic and parasympathetic nerves, and each also has an intramural plexus of neurons and nerve fibers that extends along its entire length.

As urine collects in the pelvis, the pressure in

the pelvis increases and initiates a peristaltic contraction beginning in the pelvis and spreading down along the ureter to force urine toward the bladder. A peristaltic wave occurs from once every 10 seconds to once every two to three minutes. Parasympathetic stimulation increases, and sympathetic stimulation decreases, the frequency. Transmission of the peristaltic wave is probably caused mainly by nerve impulses passing along the intramural plexus in the same manner that the intramural plexus functions in the gut.

At its lower end, the ureter penetrates the bladder obliquely through the *trigone,* as illustrated in Figure 38–1. The ureter courses for several centimeters under the bladder epithelium so that pressure in the bladder compresses the ureter, thereby preventing backflow of urine.

Pain Sensations from the Ureters, and the Ureterorenal Reflex. The ureters are well supplied with pain nerve fibers. Any time the ureter becomes blocked, intense stretch reflexes caused by ureteral dilatation elicit one of the most severe types of pain that one can experience. In addition, the pain impulses cause a sympathetic reflex back to the respective kidney to constrict the renal arterioles, thereby greatly decreasing urinary output from that kidney. This effect is called the *ureterorenal reflex;* it obviously is important to prevent excessive flow of fluid into the pelvis of a kidney whose ureter is blocked.

TONUS OF THE BLADDER, AND THE CYSTOMETROGRAM DURING BLADDER FILLING

The solid curve of Figure 38–2 is called the *cystometrogram* of the bladder. It shows the changes in intravesical pressure as the bladder fills with urine. When no urine at all is in the bladder, the intravesical pressure is approximately zero, but, by the time 100 ml. of urine has collected, the pressure will have risen to about 10 cm. water. Additional urine up to 300 to 400 ml. can collect without significant further rise in pressure; the pressure remains only 10 cm. water. Beyond this point, collection of more urine causes the pressure to rise very rapidly.

The initial portion of the cystometrogram—that is, the rise in pressure to 10 cm. of water—is called *segment I,* the plateau is called *segment II,* and the sharp rise above the 400 ml. level is called *segment III.*

Superimposed on the tonic pressure changes during filling of the bladder are periodic acute increases in pressure which last from a few

seconds to more than a minute. The pressure can rise only a few centimeters of water or it can rise to over 100 cm. water. These are *micturition waves* in the cystometrogram caused by the micturition reflex, which is discussed below.

Mechanism of Tone in the Bladder. The rapid rise of pressure in segment I of the cystometrogram presumably results from initial filling of the bladder to a level at which the bladder wall can exert its tonic pressure against the fluid. The pressure in segment II probably results from intrinsic tone of the detrusor muscle, but the intramural nerve plexus of the bladder wall might also play some role. The tone *is not* the result of signals from the spinal cord, because it still occurs even after the extrinsic nerves to the bladder are cut. The plateau in the pressure recording probably results mainly from the effect of the physical law called Laplace's law in the following way:

When a spherical vesicle begins to be filled, the increasing circumference of the vesicle makes it possible for less pressure than ever to stretch the vesicle still farther. Everyone has experienced this effect in blowing up a balloon. After the initial start, the pressure required to fill the balloon may be even less than that required to obtain the initial start.

The rapid rise in pressure in segment III in the normal bladder presumably results from stretching the bladder to its physical limit so that additional fluid will then be stretching connective tissue rather than muscle fibers.

THE MICTURITION REFLEX

Referring once again to Figure 38–2, it is evident that as the bladder fills to the upper

Figure 38–2 A normal cystometrogram showing also acute pressure waves (the dashed curves) caused by micturition reflexes.

end of segment II in the cystometrogram, many superimposed *micturition contractions* begin to appear. These are the result of a stretch reflex initiated by stretch receptors in the bladder wall. Sensory signals are conducted to the sacral segments of the cord through the pelvic nerves and then back again to the bladder through the parasympathetic fibers in these same nerves.

Once a micturition reflex begins, it is "self-regenerative." That is, initial contraction of the bladder further stretches the receptors to cause still further increase in afferent impulses from the bladder, which causes further increase in reflex contraction of the bladder, the cycle thus repeating itself again and again until the bladder has reached a strong degree of contraction. Then, after a few seconds to more than a minute, the reflex begins to fatigue, and the regenerative cycle of the micturition reflex ceases, allowing rapid reduction in bladder contraction. In other words, the micturition reflex is a single, complete cycle of (a) progressive and rapid increase in pressure, (b) a period of sustained pressure, and (c) return of the pressure back to the basal tonic pressure of the bladder. Once a micturition reflex has occurred, the nervous elements of this reflex usually remain in an inhibited state for at least a few minutes or sometimes as long as an hour or more before another micturition reflex occurs. However, as the bladder becomes more and more filled, micturition reflexes occur more and more often and more and more powerfully.

Control of Micturition by the Brain. The micturition reflex is a completely automatic cord reflex, but it can be inhibited or facilitated by centers in the brain. These include: (1) a strong *facilitatory center in the upper pons,* (2) a strong *facilitatory center in the hypothalamus* (in the septum pellucidum and medial preoptic area), (3) a moderately strong *inhibitory center in the rostral midbrain* at approximately the level of the superior colliculus, and (4) several *centers located in the cerebral cortex* that are mainly inhibitory but can at times become excitatory.

The fiber pathways of the cord through which micturition signals are transmitted to and from the brain are not known. However, some studies have indicated that the afferent fibers travel in or in close association with the lateral spinothalamic tracts and that the efferent fibers lie immediately deep to the lateral corticospinal tract.

The micturition reflex is the basic cause of micturition, but the higher centers normally exert final control of micturition by the following means:

1. The higher centers keep the micturition reflex partially inhibited all the time except when it is desired to micturate.

2. The higher centers prevent micturition by continual tonic contraction of the external urinary sphincter until a convenient time presents itself.

3. When the time to urinate arrives, the cortical centers can (a) facilitate the sacral micturition centers to initiate a micturition reflex and (b) inhibit the external urinary sphincter so that urination can occur.

ABNORMALITIES OF MICTURITION

The Atonic Bladder. Destruction of the sensory nerve fibers from the bladder to the spinal cord prevents transmission of stretch signals from the bladder and, therefore, also prevents micturition reflex contractions. Therefore, the person loses all control of his bladder despite intact efferent fibers from the cord to the bladder and despite intact neurogenic connections with the brain. Instead of emptying periodically, the bladder fills to capacity and overflows a few drops at a time through the urethra. This is called *overflow dribbling.*

The atonic bladder was a common occurrence when syphilis was widespread, because syphilis frequently causes constrictive fibrosis around the dorsal nerve root fibers where they enter the spinal cord and, subsequently, destroys these fibers. This condition is called *tabes dorsalis,* and the resulting bladder condition is called a *tabetic bladder.* Another common cause of this condition is crushing injuries to the sacral region of the cord.

The Automatic Bladder. If the spinal cord is damaged above the sacral region but the sacral segments are still intact, typical micturition reflexes still occur. However, they are no longer controllable by the brain. During the first few days to several weeks after the damage to the cord has occurred, the micturition reflexes are completely suppressed because of sudden loss of facilitatory impulses from the brain stem and cerebrum. However, the excitability of the micturition centers in the sacral cord gradually increases until typical micturition reflexes return. Sometimes these reflexes are elicited by approximately normal amounts of urine in the bladder, whereas at other times the threshold for eliciting the reflex may be low or high. The factor that causes this difference is the treatment the patient receives during the early stages of the convalescent period. If the bladder is kept completely empty by continuous drainage, contracture of the bladder causes a low threshold for development of micturition reflexes. But if the bladder is allowed to remain overly filled for long periods of time, it becomes progressively stretched, and a large quantity of urine is then necessary to elicit micturition reflexes.

It is especially interesting that scratching or otherwise stimulating the skin in the genital region can sometimes elicit a micturition reflex in this condition, thus providing a means by which some patients can still control urination.

The Uninhibited Neurogenic Bladder. Another common abnormality of micturition is the so-called "uninhibited neurogenic bladder," which results in frequent and relatively uncontrollable micturition. This condition derives from damage in the spinal cord or brain stem that interrupts most of the inhibitory signals. Therefore, facilitatory impulses passing continually down the cord keep the sacral centers so excitable that even a very small quantity of urine will elicit an uncontrollable micturition reflex and thereby promote urination.

RENAL DISEASE

The physiology of renal disease can be classified into four different physiological categories: (1) *acute renal shutdown,* in which the kidneys stop working entirely or almost entirely, (2) *chronic renal insufficiency,* in which progressively more nephrons are destroyed until the kidneys simply cannot perform all the necessary functions, (3) the *nephrotic syndrome,* in which the glomeruli have become far more permeable than normal so that large amounts of protein are lost into the urine, and (4) *specific tubular abnormalities* that cause abnormal reabsorption or lack of reabsorption of certain substances by the tubules.

ACUTE RENAL SHUTDOWN

Any condition that suddenly interferes with kidney function can cause acute renal shutdown, including acute pyelonephritis, acute glomerulonephritis, or acute damage to the tubules. Two of these, acute glomerulonephritis and tubular necrosis, deserve special consideration because of their frequent occurrence. Pyelonephritis will be considered later in relation to renal insufficiency.

Renal Shutdown Caused by Acute Glomerular Nephritis. Acute glomerular nephritis is a disease that results from an antigen-antibody reaction and in which the glomeruli become markedly inflamed. Large numbers of white blood cells collect in the inflamed glomeruli, and the endothelial cells on the vascular side of the glomerular membrane, as well as the epithelial cells on the Bowman's capsule side of the membrane, proliferate, sometimes completely filling the glomeruli and the capsule. These inflammatory reactions cause total or partial blockage of large numbers of glomeruli, and those glomeruli that are not blocked develop greatly increased permeability of the glomerular membrane, allowing large amounts of protein to leak into the glomerular filtrate. Also, rupture of the membrane in severe cases often allows large numbers of red blood cells to pass into the glomerular filtrate. Therefore, all degrees of glomerular malfunction can occur in acute glomerular nephritis, including total renal shutdown.

The inflammation of acute glomerular nephritis occurs one to three weeks following, elsewhere in the body, an infection caused by group A beta streptococci, such as a streptococcal sore throat, streptococcal tonsillitis, scarlet fever, or even streptococcal infection of the skin. The infection itself does not cause damage to the kidneys, but when antibodies develop following the infection, the glomeruli are damaged. Some clinicians believe that this effect results from precipitation of streptococcal antigen with its specific antibody and then subsequent entrapment of the precipitate in the glomerular membrane. But others believe that the streptococcal antigen causes antibodies to be formed that in turn attack the proteins of the glomerular membrane, thus resulting in a type of autoimmunity discussed in Chapter 10. Regardless of the specific cause, it is clear that the damage to the glomeruli results from antibodies that develop following a streptococcal infection and not from the streptococcal toxins themselves. Rheumatic fever is another condition that develops in a similar manner, except that the damage in this disease occurs in the valves of the heart. This condition was discussed in Chapter 32.

The acute inflammation of the glomeruli most often subsides in 10 days to two weeks, and the nephrons may return to normal function. Sometimes, however, the inflammatory reactions are so severe that many or most of the glomeruli will have been destroyed permanently. Furthermore, acute glomerulonephritis frequently becomes *chronic glomerulonephritis,* in which mild episodes of recurring glomerulonephritis return periodically, and with each attack progressively more nephrons are destroyed. Therefore, glomerulonephritis can result in chronic renal insufficiency as well as acute shutdown.

Tubular Necrosis. Another common cause of acute renal shutdown is *tubular necrosis,* which means destruction of epithelial cells in the tubules, as illustrated in Figure 38–3. The different causes of tubular necrosis are (1) *various poisons* that destroy the tubular epithelial cells, (2) *severe acute ischemia* of the kidneys, and (3) transfusion reactions.

RENAL POISONS. Among the different renal poisons are *carbon tetrachloride* and the heavy metals, such as the *mercuric ion.* These substances have specific nephrotoxic action on the tubular epithelial cells, causing death of many of them. As a result, the epithelial cells slough away from the basement membrane and plug the tubules. In some instances the basement membrane also is destroyed, but, if not, new tubular epithelial cells can usually grow along the surface of the membrane so that the tubule becomes repaired within 10 to 20 days.

SEVERE ACUTE RENAL ISCHEMIA. Severe ischemia of the kidney is likely to result from *severe circulatory shock.* In shock, the heart simply fails

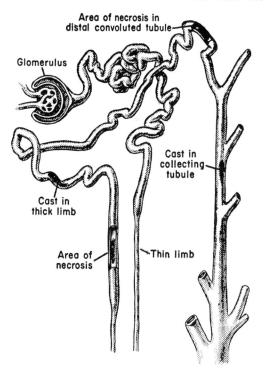

Area of necrosis in
distal convoluted tubule

Glomerulus

Cast in
collecting
tubule

Cast in
thick limb

Area of
necrosis

Thin limb

Figure 38-3 Damage to the distal tubules as a result of shock. (Modified from MacLean: Acute Renal Failure. Charles C Thomas.)

to pump sufficient amounts of blood to supply adequate nutrition to the different parts of the body, and renal blood flow is particularly likely to suffer because of strong sympathetic constriction of the renal vessels. Therefore, lack of adequate nutrition often destroys many tubular epithelial cells and thereby plugs many of the nephrons.

TRANSFUSION REACTION. A transfusion reaction normally results in hemolysis of large amounts of red blood cells with release of hemoglobin into the plasma. The size of the hemoglobin molecule is slightly less than that of the pores in the glomerular membrane so that much of the hemoglobin passes through the membrane into the glomerular filtrate. Therefore, after a transfusion reaction the tubular load of hemoglobin becomes far greater than can be reabsorbed by the proximal tubules. For years it has been taught that the excess hemoglobin is likely to precipitate in the nephron and in this way cause blockage. However, hemolysis of red blood cells also releases a vasoconstrictor agent into the blood stream, and it is believed that this vasoconstrictor can promote such poor blood supply to the tubules that this might be a further cause of tubular damage. For the present, therefore, the precise cause of blockage of the nephron is not determined, though it is possibly both tubular necrosis and hemoglobin blockage.

THE "CRUSH SYNDROME." A particularly severe type of renal shutdown frequently occurs in the so-called *crush syndrome,* in which a person has been so severely traumatized that he has both circulatory shock and severely crushed muscles

that have released large amounts of *myoglobin* into the blood stream. Myoglobin, like hemoglobin, filters readily into the glomerular filtrate and, also like hemoglobin, can block the tubules. However, here again the damage might result from vasoconstriction and tubular necrosis rather than from myoglobin plugs in the tubules. Nevertheless, the combination of severe muscle injury and circulatory shock is much more likely to cause tubular shutdown than is circulatory shock alone.

CHRONIC RENAL INSUFFICIENCY— DECREASED NUMBERS OF NEPHRONS

Even though the nephrons in a human kidney may be completely normal or almost normal, their total number may be greatly reduced. This leads to insufficient quantity of renal tissue to perform the normal functions of the kidneys. The different causes of this include *chronic glomerulonephritis, traumatic loss of a kidney, congenital absence of a kidney, congenital polycystic disease* (in which large cysts develop in the kidneys and destroy surrounding nephrons by compression), *urinary tract obstruction* resulting from renal stones, *pyelonephritis,* and *arteriosclerosis.*

Chronic glomerulonephritis was discussed previously in relation to acute glomerulonephritis. It usually results from successive episodes of mild acute glomerulonephritis that destroys a few more nephrons each time.

Probably the two other most common causes of chronic renal insufficiency are arteriosclerosis and pyelonephritis.

Arteriosclerosis as a Cause of Renal Insufficiency. Arteriosclerosis is a disease of the blood vessels in which fatty areas develop in the arterial walls, followed by fibrosis and calcification of the fatty areas, and then by total blockage of many of the vessels. This disease process will be presented much more fully in Chapter 68 in connection with fat metabolism. As a consequence of arteriosclerosis, many arteries, especially small ones, supplying different portions of the kidneys become blocked, and the respective areas of the kidney die as a result of ischemia (lack of blood flow). Figure 38-4 illustrates the Diodrast clearance of the kidneys of normal persons at different ages, showing that even in the normal person, kidney blood flow decreases to about 45 per cent of normal by the age of 80; and there is a corresponding loss of excretory function. Most of this loss of function results from progressive arteriosclerosis.

Pyelonephritis as a Cause of Renal Insufficiency. Pyelonephritis is an infectious and inflammatory process that usually begins in the renal pelvis but extends progressively into the renal parenchyma. The infection can result from many different types of bacteria, including especially the colon bacillus and staphylococci. Invasion of the kidneys by these bacteria results in progressive destruction of renal tubules, glomeruli, and any other structures in the path of the invading organisms. Consequently,

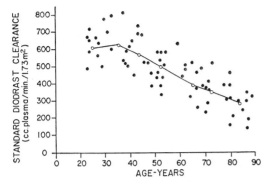

Figure 38–4   Effect of aging on the Diodrast clearance of the kidneys. (Modified from Wolstenholme et al.: Ciba Foundation Colloquia on Ageing. Little, Brown and Co.)

large portions of the functional renal tissue are lost.

A particularly interesting feature of pyelonephritis is that the invading infection usually affects the medulla of the kidney before it affects the cortex. Since one of the primary functions of the medulla is to provide the counter-current mechanism for concentrating the urine, patients with pyelonephritis frequently have reasonably normal renal function except for inability to concentrate their urine.

Inability of Insufficient Kidneys to Excrete Excess "Loads" of Excretory Products.   The most important effect of renal insufficiency is the inability of the kidneys to cope with large "loads" of electrolytes or other substances that must be excreted. Normally, one-fourth of the nephrons can eliminate essentially all the normal load of waste products from the body. However, man sometimes engages in voracious ingestion of salts, meats, and so forth, which is the reason for the excess renal functional capacity.

Function of the Remaining Nephrons in Renal Insufficiency.   In renal insufficiency, the still functioning nephrons usually become greatly overloaded in several different ways. First, for reasons not completely understood, the blood flow through the glomerulus and the amount of glomerular filtrate formed each minute by each nephron often increase to as much as double normal. Second, large amounts of excretory substances, such as urea, phosphates, sulfates, uric acid, and creatinine, accumulate in the extracellular fluid. On entering the glomerular filtrate, these constitute markedly increased tubular loads of substances that are poorly reabsorbed. These increasing loads are partially compensated by as much as 50 per cent increase in reabsorptive power of each tubule, but even this increase is often far too little to keep pace with the loads, which may be increased 1000-fold. Therefore, only a small fraction of the solutes are reabsorbed, and this also prevents reabsorption of water, resulting in rapid flushing of tubular fluid through the tubules. Consequently, the volume of urine formation by each nephron can rise to as much as 20 times normal, and a person will

occasionally have *as much as three times normal urine output* despite the fact that he has significant renal insufficiency. This paradoxical situation is caused by a greater increase in urine volume output per nephron than reduction in numbers of nephrons.

ISOSTHENURIA.   Another effect of rapid flushing of fluid through the tubules is that the normal concentrating and diluting mechanisms of the kidney do not have time to function properly, especially the concentrating mechanism. Therefore, as progressively more nephrons are destroyed, the specific gravity of the urine approaches that of the glomerular filtrate, approximately 1.008. These effects are illustrated in Figure 38–5, which gives the approximate upper and lower limits of urine specific gravity as the number of nephrons decreases. Since the concentrating mechanism is impaired more than the diluting mechanism, an important renal function test is to determine how well the kidney can concentrate urine.

Hypertension in Renal Insufficiency.   Either acute renal shutdown or chronic renal insufficiency can cause hypertension, but it more frequently occurs in the chronic condition. The most likely cause of the hypertension is retention of large quantities of water and salt, which increases the extracellular fluid volume, thereby causing a hyperactive circulation. This theory and others were discussed in detail in Chapter 26.

Anemia in Renal Insufficiency.   In addition to having hypertension, the patient with chronic renal insufficiency of severe degree also develops severe *anemia*. The probable cause of this is the following: The kidneys normally secrete *erythropoietin,* which stimulates the bone marrow to produce red blood cells. Obviously, if the kidneys are seriously damaged they are unable to produce adequate quantities of this substance, which leads to diminished red blood cell production and consequent anemia. However, other factors, such as the high concentration of urea, hydrogen ions, and potassium, might also play important roles in causing the anemia.

Osteomalacia in Renal Insufficiency.   Prolonged renal insufficiency also causes osteomalacia, a con-

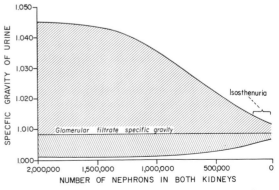

Figure 38–5   Development of *isosthenuria* in patients with decreased numbers of active nephrons.

dition in which the bones are partially absorbed and therefore greatly weakened, as will be explained in relation to the physiology of bone in Chapter 79. There are two causes of the bone absorption: first, the chronic acidosis in renal insufficiency increases the solubility of bone salts and therefore increases their removal from the bones. Second, the parathyroid glands secrete large quantities of parathyroid hormone, which in turn stimulates the osteoclasts to absorb the bone. The increased hormone secretion is believed to result secondarily from increased phosphate ions in the body fluids for reasons that will be discussed in Chapter 79.

Effects of Renal Insufficiency on the Body Fluids—Uremia

The effect of acute renal shutdown, or of severe renal insufficiency caused in any other way, on the body fluids depends to a great extent on the water and food intake of the person after shutdown has occurred. Assuming that the person continues to ingest small amounts of water and food, the concentration changes of different substances in the extracellular fluid are approximately those shown in Figure 38–6. The most important effects are (1) *generalized edema* resulting from water retention, (2) *acidosis* resulting from failure of the kidneys to rid the body of normal acidic products, (3) *high potassium concentration* resulting from failure of potassium excretion, and (4) *high concentrations of the nonprotein nitrogens,* especially *urea,* resulting from failure of the body to excrete the metabolic end-products. This condition is called *uremia* because of the high concentrations of normal urinary excretory products that collect in the body fluids.

Water Retention and Edema. If treatment is begun immediately after acute renal shutdown, the total body fluid content may not change at all. However, when the patient drinks water in response to his normal desire, the body fluids begin increasing immediately and rapidly. If he takes in

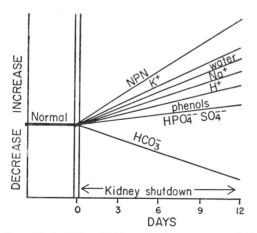

Figure 38–6 Effect of kidney shutdown on extracellular fluid constituents.

no electrolytes at the same time, approximately five-eighths of the water enters the cells and three-eighths remains in the extracellular fluids. Thus, a patient with edema in renal shutdown may have both *intracellular* and *extracellular edema.* On the other hand, if the patient should ingest large amounts of salt along with the water, the edema might be only extracellular because of greatly increased amounts of osmotically active substances in the extracellular compartment.

Acidosis in Renal Shutdown. Each day the metabolic processes of the body normally produce 50 to 100 millimols more metabolic acid than metabolic alkali. Therefore, any time the kidneys fail to function, acid begins to accumulate in the body fluids. Normally, the buffers of the fluids can buffer up to a total of 500 to 1000 millimols of acid without severe depression of the extracellular fluid pH, but after one to two weeks of renal shutdown all this buffering power is used up so that the pH then falls drastically. The patient becomes *comatose* at about this same time, and it is believed that this is caused mainly by the acidosis, as is discussed below.

Elevated Potassium Concentration in Uremia. The amount that the potassium concentration increases depends principally on the rate of protein metabolism after shutdown has occurred. Breakdown of the cellular proteins releases potassium from combination with the proteins, and the extra potassium then passes into the extracellular fluid. Obviously, failure of the kidneys to excrete the potassium as it is released causes an elevated potassium concentration in the extracellular fluid. When the potassium concentration rises to 8 mEq./liter, it begins to have a cardiotoxic effect, resulting in dilatation of the heart, and at about 12 to 15 mEq./liter it is likely to stop the heart in diastole. This is a common occurrence in uremia.

Increase in Urea and Other Nonprotein Nitrogens (Azotemia) in Uremia. The nonprotein nitrogens include urea, uric acid, creatinine, and a few less important compounds. These, in general, are the end-products of protein metabolism and must be removed from the body continually to insure continued protein metabolism in the cells. The concentrations of these, particularly of urea, can rise to as high as 10 times normal during one to two weeks of renal shutdown. However, even these high levels do not affect physiological function nearly so much as the high concentrations of hydrogen and potassium ions. Yet one of the most important means for assessing the degree of renal shutdown is to measure the concentrations of the nonprotein nitrogens.

Uremic Coma. After a week or more of renal shutdown the sensorium of the patient becomes clouded, and he soon progresses into a state of coma. The acidosis is believed to be the principal factor responsible for the coma because acidosis caused by other conditions, such as severe diabetes mellitus, also causes coma. However, many other abnormalities could also be contributory—the generalized edema, the high potassium concentration, and possibly even the high nonprotein nitrogen concentration.

The respiration usually is deep and rapid in coma, which is a respiratory attempt to compensate for the metabolic acidosis. In addition to this, during the last day or so before death, the arterial pressure falls progressively, then rapidly in the last few hours. Death occurs usually when the pH of the blood falls to about 7.0.

Dialysis of Uremic Patients with the Artificial Kidney

Artificial kidneys have now been used for about 25 years to treat patients with severe renal insufficiency. In certain types of acute renal shutdown, the artificial kidney is used simply to tide the patient over for a few weeks until the renal damage heals so that the kidneys can resume function. Also, the artificial kidney has now been developed to the point that several thousand persons with permanent renal insufficiency or even total kidney removal are being maintained in health for years at a time, their lives depending entirely on the artificial kidney.

The basic principle of the artificial kidney is to pass blood through very minute blood channels bounded by thin membranes. On the other sides of the membranes is a *dialyzing fluid* into which unwanted substances in the blood pass by diffusion.

Figure 38–7 illustrates an artificial kidney in which blood is pumped continually between two thin sheets of cellophane; on the outside of the sheets is the dialyzing fluid. The cellophane is porous enough to allow all constituents of the plasma except the plasma proteins to diffuse freely in both directions—from plasma into the dialyzing fluid and from the dialyzing fluid back into the plasma. If the concentration of a substance is greater in the plasma than in the dialyzing fluid, there will be net transfer of the substance from the plasma into the dialyzing fluid. The amount of the substance that transfers depends on: (1) the difference between the concentrations on the two sides of the membrane, (2) molecular size, the smaller molecules diffusing more rapidly than larger ones, and (3) the length of time that the blood and the fluid remain in contact with the membrane.

In normal operation of the artificial kidney, blood is continually removed from an artery, pumped through the kidney, and returned to a vein. The total amount of blood in the artificial kidney at any one time is usually less than 500 ml., the rate of flow may be several hundred ml. per minute, and the total diffusing surface is usually between 10,000 and 20,000 square centimeters. To prevent coagulation of blood in the artificial kidney, heparin is infused into the blood as it enters the "kidney." Then, to prevent bleeding in the patient as a result of the heparin, an anti-heparin substance, such as protamine, is infused into the blood as it is returned to the patient.

The Dialyzing Fluid. Table 38–1 compares the constituents in a typical dialyzing fluid with those in normal plasma and uremic plasma. Note that sodium, potassium, and chloride concentrations in the dialyzing fluid and in normal plasma are identical, but in uremic plasma the potassium and chloride concentrations are considerably greater. These two ions diffuse through the dialyzing membrane so rapidly that their concentrations fall to equal those in the dialyzing fluid within only three to four hours, exposure to the dialyzing fluid.

On the other hand, there is no phosphate, urea, urate, or creatinine in the dialyzing fluid. There-

Figure 38–7 Schematic diagram of the Skeggs-Leonards artificial kidney. (Redrawn from Bluemle et al.: *Trans. Stud. Coll. Physcns. Philad., 20:*157, 1953.)

TABLE 38–1. Comparison of Dialyzing Fluid with Normal and Uremic Plasma

Constituent	Normal Plasma	Dialyzing Fluid	Uremic Plasma
Electrolytes (mEq./liter)			
Na$^+$	142	142	142
K$^+$	5	5	8
Ca^{++}	3	3	2
Mg^{++}	1.5	1.5	1.5
Cl$^-$	107	107	117
HCO$_3^-$	27	27	14
Lactate$^-$	1.2	1.2	1.2
HPO$_4^{--}$	3	0	6
Urate$^-$	0.3	0	2
Nonelectrolytes (mg. %)			
Glucose	100	125	100
Urea	26	0	200
Creatinine	1	0	6

fore, when the uremic patient is dialyzed, these substances are lost in large quantities into the dialyzing fluid, thereby removing major proportions of them from the plasma.

Thus, the constituents of the dialyzing fluid are chosen so that those substances in excess in the extracellular fluid in uremia can be removed at rapid rates, while the normal electrolytes remain essentially normal.

Effectiveness of the Artificial Kidney. The effectiveness of an artificial kidney is expressed in terms of the amount of plasma that can be cleared of different substances each minute, which, as will be recalled from Chapter 35, is also the principal means for expressing the functional effectiveness of the kidneys themselves. Most artificial kidneys can clear 100 to 200 ml. of plasma per minute of urea, which shows that, at least in the excretion of this substance, the artificial kidney can function about two times as rapidly as the two normal kidneys together, whose urea clearance is only 70 ml. per minute. Yet the artificial kidney can be used for no more than 12 hours every three to four days because of danger from excess heparin. Therefore, the overall plasma clearance is still somewhat limited when the artificial kidney replaces the normal kidneys.

THE NEPHROTIC SYNDROME — INCREASED GLOMERULAR PERMEABILITY

Large numbers of patients with renal disease develop a so-called *nephrotic syndrome,* which is characterized especially by *loss of large quantities of plasma proteins into the urine.* In some instances this occurs without evidence of any other abnormality of renal function, but more often it is associated with some degree of renal insufficiency.

The cause of the protein loss in the urine is increased permeability of the glomerular membrane. Therefore, any disease condition that can increase the permeability of this membrane can cause the nephrotic syndrome. Such diseases include the subacute phase of *chronic glomerulonephritis* (in the previous discussion, it was noted that this disease primarily affects the glomeruli and causes a greatly increased permeability of the glomerular membrane), *amyloidosis,* which results from the deposition of an abnormal protenoid substance in the walls of blood vessels and thereby seriously damages the basement membrane, *syphilis,* which often results in fibrotic reactions around the blood vessels, and *disseminated lupus,* which is an abnormality distantly related to glomerulonephritis.

Protein Loss. In the nephrotic syndrome, as much as 30 to 40 grams of plasma proteins can be lost into the urine each day. Though the resulting low plasma protein concentration stimulates the liver to produce far more plasma proteins than usual, nevertheless, the liver often cannot keep up with the loss. Therefore, in severe nephrosis the colloid osmotic pressure sometimes falls extremely low, often from the normal level of 28 mm. Hg to as low as 6 to 8 mm. Hg.

Edema. The low colloid osmotic pressure in turn allows large amounts of fluid to filter into the interstitial spaces and also into the potential spaces of the body, thus causing serious *edema.* The nephrotic person has been known on occasion to develop as much as 40 liters of excess extracellular fluid, and as much as 15 liters of this has been *ascites* in the abdomen. Also, the joints swell, and the pleural cavity and the pericardium become partially filled with fluid.

In severe nephrosis, the loss of fluid into the interstitial spaces causes the blood volume to diminish, and this in turn elicits a sequence of events described in Chapter 76 that leads to increased aldosterone secretion by the adrenal glands. The aldosterone then causes the kidneys to reabsorb excessive quantities of sodium chloride and water, which obviously further compounds the seriousness of the edema.

A nephrotic person can be greatly benefited by intravenous infusion of large quantities of concentrated plasma proteins. Yet this is of only temporary benefit because enough protein can be lost into the urine in only a day to return the person to his original predicament. Fortunately, in some conditions, such as in glomerulonephritis, the basic disease condition can pass into a remission, and the nephrotic condition disappears.

Lipid Nephrosis. In the nephrotic syndrome, especially in nephrotic children, very high concentrations of lipids often appear in the plasma, including especially a very high concentration of *cholesterol.* Large amounts of lipids also collect in the tubular epithelium. This condition was originally labeled *lipid nephrosis,* but it is now regarded in most cases as a special manifestation of chronic glomerulonephritis.

The high concentration of plasma lipids that occurs in lipid nephrosis results from an unexplained failure of the tissues to remove lipoproteins from the plasma in nephrotic patients. Because of this failure, lipoproteins continue to accumulate.

SPECIFIC TUBULAR DISORDERS

In the discussion of active reabsorption and secretion by the tubules in Chapter 34, it was pointed out that the active transport processes are carried out by various carriers and enzymes in the tubular epithelial cells. Furthermore, there are a number of different carrier mechanisms for the different individual substances. In Chapter 3 it was also pointed out that each cellular enzyme and probably also each carrier substance are formed in response to respective genes in the nucleus. If any of the respective genes happens to be absent or abnormal, the tubules might be deficient in one of the appropriate enzymes or carriers. For this reason many different specific tubular disorders are known to occur in the transport of individual or special groups of substances through the tubular membrane. Essentially all these are hereditary disorders. Some of the more important ones are:

Renal Glycosuria. In this condition the transport mechanism for reabsorbing glucose is either deficient or lacking. The blood glucose concentration may be completely normal, but the transport maximum for reabsorption of glucose each minute is greatly limited. Consequently, despite the normal blood glucose level, large quantities of glucose pass into the urine each day. Because one of the tests for diabetes mellitus (which results from lack of insulin secretion by the pancreas) is the presence of glucose in the urine, renal glycosuria (a benign condition which causes essentially no dysfunction of the body) must always be ruled out before making a diagnosis of diabetes mellitus.

Nephrogenic Diabetes Insipidus. Occasionally the renal tubules do not respond completely to antidiuretic hormone secreted by the supraoptico-hypophyseal system, and as a consequence large quantities of dilute urine are continually excreted. As long as the person is supplied with plenty of water, this condition rarely causes any severe difficulty. However, when he fails to receive adequate quantities of water, he rapidly becomes dehydrated.

Renal Tubular Acidosis. In this condition the person is unable to secrete adequate quantities of hydrogen ions, and, as a result, large amounts of sodium bicarbonate are continually lost into the urine for reasons discussed in the preceding chapter. This causes a continual state of metabolic acidosis. However, adequate replacement therapy by continual administration of alkali can maintain normal bodily function.

Renal Hypophosphatemia. In renal hypophosphatemia the renal tubules fail to reabsorb adequate quantities of phosphate ions even when the phosphate concentration of the body fluids falls very low. This condition does not cause any serious immediate abnormalities, because the phosphate level of the extracellular fluids can vary widely without significant cellular dysfunction. However, over a long period of time the low phosphate level results in diminished calcification of the bones and causes the person to develop rickets. Furthermore, this type of rickets is refractory to vitamin D therapy in contrast to the rapid response of the usual type of rickets, which is discussed in Chapter 79.

Amino-Acidurias. Some amino acids share mutual carrier systems for reabsorption, while other amino acids have distinct carrier systems of their own. Rarely, a condition called *generalized aminoaciduria* results from deficient reabsorption of all amino acids, but, more frequently, deficiencies of specific carrier systems may result in (1) *essential cystinuria*, in which large amounts of cystine fail to be reabsorbed and often crystallize in the urine to form renal stones, (2) simple *glycinuria*, in which glycine fails to be reabsorbed, or (3) *beta-aminoisobutyric aciduria*, which occurs in about 5 per cent of all people but apparently has no clinical significance.

RENAL FUNCTION TESTS

The renal function tests can be divided into three categories: (1) determination of renal clearances, (2) measurement of substances in the blood that are normally excreted by the kidneys, and (3) chemical and physical analyses of the urine.

Renal Clearance Tests. Any of the renal clearance tests, including clearance of paraaminohippuric acid, Diodrast, inulin, mannitol, or other substances, as described in Chapter 35, can be used as a renal function test. Indeed, if all these are run, one can determine the glomerular filtration rate, the effective blood flow through the kidney per minute, the filtration fraction, and many other characteristics of renal function. However, most of these clearance tests are not satisfactory for routine clinical use because of their complexities. Instead, three other types of renal clearance tests are normally used. These are urea clearance, phenolsulfonphthalein clearance, and x-ray or radioactive measurements of Diodrast clearance into the renal pelves.

UREA CLEARANCE. Urea clearance is especially easily measured because all one needs to measure is the blood urea concentration and the total amount of urea entering the urine each minute. Employing the plasma clearance formula of Chapter 35, *urea clearance* equals *quantity of urea entering the urine each minute* divided by *plasma urea concentration.* Normal urea clearance is 70 ml. of plasma per minute for the average adult. Values less than this indicate decreased renal function. Clinical symptoms of renal insufficiency usually begin to show when the urea clearance falls below approximately 20 ml. per minute.

PHENOLSULFONPHTHALEIN CLEARANCE. Phenolsulfonphthalein, an alkaline dye, is injected either intravenously or subcutaneously, and successive samples of urine are collected. This dye is cleared into the urine both by the glomeruli and by tubular secretion. When injected intravenously, at least 60 per cent, and usually much more, of the dye is returned to the urine during the first hour if the kidneys are normal. The quantity of dye in the urine is determined by alkalinizing the urine with sodium hydroxide and checking it against a color standard. In severe cases of renal insufficiency the amount cleared into the urine during the first hour is often depressed to as low as 5 to 10 per cent of the quantity originally injected.

INTRAVENOUS PYELOGRAPHY. Several substances containing large quantities of iodine in their molecules—Diodrast, Hippuran, and Iopax—are excreted into the urine either by glomerular filtration or by active tubular secretion. Consequently, the concentration in the urine becomes very high within a few minutes after intravenous injection. Also, the iodine in the compounds makes them relatively opaque to x-rays. Therefore, x-ray pictures can be made showing shadows of the renal pelves, of the ureters, and even of the urinary bladder. Ordinarily, a sufficient quantity is excreted within five minutes after injection to give good shadows of the kidney pelves. Failure to show a distinct shadow within this time indicates depressed renal function.

RADIOACTIVE PYELOGRAPHY. If any of the above substances is prepared with *radioactive* iodine, one can measure the radioactivity from both renal pelves by placing appropriate radioactivity counters over the kidneys. Only a trace amount of the substance need be injected intravenously and the degree of radioactivity recorded during the following few minutes. Simultaneous records from the two kidneys demonstrate the functional status of the two kidneys.

The value of both x-ray and radioactive pyelography is that they measure function of each kidney independently of the other rather than total function of both kidneys together, as is measured by the other renal function tests.

Blood Analyses as Tests of Renal Function. One can also estimate how well the kidneys are performing their functions by measuring the concentrations of various substances in the blood. For instance, the normal concentration of urea in the blood is 26 mg./100 ml., but in severe cases of renal insufficiency this can rise to as much as 200 mg. per cent. The normal concentration of creatinine in the blood is 1.3 mg. per cent, but this, too, can increase as much as 10-fold. To determine the degree of metabolic acidosis resulting from renal dysfunction, one can measure the bicarbonate ion content, as discussed in the preceding chapter. Though these different tests are not so satisfactory as clearance tests for determining the functional capabilities of the kidneys, they are easy to perform, and they do tell the physician how seriously the internal environment has been disturbed.

Physical Measurements of the Urine as Renal Function Tests. Obviously, one of the most important urinary measurements is the *volume of urine* formed each day. In acute renal shutdown this can fall to zero, and in severe renal insufficiency it usually is diminished. On the other hand, moderate renal insufficiency may actually increase the urinary output, as was described above, because of overfunction of the remaining nephrons when the majority have been destroyed.

A second factor frequently measured is the *specific gravity* of the urine. Depending on the types of substances being cleared, this can vary tremendously; its upper extreme can go as high as 1.045 and it can fall to as little as 1.002. To test the ability of the kidneys to dilute the urine, the patient drinks large quantities of water, and measurements are made of the minimal specific gravity that can be attained. Then, at another time, he goes without water for 12 to 24 hours, and the maximum concentration of the urine is measured. Referring back to Figure 38–5, note that the concentrating ability of the kidneys is especially compromised as the number of nephrons decreases.

DIURETICS AND MECHANISMS OF THEIR ACTIONS

A diuretic is a substance that increases the rate of urine output. Diuretics can act either by increasing the glomerular filtration rate or by decreasing the rate of reabsorption of fluid from the tubules.

The principal use of diuretics is to reduce the total amount of fluid in the body. They are especially important in treating edema resulting from conditions other than kidney disease, such as the edema of congestive heart disease and that caused by diminished plasma proteins. Diuretics normally are of little value in treating the edema caused by kidney disease, because the abnormal kidneys rarely respond.

When using a diuretic it is usually important that the rate of sodium loss in the urine also be increased as well as the rate of water loss. The reason for this is the following: Whenever a person has edema, his extracellular fluids contain not only water but also greatly increased quantities of sodium chloride. If water alone were removed from these fluids, the fluids would become hypertonic and elicit an osmoreceptor response, followed by marked secretion of antidiuretic hormone. Consequently, large amounts of water would be reabsorbed immediately from the tubules, which would nullify the effect of the diuretic. However, if sodium is lost along with the water, this nullification will not result. Therefore, all valuable diuretics cause marked *natriuresis* (sodium loss) as well as diuresis.

Diuretics can be classified into three different physiologic types: (1) those that increase glomerular filtration, (2) those that increase the tubular osmotic load, and (3) those that inhibit the secretion of antidiuretic hormone.

DIURETICS THAT INCREASE GLOMERULAR FILTRATION

The glomerular filtration rate can be increased by *increasing the arterial pressure,* by *dilating the afferent arterioles,* by *constricting the efferent arterioles,* or by *decreasing the colloid osmotic pressure.* Some of the different agents that can increase the glomerular filtration rate in one of these ways are the following:

Agents that increase the arterial pressure include *norepinephrine,* other sympathomimetic drugs, and sometimes large infusions of fluid. However, the pressor drugs, such as norepinephrine, usually have a dual effect on fluid output.

The increase in arterial pressure caused by nor-epinephrine tends to increase the output, while *direct constriction of the afferent arterioles tends to reduce urine output.* In low and moderate pressure ranges the pressure effect is more important, so that norepinephrine normally causes diuresis, whereas in high pressure ranges, above about 160 mm. Hg, the urine output begins to fall.

Administration of *digitalis* to a patient with congestive heart failure frequently benefits the circulation so greatly that the glomerular pressure increases. And this increases the glomerular filtration rate, resulting in diuresis.

Certain chemicals directly dilate the afferent arterioles, including *theophylline* and *caffeine,* which increase the glomerular filtration rate and result in increased urinary output.

In general, the agents that increase glomerular filtration rate can increase the rate of urinary output only about three- to five-fold.

DIURETICS THAT INCREASE THE TUBULAR OSMOTIC LOAD

The Osmotic Diuretics. Injection into the blood stream of *urea, sucrose, mannitol,* or any other substance not easily reabsorbed by the tubules causes a great increase in the osmotically active substances in the tubules. The osmotic pressure of these prevents water reabsorption, so that large amounts of tubular fluid flush on into the urine.

The same effect occurs when the glucose concentration of the blood rises to very high levels in diabetes mellitus. Above a glucose concentration of about 250 mg. per cent, very little glucose is reabsorbed by the tubules; instead it acts as an osmotic diuretic and causes rapid loss of fluid into the urine. The name "diabetes" refers to the prolific urine flow.

Diuretics that Diminish Active Reabsorption — Mercurials, Carbonic Anhydrase Inhibitors, and Chlorothiazide. Any substance that inhibits carrier systems in the tubular epithelial cells and thereby diminishes active reabsorption of tubular solutes increases the tubular osmotic pressure and causes osmotic diuresis. The three most common drugs of this type are the mercurials, Diamox (acetazolamide), and chlorothiazide.

The *mercurial diuretics* are divalent organic mercurial compounds, and they have the property of combining with and inactivating sulfhydryl-containing enzymes which are important in several of the tubular transport systems. As a consequence, large doses of mercurial diuretics can decrease active reabsorption and active secretion of a number of different substances by the tubules. They have an especially important effect on active reabsorption of sodium in the distal portions of the proximal tubules; 20 to 30 per cent of the total tubular sodium load may fail to be reabsorbed. This leaves a tremendous amount of osmotically active sodium in the tubules, which prevents water reabsorption as well. Therefore, large quan-

tities of water and sodium are excreted. Because of this very high degree of effectiveness of mercurials, they are the most valuable of all the different diuretics used clinically. However, excessive usage of mercurial diuretics can sometimes destroy tubular epithelial cells, causing more harm than good.

Diamox *inhibits carbonic anhydrase,* and, therefore, when given in sufficient quantity can block hydrogen ion secretion by the tubular epithelial cells. This in turn decreases the reabsorption of both bicarbonate and sodium ions, again increasing the quantity of osmotically active substances remaining in the tubules and, therefore, causing diuresis — but also causing some degree of acidosis because of failure to excrete hydrogen ions.

Chlorothiazide inhibits both carbonic anhydrase and sodium reabsorption. However, it is not as powerful a carbonic anhydrase inhibitor as is Diamox, neither is it as powerful as the mercurials in inhibiting sodium reabsorption. It is a valuable diuretic when the combined qualities of both the other diuretics are desired. Furthermore, chlorothiazide seems to act at a different point in the tubules from the mercurials for reabsorption of sodium, because administration of mercurials and chlorothiazide at the same time causes additive effects.

DIURETICS THAT INHIBIT THE SECRETION OF ANTIDIURETIC HORMONE

The single diuretic that is most important in inhibiting the secretion of antidiuretic hormone (ADH) is *water.* This was discussed in Chapter 36. When large amounts of water are ingested, the body fluids become diluted, and ADH is no longer secreted by the supraoptico-posterior pituitary system. As a result, water fails to be reabsorbed by the distal tubules and collecting ducts, and large amounts of dilute urine flow. In addition to water, various psychic factors as well as certain drugs that affect the central nervous system, such as *narcotics, hypnotics,* and *anesthetics,* can inhibit ADH secretion. Also, *alcohol* inhibits ADH secretion. For this reason, these can all cause increases in urinary output.

REFERENCES

Bernstein, L. M., et al.: Renal Function and Renal Failure: A Programmed Text Set. 2 volumes. Baltimore, The Williams & Wilkins Co., 1965.
Black, D. A. K. (ed.): Renal Disease. Philadelphia, F. A. Davis Co., 1962.
Boyarsky, S.: The Neurogenic Bladder. Baltimore, The Williams & Wilkins Co., 1967.
Brest, A. N., and Moyer, J. H.: Renal Failure. Philadelphia, J. B. Lippincott Co., 1968.
Cafruny, E. J.: The site and mechanism of action of mercurial diuretics. *Pharm. Rev.,* 20:89, 1968.
Dahlberg, W. H.; Sreebyn, L. M.; and King, B.: Studies of parotid

saliva and blood in hemodialysis patients. *J. Appl. Physiol.,* 23:100, 1967.

Hampers, C. L., and Schupak, E.: Long-Term Hemodialysis: The Management of the Patient with Chronic Renal Failure. New York, Grune & Stratton, Inc., 1967.

Heptinstall, R. H.: Pathology of the Kidney. Boston, Little, Brown and Company, 1966.

Huang, K. C., and Lin, D. S. T.: Kinetic studies on transport of PAH and other organic acids in isolated renal tubules. *Amer. J. Physiol.,* 208:391, 1965.

Janicki, R. H., and Goldstein, L.: Glutamine synthetase and renal ammonia metabolism. *Amer. J. Physiol.,* 216:1107, 1969.

Knochel, J. P.: Formation of peritoneal fluid hypertonicity during dialysis with isotonic glucose solutions. *J. Appl. Physiol.,* 27: 233, 1969.

Kolff, W. J.: The artificial kidney. *Amer. Pract.,* 13:219, 1962.

Kuru, M.: Nervous control of micturition. *Physiol. Rev.,* 45:425, 1965.

LeBrie, S. J.: Renal lymph and osmotic diuresis. *Amer. J. Physiol.,* 215:116, 1968.

Merrill, J. P.: The Treatment of Renal Failure. 2nd ed., New York, Grune & Stratton, Inc., 1965.

Metcoff, J.: Acute Glomerulonephritis. Boston, Little, Brown and Company, 1967.

Mostofi, F. K.: The Kidney. International Academy of Pathology, Monograph No. 6. Baltimore, The Williams & Wilkins Co., 1966.

O'Brien, J. P.; Harlan, W. R., Jr.; and White, H. J. O.: Early metabolic alterations in renal allografts. *Amer. J. Physiol.,* 218:714, 1970.

Pitts, R. F.: The Physiologic Basis of Diuretic Therapy. Springfield, Ill., Charles C Thomas, Publishers, 1959.

Schreiner, G. E.: Diseases of the kidney. *In* Duncan, G. G.: Disease of Metabolism. 5th ed., Philadelphia, W. B. Saunders Company, 1964, p. 1106.

Strauss, M. B., and Walt, L. G.: Diseases of the Kidney. Boston, Little, Brown and Co., 1963.

Vander, A. J., and Carlson, J.: Mechanism of the effects of furosemide on renin secretion in anesthetized dogs. *Circ. Res.,* 25:145, 1969.

Vander, A. J.; Malvin, R. L.; Wilde, W. S.; and Sullivan, L. P.: Localization of the site of action of chlorothiazide by stop-flow analysis. *J. Pharmacol. Exp. Therap.,* 125:19, 1959.

Vander, A. J.; Malvin, R. L.; Wilde, W. S.; and Sullivan, L. P.: Localization of the site of action of mercurial diuretics by stop flow analysis. *Amer. J. Physiol.,* 195:558, 1958.

Winter, C. C.: Correctable Renal Hypertension. Philadelphia, Lea & Febiger, 1964.

Winter, C. C.: Radioisotope Renography. Baltimore, The Williams & Wilkins Co., 1963.

PART VI

RESPIRATION

CHAPTER 39

Pulmonary Ventilation

Respiration means the transport of oxygen from the atmosphere to the cells and, in turn, the transport of carbon dioxide from the cells back to the atmosphere. This process can be divided into four major stages: (1) pulmonary ventilation, which means the actual inflow and outflow of air between the atmosphere and the alveoli, (2) diffusion of oxygen and carbon dioxide between the alveoli and the blood, (3) transport of oxygen and carbon dioxide in the blood and body fluids to and from the cells, and (4) regulation of ventilation and other aspects of respiration. The present chapter and the three following discuss, respectively, these four major aspects of respiration. In subsequent chapters various pulmonary disorders and special respiratory problems related to aviation medicine and deep sea diving physiology are discussed to illustrate some of the basic principles of respiratory physiology.

MECHANICS OF RESPIRATION

BASIC MECHANISMS OF LUNG EXPANSION AND CONTRACTION

The lungs can be expanded and contracted by (1) downward and upward movement of the diaphragm to lengthen or shorten the chest cavity and (2) elevation and depression of the ribs to increase and decrease the anteroposterior diameter of the chest cavity. Figure 39–1 illustrates these two methods.

It is readily evident that contraction of the diaphragm pulls the lower boundary of the chest cavity downward and therefore increases its longitudinal length. Upward movement of the diaphragm can be caused either by simple relaxation of the diaphragm or by active contraction of the abdominal muscles, which forces the abdominal contents upward against the bottom of the diaphragm.

Elevation of the anterior portion of the chest cage causes the anteroposterior dimension of the chest cavity to increase by the following mechanism: During expiration the ribs extend in a downward direction from the spinal column, but when the sternum is lifted upward, the ribs then extend almost directly forward rather than downward. This makes the anteroposterior diameter of the chest about 20 per cent greater during inspiration than during expiration. Therefore, muscles that elevate the chest cage can be classified as muscles of inspiration, and muscles that depress the chest cage as muscles of expiration.

Normal quiet breathing is accomplished almost entirely by movement of the diaphragm; but during maximal breathing, increase in chest thickness might account for more than half of the lung enlargement.

THE MUSCLES OF INSPIRATION AND EXPIRATION

The different muscles of inspiration and expiration are the following:

MUSCLES OF INSPIRATION
    Diaphragm
    External intercostals
    Sternocleidomastoids
    Scapular elevators plus anterior serrati
    Scaleni
    Erectus muscles of the spine

MUSCLES OF EXPIRATION
    Abdominals (major muscles of expiration)
    Internal intercostals
    Posterior inferior serrati

The Diaphragm and the Abdominal Muscles. Normal *inspiration* is caused principally by contraction of the diaphragm. This muscle is bell shaped so that contraction of any of its muscle fibers pulls it downward to cause inspiration.

Ordinarily, expiration is an entirely passive process; that is, when the diaphragm relaxes, the elastic structures of the lung, chest cage, and abdomen, as well as the tone of the abdominal muscles, force the diaphragm upward. However, if forceful expiration is required, the diaphragm can

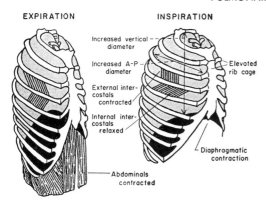

EXPIRATION INSPIRATION

Increased vertical diameter
Increased A-P diameter
External intercostals contracted
Internal intercostals relaxed
Elevated rib cage
Diaphragmatic contraction
Abdominals contracted

Figure 39–1 Expansion and contraction of the thoracic cage during expiration and inspiration, illustrating especially diaphragmatic contraction, elevation of the rib cage, and function of the intercostals.

also be pushed upward powerfully by active contraction of the abdominal muscles against the abdominal contents. Thus, all the abdominal muscles combined together represent the major muscles of expiration.

Muscles that Raise and Lower the Chest Cage. Three different groups of muscles cause inspiration by elevating the entire chest cage. The *sternocleidomastoid muscles* lift upward on the sternum; the *anterior serrati* lift many of the ribs; and the *scaleni* lift the first two ribs. Because the anterior margins of most of the ribs are connected to the sternum, lifting one portion of the anterior chest cage lifts it all.

To cause expiration, the abdominal recti, in addition to helping to compress the abdominal contents upward against the diaphragm, also pull downward on the lower ribs, thereby decreasing the anteroposterior diameter of the chest. Thus, these muscles act as muscles of expiration both by depressing the rib cage and by compressing the abdominal contents upward.

The external and internal intercostals are also important muscles of respiration despite their seemingly small size. Figure 42–1 illustrates the mechanism by which these act to cause inspiration or expiration. To the left, the ribs during expiration are angled downward and the external intercostals are stretched in a forward and downward direction. As they contract, they pull the upper ribs forward in relation to the lower ribs, and this causes leverage on the ribs to raise them upward. Conversely, in the inspiratory position, the internal intercostals are stretched, and their contraction pulls the upper ribs backward in relation to the lower ribs. This causes leverage in the opposite direction and lowers the chest cage.

RESPIRATORY PRESSURES

Intra-alveolar Pressure. The respiratory muscles cause pulmonary ventilation by alternatively compressing and distending the lungs, which in turn causes the pressure in the alveoli

to rise and fall. During inspiration the intra-alveolar pressure becomes slightly negative *with respect to atmospheric pressure,* normally about −3 mm. Hg, and this causes air to flow inward through the respiratory passageways. During normal expiration, on the other hand, the intra-alveolar pressure rises to approximately +3 mm. Hg, which causes air to flow outward through the respiratory passageways.

During maximum expiratory effort the intra-alveolar pressure can usually be increased to well over 100 mm. Hg, and during maximum inspiratory effort it can usually be reduced to as low as −80 mm. Hg.

Fluid Pressure in the Intrapleural Cavity. The lungs are physically attached to the body only at their hila, while their outer surfaces have no attachment whatsoever to the chest wall. However, the membranes of the intrapleural space constantly absorb any gas or fluid that enters this space; this absorption creates a partial vacuum that makes the lungs expand.

The normal pressure *of the fluid* in the intrapleural space is between −10 and −12 mm. Hg. This negative pressure acts as a force to hold the visceral pleura of the lungs tightly against the parietal pleura of the chest wall. When the chest cavity enlarges, this negative pressure causes the lungs also to enlarge. And when the chest cavity becomes smaller, the lungs likewise become smaller. The lungs slide up and down in the chest cavity during respiration, the visceral pleura sliding over the parietal pleura, lubricated by the few milliliters of mucopolysaccharide-containing fluid in the intrapleural space.

The cause of the very negative intrapleural *fluid* pressure is the continual tendency of the pleural capillaries to absorb fluid from the intrapleural spaces. This is particularly true of the visceral pleural capillaries because these are part of the pulmonary circulatory system and have a very low capillary pressure of about 7 mm. Hg that causes rapid absorption of fluid.

Recoil Tendency of the Lungs and the Intrapleural Pressure. The lungs have a continual tendency to collapse and therefore to recoil away from the chest wall. This tendency is caused by two different factors. First, throughout the lungs are many *elastic fibers* that are constantly stretched and are attempting to shorten. Second, and even more important, the *surface tension* of the fluid lining the alveoli causes a continual tendency for the alveoli to collapse. This effect is caused by intermolecular attraction between the surface molecules of the fluid that tends continually to reduce the surface areas of the individual alveoli; all these minute forces added together tend to collapse the whole lung and therefore to cause its recoil away from the chest wall.

Ordinarily, the elastic fibers in the lungs account for about one-third of the recoil tendency, and the surface tension phenomenon accounts for about two-thirds.

The total recoil tendency of the lungs can be measured by the amount of negative pressure in the intrapleural spaces required to prevent collapse of the lungs, and this pressure is called the *intrapleural pressure* or, occasionally, the *recoil pressure*. It is normally about −4 mm. Hg, a value quite different from the intrapleural *fluid* pressure discussed previously. That is, when the alveolar spaces are open to the atmosphere through the trachea so that their pressure is at atmospheric pressure, a pressure of −4 mm. Hg in the intrapleural space is required to keep the lungs expanded to normal size. When the lungs are stretched to very large size, such as at the end of deep inspiration, the intrapleural pressure required then to expand the lungs may be as great as −9 to −12 mm. Hg.

It is fortunate that the intrapleural *fluid* pressure is normally as much as −10 to −12 mm. Hg and therefore considerably more negative than the intrapleural pressure required to overcome the recoil tendency of the lungs. This allows the intrapleural pressure to maintain the lungs in an expanded state.

EXPLANATION OF THE DIFFERENCE BETWEEN INTRAPLEURAL FLUID PRESSURE AND INTRA-PLEURAL PRESSURE. The recent discovery that the fluid pressure in the intrapleural space is considerably more negative than the so-called intrapleural pressure has been confusing to many students as well as to some physiologists. However, if the student will return to Chapter 21 and study the discussion of interstitial fluid pressure, solid tissue pressure, and total tissue pressure, he will see that the pressure dynamics of the intrapleural space are essentially the same as those in the usual interstitial fluid spaces. The intrapleural fluid pressure is analogous to the interstitial fluid pressure, and the intrapleural pressure is analogous to the total tissue pressure. The contact pressure of the visceral pleura against the parietal pleura is analogous to solid tissue pressure. This contact pressure calculates to be about 7 mm. Hg under normal resting conditions.

"SURFACTANT" IN THE ALVEOLI, AND ITS EFFECT ON THE COLLAPSE TENDENCY. A lipo-protein mixture called "surfactant" is secreted by the alveolar epithelium into the alveoli and respiratory passages. This mixture, containing especially the phospholipid *dipalmityl lecithin,* acts in much the same way that a detergent acts. That is, it decreases the surface tension of fluids lining the alveoli and respiratory passages. In the absence of surfactant, lung expansion is extremely difficult, which illustrates that the presence of surfactant is exceedingly important to minimize the effect of surface tension in causing collapse of the lungs.

A few newborn babies do not secrete adequate quantities of surfactant, which makes lung expansion difficult. Some of these die at birth because of inadequate respiration. Others, with less severe lack of surfactant, often live only with difficulty and frequently die in early infancy. This condition is called *hyaline membrane disease* or *respiratory distress syndrome.*

Surfactant acts by forming a monomolecular layer at the interface between the fluids lining the alveoli and the air in the alveoli. This prevents the development of a water-air interface, which has 7 to 14 times as much surface tension as does the surfactant-air interface. The surface tension effect of surfactant is several times as great when the alveoli become small during expiration as it is during inspiration. This is very fortunate because a small radius of curvature of an alveolus greatly increases the tendency for surface tension to cause its collapse.

EXPANSIBILITY OF THE LUNGS AND THORAX: "COMPLIANCE"

The lungs are viscoelastic structures. Therefore, a small amount of intra-alveolar pressure causes them to expand to a certain volume, more pressure causes them to expand to a still greater volume, and so forth. Likewise, the thorax has viscoelastic properties so that the greater the pressure in the lungs, the greater becomes the expansion of the chest. The elastic properties of the lungs, as pointed out previously, are caused, first, by the surface tension of the fluids lining the alveoli and, second, by elastic fibers throughout the lung tissue itself. The elastic properties of the thorax are caused by the natural elasticity of the muscles, tendons, and connective tissue of the chest. Therefore, part of the effort expended by the inspiratory muscles during breathing is simply to stretch the elastic structures of the lungs and thorax.

The expansibility of the lungs and thorax is called compliance. This is expressed as the *volume increase in the lungs for each unit increase in intra-alveolar pressure.* The compliance of the normal lungs and thorax combined is 0.13 liter per centimeter of water pressure. That is, every time the alveolar pressure is increased by 1 cm. water, the lungs expand 130 ml.

Compliance of the Lungs Alone. The lungs alone, when removed from the chest, are almost twice as distensible as the lungs and thorax together because the thoracic cage itself must also be stretched when the lungs are expanded in situ. Thus, the compliance of the normal lungs when removed from the thorax is about 0.22 liter per cm. water. This illustrates that the muscles of inspiration must expend energy not only to expand the lungs but also to expand the thoracic cage around the lungs.

Measurement of Lung Compliance. Compliance of the lungs is measured in the following way: The person's glottis must first be completely open and remain so. Then air is inspired in steps of approximately 50 to 100 ml. at a time, and pressure measurements are made from an intra-esophageal balloon (which measures almost exactly the intrapleural pressure) at the end of each step until the total volume of air in the lungs is equal to the normal tidal volume of the person. Then the air is expired also in steps, until the lung volume returns to the expiratory resting level. The relationship of lung volume to pressure is then plotted as illustrated in Figure 39–2. This graph shows that the plot during inspiration is a different curve from the return during expiration. A diagonal line is drawn between the two ends of the curves, and the compliance of the lungs is determined from this diagonal line. Thus, in the figure the lung volume increases 600 ml. for a change in trans-lung pressure (atmospheric pressure in the lung minus intra-esophageal pressure) of 2.0 cm. water. The compliance in this instance is 0.3 liter per cm. water.

Slight modifications of this procedure can be used to measure the compliance of the lungs and thoracic cage together.

Factors that Cause Abnormal Compliance. Any condition that destroys lung tissue, causes it to become fibrotic or edematous, blocks the alveoli, or in any other way impedes lung expansion and contraction causes decreased lung compliance. When considering the compliance of both the lung and thorax together, we must also include any abnormality that reduces the expansibility of the thoracic cage. Thus, deformities of the chest cage, such as kyphosis, severe scoliosis, and other restraining conditions, such as fibrotic pleurisy or paralyzed and fibrotic muscles, can all reduce the expansibility of the lungs and thereby reduce the total pulmonary compliance.

THE "WORK" OF BREATHING

When the lungs are stretched, energy is expended by the respiratory muscles to cause the stretching. This is part of the "work" of breathing. But, in addition to the work required simply to expand the lungs, work is needed to overcome two other factors that impede the expansion and contraction of the lungs. These are (1) *viscosity* of the pulmonary tissues, which is also called *nonelastic tissue resistance,* and (2) *airway resistance.*

Nonelastic Tissue Resistance (Viscosity). Nonelastic tissue resistance simply means that a certain amount of energy (or work) is required to rearrange the molecules in the tissues of the lungs and thoracic cage in order to change them to new dimensions. This requires the slipping of molecules over each other, which constitutes a type of viscous resistance.

The nonelastic tissue resistance of the lungs alone can be measured by using the compliance diagram of Figure 39–2. As was just pointed out, the inspiratory compliance curve follows a different path from the expiratory curve. For mathematical reasons that cannot be discussed in detail at the present time, the *area* between these curves is proportional to the work required to overcome the nonelastic tissue resistance. When the lungs are edematous or when they have lost most of their elastic properties because of pulmonary fibrotic changes, this area becomes greatly widened,

Figure 39–2 Compliance diagram in a large subject. This diagram shows the compliance of the lungs alone.

indicating that the nonelastic tissue resistance has also greatly increased.

Airway Resistance. A certain amount of energy is also required to move the air along the respiratory passageways. In normal quiet breathing, this amount of energy is slight, but, when the airway becomes obstructed as a result of asthma, obstructive emphysema, diphtheria, or so forth, the airway resistance may become so greatly increased that a large amount of extra energy must be expended by the respiratory muscles simply to force the air back and forth through the narrow passageways.

Airway resistance can be measured from a modified compliance diagram recorded dynamically during the course of respiration. In Figure 39–3, the outer loop shows a curve made while air was actually flowing in and out of the lungs during the recording, and it will be noted that the area between the curves is greatly increased. The extra area over and above that recorded in Figure 39–2, denoted by the dotted shading, is a measure of the work expended to overcome the airway resistance; from this area one can calculate the actual amount of resistance in the airway. In Chapter 43 several pulmonary diseases will be discussed, especially asthma, in which airway resistance is the principal pulmonary problem.

THE PULMONARY VOLUMES AND CAPACITIES

Figure 39–4 gives a graphical representation of changes in lung volume under different conditions of breathing. For ease in describing the events of pulmonary ventilation, the air in the lungs has been subdivided at different points on this diagram into four different *volumes* and four different *capacities*, which are as follows:

THE PULMONARY "VOLUMES"

To the left in Figure 39–4 are listed four different pulmonary lung "volumes" which, when added together, equal the maximum volume to which the lungs can be expanded. The significance of each of these volumes is the following:

1. The *tidal volume* is the volume of air inspired and expired with each normal breath, and it amounts to about 500 ml. in the normal young male adult.

2. The *inspiratory reserve volume* is the extra volume of air that can be inspired over and beyond the normal tidal volume, and it is usually equal to approximately 3000 ml. in the young male adult.

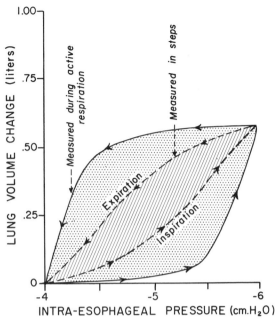

Figure 39–3 *Outer loop:* Pressure-volume diagram recorded continuously during a cycle of respiration. *Inner loop:* Static pressure-volume curve of Figure 39–2. The stippled area represents the increased work over that shown in Figure 39–2; it is caused by airway resistance.

3. The *expiratory reserve volume* is the amount of air that can still be expired by forceful expiration after the end of a normal tidal expiration; this normally amounts to about 1100 ml. in the young male adult.

4. The *residual volume* is the volume of air remaining in the lungs even after the most forceful expiration. This volume averages about 1200 ml. in the young male adult.

THE PULMONARY "CAPACITIES"

In describing events in the pulmonary cycle, it is sometimes desirable to consider two or more of the above volumes together. Such

Figure 39–4 Diagram showing respiratory excursions during normal breathing and during maximal inspiration and maximal expiration.

combinations are called pulmonary capacities. To the right in Figure 39–4 are listed the different pulmonary capacities, which can be described as follows:

1. The *inspiratory capacity* equals the *tidal volume* plus the *inspiratory reserve volume*. This is the amount of air (about 3500 ml.) that a person can breathe beginning at the normal expiratory level and distending his lungs to the maximum amount.

2. The *functional residual capacity* equals the *expiratory reserve volume* plus the *residual volume*. This is the amount of air remaining in the lungs at the end of normal expiration (about 2300 ml.).

3. The *vital capacity* equals the *inspiratory reserve volume* plus the *tidal volume* plus the *expiratory reserve volume*. This is the maximum amount of air that a person can expel from his lungs after first filling his lungs to their maximum extent and then expiring to the maximum extent (about 4600 ml.).

4. The *total lung capacity* is the maximum volume to which the lungs can be expanded with the greatest possible inspiratory effort (about 5800 ml.).

All pulmonary volumes and capacities are about 20 to 25 per cent less in the female than in the male, and they obviously are greater in large and athletic persons than in small and asthenic persons.

Resting Expiratory Level. Normal pulmonary ventilation is accomplished almost entirely by the muscles of inspiration. On relaxation of the inspiratory muscles the elastic properties of the lungs and thorax cause the lungs to contract passively. Therefore, when all inspiratory muscles are completely relaxed the lungs return to a relaxed state, and the volume of air then in the lungs is called the resting expiratory level. This is equal to the functional residual capacity, or about 2300 ml. in the young male adult.

SIGNIFICANCE OF THE PULMONARY VOLUMES AND CAPACITIES

In normal persons the volume of air in the lungs depends primarily on body size and build. Furthermore, the different "volumes" and "capacities" change with the position of the body, most of them decreasing when the person lies down and increasing when he stands. This change with position is caused by two major factors: first, a tendency for the abdominal contents to press upward against the diaphragm in the lying position, and, second, increase in the pulmonary blood volume in the lying position, which correspondingly decreases the space available for pulmonary air.

Significance of the Residual Volume. The residual volume represents the air that cannot be removed from the lungs even by forceful expiration. This is important because it provides air in the alveoli to aerate the blood even between breaths. Were it not for the residual air, the concentrations of oxygen and carbon dioxide in the blood would rise and fall markedly with each respiration, which would certainly be disadvantageous to the respiratory process.

Significance of the Vital Capacity. Other than the anatomical build of a person, the major factors that affect vital capacity are (1) the position of the person during the vital capacity measurement, (2) the strength of the respiratory muscles, and (3) the distensibility of the lungs and chest cage, which is called "pulmonary compliance."

The average vital capacity in the young adult male is about 4.6 liters, and in the young adult female about 3.1 liters, though these values are much greater in some persons of the same weight than in others. A tall, thin person usually has a higher vital capacity than an obese person, and a well-developed athlete may have a vital capacity as great as 30 to 40 per cent above normal—that is, 6 to 7 liters.

VITAL CAPACITY FOLLOWING PARALYSIS OF THE RESPIRATORY MUSCLES. Paralysis of the respiratory muscles, which occurs often following spinal cord injuries or poliomyelitis, can cause a great decrease in vital capacity, to as low as 500 to 1000 ml.—barely enough to maintain life.

DECREASED VITAL CAPACITY CAUSED BY DIMINISHED PULMONARY COMPLIANCE. Obviously, any factor that reduces the ability of the lungs to expand also reduces the vital capacity. Thus, tuberculosis, emphysema, chronic asthma, lung cancer, chronic bronchitis, and fibrotic pleurisy can all reduce the pulmonary compliance and thereby decrease the vital capacity. For this reason vital capacity measurements are among the most important of all clinical respiratory measurements for assessing the progress of different types of pulmonary fibrotic diseases.

CHANGES IN VITAL CAPACITY RESULTING FROM PULMONARY CONGESTION. In left heart disease or any other disease that causes pulmonary vascular congestion, the vital capacity becomes reduced because excess fluid in the tissues of the lungs decreases lung compliance.

Vital capacity measurements made periodically in left-sided heart disease are one of the best means for determining whether the person's condition is progressing or getting better, for these measurements can indicate the degree of pulmonary edema.

MEASUREMENT OF PULMONARY VOLUMES AND CAPACITIES

Spirometry. A simple method by which most of the pulmonary volumes and capacities can be measured is spirometry. A typical spirometer is illustrated in Figure 39–5. This consists of a drum inverted over a chamber of water, the drum counterbalanced by a weight. In the drum is a breathing mixture of gases, usually air or oxygen, and a tube connects the mouth with this gas chamber. On breathing in and out of the chamber the drum rises and falls, and an appropriate recording is made on another drum. When using this spirometer only a few breaths can be recorded because of the buildup of carbon dioxide and loss of oxygen from the gas chamber. However, a *respirometer* such as that illustrated in Figure 70–4 of Chapter 70 can be used for continual recording of the spirogram. This more elaborate apparatus chemically removes the carbon dioxide as it is formed.

Figure 39–4, discussed earlier in this chapter, shows a typical *spirogram* recorded by use of the spirometer. From this figure it is obvious that the following lung volumes and capacities can be measured by simple spirometry: tidal volume, inspiratory reserve volume, expiratory reserve volume, inspiratory capacity, and vital capacity. The volumes that *cannot* be measured this way are: residual volume, functional residual capacity, and total lung capacity. Failure of the spirometer to measure these results from its inability to measure the residual volume still remaining in the lungs after a forceful expiration.

Determination of Residual Volume, Functional Residual Capacity, and Total Lung Capacity by the Nitrogen Washout Method. The functional residual capacity can be measured by the following indirect procedure:

At the end of a normal expiration, the subject suddenly switches from breathing normal air to breathing pure oxygen, and he continues to breathe the pure oxygen for several minutes. During this period of time all the nitrogen in his lungs is "washed out" into the expired air. By collecting the expired air and analyzing the total amount of nitrogen that has been washed out of the lungs,

one can calculate the amount of air in the lungs at the beginning of the test, which equals the functional residual capacity, by the following formula:

Functional residual capacity =

$$\text{Volume of nitrogen washed out} \times \frac{100}{78}$$

This formula is based on the fact that 78 per cent of the gases in the lungs is nitrogen and 22 per cent is oxygen, carbon dioxide, and water vapor. (To be completely accurate in the measurement, additional minor corrections must be made for nitrogen diffusion out of the blood into the lungs.)

Once the functional residual capacity has been determined, the residual volume can be determined by subtracting the expiratory reserve volume from the functional residual capacity. Also, the total lung capacity can be determined by adding the inspiratory capacity to the functional residual capacity.

THE MINUTE RESPIRATORY VOLUME — RESPIRATORY RATE AND TIDAL VOLUME

The *minute respiratory volume* is the total amount of new air moved into the respiratory passages each minute, and this is equal to the *tidal volume* × the *respiratory rate*. The normal tidal volume of a young adult male, as pointed out above, is 500 ml., and the normal respiratory rate is approximately 12 breaths per minute. Therefore, the *minute respiratory volume averages about 6 liters per minute.* A person can occasionally live for short periods of time with a minute respiratory volume as low as 1.5 liters per minute and with a respiratory rate as low as two to four breaths per minute. Obviously, a greatly increased tidal volume can compensate for a markedly reduced respiratory rate.

The respiratory rate occasionally rises to as high as 40 to 50 per minute, and the tidal volume can become as great as the vital capacity, averaging about 4600 ml. in the young adult male. However, at very rapid breathing rates, a person usually cannot sustain a tidal volume greater than about one-half the vital capacity.

Maximum Breathing Capacity. A young male adult forcing himself to breathe as much volume of air as possible can usually breathe 150 to 170 liters per minute for about 15 seconds. This is called the maximum breathing capacity. On the average, this same person can maintain for long periods of strenuous exercise a minute respiratory volume as high as 100 to 120 liters per minute. It is evident, then, that the respiratory system has a marked

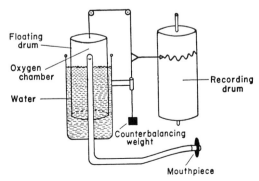

Floating drum
Oxygen chamber
Water
Recording drum
Counterbalancing weight
Mouthpiece

Figure 39–5   A spirometer.

reserve, being capable of increasing its minute respiratory volume as much as 25-fold for short periods of time and as much as 20-fold for prolonged periods of time. Yet, despite this tremendous reserve, pathologic conditions of the lungs can often so decrease the breathing capacity that a person becomes a respiratory invalid. Some of these pathologic conditions will be discussed in Chapter 43.

Maximum Expiratory Flow Rate. When a person expires with increasing force, there comes a point at which the rate of expiration reaches a maximum despite still further increase in expiratory force. This is because pressure on the lung by the chest cage compresses not only the alveoli but also the respiratory passages. Therefore, beyond a certain compressive force, the passageways collapse, which prevents further increase in flow.

In the normal person, the maximum expiratory flow rate is far greater than is required to maintain respiration at maximum breathing capacity. However, in persons with emphysema or long-standing asthma, the airways have been reduced to much smaller sizes than normal so that expiration is often extremely difficult, as we shall see in the discussion of these diseases in Chapter 43.

VENTILATION OF THE ALVEOLI

The truly important factor of the entire pulmonary ventilatory process is the rate at which the alveolar air is renewed each minute by atmospheric air; this is called *alveolar ventilation.* One can readily understand that alveolar ventilation per minute is not equal to the minute respiratory volume because a large portion of the inspired air goes to fill the respiratory passageways, the membranes of which are not capable of significant gaseous exchange with the blood.

THE DEAD SPACE

Effect of Dead Space on Alveolar Ventilation. The air that goes to fill the respiratory passages with each breath is called *dead space air.* On inspiration, much of the new air must first fill the different dead space areas—the nasal passageways, the pharnyx, the trachea, and the bronchi—before any reaches the alveoli. Then, on expiration, all the air in the dead space is expired first before any of the air from the alveoli reaches the atmosphere. *The volume of air that enters the alveoli with each breath, therefore, is equal to the tidal volume minus the dead space volume.*

Measurement of the Dead Space Volume. A simple method for measuring dead space vol-

Figure 39–6 Continuous record of the changes in nitrogen concentration in the expired air following a previous inspiration of pure oxygen. This record can be used to calculate dead space as discussed in the text.

ume is illustrated by Figure 39–6. In making this measurement the subject first breathes normal air and then suddenly takes a deep inspiration of oxygen. This, obviously, fills his entire dead space with pure oxygen, and some of the oxygen also mixes with the alveolar air. Then he expires through a rapidly recording nitrogen meter, which makes the record shown in the figure. The first portion of the expired air contains only pure oxygen, and the per cent nitrogen is zero, but toward the middle of the expiration, as the alveolar air reaches the nitrogen meter, the nitrogen concentration rises, in this instance up to 60 per cent, and then levels off. The total volume of air expired in this instance was 500 ml., and it can readily be seen that the area covered by the dots represents the air that has no nitrogen in it. Therefore, this area also represents the dead space portion of the expired air. The area covered by the hatching represents the air containing nitrogen, and therefore is the alveolar portion of the expired air. Thus, one can determine the amount of dead space air from the following equation:

Dead space air =
$$\frac{\text{Area of the dots} \times \text{Total air expired}}{\text{Area of the hatching} + \text{Area of the dots}}$$

Let us assume, for instance, that the area of the dots on the graph is 30 cm.², and the area of the hatching is 70 cm.², and the total volume expired is 500 ml. The dead space then would be

$$\frac{30}{30 + 70} \times 500, \text{ or } 150 \text{ ml.}$$

Normal Dead Space Volume. The normal dead space air in the young male adult is about

150 ml. This increases slightly with age, and it also changes greatly in different physiological states. For instance, it is reduced when a person lies down, and it is reduced in maximal expiration when the lungs are collapsed to their smallest possible size, for this causes the bronchi and trachea also to collapse to a certain extent. On the other hand, the dead space increases as much as 50 per cent when the lungs are excessively expanded. For instance, during the hyperventilation of exercise the *mean volume* of the lungs is usually considerably greater than normal, and the dead space is about 30 to 40 per cent greater than normal.

Anatomical versus Physiological Dead Space. The method just described for measuring the dead space measures the volume of all the spaces of the respiratory system besides the gas exchange areas: the alveoli and terminal ducts; this is called the *anatomical dead space.* On occasion, however, some of the alveoli themselves are not functional and, therefore, must be considered also to be dead space because there is no blood flow through the adjacent pulmonary vessels. Also, at other times the *ratio of pulmonary blood flow to ventilation* in certain alveoli is so low that these alveoli are only partially functional, so that they, too, can be considered to be partially dead space. When the alveolar dead space is included in the total measurement of dead space this is then called *physiological dead space* in contradistinction to the anatomical dead space. In the normal person, the anatomical and the physiological dead spaces are essentially equal because all alveoli are functional in the normal lung, but in persons with nonfunctional alveoli or with abnormal ratios of blood flow to alveolar ventilation in some parts of the lungs, the physiological dead space is sometimes as much as 10 times the anatomical dead space, or as much as 1 to 2 liters. These problems will be discussed further in Chapter 43 in relation to certain pulmonary diseases.

RATE OF ALVEOLAR VENTILATION

Alveolar ventilation per minute is the total volume of new air entering the alveoli each minute. It is equal to the respiratory rate times the amount of new air that enters the alveoli with each breath:

Alveolar ventilation per minute = Respiratory rate × (Tidal volume − Dead space volume)

Thus, with a normal tidal volume of 500 ml., a normal dead space of 150 ml., and a respira-

tory rate of 12 times per minute, alveolar ventilation equals $12 \times (500 - 150)$, or 4200 ml. per minute.

Theoretically, when the tidal volume falls to equal the dead space volume, no new air at all enters the alveoli with each breath, and the alveolar ventilation per minute becomes zero however rapidly the person breathes. (This relationship is not entirely true because all the dead space air is never completely expired before some of the alveolar air begins to be expired, and the same is true for inspiration. Therefore, there can be a slight amount of alveolar ventilation even with tidal volumes as little as 60 to 75 ml.)

On the other hand, when the tidal volume is several liters, the effect of dead space volume on alveolar ventilation is almost insignificant, and under these conditions the alveolar ventilation per minute almost equals the minute respiratory volume. For instance, if the tidal volume is 4 liters and the dead space volume is 200 ml., the alveolar ventilation per minute is 95 per cent of the minute respiratory volume.

Alveolar ventilation is one of the major factors determining the concentrations of oxygen and carbon dioxide in the alveoli. Therefore, almost all discussions of gaseous exchange problems in the following chapters emphasize alveolar ventilation. *The respiratory rate, the tidal volume, and the minute respiratory volume are of importance only in so far as they affect alveolar ventilation.*

FUNCTIONS OF THE RESPIRATORY PASSAGEWAYS

FUNCTIONS OF THE NOSE

As air passes through the nose, three distinct functions are performed by the nasal cavities: First, the *air is warmed* by the extensive surfaces of the turbinates and septum, which are illustrated in Figure 39-7. Second, the *air is moistened* to a considerable extent even before it passes beyond the nose. Third, the *air is filtered*. All these functions together are called the *air conditioning function* of the upper respiratory passageways. Ordinarily, the air rises to within 2 to 3 per cent of body temperature and within 2 to 3 per cent of full saturation with water vapor before it reaches the lower trachea. When a person breathes air through a tube directly into his trachea (as through a tracheostomy), the cooling and, especially, the drying effect in the lower lung can lead to lung infection.

Filtration Function of the Nose. The hairs at the entrance to the nostrils are important because

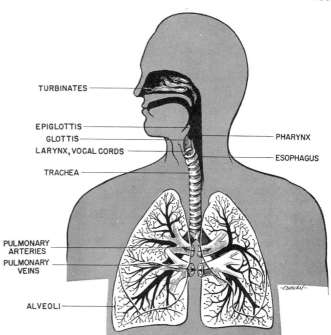

TURBINATES

EPIGLOTTIS
GLOTTIS
LARYNX, VOCAL CORDS

PHARYNX

ESOPHAGUS

TRACHEA

Figure 39–7   The respiratory passages.

PULMONARY
ARTERIES
PULMONARY
VEINS

ALVEOLI

they remove large particles. Much more important, though, is the removal of particles by *turbulent precipitation*. That is, the air passing through the nasal passageways hits many obstructing vanes, such as the turbinates, the septum, and the pharyngeal wall. Each time air hits one of these obstructions it must change its direction of movement; and the particles suspended in the air, having far more mass and momentum than air, cannot change their direction of travel as rapidly as can the air. Therefore, they continue forward, striking the surfaces of the obstructions.

All the surfaces of the nose are coated with a thin layer of *mucus,* which is secreted by the *mucous membrane* covering the surfaces. Furthermore, the epithelium of the nasal passageways is ciliated, and these *cilia* constantly beat toward the pharynx. Therefore, after particles are entrapped in the mucus, this mucus in turn is mobilized toward the pharynx and finally either expectorated or swallowed.

SIZE OF PARTICLES ENTRAPPED IN THE RESPIRATORY PASSAGES. The nasal turbulence mechanism for removing particles from air is so effective that almost no particles larger than 4 to 6 microns in diameter enter the lungs through the nose. This size is smaller than the size of red blood cells. Consequently, by far the greater proportion of dust and other particles in air is removed before the air finally reaches the lungs. Many of the remaining particles precipitate against the walls of the respiratory passages or alveoli, adhering to the alveolar fluid, but most of the particles smaller than 0.5 micron in diameter usually remain suspended even in the alveolar air and are expelled from the lungs during expiration. The smoke particles of cigarette smoke, for example, having a

particle size of approximately 0.3 micron, pass to the alveoli with almost no loss of particles, and most of them then return from the lungs to be expelled along with the expired air.

Particles that do become entrapped in the alveoli are slowly removed mainly by enlarged monocytes that have become macrophages. An excess of particles causes growth of fibrous tissue in the alveolar septa, leading to permanent debility.

THE COUGH REFLEX

The cough reflex is almost essential to life, for the cough is the means by which the passageways of the lungs are maintained free of foreign matter.

The bronchi and the trachea are so sensitive that any foreign matter or other cause of irritation initiates the cough reflex. The larynx and carina (the point where the trachea divides into the bronchi) are especially sensitive, and the terminal bronchioles and alveoli are especially sensitive to corrosive chemical stimuli, such as sulfur dioxide gas and chlorine. Afferent impulses pass from the respiratory passages mainly through the vagus nerve to the medulla. There, an automatic sequence of events is triggered by the neuronal circuits of the medulla, causing the following effects:

First, about 2.5 liters of air is inspired. Second, the epiglottis closes, and the vocal cords shut tightly to entrap the air within the lungs. Third, the abdominal muscles contract forcefully, pushing against the diaphragm while other expiratory muscles, such as the internal intercostals, also contract forcefully. Consequently, the pressure in the lungs rises usually to 100 or more mm. Hg. Fourth, the vocal cords and the epiglottis suddenly open widely so that the air under pressure in the lungs *explodes* outward. Indeed, this air is sometimes expelled at velocities as high as 75 to 100 miles an hour. Furthermore, and very important, the strong compression of the lungs also collapses the bronchi and trachea (the noncartilaginous part of the trachea invaginating inward) so that the exploding air actually passes through bronchial and tracheal slits. The rapidly moving air usually carries with it any foreign matter that is present in the bronchi or trachea.

ACTION OF THE CILIA TO CLEAR RESPIRATORY PASSAGEWAYS

In addition to the cough mechanism, the respiratory passageways of the trachea and lungs are lined with a ciliated, mucus-coated epithelium that aids in clearing the passages, for the cilia beat toward the pharynx. Thus, small foreign particles and mucus are mobilized at a velocity of as much as a centimeter per minute along the surface of the trachea toward the pharynx, in the same manner that foreign matter in the nasal passageways is also mobilized toward the pharynx. Once this material enters the pharynx, either as the result of a cough or as the result of action by the cilia, it is either expectorated or swallowed. For instance, in persons with tuberculosis, tubercle bacilli often cannot be found in the sputum from the lungs but can be found in stomach washings, for the constant movement of foreign matter from the lungs into the throat and then into the stomach provides a ready source of tubercle bacilli.

THE SNEEZE REFLEX

The sneeze reflex is very much like the cough reflex except that it applies to the nasal passageways instead of to the lower respiratory passages. The initiating stimulus of the sneeze reflex is irritation in the nasal passageways, the afferent impulses passing in the fifth nerve to the medulla where the reflex is triggered. A series of reactions similar to those for the cough reflex takes place; however, the uvula is depressed so that large amounts of air pass rapidly through the nose, as well as through the mouth, thus helping to clear the nasal passages of foreign matter.

VOCALIZATION

Speech involves the respiratory system particularly, but it also involves (1) specific speech control centers in the cerebral cortex, which will be discussed in Chapter 61, (2) respiratory centers of the brain stem, and (3) the articulation and resonance structures of the mouth and nasal cavities. Basically, speech is composed of two separate mechanical functions: (1) *phonation,* which is achieved by the larynx, and (2) *articulation,* which is achieved by the structures of the mouth.

Phonation. The larynx is specially adapted to act as a vibrator. The vibrating element is the *vocal cords,* which are folds along the lateral walls of the larynx that are stretched and positioned by several specific muscles within the confines of the larynx itself.

Figure 39–8A illustrates the basic structure of the larynx, showing that each vocal cord is stretched between the *thyroid cartilage* and an *arytenoid cartilage.* The specific muscles within the larynx that position and control the degree of stretch of the vocal cords are also shown. Thus, one can see in the figure that contraction of the *posterior cricoarytenoid* muscles pulls the arytenoid cartilages away from the thyroid cartilage and thereby stretches the vocal cords. The *transverse arytenoid muscle* pulls the arytenoid cartilages together and, therefore, approximates the two vocal cords so that they vibrate in a stream of expired air. Conversely, contraction of the *lateral cricoarytenoid muscles* pulls the arytenoid cartilages forward and apart to allow normal respiration.

The *thyroarytenoid muscles* are made up of many small slips of muscle controlled separately by different nerve fibers. The slips of muscle adjacent to the edges of the vocal cord can contract separately from those adjacent to the wall of the larynx, and other individual portions of these muscles can also contract independently of each other. These contractions control the *shapes* of the vocal cords during different types of phonation.

VIBRATION OF THE VOCAL CORDS. One might suspect that the vocal cords would vibrate in the direction of the flowing air. However, this is not the case. Instead, they vibrate laterally. The cause of the vibration is the following: When the vocal cords are approximated and air is expired, pressure of the air from below first pushes the vocal cords

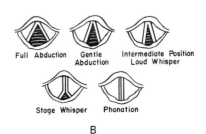

Figure 39–8   Laryngeal function in phonation. (Modified from Greene: The Voice and Its Disorders. Pitman Medical Publishing Co.)

apart, which allows rapid flow of air between their margins. The rapid flow of air then immediately creates a partial vacuum between the vocal cords, which pulls them once again toward each other. This stops the flow of air, pressure builds up behind the cords, and the cords open once more, thus continuing in a vibratory pattern.

FREQUENCY OF VIBRATION. The pitch of the sound emitted by the larynx can be changed in two different ways. First, a change can be achieved by *stretching or relaxing the vocal cords*. The mechanisms involved were partly explained above in the discussion of the intrinsic laryngeal muscles; but, in addition to the effects of the intrinsic muscles, the muscles attached to the external surfaces of the larynx can also pull against the cartilages and thereby help to stretch or relax the vocal cords. For instance, the entire larynx is moved upward by the external laryngeal muscles, which helps to stretch the vocal cords, when one emits a very high frequency sound, and the larynx is moved downward, with corresponding loosening of the vocal cords, when one emits a very bass sound.

The second means for changing the sound frequency is *to change the shape and mass of the vocal cord edges*. When very high frequency sounds are emitted, different slips of the thyroarytenoid muscles contract in such a way that the edges of the vocal cords are sharpened and thinned, whereas when bass frequencies are emitted, the thyroarytenoid muscles contract in a different pattern so that broad edges with a large mass are approximated. Figure 39–8B shows some of the positions and shapes of the vocal cords during different types of phonation.

Articulation and Resonance. The three major organs of articulation are the *lips*, the *tongue*, and the *soft palate*. These need not be discussed in detail because all of us are familiar with their movements during speech and other vocalizations.

The resonators include the *mouth*, the *nose and associated nasal sinuses*, the *pharynx*, and even the *chest cavity* itself. Here again we are all familiar with the resonating qualities of these different structures. For instance, the function of the nasal resonators is illustrated by the change in quality of the voice when a person has a severe cold.

ARTIFICIAL RESPIRATION

MANUAL METHODS OF ARTIFICIAL RESPIRATION — THE BACK-PRESSURE ARM-LIFT METHOD

A dozen or more different manual methods of artificial respiration have been advocated and practiced for many years. In general, all depend on a few seconds' compression of the chest to cause expiration followed in some instances by a few seconds' lifting of some portion of the body to help inspiration. Even with the best of these manual methods a tidal volume of no greater than 500 to 700 ml. can usually be sustained for long periods of time.

The manual method now advocated by the American Red Cross is the back-pressure arm-lift method, which is also called the *Holger Nielsen method*. This consists of the following maneuvers: The patient is placed on his face, and his arms are folded in front of his head. The operator kneels in front of the patient's head and performs two operations in rhythmical sequence approximately 12 to 15 times per minute. First, he places his hands on the back of the patient over approximately the lower half of the scapulae and compresses with about 30 to 40 pounds pressure. Second, after two to three seconds he removes this pressure and then grasps the subject under his upper arms and lifts. These two operations cause, first, expiration and, second, inspiration. The cycle is repeated again and again.

Especially important in any method of artificial respiration is *maintenance of a free airway* through the mouth or nose to the trachea. The tongue of the patient has a tendency to fall into the back of the throat and obstruct the airway. Sometimes it is necessary to pull the tongue forward, and, if available, a special airway device – a curved hollow tube – is inserted through the mouth.

MOUTH-TO-MOUTH BREATHING

A very successful method of artificial respiration is mouth-to-mouth breathing, in which the operator rapidly inspires a deep breath and then breathes into the mouth of the subject. This method has often been shunned in the past because of the belief that the expired air of the operator would not be beneficial to the subject. This is not true, because normal expired air usually still has an adequate amount of oxygen to sustain life in almost anyone. Furthermore, the carbon dioxide in the expired air is sometimes actually desirable because it helps to stimulate the respiratory center of the subject.

MECHANICAL METHODS OF ARTIFICIAL RESPIRATION

The Resuscitator. Many types of resuscitators are available, and each has its own characteristic principles of operation. Basically, the resuscitator, which is illustrated in Figure 39–9A, consists of a supply of oxygen or air, a mechanism for applying intermittent positive and sometimes negative pressure, and a mask which fits over the face of the patient. This apparatus forces air through the mask into the lungs of the patient during the positive pressure cycle and then either allows the air to flow out of the lungs during the remainder of the cycle or pulls the air out by negative pressure.

Earlier resuscitators often caused such severe damage to the lungs because of excessive positive pressures that their usage was at one time greatly

Figure 39-9   (A) The resuscitator. (B) Tank respirator.

depreciated. However, all resuscitators now have safety valves which prevent the positive pressure from rising usually above +14 mm. Hg and the negative pressure from falling below −9 mm. Hg. These pressure limits are adequate to cause excellent artificial respiration of *normal lungs* and yet are slight enough to prevent damage.

The Tank Respirator. Figure 39-9B illustrates the usual tank respirator with a patient's body inside the tank and his head protruding through a flexible but airtight collar. At the end of the tank opposite to the patient's head is a motor driven leather diaphragm that moves back and forth with sufficient excursion to raise and lower the pressure inside the tank. As the leather diaphragm moves inward, positive pressure develops around the body and causes expiration; and as the diaphragm moves outward, negative pressure causes inspiration. Check valves on the respirator control the positive and negative pressures. Ordinarily these pressures are adjusted so that the negative pressure that causes inspiration falls to −10 to −20 cm. water and the positive pressure rises to 0 to +5 cm. water.

Effect of the Resuscitator and the Tank Respirator on Venous Return. When air is forced into the lungs under positive pressure, or when the pressure around the patient's body is greatly reduced, as in the case of the tank respirator, the pressure in the entire chest cavity becomes greater than the pressure everywhere else in the body. Therefore, the flow of blood into the chest from the peripheral veins becomes impeded. As a result, use of excessive pressures with either the resuscitator or the tank respirator can reduce the cardiac output— sometimes to lethal levels. A person can usually survive as much as a 20 mm. Hg difference between intrapulmonary pressure and surrounding body pressure, but prolonged exposure to greater than a 30 mm. Hg difference usually causes death.

REFERENCES

Agostoni, E.: Thickness and pressure of the pleural liquid. *In* Fishman, A. P., and Hecht, H. H. (eds.): The Pulmonary Circulation and Interstitial Space. Chicago, University of Chicago Press, 1969, p. 65.

Banchero, N.; Schwartz, P. E.; Tsakiris, A. G.; and Wood, E. H.:

Pleural and esophageal pressures in the upright body position. *J. Appl. Physiol.,* 23:228, 1967.

Bernstein, L.: Respiration. *Ann. Rev. Physiol.,* 29:113, 1967.

Burger, E. J., Jr.; Peterson, N.; Mead, J.; and Wisnieski, K.: Method for measuring lung gas volume. *J. Appl. Physiol.,* 27:721, 1969.

Campbell, E. J. M.: Respiration. *Ann. Rev. Physiol.,* 30:105, 1968.

Clements, J. A.: Surface phenomena in relation to pulmonary function. *Physiologist,* 5:11, 1962.

Clements, J. A., and Tierney, D. F.: Alveolar instability associated with altered surface tension. *In* Handbook of Physiology. Baltimore, The Williams & Wilkins Co., 1965, Sec. III, Vol. II, p. 1565.

Comroe, J. H., Jr.: The Physiology of Respiration. Chicago, Year Book Medical Publishers, Inc., 1965.

Comroe, J. H., Jr., et al.: The Lung: Clinical Physiology and Pulmonary Function Tests. 2nd ed., Chicago, Year Book Medical Publishers, Inc., 1963.

Cumming, G.; Henderson, R.; Horsfield, K.; and Singhal, S. S.: The functional morphology of the pulmonary circulation. *In* Fishman, A. P., and Hecht, H. H. (eds.): The Pulmonary Circulation and Interstitial Space. Chicago, University of Chicago Press, 1969, p. 327.

Dubois, A. B.: Resistance to breathing. *In* Handbook of Physiology. Baltimore, The Williams & Wilkins Co., 1964, Sec. III, Vol. I, p. 451.

Froeb, H. F., and Mead, J.: Relative hysteresis of the dead space and lung in vivo. *J. Appl. Physiol.,* 25:244, 1968.

Greene, M.: The Voice and Its Disorders. 2nd ed., Philadelphia, J. B. Lippincott Co., 1965.

Guyton, A. C.: Analysis of respiratory patterns in laboratory animals. *Amer. J. Physiol.,* 150:78, 1947.

Guyton, A. C.: Electronic counting and size determination of particles in aerosols. *J. Indust. Hyg. & Toxicol.,* 28:133, 1946.

Guyton, A. C.: Measurement of the respiratory volumes of laboratory animals. *Amer. J. Physiol.,* 150:70, 1947.

Heinemann, H. O., and Fishman, A. P.: Nonrespiratory functions of mammalian lung. *Physiol. Rev.,* 49:1, 1969.

Hong, S. K.; Cerretelli, P.; Cruz, J. C.; and Rahn, H.: Mechanics of respiration during submersion in water. *J. Appl. Physiol.,* 27:535, 1969.

Hoppin, G. G., Jr.; Green, I. D.; and Mead, J.: Distribution of pleural surface pressures in dogs. *J. Appl. Physiol.,* 27:863, 1969.

Horsfield, K., and Cumming, G.: Functional consequences of airways morphology. *J. Appl. Physiol.,* 24:384, 1968.

Krahl, V. E.: Anatomy of the mammalian lung. *In* Handbook of Physiology. Baltimore, The Williams & Wilkins Co., 1964, Sec. III, Vol. I, p. 213.

Leith, D. E., and Mead, J.: Mechanisms determining residual volume of the lungs in normal subjects. *J. Appl. Physiol.,* 23:221, 1968.

Luchsinger, R., and Arnold, G.: Voice-Speech-Language: Clinical Communicology—Its Physiology and Pathology. Wadsworth Publishing Co., Inc., 1965.

Macklem, P. T., and Mead, J.: Factors determining maximum expiratory flow in dogs. *J. Appl. Physiol.,* 25:159, 1968.

Magilton, J. H., and Swift, C. S.: Response of veins draining the nose to alar-fold temperature changes in the dog. *J. Appl. Physiol.,* 27:18, 1969.

Maio, D. A., and Farhi, L. E.: Effect of gas density on mechanics of breathing. *J. Appl. Physiol.,* 23:687, 1967.

Marin, M. G., and Morrow, P. E.: Effect of changing inspired O_2

and CO_2 levels on tracheal mucociliary transport rate. *J. Appl. Physiol.,* 27:385, 1969.

McMahon, S. M.; Permutt, S.; and Proctor, D. F.: Pleural surface pressure in dogs. *J. Appl. Physiol.,* 27:881, 1969.

Mead, J.: The distribution of gas flow in lungs. *In* Wolstenholme, G. E. W., and Knight, J. (eds.): CIBA Foundation Symposium on Circulatory and Respiratory Mass Transport. London, J. & A. Churchill Ltd., 1969, p. 204.

Mead, J., and Agostoni, E.: Dynamics of breathing. *In* Handbook of Physiology. Baltimore, The Williams & Wilkins Co., 1964, Sec. III, Vol. I, p. 411.

Millette, B.; Robertson, P. C.; Ross, W. R. D.; and Anthonisen, N. R.: Effect of expiratory flow rate on emptying of lung regions. *J. Appl. Physiol.,* 27:587, 1969.

Otis, A. B.: The work of breathing. *In* Handbook of Physiology. Baltimore, The Williams & Wilkins Co., 1964, Sec. III, Vol. I, p. 463.

Permutt, S.: Respiration. *Ann. Rev. Physiol.,* 28:177, 1966.

Permutt, S.; Caldini, P.; Bane, H. N.; Howard, P.; and Riley, R. L.: Liquid pressure versus surface pressure on the esophagus. *J. Appl. Physiol.,* 23:927, 1968.

Pride, N. B.; Permutt, S.; Riley, R. L.; and Bromberger-Barnea, B.: Determinants of maximal expiratory flow from the lungs. *J. Appl. Physiol.,* 23:646, 1967.

Proctor, D. F.: Physiology of the upper airway. *In* Handbook of Physiology. Baltimore, The Williams & Wilkins Co., 1964, Sec. III, Vol. I, p. 309.

Rahn, H.; Otis, A. B.; Chadwick, L. F.; and Fenn, W. O.: The pressure-volume diagram of the thorax and lung. *Amer. J. Physiol.,* 146:161, 1946.

Scarpelli, E. M.; Clutario, B. C.; and Taylor, F. A.: Preliminary identification of the lung surfactant system. *J. Appl. Physiol.,* 23:880, 1967.

Staub, N. C.: Respiration. *Ann. Rev. Physiol.,* 31:173, 1969.

Whittenberger, J. L. (ed.): Artificial Respiration: Theory and Applications. New York, Paul B. Hoeber, 1962.

Widdicombe, J. G.: Regulation of tracheobronchial smooth muscle. *Physiol. Rev.,* 43:1, 1963.

Wood, L. D. H., and Bryan, A. C.: Effect of increased ambient pressure on flow-volume curve of the lung. *J. Appl. Physiol.,* 27:4, 1969.

Wunder, C. C.; Reed, J. H., Jr.; McConahey, W. M., III; Cronin, L.; and Wood, E. H.: Nature of pleural space of dogs. *J. Appl. Physiol.,* 27:637, 1969.

CHAPTER 40

Physical Principles of Gaseous Exchange; Diffusion of Oxygen and Carbon Dioxide Through the Respiratory Membrane

After the alveoli are ventilated with fresh air, the next step in the respiratory process is *diffusion* of oxygen from the alveoli into the pulmonary blood and diffusion of carbon dioxide in the opposite direction—from the pulmonary blood into the alveoli. The process of diffusion is simple, involving merely random molecular motion of molecules, these intertwining their ways back and forth through the respiratory membrane. However, in respiratory physiology we are concerned not only with the basic mechanism by which diffusion occurs but also with the *rate* at which it occurs, and this is a much more complicated problem, requiring a rather deep understanding of the physics of gases.

The major problems discussed in this chapter, therefore, are, first, the physical factors that determine the alveolar concentrations of gases, particularly of oxygen and carbon dioxide, and, second, the factors that affect the rate at which these gases can diffuse through the respiratory membrane. Since the following discussions entail many basic physical principles of gases, a brief review of this subject is presented as a prelude to the main text of the chapter. The student who is already well versed in these basic principles should go directly to the discussion of the composition of alveolar air.

THE PHYSICS OF GASES

Relationship of Pressure to Volume—Boyle's Law. If the mass and temperature of a gas in a chamber remain constant but the pressure is increased or decreased, the volume of the gas varies inversely with the pressure. That is,

$$\text{Volume} = \frac{\text{Constant}}{\text{Pressure}}$$

Thus, as the pressure increases from 1 to 2 atmospheres (760 to 1520 mm. Hg), the volume of 1 gram-mol of gas at 0° C. is initially 22.4 liters but then decreases to 11.2 liters. On the other hand, if the pressure of this same gram-mol of gas is decreased from 760 to 380 mm. Hg, the total volume increases from 22.4 to 44.8 liters.

Relationship of Temperature to Volume—Gay-Lussac's Law (Charles' Law). If the pressure of a given quantity of gas remains constant but the temperature is varied, the volume of the gas increases directly in proportion to the increase in temperature. When the temperature is expressed in absolute (Kelvin) degrees, the following relationship holds:

$$\text{Volume} = \text{Constant} \times \text{Temperature (K.)}$$

For example, at one atmosphere pressure, 1 gram-mol of gas has a volume of 22.4 liters at 273° K., which is 0° C. At body temperature of 37° C. (310° K.) 1 gram-mol of gas has a volume of 25.4 liters.

The Gas Law. Combining Boyle's law and Gay-Lussac's law, the following gas law applies:

$$PV = nRT$$

in which P is pressure, V is volume, n is the quantity of gas, R is a constant depending on the units of measure used for the other factors in the formula, and T is temperature. When P is expressed in millimeters Hg, V in liters, n in gram-mols, and T in absolute degrees, the value of the constant R is 62.36.

THE VAPOR PRESSURE OF WATER

All gases in the body are in direct contact with water. Therefore, all gaseous mixtures in the body are saturated with water vapor, and this must always be considered when the dynamics of gaseous exchange are discussed.

Figure 40–1 illustrates to the left a chamber con-

Figure 40–1 Demonstration of 47 mm. Hg vapor pressure in the chamber to the right.

taining normal dry air that has in it only nitrogen and oxygen. The pressure in this chamber is the normal atmospheric pressure of 760 mm. Hg, which is illustrated on the scale of the manometer. In the second chamber a few drops of water are added. Immediately, some of the water vaporizes, this vaporization increasing the mass of gas in the chamber; if the temperature in the chamber is body temperature (37° C.), the pressure increases 47 mm. Hg (the vapor pressure of water at 37° C.) to 807 mm. Hg, as shown to the right in Figure 40–1.

The vapor pressure of water depends entirely on the temperature. The greater the temperature, the greater is the activity of the molecules in the water, and the greater is the likelihood these molecules will escape from the surface of the water into the gaseous phase. On the other hand, the lower the temperature, the greater is the condensation of water from the gaseous phases. When dry air is suddenly mixed with water, the water vapor pressure is zero at first, but water molecules immediately begin escaping from the surface of the water into the air. As the air becomes progressively more

and more humidified, an equilibrium vapor pressure is approached at which the rate of condensation of water becomes equal to the rate of water vaporization. Because the kinetic activity of molecules in water increases with increased temperature, the equilibrium pressure (the *water vapor pressure*) also increases with increased temperatures. The water vapor pressure at various temperatures from 0° to 100° C. is given in Table 40–1. The most important value to remember is the vapor pressure at body temperature, 47 mm. Hg; this value will appear in most of our subsequent discussions.

THE SOLUTION OF GASES IN WATER (HENRY'S LAW)

Figure 40–2 illustrates three chambers, with chamber A having oxygen in the top of the chamber at a pressure of 100 mm. Hg and pure water in the bottom of the chamber. Large numbers of oxygen molecules continually strike the surface of the water and some of them enter the water to become dissolved. However, the quantity of dissolved oxygen in chamber A is zero because the process is just beginning. In chamber B a fairly large number of molecules of oxygen has already been dissolved in the water so that the number of molecules striking the surface of the water from below upward and leaving the water to enter the gaseous phase is equal to approximately half the number of molecules striking the surface from above and entering the dissolved state. Therefore, greater numbers of oxygen molecules are still entering the dissolved state than are leaving the dissolved state. In chamber C the amount of oxygen in the dissolved state has become great enough that the number of molecules leaving the surface of the water to enter the gaseous phase is equal to the number of molecules entering the water to become dissolved. Thus, at this point a state of equilibrium exists.

Two factors determine the quantity of a gas that will be dissolved in water or other fluid when the equilibrium state has been reached. These factors are (1) the pressure of the gas surrounding the water and (2) the solubility coefficient of the gas in water at the temperature of the water. When the gas has reached the equilibrium state, as in Figure

TABLE 40–1. Vapor Pressure of Water

Temp. (°C.)	Vapor pressure (mm.)	Temp. (°C.)	Vapor pressure (mm.)
0	4.6	39	52.0
5	6.5	40	54.9
10	9.1	41	57.9
14	11.9	42	61.0
16	13.5	43	64.3
18	15.3	44	67.8
20	17.4	46	75.1
22	19.6	48	83.2
24	22.2	50	92.0
26	25.0	55	117.5
28	28.1	60	148.9
30	31.5	65	187.1
31	33.4	70	233.3
32	35.3	75	288.8
33	37.4	80	354.9
34	39.5	85	433.2
35	41.8	90	525.5
36	44.2	95	633.7
37	46.6	100	760.0
38	49.3		

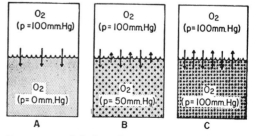

Figure 40–2 Solution of oxygen in water: (A) when the oxygen first comes in contact with pure water, (B) after the dissolved oxygen is halfway to equilibrium with the gaseous oxygen, and (C) after equilibrium has been established.

40-2C, the total quantity of dissolved gas at any particular temperature may be expressed by the following formula, which is Henry's law:

$$\text{Volume} = \text{Pressure} \times \text{Solubility coefficient}$$

When volume is expressed in volumes of gas dissolved in each volume of water at $0°$ C. and pressure is expressed in atmospheres, the solubility coefficients for important respiratory gases at body temperatures are the following:

Oxygen	0.024
Carbon dioxide	0.57
Carbon monoxide	0.018
Nitrogen	0.012
Helium	0.008

PRESSURE OF DISSOLVED GASES

If the oxygen should be removed suddenly from the chamber above the water in Figure 40-2A, no oxygen molecules would be striking the surface of the water from beneath and passing into the gaseous phase. Therefore, there would be no upward pressure against the surface of the water exerted by oxygen. In chamber B of Figure 40-2, half as many molecules of oxygen are pushing upward from the surface of the water as are striking the surface of the water from above. Because pressure is the instantaneous sum of the forces of impact of molecules against a unit area of surface, if half as many molecules are striking upward against the surface of the fluid as are striking downward against the surface of the fluid, the pressure exerted upward is equal to one-half the pressure downward. In other words, in Figure 40-2B the pressure of the oxygen in the gaseous phase is 100 mm. Hg and in the dissolved phase 50 mm. Hg. In chamber C the total force of oxygen molecules striking upward is equal to the total force of molecules striking downward, which means that the pressure exerted by the dissolved oxygen is equal to the pressure exerted by the gaseous oxygen. At this point the pressure of the oxygen in the gaseous phase is in equilibrium with the pressure of the oxygen in the dissolved phase, and both are equal to 100 mm. Hg.

The pressure exerted by a gas as it attempts to pass outward from the surface of a liquid is generally written as P_{O_2}, P_{CO_2}, P_{N_2}, P_{He}, etc., for each of the gases dissolved in the liquid.

The pressure of a dissolved gas is proportional to the quantity of gas dissolved in the fluid divided by the solubility coefficient of the gas in that particular fluid. Molecules of gases that are highly soluble are attracted by the fluid molecules so that, as they strike the surface of the fluid, this attraction prevents many of these highly soluble molecules from leaving the surface. Consequently, the total number of molecules that must be dissolved to exert a given pressure is considerably greater for very soluble gases than for gases not so soluble.

PARTIAL PRESSURES

An understanding of partial pressures is an important preliminary to understanding gaseous diffusion from the alveoli to the pulmonary blood, for it is the partial pressure of a gas that determines the force it exerts in attempting to diffuse through the pulmonary membrane.

Figure 40-3 illustrates four separate chambers, each of which has a capacity of 100 volumes of gas. The second chamber is divided by a partition, and 79 volumes of nitrogen at 760 mm. Hg pressure (1 atmosphere of pressure) are placed above the partition and 21 volumes of oxygen at 760 mm. Hg below. Then all the nitrogen in the upper part of chamber 2 is moved to chamber 1, which has a total capacity of 100 volumes. In other words, 79 volumes of nitrogen are expanded to 100 volumes. As this nitrogen is expanded, it can be shown from Boyle's law that the pressure falls to 79/100 of 760 mm. Hg — that is, to 600 mm. Hg.

The 21 volumes of oxygen at 760 mm. Hg are moved from the second chamber to the third chamber. Here again it can be shown from Boyle's law that this expansion causes the pressure of the oxygen to fall to 21/100 of 760 mm. Hg — that is, to 160 mm. Hg.

Finally, the nitrogen in the first chamber and the oxygen in the third chamber are mixed together in the fourth chamber. The original pressure of the nitrogen in 600 mm. Hg, and the pressure of the oxygen is 160 mm. Hg, but the total pressure in chamber 4 after mixing is 760 mm. Hg — the normal atmospheric pressure level. It is obvious that 600 mm. of this 760 mm. Hg results from the presence of nitrogen in the chamber, and 160 mm. results from the presence of oxygen in the chamber. The 600 mm. Hg pressure exerted by nitrogen and the 160 mm. Hg pressure exerted by the oxygen are known as the *partial pressures* of the respective gases in the mixture.

The partial pressures of the gases in a mixture are designated by the same terms used to designate pressures in liquids; i.e., P_{O_2}, P_{CO_2}, P_{N_2}, P_{He}, etc.

From the kinetic theory of gases we know that pressure against any membrane or against any other surface is determined by the number of molecules striking a unit area of the membrane

Figure 40-3 Relationship of partial pressures to gaseous percentages in mixtures of gases.

at any given instant times the average kinetic energy of the molecules. Therefore, the partial pressure of a gas in a mixture is in reality the sum of the force of impact of all the molecules of that particular gas against the surface. In other words, in chamber 4 the total force of the nitrogen molecules striking against the chamber walls is sufficient to elevate the mercury in a manometer to a level of 600 mm. The total force exerted by the oxygen molecules striking against the walls of the chamber is sufficient to elevate the mercury in a manometer to a level of 160 mm. With both these gases exerting force against the walls of the chamber at the same time, the mercury is elevated to a total level of 760 mm. Hg.

Molecules in a mixture of gases are constantly striking each other and thereby imparting energy to each other. As a result, each molecule of the mixture, regardless of its molecular weight, has the same mean kinetic energy as the next molecule. Therefore, *the partial pressure of a gas in a mixture is also a direct measure of the number of molecules of that particular gas striking a unit surface area in a given period of time.*

Pressures of Dissolved Gases in Equilibrium with a Mixture of Gases. Because each gas in a mixture exerts its own partial pressure in proportion to the concentration of its molecules, when the gases of a mixture dissolve in a liquid and come to equilibrium with the gaseous phase of the mixture, the pressure of each dissolved gas is equal to the partial pressure of the same gas in the gaseous mixture. In other words, *each gas is independent of the others in its ability to dissolve in a liquid;* obviously, this principle also applies to the solution of gases in blood. Thus, an increase in quantity of carbon dioxide dissolved in the body fluids does not significantly affect the quantity of oxygen that can be dissolved in the same fluids. (However, the various dissolved gases often do interfere with the chemical reactions of each other, as is discussed in the following chapter in connection with the reactions of oxygen and carbon dioxide with hemoglobin.)

DIFFUSION OF GASES THROUGH LIQUIDS

Figure 40–4 illustrates a chamber filled with water. The water at one end of this chamber contains a relatively large amount of dissolved oxygen. Because of the kinetic energy of matter, the dissolved gaseous molecules are constantly undergoing molecular motion, bouncing among the molecules of the solvent. Thus, in Figure 40–4 the oxygen molecules are bouncing in all directions, and this process is called *diffusion*. More of them bounce from the area of high concentration toward the area of low concentration for the following reasons:

The Pressure Gradient for Diffusion. The molecules in the area of high pressure in Figure 40–4, because of their greater number, have more *statistical chance* of bouncing into the area of low

Figure 40–4 Net diffusion of oxygen from one end of a chamber to the other.

pressure than do molecules attempting to go in the other direction. However, some molecules do bounce from the area of low pressure toward the area of high pressure. Therefore, the *net diffusion* of gas from the area of high pressure to the area of low pressure is equal to the number of molecules bouncing in this direction minus the number bouncing in the opposite direction, and this in turn is proportional to the gas pressure difference between the two areas. The pressure in area A of Figure 40–4 minus the pressure in area B divided by the distance of diffusion is known as the *pressure gradient for diffusion* or simply the *diffusion gradient*. The rate of net gas diffusion from area A to area B is directly proportional to this gradient.

This principle of diffusion from an area of high pressure to an area of low pressure holds true for diffusion of gases in a gaseous mixture, diffusion of dissolved gases in a solution, and even diffusion of gases from the gaseous phase into the dissolved state in liquids. That is, *there is always net diffusion from areas of high pressure to areas of low pressure.*

As more and more gas diffuses from area A to area B in Figure 40–4, the pressure in area B approaches the pressure in area A; as a result, the net rate of diffusion becomes less and less. After a reasonable length of time, the gaseous pressures in both ends of the chamber become essentially equal, and, thereafter, no net diffusion of gas occurs from one end to the other end. This does not mean that no molecules of gas diffuse, but merely that as many molecules then diffuse backward as forward.

If at any time the gaseous pressure at one end of the chamber falls below that at the other end, an excess of gas molecules immediately begins anew to diffuse from the high pressure area to the low pressure area until the two pressures again approach each other and eventually reach a state of equilibrium.

Rate of Diffusion. In addition to the pressure difference, several other factors affect the rate of gas diffusion in a fluid. These are (1) the solubility of the gas in the fluid, (2) cross-sectional area of the fluid, (3) the distance through which the gas must diffuse, (4) the molecular weight of the gas, (5) the viscosity of the fluid, and (6) the temperature of the fluid. In the body, the last two of these factors remain reasonably constant and usually need not be considered.

Obviously, the greater the solubility of the gas,

the greater will be the number of molecules available to diffuse for any given pressure gradient. Also, the greater the cross-sectional area of the chamber, the greater will be the total number of molecules to diffuse. On the other hand, the longer the distance that the molecules must diffuse, the longer it will take the molecules to diffuse the entire distance. Finally, the greater the velocity of kinetic movement of the molecules, which at any given temperature is inversely proportional to the square root of the molecular weight, the greater is the rate of diffusion of the gas. All of these factors can be expressed in a single formula, as follows:

$$DR \propto \frac{PD \times A \times S}{D \times MW}$$

in which DR is the diffusion rate, PD is the pressure difference between the two ends of the chamber, A is the cross-sectional area of the chamber, S is the solubility of the gas, D is the distance of diffusion, and MW is the molecular weight of the gas.

It is obvious from this formula that the characteristics of the gas itself determine two factors of the formula: solubility and molecular weight. Therefore, the *diffusion coefficient*—that is, the rate of diffusion through a given area for a given distance—for any given gas is proportional to S/MW. Considering the diffusion coefficient for oxygen to be 1, the diffusion coefficients for different gases of respiratory importance in the body fluids are:

Oxygen	1.0
Carbon dioxide	20.3
Carbon monoxide	0.81
Nitrogen	0.53
Helium	0.95

DIFFUSION OF GASES THROUGH TISSUES

The gases that are of respiratory importance are highly soluble in lipids and, consequently, are also highly soluble in cell membranes. Because of this, these gases diffuse through the cell membranes with very little impediment. Instead, the major limitation to the movement of gases in tissues is the rate at which the gases can diffuse through the

tissue fluids instead of through the cell membranes. Therefore, diffusion of gases through the tissues, including the pulmonary membrane, is almost equal to the diffusion of gases through water, as given in the above list of diffusion rates for the important respiratory gases.

COMPOSITION OF ALVEOLAR AIR

Alveolar air does not have the same concentration of gases as atmospheric air by any means, which can readily be seen by comparing the alveolar air composition in column 3 of Table 40–2 with the composition of atmospheric air in column 1. There are several reasons for the differences. First, the alveolar air is only partially replaced by atmospheric air with each breath. Second, oxygen is constantly being absorbed from the alveolar air. Third, carbon dioxide is constantly diffusing from the pulmonary blood into the alveoli. And, fourth, dry atmospheric air that enters the respiratory passages is humidified even before it reaches the alveoli.

Humidification of the Air as It Enters the Respiratory Passages. Column 1 of Table 40–2 shows that atmospheric air is composed almost entirely of nitrogen and oxygen; it normally contains almost no carbon dioxide and little water vapor. However, as soon as the atmospheric air enters the respiratory passages, it is exposed to the fluids covering the respiratory surfaces. Even before the air enters the alveoli, it becomes totally humidified.

The partial pressure of water vapor at normal body temperature of 37° C. is 47 mm. Hg, which, therefore, is the partial pressure of water in the alveolar air. Since the total pressure in the alveoli cannot rise to more than the atmospheric pressure, this water vapor simply expands the volume of the air and thereby *dilutes* all the other gases in the inspired air. In column 2 of Table 40–2 it can be seen that humidification of the air reduces the oxygen

TABLE 40–2. Partial Pressures of Respiratory Gases as They Enter and Leave the Lungs (at Sea Level) — Per Cent Concentrations Are Given in Parentheses

	Atmospheric Air* (mm. Hg)		Humidified Air (mm. Hg)		Alveolar Air (mm. Hg)		Expired Air (mm. Hg)	
N_2	597.0	(78.62%)	563.4	(74.09%)	569.0	(74.9%)	566.0	(74.5%)
O_2	159.0	(20.84%)	149.3	(19.67%)	104.0	(13.6%)	120.0	(15.7%)
CO_2	0.3	(0.04%)	0.3	(0.04%)	40.0	(5.3%)	27.0	(3.6%)
H_2O	3.7	(0.50%)	47.0	(6.20%)	47.0	(6.2%)	47.0	(6.2%)
TOTAL	760.0	(100.0%)	760.0	(100.0%)	760.0	(100.0%)	760.0	(100.0%)

* On an average cool, clear day.

partial pressure at sea level from an average of 159 mm. Hg in atmospheric air to 149 mm. Hg in the humidified air, and it reduces the nitrogen partial pressure from 597 to 563 mm. Hg.

RATE AT WHICH ALVEOLAR AIR IS RENEWED BY ATMOSPHERIC AIR

In the preceding chapter it was pointed out that the *functional residual capacity* of the lungs, which is the amount of air remaining in the lungs at the end of normal expiration, measures approximately 2300 ml. Furthermore, only 350 ml. of new air is brought into the alveoli with each normal respiration. Therefore, the amount of alveolar air replaced by new atmospheric air with each breath is only one seventh of the total, so that many breaths are required to exchange most of the alveolar air. Figure 40–5 illustrates the effect of this on the rate of renewal of the alveolar air. In the first alveolus of the figure is a highly concentrated foreign gas that has been placed momentarily in all the alveoli. The second alveolus shows slight dilution of this foreign gas with the first breath; the next alveolus shows still further dilution with the second breath, and so forth for the third, fourth, eighth, twelfth, and sixteenth breaths. Note that even at the end of 16 breaths the foreign gas still has not been completely removed from the alveoli. Figure 40–6 illustrates graphically the rate at which a foreign gas in the alveoli is normally removed, showing that with normal alveolar ventilation approximately half the foreign gas is removed in 17 seconds. When a person's rate of alveolar ventilation is only half normal, half the gas is removed in 34 seconds, and, when his rate of ventilation is two times normal, half is removed in about 8 seconds.

Value of the Slow Replacement of Alveolar Air. This slow replacement of alveolar air is of particular importance in preventing sudden changes in gaseous concentrations in the blood.

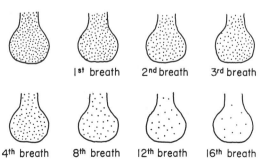

Figure 40–5 Expiration of a foreign gas from the alveoli with successive breaths.

1st breath   2nd breath   3rd breath
4th breath   8th breath   12th breath   16th breath

Figure 40–6 Rate of removal of a foreign gas from the alveoli.

This makes the respiratory control mechanism much more stable than it would otherwise be and helps to prevent excessive increases and decreases in tissue oxygenation, tissue carbon dioxide concentration, and tissue pH when the respiration is temporarily interrupted.

OXYGEN CONCENTRATION IN THE ALVEOLI

Oxygen is continually being absorbed into the blood of the lungs, and new oxygen is continually entering the alveoli from the atmosphere. The more rapidly oxygen is absorbed, the lower becomes its concentration in the alveoli; on the other hand, the more rapidly new oxygen is brought into the alveoli from the atmosphere, the higher becomes its concentration. Therefore, oxygen concentration in the alveoli is controlled by, first, the rate of absorption of oxygen into the blood and, second, the rate of entry of new oxygen into the lungs.

Figure 40–7 illustrates the effect both of alveolar ventilation and of rate of oxygen absorption into the blood on the alveolar partial pressure of oxygen (Po_2). The dark curve represents oxygen absorption at a rate of 250 ml. per minute, and the dotted curve represents 1000 ml. per minute. At a normal ventilatory rate of 4.2 liters per minute and an oxygen consumption of 250 ml. per minute, the normal operating point in Figure 40–7 is point A. However, during exercise, the rate of oxygen utilization is increased in proportion to the intensity of the exercise. From Figure 40–7 it can be seen that when 1000 ml. of oxygen is being absorbed each minute, the rate of alveolar ventilation must increase four-fold to maintain the alveolar Po_2 at the normal value of 104 mm. Hg, and at still higher rates of oxygen absorption, the rate of ventilation must rise proportionately to maintain normal alveolar Po_2.

Figure 40–7 Effect of alveolar ventilation and of rate of oxygen absorption from the alveoli on the alveolar P_{O_2}.

Another effect illustrated by Figure 40–7 is that an extremely marked increase in alveolar ventilation can never increase the alveolar P_{O_2} above 149 mm. Hg as long as the subject is breathing normal atmospheric air, for this is the maximum content of oxygen in humidified air. However, if the subject breathes gases containing concentrations of oxygen higher than 149 mm. Hg, the alveolar P_{O_2} can approach these higher concentrations when alveolar ventilation approaches maximum.

CO_2 CONCENTRATION IN THE ALVEOLI

Carbon dioxide is continually being formed in the body, then discharged into the alveoli; and it is continually being removed from the alveoli by the process of ventilation. Therefore, the two factors that determine carbon dioxide partial pressure (P_{CO_2}) in the lungs are (1) the rate of excretion of carbon dioxide from the blood into the alveoli and (2) the rate at which carbon dioxide is removed from the alveoli by alveolar ventilation.

Figure 40–8 illustrates the effects on the alveolar P_{CO_2} of both alveolar ventilation and the rate of carbon dioxide excretion. The dark curve represents normal rate of carbon dioxide excretion of 200 ml. per minute. At the

Figure 40–8 Effect on alveolar P_{CO_2} of alveolar ventilation and rate of carbon dioxide excretion from the blood.

normal rate of alveolar ventilation of 4.2 liters per minute, the operating point for alveolar P_{CO_2} is at point A in Figure 40–8 – that is, 40 mm. Hg.

Two other facts are also evident from Figure 40–8: First, *the alveolar P_{CO_2} increases directly in proportion to the rate of carbon dioxide excretion,* as represented by the dotted curve for 800 ml. CO_2 excretion per minute. Second, *the alveolar P_{CO_2} decreases in inverse proportion to alveolar ventilation.* Therefore, the concentration of both oxygen and carbon dioxide in the alveoli is determined by the rate of absorption or excretion, respectively, of these two gases and also by the alveolar ventilation.

EXPIRED AIR

Expired air is a combination of dead space air and alveolar air, and its overall composition is, therefore, determined by, first, the proportion of the expired air that is dead space air and the proportion that is alveolar air; second, the concentrations of the gases in the dead space air; and, third, the concentrations of the gases in the alveolar air. Figure 40–9 shows the progressive changes in oxygen and carbon dioxide concentrations in the expired air during the course of expiration. The very first portion of this air, the dead space air, is typical humidified air as shown in column 2 of Table 40–2. Then, progressively more and more alveolar air becomes mixed with the dead space air until all the dead space air has finally been washed out and nothing but alveolar air remains. Thus, the oxygen concentration in the first portion of the expired air is 149 mm. Hg and of the latter portion approximately 104 mm. Hg, the value in normal alveolar air. On the other hand, the early portion of the expired air has essentially no carbon dioxide, but the final portion, the alveolar air portion, has a carbon dioxide pressure of 40 mm. Hg. Indeed, one of the means for collecting alveolar air for study is simply to collect a sample of the last portion of expired air.

Normal expired air, containing both dead space air and alveolar air, has gaseous concentrations approximately as shown in column 4 of Table 40–2 – that is, concentrations somewhere between those of humidified atmospheric air and alveolar air, under normal resting conditions about two-thirds alveolar air and one-third dead space air.

DIFFUSION OF GASES THROUGH THE RESPIRATORY MEMBRANE

The Respiratory Unit. Figure 40–10 illustrates the respiratory unit, which is comprised of a *respiratory bronchiole, alveolar ducts, atria,* and *alveolar sacs* or *alveoli* (of which there are about 250 million in the two lungs, each al-

Figure 40-9 Oxygen and carbon dioxide concentrations in the various portions of expired air.

veolus having an average diameter of about 0.1 mm.). The epithelium of these structures is a very thin membrane, and the alveolar gases are in close proximity to the blood of the capillaries. Consequently, gaseous exchange between the alveolar air and the pulmonary blood occurs through the membranes of all these terminal portions of the lungs. These membranes are collectively known as the *respiratory membrane,* also called the *pulmonary membrane.*

The Respiratory Membrane. Figure 40-11 illustrates the ultrastructure of the respiratory membrane. It shows also the diffusion of oxygen from the alveolus into the red blood cell and diffusion of carbon dioxide in the opposite direction. Note the following different layers of the respiratory membrane:

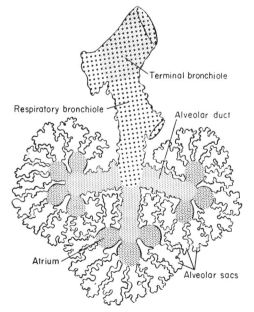

Figure 40-10 The respiratory lobule. (From Miller: The Lung. Charles C Thomas.)

1. A monomolecular layer of surfactant, a mixture of phospholipids and perhaps other substances, that spreads out over the surface of the fluid lining the alveolus.

2. A very thin layer of fluid lining the alveolus.

3. The alveolar epithelium comprised of very thin epithelial cells.

4. A very thin interstitial space between the alveolar epithelium and capillary membrane.

5. A capillary basement membrane.

6. The capillary endothelial membrane.

Despite the large number of layers, the overall thickness of the respiratory membrane in some areas is less than 0.1 micron and in essentially all areas is less than 1 micron.

From histologic studies it has been estimated that the total surface area of the respiratory membrane is approximately 70 square meters in the normal adult. This is equivalent to the floor area of a room approximately 30 feet long by 25 feet wide. The total quantity of blood in the capillaries of the lung at any given instant is about 60 to 100 ml. If this small amount of blood were spread over the entire surface of a 25 by 30 foot floor, one could readily understand how respiratory exchange of gases occurs as rapidly as it does.

The average diameter of the pulmonary capillaries is only 7 microns, which means that red blood cells must actually squeeze through them. Therefore, the red blood cell membrane usually touches the capillary wall so that oxygen and carbon dioxide need not pass through plasma as they diffuse between the red cell and alveolus. Obviously, this aids the rapidity of diffusion.

PERMEABILITY OF THE RESPIRATORY MEMBRANE. The limiting factor in the permeability of the respiratory membrane to gases is the rate at which the gases can diffuse through the water in the membrane. All gases of respiratory importance are highly soluble in the lipid substances of the cell membranes and for this reason can diffuse through these with great ease. Earlier in the chapter it was pointed out that the rate of carbon dioxide diffusion in water is about 20 times as rapid as the rate of oxygen diffusion. Therefore, it follows that the rate of diffusion of carbon dioxide through the respiratory membrane is also about 20 times as rapid as the diffusion of oxygen, as would be predicted.

Permeability of the Respiratory Membrane to Solid Solutes. In contrast to the extreme permeability of the respiratory membrane for respiratory gases, this membrane is only slightly permeable to most dissolved crystalloid substances in the interstitial

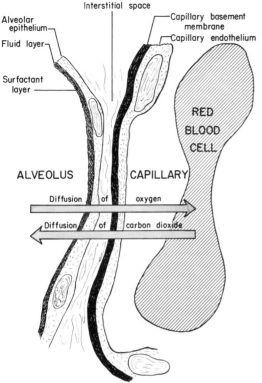

Interstitial space

Alveolar epithelium

Fluid layer

Surfactant layer

Capillary basement membrane

Capillary endothelium

RED BLOOD CELL

ALVEOLUS   CAPILLARY

Diffusion of oxygen

Diffusion of carbon dioxide

Figure 40–11   Ultrastructure of the respiratory membrane.

fluid. The pulmonary *capillary* has the same high permeability as capillary membranes elsewhere in the body, but the alveolar epithelial layer is only about 1/100 as permeable, except for lipid-soluble substances, such as alcohol and dinitrophenol, that can pass directly through the cells. Likewise, the membrane is very permeable to water and reasonably permeable to urea and potassium but very poorly permeable to sodium and glucose. By referring to the discussion of cell membrane permeability in Chapter 4, one can see that these are characteristics similar to those exhibited by the red cell membrane. In other words, the alveolar epithelial layer of the respiratory membrane seems to be a continuous layer of epithelial cells with only sparse pores between their borders. This is in contrast to the pulmonary capillary membrane, which has large pores between the individual endothelial cells through which sodium, glucose, urea, and other water-soluble substances diffuse at rates a hundred or more times as rapidly as they diffuse through the alveolar epithelium.

FACTORS AFFECTING GASEOUS DIFFUSION THROUGH THE RESPIRATORY MEMBRANE

Referring to the above discussion of diffusion through water, one can apply the same principles and same formula to diffusion of gases through the respiratory membrane. Thus, (1) the *thickness of the membrane,* (2) the *surface area of the membrane,* (3) the *diffusion coefficient* of the gas in the substance of the membrane—that is, in water, and (4) the *pressure difference* between the two sides of the membrane, all combined, determine how rapidly a gas will pass through the membrane.

The *thickness of the respiratory membrane* occasionally increases markedly, usually as a result of the presence of edema fluid in the interstitial space of the membrane. Also, fluid may collect in the alveoli, so that the respiratory gases must diffuse not only through the membrane but also through this fluid. Finally, some pulmonary diseases cause fibrosis of the lungs, which can increase the thickness of some portions of the respiratory membrane. Because the rate of diffusion through the membrane is inversely proportional to the thickness of the membrane, any factor that increases the thickness more than two to three times above normal can interfere markedly with normal respiratory exchange of gases.

The *surface area of the respiratory membrane* may be greatly decreased by many different conditions. For instance, removal of an entire lung decreases the surface area to half normal. Also, in the condition known as *emphysema* many of the alveoli coalesce, with dissolution of the septa between the alveoli. Therefore, the new chambers are much larger than the original alveoli, but the total surface area of the respiratory membrane is considerably decreased because of loss of the alveolar septa. When the total surface area is decreased to approximately a third to one-fourth normal, exchange of gases through the membrane is impeded to a significant degree even under resting conditions. And, during heavy exercise, even the slightest decrease in surface area of the lungs can be a detriment to respiratory exchange of gases.

The *diffusion coefficient* for the transfer of each gas through the respiratory membrane depends on its *solubility* in the membrane and on its *molecular weight.* The rate of diffusion in the respiratory membrane is almost exactly the same as that in water, for reasons explained above. Therefore, carbon dioxide diffuses through the membrane about 20 times as rapidly as oxygen. Oxygen in turn diffuses about two times as rapidly as nitrogen and at almost the same rate as helium.

The *pressure difference* across the respiratory membrane is the difference between the partial pressure of the gas in the alveoli and the pressure of the gas in the blood. The partial pressure represents a measure of the total number of molecules of a particular gas striking a unit area of the alveolar surface of the membrane, and the pressure of the same gas in the blood represents the number of molecules striking the same area of the membrane from the opposite side. Therefore, the differences between these two pressures, the pressure difference is a measure of the *net tendency* for the particular gas to pass through the membrane. Obviously, when the partial pressure of a gas in the alveoli is greater than the pressure of the gas in the blood, as is true of oxygen, net diffusion from the alveoli into the blood occurs, but, when the pressure of the gas in the blood is greater than the partial pressure in the alveoli, as is true of carbon dioxide, net diffusion from the blood into the alveoli occurs.

DIFFUSING CAPACITY OF THE RESPIRATORY MEMBRANE

The overall ability of the respiratory membrane to exchange a gas between the alveoli and the pulmonary blood can be expressed in terms of its *diffusing capacity*, which is defined as the *volume of a gas that diffuses through the membrane each minute for a pressure difference of 1 mm. Hg.*

Obviously, all the factors discussed above that affect diffusion through the respiratory membrane can affect the diffusing capacity.

The Diffusing Capacity for Oxygen. In the average young male adult the diffusing capacity for oxygen under resting conditions averages 21 ml. per minute. Since the oxygen pressure difference across the respiratory membrane during quiet breathing averages 11 mm. Hg, this gives a total of about 250 ml. of oxygen diffusing through the respiratory membrane each minute.

CHANGE IN OXYGEN-DIFFUSING CAPACITY DURING EXERCISE. During strenuous exercise, or during other conditions that greatly increase pulmonary activity, the diffusing capacity for oxygen increases in young male adults to a maximum of about 65 ml. per minute, which is three times the diffusing capacity under resting conditions. This increase is caused by three different factors: (1) opening up of a number of previously dormant pulmonary capillaries, thereby increasing the surface area of the blood into which the oxygen can diffuse, (2) dilatation of all the pulmonary capillaries that were already open, thereby further increasing the surface area, and (3) stretching of the alveolar membranes, increasing their surface area and decreasing their

thickness. Therefore, during exercise, the oxygenation of the blood is increased not only by increased alveolar ventilation but also by a greater capacity of the respiratory membrane for transmitting oxygen into the blood.

Diffusing Capacity for Carbon Dioxide. The diffusing capacity for carbon dioxide has never been measured because of the following technical difficulty: Carbon dioxide diffuses through the respiratory membrane so rapidly that the average P_{CO_2} in the pulmonary blood is not far different from the P_{CO_2} in the alveoli—the difference is less than 1 mm. Hg—and with the available techniques, this difference is too small to be measured accurately.

Nevertheless, measurements of diffusion of other gases have shown that the diffusing capacity varies directly with the diffusion coefficient of the particular gas. Since the diffusion coefficient of carbon dioxide is 20 times that of oxygen, one would expect a diffusing capacity for carbon dioxide under resting conditions of about 400 to 450 ml. and during exercise of about 1200 to 1300 ml. per minute.

The importance of these high diffusing capacities for carbon dioxide is this: When the respiratory membrane becomes progressively damaged, its capacity for transmitting oxygen into the blood is always impaired enough to cause death of the person long before significant impairment of carbon dioxide diffusion occurs. The only time that a low diffusing capacity for carbon dioxide causes any significant difficulty is when lung damage is far beyond that which ordinarily causes death but when the person's life is being maintained by intensive oxygen therapy that overcomes the reduction in oxygen-diffusing capacity.

Figure 40–12 compares the diffusing capacities of oxygen, carbon dioxide, and carbon monoxide at rest and during exercise, showing the extreme diffusing capacity of carbon dioxide and also the effect of exercise on the diffusing capacities of all the gases.

Measurement of Diffusing Capacity. The oxygen-diffusing capacity can be estimated from measurements of (1) alveolar P_{O_2}, (2) P_{O_2} in the pulmonary capillary blood, and (3) the rate of oxygen utilization. To measure the P_{O_2} in the pulmonary blood is a difficult procedure and is usually accomplished by measuring the P_{O_2} in the pulmonary arterial blood and the P_{O_2} in the aortic arterial blood and then mathematically obtaining a reasonable average for pulmonary capillary blood. This average must be a time-integrated average rather than a simple average, because pulmonary arterial blood on entering the capillaries becomes rapidly oxygenated during the first fraction of its transit through the capillaries and then progressively

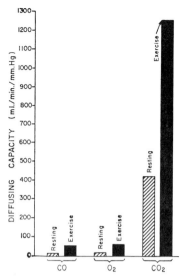

Figure 40–12 Diffusing capacities of carbon monoxide, oxygen, and carbon dioxide in the normal lungs.

per minute, and the diffusing capacity for oxygen is 1.23 times this, or 21 ml. per minute.

REFERENCES

Cerretelli, P.; Sikand, R. S.; and Farhi, L. E.: Effect of increased airway resistance on ventilation and gas exchange during exercise. *J. Appl. Physiol.*, 27:597, 1969.

Forster, R. E.: Diffusion of gases. *In* Handbook of Physiology. Baltimore, The Williams & Wilkins Co., 1964, Sec. III, Vol. I, p. 839.

Forster, R. E.: Interpretation of measurements of pulmonary diffusing capacity. *In* Handbook of Physiology. Baltimore, The Williams & Wilkins Co., 1965, Sec. III, Vol. II, p. 1453.

Guyton, A. C.; Nichols, R. J.; and Farish, C. A.: An arteriovenous oxygen difference recorder. *J. Appl. Physiol.*, 10:158, 1957.

Holloman, G. H., Jr.; Milhorn, H. T., Jr.; and Coleman, T. G.: A sampled-data regulator for maintaining a constant alveolar CO_2. *J. Appl. Physiol.*, 25:463, 1968.

La Force, R. C., and Lewis, B. M.: Diffusional transport in the human lung. *J. Appl. Physiol.*, 28:291, 1970.

Lloyd, T. C., Jr.: Hypoxic pulmonary vasoconstriction: role of perivascular tissue. *J. Appl. Physiol.*, 25:560, 1968.

Milhorn, H. T., Jr., and Pulley, P. E., Jr.: A theoretical study of pulmonary capillary gas exchange and venous admixture. *Biophys. J.*, 8:337, 1968.

Otis, A. B.: Quantitative relationships in steady-state gas exchange. *In* Handbook of Physiology. Baltimore, The Williams & Wilkins Co., 1964, Sec. III, Vol. I, p. 681.

Overfeld, E. M., and Kylstra, J. A.: Distribution component of alveolar-arterial oxygen pressure difference in man. *J. Appl. Physiol.*, 27:634, 1969.

Pattle, R. E.: Surface lining of lung alveoli. *Physiol. Rev.*, 45:48, 1965.

Power, G. G.: Gaseous diffusion between airways and alveolin in the human lung. *J. Appl. Physiol.*, 27:701, 1969.

Power, G. G., and Bradford, W. C.: Measurement of pulmonary diffusing capacity during blood-to-gas exchange in humans. *J. Appl. Physiol.*, 27:61, 1969.

Radford, E. P., Jr.: The physics of gases. *In* Handbook of Physiology. Baltimore, The Williams & Wilkins Co., 1964, Sec. III, Vol. I, p. 125.

Rahn, H., and Farhi, L. E.: Ventilation, perfusion, and gas exchange—the Va/Q concept. *In* Handbook of Physiology. Baltimore, The Williams & Wilkins Co., 1964, Sec. III, Vol. I, p. 735.

Rahn, H., and Fenn, W. O.: A Graphical Analysis of the Respiratory Gas Exchange. Washington, American Physiological Society, 1955.

Riley, R. L., and Permutt, S.: The four-quadrant diagram for analyzing the distribution of gas and blood in the lung. *In* Handbook of Physiology. Baltimore, The Williams & Wilkins Co., 1965, Sec. III, Vol. II, p. 1413.

Staub, N. C.: Alveolar-arterial oxygen tension gradient due to diffusion. *J. Appl. Physiol.*, 18:673, 1963.

Weibel, E. R.: The ultrastructure of the alveolar-capillary membrane or barrier. *In* Fishman, A. P., and Hecht, H. H. (eds.): The Pulmonary Circulation and Interstitial Space. Chicago, University of Chicago Press, 1969, p. 9.

West, J. B.: Topographical distribution of blood flow in the lung. *In* Handbook of Physiology. Baltimore, The Williams & Wilkins Co., 1965, Sec. III, Vol. II, p. 1437.

West, J.: Ventilation: Blood Flow and Gas Exchange. Philadelphia, F. A. Davis Co., 1965.

Whipp, B. J., and Wasserman, K.: Alveolar-arterial gas tension differences during graded exercise. *J. Appl. Physiol.*, 27:361, 1969.

more slowly during the latter stages of transit. Obviously, this mathematical procedure is complicated, and one never knows exactly how accurate his results are.

To obviate the difficulties encountered in measuring oxygen-diffusing capacity, physiologists usually measure carbon monoxide-diffusing capacity instead, and calculate the oxygen-diffusing capacity from this. The principle of the carbon monoxide method is the following: A small amount of carbon monoxide is breathed into the alveoli, and the partial pressure of the carbon monoxide in the alveoli is measured from appropriate alveolar air samples. Under most conditions it can be assumed that the carbon monoxide pressure in the blood is essentially zero because the hemoglobin combines with this so rapidly that the pressure never has time to build up. Therefore, the pressure difference of carbon monoxide across the respiratory membrane is equal to its partial pressure in the alveoli. Then, by measuring the volume of carbon monoxide absorbed in a short period of time and dividing this by the carbon monoxide partial pressure, one can determine accurately the carbon monoxide-diffusing capacity.

To convert carbon monoxide-diffusing capacity to oxygen-diffusing capacity, the value is multiplied by a factor of 1.23 because the diffusion coefficient for oxygen is 1.23 times that for carbon monoxide. Thus, the average diffusing capacity for carbon monoxide in young male adults is 17 ml.

CHAPTER 41

Transport of Oxygen and Carbon Dioxide in the Blood and Body Fluids

Once oxygen has diffused from the alveoli into the pulmonary blood, it is transported principally in combination with hemoglobin to the tissue capillaries where it is released for use by the cells. The presence of hemoglobin in the red cells of the blood allows the blood to transport 30 to 100 times as much oxygen as could be transported simply in the form of dissolved oxygen in the water of the blood.

In the tissue cells oxygen reacts with various foodstuffs to form large quantities of carbon dioxide. This in turn enters the tissue capillaries and is transported by the blood back to the lungs. Carbon dioxide, like oxygen, also combines with chemical substances in the blood that increase carbon dioxide transport 20- to 30-fold.

The purpose of the present chapter, therefore, is to present both qualitatively and quantitatively the physical and chemical principles of oxygen and carbon dioxide transport in the blood and body fluids.

PRESSURE DIFFERENCES OF OXYGEN AND CARBON DIOXIDE FROM THE LUNGS TO THE TISSUES

In the discussions of the preceding chapter it was pointed out that gases can move from one point to another by diffusion and that the cause of this movement is always a pressure difference from the first point to the other. Thus, oxygen diffuses from the alveoli into the pulmonary capillary blood because of a pressure difference—that is, the oxygen pressure (Po_2) in the alveoli is greater than the Po_2 in the pulmonary blood. Then the pulmonary blood is transported by way of the circulation to the peripheral tissues. There the Po_2 is still lower in the cells than in the arterial blood entering the capillaries. Here again, the much higher Po_2 in the capillary blood causes oxygen to diffuse out of the capillaries and through the interstitial spaces to the cells.

Conversely, when oxygen is metabolized with the foods in the cells to form carbon dioxide, the carbon dioxide pressure (Pco_2) in the cells rises to a high value, which causes carbon dioxide to diffuse into the tissue capillaries. Once in the blood the carbon dioxide is transported to the pulmonary capillaries where it diffuses out of the blood into the alveoli because the Pco_2 in the alveoli is lower than that in the pulmonary capillary blood.

Basically, then, the transport of oxygen and carbon dioxide by the blood depends on both diffusion and the movement of blood. We now need to consider quantitatively the factors responsible for these effects as well as their significance in the overall physiology of respiration.

UPTAKE OF OXYGEN BY THE PULMONARY BLOOD

The top part of Figure 41–1 illustrates a pulmonary alveolus adjacent to a pulmonary capillary, showing the diffusion of oxygen molecules between the alveolar air and the pulmonary blood. However, the Po_2 of the venous blood entering the capillary is only 40 mm. Hg because a large amount of oxygen was removed from this blood when it passed through the tissue capillaries. The Po_2 in the alveolus is 104 mm. Hg, giving an initial pressure difference for diffusion of oxygen into the pulmonary capillary of 64 mm. Hg. Therefore, far more oxygen diffuses into the pulmonary capillary than in the opposite direction. The graph below the capillary shows the progressive rise in blood Po_2 as the blood passes through the

Figure 41–1   Uptake of oxygen by the pulmonary capillary blood. (The curve in this figure was constructed from data in Milhorn and Pulley: *Biophys. J.,* 8:337, 1968.)

capillary. This illustrates that the P_{O_2} rises essentially to equal that of the alveolar air before reaching the midpoint of the capillary, becoming approximately 104 mm. Hg.

The average pressure difference for oxygen diffusion through the pulmonary capillary during normal respiration is about 11 mm. Hg. This is a "time-integrated" average and not simply an average of the 64 mm. Hg pressure difference at the origin of the capillary and the final zero pressure difference at the end of the capillary, because the initial pressure difference lasts for only a fraction of the transit time in the pulmonary capillary, while the low pressure difference lasts for a long time.

Uptake of Oxygen by the Pulmonary Blood During Exercise.   During strenuous exercise, a person's body may require as much as 20 times the normal amount of oxygen. However, because of the increased cardiac output, the time that the blood remains in the capillary is greatly reduced. Therefore, oxygenation of the blood might suffer for two reasons: First, the blood remains in contact with the alveoli for short periods of time and, second, far larger quantities of oxygen are needed to oxygenate the blood. Yet, because of a great *safety factor* for diffusion of oxygen through the pulmonary membrane, the blood is still *almost completely saturated* with oxygen when it leaves the pulmonary capillaries. The reasons for this are the following:

First, it was pointed out in the previous chapter that the diffusing capacity for oxygen increases about three-fold during exercise; this results mainly from increased numbers of capillaries participating in the diffusion and

also from dilatation of both the alveoli and the capillaries.

Second, even if the diffusing capacity should not increase, quantities of oxygen far greater than normal can still enter the blood because of the special nature of diffusion in the pulmonary capillaries. This can be explained by referring once again to Figure 41–1. Note that during normal respiration the blood becomes completely saturated with oxygen by the time it has passed through one-third of the pulmonary capillary, and little additional oxygen enters the capillary blood during the latter two-thirds of its transit. That is, the blood normally stays in the lungs far longer than necessary to cause adequate oxygenation anyway. Therefore, even with the shortened time of exposure in exercise, the blood still can become fully oxygenated.

Effects of Pulmonary Shunts on the P_{O_2} of Arterial Blood.   A small amount of blood, usually 1 to 2 per cent of the total cardiac output, fails to pass through the pulmonary capillaries but instead is shunted through nonaerated vessels either in the lungs themselves or in the heart. This blood mixes with the aerated blood in the left heart and slightly reduces the P_{O_2} of the blood before it enters the arterial tree. This is called *venous admixture* of blood, and its effect is illustrated in Figure 41–2. Note that the blood leaving the pulmonary capillaries has a P_{O_2} of approximately 104 mm. Hg, but the venous admixture reduces this to approximately 95 mm. Hg in the systemic arterial blood. It seems strange that only 1 to 2 per cent venous admixture could cause this much decrease in P_{O_2}, but later in the chapter we will see that pressures above 90 mm. Hg are also mainly above the level at which oxygen combines with hemoglobin. In this range even slight changes in oxygenation can change the P_{O_2} greatly even though it does not change the *total quantity* of oxygen in the blood a significant amount.

In some diseases of the pulmonary circulation the shunted blood amounts to more than 50 per cent of the total cardiac output. In these instances

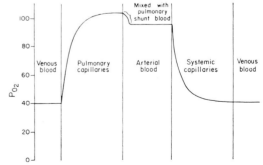

Figure 41–2   Changes in P_{O_2} in the pulmonary capillary blood, the arterial blood, and the systemic capillary blood, illustrating the effect of "venous admixture."

the arterial P_{O_2} can be decreased to as low as 40 to 50 mm. Hg.

DIFFUSION OF OXYGEN FROM THE CAPILLARIES TO THE CELLS

In the tissue capillaries, oxygen diffuses into the tissues by a process essentially the reverse of that which takes place in the lungs. That is, the P_{O_2} in the interstitial fluid immediately outside a capillary is low and very variable but averaging about 40 mm. Hg, while that in the arterial blood is high, about 95 mm. Hg. Therefore, at the arterial end of the capillary, a pressure difference of 55 mm. Hg exists for diffusion of oxygen. But, as illustrated in Figure 41–3, before the blood has gone more than a short distance into this capillary a large portion of the oxygen has diffused into the tissues, and the capillary P_{O_2} has approached the 40 mm. Hg oxygen pressure in the tissue fluids. Consequently, the venous blood leaving the tissue capillaries contains oxygen at essentially the same pressure as that immediately outside the tissue capillaries, 40 mm. Hg.

Effect of Rate of Blood Flow on Interstitial Fluid P_{O_2}. If the blood flow through a particular tissue becomes greatly increased, far greater quantities of oxygen are transported into the area in a given period of time, and the tissue P_{O_2} becomes correspondingly increased. However, the upper limit to which it can rise, even with maximal blood flow, is 95 mm. Hg, which is the oxygen pressure in the arterial blood. This is illustrated in Figure 41–4, which gives the approximate tissue P_{O_2} at different rates of blood flow.

Effect of Rate of Tissue Metabolism on Interstitial Fluid P_{O_2}. If the cells utilize far more oxygen for metabolism than normally, this tends to reduce the interstitial fluid P_{O_2}; or, conversely, the P_{O_2} rises and approaches the P_{O_2} of arterial blood when tissue metabolism ceases. Figure 41–4 also illustrates these effects, showing that increasing tissue oxygen consumption greatly reduces the interstitial fluid P_{O_2}, whereas reducing the consumption greatly increases the P_{O_2}.

Effect of Hemoglobin Concentration on Interstitial Fluid P_{O_2}. Because approximately 97 per cent of the oxygen transported in the blood is carried by hemoglobin, a decrease in hemoglobin concentration has the same effect on interstitial fluid P_{O_2} as does a decrease in the blood flow. Thus, reducing

Figure 41–4 Effect of blood flow and rate of oxygen consumption on tissue P_{O_2}.

the hemoglobin concentration to one quarter normal while maintaining normal blood flow reduces the interstitial fluid P_{O_2} to point C in Figure 41–4 —that is, to about 13 mm. Hg.

In summary, tissue P_{O_2} is determined by a balance between (a) the rate of oxygen transport to the tissues in the blood and (b) the rate at which the oxygen is utilized by the tissues.

DIFFUSION OF OXYGEN FROM THE INTERSTITIAL FLUIDS INTO THE CELLS

Since oxygen is always being used by the cells, the intracellular P_{O_2} remains lower than the interstitial fluid P_{O_2}; and when the interstitial fluid P_{O_2} falls, the intracellular P_{O_2} also falls, or, conversely, an increase in interstitial P_{O_2} causes a similar increase in intracellular P_{O_2}.

As pointed out in Chapter 4, oxygen diffuses through cell membranes extremely rapidly. Therefore, the intracellular P_{O_2} is almost equal to that of the interstitial fluids. Yet, in many instances, there is considerable distance between the capillaries and the cells. Therefore, the intracellular P_{O_2} ranges from as low as zero to as high as 40 mm. Hg, averaging (by direct measurement in the dog) 6 mm. Hg, which is the value given for the cell in Figure 41–3. Since only 1 to 5 mm. Hg oxygen pressure is required for full support of the metabolic processes of the cell, one can see that even this low cellular P_{O_2} is adequate.

DIFFUSION OF CARBON DIOXIDE FROM THE CELLS TO THE TISSUE CAPILLARIES

Because of the continual large quantities of carbon dioxide formed in the cells, the intracellular P_{CO_2} tends to rise. However, carbon dioxide diffuses about 20 times as easily as oxygen, diffusing from the cells extremely rapidly into the interstitial fluids and thence

Figure 41–3 Diffusion of oxygen from a tissue capillary to the cells.

Figure 41-5 Uptake of carbon dioxide by the blood in the capillaries.

into the capillary blood. Thus, in Figure 41-5 the intracellular P_{CO_2} is shown to be 46 mm. Hg, while that in the interstitial fluids immediately adjacent to the capillaries is about 45 mm. Hg, a pressure differential of only 1 mm. Hg.

Arterial blood entering the tissue capillaries contains carbon dioxide at a pressure of approximately 40 mm. Hg. As the blood passes through the capillaries, the blood P_{CO_2} rises to approach the 45 mm. Hg P_{CO_2} of the interstitial fluid. Because of the very large diffusion coefficient for carbon dioxide, the P_{CO_2} of venous blood is also about 45 mm. Hg, within a fraction of a millimeter of reaching complete equilibrium with the P_{CO_2} of the interstitial fluid.

Effect of Tissue Metabolism and Blood Flow on the Interstitial Fluid P_{CO_2}. Blood flow and tissue metabolism affect tissue P_{CO_2} in exactly the same way that they affect tissue P_{O_2}. Figure 41-6 shows the effects of normal metabolism, one-quarter normal metabolism, and 10 times normal metabolism on interstitial fluid P_{CO_2} at different rates of blood flow. Note that the lower limit to which the interstitial fluid P_{CO_2} can possibly fall is the P_{CO_2} of the arterial blood entering the tissue capillaries—normally about 40 mm. Hg. The elevation of tissue

Figure 41-6 Effect of blood flow and metabolic rate on tissue P_{CO_2}.

P_{CO_2} above this value depends on both blood flow and metabolic rate. Point A illustrates the P_{CO_2} when both the blood flow and metabolism are normal. Point B shows that a decrease in blood flow to one-quarter normal increases the tissue P_{CO_2} to 60 mm., while point C shows that an increase in blood flow to six times normal reduces the tissue P_{CO_2} almost to its lower limit of 40 mm. Hg. It can also be seen that increasing the rate of metabolism, especially at low blood flow rates, can increase the tissue P_{CO_2} to extremely high levels, while reducing the rate of metabolism correspondingly reduces the tissue P_{CO_2}.

REMOVAL OF CARBON DIOXIDE FROM THE PULMONARY BLOOD

On arriving at the lungs, the P_{CO_2} of the venous blood is about 45 mm. Hg while that in the alveoli is 40 mm. Hg. Therefore, as illustrated in Figure 41-7, the initial pressure difference for diffusion is only 5 mm. Hg, which is far less than that for diffusion of oxygen across the membrane. Yet, even so, because of the 20 times greater diffusion coefficient for carbon dioxide than for oxygen, the excess carbon dioxide in the blood is rapidly transferred into the alveoli. Indeed, the figure shows that the P_{CO_2} of the pulmonary capillary blood becomes almost equal to that of the alveoli within the first four-tenths of the blood's transit through the pulmonary capillary.

TRANSPORT OF OXYGEN IN THE BLOOD

Normally, about 97 per cent of the oxygen transported from the lungs to the tissues is carried in chemical combination with hemoglobin in the red blood cells, and the remaining 3 per cent is carried in the dissolved state in the water of the plasma and cells. However, when a person breathes oxygen at very high pressures, as much oxygen can sometimes be transported in the dissolved state as in chemical combination with hemoglobin. Therefore, the present discussion takes up the transport of oxygen first in combination with hemoglobin and then in the dissolved state.

THE REVERSIBLE COMBINATION OF OXYGEN WITH HEMOGLOBIN

The chemistry of hemoglobin was presented in Chapter 8, where it was pointed out that the

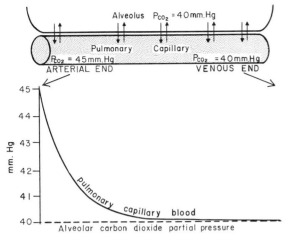

Figure 41-7 Diffusion of carbon dioxide from the pulmonary blood into the alveolus. (This curve was constructed from data in Milhorn and Pulley: *Biophys. J.,* 8:337, 1968.)

oxygen molecule combines loosely and reversibly with the heme portion of the hemoglobin. When the P_{O_2} is high, as in the pulmonary capillaries, oxygen binds with the hemoglobin, but, when the P_{O_2} is low, as in the tissue capillaries, oxygen is released from the hemoglobin. This is the basis for oxygen transport from the lungs to the tissues.

The Oxygen-Hemoglobin Dissociation Curve. Figure 41-8 illustrates the oxygen-hemoglobin dissociation curve, which shows the progressive increase in the per cent of the hemoglobin that is bound with oxygen as the P_{O_2} increases. This is called the *per cent saturation of the hemoglobin.* Since the blood leaving the lungs usually has a P_{O_2} of about 100 mm. Hg, one can see from the dissociation curve that the *usual oxygen saturation of arterial blood is about 97 per cent.* On the other hand, the P_{O_2} is about 40 mm. Hg and *the per cent saturation of the hemoglobin is about 70 per cent in normal venous blood.*

MAXIMUM AMOUNT OF OXYGEN THAT CAN COMBINE WITH THE HEMOGLOBIN OF THE BLOOD. The blood of a normal person contains approximately 15 grams of hemoglobin in each 100 ml. of blood, and each gram of hemoglobin can bind with a maximum of about 1.34 ml. of oxygen. Therefore, on the average, the hemoglobin in 100 ml. of blood can combine with a total of about 20 ml. of oxygen. This is usually expressed as 20 *volumes per cent.* The oxygen-hemoglobin dissociation curve for the normal person, therefore, can be expressed in terms of volume per cent of oxygen, as shown in Figure 41-9, rather than per cent saturation of hemoglobin.

Amount of Oxygen Released from the Hemoglobin in the Tissues. Note in Figure 41-9 that the total quantity of oxygen bound with hemoglobin in the normal arterial blood (P_{O_2} of 95 mm. Hg, 97 per cent saturated) is approximately 19.4 ml. On passing through the tissue capillaries, this amount is reduced to 14.4 ml. (P_{O_2} of 40 mm. Hg, 75 per cent saturated), or a total loss of about 5 ml. of oxygen from each 100 ml. of blood. Then, when the blood returns to the lungs, the same quantity of oxygen diffuses from the alveoli to the hemoglobin, and this too is carried to the tissues. Thus, *under normal conditions about 5 ml. of oxygen is transported by each 100 ml. of blood during each cycle through the tissues.*

Transport of Oxygen during Strenuous Exercise. In heavy exercise the muscle cells utilize oxygen at a rapid rate, which causes the interstitial fluid P_{O_2} to fall as low as 15 mm. Hg. At this pressure only 4.4 ml. of oxygen remains bound with the hemoglobin in each 100 ml. of blood, as shown in Figure 41-9. Thus, 19.4 − 4.4, or 15 ml., is the total quantity of oxygen transported by each 100 ml. of blood in each cycle through the tissues. This, obviously, is

Figure 41-8 The oxygen-hemoglobin dissociation curve.

Figure 41-9 Quantity of oxygen in each 100 ml. of normal blood (a) bound with hemoglobin and (b) dissolved in the water of the blood; (c) total oxygen in both forms.

three times as much as that normally transported by the same amount of blood, illustrating that simply an increase in rate of oxygen utilization by the tissues causes an automatic increase in the rate of oxygen release from the hemoglobin.

The Utilization Coefficient. The fraction of the hemoglobin that gives up its oxygen as it passes through the tissue capillaries is called the *utilization coefficient.* Normally, this is approximately 25 per cent of the hemoglobin. That is, *the normal utilization coefficient is approximately one-fourth.* During strenuous exercise, as high as 75 per cent of the hemoglobin can give up its oxygen; the utilization coefficient is then approximately three-quarters. This value, three-quarters, is about the highest utilization coefficient that can be attained in the overall body even when the tissues are in extreme need of oxygen. However, in local tissue areas where the blood flow is very slow or the metabolic rate very high, utilization coefficients approaching 100 per cent have been recorded —that is, essentially all the oxygen is removed.

TOTAL RATE OF OXYGEN TRANSPORT FROM THE LUNGS TO THE TISSUES

If under resting conditions about 5 ml. of oxygen is transported by each 100 ml. of blood, and if the normal cardiac output is approximately 5000 ml. per minute, the calculated total quantity of oxygen delivered to the tissues each minute is about 250 ml. This is also the amount measured by a respirometer.

This rate of oxygen transport to the tissues can be increased to about 15 times normal in heavy exercise and in other instances of excessive need for oxygen. It can be increased to three times normal simply by an increase in utilization coefficient, and it can be increased another five-fold as a result of increased cardiac output, thus accounting for the total 15-fold increase. Therefore, the maximum rate of oxygen transport to the tissues is about 15 × 250 ml., or 3750 ml., per minute in the normal young adult. Special adaptations in athletic training, such as an increase in blood hemoglobin concentration and an increase in maximum cardiac output, can sometimes increase this value to as high as 4.5 to 5 liters per minute.

Effect of Hematocrit on Oxygen Transport to the Tissues. An increase in the blood hematocrit much above the normal level of 40 reduces the cardiac output because of increased blood viscosity, and the reduction in cardiac output is more than the increase in oxygen-carrying

Figure 41–10 Depressant effect on oxygen transport of either an elevated or a decreased hematocrit. (Courtesy of Drs. J. W. Crowell and E. E. Smith.)

capacity of the blood. Since the total amount of oxygen that can be carried to the tissues each minute is the product of these two, the rate of oxygen transport is actually reduced by an excessive rise in hematocrit.

On the other hand, in anemia the oxygen-carrying capacity of the blood is reduced in proportion to the decrease in hematocrit; there is some compensatory increase in cardiac output, but this is not so great as the decrease in oxygen-carrying capacity. Here again, there is an overall reduction in the rate of oxygen transport to the tissues. The solid curve in Figure 41–10 illustrates this effect, showing that *maximum oxygen transport occurs at a hematocrit of about 40* and that *oxygen transport falls off at either high or low hematocrits.* Therefore, it is fortunate that the normal hematocrit of 40 is approximately optimal for oxygen transport.

The "optimum hematocrit" for oxygen transport rises when an animal develops a high hematocrit during acclimatization to high altitudes, as shown by the dashed curve in Figure 41–10. In this figure, the optimum hematocrit has risen from 40 to 46. The probable cause of this increase in optimal hematocrit is increase in the numbers and sizes of the peripheral small vessels, which prevents a decrease in cardiac output when the hematocrit rises.

THE OXYGEN BUFFER FUNCTION OF HEMOGLOBIN

Though hemoglobin is necessary for transport of oxygen to the tissues, it performs still another major function essential to life. This is the function of hemoglobin as an "oxygen buffer" system. That is, the hemoglobin in the blood is mainly responsible for controlling the oxygen pressure in the tissues. Were it not for the hemoglobin oxygen buffer system, extreme variations in the P_{O_2} of the tissues would occur with exercise and with every change in metabolism, blood flow, and atmospheric oxygen concentration.

Value of Hemoglobin for Maintaining Constant P_{O_2} in the Tissue Fluids. Even under basal conditions the tissues require about 5 ml. of oxygen from every 100 ml. of blood passing through the tissue capillaries. Referring back to the oxygen-hemoglobin dissociation curve in Figure 41–9, one can see that for 5 ml. of oxygen to be released from each 100 ml. of blood, the P_{O_2} must fall to about 40 mm. Hg. Therefore, tissue capillary P_{O_2} ordinarily does not rise above approximately 40 mm. Hg, for if such should occur, the oxygen needed by the tissues could not be released from the hemoglobin. In this way, the hemoglobin normally sets an upper limit on the gaseous pressure in the tissues at approximately 40 mm. Hg.

On the other hand, in heavy exercise the P_{O_2} in the tissue capillaries falls, causing the hemoglobin to release extra quantities of oxygen to the tissues. This in turn helps to prevent the capillary P_{O_2} from falling below about 15 mm. Hg.

It can be seen, then, that hemoglobin automatically delivers oxygen to the tissues at a pressure between approximately 15 and 40 mm. Hg. This seems to be a wide range of P_{O_2} in the interstitial fluid, but, when one considers how much the interstitial fluid P_{O_2} might possibly change during exercise and other types of stress, this range of 15 to 40 mm. Hg is relatively narrow.

Value of Hemoglobin for Automatically Adjusting the Tissue P_{O_2} to Different Concentrations of Atmospheric Oxygen. The normal P_{O_2} in the alveoli is approximately 104 mm. Hg, but, as one ascends a mountain or goes high in an airplane, the P_{O_2} falls considerably, or, when one enters areas of compressed air, such as deep below the sea or in tunnels, the P_{O_2} may rise to very high values. It will be seen from the oxygen-hemoglobin dissociation curve of Figure 41–8 that, when the P_{O_2} is decreased to as low as 60 mm. Hg, the hemoglobin is still 89 per cent saturated, only 8 per cent below the normal saturation of 97 per cent. Furthermore, the tissues still remove approximately 5 ml. of oxygen from every 100 ml. of blood passing through the tissues; to remove this oxygen, the P_{O_2} of the venous blood falls to only slightly less than 40 mm. Hg. Thus, the tissue P_{O_2} hardly changes despite the marked fall in alveolar P_{O_2} from 104 to 60 mm. Hg.

On the other hand, when the alveolar P_{O_2} rises far above the normal value of 104 mm. Hg, the maximum oxygen saturation of hemoglobin can never rise above 100 per cent. Therefore, even though the oxygen in the alveoli should rise to a partial pressure of 500 mm. Hg, or even more, the increase in the saturation of hemoglobin would be only 3 per cent because, even at 104 mm. Hg P_{O_2}, 97 per cent of the hemoglobin is already combined with oxygen; and only a small amount of oxygen dissolves in the fluid of the blood, as will be discussed subsequently. When the hemoglobin that has been subjected to high P_{O_2}'s in the alveoli passes to the tissue capillaries, it still loses several milliliters of oxygen to the tissues, and this loss automatically reduces the P_{O_2} of the blood to a value only a few millimeters greater than the normal 40 mm. Hg.

Consequently, the atmospheric oxygen content may vary greatly—from 60 to more than 500 mm. Hg P_{O_2}—and still the P_{O_2} of the tissue capillaries does not vary more than a few millimeters from normal.

METABOLIC USE OF OXYGEN BY THE CELLS

Relationship Between Intracellular P_{O_2} and Rate of Oxygen Usage. Only minute oxygen pressure is required in the cells for normal intracellular chemical reactions to take place. The reason for this is that the respiratory enzyme systems of the cell, which will be discussed in Chapter 67, are geared so that when the cellular P_{O_2} is more than 5 mm. Hg, oxygen availability is no longer a limiting factor in the rates of the chemical reactions. Instead, the main limiting factor then is the *concentration of adenosine diphosphate* (ADP) in the cells, as was explained in Chapter 3. This effect is illustrated in Figure 41–11, which shows the relationship between intracellular P_{O_2} and rate of oxygen usage. Note that whenever the intracellular P_{O_2} is above 5 mm. Hg the rate of oxygen usage becomes constant for any given concentration of ADP in the cell. On the other hand, when the ADP concentration is altered, the rate of oxygen usage changes in proportion to the change in ADP concentration.

It will be recalled from the discussion in Chapter 3 that when adenosine triphosphate (ATP) is utilized in the cells to provide energy, it is converted into ADP. The increasing concentration of ADP then increases the metabolic usage of both oxygen and the various nutrients that combine with oxygen to release energy. Therefore, under normal operating conditions

Figure 41–11 Effect of intracellular Po_2 on rate of oxygen usage by the cells. Note that increasing the intracellular concentration of adenosine diphosphate (*ADP*) increases the rate of oxygen usage.

Figure 41–12 Effect on oxygen consumption of altering the cardiac output. (Extrapolated to the human being from data obtained in dogs by Wigert; unpublished observations.)

the rate of oxygen utilization by the cells is controlled by the rate of energy expenditure within the cells and not by the rate of delivery of oxygen to the cells.

Effect on Metabolic Use of Oxygen of the Diffusion Distance from the Capillary to the Cell. Cells are rarely more than 50 microns away from the capillary, in which case oxygen can diffuse readily enough from the capillary to the cell to supply all the required amounts of oxygen for metabolism. However, occasionally, cells are located farther than this from the capillaries, and the rate of oxygen diffusion to these cells is so low that the intracellular Po_2 falls below the critical level of 1 to 5 mm. Hg required to maintain maximum intracellular metabolism. Thus, under these conditions, oxygen utilization by the cells is *diffusion limited* and is no longer determined by the amount of ADP formed in the cells.

Effect of Blood Flow on Metabolic Use of Oxygen. Earlier in the chapter it was pointed out that the total amount of oxygen available each minute for use in any given tissue is determined by (1) the quantity of oxygen transported in each 100 ml. of blood and (2) the rate of blood flow. If the rate of blood flow falls to zero, the amount of oxygen available each minute for metabolism obviously also falls to zero. Thus, there are times when the rate of blood flow through a tissue can be so low that the tissue Po_2 falls below the critical 1 to 5 mm. Hg required for maximal intracellular metabolism. Under these conditions, the rate of tissue utilization of oxygen is *blood flow limited.*

Effect of Cardiac Output on Metabolic Use of Oxygen. Figure 41–12 illustrates the effect of changes in cardiac output on the total body oxygen consumption. If we observe the "normal value" for oxygen utilization we see that increasing the cardiac output by as much as 60 per cent increases the rate of oxygen consumption by less than 10 per cent. In other words, in the normal human being there is almost no "blood flow limitation" in the consumption of oxygen. That is, blood flow to essentially all the tissues of the body is more than enough to supply the required oxygen.

However, in studying Figure 41–12 further, one sees that *decreasing* the cardiac output below normal causes a progressive decrease in oxygen consumption. Once the cardiac output has fallen below about one-half normal, oxygen consumption thereafter decreases almost in direct proportion to the decrease in output. Thus, whenever the cardiac output falls only slightly below normal, oxygen consumption by the body begins to be blood flow limited.

It is especially interesting that the cardiac output is normally controlled at a level just barely high enough for there to be almost no flow limitation of oxygen consumption. A higher cardiac output than this would place an unneeded load on the heart, and a lower cardiac output would limit the tissue use of oxygen. The question that must be asked is: How is the cardiac output regulated at such a critical point? The answer to this was discussed in Chapter 24, where it was pointed out that the blood flow through each local tissue is usually regulated approximately in proportion to that tissue's rate of oxygen utilization.

TRANSPORT OF OXYGEN IN THE DISSOLVED STATE

At the normal arterial Po_2 of 95 mm. Hg, approximately 0.29 ml. of oxygen is dissolved in every 100 ml. of water in the blood. When the Po_2 of the blood falls to 40 mm. Hg in the tissue capillaries, 0.12 ml. of oxygen remains dissolved. In other words, 0.17 ml. of oxygen is transported to the tissues by each 100 ml. of blood water. This compares with about 5.0 ml. transported by the hemoglobin. Therefore, the amount of oxygen transported to the tissues in the dissolved state is normally slight, only about 3 per cent of the total as compared with the 97 per cent transported by the hemoglobin. During heavy exercise, when the utilization coefficient rises, the quantity transported in the dissolved state falls to as little as 1.5 per cent. In Figure 41–9 the quantitative relationships of the amount of oxygen that can be transported in the dissolved state versus that bound with hemoglobin can be seen. This figure shows that the total oxygen in the blood under normal conditions is accounted for almost entirely by that bound with hemoglobin, while only a minute portion is dissolved in the fluids. Yet under abnormal conditions the amount transported in the dissolved state can become tremendous as follows:

Effect of Extremely High Po₂'s on Oxygen Transport. When the Po_2 in the blood rises much above 100 mm. Hg, the amount of oxygen dissolved in the water of the blood continues to increase. This effect is illustrated in Figure 41–13, which depicts the same data as Figure 41–9 except that the Po_2's in the lungs are extended to over 3000 mm. Hg in this figure. Note that in the normal range of alveolar Po_2's almost none of the total oxygen in the blood is accounted for by dissolved oxygen, but as the pressure rises progressively into the thousands of mm. Hg, a significant portion of the total oxygen is dissolved rather than bound with hemoglobin.

Effect of High Pulmonary Po₂ on Tissue Po₂. Let us assume that the Po_2 in the lungs is about 3000 mm. Hg (4 atmospheres pressure). Referring to Figure 41–13, one finds that this represents a total oxygen content in each 100 ml. of blood of about 29 volumes per cent, as illustrated by point A in

the figure. As this blood passes through the tissue capillaries and the tissues utilize their normal amount of oxygen, about 5 ml. from each 100 ml. of blood, the oxygen content on leaving the tissue capillaries is 24 volumes per cent (point B in the figure). At this point, the Po_2 in the blood leaving the tissue capillaries is still about 1200 mm. Hg, which means that oxygen is delivered to the tissues at this extremely high pressure instead of at the normal value of 40 mm. Hg. Thus, very high Po_2's breathed by the lungs can also cause very high tissue Po_2's, but only if these pressures are high enough for transport of excessive quantities of oxygen in the dissolved state.

Oxygen Poisoning. When the Po_2 in the tissues rises too high, it has a tendency to change the rates of many of the chemical reactions within the cells, sometimes to such an extent that it actually "poisons" the tissues. This will be discussed in Chapter 45 in relation to the effect of high pressures on the body.

Experimental Treatment of Anemia with Very High Concentrations of Oxygen in the Lungs. An obvious implication of this discussion is that a person with severe anemia whose blood could not carry sufficient quantities of oxygen to the tissues in combination with hemoglobin would be aided greatly by breathing pure oxygen or by being placed in a pressure chamber containing a high Po_2, for then a large quantity of oxygen could be carried to the tissues in the dissolved state rather than combined with hemoglobin. Indeed, in human beings treated in such pressure chambers, it has been possible to transport enough dissolved oxygen to supply more than two-thirds the quantity of oxygen needed by the tissues. Yet, whenever more oxygen than that used by the tissues is supplied in the dissolved state, no oxygen is removed from the hemoglobin, and the oxygen buffer function of the hemoglobin is lost. In order for hemoglobin to control the level of oxygen in the tissues, at least several per cent of the hemoglobin must lose its oxygen. Then the oxygen-hemoglobin dissociation curve operates along the steep portion of the curve, thus buffering the Po_2 of the tissue fluids as explained above. Therefore, administration of oxygen under extreme pressure may lead to death, not because of failure to transport enough oxygen but because of failure of the oxygen buffer function of hemoglobin.

COMBINATION OF HEMOGLOBIN WITH CARBON MONOXIDE

Carbon monoxide combines with hemoglobin at the same point on the hemoglobin molecule as does oxygen. Furthermore, it binds with approximately 210 times as much tenacity as oxygen, which is illustrated by the carbon monoxide-hemoglobin dissociation curve in Figure 41–14. This curve is almost identical with the oxygen-hemoglobin dissociation curve, except that the pressures of the carbon monoxide shown on the abscissa are at a level 210 times less than those in the oxygen-

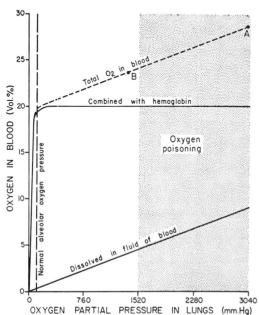

Figure 41–13 Quantity of oxygen dissolved in the water of the blood and in combination with hemoglobin at very high Po₂'s.

Figure 41–14 The carbon monoxide-hemoglobin dissociation curve.

hemoglobin dissociation curve of Figure 41–8. Therefore, a carbon monoxide pressure of only 0.5 mm. Hg in the alveoli, 210 times less than that of the alveolar oxygen, allows the carbon monoxide to compete equally with the oxygen for combination with the hemoglobin and causes half the hemoglobin in the blood to become bound with carbon monoxide instead of with oxygen. A carbon monoxide pressure of 0.7 mm. Hg (a concentration of about 0.1 per cent) is lethal.

A patient severely poisoned with carbon monoxide can be advantageously treated by administering pure oxygen, for oxygen at high alveolar pressures removes carbon monoxide from its combination with hemoglobin far more rapidly than can be achieved at the low pressure of atmospheric oxygen. That is, the high pressure of the therapeutic oxygen allows the oxygen actually to force its way into the hemoglobin molecule and thereby to displace the carbon monoxide.

The patient can also be benefited by simultaneous administration of a few per cent carbon dioxide because this strongly stimulates the respiratory center, as is discussed in the following chapter. The resulting increase in alveolar ventilation reduces the alveolar carbon monoxide concentration and thereby allows increased carbon monoxide release from the blood.

With intensive oxygen and carbon dioxide therapy, carbon monoxide can be removed from the blood 10 to 20 times as rapidly as without therapy.

TRANSPORT OF CARBON DIOXIDE IN THE BLOOD

Transport of carbon dioxide is not nearly so great a problem as transport of oxygen, because even in the most abnormal conditions carbon dioxide can usually be transported by the blood in far greater quantities than can oxygen. However, the amount of carbon dioxide in the blood does have much to do with acid-base balance of the body fluids, which was discussed in detail in Chapter 37. Under nor-

mal resting conditions *an average of 4 ml. of carbon dioxide is transported from the tissues to the lungs in each 100 ml. of blood.*

CHEMICAL FORMS IN WHICH CARBON DIOXIDE IS TRANSPORTED

To begin the process of carbon dioxide transport, carbon dioxide diffuses out of the tissue cells in the gaseous form (but not to a significant extent in the bicarbonate form because the cell membrane is far less permeable to bicarbonate than to the dissolved gas). On entering the capillary, the chemical reactions illustrated in Figure 41–15 occur immediately; the quantitative aspects of these can be described as follows:

Transport of Carbon Dioxide in the Dissolved State. A small portion of the carbon dioxide is transported in the dissolved state to the lungs. It will be recalled that the P_{CO_2} of venous blood is 45 mm. Hg and that of arterial blood is 40 mm. Hg. The amount of carbon dioxide dissolved in the fluid of the blood at 45 mm. Hg is about 2.7 ml. per 100 ml. (2.7 volumes per cent). The amount dissolved at 40 mm. Hg is about 2.4 ml., or a difference of 0.3 ml. Therefore, only about 0.3 ml. of carbon dioxide is transported in the form of dissolved carbon dioxide by each 100 ml. of blood. This is about 7 per cent of all the carbon dioxide transported.

Transport of Carbon Dioxide in the Form of Bicarbonate Ion. REACTION OF CARBON DIOXIDE WITH WATER IN THE RED BLOOD CELLS— EFFECT OF CARBONIC ANHYDRASE. The dissolved carbon dioxide in the blood reacts with water to form carbonic acid. However, this reaction would occur too slowly to be of importance were it not for the fact that inside the red blood cells the enzyme called *carbonic anhydrase* catalyzes the reaction between carbon dioxide and water, accelerating its rate many thousand-fold. Therefore, instead of requiring many seconds to occur, as is true in the plasma, the reaction occurs so rapidly in the red blood cells that it reaches almost complete equilibrium within a fraction of a second. This allows tremendous amounts of carbon dioxide to react with the red cell water even before the blood leaves the tissue capillaries.

DISSOCIATION OF CARBONIC ACID INTO BICARBONATE IONS. In another small fraction of a second the carbonic acid formed in the red cells dissociates into hydrogen and bicarbonate ions. This dissociation is about 99.9 per cent complete, so that only an infinitesimal fraction of the carbonic acid remains in the undissociated form.

As illustrated in Figure 41–15, most of the hydrogen ions formed inside the red blood

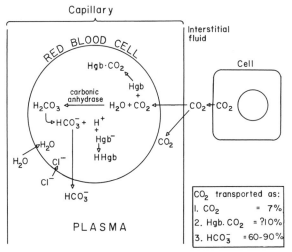

Figure 41-15   Transport of carbon dioxide in the blood.

cells react rapidly with hemoglobin, which is a powerful acid-base buffer. In reacting with hemoglobin the hydrogen ions are removed from the fluid and leave, dissolved in the fluid, a large quantity of *bicarbonate ions*. Many of these ions then diffuse into the plasma while chloride ions diffuse into the red cells. Thus, the chloride content of venous red blood cells is greater than that of arterial cells, a phenomenon called the *chloride shift*.

The reversible combination of carbon dioxide with water in the red blood cells under the influence of carbonic anhydrase accounts for 60 to 90 per cent of all the carbon dioxide transported from the tissues to the lungs. Thus, this means of transporting carbon dioxide is by far the most important of all the methods for transport. Indeed, when a carbonic anhydrase inhibitor (acetazolamide) is administered to an animal to block the action of carbonic anhydrase in the red blood cells, carbon dioxide transport from the tissues becomes very poor—so poor that the tissue Pco_2 rises to as high as 80 mm. Hg instead of the normal 45 mm. Hg.

Transport of Carbon Dioxide in Combination with Hemoglobin and Plasma Proteins—Carbaminohemoglobin. In addition to reacting with water, carbon dioxide also reacts directly with hemoglobin. The combination of carbon dioxide with hemoglobin is a reversible reaction that occurs with a very loose bond. The compound formed by this reaction is known as "carbaminohemoglobin." A small amount of carbon dioxide reacts in this same way with the plasma proteins, but this is much less significant because of the one-fourth as great quantity of these proteins in the blood.

The *maximum theoretical* quantity of carbon dioxide that can be carried from the tissues to the lungs in combination with hemoglobin and plasma proteins is approximately 30 per cent of the total

quantity transported—that is, about 1.5 ml. of carbon dioxide in each 100 ml. of blood. However, this reaction is much slower than the reaction of carbon dioxide with water inside the red blood cells. Therefore, it is doubtful that this mechanism provides transport of more than a few per cent of the total quantity of carbon dioxide.

Carbon dioxide does not combine with hemoglobin at the same point on the hemoglobin molecule as oxygen; therefore, hemoglobin can combine with both oxygen and carbon dioxide at the same time.

THE CARBON DIOXIDE DISSOCIATION CURVE

It is now apparent that carbon dioxide can exist in blood in many different forms, (1) as free carbon dioxide and (2) in chemical combination with water, hemoglobin, and plasma protein. The total quantity of carbon dioxide combined with the blood in all these forms depends on the Pco_2. The curve shown in Figure 41–16 depicts this dependence of total blood CO_2 in all its forms on Pco_2; this curve is called the *carbon dioxide dissociation curve.*

Note that the normal resting Pco_2 ranges between the limits of 40 and 45 mm. Hg, which is a very narrow range. Note also that the normal concentration of carbon dioxide in the blood is about 50 volumes per cent but that only 4 volumes per cent of this is actually exchanged in the process of transporting carbon dioxide from the tissues to the lungs. That is, the concentration rises to about 52 volumes per cent as the blood passes through the tissues, and falls to about 48 volumes per cent as it passes through the lungs.

MUTUAL INTERFERENCE BETWEEN OXYGEN AND CARBON DIOXIDE TRANSPORT IN THE BLOOD

Hemoglobin can combine with both oxygen and carbon dioxide at the same time and therefore can

Figure 41-16   The carbon dioxide dissociation curve.

transport both of these simultaneously. However, there is some degree of mutual interference, as follows:

Effect of Changes in P_{O_2} on Carbon Dioxide Content of the Blood — The "Haldane" Effect. An increase in P_{O_2} tends to displace carbon dioxide from blood, which is called the Haldane effect. There are two causes of this effect: First, the combination of oxygen with hemoglobin makes the hemoglobin a stronger acid, which decreases the affinity of the hemoglobin for combining with carbon dioxide to form carbaminohemoglobin. Second, the increased acidity of the hemoglobin increases the acidity of the fluids both in the plasma and in the red blood cells. This converts much of the bicarbonate ions into carbonic acid, which then dissociates and releases carbon dioxide from the blood.

Figure 41–17 illustrates the significance of the Haldane effect on the transport of carbon dioxide from the tissues to the lungs. This figure shows small portions of two separate carbon dioxide dissociation curves, the solid curve when the P_{O_2} is 100 mm. Hg, which is the case in the lungs, and the dashed curve when the P_{O_2} is 40 mm. Hg, which is the case in the tissues. Point A shows that the normal P_{CO_2} of 45 mm. Hg in the tissues causes 52 volumes per cent of carbon dioxide to combine with the blood. On entering the lungs the P_{CO_2} falls to 40 mm. Hg while the P_{O_2} rises to 100 mg. Hg. If the carbon dioxide dissociation curve did not change, the carbon dioxide content of the blood would fall only to 50 volumes per cent, which would be a loss of only 2 volumes per cent of carbon dioxide. However, the increase in P_{O_2} in the lungs decreases the carbon dioxide dissociation curve to the solid curve of the figure, so that the carbon dioxide content falls to 48 volumes per cent (point B). This represents an additional 2 volumes per cent loss of carbon dioxide. Thus, the Haldane effect approximately doubles the amount of carbon dioxide released from the blood in the lungs and approximately doubles the pickup of P_{CO_2} in the tissues.

Effect of P_{CO_2} on the Transport of Oxygen — The "Bohr" Effect. The Bohr effect is essentially opposite to the Haldane effect. That is, when the P_{CO_2} rises, the pH of the blood decreases and there is increased combination of carbon dioxide with the hemoglobin, both of which cause the hemoglobin to have considerably less affinity for oxygen than does carbon dioxide–free hemoglobin. Therefore, in effect, the carbon dioxide causes oxygen to dissociate from the hemoglobin.

However, from a quantitative point of view, the Bohr effect plays far less a role in affecting oxygen transport by the blood than the Haldane effect plays in affecting carbon dioxide transport. Therefore, the Bohr effect is more of academic interest than of practical significance.

CHANGE IN BLOOD ACIDITY DURING CARBON DIOXIDE TRANSPORT

The carbonic acid formed when carbon dioxide enters the blood in the tissues decreases the blood pH. Fortunately, though, the reaction of this acid with the buffers of the blood prevents the hydrogen ion concentration from rising greatly. Ordinarily, arterial blood has a pH of approximately 7.40, and, as the blood acquires carbon dioxide in the tissues, the pH falls to approximately 7.36. In other words, a pH change of 0.04 unit takes place. The reverse occurs when carbon dioxide is released from the blood in the lungs, the pH rising to the arterial value once again. In exercise, in other conditions of high metabolic activity, or when the blood flow through the tissues is sluggish, the decrease in pH in the tissue blood (and in the tissues themselves) can be as much as 0.5, or on occasion even more than this, thus causing tissue acidosis.

RELEASE OF CARBON DIOXIDE IN THE LUNGS

In the lungs the P_{CO_2} in the alveoli is slightly less than that of the blood, as explained earlier in the chapter, which causes carbon dioxide to diffuse from the blood into the alveoli. This decreases the P_{CO_2} of the red blood cells so that the carbonic acid of the cells, under the influence of carbonic anhydrase, changes back into water and carbon dioxide, and carbaminohemoglobin also releases its carbon dioxide. In other words, exactly the reverse reactions occur in a split second in the lung capillaries as in the systemic capillaries, because all the reactions for transport of carbon dioxide are reversible.

Figure 41–17 Portions of the carbon dioxide dissociation curves when the P_{O_2} is 100 mm. Hg and 40 mm. Hg, respectively. The arrow represents the Haldane effect on the transport of carbon dioxide, as explained in the text.

THE RESPIRATORY EXCHANGE RATIO

The discerning student will have noted that normal transport of oxygen from the lungs to the tissues by each 100 ml. of blood is about 5 ml., while normal transport of carbon dioxide from the tissues to the lungs is about 4 ml. Thus, under normal resting conditions only about 80 per cent as much carbon dioxide is expired from the lungs as there is oxygen uptake by the lungs. The ratio of carbon dioxide output to oxygen uptake is called the respiratory exchange ratio (R). That is,

$$R = \frac{\text{Rate of carbon dioxide output}}{\text{Rate of oxygen uptake}}$$

The value for R changes under different metabolic conditions. When a person is utilizing entirely carbohydrates for body metabolism, R rises to 1.00. On the other hand, when the person is utilizing fats almost entirely for his metabolic energy, the level falls to as low as 0.7. The reason for this difference is that when oxygen is metabolized with carbohydrates one molecule of carbon dioxide is formed for each molecule of oxygen consumed, while when oxygen reacts with fats a large share of the oxygen combines with hydrogen atoms to form water instead of carbon dioxide. Therefore, the quantity of carbon dioxide released from these reactions is far less than the quantity of oxygen consumed. In other words, the *respiratory quotient of the chemical reactions* in the tissues is now about 0.70 instead of 1.00 as is the case when carbohydrates are being utilized. The tissue respiratory quotient will be discussed in Chapter 72.

For a person on a normal diet consuming average amounts of carbohydrates, fats, and proteins, the average value for R is considered to be 0.825.

REFERENCES

Antonini, E.: Interrelationship between structure and function in hemoglobin and myoglobin. *Physiol. Rev., 45*:123, 1965.

Baumberger, J. P.; Leong, H. C.; and Neville, J. R.: Oxygen delivery rate of human blood. *J. Appl. Physiol., 23*:40, 1967.

Brown, E. B., Jr., and Michel, C. C.: Whole body CO_2 exchange. *Proc. Int. Union of Physiol. Sciences, 6*:185, 1968.

Chance, B.: Regulation of intracellular oxygen. *Proc. Int. Union of Physiol. Sciences, 6*:13, 1968.

Chinard, F. P.; Enns, T.; and Nolan, M. F.: Contributions of bicarbonate ion and of dissolved CO_2 to expired CO_2 in dogs. *Amer. J. Physiol., 198*:78, 1960.

Christoforides, C., and Hedley-Whyte, J.: Effect of temperature and hemoglobin concentration on solubility of O_2 in blood. *J. Appl. Physiol., 27*:592, 1969.

Crowell, J. W., and Smith, E. E.: Determinants of the optimal hematocrit. *J. Appl. Physiol., 22*:501, 1967.

Duvelleroy, M. A.; Buckles, R. G.; Rosenkaimer, S.; Tung, C.; and Laver, M. B.: An oxyhemoglobin dissociation analyzer. *J. Appl. Physiol., 28*:227, 1970.

Forster, R. E.: Rate of gas uptake of red cells. *In* Handbook of Physiology. Baltimore, The Williams & Wilkins Co., 1964, Sec. III, Vol. I, p. 827.

Gore, R. W., and Whalen, W. J.: Relations among tissue P_{O_2}, Q_{O_2}, and resting heat production of frog sartorius muscle. *Amer. J. Physiol., 214*:277, 1968.

Haldane, J. S., and Priestley, J. G.: Respiration. New Haven, Yale University Press, 1935.

Hemmingsen, E. A.: Accelerated transfer of oxygen through solutions of heme pigments. *Acta Physiol. Scand.,* Suppl. 246, 1965.

Jones, C. E.; Crowell, J. W.; and Smith, E. E.: Determination of mean tissue oxygen tensions by implanted perforated capsules. *J. Appl. Physiol., 26*:630, 1969.

Longmuir, I. S., and Chow, J.: Rapid method for determining effect of agents on oxyhemoglobin dissociation curves. *J. Appl. Physiol., 28*:343, 1970.

Maren, T. H.: Carbonic anhydrase: chemistry, physiology, and inhibition. *Physiol. Rev., 47*:595, 1967.

Riggs, A.: Functional properties of hemoglobins. *Physiol. Rev., 45*:619, 1965.

Root, W. S.: Carbon monoxide. *In* Handbook of Physiology. Baltimore, The Williams & Wilkins Co., 1965, Sec. III, Vol. II, p. 1087.

Roughton, F. J. W.: Transport of oxygen and carbon dioxide. *In* Handbook of Physiology. Baltimore, The Williams & Wilkins Co., 1964, Sec. III, Vol. I, p. 767.

Severinghaus, J. W.: Blood gas concentrations. *In* Handbook of Physiology. Baltimore, The Williams & Wilkins Co., 1965, Sec. III, Vol. II, p. 1475.

Sloviter, H. A.; Petkovic, M.; Ogoshi, S.; and Yamada, H.: Dispersed fluorochemicals as substitutes for erythrocytes in intact animals. *J. Appl. Physiol., 27*:666, 1969.

Van Liew, H. D.: Oxygen and carbon dioxide tensions in tissue and blood of normal and acidotic rats. *J. Appl. Physiol., 25*:575, 1968.

CHAPTER 42

Regulation of Respiration

During exercise and other physiological states that increase the metabolic activity of the body, the respiratory system is called upon to supply increased quantities of oxygen to the tissues and to remove increased quantities of carbon dioxide. The *respiratory center* in the brain stem adjusts the rate of alveolar ventilation almost exactly to the demands of the body so that, as a result, the blood oxygen pressure (Po_2) and the carbon dioxide pressure (Pco_2) are hardly altered even during strenuous exercise or other types of respiratory stress.

The present chapter describes the operation of this neurogenic system for regulation of respiration.

THE RESPIRATORY CENTER

The so-called "respiratory center" is a widely dispersed group of neurons located bilaterally in the reticular substance of the medulla oblongata and pons, as illustrated in Figure 42–1A. It is divided into three major areas: (1) the medullary rhythmicity area, (2) the apneustic area, and (3) the pneumotaxic area. The medullary rhythmicity area seems to be by far the most important. Therefore, it deserves particular attention.

The Medullary Rhythmicity Area. The medullary rhythmicity area, which is also frequently referred to simply as the *medullary respiratory center,* is located beneath the lower part of the floor of the fourth ventricle in the medial half of the medulla.

Microelectrodes inserted into this center detect some neurons throughout the center that discharge during inspiration and others that discharge during expiration. There have been some attempts to show that part of this center is primarily related to inspiration and part primarily related to expiration; however, most efforts in this direction have not been really successful. Thus, in general, *inspiratory neurons* and *expiratory neurons* intermingle in the medullary rhythmicity center. When respiration increases, both inspiratory and expiratory neurons become excited far above nor-mal, transmitting a greatly increased number of inspiratory signals to the respiratory muscles during inspiration and a greatly increased number of expiratory signals during expiration. Most of this chapter is concerned with the factors that increase and decrease the degree of activity of these neurons.

BASIC RHYTHMICITY IN THE MEDULLARY RHYTHMICITY AREA. It is in the medullary rhythmicity area that the basic rhythm of respiration is established. In the normal resting person, inspiration usually lasts for about two seconds and expiration for about three seconds. Both of these are correspondingly shortened during increased respiration and lengthened during decreased respiration.

When the medulla is transected immediately above the medullary rhythmicity area and also immediately below this area, one still finds a basic rhythmicity occurring in at least some of the inspiratory neurons. Therefore, it is certain that there is a basic oscillatory circuit located in this medullary rhythmicity area and that it is capable of causing repetitive inspiration and expiration. However, the medullary rhythmicity area is not by itself capable of giving a normal smooth pattern of respiration. If the medulla is transected immediately above the rhythmicity area but with this area still connected to the spinal cord, respiration occurs in gasps rather than in normal smooth inspirations and expirations. Furthermore, the rhythmical activity of the medullary rhythmicity center is very weak when afferent signals do not reach it from other sources. Thus, in Figure 42–1A, signals are shown entering the medullary rhythmicity area from the spinal cord, from the cerebral cortex and midbrain, from the pneumotaxic area in the upper pons, and from the apneustic area in the lower pons. All of these signals modify the rhythm of respiration and contribute to the normal smooth pattern of respiration.

Function of the Apneustic and Pneumotaxic Areas. The apneustic and pneumotaxic areas are located in the reticular substance of the

494

A.

B.

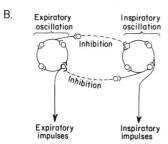

Figure 42-1 (A) The respiratory center located bilaterally in the lateral reticular substance of the medulla and lower pons. (B) Theoretical mechanism for the rhythmicity of the respiratory center.

pons. They are not necessary for maintenance of the basic rhythm of respiration. However, when the apneustic area is still connected to the medullary rhythmicity area but the pons has been transected between the apneustic and pneumotaxic areas, the animal breathes with the pattern of prolonged inspiration and very short expiration, which is exactly opposite to the pattern that occurs when breathing is accomplished entirely by the medullary rhythmicity area alone. This apneustic pattern becomes especially marked when the afferent nerve fibers of the vagus and of the glossopharyngeal nerves have been transected.

Finally, if the pneumotaxic area is also connected to the rhythmicity center, the pattern of respiration becomes essentially that normally observed, having a reasonable balance between inspiration and expiration. Also, stimulation of the pneumotaxic area can change the rate of respiration, which is the reason for its name, pneumotaxic area.

Basic Mechanism of the Rhythmicity Observed in the Medullary Rhythmicity Area. For years, physiologists have been attempting to explain the mechanism that causes the inspiratory neurons to become active during inspiration and the expiratory neurons to become active during expiration. Though the final answer to this is unknown, the following is a recent theory that is at least partially substantiated by experimental studies:

Figure 42-1B shows four expiratory neurons to the left and four inspiratory neurons to the right. Each of these neuronal groups is arranged in an oscillating network that functions in the following manner: Let us first consider the inspiratory neurons. If one of the neurons becomes excited, it excites the next one, which in turn excites the third one, and so forth. This excitation continues around and around the "inspiratory" circuit, and as it does so, collateral inspiratory impulses are transmitted to the inspiratory muscles. This process continues until the oscillation stops after about 2 seconds. The factor that stops the oscillation is perhaps fatigue of the neurons themselves, for we know that neurons cannot transmit large numbers of impulses in rapid succession without decreasing their excitability.

In addition to transmitting impulses to the inspiratory muscles, the inspiratory neurons also transmit *inhibitory* impulses into the expiratory network. This keeps the expiratory network from oscillating so long as the inspiratory network is oscillating. But just as soon as inspiratory oscillation ceases, there is no longer inhibition of the expiratory network. In the absence of this inhibition, the natural excitability of the expiratory neurons causes the expiratory network to begin oscillating. It then oscillates for about 3 seconds until its neurons likewise fatigue. During this oscillation, expiratory impulses are transmitted to the expiratory muscles, and *inhibitory* impulses are transmitted into the inspiratory network to keep it from oscillating. However, when the expiratory network has fatigued and stopped oscillating, the inhibitory impulses to the inspiratory network cease, and the inspiratory network begins to oscillate once again.

In summary, two oscillating circuits are postulated: one for inspiration and one for expiration. However, the two circuits cannot oscillate simultaneously because they inhibit each other. Thus, when the inspiratory neurons are active the expiratory neurons are inactive, and the reverse effects occur when the expiratory neurons become active. Therefore, alternation occurs back and forth between inspiratory signals and expiratory signals, this process continuing indefinitely and causing the act of respiration.

Effect on the "Respiratory Center" of Nervous Signals from Other Parts of the Nervous System

All of us are familiar with the fact that respiration is altered (a) when a person speaks,

(b) when he suddenly enters a cold shower, (c) when he is excited by a pin prick, or (d) when almost any other stressful condition suddenly occurs. The respective changes in respiration result from nerve impulses arriving in the respiratory center from other parts of the nervous system. Several special types of nervous activity that play major roles in control of respiration are described here, and still another is described later in the chapter in relation to the regulation of respiration during muscular exercise.

Facilitatory Effect of Impulses from the Spinal Cord. When the spinal cord is cut shortly below the medulla, the respiratory center becomes only very weakly active. Yet stimulation of the cut fiber tracts leading from the transected cord to the medulla will re-establish normal respiratory rhythmicity. Therefore, impulses from the spinal cord play an important facilitatory role in keeping the respiratory center alive. For this reason, it is believed that the overall degree of stimulation of different peripheral sensory receptors throughout the body play an important role in maintaining normal respiration. It is interesting that when a person has a depressed respiratory center, depressed to the point that respiration has ceased, almost any peripheral stimulus, such as slapping the skin or applying cold water to the skin, can often start respiration once again.

The Hering-Breuer Reflexes — The Inflation Reflex and the Deflation Reflex. In the lungs are many receptors that detect stretch or compression of the lungs. Some of these are located in the visceral pleura, and others are in the bronchi, bronchioles, and even the alveoli. When the lungs become stretched, the stretch receptors, especially those of the bronchioles, transmit impulses through the vagus nerves into the *tractus solitarius* of the brain stem and thence into the respiratory center where they inhibit inspiration and thereby prevent further inflation. The effect is called the *Hering-Breuer inflation reflex.* This reflex prevents overdistention of the lungs.

A *Hering-Breuer deflation reflex* also occurs during expiration: that is, as the stretch receptors become unstretched the impulses from these receptors cease, which allows inspiration to begin again. Also, it is likely that compression receptors (not well defined as yet but possibly in the walls of the alveoli) transmit impulses that inhibit expiration in the same manner that the stretch receptors inhibit inspiration. Regardless of the precise mechanism, the degree of lung deflation is reduced by this reflex. However, this deflation reflex

is normally much less active than the inflation reflex.

The major effects of the Hering-Breuer reflexes are (1) a decrease in tidal volume and (2) a compensatory increase in respiratory rate. These effects result as follows: The inspiratory phase of respiration is cut off at an earlier point than would occur were it not for the inflation reflex. Likewise the expiratory phase is shortened by the deflation reflex. Thus, the depth of respiration is greatly decreased, but there is a compensatory increase in respiratory rate.

IMPORTANCE OF THE HERING-BREUER REFLEXES IN MAINTAINING RESPIRATORY RHYTHMICITY. In addition to altering respiratory rhythmicity, the Hering-Breuer reflexes also help to maintain the respiratory rhythm. The signals from the Hering-Breuer receptors feed into the respiratory center and help to inhibit inspiration as the lungs fill and to inhibit expiration as the lungs empty. In this way these reflexes help to shift activity from the inspiratory network to the expiratory network and vice versa. Therefore, when the Hering-Breuer reflexes are blocked by cutting the vagus nerves, it is considerably easier to cause respiratory arrest than when the Hering-Breuer reflexes are intact.

Influence of the Vasomotor Center on Respiration. Most factors that increase the activity of the vasomotor center, thereby increasing blood pressure, also elevate the pulmonary ventilatory rate. Perhaps the vasomotor center itself excites the respiratory center, or it is possible — if not probable — that the same factors excite both the vasomotor center and the respiratory center at the same time — particularly since these centers are actually intermingled together in the reticular substance of the brain stem. For instance, inhibition of the baroreceptors increases the rate of pulmonary ventilation as well as increases vasomotor activity. Therefore, in many if not most instances, parallel degrees of activity occur in the vasomotor and respiratory centers at the same time.

Emotional Effects on Respiratory Control. Many emotional states can affect the character of respiration. Fear, for instance, usually increases the ventilatory rate while at the same time increasing the heart rate. This supposedly prepares the body for action and gives the animal extra capability to oppose impending danger. Because many of the different emotional states affect the sympathetic and parasympathetic centers of the hindbrain, it is easy to understand how anxiety, worry, rage, etc., can all modify the activity of the respiratory centers.

Such a consistent relationship exists between some types of emotion and their effects on respiratory regulation that psychiatrists often record the spirographic "pattern" of respiration for diagnostic purposes. In some emotional states, such as when a person is telling a lie, the spirogram becomes very irregular; in others, the spirogram contains intermittent sighs or occasionally even gasps. Though it is difficult for a physiologist to explain the meaning of the different patterns of respiration, psychiatrists have occasionally found them important for diagnosing specific psychoneurotic conditions.

Effect of Body Heat on Respiratory Regulation.
PANTING. Lower animals have very little ability to lose heat from the surfaces of their bodies for two reasons: First, the surfaces of their bodies are usually covered with fur, which prevents rapid loss of heat, and, second, most of the skin of lower animals is not supplied with sweat glands, which prevents evaporative loss of heat from the body surfaces. Therefore, the specific mechanism of panting is used by many lower animals as a means for dissipating heat.

The phenomenon of panting is probably initiated by the hypothalamic thermoregulator centers, which will be discussed in detail in Chapter 71. In essence, when the blood becomes overheated, the hypothalamus initiates a series of neurogenic reactions to decrease the body temperature. One of these in lower animals is to initiate panting. The actual panting process is then probably controlled by centers closely related to the pneumotaxic center of the pons.

When an animal pants, he breathes in and out rapidly so that large quantities of new air from the exterior come in contact with all the respiratory passages, thus cooling these passages by water evaporation from their surfaces. Yet, panting does not greatly increase the alveolar ventilation because each breath is extremely shallow, so that most of the air that passes into the alveoli is dead space air. This prevents excess loss of carbon dioxide from the blood and thereby prevents serious respiratory alkalosis.

HUMORAL REGULATION OF RESPIRATION

The ultimate goal of respiration is to maintain proper concentrations of oxygen, carbon dioxide and hydrogen ions in the body fluids. It is fortunate, therefore, that respiratory activity is highly responsive to even slight changes in any one of these in the fluids. Therefore, when one speaks of *humoral* regulation of respiration, he is referring primarily to the regulation of respiratory activity by changes in concentrations of oxygen, carbon dioxide, or hydrogen ions in the body fluids.

Carbon dioxide and hydrogen ions exert their effects primarily on the respiratory center in the brain, whereas oxygen has its effect almost entirely on the peripheral chemoreceptors, and these in turn affect the respiratory center.

EFFECTS OF CARBON DIOXIDE AND HYDROGEN ION CONCENTRATIONS ON RESPIRATORY CENTER ACTIVITY

The respiratory center, meaning by this all of the respiratory areas of the brain stem, is greatly stimulated by an increase in either carbon dioxide or hydrogen ion concentration in the fluids of the respiratory center. These effects are illustrated by two of the curves in Figure 42–2, which show, respectively, (1) that an increase in carbon dioxide concentration (while all other factors remain constant) can increase the rate of alveolar ventilation as much as six- to seven-fold and (2) that a decrease in pH, which means an increase in hydrogen ion concentration, can also increase the rate of alveolar ventilation almost exactly as much as can an excess of carbon dioxide (provided that all other factors are held constant). Later we shall see that under normal operating conditions of respiratory regulation, when all other factors are not held constant, carbon dioxide concentration affects respiration much more than does hydrogen ion concentration.

Stimulation of Respiration by Carbon Dioxide and Hydrogen Ions in Cerebrospinal Fluid — The Brain Stem Chemosensitive Areas. Recent experiments have shown the existence of a very small chemosensitive respiratory area on the surface of each side of the medulla oblongata at approximately the entry point of the eighth, ninth, and tenth cranial nerves; this area is especially sensitive to changes in hydrogen ion concentration in the cerebrospinal fluid. An increase in hydrogen ions greatly increases

Figure 42–2 Approximate effects on alveolar ventilation of changing the concentrations of carbon dioxide, hydrogen ions, and oxygen in the arterial blood when only one of the humoral factors is changed at a time and the other two are maintained at absolutely normal levels.

respiratory activity, and a decrease in hydrogen ions decreases respiratory activity.

Even though these chemosensitive areas are primarily responsive to hydrogen ions, it is mainly carbon dioxide in the blood that causes the hydrogen ions in the cerebrospinal fluids to increase. The reason for this is that carbon dioxide diffuses from the blood into the cerebrospinal fluid extremely rapidly, but hydrogen ions do not. The carbon dioxide immediately combines with water of the cerebrospinal fluid to form carbonic acid. This in turn dissociates to increase the hydrogen ion concentration. Therefore, the cerebrospinal fluid mechanism for exciting respiration is primarily responsive to changes in carbon dioxide concentration in the blood.

Unfortunately, we still do not know how important the chemosensitive areas on the surface of the medulla are to overall respiratory regulation. Some respiratory physiologists believe that they account for as much as one-half of the respiratory drive caused by carbon dioxide, whereas others believe that the respiratory drive caused by this mechanism is of almost no importance.

A Unified Concept of Humoral Stimulation of Respiratory Neurons—Role of the Hydrogen Ion. When carbon dioxide diffuses to the interior of a cell, it reacts with water inside the cell to form carbonic acid, and this in turn dissociates rapidly to increase the hydrogen ion concentration. Therefore, either an increase in hydrogen ion concentration in the extracellular fluids or an increase in extracellular carbon dioxide concentration increases the hydrogen ion concentration inside the cell. For this reason, it has been postulated that the hydrogen ion concentration inside the respiratory neurons is the common factor that actually determines their levels of activity. Though this is only a theory, it could easily explain how carbon dioxide and hydrogen ions have almost exactly the same effect in stimulating the respiratory center.

THE CHEMORECEPTOR SYSTEM FOR CONTROL OF RESPIRATORY ACTIVITY—ROLE OF OXYGEN IN RESPIRATORY CONTROL

Aside from the direct sensitivity of the respiratory center itself to humoral factors, special chemical receptors located outside the central nervous system and called *chemoreceptors* are responsive to changes in oxygen, carbon dioxide, and hydrogen ion concentrations. These transmit signals to the respiratory center to help regulate respiratory activity.

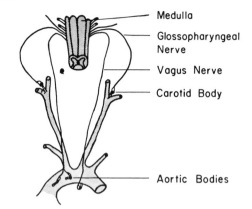

Figure 42–3   Respiratory control by the carotid and aortic bodies.

These chemoreceptors are located primarily in association with the large arteries of the thorax and neck; most of them are in the *carotid* and *aortic bodies,* which are illustrated in Figure 42–3 along with their afferent nerve connections to the respiratory center. The *carotid bodies* are located bilaterally in the bifurcations of the common carotid arteries, and their afferent nerve fibers pass through Hering's nerves to the glossopharyngeal nerves and thence to the medulla. The *aortic bodies* are located along the arch of the aorta; their afferent nerve fibers pass to the medulla through the vagi. Each of these bodies receives a special blood supply through a minute artery directly from the adjacent arterial trunk.

Stimulation of the Chemoreceptors by Decreased Arterial Oxygen.   Changes in arterial oxygen concentration have almost no direct effect on the respiratory center itself, but when the oxygen concentration in the arterial blood falls below normal, the chemoreceptors become strongly stimulated. This effect is illustrated in Figure 42–4, which shows the relationship between arterial P_{O_2} and rate of nerve impulse transmission from a carotid body.

Figure 42–4   Effect of arterial P_{O_2} on impulse rate from the carotid body of a cat. (Curve drawn from data from several sources but primarily from Von Euler.)

Note that the impulse rate is particularly sensitive to changes in arterial P_{O_2} in the range between 60 and 30 mm. Hg, which is the range in which the arterial hemoglobin saturation with oxygen decreases rapidly.

Effect of Carbon Dioxide and Hydrogen Ion Concentration on Chemoreceptor Activity. An increase in either carbon dioxide concentration or hydrogen ion concentration also excites the chemoreceptors and in this way indirectly increases respiratory activity. However, the direct effects of both of these factors in the respiratory center itself are so much more powerful than their effects mediated through the chemoreceptors that for most practical purposes one can disregard the indirect effects through the chemoreceptors. In the case of oxygen, on the other hand, this is not true, because diminished oxygen in the arterial blood can affect the respiration significantly *only* by acting through the chemoreceptors.

Net Effect of Arterial P_{O_2} Changes on Alveolar Ventilation. If care is taken to prevent simultaneous changes in either pH or P_{CO_2}, changes in arterial P_{O_2} have the effect illustrated by the dotted curve of Figure 42-2 on the rate of alveolar ventilation. This shows that so long as arterial P_{O_2} is above 100 mm. Hg there is relatively little effect on alveolar ventilation, but as the alveolar P_{O_2} falls progressively below this normal level, alveolar ventilation can increase to as much as five times normal. That is, a decrease in arterial P_{O_2} stimulates the chemoreceptors, which then stimulate the respiratory center to increase alveolar ventilation. Thus, the chemoreceptor system provides a very valuable feedback mechanism to help maintain normal arterial oxygen concentration. However, for reasons that are discussed later in the chapter, this mechanism does not operate very powerfully under normal respiratory conditions, but only under certain abnormal conditions.

INTERRELATIONSHIPS BETWEEN THE EFFECTS OF P_{CO_2}, pH, AND P_{O_2} ON RESPIRATORY ACTIVITY

The individual effects of changing P_{CO_2}, pH, or P_{O_2} while keeping the other respective factors constant were presented in Figure 42-2. However, the effects of these different factors interfere with each other in some instances and at other times support each other. One of the best attempts to show how these factors interact with each other is illustrated in Figure 42-5, which shows two separate "families" of curves relating alveolar P_{CO_2} to pulmonary ventilation. The "family" of curves represented by the solid lines is that determined in a person at different alveolar P_{O_2}'s from 40 to 100 mm. Hg and with the pH constant at the normal level of 7.4. The dashed "family" of curves was determined for

Figure 42–5 A composite diagram showing the interrelated effects of P_{CO_2}, P_{O_2}, and pH on alveolar ventilation. (Drawn from data presented in Cunningham and Lloyd: The Regulation of Human Respiration. F. A. Davis Co.)

the same person after he had developed metabolic acidosis, with a shift of his blood pH from 7.4 to 7.3. Both these families of curves show that the *slope* of the curve relating P_{CO_2} to pulmonary ventilation increases markedly as the P_{O_2} decreases. To express this another way, a decrease in P_{O_2} in the alveoli *multiplies* the effects on alveolar ventilation of carbon dioxide concentration *changes*.

Figure 42-5 shows also that acidosis shifts the entire family of curves to the left. This means that increased hydrogen ion concentration decreases the carbon dioxide concentration required to stimulate the respiratory center.

The curves in Figure 42-5 are of importance when one wishes to analyze respiratory control from a quantitative point of view because, in most instances in which respiratory control becomes altered, more than one of the individual humoral factors is changed at the same time.

QUANTITATIVE ASPECTS OF HUMORAL CONTROL OF ALVEOLAR VENTILATION

The effects of the three individual humoral factors, P_{CO_2}, P_{O_2}, and pH, on alveolar ventilation *when the other two respective factors are kept artificially at absolutely normal levels* were illustrated in Figure 42-2. This figure showed that the effects of the individual factors are all very powerful and that each of them can be the basis for an important feedback regulatory system to maintain proper concentration of itself in the body fluids. However, Figure 42-6 illustrates the effect of changing each factor when the other two respective factors *are not* artificially maintained at constant nor-

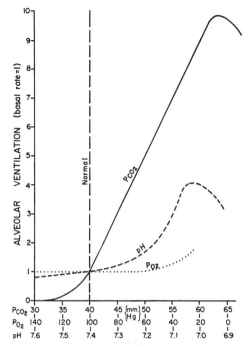

Figure 42–6 Effects of increased arterial P_{CO_2}, decreased arterial P_{O_2}, and decreased arterial pH on the rate of alveolar ventilation.

mal values. Under these conditions, one sees that the effect of P_{CO_2} on alveolar ventilation is tremendous, the effect of pH is intermediate, and the effect of P_{O_2} is slight. The purpose of this section, therefore, is to explain these differences.

CARBON DIOXIDE, THE MAJOR CHEMICAL FACTOR REGULATING ALVEOLAR VENTILATION

Note in Figure 42–6 that an increase in alveolar P_{CO_2} from the normal level of 40 mm. Hg to 63 mm. Hg causes a 10-fold increase in alveolar ventilation; this is about one and a half times as great as that which occurs when pH and P_{O_2} are controlled. The cause of this difference is that an increase in P_{CO_2} causes an almost proportional increase in hydrogen ion concentration in all the body fluids, as was explained in Chapter 37. Therefore, an increase in P_{CO_2} stimulates alveolar ventilation directly and also indirectly through its effect on hydrogen ion concentration.

Value of Carbon Dioxide as a Regulator of Alveolar Ventilation. Since carbon dioxide is one of the end-products of metabolism, its concentration in the body fluids greatly affects the chemical reactions of the cells. For this reason, the tissue fluid P_{CO_2} must be regulated exactly. In the preceding chapter, it was

pointed out that blood and interstitial fluid P_{CO_2} are determined to a great extent by the rate of alveolar ventilation. Therefore, stimulation of the respiratory center by carbon dioxide provides an important feedback mechanism for regulation of the concentration of carbon dioxide throughout the body. That is, (1) an increase in P_{CO_2} stimulates the respiratory center; (2) this increases alveolar ventilation and reduces the alveolar carbon dioxide; (3) as a result, the tissue P_{CO_2} returns to normal. In this way, the respiratory center maintains the P_{CO_2} of the tissue fluids at a relatively constant level and, therefore, might well be called a "carbon dioxide pressostat."

Effect on Respiration Caused by Breathing Progressively Increasing Concentrations of Carbon Dioxide. When one breathes air containing carbon dioxide, the alveolar and tissue P_{CO_2}'s rise above normal, and as a consequence alveolar ventilation increases. The increased ventilation, in turn, keeps the alveolar P_{CO_2} from rising as high as it otherwise would. Therefore, 1 to 2 per cent carbon dioxide in the inspired air hardly changes the carbon dioxide concentration of the tissue fluids, but as the carbon dioxide concentration rises above 5 per cent, even tremendous alveolar ventilation will not keep the tissue fluid P_{CO_2} from rising excessively. Maximum ventilation is reached at approximately 9 per cent carbon dioxide. Beyond this level, a further increase in P_{CO_2} begins to depress the respiratory center, causing progressive reduction in respiratory activity rather than a further increase. The person begins to pass into coma at 20 to 30 per cent carbon dioxide, becomes completely anesthetized at 30 to 40 per cent, and dies at 40 to 50 per cent.

CONTROL OF ALVEOLAR VENTILATION BY THE HYDROGEN ION CONCENTRATION OF THE EXTRACELLULAR FLUIDS

Figure 42–6 illustrates that changes in pH have considerably less effect on alveolar ventilation when the other humoral factors are not controlled than when they are controlled. The reason for this is that an increase in hydrogen ion concentration causes the person to breathe more rapidly than usual, which decreases the carbon dioxide concentration in the body fluids and at the same time increases the oxygen concentration. The decreased carbon dioxide *inhibits* the respiratory center and the increased oxygen *inhibits* the chemoreceptors. Therefore, opposing effects are caused by these other two factors, thus causing a "braking" effect on the hydrogen ion stimulation of respiration.

Nevertheless, despite the "braking" effect of the P_{CO_2} and P_{O_2} changes, a decrease in pH to approximately 7.1 can increase alveolar ventilation about four-fold, and an increase in pH to 7.6 can decrease ventilation to about 80 per cent of normal.

CONTROL OF ALVEOLAR VENTILATION BY ARTERIAL OXYGEN SATURATION

Unimportance of Oxygen Regulation of Respiration Under Normal Conditions — The "Braking" Effect of the P_{CO_2} and pH Regulatory Mechanisms. Figure 42–6 illustrates that changes in alveolar P_{O_2} have extremely little effect on alveolar ventilation when P_{CO_2} and pH are not controlled. This is in marked contrast to the effect when these other two factors are controlled, as was illustrated in Figure 42–2. This lack of importance of oxygen in normal regulation of respiration is caused by two effects, as follows:

First, even normally, the arterial P_{O_2} must fall below 60 mm. Hg before a marked effect on alveolar ventilation occurs, because arterial blood does not develop any significant degree of desaturation until the P_{O_2} falls below 60. Since it is the failure of blood to *deliver* oxygen to the chemoreceptors that stimulates them, one can readily see that decreasing the alveolar P_{O_2} from 100 to 60, in which range the saturation of the arterial blood hardly changes, has little effect in stimulating the chemoreceptors, as was illustrated in both Figures 41–2 and 42–4.

The second cause of the poor effect of P_{O_2} changes on respiratory control is a "braking" effect caused by both the carbon dioxide and hydrogen ion control mechanisms. The cause of this effect is illustrated in Figure 42–7, which shows a slight increase in ventilation as the alveolar P_{O_2} is decreased. The increase in ventilation blows off carbon dioxide from the

blood and therefore decreases the P_{CO_2} which is illustrated in the figure; at the same time it also decreases the hydrogen ion concentration. Therefore, two powerful respiratory inhibitory effects are caused by (a) diminished carbon dioxide and (b) diminished hydrogen ions. These two exert an inhibitory braking effect that opposes the excitatory effect of the diminished oxygen. As a result, they keep the decreased oxygen from causing marked increase in ventilation. Therefore, the maximum effect of decreased alveolar oxygen on alveolar ventilation is normally only a 66 per cent increase. This is in contrast to about a 400 per cent increase caused by decreased pH and about a 1000 per cent increase caused by increased carbon dioxide.

Yet, despite its weak response to diminished oxygen, the chemoreceptor mechanism is very important, for it is the only mechanism by which low oxygen concentration in the arterial blood can increase alveolar ventilation, and even this slight increase can sometimes be a life saver.

REASON OXYGEN REGULATION OF RESPIRATION IS NOT NORMALLY NEEDED. On first thought, it seems strange that oxygen should play so little role in the normal regulation of respiration, particularly since one of the primary functions of the respiratory center is to provide adequate intake of oxygen. However, oxygen control of respiration is not needed under most normal circumstances for the following reasons:

The respiratory system ordinarily maintains an alveolar P_{O_2} high enough to saturate fully the hemoglobin of the arterial blood. It does not matter whether alveolar ventilation is normal or 10 times normal, the blood will still be essentially fully saturated. Also, alveolar ventilation can decrease to as low as one-half normal, and the blood still remains within 10 per cent of complete saturation. Therefore, one can see that alveolar ventilation can change tremendously without significantly affecting oxygen transport to the tissues. Instead, the factors that play the major role in controlling P_{O_2} in the tissues under normal conditions are the hemoglobin-oxygen buffer system and the blood flow to the tissues, which were discussed in the preceding chapter.

On the other hand, the factor that controls carbon dioxide concentration in the tissues is primarily alveolar ventilation itself, as was explained earlier in the chapter.

Conditions Under Which Diminished Oxygen Plays a Major Role in the Regulation of Respiration. PNEUMONIA. In pneumonia or other lung ailments in which gases are not

Figure 42–7 Effect of arterial P_{O_2} on alveolar ventilation and on the subsequent decrease in arterial P_{CO_2}. (From Gray: Pulmonary Ventilation and Its Physiological Regulation. Charles C Thomas.)

readily exchanged between the atmosphere and the pulmonary blood, the oxygen regulatory system *does* then play a major role in the regulation of respiration. Contrary to the normal effect, the increased ventilation caused by oxygen lack does not now reduce the arterial Pco_2 and hydrogen ion concentration as much as normally because the pulmonary disease also diminishes carbon dioxide exchange as well as oxygen exchange. Therefore, the "braking" effects of these other two control systems on the oxygen lack system is not present. As a result, the oxygen lack system develops much of its full power and can increase alveolar ventilation as much as five- to sevenfold.

EFFECT OF THE OXYGEN LACK MECHANISM AT HIGH ALTITUDES. When a person first ascends to high altitudes or in any other way is exposed to a rarefied atmosphere, the diminished oxygen in the air stimulates his oxygen lack control system of respiration. The respiration at first increases to a maximum of about two-thirds above normal, which is a comparatively slight increase. Once again, the cause of this slight increase is the tremendous "braking" effect of the carbon dioxide and hydrogen ion control mechanisms on the oxygen lack mechanism.

However, over about a week, the respiratory center gradually becomes "adapted" to the diminished carbon dioxide and hydrogen ion concentrations in the blood so that these no longer depress the respiratory center at all. Thus, the "braking" effect on the oxygen control is gradually lost, and alveolar ventilation then rises to as high as seven times normal. This is part of the acclimatization that occurs as a person slowly ascends a mountain, thus allowing the person to adjust his respiration gradually to a level fitted for the higher altitudes. The cause of the adaptation is not known, but it has been suggested that HCO_3^- is pumped from the cerebrospinal fluid by the linings of the arachnoid space. This increases the degree of acidity in the fluid and supposedly stimulates respiration by means of the chemosensitive areas on the surface of the medulla.

REGULATION OF RESPIRATION DURING EXERCISE

In strenuous exercise, oxygen utilization and carbon dioxide formation can increase as much as 20-fold (see Fig. 42–8). Yet, except in very heavy exercise, alveolar ventilation ordinarily increases almost the same amount so that the blood Po_2 and Pco_2 remain almost exactly normal.

In trying to analyze the factors that cause increased ventilation during exercise, one is tempted immediately to ascribe this to the chemical alterations in the body fluids during exercise, including increase of carbon dioxide and hydrogen ions and decrease of oxygen. However, this is not valid, for measurements of arterial Pco_2, pH, and Po_2 show that none of these changes sufficiently in light to moderate exercise to account for the increase in ventilation. Indeed, even if a very high Pco_2 should develop during exercise, this still would be sufficient to account for only one-half the increased ventilation of heavy muscular exercise, for, as shown in Figure 42–9, the minute respiratory volume in exercise is about twice that which can be effected by maximal carbon dioxide stimulation.

Therefore, the question must be asked: What is it during exercise that causes the intense respiration? This question has not been answered, but at least two different effects seem to be predominantly concerned as follows:

1. The motor cortex, on transmitting impulses to the contracting muscles, is believed to transmit collateral impulses into the reticular substance of the brain stem to excite the respiratory center. This is analogous to the stimulatory effect that causes the arterial pressure to rise during exercise when similar collateral impulses pass to the vasomotor center.

2. During exercise, the body movements, especially of the limbs, are believed to increase pulmonary ventilation by exciting joint proprioceptors that then transmit excitatory impulses to the respiratory center. The reason for believing this is that even passive move-

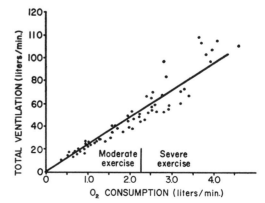

Figure 42–8 Effect of exercise on oxygen consumption and ventilatory rate. (From Gray: Pulmonary Ventilation and Its Physiological Regulation. Charles C Thomas.)

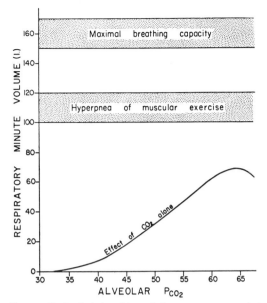

Figure 42–9 Relationship of hyperpnea caused by muscular exercise to that caused by increased alveolar P_{CO_2}. (Modified from Comroe: The Lung. Year Book Publishers.)

ments of the limbs often increase pulmonary ventilation several-fold.

It is possible also that still other undiscovered factors are even more important than all the above in regulating respiration during exercise, but since the increase in ventilation begins immediately upon the initiation of exercise, most of the increase in respiration probably results from the two neurogenic factors noted above, namely *stimulatory impulses from the motor cortex* and *proprioceptive stimulatory reflexes.*

Interrelationship Between Humoral Factors and Nervous Factors in the Control of Respiration During Exercise. Figure 42–10 illustrates diagrammatically the different factors that operate in the control of respiration during exercise, showing two neurogenic factors, (1) direct stimulation of the respiratory center by the motor cortex and (2) indirect stimulation by proprioceptors, and showing also the three humoral factors, (1) carbon dioxide, (2) hydrogen ions, and (3) oxygen. Most times when a person exercises, the nervous factors stimulate the respiratory center almost exactly the proper amount to supply the extra oxygen requirements for the exercise and to blow off the extra carbon dioxide. But, occasionally, the nervous signals are either too strong or too weak in their stimulation of the respiratory center. Then, the humoral factors play a very significant role in bringing about the final adjustment in respiration required to keep the carbon dioxide and hydrogen ion concentrations of the body fluids as nearly normal as possible. Thus, the neurogenic factors initiate the increase in alveolar ventilation during exercise and the humoral factors make additional adjustments, either upward or downward, to balance the rate of alveolar ventilation with the rate of metabolism in the body.

ABNORMALITIES OF RESPIRATORY CONTROL

RESPIRATORY DEPRESSION

Brain Edema. The activity of the respiratory center may be depressed or totally inactivated by a number of different factors, one of the most common of which is brain edema resulting from brain concussion. For instance, the head might be struck against some solid object, following which the damaged brain tissues swell, compressing the cerebral arteries against the cranial vault and thus totally or partially blocking the cerebral blood supply. As a result, the neurons of the respiratory center first become inactive and later die. In this manner brain edema may either depress or totally inactivate the respiratory center.

Occasionally, respiratory depression resulting from brain edema can be relieved by intravenous injection of hypertonic solutions. These solutions osmotically remove some of the intracellular fluids of the brain, thus relieving intracranial pressure and sometimes re-establishing respiration within a few minutes.

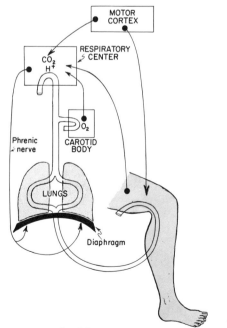

Figure 42–10 The different factors that enter into regulation of respiration during exercise.

Pressure Conus. A special condition that occurs frequently in brain injuries and in brain tumors is known as "pressure conus." That is, pressure in the cranial vault forces the brain downward through the foramen magnum, compressing the large upper portion of the medulla into the small opening of the foramen. As a result, blood flow into the medulla is partially or totally occluded so that the respiratory center becomes partially or totally inactive. Many patients with brain tumors die precipitously because of the development of such a conus. Such a situation occasionally occurs following removal of fluid from the spinal canal in patients with high cerebrospinal fluid pressure, for as soon as fluid is removed from the canal, the fluid in the cranial vault pushes the medulla into the foramen magnum, and the patient dies a few minutes later.

Anesthesia. Perhaps the most prevalent cause of respiratory depression and respiratory arrest is overdosage of anesthetics or narcotics. The best agent to be used for anesthesia is one that depresses the respiratory center the least while depressing the cerebral cortex the most. Such an anesthetic allows the anesthetist to block completely the conscious elements of the brain and yet not disturb the automatic function of respiration. Ether is probably the best of the anesthetics by these criteria, though halothane, cyclopropane, ethylene, nitrous oxide, and a few others have almost the same value. On the other hand, sodium pentobarbital is a poor anesthetic because it depresses the respiratory center considerably more than the above agents. In the early days morphine was used as an anesthetic, but this drug is now used only as an adjunct to anesthetics because it greatly depresses the respiratory center.

Respiratory Stimulants to Offset Depression. A few drugs, like picrotoxin and Metrazol, directly stimulate the respiratory center and can to a certain extent overcome the depressant effects of anesthetics, narcotics, and neuronal damage. Coramine is still another drug that stimulates respiration, but it does so indirectly. This drug excites the carotid and aortic chemoreceptors, which in turn stimulate the respiratory center. Likewise, caffeine, Benzedrine, theophylline, and many other general central nervous system stimulants can aid in stimulating the depressed respiratory center.

PERIODIC BREATHING

An abnormality of respiration called periodic breathing occurs in a number of different disease conditions. The person breathes deeply for a short interval of time and then breathes slightly or not at all for an additional interval, the cycle repeating itself over and over again.

The most common type of periodic breathing, *Cheyne-Stokes breathing,* is characterized by slow waxing and waning of respiration, occurring over and over again every 45 seconds to 3 minutes. A second important type of periodic breathing occasionally seen is *Biot's breathing,* which is charac-

terized by "runs" of several normal respirations— 1, 2, 3, 4, or more at a time—followed suddenly by a period of complete cessation of respiration, then another series of normal respirations, and so forth, the cycle repeating itself again and again. The duration of the cycle is extremely variable, sometimes as short as 10 seconds and sometimes as long as a minute.

Basic Mechanism of Cheyne-Stokes Breathing. Let us assume that the respiration becomes much more rapid and deeper than usual. This causes the P_{CO_2} in the pulmonary blood to decrease. A few seconds later the pulmonary blood reaches the brain, and the decreased P_{CO_2} inhibits respiration. As a result, the pulmonary blood P_{CO_2} gradually increases. After another few seconds the blood carrying the increased CO_2 arrives at the respiratory center and stimulates respiration again, thus making the person overbreathe once again and initiating a new cycle that continues on and on, causing Cheyne-Stokes periodic breathing.

The successive changes in pulmonary and respiratory center P_{CO_2} during Cheyne-Stokes breathing are illustrated in Figure 42–11.

Obviously, an increase and decrease in the blood concentration of oxygen could also account partially for the periodic feedback required to perpetuate Cheyne-Stokes oscillations. However, Cheyne-Stokes breathing can occur even after denervation of the chemoreceptors, illustrating that stimulation caused by oxygen lack is not a necessary feature of the condition.

Since the basic feedback mechanism for causing Cheyne-Stokes respiration is present in every person, the real question that must be answered is not why Cheyne-Stokes breathing develops under certain pathological conditions but, instead, why is it not present all the time in everyone? The reason for this is that the feedback is highly "damped," which may be explained as follows:

Damping of the Normal Respiratory Control Mechanism. In order for Cheyne-Stokes oscillation to occur, sufficient time must occur during the hyperpneic phase for the body fluid P_{CO_2} to fall considerably below the mean value. Then this decrease in P_{CO_2} initiates the apneic phase, which also must last long enough for the tissue P_{CO_2} to rise well above the mean value. To express this another

Figure 42–11 Cheyne-Stokes breathing, showing the changing P_{CO_2} in the pulmonary blood (solid line) and the delayed changes in P_{CO_2} of the fluids of the respiratory center (dashed line).

way, the total body fluid contains tremendous quantities of stored gases, and it takes a long period of hyperventilation or hypoventilation to change the concentrations of these gases significantly. In the normal person, before significant changes can occur the respiratory center ordinarily readjusts the breathing back toward normal, thereby preventing the extreme excursions of P_{CO_2} in the tissue fluids that are necessary to cause Cheyne-Stokes breathing. Thus, the tremendous storage capacity of the tissues for carbon dioxide functions as an important "damping" factor to prevent Cheyne-Stokes respiration in the normal person.

Yet even in the normal person the feedback is not completely damped. Therefore, when a person purposely overbreathes for a minute or more and then allows his involuntary respiratory control mechanisms to take over, he will first go into apnea and then go through one to two highly damped cycles of Cheyne-Stokes breathing.

A number of different conditions can overcome the damping of the feedback mechanism and cause it to oscillate spontaneously. Two of these are: (1) increased delay time in the flow of blood from the lungs to the brain and (2) increased feedback gain of the respiratory center mechanisms for control of respiration.

CHEYNE-STOKES BREATHING CAUSED BY INCREASED CIRCULATORY DELAY FROM THE LUNGS TO THE BRAIN. If the blood is long delayed in its transmission from the lungs to the brain, the damping factor discussed above cannot operate satisfactorily to block the feedback, for the body fluid gases can now change drastically before the respiratory center detects the altered blood gas concentration. As a result, the net amplification or "gain" of the feedback circuit is greatly enhanced, and the system oscillates spontaneously. Experiments have shown that a one- to two-minute delay in the passage of blood from the lungs to the brain causes most experimental animals to go into spontaneous Cheyne-Stokes respiration.

CHEYNE-STOKES BREATHING CAUSED BY INCREASED FEEDBACK GAIN IN THE RESPIRATORY CENTER. Occasionally, damage to the brain stem increases the feedback gain of the respiratory center in response to blood P_{CO_2} and blood pH. This does not mean that the sensitivity of these centers to these two humoral factors necessarily increases but instead that minute concentration *change* in one of the two humoral factors causes a drastic *change* in ventilation. Indeed, the respiratory center is often severely depressed even though the gain is very high. An example of this would be complete cessation of ventilation with a P_{CO_2} of 30 mm. Hg and yet one-half normal ventilation at a P_{CO_2} of 40 mm. Hg. In this case the *change* in ventilation is infinite. Therefore, the gain is infinite even though the ventilation is only one-half normal. Consequently, the respiratory control system can oscillate back and forth between apnea and breathing, which is the typical pattern of Cheyne-Stokes breathing. This effect probably explains why many patients with brain damage develop a type of Cheyne-Stokes breathing.

Cheyne-Stokes Respiration in Cardiac Failure. The most common cause of Cheyne-Stokes breathing is cardiac failure, which sometimes causes this type of breathing continually for many months at a time. There seem to be several factors operative in causing the Cheyne-Stokes breathing in this condition:

First, the slowed circulation in cardiac failure increases the delay time for transmission of blood from the lungs to the brain, thus reducing the effectiveness of the damping system as discussed above.

Second, the left heart volume is sometimes greatly increased. This, too, increases the circulation time from the lungs to the brain. Combining this factor with the sluggish rate of blood flow, the delay time from the lungs to the brain can increase as much as six-fold, the total interval increasing to 30 to 60 seconds instead of the normal 5 to 10 seconds. This is within the range that causes spontaneous oscillations in experimental dogs.

Third, another factor that perhaps enters into the cardiac failure type of Cheyne-Stokes breathing is impaired oxygen uptake by the edematous lungs. As a result, the arterial oxygen saturation sometimes operates in a range low enough to allow significant feedback by the oxygen-lack-chemoreceptor mechanism, thus greatly increasing the net gain of the feedback in the Cheyne-Stokes oscillatory system. One of the special reasons for believing this to be true is that cardiac Cheyne-Stokes breathing can often be stopped by oxygen therapy.

Periodic Breathing Following Brain Damage — Biot's Breathing. Many patients with very high cerebrospinal fluid pressure, with contusion of the brain tissues, with direct compression of the brain tissues, or with destructive diseases of the brain often develop periodic breathing which usually lasts only a short time before death supervenes. This type of breathing is usually of the Biot's type, though occasionally it is of the Cheyne-Stokes type. That is, the respirations usually come in couples, triples, quadruples, and so forth rather than waxing and waning.

The cause of Biot's breathing is unknown, but one would suspect that it results from direct abnormality of the basic rhythmic mechanism of the respiratory center itself. The short periodicity of Biot's breathing makes it almost certain that it is not caused by the same feedback system that causes Cheyne-Stokes breathing.

REFERENCES

American Physiological Society: Handbook of Physiology. Sec. III, Respiration. Vol. I, 1964; Vol. II, 1965. Baltimore, The Williams & Wilkins Co.

Beaver, W. L., and Wasserman, K.: Transients in ventilation at start and end of exercise. *J. Appl. Physiol.,* 25:390, 1968.

Bing, O. H. L.; Keefe, J. F.; Wolk, M. J.; Lipana, J. G.; McIntyre, K. M.; and Levine, H. J.: Cardiovascular responses to hypoxia and varying P_{CO_2} in the awake dog. *J. Appl. Physiol., 27*:204, 1969.

Burns, B. D.: The central control of respiratory movements. *Brit. Med. Bull., 19*:7, 1963.

Cohen, M. I.: Discharge patterns of brain-stem respiratory neurons during Hering-Breuer reflex evoked by lung inflation. *J. Neurophysiol., 32*:356, 1969.

Cohen, M. I.: Discharge patterns of brain-stem respiratory neurons in relation to carbon dioxide tension. *J. Neurophysiol., 31*:142, 1968.

Coleridge, H. M.; Coleridge, J. C. G.; and Howe, A.: Thoracic chemoreceptors in the dog: a histological and electrophysiological study of the location, innervation and blood supply of the aortic bodies. *Circ. Res., 26*:235, 1970.

Comroe, J. H., Jr.: The peripheral chemoreceptors. *In* Handbook of Physiology. Baltimore, The Williams & Wilkins Co., 1964, Sec. III, Vol. I, p. 557.

Cruz, J. C.; Rahn, H.; and Farhi, L. E.: Mixed venous P_{O_2}, P_{CO_2}, pH, and cardiac output during exercise in trained subjects. *J. Appl. Physiol., 27*:431, 1969.

Cunningham, D. J. C., and Lloyd, B. B. (eds.): The Regulation of Human Respiration. Philadelphia, F. A. Davis Co., 1963.

Dejours, P.: Chemoreflex hypoxic drive of ventilation in lowlanders and highlanders. *Proc. Int. Union of Physiol. Sciences, 6*:59, 1968.

Dutton, R. E.; Hodson, W. A.; Davies, D. G.; and Chernick, V.: Ventilatory adaptation to a step change in P_{CO_2} at the carotid bodies. *J. Appl. Physiol., 23*:195, 1967.

Edwards, M. W., Jr., and Mills, E.: Arterial chemoreceptor oxygen utilization and oxygen tension. *J. Appl. Physiol., 27*:291, 1969.

Eyzaguirre, C., and Lewin, J.: Chemoreceptor activity of the carotid body of the cat. *J. Physiol. (Lond.), 159*:222, 1961.

Fay, F. S.: Oxygen consumption of the carotid body. *Amer. J. Physiol., 218*:518, 1970.

Grodins, F. S.; Gray, J. S.; Schroeder, K. R.; Norins, A. L.; and Jones, R. W.: Respiratory responses to CO_2 inhalation; a theoretical study of a nonlinear biological regulator. *J. Appl. Physiol., 7*:283, 1954.

Guyton, A. C.; Crowell, J. W.; and Moore, J. W.: Basic oscillating mechanism of Cheyne-Stokes breathing. *Amer. J. Physiol., 187*:395, 1956.

Holloman, G. H., Jr.; Milhorn, H. T., Jr.; and Colman, T. G.: A sampled-data regulator for maintaining a constant alveolar CO_2. *J. Appl. Physiol., 25*:463, 1968.

Kahn, N., and Wang, S. C.: Electrophysiologic basis for pontine apneustic center and its role in integration of the Hering-Breuer reflex. *J. Neurophysiol., 30*:301, 1967.

Lloyd, B. B.: The chemical stimulus to breathing. *Brit. Med. Bull., 19*:10, 1963.

McKean, T. A.: A linear approximation of the transfer function of pulmonary mechanoreceptors of the frog. *J. Appl. Physiol., 27*:775, 1969.

Milhorn, H. T., Jr.; Benton, R.; Ross, R.; and Guyton, A. C.: A mathematical model of the human respiratory control system. *Biophys. J., 5*:27, 1965.

Milhorn, H. T., Jr., and Guyton, A. C.: An analog computer analysis of Cheyne-Stokes breathing. *J. Appl. Physiol., 20*:328, 1965.

Mitchell, R. A.: Respiration. *Ann. Rev. Physiol., 32*:415, 1970.

Pappenheimer, J. R.: The ionic composition of cerebral extracellular fluid and its relation to the control of breathing. Harvey Lectures, 1965, p. 71.

Rusher, J. L.; Stoll, P. J.; and Lenfant, C.: Exercise-induced hyperpnea and uniformity and efficiency of pulmonary ventilation. *J. Appl. Physiol., 28*:63, 1970.

Sorensen, S. C., and Cruz, J. C.: Ventilatory response to a single breath of CO_2 in normal man at sea level and high altitude. *J. Appl. Physiol., 27*:186, 1969.

Sorensen, S. C., and Severinghaus, J. W.: Irreversible respiratory insensitivity to acute hypoxia in man born at high altitude. *J. Appl. Physiol., 25*:217, 1968.

Widdicombe, J. G.: Respiratory reflexes. *In* Handbook of Physiology. Baltimore, The Williams & Wilkins Co., 1964, Sec. III, Vol. I, p. 585.

CHAPTER 43

Respiratory Insufficiency

The diagnosis and treatment of most respiratory disorders has come to depend highly on an understanding of the basic physiological principles involved. Some diseases of respiration result from inadequate ventilation, while others result from abnormalities of diffusion through the pulmonary membrane or of transport from the lungs to the tissues. In each of these instances, the therapy is often entirely different so that it is no longer satisfactory simply to make a diagnosis of "respiratory insufficiency."

Definitions. In describing different respiratory disorders the following common descriptive terms are used:

Eupnea means normal breathing; *tachypnea* means rapid breathing; and *bradypnea* means slow breathing.

Hyperpnea means a rate of alveolar ventilation great enough to cause over-respiration. However, this term is commonly used also to indicate simply a very high level of alveolar ventilation without necessarily implying over-respiration. *Hypopnea* is the opposite of hyperpnea, indicating under-respiration.

Anoxia means total lack of oxygen, but this term is more frequently used to mean simply decreased oxygen in the tissues. A more correct term for decreased oxygen in the tissues is *hypoxia.*

Anoxemia means lack of oxygen in the body fluids, but the term is generally used to mean simply reduced oxygen in the body fluids. A better term for this is *hypoxemia.*

Hypercapnia means excess carbon dioxide in the body fluids, and *hypocapnia* means depressed carbon dioxide. The term *acapnia* is also frequently used to imply hypocapnia, but, as is true with anoxemia, during life the absolute state of acapnia, with no carbon dioxide at all in the body fluids, never exists.

METHODS FOR STUDYING RESPIRATORY ABNORMALITIES

Study of Respiratory Ventilation. The measurement of vital capacity, tidal air, functional residual capacity, and other lung volumes and capacities was discussed in Chapter 39. From these, two calculations are frequently made, (1) the *breathing reserve* and (2) the *ventilatory reserve.*

THE BREATHING RESERVE. The breathing reserve is the difference between the normal minute respiratory volume and the maximum breathing capacity. Thus, if a person has a respiratory minute volume of 6 liters/minute and a maximum breathing capacity of 150 liters/minute, his breathing reserve is 144 liters/minute. In respiratory disease, the normal respiratory minute volume might be 20 liters/minute and his maximum breathing capacity 40 liters/minute. In this instance the breathing reserve is 20 liters/minute.

THE VENTILATORY RESERVE. The ventilatory reserve is the maximum percentage increase in alveolar ventilation that a person can achieve above his normal alveolar ventilation. Thus, the normal person with a normal alveolar ventilation of 4.2 liters/minute and a maximum alveolar ventilation of about 140 liters/minute has a ventilatory reserve of more than 3000 per cent. If, in the course of a disease, the normal alveolar ventilation becomes 20 liters/minute and the maximum alveolar ventilation 40 liters/minute, the ventilatory reserve is then only 100 per cent. Ordinarily, when the ventilatory reserve falls below 100 per cent the person experiences dyspnea or "air hunger," which is described later in the chapter.

Study of Blood Gas Concentrations. The physician often assesses the quantity of oxygen and carbon dioxide in the blood by observing various signs in the patient. A decrease in blood oxygen to a sufficiently low level causes the skin to become bluish because deoxygenated hemoglobin is blue. This condition, called *cyanosis,* is discussed in greater detail later in the chapter. When the blood concentration of carbon dioxide or hydrogen ions increases, the rate and depth of respiration usually also increase, which is a useful indicator in respiratory diseases. In addition to these clinical signs, blood gas analyses can be made as follows:

DETERMINATION OF BLOOD OXYGEN SATURATION. The concentration of oxygen in the blood is determined in a Van Slyke gas analysis apparatus in which a special reagent causes the hemoglobin to release its oxygen. Then the amount of released oxygen is measured.

To determine the *per cent oxygen* saturation of the blood, a similar blood sample is exposed to air until it becomes 100 per cent saturated with oxygen. The oxygen concentration of this blood is then measured as described above. Dividing this value into the initial concentration of oxygen and multiplying by 100 gives the percentage of the hemoglobin in the original blood sample that is bound with oxygen.

The normal concentration of oxygen in arterial blood is approximately 20 volumes per cent, and the normal arterial oxygen saturation is 97 per cent.

THE EAR OXIMETER. The ear oximeter, illustrated in Figure 43–1, measures the light absorption by the hemoglobin in the blood of the ear lobe at two different light frequencies. Then by appropriate combination of these two measurements it gives a direct reading of the per cent saturation of the blood. This apparatus has proved to be of great clinical value in studying blood oxygenation in patients under anesthesia, patients who have considerable respiratory paralysis, and patients who have other types of poor blood oxygenation. For instance, in polio patients in the iron lung it is essential that sufficient ventilation be maintained to assure proper oxygenation of the blood; and the ear oximeter is an excellent means for assessing this.

The light bulb of the ear oximeter heats the ear to the extent that the arterioles dilate greatly. As a result, the blood flow through the ear becomes so rapid that little deoxygenation occurs in the capillaries. Consequently, the ear oximeter measures essentially the *arterial* oxygen saturation.

MEASUREMENT OF BLOOD AND TISSUE OXYGEN PRESSURES BY POLAROGRAPHY. As explained in Chapter 41, the per cent saturation of hemoglobin is different from the actual pressure of the oxygen in the blood. A particularly valuable method has been developed for measuring the oxygen pressure (Po_2) of solutions, based on an electrochemical procedure called "polarography." Basically, this apparatus applies a negative electrical voltage to a minute platinum electrode, and the electrical current flowing from the electrode to the solution reduces the gaseous oxygen in contact with the electrode to an ionic form. Since the rate of current flow depends on the Po_2 of the solution, the Po_2 can be read directly from an electrical meter in the circuit.

When this apparatus is used for measuring blood or tissue Po_2, a very thin membrane that allows oxygen diffusion but no diffusion of the plasma proteins is usually placed over the electrode to prevent difficulties with the proteins, red cells, or tissue elements. Such electrodes have been made as small as 1 micron in diameter, small enough to measure Po_2 inside individual cells.

MEASUREMENT OF CARBON DIOXIDE CONCENTRATION AND Pco_2 OF THE BLOOD. By using different reagents in the Van Slyke gas analysis apparatus, one can liberate carbon dioxide instead of oxygen from the blood, and its concentration in the blood can thereby be measured. A modified glass electrode pH meter can be used to measure the carbon dioxide pressure (Pco_2) in the following way: A very thin membrane permeable to carbon dioxide but impermeable to bicarbonate ion is stretched over the glass electrode, and a solution of bicarbonate ions is placed in the thin space between the membrane and the electrode. The pH of this bicarbonate solution will vary as a function of the concentration of the carbon dioxide that diffuses through the membrane, and this in turn is determined by the Pco_2.

PHYSIOLOGIC TYPES OF RESPIRATORY INSUFFICIENCY

In general, the different types of abnormalities that cause respiratory insufficiency can be divided into three major categories: (1) those that cause inadequate ventilation of the alveoli, (2) those that reduce gaseous diffusion through the respiratory membrane, and (3) those that decrease oxygen transport from the lungs to the tissues.

ABNORMALITIES THAT CAUSE ALVEOLAR HYPOVENTILATION

Paralysis of the Respiratory Muscles. It is obvious how paralysis of the respiratory muscles can decrease pulmonary ventilation. Fortunately, paralysis of any single muscle of respiration is compensated by overactivity of those that are still normal. Indeed, patients with polio or cervical transection of the spinal cord resulting from a broken spinal column can sometimes maintain respiration despite complete paralysis of the diaphragm and abdominal muscles, utilizing only the neck muscles, or even helping respiration by forcing air into the lungs with the cheek muscles. This illustrates the great reserve of the respiratory muscular system.

Diseases That Increase the "Work" of Ventilation. Much more common than paralysis of the respiratory muscles are the many pulmonary diseases that increase the work required to move air in and

Figure 43–1   The ear oximeter. (Courtesy of The Waters Corporation.)

out of the lungs. These can be divided into three different types of abnormalities: (1) those that increase the airway resistance, (2) those that increase the tissue resistance, and (3) those that decrease the compliance of the lungs and thorax.

INCREASED AIRWAY RESISTANCE. One can readily understand how obstruction of the respiratory passageway can reduce a person's maximum breathing capacity. But, in addition, airway resistance is especially increased in two diseases that affect the bronchioles, *asthma* and *emphysema*, which are discussed in further detail later in the chapter. In both of these, air flow is impeded in the smaller bronchioles, which causes far greater breathing effort during expiration than inspiration for the following reasons:

When the muscles of respiration expand the lungs, the expansion process not only opens the alveoli but also expands the small bronchioles at the same time. Consequently, air flows easily into the alveoli. On the other hand, during expiration the pressure of the thoracic cage against the lungs not only compresses the alveoli but also compresses the small bronchi, in this way reducing their diameters and thereby increasing the airway resistance. As a result, the airway resistance sometimes becomes several times as great during expiration as during inspiration, and expiration is correspondingly prolonged. Also, because of this differential between expiratory and inspiratory resistance, the lungs become greatly distended, with a marked increase in functional residual capacity and residual volume.

INCREASED TISSUE RESISTANCE. Another factor that increases the work of respiration is often loss of normal pulmonary elasticity. In normal respiration, the inspiratory muscles stretch the lungs and thorax, and then the elastic recoil of the lungs and thorax causes the lungs to deflate during expiration. In this way, no effort is usually required for expiration. Many diseases of the lungs, particularly such destructive diseases as *emphysema, pulmonary fibrosis, tuberculosis,* and *various infections,* often destroy much of the elasticity of the lungs so that little recoil occurs. As a result, effort must then be expended by the respiratory muscles to deflate the lungs as well as to inflate them.

DECREASED COMPLIANCE OF THE LUNGS AND CHEST WALL. The term "compliance" means the change in lung volume caused by a unit change in distending pressure, as explained in Chapter 39. Normally, a rise in intra-alveolar pressure of 1 cm. water increases the lung volume 0.13 liter. But any disease that makes the tissues of the lungs or of the chest wall less distensible than usual—diseases such as *silicosis, sarcoidosis, tuberculosis, cancer, pneumonia,* or even *skeletal abnormalities of the chest*—reduces the compliance and, therefore, increases the work that must be expended by the respiratory muscles to expand the lungs. Also, any disease that decreases the total amount of functional lung tissue reduces the compliance and, therefore, increases the effort of respiration.

DISEASES THAT DECREASE PULMONARY DIFFUSING CAPACITY

Three different types of abnormalities can decrease the diffusing capacity of the lungs. These are: (1) decreased area of the respiratory membrane, (2) increased thickness of the respiratory membrane, called *alveolocapillary block,* and (3) abnormal ventilation-perfusion ratio in some parts of the lungs.

Decreased Area of the Respiratory Membrane. Diseases or abnormalities that decrease the area of the respiratory membrane include *removal of part or all of one lung, tuberculous destruction of the lung, cancerous destruction,* and *emphysema,* which causes gradual destruction of alveolar septa.

Also, any acute condition that fills the alveoli with fluid or otherwise prevents air from coming in contact with the alveolar membrane, such as *pneumonia, pulmonary edema,* and *atelectasis,* can temporarily reduce the surface area of the respiratory membrane.

Increased Thickness of the Respiratory Membrane — Alveolocapillary Block. The most common acute cause of increased thickness of the respiratory membrane is *pulmonary edema* resulting from left heart failure or pneumonia. However, *silicosis, tuberculosis,* and *many other fibrotic conditions* can cause progressive deposition of fibrous tissue in the interstitial spaces between the alveolar membrane and the pulmonary capillary membrane, thereby increasing the thickness of the respiratory membrane. This is usually called *alveolocapillary block,* or, occasionally, *interstitial fibrosis.* Since the rate of gaseous diffusion through the respiratory membrane is inversely proportional to the distance that the gas must diffuse, it is readily understood how alveolocapillary block can reduce the diffusing capacity of the lungs.

Abnormal Ventilation-Perfusion Ratio. Abnormal ventilation-perfusion ratio of the lungs can cause decreased pulmonary diffusing capacity. This is one of the abnormalities that is more difficult to understand, but it can be explained as follows: If the blood flow to many of the alveoli is blocked or partially blocked, the ventilation of these alveoli is of little or no benefit in the aeration of the blood, and yet, these alveoli take ventilation away from the normal alveoli.

Likewise, when blood flow is normal but ventilation of many alveoli is blocked, the blood flowing past the blocked alveoli fails to be aerated.

Therefore, it is obvious that optimum aeration of blood requires approximately equal distribution of ventilation and blood flow to the respective alveoli. When ventilation and blood flow are equally distributed, the ventilation-perfusion ratio is 1; when these are not equally distributed, the ratio can be as low as 0.2 or as high as 4.

Diseases that cause abnormal ventilation-perfusion ratios include *thrombosis of a pulmonary artery, excessive airway resistance to some alveoli (emphysema), reduced compliance of one lung without concomitant abnormality of the other lung,* and many other conditions that cause diffuse damage throughout the lungs.

ABNORMALITIES OF OXYGEN TRANSPORT FROM THE LUNGS TO THE TISSUES

Different conditions that reduce oxygen transport from the lungs to the tissues include *anemia,* in which the total amount of hemoglobin available to transport the oxygen is reduced, *carbon monoxide poisoning,* in which a large proportion of the hemoglobin becomes unable to transport oxygen, and *decreased blood flow to the tissues* caused by either low cardiac output or localized tissue ischemia. These have been discussed in Chapters 8, 41, and 28, respectively.

PHYSIOLOGIC PECULIARITIES OF SPECIFIC PULMONARY ABNORMALITIES

TUBERCULOSIS

In tuberculosis the tubercle bacilli cause a peculiar tissue reaction in the lungs including, first, invasion of the infected region by macrophages and, second, walling off of the lesion by fibrous tissue to form the so-called "tubercle." This walling-off process helps to limit further transmission of the tubercle bacilli in the lungs and, therefore, is part of the protective process against the infection. However, in approximately 3 per cent of all persons who contract tuberculosis (perhaps half the population becomes infected at some time or other), the walling-off process fails and tubercle bacilli spread throughout the lungs, causing fibrotic tubercles in many areas. In the late stages of tuberculosis, secondary infection by other bacteria causes extensive destruction of the lungs. Thus, tuberculosis causes many areas of fibrosis throughout the lungs, and, secondly, it reduces the total amount of functional lung tissue. These effects cause (1) increased effort on the part of the respiratory muscles to cause pulmonary ventila-

tion and therefore *reduced vital capacity and maximum breathing capacity,* (2) *reduced total respiratory membrane surface* and *increased thickness of the respiratory membrane,* these causing progressively diminished pulmonary diffusing capacity, and (3) *abnormal ventilation-perfusion ratio* in the lungs, further reducing the pulmonary diffusing capacity.

PULMONARY EDEMA

The circulatory aspects of pulmonary edema were discussed in Chapter 23. Briefly, any abnormality that reduces the pumping capability of the left side of the heart causes blood to dam up in the pulmonary circulation. And, if the pulmonary capillary pressure rises above approximately 28 mm. Hg, which is the amount of the colloid osmotic pressure of the plasma proteins, fluid immediately begins to transude out of the capillaries into the pulmonary interstitial spaces and even into the alveoli. This results in acute thickening of the respiratory membrane, thereby reducing the diffusing capacity of the lungs. It also causes actual blockage of many of the alveoli as a result of total filling with fluid. This further reduces the diffusing capacity by reducing the total area of the respiratory membrane. Finally, the presence of large quantities of fluid in the lungs increases the *tissue resistance* of the lungs so much that they cannot be expanded and contracted with ease; this factor increases the work load of the respiratory muscles, though it is not so detrimental an effect as the greatly reduced pulmonary diffusing capacity.

PNEUMONIA

The term pneumonia describes any lung condition in which the alveoli become filled with fluid and/or blood cells, as shown in Figure 43–2. Thus, the filling of the alveoli with fluid in severe pulmonary edema is a type of pneumonia, but another even more common type of pneumonia is *bacterial pneumonia* caused most frequently by pneumococci. This disease begins with infection in the alveoli of one part of the lungs; the alveolar mem-

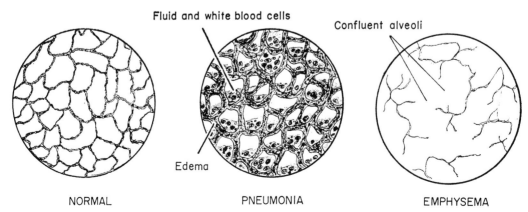

Fluid and white blood cells

Confluent alveoli

Edema

NORMAL   PNEUMONIA   EMPHYSEMA

Figure 43–2   Pulmonary changes in pneumonia and emphysema.

brane becomes edematous and highly porous so that fluid and often even red and white blood cells pass out of the blood into the alveoli. Thus, the infected alveoli become progressively filled with fluid and cells, and the infection spreads by extension of bacteria from alveolus to alveolus. Eventually, large areas of the lungs, sometimes whole lobes or even a whole lung, become "consolidated," which means that they are filled with fluids and cellular debris.

The pulmonary function of the lungs during pneumonia changes in different stages of the disease. In the early stages, the pneumonia process might well be localized to only one lung, and alveolar ventilation may be reduced even though blood flow through the lung continues normally. This results in two major pulmonary abnormalities: (1) reduction in the total available surface area of the respiratory membrane and (2) abnormal ventilation-perfusion ratio. Both these effects cause reduced diffusing capacity, which results in hypoxemia. Figure 43-3 illustrates the effect of an abnormal ventilation-perfusion ratio in pneumonia, showing that the blood passing through the aerated lung becomes 97 per cent saturated while that passing through the unaerated lung remains only 60 per cent saturated, causing the mean saturation of the aortic blood to be about 78 per cent, which is far below normal.

In other stages of pneumonia the blood flow through the diseased areas decreases concurrently with the decrease in ventilation. This gives much less debility than that resulting when the blood flow through the unventilated lung is normal.

ATELECTASIS

Atelectasis means collapse of the alveoli. It can occur in a localized area of a lung, in an entire lobe, or in an entire lung. Its causes are three-fold: (1) obstruction of the airway, (2) external compression of the lung, or (3) lack of surfactant in the fluids lining the alveoli.

Airway Obstruction. The airway obstruction type of atelectasis usually results from blockage of many small bronchi with mucus or obstruction of a major bronchus by either a large mucous plug or some solid object. The air entrapped beyond the

Figure 43-4 Effect of atelectasis on arterial blood oxygen saturation.

block is gradually absorbed by the blood flowing in the pulmonary capillaries, which pulls the walls of the alveoli together.

The effect on overall pulmonary function caused by *massive collapse* (atelectasis) of an entire lung are illustrated in Figure 43-4. Collapse of the lung tissue not only occludes the alveoli but also increases the resistance to blood flow through the pulmonary vessels as much as five-fold. This resistance increase occurs instantaneously with the collapse because the collapse compresses and folds the vessels as the volume of the lung decreases. Anoxia in the collapsed alveoli might also cause additional vasoconstriction, but this occurs slowly, as was explained in Chapter 23.

Because of the vascular constriction, blood flow through the atelectatic lung becomes slight. Essentially all the blood is routed through the ventilated lung and therefore becomes well aerated. In the situation shown in Figure 43-4, five-sixths of the blood passes through the aerated lung and only one-sixth through the unaerated lung. As a result, the aortic blood becomes almost fully saturated with oxygen despite the massive collapse of an entire lung.

Compressed Lung Tissue. If one portion of a lung is compressed more than other portions, the alveoli of this region will be smaller than the other alveoli. One can show from the physics of fluid lined bubbles, which are analogous to the alveoli, that the smaller a bubble becomes the more likely will the surface tension of the fluid cause it to collapse. Therefore, in an area partially compressed by outside pressure, this surface tension effect sometimes becomes great enough to cause the alveoli to collapse.

Lack of Surfactant—Hyaline Membrane Disease. In Chapter 39 it was pointed out that the substance called surfactant is secreted by the alveolar membrane into the fluids that line the alveoli. This substance decreases the surface tension in the alveoli 5- to 10-fold. In the normal lung this plays a major role in preventing alveolar collapse. However, a few newborn babies do not secrete an adequate quantity of surfactant. As a result, it is sometimes impossible for their respiratory forces to overcome the surface tension factor enough to expand the lungs. As a result these infants die of suffocation. This condition is sometimes called *hyaline membrane disease,* because the fluid in the

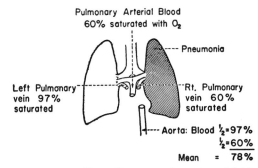

Figure 43-3 Effect of pneumonia on arterial blood oxygen saturation.

collapsed alveoli appears microscopically to be hyaline in nature.

The lungs also often lose their ability to secrete surfactant after extracorporeal perfusion of the lung during open heart surgery. As a result, pulmonary atelectasis frequently follows this type of surgical operation.

BRONCHIAL ASTHMA

Bronchial asthma is usually caused by allergic hypersensitivity of the person to foreign substances in the air — especially to plant pollens. The allergic reaction causes localized edema in the walls of the small bronchioles, secretion of thick mucus into their lumens, and spasm of their smooth muscle walls. These effects greatly increase the airway resistance to the alveoli. For reasons discussed earlier in the chapter, the bronchiolar diameter becomes even further reduced during expiration and then opens up to a certain extent during inspiration. Therefore, the asthmatic person usually can inspire quite adequately but has great difficulty expiring. This results in dyspnea, or "air hunger," which is discussed later in the chapter.

The functional residual capacity and the residual volume of the lung become greatly increased during the asthmatic attack because of the difficulty in expiring air from the lungs. Over a long period of time the chest cage becomes permanently enlarged, causing a "barrel chest," and the functional residual capacity and residual volume also become permanently increased.

CHRONIC EMPHYSEMA

A disease called chronic emphysema is becoming highly prevalent, partly because it is a disease of old age and the average age of our population is increasing each year, but mainly because of the effects of smoking. It results from two major pathophysiological changes in the lungs. First, air flow through many of the terminal bronchioles is obstructed. Second, much of the lung parenchyma is destroyed.

Many clinicians believe that chronic emphysema begins with chronic infection in the lung that causes *bronchiolitis,* which means inflammation of the small air passages of the lungs. It is likely that this inflammation involves alveolar septa and destroys many of these or that obstruction to expiration causes excess expiratory pressure in the alveoli and that this pressure ruptures the alveolar septa. At any rate, many bronchioles become irreparably obstructed, and the total surface of the respiratory membrane becomes greatly decreased as illustrated in Figure 43–2, sometimes to less than one-quarter normal.

The physiological effects of chronic emphysema are extremely varied, depending on the severity of the disease and on the relative degree of bronchiolar obstruction versus lung parenchymal destruction. However, among the different abnormalities are the following:

First, the bronchiolar obstruction greatly *increases airway resistance* and results in greatly increased work of breathing. It is especially difficult for the person to move air through the bronchioles during expiration because the compressive force on the outside of the lung not only compresses the alveoli but also compresses the bronchioles, which further increases their resistance.

Second, the marked loss of lung parenchyma greatly *decreases the diffusing capacity* of the lung, which reduces the ability of the lungs to oxygenate the blood.

Third, the obstructive process is frequently much worse in some parts of the lungs than in others so that some portions of the lungs are well ventilated while other portions are poorly ventilated. This causes an *abnormal ventilation-perfusion ratio,* which also causes poor aeration of the blood as was explained earlier in the chapter.

Fourth, loss of large portions of the lung parenchyma also decreases the number of pulmonary capillaries through which blood can pass. As a result, the pulmonary vascular resistance increases markedly, causing pulmonary hypertension. This in turn overloads the right heart and frequently causes right-heart failure.

Chronic emphysema usually progresses slowly over many years. The person develops hypoxia and hypercapnia because of hypoventilation of the alveoli and because of loss of lung parenchyma. The net result of all of these effects is severe and prolonged air hunger that can last for years until the hypoxia and hypercapnia cause death.

HYPOXIA

The term "hypoxia," as usually used, means diminished availability of oxygen to the cells of the body. The term *hypoxemia* means reduced oxygen in the body fluids, and refers especially to the oxygen in the arterial blood. Obviously, almost any of the conditions discussed in the past few sections of this chapter can cause serious degrees of hypoxia. In some of these, oxygen therapy is of great value; in others it is of moderate value, while in still others it is of almost no value. Therefore, it is important to classify the different types of hypoxia; then from this classification we can readily discuss the physiological principles of therapy. The following is a descriptive classification of the different causes of hypoxia:

1. Inadequate oxygenation of the lungs because of extrinsic reasons
   a. Deficiency of oxygen in atmosphere
   b. Hypoventilation (neuromuscular disorders)
2. Pulmonary disease
   a. Hypoventilation due to airway or pulmonary disease
   b. Uneven alveolar ventilation-perfusion ratio
   c. Diminished diffusing capacity

3. Venous-to-arterial shunts (intrapulmonary or intracardiac)
4. Inadequate transport and delivery of oxygen
   a. Anemia, abnormal hemoglobin
   b. General circulatory deficiency
   c. Localized circulatory deficiency (peripheral, cerebral, coronary vessels)
5. Inadequate tissue oxygenation or oxygen use
   a. Tissue edema
   b. Abnormal tissue demand
   c. Poisoning of cellular enzymes

This classification of the different types of hypoxia is mainly self-evident from the discussions earlier in the chapter. Only two of the types of hypoxia in the above classification need further elaboration; these are the hypoxia caused by venous-to-arterial shunts and inadequate tissue oxygenation.

Hypoxia Caused by Venous-to-Arterial Shunts. A venous-to-arterial shunt means a direct pathway for blood to flow from the systemic veins to the systemic arteries without going through the lungs. This occurs most frequently in the congenital heart disease called tetralogy of Fallot, which was discussed in Chapter 30. In this condition, the right heart pumps as much as 80 per cent of the venous blood directly into the aorta, without its having gone through the lungs at all.

Inadequate Tissue Oxygenation. The most classic type of inadequate tissue oxygenation is caused by cyanide poisoning in which the action of cytochrome oxidase is completely blocked — to such an extent that the tissues simply cannot utilize the oxygen even though plenty is available. This type of hypoxia is frequently also called *histotoxic hypoxia*.

A common type of tissue hypoxia occurs in tissue edema, which causes increased distances through which the oxygen must diffuse before it can reach the cells. This type of hypoxia can become so severe that the tissues in edematous areas actually die, as is often illustrated by serious ulcers in edematous skin.

Finally, tissues can become hypoxic when the cells themselves demand more oxygen than can be made available to them by the normal respiratory and oxygen transport systems. For instance, in strenuous exercise, one of the major limiting factors to the degree of exercise that can be performed is the tissue hypoxia that develops.

Effects of Hypoxia on the Body. Hypoxia, if severe enough, can actually cause death of the cells, but in less severe degrees it results principally in (1) depressed mental activity, sometimes culminating in coma and (2) reduced work capacity of the muscles. These effects are discussed in the following chapter in relation to aviation medicine and, therefore, are only mentioned here.

CYANOSIS

The term "cyanosis" means blueness of the skin, and its cause is excessive amounts of deoxygenated hemoglobin in the skin blood vessels, especially in the capillaries. This deoxygenated hemoglobin has an intense dark blue color that is transmitted through the skin.

The presence of cyanosis is one of the most common clinical signs of different degrees of respiratory insufficiency, and for this reason it is important to understand the factors that determine the degree of cyanosis. These are the following:

1. One of the most important factors determining the degree of cyanosis is the *quantity of deoxygenated hemoglobin in the arterial blood*. It is not the percentage deoxygenation of the hemoglobin that causes the bluish hue of the skin, but principally the *concentration of deoxygenated hemoglobin without regard to the concentration of oxygenated hemoglobin*. The reason for this is that the red color of oxygenated blood is weak in comparison with the dark blue color of deoxygenated blood. Therefore, when the two are mixed together, the oxygenated blood has relatively little coloring effect in comparison with that of the deoxygenated blood.

In general, definite cyanosis appears whenever the arterial blood contains more than 5 grams per cent of deoxygenated hemoglobin, and mild cyanosis can frequently be discerned when as little as 3 to 4 grams per cent deoxygenated hemoglobin is present. In polycythemia, cyanosis is very common because of the large amount of hemoglobin in the blood, whereas in anemia, cyanosis is rare because it is difficult for there to be enough deoxygenated hemoglobin to produce the blue color.

2. Another important factor that affects the degree of cyanosis is the *rate of blood flow through the skin,* for the following reasons: Principally, the blood in the capillaries determines skin color; the blueness of this blood is determined by two factors: (1) the concentration of deoxygenated hemoglobin in the arterial blood entering the capillaries and (2) the amount of deoxygenation that occurs as the blood passes through the capillaries. Ordinarily, the metabolism of the skin is relatively low so that little deoxygenation occurs as the blood passes through the skin capillaries. However, if the blood flow becomes extremely sluggish, even a low metabolism can cause marked desaturation of the blood and therefore can cause cyanosis. This explains the cyanosis that appears in very cold weather, particularly in children who have thin skins.

3. A final factor that affects the blueness of the skin is *skin thickness*. For instance, in newborn babies, who have very thin skin, cyanosis occurs readily, particularly in highly vascular portions of the body, such as the heels. Also, in adults (as well as babies) the lips and fingernails often appear cyanotic before the remainder of the body shows any blueness.

DYSPNEA

Dyspnea means primarily a desire for air or mental anguish associated with the act of ventilating enough to satisfy the air demand. A common synonym is "air hunger."

At least three different factors often enter into the development of the sensation of dyspnea. These are: (1) abnormality of the respiratory gases in the body fluids, especially hypercapnia and to a much lesser extent hypoxia, (2) the amount of work that must be performed by the respiratory muscles to provide adequate ventilation, and (3) the state of the mind itself. Thus, a person normally becomes very dyspneic when his muscles cannot allow him to breathe adequately to obtain oxygen and blow off carbon dioxide. This dyspnea results from excess buildup of carbon dioxide and reduction of oxygen in the body fluids.

At times, however, the levels of both carbon dioxide and oxygen in the body fluids are completely normal, but to attain this the person has to breathe forcefully. In these instances the forceful activity of the respiratory muscles themselves gives the person a sensation of air hunger. Indeed, the dyspnea can be so intense, despite normal gaseous concentrations in the body fluids, that clinicians often overemphasize this cause of dyspnea while forgetting that dyspnea can also result from abnormal gaseous concentrations.

Finally, the person's respiratory functions may be completely normal, and still he experiences dyspnea because of an abnormal state of mind. This is called *neurogenic dyspnea* or, sometimes, *emotional dyspnea.* For instance, almost anyone momentarily thinking about his act of breathing will suddenly find himself taking breaths a little more deeply than ordinarily because of a feeling of mild dyspnea. This feeling is greatly enhanced in persons who have a psychic fear of not being able to receive a sufficient quantity of air. For example, many persons on entering small or crowded rooms immediately experience emotional dyspnea, and patients with "cardiac neurosis" who have heard that dyspnea is associated with heart failure frequently experience severe psychic dyspnea even though the blood gases are completely normal. Neurogenic dyspnea has been known to be so intense that the person over-respires and causes alkalotic tetany.

HYPERCAPNIA

Hypercapnia means excess carbon dioxide in the body fluids and especially refers to excess carbon dioxide at the cellular level.

One might suspect at first thought that any respiratory condition that causes hypoxia also causes hypercapnia. However, hypercapnia usually occurs in association with hypoxia only when the hypoxia is caused by *hypoventilation* or by *circulatory deficiency.* The reasons for this are the following:

Obviously, hypoxia caused by *too little oxygen in the air,* by *too little hemoglobin,* or by *poisoning of the oxidative enzymes* has to do only with the availability of oxygen or use of oxygen by the tissues. Therefore, it is readily understandable that hyper-

capnia is *not* a concomitant of these types of hypoxia.

Also, in hypoxia resulting from poor diffusion through the pulmonary membrane or through the tissues, serious hypercapnia usually does not occur because carbon dioxide diffuses 20 times as rapidly as oxygen. Therefore, even when oxygen diffusion is greatly depressed, carbon dioxide diffusion usually is still sufficient for adequate carbon dioxide transfer.

However, in hypoxia caused by hypoventilation, including hypoventilation of only some of the alveoli (abnormal ventilation-perfusion ratio), carbon dioxide transfer between the alveoli and the atmosphere is affected as much as is oxygen transfer. Therefore, hypercapnia always results along with hypoxia. And in circulatory deficiency, diminished flow of blood decreases the removal of carbon dioxide from the tissues, resulting in tissue hypercapnia. However, the transport capacity of the blood for carbon dioxide is about three times that for oxygen, so that even here the hypercapnia is much less than the hypoxia.

EFFECTS OF HYPERCAPNIA ON THE BODY

When the alveolar P_{CO_2} rises above approximately 60 to 75 mm. Hg, dyspnea usually becomes intolerable, and as the P_{CO_2} rises to 70 to 80 mm. Hg, the person becomes lethargic and sometimes even semi-comatose. Total anesthesia and death result when the P_{CO_2} rises to 100 to 150 mm. Hg.

OXYGEN THERAPY IN THE DIFFERENT TYPES OF HYPOXIA

Oxygen therapy can be administered by (1) placing the patient's head in a "tent" which contains air fortified with oxygen, (2) allowing the patient to breathe either pure oxygen or high concentrations of oxygen from a mask, or (3) administering oxygen through an intranasal tube.

Oxygen therapy is of great value in certain types of hypoxia but of almost no value at all in other types. However, if one will simply recall the basic physiological principles of the different types of hypoxia, he can readily decide when oxygen therapy is of value and, if so, how valuable. For instance:

In *atmospheric hypoxia,* oxygen therapy can obviously completely correct the depressed oxygen level in the inspired gases and, therefore, provide 100 per cent effective therapy.

In *hypoventilation hypoxia,* a person breathing 100 per cent oxygen can move five times as much oxygen into the alveoli with each breath as he can when breathing normal air. Therefore, here again oxygen therapy can be extremely beneficial, increasing the available oxygen to as much as 400 per cent above normal.

In *hypoxia caused by impaired diffusion,* essentially the same result occurs as in hypoventilation

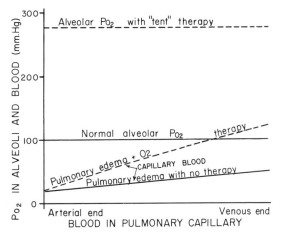

Figure 43–5 Absorption of oxygen into the pulmonary capillary blood in pulmonary edema with and without oxygen therapy.

hypoxia, for oxygen therapy can increase the P_{O_2} in the lungs from a normal value of about 100 mm. Hg to as high as 600 mm. Hg. This causes a greatly increased diffusion gradient between the alveoli and the blood, the gradient rising from a normal value of 60 mm. Hg to as high as 560 mm. Hg, or an increase of more than 800 per cent. This highly beneficial effect of oxygen therapy in diffusion hypoxia is illustrated in Figure 43–5, which shows that the pulmonary blood in a patient with pulmonary edema picks up oxygen up to eight times as rapidly as it would with no therapy.

In *hypoxia caused by anemia, carbon monoxide poisoning,* or *any other abnormality of hemoglobin transport,* oxygen therapy is of slight to moderate value because the amount of oxygen transported in the dissolved form in the fluids of the blood can still be increased above normal even though that transported by the hemoglobin is hardly altered. This is illustrated in Figure 43–6, which shows that an increase in alveolar P_{O_2} from a normal value of 100 mm. Hg to 600 mm. Hg increases the total oxygen in the blood (both combined with hemoglobin and dissolved) from 5 volumes per cent

to 6.5 volumes per cent. This represents a 30 per cent increase in the amount of oxygen transported to the tissues, and 30 per cent is often the difference between life and death. However, Figure 43–6 also shows that oxygen therapy in a pressure chamber at pressures above atmospheric pressure (760 mm. Hg) can be of still much greater value.

In *hypoxia caused by circulatory deficiency,* also called *ischemic hypoxia,* the value of oxygen therapy is usually very slight, because the problem here is sluggish flow of blood and not insufficient oxygen. Figure 43–7 illustrates, however, that even normal blood can carry a small amount of extra oxygen to the tissues (about 10 per cent extra) when the alveolar oxygen concentration is increased to 600 mm. Hg. Here again, this 10 per cent difference may save the life of the patient, as, for example, following an acute heart attack which causes the cardiac output to fall very low.

In the different types of *hypoxia caused by inadequate tissue use of oxygen,* there is no abnormality of oxygen pickup by the lungs or of transport to the tissues. Instead, the tissues simply need more oxygen than can reach them or than the enzymes can utilize. Therefore, oxygen therapy is usually of almost no benefit, though perhaps in a few instances high pressure oxygen therapy increases the P_{O_2} in the tissues sufficiently to increase oxygen utilization by as much as 1 to 5 per cent.

DANGER OF HYPERCAPNIA DURING OXYGEN THERAPY

Much of the stimulus that helps to maintain ventilation in hypoxia results from hypoxic stimulation of the aortic and carotid chemoreceptors. In the preceding chapter is was noted that in chronic hypoxia, oxygen lack becomes a far more powerful stimulus to respiration than usual, sometimes increasing the ventilation as much as 5- to 7-fold. Therefore, during oxygen therapy, relief of the hypoxia occasionally causes pulmonary ventilation to decrease so low that lethal levels of hypercapnia develop. For this reason, oxygen therapy in hypoxia is sometimes contraindicated.

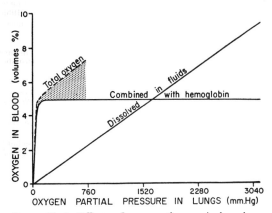

Figure 43–6 Effects of oxygen therapy in hypohemoglobinemic hypoxia.

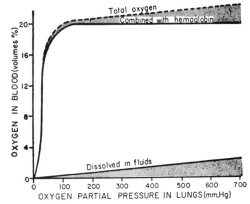

Figure 43–7 Effects of oxygen therapy in ischemic hypoxia.

ABSORPTION OF ENTRAPPED AIR

Whenever air becomes entrapped anywhere in the body, whether in the alveoli as a result of bronchial obstruction, in the gastrointestinal tract, in one of the nasal or auditory cavities, or simply as air injected beneath the skin, it is usually absorbed in a few hours or days.

Physical Principles of Air Absorption. Figure 43–8 shows the progressive stages of air absorption from a cavity. The cavity to the left is filled with recently injected air which has become humidified almost instantaneously by the fluids surrounding the cavity. Therefore, this cavity contains the normal concentrations of oxygen, nitrogen, and water vapor found in humidified air at sea level. Note that the total pressure of these three gases added together equals 760 mm. Hg. Furthermore, the pressure in the cavity remains exactly the same during the entire process of absorption because the atmospheric pressure continually presses against the body, and the body's tissues are flexible enough that the atmospheric pressure is transmitted also to almost any gas cavity beneath the skin.

At first, the P_{O_2} in the injected air is much greater than the P_{O_2} in the interstitial fluid, so that oxygen begins to be absorbed rapidly. On the other hand, the nitrogen pressure (P_{N_2}) in the cavity is actually less than that in the interstitial fluid because nitrogen, unlike oxygen, is not metabolized by the tissues of the body. As a result, minute amounts of nitrogen actually diffuse into the cavity at first. After large amounts of oxygen have been absorbed, the P_{O_2} falls to as little as 90 mm. Hg, as shown in the second cavity of Figure 43–8. This absorption of oxygen makes the cavity become smaller, thus increasing the P_{N_2} to a value greater than that in the interstitial fluid. Also, carbon dioxide diffuses into the cavity.

The final equilibrium state between the gases in the cavity and in the interstitial fluid is shown in the third cavity; the P_{O_2} has decreased to about 43 mm. Hg, the P_{CO_2} has increased to 40 mm. Hg, and the P_{N_2} has increased to 630 mm. Hg. From then on, both oxygen and nitrogen are absorbed continually. As this reduces the size of the cavity, the carbon dioxide and water pressures increase temporarily slightly above 40 and 47 mm. Hg, respectively, which are higher than the pressures of the surrounding fluids. As a consequence, the carbon dioxide and water vapor are also absorbed. This process continues until all the gases leave the cavity and the cavity collapses totally.

REFERENCES

Banyai, A. L., and Levine, E. R.: Dyspnea: Diagnosis and Treatment. Philadelphia, F. A. Davis Co., 1963.
Bendixen, H., et al.: Respiratory Care. 2nd ed., St. Louis, The C. V. Mosby Co., 1969.
Bicher, H. I., and Knisely, M. H.: Brain tissue reoxygenation time, demonstrated with a new ultramicro oxygen electrode. *J. Appl. Physiol.,* 28:387, 1970.
Brown, R.; Woolcock, A. J.; Vincent, N. J.; and Macklem, P. T.: Physiological effects of experimental airway obstruction with beads. *J. Appl. Physiol.,* 27:328, 1969.
Ciba Symposium on Oxygen Measurements in Blood and Tissues and Their Significance. Boston, Little, Brown and Company, 1966.
Comroe, J. H., Jr.; Forster, R. E., II; Dubois, A. B.; Briscoe, W. A.; and Carlsen, E.: The Lung: Clinical Physiology and Pulmonary Function Tests. 2nd ed., Chicago, Year Book Medical Publishers, Inc., 1962.
Cotes, J. E.: Lung Function: Assessment and Application in Medicine. Philadelphia, F. A. Davis Co., 1965.
Crossman, P. F.; Bushnell, L. S.; and Hedley-Whyte, J.: Dead space during artificial ventilation: gas compression and mechanical dead space. *J. Appl. Physiol.,* 28:94, 1970.
Fabian, L.: Anesthesia and the Circulation. Philadelphia, F. A. Davis Co., 1964.
Fatt, I., and St. Helen, R.: A multicathode polarographic oxygen sensor and its performance. *J. Appl. Physiol.,* 27:435, 1969.
Filley, G.: Pulmonary Insufficiency and Respiratory Failure. Philadelphia, Lea & Febiger, 1967.
Gray, F. D., Jr.: Pulmonary embolism. Philadelphia, Lea & Febiger, 1966.
Greene, D. G.: Pulmonary edema. In Handbook of Physiology. Baltimore, The Williams & Wilkins Co., 1965, Sec. III, Vol. II, p. 1585.
Guyton, A. C., and Farish, C. A.: A rapidly responding continuous oxygen consumption recorder. *J. Appl. Physiol.,* 14:143, 1959.
Hatzfeld, C.; Wiener, F.; and Briscoe, W. A.: Effects of uneven ventilation-diffusion ratios on pulmonary diffusing capacity in disease. *J. Appl. Physiol.,* 23:1, 1967.
Levy, S. E.; Shapiro, B. J.; and Simmons, D. H.: Pulmonary hemodynamics after autologous in vivo pulmonary thromboembolism. *J. Appl. Physiol.,* 27:53, 1969.
Nilsson, N. J.: Oximetry. *Physiol. Rev.,* 40:1, 1960.
Ouellet, Y.; Poh, S. C.; and Becklake, M. R.: Circulatory factors limiting maximal aerobic exercise capacity. *J. Appl. Physiol.,* 27:874, 1969.
Sobol, B. J.: Methods for simplifying constant-volume body plethysmograph. *J. Appl. Physiol.,* 27:295, 1969.
Suwa, K., and Bendixen, H. H.: Change in PaCO2 with mechanical dead space during artificial ventilation. *J. Appl. Physiol.,* 24:556, 1968.
Van Liere, E. J., and Stickney, J. C.: Hypoxia. Chicago, University of Chicago Press, 1963.

Figure 43–8 Absorption of air from an occluded cavity of the body.

PART VII

AVIATION, SPACE, AND DEEP SEA DIVING PHYSIOLOGY

CHAPTER 44

Aviation, High Altitude, and Space Physiology

As man has ascended to higher and higher altitudes in aviation, in mountain climbing, and in space vehicles, it has become progressively more important to understand the effects of altitude and low environmental gas pressures on the human body. In the early days of aviation only two factors were of concern: (1) the effects of hypoxia on the body and (2) the effects of physical factors of high altitude, such as temperature and ultraviolet radiation. When airplanes were made to go still higher and to make sharper turns, the problems of hypoxia became compounded, and, in addition, it was soon learned that airplanes could be built to withstand acceleratory forces far greater than the human body can stand. Now, with the space age at hand, all these problems have become multiplied to the point that the physical conditions in the spacecraft must be created artificially.

The present chapter deals with all these problems: first, the hypoxia at high altitudes, second, the other physical factors affecting the body at high altitudes, and, third, the tremendous acceleratory forces that occur in both aviation and space physiology.

EFFECTS OF LOW OXYGEN PRESSURE ON THE BODY

Barometric Pressures at Different Altitudes. As a prelude to discussing the effects of low oxygen pres-

sure on the body, we must recall that the total pressure of all the gases in the air, the *barometric pressure,* decreases as one rises to progressively higher and higher altitudes. Table 44–1 gives the pressures at different altitudes, showing that at sea level the pressure is 760 mm. Hg, while at 10,000 feet it is only 523 mm. Hg, and at 50,000 feet, 87 mm. Hg. This decrease in barometric pressure is the basic cause of all the hypoxia problems in high altitude physiology, for any time the barometric pressure decreases, the oxygen pressure decreases proportionately, remaining at all times slightly less than 21 per cent of the total barometric pressure.

Oxygen Content of the Air at Different Elevations. Table 44–2 shows that the partial pressure of oxygen (Po_2) in dry air at sea level is approximately 159 mm. Hg, though this can fall as much as 10 mm. when much water vapor exists in the air. The Po_2 at 10,000 feet is approximately 110 mm. Hg, at 20,000 feet 73 mm. Hg, and at 50,000 feet 18 mm. Hg.

ALVEOLAR Po_2 AT DIFFERENT ELEVATIONS

Obviously, when the Po_2 in the atmosphere decreases at higher elevations, a decrease in alveolar Po_2 is also to be expected. At low altitudes the alveolar Po_2 does not decrease quite so much as the atmospheric Po_2 because increased pulmonary ventilation helps to compensate for the diminished

TABLE 44–1.   Relationship of Barometric Pressure to Altitude

Altitude (ft.)	Pressure (mm. Hg)	Altitude (ft.)	Pressure (mm. Hg)	Altitude (ft.)	Pressure (mm. Hg)
0	760	22,000	321	44,000	116
2,000	707	24,000	294	46,000	106
4,000	656	26,000	270	48,000	96
6,000	609	28,000	247	50,000	87
8,000	564	30,000	226	52,000	79
10,000	523	32,000	206	54,000	72
12,000	483	34,000	187	56,000	66
14,000	446	36,000	170	58,000	60
16,000	412	38,000	155	60,000	54
18,000	379	40,000	141	62,000	49
20,000	349	42,000	128	64,000	45

TABLE 44-2. Effects of Low Atmospheric Pressures on Alveolar Gas Concentrations and on Arterial Oxygen Saturation

Altitude (ft.)	Barometric Pressure (mm. Hg)	P_{O_2} in Air (mm. Hg)	Breathing Air			Breathing Pure Oxygen		
			P_{CO_2} in Alveoli (mm. Hg)	P_{O_2} in Alveoli (mm. Hg)	Arterial Oxygen Saturation (%)	P_{CO_2} in Alveoli (mm. Hg)	P_{O_2} in Alveoli (mm. Hg)	Arterial Oxygen Saturation (%)
0	760	159	40	104	97	40	673	100
10,000	523	110	36	67	90	40	436	100
20,000	349	73	24	40	70	40	262	100
30,000	226	47	24	21	20	40	139	99
40,000	141	29	24	8	5	36	58	87
50,000	87	18	24	1	1	24	16	15

atmospheric oxygen. But at higher altitudes the alveolar P_{O_2} decreases even more than atmospheric P_{O_2} for peculiar reasons that are explained as follows:

Effect of Carbon Dioxide and Water Vapor on Alveolar Oxygen. Even at high altitudes carbon dioxide is continually excreted from the pulmonary blood into the alveoli. Also, water vaporizes into the alveolar space from the respiratory surfaces. Therefore, these two gases dilute the oxygen and nitrogen already in the alveoli, reducing the amount of space available in the alveoli for oxygen.

The presence of carbon dioxide and water vapor in the alveoli becomes exceedingly important at high altitudes because the total barometric pressure falls to low levels while the pressures of carbon dioxide and water vapor do not fall comparably. For instance, water vapor pressure remains 47 mm. Hg as long as the body temperature is normal, regardless of altitude; and the pressure of carbon dioxide falls from about 40 mm. Hg at sea level to about 24 mm. Hg at extremely high altitudes because of increased respiration.

Now let us see how the pressures of these two gases affect the available space for oxygen. Let us assume that the total barometric pressure falls to 100 mm. Hg; 47 mm. Hg of this must be water vapor, leaving only 53 mm. Hg for all the other gases. Under acute exposure to high altitude, 24 mm. Hg of the 53 mm. Hg must be carbon dioxide, leaving a remaining space of only 29 mm. Hg. If there were no use of oxygen by the body, one-fifth of this 29 mm. Hg would be oxygen, and four-fifths would be nitrogen; or, the P_{O_2} in the alveoli would be 6 mm. Hg. However, by this time the person's tissues would be almost totally anoxic, for which reason even this last vestige of oxygen would be absorbed into the blood, leaving not more than 1 mm. Hg oxygen pressure in the alveoli. Therefore, at a barometric pressure of 100 mm. Hg, the person could not possibly survive. This effect is somewhat different if the person is breathing pure oxygen, as we shall see in the following discussions.

A simple formula for calculating alveolar P_{O_2} is the following:

$$\text{Alveolar } P_{O_2} = \frac{P_B - P_{CO_2} - 47}{5} - P_{O_2} \text{ LOSS}$$

In this formula P_B is barometric pressure, the value 47 is the vapor pressure of water, and P_{O_2} LOSS is the oxygen pressure decrease caused by oxygen uptake into the blood.

Alveolar P_{O_2} at Different Elevations. Table 44-2 shows the P_{O_2}'s in the alveoli at different elevations when one is breathing air and when one is breathing pure oxygen. When breathing air, the alveolar P_{O_2} is 104 mm. Hg at sea level; it falls to approximately 67 mm. Hg at 10,000 feet and to only 1 mm. Hg at 50,000 feet.

Figure 44-1 illustrates the partial pressures and concentrations of the various gases in the alveoli at sea level and at 30,000 feet. These values were calculated in the manner just described.

SATURATION OF HEMOGLOBIN WITH OXYGEN AT DIFFERENT ALTITUDES

Figure 44-2 illustrates arterial oxygen saturation at different altitudes when breathing air and when breathing oxygen, and the actual per cent saturations at certain altitudes are given in Table 44-2. Up to an elevation of approximately 10,000 feet, even when breathing air, the arterial oxygen saturation remains at least as high as 90 per cent. However, above 10,000 feet the arterial oxygen saturation falls progressively, as illustrated by the left-hand curve of the figure, until it is only 70 per cent at 20,000 feet elevation and still less at higher altitudes.

EFFECT OF BREATHING PURE OXYGEN ON THE ALVEOLAR P_{O_2} AT DIFFERENT ALTITUDES

Referring once again to Figure 44-1, note that when a person breathes air at 30,000 feet, a considerable proportion of his alveolar air is nitrogen. But if he breathes pure oxygen instead of air, most of the space in the alveoli formerly occupied by nitrogen now becomes occupied by oxygen instead. Theoretically, at 30,000 feet the aviator

AIR AT O FEET:
$O_2 = 159$ mm. Hg (21 %)
$N_2 = 601$ (79%)
Total = 760   (100%)

AIR AT 30,000 FEET:
$O_2 = 47$ mm.Hg (21 %)
$N_2 = 179$ (79%)
Total = 226   (100%)

ALVEOLAR AIR AT O FEET:
$H_2O = 47$ mm. Hg (6%)
$CO_2 = 40$ (5%)
$N_2 = 569$ (75%)
        656   (86%)

ALVEOLAR AIR AT 30,000 FEET:
$H_2O = 47$ mm. Hg (21 %)
$CO_2 = 24$ (11%)
$N_2 = 134$ (59%)
        205   (91%)

Total = 760   (100%)
       −656   (86%)
$O_2 = 104$ mm. Hg (14%)

Total = 226   (100 %)
       −205   (91 %)
$O_2 = 21$ mm. Hg (9%)

Figure 44-1   Calculation of alveolar gaseous partial pressures at sea level and at 30,000 feet.

could have an alveolar Po_2 of 139 mm. Hg instead of the 21 mm. Hg that he has when he breathes air.

The second curve of Figure 44–2 illustrates the arterial oxygen saturation at different altitudes when one is breathing pure oxygen. Note that the saturation remains above 90 per cent until the aviator ascends to approximately 39,000 feet; then it falls rapidly to approximately 50 per cent at about 47,000 feet.

THE "CEILING" WHEN BREATHING AIR AND WHEN BREATHING OXYGEN IN AN UNPRESSURIZED AIRPLANE

Comparing the two arterial oxygen saturation curves in Figure 44–2, one notes that an aviator breathing oxygen can ascend to far higher altitudes than one not breathing oxygen. For instance, the arterial saturation at 47,000 feet when breathing oxygen is about 50 per cent and is equivalent to the arterial oxygen saturation at 23,000 feet when breathing air. And, because a person ordinarily can remain conscious until the arterial oxygen saturation falls to 40 to 50 per cent, the ceiling for an unacclimatized aviator in an unpressurized air-

plane when breathing air is approximately 23,000 feet and when breathing pure oxygen about 47,000 feet, provided the oxygen-supplying equipment operates perfectly.

EFFECTS OF HYPOXIA

Probably the earliest effect of hypoxia on bodily function is decreased proficiency of night vision. The amount of light needed for an aviator to see his surroundings must be increased approximately 23 per cent above normal at 5000 feet, 59 per cent at 10,000 feet, and 140 per cent at 16,000 feet. Thus, even the slightest decrease in oxygen saturation of the arterial blood depresses the function of the rods in the retina.

The rate of pulmonary ventilation ordinarily does not increase significantly until one has ascended to about 8000 feet. At this height the arterial oxygen saturation has fallen to approximately 93 per cent, at which level the chemoreceptors respond significantly. Above 8000 feet the chemoreceptor stimulatory mechanism progressively increases the ventilation until one reaches approximately 16,000 to 20,000 feet, at which elevation the ventilation has reached a maximum of approximately 65 per cent above normal. Further increase in altitude does not further activate the chemoreceptors.

Other effects of hypoxia, beginning at an elevation of approximately 12,000 feet, are drowsiness, lassitude, mental fatigue, sometimes headache, occasionally nausea, and sometimes euphoria. Most of these symptoms increase in intensity at still higher altitudes, the headache often becoming especially prominent and the cerebral symptoms sometimes progressing to the stage of twitchings or convulsions and ending, above 23,000 feet, in coma.

One of the most important effects of hypoxia is decreased mental proficiency for judgment, for achieving discrete motor movements, and for memory. Ordinarily these abilities remain absolutely normal up to approximately 9000 feet, and

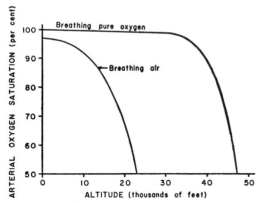

Figure 44-2   Effect of low atmospheric pressure on arterial oxygen saturation when breathing air and when breathing pure oxygen.

may be completely normal for a short time up to elevations of 15,000 feet. But, if the aviator is exposed to hypoxia for a long time, his mental proficiency, as measured by reaction times, handwriting, and other psychological tests, may decrease to 20 per cent below normal even at altitudes as low as 11,000 feet. If an aviator stays at 15,000 feet for one hour his mental proficiency ordinarily will have fallen to approximately 50 per cent below normal, and after 18 hours at this level to approximately 80 per cent below normal.

Respiratory Center Depression as the Cause of Death in Hypoxia. When a person develops severe hypoxia—that is, to the stage of coma—the respiratory center itself often becomes depressed after another few minutes because of metabolic deficit in its neuronal cells. This counteracts the stimulatory effect of the chemoreceptor mechanism, and the respiration, instead of increasing further, decreases precipitously until it actually ceases.

Effects of Sudden Exposure to Low P_{O_2}. At times the aviator who has been flying at a very high altitude with special oxygen equipment suddenly becomes detached from his oxygen supply. In this instance, the alveolar P_{O_2} falls within a few seconds to a low value, but because of oxygen stored in his body fluids (combined with hemoglobin and with various oxygen carriers of the tissues), a short time elapses before the body suffers dire results from oxygen lack.

Figure 44–3 illustrates graphically the time that must ordinarily elapse, first, before the aviator shows any signs of diminished consciousness, and, second, before actual coma results. Note from this figure that at 38,000 feet diminished consciousness begins in approximately 30 seconds, and coma results in approximately one minute, whereas at 28,000 feet diminished consciousness begins in approximately one minute, and coma occurs in approximately two minutes.

Figure 44–3 Time of exposure to low oxygen concentrations required to cause diminished consciousness or coma. (From Armstrong: Principles and Practice of Aviation Medicine. The Williams & Wilkins Co.)

ACCLIMATIZATION TO LOW P_{O_2}

If a person remains at high altitudes for days, weeks, or years, he gradually becomes acclimatized to the low P_{O_2} so that it causes fewer and fewer deleterious effects to his body and also so that it becomes possible for him to work harder or to ascend to still higher altitudes. The five principal means by which acclimatization comes about are: (1) increased pulmonary ventilation, (2) increased hemoglobin in the blood, (3) increased diffusing capacity of the lungs, (4) increased vascularity of the tissues, and (5) increased ability of the cells to utilize oxygen despite the low P_{O_2}.

Increased Pulmonary Ventilation. On immediate exposure to low P_{O_2}, the hypoxic stimulation of the chemoreceptors increases alveolar ventilation to a maximum of about 65 per cent. This is an immediate compensation for the high altitude, and it alone allows the person to rise several thousand feet higher than would be possible without the increased ventilation. Then, if he remains at a very high altitude for several days, his ventilation gradually increases to as much as five to seven times normal. The basic cause of this gradual increase is the following:

The immediate 65 per cent increase in pulmonary ventilation on rising to a high altitude blows off large quantities of carbon dioxide, reducing the P_{CO_2} and increasing the pH of the body fluids. Both of these changes *inhibit* the respiratory center and thereby *oppose the stimulation by the hypoxia.* However, during the ensuing three to five days, this inhibition fades away, allowing the respiratory center now to respond with full force to the chemoreceptor stimuli resulting from hypoxia, and the ventilation increases to about five to seven times normal. The cause of this fading inhibition is unknown, but there is some evidence that it results from reduced CSF bicarbonate ion and perhaps also from reduced brain tissue bicarbonate ion, and this in turn decreases the pH. (See Chap. 42.)

Rise in Hemoglobin During Acclimatization. It will be recalled from Chapter 8 that hypoxia is the principal stimulus for causing an increase in red blood cell production. Ordinarily, in full acclimatization to low oxygen the hematocrit rises from a normal value of 40 to 45 to an average of 60 to 65, with an average increase in hemoglobin concentration from the normal of 15 gm. per cent to about 22 gm. per cent.

In addition, the blood volume also increases, often by as much as 20 to 30 per cent, resulting in a total increase in circulating hemoglobin of as much as 50 to 90 per cent.

Unfortunately, this increase in hemoglobin and blood volume is a slow one, having almost no effect until after two to three weeks, reaching half development in a month or so, and becoming fully developed only after several years.

Increased Diffusing Capacity During Acclimatization. It will be recalled that the normal diffusing capacity for oxygen through the pulmonary membrane is approximately 20 ml. per mm. Hg pressure gradient per minute, and this diffusing capacity can increase as much as three-fold during exercise. A similar increase in diffusing capacity occurs at high altitude. Part of the increase probably results from greatly increased pulmonary capillary blood volume, which expands the capillaries and increases the surface through which oxygen can diffuse into the blood. Another part probably results from an increase in lung volume, which presumably expands the surface area of the alveolar membrane.

The Circulatory System in Acclimatization — Increased Vascularity. The cardiac output often increases as much as 20 to 30 per cent immediately after a person ascends to a high altitude, but it usually falls back to normal within a few days. In the meantime, though, the blood flow through certain organs, such as the skin and kidneys, decreases, while the flow through the muscles, heart, brain, and other organs that normally require large quantities of oxygen increases. Furthermore, histologic studies of animals that have been exposed to low oxygen levels for months or years show *greatly increased vascularity* (increased numbers and sizes of capillaries) of the hypoxic tissues. This helps to explain what happens to the 20 to 30 per cent increase in blood volume, and it means that the blood comes into much closer contact with the tissue cells than normally.

Cellular Acclimatization. In human beings and animals that are native to altitudes of 13,000 to 17,000 feet, certain cellular oxidative enzyme systems are more plentiful than in sea level inhabitants. Therefore, it is presumed that these human beings and animals can utilize oxygen that is available more effectively than can their sea level counterparts.

However, a similar increase in oxidative enzymes does not occur significantly in non-native animals or human beings exposed to high altitudes for as long as a year. Therefore, it is likely that the elevated oxidative enzyme systems of "naturally acclimatized" persons results from either many years of existence at high altitudes or from genetic differences that have come about through generations of selection at these altitudes. It is particularly significant that naturally acclimatized natives can survive for several hours at altitudes as high as 30,000 feet without breathing supplemental oxygen, whereas this would be lethal to even the best acclimatized lowlander.

DECOMPRESSION EFFECTS AT HIGH ALTITUDES

Decompression Sickness. In addition to the detrimental effects of hypoxia at high altitudes, a rapid decrease in barometric pressure can directly harm the body by causing gas bubbles in the body fluids. All degrees of damage can occur, ranging from mild pain to death. This condition is most frequently called decompression sickness, though it is sometimes also known as *dysbarism*.

Decompression sickness occurs much more frequently and much more severely in deep sea divers than in aviators. Therefore, its basic causes and effects are discussed in detail in the following chapter. Briefly, it is caused by a decrease in the barometric pressure below the total pressure of all the gases dissolved in the body fluids, and this allows the gases to escape from the dissolved state to the gaseous state, as will be explained in the next chapter.

The usual cause of decompression sickness in aviation is rapid ascent in the unpressurized airplane — ascending to over 25,000 to 30,000 feet in a few minutes. Slower ascents rarely cause the sickness.

Explosive Decompression. Experimentally, human beings have been decompressed explosively from sea level pressure to barometric pressures equal to those above 50,000 feet, and animals have been explosively decompressed to an almost complete vacuum. Essentially no harm results from the suddenness of the change in pressure. Indeed, even the ears ordinarily do not suffer any damage because the increasing volume of air in the middle ear automatically opens the eustachian tube so that the air flows outward without damage to the tympanum. Perhaps the most significant effect is the sudden increase in volume of gastrointestinal gases, for any gas in the gastrointestinal tract is expanded several-fold when one is explosively decompressed at a level of 50,000 ft. Ordinarily, most of these gases can be expelled as flatus, for most gastrointestinal gases are in the colon, which is in direct communication with the anus, and the remainder, which is ordinarily in the stomach, can be released by belching.

Boiling of the Body Fluids. The barometric pressure falls to 47 mm. Hg at an altitude of approximately 63,000 feet. Therefore, if an aviator should be subjected to the atmosphere above 63,000 feet, his total alveolar pressure would be less than the vapor pressure of water in the alveoli. Consequently, the water in the alveoli would immediately boil. Also, water vapor bubbles would appear all through the body fluids. In fact, in an animal suddenly subjected to a pressure equal to that at 70,000 feet, approximately 2.5 per cent of his body weight boils away in the three minutes that he remains alive. This means that in a human being explosively decompressed to an elevation of 70,000 feet, approximately 4 pounds of water would probably boil away from the lungs before death ensued in approximately three minutes.

EFFECTS OF ACCELERATORY FORCES ON THE BODY IN AVIATION AND SPACE PHYSIOLOGY

Because of rapid changes in velocity and direction of motion in airplanes and space ships, several types of acceleratory forces often affect the body during flight. At the beginning of flight, simple linear acceleration occurs; at the end of flight, deceleration; and, every time the airplane turns, angular acceleration occurs. In aviation physiology it is usually angular acceleration that demands greatest consideration because the structure of the airplane is capable of withstanding much greater angular acceleration than is the human body.

ANGULAR ACCELERATORY FORCES

When an airplane makes a turn, the force of angular acceleration is determined by the following relationship;

$$f = \frac{mv^2}{r}$$

in which f is the angular acceleratory force, m is the mass of the object, v is the velocity of travel, and r is the radius of curvature of the turn. From this formula it is obvious that as the velocity increases, the force of angular acceleration increases in proportion to the square of the velocity. It is also obvious that the force of acceleration is directly proportional to the sharpness of the turn.

Measurement of Angular Acceleratory Force—"G." When a subject is simply sitting in his seat, the force with which he is pressing against the seat results from the pull of gravity, and it is equal to his weight. The intensity of this force is 1 "G" because it is equal to the pull of gravity. If the force with which he presses against his seat becomes five times his normal weight during a pull-out from a dive, the force acting upon the seat is 5 G.

If the airplane goes through an outside loop so that the pilot is held down by his seat belt, *negative G* is applied to his body, and, if the force with which he is thrown against his belt is equal to the weight of his body, the negative force is −1 G.

Effects of Angular Acceleratory Force on the Body. EFFECTS ON THE CIRCULATORY SYSTEM. The most important effect of angular acceleration is on the circulatory system because blood is mobile and can be translocated by angular forces. Angular forces also tend to displace the tissues, but, because of their more solid structure, they only sag—ordinarily not enough to cause abnormal function.

When the aviator is subjected to *positive G* his blood is centrifuged toward the lower part of his body. Thus, if the angular acceleratory force is 5 G and the subject is in a standing position, the hydrostatic pressure in the veins of the feet is five times normal, or approximately 450 mm. Hg, and even in the sitting position this pressure is nearly 300 mm. Hg. As the pressure in the vessels of the lower part of the body increases, the vessels passively dilate, and a major proportion of the blood from the upper part of the body is translocated into these lower vessels. Because the heart cannot pump unless blood returns to it, the greater the quantity of blood "pooled" in the lower body the less becomes the cardiac output.

Figure 44–4 illustrates the effect of different degrees of acceleration on systemic arterial pressure in an aviator in the sitting position, showing that when the acceleration rises to 4 G the systemic arterial pressure at the level of the heart falls to approximately 40 mm. Hg. As a result, positive G diminishes blood flow to the brain. Acceleration greater than 4 to 6 G ordinarily causes "blackout" of vision within a few seconds and then unconsciousness shortly thereafter. The time required for these symptoms to appear is shown in Figure 44–5. The upper part of this figure shows that the arterial pressure recovers considerably after about 15 seconds of exposure to 3.3 G of acceleration; this recovery is caused by activation of the baroreceptor reflexes.

EFFECTS ON THE VERTEBRAE. Extremely high acceleratory forces for even a fraction of a second can fracture the vertebrae. The degree of positive acceleration that the average person can withstand in the sitting position before vertebral fracture occurs is approximately 20 G. This force also approaches the limit of safety for the structural elements of most airplanes. Therefore, any attempt to utilize such intense positive G as this in maneuvers would be extremely dangerous for reasons other than simply the effects on the body.

Transverse G. The human body can withstand tremendous transverse acceleratory forces (forces applied along the antero-posterior axis of the body —lying down in the airplane, for instance). If the transverse acceleratory forces are applied rather uniformly over large areas of the body, as much as 100 G can be withstood for a fraction of a second, and as much as 15 to 25 G can be withstood for many seconds without serious effects other than occasional collapse of a lung, which is not a lethal

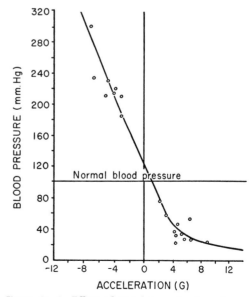

Figure 44–4 Effect of angular acceleratory forces on arterial pressure measured at the heart. (From Armstrong: Principles and Practice of Aviation Medicine. The Williams & Wilkins Co.)

Figure 44–5 *Lower curve:* Time required at different G to cause blackout. *Top curve:* Changes in systolic and diastolic arterial pressures following abrupt and continuing exposure to an angular acceleratory force of 3.3 G. (Modified from Martin and Henry: *J. Aviat. Med.,* 22:382, 1951.)

effect. Therefore, when very large acceleratory forces are involved, the aviator or astronaut flies in the semi-reclining or lying position.

Negative G. The effects of negative G on the body are less dramatic acutely but possibly more damaging permanently than the effects of positive G. An aviator can usually go through outside loops up to negative acceleratory forces of −4 to −5 G without harm except for intense momentary hyperemia of the head, though occasionally psychotic disturbances lasting for 15 to 20 minutes occur thereafter as a result of brain edema.

Figure 44–4 shows that the arterial pressure at the level of the heart increases greatly, as would be expected. An interesting but paradoxical effect is caused by this very high pressure: it elicits such an extreme baroreceptor reflex that severe vagal slowing of the heart occurs, sometimes actually stopping the heart for 5 to 10 seconds.

If the arterial pressure becomes great enough, it can cause some of the small vessels on the surface of the head and possibly in the brain to rupture. However, the vessels inside the cranium show less tendency for rupture than would be expected, for the cerebrospinal fluid is centrifuged toward the head at the same time that blood is centrifuged toward the cranial vessels, and the greatly increased pressure of this fluid acts as a cushioning buffer on the outside of the brain to prevent vascular rupture. Even so, animals exposed to negative acceleratory forces of 20 to 40 G have developed subarachnoid hemorrhages, and undoubtedly such effects could occur in the human being.

Because the eyes are not protected by the cranium, intense hyperemia occurs in them during negative G. As a result, the eyes often become temporarily blinded with "redout."

Protection of the Body Against Angular Acceleratory Forces. Specific procedures and apparatus have been developed to protect aviators against the circulatory collapse that occurs during positive G. First, if the aviator tightens his abdominal muscles to an extreme degree and leans forward to compress his abdomen, he can prevent some of the pooling of blood in the large vessels of the abdomen, thereby delaying the onset of blackout. Also, special "anti-G" suits have been devised to prevent pooling of blood in the lower abdomen and legs. The simplest of these applies positive pressure to the legs and abdomen by inflating compression bags as the G increases. Theoretically, a pilot submerged in a tank or suit of water could withstand very high degrees of acceleration, both positive and negative, for the pressures developed in the water during angular acceleration almost exactly balance the forces acting on the body. Unfortunately, however, the presence of air in the lungs still allows displacement of the heart, the lung tissues, and the diaphragm into seriously abnormal positions despite submersion in water. Therefore, even when this procedure is used, the limits of safety are about 15 to 20 G.

EFFECTS OF LINEAR ACCELERATORY FORCES ON THE BODY

Acceleratory Forces in Space Travel. In contrast to aviation physiology, the spacecraft cannot make rapid turns; therefore, angular acceleration is of little importance except when the spacecraft goes into abnormal gyrations. On the other hand, take-off acceleration and landing deceleration might be tremendous; both of these are types of linear acceleration.

Figure 44–6 illustrates a typical profile of the acceleration during take-off in a three-stage spacecraft, showing that the first stage booster causes

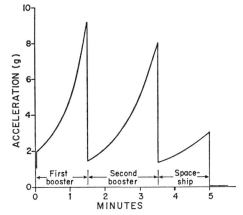

Figure 44–6 Acceleratory forces during the take-off of a spacecraft.

acceleration as high as 9 G and the second stage booster as high as 8 G. In the standing position the human body could not withstand this much acceleration, but in a semi-reclining position *transverse to the axis of acceleration,* this amount of acceleration can be withstood with ease despite the fact that the acceleratory forces continue for as long as five minutes at a time. Therefore, we see the reason for the reclining seats used by the astronauts.

Problems also occur during deceleration when the spacecraft re-enters the atmosphere. A person traveling at Mach 1 (the speed of sound and of fast airplanes) can be safely decelerated in a distance of approximately 0.12 mile, whereas a person traveling at a speed of Mach 100 (a speed possible in interplanetary space travel) requires a distance of about 10,000 miles for safe deceleration. The principal reason for this difference is that the total amount of energy that must be dispelled during deceleration is proportional to the *square* of the velocity, which alone increases the distance 10,000-fold. But, in addition to this, a human being can withstand far less deceleration if it lasts for a long time than for a short time. Therefore, deceleration must be accomplished much more slowly than is necessary at the slower initial velocities.

Deceleratory Forces Associated with Parachute Jumps. When the parachuting aviator leaves the airplane, his velocity of fall is exactly 0 feet per second at first. However, because of the acceleratory force of gravity, within 1 second his velocity of fall is 32 feet per second; in two seconds it is 64 feet per second, etc. However, as his rate of fall increases, the air resistance tending to slow his fall also increases. Finally, the air resistance is equal to the acceleratory force of gravity so that by the time he has fallen for approximately 12 seconds and a distance of 1400 feet, he will be falling at a "terminal velocity" of 109 to 119 miles per hour (175 feet per second). This is illustrated by the curve in Figure 44–7. However, if the pilot jumps from a very high altitude, where the atmosphere offers little resistance, the terminal velocity is much greater than 175 feet per second, but it slows to 175 feet per second as the body reaches the higher density atmosphere close to earth.

If the parachutist has already reached the terminal velocity of fall before he opens his parachute, an "opening shock load" of approximately 1200 pounds occurs on the parachute strands.

The usual size parachute slows the fall of the parachutist to approximately one-ninth the terminal velocity. In other words, the speed of landing is approximately 20 feet per second, and the force of impact against the earth is approximately $\frac{1}{81}$ the force of impact without a parachute. Even so, the force of impact is still great enough to cause considerable damage to the body unless the parachutist is properly trained in landing, for it is not evident as one descends how rapidly the earth is approaching. Actually, the force of impact with the earth is approximately the same as that which

Figure 44–7 Velocity of fall of a human body from a high altitude, showing the attainment of a "terminal velocity." (From Armstrong: Principles and Practice of Aviation Medicine. The Williams & Wilkins Co.)

would be experienced from jumping from a height of about 7 feet. If the parachutist is not careful, his senses will allow him to strike the earth with his legs still extended, and this will result in tremendous deceleratory forces along the skeletal axis of his body, resulting in fracture of the pelvis, of a vertebra, or a leg. Consequently, the trained parachutist strikes the earth with knees bent to cushion the shock of landing.

PERCEPTIONS OF EQUILIBRIUM AND ANGULAR ACCELERATION IN BLIND FLYING

The various sensations of equilibrium will be discussed in Chapter 57, but, in essence, any time the position of the head is not along the vertical axis of forces applied to the body, the otoliths in the utricles of the labyrinths apprise the psyche of this lack of equilibrium. However, the degree of proficiency of equilibrium perceptions has certain limits. For instance, the body must lean forward as much as 5 to 10 degrees before the utricles perceive this forward leaning. When the aviator is not flying blind, he can perceive forward leaning of the airplane by observing the ground, but, when he is flying blind, it is necessary that appropriate instruments be available to indicate the rate of descent. Also, the utricle may fail to apprise the pilot of ascent up to an error of as much as 24 degrees. Obviously, then, blind flying often results in stall of the airplane unless appropriate instruments are available.

Perhaps the least effective of the perceptive organs in blind flying are those for angular acceleration, because the organs that perceive angular acceleration, the semicircular canals, are excited only for the first few seconds after angular acceleration begins and only for the first few seconds after angular acceleration ends. Consequently, an airplane may gradually go into a turn and the aviator might perceive that he is going into the turn, but once in it he usually loses all perception of continuing to turn. Furthermore, if he goes into the turn slowly — at a rate less than 2 degrees per second per second — he may not even perceive the fact that he enters the turn. Consequently, of especial importance among the instruments for blind flying are those that apprise the aviator of the direction of travel of the airplane and the rate of turn.

PROBLEMS OF TEMPERATURE IN AVIATION AND SPACE PHYSIOLOGY

The cold temperatures of the upper atmosphere involve essentially the same physiological problems as cold temperatures on the surface of the earth; these problems will be discussed in Chapter 71. Therefore, for the present it shall only be pointed out how the temperatures change as one ascends to higher and higher elevations. If the temperature at the earth's surface is approximately 20° C., the temperatures at different altitudes are approximately the following:

feet	° C.
0	20
10,000	0
20,000	−22
30,000	−44
40,000	−55

It is obvious from these values that special clothing or special heating apparatus must be designed for flying at high altitudes.

Temperature in Space. Though the temperature of the air falls to about −55° C. several miles above the earth, the temperature rises to a very high value in space. In the ionosphere several hundred miles above the earth and in space, the kinetic energy of the few molecules, atoms, and ions is extreme. The reason for this is that any particles that can escape this far away from the gravitational pull of the earth must have a great velocity. As a result, by the time the spacecraft has reached an altitude above 350 miles, the temperature of the surrounding particles is about 3000° C. Yet, strangely enough, this has almost no effect on the temperature of the spacecraft because of the *sparsity* of these particles — that is, they are far too few to impart any significant amount of heat to the spacecraft. Instead, the temperature of the spacecraft is determined by the relative *absorption of radiant energy* from the sun versus the *re-radiation of energy* away from the spacecraft. Different coatings for the spacecraft have different absorptive and radiation characteristics. Therefore, with appropriate engineering design, the spacecraft can be constructed so that the temperature inside is almost exactly the desired value.

RADIATION AT HIGH ALTITUDES AND IN SPACE

Radiation in the Atmosphere. The electromagnetic radiations change considerably between the surface of the earth and at high altitudes because the atmosphere filters some of the radiations from the sun's rays before they reach the earth's surface. Ordinarily most of the ultraviolet light is absorbed before it reaches the earth's surface. Consequently, a person exposed to the sun in the upper atmosphere is many times as likely to develop sunburn as on the earth's surface.

Approximately 18 per cent of the visible light is filtered from the sun's rays before they strike the surface of the earth even on a perfectly clear day. Therefore, the brightness of the sun is 1.2 times as great in the upper atmosphere as on earth. This makes the earth less distinct to the aviator than otherwise for two reasons: first, he is looking from an area of greater brightness toward an area of lesser brightness, and, second, light is reflected from the atmosphere into his eyes. This reflected light especially blocks his vision of the horizon.

In aviation there is no significant hazard from gamma and x-rays, but this does become a problem in space physiology.

Radiation Hazards in Space Physiology. Large quantities of cosmic particles are continually bombarding the earth, some originating from the sun and some from outer space. The magnetic field of the earth traps many of these cosmic particles in two major belts around the earth called *Van Allen radiation belts*, illustrated in Figure 44–8. The inner belt begins at an altitude of about 300 miles and extends to about 3000 miles. The outer belt begins at about 6000 miles and extends to 20,000 miles. As also illustrated in Figure 44–8, the inner belt extends only 30 degrees on each side of the equatorial plane, and the outer belt 70 to 75 degrees.

The types of radiation in the two Van Allen belts are almost entirely high energy electrons and protons, the outer belt being comprised almost entirely of electrons and the inner belt of both. The energy level in the inner belt is extremely high, many of the particles having energies as high as 40 Mev., which makes it almost impossible to shield a spacecraft adequately against this radiation. Even with best possible shielding, a person traversing these two belts in an interplanetary space trip would be expected to receive as much as 10 roentgens of radiation, which is about one-fortieth the lethal dose; and a person in a spacecraft orbiting the earth within one of these two belts could receive enough radiation in only a few hours to cause death. During solar flares, the intensity of the outer Van Allen belt increases tremendously, which could make space travel through this belt at this time extremely dangerous.

Thus, it is important to orbit spacecraft below an altitude of 200 to 300 miles, an altitude at which the radiation hazard is slight. Also, it is possible

Figure 44-8   The Van Allen belts of radiation around the earth. (From Newell: *Science 131*:385, 1960.)

to minimize the radiation hazard during inter-planetary space travel by leaving the earth or returning to earth near one of the poles rather than near the equator.

"ARTIFICIAL CLIMATE" IN THE SEALED SPACECRAFT

Since there is no atmosphere in outer space, an artificial atmosphere and other artificial conditions of climate must be provided. The ability of a person to survive in this artificial climate depends entirely on appropriate engineering design.

Most important of all, the oxygen concentration must remain high enough and the carbon dioxide concentration low enough. The gas mixture most suitable for spacecraft travel is one of almost pure oxygen, but one having just enough nitrogen to prevent fire hazards. The total pressure is maintained as low as 200 to 400 mm. Hg. This allows more than adequate oxygenation of the astronauts, and yet the low pressure minimizes the hazard of explosive decompression of the spacecraft.

For space travel lasting more than several weeks, it will be impractical to carry along an adequate oxygen supply and enough carbon dioxide absorbent. For this reason, "recycling techniques" have been developed for use over and over again of the same oxygen. These techniques also frequently include re-use of the same food and water. Basically, they involve (1) a method for removing oxygen from carbon dioxide, (2) a method for removing water from the human excreta, and (3) use of the human excreta for resynthesizing or regrowing an adequate food supply.

Large amounts of energy are required for these processes, and the real problem at present is to derive enough energy from the sun's radiation to energize the necessary chemical reactions. Some recycling processes depend on purely physical procedures, such as distillation, electrolysis of water, and capture of the sun's energy by solar batteries, whereas others depend on biologic methods, such as use of algae, with its large store of chlorophyll, to generate foodstuffs by photosynthesis. Unfortunately, a completely practical system for recycling is yet to be achieved. The problem is the weight of the equipment that must be carried.

WEIGHTLESSNESS IN SPACE

A person in an orbiting satellite or in any non-propelled spacecraft experiences a feeling of weightlessness. That is, he is not drawn toward the bottom, sides, or top of the spacecraft but simply floats inside its chambers. The cause of this is not failure of gravity to pull on the body, because gravity from any nearby heavenly body is still active. However, the gravity acts on both the spacecraft and the person at the same time, and since there is no resistance to movement in space, both are pulled with exactly the same acceleratory forces and in the same direction. For this reason, the person simply is not attracted toward any wall of the spacecraft.

Weightlessness fortunately has not proved to be a physiological problem but rather an engineering problem in regard to providing special techniques for eating and drinking (since food and water will not stay in open plates or glasses), special waste disposal systems, and adequate hand holds or other means for stabilizing the person in the spacecraft so that he can adequately control the operation of the ship.

REFERENCES

Allen, T. H.; Maio, D. A.; Beard, S. E.; and Bancroft, R. W.: Space-cabin and suit pressures for avoidance of decompression sickness and alleviation of fire hazard. *J. Appl. Physiol., 27*:13, 1969.

Altland, P. D.; Highman, B.; and Nelson, B. D.: Serum enzyme and tissue changes in rats exercised repeatedly at altitude: effects of training. *Amer. J. Physiol., 214*:28, 1968.

Armstrong, H. G.: Aerospace Medicine. Baltimore, The Williams & Wilkins Co., 1961.

Banchers, N.; Rutishauser, W. J.; Tsakiris, A. G.; and Wood, E. H.: Pericardial pressure during transverse acceleration in dogs without thoracotomy. *Circ. Res., 20*:65, 1967.

Brown, J. H. U.: Physiology of Man in Space. New York, Academic Press., Inc., 1963.

Clamann, H. G.: Space physiology. *In* Handbook of Physiology. Baltimore, The Williams & Wilkins Co., 1965, Sec. III, Vol. II, p. 1147.

Dill, D. B.; Horvath, S. M.; Dahns, T. E.; Parker, R. E.; and Lynch, J. R.: Hemoconcentration at altitude. *J. Appl. Physiol., 27*:514, 1969.

Dill, D. B.; Myhre, L. G.; Brown, D. K.; Burrus, K.; and Gehlsen, G.: Work capacity in chronic exposures to altitude. *J. Appl. Physiol., 23*:555, 1967.

Eger, E. I., II; Kellogg, R. H.; Mines, A. H.; Lima-Ostos, M.; Morrill, C. G.; and Kent, D. W.: Influence of CO_2 on ventilatory acclimatization to altitude. *J. Appl. Physiol., 24*:607, 1968.

Hanson, P. G.: Pressure dynamics in thoracic aorta during linear deceleration. *J. Appl. Physiol., 28*:23, 1970.

Hornbein, T. F., and Severinghaus, J. W.: Carotid chemoreceptor response to hypoxia and acidosis in cats living at high altitude. *J. Appl. Physiol., 27*:837, 1969.

Hornbein, T. F., and Sorensen, S. C.: Ventilatory response to hypoxia and hypercapnia in cats living at high altitude. *J. Appl. Physiol., 27*:834, 1969.

Hurtado, A.: Animals in high altitudes: resident man. *In* Handbook of Physiology. Baltimore, The Williams & Wilkins Co., 1964, Sec. IV, p. 843.

Kellogg, R. H.: Altitude acclimatization, a historical introduction emphasizing the regulation of breathing. *Physiologist, 11*:37, 1968.

Koestler, A. G., and Reynolds, H. H.: Rapid decompression of chimpanzees to a near vacuum. *J. Appl. Physiol., 25*:153, 1968.

Kollias, J.; Buskirk, E. R.; Akers, R. F.; Prokop, E. K.; Baker, P. T.; and Picon-Reategui, E.: Work capacity of long-time residents and newcomers to altitude. *J. Appl. Physiol., 24*:792, 1968.

Korner, P. I.: Circulatory adaptations in hypoxia. *Physiol. Rev., 39*:687, 1959.

Luft, U. C.: Aviation physiology – the effects of altitude. *In* Handbook of Physiology. Baltimore, The Williams & Wilkins Co., 1965, Sec. III, Vol. II, p. 1099.

McCally, M.: Hypodynamics and Hypogravics, The Physiology of Inactivity and Weightlessness. New York, Academic Press, Inc., 1969.

Moncloa, F.; Carcelen, A.; and Beteta, L.: Physical exercise, acid-base balance, and adrenal function in newcomers to high altitude. *J. Appl. Physiol., 28*:151, 1970.

Pugh, L. G. C. E.: Animals in high altitudes: man above 5,000 meters – mountain exploration. *In* Handbook of Physiology. Baltimore, The Williams & Wilkins Co., 1964, Sec. IV, p. 861.

Reeves, J. T.; Halpin, J.; Cohn, J. E.; and Daoud, F.: Increased alveolar-arterial oxygen difference during simulated high-altitude exposure. *J. Appl. Physiol., 27*:658, 1969.

Roskamm, H.; Landry, F.; Samek, L.; Schlager, M.; Weidemann, H.; and Reindell, H.: Effects of a standardized ergometer training program at three different altitudes. *J. Appl. Physiol., 27*:840, 1969.

Schaefer, H. J.: Man and radiant energy: ionizing radiation. *In* Handbook of Physiology. Baltimore, The Williams & Wilkins Co., 1964, Sec. IV, p. 989.

Smith, E. E., and Crowell, J. W.: Influence of hematocrit ratio on survival of unacclimatized dogs at simulated high altitude. *Amer. J. Physiol., 205*:1172, 1963.

Smith, E. E., and Crowell, J. W.: Role of the hematocrit in altitude acclimatization. *Aerospace Med., 38*:39, 1966.

CHAPTER 45

Physiology of Deep Sea Diving and Other High Pressure Operations

When a person descends beneath the sea, the pressure around him increases tremendously. To keep his lungs from collapsing, air must be supplied to him also under high pressure, which exposes the blood in his lungs to extremely high alveolar gas pressures. Beyond certain limits these high pressures can cause tremendous alterations in the physiology of the body, which explains the necessity for the present discussion.

Also exposed to high atmospheric pressures are caisson workers who, in digging tunnels beneath rivers or elsewhere, often must work in a pressurized area to keep the tunnel from caving in. Here again, the same problems of excessively high gas pressures in the alveoli occur.

Before explaining the effects of high alveolar gas pressures on the body, it is necessary to review some physical principles of pressure and volume changes at different depths beneath the sea.

Relationship of Sea Depth to Pressure. A column of fresh water 33 feet high (sea water, 32 feet) exerts the same pressure at its bottom as all the atmosphere above the earth. Therefore, a person 33 feet beneath the water surface is exposed to a pressure of 2 atmospheres, 1 atmosphere of pressure caused by the air above the water and the other by the weight of the water itself. At 66 feet the pressure is 3 atmospheres, and so forth, approximately in accord with the table in the opposite column.

Effect of Depth on the Volume of Gases. Another important effect of depth is the compression of gases to smaller and smaller volumes. Figure 45–1 illustrates a bell jar at sea level containing 1 liter of air. At 33 feet beneath the sea where the pressure is 2 atmospheres, the volume has been compressed to only one-half liter. At 100 feet, where the pressure is 4 atmospheres, the volume has been compressed to one-fourth liter, and at 8 atmospheres (233 feet) to one-eighth liter. This is an extremely important effect in diving because it can cause the air chambers of the diver's body, including the lungs, to become so small in some instances that serious damage results, as is discussed later in the chapter.

Many times in this chapter it is necessary to refer to *actual volume* versus *sea level volume*. For instance, we might speak of an actual volume of 1 liter at a depth of 300 feet; this is the same quantity of air as a sea level volume of 10 liters. Therefore, these two terms must be kept in mind while studying this chapter.

Depth (feet)	Atmosphere(s)
Sea level	1
33	2
66	3
100	4
133	5
166	6
200	7
300	10
400	13
500	16

Figure 45–1 Effect of depth on gas volumes.

EFFECT OF HIGH PARTIAL PRESSURES OF GASES ON THE BODY

The three gases to which a diver breathing air is normally exposed are nitrogen, oxygen, and carbon dioxide. However, helium is often substituted for nitrogen in the diving mixture; therefore, the effects of this gas under high pressure must also be considered.

Nitrogen Narcosis at High Nitrogen Pressures. Approximately four-fifths of the air is nitrogen. At sea level pressure this has no known effect on bodily function, but at high pressures it can cause varying degrees of narcosis. When the diver is breathing air, the depth at which the first symptoms of mild narcosis appear is approximately 130 to 150 feet, at which level he begins to exhibit joviality and to lose many of his cares. At 150 to 200 feet, he becomes drowsy. At 200 to 250 feet, his strength wanes considerably, and he often becomes too clumsy to perform the work required of him. Beyond 300 feet (10 atmospheres pressure), the diver usually becomes almost useless as a result of nitrogen narcosis, and he becomes unconscious at 350 to 400 feet. It should be noted, however, that *an hour or more of the high pressure is usually required* before enough nitrogen dissolves in the body to cause these effects.

Nitrogen narcosis has characteristics very similar to those of alcohol intoxication, and for this reason it has frequently been called "raptures of the depths."

The mechanism of the narcotic effect is believed to be the same as that of essentially all the gas anesthetics. That is, nitrogen dissolves freely in the fats of the body, and it is presumed that it, like most other anesthetic gases, dissolves in the membranes or other lipid structures of the neurons and thereby reduces their excitability.

Oxygen Toxicity at High Pressures. Breathing oxygen under very high partial pressures can be detrimental to the central nervous system, sometimes resulting in epileptic convulsions followed by coma. Indeed, exposure to 3 atmospheres pressure of oxygen (P_{O_2} = 2280 mm. Hg) will cause convulsions and coma in most persons after about one hour. These convulsions often occur without any warning, and they obviously are likely to be lethal to a diver submerged beneath the sea. In an extensive study of persons who have developed oxygen toxicity, the following frequencies of different symptoms were encountered:

	per cent
Nausea	40
Muscular twitchings	21
Dizziness	17
Disturbances of vision	6
Restlessness and irritability	6
Numbness and pins-and-needles sensations	6
Convulsive seizures and coma	4

Exercise greatly increases the diver's susceptibility to oxygen toxicity at high pressures, causing symptoms to appear much earlier and with far greater severity than in the resting person.

Figure 45–2 gives the so-called *oxygen safe tolerance curve* for persons performing moderate amounts of work at different depths under the sea while breathing 100 per cent oxygen. This shows that a person performing work at a depth of only 40 feet is reasonably safe for a maximum period of 23 minutes. At a depth of 30 feet, he is safe for about 45 minutes and at 20 feet for about 1½ hours. However, there is tremendous variability in the tolerance of different persons and in the tolerance of the same person on different days. Therefore, since a convulsion underneath the sea might be lethal, even this tolerance curve is possibly too liberal for absolute safety.

The cause or causes of oxygen toxicity are yet unknown, but some of the experimental data and suggested causes are: (1) Following severe oxygen toxicity the concentrations of some of the oxidative enzymes in the tissues are considerably depleted. Therefore, it has been postulated that excess oxygen inactivates oxidative enzymes and causes toxicity by decreasing the ability of the tissues to form high energy phosphate bonds. (2) Blood flow through the brain decreases 25 to 50 per cent when a person breathes high concentrations of oxygen. This presumably results from the local tissue blood flow mechanism explained in Chapter 24. That is, the greater the amount of oxygen available in the tissues, the greater is the degree of constriction of the blood vessels, which normally helps to regulate the amount of oxygen delivered to the tissues. Yet, the decrease in blood flow can decrease the availability of other nutrients required by the cerebral tissues or can decrease the removal of excreta, such as carbon dioxide and nitrogenous end-products. Therefore, it has been postulated that either lack of certain nutrients or buildup of metabolic end-products can cause the convulsions of oxygen toxicity. (3) Still other experiments have shown that excess oxygen in the tissues can cause the development of large

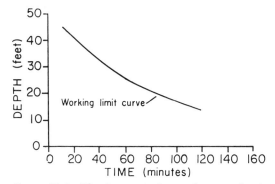

Figure 45–2 The "oxygen tolerance" curve, showing the length of time that a person can remain without danger at different depths when breathing pure oxygen. (Modified from Submarine Medicine Practice. U.S. Navy Bureau of Medicine and Surgery.)

concentrations of oxidizing free radicals that could cause oxidation of many essential elements of the cells, thereby damaging the cells' metabolic systems.

CHRONIC OXYGEN POISONING. A person can be exposed to 100 per cent oxygen at normal atmospheric pressure almost indefinitely without developing the *acute* oxygen toxicity described above. However, after a day or so of this exposure, he begins to develop *pulmonary edema* caused by damage to the linings of the bronchi and alveoli. Here again, it appears that these local tissues of the lungs are damaged as a result of oxidation of some of the essential elements of the tissues. The reason this effect occurs in the lungs and not in the other tissues is that the lungs are directly exposed to the high oxygen pressure (Po_2), whereas oxygen is delivered to the other tissues at essentially normal Po_2 because of the hemoglobin oxygen buffer system described in Chapter 41. When the Po_2 in the air rises above about 1500 mm. Hg, this buffer system fails, which then allows the Po_2 of all tissues to rise and to cause acute oxygen poisoning.

Carbon Dioxide Toxicity at Great Depths. If the diving gear is properly designed and functions properly, the diver has no problem from carbon dioxide toxicity, for depth alone does not increase the carbon dioxide partial pressure in the alveoli. This is true because carbon dioxide is manufactured in the body, and as long as the diver continues to breathe a normal tidal volume, he continues to expire the carbon dioxide as it is formed, maintaining his alveolar carbon dioxide concentration at a normal value.

Unfortunately, though, in certain types of diving gear, such as the diving helmet and the different types of rebreathing apparatuses, carbon dioxide can frequently build up in the air breathed by the diver. Up to an air concentration of 10 per cent carbon dioxide, the diver tolerates this buildup, his minute respiratory volume increasing up to a maximum of 6- to 10-fold to compensate for the increased carbon dioxide. However, beyond the 10 per cent level the situation becomes intolerable, and the respiratory center begins to be depressed rather than excited; the diver's respiration then actually begins to fail rather than to compensate. As a result, he develops varying degrees of lethargy, narcosis, and, finally, anesthesia, as was discussed in Chapter 42.

Effects of Helium at High Pressures. In deep dives, helium is used to replace the nitrogen because it has only one-fourth to one-fifth the narcotic effect of nitrogen, exhibiting essentially no narcotic effect to a depth of about 500 feet. Furthermore, three other properties make it desirable in the diving gas mixture under some conditions: (1) Because of its low atomic weight, its density is slight, which reduces the airway resistance of the diver. (2) Also because of its low atomic weight, helium diffuses through the tissues much more rapidly than nitrogen, which allows more rapid removal of helium than nitrogen from the body fluids under some conditions. (3) Helium is less soluble in the

body fluids than nitrogen, which reduces the quantity of bubbles that can form in his tissues when the diver is decompressed after a prolonged dive. Yet there are problems in the use of helium that make it undesirable under some conditions, as is discussed below in relation to decompression.

DECOMPRESSION OF THE DIVER AFTER EXPOSURE TO HIGH PRESSURES

When a person breathes air under high pressure for a long time, the amount of nitrogen dissolved in his body fluids becomes great. The reason for this is the following: The blood flowing through the pulmonary capillaries becomes saturated with nitrogen to the same pressure as that in the breathing mixture. Over several hours, nitrogen is carried to all the tissues of the body to saturate them also with dissolved nitrogen. And, since nitrogen is not metabolized by the body, it remains dissolved until the nitrogen pressure in the lungs decreases, at which time the nitrogen is then removed by the respiratory process.

Volume of Nitrogen Dissolved in the Body Fluids at Different Depths. At sea level almost exactly 1 liter of nitrogen is dissolved in the entire body. A little less than half of this is dissolved in the water of the body, and a little more than half is dissolved in the fat of the body. This is true despite the fact that fat constitutes only 15 per cent of the normal body, and it is explained by the fact that nitrogen is five times as soluble in fat as in water.

After the diver has become totally saturated with nitrogen the *sea level volume of nitrogen* dissolved in his body fluids at the different depths is:

feet	liters
33	2
100	4
200	7
300	10

However, several hours are required for the gaseous pressures of nitrogen in all the body tissues to come to equilibrium with the gas pressure of nitrogen in the alveoli simply because the blood does not flow rapidly enough and the nitrogen does not diffuse rapidly enough to cause instantaneous equilibrium. The nitrogen dissolved in the water of the body comes to almost complete saturation in about one hour, but the fat, requiring much more nitrogen for saturation and also having a relatively poor blood supply, reaches saturation only after several hours. For this reason, if a person remains at deep levels for only a few minutes not much nitrogen dissolves in his fluids and tissues, whereas if a person remains at a deep level for several hours his fluids and tissues become almost completely saturated with nitrogen.

Decompression Sickness (Synonyms: Compressed Air Sickness, Bends, Caisson Disease, Diver's Paralysis, Dysbarism). If a diver has been beneath the sea long enough so that large amounts of nitrogen dissolve in his body, and then he suddenly comes back to the surface of the sea, significant quantities of nitrogen bubbles can develop in his body fluids either intracellularly or extracellularly, and these can cause minor or serious damage in almost any area of the body, depending on the amount of bubbles formed.

The principles underlying this effect are shown in Figure 45–3. To the left, the diver's tissues have become equilibrated to a very high nitrogen pressure. However, as long as the diver remains deep beneath the sea, the pressure against the outside of his body compresses all the body tissues sufficiently to keep the dissolved gases in solution. Then, when the diver suddenly rises to sea level, the pressure on the outside of his body becomes only 1 atmosphere (760 mm. Hg), while the pressure inside the body fluids is the sum of the pressures of water vapor, carbon dioxide, oxygen, and nitrogen, or a total of 4045 mm. Hg, which is far greater than the pressure on the outside of the body. Therefore, the gases can escape from the dissolved state and form actual bubbles inside the tissues.

Exercise hastens the formation of bubbles during decompression because of increased agitation of the tissues and fluids. This is an effect analogous to that of shaking an opened bottle of soda pop to release the bubbles.

Fortunately, the phenomenon of "supersaturation" normally allows nitrogen to remain dissolved and not to form significant quantities of bubbles if the nitrogen pressure in the body fluids does not rise to more than 3.0 times the pressure on the outside of the body. Therefore, a diver can theoretically be brought immediately from a depth of 66 feet beneath the sea (3 atmospheres pressure) to sea level (1 atmosphere pressure) without significant bubble formation and without developing decompression sickness, even though on arrival at sea level the gas pressure in his body fluids is almost three times the pressure on the outside of his body. Yet for safety's sake the diver is rarely allowed to push this theoretical limit in his ascent from beneath the sea.

SYMPTOMS OF DECOMPRESSION SICKNESS. In persons who have developed decompression sickness, symptoms have occurred with the following frequencies:

	per cent
Local pain in the legs or arms	89
Dizziness	5.3
Paralysis	2.3
Shortness of breath ("the chokes")	1.6
Extreme fatigue and pain	1.3
Collapse with unconsciousness	0.5

From the above list of symptoms of decompression sickness it can be seen that the most serious problems are usually related to bubble formation in the nervous system. Bubbles sometimes actually disrupt important pathways in the brain or spinal cord, and bubbles in the peripheral nerves can cause severe pain. Unfortunately, formation of large bubbles in the central nervous system occa-

Figure 45–3 Gaseous pressures responsible for bubble formation in the body tissues.

sionally even leads to permanent paralysis or permanent mental disturbances.

But the nervous system is not the only locus of damage in decompression sickness, for bubbles can also form in the blood and become caught in the capillaries of the lungs; these bubbles block pulmonary blood flow and cause "the chokes," characterized by serious shortness of breath. This is often followed by severe pulmonary edema, which further aggravates the condition and can cause death.

The symptoms of decompression sickness usually appear within a few minutes to an hour after sudden decompression. However, occasional symptoms of decompression sickness develop as long as six or more hours after decompression.

Rate of Nitrogen Elimination from the Body. Decompression Tables. Fortunately, if a diver is brought to the surface slowly, the dissolved nitrogen is eliminated through his lungs rapidly enough to prevent decompression sickness. Figure 45–4 illustrates the rate at which nitrogen is liberated from the water of the body, from the fat of the body, and from both these sources when a person is saturated to a depth of 33 feet and then brought to sea level suddenly. Approximately two-thirds of the total nitrogen is liberated in one

Figure 45–4 Rate of nitrogen liberation from the body when a person has come to sea level from prolonged exposure to compressed air at 33 feet depth, showing separately the rate of nitrogen release from the whole body, from the water of the body, and from the fat. (From Armstrong: Principles and Practice of Aviation Medicine. The Williams & Wilkins Co.)

hour and about 90 per cent in six hours. However, some excess nitrogen is still present in the body fluids for many more hours, and the diver is not completely safe for as long as 9 to 12 hours. Therefore, a diver must be decompressed sometimes for many hours if he has been deep beneath the sea for a long time.

The rate at which a diver can be brought to the surface depends on, first, the *depth* to which he has descended and, second, the *amount of time* he has been there. If he remains at deep levels for only a short time, the body fluids do not become saturated, and, therefore, the decompression time can be reduced accordingly.

Table 45-1 gives a typical decompression table used by the U.S. Navy when the diver breathes compressed air. Note that only 20 minutes at a depth of 300 feet requires over two and a half hours decompression time, and 45 minutes at 300 feet requires over five hours. On the other hand, a person can remain at 50 feet for as long as three hours and yet be decompressed in only 12 minutes.

The "optimal time on the bottom," as given in Table 45-1, represents the optimum exposure time at each depth for the best balance between length of work period and amount of useful work the average diver can perform. Note how short these times are at great depths; this is caused principally by the nitrogen narcosis effect and, to a lesser extent, by the labored breathing that results from increased density of gases in the lungs.

OXYGEN ADMINISTRATION FOR MORE RAPID DECOMPRESSION. If oxygen is pumped to the diver in higher than normal concentrations as he ascends closer to the surface of the sea, the nitrogen partial pressure in his alveoli is considerably reduced, and, as a consequence, the rate of nitrogen removal from his body fluids is correspondingly increased. Therefore, a diver can be brought to the surface far more rapidly when oxygen is pumped to him once he has come close enough to the surface to tolerate the necessary oxygen partial pressures. Different decompression tables are used when oxygen is so supplied.

Decompression in a Tank and Treatment of Decompression Sickness. Another procedure for decompression, used especially in heavily polluted waters and when climatic situations require it, involves bringing the diver to the surface immediately and then placing him in a decompression tank within five minutes after arriving at the surface. Pressure is reapplied, and an appropriate decompression table that prevents bubble formation is used.

A person who begins to develop symptoms of decompression sickness can also be treated by placing him in such a decompression tank for a long time, several times as long as the usual decompression times, and allowing the nitrogen to be released from his body slowly.

Use of Helium-Oxygen Mixtures in Deep Dives. In deep dives helium has advantages over nitrogen, including (1) decreased decompression time, (2) lack of narcotic effect, and (3) decreased airway resistance in the lungs. The decreased decompression time results from two of its properties: (a) Only 40 per cent as much helium dissolves in the body as does nitrogen. (b) Because of its small

TABLE 45-1.

1	2	3									4
Depth of dive (feet)	Optimal time on bottom (minutes)	Stops (feet and minutes)									Approximate total decompression time (minutes)
		Feet 90	Feet 80	Feet 70	Feet 60	Feet 50	Feet 40	Feet 30	Feet 20	Feet 10	
40	240									4	6
50	190									9	12
60	150								5	15	24
70	120								13	16	33
80	115								22	26	53
90	95							2	27	21	56
100	85							6	28	21	61
110	75							14	27	37	84
120	65							13	28	32	80
130	60							13	28	28	76
140	55							15	28	32	82
150	50							16	28	32	84
160	45							17	28	43	96
170	40							19	28	46	102
185	35							19	28	46	102
200	35							22	28	46	106
210	30						5	16	28	40	100
225	27						22	26	35	48	143
250	25					2	23	26	35	51	150
300	20					9	23	26	35	51	159

atomic size it diffuses through the tissues at a velocity about 2½ times that of nitrogen.

However, helium has not proved to be as advantageous as was once thought because of another property that is different from nitrogen: bubbles begin to form when the pressure of helium in the body fluids is only 1.7 times the pressure on the outside of the body. This compares with 3.0 for nitrogen. Therefore, a diver cannot be brought up as far at a time with helium as with nitrogen. And still another factor makes nitrogen better than helium for shallow dives: the rapid diffusion of helium allows far more helium than nitrogen to become dissolved in the body fluids in a short time. Therefore, for short dives at moderate depths, nitrogen is still preferable.

If one calculates the relative advantages of helium versus nitrogen, he finds that long, deep dives favor the use of helium while short, shallow dives favor the use of nitrogen. Figure 45–5 illustrates the dividing line for most effective use of the two types of gas mixtures. For instance, point A illustrates a dive of 150 feet for 120 minutes. This is a long deep dive, and in this instance helium is far more satisfactory than nitrogen. On the other hand, point B shows a dive of 200 feet for 10 minutes, and here the person can be decompressed for a shorter time when using nitrogen than helium. Beyond 300 feet, nitrogen cannot be used at all because of nitrogen narcosis, which can develop rapidly.

SOME PHYSICAL PROBLEMS OF DIVING

Aside from the effects of high gas pressures on the body, still other physical factors place limitations on diving. These are based principally on changes in gas volumes from sea level to greater depths and include the following:

Volume of Air that Must Be Pumped to the Diver—Relationship to Rate of Carbon Dioxide Elimination. To blow off carbon dioxide from the lungs, the tidal

volume of air flowing in and out of the lungs with each breath must remain the same regardless of the depth of the dive. A tidal volume of 0.5 liter at 300 feet depth (10 atmospheres pressure) would be a sea level volume of 5 liters. Therefore, a compressor operating at sea level must pump 5 liters of air to the diver at 300 feet depth for each breath that the diver takes in order to wash the carbon dioxide out of his lungs. Stating this another way, the amount of air that must be pumped to the diver to keep his alveolar carbon dioxide normal is directly proportional to the pressure under which he is operating. At sea level the working diver requires about 1.5 cubic feet of air per minute for adequate carbon dioxide washout from his diving helmet. Therefore, the sea level volumes of air that must be pumped each minute for different depths of operation are the following:

feet	cubic feet
Sea level	1.5
33	3
66	4.5
100	6
200	10.5
300	15

Change in Density of the Air—Effect on Maximum Breathing Capacity. The density of the air increases in proportion to the pressure, which means that the density is four times as great at 100 feet depth as at sea level and seven times as great at 200 feet.

The resistance to air flow through the respiratory passageways increases directly in proportion to the density of the breathing mixture. Therefore, one can readily see that the increased density of the air will increase the work of breathing and, as a corollary, will decrease the maximum breathing capacity (the amount of air that one can breathe each minute). The following table gives the maximum breathing capacity in per cent of normal at different depths when breathing air and when breathing a mixture in which helium replaces the nitrogen of the air.

Depth (ft.)	Air (% of normal)	Oxygen-helium (% of normal)
25	75	100+
50	60	100+
100	50	86
200	35	63
400	24	48

Effect of Rapid Descent—"The Squeeze." On rapid descent, the volumes of all gases in the body become greatly reduced because of increasing pressure applied to the outside of the body. If additional quantities of air are supplied to the gas cavities during descent—especially to the lungs—no harm is done, but, if the person continues to descend without addition of gas to his cavities, the volume becomes greatly reduced, and serious

Figure 45–5 Depths and times beneath the sea at which it is more advantageous to use helium versus the depths and times at which it is more advantageous to use nitrogen. (Modified from Submarine Medicine Practice. U.S. Navy Bureau of Medicine and Surgery.)

physical damage results; this is called "the squeeze."

The most damaging effects of the squeeze occur in the lungs, for the smallest volume that the lungs can normally achieve is approximately 1.5 liters. Even if the diver inspires a maximal breath prior to descending, he can go down no farther than 100 feet before his chest begins to cave in. Therefore, to prevent lung squeeze, the diver must inspire additional air as he descends.

When air becomes entrapped in the middle ear during descent, the squeeze can cause a ruptured eardrum, and when air is entrapped in one of the nasal sinuses intense pain results. Occasionally, also, when a diver loses air pressure to his helmet his body is literally squeezed upward into the helmet, and, when air volume is lost from the mask of a free diving apparatus, the eyes can actually pop out into the mask and the face can become greatly distorted to fill the mask.

Overexpansion of the Lungs on Rapid Ascent. Exactly the opposite pulmonary effects occur on rapid ascent if the person fails to expel air from his lungs on the way up. Unfortunately, panic can frequently cause a person to close his glottis spastically, which can result in serious damage to the lungs. When the lungs become expanded to their limit the pressure continues to rise, and above an alveolar pressure of 80 to 100 mm. Hg, air is forced into the pulmonary capillaries, causing air embolism in the circulation and often resulting in death. Also, increased pressure in the lungs frequently blows out large blebs on the surfaces of the lungs or ruptures the lungs to cause pneumothorax.

In rare instances, in a diver who has been deep below the sea for a long time and in whom large amounts of gas have accumulated in the abdomen, rapid ascent can also cause serious trauma in the gastrointestinal tract.

Rapid ascent can be especially serious when a person is attempting to escape from a submarine and must rise to the surface rapidly without appropriate diving gear. It also occurs frequently when the diver loses control of his diving gear and his suit balloons up so greatly that he "blows up" to the surface.

SCUBA DIVING (SELF-CONTAINED UNDERWATER BREATHING APPARATUS)

In recent years a diving apparatus that does not require connections with the surface has been perfected and is probably best known under the trade name "Aqualung." The two basic types of self-contained underwater breathing apparatuses from which various modifications have been made are (1) the open circuit demand system and (2) the closed circuit system.

The Open Circuit Demand System. Figure 45–6 illustrates an open circuit demand type of underwater breathing apparatus showing the following components: (1) tanks of compressed air or other breathing mixture, (2) a regulator valve for reducing the pressure from the tanks, (3) a "demand" valve which allows air to be pulled into the mask with slight negative pressure in the system, (4) a mask and tube system with small "dead space," and (5) an exhalation valve located in close contiguity with the demand valve.

Basically, the demand system operates as follows: With each inspiration, slight negative pressure in the mask pulls the diaphragm of the demand valve inward, and this automatically releases air from the compressed air containers into the mask. In this way only the amount of air needed for inhalation enters the system. Then, on expiration, the air cannot go back into the tank but instead is expired through the expiration valve.

Several problems exist in relation to the open circuit demand system. First, the demand valve itself must be located as closely as possible to the level of the person's lungs. Otherwise, the hydrostatic pressure of the water pressing against the diaphragm of the demand valve would either cause too much air to flow into the mask (if the demand valve should be located deeper in the sea than the lungs) or make it difficult to inspire (if the demand valve should be located at a shallower level than the lungs).

Second, it is essential that the expiration valve be located in close contiguity with the demand valve because the difference between the level of the expiration valve and the level of the demand valve determines the least pressure difference that is required to activate the demand valve.

Figure 45–6 An open circuit demand system type of self-contained underwater breathing apparatus. (Modified from Submarine Medicine Practice. U.S. Navy Bureau of Medicine and Surgery.)

Third, the most important problem in use of the self-contained underwater breathing apparatus is the time limitation that one can remain beneath the surface; only a few minutes are possible at great depths because tremendous airflow from the tanks is required to wash carbon dioxide out of the lungs — the greater the depth, the greater the airflow required, as discussed earlier.

The Closed Circuit System. In the simplest type of closed circuit system a person breathes pure oxygen. This system contains the following elements: (1) a tank of pure oxygen, (2) a rubber bellows into which the diver can breathe back and forth, (3) a valve system for allowing oxygen to flow from the oxygen tank into the bellows as needed to keep it moderately filled, (4) a canister containing soda lime through which the rebreathed gas passes to absorb the carbon dioxide, and (5) an appropriate mask system with valves to keep the gaseous mixture flowing through the canister for carbon dioxide removal. Thus, the closed circuit system is similar to a standard anesthetic machine in which oxygen is continually rebreathed — except that no anesthetic is used.

The most important problem in use of this closed circuit system is limitation in depth that a person can remain beneath the sea because of poor oxygen tolerance. Figure 45–2 gives these tolerances for depth and time. A safe working rule is no more than 30 minutes at 30 feet depth.

SPECIAL PHYSIOLOGICAL PROBLEMS OF SUBMARINES

Escape from Submarines. Essentially the same problems as those of deep sea diving are often met in relation to submarines, especially when it is necessary to escape from a submerged submarine. Escape is possible from as deep as 300 feet even without using any special type of apparatus. Proper use of rebreathing devices using helium or hydrogen could theoretically allow escape from as deep as 600 feet.

One of the major problems of escape is prevention of air embolism. As the person ascends, the gases in his lungs expand and sometimes rupture a major pulmonary vessel, allowing the gases to enter into the pulmonary vascular system to cause embolism of the circulation. Therefore, as the person ascends, he must exhale continually.

Expansion and exhalation of gases from the lungs during ascent, even without breathing, is often rapid enough to blow off the accumulating carbon dioxide in the lungs. This keeps the concentration of carbon dioxide from building up in the blood and keeps the person from having the desire to breathe. Therefore, he can hold his breath for an extremely long time during ascent.

Health Problems in the Submarine Internal Environment. Except for escape, submarine medicine generally centers around several engineering problems to keep hazards out of the internal en-

vironment of the submarine. In atomic submarines there exists the problem of radiation hazards, but, with appropriate shielding, the amount of radiation received by the crew submerged beneath the sea has actually been less than the normal radiation received above the surface of the sea from cosmic rays. Therefore, no essential hazard results from this unless some failure in the apparatus causes unexpected release of radioactive materials.

Second, poisonous gases on occasion escape into the atmosphere of the submarine and must be controlled exactly. For instance, during several weeks' submergence, cigarette smoking by the crew can liberate sufficient amounts of carbon monoxide, if it is not removed from the air, to cause carbon monoxide poisoning, and on occasion even Freon gas has been found to diffuse through the tubes in refrigeration systems in sufficient quantity to cause dangerous toxicity. Finally, it is well known that chlorine and other poisonous gases are released when salt water comes in contact with batteries in the old type submarines.

A highly publicized factor of submarine medicine has been the possibility of psychological problems caused by prolonged submergence. Fortunately, this has turned out to be more a figment of the public's imagination than truth, for the problems here are the same as those relating (1) to any other confinement or (2) to any other type of danger. Psychological screening has been used to great advantage to keep such problems almost to zero even in month-long submergence.

DROWNING

Asphyxia in Drowning. Approximately 30 per cent of all persons who drown do not inhale water, and even after recovery of the body from the water the lungs are still dry. These persons simply die from asphyxiation. The reason they do not inhale water is that water, in attempting to enter the trachea, elicits a powerful laryngeal reflex that causes spastic closure of the vocal cords.

Inhalation of Fresh Water — Cardiac Fibrillation. When a drowning person inhales fresh water, water is absorbed through the alveolar membrane into the blood extremely rapidly by osmosis because of the total osmotic pressure of the pulmonary capillary blood (which gives an absorption gradient of about 5400 mm. Hg). Several liters of water are absorbed into the blood within one to three minutes. This greatly dilutes the electrolytes of the blood and even causes hemolysis of many red cells, with spillage of large quantities of potassium into the plasma. The resulting changes in electrolyte composition of the plasma, plus anoxia, cause the heart to fibrillate within one to three minutes after the person first inhales the water, causing death much earlier than would have occurred had the water not been inhaled.

Effect of Inhaling Salt Water. If a drowning person inhales salt water instead of fresh water, osmosis of water occurs in the opposite direction through the pulmonary membrane because the total osmotic pressure of salt water is several times that of blood. Loss of water out of the blood causes marked hemoconcentration but not hemolysis or cardiac fibrillation. Instead, the person dies of asphyxia in five to eight minutes.

REFERENCES

Andersen, H. T.: Physiological adaptations in diving vertebrates. *Physiol. Rev., 46*:212, 1966.

Behnke, A. R., Jr., and Lanphier, E. H.: Underwater physiology. *In* Handbook of Physiology. Baltimore, The Williams & Wilkins Co., 1965, Sec. III, Vol. II, p. 1159.

Bennett, P. B.: The Aetiology of Compressed Air Intoxication and Inert Gas Narcosis. International Series of Monographs in Pure and Applied Biology. Zoology Div., Vol. 31, New York, Pergamon Press, Inc., 1966.

Bennett, P. B.: Physiological problems of deep diving. *Proc. Int. Union of Physiol. Sciences, 6*:49, 1968.

Cross, M. H., and Houlihan, R. T.: Sympathoadrenomedullary response of the rat to high oxygen exposures. *J. Appl. Physiol., 27*:523, 1969.

Dubois, A. B.; Bond, G. F.; and Schaefer, K. E.: Alveolar gas exchange during submarine escape. *J. Appl. Physiol., 18*:509, 1963.

Fisher, A. B.; Dubois, A. B.; Hyde, R. W.; Knight, C. J.; and Lambertsen, C. J.: Effect of 2 months' exposure to N_2-O_2 at 2.2 Ata on lung function. *J. Appl. Physiol., 28*:70, 1970.

Fisher, A. B.; Hyde, R. W.; Puy, R. J. M.; Clark, J. M.; and Lambertsen, C. J.: Effect of oxygen at 2 atmospheres on the pulmonary mechanics of normal man. *J. Appl. Physiol., 24*:529, 1968.

Greene, D. G.: Drowning. *In* Handbook of Physiology. Baltimore, The Williams & Wilkins Co., 1965, Sec. III, Vol. II, p. 1195.

Groom, A. C.; Morin, R.; and Farhi, L. E.: Determination of dissolved N_2 in blood and investigation of N_2 washout from the body. *J. Appl. Physiol., 23*:706, 1967.

Haugaard, N.: Cellular mechanisms of oxygen toxicity. *Physiol. Rev., 48*:311, 1968.

Lambertsen, C. J.: Advance in manned undersea activity. *Proc. Int. Union of Physiol. Sciences, 6*:115, 1968.

Lambertsen, C. J.: Effects of oxygen at high partial pressure. *In* Handbook of Physiology. Baltimore, The Williams & Wilkins Co., 1965, Sec. III, Vol. II, p. 1027.

Lambertsen, C. J.: Proceedings of the Third Symposium on Underwater Physiology. Baltimore, The Williams & Wilkins Co., 1967.

Lanphier, E. H.: Man in high pressures. *In* Handbook of Physiology. Baltimore, The Williams & Wilkins Co., 1964, Sec. IV, p. 893.

Miles, S.: Underwater Medicine. 2nd ed., Philadelphia, J. B. Lippincott Co., 1966.

Overfield, E. M.; Saltzman, H. A.; Kylstra, J. A.; and Salzano, J. V.: Respiratory gas exchange in normal men breathing 0.9 per cent oxygen in helium at 31.3 Ata. *J. Appl. Physiol., 27*:471, 1969.

Puy, R. J. M.; Hyde, R. W.; Fisher, A. B.; Clark, J. M.; Dickson, J.; and Lambertsen, C. J.: Alterations in the pulmonary capillary bed during early O_2 toxicity in man. *J. Appl. Physiol., 24*:537, 1968.

Salzano, J.; Rausch, D. C.; and Saltzman, H. A.: Cardiorespiratory responses to exercise at a simulated seawater depth of 1,000 feet. *J. Appl. Physiol., 28*:34, 1970.

Submarine Medicine Practice, Department of the Navy. Washington, D.C., U.S. Government Printing Office, 1956.

Winter, P. M.; Gupta, R. K.; Michalski, A. H.; and Lanphier, E. H.: Modification of hyperbaric oxygen toxicity by experimental venous admixture. *J. Appl. Physiol., 23*:954, 1967.

Wyndam, C. H.: Physiological problems of deep level mining. *Proc. Int. Union of Physiol. Sciences, 6*:45, 1968.

PART VIII

THE NERVOUS SYSTEM AND SPECIAL SENSES

CHAPTER 46

Organization of the Nervous System

The nervous system, along with the endocrine system, provides the control functions for the body. In general, the nervous system controls the rapid activities of the body, such as muscular contractions, rapidly changing visceral events, and even the rates of secretion of some endocrine glands. The endocrine system regulates principally the metabolic functions of the body.

The nervous system is unique in the vast complexity of the control actions that it can perform. It can receive literally thousands of bits of information from the different sensory organs and then integrate all these to determine the response to be made by the body. The purpose of this chapter is to present a general outline of the basic mechanisms by which the nervous system performs such functions, and succeeding chapters analyze in detail the functions of the individual parts of the nervous system. Before beginning this discussion, however, the reader should refer to Chapters 5 and 6, which present, respectively, the principles of membrane potentials and transmission of impulses through synapses.

GENERAL DESIGN OF THE NERVOUS SYSTEM

THE SENSORY DIVISION — SENSORY RECEPTORS

Most activities of the nervous system are originated by sensory experience emanating from *sensory receptors*, whether these be visual receptors, auditory receptors, tactile receptors on the surface of the body, or other kinds of receptors. This sensory experience can cause an immediate reaction, or its memory can be stored in the brain for minutes, weeks, or years and then can help to determine the bodily reactions at some future date.

Figure 46–1 illustrates a minute portion of the *somatic* sensory system, which transmits sensory information from the receptors of the entire surface of the body and deep structures. This information enters the nervous system through the spinal nerves and is conducted

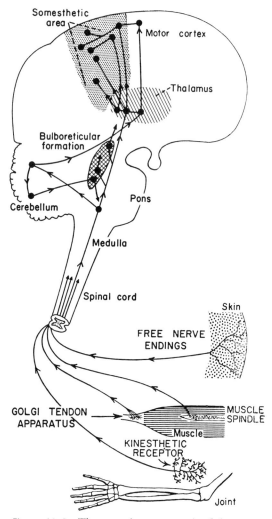

Figure 46–1 The somatic sensory axis of the nervous system.

into (a) the spinal cord at all levels, (b) the reticular substance of the medulla, pons, and mesencephalon, (c) the cerebellum, (d) the thalamus, and (e) the somesthetic areas of the cerebral cortex. But in addition to these "primary sensory" areas, signals are then relayed to essentially all other segments of the nervous system.

THE MOTOR DIVISION—THE EFFECTORS

The most important ultimate role of the nervous system is control of bodily activities. This is achieved by controlling (a) contraction of skeletal muscles throughout the body, (b) contraction of smooth muscle in the internal organs, and (c) secretion by both exocrine and endocrine glands in many parts of the body. These activities are collectively called *motor functions* of the nervous system, and the muscles and glands are called *effectors* because they perform the functions dictated by the nerve signals. That portion of the nervous system directly concerned with transmitting signals to the muscles and glands is called the motor division of the nervous system.

Figure 46–2 illustrates the *motor axis* of the nervous system for controlling skeletal muscle contraction. Operating parallel with this axis is another similar system for control of the smooth muscles and glands; it is called the *autonomic nervous system.* Note in Figure 46–2 that the skeletal muscles can be controlled from many different levels of the nervous system, including (a) the spinal cord, (b) the substance of the medulla, pons, and mesencephalon, (c) the basal ganglia, and (d) the motor cortex. Each of these different areas plays its own specific role in the control of body movements, the lower regions being concerned primarily with automatic, instantaneous responses of the body to sensory stimuli and the higher regions with deliberate movements controlled by the thought processes of the cerebrum.

PROCESSING OF INFORMATION

The nervous system would not be at all effective in controlling bodily functions if each bit of sensory information caused some motor reaction. Therefore, one of the major functions of the nervous system is to process incoming information in such a way that *appropriate* motor responses occur. Indeed, more than 99 per cent of all sensory information is continually discarded by the brain as unimportant. For instance, one is ordinarily totally unaware of the parts of his body that are in contact with his clothes and is also unaware of the pressure on his seat when he is sitting. Likewise, his attention is drawn only to an occasional object in his field of vision, and even the perpetual noise of his surroundings is usually relegated to the background.

After the important sensory information has been selected, it must be channeled into proper motor regions of the brain to cause the desired response. Thus, if a person places his hand on a hot stove, the desired response is to lift the hand, plus other associated responses, such as moving the entire body away from the stove and perhaps even shouting with pain. Yet even these responses represent activity by only a small fraction of the total motor system of the body.

Function of the Synapse in Processing Information. Transmission of impulses from one neuron to another through the synapse was discussed in Chapter 6. It will be recalled that different synapses have widely varying properties: Some transmit signals from one neuron to the next with ease while others transmit signals only with difficulty. Some postsynaptic neurons respond with large numbers of impulses while others respond with only a few. Thus, the synapses perform a selective action, often blocking the weak signals while allowing the strong signals to pass, often selecting and

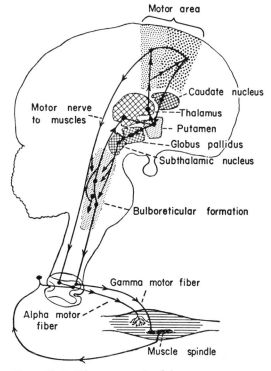

Figure 46–2   The motor axis of the nervous system.

amplifying certain weak signals, and often channeling the signal in many different directions rather than simply in one direction. The basic principles of this processing of information by the synapses are so important that they are discussed in detail in the following chapter.

STORAGE OF INFORMATION—MEMORY

Only a small fraction of the important sensory information causes an immediate motor response. The remainder is stored for future control of motor activities and for use in the thinking processes. Most of this storage of information occurs in the cerebral cortex, but not all, for even the basal regions of the brain and perhaps even the spinal cord can store small amounts of information.

The storage of information is the process we call *memory,* and this too is a function of the synapses. That is, each time a particular sensory signal passes through a sequence of synapses, the respective synapses become more capable of transmitting the same signal the next time, which process is called *facilitation.* After the sensory signal has passed through the synapses a large number of times, the synapses become so facilitated that signals from other parts of the brain can also cause transmission of impulses through the same sequence of synapses even though the sensory input has not been excited. This gives the person a feeling of experiencing the original sensation, though in effect it is only a memory of the sensation.

Unfortunately, we do not know the precise mechanism by which facilitation of synapses occurs in the memory process, but what is known about this and other details of the memory process will be discussed in Chapter 61.

Once memories have been stored in the nervous system, they become part of the processing mechanism. The thought processes of the brain compare new sensory experiences with the stored memories; the memories help to select the important new sensory information and to channel this into appropriate storage areas for future use or into motor areas to cause bodily responses.

THE THREE MAJOR LEVELS OF NERVOUS SYSTEM FUNCTION

The human nervous system has inherited specific characteristics from each stage of evolutionary development. From this heritage,

there remain three major levels of the nervous system that have special functional significance: (1) the spinal cord level, (2) the lower brain level, and (3) the higher brain or cortical level.

THE SPINAL CORD LEVEL

The spinal cord of the human being still retains many functions of the multisegmental animal. Sensory signals are transmitted through the spinal nerves into each *segment* of the spinal cord, and these signals can cause localized motor responses either in the segment of the body from which the sensory information is received or in adjacent segments. Essentially all the spinal cord motor responses are automatic and occur almost instantaneously in response to the sensory signal. In addition, they occur in specific patterns of response called *reflexes.*

Figure 46–3 illustrates two of the simpler cord reflexes. To the left is the neural control of the *muscle stretch reflex.* If a muscle suddenly becomes stretched, a sensory nerve receptor in the muscle called the *muscle spindle* becomes stimulated and transmits nerve impulses through a sensory nerve fiber into the spinal cord. This fiber synapses directly with a motoneuron in the anterior horn of the cord gray matter, and the motoneuron in turn transmits impulses back to the muscle to cause the muscle, the effector, to contract. The muscle contraction opposes the original muscle stretch. Thus, this reflex acts as a *feedback* mechanism, operating from a receptor to an effector, to prevent sudden changes in length of the muscle. This allows a person to maintain his limbs and other parts of his body in desired positions despite sudden outside forces that tend to move the parts out of position.

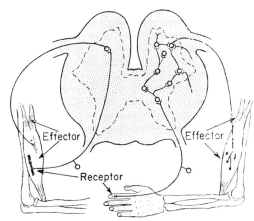

Figure 46–3 *Left:* The simple stretch reflex. *Right:* A withdrawal reflex.

To the right in Figure 46–3 is illustrated the neural control of another reflex called the *withdrawal reflex.* This is a protective reflex that causes withdrawal of any part of the body from an object that is causing pain. For instance, let us assume that the hand is placed on a sharp object. Pain signals are transmitted into the gray matter of the spinal cord, and, after appropriate selection of information by the synapses, signals are channeled to the appropriate motoneurons to cause flexion of the biceps muscle. This obviously lifts the hand away from the sharp object.

We see, then, that the withdrawal reflex is much more complex than the stretch reflex, for it involves many neurons in the gray matter of the cord and signals are transmitted to many adjacent segments of the cord to cause contraction of the appropriate muscles.

Cord Functions After the Brain Is Removed. The many reflexes of the spinal cord will be discussed in Chapter 56; however, the following list of a few of the important cord reflex functions that occur even after the brain is removed illustrates the many capabilities of the spinal cord:

1. The animal can under certain conditions be made to stand up. This is caused primarily by reflexes initiated from the pads of the feet. Sensory signals from the pads cause the extensor muscles of the limbs to tighten, which in turn allows the limbs to support the animal's body.

2. A spinal animal held in a sling so that his feet hang downward often begins walking or galloping movements involving one, two, or all his legs. This illustrates that the basic patterns for causing the limb movements of locomotion are present in the spinal cord.

3. A flea crawling on the skin of a spinal animal causes reflex to-and-fro scratching by the paw, and the paw can actually localize the flea on the surface of the body.

4. Cord reflexes exist to cause emptying of the urinary bladder or of the rectum.

5. Segmental temperature reflexes are present throughout the body. Local cooling of the skin causes vasoconstriction, which helps to conserve heat in the body. Conversely, local heating in the skin causes vasodilatation, resulting in loss of heat from the body.

This list of some of the segmental and multisegmental reflexes of the spinal cord demonstrates that many of our day-by-day and moment-by-moment activities are controlled locally by the respective segmental levels of the spinal cord, and the brain plays only a modifying role in these local controls.

THE LOWER BRAIN LEVEL

Many if not most of what we call subconscious activities of the body are controlled in the lower areas of the brain—in the medulla, pons, mesencephalon, hypothalamus, thalamus, and basal ganglia. Subconscious control of arterial blood pressure and respiration is achieved primarily in the reticular substance of the medulla and pons. Control of equilibrium is a combined function of the older portions of the cerebellum and the reticular substance of the medulla, pons, and mesencephalon. The coordinated turning movements of the head, of the entire body, and of the eyes are controlled by specific centers located in the mesencephalon, paleocerebellum, and lower basal ganglia. Feeding reflexes, such as salivation in response to taste of food and licking of the lips, are controlled by areas in the medulla, pons, mesencephalon, amygdala, and hypothalamus. And many emotional patterns, such as anger, excitement, sexual activities, reactions to pain, or reactions of pleasure, can occur in animals without a cerebral cortex.

In short, the subconscious but coordinate functions of the body, as well as many of the life processes themselves—arterial pressure and respiration, for instance—are controlled by the lower regions of the brain, regions that usually, but not always, operate below the conscious level.

THE HIGHER BRAIN OR CORTICAL LEVEL

We have seen from the above discussion that many of the intrinsic life processes of the body are controlled by subcortical regions of the brain or by the spinal cord. What then is the function of the cerebral cortex? The cerebral cortex is primarily a vast information storage area. Approximately three quarters of all the neuronal cell bodies of the entire nervous system are located in the cerebral cortex. It is here that most of the memories of past experiences are stored, and it is here that many of the patterns of motor responses are stored, which information can be called forth at will to control motor functions of the body.

Relation of the Cortex to the Thalamus and Other Lower Centers. The cerebral cortex is actually an outgrowth of the lower regions of the brain, particularly of the thalamus. For each area of the cerebral cortex there is a corresponding and connecting area of the thalamus, and activation of a minute portion of the thalamus activates the corresponding and much larger portion of the cerebral cortex. It is presumed that in this way the thalamus can call forth cortical activities at will. Also, activation of regions in the mesencephalon transmits diffuse signals to the cerebral cortex, partially through the thalamus and partially directly, to activate the entire cortex. This is the process that we call *wakefulness.* On the other hand, when these areas of the mesencephalon become inactive, the thalamic and cortical regions also become inactive, which is the process we call *sleep.*

Some areas of the cerebral cortex are not directly concerned with either sensory or motor functions of the nervous system—for example, the prefrontal lobe and large portions of the temporal and parietal lobes. These areas are set aside for the more abstract processes of thought, but even they also have direct nerve connections with the lower regions of the brain.

Large areas of the cerebral cortex can be destroyed without blocking the subconscious, and even some of the involuntary conscious, activities of the body. For instance, destruction of the somesthetic cortex does not destroy one's ability to feel objects touching his skin, but it does destroy his ability to distinguish the shapes of objects, their character, and the precise points on the skin where the objects are touching. Thus, the cortex is not required for perception of sensation, but it does add immeasurably to its depth of meaning. Likewise, destruction of the prefrontal lobe does not destroy one's ability to think, but it does destroy his ability to think in abstract terms. In other words, each time a portion of the cerebral cortex is destroyed, a vast amount of information is lost to the thinking process and some of the mechanisms for processing this information are also lost. The result is a vegetative type of existence rather than a "living" existence.

Telencephalization. In the process of evolution, the higher regions of the human nervous system have taken over many sensory and motor functions performed by lower regions of the brain in lower animals. For instance, if the spinal cord of an opossum is cut in the midthoracic region, the opossum can still walk perfectly well on both his forelimbs and hindlimbs, except that the hindlimbs are then unsynchronized with the forelimbs. A similar transection of the spinal cord in the human being causes complete loss of ability to use the legs for locomotion.

This process by which the progressively higher centers of the brain have taken over more and more of the function of the lower centers is called telencephalization. Yet, despite telencephalization, some of the very basic functions of the lower centers still remain active or at least partially

active in the human being, such as many of the cord reflexes and the stereotyped control systems of the basal regions of the brain that were just discussed.

COMPARISON OF THE NERVOUS SYSTEM WITH AN ELECTRONIC COMPUTER

When electronic computers were first designed in many different laboratories of the world by as many different engineers, it soon became apparent that all these machines had many common features. First, they all had input circuits which were comparable to the sensory portion of the nervous system and output circuits which were comparable to the motor portion of the nervous system. In the conducting pathway between the inputs and the outputs were the mechanisms for performing the different types of computations.

In simple computers, the output signals were controlled directly by the input signals, operating in a manner similar to that of the simple reflexes of the spinal cord. But, in the more complex computers, the output was determined both by the input signals and by information that had already been stored in the computer, which is analogous to the more complex reflex and processing mechanisms of our nervous system. Furthermore, as the computers became even more complex it became necessary to add still another unit, called the *programming unit,* which determines the sequence of computational operations. This unit is analogous to the mechanism in our brain that allows us to direct our attention first to one thought, then to another, and so forth, until complex sequences of thought take place.

Figure 46–4 illustrates a simple block diagram of a modern computer. Even a rapid study of this diagram will demonstrate its similarity to the nervous system. Furthermore, general purpose computers designed and built by many different scientific groups have ended up with almost identically the same basic components. The fact that these components are analogous to those of the human nervous system demonstrates that the brain is basically a computer that continuously collects sensory information and uses this along with stored information to compute the daily course of bodily activity.

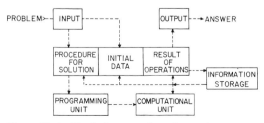

Figure 46–4 Block diagram of an electronic computer, showing the basic components and their interrelationships.

REFERENCES

Ashby, W. R.: Mathematical models and computer analysis of the function of the central nervous system. *Ann. Rev. Physiol.,* 28:89, 1966.

Asratian, E. A.: Compensatory Adaptations, Reflex Activity, and The Brain. New York, Pergamon Press, Inc., 1965.

Corning, W. C., and Balaban, M. W. (eds.): The Mind: Biological Approaches to Its Functions. New York, John Wiley & Sons, Inc., 1968.

Eldred, E., and Buchwald, J.: Central nervous system. *Ann. Rev. Physiol.,* 29:573, 1967.

Gerard, R. W., and Duyff, J. W. (eds.): Processing of Information in the Nervous System. *Proc. Int. Union of Physiol. Sci.,* Vol. III, 1962.

Konorski, J.: Integrative Activity of the Brain. Chicago, University of Chicago Press, 1967.

Merritt, H.: Textbook of Neurology. 4th ed., Philadelphia, Lea & Febiger, 1967.

Miller, R., and Burack, E.: Atlas of the Central Nervous System in Man. Baltimore, The Williams & Wilkins Co., 1968.

Noback, C. R.: The Human Nervous System: Basic Elements of Structure and Function. New York, McGraw-Hill Book Company, 1967.

Ochs, S.: Elements of Neurophysiology. New York, John Wiley & Sons, Inc., 1965.

Sarkisov, S. A.: Structure and Function of the Brain. Bloomington, Indiana, Indiana University Press, 1966.

Walshe, F.: Further Critical Studies in Neurology. Baltimore, The Williams & Wilkins Co., 1965.

CHAPTER 47

Transmission and Processing of Information in the Nervous System

INFORMATION, SIGNALS, AND IMPULSES

The term *information,* as it applies to the nervous system, means a variety of different things, such as knowledge, facts, quantitative values, intensity of pain, intensity of light, temperature, and any other aspect of the body or its immediate surroundings that has meaning. Thus, pain from a pin prick is information, pressure on the bottom of the feet is information, degree of angulation of the joints is information, and a stored memory in the brain is information.

The primary function of the brain is to transmit information from one point to another and to process this information so that it can be used advantageously or so that its meaning can become clear to the mind. However, information cannot be transmitted in its original form but only in the form of nerve impulses. Thus, a part of the body that is subjected to pain must first convert this information into *nerve impulses;* and specific areas of the brain convert abstract thoughts also into nerve impulses that are then transmitted either elsewhere in the brain or into peripheral nerves to motor effectors throughout the body. The retina of the eye converts vision into nerve impulses, the nerve endings of the joints convert degree of angulation of the joints into nerve impulses, and so forth. The nerve impulses in turn are transmitted to the opposite ends of the neurons where they cause subsequent effects either at the synapses of other neurons or at the ends of motor nerves. This is what is meant by the transmission of information.

Signals. In the transmission of information, it is frequently not desirable to speak in terms of the individual impulses but instead in terms of the overall pattern of impulses; this pattern is called a *signal.* As an example, when pressure is applied to a large area of skin, impulses are transmitted by large numbers of parallel nerve fibers, and the total pattern of impulses transmitted by all these fibers is called a signal. Thus, we can speak of visual signals, auditory signals, somesthetic sensory signals, motor signals, and so forth.

TRANSMISSION OF SIGNALS IN NERVE TRACTS

A signal is almost never transmitted only in a single nerve fiber but instead in a group of parallel fibers bound together in nerves or in *nerve tracts* in the central nervous system.

TRANSMISSION OF SIGNAL STRENGTH

Spatial Summation (Multiple Fiber Summation). Figure 47–1 illustrates a sensory nerve leading from a segment of skin, showing pain nerve fibers arborizing into an extensive plexus. A pin or other object stimulates nerve endings from a large number of different fibers, because each separate fiber divides and spreads over an area of skin sometimes as large

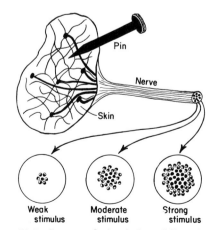

Figure 47–1   Pattern of stimulation of fibers in a nerve trunk leading from an area of skin pricked by a pin.

as 5 cm. in diameter called the *receptive field*. The number of nerve endings from each fiber is large in the very center of its "field" but becomes less and less toward the periphery. Therefore, in Figure 47–1, a pin pricking the skin stimulates a nerve fiber strongly if the point of stimulation is in the center of its field of reception, but it stimulates the fiber only weakly if the point of stimulus is in the periphery of the field.

In the lower part of Figure 47–1 is shown a cross-section of the nerve leading from the skin. Note that if the pinprick is slight, only a single fiber is stimulated strongly (the black fiber in the center), and several immediately adjacent fibers are stimulated moderately, while all other fibers are not stimulated at all. When the stimulus becomes more intense, the number of fibers strongly stimulated becomes progressively greater. Thus, the signal spreads to more and more fibers. This phenomenon is called *spatial summation*, which means simply that one of the means by which signals of increasing strength are transmitted in the nervous system is by utilization of progressively greater numbers of fibers. The increase in number of fibers as the strength of signal increases is called *recruitment* of the additional fibers.

Temporal Summation (Frequency Modulation). A second means by which signals of increasing strength are transmitted through nerve tracts is by increasing the *frequency* of nerve impulses in each fiber. This occurs in addition to the increase in number of excited fibers. This process is called *temporal summation*, which means simply that during any given interval of time, the number of impulses transmitted is directly related to the strength of the signal.

Figure 47–2 illustrates temporal summation, showing in the upper part a changing strength of signal and in the lower part the actual impulses transmitted by the nerve fiber. This figure demonstrates that the frequency increases and decreases with the strength of the signal. Thus, this is actually a process of frequency modulation, the frequency of impulses being "modulated" in proportion to the strength of the signal.

DETECTION OF A SIGNAL AT THE TERMINUS OF THE NERVE PATHWAY

The nerve impulses obviously are nothing more than a series of pulses rather than a smooth signal. However, all terminal points in the nervous system have means for averaging impulses and thereby smoothing out the

Figure 47–2 Translation of signal strength into a frequency-modulated series of nerve impulses, showing *above* the strength of signal and *below* the separate nerve impulses.

signal. Let us consider as an example skeletal muscles. Impulses controlling a skeletal muscle are transmitted through large numbers of parallel nerve fibers. If 50 nerve fibers innervate a muscle and one impulse is transmitted per second through each of these, the 50 separate impulses cause a weak but smooth muscle contraction because the pulses in the different nerve fibers are spread out over a period of time. If each nerve fiber transmits 200 impulses per second, making a total of 10,000 pulses per second, the 10,000 pulses give a strong and still smooth contraction.

Similar averaging processes occur in the central nervous system; each postsynaptic neuron in the nervous system receives literally hundreds to thousands of impulses each second from the presynaptic terminals, and the synaptic membrane is capable of summating and averaging these impulses to give relatively smooth changes in the excitatory state of the neuron, as was explained in Chapter 6.

"Granularity" of the Detected Signal. Despite the ability of the muscles and synapses in the central nervous system to average impulse rates, under some conditions the detected signal has unwanted ripples, which is called *granularity* of the signal. Two causes of this are, first, slow impulse rates and, second, synchrony of impulses in the parallel fibers through which the signal is transmitted.

A slow impulse rate causes granularity of the signal because of the nature of the averaging system itself. The synapses in the nervous system, the skeletal muscle, and the other signal detectors react to an impulse for a given period of time. Thus, the contractile twitch of a typical muscle lasts about one-thirtieth second. If a second impulse comes along before this twitch is over, it causes the process of contraction to continue. Therefore, rapidly successive impulses cause smooth contraction of the muscle called *tetanization*. On the other hand, if the impulse rate is less than 30 per second, the muscle contracts in a series of twitches, and the signal is said to be granular. The same principles apply to the averaging mechanism of the central nervous system synapses.

The second cause of granularity is *synchrony* of impulses in the parallel fibers transmitting the signal. For instance, if 100 fibers are transmitting one impulse per second, the signal will be smooth despite the low impulse frequency if the impulses are spread out in time in the different fibers. But if all these impulses are transmitted in synchrony, a very granular signal will reach the detector at a frequency of one per second.

Fidelity of the Temporal Pattern of a Signal. The "temporal pattern" of a signal means the changes in signal strength with time. Unfortunately, extremely rapid changes in signal strengths cannot be transmitted through nerve pathways. The reason for this is that the averaging mechanisms of the detectors require a considerable number of impulses before the average can be determined. Therefore, the change in signal strength in the detector always *lags* a few milliseconds behind the change in signal strength in the nerve tract itself.

The fidelity of the temporal pattern is different for different types of nerve fibers. The large type A nerve fibers that can transmit as many as 2000 impulses per second are usually associated with rapidly responding sensory receptors and rapidly responding detectors. Therefore, they can transmit changes in signal strengths that occur as rapidly as 700 cycles per second (particularly those nerve fibers associated with the pacinian corpuscles). On the other hand, the minute type C fibers that transmit only a few impulses per second are usually associated with slowly responding receptors and detectors. The frequency of signals transmitted through these pathways is often limited to as little as one or less than one cycle per second (as is the case for autonomic signals controlling the internal organs).

SPATIAL ORIENTATION OF SIGNALS IN FIBER TRACTS

How does the brain detect the precise position on the body that is receiving a sensory stimulus, and how does the brain transmit impulses precisely to individual skeletal muscle bundles? The answer to this is by transmitting their signals in a precise spatial pattern through the nerve tracts. All the different nerve tracts, both in the peripheral nerves and in the fiber tracts of the central nervous system, are spatially organized. For instance, in the dorsal columns of the spinal cord the sensory fibers from the feet lie toward the midline, while those fibers entering the dorsal columns at higher levels of the body lie progressively more toward the lateral sides of the dorsal columns. This spatial organization is maintained with precision throughout the sensory pathway all the way to the somesthetic cortex. Likewise, the fiber tracts within the brain and those extending into motor nerves are spatially oriented in the same way.

As an example, Figure 47–3 illustrates to the left three separate pins stimulating the skin; to the right the spatial orientation of the stimulated fibers in the nerve are shown. Each pin in this example stimulates a single fiber strongly and adjacent fibers less strongly. This spatial orientation of the three separate bundles of fibers is maintained all the way to the cerebral cortex and obviously allows the brain to localize the three different pinpricks to their points of stimulation.

TRANSMISSION AND PROCESSING OF SIGNALS IN NEURONAL POOLS

The central nervous system is made up of literally hundreds of separate neuronal pools, some of which are extremely small and some very large. For instance, the entire cerebral cortex could be considered to be a single large neuronal pool. If all the surface area of the cerebral cortex were flattened out, including the surfaces of the penetrating folds, the total area of this large flat pool would be several square feet. It has many separate fiber tracts coming to it (afferent fibers) and others leaving it (efferent fibers). Furthermore it maintains the same quality of spatial orientation as that found in the nerve bundles, individual points of the cortex connecting with specific points elsewhere in the nervous system or connecting through the peripheral nerves with specific points in the body. However, within this pool of neurons are large numbers of short nerve fibers whereby signals spread horizontally from neuron to neuron within the pool itself.

Other neuronal pools include the different basal ganglia and the specific nuclei in the cerebellum, mesencephalon, pons, and medulla. Also, the entire dorsal gray matter of the spinal cord could be considered to be one long pool of neurons, and the entire anterior gray matter

Figure 47–3 Spatial pattern of nerve fiber stimulation in a nerve trunk following stimulation of the skin by three separate but simultaneous pinpricks.

another long neuronal pool. Each pool has its own special characteristics of organization which cause it to process signals in its own special way. It is these special characteristics of the different pools that allow the multitude of functions of the nervous system. Yet, despite their differences in function, the pools also have many similarities of function which are described in the following pages.

TRANSMISSION OF SIGNALS THROUGH NEURONAL POOLS

Organization of Neurons in the Neuronal Pools. Figure 47–4 is a schematic diagram of the organization of neurons in a neuronal pool, showing "input" fibers to the left and "output" fibers to the right. Each input fiber divides as many as several hundred times and provides an average of several hundred terminal fibrils that spread over a large area in the pool and that finally synapse with the dendrites or cell bodies of the neurons in the pool. The area into which the endings of each incoming nerve fiber spread is called its *stimulatory field*. Note that each input fiber arborizes so that large numbers of its presynaptic terminals lie on the centermost neurons in its "field," but progressively fewer terminals lie on the neurons farther from the center of the field.

Threshold and Subthreshold Stimuli — Facilitation. Going back to the discussion of synaptic function in Chapter 6, it will be recalled that stimulation of a single excitatory presynaptic terminal almost never stimulates the postsynaptic neuron. Instead, large numbers of presynaptic terminals must discharge on the same neuron either simultaneously or in rapid succession to cause excitation. For instance, let us assume that six separate presynaptic terminals must discharge simultaneously or in rapid succession to excite any one of the neurons in Figure 47–4. Note that input fiber 1 contributes a total of 10 presynaptic terminals to neuron *a*. Therefore, an incoming impulse in fiber 1 will cause neuron *a* to discharge. This same incoming fiber has two presynaptic terminals on neuron *b* and three on neuron *c*. Neither of these two neurons will fire.

Incoming nerve fiber 2 has 12 presynaptic terminals on neuron *d*, three on neuron *c*, and two on neuron *b*. In this case, an impulse in fiber 2 excites only neuron *d*.

Though fiber 1 fails to stimulate neurons *b* and *c*, discharge of the presynaptic terminals changes the membrane potentials of these neurons so that they can be more easily excited by other incoming signals. Thus, there are two types of stimuli that enter a neuronal pool, *threshold stimuli* and *subthreshold stimuli*. A threshold stimulus causes actual excitation. A subthreshold stimulus fails to excite the neuron but does make the neuron more excitable to impulses from other sources. The neuron that is made more excitable but does not discharge is said to be *facilitated*.

"CONVERGENCE" OF SUBTHRESHOLD STIMULI TO CAUSE EXCITATION. Let us now assume that both fibers 1 and 2 in Figure 47–4 transmit concurrent impulses into the neuronal pool. In this case, neuron *c* is stimulated by six presynaptic terminals simultaneously, which is the required number to cause excitation. Thus, subthreshold stimuli can *converge* from several sources and *summate* at the neuron to cause a threshold stimulus.

SUMMATION OF FACILITATION. Now, let us consider neuron *b* of Figure 47–4 when both input nerve fibers are stimulated. In this case each fiber excites two presynaptic terminals on neuron *b*, but the total of four excited presynaptic terminals still is not up to the threshold value required to excite the neuron. However, the degree of facilitation does become greatly increased so that now only two additional presynaptic terminals excited from another source need fire to cause discharge.

THE FIELD OF TERMINALS. It must be recognized that Figure 47–4 represents a highly condensed version of the neuronal pool, for each input nerve fiber gives off ter-

Figure 47–4   Basic organization of a neuronal pool.

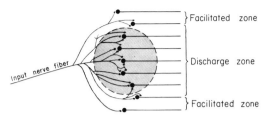

Figure 47-5 "Discharge' and "facilitated" zones of a neuronal pool.

Figure 47-7 Inhibitory circuit. Neuron #2 is an inhibitory neuron.

minals that spread to perhaps 100 or more separate neurons, and the surface of each neuron is covered by many hundred synapses. In the center of the *field* of terminals, almost all the neurons are stimulated by the incoming fiber; whereas farther toward the periphery of the field, the neurons are facilitated but do not discharge. Figure 47-5 illustrates this effect, showing the field of a single input nerve fiber. The area in the neuronal pool in which all the neurons discharge is called the *discharge* or *threshold* or *liminal zone,* and the area to either side in which the neurons are facilitated but do not discharge is called the *facilitated* or *subthreshold* or *subliminal* zone.

Facilitation of the Neuronal Pool by Signals from Accessory Sources. A neuronal pool frequently receives input nerve fibers from many different sources. Thus, in Figure 47-6 a neuron in a given pool receives impulses from a *primary source* and from two *accessory sources.* The nerve fiber from accessory source #1 is an excitatory fiber which secretes excitatory transmitter at its presynaptic terminals. If enough presynaptic terminals from this source are stimulated, the postsynaptic neuron discharges. However, in many parts of the nervous system the accessory sources supply only a few nerve fibers into the pool, not enough usually to cause excitation but yet enough to *facilitate* the neurons. Thus, in Figure 47-6, if a facilitory signal is entering the pool from

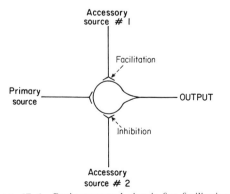

Figure 47-6 Basic neuronal circuit for facilitation or inhibition.

accessory source #1, a much weaker than usual signal from the primary source is able to excite the postsynaptic neuron. In this way the signal from accessory source #1 controls the ease with which signals pass from the primary source to the output, which explains one of the means by which different parts of the nervous system control the degree of activity of other parts.

Inhibition of a Neuronal Pool by an Accessory Source. It will be recalled from the discussions in Chapter 6 that some neurons of the central nervous system secrete an inhibitory transmitter substance instead of an excitatory transmitter substance. Thus, in Figure 47-6, stimulation of the inhibitory fibers from accessory source #2 strongly inhibits the neuronal pool so that a strong signal from the primary source is required to cause normal output.

MECHANISM BY WHICH A SINGLE INPUT SIGNAL CAN CAUSE BOTH EXCITATION AND INHIBITION—THE INHIBITORY CIRCUIT. Figure 47-7 illustrates the so-called inhibitory circuit which can change an excitatory signal into an inhibitory signal. In this figure the input fiber divides and secretes excitatory transmitter at both its endings. This causes excitation of both neurons 1 and 2. However, neuron 2 is an inhibitory neuron that secretes inhibitory transmitter at its terminal nerve endings. Excitation of this neuron therefore inhibits neuron 3.

In short, one way to cause inhibition is for a signal *to be transmitted through an inhibitory neuron,* which then secretes the inhibitory transmitter substance.

Convergence. The term "convergence" means control of a single neuron by two or more separate input nerve fibers. One type of convergence was illustrated in Figure 47-4 in which two excitatory input nerve fibers from the same source converged upon several separate neurons to stimulate them. This type of convergence from a single source is illustrated again in Figure 47-8A.

However, convergence can also result from input signals (excitatory or inhibitory) from several different sources, which is illustrated

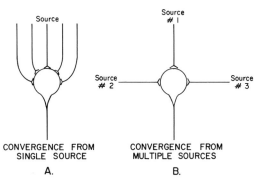

Figure 47–8 "Convergence" of multiple input fibers on a single neuron: (A) Input fibers from a single source. (B) Input fibers from multiple sources.

in Figure 47–8B. For instance, the internuncial cells of the spinal cord receive converging signals from (a) peripheral nerve fibers entering the cord, (b) propriospinal fibers passing from one segment of the cord to another, (c) corticospinal fibers from the cerebral cortex, and (d) probably several other long pathways descending from the brain into the spinal cord.

Such convergence allows summation of information from several different sources, and the resulting response is a summated effect of all these different types of information. Obviously, therefore, convergence is one of the important means by which the central nervous system correlates, summates, and sorts different types of information.

Divergence. Divergence means that excitation of a single input nerve fiber stimulates multiple output fibers from the neuronal pool. The two major types of divergence are illustrated in Figure 47–9 and may be described as follows:

An *amplifying* type of divergence often occurs, illustrated in Figure 47–9A. This means simply that an input signal spreads to an increasing number of neurons as it passes through successive pools of a nervous pathway. This type of divergence is characteristic of the corticospinal pathway in its control of skeletal muscles as follows: Stimulation of a single large pyramidal cell in the motor cortex transmits a single impulse into the spinal cord. Yet, under appropriate conditions, this impulse can stimulate perhaps 15 to 20 internuncial cells, each of which in turn stimulates perhaps as many as several hundred anterior motoneurons. Each of these then stimulates as many as 100 to 300 muscle fibers. Thus, there is a total divergence, or amplification, of as much as 50,000-fold.

The second type of divergence, illustrated in Figure 47–9B, is *divergence into multiple tracts*. In this case, the signal is transmitted in

two separate directions from the pool. This allows the same information to be transmitted to several different parts of the nervous system where it is needed. For instance, information transmitted in the dorsal columns of the spinal cord takes two courses in the lower part of the brain, (1) into the cerebellum and (2) on through the lower regions of the brain to the thalamus and cerebral cortex. Likewise, in the thalamus almost all sensory information is relayed both into deep structures of the thalamus and to discrete regions of the cerebral cortex.

TRANSMISSION OF SPATIAL PATTERNS THROUGH SUCCESSIVE NEURONAL POOLS

Most information is transmitted from one part of the nervous system to another through several successive neuronal pools. For instance, sensory information from the skin passes first through the peripheral nerve fibers, then through second order neurons that originate either in the spinal cord or in the cuneate and gracile nuclei of the medulla, and finally through third order neurons originating in the thalamus to the cerebral cortex. Such a pathway is illustrated at the top of Figure 47–10. Note that the sensory nerve endings in the skin overlap each other tremendously; and the terminal fibrils of each nerve fiber, on entering each neuronal pool, spread to many adjacent neurons, innervating perhaps 100 or more separate neurons. On first thought, one would expect signals from the skin to become completely mixed up by this haphazard arrangement of terminal fibrils in each neuronal pool. For statistical reasons, however, this does not occur, which can be explained as follows:

First, if a single point is stimulated in the skin, the nerve fiber with the most nerve end-

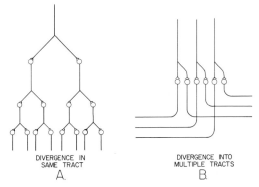

Figure 47–9 "Divergence" in neuronal pathways: (A) Divergence within a pathway to cause "amplification" of the signal. (B) Divergence into multiple tracts to transmit the signal to separate areas.

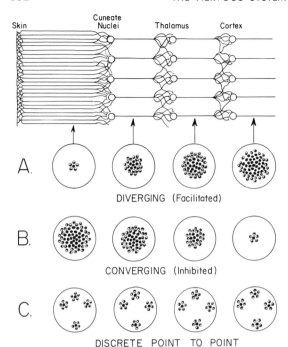

Figure 47-10 Typical organization of a sensory pathway from the skin to the cerebral cortex. *Below:* The patterns of fiber stimulation at different points in the pathway following stimulation by a pinprick when the pathway is (A) facilitated, (B) inhibited, and (C) normally excitable.

ings in that particular spot becomes stimulated to the strongest extent, while the immediately adjacent nerve fibers become stimulated less strongly, and the nerve fibers still farther away become stimulated only weakly. When this signal arrives at the first neuronal pool, the stimulus spreads in many directions in the terminal fibrils of the neuronal pool. Yet the *greatest number* of *excited presynaptic terminals* lies very near the point of entry of the most excited input nerve fiber. Therefore, the neuron closest to this central point is the one that becomes stimulated to the greatest extent. Exactly the same effect occurs in the second neuronal pool in the thalamus and again when the signal reaches the cerebral cortex.

Yet, it is true that a signal passing through *highly facilitated neuronal pools* could diverge so much that the spatial pattern at the terminus of the pathway would be completely obscured. This effect is illustrated in Figure 47-10A, which shows successive spatial patterns of neuron stimulation in such a facilitated, diverging pathway.

However, the degree of facilitation of the different neuronal pools varies from time to time. Under some conditions the degree of facilitation is so low that the pathway becomes converging, as illustrated in Figure 47-10B.

In this case, a broad area of the skin is stimulated, but the signal loses part of its fringe stimuli as it passes through each successive pool until the breadth of the stimulus becomes contracted at the opposite end. One can achieve this type of stimulation by pressing ever so lightly with a flat object on the skin. The signal converges to give the person a sensation of almost a point contact.

In Figure 47-10C, four separate points are simultaneously stimulated on the skin, and the degree of excitability in each neuronal pool is exactly that amount required to prevent either divergence or convergence. Therefore, a reasonably true spatial pattern of each of the four points of stimulation is transmitted through the entire pathway.

Centrifugal Control of Neuronal Facilitation in the Sensory Pathways. It is obvious from the above discussion that the degree of facilitation of each neuronal pool must be maintained at exactly the proper level if faithful transmission of the spatial pattern is to occur. Recent discoveries have demonstrated that the degrees of facilitation of most—indeed, probably all—neuronal pools in the different pathways are controlled by *centrifugal nerve fibers* that pass from the respective sensory areas of the cortex downward to the separate neuronal pools. Thus, these nerve fibers undoubtedly help to control the faithfulness of signal transmission.

Inhibitory Circuits to Provide Contrast in the Spatial Pattern. When a single point of the skin or other sensory area is stimulated, not only is a single fiber excited but a number of "fringe" fibers are excited less strongly at the same time, as already explained above. Therefore, the spatial pattern is blurred even before the signal begins to be transmitted through the pathway. However, in many pathways—if not all—such as in the visual pathway and in the somesthetic pathway, lateral *inhibitory circuits* inhibit the fringe neurons and re-establish a truer spatial pattern.

Figure 47-11A illustrates this circuit, showing that the nerve fibers of a pathway give off collateral fibers that excite inhibitory neurons. These inhibitory neurons in turn inhibit the less excited fringe neurons in the signal pathway. The effect of this on transmission of the spatial pattern is illustrated in Figure 47-11B, which shows the same point-to-point transmission pattern that was illustrated in Figure 47-10C. The left-hand pattern illustrates four strongly excited fibers; penumbras of fringe excitation surround each of these. To the right is illustrated removal of the penumbras by the lateral inhibitory circuits; and obviously, this increases the contrast in the signal and helps

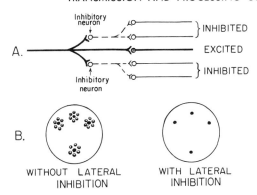

WITHOUT LATERAL
INHIBITION

WITH LATERAL
INHIBITION

Figure 47–11 (A) Lateral inhibitory circuit by which an excited fiber of a neuronal pool can cause inhibition of adjacent fibers. (B) Increase in contrast of the stimulus pattern caused by the inhibitory circuit.

the faithfulness of transmission of the spatial pattern. Unfortunately, we still know too little about this lateral inhibitory mechanism to be sure how effective it actually is. Furthermore, much of the lateral inhibition results from presynaptic inhibition, which does not require the inhibitory interneuron of Figure 47–11A; this was discussed in Chapter 6. The presynaptic type of lateral inhibition is probably the more prevalent mechanism in more peripheral areas of the nervous system, whereas the inhibitory interneuron is the more prevalent in the thalamic and cortical areas.

"Plasticity" of Terminal Nerve Endings in the Neuronal Pool. Some physiologists believe that terminal nerve endings are continually degenerating while new ones are being formed, which is called plasticity of the endings. This makes one wonder how it is possible for discrete spatial patterns to be transmitted through neuronal pools. However, if one will think for a moment, he will see that it matters little whether old terminals disappear and new ones take their place or not, for the *relative positions* of the individual signal points in the spatial pattern still remain the same.

If one wishes to delve into the philosophy of central nervous system organization, he can see that this concept of plasticity is exceedingly important in understanding how the nervous system develops. For instance, it would be impossible during the development of the embryo for neurons of one order always to connect with precisely given neurons of the next order. However, since it does not matter exactly which neurons the terminal fibers connect with—that is, since it is only the *relative positions of synapses in the neuronal pools* that are important—it becomes much simpler to comprehend how our complex nervous system could have evolved.

Relationship of Numbers of Fibers in a Pathway to the Total Amount of Spatial Information that Can Be Transmitted. The upper part of Figure 47–10 illustrates that far more sensory nerve receptors trans-

mit information into the spinal cord than there are nerve fibers transmitting the same information upward from the spinal cord to the cortex. Furthermore, the total number of points that can be recognized as individual and separate areas on the surface of the body is many times the number of nerve fibers in the sensory pathways ascending the spinal cord. The question must be asked, can a transmission system carry more discrete bits of spatial information than there are nerve fibers themselves?

In considering once again the organization of the transmission pathway shown in the top of Figure 47–10, we see that each sensory nerve fiber leads to a particular *place* in the first neuronal pool. Some fibers end almost directly on a second order neuron while others end *between* second order neurons. If a fiber lies directly on the neuron, that neuron is stimulated far more than the adjacent one. If the fiber lies between two neurons, both neurons are stimulated approximately equally. If the fiber lies three-quarters the way toward one neuron but one-quarter toward the second, the first neuron is stimulated perhaps three times as much as the second neuron. Thus, the spatial position of the incoming neuron can be transmitted by the second order neurons by the *ratios* of signal strengths in the output fibers. One ratio means one position, another ratio another position, and so forth. In this way, a large amount of information from many different sensory receptors can be condensed into much fewer nerve fibers. Here again, we do not know how highly this principle of transmission is developed. But, to give an example, each eye has about 125,000,000 receptors, and it can detect about 5,000,000 individual point areas in the visual field. Yet the total number of nerve fibers transmitting information in the optic nerve is believed to be only one million, many of which are probably concerned with other functions besides transmission of the spatial pattern. Therefore, we can guess that the fibers in this tract can carry 10 or more times as many bits of spatial information as there are fibers.

PROLONGATION OF A SIGNAL BY A NEURONAL POOL—"AFTER-DISCHARGE"

Thus far, we have considered signals that are transmitted instantaneously through a neuronal pool. However, in many instances, a signal entering a pool causes a prolonged output discharge, called after-discharge, even after the incoming signal is over. The three basic mechanisms by which after-discharge occurs are as follows:

Synaptic After-Discharge. When presynaptic terminals discharge on the surfaces of dendrites or the soma of a neuron, a postsynaptic potential develops in the neuron and lasts for many milliseconds—in the anterior motoneuron for about 15 milliseconds, though perhaps much longer in other neurons. As long as this potential lasts it can excite the neuron, causing it to transmit output impulses as was explained in Chapter 6. Thus, as a result of this synaptic after-discharge mechanism alone, it is possible for a single instantaneous input to cause a sustained signal output lasting as long as 15 milliseconds.

The Parallel Circuit Type of After-Discharge. Figure 47–12 illustrates a second type of neuronal circuit that can cause short periods of after-discharge. In this case, the input signal spreads through a series of neurons in the neuronal pool, and from many of these neurons impulses keep converging on an output neuron. It will be recalled that a signal is delayed at each synapse for about 0.5 millisecond, which is called the *synaptic delay.* Therefore, signals that pass through a succession of intermediate neurons reach the output neuron one by one after varying periods of delay. Therefore, the output neuron continues to be stimulated for many milliseconds.

It is doubtful that more than a few dozen successive neurons ordinarily enter into a parallel after-discharge circuit. Therefore, one would suspect that this type of after-discharge circuit could cause after-discharges that last for no more than perhaps 25 to 50 milliseconds. Yet, this circuit does represent a means by which a single input signal, lasting less than 1 millisecond, can be converted into a sustained output signal lasting many milliseconds.

The Reverberating (Oscillatory) Circuit as a Cause of After-Discharge. One of the most important of all circuits in the entire nervous system is the reverberating, or oscillatory, circuit, several different varieties of which are illustrated in Figure 47–13. The simplest theoretical reverberating circuit, even though such may not actually exist in the nervous system, is that illustrated in Figure 47–13A, which involves a single neuron. In this case, the output neuron simply sends a collateral nerve fiber back to its own dendrites or soma to restimulate itself; therefore, once the neuron should discharge, the feedback stimuli could

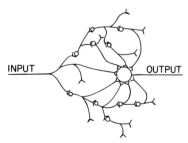

Figure 47–12   The parallel after-discharge circuit.

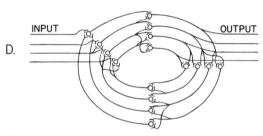

Figure 47–13   Reverberatory circuits of increasing complexity.

theoretically keep the neuron discharging for a long time thereafter.

Figure 47–13B illustrates a few additional neurons in the feedback circuit, which would give a longer period of time between the initial discharge and the feedback signal. Figure 47–13C illustrates a still more complex system in which both facilitatory and inhibitory fibers impinge on the reverberating pool. A facilitatory signal increases the ease with which reverberation takes place, while an inhibitory signal decreases the ease with which reverberation takes place.

Figure 47–13D illustrates that most reverberating pathways are constituted of many parallel fibers, and at each cell station the terminal fibrils diffuse widely. In such a system the total reverberating signal can be either weak or strong, depending on how many parallel nerve fibers are momentarily involved in the reverberation.

Finally, reverberation need not occur only in a single neuronal pool, for it can occur through a circuit of successive pools.

CHARACTERISTICS OF AFTER-DISCHARGE FROM A REVERBERATING CIRCUIT. Figure 47–14 illustrates postulated output signals from a

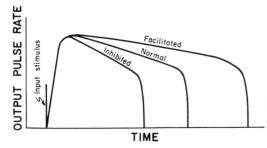

Figure 47–14   Typical pattern of the output signal from a reverberatory circuit following a single input stimulus, showing the effects of facilitation and inhibition.

reverberating after-discharge circuit. The input stimulus need last only 1 millisecond or so, and yet the output can last for many milliseconds or even minutes. The figure demonstrates that the intensity of the output signal increases to a reasonably high value early in the reverberation, then decreases to a critical point, and suddenly ceases entirely. Furthermore, the duration of the after-discharge is determined by the degree of inhibition or facilitation of the neuronal pool. In this way, signals from other parts of the brain can control the reaction of the pool to the input stimulus.

Almost these exact patterns of output signals can be recorded from the motor nerves exciting a muscle involved in the flexor reflex (discussed in Chap. 56), which is believed to be caused by a reverberating type of after-discharge following stimulation of a pain fiber.

IMPORTANCE OF SYNAPTIC FATIGUE IN DETERMINING THE DURATION OF REVERBERATION. It was pointed out in Chapter 6 that synapses fatigue if stimulated for prolonged periods of time. Therefore, one of the most important factors that determines the duration of the reverberatory type of after-discharge is probably the rapidity with which the involved synapses fatigue. Rapid fatigue would obviously tend to shorten the period of after-discharge and slow fatigue to lengthen it.

Furthermore, the greater the number of neurons in the reverberatory pathway and the greater the number of collateral feedback fibrils, the easier it would be to keep the reverberation going. Therefore, it is to be expected that longer reverberating pathways would in general sustain after-discharges for longer periods of time.

DURATION OF REVERBERATION. Typical after-discharge patterns of different reverberatory circuits have durations from as short as 10 milliseconds to as long as several minutes, or perhaps even hours. Indeed, as will be explained in Chapter 60, wakefulness is perhaps

an example of reverberation of neuronal circuits in the basal region of the brain. In this case "arousal impulses" are postulated to set off the wakefulness reverberation at the beginning of each day and thereby to cause sustained excitability of the brain, this excitability lasting 14 or more hours.

CONTINUOUS SIGNAL OUTPUT FROM NEURONAL POOLS

Some neuronal pools emit output signals continuously even without excitatory input signals. At least two different mechanisms can cause this effect, (1) intrinsic neuronal discharge and (2) reverberatory signals.

Continuous Discharge Caused by Intrinsic Neuronal Excitability. Neurons, like other excitable tissues, discharge repetitively if their membrane potentials fall below certain threshold levels. There is reason to believe that the membrane potentials of some neurons are even normally low enough to cause these neurons to emit impulses continually. The rate at which they emit their impulses can be increased by facilitatory signals or decreased by inhibitory signals; the latter can sometimes decrease the rate to extinction.

Continuous Signals Emitted from Reverberating Circuits. Obviously, a reverberating circuit that never fatigues could also be a source of continual impulses. Here again, facilitatory impulses entering the reverberating pool, as illustrated in Figure 47–13C, could increase the output signal level, and inhibition could decrease or even extinguish the output signal.

Figure 47–15 illustrates a continual output signal from a pool of neurons, whether it be a pool emitting impulses in response to intrinsic neuronal excitability or as a result of reverberation. Note that an excitatory input signal greatly increases the output signal, whereas an inhibitory input signal greatly decreases the

Figure 47–15   Continuous output from either a reverberating circuit or from a pool of intrinsically discharging neurons. This figure also shows the effect of excitatory or inhibitory input signals.

output. Those students who are familiar with radio transmitters will recognize this to be a *carrier wave* type of information transmission. That is, the excitatory and inhibitory control signals are not the *cause* of the output signal, but they do *control* it. Note that this carrier wave system allows decrease in signal intensity as well as increase, whereas, up to this point, the types of information transmission that we have discussed have been only positive information rather than negative information. This type of information transmission is used by the autonomic nervous system to control such functions as vascular tone, gut tone, degree of constriction of the iris, heart rate, and others.

RHYTHMIC SIGNAL OUTPUT

Many neuronal circuits emit rhythmic output signals — for instance, the rhythmic respiratory signal originating in the reticular substance of the medulla and pons. This repetitive rhythmic signal continues throughout life, while other rhythmic signals, such as those that cause scratching movements by the hind leg of a dog or the walking movements in an animal, require input stimuli into the respective circuits to initiate the signals.

Rhythmic signals probably result from reverberating pathways. One can readily understand that each time a signal passes around a reverberatory loop, collateral impulses could be transmitted into the output pathway. Thus, the rhythmic respiratory signals could result from such a reverberatory mechanism, as was suggested in the description of respiratory center function in Chapter 42.

Obviously, facilitatory or inhibitory signals can affect rhythmic signal output in the same way that they can affect continual signal outputs. Figure 47–16, for instance, illustrates the rhythmic respiratory signal in the phrenic nerve. However, when the carotid body is stimulated by arterial oxygen deficiency, the frequency and amplitude of the rhythmic signal pattern increase progressively.

REFERENCES

Anokhin, P. K.: Convergence of excitations on a neuron as the basis of integrative brain activity. *Proc. Int. Union of Physiol. Sciences,* 6:3, 1968.

Armett, C. J.; Gray, J. A. B.; Hunsperger, R. W.; and Lal, S.: The transmission of information in primary receptor neurones and second-order neurones of a phasic system. *J. Physiol. (Lond.),* 164:395, 1962.

Bennett, M. V. L.; Pappas, G. D.; Aljure, E.; and Nakajima, Y.: Physiology and ultrastructure of electrotonic junctions. II. Spinal and medullary electromotor nuclei in Mormyrid fish. *J. Neurophysiol.,* 30:180, 1967.

Blankenship, J. E., and Kuno, M.: Analysis of spontaneous subthreshold activity in spinal motoneurons of the cat. *J. Neurophysiol.,* 31:195, 1968.

Brooks, V. B., and Asanuma, H.: Pharmacological studies of recurrent cortical inhibition and facilitation. *Amer. J. Physiol.,* 208:674, 1965.

Bullock, T. H.: Comparative physiology of sensory systems: problems of meaning and coding. *Proc. Int. Union of Physiol. Sciences,* 6:55, 1968.

Bunge, R. P.: Glial cells and the central myelin sheath. *Physiol. Rev.,* 48:197, 1968.

Burke, R. E.: Composite nature of the monosynaptic excitatory postsynaptic potential. *J. Neurophysiol.,* 30:1114, 1967.

Harmon, L. D., and Lewis, E. R.: Neural modeling. *Physiol. Rev.,* 46:513, 1966.

Kandel, E. R.; Frazier, W. T.; and Wachtel, H.: Organization of inhibition in abdominal ganglion of *Aplysia.* I. Role of inhibition and disinhibition in transforming neural activity. *J. Neurophysiol.,* 32:496, 1969.

Kuffler, S. W.: Studies on the physiology of neuroglial cells. *Proc. Int. Union of Physiol. Sciences,* 6:79, 1968.

Moore, G. P.; Perkel, D. H.; and Segundo, J. P.: Statistical analysis and functional interpretation of neuronal spike data. *Ann. Rev. Physiol.,* 28:493, 1966.

Nicholls, J. G., and Baylor, D. A.: The after-effects of nerve impulses on glial cells and integrative processes in leech ganglia. *Proc. Int. Union of Physiol. Sciences,* 6:117, 1968.

Otsuka, M.; Kravitz, E. A.; and Potter, D. D.: Physiological and chemical architecture of a lobster ganglion and particular reference to gamma-aminobutyrate and glutamate. *J. Neurophysiol.,* 30:725, 1967.

Rall, W.: Distinguishing theoretical synaptic potentials computed for different soma-dendrite distributions of synaptic input. *J. Neurophysiol.,* 30:1138, 1967.

Rall, W.; Burke, R. E.; Smith, T. G.; Nelson, P. G.; and Frank, K.: Dendritic location of synapses and possible mechanisms for the monosynaptic EPSP in motoneurons. *J. Neurophysiol.,* 30:1169, 1967.

Redman, S. J., and Lampard, D. G.: Monosynaptic stochastic stimulation of cat spinal motoneurons. I. Response of motoneurons to sustained stimulation. *J. Neurophysiol.,* 31:485, 1968.

Rudomin, P.; Dutton, H.; and Munoz-Martinez, J.: Changes in correlation between monosynaptic reflexes produced by conditioning afferent volleys. *J. Neurophysiol.,* 32:759, 1969.

Stefanis, C., and Jasper, H.: Recurrent collateral inhibition in pyramidal tract neurons. *J. Neurophysiol.,* 27:855, 1964.

Tauc, L.: New types of neuronal interaction as studied in a mollusc. *Proc. Int. Union of Physiol. Sciences,* 6:93, 1968.

Tsukahara, N., and Fuller, D. R. G.: Conductance changes during pyramidally induced postsynaptic potentials in red nucleus neurons. *J. Neurophysiol.,* 32:35, 1969.

Waziri, R., and Kandel, E. R.: Organization of inhibition in abdominal ganglion of *Aplysia.* III. Interneurons mediating inhibition. *J. Neurophysiol.,* 32:520, 1969.

INCREASING CAROTID BODY STIMULATION

Figure 47–16 The oscillatory output from the respiratory center, showing that progressively increasing stimulation of the carotid body increases both the intensity and frequency of oscillation.

CHAPTER 48

Sensory Receptors and Their Basic Mechanisms of Action

Input to the nervous system is provided by the sensory receptors that detect such sensory stimuli as touch, sound, light, cold, warmth, and so forth. The purpose of the present chapter is to discuss the basic mechanisms by which these receptors change sensory stimuli into nerve signals and, also, how both the type of sensory stimulus and its strength are detected by the brain.

TYPES OF SENSORY RECEPTORS AND THE SENSORY STIMULI THEY DETECT

The student of medical physiology will have already studied many different anatomical types of nerve endings, and Table 48–1 gives a list and classification of most of the body's sensory receptors. This table shows that there are basically five different types of sensory receptors: (1) *mechanoreceptors*, which detect mechanical deformation of the receptor or of cells adjacent to the receptor; (2) *thermoreceptors*, which detect changes in temperature, some receptors detecting cold and others, warmth; (3) *nociceptors*, which detect damage in the tissues, whether it be physical damage or chemical damage; (4) *electromagnetic receptors*, which detect light on the retina of the eye; and (5) *chemoreceptors*, which detect taste in the mouth, smell in the nose, oxygen level in the arterial blood, osmolality of the body fluids, carbon dioxide concentration, and perhaps other factors that make up the chemistry of the body.

This chapter will discuss especially the function of specific types of receptors, primarily the mechanoreceptors, to illustrate some of the basic principles by which receptors in general operate. Other receptors will be discussed in relation to the sensory systems that they subserve, which will be presented mainly in the next few chapters. Figure 48–1 illustrates some of the different types of mechanoreceptors found in the skin or in the deep structures of the body.

DIFFERENTIAL SENSITIVITY OF RECEPTORS

The first question that must be answered is how do two types of sensory receptors detect different types of sensory stimuli? The answer to this: By virtue of differential sensitivities. That is, each type of receptor is very highly sensitive to one type of stimulus for which it is designed and yet is almost nonresponsive to normal intensities of the other types of sensory stimuli. Thus, the rods and cones are highly responsive to light but are almost completely nonresponsive to heat, cold, pressure on the eyeballs, or chemical changes in the blood. The

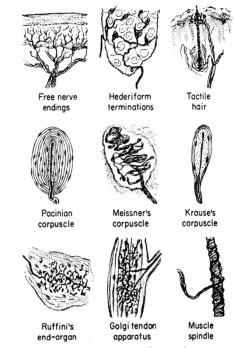

Figure 48–1 Several types of somatic sensory nerve endings. (Modified from Ramon y Cajal: Histology. William Wood and Co.)

TABLE 48–1.   Classification of Sensory Receptors

Mechanoreceptors

Skin tactile sensibilities (epidermis and dermis)
  Free nerve endings
  Expanded tip endings
    Ruffini's endings
    Merkel's discs
    Hederiform terminations
    Plus several other variants
  Encapsulated endings
    Meissner's corpuscles
    Krause's corpuscles
    Pacinian corpuscles
  Hair end-organs
Deep tissue sensibilities
  Free nerve endings
  Expanded tip endings
    Ruffini's endings
    Plus a few other variants
  Encapsulated endings
    Pacinian corpuscles
    Plus a few other variants
  Specialized endings
    Muscle spindles
    Golgi tendon receptors
Hearing
  Sound receptors of cochlea
Equilibrium
  Vestibular receptors
Arterial pressure
  Baroreceptors of carotid sinuses and aorta

Thermoreceptors

Cold
  Probably free nerve endings
Warmth
  Probably free nerve endings

Nociceptors

Pain
  Free nerve endings

Electromagnetic Receptors

Vision
  Rods
  Cones

Chemoreceptors

Taste
  Receptors of taste buds
Smell
  Receptors of olfactory epithelium
Arterial oxygen
  Receptors of aortic and carotid bodies
Osmolality
  Probably neurons of supraoptic nuclei
Blood CO_2
  Receptors in or on surface of medulla and in aortic and
    carotid bodies
Perhaps blood glucose, amino acids, fatty acids
  Receptors in hypothalamus

osmoreceptors of the supraoptic nuclei in the hypothalamus detect minute changes in the osmolality of the body fluids but yet have never been known to respond to sound. Finally, pain receptors in the skin are almost never stimulated by usual touch or pressure stimuli but do become highly active the moment tactile stimuli become severe enough to damage the tissues.

Modality of Sensation—Law of Specific Nerve Energies.   Each of the principal types of sensation that we can experience—pain, touch, sight, sound, and so forth—is called a *modality* of sensation. Yet, despite the fact that we experience these different modalities of sensation, nerve fibers transmit only impulses. Therefore, how is it that different nerve fibers transmit different modalities of sensation?

The answer to this is that each nerve tract terminates at a specific point in the central nervous system, and the type of sensation felt when a nerve fiber is stimulated is determined by the specific area in the nervous system to which the fiber leads. For instance, if a pain fiber is stimulated, the person perceives pain regardless of what type of stimulus excites the fiber. This stimulus could be electricity, heat, crushing, or stimulation of the pain nerve ending by damage to the tissue cells. Yet, whatever the means of stimulation, the person still perceives pain. Likewise, if a touch fiber is stimulated by exciting a touch receptor or in any other way, the person perceives touch because touch fibers lead to specific touch areas in the brain. Similarly, fibers from the retina of the eye terminate in the vision areas of the brain, fibers from the ear terminate in the auditory areas of the brain, and temperature fibers terminate in the temperature areas of the brain.

This specificity of nerve fibers for transmitting only one modality of sensation is called the *law of specific nerve energies.*

TRANSDUCTION OF SENSORY STIMULI INTO NERVE IMPULSES

LOCAL CURRENTS AT NERVE ENDINGS — GENERATOR POTENTIALS AND RECEPTOR POTENTIALS

All sensory receptors studied thus far have one feature in common. Whatever the type of stimulus that excites the ending, it first causes a *local flow of current* in the neighborhood of the nerve ending, and it is this current that in turn excites action potentials in the nerve fiber.

There are two different ways in which local current can be elicited. One of these is to deform or chemically alter the terminal nerve ending itself. This causes ions to diffuse through the nerve membrane, thereby setting up local flow of current. Electrical voltages caused in this manner are called *generator potentials* because the nerve fiber itself generates the current and the voltage.

The second method for causing local current involves specialized receptor cells that lie adjacent to the nerve endings. For instance, when sound enters the cochlea of the ear, the hair cells that lie on the basilar membrane emit electrical currents that in turn stimulate the terminal nerve fibrils entwining the hair cells. Potentials resulting from these currents are called *receptor potentials* because they are caused by specialized, non-neuronal "receptor" cells. However, in some instances it is difficult to tell whether the potential comes from the terminal nerve fibril itself or from a specialized receptor cell. Therefore, many physiologists use the terms generator potential and receptor potential interchangeably, meaning simply the local potential created around a terminal nerve fibril in response to a specific sensory stimulus.

The Generator Potential of the Pacinian Corpuscle. The pacinian corpuscle is a very large and easily dissected sensory receptor. For this reason, one can study in detail the mechanism by which tactile stimuli excite it and by which it causes action potentials in the sensory fiber leading from it. Note in Figure 48–1 that the pacinian corpuscle has a long central nonmyelinated nerve fiber extending through its core. Surrounding this fiber are several capsule layers so that compression on the outside of the corpuscle tends to elongate or shorten or otherwise deform the central core of the fiber, depending on how the compression is applied. The deformation causes a sudden change in membrane potential, as illustrated in Figure 48–2. This perhaps results from stretching the nerve fiber membrane, thus increasing its permeability and allowing positively charged sodium ions to leak to the interior of the fiber. This change in local potential causes a local circuit of current flow that spreads along the nerve fiber to its myelinated portion. At the first node of Ranvier, which itself lies inside the capsule of the pacinian corpuscle, the local circuit of current flow initiates action potentials in the nerve fiber. That

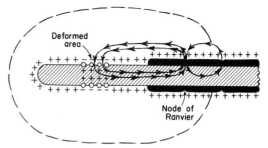

Figure 48–2 Excitation of a sensory nerve fiber by a generator potential produced in a pacinian corpuscle.

is, outward flow of current through the node depolarizes it and this then sets off a typical saltatory transmission of an action potential along the nerve fiber toward the central nervous system, as was explained in Chapter 5.

Relationship of Generator Potential to Stimulus Strength. Figure 48–3 illustrates the effect on the amplitude of the generator potential caused by progressively stronger stimuli applied to the central core of the pacinian corpuscle. Note that the amplitude increases rapidly at first but then progressively less rapidly at high stimulus strengths. The maximum amplitude that can be achieved by generator potentials is probably around 100 millivolts, though this has not been determined accurately yet. That is, a generator potential can have almost as high a voltage as an action potential.

Generator or Receptor Potentials Recorded from Other Sensory Receptors. Generator or receptor potentials have been recorded from many other sensory receptors, including most notably the muscle spindles, the hair cells of the ear, and recently the rods and cones of the

Figure 48–3 Relationship of amplitude of generator potential to strength of a stimulus applied to a pacinian corpuscle. (From Loewenstein: *Ann. N.Y. Acad. Sci.*, 94:510, 1961.)

eyes. In all of these, the amplitude of the potential increases as the strength of stimulus increases, but the additional response usually becomes progressively less as the strength of stimulus increases.

Several different modalities of sensation, such as pain and temperature, are subserved by sensory nerve endings called *free nerve endings* that appear not to have a specific receptor structure. Generator or receptor potentials have not yet been recorded in these, but their function is so similar to that of the specialized receptors that these potentials are believed to occur even in these free nerve endings.

Yet, the mechanism for causing the generator or receptor potential is not the same in different receptors. For instance, in the rods and cones of the eye it is known that changes in certain intracellular chemicals on exposure to light alter the membrane potential, resulting in the receptor potential. In this case, the basic mechanism causing the receptor potential is a chemical one in contrast to mechanical deformation that causes the generator potential in the pacinian corpuscle. In the case of thermal receptors, it is believed that changes in rates of chemical reaction at or near the membrane alter the membrane potential and thereby create a generator potential. In the case of the hair cells of the ear, movement of the hairs probably causes the receptor potentials. Thus, the mechanisms for eliciting generator or receptor potentials are individualized for each type of receptor.

Relationship of Amplitude of Generator Potential to Nerve Impulse Rate. Referring once again to Figure 48-2, we see that the generator potential generated in the core of the pacinian corpuscle causes a local circuit of current flow through the first node of Ranvier. When an action potential occurs at the node, this does not affect the generator potential being emitted by the core of the pacinian corpuscle. Instead, the core continues to emit its current as long as an effective mechanical stimulus is applied. As a result, when the node of Ranvier repolarizes after its first action potential is over, it discharges once again, and action potentials continue as long as the generator potential persists.

The rate of transmission of action potentials in the nerve fiber (impulse rate) is almost directly proportional to the amplitude of the generator or receptor potential. This relationship is illustrated in Figure 48-4, which shows the impulse rate corresponding to different voltages of generator potential recorded from a muscle spindle; there is an almost exact proportional relationship. This same relationship between generator potential and impulse rate is approximately true for most if not all sensory receptors.

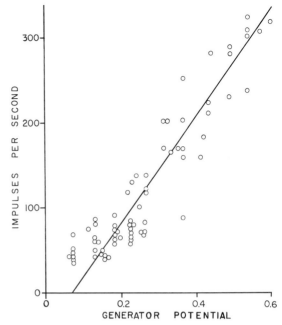

Figure 48-4 Relationship between the generator potential of a muscle spindle and the frequency of sensory impulses transmitted from the spindle. [From Katz: *J. Physiol. (Lond.), 111*:161, 1950.]

ADAPTATION OF RECEPTORS

A special characteristic of all sensory receptors is that they *adapt* either partially or completely to their stimuli after a period of time. That is, when a continuous sensory stimulus is first applied, the receptors respond at a very high impulse rate at first, then progressively less rapidly until finally many of them no longer respond at all.

Figure 48-5 illustrates typical adaptation of certain types of receptors. Note that the pacinian corpuscle and the receptor at the hair bases adapt extremely rapidly, while joint capsule and muscle spindle receptors adapt very slowly.

Furthermore some sensory receptors adapt to a far greater extent than others. For example, the pacinian corpuscles adapt to "extinction" within a fraction of a second, and the hair base receptors adapt to extinction within a second or more.

Mechanisms by which Receptors Adapt. Adaptation of receptors seems to be an individual property of each type of receptor in much the same way that development of a generator or receptor potential is an individual property. For instance, in the eye, the rods and cones adapt by changing their chemical compositions (which will be discussed in Chapter 53). In the case of the mechanoreceptors, the receptor that has been studied in greatest detail

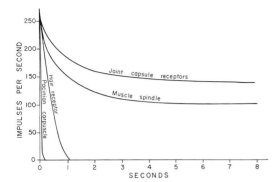

Figure 48-5 Adaptation of different types of receptors, showing rapid adaptation of some receptors and slow adaptation of others.

is the pacinian corpuscle. Adaptation occurs in this receptor in two ways. First, the corpuscular structure itself very rapidly adapts to the deformation of the tissue. That is, when the corpuscle is first compressed, it instantaneously stretches or otherwise modifies the central core nerve fiber. However, within a very small fraction of a second, fluid appears to leak from one of the encircling capsules to the next, until the surface deformation of the corpuscle no longer affects the central core fiber. For this reason, the pacinian corpuscle usually detects tissue deformation only at the very onset of the stimulus but then immediately loses its excitation. However, if the tissue continues to be deformed, the ability of the central core of the pacinian corpuscle to emit local current also becomes depressed, so that the rate of firing of the nerve fiber becomes decreased. Presumably these same two principles apply to other types of receptor potentials; that is, part of the adaptation is often in the structure of the receptor itself, and part is in the terminal nerve fibril of the receptor.

Function of the Poorly Adapting Receptors — The "Tonic" Receptors. The poorly adapting receptors, those that adapt very slowly and do not adapt to extinction, continue to transmit impulses to the brain as long as the stimulus is present (or at least for many minutes or hours). Therefore, they keep the brain constantly apprised of the status of the body and its relation to its surroundings. For instance, impulses from the joint capsule receptors allow the person to "know" the positions of the different parts of his body at all times, while impulses from the muscle spindles and Golgi tendon apparatuses allow the central nervous system to know respectively the status of muscle contraction and the load on the muscle tendon at each instant.

Other types of poorly adapting receptors include the receptors of the macula in the vestibular apparatus, the sound receptors of the ear, the pain receptors, the pressoreceptors of the arterial tree, and the chemoreceptors of the carotid and aortic bodies.

Because the poorly adapting receptors can continue to transmit information for many hours, they are called tonic receptors. It is probable that many of these poorly adapting receptors would adapt to extinction if the intensity of the stimulus should remain absolutely constant over many days. Fortunately, because of our continually changing bodily state, the tonic receptors could almost never reach a state of complete adaptation.

Function of the Rapidly Adapting Receptors — The Rate Receptors or Movement Receptors or Phasic Receptors. Obviously, receptors that adapt rapidly cannot be used to transmit a continuous signal because these receptors are stimulated only when the stimulus strength changes. Yet they react strongly *while a change is actually taking place*. Furthermore, the number of impulses transmitted is directly related to the *rate at which the change takes place*. Therefore, these receptors are called *rate* receptors, *movement* receptors, or *phasic* receptors. Thus, in the case of the pacinian corpuscle, sudden pressure applied to the skin excites this receptor for a few milliseconds, and then its excitation is over even though the pressure continues. In other words, the pacinian corpuscle is exceedingly important in transmitting information about rapid changes in pressure against the body, but it is useless in transmitting information about constant pressure applied to the body.

IMPORTANCE OF THE RATE RECEPTORS — THEIR PREDICTIVE FUNCTION. If one knows the rate at which some change in his bodily status is taking place, he can predict ahead to the state of the body a few seconds or even a few minutes later. For instance, the receptors of the semicircular canals in the vestibular apparatus of the ear detect the rate at which the head begins to turn when one runs around a curve. Using this information, a person can predict that he will turn 10, 30, or some other number of degrees within the next 10 seconds, and he can adjust the motion of his limbs *ahead of time* to keep from losing his balance. Likewise, pacinian corpuscles located in the joint capsules help to detect the rates of movement of the different parts of the body. Therefore, when one is running, information from these receptors allows the nervous system to predict ahead of time where the feet will be during any precise fraction of a second, and appropriate motor signals can be transmitted to the muscles of the legs to make any necessary anticipatory corrections in limb position so that the person will not fall. Loss of this predictive function makes it impossible for the person to run.

PSYCHIC INTERPRETATION OF STIMULUS STRENGTH

The ultimate goal of most sensory stimulation is to apprise the psyche of the state of the body and its surroundings. Therefore, it is important that we discuss briefly some of the principles related to the transmission of sensory stimulus strength to the higher levels of the nervous system.

The first question that comes to mind is, "How is it possible for the sensory system to transmit sensory experiences of tremendously varying intensities?" For instance, the auditory system can detect the weakest possible whisper but can also discern the meanings of an explosive sound only a few feet away, even though the sound intensities of these two experiences can vary as much as a trillion-fold; the eyes can see visual images with light intensities that vary as much as a million-fold; or the skin can detect pressure differences of ten thousand- to one hundred thousand-fold.

As a partial explanation of these effects, note in Figure 48–3 the relationship of the generator potential produced by the pacinian corpuscle to the strength of stimulus. At low stimulus strength, very slight changes in stimulus strength increase the generator potential markedly; whereas at high levels of stimulus strength, further increases in generator potential are very slight. Thus, the pacinian corpuscle is capable of accurately measuring extremely minute changes in stimulus strength at low intensity levels and is also capable of detecting much larger changes in stimulus strength at high intensity levels.

The transduction mechanism for detecting sound by the cochlea of the ear illustrates still another method for separating gradations of stimulus intensity. When sound causes vibration at a specific point on the basilar membrane, weak vibration stimulates only those hair cells in the very center of the vibratory point. But, as the vibration intensity increases, not only do the centermost hair cells become more intensely stimulated, but hair cells in each direction farther away from the central point also become stimulated. Signals transmitted over progressively increasing numbers of cochlear nerve fibers and the increasing intensity of signal strength in each nerve fiber are two mechanisms by which stimulus strength is transmitted into the central nervous system. These two mechanisms, which multiply each other, make it possible for the ear to operate faithfully at stimulus intensity levels changing as much as a trillion-fold.

Importance of the Tremendous Intensity Range of Sensory Reception. Were it not for the tremendous intensity range of sensory reception that we can

experience, the various sensory systems would more often than not operate in the wrong range. This is illustrated by the attempt of most persons to adjust the light exposure on a camera without using a light meter. Left to intuitive judgment of light intensity, a person almost always overexposes the film on very bright days and greatly underexposes the film at twilight. Yet, his eyes are perfectly capable of discriminating with great detail the objects around him in both very bright sunlight and at twilight; the camera cannot do this because of the narrow critical range of light intensity required for proper exposure of film.

JUDGMENT OF STIMULUS STRENGTH

Physiopsychologists have evolved numerous methods for testing one's judgment of sensory stimulus strength, but only rarely do the results from the different methods agree with each other. For instance, one testing method requires a person to select a weight that is exactly 100 per cent heavier than another. But he usually selects a weight that is about 50 per cent heavier instead of 100 per cent heavier. Thus, the weak stimulus is underestimated and the strong stimulus is overestimated. In still another test procedure, a person is given a weight to hold and is then required to select the minimum amount of additional weight that must be added for him to detect a difference. In this case he might be holding a 50 gram weight and find out that an additional 5 grams are necessary to detect a difference. Then he holds a 500 gram weight and finds that 50 grams of additional weight are required. In this instance the discriminatory ability is far greater at the low intensity level than at the high intensity level. Thus, the results of these two different types of tests are exactly opposite to each other, which means that the real argument lies in the meaning of the tests themselves. Therefore, at present no real agreement exists as to a proper method for measuring one's judgment of stimulus strength. Yet, two principles are widely discussed in the physiopsychology field of sensory interpretation: the *Weber-Fechner principle* and the *power principle.*

History and Significance of the Weber-Fechner Principle. In the mid eighteen hundreds, Weber first and Fechner later proposed the principle that gradations of stimulus strength are discriminated approximately in proportion to the logarithm of stimulus strength. This law is based primarily on one's ability to judge minimal changes in stimulus strength that can be detected, utilizing the second test described in the previous section; that is, in this example just given, a person can barely detect a 5 gram increase in weight when holding 50 grams, or a 50 gram increase when holding 500 grams. Thus, the *ratio* of the change in stimulus strength required for perception of a change remains essentially constant, which is what the logarithmic principle means.

Because the Weber-Fechner principle offers a

ready explanation for the tremendous range of stimulus strength that our nervous system can discern, it unfortunately became widely accepted for all types of sensory experience and for all levels of background sensory intensity. More recently it has become evident that this principle applies mainly to higher levels of visual, auditory, and cutaneous sensory experience and that it applies only poorly to most other types of sensory experience.

Yet, the Weber-Fechner principle is still a good one to remember because it emphasizes that the greater the background sensory stimulus, the greater also must be the additional change in stimulus strength in order for the psyche to detect the change.

The Power Law. Another attempt by physiopsychologists to find a good mathematical relationship between actual stimulus strength and interpretation of stimulus strength is the following formula, known as the power law:

$$\text{Interpreted strength} = K \, (\text{Stimulus strength})^x$$

In this formula K is a constant, and x is the power to which the stimulus strength is raised. The exponent x and the constant K are different for each type of sensation.

When this power law relationship is plotted on a graph using double logarithmic coordinates, as illustrated in Figure 48-6, a linear curve can be attained between interpreted stimulus strength and actual stimulus strength over a large range. However, as illustrated in the figure, even this power law relationship fails to hold satisfactorily at both low and high stimulus strengths.

Unfortunately, the power law is not of great philosophical importance for a very simple mathematical reason. Almost any curve (even the curve

Figure 48-6 Graphical demonstration of the "power law" relationship between actual stimulus strength and strength that the psyche interprets it to be. Note that the power law does not hold at either very weak or very strong stimulus strengths.

of a hill) that has a progressive change in slope in the same direction can be fitted over most of its range to the power law equation, provided one simply finds proper values for the exponent x and constant K. Therefore, use of the power law is more an exercise in mathematical curve fitting than it is a valuable tool in understanding sensory experience.

PHYSIOLOGICAL CLASSIFICATION OF NERVE FIBERS

Unfortunately, two separate classifications of nerve fibers are in general use. The first of these is a general classification given in Table 48-2. The fibers are divided into types A, B, and C; and the type A fibers are further subdivided into α, β, γ, and δ fibers.

Type A fibers are the typical myelinated fibers of spinal nerves. The type B fibers differ from very small type A fibers only in the fact that they do not display a negative after-potential following stimulation. However, they also are myelinated like type A fibers. They are the preganglionic autonomic nerve fibers.

Type C fibers are the very small unmyelinated nerve fibers that conduct impulses at low velocities. These constitute more than half the number of sensory fibers and also all of the postganglionic autonomic fibers.

The sizes, velocities of conduction, and functions of the different nerve fibers are given in Table 48-2. Note that the very large fibers can transmit impulses as rapidly as 120 meters per second, a distance longer than a football field. On the other hand, the smallest fibers transmit impulses as slowly as 0.5 meter per second, which is the distance from the foot to the knee.

Over two-thirds of all the nerve fibers in peripheral nerves are type C fibers. Because of their great number, these can transmit tremendous amounts of information from the surface of the body, even though their velocities of transmission are very slow. Utilization of type C fibers for transmitting this great mass of information represents an important economy of space in the nerves, for use of type A fibers would require peripheral nerves the sizes of large ropes and a spinal cord almost as large as the body itself.

Alternate Classification Used by Sensory Physiologists. Certain recording techniques have made it possible to separate the type Aα fibers into two subgroups; and, yet, these same recording techniques cannot distinguish easily between Aβ and Aγ fibers. Therefore, the following classification is frequently used by sensory physiologists:

GROUP I A. Fibers from the annulospiral endings of muscle spindles. (About 17 microns in diameter. These are alpha type A fibers in the classification of Table 48-2.)

GROUP I B. Fibers from the Golgi tendon apparatuses. (About 16 microns in diameter; these also are alpha type A fibers.)

TABLE 48–2. Properties of Different Mammalian Nerve Fibers

Type of Fiber	Diameter of Fiber (μ)	Velocity of Conduction (meters/sec.)	Duration of Spike (msec.)	Duration of Negative After-potential (msec.)	Duration of Positive After-potential (msec.)	Function
A (α)	13–22	70–120	0.4–0.5	12–20	40–60	Motor, muscle proprioceptors
A (β)	8–13	40–70	0.4–0.6	(?)	(?)	Touch, kinesthesia
A (γ)	4–8	15–40	0.5–0.7	(?)	(?)	Touch, excitation of muscle spindles, pressure
A (δ)	1–4	5–15	0.6–1.0	(?)	(?)	Pain, heat, cold, pressure
B	1–3	3–14	1.2	None	100–300	Preganglionic autonomic
C	0.2–1.0	0.2–2	2.0	50–80	300–1000	Pain, heat(?), cold(?), pressure(?), postganglionic autonomic, smell

Compiled from various sources but mainly from Grundfest.

GROUP II. Fibers from the discrete cutaneous tactile receptors and also from the flower-spray endings of the muscle spindles. (Average about 8 microns in diameter; these are beta and gamma type A fibers in the other classification.)

GROUP III. Fibers carrying temperature, crude touch, and pain sensations. (Average about 3 microns in diameter; these are delta type A fibers in the other classification.)

GROUP IV. Unmyelinated fibers carrying pain, crude pressure, and crude touch sensations. (0.2 to 1 micron in diameter; called type C fibers in the other classification.)

REFERENCES

Davis, H.: Some principles of sensory receptor action. *Physiol. Rev.,* 41:391, 1961.

Goldberg, J. M., and Lavine, R. A.: Nervous system: afferent mechanisms. *Ann. Rev. Physiol.,* 30:319, 1968.

Granit, R.: Receptors and Sensory Perception. New Haven, Yale University Press, 1955.

Gray, J. A. B.: Initiation of impulses at receptors. *In* Handbook of Physiology. Baltimore, The Williams & Wilkins Co., 1959, Sec. I, Vol. I, p. 123.

Hensel, H.; Iggo, A.; and Witt, I.: A quantitative study of sensitive cutaneous thermoreceptors with C afferent fibres. *J. Physiol. (Lond.),* 153:113, 1960.

Hunt, C. C.: On the nature of vibration receptors in the hind limb of the cat. *J. Physiol. (Lond.),* 155:175, 1961.

Hunt, C. C., and McIntyre, A. K.: Properties of cutaneous touch receptors in cat. *J. Physiol. (Lond.),* 153:88, 1960.

Hunt, C. C., and Takeuchi, A.: Responses of the nerve terminal of the pacinian corpuscle. *J. Physiol. (Lond.),* 160:1, 1962.

Loewenstein, W. R.: Biological transducers. *Sci. Amer.,* 203(2): 98, 1960.

Loewenstein, W. R.: Excitation and inactivation in a receptor membrane. *Ann. N.Y. Acad. Sci.,* 94:510, 1961.

Loewenstein, W. R.: On the 'specificity' of a sensory receptor. *J. Neurophysiol.,* 24:150, 1961.

Loewenstein, W. R.: The generation of electric activity in a nerve ending. *Ann. N.Y. Acad. Sci.,* 81:367, 1959.

Miller, W. H., et al.: How cells receive stimuli. *Sci. Amer.,* 205(3):222, 1961.

Mountcastle, V. B.: Neural dynamics in somesthesis: peripheral encoding, central neural transformation and sensory psychophysics of flutter-vibration. *Proc. Int. Union of Physiol. Sciences,* 6:260, 1968.

Neff, W. D.: Contributions to Sensory Physiology. New York, Academic Press, Inc., 1965, 1967, Vols. 1 and 2.

Terzuolo, C. A., and Washizu, Y.: Relation between stimulus strength, generator potential and impulse frequency in stretch receptor of *Crustacea. J. Neurophysiol.,* 25:56, 1962.

Winkelmann, R. K.: Nerve Endings in Normal and Pathologic Skin. Springfield, Illinois, Charles C Thomas, Publisher, 1960.

Zotterman, Y.: Sensory Mechanisms. Progress in Brain Research. New York, American Elsevier Publishing Co., Inc., 1967, Vol. 23.

CHAPTER 49

Somatic Sensations: I. The Mechanoreceptive Sensations

The *somatic senses* are the nervous mechanisms that collect sensory information from the body. These senses are in contradistinction to the *special senses*, which mean specifically sight, hearing, smell, taste, and equilibrium.

CLASSIFICATION OF SOMATIC SENSATIONS

The somatic senses can be classified into three different physiological types: (1) The *mechanoreceptive somatic senses*, stimulated by mechanical displacement of some tissue of the body, (2) The *thermoreceptive senses*, which detect heat and cold, and (3) the *pain sense*, which is activated by any factor that damages the tissues. The present chapter deals with the mechanoreceptive somatic senses, and the following chapter deals with the thermoreceptive and pain senses.

The mechanoreceptive senses include *touch, pressure*, and *vibration* senses (which are frequently called the *tactile senses*) and the *kinesthetic sense*, which determines the relative positions and rates of movement of the different parts of the body.

Other Classifications of Somatic Sensations. Different physiological types of somatic sensations are grouped together in special classes that are not necessarily mutually exclusive, as follows:

Exteroceptive sensations are those from the surface of the body.

Proprioceptive sensations are those having to do with the physical state of the body, including kinesthetic sensations, possible tendon and muscle sensations, pressure sensations from the bottom of the feet, and even the sensation of equilibrium, which is generally considered to be a "special" sensation rather than a somatic sensation.

Visceral sensations are those from the viscera of the body; in using this term one usually refers specifically to sensations from the internal organs.

The *deep sensations* are those that come from the deep tissues, such as from the bone, fasciae, and so forth. These include mainly "deep" pressure, pain, and vibration.

DETECTION AND TRANSMISSION OF TACTILE SENSATIONS

Interrelationship Between the Tactile Sensations of Touch, Pressure, and Vibration. Though touch, pressure, and vibration are frequently classified as separate sensations, they are all detected by the same types of receptors. The only differences among these three types of sensations are (1) touch sensation generally results from stimulation of tactile receptors in the skin or in tissues immediately beneath the skin, (2) pressure sensation generally results from deformation of deeper tissues, and (3) vibration sensation results from rapidly repetitive sensory signals, but the same types of receptors as those for both touch and pressure are utilized.

The Tactile Receptors. At least six entirely different types of tactile receptors are known, but many more similar to these probably also exist. Some of these receptors were illustrated in Figure 48–1 of Chapter 48, and their special characteristics are the following:

First, some *free nerve endings*, which are found everywhere in the skin and in many other tissues, can detect touch and pressure. For instance, even light contact with the cornea of the eye, which contains no other type of nerve ending besides free nerve endings, can nevertheless elicit touch and pressure sensations.

Second, a touch receptor of special sensitivity is *Meissner's corpuscle*, an encapsulated nerve ending that excites a large myelinated sensory nerve fiber. Inside the capsulation are many whorls of terminal nerve filaments. These receptors are particularly abundant in the fingertips, lips, and other areas of the skin where one's ability to discern spatial characteristics of touch sensations is highly developed. These receptors, along with the expanded tip receptors described subsequently, are probably responsible for the ability to recognize exactly what point of the body is touched and to recognize the texture of objects touched. Meissner's corpuscles probably adapt within a second or more after they are stimulated, which means that they are particularly sensitive to movement of very light objects over the surface of the skin.

Third, the fingertips and other areas that contain large numbers of Meissner's corpuscles also contain *expanded tip tactile receptors*, one type of which is *Merkel's discs*. These receptors differ from Meissner's corpuscles in that they transmit long-continuing signals rather than adapting rapidly. Therefore, they are probably responsible for giving steady state signals that allow one to determine continuous touch of objects against the skin. The hairy parts of the body contain almost no Meissner's corpuscles but do contain a few expanded tip receptors.

Fourth, slight movement of any hair on the body stimulates the nerve fiber entwining its base. Thus, each hair and its basal nerve fiber, called the *hair end-organ*, is also a type of touch receptor. This receptor adapts readily and, therefore, like Meissner's corpuscles, detects mainly movement of objects on the surface of the body.

Fifth, located in the deeper layers of the skin and also in deeper tissues of the body are many *Ruffini's end-organs*, which are multi-branched endings, as described and illustrated in the previous chapter. These endings do not adapt rapidly and, therefore, are important for signaling continuous states of deformation of the deeper tissues, such as heavy and continuous touch signals and pressure signals.

Sixth, many *pacinian corpuscles*, which were discussed in detail in Chapter 48, lie both immediately beneath the skin and also deep in the tissues of the body. These are stimulated only by very rapid movement of the tissues because these receptors adapt in a small fraction of a second. Therefore, they are particularly important for detecting tissue vibration or other extremely rapid changes in the mechanical state of the tissues.

Transmission of Tactile Sensations in Peripheral Nerve Fibers. The specialized sensory receptors, such as Meissner's corpuscles, expanded tip endings, pacinian corpuscles, and Ruffini's endings, all transmit their signals in beta type A nerve fibers that have transmission velocities of 30 to 60 meters per second. On the other hand, free nerve ending receptors and probably some of the hair end-organs transmit signals via very small delta type A nerve fibers and perhaps via some type C fibers that conduct at velocities of 1 to 8 meters per second. Thus, the more critical types of sensory signals — those that help to determine precise localization on the skin, minute gradations of intensity, or rapid changes in sensory signal intensity — are all transmitted in the rapidly conducting types of sensory never fibers. On the other hand, the cruder types of signals, such as crude deep pressure and poorly localized touch, are transmitted via much slower nerve fibers, fibers that also require much less space in the nerves.

DETECTION OF VIBRATION

All the different tactile receptors are involved in detection of vibration, though different receptors detect different frequencies of vibration. Pacinian corpuscles can discern vibrations up to as high as 700 cycles per second, because they respond extremely rapidly to minute and rapid deformations of the tissues, and they also transmit their signals over beta type A nerve fibers, which can transmit more than 1000 impulses per second.

Low frequency vibrations up to 100 cycles per second, on the other hand, stimulate other tactile receptors — such receptors as Meissner's corpuscles, expanded tip receptors, and even Ruffini's endings in the deeper tissues — in preference to pacinian corpuscles.

POSSIBILITY OF A "MUSCLE SENSE"

Special sensory receptors are found in both skeletal muscles and tendons, *muscle spindles* in muscles and *Golgi tendon apparatuses* in tendons. These receptors transmit their signals into the spinal cord and cerebellum to control reflex contraction of the muscles. But, from what is known at present, these signals operate entirely at a subconscious level. These specialized receptors and the reflexes that they subserve will be considered in detail in relation to spinal cord reflexes and control of muscle contraction in Chapter 56.

However, some psychological effects have pointed to the existence of a conscious "muscle sense." For instance, one is capable of determining how heavy an object is, and this has long been believed to be detected by the specialized muscle spindles or tendon organs; but critical experiments show this not to be true. Instead, this type of information is probably signaled in two ways: (1) by usual tactile signals in the deep tissues and (2) by signals from the motor cortex indicating intensity of motor signals to the muscles required to lift an object.

KINESTHETIC SENSATIONS

The term "kinesthesia" means conscious recognition of the orientation of the different parts of the body with respect to each other as well as of the rates of movement of the different parts of the body. These functions are subserved principally by extensive sensory endings in the joint capsules and ligaments.

The Kinesthetic Receptors. Three major types of nerve endings have been described in the joint capsules and ligaments about the joints. (1) By far the most abundant of these is a spray type of ending similar to *Ruffini's end-organ*, which was illustrated in Figure 48–1 of Chapter 48. These endings are stimulated strongly when the joint is suddenly moved; they adapt slightly at first but then transmit a steady signal thereafter. (2) A second type of ending, resembling the stretch receptors found in muscle tendons (called Golgi tendon receptors), is found particularly in the ligaments about the joints. Though far less numerous than the Ruffini end-organs, they have essentially the same response properties. (3) A few pacinian corpuscles are also found in the tissues around the joints. These adapt extremely rapidly and presumably help to detect *rate of movement* at the joint.

STIMULATION OF THE JOINT RECEPTORS. Figure 49–1 illustrates the excitation of seven different nerve fibers leading from separate joint receptors in the capsule of a cat's knee joint. Note that at 180 degrees angulation one of the receptors is stimulated; then at 150 degrees still another is stimulated; at 140 degrees two are stimulated, and so forth. The information from these joint receptors continually apprises the central nervous system of the momentary angulation of the joint. That is, the angulation determines *which* receptor is stimulated and how much it is stimulated, and from this the brain knows how far the joint is bent.

Figure 49–1 Responses of seven different nerve fibers from knee joint receptors in a cat. (Modified from Skogland: *Acta Physiol. Scand*, Suppl. 124, *36*:1, 1956.)

DETECTION OF RATE OF MOVEMENT AT THE JOINT. Because pacinian corpuscles are especially adapted for detecting movement of tissues, it is tempting to suggest that rate of movement at the joints is detected by the pacinian corpuscles. However, the number of pacinian corpuscles in the joint tissues is small, for which reason rate of movement at the joint is probably detected mainly in the following way: The Ruffini and Golgi endings in the joint tissues are stimulated very strongly at first by the process of joint movement, but within a fraction of a second this strong level of stimulation fades to a lower, steady state rate of firing. Nevertheless, this early over-shoot in receptor stimulation is directly proportional to the rate of joint movement and is believed to be the signal used by the brain to discern the rate of movement. However, it is likely that the few pacinian corpuscles also play at least some role in this process.

Transmission of Kinesthetic Signals in the Peripheral Nerves. Kinesthetic signals, like those from the critical tactile sensory receptors, are transmitted almost entirely in the beta type A sensory nerve fibers, which carry signals very rapidly to the cord and, thence, to the brain. This rapid transmission of kinesthetic signals is particularly important when parts of the body are being moved rapidly, because it is essential for the central nervous system to "know" at each small fraction of a second the exact locations of the different parts of the body; otherwise one would not be capable of controlling further movements.

THE DUAL SYSTEM FOR TRANSMISSION OF MECHANORECEPTIVE SOMATIC SENSATIONS IN THE CENTRAL NERVOUS SYSTEM

All sensory information from the somatic segments of the body enters the spinal cord through the posterior roots. On entering the cord, the fibers separate into medial and lateral divisions. The medial fibers immediately enter the *dorsal columns* of the cord and ascend the entire length of the cord, while the lateral fibers travel upward for one to six segments and downward for one to two segments, and then synapse with dorsal horn cells that give rise to the *ventral* and *lateral spinothalamic tracts*. These tracts ascend to the brain in, respectively, the anterior and lateral columns of the spinal cord. This separation of the fibers at the dorsal roots represents a separation of the pathways for transmission of sensory impulses: the dorsal column pathway gives rise to the *dorsal column system*, and the spinothalamic tracts give rise to the *spinothalamic system*.

Comparison of the Dorsal Column and Spinothalamic Systems. The dorsal column system is composed of large, heavily myelinated nerve fibers and transmits signals to the brain at velocities of 35 to 70 meters per second. Also, there is a high degree of spatial orientation of the nerve fibers with respect to their origin on the surface of the body. On the other hand, the spinothalamic system is composed mainly of small fibers, some of which are not myelinated at all, or are poorly myelinated, and which transmit impulses at low velocities. These fibers are spatially oriented but only poorly so.

These differences in the two systems immediately characterize the types of sensory information that are transmitted by the two pathways. First, sensory information that must be transmitted rapidly and with temporal fidelity is transmitted in the dorsal column system, while that which does not need to be transmitted rapidly is transmitted mainly in the spinothalamic system. Second, those sensations that detect fine gradations of intensity are transmitted in the dorsal column system, while those that lack the fine gradations are transmitted in the spinothalamic system. And, third, sensations that are discretely localized to exact points in the body are transmitted in the dorsal column system, while those transmitted in the spinothalamic system can be localized much less exactly. With this differentiation in mind we can now list the types of sensations transmitted in the two systems.

The Dorsal Column System

1. Touch sensations requiring a high degree of localization of the stimulus and requiring the transmission of fine gradations of intensity.
2. Phasic sensations, such as vibratory sensations.
3. Kinesthetic sensations.
4. Pressure sensations having to do with fine degrees of judgment of pressure intensity.

The Spinothalamic System

1. Pain.
2. Thermal sensations, including both warm and cold sensations.
3. Crude touch sensations capable of much

less localizing ability on the surface of the body and perhaps requiring slightly increased intensity of stimulus to elicit.
4. Pressure sensations of a somewhat cruder nature than those transmitted by the dorsal column system.
5. Tickle and itch sensations.
6. Sexual sensations.

TRANSMISSION IN THE DORSAL COLUMN SYSTEM

Because the dorsal column system is large and its fibers numerous, we know much about transmission of sensory impulses in this system in comparison with very little about transmission of impulses in the spinothalamic system. Therefore, the dorsal column system will be discussed at length and then the known differences between this and the spinothalamic system will be pointed out.

ANATOMY OF THE DORSAL COLUMN PATHWAY

The dorsal column system is illustrated in Figure 49-2. The nerve fibers entering the *dorsal columns* pass all the way up these columns

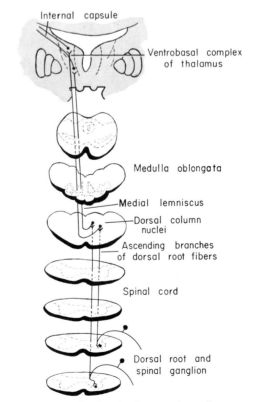

Figure 49-2 The dorsal column pathway for transmitting critical types of mechanoreceptive signals. (Modified from Ranson and Clark: Anatomy of the Nervous System.)

to the medulla, where they synapse in the *dorsal column nuclei* (the *cuneate* and *gracile nuclei*). From here, *second order neurons* decussate immediately to the opposite side and then pass upward to the thalamus through bilateral pathways called the *medial lemnisci*. Each medial lemniscus terminates in a *ventrobasal complex of nuclei* located posteriorly and ventrally in each side of the *thalamus*. In its pathway through the hindbrain, the medial lemniscus is joined by additional fibers from the *main sensory nucleus of the trigeminal nerve and from the upper portion of its descending nuclei*; these fibers subserve the same sensory functions for the head that the dorsal column fibers subserve for the body.

From the ventrobasal complex, *third order neurons* project, as shown in Figure 49-3, mainly to the *postcentral gyrus* of the *cerebral cortex*. But, in addition, neurons also project to closely associated regions of the cortex behind and in front of the postcental gyrus.

Collateral Fibers from the Dorsal Column Pathway to Centers in the Cord, Brain Stem, and Thalamus. Collateral fibers deviate from the dorsal column system to several distinct areas: First, a large number of collaterals spread from the dorsal column fibers to the gray matter of the spinal cord in the *initial segment* where each respective dorsal root enters. Second, collateral fibers pass from the dorsal column nuclei to the anterior cerebellum. Third, *very* few collateral fibers pass to the brain stem nuclei, particularly to those in the tegmentum of the mesencephalon. And, fourth, a few collaterals project to additional nuclei in the thalamus besides those of the ventrobasal complex. These collateral fibers subserve the functions of (a) localized segmental reflexes in the cord, (b) cerebellar reflexes that help to coordinate motor movements in the body, and (c) tegmental and thalamic responses that are probably at least partially responsible for the conscious perception of sensations.

It is particularly notable that the lower portions of the hindbrain receive only few collaterals from the dorsal column system. But, as will be pointed out later, large numbers of collaterals enter this area from the spinothalamic system.

Spatial Orientation of the Nerve Fibers in the Dorsal Column Pathway. All the way from the origin of the dorsal columns to the cerebral cortex, a distinct spatial orientation of the fibers from individual parts of the body is maintained. The fibers from the lower parts of the body lie toward the center of the dorsal columns, while those that enter the dorsal columns at progressively higher and higher levels form successive layers on the lateral sides of the dorsal columns.

In the thalamus, the distinct spatial orientation is still maintained, with the tail end of the body represented by the most lateral portions of the ventrobasal complex and the head and face represented in the medial component of the complex. However, note that because of the crossing of the medial lemnisci in the medulla, the left side of the body is represented in the right side of the thalamus and the right side of the body is represented in the left side of the thalamus. In a similar manner, the fibers passing to the cerebral cortex also are accurately spatially oriented so that a single part of the cortex receives signals from a discrete area of the body, as is described below.

THE SOMESTHETIC CORTEX

The area of the cerebral cortex to which the primary sensory impulses are projected is called the *somesthetic cortex*. In the human being, this area lies mainly in the anterior portions of the parietal lobes. Two distinct and separate areas are known to receive direct afferent nerve fibers from the relay nuclei of the thalamus; these, called *somatic sensory area I* and *somatic sensory area II*, are illustrated in Figure 49-4.

Though the term "somesthetic cortex" actually means all the areas of the cerebral cortex that receive primary sensory information from the body, somatic sensory area I is so much more important to the sensory functions of the body than is somatic sensory area II that in popular usage the term somesthetic cortex is almost always used to designate area I exclusive of area II. Yet, to keep these two areas separated, we will henceforth refer to

Figure 49-3 Projection of the dorsal column system from the thalamus to the somesthetic cortex. (Modified from Brodal: Neurological Anatomy in Relation to Clinical Medicine. Oxford University Press.)

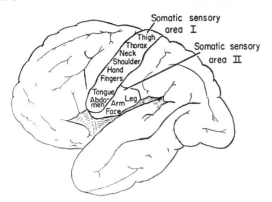

Figure 49-4 The two somesthetic cortical areas, somatic sensory areas I and II.

them separately as somatic sensory area I and somatic sensory area II.

Projection of the Body in Somatic Sensory Area I. Somatic sensory area I lies in the postcentral gyrus of the human cerebral cortex. A distinct spatial orientation exists in this area for reception of nerve signals from the different areas of the body. Figure 49–5 illustrates a cross-section through the brain at the level of the postcentral gyrus, showing the representations of the different parts of the body in separate regions of somatic sensory area I. Note, however, that each side of the cortex receives sensory information almost exclusively from the opposite side of the body (except for a small amount of sensory information from the same side of the face).

Some areas of the body are represented by large areas in the somatic cortex—the lips by

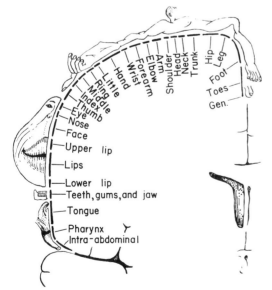

Figure 49-5 Representation of the different areas of the body in the somatic sensory area I of the cortex. (From Penfield and Rasmussen: The Cerebral Cortex in Man. The Macmillan Co.)

far the greatest area of all, followed by the face and thumb—while the entire trunk and lower part of the body are represented by relatively small areas. The sizes of these areas are directly proportional to the number of specialized sensory receptors in each respective peripheral area of the body. For instance, a great number of specialized nerve endings are found in the lip and thumb, while only a few are present in the skin of the trunk.

Note also that the head is represented in the lower or lateral portion of the postcentral gyrus, while the lower part of the body is represented in the medial or upper portion of the postcentral gyrus.

Modality Differentiation in Somatic Sensory Area I. Not only is there spatial projection of the body in somatic sensory area I, but there is also a moderate degree of modality separation between tactile signals and kinesthetic signals. Tactile signals stimulate neurons mainly in the anteriormost portion of the postcentral gyrus where it dips into the central sulcus. On the other hand, kinesthetic signals stimulate neurons mainly in the posteriormost portion of the postcentral gyrus. This separation of modalities is important because it helps us to understand how the brain dissects different types of information from incoming sensory signals, a subject that will be discussed in more detail later in the chapter.

Somatic Sensory Area II. The second cortical area to which somatic afferent fibers project, somatic sensory area II, lies posterior and inferior to the lower end of the postcentral gyrus and on the upper wall of the lateral fissure, as shown in Figure 49–4. The degree of localization of the different parts of the body is far less acute in this area than in somatic sensory area I. The face is represented anteriorly, the arms centrally, and the legs posteriorly.

So little is known about the function of somatic sensory area II that it cannot be discussed intelligently. It is known that signals enter this area from both the dorsal column system and the spinothalamic system, though perhaps more from the latter. In fact, it has been suggested that somatic sensory area II might be the cortical terminus of pain information, though this is questionable. Also, stimulation of somatic sensory area II in some instances causes complex body movements, for which reason it possibly plays a role in sensory control of motor functions.

Excitation of Vertical Columns of Neurons in the Somesthetic Cortex.   The cerebral cortex contains six separate layers of neurons, and, as would be expected, the neurons in each layer perform functions different from those in other layers. Unfortunately, we still know little about these functions except that the incoming sensory signal excites neuron layers 3 and 4 first. Then the signal spreads outward toward the surface of the cortex, and also inward toward the deeper layers, along a *vertical column of neurons*, continuing through all of the six cortical layers. The nature of the signal changes slightly as it passes through the different layers, and accessory signals are transmitted laterally from this column of neurons to other cells as well as to other areas of the brain. Presumably, special characteristics of the sensory signal are discerned from the parent signal as it traverses the vertical column of neurons, though how this occurs is still a mystery.

Functions of Somatic Sensory Area I. The functional capabilities of different areas of the somesthetic cortex have been determined by selective excision of the different portions. Widespread excision of somatic sensory area I causes loss of the following types of sensory judgment:

1. The person is unable to localize discretely the different sensations in the different parts of the body. However, he can localize these sensations crudely, which indicates that the thalamus or parts of the cerebral cortex not normally considered to be concerned with somatic sensations can perform some degree of localization.

2. He is unable to judge critical degrees of pressure against his body.

3. He is unable to judge exactly the weights of objects.

4. He is unable to judge shapes or forms of objects. This is called *astereognosis*.

5. He is unable to judge texture of materials, for this type of judgment depends on highly critical sensations caused by movement of the skin over the surface to be judged.

6. He is unable to judge fine gradations in temperature.

7. He is unable to recognize the relative orientation of the different parts of his body with respect to each other.

CHARACTERISTICS OF TRANSMISSION IN THE DORSAL COLUMN PATHWAY

Faithfulness of Transmission. The most important functional characteristic of the dorsal column pathway is its *faithfulness* of transmission. That is, each time a point in the periphery is stimulated, a signal ordinarily is transmitted all the way to the somesthetic cortex. Also, if this peripheral stimulus increases in intensity, the intensity of the signal at the cerebral cortex increases proportionately. And, finally, when a discrete area of the body is stimulated, the signal from this area is transmitted to a discrete area of the cerebral cortex. Thus, the dorsal column pathway is adequately organized for transmission of accurate information from the periphery to the sensorium. Furthermore, the responsiveness of this system can be altered only moderately by stimuli from other areas of the nervous system, and it cannot be depressed to a significant extent by extraneous agents, such as moderate degrees of anesthesia.

Basic Neuronal Circuit and Discharge Pattern in the Dorsal Column System. The lower part of Figure 49–6 illustrates the basic organization of the neuronal circuit of the dorsal column system, showing that at each synaptic stage a moderate degree of divergence occurs. However, the upper part of Figure 49–6 shows that a single receptor stimulus on the skin does not cause all the

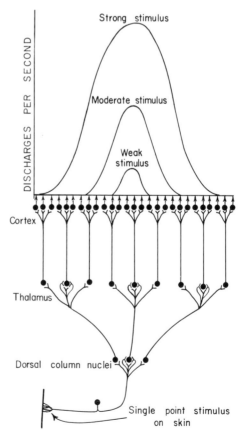

Figure 49–6   Transmission of a pinpoint stimulus signal to the cortex.

cortical neurons with which that receptor connects to discharge at the same rate. Instead, the cortical neurons that discharge to the greatest extent are those in a central part of the cortical "field" for each respective receptor. Thus, a weak stimulus causes only central neurons to fire. A moderate stimulus causes still more neurons to fire, but those in the center discharge at a considerably more rapid rate than do those farther away from the center. Finally, a strong stimulus causes widespread discharge in the cortex but again a much more rapid discharge of the central neurons in comparison with the peripheral neurons.

Point-to-Point Transmission in the Dorsal Column Pathway. Referring once again to Figure 49–6, it is evident that, when a single receptor is stimulated in the skin, a particular point in the somesthetic cortex is excited more intensively than are all the surrounding areas. This does not mean that adjacent areas of the somesthetic cortex are not also stimulated. However, these adjacent areas are stimulated *less* intensively than a central point in the cortical field of the stimulated receptor. Thus, by this means, the sensory cortex is capable of detecting the precise localization of signals from the skin, despite the spread of excitation in the cortex.

TWO-POINT DISCRIMINATION. A method frequently used to test a person's tactile capabilities is to determine his so-called "two-point discriminatory ability." In this test, two needles are pressed against the skin, and the subject determines whether he feels two points of stimulus or one point. On the tips of the fingers a person can distinguish two separate points even when the needles are as close together as 1 to 2 mm. However, on the back, the needles must usually be as far apart as 10 to 15 mm. before one can detect two separate points. The reason for this is that there are many specialized tactile receptors in the tips of the fingers in comparison with a small number in the skin of the back. Referring back to Figure 49–5, we can see also that the portions of the body that have a high degree of two-point discrimination have a correspondingly large cortical representation in somatic sensory area I.

Figure 49–7 illustrates the probable mechanism by which the dorsal column pathway transmits two-point discriminatory information. This shows two adjacent points on the skin that are strongly stimulated, and it shows the small area of the somesthetic cortex (greatly enlarged) that is excited by signals

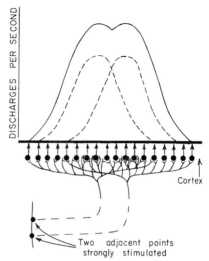

Figure 49–7 Transmission of signals to the cortex from two adjacent pinpoint stimuli.

from the two stimulated points in the skin. The two dashed curves show the individually excited cortical fields, and the solid curve shows the resultant cortical excitation when both the skin points are stimulated simultaneously. Note that the resultant zone of excitation has two separate peaks. It is believed to be these two peaks, separated by a valley, that allow the sensory cortex to detect the presence of two stimulatory points rather than a single stimulatory point.

INCREASE IN CONTRAST IN THE SPATIAL PATTERN CAUSED BY LATERAL INHIBITION. In Chapter 47 it was pointed out that contrast in sensory patterns is increased by inhibitory signals transmitted laterally in the sensory pathway. This effect was illustrated in Figure 47–11 of that chapter.

In the case of the dorsal column system, an excited sensory receptor in the skin transmits not only excitatory signals to the somesthetic cortex but also inhibitory signals laterally to adjacent fiber pathways. These inhibitory signals help to block lateral spread of the excitatory signal. As a result, the peak of excitation stands out, and much of the surrounding diffuse stimulation is blocked. Obviously, this mechanism accentuates the contrast between the areas of peak stimulation and the surrounding areas, thus greatly increasing the contrast or sharpness of the perceived spatial pattern.

Transmission of Rapidly Changing and Repetitive Sensations. The dorsal column system is of particular value for apprising the sensorium of rapidly changing peripheral conditions. This system can "follow" changing stimuli up to at least 100 cycles per second and can "detect" changes as high as 700 cycles. Since most of the tactile impulses are elicited by rapidly changing conditions, one readily understands the importance of faithful transmission of the resultant signals. This allows one to direct his attention immediately to any point of contact, which, in turn, allows him to make necessary corrections before damage can be done.

VIBRATORY SENSATION. Because vibratory signals are rapidly repetitive (can be detected up to 700 cycles per second), they can be transmitted only in the dorsal column system and not at all in the slowly transmitting spinothalamic system. For this reason, application of vibration with a tuning fork to different peripheral parts of the body is an important tool used by the neurologist for testing functional integrity of the dorsal column system.

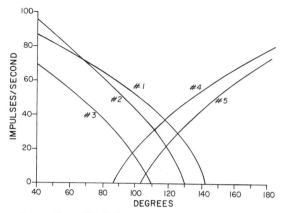

Figure 49–8 Typical responses of five different neurons in the knee joint receptor field of the ventrobasal complex when the knee joint is moved through its range of motion. (The curves were constructed from data in Mountcastle: *J. Neurophysiol.,* 26:807, 1963.)

PROCESSING OF INFORMATION IN THE DORSAL COLUMN SYSTEM

Despite the faithfulness of transmission of signals from the periphery to the sensory cortex through the dorsal column system, there is also some processing of sensory information at lower synaptic levels before it reaches the cerebral cortex. For instance, the signal pattern from kinesthetic receptors changes as it passes progressively up the dorsal column system. Figure 49–1 showed that individual joint receptors are stimulated maximally at specific degrees of angulation of the joint, with the intensity of stimulation decreasing on either side of the maximal point. However, the kinesthetic signal is quite different at the level of the ventrobasal complex of the thalamus, as can be seen by referring to Figure 49–8. This figure shows that the ventrobasal neurons are of two types: (1) those that are maximally stimulated when the joint is at full angulation and (2) those that are maximally stimulated when the joint is at minimal angulation. In each case, as the degree of angulation changes, the rate of stimulation of the neuron either decreases or increases depending on the direction in which the joint is being angulated. Furthermore, the intensity of neuronal excitation changes over angles of 40 to 60 degrees of angulation in contrast to 20 to 30 degrees for the individual receptors, as was illustrated in Figure 49–1. Thus, the signals from the individual joint receptors are integrated in the space domain by the thalamic neurons, giving a progressively stronger signal as the joint moves in only one

direction rather than giving a peaked signal as occurs in stimulation of individual receptors.

Another instance of information processing in the dorsal column system is physical separation of signals of different modalities, with cutaneous sensibilities terminating anteriorly in somatic area I and kinesthetic sensibilities terminating posteriorly. It is presumably these types of changes in signal patterns and separation of modalities that allow the sensorium to separate out all the ingredients of sensory information.

TRANSMISSION IN THE SPINOTHALAMIC SYSTEM

It was pointed out earlier in the chapter that the spinothalamic system transmits sensory signals that do not require rapid transmission nor highly discrete localization in the body. These include pain, heat, cold, crude touch, crude pressure, and sexual sensations. In the following chapter, pain and temperature sensations are discussed specifically, but the present chapter is concerned principally with transmission of crude touch and pressure in the spinothalamic system.

THE SPINOTHALAMIC PATHWAY

Ventral Spinothalamic Tract. Most of the spinothalamic fibers subserving the senses of touch and pressure enter the dorsal columns along with the fibers of the dorsal column system, as shown in Figure 49–9, and then travel upward as much as six segments or downward as much as two segments before ending on second order neurons in the dorsal horns. Then fibers from these neurons cross through the anterior commissure to the

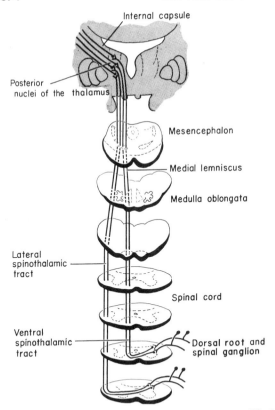

Internal capsule

Posterior
nuclei of the thalamus

Mesencephalon

Medial lemniscus

Medulla oblongata

Lateral
spinothalamic
tract

Spinal cord

Ventral
spinothalamic
tract

Dorsal root and
spinal ganglion

Figure 49–9 The spinothalamic pathways. (Modified from Ranson and Clark: Anatomy of the Nervous System.)

opposite anterior column and form the *ventral spinothalamic tract* that passes all the way to the thalamus.

The ventral spinothalamic tract terminates, along with the terminations of the dorsal column system, mainly in the ventrobasal nuclear complex of the thalamus, though a few of the fibers of the ventral spinothalamic tract terminate in the bulbar and mesencephalic reticular areas.

The fibers of the ventral spinothalamic tract are mainly larger size delta type A nerve fibers in contrast to smaller fibers found in the lateral spinothalamic tract. Furthermore, there is crude localization of touch and pressure sensations transmitted through this tract, indicating a moderate degree of *spatial orientation* of the fibers in the tract. This tract is sometimes called the *neospinothalamic* system, in contradistinction to the *paleospinothalamic* system of the lateral spinothalamic tract.

Lateral Spinothalamic and Spinotectal Tracts. On entering the cord through the posterior roots, the pain and temperature fibers travel upward in the *tract of Lissauer* for one to three segments and then terminate on *second order neurons* in the gray matter of the dorsal horns. The fibers from these neurons then cross through the anterior commissure of the cord gray matter, pass to the opposite lateral column of the spinal cord where they form the *lateral spinothalamic tract,* and eventually

terminate mainly in the intralaminar nuclei of the thalamus, a different terminus from that of the ventral spinothalamic tract.

Many second order neurons of the lateral spinothalamic tract terminate in the reticular substance of the bulbar and mesencephalic areas or send collaterals into these areas. These fibers, though often considered to be part of the spinothalamic system, are also frequently designated as the *spinotectal* or the *spinobulbar tract.* After terminating in the reticular substance, higher order neurons transmit the signals to generally the same areas in the thalamus as does the lateral spinothalamic tract.

The fibers of the spinothalamic and spinotectal tracts are mainly very small delta type A or type C fibers.

Fibers also enter both the lateral and the ventral spinothalamic tracts from the spinal nucleus of the fifth nerve. These fibers transmit respective types of sensory information from the head.

Projection of Spinothalamic Signals from the Thalamus to the Cortex. Third order neurons from both the ventrobasal complex of the thalamus and the intralaminar nuclei probably project primarily to somatic area I and, to a lesser extent, to somatic area II. However, on the basis of inadequate information, it has been suggested that the lateral spinothalamic system, which transmits pain and temperature sensations, might project more to somatic area II. On the other hand, this is disputed by the fact that destruction of somatic area I considerably reduces one's ability to appreciate gradations of temperature, whereas destruction of somatic area II does not have this effect.

Separation of Modalities in the Spinothalamic Tracts. It has already been pointed out in the above description of the spinothalamic pathway that the nerve fibers that subserve the senses of touch and pressure are transmitted in the ventral spinothalamic tract, and those that subserve pain and temperature are transmitted in the lateral spinothalamic tract. In addition, the lateral spinothalamic tract is at least partially divided into *superficial* and *deep divisions*, the superficial division transmitting pain signals and the deep division transmitting temperature signals. Thus, the process of separating the different modalities of sensation actually begins in the spinal cord.

Characteristics of Transmission in the Spinothalamic Tracts. The precise nature of transmission in the spinothalamic tracts is not nearly so well known as in the dorsal column system. In general, the same principles apply to transmission in this tract as in the dorsal column system, except for the following differences: (a) the velocities of transmission in the spinothalamic tract are only one-fifth to a half those in the dorsal column system; (b) the spatial localization of signals transmitted in the spinothalamic tract is far less acute than in the dorsal column system, especially in the lateral spinothalamic tract that transmits pain and temperature signals; (c) the gradations of intensities are also far less acute, most of the sensations being recognized in 10 to 20 gradations of strength rather than the 100 or more gradations for the dorsal column system; and (d) the ability to transmit rapidly repetitive sensations is almost nil in the spinothalamic system.

Thus, it is evident that the spinothalamic tracts are by far a cruder type of transmission system than is the dorsal column system. However, it must be realized that certain types of sensations are, nevertheless, transmitted only by the spinothalamic tracts, including pain, thermal sensations, and sexual sensations. Only the mechanoreceptive sensations are transmitted by both systems, the dorsal column system having to do with the critical types of mechanoreceptive sensations, and the ventral spinothalamic tract having to do with the cruder types of mechanoreceptive sensations.

SOME SPECIAL ASPECTS OF SENSORY FUNCTION

Function of the Thalamus in Somatic Sensations. The major function of the thalamus in tactile sensation is probably to relay information to the cortex. However, since a person with his somesthetic cortex removed can redevelop a mild to moderate degree of tactile sensibility, it must be assumed that the thalamus can do much more than simply relay tactile information to the somesthetic cortex. The thalamus seems to be even more important in pain sensation, as pointed out in the following chapter.

Cortical Control of Sensory Sensitivity. Almost all sensory information that enters the cerebrum, including sensory information from the eyes, the ears, the taste receptors, and all the somatic receptors, is relayed through one or another of the thalamic nuclei. Furthermore, the conscious brain is capable of directing its attention to different segments of the sensory system. This function is believed to be mainly achieved through facilitation or inhibition of the cortical receptive areas or the relay stations in the thalamus.

But, in addition, "centrifugal" impulses are transmitted from the cortex to the still lower relay stations in the sensory pathways to facilitate or inhibit transmission. For instance, centrifugal pathways control the sensitivity of the synapses in the dorsal column nuclei and in the posterior horn relay station of the spinothalamic system. Also, similar systems are known for the visual, auditory,

and olfactory systems, which are discussed in later chapters. Each centrifugal pathway begins in the cortex where the sensory pathway that it controls terminates. Thus, a feedback loop exists for each sensory pathway.

Obviously, centrifugal control of sensory input could help the cerebral cortex determine the threshold for different sensory signals. Also, it might help the brain focus its attention on specific types of information, which is an important and necessary quality of nervous system function.

Segmental Fields of Sensation—The Dermatomes. Each spinal nerve innervates a "segmental field" of the skin called a *dermatome*. The different dermatomes are illustrated in Figure 49–10. However, these are shown as if there were distinct borders between the adjacent dermatomes, which is far from true because much overlap exists from segment to segment. Indeed, because of the great overlap, the posterior roots from an entire segment of the spinal cord can be destroyed without causing significant loss of sensation in the skin.

Figure 49–10 shows that the anal region of the body lies in the dermatome of the most distal cord segment. In the embryo, this is the tail region and is the most distal portion of the body. The legs develop from the lumbar and upper sacral segments rather than from the distal sacral segments, which is evident from the dermatomal map. Obviously, one can use a dermatomal map such as that illus-

Figure 49–10 The dermatomes. (Modified from Grinker and Bucy: Neurology. Charles C Thomas.)

trated in Figure 49–10 to determine the level in the spinal cord at which various cord injuries may have occurred when the peripheral sensations are disturbed.

REFERENCES

Adrian, E. D.: The Physical Background of Perception. Oxford, Clarendon Press, 1947.

Andersen, P.; Eccles, J. C.; Schmidt, R. F.; and Yokota, T.: Identification of relay cells and interneurons in the cuneate nucleus. *J. Neurophysiol.,* 27:1080, 1964.

Bishop, P. O.: Central nervous system: afferent mechanisms and perception. *Ann. Rev. Physiol.,* 29:427, 1967.

Bourassa, C. M., and Swett, J. E.: Sensory discrimination thresholds with cutaneous nerve volleys in the cat. *J. Neurophysiol.,* 30:515, 1967.

Darian-Smith, I.: Somatic sensation. *Ann. Rev. Physiol.,* 31:417, 1969.

Desmedt, J. E.: Feedback control in sensory systems. *Proc. Int. Union of Physiol. Sciences,* 6:264, 1968.

Gordon, G., and Paine, C. H.: Functional organization in nucleus gracilis of the cat. *J. Physiol. (Lond.),* 153:331, 1960.

Jerge, C. R.: Organization and function of the trigeminal mesencephalic nucleus. *J. Neurophysiol.,* 26:379, 1963.

Krauthamer, G.; Besson, J. M.; Guilbaud, G.; Abdelmoumene, M.; and Lim, R. K. S.: Descending control systems. *Proc. Int. Union of Physiol. Sciences,* 6:239, 1968.

Mountcastle, V. B.; Poggio, G. F.; and Werner, G.: The relation of thalamic cell response to peripheral stimuli varied over an intensive continuum. *J. Neurophysiol.,* 26:807, 1963.

Mountcastle, V. B.; Talbot, W. H.; Sakata, H.; and Hyvarinen, J.: Cortical neuronal mechanisms in flutter-vibration studied in unanesthetized monkeys. *J. Neurophysiol.,* 32:452, 1969.

Nicholls, J. G., and Baylor, D. A.: Specific modalities and receptive fields of sensory neurons in CNS of the leech. *J. Neurophysiol.,* 31:740, 1968.

Sinclair, D. C.: Cutaneous Sensation. New York, Oxford University Press, 1967.

Wagman, I. H., and Price, D. D.: Responses of dorsal horn cells of *M. mulatta* to cutaneous and sural nerve A and C fiber stimuli. *J. Neurophysiol.,* 32:803, 1969.

Wall, P. D., and Pomeranz, B.: Cord cells responding to fine myelinated afferents from viscera, muscle, and skin. *Proc. Int. Union of Physiol. Sciences,* 6:237, 1968.

Werner, G., and Mountcastle, V. B.: Neural activity in mechanoreceptive cutaneous afferents: stimulus-response relations, Weber functions, and information transmission. *J. Neurophysiol.,* 28:359, 1965.

Whitlock, D. G., and Perl, E. R.: Thalamic projects of spinothalamic pathways in monkeys. *Exp. Neurol.,* 3:240, 1961.

Whitsel, B. L.; Petrucelli, L. M.; and Werner, G.: Symmetry and connectivity in the map of the body surface in somatosensory area II of primates. *J. Neurophysiol.,* 32:170, 1969.

CHAPTER 50

Somatic Sensations:
II. Pain, Visceral Pain, Headache, and Thermal Sensations

Many, if not most, ailments of the body cause pain. Furthermore, one's ability to diagnose different diseases depends to a great extent on a knowledge of the different qualities of pain, a knowledge of how pain can be referred from one part of the body to another, how pain can spread in all directions from the painful site, and, finally, what the different causes of pain are. For these reasons, the present chapter is devoted mainly to pain and to the physiologic basis of some of the associated clinical phenomena.

The Purpose of Pain. Pain is a protective mechanism for the body; it occurs whenever any tissues are being damaged and causes the individual to react reflexly to remove the pain stimulus. Even such simple activities as sitting for a long time on the ischia can cause tissue destruction because of lack of blood flow to the skin where the skin is compressed by the weight of the body. When the skin becomes painful as a result of the ischemia, the person shifts his weight unconsciously. A person who has lost his pain sense, such as after spinal cord injury, fails to feel the pain and therefore fails to shift his weight. This eventually results in ulceration at the areas of pressure unless special measures are taken to move the person from time to time.

QUALITIES OF PAIN

Pain has been classified into three different major types: pricking, burning, and aching pain. Other terms used to describe different types of pain include throbbing pain, nauseous pain, cramping pain, and sharp pain, all of which are well known to almost everyone.

Pricking pain is felt when a needle is stuck into the skin or when the skin is cut with a knife. It is also often felt when a widespread area of the skin is diffusely but strongly irritated.

Burning pain is, as its name implies, the type of pain felt when the skin is burned. It can be excruciating and is the most likely of the pain types to cause suffering.

Aching pain ordinarily is not felt on the surface of the body, but, instead, is a deep pain with varying degrees of annoyance. Aching pain of low intensity in widespread areas of the body can summate into a very disagreeable sensation.

It is not necessary to describe these different qualities of pain in great detail because they are well known to all persons. The real problem, and one that is only partially solved, is what causes the differences in quality. Pricking pain results from stimulation of delta type A pain fibers, whereas burning and aching pain results from stimulation of the more primitive type C fibers, which will be discussed later in the chapter.

METHODS FOR MEASURING THE PERCEPTION OF PAIN

The intensity of a stimulus necessary to cause pain can be measured in many different ways, but the most used methods have been pricking the skin with a pin at measured pressures, pressing a solid object against a protruding bone with measured force, or heating the skin with measured amounts of heat. The latter method has proved to be especially accurate from a quantitative point of view.

Figure 50–1 illustrates the basic principles of a heat apparatus used for measuring pain threshold. An intense light is focused by a large condenser lens onto a black spot painted on the forehead of the subject, and the heat intensity delivered by the light is controlled by a rheostat. In determining the subject's threshold for pain, the intensity of the heat is increased in progressive steps, and the

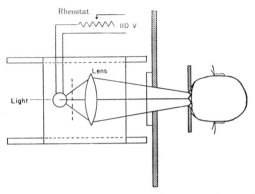

Figure 50–1 Heat apparatus for measuring pain threshold. (From Hardy, Wolff, and Goodell: *J. Clin. Invest*, 19:649, 1940.)

length of time required for the forehead to heat sufficiently to elicit pain is recorded for each heat intensity. These data are then plotted in the form of a "strength-duration curve" to express pain threshold, as follows.

Strength-Duration Curve for Expressing Pain Threshold. Figure 50–2 illustrates a typical strength-duration curve obtained by using the above procedure for measuring pain threshold. This curve is almost identical with the strength-duration curves discussed in Chapter 5 for stimulation of nerve fibers by electrical currents of increasing intensities. Note that a very intense stimulus applied for only a second elicits a sensation of pain while a stimulus of much less intensity may require many seconds. The lowest intensity of stimulus that will excite the sensation of pain when the stimulus is applied continuously is called the *pain threshold.*

The Intensity of Pain—"JND's." The increase in stimulus intensity that will barely cause a detectable difference in degree of pain is called a *just noticeable difference* (JND). By applying all different stimulus intensities between the level of no pain at all and the most intense pain that a person can distinguish, it has been found that approximately 22 JND's can be discerned by the average person.

Figure 50–3 illustrates the relationship of stimulus intensity (applied by the apparatus of Fig. 50–1) to the intensity of pain expressed in JND's.

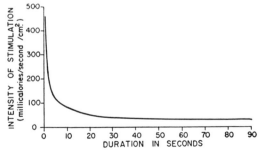

Figure 50–2 Strength-duration curve for depicting pain threshold. (From Hardy: *J. Chronic Dis.*, 4:20, 1956.)

Uniformity of Pain Threshold in Different People. Figure 50–4 shows graphically the skin temperature at which pain is first perceived by different persons. By far the greatest number of subjects perceive pain when the skin temperature reaches almost exactly 45° C., and almost everyone perceives pain before the temperature reaches 47° C. In other words, it is almost never true that some persons are unusually sensitive or insensitive to pain. Indeed, measurements in people as widely different as Eskimos, Indians, and whites have shown no significant differences in their *thresholds for pain.* However, different people do *react* very differently to pain, as is discussed below.

THE PAIN RECEPTORS AND THEIR STIMULATION

Free Nerve Endings as Pain Receptors. The pain receptors in the skin and other tissues are all free nerve endings. They are widespread in the superficial layers of the *skin* and also in certain internal tissues, such as the *periosteum,* the *arterial walls,* the *joint surfaces,* and the *falx* and *tentorium* of the cranial vault. Most of the other deep tissues are not extensively supplied with pain endings but are diffusely supplied; nevertheless, any widespread tissue damage can still summate to cause the aching type of pain in these areas.

RATE OF TISSUE DAMAGE AS THE CAUSE OF PAIN

The average critical value of 45° C. at which a person first begins to perceive pain is also the temperature at which the tissues begin to be damaged by heat; indeed, the tissues are eventually completely destroyed if the temperature remains at this level indefinitely. Therefore, it is immediately apparent that pain resulting from heat is closely correlated with the ability of heat to damage the tissues.

Furthermore, in studying soldiers who had been severely wounded in World War II, it was found that the majority of them felt little or no pain except for a short time after the severe wound had been sustained. This, too, indicates that *pain generally is not felt after damage has been done* but only *while damage is being done.*

Bradykinin and Histamine as Possible Stimulators of Pain Endings. The precise mechanism by which tissue damage stimulates pain endings is not known. However, many research

Figure 50–3 Pain intensity, expressed in JNDs, at different intensities of stimulus. (From Hardy: *J. Chronic Dis., 4:22, 1956.*)

workers have shown that extracts from damaged tissues cause intense pain when injected beneath the normal skin. Therefore, it is almost certain that some chemical substance or substances released from the cells or formed in damaged tissues excite the pain nerve endings.

There are many reasons to believe that the substance *bradykinin* or some similar product might be the principal substance that stimulates pain endings. For instance, when this substance is injected in extremely minute quantities underneath the skin, severe pain is felt. Furthermore, cell damage releases proteolytic enzymes that almost immediately split bradykinin and other similar substances from the globulins in the interstitial fluid. And, finally, bradykinin and similar substances can be found in the skin when painful stimuli are applied.

Thus, the postulated mechanism for eliciting pain: damage to cells releases proteolytic enzymes that then split bradykinin and associated

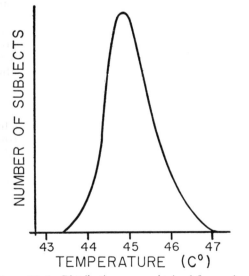

Figure 50–4 Distribution curve obtained from a large number of subjects of the minimal skin temperature that causes pain. (Modified from Hardy: *J. Chronic Dis., 4:22, 1956.*)

substances from globulin, and these in turn stimulate the nerve endings.

Another substance possibly involved in at least some types of pain is histamine, because damaged cells also release this substance and because almost infinitesimal amounts of it, too, can cause very severe pain upon its injection beneath the skin.

Tissue Ischemia and Muscle Spasm as Causes of Pain. When blood flow to a tissue is blocked, the tissue becomes very painful within a few minutes. And the greater the rate of metabolism of the tissue, the more rapidly the pain appears. For instance, if a blood pressure cuff is placed around the upper arm and inflated until the arterial blood flow ceases, exercise of the forearm muscles can cause severe muscle pain within 15 to 20 seconds. In the absence of muscle exercise, the pain will not appear for three to four minutes. Cessation of blood flow to the skin, in which the metabolic rate is very low, usually does not cause pain for about 20 to 30 minutes.

Muscle spasm is a frequent cause of pain, also. The reason for this is probably two-fold. First, the contracting muscle compresses the intramuscular blood vessels and either reduces or cuts off the blood flow. Second, muscle contraction increases the rate of metabolism of the muscle. Therefore, muscle spasm probably causes relative muscle ischemia so that typical ischemic pain results.

The cause of pain in ischemia is yet unknown; however, it is relieved by supplying oxygen to the ischemic tissue. Flow of unoxygenated blood to the tissue will not relieve the pain.

One of the suggested causes of pain in ischemia is accumulation of large amounts of lactic acid in the tissues, formed as a consequence of the anaerobic metabolism (metabolism without oxygen) that occurs during ischemia. However, it is also possible that other chemical agents, such as bradykinin and histamine, are formed in the tissues because of muscle cell damage and that these, rather than lactic acid, stimulate the pain nerve endings.

TICKLING AND ITCH

The phenomenon of tickling and itch has often been stated to be caused by very mild stimulation of pain nerve endings, since whenever pain is blocked by anesthesia of a nerve or by compressing the nerve, the phenomenon of tickling and itch also disappears. However, recent neurophysiologic studies have demonstrated the existence of a few very sensitive free nerve endings that respond to extremely light touch. These are believed to carry the sensation of tickle and itch, and their signals are transmitted over primitive type C fibers similar to those that transmit the burning type of pain.

TRANSMISSION OF PAIN SIGNALS INTO THE CENTRAL NERVOUS SYSTEM

"Fast" Pain Fibers and "Slow" Pain Fibers. Pain signals are transmitted by small delta type A fibers at velocities of between 3 and 10 meters per second and also by type C fibers at velocities between 0.5 and 2 meters per second. When the delta type A fibers are blocked by compressing the nerve trunk, the pricking type of pain disappears. On the other hand, when the type C fibers are blocked by low concentrations of local anesthetic, the burning and aching types of pain disappear.

Therefore, a sudden painful stimulus gives a "double" pain sensation: a fast pricking pain sensation followed a second or so later by a slow burning pain sensation. The pricking pain presumably apprises the person very rapidly of a damaging influence and, therefore, plays an important role in making the person react immediately to remove himself from the stimulus. On the other hand, the slow burning sensation tends to become more and more painful over a period of time. It is this sensation that gives one the intolerable feeling toward pain.

Transmission in the Spinothalamic and Spinotectal Tracts. Pain fibers enter the cord through the dorsal roots, ascend or descend one to three segments in the *tract of Lissauer,* and then end on second order neurons in the posterior horns of the cord gray matter. These neurons then give rise to fibers that cross immediately to the opposite side of the cord in the *anterior commissure* and pass upward to the brain in the *lateral spinothalamic* and *spinotectal tracts.*

The fibers of the spinotectal tract terminate in the reticular areas of the medulla, pons, and mesencephalon, and large numbers of collateral fibers from the lateral spinothalamic tract also terminate in these same areas, as shown in Figure 50–5. However, a small portion of the fibers pass all the way to the thalamus where they terminate in the intralaminar nuclei and in the posterior nuclear group. The

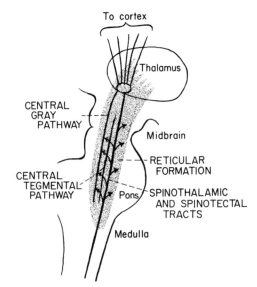

Figure 50–5   Tranmission of pain signals into the hindbrain, thalamus, and cortex through multiple diffuse pathways.

type C fibers that transmit burning and aching pain terminate in the intralaminar nuclei. Also, similar signals from the spinotectal tract are relayed through the bulbar areas of the brain to the intralaminar nuclei. On the other hand, the small myelinated fibers that are responsible for pricking pain terminate in the posterior nuclear group. It is in this same area that the ventral spinothalamic fibers for transmission of touch and pressure terminate.

Pain Pathways in the Brain.   Unfortunately, very little is known about the transmission of pain signals in the bulbar, diencephalic, and cortical regions of the brain. It is believed that the pricking pain signals that enter the posterior nuclear group are then conducted by way of third order neurons to the somesthetic areas of the cortex along much the same pathways that touch and pressure signals are transmitted.

On the other hand, the burning and aching types of pain signals seem to spread through the reticular regions of the brain stem along pathways illustrated in Figure 50–5 and, finally, pass into the intralaminar nuclei of the thalamus. What happens beyond here is very unclear. There is reason to believe that at least a portion of the signals is relayed to somatic area II, because extirpation of area II in patients with pain has in some instances, relieved the pain.

The pain signals that enter the reticular substance of the hindbrain probably are of special importance in causing the suffering quality of pain, because stimulation in these regions can apparently cause very severe suffering in animals. Also, these signals increase the general level of excitability of the entire brain by stimulating the *ascending reticular system,* as will be discussed in Chapter 60.

Figure 50–5 shows two brain stem pathways that supposedly play special roles in pain sensation: (1) The *central gray pathway,* which transmits pain signals from the reticular regions to the thalamus; its destruction supposedly reduces the intensity of pain sensations perceived. (2) The *central tegmental pathway,* which supposedly suppresses pain; when it is destroyed in an animal, the animal seems to become excessively sensitive to pain, sometimes seemingly experiencing continuous pain.

SURGICAL INTERRUPTION OF PAIN PATHWAYS. Often a person has such severe and intractable pain (this often results from rapidly spreading cancer), that it is necessary to relieve the pain. To do this the pain pathway can be destroyed at any one of several different points. If the pain is in the lower part of the body, a *cordotomy* in the upper thoracic region usually relieves the pain. To do this, the spinal cord on the side opposite to the pain is sectioned almost entirely through its anterolateral quadrant, which interrupts the spinothalamic tracts.

A cordotomy does not relieve pain in the upper part of the body because many pain fibers from the upper body do not cross to the opposite side of the spinal cord before reaching the medulla. In these patients, the lateral spinothalamic tract is sectioned in the brain stem where it passes over the inferior olive. This procedure is called a *bulbar tractotomy.*

Finally, if the pain originates in the neck or face, even a bulbar tractotomy is not satisfactory. A recent operation has been devised to destroy those posterior group nuclei of the thalamus where the pain endings synapse with third order neurons. These neurons lie slightly posterior to the nuclei of the ventrobasal complex that transmit most of the tactile sensations, so that the tactile sensations are not disturbed.

Function of the Thalamus and Cerebral Cortex in the Appreciation of Pain.   Complete removal of the sensory areas of the cerebral cortex does not destroy one's ability to perceive pain. Therefore, it is believed that pain impulses entering only the thalamus and even lower centers cause at least some conscious perception of pain. However, this does not mean that the cerebral cortex has nothing to do with normal pain appreciation; indeed, electrical stimulation of the somesthetic cortical areas causes a person to perceive mild pain in approximately 3 per cent of the stimulations. Furthermore, lesions in these areas, particularly in somatic sensory area II, at times give

rise to severe pain. Thus, there is reason to believe that somatic sensory area II may be more concerned with pain sensation than is somatic sensory area I.

Localization of Pain in the Body.   Most localization of pain probably results from simultaneous stimulation of tactile receptors along with the pain stimulation. However, the pricking type of pain, transmitted through delta type A fibers, can be localized perhaps within 10 to 20 cm. of the stimulated area. On the other hand, the burning and aching types of pain, transmitted through type C fibers, is localizable only very grossly, perhaps to a major part of the body such as a limb but certainly not to small areas. This is in keeping with the fact that these fibers terminate extremely diffusely in the hindbrain and thalamus.

THE REACTION TO PAIN

Even though the threshold for recognition of pain remains approximately equal from one person to another, the degree to which each one reacts to pain varies tremendously. Stoical persons, such as members of the American Indian race, react to pain far less intensely than do more emotional persons. In the preceding chapter, it was pointed out that conditioning impulses entering the sensory areas of the central nervous system from various portions of the central and peripheral nervous systems can determine whether incoming sensory impulses will be transmitted extensively or weakly to other areas of the brain. It is probably some such mechanism as this that determines how much one reacts to pain.

Pain causes both reflex motor reactions and psychic reactions. Some of the reflex actions occur directly from the spinal cord, for pain impulses entering the gray matter of the cord can directly initiate "withdrawal reflexes" that remove the body or a portion of the body from the noxious stimulus, as will be discussed in Chapter 56. These primitive spinal cord reflexes, though important in lower animals, are mainly suppressed in the human being by the higher centers of the central nervous system. Yet, in their place, much more complicated and more effective reflexes from the motor cortex are initiated by the pain stimuli to eliminate the painful stimulus.

The psychic reactions to pain are likely to be far more subtle; they include all the well-known aspects of pain such as anguish, anxiety, crying, depression, nausea, and excess muscular excitability throughout the body. These reactions vary tremendously from one person to another following comparable degrees of pain stimuli.

REFERRED PAIN

Often, a person feels pain in a part of his body that is considerably removed from the tissues causing the pain. This pain is called referred pain. On occasion, pain can even be referred from one surface area of the body to another, but more frequently it is initiated in one of the visceral organs and referred to an area on the body surface. Also, pain may originate in a viscus and be referred to another deep area of the body not exactly coincident with the location of the viscus producing the pain. A knowledge of these different types of referred pain is extremely important because many visceral ailments cause no other signs except referred pain.

MECHANISM OF REFERRED PAIN

Figure 50–6 illustrates the most generally accepted mechanism by which most pain is referred. In the figure, branches of visceral pain fibers are shown to synapse in the spinal cord with some of the same second order neurons that receive pain fibers from the skin. When the visceral pain fibers are stimulated intensely, pain sensations from the viscera spread into some of the neurons that normally conduct pain sensations only from the skin, and the

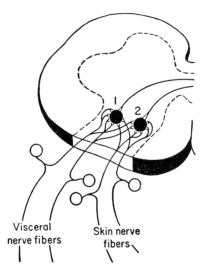

Figure 50–6 Mechanism of referred pain and referred hyperalgesia.

person has the feeling that the sensations actually originate in the skin itself. It is also possible that some referred pain results from convergence of visceral and skin impulses at the level of the thalamus rather than in the spinal cord.

REFERRED PAIN CAUSED BY REFLEX MUSCULAR SPASM

Some types of referred pain are caused secondarily by reflex muscular spasm. For instance, pain in a ureter can cause reflex spasm of the lumbar muscles. Often the pain from the ureter itself is hardly felt at all, but instead almost all the pain results from spasm of the lumbar muscles.

Many back pains and some types of headache also appear to be caused by muscular spasm, the spasm originating reflexly from much weaker pain impulses originating elsewhere in the body.

VISCERAL PAIN

In clinical diagnosis, pain from the different viscera of the abdomen and chest is one of the few criteria that can be used for diagnosing visceral inflammation, disease, and other ailments. In general, the viscera have sensory receptors for no other modalities of sensation besides pain, and visceral pain differs from surface pain in many important aspects.

One of the most important differences between surface pain and visceral pain is that highly localized types of damage to the viscera rarely cause severe pain. For instance, a surgeon can cut the gut entirely in two in a patient who is awake without causing significant pain. On the other hand, any stimulus that causes *diffuse stimulation of pain nerve endings* throughout a viscus causes pain that can be extremely severe. For instance, occluding the blood supply to a large area of gut stimulates many diffuse pain fibers at the same time and can result in extreme pain.

CAUSES OF TRUE VISCERAL PAIN

Any stimulus that excites pain nerve endings in diffuse areas of the viscera causes visceral pain. Such stimuli include ischemia of visceral tissue, chemical damage to the surfaces of the viscera, spasm of the smooth muscle in a hollow viscus, distention of a hollow viscus, or stretching of the ligaments.

Ischemia. Ischemia causes visceral pain in exactly the same way that it does in other tissues, presumably because of the formation of acidic metabolic end-products or tissue degenerative products, such as bradykinin or histamine, that stimulate the pain nerve endings.

Chemical Stimuli. On occasion, damaging substances leak from the gastrointestinal tract into the abdominal cavity. For instance, proteolytic acidic gastric juice often leaks through a ruptured gastric or duodenal ulcer. This juice causes widespread digestion of the visceral peritoneum, thus stimulating extremely broad areas of pain fibers. The pain is usually extremely severe.

Spasm of a Hollow Viscus. Spasm of the gut, the gallbladder, a bile duct, the ureter, or any other hollow viscus can cause pain in exactly the same way that spasm of the skeletal muscle causes pain. This presumably results from diminished blood flow to the muscle, combined with increased metabolic need of the muscle for nutrients. Thus, *relative* ischemia develops, which causes severe pain.

Often, pain from a spastic viscus occurs in the form of *cramps,* the pain increasing to a high degree of severity and then subsiding, this process continuing rhythmically once every few minutes. The rhythmic cycles result from rhythmic contraction of smooth muscle. For instance, each time a peristaltic wave travels along an overly excitable spastic gut, a cramp occurs. The cramping type pain frequently occurs in gastroenteritis, constipation, menstruation, parturition, gallbladder disease, or ureteral obstruction.

Overdistention of a Hollow Viscus. Extreme overfilling of a hollow viscus also results in pain, presumably because of overstretch of the tissues themselves. However, overdistention can also collapse the blood vessels that encircle the viscus, or that pass into its wall, thereby perhaps promoting ischemic pain.

"PARIETAL" PAIN CAUSED BY VISCERAL DAMAGE

In addition to true visceral pain, some pain sensations are also transmitted from the viscera through nerve fibers that innervate the parietal peritoneum, pleura, or pericardium. The parietal surfaces of the visceral cavities are supplied mainly by nerve fibers that penetrate from the surface of the body inward.

Characteristics of Parietal Visceral Pain. When a disease affects a viscus, it often spreads to the parietal wall of the visceral cavity. This wall, like the skin, is supplied with extensive innervation. Therefore, pain from the parietal wall of a visceral cavity is frequently very sharp and pricking in quality, though it can also have burning and aching qualities if the pain stimu-

lus is diffuse. Thus, a knife incision through the *parietal* peritoneum is very painful, even though a similar cut through the visceral peritoneum or through a gut is not painful.

INSENSITIVE VISCERA

A few visceral areas are almost entirely insensitive to pain of any type. These include the parenchyma of the liver and the alveoli of the lungs. Yet the liver *capsule* is extremely sensitive to both direct trauma and stretch, and the *bile ducts* are also sensitive to pain. In the lungs, even though the alveoli are insensitive, the bronchi and the parietal pleura are both very sensitive to pain.

LOCALIZATION OF VISCERAL PAIN—REFERRED VISCERAL PAIN

Pain from the different viscera is frequently difficult to localize for a number of reasons. First, the brain does not know from firsthand experience that the different organs exist, and, therefore, any pain that is localized internally can be localized only generally. Second, sensations from the abdomen and thorax are transmitted by two separate pathways to the central nervous system—the *true visceral pathway* and the *parietal pathway.* The true visceral pain is transmitted via sensory fibers of the autonomic nervous system, and the sensations are *referred* to surface areas of the body often far from the painful organ. On the other hand, parietal sensations are conducted *directly* from the parietal peritoneum, pleura, or pericardium, and the sensations are usually *localized directly over the painful area.*

The Visceral Pathway for Transmission of Pain. Most of the internal organs of the body are supplied by pain fibers that pass along the visceral sympathetic nerves into the spinal cord and thence up the lateral spinothalamic tract along with the pain fibers from the body's surface. A few visceral pain fibers—those from the distal portions of the colon, from the rectum, and from the bladder—enter the spinal cord through the sacral parasympathetic nerves, and some enter the central nervous system through various cranial nerves. These include fibers in the glossopharyngeal and vagus nerves, which transmit pain from the pharynx, trachea, and upper esophagus. And fibers from the surfaces of the diaphragm as well as from the lower esophagus are carried in the phrenic nerves.

LOCALIZATION OF REFERRED PAIN TRANSMITTED BY THE VISCERAL PATHWAYS. The position in the cord to which visceral afferent fibers pass from each organ depends on the segment of the body from which the organ developed embryologically. For instance, the heart originated in the neck and upper thorax. Consequently, the heart's visceral pain fibers enter the cord all the way from C-3 down to T-5. The stomach had its origin approximately from the seventh to the ninth thoracic segments of the embryo, and consequently the visceral afferents from the stomach enter the spinal cord between these levels. The gallbladder had its origin almost entirely in the ninth thoracic segment, so that the visceral afferents from the gallbladder enter the spinal cord at T-9.

Because the visceral afferent pain fibers are responsible for transmitting referred pain from the viscera, the location of the referred pain on the surface of the body is in the dermatome of the segment from which the visceral organ was originally derived in the embryo. Some of the areas of referred pain on the surface of the body are shown in Figure 50–7.

The Parietal Pathway for Transmission of Abdominal and Thoracic Pain. Where skeletal nerves overlie the abdomen or thorax, pain fibers penetrate inward to innervate the parietal peritoneum, parietal pleura, and parietal pericardium. Also, retroperitoneal visceral organs and perhaps portions of the mesentery are innervated to some extent by parietal pain fibers. The kidney, for instance, is supplied by both visceral and parietal fibers.

Pain from the viscera is frequently localized in two surface areas of the body because of the

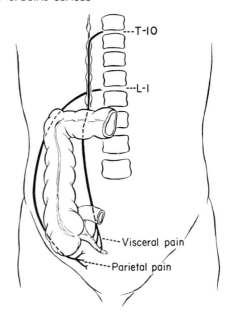

Figure 50–8 Visceral and parietal transmission of pain from the appendix.

dual pathways for transmission of pain. Figure 50–8 illustrates dual transmission of pain from an inflamed appendix. Impulses pass from the appendix through the sympathetic visceral pain fibers into the sympathetic chain and then into the spinal cord at approximately T-10 or T-11; this pain is referred to an area around the umbilicus. On the other hand, pain impulses also often originate in the parietal peritoneum where the inflamed appendix touches the abdominal wall, and these impulses pass directly through the skeletal nerves into the spinal cord at a level of approximately L-1 or L-2. This pain is localized directly over the irritated peritoneum in the right lower quadrant of the abdomen.

VISCERAL PAIN FROM VARIOUS ORGANS

Cardiac Pain. Almost all pain that originates in the heart results from ischemia secondary to coronary sclerosis, which was discussed in Chapter 31. This pain is referred mainly to the base of the neck, over the shoulders, over the pectoral muscles, and down the arms. Most frequently, the referred pain is on the left side rather than on the right – probably because the left side of the heart is much more frequently involved in coronary disease than is the right side – but occasionally mild referred pain occurs on the right side of the body as well as on the left.

The pain impulses are conducted through nerves passing to the middle cervical ganglia, to the stellate ganglia, and to the first four or five thoracic ganglia of the sympathetic chains. Then the im-

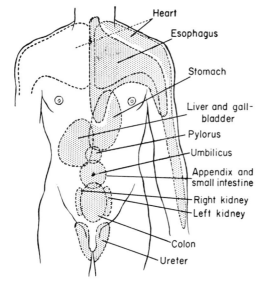

Figure 50–7 Surface areas of referred pain from different visceral organs.

pulses pass into the spinal cord through the second, third, fourth, and fifth thoracic spinal nerves.

DIRECT PARIETAL PAIN FROM THE HEART. When coronary ischemia is extremely severe, such as immediately after a coronary thrombosis, intense cardiac pain sometimes occurs directly underneath the sternum simultaneously with pain referred to other areas. This direct pain from underneath the sternum is difficult to explain on the basis of the visceral nerve connections. Therefore, it is highly probable that skeletal nerve endings passing from the heart through the pericardial reflections around the great vessels conduct this direct pain.

In addition to pain from the heart, other sensations may accompany coronary thrombosis. One of these is a tight, oppressive sensation beneath the sternum. The exact cause of this is unknown, but a possible cause is reflex spasm of blood vessels, bronchioles, or muscles in the chest region.

RELIEF OF REFERRED CARDIAC PAIN BY SYMPATHECTOMY. To interrupt pain impulses from the heart one can either cut the sympathetic nerves that pass from the heart to the sympathetic chains or, as is usually performed, cut nerve fibers as they pass through the sympathetic chains into the spinal cord. Relief of cardiac pain can frequently be accomplished simply by removing the sympathetic chain from T-2 through T-5 on only the left side, but sometimes it is necessary to remove the fibers on both sides in order to obtain satisfactory results.

Esophageal Pain. Pain from the esophagus is usually referred to the pharynx, to the lower neck, to the arms, or to midline chest regions beginning at the upper portion of the sternum and ending approximately at the lower level of the heart. Irritation of the gastric end of the esophagus may cause pain directly over the heart, though the pain has nothing whatsoever to do with the heart. Such pain may be caused by spasm of the cardia, the area where the esophagus empties into the stomach, which causes excessive dilatation of the lower esophagus, or it may result from chemical, bacterial, or other types of inflammatory irritations.

Gastric Pain. Pain arising in the fundus of the stomach — usually caused by gastritis — is referred to the anterior surface of the chest or upper abdomen from slightly below the heart to an inch or so below the xyphoid process. These pains are frequently characterized as burning pains; and they, or pains from the lower esophagus, cause the condition known as "heartburn."

Most peptic ulcers occur within 1 to 2 inches on either side of the pylorus in the stomach or in the duodenum, and pain from such ulcers is usually referred to a surface point approximately midway between the umbilicus and the xyphoid process. The origin of ulcer pain is almost undoubtedly chemical, because when the acid juices of the stomach are not allowed to reach the pain fibers in the ulcer crater the pain does not exist. This pain is characteristically intensely burning.

Biliary and Gallbladder Pain. Pain from the bile ducts and gallbladder is localized in the midepigastrium almost coincident with pains caused by peptic ulcers. Also, biliary and gallbladder pain is often burning, like that from ulcers, though cramps often occur too.

Biliary disease, in addition to causing pain on the abdominal surface, frequently refers pain to a small area at the tip of the right scapula. This pain is transmitted through sympathetic afferent fibers that enter the ninth thoracic segment of the spinal cord.

Pancreatic Pain. Lesions of the pancreas, such as in acute or chronic pancreatitis in which the pancreatic enzymes eat away the pancreas and surrounding structures, promote intense pain in areas both anterior to and behind the pancreas. It should be remembered that the pancreas is located beneath the parietal peritoneum and that it receives many sensory fibers from the posterior abdominal wall. Therefore, the pain is frequently localized directly behind the pancreas in the back and is severe and burning in character.

Renal Pain. The kidney, kidney pelvis, and ureters are all retroperitoneal structures and receive most of their pain fibers directly from skeletal nerves. Therefore, pain is usually felt directly behind the ailing structure. However, pain occasionally is referred via visceral afferents to the anterior abdominal wall below and about 2 inches to the side of the umbilicus.

Pain from the bladder is felt directly over the bladder, presumably because the bladder is well innervated by parietal pain fibers. However, pain is also frequently referred into the groin and testicles because some nerve fibers from the bladder apparently synapse in the cord in association with fibers from the genital areas.

Uterine Pain. Both parietal and visceral afferent pain may be transmitted from the uterus. The low abdominal cramping pains of dysmenorrhea are mediated through the sympathetic afferents, and an operation to cut the hypogastric nerves between the hypogastric plexus and the uterus will in almost all instances relieve this pain. On the other hand, lesions of the uterus that spread into the adnexa around the uterus, or lesions of the fallopian tubes and broad ligaments, usually cause pain in the lower back or side. This pain is conducted over parietal nerve fibers and is usually quite aching or burning in nature rather than resembling the cramping pain of true dysmenorrhea.

SOME CLINICAL ABNORMALITIES OF PAIN AND OTHER SENSATIONS

PARESTHESIA

Sometimes a sensory nerve tract becomes irritated by a lesion in a peripheral nerve or in the central nervous system. Spontaneous impulses are then transmitted from the irritated nerve fiber all the way to the brain, and a sensation is localized in a peripheral area of the body which has not been directly stimulated. This sensation is called a paresthesia. Obviously, the irritative lesion can cause any one of the different modalities of sensation, depending on which type of fiber is involved.

HYPERESTHESIA

Hyperesthesia is very similar to paresthesia except that there is no spontaneous sensation. Instead, the sensory fibers are merely facilitated so that they are excessively sensitive to sensory stimuli.

The basic causes of hyperesthesia are, first, excessive sensitivity of the receptors themselves, which is called *primary hyperesthesia,* or, second, facilitation of sensory transmission, which is called *secondary hyperesthesia.*

Hyperesthesia that involves only pain sensations is called *hyperalgesia.* An example of *primary hyperalgesia* is the extreme sensitivity of sunburned skin. An example of *secondary hyperalgesia* is extreme pain sensitivity caused by lesions in some areas of the hindbrain or thalamus.

HERPES ZOSTER

Occasionally a virus similar to (and possibly identical with) that of chickenpox infects a dorsal root ganglion, irritating and stimulating the dorsal root neurons and causing a disease known as herpes zoster. Impulses spread into the central nervous system and cause severe paresthetic pain and other types of paresthesias or hyperesthesias. Usually, only one dorsal root ganglion is infected at a time. Therefore, the paresthesias and hyperesthesias are localized in the entire half dermatome subserved by the infected dorsal root ganglion and are not present even to the slightest extent in neighboring dermatomes or on the opposite side of the body.

The impulses that radiate peripherally from the infected ganglion also cause several local physiologic effects at the sensory nerve endings in the half dermatome subserved by the dorsal root ganglion. First, vasodilatation may occur, and, second, vesicles can appear. The exact cause of these effects is not known, but impulses spreading peripherally along a sensory nerve fiber probably liberate a humoral substance at the endings that in turn causes the vasodilatation and exudation. Among the possible humoral substances that have been suggested are adenosine phosphate compounds, histamine, and acetylcholine.

THE BROWN-SÉQUARD SYNDROME

Obviously, if the spinal cord is transected entirely, all sensations and motor functions distal to the segments of transection are blocked, but if only one side of the spinal cord is transected, the so-called Brown-Séquard syndrome occurs. The following effects of such a transection occur, and these can be predicted from a knowledge of the cord fiber tracts illustrated in Figure 50–9: All motor functions are blocked on the side of the transection in all segments below the level of the transection; while only some of the modalities of sensation are lost on the transected side and others are lost on the opposite side. The sensations of pain, heat, and cold are lost on the opposite side of the body in all dermatomes two to six segments below the level of the transection. The sensations that are transmitted only in the dorsal columns—kinesthetic sensation, vibration sensation, discrete localization, and two-point discrimination—are lost entirely on the side of the transection in all dermatomes below the level of the transection. Touch is impaired on the side of the transection because the principal pathway for transmission of light touch, the dorsal columns, is transected. Yet, "crude touch," which is poorly localized, still persists because of transmission in the opposite ventral spinothalamic tract.

THE THALAMIC SYNDROME

Occasionally the posterolateral branch of the posterior cerebral artery, a small artery supplying the posteroventral portion of the thalamus, becomes blocked by thrombosis, so that the nuclei of this area of the thalamus degenerate, while the medial and anterior nuclei of the thalamus remain intact. The patient suffers a series of abnormalities, as follows: First, loss of almost all sensations from the opposite side of the body occurs because of destruction of the relay nuclei. Second, ataxia (inability to control movements precisely) may be evident because of loss of kinesthetic impulses normally relayed through the thalamus to the cortex. Third, after a few weeks to a few months some sensory perception in the opposite side of the body returns, but strong stimuli are necessary to elicit

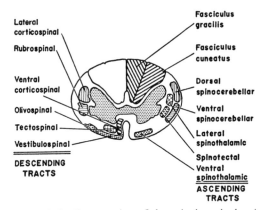

Figure 50–9 Cross-section of the spinal cord, showing principal ascending tracts on the right and principal descending tracts on the left.

this. When the sensations do occur, they are almost always very painful, sometimes lancinating, regardless of the type of stimulus applied to the body. Fourth, the person is likely to perceive many affective sensations of extreme unpleasantness or, rarely, extreme pleasantness; the unpleasant ones are often associated with emotional tirades.

It is especially difficult to explain the emotional and affective phenomena that occur in the thalamic syndrome, but the medial nuclei of the thalamus are not destroyed by thrombosis of the artery, and it is possible that these nuclei become facilitated and give rise to the affective sensations.

TIC DOULOUREUX

Lancinating pains occur in some persons over one side of the face in part of the sensory distribution area of the fifth or ninth nerves; this phenomenon is called tic douloureux. The pains feel like sudden electric shocks, and they may appear for only a few seconds at a time or they may be almost continuous. Often, they are set off by exceedingly sensitive "trigger areas" on the surface of the head, in the mouth, or in the nose. For instance, when the patient swallows a bolus of food, as the food touches a tonsil it might set off a severe lancinating pain in the mandibular portion of the fifth nerve.

The pain of tic douloureux can usually be blocked by cutting the peripheral nerve from the hypersensitive area. The fifth nerve is often sectioned immediately inside the cranium, where the motor and sensory roots of the fifth nerve can be separated so that the motor portions, which are needed for many of the jaw movements, are spared while the sensory elements are destroyed. Obviously, this operation leaves the side of the face anesthetic, which in itself may be annoying. Therefore, still another operation is often performed in the medulla for relief of this pain, as follows:

The sensory nuclei of the fifth nerve divide into the main sensory nucleus, which is located in the pons, and the spinal sensory nucleus, which descends into the upper part of the spinal cord. It is the spinal nucleus that subserves the function of pain. Therefore, the spinal tract of the fifth nerve can be cut in the medulla as it passes to the spinal nucleus. This blocks pain sensations from the side of the face but does not block the sensations of touch and pressure.

HEADACHE

Headaches are actually referred pain to the surface of the head from the deep structures, and most headaches probably result from pain stimuli arising inside the cranium, but at least some are known to result from pain arising outside the cranium.

HEADACHE OF INTRACRANIAL ORIGIN

Pain-Sensitive Areas in the Cranial Vault. The brain itself is almost totally insensitive to pain. Even cutting or electrically stimulating the somesthetic centers of the cortex only occasionally causes pain; instead, it causes tactile paresthesias on the area of the body represented by the portion of the somesthetic cortex stimulated. Therefore, it is obvious from the outset that much or most of the pain of headache probably is not caused by damage within the brain itself.

On the other hand, *tugging on the venous sinuses, damaging the tentorium,* or *stretching the dura at the base of the brain* can all cause intense pain that is recognized as headache. Also, almost any type of traumatizing, crushing, or stretching stimulus to the *blood vessels of the dura* can cause headache. One of the most sensitive structures of the entire cranial vault is the middle meningeal artery, and neurosurgeons are careful to remain clear of this artery as much as possible when performing brain operations under local anesthesia.

Areas of the Head to which Intracranial Headache Is Referred. Stimulation of pain receptors in the intracranial vault above the tentorium, including the upper surface of the tentorium itself, initiates impulses in the fifth nerve and, therefore, causes referred headache on the outside of the head in the area supplied by the fifth cranial nerve. This area includes the upper part of the head anterior to the ear, as outlined by the dark shaded area of Figure 50–10. Thus, pain arising above the tentorium causes what is called "frontal headache."

On the other hand, pain impulses from beneath the tentorium enter the central nervous system mainly through the second cervical nerve, which also supplies the scalp behind the ear. Therefore, subtentorial pain stimuli cause "occipital headache" referred to the posterior part of the head shown in Figure 50–10.

Types of Intracranial Headache. HEADACHE OF MENINGITIS. One of the most severe headaches

Supratentorial headaches

Subtentorial headaches

Nasal sinus and eye headaches

Figure 50–10 Areas of headache resulting from different causes.

of all is that resulting from meningitis, which causes inflammation of all the meninges, including the sensitive areas of the dura and the sensitive areas around the venous sinuses. Such intense damage as this can cause extreme headache pain referred over the entire head.

HEADACHE RESULTING FROM DIRECT MENINGEAL TRAUMA.   Following a brain operation one ordinarily has intense headache for several days to several weeks. Though part of this headache may result from the trauma of the brain itself, experiments indicate that most of it results from meningeal irritation.

Another type of meningeal trauma that almost invariably causes headache is the meningeal irritation resulting from brain tumor. Usually, tumor headache is referred to a localized area of the head, the exact area depending on the portion of the meninges affected by the tumor. Since any tumor above the tentorium refers its pain to the frontal areas and any tumor below the tentorium refers its pain to the occipital region of the skull, the general location of an intracranial tumor can often be predicted from the area of the headache.

HEADACHE CAUSED BY LOW CEREBROSPINAL FLUID PRESSURE.   Removing as little as 20 ml. of fluid from the spinal canal, particularly if the person remains in the upright position, often causes intense intracranial headache. Removing this quantity of fluid causes the pressure in the cranial vault to fall to a negative value of about −200 mm. water. This presumably stretches the various dural surfaces and thereby elicits the pain which causes the headache. Also, it is possible that the decreased fluid in the cranial vault allows the brain to rub excessively against the vault, thereby causing undue stretch or irritation of the meninges.

MIGRAINE HEADACHE.   Migraine headache is a special type of headache that is thought to result from abnormal vascular phenomena, though the exact mechanism is unknown.

Migraine headaches often begin with various prodromal sensations, such as nausea, loss of vision in part of the fields of vision, visual aura, or other types of sensory hallucinations. Ordinarily, the prodromal symptoms begin half an hour to an hour prior to the beginning of the headache itself. Therefore, any theory that explains migraine headache must also explain these prodromal symptoms.

One of the theories of the cause of migraine headaches is that prolonged emotion or tension causes reflex vasospasm of some of the arteries of the head, including arteries that supply the brain itself. The vasospasm theoretically produces ischemia of portions of the brain, and this is responsible for the prodromal symptoms. Then, as a result of the intense ischemia, something happens to the vascular wall to allow it to become flaccid and incapable of maintaining vascular tone for 24 to 48 hours. The blood pressure in the vessels causes them to dilate and pulsate intensely, and

it is supposedly the excessive stretching of the walls of the arteries that causes the actual pain of migraine headaches. However, it is possible that diffuse after-effects of ischemia in the brain itself are responsible for this type of headache.

ALCOHOLIC HEADACHE.   As many people have experienced, a headache usually follows an alcoholic binge. It is most likely that alcohol, because it is toxic to tissues, directly irritates the meninges and causes the cerebral pain.

HEADACHE CAUSED BY CONSTIPATION.   Constipation causes headache in many persons. Because it has been shown that constipation headache can occur in persons whose spinal cords have been cut, we know that this headache is not caused by nervous impulses from the colon. Therefore, it possibly results from absorbed toxic products or from changes in the circulatory system. Indeed, constipation sometimes does cause temporary loss of plasma into the wall of the gut, and a resulting poor flow of blood to the head could be the cause of the headache.

EXTRACRANIAL TYPES OF HEADACHE

Headache Resulting from Muscular Spasm.   Emotional tension often causes many of the muscles of the head, including especially those muscles attached to the scalp and the neck muscles attached to the occiput, to become moderately spastic, and it is postulated that this causes headache. The pain of the spastic head muscles supposedly is referred to the overlying areas of the head and gives one the same type of headache as do intracranial lesions.

Headache Caused by Irritation of the Nasal and Accessory Nasal Structures.   The mucous membranes of the nose and also of all the nasal sinuses are sensitive to pain, but not intensely so. Nevertheless, infection or other irritative processes in widespread areas of the nasal structures usually cause headache that is referred behind the eye or, in the case of frontal sinus infection, to the frontal surfaces of the forehead and scalp, as illustrated in Figure 50–10. Also, pain from the lower sinuses — such as the maxillary sinuses — can be felt in the face.

Headache Caused by Eye Disorders.   Difficulty in focusing one's eyes clearly may cause excessive contraction of the ciliary muscles in an attempt to gain clear vision. Even though these muscles are extremely small, tonic contraction of them can be the cause of retroorbital headache. Also, excessive attempts to focus the eyes can result in reflex spasm in the various facial and extraocular muscles. Tonic spasm in these muscles is also a possible cause of headache.

A second type of headache originating in the eyes occurs when the eyes are exposed to excessive irradiation by ultraviolet light rays. Watching the sun or the arc of an arc-welder for even a few seconds may result in headache that lasts from 24 to 48 hours. The headache sometimes results from "actinic" irritation of the conjunctivae, and the pain is referred to the surface of the head or retro-orbitally. However, focusing intense light from an arc or the sun on the retina can actually burn the retina, and this could result in headache.

THERMAL SENSATIONS

THERMAL RECEPTORS AND THEIR EXCITATION

The human being can perceive different gradations of cold and heat, progressing from *cold* to *cool* to *indifferent* to *warm* to *hot;* some persons perceive even freezing cold and burning hot.

The different thermal gradations in the body are discriminated by at least three different types of sensory end-organs, the *cold receptors,* the *warm receptors,* and the *pain receptors.* The pain receptors are stimulated only by extreme degrees of heat or cold and therefore will not be considered as thermal receptors but merely as pain receptors that become stimulated when either heat or cold becomes sufficiently severe to cause damage.

Both heat and cold can be perceived from areas of the body that contain only free nerve endings, such as the cornea. Therefore, it is known that free nerve endings can subserve these sensory functions. Thus far, no complex specialized receptors for either cold or warmth have been discovered.

Stimulation of Thermal Receptors—Sensations of Cold, Warmth, and Hot. Figure 50–11 illustrates the responses to different temperatures of three different types of nerve fibers: (1) a cold fiber, (2) a warm fiber, and (3) a pain fiber. Note especially that these fibers respond differently at different levels of temperature. For instance, in the *very* cold region only the pain fibers are stimulated (if the skin becomes even colder so that it nearly freezes or actually does freeze, even these fibers cannot be stimulated). As the temperature rises to 10 to 15° C., pain impulses cease, but the cold receptors begin to be stimulated. Then, above about 25° C. the warm end-organs become stimulated while the cold end-organs fade out at about 35° C. Finally, at around 45° C. the warm endings become nonresponsive once more; but the cold endings begin to respond again, and the pain endings also begin to be stimulated.

One can understand from Figure 50–11, therefore, that a person determines the different gradations of thermal sensations by the relative degrees of stimulation of the different types of endings. For instance, at 20° C. only cold endings are stimulated, whereas at 40° C. only warm endings are stimulated; at 33° C. both cold and warm endings are stimulated, and at 50° C. both cold and pain endings are stimulated. One can understand also from this figure why extreme degrees of cold or heat can be painful and why both these sensations, when intense enough, may give almost exactly the same quality of sensations—freezing cold and burning hot sensations feel almost alike to the person.

It is particularly interesting that a few areas of the body, such as the tip of the penis, do not contain any warmth receptors; but these areas can experience the sensations of cold with ease and of "hot," which depends on stimulating cold and pain endings, but never the sensation of warmth.

Stimulatory Effects of Rising and Falling Temperature—Adaptation of Thermal Receptors. When a cold receptor is suddenly subjected to an abrupt fall in temperature, it becomes strongly stimulated at first, but this stimulation fades rapidly during the first minute and progressively more slowly during the next half hour or more. In other words, the receptor adapts to a great extent; this is illustrated in Figure 50–12, which shows that the frequency of discharge of a cold receptor rose approximately four-fold when the temperature fell suddenly from 32° to 30° C., but in less than a minute the frequency fell about five sixths of the way back to the original control value. Later, the temperature was suddenly raised from 30° to 32° C. At this point the cold receptor stopped firing entirely for a short time, but after adaptation returned to its original control level. The same type of response occurs in warm receptors.

Thus, it is evident that thermal receptors

Figure 50–11 Frequencies of discharge of a cold receptor, a warm receptor, and a pain nerve fiber at different temperatures. (The responses of the cold and warm receptors are modified from Zotterman: *Ann Rev Physiol.,* *15:*357, 1953.)

Figure 50–12 Response of a nerve fiber from a cold receptor following, first, instantaneous change in skin temperature from 32° to 30° C. and, second, instantaneous change back to 32° C. Note the adaptation of the receptor and also the higher steady state level of discharge at 30° than at 32°.

respond markedly to *changes in temperature* in addition to being able to respond to steady states of temperature, as was depicted in Figure 50–11. This means, therefore, that when the temperature of the skin is actively falling, a person feels much colder than when the temperature remains at the same level. Conversely, if the temperature is actively rising the person feels much warmer than he would at the same temperature if it were constant.

The response to changes in temperature explains the extreme degree of heat that one feels when he first enters a hot tub of water and the extreme degree of cold he feels when he goes from a heated room to the out-of-doors on a cold day. The adaptation of the thermal receptors explains the ability of the person to become gradually accustomed to his new temperature environment.

However, since the thermal receptors do not adapt all the way to extinction, they still remain stimulated to some extent at the different steady states of temperature. Therefore, even after prolonged exposure to different temperatures, the person can still perceive cold and warmth even though the perceived intensity is not nearly so great as it is on first exposure.

Mechanism of Stimulation of the Thermal Receptors. Thermal receptors are directly responsive to (1) the temperature of the receptor itself and (2) changes in temperature. On the other hand, they do not respond, as has been claimed in the past, to temperature gradients either between different depths or different areas of the skin.

It is believed that the thermal receptors are stimulated by changes in their metabolic rates, these changes resulting from the fact that temperature alters the rates of intracellular chemical reactions about 2.3 times for each

10° C. change. In other words, thermal detection probably results not from direct physical stimulation but instead from chemical stimulation of the endings as modified by the temperature.

Spatial Summation of Thermal Sensations. The number of cold or warmth endings in any small surface area of the body is very slight, so that it is difficult to judge gradations of temperature when small areas are stimulated. However, when a large area of the body is stimulated, the thermal signals from the entire area summate. Indeed, one reaches his maximum ability to discern minute temperature variations when his entire body is subjected to a temperature change all at once. For instance, rapid changes in temperature as little as 0.01° C. can be detected if this change affects the entire surface of the body simultaneously. On the other hand, temperature changes 100 times this great might not be detected when the skin surface affected is only a square centimeter or so in size.

TRANSMISSION OF THERMAL SIGNALS IN THE NERVOUS SYSTEM

Thermal signals are transmitted by the smallest myelinated fibers of the peripheral nerves, the delta group with diameters of only 2 to 3 microns. It is also possible that some very crude thermal sensations, with very poor differentiation of temperature gradations, are transmitted by type C fibers.

On entering the central nervous system, the thermal fibers travel in the lateral spinothalamic tract. The exact point in the thalamus at which the impulses synapse is not known, but since other spinothalamic fibers terminate mainly in the posterior group nuclei of the thalamus, it is believed that the thermal tracts do too. Then, third order neurons travel to the somesthetic cortex, but the portion of the cerebral cortex to which the major share of thermal fibers radiate is also unknown.

Occasionally, a neuron in somatic sensory area I has been found by microelectrode studies to be directly responsive to either cold or warm stimuli in specific areas of the skin. Furthermore, it is known that removal of the postcentral gyrus in the human being reduces his ability to distinguish different gradations of temperature. Therefore, at present it is believed that somatic sensory area I, rather than area II, is the major somesthetic area concerned with thermal sensations.

REFERENCES

Albe-Fessard, D., and Bowsher, D.: Central pathways for painful messages. *Proc. Int. Union of Physiol. Sciences, 6*:241, 1968.

Bessou, P., and Perl, E. R.: Response of cutaneous sensory units with unmyelinated fibers to noxious stimuli. *J. Neurophysiol., 32*:1025, 1969.

Ciba Foundation Symposium: Touch, Heat and Pain. Boston, Little, Brown and Company, 1965.

Guyton, A. C., and Reeder, R. C.: Pain and contracture in poliomyelitis. *Arch. Neurol. & Psychiat., 63*:954, 1950.

Hardy, J. D.: The nature of pain. *J. Chronic Dis., 4*:22, 1956.

Hardy, J. D.; Wolff, H. C.; and Goodell, H.: Pain Sensations and Reactions. Baltimore, The Williams & Wilkins Co., 1952.

Henry Ford Hospital Symposium: Pain. Boston, Little, Brown and Company, 1966.

Lim, R. K. S.: Pain. *Ann. Rev. Physiol., 32*:269, 1970.

Perl, E. R.: Relation of cutaneous receptors to pain. *Proc. Int. Union of Physiol. Sciences, 6*:235, 1968.

Poulos, D. A., and Benjamin, R. M.: Response of thalamic neurons to thermal stimulation of the tongue. *J. Neurophysiol., 31*:28, 1968.

Segundo, J. P.; Takenaka, T.; and Encabo, H.: Somatic sensory properties of bulbar reticular neurons. *J. Neurophysiol., 30*: 1221, 1967.

Sweet, W. H.: Pain. *In* Handbook of Physiology. Baltimore, The Williams & Wilkins Co., 1959, Sec. I, Vol. I, p. 459.

Way, E. L.: New Concepts in Pain and Its Clinical Management. Philadelphia, F. A. Davis Co., 1967.

Wolff, H. G.: Headache and Other Head Pain. 2nd ed., New York, Oxford University Press, 1963.

Yokota, T., and MacLean, P. D.: Fornix and fifth-nerve interaction on thalamic units in awake, sitting squirrel monkeys. *J. Neurophysiol., 31*:358, 1968.

Zotterman, Y.: Thermal sensations. *In* Handbook of Physiology. Baltimore, The Williams & Wilkins Co., 1959. Sec. I, Vol. I, p. 431.

CHAPTER 51

The Sense of Hearing

In this chapter we begin discussion of the *special senses*, hearing, sight, taste, smell, and equilibrium. Hearing, like many somatic senses, is a mechanoreceptive sense, for the ear responds to mechanical vibration of the sound waves in the air. The purpose of the present chapter is to describe and explain the mechanism by which the ear receives sound waves, discriminates their frequencies, and finally transmits auditory information into the central nervous system.

THE TYMPANIC MEMBRANE AND THE OSSICULAR SYSTEM

TRANSMISSION OF SOUND FROM THE TYMPANIC MEMBRANE TO THE COCHLEA

Figure 51–1 illustrates the *tympanic membrane*, commonly called the *eardrum*, and the *ossicular system*, which transmits sound through the middle ear. The tympanic membrane is cone shaped, with its concavity facing downward toward the auditory canal. Attached to the very center of the tympanic membrane is the *handle* of the *malleus*. The malleus is tightly bound by ligaments at its other end with the *incus* so that whenever the malleus moves, the incus generally moves in unison with it. The opposite end of the incus in turn articulates with the stem of the *stapes*, and the *faceplate* of

Figure 51–1 The tympanic membrane, the ossicular system of the middle ear, and the inner ear.

the stapes lies in the opening of the oval window where sound waves are transmitted into the inner ear, the *cochlea.*

The ossicles of the middle ear are suspended by ligaments in such a way that the combined malleus and incus act as a single lever having its fulcrum approximately at the border of the tympanic membrane. The large *head* of the malleus, which is on the opposite side of the fulcrum from the handle, almost exactly balances the other end of the lever so that changes in position of the body will not increase nor decrease the tension on the tympanic membrane.

The articulation of the incus with the stapes causes the stapes to rock backward every time the handle of the malleus moves inward and to rock forward every time the malleus moves outward, which promotes inward and outward motion of the faceplate at the oval window.

The handle of the malleus is constantly pulled inward by ligaments and by the tensor tympani muscle, which keeps the tympanic membrane tensed. This allows sound vibrations on *any* portion of the tympanic membrane to be transmitted to the malleus, which would not be true if the membrane were lax.

Impedance Matching by the Ossicular System. The amplitude of movement of the stapes faceplate with each sound vibration is only three-fourths as much as the amplitude of the handle of the malleus. Therefore, the ossicular lever system does not amplify the movement distance of the stapes, as is commonly believed, but instead the system increases the *force* of movement about 1.3 times. However, the surface area of the tympanic membrane is approximately 55 sq. mm., whereas the surface area of the stapes averages 3.2 sq. mm. This 17-fold difference times the 1.3-fold ratio of the lever system, allows all the energy of a sound wave impinging on the tympanic membrane to be applied to the small faceplate of the stapes, causing approximately 22 times as much pressure on the fluid of the cochlea as is exerted by the sound wave against the tympanic membrane. Since fluid has far greater inertia than air, it is easily understood that increased amounts of pressure are needed to cause vibration in the fluid. Therefore, the tympanic membrane and ossicular system provide *impedance matching* between the sound waves in air and the sound vibrations in the fluid of the cochlea. Indeed, the impedance matching is probably 50 to 75 per cent of perfect for sound frequencies between 300 to 3000 cycles per second, which allows almost full utilization of the energy in the incoming sound waves. In the absence of this impedance matching, it is calculated that 97 per cent of the sound energy would be reflected from the tympanic membrane back into the air and, therefore, would be lost, thus decreasing the sensitivity of hearing about 13 decibels (a decrease from a loud voice level to a very soft voice level).

Transmission Characteristics of the Ossicular System. Every vibrating system that has inertia and that has an elastic component also has a natural frequency at which it can vibrate back and forth most easily. This is called its *resonant frequency.* Since the ossicular system does have inertia and since it is suspended by elastic ligaments, it has a broadly tuned natural resonating frequency between 700 and 1400 cycles per second. However, the ligaments and other structures attached to the ossicles are also viscous, which prevents excessive resonance; this is called *damping of the vibrations.* Because of the slight amount of resonance that does occur, sound waves of approximately 1200 cycles per second can be transmitted through the ossicular system with greater ease than can sound waves of other frequencies.

The external auditory canal, because of its dimensions, acts as an air column resonator and has a natural resonating frequency of about 3000 cycles per second. But here again, the degree of resonance is slight and potentiates the 3000 cycle sound only a minute amount.

Combining the resonant effects of the ossicular system and of the auditory canal, transmission of sound from the air to the cochlea is excellent between the limits of 600 and 6000 cycles per second and fades both above and below these limits.

Attenuation of Sound by Contraction of the Stapedius and Tensor Tympani Muscles. When loud sounds are transmitted through the ossicular system into the central nervous system, a reflex occurs after a latent period of only 10 milliseconds to cause contraction of both the stapedius and tensor tympani muscles. The tensor tympani muscle pulls the handle of the malleus inward while the stapedius muscle pulls the stapes outward. These two forces oppose each other and thereby cause the entire ossicular system to develop a high degree of rigidity, thus greatly reducing the transmission of low frequency sound, frequencies below 1000 cycles per second, to the cochlea.

This *attenuation reflex* can reduce the intensity of sound transmission by as much as 30 to 40 decibels, which is about the same difference as that between a whisper and the sound emitted by a loud radio speaker. The function of this mechanism is partly to allow adaptation of the ear to sounds of different intensities, but even more, to protect the cochlea from damaging vibrations caused by excessively loud sounds. It is mainly low frequency sounds, the ones that are attenuated, that frequently are loud enough to damage the basilar membrane of the cochlea. Also, when a person talks, collateral signals are sent to his muscles to decrease his own hearing sensitivity so that his speech will not overstimulate the hearing mechanism.

TRANSMISSION OF SOUND THROUGH THE BONE

Because the inner ear, the *cochlea,* is embedded in a bony cavity in the temporal bone, vibrations of the entire skull can cause fluid vibrations in the cochlea itself. Therefore, under appropriate conditions, a tuning fork or an electronic vibrator placed on any bony protuberance of the skull causes the person to hear the sound if it is intense enough.

THE COCHLEA

FUNCTIONAL ANATOMY OF THE COCHLEA

The cochlea is a system of coiled tubes, shown in Figure 51–2, with three different tubes coiled side by side, the *scala vestibuli,* the *scala media,* and the *scala tympani.* The scala vestibuli and scala media are separated from each other by the *vestibular membrane,* and the scala tympani and scala media are separated from each other by the *basilar membrane.* On the surface of the basilar membrane lies a structure, the *organ of Corti,* which contains a series of mechanically sensitive cells, the *hair cells.* These are the receptive end-organs that generate nerve impulses in response to sound vibrations.

Figure 51–3 illustrates schematically the functional parts of the uncoiled cochlea for transmission of sound vibrations. First, note that the vestibular membrane is missing from this figure. This membrane is so thin and so easily moved that it does not obstruct the pas-

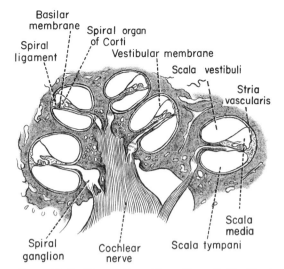

Figure 51–2   The cochlea. (From Goss, C. M.: Gray's Anatomy of the Human Body. Lea & Febiger.)

sage of sound vibrations from the scala vestibuli into the scala media at all. Therefore, so far as the transmission of sound is concerned, the scala vestibuli and scala media are considered to be a single chamber. The importance of the vestibular membrane is to maintain a special fluid in the scala media that is required for normal function of the sound receptive hair cells, as discussed later in the chapter.

Sound vibrations enter the scala vestibuli from the faceplate of the stapes at the oval window. The faceplate covers this window and is connected with the window's edges by a relatively loose annular ligament so that it can move inward and outward with the sound vibrations. Inward movement causes the fluid to move into the scala vestibuli and scala media, which immediately increases the pressure in the entire cochlea and causes the round window to bulge outward.

Note from Figure 51–3 that the distal end of the scala vestibuli and scala tympani are continuous with each other by way of the *helicotrema*. If the stapes moves inward *very slowly*,

Figure 51–3   Movement of fluid in the cochlea following forward thrust of the stapes.

fluid from the scala vestibuli is pushed through the helicotrema into the scala tympani, and this causes the round window to bulge outward. However, if the stapes vibrates inward and outward rapidly, the fluid simply does not have time to pass all the way to the helicotrema, then to the round window, and back again to the oval window between each two successive vibrations. Instead, the fluid wave takes a short-cut through the basilar membrane, causing it to bulge back and forth with each sound vibration. We shall see later that each frequency of sound causes a different "pattern" of vibration in the basilar membrane and that this is one of the important means by which the sound frequencies are discriminated from each other.

The Basilar Membrane and Its Resonating Qualities.   The basilar membrane contains about 20,000 or more *basilar fibers* that project from the bony center of the cochlea, the *modiolus,* toward the outer wall. These fibers are stiff elastic hairlike structures that are not fixed at their distal ends except that they are embedded in the membrane. Because they are stiff and because they are free at one end, they can vibrate like reeds of a harmonica.

The lengths of the basilar fibers increase progressively from the base of the cochlea to the helicotrema, from approximately 0.04 mm. at the base to 0.5 mm. at the helicotrema, a 12-fold increase in length. This increase in length helps the shorter fibers near the base of the cochlea to vibrate at a high frequency, while those near the helicotrema vibrate more easily at a low frequency.

In addition to the differences in lengths of the basilar fibers, they are also differently "loaded" by the fluid of the cochlea. That is, when a fiber vibrates back and forth, all the fluid between the vibrating fiber and the oval and round windows must also vibrate back and forth at the same time. For a fiber vibrating near the base of the cochlea, the total mass of the fluid is slight in comparison with that for a fiber vibrating near the helicotrema. This difference, too, favors high frequency vibration near the windows and low frequency vibration near the tip of the cochlea.

Thus, high frequency resonance of the basilar fibers occurs near the base and low frequency resonance occurs near the apex because of (1) difference in lengths of the fibers and (2) difference in "loading." The product of these two factors multiplied together is called the *volume elasticity* of the fibers. The volume elasticity of the fibers near the helicotrema is at least 100 times as great as that of the fibers near the stapes, which corresponds to a difference in resonating frequencies between the two extremes of the cochlea of about 7 octaves.

TRANSMISSION OF SOUND WAVES IN THE COCHLEA—THE "TRAVELING WAVE"

If the foot of the stapes moves inward instantaneously, the round window must also bulge outward instantaneously. Since the fluid wave will not have time to move all the way from the oval window to the helicotrema and back to the round window, the initial effect is to cause the basilar membrane at the very base of the cochlea to bulge in the direction of the round window. However, the elastic tension that is built up in the basilar fibers as they bend toward the round window initiates a wave that "travels" downward along the basilar membrane, as illustrated in Figure 51–4. Figure 51–4A shows movement of a high frequency wave down the basilar membrane, Figure 51–4B a medium frequency wave, and Figure 51–4C a very low frequency wave. Movement of the wave along the basilar membrane is comparable to the movement of a pressure wave along the arterial walls, which was discussed in Chapter 19, or it is also comparable to the wave that travels along the surface of a pond.

Pattern of Vibration of the Basilar Membrane for Different Sound Frequencies. Note in Figure 51–4 the different patterns of transmission for sound waves of different frequencies. Each wave is relatively weak at the outset but be-

comes strong when it reaches that portion of the basilar membrane that has a natural resonant frequency equal to the respective sound frequency. At this point the basilar membrane can vibrate back and forth with such great ease that the energy in the wave is completely dissipated. Consequently, the wave ceases at this point and fails to travel the remaining distance along the basilar membrane. Thus, a high frequency sound wave travels only a short distance along the basilar membrane before it reaches its resonant point and dies out; a medium frequency sound wave travels about halfway and then dies out; and, finally, a very low frequency sound wave travels the entire distance along the membrane before it dies.

Another feature of the traveling wave is that it travels fast along the initial portion of the basilar membrane but progressively more slowly as it goes farther and farther into the cochlea. The cause of this is the high coefficient of elasticity of the basilar fibers near the stapes but a progressively decreasing coefficient farther along the membrane. This rapid initial transmission of the wave allows the high frequency sounds to travel far enough into the cochlea to spread out and separate from each other on the basilar membrane. Without this spread, all the high frequency waves would be bunched together within the first millimeter or so of the basilar membrane, and their frequencies could not be discriminated one from the other.

AMPLITUDE PATTERN OF VIBRATION OF THE BASILAR MEMBRANE. The dashed curves of Figure 51–5A show the position of a sound wave on the basilar membrane when the stapes (a) is all the way inward, (b) has moved back to the neutral point, (c) is all the way outward, and (d) has moved back again to the neutral point but is moving inward. The shaded area around these different waves shows the maximum extent of vibration of the basilar membrane during a complete vibratory cycle. This is the amplitude pattern of vibration of the basilar membrane for this particular sound frequency.

Figure 51–5B shows the amplitude patterns of vibration for different frequencies, showing that the maximum amplitude for 8000 cycles occurs near the base of the cochlea, while that for frequencies of 50 to 100 cycles per second occurs near the helicotrema.

Note in Figure 51–5B that the basal end of the basilar membrane vibrates at least weakly for all frequencies. However, beyond the resonant area for each given frequency, the

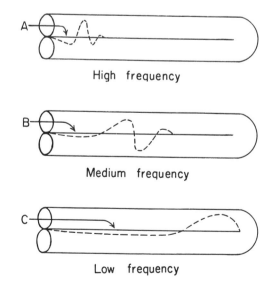

High frequency

Medium frequency

Low frequency

Figure 51–4 Diagrammatic representation of "traveling waves" along the basilar membrane for high, medium, and low frequency sounds.

Figure 51-5 (A) Amplitude pattern of vibration of the basilar membrane for a medium frequency sound. (B) Amplitude patterns for sounds of all frequencies between 50 and 8000 per second, showing the points of maximum amplitude (the resonance points) on the basilar membrane for the different frequencies.

vibration of the basilar membrane cuts off sharply. The principal method by which sound frequencies are discriminated from each other is based on the pattern of stimulation of the nerve fibers from the basilar membrane as follows.

FUNCTION OF THE ORGAN OF CORTI

The organ of Corti, illustrated in Figure 51-6, is the receptor organ that generates nerve impulses in response to vibration of the basilar membrane. Note that the organ of Corti lies on the surface of the basilar fibers and that the actual sensory receptors in the organ of Corti are two types of *hair cells,* a

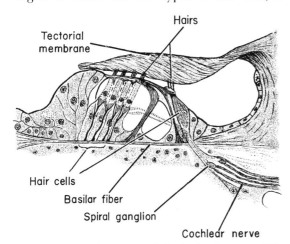

Figure 51-6 The organ of Corti, showing especially the hair cells and the tectorial membrane against the projecting hairs.

single row of *internal hair cells,* numbering about 3500 and measuring about 12 microns in diameter, and three to four rows of *external hair cells,* numbering about 20,000 and having diameters of only about 8 microns. The bases of the hair cells are enmeshed by a network of cochlear nerve endings. These lead to the *spiral ganglion of Corti,* which lies in the modiolus of the cochlea. The spiral ganglion in turn sends axons into the *cochlear nerve* and thence into the central nervous system at the level of the upper medulla. The relationship of the organ of Corti to the spiral ganglion is illustrated in Figure 51-2.

Excitation of the Hair Cells. Note in Figure 51-6 that minute hairs project upward from the hair cells and are embedded in the surface of the *tectorial membrane,* which lies above in the scala media. These hair cells are almost identical with the hair cells found in the macula and cristae ampullaris of the vestibular apparatus, which will be discussed in Chapter 57. Bending of the hairs excites the hair cells, and this in turn excites the nerve fibers enmeshing their bases.

Figure 51-7 illustrates the mechanism by which vibration of the basilar membrane excites the hair endings. This shows that the upper ends of the hair cells are fixed tightly in a structure called the *reticular lamina.* Furthermore, the reticular lamina is very rigid and is continuous with a rigid triangular structure called the *rods of Corti* that rests on each basilar fiber. Therefore, the basilar fiber, the rods of Corti, and the reticular lamina all move as a unit.

Upward movement of the basilar fiber rocks the reticular lamina upward and *inward.* Then, when the basilar membrane moves downward, the reticular lamina rocks downward and *outward.* The inward and outward motion causes the hairs to sheer back and forth in the tectorial membrane, thus exciting the cochlear nerve fibers whenever the basilar membrane vibrates.

Mechanism by which the Hair Cells Excite the Nerve Fibers—Receptor Potentials. Back-and-forth bending of the hairs causes alternate changes in the electrical potential across the surface of the hair cells. This alternating potential is the *receptor potential* of the hair cell; and in some way not yet understood, it stimulates the nerve filaments that surround the bases of the hair cells. Many physiologists believe that the receptor potential stimulates the filaments by direct electrical excitation, though it is possible that a chemical transmitter system is involved.

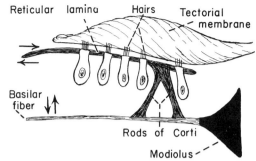

Reticular lamina   Hairs   Tectorial membrane

Basilar fiber

Rods of Corti

Modiolus

Figure 51-7 Stimulation of the hair cells by the to-and-fro movement of the hairs in the tectorial membrane.

When the basilar fiber bends toward the scala vestibuli (in the upward direction in Fig. 51-7), the hair cell membrane potential decreases; that is, the cell becomes depolarized, and it is this depolarization that excites action potentials in the nerve fiber. When the basilar fiber moves in the opposite direction, the hair cell becomes hyperpolarized, and the action potentials cease.

THE ENDOCOCHLEAR POTENTIAL. To explain even more fully the electrical potentials generated by the hair cells, we need to explain still another electrical phenomenon called the endocochlear potential: The scala media is filled with a fluid called *endolymph* in contradistinction to the *perilymph* present in the scala vestibuli and scala tympani. The scala vestibuli and scala tympani in most young children and in some adults communicate directly with the subarachnoid space around the brain, and the perilymph is almost identical with cerebrospinal fluid. On the other hand, the endolymph that fills the scala media is an entirely different fluid secreted by the *stria vascularis,* a highly vascular area on the outer wall of the scala media. Endolymph contains a very high concentration of potassium and a very low concentration of sodium, which is exactly opposite to the perilymph.

An electrical potential of approximately 80 mv. exists all the time between the endolymph and the perilymph, with positivity inside the scala media and negativity outside. This is called the *endocochlear potential,* and it is generated by continual secretion of positive potassium ions into the scala media by the stria vascularis.

The importance of the endocochlear potential is that the tops of the hair cells project through the reticular lamina into the endolymph of the scala media while perilymph bathes the lower bodies of the hair cells. Furthermore, the hair cells have a negative intracellular potential −70 mv. with respect to the perilymph, which means that the upper border of the hair cells has a total membrane potential between the endolymph and their intracellular fluids of 150 mv. (70 + 80). It is believed that this high electrical potential at the hair border of the cell greatly sensitizes the cell, thereby increasing its ability to respond to slight movement of the hairs.

DETERMINATION OF PITCH — THE "PLACE" PRINCIPLE

From earlier discussions in this chapter it is already apparent that low pitch (or low frequency) sounds cause maximal activation of the basilar membrane near the apex of the cochlea, sounds of high pitch (or high frequency) activate the basilar membrane near the base of the cochlea, and intermediate frequencies activate the membrane at intermediate distances between these two extremes. Furthermore, there is spatial organization of the cochlear nerve fibers from the cochlea to the cochlear nuclei in the brain stem, the fibers from each respective area of the basilar membrane terminating in a corresponding area in the cochlear nuclei. We shall see later that this spatial organization continues all the way up the brain stem to the cerebral cortex. The recording of signals from the auditory tracts in the brain stem and from the auditory receptive fields in the cerebral cortex shows that specific neurons are activated by specific pitches. Therefore, the primary method used by the nervous system to detect different pitches is to determine the position along the basilar membrane that is most stimulated. This is called the *place principle* for determination of pitch.

Yet, pitches probably can be discriminated much less accurately in other ways as well. For instance, destruction of the apical half of the cochlea destroys the basilar membrane where low frequency sounds are normally detected, but low frequency sounds can still be discriminated from high frequency sounds to a slight degree. Presumably, when the low frequency wave travels over the basal part of the basilar membrane, signals are transmitted into the neural system, and these signals are detected as low frequency sound. It is believed that the low frequency sounds cause synchronized low frequency volleys of impulses that can be distinguished from high frequency signals. This is pointed out not to decrease the importance of the place mechanism for pitch determination but to show that another, much cruder, backup mechanism probably exists in case portions of the cochlea are destroyed.

DETERMINATION OF LOUDNESS

Loudness is determined by the auditory system in at least three different ways: First, as the sound becomes louder, the amplitude of vibration of the basilar membrane and hair cells also increases so that the hair cells excite the nerve endings at faster rates. Second, as the amplitude of vibration increases, it causes more and more of the hair cells on the fringes of the vibrating portion of the basilar membrane to become stimulated, thus causing *spatial summation* of impulses—that is, transmission through many nerve fibers rather than through a few. Third, certain hair cells do not become stimulated until the vibration of the basilar membrane reaches a relatively high intensity, and it is believed that stimulation of these cells in some way apprises the nervous system that the sound is then very loud.

Detection of Changes in Loudness—The Power Law. It was pointed out in Chapter 48 that a person interprets changes in intensity of sensory stimuli approximately in proportion to a power function of the actual intensity. In the case of sound, the interpreted sensation changes approximately in proportion to the cube root of the actual sound intensity. To express this another way, the ear can discriminate changes in sound intensity from the softest whisper to the loudest possible noise of *approximately one trillion times* as much sound energy. Yet the ear interprets this much difference in sound level as approximately a 10,000-fold change. Thus, the scale of intensity is greatly "compressed" by the sound perception mechanisms of the auditory system. This obviously allows a person to interpret differences in sound intensities over an extremely wide range, a far broader range than would be possible were it not for compression of the scale.

The Decibel Unit. Because of the extreme changes in sound intensities that the ear can detect and discriminate, sound intensities are usually expressed in terms of the logarithm of their actual intensities. A 10-fold increase in intensity is called 1 *bel*, and one-tenth bel is called 1 *decibel*. One decibel represents an actual increase in intensity of 1.26 times.

Another reason for using the decibel system in expressing changes in loudness is that, in the usual sound range for communication, the ears can detect approximately a 1 decibel change in sound intensity.

Threshold of Hearing for Sounds of Different Frequencies. The intensity of sound is measured by the amount of energy that is dissipated when the sound is absorbed at a surface, and sound engineers arbitrarily consider *1 microwatt of sound energy per cm.² of surface area to have "unit" intensity, which is zero decibels when converted to the decibel scale* because the logarithm of unity is zero. If the intensity of a sound is +10 decibels intensity, the actual intensity is 10 microwatts/cm.²; +30 decibels is 1000 microwatts/cm.²; and, conversely, −30 decibels is 1/1000 microwatt/cm.².

Figure 51–8 Relationship of the threshold of hearing and the threshold of somesthetic perception to the sound energy level at each sound frequency. (Modified from Stevens and Davis: Hearing. John Wiley & Sons.)

Figure 51–8 shows the energy threshold at which sounds of different frequencies can be heard by the ear. This figure illustrates that a 2000 cycle per second sound can be heard even when its intensity is as low as −70 decibels, which is one ten-millionth microwatt/cm.². On the other hand, a 100 cycle per second sound can be detected only if its intensity is 10,000 times as great as this—that is, an intensity of −30 decibels.

FREQUENCY RANGE OF HEARING. The frequencies of sound that a young person can hear, before aging has occurred in the ears, is generally stated to be between 30 and 20,000 cycles per second. However, referring again to Figure 51–8, we see that the sound range depends to a great extent on intensity. If the intensity is only −60 decibels, the sound range is 500 to 5000 cycles per second, but, if the sound intensity is −20 decibels, the frequency range is about 70 to 15,000 cycles per second, and only with intense sounds can the complete range of 30 to 20,000 cycles be achieved. In old age, the frequency range falls to 50 to 8,000 cycles per second or less, as is discussed later in the chapter.

CENTRAL AUDITORY MECHANISMS

THE AUDITORY PATHWAY

Figure 51–9 illustrates the major auditory pathways. It shows that nerve fibers from the *spiral ganglion of the organ of Corti* enter the *dorsal* and *ventral cochlear nuclei* located in the upper part of the medulla. At this point, all the fibers synapse, and second order neurons pass mainly to the opposite side of the brain stem through the *trapezoid body* to end in the *superior olivary nucleus*. However, some of the second order fibers pass ipsilaterally to the superior olivary nucleus on the same side. Most of the fibers entering the superior olivary nucleus on either side terminate here, but some pass on through this nucleus. From the superior olivary nucleus the auditory pathway then

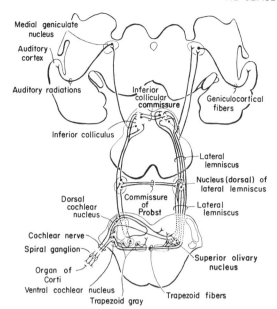

Figure 51-9   The auditory pathway. (Modified from Crosby, Humphrey, and Lauer: Correlative Anatomy of the Nervous System. The Macmillan Co.)

passes upward through the *lateral lemniscus,* and most of the fibers end in the *inferior colliculus,* but some terminate at a lower level in the *nucleus of the lateral lemniscus.* A few fibers cross from the nucleus of the lateral lemniscus through the *commissure of Probst* to the contralateral nucleus, and still other fibers cross through the *inferior collicular commissure* from one inferior colliculus to the other. From the inferior colliculus, the pathway then passes through the *peduncle of the inferior colliculus* to the *medial geniculate nucleus,* where all the fibers synapse. From here, the auditory tract spreads by way of the *auditory radiation* to the *auditory cortex* located mainly in the superior temporal gyrus.

Several points of importance in relation to the auditory pathway should be noted. First, impulses from either ear are transmitted through the auditory pathways of both sides of the brain stem with only slight preponderance of transmission in the contralateral pathway. In at least three different places in the brain stem crossing-over occurs between the two pathways: (a) in the trapezoid body, (b) in the commissure of Probst, and (c) in the commissure connecting the two inferior colliculi.

Second, many collateral fibers from the auditory tracts pass directly into the reticular activating system of the brain stem. This system projects diffusely upward into the cerebral cortex and downward into the spinal cord.

Third, the pathway for transmission of sound impulses from the cochlea to the cortex consists of at least four neurons and sometimes as many as six. Neurons *may* or *may not*

synapse in the superior olivary nuclei, in the nuclei of the lateral lemniscus, and in the inferior colliculi. Therefore, some of the tracts are more direct than others, which means that some impulses arrive at the cortex well ahead of others even though they might have originated at exactly the same time.

Fourth, several important pathways also exist from the auditory system into the cerebellum: (a) directly from the cochlear nuclei, (b) from the inferior colliculi, (c) from the reticular substance of the brain stem, and (d) from the cerebral auditory areas. These activate the *cerebellar vermis* instantaneously in the event of a sudden noise.

Firing Rates at Different Levels of the Auditory Tract.   Single nerve fibers entering the cochlear nuclei from the eighth nerve can fire at rates up to 1000 per second, the rate being determined mainly by the loudness of the sound. At low sound frequencies, the nerve impulses are usually synchronized with the sound waves.

In the auditory tracts of the brain stem, the firing is usually no longer synchronized with the sound frequency except at sound frequencies below 200 cycles per second. Furthermore, the firing rates are often considerably different from those in the eighth nerve. These findings demonstrate that the sound signals are not transmitted unchanged directly from the ear to the higher levels of the brain; instead, information from the sound signals begins to be dissected from the impulse traffic at levels as low as the cochlear nuclei. We will have more to say about this later, especially in relation to perception of direction from which sound comes.

Another significant feature of the auditory pathways is that low rates of impulse firing continue even in the absence of sound all the way from the cochlear nerve fibers to the auditory cortex. When the basilar membrane moves toward the scala vestibuli, the impulse traffic increases; and when the basilar membrane moves toward the scala tympani, the impulse traffic decreases. Thus, the presence of this background signal allows information to be transmitted from the basilar membrane when the membrane moves in either direction: positive information in one direction and negative information in the opposite direction. Were it not for the background signal, only the positive half of the information could be transmitted. This type of so-called "carrier wave" method for transmitting information is utilized in many parts of the brain, as will be discussed in several of the following chapters.

Integration of Auditory Information in the Relay Nuclei.   Very little is known about the function of the different nuclei in the auditory pathway. However, a lower animal can still hear and can even detect normal threshold auditory signals when the cerebral cortex is removed bilaterally, which indicates that the nuclei in the brain stem and thalamus can perform many auditory functions even without the cerebral cortex. In man, *bilateral* destruction of the cortical auditory centers gives a different picture: it causes almost total deafness. This indicates that in man the conscious elements of the lower centers are almost completely suppressed.

One of the important features of auditory transmission through the relay nuclei is the spatial orientation of the pathways for sounds of different frequencies. For instance, in the dorsal cochlear nucleus, high frequencies are represented along the medial edge while low frequencies are represented along the lateral edge, and a similar type of spatial orientation occurs throughout the auditory pathway as it travels upward to the cortex.

FUNCTION OF THE CEREBRAL CORTEX IN HEARING

The projection of the auditory pathway to the cerebral cortex is illustrated in Figure 51–10, which shows that the auditory cortex lies principally on the *supratemporal plane of the superior temporal gyrus* but also extends over the *lateral border of the temporal lobe,* over much of the *insular cortex,* and even into the most lateral portion of the *parietal operculum.*

Two separate areas are shown in Figure 51–10, a *short latency area* and a *long latency*

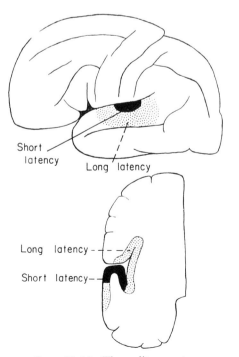

Figure 51–10   The auditory cortex.

area. The short latency area is directly excited by projections from the medial geniculate body, while the long latency area is usually excited secondarily by impulses from the short latency area and by projections from the medial geniculate body. The short latency area is frequently called the *primary auditory cortex* and the long latency area the *secondary auditory cortex.*

Locus of Sound Frequency Perception in the Auditory Cortex.   Certain parts of the auditory cortex are known to respond to high frequencies and other parts to low frequencies. In monkeys, the posterior part of the supratemporal plane responds to high frequencies, while the anterior part responds to low frequencies. Unfortunately, the frequency localization in the human cortex is yet unknown.

The frequency range to which each individual neuron in the auditory cortex responds is much narrower than that in the cochlear and brain stem relay nuclei. Referring back to Figure 51–5B, we note that the nerve fibers near the base of the cochlea can be stimulated by all frequency sounds, and in the cochlear nuclei this same breadth of sound representation is found. Yet by the time the excitation has reached the cerebral cortex, each neuron responds to only a narrow range of frequencies rather than to a broad range. Therefore, somewhere along the pathway integrative mechanisms in some way "sharpen" the frequency response. It is believed that this sharpening effect is caused mainly by the phenomenon of lateral inhibition, which was discussed in Chapter 47 in relation to mechanisms for transmitting information in nerves. That is, stimulation of the cochlea at one frequency causes inhibition of signals caused by sounds a few cycles per second on either side of the stimulated frequency, this effect resulting from collateral fibers angling off the primary signal pathway and exerting inhibitory influences on adjacent pathways. The same effect has also been demonstrated to be important in sharpening patterns of somesthetic images, visual images, and other types of sensations.

Some of the neurons in the auditory cortex do not respond at all to sounds in the ear. It is believed that these neurons "associate" different sound frequencies with each other or associate sound information with information from other sensory areas of the cortex. Indeed, the parietal portion of the auditory cortex partly overlaps somatic sensory area II, which could provide easy opportunity for association of auditory information with somatic sensory information.

Discrimination of Sound Patterns by the Auditory Cortex. Complete bilateral removal of the auditory cortex does not prevent an animal from detecting sounds or reacting in a crude manner to the sounds. However, it does greatly reduce or sometimes even abolish his ability to discriminate different *patterns of sound.* For instance, an animal that has been trained to recognize a combination or a sequence of tones, one following the other in a particular pattern, loses this ability when the cerebral cortex is destroyed, and, furthermore, he cannot relearn this type of response. Therefore, the cerebral cortex is important in the discrimination of *tonal patterns.*

In the human being, direct electrical stimulation in the short latency area of the auditory cortex causes the person to hear (a) a tingling sound, (b) a loud roar, or (c) some other single discrete type of sound; whereas stimulation lower down on the side of the temporal lobe occasionally causes the person to perceive intelligible sounds that even include whole sentences. Thus, it would be possible to divide the auditory cortex into a short latency *primary receptive area* and a long latency *interpretive area.*

DISCRIMINATION OF DIRECTION FROM WHICH SOUND EMANATES

A person determines the direction from which sound emanates by at least two different mechanisms: (1) by the time lag between the entry of sound into one ear and into the opposite ear and (2) by the difference between the intensities of the sounds in the two ears. The first mechanism functions best for frequencies below 3000 cycles per second, and the intensity mechanism operates best at higher frequencies because the head acts as a sound barrier at these frequencies. The time lag mechanism discriminates direction much more exactly than the intensity mechanism, for the time lag mechanism does not depend on extraneous factors but only on an exact interval of time between two acoustical signals. If a person is looking straight toward the sound, the sound reaches both ears at exactly the same instant, while, if the right ear is closer to the sound than the left ear, the sound impulses from the right ear are perceived ahead of those from the left ear.

If a sound emanates from the right side and 45 degrees from in front and if still a second sound emanates also from the right but 45 degrees from behind, the difference in time of arrival of the two sounds at the two ears will be exactly the same. For this reason, it is diffi-cult to distinguish whether the sound is originating from the frontal or the posterior quadrant. The principal method by which a person determines this is to rotate his head quickly. If he turns his head toward the right when the sound is coming from the frontal quadrant, the time lag between the two ears becomes smaller. If the sound is originating from the posterior quadrant, the time lag becomes greater. A sudden sound that occurs so rapidly that the person does not have time to move his head often cannot be properly localized to the frontal or posterior quadrant.

Neural Mechanisms for Detecting Sound Direction. Destruction of the auditory cortex on both sides of the brain, in either man or lower mammals, causes loss of almost all ability to detect the direction from which sound comes. Yet, the mechanism for this detection process begins in the superior olivary nuclei, even though it requires the neural pathways all the way from these nuclei to the cortex for interpretation of the signals. The mechanism is believed to be the following:

When sound enters one ear slightly before it enters the other ear, it excites the neurons in the contralateral superior olivary nucleus but, at the same time, *inhibits* the neurons in the ipsilateral superior olivary nucleus, this inhibition lasting for a fraction of a millisecond. Therefore, for a short period of time after the sound reaches the first ear, the pathway for sound coming from the second ear is in an inhibited state. Furthermore, certain portions of the superior olivary nuclei seem to have longer time lags of inhibition than do other portions. Therefore, when the sound hits the second ear and when its signal enters the inhibited superior olivary nucleus, the signal will pass up the auditory pathway through some of the neurons but not through others. Thus, a spatial pattern of neuronal stimulation develops, with the short lagging sounds stimulating one set of neurons maximally and the long lagging sounds stimulating another set of neurons maximally. This spatial orientation of signals is then transmitted all the way to the auditory cortex where sound direction is determined by the locus in the cortex that is stimulated maximally. It is believed that the signals for determining sound direction are transmitted through a slightly different pathway and that this pathway terminates in the cerebral cortex in a slightly different locus compared with the transmission pathway and the termination locus of the tonal patterns of sound.

This postulated mechanism for detection of sound direction indicates again how sensory signals are dissected as they pass through different levels of neuronal activity. In this case, the quality of sound direction is believed to be separated from the quality of sound tones at the level of the superior olivary nuclei.

CENTRIFUGAL CONDUCTION OF IMPULSES FROM THE CENTRAL NERVOUS SYSTEM

Retrograde pathways have been demonstrated at each level of the central nervous system all the way from the auditory cortex to the cochlea. The final pathway is from the olivary nucleus to the organ of Corti.

These retrograde fibers are mainly inhibitory. Indeed, direct stimulation of discrete points in the olivary nucleus have been shown to inhibit specific areas of the organ of Corti, reducing their sound sensitivities as much as 15 to 20 decibels. One can readily understand how this could allow a person to direct his attention to sounds of particular qualities while rejecting sounds of other qualities. This is readily demonstrated when one listens to a single instrument in a symphony orchestra.

HEARING ABNORMALITIES

TYPES OF DEAFNESS

Deafness is usually divided into two types: first, that caused by impairment of the cochlea or auditory nerve, which is usually classed under the heading "nerve deafness," and, second, that caused by impairment of the middle ear mechanisms for transmitting sound into the cochlea, which is usually called "conduction deafness." Obviously, if either the cochlea or the auditory nerve is completely destroyed the person is permanently deaf. However, if the cochlea and nerve are still intact but the ossicular system has been destroyed or ankylosed, sound waves can still be conducted into the cochlea by means of bone conduction.

Tuning Fork Test for Differentiation Between Nerve Deafness and Conduction Deafness. To test an ear for bone conduction with a tuning fork, one places a vibrating fork in front of the ear and the subject listens to the tone of the fork until he can no longer hear it. Then the butt of the still weakly vibrating fork is immediately placed against the mastoid process. If his bone conduction is better than his air conduction, he will again hear the sound of the tuning fork. If this occurs, his deafness may be considered to be conduction deafness. However, if on placing the butt of the fork against the mastoid process he cannot hear the sound of the tuning fork, his bone conduction is probably decreased as much as his air conduction, and any deafness that exists in the ear is presumably due to damage in the cochlea or in the nervous system rather than in the ossicular system—that is, it is nerve deafness.

Figure 51–11   Audiogram of the old age type of nerve deafness.

The Audiometer. To determine the nature of hearing disabilities more exactly than can be accomplished by the above method, the "audiometer" is used. This is simply an earphone connected to an electronic oscillator capable of emitting pure tones ranging from low frequencies to high frequencies. Based on previous studies of normal persons, the instrument is calibrated so that the zero intensity level of sound at each frequency is the loudness that can barely be heard by the normal person. However, the calibrated volume control can be changed to increase or decrease the loudness of each tone above or below the zero level. If the loudness of a tone must be increased to 30 decibels above normal before the subject can hear it, he is said to have a *hearing loss* of 30 decibels for that particular tone.

In performing a hearing test using an audiometer, one tests approximately 8 to 10 tones covering the auditory spectrum one at a time, and the hearing loss is determined for each of these tones. Then the so-called "audiogram" is plotted as shown in Figures 51–11 and 51–12, depicting the hearing loss for each of the tones in the auditory spectrum.

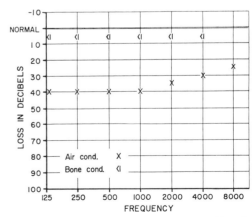

Figure 51–12   Audiogram of deafness resulting from middle ear sclerosis.

The audiometer, in addition to being equipped with an earphone for testing air conduction by the ear, is also equipped with an electronic vibrator for testing bone conduction from the mastoid process into the cochlea.

THE AUDIOGRAM IN NERVE DEAFNESS. If a person has nerve deafness—this term including damage to the cochlea, to the auditory nerve, or to the central nervous system circuits from the ear—he has lost the ability to hear sound as tested by both the air conduction apparatus and the bone conduction apparatus. An audiogram depicting nerve deafness is illustrated in Figure 51–11. In this figure the deafness is mainly for high frequency sound. Such deafness could be caused by damage to the base of the cochlea. This type of deafness occurs to some extent in almost all older persons.

THE AUDIOGRAM IN CONDUCTION DEAFNESS. A second and frequent type of deafness is that caused by fibrosis of the middle ear following infection in the middle ear. In this instance the sound waves cannot be transmitted easily to the oval window. Figure 51–12 illustrates an audiogram from a person with "middle ear deafness" of this type. In this case the bone conduction is essentially normal, but air conduction is greatly depressed at all frequencies, though more so at the low frequencies. In this type of deafness, the faceplate of the stapes frequently becomes "ankylosed" by bony overgrowth, to the edges of the oval window. In this case, the person becomes totally deaf for air conduction; but he can be made to hear again, with only 15 decibels hearing loss, by removing the stapes so that sound waves can strike the oval window directly.

REFERENCES

Aitkin, L. M., and Dunlop, C. W.: Interplay of excitation and inhibition in the cat medial geniculate body. *J. Neurophysiol.,* 31:44, 1968.
Brugge, J. F.; Anderson, D. J.; Hind, J. E.; and Rose, J. E.: Time structure of discharges in single auditory nerve fibers of the squirrel monkey in response to complex periodic sounds. *J. Neurophysiol.,* 32:386, 1969.
Brugge, J. F.; Dubrovsky, N. A.; Aitkin, L. M.; and Anderson, D. J.: Sensitivity of single neurons in auditory cortex of cat to binaural tonal stimulation: effects of varying interaural time and intensity. *J. Neurophysiol.,* 32:1005, 1969.
Carmel, P. W., and Starr, A.: Acoustic and nonacoustic factors modifying middle-ear muscle activity in waking cats. *J. Neurophysiol.,* 26:598, 1963.
Davis, H.: Biophysics and physiology of the inner ear. *Physiol. Rev.,* 37:1, 1957.
Dewson, J. H., III: Efferent olivocochlear bundle: some relationships to noise masking and to stimulus attenuation. *J. Neurophysiol.,* 30:817, 1967.
Dewson, J. H., III: Efferent olivocochlear bundle: some relationships to stimulus discrimination in noise. *J. Neurophysiol.,* 31:122, 1968.
Geisler, C. D.; Rhode, W. S.; and Hazelton, D. W.: Responses of inferior colliculus neurons in the cat to binaural acoustic stimuli having wide-band spectra. *J. Neurophysiol.,* 32:960, 1969.
Gerstein, G. L.; Butler, R. A.; and Erulkar, S. D.: Excitation and inhibition in cochlear nucleus. I. Tone-burst stimulation. *J. Neurophysiol.,* 31:526, 1968.
Goldberg, J. M., and Brown, P. B.: Functional organization of the dog superior olivary complex: an anatomical and electrophysiological study. *J. Neurophysiol.,* 31:639, 1968.
Goldberg, J. M., and Brown, P. B.: Response of binaural neurons of dog superior olivary complex to dichotic tonal stimuli: some physiological mechanisms of sound localization. *J. Neurophysiol.,* 32:613, 1969.
Grinnell, A. D.: Comparative physiology of hearing. *Ann. Rev. Physiol.,* 31:545, 1969.
Hawkins, J. E., Jr.: Hearing. *Ann. Rev. Physiol.,* 26:453, 1964.
Horvath, R. S.: Variability of cortical auditory evoked response. *J. Neurophysiol.,* 32:1056, 1969.
Katsuki, Y.: Comparative neurophysiology of hearing. *Physiol. Rev.,* 45:380, 1965.
Kiang, N. Y. S.: Auditory neurophysiology: peripheral aspects. *Proc. Int. Union of Physiol. Sciences,* 6:254, 1968.
Masterton, R. B.; Jane, J. A.; and Diamond, I. T.: Role of brainstem auditory structures in sound localization. 1. Trapezoid body, superior olive, and lateral lemniscus. *J. Neurophysiol.,* 30:341, 1967.
Masterton, R. B.; Jane, J. A.; and Diamond, I. T.: Role of brainstem auditory structures in sound localization. II. Inferior colliculus and its brachium. *J. Neurophysiol.,* 31:96, 1968.
Rose, J. E.: Anatomical and physiological organization of the auditory system in the mid-brain. *Proc. Int. Union of Physiol. Sciences,* 6:256, 1968.
Rose, J. E.; Brugge, J. F.; Anderson, D. J.; and Hind, J. E.: Some possible neural correlates of combination tones. *J. Neurophysiol.,* 32:402, 1969.
Schwartzkopff, J.: Hearing. *Ann. Rev. Physiol.,* 29:485, 1967.
Thompson, R. F.: Function of auditory cortex of cat in frequency discrimination. *J. Neurophysiol.,* 23:321, 1960.
Wever, E. G.: Electrical potentials of the cochlea. *Physiol. Rev.,* 46:102, 1966.
Wever, E. G., and Lawrence, M.: Physiological Acoustics. Princeton, Princeton University Press, 1954.
Whitfield, I. C.: Auditory Pathway. Baltimore, The Williams & Wilkins Co., 1968.

CHAPTER 52

The Eye:
I. Optics of Vision

PHYSICAL PRINCIPLES OF OPTICS

Before it is possible to understand the optical systems of the eye, the student must be thoroughly familiar with the basic physical principles of optics, including the physics of refraction, and have a knowledge of focusing, depth of focus, and so forth. Therefore, in the present study of the optics of the eye, a brief review of these physical principles is first presented, and then the optics of the eye are discussed.

REFRACTION OF LIGHT

The Refractive Index of a Transparent Substance. Light rays travel through a vacuum at a velocity of approximately 300,000 kilometers per second and practically as fast through air and other gaseous media but much slower through solids and liquids. The refractive index of a transparent substance is the *ratio* of the velocity of light in air to that in the substance. Obviously, the refractive index of air itself is 1.00

If light travels through a particular type of glass at a velocity of 200,000 kilometers per second, the refractive index of this glass is 300,000 divided by 200,000, or 1.50.

Refraction of Light Rays at an Interface Between Two Media with Different Refractive Indices. When light waves traveling forward in a beam, as shown in the upper part of Figure 52–1, strike an interface that is perpendicular to the beam, the waves enter the second refractive medium without deviating in their course. The only effect that occurs is decreased velocity of transmission. On the other hand, if, as illustrated in the lower part of Figure 52–1, the light waves strike an angulated interface, the light rays bend if the refractive indices of the two media are different from each other. In this particular figure the light waves are leaving air, which has a refractive index of 1.00, and are entering a block of glass having a refractive index of 1.50. When the beam first strikes the angulated interface, the lower edge of the beam enters the glass ahead of the upper edge. The wave front in

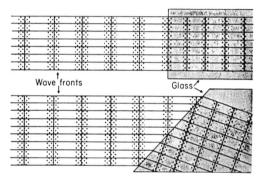

Figure 52–1 Wave fronts entering (*top*) a glass surface perpendicular to the light rays and (*bottom*) a glass surface angulated to the light rays. This figure illustrates that the distance between waves after they enter the glass is shortened to approximately two-thirds that in air. It also illustrates that light rays striking an angulated glass surface are refracted.

the upper portion of the beam continues to travel at a velocity of 300,000 kilometers per second while that which has entered the glass travels at a velocity of 200,000 kilometers per second. This causes the upper portion of the wave front to move ahead of the lower portion so that it is no longer vertical but is angulated to the right. Because the direction in which light travels is always perpendicular to the wave front, the direction of travel of the light beam now bends downward.

The bending of light rays at an angulated interface is known as *refraction*. Note particularly that the degree of refraction increases as a function of (1) the ratio of the two refractive indices of the two transparent media and (2) the degree of angulation between the interface and the wave front of the beam.

APPLICATION OF REFRACTIVE PRINCIPLES TO LENSES

The Convex Lens. Figure 52–2 shows parallel light rays entering a convex lens. The light rays passing through the center of the lens strike the lens exactly perpendicular to the lens surfaces and therefore pass through the lens without being re-

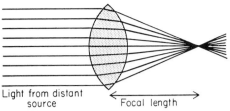

Figure 52–2 Bending of light rays at each surface of a convex spherical lens, showing that parallel light rays are focused to a point focus.

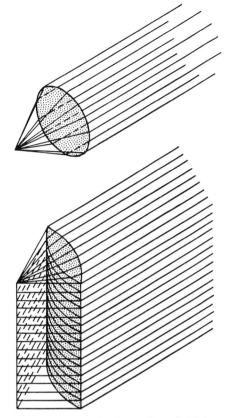

fracted at all. Toward either edge of the lens, however, the light rays strike a progressively more angulated interface. Therefore, the outer rays bend more and more toward the center.

When the rays have passed through the glass, the outer rays bend still more toward the center. Thus, parallel light rays entering an appropriately formed convex lens come to a single point focus at some distance beyond the lens.

The Concave Lens. Figure 52–3 shows the effect of a concave lens on parallel light rays. The rays that enter the very center of the lens strike an interface that is absolutely perpendicular to the beam and, therefore, do not refract at all. The rays at the edge of the lens enter the lens ahead of the rays toward the center. This is opposite to the effect in the convex lens, and it causes the peripheral light rays to *diverge* away from the light rays that pass through the center of the lens.

Thus, the concave lens *diverges* light rays, whereas the convex lens *converges* light rays.

Spherical versus Cylindrical Lenses. Figure 52–4 illustrates both a convex *spherical* lens and a convex *cylindrical* lens. Note that the convex cylindrical lens bends light rays from the two sides of the lens but not from either the top or the bottom. Therefore, parallel light rays are bent to a focal *line*. On the other hand, the light rays that pass through the spherical lens are refracted at all edges of the lens toward the central ray, and all the rays come to a focal *point*.

The cylindrical lens is well illustrated by a test tube full of water. If the test tube is placed in a beam of light and a piece of paper is brought progressively closer to the tube, a certain distance

Figure 52–4 *Top:* Point focus of parallel light rays by a spherical convex lens. *Bottom:* Line focus of parallel light rays by a cylindrical convex lens.

will be found at which the light rays come to a focal line. On the other hand, the spherical lens is illustrated by an ordinary magnifying glass. If such a lens is placed in a beam of sunlight and a piece of paper is brought progressively closer to the lens, the light rays will impinge on a common focal point at an appropriate distance.

Concave cylindrical lenses *diverge* light rays in only one plane in the same manner that convex cylindrical lenses converge light rays in one plane.

Figure 52–5 shows two convex cylindrical lenses at right angles to each other. The vertical cylindrical lens causes convergence of the light rays that pass through the two sides of the lens. The horizontal lens converges the top and bottom rays. Thus, all the light rays come to a single point focus. In other words, *two cylindrical lenses crossed at right angles to each other perform the same function as one spherical lens of the same refractive power.*

FOCUSING BY CONVEX LENSES

The distance from a convex lens at which parallel light rays converge to a common focal point is the *focal length* of the lens. The diagram at the top of Figure 52–6 illustrates this focusing of parallel light rays. In the middle diagram of

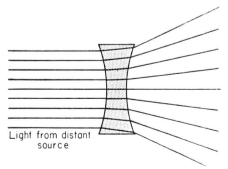

Figure 52–3 Bending of light rays at each surface of a concave spherical lens, illustrating that parallel light rays are diverged by a concave lens.

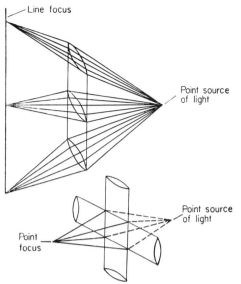

Figure 52–5 Two cylindrical convex lenses at right angles to each other, illustrating that one lens converges light rays in one plane and the other lens converges light rays in the plane at right angles. The two lenses combined give the same point focus as that obtained with a spherical convex lens.

Figure 52–6, the light rays that enter the convex lens are not parallel but are diverging because the origin of the light is a point source not far away from the lens itself. The rays striking the center of the lens are perpendicular to the lens surface and pass through the lens without any refraction. In the discussion above it was shown that the degree of bending of light rays is a function of (a) the ratio of refractive indices at the interface and (b) the angle of incidence of the light rays. Therefore, the edges of the convex lens bend the

diverging rays approximately the same amount as they do parallel rays because the greater refraction at one interface is approximately balanced by less refraction at the other interface. However, because these rays are already diverging, even though they are bent the same extent as parallel rays, it can be seen from the diagram that they do not come to a point focus at the same distance away from the lens as do the parallel rays. In other words, when rays of light that are already diverging enter a convex lens, the distance of focus on the other side of the lens is farther from the lens than is the case when the entering rays are parallel to each other.

In the lower diagram of Figure 52–6 are shown light rays that also are diverging toward a convex lens that has far greater curvature than that of the upper two lenses of the figure. In this diagram the distance from the lens at which the light rays come to a focus is exactly the same as that from the lens in the first diagram, in which the lens was convex but the light rays entering it were parallel. This illustrates that both parallel rays and diverging rays can be focused at the same distance behind a lens provided that the lens changes its convexity. It is obvious that the nearer the point source of light is to the lens, the greater is the divergence of the light rays and the greater must be the curvature of the lens to cause focusing at the same distance.

The relationship of focal length of the lens, distance of the point source of light, and distance of focus is expressed by the following formula:

$$\frac{1}{f} = \frac{1}{a} + \frac{1}{b}$$

in which f is the focal length of the lens, a the distance of the point source of light from the lens, and b the distance of focus from the lens.

FORMATION OF AN IMAGE BY A CONVEX LENS

The upper drawing of Figure 52–7 illustrates a convex lens with two point sources of light to the

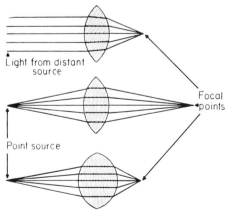

Figure 52–6 The upper two lenses of this figure have the same strength, but the light rays entering the top lens are parallel, while those entering the second lens are diverging; the effect of parallel versus diverging rays on the focal distance is illustrated. The bottom lens has far more refractive power than either of the other two lenses, illustrating that the stronger the lens the nearer to the lens is the point focus.

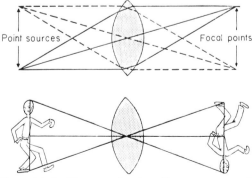

Figure 52–7 The top drawing illustrates two point sources of light focused at two separate points on the opposite side of the lens. The lower drawing illustrates formation of an image by a convex spherical lens.

left. Because light rays from any point source pass through the center of a convex lens without being refracted in either direction, the light rays from both point sources of light are shown to pass straight through the lens center. Furthermore, the other light rays from each point source of light, whether they pass through the upper edge of the lens, the sides of the lens, or the lower edge, all come to the same point focus behind the lens *directly in line with the point source of light and the center of the lens.*

Any object in front of the lens is in reality a mosaic of point sources of light. Some of these points are very bright, some are very weak, and they vary in color. The light rays from each point source of light that enter the very center of the convex lens pass directly through this lens without any of the rays bending. Furthermore, the light rays that enter the edges of the lens finally come to a focal point behind the lens in line with the rays that pass through the center. Therefore, every point source of light on the object comes to a separate point focus on the opposite side of the lens. If all portions of the object are the same distance in front of the lens, all the focal points behind the lens will fall in a common plane a certain distance behind the lens. If a white piece of paper is placed at this distance, one can see an image of the object, as is illustrated in the lower portion of Figure 52-7. However, this image is upside down with respect to the original object, and the two lateral sides of the image are reversed with respect to the original object. This is the method by which the lens of a camera focuses light rays on the camera film.

MEASUREMENT OF THE REFRACTIVE POWER OF A LENS—THE DIOPTER

The more a lens bends light rays, the greater is its "refractive power." This refractive power is measured in terms of *diopters.* The refractive power of a convex lens is equal to 1 meter divided by its focal length. Thus a lens has a refractive power of +1 diopter when it is capable of converging parallel light rays to a focal point 1 meter beyond the lens, as illustrated in Figure 52-8.

If the lens is capable of bending parallel light rays twice as much as a lens with a power of +1 diopter, it is said to have a strength of +2 diopters, and, obviously, the light rays come to a focal point 0.5 meter beyond the lens. A lens capable of converging parallel light rays to a focal point only 10 cm. beyond the lens has a refractive power of +10 diopters.

The refractive power of concave lenses cannot be stated in terms of the focal distance beyond the lens because the light rays do not focus to a point. Therefore, the power of a concave lens is stated in terms of its ability to diverge light rays in comparison with the ability of convex lenses to converge light rays. That is, if a concave lens diverges light rays the same amount that a 1 diopter convex lens converges them, the concave lens is said to have a dioptric strength of −1. Likewise, if the concave lens diverges the light rays as much as a +10 diopter lens converges them, it is said to have a strength of −10 diopters.

Note particularly that concave lenses can "neutralize" the refractive power of convex lenses. Thus, placing a 1 diopter concave lens immediately in front of a 1 diopter convex lens results in a lens system with zero refractive power.

The strengths of cylindrical lenses are computed in the same manner as the strengths of spherical lenses. If a cylindrical lens focuses parallel light rays to a line focus 1 meter beyond the lens, it has a strength of +1 diopter. On the other hand, if a cylindrical lens of a concave type *diverges* light rays as much as a +1 diopter cylindrical lens *converges* them, it has a strength of −1 diopter.

THE OPTICS OF THE EYE

THE EYE AS A CAMERA

The eye, as illustrated in Figure 52-9 is optically equivalent to the usual photographic camera, for it has a lens system, a variable aperture system, and a retina that corresponds to the film. The lens system of the eye is composed of (1) the interface between air and the anterior surface of the cornea, (2) the interface between the posterior surface of the

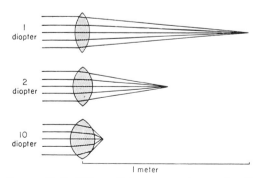

Figure 52-8 Effect of lens strength on the focal distance.

Figure 52-9 The eye as a camera. The numbers are the refractive indices.

cornea and the aqueous humor, (3) the interface between the aqueous humor and the anterior surface of the lens, and (4) the interface between the posterior surface of the lens and the vitreous humor. The refractive index of air is 1; the cornea, 1.38; the aqueous humor, 1.33; the lens (on the average), 1.40; and the vitreous humor, 1.34.

The total refractive power of the eye lens when it is surrounded by fluid on each side is only about 18 diopters; whereas, if this lens were removed from the eye and then surrounded by air, its refractive power would be about 150 diopters. Thus, it can be seen that the lens inside the eye is not nearly so powerful as it is outside the eye. The reason for this is that the fluids surrounding the lens have refractive indices not greatly different from the refractive index of the lens itself, the smallness of the differences greatly decreasing the amount of light refraction at the lens interfaces.

The curvature of the anterior surface of the cornea is reasonably great, but more important than this are the facts that (1) the refractive index of the cornea is markedly different from that of air and (2) the cornea is further away from the retina. This interface accounts for about 46 diopters of the refractive power of the eye.

The posterior surface of the cornea is concave and actually acts as a concave lens, but, because the difference in refractive index of the cornea and the aqueous humor is slight, this posterior surface of the cornea has a refractive power of only about −5 diopters, which neutralizes part of the refractive power of the other refractive surfaces of the eye.

THE REDUCED EYE. If all the refractive surfaces of the eye are algebraically added together and then considered to be one single lens, the optics of the normal eye may be simplified and represented schematically as a "reduced eye." This is useful in simple calculations. In the reduced eye, a single lens is considered to exist with its central point 17 mm. in front of the retina and have a total refractive power of approximately 59 diopters when the lens is accommodated for distant vision.

Formation of an Image on the Retina. In exactly the same manner that a glass lens can focus an image on a sheet of paper, the lens system of the eye can also focus an image on the retina. The image is inverted and reversed with respect to the object. However, the mind perceives objects in the upright position despite the upside down orientation on the retina because the brain is trained to consider an inverted image as the normal.

THE MECHANISM OF ACCOMMODATION

The refractive power of the crystalline lens of the eye can be voluntarily increased from 18 diopters to approximately 32 diopters in young children; this is a total "accommodation" of 14 diopters. To do this, the shape of the lens is changed from that of a moderately convex lens to that of a very convex lens. The mechanism of this is the following:

Normally, the lens is composed of a strong elastic capsule filled with viscous proteinaceous, but transparent fibers. When the lens is in a relaxed state, with no tension on its capsule, it assumes a spherical shape, owing entirely to the elasticity of the lens capsule. However, as illustrated in Figure 52–10, approximately 70 ligaments attach radially around the lens, pulling the lens edges toward the ciliary body. These ligaments are constantly tensed by the elastic pull of their attachments at the ciliary body, and the tension on the ligaments causes the lens to remain relatively flat under normal resting conditions of the eye. At the insertions of the tendons in the ciliary body is the ciliary muscle, which has two sets of smooth muscle fibers, the *meridional fibers* and the *circular fibers*. The meridional fibers extend from the corneoscleral junction to the insertions of the ligaments in the ciliary body approximately 2 to 3 mm. behind the corneoscleral junction. When these muscle fibers contract, the insertions of the ligaments are pulled forward, thereby releasing a certain amount of tension on the crystalline lens. The circular fibers are arranged circularly all the way around the eye so that when they contract a sphincter-like action occurs, decreasing the diameter of the circle of ligament attachments and allowing

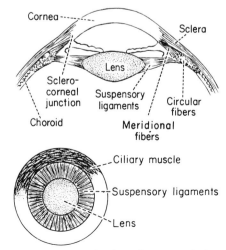

Figure 52–10 Mechanism of accommodation.

the ligaments to pull less on the lens capsule.

Thus, contraction of both sets of smooth muscle fibers in the ciliary muscle relaxes the ligaments to the lens capsule, and the lens assumes a more spherical shape, like that of a balloon, because of elasticity of its capsule. When the ciliary muscle is completely relaxed, the dioptric strength of the lens is as weak as it can become. On the other hand, when the ciliary muscle contracts as strongly as possible, the dioptric strength of the lens becomes maximal.

Parasympathetic Control of Accommodation. The ciliary muscle is controlled by the parasympathetic nervous system which will be discussed in Chapter 79. Stimulation of the parasympathetic fibers to the eye contracts the ciliary muscle, which in turn relaxes the ligaments of the lens and increases its refractive power. With an increased refractive power, the eye is capable of focusing on objects that are nearer to it than is an eye with less refractive power. Consequently, as a distant object moves toward the eye, the number of parasympathetic impulses impinging on the ciliary muscle must be progressively increased for the eye to keep the object constantly in focus.

Presbyopia. As a person grows older, his lens loses its elastic nature and becomes a relatively solid mass, probably because of progressive denaturation of the proteins. Therefore, the ability of the lens to assume a spherical shape progressively decreases, and the power of accommodation decreases from approximately 14 diopters shortly after birth to approximately 2 diopters at the age of 45 to 50. Thereafter, the lens of the eye may be considered to be almost totally nonaccommodating, which condition is known as "presbyopia."

Once a person has reached the state of presbyopia, each eye remains focused permanently at an almost constant distance; this distance depends on the physical characteristics of each individual's eyes. Obviously, the eyes can no longer accommodate for both near and far vision. Therefore, for an older person to see clearly both in the distance and nearby, he must wear glasses for far-seeing or near-seeing—for one or the other or for both.

THE PUPILLARY APERTURE

A major function of the iris is to increase the amount of light that enters the eye during darkness and to decrease the light that enters the eye in bright light. The reflexes for con-

trolling this mechanism will be considered in the discussion of the neurology of the eye in Chapter 59. The amount of light that enters the eye through the pupil is proportional to the area of the pupil or to the *square of the diameter* of the pupil. The pupil of the human eye can become as small as approximately 1.5 mm. and can become almost as large as 8 mm. in diameter. Therefore, the quantity of light entering the eye may vary approximately 30 times as a result of changes in pupillary aperture size.

Depth of Focus of the Lens System of the Eye. Figure 52–11 illustrates two separate eyes that are exactly alike except that the diameters of the pupillary apertures are different. In the upper eye the pupillary aperture is small, and in the lower eye the aperture is large. In front of each of these two eyes are two small point sources of light, and light from each passes through the pupillary apertures and focuses on the retina. Consequently, in both eyes the retina sees two spots of light in perfect focus. It is evident from the diagrams, however, that if the retina is moved forward or backward, the size of each spot will not change much in the upper eye, but in the lower eye the size of the spot will increase greatly, and it becomes a "blur circle." In other words, the upper lens system has far greater *depth of focus* than the bottom lens system. When a lens system has great depth of focus, the retina can be considerably displaced from the focal plane and still discern the various points of an image rather distinctly; whereas, when a lens system has shallow depth of focus, having the retina only slightly away from the focal plane causes extreme blurring of the image.

The greatest possible depth of focus occurs when the pupil is extremely small. The reason for this is that with a very small aper-

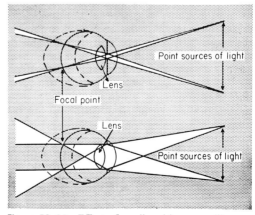

Figure 52–11 Effect of small and large pupillary apertures on the depth of focus.

ture the eye acts more or less as a pinhole camera. It will be recalled from the study of physics that light rays coming from an object and passing through a pinhole are in focus on any surface at all distances beyond the pinhole. In other words, the depth of focus of a pinhole camera is infinite.

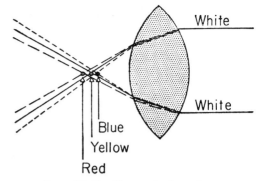

Figure 52–13 Chromatic aberration.

"NORMAL" ABERRATIONS OF VISION

Spherical Aberration. The crystalline lens of the eye is not nearly so regularly formed as lenses made by good opticians. Indeed, the light rays passing through the peripheral edges of the eye lens are never brought to really sharp focus with the other light rays, as illustrated in Figure 52–12. This effect is known as "spherical aberration," and the lens system of the human eye is quite subject to such an error. Therefore, increasing the aperture of the pupil progressively decreases the sharpness of focus. This partially explains why visual acuity is decreased at low illumination levels, which cause the pupil to dilate.

Chromatic Aberration. The lens of the eye is also subject to "chromatic aberration," which is illustrated in Figure 52–13. This means that the refractive power of the lens is different for the different colors, so that different colors focus at different distances behind the lens. Furthermore, the greater the aperture of the lens, the greater are the errors of chromatic aberration, for those light rays passing through the center of the lens are proportionately less affected than are the rays passing through the periphery of the lens.

Diffractive Errors of the Eye. Still another error in the optical system of all eyes is "diffraction" of light rays. Diffraction means bending of light rays when they pass over sharp edges; this obviously occurs as the rays pass over the edges of the pupil. Diffractive errors occur especially when the pupil becomes very small, because "interference" patterns then appear on the retina. A thorough consideration of diffraction at this point is impossible, and this phenomenon is simply noted to explain that, as the pupil becomes 1.5 mm. in size, the sharpness of vision becomes less than it is when the pupil is approximately 2.5 mm. in size despite the fact that the depth of focus is better the smaller the pupillary diameter.

Cataracts. Cataracts are an especially common eye abnormality that occurs in older people. A cataract is a cloudy or opaque area in the lens. In

the early stage of cataract formation the proteins in the lens fibers immediately beneath the capsule become denatured. Later, these same proteins coagulate to form opaque areas in place of the normal transparent protein fibers of the lens. Finally, in still later stages, a considerable amount of calcium is deposited in the coagulated proteins, thus further increasing the opacity.

When a cataract has obscured light transmission so greatly that it seriously impairs vision, the condition can be corrected by surgical removal of the entire lens. When this is done, however, the eye loses a large portion of its refractive power, which must be replaced by a powerful convex lens in front of the eye, as will be explained in the following sections.

ERRORS OF REFRACTION

Emmetropia. As shown in Figure 52–14, the eye is considered to be normal or "emmetropic" if, when the ciliary muscle is completely relaxed, parallel light rays from distant objects are in sharp focus on the retina. This means that the

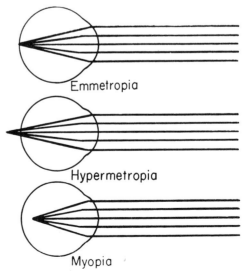

Figure 52–14 Parallel light rays focus on the retina in emmetropia, behind the retina in hypermetropia, and in front of the retina in myopia.

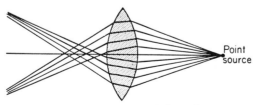

Figure 52–12 Spherical aberration.

emmetropic eye can, with its ciliary muscle completely relaxed, see all distant objects clearly, but to focus objects at close range it must contract its ciliary muscle and thereby provide various degrees of accommodation.

Hypermetropia (Hyperopia). Hypermetropia, which is also known as "far-sightedness," is due either to an eyeball that is too short or to a lens system that is too weak when the ciliary muscle is completely relaxed. In this condition, parallel light rays are not bent sufficiently by the lens system to come to a focus by the time they reach the retina. In order to overcome this abnormality, the ciliary muscle may contract to increase the strength of the lens. Therefore, the far-sighted person is capable, by using his mechanism of accommodation, of focusing distant objects on his retina. If he has used only a small amount of strength in his ciliary muscle to accommodate for the distant objects, then he still has much accommodative power left, and objects closer and closer to the eye can also be focused sharply until the ciliary muscle has contracted to its limit. The distance of the object away from the eye at this point is known as the "near point" of vision.

In old age, when the lens becomes presbyopic, the far-sighted person often is not able to accommodate his lens sufficiently to focus even distant objects, much less to focus near objects.

Myopia. In myopia, or "near-sightedness," even when the ciliary muscle is completely relaxed, the strength of the lens is still so great that light rays coming from distant objects are focused in front of the retina. This is usually due to too long an eyeball, but it can rarely result from too much power of the lens system of the eye.

No mechanism exists by which the eye can decrease the strength of its lens beyond that which exists when the ciliary muscle is completely relaxed. Therefore, the myopic person has no mechanism by which he can ever focus distant objects sharply on his retina. However, as an object comes nearer and nearer to his eye it finally comes near enough that its image is focused on the retina. Then, when the object comes still closer to the eye, the person can use his mechanism of accommodation to keep the image focused clearly. Therefore, a myopic person has a definite limiting "far point" for acute vision as well as a "near point"; when an object comes inside the "far point," he can use his mechanism of accommodation to keep the object in focus until the object reaches the "near point."

CORRECTION OF MYOPIA AND HYPERMETROPIA BY USE OF LENSES. It will be recalled that light rays passing through a concave lens diverge. Therefore, if the refractive surfaces of the eye have too much refractive power, as in myopia, some of this excessive refractive power can be neutralized by placing in front of the eye a concave spherical lens, which will diverge the rays. On the other hand, in a person who has hypermetropia—that is, one who has too weak a lens for the distance of the retina away from the lens—the abnormal vision can be

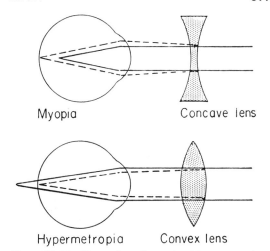

Figure 52-15 Correction of myopia with a concave lens and correction of hypermetropia with a convex lens.

corrected by adding refractive power with a convex lens in front of the eye. These corrections are illustrated in Figure 52–15. One usually determines the strength of the concave or convex lens needed for clear vision by "trial and error"—that is, by trying first a strong lens and then a stronger or weaker lens until the one that gives the best visual acuity is found.

Astigmatism. Astigmatism is a refractive error of the lens system of the eye caused usually by an oblong shape of the cornea or, rarely, by an oblong shape of the lens. A lens surface like the side of a football lying edgewise to the incoming light, for instance, would be an example of an astigmatic lens. The degree of curvature in a plane through the long axis of the football is not nearly so great as the degree of curvature in a plane through the short axis. The same is true of an astigmatic lens. Because the curvature of the astigmatic lens along one plane is less than the curvature along the other plane, light rays striking the peripheral portions of the lens in one plane are not bent nearly so much as are rays striking the peripheral portions of the other plane.

This is illustrated in Figure 52–16, which shows

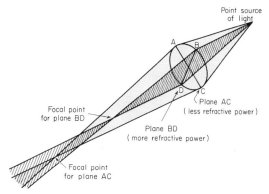

Figure 52-16 Astigmatism, illustrating that light rays focus at one focal distance in one focal plane and at another focal distance in the plane at right angles.

what happens to rays of light emanating from a point source and passing through an astigmatic lens. The light rays in the vertical plane, which is indicated by plane BD, are refracted greatly by the astigmatic lens because of the greater curvature in the vertical direction than in the horizontal direction. However, the light rays in the horizontal plane, indicated by plane AC, are bent not nearly so much as the light rays in the vertical plane. It is obvious, therefore, that the light rays passing through an astigmatic lens do not all come to a common focal point because the light rays passing through one plane of the lens focus far in front of those passing through the other plane.

Placing an appropriate *spherical* lens in front of an astigmatic eye can bring the light rays that pass through *one plane* of the lens into focus on the retina, but spherical lenses can never bring *all* the light rays into complete focus at the same time. This is the reason why astigmatism is a very undesirable refractive error of the eyes. Furthermore, the accommodative powers of the eyes cannot compensate for astigmatism for the same reasons that spherical lenses placed in front of the eyes cannot correct the condition.

CORRECTION OF ASTIGMATISM WITH A CYLINDRICAL LENS. In correcting astigmatism with lenses it must always be remembered that two cylindrical lenses of equal strength may be crossed at right angles to give the same refractive effects as a spherical lens. However, if one of the crossed cylindrical lenses has a different strength from that of the second lens, the light rays in one plane may not be brought to a common focal point with the light rays in the opposite plane. This is the situation that one usually finds in the astigmatic eye. In other words, one may consider an astigmatic eye as having a lens system made up of two cylindrical lenses of slightly different strengths. Another way of looking upon the astigmatic lens system of the eye is that this system is a spherical lens with a superimposed cylindrical lens.

To correct the focusing of an astigmatic lens system, it is necessary to determine both the *strength* of the cylindrical lens needed to neutralize the excess cylindrical power of the eye lens and the *axis* of this abnormal cylindrical lens.

There are several methods for determining the axis of the abnormal cylindrical component of the lens system of an eye. One of these methods is based on the use of parallel black bars, as shown in Figure 52–17. Some of these parallel bars are vertical, some horizontal, and some at various angles to the vertical and horizontal axes. After placing by trial and error various spherical lenses in front of the astigmatic eye, one set of these parallel bars finally will be brought into sharp focus on the retina of the astigmatic eye. If there is no accompanying myopia or hypermetropia, the spherical lenses may not be necessary.

It can be shown from the physical principles of optics discussed earlier in this chapter that the axis of the *out-of-focus* cylindrical component of the optical system is parallel to the black bars that

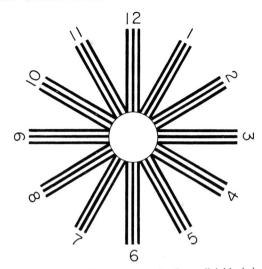

Figure 52–17 Chart composed of parallel black bars for determining the axis (meridian) of astigmatism.

are fuzzy in appearance. Once this axis is found, the examiner tries progressively stronger and weaker positive or negative cylindrical lenses, the axes of which are placed parallel to the out-of-focus bars until the patient sees all the crossed bars with equal clarity. When this has been accomplished, the examiner directs the optician to grind a special lens having both the spherical correction plus the cylindrical correction at the appropriate axis.

SIZE OF THE IMAGE ON THE RETINA AND VISUAL ACUITY

If the distance from an object to the eye lens is 17 meters and the distance from the center of the lens to the image is 17 millimeters, the ratio of the object size to image size is 1000 to 1. Therefore, an object 17 meters in front of the eye and 1 meter in size produces an image on the retina 1 millimeter in size.

Theoretically, a point of light from a distant point source, when focused on the retina, should be infinitely small. However, since the lens system of the eye is not absolutely perfect, such a retinal spot ordinarily has a total diameter of about 11 microns even with maximum resolution of the optical system. However, it is brightest in its very center and shades off gradually toward the edges.

The average diameter of cones *in the fovea* of the retina is approximately 1.5 microns, which is one-seventh the diameter of the spot of light. Nevertheless, since the spot of light has a bright center point and shaded edges, a person can distinguish two separate points if their centers lie approximately 2 microns apart on the retina, which is slightly greater than the width of a foveal cone. This discrimination between points is illustrated in Figure 52–18.

The maximum visual acuity of the human eye for discriminating between point sources of light is 26 seconds. That is, when light rays from two separate points strike the eye with an angle of at least 26 seconds between them, they can usually be recognized as two points instead of one. This means that a person with maximal acuity looking at two bright pinpoint spots of light 10 meters away can barely distinguish the spots as separate entities when they are 1 millimeter apart.

The visual acuity is no more than one-tenth as good in the retina outside the foveal region, becoming poorer and poorer as the periphery is approached. This is caused by the connection of many rods and cones to the same nerve fiber, as discussed in Chapter 54.

Clinical Method for Stating Visual Acuity. Usually the test chart for testing eyes is placed 20 feet away from the tested person, and if the person can see the letters of the size that he should be able to see at 20 feet, he is said to have 20/20 vision: that is, normal vision. If he can see only letters that he should be able to see at 200 feet, he is said to have 20/200 vision. On the other hand, if he can see at 20 feet letters that he should be able to see only at 15 feet, he is said to have 20/15 vision. In other words, the clinical method for expressing visual acuity is to use a mathematical fraction that expresses the ratio of two distances, which is also the ratio of one's visual acuity to that of the normal person.

DETERMINATION OF DISTANCE OF AN OBJECT FROM THE EYE—DEPTH PERCEPTION

There are three major means by which the visual apparatus normally perceives distance, a phenomenon that is known as depth perception. These are (1) relative sizes of objects, (2) moving parallax, and (3) stereopsis.

Determination of Distance by Relative Sizes. If a person knows that a man is six feet tall and then he sees this man even with only one eye, he can determine how far away the man is simply by the size of the man's image on his retina. He does not consciously think about the size of this image, but his brain has learned to determine automatically from the image sizes the distance of objects from the eye when the dimensions of these objects are already known.

Determination of Distance by Moving Parallax. Another important means by which the eyes determine distance is that of moving parallax. If a person looks off into the distance with his eyes completely still, he perceives no moving parallax, but, when he moves his head to one side or the other, the images of objects close to him move rapidly across his retinae while the images of distant objects remain rather stationary. For instance, if he moves his head 1 inch and an object is only 1 inch in front of his eye, the image moves almost all the way across his retinae, whereas the image of an object 200 feet away from his eyes does not move perceptibly. Thus, by this mechanism of moving parallax, one can tell the *relative distances* of different objects even though only one eye is used.

Determination of Distance by Stereopsis. Another method by which one perceives parallax is that of binocular vision. Because one eye is a little more than 2 inches to one side of the other eye, the images on the two retinae are different one from the other—that is, an object that is 1 inch in front of the bridge of the nose forms an image on the temporal portion of the retina of each eye, whereas a small object 20 feet in front of the nose has its image at closely corresponding points in the middle of the eye. This type of parallax is illustrated in Figure 52–19, which shows the images of a black spot

Figure 52–18 Maximal visual acuity for two point sources of light.

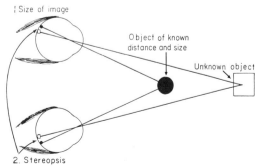

Figure 52–19 Perception of distance (1) by the size of the image on the retina and (2) as a result of stereopsis.

and a square actually reversed on the retinae because they are at different distances in front of the eyes. This gives a type of parallax that is present all the time when both eyes are being used. It is almost entirely this binocular parallax (or stereopsis) that gives a person with two eyes far greater ability to judge relative distances *when objects are nearby* than a person who has only one eye. However, stereopsis is virtually useless for depth perception at distances beyond 200 feet.

OPTICAL INSTRUMENTS

THE OPHTHALMOSCOPE

The ophthalmoscope is an instrument designed so that an observer can look through it into another person's eye and see the retina with clarity. Though the ophthalmoscope appears to be a relatively complicated instrument, its principles are simple. The basic portions of such an instrument are illustrated in Figure 52–20 and may be explained as follows:

If a bright spot of light is on the retina of an emmetropic eye, light rays from the spot diverge toward the lens system of the eye, and, after they pass through the lens system, they are parallel with each other because the retina is located in the focal plane of the lens. When these parallel rays pass into an emmetropic eye of another person, they focus to a point on the retina of the second person because his retina is in the focal plane of his lens. Therefore, any spot of light on the retina of the observed eye comes to a focal spot on the retina of the observing eye. Likewise, when the bright spot of light is moved to different portions of the observed retina, the focal spot on the retina of the observer also moves. Thus, if the retina of one person is made to emit light, the image of his retina will be focused on the retina of the observer provided that the two eyes are simply looking into each other. These principles, of course, apply only to completely emmetropic eyes.

To make an ophthalmoscope, one need only devise a means for illuminating the retina to be examined. Then, the reflected light from that retina can be seen by the observer simply by putting the two eyes close to each other. To illuminate the retina of the observed eye, an angulated mirror or a segment of a prism is placed in front of the observed eye in such a manner that light from a bulb is reflected into the observed eye. Thus, the retina is lighted through the pupil, and the observer can also see into the subject's pupil by looking over the edge of the mirror or prism.

It was noted above that these principles apply only to persons with completely emmetropic eyes. If the refractive power of either eye is abnormal, it is necessary to correct this refractive power in order for the observer to see a sharp image of the observed retina. Therefore, the usual ophthalmoscope has a series of about 20 lenses mounted on a turret so that the turret can be rotated from one lens to another, and the correction for abnormal refractive power of either or both eyes can be made at the same time by selecting a single lens of appropriate strength. In normal young persons, when the two eyes come close together, a natural accommodative reflex occurs that causes approximately +2 diopters increase in the strength of the lens of each eye. To correct for this, it is necessary that the lens turret be rotated to approximately −4 diopters correction.

THE RETINOSCOPE

The retinoscope, illustrated in Figure 52–21, is an instrument that can be used to determine the refractive power of an eye even though the subject cannot converse with the observer. Such a procedure is valuable for fitting glasses to an infant.

To use the retinoscope, one places a bright spot of light behind and to one side of the observed eye, and stands 1 meter away, looking through a hole in the middle of a mirror. The observer then rotates this mirror from side to side, casting a reflected beam of light into the pupil of the observed eye while the subject keeps his gaze intently on the observer's eye. If the observed eye is normal, when this beam of light first hits the edge of the pupil the entire pupil suddenly glows red. If the eye has abnormal refractive powers, the red glow appears either on the side of the pupil into which the light shines or on the opposite side of the pupil. *In hyperopia the first glow appears on the side of the pupil from which the light beam is being moved.*

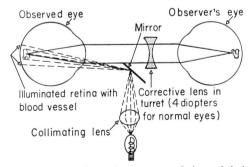

Observed eye   Observer's eye

Mirror

Illuminated retina with blood vessel   Corrective lens in turret (4 diopters for normal eyes)

Collimating lens

Figure 52–20 The optical system of the ophthalmoscope.

(1) Glows on this side in myopia

Mirror

1 diopter lens

Observer's eye

(2) Glows on this side in hypermetropia

(3) Glows over entire pupil in emmetropia

Figure 52–21 The retinoscope.

In myopia the first glow appears on the opposite side of the pupil. The cause of this difference is that the lens of the hyperopic eye is too weak and that of the myopic eye too strong to focus the light from the edge of the pupil on the retina exactly in line with the observer's eye. For a fuller understanding of this effect, however, the student is referred to texts on physiologic optics.

One can fit glasses to a patient by placing selected lenses in front of the observed eye one at a time until the glow suddenly covers the pupil over its entire extent rather than spreading from one side of the pupil to the other. However, it should be noted that in retinoscopy one tests an eye that is focused on the observer's eye at 1 meter's distance. This must be taken into consideration in prescribing glasses; 1 diopter strength must be subtracted for far vision.

REFERENCES

Adler, F. H.: Physiology of the Eye. 4th ed., St. Louis, The C. V. Mosby Co., 1965.

Armington, J. C.: Vision. *Ann. Rev. Physiol., 27*:163, 1965.

Brecher, G. A.; Lewis, D.; and Eastman, A. A.: Test comparing a new with a conventional astigmatic chart. *Amer. J. Ophthal., 48*:118, 1959.

Davson, H.: Physiology of the Eye. 2nd ed., Boston, Little, Brown and Co., 1962.

Davson, H. (ed.): The Eye. 4 volumes. New York, Academic Press, Inc., 1962.

Duke-Elder, S., and Perkins, E. S. (eds.): The Transparency of the Cornea. Springfield, Illinois, Charles C Thomas, Publishers, 1960.

Fry, G. A.: The image-forming mechanism of the eye. *In* Handbook of Physiology. Baltimore, The Williams & Wilkins Co., 1959, Sec. I, Vol. I, p. 647.

Girard, L. J.: Corneal and Scleral Contact Lenses: Proceedings of the International Congress. St. Louis, The C. V. Mosby Co., 1968.

Harris, J. D.: Symposium on the Lens. St. Louis, The C. V. Mosby Co., 1965.

Linksz, A.: Physiology of the Eye: Vol. I. Optics. New York, Grune & Stratton, Inc., 1950.

Paterson, C. A.: Extracellular space of the crystalline lens. *Amer. J. Physiol., 218*:797, 1970.

Polyak, S.: The Vertebrate Visual System. Chicago, University of Chicago Press, 1957.

Sloane, A. E.: Manual of Refraction. Boston, Little, Brown and Co., 1961.

Smelser, G. K. (ed.): The Structure of the Eye. New York, Academic Press, Inc., 1961.

CHAPTER 53

The Eye:
II. Receptor Functions of
the Retina

The retina is the light-sensitive portion of the eye, containing the cones, which detect specific colors, and the rods, which detect light of any color besides deep red. When the rods and cones are excited, signals are transmitted through successive neurons in the retina itself and finally into the optic nerve fibers and cerebral cortex. The purpose of the present chapter is to explain specifically the mechanisms by which the rods and cones detect both white and colored light.

ANATOMY AND FUNCTION OF THE STRUCTURAL ELEMENTS OF THE RETINA

The Layers of the Retina. Figure 53–1 shows the functional components of the retina arranged in

Figure 53–1 Plan of the retinal neurons. (From Polyak: The Retina. University of Chicago Press.)

layers from the outside to the inside as follows: pigment layer, layer of rods and cones projecting into the pigment, outer limiting membrane, outer nuclear layer, outer plexiform layer, inner nuclear layer, inner plexiform layer, ganglion layer, layer of optic nerve fibers, and inner limiting membrane.

Since the rods and cones are the light-sensitive portion of the retina, light must pass through all the other layers besides the pigment layer before reaching the excitable cells, and this obviously decreases the visual acuity. However, in the central region of the retina, as will be discussed more fully shortly, the inner layers are pulled aside to prevent this loss of acuity.

The Rods and Cones. The nuclei of the rods and cones are located in the outer nuclear layer while the light-sensitive portions of both these receptors project through the outer limiting membrane and protrude part way into the pigment layer of the retina. The rods are relatively slender, having a diameter of about 2 microns in the more central portions of the retina and 4 to 5 microns in the more peripheral portions. The cones, on the other hand, have a diameter of about 1.5 microns in the *fovea,* the centralmost portion of the retina, and about 5 to 8 microns in the periphery.

The sensitive portions of both the rods and cones contain light-sensitive chemicals which decompose on exposure to light. The decomposition products in turn stimulate the cell membranes of the rods and cones, eliciting signals that are then transmitted into the nervous system. Indeed, the rods and cones themselves are modified neurons and can actually be considered to be part of the nervous system.

The Foveal Region of the Retina and Its Importance in Acute Vision. A minute area in the center of the retina called the *macula* and occupying a total area of about 1 square millimeter is especially capable of acute and detailed vision. This area is composed entirely of cones, but the cones are very much elongated and have a diameter of only 1.5 microns in contradistinction to the very large cones located farther peripherally in the retina. The central portion of the macula is called the *fovea;* in this region

The figure labels read:
- Pigmented layer
- Layer of rods and cones
- Outer limiting membrane
- Outer nuclear layer
- Outer plexiform layer
- Horizontal cell
- Inner nuclear layer
- Fiber of Müller
- Amacrine cell
- Inner plexiform layer
- Ganglion cells
- Ganglionic layer
- Stratum opticum
- Inner limiting membrane

the blood vessels, the ganglion cells, the inner nuclear layer of cells, and the plexiform layers are all displaced to one side rather than resting directly on top of the cones. This allows light to pass unimpeded to the cones rather than through several layers of retina, which probably aids in the acuity of visual perception by this region of the retina.

The Pigment Layer of the Retina. The black pigment *melanin* in the pigment layer, and still more melanin in the choroid, prevents light reflection throughout the globe of the eyeball, and this is extremely important for acute vision. This pigment performs the same function in the eye as the black paint inside the bellows of a camera. Without it, light rays would be reflected in all directions within the eyeball and would cause diffuse lighting of the retina rather than the contrasts between dark and light spots required for the formation of precise images.

The importance of melanin in the pigment layer and choroid is well illustrated by its absence in *albinos,* persons hereditarily lacking in melanin pigment in all parts of their bodies. When an albino enters a bright area, light that impinges on the retina is reflected in all directions by the white surface of the unpigmented choroid so that a single discrete spot of light that would normally excite only a few rods or cones is reflected everywhere and excites many of the receptors. Also, a large amount of diffuse light enters the eye through the unpigmented iris. Thus, the lack of melanin causes the visual acuity of albinos, even with the best of optical correction, to be rarely better than 20/100 to 20/200.

Relationship of the Retina to the Choroid. The retina is adhered to the choroid, which is a highly vascular layer of the eye underlying the sclera. The nutrient blood supply to the inside layers of the retina is derived from the central retinal artery, which enters the inside of the eye along with the optic nerve and then divides to supply the entire inner retinal surface. Thus, to a great extent the retina has its own blood supply independent of the underlying vascular choroid. However, the outer layers of the retina, including especially the rods and cones, are at least partially dependent on diffusion of nutrients from the choroid. For instance, most of the vitamin A utilized by the rods and cones for synthesis of their light-sensitive chemicals is derived from the choroid and stored in the pigment layer of retina until needed.

RETINAL DETACHMENT. The neural retina occasionally detaches from the pigment epithelium. In a few instances the cause of such detachment is injury to the eyeball that allows fluid or blood to collect between the retina and the pigment epithelium, but more often it is caused by contracture of the fine collagenous fibrils in the vitreous humor, which pull the retina unevenly toward the interior of the globe.

Fortunately, partly because of diffusion across the detachment gap and partly because of the independent blood supply to the retina through the retinal artery, the retina can resist degeneration for many days and can become functional once again if surgically replaced in its normal relationship with the pigment epithelium. But, if not replaced soon, the retina finally degenerates and then is unable to function even after surgical repair.

PHOTOCHEMISTRY OF VISION

Both the rods and cones contain chemicals that decompose on exposure to light and, in the process, excite the nerve fibers leading from the eye. The chemical in the *rods* has been isolated and is called *rhodopsin.* Unfortunately, the light-sensitive chemical in the *cones* of human beings has not been isolated. However, in some lower animals these chemicals have been isolated and in all instances have chemical compositions only slightly different from that of rhodopsin.

In the present section we will discuss principally the photochemistry of rhodopsin, but it should be recognized that almost exactly the same principles apply to the photochemistry of the cones.

THE RHODOPSIN-RETINENE VISUAL CYCLE, AND EXCITATION OF THE RODS

Rhodopsin and Its Decomposition by Light Energy. The portion of the rod that projects into the pigment layer of the retina has a concentration of about 40 per cent of the light-sensitive pigment called *rhodopsin* or *visual purple.* This substance is a combination of the protein *scotopsin* and the carotenoid pigment *retinene.* Furthermore, the retinene is a particular type called neo-b retinene, which is a type of 11-*cis* retinene. This *cis* form of the retinene is important because only this type of retinene can combine with scotopsin to form rhodopsin.

When light energy is absorbed by rhodopsin, the rhodopsin immediately begins to decompose as shown in Figure 53–2. The cause of this is an instantaneous change of the *cis* form of retinene into an all-*trans* form, which still has the same chemical structure as the *cis* form but has a different physical structure. Yet, the reactive sites of the all-*trans* retinene no longer fit with the reactive sites of the protein scotopsin, and it begins to pull away from the scotopsin. The immediate product is *lumirhodopsin,* which is a partially split combination of the all-*trans* retinene and scotopsin. How-

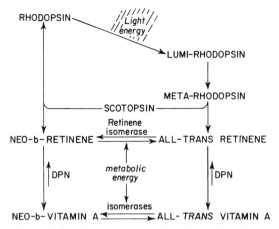

Figure 53–2 Photochemistry of the rhodopsin-retinene-vitamin A visual cycle.

ever, lumi-rhodopsin is an extremely unstable compound and decays in a small fraction of a second to meta-rhodopsin, which is still another loose combination of the all-*trans* retinene and scotopsin. This compound, too, is unstable, and it slowly decomposes during the next few seconds into completely split products—scotopsin and all-*trans* retinene. During the process of splitting, the rods become excited, and signals are transmitted into the central nervous system.

Reformation of Rhodopsin. The first stage in reformation of rhodopsin, as shown in Figure 53–2, is to reconvert the all-*trans* retinene into neo-b retinene, the *cis* form of this pigment. This process is catalyzed by the enzyme *retinene isomerase*. However, this process also requires active metabolism and energy transfer to the neo-b retinene; once the neo-b retinene is formed, it automatically recombines with the scotopsin to reform rhodopsin, an exergonic process (which means that it gives off energy). The product, rhodopsin, is a stable compound until its decomposition is again triggered by absorption of light energy.

Excitation of the Rod as Rhodopsin Decomposes. The exact method by which rhodopsin decomposition excites the rod is still speculative. However, it is known that when the all-*trans* retinene begins to split away from the scotopsin, several ionized radicals are momentarily exposed, and this could excite the rod in one of several ways. One suggestion has been that the instantaneous ionic field of the radicals causes the membrane potential of the rod to change and that this in turn causes excitation. The physical construction of the pigment portion of the rod is compatible with this concept because the cell membrane folds inward to form tremendous numbers of discs, lying

one on top of the other, and each one is surfaced on the inside with aggregates of rhodopsin. This extensive relationship between rhodopsin and the cell membrane could explain why the rod is so exquisitely sensitive to light, being capable of detectable excitation following absorption of only one quantum of light energy.

A second theory for excitation of the rod is that the momentarily ionized all-*trans* retinene–scotopsin complex acts as an enzyme to promote some yet unspecified chemical reaction within the outer segment of the rod. This reaction in turn supposedly causes changes in the rod membrane potential.

THE RECEPTOR POTENTIAL. Regardless of the precise mechanism by which the rod is excited, the membrane potential of the rod changes during exposure to light, and the degree of change is a direct function of the intensity of the light. Thus, a *receptor potential* is generated in the rod.

However, this receptor potential is peculiar in that its polarity is opposite to the polarity of receptor potentials recorded in other types of receptors. Furthermore, the receptor potential has two phases: an instantaneous very low voltage phase, the electrical potential of which is directly proportional to the light energy; and a secondary phase of much greater voltage, the potential of which is proportional to the logarithm of the light energy. The first potential is believed to result from the immediate interaction of light with the rhodopsin. The second potential is believed to result from subsequent chemical events in the rod or its membrane—chemical events that could or could not be separate from the rhodopsin cycle.

This transduction of light energy into a logarithmic membrane potential seems to be an exceedingly important one for function of the eye, because it allows the eye to discriminate light intensities through a range many thousand times as great as would otherwise be possible. In general, regardless of the original light intensity, the eye can detect an increase in light intensity when this increase is about 0.5 to 1 per cent, which again illustrates a logarithmic type of response of the retina.

Relationship Between Retinene and Vitamin A. The lower part of the scheme in Figure 53–2 illustrates that each of the two types of retinene can be converted into a corresponding type of vitamin A, and in turn the two types of vitamin A can be reconverted into the two types of retinene. Thus, the two retinenes are in equilibrium with vitamin A. Most of the vitamin A of the retina is stored in the pigment layer of the retina rather than in the rods themselves, but this vitamin A is readily available to the rods.

Retinene is formed from vitamin A by dehydrogenation in accordance with the following reaction:

$$C_{19}H_{27}CH_2OH + DPN^+ \xrightarrow{\text{Alcohol dehydrogenase}}$$
$$\text{Vitamin A}$$
$$C_{19}H_{27}CHO + DPNH + H^+$$
$$\text{Retinene}$$

The equilibrium of this equation is strongly in favor of vitamin A. Therefore, vitamin A concentration is normally far greater than free retinene concentration in the retina.

Another important feature of the conversion of retinene into vitamin A, or the converse conversion of vitamin A into retinene, is that these processes require a much longer time to approach equilibrium than it takes for conversion of retinene and scotopsin into rhodopsin, or for conversion (under the influence of strong light energy) of rhodopsin into retinene and scotopsin. Therefore, all the reactions of the upper part of Figure 53–2 can take place relatively rapidly in comparison with the slow interconversions between retinene and vitamin A.

Yet, if the retina remains exposed to strong light for a long time, most of the stored rhodopsin will eventually be converted into retinene, which causes a continuously high concentration of retinine in the rods, resulting in conversion of much of the retinene into vitamin A. Therefore, gradually a larger portion of the rhodopsin of the rods becomes converted first into retinene and thence into vitamin A, thereby decreasing the concentration of all the photochemicals in the rods much more than would be true were it not for this subsequent interconversion from retinene to vitamin A.

Conversely, during total darkness, essentially all the retinene in the rods becomes converted into rhodopsin, thereby reducing the concentration of retinene almost to zero. This then allows much of the vitamin A to be converted into retinene, which becomes rhodopsin within another few minutes. Thus, in complete darkness not only is almost all the retinene of the rods converted into rhodopsin but also much of the vitamin A stored in the pigment layer of the retina is absorbed by the rods and also converted into rhodopsin.

NIGHT BLINDNESS. Night blindness is associated with vitamin A deficiency, for, when the total quantity of vitamin A in the blood becomes greatly reduced, the quantities of vitamin A, retinene, and rhodopsin in the rods, as well as the color photosensitive chemicals in the cones, are all depressed, thus decreasing the sensitivities of the rods and cones. This condition is called night blindness because at night the amount of available light is far too little to permit adequate vision, whereas in daylight, sufficient light is available to excite the rods and cones despite their reduction in photochemical substances.

For night blindness to occur, a person often must remain on a vitamin A deficient diet for weeks or months, because large quantities of vitamin A are normally stored in the liver and are made available to the rest of the body in times of need. However, once night blindness does develop it can sometimes be completely cured in a half hour or more by intravenous injection of vitamin A. This results from the ready conversion of vitamin A into retinene and thence into rhodopsin.

PHOTOCHEMISTRY OF COLOR VISION BY THE CONES

It was pointed out at the outset of this discussion that the photochemicals in the cones have almost exactly the same chemical compositions as rhodopsin in the rods. The only difference is that the protein portions, the opsins, called *photopsins* in the cones, are different from the scotopsin of the rods. The retinene portions are exactly the same in the cones as in the rods. The color-sensitive pigments of the cones, therefore, are combinations of retinene and photopsins.

In the discussion of color vision later in the chapter, it will become evident that three different types of pigments are present in different cones, thus making these cones selectively sensitive to the different colors, red, green, and blue. The absorption characteristics of the pigments in the three types of cones show peak absorbences at, respectively, 430, 535, and 575 millimicrons. The approximate absorption curves for these three pigments are shown in Figure 53–3. The peak absorption for the rhodopsin of the rods, on the other hand, occurs at 505 millimicrons.

Note that the peak absorption of the so-called "red" cone is actually in the orange color band. It is called the "red" cone because it responds to red much more intensely than do any of the other cones.

AUTOMATIC REGULATION OF RETINAL SENSITIVITY—DARK AND LIGHT ADAPTATION

Relationship of Sensitivity to Pigment Concentration. The sensitivity of rods is approxi-

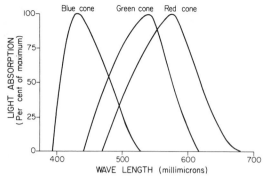

Figure 53–3 Light absorption by the respective pigments of the three color receptive cones of the human retina. (Drawn from curves recorded by Marks, Dobelle, and MacNichol, Jr.: *Science, 143:*1181, 1964, and by Brown and Wald: *Science, 144:*45, 1964.)

Figure 53–4 Dark adaptation, illustrating the relationship of cone adaptation to rod adaptation.

mately proportional to the antilogarithm of the rhodopsin concentration, and it is assumed that this relationship also holds true in the cones. This means that a minute decrease in rhodopsin reduces the sensitivity of rods tremendously. For instance, the sensitivity of a rod is reduced about 8.5 times when its concentration of rhodopsin is reduced from maximum by only 0.006 per cent, and the sensitivity decreases over 3000 times when the rhodopsin concentration is reduced 0.6 per cent. This antilog effect is postulated to result from the fact that the first rhodopsin to be decomposed on exposure to light lies near the surface of the membrane discs in the rod. After this portion has decomposed, far greater decomposition of the deeper layers of rhodopsin is required to excite the rods.

In summary, the sensitivity of the rods and cones can be altered up or down tremendously by only slight changes in concentrations of the photosensitive chemicals.

Light and Dark Adaptation. If a person has been in bright light for a long time, large proportions of the photochemicals in both the rods and cones are reduced to retinene and opsins. Furthermore, much of the retinene in both the rods and cones is converted into vitamin A. Because of these two effects, the concentrations of the photochemicals are considerably reduced, and the sensitivity of the eye to light is even more reduced. This is called *light adaptation.*

On the other hand, if the person remains in the darkness for a long time, essentially all the retinene and opsins in the rods and cones become converted into light-sensitive pigments. Furthermore, large amounts of vitamin A are converted into retinene, which is then changed into additional light-sensitive pigments, the final limit being determined by the amount of

opsins in the rods and cones. Because of these two effects, the visual receptors gradually become so sensitive that even the minutest amount of light causes excitation. This is called *dark adaptation.*

Figure 53–4 illustrates the course of dark adaptation when a person is exposed to total darkness after having been exposed to bright light for several hours. Note that retinal sensitivity is very low when one first enters the darkness, but within 1 minute the sensitivity has increased 10-fold—that is, the retina can respond to light of one-tenth intensity. At the end of 20 minutes the sensitivity has increased about 6000-fold, and at the end of 40 minutes it has increased about 25,000-fold.

The resulting curve of Figure 53–4 is called the *dark adaptation curve.* Note, however, the inflection in the curve. The early portion of the curve is caused by adaptation of the cones, for these adapt much more rapidly than the rods because of a basic difference in the rate at which they resynthesize their photosensitive pigments. On the other hand, the cones do not achieve anywhere near the same degree of sensitivity as the rods. Therefore, despite rapid adaptation by the cones, they cease adapting after only a few minutes, while the slowly adapting rods continue to adapt for many minutes or even hours, their sensitivity increasing tremendously. However, a large share of the greater sensitivity of the rods is caused by convergence of as many as 100 rods onto a single ganglion cell in the retina; these rods summate to increase their sensitivity, as will be discussed in the following chapter.

Figure 53–5 illustrates dark adaptation, but in this figure it is shown that the rate of dark adaptation differs, depending on the previous degree of exposure of the retina to light. Curve number 1 shows the dark adaptation after a person had been in bright light for approximately 20 minutes, curve number 2 shows the rate after he had been in bright light for only a few minutes, and curve number 3 after he had been in bright light for many hours. Note that the eye exposed to bright light for a long time adapts slowly in comparison with the eye that has been in bright light for only a few minutes. The cause of this difference is probably the slowness of the interconversions between vitamin A and retinene. The eye that has remained in bright light for only a few minutes has not converted significant quantities of its retinene into vitamin A. Therefore, this eye can resynthesize its photosensitive pigments in a short time. Conversely, the eye that has been exposed to prolonged bright light will have had time to convert most of its retinene into vitamin A, and since the reconversion from vitamin A to retinene is very slow, dark adaptation of the eye also is slow.

To the right in Figure 53–5 is shown a typical light adaptation curve, illustrating that light adaptation occurs much more rapidly than dark adaptation. Also, the slow component caused by retinene-vitamin A interconversion does not play as prominent a role in light adaptation as in dark adaptation. Therefore, a person exposed to extremely bright light after having been in prolonged darkness becomes adjusted to the new light conditions in only a few minutes. This is in contrast to dark adaptation, which requires 10 to 20 minutes for moderate adaptation and 10 to 18 hours for maximal adaptation.

Value of Light and Dark Adaptation in Vision. Between the limits of maximal dark adaptation and maximal light adaptation, the retina of the eye can change its sensitivity to light by as much as 500,000 to 1,000,000 times, the sensitivity automatically adjusting to changes in illumination.

Since the registration of images by the retina requires detection of both dark and light spots in the image, it is essential that the sensitivity of the retina always be adjusted so that the receptors respond to the lighter areas and not to the darker areas. An example of maladjust-

ment of the retina occurs when a person leaves a movie theater and enters the bright sunlight, for even the dark spots in the images then seem exceedingly bright, and, as a consequence, the entire visual image is bleached, having little contrast between its different parts. Obviously, this is poor vision, and it remains poor until the retina has adapted sufficiently that the dark spots of the image no longer stimulate the receptors excessively.

Conversely, when a person enters darkness, the sensitivity of the retina is usually so slight that even the light spots in the image cannot excite the retina. But, after dark adaptation, the light spots begin to register. As an example of the extremes of light and dark adaptation, the light intensity of the sun is approximately 30,000 times that of the moon; yet the eye can function well both in bright sunlight and in bright moonlight.

Negative After-Images. If one looks steadily at a scene for a while, the bright portions of the image cause light adaptation of the retina while the dark portions of the image cause dark adaptation. In other words, areas of the retina that are stimulated by light become less sensitive while areas that are exposed only to darkness gain in sensitivity. If the person then moves his eyes away from the scene and looks at a bright white surface he sees exactly the same scene that he had been viewing, but the light areas of the scene now appear dark, and the dark areas appear light. This is known as the negative after-image, and it is a natural consequence of light and dark adaptation.

The negative after-image persists as long as any degree of light and dark adaptation remains in the respective portions of the retina. Referring back to Figure 53–5, one can see from the dark adaptation curves that a negative after-image could possibly persist as long as an hour under favorable conditions.

FUSION OF FLICKERING LIGHTS BY THE RETINA

A flickering light is one whose intensity alternately increases and decreases rapidly. An instantaneous flash of light excites the visual receptors for as long as $1/10$ to $1/5$ second, and because of this *persistence* of excitation, rapidly successive flashes of light become *fused* together to give the appearance of being continuous. This effect is well known when one observes motion pictures or television. The images on the motion picture screen are flashed at a rate of 24 frames per second, while those of the television screen are flashed at a rate of 60 frames per second. As a result, the images fuse together, and continuous motion can be observed.

Figure 53–5 Dark and light adaptation. (The retina is considered to have a sensitivity of 1 when maximally light adapted.)

Figure 53–6 illustrates that the critical frequency at which flicker fusion occurs varies with the light intensity. At a low intensity, fusion results even when the rate of flicker is as low as 5 or 6 per second. However, in bright illumination, the critical frequency for fusion rises to as great as 60 flashes per second. This difference results at least partly from the fact that the cones, which operate mainly at high levels of illumination, can detect much more rapid alterations in illumination than can the rods, which are the important receptors in dim light. Also, it is likely that intense light causes such intense nerve signals that it is easier for the brain to detect alternation in the signals.

COLOR VISION

From the preceding sections, we know that the cones contain light-sensitive pigments that are different from the rhodopsin in the rods and also that different cones are sensitive to different colors of light. The present section is a discussion of the mechanisms by which the retina detects the different gradations of color in the visual spectrum.

THE TRI-COLOR THEORY OF COLOR PERCEPTION

Many different theories have been proposed to explain the phenomenon of color vision, and they are all based on the well-known observation that the human eye can detect almost all gradations of colors when red, green, and blue monochromatic lights are appropriately mixed in different combinations.

The first important theory of color vision

Figure 53–6   Relationship of intensity of illumination to the critical frequency for fusion.

was that of Young, which was later expanded and given more experimental basis by Helmholtz. Therefore, the theory is known as the *Young-Helmholtz theory*. According to this theory, there are three different types of cones, each of which responds maximally to a different color, one type responding maximally to red, another to green, and another to blue.

As time has gone by, the Young-Helmholtz theory has been expanded and more details have been worked out. It now is generally accepted as the mechanism of color vision.

Spectral Sensitivities of the Three Types of Cones.   On the basis of psychological tests, the spectral sensitivities of the three different types of cones in human beings are essentially the same as the light absorption curves for the three types of pigment found in the respective cones. These were illustrated in Figure 53–3, and they can readily explain almost all the phenomena of color vision.

Interpretation of Color in the Nervous System. Referring again to Figure 53–3, one can see that a red monochromatic light with a wavelength of 610 millimicrons stimulates the red cones to a stimulus value of approximately 0.75 while it stimulates the green cones to a stimulus value of approximately 0.13 and the blue cones not at all. Thus, the ratios of stimulation of the three different types of cones in this instance are 75:13:0. The nervous system interprets this set of ratios as the sensation of red. On the other hand, a monochromatic blue light with a wavelength of 450 millimicrons stimulates the red cones to a stimulus value of 0, the green cones to a value of 0.14 and the blue cones to a value of 0.86. This set of ratios—0:14:86—is interpreted by the nervous system as blue. Likewise, ratios of 100:50:0 are interpreted as orange-yellow and 50:85:15 as green.

This scheme also shows how it is possible for a person to perceive a sensation of yellow when a red light and a green light are shone into the eye at the same time, for this stimulates the red and green cones approximately equally, which gives a sensation of yellow even though no wavelength of light corresponding to yellow is present.

PERCEPTION OF WHITE LIGHT. Approximately equal stimulation of all the red, green, and blue cones gives one the sensation of seeing white. Yet there is no wavelength of light corresponding to white; instead, white is a combination of all the wavelengths of the spectrum. Furthermore, the sensation of white can be achieved by stimulating the retina with a proper combination of only three chosen

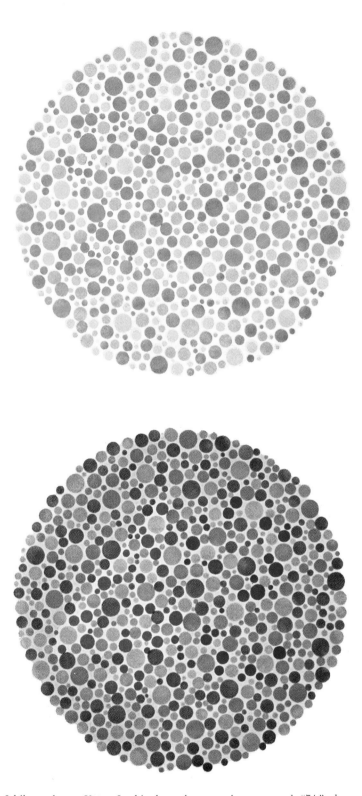

Figure 53–7   Two Ishihara charts: *Upper:* In this chart, the normal person reads "74," whereas the red-green color blind person reads "21." *Lower:* In this chart, the red-blind person (protanope) reads "2," while the green-blind person (deuteranope) reads "4." The normal person reads "42." (From Ishihara: Tests for Colour-Blindness. Tokyo, Kanehara and Co.)

colors that stimulate the respective types of cones.

Integration of Color Sensations by the Retina and the Cerebral Cortex. From psychological studies, we know that the interpretation of color is performed partly by the retina and partly by the cerebral cortex. If a person places a monochromatic green filter in front of his left eye and a monochromatic red filter in front of his right eye the visual object appears mainly yellow. This integration of color sensations obviously could not be occurring in the retina because one retina is allowed to respond only to green light and the other only to red light. However, sensations perceived in this way are not as clearly mixed as those perceived when the two monochromatic lights are mixed in the same retina. Therefore, at least some degree of the interpretation of color occurs in the retina itself even before the light information is transmitted into the brain.

COLOR BLINDNESS

Red-Green Color Blindness. When a single group of color receptive cones is missing from the eye, the person is unable to distinguish some colors from others. As can be observed by studying Figure 53–3, if the red cones are missing, light of 525 to 625 millimicrons wavelength can stimulate only the green-sensitive cones, so that the *ratio* of stimulation of the different cones does not change as the color changes from green all the way through the red spectrum. Therefore, within this wavelength range, all colors appear to be the same to this "color blind" person.

On the other hand, if the green-sensitive cones are missing, the colors in the range from green to red can stimulate only the red-sensitive cones, and the person also perceives only one color within these limits. Therefore, when a person lacks either the red or green types of cones, he is said to be "red-green" color blind. However, when one or more types of cones are abnormal but still function partially, a person has "color weakness," instead of color blindness.

The person with loss of red cones is called a *protanope;* his overall visual spectrum is noticeably shortened at the long wavelength end because of lack of the red cones. The color blind person who lacks green cones is called a *deuteranope;* this person has a perfectly normal visual spectral width because the green cones operate in the middle of the spectrum.

Blue Weakness. Occasionally, a person has "blue weakness," which results from diminished or absent blue receptors. If we observe Figure 53–3 once again, we can see that the blue cones are sensitive to a spectral range almost entirely different from that of both the red and green cones. Therefore, if the blue receptors are completely absent, the person has a greater preponderance of green,

yellow, orange, and red in his visual spectrum than he has of blue, thus producing this rarely observed type of color weakness or blindness.

Tests for Color Blindness. Tests for color blindness depend on the person's ability to distinguish various colors from each other and also on his ability to judge correctly the degree of contrast between colors. For instance, to determine whether or not a person is red-green color blind, he may be given many small tufts of wool whose colors encompass the entire visual spectrum. He is then asked to place those tufts that have the same colors in the same piles. If he is not color blind, he recognizes immediately that all the tufts have slightly different colors; however, if he is red-green color blind, he places the red, orange, yellow, and yellow-green colors all together as having essentially the same color.

STILLING AND ISHIHARA TEST CHARTS. A rapid method for determining color blindness is based on the use of spot-charts such as those illustrated in Figure 53–7. These charts are arranged with a confusion of spots of several different colors. In the top chart, the normal person reads "74," while the red-green color blind person reads "21." In the bottom chart, the normal person reads "42," while the red blind protanope reads "2," and the green blind deuteranope reads "4."

If one will study these charts while at the same time observing the spectral sensitivity curves of the different cones in Figure 53–3, he can readily understand how excessive emphasis can be placed on spots of certain colors by color blind persons in comparison with normal persons.

Genetics of Color Blindness. Color blindness is sex-linked and probably results from absence of appropriate color genes in the X chromosomes. This lack of color genes is a recessive trait; therefore, color blindness will not appear as long as another X chromosome carries the genes necessary for development of the respective color-receptive cones.

Because the male human being has only one X chromosome, all three color genes must be present in this single chromosome if he is not to be color blind. In approximately 1 out of every 50 times, the X chromosome lacks the red gene; in approximately 1 out of every 16 it lacks the green gene; and rarely it lacks the blue gene. This means, therefore, that 2 per cent of all men are red color blind (protanopes) and that 6 per cent are green color blind (deuteranopes), making a total of approximately 8 per cent who are red-green color blind. Because a female has two X chromosomes, red-green color blindness is a rare abnormality in the female, occurring in only 1 out of every 250 women.

REFERENCES

Alpern, M.: Distal mechanisms of vertebrate color vision. *Ann. Rev. Physiol.,* 30:279, 1968.
Baker, H. D., and Rushton, W. A. H.: The red-sensitive pigment in normal cones. *J. Physiol. (Lond.),* 176:56, 1965.

Ballantyne, A. J., and Michaelson, I. C.: Textbook of the Fundus of the Eye. Baltimore, The Williams & Wilkins Co., 1962.

Bishop, P. O.: The visual system: the neural basis of visual perception. *Proc. Int. Union of Physiol. Sciences, 6*:262, 1968.

Brindley, G. S.: Physiology of the Retina and Visual Pathway. Baltimore, The Williams & Wilkins Co., 1960.

Brown, P. K., and Wald, G.: Visual pigments in single rods and cones of the human retina. *Science, 144*:45, 1964.

Ciba Foundation: Colour Vision: Physiology and Experimental Psychology. Boston, Little, Brown and Company, 1965.

Dartnall, H. J. A., and Tansley, K.: Physiology of vision: retinal structure and visual pigments. *Ann. Rev. Physiol., 25*:433, 1963.

Davson, H. (ed.): The Visual Process. *In* The Eye, New York, Academic Press, Inc., 1962, Vol. 2.

Dowling, J. E.: Night blindness, *Sci. Amer., 215*:78(4), 1966.

Duke-Elder, S.: Diseases of the Retina. *In* Systems of Ophthalmology, St. Louis, The C. V. Mosby Co., 1967, Vol. 10.

Hartline, H. K.: Excitation and inhibition in the retina. *Proc. Int. Union of Physiol. Sciences, 6*:73, 1968.

Linksz, A.: An Essay on Color Vision and Clinical Color Vision Tests. New York, Grune & Stratton, Inc., 1964.

Linksz, A.: Physiology of the Eye. Vol. II. Vision. New York, Grune & Stratton, Inc., 1952.

MacNichol, E. F., Jr.: Three-pigment color vision. *Sci. Amer., 211*:48(6), 1964.

Marks, W. B.; Dobelle, W. H.; and MacNichol, E. E., Jr.: Visual pigments of single primate cones. *Science, 143*:1181, 1964.

Neisser, U.: The processes of vision. *Sci. Amer., 219*:204(3), 1968.

Rushton, W. A. H.: A foveal pigment in the deuteranope. *J. Physiol. (Lond.), 176*:24, 1965.

Rushton, W. A. H.: Cone pigment kinetics in the deuteranope. *J. Physiol. (Lond.), 176*:38, 1965.

CHAPTER 54

The Eye:
III. Neurophysiology of Vision

THE VISUAL PATHWAY

Figure 54-1 illustrates the visual pathway from the two retinae back to the *visual cortex.* After impulses leave the retinae they pass backward through the *optic nerves.* At the *optic chiasm* all the fibers from the nasal halves of the retinae cross to the opposite side where they join the fibers from the opposite temporal retinae to form the *optic tracts.* The fibers of each optic tract synapse in the *lateral geniculate body,* and from here, the *geniculocalcarine fibers* pass through the *optic radiation,* or *geniculocalcarine tract,* to the *optic* or *visual cortex* in the calcarine area of the occipital lobe.

In addition, visual fibers pass to three other areas of the brain: (1) from the lateral geniculate body to the lateral thalamus, the ventral thalamus, the superior colliculi, and the pretectal nuclei; (2) from the optic tracts directly to the superior colliculi; and (3) from the optic tracts directly into the pretectal nuclei of the brain stem.

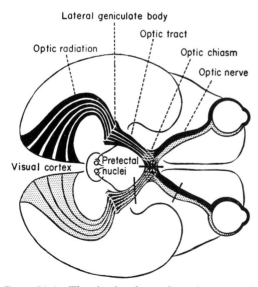

Figure 54-1 The visual pathways from the eyes to the visual cortex. (Modified from Polyak: The Retina. University of Chicago Press.)

NEURAL FUNCTION OF THE RETINA

NEURAL ORGANIZATION OF THE RETINA

The anatomy of the retina was illustrated in Figure 53-1 of the preceding chapter, and Figure 54-2 illustrates the basic essentials of the retina's neural connections; to the left is the general organization of the neural elements in a peripheral retinal area and to the right the organization in the foveal area. Note that in the peripheral region both rods and cones converge on *bipolar cells* which in turn converge on *ganglion cells.* In the fovea, where only cones exist, there is little or no convergence; instead, the cones are represented by approximately equal numbers of bipolar and ganglion cells.

Not shown in the figure is the fact that in both the outer and inner plexiform layers (that is, where the neurons synapse with each other) there are many lateral connections between the different neural elements. Many of these lateral connections are collateral branches of the dendrites and axons of the bipolar and ganglion cells. But, in addition, two special types of cells are present that transmit impulses laterally, (1) the *horizontal cells* in the outer plexiform layer and (2) the *amacrine cells* in the inner plexiform layer.

Each retina contains about 125 million rods and 5.5 million cones; yet, as counted with the light microscope, only 900,000 optic nerve fibers lead from the retina to the brain. Thus, an average of 140 rods plus 6 cones converge on each optic nerve fiber. However, there are major differences between the peripheral retina and the central retina, for nearer the fovea, fewer and fewer rods and cones converge on each optic fiber, and the rods and cones both become slenderer. These two effects progressively increase the acuity of vision toward the central retina. In the very central portion, in the fovea, there are no rods at all. Also, the number of optic nerve fibers leading

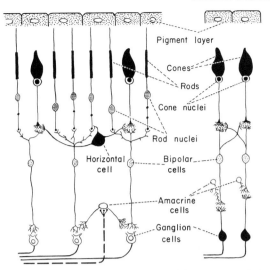

Figure 54-2 Neural organization of the retina: peripheral area to the left, foveal area to the right.

from this part of the retina is almost equal to the number of cones in the fovea, as shown to the right in Figure 54-2. This mainly explains the high degree of visual acuity in the central portion of the retina in comparison with the very poor acuity in the peripheral portions.

Another difference between the peripheral and central portions of the retina is that there is considerably greater sensitivity of the peripheral retina to weak light. This is believed to result mainly from the fact that as many as 600 rods converge on the same optic nerve fiber in the most peripheral portions of the retina; the signals from these rods summate to give more intense stimulation of the peripheral ganglion cells. The rods are probably also more sensitive to weak light than are the cones, but this is not certain.

STIMULATION OF THE RODS AND CONES — THE RECEPTOR POTENTIAL

Stimulation of the rods and cones was discussed at length in the preceding chapter. Briefly, when light strikes either of these receptors, it decomposes a photochemical that in turn acts on the membrane of the *outer segment* of the receptor (the part that protrudes into the pigment layer) to cause a sustained *receptor potential* that lasts as long as the light continues. However, as pointed out in the previous chapter, this receptor potential is different from all other receptor potentials that have been recorded, for it is positive rather than negative, and its intensity is pro-

portional to the logarithm of light energy in contrast to the more linear response of many other receptors. This logarithmic response seems to be very important to vision, because it allows the eyes to detect contrasts in the image even when light intensities vary many thousand-fold, an effect that would not be possible if there were linear transduction of the signal.

Receptor potentials are transmitted unchanged through the bodies of the rods and cones. It is believed that neither the rods nor the cones generate action potentials. Instead, the receptor potentials themselves, acting at the ends of the axons, are believed to induce signals directly in successive neurons.

STIMULATION OF THE BIPOLAR AND HORIZONTAL CELLS

Very little is known about function of the neurons in the inner nuclear layer—the bipolar cells and horizontal cells—except that the horizontal cells transmit signals laterally in the retina and the bipolar cells transmit signals from the rods and cones to the ganglion cells. The axons of the rods and cones make extremely intimate contact with the dendrites of both the bipolar cells and the horizontal cells. Recordings of intracellular signals in the horizontal cells indicate that these cells, like the rods and cones, perhaps also transmit no action potentials but instead transmit signals of the electrotonic variety. On the other hand, action potentials have been recorded from neurons in the inner nuclear layer that are believed to be the bipolar cells.

STIMULATION OF THE GANGLION CELLS

Spontaneous, Continuous Discharge of the Ganglion Cells. Though what happens in the transmission of visual signals from the rods and cones to the ganglion cells is not clear, it is known that the visual signals have already been changed tremendously by the time they reach the ganglion cells. The first important effect is that the ganglion cells transmit continuous nerve impulses at a rate of about 5 per second even when the retina is in total darkness. Then, the visual signal is superimposed onto this basic level of ganglion cell stimulation.

Effect of Light on Ganglion Cell Discharge. Figure 54-3 illustrates a minute spot of light shining on the very center of the *retinal field* of a peripheral ganglion cell. The light stimulates the rods and cones in this center, and the

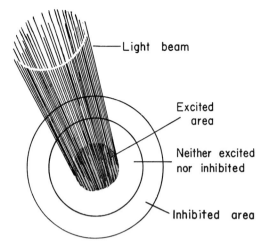

Figure 54-3 Excitation and inhibition of a retinal area caused by a small beam of light.

Figure 54-4 Responses of ganglion cells to light in (1) an area excited by a spot of light and (2) an area immediately adjacent to the excited spot; the ganglion cells in this area are inhibited by the mechanism of lateral inhibition. (Modified from Granit: Receptors and Sensory Perception. Yale University Press.)

signals from these converge on the ganglion cell. The top record of Figure 54-4 illustrates the effect of turning the spot of light on. This record shows to the left the normal background rate of discharge of the ganglion cell and then shows that the rate of discharge increases markedly a fraction of a second after the light is turned on. Within another fraction of a second, the rate of impulses decreases almost all the way back to the background rate of discharge; in the case of at least some of the ganglion cells, the rate does not decrease quite all the way back. From this record, one can see two aspects of ganglion cell function. First, there is an intense transient discharge when the light is turned on. Second, there is a much weaker increase in discharge of some of the cells that lasts probably as long as the light remains on.

When the light is turned off, exactly opposite effects occur. For a fraction of a second, the ganglion cell transmits no impulses at all, but the rate gradually picks back up over another few seconds until the normal background rate of discharge returns.

Lateral Inhibition of Surrounding Ganglion Cells. Referring once again to Figure 54-3, one sees a concentric area surrounding the spot of light on the retina labeled "inhibited area." The lower record in Figure 54-4 shows the effect of the spot of light on discharge of a ganglion cell whose field center is in this area. Note that when the light is turned on, these ganglion cells become completely inhibited and stop transmitting even their normal background impulses. Then, when the light is turned off, the impulse rate increases at first to a level considerably greater than the normal rate of discharge, this effect lasting for a frac-

tion of a second until the cells readapt to their normal level of discharge.

MECHANISM OF LATERAL INHIBITION. It is believed that lateral inhibition is caused by signals carried in the *horizontal cells* that have extensive horizontal connections in the outer plexiform layer of the retina. One of the reasons for believing this is that these cells have far more extensive fiber networks in the peripheral retina than in the central retina, and it is found experimentally that the area of inhibition is great in the periphery but circumscribed in the central retina, which is consonant with the more discrete visual functions of the central retina than of the peripheral retina.

SIGNIFICANCE OF LATERAL INHIBITION—ENHANCEMENT OF "CONTRAST" IN THE VISUAL SIGNAL. In the above description of the neural elements of the retina we noted that the dendrites and axons in the retina spread laterally a moderate amount within the plexiform layers. This obviously causes the signals from adjacent receptors of the retina to spread into each other. Counteracting this lateral mixing of signals is the phenomenon of lateral inhibition. That is, the excitatory signals that spread laterally are opposed and nullified by the lateral inhibitory signals. This obviously helps to maintain sharp boundaries between the signals from excited and nonexcited areas of the retina, which enhances the *contrast* of the visual signal.

The "On-Off" Responses of Ganglion Cells. Note in Figure 54-4 that the ganglion cell receiving signals from an excited area of the retina transmits large numbers of impulses when the light is turned on. This is called the "on" response of the ganglion cell. On the other hand, note that the ganglion cell in the inhibited area of the retina adjacent to the excited area transmits large numbers of impulses immediately after the light is cut off. This is called the "off" response of a ganglion cell. In the intermediate zone between the excited and the inhibited areas, the ganglia give an "on-off" response in which they fire both when the light is turned on and when it is turned off. Also, some ganglia give an "off" response when they are in the center of the light spot and an "on" response when slightly to one side of the spot. These types of responses are described here because they are used by neurophysiologists to track patterns of excitation or inhibition through the visual pathway.

REGULATION OF RETINAL FUNCTION BY CENTRIFUGAL FIBERS FROM THE CENTRAL NERVOUS SYSTEM

The sensitivity of the retina is at least partially controlled by signals from the central nervous system. Centrifugal fibers pass in the retrograde direction in the optic nerve from the brain to the retina and synapse in the retina with the *amacrine cells* and perhaps to some extent directly with the ganglion cells. Stimulation of specific areas in the brain stem can enhance in some instances and diminish in other instances the sensitivities of specific areas of the retina. It is supposed that this represents a mechanism by which the central nervous system can direct one's attention to specific portions of the visual field.

FUNCTION OF THE LATERAL GENICULATE BODY

Anatomical Organization of the Lateral Geniculate Nuclei. Each lateral geniculate body is composed of six nuclear layers. Layers 2, 3, and 5 (from the surface inward) receive signals from the temporal portion of the ipsilateral retina, while layers 1, 4, and 6 receive signals from the nasal retina of the opposite eye.

All layers of the lateral geniculate body relay visual information to the *visual cortex*. It is from here that the *geniculocalcarine fibers* originate.

The pairing of layers from the two eyes probably plays a major role in *fusion of vision,* because corresponding retinal fields in the two eyes connect with respective neurons that are approximately superimposed over each other in the successive layers. Also, with a little imagination, one can postulate that interaction between the successive layers could be part of the mechanism by which stereoscopic visual depth perception occurs, because this depends on comparing the visual images of the two eyes and determining their slight differences, as was discussed in Chapter 52.

Characteristics of the Visual Signal in the Lateral Geniculate Body. The signals recorded in the relay neurons of the lateral geniculate body are almost the same as those recorded in the ganglion cells of the retinae. That is, a

lateral geniculate neuron representing an excited area of the retina discharges at a rapid rate immediately after the light is turned on, and then this rate decreases with time. Likewise, neurons representing areas of the retina immediately adjacent to the excited area are inhibited in the same way that corresponding ganglion cells are inhibited when the light is turned on. In other words, the lateral geniculate body appears to relay information from the optic nerve fibers into the geniculocalcarine tract without altering it greatly.

However, there is one important difference between the visual signals in the neurons of the lateral geniculate body and those in the ganglion cells: the boundaries between excited and nonexcited areas of the visual signals are considerably sharper in the lateral geniculate neurons than in the ganglion cells of the retina. This indicates that additional lateral inhibition probably occurs in the lateral geniculate body in addition to that which occurs in the retina, this lateral inhibition further enhancing the degree of contrast in the visual pattern.

FUNCTION OF THE PRIMARY VISUAL CORTEX

The ability of the visual system to detect spatial organization of the visual scene—that is, to detect the forms of objects, brightness of the individual parts of the objects, shading, and so forth—is dependent on function of the *primary visual cortex,* which is illustrated in Figure 54–5. This area lies mainly in the calcarine fissure located bilaterally on the medial aspect of each occipital cortex. Specific points of the retina connect with specific points of the visual cortex, the right halves of the two respective retinae connecting with the right visual cortex and the left halves connecting with the left visual cortex. The macula is repre-

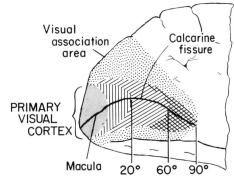

Figure 54–5 The visual cortex.

Retinal Image   Cortical Stimulation

Figure 54–6   Pattern of excitation occuring in the visual cortex in response to a retinal image of a black cross.

sented at the occipital pole of the visual cortex and the peripheral regions of the retina are represented in concentric circles farther and farther forward from the occipital pole. The upper portion of the retina is represented superiorly in the visual cortex and the lower portion inferiorly. Note the large area of cortex receiving signals from the macular region of the retina. It is in the center of this region that the fovea is represented, which gives the highest degree of visual acuity.

DETECTION OF LINES AND BORDERS BY THE PRIMARY VISUAL CORTEX

If a person looks at a blank wall, very little stimulation occurs in the neurons of the primary visual cortex, whether the illumination of the wall be bright or weak. Therefore, the question must be asked: What does the visual cortex do? To answer this question, let us now place on the wall a large black cross such as that illustrated to the left in Figure 54–6. To the right is illustrated the spatial pattern of excitation that one finds in the visual cortex. *Note that the areas of excitation occur along the sharp borders of the visual pattern.* Thus, by the time the visual signal is recorded in the primary visual cortex, it is concerned almost entirely with the *contrasts* in the visual scene rather than with the flat areas. At each point in the visual scene where there is a change from dark to light or light to dark, the corresponding area of the primary visual cortex becomes stimulated. The intensity of stimulation is determined by the *gradient of contrast.* That is, the greater the sharpness in the contrast border and the greater the difference in intensities between the light and dark areas, the greater is the degree of stimulation.

Thus, the *pattern of contrasts* in the visual scene is. impressed upon the neurons of the visual cortex, and this pattern has a spatial orientation roughly the same as that of the retinal image.

Detection of Orientation of Lines and Borders. Not only does the visual cortex detect the existence of lines and borders in the different areas of the retinal image, but it also detects the orientation of each line or border—that is, whether it is a vertical or horizontal border, or lies at some degree of inclination. The way in which the visual cortex performs this function is the following:

The visual cortex is composed of thousands of minute columns of neurons, each column having a diameter of about 0.5 mm. and extending from the surface of the cortex down through its six layers. Each column of cells is excited by a border or line that is oriented only in one particular direction. Thus, a column of cells that will respond to a horizontal border will not respond to a vertical border. However, immediately adjacent to this horizontal column of cells is another column that responds to a border slightly inclined from the horizontal, then another adjacent column that responds to a border with still more inclination, and still another column that responds to a vertical border. Thus, each minute area of the visual cortex has multiple neuron columns, each of which responds specifically to a border or line oriented in a specific direction.

DETECTION OF LINES AND BORDERS BY NEURONS IN THE RETINA AND LATERAL GENICULATE BODY. In both the retina and the lateral geniculate body, some neurons become excited when a line or a border appears on the corresponding retinal receptors. It is believed that the phenomenon of lateral inhibition causes this effect as follows: A particular ganglion cell is excited by light in one area and yet inhibited by light in a contiguous area. Thus, if a contrast border occurs between these two areas so that light falls on the excitable field of the ganglion cell while the dark area falls on the inhibitory field, the ganglion cell will be maximally excited. Yet, flat white light would cause almost no excitation because the ganglion cell would be inhibited by one retinal field and excited by the other.

Similar mechanisms of interaction are believed to occur at the lateral geniculate level and again in the visual cortex, the recognition of lines and borders becoming progressively more acute the higher into the visual nervous system one goes.

NEURAL ANALYSIS OF COLOR

Almost nothing is known at present about the mechanisms by which the visual nervous system decodes color information from the retina. However, decoding seems to begin in the retina itself by interaction among signals from different cones. A few ganglion cells transmit signals from blue receptive cones, others from green cones, and still others from red cones, representing the three primary colors. But most ganglion cells transmit combined color information from several types of cones. Two of the combinations are the following:

1. Some ganglion cells combine information from all three types of cones, supposedly giving no color differentiation at all but only an overall luminosity signal.

2. Some ganglion cells give positive responses to blue and negative responses to yellow, whereas others give positive responses to yellow and negative to blue. Still other cells give positive responses to red and negative to blue, and so forth.

These interactions are much the same as those that result at contrast borders in the visual image and presumably are caused by essentially the same mechanism. That is, cones responding to one color send an inhibitory signal to the ganglion cell while cones stimulated by another color send an excitatory signal.

It is believed that further interactions of a similar nature but of a higher order occur at higher levels in the visual nervous system—in the lateral geniculate body and the visual cortex. That such interactions can occur at these higher levels is proved by the fact that looking at an object through a red filter with one eye and a green filter with the other eye will give the impression of seeing an image that has neither a red nor green color but some intermediate color such as yellow or orange.

PERCEPTION OF LUMINOSITY

One of the most important features of any visual image is its level of light intensity, or its *luminosity*. Yet, it has been very difficult to understand how luminosity is perceived by the visual neural apparatus because most ganglion cells in the retina respond primarily to *changes* in light intensities or to contrast borders in the retinal image. Yet, a few ganglion cells do seem to respond weakly at continuous rates proportional to the overall luminosity rather than simply to changes or to contrast. These signals are believed to carry luminosity information to the brain.

Yet, once the information reaches the brain, it is still a question which parts of the brain are responsible for perception of luminosity. In the human being, destruction of the primary

visual cortex causes total blindness, including loss of ability to determine luminosity. However, some lower animals can still detect changes in luminosity almost as well as previously, and this suggests that lower centers could be very important in this function of the brain.

Perception of Rapid Changes in Light Intensity and of Movement. The primary visual cortex is very strongly excited by (1) sudden changes in light intensity, particularly when these changes occur along sharp borders, and (2) movement of an image across the retina. Both of these effects cause rapid discharge of ganglion cells for a fraction of a second after the initial stimulation. These effects emphasize again that the visual cortex is especially well adapted for perception of contrast signals in contradistinction to perception of steady signals.

TRANSMISSION OF VISUAL INFORMATION INTO OTHER REGIONS OF THE CEREBRAL CORTEX

Even when a single spot of light is applied to the retina, impulses appear within a few milliseconds in widespread areas of the cerebral cortex adjacent to the primary visual cortex. These areas are called the *visual association areas,* some of which were illustrated in Figure 54–5. Figure 54–7 shows the association areas on the lateral surface of the cortex that are concerned with vision. Area 17 is the primary visual cortex, which lies almost entirely on the medial aspect of the cerebral hemisphere but extends barely out of the longitudinal fissure onto the outer surface of the occipital pole. Area 17 is also called the *striate area* because of its striped appearance to the naked eye. Area 18 lies immediately above and lateral to the visual cortex, and area 19 lies still farther above and lateral to area 18.

The retinal image is spatially oriented in both area 18 and area 19, just as occurs in the primary visual cortex. Area 18 receives signals directly from the primary visual cortex via subcortical fibers. Area 19 receives signals from area 18 and

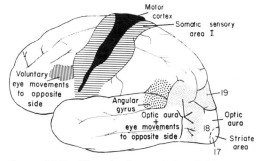

Figure 54–7 The visual association fields and the cortical areas for control of eye movements.

also directly from the lateral pulvinar of the thalamus, which lies immediately adjacent to the lateral geniculate body and receives signals from it.

The visual projection images in areas 18 and 19 are organized into columns of cells in the same manner as that described earlier for the primary visual cortex. However, these columns of cells respond to more complex patterns than do those in the primary visual cortex. For instance, some columns of cells respond to lines or borders only of particular lengths while other lengths fail to stimulate. Other columns are stimulated by simple geometric patterns, such as right angles, curving borders, or so forth, and others by specific combinations of colors. It is presumably these progressively more complex interpretations of form, colors, and probably also temporal changes that eventually decode the visual information, giving the person his overall impression of the visual scene that he is observing.

Effects of Electrical Stimulation or Destruction of Visual Cortical Areas in the Human Being.

Electrical stimulation in the primary visual cortex, area 18, or area 19 causes a person to have optic auras— that is, flashes of light, colors, or simple forms such as lines, stars, discs, triangles, and so forth—but he does not see complicated forms.

Stimulation of the lower border of the *temporal cortex,* on the other hand, often elicits complicated visual perceptions, sometimes causing the person to "see" a scene that he had known many years before.

Widespread destruction of areas 18 and 19 decreases one's ability to interpret the shapes of objects, their sizes, or their meanings, and can cause particularly an abnormality known as *alexia,* or *word blindness,* which means that the person cannot identify the meanings of words that he sees. Destruction of the cerebral cortex in the angular gyrus region where the temporal, parietal, and occipital lobes all come together usually makes it difficult for a person to correlate visual images with the motor functions. For instance, he is able to see his plate of food perfectly well but is unable to utilize the visual information to direct his fork toward the food. Yet, if he feels the plate with his other hand, he can use stereotaxic information from his somesthetic cortex to direct the fork accurately. Other aspects of visual interpretation in relation to overall function of the cerebral cortex will be discussed in Chapter 61, and later in this chapter we will return to the cortical control of eye movements.

THE FIELDS OF VISION; PERIMETRY

The *field of vision* is the area seen by an eye at a given instant. The area seen to the nasal side is called the *nasal field of vision,* and the area seen to the lateral side is called the *temporal field of vision.*

To diagnose blindness in specific areas of the eyes, one charts the field of vision for each eye by a process known as *perimetry.* This is done by having the subject look with one eye toward a central spot directly in front of the eye. Then a small dot of light or a small object is moved back and forth in all areas of the field of vision, both laterally and nasally and upward and downward, and the person indicates when he can see the spot of light or object and when he cannot see it. At the same time a chart (Fig. 54-8) is made for the eye, showing the areas in which the subject can see the spot and in which he cannot see it. Thus, the field of vision, with both its blind and normal areas, is plotted.

In all perimetry charts, a blind spot caused by lack of rods and cones in the retina over the optic disc is found approximately 15 degrees lateral to the central point of vision, as illustrated in the figure.

Abnormalities in the Fields of Vision. Occasionally blind spots are found in other portions of the field of vision besides the optic disc area. Such blind spots are called *scotomata;* they frequently result from allergic reactions in the retina or from toxic conditions, such as lead poisoning or even excessive use of tobacco.

Still another condition that can be diagnosed by perimetry is *retinitis pigmentosa.* In this disease, portions of the retina degenerate and excessive melanin pigment deposits in the degenerated areas. Retinitis pigmentosa generally causes blindness in the peripheral field of vision first and then gradually encroaches on the central areas.

EFFECT OF LESIONS IN THE OPTIC PATHWAY ON THE FIELDS OF VISION. Perhaps the most important use of perimetry is in localization of lesions in the brain. Lesions in the optic nerve, in the optic chiasm, in the optic tract, and in the geniculocalcarine tract all cause blind areas in the visual fields, and the "patterns" of these blind areas indicate the location of the lesion in the optic pathway.

Destruction of an entire *optic nerve* obviously causes blindness of the respective eye. Destruction of the *optic chiasm,* as shown by the longitudinal line across the chiasm in Figure 54-1, prevents the

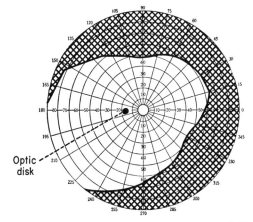

Figure 54-8 A perimetry chart, showing the field of vision for the left eye.

passage of impulses from the nasal halves of the two retinae to the opposite optic tracts. Therefore, the nasal halves are both blinded, which means that the person is blind in both temporal fields of vision; this condition is called *bitemporal hemianopsia.* Such lesions frequently result from tumors of the adenohypophysis pressing upward on the optic chiasm.

Interruption of an *optic tract,* which is also shown by a line in Figure 54–1, denervates the corresponding half of each retina on the same side as the lesion, and, as a result, neither eye can see objects to the opposite side. This condition is known as *homonymous hemianopsia.* Destruction of the *optic radiation* or the *visual cortex* of one side also causes homonymous hemianopsia. A common condition that destroys the visual cortex is thrombosis of the posterior cerebral artery, which infarcts the occipital cortex except for the foveal area, thereby sparing central vision.

One can differentiate a lesion in the optic tract from a lesion in the geniculocalcarine tract or visual cortex by determining whether impulses can still be transmitted into the pretectal nuclei to initiate a pupillary light reflex. To do this, light is shown onto the blinded half of one of the retinae, and, if a pupillary light reflex can still occur, it is known that the lesion is at or beyond the lateral geniculate body, that is, in the geniculocalcarine tract or visual cortex. However, if the light reflex is also lost, then the lesion is in the optic tract itself.

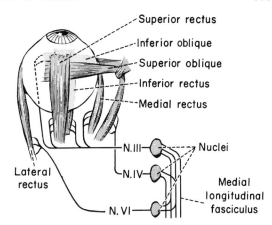

Figure 54–9 The extraocular muscles of the eye and their innervation.

EYE MOVEMENTS AND THEIR CONTROL

To make use of the abilities of the eye, it has been almost equally as important to develop a cerebral control system for directing the eyes toward the object to be viewed as to develop a system for interpretation of the visual signals from the eyes.

Muscular Control of Eye Movements. The eye movements are controlled by three separate pairs of muscles shown in Figure 54–9: (1) the medial and lateral recti, (2) the superior and inferior recti, and (3) the superior and inferior obliques. The medial and lateral recti contract reciprocally to move the eyes from side to side. The superior and inferior recti contract reciprocally to move the eyes upward or downward. And the oblique muscles function mainly to rotate the eyeballs to keep the visual fields in the upright position.

Neuronal Pathways for Control of Eye Movements. Figure 54–9 also illustrates the innervation of the ocular muscles by the third, fourth, and sixth cranial nerves, showing their respective nuclei in the brain stem. Figure 54–10 illustrates some of the interconnections between these nuclei and their control by the cerebrum.

Each of the three sets of muscles to each eye is reciprocally innervated so that one muscle of the pair relaxes while the other contracts.

Simultaneous movement of both eyes in the same direction is called *conjugate movement* of the eyes. Though the precise brain centers that control conjugate movements are still unknown, it is believed that most of these are controlled by the

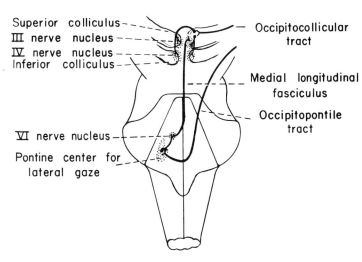

Figure 54–10 Neural pathways for control of the conjugate movements of the eyes.

superior colliculi. The association areas of the visual cortex, areas 18 and 19 in particular, transmit impulses through the *occipitocollicular tracts* to control the activities of the superior colliculi, and it is believed that these nuclei then transmit signals through the medial longitudinal fasciculus to the third, fourth, and sixth nuclei that control the eye muscles.

Some physiologists believe that there are additional centers located bilaterally in the pons, called the *pontile centers for lateral gaze,* that control the lateral conjugate movements while the superior colliculi control the vertical conjugate movements.

It must be pointed out, however, that bilateral destruction of the superior colliculi or of the pontile centers for lateral gaze does not prevent all vertical and lateral conjugate movements. Therefore, other centers of the brain stem are perhaps equally as important, or even more important, in the control of conjugate movements.

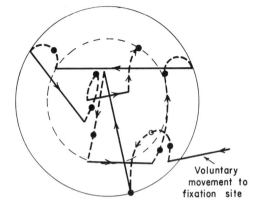

Voluntary
movement to
fixation site

Figure 54–11   Movements of a spot of light on the fovea, showing sudden "flicking" movements to move the spot back toward the center of the fovea whenever it drifts to the foveal edge. (The dashed lines represent slow drifting movements, and the solid lines represent sudden flicking movements.) (Modified from Whitteridge: Handbook of Physiology, Sec. I, Vol. II, p. 1089, 1960. Baltimore, The Williams & Wilkins Co.)

FIXATION MOVEMENTS OF THE EYES

Perhaps the most important movements of the eyes are those that cause the eyes to "fix" on a discrete portion of the field of vision.

Fixation movements are controlled by two entirely different neuronal mechanisms. The first of these allows the person to move his eyes voluntarily to find the object upon which he wishes to fix his vision; this is called the *voluntary fixation mechanism.* The second is an involuntary mechanism that holds the eyes firmly on the object once it has been found; this is called the *involuntary fixation mechanism.*

The voluntary fixation movements are controlled by a small cortical field located bilaterally in the premotor cortical regions of the frontal lobes, as illustrated in Figure 54–7. Bilateral dysfunction or destruction of these areas makes it difficult or almost impossible for the person to "unlock" his eyes from one point of fixation and then move them to another point. It is usually necessary for him to blink his eyes or put his hand over his eyes for a short time, which then allows him to move the eyes.

On the other hand, the fixation mechanism that causes the eyes to "lock" on the object of attention once it is found is controlled by the *eye fields of the occipital cortex* — mainly area 19 — which are also illustrated in Figure 54–7. When these areas are destroyed bilaterally, the person becomes completely unable to keep his eyes directed toward a given fixation point.

To summarize, the posterior eye fields automatically "lock" the eyes on a given spot of vision and thereby prevent movement of the image across the retina. To unlock this visual fixation, voluntary impulses must be transmitted from the "voluntary" eye fields located in the frontal areas.

Mechanism of Fixation.   Visual fixation results from a negative feedback mechanism that prevents the object of attention from leaving the foveal portion of the retina. The eyes even normally have three types of continuous but almost imperceptible movements: (1) a *continuous tremor* at a rate of 30 to 80 cycles per second caused by successive contractions of the motor units in the ocular muscles, (2) a *slow drift* of the eyeballs in one direction or another, and (3) sudden *flicking movements* which are controlled by the involuntary fixation mechanism. When a spot of light has become fixed on the foveal region of the retina, the tremorous movements cause the spot to move back and forth at a rapid rate across the cones, and the drifting movements cause it to drift slowly across the cones. However, each time the spot of light approaches the edge of the fovea, a sudden flicking movement moves the spot away from this edge back toward the center, which is an automatic response to move the image back toward the central portion of the fovea. These drifting and flicking motions are illustrated in Figure 54–11, which shows by the dashed lines the slow drifting across the retina and by the solid lines the flicks that keep the image from leaving the foveal region.

Fixation on Moving Objects—"Pursuit Movements." The eyes can also remain fixed on a moving object, which is called *pursuit movement.* A highly developed cortical mechanism automatically detects the course of movement of an object and then gradually develops a similar course of movement of the eyes. For instance, if an object is moving up and down in a wavelike form at a rate of several times per second, the eyes at first may be completely unable to fixate on it. However, after a second or so the eyes begin to jump coarsely in approximately the same pattern of movement as that of the object. Then after a few more seconds, the eyes develop progressively smoother and smoother movements and finally follow the course of movement almost exactly. This represents a high degree of calculating ability by the cerebral cortex.

Nystagmus. OPTIKOKINETIC NYSTAGMUS. A particularly important type of pursuit movement is that called *nystagmus,* which allows the eyes to fixate on successive points in a continuously moving scene. For instance, if a person is looking out the window of a train, his eyes fix on successive points in the visual scene. To do this, the eyes fixate on some object and move slowly backward as the object also moves backward. When the eyes have moved far to the side, they automatically jump forward to fix on a new object, which is followed again by slow movement in the backward direction. This type of movement is called *optikokinetic nystagmus.* The slow movement in one direction is called the *slow component* of the nystagmus, while the rapid movement in the other direction is called the *fast component.*

Optikokinetic nystagmus obviously is a type of pursuit movement. That is, after an initial second or so of orientation, the visual system automatically calculates the course and rate of movement of the visual scene, then follows this exactly until the eyes reach a lateral limit, at which time they jump to a new point in the scene.

VESTIBULAR CONTROL OF EYE MOVEMENTS. Another type of eye movement is elicited by stimulation of the vestibular apparatus. The vestibular nuclei are connected directly with the brain stem nuclei that control ocular movements, and, any time the head is accelerated in a vertical, longitudinal, lateral or angular direction, an immediate compensatory motion of the eyes occurs in the opposite direction. This allows the eyes to remain fixed on an object of attention despite rapid movements of the body or head.

Vestibular control of the eyes is especially valuable when a person is subjected to jerky motions of his body. For instance, when a person with bilateral destruction of his vestibular apparatuses rides over rough roads, he has extreme difficulty fixing his eyes on the road or on any horizontal scene. The optikokinetic type of pursuit movement is not capable of keeping the eyes fixed under such conditions because it has a latent period of about one-fifth second before the direction of movement can be detected and followed by the eyes. On the other hand, the vestibular type of eye movement has a short latent period, measured in hundredths of a second rather than one-fifth second.

Also when a person begins to rotate, his vestibular apparatuses cause a vestibular type of nystagmus. That is, the eyes lag behind the rotating head, then jump forward, then lag behind again, jump again, and so forth. This type of nystagmus is discussed in Chapter 57 in relation to vestibular function.

Pathologic Types of Nystagmus. Occasionally, abnormalities occur in the control system for eye movements that cause continuous nystagmus despite the fact that neither the visual scene nor the body is moving. This is likely to occur when one of the vestibular apparatuses is damaged or when severe damage is sustained in the deep nuclei of the cerebellum. This will be discussed further in Chapter 58.

Another pathologic type of eye movement that is sometimes called nystagmus occurs when the foveal regions of the two eyes have been destroyed or when the vision in these areas is greatly weakened. In such a condition the eyes attempt to fix the object of attention on the foveae but always overshoot the mark because of foveal insensitivity. Therefore, they oscillate back and forth but never achieve foveal fixation. Even though this condition is known clinically as a type of nystagmus, physiologically it is completely different from the nystagmus that keeps the eyes fixed on a moving scene.

FUSION OF THE VISUAL IMAGES

To make the visual perceptions more meaningful and also to aid in depth perception by the mechanism of stereopsis, which was discussed in Chapter 52, the visual images in the two eyes normally *fuse* with each other on "corresponding points" of the two retinae. Furthermore, three different types of fusion are required: lateral fusion, vertical fusion, and torsional fusion (same rotation of the two eyes about their optical axes).

Fusion of the images of the two eyes probably results from two mechanisms in the visual system. First, crude fusion is inherent in the newborn baby because of inherited conjugate movements of the eyes. Second, as the visual system becomes more and more developed, the ability of the two eyes to fixate on the same object of attention causes even greater accuracy of fusion. Then, as the visual system develops its pursuit abilities, the pathways for controlling the conjugate movements of the two eyes normally develop equally. And the more the two eyes are moved in unison with each other the more "set" become the patterns of movement for the eyes, and, consequently, the more exact becomes the degree of fusion that can be maintained regardless of the position or rapidity of movement of the eyes.

It was pointed out earlier in the chapter that the lateral geniculate body probably plays a major role in fusion of visual images, because corresponding points of the two retinae transmit visual signals respectively to the successive nuclear layers of the lateral geniculate body. When the eyes are appropriately oriented, the retinal images will be appropriately superimposed upon each other in the respective layers.

Strabismus. Strabismus, which is also called *squint* or *cross-eyedness,* means lack of fusion of the eyes in one or more of the coordinates described above. Three basic types of strabismus are illustrated in Figure 54–12: *horizontal strabismus, vertical strabismus,* and *torsional strabismus.* However, combinations of two or even of all three of the different types of strabismus often occur.

Strabismus is believed to be caused by an abnormal "set" of the fusion mechanism of the visual system. That is, in the early efforts of the child to fixate the two eyes on the same object, one of the eyes fixates satisfactorily while the other fails to fixate, or they both fixate satisfactorily but never simultaneously. Soon, the patterns of conjugate movements of the eyes become abnormally "set" so that the eyes never fuse.

Frequently, some abnormality of the eyes contributes to the failure of the two eyes to fixate on the same point. For instance, if at birth one eye has poor vision in comparison with the other, the good eye tends to fixate on the object of attention while the poor eye might never do so. Also, in hyperopic infants, intense impulses must be transmitted to the ciliary muscles to focus the eyes, and some of these impulses overflow into the oculomotor nuclei to cause simultaneous convergence of the eyes, as will be discussed below. As a result, the child's fusion mechanism becomes "set" for continual inward deviation of the eyes.

SUPPRESSION OF VISUAL IMAGE FROM A REPRESSED EYE. In most patients with strabismus the eyes alternate in fixing on the object of attention. However, in some patients, one eye alone is used all the time while the other eye becomes repressed and is never used for vision. The vision in the repressed eye develops only slightly, usually remaining 20/400 or less. If the dominant eye then becomes blinded, vision in the repressed eye can develop slowly in the adult and far more rapidly in children. This illustrates that visual acuity is highly dependent on the proper development of the central synaptic connections from the eyes.

AUTONOMIC CONTROL OF ACCOMMODATION AND PUPILLARY APERTURE

The Autonomic Nerves to the Eyes. The eye is innervated by both parasympathetic and sympathetic fibers. The parasympathetic fibers arise in the *Edinger-Westphal nucleus* and then pass in the *third nerve* to the *ciliary ganglion,* which lies about 1 cm. behind the eye. Here the fibers synapse with postganglionic parasympathetic neurons that pass through the *ciliary nerves* into the eyeball. These nerves excite the ciliary muscle and the sphincter of the iris.

The sympathetic innervation of the eye originates in the *intermediolateral horn cells* of the first thoracic segment of the spinal cord. From here, sympathetic fibers enter the sympathetic chain and pass upward to the *superior cervical ganglion* where they synapse with postganglionic neurons. Fibers from these spread along the carotid artery and along successively smaller arteries until they reach the eyeball. There the sympathetic fibers innervate the radial fibers of the iris as well as several extraocular structures around the eye, which are discussed shortly in relation to Horner's syndrome.

CONTROL OF ACCOMMODATION

The accommodation mechanism—that is, the mechanism which focuses the lens system of the eye—is essential to a high degree of visual acuity. Accommodation results from contraction or relaxation of the ciliary muscle, contraction causing increased strength of the lens system, as explained in Chapter 52, and relaxation causing decreased strength. The question that must be answered now is: How does one adjust his accommodation to keep his eyes in focus all the time?

Accommodation of the lens is regulated by a negative feedback mechanism that automatically adjusts the focal power of the lens for the highest degree of visual acuity. Though the precise mechanism by which this negative feedback operates is not known, there seem to be two specific processes involved:

First, if a visual image is not in focus, the ciliary muscle automatically begins to contract or relax. If the focus improves, the accommodation continues to change in the same direction, but, if it worsens, the accommodation automatically turns around to the other direction and changes until the proper focus is attained. Thus, the accommodation mechanism continually searches back and forth on the two sides of appropriate focus, thereby keeping the focus always near the point of maximal visual acuity.

Second, and perhaps even more important than the above mechanism, is one that results from the fact that the center of the fovea is depressed in relation to the remainder of the retina. Because of this, an image might well be

| Horizontal strabismus | Torsional strabismus | Vertical strabismus |

Figure 54–12   The three basic types of strabismus.

in better focus in the center of the fovea than along the edges. If this is true, the visual system theoretically could detect that the focal plane of the lens system lies too far posteriorly. On the other hand, if the focus on the edges of the fovea is better than in the very center, then it could be detected that the focal plane lies too far anteriorly. From this information, appropriate integrative impulses could adjust the degree of accommodation to maintain a continual state of precise focus halfway between the depth and the sides of the fovea.

It is presumed that the cortical areas that control accommodation closely parallel those that control fixation movements of the eyes, with integration of visual signals in areas 18 and 19 and transmission of motor signals to the ciliary muscle through the pretectal area and Edinger-Westphal nucleus.

Association of Accommodation and Convergence. When the eyes fixate on a distant object and then on another object nearby, the degree of accommodation of the lens must change to maintain proper focus. Fortunately, when the eyes converge while fixating on closer objects, an automatic association between convergence and accommodation causes increased accommodation of the lens at the same time. Likewise, accommodation of the lens for near vision automatically causes convergence, a process called *accommodation-convergence synkinesis.*

CONTROL OF THE PUPILLARY APERTURE

Stimulation of the parasympathetic nerves excites the pupillary sphincter, thereby decreasing the pupillary aperture; this is called *miosis*. On the other hand, stimulation of the sympathetic nerves excites the radial fibers of the iris and causes pupillary dilatation, which is called *mydriasis*.

The Pupillary Light Reflex. When light is shone into the eyes the pupils constrict, a reaction that is called the pupillary light reflex. The neuronal pathway for this reflex is illustrated in Figure 54–13. When light impinges on the retina, the resulting impulses pass through the optic nerves and optic tracts to the pretectal nuclei. From here, impulses pass to the *Edinger-Westphal nucleus* and finally back through the *parasympathetic nerves* to constrict the sphincter of the iris. In darkness, the Edinger-Westphal nucleus becomes inhibited, which results in dilatation of the pupil.

The function of the light reflex is to help the eye adapt extremely rapidly to changing light conditions, the importance of which was explained in relation to retinal adaptation in the previous chapter. The limits of pupillary diameter are about 1.5 mm. on the small side and 8 mm. on the large side. Therefore, the range of light adaptation that can be effected by the pupillary reflex is about 30 to 1.

Association Between Accommodation and Pupillary Constriction. In almost the same way that accommodation and convergence are associated, which was discussed above, the pupil of the eye also constricts when the eye accommodates for near vision. This effect presumably results from a spillover of impulses from the accommodative system into the pupillary control system. This association is used clinically to determine whether the parasympathetic nerves to the eye are intact. Its function is believed to be two-fold: (1) It helps to compensate for the increased light focused on the retina by the shorter focal length lens. (2) It increases the depth of focus of the lens system, which is greatly needed when one is focusing his eyes on near objects.

Pupillary Reflexes in Syphilis. Central nervous system syphilis almost always eventually blocks the light reflex pathway between the retina and the Edinger-Westphal nucleus. It is believed that this block most likely occurs in the pretectal region of the brain stem, though it could result from destruction of small afferent fibers in the optic nerves. One of the distinguishing features of the loss of the light reflex in syphilis is that the pupillary constriction associated with accommodation is not lost at the same time because the parasympathetic pathways to the eyes are not involved in syphilis.

When the impulses to the Edinger-Westphal nuclei are blocked by syphilis, these nuclei are also released from inhibitory impulses arriving from some outside source. As a result, the nuclei be-

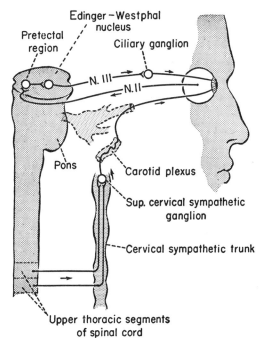

Figure 54–13 Autonomic innervation of the eye, showing also the reflex arc of the light reflex. (Modified from Ranson and Clark: Anatomy of the Nervous System.)

come tonically active, causing the pupils thereafter to remain constricted in addition to their failure to respond to light. Such a pupil that fails to respond to light and also is very small (an *Argyll Robertson pupil*) is an important diagnostic sign of central nervous system syphilis. However, a few other conditions, including central nervous system damage from alcoholism, encephalitis, and so forth, can occasionally also cause an Argyll Robertson pupil.

Horner's Syndrome. The sympathetic nerves to the eye are occasionally interrupted, and this interruption frequently occurs in the cervical chain. This results in Horner's syndrome, which consists of the following effects: First, because of interruption of fibers to the pupillary dilator muscle, the pupil remains persistently constricted to a smaller diameter than that of the pupil of the opposite eye. Second, the superior eyelid droops because this eyelid is normally maintained in an open position during the waking hours partly by contraction of a smooth muscle, the *superior palpebral muscle,* which is innervated by the sympathetics. Therefore, destruction of the sympathetics makes it impossible to open the superior eyelid nearly as widely as normally. Third, the blood vessels on the corresponding side of the face and head become persistently dilated. And, fourth, sweating cannot occur on the side of the face and head affected by Horner's syndrome.

REFERENCES

Brindley, G. S.: Central pathways of vision. *Ann. Rev. Physiol., 32*:259, 1970.

Brown, K. T., and Wiesel, T. N.: Localization of origins of electroretinogram components by intraretinal recording in the intact cat eye. *J. Physiol. (Lond.), 158*:257, 1961.

Cohen, B.; Feldman, M.; and Diamond, S. P.: Effects of eye movement, brain-stem stimulation, and alertness on transmission through lateral geniculate body of monkey. *J. Neurophysiol., 32*:583, 1969.

Creutzfeldt, O., and Sakmann, B.: Neurophysiology of vision. *Ann. Rev. Physiol., 31*:499, 1969.

Dow, B. M., and Dubner, R.: Visual receptive fields and responses to movement in an association area of cat cerebral cortex. *J. Neurophysiol., 32*:773, 1969.

Guyton, J. S., and Kirkman, N.: Ocular movement: I. Mechanics, pathogenesis, and surgical treatment of alternating hypertropia (dissociated vertical divergence, double hypertropia) and some related phenomena. *Amer. J. Ophthal., 41*:438, 1956.

Hubel, D. H.: Eleventh Bowditch Lecture. *Physiologist, 10*:17, 1967.

Hubel, D. H.: The visual cortex of the brain. *Sci. Amer., 209*(5):54, 1963.

Hubel, D. H., and Wiesel, T. N.: Cortical and callosal connections concerned with the vertical meridian of visual fields in the cat. *J. Neurophysiol., 30*:1561, 1967.

Hubel, D. H., and Wiesel, T. N.: Receptive fields and functional architecture in two nonstriate visual areas (18 and 19) of the cat. *J. Neurophysiol., 28*:229, 1965.

Hubel, D. H., and Wiesel, T. N.: Receptive fields of cells in striate cortex of very young, visually inexperienced kittens. *J. Neurophysiol., 26*:994, 1963.

Hubel, D. H., and Wiesel, T. N.: Shape and arrangement of columns in cat's striate cortex. *J. Physiol. (Lond.), 165*:559, 1963.

Hyde, J. E., and Slusher, M. A.: Functional role of median longitudinal fasciculus in evoked conjugate ocular deviations in cats. *Amer. J. Physiol., 200*:919, 1961.

Ogle, K. M.; Martens, T. G.; and Dyer, T. A.: Oculomotor Imbalance in Binocular Vision and Fixation Disparity. Philadelphia, Lea & Febiger, 1967.

Poggio, G. F.; Baker, F. H.; Lamarre, Y.; and Sanseverino, E. R.: Afferent inhibition at input to visual cortex of the cat. *J. Neurophysiol., 32*:892, 1969.

Polyak, S.: The Retina. Chicago, University of Chicago Press, 1941.

Robinson, D. A., and Fuchs, A. F.: Eye movements evoked by stimulation of frontal eye fields. *J. Neurophysiol., 32*:637, 1969.

Snyder, M.; Killackey, H.; and Diamond, I. T.: Color vision in the tree shrew after removal of posterior neocortex. *J. Neurophysiol., 32*:554, 1969.

Stark, L.: Environmental clamping of biological systems: pupil servomechanism. *J. Opt. Soc. Amer., 52*:925, 1962.

Sterling, P., and Wickelgren, B. G.: Visual receptive fields in the superior colliculus of the cat. *J. Neurophysiol., 32*:1, 1969.

Thomas, E. L.: Movements of the eye. *Sci. Amer., 219*:88(2), 1968.

Walsh, F. B.: Clinical Neuro-ophthalmology. 3rd ed., Baltimore, The Williams & Wilkins Co., 1968.

Wolbarsht, M. L., and Yeandle, S. S.: Visual processes in the Limulus eye. *Ann. Rev. Physiol., 29*:513, 1967.

Wurtz, R. H.: Visual receptive fields of striate cortex neurons in awake monkeys. *J. Neurophysiol., 32*:727, 1969.

CHAPTER 55

The Chemical Senses — Taste and Smell

THE SENSE OF TASTE

Taste is a function of the *taste buds* in the mouth, and its importance lies in the fact it allows the person to select his food in accord with his desires and perhaps also in accord with the needs of the tissues for specific nutritive substances.

On the basis of psychologic studies, there are generally believed to be at least four *primary* sensations of taste: *sour, salty, sweet,* and *bitter*. Yet we know that a person can perceive literally hundreds of different tastes. These are all supposed to be combinations of the four primary sensations in the same manner that all the colors of the spectrum are combinations of three primary color sensations, as described in Chapter 53. However, there might be other less conspicuous classes or subclasses of primary sensations. Nevertheless, the following discussion is based on the usual classification of only four primary tastes.

THE PRIMARY SENSATIONS OF TASTE

The Sour Taste. The sour taste is caused by acids, and the intensity of the taste sensation is approximately proportional to the loga-

rithm of the *hydrogen ion concentration*. That is, the more acidic the acid, the stronger becomes the sensation.

The Salty Taste. The salty taste is elicited by ionized salts. The quality of the taste varies somewhat from one salt to another because the salts also elicit other taste sensations besides saltiness. The cations of the salts are mainly responsible for the salty taste, but the anions also contribute at least to some extent.

The Sweet Taste. The sweet taste is not caused by any single class of chemicals. A list of some of the types of chemicals that cause this taste includes: sugars, glycols, alcohols, aldehydes, ketones, amides, esters, amino acids, sulfonic acids, halogenated acids, and inorganic salts of lead and beryllium. Note specifically that almost all the substances that cause a sweet taste are organic chemicals; the only inorganic substances that elicit the sweet taste at all are certain salts of lead and beryllium.

Table 55–1 shows the relative intensities of taste of certain substances that cause the sweet taste. *Sucrose*, which is common table sugar, is considered to have an index of 1. Note that one of the substances has a sweet index 5000 times as great as that of sucrose.

TABLE 55–1. Relative Taste Indices of Different Substances

Sour Substances	Index	Bitter Substances	Index	Sweet Substances	Index	Salty Substances	Index
Hydrochloric acid	1	Quinine	1	Sucrose	1	NaCl	1
Formic acid	1.1	Brucine	11	1-propoxy-2-amino-		NaF	2
Chloracetic acid	0.9	Strychnine	3.1	4-nitrobenzene	5000	CaCl₂	1
Acetyllactic acid	0.85	Nicotine	1.3	Saccharin	675	NaBr	0.4
Lactic acid	0.85	Phenylthiourea	0.9	Chloroform	40	NaI	0.35
Tartaric acid	0.7	Caffeine	0.4	Fructose	1.7	LiCl	0.4
Malic acid	0.6	Veratrine	0.2	Alanine	1.3	NH₄Cl	2.5
Potassium H tartrate	0.58	Pilocarpine	0.16	Glucose	0.8	KCl	0.6
Acetic acid	0.55	Atropine	0.13	Maltose	0.45		
Citric acid	0.46	Cocaine	0.02	Galactose	0.32		
Carbonic acid	0.06	Morphine	0.02	Lactose	0.3		

From Derma: *Proc. Oklahoma Acad. Sc.,* **27:9, 1947**; and Pfaffman: Handbook of Physiology, Sec. I, Vol. I, p. 507, 1959. Baltimore, The Williams & Wilkins Co.

However, this extremely sweet substance, known as *P-4000*, is unfortunately extremely toxic and therefore cannot be used as a sweetening agent. *Saccharin*, on the other hand, is also more than 600 times as sweet as common table sugar, and since it is not toxic it can be used with impunity as a sweetening agent.

The Bitter Taste. The bitter taste, like the sweet taste, is not caused by any single type of chemical agent, but, here again, the substances that give the bitter taste are almost entirely organic substances. Two particular classes of substances are especially likely to cause bitter taste sensations, (1) long chain organic substances and (2) alkaloids. The alkaloids include many of the drugs used in medicines, such as quinine, caffeine, strychnine, and nicotine.

Some substances that at first taste sweet have a bitter after-taste. This is true of saccharin, which makes this substance objectionable to some people. Some substances have a sweet taste on the front of the tongue, where taste buds with special sensitivity to the sweet taste are principally located, and a bitter taste on the back of the tongue, where taste buds more sensitive to the bitter taste are located.

The bitter taste, when it occurs in high intensity, usually causes the person or animal to reject the food. This is undoubtedly an important purposive function of the bitter taste sensation because many of the deadly toxins found in poisonous plants are alkaloids, and these all cause an intensely bitter taste.

Threshold for Taste

The threshold for stimulation of the sour taste by hydrochloric acid averages 0.0009 N, for stimulation of the salty taste by sodium chloride: 0.01 M; for the sweet taste by sucrose: 0.01 M; and for the bitter taste by quinine: 0.000008 M. Note especially how much more sensitive is the bitter taste sense to stimuli than all the others, which would be expected since this sensation provides an important protective function.

Table 55-1 gives the relative taste indices (the reciprocals of the taste thresholds) of different substances. In this table, the intensities of the four different primary sensations of taste are referred, respectively, to the intensities of taste of hydrochloric acid, quinine, sucrose, and sodium chloride, each of which is considered to have a taste index of 1.

Taste Blindness. Many persons are taste blind for certain substances, especially for different types of thiourea compounds. A substance used frequently by psychologists for demonstrating taste blindness is *phenylthiocarbamide*, for which approximately 15 to 30 per cent of all people exhibit taste blindness, the exact percentage depending on the method of testing.

THE TASTE BUD AND ITS FUNCTION

Figure 55-1 illustrates a taste bud, which has a diameter of about $1/30$ millimeter and a length of about $1/16$ millimeter. The taste bud is composed of about 20 modified epithelial cells called *taste cells*. These cells are continually being replaced so that some are young cells and others mature cells; the mature cells lie toward the center of the bud. The life span of each taste cell is about seven days.

The outer tips of the taste cells are arranged around a minute *taste pore*, shown in Figure 55-1. From the tip of each taste cell, several *microvilli*, or *taste hairs*, about 3 microns in length and 0.2 microns in width, protrude outward through the taste pore into the cavity of the mouth. These microvilli are believed to provide the receptor surface for taste.

Interweaving among the taste cells is a branching terminal network of two or three *taste nerve fibers* that are stimulated by the taste cells. These fibers invaginate deeply into folds of the taste cell membranes, so that there is extremely intimate contact between the taste cells and the nerves.

An interesting feature of the taste buds is that they completely degenerate when the taste nerve fibers are destroyed. Then, if the taste fibers regrow to the epithelial surface of the mouth, the local epithelial cells regroup themselves to form new taste buds. This illustrates the important principle of "trophic"

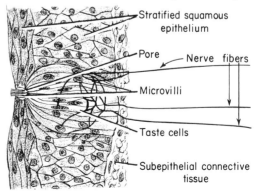

Figure 55-1   The taste bud.

function of nerve fibers in certain parts of the body.

Location of the Taste Buds. The taste buds are found on three out of four of the different types of papillae of the tongue, as follows: (1) A large number of taste buds are on the circumvallate papillae which form a **V** line toward the posterior of the tongue. (2) Moderate numbers of taste buds are on the fungiform papillae all over the surface of the tongue. (3) Moderate numbers are on the foliate papillae located in folds along the posterolateral surfaces of the tongue. Additional taste buds are located on the tonsillar pillars and at other points around the nasopharynx, including even a few on the posterior pharyngeal wall itself. Adults have approximately 10,000 taste buds, and children a few more. Beyond the age of 45 many taste buds begin to degenerate, causing the taste sensation to become progressively less critical.

Especially important in relation to taste is the tendency for taste buds subserving particular primary sensations of taste to be localized in special areas. The sweet taste is localized *principally* on the anterior surface and tip of the tongue, the sour taste on the two lateral sides of the tongue, and the bitter taste on the circumvallate papillae on the posterior surface of the tongue. The salty taste is present over the entire tongue.

Specificity of Taste Buds for the Primary Taste Stimuli. In the foregoing paragraphs we have discussed taste buds as if each responded to a particular type of taste stimulus and not to other taste stimuli. In a statistical sense this is true, but so far as any single taste bud is concerned, it is not true, for most taste buds respond to varying extents to three and sometimes to all four of the primary taste stimuli.

Figure 55–2 illustrates the responsiveness of four different taste buds to the different primary tastes. Figure 55–2A illustrates a bud responsive to all four types of taste stimuli, but especially to saltiness. Figure 55–2B illustrates a taste bud strongly responsive to sourness and saltiness but also responsive to a moderate degree to both bitterness and sweetness. Likewise, Figures 55–2C and 55–2D illustrate response characteristics of two other types of taste buds.

DETECTION OF DIFFERENT SENSATIONS OF TASTE BY THE TASTE BUDS. Since each taste bud responds to multiple primary taste stimuli, it is difficult to understand how a person perceives the different primary taste sensations independently of each other. However, a theory attempting to explain this is the following: Some area in the brain presumably is capable of detecting the *ratios* of stimulation of the different types of taste buds. For the sweet taste, a taste bud that

Figure 55–2 Specific responsiveness of four different types of taste buds, showing multiple stimulation by the different primary sensations of taste in the case of each of the taste buds.

responds strongly to sweet stimuli will transmit a stronger signal to the brain than will a taste bud that responds only weakly to the sweet taste. Thus, the ratio of the signal strength from the first bud to the signal strength from the second is high, and it is this *ratio* that elicits the sweet taste. Let us now assume that the second taste bud has a strong response to the bitter taste while the first has a weak response. The ratio will be exactly the reverse when a bitter substance is placed in the mouth. Likewise, other ratios, often involving many taste buds rather than simply two taste buds, will theoretically be perceived as the salty or sour sensation.

Tastes besides the four primary tastes obviously would give still other ratios of stimulation of the different taste buds and thus give all the different gradations of taste sensations that are known to occur.

Mechanism of Stimulation of Taste Buds. THE RECEPTOR POTENTIAL. The membrane of the taste cell, like that of other sensory receptor cells, normally is negatively charged on the inside with respect to the outside. Application of a taste substance to the taste hairs causes partial loss of this negative potential, and the decrease in potential is approximately proportional to the logarithm of concentration of the stimulating substance. This change in potential in the taste cell is the *receptor potential* for taste.

The mechanism by which the stimulating substance reacts with the taste hairs to initiate the receptor potential is unknown. It is believed by some physiologists that the substance is simply adsorbed to the surface of the taste hair and that this adsorption changes the physical characteristics of the hair membrane sufficiently to alter its electrical potential. The substance supposedly is gradually washed away from the taste hair by the saliva, thus removing the taste stimulus.

Another theory suggests that the substance combines chemically with a specific receptor substance in the taste hair and that the combination alters the membrane permeability. Supposedly, the type of receptor substance or substances in each taste hair determine the types of taste substances that will elicit a response.

GENERATION OF NERVE IMPULSES BY THE TASTE BUD. The taste nerve fiber endings are encased by folds of the taste cell membranes. In some way not understood the receptor potentials of the taste cells generate impulses in the taste fibers. On first application of the taste stimulus, the rate of discharge of the nerve fibers rises to a peak in a small fraction of a second, but then it adapts within the next 2 seconds back to a much lower steady level. Thus, a strong immediate signal is transmitted by the taste nerve and a much weaker continuous signal is transmitted as long as the taste bud is exposed to the taste stimulus.

TRANSMISSION OF TASTE SIGNALS INTO THE CENTRAL NERVOUS SYSTEM

Figure 55–3 illustrates the neuronal pathways for transmission of taste sensations from the tongue and pharyngeal region into the central nervous system. Taste impulses from the anterior two thirds of the tongue pass first into the *fifth nerve* and then through the *chorda tympani* into

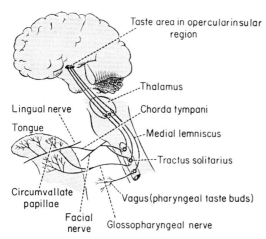

Figure 55–3 Transmission of taste impulses into the central nervous system.

the *seventh nerve*, thence into the *tractus solitarius* in the brain stem. Taste sensations from the circumvallate papillae on the back of the tongue and from other posterior regions of the mouth are transmitted through the *ninth nerve* also into the *tractus solitarius* but at a slightly lower level. Finally, a few taste impulses are transmitted into the *tractus solitarius* from the base of the tongue and other parts of the pharyngeal region by way of the *vagus nerve*. All taste fibers synapse in the *nuclei of the tractus solitarius* and send second order neurons to a small area of the *thalamus* located slightly medial to the thalamic terminations of the facial regions of the dorsal column-medial lemniscal system. From the thalamus, third order neurons are believed to be transmitted to the *parietal opercular-insular area* of the cerebral cortex. This lies at the very lateral margin of the postcentral gyrus in the sylvian fissure in close association with, or even superimposed on, the tongue area of somatic sensory area I.

From this description of the taste pathways, it immediately becomes evident that they parallel closely the somatic pathways from the tongue.

Taste Reflexes. From the tractus solitarius a large number of impulses are transmitted directly into the *superior* and *inferior salivatory nuclei,* and these in turn transmit impulses to the submaxillary and parotid glands to help control the secretion of saliva during the ingestion of food.

Adaptation of Taste. Everyone is familiar with the fact that taste sensations adapt rapidly. For instance, the salty taste sensation adapts almost to extinction even when concentrations of salt solutions as high as 5 to 10 per cent are placed directly on a salty taste bud, and similar high degrees of adaptation are also known for the other taste buds.

Yet, from electrophysiological studies of taste nerve fibers, it seems that the taste buds themselves do not adapt enough by themselves to account for all or most of the taste adaptation. They have a rapid period of adaptation during the first 2 to 3 seconds after contact with the taste stimulus. The first burst of impulses from the taste bud allows one to detect extremely minute concentrations of taste substances, but normal concentrations of taste substances cause prolonged discharge of the taste fibers. Therefore, the extreme adaptation that occurs in the sensation of taste has been postulated to occur in the central nervous system itself, though the mechanism and site of this is not known. If this is true—and it is still questionable—it is a mechanism that is different from that of other sensory systems, almost all of which adapt almost entirely in the receptors.

SPECIAL ATTRIBUTES OF THE TASTE SENSE

Affective Nature of Taste. *Pleasantness* and *unpleasantness* are called the "affective" attributes of a sensation. Figure 55–4 illustrates the affective effects of different types of taste at different concentrations of the stimulating substances, showing,

Figure 55-4 The affective nature of the different primary sensations of taste at progressively increasing degrees of taste stimulus. (From Engel in Woodworth: Experimental Psychology. Henry Holt and Co.)

strangely enough, that the sweet taste is likely to be unpleasant at a very low concentration but very pleasant at high concentrations. The other types of taste, on the other hand, are likely to be pleasant at low concentrations but exceedingly unpleasant at high concentrations. This is particularly true of the bitter taste.

Importance of the Sense of Smell in Taste. Persons with severe colds frequently state that they have lost their sense of taste. However, on testing the taste sensations, these are found to be completely normal. This illustrates that much of what we call taste is actually smell. Odors from the food can pass upward in the nasopharynx, often stimulating the olfactory system thousands of times as strongly as the taste system. For instance, if the olfactory system is intact, alcohol can be "tasted" in $1/50,000$ the concentration required when the olfactory system is not intact.

Taste Preference and Control of the Diet. Taste preference means simply that an animal will choose certain types of food in preference to others, and he automatically uses this to help control the type of diet that he eats. Furthermore, to a great extent his taste preferences change in accordance with the needs of the body for certain specific substances. The following experimental studies will illustrate this ability of an animal to choose food in accord with the need of his body: First, adrenal-ectomized animals automatically select drinking water with a high concentration of sodium chloride in preference to pure water, and this in many instances is sufficient to supply the needs of the body and prevent death as a result of salt depletion. Second, an animal injected with excessive amounts of insulin develops a depleted blood sugar, and he automatically chooses the sweetest food from among many samples. Third, para-thyroidectomized animals automatically choose drinking water with high concentrations of calcium chloride.

These same phenomena are also observed in many instances of everyday life. For instance, the salt licks of the desert region are known to attract animals from far and wide, and even the human being rejects any food that has an unpleasant affective sensation, which certainly in many instances protects our bodies from undesirable substances.

The phenomenon of taste preference almost certainly results from some mechanism located in the central nervous system and not from a mechanism in the taste buds themselves, because many experiments have demonstrated that taste preference can occur in animals even in the absence of changes in stimulus threshold of the taste buds for the substances involved. Another reason for believing this to be a central phenomenon is that previous experience with unpleasant or pleasant tastes plays a major role in determining one's different taste preferences. For instance, if a person becomes sick immediately after eating a particular type of food, he generally develops a negative taste preference, or taste aversion, for that particular food thereafter; the same effect can be demonstrated in animals.

THE SENSE OF SMELL

Smell is the least well understood sense. This results partly from the location of the olfactory membrane high in the nose where it is difficult to study and partly from the fact that the sense of smell is a subjective phenomenon that cannot be studied with ease in lower animals. Still another complicating problem is the fact that the sense of smell is almost rudimentary in the human being in comparison with that of some lower animals.

THE OLFACTORY MEMBRANE

The olfactory membrane lies in the superior part of each nostril, as illustrated in Figure 55-6. Medially it folds downward over the surface of the septum, and laterally it folds over the superior turbinate and even over a small portion of the upper surface of the middle turbinate. In each nostril the olfactory membrane has a surface area of approximately 2.4 square centimeters.

The Olfactory Cells. The receptor cells for the smell sensation are the olfactory cells, which are actually bipolar nerve cells derived originally from the central nervous system itself. There are on the order of 100 million of these cells in the olfactory epithelium interspersed among *sustentacular cells,* as shown in Figure 55-5. The mucosal end of the olfactory

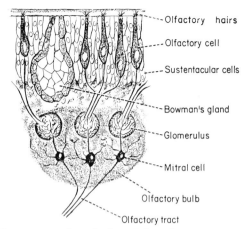

Figure 55-5 Organization of the olfactory membrane. (After Maximow and Bloom: A Textbook of Histology.)

cell forms a *knob* from which large numbers of *olfactory hairs*, or cilia, 0.3 micron in diameter and 3 microns in length, project into the mucus that coats the inner surface of the nasal cavity. These projecting olfactory hairs are believed to react to odors in the air and then to stimulate the olfactory cells, as is discussed below. Spaced among the olfactory cells in the olfactory membrane are many small *glands of Bowman* that secrete mucus onto the surface of the olfactory membrane.

STIMULATION OF THE OLFACTORY CELLS

The Necessary Stimulus for Smell. We do not know what it takes chemically to stimulate the olfactory cells. Yet we do know the physical characteristics of the substances that cause olfactory stimulation: First, the substance must be volatile so that it can be sniffed into the nostrils. Second, it must be at least slightly water soluble so that it can pass through the mucus to the olfactory cells. And, third, it must also be lipid soluble, presumably because the olfactory hairs and outer tips of the olfactory cells are composed principally of lipid materials.

Regardless of the basic mechanism by which the olfactory cells are stimulated, it is known that they become stimulated only when air blasts upward into the superior region of the nose. Therefore, smell occurs in cycles along with the inspirations, which indicates that the olfactory receptors respond in milliseconds to the volatile agents. Because smell intensity is exacerbated by blasting air through the upper reaches of the nose, a person can greatly increase his sensitivity of smell by the well-known sniffing technique.

Generator Potentials (Receptor Potentials) in

Olfactory Cells. The olfactory cells are believed to react to olfactory stimuli in the same manner that most other sensory receptors react to their specific stimuli; that is, by generating a generator potential or receptor potential, which in turn initiates nerve impulses in the olfactory nerve fibers. An experiment that demonstrates these properties of the olfactory receptors is the following: An electrode is placed on the surface of the olfactory membrane, and its electrical potential with respect to the remainder of the body is recorded. When an odorous substance is insufflated into the nostril, the potential becomes negative and remains negative as long as the odorous air continues to pass through the nostril. This electrical recording is called the *electro-olfactogram*, and it is believed to result from summation of generator potentials developed in the receptor ends of the olfactory cells.

Over a wide range, both the amplitude of the electro-olfactogram and the rate of olfactory nerve impulses are approximately proportional to the logarithm of the stimulus strength, which illustrates that the olfactory receptors tend to obey principles of transduction similar to those of other sensory receptors.

Adaptation. The olfactory receptors adapt approximately 50 per cent in the first second or so after stimulation. Thereafter, they adapt further very slowly. Yet we all know from our own experience that smell sensations adapt almost to extinction within a minute or more after one enters a strongly odorous atmosphere. Since the psychological adaptation seems to be more rapid than the adaptation of the receptors, it has been suggested that at least part of this adaptation occurs in the central nervous system, as has also been postulated for adaptation of taste sensations.

Search for the Primary Sensations of Smell. All physiologists are convinced that the many sensations of smell are subserved by a few rather discrete primary sensations in the same way that taste is subserved by sour, sweet, bitter, and salty sensations. Thus far, only minor success has been achieved in classifying the primary sensations of smell. Yet, on the basis of psychological tests and on the basis of action potential studies from individual olfactory cells as well as from various points in the olfactory nerve pathways, it is believed that about seven different primary classes of olfactory stimulants selectively excite separate olfactory cells.

Two basic theories have been postulated to explain the abilities of different receptors to respond selectively to different types of olfactory stimulants, the *chemical theory* and the *physical theory*. The chemical theory assumes that *receptor chemicals* in the membranes of the olfactory hairs react specifically with the different types of olfactory stimulants. The type of receptor chemical determines the type of stimulant that will elicit a response in the olfactory cell. The reaction between the stimulant and the receptor substance supposedly increases the permeability of the olfactory hair membrane, and this in turn creates the generator potential in the olfactory cell that generates impulses in the olfactory nerve fibers.

The physical theory assumes that differences in *physical receptor sites* on the olfactory hair membranes of separate olfactory cells allow specific olfactory stimulants to adsorb to the membranes of different olfactory cells. A fact that supports this theory is that many substances that have very different chemical properties but that have almost identical molecular shapes have the same odor. This indicates that a physical property of the stimulant determines the odor. On the basis of this theory, the following primary classes of odor have been postulated:

1. Camphoraceous   5. Ethereal
2. Musky   6. Pungent
3. Floral   7. Putrid
4. Pepperminty

It is unlikely that this list actually represents the primary sensations of smell, but it does illustrate one of the many attempts to classify them.

Affective Nature of Smell. Smell, equally as much as taste, has the affective qualities of either pleasantness or unpleasantness. Because of this, smell is as important as, if not more important than, taste in the selection of food. Indeed, a person who has previously eaten food that has disagreed with him is often nauseated by the smell of that same type of food on a second occasion. Other types of odors that have proved to be unpleasant in the past may also provoke a disagreeable feeling; on the other hand, perfume of the right quality can wreak havoc with the masculine emotions.

Threshold for Smell. One of the principal characteristics of smell is the minute quantity of the stimulating agent in the air required to effect a smell sensation. For instance, the substance *methyl mercaptan* can be smelled when only 1/25,000,000,000 milligram is present in each milliliter of air. Because of this low threshold, this substance is mixed with natural gas to give it an odor that can be detected when it leaks from a gas pipe.

MEASUREMENT OF SMELL THRESHOLD. One of the problems in studying smell has been difficulty in obtaining accurate measurements of the threshold stimulus required to induce smell. The simplest technique is simply to allow a person to sniff different substances in his usual manner of smelling. Indeed, some investigators feel that this is equally as satisfactory as almost any other procedure. However, to eliminate variations from person to person, more objective methods have been developed: First, a "blast" technique has been used, in which air containing odoriferous material of measured concentrations is blasted upward in a nostril. With this technique precisely the same degree of blast and precisely the same concentration of olfactory agent can be employed from one person to another. Even so, much difficulty has still been experienced in establishing stable thresholds, presumably because the degree of turbulence of the blasted air and other factors, as well as the concentration of the odoriferous agent, enter into the intensity of smell.

A second method that has been used to test relative thresholds has been to place the subject's head into a box containing the volatilized agent. Appropriate measures are taken to exclude odors from the person's own body. The person is allowed to breathe naturally, but the volatilized agent is distributed evenly in the air that is breathed.

Gradations of Smell Intensities. Though the threshold concentrations of substances that evoke smell are extremely slight, concentrations only 10 to 50 times above the threshold values evoke maximum intensity of smell. This is in contrast to most other sensory systems of the body, in which the ranges of detection are tremendous — for instance 500,000 to 1 in the case of an eye and almost a trillion to 1 in the case of the ears. It seems, therefore, that the smell sense is concerned more with detecting the presence or absence of odors than with quantitative detection of their intensities.

TRANSMISSION OF SMELL SENSATIONS INTO THE CENTRAL NERVOUS SYSTEM

The function of the central nervous system in olfaction is almost as vague as the function of the peripheral receptors. However, Figures 55–5 and 55–6 illustrate the general plan for transmission of olfactory sensations into the central nervous system. Figure 55–5 shows a number of separate *olfactory cells* sending axons into the *olfactory bulb* to end on *dendrites from mitral cells* in a structure called the *glomerulus*. Approximately 25,000 axons enter the glomerulus and synapse with only 25 mitral cells that in turn send signals into the brain.

It has been claimed that each glomerulus becomes excited in response to only a single type of smell.

Figure 55–6 shows the principal pathways for transmission of olfactory signals from the mitral cells into the brain. The fibers from the mitral cells

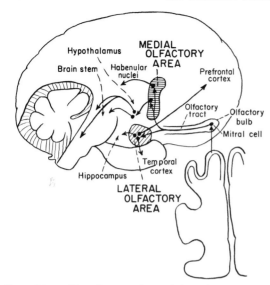

Figure 55–6 Neural connections of the olfactory system.

travel through the olfactory tract and terminate in two principal areas of the brain called the *medial olfactory area* and the *lateral olfactory area,* respectively. The medial olfactory area is composed of a large group of nuclei located in the midportion of the brain superiorly and anteriorly to the hypothalamus. This group includes the *septum pellucidum,* the *gyrus subcallosus,* the *paraolfactory area,* the *olfactory trigone,* and the *medial part of the anterior perforated substance.*

The lateral olfactory area is composed of the *prepyriform area,* the *uncus,* the *lateral part of the anterior perforated substance,* and part of the *amygdaloid nuclei.*

Secondary olfactory tracts pass from the nuclei of both the medial olfactory area and the lateral olfactory area into the *hypothalamus, thalamus, hippocampus,* and *brain stem nuclei.* These secondary areas control the automatic responses of the body to olfactory stimuli, including automatic feeding activities and also emotional responses, such as fear, excitement, and pleasure.

Secondary olfactory tracts also spread from the lateral olfactory area into the temporal cortex and prefrontal cortex. It is probably in this lateral olfactory area, especially in the amygdala and its associated cortical regions, that the more complex aspects of olfaction are integrated, such as association of olfactory sensations with somatic, visual, tactile, and other types of sensation. Complete removal of the lateral olfactory area hardly affects the primitive responses to olfaction, such as licking the lips, salivation, and other feeding responses caused by the smell of food or such as the various emotions associated with smell. On the other hand,

its removal does abolish the more complicated conditioned reflexes depending on olfactory stimuli. Therefore, this region is considered to be the primary *cerebral* area for smell. In human beings, tumors in the region of the uncus and amygdala frequently cause the person to perceive abnormal smells.

Centrifugal Control of the Olfactory Bulb by the Central Nervous System. The central nervous system also transmits impulses in a backward direction to the olfactory bulb. These originate mainly in undetermined areas in the brain stem and olfactory portion of the cerebrum and terminate in the small *granule* cells in the center of the bulb, which in turn send axons to the mitral cells to control these cells. The signals from the central nervous system are mainly inhibitory, but they are occasionally excitatory.

Mechanism of Function of the Olfactory Tracts. Electrophysiological studies of the olfactory system show that the mitral cells are continually active, and superimposed on this background activity are evoked potentials caused by olfactory stimuli. Thus, the olfactory stimuli presumably *modulate* the rate of rhythmic impulses in the olfactory system and in this way transmit the olfactory information.

REFERENCES

Amoore, J. E.; Johnston, J. W., Jr.; and Rubin, M.: The stereochemical theory of odor. *Sci. Amer.,* 210(2):42, 1964.
Biedenbach, M. A., and Stevens, C. F.: Electrical activity in cat olfactory cortex produced by synchronous orthodromic volleys. *J. Neurophysiol.,* 32:193, 1969.
Denton, D. A.: Salt appetite. *In* Handbook of Physiology. Baltimore. The Williams & Wilkins Co., 1967, Sec. VI, Vol. I, p. 433.
Freeman, W. J.: Relations between unit activity and evoked potentials in prepyriform cortex of cats. *J. Neurophysiol.,* 31:337, 1968.
Hodgson, E. S.: Taste receptors. *Sci. Amer.,* 204(5):135, 1961.
Kare, M. R., and Maller, O.: The Chemical Sense and Nutrition. Baltimore, The Johns Hopkins Press, 1967.
Lat, J.: Self-selection of dietary components. *In* Handbook of Physiology. Baltimore, The Williams & Wilkins Co., 1967, Sec. VI, Vol. I, p. 367.
Moulton, D. G., and Beidler, L. M.: Structure and function in the peripheral olfactory system. *Physiol. Rev.,* 47:1, 1967.
Oakley, B., and Benjamin, R. M.: Neural mechanisms of taste. *Physiol. Rev.,* 46:173, 1966.
O'Connell, R. J., and Mozell, M. M.: Quantitative stimulation of frog olfactory receptors. *J. Neurophysiol.,* 32:51, 1969.
Pfaffman, C.: Taste, its sensory and motivating properties. *Amer. Sci.,* 52:187, 1964.
Stevens, C. F.: Structure of cat frontal olfactory cortex. *J. Neurophysiol.,* 32:184, 1969.
Wenzel, B. M., and Sieck, M. H.: Olfaction. *Ann. Rev. Physiol.,* 28:381, 1966.
Young, P. T.: Palatability: the hedonic response to foodstuffs. *In* Handbook of Physiology. Baltimore, The Williams & Wilkins Co., 1967, Sec. VI, Vol. I, p. 353.
Zotterman, Y.: Olfaction and Taste. New York, The Macmillan Company, 1963.

CHAPTER 56

The Cord Reflexes

In the discussion of the central nervous system thus far, we have considered principally the input of sensory information. In the following chapters we will discuss the origin and output of motor signals, the signals that cause muscular and other effects throughout the body. Sensory information is integrated at all levels of the nervous system and causes corresponding motor responses, beginning in the spinal cord with relatively simple reflexes, extending into the brain stem with still more complicated responses, and, finally, extending to the cerebral cortex where the most complicated responses are controlled. The present chapter discusses the control of muscle function by means of spinal cord reflexes.

Exacerbation of the Reflexes in the Decerebrate Animal. To study many of the cord reflexes adequately, it is often necessary to increase the excitability of the cord centers that integrate the reflexes. This can be accomplished by decerebration of the animal at a level midway between the superior and inferior colliculi. Transecting the brain at this point removes inhibitory influences from above this level but leaves intact a large area of the brain stem called the *bulboreticular facilitatory area,* which will be discussed in the following chapter. This area is intrinsically excitable and transmits large numbers of impulses to the spinal cord through the *reticulospinal, tectospinal, rubro-spinal,* and *vestibulospinal tracts,* thus maintaining a high degree of facilitation in the cord. In such an animal, some of the cord reflexes become highly sensitized, and they can then be elicited with ease. Therefore, the following discussion describes principally reflexes in the decerebrate preparation. However, even in the normal human being, facilitatory signals from the brain can make the cord reflexes equally as powerful or sometimes even more powerful under natural conditions than in the decerebrate preparation.

ORGANIZATION OF THE SPINAL CORD FOR REFLEX FUNCTIONS

The cord gray matter is the integrative area for the cord reflexes. Figure 56–1 shows the typical organization of the cord gray matter in a single cord segment. Sensory signals enter the cord through the sensory roots; and once in the cord, they travel to the brain for sensory perception. They also spread through the cord itself to control the cord reflexes, the motor signals for which leave by way of the motor roots. Between these two roots are two major types of cells concerned with reflex functions of the cord, the *anterior motoneurons* and the *internuncial cells* (also called *intermediate cells* or *interneurons*).

Function of the Anterior Motoneurons. The anterior motoneurons give rise to large type A alpha nerve fibers ranging from 9 to 16 microns in diameter and passing through the spinal nerves to innervate the skeletal muscle fibers. Stimulation of a single nerve fiber excites approximately 100 to 300 skeletal muscle fibers, which are collectively called the *motor unit.* Thus, it is the anterior motoneurons that control the contractile functions of the skeletal muscle fibers. Transmission of nerve impulses

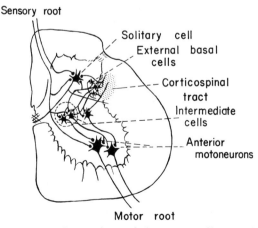

Figure 56–1 Connections of the sensory fibers and corticospinal fibers with the internuncial cells ("intermediate cells") and anterior motoneurons of the spinal cord. (Modified from Lloyd: *J. Neurophysiol., 4:*525, 1941.)

Sensory root

Solitary cell
External basal cells
Corticospinal tract
Intermediate cells
Anterior motoneurons

Motor root

into skeletal muscles and their stimulation of the muscle fibers was discussed in Chapters 6 and 7.

THE GAMMA MOTONEURONS. In addition to the large motoneurons that excite contraction of the skeletal muscle fibers, much smaller motoneurons transmit impulses through type A gamma fibers, averaging 5 microns in diameter, to special skeletal muscle fibers called *intrafusal fibers.* These are part of the *muscle spindle,* which is discussed at length later in the chapter.

The Internuncial Cells. The internuncial cells are present in the base of the dorsal horns, are spread diffusely in the anterior horn, and are also in the intermediate areas between these two. These cells are numerous— approximately 30 times as numerous as the anterior motoneurons. They are small and highly excitable, often exhibiting spontaneous activity and being capable of firing as rapidly as 1500 times per second. They have many interconnections one with the other, and many of them directly innervate the anterior motoneurons as illustrated in Figure 56–1. The interconnections between the internuncial cells and anterior motoneurons are responsible for many of the integrative functions of the spinal cord that are discussed during the remainder of this chapter.

Essentially all the different types of neuronal circuits described in Chapter 47 are found in the internuncial pool of cells of the spinal cord, including the *diverging, converging,* and *repetitive-discharge* circuits. In the following sections of this chapter we will see the many applications of these different circuits to the performance of specific reflex acts by the spinal cord.

Only a few incoming sensory signals to the spinal cord or signals from the brain terminate directly on the anterior motoneurons. Instead, most of them are transmitted first through internuncial cells where they are appropriately processed before stimulating the anterior motoneurons. Thus, in Figure 56–1, it is shown that the corticospinal tract terminates almost entirely on internuncial cells, and it is through these that the cerebrum transmits most of its signals for control of muscular function.

Sensory Input to the Motoneurons and Internuncial Cells. After entering the spinal cord, sensory fibers divide, sending one branch toward the brain through long ascending fiber tracts, while the other branch spreads locally in the cord segment of entry and in neighboring segments. Most of the terminal fibrils end

on internuncial cells, but a very small number terminate on the anterior motoneurons. Thus, there are two pathways through the spinal cord that cord reflexes can take: either directly to the anterior motoneuron itself, utilizing a monosynaptic pathway; or through one or more internuncial cells first, before passing to the anterior motoneuron. The monosynaptic pathway provides an extremely rapid reflex feedback system and is the basis of the very important muscle stretch reflex that will be discussed later in the chapter. All other cord reflexes utilize the internuncial cell pathway, a pathway that can modify the signals tremendously and that can cause complex reflex patterns. For instance, the very important protective reflex called the withdrawal reflex utilizes this pathway.

Multisegmental Connections of the Spinal Cord—The Propriospinal Fibers. About half of all the nerve fibers ascending and descending in the spinal cord are propriospinal fibers. These are fibers that run from one segment of the cord to another. In addition, the terminal fibrils of sensory fibers as they enter the cord branch both up and down the spinal cord, some of the branches transmitting signals only a segment or two in each direction, while others transmit signals many segments. These ascending and descending fibers of the cord enter into the multisegmental reflexes that will be described in this chapter, including many reflexes that coordinate movements in both the forelimbs and hindlimbs simultaneously.

MUSCLE AND TENDON RECEPTORS

In addition to a few pain and deep pressure sensory endings, muscles and tendons have an abundance of two special types of receptors: (1) *muscle spindles,* which detect momentary lengths of muscle fibers and rate of change of these lengths, and (2) *Golgi tendon organs,* which detect the tension applied to the tendon fibers during muscle contraction.

Signals transmitted to the central nervous system from these two receptors operate entirely at a subconscious level, causing no sensory perception at all. Instead, they transmit tremendous amounts of information from the muscles and tendons to (1) the motor control systems of the spinal cord and (2) the motor control systems of the cerebellum. They cause reflexes associated with "damping" of muscle movements, with equilibrium, and with posture.

THE MUSCLE SPINDLE AND ITS EXCITATION

The physiologic organization of the muscle spindle is illustrated in Figure 56–2A. Each spindle is built around 3 to 10 small *intrafusal muscle fibers* that are pointed at their ends and attach to the sheaths of the surrounding skeletal muscle fibers. Midway between the ends of each intrafusal fiber is a heavily nucleated area that has lost all cross striations and cannot contract, but instead becomes stretched whenever the spindle is stretched or whenever the ends of the intrafusal fibers contract. Entwined around the central areas of the intrafusal fibers is a nerve ending called the *primary* or *annulospiral receptor,* and from this a large type A nerve fiber, averaging 17 microns in diameter, passes into the sensory roots of the spinal cord. On each side of the annulospiral receptor is usually found another ending called the *secondary* or *flower-spray receptor* that excites another sensory nerve fiber that is much smaller than that excited by the annulospiral receptor, averaging approximately 8 microns in diameter.

The intrafusal muscle fibers of the spindle are innervated by small *gamma* motor nerve fibers, averaging 5 microns in diameter, which are distinct from the large alpha motor fibers to the extrafusal skeletal muscle fibers. Note again that the intrafusal fibers cannot contract in their middle portions where the primary and secondary receptors attach, but instead contract only in their two end portions. In fact, *contraction of these end portions actually stretches the central portion of the spindle.*

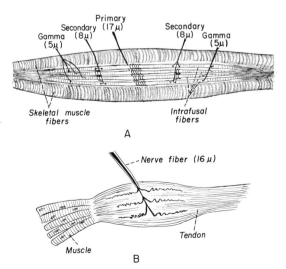

Figure 56–2 (A) Arrangement of the muscle spindle with respect to the surrounding skeletal muscle fibers. (B) Anatomy of Golgi tendon organ.

Stimulation of the Muscle Spindle. The muscle spindle can be stimulated by (1) stretch of the entire muscle belly, which also stretches the muscle spindle, or (2) contraction of the intrafusal fibers of the spindle, which contracts the two ends of the spindle but stretches the middle. In other words, the stimulus that normally excites the muscle spindle is stretch of its middle portion, caused either by stretching the entire muscle or by stimulating the intrafusal fibers.

Response Characteristics of the Primary and Secondary Receptors. The primary and secondary receptors are quite different in their responses to stretch. The primary receptors, being large and highly sensitive, respond in a fraction of a millisecond, and they transmit tremendous numbers of impulses following even very slight stretch. However, the rate of firing fades off within a fraction of a second and then settles down to a much lower, steady rate of firing. The intense excitation at the onset of stretch transmits information to the central nervous system and depicts the *rate of change* of the receptor length. On the other hand, the steady state signal transmits information depicting the actual *length* of the receptor.

In contrast, the secondary receptor requires several milliseconds to reach full response; and when it does respond, the degree of excitation is determined almost entirely by the length of the receptor alone and hardly at all by the rate of length change. Therefore, this receptor is capable of transmitting to the central nervous system a signal depicting only the length of the receptor itself.

Most functions of the muscle spindle that have been discovered thus far have been ascribed to signals from the primary receptors, as will be discussed in the following few sections.

Control of Muscle Spindle Function by the Gamma Efferent Fibers. Stimulation of the gamma efferent fibers to the muscle spindles contracts the two end portions of the intrafusal fibers. This decreases the overall length of the muscle spindle. If the surrounding skeletal muscle does not contract simultaneously, contraction of the end portions of the intrafusal fibers must obviously stretch the central portions where the sensory nerve receptors attach, thereby exciting the receptors. Conversely, reduction in the degree of gamma efferent excitation relaxes the stretch and correspondingly reduces the excitation of the spindle.

In other words, the muscle spindle functions

as a *comparator,* comparing the length of the spindle with the length of the skeletal muscle fibers that surround it. If the length of the surrounding muscle fibers is greater than the length of the muscle spindle, the sensory receptors of the spindle are excited. If the length of the surrounding muscle fibers is less than that of the spindle the excitation of the receptors is reduced. And when the spindle is shortened because of gamma efferent stimulation, its sensory receptors become excited because the spindle is then shorter than the surrounding muscle fibers.

Transmission of Information from the Muscle Spindles into the Central Nervous System. The large fibers from the primary receptors, on entering the spinal cord, send branches directly to the anterior motoneurons that supply the same muscle in which the spindles are located. In addition, collaterals from these fibers synapse in the dorsal cord and send second order fibers up the spinocerebellar tracts directly to the cerebellum. The branches that course directly to the anterior motoneurons cause instantaneous reflex effects, which are described later in this chapter. Those signals that are transmitted to the cerebellum cause more delayed cerebellar reflexes, which will be described in Chapter 58.

The fibers from the secondary receptors, on entering the spinal cord, terminate principally on the internuncial cells and elicit more delayed and more variable patterns of muscle reflexes, some of which will be discussed later.

THE GOLGI TENDON ORGAN AND ITS EXCITATION

Golgi tendon organs, illustrated in Figure 56–2B, lie within muscle tendons immediately beyond their attachments to the muscle fibers. An average of 10 to 15 muscle fibers is usually connected in series with each Golgi tendon organ, and the organ is stimulated by the tension produced by this small bundle of muscle fibers. Thus, the major difference between the function of the Golgi tendon apparatus and the muscle spindle is that the spindle detects muscle length, and the tendon organ detects muscle *tension.*

The tendon organ, like the primary receptor of the muscle spindle, responds with overexcitation to the onset of increased muscle tension, but within a small fraction of a second it settles down to a lower level of steady state firing that is almost directly proportional to the muscle tension. Thus, the Golgi tendon organs provide to the central nervous system instantaneous information of the degree of tension on each small segment of each muscle.

Transmission of Impulses from the Tendon Organ into the Central Nervous System. Signals from the tendon organ are transmitted through large, rapidly conducting type A alpha nerve fibers, fibers only slightly smaller than those from the primary receptors of the muscle spindle. These fibers, like those from the primary receptors, transmit signals both into local areas of the cord and through the spinocerebellar tracts into the cerebellum. The local signal is believed to excite a single inhibitory interneuron that in turn inhibits the anterior motoneuron. Thus, this local circuit directly inhibits the individual muscle without affecting adjacent muscles. The signals to the cerebellum will be discussed in Chapter 58.

THE STRETCH REFLEX

All reflexes require a *receptor,* an *effector,* and a *medium for transmission of signals* from the receptor to the effector. The stretch reflex, also called the myotatic reflex, employs the fewest number of neurons of any cord reflex: a single sensory neuron and a single motoneuron, as illustrated in Figure 56–3. This reflex is initiated by stretch of the primary receptor of the muscle spindle, and it causes the stretched muscle to contract.

Note again in Figure 56–2A that the ends of the muscle spindle are attached to the connective tissue between the skeletal muscle fibers. Because of this arrangement, stretch of the skeletal muscle fibers also stretches the muscle spindle and thereby excites the receptor. And impulses are transmitted directly to the anterior motoneurons to contract reflexly the stretched muscle. The net effect of this reflex, therefore, is *to oppose any stretch of the muscle* beyond its present length.

Simultaneous Inhibition of the Antagonistic Muscle. At the same time that impulses are

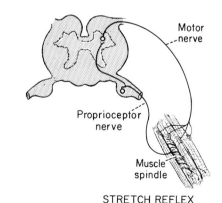

STRETCH REFLEX

Figure 56–3 Neuronal circuit of the stretch reflex.

transmitted by the stretch reflex to excite the stretched muscle, inhibitory impulses are transmitted to the motoneurons of the antagonist muscles, thus inhibiting these muscles and allowing the reflex contraction of the stretched muscle to be more effective.

It is believed that the signals transmitted to the antagonist muscles are transmitted through a single internuncial *inhibitory neuron* and that this secretes an inhibitory hormone at its synapse with the anterior motoneuron.

Static versus Dynamic Stretch Reflexes. Referring back to the discussion of muscle spindle excitation earlier in the chapter, one sees that the spindle can be stimulated many times as strongly by sudden increase in the degree of stretch as it can by continuous stretch. Therefore, as would be expected, the intensity of the reflex caused by a sudden increase in stretch is extremely powerful in comparison with the reflex caused by slow increase in stretch. The relatively weak reflex caused by continuous stretch is called the *static stretch reflex,* while the strong reflex that occurs following sudden increase in stretch is called the *dynamic stretch reflex.*

The Negative Stretch Reflex. Even under normal resting conditions the primary receptor of the muscle spindle transmits a tonic discharge of nerve impulses. Therefore, the signal output from the muscle spindle can be *decreased* as well as increased. Thus, when a muscle is shortened, the rate of impulses becomes reduced, particularly so when the muscle is shortened rapidly. This decrease in impulses immediately inhibits muscle tone in the shortened muscle and, simultaneously, automatically excites the antagonist muscles. Obviously, therefore, this negative stretch reflex opposes the shortening of the muscle length, particularly if the shortening occurs rapidly.

Damping Function of the Stretch Reflex. In the above discussion of the different manifestations of the stretch reflex, it has been obvious that one of the primary functions of the stretch reflex is to *oppose changes in muscle length*—especially *sudden* changes in length because of the extreme power of the *dynamic* stretch reflex. This powerful dynamic reflex effect has a very important function to "damp" muscle contractions, that is, to remove jerkiness in the contractions. For instance, impulses transmitted down the corticospinal tract to the anterior motoneurons often arrive in fairly irregular packets, and in animals whose posterior roots to a respective muscle have been sectioned and the stretch reflex thereby made nonfunctional, the muscle contraction is jerky

rather than smooth. However, with the dynamic stretch reflex intact, the movements are smooth, and no jerkiness is perceptible.

CONTROL OF MUSCLE CONTRACTION BY THE GAMMA EFFERENT MOTOR FIBERS — A SERVO CONTROL SYSTEM

Until recently it was believed that skeletal muscle contraction was controlled almost entirely by signals passing from the brain and thence through the large anterior motoneurons to the skeletal muscle fibers. It is now known that impulses transmitted to the muscle through the small *gamma nerve fibers* can also control muscle contraction. However, they control muscle contraction in an indirect manner, as follows:

Figure 56–4A shows a normal muscle with an inlying muscle spindle. Figure 56–4B shows the same spindle after contraction of the intrafusal fibers. It will be recalled that the intrafusal fibers contract only at the two ends, and contraction of these ends *stretches the primary receptor* in the central portion of the spindle. Therefore, in Figure 56–4B the receptor transmits impulses that excite the anterior motoneurons of this same muscle, and these in turn cause the large muscle fibers surrounding the spindle to contract as shown in Figure 56–4C until the central portion of the spindle returns to its normal length. When this has occurred, the receptor stops sending excess impulses.

From the above sequence, one can see that contraction of the spindle *causes a secondary*

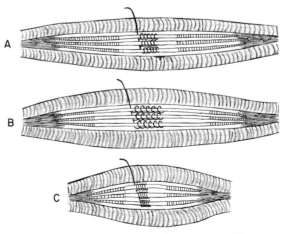

Figure 56–4 (A) The resting muscle spindle. (B) Contraction of the intrafusal fibers of the muscle spindle, resulting in stretch of the central receptor portion of the spindle. (C) Reflex contraction of the skeletal muscle fibers until the central portion of the spindle returns to its resting length.

contraction of the surrounding muscle until its contraction is equal to the contraction of the spindle. Furthermore, this effect is accomplished without any nerve impulses being transmitted directly from the brain to the anterior motoneurons. Such a mechanism as this is called a *servo mechanism,* the principles of which were discussed in Chapter 1.

To recapitulate, the gamma motor nerves can change the length of the muscle spindle. And the automatic feedback system of the stretch reflex causes the length of the surrounding skeletal muscle to adjust automatically to equal the length of the spindle.

Advantages of the Servo Type of Muscle Contraction. On first thought, it is difficult to understand why the nervous system would have two different types of muscle control mechanisms. However, these two mechanisms control two entirely different types of muscle functions. Direct stimulation of the muscle through the large anterior motoneurons causes instantaneous and rapid contraction of the muscle without any limit to the degree of shortening that will occur. On the other hand, the gamma system causes the muscle to contract to a *predetermined length.* The muscle will continue to contract until it has shortened to equal the degree of contraction of the spindle. Then it will stop contracting at that point.

But even more important than simply contracting the muscle to a fixed length is the fact that any extraneous force that tries to lengthen or shorten the muscle from the desired position will be automatically opposed by the muscle spindle servo feedback. That is, if an excessive load is put onto the muscle, it will stretch the muscle to a length greater than that of the spindle. As a result, impulses from the spindle cause increased contraction of the muscle to oppose the increased load and thereby return the length of the muscle back once again to the desired value.

Therefore, the real importance of gamma efferent control of muscle contraction is that the degree of shortening can be controlled *irrespective of the amount of load that the muscle is operating against.*

Function of the Gamma Efferent System in Postural Muscle Contractions. Certain muscles of the body, primarily the extensor muscles of the legs, hips, and back, are specifically adapted for support of the body against gravity. These muscles have different characteristics of contraction from the other muscles, exhibiting much slower contractions and being capable of contracting for prolonged periods of time.

Another characteristic of the postural muscles is that they usually contract to the appropriate lengths to maintain proper posture and remain contracted for long periods of time at precisely the same lengths. Therefore, many physiologists believe that the usual postural contractions of the body are controlled primarily by the gamma efferent servo control mechanism instead of by the alpha control mechanism. The advantage of the servo control mechanism would be that one could set the different parts of the body to exact positions and these parts would remain in the same positions regardless of outside disturbing forces. Thus, if a person should lift 50 pounds in front of him, the servo positioning system would automatically contract the extensor muscles of the hips to offset the tendency of the weight to flex the hips forward. Such a system would obviously be very simple compared to readjustment of each hip muscle contraction by signals from the brain, which signals would have to be changed every time the person should make even a simple movement or lift a weight.

Muscle Tone and Its Relation to the Gamma Efferent System. Skeletal muscle tone is caused by a continual slow rate of discharge of the alpha nerve fibers to each respective muscle. The intensity of the tone is increased by increasing this basal rate of nerve discharge, and it is decreased by decreasing the rate of discharge. One factor that causes the anterior motoneuron to emit continuous discharges is its facilitation by afferent impulses from the muscle spindle. When the rate of discharge of the spindle impulses increases, the tone of the muscle also increases. When the rate of discharge decreases, the tone of the muscle also decreases. And if the posterior roots to the spinal cord are cut, the tone of the muscle disappears either entirely or almost entirely because of loss of the facilitatory impulses from the spindles. Therefore, a large portion of even the basal tone of skeletal muscles is caused by the feedback arc from the muscle spindles.

EFFECT OF GAMMA FIBER STIMULATION ON TONE. Let us assume that the gamma fibers to muscles on both sides of a joint are stimulated. This shortens the muscle spindles, and the spindle feedback attempts to shorten the muscles as well. Yet the contractions of the two muscles on the two sides of the joint oppose each other and prevent shortening of either muscle. As a consequence, the shortened muscle spindles continue to discharge and continue to excite the respective muscles, so that the tone of each muscle increases. In this way, the gamma efferent system can increase or decrease the tone of the skeletal muscles.

Brain Areas for Control of the Gamma Efferent System. The gamma efferent system is excited to an extensive degree by the *bulboreticular facilitatory* region of the brain stem and secondarily by impulses transmitted into the bulboreticular facilitatory area from (a) the *cerebellum,* (b) the *basal ganglia,* and (c) even the *cerebral cortex.* Unfortunately, little is known about the precise mechanisms of control of the gamma efferent system. However, the bulbo-reticular facilitatory area is particularly concerned with postural contractions, which emphasizes the possible or probable important role of the gamma efferent mechanism in the control of muscular contraction for positioning the different parts of the body.

To emphasize the importance of the gamma system for control of muscles, it is particularly interesting that 31 per cent of all motor fibers to the muscles are gamma efferents rather than alpha efferents.

Shift from Direct Control of Muscles by the Alpha System to Servo Control by the Gamma System. At times it is important for the muscles to operate without the muscle spindle system opposing changes in muscle length. For instance, if a person wishes to throw his arm forward at a very rapid rate, a very active muscle spindle feedback system would prevent this from occurring. Neurophysiologic studies have shown that signals from the brain to cause rapid movement of parts of the body often excite both the alpha motoneurons and the gamma neurons simultaneously. This allows both the large skeletal muscle fibers and the muscle spindle fibers to contract in unison. As a result, no interfering feedback signals from the muscle spindles occur.

Another condition in which the alpha system can control muscles almost independently of the gamma system occurs when the muscle spindle intrafusal fibers are almost entirely in a flaccid state. Under these conditions the muscle spindle can be stretched without significant stimulation of the spindle receptors, because the flaccidity of the fibers prevents any tension on the spindle.

On the other hand, there are many other times when it is extremely important for the muscle spindles to operate mainly in the servo control mode. An example of this is the positioning contractions for different parts of the body when one is attempting to perform a very delicate task. Thus, if the person is replacing a small balance wheel on a watch, it is essential that the entire upper arm be positioned very exactly during this process. Furthermore, any force that would tend to move the arm out of position should be corrected instantly and without the necessity of conscious thought. It is believed that this is accomplished by transmitting signals from the brain to increase the *gain* of the muscle spindle-servo feedback mechanism. To do this, the gamma fibers to both agonist and antagonist muscles are excited, thus tightening the muscle spindles and increasing the numbers of impulses trans-

mitted by their receptors. As a result, the tone of all the involved muscle becomes greatly increased. But, in addition, the dynamic stretch reflex also becomes very highly sensitized because of the new rigidity of the intrafusal muscle fibers of the spindle. Therefore, even the minutest force tending to push an arm or other part of the body out of position will be resisted instantly by the servo control system.

Though much of the preceding is still theory, it possibly explains how we are capable of performing, on the one hand, exceedingly rapid and intense muscle functions without inhibitory effects from the stretch reflex; and yet, on the other hand, how it is possible for us to "set" the different parts of our body into very exact positions that will be held irrespective of changes in load.

CLINICAL APPLICATIONS OF THE STRETCH REFLEX

The Knee Jerk and Other Muscle Jerks. Clinically, a method used to determine the functional integrity of the stretch reflexes is to elicit the knee jerk and other muscle jerks. The knee jerk can be elicited by simply striking the patellar tendon with a reflex hammer; this stretches the quadriceps muscle and initiates a dynamic stretch reflex to cause the lower leg to jerk forward. The upper part of Figure 56–5 illustrates a myogram from the quadriceps muscle recorded during a knee jerk.

Similar reflexes can be obtained from almost any muscle of the body either by striking the tendon of the muscle or by striking the belly of the muscle itself. In other words, sudden stretch of muscle spindles is all that is required to elicit a stretch reflex.

The muscle jerks are used by neurologists to assess the degree of facilitation of spinal cord centers. When large numbers of facilitatory impulses are being transmitted from the upper regions of the central nervous system into the cord,

Figure 56–5 Myograms recorded from the quadriceps muscle during elicitation of the knee jerk and from the gastrocnemius muscle during ankle clonus.

the muscle jerks are greatly exacerbated. On the other hand, if the facilitatory muscles are depressed or abrogated, the muscle jerks are considerably weakened or completely absent. These reflexes are used most frequently to determine the presence or absence of muscle spasticity following lesions in the motor areas of the brain. Ordinarily, diffuse lesions in the contralateral motor areas of the cerebral cortex cause greatly exacerbated muscle jerks.

CLONUS. Under appropriate conditions, the muscle jerks can oscillate, a phenomenon called clonus (see lower myogram, Fig. 56–5).

Oscillation can be explained particularly well in relation to ankle clonus as follows:

If a person is standing on his tiptoes and suddenly drops his body downward to stretch one of his gastrocnemius muscles, impulses are transmitted from the muscle spindles into the spinal cord. These reflexly excite the stretched muscle, which lifts the body back up again. After a fraction of a second, the reflex contraction of the muscle dies out and the body falls again, thus stretching the spindles a second time. Again a dynamic stretch reflex lifts the body, but this too dies out after a fraction of a second, and the body falls once more to elicit still a new cycle. In this way, the stretch reflex of the gastrocnemius muscle continues to oscillate, often for long periods of time; this is clonus.

Clonus ordinarily occurs only if the stretch reflex is highly sensitized by facilitatory impulses from the cerebrum. For instance, in the decerebrate animal, in which the stretch reflexes are highly facilitated, clonus develops readily. Therefore, to determine the degree of facilitation of the spinal cord, neurologists test patients for clonus by suddenly stretching a muscle and keeping a steady stretching force applied to the muscle. If clonus occurs, the degree of facilitation is certain to be very high.

THE TENDON REFLEX

Inhibitory Nature of the Tendon Reflex. The Golgi tendon organ detects tension applied to the tendon by muscle contraction, as was discussed earlier in the chapter. Signals from the tendon organ are then transmitted into the spinal cord to cause reflex effects in the respective muscle. However, this reflex is entirely inhibitory, the exact opposite to the muscle spindle reflex. The signal from the tendon organ supposedly excites inhibitory internuncial cells, and these in turn inhibit the anterior motoneurons to the respective muscle.

When tension on the muscle and, therefore, on the tendon becomes extreme, the inhibitory effect from the tendon organ can be so great that it causes sudden relaxation of the entire muscle. This effect is called the *lengthening reaction;* it is probably a protective mechanism to prevent tearing of the muscle or avulsion of the tendon from its attachments to the bone. We know, for instance, that direct electrical stimulation of muscles in the laboratory can frequently cause such destructive effects.

However, much more important than this protective reaction is probably the function of the tendon reflex as a part of the overall servo control of muscle contraction in the following manner:

The Tendon Reflex as a Servo Control Mechanism for Muscle Tension. In the same way that the stretch reflex operates as a feedback mechanism to control length of a muscle, the tendon reflex theoretically can operate as a servo feedback mechanism to control muscle tension. That is, if the tension on the muscle becomes too great, inhibition from the tendon organ decreases this tension back to a lower value. On the other hand, if the tension becomes too little, impulses from the tendon organ cease; and this loss of inhibition allows the anterior motoneurons to become active again, thus increasing muscle tension back toward a higher level.

Very little is known at present about the function of or control of this tension servo feedback mechanism, but it is postulated to operate in the following basic manner: Signals from the brain are presumably transmitted to the cord centers to set the gain of the tendon feedback system. This can be done by changing the degree of facilitation of the neurons in the feedback loop. If the gain is high, then this system will be extremely sensitive to signals coming from the tendon organs; on the other hand, lack of excitatory signals from the brain could make the system very insensitive to the signals from the tendon organ. In this way, control signals from higher nervous centers could automatically set the level of tension at which the muscle would be maintained. If the required tension is high, then the muscle tension would be set by the servo feedback mechanism to this high level of tension. On the other hand, if the desired tension level is low, the muscle tension would automatically adjust to this low level.

This description of the tension servo system is obviously very incomplete and almost entirely theoretical at the present time, but it depicts the approximate state of present speculation on this effect.

Value of a Servo Mechanism for Control of Tension. An obvious value of a mechanism for setting degree of muscle tension would be to allow the different muscles to apply a certain amount of force irrespective of how far the muscles contract. An example of this would be paddling a boat, during which a person sets the amount of force that he pulls backwards on the paddle and maintains that degree of force throughout the entire movement. Were it not for some type of tension feedback mechanism, the same amount of force would not be maintained throughout the stroke because uncontrolled muscles change their force of contraction as their lengths change. One can imagine hundreds of different patterns of muscle function that require maintenance of constant tension rather than maintenance of constant lengths of the muscles.

THE FLEXOR REFLEX

In the spinal or decerebrate animal, almost any type of sensory stimulus to a limb is likely to cause the flexor muscles of the limb to contract strongly, thereby withdrawing the limb from the stimulus. This is called the flexor reflex.

In its classical form the flexor reflex is elicited most frequently by stimulation of pain endings, such as by pinprick, heat, or some other painful stimulus, for which reason it is frequently called a *nociceptive reflex*. However, stimulation of the light touch receptors can also elicit a weaker and less prolonged flexor reflex.

If some other part of the body besides one of the limbs is stimulated, this part also will be withdrawn from the stimulus, but the reflex may not be confined entirely to flexor muscles even though it is basically the same reflex. Therefore, the reflex is frequently called a *withdrawal reflex*, too.

Neural Mechanism of the Flexor Reflex. The left-hand portion of Figure 56–6 illustrates the neuronal pathways for the flexor reflex. In this instance, a painful stimulus is applied to the hand; as a result, the flexor muscles of the upper arm become reflexly excited, thus withdrawing the hand from the painful stimulus.

The pathways for eliciting the flexor reflex do not pass directly to the anterior motoneurons but, instead, pass first into the internuncial pool of neurons and then to the motoneurons. The shortest possible circuit is a three- or four-neuron arc; however, most of the signals of the reflex traverse many more neurons than this and involve the following basic types of

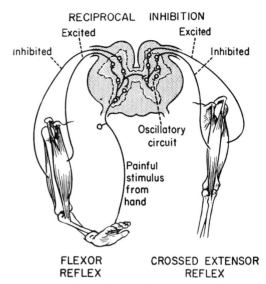

Figure 56–6 The flexor reflex, the crossed extensor reflex, and reciprocal inhibition.

circuits: (1) diverging circuits to spread the reflex to the necessary muscles for withdrawal, (2) circuits to inhibit the antagonist muscles, called *reciprocal inhibition circuits,* and (3) circuits to cause a prolonged repetitive after-discharge even after the stimulus is over.

Figure 56–7 illustrates a typical myogram from a flexor muscle during a flexor reflex. Within a few milliseconds after a pain nerve is stimulated, the flexor response occurs. Then, in the next few seconds the reflex begins to *fatigue*, which is characteristic of essentially all integrative reflexes of the spinal cord. Then, soon after the stimulus is over, the contraction of the muscle begins to return toward the base line, but, because of *after-discharge*, will not return all the way for many milliseconds. The duration of the after-discharge depends on the intensity of the sensory stimulus that had elicited the reflex; a weak stimulus causes almost no after-discharge, in contrast to an after-discharge lasting for several seconds following a strong stimulus. Furthermore, a flexor reflex

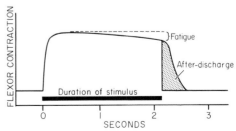

Figure 56–7 Myogram of a flexor reflex, showing rapid onset of the reflex, an interval of fatigue, and finally after-discharge after the stimulus is over.

initiated by nonpainful stimuli and transmitted through the large sensory fibers causes essentially no after-discharge, whereas nociceptive impulses transmitted through the small type A fibers and type C fibers cause prolonged after-discharge.

The after-discharge that occurs in the flexor reflex almost certainly results from both types of repetitive-discharge circuits that were discussed in Chapter 47. Electrophysiological studies show that the immediate after-discharge, lasting for about 6 to 8 milliseconds, results from the parallel type of circuit, with impulses being transmitted from one interneuron to another to another and all these in turn transmitting their signals successively to the anterior motoneuron. However, the prolonged after-discharge that occurs following strong pain stimuli almost certainly involves reverberating circuits in the interneurons, these transmitting impulses to the anterior motoneurons sometimes for several seconds after the incoming sensory signal is completely over.

Thus, the flexor reflex is appropriately organized to withdraw a pained or otherwise irritated part of the body away from the stimulus. Furthermore, because of the after-discharge it will hold the irritated part away from the stimulus for as long as 1 to 3 seconds even after the irritation is over. During this time, other reflexes and other actions of the central nervous system can move the entire body away from the painful stimulus.

The Pattern of Withdrawal. The pattern of withdrawal that results when the flexor (or withdrawal) reflex is elicited depends on the sensory nerve that is stimulated. Thus, a painful stimulus on the inside of the arm not only elicits a flexor reflex in the arm but also contracts the abductor muscles to pull the arm outward. In other words, the integrative centers of the cord cause those muscles to contract that can most effectively remove the pained part of the body from the object that causes pain. This same principle applies to any part of the body but especially to the limbs, for they have highly developed flexor reflexes.

The Principles of "Local Sign" and "Final Position" in Relation to the Flexor Reflex. The position to which a limb is withdrawn following a painful stimulus is determined entirely by the point on the skin that is stimulated, and, if another point only a few centimeters away is stimulated, the withdrawal position will be correspondingly different. This is called the phenomenon of *local sign*. That is, the response depends on the locus of the stimulus.

Another extremely important principle in relation to the flexor reflex and to many other motor responses of the spinal cord is that the original position of the limb has nothing to do with the final position of the limb following the stimulus. For instance, a limb that is already flexed will be withdrawn to exactly the same position following stimulation of a particular pain nerve as it would if the original position of the limb had been in full extension. This is called the principle of *final position*. It results from a servo control type of organization of the interneurons between the incoming sensory nerve and the outgoing motor nerves. Thus, each incoming sensory nerve is connected to a particular complex of interneurons to give a different final position from that of any other incoming sensory nerve.

The Stretch Flexor Reflex. Another type of stretch reflex is that which results from stimulation of the *secondary receptors* of the muscle spindles. Such stimulation usually results in a flexor pattern of withdrawal; but in some instances, for unknown reasons, it can also cause an extensor response pattern. The precise function of this type of flexor reflex is not yet known, though it has been suggested that it might operate in the oscillating feedback circuits associated with walking, which are discussed later in the chapter.

THE CROSSED EXTENSOR REFLEX

Approximately 0.2 to 0.5 second after a stimulus elicits a flexor reflex in one limb, the opposite limb begins to extend. This is called the crossed extensor reflex. Extension of the opposite limb obviously can push the entire body away from the object causing the painful stimulus.

Neuronal Mechanism of the Crossed Extensor Reflex. The right-hand portion of Figure 56–6 illustrates the neuronal circuit responsible for the crossed extensor reflex, showing that signals from the sensory nerves cross to the opposite side of the cord to cause exactly opposite reactions to those that cause the flexor reflex. Because the crossed extensor reflex usually does not begin until 200 to 500 milliseconds following the initial pain stimulus, it is certain that many internuncial neurons are in the circuit between the incoming sensory neuron and the motoneurons of the opposite side of the cord responsible for the crossed extension. Furthermore, after the painful stimulus is removed, the crossed extensor reflex continues for an even longer period of after-discharge than that for the flexor reflex. Therefore, again, it is almost certain that this prolonged after-discharge results from reverberatory circuits among the internuncial cells.

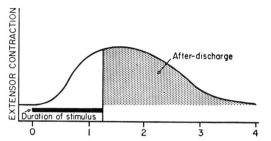

Figure 56–8 Myogram of a crossed extensor reflex, showing slow onset but prolonged after-discharge.

Figure 56–9 Myogram of a flexor reflex, illustrating crossed inhibition caused by a stronger flexor reflex in the opposite limb.

Figure 56–8 illustrates a myogram recorded from a muscle involved in a crossed extensor reflex. This shows the relatively long latency before the reflex begins and also the very long after-discharge following the end of the stimulus. The prolonged after-discharge obviously would be of benefit in holding the entire body away from a painful object until other neurogenic reactions should cause the body to move away.

The Principle of "Final Position" in the Crossed Extensor Reflex. In exactly the same way that a sensory stimulus from a given local area of the body causes a certain "final position of withdrawal," it also causes a "final position of extension" on the opposite side of the body. This final position is different for each sensory nerve stimulated.

RECIPROCAL INNERVATION

In the foregoing paragraphs we have pointed out several times that excitation of one group of muscles is often associated with inhibition of another group. For instance, when a stretch reflex excites one muscle, it simultaneously inhibits the antagonist muscles. This is the phenomenon of *reciprocal inhibition,* and the neuronal mechanism that causes this reciprocal relationship is called *reciprocal innervation.* Likewise, reciprocal relationships exist between the two sides of the cord as exemplified by the flexor and extensor reflexes as described above.

We will see below that the principle of reciprocal innervation is also important in most of the cord reflexes that subserve locomotion, for it helps to cause forward movement of one limb while causing backward movement of the opposite limb.

Figure 56–9 illustrates a typical example of reciprocal inhibition. In this instance, a moderate but prolonged flexor reflex is elicited from one limb of the body, and while this reflex is still being elicited a still stronger flexor reflex is elicited from the opposite limb, causing reciprocal inhibition of the first limb. Then removal of the strong flexor reflex allows the original reflex to reassume its previous intensity.

FATIGUE OF REFLEXES; REBOUND

Figure 56–7 illustrated that the flexor reflex begins to *fatigue* within a few seconds after its initiation. This is a common effect in most of the cord reflexes as well as many other reflexes of the central nervous system, and it presumably results from progressive diminishment of excitability of some neurons in the reflex circuits.

Another effect closely allied to fatigue is *rebound.* This means that, immediately after a reflex is over, a second reflex of the same type is much more difficult to elicit for a given time thereafter. However, because of reciprocal innervation, reflexes of the antagonist muscles become even more easily elicited during that same period of time. For instance, if a flexor reflex occurs in a left limb, a second flexor reflex is more difficult to establish for a few seconds thereafter, but a crossed extensor reflex in this same limb will be greatly exacerbated. Rebound is probably one of the important mechanisms by which the rhythmic to-and-fro movements required in locomotion are effected, which are described in more detail later in the chapter.

THE REFLEXES OF POSTURE AND LOCOMOTION

THE POSTURAL AND LOCOMOTIVE REFLEXES OF THE CORD

The Positive Supportive Reaction. Pressure on the footpad of a decerebrate animal causes the limb to extend against the pressure that is being applied to the foot. Indeed, this reflex is so strong that a decerebrate animal can be placed on its feet, and the pressure on the foot-pads will reflexly stiffen the limbs sufficiently that they can support the weight of the body—and the animal will stand in a rigid position. This reflex is called the positive supportive reaction.

The positive supportive reaction involves a complex circuit in the internuncial cells similar to those responsible for the flexor and the crossed-extensor reflexes. Furthermore, the principle of "final position" applies to this reflex equally as much as to the flexor and crossed extensor reflex. That is, the locus of the pressure on the pad of the foot determines the final position to which the limb is extended.

THE MAGNET REACTION.   If pressure is applied on one side of the foot, the foot moves in that direction, or, if the pressure is moved forward, the foot moves forward, or, if it is moved backward, the foot moves backward. In this way, the bottom of the foot remains continuously extended toward the pressure that is being applied. This is called the magnet reaction, and it obviously could help in the maintenance of equilibrium, for, if an animal should tend to fall to one side, that side of the foot-pad would be stimulated, and the limb would automatically extend toward the falling side to push the body of the animal in the opposite direction.

The Cord "Righting" Reflexes.   When a spinal cat or even a well-recovered young spinal dog is laid on its side, it will make incoordinate movements that indicate that it is trying to raise itself to the standing position. This is called a cord righting reflex, and it illustrates that relatively complicated reflexes associated with posture are at least partially integrated in the spinal cord. Indeed, a puppy with a well-healed transected thoracic cord can completely right itself from the lying position and can even walk on its hindlimbs. And, in the case of the opossum with a similar transection of the thoracic cord, the walking movements of the hindlimbs are hardly different from those in the normal opossum—except that the hindlimb movements are not synchronized with those of the forelimbs as is normally the case.

The Rhythmic Stepping Reflex of the Single Limb.   Rhythmic stepping movements are frequently observed in the limbs of spinal animals. Indeed, even when the lumbar portion of the spinal cord is separated from the remainder of the cord and a longitudinal section is made down the center of the cord to block neuronal connections between the two limbs, each hind limb can still perform step-

ping functions. Forward flexion of the limb is followed a second or so later by backward extension. Then flexion occurs again, and the cycle is repeated over and over.

This oscillation back and forth between the flexor and extensor muscles seems to result mainly from reciprocal inhibition and rebound. That is, the forward flexion of the limb causes reciprocal inhibition of extensor muscles, but shortly thereafter the flexion begins to die out; as it does so, *rebound* excitation of the extensors causes the leg to move downward and backward. After extension has continued for a time, it, too, begins to die and is followed by rebound excitation of the flexor muscles.

Another reflex that might enter into the rhythmic stepping pattern of the single limb is the stretch flexor reflex initiated, as noted earlier in the chapter, by stretch of the secondary receptors of the muscle spindles. Downward extension of the limb can elicit such a reflex and cause the limb to lift again. However, this reflex is not essential to the stepping movements because these can still occur when all feedback signals from the muscles are blocked.

Figure 56–10 shows myograms of alternate extensor and flexor contractions in a limb during the course of stepping movements.

Reciprocal Stepping of Opposite Limbs.   If the lumbar spinal cord is not sectioned down its center as noted above, every time stepping occurs in the forward direction in one limb, the opposite limb ordinarily steps backward. This effect results from reciprocal innervation between the two limbs.

Diagonal Stepping of All Four Limbs—The "Mark Time" Reflex.   If a decerebrate animal is held up from the table and its legs are allowed to fall downward as illustrated in Figure 56–11, the stretch on the limbs occasionally elicits stepping reflexes that involve all four limbs. In general, stepping occurs diagonally

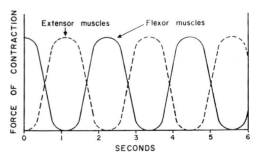

Figure 56–10   Alternate contraction of the extensor and flexor muscles in the limb of a spinal animal when a stepping reflex was elicited.

Figure 56–11 Diagonal stepping movements exhibited by a spinal animal.

between the fore- and hindlimbs. That is, the right hindlimb and the left forelimb move backward together while the right forelimb and left hindlimb move forward. This diagonal response is another manifestation of reciprocal innervation, this time occurring the entire distance up and down the cord between the fore- and hindlimbs. Such a walking pattern is often called a mark time reflex.

The Galloping Reflex. Another type of reflex that occasionally develops in the spinal animal is the galloping reflex, in which both forelimbs move backward in unison while both hindlimbs move forward. If stretch or pressure stimuli are applied almost exactly equally to opposite limbs at the same time, a galloping reflex will likely result, whereas unequal stimulation of one side versus the other elicits the diagonal walking reflex. This is in keeping with the normal patterns of walking and of galloping, for, in walking, only one limb at a time is stimulated, and this would predispose to continued walking. Conversely, when the animal strikes the ground during galloping, the limbs on both sides are stimulated approximately equally; this obviously would predispose to further galloping and, therefore, would continue this pattern of motion in contradistinction to the walking pattern.

THE SCRATCH REFLEX

An especially important cord reflex in lower animals is the scratch reflex, which actually involves two different functions: (1) a *position sense* that allows the paw to find the exact point of irritation on the surface of the body and (2) a *to-and-fro scratching movement.*

Obviously, the scratch reflex, like the stepping movements of locomotion, involves reciprocal in-

nervation circuits that cause oscillation. One of the important discoveries in relation to the to-and-fro movement of the scratch reflex is that it can still occur even when all the sensory roots from the oscillating limb are sectioned. In other words, feedback from the limb itself is not necessary to maintain the neuronal oscillation, which almost certainly means that the oscillation can occur intrinsically as a result of oscillating circuits within the spinal internuncial cells themselves.

The *position sense* of the scratch reflex is also a highly developed function, for even though a flea might be crawling as far forward as the shoulder of a spinal animal, the hind paw can find its position. Furthermore, this can be accomplished even though 19 different muscles in the limb must be contracted simultaneously in a precise pattern to bring the paw to the exact position of the crawling flea. To make the reflex even more complicated, when the flea crosses the midline, the paw stops scratching, but the opposite paw suddenly begins the to-and-fro motion and finds the flea immediately.

THE SPINAL CORD REFLEXES THAT CAUSE MUSCLE SPASM

In human beings, local muscle spasm is often observed. The mechanism of this has not been elucidated to complete satisfaction even in experimental animals, but it is known that pain stimuli can cause reflex spasm of local muscles, which presumably is the cause of much if not most of the muscle spasm observed at localized regions of the human body.

Muscle Spasm Resulting from a Broken Bone. One type of clinically important spasm occurs in muscles surrounding a broken bone. This seems to result from the pain impulses initiated from the broken edges of the bone, which cause the muscles surrounding the area to contract powerfully and tonically. Relief of the pain by injection of a local anesthetic relieves the spasm; a general anesthetic also relieves the spasm. One of these procedures is often necessary before the spasm can be overcome sufficiently for the two ends of the bone to be set back into appropriate positions.

Abdominal Spasm in Peritonitis. Another type of local spasm caused by a cord reflex is the abdominal spasm resulting from irritation of the parietal peritoneum by peritonitis. Here, again, relief of the pain caused by the peritonitis allows the spastic muscles to relax. Almost the same type of spasm occurs during surgical operations; pain impulses from the parietal peritoneum cause the abdominal muscles to contract extensively and sometimes actually to extrude the gut through the surgical wound. For this reason deep surgical anesthesia is usually required for intraabdominal operations.

Muscle Cramps. Still another type of local spasm is the typical muscle cramp. In the past this has been ascribed to local contraction of the muscle in response to abnormal metabolic end-products in the muscle itself. However, electromyographic studies indicate that the cause of a muscle cramp is the following:

Any local irritating factor or metabolic abnormality of a muscle – such as severe cold, lack of blood flow to the muscle, or overexercise of the muscle – can elicit pain or other types of sensory impulses that are transmitted from the muscle to the spinal cord, thus causing reflex muscle contraction. The contraction in turn stimulates the same sensory receptors still more, which causes the spinal cord to increase the intensity of contraction still further. Thus, a positive feedback mechanism occurs so that a small amount of initial irritation causes more and more contraction until a full-blown muscle cramp ensues. Reciprocal inhibition of the muscle can sometimes completely relieve the cramp. That is, if a person purposefully contracts the muscle on the opposite side of the joint from the cramped muscle while at the same time using another hand or leg to prevent movement of the joint, the reciprocal inhibition that occurs in the cramped muscle can at times relieve the cramp immediately.

THE AUTONOMIC REFLEXES IN THE SPINAL CORD

Many different types of segmental autonomic reflexes occur in the spinal cord, most of which are discussed in other chapters. Briefly, these include: (1) changes in vascular tone resulting from heat and cold (Chap. 71), (2) sweating, which results from localized heat at the surface of the body (Chap. 71), (3) intestino-intestinal reflexes that control some motor functions of the gut (Chap. 63), (4) peritoneo-intestinal reflexes that inhibit gastric motility in response to peritoneal irritation (Chap. 66), and (5) evacuation reflexes for emptying the bladder (Chap. 38) and the colon (Chap. 63). In addition, all the segmental reflexes can at times be elicited simultaneously in the form of the so-called mass reflex as follows:

The Mass Reflex. In a spinal animal or human being, the spinal cord sometimes suddenly becomes excessively active, causing massive discharge of large portions of the cord. The usual stimulus that causes this is a strong nociceptive stimulus to the skin or excessive filling of a viscus, such as overdistention of the bladder or of the gut. Regardless of the type of stimulus, the resulting reflex, called the mass reflex, involves large portions or even all of the cord, and its pattern of reaction is the same. The effects of the mass reflex are: (1) a major portion of the body goes into strong flexor spasm, (2) the colon and bladder are likely to evacuate, (3) the arterial pressure often

rises to maximal values – sometimes to a mean pressure well over 200 mm. Hg, and (4) large areas of the body break out into profuse sweating. The mass reflex might be likened to the epileptic attacks that involve the central nervous system in which large portions of the brain become massively activated.

The precise neuronal mechanism of the mass reflex is unknown. However, since its duration is prolonged, it presumably results from great masses of reverberating circuits that excite large areas of the cord at once.

SPINAL CORD TRANSECTION AND SPINAL SHOCK

When the spinal cord is suddenly transected, essentially all cord functions immediately become depressed, a reaction called *spinal shock*. The reason for this is that normal activity of the cord neurons depends to a great extent on continual tonic discharges from higher centers, particularly discharges transmitted through the vestibulospinal tract and the excitatory portion of the reticulospinal tracts.

After a few days to a few weeks of spinal shock, the spinal neurons gradually regain their excitability. This seems to be a natural characteristic of neurons everywhere in the nervous system – that is, after they lose their source of facilitatory impulses, they usually increase their own natural degree of excitability to make up for the loss. In most nonprimates, the excitability of the cord centers returns essentially to normal within a few hours to a few weeks, but in man the return is often delayed for several months and occasionally is never complete; or, on the other hand, recovery is sometimes excessive, with resultant hyperexcitability of all or most cord functions.

Some of the spinal functions specifically affected during or following spinal shock are: (1) The arterial blood pressure falls immediately – sometimes to as low as 40 mm. Hg – thus illustrating that sympathetic activity becomes blocked almost to extinction. However, the pressure ordinarily returns to normal within a few days. (2) All skeletal muscle reflexes integrated in the spinal cord are completely blocked during the initial stages of shock. In lower animals, a few hours to 3 weeks are required for these reflexes to return to normal, and in man they rarely return to normal. Yet sometimes, both in animals and man, some reflexes eventually become hyperexcitable, particularly if a few facilitatory pathways remain intact between the brain and the cord while the remainder of the spinal cord is transected. (3) The sacral reflexes for control of bladder and colon evacuation are completely suppressed in man for the first few weeks following cord transection, but they eventually return. These effects are discussed in Chapters 38 and 66.

REFERENCES

Creed, R. S.; Denny-Brown, D.; Eccles, J. C.; Liddell, E. G. T.; and Sherrington, C. S.: Reflex Activity of the Spinal Cord. New York, Oxford University Press, Inc., 1932.

Emonet-Denand, F., and Houk, J.: Some effects of polarizing current on discharges from muscle spindle receptors. *Amer. J. Physiol., 216*:404, 1969.

Eyzaguirre, C.: The electrical activity of mammalian intrafusal fibres. *J. Physiol. (Lond.), 150*:169, 1960.

Fields, H. L.; Evoy, W. H.; and Kennedy, D.: Reflex role played by efferent control of an invertebrate stretch receptor. *J. Neurophysiol., 30*:859, 1967.

Gillary, H. L., and Kennedy, D.: Neuromuscular effects of impulse pattern in a crustacean motoneuron. *J. Neurophysiol., 32*:607, 1969.

Gillary, H. L., and Kennedy, D.: Pattern generation in a crustacean motoneuron. *J. Neurophysiol., 32*:595, 1969.

Gilman, S., and McDonald, W. I.: Relation of afferent fiber conduction velocity to reactivity of muscle spindle receptors after cerebellectomy. *J. Neurophysiol., 30*:1513, 1967.

Granit, R.: Muscular tone. *J. Sport Med., 2*:46, 1962.

Granit, R.; Pompeiano, O.; and Waltman, B.: Fast supraspinal control of mammalian muscle spindles: extra- and intrafusal co-activation. *J. Physiol. (Lond.), 147*:385, 1959.

Henneman, E.: Functional relations between components of the stretch reflex circuit. *Proc. Int. Union of Physiol. Sciences, 6*:113, 1968.

Houk, J., and Henneman, E.: Responses of Golgi tendon organs to active contractions of the soleus muscle of the cat. *J. Neurophysiol., 30*:466, 1967.

Houk, J., and Simon, W.: Responses of Golgi tendon organs to forces applied to muscle tendon. *J. Neurophysiol., 30*:1466, 1967.

Hunt, C. C., and Perl, E. R.: Spinal reflex mechanisms concerned with skeletal muscle. *Physiol. Rev., 40*:538, 1960.

Laporte, Y.: Muscle spindles. *Proc. Int. Union of Physiol. Sciences, 6*:81, 1968.

Liddell, E. G. T.: The Discovery of Reflexes. New York, Oxford University Press, Inc., 1960.

Ottoson, D.; McReynolds, J. S.; and Shepherd, G. M.: Sensitivity of isolated frog muscle spindle during and after stretching. *J. Neurophysiol., 32*:24, 1969.

Partridge, L. D., and Glaser, G. H.: Adaptation in regulation of movement and posture. A study of stretch responses in spastic animals. *J. Neurophysiol., 23*:257, 1960.

Phillips, C. G.: Neurophysiology: motor mechanisms. *Proc. Int. Union of Physiol. Sciences, 6*:119, 1968.

Roberts, Tristan D. M.: Neurophysiology of Postural Mechanisms. New York, Plenum Publishing Corporation, 1967.

Rudomin, P., and Dutton, H.: Effects of conditioning afferent volleys on variability of monosynaptic responses of extensor motorneurons. *J. Neurophysiol., 32*:140, 1969.

Rudomin, P., and Dutton, H.: Effects of muscle and cutaneous afferent nerve volleys on excitability fluctuations of Ia terminals. *J. Neurophysiol., 32*:158, 1969.

Sherrington, C. S.: The Integrative Action of the Nervous System. New Haven, Yale University Press, 1911.

Shimamura, M.; Mori, S.; and Yamauchi, T.: Effects of spino-bulbo-spinal reflex volleys on extensor motoneurons of hindlimb in cats. *J. Neurophysiol., 30*:319, 1967.

Thomas, R. C., and Wilson, V. J.: Recurrent interactions between motoneurons of known location in the cervical cord of the cat. *J. Neurophysiol., 30*:661, 1967.

CHAPTER 57

Motor Functions of the Brain Stem and Basal Ganglia—Reticular Formation, Vestibular Apparatus, Equilibrium, and Brain Stem Reflexes

The brain stem is a complex extension of the spinal cord. Collected in it are numerous neuronal circuits to control respiration, cardiovascular function, gastrointestinal function, eye movement, equilibrium, support of the body against gravity, and many special stereotyped movements of the body. Some of these functions—such as control of respiration and cardiovascular functions—are described in special sections of this text. The present chapter deals primarily with the motor functions of the brain stem.

THE RETICULAR FORMATION, AND SUPPORT OF THE BODY AGAINST GRAVITY

Throughout the entire extent of the brain stem—in the medulla, pons, mesencephalon, and even in portions of the diencephalon—are areas of diffuse neurons collectively known as the *reticular formation*. Figure 57–1 illustrates the extent of the reticular formation, showing it to begin at the upper end of the spinal cord and to extend (a) into the hypothalamus, (b) to the sides of the thalamus, and (c) even upward through the central portion of the thalamus. The lower end of the reticular formation is continuous with the internuncial cells of the spinal cord, and, indeed, the reticular formation of the brain stem functions in a manner quite analogous to many of the functions of the internuncial system of the cord.

Interspaced in the reticular formation are many small special nuclei, some of which are motor in function and some of which are sensory in function. A few of the motor nuclei are illustrated in Figure 57–1. These operate in close conjunction with the diffuse reticular neurons to provide many of the subconscious motor activities of the body.

EXCITATORY AND INHIBITORY REGIONS OF THE RETICULAR FORMATION

By far the majority of the reticular formation is excitatory, including especially the

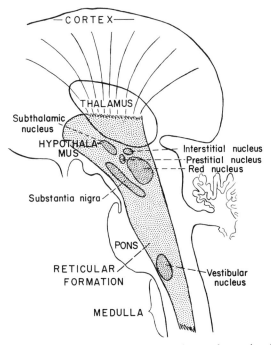

Figure 57–1 The reticular formation and associated nuclei.

reticular formation in the pons, mesencephalon, and diencephalon. In addition, the lateral portions of the reticular formation in the medulla are also excitatory. These areas are known collectively as the *bulboreticular facilitatory area.* Diffuse stimulation in this facilitatory area causes a general increase in muscle tone either throughout the body or in localized areas.

A small portion of the reticular formation, located in the ventromedial portion of the medulla, has mainly inhibitory functions and is known as the *bulboreticular inhibitory area.* Diffuse stimulation in this area causes decreased tone of the musculature throughout the body.

The bulboreticular facilitatory area is intrinsically excitable. That is, if it is not inhibited by signals from other parts of the nervous system, it has a natural tendency to transmit continuous nerve impulses. Thus, when the facilitatory portion of the reticular formation is uninhibited by signals from other sources, it transmits continual impulses to the skeletal muscles throughout the body. But, in the normal animal, inhibitory signals are continually available from the basal ganglia and cerebral cortex to keep the reticular facilitatory area from becoming overactive. As will be noted later in the chapter in the discussion of the decerebrate animal, removal of the higher portions of the nervous system removes this inhibition and allows the reticular facilitatory area to become tonically active, which causes rigidity of the skeletal muscles throughout the body.

On the other hand, when the brain stem is sectioned at a level slightly lower than the vestibular nuclei, most function of the reticular facilitatory area is then blocked, which causes almost all the musculature throughout the body to become totally flaccid.

Discrete Motor Functions Elicited from the Reticular Formation. Stimulation of discrete points in the reticular formation will at times cause discrete muscle contraction or inhibition in specific parts of the body. It is particularly interesting that stimulation in the reticular nuclei of the pons or mesencephalon frequently causes excitation of a protagonist muscle along with simultaneous inhibition of an antagonist muscle. Unfortunately, we know very little about the ability of the discrete neurons in the reticular formation to control muscle movements, but the discussions in the following sections will show that the reticular formation does not necessarily act as one large unit but, instead, has the capability of controlling many discrete motor functions as well. Among the most important of these motor functions are the control

of equilibrium and the support of the body against gravity.

SUPPORT OF THE BODY AGAINST GRAVITY

When a person or an animal is in a standing position, continuous impulses are transmitted from the reticular formation and from closely allied nuclei, particularly from the vestibular nuclei, into the spinal cord and thence to the extensor muscles to stiffen the limbs. This allows the limbs to support the body against gravity. These impulses are transmitted mainly by way of the reticulospinal and vestibulospinal tracts.

The normal excitatory nature of the upper portion of the reticular formation provides much of the intrinsic excitation required to maintain tone in the extensor muscles. However, the degree of activity in the individual extensor muscles is determined by the equilibrium mechanisms. Thus, if an animal begins to fall to one side, the extensor muscles on that side stiffen while those on the opposite side relax. And similar effects occur when it tends to fall forward or backward.

In essence, then, the reticular formation provides the nervous energy to support the body against gravity. But other factors, particularly the vestibular apparatuses, control the relative degree of extensor contraction in the different parts of the body, which provides the function of equilibrium.

VESTIBULAR SENSATIONS AND THE MAINTENANCE OF EQUILIBRIUM

THE VESTIBULAR APPARATUS

The vestibular apparatus is the sensory organ that detects sensations concerned with equilibrium. It is composed of a *bony labyrinth* containing the *membranous labyrinth,* the functional part of the apparatus. The top of Figure 57–2 illustrates the membranous labyrinth; it is composed mainly of the *cochlear duct,* the three *semicircular canals,* and the two large chambers known as the *utricle* and the *saccule.* The cochlear duct is concerned with hearing and has nothing to do with equilibrium; the saccule is concerned with either hearing, detection of vibration, or perhaps equilibrium in some lower animals but probably is nonfunctional in the human being. However, the *utricle* and the *semicircular canals* are especially important for maintaining equilibrium.

The Utricle. Located in the utricle is the

MEMBRANOUS LABYRINTH

CRISTA AMPULLARIS AND MACULA

Figure 57–2 The membranous labyrinth, and organization of the crista ampullaris and the macula. (From Goss: Gray's Anatomy of the Human Body. Lea & Febiger; and modified from Kolmer by Buchanan: Functional Neuroanatomy. Lea & Febiger.)

macula, which is covered by a gelatinous layer in which many small calcified masses known as *otoconia* are embedded. Also, in the macula are many *hair cells,* which project *cilia* or *hairs* up into the gelatinous layer. Around these hair cells are entwined sensory axons of the vestibular nerve. Bending the hairs to one side transmits impulses to apprise the central nervous system of the relative position of the otoconia in the gelatinous mass over the macula. Therefore, the weight of the otoconia compressing, bending, or pulling the hairs provides signals from the utricle, and these signals in turn are transmitted by appropriate nerve tracts to the brain, thence controlling equilibrium.

The Semicircular Canals. The three semicircular canals in each vestibular apparatus, known respectively as the *superior, posterior,* and *external* (or *horizontal*) *semicircular canals,* are arranged at right angles to each other so that they represent all three planes in space. When the head is bent forward at approximately 30 degrees, the two external semicircular canals are located approximately horizontal with respect to the surface of the earth. The superior

canals are then located in vertical planes that project *forward and 45 degrees outward,* and the posterior canals are also in vertical planes but project *backward and 45 degrees outward.* Thus, the superior canal on each side of the head is in a plane parallel to that of the posterior canal on the opposite side of the head, whereas the two external canals on the two sides are located in approximately the same plane.

In the *ampullae* of the semicircular canals, as illustrated in Figure 57–2, are small crests, each called a *crista ampullaris,* and on top of the crest is a gelatinous mass similar to that in the utricle and known as the *cupula.* Into the cupula are projected hairs from hair cells located along the ampullary crest, and these hair cells in turn are connected to sensory nerve fibers that pass into the *vestibular nerve.* Bending the cupula to one side, caused by flow of fluid in the canal, stimulates the hair cells, while bending in the opposite direction inhibits them. Thus, appropriate signals are sent through the vestibular nerve to apprise the central nervous system of fluid movement in the respective canal.

Directional Sensitivity of the Hair Cells—The Kinocilium. Each hair cell has a large number of very small cilia plus one very large cilium called the *kinocilium.* This kinocilium is located to one side of the hair cell, always on the same side of the cell with respect to its orientation on the ampullary crest. This presumably is the cause of the directional sensitivity of the hair cells: namely, stimulation when the hairs are bent in one direction and inhibition when bent in the opposite direction.

In the macula of the utricle, the different hair cells have varying orientations, and this allows a different concert of hair cells to be stimulated for each respective position of the head in space.

Neuronal Connections of the Vestibular Apparatus with the Central Nervous System. Figure 57–3 illustrates the central connections of the vestibular nerve. Most of the vestibular nerve fibers end in the vestibular nuclei, which are located approximately at the junction of the medulla and the pons, but some fibers pass without synapsing to the flocculonodular lobes of the cerebellum and to other regions of the brain stem. The fibers that end in the vestibular nuclei synapse with second order neurons that in turn send fibers into the flocculonodular lobe of the cerebellum, to the cortex of other portions of the cerebellum, into the vestibulospinal tract, into the medial longitudinal fasciculus, and to other areas of the brain stem, particularly the reticular nuclei.

The primary pathway for the reflexes of equilibrium begins in the *vestibular nerves,* passes next to the *vestibular nuclei,* and then to the *reticular*

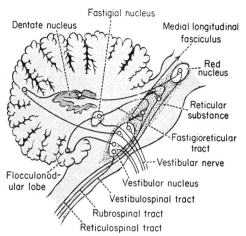

Figure 57–3 Connections of vestibular nerves in the central nervous system.

nuclei of the brain stem. The reticular nuclei in turn control the interplay between facilitation and inhibition of the extensor muscles, thus automatically controlling equilibrium.

However, destruction of the *flocculonodular lobes* of the cerebellum causes essentially the same effects on equilibrium as loss of the semicircular canals. Therefore, equilibrium reflexes initiated in the semicircular canals undoubtedly require function of the flocculonodular lobes of the cerebellum as well as of the reticular nuclei of the brain stem. This is in keeping with the usual function of the cerebellum to coordinate rapid body movements, for the semicircular canals are also concerned with equilibrium during *rapid changes in direction* of motion and not with equilibrium under static conditions, as is pointed out in subsequent sections.

FUNCTION OF THE UTRICLE IN THE MAINTENANCE OF STATIC EQUILIBRIUM

Gravitational pull on the *otoconia* causes them to press continually on the *macular hair tufts* in the utricles, and this apprises the central nervous system of the position of the head with respect to the direction of gravitational pull. The density of the otoconia is almost three times that of the surrounding fluid and tissues, and because of this they bend the hair tufts forward when the head leans forward, backward when the head leans backward, and to one side when the head leans to that side.

It is especially important that the different hair cells are oriented in several different directions in the maculae so that at different positions of the head, different hair cells become stimulated. The "patterns" of stimulation of the different hair cells apprise the nervous system of the position of the head with respect

to the pull of gravity. In turn, the reticular nuclei reflexly excite the appropriate muscles to maintain proper equilibrium.

Detection of Linear Acceleration by the Utricle. When the body is suddenly thrust forward, the otoconia, which have greater inertia than the surrounding fluids, fall backward on the hair tufts, and information of mal-equilibrium is sent into the nervous centers, causing the individual to feel as if he were falling backward. This automatically causes him to lean his body forward; the forward lean then shifts the otoconia anteriorly. Thus, the body automatically leans farther and farther forward until the anterior shift of the otoconia exactly equals the tendency for the otoconia to fall backward because of the linear acceleration. At this point, the nervous system detects a state of proper equilibrium and therefore shifts the body no farther forward. As long as the degree of linear acceleration remains constant and the body is maintained in this forward leaning position, the person falls neither forward nor backward. Thus, the otoconia operate to maintain equilibrium during linear acceleration in exactly the same manner as they operate in static equilibrium.

The otoconia *do not* operate for the detection of linear *motion.* When a runner first begins to run, he must lean far forward to keep from falling over backward because of acceleration, but once he has achieved running speed, he would not have to lean forward at all if he were running in a vacuum. When running in air he leans forward to maintain equilibrium only because of the air resistance against his body, and in this instance it is not the otoconia that make him lean but the pressure of the air acting on pressure end-organs in the skin, which initiate the appropriate equilibrium adjustments to prevent falling.

THE SEMICIRCULAR CANALS AND THEIR DETECTION OF ANGULAR ACCELERATION

When the head suddenly *begins* to rotate in any direction, the endolymph in the membranous semicircular canals, because of its inertia, tends to remain stationary while the semicircular canals themselves turn. This causes relative fluid flow in the canals in a direction opposite to the rotation of the head. Figure 57–4 illustrates an *ampulla* of one of the semicircular canals, showing the *cupula* and its embedded hairs bending in the direction of fluid movement. And Figure 57–5 illustrates the discharge from a single hair cell in the crista ampullaris when an animal is rotated for

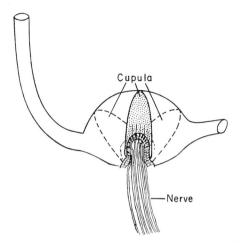

Figure 57–4 Movement of the cupula and its embedded hairs during rotation first in one direction and then in the opposite direction.

40 seconds, showing that (1) even when the cupula is in its resting position the hair cell emits a tonic discharge of approximately 15 impulses per second; (2) when the animal is rotated, the hairs bend to one side and the rate of discharge increases greatly; and (3) with continued rotation, the excess discharge of the hair cell gradually subsides back to the resting level in about 20 seconds.

The reason for this adaptation of the receptor is that within a second or more of rotation, friction in the semicircular canal causes the endolymph to rotate as rapidly as the semicircular canal itself; then in an additional 15 to 20 seconds the cupula slowly returns to its resting position in the middle of the ampulla.

When the rotation suddenly stops, exactly the opposite effects take place: the endolymph continues to rotate while the semicircular canal stops. This time the cupula is bent in the opposite direction, causing the hair cell to stop

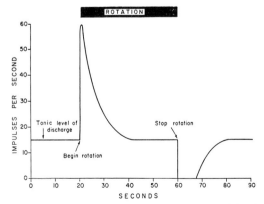

Figure 57–5 Response of a hair cell when a semicircular canal is stimulated first by rotation and then by stopping rotation.

discharging entirely. After another few seconds, the endolymph stops moving, and the cupula returns gradually to its resting position in about 20 seconds, thus allowing the discharge of the hair cells to return to its normal tonic level as shown to the right in Figure 57–5.

Thus, the semicircular canals transmit a signal when the head *begins* to rotate or when it *stops* rotating. Furthermore, this organ responds to rotation in any plane – horizontal, sagittal, or coronal – for fluid movement always occurs in at least one semicircular canal. In summary, the semicircular canals detect the *rate of change* of rotation of the head in any plane, which is called *angular acceleration*.

Rate of Angular Acceleration Required to Stimulate the Semicircular Canals. The angular acceleration required to stimulate the semicircular canals in the human being averages about 1 degree per second per second. In other words, when one begins to rotate, his velocity of rotation must be as much as 1 degree per second by the end of the first second, 2 degrees per second by the end of the second second, 3 degrees per second by the end of the third second, and so forth, in order for him barely to detect that his rate of rotation is increasing.

"Predictive" Function of the Semicircular Canals in the Maintenance of Equilibrium. Since the semicircular canals do not detect that the body is off balance in the forward direction, in the side direction, or in the backward direction, one might at first ask: What is the function of the semicircular canals in the maintenance of equilibrium? All they detect is that the person's head is beginning to rotate or is stopping rotation in one direction or another. Therefore, the function of the semicircular canals is not likely to be to maintain static equilibrium or to maintain equilibrium during linear acceleration or when the person is exposed to steady centrifugal forces. Yet loss of function of the semicircular canals causes a person to have poor equilibrium, especially when he attempts performing *rapid* and *intricate* body movements.

We can explain the function of the semicircular canals best by the following illustration. If a person is running forward rapidly, and then suddenly begins to turn to one side, he falls off balance a second or so later unless appropriate corrections are made *ahead of time*. But, unfortunately, the utricle cannot detect that he is off balance until *after* this has occurred. On the other hand, the semicircular canals will have already detected that the person is beginning to turn, and this information can easily apprise the central nervous system of the fact that the person *will* fall off balance within the next second or so unless some correction is made. In other words, the semicircular canal mechanism *predicts ahead of time* that mal-equilibrium is going to occur even before it occurs and thereby causes the equilibrium centers to make appropriate adjustments. In this way, the person need not fall off balance before he begins to correct the situation.

Removal of the flocculonodular lobes of the cerebellum prevents normal function of the semicircular canals but does not prevent normal function of the macular receptors. It is especially interesting in this connection that the cerebellum serves as a "predictive" organ for most of the other rapid movements of the body as well as those having to do with equilibrium. These other functions of the cerebellum are discussed in the following chapter.

The Vestibular Phasic Reflexes. Sudden changes in the orientation of an animal in space elicits reflexes that help to maintain equilibrium and posture. For instance, if an animal is suddenly pushed to the right, even before he can fall more than a few degrees his right legs extend instantaneously. In other words, this mechanism *anticipates* that the animal will be off balance in a few seconds and makes appropriate adjustments to prevent this.

Another type of vestibular phasic reflex occurs when the animal's head suddenly falls downward. When this occurs, the forepaws extend forward, the extensor muscles tighten, and the muscles in the back of the neck stiffen to prevent the animal's head from striking the ground. This reflex is probably also of importance in locomotion, for, in the case of the galloping horse, the downward thrust of the head could automatically provide reflex thrust of the forelimbs to move the animal forward for the next gallop.

The Vestibular "Righting" Reflex and Equilibrium. In the previous chapter it was pointed out that some of the crude motions required to move an animal from a lying to a standing position can be elicited by circuits in the spinal cord. However, these *cord* "righting" reflexes cannot perform the full function of righting an animal to the upright position without signals from the brain stem. The static equilibrium organ, the macula of the utricle, transmits impulses into the brain stem to apprise the nervous system of the status of the animal with respect to the pull of gravity, and, if the animal is in a lying position, these can elicit appropriate reflexes from the vestibular and reticular nuclei to cause the animal to climb to his feet. This is called the vestibular righting reflex.

If the standing decerebrate animal begins to fall to one side, the legs on that side extend automatically, or, if he begins to fall to the opposite side, the legs on that side become extended. These vestibular reflexes require neuronal centers in all parts of the brain stem up to as high as the mesencephalon.

NYSTAGMUS CAUSED BY STIMULATION OF THE SEMICIRCULAR CANALS

When a person rotates his head, his eyes must rotate in the opposite direction if they are to remain "fixed" on any one object long enough to gain a clear image. After they have rotated far to one side, they jump suddenly in the direction of rotation of the head to "fix" on a new object and then rotate slowly backward again. This sudden jumping motion forward and then slow backward motion is known as nystagmus. The jumping motion is called the *fast component* of the nystagmus, and the slow movement is called the *slow component*.

Nystagmus always occurs automatically when the semicircular canals are stimulated. For instance, if a person's head begins to rotate to the right, backward movement of fluid in the left horizontal canal and forward movement in the right horizontal canal cause the eyes to move slowly to the left; thus, the slow component of nystagmus is initiated by the vestibular apparatuses. But, when the eyes have moved as far to the left as they reasonably can, centers located in the brain stem in close approximation with the nuclei of the abducens nerves cause the eyes to jump suddenly to the right; then the vestibular apparatuses take over once more to move the eyes again slowly to the left.

The precise pathways that transmit nystagmus information from the vestibular apparatus to the ocular nuclei are not known. A probable pathway is through the *medial longitudinal fasciculus,* which connects the vestibular nuclei directly with these nuclei. However, complete transection of this fasciculus does not prevent vestibular nystagmus, which indicates that other pathways are undoubtedly involved as well.

CLINICAL TESTS FOR INTEGRITY OF VESTIBULAR FUNCTION

Balancing Test. One of the simplest clinical tests for integrity of the equilibrium mechanism is to have the individual stand perfectly still with his eyes closed. If he no longer has a functioning static equilibrium system of the utricles, he will waver to one side or the other and possibly even fall. However, as noted below, some of the proprioceptive mechanisms of equilibrium are occasionally sufficiently well developed to maintain balance even with the eyes closed.

Barany Test. A second test that is frequently performed determines the integrity of the semicircular canals. In this instance the individual is placed in a "Barany chair" and rotated rapidly while he places his head respectively in various planes—first forward, then angulated to one side or the other. By such positioning, each pair of semicircular canals is successively placed in the horizontal plane of rotation. When the rotation of the chair is stopped suddenly, the endolymph, because of its momentum, continues to rotate in the pair of semicircular canals that has been placed in the horizontal plane, this flow of endolymph causing the cupula to bend in the direction of rotation. As a result, nystagmus occurs, with the slow component in the direction of rotation and the fast component in the opposite direction. Also, as long as the nystagmus lasts (about 10 to 20 seconds) the individual has the sensation that he is rotating in the direction *opposite* to that in which he was actually rotated in the chair. Obviously, this test checks the semicircular canals on both sides of the head at the same time.

Ice Water Test. A clinical test for testing one vestibular apparatus separately from the other depends on placing ice water in one ear. The external semicircular canal lies adjacent to the ear, and cooling the ear can transfer a sufficient amount of heat from this canal to cool the endolymph. This increases the density of the endolymph, thereby causing it to sink downward and resulting in slight movement of fluid around the semicircular canal. This stimulates the canal, giving the individual a sensation of rotating and also initiating nystagmus. From these two findings one can determine whether the respective semicircular canals are functioning properly. When the semicircular canals are normal, the utricles are usually normal also, for disease usually destroys the function of both at the same time.

OTHER FACTORS CONCERNED WITH EQUILIBRIUM

The Neck Proprioceptors. The vestibular apparatus detects the orientation and movements *only of the head.* Therefore, it is essential that the nervous centers also receive appropriate information depicting the orientation of the head with respect to the body as well as the orientation of the different parts of the body with respect to each other. This information is transmitted from the proprioceptors of the neck and body either directly into the reticular nuclei of the brain stem or by way of the cerebellum and thence into the reticular nuclei.

By far the most important proprioceptive information needed for the maintenance of equilibrium is that derived from the *joint receptors of the neck,* for this apprises the nervous system of the orientation of the head with respect to the body. When the head is bent in one direction or the other, impulses from the neck proprioceptors keep the vestibular apparatuses from giving the person a sense of malequilibrium. They do this by exactly opposing the impulses transmitted from the vestibular apparatuses. However, *when the entire body* is changed to a new position with respect to gravity, the impulses from the vestibular apparatuses *are not opposed* by the neck proprioceptors; therefore, the person in this instance does perceive a change in equilibrium status.

The Neck Reflexes. In an animal *whose vestibular apparatuses have been destroyed,* bending the neck causes immediate muscular reflexes called *neck reflexes* occurring especially in the forelimbs.

For instance, bending the head forward causes both forelimbs to relax. However, when the vestibular apparatuses are *intact,* this effect does *not* occur because the vestibular reflexes function almost exactly oppositely to the neck reflexes. Thus, if the head is flexed downward, the vestibular reflex tends to extend the forelimbs while the neck reflexes tend to relax them. Since the equilibrium of the entire body and not of the head alone must be maintained, it is easy to understand that the vestibular and neck reflexes must function oppositely. Otherwise, each time the neck should bend, the animal would immediately fall off balance.

Proprioceptive and Exteroceptive Information from Other Parts of the Body. Proprioceptive information from other parts of the body besides the neck is also necessary for maintenance of equilibrium because appropriate equilibrium adjustments must be made whenever the body is angulated in the chest or abdomen region or elsewhere. Presumably, all this information is algebraically added in the reticular substance of the brain stem, thus causing appropriate adjustments in the postural muscles.

Also important in the maintenance of equilibrium are several types of exteroceptive sensations. For instance, pressure sensations from the footpads can tell one (a) whether his weight is distributed equally between his two feet and (b) whether his weight is more forward or backward on his feet.

Another instance in which exteroceptive information is necessary for maintenance of equilibrium occurs when a person is running. The air pressure against the front of his body signals that a force is opposing the body in a direction different from that caused by gravitational pull; as a result, the person leans forward to oppose this.

Importance of Visual Information in the Maintenance of Equilibrium. After complete destruction of the vestibular apparatuses, and even after loss of most proprioceptive information from the body, a person can still use his visual mechanisms effectively for maintaining equilibrium. Visual images help the person maintain equilibrium simply by visual detection of the upright stance. Also, slight linear or angular movement of the body instantaneously shifts the visual images on the retina, and this information is relayed to the equilibrium centers. In this respect the optic information is similar to that from the semicircular canals and can help the equilibrium centers predict that the person will fall off balance before this actually occurs if appropriate adjustments are not immediately made. Many persons with complete destruction of the vestibular apparatus have almost normal equilibrium as long as their eyes are open and as long as they perform all motions slowly. But, when moving rapidly or when the eyes are closed, equilibrium is immediately lost.

Conscious Perception of Equilibrium. A cortical center for conscious perception of the state of equilibrium has been found in man to lie in the upper portion of the temporal lobe in close association with the primary cortical area for hearing. The sensations from the vestibular apparatuses, from the neck proprioceptors, and from most of the other proprioceptors are probably first integrated in the equilibrium centers of the brain stem before being transmitted to the cerebral cortex. Various pathologic processes in the vestibular apparatuses or in the vestibular neuronal circuits often affect the equilibrium sensations. Thus, loss of one or both flocculonodular lobes or one or both fastigial nuclei of the cerebellum gives the person a sensation of constant mal-equilibrium; this probably results because impulses from these cerebellar areas are normally integrated into the subconscious sense of equilibrium by the reticular nuclei even before equilibrium information is sent to the cerebral cortex.

FUNCTIONS OF THE RETICULAR FORMATION AND SPECIFIC BRAIN STEM NUCLEI IN CONTROLLING SUBCONSCIOUS, STEREOTYPED MOVEMENTS

Rarely, a child is born without brain structures above the mesencephalic region, and some of these children have been kept alive for many months. Such a child is able to perform essentially all the functions of feeding, such as sucking, extrusion of unpleasant food from the mouth, and moving his hands to his mouth to suck his fingers. In addition, he can yawn and stretch. He can cry and follow objects with his eyes and by movements of his head. Also, placing pressure on the upper anterior parts of his legs will cause him to pull to the sitting position.

Therefore, it is obvious that many of the stereotyped motor functions of the human being are integrated in the brain stem. Unfortunately, the loci of most of these different motor control systems have not been found except for the following:

Stereotyped Body Movements. Most movements of the trunk and head can be classified into several simple movements, such as forward flexion, extension, rotation, and turning movements of the entire body. These types of movements are controlled by special nuclei located mainly in the mesencephalic and lower diencephalic region. For instance, *rotational movements* of the head and eyes are controlled by the *interstitial nucleus,* which is illustrated in Figure 57–1. This nucleus lies in close approximation to the *medial longitudinal fasciculus,* through which it transmits a major portion of its control impulses. The *raising movements* of the head and body are controlled by the *prestitial nucleus,* which is located approximately at the juncture of the diencephalon and mesencephalon. On the other hand, the *flexing movements* of the head and body are controlled by the *nucleus precommissuralis* located at the level of the posterior commissure. Finally, the *turning movements,* which are much more complicated, involve both the pontile and mesencephalic reticular formation. However, for full expression of the turning movements, the caudate nucleus and the cingulate gyrus of the cerebral cortex are also required. The turning movements can cause an animal or person to continue circling around and around in one direction or the other.

Function of the Red Nucleus. The red nucleus is composed of two distinct parts, the *magnocellular portion* and a *small cellular portion.*

As illustrated in Figure 57–6, the magnocellular nuclei receive impulses both from the *caudate nucleus* and the *putamen* as well as from the *prestitial nucleus* and the *nucleus commissuralis.* In turn it transmits signals by way of the *rubrospinal tract* into the spinal cord and by way of collaterals from the rubrospinal tract into the *reticular nuclei* of the brain stem. Stimulation of this portion of the red nucleus causes the head and upper trunk to extend backward. Thus, this portion of the red nucleus enters into a particular type of gross body movement concerned with forward and backward deviation of the body axis.

The small cells of the red nucleus are excited principally by impulses from the *cerebellum* through the *cerebellorubral tracts.* In turn, they send impulses mainly into the reticular formation of the brain stem. The functions of these tracts in the regulation of motor movements are discussed in the following chapter in relation to function of the cerebellum.

Function of the Substantia Nigra. The precise function of the substantia nigra, also shown in Figure 57–6, is unknown. However, it is known to receive considerable numbers of fibers from the caudate nucleus and putamen as well as short fibers from the upper reticular

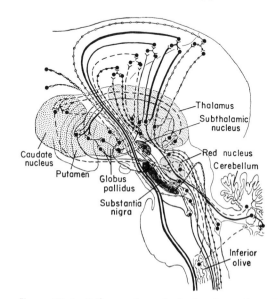

Figure 57–6 Pathways through the basal ganglia and related structures of brain stem, the thalamus, and cerebral cortex. (From Jung and Hassler: Handbook of Physiology, Sec. I, Vol. II, 1960. Baltimore, The Williams & Wilkins Co.)

formation. Destruction of the substantia nigra causes serious loss of control of the gamma activating system of the muscle spindles; therefore, it has been suggested that this area is the major brain stem control center for excitation of the gamma efferent nerve fibers.

When voluntary muscular functions are to be performed, activation of the substantia nigra, presumably by way of the basal ganglia, supposedly can activate the gamma efferent system at the same time as or even before the alpha motor neurons to the muscles are activated. This is believed to provide background muscular tone and to help "position" the parts of the body with respect to each other so that discrete motor functions of the hands or feet can be performed.

Function of the Subthalamic Areas—Forward Progression. Much less is known about function of higher brain stem centers in posture and locomotion than of the lower centers, principally because of the complexity of the neuronal connections. However, it is known that stimulation of centers in or around the subthalamic nuclei can cause rhythmic motions, including crude forward walking reflexes. This does not mean that the individual muscles of walking are necessarily controlled from this region but simply that excitation of this region excites the appropriate patterns of action in the brain stem and cord to cause the walking movements.

A cat with its brain transected beneath the thalamus but above the subthalamus can walk in an almost completely normal fashion—so much so that the observer cannot tell the difference. However, when the animal comes to an obstruction he simply butts his head against it and tries to keep on walking. Thus, he lacks *purposefulness of locomotion.*

The function of the subthalamic region in walking is frequently described as that of controlling *forward progression.* In the foregoing discussions we have already noted that a decerebrate animal can stand perfectly well and that he can make attempts to right himself from the lying position. Unfortunately, though, the decerebrate animal cannot force himself to move forward in a normal walking pattern. This seems to be the function of areas either in or somewhere close to the subthalamic nuclei.

MOTOR FUNCTIONS OF THE BASAL GANGLIA

Physiologic Anatomy of the Basal Ganglia. The anatomy of the basal ganglia is so complex and is so poorly known in its details that it would be pointless to attempt a complete descripion at this time. However, Figure 57–6 illustrates the principal structures of the basal ganglia and their neural connections with other parts of the nervous system. Anatomically, the basal ganglia are the *caudate nucleus, putamen, globus pallidus, amygdaloid nucleus,* and *claustrum.* The amygdaloid nucleus, which will be discussed in Chapter 62, and the claustrum are not concerned directly with motor functions of the central nervous system. On the other hand, the *thalamus, subthalamus, substantia nigra,* and *red nucleus* all operate in close association with the caudate nucleus, putamen, and globus pallidus and are considered to be part of the basal ganglia system for motor control.

Some important features of the different pathways illustrated in Figure 57–6 are the following:

1. Numerous nerve pathways pass from the motor portions of the cerebral cortex to the caudate nucleus and putamen. In turn, the caudate nucleus and putamen send numerous fibers to the globus pallidus, the globus pallidus to the ventrolateral nucleus of the thalamus, and this nucleus back to the motor areas of the cerebral cortex. Thus, circular pathways are established from the motor cortical regions to the basal ganglia, the thalamus, and back to the same motor regions from which the pathways begin. These circuits obviously could operate as a feedback system of the servo control type, as will be discussed later.

2. In the following chapter it will be noted that signals also pass from the motor regions of the cerebral cortex to the pons and then into the cerebellum. In turn, signals from the cerebellum are transmitted back to the motor cortex by way of the ventrolateral nucleus of the thalamus, the same thalamic nucleus through which the basal ganglial signals are transmitted. This circuit could allow integration between the basal ganglial feedback signals and the feedback signals from the cerebellum.

3. The basal ganglia have numerous short neuronal connections among themselves. Also, the lower basal ganglia, such as the globus pallidus, send tremendous numbers of nerve fibers into the lower brain stem, projecting especially onto the reticular nuclei of the mesencephalon and onto the red nucleus. It is presumably through these pathways that many of the so-called extrapyramidal signals for motor control are transmitted.

4. Large numbers of nerve fibers pass directly from the motor cortex to the reticular nuclei and other nuclei of the mesencephalic and pontine areas. These by-pass the basal ganglia but are integrated in the reticular nuclei with those that come from the basal ganglia.

FUNCTIONS OF THE DIFFERENT BASAL GANGLIA

Before attempting to discuss the functions of the basal ganglia in man, we should speak briefly of the better known functions of these ganglia in lower animals. In birds, for instance, the cerebral cortex is poorly developed, whereas the basal ganglia are highly developed. These ganglia perform essentially all the motor functions, even controlling the voluntary movements in much the same manner that the motor cortex of the human being controls voluntary movements. Furthermore, in the cat, and, to a less extent, in the dog, decortication removes only the discrete types of motor functions and does not interfere with the animal's ability to walk, eat, fight, develop rage, have periodic sleep and wakefulness, and even participate naturally in sexual activities. However, if a major portion of the basal ganglia is destroyed, only gross stereotyped movements remain, which were discussed above in relation to the mesencephalic animal.

Finally, in the human being, decortication of very young individuals destroys the discrete movements of the body, particularly of the hands and distal portions of the lower limbs, but does not destroy the person's ability to walk, to control his equilibrium, or to perform many other subconscious types of movements. However, simultaneous destruction of a major portion of the caudate nuclei almost totally paralyzes the opposite side of the body except for a few stereotyped reflexes integrated in the cord or brain stem.

With this brief background of the overall function of the basal ganglia, we can attempt to dissect the functions of the individual portions of the basal ganglia system, realizing that the system actually operates as a total unit and that individual functions cannot be ascribed completely to the different parts of the basal ganglia.

Inhibition of Motor Tone by the Basal Ganglia. Though it is wrong to ascribe a single function to all the basal ganglia, nevertheless, one of the general effects of basal ganglial excitation is to inhibit muscle tone throughout the body. This effect results from inhibitory signals transmitted from the basal ganglia to the bulboreticular facilitatory area and excitatory signals to the bulboreticular inhibitory area. Therefore, whenever widespread destruction of the basal ganglia occurs, the facilitatory area becomes overactive and the inhibitory area underactive, and this causes muscle rigidity throughout the body. For instance, when the brain stem is transected at the mesencephalic level, which removes the inhibitory effects of the basal ganglia, the phenomenon of decerebrate rigidity occurs.

Yet, despite this general inhibitory effect of the basal ganglia, stimulation of specific areas within the basal ganglia can at times elicit positive muscle contractions and at times even complex patterns of movements.

Function of the Caudate Nucleus and Putamen —The Striate Body. The caudate nucleus and putamen, because of their gross appearance on sections of the brain, are together called the *striate body.* They seem to function together to initiate and regulate gross intentional movements of the body. To perform this function they transmit impulses through two different pathways: (1) into the *globus pallidus,* thence by way of the *thalamus* to the *cerebral cortex,* and finally downward into the spinal cord through the *corticospinal* and *extracorticospinal pathways;* (2) downward through the *globus pallidus* and the *substantia nigra* by way of short pathways into the *reticular formation* and finally into the spinal cord mainly through the *reticulospinal tracts.*

In summary, the striate body helps to control gross intentional movements that we normally perform unconsciously. However, this control also involves the motor cortex, with which the striate body is very closely connected.

Function of the Globus Pallidus. It has been suggested that the principal function of the globus pallidus is to provide background muscle tone for intended movements, whether these be initiated by impulses from the cerebral cortex or from the striate body. That is, if a person wishes to perform an exact function with one hand, he positions his body appropriately and then tenses the muscles of the upper arm. These associated tonic contractions are supposedly initiated by a circuit that strongly involves the globus pallidus. Destruction of the globus pallidus removes these associated movements and, therefore, makes it difficult for the distal portions of the limbs to perform their more discrete activities.

The globus pallidus is believed to function through two pathways: first, through feedback circuits to the thalamus, to the cerebral cortex, and thence by way of corticospinal and extracorticospinal tracts to the spinal cord; and, second, by way of short pathways to the reticular formation of the brain stem and thence mainly by way of the reticulospinal tracts into the spinal cord.

Electrical stimulation of the globus pallidus while an animal is performing a gross body movement will often stop the movement in a static position, the animal holding that position for many seconds while the stimulation continues. This reaction fits with the concept that the globus pallidus is involved in some type of servo feedback motor control system that is capable of locking the different parts of the body into specific positions. Obviously, such a circuit could be extremely important in providing the background body movements and upper limb movements when a person performs delicate tasks with his hands.

CLINICAL SYNDROMES RESULTING FROM DAMAGE TO THE BASAL GANGLIA

Much of what we know about the function of the basal ganglia comes from study of patients with basal ganglia lesions whose brains have undergone pathologic studies after death. Among the different clinical syndromes are:

Chorea. Chorea is a disease in which random uncontrolled contractions of different muscle groups of the body occur continuously. Normal progression of movements cannot occur; instead, the person may perform a normal sequence of movements for a few seconds and then suddenly begin another sequence of movements; then still another sequence begins after a few seconds. Because of this peculiar progression of movements, one type of chorea is frequently called *St. Vitus' dance.*

Pathologically, chorea results from *diffuse and widespread damage of the striate body,* but the manner in which this causes the choreiform movements is not known.

Athetosis. In this disease, slow, writhing movements of the peripheral parts of one or more limbs occur continually. The movements are likely to be wormlike, first, with overextension of the hands and fingers, then flexion, then rotatory twisting to the side — all these continuing in a slow, rhythmic pattern. The contracting muscles exhibit a high degree of spasm, and the movements are enhanced by emotions or by excessive signals from the sensory organs. Furthermore, voluntary movements in the affected area are greatly impaired or sometimes even impossible.

The damage in athetosis is usually found in the *outer portion of the globus pallidus* or in this area and the striate body. Athetosis is usually attributed to the interruption of feedback circuits among the basal ganglia, thalamus, and cerebral cortex. The normal feedback circuits presumably allow a constant and rapid interplay between antagonistic muscle groups so that finely controlled movements can take place. However, if the feedback circuits are blocked, it is supposed that the detouring impulses may take devious routes through the basal ganglia, thalamus, and motor cortex, causing a succession of abnormal movements.

Hemiballismus. Hemiballismus is an uncontrollable succession of violent movements of large areas of the body. These may occur once every few seconds or sometimes only once in many minutes.

For instance, an entire leg might suddenly jerk uncontrollably to full flexion, or the entire trunk might go through an extreme torsion movement, or an arm might be pulled upward suddenly with great force. Hemiballismus of the legs or trunk causes the person to fall to the ground if he is walking, and even when the person is in bed, he tosses violently when affected by these powerful and strong intermittent movements. Furthermore, attempts to perform voluntary movements frequently invoke ballistic movements in place of the normal movements.

Hemiballismus of one side of the body results from a *large lesion in the opposite subthalamus.* The smooth, rhythmic movements of the limbs or other parts of the body normally integrated in this area can no longer occur, but excitatory impulses attempting to evoke such movements elicit instead uncontrollable ballistic movements.

Parkinson's Disease. Parkinson's disease, which is also known as *paralysis agitans,* results almost invariably from *widespread destruction of the substantia nigra,* the *globus pallidus,* or *closely associated areas.* It is characterized by (1) *rigidity* of the musculature in either widespread areas of the body or in isolated areas, (2) *tremor at rest* of the involved areas in most but not all instances, and (3) *loss of involuntary and associated movements.* Also, there may be enhancement of salivation, sweating, and seborrheic secretion. And, finally, the gamma activating system for exciting the muscle spindles becomes completely or almost completely inactivated, presumably because the substantia nigra is one of the major areas of the brain stem for controlling the gamma efferent system.

The rigidity in this disease is quite different from that which occurs in decerebrate rigidity, for decerebrate rigidity results from hyperactivity of the muscle spindle system, and even slight movement of a muscle is met with rather extreme reflex resistance resulting from feedback through the stretch reflex mechanism. Parkinsonian rigidity, on the other hand, is of a "plastic" type. That is, sudden movement is not met by intense resistance from the stretch reflexes as in the decerebrate type of rigidity, but instead both protagonist and antagonist muscles remain tightly contracted throughout the movement. Furthermore, the rigidity of Parkinson's disease is known to result from excess impulses transmitted in the corticospinal system, thus activating the alpha motor fibers to the muscles, whereas decerebrate rigidity results from excess activation of the gamma efferents to the muscle spindles.

Tremor usually, though not always, occurs in Parkinson's disease. Its frequency normally is 4 to 8 cycles per second. When the person performs voluntary movements, weak tremors become temporarily blocked, possibly because cortical signals to the spinal cord excite the gamma efferent fibers to the muscle spindles at the same time that they excite the alpha fibers; the gamma stimulation sensitizes the muscle spindles which theoretically could then damp out the tremor.

The mechanism of the tremor in Parkinson's disease is not known, but at least three different theories must be considered: (1) Loss of inhibitory influences from the substantia nigra allows the control nuclei of antagonistic muscles of the *reticular substance of the brain stem* to oscillate. (2) Loss of inhibition of the *spinal cord* or loss of the damping function of the gamma efferent muscle spindle system allows oscillation back and forth between the motoneurons controlling antagonistic muscle groups. (3) Lack of impulses in the lower basal ganglia enhances feedback in the globus pallidus–thalamus–cortical circuit, causing it to oscillate.

Another abnormality of Parkinson's disease is a high degree of "motor stiffness" and loss of automatisms. When a person begins to perform a discrete voluntary movement with his hands, the automatic "associated" adjustments of the trunk of his body and the upper arm segments do not occur. Instead, he must voluntarily adjust these segments before he can use his hands. Furthermore, a tremendous amount of nervous effort must be made by the voluntary motor control system to overcome the motor stiffness of his musculature. Thus, the person with Parkinson's disease has a masklike face, showing almost no automatic, emotional facial expressions; he is usually bent forward because of his muscle rigidity, and all his movements of necessity are highly deliberate rather than the many casual subconscious movements that are normally a part of our everyday life. This loss of automatisms is believed to result from diminished function of the gamma efferent system. It should be remembered in this connection that many of the automatic movements of the body are probably effected not by transmission of signals directly through the anterior motoneurons but instead through gamma efferent neurons.

Coagulation of the Ventrolateral Nucleus of the Thalamus for Treatment of Parkinson's Disease. In recent years neurosurgeons have treated Parkinson's disease patients, with varying success, by destroying portions of the basal ganglia or of the thalamus. The most prevalent treatment at present is widespread destruction of the ventrolateral nucleus of the thalamus, either by electrocoagulation or by injecting a sclerosing chemical. Most fiber pathways from the basal ganglia and cerebellum to the cerebral cortex pass through this nucleus so that its destruction blocks many or most of the feedback functions of the basal ganglia and cerebellum. It is presumed that blockage of some of these feedbacks limits the functions of at least certain basal ganglia and thereby removes the factors that cause the rigidity and tremor of Parkinson's disease. For instance, recent experiments in monkeys show that destruction of the ventrolateral nucleus allows increased gamma efferent stimulation of the muscle spindles. It has been suggested that this increases the degree of damping afforded by the muscle spindles, as was explained in the preceding chapter, thereby blocking the tremor that is characteristic of Parkinson's disease.

Treatment with L-Dopa. Administration of L-dopa to patients with Parkinson's disease ameliorates most of the symptoms in about two-thirds of patients. It is believed by many investigators that a derivative of L-dopa, perhaps dopamine, is secreted by the nerve endings of the neurons whose cell bodies are in the substantia nigra. These endings are found especially in the basal ganglia. Therefore, the L-dopa administered to a patient with Parkinson's disease is believed to substitute for the transmitter substance not secreted by the destroyed neurons.

FUNCTIONS OF THE CEREBRAL CORTEX IN POSTURE AND LOCOMOTION – PURPOSEFULNESS OF LOCOMOTION

The principal function of the cerebral cortex in locomotion is to add *purposefulness* to the locomotion, as illustrated above by the subthalamic cat that would not go around an obstruction but would simply butt its head against the obstruction and continue its attempts to move forward. Such an effect obviously would not occur in an animal with a cerebral cortex. However, in addition to the purposefulness added by cortical function, several postural and righting reflexes are also integrated in the cortex. These include the *visual righting reflex,* certain aspects of *exteroceptive righting reflexes,* and the *placing reaction.*

The Visual and Exteroceptive Righting Reflexes. The visual righting reflexes depend on function of the visual cortex. That is, an animal can tell by images from the periphery whether his body is in the appropriate position with respect to its surroundings. From this information, he can raise himself from the lying position to the standing position even in the absence of the vestibular righting reflexes.

Similarly, the animal's sensory and motor cortex causes righting in response to exteroceptive information that is transmitted into the cerebrum. It has been pointed out that sensations from the skin can cause crude righting responses even in the spinal animal and even much better responses in the decerebrate animal. However, for full expression of the exteroceptive righting reflexes, the cerebral cortex is also required.

THE PLACING REACTION. A blindfolded animal held in the air and then brought in contact with a surface so that almost any part of his body makes contact with the surface immediately brings his paws toward the point of contact, thus placing his feet in an advantageous position to gain support. This is called the placing reaction. Though some expression of the placing reaction can occur even in animals without the cerebral cortex, here again the complete reaction can occur only when the cortex is available to provide stereognostic orientation of the different parts of the body in relation to each other, thus allowing the animal to "know" where his feet need to be moved in order to make appropriate contact.

THE HOPPING REACTION. The hopping reaction is another reaction that requires the cerebral cortex for full attainment. This reaction probably results mainly from direct transmission of proprioceptive and vestibular information to the somesthetic area in the postcentral gyrus, where direct connections are made with the motor cortex to cause the hopping.

SUMMARY OF THE DIFFERENT FUNCTIONS OF THE CENTRAL NERVOUS SYSTEM IN POSTURE AND LOCOMOTION

From the discussions in the previous chapter, we could see that almost all the discrete "patterns" of muscle movement required for posture and locomotion can be elicited by the spinal cord alone. However, coordination of these patterns to provide equilibrium, progression, and purposefulness of movement requires neuronal function at progressively higher levels of the central nervous system. Centers in the brain stem provide most of the nervous energy required to maintain postural tone and therefore to support the body against gravity. But, in addition, the brain stem centers provide especially the equilibrium reflexes. Then moving still higher we find a control center for "progression" in the subthalamic region, a center that makes the animal move forward in a normal rhythmic pattern of walking and also in a straight line. Even so, this animal still does not have purposefulness in his motion. For this the basal ganglia, the thalamus, and the cerebral cortex are required.

With this background, we can now describe the characteristics of animals with transections through their nervous systems at the different levels shown in Figure 57-7.

The Spinal Animal. In the spinal animal, the flexor reflex, the crossed extensor reflex, the stretch reflex, and the reflexes that control the basic rhythm and pattern of locomotion are intact after the stage of spinal shock is over. Furthermore, under the most favorable conditions the animal can even stand for a few seconds because of the positive supportive reaction.

Transection of the Medulla Below the Vestibular Nuclei. If the medulla is cut below the vestibular nuclei, the lower portion of the reticular substance of the brain stem will be intact; this, if anything,

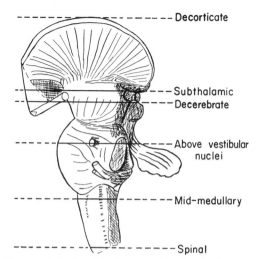

Figure 57–7 Levels of transection in the brain, the descriptions of which are given in the text.

transmits *inhibitory* impulses into the spinal cord. Therefore, few cord reflexes can be elicited, and the animal certainly cannot stand.

Transection of the Brain Stem Above the Vestibular Nuclei. If the brain stem is transected above the vestibular nuclei, the vestibular nuclei and the upper reticular substance of the brain stem transmit *facilitatory* impulses into the spinal cord and actually provide far more facilitation than normal. If the transection is well above the vestibular nuclei, sufficient postural tone can be attained for the animal to stand for minutes at a time.

The Classical Decerebrate Animal—Decerebrate Rigidity. The classical decerebrate animal has a transection between the superior and the inferior colliculi. In this animal the inhibitory signals from the basal ganglia and cerebral cortex to the reticular facilitatory area are removed, which allows the bulboreticular facilitatory area to become intrinsically very active, as explained earlier in the chapter. Impulses are transmitted downward to cause muscle rigidity called *decerebrate rigidity*. Especially the extensor muscles, but to some extent the flexor muscles as well, are maintained in a state of strong tonic contraction. The animal can occasionally stand, supporting his body against gravity. He also can make crude equilibrium adjustments; but for full expression of equilibrium, he needs a completely intact mesencephalon.

A major part of the increased postural tone in decerebrate rigidity results from stimulation of the *gamma activating system of the muscle spindles* and not from direct activation of the alpha motor neurons.

The Subthalamic Animal. With the transection immediately above the superior colliculi, preserving the subthalamus as well as the lower portions of the hypothalamus, the animal no longer has severe rigidity and is capable of walking with normal progression.

The Decorticate Animal. In a decorticate animal the cortex has been removed but the thalamus and basal ganglia are still intact. The capabilities of this type of animal depend on its degree of cerebral development. The cat, for instance, is capable of almost any type of motion, even to the extent of going around obstructions in his way, but, on the other hand, he still lacks much of the purposefulness of locomotion and frequently sits very still for hours at a time. The human being loses essentially all purposeful or voluntary motor functions on decortication.

FUNCTION OF THE CORD AND BRAIN STEM REFLEXES IN MAN

Though the foregoing discussions have concerned principally reflexes in animals rather than man, practically all the same principles of cord and brain stem function apply almost as well to man. Yet there are quantitative differences that must always be remembered.

First, man walks in an upright position and his limbs have become straightened to the point that almost no muscular strength is required to maintain the weight of the body against gravity. For instance, the direct line between the center of mass of the body and the direction of gravitational pull runs slightly behind the axes of the hip joints so that gravity tends to extend the hips and so that the ligaments of the hip joints, rather than the muscles, support the body against gravity. Likewise, the line of gravitational pull runs slightly in front of the mid-position of the knee joints and well in front of the ankle joints. Here again, it is mainly tension in the ligaments and uncontracted muscles that maintains the weight of the body against gravity rather than postural tone.

Therefore, in man, contraction of postural muscles occurs principally to maintain equilibrium rather than to support the body against gravity. Yet, if man crouches into an animal-like position, postural tone subserves essentially the same functions as in lower animals.

In man with a spinal cord transection, most of the same reflexes as those that occur in lower animals can be elicited either wholly or partially. The only reflexes that are rarely if ever elicited in spinal man are the rhythmic reflexes; yet the reciprocal innervation that is required for rhythmic reflexes can be demonstrated. For instance, movement of an arm in one direction often causes simultaneous and directionally similar movement of the opposite leg, thus illustrating that appropriate reciprocal innervation is still present in the cord of man to cause at least crude types of diagonal movements necessary for the mark time reflex.

Also, stimulation in the subthalamic region in man causes rhythmic walking movements similar to those that occur in animals, illustrating that this same region of the brain subserves similar functions in man as in lower animals. However, a man who has lost large portions of the motor areas of his cortex cannot use these lower regions of the brain stem effectively enough actually to provide locomotion. This is a manifestation of the process of *encephalization* that has occurred in higher orders of animals, which suppresses many functions of the lower centers while the same functions are taken over, usually in a still more complicated manner, by the higher centers. Thus, in man, the motor areas of the cortex are needed to provide progression in the act of locomotion and also to provide purposefulness of locomotion.

Nevertheless, in our subsequent discussions of the motor functions of the brain in man, we must always keep in mind the basic activities of the lower centers. The cord patterns of response are integrated into the overall control of muscular activity even when this control is initiated in the cerebral cortex. It is still the anterior motoneurons that have the final decisive power in the contraction of given muscles, and it is still the pattern of organization of the internuncial cells in the gray matter of the cord that provides many of the patterns of muscular contraction. Impulses from above ordinarily control only the sequence in which these patterns of contraction occur rather than the contractions of the individual muscles.

REFERENCES

Barbeau, A., et al.: Parkinson's Disease: Trends in Research and Treatment. New York, Grune & Stratton, Inc., 1965.

Brookhart, J. M.; Mori, S.; and Reynolds, P. J.: Postural reactions to two directions of displacement in dogs. *Amer. J. Physiol.,* 218:719, 1970.

Carman, J. B.: Anatomic basis of surgical treatment of Parkinson's disease. *New Eng. J. Med.,* 279:919, 1968.

Cohen, L. A.: Role of eye and neck proprioceptive mechanisms in body orientation and motor coordination. *J. Neurophysiol.,* 24:1, 1961.

Denny-Brown, D.: The Basal Ganglia: And Their Relation to Disorders of Movement. New York, Oxford University Press, Inc., 1962.

Fischer, J. H.: The Labyrinth: Physiology and Functional Tests. New York, Grune & Stratton, Inc., 1956.

French, J. D.: The reticular formation. *In* Handbook of Physiology. Baltimore, The Williams & Wilkins Co., 1960, Sec. I, Vol. II, p. 1281.

Gernandt, B. E.: Vestibular mechanisms. *In* Handbook of Physiology. Baltimore, The Williams & Wilkins Co., 1959, Sec. I, Vol. I, p. 549.

Higgins, D. C.: Vestibular component in dynamic muscle stretch. *Amer. J. Physiol.,* 214:482, 1968.

Kidokoro, Y.; Shuto, S.; Kubota, K.; and Sumino, R.: Reflex organization of cat masticatory muscles. *J. Neurophysiol.,* 31:695, 1968.

Kim, J. H., and Partridge, L. D.: Observations on types of response to combinations of neck, vestibular, and muscle stretch signals. *J. Neurophysiol.,* 32:239, 1969.

Martin, J. P.: Basal Ganglia and Posture. Philadelphia, J. B. Lippincott Co., 1967.

Mori, S., and Brookhart, J. M.: Characteristics of the postural reactions of the dog to a controlled disturbance. *Amer. J. Physiol.,* 215:339, 1968.

Mori, S.; Reynolds, P. J.; and Brookhart, J. M.: Contribution of pedal afferents to postural control in the dog. *Amer. J. Physiol.,* 218:726, 1970.

Partridge, L. D., and Kim, J. H.: Dynamic characteristics of response in a vestibulomotor reflex. *J. Neurophysiol.,* 32:485, 1969.

Segundo, J. P.; Takenaka, T.; and Encabo, H.: Electrophysiology of bulbar reticular neurons. *J. Neurophysiol.,* 30:1194, 1967.

Sherrington, C. S.: Decerebrate ridigity and reflex coordination of movements. *J. Physiol. (Lond.), 22*:319, 1898.

Stiles, R. N., and Randall, J. E.: Mechanical factors in human tremor frequency. *J. Appl. Physiol., 23*:324, 1967.

Valdman, A. V. (ed.): Pharmacology and Physiology of the Reticular Formation. Progress in Brain Research. New York, American Elsevier Publishing Co., Inc., 1967, Vol. 20.

Wilson, V. J.; Wylie, R. M.; and Marco, L. A.: Organization of the medial vestibular nucleus. *J. Neurophysiol., 31*:166, 1968.

Wilson, V. J.; Wylie, R. M.; and Marco, L. A.: Synaptic inputs to cells in the medial vestibular nucleus. *J. Neurophysiol., 31*:176, 1968.

Wilson, V. J., and Yoshida, M.: Comparison of effects of stimulation of Deiters' nucleus and medial longitudinal fasciculus on neck, forelimb, and hindlimb motoneurons. *J. Neurophysiol., 32*:743, 1969.

Wolfson, R. J.: The Vestibular System and Its Diseases. Philadelphia, University of Pennsylvania Press, 1966.

CHAPTER 58

Cortical and Cerebellar Control of Motor Functions

In preceding chapters we have been concerned with many of the subconscious motor activities integrated in the spinal cord and brain stem, especially those responsible for locomotion. In the present chapter we will discuss the control of motor function by the cerebral cortex and cerebellum, much of which is "voluntary" control in contradistinction to the subconscious control effected by the lower centers. Yet we will also see that at least some motor functions of the cerebral cortex and cerebellum are not entirely "voluntary."

PHYSIOLOGIC ANATOMY OF THE MOTOR AREAS OF THE CORTEX AND THEIR PATHWAYS TO THE CORD

Those parts of the cortex which, when stimulated electrically, cause movements somewhere in the body are considered to be motor areas. Figure 58–1 illustrates the principal motor area of the cortex from which most motor responses are elicited, showing that this area is located in the posterior part of the frontal lobe. The area immediately in front of the central sulcus, designated in the figure by the darkest shading, contains large

numbers of *giant Betz cells,* or *pyramidal cells,* for which reason it is also called the *area pyramidalis.* This area causes motor movements following the least amount of electrical excitation.

The Corticospinal Tract (Pyramidal Tract). One of the major pathways by which motor signals are transmitted from the motor areas of the cortex to the anterior motoneurons of the spinal cord is through the corticospinal tract, or pyramidal tract, which is illustrated in Figure 58–2. This tract originates in all the shaded areas in Figure 58–1, including both the motor and the somesthetic areas, about three-quarters from the motor area and about one-quarter from the somesthetic regions posterior to the central sulcus. The func-

Figure 58–1 Relationship of the motor cortex to the somesthetic cortex.

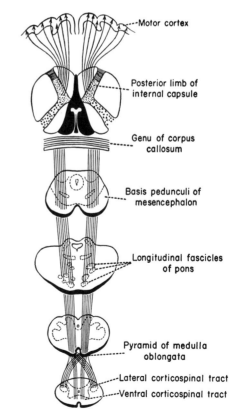

Motor cortex

Posterior limb of internal capsule

Genu of corpus callosum

Basis pedunculi of mesencephalon

Longitudinal fascicles of pons

Pyramid of medulla oblongata

Lateral corticospinal tract

Ventral corticospinal tract

Figure 58–2 The corticospinal tract. (From Ranson and Clark: Anatomy of the Nervous System.)

tion of the fibers from the somesthetic cortex is not clear, for only rarely does stimulation of this area cause motor movements.

The most impressive fibers in the corticospinal tract are the large myelinated fibers that originate in the giant Betz cells of the motor area. These account for approximately 34,000 large fibers (mean diameter of about 16 microns) in the cortico-spinal tract from each side of the cortex. However, since each corticospinal tract contains more than a million fibers, only 3 per cent of the total number of corticospinal fibers are of this large type. The other 97 per cent are mainly fibers smaller than 4 microns diameter, 60 per cent of which are mye-linated and 40 per cent unmyelinated.

The corticospinal tract passes downward through the *posterior limb of the internal capsule* and through the *brain stem;* then it decussates mainly to the opposite side in the *pyramids of the medulla.* By far the majority of the corticospinal fibers de-scend in the *lateral corticospinal tracts* of the cord and terminate principally on internuncial cells at the bases of the dorsal horns of the cord gray matter. A few fibers, however, do not cross to the opposite side but pass on down the cord ipsilater-ally. Another small percentage passes ipsilaterally down the cord in the *ventral corticospinal tracts* but then mainly decussates to the opposite sides farther down the cord.

COLLATERALS FROM THE CORTICOSPINAL TRACT IN THE BRAIN. Even before the corticospinal tract leaves the brain, many collaterals are given off as follows:

1. The axons from the giant Betz cells send short collaterals back to the cortex itself. It is believed that these collaterals either activate or inhibit adjacent regions of the cortex when the Betz cells discharge.

2. A large number of collateral fibers pass into the *striate body,* which is composed of the *caudate nucleus* and *putamen.* From here additional path-ways extend down the brain stem and then into the spinal cord through the *extracorticospinal tracts,* which are discussed later in the chapter.

3. A moderate number of collaterals pass from the corticospinal tract into the *substantia nigra* and *red nuclei.* From these, additional pathways pass down the cord into the *reticular substance of the brain stem,* and thence through reticulospinal tracts into the spinal cord.

4. A moderate number of collaterals also deviate from the corticospinal tract directly into the *reticular substance* of the brain stem.

5. A tremendous number of collaterals synapse in the pontile nuclei, which give rise to the *pontocerebellar fibers.* Thus, each time signals are transmitted from the motor cortex through the corticospinal tract, simultaneous signals are trans-mitted into the cerebellar hemispheres.

Importance of the Area Pyramidalis. The giant pyramidal cells of the *area pyramidalis* are very excitable. A much less degree of electrical stimu-lation is required in this area to cause motor movement than in any other area of the brain.

Therefore, it is frequently stated that the area pyramidalis is the *primary motor area* of the cor-tex. However, we shall see in subsequent sections that it is difficult to make this exact distinction, because areas located anterior to the area pyram-idalis also seem to have direct motor excitatory tracts to the spinal cord.

A single weak shock applied to a pyramidal area is sometimes strong enough to cause a definite motor movement somewhere in the body—for instance, a flick of a finger, deviation of the lip, or some other discrete movement; and a short train of stimuli almost always elicits a movement in some discrete part of the body. In almost no other region of the cortex can one be certain of achieving a motor response with such weak stimuli.

The Extracorticospinal Tracts (Extrapyramidal Tracts). The extracorticospinal tracts, or extra-pyramidal tracts, are collectively all the tracts, besides the corticospinal tract itself, that transmit motor signals from the cortex to the spinal cord. In the above discussion of the collaterals from the corticospinal tract, we have noted some of these extracorticospinal pathways that pass through the striate nuclei, substantia nigra, red nuclei, and reticular substance of the brain stem. In addition to these collateral pathways from the corticospinal tract, still many other pathways course directly from the cortex to these intermediate nuclei. For instance, large numbers of neurons project di-rectly from other motor areas of the cortex, in addition to the area pyramidalis, into the *caudate nucleus* and then through the *putamen, globus pallidus, subthalamic nucleus, red nucleus, sub-stantia nigra,* and *reticular substance of the brain stem* before passing into the spinal cord. The multi-plicity of connections within these intermediate nuclei of the basal ganglia and reticular substance was presented in the preceding chapter.

The pathways for transmission of extracortico-spinal signals into the cord are the *reticulospinal tracts,* which lie in both the ventral and lateral columns of the cord; and, to a less extent, the *rubrospinal, tectospinal,* and *vestibulospinal tracts.*

SPATIAL ORGANIZATION OF THE MOTOR CORTEX

Figure 58–3 shows the areas of the body in which muscle contractions occur when different parts of the motor cortex are stimu-lated. Thus, stimulation of the most lateral portions of the cortex causes swallowing or chewing movements, while stimulation of the midline portion of the motor cortex, where the motor cortex bends over into the longi-tudinal sulcus, causes contraction in the legs, feet, or toes. The spatial organization is similar to that of the sensory cortex, which was pre-sented in Figure 49–4 of Chapter 49.

Degree of Representation of Different Muscle Groups in the Motor Cortex. The different

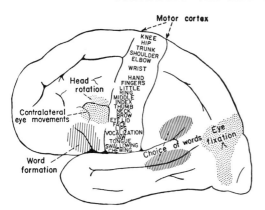

Figure 58–3   Representation of the different muscles of the body in the motor cortex and location of other cortical areas responsible for certain types of motor movements.

muscle groups of the body are not represented equally in the motor cortex. In general, the degree of representation is proportional to the discreteness of movement required of the respective part of the body. Thus, the thumb and fingers have large representations, as is true also of the lips, tongue, and vocal cords. The relative degrees of representation of the different parts of the body are illustrated in Figure 58–4, a figure constructed by Penfield and Rasmussen on the basis of stimulatory charts made of the human motor cortex during brain operations performed under local anesthesia.

When using barely threshold stimuli, only small segments of the peripheral musculature ordinarily contract at a time. In the "finger" and "thumb" regions, which have tremendous

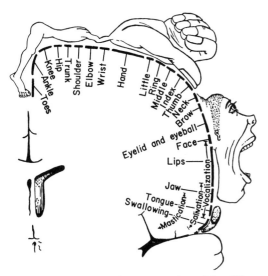

Figure 58–4   Degree of representation of the different muscles of the body in the motor cortex. (From Penfield and Rasmussen: The Cerebral Cortex of Man. The Macmillan Co.)

representation in the cerebral cortex, threshold stimuli can sometimes cause single muscles or, at times, even single fasciculi of muscles to contract, thus illustrating that a high degree of control is exercised by this portion of the motor cortex over discrete muscular movement.

On the other hand, threshold stimuli in the trunk region of the body might cause as many as 30 to 50 small trunk muscles to contract simultaneously, thus illustrating that the motor cortex does not control discrete trunk muscles but instead controls *groups* of muscles. Similarly, threshold stimuli in the lip, tongue, and vocal cord regions of the motor cortex cause contraction of minute muscular areas, whereas stimulation in the leg region ordinarily excites several synergistic muscles at a time, causing some gross movement of the leg. Also, it should be recognized that the finger muscles have far more representation in the motor cortex than do the muscles of the upper arm.

Coordinate Movements Elicited by Stimulation of the Motor Cortex — The Principle of "Final Position."   Even though bare threshold stimulation of discrete points in the motor cortex can cause relatively minute portions of the musculature to contract individually, more moderate stimulation almost always causes coordinate contraction of several different muscles simultaneously. Furthermore, stimulation of an area of the motor cortex often causes a limb or other part of the body to move to essentially the same "final position" regardless of its original position. This has been demonstrated in normal unanesthetized monkeys in which electrodes had been implanted previously in a limb area of the motor cortex. Each time a stimulus was applied, regardless of the initial position of the limb, it would always move to a given position. Thus, stimulation of one point in the motor cortex would cause the forearm to lift upward and the hands to contort outward. Shifting the stimulus to a slightly different point would cause a slightly different "final position," involving mainly the same musculature but recruiting additional musculature for the new position and dropping some of the original musculature. In this way, stimulation of a succession of points across the surface of the motor cortex could theoretically cause a limb to move from one final position to the next, to the next, to the next, each final position being slightly different from the previous one. In this way, it would be possible to achieve a sequence of movements.

The mechanism by which the different "final positions" are achieved probably involves special circuits in the basal ganglia and in the internuncial system of the spinal cord. It is reasonable to believe that the final positions elicited by stimulation of the motor cortex are achieved by sending appropriate stimuli into the spinal cord to excite given groups of internuncial cells, each of which in turn is organized to give a particular final position.

Sequential Movements and Rhythmic Movements Elicited by Stimulating the Motor Cortex. Stimulation of a given area of the human cortex occasionally also causes rhythmic types of contraction, with alternate contractions between agonist and antagonist muscles. From the mouth region of the motor cortex, normal chewing or swallowing movements can be elicited. In lower animals stimulation of the foreleg region can occasionally cause rhythmic forward and backward movement of the limb in a stepping fashion, and stimulation of another immediately adjacent area might cause the limb to perform a batting type of movement instead of the stepping movement.

SKILLED MOVEMENTS, AND THE CONCEPT OF A "PREMOTOR AREA"

As one proceeds farther forward in the motor cortex, particularly into the area 1 to 3 centimeters anterior to the area pyramidalis, the patterns of elicited motor movements change somewhat from those observed upon stimulation of the pyramidal cells in the area pyramidalis. Stimulation in this forward region requires considerably greater electrical current and longer trains of stimuli to elicit responses. When responses do occur, they are likely to be slower to develop and also to be gross movements involving many muscles in contradistinction to the more discrete movements elicited from the area pyramidalis. Thus, the portion of the motor cortex anterior to the area pyramidalis seems to be concerned with movements that require coordination between many muscles rather than contraction of isolated muscles as elicited from the area pyramidalis. Therefore, some physiologists have termed this anterior portion of the motor cortex the *premotor area* and attribute to it special characteristics that are different from those of the posterior part of the motor cortex called the primary motor area.

The distinction between premotor and primary motor areas is a very inexact one because both of these areas transmit signals into both the corticospinal and extracorticospinal tracts, and their functions are tightly interwoven. Yet, the facts that slightly more coordinated movements are elicited from the anterior portions of the motor cortex and that more discrete movements are elicited from the posterior portions (the area pyramidalis) do give some justification to the term "premotor area."

The premotor area performs its functions by sending signals through two main routes: (1) the area pyramidalis and (2) the extracorticospinal tracts. For instance, removal of the finger portion of the area pyramidalis prevents fine movements of the fingers when the premotor area is stimulated, indicating that the premotor area must work through the area pyramidalis to give fine movements. Yet, stimulation of the premotor area, even after removal of its corresponding area pyramidalis, can still cause postural movements such as movements of an entire arm, head rotation, movement of the trunk, or "fixation" of the upper arms or legs.

Control of Skilled Movements. In human beings in whom small areas of the premotor region have been removed, the coordination of skilled movements becomes impaired. The discrete movements of the hands are still intact, but the organized movements needed for performance of skilled acts are gone. Therefore, it is believed that the premotor region is specifically concerned with the *acquisition of specialized motor skills.*

Electrical stimulation of the premotor region and other closely associated areas of the motor cortex in human beings will often cause complex motor movements such as turning the hand, moving the thumb and fingers toward each other, and so forth. This also fits with the concept that the more anterior portions of the motor cortex are concerned with skilled movements.

In addition, complex eye movements for fixation of the eyes and complex movements of the larynx and mouth for formation of words can be elicited by stimulating the eye, laryngeal, and mouth portions of the premotor regions, as discussed elsewhere in the text.

MOTOR FUNCTIONS OF THE SOMATIC SENSORY CORTEX

If we will now refer back to the earlier discussion of the corticospinal and extracorticospinal tract, we will recall that as many as one-quarter of the fibers in these tracts originate from areas that are normally considered to be sensory in function and which lie posterior to the central sulcus. Especially in the anterior part of the postcentral gyrus is there intense overlap between the somatic sensory and motor areas. Electrical stimulation in this region causes somatic sensations and, occasionally, also simultaneous muscular contractions in the same general area of the body where the sensations are felt.

From Chapter 49, it will also be recalled that there is a *second somatic sensory cortex* located at the lower tip of the postcentral gyrus. From this region, motor responses can also be elicited, and the representation of the muscles of the body in this area is very much the same as the sensory representation, for which reason it is believed that the second somatic area might be specifically organized to associate the sensory system with the motor system. However, ablation of this region does not cause measurable impairment of motor function.

EFFECTS OF ABLATING PORTIONS OF THE MOTOR CORTEX

Ablation of the Area Pyramidalis. Removal of a very small portion of the area pyramidalis in a monkey usually causes flaccid paralysis of the represented muscles. If the area removed is principally in the posterior part of the motor cortex, the flaccidity continues indefinitely, but, if the ablation extends anteriorly as far as the anterior edge of the precentral gyrus, the flaccidity is replaced by spasticity after a few weeks. Because of this, some neurophysiologists have postulated a *suppressor zone* that normally transmits signals through the basal ganglia to suppress muscular tone in a small strip of cortex lying immediately anterior to the area pyramidalis. It has been supposed that removal of this suppressor region allows the muscle tone throughout the body to increase markedly.

If the caudate nucleus is left intact, gross postural and limb "fixation" movements can still be performed, but discrete movements of the distal segments of the limbs — the hands and fingers, for instance — are lost.

Ablation of the Premotor Area. If the premotor area is removed without destroying any of the adjacent area pyramidalis, the affected region of the body is likely to become spastic, but control of the discrete movements from the area is not affected. On the other hand, some of the skilled movements that had previously been performed by the involved muscles seem no longer to be performed in a coordinate manner.

The Babinski Sign. Destruction of the foot region of the area pyramidalis or transection of the foot portion of the corticospinal tract causes a peculiar response of the foot called the Babinski sign, which is demonstrated when a firm tactile stimulus is applied to the sole — the great toe extends and the smaller toes fan out. This is in contradistinction to the normal effect in which all the toes bend

downward. Therefore, the Babinski sign is often used clinically to detect abnormalities in the corticospinal portion of the motor control system. Yet, even so, the physiological mechanism of the Babinski sign is not known.

Effect of Hemidecortication. In rare instances it has been necessary to remove the cortex of an entire hemisphere. When this is done in a child and the basal ganglia and thalamus are left intact, the child usually learns to walk adequately and to perform postural and other crude subconscious movements. However, he can never learn to perform discrete movements with his hand on the contralateral side of his body or to perform complicated "voluntary" movements involving any of his contralateral muscles.

If the surgical incision involves significant portions of the caudate nuclei or of the thalamus, the person never learns to walk or to perform any satisfactory purposeful movements on the contralateral side of his body. This illustrates forcefully the importance of the diencephalic nuclei and basal ganglia in the control of motor functions.

EFFECTS OF PYRAMIDAL TRANSECTION

Since transection of a pyramid in the medulla destroys all or almost all the corticospinal fibers from one cerebral hemisphere, such a procedure in an experimental animal discloses much information about the overall function of the corticospinal tract as distinguished from the extracorticospinal tracts.

Transection of a pyramid on one side of the medulla causes *loss of all discrete movements on the opposite side of the body,* which illustrates that the corticospinal tract, and not the extracorticospinal tracts, controls the fine movements — especially in the hands, fingers, feet, and toes.

A second effect of pyramidal transection is *hypotonia* throughout the body; this results from loss of the normal facilitation of the corticospinal tract on the spinal motor centers. Associated with the hypotonia is gradual atrophy of the muscles everywhere in the body, especially the distal musculature of the limbs.

Functions of the Extracorticospinal Pathways That Are Retained After Pyramidal Transection. After pyramidal transection, the motor functions not lost are mainly the postural contractions and other gross movements of the body. For instance, the different movements required for equilibrium and support of the body against gravity are still completely intact. Furthermore, stimulation of motor cortical regions still causes gross movements of the trunk and upper portions of the limbs. These movements are sometimes called "fixation movements," and their function is the following: when a discrete motor function is to be performed by a hand, for instance, the trunk and upper arm must be "fixed" in appropriate positions before the hand can perform its required function. These fixation movements still occur even after complete transection of the pyramidal tract. In other words, the extracorticospinal pathways are responsible for almost all the postural and fixation movements, while the corticospinal tracts are responsible mainly for the superimposed discrete movements, especially those of the distal portions of the limbs.

FUNCTION OF THE INTERNUNCIAL SYSTEM OF THE SPINAL CORD

As illustrated in Figure 58–5, almost all corticospinal fibers end on small *internuncial cells in the base of the dorsal horns.* From these, impulses are transmitted mainly through *intermediate internuncial cells* to the anterior motoneurons. Thus, usually, at least two internuncial cells must relay the signals from the corticospinal tract to the anterior motoneurons, and sometimes three or even more. On the other hand, a few corticospinal fibers end directly on the anterior motoneurons.

FUNCTIONS OF THE INTERNUNCIAL CELLS IN THE MOTOR TRANSMISSION SYSTEM

In Chapter 56, we noted that the internuncial cells are the loci of "patterns of movement" in the spinal cord. These can be either fixed movements with "final positions" or rhythmic movements, such as stepping or scratching movements. Therefore, we must presume that these same patterns of movement can be activated by impulses from the corticospinal and extracorticospinal tracts. In addition, "patterns of movement" not normally found in the spinal animal can almost certainly be elicited by impulses from the brain. It might be sup-

posed that stimulation of successive points in the motor cortex causes a sequence of "patterns of movement," each of which is actually an intrinsic function of the internuncial system. In this way, the impulses from the motor cortex would not be stimulating the individual muscles directly but would be stimulating successively different complexes of internuncial cells of the spinal cord. Thus, *spread of excitation across a minute area of the motor cortex could reasonably be expected to cause a sequential movement of a limb, perhaps involving 100 or more successive cord integrated patterns of movement.*

Convergence of Signals in the Internuncial System. Another important function of the internuncial system is convergence of motor stimulatory signals from many different sources. For instance, sensory stimuli from the peripheral nerves can cause certain specific patterns of movement as described in relation to the cord reflexes in Chapter 56. Then the motor cortex can elicit its patterns of movement, and stimulation of other cortical motor areas as well as areas of the basal ganglia and the brain stem elicits still other patterns of movement mainly concerned with posture, equilibrium, and fixation of limbs. Since all these different types of patterns of movement might be elicited simultaneously, the internuncial cells and anterior motoneurons must algebraically add all the different stimuli and come out with the net result. Therefore, the internuncial system, along with the motoneurons themselves, constitutes a final level for integration of motor signals from all sources.

STIMULATION OF THE ANTERIOR MOTONEURON AND ITS OUTPUT TO THE PERIPHERY

Stimulation of the anterior motoneuron has been discussed in detail in Chapters 6 and 56. We can summarize the stimulation of the anterior motoneurons by pointing out that both excitatory and inhibitory impulses impinge on its dendrites and soma. The impulses from the corticospinal pathway, from the vestibulospinal pathway, and from parts of the reticulospinal tracts are all excitatory while impulses from another portion of the reticulospinal tract are mainly inhibitory. Also, many of the internuncial cells that help to form the different patterns of movement transmit inhibitory impulses.

Though most of the impulses to the anterior motoneurons are routed through the internuncial cells, a few pass directly to the anterior motoneurons. These presumably perform much less complicated functions than do those that pass through the internuncial system, including such functions as

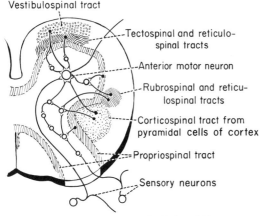

Vestibulospinal tract

Tectospinal and reticulo-spinal tracts

Anterior motor neuron

Rubrospinal and reticulo-lospinal tracts

Corticospinal tract from pyramidal cells of cortex

Propriospinal tract

Sensory neurons

Figure 58–5 Convergence of all the different motor pathways on the anterior motoneurons.

exciting highly discrete muscular contractions instead of patterns of movement, or providing general facilitation or general inhibition.

MOTOR FUNCTIONS OF THE CEREBELLUM

The cerebellum has long been called a silent area of the brain, principally because electrical excitation of the cerebellum does not cause any sensation and rarely any motor movement. However, as we shall see, removal of the cerebellum does make motor movements highly abnormal. Furthermore, even though electrical stimulation of the cerebellum normally does not cause motor movements, it can inhibit or, rarely, enhance motor movements that are being elicited by the motor cortex.

Physiologic Anatomy. The human cerebellum is generally divided into two major lobes, the *anterior cerebellum,* which is small, and the *posterior cerebellum,* which is much larger. The posterior cerebellum is greatly enlarged bilaterally to form the *cerebellar hemispheres,* which are also known as the *neocerebellum* (new cerebellum) because they represent a phylogenetically new portion of the cerebellum. The older portions of the cerebellum, including the small midline section of the posterior cerebellum and the entire anterior cerebellum, are called the *paleocerebellum* (old cerebellum).

The afferent pathways. The basic afferent pathways to the cerebellum are illustrated in Figures 58–6 and 58–7. An extensive and important afferent pathway is the *corticocerebellar pathway,* which originates mainly in the *motor cortex* (and, to a less extent, in the sensory cortex as well) and then passes by way of the *pontile nuclei* and *pontocerebellar tracts* directly to the cortex of the

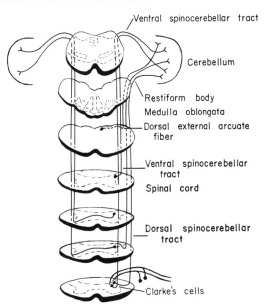

Figure 58–7 The spinocerebellar tracts.

cerebellum. In addition, important afferent tracts originate in the brain stem; they include: (a) an extensive *olivocerebellar* tract, which passes from the *inferior olive* to all parts of the cerebellum; this tract is excited by fibers from the *striate body* and from widespread areas of the *reticular formation;* (b) *vestibulocerebellar fibers,* some of which originate in the vestibular nuclei and some of which originate in the vestibular apparatus itself; most of these terminate in the *flocculonodular lobe* and *fastigial nucleus* of the cerebellum; (c) *reticulocerebellar fibers,* which originate in different portions of the reticular formation and terminate mainly in the midline structures of the paleocerebellum.

The cerebellum receives still other important signals from the peripheral parts of the body, which reach the cerebellum by way of the *ventral* and *dorsal spinocerebellar* tracts illustrated in Figures 58–6 and 58–7. The signals transmitted in these tracts originate almost entirely in the muscle spindles and Golgi tendon organs and apprise the cerebellum of the momentary status of muscle contraction and momentary degree of tension on the muscle tendons.

Note in Figure 58–7 that the nerve fibers from the muscle spindles and Golgi tendon organs terminate on Clarke's cells at the bases of the dorsal horns and that most of these fibers pass ipsilaterally up to the cerebellum. These pathways can transmit impulses at velocities of over 100 meters per second, which is the most rapid conduction of any pathway in the entire central nervous system. This extremely rapid conduction allows instantaneous apprisal of the cerebellum of changes that take place in the status of the muscles.

In addition to the signals in the spinocerebellar tracts, other signals are transmitted through the dorsal columns to the medulla and then relayed

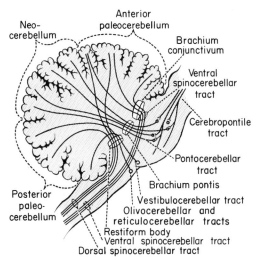

Figure 58–6 The principal afferent tracts to the cerebellum.

Figure 58-8 The sensory projection areas, called "homunculi," on the cortex of the cerebellum. (From Snider: *Sci. Amer., 199*(6):4, 1958.)

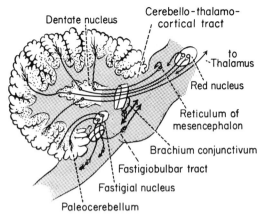

Figure 58-9 Principal efferent tracts from the cerebellum.

from there to the cerebellum. Likewise, signals are transmitted through the spinotectal pathway to the reticular substance of the brain stem and then relayed to the cerebellum. These pathways transmit information from the joints, the pressure receptors of the deep tissues, and even from the cutaneous receptors. Thus, the cerebellum collects continual information about the instantaneous physical status of all parts of the body even though the cerebellum is operating at a subconscious level.

Afferent fibers entering the cerebellum terminate in distinct spatially oriented areas of the cerebellar cortex. Figure 58-8 illustrates the locus of two different "homunculi" that represent the terminations in the cerebellum of the sensory fibers from all parts of the body. One of these lies in the anterior cerebellum, and the other is split in half and lies in the posterior cerebellum. Also, *fibers from the motor cortex project to the same corresponding points in both these areas of the cerebellum.*

The efferent pathways. The efferent tracts that leave the cerebellum, illustrated in Figure 58-9, arise in four different nuclei, the *dentate, globose, emboliform,* and *fastigial nuclei.* The first three of these send their fibers upward through the brachium conjunctivum to the *red nucleus,* to the upper portion of the reticular formation, and to the *ventrolateral nucleus of the thalamus* and thence to the motor cortex. The *fastigial nuclei,* on the other hand, send their efferent fibers through the restiform bodies into the lower portion of the reticular formation. These nuclei represent a major pathway for transmission of impulses from the paleocerebellum into the reticular formation.

FUNCTION OF THE CEREBELLUM IN VOLUNTARY MOVEMENTS

The cerebellum functions only in association with motor activities initiated elsewhere in the central nervous system. These activities may be initiated originally in the spinal cord, in the reticular formation, in the basal ganglia, or in the motor areas of the cerebral cortex. We will discuss, first, the operation of the cerebellum in association with the motor cortex for the control of voluntary movements.

Figure 58-10 illustrates the basic cerebellar pathways involved in cerebellar control of voluntary movements. When motor impulses are transmitted from the cerebral cortex downward through the corticospinal tract to excite the voluntary muscles, collateral impulses are transmitted simultaneously into the cerebellum through the pontocerebellar tracts. Therefore, for every motor movement that is

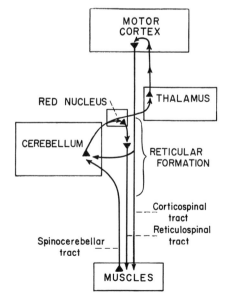

Figure 58-10 Pathways for cerebellar "error" control of voluntary movements.

performed, not only do the muscles receive activating signals, but the cerebellum receives similar signals at the same time.

When the muscles respond to the motor signals, the muscle spindles, Golgi tendon apparatuses, joint receptors, and other peripheral receptors transmit impulses upward mainly through the spinocerebellar and dorsal column pathways to terminate in the anterior cerebellum. Furthermore, these impulses go to exactly the same portion of the anterior cerebellum that is stimulated by the downcoming impulses from the motor cortex.

After the signals from the periphery and those from the motor cortex are integrated, efferent impulses are transmitted from the cerebellar cortex to the dentate nuclei and thence upward through the ventrolateral nuclei of the thalamus back to the motor cortex where the stimulus first originated.

"Error Control" by the Cerebellum. One will readily recognize that the circuit described above represents a complicated feedback circuit beginning in the motor cortex and also returning to the motor cortex. Furthermore, experiments have shown that the cerebellum operates in much the same manner as a servomechanism such as that used in (a) industrial control systems, (b) the control system of an anti-aircraft gun, or (c) the control system of an automatic pilot. That is, the cerebellum seems to compare the "intentions" of the cortex with the "performance" by the parts of the body, and, if the parts have not moved according to the intentions of the cortex, the "error" between these two is calculated by the cerebellum so that appropriate and immediate corrections can be made. Thus, if the cortex has transmitted a signal intending the limb to move to a particular point, but the limb begins to move too fast and obviously is going to overshoot the point of intention, the cerebellum can initiate "braking impulses" to slow down the movement of the limb and stop it at the point of intention.

Ordinarily, the motor cortex transmits far more impulses than are needed to perform each intended movement, and the cerebellum therefore must act to inhibit the motor cortex at the appropriate time after the muscle has begun to move. The cerebellum automatically assesses the rate of movement and calculates the length of time that will be required to reach the point of intention. Then appropriate inhibitory impulses are transmitted to the motor cortex to inhibit the agonist muscle and to excite the antagonist muscle. In this way, appropriate "brakes" are applied to stop the movement at the precise point of intention.

Experiments performed recently by Smith have demonstrated two important characteristics of the cerebellar feedback system:

1. When a person performs a particular act, such as moving a heavy weight rapidly to a new position, at first he is unable to judge the degree of inertia that will be involved in the movement. Therefore, he almost always overshoots the point of intention on his first trial. However, after several trials, he can stop the movement at the precise point. This illustrates that learned knowledge of the inertia of the system is an important feature of the cerebellar feedback mechanism.

2. When a rapid movement is made toward a point of intention, the agonist muscle contracts strongly throughout the early course of movement. Then, suddenly, shortly before the point of intention is reached, the agonist muscle becomes completely inhibited while the antagonist muscle becomes strongly excited. Furthermore, the point at which this reversal of excitation occurs depends on the rate of movement and on the previously learned knowledge of the inertia of the system. The faster the movement and the greater the inertia, the earlier the reversal point appears in the course of movement.

Since all these events transpire much too rapidly for the motor cortex to reverse the excitation "voluntarily," it is evident that the excitation of the antagonist muscle toward the end of a movement is an entirely automatic and subconscious function and is not a "willed" contraction of the same nature as the original contraction of the agonist muscle. We shall see below that in patients with serious cerebellar damage, excitation of the antagonist muscles does not occur at the appropriate time but instead always too late. Therefore, it is almost certain that one of the major functions of the cerebellum is automatic excitation of antagonist muscles at the end of a movement while at the same time inhibiting agonist muscles that have started the movement.

The "Damping" Function of the Cerebellum. One of the by-products of the cerebellar feedback mechanism is its ability to "damp" muscular movements. To explain the meaning of "damping" we must first point out that essentially all movements of the body are "pendular." For instance, when an arm is moved, momentum develops, and the momentum must be overcome before the movement can be stopped. And, because of the momentum, all pendular movements have a tendency to overshoot. If overshooting does occur in a person whose cerebellum has been destroyed, the conscious centers of the cerebrum eventually recognize this and initiate a movement in the opposite direction to bring the arm to its intended position. But again the arm, by virtue of its momentum, overshoots, and appropriate corrective signals must again be instituted. Thus the arm oscillates back and forth past its intended point for several cycles before it finally reaches its mark. This effect is called an *action tremor,* or *intention tremor.*

However, if the cerebellum is intact, appropriate subconscious signals stop the movement precisely at the intended point, thereby preventing the overshoot and also the tremor. This is the basic characteristic of a damping system. All servocontrol systems regulating pendular elements that have inertia must have damping circuits built into the servomechanisms. In the motor control system of our central nervous system, the cerebellum seems to provide much of this damping function.

RELATIONSHIP OF CEREBELLAR DAMPING TO DAMPING CAUSED BY THE STRETCH REFLEX. The stretch reflex, which is strictly a cord reflex, also provides much of the damping of body movements, as pointed out in Chapter 56. The cerebellar damping mechanism seems to be superimposed onto the basic cord damping mechanism. The muscle spindles excite both these functions. Rapid stretch of a muscle transmits spindle signals both directly to the anterior motoneuron to cause the damping action of the stretch reflex and simultaneously to the cerebellum to cause additional damping. The cerebellum probably provides its damping action by transmitting reflex signals into the reticular formation and thence back down the cord to stimulate the gamma efferent neurons. These in turn alter the degree of contraction of the intrafusal fibers of the muscle spindles and thereby change their degree of excitation.

The cerebellar damping mechanism at times probably supports the stretch reflex damping system and at other times inhibits it, thus illustrating that the cerebellar system has considerable latitude of control that makes it adaptable to more complex motor activities than is true of the stretch reflex.

Function of the Cerebellum in Prediction. Another important by-product of the cerebellar feedback mechanism is its ability to help the central nervous system predict the future positions of moving parts of the body. Without the cerebellum a person "loses" his limbs when they move rapidly, indicating that feedback information from the periphery must be analyzed by the cerebellum if the brain is to keep up with the motor movements. Thus, somewhere in the circuit of the cerebellum there seems to be an integrative system that detects from the incoming proprioceptive signals the rapidity with which the limb is mov-

ing and then predicts from this the projected time course of movement. This allows the cerebellum, operating through the cerebral cortex, to inhibit the agonist muscles and to excite the antagonist muscles when the movement approaches the point of intention.

Without the cerebellum this predictive function is so deficient that moving parts of the body move much farther than the point of intention. This failure to control the distance that the parts of the body move is called *dysmetria,* which means simply poor control of the distance of movement. As would be expected, dysmetria becomes greatly enhanced in rapid movements in comparison with slow movements.

FAILURE OF "PROGRESSION." One of the most important features of normal motor function is one's ability to progress from one movement to the next in an orderly succession. When the cerebellum becomes dysfunctional and the person loses his subconscious ability to predict ahead of time how far the different parts of his body will move in a given time, he becomes unable also to control the beginning of the next movement. As a result, the succeeding movement may begin much too early or much too late. Therefore, movements such as those required for writing, running, or even talking all become completely incoordinate, lacking completely in the ability to progress in an orderly sequence from one movement to the next.

Extramotor Predictive Functions of the Cerebellum. The cerebellum also plays a role in predicting other events besides simply movements of the body. For instance, the rates of progression of both auditory and visual phenomena can be predicted. As an example, a person can predict from the changing visual scene how rapidly he is approaching an object. A striking experiment that demonstrates the importance of the cerebellum in this ability is the effect of removing the midline portion of the paleocerebellum in monkeys. Such a monkey occasionally charges the wall of a corridor and literally bashes its brains out because it is unable to predict when it will reach the wall.

Unfortunately, we are only now beginning to learn about these extramotor predictive functions of the cerebellum. It is quite possible that the cerebellum provides a "time base" against which signals from other parts of the central nervous system can be compared. It is often stated that the cerebellum is especially important in interpreting spatiotemporal relationships in sensory information.

FUNCTION OF THE CEREBELLUM IN INVOLUNTARY MOVEMENTS

The cerebellum functions in involuntary movements in almost exactly the same manner as it does in voluntary movements except that somewhat different pathways are involved, as shown in Figure 58–11. The extracorticospinal impulses that originate in the basal ganglia or reticular formation to cause involuntary movements pass also into the inferior olive and thence into the cerebellum, primarily to the paleocerebellum. Then as the muscle movements occur, proprioceptive information from the muscles, joints, and other peripheral parts of the body passes to the cerebellum, thus providing the same type of "error" control for involuntary movements as is provided for voluntary movements, as discussed above. Impulses feed from the cerebellum directly back into the *reticular formation* by way of the *fastigial nuclei* and *fastigiobulbar tracts*. They feed back into the basal ganglia by way of the *emboliform nuclei,* then upward through the *centrum medianum* of the thalamus and thence into the *caudate nucleus* and *putamen.*

Among the principal pathways through which the corrective signals are transmitted to the muscles are the *reticulospinal tracts* that control the *gamma efferent system* to the muscle spindles. Indeed, the cerebellum is one of the most powerful of all the activators of this system, which indicates the importance of the gamma efferents in cerebellar functions.

Function of the Cerebellum for Control of Spinal Reflex Movements. Even movements elicited by spinal reflexes are modified to some extent by the cerebellum. For instance, in the absence of the cerebellum, sudden movement of a limb by some extraneous force is met with almost no resistance, but in the presence of the cerebellum the stretch reflexes oppose the movement almost instantaneously. Thus, the cerebellum, operating through a cerebellar-reticular feedback system, greatly enhances the power of the spinal stretch reflex.

On the other hand, many movements integrated in the spinal cord, such as scratching and cleaning of the face with the paws, can be performed completely coordinately in the absence of a cerebellum.

FUNCTION OF THE CEREBELLUM IN EQUILIBRIUM

In the discussion of the vestibular apparatus and of the mechanism of equilibrium in Chapter 57, it was pointed out that the flocculonodular lobes of the cerebellum are required for proper integration of the equilibratory impulses. Removal of the *nodulus* in particular causes almost complete loss of equilibrium, and removal of the *flocculus* causes temporary loss. The symptoms of mal-equilibrium are essentially the same as those that result from destruction of the semicircular canals. This indicates that these areas of the cerebellum are particularly important for integration of *changes in direction of motion* as detected by the semicircular canals, but not so important for integrating static signals of equilibrium as detected by the maculae in the utricles. This is in keeping with the other functions of the cerebellum; that is, the semicircular canals allow the central nervous system to *predict* ahead of time that rotational movements of the body are going to cause mal-equilibrium, and this predictive function causes appropriate muscles to contract to prevent the person from losing equilibrium even before this state develops. This is very much the same as the predictive function of the cerebellum for coordination of rapid voluntary movements.

ELECTRICAL ACTIVITY IN THE CEREBELLUM

The cerebellar cortex is continually active— rhythmic discharges occurring throughout the cortex at rates of 150 to 250 times per second. Signals passing through the cerebellum "modulate" the intrinsic activity of the cerebellum, sometimes increasing and at other times decreasing the activity.

The continual rhythmic impulses in the cerebellum facilitate the motor cortex. Enhanced activity of a particular region in the cerebellar cortex enhances the degree of facilitation of a corresponding point in the motor cortex. Conversely, decreased activity in a given point of the cerebellum causes corresponding decrease in activity of its representative area in the motor cortex.

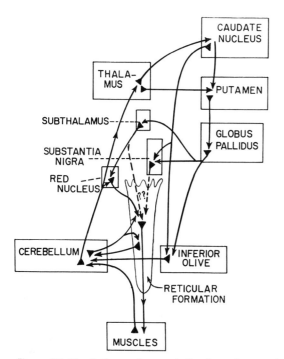

Figure 58–11 Pathways for cerebellar "error" control of involuntary movements.

CLINICAL ABNORMALITIES OF THE CEREBELLUM

Destruction of small portions of the cerebellar cortex cause no detectable abnormality in motor function. In fact, several months after as much as half the cerebellar cortex has been removed, the motor functions appear to be almost entirely normal. Thus, the remaining areas of the cerebellum compensate tremendously for loss of part of the cerebellum.

To cause serious dysfunction of the cerebellum, the cerebellar lesion must usually involve the deep cerebellar nuclei—the *dentate, emboliform, globose,* and *fastigial nuclei*—as well as the cerebellar cortex.

Dysmetria and Ataxia. Two of the most important symptoms of cerebellar disease are dysmetria and ataxia. It was pointed out above that in the absence of the cerebellum a person cannot predict ahead of time how far his movements will go. Therefore, the movements ordinarily overshoot their intended mark. This effect is called *dysmetria,* and it results in incoordinate movements which are called *ataxia*.

Dysmetria and ataxia can also result from lesions in the spinocerebellar tracts, for the feedback information from the moving parts of the body is essential for accurate control of the muscular movements.

PAST POINTING. Past pointing means that in the absence of the cerebellum a person ordinarily moves his hand or some other moving part of his body considerably beyond the point of intention. This probably results from the following effect: The motor cortex normally transmits more impulses to the muscles to perform a given motor function than are actually needed. The cerebellum automatically corrects this by inhibiting the movement after it has begun. However, if the cerebellum is not available to cause this inhibition, the movement ordinarily goes far beyond the intended point. Therefore, past pointing is actually a manifestation of dysmetria.

Failure of Progression. DYSDIADOCHOKINESIA. When the motor control system fails to predict ahead of time where the different parts of the body will be at a given time, it temporarily "loses" the parts during rapid motor movements. As a result, the succeeding movement may begin much too early or much too late so that no orderly "progression of movement" can occur. One can demonstrate this readily in a patient with cerebellar damage by having him turn one of his hands upward and downward at a rapid rate. He rapidly "loses" his hand and does not know its position during any portion of the movements. As a result, a series of jumbled movements occurs instead of the normal coordinate upward and downward motions. This is called dysdiadochokinesia.

DYSARTHRIA. Another instance in which failure of progression occurs is in talking, for the formation of words depends on rapid and orderly succession of individual muscular movements in the larynx. Lack of coordination between these and inability to predict either the intensity of the sound or the duration of each successive sound cause jumbled vocalization, with some syllables loud, some weak, some held long, some held for a short interval, and resultant speech that is almost completely unintelligible. This is called dysarthria.

Intention Tremor. When a person who has lost his cerebellum performs a voluntary act, his muscular movements are jerky; this reaction is called an *intention tremor* or an *action tremor,* and it results from failure of the cerebellar system to damp the motor movements. Tremor is particularly evident when the dentate nuclei or the brachium conjunctivum is destroyed, but it is not present when the spinocerebellar tracts from the periphery to the cerebellum are destroyed. This indicates that the feedback pathway from the cerebellum to the motor cortex is a principal pathway for damping of muscular movements.

CEREBELLAR NYSTAGMUS. Cerebellar nystagmus is a tremor of the eyeballs that occurs usually when a person attempts to fixate his eyes on a scene to the side of his head. This off-center type of fixation results in rapid tremorous movements of the eyes rather than a steady fixation, and it is probably another manifestation of the failure of damping by the cerebellum. It also occurs especially when the flocculonodular lobes are damaged; and in this instance it is associated with loss of equilibrium, presumably because of dysfunction of the pathways through the cerebellum from the semicircular canals. Nystagmus resulting from damage to the semicircular canals was discussed in Chapter 57.

Rebound. If a person with cerebellar disease is asked to contract his arm tightly while the physician holds it back at first and then lets go, the arm will fly back until it strikes the face instead of being automatically stopped. This is called rebound, and it results from loss of the cerebellar component of the stretch reflex. That is, the normal cerebellum ordinarily instantaneously and powerfully sensitizes the spinal cord stretch reflex mechanism whenever a portion of the body begins to move unexpectedly in an unwilled direction. But, without the cerebellum, activation of the antagonist muscles fails to occur, thus allowing overmovement of the limb.

HYPOTONIA. Loss of the cerebellar hemispheres causes moderate decrease in tone of the peripheral musculature, though after several months the motor cortex usually compensates for this by an increase in its intrinsic activity. The hypotonia results from loss of facilitation of the motor cortex by the cerebellocortical pathways.

LOSS OF EQUILIBRIUM. Destruction of the flocculonodular portion of the cerebellum causes loss of equilibrium, particularly when the person attempts to perform rapid movements. This loss of equilibrium is similar to that which occurs when the semicircular canals are destroyed, as was discussed in Chapter 57.

SENSORY FEEDBACK CONTROL OF MOTOR FUNCTIONS

Everyone who has ever studied the relationship of the somatic sensory areas to the motor areas of the cortex has been impressed by the close functional interdependence of the two areas. Anatomically, the characteristic pyramidal cells of the motor cortex extend backward into the anterior lip of the postcentral gyrus where they intermingle with the large numbers of granule cells of the somatic sensory cortex. Likewise, the granule cells of the sensory cortex extend anteriorly into the precentral gyrus, the *area pyramidalis*. Thus the two areas fade into each other.

Furthermore, when a portion of the somatic sensory cortex in the postcentral gyrus is removed, the muscles controlled by the motor cortex immediately anterior to the removed area often lose much of their coordination. This observation illustrates that the somatic sensory area plays a major role in the control of motor functions. The overall mechanism by which sensory feedback plays this role has been postulated to be the following:

THE SENSORY "ENGRAM" FOR MOTOR MOVEMENTS

A person performs a motor movement mainly to achieve a purpose. It is primarily in the sensory and sensory association areas that he experiences effects of motor movements and records "memories" of the different patterns of motor movements. These are called sensory engrams of the motor movements. When he wishes to achieve some purposeful act, he presumably calls forth one of these engrams and then sets the motor system of the brain into action to reproduce the sensory pattern that is laid down in the engram.

The Proprioceptor Feedback Servomechanism for Reproducing the Sensory Engram. From the earlier discussion of cerebellar function, it is clear how proprioceptor signals from the periphery can affect motor activity. However, in addition to the feedback pathways through the cerebellum, more slowly acting feedback pathways also pass from proprioceptors to the sensory areas of the cerebral cortex and thence back to the motor cortex. Each of these feedback pathways is capable of modifying the motor response. For instance, if a person learns to cut with scissors, the movements involved in this process cause a particular sequential pattern of proprioceptive impulses to pass to the somatic sensory area. Once this pattern has been "learned" by the sensory cortex, the memory engram of the pattern can be used to activate the motor system to perform the same sequential pattern whenever it is required.

To do this, the proprioceptor signals from the fingers, hands, and arms are compared with the engram, and if the two do not match each other, the difference, called the "error," supposedly initiates additional motor signals that automatically activate appropriate muscles to bring the fingers, hands, and arms into the necessary attitudes for performance of the task. Each successive portion of the engram presumably is projected according to a time sequence, and the motor control system automatically follows from one point to the next so that the fingers go through the precise motions necessary to duplicate exactly the sensory engram of the motor activity.

Thus, one can see that the motor system in this case actually acts as a servomechanism, for it is not the motor cortex itself that controls the pattern of activity to be accomplished. Instead, the pattern is located in the sensory part of the brain, and the motor system merely "follows" the pattern, which is the definition of a servomechanism. If ever the motor system fails to follow the pattern, sensory signals are fed back to the cerebral cortex immediately to apprise the sensorium of this failure, and appropriate corrective signals are transmitted to the muscles.

Other sensory signals besides somesthetic signals are also involved in motor control, particularly visual signals. However, these other sensory systems are often slower to recognize error than is the somesthetic proprioceptor system. Therefore, when the sensory engram depends on visual feedback for control purposes, the motor movements are usually considerably slowed in comparison with those that depend on somesthetic feedback.

An extremely interesting experiment that demonstrates the importance of the sensory engram for control of motor movements is one in which a monkey has been trained to perform some skilled activity and then various portions of his cortex are removed. Removal of small portions of the motor cortex that control the muscles normally used for the skilled activity does not prevent the monkey from performing the activity. Instead he automatically uses other muscles in place of the paralyzed ones to perform the same skilled activity. On the other hand, if the corresponding somesthetic cortex is removed but the motor cortex is left intact, the monkey loses all ability to perform the skilled activity. Thus, this experiment demonstrates that the motor system acts automatically as a servomechanism to use whatever muscles are available to follow the pattern of the sensory engram, and if some muscles are missing, other muscles are substituted automatically. The experiment also demonstrates forcefully that the somesthetic cortex is essential to the motor performance.

Function of Lower Brain Centers in Correlating Sensory and Motor Functions. Voluntary motor movements are almost certainly initiated in control centers located deep in the brain, such as in the reticular areas of the mesencephalon and in the diencephalon, and not in the cortex. These areas are believed to transmit signals to the sensory cortex to select the sensory engram of the motor activity to be performed. Then, the signals from this engram go back down to the thalamus and mesencephalon and thence up to the motor areas to control the motor system. Thus, the signals do not pass directly from one cortical area to another but indirectly through the lower areas. The primary evidence for this is that deep vertical slits in the cerebral cortex do not prevent transmission of most cortical signals from one cortical area to another. Indeed, complete transection of the *corpus callosum*, which is the major bundle of fibers connecting the two cerebral hemispheres, has almost no effect on motor function. And, similarly, a deep transection of the cortex in the central sulcus between the somesthetic sensory area and the motor cortex has no demonstrable effect on sensory control of motor function.

This ability of the lower centers of the brain to control cortical function and to call forth specific activity in certain areas of the cortex will be discussed in Chapter 60. Suffice it to say at the present that it is almost certainly the lower areas of the brain that initiate the original idea which sets into action the sensory and the motor cortical regions to perform specific motor tasks.

ESTABLISHMENT OF "SKILLED" MOTOR PATTERNS IN THE MOTOR CORTEX

Many motor activities are performed so rapidly that there is insufficient time for sensory feedback signals to control these activities. For instance, the movements of the fingers during typing occur much too rapidly for somesthetic sensory signals to be transmitted back to the cortex and for these then to control each discrete movement. It is believed that the patterns for control of these rapid coordinate muscular movements are established in the motor areas of the frontal lobe. The reason for believing this is that ablation of these areas of the cortex destroys one's ability to perform rapid coordinated muscular contractions, such as those required during the act of typing, talking, or writing by hand.

Role of Sensory Feedback During Establishment of the Rapid Motor Patterns. Even a highly skilled motor activity can be performed the very first time if it is performed extremely slowly — slowly enough for sensory feedback to guide the movements through each step. However, to be really useful, many skilled motor activities must be performed rapidly. This probably is achieved by successive performance of the same skilled activity until finally an engram of the skilled activity is laid down in the motor regions of the cortex as well as in the sensory region. This engram causes a precise set of muscles to go through a specific sequence of movements required to perform the skilled activity. Therefore, such an engram is called a *pattern of skilled motor function,* and the motor areas are primarily concerned with this.

After a person has performed a skilled activity many times, the motor pattern of this activity can thereafter cause the hand or arm or other part of the body to go through the same pattern of activity again and again, now entirely *without* sensory feedback control. However, even though sensory feedback control is no longer present, the sensory system still determines whether or not the act has been performed correctly. This determination is made in retrospect rather than while the act is being performed. If the pattern has not been performed correctly, information from the sensory system supposedly can help to correct the pattern the next time it is performed.

Thus, eventually, hundreds of patterns of different coordinate movements are laid down in the motor areas of the cortex, and these can be called upon one at a time in different sequential orders to perform literally thousands of complex motor activities.

An interesting experiment that demonstrates the applicability of these theoretical methods of muscular control is one in which the eyes are made to "follow" an object that moves around and around in a circle. At first, the eyes can follow the object only when it moves around the circle slowly, and even then the movements of the eyes are extremely jerky. Thus, sensory feedback is being utilized to control the eye movements for following the object. However, after a few seconds, the eyes begin to follow the moving object rather faithfully, and the rapidity of movement around the circle can be increased to many times per second, and still, the eyes continue to follow the object. Sensory feedback control of each stage of the eye movements at these rapid rates would be completely impossible. Therefore, by this time, the eyes have developed a pattern of movement that is not dependent upon step-by-step sensory feedback. Nevertheless, if the eyes should fail to follow the object around the circle, the sensory system would immediately become aware of this and presumably could make corrections in the pattern of movement.

EFFECT OF DIFFERENT MUSCULAR LOADS ON THE PERFORMANCE OF SKILLED MOTOR ACTIVITIES

All of us know that a person can move his hand around and around in a circle whether the hand be empty or loaded with a 10-pound weight. That is, the load makes little difference in the faithfulness of performance of the required activity. The reason for this is that the motor activity is probably controlled by a servo feedback mechanism that monitors the performance of the motor act constantly, and, if the performance is not according to the sensory engram of the required movement, appropriate "error" signals are automatically transmitted into the motor system to achieve the desired result. One can see, therefore, the practical significance of the servomechanism type of control of muscular activity.

REFERENCES

Asanuma, H.; Stoney, S. D., Jr.; and Abzug, C.: Relationship between afferent input and motor outflow in cat motorsensory cortex. J. Neurophysiol., 31:670, 1968.

Bell, C. C., and Grimm, R. J.: Discharge properties of Purkinje cells recorded with single and double microelectrodes. J. Neurophysiol., 32:1044, 1969.

Bloedel, J. R., and Roberts, W. J.: Functional relationship among neurons of the cerebellar cortex in the absence of anesthesia. J. Neurophysiol., 32:75, 1969.

Eccles, J. C.: Integration of information in the cerebellum. Proc. Int. Union of Physiol. Sciences, 6:19, 1968.

Eccles, R. M.; Shealy, C. N.; and Willis, W. D.: Patterns of innervation of kitten motoneurones. J. Physiol. (Lond.), 165:392, 1963.

Evarts, E. V.: Activity of pyramidal tract neurons during postural fixation. J. Neurophysiol., 32:375, 1969.

Evarts, E. V.: Relation of pyramidal tract activity to force exerted during voluntary movement. J. Neurophysiol., 31:14, 1968.

Evarts, E. V., and Thach, W. T.: Motor mechanisms of the CNS: cerebrocerebellar interrelations. Ann. Rev. Physiol., 31:451, 1969.

Fetz, E. E.: Pyramidal tract effects on interneurons in the cat lumbar dorsal horn. J. Neurophysiol., 31:69, 1968.

Fidone, S. J., and Preston, J. B.: Patterns of motor cortex control of flexor and extensor cat fusimotor neurons. J. Neurophysiol., 32:103, 1969.

Fox, C. A., and Snider, R. S.: The Cerebellum. Progress in Brain Research. New York, American Elsevier Publishing Co., Inc., 1967, Vol. 25.

Gilman, S., and McDonald, W. I.: Cerebellar facilitation of muscle spindle activity. J. Neurophysiol., 30:1494, 1967.

Higgins, D. C.; Partridge, L. D.; and Glaser, G. H.: A transient cerebellar influence on stretch responses. J. Neurophysiol., 25:684, 1962.

Hunt, C. C., and Kuno, M.: Properties of spinal interneurones. J. Physiol. (Lond.), 147:346, 1959.

Kameda, K.; Nagel, R.; and Brooks, V. B.: Some quantitative aspects of pyramidal collateral inhibition. J. Neurophysiol., 32:540, 1969.

Liles, S. L., and Davis, G. D.: Interrelation of caudate nucleus and thalamus in alteration of cortically induced movement. J. Neurophysiol., 32:564, 1969.

Llinas, R.; Bloedel, J. R.; and Hillman, D. E.: Functional characterization of neuronal circuitry of frog cerebellar cortex. J. Neurophysiol., 32:847, 1969.

Marchiafava, P. L.: Activities of the central nervous system: motor. Ann. Rev. Physiol., 30:359, 1968.

Massion, J.: The mammalian red nucleus. Physiol. Rev., 47:383, 1967.

O'Brien, J. H., and Fox, S. S.: Single-cell activity in cat motor cortex. I. Modifications during classical conditioning procedures. J. Neurophysiol., 32:267, 1969.

Oscarsson, O.: Functional organization of the spino- and cuneocerebellar tracts. Physiol. Rev., 45:495, 1965.

Paillard, J.: The patterning of skilled movements. In Handbook of Physiology. Baltimore, The Williams & Wilkins Co., 1960, Sec. I, Vol. III, p. 1679.

Partridge, L. D., and Glaser, G. H.: Adaptation in regulation of movement and posture. J. Neurophysiol., 23:257, 1960.

Penfield, W., and Rasmussen, T.: The Cerebral Cortex of Man. New York, The Macmillan Company, 1950.

Porter, R., and Hore, J.: Time course of minimal corticomotoneuronal excitatory postsynaptic potentials in lumbar motoneurons of the monkey. J. Neurophysiol., 32:443, 1969.

Stark, L.; Ida, M.; and Willis, P. A.: Dynamic characteristics of the motor coordination system in man. Biophys. J., 1:279, 1961.

Stewart, D. H., Jr., and Preston, J. B.: Functional coupling between the pyramidal tract and segmental motoneurons in cat and primate. J. Neurophysiol., 30:453, 1967.

Thach, W. T.: Discharge of Purkinje and cerebellar nuclear neurons during rapidly alternating arm movements in the monkey. J. Neurophysiol., 31:785, 1968.

Wyman, R. J.: Lateral inhibition in a motor output system. I. Reciprocal inhibition in dipteran flight motor system. J. Neurophysiol., 32:297, 1969.

CHAPTER 59

The Autonomic Nervous System; and Vegetative Functions of the Hypothalamus

The portion of the nervous system that controls the visceral functions of the body is called the *autonomic nervous system.* This system helps to control arterial pressure, gastrointestinal motility and secretion, urinary output, sweating, body temperature, and many other activities, some of which are controlled almost entirely by the autonomic nervous system and some only partially.

The *hypothalamus* is one of the major areas of the brain that in turn controls the activities of the autonomic nervous system. But in this structure are also control centers that do not operate through the autonomic system, such as body weight regulation, body fluid regulation, and regulation of many endocrine glands. These functions of the hypothalamus, along with the functions mediated through the autonomic nervous system, are collectively called the *vegetative functions* of the hypothalamus.

GENERAL ORGANIZATION OF THE AUTONOMIC NERVOUS SYSTEM

The autonomic nervous system is activated mainly by centers located in the *spinal cord, brain stem,* and *hypothalamus.* Also, portions of the cerebral cortex can transmit impulses to the lower centers and in this way effect autonomic control. Often the autonomic nervous system operates by means of *visceral reflexes.* That is, sensory signals from parts of the body send impulses into the centers of the cord, brain stem, or hypothalamus, and these in turn transmit appropriate reflex responses back to the visceral organs to control their activities.

The autonomic impulses are transmitted to the body through two major subdivisions called the *sympathetic* and *parasympathetic systems,* the characteristics and functions of which follow.

PHYSIOLOGIC ANATOMY OF THE SYMPATHETIC NERVOUS SYSTEM

Figure 59–1 illustrates the general organization of the sympathetic nervous system, showing one of the two *sympathetic chains* to the side of the spinal column and nerves extending to the different internal organs. The sympathetic nerves originate in the spinal cord between the segments T-1 and L-2.

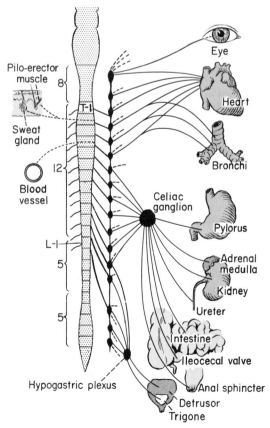

Figure 59–1   The sympathetic nervous system. Dashed lines represent postganglionic fibers in the gray rami leading into the spinal nerves for distribution to blood vessels, sweat glands, and pilo-erector muscles.

They begin in the *sympathetic motoneurons* of the *intermediolateral horns* of the spinal gray matter.

Preganglionic and Postganglionic Sympathetic Neurons. The sympathetic nerves are different from skeletal motor nerves in the following way: Each motor pathway to a skeletal muscle is comprised of a single fiber originating in the cord. Each sympathetic pathway is comprised of a *preganglionic neuron* and a *postganglionic neuron.* The cell body of the preganglionic neuron lies in the intermediolateral horn of the spinal cord, and its fiber passes, as illustrated in Figure 59–2, through an *anterior root* of the cord into a *spinal nerve* and finally through the *white ramus* from the spinal nerve to a *ganglion* of the *sympathetic chain.* Here the fiber either synapses immediately with postganglionic neurons or often passes on through the chain into one of its radiating nerves to synapse with postganglionic neurons in an outlying sympathetic ganglion. The fiber of each postganglionic neuron then travels to its destination in one of the organs.

Sympathetic Nerve Fibers in the Skeletal Nerves. Many of the fibers from the postganglionic neurons in the sympathetic chain pass back into the spinal nerves through *gray rami* (see Fig. 59–2) at all levels of the cord. These pathways are made up of type C fibers that extend to all parts of the body in the skeletal nerves. They control the blood vessels, sweat glands, and pilo-erector muscles of the hairs. Approximately 8 per cent of the fibers in the average skeletal nerve are sympathetic fibers, a fact that indicates their importance.

A few preganglionic neurons enter the spinal nerves directly from the spinal cord rather than passing first through the sympathetic chain; they synapse with postganglionic neurons located in the spinal nerves themselves. Therefore, destruction of the sympathetic chain does not remove all sympathetic activity, though it does remove at least 90 per cent of it.

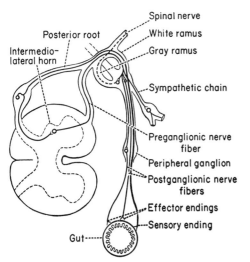

Figure 59–2 Nerve connections between the spinal cord, sympathetic chain, spinal nerves, and peripheral sympathetic nerves.

Spinal nerve
White ramus
Posterior root
Gray ramus
Intermedio-
lateral horn
Sympathetic chain
Preganglionic nerve fiber
Peripheral ganglion
Postganglionic nerve fibers
Effector endings
Sensory ending
Gut

Segmental Distribution of Sympathetic Nerves. The sympathetic pathways originating in the different segments of the spinal cord are not necessarily distributed to the same part of the body as the spinal nerve fibers from the same segments. Instead, the *sympathetic fibers from T-1 generally pass up the sympathetic chain into the head; from T-2 into the neck; T-3, T-4, T-5, and T-6 into the thorax; T-7, T-8, T-9, T-10, and T-11 into the abdomen; T-12, L-1, and L-2 into the legs.* This distribution is only approximate and overlaps greatly.

The distribution of sympathetic nerves to each organ is determined partly by the position in the embryo at which the organ originates. For instance, the heart receives many sympathetic nerves from the neck portion of the sympathetic chain because the heart originates in the neck of the embryo. Likewise, the abdominal organs receive their sympathetic innervation from the lower thoracic segments because the primitive gut originates in the lower thoracic area.

Special Nature of the Sympathetic Nerve Endings in the Adrenal Medullae. Preganglionic sympathetic nerve fibers pass, without synapsing, all the way from the intermediolateral horn cells of the spinal cord, through the sympathetic chains, through the splanchnic nerves, and finally into the adrenal medullae. There they end directly on special cells that secrete epinephrine and norepinephrine. These secretory cells are embryologically derived from nervous tissue and are analogous to postganglionic neurons; indeed, they even have rudimentary nerve fibers.

PHYSIOLOGIC ANATOMY OF THE PARASYMPATHETIC NERVOUS SYSTEM

The parasympathetic nervous system is illustrated in Figure 59–3, showing that parasympathetic fibers leave the central nervous system through several of the cranial nerves, the second and third sacral spinal nerves, and occasionally the first and fourth sacral nerves. Probably 80 per cent or more of all parasympathetic nerve fibers are in the vagus nerves, passing to the entire thoracic and abdominal regions of the body. However, the vagus nerve is not entirely parasympathetic in function because it also carries some voluntary skeletal nerve fibers from the nucleus ambiguus to the laryngeal and pharyngeal muscles and some afferent nerve fibers from the pressoreceptors of the arteries and from stretch receptors of the lungs to the medulla. Nevertheless, when a physiologist speaks of the parasympathetic nervous system, in the back of his mind he thinks mainly of the two vagus nerves. The vagus nerves supply parasympathetic nerves to the heart, the lungs, the esophagus, the stomach, the small intestine, the proximal half of the colon, the liver, the gallbladder, the pancreas, and the upper portions of the ureters.

Parasympathetic fibers in the *third nerve* flow to the pupillary sphincters and ciliary muscles of the eye. Fibers from the *seventh nerve* pass to the lacrimal, nasal, and submaxillary glands, and

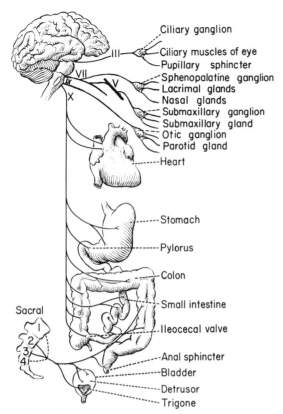

Ciliary ganglion
Ciliary muscles of eye
Pupillary sphincter
Sphenopalatine ganglion
Lacrimal glands
Nasal glands
Submaxillary ganglion
Submaxillary gland
Otic ganglion
Parotid gland
Heart
Stomach
Pylorus
Colon
Small intestine
Ileocecal valve
Anal sphincter
Bladder
Detrusor
Trigone
Sacral

Figure 59–3 The parasympathetic nervous system.

fibers from the *ninth nerve* pass to the parotid gland.

The sacral parasympathetic fibers congregate in the form of the two *nervi erigentes,* which leave the sacral plexus on each side of the cord and distribute their peripheral fibers to the descending colon, rectum, bladder and lower portions of the ureters. Also, this sacral group of parasympathetics supplies fibers to the external genitalia to cause various sexual reactions.

Preganglionic and Postganglionic Parasympathetic Neurons. The parasympathetic system, like the sympathetic, has both preganglionic and postganglionic neurons, but, except in the case of a few cranial parasympathetic nerves, the preganglionic fibers pass uninterrupted to the organ that is to be excited by parasympathetic impulses. In the wall of the organ are located the *postganglionic neurons* of the parasympathetic system. The preganglionic fibers synapse with these; and then short postganglionic fibers, 1 millimeter to several centimeters in length, leave the neurons to spread in the substance of the organ. This location of the parasympathetic postganglionic neurons in the visceral organ itself is quite different from the arrangement of the sympathetic ganglia, for the cell bodies of the sympathetic postganglionic neurons are always located in ganglia of the sympathetic chain or in various other discrete ganglia in the abdomen or thorax rather than in the excited organ itself.

BASIC CHARACTERISTICS OF SYMPATHETIC AND PARASYMPATHETIC FUNCTION

CHOLINERGIC AND ADRENERGIC FIBERS— SECRETION OF ACETYLCHOLINE AND NOREPINEPHRINE BY THE POSTGANGLIONIC NEURONS

It will be recalled from Chapter 6 that skeletal nerve endings secrete acetylcholine. This is also true of the preganglionic neurons of both the sympathetic and parasympathetic system, and it is true, too, of the parasympathetic postganglionic neurons. Therefore, all these fibers are said to be *cholinergic because they secrete acetylcholine at their nerve endings.*

A few of the postganglionic endings of the sympathetic nervous system also secrete acetylcholine, and these fibers, too, are cholinergic; but by far the majority of the sympathetic postganglionic endings secrete *norepinephrine.* These fibers are said to be *adrenergic,* a term derived from *noradrenalin,* which is another name for norepinephrine. Thus, there is a basic functional difference between the respective postganglionic neurons of the parasympathetic and sympathetic systems, one secreting acetylcholine and the other principally norepinephrine. The chemical structures of these substances are the following:

$$CH_3-\underset{\underset{O}{\|}}{C}-O-CH_2-CH_2-\underset{\underset{CH_3}{|}}{\overset{\overset{Cl}{|}}{N}}\diagdown\overset{CH_3}{\diagdown CH_3}$$

Acetylcholine chloride

Norepinephrine

Parasympathetic and Sympathetic Mediation by Acetylcholine and Norepinephrine. The acetylcholine and norepinephrine secreted by the postganglionic neurons act on the different organs to cause the respective parasympathetic or sympathetic effects. Therefore, these substances are called *parasympathetic* and *sympathetic mediators,* respectively, or sometimes *cholinergic* and *adrenergic* mediators.

Mechanism of Secretion of Acetylcholine and Norepinephrine by Nerve Endings. Electron micrographs show that small vesicles of acetylcholine are stored in parasympathetic nerve endings and vesicles of norepinephrine are stored in sympathetic nerve endings. Supposedly, nerve impulses cause short periods of increased permeability of the membrane at the fiber ending, thereby allowing sudden emptying of some of these vesicles.

Adrenergic nerves also secrete a minute amount of acetylcholine along with norepinephrine. If a drug such as guanethidine which blocks release of norepinephrine from adrenergic nerves is given to an animal, the nerve then secretes only acetylcholine when it is stimulated, and this causes parasympathetic effects rather than sympathetic effects. On the basis of this, it has been postulated that the nerve impulse causes norepinephrine release in the following manner: Supposedly the primary effect is the release of a minute amount of acetylcholine. The acetylcholine then acts on the stored vesicles of norepinephrine in the nerve endings to cause them to discharge norepinephrine into the surrounding fluids. The amount of norepinephrine released is very great in comparison with the minute amount of acetylcholine that supposedly causes its release, which means that the acetylcholine effect is normally too slight to be observed in comparison with the norepinephrine effect.

Destruction of Acetylcholine, and Duration of Its Action. The acetylcholine secreted by the parasympathetic nerve endings, like that also secreted at skeletal nerve endings, is destroyed by the enzyme *cholinesterase,* which is present in all the effector organs or the surrounding fluids. However, this destruction does not take place nearly so rapidly as it does in skeletal muscle fibers. The acetylcholine sometimes persists for as long as several seconds after its release and therefore has a long period of action.

Removal of Norepinephrine from the Tissues, and Duration of Its Action. Norepinephrine secreted by sympathetic nerve endings is removed from the tissues by (1) reabsorption into the sympathetic nerve endings themselves and (2) methylation caused by o-*methyl transferase,* an enzyme analogous to cholinesterase. Ordinarily, the action of norepinephrine secreted in the tissues lasts for only a few seconds, illustrating that its destruction or removal from the tissues is rapid. However, norepinephrine and epinephrine secreted into the blood by the adrenal medullae are not removed or destroyed significantly until they diffuse into tissues. Therefore, when secreted into the blood, these two hormones remain active for one to several minutes.

EXCITATORY AND INHIBITORY ACTIONS OF SYMPATHETIC AND PARASYMPATHETIC STIMULATION

Table 59–1 gives the effects on different visceral functions of the body of stimulating the parasympathetic and sympathetic nerves. From this table it can readily be seen that *sympathetic stimulation causes excitatory effects in some organs but inhibitory effects in others. Likewise, parasympathetic stimulation causes excitation in some organs but inhibition in others.* Also, when sympathetic stimulation excites a particular organ, parasympathetic stimulation often inhibits it, illustrating that the two systems occasionally act reciprocally to each other. However, most organs are dominantly controlled by one or the other of the two systems, so that, except in a few instances, the two systems do not actively oppose each other.

There is no generalization one can use to explain whether sympathetic or parasympathetic stimulation will cause excitation or inhibition of a particular organ. Therefore, to understand sympathetic and parasympathetic function, one must learn the functions of these two nervous systems as listed in Table 59–1. Some of these functions need to be clarified in still greater detail as follows:

EFFECTS OF SYMPATHETIC AND PARASYMPATHETIC STIMULATION ON SPECIFIC ORGANS

The Eye. Two functions of the eye are controlled by the autonomic nervous system: These are the pupillary opening and the focus of the lens. Sympathetic stimulation contracts the meridional *fibers of the iris* and, therefore, dilates the pupil, whereas parasympathetic stimulation contracts the *circular muscle of the iris* to constrict the pupil. The parasympathetics that control the pupil are reflexly stimulated when excess light enters the eyes; this reflex reduces the pupillary opening and decreases the amount of light that strikes the retina. On the other hand, the sympathetics become stimulated during periods of excitement and, therefore, increase the pupillary opening at these times.

Focusing of the lens is controlled entirely by the parasympathetic nervous system. The lens is normally held in a flattened state by tension of its radial ligaments. Parasympathetic excitation contracts the *ciliary muscle,* which releases this tension and allows the lens to become more convex.

TABLE 59–1.   Autonomic Effects on Various Organs of the Body

Organ	Effect of Sympathetic Stimulation	Effect of Parasympathetic Stimulation
Eye: Pupil	Dilated	Contracted
Ciliary muscle	None	Excited
Glands: Nasal	Vasoconstriction	Stimulation of thin, copious
Lacrimal		secretion containing many enzymes
Parotid		
Submaxillary		
Gastric		
Pancreatic		
Sweat glands	Copious sweating (cholinergic)	None
Apocrine glands	Thick, odoriferous secretion	None
Heart: Muscle	Increased rate	Slowed rate
	Increased force of beat	Decreased force of atrial beat
Coronaries	Vasodilated	Constricted
Lungs: Bronchi	Dilated	Constricted
Blood vessels	Mildly constricted	None
Gut: Lumen	Decreased peristalsis and tone	Increased peristalsis and tone
Sphincter	Increased tone	Decreased tone
Liver	Glucose released	None
Gallbladder and bile ducts	Inhibited	Excited
Kidney	Decreased output	None
Ureter	Inhibited	Excited
Bladder: Detrusor	Inhibited	Excited
Trigone	Excited	Inhibited
Penis	Ejaculation	Erection
Systemic blood vessels:		
Abdominal	Constricted	None
Muscle	Constricted (adrenergic)	None
	Dilated (cholinergic)	
Skin	Constricted (adrenergic)	Dilated
	Dilated (cholinergic)	
Blood: Coagulation	Increased	None
Glucose	Increased	None
Basal metabolism	Increased up to 150%	None
Adrenal cortical secretion	Increased	None
Mental activity	Increased	None
Piloerector muscles	Excited	None

This causes the eye to focus on objects near at hand. The focusing mechanism was discussed in Chapter 54 in relation to function of the eyes.

The Glands of the Body. The *nasal, lacrimal, salivary,* and many *gastrointestinal glands* are all strongly stimulated by the parasympathetic nervous system, resulting in copious quantities of secretion. The glands of the alimentary tract most strongly stimulated by the parasympathetics are those of the upper tract, including especially those of the mouth and stomach. The glands of the small and large intestines are controlled principally by local factors in the intestinal tract itself and not by the autonomic nerves.

Sympathetic stimulation has little direct effect on glandular secretion, but it does cause vasoconstriction of the blood vessels supplying the glands and in this way often reduces their rates of secretion.

The *sweat glands* secrete large quantities of sweat when the sympathetic nerves are stimulated, but no effect is caused by stimulating the parasympathetic nerves. However, the sympathetic fibers to most sweat glands are *cholinergic,*

in contrast to most other sympathetic fibers, which are adrenergic. Furthermore, the sweat glands are stimulated primarily by centers in the hypothalamus that are usually considered to be parasympathetic centers. Therefore, sweating could be called a parasympathetic function.

The *apocrine glands* secrete a thick, odoriferous secretion as a result of sympathetic stimulation, but they do not react to parasympathetic stimulation. Furthermore, the apocrine glands, despite their close embryological relationship to sweat glands, are controlled by adrenergic fibers rather than by cholinergic fibers and are controlled by the sympathetic centers of the central nervous system rather than by the parasympathetic centers.

The Gastrointestinal System. The gastrointestinal system has its own intrinsic set of nerves known as the *intramural plexus.* However, both parasympathetic and sympathetic stimulation can affect gastrointestinal activity—parasympathetic especially. Parasympathetic stimulation, in general, increases the overall degree of activity of the gastrointestinal tract by promoting peristalsis, thus allowing rapid propulsion of contents along the tract. This propulsive effect is associated with simultaneous increases in rates of secretion by many of the gastrointestinal glands, which was described above.

Normal function of the gastrointestinal tract is not very dependent on sympathetic stimulation. However, in some diseases, strong sympathetic stimulation inhibits peristalsis and increases the tone of the sphincters. The net result is greatly slowed propulsion of food through the tract.

The Heart. In general, sympathetic stimulation increases the overall activity of the heart. This is accomplished by increasing both the rate and force of heartbeat. Parasympathetic stimulation causes mainly the opposite effects, decreasing the overall activity of the heart. To express these effects in another way, sympathetic stimulation increases the effectiveness of the heart as a pump, whereas parasympathetic stimulation decreases its effectiveness. However, sympathetic stimulation unfortunately also greatly increases the metabolism of the heart while parasympathetic stimulation decreases its metabolism and allows the heart a certain degree of rest.

Systemic Blood Vessels. Most blood vessels, especially those of the abdominal viscera and the skin of the limbs, are constricted by sympathetic stimulation, whereas parasympathetic stimulation dilates vessels in certain restricted areas. In muscles, there are two types of sympathetic nerve fibers, adrenergic and cholinergic. The adrenergic fibers cause mild constriction of the vessels. On the other hand, the cholinergic fibers can strongly dilate the vessels during the early phases of muscular exercise. This was discussed in detail in Chapter 32.

EFFECT OF SYMPATHETIC AND PARASYMPATHETIC STIMULATION ON ARTERIAL PRESSURE. The arterial pressure in the circulatory system is caused by propulsion of blood by the heart and by resistance to flow of this blood through the vascular system. In general, sympathetic stimulation increases both propulsion by the heart and resistance to flow, which can cause the pressure to increase greatly.

On the other hand, parasympathetic stimulation decreases the pumping effectiveness of the heart, which lowers the pressure a moderate amount, though not nearly so much as the sympathetics can increase the pressure.

The Lungs. In general, the structures in the lungs do not have extensive sympathetic or parasympathetic innervation. Therefore, all effects of stimulation are mild. Sympathetic stimulation can mildly dilate the bronchi and mildly constrict the blood vessels. On the contrary, parasympathetic stimulation can cause mild constriction of the bronchi and can perhaps mildly dilate the vessels.

Effects of Sympathetic and Parasympathetic Stimulation on Other Functions of the Body. Because of the great importance of the sympathetic and parasympathetic control systems, these are discussed many times in this text in relation to a myriad of body functions that are not considered in detail

here. In general, most of the entodermal structures, such as the ducts of the liver, the gallbladder, the ureter, and the bladder, are inhibited by sympathetic stimulation but excited by parasympathetic stimulation. Sympathetic stimulation also has metabolic effects, causing release of glucose from the liver, increase in blood glucose concentrations, increase in basal metabolic rate, and increase in mental activity. Finally, the sympathetics and parasympathetics are involved in regulating the male and female sexual acts, as will be explained in Chapters 80 and 81.

FUNCTION OF THE ADRENAL MEDULLAE

Stimulation of the sympathetic nerves to the adrenal medullae causes large quantities of epinephrine and norepinephrine to be released into the circulating blood, and these two hormones in turn are carried in the blood to all tissues of the body. On the average, approximately 75 per cent of the secretion is epinephrine and 25 per cent norepinephrine, though the relative proportions of these change considerably under different physiological conditions.

The circulating norepinephrine has almost the same effects on the different organs as those caused by direct sympathetic stimulation, except that *the effects last about 10 times as long* because norepinephrine is removed from the blood slowly. For instance, norepinephrine causes constriction of essentially all the blood vessels of the body; it causes increased activity of the heart, inhibition of the gastrointestinal tract, dilatation of the pupil of the eye, and so forth.

Epinephrine, also, causes almost the same effects as those caused by norepinephrine, but the effects differ in the following respects: First, epinephrine has a greater effect on cardiac activity than norepinephrine. Second, epinephrine causes only weak constriction of the blood vessels of the muscles in comparison with a much stronger constriction that results from norepinephrine. Since the muscle vessels represent a major segment of all the vessels of the body, this difference is of special importance because norepinephrine greatly increases the total peripheral resistance and thereby greatly elevates arterial pressure, whereas epinephrine raises the arterial pressure to a much less extent but increases the cardiac output considerably because of its effect on the heart and veins.

A third difference between the action of epinephrine and norepinephrine relates to their effects on tissue metabolism. Epinephrine probably has several times as great a metabolic effect as norepinephrine. Indeed, the epinephrine secreted by the adrenal medullae increases the metabolic rate of the body often to as much as 100 per cent above normal, in this way increasing the activity and excitability of the whole body. It also increases the rate of other metabolic activities, such as glycogenolysis in the liver and glucose release into the blood.

In summary, stimulation of the adrenal medullae causes the release of hormones that have almost the same effects throughout the body as direct sympathetic stimulation, except that the effects are greatly prolonged. The only significant differences are caused by the epinephrine in the secretion, which increases the rate of metabolism and cardiac output to a greater extent than is caused by direct sympathetic stimulation.

Value of the Adrenal Medullae to the Function of the Sympathetic Nervous System. Usually, when any part of the sympathetic nervous system is stimulated, the entire system, or at least major portions of it, are stimulated at the same time. Therefore, norepinephrine and epinephrine are almost always released by the adrenal medullae at the same time that the different organs are being stimulated directly by the sympathetic nerves. Therefore, the organs are actually stimulated in two different ways simultaneously, directly by the sympathetic nerves and indirectly by the medullary hormones. The two means of stimulation support each other, and either can actually substitute for the other. For instance, destruction of the direct sympathetic pathways to the organs does not abrogate excitation of the organs because norepinephrine and epinephrine are still released into the circulating fluids and indirectly cause stimulation. Likewise, total loss of the two adrenal medullae usually has little significant effect on the operation of the sympathetic nervous system because the direct pathways can still perform almost all the necessary duties. Thus, the dual mechanism of sympathetic stimulation provides a safety factor, one mechanism substituting for the other when the second is missing.

Another important value of the adrenal medullae is the capability of epinephrine and norepinephrine to stimulate structures of the body that are not innervated by direct sympathetic fibers. For instance, the metabolic rate of every cell of the body is increased by these hormones, especially by epinephrine, even though only a small proportion of all the cells in the body are innervated by sympathetic fibers.

"Alpha" and "Beta" Adrenergic Receptors. To explain the differences in action of norepinephrine and epinephrine on different sympathetic effector organs, it has been postulated that two different types of receptor substances exist in the different effector cells, one called *alpha receptor substance* and the other *beta receptor substance. Supposedly, norepinephrine can affect only those effector cells that contain the alpha receptor, while epinephrine can affect both.*

Table 59–2 gives the postulated distribution of alpha and beta receptors in organs that are controlled by the sympathetics. Note that certain alpha functions are excitatory while others are inhibitory. Likewise, certain beta functions are excitatory and others inhibitory. Therefore, alpha and beta receptors are not necessarily associated with excitation or inhibition but simply with the affinity of the hormone for the effector organ.

Another hormone closely similar to epinephrine and norepinephrine, *isopropyl norepinephrine,* has an extremely strong action on beta functions but almost no action on alpha functions. Furthermore, small amounts of this third hormone have been reported to be present in the adrenal medullae and possibly are secreted into the blood along with norepinephrine and epinephrine.

Unfortunately, the alpha and beta receptor concept has not been able to explain all the differences between the actions of the different sympathomimetic hormones.

RELATIONSHIP OF STIMULUS RATE TO DEGREE OF SYMPATHETIC AND PARASYMPATHETIC EFFECT

A special difference between the autonomic nervous system and the skeletal nervous system is the low frequency of stimulation required for full activation of autonomic effectors. In general, only one impulse every few seconds suffices to maintain normal sympathetic or parasympathetic effect, and full activation occurs when the nerve fibers discharge 10 to 15 times per second. This compares with full activation in the skeletal nervous system at about 75 to 200 impulses per second.

SYMPATHETIC AND PARASYMPATHETIC "TONE"

The sympathetic and parasympathetic systems are continually active, and the basal rates

TABLE 59–2.   Relationships of Adrenergic Receptors to Function

Alpha Receptor	Beta Receptor
Vasoconstriction	Vasodilatation (muscle)
Cardioacceleration	Cardioacceleration
Iris dilatation	Increased myocardial strength
Intestinal relaxation	Myometrial relaxation
Pilomotor contraction	Bronchial relaxation

of activity are known, respectively, as *sympathetic tone* or *parasympathetic tone*.

The value of tone is that it allows a single nervous system to increase or to decrease the activity of a stimulated organ. For instance, sympathetic tone normally keeps almost all the blood vessels of the body constricted to approximately half their maximum diameter. By increasing the degree of sympathetic stimulation, the vessels can be constricted even more; but, on the other hand, by inhibiting the normal tone, the vessels can be dilated. If it were not for the continual sympathetic tone, the sympathetic system could cause only vasoconstriction, never vasodilatation.

Another interesting example of tone is that of the parasympathetics in the gastrointestinal tract. Surgical removal of the parasympathetics can cause serious and prolonged gastric and intestinal "atony," thus illustrating that in normal function the parasympathetic tone to the gut is strong. This tone can be decreased by the brain, thereby inhibiting gastrointestinal motility, or, on the other hand, it can be increased, thereby promoting increased gastrointestinal activity.

Tone Caused by Basal Secretion of Norepinephrine and Epinephrine by the Adrenal Medullae. The normal resting rate of secretion by the adrenal medullae is about 0.2 μgm./kg./min. of epinephrine and about 0.07 μgm./kg./min. of norepinephrine. These quantities are considerable—indeed, enough to maintain the blood pressure almost up to the normal value even if all direct sympathetic pathways to the cardiovascular system are removed. Therefore, it is obvious that much of the overall tone of the sympathetic nervous system results from basal secretion of epinephrine and norepinephrine in addition to that which results from direct sympathetic stimulation.

Effect of Loss of Sympathetic or Parasympathetic Tone Following Denervation. Immediately after a sympathetic or parasympathetic nerve is cut, the innervated organ loses its sympathetic or parasympathetic tone. In the case of the blood vessels, for instance, cutting the sympathetic nerves results immediately in almost maximal vasodilatation. However, over several days or weeks, the *intrinsic tone* in the smooth muscle of the vessels increases, usually restoring almost normal vasoconstriction.

Essentially the same events occur in any organ whenever sympathetic or parasympathetic tone is lost. That is, compensation soon develops to return the function of the organ almost to its normal basal level. However, the compensation sometimes requires many

months. For instance, loss of parasympathetic tone to the heart increases the heart rate from 90 to about 160 beats per minute in a dog, and this will still be about 120 six months later.

DENERVATION HYPERSENSITIVITY OF SYMPATHETIC AND PARASYMPATHETIC ORGANS FOLLOWING DENERVATION

During the first week or so after a sympathetic or parasympathetic nerve is destroyed, the innervated organ becomes more and more sensitive to injected norepinephrine or acetylcholine, respectively. This effect is illustrated in Figure 59–4; the blood flow in the forearm before removal of the sympathetics was 200 ml. per minute, and a test dose of norepinephrine caused only a slight depression in flow. Then the stellate ganglion was removed, and normal sympathetic tone was lost. At first, the blood flow rose markedly because of the lost vascular tone, but over a period of days to weeks the blood flow returned almost to normal because of progressive increase in intrinsic tone of the vascular musculature itself, thus compensating for the loss of sympathetic tone. Another test dose of norepinephrine was then administered and the blood flow decreased much more than before, illustrating that the blood vessels had become about two to four times as responsive to norepinephrine as previously. This phenomenon is called denervation hypersensitivity. It occurs in both sympathetic and parasympathetic organs and to a far greater extent in some organs than in others, often increasing the response as much as 10-fold.

Mechanism of Denervation Hypersensitivity. The precise means by which denervation hypersensitivity develops is not known, though several suggestions have been made. One of these is that prolonged lack of nerve stimulation allows the quantity of "receptor substances" in the membranes of the effector cells to increase, thereby increasing the responsiveness of the cells to the circulating norepinephrine, epinephrine, or acetylcholine.

THE AUTONOMIC REFLEXES

It is mainly by means of autonomic reflexes that the autonomic nervous system regulates visceral functions. Throughout this text the functions of these reflexes are discussed in detail in relation to individual organs, but, to illustrate their importance, a few are presented here briefly.

Cardiovascular Autonomic Reflexes. Several reflexes in the cardiovascular system help to control the arterial blood pressure, cardiac output, and heart rate. One of these is the *baroreceptor reflex*, which was described in Chapter 26 along with other cardiovascular reflexes. Briefly, stretch receptors called *baroreceptors* are located in the walls of the major

Figure 59–4 Effect of sympathectomy on blood flow in the arm, and the effect of a test dose of norepinephrine before and after sympathectomy, showing sensitization of the vasculature to norepinephrine.

arteries, including the carotid arteries and the aorta. When these become stretched by high pressure, impulses are transmitted to the brain stem, where they inhibit the sympathetic centers. This results in decreased sympathetic impulses to the heart and blood vessels, which allows the arterial pressure to fall back toward normal.

The Gastrointestinal Autonomic Reflexes. The uppermost part of the gastrointestinal tract and the rectum are controlled principally by autonomic reflexes. For instance, the smell of appetizing food initiates impulses from the nose to the vagal, glossopharyngeal, and salivary nuclei of the brain stem. These in turn transmit impulses through the parasympathetic nerves to the secretory glands of the mouth and stomach, causing secretion of digestive juices even before food enters the mouth. And when fecal matter fills the rectum at the other end of the alimentary canal, sensory impulses initiated by stretching the rectum are sent to the sacral portion of the spinal cord, and a reflex signal is retransmitted through the parasympathetics to the distal parts of the colon; these result in strong peristaltic contractions that empty the bowel.

Other Autonomic Reflexes. Emptying of the bladder is also controlled in the same way as emptying of the rectum; stretching the bladder sends impulses to the sacral cord, and this in turn causes contraction of the bladder as well as relaxation of the urinary sphincters, thereby promoting micturition.

Also important are the sexual reflexes which are initiated both by psychic stimuli from the brain and stimuli from the sexual organs. Impulses from these sources converge on the sacral cord and, in the male, result, first, in erection and then in ejaculation.

Other autonomic reflexes include reflex regulation of pancreatic secretion, gallbladder emptying, urinary excretion, sweating, blood glucose concentration, and many other visceral functions, all of which are discussed in detail at many points throughout this text.

MASS DISCHARGE OF SYMPATHETIC SYSTEMS VERSUS DISCRETE CHARACTERISTICS OF PARASYMPATHETIC REFLEXES

In general, large portions of the sympathetic nervous system often become stimulated simultaneously, a phenomenon called *mass discharge*. This characteristic of sympathetic action is in keeping with the usually diffuse nature of sympathetic function, such as overall regulation of arterial pressure or of metabolic rate.

However, in a few instances sympathetic activity does occur in isolated portions of the system without involving the entire system. The most important of these are: (1) In the process of heat regulation, the sympathetics control sweating and blood flow in the skin without affecting other organs innervated by the sympathetics. (2) During muscular activity, the cholinergic vasodilator fibers of the skeletal muscles are stimulated independently of all the remainder of the sympathetic system. (3) Many "local reflexes" involving the spinal cord but not the higher nervous centers affect local areas. For instance, heating a local skin area causes local vasodilatation and sweating while cooling causes vasoconstriction.

In contrast to the sympathetic system, most reflexes of the parasympathetic system are very specific. For instance, parasympathetic cardiovascular reflexes usually act only on the heart to increase or decrease its activity. Likewise, parasympathetic reflexes frequently cause secretion only in the mouth or, in other instances, secretion only by the stomach glands. Finally, the rectal emptying reflex does not affect other parts of the bowel to a major extent.

Yet there is often association between closely allied parasympathetic functions. For instance, though salivary secretion can occur independently of gastric secretion, these two often also occur together, and pancreatic secretion frequently occurs at the same time. Also, the rectal emptying reflex often initiates a bladder emptying reflex, resulting in simultaneous emptying of both the bladder and rectum. Conversely, the bladder emptying reflex can help to initiate rectal emptying.

"ALARM" OR "STRESS" FUNCTION OF THE SYMPATHETIC NERVOUS SYSTEM

From the preceding discussion of the functions of the sympathetic nervous system one can readily understand that mass sympathetic discharge results in increased activity of many functions of the body, including increase in arterial pressure, increase in blood supply to the tissues, in rate of cellular metabolism throughout the body, blood glucose concentration, rate of blood coagulation, and even mental activity. The sum of these effects permits the person to perform strenuous physical activity to a far greater extent than would otherwise be possible. And, since it is physical stress that usually excites the sympathetic system, it is frequently said that the purpose of the sympathetics is to provide extra energy for the body in states of stress, and this is often called the sympathetic *alarm reaction,* or *stress reaction.*

PHARMACOLOGY OF THE AUTONOMIC NERVOUS SYSTEM

DRUGS THAT ACT ON ADRENERGIC EFFECTOR ORGANS — THE SYMPATHOMIMETIC DRUGS

From the foregoing discussion, it is obvious that intravenous injection of norepinephrine causes essentially the same effects throughout the body as sympathetic stimulation. Therefore, norepinephrine is called a *sympathomimetic, or adrenergic, drug.* Other sympathomimetic drugs include epinephrine, phenylephrine, methoxamine, and many others. These differ from each other in the degree to which they stimulate different sympathetic effector organs and in their duration of action. Norepinephrine and epinephrine have actions as short as one to three minutes, while the actions of most other commonly used sympathomimetic drugs last 30 minutes to 2 hours.

Drugs that Cause Release of Norepinephrine from Nerve Endings. Certain drugs have a sympathomimetic action in an indirect manner rather than by directly exciting adrenergic effector organs. These drugs include ephedrine, tyramine, and amphetamine. Their effect is to cause release of norepinephrine from its storage vesicles in the sympathetic nerve endings. The norepinephrine in turn causes the sympathetic effects.

Drugs that Block Adrenergic Activity. Adrenergic activity can be blocked at several different points in the stimulatory process as follows:

1. The synthesis and storage of norepinephrine in the sympathetic nerve endings can be prevented. The best known drug that causes this effect is *reserpine.*

2. Release of norepinephrine from the sympathetic endings can be blocked. This is caused by *guanethidine.*

3. The alpha receptors can be blocked. Two drugs that cause this effect are *Dibenamine and phenoxybenzamine.*

4. The beta receptors can be blocked. Drugs that do this are *dichloroisoproterenol* and *nethalide.*

5. Sympathetic activity can be blocked by drugs that block transmission of nerve impulses through the autonomic ganglia. These are discussed in the following section, but the most important drug for blockade of both sympathetic and parasympathetic transmission through the ganglia is *hexamethonium.*

DRUGS THAT ACT ON CHOLINERGIC EFFECTOR ORGANS

Parasympathomimetic Drugs (Muscarinic Drugs). Acetylcholine injected intravenously usually does not cause exactly the same effects throughout the body as parasympathetic stimulation because the acetylcholine is destroyed by cholinesterase in the blood and body fluids before it can reach all the effector organs. Yet a number of other drugs that are not so rapidly destroyed can produce typical parasympathetic effects, and these are called parasympathomimetic drugs. They are also called muscarinic drugs, because *muscarine,* which is derived from the toadstool, was among the first drugs of this type to be discovered.

Three commonly used parasympathomimetic drugs are *pilocarpine, methacholine,* and *carbamylcholine.* Although these drugs have almost the same effects on the different effector organs as does parasympathetic stimulation, there are slight variations. For instance, carbamylcholine is especially active on the bladder and gastrointestinal tract, while methacholine is especially active on the cardiovascular system and eyes.

Parasympathomimetic drugs act on the effector organs of cholinergic *sympathetic* fibers also. For instance, these drugs cause profuse sweating. Also, they cause vascular dilatation, this effect occurring even in vessels not innervated by cholinergic fibers.

Drugs that Have a Parasympathetic Potentiating Effect. Some drugs do not have a direct effect on parasympathetic effector organs but do potentiate the effects of the naturally secreted acetylcholine at the parasympathetic endings. These are the same drugs as those listed in Chapter 6 that potentiate the effect of acetylcholine at the neuromuscular junction — that is, *neostigmine, physostigmine,* and *diisopropyl fluorophosphate.* These inhibit cholinesterase, thus preventing rapid destruction of the acetylcholine liberated by the parasympathetic nerve endings. As a consequence, the quantity of acetylcholine acting on the effector organs progressively increases with successive stimuli, and the degree of action also increases.

Drugs that Block Cholinergic Activity at Effector Organs. *Atropine* and similar drugs, such as *homatropine* and *scopolamine,* block the action of acetylcholine on cholinergic effector organs. However, these drugs do not affect the action of acetylcholine on the postganglionic neurons or on skeletal muscle.

DRUGS THAT STIMULATE THE POST-GANGLIONIC NEURONS—"NICOTINIC DRUGS"

The preganglionic neurons of both the parasympathetic and sympathetic nervous systems secrete acetylcholine at their endings, and the acetylcholine in turn stimulates the postganglionic neurons. Therefore, injected acetylcholine can also stimulate the postganglionic neurons of both systems, thereby causing both sympathetic and parasympathetic effects in the body. *Nicotine* is a drug that can also stimulate postganglionic neurons in the same manner as acetylcholine, but it cannot directly stimulate the autonomic effector organs. Therefore, drugs that cause autonomic effects by stimulating the postganglionic neurons are frequently called nicotinic drugs. Some drugs, such as *acetylcholine* itself, *carbamylcholine* and *methacholine,* have both nicotinic and muscarinic actions.

Nicotine excites both the sympathetic and parasympathetic systems at the same time, resulting in strong sympathetic vasoconstriction in the abdominal organs and limbs, but at the same time resulting in parasympathetic effects, such as increased gastrointestinal activity and, sometimes, slowing of the heart.

Nicotinic drugs also stimulate skeletal muscle as well as postganglionic neurons. Because of this, it is believed that the receptor system in the postganglionic neurons is similar to that in skeletal muscle fibers.

Ganglionic Blocking Drugs. Many important drugs that block impulse transmission from the preganglionic neurons to the postganglionic neurons have been discovered in the past few years, including *tetraethyl ammonium ion,* *hexamethonium ion,* and *pentolinium.* These inhibit impulse transmission in both the sympathetic and parasympathetic systems simultaneously. They are often used for blocking sympathetic activity but rarely for blocking parasympathetic activity because the sympathetic blockade usually far overshadows the effects of parasympathetic blockade. The ganglionic blocking drugs have been especially important in reducing arterial pressure in patients with hypertension.

MEDULLARY, PONTINE, AND MESENCEPHALIC CONTROL OF THE AUTONOMIC NERVOUS SYSTEM

Many areas in the reticular substance of the medulla, pons, and mesencephalon, as well as many special nuclei, (see Fig. 59–5) control different autonomic functions such as arterial pressure, heart rate, glandular secretion in the upper part of the gastrointestinal tract, gastrointestinal peristalsis, the degree of contraction of the urinary bladder, and many others. The control of each of these is discussed in detail at appropriate points in this text. Suffice it to point out here that the most

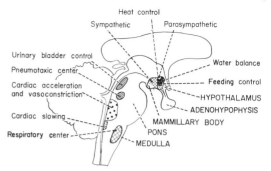

Figure 59–5 Autonomic control centers of the brain stem.

important factors controlled in the lower brain stem are arterial pressure and heart rate. Indeed, transection of the brain stem at the midpontile level allows normal basal control of arterial pressure to continue as before, though transection immediately below the medulla causes the arterial pressure to fall to about one-half normal. Closely associated with the cardiovascular regulatory centers in the medulla is the medullary center for regulation of respiration. Though this is not considered to be an autonomic function, it is one of the *involuntary* functions of the body.

Control of Lower Brain Stem Autonomic Centers by Higher Areas. Signals from the hypothalamus and even from the cerebrum can affect the activities of almost all the lower brain stem autonomic control centers. For instance, stimulation in appropriate areas of the hypothalamus can activate the medullary cardiovascular control centers strongly enough to increase the arterial pressure to more than double normal. Likewise, higher centers can transmit signals into the lower centers to increase or decrease salivation, gastrointestinal activity, or urinary bladder contraction. Therefore, to a great extent the autonomic centers in the lower brain stem are relay stations for control activities integrated at higher levels of the brain.

VEGETATIVE FUNCTIONS OF THE HYPOTHALAMUS

The hypothalamus is one of the most important areas of the brain for control of autonomic and other involuntary functions of the body that are collectively called vegetative functions, meaning by this term the involuntary functions of the body that are necessary for living. The different hypothalamic centers are so important in the control of bodily activities that they are discussed in detail throughout the text in association with specific functional systems, such as in connection with arterial pressure regulation in Chapter 26, thirst and water conservation in Chapter 36, and temperature regulation in Chapter 71. However, to illustrate the organization of the hypothalamus as a functioning unit, we will summarize some of its vegetative functions here.

Figure 59-6 summarizes most of the vegetative functions of the hypothalamus and shows the major nuclei or major areas which when stimulated control respective vegetative activities. A word of caution must be issued in studying this diagram, however, for the areas that cause specific activities are not nearly so discrete as shown in the figure. The illustrated areas are merely those from which the functions are likely to be elicited. Also, it is not known whether the effects noted in the figure result from activation of specific control nuclei or whether they result merely from activation of fiber tracts leading from control nuclei located elsewhere. For instance, stimulation of the posterior hypothalamus, through which most of the fiber pathways from the anterior hypothalamus must pass, can at times elicit many or most functions believed to be controlled by nuclei in the anterior hypothalamus. With this caution in mind, we can give the following general description of the vegetative control functions of the hypothalamus.

Cardiovascular Regulation. Stimulation of widespread areas throughout the hypothalamus can cause every known type of neurogenic effect in the cardiovascular system, including increased arterial pressure, decreased arterial pressure, increased heart rate, and decreased heart rate. In general, stimulation in the *posterior* and *lateral hypothalamus* increases the arterial pressure and heart rate, while stimulation in the *preoptic area* in the anterior hypothalamus has exactly the opposite effects, causing marked decrease in both heart rate and arterial pressure. These effects are transmitted mainly through the cardiovascular control centers in the reticular substance of the medulla and pons.

Regulation of Body Temperature. Large areas in the anterior hypothalamus, including especially the *preoptic area,* are concerned with regulation of body temperature. An increase in temperature of the blood flowing through these areas increases their activity, while a decrease in temperature decreases their activity. In turn, these areas control the mechanisms for increasing or decreasing body temperature, as discussed in Chapter 71.

Regulation of Body Water. The hypothalamus regulates body water in two separate ways, by (1) creating the sensation of thirst, which makes an animal drink water, and (2) controlling the excretion of water into the urine. A small area called the *thirst center* is located in the vicinity of the *paraventricular nucleus* and the *ventromedial nucleus.* When the electrolytes inside the neurons of this small center become too concentrated, the animal develops an intense desire to drink water; he will search out the nearest source of water and drink enough to return the electrolyte concentration of the thirst center to normal.

Control of renal excretion of water is vested mainly in the *supraoptic nuclei* and to a lesser extent in the neurons of the *paraventricular nucleus.* When the body fluids become too concentrated, the neurons of these areas become stimulated and secrete a hormone called *antidiuretic hormone.* This hormone acts on the distal tubules and collecting ducts of the kidneys to cause massive reabsorption of water, thereby decreasing the loss of water into the urine.

Gastrointestinal and Feeding Regulation. Excitation of several areas of the hypothalamus causes an animal to experience extreme hunger, a voracious appetite, and an intense desire to search for food. The two areas that, when excited, are most associated with hunger are the *perifornical nucleus* and the *lateral hypothalamic area.* On the other hand, damage to these areas of the hypothalamus causes the animal to lose all desire for food; he literally wastes away.

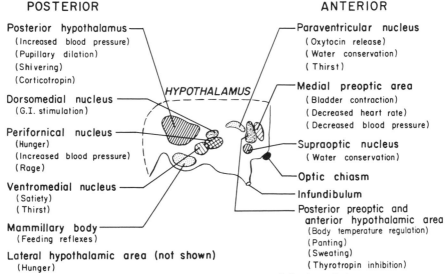

Figure 59-6 Autonomic control centers of the hypothalamus.

A center that opposes the desire for food, called the *satiety center,* is located in the *ventromedial nucleus.* When this center is stimulated, an animal that is eating food suddenly stops eating and shows complete indifference to the food. On the other hand, if this area is destroyed, the animal can never be satiated, but, instead, his hypothalamic hunger centers become overactive so that he has a voracious appetite at all times, resulting in tremendous obesity.

Another area of the hypothalamus that enters into the overall control of gastrointestinal activity is the *dorsomedial nucleus* which, when stimulated, increases the degree of peristalsis in the gut and also increases glandular secretion. Also, the *mammillary body* enters into many feeding reflexes, such as licking the lips and swallowing.

Excitement and Rage Mechanism of the Hypothalamus. Stimulation of certain areas of the hypothalamus, particularly in the region of the *perifornical nucleus,* causes an animal to become extremely excited, plus development of elevated blood pressure, dilated pupils, and symptoms of rage, such as hissing, arching his back, and assuming a stance ready to attack. Thus, the hypothalamus plays a major role in the overall behavior of the animal, a subject that will be discussed in much more detail in Chapter 62.

Hypothalamic Control of Endocrine Functions. Stimulation of certain areas of the hypothalamus also causes the adenohypophysis to secrete its hormones. This subject will be discussed in detail in Chapter 75 in relation to the overall control of the endocrine glands, but, briefly, the basic mechanism of the control of the adenohypophysis is the following:

The adenohypophysis receives its blood supply mainly from veins that flow into the adenohypophyseal sinuses from the lower part of the hypothalamus. As the blood courses through the hypothalamus before reaching the adenohypophysis, special *neurosecretory substances* are secreted into the blood by various hypothalamic nuclei. They are then transported in the blood to the adenohypophysis where they act on the glandular cells to cause release of the adenohypophyseal hormones.

Stimulation of the following areas of the hypothalamus causes secretion of specific hormones by the adenohypophysis: (1) preoptic and anterior hypothalamic areas, thyrotropin; (2) posterior hypothalamus and median eminence of the infundibulum, corticotropin; (3) median eminence, follicle stimulating hormone; and (4) anterior hypothalamus, luteinizing hormone.

Summary. A number of discrete areas of the hypothalamus have now been found that control specific vegetative functions. However, these areas overlap each other tremendously, so much so that the above separation of different areas for different hypothalamic functions is partially artificial.

REFERENCES

Adams, D. B.; Baccelli, G.; Mancia, G.; and Zanchetti, A.: Cardiovascular changes during naturally elicited fighting behavior in the cat. *Amer. J. Physiol., 216*:1226, 1969.

Axelrod, J.: The fate of norepinephrine and the effect of drugs. *Physiologist, 11*:63, 1968.

Bleier, R.: The Hypothalamus of the Cat. Baltimore, The Johns Hopkins Press, 1961.

Brown-Grant, K., and Cross, B. A.: Recent studies on the hypothalamus. *Brit. Med. Bull., 22*:195, 1966.

Burn, J. H.: The Autonomic Nervous System: For Students of Physiology and of Pharmacology. 2nd ed., Philadelphia, F. A. Davis Co., 1965.

Burnstock, G.: The autonomic neuromuscular junction. *Proc. Int. Union of Physiol. Sciences, 6*:7, 1968.

Burnstock, G., and Holman, M. E.: Smooth muscle: autonomic nerve transmission. *Ann. Rev. Physiol., 25*:61, 1963.

Cross, B. A.: The hypothalamus in mammalian homeostasis (symposium). *Soc. Exp. Biol., 18*:157, 1964.

Emmelin, N.: Nervous control of salivary glands. In Handbook of Physiology. Baltimore, The Williams & Wilkins Co., 1967, Sec. VI, Vol. II, p. 595.

Engelman, K., and Portnoy, B.: A sensitive double-isotope derivative assay for norepinephrine and epinephrine: normal resting human plasma levels. *Circ. Res., 26*:53, 1970.

Ferry, C. B.: Cholinergic link hypothesis in adrenergic neuro-effector transmission. *Physiol. Rev., 46*:420, 1966.

Gebber, G. L., and Snyder, D. W.: Hypothalamic control of baroreceptor reflexes. *Amer. J. Physiol., 218*:124, 1970.

Gillespie, J. S.: Adrenergic mechanisms. *Proc. Int. Union of Physiol. Sciences, 6*:107, 1968.

Guyton, A. C., and Gillespie, W. M., Jr.: Constant infusion of epinephrine: rate of epinephrine secretion and destruction in the body. *Amer. J. Physiol., 165*:319, 1951.

Guyton, A. C., and Reeder, R. C.: Quantitative studies on the autonomic actions of curare. *J. Pharmacol. Exp. Therap., 98*: 188, 1950.

Hess, W. R.: The Functional Organization of the Diencephalon. New York, Grune & Stratton, Inc., 1958.

Ingram, R. H., Jr.; Szidon, J. P.; and Fishman, A. P.: Response of the main pulmonary artery of dogs to neuronally released versus blood-borne norepinephrine, *Circ. Res., 26*:249, 1970.

Kosterlitz, H. W.: Intrinsic and extrinsic nervous control of motility of the stomach and the intestines. In Handbook of Physiology. Baltimore, The Williams & Wilkins Co., 1968, Sec. VI, Vol. IV, p. 2147.

Malmejac, J.: Activity of the adrenal medulla and its regulation. *Physiol. Rev., 44*:186, 1964.

Mauskopf, J. M.; Gray, S. D.; and Renkin, E. M.: Transient and persistent components of sympathetic cholinergic vasodilation. *Amer. J. Physiol., 216*:92, 1969.

Ninomiya, I.; Judy, W. V.; and Wilson, M. F.: Hypothalamic stimulus effects on sympathetic nerve activity. *Amer. J. Physiol., 218*:453, 1970.

Nishi, S.; Soeda, H.; and Koketsu, K.: Release of acetylcholine from sympathetic preganglionic nerve terminals. *J. Neurophysiol., 30*:114, 1967.

Priola, D. V., and Fulton, R. L.: Positive and negative inotropic responses of the atria and ventricles to vagosympathetic stimulation in the isovolumic canine heart. *Circ. Res., 25*:265, 1969.

Root, W. S., and Hofmann, F. G. (eds.): Physiological Pharmacology. New York, Academic Press, Inc., 1967, Vol. 3, Pts. C and D.

Soderberg, U.: Neurophysiological aspects of homeostasis. *Ann. Rev. Physiol., 26*:271, 1964.

Taylor, D. B., and Nedergaard, O. A.: Relation between structure and action of quaternary ammonium neuromuscular blocking agents. *Physiol. Rev., 45*:523, 1965.

von Euler, U. S.: Adrenergic transmitter granules. *Proc. Int. Union of Physiol. Sciences, 6*:95, 1968.

von Euler, U. S.: Neurotransmission in the adrenergic nervous system. *Harvey Lect., 55*:43, 1961.

von Euler, U. S.: Noradrenaline. Springfield, Illinois, Charles C Thomas, Publisher, 1956.

CHAPTER 60

The Reticular Activating System: Wakefulness, Sleep, Attention, Brain Waves, and Epilepsy

In the previous chapters of this section on neurophysiology we have discussed, first, the sensory mechanisms of the nervous system and, second, the motor mechanisms of the nervous system. In the present chapter we will consider the functions of the *reticular activating system,* a system that controls the overall degree of central nervous system activity, including control of wakefulness and sleep, and control of at least part of our ability to direct attention toward specific areas of our conscious minds.

Figure 60–1A illustrates the extent of this system, showing that it begins in the lower brain stem and extends upward through the mesencephalon and thalamus to be distributed throughout the cerebral cortex. Impulses are transmitted from the ascending reticular activating system to the cortex by two different pathways. One pathway passes upward from the brain stem portion of the reticular formation to the *subthalamus* and adjacent structures and thence to the cortex. The other pathway passes upward from the mesencephalon mainly through the intralaminar and midline nuclei of the thalamus to the anteroventral nuclei and reticular nuclei of the thalamus and finally through diverse pathways to essentially all parts of the cerebral cortex.

FUNCTION OF THE RETICULAR ACTIVATING SYSTEM IN WAKEFULNESS

Diffuse electrical stimulation in the *mesencephalic and upper pontile portion of the reticular formation*—an area also called the *bulboreticular facilitatory area* and discussed in Chapter 57 in relation to the motor functions of the nervous system—causes immediate and marked activation of the cerebral cortex and even causes a sleeping animal to awaken instantaneously. Furthermore, when this brain stem portion of

the reticular formation is damaged severely, as occurs (a) when a *brain tumor* develops in this region, (b) when serious *hemorrhage* occurs, or (c) in diseases such as *encephalitis lethargica* (sleeping sickness), the person passes into coma and is completely nonsusceptible to normal awakening stimuli.

Function of the Brain Stem Portion of the Reticular Activating System. Electrical stimuli applied to different portions of the reticular activating system have shown that the brain stem portion functions quite differently from the thalamic portion. Electrical stimulation of the brain stem portion causes generalized activation of the entire brain, including activation of the cerebral cortex, thalamic nuclei, basal ganglia, hypothalamus, other portions of the brain stem, and even the spinal cord. Furthermore, once the brain stem portion is stimulated, the degree of activation throughout the nervous system remains high for as long as a half minute or more after the stimulation is over. Therefore, *it is believed that the brain stem portion of the reticular activating system is basically responsible for normal wakefulness of the brain.*

The brain stem portion of the reticular activating system can transmit signals to the cerebral cortex through two separate pathways: (1) a more direct pathway through the region of the subthalamus, and (2) an indirect pathway through many successive synapses in the thalamus. It is assumed that the direct pathway allows immediate activation of cortical function. On the other hand, transmission of signals through the indirect thalamic pathway is considerably delayed, and the signals are believed to stimulate some areas of the cortex more than others because of intrathalamic selection of pathways, as is explained in the following paragraphs.

A

B

Figure 60–1 (A) The ascending reticular activating system schematically projected on a monkey brain. (From Lindsley: Reticular Formation of the Brain. Little, Brown and Co.) (B) Convergence of pathways from the cerebral cortex and from the spinal afferent systems on the reticular activating system. (From French, Hernandez-Peon, and Livingston: *J. Neurophysiol., 18:*74, 1955.)

Function of the Thalamic Portion of the Reticular Activating System. Electrical stimulation in different areas of the thalamic portion of the reticular activating system activates specific regions of the cerebral cortex more than others. This is distinctly different from stimulation in the mesencephalic portion which activates all the brain at the same time. Therefore, it is believed that the thalamic portion of the reticular activating system has two specific functions: first, it relays some of the diffuse facilitatory signals from the mesencephalic portion to all parts of the cerebral cortex to cause generalized activation of the cerebrum, and, second, stimulation of selected points in the thalamic activating system causes specific activation of certain areas of the cerebral cortex in distinction to the other

areas. This selective activation of specific cortical areas probably plays an important role in our ability to direct our attention to certain parts of our mental activity, which is discussed later in the chapter.

THE AROUSAL REACTION — SENSORY ACTIVATION OF THE RETICULAR ACTIVATING SYSTEM

When an animal is asleep, the reticular activating system is mainly dormant. Yet almost any type of sensory signal can immediately activate the system. For instance, proprioceptive signals from the joints and possibly the muscle receptors as well, pain impulses from the skin, visual signals from the eyes, auditory signals from the ears, or even visceral sensations from the gut can all cause sudden activation of the reticular activating system and therefore arouse the animal. This is called the *arousal reaction.*

Some types of sensory stimuli are more potent than others in eliciting the arousal reaction, the most potent being pain and proprioceptive somatic impulses. This importance of somatic impulses in maintaining an activated state in the reticular activating system is borne out by the fact that progressive destruction of the sensory input tracts to this system progressively decreases the degree of wakefulness. Indeed, an animal falls asleep permanently when the brain stem is transected slightly above the point of entrance of the sensory impulses from the fifth nerve into the pontine region of the reticular formation. Such a transection causes all somatic sensory impulses to be lost. However, even after this procedure, electrical stimulation in the remaining reticular activating system above the transection can still awaken the animal. This experiment illustrates the importance of the arousal reaction and of somatic impulses in maintaining the wakefulness state of the animal.

Anatomically the reticular formation of the brain stem is admirably constructed to perform the arousal functions. It receives tremendous numbers of collateral fibers from the *spinothalamic tracts,* the *spinocerebellar tracts,* the *spinotectal tracts,* the *auditory tracts,* the *visual tracts,* and others, so that almost any sensory stimulus in the body can activate it. The reticular formation in turn can transmit signals both upward into the brain and downward into the spinal cord. Indeed, many of the fibers originating from cells in the reticular formation divide, with one branch of the fiber passing upward and the other branch passing downward.

CEREBRAL ACTIVATION OF THE RETICULAR ACTIVATING SYSTEM

In addition to activation of the reticular activating system by sensory impulses, the cerebral cortex can also stimulate this system. Direct fiber pathways pass into the reticular activating system as shown in Figure 60–1B from almost all parts of the cerebrum but particularly from (1) the *sensorimotor cortex* of the pre- and postcentral gyri, (2) the *frontal cortex,* (3) the *cingulate gyrus,* (4) the *hippocampus* and *other structures of the rhinencephalon,* and (5) the *basal ganglia.* Because of an exceedingly large number of nerve fibers that pass from the motor regions of the cerebral cortex to the

reticular formation, motor activity in particular is associated with a high degree of wakefulness, which partially explains the importance of movement to keep a person awake. However, intense activity of any other part of the cerebrum can also activate the reticular activating system and consequently can cause a high degree of wakefulness.

THE FEEDBACK THEORY OF WAKEFULNESS AND SLEEP

From the above discussion we can see that activation of the reticular activating system greatly intensifies the degree of activity in the cerebral cortex. But, in turn, increased activity in the cerebral cortex increases the degree of activity of the reticular activating system. Thus, there exists a *positive* "feedback" that helps to keep the ascending reticular activating system active once it becomes excited.

Likewise, increased activity of this system increases the degree of muscle tone throughout the body, which was discussed in Chapter 57 in relation to the gamma efferent system. Also, stimulation of the reticular activating system increases many autonomic activities throughout the body. In turn, these peripheral effects cause increased somatic impulses to be transmitted into the central nervous system— particularly important are the proprioceptor impulses, for these have a high degree of arousal activity. Thus, here again, we find another *positive* "feedback." That is, activity in the reticular activating system causes increased peripheral activity and this in turn feeds back to the reticular activating system to promote further increase in excitation there.

Finally, still a *third* "positive" feedback results from stimulation of the sympathetic nerves by the reticular activating system, for this causes release of epinephrine that then has a direct effect on the brain stem portion of the reticular activating system to enhance further its activity, though it is still not known whether this effect is a significant one.

On the basis of these three positive feedback loops—(1) to the cerebral cortex and back to the reticular formation, (2) to the peripheral muscles and back, and (3) possibly to the adrenal medullae and back via the hormone epinephrine—a feedback theory of wakefulness and sleep can be formulated as follows:

It is assumed that once the reticular activating system becomes activated, the feedback impulses from both the cerebral cortex and from the periphery tend to maintain its excitation. Thus, after the person has become awakened, he tends to remain awake at least for the time being.

After prolonged wakefulness, it can also be assumed that many of the neuronal cells of the feedback loops, particularly the neuronal cells in the reticular activating system itself, gradually become fatigued or less excitable for other reasons. When this happens, the degree of activation of the entire system begins to decrease. As a result, the intensity of the cortical and peripheral feedbacks also decreases, which results in further depression of the reticular activating system. The cortex and peripheral functions are further depressed, followed by still further depression of the reticular activating system. Thus, once the degree of excitability of the neurons falls to a critical level, a vicious cycle of depression ensues until all or most components of the feedback loops fade into decreased activity, and this then represents the state of sleep.

The next element of the feedback theory of wakefulness and sleep assumes that, after the reticular activating system has been dormant for a time, the neuronal cells involved in the feedback loops gradually regain excitability. Yet wakefulness still does not occur until some arousal signal initiates activity in the system. Once such has occurred, the feedback loops immediately come into play, and the person passes from the state of sleep into the state of wakefulness.

One might immediately ask: How is it that a person can also have various degrees of wakefulness and sleep? To answer this, we need only realize that the feedback loops contain literally millions of parallel pathways. If the feedback system is operating only through a few of these, a person would theoretically have a slight degree of wakefulness, but if the feedback is operating through tremendous numbers of pathways simultaneously, the degree of wakefulness obviously would be very great. Furthermore, the numbers of pathways involved at any one time would depend on the various stimulatory influences entering the reticular activating system. That is, the greater the intensity of signals from the sensory systems and the greater the intensity of signals from activated regions of the cerebral cortex, the greater also would be the degree of wakefulness. Indeed, these effects are precisely those known to occur in everyone.

One of the most important bits of evidence in favor of the feedback theory of wakefulness and sleep is the fact that complete relaxation of the peripheral musculature throughout the body, even in the absence of muscular or neuronal fatigue, can frequently cause a person to go to sleep. Obviously, relaxation of the musculature would break one of the major feedback loops for maintenance of wakefulness.

Those students who are familiar with the theory of oscillating mechanisms will immediately recognize that the oscillation between sleep and wakefulness as just described is the same as that which occurs in the so-called relaxation oscillator. Oscillation of this type is characterized by rapid phases of positive feedback and slow phases of negative feedback. In the case of the sleep-wakefulness cycle, rapid positive feedback results from the three positive feedback loops. That is, once excitation of the reticular activating system begins, positive feedback causes the activation to become progressively greater and greater. Also, positive feedback occurs when the reticular activating system begins to be depressed, progressive depression of more and more of the feedback loops causing the depression to proceed rapidly to the stage of sleep.

On the other hand, the slow negative feedback components of the oscillation are (a) slow fatigue of the neurons during the wakefulness cycle and (b) slow re-establishment of excitability of the neurons during the sleep phase. Once the slow negative feedback has proceeded beyond a critical point, the rapid positive feedback then (theoretically) shifts the person rapidly from one state to the other—from sleep to wakefulness or from wakefulness to sleep.

Active Production of Sleep by Some Neuronal Centers. Though sleep is generally associated with decreased activity of the reticular activating system and, therefore, also with decreased activity of the cerebrum as well, electrical stimulation of a few centers in the brain can actively cause sleep. The most important of these is located in the upper medullary and lower pontine area of the reticular formation, an area closely associated with the upper regions of the solitary fasciculus. Electrical stimulation in this region causes changes in the electroencephalogram that are typical of those changes in the early stages of sleep. On the contrary, when this area is destroyed bilaterally, the animal develops insomnia.

Therefore, it is almost certain that the excitatory portion of the reticular activating system is at least partially balanced by an inhibitory area that, when stimulated, can actively produce sleep or at least tend to produce sleep. This reticular inhibitory area receives a large inflow of sensory signals from the cutaneous mechanoreceptors; and experiments have demonstrated that slow stimulation of these mechanoreceptors can have a soporific effect on a person.

REM Sleep and the Process of Dreaming. Periodically during sleep, a person usually passes through a state of dreaming associated with mild involuntary muscle jerks and with rapid eye movements. The rapid eye movements have given this stage of sleep the name

REM sleep. Electroencephalograms during these periods are typical of those of light sleep but not of deep sleep. Yet, strangely enough, muscle tone throughout the body is diminished almost to zero, the heart rate may be as low as 20 beats below normal, and the arterial pressure may be as low as 30 mm. Hg below normal. Thus, the person seems physiologically to be in very deep sleep, rather than light sleep, despite the fact that he is dreaming and has uncontrolled muscle movements.

REM sleep usually occurs three to four times during each night at intervals of 80 to 120 minutes, each occurrence lasting from 5 minutes to more than an hour. As much as 50 per cent of the infant's sleep cycle is composed of REM sleep; and in the adult, approximately 20 per cent is REM sleep.

The causation of REM sleep and its cyclic appearance is still unknown. However, it seems to be especially important, because the degree of restfulness of sleep is associated with the amount of REM sleep that a person receives during the night. For instance, in persons who have been awakened each time that the electroencephalogram demonstrated the onset of REM sleep, extreme tiredness and serious neurotic tendencies have developed.

PHYSIOLOGICAL EFFECTS OF SLEEP

Sleep causes two major types of physiological effects: first, effects on the nervous system itself and, second, effects on other structures of the body. The first of these seems to be by far the more important, for any person who has a transected spinal cord in the neck shows no physiological effects in the body beneath the level of transection that can be attributed to a sleep and wakefulness cycle; this lack of sleep and wakefulness causes neither significant harm to the bodily organs, nor even any deranged function. On the other hand, lack of sleep certainly does affect the functions of the central nervous system.

Prolonged wakefulness is often associated with progressive malfunction of the mind and behavioral activities of the nervous system. We are all familiar with the increased sluggishness of thought that occurs toward the end of a prolonged wakeful period, but in addition, a person can become irritable or even psychotic following forced wakefulness for prolonged periods of time or even following loss of the REM portion of his sleep. Therefore, we can assume that sleep restores normal "balance" between the different parts of the central nervous system. This might be likened to the "rezeroing" of electronic analog computers after prolonged use, for all computers of this type gradually lose their "base line" of operation; it is reasonable to assume that the same effect occurs in the central nervous system, since it, too, is an analog computer (though using a digital method of information transmission). Therefore, in the absence of any definitely demonstrated functional value of sleep, we might postulate, on the basis of known psychological changes that occur with wakefulness and sleep, that sleep performs this rezeroing function for the nervous system. Yet this is but a suggestion based purely on psychological data and on analogy with computers.

Even though, as pointed out above, wakefulness and sleep have not been shown to be necessary for somatic functions of the body, the cycle of enhanced and depressed nervous excitability that follows along with the cycle of wakefulness and sleep does have moderate effects on the peripheral body. For instance, there is enhanced sympathetic activity during wakefulness and also enhanced numbers of impulses to the skeletal musculature to increase muscular tone. Conversely, during sleep sympathetic activity decreases while parasympathetic activity occasionally increases, and the muscular tone becomes almost nil. Therefore, during sleep, arterial blood pressure falls, pulse rate decreases, skin vessels dilate, activity of the gastrointestinal tract sometimes increases, muscles fall into a completely relaxed state, and overall basal metabolic rate of the body falls by about 10 to 20 per cent.

ATTENTION

So long as a person is awake, he has the ability to direct his attention to specific aspects of his mental environment. Furthermore, his degree of attention can change remarkably from (a) almost no attention at all to (b) broad attention to almost everything that is going on, or to (c) intense attention to a minute facet of his momentary mental experience.

Unfortunately, the basic mechanisms by which the brain accomplishes its diverse acts of attention are not known. However, a few clues are beginning to fall into place, as follows:

Brain Stem Control of Overall Attentiveness. In exactly the same way that a person can change from a state of sleep to a state of wakefulness, there can be all degrees of wakefulness, from wakefulness in which a person is nonattentive to almost all his surroundings to an extremely high degree of wakefulness in which the person reacts instantaneously to almost any sensory experience. These changes in degree of *overall attentiveness* seem to be caused primarily by changes in activity of the brain stem portion of the reticular activating system. Thus, control of the general level of attentiveness is probably exerted by the same mechanism that controls wakefulness and sleep, the control center for which is located in the mesencephalon and upper pons.

Function of the Thalamus in Attention. Earlier in the chapter it was pointed out that stimulation of a single specific area in the thalamic portion of the reticular activating system generally activates only a specific area of the cerebral cortex. Since the cerebral cortex is one of the most important areas of the brain for conscious awareness of our surroundings, one can surmise that the ability of specific thalamic areas to excite specific cortical regions might be one of the mechanisms by which a person can direct his attention to specific aspects of his mental environment, whether these be immediate sensory experiences or stored memories.

Relation of Centrifugal Control of Sensory Information to Attention. Nervous pathways extend centrifugally from almost all sensory areas of the brain toward the lower centers to control the intensity of sensory input to the brain. For instance, the auditory cortex can either inhibit or facilitate signals from the cochlea, the visual cortex can control the signals from the retina in the same way, and the somesthetic cortex can control the intensities of signals from the somatic areas of the body.

Thus, it is likely that activated regions of the cortex control their own sensory input under some conditions. This is another means by which the brain might direct its attention to specific phases of its mental activity.

COORDINATION OF DIFFERENT CORTICAL FUNCTIONS BY THE RETICULAR ACTIVATING SYSTEM

Though many fiber pathways interconnect different areas of the cerebral cortex directly, these have only limited function in coordinating the activities of different parts of the cerebral cortex with each other. For instance, the entire corpus callosum can be cut, which removes almost all direct pathways between one cerebral hemisphere and the other, and yet it is difficult, when using usual psychological tests, to determine that there is any decrease in cerebral capability. Likewise, a deep transection can be made in the cerebral cor-

tex in the bottom of the central sulcus, separating the parietal cortex from the frontal cortex, and yet the sensory information stored in the parietal cortex can still be used to control motor activities of the frontal cortex in such a way that the experimental observer cannot tell the difference. Thus, it is evident that the more important routes of communication from one major part of the cortex to another are mainly through deep areas of the brain.

Anatomically, many pathways exist between the different thalamic nuclei, which suggests that much of the coordination between the different cortical areas could occur in the thalamus. However, even in the thalamus only insignificant pathways exist *from one side of the brain to the other.* Therefore, it is likely that most of the coordination between one cerebral hemisphere and the other is controlled by the brain stem reticular activating area and that many of the important correlations even within a single hemisphere are also controlled through this area. Yet, it is also likely that many, if not most, correlations between closely associated cortical areas of the same hemisphere are controlled primarily by the thalamic nuclei.

POSSIBLE "SCANNING" AND "PROGRAMMING" FUNCTIONS OF THE RETICULAR ACTIVATING SYSTEM

At this point it is important to refer back to the analogy drawn in Chapter 46 between the functions of the central nervous system and of a general purpose computer. Almost all complex computers, particularly those that are capable of storing information and then recalling this information at later times, have a control center called a "programming unit" that directs the attention of different parts of the computer to specific information stored in other parts of the computer. It is quite likely that the reticular activating system, particularly the thalamic part, operates in much the same way as the programming unit of a computer for the following reasons:

1. The ability of the reticular activating system to direct one's attention to specific mental activities is analogous to the ability of the programming unit to call forth information that has been stored in the memory of the computer or to call forth information that is given to the computer at its input, information that is comparable to sensory information in the human being.

2. The programming unit can "find information" when its exact locus is not known. In the computer, this is achieved by dictating certain qualities required of the information and then searching through hundreds or thousands of stored bits of information until the appropriate information is found, a function called *scanning.* We know from psychological tests that our own brain can perform this same function, though we do not know how it does so. Yet, since the thalamus perhaps plays a role in directing our attention to stored information in specific areas of the cortex, it is reasonable

to postulate that the thalamus, probably operating in conjunction with other basal areas of the brain, plays a major role in this scanning operation.

3. The programming unit of a computer determines the sequence of processing of information. Our brain can also control its thoughts in orderly sequence. Here again, mainly on the basis of anatomical considerations and on the basis of the fact that the reticular activating system is the primary controller of cerebral activity, it can be surmised, until we know more about the subject, that the reticular activating system is the primary controller of our orderly sequence of thoughts.

EFFECT OF BARBITURATE ANESTHESIA ON THE RETICULAR ACTIVATING SYSTEM

The barbiturates have a specific depressant effect on the brain stem portion of the reticular activating system. Therefore, barbiturates obviously can either depress brain activity or even cause sleep. Yet it is especially interesting that barbiturate anesthesia does not block transmission in most of the specific sensory systems and also does not block function of the thalamic portion of the reticular activating system. It is probable that many other clinically used anesthetics also have specific depressant effects on the brain stem portion of the reticular activating system and in this way cause general anesthesia.

BRAIN WAVES

Electrical recordings from the surface of the brain or from the outer surface of the head demonstrate continuous electrical activity in the brain. Both the intensity and patterns of this electrical activity are determined to a great extent by the overall excitation of the brain resulting from functions in the reticular activating system. The undulations in the recorded electrical potentials, shown in Figure 60–2, are called *brain waves,* and the entire record is called an *electroencephalogram* (EEG).

The intensities of the brain waves on the surface of the scalp range from zero to 300 microvolts, and their frequencies range from once every few seconds to 50 or more per second. The character of the waves is highly dependent on the degree of activity of the cerebral cortex, and the waves change markedly between the states of wakefulness and sleep.

Much of the time, the brain waves are irregular and no general pattern can be discerned in the EEG. However, at other times, distinct patterns do appear. Some of these are characteristic of specific abnormalities of the brain, such as epilepsy, which is discussed later. Others occur even in normal persons and can be classified into *alpha, beta, theta,* and *delta waves,* which are all illustrated in Figure 60–2.

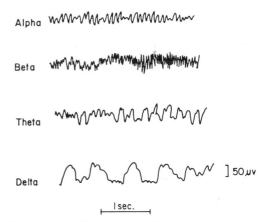

Figure 60–2 Different types of normal electroencephalographic waves.

Alpha waves are rhythmic waves occurring at a frequency between 8 and 13 per second and are found in the EEG's of almost all normal persons when they are awake in a quiet, resting state of cerebration. These waves occur most intensely in the occipital region but can also be recorded at times from the parietal and frontal regions of the scalp. Their voltage usually is about 50 microvolts. During sleep the alpha waves disappear entirely, and when the awake person's attention is directed to some specific type of mental activity, the alpha waves are replaced by asynchronous, higher frequency but lower voltage waves. Figure 60–3 illustrates the effect on the alpha waves of simply opening the eyes in bright light and then closing the eyes again. Note that the visual sensations cause immediate cessation of the alpha waves and that these are replaced by low voltage, asynchronous waves.

Beta waves occur at frequencies of more than 14 cycles per second and sometimes as high as 50 cycles per second. These are most frequently recorded from the parietal and frontal regions of the scalp, and they can be divided into two major types, *beta I* and *beta II.* The beta I waves, which are illustrated in Figure 60–2, have a frequency about twice that of the alpha waves, and these are affected by mental activity in very much the same way as the alpha waves—that is, they disappear and in their place appears an asynchronous but low voltage recording. The beta II waves, on the contrary, appear during *intense* activation of the central nervous system or during tension. Thus, one type of beta wave is inhibited by cerebral activity while the other is elicited.

Theta waves have frequencies between 4 and 7 cycles per second. These occur mainly in the parietal and temporal regions in children, but

Figure 60–3 Replacement of the alpha rhythm by an asynchronous discharge on opening the eyes.

they also occur during emotional stress in some adults, particularly during disappointment and frustration. They can often be brought out in the EEG of a frustrated person by allowing him to enjoy some pleasant experience and then suddenly removing this element of pleasure; this causes approximately 20 seconds of theta waves.

Delta waves include all the waves of the EEG below 3½ cycles per second and sometimes as low as 1 cycle every 2 to 3 seconds. These occur in deep sleep, in infancy, and in serious organic brain disease. And they also occur in the cortex of animals that have had subcortical transections separating the cerebral cortex from the reticular activating system. Therefore, delta waves can occur strictly in the cortex independently of activities in lower regions of the brain.

ORIGIN OF THE DIFFERENT TYPES OF BRAIN WAVES

The discharge of a single neuron or single nerve fiber in the brain cannot be recorded from the surface of the head. Instead, for an electrical potential to be recorded all the way through the skull, large portions of nervous tissue must emit electrical current simultaneously. There are two ways in which this can occur. First, tremendous numbers of nerve fibers can discharge in synchrony with each other, thereby generating very strong electrical currents. Second, large numbers of neurons can partially discharge, though not emit action potentials; furthermore, these partially discharged neurons can give prolonged periods of current flow that can undulate slowly with changing degrees of excitability of the neurons. Simultaneous electrical measurements within the brain while recording brain waves from the scalp indicate that it is the second of these that causes the usual brain waves.

To be more specific, the surface of the cerebral cortex is composed almost entirely of a mat of dendrites from neuronal cells in layers one and two of the cortex (there are four more layers of cortex below these). When signals impinge on these dendrites, they become partially discharged, emitting negative potentials characteristic of excitatory postsynaptic potentials as discussed in Chapter 6. This partially discharged state makes the neurons of the outer two layers of the cortex highly excitable, and the negative potential is simultaneously recorded from the surface of the scalp, indicating this high degree of excitability.

One of the important sources of signals to excite the outer dendritic layer of the cerebral cortex is the ascending reticular activating system. Therefore, brain wave intensity is closely related to the degree of activity in either the brain stem or the thalamic portions of the reticular activating system.

Origin of Delta Waves. Transection of the fiber tracts from the thalamus to the cortex, which blocks the reticular activating system fibers, causes delta waves in the cortex. This indicates that some synchronizing mechanism can occur in the cortical neurons themselves — entirely independently of lower structures in the brain — to cause the delta waves.

Delta waves also occur in very deep non-REM sleep; and this suggests that the cortex is then released from the activating influences of the reticular activating system, as was explained earlier in the chapter.

Origin of the Alpha Waves; and the "Recruiting Response." Alpha waves will *not* occur in the cortex without connections with the reticular activating system. It is possible that this system enters into the genesis of the alpha waves in the following manner:

Electrical stimulation of the thalamic portion of the reticular activating system causes slow development of an electrical potential in the cerebral cortex called a *recruiting response*. This is illustrated in Figure 60–4, which shows comparative excitation of the visual cortex following stimulation of (A) the lateral geniculate body that directly excites the visual cortex and (B) the thalamic reticular activating system that indirectly excites the visual associative cortex. Note the prolonged delay and slow buildup of the electrical potential following stimulation of the reticular activating system; this is the recruiting response.

Another feature of interest regarding the recruiting response is that repetitive stimuli in the thalamus at approximately 10 cycles per second cause the response to become more and more intense with each repetition during the first few repetitive stimuli, which is the origin of the term "recruiting." And a final point of importance is that strong electrical stimulation of the cortex causes reverberation of impulses among the nuclei of the thalamus at frequencies ranging between 8 and 13 per second, the frequency of the alpha waves.

On the basis of these facts, one might postulate that the recruiting response of the thalamus combined with reverberation within the thalamus could cause both the periodicity of the alpha waves and the synchronous activation of literally millions of cortical neurons at the same time.

Figure 60–4 Response of the visual cortex to stimulation of the lateral geniculate body (which causes direct excitation of the cortex), and the "recruiting response" in the visual cortex following stimulation of the thalamic portion of the reticular activating system. Note the long latent period before appearance of the recruiting response.

EFFECT OF VARYING DEGREES OF CEREBRAL ACTIVITY ON THE BASIC RHYTHM OF THE ELECTROENCEPHALOGRAM

There is a general relationship between the degree of cerebral activity and the average frequency of the electroencephalographic rhythm, the frequency increasing progressively with higher and higher degrees of activity. This is illustrated in Figure 60–5, which shows the existence of delta waves in stupor, surgical anesthesia, and sleep; theta waves in psychomotor states and in infants; alpha waves during relaxed states; and beta waves during periods of intense mental activity. However, during periods of mental activity the waves usually become asynchronous rather than synchronous so that the voltage falls considerably, despite increased cortical activity, as illustrated in Figure 60–3.

CLINICAL USE OF THE ELECTROENCEPHALOGRAM

One of the most important uses of the EEG is to diagnose different types of epilepsy and to find the focus in the brain causing the epilepsy. This is discussed further below. But, in addition, the EEG can be used to localize brain tumors or other space-occupying lesions of the brain and to diagnose certain types of psychopathic disturbances.

There are two means by which brain tumors can be localized. Some brain tumors are so large that they block electrical activity from a given portion of the cerebral cortex, and when this occurs the intensity of the brain waves is considerably reduced in the region of the tumor. However, more frequently a brain tumor compresses the surrounding neuronal tissue and thereby causes abnormal electrical excitation of these surrounding areas, this in turn leading to synchronous discharges of very high voltage waves in the EEG, as shown in the middle two records of Figure 60–6. Localization of the origin of these spikes on the surface of the scalp is a valuable means for locating the brain tumor.

The upper part of Figure 60–6 shows the placement of 16 different electrodes on the scalp, and the lower part of the figure shows the brain waves from four of these electrodes marked in the figure by X's. Note that in two of these intense brain waves are recorded and, furthermore, that the two waves are essentially of reverse polarity to each other. This reverse polarity means that the origin of the spikes is somewhere in the area *between* the two respective electrodes. Thus, the excessively excitable area of the brain has been located, and this is a lead to the location of the brain tumor.

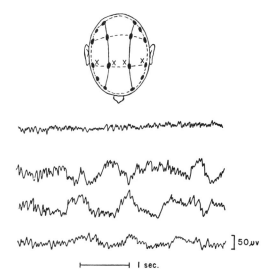

Figure 60–6 Localization of a brain tumor by means of the EEG, illustrating above the placement of electrodes and below the records from the four electrodes designated by X's.

Use of brain waves in diagnosing psychopathic abnormalities is generally not very satisfactory because only a few of these cause distinct brain wave patterns. Yet by observing combinations of different types of basic rhythms, reactions of the rhythms to attention, changes of the rhythms with forced breathing to cause alkalosis, the appearance of particular characteristics in the brain waves (such as "spindles" of alpha waves), and so forth, an experienced electroencephalographer can detect at least certain types of psychopathic disturbances. For instance, it was pointed out above that theta waves are frequently found in persons under severe tension.

EPILEPSY

Epilepsy is characterized by uncontrolled excessive activity of either a part of the central nervous system or all of it. A person who is predisposed to epilepsy has attacks when the basal level of excitability of his nervous system (or of the part that is susceptible to the epileptic state) rises above a certain critical threshold. But, as long as the degree of excitability is held below this threshold, no attack occurs.

Basically, the two different types of epilepsy are: *generalized epilepsy* and *partial epilepsy*. Generalized epilepsy involves essentially all parts of the brain at once, whereas partial epilepsy involves only a portion — sometimes only a minute focal spot and other times a moderate portion of the brain. Generalized epilepsy can be divided into *grand mal epilepsy* and *petit mal epilepsy*.

GENERALIZED EPILEPSY

Grand Mal. Grand mal epilepsy is characterized by extreme neuronal discharges originating in the

Figure 60–5 Effect of varying degrees of cerebral activity on the basic rhythm of the EEG. (From Gibbs and Gibbs: Atlas of Encephalography. The Authors.)

brain stem portion of the reticular activating system. These then spread throughout the entire central nervous system, to the cortex, to the deeper parts of the brain, and even into the spinal cord to cause generalized *tonic convulsions* of the entire body followed toward the end of the attack by alternating muscular contractions called *clonic convulsions*. The grand mal seizure lasts from a few seconds to as long as three to four minutes and is characterized by post-seizure depression of the entire nervous system; the person might remain in a stupor for one minute to as long as a day or more after the attack is over.

The middle recording of Figure 60-7 illustrates a typical electroencephalogram from almost any region of the cortex during a grand mal attack. This illustrates that high voltage synchronous discharges occur over the entire cortex having almost the same periodicity as the normal alpha waves. Furthermore, the same type of discharge occurs on both sides of the brain at the same time, illustrating that the origin of the abnormality is in the lower centers of the brain that control the activity of the cerebral cortex and not in the cerebral cortex itself.

In experimental animals or even in human beings, grand mal attacks can be initiated by administering neuronal stimulants, such as the well-known drug Metrazol, or they can be caused by insulin hypoglycemia or by the passage of alternating electrical current directly through the brain. Electrical recordings from the thalamus and also from the reticular formation of the brain stem during the grand mal attack show typical high voltage activity in both of these areas similar to that recorded from the cerebral cortex. Furthermore, in an experimental animal, even after transecting the brain stem as low as the mesencephalon, a typical grand mal seizure can still be induced in the portion of the brain stem beneath the transection.

Presumably, therefore, a grand mal attack is caused by intrinsic overexcitability of the neurons

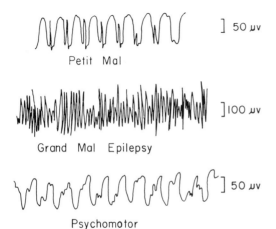

] 50 μv

Petit Mal

] 100 μv

Grand Mal Epilepsy

] 50 μv

Psychomotor

Figure 60-7 Electroencephalograms in different types of epilepsy.

that make up the reticular activating structures or from some abnormality of the local neuronal pathways. The synchronous discharges from this region could result from local reverberating circuits.

One might ask: What stops the grand mal attack after a given time? This is believed to result from (1) *fatigue of the neurons* involved in precipitating the attack and (2) *active inhibition* by certain structures of the brain. The change from the tonic type of convulsion to the clonic type during the latter part of the grand mal attack has been suggested to result from partial fatigue of the neuronal system so that some of the excited neurons fade out for a moment, then return to activity after a brief rest, only to fatigue a second time and then a third time, and so forth, until the entire seizure is over. And the stupor that lasts for a few minutes to many hours after a grand mal seizure is over is believed to result from the intense fatigue of the neurons following their intensive activity during the grand mal attack.

The *active inhibition* that helps to stop the grand mal attack is believed to result from feedback circuits through inhibitory areas of the brain. The grand mal attack undoubtedly excites such areas as the basal ganglia, which in turn emit many inhibitory impulses into the reticular formation of the brain stem. But the nature of such active inhibition is very much a matter of speculation.

Petit Mal Epilepsy. Petit mal epilepsy is closely allied to grand mal epilepsy, and it occurs in two forms, the *myoclonic form* and the *absence form*.

In the *myoclonic form of epilepsy*, a burst of neuronal discharges lasting a fraction of a second occurs throughout the nervous system. These are similar to those that occur at the beginning of a grand mal attack, and the person exhibits a single violent muscular jerk usually involving his arms or head. However, the entire process stops immediately and is usually over before he loses consciousness or even before he stops what he is doing. Yet this myoclonic type of attack often becomes more and more severe until the person begins to have grand mal attacks. Therefore, from an etiological point of view a myoclonic attack is essentially the same as a grand mal attack except that some inhibitory influence stops it immediately.

The *absence type of petit mal epilepsy* is characterized by 5 to 20 seconds of unconsciousness during which the person has several twitchlike contractions of the muscles, usually in the head region—especially blinking of the eyes; this is followed by return of consciousness and continuation of previous activities.

This type of epilepsy also is closely allied to grand mal epilepsy, and, in rare instances, it initiates grand mal attacks. Furthermore, persons who have petit mal attacks during early life may become progressively disposed to grand mal attacks in later life. The first recording of Figure 60-7 illustrates a typical *spike and dome* pattern that is recorded during the absence type of petit mal epilepsy. The spike portion of this recording is almost identical to the spikes that occur in grand mal epilepsy, but the dome portion is distinctly different. The spike and dome can be recorded over the entire cerebral cortex, illustrating that the seizure originates in the reticular activating system of the brain.

Since petit mal and grand mal epilepsy both originate in essentially the same locus of the brain stem, it is believed that they have the same common cause but that some influence inhibits neuronal activity during the petit mal attack so that it will not progress into the grand mal seizure.

PARTIAL EPILEPSY

Partial epilepsy can involve almost any part of the brain, either localized regions of the cerebral cortex or deeper structures of both the cerebrum and brain stem. And almost always, partial epilepsy results from some localized organic lesion of the brain, such as a scar that pulls on the neuronal tissue, a tumor that compresses an area of the brain, or a destroyed area of brain tissue. Lesions such as these can promote extremely rapid discharges in the local neurons, and when the discharge rate rises above approximately 1000 per second, synchronous waves begin to spread over the adjacent cortical regions. These waves presumably result from *localized reverberating circuits* that gradually recruit adjacent areas of the cortex into the discharge zone. The process spreads to adjacent areas at a rate as slow as a few millimeters a minute to as fast as several centimeters per second. When such a wave of excitation spreads over the motor cortex, it causes a progressive "march" of muscular contractions throughout the opposite side of the body, beginning perhaps in the leg region and marching progressively upward to the head region, or at other times marching in the opposite direction. This is called *jacksonian epilepsy.*

A partial epileptic attack may be confined to a single area of the brain, but in many instances it excites the brain stem portion of the reticular activating system so greatly that it initiates a grand mal epileptic attack.

One type of partial epilepsy is the so-called *psychomotor seizure,* which may cause (1) a short period of amnesia, (2) an attack of abnormal rage, (3) sudden anxiety, discomfort, or fear, (4) a moment of incoherent speech or mumbling of some trite phrase, or (5) a motor act to attack someone, to rub the face with the hand, or so forth. Sometimes the person cannot remember his activities during the attack, but at other times he will have been conscious of everything that he had been doing but unable to control it.

The lower tracing of Figure 60–7 illustrates a typical electroencephalogram during a psychomotor attack, showing a low frequency rectangular wave with a frequency between 2 and 4 per second and with superimposed 14 per second waves.

The electroencephalogram can often be used to localize abnormal spiking waves originating in areas of organic brain disease that might predispose to epileptic attacks. Once such a focal point is found, surgical excision of the focus frequently prevents future epileptic attacks.

REFERENCES

Brazier, M. A. B.: Electrical Activity of the Brain. New York, The Macmillan Company, 1962.

Ciba Foundation Symposium: The Nature of Sleep. Boston, Little, Brown and Company, 1961.

Fessard, A.: Brain potentials and rhythms—introduction. *In* Handbook of Physiology. Baltimore, The Williams & Wilkins Co., 1959, Sec. I, Vol. I, p. 255.

Jasper, H. H.: Unspecific thalamocortical relations. *In* Handbook of Physiology. Baltimore, The Williams & Wilkins Co., 1960, Sec. I, Vol. II, p. 1307.

Jouvet, M.: Experimental evidence for the participation of monoamines in sleep mechanisms. *Proc. Int. Union of Physiol. Sciences,* 6:33, 1968.

Jouvet, M.: Neurophysiology of the states of sleep. *Physiol. Rev.,* 47:117, 1967.

Kleitman, N.: Sleep and Wakefulness. Chicago, University of Chicago Press, 1963.

Laidlaw, J. P., and Stanton, J. B.: The EEG in Clinical Practice. Baltimore, The Williams & Wilkins Co., 1967.

Lindsley, D. B.: Attention, consciousness, sleep and wakefulness. *In* Handbook of Physiology. Baltimore, The Williams & Wilkins Co., 1960, Sec. I, Vol. III, p. 1553.

Magoun, H. W.: The Waking Brain. 2nd ed., Springfield, Illinois, Charles C Thomas, Publisher, 1962.

Matsumoto, H.; Ayala, G. F.; and Gumnit, R. J.: Neuronal behavior and triggering mechanism in cortical epileptic focus. *J. Neurophysiol.,* 32:688, 1969.

Mills, J. N.: Human circadian rhythms. *Physiol. Rev., 46:*128, 1966.

Moruzzi, G.: Ascending reticular system and sleep physiology. *Proc. Int. Union of Physiol. Sciences,* 6:87, 1968.

Olds, J.: Higher functions of the nervous system. *Ann. Rev. Physiol.,* 21:381, 1959.

O'Leary, J. L., and Goldring, S.: D-C potentials of the brain. *Physiol. Rev.,* 44:91, 1964.

Oswald, I.: Sleeping and Waking. Physiology and Psychology. New York, American Elsevier Publishing Co., Inc., 1962.

Owen, M., and Bliss, E. L.: Sleep loss and cerebral excitability. *Amer. J. Physiol.,* 218:171, 1970.

Skultety, F. M.: Circus movements in cats following midbrain stimulation through chronically implanted electrodes. *J. Neurophysiol.,* 25:152, 1962.

Steriade, M.; Iosif, G.; and Apostol, V.: Responsiveness of thalamic and cortical motor relays during arousal and various stages of sleep. *J. Neurophysiol.,* 32:251, 1969.

Valdman, A. V.: Pharmacology and Physiology of the Reticular Formation: Progress in Brain Research. New York, American Elsevier Publishing Co., Inc., 1967, Vol. 20.

CHAPTER 61

The Cerebral Cortex and Intellectual Functions of the Brain

It is ironic that we know least about the mechanisms of the cerebral cortex of all parts of the brain, even though it is by far the largest portion of the nervous system. Yet we do know the effects of destruction or specific stimulation of various portions of the cortex, and still more information has been gained from electrical recordings from the cortex or from the surface of the scalp. In the early part of the present chapter the known facts about cortical functions are discussed and then some basic theories of the neuronal mechanisms involved in thought processes, memory, analysis of sensory information, and so forth, are presented briefly.

PHYSIOLOGIC ANATOMY OF THE CEREBRAL CORTEX

The functional part of the cerebral cortex is comprised mainly of a thin layer of neurons 1.5 to 4 mm. in thickness, covering the surface of all the convolutions of the cerebrum and having a total area of about 2300 sq. cm. The total cerebral cortex contains approximately 9.3 billion neurons.

Figure 61–1 illustrates the typical structure of the cerebral cortex, showing successive layers of different types of cells. Most of the cells are either *granule cells* or *pyramical cells,* the latter named for their characteristic pyramidal shape. To the right in Figure 61–1 is illustrated the typical organization of nerve fibers within the different layers of the cortex. Note particularly the large number of horizontal fibers extending between adjacent areas of the cortex, but note also the vertical fibers that extend to and from the cortex to lower areas of the brain stem or to distant regions of the cerebral cortex through long association bundles of fibers.

Neurohistologists have divided the cerebral cortex into almost 100 different areas which have slightly different architectural characteristics. Yet in all these different areas there still persist at

Figure 61–1 Structure of the cerebral cortex, illustrating: *I,* molecular layer; *II,* external granular layer; *III,* layer of pyramidal cells; *IV,* internal granular layer; *V,* ganglionic layer; *VI,* layer of fusiform or polymorphic cells. (From Ranson and Clark (after Brodmann): Anatomy of the Nervous System.)

least crude representations of all the six major layers of the cortex. Furthermore, to the untrained histologist, not more than five major architectural types of cortex can be distinguished as follows: Type 1 is composed almost entirely of pyramidal cells with few granular cells and, therefore, is often called the *agranular cortex.* At the other extreme, type 5 contains almost no pyramidal cells but is filled with closely packed granule cells and is called the *granular cortex.* Types 2, 3, and 4 have intermediate characteristics containing gradated proportions of pyramidal and agranular cells.

The agranular cortex, which contains large

Figure 61–2 Structurally distinctive areas of the human cerebral cortex. (From Brodmann, modified by Buchanan: Functional Neuroanatomy. Lea & Febiger.)

numbers of pyramidal cells, is characteristic of the motor areas of the cerebral cortex; while the granular cortex, containing almost no pyramidal cells, is characteristic of the primary sensory areas. The other three intermediate types of cortex are characteristic of the association areas between the primary sensory and motor regions.

Figure 61–2 shows a map of cortical areas classified on the basis of histological studies by Brodmann. This classification provides a basis for discussion of functional areas of the brain, even though different architectural types of cortex often have similar functions.

Anatomical Relationship of the Cerebral Cortex to the Thalamus and Other Lower Centers. All areas of the cerebral cortex have direct afferent and efferent connections with the thalamus. Figure 61–3 shows the areas of the cerebral cortex connected with specific parts of the thalamus. These connections are always in *two* directions, both from the thalamus to the cortex and then from the cortex back to the same area of the thalamus. Furthermore, when the thalamic connections are cut, the functions of the corresponding cortical area become entirely or almost entirely abrogated. Therefore,

the cortex operates in close association with the thalamus and can almost be considered both anatomically and functionally a large outgrowth of the thalamus. Also all pathways from the sensory organs to the cortex pass through the thalamus, with the single exception of the sensory pathways of the olfactory tract.

FUNCTIONS OF CERTAIN SPECIFIC CORTICAL AREAS

In early attempts to analyze the function of the cerebral cortex it was assumed that different parcels of the cortex performed discrete functions, all independent of each other and to a great extent independent of the deeper structures of the cerebrum and brain stem. We now know that this is not true, and we know also that, except for a few specific areas of the cerebral cortex, when any moderate-sized portion of the cortex is destroyed, other segments of the cortex can take over the lost functions. Yet, even so, studies in human beings by neurosurgeons have shown that at least some specific functions are localized to certain general areas of the cerebral cortex.

Figure 61–4 gives a map of these areas as determined by Penfield and Rasmussen from electrical stimulation of the cortex or by neurological examination of patients after portions of the cortex had been removed. The lightly shaded areas are primary sensory areas, while the darkly shaded area is the voluntary motor area from which muscular movements can be elicited with relatively weak electrical stimuli. These primary sensory and motor areas have highly specific functions, but all the other areas of the cortex perform the more general functions that we commonly call cerebration.

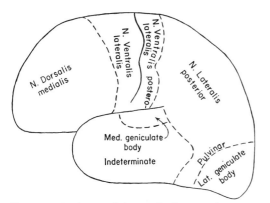

Figure 61–3 Areas of the cerebral cortex that connect with specific portions of the thalamus. (Modified from Elliott: Textbook of the Nervous System. J. B. Lippincott Co.)

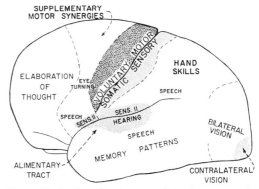

Figure 61–4 Functional areas of the human cerebral cortex as determined by electrical stimulation of the cortex during neurosurgical operations and by neurological examinations of patients with destroyed cortical regions. (From Penfield and Rasmussen: The Cerebral Cortex of Man. The Macmillan Co.)

SPECIFIC FUNCTIONS OF THE PRIMARY SENSORY AREAS

The primary sensory areas all have certain functions in common. For instance, somatic sensory areas, visual sensory areas, and auditory sensory areas all have spatial localizations of impulses from the peripheral receptors (which has already been discussed in detail in previous chapters).

Electrical stimulation of the primary sensory areas in awake patients gives relatively uncomplicated sensations. For instance, in the somatic sensory area the patient states that he feels tingling in the skin, numbness, mild electric shock, or, rarely, mild degrees of temperature sensations. And these sensations are localized to discrete areas of the body in accord with the spatial representation in the somatic sensory cortex, as described in Chapter 49. Therefore, it is believed that the primary somatic sensory cortex analyzes only the simple aspects of sensations and that analysis of intricate patterns of sensory experience requires other parts of the brain operating in association with this area.

Electrical stimulation of the primary visual cortex causes the person to see flashes of light, bright lines, colors, or other simple visions. Here again, the visual images are localized to specific regions of the visual fields in accord with the portion of the primary visual cortex stimulated, as described in Chapter 54. But the visual cortex alone is not capable of analyzing complicated visual patterns; for this, the visual cortex must operate in association both with other regions of the cerebral cortex and with lower brain centers.

Electrical stimulation of the auditory cortex causes a person to hear a simple sound which may be weak or loud, have low or high frequency, or have other uncomplicated characteristics, such as a squeak or even an undulation. But never are words or any other fully intelligible sound heard. Thus, the primary auditory cortex, like the other primary sensory areas, can detect the individual elements of auditory experience but cannot analyze complicated sounds. Therefore, the primary auditory cortex alone is not sufficient to give one even the usual auditory experiences.

Despite the inability of the primary sensory areas to analyze the incoming sensations fully, when these areas are destroyed, the ability of the person to utilize the respective sensations usually suffers drastically. For instance, loss of the primary visual cortex in one occipital cortex causes a person to become blind in the ipsilateral halves of both retinae, and loss of the primary visual cortices in both hemispheres causes total blindness. Likewise, loss of both primary auditory cortices causes almost total deafness. Loss of the postcentral gyri causes *depression* of somatic sensory sensations, though not total loss presumably because of additional cortical representation of these sensations in somatic sensory area II. (In animals far down the phylogenetic scale, loss of the visual and auditory cortices may have little effect on vision and hearing, and even an anencephalic human infant detects some visual scenes in the absence of the visual cortex, for such a child can observe a moving object and even follow it by movement of his eyes and head.)

Therefore, we can summarize the functions of the primary sensory areas of the human cerebral cortex in the following way: The lower centers of the brain relay a large part of the sensory signals to the cerebral cortex for analysis. In turn, the primary sensory areas transmit the results of their analyses back to the lower centers and to other regions of the cerebral cortex, as is discussed later in the chapter.

THE SECONDARY SENSORY AREAS

Around the borders of the primary sensory areas are regions which are variously called *secondary sensory areas,* or *sensory association areas.* In general these areas extend for 1 to 5 cm. in all directions from the primary sensory areas; and each time a primary area receives a sensory signal, secondary signals immediately spread into these secondary areas as well. Electrical stimulation of a secondary sensory area causes almost the same types of sensory experiences in awake patients as stimulation in the respective primary sensory areas. Therefore, it is doubtful, in contradistinction to former beliefs, that even the secondary sensory areas are capable by themselves of analyzing the complex types of sensory experiences. Yet destruction of these secondary sensory areas greatly reduces the capability of the brain to analyze different characteristics of sensory experiences. For instance, damage in the superior temporal gyrus below and behind the primary auditory area in the "dominant hemisphere" of the brain often causes a person to lose all understanding of the meanings of words or of other auditory experiences.

Likewise destruction of areas 18 and 19 or the presence of a brain tumor or other lesion in these areas, which represent the secondary visual areas, does not cause blindness or prevent normal activation of the primary visual cortex but greatly reduces the person's ability to interpret what he sees. Such a person often loses his ability to recognize the meanings of words, a condition that is called *word blindness.*

Finally, destruction of the secondary somatic sensory area in the parietal cortex posterior to primary somatic area I causes the person to lose his spatial perception for location of the different parts of his body. In the case of the hand that has been "lost," the skills of the hand are greatly reduced. Thus, this area of the cortex seems to be necessary for interpretation of somatic sensory experiences.

INTERPRETATIVE FUNCTIONS OF THE ANGULAR GYRUS AND TEMPORAL LOBE REGIONS

The temporal lobe and the angular gyrus — the gyrus that lies at the posterior tip of the lateral fissure where the temporal, parietal, and occipital lobes all come together — are especially important for many of the intellectual functions of the cerebral cortex, particularly for interpretation of the different sensory experiences, whether these be visual, auditory, or somatic. Following severe damage in these regions, a person might hear perfectly well and even recognize different words but still be unable to arrange these words into a coherent thought. Likewise, he may be able to read words from a printed page but again be unable to recognize the thought that is conveyed.

Electrical stimulation in widespread areas of the temporal lobe and angular gyrus of a conscious patient causes highly complex thoughts, including memories of complicated visual scenes or auditory hallucinations, such as specific musical pieces or a discourse by a specific person. For this reason it is believed that complicated memory patterns are stored in the temporal lobes and angular gyrus. This is in accord with the importance of these areas in the interpretation of complicated meanings of different sensory experiences.

The Dominant Hemisphere. The interpretative functions of the temporal lobe and angular gyrus are usually highly developed in only one cerebral hemisphere, which is called the dominant hemisphere. At birth these regions in both of the two hemispheres have almost the same capability of developing; we know this because removal of either hemisphere in early childhood causes the opposite to develop full dominant characteristics. Yet a theory that might explain dominance is the following: The attention of the "mind" is usually directed to one portion of the brain at a time, as was explained in the preceding chapter. Presumably, one angular gyrus region begins to be used to a greater extent than the other, and, thenceforth, because of the tendency to direct one's attention to the better developed region, the rate of learning in the cerebral hemisphere that gains the first start increases rapidly while that in the opposite side remains slight. Therefore, in the normal human being, one side becomes dominant over the other.

In more than 9 out of 10 persons the left temporal lobe and angular gyrus become dominant, and in the remaining one tenth of the population either both sides develop simultaneously to have dual dominance, or, more rarely, the right side alone becomes highly developed.

Usually associated with the dominant temporal lobe and angular gyrus is dominance of certain portions of the somesthetic cortex and motor cortex for control of voluntary motor functions. For instance, as is discussed later in the chapter, the premotor speech area, located far laterally in the intermediate frontal area, is almost always dominant on the same side of the brain as the dominant temporal lobe and angular gyrus. This speech area causes the formation of words by exciting simultaneously the laryngeal muscles, the respiratory muscles, and the muscles of the mouth.

Though the interpretative areas of the temporal lobe and angular gyrus, as well as many of the motor areas, are highly developed in only a single hemisphere, they are capable of receiving sensory information from both hemispheres and are also capable of controlling motor activities in both hemispheres. This unitary, cross-feeding organization prevents interference between the two sides of the brain; such interference, obviously, could create havoc with both thoughts and motor responses.

EFFECT OF TEMPORAL LOBE OR ANGULAR GYRUS DESTRUCTION IN THE DOMINANT HEMISPHERE. Destruction of the dominant temporal lobe and angular gyrus in an adult person normally leaves a great void in his intellect because of his inability to interpret the meanings of sensory experiences. Therefore, the neurosurgeon assiduously avoids surgery in this region. The adult can only gradually develop the interpretative functions of the nondominant temporal lobe and angular gyrus, and even then only to a slight extent. Thus, only a small amount of intellect can return. However, if this area is destroyed in a child under six years of age, the opposite side can usually develop to full extent, thus eventually returning the capabilities of the child essentially to normal.

Function of the Temporal Lobes in Recent Memory. Another function of the temporal lobes is to store recent memories. When both temporal lobes are completely destroyed, a person can recall only the simplest recent events, though he can recall facts that had previously been deeply engrained in his mind by long experience. It seems, therefore, that *the temporal lobes are highly impressionistic areas that rapidly and easily develop memory engrams and can also use these easily and rapidly for comparison with new experience, thereby interpreting the new in the light of the old.*

THE PREFRONTAL AREAS

The prefrontal areas are those portions of the frontal lobes that lie anterior to the motor regions. For years, this part of the cortex has been considered to be the locus of the higher intellect of the human being, principally because the main difference between the brain of monkeys and that of man is the great prominence of man's prefrontal lobes. Yet efforts to show that the prefrontal cortex is more important in higher intellectual functions than other portions of the cortex have not been successful. Indeed, destruction of the angular gyrus or posterior temporal lobe in the dominant hemisphere causes infinitely more harm to the intellect than does destruction of both prefrontal areas.

Psychological studies in lower animals have shown that all portions of the cortex not immediately associated with either sensory or motor functions are important in the ability of an animal to learn complicated information — such as a rat's learning to run through a maze — and that all the different areas are approximately equipotent in this regard. Therefore, the importance of the prefrontal areas to the human being is perhaps not that these areas perform some specific function different from those of other cortical areas, but primarily that they supply much additional cortical area in which cerebration can take place.

The prefrontal areas also control some of the autonomic functions of the body by transmitting impulses to the hypothalamus either directly or indirectly through the thalamus. These functions are described in the following chapter.

Prevention of Distractibility by the Prefrontal Areas. One of the outstanding characteristics of a person who has lost his prefrontal areas is the ease with which he can be *distracted* from a sequence of thoughts. Likewise, in lower animals whose prefrontal areas have been removed, the ability to concentrate on psychological tests is almost completely lost. The human being without prefrontal areas is still capable of performing many intellectual tasks, such as answering short questions and performing simple arithmetic computations (such as $9 \times 6 = 54$), thus illustrating that the basic intellectual activities of the cerebral cortex are still intact without the prefrontal areas. Yet if concerted *sequences* of cerebral functions are required of the person, he becomes completely disorganized. Therefore, the prefrontal areas seem to be important in keeping the mental functions directed toward goals.

Elaboration of Thought by the Prefrontal Areas. Another function that has been ascribed to the prefrontal areas by psychologists is called *elaboration of thought.* This means simply an increase in depth and abstractness of the different thoughts. It can be shown by psychological tests that lower animals presented with successive bits of sensory information forget these bits, however important they might be, within a second or more if the prefrontal areas have been removed. Yet if the prefrontal areas are intact, many successive bits of information can be remembered for many seconds. Therefore, there is much reason to believe that the prefrontal areas are especially capable of storing many bits of information *temporarily* and then correlating these into thoughts of a high order. This ability to hold and correlate many types of information simultaneously in the prefrontal areas could well explain the many functions of the brain that we associate with higher intelligence, such as abilities to (1) plan for the future, (2) delay action in response to incoming sensory signals so that the sensory information can be weighed until the best course of response is decided, (3) consider the consequences of motor actions even before these are performed, (4) solve complicated mathematical, legal, or philosophical problems, (5) correlate all avenues of information in diagnosing rare diseases, and (6) control one's activities in accord with moral laws.

Effects of Destruction of the Prefrontal Areas

The person without prefrontal areas ordinarily acts precipitously in response to incoming sensory impulses, such as striking an adversary too large to be beaten instead of pursuing the more judicious course of running away. Also, he is likely to lose many or most of his morals; he has little embarrassment in relation to his excretory, sexual, and social activities; and he is prone to quickly changing moods of sweetness, hate, joy, sadness, exhilaration, and rage. In short, he is a highly *distractible* person with lack of ability to pursue long and complicated thoughts.

THOUGHTS, MEMORY, LEARNING, AND CONSCIOUSNESS

Though all of us know what a *thought* is, any attempt to describe it in abstract terms is almost impossible. The comprehension of a visual scene at a given instant, for instance, would be a single thought; or overall comprehension of one's surroundings in response to other sensory information would be another thought; projection in one's mind of a mathematical equation is still a third thought. Thus, each instant of awareness could be defined as a thought. And the awareness itself can be defined as *consciousness.*

With the above definitions of thoughts and consciousness in mind, it now becomes easy to define memory and learning. *Memory* is the capability of recalling a thought at least once and usually again and again, and *learning* is the capability of the nervous system to store memories. Therefore, thoughts, consciousness, memory, and learning all go hand in hand and, in so far as the central nervous system is concerned, are almost inseparable.

Our most difficult problem in discussing consciousness, thoughts, memory, and learning is that we do not know the neural mechanism of a thought. We know that destruction of large portions of the cerebral cortex does not prevent a person from having thoughts, but it does reduce his *degree* of awareness of his surroundings. On the other hand, destruction of far smaller portions of the thalamus, or especially of the mesencephalic tegmentum, can cause tremendously decreased awareness or even complete unconsciousness.

Each thought almost certainly involves simultaneous action in portions of the cerebral cortex, thalamus, rhinencephalon, and reticular formation of the brain stem. Some crude thoughts probably depend almost entirely on lower centers; the thought of pain is probably a good example, for electrical stimulation of

the human cortex rarely elicits anything more than the mildest degrees of pain, while stimulation of certain areas of the hypothalamus and mesencephalon in animals apparently causes excruciating pain. On the other hand, a type of thought pattern that requires mainly the cerebral cortex is that involving vision, because loss of the visual cortex causes complete inability to perceive visual form or color.

Therefore, we might formulate a definition of a thought in terms of neural activity as follows: A thought probably results from the momentary "pattern" of stimulation of many different parts of the nervous system at the same time, probably involving most importantly the cerebral cortex, the thalamus, the rhinencephalon, and the upper reticular formation of the brain stem. The stimulated areas of the rhinencephalon, thalamus, and reticular formation probably determine the crude nature of the thought, giving it such qualities as pleasure, displeasure, pain, comfort, crude modalities of sensation, localization to general parts of the body, and other gross characteristics. On the other hand, the stimulated area of the cortex probably determines the discrete characteristics of the thought (such as specific localization of sensations in the body and of objects in the fields of vision), discrete patterns of sensation (such as the rectangular pattern of a concrete block wall or the texture of a rug), and other individual characteristics that enter into the overall awareness of a particular instant.

THE BASIS OF MEMORY

If we accept the above approximation of what constitutes a thought, we can see immediately that the mechanism of memory must be equally as complex as the mechanism of a thought, for, to provide memory, the nervous system must create the same spatial and temporal pattern of stimulation in the central nervous system at some future date. Though we cannot explain in detail what a memory is, we do know some of the basic neuronal processes that probably lead to the process of memory.

All of us know that all degrees of memory occur, some memories lasting only a few seconds and others lasting hours, days, months, or years. Possibly all of these types of memory are caused by the same mechanism operating to different degrees of fulfillment. Yet, it is also possible that different mechanisms of memory do exist. Indeed, most physiologists distinguish either two or three types of memory. For purposes of the following discussion, we will divide memory into (1) *instantaneous memory* and (2) *fixed memory,* which includes two major subdivisions (a) *short-term memory* (also called *recent memory*) that lasts for a few days and (b) *long-term memory* that can last for any period of time up to the entire life span of the person.

Instantaneous Memory.   Instantaneous memory refers to one's ability to recall tremendous amounts of information from one second to the next or from minute to minute. For instance, if one is walking through a large industrial plant, literally thousands or perhaps millions of small bits of information are passed into the nervous system from the sensory receptors. Most of these bits are stored without one's thinking about them, but for periods of seconds or minutes thereafter, the information can still be recalled. Unfortunately, we do not know the basic mechanism of this type of memory, but the following three theories are representative of possible mechanisms.

REVERBERATING CIRCUIT THEORY OF INSTANTANEOUS MEMORY.   When a tetanizing electrical stimulus is applied directly to the surface of the cerebral cortex and then removed after a second or more, the local area excited by this stimulus continues to emit rhythmic action potentials for minutes or, under favorable conditions, for as long as one hour. This effect results from local reverberating circuits, the impulses passing through a multiconnected circuit of neurons in the local area of the cortex itself or back and forth between the cortex and the thalamus.

It is presumed that sensory signals reaching the cerebral cortex can set up similar reverberating oscillations that continue for many minutes. Then gradually, as the reverberating circuits fatigue and the oscillation ceases, the instantaneous or temporary memory fades away.

One of the principal observations in support of this theory of instantaneous or temporary memory is that any factor that causes a general disturbance of brain function, such as sudden concussion of the brain, immediately erases all instantaneous or temporary memory, and the memory cannot be recalled when the disturbance is over.

POST-TETANIC POTENTIATION THEORY OF INSTANTANEOUS OR TEMPORARY MEMORY.   In most areas of the nervous system, including even the anterior motoneurons of the spinal cord, tetanic stimulation of a neuron for a few seconds causes a few seconds to a few hours of increased excitability of the neuron. If during this period the neuron is stimulated again, it responds much more vigorously than normally, a phenomenon called *post-tetanic potentiation.* This is obviously a type of memory that depends on changes in excitability of the involved neurons. It is likely that this phenomenon results from some temporary change in the synapses or in the neurons themselves.

DC POTENTIAL THEORY OF INSTANTANEOUS OR TEMPORARY MEMORY.   Another change that occurs in neurons following a period of excitation is a change in the membrane potential of the neuron, lasting for seconds to hours. Because this changes the excitability of the neuron, it therefore could be the basis of a memory. These changes in membrane potential are called *DC potentials* or, sometimes, *electronic potentials.* Measurements in the cerebral cortex show that such potentials do indeed occur, especially in the dendritic layers of the cortex.

Fixed Memory—The Time Factor in the Fixation of Memories.   If a memory is to last in the brain so that it can be recalled days later, it must become "fixed" in the neuronal circuits. This process requires 15 to 20 minutes for minimal fixation and an hour or more for maximal fixation. For instance, if a strong sensory impression is made on the brain but is then followed by strong electrical stimulation of the brain, the sensory experience will not be remembered at all. However, if the same sensory stimulus is impressed on the brain and the strong electrical shock is delayed for as long as an hour, the memory "engram" will have become established. If we hold to the reverberation theory of instantaneous or temporary memory, one could explain these differences in the following way: If the reverberation is disrupted soon after it begins by strong electrical stimulation, significant physical or chemical changes in the synapses will not occur to develop a fixed memory. But if the thought pattern is not disrupted, sufficient reverberation can take place to establish a deeply trodden memory pathway.

Similarly, if many strong sensory experiences occur one after another, it usually will not be possible to remember any one of them well, presumably because each successive experience disrupts the previous reverberating circuit (or previous electrotonic potentials in the case of the electrotonic theory) before an engram of the memory can be established.

MECHANISM OF FIXED MEMORY—ALTERATION OF TRANSMISSION FACILITY AT THE SYNAPSES.   Fixed memory means the ability of the nervous system to recall thoughts long after initial excitation of the thoughts is over. We know that fixed memory does not depend on continued activity of the nervous system, because the brain can be totally inactivated by cooling, by general anesthesia, by hypoxia, by ischemia, or by any other method and yet memories that have been previously stored are still retained when the brain becomes active once again. Therefore, it is assumed that fixed memory must result from some actual alterations of the synapses, either physical or chemical.

Many different theories have been offered to explain the synaptic changes that cause fixed memory. Among the most important of these are:

1. *Anatomical changes in the synapses.* Cajal, more than half a century ago, discovered that the number of terminal fibrils ending on neuronal cells and dendrites in the cerebral cortex increases with age. Conversely, physiologists have shown that inactivity of regions of the cortex causes thinning of the cortex: for instance, thinning of the primary visual cortex in animals that have lost their eyesight. Also, intense activity of a particular part of the cortex can cause excessive thickening of the cortical shell in that area alone. This has been demonstrated especially in the visual cortex of animals subjected to repeated visual experiences. Finally, some neuroanatomists believe that they can observe electronmicrographic changes in presynaptic terminals that have been subjected to intense and prolonged activity.

All of these observations have led to a widely held belief that fixation of memories in the brain results from physical changes in the synapses themselves: perhaps changes in numbers of presynaptic terminals, perhaps in sizes of the terminals, or perhaps in their chemical compositions. Also, it has been suggested that permanent changes might occur in excitability of the postsynaptic neurons. Such physical changes could cause permanent or semipermanent increase in the degree of facilitation of the synapses, thus allowing signals to pass through the synapses with progressive ease the more often the memory engram is used. This obviously would explain the tendency for memories to become more and more deeply fixed in the nervous system the more often they are recalled or the more often the person repeats the sensory experience that leads to the memory engram.

2. *Theoretical function of RNA in memory.* The discovery that DNA and RNA can act as codes to control reproduction, which in itself is a type of memory from one generation to another, plus the fact that these substances once formed in a cell tend to persist for the lifetime of the cell, has led to the theory that nucleic acids might be involved in memory changes of the neurons, changes that could last for the lifetime of the person. In addition, biochemical studies have shown an increase in RNA in some active neurons. Yet, a mechanism by which RNA could cause facilitation of synaptic transmission has never been found, and more recently it has been shown that memories can be fixed in neurons even when the chemical processes for formation of RNA are blocked. Therefore, this theory seems to have been based almost entirely on analogy rather than on factual evidence.

3. *The glial theory and other extraneuronal theories.* Anatomists have demonstrated that the structures of glial cells surrounding neurons change markedly under different functional conditions. This obviously has led to the theory that glial changes might increase the facilitation of synapses and, therefore, be the basis of memory. Also, chemists have demonstrated that the quantity and chemical composition of mucopolysaccharides surrounding synapses change under different functional conditions, and this leads to still another theory for memory. However, in both of these instances there has been little more than circumstantial evidence.

In summary, the theory that seems most likely at present for explaining the fixation of memory is that some actual physical or chemical change occurs in the presynaptic terminals themselves or possibly in the postsynaptic neuron, these changes permanently facilitating the transmission of impulses at the synapses. If all the synapses are thus facilitated in a thought circuit, this circuit can be re-excited by any one of many diverse incoming signals at later dates, thereby causing memory. The overall facilitated circuit is called a *memory engram.*

SHORT-TERM VERSUS LONG-TERM FIXED MEMORY. Physiologists also differentiate between memories that can be remembered for only a few days versus those that can be recalled many months or years later. There are several reasons for making this differentiation, as follows: (1) Destruction of the temporal lobes in a person causes him to lose his memory for recent events that have occurred over a period of about two to six days prior to the destruction. However, this destruction does not cause loss of long-term, well-ingrained memories. (2) Electroshock therapy in mental patients, or the occurrence of convulsions in epileptic patients, frequently causes loss of memories of the past few days but not loss of long-term memories.

Therefore, it is believed that certain areas of the brain can support memories for much longer periods of time than can other areas, some areas perhaps being able to store memories for only a few seconds, a few minutes, or a few days, while others can store memories for years or decades. It is also possible that memories are first stored in temporary areas, such as in the temporal lobes, and then are transferred over a period of time to other more permanent storage areas. This process of transferring memory storage has been proved to occur, for memories that utilize primarily one cerebral cortex can be transferred to the other cerebral cortex through the corpus callosum, as will be discussed in the following paragraph.

FUNCTION OF THE CORPUS CALLOSUM AND ANTERIOR COMMISSURE TO TRANSFER MEMORIES TO THE OPPOSITE HEMISPHERE

Fibers in the *corpus callosum* connect the respective cortical areas of the two hemispheres with each other except for the cortices of the anterior portions of the temporal lobes; these temporal cortical areas are interconnected by fibers that pass through the *anterior commissure.* Because of the tremendous number of fibers in the corpus callosum, it was very early believed that this massive structure must have some important function to correlate activities of the two cerebral hemispheres. However, cutting the corpus callosum in experimental animals, or in occasional instances even in the human being, caused no change in the function of the brain that could be discerned by usual psychological tests. Therefore, for a long time the function of the corpus callosum was a complete mystery.

Recently, improved psychological experiments have demonstrated the function of the corpus callosum and the anterior commissure. This can be explained best by recounting one of the experiments. A monkey is first prepared by cutting the corpus callosum and splitting the optic chiasm longitudinally. Then he is taught to recognize different types of objects with his right eye while his left eye is covered. Then the right eye is covered and he is tested to determine whether or not his left eye can recognize the same object. The answer to this is that the left eye *can not* recognize the object. Yet, on repeating the same experiment with the optic chiasm still split but the corpus cal-

losum intact, it is found invariably that recognition in one hemisphere of the brain creates recognition also in the opposite hemisphere.

Thus, the function of the corpus callosum and the anterior commissure is to associate respective cortical areas of the two hemispheres in such a way that a memory engram impressed on one hemisphere will become simultaneously impressed on the opposite hemisphere. Yet, it should be recalled once again that many other types of correlation between different cortical areas normally takes place through lower brain centers, particularly through centers in the reticular formation and thalamus. This explains why it is possible to transect the corpus callosum and other major association bundles in the cerebrum without affecting most of our sensory and motor functions of the brain.

RELATIONSHIP OF CONDITIONED REFLEXES TO LEARNING

Since the time of Pavlov, it has often been taught that learning depends entirely or almost entirely on alteration of the *natural,* or *unconditioned, reflexes.* The unconditioned reflexes result from inherent properties of the nervous system that are present at the time of birth, and they include such reflexes as the sucking act in response to oral stimuli, salivation in response to the smell of food, jerk of the arms in response to loud sounds or flashes of light, crying reflexes, reflexes that cause walking motions, flexion, extension, and so forth.

A *conditioned reflex* can be established by associating a new sensory stimulus with a natural stimulus that elicits an unconditioned reflex. For instance, in one of Pavlov's classical experiments, food given to an animal caused the unconditioned reflex of salivation. A bell was rung each time before food was offered to the animal. After this had been done a number of times, the bell was rung but food was not offered. Nevertheless, the dog still salivated a short interval after the bell had been rung. This reaction is called a conditioned reflex.

A *secondary* conditioned reflex can then be built onto the primary conditioned reflex. For instance, a flash of light before the bell ringing and reflex salivation response will eventually establish a secondary conditioned reflex so that flashing the light alone is followed by salivation.

One can easily see that many successive reflexes could theoretically be built one on top of the other until finally the central nervous system could perform a multitude of reflex functions.

Learning by Trial and Error. Conditioned reflexes by themselves, however, are not sufficient to explain most learning that occurs in the brain, for many chance happenings can lead to learning—for instance, chance movement of the head to a position that provides a particularly interesting scene can elicit a memory that will cause the person to move his head to that same position again to see the same scene. Therefore, trial and error happenings that provide motor and sensory associations are another mechanism of learning—and perhaps an even more extensively used mechanism—in addition to that of conditioned reflexes.

Learning by Precept. Another type of learning occurs secondarily to our thought processes. That is, even in the absence of new sensory experience, simply thinking about previous sensory experiences can give rise to new thought patterns that in turn become memories. For instance, let us assume that one has never seen an automobile but is completely familiar with the wagon and with the gasoline engine. Only a slight amount of thought can then generate the concept of the automobile. Similarly, multiple induction from our wide variety of sensory experiences can lead to abstract thoughts, and correlation of abstract thoughts can in turn lead to still other abstract thoughts, the process continuing on and on.

INTELLECTUAL OPERATIONS OF THE BRAIN

Thus far, we have considered the approximate nature of thoughts and possible mechanisms by which memory and learning can occur. Now we need to consider the mechanisms by which the brain performs complex intellectual operations, such as the analysis of sensory information and the establishment of abstract thoughts. About these mechanisms we know almost nothing, but experiments along these lines have established the following important facts: First, the brain can focus its attention on specific types of information (which was discussed in the preceding chapter). Second, the different qualities of each set of information signals are split away from the central signal and are transmitted to multiple areas of the brain. Third, the brain compares new information with old information in its memory loci. And, fourth, the brain determines patterns of stimulation.

Analysis of Information by Splitting Its Qualities. When sensory information enters the nervous system, one of the first steps in its analysis is to transmit the signal to separate parts of the brain that are selectively adapted for detecting specific qualities. For instance, if the hand is placed on a hot stove, (a) pain information is sent through the spinotectal and spinothalamic tracts into the reticular formation of the brain stem and into specific thalamic and hypothalamic nuclei, (b) tactile information is sent through the dorsal column system into the somesthetic cortex, giving the cortex a detailed description of the part of the hand that is touching the hot stove, and (c) kinesthetic information from the muscles and joints is sent to the cerebellum, the reticular formation, and to higher centers of the brain to give momentary information of the position of the hand. Thus, the different qualities of the overall information are dissected and transmitted to different parts of the brain. It is on the basis of analyses in all these parts of the brain that motor responses to the incoming sensory information are formulated.

One can see that different centers of the brain react to specific qualities of information—the reticular formation and certain regions of the thalamus and hypothalamus to pain, regions of the mesencephalon and hypothalamus to the affective nature of the sensation (that is, whether it is pleasant or unpleasant), the somatic cortex to the localization of sensation, other areas to kinesthetic activities, and others to visual, auditory, gustatory, olfactory, and vestibular information.

Analysis of Information by Comparison with Memories. We all know that new sensory experiences are immediately compared with previous experiences of the same or similar types. For instance, this is the way we recognize a person whom we know. Yet, it still remains a mystery how we can make these comparisons. One theory suggests that the memory engram might be stored in one layer of the cerebral cortex and that the new incoming sensory experience might be impressed on another layer. If the new pattern of stimulation matches the memory engram, some interaction between the two gives the person a sense of recognition.

Analysis of Patterns. In analyzing information, the brain depends to a great extent on "patterns" of stimulation. For instance, a square is detected as a square regardless of its position or angle of rotation in the visual field. Likewise, a series of parallel bars is detected as parallel bars regardless of their orientation, or a fly is detected as a fly whether it is seen in the peripheral or central field of vision. We can extend the same logic to the somatic sensory areas, for a person can detect a square simply by feeling it whether it be in an upright, horizontal, or angulated position. Also, it can be detected even by the feet even though the feet may have never felt a square before.

It is believed that one of the major methods for analyzing information is to dissect it into its component patterns. For instance, repetitive auditory sensations and visual sensations of a series of lights have at least one characteristic in common: both have the pattern of repetitiveness. Yet we cannot say how patterns of sensations are detected. One theory holds that their detection results from traveling waves passing through the cortex. As these cross the brain they might cause a signal to be transmitted every time they interact with an excited point of the cortex. For instance, if a traveling wave were passing over a visual cortex that is being excited by a regular series of lights, every time this wave should come in contact with an activated point, a signal would theoretically be transmitted away from the cortex to some *analyser portion* of the brain. The analyser would receive a train of rhythmic impulses, each impulse denoting a separate light, and this train of impulses could then theoretically give the person the sensation of repetitiveness. Likewise, repetitive sounds entering the ears could transmit a train of impulses to the same analyzer which would also give a sensation of repetitiveness, or movement of the hand over a corrugated surface could give the analyser the same sensation of repetitiveness.

If we use our imaginations, we can see that only a few different types of analytical patterns can go a long way toward explaining almost all types of sensory impressions.

PROCESSING OF INFORMATION SO THAT PATTERNS CAN BE DETERMINED. Beautiful examples of the ability of the brain to process information for determination of patterns have been discovered in relation to the visual system. In Chapter 54 we pointed out that the retina itself processes visual information to a great extent even before it is transmitted into the brain. For instance, the retina is especially responsive to rapidly changing light intensities or to movement of images across the receptors, and the signals sent to the brain are appropriately modified to transmit these qualities. Then, when the image is transmitted through the lateral geniculate body and thence into the primary visual cortex, the visual signals are altered again in such a way that the cortex detects mainly boundaries between light and dark areas. That is, the visual scene is "differentiated" by this mechanism, bringing out strongly the contrasts of the scene but deemphasizing the flat areas. This explains why a few lines drawn on a paper can give one the impression of a person's appearance. Certainly the lines do not represent the actual picture of a person, but they do give the visual cortex the same pattern of contrasts that his own visual system would give.

One can see, therefore, that the visual cortex utilizes transformed types of information to point up the most striking characteristics of the visual scene while neglecting unimportant information. Then, the patterns of the visual scene can be extracted from this "pre-selected" information.

Conversion of Information in Association Areas of the Brain. Certain areas of the brain perform specific functions, such as the primary sensory areas and the motor area. On the other hand, some of the remainder of the brain, called *association areas,* do not necessarily perform functions associated with a particular modality of sensation or with a specific motor function. These areas often receive information from different sensory inputs and also from other association areas of the brain. Some association areas are located in the cortex and others in lower brain regions, such as in the *thalamus* and *caudate nucleus.* Indeed, for every association area in the cortex, there is a reciprocally connected association area of the thalamus. In addition, other areas of the thalamus, such as the *centrum medianum,* interconnect widely within the thalamus itself and presumably act as association areas between the different thalamic nuclei.

It is in the association areas that information from many different sources interact. Thus, the angular gyrus of the cerebral cortex, where the temporal, parietal, and occipital lobes all come together, is an association area that receives auditory, visual, and tactile information. Neurophysiological experiments have demonstrated that these interacting types of information can at times inhibit each other and at other times facilitate each other. Yet, how the interactions in the association areas are analyzed beyond this point is still a mystery.

INTELLECTUAL ASPECTS OF MOTOR CONTROL

Control of muscular movements by the cerebral cortex and other areas of the brain was discussed in Chapter 58, but we still need to consider the manner in which the intellectual operations of the brain elicit motor events. Almost all sensory and even abstract experiences of the mind are eventually expressed in some type of motor activity, such as actual muscular movements of a directed nature, tenseness of the muscles, total relaxation of the muscles, attainment of certain postures, tapping of the fingers, grimaces of the face, or speech.

Psychological tests show that the analytical portions of the brain control motor activities in the following sequence of three stages: (1) origin of the thought of the motor activity to be performed, (2) determination of the sequence of movements necessary to perform the overall task, and (3) control of the muscular movements themselves.

Origin of the Thought of a Motor Activity. A large number of motor movements, which we call subconscious movements, are conceived in the lower regions of the brain, but complex learned tasks can be performed only when the cerebral cortex is present. The most important cortical areas for this purpose are the posterior part of the superior temporal gyrus and the angular gyrus of the dominant hemisphere. The student will immediately recognize this area to be the *gnostic region* of the brain. Thus, the sensory interactions that take place in this all-important sensory part of the dominant hemisphere not only play a major role in interpreting sensory experiences but also in determining the person's course of motor activities.

Determination of Sequences of Movement. In our earlier discussion of cortical control of motor activities, it was pointed out that the somatic association area is necessary for performance of complex learned movements; destruction of this area both in animals and human beings causes the motor activities to lose much of their purposefulness. Therefore, it is postulated that once the gnostic area of the brain has determined the motor activity to be performed, the actual sequences of movement are determined in the somatic association areas as discussed in Chapter 58. Signals are transmitted from here to the motor portion of the brain to control the actual motor movements, as also discussed in Chapter 58.

It is particularly significant that destruction of the most lateral portion of the somatic association area in the dominant hemisphere, the area most concerned with facial and laryngeal movements, almost completely destroys one's capability of organizing his speech into phrases or sentences that express a thought. This is discussed further in the following section in relation to the function of the brain in communication.

FUNCTION OF THE BRAIN IN COMMUNICATION

One of the most important differences between the human being and lower animals is the facility with which human beings can communicate with one another. Furthermore, because neurological tests can easily assess the ability of a person to communicate with others, we know perhaps more about the sensory and motor systems related to communication than about any other segment of cortical function. Therefore, we will review rapidly the function of the cortex in communication, and from this one can see immediately how the principles of sensory analysis and motor control that were just discussed apply to this art.

There are two aspects to communication: first, the *sensory aspect,* involving the ears and eyes, and, second, the *motor aspect,* involving vocalization and its control.

Sensory Aspects of Communication. We noted earlier in the chapter that destruction of portions of the *secondary auditory* and *visual areas* of the cortex can result in inability to understand the spoken word or the written word. These effects are called respectively *auditory agnosia* and *visual agnosia* or, more commonly, *word deafness* and *word blindness.* On the other hand, some persons are perfectly capable of understanding either the spoken word or the written word but are unable to understand their meanings when used to express a thought. This results most frequently when the *angular gyrus* or *posterior portion of the superior temporal gyrus* in the dominant hemisphere is damaged or destroyed. This is considered to be *general agnosia.*

Motor Aspects of Communication—Aphasia. SENSORY APHASIAS. The process of speech involves (a) formation in the mind of thoughts to be expressed and choice of the words to be used and (b) the actual act of vocalization. The formation of thoughts and choice of words is principally the function of the sensory areas of the brain, for we find that a person who has a destructive lesion in the same temporal and angular gyrus region that is involved in general agnosia also has inability to formulate intelligible thoughts to be communicated. And at other times the thoughts can be formulated, but the person is unable to put together the appropriate sequence of words to express the thought; this results most frequently from destruction of the lowermost tip of the somatic sensory association area where it dips into the sylvian fissure. These inabilities to formulate thoughts and word sequences are called *sensory aphasias.*

MOTOR APHASIA. Often a person is perfectly capable of deciding what he wishes to say, and he is capable of vocalizing, but he simply cannot make his vocal system emit

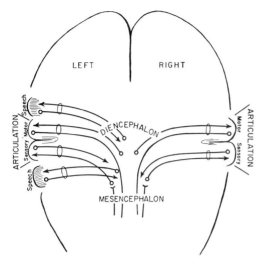

and mesencephalic areas, as illustrated in Figure 61–5, rather than directly from one area of the cortex to the next.

REFERENCES

Bitterman, M. E.: The evolution of intelligence. *Sci. Amer., 212*: 92(1), 1965.
Bogoch, S.: The Biochemistry of Memory: With an Inquiry into the Function of the Brain Mucoids. New York, Oxford University Press, Inc., 1968.
Boycott, B. B.: Learning in the octopus. *Sci. Amer., 212*:42(3), 1965.
Brain, R.: Speech Disorders. London, Butterworth & Co., 1961.
Ciba Foundation Study Group No. 20: Function of the Corpus Callosum. Boston, Little, Brown and Company, 1965.
Ciba Symposium on Disorders of Language. Boston, Little, Brown and Company, 1964.
Creutzfeldt, O. D.: The general physiology of the cortex. *Proc. Int. Union of Physiol. Sciences, 6*:258, 1968.
Darley, F. L.: Brain Mechanisms Underlying Speech and Language. New York, Grune & Stratton, Inc., 1967.
Ferster, C. B.: Arithmetic behavior in chimpanzees. *Sci. Amer., 210*:98(5), 1964.
Galambos, R.: Neurophysiological studies on learning and motivation. *Fed. Proc., 20*:603, 1961.
Gerard, R. W.; Fessard, A.; and Konorski, J. (eds.): Brain Mechanisms and Learning. Philadelphia, F. A. Davis Co., 1961.
Hernandez-Peon, R.: The Physiological Basis of Mental Activity. New York, American Elsevier Publishing Co., Inc., 1963.
Hess, W. R.: The Biology of Mind. Chicago, University of Chicago Press, 1964.
Kandel, E. R., and Spencer, W. A.: Cellular neurophysiological approaches in the study of learning. *Physiol. Rev., 48*:65, 1968.
Laursen, A. M.: Higher functions of the central nervous system. *Ann. Rev. Physiol., 29*:543, 1967.
McConnell, J. V.: Comparative physiology: learning in invertebrates. *Ann. Rev. Physiol., 28*:107, 1966.
Penfield, W.: The Excitable Cortex in Conscious Man. Springfield, Illinois, Charles C Thomas, Publisher, 1958.
Peterson, L. R.: Short-term memory. *Sci. Amer., 215*:90(1), 1966.
Schaltenbrand, G., and Woolsey, C. N.: Principles of Cerebral Localization and Organization. Madison, University of Wisconsin Press, 1964.
Scher, J. M. (ed.): Theories of the Mind. New York, The Macmillan Company, 1962.
Sperry, R. W.: Mental unity following surgical disconnection of the cerebral hemispheres. *Harvey Lectures, 1966–1967*, p. 293.
Sperry, R. W.: The great cerebral commissure. *Sci. Amer., 210*: 42(1), 1964.
Wickelgren, B. G.: Habituation of spinal interneurons. *J. Neurophysiol., 30*:1424, 1967.
Wiener, N., and Schade, J. P.: Progress in Brain Research. Nerve, Brain and Memory Models. New York, American Elsevier Publishing Co., Inc., 1963, Vol. II.
Young, J. Z.: The Memory System of the Brain. Berkeley, University of California Press, 1966.

Figure 61–5 Centers of the cortex for speech control. (From Penfield and Rasmussen: The Cerebral Cortex of Man. The Macmillan Co.)

words instead of noises. This effect, called *motor aphasia,* almost always results from damage to *Broca's area,* which lies in the *anterior* facial region of the motor cortex in the dominant hemisphere. Therefore, we assume that the *patterns* for control of the larynx, lips, mouth, respiratory system, and other accessory muscles of articulation are all controlled in this area.

Finally, we have the act of *articulation* itself, which means the muscular movements of the mouth, tongue, larynx, and so forth, that are responsible for the actual emission of sound. The *posterior* facial and laryngeal regions of the motor cortex activate these muscles, and the sensory cortex helps to control the muscle contractions by feedback servomechanisms described in Chapter 58. Destruction of these regions can cause either total or partial inability to speak distinctly.

Finally, note that the pathways from one part of the cortex to another that control communication pass mainly through the thalamic

CHAPTER 62

Behavioral Functions of the Brain: The Limbic System, and Role of the Hypothalamus

Behavior is a function of the entire nervous system, not of any particular portion. Even the discrete cord reflexes are an element of behavior, and the wakefulness and sleep cycle discussed in Chapter 60 is certainly one of the most important of our behavioral patterns. However, in this chapter we will deal with those special types of behavior associated with emotions, subconscious motor and sensory drives, and the intrinsic feelings of pain and pleasure. These functions of the nervous system are performed mainly by subcortical structures, such as the hypothalamus and adjacent areas, but older portions of the cerebral cortex that are located on the medial and ventral portions of the cerebral hemispheres also play a role. This overall group of brain structures is called the *limbic system.*

THE LIMBIC SYSTEM

Figure 62–1 illustrates most of the important structures of the limbic system, showing the *hypothalamus* as one of the central elements of the system; surrounding it are other subcortical structures including the *preoptic area,* the *septum,* the *paraolfactory area,* the *epithalamus,* the *anterior nuclei of the thalamus, portions of the basal ganglia,* the *hippocampus,* and the *amygdala.* Surrounding the subcortical areas is the *limbic cortex* composed of a ring of cerebral cortex beginning (a) in the *orbitofrontal area* on the ventral surface of the frontal lobes, extending (b) upward in front of and over the corpus callosum onto the medial aspect of the cerebral hemisphere to the *cingulate gyrus,* and finally (c) downward posterior to the corpus callosum onto the ventromedial surface of the temporal lobe in the *hippocampal gyrus, pyriform area,* and *uncus.* Thus, on the medial and ventral surfaces of each cerebral hemisphere is a ring of

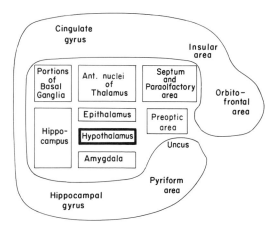

Figure 62–1   The limbic system.

paleocortex that surrounds a group of deep structures intimately associated with overall behavior and with emotions. This entire system of structures is the *limbic system,* and the cortical regions of the system are called the *limbic cortex.*

In the past, many neurophysiologists have lumped the entire limbic system together as a unit structure that controls the different emotional and behavioral patterns of the nervous system. As we learn more about the various portions of this system, it is becoming evident that different areas of the system perform very specific functions.

BEHAVIORAL FUNCTIONS OF THE HYPOTHALAMUS AND RELATED AREAS

The vegetative functions of the hypothalamus—that is, the subconscious control of many of the body's internal activities by the hypothalamus—were discussed in Chapter 59 in relation to the autonomic nervous system. Some of these functions include regulation of arterial pressure, body fluid balance, electrolyte content of the body fluids, feeding, gastrointestinal activity, and many internal secretions of the endocrine glands.

In addition to these vegetative functions, the hypothalamus and closely associated areas of the diencephalon and mesencephalon also have important behavioral functions, as follows:

PLEASURE AND PAIN; REWARD AND PUNISHMENT

In the past several years, it has been learned that many hypothalamic and closely related structures are particularly concerned with the affective nature of sensory sensations—that is, with whether the sensations are *pleasant* or *painful*. These affective qualities are also called *reward* and *punishment*. Electrical stimulation of certain regions soothes the animal, whereas electrical stimulation of other regions causes extreme pain, fear, defense, escape reactions, and all the other elements of punishment. Obviously, these two oppositely responding systems greatly affect the behavior of the animal.

Figure 62–2 illustrates a technique that has been used for localizing the specific reward and punishment areas of the brain. In this figure a lever is placed at the side of the cage and is arranged so that depressing the lever makes electrical contact with a stimulator. Electrodes are placed successively at different areas in the brain so that the animal can stimulate the area

Figure 62–2   Technique for localizing reward and punishment centers in the brain of a monkey.

by pressing the lever. If stimulating the particular area gives the animal a sense of reward, then he will press the lever again and again, sometimes as much as 5000 times per hour. Furthermore, when offered the choice of eating some delectable food as opposed to the opportunity to stimulate the reward center, he often chooses the electrical stimulation.

By using this procedure, the major reward centers have been found to be located in the hypothalamus, primarily *in the ventromedial nuclei of the hypothalamus*. Less potent reward centers, which are probably secondary to the major ones in the hypothalamus, are found in the amygdala, septum, certain areas of the thalamus and basal ganglia, and finally extending downward into the basal tegmentum of the mesencephalon.

The apparatus illustrated in Figure 62–2 can also be connected so that pressing the lever turns off rather than turns on an electrical stimulus. In this case, the animal will not turn the stimulus off when the electrode is in one of the reward areas, but when it is in certain other areas he immediately learns to turn it off. Stimulation in these areas causes the animal to show all the signs of pain and displeasure. Furthermore, prolonged stimulation for 24 hours or more causes the animal to become severely sick and actually leads to death.

By means of this technique, the principal centers for pain, punishment, and escape tendencies have been found in the *midbrain periventricular area of the dorsomedial tegmentum* and extending upward into the *periventricular structures of the hypothalamus and thalamus*. The *perifornical nucleus* of the hypothalamus is the most reactive of all the hypothalamic areas. Another area that causes intense pain and punishment tendencies is located in the central gray area of the mesencephalon.

It is particularly interesting that stimulation in the pain and punishment centers can frequently inhibit the reward and pleasure centers completely, illustrating that pain can take precedence over pleasure and reward.

Importance of Reward and Punishment in Behavior.   Almost everything that we do depends on reward and punishment. If we are doing something that is rewarding, we continue to do it; if it is punishing, we cease to do it. Therefore, the reward and punishment centers undoubtedly constitute one of the most important of all the controllers of our bodily activities, our motivations, and so forth.

Importance of Reward and Punishment in Learning—Habituation and Reinforcement.   It has been found in animal experiments that a sensory experience that causes neither reward nor punishment is remembered hardly at all. Also, electrical recordings have shown that new and novel sensory stimuli always excite the cerebral cortex. But repetition of the stimulus over a period of time leads to almost complete extinction of the cortical response if this stimulus does not excite either reward or punishment centers. Thus, the animal becomes *habituated* to the sensory stimulus. But if the stimulus causes either reward or punishment rather than indifference, the cortical response becomes progressively more and more intense with repetitive stimulation instead of fading away, and the response is said to be *reinforced.* Therefore, an animal builds up strong memory engrams for sensations which are either rewarding or punishing, but, on the other hand, develops complete habituation to indifferent sensory stimuli. Therefore, it is evident that the reward and punishment centers of the midbrain have much to do with controlling the type of information that we learn.

Effect of Tranquilizers on the Reward and Punishment Centers. Administration of a tranquilizer, such as chlorpromazine, inhibits both the reward and punishment centers, thereby greatly decreasing the affective reactivity of the animal. Therefore, it is presumed that tranquilizers function in psychotic states by suppressing many of the important behavioral areas of the hypothalamus and its associated regions of the brain.

THE AFFECTIVE-DEFENSIVE PATTERN—RAGE

An emotional pattern that involves the hypothalamus and has been well characterized by Hess is the affective-defensive pattern. This can be described as follows:

Stimulation of the *perifornical nuclei of the hypothalamus*, which are also the hypothalamic regions that give the most intense sensation of punishment, causes the animal to (1) develop a defense posture, (2) extend his claws, (3) lift his tail, (4) hiss, (5) spit, (6) growl, and (7) develop pilo-erection, wide-open eyes, and dilated pupils. Furthermore, even the slightest provocation causes an immediate savage attack. This is approximately the behavior that one would expect from an animal being severely punished, and it is a pattern of behavior that has also been called simply *rage*. It can even occur in decorticated animals, illustrating that the basic behavioral patterns for protective activities are present in the lower regions of the brain.

Stimulation of the more rostral areas of the punishment areas—that is, in the midline preoptic and septal areas—causes mainly fear and anxiety, associated with a tendency for the animal to run away.

Exactly the opposite emotional behavioral patterns occur when the reward centers are stimulated: namely, docility and tameness.

FUNCTION OF THE HYPOTHALAMUS IN SLEEP, WAKEFULNESS, ALERTNESS, AND EXCITEMENT

Stimulation of regions of the hypothalamus dorsolateral and caudal to the mammillary bodies greatly excites the reticular activating system and therefore causes wakefulness, alertness, and excitement. In addition, stimulation of the dorsal regions of the hypothalamus excites the sympathetic nervous system in general, increasing the arterial pressure, causing pupillary dilatation, and enhancing other activities associated with sympathetic activity. On the other hand, stimulation in a few isolated areas of the lateral hypothalamus or anterior hypothalamus or in isolated points of the thalamic portions of the reticular activating system often inhibits the mesencephalic portion of the reticular activating system, causing somnolence and sometimes actual sleep. Thus, the hypothalamus indirectly contributes much to the control of the degree of excitement and alertness.

PSYCHOSOMATIC EFFECTS OF THE HYPOTHALAMUS AND RETICULAR ACTIVATING SYSTEM

We are all familiar with the fact that abnormal function in the central nervous system can frequently lead to serious dysfunction of the different somatic organs of the body. This is also true in experimental animals, for, as pointed out above, prolonged electrical stimulation in the punishment regions of the brain can actually lead to severe sickness of the animal, culminating in death within 24 to 48 hours. We need, therefore, to understand briefly the mechanisms by which stimulatory effects in the brain can affect the peripheral organs. Ordinarily, this occurs through three routes: (1) through the motor nerves to the skeletal muscles throughout the body, (2) through the autonomic nerves to the different internal organs of the body, and (3) through the hormones secreted by the pituitary gland in response to nervous activity in the hypothalamus, as will be explained in Chapter 75.

Psychosomatic Disorders Transmitted Through the Skeletal Nervous System. Abnormal psychic states can greatly alter the degree of nervous stimulation to the skeletal musculature throughout the body and thereby increase or decrease the skeletal muscular tone. During states of attention the general skeletal muscular tone as well as sympathetic tone normally increases, whereas during somnolent states skeletal muscular and sympathetic activity both decrease greatly. In neurotic and psychotic states, such as anxiety, tension, and mania, generalized overactivity of both the muscles and sympathetic system often occur throughout the body. This in turn results in intense feedback from the muscle proprioceptors to the reticular activating system, and the epinephrine circulating in the blood as a result of the sympathetic activity directly excites the reticular activating system, both of which undoubtedly help to maintain an extreme degree of wakefulness and alertness that characterizes these emotional states. Unfortunately, though, the wakefulness prevents adequate sleep and also leads to progressive bodily fatigue but still inability to go to sleep.

Transmission of Psychosomatic Effects Through the Autonomic Nervous System. Many psychosomatic abnormalities result from hyperactivity of either the sympathetic or parasympathetic system. In general, hyperactivity of the sympathetic system occurs everywhere in the body at the same time rather than in focal areas, and the usual effects are (1) increased heart rate—sometimes with palpitation of the heart, (2) increased arterial pressure, (3) constipation, and (4) increased metabolic rate. On the other hand, parasympathetic signals are likely to be much more focal. For instance, stimulation of specific areas in the dorsal motor nuclei of the vagus nerves can cause more or less specifically (1) increased or decreased heart rate and palpitation of the heart, (2) esophageal spasm, (3) increased peristalsis in the upper gastrointestinal tract, or (4) increased hyperacidity of the stomach with resultant development of peptic ulcer. Stimulation of the sacral region of the parasympathetic system, on the other hand, is likely to cause extreme colonic glandular secretion and peristalsis with resulting diarrhea. One can readily see, then, that emotional patterns controlling the sympathetic and parasympathetic centers of the hypothalamus can cause wide varieties of peripheral psychosomatic effects.

Psychosomatic Effects Transmitted Through the Anterior Pituitary Gland. Electrical stimulation of the posterior hypothalamus increases the secretion of corticotropin by the anterior pituitary gland (as explained in Chapter 59) and therefore indirectly increases the output of adrenocortical hormones. One of the effects of this is a gradual increase in stomach hyperacidity because of the effect of glucocorticoids on stomach secretion. Over a prolonged period of time this obviously could lead to peptic ulcer, which is a well-known effect of hypersecretion of the adrenal cortex. Likewise, activity in the anterior hypothalamus increases the pituitary secretion of thyrotropin, which in turn increases the output of thyroxine and leads to an elevated basal metabolic rate. It is well known that differ-

ent types of emotional disturbances can lead to thyrotoxicosis (as will be explained in Chapter 77), this presumably resulting from overactivity in the anterior hypothalamus.

From these examples, therefore, it is evident that many types of psychosomatic diseases of the body can be caused by abnormal control of anterior pituitary secretion. The precise mechanisms by which stimulation in the central nervous system can cause increased pituitary secretion will be discussed in Chapter 75. Basically, it results from the formation in the hypothalamus of *neurosecretory substances* that are transported through a *hypophyseal portal system* to the anterior pituitary gland, there exciting the secretion of the pituitary hormones.

FUNCTIONS OF OTHER PARTS OF THE LIMBIC SYSTEM

A large portion of the signals originating in other parts of the limbic system besides the hypothalamus are funneled through the hypothalamus, giving rise to various vegetative or behavioral effects that we know to result when the hypothalamus is stimulated. Other signals are funneled into the reticular activating system to control the degree of wakefulness, alertness, and attention. And still other behavioral signals are transmitted directly from the limbic structures into the basal ganglia and into nonlimbic portions of the cerebral cortex to control such diverse activities as its analytical function, its sensory functions, and even at times its motor functions.

Yet, unfortunately, the effects caused by the different limbic structures are mainly subjective, which prevents adequate experimental study in animals. Therefore, our knowledge of the functions of the limbic structures is very poor indeed. About the best that we can do at present is to catalogue some of the overt effects known to occur when different parts of the limbic system are stimulated.

FUNCTIONS OF THE AMYGDALA

The amygdala is a complex of nuclei located immediately beneath the ventral surface of the cerebral cortex in the pole of each temporal lobe. In lower animals, this complex is concerned primarily with association of olfactory stimuli with stimuli from other parts of the brain. It was pointed out in Chapter 55 that one of the major divisions of the olfactory tract leads directly to a portion of the amygdala called the *corticomedial nuclei* that lie immediately beneath the cortex in the pyriform area of the temporal lobe. However, in the human being, another portion of the amygdala, the *basolateral nuclei,* has become much more highly developed than the olfactory portion and plays very important roles in many behavioral activities not in any way associated with olfactory stimuli.

The amygdala receives impulses from all portions of the limbic cortex, from the orbital surfaces of the frontal lobes, from the cingulate gyrus, and from the hippocampal gyrus. In turn, it transmits signals (a) back into these same cortical areas, (b) into the hippocampus, and (c) especially into the hypothalamus.

Effects of Stimulating the Amygdala. In general, stimulation in the amygdala can cause almost all the same effects as those elicited by stimulation of the hypothalamus, plus still other effects. The effects that are mediated through the hypothalamus include (1) increases or decreases in arterial pressure, (2) increases or decreases in heart rate, (3) increases or decreases in gastrointestinal motility and secretion, (4) defecation and micturition, (5) pupillary dilatation or, rarely, constriction, (6) pilo-erection, (7) secretion of the various anterior pituitary hormones, including especially the gonadotropins and corticotropin.

Aside from these effects mediated through the hypothalamus, amygdala stimulation can also cause different types of involuntary movement. These include (1) tonic movements, such as raising the head or bending the body, (2) circling movements, (3) occasionally clonic, rhythmic movements, and (4) different types of movements associated with olfaction and eating, such as licking, chewing, and swallowing. And stimulation can inhibit respiration or at other times stop all spontaneous movements of an animal. On rare occasions, it can also accelerate respiration.

In addition, stimulation of certain amygdaloid nuclei can cause a pattern of rage, escape, punishment, and pain similar to the affective-defense pattern elicited from the hypothalamus as described above. And stimulation of other nuclei can give reactions of reward and pleasure.

Finally, excitation of still other portions of the amygdala can cause sexual activities that include erection, copulatory movements, ejaculation, ovulation, uterine activity, and premature labor.

In short, stimulation of appropriate portions of the amygdaloid nuclei can give almost any pattern of behavior. It is believed that the normal function of the amygdaloid nuclei is to help control the overall pattern of behavior demanded for each occasion.

FUNCTIONS OF THE HIPPOCAMPUS

The hippocampus is an elongated structure composed of a modified type of cerebral cortex. It folds inward to form the ventral surface of the inferior horn of the lateral ventricle. One end of the hippocampus lies immediately adjacent to the amygdaloid nuclei, and it also fuses along one of its borders with the hippocampal gyrus, which is the cortex of the ventromedial surface of the temporal lobe. The hippocampus has numerous connections with almost all parts of the limbic system, including especially the amygdala, hippocampal gyrus, cingulate gyrus, hypothalamus, and other areas closely related to the hypothalamus.

Unfortunately, as is true of most limbic structures, we know very little about the overall function of the hippocampus and, therefore, for the present can only describe some overt effects that occur when this area is stimulated. These are the following:

Stimulation of different regions of the hippocampus can cause *involuntary tonic or clonic movements* in different parts of the body. At times it causes *rage or other emotional reactions*. And stimulation sometimes causes different types of *sexual phenomena*. One of the most remarkable effects of stimulation in conscious man is immediate *loss of contact* with any person with whom he might be talking; indicating that the hippocampus can play a role in determining a person's attention.

A special feature of the hippocampus is that weak electrical stimuli can cause local epileptic seizures in this region. These seizures cause various psychomotor effects including olfactory, visual, auditory, tactile, and other types of hallucinations that are uncontrollable even though the person has not lost consciousness and even though he knows them to be unreal. One of the reasons for this hyperexcitability of the hippocampus is probably that it is composed of a different type of cortex from that elsewhere in the cerebrum, having only three layers instead of the six layers found elsewhere.

Theoretical Function of the Hippocampus. Almost any type of sensory experience causes instantaneous activation of different parts of the hippocampus, and the hippocampus in turn distributes many outgoing signals to the hypothalamus and other parts of the diencephalon through the *fornix*. Thus, one of the major sources of input signals to the hypothalamus and its related structures is the hippocampus. On the basis of this, it is often stated that the primary role of the hippocampus is to provide a channel through which different incoming sensory signals can excite appropriate hypothalamic reactions.

FUNCTION OF THE HIPPOCAMPUS IN LEARNING. Earlier in the chapter it was pointed out that reward and punishment centers play a major role in determining which factors will be learned. A person becomes rapidly habituated to indifferent stimuli but learns assiduously any sensory experience that causes either pain or pleasure. Yet what is the route by which reward and punishment centers receive their original information? Much of this information is transmitted directly into these centers, but, on the other hand, much of it also depends on correlation of sensory experiences from several different sources. It has been suggested that the hippocampus plays the role of associating the affective characteristics of different sensory signals and then in turn transmitting the correlated information into the reward or punishment areas of the hypothalamus and other centers to help control the information that a person will or will not learn.

FUNCTION OF THE HIPPOCAMPAL, CINGULATE, AND ORBITOFRONTAL CORTEX

Probably the most poorly understood portion of the entire limbic system is the ring of cerebral cortex, called the *limbic cortex,* that surrounds the subcortical limbic structures. The limbic cortex is among the oldest of all parts of the cerebral cortex. In lower animals it plays a major role in various olfactory, gustatory, and feeding phenomena. However, in the human being, these functions of the limbic cortex are of minor importance. Instead, the limbic cortex of the human being is believed to be the cerebral association cortex for control of the lower centers that have to do primarily with behavior. For instance, the hippocampal gyrus has close interconnections with the hippocampus and thence through the fornix with almost all areas of the diencephalon, especially with the hypothalamus. The cingulate gyrus likewise connects directly with the hippocampus and also connects through the thalamus with almost all diencephalic areas. Finally, the orbitofrontal cortex interconnects profusely with the septal area, hypothalamus, preoptic area, and other associated regions.

Effect of Stimulation. Stimulation of the different regions of the limbic cortex has failed to give any real idea of their functions, because the different types of effects are just as diverse as those already recounted for stimulation of the hypothalamus, amygdala, or hippocampus, including such effects as (a) inhibition or acceleration of respiration, (b) facilitation or inhibition of spontaneous or cortical induced movements, (c) facilitation of clonic or tonic movements of the body, (d) vocalization, (e) chewing, (f) licking, (g) swallowing, (h) cardiovascular effects, such as increases or decreases in arterial pressure and increases or decreases in heart rate, (i) gastric effects, such as increases or decreases in gastrointestinal motility, (j) increases or decreases in gastrointestinal secretions, and (k) affective reactions including rage, docility, excitement, alertness, etc.

Effect of Ablation of Different Portions of the Limbic Cortex. Ablation experiments also have failed to give much idea of limbic cortical function, because the effects observed by different experimenters have often been quite different from each other. Also, the effects of ablation have sometimes been so inconsequential that no changes have been observed in the animal's behavior. However, the following general effects have been observed following ablation of certain specific areas.

ABLATION OF THE ANTERIOR PORTION OF THE CINGULATE GYRI. Bilateral removal of the anterior portion of cingulate gyri causes increased tameness of an animal to the extent that he no longer has fear of a man even though this fear might have existed preoperatively. Furthermore,

previous rage reactions are suppressed, and the animal seems to lack "social consciousness."

ABLATION OF THE POSTERIOR ORBITAL CORTEX. Bilateral removal of the posterior portion of the fronto-orbital cortex often causes an animal to develop an intense degree of motor restlessness, becoming unable to sit still, but instead moving continually.

ABLATION OF THE ANTERIOR TEMPORAL REGION— THE KLUVER-BUCY SYNDROME. When the anterior temporal cortex is removed, many important subcortical structures are also frequently damaged or removed, including especially the amygdala. Therefore, it is doubtful that what has been learned from anterior temporal ablation is of particular value in determining the function of the temporal limbic *cortex.* However, removal of the entire anterior tip of the temporal lobe in the dominant hemisphere of a monkey, and sometimes in man, causes intense changes in behavior, including (1) excessive tendency to examine objects, (2) loss of fear, (3) decreased aggressiveness, (4) tameness, (5) changes in dietary habits such that a herbivorous animal sometimes even becomes carnivorous, and (6) sometimes psychic blindness. This combination of effects is called the Kluver-Bucy syndrome.

In cats, opposite effects are sometimes observed, such as extreme savagery and rage. Yet, in other cats, removal of the anterior temporal poles causes extreme docility. Since different experimenters remove or damage different amounts of tissue in their experiments, these results show primarily that the anterior temporal poles, including the sublying amygdala, can have diverse effects on behavior, depending on the precise areas destroyed. This is to be expected because of the great diversity of effects that can be caused by stimulation of different areas of the amygdala and other closely associated regions.

Overall Function of the Hippocampal Gyrus, Cingulate Gyrus, and Orbitofrontal Cortex. After the above discussion, we find ourselves immediately perplexed regarding the function of the cortical regions of the limbic system. The reason for this probably is that these regions are actually association areas that correlate information from many other regions of the brain but cause no direct, overt effects that can be observed objectively. Thus, in the insular and anterior temporal cortex we find gustatory and olfactory associations in particular. In the hippocampal gyrus, there is a tendency for complex auditory associations. In the cingulate cortex, there is some reason to believe that sensorimotor associations occur. And, finally, the orbitofrontal cortex presumably aids in the analytical functions of the prefrontal lobes.

Therefore, until further information is available, it is perhaps best to state that the cortical regions of the limbic system occupy intermediate associative positions between the functions of the remainder of the cerebral cortex and the functions of the lower centers for control of behavioral patterns.

REFERENCES

Adey, W. R., and Tokizane, T.: Structure and Function of the Limbic System. Progress in Brain Research. New York, American Elsevier Publishing Co., Inc., 1967, Vol. 27.

Anand, B. K.: Limbic system in innate animal behavior. *Proc. Int. Union of Physiol. Sciences, 6*:218, 1968.

Baldwin, M.: The limbic system in man. *Proc. Int. Union of Physiol. Sciences, 6*:226, 1968.

Berlyne, D. E.: Conflict and arousal. *Sci. Amer., 215*:82(2), 1966.

Boyd, E. S., and Gardner, L. C.: Effect of some brain lesions on intracranial self-stimulation in the rat. *Amer. J. Physiol., 213*:1044, 1967.

Brady, J. V.: Emotional behavior. *In* Handbook of Physiology. Baltimore, The Williams & Wilkins Co., Sec. I, Vol. III, 1960, p. 1529.

Brutkowski, S.: Functions of prefrontal cortex in animals. *Physiol. Rev., 45*:721, 1965.

Delgado, J. M. R.: Electrical stimulation of the limbic system. *Proc. Int. Union of Physiol. Sciences, 6*:222, 1968.

Egger, D. M., and Flynn, J. P.: Effects of electrical stimulation of the amygdala on hypothalamically elicited attack behavior in cats. *J. Neurophysiol., 26*:705, 1963.

Gellhorn, E., and Loofbourrow, G. N.: Emotions and Emotional Disorders. New York, Paul B. Hoeber, 1962.

Green, J. D.: The hippocampus. *Physiol. Rev., 44*:561, 1964.

Grossman, S. P.: Behavioral responses to microinjections of drugs into the limbic system. *Proc. Int. Union of Physiol. Sciences, 6*:224, 1968.

Hilton, S. M., and Zbrozyna, A. W.: Amygdaloid region for defence reactions and its efferent pathway to the brain stem. *J. Physiol. (Lond.), 165*:160, 1963.

Kaada, B. R.: Cingulate, posterior orbital, anterior insular and temporal pole cortex. *In* Handbook of Physiology. Baltimore, The Williams & Wilkins Co., 1960, Sec. I, Vol. II, p. 1345.

Levi, L.: Emotional Stress. New York, American Elsevier Publishing Co., Inc., 1967.

Olds, J.: Hypothalamic substrates of reward. *Physiol. Rev., 42*:554, 1962.

Olds, J.; Yuwiler, A.; Olds, M. E.; and Yun, C.: Neurohumors in hypothalamic substrates of reward. *Amer. J. Physiol., 207*:242, 1964.

Simon, A.; Herbert, C. C.; and Straus, R. (eds.): The Physiology of Emotions. Springfield, Illinois, Charles C Thomas, Publisher, 1961.

Summers, T. B., and Kaelber, W. W.: Amygdalectomy: effects in cats and a survey of its present status. *Amer. J. Physiol., 203*: 1117, 1962.

Weiss, J. M.; McEwen, B. S.; Silva, M. T.; and Kalkut, M.: Pituitary-adrenal alterations and fear responding. *Amer. J. Physiol., 218*:864, 1970.

Wickelgren, W. O.: Effect of acoustic habituation on click-evoked responses in cats. *J. Neurophysiol., 31*:777, 1968.

PART IX

THE GASTROINTESTINAL TRACT

CHAPTER 63

Movement of Food Through the Alimentary Tract

The primary function of the alimentary tract is to provide the body with a continual supply of water, electrolytes, and nutrients, but before this can be achieved food must be moved along the alimentary tract at an appropriate rate for the digestive and absorptive functions to take place. Therefore, discussion of the alimentary system is presented in three different phases in the next three chapters: (1) movement of food through the alimentary tract, (2) secretion of the digestive juices, and (3) absorption of the digested foods, water, and the various electrolytes.

Figure 63–1 illustrates the entire alimentary tract, showing major anatomical differences between its parts. Each part is adapted for specific functions, such as: (1) simple passage of food from one point to another, as in the esophagus, (2) storage of food or fecal matter in the body of the stomach and in the descending colon, respectively, (3) digestion of food in the stomach, duodenum, jejunum, and ileum, and (4) absorption of the digestive end-products in the entire small intestine and proximal half of the colon. One of the most important features of the gastrointestinal tract that is discussed in the present chapter is the myriad of autoregulatory processes in the gut that keeps the food moving at an appropriate pace — slow enough for digestion and absorption to take place but fast enough to provide the nutrients needed by the body.

GENERAL PRINCIPLES OF GASTRO-INTESTINAL MOTILITY

CHARACTERISTICS OF THE INTESTINAL WALL

Figure 63–2 illustrates a typical section of the intestinal wall, showing the following layers

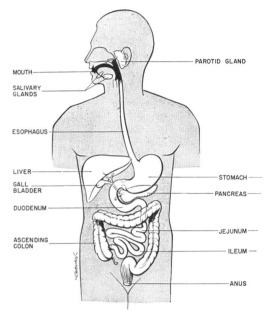

Figure 63–1 The alimentary tract.

from the outside inward: (1) the *serosa*, (2) a *longitudinal muscle layer*, (3) a *circular muscle layer*, (4) the *submucosa*, and (5) the *mucosa*. In addition, a sparse layer of smooth muscle fibers, the *muscularis mucosae*, lies in the deeper layers of the mucosa. The motor functions of the gut are performed by the different layers of smooth muscle.

The smooth muscle of each layer of the gastrointestinal tract forms a functional *syncytium* — that is, the individual muscle fibers lie in such close contact with each other that stimulatory impulses initiated in any part of the muscle tend to spread to adjacent areas. The conduction rate in the muscle itself is approximately 4 mm. per second. As this excitatory wave travels through the muscle, a prolonged contraction lasting 1 to 7 seconds occurs, the duration depending on the char-

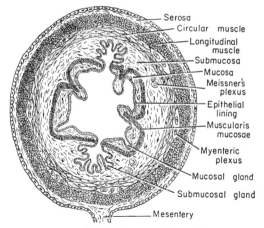

Figure 63-2 Typical cross-section of the gut.

acteristics of the specific segment of the gut stimulated.

The smooth muscle of the gastrointestinal tract exhibits both tonic contraction and rhythmic contraction which are characteristic of most types of smooth muscle, as discussed in Chapter 7.

Tonic contraction is continuous, lasting minute after minute or even hour after hour, sometimes increasing or decreasing in intensity but nevertheless continuing. The degree of tonic contraction in each segment of the gut determines the amount of steady pressure in the segment, and tonic contraction of the sphincters determines the amount of resistance to movement of food offered by each sphincter. In this way the *pyloric, ileocecal,* and *anal sphincters* all help to regulate food movement in the gut.

The *rhythmic contractions* of the gastrointestinal smooth muscle occur at rates as rapid as 15 to 20 times per minute or as slow as 2 to 3 times per minute. These contractions are responsible for the phasic functions of the gastrointestinal tract, such as mixing of the food and peristaltic propulsion of food, as is discussed below.

INNERVATION OF THE GUT — THE INTRAMURAL PLEXUS

Beginning in the esophageal wall and extending all the way to the anus is an *intramural nerve plexus.* This is composed principally of two layers of neurons and appropriate connecting fibers: the outer layer, called the *myenteric plexus* or *Auerbach's plexus,* lies between the longitudinal and circular muscular layers; and the inner layer, called the *submucosal plexus* or *Meissner's plexus,* lies in the submucosa. The myenteric plexus is far more extensive than the submucosal plexus, for which reason some physiologists often refer to the entire intramural plexus as simply the myenteric plexus.

In general, stimulation of the myenteric plexus increases the activity of the gut, causing four principal effects: (1) increased tonic contraction, or "tone," of the gut wall, (2) increased intensity of the rhythmic contractions, (3) increased rate of rhythmic contraction, and (4) increased velocity of conduction of excitatory waves along the gut wall. On the other hand, a few myenteric plexus fibers are inhibitory rather than excitatory; these are adrenergic fibers, in contrast to the excitatory fibers, which are cholinergic.

The intramural plexus is especially responsible for many neurogenic reflexes that occur locally in the gut, such as reflexes from the mucosal epithelium to increase the excitability of the muscle layers or to cause localized secretion of digestive juices by the submucosal glands. This plexus is also intimately involved in coordination of the motor movements of the gastrointestinal tract, which are discussed later in the chapter.

Autonomic Control of the Gastrointestinal Tract. The entire gastrointestinal tract receives extensive parasympathetic and sympathetic innervation that is capable of altering the overall activity of the entire gut or of specific parts of it.

PARASYMPATHETIC INNERVATION. The parasympathetic supply to the gut is divided into *cranial* and *sacral divisions,* which were discussed in Chapter 59. Except for a few parasympathetic fibers to the mouth and pharyngeal regions of the alimentary tract, the cranial parasympathetics are transmitted almost entirely in the *vagus nerves.* These fibers provide extensive innervation to the esophagus, stomach, and pancreas, and, to a much less extent, to the small intestine, gallbladder, and first half of the large intestine. The sacral parasympathetics originate in the second, third, and fourth sacral segments of the spinal cord and pass through the *nervi erigentes* to the distal half of the large inestine. The sigmoidal, rectal, and anal regions of the large intestine are considerably better supplied with parasympathetic fibers than are the other portions. These fibers function especially in the defecation reflexes, which are discussed later in chapter.

The postganglionic neurons of the parasympathetic system are part of the myenteric plexus, so that stimulation of the parasympathetic nerves causes a general increase in activity of this plexus. This in turn excites the gut wall and facilitates most of the intrinsic nervous reflexes of the gastrointestinal tract.

SYMPATHETIC INNERVATION. The sympathetic fibers to the gastrointestinal tract originate in the spinal cord between the segments T-8 and L-3. The preganglionic fibers, after leaving the cord, enter the sympathetic chains and pass through the chains to outlying ganglia, such as the *celiac ganglia* and various *mesenteric ganglia*. Here, the postganglionic neuron bodies are located, and postganglionic fibers spread from them to all parts of the gut. The sympathetics innervate essentially all portions of the gastrointestinal tract rather than being more extensively supplied to the most orad and most analward portions as is true of the parasympathetics.

In general, stimulation of the sympathetic nervous system inhibits activity in the gastrointestinal tract, causing effects essentially opposite to those of the parasympathetic system. However, the sympathetic system elicits excitatory effects in at least two instances: it excites (1) the ileocecal sphincter and (2) the internal anal sphincter. Thus, strong stimulation of the sympathetic system can totally block movement of food through the gastrointestinal tract, both by inhibition of the gut wall and by excitation of at least two major sphincters of the gastrointestinal tract.

EFFECTS OF PARASYMPATHETIC AND SYMPATHETIC DENERVATION. Denervating the sympathetic nervous system normally has no more than transient effects on the function of the gastrointestinal tract, which illustrates that sympathetic effects on gastrointestinal function are usually minimal.

On the other hand, denervation of the parasympathetic system almost always decreases the activity of the denervated portions of the gut for weeks, months, or years. For instance, destruction of the vagus nerves markedly reduces the tone and degree of peristaltic activity in the terminal esophagus, stomach, and upper intestine, while destruction of the sacral parasympathetics decreases the tone of the descending colon, sigmoid, and rectum, thereby seriously impairing defecation.

It is evident, then, that the gastrointestinal tract is usually tonically excited by the parasympathetics, and loss of this tone reduces gastrointestinal functions. Fortunately, after several months to a year or more, the intramural nervous system of the gut usually compensates at least partially by gradually

increasing its intrinsic excitability so that the excitability usually returns almost to normal.

Afferent Nerve Fibers from the Gut. Afferent nerve fibers arise in the gut, also. Some of these have their cell bodies in the submucosa and terminate within the intramural plexus. The nerve endings can be stimulated by (a) irritation of the gut mucosa, (b) excessive distention of the gut, or (c) presence of specific chemical substances in the gut. Signals transmitted through these fibers control excitation or inhibition of intestinal movements or intestinal secretion.

In addition to the afferent fibers that terminate in the intramural plexus, other afferent fibers transmit signals all the way to the central nervous system, traveling along with the sympathetic or parasympathetic nerves. For example, pain nerve fibers, except those from the esophagus, course along with the sympathetic nerves to the spinal cord. Also, about 80 per cent of the nerve fibers in the vagus nerves are afferent rather than efferent. These fibers transmit signals into the medulla to help initiate vagal signals that in turn control many functions of the gastrointestinal tract, as described in this and the following chapter.

FUNCTIONAL TYPES OF MOVEMENTS IN THE GASTROINTESTINAL TRACT

Two basic types of movements occur in the gastrointestinal tract: (1) *mixing movements,* which keep the intestinal contents thoroughly mixed at all times, and (2) *propulsive movements,* which cause food to move forward along the tract at an appropriate rate for digestion and absorption.

THE MIXING MOVEMENTS

In most parts of the alimentary tract, the mixing movements are caused by either *peristaltic contractions or local contractions of small segments of the gut wall.* These movements are modified in different parts of the gastrointestinal tract for proper performance of the respective activities of each part, as discussed separately later in the chapter.

THE PROPULSIVE MOVEMENTS — PERISTALSIS

The basic propulsive movement of the gastrointestinal tract is peristalsis, which is illustrated in Figure 63–3. A contractile ring appears around the gut and then moves forward; this is analogous to putting one's fingers around a thin distended tube, then constricting the fingers and moving them forward

Figure 63–3 Peristalsis.

along the tube. Obviously, any material in front of the contractile ring is moved forward.

Peristalsis is an inherent property of any syncytial smooth muscle tube, and stimulation at any point causes a contractile ring to spread in both directions. Thus, peristalsis occurs in (a) the gastrointestinal tract, (b) the bile ducts, (c) other glandular ducts throughout the body, (d) the ureters, and (e) most other smooth muscle tubes of the body.

The usual stimulus for peristalsis is *distention*. That is, if a large amount of food collects at any point in the gut, the distention stimulates the gut wall 2 to 3 cm. above this point, and a contractile ring appears and initiates a peristaltic movement.

Function of the Intramural Nerve Plexus in Peristalsis. Even though peristalsis is a basic characteristic of all tubular smooth muscle structures, it occurs only weakly or not at all in portions of the gastrointestinal tract that have congenital absence of the intramural nerve plexus. Also, it is greatly depressed or completely blocked in the entire gut when the person is treated with atropine to paralyze the intramural plexus. Finally, the intramural plexus causes the velocity of transmission of peristalsis along the gut to be 1 to 10 cm. per second, which is some 3 to 25 times as great as the natural conductive velocity of impulses in the smooth muscle itself (4 mm./second). Furthermore, since the intramural plexus is principally under the control of the parasympathetic nerves, the intensity of peristalsis and its velocity of conduction can be altered by parasympathetic stimulation.

Therefore, *even though the basic phenomenon of peristalsis is not dependent on the intramural nerve plexus, effectual peristalsis does require an active intramural plexus.*

Analward Peristaltic Movements. Peristalsis, theoretically, can occur in either direction from a stimulated point, but it normally dies out rapidly in the orad direction while continuing for a considerable distance analward. The cause of this directional transmission of peristalsis has never been ascertained, though several suggestions have been offered as follows:

RECEPTIVE RELAXATION AND THE "LAW OF THE GUT." Some physiologists believe the directional movement of peristalsis to be caused by a special organization of the intramural nerve plexus (principally the myenteric portion) which allows preferential transmission of analward signals. One reason for believing this is the effect of electrical stimulation on the gut: An electrical stimulus causes a contractile ring to appear near the point of the stimulus but at the same time sometimes causes relaxation, called "receptive relaxation," several centimeters down the gut toward the anus. It is believed that this relaxation could occur only as a result of conduction in the myenteric plexus. Obviously, a leading wave of receptive relaxation could allow food to be propelled more easily analward than in the orad direction.

This response to electrical stimulation is also called the "law of the gut" or sometimes simply the "myenteric reflex." It can be particularly well demonstrated in the esophagus and in the pyloric region of the stomach.

THE GRADIENT THEORY FOR FORWARD PROPULSION. Another theory for forward propulsion of intestinal contents is the gradient theory. This is based on the fact that the upper part of the gastrointestinal tract usually displays a higher degree of activity than the lower part. For instance, the basic contractile rhythm in the duodenum is approximately 8 to 9 contractions per minute, while that in the ileum is 5 to 6 contractions per minute. Also, in the upper part of the gastrointestinal tract far greater quantities of secretions are formed, and these, by distending the gut, initiate far more peristaltic waves than are initiated in the lower gut. Therefore, it is postulated that a greater number of peristaltic impulses originate in the orad portions of the gut than in the more distal regions, thereby causing peristaltic waves generally to travel down rather than up the gut. This is analogous to the *pacemaker* function of the S-A node in the heart.

INGESTION OF FOOD

The amount of food that a person ingests is determined principally by the intrinsic desire for food called *hunger,* and the type of food that he preferentially seeks is determined by his *appetite.* These mechanisms in themselves are extremely important automatic regulatory systems for maintaining an adequate nutritional supply for the body, and they will be discussed in detail in Chapter 72 in relation to nutrition of the body. The present discussion is confined to the actual mechanical aspects of food ingestion, including especially *mastication* and *swallowing*.

MASTICATION (CHEWING)

The teeth are admirably designed for chewing, the anterior teeth (incisors) providing a strong cutting action and the posterior teeth (molars) a grinding action. All the jaw muscles working together can close the teeth with a force as great as 55 pounds on the incisors and 200 pounds on the molars. When this is applied to a small object, such as a small seed between the molars, the actual force *per square inch* may be several thousand pounds.

Most of the muscles of chewing are innervated by the motor branch of the 5th cranial nerve, and the chewing process is controlled by nuclei in the hindbrain. Stimulation of the reticular formation near the hindbrain centers for taste can cause continual rhythmic chewing movements. Also, stimulation of areas in the hypothalamus, amygdaloid nuclei, and even in the cerebral cortex near the sensory areas for taste and smell can cause chewing.

Much of the chewing process is caused by the *chewing reflex,* which may be explained as follows: The presence of a bolus of food in the mouth causes reflex inhibition of the muscles of mastication, which allows the lower jaw to drop. The sudden drop in turn initiates a stretch reflex of the jaw muscles that leads to *rebound* contraction. This automatically raises the jaw to cause closure of the teeth, but it also compresses the bolus again against the linings of the mouth, which inhibits the jaw muscles once again, allowing the jaw to drop and rebound, and this is repeated again and again.

Chewing of the food is especially important for digestion of most fruits and raw vegetables, because these have undigestible cellulose membranes around their nutrient portions which must be broken before the food can be utilized. However, chewing also aids in the utilization of all other types of food for the following simple reason: Since the *digestive enzymes act only on the surfaces of food particles,* the rate of digestion is highly dependent on the total surface area exposed to the intestinal secretions. Also, grinding the food to a very fine particulate consistency prevents excoriation of the gastrointestinal tract and increases the ease with which food is emptied from the stomach into the small intestine and thence into all succeeding segments of the gut.

SWALLOWING (DEGLUTITION)

Swallowing is a complicated mechanism, principally because the pharynx most of the time subserves several other functions besides swallowing and is converted for only a few

seconds at a time into a tract for propulsion of food. Especially is it important that respiration not be seriously compromised during swallowing.

In general, swallowing can be divided into: (1) the *voluntary stage,* which initiates the swallowing process, (2) the *pharyngeal stage,* which is involuntary and constitutes the passage of food through the pharynx into the esophagus, and (3) the *esophageal stage,* another involuntary phase which promotes passage of food from the pharynx to the stomach.

Voluntary Stage of Swallowing. When the food is ready for swallowing, it is "voluntarily" squeezed or rolled posteriorly by pressure of the tongue upward and backward against the palate, as shown in Figure 63–4. Thus, the tongue forces the bolus of food into the pharynx. From here on, the process of swallowing becomes entirely, or almost entirely, automatic and ordinarily cannot be stopped.

Pharyngeal Stage of Swallowing. When the bolus of food is pushed backward in the mouth, it stimulates *swallowing receptor areas* all around the opening of the pharynx, especially on the tonsillar pillars, and impulses from these pass to the brain stem to initiate a series of automatic pharyngeal muscular contractions as follows:

1. The soft palate is pulled upward to close the posterior nares, in this way preventing reflux of food into the nasal cavities.

2. The palatopharyngeal folds on either side of the pharynx are pulled medialward to approximate each other. In this way these

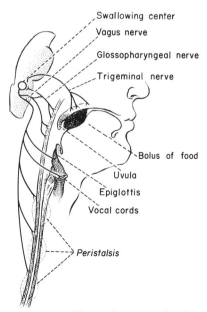

Figure 63–4 The swallowing mechanism.

folds form a sagittal slit through which the food must pass into the posterior pharynx. This slit performs a selective action, allowing food that has been masticated properly to pass with ease while impeding the passage of large objects. Since this stage of swallowing lasts only 1 second, any large object is usually impeded too much to pass through the pharynx into the esophagus.

3. The vocal cords of the larynx are strongly approximated, and the epiglottis swings backward over the superior opening of the larynx. Both of these effects prevent passage of food into the trachea. Especially important is the approximation of the vocal cords, but the epiglottis helps to prevent food from ever getting as far as the vocal cords. Destruction of the vocal cords or of the muscles that approximate them can cause strangulation. On the other hand, removal of the epiglottis usually does not cause serious debility in swallowing.

4. The entire larynx is pulled upward and forward by muscles attached to the hyoid bone; this movement of the larynx stretches the opening of the esophagus. At the same time, the *hypopharyngeal sphincter* around the esophageal entrance, which normally prevents air from going into the esophagus during respiration, relaxes, thus allowing food to move easily and freely from the posterior pharynx into the upper esophagus. The upward movement of the larynx also lifts the glottis out of the main stream of food flow so that the food usually passes on either side of the epiglottis rather than over its surface; this adds still another protection against passage of food into the trachea.

5. At the same time that the larynx is raised and that the hypopharyngeal sphincter is relaxed, the superior constrictor muscle of the pharynx contracts, giving rise to a rapid peristaltic wave passing downward over the pharyngeal muscles and into the esophagus.

To summarize the mechanics of the pharyngeal state of swallowing—the trachea is closed, the esophagus is opened, and a fast peristaltic wave originating in the pharynx then forces the bolus of food into the upper esophagus, the entire process occurring in 1 to 2 seconds.

NERVOUS CONTROL OF THE PHARYNGEAL STAGE OF SWALLOWING. The most sensitive tactile areas of the pharynx for initiation of the pharyngeal stage of swallowing lie in a ring around the pharyngeal opening, with greatest sensitivity in the tonsillar pillars. Impulses are transmitted from these areas through the sensory portions of the trigeminal and glosso-pharyngeal nerves into a region of the medulla oblongata closely associated with the *tractus solitarius,* which receives essentially all sensory impulses from the mouth.

The successive stages of the swallowing process are then automatically controlled in orderly sequence by neuronal areas distributed throughout the reticular substance of the medulla and lower portion of the pons. The sequence of the swallowing reflex remains the same from one swallow to the next, and the timing of the entire cycle also remains constant from one swallow to the next. The areas in the medulla and lower pons that control swallowing are collectively called the *deglutition* or *swallowing center.*

The motor impulses from the swallowing center to the pharynx and upper esophagus that cause swallowing are transmitted by the 5th, 9th, 11th, and 12th cranial nerves and even a few of the superior cervical nerves.

In summary, the pharyngeal stage of swallowing is principally a reflex act. It is almost never initiated by direct stimuli to the swallowing center from higher regions of the central nervous system. Instead, it is almost always initiated by voluntary movement of food into the pharynx, which, in turn, elicits the swallowing reflex.

EFFECT OF THE PHARYNGEAL STAGE OF SWALLOWING ON RESPIRATION. The entire pharyngeal stage of swallowing occurs in less than 1 to 2 seconds, thereby interrupting respiration for only a fraction of a usual respiratory cycle. The swallowing center specifically inhibits the respiratory center of the medulla during this time, halting respiration at any point in its cycle to allow swallowing to proceed. Yet, even while a person is talking, swallowing interrupts respiration for such a short time that it is hardly noticeable.

Esophageal Stage of Swallowing. The esophagus functions primarily to conduct food from the pharynx to the stomach, and its movements are organized specifically for this function.

Normally the esophagus exhibits two types of peristaltic movements—*primary peristalsis* and *secondary peristalsis.* Primary peristalsis is simply a continuation of the peristaltic wave that begins in the pharynx and spreads into the esophagus during the pharyngeal stage of swallowing. This wave passes all the way from the pharynx to the stomach in approximately 5 to 10 seconds. However, food swallowed by a person who is in the upright position is usually transmitted to the lower end of the esophagus even more rapidly than the peristaltic wave itself, in about 4 to 8 seconds, because of the additional effect of gravity pulling the food downward. If the primary peristaltic wave fails to move all the food that has entered the esophagus on into the stomach, secondary peristaltic waves result from distention of the esophagus by the retained food. These waves are essentially the same as the primary peristaltic waves, except that they originate in the esophagus itself rather than in the pharynx. Secondary peristaltic waves continue to be initiated until all the food has emptied into the stomach.

The peristaltic waves of the esophagus are controlled almost entirely by vagal reflexes that are part of the overall swallowing mechanism. These reflexes are transmitted through *vagal afferent fibers* from the esophagus to the medulla and then back again to the esophagus through *vagal efferent fibers.*

The musculature of the pharynx and the upper third of the esophagus is skeletal muscle, and, therefore, the peristaltic waves in these regions are controlled only by skeletal nerve impulses. In the lower two thirds of the esophagus, the musculature is smooth, but even this portion of the esophagus is normally under the control of the vagus nerve. However, when the vagus nerves to the esophagus are sectioned, the myenteric nerve plexus of the esophagus becomes excitable enough after several hours to several days to cause weak secondary peristaltic waves even without support from the vagal reflexes. Therefore, following paralysis of the swallowing reflex, food forced into the upper esophagus and then pulled by gravity to the lower esophagus still passes readily into the stomach.

RECEPTIVE RELAXATION OF THE STOMACH. As the esophageal peristaltic wave passes toward the stomach, a wave of relaxation precedes the constriction. Furthermore, the entire stomach and, to a less extent, even the duodenum become relaxed as this wave reaches the lower end of the esophagus. Especially important, also, is relaxation of the gastroesophageal constrictor at the juncture between the esophagus and the stomach. In other words, the constrictor and the stomach are prepared ahead of time to receive food being propelled down the esophagus during the swallowing act.

FUNCTION OF THE GASTROESOPHAGEAL CONSTRICTOR

At the lower end of the esophagus, about 5 cm. above its juncture with the stomach at the *cardia,* the circular muscle is slightly hyper-trophied to form a so-called *gastroesophageal constrictor.* Anatomically this constrictor is no different from the remainder of the esophagus except for the hypertrophy. However, physiologically, it remains tonically constricted in contrast to the remainder of the esophagus which normally remains completely relaxed. However, when a peristaltic wave of swallowing passes down the esophagus, "receptive relaxation" relaxes the gastroesophageal sphincter ahead of the peristaltic wave, and allows propulsion of the swallowed food on into the stomach after a delay of only 1 to 3 seconds. Rarely, the gastroesophageal constrictor does not relax satisfactorily, resulting in the condition called *achalasia,* which will be discussed in detail in Chapter 66.

A principal function of the gastroesophageal constrictor is to prevent reflux of stomach contents into the upper esophagus. The stomach contents are highly acidic and contain many proteolytic enzymes. The esophageal mucosa, except in the lower eighth of the esophagus, is not capable of resisting for long the digestive action of gastric secretions in the esophagus. Fortunately, the tonic constriction of the gastroesophageal constrictor prevents significant reflux of stomach contents into the upper esophagus except under abnormal conditions.

Another factor that prevents reflux is a valvelike mechanism of that portion of the esophagus that lies immediately beneath the diaphragm. Greatly increased intra-abdominal pressure caves the esophagus inward at this point and therefore prevents high pressure in the stomach from forcing stomach contents into the esophagus. Otherwise, every time we should walk, cough, or breathe hard, we would expel acid into the esophagus.

MOTOR FUNCTIONS OF THE STOMACH

The motor functions of the stomach are three-fold: (1) storage of large quantities of food until it can be accommodated in the lower portion of the gastrointestinal tract, (2) mixing of this food with gastric secretions until it forms a semifluid mixture called *chyme,* and (3) slow emptying of the food from the stomach into the small intestine at a rate suitable for proper digestion and absorption by the small intestine.

Figure 63–5 illustrates the basic anatomy of the stomach. Physiologically, the stomach can be divided into two major parts: (1) the *corpus,* or *body,* and (2) the *antrum.* The *fundus,* located

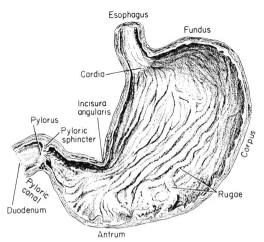

Figure 63-5   Physiologic anatomy of the stomach.

at the upper end of the body of the stomach, is often considered by anatomists to be a separate entity from the body, but from a physiological point of view, the fundus is actually a functional part of the body.

STORAGE FUNCTION OF THE STOMACH

As food enters the stomach, it forms concentric circles in the body of the stomach, the newest food lying closest to the *cardia* (where the esophagus empties into the stomach) and the oldest food lying nearest the wall of the stomach. Normally the body of the stomach has relatively little tone in its muscular wall so that it can bulge progressively outward, thereby accommodating greater and greater quantities of food up to a limit of about 1 liter. The pressure in the stomach remains low until this limit is approached, for two reasons: First, the smooth muscle in the wall of the stomach exhibits a quality of *plasticity*, which means that it can increase its length greatly without significantly changing its tone. Second, the greater the diameter of the stomach, the greater also becomes the radius of curvature of the walls. At a given pressure, the stretching force on the walls increases in direct proportion to this radius of curvature (an effect called the *law of Laplace*), for which reason the pressure inside the stomach increases only slightly despite marked distention.

MIXING IN THE STOMACH

The digestive juices of the stomach are secreted by the *gastric glands*, which cover almost the entire wall of the body and fundus of the stomach. These secretions come immediately

into contact with the stored food lying against the inner surface of the stomach; and when the stomach is filled, weak waves called *tonus waves,* or *mixing waves,* move along the stomach wall approximately once every 20 seconds. These waves begin near the cardia and spread for variable distances toward the pylorus. In general, they become more intense as they approach the antral portion of the stomach.

The mixing waves tend to move the gastric secretions and the outermost layer of food gradually toward the antral part of the stomach. On entering the antrum the waves become stronger, and the food and gastric secretions become progressively mixed to a greater and greater degree of fluidity.

In addition to the mixing caused by the tonus waves, mixing is also caused by intense peristaltic movements in the antral portion of the stomach. These movements cause mixing in the following way: Each time a peristaltic wave passes over the antrum toward the pylorus, it digs deeply into the contents of the antrum. Yet the opening of the pylorus is so small that only a few milliliters of antral contents are expelled into the duodenum with each peristaltic wave. Instead, the great mass of antral contents is squeezed or squirted backward through the peristaltic ring toward the body of the stomach. Thus, the moving peristaltic constrictive ring, combined with the squirting action, is an exceedingly important mixing mechanism of the stomach.

Chyme.   After the food has become mixed with the stomach secretions, the resulting mixture that passes on down the gut is called chyme. The degree of fluidity of chyme depends on the relative amounts of food and stomach secretions and on the degree of digestion that has occurred. The appearance of chyme is that of a murky, milky, semifluid or paste.

Propulsion of Food Through the Stomach. Strong peristaltic waves occur about 20 per cent of the time in the antrum of the stomach. These waves, like the mixing waves, occur about once every 20 seconds. In fact, they are probably extensions of the mixing waves, which become potentiated as they spread from the body of the stomach into the antrum. They become intense approximately at the incisura angularis, from which they spread through the antrum, then over the pylorus, and often even into the duodenum. As the stomach becomes progressively more and more empty, these intense waves begin farther and farther up the body of the stomach, gradually pinching off the lowermost portions of stored food, adding this food to the chyme in the antrum.

The peristaltic waves often exert as much as 50 to 70 cm. of water pressure, which is about six times as powerful as the usual mixing waves.

Hunger Contractions. In addition to the mixing and peristaltic contractions of the stomach, a third type of intense contractions, called hunger contractions, often occurs when the stomach has been empty for a long time. These are usually rhythmic peristaltic contractions probably representing exacerbated mixing waves in the *body* of the stomach. However, when they become extremely strong, they often fuse together to cause a continuing tetanic contraction lasting for as long as two to three minutes.

Hunger contractions are usually most intense in young healthy persons with high degrees of gastrointestinal tonus, and they are also greatly increased by a low level of blood sugar.

When hunger contractions occur in the stomach, the person sometimes experiences a sensation of pain in the pit of his stomach called *hunger pangs.* Hunger pangs usually do not begin until 12 to 24 hours after the last ingestion of food; in starvation they reach their greatest intensity in three to four days, and then gradually weaken in succeeding days.

Hunger contractions are often associated with a feeling of hunger and therefore are perhaps an important means by which the alimentary tract intensifies the desire for food when a person is in a state of incipient starvation.

Reflex Regulation of Stomach Contractions. Distention of the stomach by food initiates vagal afferent signals that pass to the medulla oblongata and reflexly inhibit stomach contractions but at the same time increase the rate of stomach secretion. This obviously allows prolonged digestion of the food prior to emptying it into the duodenum.

On the other hand, irritation of the stomach mucosa causes increased stomach contractions. Thus, accumulation of stomach secretions containing large quantities of hydrochloric acid increases the intensity of stomach contractions. Therefore, when the food becomes oversaturated with stomach secretions, it is automatically propelled toward the duodenum.

EMPTYING OF THE STOMACH

The Pyloric Pump. Basically, stomach emptying is *promoted by peristaltic waves* traveling from the stomach to the duodenum, and it is *opposed by resistance of the pylorus* to the passage of food. The pylorus normally remains almost but not completely closed because of mild tonic contraction. A pressure gradient of 3 to 5 cm.

water normally exists from the stomach to the pylorus, but this is great enough to cause emptying of nothing more than water and other highly fluid contents. Yet when peristaltic waves intermittently increase the pressure on the antral side of the pylorus, chyme of all degrees of consistency can be emptied. Therefore, for all practical purposes *the rate of emptying of the stomach is determined principally by the degree of activity of the antral peristaltic waves.*

The antral peristaltic waves, when active as just explained, characteristically occur almost exactly three times per minute, becoming very strong near the incisura angularis and moving over the antrum, then over the pylorus, and finally into the duodenum. As the wave moves forward, the pyloric sphincter and the proximal portion of the duodenum are inhibited, which is an instance of "receptive relaxation." With each peristaltic wave, several milliliters of chyme are forced into the duodenum. This "pumping" action of the antral portion of the stomach is sometimes also called the "pyloric pump."

Regulation of Emptying. The rate at which food is emptied from the stomach into the duodenum by the pyloric pump is regulated by two principal factors: (1) the fluidity of the chyme in the stomach and (2) the receptivity of the small intestine—especially the duodenum—to the chyme.

The *degree of fluidity of the chyme* is determined by (1) the type of food that has been eaten, (2) the degree of mastication of the food, (3) the length of time the food has been digested and mixed with secretions in the stomach, and (4) the intensity of the mixing and peristaltic waves of the stomach.

On the other hand, the *receptivity of the small intestine to chyme* is determined principally by (1) the amount of chyme already present in the duodenum and other parts of the upper intestine, (2) the acidity of the chyme, and (3) the type of food in the chyme. These three factors control the activity of the pyloric pump by means of (a) an *enterogastric reflex* from the duodenum to the stomach which inhibits the degree of antral peristalsis or (b) release from the small intestine of a hormone, *enterogastrone,* which passes by way of the blood to the stomach also to inhibit the degree of antral peristalsis.

Some of the factors that can elicit either the enterogastric reflex or the release of enterogastrone are the following:

EXCITATION OF THE ENTEROGASTRIC REFLEX

by Excessive Pressure, Irritants, Proteins, and Acid in the Duodenum, and the Effect on Gastric Emptying. When large quantities of chyme have been emptied into the duodenum, thereby raising the pressure in the duodenum, or whenever an obstruction in the small intestine prevents forward movement of the intestinal contents, the high pressure in the upper portion of the small bowel elicits the *enterogastric reflex,* which inhibits gastric peristalsis. Various aspects of this reflex are transmitted by three separate routes to the stomach, (1) by the myenteric plexus, (2) by vagal afferents to the medulla to inhibit vagal efferents, and (3) possibly by way of the celiac ganglion and then through sympathetic nerves to the stomach. Thus, the small intestine protects itself from becoming overloaded because of too rapid emptying of the stomach contents.

Irritants in the small intestine, excessive *acidity* of the chyme, excessive *protein* breakdown products, and even *hypo-* or *hypertonicity* of the chyme can also elicit the enterogastric reflex. For instance, any time the pH of the chyme entering the small intestine falls below approximately 3.5 to 4, the enterogastric reflex is immediately elicited, which inhibits gastric peristalsis and reduces or blocks further release of acidic stomach contents into the duodenum until the duodenal chyme can be neutralized by pancreatic and other secretions entering the duodenum.

The breakdown products of protein digestion can also elicit this reflex; by slowing the rate of stomach emptying, sufficient time is insured for adequate protein digestion in the upper portion of the small intestine.

Finally, the fact that either hyper- or hypotonic fluids (especially hypertonic) can elicit the enterogastric reflex prevents too rapid flow of these types of fluids into the small intestine, thereby preventing rapid changes in electrolyte balance of the body fluids during absorption of the intestinal contents.

Enterogastrone Inhibition of the Stomach Following Ingestion of Fats and Carbohydrates. When fatty foods, especially fatty acids, are present in the chyme, the hormone *enterogastrone* is extracted by the chyme from the mucosa of the duodenum and jejunum. This in turn is absorbed into the blood and within a few minutes inhibits the motility of the stomach, thereby slowing the rate of stomach emptying to as little as half to one-third the previous rate. This obviously allows prolonged time for digestion of fats in the small intestine, an important effect be-cause fats are the slowest of all foods to be digested.

To a much less extent, carbohydrates and acid in the small intestine extract small quantities of enterogastrone and therefore slightly inhibit the rate of stomach emptying.

Summary. The duodenum and other portions of the upper intestine control the rate of stomach emptying by (1) the enterogastric reflex and (2) the enterogastrone mechanism. These two work together to slow down the rate of emptying when (a) too much food is already in the small intestine or (b) when the chyme is excessively acid, contains too much fat, is hypotonic or hypertonic, or is irritating. In this way, the rate of stomach emptying is limited to that amount of chyme that the small intestine can process.

Among the different foods that affect the rate of emptying, fats inhibit emptying the greatest, sometimes delaying the emptying of a fatty meal for as long as three to six hours; proteins have an intermediate effect, and carbohydrates only a mild delaying effect.

Role of Pyloric Contraction in the Regulation of Emptying. *Ordinarily, the degree of pyloric contraction is a minor factor in the control of gastric emptying.* But, when irritant factors in the small intestine become extreme —much beyond the normal physiological range—it is possible for the pylorus to constrict and help the usual peristalsis inhibitory regulatory mechanisms to prevent further emptying.

MOVEMENTS OF THE SMALL INTESTINE

MIXING CONTRACTIONS (SEGMENTATION CONTRACTIONS)

Small ringlike contractions occur most of the time throughout the entire small intestine. These often occur irregularly, but at other times they are rhythmic, occurring at rates of 8 to 9 per minute in the duodenum and at slower rates progressively farther down the small intestine. These contractions cause "segmentation" of the small intestine as illustrated in Figure 63–6, dividing the intestine at times into regularly spaced segments that have the appearance of a chain of sausages. The next contractions occur at new points. Therefore, the segmentation contractions "chop" the chyme many times per minute, in this way

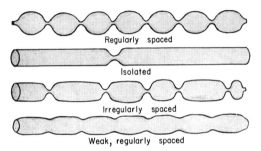

Figure 63-6 Segmentation movements of the small intestine.

keeping the solid suspension of food mixed with the secretions of the small intestine.

These mixing movements are only partially dependent on the intramural nervous system of the gut, for weak contractions still occur even when this system is blocked by atropine. However, their intensity can be increased by parasympathetic action or decreased by sympathetic action.

Chyme is also rotated counterclockwise as it passes along the gut. This results from the fact that the longitudinal muscle of the gut is arranged in a long counterclockwise spiral, while the circular muscle is arranged in a tight spiral.

In addition to mixing the chyme continually with new secretions from the intestinal wall, the mixing movements perform an equally important function in keeping all portions of the chyme continually exposed to the mucosa so that adequate absorption can take place.

Pendular Movements. Another type of mixing movement that can be seen only by close observation is pendular movement. These movements consist of small constrictive waves that sweep forward, then backward, up and down a few centimeters of the gut. By this process, the intestinal contents are moved back and forth, which obviously could play a major role in keeping the chyme thoroughly mixed.

PROPULSIVE MOVEMENTS

Chyme is propelled through the small intestine by *peristaltic waves.* These occur in any part of the small intestine, and they always move analward at a velocity of 1 to 2 cm. per second. However, they are normally weak and usually die out after traveling only a few centimeters. As a result, the movement of the chyme along the small intestine is slow, averaging only 1 cm. per minute. This means that 3 to 10 hours are normally required for pas-

sage of chyme from the pylorus to the ileocecal valve.

Peristaltic activity of the small intestine is greatly increased immediately after a meal. This is caused by the so-called *gastroenteric reflex* that is initiated by distention of the stomach and conducted principally through the myenteric plexus from the stomach down along the wall of the small intestine. This reflex increases the overall degree of excitability of the small intestine, including both increased motility and secretion.

Also, intense irritation of the intestinal mucosa or extreme distention of the intestine can elicit a so-called *peristaltic rush*, which is a powerful peristaltic wave that begins in the duodenum and passes the entire distance down the small intestine to the ileocecal valve in a few minutes. This obviously sweeps the contents of the intestine into the colon and thereby relieves the small intestine of either irritants or excess distention.

REGULATION OF RATE OF MOVEMENT OF CHYME THROUGH THE SMALL INTESTINE. The rate of movement of chyme through the small intestine is regulated mainly by the rate of emptying of the stomach into the duodenum. As the duodenum becomes distended, the intensity of peristaltic waves in the duodenum increases, and the duodenal contents are propelled rapidly along the small intestine to reduce the duodenal distention. The lack of distention then decreases the intensity of peristalsis until additional stomach emptying occurs.

In other words, the function of the peristaltic waves in the small intestine is to spread out the chyme along the intestinal mucosa soon after it is emptied from the stomach. On reaching the ileocecal valve the chyme is blocked sometimes for several hours until the person eats another meal, at which time a gastroenteric (gastroileal) reflex intensifies the peristalsis in the ileum and forces the remaining chyme through the ileocecal valve into the cecum.

Movements Caused by the Muscularis Mucosae and Movements of the Villi. The muscularis mucosae, which is stimulated by local nervous reflexes in the intramural plexus and by the sympathetic nervous system but not by the parasympathetic system, can cause short or long folds to appear in the intestinal mucosa. Also, individual fibers from this muscle extend upward into the intestinal villi and cause them to contract intermittently. The mucosal folds increase the surface area exposed to the chyme, thereby increasing the rate of absorption. The contractions of the villi — shortening, elongating, and shortening again — "milk" the villi so that lymph flows freely from the central lacteals into the lymphatic system. And both these types of contraction also agitate the fluids surrounding the villi so that progressively new areas of fluid become exposed to absorption.

Villus contractions are initiated by local nervous reflexes that occur in response to chyme in the small intestine. It has been claimed, though not proved, that the villi can also be stimulated by a hormone called *villikinin*. Supposedly, the chyme in the intestine extracts villikinin from the mucosa, and this in turn is absorbed into the blood to excite the villi.

CONTRACTION AND EMPTYING OF THE GALLBLADDER — CHOLECYSTOKININ

The functions of the gallbladder will be discussed specifically in relation to the overall function of the liver in Chapter 74. However, its activity is closely associated with that of the small intestine.

The liver continually secretes bile, which is then stored and concentrated in the gallbladder. The bile does not enter the small intestine until a specific stimulus causes the gallbladder to contract; this stimulus in general is initiated by the presence of fat in the small intestine as follows:

When fat enters the small intestine, within a few minutes it extracts from the mucosa a hormone called *cholecystokinin*. This in turn passes by way of the blood to the gallbladder and causes it to contract rhythmically. However, these contractions alone also are not sufficient to cause emptying of the gallbladder, because the sphincter of Oddi, which constricts the common bile duct where it empties into the small intestine, normally remains tonically contracted despite gallbladder contraction. But, every time a peristaltic wave passes over the duodenum, the wave of relaxation immediately ahead of the peristaltic contraction relaxes the sphincter of Oddi at the same time that it relaxes the gut wall, thus allowing small squirts of bile to flow into the duodenum. Therefore, the emptying of the gallbladder results from a combined action of cholecystokinin and peristalsis.

Regulation of gallbladder emptying by fats is undoubtedly a purposive effect, for bile salts help to emulsify the fats in the gut so that they can be digested by the intestinal lipases, and they also help to transport the digested fats through the intestinal epithelium.

FUNCTION OF THE ILEOCECAL VALVE

A principal function of the ileocecal valve is to prevent backflow of fecal contents from the colon into the small intestine. As illustrated in Figure 63–7, the lips of the ileocecal valve, which protrude into the lumen of the cecum, are admirably adapted for this function. Usually the valve can resist reverse pressure of as much as 50 to 60 cm. water.

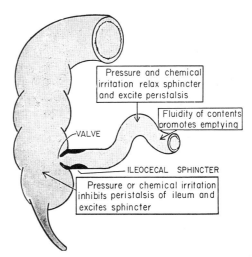

Pressure and chemical irritation relax sphincter and excite peristalsis

Fluidity of contents promotes emptying

VALVE

ILEOCECAL SPHINCTER

Pressure or chemical irritation inhibits peristalsis of ileum and excites sphincter

Figure 63–7   Emptying at the ileocecal valve.

The wall of the ileum for several centimeters immediately preceding the ileocecal valve has a thickened muscular coat called the *ileocecal sphincter*. This normally remains mildly constricted and prevents emptying of the ileal contents into the cecum except immediately following a meal when a gastroileal reflex (described above) intensifies the peristalsis in the ileum. When this occurs, about 4 ml. of chyme empties from the ileum into the cecum with each peristaltic wave. Yet, even so, only about 450 ml. of chyme empties into the cecum each day. The resistance to emptying at the ileocecal valve prolongs the stay of chyme in the ileum and, therefore, facilitates absorption.

Control of the Ileocecal Sphincter. The degree of contraction of the ileocecal sphincter is controlled primarily by reflexes from the cecum. Whenever the cecum is distended, the degree of contraction of the ileocecal sphincter is intensified, which greatly delays emptying of additional chyme from the ileum. Also, any irritant in the cecum causes constriction of the ileocecal sphincter. For instance, when a person has an inflamed appendix, the irritation of this vestigial remnant of the cecum can cause such intense spasm of the ileocecal sphincter that it completely blocks emptying of the ileum. These reflexes from the cecum to the ileocecal sphincter are mediated by way of the myenteric plexus.

In addition, various viscero-sympathetic reflexes initiated by irritation of other portions of the gastrointestinal tract, of the kidneys, or of the peritoneum can also cause intense contraction of the sphincter and thereby delay, or even completely stop, movement of intestinal contents through the ileocecal valve.

MOVEMENTS OF THE COLON

The functions of the colon are (1) absorption of water and electrolytes from the chyme and (2) storage of fecal matter until it can be expelled. The proximal half of the colon, illustrated in Figure 63–8, is concerned principally with absorption, and the distal half with storage; since intense movements are not required for these functions, the movements of the colon are normally sluggish. Yet in a sluggish manner, the movements still have characteristics similar to those of the small intestine and can be divided once again into mixing movements and propulsive movements.

Mixing Movements—Haustrations. In the same manner that segmentation movements occur in the small intestine, large circular constrictions also occur in the large intestine. At each of these constriction points, about 2.5 cm. of the circular muscle contracts, sometimes constricting the lumen of the colon to almost complete occlusion. At the same time, the longitudinal muscle of the colon, which is aggregated into three longitudinal stripes called the *tineae coli*, contract. These combined contractions of the circular and longitudinal smooth muscle cause the unstimulated portion of the large intestine to bulge outward into baglike sacs called *haustrations*. The haustral contractions, once initiated, usually reach peak intensity in about 30 seconds and then disappear during the next 60 seconds. They at times also move slowly analward during their period of contraction. After another few minutes, new haustral contractions occur in nearby areas but not in the same areas. Therefore, the fecal material in the large intestine is slowly "dug" into and rolled over in much the same manner that one spades the earth. In this way, all the fecal material is gradually exposed to the surface of the large intestine, and fluid is progressively absorbed until only 80 ml. of the 450 ml. daily load of chyme is lost in the feces. These contractions also help to move the fecal contents of the cecum and ascending colon into the transverse colon; and in this function, they act as very weak propulsive, as well as mixing, movements.

Propulsive Movements—"Mass Movements." Peristaltic waves of the type seen in the small intestine do not occur in the colon. Instead, another type of movement, called mass movements, propels the fecal contents toward the anus. These movements usually occur only a few times each day, most abundantly for about 10 minutes during the first hour or so after eating breakfast.

A mass movement is characterized by the following sequence of events: First, a constrictive point occurs at a distended or irritated point in the colon, and then rapidly thereafter the 20 or more cm. of colon *distal* to the constriction contracts almost as a unit, forcing the fecal material in this segment *en masse* down the colon. The initiation of contraction is complete in about 30 seconds, and relaxation then occurs during the next two to three minutes. Mass movements can occur in any part of the colon, though most often they occur in the transverse or descending colon. When they have forced a mass of feces into the rectum, the desire for defecation is felt.

INITIATION OF MASS MOVEMENTS BY THE DUODENOCOLIC REFLEX. The appearance of mass movements after meals is caused principally by the *duodenocolic reflex*. This results from filling of the duodenum, which initiates a reflex from the duodenum to the colon and thereby increases the excitability of the entire colon. To a less extent, a *gastrocolic reflex* also occurs.

These reflexes still take place, though with decreased intensity, when the autonomic nerves are removed; therefore, it is evident that the reflexes are basically transmitted through the myenteric plexus all the way from the upper alimentary tract to the colon. It is probable, though, that reflexes conducted in the autonomic nervous system also help to reinforce this direct route of transmission.

Irritation in the colon can also initiate intense mass movements. For instance, a person who has an ulcerated condition of the colon (*ulcerative colitis*) frequently has mass movements that persist almost all of the time.

Mass movements are also initiated by intense stimulation of the parasympathetic nervous system or simply by overdistention of a segment of the colon.

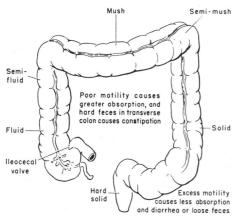

Figure 63–8 Absorptive and storage functions of the large intestine.

DEFECATION

Continual dribble of fecal matter through the anus is prevented by tonic constriction of (1) the *internal anal sphincter*, a circular mass of smooth muscle that lies immediately inside the anus, and (2) the *external anal sphincter*, composed of striated voluntary muscle and controlled by the somatic nervous system.

Ordinarily, defecation results from the *defecation reflex*, which can be described as follows: When the feces enter the rectum, distention of the rectal wall initiates afferent signals that spread through the *myenteric plexus* to initiate peristaltic waves in the descending colon and sigmoid, forcing feces toward the anus. As the peristaltic wave approaches the anus, the internal anal sphincter is inhibited by the usual phenomenon of "receptive relaxation," and if the external anal sphincter is relaxed, defecation will occur. This overall effect is called the defecation reflex.

However, the defecation reflex itself is extremely weak, and to be effective in causing defecation it must be fortified by another reflex that involves the sacral segments of the spinal cord, as illustrated in Figure 63-9. When the afferent fibers in the rectum are stimulated, signals are transmitted into the spinal cord and thence, reflexly, back to the descending colon, sigmoid, rectum, and anus by way of parasympathetic nerve fibers in the *nervi erigentes*. These parasympathetic signals greatly intensify the peristaltic waves and convert the defecation reflex from an ineffectual weak movement into a powerful process of defecation that is sometimes effective in emptying the large bowel all the way from the splenic flexure to the anus. Also, the afferent signals entering the spinal cord initiate other effects,

such as taking a deep breath, closure of the glottis, and contraction of the abdominal muscles to force downward on the fecal contents of the colon while at the same time causing the pelvic floor to pull outward and upward on the anus to evaginate the feces downward in the same manner that a person would squeeze sausage out of a bag.

However, the normal person prevents defecation until a socially acceptable time presents itself. This is achieved by voluntary contraction of the external sphincter, a skeletal muscle that is kept tonically contracted when a person wishes to prevent defecation. When this is done, the defecation reflex dies out after a few minutes and usually will not return until many hours thereafter.

When it becomes convenient for the person to defecate, defecation reflexes can usually be initiated by taking a deep breath to move the diaphragm downward and then contracting the abdominal muscles to increase the pressure in the abdomen, thus forcing fecal contents into the rectum to elicit new reflexes. Unfortunately, reflexes initiated in this way are never as effective as those that arise naturally, for which reason people who inhibit their natural reflexes too often become severely constipated.

In the newborn baby or in a person with a transected spinal cord, the defecation reflex causes automatic emptying of the lower bowel without the normal control exercised through contraction of the external anal sphincter.

OTHER AUTONOMIC REFLEXES AFFECTING BOWEL ACTIVITY

Aside from the duodenocolic, gastrocolic, gastroileal, enterogastric, and defecation reflexes which have been discussed in this chapter, several other important autonomic reflexes can affect the overall degree of bowel activity. These are the intestino-intestinal reflex, peritoneo-intestinal reflex, reno-intestinal reflex, vesico-intestinal reflex, and somato-intestinal reflex. All these reflexes are initiated by sensory signals that pass to the spinal cord and then are transmitted through the sympathetic nervous system back to the gut. And they always *inhibit* gastrointestinal activity. Thus, the *intestino-intestinal reflex* occurs when one part of the intestine becomes overdistended or its mucosa becomes excessively irritated; this blocks activity in other parts of the intestine while the local distention or irritability increases activity in the localized region and moves the intestinal contents away from the distended or irritated area.

The *peritoneo-intestinal reflex* is very much like the intestino-intestinal reflex, except that it re-

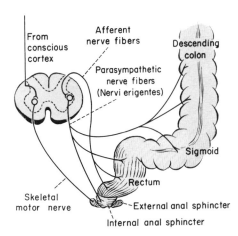

Figure 63-9 The afferent and efferent pathways of the parasympathetic mechanism for enhancing the defecation reflex.

From conscious cortex

Afferent nerve fibers

Descending colon

Parasympathetic nerve fibers (Nervi erigentes)

Sigmoid

Rectum

Skeletal motor nerve

External anal sphincter

Internal anal sphincter

sults from irritation of the peritoneum; it causes intestinal paralysis. The *reno-intestinal* and *vesico-intestinal reflexes* inhibit intestinal activity as a result of kidney or bladder irritation. Finally, the *somato-intestinal reflex* causes intestinal inhibition when the skin over the abdomen is irritatingly stimulated. Some of these reflexes will be discussed further in Chapter 66 in relation to abnormalities of gastrointestinal function.

REFERENCES

Bortoff, A., and Sachs, F.: Electrotonic spread of slow waves in circular muscle of small intestine. *Amer. J. Physiol., 218*:576, 1970.

Bosma, J. F.: Deglutition: pharyngeal stage. *Physiol. Rev., 37*:275, 1957.

Code, C. F., and Carlson, H. C.: Motor activity of the stomach. *In* Handbook of Physiology. Baltimore, The Williams & Wilkins Co., 1968, Sec. VI, Vol. IV, p. 1903.

Creamer, B.: Motor disturbances of the esophagus. *In* Handbook of Physiology. Baltimore, The Williams & Wilkins Co., 1968, Sec. VI, Vol. IV, p. 2331.

Daniel, E. E.: Digestion: motor function. *Ann. Rev. Physiol., 31*:203, 1969.

Davenport, H.: Physiology of the Digestive Tract. 2nd ed., Chicago, Year Book Publishers, Inc., 1966.

Diamant, N. E., and Bortoff, A.: Nature of the intestinal slow-wave frequency gradient. *Amer. J. Physiol., 216*:301, 1969.

Edwards, D. A. W., and Rowlands, E. N.: Physiology of the gastro-duodenal junction. *In* Handbook of Physiology. Baltimore, The Williams & Wilkins Co., 1968, Sec. VI, Vol. IV, p. 1985.

Grossman, M. I.: The digestive system. *Ann. Rev. Physiol., 25*: 165, 1963.

Hendrix, T. R., and Bayless, T. M.: Digestion: intestinal secretion. *Ann. Rev. Physiol., 32*:139, 1970.

Hightower, N. C., Jr.: Motor action of the small bowel. *In* Handbook of Physiology. Baltimore, The Williams & Wilkins Co., 1968, Sec. VI, Vol. IV, p. 2001.

Hunt, J. N., and Knox, M. T.: Regulation of gastric emptying. *In* Handbook of Physiology. Baltimore, The Williams & Wilkins Co., 1968, Sec. VI, Vol. IV, p. 1917.

Jenkins, G. N.: The Physiology of the Mouth. 3rd ed., Philadelphia, F. A. Davis Co., 1965.

Kelley, M. L., Jr., and DeWeese, J. A.: Effects of eating and intraluminal filling on ileocolonic junctional zone pressures. *Amer. J. Physiol., 216*:1491, 1969.

Malhotra, S. L.: Effect of saliva on gastric emptying. *Amer. J. Physiol., 213*:169, 1967.

Nelsen, T. S., and Becker, J. C.: Simulation of the electrical and mechanical gradient of the intestine. *Amer. J. Physiol., 214*: 749, 1968.

Schuster, M. M.: Motor action of rectum and anal sphincters in continence and defecation. *In* Handbook of Physiology. Baltimore, The Williams & Wilkins Co., 1968, Sec. VI, Vol. IV, p. 2121.

Shehadeh, Z.; Price, W. E.; and Jacobson, E. D.: Effects of vasoactive agents on intestinal blood flow and motility in the dog. *Amer. J. Physiol., 216*:386, 1969.

Truelove, S. C.: Movements of the large intestine. *Physiol. Rev., 46*:457, 1966.

Vagne, M., and Grossman, M. I.: Cholecystokinetic potency of gastrointestinal hormones and related peptides. *Amer. J. Physiol., 215*:881, 1968.

Wood, J. D., and Perkins, W. E.: Mechanical interaction between longitudinal and circular axes of the small intestine. *Amer. J. Physiol., 218*:762, 1970.

Youmans, W. B.: Innervation of the gastrointestinal tract. *In* Handbook of Physiology. Baltimore, The Williams & Wilkins Co., 1968, Sec. VI, Vol. IV, p. 1655.

CHAPTER 64

Secretory Functions of the Alimentary Tract

Throughout the gastrointestinal tract secretory glands subserve two primary functions: First, digestive enzymes are secreted in all areas from the mouth to the distal end of the ileum. Second, mucous glands, present from the mouth to the anus, provide mucus for lubrication and protection of all parts of the alimentary tract.

Most digestive secretions are formed only in response to the presence of food in the alimentary tract, and the quantity secreted in each segment of the tract is almost exactly the amount needed for proper digestion. Furthermore, in some portions of the gastrointestinal tract even the types of enzymes and other constituents of the secretions are varied in accordance with the types of food present. The purpose of the present chapter, therefore, is to describe the different alimentary secretions, their functions, and regulation of their production.

GENERAL PRINCIPLES OF GASTROINTESTINAL SECRETION

ANATOMICAL TYPES OF GLANDS

Several types of glands provide the different types of secretions in the gastrointestinal tract. First, on the surface of the epithelium in most parts of the gastrointestinal tract are literally billions of *single cell mucous glands* called *goblet cells.* These function entirely by themselves without the necessity of coordination with other goblet cells, and they simply extrude their mucus directly into the lumen of the intestine.

Second, the most abundant type of gland in the gastrointestinal tract is the tubular gland, two examples of which are illustrated, in Figure 64–1A and 64–1B. Figure 64–1A shows a *crypt of Lieberkühn,* found throughout the small intestine and in modified form in the large intestine. These are simple pits lined with goblet cells to produce mucus

Figure 64–1 The anatomical types of gastrointestinal glands: (A) A simple tubular gland represented by a crypt of Lieberkühn. (B) A more elongated tubular gland represented by a gastric gland. (C) A compound acinous gland represented by the pancreas.

and epithelial Paneth cells to produce mainly serous fluids but also a small amount of enzymes. Figure 64–1B illustrates a *gastric gland* in the main body of the stomach. This is a considerably deeper tubular gland, and it is occasionally branched.

Also associated with the gastrointestinal tract are several complex glands—the *salivary glands,* the *pancreas,* and the *liver*—which provide secretions for digestion or emulsification of food. The liver has a highly specialized structure that will be discussed in Chapter 74. The salivary glands and the pancreas are compound acinous glands of the type illustrated in Figure 64–1C. These glands lie completely outside the walls of the gastrointestinal tract and, in this, differ from all other gastrointestinal glands. The secreting structures, called *acini,* are lined with secreting glandular cells; these feed into a system of ducts that finally empty through one or more portals into the intestinal tract itself.

BASIC MECHANISMS OF STIMULATION OF THE GASTROINTESTINAL GLANDS

Effect of Local Stimuli. The mechanical presence of food in a particular segment of the gastrointestinal tract usually causes the glands of that region and often of adjacent regions to secrete moderate to large quantities of digestive juices. Part of this local effect results from direct stimulation of the surface glandular cells themselves by contact with the food. For instance, the goblet cells on the surface of the epithelium are stimulated mainly in this way. However, most local stimulation of the intestinal glands results from one of the following three methods of stimulation:

(1) Tactile stimulation or chemical irritation of the mucosa can elicit reflexes that pass through the intramural nervous system of the intestinal wall to stimulate either the mucous cells on the surface or the deeper glands of the mucosa. (2) Distention of the gut can also elicit nervous reflexes that stimulate secretion. (3) Either tactile stimuli or distention can result in increased motility of the gut, as described in the preceding chapters, and the motility in turn can then increase the rate of secretion.

Autonomic Stimulation of Secretion. PARA-SYMPATHETIC STIMULATION. Stimulation of the parasympathetic nerves to the alimentary tract almost invariably increases the rates of glandular secretion. This is especially true of the glands in the upper portion of the tract, including the salivary, esophageal, and gastric glands, the pancreas, and Brunner's glands in the duodenum. Secretion in the remainder of the intestinal tract more frequently occurs in response to local stimuli. However, even the glands of the small intestine and the mucous glands of the large intestine can be activated to a slight extent by parasympathetic impulses.

SYMPATHETIC STIMULATION. Stimulation of sympathetic nerves in most parts of the gastrointestinal tract causes a slight increase in secretion by the respective glands. On the other hand, sympathetic stimulation also results in constriction of the blood vessels supplying the glands. Therefore, sympathetic stimulation can have a dual effect: First, sympathetic stimulation alone usually slightly increases secretion. But, second, if parasympathetic stimulation is causing copious secretion by the glands, superimposed sympathetic stimulation often reduces the secretion because of reduced blood supply.

A few intestinal glands are specifically inhibited, rather than excited, by sympathetic stimulation. Especially important is the inhibition of Brunner's glands, for this often leaves the initial portion of the duodenum unprotected against the acidic digestive juices from the stomach, thus allowing development of *peptic ulcer*.

Regulation of Glandular Secretion by Hormones. In the stomach and intestine several different *gastrointestinal hormones* help to regulate the volume and character of the secretions. These hormones are liberated from the gastrointestinal mucosa in response to the presence of foods in the lumen of the gut. They then are absorbed into the blood and are carried to the respective glands which they stimulate. This type of stimulation is particularly valuable in increasing the output of gastric juice and pancreatic juice when food enters the stomach or duodenum. Also, hormonal stimulation of the gallbladder wall causes it to empty its stored bile into the duodenum, as discussed in the preceding chapter. Other hormones that are still of doubtful value have also been postulated to stimulate secretion by the glands of the small intestine.

Chemically, the gastrointestinal hormones are either proteins or polypeptides, or derivatives of these.

BASIC MECHANISM OF SECRETION BY GLANDULAR CELLS

Though all the basic mechanisms by which glandular cells form different secretions and then extrude these to the exterior are not known, experimental evidence points to the following basic principles of secretion by glandular cells, as illustrated in Figure 64–2. (1) The nutrient material needed for formation of the secretion must diffuse or be actively transported from the capillary into the base of the glandular cell. (2) Many *mitochondria* located inside the cell near its base provide oxidative energy for formation of adenosine triphosphate. (3) Energy from the adenosine triphosphate (see Chap. 67), along with appropriate nutrients, is then used for synthesis of the organic substances, this synthesis occurring almost entirely in the *endoplasmic reticulum*. The *ribosomes* adherent to this reticulum are specifically responsible for formation of proteins that are to be secreted. (4) The secretory materials flow through the tubules of the endoplasmic reticulum into the vesicles of the Golgi apparatus, which lies near the secretory ends of the cells. (5) The materials then are concentrated and discharged into the cytoplasm in

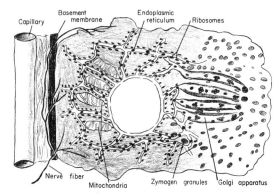

Figure 64–2 Basic mechanism of secretion by a glandular cell.

the form of *secretory granules.* (6) These granules are then extruded through the secretory surface into the lumen of the gland.

Water and Electrolyte Secretion in Response to Nervous Stimulation. A second necessity for glandular secretion is sufficient water and electrolytes to be secreted along with the organic substances. The following is a postulated method by which nervous stimulation causes water and salts to pass through the glandular cells in great profusion, which washes the organic substances through the secretory border of the cells at the same time: (1) Nerve stimulation has a specific effect on the *basal* portion of the cell membrane to cause active transport of chloride ions to the interior. (2) The resulting increase in electronegativity inside the cell then causes positive ions also to move to the interior of the cell. (3) The excess of both of these ions inside the cell creates an osmotic force which pulls water to the interior, thereby increasing the hydrostatic pressure inside the cells and causing the cell itself to swell. (4) The pressure in the cell then results in minute ruptures of the secretory border of the cell and causes flushing of water, electrolytes, and organic materials out of the secretory end of the glandular cell and into the lumen of the gland.

In support of this theory have been the following findings: First, the nerve endings on glandular cells are principally on the bases of the cells. Second, microelectrode studies show that the electrical potential across the membrane at the base of the cell is between 30 and 40 mv., with negativity on the interior and positivity on the exterior. Nerve stimulation increases this polarization voltage to values some 10 to 20 mv. greater than normal, parasympathetic stimulation normally having a considerably greater effect than sympathetic stimulation. This increase in polarization occurs a second or more after the nerve impulse has arrived, indicating that it is caused by movement of negative ions through the membrane to the interior of the cell.

Though this mechanism for secretion is still partly theoretical, it does explain how it would be possible for nerve impulses to regulate secretion. Obviously, hormonal effects on the cell membrane could cause similar results.

LUBRICATING AND PROTECTIVE PROPERTIES OF MUCUS AND ITS IMPORTANCE IN THE GASTROINTESTINAL TRACT

Mucus is a thick secretion composed of water, electrolytes, and a mixture of several mucopolysaccharides. Mucus is slightly different in different parts of the gastrointestinal tract, but everywhere it has several important characteristics that make it both an excellent lubricant and a protectant for the wall of the gut. *First,* mucus has adherent qualities that make it adhere tightly to the food or other particles and spread as a thin film over the surfaces. *Second,* it has sufficient *body* that it coats the wall of the gut and prevents actual contact of food particles with the mucosa. *Third,* mucus has a low resistance to shear or slippage so that the particles can slide along the epithelium with great ease. *Fourth,* mucus causes fecal particles to adhere to each other to form the fecal masses that are expelled during a bowel movement. *Fifth,* mucus is strongly resistant to digestion by the gastrointestinal enzymes. And, *sixth,* the mucopolysaccharides of mucus have amphoteric properties (as is true of all proteins) and are therefore capable of buffering small amounts of either acids or alkalies.

In summary, mucus has the ability to allow easy slippage of food along the gastrointestinal tract and also to prevent excoriative or chemical damage to the epithelium. One becomes acutely aware of the lubricating qualities of mucus when his salivary glands fail to secrete, for under these circumstances it is extremely difficult to swallow solid food even when it is taken with large amounts of water.

SECRETION OF SALIVA

The Salivary Glands; Characteristics of Saliva. The principal glands of salivation are the *parotid, submaxillary,* and *sublingual glands,* but in addition to these are many small *buccal glands.* The daily secretion of saliva normally ranges between 1000 and 1500 milliliters, as shown in Table 64–1.

TABLE 64–1.   Daily Secretion of Intestinal Juices

	Daily volume (ml.)	pH
Saliva	1200	6.0–7.0
Gastric secretion	2000	1.0–3.5
Pancreatic secretion	1200	8.0–8.3
Bile	700	7.8
Succus entericus	3000	7.8–8.0
Brunner's gland secretion	50(?)	8.0–8.9
Large intestinal secretion	60	7.5–8.0
Total	8210	

Saliva contains two different types of secretion: (1) a *serous secretion* containing *ptyalin* (an α-amylase), which is an enzyme for digesting starches, and (2) *mucous secretion* for lubricating purposes. The parotid glands secrete entirely the serous type, and the submaxillary glands secrete mainly the serous type but also a large amount of mucus as well. The sublingual glands secrete mainly mucus, and the buccal glands secrete only mucus. Saliva has a pH between 6.0 and 7.0, a favorable range for the digestive action of ptyalin.

Nervous Regulation of Salivary Secretion. Figure 64-3 illustrates the nervous pathways for regulation of salivation, showing that the submaxillary and sublingual glands are controlled principally by nerve impulses from the superior portions of the *salivatory nuclei* and the parotid gland by impulses from the inferior portions of these nuclei. The salivatory nuclei are located approximately at the juncture of the medulla and pons and are excited by both taste and tactile stimuli from the tongue and other areas of the mouth. Taste stimuli that are agreeable result in copious secretion of saliva. Also, certain types of tactile stimuli, such as the presence of smooth objects in the mouth (a pebble, for instance), cause marked salivation, whereas rough objects cause less salivation and occasionally even inhibit salivation.

Salivation can also be stimulated or inhibited by impulses arriving in the salivatory nuclei from higher centers of the central nervous system. For instance, when a person eats food that he particularly likes, salivation is far greater than when he eats food that he detests.

The *appetite area* of the brain that partially regulates these effects is located in close proximity to the parasympathetic centers of the anterior hypothalamus, and it functions to a great extent in response to signals from the taste and smell areas of the cerebral cortex or amygdala.

Finally, salivation also occurs in response to reflexes originating in the stomach and upper intestines—particularly when very irritating foods are swallowed or when a person is nauseated because of some gastrointestinal abnormality. The swallowed saliva presumably helps to remove the irritating factor in the gastrointestinal tract by diluting or neutralizing the irritant substances.

ESOPHAGEAL SECRETION

Briefly, the esophageal secretions are entirely mucoid in character and function principally to provide lubrication for swallowing. The main body of the esophagus is lined with many simple mucous glands, but at the gastric end and, to a less extent, in the initial portion of the esophagus there are many compound mucous glands. The mucus secreted by the compound glands in the upper esophagus prevents mucosal excoriation by the newly entering food, while the compound glands near the cardia (the esophagogastric juncture) protect the esophageal wall from being digested by gastric juices that reflux into the lower esophagus. Despite this protection, a peptic ulcer at times still occurs at the gastric end of the esophagus.

GASTRIC SECRETION

CHARACTERISTICS OF THE GASTRIC SECRETIONS

In addition to the mucus secreting cells that line the surface of the stomach, the stomach mucosa has two entirely different types of tubular glands: the *gastric* or *fundic* glands and the *pyloric glands*. The gastric glands secrete the digestive juices, and the pyloric glands secrete almost entirely mucus for the protection of the pyloric mucosa. The gastric glands are located everywhere in the mucosa of the body and fundus of the stomach, and the pyloric glands are located in the antral portion of the stomach. In addition, a few *cardiac glands*, which are almost identical to the pyloric glands, are located about 1 cm. immediately surrounding the cardia.

Figure 64-3   Nervous regulation of salivary secretion.

The Digestive Secretions from the Gastric Glands. A typical gastric gland is shown in Figure 64–1B, which is composed of three different types of cells: the *mucous neck cells,* which secrete mucus; the *chief cells,* which secrete the digestive enzymes (*pepsin* in particular); and the *parietal cells,* which secrete hydrochloric acid and which lie mainly behind the mucous neck cells or, less often, behind the chief cells. A postulated mechanism for secretion of the enzymes was given earlier in the chapter and was depicted in Figure 64–2. However, secretion of hydrochloric acid by the parietal cells involves particular problems which require further consideration as follows:

BASIC MECHANISM OF HYDROCHLORIC ACID SECRETION. The parietal cells secrete an electrolytic solution containing 150 millimols of hydrochloric acid per liter, which is almost exactly isotonic with the body fluids. The pH of this acid solution is approximately 0.8, thus illustrating its extreme acidity. At this pH the hydrogen ion concentration is about four million times that of the arterial blood. And to concentrate the hydrogen ions this tremendous amount requires over 1500 calories of energy per liter of gastric juice, as discussed in Chapter 4 in relation to membrane transport mechanisms.

Figure 64–4 illustrates the basic structure of a parietal cell, showing that it contains a system of *intracellular canaliculi.* This system is a modification of the endoplasmic reticulum. Presumably the hydrochloric acid is formed at the membranes of these canaliculi and then conducted to the exterior.

Different suggestions for the precise mechanism of hydrochloric acid formation have been offered, a simple one of which is illustrated in Figure 64–5. This diagram shows that *carbon dioxide,* either formed during metabolism in the cell or entering the cell from the blood,

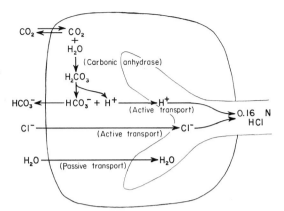

Figure 64–5 Postulated mechanism for the secretion of hydrochloric acid.

combines with water under the influence of *carbonic anhydrase* to form *carbonic acid.* This, in turn, dissociates into *bicarbonate ion* and *hydrogen ion.* By some yet unknown active transport process, the hydrogen ion is transported through the wall of the canaliculus and into its lumen. The bicarbonate ion, in turn, diffuses backward into the blood.

Chloride ion, also, is actively transported from the blood into the canaliculi, though the basic process for this, too, is unknown. Finally, water passes into the canaliculi by passive diffusion.

The importance of carbon dioxide in the chemical reactions for formation of hydrochloric acid is illustrated by the fact that inhibition of carbonic anhydrase by the drug *acetazolamide* almost completely blocks the formation of hydrochloric acid.

SECRETION OF PEPSIN. The principal enzyme secreted by the chief cells is pepsin. This is formed inside the cells in the form of *pepsinogen,* which has no digestive activity. However, once pepsinogen is secreted and comes in contact with previously formed pepsin in the presence of hydrochloric acid, it is immediately activated to form active pepsin. In this process, the pepsinogen molecule having a molecular weight of 42,500 is split to the pepsin molecule having a molecular weight of 35,000.

Pepsin is an active proteolytic enzyme in a highly acid medium (optimum pH = 2.0), but above a pH of about 5 it has little proteolytic activity and soon becomes completely inactivated. Therefore, hydrochloric acid secretion is equally as necessary as pepsin secretion for protein digestion in the stomach.

(Several other types of pepsin are secreted in smaller amounts, the most important of which is called *gastricsin* or pepsin I. The peptic activity of gastricsin in the human stomach is approximately one-fourth that of pepsin. It has optimal activity

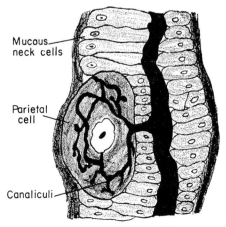

Mucous neck cells

Parietal cell

Canaliculi

Figure 64–4 Anatomy of the canaliculi in a parietal cell.

at a pH of 3.0 in contrast to an optimal pH of 2.0 for the more abundant type of pepsin, known as pepsin II.)

SECRETION OF OTHER ENZYMES. Small quantities of other enzymes are also secreted in the stomach juices, including *gastric lipase, gastric amylase,* and *gastric urease.* Gastric lipase is actually a *tributyrase,* for its principal activity is on tributyrin, which is butterfat, but it has almost no lipolytic activity on the usual fats. Gastric amylase plays a very minor role in digestion of starches. Gastric urease is not formed by the gastric mucosa but instead by bacterial contaminants in the stomach. It splits urea in the food or in the gastric juice to form ammonia, which in turn becomes ammonium hydroxide and neutralizes a small portion of the gastric acid. In uremia, considerable amounts of ammonia can be belched from the stomach.

Secretion of Mucus in the Stomach. The pyloric and cardiac glands are structurally similar to the gastric glands, but contain almost no chief and parietal cells. Instead, they contain almost entirely mucous cells that are identical with the mucous neck cells of the gastric glands. All these cells secrete a thin mucus, which protects the stomach wall from digestion by the gastric enzymes.

In addition, the surface of the stomach mucosa between glands has a continuous layer of mucous cells that secrete large quantities of a far more *viscid and alkaline mucus* that coats the mucosa with a mucous gel layer over 1 mm. thick, thus providing a major shell of protection for the stomach wall as well as contributing to lubrication of food transport. Even the slightest irritation of the mucosa directly stimulates the mucous cells to secrete copious quantities of this thick, viscid mucus.

REGULATION OF GASTRIC SECRETION BY NERVOUS AND HORMONAL MECHANISMS

Gastric secretion is regulated by both nervous and hormonal mechanisms, nervous regulation being effected through the parasympathetic fibers of the vagus nerves and hormonal regulation taking place by means of the hormone *gastrin.* Thus, regulation of gastric secretion is different from the regulation of salivary secretion, which is effected entirely by nervous mechanisms.

Vagal Stimulation of Gastric Secretion. Nervous signals to cause gastric secretion originate in the dorsal motor nuclei of the vagi and pass via the vagus nerves to the myenteric plexus of the stomach and thence to the gastric glands. In response, these glands secrete vast quantities of both pepsin and acid. Also, vagal

signals to the pyloric glands, the cardiac glands, and the mucous neck cells of the gastric glands cause greatly increased secretion of mucus as well.

Still another effect of vagal stimulation is to cause the antral part of the stomach mucosa to secrete the hormone gastrin. As will be explained in the following paragraphs, this hormone then acts on the gastric glands to cause additional flow of highly acid gastric juice. Thus, vagal stimulation excites stomach secretion both directly by stimulation of the gastric glands and indirectly through the gastrin mechanism.

Stimulation of Gastric Secretion by Gastrin. When food enters the stomach, it causes the antral portion of the stomach mucosa to secrete the hormone gastrin, a heptadecapeptide. The food causes release of this hormone in two ways: (1) The actual bulk of the food distends the stomach, and this in some way not yet understood causes the hormone gastrin to be released from the antral mucosa. (2) Certain substances called secretagogues—such as food extractives, partially digested proteins, alcohol (in low concentration), caffeine, and so forth—also cause gastrin to be liberated from the antral mucosa.

Gastrin is absorbed into the blood and carried to the gastric glands where it stimulates mainly the parietal cells but to a less extent also the chief cells. The parietal cells increase their rate of hydrochloric acid secretion as much as eight-fold, and the chief cells increase their rate of enzyme secretion two- to four-fold.

The rate of secretion in response to gastrin is somewhat less than to vagal stimulation, 200 ml. per hour in contrast to about 500 ml. per hour, indicating that the gastrin mechanism is a less potent mechanism for stimulation of stomach secretion than is vagal stimulation. However, the gastrin mechanism usually continues for several hours in contrast to a much shorter period of time for vagal stimulation. Therefore, as a whole, it is likely that the gastrin mechanism is equally as important as, if not more important than, the vagal mechanism for control of gastric secretion. Yet, when both of these work together, the total secretion is much greater than the sum of the individual secretions caused by each of the two mechanisms. In other words, the two mechanisms multiply each other rather than simply add to each other.

Histamine, an amino acid derivative, also stimulates gastric secretion in almost exactly the same way as gastrin, except that it is only about 80 per cent as potent as gastrin.

FEEDBACK INHIBITION OF GASTRIC ACID SECRETION. When the acidity of the gastric juices increases to a pH of 2.0, the gastrin mechanism for stimulating gastric secretion becomes totally blocked. This effect probably results from two different factors. *First*, greatly enhanced acidity depresses or blocks the extraction of gastrin itself from the antral mucosa. *Second*, the acid seems to extract an inhibitory hormone from the gastric mucosa or to cause an inhibitory reflex that inhibits gastric acid secretion.

Obviously, this feedback inhibition of the gastric glands plays an important role in protecting the stomach against excessively acid secretions, which would readily cause peptic ulceration. However, in addition to this protective effect, the feedback mechanism is also important to maintain optimal pH for function of the peptic enzymes in the digestive process.

The Three Phases of Gastric Secretion. Gastric secretion is said to occur in three separate phases (as illustrated in Fig. 64–6): a *cephalic phase,* a *gastric phase,* and an *intestinal phase.* However, as will be apparent in the following discussion, these three phases in reality fuse together.

THE CEPHALIC PHASE. The cephalic phase of gastric secretion occurs even before food enters the stomach. It results from the sight, smell, thought, or, taste of food; and the greater the appetite, the more intense is the stimulation. Neurogenic signals causing the cephalic phase of secretion can originate in the cerebral cortex or in the appetite centers of the amygdala or hypothalamus. They are transmitted through the dorsal motor nuclei of the vagi to the stomach. This phase of secretion accounts for one-tenth to one-fifth of the gastric secretion normally associated with eating a meal.

THE GASTRIC PHASE. Once food enters the stomach, it excites the gastrin mechanism, which in turn causes a low rate of secretion of gastric juice that continues throughout the several hours that the food remains in the stomach.

In addition, the presence of food in the stomach also causes (a) local reflexes in the intramural plexus of the stomach and (b) vagovagal reflexes that pass all the way to the brain stem and back to the stomach; both of these reflexes cause parasympathetic stimulation of the gastric glands and add to the secretion caused by the gastrin mechanism.

The gastric phase of secretion accounts for more than two-thirds of the total gastric secretion associated with eating a meal and, therefore, accounts for most of the total daily gastric secretion of about 2000 ml.

THE INTESTINAL PHASE. Even after the food has left the stomach, small amounts of gastric juice continue to be secreted for six to eight hours; that is, as long as chyme remains in the small intestine. Furthermore, this secretion will still occur even when all nerve connections are cut. Therefore, it is presumed that this phase of secretion results from one or more hormones liberated by the mucosa of the lower intestine and carried by the blood back to the stomach, but no specific hormone has been proved to cause this effect. Only about 5 per cent

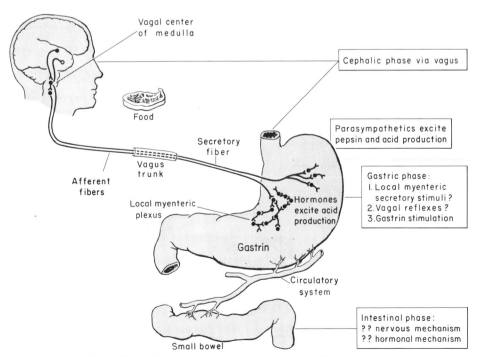

Figure 64–6   The phases of gastric secretion and their regulation.

of the total gastric secretion occurs during the intestinal phase.

Inhibition of Gastric Secretion by Intestinal Factors. Though chyme stimulates gastric secretion during the intestinal phase of secretion, it paradoxically often partially inhibits secretion during the gastric phase. This results from at least two different influences:

1. The presence of food in the small intestine initiates an *enterogastric reflex,* transmitted through the myenteric plexus, the sympathetic nerves, and the vagus nerves, that inhibits stomach secretion. This reflex can be initiated by distention of the small bowel, the presence of acid in the upper intestine, the presence of protein breakdown products, or irritation of the mucosa. This is part of the complex mechanism discussed in the preceding chapter for slowing down stomach emptying when the intestines are already filled.

2. The presence of fat in the small intestine causes the hormone *enterogastrone* to be released from the intestinal mucosa, and this also inhibits gastric secretion as well as gastric motility; this, too, was discussed in the preceding chapter.

SECRETION DURING THE INTERDIGESTIVE PERIOD. The stomach secretes only a few milliliters of gastric juice per hour during the "interdigestive period" when no digestion is occurring anywhere in the gut. Fortunately, excessive secretion is usually inhibited during this time by the acid inhibitory feedback noted above. However, strong emotional stimuli frequently increase this secretion to 50 or more ml. per hour in very much the same manner that the cephalic phase excites secretion at the onset of a meal. This increase of secretion during emotions is believed to be one of the factors in the development of peptic ulcers, as will be discussed in Chapter 66.

PANCREATIC SECRETION

Characteristics of Pancreatic Juice. Pancreatic juice contains enzymes for digesting all three major types of food: proteins, carbohydrates, and fats.

The proteolytic enzymes are *trypsin,* two different *chymotrypsins, carboxypolypeptidase, ribonuclease,* and *deoxyribonuclease.* The first two of these split whole and partially digested proteins, while carboxypolypeptidase splits a particular peptide linkage of small peptides. The nucleases split the two types of nucleic acids: ribonucleic and deoxyribonucleic acids.

The digestive enzyme for carbohydrates is *pancreatic amylase,* which hydrolyzes starches, glycogen and most other carbohydrates, except cellulose, to form disaccharides.

The enzyme for fat digestion, *pancreatic lipase,* is capable of hydrolyzing neutral fat into glycerol and fatty acids.

The proteolytic enzymes as synthesized in the pancreatic cells are in the inactive forms *trypsinogen, chymotrypsinogen,* and *procarboxypolypeptidase,* which are all enzymatically inactive. These become activated only after they are secreted into the intestinal tract. Trypsinogen is activated by an enzyme called *enterokinase,* which is released from the intestinal mucosa whenever chyme comes in contact with the mucosa. Also, trypsinogen can be activated to a slight extent by trypsin that has already been formed. Chymotrypsinogen is activated by trypsin to form chymotrypsin, and procarboxypolypeptidase is presumably activated in some similar manner.

Secretion of Trypsin Inhibitor. It is important that the proteolytic enzymes of the pancreatic juice not become activated until they have been secreted into the intestine, for the trypsin and other enzymes would digest the pancreas itself. Fortunately, the same cells that secrete the proteolytic enzymes into the acini of the pancreas secrete simultaneously another substance called *trypsin inhibitor.* This substance is stored in the cytoplasm of the glandular cells surrounding the enzyme granules, and it prevents activation of trypsin both inside the secretory cells and in the acini and ducts of the pancreas. Since it is trypsin that activates the other pancreatic proteolytic enzymes, trypsin inhibitor also prevents the subsequent activation of all these.

However, when the pancreas becomes severely damaged or when a duct becomes blocked, large quantities of pancreatic secretion become pooled in the damaged areas of the pancreas. Under these conditions, the effect of trypsin inhibitor is sometimes overwhelmed, in which case the pancreatic secretions rapidly become activated and literally digest the entire pancreas within a few hours, giving rise to the condition called *acute pancreatitis.* This usually is lethal because of accompanying shock, and even if not lethal it leads to a lifetime of pancreatic insufficiency.

REGULATION OF PANCREATIC SECRETION

Pancreatic secretion, like gastric secretion, is regulated by both nervous and hormonal mechanisms. However, in this case, hormonal regulation is by far the more important of the two.

Nervous Regulation. When the cephalic and gastric phases of stomach secretion occur, parasympathetic impulses are simultaneously transmitted along the vagus nerves to the pancreas, resulting in secretion of moderate amounts of enzymes into the pancreatic acini. However, little secretion flows through the pancreatic ducts to the intestine because little water and electrolytes are secreted along with the enzymes. Therefore, most of the enzymes are temporarily stored in the acini.

Hormonal Regulation.   After food enters the small intestine, pancreatic secretion becomes copious, mainly in response to the hormone *secretin.* In addition, a second hormone, *pancreozymin,* causes greatly increased secretion of enzymes.

Secretin Stimulation of Pancreatic Secretion — "Hydrelatic" Stimulation.   Secretin is a small polypeptide (molecular weight of about 3400) that is present in the mucosa of the upper small intestine in the inactive form *prosecretin.* When chyme enters the intestine, it causes the release and activation of secretin, which is subsequently absorbed into the blood. The one constituent of chyme that causes greatest secretin release is hydrochloric acid, though almost any type of food will cause at least some release.

Secretin causes the pancreas to secrete large quantities of fluid containing a high concentration of bicarbonate ion (up to 145 mEq./liter) but a low concentration of chloride ion. This copious flow of fluid is called "hydrelatic secretion" because the fluid is composed principally of a thin watery solution containing almost no enzymes.

The secretin mechanism is especially important for two reasons: *First,* secretin is released in especially large quantities from the mucosa of the small intestine any time the pH of the duodenal contents falls below approximately 4.0. This immediately causes large quantities of pancreatic juice containing abundant amounts of sodium bicarbonate to be secreted, which results in the following reaction in the duodenum:

$$HCl + NaHCO_3 \rightarrow NaCl + H_2CO_3$$

The carbonic acid immediately dissociates into carbon dioxide and water, and the carbon dioxide is absorbed into the body fluids, thus leaving a neutral solution of sodium chloride in the duodenum. In this way, the acid contents emptied into the duodenum from the stomach become neutralized, and the peptic activity of the gastric juices is immediately blocked. Since the mucosa of the small intestine cannot withstand the intense digestive properties of gastric juice, this is a highly important protective mechanism against the

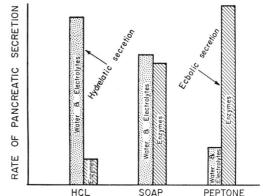

Figure 64–7 Hydrelatic and ecbolic secretion by the pancreas caused respectively by the presence of acids or peptone solutions in the duodenum.

development of duodenal ulcers, which will be discussed in further detail in Chapter 66.

A *second* importance of hydrelatic secretion by the pancreas is to provide an appropriate pH for action of the pancreatic enzymes. All of these function optimally in a slightly alkaline or neutral medium. The pH of the hydrelatic secretion averages 8.0.

Pancreozymin — "Ecbolic" Secretion by the Pancreas.   The presence of food in the upper small intestine also causes a second hormone, pancreozymin, to be released from the mucosa. This results especially from the presence of proteoses and peptones, which are products of partial protein digestion. Pancreozymin, like secretin, passes by way of the blood to the pancreas but, instead of causing hydrelatic secretion, causes secretion of large quantities of digestive enzymes, which is similar to the effect of vagal stimulation. This type of secretion is called *ecbolic secretion* in contradistinction to hydrelatic secretion. However, pancreozymin has yet to be isolated in a pure form.

The differences between the effects of secretin and pancreozymin are shown in Figure 64–7, which illustrates (a) intense hydrelatic secretion in response to acid in the duodenum, (b) a dual effect in response to soap, and (c) intense ecbolic secretion in response to peptones.

Figure 64–8   Regulation of pancreatic secretion.

Figure 64–8 summarizes the overall regulation of pancreatic secretion. The total amount secreted each day is about 1200 ml.

SECRETION OF BILE BY THE LIVER

The secretion of bile by the liver will be discussed in more detail in relation to the overall function of the liver in Chapter 74. Basically, bile contains no digestive enzyme and is important for digestion only because of the presence of bile salts which (1) help to emulsify fat globules so that they can be digested by the intestinal lipases and (2) help render the end-products of fat digestion soluble so that they can be absorbed through the gastrointestinal mucosa into the lymphatics.

Bile is secreted continually by the liver rather than intermittently as in the case of most other gastrointestinal secretions, but the bile is stored in the gallbladder until it is needed in the gut. The gallbladder then empties the bile into the intestine in response to *cholecystokinin,* a hormone extracted from the intestinal mucosa by fats in the intestine. Therefore, bile becomes available to aid in the processes of fat digestion. This mechanism of gallbladder emptying was discussed in the preceding chapter.

The rate of bile secretion can be altered in response to four different effects: (1) vagal stimulation can sometimes more than double the secretion, (2) secretin has a moderate effect on liver secretion, effecting as much as 80 per cent increase in bile production (but without any increase in bile acid production) at the same time that it also causes the hydrelatic response of the pancreas, (3) the greater the liver blood flow (up to a point), the greater the secretion, and (4) the presence of large amounts of bile salts in the blood increases the rate of liver secretion proportionately. Most of the bile salts secreted in the bile are reabsorbed by the intestines and then resecreted by the liver over and over again, performing their actions in relation to fat digestion and absorption many times before being lost in the feces. This continual recirculation of bile salts is important for maintaining even the normal daily flow of bile. When the salts are lost to the exterior through a bile fistula rather than being reabsorbed, the rate of bile secretion becomes reduced several fold.

The daily volume of bile production averages 600 to 700 ml.

SECRETIONS OF THE SMALL INTESTINE

SECRETION OF MUCUS BY BRUNNER'S GLANDS AND BY MUCOUS CELLS OF THE INTESTINAL SURFACE

An extensive array of compound mucous glands, called *Brunner's glands,* is located in the first few centimeters of the duodenum, mainly between the pylorus and the papilla of Vater where the pancreatic juices and bile empty into the duodenum. These glands secrete large quantities of mucus in response to: (a) direct tactile stimuli or irritating stimuli of the overlying mucosa, (b) vagal stimulation, which causes secretion concurrently with increase in stomach secretion, and (c) intestinal hormones, the exact natures of which have not been determined. The function of the mucus secreted by Brunner's glands is to protect the duodenal wall from digestion by the gastric juice, and their rapid and intense reponse to irritating stimuli is especially geared to this purpose.

Brunner's glands are strongly inhibited by sympathetic stimulation; therefore, such stimulation is likely to leave the duodenal bulb almost totally unprotected and is perhaps one of the factors that cause this area of the gastrointestinal tract to be the site of peptic ulcers in about 50 per cent of the cases.

Mucus is also secreted in large quantities by goblet cells located extensively over the surface of the intestinal mucosa. This secretion results principally from direct tactile or chemical stimulation of the mucosa by the chyme. Additional mucus is also secreted by the goblet cells in the intestinal glands. This secretion is probably controlled mainly by local nervous reflexes.

SECRETION OF THE INTESTINAL DIGESTIVE JUICES—THE CRYPTS OF LIEBERKÜHN

Located on the entire surface of the small intestine, with the exception of the Brunner's gland area of the duodenum, are small crypts called crypts of Lieberkühn, one of which is illustrated in Figure 64–1. The intestinal secretions are believed to be formed by the epithelial cells in these crypts at a rate of about 3000 ml. per day. The secretions are almost pure extracellular fluid, and they have a neutral pH in the range of 6.5 to 7.5. The secretions are rapidly reabsorbed by the villi. This circulation of fluid from the crypts to the villi obviously supplies a watery vehicle for absorption of substances from the small intestine, which is one of the primary functions of the small intestine.

Enzymes in the Small Intestinal Secretion. When secretions of the small intestine are collected without cellular debris, they have almost no enzymes, the only two enzymes being *enterokinase,* which activates trypsin as explained earlier in the chapter, and a small amount of *amylase.*

However, the epithelial cells of the mucosa contain large quantities of digestive enzymes and presumably digest food substances *while* they are being absorbed through epithelium. These enzymes are the following: (1) several different *peptidases* for splitting polypeptides into amino acids, (2) four enzymes for splitting disaccharides into monosaccharides—*sucrase, maltase, isomaltase,* and *lactase,* (3) *intestinal lipase* for splitting neutral fats into glycerol and fatty acids, and (4) a small amount of intestinal amylase for splitting carbohydrates into disaccharides.

The epithelial cells deep in the crypts of Lieberkühn continually undergo mitosis, and the new cells gradually migrate along the basement membrane upward out of the crypts toward the tips of the villi where they are finally shed into the intestinal secretions. The life cycle of an intestinal epithelial cell is approximately 48 hours. This rapid growth of new cells allows immediate repair of any excoriation that occurs in the mucosa.

REGULATION OF SMALL INTESTINAL SECRETION

Local Stimuli. By far the most important means for regulating small intestinal secretion is various local intramural nervous reflexes or direct stimuli. Especially important is distention of the small intestine, which causes copious secretion from the crypts of Lieberkühn. In addition, tactile or irritative stimuli can result in intense secretion. Therefore, for the most part, secretion in the small intestine occurs simply in response to the presence of food in the intestine—the greater the amount, the greater the secretion.

Extrinsic Nervous Regulation. Parasympathetic stimulation can increase basal secretion of the small intestine by two- to three-fold, but this is actually a small increase in comparison with the effects of local reflexes resulting from distention or irritation of the mucosa. Nevertheless, parasympathetic stimulation can aid in the overall regulation of secretion by the intestine.

Hormonal Regulation. Crude experiments have indicated that a hormone (or hormones) called *enterocrinin* extracted from the mucosa of the small intestine by the chyme might control intestinal secretion and even the types of enzymes secreted. However, more refined experiments have thus far failed to prove that this is an important mechanism.

SECRETIONS OF THE LARGE INTESTINE

Mucus Secretion. The mucosa of the large intestine, like that of the small intestine, is lined with crypts of Lieberkühn, but the epithelial cells contain almost no enzymes. Instead, they are lined almost entirely by goblet cells. Also, on the surface epithelium of the large intestine are large numbers of goblet cells dispersed among the other epithelial cells.

Therefore, the only significant secretion in the large intestine is mucus. Its rate of secretion is regulated principally by direct, tactile stimulation of the goblet cells on the surface of the mucosa and perhaps by local nervous reflexes to the goblet cells in the crypts of Lieberkühn. However, stimulation of the nervi erigentes, which carry the parasympathetic innervation to the distal half of the large intestine, also causes marked increase in the secretion of mucus. This occurs along with an increase in motility, which was discussed in the preceding chapter. Therefore, during extreme parasympathetic stimulation, often caused by severe emotional disturbances, so much mucus may be secreted into the large intestine that the person has a bowel movement of ropy mucus as often as every 30 minutes; the mucus contains little or no fecal material.

Mucus in the large intestine obviously protects the wall against excoriation, but, in addition, it provides the adherent qualities for holding fecal matter together. Furthermore, it protects the intestinal wall from the great amount of bacterial activity that takes place inside the feces, and it, plus the alkalinity of the secretion (pH of 8.0), also provides a barrier to keep acids formed deep in the feces from attacking the intestinal wall.

Secretion of Water and Electrolytes in Response to Irritation. Whenever a segment of the large intestine becomes intensely irritated, such as occurs when bacterial infection becomes rampant during *gastroenteritis,* the mucosa then secretes large quantities of water and electrolytes in addition to the normal viscid solution of mucus. This acts to dilute the irritating factors and to cause rapid movement of the feces toward the anus. The usual result is *diarrhea* with loss of large quantities of water and electrolytes but also earlier recovery from the disease than would otherwise occur. The lost electrolytes contain an especially large amount of potassium, the loss of which sometimes causes hypokalemic paralysis.

REFERENCES

Altamirano, M.: Back diffusion of H⁺ during gastric secretion. *Amer. J. Physiol.,* 218:1, 1970.

Burgen, A. S. V.: Secretory processes in salivary glands. *In* Handbook of Physiology. Baltimore, The Williams & Wilkins Co., 1967, Sec. VI, Vol. II, p. 561.

Ciba Foundation Symposium: The Exocrine Pancreas: Normal and Abnormal Function. Boston, Little, Brown and Company, 1962.

Code, C. F.: Hormonal inhibition of gastric secretion. *Proc. Int. Union of Physiol. Sciences,* 6:194, 1968.

Cooke, A. R.: Potentiation of acid secretion in dogs. *Amer. J. Physiol.,* 216:968, 1969.

Cooke, A. R.: The glands of Brunner. *In* Handbook of Physiology. Baltimore, The Williams & Wilkins Co., 1967, Sec. VI, Vol. II, p. 1087.

Cooke, A. R., and Grossman, M. I.: Comparison of stimulants of antral release of gastrin. *Amer. J. Physiol.,* 215:314, 1968.

Cowley, D. J., and Code, C. F.: Effects of secretory inhibitors on mucosal blood flow in nonsecreting stomach of conscious dogs. *Amer. J. Physiol.,* 218:270, 1970.

Davenport, H. W.: Physiological structure of the gastric mucosa. *In* Handbook of Physiology. Baltimore, The Williams & Wilkins Co., 1967, Sec. VI, Vol. II, p. 759.

Dombro, R.; Huang, M.; Dittbrenner, M.; and Ragins, H.: Metabolic pathway of histamine in canine gastric mucosa in relation to acid secretion. *Amer. J. Physiol.,* 216:988, 1969.

Farrar, G. E., Jr., and Bower, R. J.: Gastric juice and secretion: physiology and variations in disease. *Ann. Rev. Physiol.,* 29:141, 1967.

Fritz, M. E., and Botelho, S. Y.: Membrane potentials in unstimulated parotid gland of the cat. *Amer. J. Physiol.,* 216:1180, 1969.

Fritz, M. E., and Botelho, S. Y.: Role of autonomic nerve impulses in secretion by the parotid gland of cat. *Amer. J. Physiol.,* 216:1392, 1969.

Gregory, R. A.: Chemistry of gastrin. *Proc. Int. Union of Physiol. Sciences,* 6:188, 1968.

Gregory, R. A.: Secretory mechanisms of the digestive tract. *Ann. Rev. Physiol.,* 27:395, 1965.

Grossman, M. I.: Gastrin: new life for an old hormone. *Gastroenterology,* 40:149, 1961.

Grossman, M. I.: Hormonal regulation of gastrointestinal function. *Proc. Int. Union of Physiol. Sciences,* 6:187, 1968.

Grossman, M. I.: Neuronal and hormonal stimulation of gastric secretion of acid. *In* Handbook of Physiology. Baltimore, The Williams & Wilkins Co., 1967, Sec. VI, Vol. II, p. 835.

Harper, A. A.: Hormonal control of pancreatic secretion. *In* Handbook of Physiology. Baltimore, The Williams & Wilkins Co., 1967, Sec. VI, Vol. II, p. 969.

Hutchison, G. A., and Hirschowitz, B. I.: Two models of gastric H, Na, K, and Cl secretion in the dog using histamine and pentagastrin. *Amer. J. Physiol.,* 216:487, 1969.

Jacobson, E. D.; Swan, K. G.; and Grossman, M. I.: Blood flow and secretion in the stomach. *Gastroenterology,* 52:414, 1967.

Johnson, L. R., and Grossman, M. I.: Analysis of inhibition of acid secretion by cholecystokinin in dogs. *Amer. J. Physiol.,* 218:550, 1970.

Johnson, L. R., and Grossman, M. I.: Effects of fat, secretin and cholecystokinin on histamine-stimulated gastric secretion. *Amer. J. Physiol.,* 216:1176, 1969.

Johnson, L. R., and Grossman, M. I.: Secretin: the enterogastrone released by acid in the duodenum. *Amer. J. Physiol.,* 215:885, 1968.

Jones, R. S., and Grossman, M. I.: Dose response relationships of the choleretic effect of histamine. *Amer. J. Physiol.,* 216:335, 1969.

Kasbekar, D. K.; Ridley, H. A.; and Forte, J. G.: Pentagastrin and acetylcholine relation to histamine in H⁺ secretion by gastric mucosa. *Amer. J. Physiol.,* 216:961, 1969.

Lee, Y. H., and Thompson, J. H.: Gastric electrolyte secretion during nonsteady-state in gastric fistula rats. *Amer. J. Physiol.,* 216:1481, 1969.

Morris, G. C. R.: Pathological physiology of salivary glands. *In* Handbook of Physiology. Baltimore, The Williams & Wilkins Co., 1967, Sec. VI, Vol. II, p. 679.

Nakajima, S., and Magee, D. F.: Influences of duodenal acidification on acid and pepsin secretion of the stomach in dogs. *Amer. J. Physiol.,* 218:545, 1970.

Nakajima, S.; Nakamura, M.; and Magee, D. F.: Effect of secretin on gastric acid and pepsin secretion in response to various stimuli. *Amer. J. Physiol.,* 216:87, 1969.

Rothman, S. S.: Subcellular distribution of trypsinogen and chymotrypsinogen in rabbit pancreas. *Amer. J. Physiol.,* 218:372, 1970.

Schachter, M., and Beilenson, S.: Kallikrein and vasodilation in the submaxillary gland. *Gastroenterology,* 52:401, 1967.

Schneyer, L. H., and Schneyer, C. A.: Inorganic composition of saliva. *In* Handbook of Physiology. Baltimore, The Williams & Wilkins Co., 1967, Sec. VI, Vol. II, p. 497.

Taylor, W. H.: Proteinases of the stomach in health and disease. *Physiol. Rev.,* 42:519, 1962.

Uvnas, B.: Gastrin and vagus. *Proc. Int. Union of Physiol. Sciences,* 6:189, 1968.

CHAPTER 65

Digestion and Absorption in the Gastrointestinal Tract

The foods on which the body lives, with the exception of small quantities of vitamins and minerals, can be classified as carbohydrates, fats, and proteins. However, these generally cannot be absorbed in their natural form through the gastrointestinal mucosa and, for this reason, are useless as nutrients without the preliminary process of digestion. Therefore, the present chapter discusses, first, the processes by which carbohydrates, fats, and proteins are digested into small enough compounds for absorption and, second, the mechanisms by which the digestive end-products as well as water, electrolytes, and other substances are absorbed.

DIGESTION OF THE VARIOUS FOODS

Hydrolysis as the Basic Process of Digestion. Almost all the carbohydrates of the diet are large *polysaccharides* or *disaccharides,* which are combinations of many *monosaccharides* bound to each other by the process of *condensation*. This means that a hydrogen ion is removed from one of the monosaccharides while a hydroxyl ion is removed from the next one; the two monosaccharides then combine with each other at these sites of removal, and the hydrogen and hydroxyl ions combine to form water. When the carbohydrates are digested back into monosaccharides, specific enzymes return the hydrogen and hydroxyl ions to the polysaccharides and thereby separate the monosaccharides from each other. This process, called hydrolysis, is the following:

$$R_1 - R_2 + H_2O \rightarrow R_1OH + R_2H$$

Almost the entire fat portion of the diet consists of triglycerides (neutral fats), which are combinations of three *fatty acid* molecules condensed with a single *glycerol* molecule. In the process of condensation, three molecules of water had been removed. Digestion of the triglycerides consists of the reverse process, the fat-digesting enzymes returning the three molecules of water to each molecule of neutral fat and thereby splitting the fatty acid molecules away from the glycerol. Here again, the process is one of hydrolysis.

Finally, proteins are formed from *amino acids* that are bound together by means of *peptide linkages*. In this linkage a hydroxyl ion is removed from one amino acid, while a hydrogen ion is removed from the succeeding one; thus, the amino acids also combine together by a process of condensation while losing a molecule of water. Digestion of proteins, therefore, also involves a process of hydrolysis, the proteolytic enzymes returning the water to the protein molecules to split them into their constituent amino acids.

Therefore, the chemistry of digestion is really simple, for in the case of all three major types of food, the same basic process of *hydrolysis* is involved. The only difference lies in the enzymes required to promote the reactions for each type of food.

All the digestive enzymes are proteins. Their secretion by the different gastrointestinal glands is discussed in the preceding chapter.

DIGESTION OF CARBOHYDRATES

The Carbohydrate Foods of the Diet. Only three major sources of carbohydrates exist in the normal human diet. These are sucrose, which is the disaccharide known popularly as cane sugar; lactose, which is a disaccharide in milk; and starches, which are large polysaccharides present in almost all foods and particularly in the grains. Other types of carbohydrates ingested to a slight extent are glycogen, alcohol, lactic acid, pyruvic acid, pectins, dextrins, and minor quantities of other carbohydrate derivatives in meats. The diet also contains a large amount of cellulose, which is a carbohydrate. However, no enzymes capable of hydrolyzing cellulose are secreted by the human digestive tract. Consequently, cellulose cannot be considered to be a food.

Because starches make up the largest proportion of the carbohydrate intake of the human being, the digestive system is especially geared for their digestion. However, other polysaccharides, such as the pectins, glycogen, and the dextrins, are digested in essentially the same manner as the starches.

Digestion of Carbohydrates in the Mouth. When food is chewed, it is mixed with the saliva, which contains the enzyme *ptyalin* (α-amylase) secreted mainly by the parotid glands. This enzyme hydrolyzes starch into the disaccharide *maltose* (and another disaccharide, *isomaltose*), as shown in Figure 65–1; but the food remains in the mouth only a short time, and probably not more than 3 to 5 per cent of all the starches that are eaten will have become hydrolyzed into maltose by the time the food is swallowed. One can demonstrate the digestive action of ptyalin in the mouth by chewing a piece of bread for several minutes; after this time, the bread tastes sweet because of the maltose that has been liberated from the starches of the bread.

Most starches in their natural state, unfortunately, are present in the food in small globules, each of which has a thin protective cellulose covering. Therefore, most naturally occurring starches cannot be digested by ptyalin unless the food is cooked to destroy the protective membrane.

Digestion of Carbohydrates in the Stomach. Even though food does not remain in the mouth long enough for ptyalin to complete the breakdown of starches into maltose, the action of this enzyme continues for as long as 30 to 50 minutes after the food has entered the stomach; that is, until the contents of the fundus are mixed with the stomach secretions. Then the activity of the salivary amylase is blocked by the acid of the gastric secretions, for it is essentially nonactive as an enzyme once the pH of the medium falls below approximately 4.0. Nevertheless, on the average, before the food becomes completely mixed with the gastric secretions, as much as 30 to 40 per cent of the starches will have been changed into maltose.

HYDROLYSIS OF STARCHES AND DISACCHARIDES BY THE ACID OF THE STOMACH. The acid of the stomach juices can, to a slight extent, hydrolyze starches and disaccharides. However, quantitatively this reaction occurs to such a slight extent that it is usually ignored as an unimportant effect.

Digestion of Carbohydrates in the Small Intestine.

DIGESTION BY PANCREATIC AND INTESTINAL AMYLASE. Pancreatic secretion, like saliva, contains a large quantity of α-amylase which is capable of splitting starches into *maltose* and *isomaltose*. Therefore, immediately after the chyme empties from the stomach into the duodenum and mixes with pancreatic juice, the starches that have not already been split are digested by amylase. In general, the starches are almost totally converted into maltose and isomaltose before they have passed beyond the jejunum. Indeed, pancreatic amylase is a more powerful enzyme than salivary amylase, for pancreatic amylase can actually digest much of the uncooked starch (protective cellulose membrane and all), which is not true of salivary amylase.

Hydrolysis of Disaccharides into Monosaccharides by the Intestinal Epithelial Enzymes. The epithelial cells of the small intestinal epithelium contain the four enzymes *lactase, sucrase, maltase,* and *isomaltase,* which are capable of splitting the disaccharides lactose, sucrose, maltose, and isomaltose, respectively, into their constituent monosaccharides. These enzymes are located in the brush border of the cells lining the lumen of the intestine, and the disaccharides are digested as they come in contact with this border or as they diffuse into the microvilli that constitute the brush. The digested products are then immediately absorbed into the portal blood. Lactose splits into a molecule of *galactose* and a molecule of *glucose*. Sucrose splits into a molecule of *fructose* and a molecule of *glucose*. Maltose and isomaltose split into *two molecules of glucose*. Thus, the final products of carbohydrate digestion that are absorbed into the blood are all monosaccharides.

In the ordinary diet, which contains far more starches than either sucrose or lactose, glucose represents about 80 per cent of the final products of carbohydrate digestion, and galactose and fructose each represent, on the average, about 10 per cent of the products of carbohydrate digestion.

DIGESTION OF FATS

The Fats of the Diet. By far the most common fats of the diet are the neutral fats, also known as

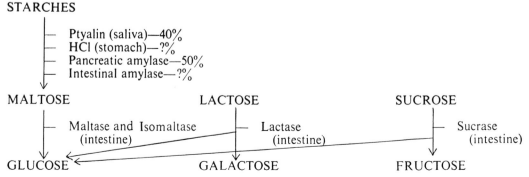

Figure 65–1 Digestion of carbohydrates.

triglycerides, each molecule of which is composed of a glycerol nucleus and three fatty acids, as illustrated in Figure 65–2. Neutral fat is found in food of both animal origin and plant origin.

In the usual diet are also small quantities of phospholipids, cholesterol, and cholesterol esters. The phospholipids and cholesterol esters contain fatty acid and, therefore, can be considered to be fats themselves. Cholesterol, on the other hand, is a sterol compound containing no fatty acid, but it does exhibit some of the physical and chemical characteristics of fats; it is derived from fats, and it is metabolized similarly to fats. Therefore, cholesterol is considered from a dietary point of view to be a fat.

Digestion of Fats in the Intestine. A small amount of short chain triglycerides of butter fat origin is digested in the stomach by gastric lipase (*tributyrase*). However, the amount of digestion is so slight that it is unimportant. Instead, essentially all fat digestion occurs in the small intestine as follows:

EMULSIFICATION OF FAT BY BILE ACIDS. The first step in fat digestion is to break the fat globules into small sizes so that the water-soluble digestive enzymes can act on the globule surfaces. This process is called emulsification of the fat, and it is achieved under the influence of *bile,* the secretion of the liver that does not contain any digestive enzymes. However, it does contain a large quantity of *bile salts* mainly in the form of ionized sodium salts which are extremely important for the emulsification of fat. The carboxyl part of the bile salt is highly soluble in water, whereas the sterol portion of the bile salt is highly soluble in fat. Therefore, bile salts aggregate at the surfaces of the fat globules in the intestinal contents with the carboxyl portion of the bile salt projecting outward and soluble in the surrounding fluids, while the

sterol portion is dissolved in the fat itself; this effect greatly decreases the interfacial tension of the fat.

When the interfacial tension of a globule of nonmiscible fluid is low, this nonmiscible fluid on agitation can be broken up into many minute particles far more easily than it can when the interfacial tension is great. Consequently, the function of the bile salts is to make the fat globules readily fragmentable by agitation in the small bowel. This action is the same as that of many detergents that are used widely in most household cleansers for removing grease.

Each time the diameters of the fat globules are decreased by a factor of 2 as a result of agitation in the small intestine, the total surface area of the fat increases 2 times. In other words, the total surface area of the fat particles in the intestinal contents is inversely proportional to the diameter of the particles.

The lipases are water-soluble compounds and can attack the fat globules only on their surfaces. Consequently, it can be readily understood how important bile salts are for the digestion of fats.

Digestion of Fats by Pancreatic Lipase and Enteric Lipase. By far the most important enzyme for the digestion of fats is *pancreatic lipase* in the pancreatic juice. However, the epithelial cells of the small intestine also contain a small quantity of lipase known as *enteric lipase.* Both of these act alike to cause hydrolysis of fat.

END-PRODUCTS OF FAT DIGESTION. Some research workers believe that almost all the triglycerides of the diet are finally broken into fatty acids and glycerol, though most believe that many of these are digested only to the mono- and diglyceride stages, as shown in Figure 65–3. From a practical point of view, the farther the process of fat hydrolysis proceeds in the intestinal tract, the better is the absorption of the fats. Furthermore, diseases that prevent secretion of pancreatic lipase invariably lead to poor digestion of fats and, consequently, to almost no absorption of fats.

DIGESTION OF PROTEINS

The Proteins of the Diet. The dietary proteins are derived almost entirely from meats and vegetables. These proteins in turn are formed of long chains of amino acids bound together by *peptide linkages.* A typical linkage is the following:

$$CH_3-(CH_2)_{16}-\overset{\overset{\text{O}}{\|}}{C}-O-CH_2$$

$$CH_3-(CH_2)_{16}-\overset{\overset{\text{O}}{\|}}{C}-O-CH \ + \ 3H_2O \ \xrightarrow{\text{lipase}}$$

$$CH_3-(CH_2)_{16}-\overset{\overset{\text{O}}{\|}}{C}-O-CH_2$$

(Tristearin)

$$HO-CH_2$$
$$HO-CH \ + \ 3CH_3-(CH_2)_{16}-\overset{\overset{\text{O}}{\|}}{C}-OH$$
$$HO-CH_2$$

(Glycerol)   (Stearic acid)

Figure 65–2 Hydrolysis of neutral fat catalyzed by lipase.

$$R-\overset{\overset{\text{NH}_2}{|}}{C}H-\overset{\overset{\|}{\text{O}}}{C}-OH \ + \ H-\overset{\overset{\text{H}}{|}}{N}-\overset{}{C}H-COOH \ \rightarrow$$
$$\qquad\qquad\qquad\qquad\qquad R$$

$$R-\overset{\overset{\text{NH}_2}{|}}{C}H-\overset{\overset{\|}{\text{O}}}{C}-\overset{\overset{\text{H}}{|}}{N}-CH-COOH \ + \ H_2O$$
$$\qquad\qquad\qquad\qquad\qquad R$$

$$\text{Fat} \xrightarrow{\text{(Bile + Agitation)}} \text{Emulsified fat}$$

$$\text{Emulsified fat} \xrightarrow[\text{(Enteric lipase)}]{\text{(Pancreatic lipase)}} \begin{cases} \text{Fatty acids} \\ \text{Glycerol} \\ \text{Glycerides} \end{cases} \begin{matrix} 60\% \text{ (?)} \\ 40\% \text{ (?)} \end{matrix}$$

Figure 65–3   Digestion of fats.

The characteristics of each type of protein are determined by the types of amino acids in the protein molecule and by the arrangement of these amino acids. The physical and chemical characteristics of the different proteins will be discussed in Chapter 69.

Digestion of Proteins in the Stomach. Pepsin II, and *gastricsin* (also called pepsin I), the two most important peptic enzymes of the stomach, are most active at pH's of about 2 and 3, respectively, and are completely inactive at a pH above approximately 5. Consequently, for these enzymes to cause any digestive action on protein, the stomach juices must be acidic. It will be recalled from Chapter 64 that the gastric glands secrete a large quantity of hydrochloric acid. This hydrochloric acid is secreted at a pH of about 0.8, but, by the time it is mixed with the stomach contents, the pH ranges around 2 to 3, a highly favorable range of acidity for pepsin and gastricsin activity.

Pepsin and gastricsin are capable of digesting essentially all the different types of proteins in the diet. One of the important features of pepsin digestion is its ability to digest collagen, an albuminoid that is affected little by other digestive enzymes. Collagen is a major constituent of the intercellular connective tissue of meats, and for the digestive enzymes of the digestive tract to penetrate meats and digest the cellular proteins it is first necessary that the collagen fibers be digested. Consequently, in persons lacking peptic activity in the stomach, the ingested meats are poorly penetrated by the digestive enzymes and, therefore, are poorly digested.

As illustrated in Figure 65–4, pepsin usually does not complete the process of protein digestion all the way to amino acids but simply splits the proteins into proteoses, peptones, and large polypeptides. This splitting of proteins is a process of "hydrolysis" occurring at the peptide linkages between the amino acids.

Digestion of Proteins by Pancreatic Secretions. TRYPSIN. When the proteins leave the stomach, they ordinarily are in the form of proteoses, peptones, large polypeptides, and about 15 per cent amino acids. Immediately upon entering the small intestine, the partial breakdown products are attacked by the pancreatic enzyme trypsin. As illustrated in Figure 65–4, trypsin is capable of hydrolyzing all the partial breakdown products of protein to peptides, and it even hydrolyzes some of them to the final stage of amino acids; but most of the products of trypsin digestion are dipeptides or other rather small polypeptides.

CHYMOTRYPSIN AND CARBOXYPOLYPEPTIDASES. Pancreatic secretions also contain two types of chymotrypsin and a carboxypolypeptidase. Chymotrypsin acts very much like trypsin except that it catalyzes hydrolysis of different types of peptide linkages. The carboxypolypeptidase is capable of hydrolyzing certain polypeptides to amino acids.

Digestion of Peptides by the Epithelium of the Small Intestine. The epithelial cells of the small intestine contain several different enzymes for hydrolyzing the final peptide linkages of the different dipeptides and other small polypeptides as they are absorbed through the epithelium into the portal blood. Because some of these linkages are slightly different from each other in their quantity of bond energy and other factors, the linkages between specific types of amino acids require separate enzymes for hydrolysis. The enzymes re-

$$\left.\begin{matrix} \text{Proteins} \\ \text{Collagen} \\ \text{Nucleoproteins} \end{matrix}\right\} \xrightarrow[\text{(gastricsinogen + HCl} \to \text{gastricsin)}]{\text{(pepsinogen + HCl} \to \text{pepsin)}} \left\{\begin{matrix} \text{Proteoses} \\ \text{Peptones} \\ \text{Polypeptides} \end{matrix}\right\}$$

$$\text{(trypsinogen + enterokinase} \to \text{trypsin)}$$

$$\text{(chymotrypsinogen + trypsin} \to \text{chymotrypsin)}$$
$$\text{(carboxypolypeptidases)}$$

$$\left\{\begin{matrix} \text{Polypeptides} \\ + \\ \text{Amino acids} \end{matrix}\right\} \xrightarrow[\text{(dipeptidases)}]{\text{(aminopolypeptidases)}} \text{Amino acids}$$

Figure 65–4   Digestion of proteins.

sponsible for final hydrolysis of the peptides into amino acids are *amino polypeptidase* and the *dipeptidases.*

When food has been properly masticated and is not eaten in too large a quantity at any one time, about 98 per cent of all the proteins finally become amino acids. A few molecules of protein are never digested at all, and some remain in the stages of proteoses, peptones, and varying sizes of polypeptides.

BASIC PRINCIPLES OF GASTROINTESTINAL ABSORPTION

ANATOMICAL BASIS OF ABSORPTION

The total quantity of fluid that must be absorbed each day is equal to the ingested fluid (about 1.5 liters) plus that secreted in the various gastrointestinal secretions (about 8.5 liters). This comes to a total of approximately 10 liters. About 9.5 liters of this is absorbed in the small intestine, leaving only 0.5 liter to pass through the ileocecal valve into the colon each day.

The stomach is a poor absorptive area of the gastrointestinal tract. Only a few highly lipid-soluble substances, such as alcohol and some drugs, can be absorbed in small quantities.

The Absorptive Surface of the Intestinal Mucosa — The Villi. Figure 65–5 illustrates the absorptive surface of the intestinal mucosa, showing many folds called *valvulae conniventes,* which increase the surface area of the absorptive mucosa about three-fold. These folds extend circularly all the way around the intestine and are especially well developed in the duodenum and jejunum, where they often protrude as much as 8 mm. into the lumen.

Located over the entire surface of the small intestine, from approximately the point at which the common bile duct empties into the duodenum down to the ileocecal valve, are literally millions of small *villi,* which project about 1⁻mm. from the surface of the mucosa, as shown on the surfaces of the valvulae conniventes in Figure 65–5 and in detail in Figure 65–6. These villi lie so close to each other in the upper small intestine that they actually touch in most areas, but their distribution is less profuse in the distal small intestine. The presence of villi on the mucosal surface enhances the absorptive area another 10-fold.

The intestinal epithelial cells are characterized by a brush border, consisting of about 600 *microvilli* 1μ in length and 0.1μ in diameter, protruding from each cell; these are illustrated in the electron micrograph in Figure 65–7. This increases the surface area exposed to the intestinal materials another 20-fold. Thus, the combination of the valvulae conniventes, the villi, and the microvilli increases the absorptive area of the mucosa about 600-fold, making a tremendous total area of about 550 square meters for the entire small intestine.

Figure 65–6A illustrates the general organization of a villus, emphasizing especially the advantageous arrangement of the vascular system for absorption of fluid and dissolved material into the portal blood, and the arrangement of the *central lacteal* for absorption into the lymphatics. Figure 65–6B shows the cross-section of a villus, and Figure 65–7 shows many small *pinocytic* vesicles, which are pinched off portions of infolded epithelium surrounding extracellular materials that have been entrapped inside the cells. Some substances are absorbed by this physical process of *pinocytosis,* though, as noted later in the chapter, most absorption occurs by means of single molecular transfer. Located near the brush border of the epithelial cell are many *mitochondria,* which supply the cell with oxidative energy needed for "active transport" of materials through the intestinal epithelium; this also is discussed later in the chapter.

BASIC MECHANISMS OF ABSORPTION

Absorption through the gastrointestinal mucosa occurs by *active transport* and by *diffusion,* as is also true for other membranes. The physical principles of these processes were explained in Chapter 4.

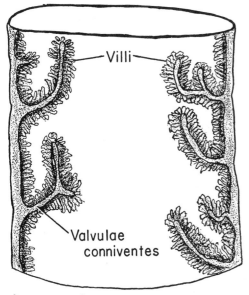

Figure 65–5 A longitudinal section of the small intestine, showing the valvulae conniventes covered by villi.

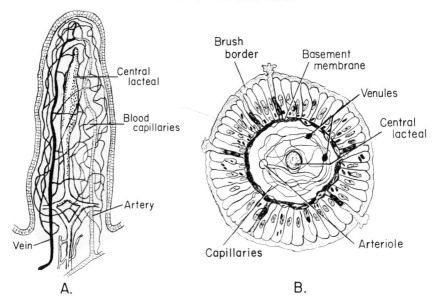

Figure 65–6 Functional organization of the villus: (A) Longitudinal section. (B) Cross-section showing the epithelial cells and basement membrane.

Briefly, active transport imparts energy to the substance as it is being transported for the purpose of concentrating it on the other side of the membrane or for moving it against an electrical potential. On the other hand, the term "diffusion" means simply transport of substances through the membrane as a result of molecular movement *along,* rather than against, an electrochemical gradient.

ABSORPTION IN THE SMALL INTESTINE

ABSORPTION OF NUTRIENTS

Absorption of Carbohydrates. Essentially all the carbohydrates are absorbed in the form of monosaccharides, only a small fraction of a per cent being absorbed as disaccharides and almost none as larger carbohydrate compounds. Furthermore, little carbohydrate absorption results from diffusion, for the pores of the mucosa through which diffusion occurs are essentially impermeable to water-soluble solutes with molecular weights greater than 100.

That the transport of monosaccharides is an active process is demonstrated by several important experimental observations:

1. Their transport can be blocked by metabolic inhibitors, such as iodoacetic acid, cyanides, and phlorhizin.

2. The transport is selective, specifically transporting certain monosaccharides without

transporting others. The order of preference for transporting different monosaccharides and their relative rates of transport in comparison with glucose are:

> Galactose 1.1
> Glucose 1.0
> Fructose 0.4
> Mannose 0.2
> Xylose 0.15
> Arabinose 0.1

3. There is a maximum rate of transport for each type of monosaccharide. The most rapidly transported monosaccharide is galactose, with glucose running a close second. Fructose, which is also one of the three important monosaccharides for nutrition, is ab-

Figure 65–7 Brush border of the gastrointestinal epithelial cell, showing, also, pinocytic vesicles, mitochondria, and endoplasmic reticulum lying immediately beneath the brush border. (Courtesy of Dr. William Lockwood.)

sorbed less than half as rapidly as either galactose or glucose, because it must first diffuse into the cell and then be converted to glucose before the active carrier system will transport it.

4. There is competition between certain sugars for the carrier system. For instance, if large amounts of galactose are being transported, the amount of glucose that can be transported simultaneously is considerably reduced.

These are identically the same principles demonstrated for active transport of monosaccharides through the tubular membranes in the kidneys. Therefore, it is reasonable to believe that carbohydrate transport through the tubular epithelium and through the gastrointestinal mucosa involve essentially the same processes.

MECHANISM OF MONOSACCHARIDE ABSORPTION. We still do not know the precise mechanism of monosaccharide absorption, but we do know the following facts about the mechanism:

First, the chemical nature of monosaccharide molecules that are actively transported, in contradistinction to those that are not actively transported, or that are poorly transported is: (a) six or more carbon atoms in their chain, (b) a D-pyranose ring structure, and (c) an intact hydroxyl group at carbon 2.

Second, glucose transport becomes blocked whenever sodium transport is blocked. Therefore, it is assumed that the energy required for glucose transport is actually provided by the sodium transport system. That is, sodium is actively transported out of the mucosal cells into the subepithelial spaces, and some coupled reaction between sodium transport and glucose transport seems to provide the energy for moving glucose through the intestinal wall. Because of this dependence of glucose transport on sodium transport, the mechanism for glucose transport is said to be *secondary active transport*, in contradistinction to *primary active transport*, the term applied to the sodium type of transport mechanism.

Absorption of Proteins. Essentially all proteins are absorbed in the form of amino acids. However, minute quantities of dipeptides are also absorbed, and extremely minute quantities of whole proteins can be absorbed at times, probably by the process of pinocytosis described above though not by the usual absorptive mechanisms.

The absorption of amino acids also obeys the general principles listed above for active absorption of glucose; that is, certain types of amino acids are absorbed selectively and certain ones interfere with the absorption of others, illustrating that common carrier systems exist. Finally, metabolic poisons block the absorption of amino acids in the same way that they block the absorption of glucose.

Absorption of amino acids through the intestinal mucosa can occur far more rapidly than can protein digestion in the lumen of the intestine. As a result, the normal rate of absorption is determined not by the rate at which they can be absorbed but by the rate at which they can be released from the proteins during digestion. For these reasons, essentially no free amino acids can be found in the intestine during digestion—that is, they are absorbed as rapidly as they are formed.

BASIC MECHANISMS OF AMINO ACID TRANSPORT. As is true for monosaccharide absorption, very little is known about the basic mechanisms of amino acid transport. However, at least three different systems transport different amino acids—one transports *neutral amino acids,* a second transports *basic amino acids,* and a third has specificity for the two amino acids *proline* and *hydroxyproline.* Also, the transport mechanisms have far greater affinity for transporting L-stereoisomers of amino acids than D-stereoisomers. And experiments have demonstrated that *pyridoxal phosphate,* a derivative of the vitamin pyridoxine, is required for transport of many amino acids.

Amino acid transport, like monosaccharide transport, also ceases when sodium transport is blocked. Therefore, amino acid transport is also a type of secondary active transport, the energy for which is provided by the transport of sodium through the membrane.

Absorption of Fats. Fats are believed to be absorbed through the intestinal membrane principally in the form of fatty acids and monoglycerides, though a few diglycerides and triglycerides are also absorbed.

The mechanism for absorption of fatty acids appears to be the following: First, the fatty acid molecule, being highly lipid-soluble, becomes dissolved in the membrane of the brush border of the epithelial cell and diffuses to the interior of the cell. There it comes in contact with the endoplasmic reticulum, which uses fatty acid molecules to resynthesize triglycerides similar to those from which the fatty acid had been derived by digestion in the intestine. These triglycerides are then conducted through the tubules of the endoplasmic reticulum to the sides of the epithelial cell, where they are discharged through the intercellular channel into the submucosal fluids.

Figure 65-8 Chemical steps in the resynthesis of triglycerides by the endoplasmic reticulum.

The chemical mechanism for resynthesis of triglycerides by the endoplasmic reticulum is illustrated in Figure 65-8. This shows that the fatty acids first combine with coenzyme A and then through a succession of stages with the substance glycerophosphate, from which the glycerol portion of the triglyceride is derived.

The absorption of monoglycerides, and perhaps small quantities of diglycerides, occurs in very much the same way as the absorption of fatty acids—that is, by diffusion through the brush border of the membrane and then synthesis of triglycerides by the endoplasmic reticulum. However, the chemical steps for the synthesis of triglycerides are different because the glycerol portion of the triglyceride is already present. It is believed that additional fatty acid radicals are added to the mono- or diglycerides by direct combination of fatty acid coenzyme A as illustrated by the last step in Figure 65-8.

One of the important features of this mechanism for transporting fatty acids and glycerides is that the active chemical process occurs at the membrane of the endoplasmic reticulum rather than at the membrane of the epithelial cell itself where active transport of monosaccharides and amino acids occurs.

CHYLOMICRONS, AND TRANSPORT OF FAT IN THE LYMPH. The triglycerides are discharged by the epithelial cells into the submucosal fluids in the form of minute fat droplets about 0.5 micron in diameter. Small quantities of phospholipids and cholesterol are present in these droplets along with the triglycerides, and the surfaces of the droplets are coated with a layer of protein, which makes them hydrophilic, allowing a reasonable degree of suspension stability in the extracellular fluids. These small particles are called chylomicrons.

From beneath the epithelial cells the chylomicrons wend their way into the central lacteals of the villi and from here are pumped by the lymphatic pump upward through the thoracic duct to be emptied into the great veins of the neck. Between 80 and 90 per cent of all fat absorbed from the gut is absorbed in this manner and transported to the blood by way of the thoracic lymph in the form of chylomicrons.

DIRECT ABSORPTION OF FATTY ACIDS INTO THE PORTAL BLOOD. Much smaller quantities of fatty acids, not over 10 to 20 per cent of the total, are absorbed directly into the portal blood rather than being converted into triglycerides and absorbed into the lymphatics. In general, the shorter the fatty acid chain, the greater is the proportion absorbed in this way; thus, a major portion of the short chain fatty acids in butterfat is absorbed directly into the portal blood rather than into the lymphatics. The cause of this difference between short and long chain fatty acid absorption is presumably that the shorter chain fatty acids are more diffusible than the longer ones and are also more water-soluble, which allows direct diffusion of fatty acids from the epithelial cells into the capillary blood of the villus.

HYDROTROPIC ACTION OF BILE SALTS. Bile salts are known to increase the rate of absorption of both fatty acids and glycerides. This effect results from the ability of bile salts to promote emulsification of the fatty acid and glyceride particles to extremely minute sizes so that they can become entrapped by the membrane of the brush border of the epithelial cells. In the absence of bile salts, fat absorption is seriously reduced.

ABSORPTION OF WATER AND ELECTROLYTES

Absorption of Electrolytes. The rate of absorption of electrolytes is greater in the upper small intestine than in the distal portions, which primarily results from the greater surface area in the upper regions caused by the valvulae conniventes and the profuse villi.

The monovalent electrolytes—sodium, potassium, chloride, nitrate, and bicarbonate—are easily absorbed through the intestinal membrane, while most of the polyvalent electrolytes, such as calcium, magnesium, and sulfate, are poorly absorbed.

Even though the divalent ions are not readily absorbed through the gastrointestinal mucosa, these also are usually not required to nearly so great an extent by the body as the monovalent ions.

ACTIVE ABSORPTION OF SODIUM, CHLORIDE, AND OTHER IONS. Sodium is actively absorbed in large quantity by the mucosa of the small intestine, and at least some chloride is also actively absorbed. However, in addition, transport of sodium across the epithelial membrane creates an electrical potential that pulls additional chloride ions through the membrane simultaneously.

The transport of sodium occurs in the following manner: First, the sodium is actively transported out of the epithelial cells through their sides, the sodium entering the intercellular spaces and flowing along with absorbed water to the center of the villus. Second, the resultant depletion of sodium in the epithelial cells causes a large concentration difference between the lumen of the gut and the cell. The concentration difference causes rapid diffusion of sodium inward through the brush border; and this sodium, too, is subsequently transported into the subepithelial spaces.

Calcium ions are actively absorbed, especially from the duodenum and jejunum, and calcium ion absorption is exactly controlled in relation to the need of the body for calcium. One important factor controlling calcium absorption is parathyroid hormone secreted by the parathyroid glands. In the presence of this hormone, calcium can be absorbed against approximately twice as great a concentration gradient as is true in the absence of parathyroid homone. Also, vitamin D enhances calcium absorption. These effects are discussed in Chapter 79.

Iron ions are also actively absorbed from the small intestine. The principles of iron absorption and the regulation of its absorption in proportion to the body's need for iron were discussed in Chapter 8.

It is probable that hydrogen, potassium, magnesium, phosphate, and bicarbonate ions can also be actively absorbed or secreted in varying amounts through the mucosa of the small intestine.

Absorption of Water.   Water is absorbed by simple diffusion—that is, by osmosis. Some of the reasons for coming to this conclusion are:

(1) When a hyper- or hypotonic solution is placed in the small intestine, it soon becomes isotonic, indicating rapid diffusion across the membrane to establish osmotic equilibrium between the body fluids and the intestinal fluids. (2) When glucose or electrolytes are actively absorbed through the membrane, thereby reducing the osmolality of the intestinal fluids and increasing the osmolality on the opposite side of the membrane, water "follows" through the membrane to maintain isotonicity between the two fluids.

Thus, the simple principles of diffusion can adequately explain the absorption of water—that is, as the electrolytes and nutrients are absorbed, the intestinal fluids become hypotonic, which then causes water to be absorbed by osmosis.

ABSORPTION IN THE LARGE INTESTINE; FORMATION OF THE FECES

Approximately 500 ml. of chyme pass through the ileocecal valve into the large intestine each day. Most of the water and electrolytes in this are absorbed in the colon, leaving only about 100 ml. of fluid to be excreted in the feces.

Essentially all the absorption in the large intestine occurs in the proximal half of the colon, giving this portion the name *absorbing colon,* while the distal colon functions principally for storage and is therefore called the *storage colon.*

Active Absorption of Sodium; Diffusion of Chloride and Water.   The mucosa of the large intestine is capable of absorbing sodium actively. Loss of sodium from the lumen of the gut creates a strong electrical potential across the membrane, which results in the movement of chloride and other negative ions out of the colon into the interstitial fluids. Because of this mechanism, little sodium chloride is lost in the feces. The absorption of sodium and chloride then causes osmotic absorption of water as explained above so that little water also is lost in the feces. Left behind as the most abundant salt in the feces is generally a moderate amount of almost insoluble calcium phosphate. Also, the stools tend to have a high bicarbonate concentration because bicarbonate ion is absorbed through the membrane poorly and also because it is actively secreted in small amounts. The bicarbonate causes alkalinity of the outer surfaces of the fecal boluses, while deep in the feces bacterial activity usually creates acidity.

Bacterial Action in the Colon.   Numerous bacteria, especially colon bacilli, are present in the absorbing colon. These are capable of digesting small amounts of cellulose, in this way providing a few calories of nutrition to the body each day. In herbivorous animals this source of energy is sometimes very significant though it is of negligible importance in the human being. Other substances formed as a result of bacterial activity are vitamin K, vitamin B_{12}, thiamin, riboflavin, and various gases that contribute to *flatus* in the colon. Vitamin K is especially important, for the amount of this vitamin in the ingested foods is normally insufficient to maintain adequate blood coagulation.

Composition of the Feces. The feces normally are about three-fourths water and one-fourth solid matter composed of about 30 per cent dead bacteria, 10 to 20 per cent fat, 10 to 20 per cent inorganic matter, 2 to 3 per cent protein, and 30 per cent undigested roughage of the food and dried constituents of digestive juices, such as bile pigment and sloughed epithelial cells. The large amount of fat derives from unabsorbed fatty acids from the diet, fat formed by bacteria, and fat in the sloughed epithelial cells.

The brown color of feces is caused by *stercobilin* and *urobilin,* which are derivatives of bilirubin. The odor is caused principally by the products of bacterial action; these vary from one person to another, depending on each person's colonic bacterial flora and on the type of food he has eaten. The actual odoriferous products include *indole, skatol, mercaptans,* and *hydrogen sulfide.*

REFERENCES

Altamirano, M.: Action of concentrated solutions of nonelectrolytes on the dog gastric mucosa. *Amer. J. Physiol.,* 216:33, 1969.

Altamirano, M.: Action of solutions of reduced osmotic concentration on the dog gastric mucosa. *Amer. J. Physiol.,* 216:25, 1969.

Annegers, J. H.: Intestinal absorption of amino acids in the dog. *Amer. J. Physiol.,* 216:1, 1969.

Benson, J. A., Jr., and Rampone, A. J.: Gastrointestinal absorption. *Ann. Rev. Physiol.,* 28:201, 1966.

Booth, C. C.: Absorption from the small intestine. *Scient. Basis Med. Ann. Rev.,* 171, 1963.

Cook, D. A.; McGilliard, A. D.; Young, J. W.; Jacobson, N. L.; and Allen, R. S.: Lymphatic absorption of palmitate-1-¹⁴C and aleate-1-¹⁴C in the bovine. *Amer. J. Physiol.,* 216:41, 1969.

Crane, R. K.: Absorption of sugars. *In* Handbook of Physiology. Baltimore, The Williams & Wilkins Co., 1968, Sec. VI, Vol. III, p. 1323.

Curran, P. F., and Schultz, S. G.: Transport across membranes: general principles. *In* Handbook of Physiology. Baltimore, The Williams & Wilkins Co., 1968, Sec. VI, Vol. III, p. 1217.

Hagemann, R. F.; Sigdestad, C. P.; and Lesher, S.: Quantitation of intestinal villus cells on a weight basis: values for C57BL/6 mice. *Amer. J. Physiol.,* 218:637, 1970.

Hakim, A. A., and Lifson, N.: Effects of pressure on water and solute transport by dog intestinal mucosa in vitro. *Amer. J. Physiol.,* 216:276, 1969.

Heaton, J. W., Jr., and Code, C. F.: Sodium-glucose relationships during intestinal sorption in dogs. *Amer. J. Physiol.,* 216:749, 1969.

Johnston, J. M.: Mechanism of fat absorption. *In* Handbook of Physiology. Baltimore, The Williams & Wilkins Co., 1968, Sec. VI, Vol. III, p. 1353.

Lifson, N.; Gruman, L. M.; and Levitt, D. G.: Diffusive-convective models for intestinal absorption of D_2O. *Amer. J. Physiol.,* 215:444, 1968.

Martin, D. L., and DeLuca, H. F.: Influence of sodium on calcium transport by the rat small intestine. *Amer. J. Physiol.,* 216:1351, 1969.

Matthews, D. M., and Laster, L.: Competition for intestinal transport among five neutral amino acids. *Amer. J. Physiol.,* 208:601, 1965.

Reiser, S., and Christiansen, P. A.: Intestinal transport of amino acids as affected by sugars. *Amer. J. Physiol.,* 216:915, 1969.

Schedl, H. P.; Miller, D. L.; Wilson, H. D.; and Flores, P.: a-Aminoisobutyric acid transport and tissue concentration at various intestinal sites. *Amer. J. Physiol.,* 216:1131, 1969.

Schultz, S. G., and Curran, P. F.: Intestinal absorption of sodium chloride and water. *In* Handbook of Physiology. Baltimore, The Williams & Wilkins Co., 1968, Sec. VI, Vol. III, p. 1245.

Smyth, D. H.: Intestinal absorption. *Brit. Med. Bull.,* 23:205, 1967.

Stevens, C. E.; Dobson, A.; Mammano, J. H.: A transepithelial pump for weak electrolytes. *Amer. J. Physiol.,* 216:983, 1969.

Taylor, A. E.; Wright, E. M.; Schultz, S. G.; and Curran, P. F.: Effect of sugars on ion fluxes in intestine. *Amer. J. Physiol.,* 214:836, 1968.

Wilson, T. H.: Intestinal Absorption. Philadelphia, W. B. Saunders Company, 1962.

Wiseman, G.: Absorption of amino acids. *In* Handbook of Physiology. Baltimore, The Williams & Wilkins Co., 1968, Sec. VI, Vol. III, p. 1277.

CHAPTER 66

Gastrointestinal Disorders

The logical treatment of most gastrointestinal disorders depends on a basic knowledge of gastrointestinal physiology. The purpose of this chapter, therefore, is to discuss certain types of malfunction that have special physiologic consequences.

DISORDERS OF SWALLOWING AND OF THE ESOPHAGUS

Paralysis of the Swallowing Mechanism. Damage to the 5th, 9th, or 10th nerve can cause paralysis of significant portions of the swallowing mechanism. Also, a few diseases, most commonly poliomyelitis, prevent normal swallowing by damaging the swallowing center in the brain stem. Finally, malfunction of the swallowing muscles, as occurs in *muscle dystrophy* or in failure of neuromuscular transmission in *myasthenia gravis* or *botulism,* can also prevent normal swallowing.

When the swallowing mechanism is partially or totally paralyzed, the abnormalities that can occur include: (1) complete abrogation of the swallowing act so that swallowing cannot occur at all, (2) failure of the glottis to close so that food passes into the trachea as well as into the esophagus, (3) failure of the soft palate and uvula to close the posterior nares so that food refluxes into the nose during swallowing, or (4) failure of the cricoid esophageal sphincter to remain closed during normal breathing, thus allowing large quantities of air to be sucked into the esophagus.

One of the most serious instances of paralysis of the swallowing mechanism occurs when a person is under deep anesthesia. Often he vomits large quantities of materials from the stomach into the pharynx; then instead of swallowing the materials again he simply sucks them into his trachea because the anesthetic has blocked the reflex mechanism of swallowing. As a result, such patients often choke to death on their own vomitus.

Achalasia. Achalasia is a condition in which the lower few centimeters of the esophagus fail to relax during the swallowing mechanism. As a result, food transmission from the esophagus into the stomach is impeded or is sometimes completely blocked. Pathological studies have shown the cause of this condition to be either damaged or absent myenteric plexus in the lower portion of the esophagus. The musculature of the lower end of the esophagus is still capable of contracting, and even exhibits incoordinate movements, but it has lost the ability to conduct a peristaltic wave and has lost the ability of "receptive relaxation" of the gastroesophageal sphincter as food approaches this area during the swallowing process.

Mega-esophagus. Mega-esophagus means a greatly enlarged esophagus, and it ordinarily results from achalasia. Food is dammed up behind the aperistaltic portion of the esophagus, causing the middle and upper esophagus to dilate progressively until it sometimes becomes 3 to 4 inches in diameter. In severe cases of achalasia, food sometimes stays in the esophagus as long as 24 hours before passing into the stomach. This prolonged storage of food allows extreme putrefaction of the food and often also allows bacterial penetration of the esophageal mucosa, which results in ulceration of the esophageal wall and, possibly, death.

DISORDERS OF THE STOMACH

Gastritis. Gastritis means inflammation of the gastric mucosa. This can result from (a) action of irritant foods on the gastric mucosa, (b) excessive excoriation of the stomach mucosa by the stomach's own peptic secretions, or (c) occasionally, bacterial inflammation. One of the most frequent causes of gastritis is irritation of the mucosa by alcohol.

The inflamed mucosa in gastritis is often painful, causing a diffuse burning pain referred to the high epigastrium. Reflexes initiated in the stomach mucosa cause the salivary glands to salivate intensely, and the frequent swallowing of foamy saliva makes air accumulate in the stomach. As a result, the person usually belches profusely, a burning sensation occurring in his throat with each belch.

Gastric Atrophy. In about one third of all persons who have gastritis, the mucosa gradually becomes atrophic until little or no gastric gland activity remains. Loss of the stomach secretions then leads to *anacidity* and, occasionally, to *pernicious anemia.*

ANACIDITY (AND HYPOACIDITY). Anacidity means simply that the stomach fails to secrete hydrochloric acid, and hypoacidity means diminished acid secretion. Usually, when acid is not secreted, pepsin also is not secreted, and, even if it were, the lack of acid would prevent it from functioning because pepsin requires an acid medium for activity. Obviously, then, essentially all digestive function in the stomach is lost when anacidity is present.

The method usually used to determine the degree of hypoacidity is to inject 0.5 milligram of histamine and then to aspirate the stomach secretions through a tube for the following hour. Each 10-minute sample of stomach secretion is titrated against sodium hydroxide to a pH of 8.5, which gives a measure of *total acid* secreted. The maximum rate of acid secretion during any 10-minute interval in the normal person rises to almost exactly 1 milliequivalent of hydrochloric acid per minute, and the total hydrochloric acid secreted during the entire hour after the injection averages about 18 milliequivalents. All degrees of hypoacidity – down to no acid whatsoever – occur.

The acid secretion is frequently divided into *free acid* and *combined acid*. The amount of free acid is determined by titrating the gastric secretions to a pH of 3.5, using *dimethylamino-azobenzene* as an indicator. After this titration has been performed, the same secretions are titrated to a pH of 8.5, using *phenolphthalein* as an indicator; this measures the combined acid. Gastric secretions mixed with food in the stomach usually show little or no free acid but a large amount of combined acid. On the other hand, when the stomach secretes large quantities of gastric juice while it is almost empty of food, the larger portion of the acid may then be free acid and only a small amount is combined.

Though anacidity is associated with depressed or even no digestive capability by the stomach, the overall digestion of food in the entire gastrointestinal tract is still almost normal. This is because trypsin and other enzymes secreted by the pancreas are capable of digesting most of the protein in the diet – particularly if the food is well chewed so that no portion of the protein is protected by collagen fibers, which need peptic activity for most effective digestion.

Pernicious Anemia in Gastric Atrophy. Pernicious anemia, which was discussed in Chapter 8, is a common accompaniment of gastric atrophy. The normal gastric secretions contain a mucopolypeptide called *intrinsic factor,* which is secreted by the parietal cells (the HCl-producing cells) and which must be present for adequate absorption of vitamin B_{12} from the small intestine. The intrinsic factor combines with vitamin B_{12}, making it resistant to digestion and also aiding in its absorption. In the absence of intrinsic factor, an adequate amount of vitamin B_{12} is not made available from the foods. As a result, maturation failure occurs in the bone marrow, resulting in pernicious anemia. This subject was discussed in more detail in Chapter 8.

Pernicious anemia also occurs frequently when

most of the stomach has been removed for treatment of either stomach ulcer or gastric cancer.

PEPTIC ULCER

A peptic ulcer is an excoriated area of the mucosa caused by the digestive action of gastric juice. Figure 66–1 illustrates the points in the gastrointestinal tract at which peptic ulcers frequently occur, showing that by far the most frequent site of peptic ulcers is in the pyloric region of the duodenum. In addition, peptic ulcers frequently occur along the lesser curvature of the stomach or, more rarely, in the lower end of the esophagus where stomach juices frequently reflux. A peptic ulcer called a *marginal ulcer* also frequently occurs wherever an abnormal opening, such as a gastrojejunostomy, is made between the stomach and some portion of the small intestine.

Basic Cause of Peptic Ulceration. The usual cause of peptic ulceration is too much secretion of gastric juice in relation to the degree of protection afforded the mucosa by the mucus that is also secreted. It will be recalled that all areas normally exposed to gastric juice are well supplied with mucous glands, beginning with the compound mucous glands of the lower esophagus, then including the mucous cell coating of the stomach mucosa, the mucous neck cells of the gastric glands, the deep pyloric glands that secrete almost nothing but mucus, and, finally, the coiled tubular glands of Brunner of the upper duodenum, which secrete a highly alkaline mucus.

In addition to the mucus protection of the mucosa, the duodenum is also protected by the alkalinity of the small intestinal secretions. Especially important is pancreatic secretion, which contains large quantities of sodium bicarbonate that neutralize the hydrochloric acid of the gastric juice, thus inactivating the pepsin to prevent digestion of the mucosa. Two additional mechanisms insure that this neutralization of gastric juices is complete:

1. When excess acid enters the duodenum, it reflexly inhibits antral peristalsis in the stomach, thereby decreasing the rate of gastric emptying. This allows increased time for pancreatic secretion to enter the duodenum to neutralize the acid

CAUSES:
1. high acid and peptic content
2. irritation
3. poor blood supply
4. poor secretion of mucous
5. ? infection

cardia

ulcer sites

pylorus

marginal ulcer

Figure 66–1   Peptic ulcer.

already present. After neutralization has taken place, the reflex subsides and more stomach contents are emptied.

2. The presence of acid in the small intestine liberates secretin from the intestinal mucosa, which then passes by way of the blood to the pancreas to promote rapid secretion of pancreatic juice. Secretion also stimulates the pancreas to secrete a juice containing a high concentration of sodium bicarbonate, thus making more sodium bicarbonate available for neutralization of the acid. These mechanisms were discussed in detail in Chapters 63 and 61 in relation to gastrointestinal motility and secretion.

Experimental Peptic Ulcer. Experimental peptic ulcers have been created in dogs and other animals in the following ways: (a) *Feeding ground glass* to an animal causes excoriation of the pyloric wall and allows the peptic juices to begin digesting the deeper layers of the mucosa. (b) *Transplantation of the pancreatic duct to the ileum* removes the normal neutralizing effect of pancreatic secretion in the duodenum and, therefore, allows the gastric juice to attack the mucosa of the upper duodenum. (c) *Repeated injection of histamine* causes excessive secretion of gastric juice. (d) *Continual infusion of hydrochloric acid* through a tube into the stomach causes direct irritation of the stomach mucosa and also prevents full neutralization of the gastric juices by the pancreatic and other secretions of the small intestine. Therefore, peptic activity is enhanced, and ulcers appear in either the stomach or duodenum. (e) A portion of the *stomach is anastomosed directly with the small intestine* so that gastric juice can pass directly and rapidly into the small intestine. The mucosa of the small intestine, except at the uppermost part of the duodenum, is not sufficiently resistant to gastric juice to prevent peptic digestion. (f) *Obstruction of the blood flow* or even reduction of the blood flow to an area of the stomach or upper duodenum will cause ulcers to develop because the local area cannot produce appropriate protective secretions.

In summary, any factor that (1) *increases the rate of production of gastric juice* or (2) *blocks the normal protective mechanisms* against this juice can produce peptic ulcers. The same general principles apply to the development of peptic ulcers in the human being.

Causes of Peptic Ulcer in the Human Being. Peptic ulcer occurs much more frequently in the white collar worker than in the laborer, and persons subjected to extreme anxiety for a long time seem particularly prone to peptic ulcer. For instance, the number of persons who developed peptic ulcer increased greatly during the air raids of London. Therefore, it is believed that most instances of peptic ulcer in the human being result from excessive stimulation of the dorsal motor nucleus of the vagus by impulses originating in the cerebrum. The ulcers develop in the least protected areas of the gastrointestinal tract—especially immediately beyond the pylorus—as illustrated in Figure 66–1. Supporting this theory is the fact that ulcer patients have a high rate of gastric secretion during the interdigestive period between meals when the stomach is empty. The normal stomach secretes a total of approximately 18 milliequivalents of hydrochloric acid during the 12-hour interdigestive period through the night, while ulcer patients occasionally secrete as much as 300 milliequivalents of hydrochloric acid during this same time.

Physiology of Treatment. A usual medical treatment for peptic ulcer is a diet of six or more small meals per day rather than the normal three large meals. In this way food is kept in the stomach most of the time, and the food neutralizes much if not most of the acid and dilutes the gastric juice so that its digestive action on the mucosa is minimized. Also, the composition of the diet is prescribed in accordance with the following two essential principles: (1) The diet should be bland, containing as few secretagogues as possible—that is, no alcohol, few spices, and meats boiled until the extractives have been washed away. Lack of the secretagogues prevents the extraction of gastrin from the pyloric mucosa and therefore reduces the hormonal stimulation of the gastric glands during the gastric phase of stomach secretion. This mechanism was discussed in Chapter 64. (2) The diet is formulated to contain large quantities of fat. On entering the small intestine, the fat activates the enterogastrone mechanism that inhibits gastric secretion and also slows the rate of gastric emptying. This reduces the quantity of gastric juice secreted and also allows prolonged stasis of food in the stomach to buffer the gastric juice that is secreted.

Surgical treatment of peptic ulcer is effected by one or both of two procedures: (1) removal of a large portion of the stomach or (2) vagotomy. When a peptic ulcer is surgically removed, usually at least the lower three fourths to four fifths of the stomach is also removed and the upper stump of the stomach then anastomosed to the jejunum. If less of the stomach than this is removed, far too much gastric juice continues to be secreted, and a marginal ulcer soon develops where the stomach is anastomosed to the intestine. Since a marginal ulcer is frequently equally as debilitating as, or sometimes even more debilitating than, the original ulcer, nothing will have been accomplished. However, if only one fifth of the stomach remains, the rate of gastric secretion is usually low enough that a marginal ulcer does not develop, and yet the stomach is still large enough to hold small meals. Furthermore, the presence of at least some functional stomach mucosa usually prevents the development of pernicious anemia.

Vagotomy temporarily blocks almost all secretion by the stomach and usually cures a peptic ulcer within less than a week after the operation is performed. The vagus nerves may be cut as they pass downward over the surface of the esophagus in the thorax or as they pass through the diaphragm to spread over the surface of the stomach. Unfortunately, though, a large amount of basal stomach secretion returns three to six months after the vagotomy, and in many patients the ulcer itself also returns. Even more distressing is the *gastric atony* that usually follows vagotomy, for the motility of the stomach is often reduced to such a low level that almost no gastric emptying occurs after vagotomy. For this reason, vagotomy alone is rarely performed in the treatment of peptic ulcer. However, it is performed frequently in association with simultaneous plastic procedures to increase the size of the opening of the pylorus from the stomach into the small intestine.

DISORDERS OF THE SMALL INTESTINE

In comparison with disorders of the stomach and colon, disorders of the small intestine are surprisingly few. Perhaps the most common disorder of the small intestine and its associated structures is the development of gallstones in the gallbladder and bile ducts. (This will be discussed in Chap. 74 in relation to the function of the liver and the bile ducts.) The presence of stones in the gallbladder or bile ducts does not alter the function of the small intestine unless a stone blocks the flow of bile into the intestine. Lack of bile then prevents adequate absorption of fatty acids from the small intestine with resultant loss of large quantities of fats in the feces.

Abnormal Digestion of Food in the Small Intestine; Pancreatic Failure. By far the greater portion of all digestion in the alimentary tract occurs in the upper third of the small intestine. Only rarely is digestion impaired enough to cause malnutrition. Indeed, as much as three-fourths of the small intestine have been removed without the patient's suffering serious malnutrition.

Perhaps the commonest cause of abnormal digestion is failure of the pancreas to secrete its juice into the small intestine. Lack of pancreatic secretion frequently occurs (a) in *pancreatitis,* which is discussed below, (b) when the *pancreatic duct is blocked* by a gallstone at the papilla of Vater, or (c) after the *head of the pancreas has been removed* because of malignancy. Loss of pancreatic juice means loss of trypsin, chymotrypsin, carboxypolypeptidase, pancreatic amylase, pancreatic lipase, and still a few other digestive enzymes. Without these enzymes, as much as three quarters of the fat entering the small intestine may go undigested and as much as a third to one half of the proteins and starches. As a result, large portions of the ingested food are not utilized for nutrition; and copious, fatty feces are excreted.

Pancreatitis. Pancreatitis means inflammation of the pancreas, and this can occur either in the form of *acute pancreatitis* or *chronic pancreatitis.* The commonest cause of acute pancreatitis is blockage of the papilla of Vater by a gallstone; this blocks the main secretory duct from the pancreas as well as the common bile duct. The pancreatic

enzymes are then dammed up in the ducts and acini of the pancreas. Eventually, so much trypsinogen accumulates that it overcomes the *trypsin inhibitor* in the secretions, and a small quantity of trypsinogen becomes activated to form trypsin. Once this happens the trypsin activates still greater quantities of trypsinogen as well a chymotrypsinogen and carboxypolypeptidase, resulting in a vicious cycle until all the proteolytic enzymes in the pancreatic ducts and acini become activated. Rapidly these digest large portions of the pancreas itself, sometimes completely and permanently destroying the ability of the pancreas to secrete digestive enzymes.

Often in acute pancreatitis the proteolytic enzymes eat their way all the way to the surface of the pancreas, and pancreatic juice then empties into the peritoneal cavity where additional proteolytic activity occurs, causing *chemical peritonitis.* However, as soon as large portions of the pancreas have been destroyed, the acute stage of pancreatitis subsides, leaving the person with diminished or sometimes totally absent pancreatic secretion into the gut.

Chronic pancreatitis usually occurs in much the same manner as does acute pancreatitis, beginning with blockage of the pancreatic duct but progressing slowly until pancreatic function becomes greatly diminished.

Ordinarily, the islets of Langerhans are not adversely affected by pancreatitis, so that insulin still continues to be secreted by the pancreas even though secretion into the intestine is markedly reduced.

Malabsorption from the Small Intestine — "Sprue." Occasionally, nutrients are not adequately absorbed from the small intestine even though the food is well digested. Several different diseases can cause decreased absorbability by the mucosa; these are often classified together under the general heading of sprue, or in young children the condition is frequently called *celiac disease.* Obviously, also, malabsorption can occur when large portions of the small intestine have been removed.

The cause of celiac disease in children or so-called "idiopathic sprue" in adults is not known, though removal of wheat and rye flour from the food — that is, removal of the protein *gliadin* (a type of gluten) from the diet — will in many instances effect a cure or at least aid in the absorption of nutrients from the intestinal tract. It is believed that gliadin is not fully digested and that remaining glutamine-containing polypeptides are toxic to the mucosa. The newly forming epithelial cells in the crypts of Lieberkühn die before they can migrate upward onto the villi. As a result, the villi become blunted or disappear altogether, thus greatly reducing the absorptive area of the gut.

A different type of sprue called "tropical sprue" frequently occurs in the tropics and can often be treated with antibacterial agents. Even though no specific bacterium has ever been implicated as the cause of this tropical sprue, it is believed that this variety is often caused by inflammation of the intestinal mucosa resulting from a yet unidentified infectious agent.

In the early stages of sprue, the absorption of fats is more impaired than the absorption of other digestive products. The fat appearing in the stools is almost entirely in the form of soaps rather than undigested neutral fat, illustrating that the problem is one of absorption and not of digestion. In this stage of sprue, the condition is frequently called *idiopathic steatorrhea*, which means simply excess fats in the stools as a result of unknown causes.

In more severe cases of sprue the absorption of proteins, carbohydrates, calcium, vitamin K, folic acid, and vitamin B$_{12}$, as well as many other important substances, becomes greatly impaired. As a result, the person suffers (1) severe nutritional deficiency, often developing severe wasting of the tissues, (2) osteomalacia (demineralization of the bones because of calcium lack), (3) inadequate blood coagulation due to lack of vitamin K, and (4) macrocytic anemia of the pernicious anemia type, owing to diminished vitamin B$_{12}$ and folic acid absorption.

Regional Ileitis and Appendicitis. Regional ileitis means inflammation of a portion of the ileum, occurring usually at its distal end, and appendicitis means inflammation of the appendix. Appendicitis is usually caused by an actual infection in the appendix, while the cause of regional ileitis is not known but could also result from a low grade infectious process. Either of these two conditions causes pain that is referred to the mid-abdomen. Simultaneously, they also elicit intense enterointestinal reflexes resulting in severe inhibition of gastrointestinal mobility; as a result, functional obstruction often occurs in the small bowel, causing symptoms of acute intestinal obstruction, which will be discussed later in the chapter. These are mainly distention of the upper small intestine associated with protracted cramping pains and vomiting.

DISORDERS OF THE LARGE INTESTINE

CONSTIPATION

Constipation means slow movement of feces through the large intestine, and it is usually associated with large quantities of dry, hard feces in the descending colon which accumulate because of the long time allowed for absorption of fluid.

The most frequent cause of constipation is irregular bowel habits that have developed through a lifetime of inhibition of the normal defecation reflexes. The newborn child is rarely constipated, but part of his training in the early years of life requires that he learn to control defecation, and this control is effected by inhibiting the natural defecation reflexes. Clinical experience shows that if one fails to allow defecation to occur when the defecation reflexes are excited, the reflexes themselves become progressively less strong over a period of time. For this reason, if a person establishes regular bowel habits early in life, usually defecating in the morning after breakfast when the gastrocolic and duodenocolic reflexes cause mass movements in the large intestine, he can generally prevent the development of constipation in later life.

Constipation can also result from *atony* of the large intestine, or at times it results from *spasm of a small segment of the sigmoid,* which blocks movement of feces from the proximal portions of the large intestine. It should be recalled that motility in the large intestine even normally is weak, so that even a slight degree of spasm in any portion of the descending colon or sigmoid is often capable of causing serious constipation.

Megacolon. Occasionally, constipation becomes so severe that bowel movements occur only once every several weeks or, occasionally, only once a month. Obviously, this allows tremendous quantities of fecal matter to accumulate in the colon, causing the colon sometimes to distend to a diameter as great as 3 to 4 inches. The condition is called megacolon, or *Hirschsprung's disease.*

The most frequent cause of megacolon is congenital lack of a myenteric plexus in a segment of the sigmoid. This causes the sigmoid to become relatively amotile so that normal defecation reflexes or propulsive motility cannot occur through this area of the large intestine. As a consequence, the sigmoid itself becomes small and almost spastic while feces accumulate proximal to this area, causing megacolon. Thus, the diseased portion of the large intestine appears normal by x-ray, while the originally normal portion of the large intestine appears greatly enlarged.

DIARRHEA

Diarrhea, which is the opposite of constipation, results from rapid movement of fecal matter through the large intestine. The major causes of diarrhea are: (1) an infection in the gastrointestinal tract called *gastroenteritis* and (2) excessive parasympathetic stimulation of the large intestine, which is called *neurogenic diarrhea.*

Gastroenteritis. Gastroenteritis means infection at any point in the gastrointestinal tract or throughout the entire extent of the tract. In usual infectious diarrhea, the infection is most extensive in the large intestine, though sometimes it extends into the small intestine as well. Everywhere that the infection is present, the mucosa becomes extensively irritated, and its rate of secretion becomes greatly enhanced. In addition, the motility of the intestinal wall usually increases many fold. As a result, large quantities of fluid are made available for washing the infectious agent toward the anus, and at the same time strong mass movements propel this fluid forward. Obviously, this is an important mechanism for ridding the intestinal tract of the debilitating infection.

Neurogenic Diarrhea. Everyone is familiar with the diarrhea that accompanies periods of nervous tension, such as (a) during examination time or (b) when a soldier is about to go into battle. This type of diarrhea, called neurogenic diarrhea, is caused by excessive stimulation of the parasympathetic nervous system, which greatly excites both motility and secretion of mucus in the colon. These two effects added together can cause marked diarrhea. Indeed, in tense states, even the chyme in the small intestine can be flushed into the large intestine and then rapidly on through to the anus, causing at times loss of large quantities of water and electrolytes.

Ulcerative Colitis. Ulcerative colitis is a disease in which extensive areas of the walls of the large intestine become ulcerated. This, like neurogenic diarrhea, is frequently associated with different states of nervous tension. The motility of the ulcerated colon is often so great that mass movements occur most of the time, rather than the usual 10 to 20 minutes per day. Also, the colon's secretions are greatly enhanced. As a result, the patient has repeated diarrheal bowel movements.

Though ulcerative colitis is associated with neurogenic disorders, the precise cause of the actual ulcers is unknown. Some clinicians believe that these are caused by infectious bacteria invading the mucosa, while others believe that they result from excoriation or digestion of the mucosa by the contents of the large intestine. Regardless of the cause, if the condition has not progressed too far, removal of factors that have been causing nervous tension frequently leads to a cure. On the other hand, if the condition has progressed extremely far, the ulcers usually will not heal until an ileostomy is performed to allow the intestinal contents to drain to the exterior rather than to flow through the colon. Even then the ulcers sometimes fail to heal, and the only solution is complete removal of the colon itself.

PARALYSIS OF DEFECATION IN SPINAL CORD INJURIES

From Chapter 63 it will be recalled that defecation is normally initiated by the movement of feces into the rectum, which causes a reflex to pass from the rectum to the spinal cord and then back to the descending colon, sigmoid, rectum, and anus to intensify greatly the intrinsic defecation reflex of the myenteric plexus. Frequently, this *cord* defecation reflex is blocked or altered by spinal cord injuries. For instance, destruction of the *conus medullaris* of the spinal cord destroys the sacral centers in which the cord reflex is integrated and therefore almost completely paralyzes defecation. In such instances defecation requires extensive supportive measures, such as cathartics and large enemas.

But, more frequently, the spinal cord is injured somewhere between the conus medullaris and the brain, in which case the voluntary portion of the defecation act is blocked while the basic cord reflexes for defecation are still intact. Nevertheless, loss of the voluntary aid to defecation — that is, loss of the increased abdominal pressure, the lifting of the pelvic floor, and the stretching of the anal ring by the pelvic muscles — often makes defecation a difficult process in the person with this type of cord injury. Yet, since cord defecation reflexes can still occur, a small enema given to potentiate the action of these reflexes, usually in the morning shortly after a meal, can often cause adequate defecation. In this way persons with spinal cord injuries can usually control their bowel movements each day.

GENERAL DISORDERS OF THE GASTROINTESTINAL TRACT

VOMITING

Vomiting is the means by which the upper gastrointestinal tract rids itself of its contents when the gut becomes excessively irritated, overdistended, or even overexcitable. The stimuli that cause vomiting can occur in any part of the gastrointestinal tract, though distention or irritation of the stomach or duodenum provides the strongest stimulus. Impulses are transmitted, as illustrated in Figure 66–2, by both vagal and sympathetic afferents to the bilateral *vomiting center* of the medulla, which lies near the tractus solitarius at approximately the level of the dorsal motor nucleus of the vagus. Appropriate motor reactions are then instituted to cause the vomiting act, and the motor impulses that cause the actual vomiting are transmitted from the vomiting center through the fifth, seventh, ninth, tenth, and twelfth cranial nerves to the upper gastrointestinal tract and through the spinal nerves to the diaphragm and abdominal muscles.

The Vomiting Act. Once the vomiting center has been sufficiently stimulated and the vomiting act instituted, the first effects are (1) a deep breath, (2) raising of the hyoid bone and the larynx to pull the cricoesophageal sphincter open, (3) closing of the glottis, and (4) lifting of the soft palate to close the posterior nares. Next comes a strong downward contraction of the diaphragm along with simultaneous contraction of all the abdominal muscles. This obviously squeezes the stomach between the two sets of muscles, building the intragastric pressure to a high level. At the same time, strong reverse peristalsis begins in the antral region of the stomach and passes over the body of the stomach, thus forcing the stomach contents toward the esophagus. Finally, the gastroesophageal sphincter relaxes, allowing expulsion of the gastric contents upward through the esophagus.

Thus, the vomiting act results from a squeezing action of the muscles of the stomach and abdomen associated with opening of the esophageal sphincters so that the gastric contents can be expelled.

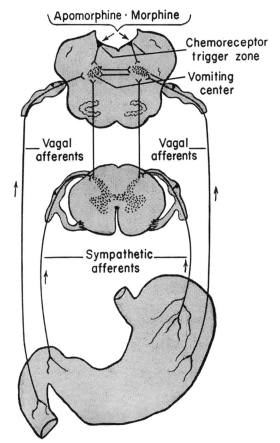

Figure 66-2 The afferent connections of the vomiting center.

The Chemoreceptor Trigger Zone of the Medulla for Initiation of Vomiting by Drugs or by Motion Sickness. Aside from the vomiting initiated by irritative stimuli in the gastrointestinal tract itself, vomiting can also be caused by impulses arising in areas of the brain outside the vomiting center. This is particularly true of a small area located bilaterally on the floor of the fourth ventricle above the area postrema and called the *chemoreceptor trigger zone.* Electrical stimulation of this area will initiate vomiting, but, more importantly, administration of certain drugs, including apomorphine, morphine, copper sulfate, and some of the digitalis derivatives, can directly stimulate the chemoreceptor trigger zone and initiate vomiting. Destruction of this area blocks this type of vomiting but does not block vomiting resulting from irritative stimuli in the gastrointestinal tract itself.

Also, it is well known that rapidly changing motions of the body cause certain people to vomit. The mechanism for this is the following: The motion stimulates the receptors of the labyrinth, and impulses are transmitted either directly or by way of the vestibular nuclei into the cerebellum. After passing through the uvula and nodule of the cerebellum, the signals are believed to be transmitted to the chemoreceptor trigger zone and thence to the vomiting center to cause vomiting.

Cortical Excitation of Vomiting. Various psychic stimuli, including disquieting scenes, noisome odors, and other similar psychological factors, can also cause vomiting. Stimulation of certain areas of the hypothalamus also causes vomiting. The precise neuronal connections for these effects are not known, though it is probable that the impulses pass directly to the vomiting center and do not involve the chemoreceptor trigger zone.

NAUSEA

Everyone has experienced the sensation of nausea and knows that it is often a prodrome of vomiting. Nausea is the conscious recognition of subconscious excitation in an area of the medulla closely associated with or part of the vomiting center, and it can be caused by irritative impulses coming from the gastrointestinal tract, impulses originating in the brain associated with motion sickness, or impulses from the cortex to initiate vomiting. However, vomiting occasionally occurs without the prodromal sensation of nausea, indicating that only certain portions of the vomiting centers are associated with the sensation of nausea.

A common cause of nausea is distention or irritation of the duodenum or lower small intestine. When this occurs, the intestine contracts forcefully while the stomach relaxes, thus allowing the intestinal contents to reflux into the stomach. This is preliminary to the vomiting that often follows.

GASTROINTESTINAL OBSTRUCTION

The gastrointestinal tract can become obstructed at almost any point along its course, as illustrated in Figure 66-3; some common causes of obstruction are: *cancer, fibrotic constriction resulting from ulceration or from peritoneal adhesions, spasm of a segment of the gut,* or *paralysis of a segment of the gut.*

The abnormal consequences of obstruction depend on the point in the gastrointestinal tract that becomes obstructed. If the obstruction occurs at the pylorus, which results often from fibrotic constric-

Figure 66-3 Obstruction in different parts of the gastrointestinal tract.

tion following peptic ulceration, persistent vomiting of stomach contents occurs. This obviously depresses bodily nutrition, and it also causes excessive loss of hydrogen ions from the body and can result in various degrees of alkalosis.

If the obstruction is beyond the stomach, reflux from the small intestine causes the intestinal juices to flow backward into the stomach, and these are vomited along with the stomach secretions. In this instance the person loses large amounts of water and electrolytes so that he becomes severely dehydrated, but the loss of acids and bases may be approximately equal so that little change in acid-base balance occurs. If the obstruction is near the lower end of the small intestine, then it is actually possible to vomit more basic than acid substances; in this case acidosis may result.

Also important in obstruction of the small intestine is marked distention of the intestine proximal to the obstructed point. Large quantities of fluid and electrolytes continue to be secreted into the lumen of the small intestine and even large amounts of proteins are lost from the blood stream, partly into the intestinal lumen and partly into the gut wall, which becomes edematous as a result of the excessive distention. The plasma volume diminishes because of the protein loss, and severe circulatory shock often ensues. One might immediately ask: Why does the small intestine not absorb these fluids and electrolytes? The answer seems to be that distention of the gut greatly stimulates the secretory activity of the gut but does not correspondingly increase the rate of absorption. Normally this would act to flush the chyme farther down the small intestine and therefore relieve the distention. But, if obstruction is present, obviously this normal mechanism backfires and simply causes a vicious cycle of more and more distention.

If the obstruction is near the distal end of the large intestine, feces can accumulate in the colon for several weeks. The patient develops an intense feeling of constipation, but in the first stages of the obstruction vomiting is not severe. After the large intestine has become completely filled and it finally becomes impossible for additional chyme to move from the small intestine into the large intestine, vomiting begins severely. Reflux of the large intestinal contents can sometimes occur through the ileocecal valve all the way to the stomach, and the character of the vomitus becomes fecal. Some physiologists believe that this reflux is caused by reverse peristalsis, an effect not normal to the intestinal tract. Prolonged obstruction of the large intestine can finally cause rupture of the intestine itself or dehydration and circulatory shock resulting from the severe vomiting.

GASES IN THE GASTROINTESTINAL TRACT (FLATUS)

Gases can enter the gastrointestinal tract from three different sources: (1) swallowed air, (2) gases formed as a result of bacterial action, and (3) gases that diffuse from the blood into the gastrointestinal tract.

Most gases in the stomach are nitrogen and oxygen derived from swallowed air, and a large proportion of these are expelled by belching.

Only small amounts of gas are usually present in the small intestine, and these are composed principally of air that passes from the stomach into the intestinal tract. In its transport through the small intestine, only 5 to 15 per cent of the air is absorbed, and a considerable amount of carbon dioxide actually diffuses from the blood into the air to bring it into equilibrium with the carbon dioxide of the tissue fluids.

In the large intestine, the greater proportion of the gases is derived from bacterial action; these gases include especially *carbon dioxide, methane,* and *hydrogen.* When the methane and hydrogen become suitably mixed with oxygen from swallowed air, an actual explosive mixture is occasionally formed.

Essentially all the gases in the large intestine are highly diffusible through the intestinal mucosa. Therefore, if the gases remain in the large intestine for many hours the final mixture contains approximately 75 per cent or more of nitrogen and little of the other gases. The reason for this is that nitrogen in the gut cannot easily be absorbed into the blood because of the high P_{N_2} already in the blood, as explained in Chapter 43. However, if the gases are passed on through the colon rapidly, the composition of the expelled flatus may be as little as 20 per cent nitrogen, with the remaining 80 per cent composed mainly of carbon dioxide, methane, and hydrogen.

Certain foods are known to cause greater expulsion of flatus from the large intestine than others — beans, cabbage, onions, cauliflower, corn, and certain highly irritant foods such as vinegar. Some of these foods serve as a suitable medium for gas-forming bacteria, but in other instances excess gas results from irritation of the large intestine which promotes rapid expulsion of the gases before they can be absorbed.

The amount of gases entering or forming in the large intestine each day averages 7 to 10 liters, whereas the average amount expelled is usually only about 0.5 liter. The remainder is absorbed through the intestinal mucosa. Most often, a person expels large quantities of gases not because of excessive bacterial activity but because of excessive motility of the large intestine, the gases being moved on through the large intestine before they can be absorbed.

REFERENCES

Bockus, H. L. (ed.): Gastroenterology. 3 volumes. 2nd ed., Philadelphia, W. B. Saunders Company, 1963–1965.
Cleave, T. L.: Peptic Ulcer: Its Cause, Prevention, and Arrest. Baltimore, The Williams & Wilkins Co., 1963.
Code, C. F.: The peptic ulcer problem. A physiologic appraisal. *Amer. J. Dig. Dis.,* 5:288, 1960.

Code, C. F.: The physiologic basis of some disorders of motor function of the esophagus. *Postgrad. Med. J., 26*:265, 1959.

Grossman, M. I., and Ventzke, L. E.: Response of patients with duodenal ulcer to augmented histamine test as related to blood groups and to secretor status. *Gastroenterology, 42*:292, 1962.

Harkins, H. N., and Nyhus, L. M.: Surgery of the Stomach and Duodenum. Boston, Little, Brown and Company, 1968.

Jarnum, S.: Protein-Losing Gastroenteropathy. Philadelphia, F. A. Davis Co., 1963.

Johnson, L. R.; Jones, R. S.; Aures, D.; and Hakanson, R.: Effect of antrectomy on gastric histidine decarboxylase activity in the rat. *Amer. J. Physiol., 216*:1051, 1969.

Jones, F. A., and Gummer, J. W.: Clinical Gastroenterology. 2nd ed., Philadelphia, F. A. Davis Co., 1965.

Levine, R. J., and Senay, E. C.: Histamine in the pathogenesis of stress ulcers in the rat. *Amer. J. Physiol., 214*:892, 1968.

Naish, J. M., and Read, A. E.: Basic Gastroenterology. Baltimore, The Williams & Wilkins Co., 1966.

Palmer, E. D.: Functional Gastrointestinal Disease. Baltimore, The Williams & Wilkins Co., 1967.

Schiff, L.: The large intestine. *In* Sodeman, W. A., and Sodeman, W. A., Jr. (eds.): Pathologic Physiology. 4th ed., Philadelphia, W. B. Saunders Company, 1967, p. 637.

Silber, W.: Pressure studies in the diagnosis of esophageal disease. *Amer. J. Dig. Dis., 13*:356, 1968.

Symposium on gastrointestinal gas. *Ann. N. Y. Acad. Sci., 150*:1, 1968.

Volwiler, W., and McGuigan, J. E.: Metabolic considerations in functions and disorders of the digestive tract. *In* Duncan, G. G. (ed.): Diseases of Metabolism. 5th ed., Philadelphia, W. B. Saunders Company, 1964, p. 1331.

PART X

METABOLISM AND TEMPERATURE REGULATION

CHAPTER 67

Metabolism of Carbohydrates and Formation of Adenosine Triphosphate

The next few chapters deal with metabolism in the body, which means the chemical processes that make it possible for the cells to continue living. It is not the purpose of this textbook, however, to present the chemical details of all the various cellular reactions, for this lies in the discipline of biochemistry. Instead, these chapters are devoted to: (1) a review of the principal chemical processes of the cell and (2) an analysis of their physiological implications, especially in relation to the manner in which they fit into the overall concept of homeostasis.

RELEASE OF ENERGY FROM FOODS AND THE CONCEPT OF "FREE ENERGY"

The greater proportion of the chemical reactions in the cells is concerned with making the energy in foods available to the various physiological systems of the cell. For instance, energy is required for (a) muscular activity, (b) secretion by the glands, (c) maintenance of membrane potentials by the nerve and muscle fibers, (d) synthesis of substances in the cells, and (e) absorption of foods from the gastrointestinal tract.

Coupled Reactions. All the energy foods — carbohydrates, fats, and proteins — can be oxidized with oxygen in the cells, and in this process large amounts of energy are released. These same foods can also be burned with pure oxygen outside the body in an actual fire, again releasing large amounts of energy. However, this time the energy is released suddenly, all in the form of heat. The energy needed by the physiological processes of the cells is not heat but instead energy (a) to cause mechanical movement in the case of muscle function, (b) to concentrate solutes in the case of glandular secretion, and (c) to effect other functions. To provide this energy, the chemical reactions must be "coupled" with the systems responsible for these physiological functions. This coupling is accomplished by special cellular enzyme and energy transfer systems, some of which will be explained in this and subsequent chapters.

"FREE ENERGY." The amount of energy liberated by complete oxidation of a food is called the free energy of the food, and this is generally represented by the symbol ΔF. Free energy is usually expressed in terms of calories per mol of food substance. For instance, the amount of free energy liberated by oxidation of 1 mol of glucose (180 grams of glucose) is 686,000 calories.

ROLE OF ADENOSINE TRIPHOSPHATE (ATP) IN METABOLISM

ATP is a labile chemical compound that is present in all cells and has the following chemical structure:

From this formula it can be seen that ATP is a combination of adenine, ribose, and three phosphate radicals. The last two phosphate radicals are connected with the remainder of the molecule by so-called *high energy bonds*, which are indicated by the symbol ~. The amount of free energy in each of these high energy bonds per mol of ATP is approximately 7000 calories under standard conditions and 8000 calories under the conditions of temperature and concentrations of the reactants in the body. Therefore, removal of each phosphate radical liberates 8000 calories of energy. After loss of one phosphate radical from ATP, the compound becomes *adenosine diphosphate* (ADP), and after loss of the second phosphate radical the compound becomes *adenosine monophosphate* (AMP). The interconversions between ATP, ADP, and AMP are the following:

$$ATP \underset{+8000 \text{ cal. } PO_4}{\overset{-8000 \text{ cal. } ADP}{\rightleftharpoons}} + \underset{+8000 \text{ cal. } 2PO_4}{\overset{-8000 \text{ cal. } AMP}{\rightleftharpoons}} +$$

ATP is present everywhere in the cytoplasm and nucleoplasm of all cells, and essentially all the physiological mechanisms that require energy for operation obtain this directly from the stored ATP. In turn, the food in the cells is gradually oxidized, and the released energy is used to re-form the ATP, thus always maintaining a supply of this substance; all of these energy transfers take place by means of coupled reactions.

In summary, ATP is an intermediary compound that has the peculiar ability of entering into many coupled reactions—reactions with the food to extract energy, and reactions in relation to many physiological mechanisms to provide energy for their operation. For this reason, ATP has frequently been called the energy *currency* of the body that can be gained and spent again and again.

The principal purpose of the present chapter is to explain how the energy from carbohydrates can be used to form ATP in the cells. At least 99 per cent of all the carbohydrates utilized by the body is used for this purpose.

TRANSPORT OF MONOSACCHARIDES THROUGH THE CELL MEMBRANE

From Chapter 65 it will be recalled that the final products of carbohydrate digestion in the alimentary tract are almost entirely glucose, fructose, and galactose, glucose representing by far the major share of these. These three monosaccharides are absorbed into the portal blood, and, after passing through the liver, are carried everywhere in the body by the circulatory system. But, before they can be used by the cells, they must be transported through the cell membrane into the cellular cytoplasm.

Monosaccharides cannot diffuse through the pores of the cell membrane, for the maximum molecular weight of particles that can do this is about 100, whereas glucose, fructose, and galactose all have molecular weights of 180. Yet these monosaccharides do pass to the interior of the cells with a reasonable degree of freedom. Therefore, it is almost certain that they are transported through the membrane by a carrier system. They presumably combine with some carrier substance that makes them soluble in the membrane and then after passing through the membrane become dissociated from this carrier. Thus far, the nature of the carrier is uncertain, though it is believed to be a protein of small molecular weight, as was discussed in Chapter 4. It is known that the carrier mechanism works in both directions, both into and out of the cell. Furthermore, the rate of transport is approximately proportional to the concentration differences of the monosaccharides on the two sides of the membrane, and the monosaccharides are *not* transported through the usual cell membrane against a concentration difference. Therefore, the transport mechanism is one of *facilitated diffusion* and not of true active transport. These concepts were discussed in more detail in Chapter 4.

FACILITATION OF GLUCOSE TRANSPORT BY INSULIN

The rate of glucose transport and also transport of some other monosaccharides is greatly increased by insulin. When large amounts of insulin are secreted by the pancreas, the rate of glucose transport into some cells increases to as much as 10 times the rate of transport when no insulin at all is secreted. And, for practical considerations, the amounts of glucose that can diffuse to the insides of most cells of the body in the absence of insulin are far too little to supply anywhere near the amount of glucose normally required for energy metabolism. Therefore, in effect, the rate of carbohydrate utilization by the cells is controlled by the rate of insulin secretion in the pancreas. The functions of insulin and its control of carbohydrate metabolism will be discussed in detail in Chapter 78.

Failure of Disaccharides To Be Transported. Small amounts of disaccharides are absorbed into the blood from the gastrointestinal tract, but none of these can be transported into the cells. Therefore, no disaccharides or larger polysaccharides are utilized for cellular metabolism. Instead, they are excreted completely in the urine.

PHOSPHORYLATION OF THE MONOSACCHARIDES

Immediately upon entry into the cells, the monosaccharides combine with a phosphate radical in accordance with the following reaction:

$$\text{Glucose} \xrightarrow[\text{+ATP}]{\text{glucokinase}} \text{Glucose 6-phosphate}$$

This phosphorylation is promoted by enzymes called *hexokinases*, which are specific for each particular type of monosaccharide; thus, *glucokinase* promotes glucose phosphorylation, *fructokinase* promotes fructose phosphorylation, and *galactokinase* promotes galactose phosphorylation.

The phosphorylation of monosaccharides is almost completely irreversible except in the liver cells, the renal tubular epithelium, and the intestinal epithelial cells in which specific phosphatases are available for reversing the reaction. Therefore, in most tissues of the body phosphorylation serves to *capture* the monosaccharide in the cell—once *in* the cell the monosaccharide will not diffuse back out except in special cells that have the necessary phosphatases.

Conversion of Fructose and Galactose Into Glucose. In liver cells, appropriate enzymes are available to promote interconversions between the monosaccharides, as shown in Figure 67–1; and the dynamics of the reactions are such that the final product of these interconversions is almost entirely glucose. In essentially all other cells glucose and fructose can be reversibly interconverted, but some of the enzymes required for conversion of galactose into the other two monosaccharides are missing. Later in the chapter it is noted that the monosaccharides must all become either *glucose 6-phosphate* or *fructose 6-phosphate* before they can be used for energy by the cells. For this reason, before galactose can be utilized by the tissues, it must be converted by the liver cells into glucose, which is then transported by the blood to the other cells. Much of the fructose also is converted to glucose by the liver and then transported to the other cells, though this is not a necessary step.

STORAGE OF GLYCOGEN IN LIVER AND MUSCLE

After absorption into the cells, glucose can be used immediately for release of energy to the cells or it can be stored in the form of glycogen, which is a large polymer of glucose. Also, after the other monosaccharides are converted into glucose, they too can all be polymerized into glycogen.

All cells of the body are capable of storing at

Figure 67-1   Interconversions of the three major monosaccharides—glucose, fructose, and galactose—in liver cells.

least some glycogen, but certain cells can store large amounts, especially the liver cells, which can store up to 8 per cent of their weight as glycogen, and muscle cells, which can store up to 1 per cent glycogen. The glycogen molecules can be polymerized to almost any molecular weight, the average molecular weight being 5,000,000 or greater, and most of the glycogen precipitates in the form of solid granules. This conversion of the monosaccharides into a high molecular weight, precipitated compound makes it possible to store large quantities of carbohydrates without significantly altering the osmotic pressure of the intracellular fluids. Obviously, high concentrations of low molecular weight, soluble monosaccharides would play havoc with the osmotic relationships between intracellular and extracellular fluids.

Some physiologists believe that the glycogen stores exist in two forms, one a readily released and labile form called *free* glycogen and another, called *fixed* glycogen, that can be released only slowly.

GLYCOGENESIS

Glycogenesis is the process of glycogen formation, the chemical reactions for which are illustrated in Figure 67-2. From this figure it

can be seen that *glucose 6-phosphate* first becomes *glucose 1-phosphate;* then this is converted to *uridine diphosphate glucose,* which is then converted into glycogen. Several specific enzymes are required to cause these conversions, and any monosaccharide that can be converted into glucose obviously can enter into the reactions. Certain smaller compounds, including *lactic acid, glycerol, pyruvic acid,* and *some deaminated amino acids,* can also be converted into

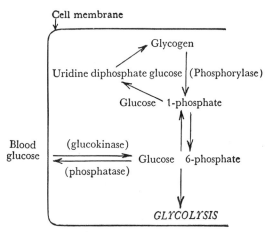

Figure 67-2   The chemical reactions of glycogenesis and glycogenolysis, showing also the interconversions between blood glucose and liver glycogen. (The phosphatase required for release of glucose from the cell is absent in muscle cells.)

glucose or closely allied compounds and thence into glycogen.

GLYCOGENOLYSIS

Glycogenolysis means the breakdown of glycogen to re-form glucose in the cells. Glycogenolysis does not occur by reversal of the same chemical reactions that serve to form glycogen; instead, each succeeding glucose molecule in the glycogen polymer is split away by a process of *phosphorylation*, catalyzed by the enzyme *phosphorylase.*

Under resting conditions, the phosphorylase of most cells is in an inactive form so that glycogen can be stored but not reconverted into glucose. When it is necessary to re-form glucose from glycogen, therefore, the phosphorylase must first be activated. This is accomplished in the following two ways:

Activation of Phosphorylase by Epinephrine and Glucagon.   Two hormones, epinephrine and glucagon, can specifically activate phosphorylase and thereby cause rapid glycogenolysis. The initial effect of each of these hormones is to increase the formation of cyclic adenylate in the cells, and it is this substance that activates the phosphorylase.

Epinephrine is released by the adrenal medullae when the sympathetic nervous system is stimulated. Therefore, one of the functions of the sympathetic nervous system is to increase the availability of glucose for rapid metabolism. This function of epinephrine occurs markedly both in liver cells and in muscle, thereby contributing, along with other effects of sympathetic stimulation, to preparation of the body for action, as was discussed in Chapter 59.

Glucagon is a hormone secreted by the *alpha cells* of the pancreas when the blood glucose concentration falls disastrously low. It stimulates the formation of cyclic adenylate mainly in the liver rather than elsewhere in the body. Therefore, its effect is primarily to dump glucose out of the liver into the blood, thereby elevating blood glucose concentration. The function of glucagon in blood glucose regulation is discussed in Chapter 78.

Transport of Glucose Out of the Cells.   The cells of the liver, of the kidney tubules, and of the intestinal mucosa contain *phosphatases* that can split phosphate away from glucose 6-phosphate and therefore make the glucose available for retransport out of the cell into the interstitial fluids. Therefore, in these cells glucose fails to be "captured" by the usual phosphorylation mechanism of the cells but instead can diffuse rather freely in both directions through the cell membrane.

Since glucose can pass out of as well as into liver cells, when it is formed in the liver as a result of glycogenolysis, most of it immediately passes into the blood. Therefore, liver glycogenolysis causes an immediate rise in blood glucose concentration. Glycogenolysis in most other cells of the body, especially in the muscle cells, simply makes increased amounts of glucose 6-phosphate available inside the cells and increases the local rate of glucose utilization but does not release the glucose into the extracellular fluids because the glucose 6-phosphate cannot be dephosphorylated.

RELEASE OF ENERGY FROM THE GLUCOSE MOLECULE BY THE GLYCOLYTIC PATHWAY

Since complete oxidation of 1 gram-mol of glucose releases 686,000 calories of energy, and only 8000 calories of energy are required to form 1 gram-mol of adenosine triphosphate (ATP), it would be extremely wasteful of energy if glucose should be decomposed all the way into water and carbon dioxide at once while forming only a single ATP molecule. Fortunately, cells contain an extensive series of different protein enzymes that cause the glucose molecule to split a little at a time in many successive steps with its energy released in small packets to form one molecule of ATP at a time, forming a total of 38 mols of ATP for each mol of glucose utilized by the cells.

The purpose of the present section is to describe the basic principles by which the glucose molecule is progressively dissected and its energy released to form ATP.

GLYCOLYSIS AND THE FORMATION OF PYRUVIC ACID

By far the most important means by which energy is released from the glucose molecule is by the processes of *glycolysis* and then *oxidation of the end-products of glycolysis.* Glycolysis means splitting of the glucose molecule to form two molecules of pyruvic acid. This occurs by 10 successive steps of chemical reactions, as illustrated in Figure 67–3. Each step is catalyzed by at least one specific protein enzyme. Note that glucose is first converted into fructose 1,6-phosphate and then is split into two three-carbon atom molecules, each of which is then converted through five successive steps into pyruvic acid.

Figure 67–3   The sequence of chemical reactions responsible for glycolysis.

Net reaction:

$$\text{Glucose} + 2\text{ADP} + 2\text{PO}_4^{---} \longrightarrow 2 \text{ Pyruvic acid} + 2\text{ATP} + 4\text{H}$$

Formation of Adenosine Triphosphate (ATP) During Glycolysis. Despite the many chemical reactions in the glycolytic series, little energy is released, and most of the energy that is released simply becomes heat and is lost to the metabolic systems of the cells. However, between the 1,3-diphosphoglyceric acid and the 3-phosphoglyceric acid stages, and again between the phosphopyruvic acid and the pyruvic acid stages, the packets of energy released are greater than 8000 calories per mol, the amount required to form ATP, and the reactions are coupled in such a way that ATP is formed. Thus, a total of 4 mols of ATP is formed for each mol of fructose 1,6-phosphate that is split to pyruvic acid.

Yet two mols of ATP had been required to phosphorylate the original glucose to form fructose 1,6-phosphate before glycolysis could begin. Therefore, the net gain in ATP molecules by the entire glycolytic process is only two mols for each mol of glucose utilized. This amounts to 16,000 calories of energy stored in the form of ATP, but during glycolysis a total of 56,000 calories of energy is lost from the original glucose, giving an overall *efficiency* for ATP formation of 29 per cent. The remaining 71 per cent of the energy is lost in the form of heat.

Release of Hydrogen Atoms During Glycolysis. Note in Figure 67–3 that two hydrogen atoms are released from each molecule of glyceraldehyde during its conversion to 1,3-diphosphoglyceric acid. And, since two molecules of

glyceraldehyde are formed from each glucose molecule, a total of four hydrogen atoms is released. Later in the chapter we will see that still many more hydrogen atoms are released when pyruvic acid is split into its component parts and that these hydrogen atoms plus the four released during glycolysis are oxidized to provide most of the energy used in the synthesis of ATP.

CONVERSION OF PYRUVIC ACID TO ACETYL COENZYME A

The next stage in the degradation of glucose is conversion of the two pyruvic acid molecules into two molecules of *acetyl coenzyme A* (acetyl Co-A) in accordance with the following reaction:

$$2 \underset{\text{(Pyruvic acid)}}{CH_3-\overset{\overset{O}{\|}}{C}-COOH} + 2 \underset{\text{(Coenzyme A)}}{Co\text{-}A-SH} \longrightarrow$$

$$2 \underset{\text{(Acetyl Co-A)}}{CH_3-\overset{\overset{O}{\|}}{C}-S-Co\text{-}A} + 2CO_2 + 4H$$

From this reaction it can be seen that two carbon dioxide molecules and four hydrogen atoms are released, while the remainders of the pyruvic acid molecules combine with coenzyme A, a derivative of the vitamin pantothenic acid, to form two molecules of acetyl Co-A. In this conversion, no ATP is formed, but six molecules of ATP are formed when the four hydrogen atoms are later oxidized, as will be discussed in a later section.

THE CITRIC ACID CYCLE

The next stage in the degradation of the glucose molecule is called the *citric acid cycle* (*tricarboxylic acid cycle*, or *Krebs cycle*). This is a sequence of chemical reactions in which the acetyl portion of acetyl Co-A is degraded to carbon dioxide and hydrogen atoms. Then the hydrogen atoms are subsequently oxidized, releasing still more energy to form ATP.

The enzymes responsible for the citric acid cycle and also for the subsequent oxidation of the hydrogen to form ATP are all contained in the *mitochondria*. It is believed that the enzymes are attached to the surfaces of the intramitochondrial "shelves" and that they are arranged in a sequential order so that successive products of the reactions can be shuttled from enzyme to enzyme. This obviously would greatly enhance the speed of the chemical reactions and would promote increased rate of ATP formation. Because of the special function of the mitochondria in providing ATP to the remainder of the cell, they are frequently called the "powerhouses" of the cell.

Figure 67–4 shows the different stages of the chemical reactions in the citric acid cycle. The substances to the left are added during the chemical reactions, and the products of the chemical reactions are shown to the right. Note at the top of the column that the cycle begins with *oxaloacetic acid*, and then at the bottom of the chain of reactions *oxaloacetic acid* is formed once more. Thus, the cycle can continue over and over again.

In the initial stage of the citric acid cycle, *acetyl Co-A* combines with *oxaloacetic acid* to form *citric acid*. The coenzyme A portion of the acetyl Co-A is released and can be used again and again for the formation of still more quantities of acetyl Co-A from pyruvic acid. The acetyl portion, however, becomes an integral part of the citric acid molecule. During the successive stages of the citric acid cycle, several molecules of water are added; and *carbon dioxide* and *hydrogen atoms* are released at various stages in the cycle, as shown on the right in the figure.

The net results of the entire citric acid cycle are shown at the bottom of Figure 67–4, illustrating that for each molecule of glucose originally metabolized, 2 acetyl Co-A molecules enter into the citric acid cycle along with 6 molecules of water. These then are degraded

Net reaction per molecule of glucose:

$$2 \text{ Acetyl Co-A} + 6H_2O + 2ADP \longrightarrow$$
$$4CO_2 + 16H + 2Co\text{-}A + 2ATP$$

Figure 67–4 The chemical reactions of the citric acid cycle, showing the release of carbon dioxide and an especially large number of hydrogen atoms during the cycle.

into 4 carbon dioxide molecules, 16 hydrogen atoms, and 2 molecules of coenzyme A.

Formation of ATP in the Citric Acid Cycle. Not a great deal of energy is released during the citric acid cycle itself, and in only one of the chemical reactions—during the change from α-ketoglutaric acid to succinic acid—is a molecule of ATP formed. Thus, for each molecule of glucose metabolized, two acetyl Co-A molecules pass through the citric acid cycle, each forming a molecule of ATP; or, a total of two molecules of ATP is formed.

Function of the Dehydrogenases and Diphosphopyridine Nucleotide (DPN) To Cause Release of Hydrogen Atoms. As noted at several points in this discussion, hydrogen atoms are released during different chemical reactions— 4 hydrogen atoms during glycolysis, 4 during the formation of acetyl Co-A from pyruvic acid, and 16 in the citric acid cycle; this makes a total of 24 hydrogen atoms. However, the hydrogen atoms are not simply turned loose in the intracellular fluid. Instead, they are released in packets of two, and in each instance the release is catalyzed by a specific protein enzyme called a *dehydrogenase*. Twenty of the twenty-four hydrogen atoms immediately combine with diphosphopyridine nucleotide (DPN), a derivative of the vitamin nicotinamide, in accordance with the following reaction:

$$DPNH + H^+ + Substrate$$

This reaction will not occur without the initial intermediation of the dehydrogenase nor without the availability of DPN to act as a carrier for a hydrogen atom. Both the hydrogen ion and the hydrogen bound with DPN subsequently enter into the oxidative chemical reactions that form tremendous quantities of ATP, as will be discussed later.

The remaining four hydrogen atoms released during the breakdown of glucose—the four released during the citric acid cycle between the succinic and fumaric acid stages— combine with a specific dehydrogenase but are not then subsequently released to DPN. Instead, they pass directly from the dehydrogenase into the oxidative processes.

Function of Decarboxylases To Cause Release of Carbon Dioxide. Referring again to the chemical reactions of the citric acid cycle and also to those for formation of acetyl Co-A

from pyruvic acid, we find that there are three stages in which carbon dioxide is released. To cause carbon dioxide release, other specific protein enzymes, called *decarboxylases*, split the carbon dioxide away from the substrate. The carbon dioxide in turn becomes dissolved in the fluids of the cells and is transported to the lungs where it is expired from the body (see Chap. 41).

OXIDATION OF HYDROGEN ATOMS BY THE OXIDATIVE ENZYMES

Despite all the complexities of (a) glycolysis, (b) the citric acid cycle, (c) dehydrogenation, and (d) decarboxylation, pitifully small amounts of ATP are formed during all these processes. Instead, about 90 per cent of the final ATP is formed during subsequent oxidation of the hydrogen atoms that are released during these earlier stages of glucose degradation. Indeed, the principal function of all these earlier stages is to make the hydrogen of the glucose molecule available in a form that can be utilized for oxidation.

Oxidation of hydrogen is accomplished by a series of enzymatically catalyzed reactions that (a) change the hydrogen atoms into hydrogen ions and (b) change the dissolved oxygen of the fluids into hydroxyl ions. Then these two products combine with each other to form water. During the sequence of oxidative reactions, tremendous quantities of energy are released to form ATP. The oxidative enzymes, like the dehydrogenases and decarboxylases of the citric acid cycle, are believed to be arranged in an orderly fashion on the inner surfaces of the mitochondria, thus allowing rapid procession of the chemical reactions.

Ionization of Hydrogen and Oxygen by the Oxidative Enzymes. Figure 67–5 illustrates a schema of the reactions that cause ionization of both hydrogen and oxygen, with subsequent formation of water. In this schema, the two hydrogens removed from each substrate during the degradation of glucose are shuttled from compound to compound until finally they are released into the body fluids at the cytochrome b stage in the form of two free hydrogen ions. To change a hydrogen atom into a hydrogen ion, an electron must be removed from it. This is performed by cytochrome b. Cytochrome b then passes the electron to cytochrome c, and cytochrome c in turn passes the electron to cytochrome a, which is also known as cytochrome oxidase. At this point, two electrons are transferred from two cytochrome a molecules to one-half molecule

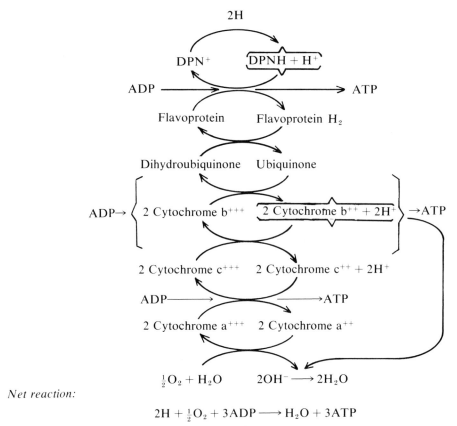

Net reaction:

$$2H + \tfrac{1}{2}O_2 + 3ADP \longrightarrow H_2O + 3ATP$$

Figure 67–5 The chemical processes of oxidative phosphorylation, showing ionization of hydrogen and oxygen and formation of water and adenosine triphosphate.

of dissolved oxygen plus a molecule of water to form two hydroxyl ions. Thus, hydrogen ions are formed by removing electrons from hydrogen atoms, and hydroxyl ions are formed by transferring these electrons to the dissolved oxygen in the water. Once both of these ions are present in the same fluid medium, they immediately react to form water, as shown in Figure 65–5.

Formation of ATP During Oxidation of Hydrogen—"Oxidative Phosphorylation." The net reaction for the oxidation of each two hydrogen atoms is shown at the bottom of Figure 67–5. This illustrates that three molecules of ATP are formed (but occasionally as many as four or five are formed) during this oxidation: one between the DPN and flavoprotein stages, one between the ubiquinone and cytochrome c stages, and the third between the cytochrome c and cytochrome a stages. This formation of ATP during the oxidation of hydrogen is called *oxidative phosphorylation*. The precise chemical means by which the oxidation of hydrogen is coupled with the conversion of adenosine diphosphate (ADP) into ATP is almost completely unknown, representing one of the most important voids in our understanding of the body's metabolic processes.

SUMMARY OF ATP FORMATION DURING THE BREAKDOWN OF GLUCOSE

We can now determine the total number of ATP molecules formed by the energy from one molecule of glucose.

First, during glycolysis, four molecules of ATP are formed, while two are expended to cause the initial phosphorylation of glucose to start the process going. This gives a net gain of *two molecules of ATP.*

Second, during each revolution of the citric acid cycle, one molecule of ATP is formed. However, because each glucose molecule splits into two pyruvic acid molecules, there are two revolutions of the cycle for each molecule of glucose metabolized, giving a net production of *two molecules of ATP.*

Third, during the entire schema of glucose breakdown, a total of 24 hydrogen atoms is released during glycolysis and during the citric acid cycle. Twenty of these atoms are oxidized by the schema of oxidative reactions shown in Figure 67–5, with the release of three (but occasionally four or five) ATP molecules per two atoms of hydrogen metabolized. This gives a minimum production of *30 ATP molecules.*

Fourth, the remaining four hydrogen atoms are released by their dehydrogenase into the oxidative schema beyond the first stage of Figure 67–5, so that for these four hydrogen atoms only two ATP molecules are usually released for each two hydrogen atoms oxidized, giving a total of *four ATP molecules.*

Now, adding all the ATP molecules formed, we find a minimum of *38 ATP molecules* formed for each molecule of glucose degraded to carbon dioxide and water. Thus, 304,000 calories of energy are stored in the form of ATP, while 686,000 calories are released during the complete oxidation of each gram-mol of glucose. This represents an overall *efficiency* of energy transfer of at least 44 per cent. The remaining 56 per cent of the energy becomes heat and therefore cannot be used by the cells to perform specific functions.

CONTROL OF GLYCOLYSIS AND OXIDATION BY ADENOSINE DIPHOSPHATE (ADP)

Continual release of energy from glucose when the energy is not needed by the cells would be an extremely wasteful process. Fortunately, glycolysis and the subsequent oxidation of hydrogen atoms is continually controlled in accordance with the needs of the cells for ATP. This control is accomplished simply in the following manner:

Referring back to the various chemical reactions of the glycolytic series, the citric acid cycle, and the oxidation of hydrogen, we see that at different stages *ADP is converted into ATP. If ADP is not available at each of these stages, the reactions cannot occur, thus stopping the degradation of the glucose molecule.* Therefore, once all the ADP in the cells has been converted to ATP, the entire glycolytic and oxidative process stops. Then, when more ATP is used to perform different physiological functions in the cell, new ADP is formed, which automatically allows glycolysis and oxidation to proceed once more. In this way, essentially a full store of ATP is automatically maintained all the time, except when the activity of the cell becomes so great that ATP is used up more rapidly than it can be formed.

GLUCOSE CONVERSION TO GLYCOGEN OR FAT

When glucose is not immediately required for energy, the extra glucose that continually enters the cells either is stored as glycogen or is converted into fat and stored in this form, as discussed in the following chapter. Glucose is preferentially stored as glycogen until the cells have stored as much glycogen as they can—an amount sufficient to supply the energy needs of the body for only a few hours. When the cells (primarily liver and muscle cells) approach saturation with glycogen, the additional glucose is converted into fat; the chemistry of this conversion is discussed in the following chapter.

ANAEROBIC RELEASE OF ENERGY— "ANAEROBIC GLYCOLYSIS"

Occasionally, oxygen becomes either unavailable or insufficient so that cellular oxidation of glucose cannot take place. Yet, even under these conditions, a small amount of energy can still be released to the cells by glycosis, for the chemical reactions in the glycolytic breakdown of glucose to pyruvic acid do not require oxygen. Unfortunately, this process is extremely wasteful of glucose because only 16,000 calories of energy are used to form ATP for each molecule of glucose utilized, which represents only a little over 2 per cent of the total energy in the glucose molecule. Nevertheless, this release of glycolytic energy to the cells can be a lifesaving measure for a few minutes when oxygen becomes unavailable.

Formation of Lactic Acid During Anaerobic Glycolysis. The *law of mass action* states that as the end-products of a chemical reaction build up in the reacting medium, the rate of the reaction approaches zero. The two end-products of the glycolytic reactions (see Fig. 67–3) are (1) pyruvic acid and (2) hydrogen atoms, which are combined with DPN to form DPNH and H^+. The build-up of either or both of these would stop the glycolytic process and prevent further formation of ATP. Fortunately, when their quantities begin to be excessive these two end-products react with each other to form lactic acid in accordance with the following equation:

$$\underset{\text{(Pyruvic acid)}}{CH_3-\overset{\overset{\text{O}}{\|}}{C}-COOH} + DPNH + H^+ \xrightarrow[\longleftarrow]{\text{lactic dehydrogenase}}$$

$$\underset{\underset{\text{(Lactic acid)}}{\overset{|}{H}}}{CH_3-\overset{\overset{\text{OH}}{|}}{C}-COOH} + DPN^+$$

Thus, under anaerobic conditions, by far the major proportion of the pyruvic acid is converted into lactic acid, which diffuses readily out of the cells into the extracellular fluids and even into the intracellular fluids of other less active cells. Therefore, lactic acid represents a type of "sinkhole" into which the glycolytic end-products can disappear, thus allowing glycolysis to proceed far longer than would be possible if the pyruvic acid and hydrogen were not removed from the reacting medium. Indeed, glycolysis could proceed for only a few seconds without this conversion. Instead, it can proceed for several minutes, supplying the body with considerable quantities of ATP even in the absence of respiratory oxygen.

Reconversion of Lactic Acid and Pyruvic Acid to Glucose in the Presence of Oxygen. The entire schema of glycolysis, all the way from glucose to lactic acid, is theoretically reversible, but in practice the reverse change of pyruvic acid into phosphoenolpyruvic acid is extremely slow. However, other enzymes cause the pyruvic acid to be converted first into oxaloacetic acid and then into phosphoenolpyruvic acid, after which the scheme of reconversion can proceed. Unfortunately, however, this reverse process consumes six ATP molecules in comparison with the two ATP molecules formed during glycolysis.

When a person begins to breathe oxygen again after a period of anaerobic metabolism, the extra DPNH and H$^+$ that have built up in the body fluids are rapidly oxidized, thereby greatly reducing their concentrations. As a result, the chemical reaction for formation of lactic acid immediately reverses itself, the lactic acid once again becoming pyruvic acid. Large portions of this are immediately utilized by the citric acid cycle to provide additional oxidative energy, and large quantities of ATP are formed. This excess ATP then causes as much as three-fourths of the remaining excess pyruvic acid to be converted back into glucose.

Thus, the great amount of lactic acid that forms during anaerobic glycolysis does not become lost to the body, for when oxygen is again available, the lactic acid either can be reconverted to glucose or can be used directly for energy. By far the greater proportion of this reconversion occurs in the liver, but a small amount can also occur in some other tissues.

USE OF LACTIC ACID BY THE HEART FOR ENERGY. Heart muscle is especially capable of converting lactic acid to pyruvic acid and then utilizing this for energy. This occurs especially in very heavy exercise, during which large amounts of lactic acid are released into the blood from the skeletal muscles.

RELEASE OF ENERGY FROM GLUCOSE BY THE PHOSPHOGLUCONATE PATHWAY

Though 95 per cent or more of all the carbohydrates utilized by the muscles are degraded to pyruvic acid by glycolysis and then oxidized, the glycolytic schema is not the only means by which glucose can be degraded and then oxidized to provide energy. A second important schema for breakdown and oxidation of glucose is called the *phosphogluconate pathway,* which is perhaps responsible for as much as 30 per cent of the glucose breakdown in the liver and even more than this in fat cells. On the other hand, it occurs hardly at all in muscle cells. It is especially important because it can provide energy independently of all the enzymes of the tricarboxylic acid cycle and therefore is an alternate pathway for energy metabolism in case of some enzymatic abnormality of the cells.

Release of Carbon Dioxide and Hydrogen by Means of the Phosphogluconate Pathway. Figure 67–6 illustrates most of the basic chemical reactions in the phosphogluconate pathway. This shows that glucose, after several stages of conversion, releases one molecule of carbon dioxide and four atoms of hydrogen, with resultant formation of a five-carbon sugar, d-ribulose. This substance in turn can change progressively into several other five-, four-, seven-, and three-carbon sugars. Finally, various combinations of these sugars can resynthesize glucose. However, *only five molecules of glucose are resynthesized for every six molecules of glucose that initially enter into the reactions.* That is, the phosphogluconate pathway is a cyclic proc-

Glucose   6-phosphate   → 2H
6-Phosphoglucono-δ-lactone
6-Phosphogluconic acid   → 2H
3-Keto-6-phosphogluconic acid   → CO$_2$
D-Ribulose 5-phosphate

H$_2$O

{D-Xylulose 5-phosphate + D-Ribose 5-phosphate}

{D-Sedoheptulose 7-phosphate + D-Glyceraldehyde 3-phosphate}

{Fructose 6-phosphate + Erythrose 4-phosphate}

Net reaction:
Glucose + 12TPN + 6H$_2$O ⟶ 6CO$_2$ + 12H + 12TPNH

Figure 67–6 The phosphogluconate pathway for glucose metabolism.

ess in which one molecule of glucose is metabolized for each revolution of the cycle. As illustrated at the bottom of Figure 67–6, the net reaction is conversion of the single molecule of glucose plus six molecules of water into six carbon dioxide molecules and 24 hydrogen atoms. Thus, by revolution of the cycle again and again, all the glucose can eventually be converted into carbon dioxide and hydrogen, and the hydrogen in turn can then be utilized for energy.

Use of Hydrogen; The Function of Triphosphopyridine Nucleotide (TPN). The hydrogen released during the phosphogluconate cycle does not combine with DPN, as in the glycolytic pathway, but combines with TPN, which is almost identical with DPN except for an extra phosphate radical. This difference is extremely important because only hydrogen bound with TPN in the form of TPNH can be used for synthesis of fats from carbohydrates, which is discussed in the following chapter. When glycolysis becomes slowed down because of cellular inactivity, the phosphogluconate pathway still remains operative (mainly in the liver) to break down any excess glucose that continues to be transported into the cells, and TPNH becomes abundant to help convert acetyl radicals into long fatty acid chains.

FORMATION OF CARBOHYDRATES FROM PROTEINS AND FATS— "GLUCONEOGENESIS"

When the body's stores of carbohydrates decrease below normal, moderate quantities of glucose can be formed from *amino acids* and from the *glycerol* portion of fat. This process is called *gluconeogenesis.* Approximately 60 per cent of the amino acids in the body proteins can be converted into carbohydrates, while the remaining 40 per cent have chemical configurations that make this difficult. Each amino acid is converted into glucose by a slightly different chemical process. For instance, alanine can be converted directly into pyruvic acid simply by deamination; the pyruvic acid then is converted into glucose, as explained previously. Several of the more complicated amino acids can be converted into different sugars containing three-, four-, five-, or seven-carbon atoms; these can then enter the phosphogluconate pathway and eventually form glucose. Thus, by means of deamination plus several simple interconversions, many of the amino acids become glucose. Similar interconversions can change glycerol into carbohydrates.

Regulation of Gluconeogenesis. Diminished carbohydrates in the cells and decreased blood sugar are the basic stimuli that set off an increase in the rate of gluconeogenesis. The diminished carbohydrates can directly cause reversal of many of the glycolytic and phosphogluconate reactions, thus allowing conversion of deaminated amino acids and glycerol into carbohydrates. However, in addition, several of the hormones secreted by the endocrine glands are especially important in this regulation.

EFFECT OF CORTICOTROPIN AND GLUCOCORTICOIDS ON GLUCONEOGENESIS. When normal quantities of carbohydrates are not available to the cells, it is believed that the adenohypophysis, for reasons not completely understood at present, begins to secrete increased quantities of corticotropin, which stimulate the adrenal cortex to produce large quantities of *glucocorticoid hormones,* especially cortisol. In turn, cortisol mobilizes proteins from essentially all cells of the body, making these available in the form of amino acids in the body fluids. A high proportion of these immediately becomes deaminated in the liver and therefore provides ideal substrates for conversion into glucose. Thus, one of the most important means by which gluconeogenesis is promoted is through the release of glucocorticoids from the adrenal cortex.

EFFECT OF THYROXINE ON GLUCONEOGENESIS. Thyroxine, secreted by the thyroid gland, also increases the rate of gluconeogenesis. This, too, is believed to result principally from mobilization of proteins from the cells. However, it might result to some extent also from the mobilization of fats from the fat depots, the glycerol portion of the fats being converted into glucose.

BLOOD GLUCOSE

Except immediately after a meal, glucose is the only monosaccharide present in significant quantities in the blood and interstitial fluids. There are two reasons for this: (1) Usually 80 to 100 per cent of the monosaccharides absorbed from the gastrointestinal tract is glucose, and only rarely is more than 20 per cent of these fructose and galactose together. (2) Within less than an hour after absorption from the gut, essentially all the fructose and galactose will have entered the cells and been converted into glucose.

The normal blood glucose concentration in a person who has not eaten a meal within the past three to four hours is approximately 90 mg. per cent, and, even after a meal containing large amounts of carbohydrates, this rarely rises above 140 mg. per cent unless the person has diabetes mellitus, which will be discussed in Chapter 78.

The regulation of blood glucose concentration is so intimately related to insulin that this subject will be discussed in detail in Chapter 78 in relation to the function of insulin.

REFERENCES

Awapara, J., and Simpson, J. W.: Comparative physiology: metabolism. *Ann. Rev. Physiol., 29*:87, 1967.

Baxter, J. H.: Routes of glyceride delivery from liver to plasma in the rat. *Amer. J. Physiol., 218*:790, 1970.

Boerth, R. C.; Covell, J. W.; Seagren, S. C.; and Pool, P. E.: High-energy phosphate concentrations in dog myocardium during stress. *Amer. J. Physiol., 216*:1103, 1969.

Ciba Symposium on Control of Glycogen Metabolism. Boston, Little, Brown and Company, 1964.

Corsi, A.; Midrio, M.; and Granata, A. L.: In situ utilization of glycogen and blood glucose by skeletal muscle during tetanus. *Amer. J. Physiol., 216*:1534, 1969.

Crane, F. L., and Low, H.: Quinones in energy-coupling systems. *Physiol. Rev., 46*:662, 1966.

Dixon, M., and Webb, E. C.: Enzymes. New York, Academic Press, Inc., 1964.

Bondy, P. K. (ed.): Duncan's Diseases of Metabolism. 6th ed., Philadelphia, W. B. Saunders Company, 1964.

Glinsmann, W. H., and Mortimore, G. E.: Influence of glucagon and 3',5'-AMP on insulin responsiveness of the perfused rat liver. *Amer. J. Physiol., 215*:553, 1968.

Kaden, M.; Oakley, N. W.; and Field, J. B.: Effect of alcohol on gluconeogenesis using the isolated rat liver perfusion technique. *Amer. J. Physiol., 216*:756, 1969.

Krebs, H. A.: The tricarboxylic acid cycle. *Harvey Lectures, 44*:165, 1948–1949.

Lemberg, M. R.: Cytochrome oxidase. *Physiol. Rev., 49*:48, 1969.

Levine, R.: Carbohydrate metabolism. In Duncan, G. G. (ed.): Diseases of Metabolism. 5th ed., Philadelphia, W. B. Saunders Company, 1964, p. 105.

Levine, R.; Goldstein, M. S.; Huddleston, B.; and Klein, S. P.: Action of insulin on "permeability" of cells to free hexoses, as studied by its effect on distribution of galactose. *Amer. J. Physiol., 163*:79, 1950.

Mommaerts, W. F. H. M.: Energetics of muscular contraction. *Physiol. Rev., 49*:427, 1969.

Neely, J. R.; Bowman, R. H.; and Morgan, H. E.: Effects of ventricular pressure development and palmitate on glucose transport. *Amer. J. Physiol., 216*:804, 1969.

Omachi, A.; Scott, C. B.; and Parry, T. E.: Influence of glycolysis on NADH content in human erythrocytes. *Amer. J. Physiol., 216*:527, 1969.

Park, C. R.; Johnson, L. H.; Wright, J. H., Jr.; and Batsel, H.: Effect of insulin on transport of several hexoses and pentoses into cells of muscle and brain. *Amer. J. Physiol., 191*:13, 1957.

Pyke, D. (ed.): Disorders of Carbohydrate Metabolism. Philadelphia, J. B. Lippincott Co., 1963.

Reynafarje, B.; Oyola, L.; Cheesman, R.; Marticorena, E.; and Jimenez, S.: Fructose metabolism in sea-level and high-altitude natives. *Amer. J. Physiol., 216*:1542, 1969.

Ricketts, H. T., and Krahl, M. E.: Carbohydrate metabolism. In Sodeman, W. A., and Sodeman, W. A., Jr. (eds.): Pathologic Physiology. 4th ed., Philadelphia, W. B. Saunders Company, 1967, p. 94.

Rivlin, R. S.: Perinatal development of enzymes synthesizing FMN and FAD. *Amer. J. Physiol., 216*:979, 1969.

Ruderman, N. B., and Herrera, M. G.: Glucose regulation of hepatic gluconeogenesis. *Amer. J. Physiol., 214*:1346, 1968.

Stetten, D., Jr., and Stetten, M. R.: Glycogen metabolism. *Physiol. Rev., 40*:505, 1960.

Weintraub, B.; Sarcione, E. J.; and Sokal, J. E.: Effect of glucagon on phosphorylase activity of the isolated perfused liver. *Amer. J. Physiol., 216*:521, 1969.

CHAPTER 68

Lipid Metabolism

A number of different chemical compounds of the food and body are classified together as *lipids*. These include (1) *neutral fat,* known also as *triglycerides,* (2) the *phospholipids,* (3) *cholesterol,* and (4) a few others of less importance. These substances have certain similar physical and chemical properties, especially the fact that they are miscible with each other. Chemically, the basic lipid moiety of both the triglycerides and the phospholipids is *fatty acids,* which are simply long chain hydrocarbon organic acids. A typical fatty acid, palmitic acid, is the following:

$$CH_3(CH_2)_{14}COOH$$

Though cholesterol does not contain fatty acid, its sterol nucleus, as pointed out later in the chapter, is synthesized from degradation products of fatty acid molecules, thus giving it many of the physical and chemical properties of other lipid substances.

The triglycerides are used in the body mainly to provide energy for the different metabolic processes; this function they share almost equally with the carbohydrates. However, some lipids, especially cholesterol, the phospholipids, and derivatives of these, are used throughout the body to provide other intracellular functions.

Basic Chemical Structure of Triglycerides (Neutral Fat). Since most of this chapter deals with utilization of triglycerides for energy, the following basic structure of the triglyceride molecule must be understood:

$$
\begin{array}{l}
CH_3-(CH_2)_{16}-COO-CH_2 \\
CH_3-(CH_2)_{16}-COO-CH \\
CH_3-(CH_2)_{16}-COO-CH_2 \\
\quad\quad\quad\text{Tri-stearin}
\end{array}
$$

Note that three long chain fatty acid molecules are bound with one molecule of glycerol. In the human body, the three fatty acids most commonly present in neutral fat are (1) *stearic acid,* which has an 18-carbon chain and is fully saturated with hydrogen atoms, (2) *oleic acid,* which also has an 18-carbon chain but has one double bond in the middle of the chain, and (3) *palmitic acid,* which has 16-carbon atoms and is fully saturated. These or closely similar fatty acids are also the major constituents of the fats in the food.

TRANSPORT OF LIPIDS IN THE BLOOD

TRANSPORT FROM THE GASTROINTESTINAL TRACT — THE "CHYLOMICRONS"

It will be recalled from Chapter 65 that essentially all the fats of the diet are absorbed into the lymph (with the exception of the short chain fatty acids that can be absorbed directly into the portal blood but that normally represent only a small fraction of the fat in the diet). In the digestive tract, most triglycerides are split into glycerol and fatty acids or into monoglycerides and fatty acids. Then, on passing through the intestinal epithelial cells, these are resynthesized into new molecules of triglycerides which aggregate and enter the lymph as minute, dispersed droplets called *chylomicrons,* having a size about 0.5 micron. A small amount of protein adsorbs to the outer surfaces of the chylomicrons; this increases their suspension stability in the fluid of the lymph and prevents their adhering to the lymphatic vessel walls.

Most of the cholesterol and phospholipids absorbed from the gastrointestinal tract, as well as small amounts of phospholipids that are continually synthesized by the intestinal mucosa, also enter the chylomicrons. Thus, chylomicrons are composed principally of triglycerides, but they also contain small amounts of phospholipids, cholesterol, and protein, as indicated in Table 68–1. The chylomicrons are then transported up the thoracic duct and emptied into the venous blood at the juncture of the jugular and subclavian veins.

TABLE 68–1. The Lipoproteins

	Density	Diameter μ	Sf	Triglycerides %	Phospholipids %	Cholesterol %	Protein %
Chylomicrons	0.94	0.700	4000–40,000	85	7	7	1
Low density lipoportein							
β_1	0.98	0.045	10–400	50	18	23	9
β_2	1.03	0.023	2–10	10	22	47	21
High density lipoprotein							
α_1	1.09	0.013	−1	8	29	30	33
α_2	1.14	0.010	−2	5	21	17	57

Removal of the Chylomicrons from the Blood.
Immediately after a meal that contains large quantities of fat, the chylomicron concentration in the plasma may rise to as high as 1 to 2 per cent, and, because of the large sizes of the chylomicrons, the plasma appears turbid and sometimes yellow. However, the chylomicrons are gradually removed within two to three hours, and the plasma becomes clear once again. The fat of the chylomicrons is removed in two different ways as follows:

HYDROLYSIS OF THE CHYLOMICRON TRIGLYCERIDES BY LIPOPROTEIN LIPASE. The major part of the triglycerides in the chylomicrons is probably hydrolyzed into glycerol and fatty acids under the influence of an enzyme in the blood called *lipoprotein lipase*. The glycerol is then metabolized in much the same way as glucose, as explained in the preceding chapter, while the fatty acid molecules are transported, combined with albumin, to the various cells of the body, as explained below. The fatty acid molecules can there be directly oxidized for energy, or they can be used by the fat cells of adipose tissue to resynthesize triglycerides, which are stored in the fat cells.

Only small amounts of lipoprotein lipase are normally present in the blood. However, certain tissues of the body, especially adipose tissue, heart muscle, and, to a lesser extent, skeletal muscle, contain large quantities of this enzyme. It is believed that as the chylomicrons pass through the blood vessels of these tissues, the enzyme enters the capillaries to cause immediate release of fatty acids, and since fatty acids are lipid soluble, they diffuse extremely rapidly through the capillary and cell membranes to be used by the local cells. Or they can be transported in combination with albumin to any other part of the body, as explained below.

ABSORPTION OF WHOLE CHYLOMICRONS BY THE LIVER CELLS. The second means by which the chylomicrons are removed from the blood is by transport of whole chylomicrons through the capillary wall directly into liver cells, which then either use the lipids of the chylomicrons for energy or convert them into other lipid substances.

Lipoprotein Lipase as a "Clearing Factor"; Effect of Heparin. When heparin is injected into a person who has large quantities of chylomicrons in his plasma, the chylomicrons disappear rapidly. This effect probably comes about in the following way: Heparin seems to activate the lipoprotein lipase in the plasma or to cause lipoprotein lipase to be released into the blood from the adipose tissue, muscles, and other tissues. This effect is probably related to the fact that lipoprotein lipase itself contains heparin or some heparin-like substance in its molecular complex. The lipoprotein lipase in turn hydrolyzes the triglycerides of the chylomicrons, thereby "clearing" the plasma of its opalescence or turbidity. Therefore, lipoprotein lipase has been called "clearing factor."

TRANSPORT OF FATTY ACIDS IN COMBINATION WITH ALBUMIN—"FREE FATTY ACID"

A small and highly variable quantity of fatty acids is always present in the blood combined with the albumin of the plasma proteins. Remember that fatty acids ionize strongly in water; therefore, immediately upon release of fatty acids into the blood, either by the action of lipoprotein lipase on chylomicrons or by the action of this same enzyme on triglycerides stored in the adipose tissue, the ionized fatty acid combines almost immediately with albumin of the plasma proteins. The fatty acid bound with proteins in this manner is called *free fatty acid* or *nonesterified fatty acid* (or simply *NEFA*) to distinguish it from essentially all the remainder of the fatty acid in the plasma which is transported in the form of esters of glycerol, cholesterol, or other substances.

The concentration of free fatty acid in the plasma under resting conditions is about 15 mg. per 100 ml. of blood, which is a total of only 0.75 gram of fatty acids in the entire circulatory system. Yet, strangely enough, even this small amount accounts for most of the transport of lipids from one part of the body to another for the following reasons:

(1) Despite the minute amount of free fatty acid in the blood, its rate of "turnover" is extremely rapid, *half the fatty acid being replaced by new fatty acid every two to three minutes.* One can calculate that at this rate over half of all the energy required by the body can be provided in this way even without increasing the free fatty acid concentration. (2) All conditions that increase the rate of utilization of fat for cellular energy also increase the free fatty acid concentration in the blood; this sometimes increases as much as 10-fold. Especially does this occur in starvation and in diabetes when a person is not or cannot be using carbohydrates for energy.

Under normal conditions about three molecules of fatty acid combine with each molecule of albumin, but as many as 30 fatty acid molecules can at times combine with a single molecule of albumin when the need for fatty acid transport is extreme. This shows how variable the rate of lipid transport can be under different physiological needs.

THE LIPOPROTEINS

In the postabsorptive state — that is, when no chylomicrons are in the blood — over 95 per cent of all the lipids in the plasma are in the form of lipoproteins, which are mixtures of *triglycerides, phospholipids, cholesterol,* and *protein.* The protein in the mixture averages about a fourth to one-third the total constituents, and the remainder is lipids. The total concentration of lipoproteins in the plasma averages about 700 mg. per 100 ml., and this can be broken down into the following average concentrations of the individual constituents:

	mg./100 ml. of plasma
Cholesterol	180
Phospholipids	160
Triglycerides	160
Lipoprotein protein	200

Table 68–1 illustrates the two major groups of lipoprotein: (1) the *low density lipoproteins,* which contain high concentrations of triglycerides and low concentrations of protein, and (2) the *high density lipoproteins,* which contain low concentrations of triglycerides and high concentrations of protein. The lipoproteins can also be classified on the basis of their elec-

trophoretic mobility in an electrophoresis apparatus as (a) *alpha lipoproteins,* which are high density lipoproteins and which migrate with the alpha globulins in the electrophoresis apparatus, and (b) *beta lipoproteins,* which are low density lipoproteins and which migrate with the beta globulins.

Another special classification of the lipoproteins is the S_f classification, which is a classification of the rate at which they either float to the surface during centrifugation of a sample of plasma or the rate at which they settle to the bottom. The low density lipoproteins have positive S_f values, which means that they float to the surface, and high density lipoproteins have negative S_f values, which means that they are centrifuged downward.

Function of the Lipoproteins. The function of the lipoproteins in the plasma is poorly known, though they are believed to be a means by which lipid substances can be transported from the liver to other parts of the body. For instance, the turnover of triglycerides in the lipoproteins is as much as 1.5 grams per hour, which could account for as much as 10 to 20 per cent of the total lipids utilized by the body for energy. It has especially been suggested that fats synthesized from carbohydrates in the liver are transported to the adipose tissue in lipoproteins because almost no free fatty acids are present in the blood when fat synthesis is occurring, while plenty of lipoproteins are present. Even more important might be the transport of cholesterol and phospholipids by the lipoproteins, because these substances are not known to be transported to any significant extent in any other form.

The lipoproteins are formed either entirely or almost entirely in the liver, which is in keeping with the fact that most plasma phospholipids, cholesterol, and triglycerides (except those in the chylomicrons) are synthesized in the liver.

THE FAT DEPOSITS

ADIPOSE TISSUE

Large quantities of fat are frequently stored in two major tissues of the body, the adipose tissue and the liver. The adipose tissue is usually called the *fat deposits,* or simply the *fat depots.* The factors that cause storage and removal of fat from adipose tissue are quite different from those that affect the fat in the liver, as will become evident in the succeeding paragraphs.

The major function of adipose tissue is storage of triglycerides until these are needed to provide energy elsewhere in the body. However, a subsidiary function is to provide heat insulation for the body, as will be discussed in Chapter 71.

The Fat Cells. The fat cells of adipose tissue are modified fibroblasts that are capable of storing almost pure triglycerides in quantities equal to 80 to 95 per cent of their volume. The triglycerides are generally in a liquid form, and when the tissues of the skin are exposed to prolonged cold, the fatty acid chains of the triglycerides, over a period of weeks, become either shorter or more unsaturated to decrease their melting point, thereby always allowing the fat in the fat cells to remain in a liquid state. This is particularly important because only liquid fat can be hydrolyzed and then transported from the cells.

Fat cells can also synthesize fatty acids and triglycerides from carbohydrates, this function supplementing the synthesis of fat in the liver, as discussed later in the chapter.

Exchange of Fat Between the Adipose Tissue and the Blood—Tissue Lipase. Large quantities of lipase are present in adipose tissue. This enzyme catalyzes rapid exchange between the lipoprotein triglycerides and free fatty acids of the plasma and the triglycerides of the adipose tissue. Because of this rapid exchange, the triglycerides in the fat cells are renewed approximately once every two to three weeks, which means that the fat stored in the tissues one month, even of the obese person, is not the same fat that was stored the preceding month, thus emphasizing the dynamic state of the storage fat.

THE LIVER LIPIDS

The principal functions of the liver in lipid metabolism are: (1) to degrade the triglycerides into small compounds that can be used for energy, (2) to synthesize triglycerides mainly from carbohydrates and, to a lesser extent, from proteins, and (3) to synthesize other lipids from triglycerides.

Large quantities of triglycerides appear in the liver (a) during starvation, (b) in diabetes mellitus, or (c) in any other condition in which fat is being utilized rapidly for energy. In these conditions, the triglycerides are mobilized from the adipose tissue and transported as free fatty acids in the blood, and then are redeposited as triglycerides in the liver, where the initial stages of much of the fat degradation begin. Thus, under normal physiological conditions the total amount of triglycerides in the liver is controlled to a great extent by the overall rate at which lipids are being utilized for energy.

The liver cells, in addition to containing triglycerides, contain large quantities of phospholipids and cholesterol, which are continually synthesized by the liver. Also, the liver cells are much more capable than other tissues of desaturating fatty acids so that the liver triglycerides normally are much more unsaturated than the triglycerides of the adipose tissue. This capability of the liver to desaturate fatty acids seems to be functionally important to all the tissues of the body, because many of the structural members of all cells contain reasonable quantities of desaturated fats, and their principal source seems to be the liver. This desaturation is accomplished by a dehydrogenase in the liver cells, as discussed later.

USE OF TRIGLYCERIDES FOR ENERGY, AND FORMATION OF ADENOSINE TRIPHOSPHATE (ATP)

Approximately 40 to 45 per cent of the calories in the normal American diet are derived from fats, which is about equal to the calories derived from carbohydrates. Therefore, the use of fats by the body for energy is equally as important as the use of carbohydrates. In addition, an average of 30 to 50 per cent of the carbohydrates ingested with each meal is converted into triglycerides, then stored, and later utilized as triglycerides for energy. Therefore, as much as two thirds to three quarters of all the energy derived directly by the cells might be supplied by triglycerides rather than by carbohydrates. For this reason it is equally as important, if not more important, to understand the principles of triglyceride oxidation as to understand carbohydrate oxidation.

Hydrolysis of the Triglycerides. The first stage in the utilization of triglycerides for energy is hydrolysis into fatty acids and glycerol. The glycerol is immediately changed by intracellular enzymes into glyceraldehyde, which enters the phosphogluconate pathway of glucose breakdown and is used for energy. But before the fatty acids can be used for energy, they must be degraded further in the following way:

Degradation of Fatty Acid to Acetyl Coenzyme A by "Beta Oxidation." The fatty acid molecule is degraded by progressive release of two-carbon segments in the form of acetyl coenzyme A (acetyl Co-A). This process is illustrated in Figure 68–1, and it is called the *beta oxidation* process for degradation of fatty acids. The successive stages are the following:

1. The fatty acid molecule first combines with coenzyme A to form a *fatty acyl Co-A* molecule.

2. Fatty acyl Co-A loses two hydrogen atoms from the alpha and beta carbons, leaving a double bond at this point. The hydrogen atoms that are removed become attached to a flavoprotein (FAD) and later are oxidized, as discussed below.

3. A water molecule reacts at the site of the double bond so that a hydrogen atom from the water attaches to the alpha carbon and the remaining hydroxyl radical attaches to a beta carbon.

4. Two additional hydrogen atoms are removed, one from the beta carbon and one from the hydroxyl radical. The hydrogen atoms removed in this process combine with DPN and are also oxidized, as discussed below.

5. The compound splits between the alpha and beta carbons, the long portion of the chain combining with a new molecule of coenzyme A, while the shorter acetyl portion remains combined with the original coenzyme A in the form of *acetyl Co-A.*

The new fatty acyl Co-A, which now has two carbon atoms less than the original fatty acyl Co-A, re-enters reaction number 2 in Figure 68–1 and proceeds through the four stages of chemical reactions until another acetyl Co-A molecule is released. This process is repeated again and again until the entire fatty acid molecule is split into acetyl Co-A. For instance, from each molecule of stearic acid, nine molecules of acetyl Co-A are formed.

Oxidation of Acetyl Co-A. The acetyl Co-A molecules formed by beta oxidation of fatty acids enter into the citric acid cycle as explained in the preceding chapter, combining first with oxaloacetic acid to form succinic acid, which then is degraded into carbon dioxide and hydrogen atoms. The hydrogen is subsequently oxidized by the oxidative enzymes of the cells, which was also explained in the preceding chapter. The net reaction for each molecule of acetyl Co-A is the following:

$$CH_3COCo\text{-}A + Oxaloacetic\ acid + 3H_2O + ADP$$
$$\xrightarrow{\textit{Citric acid cycle}}$$
$$2CO_2 + 8H + HCo\text{-}A + ATP + Oxaloacetic\ acid$$

Thus, after the initial degradation of fatty acids to acetyl Co-A, their final breakdown is precisely the same as that of the acetyl Co-A formed from pyruvic acid during the metabolism of glucose.

ATP Formed by Oxidation of Fatty Acid. In Figure 68–1 note also that 4 hydrogen atoms are released each time a molecule of acetyl Co-A is formed from the fatty acid chain. Therefore, for every stearic acid molecule that is split, a total of 32 hydrogen atoms are removed. In addition, for each acetyl Co-A degraded by the tricarboxylic acid cycle, 8 hydrogen atoms are removed, making an additional 72 hydrogens for each molecule of stearic acid metabolized. This added to the above 32 hydrogen atoms makes a total of 104 hydrogen atoms. Of this group 34 are removed from the degrading fatty acid by flavoproteins and 70 are removed by DPN. These two groups of hydrogen atoms are oxidized by the cells, as discussed in the preceding chapter, but they enter the oxidative system at different points, so that 1 molecule of ATP is synthesized for each of the 34 flavoprotein hydrogens and $1\frac{1}{2}$ molecules of ATP are synthesized for each of the 70 DPN hydrogens. This makes 34 plus 105, or a total of 139 molecules of ATP formed by the oxidation of hydrogen derived from each molecule of stearic acid. And another 9 molecules of ATP are formed in the citric acid cycle, one for each of the 9 acetyl Co-A molecules metabolized. Thus, a total of 148 molecules of ATP are formed during the complete oxidation of one molecule of stearic acid. However, 2 high energy bonds are consumed in the initial combination of coenzyme A with the fatty acid molecule, making a net gain of 146 molecules of ATP.

(1) $RCH_2CH_2CH_2COOH + Co\text{-}A + ATP \xrightleftharpoons{\text{thiokinase}} RCH_2CH_2CH_2COCo\text{-}A + AMP + Pyrophosphate$
    (Fatty acid)   (Fatty acyl Co-A)

(2) $RCH_2CH_2CH_2COCo\text{-}A + FAD \xrightarrow{\text{acyl dehydrogenase}} RCH_2CH=CHCOCo\text{-}A + FADH_2$
    (Fatty acyl Co-A)

(3) $RCH_2CH=CHCOCo\text{-}A + H_2O \xrightleftharpoons{\text{enoyl hydrase}} RCH_2CHOHCH_2COCo\text{-}A$

(4) $RCH_2CHOHCH_2COCo\text{-}A + DPN^+ \xrightleftharpoons[\text{dehydrogenase}]{\beta\text{-hydroxyacyl}} RCH_2COCH_2COCo\text{-}A + DPNH + H^+$

(5) $RCH_2COCH_2COCo\text{-}A + Co\text{-}A \xrightleftharpoons{\text{thiolase}} RCH_2COCo\text{-}A + CH_3COCo\text{-}A$
                                                             (Fatty acyl Co-A)   (Acetyl Co-A)

Figure 68–1 Beta oxidation of fatty acids to yield acetyl coenzyme A.

FORMATION OF ACETOACETIC ACID IN THE LIVER AND ITS TRANSPORT IN THE BLOOD

Over 50 per cent of the initial degradation of fatty acids occurs in the liver. However, the liver cannot use anything like this amount of fatty acids for its own intrinsic metabolic processes. Instead, when the fatty acid chains have been split into acetyl Co-A, two molecules of acetyl Co-A condense to form one molecule of acetoacetic acid as follows:

$$2CH_3COCo\text{-}A + H_2O \underset{\text{other cells}}{\overset{\text{liver cells}}{\rightleftharpoons}}$$
Acetyl Co-A

$$CH_3COCH_2COOH + 2HCo\text{-}A$$
Acetoacetic acid

The acetoacetic acid then freely diffuses through the liver cell membranes and is transported by the blood to the peripheral tissues. Here it again diffuses freely into the cells where exactly the reverse reactions occur and two acetyl Co-A molecules are formed. These in turn enter the citric acid cycle and are oxidized for energy as explained above.

Normally, the acetoacetic acid that enters the blood is transported so rapidly to the tissues that its concentration in the plasma rarely rises above 3 mg. per cent. Yet despite the small quantities in the blood, tremendous amounts are actually transported; this is analogous to the tremendous rate of fatty acid transport in the free form. The rapid transport of both these substances probably depends on their high degree of lipid solubility, which allows rapid diffusion through the cell membranes.

"Ketosis" and Its Occurrence in Starvation, Diabetes, and Other Diseases. Large quantities of acetoacetic acid occasionally accumulate in the blood and interstitial fluids; this condition is called *ketosis* because acetoacetic acid is a keto acid. It occurs especially in starvation, in diabetes mellitus, or even when a person's diet is composed almost entirely of fat. In all these states, essentially no carbohydrates are metabolized—in starvation and following a high fat diet because carbohydrates are not available and in diabetes because insulin is not available to cause glucose transport into the cells.

When carbohydrates are not utilized for energy, essentially all the energy of the body must come from metabolism of fats. We shall see later in the chapter that lack of availability of carbohydrates automatically increases the rate of removal of fatty acids from adipose tissues, and in addition, several hormonal responses—such as increased secretion of corticotropin by the adenohypophysis, increased secretion of glucocorticoids by the adrenal cortex, and decreased secretion of insulin by the pancreas—all further enhance the removal of fatty acids from the fat tissues. As a result, tremendous quantities of fat become available to the liver for degradation into acetoacetic acid. The acetoacetic acid in turn pours out of the liver to be carried to the cells. Yet the cells are limited in the amount of acetoacetic acid that can be oxidized for two reasons: First, acetoacetic acid can be oxidized only as rapidly as adenosine diphosphate is formed in the tissues to initiate the oxidative reactions. Second, lack of carbohydrate intermediates of the glycolytic series depresses the activity of citrate lyase (one of the important rate limiting enzymes controlling the citric acid cycle), which is mainly responsible for degradation of acetoacetic acid, as explained previously. Thus, because of this limit and because of the simultaneous outpouring of tremendous quantities of acetoacetic acid from the liver, the blood concentration of acetoacetic acid sometimes rises to as high as 30 times normal, thus leading to extreme acidosis, as explained in Chapter 37.

THE KETONE BODIES. Acetoacetic acid is readily converted in the body fluids into two other compounds, *beta-hydroxybutyric acid* and *acetone,* in the following manner:

These two substances, along with the acetoacetic acid itself, are called *ketone bodies,* and clinical procedures for analyzing the total concentration of these are often used as a measure of the degree of ketosis in a person. Beta-hydroxybutyric acid has almost the same properties as acetoacetic acid, while acetone is a volatile substance that is blown off in small quantities in the expired air of the lungs, often giving the breath an acetone smell. This smell, too, is frequently used as a diagnostic criterion of ketosis.

ADAPTATION TO A HIGH FAT DIET. If a person changes slowly from a carbohydrate diet to an almost completely fat diet, his body adapts to the utilization of far more acetoacetic acid than usual, and, in this instance, ketosis does not occur. For instance, the Eskimos, who sometimes live almost entirely on a fat diet, do not develop ketosis. Presumably some factor in these people enhances the rate of acetoacetic acid metabolism by the cells.

SYNTHESIS OF TRIGLYCERIDES FROM CARBOHYDRATES

Whenever a greater quantity of carbohydrates enters the body than can be used immediately for energy or stored in the form of glycogen, the excess is rapidly converted into triglycerides and then stored in this form in the adipose tissue. In the past it has been generally thought that most of this fat synthesis occurs in the liver, though recent experiments demonstrate that most triglyceride synthesis occurs in the adipose tissue itself. And even when the synthesis does occur in the liver, the triglycerides are then mainly transported to the adipose tissue where they are stored until needed for energy.

Conversion of Acetyl Co-A into Fatty Acids. All the basic processes for synthesis of triglycerides from carbohydrates are not completely understood. However, the first step is conversion of carbohydrates into acetyl Co-A. It will be recalled from the preceding chapter that this occurs during the normal degradation of glucose by the glycolytic system. It will also be remembered from earlier in this chapter that fatty acids are actually large polymers of acetic acid. Therefore, it is easy to understand how acetyl Co-A could be converted into fatty acids.

However, the synthesis of triglycerides from acetyl Co-A is not simply the reverse of the oxidative degradation that was described above. Instead, its first step is conversion of acetyl Co-A into *malonyl Co-A* in accordance with step 1 of Figure 68–2. A large amount of energy is transferred from ATP to malonyl Co-A, and it is this energy that is utilized to cause the subsequent reactions that are required in the formation of the fatty acid molecule. Step 2 in Figure 68–2 gives the net reaction in the formation of a stearic acid molecule, showing that one acetyl Co-A molecule and eight malonyl Co-A molecules combine with TPNH and hydrogen ions to form the fatty acid molecule. Carbon dioxide and Co-A are both liberated, and these are utilized again and again in the formation of malonyl Co-A. Un-

fortunately, all the steps that lead to the net reaction illustrated at the bottom of Figure 68–2 are not known. However, the acetyl Co-A that is converted into fatty acid molecules is derived mainly from the *glycolytic* breakdown of glucose, and the TPNH required for fatty acid synthesis is derived from the *phosphogluconate pathway* of glucose degradation, which emphasizes the importance of both these pathways in fat synthesis. These two mechanisms of glucose degradation occur side-by-side in the fat cell, contributing the appropriate proportions of acetyl Co-A and TPNH required for fatty acid synthesis.

In addition to the above major pathway for fat synthesis, a small amount of fatty acid synthesis occurs in the mitochondria, utilizing a schema that is essentially the reverse of the normal degradation of fatty acids, with the exception of one or two steps that are necessary to provide energy for polymerization.

Combination of Fatty Acids with α-Glycerophosphate to Form Triglycerides. Once the synthesized fatty acid chains have grown to contain 14 to 18 carbon atoms, they are converted to triglycerides. This occurs because the enzymes that cause this conversion are highly specific for fatty acids with chain lengths of 14 carbon atoms or greater. This specificity of the enzymes actually controls the physical quality of the triglycerides that are stored in the body.

As illustrated in Figure 68–3, the glycerol portion of the triglyceride is furnished by α-glycerophosphate, which is a product derived from the glycolytic schema of glucose degradation. As illustrated in Figure 65–8 of Chapter 65, fatty acids first combine with glycerophosphate to form phosphatidic acid. Then the phosphate radical is removed from this to give a diglyceride. Finally, a third fatty acid combines with this to form the triglyceride.

The real importance of this mechanism for

Step 1:

$$CH_3COCo\text{-}A + CO_2 + ATP$$

$\updownarrow$ (*Acetyl Co-A carboxylase*)

$$\begin{array}{l} COOH \\ | \\ CH_2 \qquad + ADP + PO_4^{---} \\ | \\ O{=}C{-}Co\text{-}A \end{array}$$

Malonyl Co-A

Step 2:

$$1 \text{ Acetyl Co-A} + 8 \text{ Malonyl Co-A} + 16TPNH^+ + 16H^+$$
$$\downarrow$$
$$1 \text{ Stearic acid} + 8CO_2 + 9Co\text{-}A + 16TPN + 7H_2O$$

Figure 68–2 Synthesis of fatty acids.

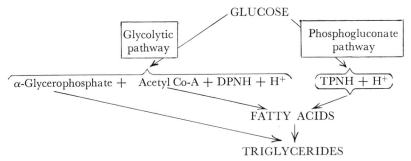

Figure 68–3   An overall schema for synthesis of triglycerides from glucose.

formation of triglycerides is that the whole process is controlled to a great extent by the concentration of α-glycerophosphate. Therefore, when carbohydrates are not available to form α-glycerophosphate, the equilibrium between fatty acids and triglycerides shifts to cause the triglycerides to split into fatty acids, and the fatty acids are then degraded to form acetyl Co-A. This acetyl Co-A then becomes the source of energy in place of acetyl Co-A derived from glucose.

Efficiency of Carbohydrate Conversion into Fat.   During triglyceride synthesis, only about 15 per cent of the original energy in the glucose is lost in the form of heat, while the remaining 85 per cent is transferred to the stored triglycerides.

Importance of Fat Synthesis.   Fat synthesis from carbohydrates is especially important for two reasons: (1) The ability of the different cells of the body to store carbohydrates in the form of glycogen is generally slight; only a few hundred grams of glycogen are stored in the liver, the skeletal muscles, and all other tissues of the body put together. Therefore, fat synthesis provides a means by which the energy of excess ingested carbohydrates (and proteins, too) can be stored for later use. Indeed, the average person has about 60 times as much energy stored in the form of fat as energy stored in the form of carbohydrate. (2) Each gram of fat contains approximately 2¼ times as many calories of energy as each gram of glycogen. Therefore, for a given weight gain, a person can store far more energy in the form of fat than in the form of carbohydrate, which would be important when an animal must be highly motile to survive.

Failure to Synthesize Fats from Carbohydrates in the Absence of Insulin.   When insulin is not available, as in diabetes mellitus, fats are poorly, if at all, synthesized by the cells. This seems to result from the following effects: First, when insulin is not available, glucose cannot enter the adipose cells satisfactorily, so that little of the acetyl Co-A needed for fat synthesis can be derived from glucose. Second, lack of glucose in the cells greatly reduces the availability of α-glycerophosphate, which also makes it difficult for the tissues to form triglycerides.

SYNTHESIS OF TRIGLYCERIDES FROM PROTEINS

Many amino acids can be converted into acetyl Co-A, as will be discussed in the following chapter. Obviously, this can be synthesized into triglycerides. Therefore, when a person has more proteins in his diet than his tissues can use as proteins, a large share of the excess is stored as fat.

REGULATION OF ENERGY RELEASE FROM TRIGLYCERIDES

Regulation of Energy Release by Formation of Adenosine Diphosphate (ADP) in the Tissues. The primary factor that causes energy release from all foodstuffs is the rate of formation of ADP in the tissues. As explained at many points in this text, essentially all functions of the body are energized by the high energy phosphate bonds of ATP. In the process of liberating this energy the ATP becomes ADP, which in turn is a necessary substrate for the oxidation reactions responsible for energy release from essentially all foods. For instance, in the absence of ADP, acetyl Co-A cannot be oxidized, and it accumulates in the tissues. Then, in accordance with the *law of mass action*, this accumulation of acetyl Co-A brings the degradation of fatty acids to a halt. Likewise, failure to oxidize DPNH builds up this end-product of fat degradation, which likewise slows fatty acid degradation.

Conversely, when the activity of the tissues accelerates so that increased quantities of

ADP are formed, all the oxidative processes accelerate, and the degradation and utilization of fatty acids and other foodstuffs for energy proceeds apace.

Fat-Sparing Effect of Carbohydrates. When adequate quantities of carbohydrates are available in the body, the utilization of triglycerides for energy is greatly depressed even though large quantities of ADP are formed. In place of fat utilization, the carbohydrates are utilized preferentially. There are several different reasons for this "fat sparing" effect of carbohydrates, but probably the most important is the following: The fats in adipose tissue cells are present in two different forms, triglycerides and small quantities of free fatty acids. These are in constant equilibrium with each other. When excess quantities of α-*glycerophosphate* are present, the equilibrium between free fatty acids and triglycerides shifts toward the triglycerides as explained earlier in the chapter; and as a result, only minute quantities of fatty acids are then available to be utilized for energy. Since α-glycerophosphate is one of the important degradation products of glucose, the availability of adequate amounts of glucose provides this substance in excess and thereby automatically inhibits the use of fatty acids for energy.

Indeed, when carbohydrates are available in excess, fats are synthesized instead of being degraded. This effect is caused partially by the large quantities of acetyl Co-A formed from the carbohydrates and by the low concentration of free fatty acids in the adipose tissue, thus creating conditions appropriate for conversion of acetyl Co-A into fatty acids. However, a second effect that promotes conversion of carbohydrates to fats is the following: The first step, and the rate-limiting step, in the synthesis of fatty acids is carboxylation of acetyl Co-A to form malonyl Co-A. The rate of this reaction is controlled primarily by the enzyme *acetyl Co-A carboxylase,* whose own activity is accelerated in the presence of the intermediates of the citric acid cycle. Therefore, when excess carbohydrates are being utilized in the citric acid cycle, this automatically causes simultaneous synthesis of fatty acids. Thus, an excess of carbohydrates in the diet not only spares the fat already in the fat stores but also increases these.

Hormonal Regulation of Fat Metabolism. At least seven of the hormones secreted by the endocrine glands also have marked effects on fat metabolism as follows:

As already pointed out, *insulin lack* causes depressed glucose utilization, which in turn decreases fat synthesis and promotes fat mobilization from the tissues as well as increased rate of fat utilization. In severe diabetes, a person can become extremely emaciated because of depletion of the fat stores. Conversely, an excess of insulin greatly enhances the availability of glucose to the cells, which inhibits fat utilization and enhances fat synthesis. In addition, insulin directly enhances the entry of triglycerides into fat cells, thus further increasing the fat stores while decreasing the use of fats for energy.

Glucocorticoids secreted by the adrenal cortex have a direct effect on the fat cells to increase the rate of fat mobilization. It has been postulated that this results from an effect of glucocorticoids to increase the cell membrane permeability of fat cells. In the absence of glucocorticoids, mobilization of fat is depressed, causing considerable depression in fat utilization. Furthermore, glucocorticoids must be available before sufficient fat can ever be mobilized to cause ketosis. Therefore, glucocorticoids are frequently said to have a *ketogenic effect.*

Corticotropin, especially, and *growth hormone,* to a lesser extent, from the adenohypophysis both have a fat-mobilizing effect similar to that of the adrenocortical glucocorticoids, though the precise causes of the mobilization are not known. The corticotropin also stimulates secretion of glucocorticoids which mobilize still more fat, as explained above.

Thyroid hormone causes rapid mobilization of fats, which is believed to result indirectly from an increased rate of energy metabolism in all cells of the body under the influence of this hormone. The resulting reduction in acetyl Co-A in the cells would then be a stimulus to cause fat mobilization.

Finally, *epinephrine* and *norepinephrine* have direct effects on fat cells to increase their rate of fat mobilization. In times of stress, the release of epinephrine from the adrenal medullae is an important means by which fatty acids are made available for metabolism, sometimes increasing the nonesterified fatty acids in the blood as much as 10- to 15-fold.

The effects of the different hormones on metabolism are discussed further in the chapters dealing with each of them.

OBESITY

Obesity means deposition of excess fat in the body. This subject will be discussed in detail in relation to dietary balances in Chapter 72, but briefly it is caused by ingestion of greater amounts of food than can be utilized by the body for energy. The excess food, whether fats, carbohydrates, or proteins, is then stored as fat in the adipose tissue to be used later for energy. Strains of rats have been found in which *hereditary obesity* occurs. In at least one of these strains, the obesity is caused by ineffective mobilization of fat from the adipose tissue while synthesis and storage of fat continues normally. Obviously, such a one-way process causes progressive enhancement of the fat stores, resulting in severe obesity.

PHOSPHOLIPIDS AND CHOLESTEROL

PHOSPHOLIPIDS

The three major types of body phospholipids are the *lecithins,* the *cephalins,* and the *sphingomyelins,* typical examples of which are shown in Figure 68-4.

Phospholipids always contain one or more fatty acid molecules and one phosphoric acid radical, and they usually contain a nitrogenous base. Though the chemical structures of phospholipids are somewhat variant, their physical properties are similar, for they are all lipid soluble, are trans-

$$H_2C-O-\overset{\overset{\displaystyle O}{\|}}{C}-(CH_2)_7-CH=CH-(CH_2)_7-CH_3$$

$$HC-O-\overset{\overset{\displaystyle O}{\|}}{C}-(CH_2)_{16}-CH_3$$

$$H_2C-O-\overset{\overset{\displaystyle O}{\|}}{P}-O-CH_2-CH_2-N\overset{CH_3}{\underset{CH_3}{\overset{CH_3}{<}}}$$
OH   OH

A lecithin

$$H_2C-O-\overset{\overset{\displaystyle O}{\|}}{C}-(CH_2)_7-CH=CH-(CH_2)_7-CH_3$$

$$HC-O-\overset{\overset{\displaystyle O}{\|}}{C}-(CH_2)_{16}-CH_3$$

$$H_2C-O-\overset{\overset{\displaystyle O}{\|}}{P}-O-CH_2-CH_2-NH_2$$
OH

A cephalin

CH_3
|
(CH_2)_{12}
|
CH
‖
CH
|
HO—C—H   O
| ‖
HC—NH—C—(CH_2)_{16}—CH_3
|
$$HC-O-\overset{\overset{\displaystyle O}{\|}}{P}-O-CH_2-CH_2-N\overset{CH_3}{\underset{CH_3}{\overset{CH_3}{<}}}$$
H   OH   OH

Sphingomyelin

Figure 68-4 Typical phospholipids.

ported together in lipoproteins in the blood, and seem to be utilized similarly throughout the body for various structural purposes.

Formation of Phospholipids. Phospholipids are formed in essentially all cells of the body, though certain cells have a special ability to form them. Probably 90 per cent or more of the phospholipids that enter the blood are formed in the liver cells, though reasonably large quantities can also be formed by the intestinal mucosa.

The rate of phospholipid formation is governed to some extent by the usual factors that control the rate of fat metabolism, for when triglycerides are deposited in the liver, the rate of phospholipid formation increases. Also, certain specific chemical substances are needed for formation of some phospholipids. For instance, *choline,* either in the diet or synthesized in the body, is needed for the formation of lecithin because choline is the nitrogenous base of the lecithin molecule. Also, *inositol* is needed for the formation of some cephalins.

Specific Uses of Phospholipids. Several isolated functions of the phospholipids are the following: (1) Phospholipids possibly help to transport fatty acids through the intestinal mucosa into the lymph, possibly by making the fat particles more miscible with water. (2) Thromboplastin, which is necessary to initiate the clotting process, is composed mainly of one of the cephalins. (3) Large quantities of sphingomyelin are present in the nervous system; this substance acts as an insulator in the myelin sheath around nerve fibers. (4) Phospholipids are donors of phosphate radicals when these are needed for different chemical reactions in the tissues. (5) Perhaps the most important of all the functions of phospholipids is the formation of structural elements within cells throughout the body, as is discussed below in connection with cholesterol.

CHOLESTEROL

Cholesterol, the formula of which is illustrated, is present in the diet of all persons, and it can be absorbed slowly from the gastrointestinal tract into the intestinal lymph without any previous digestion. It is highly fat soluble, but only slightly soluble in water, and it is capable of forming esters with fatty acids. Indeed, approximately 70 per cent of the cholesterol of the plasma is in the form of cholesterol esters.

Formation of Cholesterol. Besides the cholesterol absorbed each day from the gastrointestinal tract, which is called *exogenous cholesterol,* a large quantity, called *endogenous cholesterol,* is formed in the cells of the body. Essentially all the endogenous cholesterol that circulates in the lipoproteins of the plasma is formed by the liver, but all the other cells of the body form at least some cholesterol, which is consistent with the fact that many of the membranous structures of all cells are partially composed of this substance.

Cholesterol

As illustrated by the formula of cholesterol, its basic structure is a sterol nucleus. This is synthesized entirely from acetyl Co-A. In turn, the sterol nucleus can be modified by means of various side chains to form (a) cholesterol, (b) cholic acid, which is the basis of the bile acids formed in the liver, and (c) several important steroid hormones secreted by the adrenal cortex, the ovaries, and the testes (these are discussed in later chapters).

Factors that Affect the Plasma Cholesterol Concentration—Feedback Control of Body Cholesterol. Among the important factors that affect plasma cholesterol concentration are the following:

1. An increase in the amount of cholesterol ingested each day increases the plasma concentration slightly. However, when cholesterol is ingested, an intrinsic feedback control system for control of body cholesterol causes the liver to compensate almost completely for this by synthesizing smaller quantities of endogenous cholesterol. As a result, a diet containing large quantities of cholesterol usually increases the blood cholesterol concentration no more than 15 per cent.

2. A *saturated* fat diet increases blood cholesterol concentration as much as 15 to 25 per cent. This presumably results from increased fat deposition in the liver, which then provides increased quantities of acetyl Co-A in the liver cells for production of cholesterol. Therefore, to decrease the blood cholesterol concentration, it is equally as important to maintain a diet low in saturated fat as to maintain a diet low in cholesterol concentration.

3. Ingestion of fat containing highly unsaturated fatty acids usually depresses the blood cholesterol concentration a slight to moderate amount.

4. Lack of thyroid hormone increases the blood cholesterol concentration, whereas excess thyroid hormone decreases the concentration. This effect is believed to be related to the increased metabolism of all lipid substances under the influence of thyroxine.

5. The blood cholesterol also rises greatly in diabetes mellitus. This is believed to result from the general increase in lipid mobilization in this condition.

6. The female sex hormones, the *estrogens,* decrease blood cholesterol; whereas the male sex hormones, the *androgens,* increase blood cholesterol. Unfortunately, the mechanisms of these effects are unknown, but the sex effects are very important because the higher cholesterol in the male is associated with a higher incidence of heart attacks.

7. In renal retention diseases, the blood cholesterol rises greatly, along with similar increases in blood triglycerides and phosphatides. This is believed to result from inhibition of plasma lipoprotein lipase, resulting in diminished removal of lipoproteins from the plasma, and thus causing their concentration to increase markedly.

Specific Uses of Cholesterol. By far the most abundant use of cholesterol in the body is to form cholic acid in the liver. As much as 80 per cent of the cholesterol is converted into cholic acid. This, in turn, is conjugated with other substances to form bile salts, which have already been discussed in connection with fat digestion.

A small quantity of cholesterol is used (a) by the adrenal glands to form adrenocortical hormones, (b) by the ovaries to form progesterone and estrogen, and (c) by the testes to form testosterone. However, these glands can also synthesize their own sterols and then form their hormones from these, as is discussed in the chapters on endocrinology later in the text.

A large amount of cholesterol is precipitated in the corneum of the skin. This, along with other lipids, makes the skin highly resistant to the absorption of water-soluble substances and also to the action of many chemical agents, for cholesterol and the other lipids are highly inert to such substances as acids and different solvents that might otherwise easily penetrate the body. Also, these lipid substances help to prevent water evaporation from the skin; without this protection the amount of evaporation would probably be 15 to 20 liters per day instead of the usual 300 to 400 ml.

STRUCTURAL FUNCTIONS OF PHOSPHOLIPIDS AND CHOLESTEROL

The specific uses of phospholipids and cholesterol are probably of only minor importance in comparison with their importance for general structural purposes throughout the cells of the body.

In Chapter 2 it was pointed out that reasonably large quantities of phospholipids and cholesterol are present in the cell membrane and in the membranes of the internal organelles of all cells. It has even been claimed that cholesterol and possibly also the phospholipids have a controlling effect over the permeability of cell membranes.

For membranes to develop, substances that are not soluble in water must be available, and, in general, the only substances in the body that are not soluble in water (besides the inorganic substances of bone) are the lipids and some proteins. Thus, the physical integrity of cells throughout the body is based mainly on phospholipids, triglycerides, cholesterol, and certain insoluble proteins. Some phospholipids are somewhat water soluble as well as lipid soluble, which gives them the important property of helping to decrease the interfacial tension between the membranes and the surrounding fluids.

Another fact that indicates phospholipids and cholesterol to be mainly concerned with the formation of structural elements of the cells is the slow turn-over rate of these substances. For instance, radioactive phospholipids formed in the brain of mice remain in the brain several months after they are formed. Thus, these phospholipids are only slowly metabolized, and the fatty acid is not split away from them to any major extent. Consequently, the purpose of their being in the cells of the brain is presumably related to their indestructible physical properties rather than to their chemical properties—in other words, for the formation of actual physical structures within the cells of the brain.

ATHEROSCLEROSIS

Atherosclerosis is principally a disease of the large arteries in which lipid deposits called *atheromatous plaques* appear in the subintimal layer of the arteries. These plaques contain an especially large amount of cholesterol and often are simply called cholesterol deposits. They usually are also associated with degenerative changes in the arterial wall. In a later stage of the disease, fibroblasts infiltrate the degenerative areas and cause progressive sclerosis of the arteries. In addition, calcium often precipitates with the lipids to develop *calcified plaques*. When these two reactions occur, the arteries become extremely hard, and the disease is then called *arteriosclerosis,* or simply "hardening of the arteries."

Obviously, arteriosclerotic arteries lose most of their distensibility, and, because of the degenerative areas, they are easily ruptured. Also, the atheromatous plaques often protrude through the intima into the flowing blood, and the roughness of their surfaces causes blood clots to develop, with resultant thrombus or embolus formation (see Chapter 12). Almost half of all human beings die of arteriosclerosis; approximately two thirds of the deaths are caused by thrombosis of one or more coronary arteries and the remaining one third by thrombosis or hemorrhage of vessels in other organs of the body—especially the brain, kidneys, liver, gastrointestinal tract, limbs, and so forth.

Despite the extreme prevalence of atherosclerosis, little is known about its cause. Therefore, it is necessary to outline the general trends of the experimental studies rather than to present a definitive description of the mechanisms that cause atherosclerosis.

EXPERIMENTAL PRODUCTION OF ATHEROSCLEROSIS IN ANIMALS

Severe atherosclerosis can be produced easily in rabbits by simply feeding them large quantities of cholesterol. Rabbits normally have no cholesterol in their diet, for cholesterol is not present in plants, which constitute their entire dietary intake. Therefore, the metabolic processes of the rabbit are not adapted to utilize the ingested cholesterol. Instead, the cholesterol precipitates in many areas of the body, especially in the liver and in the subintimal layer of the arteries.

In carnivorous animals, such as the dog, the induction of atherosclerosis is much more difficult, but it can be caused by excessive administration of cholesterol after the thyroid gland has been removed or inhibited by propylthiouracil. Loss of thyroid secretion greatly depresses cholesterol utilization, thus allowing much of the ingested cholesterol to deposit in the arterial walls.

As pointed out above in relation to the metabolism of cholesterol, the plasma cholesterol concentration increases in either a human being or an experimental animal that has a very high fat diet. On the other hand, substitution of unsaturated fats for saturated fats usually decreases the cholesterol level and, in some experimental animals at least, also inhibits the development of atherosclerosis.

ATHEROSCLEROSIS IN THE HUMAN BEING

Effect of Age, Sex, and Heredity on Atherosclerosis. Atherosclerosis is mainly a disease of old age, but small atheromatous plaques can almost always be found in the arteries of young adults. Therefore, the full-blown disease seems to culminate from a lifetime of lipid deposition rather than deposition over a few years.

Far more men die of atherosclerotic heart disease than do women. This is especially true of men younger than 50. For this reason, it is possible that the male sex hormone accelerates the development of atherosclerosis, or that the female sex hormone *protects* a person from atherosclerosis. Indeed, administration of estrogens to men who have already had coronary thromboses seems to decrease the number of secondary coronary attacks. Furthermore, administration of estrogens to chickens with atheromatous plaques in their coronaries has in some instances decreased the degree of the disease.

Atherosclerosis and atherosclerotic heart disease are highly hereditary in some families. In some instances, this is related to an inherited hypercholesterolemia, but in other instances the blood cholesterol level is completely normal. In this second group, the cholesterol is mainly present in large β-lipoproteins rather than in smaller ones as is usual. These large lipoproteins have a high cholesterol to phospholipid ratio, a factor that supposedly leads to decreased suspension stability of the lipoproteins in the plasma and increased lipid deposition in the arterial wall. Inheritance of the tendency to atherosclerosis is sometimes caused by dominant genes, which means that once this trait enters a family a high incidence of the disease occurs among the offspring.

Other Diseases that Predispose to Atherosclerosis. Human beings with severe *diabetes* or severe *hypothyroidism* frequently develop premature and severe atherosclerosis. In both these conditions the blood cholesterol is greatly elevated, which seems to be the cause of the atherosclerosis.

Another disease associated with atherosclerosis, in human beings, as well as in experimental animals, is *hypertension;* the incidence of atherosclerotic coronary heart disease is about twice as great in hypertensives as in normal persons. Though the cause of this is not known, it possibly results from pressure damage to the arterial walls with subsequent deposition of cholesterol plaques.

Relationship of Dietary Fat to Atherosclerosis in the Human Being. A high fat diet, especially one containing cholesterol and saturated fats, greatly increases one's chances of developing atherosclerosis. Therefore, decreasing the fat can help greatly in protecting against atherosclerosis, and some experiments indicate that this can benefit even patients who have already had coronary heart attacks. Also, life insurance statistics show that the rate of mortality—mainly from coronary disease—of normal weight older age persons is about half the mortality rate of overweight subjects of the same age.

SUMMARY OF FACTORS CAUSING ATHEROSCLEROSIS

Atherosclerosis is almost certainly caused by an abnormality of lipid metabolism, but it is also exacerbated by almost any factor that injures the arterial wall. Especially is elevated blood cholesterol often related to atherosclerosis. Finally, perhaps more important than any of these might be some undiscovered third factor that is inherited from generation to generation to cause an in-

creased rate of cholesterol production or deposition in the arterial walls themselves, irrespective of the blood cholesterol concentration.

REFERENCES

Adipose tissue and metabolism (conference). *Ann. N. Y. Acad. Sci.,* 131:1, 1965.

DiLuzio, N. R.: Reticuloendothelial involvement in lipid metabolism. *Ann. N.Y. Acad. Sci.,* 88:244, 1960.

Fredrickson, D. S., and Gordon, R. S., Jr.: Transport of fatty acids. *Physiol. Rev.,* 38:585, 1958.

Freidman, M., and Byers, S. O.: Epinephrine-induced reduction of postprandial lipemia in the rat. *Amer. J. Physiol.,* 213:829, 1967.

Frohman, L. A.; Bernardis, L. L.; Schnatz, J. D.; and Burek, L.: Plasma insulin and triglyceride levels after hypothalamic lesions in weanling rats. *Amer. J. Physiol.,* 216:1496, 1969.

Goodman, DeW. S.: Cholesterol ester metabolism. *Physiol. Rev.,* 45:747, 1965.

Gurin, S., and Marsh, J. B.: Lipid metabolism. *In* Duncan, G. G. (ed.): Diseases of Metabolism. 5th ed., Philadelphia, W. B. Saunders Company, 1964, p. 191.

Havel, R. J.; Pernow, B.; and Jones, N. L.: Uptake and release of free fatty acids and other metabolites in the legs of exercising men. *J. Appl. Physiol.,* 23:90, 1967.

Jones, N. L., and Havel, R. J.: Metabolism of free fatty acids and chylomicron triglycerides during exercise in rats. *Amer. J. Physiol.,* 213:819, 1967.

Khachadurian, A. K.; Kamelian, M.; and Adrouni, B.: Metabolism of sheep adipose tissue in vitro. *Amer. J. Physiol.,* 213:1385, 1967.

Kirk, J. E.: Arteriosclerosis and arterial metabolism. *In* Bittar, E. E., and Bittar, N. (eds.): The Biological Basis of Medicine. New York, Academic Press, Inc., 1968, Vol. 1, p. 493.

Lee, D. S.; Hashim, S. A.; and Van Italie, T. B.: Effect of long chain triglyceride on chylous transport of medium chain fatty acids. *Amer. J. Physiol.,* 214:294, 1968.

Malinow, M. R.; Perley, A.; and McLaughlin, P.: Muscular exercise and cholesterol degradation: mechanisms involved. *J. Appl. Physiol.,* 27:662, 1969.

Morgan, T. E.; Short, F. A.; and Cobb, L. A.: Effect of long-term exercise on skeletal muscle lipid composition. *Amer. J. Physiol.,* 216:82, 1969.

Myant, N. B.: Hormonal control of cholesterol metabolism. *In* Bittar, E. E., and Bittar, N. (eds.): The Biological Basis of Medicine. New York, Academic Press, Inc., 1968, Vol. 2, p. 193.

Myant, N. B.: Hormonal control of fatty acid metabolism. *In* Bittar, E. E., and Bittar, N. (eds.): The Biological Basis of Medicine. New York, Academic Press, Inc., 1968, Vol. 2, p. 151.

Olson, R. E., and Vester, J. W.: Nutrition-endocrine interrelationships in the control of fat transport in man. *Physiol. Rev.,* 40:677, 1960.

Paul, P.: FFA metabolism of normal dogs during steady-state exercise at different work loads. *J. Appl. Physiol.,* 28:127, 1970.

Perl, W., and Sanuel, P.: Input-output analysis for total input rate and total traced mass of body cholesterol in man. *Circ. Res.,* 25:191, 1969.

Schonfeld, G., and Kipnis, D. M.: Effects of fatty acids on carbohydrate and fatty acid metabolism of rat diaphragm. *Amer. J. Physiol.,* 215:513, 1968.

Spain, D. M.: Atherosclerosis. *Sci. Amer.,* 215:48(2), 1966.

Stauffacher, W., and Renold, A. E.: Effect of insulin in vivo on diaphragm and adipose tissue of obese mice. *Amer. J. Physiol.,* 216:98, 1969.

Tubbs, P. K.: Membranes and fatty acid metabolism. *Brit. Med. Bull.,* 24:158, 1968.

Wirsen, C.: Studies in lipid mobilization. *Acta Physiol. Scand.,* 65(Suppl. 252), 1965.

Zilversmit, D. B.; McCandless, E. L.; Jordan, P. H.; Henly, W. S.; and Ackerman, R. F.: The synthesis of phospholipids in human atheromatous lesions. *Circulation,* 23:370, 1961.

CHAPTER 69

Protein Metabolism

About three quarters of the body solids are proteins. These include *structural proteins, enzymes, genes, proteins that transport oxygen, proteins of the muscle that cause contraction,* and many other types that perform specific functions both intracellularly and extracellularly throughout the body.

The basic chemical properties of proteins that explain their diverse functions are so extensive that they are a major portion of the entire discipline of biochemistry. For this reason, the present discussion is confined to the general aspects of protein metabolism.

BASIC PROPERTIES OF PROTEINS

THE AMINO ACIDS

The principal constituents of proteins are amino acids, 21 of which are present in the body in significant quantities. Figure 69–1 illustrates the chemical formulas of these 21 amino acids, showing that they all have two features in common: Each amino acid has an acidic group ($-COOH$) and a nitrogen radical that lies in close association with the acidic radical, usually represented by the amino group ($-NH_2$).

Peptide Linkages and Peptide Chains. In proteins, the amino acids are aggregated into long chains by means of so-called *peptide linkages,* one of which is illustrated by the following reaction:

$$R-\underset{\overset{|}{NH_2}}{CH}-COOH + R'-\underset{\overset{|}{NH_2}}{CH}-COOH \longrightarrow$$

$$R-\underset{\overset{|}{NH_2}}{CH}-CO$$
$$\underset{\overset{|}{NH}}{} + H_2O$$
$$R'-CH-COOH$$

Note in this reaction that the amino radical of one amino acid combines with the carboxyl radical of the other amino acid. A hydrogen atom is released from the amino radical, and a hydroxyl radical is released from the carboxyl radical; and these two combine to form a molecule of water. Note that after the peptide linkage has been formed, an amino radical and a carboxyl radical are still in the new molecule, both of which are capable of combining with additional amino acids to form a *peptide chain.* Some complicated protein molecules have as many as a hundred thousand amino acids combined together principally by peptide linkages, and even the smallest protein usually has more than 20 amino acids combined together by peptide linkages.

Other Linkages in Protein Molecules. Some protein molecules are composed of several peptide chains rather than a single chain, and these in turn are bound with each other by other linkages, principally by *hydrogen bonds* between the CO and NH radicals of the peptides as follows:

Also, many peptide chains are coiled, and the successive coils are held in a tight spiral by similar hydrogen bonding. But, in addition to hydrogen bonds, separate peptide chains can be held together by sulfhydryl, phenolic, and salt linkages, as well as by others.

PHYSICAL CHARACTERISTICS OF PROTEINS

GLOBULAR PROTEINS

With the exception of the few fibrous proteins, which are discussed subsequently, most proteins of the body assume either a globular or an elliptical shape and are called *globular proteins.* These, in general, are soluble in water or salt solutions,

AMINO ACIDS

Figure 69–1   The amino acids, showing the 10 essential amino acids, which cannot be synthesized at all or in sufficient quantity in the body.

and they are held in a globular shape by continuous coiling of the peptide chains.

Important Types of Globular Proteins in the Body. There is no simple functional classification of the globular proteins for two reasons: First, these proteins perform literally thousands of different functions in the body. Second, proteins of widely varying chemical and physical characteristics many times perform very much the same function. Yet for purposes of reference the different globular proteins can be classified mainly on the basis of their chemical properties as follows:

The *albumins* are low molecular weight simple proteins that are readily soluble in water. They are the major proteins of the plasma but are also present in small quantities inside the cell.

Globulins are also simple proteins that are soluble in salt solution but poorly soluble in water. These, like albumin, are also present in large quantities in the plasma and are extensively present inside the cells. Conjugated products of globulins constitute most cellular enzymes.

Histones and *protamines* are highly basic proteins because of large quantities of basic amino acids in their molecules. These are the principal proteins bound with nucleic acids to form nucleoproteins.

FIBROUS PROTEINS

Many of the highly complex proteins are fibrillar and are called fibrous proteins. In these the peptide chains are elongated, and many separate chains are held together in parallel bundles by cross-linkages. The three major types of fibrous proteins are (1) *collagens,* which are the basic structural proteins of connective tissue, tendons, cartilage, and bone; (2) *elastins,* which are the elastic fibers of tendons, arteries, and connective tissue; and (3) *keratins,* which are the structural proteins of hair and nails.

Since fibrillar proteins, the collagen fibers in particular, are the principal structural proteins of the body, it is important to know about their physical characteristics. In general, the fibrils are extremely strong and are capable of being stretched and then recoiling to their natural length; these are properties of typical *elastomers.* Another elastomeric characteristic of the fibrillar proteins is their tendency to *creep;* that is, if stretched for a long time, their basic length gradually becomes the stretched length, but, on the other hand, if the tension on the two ends of the fibrils is relaxed, the fibrils creep to a shorter and shorter length. For instance, a large scar that forms shortly after a severe wound gradually creeps to a smaller and smaller size if tension is not placed on the scar, but, on the other hand, such a scar can creep to a larger and larger size if it is in an area where tension is high, as occurs frequently in the skin of a person who becomes progressively more obese.

CONJUGATED PROTEINS

In addition to the simple globular and fibrous proteins, many proteins are combined as conjugated proteins with nonprotein substances. These include the following:

Nucleoproteins are combinations of simple proteins and nucleic acid. The deoxyribose nucleoproteins are the principal constituents of genes, and ribose nucleoproteins perform essential functions in the cytoplasm necessary for the synthesis of proteins, which was discussed in detail in Chapter 3.

Mucoproteins are a type of conjugated protein containing large quantities of complex polysaccharides. One of the most important mucoproteins, that found in mucus, is highly resistant to chemical destruction and also has exceptionally important properties that were discussed in Chapter 63 in relation to gastrointestinal physiology.

In addition to these important types of proteins are the *lipoproteins,* which contain lipid materials, *chromoproteins* (such as hemoglobin and cytochromes), which contain coloring agents, *phosphoproteins,* which contain phosphorus, and *metalloproteins,* which contain magnesium, copper, iron, zinc, or other metallic ions and which constitute many of the enzymes.

TRANSPORT AND STORAGE OF AMINO ACIDS

THE BLOOD AMINO ACIDS

The normal concentration of amino acids in the blood is between 35 and 65 mg. per cent. This is an average of about 2 mg. per cent for each of the 21 amino acids, though some are present in far greater concentrations than others. Since the amino acids are relatively strong acids, they exist in the blood principally in the ionized state and account for 2 to 3 milliequivalents of the negative ions in the blood. The precise distribution of the different amino acids in the blood depends to some extent on the types of proteins ingested, but the concentrations of some individual amino acids are also regulated to a certain extent by (a) selective synthesis in the different cells and (b) selective excretion by the kidneys.

Fate of Amino Acids Absorbed from the Gastrointestinal Tract.   It will be recalled from Chapter 65 that the end-products of protein digestion in the gastrointestinal tract are almost entirely amino acids and that only rare polypeptide or protein molecules are absorbed. Immediately after a meal, the amino acid concentration in the blood rises, but the rise is usually only a few milligrams per cent for two reasons: First, protein digestion and absorption is usually extended over two to three hours, which allows only small quantities of amino acids to be absorbed at a time. Second, after entering the blood, the excess amino acids are absorbed within 5 to 10 minutes by cells throughout the entire body. Therefore, almost never do large concentrations of amino acids accumulate in the blood. Nevertheless, the turnover rate of the amino acids is so rapid that many grams of proteins can be carried from one part of the body to another in the form of amino acids each hour.

Active Transport of Amino Acids into the Cells. The molecules of essentially all the amino acids are much too large to diffuse through the pores of the cell membranes. Small quantities perhaps can dissolve in the matrix of the cell membrane and diffuse to the interior of the cells in this manner, but large quantities of amino acids can be transported through the membrane only by active transport utilizing carrier mechanisms. The nature of the carrier mechanisms is still poorly understood, but some of the theories are discussed in Chapter 4.

RENAL THRESHOLD FOR AMINO ACIDS. One of the special functions of carrier transport of amino acids is to prevent loss of these in the urine. All the different amino acids can be *actively transported* through the proximal tubular epithelium, thus removing them from the glomerular filtrate and returning them to the blood. However, as is true of other active transport mechanisms in the renal tubules, there is an upper limit to the rate at which each type of amino acid can be transported. For this reason, when a particular type of amino acid rises to too high a concentration in the plasma and glomerular filtrate, the excess above that which can be actively reabsorbed is lost into the urine.

In Chapter 38 it was pointed out that appropriate enzyme systems for active reabsorption of certain amino acids are often deficient or lacking from the renal tubular epithelium. In these conditions the plasma threshold for the respective amino acids is greatly reduced. However, in the normal person, the loss of amino acids in the urine each day is insignificant, thus allowing essentially all amino acids absorbed from the gastrointestinal tract to be utilized by the cells.

STORAGE OF AMINO ACIDS AS PROTEINS IN THE CELLS

Almost immediately after entry into the cells, amino acids are conjugated under the influence of intracellular enzymes into cellular proteins so that the concentration of amino acids inside the cells probably always remains low. Thus, so far as is known, storage of large quantities of amino acids as such probably does not occur in the cells; instead, they are stored in the form of actual proteins. Yet many intracellular proteins can be rapidly decomposed again into amino acids under the influence of intracellular enzymes called *kathepsins,* and these amino acids in turn can be transported back out of the cell into the blood. The proteins that can be thus decomposed include many cellular enzymes as well as some other functioning proteins. However, the genes of the nucleus and structural proteins, such as collagen and muscle contractile proteins, do not participate significantly in this reversible storage of amino acids.

Some tissues of the body participate in the storage of amino acids to a greater extent than others, particularly tissues that have high metabolic rates. Thus, the liver, which is a large organ and also has a high metabolic rate, stores large quantities of proteins; this is also true of the kidney and the intestinal mucosa.

Release of Amino Acids from the Cells and Regulation of Plasma Amino Acid Concentration. Whenever the plasma amino acid concentration falls below its normal level, amino acids are transported out of the cells to replenish the supply in the plasma. Simultaneously, intracellular proteins are degraded back into amino acids.

The plasma concentration of each type of amino acid is maintained at a reasonably constant value. Later it will be noted that the various hormones secreted by the endocrine glands are able to alter the balance between tissue proteins and circulating amino acids; growth hormone and insulin increase the formation of tissue proteins, while the adrenocortical glucocorticoid hormones increase the concentration of circulating amino acids.

Reversible Equilibrium Between the Proteins of Different Parts of the Body. Since cellular proteins can be synthesized rapidly from plasma amino acids and many of these in turn can be degraded and returned to the plasma almost equally as rapidly, there is constant equilibrium between the plasma amino acids and most of the proteins in the cells of the body. Therefore, it follows that there is also equilibrium between the proteins from one type of cell to the next. For instance, if any particular tissue loses proteins, it can synthesize new proteins from the amino acids of the blood; in turn, these are replenished by degradation of proteins from other cells of the body. These effects are particularly noticeable in relation to protein synthesis in cancer cells. Cancer cells are prolific users of amino acids, and, simultaneously, the proteins of the other tissues become markedly depleted.

Upper Limit to the Storage of Proteins. Each particular type of cell has an upper limit to the amount of proteins that it can store. After all the cells have reached their limit, the excess amino acids in the circulation are then degraded into other products and used for energy, as is discussed subsequently, or converted to fat and stored in this form.

THE PLASMA PROTEINS

The three major types of protein present in the plasma are *albumin*, *globulin*, and *fibrinogen*. The principal function of albumin is to provide *colloid osmotic pressure*, which in turn prevents plasma loss from the capillaries, as discussed in Chapter 20. The globulins perform a number of enzymatic functions in the plasma itself, but, more important than this, they are principally responsible for both the natural and acquired immunity that a person has against invading organisms, which was discussed in Chapter 10. The fibrinogen polymerizes into long fibrin threads during blood coagulation, thereby forming blood clots that help to repair leaks in the circulatory system, which was discussed in Chapter 12.

Formation of the Plasma Proteins. Essentially all the albumin and fibrinogen of the plasma proteins, as well as 50 per cent or more of the globulins, are formed in the liver. The remainder of the globulins are formed in the lymphoid tissues and other cells of the reticuloendothelial system. These are mainly the gamma globulins that constitute the antibodies.

The rate of plasma protein formation by the liver can be extremely high, as great as 4 grams per hour or as much as 100 grams per day. Certain disease conditions often cause rapid loss of plasma proteins; severe burns that denude large surface areas cause loss of many liters of plasma through the denuded areas each day. The rapid production of plasma proteins by the liver is obviously valuable in preventing death in such states. Furthermore, occasionally, a person with severe renal disease loses as much as 20 to 30 grams of plasma protein in the urine each day for years. In some of these patients the plasma protein concentration may remain normal or almost normal throughout the entire illness.

Use of Plasma Proteins by the Tissues. When the tissues become depleted of proteins, the plasma proteins can act as a source for rapid replacement of the tissue proteins. Indeed whole plasma proteins can be imbibed *in toto* by the reticuloendothelial cells; then, once in the cells, these are split into amino acids that are transported back into the blood and utilized throughout the body to build cellular proteins. In this way, then, the plasma proteins function as a labile protein storage medium and represent a rapidly available source of amino acids whenever a particular tissue requires these.

Reversible Equilibrium Between the Plasma Proteins and the Tissue Proteins. The rate of synthesis of plasma proteins by the liver is dependent on the concentration of amino acids in the blood, which means that the concentration of plasma proteins becomes reduced whenever an appropriate supply of amino acids is not available. On the other hand, whenever excess proteins are available in the plasma but insufficient proteins are present in the cells, the plasma proteins are used to form tissue proteins. Thus, there is a constant state of equilibrium, as illustrated in Figure 69–2, between the plasma proteins, the amino acids of the blood, and the tissue proteins. It has been estimated from radioactive tracer studies that about 400 gms. of body protein are synthesized and degraded each day as part of the continual state of flux of amino acids. This illustrates once again the general principle of reversible exchange of amino acids among the different proteins of the body. Even during starvation or during severe debilitating diseases, the ratio of total tissue proteins in the body to total plasma proteins remains relatively constant at about 33 to 1.

Because of this reversible equilibrium between plasma proteins and the other proteins of the body, one of the most effective of all therapies for severe acute protein deficiency is intravenous administration of plasma protein. Within hours, the protein becomes distributed throughout the cells of the body where it is needed.

CHEMISTRY OF PROTEIN SYNTHESIS

Proteins are synthesized in all cells of the body, and the functional characteristics of each cell are dependent upon the types of protein that it can form. Basically, the genes of the cells control the protein types and thereby control the functions of the cell. This regulation of cellular function by the genes was discussed in detail in Chapter 3. Chemically, two basic processes must be accomplished for the synthesis of proteins; these are: (1) synthesis of the amino acids and (2) appropriate conjugation of the amino acids to form the respective types of whole proteins in each individual cell.

Essential and Nonessential Amino Acids. Eleven of the 21 amino acids normally present in animal proteins can be synthesized in the cells, while the other 10 either cannot be synthesized at all or are synthesized in too small a quantity to supply the body's needs. The first group of amino acids is

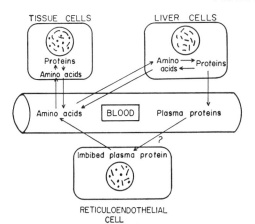

Figure 69-2 Reversible equilibrium between the tissue proteins, plasma proteins, and plasma amino acids.

called *nonessential,* while the second group is called *essential amino acids.* The essential amino acids obviously must be present in the diet if protein formation is to take place in the body. Use of the word "essential" does not mean that the other 11 amino acids are not equally as essential in the formation of the proteins, but only that these others are not essential in the diet.

Synthesis of the nonessential amino acids depends on the formation first of appropriate α-keto acids, which are the precursors of the respective amino acids. For instance, *pyruvic acid,* which is formed in large quantities during the glycolytic breakdown of glucose, is the keto acid precursor of the amino acid *alanine.* Then, by the simple process of *transamination,* an amino radical is transferred to the α-keto acid while the keto oxygen is transferred to the donor of the amino radical. This reaction is illustrated in Figure 69-3. Note in this reaction that the amino radical is transferred to the pyruvic acid from a substance called *glutamine.* Glutamine is present in the tissues in large quantities, and it functions specifically as an amino radical storehouse. In addition, amino radicals can be transferred from *asparagine* and from some amino acids of the body fluids, particularly from glutamic acid and aspartic acid.

Transamination is promoted by enzymes called *transaminases,* all of which are derivatives of pyridoxine, one of the B vitamins. Without this vitamin, the nonessential amino acids cannot be synthesized, and, therefore, protein formation cannot proceed normally.

Formation of Proteins from Amino Acids. Once the appropriate amino acids are present in a cell,

whole proteins are synthesized rapidly. However, each peptide linkage requires from 500 to 4000 calories of energy, and this must be supplied from ATP in the cell. Protein formation proceeds through two steps: (1) "activation" of each amino acid, during which the amino acid is "energized" by energy derived from ATP, and (2) alignment of the amino acids into the peptide chains, a function that is under control of the genetic system of each individual cell. Both of these processes were discussed in detail in Chapter 3. Indeed, the formation of cellular proteins is the basis of life itself and is so important that the reader would do well to review Chapter 3.

USE OF PROTEINS FOR ENERGY

It was pointed out earlier that there is an upper limit to the amount of protein that can accumulate in each particular type of cell. Once the cells are filled to their limits, any additional amino acids in the body fluids are degraded and used for energy or stored as fat. This degradation occurs almost entirely in the liver, and it begins with the process known as deamination.

Deamination. Deamination means removal of the amino groups from the amino acids. This can occur by several different means, two of which are especially important: (1) transamination, which means transfer of the amino group to some acceptor substance as explained above in relation to the synthesis of amino acids, and (2) oxidative deamination.

The greatest amount of deamination occurs by the following transamination schema:

Note from this schema that the amino group from the amino acid is transferred to α-ketoglutaric acid, which then becomes glutamic acid. The glutamic acid can then transfer the

$$NH_2-C-CH_2-CH_2-CH-COOH + CH_3-C-COOH \xrightarrow{\text{Transaminase}}$$

$$\begin{matrix} \| & & | & & \| \\ O & & NH_2 & & O \end{matrix}$$

(Glutamine)   (Pyruvic acid)

$$NH_2-C-CH_2-CH_2-C-COOH + CH_3-CH-COOH$$

$$\begin{matrix} \| & & \| & & | \\ O & & O & & NH \end{matrix}$$

(α—Ketoglutamic acid)   (Alanine)

Figure 69-3 Synthesis of alanine from pyruvic acid by transamination.

amino group to still other substances or can release it in the form of ammonia. In the process of losing the amino group, the glutamic acid once again becomes α-ketoglutaric acid, so that the cycle can be repeated again and again.

Oxidative deamination occurs to a much less extent and is catalyzed by amino acid oxidases. In this process the amino acid is oxidized at the point where the amino radical attaches, which causes the amino radical to be released.

UREA FORMATION BY THE LIVER. The ammonia released during deamination is removed from the blood almost entirely by conversion into urea, two molecules of ammonia and one molecule of carbon dioxide combining in accordance with the following net reaction:

$$2NH_3 + CO_2 \rightarrow H_2N-\underset{\underset{O}{\|}}{C}-NH_2 + H_2O$$

Essentially all urea formed in the human body is synthesized in the liver. In the absence of the liver or in serious liver disease, ammonia accumulates in the blood. This in turn is extremely toxic, especially to the brain, often leading to a state called *hepatic coma*.

The stages in the formation of urea from ammonia and carbon dioxide are not all known, but the following represents the general net mechanism:

The reaction begins with the amino acid derivative *ornithine*, which combines with one molecule of carbon dioxide and one molecule of ammonia to form a second substance, *citrulline*. This in turn combines with still another molecule of ammonia to form *arginine*, which then splits into *ornithine* and *urea*. The urea diffuses from the liver cells into the body fluids and is excreted by the kidneys, while the ornithine is reused in the cycle again and again.

Oxidation of Deaminated Amino Acids. Once amino acids have been deaminated, the resulting keto acid products can in most instances be oxidized to release energy for metabolic purposes. This usually involves two processes: (1) the keto acid is changed into an appropriate chemical substance that can combine with one or more of the active compounds of the tricarboxylic acid cycle and (2) this substance is then degraded by this cycle in the same manner that acetyl Co-A derived from carbohydrate and lipid metabolism is degraded.

In general, the amount of adenosine triphosphate formed for each gram of protein that is oxidized is approximately the same as that formed for each gram of glucose oxidized.

Gluconeogenesis and Ketogenesis. Certain deaminated amino acids are similar to the breakdown products that result from glucose and fatty acid metabolism. For instance, deaminated alanine is pyruvic acid. Obviously, this can be converted into glycose or glycogen. Or it can be converted into acetyl Co-A, which can then be polymerized into fatty acids. Also, two molecules of acetyl Co-A can condense to form acetoacetic acid, which is one of the ketone bodies, as explained in the preceding chapter.

The conversion of amino acids into glucose or glycogen is called *gluconeogenesis*, and the conversion of amino acids into keto acids (or fatty acids) is called *ketogenesis*. Eighteen out of 21 of the deaminated amino acids have chemical structures that allow them to be converted into glucose, and 19 can be converted into fats —5 directly and the other 14 by becoming carbohydrate first and then becoming fat.

OBLIGATORY DEGRADATION OF PROTEINS

When a person is on a starvation diet or eats no proteins, a certain proportion of his own body proteins continues to be degraded into amino acids, then deaminated and oxidized. This involves about 30 grams of protein each day; which is called the *obligatory loss* of proteins. Therefore, to prevent a net loss of protein from the body, one must ingest at least 30 grams of protein each day, and to be on the safe side as much as 75 grams is usually recommended.

The ratios of the different amino acids in the dietary protein must be about the same as the ratios in the body tissues for the entire protein to be usable. If one particular type of essential amino acid is low in concentration, the others might be unusable, because cells form either whole proteins or none at all, as explained in Chapter 3 in relation to protein synthesis. The unusable amino acids are then deaminated and oxidized. A protein that has a ratio of amino acids different from that of the average body protein is called a *partial protein,* and such a protein is obviously less valuable for nutrition than is the *complete protein.*

Effect of Starvation on Protein Degradation. Except for the excess protein in the diet or the 30 grams of obligatory protein degradation each day, the body uses almost entirely carbohydrates or fats for energy as long as these are available. However, after several weeks of starvation, when the quantity of stored fats begins to run out, the amino acids of the blood begin to be rapidly deaminated and oxidized for energy. From this point on, the proteins of the tissues degrade rapidly, and the cellular functions deteriorate precipitously.

Because carbohydrate and fat utilization for energy occurs in preference to protein utilization, carbohydrates and fats are called *protein sparers.*

HORMONAL REGULATION OF PROTEIN METABOLISM

Growth Hormone. Growth hormone increases the rate of synthesis of cellular proteins, causing the tissue proteins to increase while the plasma amino acid concentration falls. The precise mechanism by which growth hormone increases the rate of protein synthesis is not known, but growth hormone is believed to act in some direct manner either to enhance the transport of amino acids through the cell membranes or to accelerate the actual chemical processes of protein synthesis. Part of the action might also result from the effect of growth hormone on fat metabolism, for this hormone causes an increased rate of fat liberation from the fat depots, making this available for energy. This in turn reduces the rate of oxidation of amino acids and consequently makes increased quantities of amino acids available to the tissues to be synthesized into proteins.

Insulin. Lack of insulin reduces protein synthesis almost to zero. The mechanism by which this hormone affects protein metabolism is also unknown, but insulin does slightly accelerate amino acid transport into the cells, which could be a stimulus to protein synthesis. Also, insulin increases the availability of glucose to the cells so that the use of amino acids for energy becomes correspondingly reduced. This, too, undoubtedly makes far larger quantities of amino acids available to the tissues for protein synthesis.

Glucocorticoids. The glucocorticoids secreted by the adrenal cortex *decrease* the quantity of protein in most tissues while increasing the amino acid concentration in the plasma. However, contrary to elsewhere in the body, these hormones *increase* both the liver proteins and the plasma proteins. It is believed that the glucocorticoids act by increasing the rate of breakdown of extrahepatic proteins, thereby making increased quantities of amino acids available in the body fluids. This in turn supposedly allows the liver to synthesize increased quantities of liver and plasma proteins.

The effects of glucocorticoids on protein metabo-

lism are especially important in ketogenesis and gluconeogenesis, for in the absence of these hormones insufficient quantities of amino acids are usually available in the plasma to allow either significant gluconeogenesis or ketogenesis from proteins.

Testosterone. Testosterone, the male sex hormone, causes increased deposition of protein in the tissues throughout the body, including especially an increase in the contractile proteins of the muscles. The mechanism by which this effect comes about is unknown, but it is definitely different from the effect of growth hormone in the following way: Growth hormone causes tissues to continue growing almost indefinitely, while testosterone causes the muscles and other protein tissues to enlarge only for several months; beyond that time, despite continued administration of testosterone, further protein deposition ceases.

Estrogens, the principal female sex hormones, also cause slight deposition of protein, but their effect is relatively insignificant in comparison with that of testosterone.

Thyroxine. Thyroxine increases the rate of metabolism of all cells, and, as a result, indirectly affects protein metabolism. If insufficient carbohydrates and fats are available for energy, thyroxine causes rapid degradation of proteins to be used for energy. On the other hand, if adequate quantities of carbohydrates and fats are available and excesses of amino acids are also available in the extracellular fluid, thyroxine can actually increase the rate of protein synthesis. Conversely, in growing animals deficiency of thyroxine causes growth to be greatly inhibited because of lack of protein synthesis. In essence, it is believed that thyroxine has no specific effect on protein metabolism but does have an important general effect to increase the rates of both normal anabolic and normal catabolic protein reactions.

REFERENCES

Borasky, R.: Ultrastructure of Protein Fibers. New York, Academic Press, Inc., 1963.

Cohen, J. R.; Feldman, E.; and Whitbeck, A. A.: Effects of energy availability on protein synthesis in isolated rat atria. *Amer. J. Physiol., 216:*76, 1969.

Coles, B. L., and MacDonald, I.: The influence of dietary carbohydrate intake on serum protein levels. *J. Physiol. (Lond.), 165:* 327, 1963.

Coulson, R. A., and Hernandez, T.: Amino acid catabolism in the intact rat. *Amer. J. Physiol., 215:*741, 1968.

Davis, M. M., and Richmond, J. E.: Effect of dietary protein on serum proteins. *Amer. J. Physiol., 215:*366, 1968.

Elwyn, D. H.; Parikh, H. C.; and Schoemaker, W. C.: Amino acid movements between gut, liver, and periphery in unanesthetized dogs. *Amer. J. Physiol., 215:*1260, 1968.

Goldberg, A. L., and Goodman, H. M.: Amino acid transport during work-induced growth of skeletal muscle. *Amer. J. Physiol., 216:*1111, 1969.

Goldberg, A. L., and Goodman, H. M.: Effects of disuse and denervation on amino acid transport by skeletal muscle. *Amer. J. Physiol., 216:*1116, 1969.

Greenstein, J. P., and Winitz, M.: Chemistry of the Amino Acids. 3 volumes. New York, John Wiley & Sons, Inc., 1961.

Lacy, W. W.: Effect of acute uremia on amino acid uptake and urea production by perfused rat liver. *Amer. J. Physiol., 216*: 1300, 1969.

Munro, H., and Allison, J. B.: Mammalian Protein Metabolism. New York, Academic Press, Inc., 1967, Vols. 1 and 2.

Neurath, H. (ed.): The Proteins: Composition, Structure, and Function. 2nd ed., New York, Academic Press, Inc., 1963, Vol. 1.

Reeve, E. B., and McKinley, J. E.: Measurement of albumin synthetic rate with bicarbonate-^{14}C. *Amer. J. Physiol., 218*:498, 1970.

Rothschild, M. A.; Oratz, M.; Mongelli, J.; and Schreiber, S. S.: Effect of albumin concentration on albumin synthesis in the perfused liver. *Amer. J. Physiol., 216*:1127, 1969.

Salter, J. M.: Protein metabolism. *In* Duncan, G. G. (ed.): Diseases of Metabolism. 5th ed., Philadelphia, W. B. Saunders Company, 1964, p. 1.

Schedl, H. P.; Miller, D. L.; Wilson, H. D.; and Flores, P.: a-Aminoisobutyric acid transport and tissue concentration at various intestinal sites. *Amer. J. Physiol., 216*:1131, 1969.

Surks, M. I.: Effect of hypoxia on in vitro incorporation of leucine-^{14}C into rat liver protein. *Amer. J. Physiol., 218:*842, 1970.

CHAPTER 70

Energetics and Metabolic Rate

IMPORTANCE OF ADENOSINE TRIPHOSPHATE (ATP) IN METABOLISM

In the last few chapters it has been pointed out that carbohydrates, fats, and proteins can all be used by the cells to synthesize large quantities of ATP, and that in turn the ATP can be used as an energy source for many other cellular functions. For these reasons, ATP has been called an energy "currency" that can be created and expended. Indeed, the cells can transfer energy from the different foodstuffs to most functional systems of the cells only through this medium of ATP. Many of the attributes of ATP were presented in Chapter 2, but others require discussion at this point.

An attribute of ATP that makes it highly valuable as a means of energy currency is the large quantity of free energy (about 8000 calories per mol under physiological conditions) vested in each of its two high energy phosphate bonds. The amount of energy in each bond, when liberated by decomposition of one molecule of ATP, can cause almost any stage of any chemical reaction in the body to take place. If the bonds of ATP did not have this much energy, many of the chemical reactions could not be energized by this compound. Some chemical reactions that require ATP energy use only a few hundred of the available 8000 calories, and the remainder of this energy is then lost in the form of heat. Yet even this inefficiency in the utilization of energy is better than not being able to energize the necessary chemical reactions at all.

The precise methods by which ATP is used to cause the many physical, chemical, and other types of functions in the cells lie principally in the province of biochemistry. Therefore, only the more important functions of ATP are listed here.

Use of ATP for Synthesis of Important Cellular Components. Probably by far the most important intracellular process that requires ATP is formation of peptide linkages between amino acids during the synthesis of proteins. The energy from one high energy bond of ATP is used for each peptide linkage. Since an occasional protein of the body has as many as 100,000 amino acids linked together in this manner, it can readily be understood how much ATP is required for this cellular function. The different peptide linkages, depending on which types of amino acids are linked together, require from 500 to 4000 calories of energy per mol. In each instance, the amount of energy available from the ATP high energy bond, 8000 calories, is always more than sufficient.

Also, it will be recalled from the preceding chapters that ATP is utilized in the synthesis of glucose from lactic acid and in the synthesis of fatty acids from acetyl Co-A. In addition, ATP is utilized in the synthesis of cholesterol, phospholipids, the hormones, and almost all other substances of the body. Even the urea excreted by the kidneys requires ATP to cause its formation. One might wonder at the advisability of expending energy to form urea which then is simply thrown away from the body. However, if he remembers the extreme toxicity of ammonia in the body fluids, he can see the value of this reaction, which keeps the ammonia concentration of the body fluids always at a low level.

Use of ATP for Muscular Contraction. Muscular contraction will not occur without energy from ATP. Myosin, one of the important contractile proteins of the muscle fiber, acts as an enzyme to cause breakdown of ATP into adenosine diphosphate (ADP), thus causing the release of energy. However, the means by which this energy is coupled to the contractile process of the muscle fiber is still conjecture. Only a small amount of ATP is normally degraded in muscles when muscular contraction is not occurring, but this rate of ATP usage can rise to more than 100 times the resting level during maximal contraction. Postulated mechanisms by which ATP is utilized to cause muscle contraction were discussed in Chapter 7.

Use of ATP for Active Transport Across Membranes. In Chapters 4, 34, and 65, active absorption of electrolytes and various nutrients across cell membranes and from the renal tubules and gastrointestinal tract was discussed. In each instance, it was noted that active transport of most electrolytes and other substances, such as glucose, amino acids, and acetoacetate, can occur against an electrochemical gradient, even though the natural diffusion of the substances would be in the opposite direction. Obviously, to oppose the electrochemical gradient requires energy, as was discussed in Chapter 4. This energy is provided by ATP. However, the precise means by which the energy is transferred from ATP to the active transport systems are almost completely unknown.

ENERGY FOR GLANDULAR SECRETION. The same principles apply to glandular secretion as to the absorption of substances against concentration gradients, for many substances to be secreted are concentrated as they pass through the glandular cell. This concentrating process requires appropriate amounts of energy. In addition, energy is also required to synthesize the organic compounds to be secreted.

ENERGY FOR NERVE CONDUCTION. The energy utilized during propagation of a nerve impulse is derived from the concentration differences of ions across the membranes. That is, a high concentration of potassium inside the fiber and a low concentration outside the fiber constitutes a type of energy storage. Likewise, a high concentration of sodium on the outside of the membrane and a low concentration on the inside represents another store of energy. The energy needed to pass each impulse along the fiber membrane is derived from this energy storage, with potassium transferring out of the cell and sodium into the cell. However, active transport systems then retransport the ions back through the membrane to their former positions. Here, ATP is utilized in abundance to retransfer the sodium and potassium ions after nerve impulses have been conducted, and the rate of ATP usage increases in proportion to the number of nerve impulses transmitted.

CREATINE PHOSPHATE AS A STORAGE DEPOT FOR ENERGY

Despite the paramount importance of ATP as a coupling agent for energy transfer, this substance is not the most abundant store of high energy phosphate bonds in the cells. On the contrary, *creatine phosphate*, which also contains high energy phosphate bonds, is many times more abundant. The high energy bond of creatine phosphate contains about 8500 calories per mol under standard conditions, or 9500 calories per mol under conditions in the body (38° C. and low concentrations of the reactants). This is not greatly different from the 8000 calories per mol in each of the two high energy phosphate bonds of ATP. The formula for creatine phosphate is the following:

$$HOOC-CH_2-\underset{\underset{H}{|}}{\overset{\overset{CH_3}{|}}{N}}-\underset{}{\overset{\overset{NH}{||}}{C}}-\underset{}{\overset{\overset{H}{|}}{N}} \sim \underset{\underset{H}{||}{O}}{\overset{\overset{O}{||}}{P}}-OH$$

Creatine phosphate cannot act in the same manner as ATP as a coupling agent for transfer of energy between the foods and the functional cellular systems. But it can transfer energy interchangeably with ATP. When extra amounts of ATP are available in the cell, much of its energy is utilized to synthesize creatine phosphate, thus building up this storehouse of energy. Then when the ATP begins to be used up, the energy in the creatine phosphate is transferred rapidly back to ATP and from this to the functional systems of the cells. This reversible interrelationship between ATP and creatine phosphate is illustrated by the following equation:

$$\text{Creatine phosphate} + \text{ADP}$$
$$\Updownarrow$$
$$\text{ATP} + \text{Creatine}$$

Note particularly that the higher energy level of the high energy phosphate bond in creatine phosphate, 9500 in comparison with 8000 calories per mol, causes the reaction between creatine phosphate and ATP to proceed to an equilibrium state very much in favor of ATP. Therefore, the slightest utilization of ATP by the cells calls forth the energy from the creatine phosphate to synthesize new ATP. This effect keeps the concentration of ATP at an almost constant level for the following reason: The greater energy in the creatine phosphate bond than in the ATP bond allows such rapid transfer of energy to the ATP system that almost all the creatine phosphate must be used up in the cell before the concentration of ATP will fall significantly. For this reason we can actually call creatine phosphate an "ATP sparer."

ANAEROBIC VERSUS AEROBIC ENERGY

Anaerobic energy means energy that can be derived from the foods without the simultaneous utilization of oxygen; *aerobic energy* means energy that can be derived from the foods only by oxidative metabolism. In the discussions in the preceding three chapters it was noted that carbohydrates, fats, and proteins can all be oxidized to cause synthesis of ATP. However, carbohydrates are the only significant foods that can be utilized to provide energy without utilization of oxygen; this energy release occurs during glycolytic breakdown of glucose or glycogen to pyruvic acid. For each mol of glucose that is split into pyruvic acid, 2 mols of ATP are formed. However, when glycogen is split to pyruvic acid, each mol of glucose in the glycogen gives rise to 3 mols of ATP. The reason for this difference is that free glucose entering the cell must be phosphorylated by 1 mol of ATP before it can begin to be split, whereas this is not true of glucose derived from glycogen because it comes from the glycogen in the phosphorylated state. Thus, the best source of energy under anaerobic conditions is the stored glycogen of the cells.

Anaerobic Energy During Hypoxia.   One of the prime examples of anaerobic energy utilization occurs in acute hypoxia. When a person stops breathing, he already has a small amount of oxygen stored in his lungs and an additional amount stored in the hemoglobin of his blood. However, these are sufficient to keep the metabolic processes functioning for only about two minutes. Continued life beyond this time requires an additional source of energy. This can be derived for another minute or so from glycolysis, the glycogen of the cells splitting into pyruvic acid and the pyruvic acid in turn becoming lactic acid, which diffuses out of the cells as described in Chapter 67.

Anaerobic Energy Usage in Strenuous Bursts of Activity.   It is common knowledge that muscles can perform extreme feats of strength for a few seconds but are much less capable during prolonged activity. The energy used during strenuous activity is derived from: (1) ATP already present in the muscle cells, (2) stored creatine phosphate in the cells, (3) anaerobic energy released by glycolytic breakdown of glycogen to lactic acid, and (4) oxidative energy released continuously by oxidative processes in the cells. The oxidative processes cannot operate anywhere near rapidly enough to supply all the energy demands during strenuous bursts of activity. Therefore, the other three sources of energy are called upon to their maximum extent.

The amount of ATP in cells is only about 5 millimols per liter of intracellular fluid, and this amount can maintain maximum muscle contraction for not more than a few seconds. The amount of creatine phosphate in the cells may be as much as 10 times this amount, but even by utilization of all the creatine phosphate, the amount of time that maximum contraction can be maintained is still only 20 to 30 seconds. Release of energy by glycolysis can

occur much more rapidly than can oxidative release of energy. Consequently, most of the extra energy required during strenuous activity that lasts for more than 30 seconds but less than two minutes is probably derived from anaerobic glycolysis. As a result, the glycogen content of muscles after strenuous bouts of exercise becomes almost zero, while the lactic acid concentration of the blood rises. Then, immediately after the exercise is over, oxidative metabolism is used to reconvert about four-fifths of the lactic acid into glucose, while the remainder becomes pyruvic acid and is degraded and oxidized in the citric acid cycle. The reconversion to glucose occurs principally in the liver cells, and the glucose is then transported in the blood back to the muscles where it is stored once more in the form of glycogen.

Oxygen Debt.   After a period of strenuous exercise, the oxidative metabolic processes continue to operate at a high level of activity for many minutes to (1) reconvert the lactic acid into glucose and (2) reconvert the decomposed ATP and creatine phosphate to their original states. The extra oxygen that must be used in the oxidative energy processes to rebuild these substances is called the *oxygen debt*.

The oxygen debt is illustrated in Figure 70–1 by the shaded area. This figure shows, first, the normal rate of energy expenditure by the body, and, second, the normal rate of oxidative metabolism. At the three-minute mark, the person exercises strongly for about four minutes. During this time the oxidative release of energy provides some energy expenditure, but a large share of energy expenditure occurs principally at the expense of glycolysis. At the end of exercise, the oxidative release of energy continues to be high while the rate of energy expenditure returns to normal. The excess oxidative metabolism that must occur even after the period of exercise

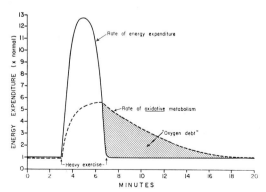

Figure 70–1   Oxygen debt occurring after a bout of strenuous exercise.

is over, as shown by the shaded area in the figure, represents the oxygen debt. This is usually expressed in terms of the oxygen that must be utilized to provide this excess metabolism.

SUMMARY OF ENERGY UTILIZATION BY THE CELLS

With the background of the past few chapters and of the preceding discussion, we can now synthesize a composite picture of overall energy utilization by the cells as illustrated in Figure 70–2. This figure shows the anaerobic utilization of glycogen and glucose to form ATP and also the aerobic utilization of compounds derived from carbohydrates, fats, proteins, and other substances for the formation of still additional ATP. In turn, ATP is in reversible equilibrium with creatine phosphate in the cells, and, since large quantities of creatine phosphate are present in the cell, much of the stored energy of the cell is in this energy storehouse.

In turn, energy from ATP can be utilized by the different functioning systems of the cells to provide for synthesis and growth, muscular contraction, glandular secretion, impulse conduction, active absorption, and other cellular activities. If greater amounts of energy are called forth for cellular activities than can be provided by oxidative metabolism, the creatine phosphate storehouse is first utilized, and this is followed rapidly by anaerobic breakdown of glycogen. Thus, oxidative metabolism cannot deliver energy to the cells nearly so rapidly as can the anaerobic processes, but in contrast it is quantitatively almost inexhaustible.

CONTROL OF ENERGY RELEASE IN THE CELL

Rate Control of Enzyme-Catalyzed Reactions. Before it is possible to discuss the control of energy release in the cell, it is necessary to consider the basic principles of *rate control* of enzymatically catalyzed chemical reactions, which are the type of reactions that occur almost universally throughout the body.

The mechanism by which an enzyme catalyzes a chemical reaction is, first to combine loosely with one of the substrates of the reaction. This alters the electrical forces on the substrate sufficiently that it can then react with other substances. Therefore, the rate of the overall chemical reaction is determined by both the concentration of the enzyme and the concentration of the substrate that binds with the enzyme. The basic equation expressing this concept is the following:

$$\text{Rate of reaction} = \frac{K_1 \cdot \text{Enzyme} \cdot \text{Substrate}}{K_2 + \text{Substrate}}$$

This is called the *Michaelis-Menten equation.* Figure 70–3 illustrates graphically the application of this equation, showing how both substrate concentration and enzyme concentration affect the rate of reaction.

ROLE OF ENZYME CONCENTRATION IN THE REGULATION OF METABOLIC REACTIONS. Figure 70–3 illustrates that when the substrate is present in excess, the rate of a chemical reaction is determined almost entirely by the concentration of the enzyme. Thus, as shown to the right in the figure, as the enzyme con-

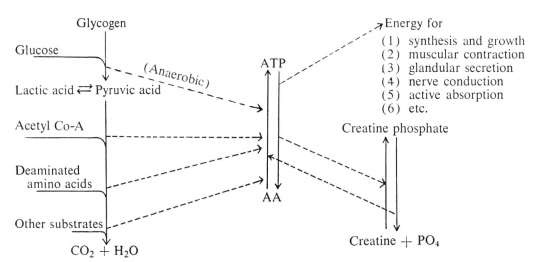

Figure 70–2 Overall schema of energy transfer from foods to the adenylic acid system and then to the functional elements of the cells. (Modified from Soskin and Levine: Carbohydrate Metabolism. University of Chicago Press.)

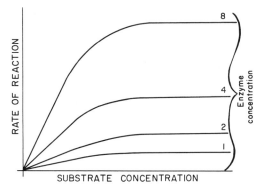

Figure 70–3 Effect of substrate and enzyme concentrations on the rate of enzyme-catalyzed reaction.

centration increases from an arbitrary value of 1 to 2, then to 4 or 8, the rate of the reaction increases proportionately. As an example of this effect, when large quantities of glucose enter the renal tubules in diabetes mellitus, the rate of reabsorption of the glucose is determined almost entirely by the concentration of the glucose transport enzymes in the proximal tubular cells because the substrate is then present in excess.

ROLE OF SUBSTRATE CONCENTRATION IN REGULATION OF METABOLIC REACTIONS. Note also from Figure 70–3 that when the substrate concentration becomes low—low enough that only small portions of the enzymes are required in the reaction—the rate of the reaction is directly proportional to the substrate concentration as well as enzyme concentration. This is the effect seen in the absorption of substances from the intestinal tract and renal tubules when their concentrations are very low. That is, the rate of absorption is then directly dependent on the concentration of the substance as well as on the concentration of the enzyme.

Rate Limitation in a Series of Reactions. It has become evident from the discussions in the preceding few chapters that almost all chemical reactions of the body occur in series, the products of one reaction acting as the substrates of the next reaction, and so on. It is immediately obvious also that the overall rate of a complex series of chemical reactions is determined by the rate of reaction of the slowest step in the series. This is called the *rate limiting reaction* for the entire series.

Adenosine Diphosphate (ADP) Concentration as the Rate Controlling Factor in Energy Release. Under normal resting conditions, the concentration of ADP in the cells is extremely slight so that the chemical reactions that depend on ADP as one of the substrates likewise are very slow. These include all the oxidative metabolic pathways as well as essentially all other pathways for release of energy in the body. Thus, *ADP is the major rate limiting factor* for almost all energy metabolism of the body.

When the cells become excessively active, regardless of the type of activity, large quantities of ATP are converted into ADP, increasing the concentration of ADP in direct proportion to the degree of activity of the cell. This automatically increases the rates of reactions of the ADP-rate-limiting steps in the metabolic release of energy. Thus, by this simple process, the amount of energy released in the cell is controlled by the degree of activity of the cell. In the absence of cellular activity, the release of energy stops, while in the presence of cellular activity the concentration of ADP increases, and all the energy-giving processes are automatically set in motion.

THE METABOLIC RATE

The *metabolism* of the body means simply all the chemical reactions in all the cells of the body, and the *metabolic rate* is normally expressed in terms of the rate of heat liberation during the chemical reactions.

Heat as the Common Denominator of All the Energy Released in the Body. In discussing many of the metabolic reactions of the preceding chapters, we have noted that not all the energy in the foods is transferred to ATP; instead, a large portion of this energy becomes heat. On the average, about 55 per cent of the energy in the foods becomes heat during ATP formation. Then still more energy becomes heat as it is transferred from ATP to the functional systems of the cells, so that not more than about 25 per cent of all the energy from the food is finally utilized by the functional systems.

Even though 25 per cent of the energy finally reaches the functional systems of the cells, the major proportion of this also becomes heat for the following reasons: We might first consider the synthesis of protein and other growing elements of the body. When proteins are synthesized, large portions of ATP are used to form the peptide linkages, and this stores energy in these linkages. But we also noted in our discussions of proteins in Chapter 69 that there is continuous turnover of proteins, some being degraded while others are being formed. When the proteins are degraded, the energy stored in the peptide linkages is released in the form of heat into the body.

Now let us consider the energy used for muscle activity. Much of this energy simply overcomes the viscosity of the muscles themselves or of the tissues so that the limbs can move. The viscous movement in turn causes friction within the tissues, which generates heat.

We might also consider the energy expended by the heart in pumping blood. This blood distends the arterial system, the distention in itself representing a reservoir of potential energy. However, as the blood flows through the peripheral vessels, the friction of the different layers of blood flowing over each other and the friction of the blood against the walls of the vessels turns this energy into heat.

Therefore we can say that essentially all the energy expended by the body is converted into heat. The only real exception to this occurs when the muscles are used to perform some form of work outside the body. For instance, when the muscles elevate an object to a height or carry the person's body up steps, a type of potential energy is thus created by raising a mass against gravity. Also, if muscular energy is used to turn a flywheel so that kinetic energy is developed in the flywheel, this, too, would be external expenditure of energy. But, when external expenditure of energy is not taking place, it is safe to consider that all the energy released by the metabolic processes eventually becomes body heat.

The Calorie. To discuss the metabolic rate and related subjects intelligently, it is necessary to use some unit for expressing the quantity of energy released from the different foods or expended by the different functional processes of the body. In general, the *Calorie* is the unit used for this purpose. It will be recalled that 1 *calorie,* spelled with a small "c," is the quantity of heat required to raise the temperature of 1 gram of water 1° C. The calorie is much too small a unit for ease of expression in speaking of energy in the body. Consequently the large Calorie, spelled with a capital "C," which is equivalent to 1000 calories, is the unit ordinarily used in discussing energy metabolism.

MEASUREMENT OF THE METABOLIC RATE

Direct Calorimetry. As pointed out above, it is only when the body performs some external work that energy expended within the body does not become heat. Since a person is ordinarily not performing any external work, his metabolic rate can be determined by simply measuring the total quantity of heat liberated from the body in a given time. This method is called *direct calorimetry.*

In determining the metabolic rate by direct calorimetry, one measures the quantity of heat liberated from the body in a large, specially constructed *calorimeter* as follows: The subject is placed in an air chamber so well insulated that no heat can leak through the walls of the chamber. As heat is formed by his body, it warms the air of the chamber. However, the air temperature within the chamber is maintained at a constant value by forcing the air through pipes in a cool water bath. The rate of heat gain by the water bath, therefore, is equal to the rate at which heat is liberated by the subject's body.

Obviously, direct calorimetry is physically difficult to perform and, therefore, is used only for research purposes.

Indirect Calorimetry. Since more than 95 per cent of the energy expended in the body is derived from reaction of oxygen with the different foods, the metabolic rate can also be calculated with a high degree of accuracy from the rate of oxygen utilization. When 1 liter of oxygen is metabolized with glucose, 5.01 Calories of energy are released; when metabolized with starches, 5.06 Calories are released; with fat, 4.70 Calories; and with protein, 4.60 Calories.

From these figures it is striking how nearly equivalent are the quantities of energy liberated per liter of oxygen regardless of the type of food that is being burned. For the average diet, the *quantity of energy liberated per liter of oxygen utilized in the body averages approximately 4.825 Calories.* Using this *energy equivalent* of oxygen, one can calculate approximately the rate of heat liberation in the body from the quantity of oxygen utilized in a given period of time.

If a person should metabolize only carbohydrates during the period of the metabolic rate determination, the calculated quantity of energy liberated based on the value for the average energy equivalent of oxygen (4.825 Calories per liter), would be approximately 4 per cent too little. On the other hand, if the person were obtaining most of his energy from protein, the calculated value would be approximately 4 per cent too great, and, if he were burning almost entirely fat during the test, the error would be insignificant.

THE METABOLATOR. Figure 70–4 illustrates the metabolator usually used for indirect calorimetry. This apparatus contains a floating drum, under which is an oxygen chamber connected to a mouthpiece through two rubber tubes. A valve in one of these rubber tubes allows air to pass from the oxygen chamber into the mouth, while air passing from the mouth back to the chamber is directed by means of another valve through the second tube. Before the expired air from the mouth enters the upper portion of the oxygen chamber, it flows through a lower chamber containing pellets of soda lime, which combine chemically with the carbon dioxide in the expired air. Therefore, as oxygen is used by the person's body and the carbon

Figure 70-4 The metabolator.

dioxide is absorbed by the soda lime, the floating oxygen chamber, which is precisely balanced by a weight, gradually sinks in the water, owing to the oxygen loss. This chamber is approximately coupled to a pen that records on a moving drum the rate at which the chamber sinks in the water and thereby records the rate at which the body utilizes oxygen.

FACTORS THAT AFFECT THE METABOLIC RATE

Any factor that increases the chemical activity in the cells increases the metabolic rate. Some of these are the following:

Exercise. The factor that causes by far the most dramatic effect on metabolic rate is strenuous exercise. Short bursts of maximal muscle contraction in any single muscle liberates as much as a hundred times its normal resting amount of heat. In considering the entire body, however, maximal muscle exercise can increase the overall heat production of the body for a few seconds at a time to about 50 times normal or for several minutes to about 20 times normal, which is an increase in metabolic rate to 2000 per cent of normal.

ENERGY REQUIREMENTS FOR DAILY ACTIVITIES. When an average man of 70 kilograms lies in bed all day, he utilizes approximately 1650 Calories of energy. The process of eating increases the amount of energy utilized each day by an additional 200 or more Calories so that the same man lying in bed and also eating a reasonable diet requires a dietary intake of approximately 1850 Calories per day. If he sits in a chair all day, his total energy requirement reaches 2000 to 2250 Calories. Therefore, in round figures, it can be assumed that the daily energy requirements simply for existing (that is, performing essential functions only) is about 2000 Calories.

EFFECTS OF DIFFERENT TYPES OF WORK ON DAILY ENERGY REQUIREMENTS. Table 70-1 illustrates the rates of energy utilization while one performs different types of activities. Note that walking up stairs requires approximately 17 times as much energy as lying in bed asleep. And, in general, a laborer can, over 24 hours, average a rate of energy utilization as great as 6000 to 7000 Calories, or, in other words, as much as 3½ times the basal rate of metabolism.

Specific Dynamic Action of Food. After a meal is ingested, the metabolic rate increases. This is believed to result partly from the different chemical reactions associated with digestion, absorption, and storage of food in the body. However, it is also known that certain of the amino acids derived from the proteins of the ingested food directly stimulate cellular chemical processes.

After a meal containing a large quantity of carbohydrates, the metabolic rate usually increases about 4 per cent, but it has been observed to increase to as high as 30 per cent above normal, this increase lasting two to five hours. After a meal containing mainly fat, the metabolic rate also normally rises about 4 per cent, but, here again, it has been observed to rise to as high as 10 to 15 per cent above normal, the effect sometimes lasting seven to nine hours. After a meal containing large quantities of proteins, the metabolic rate usually begins rising within one to two hours, reaches a maximum usually about 30 per cent above normal but sometimes as high as 50 to 70 per cent above normal, and lasts for as long as 10 to 12 hours. This effect of the different foods on the metabolic rate is called the *specific dynamic action* of food.

The much greater specific dynamic action of proteins than of carbohydrates and fats is probably caused by two factors: First, a greater number of energy-releasing chemical reactions occur during protein digestion, absorption, and metabolism than during digestion and assimilation of the other two types of food. Second, certain amino acids seem to have a direct stimulatory effect on all chemical reactions of the cell—much the same effect as epinephrine and norepinephrine on the metabolic rate, which is discussed below.

Age. The metabolic rate of the newborn child in relation to his body surface area is almost two times that of an aged person. This is illustrated in

TABLE 70-1. Energy Expenditure per Hour During Different Types of Activity for a 70 Kilogram Man

Form of Activity	Calories per hour
Sleeping	65
Awake lying still	77
Sitting at rest	100
Standing relaxed	105
Dressing and undressing	118
Tailoring	135
Typewriting rapidly	140
"Light" exercise	170
Walking slowly (2.6 miles per hour)	200
Carpentry, metal working, industrial painting	240
"Active" exercise	290
"Severe" exercise	450
Sawing wood	480
Swimming	500
Running (5.3 miles per hour)	570
"Very severe" exercise	600
Walking very fast (5.3 miles per hour)	650
Walking up stairs	1100

Extracted from data compiled by Professor M. S. Rose.

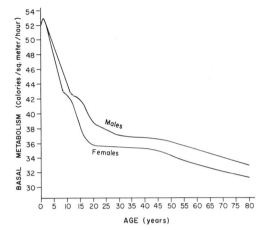

Figure 70–5 Normal basal metabolic rates at different ages for each sex.

Figure 70–5, which shows the metabolic rates of both males and females from birth until very old age. The high metabolic rate of young children results from high rates of cellular reactions, but also partly from rapid synthesis of cellular materials and growth of the body, which require moderate quantities of energy.

Thyroid Hormone. When the thyroid gland secretes maximal quantities of thyroxine, the metabolic rate sometimes rises to as much as 100 per cent above normal. On the other hand, total loss of thyroid secretion decreases the metabolic rate to as low as 50 per cent of normal. These effects can readily be explained by the basic function of thyroxine to increase the rates of activity of almost all the chemical reactions in all cells of the body. This relationship between thyroxine and metabolic rate will be discussed in much greater detail in Chapter 77 in relation to thyroid function, because one of the most useful methods for diagnosing abnormal rates of thyroid secretion is to determine the basal metabolic rate of the patient. A normal person usually has a basal metabolic rate within 10 to 15 per cent of normal, while the hyperthyroid person often has a basal metabolic rate as high as 40 to 80 per cent above normal, and a hypothyroid person can have a basal metabolic rate as low as 40 to 50 per cent below normal.

Sympathetic Stimulation. Stimulation of the sympathetic nervous system with liberation of norepinephrine and epinephrine increases the metabolic rates of essentially all the tissues of the body. These hormones have a direct effect on cells to cause glycogenolysis, and this, probably along with other intracellular effects of these hormones, increases cellular activity.

Maximal stimulation of the sympathetic nervous system can increase the metabolic rate in the rat as much as 80 per cent, but the amount of this effect in human beings is doubtful—probably 25 per cent or less.

Male Sex Hormone. The male sex hormone can increase the basal metabolic rate about 10 to 15 per cent, and the female sex hormone perhaps a few per cent but usually not enough to be of great significance. The difference in metabolic rates of males and females is illustrated in Figure 70–5.

Growth Hormone. Growth hormone can increase the basal metabolic rate as much as 15 to 20 per cent as a result of direct stimulation of cellular metabolism.

Fever. Fever, regardless of its cause, increases the metabolic rate. This is because all chemical reactions, either in the body or in the test tube, increase their rates of reaction approximately 130 per cent for every 10° C. rise in temperature. An increase in body temperature to 110° F. increases the metabolic rate almost 100 per cent.

Climate. Studies of metabolic rates of persons living in the different geographic zones have shown as much as 10 to 20 per cent lower metabolic rates in tropical regions than in arctic regions. This difference is caused to a great extent by adaptation of the thyroid gland, with increased secretion in cold climates and decreased secretion in hot climates. Indeed, far more persons develop hyperthyroidism in cold regions of the earth than in tropical regions.

Sleep. The metabolic rate falls approximately 10 to 15 per cent below normal during sleep. This fall is presumably due to two principal factors: (1) decreased tone of the skeletal musculature during sleep and (2) decreased activity of the sympathetic nervous system.

Malnutrition. Prolonged malnutrition often decreases the metabolic rate as much as 20 to 30 per cent; this decrease is presumably caused by the paucity of necessary food substances in the cells.

In the final stages of many disease conditions, the inanition that accompanies the disease frequently causes marked premortem decrease in metabolic rate, even to the extent that the body temperature may fall a number of degrees shortly before death.

THE BASAL METABOLIC RATE

The Basal Metabolic Rate as a Method for Comparing Metabolic Rates Between Individuals. It has been extremely important to establish a procedure that will measure the inherent activity of the tissues independently of exercise and other extraneous factors that would make it impossible to compare one person's metabolic rate with that of another person. To do this, the metabolic rate is usually measured under so-called *basal conditions,* and the metabolic rate then measured is called the *basal metabolic rate.*

Basal Conditions. The basal metabolic rate means the rate of energy utilization in the body during absolute rest but while the person is awake. The following basal conditions are necessary for measuring the basal metabolic rate:

1. The person must not have eaten any food for at least 12 hours because of the specific dynamic action of foods.

2. The basal metabolic rate is determined after a night of restful sleep, for rest reduces the activity of the sympathetic nervous system and of other metabolic excitants to their minimal level.

3. No strenuous exercise is performed after the night of restful sleep, and the person must remain at complete rest in a reclining position for at least 30 minutes prior to actual determination of the metabolic rate. This is perhaps the most important of all the conditions for attaining the basal state because of the extreme effect of exercise on metabolism.

4. All psychic and physical factors that cause excitement must be eliminated, and the subject must be made as comfortable as possible. These conditions, obviously, help to reduce the degree of sympathetic activity to as little as possible.

5. The temperature of the air must be comfortable and be somewhere between the limits of 65° and 80° F. Below 65° F., the sympathetic nervous system becomes progressively more activated to help maintain body heat, and above 80° F., discomfort, sweating, and other factors increase the metabolic rate.

Usual Technique for Determining the Basal Metabolic Rate. The usual method for determining basal metabolic rate is first to establish the subject under basal conditions and then to measure his rate of oxygen utilization using a metabolator of the type illustrated in Figure 70–4. Then the basal metabolic rate is calculated as shown in Figure 70–6.

In the upper portion of Figure 70–6, the quantity of heat liberated in the body of a person is calculated from the quantity of oxygen utilized. In this figure, note that 15 liters of oxygen (corrected to standard conditions) are consumed in one hour. Multiplying this times the energy equivalent for 1 liter of oxygen (4.825 Calories), the total quantity of energy liberated within the body during the hour is 72.4 Calories.

15 liters	— O_2 at standard conditions consumed in 1 hr.
$\times 4.825$	— Calories liberated per liter of O_2 burned
72.4	— Calories liberated per hour
$\div 1.5$	— Body surface area in square meters
48.3	— Calories per square meter per hour
-38.5	— Normal value for 20-year-old man
9.8	— Excess Calories above normal

$$\frac{9.8 \times 100}{38.5} = 25.5 \text{ per cent above normal}$$

$$BMR = +25.5$$

Figure 70–6 Calculation of basal metabolic rate from the rate of oxygen consumption.

Figure 70–7 Relationship of height and weight to body surface area. (From DuBois: Metabolism in Health and Disease. Lea & Febiger.)

Expressing the Basal Metabolic Rate in Terms of Surface Area. Obviously, if one subject is much larger than another, the total amount of energy utilized by the two subjects will be considerably different simply because of differences in body size. Experimentally, among normal persons the average basal metabolic rate varies approximately *in proportion to the body surface area.*

Note from Figure 70–6 that the total number of Calories liberated by the person per hour is divided by his total body surface area, 1.5 square meters (the value for the average adult is 1.73 square meters). This means that his basal metabolic rate is 48.3 Calories per square meter per hour.

METHOD FOR CALCULATING THE TOTAL SURFACE AREA. The surface area of the body varies approximately in proportion to weight$^{0.67}$. However, measurements of the body surface area have shown that it can be calculated more accurately by a complicated formula based on weight and height of the subject as follows:

Body surface area =
$$\text{Weight}^{0.425} \times \text{Height}^{0.725} \times 0.007184$$

Figure 70–7 presents a graph based on this formula. In the formula and in the figure, body surface area is expressed in *square meters,* weight in *kilograms,* and height in *centimeters.*

Expressing the Basal Metabolic Rate in Percentage Above or Below Normal. To compare the basal metabolic rate of any one subject with the normal basal metabolic rate, it is necessary to refer to a chart such as that in Figure 70–5, which gives the normal basal metabolic rate per square meter at each age for each sex. For example, in Figure 70–5 the normal metabolic rate for a 20-year-old male is shown to be 38.5 Calories per square meter per hour. Therefore, if the person represented in the calculations of Figure 70–6 is a 20-year-old male, he liberates 9.8 Calories per square meter per hour above the normal mean value. It is then determined that this is 25.5 per cent above normal. Therefore, the basal metabolic rate is expressed as *plus 25.5.* Similarly, basal metabolic rates below normal are expressed as minus values.

Constancy of the Basal Metabolic Rate in the Same Person. Basal metabolic rates have been measured in many subjects at repeated intervals for as long as 20 or more years. As long as a subject remains healthy, almost invariably his basal metabolic rate as expressed in percentage of normal does not vary more than 5 to 10 per cent.

Constancy of the Basal Metabolic Rate from Person to Person.   When the basal metabolic rate is measured in a wide variety of different persons and comparisons are made within single age, weight, and sex groups, 85 per cent of normal persons have been found to have basal metabolic rates within 10 per cent of the mean. Thus, it is obvious that measurements of metabolic rates performed under basal conditions offer an excellent means for comparing the rates of metabolism from one person to another.

REFERENCES

Barnard, R. J., and Foss, M. L.: Oxygen debt: effect of beta-adrenergic blockade on the lactacid and alactacid components. *J. Appl. Physiol., 27*:813, 1969.

Barnard, R. J., and Peter, J. B.: Effect of training and exhaustion on hexokinase activity of skeletal muscle. *J. Appl. Physiol., 27*:691, 1969.

Chance, B.: Enzymes in action in living cells: the steady state of reduced pyridine nucleotides. *Harvey Lectures, 49*:145, 1953–1954.

Consolazio, C. F.; Johnson, R.; and Pecora, L.: Physiological Measurements of Metabolic Functions in Man. New York, McGraw-Hill Book Company, 1963.

Costill, D. L.: Metabolic responses during distance running. *J. Appl. Physiol., 28*:251, 1970.

Fox, E. L.; Robinson, S.; and Wiegman, D. L.: Metabolic energy sources during continuous and interval running. *J. Appl. Physiol., 27*:174, 1969.

Gold, A. J.; Zornitzer, A.; and Samueloff, S.: Influence of season and heat on energy expenditure during rest and exercise. *J. Appl. Physiol., 27*:9, 1969.

Guyton, A. C., and Farrish, C. A.: A rapidly responding continuous oxygen consumption recorder. *J. Appl. Physiol., 14*:143, 1959.

Herrmann, H., and Tootle, M. L.: Specific and general aspects of the development of enzymes and metabolic pathways. *Physiol. Rev., 44*:289, 1964.

Hunt, L. M.: An analytic formula to instantaneously determine total metabolic rate for the human system. *J. Appl. Physiol., 27*:731, 1969.

Kleiber, M.: Respiratory exchange and metabolic rate. *In* Handbook of Physiology. Baltimore, The Williams & Wilkins Co., 1965, Sec. III, Vol. II, p. 927.

Lister, D.; Sair, R. A.; Will, J. A.; Schmidt, G. R.; Cassens, R. G.; Hoekstra, W. G.; and Briskey, E. J.: Metabolism of striated muscle of stress-susceptible pigs breathing oxygen or nitrogen. *Amer. J. Physiol., 218*:102, 1970.

Lundholm, L.; Mohme-Lundholm, E.; and Svedmyr, N.: Metabolic effects of catecholamines. *In* Bittar, E. E., and Bittar, N. (eds.): The Biological Basis of Medicine. New York, Academic Press, Inc., 1968, Vol. 2, p. 101.

Myant, N. B.: Some aspects of the control of cell metabolism. *In* Bittar, E. E., and Bittar, N. (eds.): The Biological Basis of Medicine. New York, Academic Press, Inc., 1968, Vol. 2, p. 133.

Newman, R. W.: Cold acclimation in Negro Americans. *J. Appl. Physiol., 27*:316, 1969.

Rochelle, R. H., and Horvath, S. M.: Metabolic responses to food and acute cold stress. *J. Appl. Physiol., 27*:710, 1969.

Schneider, E. G.; Robinson, S.; and Newton, J. L.: Oxygen debt in aerobic work. *J. Appl. Physiol., 25*:58, 1968.

Sodeman, W. A.: Nutritional factors: protein and fat metabolism. *In* Sodeman, W. A., and Sodeman, W. A., Jr.(eds.): Pathologic Physiology. 4th ed., Philadelphia, W. B. Saunders Company, 1967, p. 67.

Soule, R. G., and Goldman, B. F.: Energy cost of loads carried on the head, hands, or feet. *J. Appl. Physiol., 27*:687, 1969.

Wyndham, C. H., and Loots, H.: Responses to cold during a year in Antarctica. *J. Appl. Physiol., 27*:696, 1969.

CHAPTER 71

Body Temperature, Temperature Regulation, and Fever

The temperature of the inside of the body remains almost exactly constant, within ± 1° F., day in and day out except when a person develops a febrile illness. Indeed, the nude person can be exposed to temperatures as low as 55° to 60° F. or as high as 150° F. and still maintain an almost constant internal body temperature. Therefore, it is obvious that the mechanisms for control of body temperature represent a beautifully designed control system. It is the purpose of this chapter to discuss this system as it operates in health and in disease.

"Core" Temperature versus Surface Temperature. When speaking of the body temperature, one usually means the temperature in the interior, called the *core temperature,* and not the temperature of the skin or tissues immediately underlying the skin. The core temperature is accurately regulated, normally varying from the mean by not more than 1° F. On the other hand, the *surface temperature* rises and falls with the temperature of the surroundings. In speaking of body temperature regulation we almost always refer to the core temperature; when we refer to the ability of the skin to lose heat to the surroundings, we usually speak of the surface temperature; and when we wish to calculate the total amount of heat stored in the body, we use the *average body temperature.* The average body temperature can be approximated by the following formula:

Average temperature =
0.7 internal temperature + 0.3 surface temperature

The Normal Body Temperature. No single temperature level can be considered to be normal, for measurements on many normal persons have shown a *range* of normal temperatures, as illustrated in Figure 71–1, from approximately 97° F. to over 99° F. When measured by rectum, the values are approximately 1° F. greater than the oral temperatures. The average normal tempera-

Figure 71–1 Estimated range of body temperature in normal persons. (From DuBois: Fever. Charles C Thomas.)

ture is generally considered to be 98.6° F. (37° C.) when measured orally and approximately 1° F. or 0.6° C. higher when measured rectally.

The body temperature varies considerably with exercise and with the temperature of the surroundings, for the temperature regulatory mechanisms are not 100 per cent effective. When excessive heat is produced in the body by strenuous exercise, the rectal temperature can rise to as high as 101° to 104° F. On the other hand, when the body is exposed to extreme cold, the rectal temperature can often fall to values considerably below 98° F.

Relationship of Body Heat to Body Temperature— Specific Heat of the Tissues. The temperature of an object is a measure of the kinetic activity of its molecules, and this is proportional to the amount of heat stored in the object. Therefore, body temperature is directly proportional to the heat in the body.

On the average, the body temperature increases 1° C. for each 0.83 Calorie stored per kilogram of body weight. In other words, the specific heat of the tissues is 0.83 Cal./kg./degree C. For a 70 kg. man, approximately 58 Calories of heat must be added to the body to raise the body temperature 1° C., or 1.8° F.

831

1. Basal metabolism

2. Muscular activity
   — Shivering —

3. Thyroxine effect on cells

4. Sympathetic effect on cells

5. Temperature effect on cells

1. Radiation

2. Evaporation
   — Convection —

3. Conduction
   — Convection —

HEAT PRODUCTION   HEAT LOSS

Figure 71–2 Balance of heat production versus heat loss.

BALANCE BETWEEN HEAT PRODUCTION AND HEAT LOSS

Heat is continually being produced in the body as a by-product of metabolism, and body heat is also continually being lost to the surroundings. When the rate of heat production is exactly equal to the rate of loss, the person is said to be in heat balance. But when the two are out of equilibrium, the body heat, and the body temperature as well, will obviously be either increasing or decreasing.

Figure 71–2 illustrates this balance between heat production and heat loss. To the left are the important factors that play major roles in determining the rate of heat production and which were discussed in detail in Chapter 70; they may be listed simply as follows: (1) basal rate of metabolism of all the cells of the body; (2) increase in rate of metabolism caused by muscle activity, including that caused by shivering; (3) increase in metabolism caused by the effect of thyroxine on cells; (4) increase in metabolism caused by the effect of norepinephrine and sympathetic stimulation on cells; and (5) increase in metabolism caused by increased temperature of the body cells.

HEAT LOSS

The various methods by which heat is lost from the body are depicted on the right side of the balance in Figure 71–2 and are shown pictorially in Figure 71–3. These include *radiation, conduction,* and *evaporation.* However, the amount of heat lost by each of these different mechanisms varies with atmospheric conditions.

Radiation. As illustrated in Figure 71–3, a nude person in a room at normal room temperature loses about 60 per cent of his total heat loss by radiation.

Loss of heat by radiation means loss in the form of infrared heat rays, a type of electromagnetic waves. Most infrared heat rays radiating from the body have wavelengths of 5 to 20 microns, 10 to 30 times the wavelengths of light rays. All mass in the universe that is not at absolute zero temperature radiates such rays. Therefore, the human body radiates heat rays in all directions. However, heat rays are also being radiated from the walls and other objects toward the body. If the temperature of the body is greater than the temperature of the surroundings, a greater quantity of heat is radiated from the body than is radiated to the body. This is the usual situation. Yet, at times, especially in the summer, the surroundings become hotter than the human body, under which circumstances more radiant heat is transmitted to the body than from the body.

Heat loss by radiation varies directly with the temperature difference between the temperature of the body surface and the average temperature of the surroundings. Therefore, it is impossible to state categorically exactly what percentage of the body heat will be lost by radiation unless all of the conditions surrounding the body are momentarily defined.

The surface of the human body is extremely absorbent for heat rays. A body exposed to infrared heat rays from a stove absorbs approximately 97 per cent of the rays that hit it. This rate of infrared absorption is approximately equal for human beings with either white or black skin, for at these wavelengths the different colors of the skin have no effect on absorption. On the other hand, the energy from the sun is transmitted mainly in the form of light rays rather than infrared rays. Approximately 35 per cent of these waves are reflected from the white skin but only a small amount from the dark skin. Consequently, in sunlight, dark skin does absorb more heat than white skin.

Conduction. Usually, only minute quantities of heat are lost from the body by direct

Figure 71–3 Mechanisms of heat loss from the body.

conduction from the surface of the body to other objects, such as a chair or a bed. When one first sits on a chair while nude, heat is conducted from the body to the chair rapidly, but within a few minutes the temperature of the chair rises almost to equal the temperature of the body, and thereafter the chair actually becomes an insulator to prevent further loss of heat. Consequently, the loss of heat by *conduction to objects* represents only a small per cent of the total heat loss from the body, as illustrated in Figure 71–3.

On the other hand, loss of heat by *conduction to air* does represent a sizeable proportion of the body's heat loss even under normal conditions. It will be recalled that heat is actually the kinetic energy of molecular motion, and the molecules that comprise the skin of the body are continually undergoing vibratory motion. Thus, the vibratory motion of the skin molecules can cause increased velocity of motion of the air molecules that come into direct contact with the skin. However, once the temperature of the air immediately adjacent to the skin equals the temperature of the skin, no further exchange of heat from the body to the air can occur. Therefore, conduction of heat from the body to the air is self-limited unless the heated air moves away from the skin so that new, unheated air is continually brought in contact with the skin.

Convection. Movement of air is known as convection, and the removal of heat from the body by convection air currents is commonly called "heat loss by convection." Actually, the heat must first be *conducted* to the air and then carried away by the convection currents.

A small amount of convection almost always occurs around the body because of the tendency for the air adjacent to the skin to rise as it becomes heated. Therefore, a nude person seated in a comfortable room without gross air movement still loses about 12 per cent of his heat by conduction to the air and then by convection away from the body.

COOLING EFFECT OF WIND. When the body is exposed to wind, the layer of air immediately adjacent to the skin is replaced by new air much more rapidly than normally, and heat loss by convection increases accordingly. Figure 71–4 shows that the cooling effect of wind at low velocities is approximately proportional to the square root of the wind velocity. For instance, a wind of 4 miles per hour is about two times as effective for cooling as a wind of 1 mile per hour. However, when the wind velocity rises beyond a few miles per hour, additional cooling does not occur to a great extent. This is because once the wind has cooled the skin to the temperature of the air itself, the rate of heat loss cannot increase further, regardless of the wind

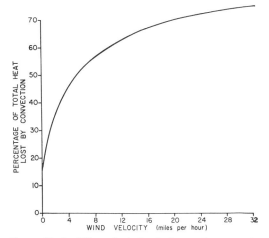

Figure 71–4 Effect of wind velocity on the percentage of heat loss that occurs by convection.

velocity. Instead, the rate at which heat can flow from the core of the body to the skin is then the factor that determines the rapidity with which heat can be lost.

CONDUCTION AND CONVECTION OF HEAT FROM A BODY EXPOSED TO WATER. Water has a specific heat several thousand times as great as that of air so that each unit portion of water adjacent to the skin can absorb far greater quantities of heat than can air. Also, the conductivity of heat through water is marked in comparison with that in air. Consequently, heating a thin layer of water next to the body does not result in the formation of an "insulator zone" as occurs in air. For this reason, the rate of heat loss from the body into nonflowing water is almost as much as the rate of heat loss into rapidly flowing water.

Because of these differences in heat conduction by air and water, the rate of heat loss to water at moderate temperatures is many times as great as the rate of heat loss to air of the same temperature. However, when the water and air are extremely cold, the rate of heat loss to air becomes almost as great as to water, for both water and air are then capable of carrying away essentially all the heat that can reach the skin from the body core.

Evaporation. When water evaporates from the body surface, 0.58 Calorie of heat is lost for each gram of water that evaporates, and water evaporates *insensibly* from the skin and lungs at a rate of about 600 ml. per day. This causes continual heat loss at a rate of 12 to 18 Calories per hour. Unfortunately, this insensible evaporation of water directly through the skin and lungs cannot be controlled for purposes of temperature regulation because it results from continual diffusion of water molecules through the skin and alveoli regardless of body temperature. However, evaporative loss of heat can be controlled by regulating the rate of sweating, which is discussed below.

EVAPORATION AS A NECESSARY REFRIGERATION MECHANISM AT HIGH AIR TEMPERATURES. In the preceding discussions of radiation and conduction it was noted that as long as the body temperature is greater than that of the surroundings, heat is lost by radiation and conduction, but when the temperature of the surroundings is greater than that of the skin, instead of losing heat, the body gains heat by radiation and conduction from the surroundings. Under these conditions, *the only means by which the body can rid itself of heat is by evaporation.* Therefore, any factor that prevents adequate evaporation when the surrounding temperatures are higher than body temperature permits the body temperature to rise. This occurs occasionally in human beings who are born with congenital absence of sweat glands. These persons can withstand cold temperatures as well as normal persons, but they are likely to die of heat stroke in tropical zones, for without the evaporative refrigeration system their body temperatures must remain at values greater than those of the surroundings.

EFFECT OF HUMID WEATHER ON EVAPORATIVE LOSS OF HEAT. It is well known that hot, muggy, summer days are extremely uncomfortable, and that sweat pours from the body far more profusely on these days than normally. This is because the air is already humidified almost to its maximal extent. As a result, the rate of evaporation may be greatly reduced or totally prevented so that the secreted sweat remains in the fluid state. Consequently, the body temperature approaches the temperature of the surroundings or rises above this temperature even though sweat continues to pour forth.

EFFECT OF CONVECTION AIR CURRENTS ON EVAPORATION. As already pointed out, a thin zone of air adjacent to the skin usually remains relatively stationary and is not exchanged for new air at a rapid rate unless convection currents are present. Lack of air movement prevents effective evaporation in the same manner that effective cooling by conduction of heat to the air is prevented — that is, the local air becomes saturated with water vapor, and further evaporation cannot occur. Convection currents cause air that has become saturated with moisture to move away from the skin while unsaturated air replaces it. Indeed, convection is probably of even more importance for heat loss from the body by evaporation than by conduction, for the instances in which one especially needs to lose heat from the body, such as on hot days, are the same times when evaporative loss of heat from the body is far greater than conductive loss. This also explains why fans are in demand on hot days.

SWEATING AND ITS REGULATION BY THE AUTONOMIC NERVOUS SYSTEM

When the body becomes overheated, large quantities of sweat are secreted onto the surface of the skin by the eccrine sweat glands to provide rapid *evaporative cooling* of the body. Stimulation of the preoptic area immediately anterior to the hypothalamus excites sweating. The impulses from this area that cause sweating are transmitted in the autonomic pathways to the cord and thence through the sympathetic outflow to the skin everywhere in the body.

It should be recalled from the discussion of the autonomic nervous system in Chapter 59 that the eccrine sweat glands are innervated by *cholinergic* nerve fibers. However, these glands can also be stimulated by epinephrine or norepinephrine circulating in the blood even though the glands themselves, in most parts of the body, do not have adrenergic innervation. It is possible that the sweat glands of the hands and feet do have some adrenergic innervation as well as cholinergic innervation, for many emotional states that excite the adrenergic portions of the sympathetic nervous system are known also to cause local sweating of the hands and feet. Also, during muscular exercise, which normally excites adrenergic activity, localized sweating of the hands and feet occurs. In this instance, it is believed that the moisture from the sweat helps the surfaces of the hands and feet to gain traction against smooth surfaces and also prevents drying of the thick, cornified layers of skin.

Rate of Sweating. In cold weather, the rate of sweat production is essentially zero but in very hot weather the maximum rate of sweat production is from 1.5 liters per hour in the unacclimatized person to 4 liters per hour in the person maximally acclimatized to heat. Thus, during maximal sweating, a person can lose more than 8 pounds of body weight per hour.

Mechanism of Sweat Secretion. The sweat glands are tubular structures consisting of two parts: (1) a deep *coiled portion* that secretes the sweat, and (2) a *duct portion* passing outward through the dermis of the skin. As is true of so many other glands, the secretory portion of the sweat gland secretes a fluid called the *precursor secretion;* then certain constituents of the fluid are believed to be reabsorbed as it flows through the duct.

The precursor secretion is an active secretory product of the epithelial cells lining the coiled portion of the sweat gland. Cholinergic sympathetic nerve fibers ending on or near the gland elicit the secretion, though the precise mechanism by which this occurs is unknown.

Since large amounts of sodium chloride are lost in the sweat, it is important especially to know how the sweat glands handle sodium and chloride during the secretory process. When the rate of sweat secretion is very low, the sodium and chloride concentrations of the sweat are very low, probably because these ions are reabsorbed from the precursor secretion before it reaches the surface of the body; their concentrations are sometimes as low as 5 mEq./liter each. On the other hand, when the rate of secretion becomes progressively greater, the rate of sodium chloride reabsorption apparently does not increase commensurately, so that then the concentration of sodium chloride in the sweat can rise almost to the level in plasma.

Other substances lost in reasonable quantities in the sweat include urea, lactic acid, and potassium ions. At low rates of sweat secretion, the concentrations of all these can be extremely high, but at high rates of secretion the concentration of urea is about two times that in the plasma, lactic acid about four times, and potassium about 1.2 times.

EFFECT OF ALDOSTERONE ON SODIUM LOSS IN THE SWEAT. Aldosterone decreases the sodium and chloride concentrations in sweat, thereby conserving these ions in the body fluids. It is believed that aldosterone functions in much the same way in the sweat glands as it functions in the renal tubules—that is, to increase the rate of active reabsorption of sodium by the ducts. The reabsorption of sodium carries chloride ions along as well because of the electrical gradient that develops when sodium is reabsorbed. The importance of this aldosterone effect is to minimize loss of sodium chloride in the sweat when the blood sodium chloride concentration is already low.

Acclimatization of the Sweating Mechanism. A person exposed to hot weather for several weeks sweats progressively more and more profusely, sweating an average maximum of about 1.5 liters per hour at first, which rises to about double this value within 10 days and to about 2½ times as much within six weeks, as illustrated in Figure 71–5. This increased effectiveness of the sweating mechanism is caused by a direct increase in sweating capa-

Figure 71–5 Acclimatization of the sweating mechanism, showing progressive increase in the maximum rate of sweating during the first few weeks of exposure to tropical weather.

bility of the sweat glands themselves. Associated with the increased sweating is usually decreased concentration of sodium chloride in the sweat, which allows progressively better conservation of salt. Most of this adaptation for conserving salt is probably caused by increased secretion of aldosterone, but it possibly also results in part from decreased sodium chloride concentration in the body fluids and from local changes in the sweat glands.

Extreme sweating can deplete the extracellular fluids of electrolytes, particularly of sodium and chloride. Consequently, extra sodium chloride usually must be supplied in the diet in tropical climates. A person who sweats profusely may lose as much as 15 to 20 grams of sodium chloride each day until he becomes acclimatized. On the other hand, after four to six weeks of acclimatization the loss of sodium chloride may be as little as 3 to 5 grams per day. This occurs at least partially because of increased aldosterone secretion resulting from depletion of the salt reserves of the body.

Long-term Acclimatization of the Sweat Apparatus. In addition to the rapid acclimatization of the sweating mechanism just described, acclimatization can occur over many years in a different way, as follows:

A person who has lived in the tropics since childhood actually has greater numbers of active sweat glands in the body. A person is born with a considerable excess of sweat glands, but if he lives in a temperate zone, many of these become permanently inactivated during childhood. However, if he lives in the tropics they remain functional throughout life. Therefore, a person who has spent his childhood in the tropics usually possesses a much more effective sweating mechanism than does a person reared elsewhere.

Panting as a Means of Evaporative Heat Loss. Many lower animals do not have sweat glands, but to offset this they lose large amounts of heat by the panting mechanism. During panting, only small volumes of air pass in and out of the lungs with each breath so that mainly dead space air enters the alveoli. Yet, because of the very rapid breathing rate, tremendous amounts of air are moved over the surfaces of the tongue, mouth, and trachea without overrespiration. Evaporation from these respiratory surfaces, especially of saliva on the tongue, is an important mechanism for heat control.

A special nervous center in the pons controls panting. This center modifies the normal respiratory pattern to provide the rapid and shallow breathing required for the panting mechanism.

THE INSULATOR SYSTEM OF THE BODY

The skin, the subcutaneous tissues, and especially the fat of the subcutaneous tissues are a heat insulator for the body. The fat is especially important because it conducts heat only *one-fourth* as rapidly as do the other tissues. When no blood is flowing from the heated internal organs to the skin, the insulating properties of the male body are approximately equal to three-quarters the insulating properties of a usual suit of clothes. In women this insulation is still better. Obviously, the degree of insulation varies from one person to another, depending to a great extent on the quantity of adipose tissue.

Because most body heat is produced in the deeper portions of the body, the insulation beneath the skin is an effective means for maintaining normal internal temperatures, even though it allows the temperature of the skin to approach the temperature of the surroundings.

FLOW OF BLOOD TO THE SKIN—THE "RADIATOR" SYSTEM OF THE BODY

Blood vessels penetrate the subcutaneous insulator tissues and are distributed profusely in the subpapillary portions of the skin. Indeed, immediately beneath the skin is a continuous venous plexus that is supplied by rapid inflow of arterial blood. Indeed, in the most exposed areas of the body—the hands, feet, and ears—blood is supplied through direct arteriovenous shunts from the arterioles to the veins. The rate of blood flow into this venous plexus can vary tremendously—from barely above zero to as great as 30 per cent of the total cardiac output. A high rate of blood flow causes heat to be conducted from the internal portions of the body to the skin with great efficiency, whereas reduction in the rate of blood flow decreases the efficiency of heat conduction from the internal portions of the body.

Obviously, therefore, the skin is an effective radiator system and the flow of blood to the skin is the mechanism of heat transfer from the body "core" to the skin. If blood flow from the internal structures to the skin is depressed, the only means by which heat produced internally can be lost to the exterior is by heat diffusion through the insulator tissues of the skin and subcutaneous areas.

Control of Heat Conduction to the Skin. Heat conduction to the skin by the blood is controlled by the degree of vasoconstriction of the arterioles that supply blood to the venous plexus of the skin, and this vasoconstriction is controlled almost entirely by the sympathetic nervous system. Ordinarily, the sympathetics remain tonically active, causing continual constriction of the arterioles supplying the skin. When the sympathetic centers of the posterior hypothalamus are stimulated, the blood vessels are constricted even more, and blood flow to the skin may almost cease, but, when these posterior centers of the hypothalamus are inhibited, decreased numbers of sympathetic impulses are transmitted to the periphery, and the blood vessels dilate.

Effect of Clothing on Heat Loss. EFFECT ON CONDUCTIVE HEAT LOSS. Clothing entraps air next to the skin and in the weave of the cloth, thereby increasing the thickness of the so-called "private zone" of air adjacent to the skin and decreasing the flow of convection air currents. Consequently, the rate of heat loss from the body by conduction is greatly depressed.

A usual suit of clothes decreases the rate of heat loss from the body to about half that from a nude body, while arctic-type clothing can decrease this heat loss to as little as one sixth that of the nude state.

EFFECT ON HEAT LOSS BY RADIATION. About half the heat transmitted from the skin to the clothing probably is radiated to the clothing instead of being conducted across the small intervening space, for sputtering the inside of clothing with a thin layer of gold, which reflects radiant heat, makes the insulating properties of clothing far more effective than otherwise. This is because the heat leaving the skin is reflected back from the clothing to the skin, and, consequently, little heat can then be transmitted to the clothing by radiation. As a result of this new technique, clothing for use in the arctic can be decreased in weight by about half.

LOSS OF HEAT THROUGH WET CLOTHING. The effectiveness of clothing in preventing heat loss is almost completely lost when it becomes wet, for there is no entrapped air to act as an insulator. Instead, the interstices of the clothing are filled with water which, because of its high conductivity for heat, increases the rate of heat transmission about 5- to 20-fold. One of the most important factors for protecting the body against cold in arctic regions is extreme precaution against wet clothing. Indeed, one must be careful not to over-heat oneself temporarily, for sweating in one's clothes makes them much less effective thereafter as an insulator.

EFFECT ON HEAT LOSS BY EVAPORATION. Clothing that is pervious to moisture allows almost normal heat loss from the body by evaporation, for when sweating occurs the sweat itself can dampen the clothing, and evaporation then occurs from the surface of the clothing. This cools the clothing, which in turn cools the skin. Consequently, in tropical regions, light clothing that is pervious to sweat but impervious to radiant heat from the sun prevents the body from gaining radiant heat while at the same time allowing it to lose heat at a rate almost as if one were not wearing clothing.

REGULATION OF BODY TEMPERATURE—FUNCTION OF THE HYPOTHALAMUS

Figure 71–6 illustrates approximately what happens to the temperature of the nude body

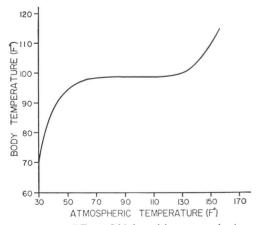

Figure 71–6 Effect of high and low atmospheric temperatures for several hours' duration on the internal body temperature, showing that the internal body temperature remains stable despite wide changes in atmospheric temperature.

after a few hours' exposure to air ranging from 30° to 170° F. Obviously, the precise dimensions of this curve vary, depending on the movement of air, the amount of moisture in the air, and even the nature of the surroundings. However, in general, between approximately 60° and 130° F., the nude body is capable of maintaining indefinitely a normal body core temperature somewhere between 98° and 100° F.

The temperature of the body is regulated almost entirely by nervous feedback mechanisms, and almost all of these operate through a *temperature regulating center* located in the hypothalamus. However, for these feedback mechanisms to operate, there must also exist temperature detectors to determine when the body temperature becomes either too hot or too cold. Some of these receptors are the following:

Temperature Receptors. Probably the most important temperature receptors for control of body temperature are many special *heat sensitive neurons* located in the preoptic area of the anterior hypothalamus. These neurons increase their impulse output as the temperature rises and decrease their output when the temperature decreases. The firing rate sometimes increases as much as 10-fold with an increase in body temperature of 10° C.

In addition to these heat sensitive neurons of the preoptic area, other receptors sensitive to temperature include: (1) a few *cold sensitive neurons* found in different parts of the hypothalamus, the septum, and the reticular substance of the midbrain, all of which increase their rate of firing when exposed to cold (the numbers of these are few and there is doubt that they play any role in the regulation of body temperature); (2) *skin temperature receptors,* including both *warmth* and *cold receptors,* that transmit nerve impulses into the spinal cord and thence to the hypothalamic region of the brain to help control body temperature, as will be discussed later; and (3) perhaps *temperature receptors in some internal organs* of the body that possibly also transmit signals to the central nervous system to help control body temperature.

THERMOSTATIC CENTER IN THE ANTERIOR HYPOTHALAMUS

In recent years experiments have been performed in which minute areas in the base of the brain have been either heated or cooled by use of a so-called *thermode.* This is a device

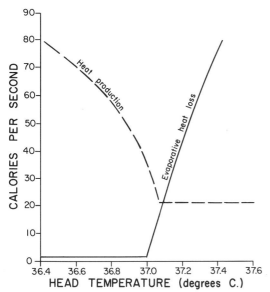

Figure 71–7 Effect of hypothalamic temperature on: (1) evaporative heat loss from the body and (2) heat production caused primarily by muscular activity and shivering. This figure demonstrates the extremely critical temperature level at which increased heat loss begins and increased heat production stops. (Drawn from data in Benzinger, Kitzinger, and Pratt: Temperature, Part 3. Ed. Hardy, p. 637. Reinhold Publishing Corp.)

that can be either heated or cooled by electrical means or by passing cold or hot water through it. The principal area in the brain where heat or cold from the thermode affects body temperature control is the preoptic and adjacent regions of the anterior hypothalamus, which is also where the heat sensitive neurons are found. Therefore, this small area is called the *thermostatic center*. This center controls body temperature by altering both the rate of heat loss from the body and the rate of heat production, as will be explained later.

Figure 71–7 demonstrates the extreme precision with which the hypothalamic thermostat can control heat loss and heat production. The solid curve shows that almost precisely at 37° C. (98.4° F.), sweating begins and then increases rapidly as the temperature rises above this value; on the other hand, it ceases at any temperature below this same critical level.

Likewise, the thermostat controls the rate of heat production, which is illustrated by the dashed curve. At any temperature above 37.1° C., the heat production remains almost exactly constant, but whenever the temperature falls below this level, the various mechanisms for increasing heat production become markedly activated, especially the increase in muscular activity which culminates in shivering.

MECHANISMS OF INCREASED HEAT LOSS WHEN THE BODY BECOMES OVERHEATED

Overheating the preoptic thermostatic area increases the rate of heat loss from the body in two principal ways: (1) by stimulating the sweat glands to cause evaporative heat loss from the body and (2) by inhibiting sympathetic centers in the posterior hypothalamus; this removes the normal vasoconstrictor tone to the skin vessels, thereby allowing vasodilatation and loss of heat from the skin.

MECHANISMS OF HEAT CONSERVATION AND INCREASED HEAT PRODUCTION WHEN THE BODY BECOMES COOLED

When the preoptic area of the hypothalamus is cooled below approximately 37° C., special mechanisms are set into play to conserve the heat that is already in the body, and still other mechanisms are set into play to increase the rate of heat production, as follows:

Heat Conservation. VASOCONSTRICTION IN THE SKIN. One of the first effects is intense vasoconstriction of the skin vessels over the entire body. This probably results from *release of the posterior hypothalamic sympathetic areas from inhibition* by the preoptic thermostatic signals. As a result, the sympathetic areas become overactive, and intense adrenergic vasoconstriction occurs throughout the body. This vasoconstriction obviously prevents the "radiator system" from transferring heat from the internal portions of the body to the skin. Consequently, with maximal vasoconstriction the only heat that can leave the body is that which can be conducted directly through the insulator layers of the skin. This effect conserves the quantity of heat in the body.

PILO-ERECTION. A second means by which heat is conserved when the hypothalamus is cooled is pilo-erection—that is, the hairs "stand on end." Obviously, this effect is not important in the human being because of the paucity of hair, but in lower animals the upright projection of the hairs in cold weather entraps a thick layer of insulator air next to the skin so that the transfer of heat to the surroundings is greatly depressed.

ABOLITION OF SWEATING. Sweating is completely abolished by cooling the preoptic thermostat below about 37° C. This obviously causes evaporative cooling of the body to cease except for that resulting from insensible evaporation.

Increased Production of Heat. Heat production is increased in three separate ways when the temperature of the hypothalamic thermostat falls below 37° C.:

HYPOTHALAMIC STIMULATION OF SHIVERING. Located in the dorsomedial portion of the posterior hypothalamus near the wall of the third ventricle is an area called the *primary motor center for shivering.* This area is normally inhibited by signals from the preoptic thermostatic area. However, when the preoptic thermostat becomes cooled and thereby loses its excitation, the normal inhibition of the primary motor center no longer exists. The self-excitation property of this center then causes it to transmit impulses through bilateral tracts down the brain stem, into the lateral columns of the spinal cord, and, finally, to the anterior motoneurons. These impulses are nonrhythmic and do not cause the actual muscle shaking. Instead, they progressively increase the tone of the skeletal muscles throughout the body. The resulting increase in muscle metabolism increases the rate of heat production, often raising total body heat production as much as 50 per cent even before shivering occurs. Then, once the tone of the muscles rises above a certain critical level, shivering begins. This probably results from feedback oscillation of the muscle spindle stretch reflex mechanism. When shivering begins, body heat production can rise to as high as 200 to 400 per cent of normal.

SYMPATHETIC "CHEMICAL" EXCITATION OF HEAT PRODUCTION. It was pointed out in Chapter 70 that either sympathetic stimulation or circulating norepinephrine and epinephrine in the blood can cause an immediate increase in the rate of cellular metabolism; this effect is called *chemical thermogenesis,* and is believed to result at least partially from the ability of norepinephrine and epinephrine to uncouple oxidative phosphorylation, as a result of which far more oxidation of foodstuffs must occur to produce the required amounts of high energy phosphate compounds required for normal function of the body. Therefore, the rate of cellular metabolism increases markedly.

The process of acclimatization greatly affects the intensity of chemical thermogenesis; an animal that has been exposed for several weeks to a very cold environment has as much as 100 to 200 per cent increase in metabolism when acutely exposed to cold, in contrast to the unacclimatized animal, which responds with an increase in metabolism of only 40 to 50 per cent.

In man, who rarely becomes completely acclimatized to cold environments, it is rare that chemical thermogenesis increases the rate of heat production more than 25 to 50 per cent, but even this amount is important in the control of body temperature.

INCREASED THYROXINE OUTPUT AS A CAUSE OF INCREASED HEAT PRODUCTION. Cooling the preoptic area of the hypothalamus also increases the production of the neurosecretory hormone *thyrotropin-releasing factor* by the hypothalamus. This hormone is carried by way of the hypothalamic portal veins to the adenohypophysis where it stimulates the secretion of *thyrotropin.* Thyrotropin, in turn, stimulates increased output of thyroxine by the thyroid gland, as will be explained in Chapter 77. The increased thyroxine increases the rate of cellular metabolism throughout the body. This increase in metabolism through the thyroid mechanism does not occur immediately but requires several weeks for the thyroid gland to hypertrophy before it reaches its new level of thyroxine secretion.

Exposure of animals to extreme cold for several weeks can cause their thyroid gland to increase in size as much as 20 to 40 per cent. Also, military personnel residing for several months in the Arctic develop basal metabolic rates 16 to 20 per cent above normal, and Eskimos have basal metabolic rates even higher than this. This continual stimulatory effect of cold on the thyroid gland probably also explains the large number of toxic thyroid goiters that occur in colder climates.

EFFECT OF THE SKIN COLD AND WARMTH RECEPTORS ON BODY TEMPERATURE REGULATION

Nerve receptors for both cold and heat are present everywhere in the skin. The function of these in transmitting cold and warmth sensations into the nervous system was discussed in Chapter 50. However, in addition to their effect in providing these sensations, these receptors probably also play a role in regulating internal body temperature in three different ways: they (1) cause a psychic desire for warmer surroundings and therefore make the person seek appropriate shelter or clothing, (2) transmit nerve impulses into the central nervous system and alter the "setting" of the hypothalamic thermostat, and (3) elicit local cord reflexes that affect skin blood flow or sweating and thereby help to maintain normal body temperature.

Shift in "Setting" of the Hypothalamic Thermostat by Cold or Warmth Sensations from the Skin. When the temperature receptors of the skin are strongly stimulated, impulses from these interact with the hypothalamic temperature regulatory mechanism to reset the control level of the hypothalamic thermostat slightly. Thus, when the skin becomes overheated, the thermostatic setting is reduced a few tenths of a degree below 37.0° C. That is, sweating begins to occur at perhaps 36.5° C. instead of at the normal 37.0° C. level, and increased heat production begins to occur at levels below 36.6° C. instead of below the normal level of 37.1° C.

Conversely, extreme cold sets the hypothalamic thermostat to a few tenths of a degree above 37.0° C.

IMPORTANCE OF PERIPHERAL STIMULI IN CONTROL OF INTERNAL BODY TEMPERATURE. It is obvious from the preceding discussion that changes in skin temperature cannot greatly alter the internal body temperature. However, these signals do make the thermostatic mechanisms more effective than usual, because through their auspices overheating of the skin causes the heat-losing activities of the thermostatic center to come into play long before the excessive heat can be transmitted from the skin to the interior of the body. Likewise, excessive cooling of the skin causes heat conservation and increased heat production to begin even before the cold can be transmitted to the interior of the body.

Yet, it would be impossible for the peripheral receptors alone to regulate internal body temperature. This fact follows from a basic understanding of control principles: *Stable operation of a control system requires that the receptors exciting the control system detect the factor that is being controlled.* In this instance the factor that is being controlled is the *internal body temperature.* Therefore, even though skin receptors are a valuable adjunct in controlling internal body temperature, they do not detect this temperature and therefore cannot substitute for the hypothalamic thermostatic control system.

Local Skin Reflexes. When one places a foot under a hot lamp and leaves it there for a short time, he finds *local vasodilatation* and mild *local sweating.* Conversely, placing a foot in cold water causes vasoconstriction and cessation of sweating. These reactions are caused by local cord reflexes conducted from the skin receptors to the spinal cord and back to the same skin area. However, their *intensity* is controlled by the hypothalamic thermostat, so that the overall effect is approximately proportional to the thermostatic signal *times* the local signal. Such reflexes can help to prevent excessive heat exchange from locally cooled or heated portions of the body.

Regulation of Internal Body Temperature After Cutting the Spinal Cord. After cutting the spinal cord in the neck above the sympathetic outflow from the cord, regulation of body temperature becomes extremely poor, for the hypothalamus can then no longer control either skin blood flow or the degree of sweating anywhere in the body. On the other hand, the local temperature reflexes originating in the skin receptors still exist. Unfortunately, these reflexes are not powerful, and, since none of the skin receptors are capable of detecting *internal* temperature, automatic regulation of internal body temperature is almost completely lost. In these persons, body temperature must be regulated principally by the patient's psychic response to cold and hot sensations in his head region. That is, if he feels himself becoming too hot or if he develops a headache from the heat, he knows to select cooler surroundings, and, conversely, if he has cold sensations, he selects warmer surroundings.

ABNORMALITIES OF BODY TEMPERATURE REGULATION

FEVER

Fever, which means a body temperature above the usual range of normal, may be caused by abnormalities in the brain itself or by toxic substances that affect the temperature regulating centers. Some causes of fever are presented in Figure 71–8. These include bacterial diseases, brain tumors, and a vicious cycle of heat production which may terminate in heat stroke.

Resetting the Hypothalamic Thermostat in Febrile Diseases — Effect of Pyrogens

Many proteins, breakdown products of proteins, and certain other substances, such as

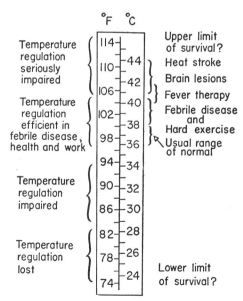

Figure 71–8 Body temperatures under different conditions. (From DuBois: Fever. Charles C Thomas.)

lipopolysaccharide toxins secreted by bacteria, can cause the "setting" of the hypothalamic thermostat to rise. Substances that cause this effect are called pyrogens. It is pyrogens secreted by toxic bacteria or pyrogens released from degenerating tissues of the body that cause fever during disease conditions. When the setting of the hypothalamic thermostat becomes increased to a higher level than normal, all the mechanisms for raising the body temperature are set into play, including heat conservation and increased heat production. Within a few hours after the thermostat has been set to a higher level, the body temperature also approaches this level.

Mechanism of Action of Pyrogens to Cause Fever—Leukocyte Pyrogen. Some physiologists believe that pyrogenic substances have a direct action on the hypothalamic thermostat to increase its setting. However, there is reason to believe that most pyrogens affect the hypothalamic thermostat indirectly in the following manner:

When a body tissue degenerates or becomes diseased, polymorphonuclear leukocytes enter the tissue; these in turn are autocatalytically destroyed as they phagocytize the degenerating tissue. And, as the leukocytes degenerate, they release a substance called *leukocyte pyrogen* from their "buffy coat." This substance has a direct effect on the hypothalamic thermostat to increase its setting to febrile levels.

Likewise, when bacterial pyrogens are injected into a person they usually do not affect the hypothalamic thermostat immediately; instead, the effect is delayed for many minutes to several hours. This delay is believed to result from the fact that the pyrogens must react with leukocytes (or perhaps plasma) to form a substance called *endogenous pyrogen*. The endogenous pyrogen is then transmitted in the blood to the hypothalamic thermostat and increases its setting to the febrile level. It is possible that this endogenous pyrogen is the same as leukocyte pyrogen.

To give one an idea of the extremely powerful effect of pyrogens in resetting the hypothalamic thermostat, as little as 1 *microgram* of lipopolysaccharide pyrogen derived from *E. coli* can cause severe fever when injected into man.

Effect of Dehydration on the Hypothalamic Thermostat. Dehydration is another factor that can cause the body temperature to rise considerably. Part of this elevation of temperature probably results from lack of available fluid for sweating, but dehydration can also cause temperature elevation even in a cold

atmosphere. Consequently, dehydration almost certainly has a direct effect on the hypothalamic centers to set the hypothalamic thermostat to febrile levels.

Characteristics of Febrile Conditions

Chills. When the setting of the thermostat is suddenly changed from the normal level to a higher-than-normal value as a result of tissue destruction, pyrogenic substances, or dehydration, the body temperature usually takes several hours to reach the new temperature setting. For instance, the temperature setting of the hypothalamic thermostat, as illustrated in Figure 71–9, might suddenly rise to 103° F. Because the blood temperature is less than the temperature setting of the hypothalamic thermostat, the usual autonomic responses to cause elevation of body temperature occur. During this period the person experiences chills, during which he feels extremely cold, even though his body temperature may be already above normal. Also, his skin is cold because of vasoconstriction, and he shakes all over because of shivering. His chills continue until his body temperature rises to the hypothalamic setting of 103° F. Then, when the temperature of the body reaches this value, he no longer experiences chills but instead feels neither cold nor hot. As long as the factor that is causing the hypothalamic thermostat to be set at a high value continues its effect, the body temperature is regulated more or less in the normal manner but at the high temperature level.

The Crisis. If the factor that is causing the high temperature is suddenly removed, the hypothalamic thermostat is suddenly set at a lower value—perhaps even back to the normal level, as illustrated in Figure 71–9. In this instance, the blood temperature is still 103° F., but the hypothalamus is attempting to regulate the body temperature at 98.6° F. This situation is analogous to excessive heating of the anterior hypothalamus, which causes intense sweating and sudden development of a hot skin because of vasodilatation everywhere. This sudden change of events in a febrile disease is known as the "crisis." In olden days, the

Figure 71–9 Effects of changing the setting of the "hypothalamic thermostat."

doctor always awaited the crisis, for once this occurred he knew immediately that the patient's temperature would soon be falling.

Heat Stroke

The limits of extreme heat that one can stand depend almost entirely on whether the heat is dry or wet. If the air is completely dry and sufficient convection air currents are flowing to promote rapid evaporation from the body, a person can withstand several hours of air temperature at 200° F. with no apparent ill effects. On the other hand, if the air is 100 per cent humidified and evaporation cannot occur or if the body is in water, the body temperature begins to rise whenever the surrounding temperature rises above approximately 94° F. If the person is performing heavy work, this critical temperature level may fall to 85° to 90° F.

Also, as the body temperature rises, the basal metabolism increases about 6 per cent for each degree F. rise, or 10 per cent for each degree C. rise, because of the intrinsic effect of heat to increase the rates of chemical reactions. Once the body temperature rises to approximately 110° F., the rate of metabolism has doubled.

Unfortunately, there is a limit to the rate at which the body can lose heat even with maximal sweating. Furthermore, when the hypothalamus becomes excessively heated, its heat regulating ability becomes greatly depressed and sweating diminishes. As a result, a vicious cycle develops: high temperature causes increased production of heat, which increases the body temperature still higher, which causes still greater production of heat and a still higher body temperature, etc. Once the body temperature rises above 107° to 110° F., the heat regulating mechanisms often can no longer dissipate the excessive heat being produced. Therefore, the temperature may then rise abruptly until it causes death, unless this rise is checked artificially.

Relationship of Heat Stroke to Dehydration. People who work in hot and humid atmospheres can often become excessively dehydrated because of sweating and, consequently, can develop *circulatory shock*. As pointed out earlier in the chapter, dehydration probably also has a direct effect on the hypothalamus to set the thermostat at a high level, in addition to its effects to diminish sweating drastically; and these effects undoubtedly add to the vicious cycle of hyperpyrexia. This explains why it is important to maintain an adequate intake of water and sodium chloride in hot surroundings if heat stroke is to be avoided.

Acclimatization to Heat. It is often extremely important to acclimatize persons to extreme heat; some examples are acclimatization of soldiers for tropical duty and acclimatization of miners for work in the two mile deep gold mines of South Africa where the temperature approaches body temperature and the humidity approaches 100 per cent. Exposure of a person to heat for several hours each day while working a reasonably heavy work load will develop increased tolerance to hot and humid conditions in about one week. Probably the most important physiologic changes that occur during this acclimatization process are an increase in plasma volume and diminished loss of salt in the sweat and urine, these effects resulting from increased secretion of aldosterone.

Harmful Effects of High Temperature

When the body temperature rises above approximately 106° F., the parenchyma of many cells usually begins to be damaged. The pathologic findings in a person who dies of hyperpyrexia are local hemorrhages and parenchymatous degeneration of cells throughout the entire body. The brain is especially likely to suffer, because neuronal cells once destroyed can never be replaced. When the body temperature rises to 110° F., the person usually has only a few hours to live unless his temperature is brought back within normal range rapidly by sponging his body with alcohol, which evaporates and cools the body, or by bathing him in ice water.

Effects of Chemicals and Drugs on Body Temperature

An extremely large number of foreign substances when injected into the body fluids can cause the body temperature to rise—that is, they are pyrogenic. Bacteria, pollens, dust, and vaccines are all pyrogenic because of their protein content. Ordinarily, human plasma is not pyrogenic, but it is possible for the plasma proteins and other proteins that normally do not cause pyrogenic reactions to be changed chemically by degradation or denaturization and thereafter to cause pyrogenic reactions. A few nonprotein chemicals are also known to elevate the body temperature; these include especially certain polysaccharides and some nitrated phenol compounds.

Antipyretics. Aspirin, antipyrine, aminopyrine, and a number of other substances known as "antipyretics" have an effect on the hypothalamic thermostat opposite to that of the pyrogens. In other words, they cause the setting of the thermostat to be lowered so that the body temperature falls, though usually not more than a degree or so. Aspirin is especially effective in lowering the hypothalamic setting when pyrogens have raised the setting, but aspirin will not lower the normal temperature. On the other hand, aminopyrine will decrease even the normal body temperature. Obviously, these drugs can be used to prevent damage to the body from excessively high body temperature.

EXPOSURE OF THE BODY TO EXTREME COLD

A person exposed to ice water for approximately 20 to 30 minutes ordinarily dies unless treated immediately. By that time, the internal body temperature will have fallen to about 77° F. Yet if he is warmed rapidly by application of external heat, his life can often be saved.

Treatment of a person whose body temperature has fallen into the seventies usually consists of application of wet heat either in the form of tub treatment or hot packs, with the water at approximately 110° F. If the temperature of the water is less than this, the rate at which heat is returned to the body is too slow for maximal benefit, and, if it is greater than this, the skin might be severely damaged because it becomes overheated while not receiving a satisfactory blood supply.

Loss of Temperature Regulation at Low Temperatures. As noted in Figure 71–8, once the body temperature has fallen below 85° F., the ability of the hypothalamus to regulate temperature is completely lost, and it is greatly impaired even when the body temperature falls below approximately 94° F. Part of the reason for this loss of temperature regulation is that the rate of heat production in each cell is greatly depressed by the low temperature. Also, sleepiness and even coma are likely to develop, which depress the activity of the central nervous system heat-control mechanisms and prevent shivering.

Frostbite. When the body is exposed to extremely low temperatures, surface areas can actually freeze; the freezing is called "frostbite." This occurs especially in the lobes of the ears and in the digits of the hands and feet. If the parts are thawed immediately, especially with water that is not above approximately 110° F., no permanent damage may result. On the other hand, prolonged freezing causes permanent circulatory impairment as well as local tissue damage. Often gangrene follows thawing, and the frost-bitten areas are lost.

Artificial Hypothermia. It is possible to decrease the temperature of a person by giving him a sedative to depress the hypothalamic thermostat and then packing him in ice until his temperature falls. His temperature can then be maintained below 90° F. for several days to a week or more by continual sprinkling of cool water or alcohol on the body. Such artificial cooling is often used during heart surgery so that the heart can be stopped for many minutes at a time. Cooling to this extent does not cause severe physiologic results.

REFERENCES

American Physiological Society: Adaptation to Environment. *In* Handbook of Physiology. Baltimore, The Williams & Wilkins Co., 1965, Sec. IV.

Atkins, E.: Pathogenesis of fever. *Physiol. Rev.,* 40:580, 1960.

Bell, C. R., and Walters, J. D.: Reactions of men working in hot and humid conditions. *J. Appl. Physiol.,* 27:684, 1969.

Benzinger, T. H.: Heat regulation: homeostasis of central temperature in man. *Physiol. Rev.,* 49:671, 1969.

Cooper, K. E.: Temperature regulation and the hypothalamus. *Brit. Med. Bull.,* 22:238, 1966.

Cunningham, D. J.; Stolwijk, J. A. J.; Murakami, N.; and Hardy, J. D.: Responses of neurons in the preoptic area to temperature, serotonin, and epinephrine. *Amer. J. Physiol.,* 213:1570, 1967.

Curtis, S. E., and Rogler, J. C.: Thermoregulatory ontogeny in piglets: sympathetic and adipokinetic responses to cold. *Amer. J. Physiol.,* 218:149, 1970.

Dill, D. B.; Horvath, S. M.; van Beaumont, W.; Gehlsen, G.; and Burrus, K.: Sweat electrolytes in desert walks. *J. Appl. Physiol.,* 23:746, 1967.

Downey, J. A.; Miller, J. M.; and Darling, R. C.: Thermoregulatory responses to deep and superficial cooling in spinal man. *J. Appl. Physiol.,* 27:209, 1969.

Eisenman, J. S.: Pyrogen-induced changes in the thermosensitivity of septal and preoptic neurons. *Amer. J. Physiol.,* 216:330, 1969.

Folk, G. E.: Introduction to Environmental Physiology: Extremes and Mammalian Survival. Philadelphia, Lea & Febiger, 1966.

Hammel, H. T.: Regulation of internal body temperature. *Ann. Rev. Physiol.,* 30:641, 1968.

Hardy, J. D.: Physiology of temperature regulation. *Physiol. Rev.,* 41:521, 1961.

Hardy, J. D. (ed.): Temperature; Its Measurement and Control in Science and Industry. Biology Medicine Level. New York, Reinhold Publishing Corp., 1963, Pt. 3.

Hemingway, A.: Shivering. *Physiol. Rev.,* 43:397, 1963.

Jackson, D. L.: A hypothalamic region responsive to localized injection of pyrogens. *J. Neurophysiol.,* 30:586, 1967.

Lind, A. R.; Humphreys, P. W.; Collins, K. J.; Foster, K.; and Sweetland, K. F.: Influence of age and daily duration of exposure on responses of men to work in heat. *J. Appl. Physiol.,* 28:50, 1970.

MacIntyre, B. A.; Bullard, R. W.; Banerjee, M.; and Elizondo, R.: Mechanism of enhancement of eccrine sweating by localized heating. *J. Appl. Physiol.,* 25:255, 1968.

Nakayama, T., and Hardy, J. D.: Unit responses in the rabbit's brain stem to changes in brain and cutaneous temperature. *J. Appl. Physiol.,* 27:848, 1969.

Robinson, S., and Robinson, A. H.: Chemical composition of sweat. *Physiol. Rev.,* 34:202, 1954.

Saltin, B.; Gagge, A. P.; and Stolwijk, J. A. J.: Body temperatures and sweating during thermal transients caused by exercise. *J. Appl. Physiol.,* 28:318, 1970.

Sams, W. M., Jr., and Winkelmann, R. K.: Temperature effects on isolated resistance vessels of skin and mesentery. *Amer. J. Physiol.,* 216:112, 1969.

Smith, R. E., and Horwitz, B. A.: Brown fat and thermogenesis. *Physiol. Rev.,* 49:330, 1969.

Snell, E. S., and Atkins, E.: The mechanisms of fever. *In* Bittar, E. E., and Bittar, N. (eds.): The Biological Basis of Medicine. New York, Academic Press, Inc., 1968, Vol. 1, p. 397.

Stolwijk, J. A. J.: Body temperature regulation. *Proc. Int. Union of Physiol. Sciences,* 6:196, 1968.

Symposium on metabolic adaptations to temperature. *Fed. Proc.,* 25:1153, 1966.

Vanhoutte, P. M., and Shepherd, J. T.: Effect of temperature on reactivity of isolated cutaneous veins of the dog. *Amer. J. Physiol.,* 218:187, 1970.

Wurster, R. D., and McCook, R. D.: Influence of rate of change in skin temperature on sweating. *J. Appl. Physiol.,* 27:237, 1969.

CHAPTER 72

Dietary Balances, Regulation of Feeding, Obesity, and Starvation

The intake of food must always be sufficient to supply the metabolic needs of the body and yet not enough to cause obesity. Also, since different foods contain different proportions of proteins, carbohydrates, and fats, appropriate balance must be maintained between these different types of food so that all segments of the body's metabolic systems can be supplied with the requisite materials. This chapter therefore discusses the problems of balance between the three major types of food and also the mechanisms by which the intake of food is regulated in accordance with the metabolic needs of the body.

DIETARY BALANCES

ENERGY AVAILABLE IN FOODS

The energy liberated from each gram of carbohydrate as it is oxidized to carbon dioxide and water is 4.1 Calories, and that liberated from fat is 9.3 Calories. The energy liberated from metabolism of the average protein of the diet as each gram is oxidized to carbon dioxide, water, and urea is 4.35 Calories. Also, these different substances vary in the average percentages that are absorbed from the gastrointestinal tract; approximately 98 per cent of the carbohydrate, 95 per cent of the fat, and 92 per cent of the protein. Therefore, in round figures the average *physiologically available energy* in each gram of the three different foodstuffs in the diet is:

	Calories
Carbohydrates	4.0
Fat	9.0
Protein	4.0

AVERAGE COMPOSITION OF THE DIET

The average American receives approximately 15 per cent of his energy from protein, about 40 per cent from fat, and 45 per cent from carbohydrates. In most other parts of the world the quantity of energy derived from carbohydrates far exceeds that derived from both proteins and fats. Indeed, in Mongolia the energy received from fats and proteins combined is said to be no greater than 15 to 20 per cent.

Daily Requirement for Protein. About 30 grams of the body proteins are degraded and used for energy daily. Therefore, all cells must continue to form new proteins to take the place of those that are being destroyed, and a supply of protein is needed in the diet for this purpose. An average man can maintain his normal stores of protein provided that his *daily intake is above 30 to 45 grams.*

PARTIAL PROTEINS. Another factor that must be considered in analyzing the proteins of the diet is whether the dietary proteins are *complete* proteins or *partial* proteins. Complete proteins have compositions of amino acids in appropriate proportion to each other so that all the amino acids can be properly used by the human body. In general, proteins derived from animal foodstuffs are more nearly complete than are proteins derived from vegetable and grain sources. This subject is more completely discussed in Chapter 69; therefore, for the present, suffice it to say that when partial proteins are in the diet an increased minimal quantity of protein is necessary in the daily rations to maintain protein balance. A particular example of this occurs in the diet of many African natives who subsist primarily on a corn meal diet. The protein of corn is almost totally lacking in tryptophan; and this means that this diet, in effect, is almost completely protein deficient. As a result, the natives, especially the children, develop the protein deficiency syndrome called *kwashiorkor,* which consists of failure to grow, lethargy, depressed mentality, and hypoprotein edema.

Necessity for Fat in the Diet. The human body is capable of desaturating fatty acids to a slight extent, especially in the liver. However, the liver can form only minute quantities of the more highly unsaturated fatty acids that have several double bonds in their fatty chains. For this reason, some of the more highly unsaturated fatty acids, such as *arachidonic, linoleic,* and *linolenic acids,* appear to be essential constituents of the diet for normal operation of the body. In lower animals, scaling and exudative skin lesions appear when these unsaturated fatty acids are totally absent from the diet; and the unsaturated fatty acids are probably also of special importance for formation of some of the structural elements of all cells throughout the body, because lack of these acids causes failure of growth. How much unsaturated fat is needed in the diet has not been determined; but because Mongolians exist on diets with fat contents as low as 10 per cent, it is presumed that only small quantities of unsaturated fats are needed. However, it was pointed out in Chapter 68 that a high *ratio* of saturated to unsaturated fats in the diet predisposes to atherogenesis, which often leads to heart attacks. The cause of this effect is yet unknown; but nevertheless, most nutritionists recommend that the fat of the diet contain a high proportion of unsaturated fatty acids.

Necessity for Carbohydrates in the Diet. In the discussion of carbohydrate metabolism in Chapter 67, it was pointed out that failure to metabolize carbohydrate in sufficient quantity causes such rapid metabolism of fats that ketosis is likely to develop. Also, fats and proteins alone usually cannot supply sufficient energy for all operations of the human body, and, consequently, carbohydrates appear to be essential to prevent weakness. This is especially true when the body undertakes a considerable work load. Finally, as already discussed,

carbohydrate is a *fat sparer* and a *protein sparer;* that is, carbohydrate is burned in preference to the burning of fat and protein, and this is especially important for preserving the functional proteins in the cells.

Composition of Different Foods. Table 72–1 presents the composition of a selected group of foods, illustrating especially the high proportions of fats and proteins in meat products and the high proportions of carbohydrates in most vegetable products.

Fat is deceptive in the diet, for it often exists as 100 per cent fat undiluted by any other substances, whereas essentially all the proteins and carbohydrates of foods are mixed in watery media and often represent less than 25 per cent of the weight of the food. Thus, the fat of one pat of butter mixed with an entire helping of potato may contain as much energy as all the potato itself. This is one of the reasons why fats are assiduously avoided in the prescription of diets for weight reduction.

STUDY OF ENERGY BALANCES

It is important in many physiological studies to compute the intake of food and also to compute at the same time the utilization of the different types of foods by the body. Measurement of the intake of

TABLE 72–1.   Protein, Fat, and Carbohydrate Content of Different Foods

Food	Protein, %	Fat, %	Carbohydrate, %	Fuel Value per 100 Grams, Calories
Apples	0.3	0.4	14.9	64
Asparagus	2.2	0.2	3.9	26
Bacon, fat	6.2	76.0	0.7	712
broiled	25.0	55.0	1.0	599
Beef, medium	17.5	22.0	1.0	268
Beets, fresh	1.6	0.1	9.6	46
Bread, white, milk	9.0	3.6	49.8	268
Butter	0.6	81.0	0.4	733
Cabbage	1.4	0.2	5.3	29
Carrots	1.2	0.3	9.3	45
Cashew nuts	19.6	47.2	26.4	609
Cheese, Cheddar, American	23.9	32.3	1.7	393
Chicken, total edible	21.6	2.7	1.0	111
Chocolate	(5.5)	52.9	(18.)	570
Corn (maize), entire	10.0	4.3	73.4	372
Haddock	17.2	0.3	0.5	72
Lamb, leg, intermediate	18.0	17.5	1.0	230
Milk, fresh whole	3.5	3.9	4.9	69
Molasses, medium	0.0	0.0	(60.)	240
Oatmeal, dry, uncooked	14.2	7.4	68.2	396
Oranges	0.9	0.2	11.2	50
Peanuts	26.9	44.2	23.6	600
Peas, fresh	6.7	0.4	17.7	101
Pork, ham, medium	15.2	31.0	1.0	340
Potatoes	2.0	0.1	19.1	85
Spinach	2.3	0.3	3.2	25
Strawberries	0.8	0.6	8.1	41
Tomatoes	1.0	0.3	4.0	23
Tuna, canned	24.2	10.8	0.5	194
Walnuts, English	15.0	64.4	15.6	702

Extracted from data compiled by Chatfield and Adams, U.S. Department of Agriculture Circular No. 549, 1940.

the different types of foods can be accomplished by simply weighing the foods before they are eaten and analyzing these foods for their relative contents of carbohydrates, fats, and protein. Then the loss of foods in the fecal and urinary excretions is subtracted from the intake to determine the *net intake*. On the other hand, special procedures must be performed for determining *utilization* of the different types of foods for energy, as follows:

Determination of the Rate of Protein Metabolism in the Body. The average protein of the diet contains approximately 16 per cent nitrogen, and the remaining 84 per cent is composed of carbon, hydrogen, oxygen, and sulfur. Numerous experiments have shown that when protein is metabolized in the body, the average person excretes about 90 per cent of the protein nitrogen in the urine in the form of urea, uric acid, creatinine, and other less important nitrogen products. The remaining 10 per cent of the nitrogen from the metabolized protein is ordinarily excreted in the feces. It is possible, therefore, to estimate relatively accurately the total quantity of protein metabolized by the body in a given time by analyzing the amount of nitrogen excreted in the urine. For instance, if 8 grams of nitrogen are excreted into the urine each day, then the total excretion, including that in the feces (an additional 10 per cent), will be about 8.8 grams. This value multiplied by 100/16 gives a calculated total protein metabolism of 55 grams per day.

"Nitrogen Balance" in the Body. Nitrogen balance experiments are frequently performed to determine the rate of protein increase or decrease in the body. These balance studies are performed by simultaneously measuring (a) the rate of protein intake and (b) the rate of protein utilization, the total protein utilization being estimated as was just described.

Ordinarily, such nitrogen balance studies are not valid unless they are continued for a week or more because of the difficulty of maintaining steady rates of protein ingestion and of nitrogen excretion.

A *negative nitrogen balance* implies greater protein utilization than protein intake, causing loss of protein from the body, and a *positive nitrogen balance* implies a net gain of protein in the body. Factors that cause negative balance include malnutrition, debilitating diseases, and glucocorticoid hormones. Factors that cause positive nitrogen balance include exercise, growth hormone, and testosterone.

Relative Utilization of Fat and Carbohydrates — The Respiratory Quotient. When one molecule of glucose is oxidized, the number of molecules of carbon dioxide liberated is exactly equal to the number of oxygen molecules necessary for the oxidative process. This fact is illustrated in Figure 72–1. Therefore the *respiratory quotient,* which is defined as the *ratio of carbon dioxide output to oxygen usage,* is 1.00. On the other hand, oxidation of triolein (the most abundant fat in the body) liberates 57

carbon dioxide molecules while 80 oxygen molecules are being utilized. Consequently, the respiratory quotient in this instance is 0.71. Finally, oxidation of alanine liberates five carbon dioxide molecules for every six oxygen molecules entering into the reaction, giving a respiratory quotient of 0.83.

The respiratory quotient for utilization of carbohydrates is always 1.00 because the quantity of oxygen in each carbohydrate molecule is always exactly sufficient to oxidize only the hydrogen within the molecule, and oxidation of each carbon atom in the molecule requires one molecule of respiratory oxygen. On the other hand, the respiratory quotient of dietary fat has been found to average 0.707, while the respiratory quotient of the average protein of the diet is 0.801.

Use of the Respiratory Quotient for Estimating Relative Rates of Carbohydrate and Fat Metabolism. As already pointed out, the average person receives only 15 per cent of his total energy from protein metabolism. Furthermore, the respiratory quotient of protein is approximately midway between the respiratory quotients of fat and carbohydrate (see preceding paragraph). Consequently, when the overall respiratory quotient of a person (which is equal to the respiratory exchange ratio, as discussed in Chap. 40) is measured by determining the total respiratory intake of oxygen and the total output of carbon dioxide from the lungs, one has a reasonable measure of the relative quantities of fat and carbohydrate being metabolized by the body during that particular time. For instance, if the respiratory quotient is approximately 0.71, the body is burning almost entirely fat to the exclusion of carbohydrates and proteins. If, on the other hand, the respiratory quotient is 1.00, the body is probably metabolizing almost entirely carbohydrate to the exclusion of fat and protein. Finally, a respiratory quotient of 0.85 indicates approximately equal utilization of carbohydrate and fat.

To be still more accurate in the calculation, one must consider the portion of the respiratory quotient that represents protein metabolism. To do this the rate of protein utilization in the body is first determined as discussed above; then the quantities of oxygen utilized and carbon dioxide released as a result of the protein metabolism are calculated and subtracted from the intake of oxygen and output of carbon dioxide, respectively, in order to establish the net respiratory quotient as it applies to fat and carbohydrate metabolism.

Variations in the Respiratory Quotient. Shortly after a meal, almost all the food metabolized is carbohydrates. Consequently, the respiratory quotient approaches 1.00 at this time. Approximately 8 to 10 hours following a meal, the quantity of carbohydrate being metabolized is relatively slight, and the respiratory quotient approaches that for fat metabolism; that is, approximately 0.71.

In diabetes mellitus, little carbohydrate is utilized by the body, and consequently, most of the

$$C_6H_{12}O_6 + 6\ O_2 \rightarrow 6\ CO_2 + 6\ H_2O$$
Glucose

Respiratory Quotient:

$$\frac{6}{6} = 1.00$$

$$C_{57}H_{104}O_6 + 80\ O_2 \rightarrow 57\ CO_2 + 52\ H_2O$$
Triolein

$$\frac{57}{80} = 0.71$$

$$2\ C_3H_7O_2N + 6\ O_2 \rightarrow (NH_2)_2CO + 5\ CO_2 + 5\ H_2O \quad \frac{5}{6} = 0.83$$
Alanine

Figure 72–1 Utilization of oxygen and release of carbon dioxide during the oxidation of carbohydrate, fat, and protein. The respiratory quotient for each of these reactions is calculated.

energy is derived from fat. Therefore, most persons with severe diabetes have respiratory quotients approaching the value for fat metabolism, 0.71.

In the human being, the respiratory quotient rarely becomes greater than 1.00, but such values have been recorded in animals in which carbohydrate is being converted rapidly into fat. When carbohydrate is being converted into fat, considerable oxygen is liberated from the carbohydrate molecule, and this oxygen becomes available for oxidation of other foodstuffs; this lessens the quantity of oxygen that must be brought into the body through the lungs. Consequently, if the animal is at the same time metabolizing only carbohydrates, the amount of carbon dioxide excreted through the lungs is greater than the intake of oxygen so that the respiratory quotient can rise to values as high as perhaps 1.10.

REGULATION OF FOOD INTAKE

Hunger. The term "hunger" means a craving for food, and it is associated with a number of objective sensations. For instance, as pointed out in Chapter 63, in a person who has not had food for many hours, the stomach undergoes intense rhythmic contractions called *hunger contractions*. These cause a tight or a gnawing feeling in the pit of the stomach and sometimes actually cause pain called *hunger pangs*. In addition to the hunger pangs, the hungry person also becomes more tense and restless than usual, and often has a strange feeling throughout his entire body that might be described by the nonphysiological term "twitterness."

Some physiologists actually define hunger as the tonic contractions of the stomach. However, even after the stomach is completely removed, the psychic sensations of hunger still occur, and equally as much craving for food still makes the person search for an adequate food supply.

Appetite. The term "appetite" is often used in the same sense as hunger except that it usually implies desire for specific types of food instead of food in general. Therefore, appetite helps a person choose the quality of food he eats.

Satiety. Satiety is the opposite of hunger. It means a feeling of fulfillment in the quest for food. Satiety usually results from a filling meal, particularly when the person's nutritional storage depots, the adipose tissue and the glycogen stores, are already filled.

NEURAL CENTERS FOR REGULATION OF FOOD INTAKE

Stimulation of the *lateral hypothalamus* causes an animal to eat voraciously, while stimulation of the *ventromedial nuclei of the hypothalamus* causes complete satiety, and, even in the presence of highly appetizing food, the animal will still refuse to eat. Conversely, a destructive lesion of the medial nuclei causes exactly the same effect as stimulation of the lateral hypothalamic nuclei—that is, voracious and continued eating until the animal becomes extremely obese. Indeed, the way in which such an animal approaches food demonstrates his extreme desire for food, for he actually *attacks* the food rather than eating calmly. Lesions of the lateral hypothalamic nuclei cause exactly the opposite effects, sometimes complete lack of desire for food and progressive inanition of the animal. Therefore, we can label the lateral nuclei of the hypothalamus as the *hunger center* or the *feeding center,* while we can label the ventromedial nuclei of the hypothalamus as a *satiety center.*

The feeding center operates by directly exciting the emotional drive to search for food. On the other hand, it is believed that the satiety center operates primarily by inhibiting the feeding center.

Other Neural Centers that Enter into Feeding. If the brain is sectioned between the hypothalamus and the mesencephalon, the animal can still perform the basic mechanical features of the feeding process. He can salivate, lick his lips, chew food, and swallow. Therefore, the actual mechanics of feeding are all controlled by centers in the brain stem and are entirely independent of the hypothalamus or cerebrum. The function of the hypothalamus in feeding, then, is to control the quantity of food intake and not the actual processes by which intake occurs.

Higher centers than the hypothalamus also play important roles in the control of feeding, particularly in the control of appetite. These centers include especially the amygdala and the cortical areas of the limbic system. It will be recalled from the discussion of the sense of smell that the amygdala is one of the major parts of the olfactory nervous system. Destructive lesions in the amygdala have demonstrated that some of its areas greatly increase feeding, while others inhibit feeding. In addition, stimulation of some areas of the amygdala elicits the mechanical act of feeding. However, the most important effect of destruction of the amygdala on both sides of the brain is a "psychic blindness" in the choice of foods. In other words, the person loses or at least partially loses the mechanism of appetite control of type and quality of food that he eats.

The cortical regions of the limbic system, including the infraorbital regions, the hippocampal gyrus, and the cingulate gyrus, all have areas that when stimulated can either increase or decrease feeding activities. These areas seem especially to play a role in the animal's drive to search for food when he is hungry. It is presumed that these centers are also responsible for determining the quality of food that is eaten. For instance, a previous unpleasant experience with almost any type of food often kills a person's appetite for that food thenceforth.

FACTORS THAT REGULATE FOOD INTAKE

We can divide the regulation of food into *long-term regulation,* which means regulation over weeks and months, and *short-term regulation,* which means regulation of hunger and feeding from minute to minute or hour to hour. Long-term regulation is concerned primarily with maintenance of normal quantities of nutrient stores in the body. Short-term regulation is concerned primarily with the immediate effects of feeding on the alimentary tract.

Long-term Regulation (Metabolic Regulation). An animal that has been starved for a long time and is then presented with unlimited food eats a far greater quantity than does an animal that has been on a regular diet. Conversely, an animal that has been force-fed for several weeks eats little when allowed to eat according to its own desires. Thus, the feeding center in the hypothalamus is geared to the nutritional status of the body. Some of the nutritional factors that control the degree of activity of the feeding center are the following:

AVAILABILITY OF GLUCOSE TO THE BODY CELLS—THE GLUCOSTATIC THEORY OF HUNGER AND OF FEEDING REGULATION. It has long been known that a decrease in blood glucose concentration is associated with development of hunger, which has led to the so-called glucostatic theory of hunger and of feeding regulation, as follows: When the blood glucose level falls too low, this automatically causes the animal to increase his feeding, which eventually returns the glucose concentration back toward normal. There are two other observations that also support the glucostatic theory: (1) An increase in blood glucose level increases the measured electrical activity in the satiety center in the ventromedial nuclei of the hypothalamus and simultaneously decreases the electrical activity in the feeding center of the lateral nuclei. (2) Chemical studies show that the ventromedial nuclei (the satiety center) concentrates glucose while other areas of the hypothalamus fail to concentrate glucose; therefore, it is assumed that glucose acts by increasing the degree of satiety.

EFFECT OF BLOOD AMINO ACID CONCENTRATION ON FEEDING. An increase in amino acid concentration in the blood also reduces feeding, and a decrease enhances feeding. In general, though, this effect is not as powerful as the glucostatic mechanism.

EFFECT OF FAT METABOLITES ON FEEDING. The overall degree of feeding varies almost inversely with the amount of adipose tissue in the body. That is, as the quantity of adipose tissue increases, the rate of feeding decreases. Therefore, many physiologists believe that *long-term regulation* of feeding is controlled mainly by fat metabolites of undiscovered nature. In support of this is the fact that the long-term average concentration of free fatty acids in the blood is directly proportional to the quantity of adipose tissue in the body. Therefore, it is likely that the free fatty acids or some other similar fat metabolites act in the same manner as glucose and amino acids to cause a negative feedback regulatory effect on feeding. It is also possible that this is by far the most important long-term regulator of feeding.

INTERRELATIONSHIP BETWEEN BODY TEMPERATURE AND FOOD INTAKE. When an animal is exposed to cold, he tends to overeat; and when exposed to heat, he tends to undereat. This is caused by interaction within the hypothalamus between the temperature regulating system (see Chap. 71) and the food intake regulating system. It is important because increased food intake in the cold animal (a) increases its metabolic rate and (b) provides increased fat for insulation, both of which tend to correct the cold state.

SUMMARY OF LONG-TERM REGULATION. Even though our information on the different feedback factors in long-term feeding regulation is imprecise, we can make the following general statement: When the nutrient stores of the body fall below normal, the feeding center of the hypothalamus becomes highly active, and the person exhibits increased hunger; on the other hand, when the nutrient stores are abundant, the person loses his hunger and develops a state of satiety.

Short-term Regulation (Nonmetabolic Factors). The degree of hunger or satiety can be temporarily increased or decreased by daily habits. For instance, the normal person has the habit of eating three meals a day, and, if he misses one, he is likely to develop a state of hunger at mealtime despite completely adequate nutritional stores in his tissues. But, in addition to habit, several other short-term physiological stimuli can alter one's desire for food for several hours at a time as follows:

GASTROINTESTINAL DISTENTION. When the gastrointestinal tract becomes distended, especially the stomach, inhibitory signals suppress the feeding center, thereby reducing the desire for food. This effect occurs at least partially even after the vagi and the sympathetic nerves from the upper gastrointestinal tract have been severed, indicating that simple overstretching of the abdominal cavity might be the effective stimulus for this feedback response. Obviously, this mechanism is of particular importance in bringing one's feeding to a halt during a heavy meal.

METERING OF FOOD BY HEAD RECEPTORS. When a person with an esophageal fistula is fed large quantities of food, even though this food is immediately lost again to the exterior, his appetite is decreased after a reasonable quantity of food has passed through his mouth. This effect occurs despite the fact that the gastrointestinal tract does not become the least bit filled. Therefore, it is postulated that various "head factors" relating to feeding, such as chewing, salivation, swallowing, and tasting, "meter" the food as it passes through the mouth, and after a certain amount has passed

through, the hypothalamic feeding center becomes inhibited.

Importance of Having Both Long- and Short-term Regulatory Systems for Feeding. The long-term regulatory system obviously helps an animal to maintain constant stores of nutrients in his tissues, preventing these from becoming too low or too high. However, an animal that forages for food must at times store excess quantities of food for future use even though he has no immediate use for the food. This explains the value of the short-term regulatory stimuli—they make the animal eat in a rhythmic pattern so that the food passes through his gastrointestinal tract fairly continuously and so that his digestive, absorptive, and storage mechanisms can all work at a steady pace rather than only when the animal needs food for energy. Indeed, the digestive, absorptive, and storage mechanisms can increase their rates of activity above normal only four- to five-fold, whereas the rate of usage of stored nutrients for energy sometimes increases to 20 times normal.

It is important, then, that feeding occur rather continuously (but at a rate that the gastrointestinal tract can accommodate), regulated principally by the short-term regulatory mechanism. However, it is also important that the intensity of the daily rhythmic feeding habits be modulated up or down by the long-term regulatory system, based principally on the level of nutrient stores in the body.

OBESITY

Energy Input versus Energy Output. When greater quantities of energy (in the form of food) enter the body than are expended, the body weight increases. Therefore, obesity is obviously caused by excess energy input over energy output. For each 9.3 Calories excess energy entering the body, 1 gram of fat is stored.

Excess energy input occurs *only during the developing phase of obesity,* and once a person has become obese, all that is required for him to remain obese is that his energy input equal his energy output. For the person to reduce, the output must be *greater* than the input. Indeed, studies of obese persons have shown that their intake of food is statistically identical to that for normal persons.

Effect of Muscular Activity on Energy Output. About half the energy used each day by the normal person goes into muscular activity, and in the laborer as much as three-fourths is used in this way. Therefore, muscular activity is by far the most important means by which energy is expended in the body. Therefore, it is frequently said that obesity results from *too high a ratio of food intake to daily exercise.*

ABNORMAL FEEDING REGULATION AS A PATHOLOGIC CAUSE OF OBESITY

The preceding discussion of the mechanisms that regulate feeding emphasized that the rate of feeding is normally regulated in proportion to the nutrient stores in the body. When these stores begin to approach an optimal level in a normal person, feeding is automatically reduced to prevent overstorage. However, in many obese persons this is not true, for feeding does not slacken until body weight is far above normal. Therefore, in effect, obesity is often caused by an abnormality of the feeding regulatory mechanism. This can result from either psychogenic factors that affect the regulation or actual abnormalities of the hypothalamus itself.

Psychogenic Obesity. Studies of obese patients show that a large proportion of obesity results from psychogenic factors. Perhaps the most common psychogenic factor contributing to obesity is the prevalent idea that healthy eating habits require three meals a day and that each meal must be filling. Many children are forced into this habit by overly solicitous parents, and the children continue to practice it throughout life. In addition, persons are known often to gain large amounts of weight during or following stressful situations, such as the death of a parent, a severe illness, or even mental depression. It seems that eating is often a means of release from tension.

Hypothalamic Abnormalities as a Cause of Obesity. In the preceding discussion of feeding regulation, it was pointed out that lesions in the ventromedial nuclei of the hypothalamus cause an animal to eat voraciously, and, therefore, to become obese. Also, many persons with hypophyseal tumors that encroach on the hypothalamus develop progressive obesity, illustrating that obesity in the human being, too, can definitely result from damage to the hypothalamus. Though in the normal obese person hypothalamic damage is almost never found, it is possible that the functional organization of the feeding center is different in the obese person from that of the nonobese person. For instance, an obese person who has made himself reduce to normal weight usually develops hunger that is demonstrably far greater than that of the normal person. This indicates that the "setting" of his feeding center is at a much higher level of nutrient storage than is that of the normal person.

Genetic Factors in Obesity. Obesity definitely runs in families. Furthermore, identical twins usually maintain weight levels within 2 pounds of each other throughout life if they live under similar conditions, or within 5 pounds of each other if their conditions of life differ markedly. This might result partly from eating habits engendered during childhood, but it is generally believed that this close similarity between twins is genetically controlled.

The genes can direct the degree of feeding in several different ways, including (1) a genetic abnormality of the feeding center to set the level of nutrient storage high or low and (2) abnormal hereditary psychic factors that either whet the appetite or cause the person to eat as a "release" mechanism.

A genetic abnormality in the *chemistry of fat storage* is also known to cause obesity in a certain strain of rats. In these rats, fat is easily stored in the adipose tissue, but the quantity of lipoprotein lipase formed in the adipose tissue is greatly reduced, so that little of the fat can be removed. This obviously results in a one-way path, the fat continually being deposited but never released. This, too, is another possible mechanism of human obesity in some human beings.

TREATMENT OF OBESITY

Treatment of obesity depends simply on decreasing energy input below energy expenditure. In other words, this means partial starvation. For this purpose, most diets are designed to contain large quantities of "bulk" which, in general, are made up of non-nutritive cellulose substances. This bulk distends the stomach and thereby partially appeases the hunger. In most lower animals such a procedure simply makes the animal increase his food intake still further, but the human being can often fool himself because his food intake is sometimes controlled as much by habit as by hunger. As pointed out below in connection with starvation, it is important to prevent vitamin deficiencies during the dieting period.

Various *drugs for decreasing the degree of hunger* have been used in the treatment of obesity. The most important of these is *amphetamine* (or amphetamine derivatives), which directly stimulates the satiety center in the ventromedial nuclei of the hypothalamus; this in turn inhibits the feeding center. However, there is danger in using this drug because it simultaneously overexcites the central nervous system, making the person nervous and elevating the blood pressure.

Finally, the more exercise one takes, the greater is his daily energy expenditure and the more rapidly his obesity disappears. Therefore, forced exercise is often an essential part of the treatment for obesity.

INANITION

Inanition is the exact opposite of obesity. In addition to inanition caused by inadequate food, both psychogenic and hypothalamic abnormalities can on occasion cause greatly decreased feeding. One such condition, *anorexia nervosa,* is an abnormal psychic state in which a person loses all desire for food and even becomes nauseated by food; as a result, severe inanition occurs. Also, destructive lesions of the hypothalamus, particularly caused by vascular thrombosis, frequently cause a condition called *cachexia;* the term simply means severe inanition.

STARVATION

Use of Food Stores During Starvation. During starvation the body preferentially uses carbohydrates rather than fat and protein (and fat rather than protein) for metabolism. However, because the carbohydrate stores are very slight (only a few hundred grams of glycogen in the liver and muscles), they are used up almost entirely within 24 to 48 hours, as shown in Figure 72–2. Indeed, the carbohydrate stores, if utilized alone without utilization of either fat or protein, would be sufficient to maintain life for only about 13 hours.

Figure 72–2 also illustrates the rates of fat and protein depletion throughout the weeks of starvation, assuming that the person has approximately 15 per cent body fat at the beginning of starvation. Under these conditions essentially all the fat is removed within approximately five to six weeks, and thereafter no fat remains to act as a protein sparer.

Unfortunately, starvation occurs more frequently in unhealthy than in healthy people, such as in cancer patients or patients with fever. In these, the basal metabolic rate is usually considerably above the mean value, and, therefore, the food stores are depleted at a much more rapid rate than would otherwise occur.

PREMORTEM DEPLETION OF PROTEINS. In the early stages of starvation, the only utilization of protein each day is the "obligatory" utilization due simply to degradation of intracellular protein as a result of their participation in cell metabolism. However, after most of the fat is gone—that is, after five to six weeks—the proteins must also be used for energy. Consequently, the quantity of protein thereafter decreases rapidly. Obviously, the disappearance of protein from the cells decreases the ability of the cells to perform normal cellular functions, and death soon ensues.

Vitamin Deficiencies in Starvation. The stores of some of the vitamins, especially the water-soluble vitamins—the vitamin B group and vitamin C—do not last long during starvation. Consequently, after a week or more of starvation mild vitamin deficiencies usually begin to appear, and over several weeks severe vitamin deficiencies can occur. Obviously, these can add to the debility that leads to death.

REFERENCES

Adolescent obesity (symposium): *Fed. Proc., 25*:1, 1966.
Albanese, A. A.: New Methods of Nutritional Biochemistry. New York, Academic Press, Inc., 1967, Vol. 3.
Anand, B. K.: Central chemosensitive mechanisms related to feeding. *In* Handbook of Physiology. Baltimore, The Williams & Wilkins Co., 1967, Sec. VI, Vol. I, p. 249.
Anand, B. K.: Nervous regulation of food intake. *Physiol. Rev., 41*:677, 1961.
Brock, J. F., and Truswell, A. S.: Biochemistry of malnutrition. *In* Bittar, E. E., and Bittar, N. (eds.): The Biological Basis of Medicine. New York, Academic Press, Inc., 1968, Vol. I, p. 415.
Crews, E. L., III; Fuge, K. W.; Oscai, L. B.; Holloszy, J. O.; and Shank, R. E.: Weight, food intake, and body composition: effects of exercise and of protein deficiency. *Amer. J. Physiol., 216*:359, 1969.
Davidson, S., and Passmore, R.: Human Nutrition and Dietetics. 3rd ed., Baltimore, The Williams & Wilkins Co., 1966.
Debons, A. F.; Krimsky, I.; Likuski, H. J.; From, A.; and Cloutier, R. J.: Gold thioglucose damage to the satiety center: inhibition in diabetes. *Amer. J. Physiol., 214*:652, 1968.
DeGroot, J.: Organization of hypothalamic feeding mechanisms. *In* Handbook of Physiology. Baltimore, The Williams & Wilkins Co., 1967, Sec. VI, Vol. I, p. 239.
De Ruiter, L.: Feeding behavior and the hypothalamus. *Proc. Int. Union of Physiol. Sciences, 6*:198, 1968.
Hunter, S. H., and Provasoli, L.: Comparative physiology: nutrition. *Ann. Rev. Physiol., 27*:19, 1965.
Kakolewski, J. W., and Deaux, E.: Initiation of eating as a function of the ingestion of hypoosmotic solutions. *Amer. J. Physiol., 218*:590, 1970.
Kennedy, G. C.: Food intake, energy balance and growth. *Brit. Med. Bull., 22*:216, 1966.
Keys, A.: Undernutrition. *In* Duncan, G. G. (ed.): Diseases of Metabolism. 5th ed., Philadelphia, W. B. Saunders Company, 1964, p. 664.
Mayer, J.: General characteristics of the regulation of food intake. *In* Handbook of Physiology. Baltimore, The Williams & Wilkins Co., 1967, Sec. VI, Vol. I, p. 3.
Peng, Y.; Benevenga, N. J.; and Harper, A. E.: Amino acid balance and food intake: effect of previous diet on plasma amino acids. *Amer. J. Physiol., 216*:1020, 1969.
Pool, R.: Suction lesions and hypothalamic hyperphagia. *Amer. J. Physiol., 213*:31, 1967.
Scharrer, E.; Baile, C. A.; and Mayer, J.: Effect of amino acids and protein on food intake of hyperphagic and recovered aphagic rats. *Amer. J. Physiol., 218*:400, 1970.
Schemmel, R.; Mikelsen, O.; and Tolgay, Z.: Dietary obesity in rats: influence of diet, weight, age, and sex on body composition. *Amer. J. Physiol., 216*:373, 1969.
Stark, P.; Totty, C. W.; Turk, J. A.; and Henderson, J. K.: A possible role of a cholinergic system affecting hypothalamic-elicited eating. *Amer. J. Physiol., 214*:463, 1968.
Steele, R.; Winkler, B.; Rathgeb, I.; Bjerknes, C.; and Altszuler, N.: Plasma glucose and free fatty acid metabolism in normal long-fasted dogs. *Amer. J. Physiol., 214*:313, 1968.
Strang, J. M.: Obesity. *In* Duncan, G. G. (ed.): Diseases of Metabolism. 5th ed., Philadelphia, W. B. Saunders Company, 1964, p. 693.
Wilmore, J. H., and Behnke, A. R.: An anthropometric estimation of body density and lean body weight in young men. *J. Appl. Physiol., 27*:25, 1969.

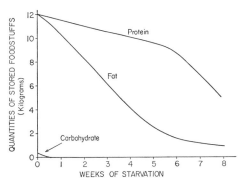

Figure 72–2 Effect of starvation on the food stores of the body.

CHAPTER 73

Vitamin and Mineral Metabolism

The study of vitamin and mineral metabolism rightfully falls in the province of biochemistry. Therefore, the present discussion of this subject is greatly abbreviated. (Refer to biochemistry texts for detailed information.) It is the purpose of this chapter to present the major aspects of vitamin and mineral metabolism as they relate to the overall physiology of the body.

VITAMINS

A vitamin is an organic compound needed in small quantities for operation of normal bodily metabolism and that cannot be manufactured in the cells of the body. Probably hundreds of such substances exist, most of which have not been discovered. However, a few have been studied extensively because they are present in the foods in relatively small quantities, and, as a result, dietary deficiency of one or more of them often occurs. Therefore, from a clinical point of view the agents that are generally considered to be vitamins are those organic compounds that occur in the diet in small quantities and, when lacking, can cause specific metabolic deficits.

Daily Requirements of Vitamins. Table 73–1 illustrates the usually recommended daily requirements of the different important vitamins. These requirements vary considerably, depending on the metabolic condition of each person. For instance, the greater the person's size, the greater is his vitamin requirement. Second, growing persons usually require greater quantities of vitamins than do others. Third, when the person performs exercise, the vitamin requirements are increased. Fourth, during disease and fevers, the vitamin requirements are ordinarily increased. Fifth, when greater than normal quantities of carbohydrates are metabolized, the requirements of thiamine and perhaps some of the other vitamins of the B complex are increased. Sixth, during pregnancy and lactation the requirement for vitamin D by the mother is greatly increased, and the requirement for vitamin D is considerable during the period of growth in children. Finally, a number of metabolic deficits may occur pathologically in which the vitamins themselves cannot be utilized properly in the body; in such conditions the requirement for one or more specific vitamins may be extreme.

Storage of Vitamins in the Body. Vitamins are stored to a slight extent in all the cells. However, some vitamins are stored to a major extent in the liver. For instance, the quantity of vitamin A stored in the liver may be sufficient to maintain a person without any intake of vitamin A for up to six months, and ordinarily the quantity of vitamin D stored in the liver is sufficient to maintain a person for three to five months without any additional intake of vitamin D.

The storage of vitamin K and of most water-soluble vitamins is relatively slight; this applies especially to the vitamin B compounds, for when a person's diet is deficient in vitamin B compounds, clinical symptoms of the deficiency can sometimes be recognized within a few days. Absence of vitamin C, another water-soluble vitamin, can cause symptoms within a few weeks and can cause death from scurvy in 20 to 30 weeks.

VITAMIN A

Vitamin A occurs in animal tissues as *retinol,* the formula of which is illustrated. This vitamin does not occur in foods of vegetable origin, but *provitamins* for the formation of vitamin A do occur in abundance in many different vegetable foods. These are the yellow and red *carotenoid pigments,* which, since they have chemical structures similar to that of vitamin A, can be changed into vitamin A in the human body, this change occurring mainly in the liver cells.

TABLE 73–1. Daily Requirements of the Vitamins

Vitamin	Daily Requirement
A	3.1 mg.
Thiamine	1.3 mg.
Riboflavin	1.8 mg.
Niacin	18 mg.
Ascorbic acid	80 mg.
D (children and during pregnancy)	11 μg.
E	unknown
K	none
Folic acid	unknown
B_{12}	unknown
Inositol	unknown
Pyridoxine	unknown
Pantothenic acid	unknown
Biotin	unknown
Para-aminobenzoic acid	unknown

The basic function of vitamin A in the metabolism of the body is not known except in relation to its use in the formation of retinal pigments, which was discussed in Chapter 53. Nevertheless, some of the other physiologic results of vitamin A lack have been well documented. Vitamin A is necessary for normal growth of most cells of the body and especially for normal growth and proliferation of the different types of epithelial cells. When vitamin A is lacking, the epithelial structures of the body tend to become stratified and keratinized. Therefore, vitamin A deficiency manifests itself by (1) scaliness of the skin and sometimes acne, (2) failure of growth of young animals, (3) failure of reproduction in many animals, associated especially with atrophy of the germinal epithelium of the testes and sometimes with interruption of the female sexual cycle, and (4) keratinization of the cornea with resultant corneal opacity and blindness.

Also, the damaged epithelial structures often become infected, for example, in the eyes, the kidneys, or the respiratory passages. Therefore, vitamin A has been called an "anti-infection" vitamin. Vitamin A deficiency also frequently causes kidney stones, probably owing to infection in the renal pelvis.

THIAMINE (VITAMIN B₁)

Thiamine operates in the metabolic systems of the body principally as *thiamine pyrophosphate;* this compound functions as a *cocarboxylase,* operating mainly in conjunction with a protein decarboxylase for decarboxylation of pyruvic acid and other α-keto acids, as discussed in Chapter 67.

Thiamine chloride

Thiamine deficiency causes decreased utilization of pyruvic acid and some amino acids by the tissues but increased utilization of fats. Thus, thiamine is specifically needed for final metabolism of carbohydrates and proteins. Probably the decreased utilization of these nutrients is the responsible factor for the debilities associated with thiamine deficiency.

Thiamine Deficiency and the Nervous System. The central nervous system depends almost entirely on the metabolism of carbohydrates for its energy. In thiamine deficiency the utilization of glucose by the central nervous system may be decreased as much as 50 to 60 per cent. Therefore, it is readily understandable how thiamine deficiency could greatly impair function of the central nervous system. The neuronal cells of the central nervous system frequently show chromatolysis and swelling during thiamine deficiency, changes that are characteristic of neuronal cells with poor nutrition. Obviously, such changes as these can disrupt communication in many different portions of the central nervous system.

Also, thiamine deficiency can cause *degeneration of myelin sheaths* of nerve fibers both in the peripheral nerves and in the central nervous system. The lesions in the peripheral nerves frequently cause these nerves to become extremely irritable, resulting in "polyneuritis" characterized by pain radiating along the course of one or more peripheral nerves. Also, in severe thiamine deficiency, the peripheral nerve fibers and fiber tracts in the cord can degenerate to such an extent that *paralysis* occasionally results; and, even in the absence of paralysis, the muscles atrophy, with resultant severe weakness.

Thiamine Deficiency and the Cardiovascular System. Thiamine deficiency also weakens the heart muscle, so that a person with severe thiamine deficiency sometimes develops *cardiac failure.* In general, the right side of the heart becomes greatly enlarged in thiamine deficiency. Furthermore, the return of blood to the heart may be increased to as much as three times normal. This indicates that thiamine deficiency causes *peripheral vasodilation* throughout the circulatory system, possibly as a result of metabolic deficiency in the smooth muscle of the vascular system itself. Therefore, the cardiac effects of thiamine deficiency are due partly to excessive return of blood to the heart and partly to primary weakness of the cardiac muscle. *Peripheral edema* and *ascites* also occur to a major extent in some persons with thiamine deficiency because of the cardiac failure.

Thiamine Deficiency and the Gastrointestinal Tract. Among the gastrointestinal symptoms of thiamine deficiency are indigestion, severe constipation, anorexia, gastric atony, and hypochlorhydria. All these effects possibly result from failure of the smooth muscle and glands of the gastrointestinal tract to derive sufficient energy from carbohydrate metabolism.

The overall picture of thiamine deficiency, including polyneuritis, cardiovascular symptoms, and gastrointestinal disorders, is frequently re-

Vitamin A

ferred to as "beriberi"—especially when the cardiovascular symptoms predominate.

NIACIN

Niacin, also called *nicotinic acid* and formerly known as vitamin G, functions in the body in the forms of *diphosphopyridine nucleotide* (DPN) and *triphosphopyridine nucleotide* (TPN), which are also known as coenzymes I and II. These coenzymes are hydrogen acceptors which combine with hydrogen atoms as they are removed from food substrates by many different types of dehydrogenases. Typical operation of both of them is presented in Chapter 67. When a deficiency of niacin exists, the normal rate of dehydrogenation presumably cannot be maintained, and, therefore, oxidative delivery of energy from the foodstuffs to the functioning elements of the cells likewise cannot occur at normal rates.

Niacin

Because coenzymes I and II apparently operate in essentially all cells of the body, it is readily understood how lack of niacin can cause multiple symptoms. Clinically, niacin deficiency causes mainly gastrointestinal symptoms, neurologic symptoms, and a characteristic dermatitis. However, it is probably much more proper to say that essentially all functions of the body are depressed.

In the early stages of niacin deficiency simple physiologic changes, such as muscular weakness of all the different types of muscles and poor glandular secretion, may occur, but in severe niacin deficiency actual death of tissues ensues. Pathologic lesions appear in many parts of the central nervous system, and permanent dementia or any of many different types of psychoses may result. Also, the skin develops a cracked, pigmented scaliness in areas that are exposed to mechanical irritation or sun irradiation; thus, it seems as if the skin were unable to repair the different types of irritative damage.

Niacin deficiency causes intense irritation and inflammation of the mucous membranes of the mouth and other portions of the gastrointestinal tract, thus instituting many digestive abnormalities. It is possible that this results from generalized depression of metabolism in the gastrointestinal epithelium and failure of appropriate epithelial repair.

The clinical entity called "pellagra" and the canine disease called "black tongue" are caused mainly by niacin deficiency. Pellagra is greatly exacerbated in persons on a corn diet (such as many of the natives of Africa) because corn is very deficient in the amino acid tryptophan, which can be converted in limited quantities to niacin in the body.

RIBOFLAVIN (VITAMIN B₂)

Riboflavin normally combines in the tissues with phosphoric acid to form two coenzymes, *flavin mononucleotide* (FMN), and *flavin adenine dinucleotide* (FAD). These in turn operate as hydrogen carriers in several of the important oxidative systems of the body. Usually, DPN, operating in association with specific dehydrogenases, accepts hydrogen removed from various food substrates and then passes the hydrogen to FMN or FAD; finally, the hydrogen is released as an ion into the surrounding fluids to become oxidized by nascent oxygen, the system for which is described in Chapter 67.

Riboflavin

Deficiency of riboflavin in lower animals causes severe *dermatitis, vomiting, diarrhea, muscular spasticity* which finally becomes muscular weakness, and then *death* preceded by coma and decline in body temperature. Thus, severe riboflavin deficiency can cause many of the same effects as lack of niacin in the diet; presumably the debilities that result in each instance are due to generally depressed oxidative processes within the cells.

In the human being riboflavin deficiency has never been known to be severe enough to cause the marked debilities noted in animal experiments, but mild riboflavin deficiency is probably common. Such deficiency causes digestive disturbances, burning sensations of the skin and eyes, cracking at the corners of the mouth, headaches, mental depression, forgetfulness, etc. Perhaps the most common characteristic lesion of riboflavin deficiency is *cheilosis*, which is inflammation and cracking at the angles of the mouth. In addition, a fine, scaly dermatitis often occurs at the angles of the nares, and keratitis of the cornea may occur with invasion of the cornea by capillaries.

Though the manifestations of riboflavin deficiency are usually relatively mild, this deficiency frequently occurs in association with deficiency of thiamine or niacin. Therefore, many deficiency syndromes, including pellagra, beriberi, sprue, and kwashiorkor, are probably due to a combined deficiency of a number of the vitamins, as well as to other aspects of malnutrition.

VITAMIN B$_{12}$

Several different *cobalamin* compounds which possess the common prosthetic group illustrated below exhibit so-called "vitamin B$_{12}$" activity.

Note that this prosthetic group contains cobalt that has four coordination bonds similar to those found in the hemoglobin molecule. It is likely that the cobalt atom functions in much the same way that the iron atom functions in hemoglobin.

Almost nothing is known about the manner in which vitamin B$_{12}$ enters into the metabolic organization of cells, though experiments indicate that it acts as a coenzyme in the conversion of amino acids and similar compounds into other substances. It also seems to be important for reducing ribonucleotides to deoxyribonucleotides, a step that is important in the formation of genes. This could explain the two major functions of vitamin B$_{12}$, (1) promotion of growth and (2) red blood cell maturation. This latter function was described in detail in Chapter 8.

A special effect of vitamin B$_{12}$ deficiency is often demyelination of the large nerve fibers of the spinal cord, especially of the posterior columns and occasionally of the lateral columns. As a result, persons with pernicious anemia (anemia caused by failure of red cell maturation) frequently have much loss of peripheral sensation and, in severe cases, even become paralyzed.

FOLIC ACID (PTEROYLGLUTAMIC ACID)

Several different pteroylglutamic acids, one of which is illustrated below, exhibit the "folic acid effect." Folic acid enters into the chemical reactions for conversion of glycine into serine and also for formation of methionine, purines, and thymine. These latter two are required for formation of deoxyribonucleic acid. Therefore, folic acid is required for reproduction of the cellular genes. This perhaps explains one of the most important functions of folic acid—that is, to promote growth.

Folic acid is an even more potent growth promoter than vitamin B$_{12}$, and, like vitamin B$_{12}$, is also important for the maturation of red blood cells, as discussed in Chapter 8. However, vitamin B$_{12}$ and folic acid each perform specific and different functions in promoting growth and maturation of red blood cells.

PYRIDOXINE (VITAMIN B$_6$)

Pyridoxine exists in the form of *pyridoxal phosphate* in the cells and functions as a coenzyme for many different chemical reactions relating to amino acid and protein metabolism. For instance, it is required as a coenzyme in (a) transamination; (b) decarboxylation; (c) synthesis of at least one amino acid, tryptophan; (d) conversion of a number of the amino acids into other substances needed by the cells; and (e) desaturation of fats in the liver to form unsaturated fats. As a result of this myriad of chemical functions, pyridoxine plays many key roles in metabolism—especially in protein metabolism. Also, it is believed to act in the transport of some amino acids across cell membranes.

Pyridoxine

Dietary lack of pyridoxine in lower animals can cause dermatitis, decreased rate of growth, development of fatty liver, anemia, and evidence of mental deterioration. In the human being, pyridoxine deficiency has been known to cause convulsions, dermatitis, and gastrointestinal disturbances such as nausea and vomiting. However, this deficiency is rare.

PANTOTHENIC ACID

Pantothenic acid mainly is incorporated in the body into coenzyme A, which has many metabolic roles in the cells. Two of these discussed at length in Chapters 67 and 68 are: (1) acetylation of decarboxylated pyruvic acid to form acetyl Co-A prior to its entry into the tricarboxylic acid cycle and (2) degradation of fatty acid molecules into multiple molecules of acetyl Co-A. Thus, lack of panto-

Folic acid (pteroylglutamic acid)

thenic acid can lead to depressed metabolism of both carbohydrates and fats.

OH   CH₃ OH
│   │   │
CH₂—C——CH—C—NH—CH₂—CH₂—COOH
│   ‖
CH₃   O
Pantothenic acid

Deficiency of pantothenic acid in lower animals can cause retarded growth, failure of reproduction, graying of the hair, dermatitis, fatty liver, and hemorrhagic adrenal cortical necrosis. In the human being, no definite deficiency syndrome has been proved, presumably because of the wide occurrence of this vitamin in almost all foods. Nevertheless, this does not mean that pantothenic acid is not of value in the metabolic systems of the body; indeed, it is perhaps as necessary as any other vitamin.

BIOTIN

Deficiency of biotin has been shown in human experiments to cause a mild dermatitis, lassitude, gastrointestinal symptoms, and hyperesthesias, including muscle pains.

O
‖
C
╱   ╲
HN   NH
│   │
HC————CH
│   │
H₂C   C
╲   ╱ │ ╲
S   H (CH₂)₄COOH
Biotin

Biotin acts in the body as a prosthetic group in several different coenzymes that cause combination of carbon dioxide with other organic compounds—for instance: (1) the combination of carbon dioxide with pyruvic acid to form oxaloacetic acid which, it will be recalled, is an important compound for operation of the tricarboxylic acid cycle; (2) combination of ammonia with carbon dioxide to form urea; and (3) utilization of carbon dioxide in the formation of purines and in the formation of fatty acids. Thus, biotin deficiency could disorganize the metabolic efficiency of the body.

INOSITOL

Lack of inositol in lower animals can cause decreased rate of growth, loss of hair, and the development of fatty liver. It is this latter effect that has received most attention from clinicians, for it has been supposed that deficiency of inositol might also cause fatty liver in human beings. Inositol is necessary for the liver to form certain types of cephalins, which could explain why inositol might be important for removing fat from the liver. Yet the fat that develops in the liver in inositol deficiency ordinarily is cholesterol rather than neutral fat,

OH   OH
│   │
HO╱ H   H   H
│
H ╲ H   OH   OH
│   │
OH   H
Inositol

which is an unexplained effect. Even though inositol is often used clinically for treatment of fatty liver, it has not been proved to be an effective measure.

CHOLINE

Deficiency of choline also results in the development of a fatty liver, but this time containing large quantities of triglycerides instead of cholesterol. Choline is utilized by the liver for formation of lecithin, and, in so doing, fatty acids supposedly are removed from the triglycerides of the liver, thereby decreasing the triglyceride content of the liver. Greater than normal quantities of choline seem to be needed for prevention of a fatty liver when the liver is already damaged, as in cirrhosis of the liver, etc. Under normal circumstances it is doubtful that choline deficiency ever exists.

CH₃   CH₃
╲   ╱
N
╱ │ ╲
CH₃   │   CH₂—CH₂OH
│
OH
Choline

Choline is also needed by the body as a source for methyl radicals, which are used for various synthetic processes and also for detoxification of many toxic compounds. Finally, choline is necessary as a precursor for formation of acetylcholine.

ASCORBIC ACID (VITAMIN C)

Ascorbic acid is a strong reducing compound, and it probably can be reversibly oxidized and reduced within the body. It is believed that ascorbic acid functions either as a reducing agent in specific metabolic processes or as an oxidation-reduction system, reversibly exchanging electrons with other oxidation-reduction systems. However, the precise function of ascorbic acid is almost totally unknown, though many of the physiologic effects of ascorbic acid deficiency have been well catalogued. For instance, oxidation of tyrosine and phenylalanine requires an adequate supply of ascorbic acid, and ascorbic acid plays a role in formation of hydroxyproline, an integral constituent of collagen. Yet, it is almost certain from the deficiency symptoms that occur when ascorbic acid is lacking that it performs far more functions than simply these.

Though the mechanisms are not clear, ascorbic acid enhances the removal of iron from cellular ferritin, thereby increasing the concentration of iron in the body fluids. Also, in some way not clarified, ascorbic acid potentiates the effects of folic acid in at least some metabolic processes.

Ascorbic acid
(vitamin C)

Physiologically, the major function of ascorbic acid appears to be maintenance of normal intercellular substances throughout the body. This includes the formation of collagen, probably because of the action of ascorbic acid in synthesis of hydroxyproline. It also enhances the intercellular cement substance between the cells, the formation of bone matrix, and the formation of tooth dentin.

Deficiency of ascorbic acid for 20 to 30 weeks, as occurred during long sailing voyages in olden days, causes *scurvy*, some effects of which are the following:

One of the most important effects of scurvy is *failure of wounds to heal*. This is caused by failure of the cells to deposit collagen fibrils and intercellular cement substances. As a result, healing of a wound may require several months instead of the several days ordinarily necessary.

Lack of ascorbic acid causes *cessation of bone growth*. The cells of the growing epiphyses continue to proliferate, but no new matrix is laid down between the cells, and the bones fracture easily at the point of growth because of failure to ossify. Also, when an already ossified bone fractures in a person with ascorbic acid deficiency, the osteoblasts cannot secrete a new matrix for the deposition of new bone. Consequently, the fractured bone does not heal.

The *blood vessel walls become extremely fragile* in scurvy, presumably because of failure of the endothelial cells to be cemented together properly. Especially are the capillaries likely to rupture, and as a result many small petechial hemorrhages occur throughout the body. The hemorrhages beneath the skin cause purpuric blotches, some-

times over the entire body. To test for ascorbic acid deficiency, one can produce such petechial hemorrhages by inflating a blood pressure cuff over the upper arm; this occludes the venous return of blood, the capillary pressure rises, and red blotches occur in the skin immediately if there is a sufficiently severe ascorbic acid deficiency.

In extreme scurvy the muscle cells sometimes fragment; lesions of the gums with loosening of the teeth occur; infections of the mouth develop; vomiting of blood, bloody stools, and cerebral hemorrhage can all occur; and, finally, high fever often develops.

VITAMIN D

Vitamin D increases calcium absorption from the gastrointestinal tract and also helps to control calcium deposition in the bone. The mechanism by which vitamin D increases calcium absorption is to promote active transport of calcium through the epithelium of the ileum. It increases the formation of a calcium-binding protein in the epithelial cells, and it is believed to be this protein that aids in calcium absorption. The specific functions of vitamin D in relation to bone formation are presented in Chapter 79.

Vitamin D₃ (activated 7-dehydrocholesterol)

Several vitamin D compounds exist, one of which is illustrated. This is the natural vitamin D, which results from ultraviolet irradiation of 7-dehydrocholesterol in the skin. A synthetic vitamin D compound, calciferol, is formed by irradiation of ergosterol and is used to a major extent in vitamin D therapy. Another synthetic compound, dehydrotachysterol, has an extremely potent action for increasing the blood calcium concentration, as will be discussed in Chapter 79, and therefore is often used for treatment of persons whose parathyroid glands have been removed.

Unfortunately, the chemical manner in which vitamin D increases calcium absorption from the gastrointestinal tract is totally unknown.

VITAMIN E

Several related compounds, one of which is illustrated below, exhibit so-called "vitamin E activity."

It is doubtful whether vitamin E deficiency occurs significantly in human beings. In lower animals, lack of vitamin E can cause degeneration of the germinal epithelium in the testis and therefore can cause male sterility. Lack of vitamin E can also cause resorption of a fetus after conception in the female. Because of these effects of vitamin E deficiency, vitamin E is sometimes called the "anti-sterility vitamin."

Vitamin E deficiency in animals can also cause paralysis of the hindquarters, and pathologic changes occur in the muscles similar to those found in the disease entity "muscular dystrophy" of the human being. However, administration of vitamin E to patients with muscular dystrophy has not proved to be of any benefit.

Vitamin E (alpha-tocopherol)

Finally, as is true of almost all the vitamins, deficiency of vitamin E prevents normal growth, and it sometimes causes degeneration of the renal tubular cells.

Vitamin E is believed to function mainly in relation to unsaturated fatty acids, providing a protective role to prevent oxidation of the unsaturated fats. In the absence of vitamin E, the quantity of unsaturated fats in the cells becomes diminished, causing abnormal structure and function of such cellular organelles as the mitochondria and the lysosomes. Indeed, the muscular dystrophy-like syndrome that occurs in the vitamin E deficiency probably results from continual rupture of lysosomes and subsequent autodigestion of the muscle.

VITAMIN K

Vitamin K is necessary for the formation by the liver of prothrombin and factor VII (proconvertin), both of which are important in blood coagulation. Therefore, when vitamin K deficiency occurs, blood clotting is retarded. The function of this vitamin and its relationships with some of the anticoagulants, such as Dicumarol, have been presented in greater detail in Chapter 12.

Several different compounds, both natural and synthetic, exhibit vitamin K activity. The chemical formula for one of the natural vitamin K compounds is illustrated below. Because vitamin K is synthesized by bacteria in the colon, a dietary source for this vitamin is not usually necessary, but, when the bacteria of the colon are destroyed by administration of large quantities of antibiotic drugs, vitamin K deficiency occurs readily because of the paucity of this compound in the normal diet.

MINERAL METABOLISM

The functions of many of the minerals, such as sodium, potassium, chloride, and so forth, have been presented at appropriate points in the text. Therefore, only specific functions of minerals not covered elsewhere are mentioned here.

The body content of the most important minerals is listed in Table 73–2, and the daily requirements of these are given in Table 73–3.

TABLE 73–2. Content in Grams of a 70 Kilogram Adult Man

Water	41,400	Mg	21
Fat	12,600	Cl	85
Protein	12,600	P	670
Carbohydrate	300	S	112
Na	63	Fe	3
K	150	I	0.014
Ca	1,160		

TABLE 73–3. Daily Requirements of Minerals

Na	3.0 grams	I	250.0 μg.
K	1.0 grams	Mg	unknown
Cl	3.5 grams	Co	unknown
Ca	0.8 gram	Cu	unknown
PO$_4$	1.5 grams	Mn	unknown
Fe	12.0 mg.	Zn	unknown

Magnesium. Magnesium is approximately one-sixth as plentiful in cells as potassium, and it undoubtedly performs at least some of the same intracellular functions. But magnesium is also

Vitamin K$_1$ (2-methyl-3-phytyl-1,4-naphthoquinone)

required as a catalyst for many intracellular enzymatic reactions, particularly those relating to carbohydrate metabolism.

The extracellular magnesium concentration is slight, only 2 to 3 mEq./liter. Increased extracellular concentration of magnesium depresses activity in the nervous system and also depresses skeletal muscle contraction. This latter effect can be blocked by administration of calcium. Low magnesium concentration causes greatly increased irritability of the nervous system, peripheral vasodilatation, and cardiac arrhythmias.

Calcium. Calcium is present in the body mainly in the form of calcium phosphate in the bone. This subject will be discussed in detail in Chapter 79, as is also the calcium content of the extracellular fluid.

Excess quantities of calcium ion in the extracellular fluids can cause the heart to stop in systole and can act as a mental depressant. On the other hand, low levels of calcium can cause spontaneous discharge of nerve fibers, resulting in tetany. This, too, will be discussed in Chapter 79.

Phosphorus. Phosphate is the major anion of intracellular fluids. Phosphates have the ability to combine reversibly with a multitude of coenzyme systems and also with a multitude of other compounds that are necessary for the operation of metabolic processes. Many important reactions of phosphates have been catalogued at other points in this text, especially in relation to the function of ATP, ADP, creatine phosphate, and so forth. Suffice it to say that phosphates are perhaps the single most important mineral constituent required for cellular activity. Also, bone contains a tremendous amount of calcium phosphate, which will be discussed in Chapter 79.

Iron. The function of iron in the body, especially in relation to the formation of hemoglobin, was discussed in Chapter 8. The major proportion of iron in the body is in the form of hemoglobin, though smaller quantities are present in other forms, especially in the liver and in the bone marrow. Electron carriers containing iron (especially the cytochromes) are present in all the cells of the body and are essential for most of the oxidation that occurs in the cells. Therefore, iron is absolutely essential both for transport of oxygen to the tissues and for maintenance of oxidative systems within the tissue cells, without which life would cease within a few seconds.

Important Trace Elements in the Body. A few elements are present in the body in such small quantities that they are called "trace elements." Usually, the amounts of these in the foods are also minute. Yet without any one of them a specific deficiency syndrome is likely to develop.

IODINE. The best known of the trace elements is iodine. This element is discussed in Chapter 77 in connection with thyroid hormone; as illustrated in Table 73–2, the entire body contains an average of only 14 milligrams. Iodine is essential for the formation of *thyroxine,* which in turn is essential for maintenance of normal metabolic rates in all the cells.

COPPER. Copper deficiency has been observed on rare occasions in infants receiving only a milk diet. The effect of the deficiency is a microcytic, normochromic anemia caused by failure of the intestines to absorb adequate quantities of iron. Lack of copper also decreases the quantities of certain copper-containing enzymes in the tissue cells.

ZINC. Zinc is an integral part of the enzyme *carbonic anhydrase,* which is present in especially high concentration in the red blood cells. This enzyme is responsible for rapid combination of carbon dioxide with water in the red blood cells of the peripheral capillary blood and for rapid release of carbon dioxide from the pulmonary capillary blood into the alveoli. Carbonic anhydrase is also present to a major extent in the gastrointestinal mucosa, in the tubules of the kidney, and in the epithelial cells of many glands of the body. Consequently, zinc in small quantities is essential for the performance of many reactions relating to carbon dioxide metabolism.

Considerable zinc is present in the pancreas, and zinc combines readily with insulin. Therefore, even though zinc is not necessary for the function of insulin, it has been suggested that much of the insulin of the pancreas is stored in this organ as a zinc compound.

Zinc is also a component of *lactic dehydrogenase* and, therefore, is important for the interconversions of pyruvic acid and lactic acid. Finally, zinc is a component part of some *peptidases* and therefore is important for digestion of proteins in the gastrointestinal tract.

COBALT. Cobalt is an essential part of vitamin B_{12}, and vitamin B_{12} is essential for maturation of red blood cells, as discussed earlier in this chapter.

Excess cobalt in the diet causes the opposite of anemia; i.e., polycythemia. However, the polycythemic cells formed contain relatively small concentrations of hemoglobin, so that the total quantity of hemoglobin in the circulating blood remains about normal. Thus, it appears that cobalt is concerned principally with formation of the red blood cell structure and not with formation of hemoglobin.

MANGANESE. Lack of manganese in the diet of animals causes testicular atrophy, though the cause of this is not known. Also, manganese is necessary in the body to activate arginase, which is the principal enzyme necessary for the formation of urea. Consequently, lack of manganese in the diet might prevent the conversion of ammonium ions into urea, and excessive quantities of ammonium compounds developing in the body fluids could cause toxicity. Finally, manganese activates many of the other metabolic enzymes, including cholinesterase and at least one of the enzymes required for glycolysis.

FLUORINE. Fluorine does not seem to be a necessary element for metabolism, but the presence of a small quantity of fluorine in the body during the period of life when the teeth are being formed subsequently protects against carious teeth. Fluorine does not make the teeth themselves stronger but, instead, has some yet unknown effect in suppressing the cariogenic process. It has been suggested that fluorine in the teeth combines with various trace metals that are necessary for activation of the bacterial enzymes, and, because the enzymes are deprived of these trace metals, they remain inactive and cause no caries.

Excessive intake of fluorine causes *fluorosis,* which is manifest in its mild state by mottled teeth and in a more severe state by enlarged bones. It has been postulated that in this condition fluorine combines with trace metals in some metabolic enzymes, including the phosphatases, so that various metabolic systems become partially inactivated. According to this theory, the mottled teeth and enlarged bones are due to abnormal enzyme systems in the odontoblasts and osteoblasts. Even though mottled teeth are highly resistant to the development of caries, the structural strength of these teeth is considerably lessened by the mottling process.

REFERENCES

B$_{12}$ coenzymes (symposium). *Fed. Proc., 25:*1623, 1966.

Biological availability of different forms of vitamins A and E (symposium). *Fed. Proc., 24:*901, 1965.

Ciba Foundation Study Group No. 11: The Mechanism of Action of Water-Soluble Vitamins. Boston, Little, Brown and Company, 1962.

Ciba Symposium on Nutrition and Infection. Boston, Little, Brown and Company, 1967.

Ciba Symposium on Thiamine Deficiency. Boston, Little, Brown and Company, 1967.

Cotzias, G. C., and Foradori, A. C.: Trace metal metabolism. *In* Bittar, E. E., and Bittar, N. (eds.): The Biologic Basis of Medicine. New York, Academic Press, Inc., 1968, Vol. I, p. 105.

Comar, C. L., and Bronner, F. (eds.): Mineral Metabolism. 2 volumes. New York, Academic Press, Inc., 1963.

Frieden, E.: The biochemistry of copper. *Sci. Amer., 218:*102(5), 1968.

Glass, G. B.: Gastric intrinsic factor and its function in the metabolism of vitamin B$_{12}$. *Physiol. Rev., 43:*529, 1963.

Goldsmith, G. A.: Vitamins and avitaminosis. *In* Duncan, G. G. (ed.): Diseases of Metabolism. 5th ed., Philadelphia, W. B. Saunders Company, 1964, p. 567.

Gunderson, F. L.; Gunderson, H. W.; and Ferguson, E. R., Jr.: Food Standards and Definitions in the United States. New York, Academic Press, Inc., 1963.

Merivale, W. H. H.: Mineral metabolism. *In* Duncan, G. G. (ed.): Diseases of Metabolism. 5th ed., Philadelphia, W. B. Saunders Company, 1964, p. 237.

Mottram, V. H.: Human Nutrition. 2nd ed., Baltimore, The Williams & Wilkins Co., 1963.

Pike, R. L., and Brown, M. L.: Nutrition: An Integrated Approach. New York, John Wiley & Sons, Inc., 1967.

Robinson, F. A.: The Vitamin Co-factors of Enzyme Systems. New York, Pergamon Press, Inc., 1967.

Stokstad, E. L. R., and Koch, J.: Folic acid metabolism. *Physiol. Rev., 47:*83, 1967.

Wohl, M. G., and Goodhart, R. S.: Modern Nutrition in Health & Disease. 4th ed., Philadelphia, Lea & Febiger, 1968.

CHAPTER 74

The Liver and Biliary System

Different functions of the liver have been presented at many points in this text because the liver has so many and such varied functions that it is impossible to separate its actions from those of the other organ systems. Therefore, the purpose of the present chapter is to coordinate the different functions of the liver and to show how the liver operates as an individual organ.

The basic functions of the liver can be divided into: (1) its vascular functions for storage and filtration of blood, (2) its secretory function for secreting bile into the gastrointestinal tract, and (3) its metabolic functions concerned with the majority of the metabolic systems of the body.

PHYSIOLOGIC ANATOMY OF THE LIVER

The basic functional unit of the liver is the liver lobule, which is usually a cylindrical structure several millimeters in length and 0.8 to 2 mm. in diameter. The human liver contains 50,000 to 100,000 individual lobules.

The liver lobule is constructed around a *central vein* that empties into the hepatic veins and thence into the vena cava. The lobule itself is composed principally of many *hepatic cellular plates* (two of which are illustrated in Figure 74–1) that radiate centrifugally from the central vein like spokes in a wheel. Each hepatic plate is usually two cells thick, and between the adjacent cells lie small *bile canaliculi* that empty into *terminal bile ducts*

lying in the septa between the adjacent liver lobules.

Also in the septa are small *portal venules* that receive their blood from the portal veins. From these venules blood flows into flat, branching *hepatic sinusoids* that lie between the hepatic plates, and thence into the central vein. Thus, the hepatic cells are exposed on one side to portal venous blood and on the other side to bile canaliculi.

In addition to the portal venules, *hepatic arterioles* are also present in the interlobular septa. These arterioles supply arterial blood to the septal tissues, and many of the small arterioles also empty directly into the hepatic sinusoids.

The venous sinusoids are lined with two types of cells: (1) typical *endothelial cells* and (2) large *Kupffer cells,* which are reticuloendothelial cells capable of phagocytizing bacteria and other foreign matter in the blood. The endothelial lining of the venous sinusoids has extremely large pores, some of which are almost 1 micron in diameter. Beneath this lining, between the endothelial cells and the hepatic cells, is a very narrow space called the *space of Disse.* Because of the large pores in the endothelium, substances in the plasma move freely into the space of Disse. Even large portions of plasma proteins diffuse freely into this space.

In the interlobular septa are also vast numbers of *terminal lymphatics.* The spaces of Disse connect directly with these lymphatics so that excess fluid in these spaces is removed through the lymphatics.

FUNCTION OF THE HEPATIC VASCULAR SYSTEM

The function of the hepatic vascular system was discussed in Chapter 32 in connection with the portal veins. Briefly, this can be summarized as follows:

Blood Flow Through the Liver. About 1000 ml. of blood flows from the portal vein through the liver sinusoids each minute, and approximately an additional 400 ml. flows into the sinusoids from the hepatic artery, the total averaging about 1400 ml. per minute.

Total hepatic blood flow per minute can be measured by a modified Fick procedure in which the

Figure 74–1 Basic structure of the liver lobule, showing the hepatic cellular plates, the blood vessels, and the bile ducts.

Labels in figure: Hepatic cell plate; Bile canaliculus; Central vein; Venous sinusoid; Kupffer cell; Terminal bile duct; Portal venule dye *Bromsulphalein* is injected into the circulatory system, and its concentration is measured in the arterial blood and in venous blood collected from the hepatic vein by means of a catheter. From these measurements, one can calculate the arterio-venous difference of the Bromsulphalein. In addition, the rate at which the Bromsulphalein disappears from the blood is determined over a period of 10 minutes or more. Since the only significant route of disappearance is through the hepatic cells into the biliary tract, one can then calculate liver blood flow by the usual Fick formula as follows:

Hepatic blood flow =
$$\frac{\text{Rate of blood clearance of Bromsulphalein}}{\text{A-V difference in Bromsulphalein}}$$

Pressures and Resistance in the Hepatic Vessels. The pressure in the hepatic vein leading from the liver into the vena cava averages almost exactly 0 mm. Hg, while the pressure in the portal vein leading into the liver averages 8 mm. Hg. This shows that the resistance to blood flow from the portal venous system to the systemic veins is normally low, especially when one considers that about 1.4 liters of blood flow by this route each minute. However, various pathologic conditions can cause the resistance to rise markedly, sometimes increasing the portal venous pressure to as high as 20 to 30 mm. Hg. The most common cause of increased hepatic vascular resistance is the disease *portal cirrhosis*, in which many of the exit pathways from the sinusoids into the central lobular venules are blocked.

Storage of Blood in the Liver; Hepatic Congestion. An increase in pressure in the veins draining the liver dams blood in the liver sinusoids and thereby causes the entire liver to swell markedly. The liver can store 200 to 400 ml. of blood in this way as the result of only 4 to 8 mm. Hg rise in hepatic venous pressure. For this reason, the liver is frequently said to be one of the *blood reservoirs*. Conversely, if a person hemorrhages so that large amounts of blood are lost from the circulatory system, much of the normal blood in the liver sinusoids drains into the remainder of the circulation to help replace the lost blood.

The most important cause of hepatic congestion is cardiac failure, which often increases the central venous pressure to as high as 10 to 15 mm. Hg. The continual stretching of the liver sinusoids that results and the stasis of blood caused by the hepatic congestion gradually leads to necrosis of many of the hepatic cells in the hepatic cellular plates.

Lymph Flow from the Liver. Because the pores in the hepatic sinusoids allow ready passage of proteins, the lymph draining from the liver has a protein concentration usually about 6 grams per cent, which is only slightly less than the protein concentration of plasma. Also, the extreme permeability of the liver sinusoids allows large quantities of lymph to form. Indeed, between one third and one half of all the lymph formed in the body under resting conditions arises in the liver.

Effects of High Hepatic Vascular Pressures on Fluid Transudation from the Liver Sinusoids and Portal Capillaries. When the hepatic venous pressure rises only 3 to 5 mm. Hg above normal, fluid begins to transude into the lymph and also to leak through the outer surface of the liver capsule directly into the abdominal cavity. This fluid is almost pure plasma, containing 80 to 90 per cent as much protein as normal plasma. At still higher hepatic venous pressures, 10 to 15 mm. Hg, liver lymph flow increases to as much as 20 times normal, and the "sweating" from the surface of the liver can be so great that it causes severe *ascites* (free fluid in the abdominal cavity).

Blockage of portal blood flow into or through the liver causes very high portal capillary pressures, resulting in edema of the gut wall and transudation of fluid through the serosa of the gut into the abdominal cavity. This, too, can cause ascites but is less likely to do so than is sweating from the liver surface, because collateral vascular channels develop rapidly from the portal veins to the systemic veins, lowering the intestinal capillary pressure back to a safe value.

The Hepatic Reticuloendothelial System. The inner surfaces of all the liver sinusoids are loaded with many *Kupffer cells,* which protrude into the flowing blood as illustrated in Figure 74–1 and as shown in detail in Figure 9–3 of Chapter 9. These cells are very phagocytic, so much so that they can remove perhaps 99 per cent of bacteria in the portal venous blood before they can pass all the way through the liver sinusoids. Since the portal blood drains from the intestines, it almost always contains a reasonable number of colon bacilli. Therefore, the importance of the Kupffer cell filtration system is readily apparent. The number of Kupffer cells in the sinusoids increases markedly when increased quantities of particulate matter or other debris are present in the blood.

SECRETION OF BILE AND FUNCTIONS OF THE BILIARY TREE

PHYSIOLOGIC ANATOMY OF BILIARY SECRETION

All the hepatic cells continually form a small amount of secretion called *bile*. This is secreted into the minute *bile canaliculi,* which lie between the double layer of cells in the hepatic plates, and the bile then flows peripherally toward the interlobular septa, where the canaliculi empty into *terminal bile ducts,* then into progressively larger ducts, finally reaching the *hepatic duct* and *common bile duct* from which the bile either empties directly into the duodenum or is diverted into the gallbladder.

Storage of Bile in the Gallbladder. In the discussion of bile and its functions in relation to gastrointestinal secretions in Chapter 64, it was pointed out that bile secreted continually by the liver cells is normally stored in the gallbladder until needed in the duodenum. The total secretion of bile each day is some 600 to 800 ml., and the maximum volume of the gallbladder is only 40 to 70 ml. Nevertheless, at least 12 hours' bile secretion can be stored in the gallbladder because water, sodium, chloride, and most other small electrolytes are continually absorbed by the gallbladder mucosa, concentrating the other bile constituents, including the bile salts, cholesterol, and bilirubin. Bile is normally concentrated about five-fold, but it can be concentrated up to a maximum of about 10-fold.

EMPTYING OF THE GALLBLADDER. Two basic conditions are necessary for the gallbladder to empty: (1) The sphincter of Oddi must relax to allow bile to flow from the common bile duct into the duodenum and (2) the gallbladder itself must contract to provide the force required to move the bile along the common duct. After a meal, particularly one that contains a high concentration of fat, both these effects take place in the following manner:

First, the fat in the food entering the small intestine extracts a hormone called *cholecystokinin* from the intestinal mucosa, especially from the upper regions of the small intestine. The cholecystokinin in turn is absorbed into the blood and on passing to the gallbladder causes specific contraction of the gallbladder muscle. This provides the pressure that forces bile toward the duodenum.

Second, vagal stimulation associated with the cephalic phase of gastric secretion or with various intestino-intestinal reflexes causes an additional weak contraction of the gallbladder, which helps to force the bile from the gallbladder into the duodenum.

Third, when the gallbladder contracts, the sphincter of Oddi becomes inhibited, this effect resulting from either a neurogenic or a myogenic reflex from the gallbladder to the sphincter of Oddi.

Fourth, the presence of food in the duodenum causes the degree of peristalsis in the duodenal wall to increase. Each time a peristaltic wave travels toward the sphincter of Oddi, this sphincter, along with the adjacent intestinal wall, momentarily relaxes because of the phenomenon of "receptive relaxation," and, if the bile in the common bile duct is under sufficient pressure, a small quantity of the bile squirts into the duodenum.

In summary, the gallbladder empties its store of concentrated bile into the duodenum mainly in response to the cholecystokinin stimulus. When fat is not in the meal, the gallbladder empties poorly, but, when adequate quantities of fat are present, the gallbladder empties completely in about one hour.

Figure 74–2 summarizes the secretion of bile, its

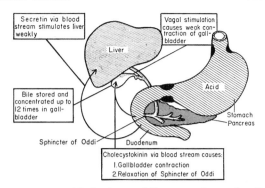

Figure 74–2 Mechanisms of liver secretion and gallbladder emptying.

storage in the gallbladder, and its release from the bladder to the gut.

Composition of Bile. Table 74–1 gives the composition of bile when it is first secreted by the liver and then after it has been concentrated in the gallbladder. This table shows that the most abundant substance secreted in the bile is the *bile salts,* but also secreted or excreted in large concentrations are *bilirubin, cholesterol, fatty acids,* and the usual *electrolytes* of plasma. In the concentrating process in the gallbladder, water, and large portions of the electrolytes are reabsorbed by the gallbladder mucosa, but essentially all the other constituents, including especially the bile salts

TABLE 74–1. Composition of Bile

	Liver Bile	Gallbladder Bile
Water	97.5 gm. %	92 gm. %
Bile salts	1.1 gm. %	3 to 10 gm. %
Bilirubin	0.2 gm. %	0.6 to 2.0 gm. %
Cholesterol	0.1 gm. %	0.3 to 0.9 gm. %
Fatty acids	0.12 gm. %	0.3 to 1.2 gm. %
Lecithin	0.04 gm. %	0.1 to 0.4 gm. %
Na^+	145 mEq./l.	130 mEq./l.
K^+	5 mEq./l.	9 mEq./l.
Ca^{++}	5 mEq./l.	12 mEq./l.
Cl^-	100 mEq./l.	75 mEq./l.
HCO_3^-	28 mEq./l.	10 mEq./l.

and lipid substances such as cholesterol, are not reabsorbed and therefore become highly concentrated in the gallbladder bile.

THE BILE SALTS AND THEIR FUNCTION

The liver cells form about 1 gram of *bile salts* daily. The precursor of the bile salts is *cholesterol,* which is either supplied in the diet or synthesized in the liver cells during the course of fat metabolism and then is converted to *cholic acid.* This then combines principally with glycine to form *glycocholic acid* and to a lesser extent with taurine to form *taurocholic acid.* The salts of these acids are secreted in the bile.

The bile salts have two important actions in the intestinal tract. First, they have a detergent action on the fat particles in the food, which decreases the surface tension of the particles and allows the agitation in the intestinal tract to break the fat globules into minute sizes. This is called the *emulsifying* or *detergent function* of bile salts. Second, and even more important than the emulsifying function, bile salts help in the absorption of fatty acids, monoglycerides, cholesterol, and other lipids from the intestinal tract, and this is called their *hydrotropic function*. It is believed that the bile salt ions become physically adsorbed to the lipids, and the electrical charges of these ions presumably then increase the solubility of the lipids, thereby allowing ready passage through the intestinal mucosa. Without the presence of bile salts in the intestinal tract, up to 40 per cent of the lipids are lost into the stools, and the person often develops a metabolic deficit due to this nutrient loss.

Also, when fats are not absorbed adequately, the fat-soluble vitamins are not absorbed satisfactorily. Therefore, in the absence of bile salts, vitamins A, D, E, and K are poorly absorbed. Though large quantities of the first three of these vitamins are usually stored in the body, this is not true of vitamin K. Within only a few days after bile secretion ceases, the person usually develops a deficiency of vitamin K. This in turn results in deficient formation by the liver of factor VII and prothrombin, thus resulting in serious impairment of blood coagulation.

Enterohepatic Circulation of Bile Salts. Approximately 94 per cent of the bile salts are reabsorbed by the intestinal mucosa along with the absorption of the fats, but, after passing through the mucosa, the bile salts separate from the fats and pass into the portal blood, while the fats travel in the lymphatics. On reaching the liver the bile salts are absorbed from the venous sinusoids into the hepatic cells and then resecreted into the bile canaliculi. In this way about 94 per cent of all the bile salts are recirculated into the bile, so that on the average these salts make the entire circuit some 18 times before being carried out in the feces. The small quantities of bile salts lost into the feces are replaced by new amounts formed continually by the liver cells. This recirculation of the bile salts is called the *enterohepatic circulation*.

The quantity of bile secreted by the liver each day is highly dependent on the availability of bile salts—the greater the quantity of bile salts in the enterohepatic circulation (usually a total of about 4 gm.), the greater is the rate of bile secretion. When a bile fistula forms so that bile is lost directly from the common bile duct to the exterior, the bile salts cannot be reabsorbed. Therefore, the total quantity of bile salts in the enterohepatic circulation becomes greatly depressed, and concurrently the volume of liver secretion is also depressed.

However, if a bile fistula continues to empty the bile salts to the exterior for several days to several weeks, the liver increases its production of bile salts as much as 10-fold, which increases the rate of bile secretion approximately back to normal. This also demonstrates that the daily rate of bile salt secretion is actively controlled, though the mechanism of this control is unknown.

EXCRETION OF BILIRUBIN IN THE BILE

In addition to secreting substances synthesized by the liver itself, the liver cells also *excrete* a number of substances formed elsewhere in the body. Among the most important of these is *bilirubin*, which is one of the major end-products of hemoglobin decomposition, as was pointed out in Chapter 8.

Briefly, when the red blood cells have lived out their life span, averaging 120 days, and have become too fragile to exist longer in the circulatory system, their cell membranes rupture, and the released hemoglobin is phagocytized by reticuloendothelial cells throughout the body. Here, the hemoglobin is first split into *globin* and *heme,* and the heme ring is opened to give a straight chain of four pyrrole nuclei that is the substrate from which the bile pigments are formed. The first pigment formed is *biliverdin,* but this is rapidly reduced to *bilirubin,* which is gradually released into the plasma. However, bilirubin itself is almost insoluble in the body fluids. Instead, it combines with some of the plasma proteins, mainly albumin, and is transported in this combination throughout the blood and interstitial fluids. This protein-bound bilirubin eventually is absorbed into the liver cells, and the bilirubin is removed from the protein and conjugated with other substances that make it become highly soluble. About 80 per cent of it conjugates with glucuronic acid to form *bilirubin glucuronide;* an additional 10 per cent conjugates with sulfate to form *bilirubin sulfate,* and the final 10 per cent conjugates with a multitude of other solubilizing substances. It is in these soluble forms that the bilirubin is excreted into the bile.

A small portion of the soluble bilirubin formed by the hepatic cells returns to the plasma. Either this is resecreted by the liver cells directly into the liver sinusoids or it is reabsorbed into the blood from the bile ducts or lymphatics. Regardless of the exact mechanism by which it re-enters the blood, this causes a small portion of the bilirubin in the extracellular fluids always to be of the soluble type rather than of the protein-bound type.

Formation and Fate of Urobilinogen. Once in the intestine, bilirubin is converted by bacterial action mainly into the substance *urobilinogen,* which is highly soluble. Some of the urobilinogen is reabsorbed through the intestinal mucosa into the blood and is re-excreted by the liver back into the gut, or the urobilinogen may be excreted by the kidneys into the urine. After exposure to air in the urine, the urobilinogen becomes oxidized to *urobilin,* or it becomes altered and oxidized in the feces to form *stercobilin.* These interrelationships of bilirubin and the other bile pigments are illustrated in Figure 74–3.

Jaundice. The word "jaundice" means a yellowish tint to the body tissues, including yellowness of the skin and also of the deep tissues. The usual cause of jaundice is large quantities of bilirubin in the extracellular fluids, either protein-bound bilirubin or soluble bilirubin. The normal plasma concentration of bilirubin, including both the protein-bound and the soluble forms, averages 0.5 mg. per 100 ml. of plasma. However, in certain abnormal conditions this can rise to as high as 40 mg. per 100 ml. The skin usually begins to appear jaundiced when the concentration rises to about three times normal — that is, above 1.5 mg. per 100 ml.

The common causes of jaundice are: (1) increased destruction of red blood cells with rapid release of bilirubin into the blood or (2) obstruction of the bile ducts or damage to the liver cells so that even the usual amounts of bilirubin cannot be excreted into the gastrointestinal tract. These two types of jaundice are called, respectively, *hemolytic jaundice* and *obstructive jaundice*. They differ from each other in the following ways:

HEMOLYTIC JAUNDICE. In hemolytic jaundice, the excretory function of the liver is not impaired in the least, but red blood cells are hemolyzed rapidly and the hepatic cells simply cannot excrete the bilirubin as rapidly as it is formed. Therefore, the plasma concentration of *protein-bound bilirubin* rises especially high. Likewise, the rate of formation of *urobilinogen* in the intestine is greatly increased, and much of this is absorbed into the blood and later excreted by the kidneys.

OBSTRUCTIVE JAUNDICE. In obstructive jaundice, caused either by obstruction of the bile ducts or by damage to the liver cells, the rate of bilirubin formation is normal, but the bilirubin formed simply cannot pass from the blood into the intestines. However, the protein-bound bilirubin usually does still enter the liver cells, the bilirubin is split from the proteins and conjugated with glucuronic acid, and this soluble form of bilirubin is returned to the blood probably mainly by rupture of the conjested bile canaliculi and direct

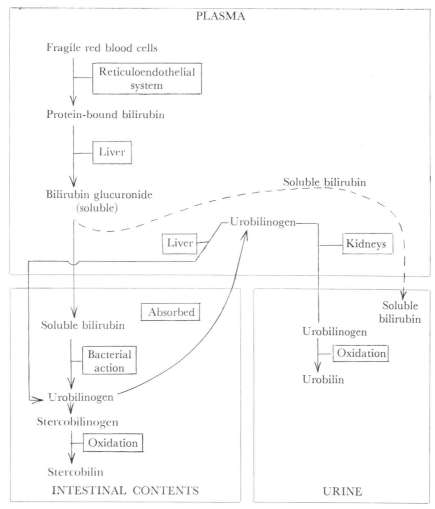

Figure 74–3 Bilirubin formation and excretion.

emptying of the bile into the lymph leaving the liver. Thus, *most of the bilirubin in the plasma becomes the soluble type* rather than the protein-bound type.

DIAGNOSTIC DIFFERENCES BETWEEN HEMOLYTIC AND OBSTRUCTIVE JAUNDICE. A simple test called the *van den Bergh test* can be used to differentiate between protein-bound and soluble bilirubin in the plasma. If an immediate reaction occurs with the van den Bergh reagent, the bilirubin is of the soluble type, and the reaction is called a "direct van den Bergh reaction." However, to demonstrate the presence of protein-bound bilirubin, one must first add alcohol to the plasma. This precipitates the protein and dissolves the formerly protein-bound bilirubin so that it can then combine with the van den Bergh reagent. Therefore, the test gives an "indirect van den Bergh reaction." Thus, *in hemolytic jaundice an indirect van den Bergh reaction occurs (increased protein-bound bilirubin) and in obstructive jaundice a direct van den Bergh reaction occurs (increased soluble bilirubin).*

When there is total obstruction of bile flow, no bilirubin at all can reach the intestines to be converted into urobilinogen by bacteria. Therefore, urobilinogen is not reabsorbed into the blood and is not excreted by the kidneys into the urine. Consequently, in *total* obstructive jaundice, tests for urobilinogen in the urine are completely negative. Also, the stools become clay colored for lack of stercobilin and other bile pigments.

Another major difference between protein-bound and soluble bilirubin is that the kidneys excrete soluble bilirubin but not protein-bound bilirubin. Therefore, in severe obstructive jaundice, large quantities of soluble bilirubin appear in the urine. This can be demonstrated simply by shaking the urine and observing the foam, which becomes colored an intense yellow.

Thus, by understanding the physiology of bilirubin excretion by the liver and by use of a few simple tests, it is possible to differentiate between obstructive and hemolytic jaundice and, often, also to determine the severity of the disease.

SECRETION OF CHOLESTEROL; GALLSTONE FORMATION

Bile salts are formed in the hepatic cells from cholesterol, and in the process of secreting the bile salts about one-tenth as much cholesterol is also secreted into the bile. No specific function is known for the cholesterol in the bile, and it is presumed that it is simply a by-product of bile salt formation and secretion.

Cholesterol is almost insoluble in pure water, but the bile salts, fatty acids, and lecithin in bile exhibit a hydrotropic action on the cholesterol to make it soluble. When the bile becomes concentrated in the gallbladder, all the hydrotropic substances become concentrated along with the cholesterol, which keeps the cholesterol in solution. Under abnormal conditions, however, the

cholesterol may precipitate, resulting in the formation of *gallstones*, as shown in Figure 74–4. The different conditions that can cause cholesterol precipitation are: (1) too much absorption of water from the bile, (2) too much absorption of bile salts, fatty acids, and lecithin from the bile, (3) too much secretion of cholesterol in the bile, and (4) inflammation of the epithelium of the gallbladder. The latter two of these require special explanation, as follows:

The amount of cholesterol in the bile is determined partly by the quantity of fat that the person eats, for the hepatic cells synthesize cholesterol as one of the products of fat metabolism in the body. For this reason, persons on a high fat diet over a period of many years are prone to the development of gallstones.

Inflammation of the gallbladder epithelium often results from low grade chronic infection; this changes the absorptive characteristics of the gallbladder mucosa, sometimes allowing excessive absorption of water, bile salts, or other substances that are necessary to keep the cholesterol in solution. As a result, cholesterol begins to precipitate, usually forming many small crystals of cholesterol on the surface of the inflamed mucosa. These, in turn, act as nidi for further precipitation of cholesterol, and the crystals grow larger and larger. Occasionally tremendous numbers of sand-like stones develop, but much more frequently they coalesce to form a few large gallstones, or even a single stone that fills the entire gallbladder.

X-ray Opaque Gallstones. Calcium sometimes precipitates in the form of calcium carbonate in the gallstones or precipitates with some of the fatty substances involved in the formation of cholesterol gallstones. In about one-fourth of all cases of gallstones, the concentration of calcium in the stones is great enough to make them x-ray opaque. That is, the gallstones show up on x-ray pictures of the abdomen. In the remaining three-fourths of

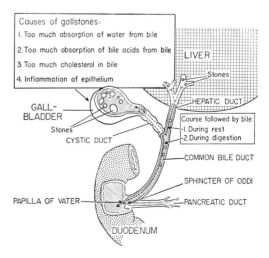

Figure 74–4 Formation of gallstones.

the cases, the stones have almost no calcium and therefore have almost the same x-ray opaqueness as the tissues of the body. Consequently, the stones cannot be seen on an x-ray film. In these instances, special iodinated drugs can be administered to the patient. These drugs concentrate in the bile surrounding the stones, and an x-ray picture shows the outlines of the stones.

METABOLIC FUNCTIONS OF THE LIVER

The metabolic functions of the liver are so numerous and intricate that they could not possibly be presented completely in this chapter. Therefore, for details, refer to the preceding chapters on the metabolism of carbohydrates, fats, and proteins. Briefly, the specific roles of the liver in the different metabolic processes are described as follows:

CARBOHYDRATE METABOLISM

In carbohydrate metabolism the liver performs the following specific functions: (1) storage of glycogen, (2) conversion of galactose and fructose to glucose, (3) gluconeogenesis, and (4) formation of many important chemical compounds from the intermediate products of carbohydrate metabolism.

Most of these functions of the liver in carbohydrate metabolism are especially important in maintaining a normal blood glucose concentration. For instance, storage of glycogen allows the liver to remove excess glucose from the blood, store it, and then return it to the blood when the blood glucose concentration begins to fall too low. This is called the *glucose buffer function* of the liver. For instance, immediately after a meal containing large amounts of carbohydrates, the blood glucose concentration rises about three times as much in a person with a nonfunctional liver as in a person with a normal liver.

Gluconeogenesis in the liver is also concerned with maintaining a normal blood glucose concentration, for gluconeogenesis occurs to a significant extent only when the glucose concentration begins to fall below normal. In such a case, large amounts of amino acids are converted into glucose, thereby helping to maintain a relatively normal blood glucose concentration.

The conversion of galactose and fructose to glucose is also important for maintaining normal blood glucose, because the greater proportion of the glucose formed in this manner is eventually returned by the liver to the blood. This function is particularly important for the metabolism of galactose because almost none of the cells of the body besides those of the liver can convert galactose into glucose. However, fructose can be used by all the cells in much the same manner as glucose.

FAT METABOLISM

Though fat metabolism can take place in almost all cells of the body, certain aspects of fat metabolism occur so much more rapidly in the liver than in the other cells that probably 50 per cent or more of the different fat interconversions occur in the liver rather than elsewhere. Some specific functions of the liver in fat metabolism are: (1) beta oxidation of fatty acids and formation of acetoacetic acid, (2) formation of the lipoproteins, (3) formation of large quantities of cholesterol and phospholipids, and (4) conversion of large quantities of carbohydrates and proteins to fat.

To derive energy from neutral fats, the fat is first split into glycerol and fatty acids; then the fatty acids are split by *beta oxidation* into two-carbon acetyl radicals which form *acetyl coenzyme A* (acetyl Co-A). This in turn can then enter the tricarboxylic acid cycle and be oxidized to liberate tremendous amounts of energy. Beta oxidation can probably take place in all cells of the body, but it occurs so rapidly in the hepatic cells in comparison with the others that about 60 per cent of all initial oxidation of fatty acids in the body occurs in the liver. Yet the liver itself cannot utilize all the acetyl Co-A that is formed; instead, this is converted by condensation of two molecules of acetyl Co-A into *acetoacetic acid,* which is a highly soluble acid that passes from the liver cells into the extracellular fluids and then is transported throughout the body to be absorbed by the other tissues. These tissues in turn reconvert the acetoacetic acid into acetyl Co-A and then oxidize it in the usual manner. In this way, therefore, the liver is responsible for the major part of the metabolism of fats.

Except for the use of cholesterol to form bile salts, the functions of the cholesterol and phospholipids formed in the liver are still in doubt. About 80 per cent of the cholesterol is converted into bile salts, but the remainder enters the blood to be transported principally in the lipoproteins. The phospholipid lecithin likewise is transported principally in the lipoproteins. It is possible that both these substances, along with the triglyceride fractions of the lipoproteins, are absorbed by cells everywhere in the body to help form the cell membranes and intracellular structures, for it is well known that most membranous structures throughout the body contain major amounts of cholesterol, phospholipids, and triglycerides.

Some of the fat synthesis in the body from carbohydrates and proteins also occurs in the liver. After fat is synthesized in the liver it is transported in the lipoproteins to the adipose tissue to be stored.

PROTEIN METABOLISM

Even though a large proportion of the metabolic processes for carbohydrates and fat metabolism occurs in the liver, the body could probably dispense with these functions of the liver and still survive. On the other hand, the body could not dispense with the services of the liver in protein metabolism for more than a few days without death ensuing. The most important functions of the liver in protein metabolism are: (1) deamination of amino acids, (2) formation of urea for removal of ammonia from the body fluids, (3) formation of plasma proteins, and (4) interconversions among the different amino acids and other compounds important to the metabolic processes of the body.

Deamination of the amino acids is required before these can be used for energy or before they can be converted into carbohydrates or fats. A small amount of deamination can occur in the other tissues of the body, especially in the kidneys; but here again, quantitatively the percentage of deamination occurring extrahepatically is so small that it is almost completely unimportant.

Formation of urea by the liver removes ammonia from the body fluids. Since large amounts of ammonia are continually formed in the gut by bacteria and then absorbed into the blood, without this function of the liver the plasma ammonia concentration rises rapidly and results in *hepatic coma* and death.

Essentially all the plasma proteins, with the exception of part of the gamma globulins, are formed by the hepatic cells. This accounts for more than 95 per cent of all the plasma proteins. The remaining gamma globulins are the immune bodies formed mainly by the plasma cells elsewhere in the body. The liver can form plasma proteins at a maximum rate of 50 grams per day. Therefore, after loss of as much as half the plasma proteins from the body, these can be replenished in approximately four to seven days. It is particularly interesting that plasma protein depletion causes rapid mitosis of the hepatic cells and actual growth of the liver to a larger size, these effects coupled with rapid output of plasma proteins until the plasma concentration returns to normal.

Among the most important functions of the liver is its ability to synthesize certain amino acids and also to synthesize other important chemical compounds from amino acids. For instance, the so-called nonessential amino acids can be synthesized in the liver. To do this, a keto acid having the same chemical composition (except at the keto oxygen) as that of the amino acid to be formed is first synthesized. Then an amino radical is transferred through several stages of *transamination* from an available amino acid to the keto acid to take the place of the keto oxygen.

MISCELLANEOUS METABOLIC FUNCTIONS OF THE LIVER

Storage of Vitamins. The liver has a particular propensity for storing vitamins and has long been known as an excellent source of certain vitamins in treating patients. The single vitamin stored to the greatest extent in the liver is vitamin A, but large quantities of vitamin D and vitamin B_{12} are normally stored as well. Sufficient quantities of vitamin A can be stored to prevent vitamin A deficiency for as long as one to two years, and sufficient vitamin D and vitamin B_{12} can be stored to prevent deficiency for as long as three to six months.

Relationship of the Liver to Blood Coagulation. The liver forms a large proportion of the blood substances utilized in the coagulation process. These are fibrinogen, prothrombin, accelerator globulin, factor VII, and several other less important coagulation factors. Vitamin K is required by the metabolic processes of the liver for the formation of prothrombin and factors VII, IX, and X. In the absence of vitamin K the concentrations of these substances fall very low and almost prevent blood coagulation.

Storage of Iron. Except for the iron in the hemoglobin of the blood, by far the greater proportion of the iron in the body is usually stored in the liver in the form of *ferritin*. The hepatic cells contain large amounts of a protein called *apoferritin*, which is capable of combining with either small or large quantities of iron. Therefore, when iron is available in the body fluids in extra quantities, it combines with the apoferritin to form ferritin and is stored in this form until needed by the body. When the iron in the circulating body fluids falls low, the ferritin releases the iron. Thus, the apoferritin-ferritin system of the liver acts as an *iron buffer* and also as an iron storage medium. Other functions of the liver in relation to iron metabolism are considered in Chapter 8.

REFERENCES

Billing, B. H.: Bile pigment metabolism. *Scient. Basis Med. Ann. Rev.,* 197, 1963.

Brauer, R. W.: Liver circulation and function. *Physiol. Rev., 43:* 115, 1963.

Cohn, J. N., and Pinkerson, A. L.: Intrahepatic distribution of hepatic arterial and portal venous flows in the dog. *Amer. J. Physiol.,* 216:285, 1969.

Di Luzio, N. R.: Comparative study of the effect of alcoholic beverages on the development of the acute ethanol-induced fatty liver. *Quart. J. Stud. Alcohol,* 23:557, 1962.

Glinsmann, W. H.; Hern, E. P.; and Lynch, A.: Intrinsic regulation of glucose output by rat liver. *Amer. J. Physiol.,* 216:698, 1969.

Goldstein, F.: The gallbladder and pancreas. *In* Sodeman, W. A., and Sodeman, W. A., Jr. (eds.): Pathologic Physiology. 4th ed., Philadelphia, W. B. Saunders Company, 1967, p. 696.

Ingelfinger, F. J.: The liver. *In* Sodeman, W. A., and Sodeman, W. A., Jr. (eds.): Pathologic Physiology. 4th ed., Philadelphia, W. B. Saunders Company, 1967, p. 661.

Kamm, D. E., and Cahill, G. F., Jr.: Effect of acid-base status on renal and hepatic gluconeogenesis in diabetes and fasting. *Amer. J. Physiol.,* 216:1207, 1969.

Katz, M. L., and Bergman, E. N.: Hepatic and portal metabolism of glucose, free fatty acids, and ketone bodies in the sheep. *Amer. J. Physiol.,* 216:953, 1969.

Klaasen, C. D., and Plaa, G. L.: Hepatic disposition of phenoldi-bromphthalein disulfonate and sulfobromophthalein. *Amer. J. Physiol.,* 215:971, 1968.

Parli, C. J.; Burns, L. C.; Meghal, S. K.; Kinneberg, K.; O'Neal, R. M.; and Koeppe, R. E.: Pyruvate metabolism in mammalian liver. *Amer. J. Physiol.,* 216:1525, 1969.

PART XI

ENDOCRINOLOGY AND REPRODUCTION

CHAPTER 75

Introduction to Endocrinology, the Hypophyseal Hormones, and the Pineal Gland

The functions of the body are regulated by two major control systems: (1) the nervous system, which has been discussed, and (2) the hormonal, or endocrine, system. In general, the hormonal system is concerned principally with the different metabolic functions of the body: controlling the rates of chemical reactions in the cells or the transport of substances through cell membranes or other aspects of cellular metabolism, such as growth and secretion. Some hormonal effects occur in seconds, while others require several days even to get started and then last for weeks, months, or even years.

Many interrelationships exist between the hormonal and nervous systems. For instance, at least two glands secrete their hormones only in response to appropriate nerve stimuli, the *adrenal medullae* and the *neurohypophysis,* and few of the adenohypophyseal hormones are secreted to a significant extent except in response to nervous activity in the hypothalamus, as is detailed later in this chapter.

NATURE OF A HORMONE

A hormone is a chemical substance that is secreted into the body fluids by one cell or a group of cells and that exerts a physiological effect on other cells of the body.

At many points in this text we have already discussed different hormones, some of which are called *local hormones* and others that are called *general hormones.* The local hormones include *acetylcholine* released at the parasympathetic and skeletal nerve endings, *secretin* released by the duodenal wall and transported in the blood to the pancreas to cause a watery pancreatic secretion, *cholecystokinin* released in the small intestine and transported to the gallbladder to cause contraction, and many others. These hormones obviously have specific local effects, whence comes the name local hormones.

On the other hand, the general hormones are secreted by specific *endocrine glands* and are transported in the blood to cause physiologic actions at distant points in the body. A few of the general hormones affect all, or almost all, cells of the body; examples are *growth hormone* from the adenohypophysis and *thyroid hormone* from the thyroid gland. On the other hand, other general hormones affect specific tissues far more than other tissues; for instance, *corticotropin* from the adenohypophysis specifically stimulates the adrenal cortex, and the *ovarian hormones* have specific effects on the uterine endometrium. The tissues affected specifically in this way are called *target tissues.* Many examples of target organs will become apparent in the following chapters on endocrinology.

The following general hormones have proved to be of major significance and are discussed in detail in this and the following chapters:

Adenohypophyseal hormones: *growth hormone, corticotropin, thyrotropin, follicle-stimulating hormone, luteinizing hormone, luteotropic hormone,* and *melanocyte-stimulating hormone.*

Neurohypophyseal hormones: *antidiuretic hormone* and *oxytocin.*

Adrenocortical hormones: especially *cortisol* and *aldosterone.*

Thyroid hormones: *thyroxine* and *calcitonin.*

Pancreatic hormones: *insulin* and *glucagon.*

Ovarian hormones: *estrogens* and *progesterone.*

Testicular hormones: especially *testosterone.*

Parathyroid hormone: *parathormone* and *calcitonin.*

Placental hormones: *chorionic gonadotropin, estrogens,* and *progesterone.*

Negative Feedback in the Control of Hormonal Secretion. At many points both in this chapter and in the succeeding few chapters we will see that once a hormone accomplishes its physiological function, its rate of secretion usually begins to decrease. This is caused by negative feedback, a phenomenon we have seen to be important in nervous control systems. In general, each gland has a basic tendency to *oversecrete* its particular hormone, but, once the normal physiological effect of the hormone has been achieved, information is transferred in some way back to the producing gland to check further secretion. On the other hand, if the gland undersecretes, the physiological effects of the hormone diminish, and the feedback decreases, thus allowing the gland to begin secreting adequate quantities of the hormone once again. In this way, the rate of secretion of each hormone is controlled in accord with the need for the hormone. The specific negative feedback mechanisms are discussed in relation to the different individual hormones.

Chemistry of the Hormones. Chemically, the basic types of hormones are: (1) proteins or derivatives of proteins or amino acids and (2) steroid hormones. For example, the hormones of the pancreas and adenohypophysis are proteins or large polypeptides, while the hormones of the neurohypophysis, thyroid, and adrenal medulla are derivatives of proteins or amino acids. The steroids are secreted by tissues derived from the mesenchymal zone of the embryo, including the adrenal cortex, the ovary, and the testis.

INTRACELLULAR HORMONAL MEDIATORS — CYCLIC AMP

Many hormones probably exert their effects on cells by first causing the substance called cyclic $3',5'$-adenosine monophosphate (cyclic AMP) to be formed in the cell. Once formed, the cyclic AMP causes the hormonal effects inside the cell. Thus, cyclic AMP is an intracellular hormonal mediator.

Figure 75–1 illustrates the function of cyclic AMP in more detail. It is believed that the stimulating hormone acts at the membrane of the target cell, presumably combining with a specific receptor for that particular type of hormone. The specificity of the receptors determines which hormones will affect which target cells. After binding with the receptor, the combination of hormone and receptor activates the enzyme *adenyl cyclase* in the membrane, and the portion of the adenyl cyclase

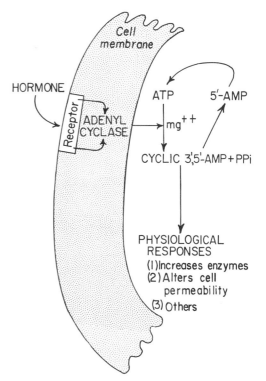

Figure 75–1 The cyclic AMP mechanism by which many hormones exert their control of cell function.

that is exposed to the cytoplasm causes immediate conversion of cytoplasmic ATP into cyclic AMP. The cyclic AMP then initiates any number of cellular functions before it itself is destroyed; functions such as increasing the number of enzymes in the cell, altering the cell permeability, initiating synthesis of specific intracellular chemicals, and so forth. The types of effects that will occur inside the cell are determined by the character of the cell itself. Thus, a thyroid cell stimulated by cyclic AMP forms thyroid hormones, whereas an adrenocortical cell forms adrenocortical hormones. On the other hand, cyclic AMP affects epithelial cells of the renal tubules by increasing their permeability to water.

The cyclic AMP mechanism has been proposed as an intracellular hormonal mediator for at least the following hormones (and perhaps there are more): thyrotropin, corticotropin, melanocyte-stimulating hormone, antidiuretic hormone, glucocorticoids, thyroid hormone, luteinizing hormone, insulin, glucagon, parathyroid hormone, catecholamines, seratonin, and estrogens.

It has been postulated that other types of intracellular hormonal mediators also exist. One other proposed type is the prostaglandins, which are a series of related compounds that frequently cause intracellular inhibition, in contrast to the activation usually caused by cyclic AMP. However, such a prostaglandin system is mainly theoretical at present.

THE HYPOPHYSIS AND ITS RELATIONSHIP TO THE HYPOTHALAMUS

The hypophysis (Fig. 75–2), also called the *pituitary gland,* is a small gland—about 1 cm. in diameter—that lies in the *sella turcica* at the base of the brain and is connected with the hypothalamus by the *hypophyseal,* or *pituitary, stalk.* Physiologically, the hypophysis is divisible into two distinct portions: the *adenohypophysis,* also known as the *anterior pituitary gland,* and the *neurohypophysis,* also known as the *posterior pituitary gland.* Between these is a small, relatively avascular zone called the *pars intermedia,* which is almost absent in the human being but is much larger and much more functional in some lower animals.

Embryologically, the two portions of the hypophysis originate from different sources, the adenohypophysis from *Rathke's pouch,* which is an embryonic invagination of the pharyngeal epithelium, and the neurohypophysis from an outgrowth of the hypothalamus. The origin of the adenohypophysis from the pharyngeal epithelium explains the epithelioid nature of its cells, while the origin of the neurohypophysis from neural tissue explains the presence of large numbers of glial type cells in this gland.

Six important hormones plus several less important ones are secreted by the *adeno*hypophysis, and two important hormones are

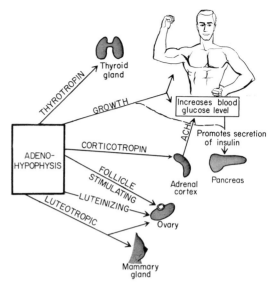

Figure 75–3 Metabolic functions of the adenohypophyseal hormones.

secreted by the *neuro*hypophysis. The hormones of the adenohypophysis play major roles in the control of metabolic functions throughout the body, as shown in Figure 75–3; thus: (1) *Growth hormone* promotes growth of the animal by affecting many metabolic functions throughout the body, especially protein formation. (2) *Corticotropin* controls the secretion of some of the adrenocortical hormones, which in turn affect the metabolism of glucose, proteins, and fats. (3) *Thyrotropin* controls the rate of secretion of thyroxine by the thyroid gland, and thyroxine in turn controls the rates of most chemical reactions of the entire body. Three separate gonadotropic hormones, (4) *follicle-stimulating hormone,* (5) *luteinizing hormone,* and (6) *luteotropic hormone,* control the growth of the gonads as well as their reproductive activities.

The two hormones secreted by the neurohypophysis play other roles: (1) *Antidiuretic hormone* controls the rate of water excretion into the urine and in this way helps to control the concentration of water in the body fluids. (2) *Oxytocin* (a) helps to deliver milk from the glands of the breast to the nipples during suckling; (b) probably helps to promote fertilization of the ovum following discharge of sperm into the vagina; and (c) probably helps in the delivery of the baby at the end of gestation.

CONTROL OF HYPOPHYSEAL SECRETION BY THE HYPOTHALAMUS

Almost all secretion by the hypophysis is controlled by signals transmitted from the

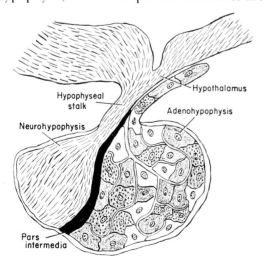

Figure 75–2 The hypophysis.

hypothalamus. Indeed, when the hypophysis is removed from its normal position beneath the hypothalamus and transplanted to some other part of the body, its rates of secretion of the different hormones fall to extremely low levels — in the case of some of the hormones, almost to zero.

Secretion from the neurohypophysis is controlled by nerve fibers originating in the hypothalamus and terminating in the neurohypophysis. In contrast to this, secretion by the adenohypophysis is controlled by *neurosecretory substances* secreted within the hypothalamus itself and then conducted to the adenohypophysis through minute blood vessels called *hypothalamic-hypophyseal portal vessels*. In the adenohypophysis these neurosecretory substances act on the glandular cells to control their secretion. This system of control will be discussed in detail later in the chapter.

The hypothalamus receives signals from almost all possible sources in the nervous system. Thus, when a person is exposed to pain, a portion of the pain signal is transmitted into the hypothalamus. Likewise, when a person experiences some powerful depressing or exciting thought, a portion of the signal is transmitted into the hypothalamus. Olfactory stimuli denoting pleasant or unpleasant smells transmit strong signal components through the amygdaloid nuclei into the hypothalamus. Even the concentrations of nutrients and electrolytes in the blood excite various portions of the hypothalamus. Thus, the hypothalamus is a collecting center for information concerned with the well-being of the body, and in turn this information is used to control secretion by the hypophysis.

THE ADENOHYPOPHYSIS AND ITS REGULATION BY HYPOTHALAMIC NEUROSECRETORY SUBSTANCES

CELL TYPES OF THE ADENOHYPOPHYSIS

The adenohypophysis is composed of three major types of cells: (1) the *chromophobes*, which account for about 50 per cent of all the cells, (2) the *acidophils*, also known as *alpha cells*, which account for approximately 40 per cent of all the cells, and (3) the *basophils*, also known as *beta cells*, which account for about 10 per cent of all the cells.

The *acidophils* are believed to produce *growth hormone* and *luteotropic hormone* (also called prolactin), whereas the *basophils* are believed to produce *luteinizing hormone, follicle-stimulating hormone, thyrotropin,* and probably *corticotropin;*

however, these beliefs are based mainly on indirect evidence.

The chromophobes have been supposed to secrete none of the adenohypophyseal hormones but instead simply to be precursor cells of either the acidophils or basophils. Yet this concept is probably too naïve because tumors composed entirely of chromophobes are often associated with hypersecretion of one or more adenohypophyseal hormones.

THE HYPOTHALAMIC-HYPOPHYSEAL PORTAL SYSTEM

The adenohypophysis is a highly vascular gland with extensive sinuses among the glandular cells. Almost all of the blood that enters these sinuses passes first through the tissue of the lower hypothalamus and then through small *hypothalamic-hypophyseal portal vessels* into the sinuses. Thus, Figure 75–4 illustrates a small artery supplying the lowermost portion of the hypothalamus called the *median eminence*. Small vascular tufts project into the substance of the median eminence and then return to its surface, coalescing to form the hypothalamic-hypophyseal portal vessels. These in turn pass downward along the hypophyseal stalk to supply the adenohypophyseal sinuses.

Secretion of Hypothalamic Neurosecretory Substances into the Median Eminence. Nerve endings have long been known to secrete acetylcholine, norepinephrine, and perhaps a few other transmitter substances for transmission of nerve signals. Now, it is rapidly becoming apparent that special types of nerve cells can also secrete still many other hormones called *neurosecretory substances.*

Neurons originating in various parts of the hypothalamus send nerve fibers into the *median eminence* and into the *tuber cinereum*, which

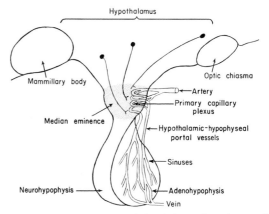

Figure 75-4 The hypothalamic-hypophyseal portal system.

is the hypothalamic tissue that surrounds the hypophyseal stalk. These nerve endings are different from most endings in the central nervous system because their function is not to transmit signals from one neuron to another but merely to secrete hormones, the *hypothalamic hormones* or *hypothalamic neurosecretory substances*, that are immediately absorbed into the hypothalamic-hypophyseal portal capillaries and carried directly to the sinuses of the adenohypophysis.

The "Releasing" Factors. The function of the hormones secreted into the median eminence and then transported to the adenohypophysis is to control release of adenohypophyseal hormones from the adenohypophyseal cells. Therefore, the hypothalamic neurosecretory hormones are also called *releasing factors.* Several small polypeptides have been shown to cause release of specific adenohypophyseal hormones, for which reason it is assumed that the releasing factors are probably small polypeptide hormones similar in nature to, but slightly larger than, the two neurohypophyseal hormones, antidiuretic hormone and oxytocin, which are discussed later in the chapter.

For each type of adenohypophyseal hormone, with the possible exception of luteotropic hormone, there is a corresponding hypothalamic-releasing factor. Thus, the releasing factor that causes secretion of thyrotropin is called *thyrotropin-releasing factor (TRF),* the factor that causes release of corticotropin is called *corticotropin-releasing factor (CRF),* and the factor that causes release of somatotropin (growth hormone) is called *somatotropin-releasing factor (SRF).* Similarly, releasing factors exist for follicle-stimulating and luteinizing hormones and perhaps for the less important hormones such as melanocyte-stimulating hormone and different lipolytic hormones secreted by the adenohypophysis. In addition, a luteotropin *inhibitory* factor has been postulated; this will be discussed in Chapter 81.

Specific Areas in the Hypothalamus that Control Secretion of Specific Adenohypophyseal Hormones. Unfortunately, we know much too little about the discrete areas of the hypothalamus that regulate secretion of different adenohypophyseal hormones, but some of these areas have been delineated by two types of experimentation: (1) by studying the effects of discrete hypothalamic *lesions* on the hormonal secretory rates and (2) by studying the effects of *stimulation* of discrete hypothalamic areas. Using these two techniques, it has been concluded that the areas shown in Figure 75–5

are probably important in controlling the secretion of some of the adenohypophyseal hormones. These areas are:

The secretion of *growth hormone (GH)* seems to be controlled primarily by the lower *paraventricular area* of the hypothalamus.

The secretion of *thyrotropin (TSH)* is controlled by an area located in the *preoptic area* of the anterior hypothalamus immediately above and posterior to the optic chiasm.

Secretion of *corticotropin (ACTH)* is probably controlled by an area located in the lowermost and posterior portions of the hypothalamus, including the median eminence itself.

Secretion of two of the *gonadotropic hormones, FSH and LH,* is controlled by an area of the inferior hypothalamus that overlaps the area for ACTH but also extends posteriorly toward the mammillary body. In addition, secretion of LH and FSH is partially controlled by an area in the *preoptic* portion of the hypothalamus.

Luteotropic hormone (LTH) secretion is controlled by a small area immediately anterior to the mammillary body.

It must be recognized, however, that the areas just described might simply be "way stations" for relay of appropriate secretory signals. Therefore, these localizations are not to be taken too seriously, other than to recognize that different parts of the hypothalamus do have specific effects on the secretion of specific adenohypophyseal hormones.

PHYSIOLOGIC FUNCTIONS OF THE ADENOHYPOPHYSEAL HORMONES

Five of the major adenohypophyseal hormones, *corticotropin, thyrotropin,* and the three *gonadotropins,* exert their effects by stimulating "target glands," the adrenal cortex, the thyroid gland, and the gonads, respectively. The functions of these hormones are so intimately con-

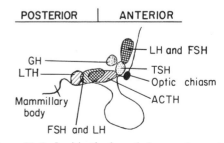

Figure 75–5 Loci in the hypothalamus where stimulation will cause secretion of different adenohypophyseal hormones.

cerned with the functions of the target glands themselves that these particular hormones will not be discussed here but, instead, in subsequent chapters along with the functions of the target gland hormones. On the other hand, *growth hormone* does not function through a target gland but acts directly on all or almost all tissues of the body.

GROWTH HORMONE

Growth hormone (GH), also called *somatotropic hormone* (SH) or *somatotropin,* causes growth of all tissues of the body that are capable of growing. It promotes both increased sizes of the cells and increased mitosis with development of increased numbers of cells.

Figure 75–6 illustrates the typical effect of daily injections of growth hormone into a growing rat in comparison with a litter-mate that did not receive growth hormone. This figure shows marked exacerbation of growth by growth hormone—both in the early days of life and even after the two rats had reached adulthood. In the early stages of development, all organs of the treated rat increased proportionately in size, but, after adulthood was reached, most of the bones ceased growing while the soft tissues continued to grow. This results from the fact that, once the epiphyses of the long bones have united with the shafts, further growth of the bones cannot occur even though most other tissues of the body can continue to grow throughout life.

Basic Metabolic Effects for Promotion of Growth. Growth hormone is known to have the following basic effects on the metabolic processes of the body:

1. Increased rate of protein synthesis in all the cells of the body.
2. Decreased rate of carbohydrate utilization throughout the body.
3. Increased mobilization of fats and use of fats for energy.

Thus, in effect, growth hormone enhances the body proteins, conserves carbohydrates, and uses up the fat stores. It is probable that the increased rate of growth results mainly from the increased rate of protein synthesis.

NECESSITY OF INSULIN AND CARBOHYDRATE FOR THE ACTION OF GROWTH HORMONE. Growth hormone fails to cause growth of a pancreatectomized animal, and it also fails to cause growth if carbohydrates are excluded from the diet. This shows that adequate insulin activity, as well as adequate availability of carbohydrates, is necessary for growth hormone to be effective. Furthermore, administration of growth hormone over several days probably has a direct effect on the pancreas to increase the rate of insulin secretion. Indeed, some physiologists feel that some of the growth-promoting effect of growth hormone is brought about by the increased secretion of insulin.

Enhancement of Amino Acid Transport Through the Cell Membrane as the Basic Mechanism for Promotion of Growth. Since the earliest studies on growth hormone, it has been known that its important effect is to cause protein accumulation in all or almost all cells of the body. Recently, it has been learned that growth hormone directly enhances transport of at least some, and perhaps most, amino acids through the cell membrane to the interior of the cell. This increases the concentration of amino acids in the cells, which is presumed to set off the cycle for protein building. Thus, this major effect of growth hormone can possibly be explained by a simple action at the cell membrane itself.

This control of amino acid transport through the cell membrane is analogous to the effect of insulin in controlling glucose transport through the membrane as discussed in Chapters 67 and 78. It is likely also that many other hormones, especially the adrenocortical hormones, exert many of their control functions by regulation of membrane transport.

Possible Effects of Growth Hormone on RNA. Some physiologists do not believe that all the effects of growth hormone in increasing protein synthesis can be ascribed to the promotion of amino acid transport into cells. There are two reasons for this: (1) In the presence of growth hormone, the quantity of messenger RNA in the cells is increased, which could in

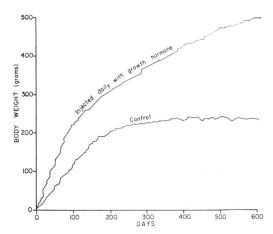

Figure 75–6 Comparison of weight gain of a rat injected daily with growth hormone with that of a normal rat of the same litter.

itself increase the rate of protein synthesis. (2) Protein synthesis can be enhanced by growth hormone even in cell homogenates in which the cell membranes have been totally disrupted. Therefore, it has been suggested that growth hormone might increase the rate of protein synthesis either by increasing the rate of formation of RNA in the nucleus or by increasing the activity of RNA after it has been formed.

Mobilization of Fatty Acids — Ketogenic Effect of Growth Hormone. Growth hormone also increases the release of fatty acids from the adipose tissue and therefore increases the fatty acid concentration in the body fluids; this in turn increases the quantity of stored fat in the liver.

One of the consequences of increased fatty acid mobilization is greatly increased use of fatty acids for supplying energy to the body. Likewise, far greater quantities of acetoacetic acid are formed in the liver and transported into the blood. Therefore, a person exposed to excessive growth hormone is much more liable to develop ketosis than is the normal person. This is called the *ketogenic effect* of growth hormone.

Effect on Carbohydrate Metabolism — Diabetogenic Effect of Growth Hormone. Growth hormone causes a simultaneous decrease in utilization of carbohydrate for energy. Instead, glycogen is stored in the cells until the cells become saturated; thereafter, the uptake of glucose by the cells becomes greatly depressed. As a result, the blood glucose concentration rises above normal, for which reason it is said that growth hormone has a *diabetogenic effect*.

Unfortunately, we do not know the precise mechanism by which growth hormone decreases the glucose utilization by the cells. This possibly results entirely secondarily to the increased mobilization of fatty acids and increased transport of amino acids into cells. That is, the fatty acids and deaminated amino acids perhaps form so much acetyl coenzyme A in the cells that the increased quantities of this substance block the glycolytic breakdown of glucose into the same substance. In addition, it is known that growth hormone diminishes the transport of glucose into the cells, which is an effect opposite to that of insulin on the transport of glucose. The answer to this question remains a fertile field for research.

Diabetogenic Effects of Other Adenohypophyseal Hormones. Growth hormone is not the only adenohypophyseal hormone that increases the blood glucose concentration, for at least three others can do the same, *cortico-*

tropin, thyrotropin, and *luteotropin.* Especially important is corticotropin, which increases the rate of cortisol secretion by the adrenal cortex. Cortisol then increases the blood glucose concentration by increasing the rate of gluconeogenesis. Indeed, this effect, quantitatively, is probably more important than the direct effect of growth hormone.

PITUITARY DIABETES. From the preceding discussion it is readily apparent that generalized increase in secretion of all the adenohypophyseal hormones causes elevated blood glucose concentration; this condition is called pituitary diabetes, and it differs from *diabetes mellitus,* which results from insulin lack, in the following ways: First, in pituitary diabetes the rate of glucose utilization by the cells is only moderately depressed, in comparison with almost no utilization in diabetes mellitus. Second, the blood glucose concentration is relatively *refractory to insulin* because adequate insulin is already available in the body to cause glucose transfer into the cell. Third, many of the side effects that result from reduced carbohydrate metabolism in diabetes mellitus are absent in pituitary diabetes.

DIABETES MELLITUS CAUSED BY ADENOHYPOPHYSEAL SECRETION. Injection of a potent extract containing all the hormones of the adenohypophysis into an experimental animal allowed to eat large quantities of carbohydrates causes the beta cells of the islets of Langerhans to degenerate within a few weeks. These cells are said to "burn out" from overstimulation. Once this has occurred, the animal develops frank diabetes mellitus.

In human beings, hypersecretion by the adenohypophysis, such as occurs in giantism (described later in the chapter), frequently causes a similar effect, the blood glucose level at first being elevated simply because of pituitary diabetes but later remaining elevated because of progressive destruction of the beta cells of the islets of Langerhans. The direct effect of growth hormone in stimulating pancreatic secretion is believed to be the major, but not the only, effect of the adenohypophyseal hormones in causing burnout of the beta cells.

Regulation of Growth Hormone Secretion. For many years it was believed that growth hormone was secreted primarily during the period of growth but then disappeared from the blood at adolescence. However, this has proved to be very far from the truth, because after adolescence, secretion continues at a rate about two-thirds as great as that in childhood. Furthermore, the rate of growth hormone secretion increases and decreases extremely rapidly in relation to the person's state of nutrition. The normal concentration of growth hormone in the plasma of an adult is about 4 millimicrograms per milliliter and in the child about 7 millimicrograms per milliliter. However, these values often increase to

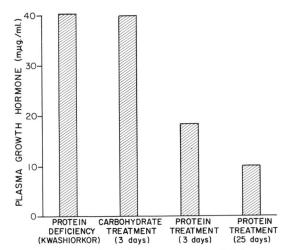

Figure 75-7 Effect of extreme protein deficiency on the concentration of growth hormone in the plasma in the disease kwashiorkor. The figure also shows the failure of carbohydrate treatment but the effectiveness of protein treatment in lowering growth hormone concentration. (Drawn from data in Pimstone: *Amer. J. Clin. Nutr.*, *21:*482, 1968.)

as high as 50 millimicrograms per milliliter after depletion of the body stores of proteins or carbohydrates. For instance, extremely high levels occur during wasting diseases or during starvation; the hormone concentration is mainly an inverse reflection of the degree of protein depletion. Figure 75-7 illustrates this relationship: the first block shows growth hormone levels in children with extreme protein deficiency during the malnutrition disease called *kwashiorkor;* the second block shows the levels in the same children after three days of treatment with more than adequate quantities of carbohydrates in their diets; block three shows the levels after treatment with protein supplement in their diet for three days; and block four shows the levels after treatment with protein supplement for 25 days. These results demonstrate that under very severe conditions of protein malnutrition, adequate calories alone are not sufficient to correct the excess production of growth hormone. Instead, the protein deficiency must also be corrected before the growth hormone concentration will return to normal.

Thus, it is almost certain that growth hormone secretion is controlled moment by moment by the nutritional status of the body, and it seems that the most important nutrient in the control of growth hormone secretion is protein, though changes in blood glucose concentration can also cause extremely rapid and dramatic alterations in growth hormone secretion. Consequently, it can be postulated that growth hormone operates in a feedback con-

trol system as follows: When the tissues begin to suffer from malnutrition, especially from poor protein nutrition, large quantities of growth hormone are secreted. Growth hormone, in turn, promotes the transport of amino acids into the cells and perhaps also promotes increased activity of the RNA system for formation of proteins, thus enhancing the protein nutritional status of the cells; and this inhibits the production of growth hormone, returning the plasma growth hormone level back to normal value.

Assay of Growth Hormone in Plasma. Our understanding of growth hormone function has been greatly retarded until recently, because of lack of an accurate method for assaying growth hormone concentrations in body fluids. However, a *radio immunologic method* has been developed that now allows accurate measurement of extremely minute quantities of growth hormone in the plasma. This method is basically the following: (1) Purified growth hormone is first labeled with a radioactive substance. (2) The labeled hormone is then combined with specific antibody for growth hormone. (3) This combination of radioactive growth hormone and antibody is mixed with a sample of plasma. Growth hormone in the plasma competes with the radioactive growth hormone for the reactive sites on the antibody and pushes some of the radioactive growth hormone off the antibody. (4) The antibody and its bound growth hormone is then separated from the solution, and the degree of radioactivity is measured. The less this degree of radioactivity, the greater is the concentration of growth hormone in the sample of plasma.

The same radio immunologic method has been applied to essentially all other protein hormones secreted in the body, including several other adenohypophyseal hormones, insulin, parathyroid hormone, and calcitonin.

ABNORMALITIES OF GROWTH HORMONE SECRETION

Panhypopituitarism. This term means decreased secretion of all the adenohypophyseal hormones. The decrease in secretion may be congenital (present from birth), or it may occur suddenly or slowly at any time during the life of the individual.

DWARFISM. Some instances of dwarfism result from deficiency of adenohypophyseal secretion during childhood—especially deficiency of growth hormone. In general, the features of the body develop in appropriate proportion to each other, but the rate of development is greatly decreased. A child who has reached the age of 10 may have the bodily development of a child of 4 to 5, whereas the same person on reaching the age of 20 may have the bodily development of a child of 7 to 10.

The dwarf usually does not exhibit specific thyroid deficiency or adrenocortical deficiency, for the entire body remains so small that only small quantities of thyrotropic and adrenocorticotropic hormones are needed. Also, there is no mental retardation. On the other hand, the pituitary dwarf does not pass through puberty and never secretes a sufficient quantity of gonadotropic hormones to develop adult sexual functions.

PANHYPOPITUITARISM IN THE ADULT. Panhypopituitarism occurring in adulthood frequently results from one of three different abnormalities: Two tumorous conditions, craniopharyngiomas and chromophobe tumors, may compress the pituitary gland until the functioning adenohypophyseal cells are totally or almost totally destroyed. The third cause is thrombosis of the hypophyseal blood vessels. This occurs occasionally when a mother develops circulatory shock following birth of a baby.

The effects of panhypopituitarism, in general, are (1) hypothyroidism, (2) depressed production of glucocorticoids by the adrenal glands, and (3) suppressed secretion of the gonadotropic hormones to the point that sexual functions are lost. Thus, the picture is that of a lethargic person (from lack of thyroxine) who is gaining weight because of lack of fat mobilization by growth, corticotropic, adrenocortical, and thyroid hormones and who has lost all sexual functions. Except for the abnormal sexual functions, the patient can usually be treated satisfactorily by administration of adrenocortical and thyroid hormones.

Giantism. Occasionally, the acidophilic cells of the adenohypophysis become excessively active, and sometimes even acidophilic tumors occur in the gland. As a result, large quantities of growth hormone and probably, to a less extent, some of the other adenohypophyseal hormones are produced. All body tissues grow rapidly, including the bones, and if the epiphyses of the long bones have not become fused with the shafts before the development of the adenohypophyseal acidophilia, height increases so that the person becomes a giant. Thus, for giantism to occur, the acidophilia must occur prior to adolescence.

The pituitary giant ordinarily has a basal metabolic rate 20 per cent or more greater than normal; this may be due to the direct effect of growth hormone on metabolism, or perhaps thyrotropic hormone is secreted in increased quantities simultaneously with the secretion of excessive growth hormone. The giant also ordinarily has hyperglycemia, and the beta cells of the islets of Langerhans in the pancreas may eventually degenerate, partially because they become overactive, owing to the hyperglycemia and partially because of a direct overstimulating effect of growth hormone on the islet cells. Consequently, many giants finally develop full-blown diabetes mellitus.

As a result of the increased metabolism, the diabetes, and other effects of excessive adenohypophyseal hormones, the life of a giant is often considerably shortened. However, once giantism is diagnosed, further development can usually be blocked by radium irradiation in the hypophysis (by means of a needle implant) to reduce the degree of eosinophilia.

Acromegaly. If adenohypophyseal eosinophilia or an eosinophilic tumor occurs after adolescence — that is, after the epiphyses of the long bones have fused with the shafts — the person cannot grow taller; but his soft tissues can continue to grow, and the bones can grow in thickness. This condition is known as "acromegaly." Enlargement is especially marked in the small bones of the hands and feet and the *membranous bones,* including the cranium, the nose, the bosses on the forehead, the supra-orbital ridges, the lower jawbone, and portions of the vertebrae, for their growth does not cease at adolescence anyway. Consequently, the jaw protrudes forward, sometimes as much as a half inch or more, the forehead slants forward because of excess development of the supra-orbital ridges, the nose increases to as much as twice normal size, the foot requires a size 14 or larger shoe, and the fingers become extremely thickened so that the hand develops a size almost twice normal. In addition to these effects, changes in the vertebrae ordinarily cause a hunched back, which is known clinically as "kyphosis." Finally, many soft tissue organs, such as the tongue and liver, become greatly enlarged. A typical acromegalic is shown in Figure 75–8.

CHEMISTRY OF THE ADENOHYPOPHYSEAL HORMONES

All the adenohypophyseal hormones thus far isolated are small proteins or polypeptides. The molecular weights of these differ for the same hormone in different types of animals, but they usually range between 25,000 and 100,000. Corticotropin, for instance, is a polypeptide containing 39 amino acids.

The molecular weight of human growth hormone is approximately 21,500. That of growth hormone secreted by most lower animals is about twice this. It is particularly significant that bovine growth hormone, or that from other animals below the level of primates, has no significant effect in the human being. For this reason, therapeutic use of growth hormone has been long impeded.

MELANOCYTE-STIMULATING HORMONE

Another hormone normally secreted in small quantities, and perhaps in large quantities under abnormal conditions, is melanocyte-stimulating hormone (MSH), also known as *intermedin.* This hormone is secreted by the *pars intermedia* of the hypothalamus. MSH stimulates the *melanocytes,*

Figure 75-8   An acromegalic. (Courtesy of Dr. Herbert Langford.)

which are cells that contain the black pigment melanin and that occur in abundance between the dermis and epidermis of the skin.

In some amphibia, melanin is collected in the melanocytes in small granules called *melanosomes*. When melanocyte-stimulating hormone is not available, the melanosomes become concentrated near the nuclei of the melanocytes in such a way that the melanocytes appear light colored. Then, when MSH is secreted, the melanosomes disperse throughout the cytoplasm of the melanocytes, and the entire cell becomes almost black within a few seconds to minutes. In this way, the color of the skin can be changed from very light in the absence of the hormone to very dark in its presence.

Exposure of the human being to melanocyte-stimulating hormone over a period of days causes intense darkening of the skin. It has a much greater effect in persons who have a genetically dark skin than in persons who have a genetically light skin. It is likely that this hormone acts in a different manner in the human being than in amphibia, for much of the melanin formed by human melanocytes actually leaves the melanocytes and becomes dispersed in the cells of the epidermis rather than remaining in the melanocytes.

Corticotropin, one of the major hormones secreted by the adenohypophysis, has about one-thirtieth the melanocyte-stimulating effect of MSH, and the two hormones are often secreted in high concentrations simultaneously. This frequently occurs in Addison's disease in which the adrenal cortex is damaged so much that it cannot secrete adrenocortical hormones; lack of these hormones causes compensatory increase in corticotropin secretion and, along with it, increased MSH secretion. Consequently, the skin becomes considerably darkened. MSH has exactly the same chemical structure as the first 13 amino acid portion of the 39 amino acid corticotropin polypeptide chain.

THE NEUROHYPOPHYSIS AND ITS RELATION TO THE HYPOTHALAMUS

The neurohypophysis is composed mainly of glial-like cells called *pituicytes*. However, the pituicytes do not secrete hormones; they act simply as a supporting structure for large numbers of *terminal nerve fibers* and *terminal nerve endings* from nerve tracts that originate in the *supraoptic* and *paraventricular nuclei* of the hypothalamus, as shown in Figure 75–9. These tracts pass to the neurohypophysis through the *hypophyseal stalk* (pituitary stalk). It is the nerve endings that secrete the neurohypophyseal hormones: (1) *antidiuretic hormone* (ADH), also called *vasopressin*, and (2) *oxytocin*.

If the hypophyseal stalk is cut near the hypophysis, leaving the entire hypothalamus intact, the neurohypophyseal hormones continue to be secreted almost normally, but they are then secreted by the cut ends of the fibers within the hypothalamus and not by the nerve

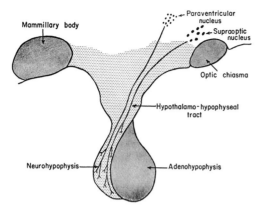

Figure 75–9 Hypothalamic control of the neurohypophysis.

endings in the neurohypophysis. Furthermore, the hormones have been demonstrated in the nerve fibers as far up as the supraoptic and paraventricular nuclei themselves. For these reasons, it is almost certain that the neurohypophyseal hormones are formed in the cell bodies of the supraoptic and paraventricular nuclei and are then transported down the nerve fibers to the nerve ending in the neurohypophysis. Indeed, movement at a rate of about 3 mm. per day of hormonal granules down the nerve fibers toward the neurohypophysis has actually been observed.

ADH is formed primarily in the supraoptic nuclei, while *oxytocin is formed primarily in the paraventricular nuclei.* However, each of these two nuclei can secrete approximately one-sixth as much of the second hormone as of its primary hormone.

Under resting conditions, large quantities of both ADH and oxytocin accumulate in the neurohypophysis, probably bound with a "carrier protein" in the terminal nerve fibers themselves. However, when nerve impulses are transmitted downward along the fibers from the supraoptic and paraventricular nuclei, the hormones are immediately released from the nerve endings and are absorbed into the local blood vessels. It is believed that the nerve impulses first release small vesicles of some other "transmitter" substance at the nerve endings and that it is this transmitter substance that dissolutes the hormones from the carrier protein and allows them to diffuse out of the nerve endings.

PHYSIOLOGIC FUNCTIONS OF ANTIDIURETIC HORMONE (VASOPRESSIN)

Extremely minute quantities of antidiuretic hormone (ADH), as little as 2 millimicrograms, when injected into a person cause antidiuresis; that is, decreased excretion of water by the kidneys. This antidiuretic effect was discussed in detail in Chapter 36. Briefly, in the absence of ADH, the distal tubules, collecting tubules, and parts of the loops of Henle are almost totally impermeable to water, which prevents significant reabsorption of water and therefore allows extreme loss of water into the urine. On the other hand, in the presence of ADH, the permeability of these tubules to water increases greatly and allows most of the water to be reabsorbed, thereby conserving water in the body.

The precise mechanism by which ADH acts on the tubules to increase their permeability is unknown. However, the hormone first becomes fixed to the epithelium of the tubules and becomes highly concentrated in these cells. It then causes these tissues to form large quantities of cyclic adenylic acid. This in turn supposedly opens up the pores in the cell membranes and allows free diffusion of water between the tubular and peritubular fluids, though the manner in which cyclic adenylic acid causes its effect on the pores is completely unknown. The water is then absorbed by osmosis, as explained in relation to the concentrating mechanism of the kidney in Chapter 36.

Regulation of ADH Production. OSMOTIC REGULATION. When a concentrated electrolyte solution is injected into the artery supplying the hypothalamus, the supraoptic nuclei immediately transmit impulses into the neurohypophysis to release large quantities of ADH into the circulating blood. Conversely, injection of pure water into this artery causes complete cessation of impulses from the supraoptic nuclei and essentially complete cessation of ADH production. The ADH that has already been produced is destroyed by the tissues at a rate of approximately one-half every seven minutes. Thus, the concentration of ADH in the body fluids can change from small amounts to large amounts, or vice versa, in only a few minutes.

It has been postulated that the neurons of the supraoptic nuclei function as *osmoreceptors,* and histologists have claimed to have observed fluid chambers in these cells which presumably could increase and decrease in size in relation to the degree of concentration of the extracellular fluids. When the extracellular fluids are highly dilute, osmosis of water into the cell supposedly increases the volume of the fluid chamber, while concentrated extracellular fluids supposedly reduce the volume. Regardless of whether these effects are true, concentrated body fluids do stimulate the supraoptic nuclei, while dilute body fluids inhibit them. Therefore, a feedback regulatory mechanism is available for regulation of the total osmotic pressure of the body fluids, operating as follows:

When the body fluids become highly concentrated, the supraoptic nuclei become excited, impulses are transmitted to the neurohypophysis, and ADH is secreted. This passes by way of the blood to the kidneys, where it increases the permeability of the collecting tubules to water. As a result, most of the water is reabsorbed from the urine, while electrolytes continue to be lost. This effect dilutes the extracellular fluids, returning them to a reasonably normal osmotic composition.

EFFECT OF HEMORRHAGE—PRESSOR EFFECT OF ADH. The most powerful stimuli of all for increasing the secretion of ADH are severe hemorrhage and circulatory shock. It is most generally believed that the decreased pressure in the various baroreceptor areas of the body causes baroreceptor signals to be transmitted to the supraoptic nuclei to increase the output of ADH. However, it is perhaps equally possible that decreased blood flow to the hypothalamus itself might be the primary stimulus to cause this greatly increased output of ADH.

The *normal* rate of ADH secretion is much too small to cause any significant pressor effect. (The name "vasopressin" was given to ADH on the basis of pharmacologic experiments in which pressor effects are observed when large quantities of ADH are injected into animals.) However, the quantities of ADH secreted following severe hemorrhage *are* large enough to cause a mild pressor effect in the circulatory system, thereby helping to return the arterial pressure toward normal. Yet because the body has so many other pressure regulating systems that are far more powerful than this one, it is doubtful that the pressor action of ADH is of any real significance to the regulation of the circulatory system.

OTHER FACTORS THAT AFFECT ADH PRODUCTION. Other factors that frequently increase the output of ADH include *trauma* to the body, *pain, anxiety,* and drugs such as *morphine, nicotine, tranquilizers,* and some *anesthetics.* Each of these factors can cause retention of water in the body. This explains the frequent accumulation of water in many emotional states, and it also explains the diuresis that occurs when the state is over.

A substance that inhibits ADH secretion is *alcohol.* Therefore, during an alcoholic bout, lack of ADH allows marked diuresis. Alcohol probably also dilates the afferent arterioles of the nephrons, which adds to the diuretic effect.

Diabetes Insipidus. Diabetes insipidus is the disease that occurs when the supraoptico-hypophyseal system for secreting ADH fails. In a person with full-blown diabetes insipidus, lack of ADH keeps his urine from ever being concentrated. The urine specific gravity remains almost constantly between 1.002 and 1.006; the urine output is usually 4 to 6 liters per day but can be as great as 12 to 15 liters per day, depending principally on how much water the person drinks. Furthermore, the rapid loss of fluid in the urine creates a constant thirst, which keeps the water flushing through his body.

The person with diabetes insipidus has a tendency to become dehydrated and also to "wash out" an excess of electrolytes from his body fluids. However, these tendencies are usually quite well offset by the increased thirst and an increased desire for salt in food. Under conditions of circulatory stress or when water and salt might not be adequately available, the fluid and electrolyte loss can become important.

Diabetes insipidus can be treated easily by simply insufflating a small amount of powdered posterior pituitary gland or synthetic vasopressin into the nose several times a day. Though only a minute amount of ADH is absorbed in this way, it takes only 0.1 microgram of the hormone to cause a maximal antidiuretic effect.

Diabetes insipidus occurs most frequently as a result of a tumor of the hypothalamus or hypophysis that destroys the portions of the hypothalamus that control ADH secretion. Also, injury to the hypophyseal stalk temporarily reduces the production of ADH. If the stalk is sectioned near the neurohypophysis, little ADH is secreted for the first few weeks, but gradually secretion returns, sometimes to normal, thus illustrating that ADH is actually synthesized in the hypothalamus rather than in the neurohypophysis.

Other Actions of ADH. Large doses of ADH can also cause contraction of almost any smooth muscle tissue in the body, including contraction of most of the intestinal musculature, the bile ducts, and the uterus. However, the concentrations required to cause these effects are far greater than that required to cause antidiuresis, and it is doubtful that these are significant physiologic effects for normal function of the body.

OXYTOCIC HORMONE

Effect on the Uterus. An "oxytocic" substance is one that causes contraction of the pregnant uterus. The hormone *oxytocin,* in accordance with its name, powerfully stimulates the pregnant uterus, especially toward the end of gestation. Therefore, many obstetricians believe that this hormone is at least partially responsible for effecting birth of the baby. This is supported by the following facts: (1) In a hypophysectomized animal, the duration of labor is considerably prolonged, thus indicating a probable effect of oxytocin during delivery. (2) The amount of oxytocin in the plasma has been reported to increase during labor. (3) Stimulation of the cervix in a pregnant animal elicits nervous signals that pass to the hypothalamus and cause increased secretion of oxytocin.

ADH also stimulates the pregnant uterus, though not nearly so strongly as oxytocin. Likewise, oxytocin excites contraction of a number of other smooth muscle structures in the body besides the uterus, though usually to a much less extent than does ADH. The fact that the functions of the two hormones partially overlap illustrates their physiologic relation to each other; as pointed out below, their chemical structures are also similar.

Possible Effect of Oxytocin in Promoting Fertilization of the Ovum. Sexual stimulation of the female during intercourse increases the secretion of oxytocin, and the increased oxytocin is probably at least partially responsible for the uterine contractions that occur during the female orgasm. For these reasons, it has been proposed that oxytocin promotes fertilization of the ovum by causing uterine propulsion of the male semen upward through the fallopian tubes.

Effect of Oxytocin on Milk Ejection. Oxytocin plays an especially important function in the process of lactation, for this hormone causes milk to be expressed from the alveoli into the ducts so that the baby can obtain it by suckling. This mechanism works as follows: The suckling stimuli on the nipple of the breast cause impulses to be transmitted through the sensory nerves to the brain, the impulses finally reaching the paraventricular nuclei in the anterior hypothalamus to cause release of oxytocin. The oxytocin then is carried by the blood to the breasts where it causes contraction of *myoepithelial cells,* which lie outside of and compress the alveoli of the mammary glands. In less than a minute after the beginning of suckling, milk begins to flow. Therefore, this mechanism is frequently called *milk letdown* or *milk ejection.* ADH also has a moderate effect in promoting milk ejection.

Possible Effect of Oxytocin on Milk Secretion by the Mammary Glands. Oxytocin release by the neurohypophysis is often associated with increased secretion of luteotropic hormone by the adenohypophysis. It has been postulated that a small portion of the oxytocin released by the neurohypophysis is transmitted directly from the neurohypophysis to the adenohypophysis through local blood vessels. Then, on reaching the glandular cells of the adenohypophysis, it supposedly promotes luteotropin production. Luteotropin in turn is one of the hormones principally responsible for milk secretion and therefore is also called *prolactin;* this will be discussed in Chapter 82.

CHEMICAL NATURE OF ANTIDIURETIC HORMONE (ADH) AND OXYTOCIN

Both oxytocin and ADH (vasopressin) are polypeptides containing eight amino acids. They have been isolated in pure form from neurohypophyseal glands, and they have also been synthesized from their basic amino acid components. The amino acid compositions of these are the following:

VASOPRESSIN	OXYTOCIN
Tyrosine	Tyrosine
Proline	Proline
Glutamic acid	Glutamic acid
Aspartic acid	Aspartic acid
Glycine	Glycine
Cystine	Cystine
Phenylalanine	Leucine
Arginine	Isoleucine

Note that these two hormones are identical except that in vasopressin phenylalanine and arginine replace leucine and isoleucine of the oxytocin molecule. The similarity of the molecules explains the occasional functional similarities between these two hormones, while the slight dissimilarities of the molecules illustrate that slight chemical changes can alter the physiological properties of hormones markedly.

Both ADH and oxytocin in the nerve fibers of the supraoptic and paraventricular neurons are adsorbed to a "carrier protein" that itself has a molecular weight of about 30,000. These hormones remain in this form until they are released from the nerve endings in the neurohypophysis.

THE PINEAL GLAND

The structure known as the pineal gland is frequently called an endocrine gland. However, thus far, it has not been proved to secrete a definite and important hormone. Literally thousands of attempts have been made to extract hormones from this "gland," and many such hormones have been claimed and later disclaimed by different research workers. The two most usual hormonal effects postulated have been *growth stimulation* and *sexual stimulation.*

In adults the pineal gland often becomes completely calcified; this does not cause any significant physiologic effect on the body, which makes it doubtful that the pineal gland has any significant function. Yet the following facts are known about this structure.

Embryologically, the pineal gland is derived from a third eye that begins to develop early in the embryo and later degenerates. However, the epithelioid cells of this structure persist into adulthood in the pineal body, and it is possible that they could secrete a hormone or hormones of importance to the body. Recently, a substance called *melatonin* has been extracted from the bovine pineal gland,

and it has been suggested that exposure of the eyes to light causes nerve signals that in turn promote melatonin release into the blood. In the frog, this substance causes melanin, the dark pigment in the *melanocytes* of the skin, to concentrate near the nuclei of the melanocytes in such a way that the skin becomes lighter than normal. Also, melatonin inhibits gonadal function in the rat, probably by inhibiting the formation of gonadotropic hormones. However, it is not known whether or not similar effects occur in man. Also, this same substance has been extracted from other tissues of the body as well, which makes it doubtful that the pineal gland is of real significance in relation to skin pigmentation and gonadal function.

Some physiologists also believe that a substance that stimulates the secretion of aldosterone by the adrenal cortex is secreted somewhere in the mid-portion of the brain—in the region of the mesencephalon or upper pons. Because of the proximity of the pineal gland to this area, it has been suggested that the pineal gland is the locus of secretion of the aldosterone-stimulating substance.

REFERENCES

Aulsebrook, L. H., and Holland, R. C.: Central regulation of oxytocin release with and without vasopressin release. *Amer. J. Physiol., 216*:818, 1969.

Bajusz, E.: Physiology and Pathology of Adaptation Mechanisms: Neural-Neuro-Endocrine-Hormonal. New York, Pergamon Press, Inc., 1968.

Berson, S. A., and Yalow, R. S.: Peptide hormones in plasma. Harvey Lectures, 1966–1967, p. 107.

Berson, S. A., and Yalow, R. S.: Radioimmunoassay of peptide hormones. *Proc. Int. Union of Physiol. Sciences, 6*:5, 1968.

Brown-Grant, K.: The action of hormones on the hypothalamus. *Brit. Med. Bull., 22*:273, 1966.

Bullough, W. S.: The control of tissue growth. *In* Bittar, E. E., and Bittar, N. (eds.): The Biological Basis of Medicine. New York, Academic Press, Inc., 1968, Vol. 1, p. 311.

Butcher, R. W.: The biological role of adenosine 3′,5′-monophosphate(cyclic AMP). *Proc. Int. Union of Physiol. Sciences, 6*:161, 1968.

Campbell, H. J.: Control of hormone secretion. *In* Bittar, E. E., and Bittar, N. (eds.): The Biological Basis of Medicine. New York, Academic Press, Inc., 1968, Vol. 2, p. 47.

Douglas, W. W.: Cellular mechanisms of secretion in the adrenal medulla and posterior pituitary gland: possible methods of stimulus-secretion coupling. *Proc. Int. Union of Physiol. Sciences, 6*:63, 1968.

Exley, D., and Hockaday, T. D. R.: Transport and metabolism of hormones. *In* Bittar, E. E., and Bittar, N. (eds.): The Biological Basis of Medicine. New York, Academic Press, Inc., 1968, Vol. 2, p. 25.

Farrell, G.; Fabre, L. F.; and Rauschkolb, E. W.: The neurohypophysis. *Ann. Rev. Physiol., 30*:557, 1968.

Fasciolo, J. C.; Totel, G. L.; and Johnson, R. E.: Antidiuretic hormone and human eccrine sweating. *J. Appl. Physiol., 27*:303, 1969.

Fraser, T. R.: Human growth hormone. *Scient. Basis Med. Ann. Rev.*, 36, 1963.

Growth hormone (symposium). *Ann. N.Y. Acad. Sci., 148*:289, 1968.

Guillemin, R.: Control of the secretions of the adenohypophysis by the central nervous system. *Proc. Int. Union of Physiol. Sciences, 6*:69, 1968.

Guillemin, R.: The adenohypophysis and its hypothalamic control. *Ann. Rev. Physiol., 29*:313, 1967.

Harris, G. W., and Donovan, B. T.: The Pituitary Gland. 3 volumes. Berkeley, University of California Press, 1966.

Harris, G. W.; Reed, M.; and Fawcett, C. P.: Hypothalamic releasing factors and the control of anterior pituitary function. *Brit. Med. Bull., 22*:266, 1967.

Knobil, E.: Growth hormone. *Proc. Int. Union of Physiol. Sciences, 6*:77, 1968.

Knobil, E., and Hotchkiss, J.: Growth hormone. *Ann. Rev. Physiol., 26*:47, 1964.

Kupperman, H. S.: Human Endocrinology. 2nd ed., Philadelphia, F. A. Davis Co., 1969.

Lissak, K., and Endroczi, E.: The Neuroendocrine Control of Adaptation. New York, Pergamon Press, Inc., 1966.

Malaisse, W. J.; Malaisse-Lagae, F.; King, S.; and Wright, P. H.: Effect of growth hormone on insulin secretion. *Amer. J. Physiol., 215*:423, 1968.

Manchester, K. L.: Hormonal control of protein biosynthesis. *In* Bittar, E. E., and Bittar, Neville (eds.): The Biological Basis of Medicine. New York, Academic Press, Inc., 1968, Vol. 2, p. 221.

McCann, S. M.: Present status of hypothalamic hypophyseal stimulating and inhibiting hormones. *Proc. Int. Union of Physiol. Sciences, 6*:202, 1968.

McCann, S. M., and Porter, J. C.: Hypothalamic pituitary stimulating and inhibiting hormones. *Physiol. Rev., 49*:240, 1969.

McCann, S. M.; Dhariwal, A. P. S.; and Porter, J. C.: Regulation of the adenohypophysis. *Ann. Rev. Physiol., 30*:589, 1968.

O'Brien, C. P., and Bach, L. M. N.: Observations concerning hypothalamic control of growth. *Amer. J. Physiol., 218*:226, 1970.

Orloff, J.: The role of adenyl cyclase in the regulation of antidiuretic hormone action in toad bladder. *Proc. Int. Union of Physiol. Sciences, 6*:163, 1968.

Pliska, V.; Rudinger, J.; Dousa, T.; and Cort, J. H.: Oxytocin activity and the integrity of the disulfide bridge. *Amer. J. Physiol., 215*:916, 1968.

Randle, P. J.: Endocrine control of metabolism. *Ann. Rev. Physiol., 25*:291, 1963.

Relkin, R.: Pineal function relation to absolute darkness and sexual maturation. *Amer. J. Physiol., 213*:999, 1967.

Robison, G. A.: Functional effects of cyclic AMP. *Proc. Int. Union of Physiol. Sciences, 6*:165, 1968.

Sloper, J. C.: Hypothalamic neurosecretion. *Brit. Med. Bull., 22*:209, 1966.

Skelton, C. L.; Levery, G. S.; and Epstein, S. E.: Positive inotropic effects of dibutyryl cyclic adenosine 3′,5′-monophosphate. *Circ. Res., 26*:35, 1970.

Tepperman, J.: Metabolic and Endocrine Physiology. 2nd ed., Chicago, Year Book Medical Publishers, Inc., 1968.

Turner, C. D.: General Endocrinology. 4th ed., Philadelphia, W. B. Saunders Company, 1966.

Venning, E. H.: Adenohypophysis and adrenal cortex. *Ann. Rev. Physiol., 27*:107, 1965.

Williams, R.: Textbook of Endocrinology. 4th ed., Philadelphia, W. B. Saunders Company, 1968.

CHAPTER 76

The Adrenocortical Hormones

The adrenal glands, which lie at the superior poles of the two kidneys, are each composed of two distinct parts, the *adrenal medulla* and the *adrenal cortex*. The adrenal medulla is functionally related to the sympathetic nervous system, and it secretes the hormones *epinephrine* and *norepinephrine* in response to sympathetic stimulation. In turn, these hormones cause almost the same effects as direct stimulation of the sympathetic nerves in all parts of the body. These effects were discussed in detail in Chapter 59 in relation to the sympathetic nervous system.

The adrenal cortex secretes an entirely different group of hormones called *corticosteroids*. As far as is known, there is no direct functional relationship between the adrenal medulla and the adrenal cortex, their anatomical association being simply a chance occurrence.

The adrenocortical hormones do not all cause exactly the same effects in the body. Two major types of hormones—the *mineralocorticoids* and the *glucocorticoids*—are secreted by the adrenal cortex. In addition to these, small amounts of *androgenic hormones*, which exhibit the same effects in the body as the male sex hormone testosterone, are also secreted. These are normally unimportant, though in certain abnormalities of the adrenal cortices extreme quantities can be secreted (which is discussed later in the chapter) and can then result in masculinizing effects.

The *mineralocorticoids* have gained this name because they especially affect the electrolytes of the extracellular fluids—sodium, potassium, and chloride in particular. The *glucocorticoids* have gained this name because they exhibit an important effect in increasing blood glucose concentration. However, the glucocorticoids have additional effects on both protein and fat metabolism which may be even more important to body function than their effects on carbohydrate metabolism.

Over 30 different steroids have been isolated from the adrenal cortex, but only two of these

are of major importance to the endocrine functions of the body—*aldosterone* and *cortisol*.

FUNCTIONS OF THE MINERALO-CORTICOIDS—ALDOSTERONE

Loss of adrenocortical secretion causes death within three days to a week unless the person receives extensive salt therapy or mineralocorticoid therapy. Without mineralocorticoids, the sodium and chloride concentrations of the extracellular fluid decrease markedly, and the total extracellular fluid volume and blood volume also become greatly reduced. The person soon develops diminished cardiac output, which proceeds to a shocklike state followed by death. This entire sequence can be prevented by the administration of aldosterone or some other mineralocorticoid. Therefore, the mineralocorticoids are said to be the "life-saving" portion of the adrenocortical hormones, while the glucocorticoids are of particular importance in helping the person resist different types of stresses, as is discussed later in the chapter.

Aldosterone exerts at least 95 per cent of the mineralocorticoid activity of the adrenocortical secretion, but two other hormones secreted by the adrenal cortex provide at least some significant mineralocorticoid activity. These are *corticosterone*, which also exerts glucocorticoid effects, and *deoxycorticosterone*, which is secreted in minute quantities and has almost the same effects as aldosterone, but a potency only one-thirtieth that of aldosterone.

The basic actions of the mineralocorticoids are illustrated in Figure 76–1, which shows two major basic effects as follows: (1) *increased renal tubular reabsorption of sodium* and (2) *increased renal excretion of potassium.* These two basic effects, however, cause many additional secondary effects also illustrated in the figure.

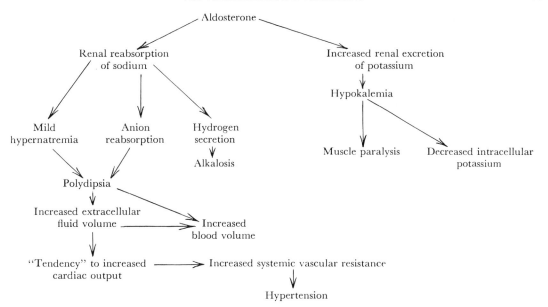

Figure 76–1 Effects of aldosterone.

RENAL EFFECTS OF ALDOSTERONE

Effect on Tubular Reabsorption of Sodium. By far the most important effect of aldosterone and other mineralocorticoids is to increase the rate of tubular reabsorption of sodium. It will be recalled from Chapter 34 that sodium is reabsorbed from the renal tubules along their entire extent. By selectively destroying different parts of the renal tubules and then studying sodium reabsorption, investigators have shown that aldosterone has an especially potent effect in the distal tubule, collecting tubule, and at least part of the loop of Henle. In addition, it increases sodium reabsorption in the proximal tubule but does not seem to play the all-powerful role here that it plays in the other tubules.

A high secretion rate of aldosterone can decrease the loss of sodium in the urine to as little as a few milligrams per day during the first few days of therapy. Beyond this time, compensatory adaptations in the body tend to return sodium excretion to normal. Conversely, total lack of aldosterone secretion can cause loss of as much as 12 grams of sodium in the urine in a day, an amount equal to one seventh of all the sodium in the body.

The increased tubular reabsorption of sodium tends to increase the *quantity* of sodium in the extracellular fluids. However, significant *hypernatremia*, which means an increase in sodium *concentration*, usually does not occur because the person becomes extremely thirsty and keeps adding water to his body fluids to dilute the sodium as it is reabsorbed, thereby maintaining almost normal sodium concentration. Rarely, the concentration does rise 3 to 8 milliequivalents.

Effect on Tubular Reabsorption of Chloride. Aldosterone also increases the reabsorption of chloride ions from the tubules. This probably occurs secondarily to the increased sodium reabsorption as follows: Absorption of the positively charged sodium ions causes an electrical potential gradient to develop between the lumen and the outside of the tubules, with positivity on the outside. This positivity in turn attracts negatively charged diffusible anions through the membrane. Since chloride ions are by far the most prevalent anions in the tubular fluids, the absorption of chloride increases considerably.

Despite a considerable increase in total *quantity* of chloride in the extracellular fluids, its *concentration* actually decreases, for, as pointed out below, a greater than normal proportion of bicarbonate is also reabsorbed. The increased quantity of chloride in the extracellular fluid results from the increased volume of extracellular fluid and not from an increase in chloride concentration.

Alkalosis Caused by Aldosterone. As pointed out in the discussion of acid-base balance in Chapter 37, a large proportion of the sodium reabsorption from the tubules results from an exchange reaction in which hydrogen ions are secreted into the tubules to take the place of the sodium that is reabsorbed.

Therefore, when the rate of sodium reabsorption is enhanced in response to aldosterone, the hydrogen ion concentration in the body fluids becomes reduced.

The secreted hydrogen ions originate from carbonic acid molecules, and when the hydrogen ions are secreted bicarbonate ions are left in the extracellular fluids. Thus, for each sodium ion reabsorbed by this exchange reaction, one bicarbonate ion enters the body fluids, which shifts the hydrogen ion concentration in the body fluids toward the alkaline side. For these reasons, increased secretion of aldosterone promotes alkalosis, while decreased secretion promotes acidosis. The changes in body fluid pH are not excessive, however, and ordinarily can be adequately compensated by the normal acid-base regulatory mechanisms.

Increased Renal Excretion of Potassium. At the same time that aldosterone causes increased tubular reabsorption of sodium, it also increases the loss of potassium into the urine. The probable cause of this is the interaction between sodium and potassium when sodium is actively absorbed by the tubular epithelium. That is, when aldosterone stimulates active reabsorption of sodium, much of the sodium is exchanged for potassium ions, similar to its exchange for hydrogen ions as discussed above. Therefore, increased numbers of potassium ions are lost in the urine in response to aldosterone.

HYPOKALEMIA AND MUSCLE PARALYSIS. The loss of potassium decreases the concentration of potassium in the extracellular fluids, which is the condition called *hypokalemia*. Thus, at the same time that the sodium and chloride ions become increased in the extracellular fluids, the potassium ions become decreased.

The low potassium ion concentration sometimes leads to muscle paralysis; this is caused by hyperpolarization of the nerve and muscle fiber membranes (see Chap. 5), which prevents transmission of action potentials.

EFFECTS OF ALDOSTERONE ON FLUID VOLUMES AND CARDIOVASCULAR DYNAMICS

Effect on Extracellular Fluid Volume — Edema. From the preceding discussion, it is evident that mineralocorticoids greatly increase the quantities of sodium, chloride, and bicarbonate ions in the extracellular fluids. The net result is a marked increase in total *quantity* of electrolytes in the extracellular fluids, even though there is only a slight increase in electrolyte *concentration*. This in turn promotes increased water reabsorption from the tubules by (a) stimulating the hypothalamic antidiuretic hormone system as explained in the preceding chapter and (b) by creating an osmotic gradient across the tubular membrane when the electrolytes are absorbed, this in turn carrying water through the membrane in the wake of the electrolyte absorption. Also the increased electrolyte concentration causes thirst, thereby making the person drink excessive amounts of water; this effect is called *polydipsia*. The final result, therefore, is an increase in extracellular fluid volume, sometimes enough to cause generalized extracellular edema.

However, moderate quantities of mineralocorticoids usually do not cause frank edema for the following reason: The polydipsia associated with excess electrolyte concentration causes *polyuria* — that is, increased urinary output. Even though this urine contains small quantities of electrolytes, its greatly increased volume keeps washing more and more electrolytes from the body fluids; this to some extent nullifies the general tendency of the aldosterone to expand the extracellular fluid volume. It will be recalled from Chapter 22 that the extracellular fluid volume must increase about 30 per cent before frank edema occurs.

EFFECT ON BLOOD VOLUME. The plasma portion of the blood is part of the extracellular fluid, and the plasma volume increases almost proportionally during the early part of the increase in extracellular fluid volume. Therefore, one of the effects of aldosterone secretion is a mild to moderate increase in blood volume.

Effect on Cardiac Output. Lack of aldosterone secretion has a profound effect on cardiac output, for the extracellular fluid and blood volumes decrease so greatly that venous return becomes markedly compromised. As a result, cardiac output actually falls to shock levels.

Conversely, when aldosterone secretion is excessive the resultant increase in extracellular fluid volume and blood volume "tends" to increase the cardiac output. However, as discussed in Chapter 24, all tissues of the body regulate their own blood flow, thereby preventing a rise in cardiac output much above that amount required to supply the tissues adequately with nutrition. For this reason, the cardiac output rarely rises more than 10 to 20 per cent.

Effect on Arterial Pressure. When aldosterone is not secreted, a shock-like state of hypotension occurs because of the greatly reduced cardiac output.

On the other hand, excess secretion of aldosterone usually causes moderate to marked hypertension. The cause of this is not easy to analyze, because the increase in cardiac output of 10 to 20 per cent is certainly not enough to account for the 50 per cent or more increase in arterial pressure that often results. Yet the answer to this, as explained in the discussion of hypertension in Chapter 26, perhaps lies in the mechanism of "long-term" local regulation of blood flow in the intrinsic tissues. That is, the increased cardiac output supplies each tissue with too much blood flow, and the local vessels constrict more and more in an attempt to return the blood flow to normal levels. This response obviously increases the total peripheral resistance and would therefore cause hypertension.

EFFECTS OF ALDOSTERONE ON SWEAT GLANDS, SALIVARY GLANDS, AND INTESTINAL GLANDS

The mineralocorticoids have almost the same effect on the sweat glands, salivary glands, and intestinal glands as on the renal tubules, greatly reducing the loss of sodium and chloride in the glandular secretions. The effect on the sweat glands is important to conserve body salt in hot environments, whereas the effect on the gastrointestinal glands is probably of importance to prevent salt loss in the gastrointestinal excretory products.

MECHANISM OF ACTION OF ALDOSTERONE

Though we have known for many years the overall effects of mineralocorticoids on the body, the means by which aldosterone acts on the tubular cells to increase transport of sodium has until recently been a complete mystery. But several experiments now point toward a possible solution, as follows:

First, aldosterone concentrates in the nuclei of the tubule epithelial cells.

Second, aldosterone causes an increase in cytoplasmic RNA, the concentration of this beginning to rise 20 to 50 minutes afer exposure to aldosterone.

Third, approximately 50 minutes after exposure to aldosterone, the rate of sodium transport begins to increase.

Fourth, poisons that block the formation of messenger RNA in the nucleus or that block the formation of protein enzymes in the ribosomes also block the effect of aldosterone on sodium transport.

Therefore, it has been reasoned that aldosterone acts by stimulating the specific DNA molecules (the genes) that control sodium reabsorption. These set into motion the synthesis of specific messenger RNA molecules that in turn cause formation of specific enzymes that catalyze one or more of the chemical reactions responsible for sodium transport through the cell membrane.

REGULATION OF ALDOSTERONE SECRETION

Primary Factors that Stimulate Aldosterone Secretion.   In the normal person the rate of aldosterone secretion is automatically regulated to the level required to maintain *normal sodium concentration in the extracellular fluids* and *normal extracellular fluid volume*. To perform these functions, the rate of aldosterone secretion is responsive to the following changes:

1. *Decreased sodium concentration in the extracellular fluids* exerts a powerful effect on aldosterone secretion; only 10 per cent reduction in sodium concentration (about 14 mEq./liter) increases the rate of aldosterone secretion about two-fold almost immediately and perhaps much more than this if the stimulus is continued for several days.

2. *Increased potassium concentration* also increases the rate of aldosterone secretion; a 10 per cent change in potassium (about 0.5 mEq./liter) is required to effect a two-fold change in secretion. Since sodium and potassium affect aldosterone secretion in opposite directions, it is frequently said that it is the *sodium to potassium ratio* that regulates aldosterone secretion.

3. Either decreased extracellular fluid volume or decreased blood volume also increases the rate of aldosterone secretion, though this effect usually does not occur to a great extent until the arterial pressure and cardiac output also begin to fall. Therefore, it is probable that *diminished circulation to the tissues* in some way causes increased aldosterone secretion.

Basic Mechanisms for Stimulating Aldosterone Secretion.   Though we know most of the basic *stimuli* that elicit aldosterone secretion, we are much less sure of the mechanisms by which these stimuli act on the adrenal cortex. Some theories relating to this were discussed in Chapter 36; but to make the story complete here, these theories are summarized briefly. In general, the theories fall into three different categories: (1) direct stimulation of the adrenal cortex to regulate its output, (2) renin-angiotensin regulation, and (3) neurosecretory regulation of adrenocortical output. Figure 76–2 illustrates the basic interrelations of these several theories. The solid lines represent excitatory effects, while the dashed lines represent inhibitory effects.

DIRECT STIMULATION OF THE ADRENAL CORTEX.   When blood containing a *low concentration of sodium ions* is infused directly into the adrenal artery, the output of aldosterone often increases 1.5- to 3-fold almost immediately. Likewise, blood containing *excess potassium ions* causes an even greater effect. This illustrates that decreased blood sodium or increased blood potassium could have a

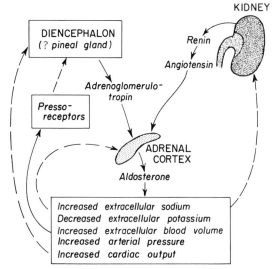

Figure 76–2 Postulated mechanisms for regulation of aldosterone secretion.

direct effect on the adrenal glands to regulate the output of aldosterone. Indeed, the quantitative changes in aldosterone output are sufficient to account for much of the regulation of aldosterone secretion.

The Renin-Angiotensin Mechanism. Infusion of angiotensin into the circulation causes marked increase in aldosterone secretion from the adrenal cortex, and other research studies indicate that either low blood sodium or diminished blood flow to the kidneys increases the formation of renin, which in turn leads to the formation of angiotensin (see Chap. 26). Therefore, it has been suggested that both of these stimuli increase aldosterone secretion through the mediation of angiotensin.

Neurosecretory Control of Adrenocortical Secretion. We shall see later in the chapter that one of the adenohypophyseal hormones, corticotropin (ACTH), is the primary regulator of glucocorticoid secretion. This hormone also increases aldosterone secretion to a slight extent but perhaps not enough to have major significance.

Destruction of the diencephalic portion of the brain has, in the hands of some research workers, a depressant effect on adrenocortical secretion of aldosterone, and a substance called *adrenoglomerulotropin* isolated from the pineal gland of animals has been claimed to restore the secretion of aldosterone. Therefore, it has been suggested that decreased sodium, increased potassium, decreased blood flow to the tissues of the diencephalon, or signals from the neck baroreceptors causes the release of adreno-

glomerulotropin, which in turn increases the output of aldosterone and returns the body fluids and circulatory conditions to normal.

Summary. The precise mechanism or mechanisms by which aldosterone secretion is regulated are unknown. However, few research workers doubt that plasma sodium and potassium changes can have direct effects on adrenal output of aldosterone. What most research workers do doubt is whether or not these direct effects are quantitatively sufficient to explain all the known aspects of regulation of aldosterone secretion.

Unfortunately, both the nervous and renal mechanisms of regulation are still doubtful because not all research workers have been able to confirm all the postulations. For instance, destruction of the diencephalon by some research workers has failed to affect aldosterone secretion. And in the case of the renal mechanism, a clamp on a renal artery to decrease blood flow to the kidney (in some experiments) actually decreases aldosterone secretion rather than increases it as the renin-angiotensin mechanism would predict.

FUNCTIONS OF THE GLUCOCORTICOIDS

Even though mineralocorticoids usually save the life of an adrenalectomized animal, the animal still is not completely normal. Instead, its metabolic systems for utilization of carbohydrates, proteins, and fats are considerably deranged. Furthermore, the animal cannot resist different types of physical or even mental stress, and minor illnesses, such as respiratory tract infections, can lead to death. Therefore, the glucocorticoids have functions just as important to the life of the animal as do the mineralocorticoids. These are explained in the following sections.

At least 95 per cent of the glucocorticoid activity of the adrenocortical secretions results from the secretion of *cortisol*, known also as *hydrocortisone* and *compound F*. In addition to this, a small amount of glucocorticoid activity is provided by *corticosterone* and a minute amount by *cortisone*.

EFFECT OF CORTISOL ON CARBOHYDRATE METABOLISM

Stimulation of Gluconeogenesis. By far the best-known metabolic effect of cortisol and other glucocorticoids on metabolism is their ability to stimulate gluconeogenesis by the liver, often increasing the rate of gluconeogenesis as much as 6- to 10-fold. However, the manner in which cortisol promotes gluconeogenesis is still not clear. Three different theories have been offered:

First, cortisol increases the transport of amino acids from the extracellular fluids into the liver cells. This obviously could increase the availability of amino acids for use in the liver cells, including conversion of the amino acids to glucose.

Second, several of the enzymes required to convert amino acids into glucose are increased in quantity in the liver cells of animals treated with glucocorticoids. Also, the concentration of RNA is increased in the liver cells. Therefore, it is assumed that glucocorticoids activate nuclear formation of messenger RNA's that in turn lead to the array of enzymes required for gluconeogenesis.

Third, the gluconeogenic effect could result secondarily from mobilization of amino acids from the extrahepatic tissues, which is another effect of *cortisol*, as explained later. As a result of protein mobilization, more amino acids become available in the plasma to enter into the gluconeogenesis process and thereby promote the formation of glucose.

INCREASED LIVER GLYCOGEN. One of the effects of increased gluconeogenesis is a marked increase in glycogen in the liver cells. On the other hand, the quantity of glycogen in other cells of the body generally is not increased or is even decreased, illustrating that there is a basic difference in the effect of cortisol on liver glycogen and cellular glycogen elsewhere.

Decreased Glucose Utilization by the Cells. Cortisol also usually decreases the rate of glucose utilization by the cells, though this effect is often slight. It is possible that the reduction in glucose utilization results secondarily from enhanced utilization of fats, which is discussed below, but most physiologists believe that somewhere between the point of entry of glucose into the cells and its final degradation cortisol directly delays the rate of glucose utilization. A suggested mechanism for this effect is based on the observation that glucocorticoids depress the oxidation of DPNH. Since DPNH must be oxidized to allow rapid glycolysis, this effect could account for the diminished utilization of glucose by the cells.

Also, it is known that glucocorticoids slightly depress glucose transport into the cells, which might be the primary means by which these hormones diminish glucose utilization.

Elevated Blood Glucose Concentration and Adrenal Diabetes. Both the increased rate of gluconeogenesis and the mild reduction in rate of glucose utilization by the cells cause the blood glucose concentration to rise. In addition, the enzyme *glucose 6-phosphatase* is increased in the liver in response to cortisol; this enzyme catalyzes the dephosphorylation of liver glucose and thereby promotes its transport into the blood, thus further increasing the blood glucose concentration.

The increased blood glucose concentration is called *adrenal diabetes*, and it has many similarities to pituitary diabetes, which was discussed in the preceding chapter. Administration of insulin lowers the blood glucose concentration only a moderate amount in adrenal diabetes, not nearly so much as it does in pancreatic diabetes. On the other hand, insulin causes greater decrease in blood glucose concentration in adrenal diabetes than in pituitary diabetes. Therefore, *pituitary diabetes is said to be weakly insulin sensitive, adrenal diabetes moderately insulin sensitive, and pancreatic diabetes strongly insulin sensitive.*

EFFECTS OF CORTISOL ON PROTEIN METABOLISM

Reduction in Cellular Protein. One of the principal effects of cortisol on the metabolic systems of the body is reduction of the protein stores in essentially all body cells except those of the liver and gastrointestinal tract. This is caused primarily by decreased protein anabolism — that is, by decreased rate of protein synthesis — but increased catabolism of proteins already in the cells also contributes partly to the decreased cellular stores. Yet the basic mechanism by which these effects take place is completely unknown; possible causes are discussed below.

INCREASED LIVER PROTEIN, GASTROINTESTINAL PROTEIN, AND PLASMA PROTEINS CAUSED BY CORTISOL. Coincidently with the reduced proteins elsewhere in the body, the liver and gastrointestinal proteins become enhanced. Furthermore, the plasma proteins (which are produced by the liver and then released into the blood) are also increased. Therefore, these are exceptions to the protein depletion that occurs elsewhere in the body. It is believed that this effect is caused both by the ability of cortisol to enhance amino acid transport into liver cells (but not into other cells) and by enhancement of the liver enzymes required for protein anabolism.

Increased Blood Amino Acids. The failure of amino acids to be incorporated into proteins by the cells under the influence of glucocorticoids, plus the increased catalysis of cellular proteins, increases the blood amino acid concentration; and this perhaps accounts for many of the important functions of glucocorticoids. For instance, the different cells continually synthesize a multitude of chemical substances from amino acids. Therefore, mobilization of amino acids from the tissues might play a valuable function in providing an available supply of amino acids in time of need.

Diminished Transport of Amino Acids into Extrahepatic Cells, and Enhanced Transport into Hepatic Cells. Recent studies in isolated tissue have demonstrated that cortisol depresses amino acid transport into muscle cells and perhaps into other extrahepatic cells. But, in contrast to this, cortisol enhances the transport of amino acids into liver cells.

Obviously, the decreased transport of amino acids into extrahepatic cells decreases their intracellular amino acid concentrations and as a consequence decreases the synthesis of protein. Yet catabolism of proteins in the cells continues to release amino acids from the already existing proteins, and these diffuse out of the cells to increase the plasma amino acid concentration. Therefore, it is said that *cortisol mobilizes amino acids from the tissues.*

The increased plasma concentration of amino acids, plus the fact that cortisol enhances transport of amino acids into the hepatic cells, could also account for enhanced utilization of amino acids by the liver in the presence of cortisol—such effects as: (1) increased rate of deamination of amino acids by the liver, (2) increased protein synthesis in the liver, (3) increased formation of plasma proteins by the liver, and (4) increased conversion of amino acids to glucose—that is, enhanced gluconeogenesis.

Thus, it is possible that many of the effects of cortisol on the metabolic systems of the body can be explained by these changes in transport of amino acids through the cell membrane; though cortisol also increases the enzymes required for these effects, possibly by causing the formation of appropriate messenger RNA's.

EFFECTS OF CORTISOL ON FAT METABOLISM

Mobilization of Fatty Acids. In much the same manner that cortisol promotes amino acid mobilization from muscle, it also promotes mobilization of fatty acids from adipose tissue. This in turn increases the concentration of unesterified fatty acids in the plasma and increases their utilization for energy. In this way, cortisol helps to shift the metabolic systems of the cells from utilization of glucose for energy to utilization of fatty acids instead.

Unfortunately, we do not know the mechanism by which cortisol promotes fatty acid mobilization, though it might cause some change in cell membrane transport of fatty acids, as seems to be the case for amino acids.

KETOGENIC AND ANTIKETOGENIC EFFECTS OF CORTISOL. It is frequently said that cortisol has a *ketogenic effect* because ketosis usually will not develop unless cortisol is available to cause fat mobilization. However, this ketogenic effect occurs only under certain conditions, such as when insulin is not available.

REGULATION OF CORTISOL SECRETION — CORTICOTROPIN (ACTH)

Effect of Corticotropin on Adrenocortical Secretion of Cortisol. Corticotropin, also known as *corticotropic hormone* or *adrenocorticotropic hormone* (ACTH), is secreted by the adenohypophysis. It in turn stimulates the adrenal cortex to secrete greatly increased quantities of the glucocorticoids, cortisol and corticosterone, and slightly increased quantities of aldosterone and adrenal androgens.

Small quantities of corticotropin are secreted continually by the adenohypophysis and will continue to be secreted even when the hypophysis is separated from the hypothalamus and transplanted to some other part of the body. However, when the hypothalamus and the hypothalamic-hypophyseal portal system are intact, corticotropin secretion can be enhanced as much as 20-fold within minutes by physiological stress. This is illustrated in Figure 76-3, which shows the adrenal response to corticotropin within a few minutes after breaking the tibia and fibula of a rat. It is believed that pain stimuli caused by the stress are first transmitted to the perifornical area of the hypothalamus. This in turn transmits impulses into the median eminence, as illustrated in Figure 76-4, where a neurosecretory product, coritcotropin-releasing factor (CRF), is secreted into the primary capillary plexus of the hypophyseal portal system and then carried by this portal system to the sinuses of the adenohypophysis. CRF then excites the glandular cells to cause corticotropin secretion, which in turn causes adrenocortical secretion

Figure 76–3   Rapid reaction of the adrenal cortex of a rat to stress caused by fracture of the tibia and fibula. (Courtesy of Drs. Guillemin, Dear, and Lipscomb.)

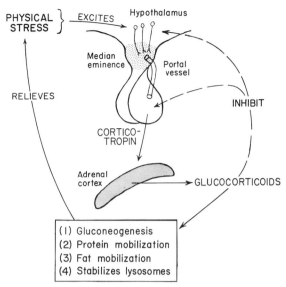

Figure 76–4   Mechanism for regulation of glucocorticoid secretion.

of cortisol. This adenohypophyseal-adrenocortical response to stress is especially important in helping the animal survive under adverse conditions, which are discussed later in the chapter.

INHIBITORY EFFECT OF CORTISOL ON CORTICOTROPIN SECRETION. An elevated concentration of cortisol in the blood causes negative feedback to the adenohypophysis to reduce the production of corticotropin. Though the exact means by which this feedback occurs is unknown, physiologists in general believe that cortisol acts both on the hypothalamus and directly on the adenohypophysis itself to inhibit adenohypophyseal secretion of corticotropin.

However, from a consideration of the theory of control systems, it is much more likely that some *effect of cortisol*, rather than the cortisol itself, is the basic feedback stimulus for control of corticotropin secretion. The reason for believing this is that no hormone is yet known in which the feedback control system

is geared primarily to control the concentration of the hormone itself. Instead, the respective control systems are geared to control some factor that the hormone affects. To express this another way, the hormones are "operators," and they vary up or down tremendously in order to cause the desired effect. Thus, it is the desired effect that is the object of the control system and not the concentration of the hormone. In the case of pancreatic secretion of insulin, insulin secretion can increase or decrease as much as 25-fold, acting as an operator that has as its object the control of blood glucose concentration. In the case of the corticotropin-cortisol system, we still do not know what metabolic factor cortisol controls, but when this factor is discovered it is almost certain that it will be this factor that in turn provides the feedback stimulus to the hypothalamus or adenohypophysis to regulate corticotropin secretion.

Function of Cortisol in Different Types of Stress. It is amazing that almost any type of stress, whether it be physical or neurogenic, will cause an immediate and marked increase in corticotropin secretion, followed within minutes by greatly increased adrenocortical secretion of cortisol. Some of the different types of stress that increase cortisol release are the following:

1. Trauma of almost any type
2. Intense heat
3. Intense cold
4. Injection of norepinephrine and other sympathomimetic drugs

5. Surgical operations
6. Injection of necrotizing substances beneath the skin
7. Restraining an animal so that he cannot move
8. Almost any debilitating disease

Thus, a wide variety of nonspecific stimuli can cause marked increase in the rate of cortisol secretion by the adrenal cortex.

Yet, even though we know that cortisol secretion often increases greatly in stressful situations, we still are not sure why this is of significant benefit to the animal. One guess, which is probably as good as any other, is that the glucocorticoids cause rapid mobilization of amino acids and fats from their cellular stores, making these available both for energy and for synthesis of other compounds needed by the different tissues of the body. Indeed, it is well known that when proteins are released from most of the tissue cells, the liver cells can use the mobilized amino acids to form new proteins. It is possible that other tissues, particularly damaged tissues which are momentarily depleted of proteins, can also utilize the newly available amino acids to form new proteins that are essential to the lives of the cells. Or perhaps the amino acids are used to synthesize such essential intracellular substances as purines, pyrimidines, and creatine phosphate, which are necessary for maintenance of cellular life.

But all this is mainly supposition. It is supported only by the fact that cortisol usually does not mobilize the basic functional proteins of the cells, such as the contractile proteins of muscle fibers, until almost all other proteins have been released. This preferential effect of cortisol in mobilizing labile proteins could make amino acids available to needy cells to synthesize substances essential to life.

Effect of Cortisol in Inflammation: Stabilization of Lysosomes. Almost all tissues of the body respond to tissue damage, whether the damage results from simple trauma or from some cellular disease, by the process called *inflammation*. There are three stages in this process: (1) leakage of large quantities of plasma-like fluid out of the capillaries into the damaged area followed by clotting of the fluid, (2) infiltration of the area by leukocytes, (3) tissue healing, which is accomplished to a great extent by ingrowth of fibrous tissue. The amount of cortisol normally secreted does not significantly affect inflammation and healing; but large amounts of cortisol or other glucocorticoids administered to a person block all stages of the process, even blocking the initial leakage of plasma from the capillaries into the damaged area.

Unfortunately, we still know very little about the basic mechanism of cortisol's *anti-inflammatory effect*. However, it has recently been discovered that cortisol and other glucocorticoids stabilize the membranes of cellular lysosomes so that they rupture only with difficulty. It will be recalled that lysosomes contain large quantities of hydrolytic enzymes that are capable of digesting intracellular proteins. This stabilizing effect on the lysosome membrane could alone account for a major share of the anti-inflammatory effect of cortisol, because it would prevent the usual destruction of tissues that occurs in inflammation because of liberation of lysosomal enzymes.

A second factor of possible importance in the anti-inflammatory effect of cortisol is its ability to decrease the formation of bradykinin, a very powerful vasodilating substance that appears in inflamed tissues. This substance is split from an alpha globulin by the proteolytic enzyme kallikrein. It is possible that stabilization of the lysosome membrane prevents release of kallikrein into the cell from the lysosome.

Finally, cortisol decreases the permeability of the capillary membrane, and this is a significant factor in preventing the usual protein leakage into inflamed tissues. Part of this decrease in capillary permeability might result from inhibition of bradykinin formation, because bradykinin vasodilatation tends to overstretch the capillary membrane.

Regardless of the precise mechanisms by which the anti-inflammatory effect occurs, this effect of cortisol perhaps plays a major role in combating certain types of diseases, such as rheumatoid arthritis, rheumatic fever, and acute glomerulonephritis. All these are characterized by severe local inflammation, and the harmful effects to the body are caused mainly by the inflammation itself and not by other aspects of the disease. When cortisol or other glucocorticoids are administered to patients with these diseases, almost invariably the inflammation subsides within 24 to 48 hours. And even though the cortisol does not correct the basic disease condition itself but merely prevents the damaging effects of the inflammatory response, this alone can often be a lifesaving measure.

Effect on Blood Cells and on Immunity. Cortisol decreases the number of eosinophils and lymphocytes in the blood; this effect begins within a few minutes after the injection of cortisol and is marked within a few hours. Indeed, a finding of lymphocytopenia or eosinopenia is an important diagnostic criterion for overproduction of cortisol by the adrenal gland.

Likewise, administration of large doses of cortisol causes significant atrophy of all the lymphoid tissue throughout the body, which in turn decreases the output of antibodies from the lymphoid tissue. As a result, the level of humoral immunity for almost all foreign invaders of the body is decreased.

On the other hand, cortisol increases the production of red blood cells, the cause of which is unknown. When excess cortisol is secreted by the adrenal glands, polycythemia results, and, conversely, when the adrenal glands secrete no cortisol, severe anemia often results.

EFFECT ON ALLERGY. Cortisol blocks the inflammatory response to allergic reactions in exactly the same way that it blocks other types of inflammatory responses. The basic allergic reaction between antigen and antibody is not affected by cortisol, and even some of the secondary effects of the allergic reaction, such as the release of histamine, still occur the same as ever. However, since the inflammatory response is responsible for many of the serious and sometimes lethal effects of allergic reactions, administration of cortisol can often be lifesaving. For instance, cortisol effectively prevents shock or death in anaphylaxis which otherwise kills many persons, as explained in Chapter 10.

CHEMISTRY OF ADRENOCORTICAL SECRETION

Locus of Hormone Formation in the Adrenal Cortex. Aldosterone is secreted by the *zona glomerulosa* of the adrenal cortex, the glucocorticoids by the *zona fasciculata,* and the adrenal androgens probably by the *zona reticularis*. These zones are illustrated in Figure 76–5. During prolonged corticotropin stimulation of the adrenal cortex, the zona fasciculata and zona reticularis both hypertrophy, whereas complete lack of corticotropin causes these two zones to atrophy almost entirely while leaving the zona glomerulosa mainly intact. On the contrary, conditions that increase the output of aldosterone cause hypertrophy of the zona glomerulosa while not affecting the other two zones.

Chemistry of the Adrenocortical Hormones. All the adrenocortical hormones are steroid compounds, and they can be formed in the adrenal cortex from acetyl coenzyme A or from cholesterol preformed in the cortical cells in the body. Figure 76–6 illustrates a schema for formation of most of the important steroids found in the adrenal cortex. From this schema it can be seen that the adrenal cortex can form hormones having at least five different types of effects, two of which, the mineralocorticoids and glucocorticoids, are extremely important because they are secreted in large quan-

ZONA GLOMERULOSA
? secretes mineralocorticoids

ZONA FASCICULATA
? secretes glucocorticoids

ZONA RETICULARIS
? secretes adrenal
sex hormones

MEDULLA   CORTEX

MAGNIFIED SECTION

Figure 76–5   Secretion of adrenocortical hormones by the different zones of the adrenal cortex.

tities, and three of which, the adrenal sex hormones, are normally of only minor importance because they are secreted in minute quantities. However, a change in even a single enzyme system somewhere in the scheme can cause vastly different types of hormones to be formed; occasionally, large quantities of masculinizing or, rarely, feminizing sex hormones are secreted by adrenal tumors, as discussed later in the chapter.

Figure 76–7 illustrates the chemical formulas of four of the adrenal steroid compounds. Of these, cortisol and aldosterone are secreted in large quantities by the adrenal cortex. Cortisone and deoxycorticosterone are secreted in only minor amounts, but both are used to a great extent in therapy because they can be synthesized readily.

In addition, several steroids that do not naturally occur in the adrenal cortex but that have even more potent actions than the natural hormones have been synthesized. Thus, *prednisolone,* which is a slight modification of cortisol, has three times as much glucocorticoid potency as cortisol but no increase in mineralocorticoid potency. Therefore, it is used clinically when one wishes maximum glucocorticoid activity and yet wishes to prevent electrolyte and water retention. Likewise, *2-methyl-9α-fluorocortisol* has three times as much mineralocorticoid activity as aldosterone and therefore can be used in minute quantities, only 25 micrograms per day, to treat total deficiency of mineralocorticoid secretion.

On the other hand, other substances similar in chemical structure but having almost no mineralocorticoid activity—the *spirolactones,* for instance—become firmly fixed in the renal tubular epithelial cells at the sites where aldosterone normally acts. Therefore, these compounds block the effect of the body's own aldosterone. As a result, the kidneys then function as if no aldosterone were being secreted in the body.

Mechanism by which Corticotropin Activates Adrenocortical Cells To Produce Steroids — Function of Cyclic AMP. It is believed that the primary effect of corticotropin on the adrenocortical cells is to cause formation of cyclic adenylic acid (cyclic AMP) in the cells by the mechanism discussed in Chapter 75. This substance, in turn, is believed to promote the intracellular reactions for formation of adrenocortical hormones. Cyclic AMP has a specific effect to activate the phosphogluconate pathway for glucose metabolism, which was discussed in Chapter 67 and which causes the formation of large quantities of TPNH (reduced triphosphopyridine nucleotide). The TPNH, in turn, has an extremely high capacity for causing hydroxylation. Therefore, it is believed that the chemical reactions for formation of the adrenal steroids begin with cholesterol that is gradually changed, mainly by a series of hydroxylations, into the different adrenal steroids.

Chemistry of Corticotropin (ACTH). ACTH has been isolated in pure form from the adenohypophysis. It is a large polypeptide having a chain length of 39 amino acids. A digested product of ACTH having a chain length of 24 amino acids has all the corticotropin effects of the total molecule.

Chemistry of Corticotropin-Releasing Factor. Several small peptides that have a chemical configuration similar to that of vasopressin and oxytocin have been isolated from the hypothalamus which, upon injection into an animal, cause rapid release of corticotropin from the adenohypophysis. Also, vasopressin itself has a weak effect to cause corticotropin release. It is not known whether one of these is the specific corticotropin-releasing factor.

Transport and Fate of the Adrenal Hormones. Cortisol combines loosely with an alpha globulin called "transcortin" and, to a less extent, with albumin. On the other hand, aldosterone combines loosely with albumin and, to a less extent, with alpha globulin. In these combined forms the hormones are transported throughout the extracellular fluid compartment. In general, the hormones become fixed in the target tissues within 10 to 20 minutes after their secretion.

The adrenal steroids are deactivated mainly in the liver and conjugated to form glucuronides and, to a less extent, sulfates. About 25 per cent of these are excreted in the feces and the remaining 75 per cent in the urine. The conjugated forms of these hormones are inactive.

The normal concentration of aldosterone in blood is about 0.08 μg./100 ml., and the secretory rate is about 150 μg. per day. Once secreted, aldosterone functions in the body for several hours.

The concentration of cortisol in the blood averages 12 μg./100 ml., and the secretory rate averages 15 mg. per day. Its half-life in the body is about 100 minutes. Thus, the effects of both aldosterone and cortisol last for only a few hours at most after secretion.

THE ADRENAL ANDROGENS

Several moderately active male sex hormones called *adrenal androgens* (two of which are illustrated in Fig. 76–6) are continually secreted by the adrenal cortex. Also, progesterone and estrogens, which are female sex hormones, have been extracted from the adrenal cortex, though these are secreted in only minute quantities.

In normal physiology of the human being, even the adrenal androgens have almost insignificant effects. However, it is possible that part of the early development of the male sex organs results from childhood secretion of adrenal androgens. The adrenal androgens possibly also exert mild effects in the female, not only before puberty but also throughout life. One of these effects is said to be increased libido. The physiologic effects of andro-

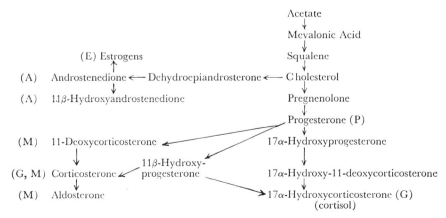

Figure 76–6   Schema showing the stages through which the different corticosteroids are synthesized. The physiologic characteristics of some of these corticosteroids are designated as follows: *A,* androgenic effects; *E,* estrogenic effects; *G,* glucocorticoid effects; *M,* mineralocorticoid effects; *P,* progesteronic effects.

Figure 76–7   Four important corticosteroids.

gens will be discussed in Chapter 80 in relation to male sexual functions.

ABNORMALITIES OF ADRENOCORTICAL SECRETION

HYPOADRENALISM — ADDISON'S DISEASE

Addison's disease results from failure of the adrenal cortices to produce adrenocortical hormones, and this in turn is most frequently caused by *primary atrophy* of the adrenal cortices but also frequently by tuberculous destruction of the adrenal glands or invasion of the adrenal cortices by cancer. Basically, the disturbances in Addison's disease are:

Mineralocorticoid Deficiency.   Lack of aldosterone secretion greatly decreases sodium reabsorption and consequently allows sodium ions, chloride ions, and water to be lost into urine in great profusion. The net result is a greatly decreased extracellular fluid volume. Furthermore, the person develops acidosis because of failure of hydrogen ions to be secreted in exchange for sodium reabsorption.

As the extracellular fluid becomes depleted, the plasma volume falls, the red blood cell concentration rises markedly, the cardiac output decreases, and the patient dies in shock, death usually occurring four to seven days after complete cessation of mineralocorticoid secretion.

Glucocorticoid Deficiency.   Loss of cortisol secretion makes it impossible for the person with Addison's disease to maintain normal blood glucose concentration between meals because he cannot synthesize significant quantities of glucose by gluconeogenesis. Furthermore, lack of cortisol reduces the mobilization of both proteins and fats from the tissues, thereby depressing many other metabolic functions of the body. This sluggishness of energy mobilization when cortisol is not available is perhaps one of the major detrimental effects of glucocorticoid lack. However, even when excess quantities of glucose and other nutrients are available, the person's muscles are still weak, indicating that glucocorticoids are also needed to maintain other metabolic functions of the tissues besides simply energy metabolism.

Lack of adequate glucocorticoid secretion also makes the person with Addison's disease highly susceptible to the deteriorating effects of different types of stress, and even a mild respiratory infection can sometimes cause death.

Melanin Pigmentation.   Another characteristic

of some persons with Addison's disease is melanin pigmentation of the skin over most of the body. This melanin is not always deposited evenly but occasionally in blotches and especially in the thin skin areas, such as the mucous membranes of the lips and the thin skin of the nipples.

The cause of the melanin deposition is believed to be the following: It is well known that the *pars intermedia* of the adenohypophysis in some lower animals, on appropriate stimulation, secretes large quantities of *melanocyte-stimulating hormone* (MSH), and that this in turn increases the melanin production by the melanocytes in the skin. Therefore, it has been assumed that in persons with Addison's disease, lack of cortisol secretion exerts a feedback effect on the adenohypophysis, causing it to secrete excessive amounts of MSH, and that this in turn causes the pigmentation. In addition, however, lack of secretion of cortisol by the adrenal cortex also causes tremendous secretion of adrenocorticotropic hormone (ACTH) by the adenohypophysis, and ACTH has one-thirtieth as much melanocyte-stimulating effect as does melanocyte-stimulating hormone (MSH). Indeed, MSH has the same peptide structure as the first 13 amino acids in the 39 amino acid chain of the ACTH molecule.

Diagnosis of Addison's Disease. A person with this disease has almost no urinary secretion of steroids derived from the adrenals. For instance, in full-blown Addison's disease, less than 1 mg. of 17-hydroxysteroids, derived almost entirely from cortisol secreted by the adrenals, is excreted in a 24 hour period. However, to be certain that the person has Addison's disease, corticotropin (ACTH) is infused into the patient over a period of 8 hours or more, and the rate of secretion of 17-hydroxysteroids is again measured. If the Addison's disease is caused by total destruction of the adrenal cortices, there will be no increase in urinary 17-hydroxysteroids. If, on the other hand, the Addison's disease is caused by failure of the adenohypophysis to stimulate the adrenal cortices, successive administration of ACTH several days in a row will cause the adrenal glands to secrete normal quantities of 17-hydroxysteroids.

Treatment of Persons with Addison's Disease. The untreated person with Addison's disease dies within a few days because of electrolyte disturbances. Yet such a person can usually live for years, though in a weakened condition, if small quantities of mineralocorticoids are administered — for instance, 25 mg. of 2-methyl-9α-fluorocortisol by mouth daily. Also implantation of small pellets of solid deoxycorticosterone in a fat pad of the back can result in satisfactory electrolyte balance for as long as 6 to 12 months.

Unfortunately, persons treated with mineralocorticoids but without glucocorticoids are still subject to stress reactions that may cause rapid demise. Therefore, full treatment of Addison's disease includes administration of glucocorticoids — usually cortisone, cortisol, or prednisolone — as well as mineralocorticoids.

The Addisonian Crisis. As noted earlier in the chapter, great quantities of glucocorticoids are occasionally secreted in response to different types of physical or mental stress. In the person with Addison's disease, the output of glucocorticoids does not increase during stress. Yet whenever he is specifically subjected to different types of trauma, disease, or other stresses, he is likely to develop an acute need for excessive amounts of glucocorticoids, and must be given many times the normal quantities of glucocorticoids in order to prevent death.

This critical need for extra corticosteroids and the associated severe debility in times of stress is called an Addisonian crisis.

HYPERADRENALISM — CUSHING'S DISEASE

One type of hypersecretion by the adrenal cortex causes a complex of hormonal effects called Cushing's disease, and this usually results from general hyperplasia of both adrenal cortices. The hyperplasia in turn is most often caused by increased secretion of corticotropin by the adenohypophysis. Most abnormalities of Cushing's disease are ascribable to abnormal amounts of cortisol and, to a less extent, of aldosterone, but in some persons the secretion of adrenal sex hormones is also of significance.

Effects on Electrolytes, Body Fluids, and the Cardiovascular System. In Cushing's disease sodium retention is moderately enhanced because of the moderately increased aldosterone secretion, and this in turn leads to mildly increased extracellular fluid volume, increased plasma volume, mild degrees of hypokalemia, and mild alkalosis. The increase in extracellular fluid volume is usually not enough to cause frank edema, but it may cause slight puffiness of the skin, especially in the facial region, as illustrated in Figure 76–8.

The changes in fluid and electrolyte balance also usually lead to slightly increased cardiac output, as explained earlier in the chapter, and also to hypertension. Thus, most persons with Cushing's disease have a moderate to severe degree of hypertension, which in most instances is relieved by subtotal removal of the hypersecreting glands.

Effects on Carbohydrate and Protein Metabolism. The abundance of glucocorticoids secreted in Cushing's disease causes increased blood glucose concentration, sometimes to values as high as 140 to 200 mg. per cent. This effect probably results mainly from enhanced gluconeogenesis. If this "adrenal diabetes" lasts for many months, the beta cells in the islets of Langerhans occasionally "burn out" because the high blood glucose greatly overstimulates them to secrete insulin. The destruction of these cells then causes frank diabetes mellitus, which is permanent for the remainder of life.

The effects of glucocorticoids on protein catabolism are often profound in Cushing's disease, causing greatly decreased tissue proteins almost

Figure 76–8 A person with Cushing's disease before subtotal adrenalectomy (left) and after subtotal adrenalectomy (right). (Courtesy of Dr. Leonard Posey.)

everywhere in the body with the exceptions of the liver, the gastrointestinal tract, and the plasma proteins. Even the collagen fibers in the subcutaneous tissue are diminished so that the subcutaneous tissues tear easily, resulting in development of large *purplish striae;* these are actually scars where the subcutaneous tissues have torn apart. In addition, lack of protein deposition in the bones causes *osteoporosis* with consequent weakness of the bones.

Effects of the Adrenal Sex Hormones. In Cushing's disease the quantities of adrenal sex hormones, the androgens principally, are usually also increased to significant levels, and these result in typical masculinizing effects in a female (or in a male child), such as growth of a beard, increased deepness of the voice, occasional baldness, growth of the clitoris to resemble a penis or growth of a child's penis to adult proportions. In the male adult with Cushing's disease, the androgenic effects of the adrenal sex hormones will not cause any distinctive changes because of the much more potent androgenic effects of the normal testicular hormone secretion.

Diagnosis and Treatment of Cushing's Disease. Diagnosis of Cushing's disease is most frequently made on the basis of such typical findings as the buffalo torso, puffiness of the face, masculinizing effects, hypertension, elevated blood glucose concentration that is moderately insulin resistant, and secretion of 3 to 5 times the normal quantities of 17-hydroxysteroids in the urine.

Treatment in Cushing's disease consists of decreasing the secretion of corticotropin, if this is possible. Hypertrophied pituitary glands or tumors

that oversecrete corticotropin can be surgically removed or destroyed by radiation. If corticotropin secretion cannot easily be decreased, the only satisfactory treatment is usually bilateral total or partial adrenalectomy followed by administration of adrenal steroids to make up for any insufficiency that develops.

HYPERSECRETION OF INDIVIDUAL HORMONES

Occasionally, an abnormality of one or more enzyme systems for synthesis of the adrenocortical hormones causes excessive secretion of one or more particular types of adrenocortical hormones. For instance, tumors of the adrenal cortex or congenital absence of one of the enzymes in the adrenal cells can cause excessive secretion of (a) one or more of the adrenal sex hormones, (b) one of the glucocorticoids, or (c) one of the mineralocorticoids. In these conditions, a person may develop any one (or more) of the effects noted previously in relation to Cushing's disease without having the other effects. Three common types of specific hormonal excesses are (1) primary aldosteronism, (2) primary glucocorticoid hypersecretion, and (3) the adrenogenital syndrome.

Primary Aldosteronism. In primary aldosteronism all the symptoms of Cushing's disease that result from excess mineralocorticoid secretion are seen without the other symptoms. These include particularly *increased extracellular fluid volume, hypertension, hypokalemia, mild alkalosis,* and a *tendency toward hypernatremia.* Ordinarily, the person has such intense polydipsia that he keeps washing sodium chloride through his kidneys despite a very low concentration of sodium chloride in his urine, and this prevents him from having frank edema in mild cases. However, in severe cases, edema does appear. Especially interesting in primary aldosteronism are occasional periods of *muscular paralysis caused by the hypokalemia,* the mechanism of which was explained in Chapter 5.

Primary Excess Glucocorticoid Secretion. Rarely, excess glucocorticoids are secreted without excess secretion of the other adrenocortical hormones. When this occurs, the person develops (a) adrenal diabetes, (b) signs of protein deficiency, including especially development of purplish striae and osteoporosis, and (c) increased lipogenesis, which, in extreme cases causes a buffalo torso.

A high percentage of the persons who secrete excess amounts of glucocorticoids is likely to secrete a reasonable percentage of androgens along with the glucocorticoids. In such an instance, the female patient develops a beard and other masculinizing effects; the condition is then frequently called *diabetes of bearded women.*

Adrenogenital Syndrome. Tumors of the adrenal cortex that secrete excess adrenal sex hormones, except in rare instances, secrete androgenic hormones that cause intense masculinizing effects throughout the body. For instance, the female develops extremely virile characteristics, including growth of a beard, development of a much deeper voice, occasionally development of baldness if she also has the genetic inheritance for baldness, development of a masculine distribution of hair on the body and on the pubis, growth of the clitoris to resemble a penis, and deposition of proteins in the skin and especially in the muscles to give typical masculine characteristics.

In the prepubertal male a virilizing adrenal tumor causes the same characteristics as in the female, plus rapid development of the male sexual organs and creation of male sexual desires. Typical development of the male sexual organs in a 4-year-old boy with the adrenogenital syndrome is shown in Figure 76–9.

In the adult male, the virilizing characteristics of the adrenogenital syndrome are usually completely obscured by the normal virilizing characteristics of the testosterone secreted by the testes. Therefore, it is often difficult to make a diagnosis of adrenogenital syndrome in the male adult. However, an occasional adrenal tumor secretes sufficient quantities of feminizing hormones that the male patient develops *gynecomastia,* which means excessive growth of the male mammary glands. Also, in the adrenogenital syndrome, the excretion of 17-ketosteroids (which are derived from androgens) in the urine may be as much as 10 to 15 times normal.

Figure 76–9   Adrenogenital syndrome in a 4-year-old boy. (Courtesy of Dr. Leonard Posey.)

REFERENCES

Adler, S.: An extrarenal action of aldosterone on mammalian skeletal muscle. *Amer. J. Physiol., 218*:616, 1970.

Appelgren, L.: Sites of steroid hormone formation. *Acta Physiol. Scand., 71,* Suppl. 301, 1967.

Blair-West, J.; Cain, M.; Catt, K.; Coghlan, J.; Denton, D.; Funder, J.; Nelson, J.; Scoggins, B.; Wintour, M.; and Wright, R.: The control of aldosterone secretion. *Proc. Int. Union of Physiol. Sciences, 6*:249, 1968.

Blair-West, J.; Coghlan, J. P.; Denton, D. A.; and Wright, R. D.: Effect of endocrines on salivary glands. *In* Handbook of Physiology. Baltimore, The Williams & Wilkins Co., 1967, Sec. VI, Vol. II, p. 633.

Bransome, E. D., Jr.: Adrenal cortex. *Ann. Rev. Physiol., 30*:171, 1968.

Ciba Foundation Study Group No. 27: The Human Adrenal Cortex. Boston, Little, Brown and Company, 1967.

Conn, J. W.: The evolution of primary aldosteronism: 1954–1967. Harvey Lectures, 1966–1967, p. 257.

Cope, C. L.: Adrenal Steroids and Disease. Philadelphia, J. B. Lippincott Co., 1965.

Davis, J. O.: The control of aldosterone secretion. *Physiologist, 5*:65, 1962.

Denton, D. A.: Evolutionary aspects of the emergence of aldosterone secretion and salt appetite. *Physiol. Rev., 45*:245, 1965.

Eisenstein, A. B.: The Adrenal Cortex. Boston, Little, Brown and Company, 1967.

Fimognari, G. M.; Fanestil, D. D.; and Edelman, I. S.: Induction of RNA and protein synthesis in the action of aldosterone in the rat. *Amer. J. Physiol., 213*:954, 1967.

Goodwin, F. J.; Knowlton, A. I.; and Laragh, J. H.: Absence of renin suppression by deoxycorticosterone acetate in rats. *Amer. J. Physiol., 216*:1476, 1969.

Hechter, O., and Halkerston, I. D. K.: Effects of steroid hormones of gene regulation and cell metabolism. *Ann. Rev. Physiol.,* *27*:133, 1965.

Howards, S. S.; Davis, J. O.; Johnston, C. I.; and Wright, F. S.: Steroidogenic response in normal dogs receiving blood from dogs with caval constriction. *Amer. J. Physiol., 214*:990, 1968.

Kostyo, J. L., and Schmidt, J. E.: Inhibitory effects of cardiac glycosides and adrenal steroids on amino acid transport. *Amer. J. Physiol., 204*:1031, 1963.

McKerns, K. W. (ed.): Functions of the Adrenal Cortex. New York, Appleton-Century-Crofts, 1968, Vols. 1 and 2.

Robb, C. A.; Davis, J. O.; Johnston, C. I.; and Hartroft, P. M.: Effects of deoxycorticosterone on plasma renin activity in conscious dogs. *Amer. J. Physiol., 216*:884, 1969.

Sharp, G. W. G., and Leaf, A.: Mechanism of action of aldosterone. *Physiol. Rev., 46*:593, 1966.

Urquhart, J., and Li, C. C.: The dynamics of adrenocortical secretion. *Amer. J. Physiol., 214*:73, 1968.

Venning, E. H.: Adenohypophysis and adrenal cortex. *Ann. Rev. Physiol., 27*:107, 1965.

Voogt, J. L.; Sar, M.; and Meites, J.: Influence of cycling, pregnancy, labor, and suckling on corticosterone-ACTH levels. *Amer. J. Physiol., 216*:655, 1969.

Yates, F. E.: The regulation of cortisol secretion. *Proc. Int. Union of Physiol. Sciences, 6*:247, 1968.

Yates, F. E., and Urquhart, J.: Control of plasma concentrations of adrenocortical hormones. *Physiol. Rev., 42*:359, 1962.

CHAPTER 77

The Thyroid Hormones

The thyroid gland, which is located immediately below the larynx on either side of and anterior to the trachea, secretes *thyroxine* and, also, much less quantities of several other closely related iodinated hormones that have a profound effect on the metabolic rate of the body. Complete lack of thyroid secretion usually causes the basal metabolic rate to fall to −40 to −50, and extreme excesses of thyroid secretion can cause the basal metabolic rate to rise as high as +60 to +100. Thyroid secretion is controlled primarily by *thyrotropin* secreted by the adenohypophysis.

The purpose of this chapter is to discuss the formation and secretion of the thyroid hormones, their functions in the metabolic schema of the body, and regulation of their secretion.

FORMATION AND SECRETION OF THE THYROID HORMONE

By far the most abundant of the hormones secreted by the thyroid glands is *thyroxine.* However, significant amounts of *triiodothyronine* are also secreted. The functions of these two hormones are qualitatively the same, but they differ in rapidity and intensity of action.

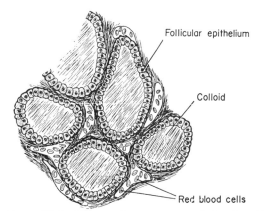

Follicular epithelium

Colloid

Red blood cells

Figure 77–1 Microscopic appearance of the thyroid gland, showing the secretion of thyroglobulin into the follicles.

In this chapter mainly the functions of thyroxine are discussed because of the great abundance of this hormone, but when necessary we will point out the differences between the actions of thyroxine and those of triiodothyronine; triiodothyronine is about five times as potent as thyroxine, even though it is present in the blood in only minute quantities.

Physiologic Anatomy of the Thyroid Gland. The thyroid gland is composed, as shown in Figure 77–1, of large numbers of encysted *follicles* filled with a substance called *colloid* and lined with *cuboidal epithelioid cells* that secrete into the interior of the follicles. Once the secretion has entered the follicles, it must be absorbed back through the follicular epithelium into the blood before it can function in the body. (The thyroid gland has a blood supply as rich as that of any other area of the body with the probable exception of the adrenal cortex.)

REQUIREMENTS OF IODINE FOR FORMATION OF THYROXINE

To form normal quantities of thyroxine, approximately 35 to 50 mg. of ingested iodine are required *each year,* or approximately *1 mg. per week.* To prevent iodine deficiency, common table salt is iodized with one part sodium iodide to every 100,000 parts sodium chloride.

Fate of Ingested Iodides. Iodides ingested orally are absorbed from the gastrointestinal tract into the blood in approximately the same manner as chlorides, but iodides do not remain in the circulatory system for a prolonged time. Within the first three days two thirds of the ingested iodides are normally lost into the urine, and the remaining one third is selectively removed from the circulating blood by the cells of the thyroid gland and used for synthesis of thyroid hormones, which are either stored in the form of *thyroglobulin* in the follicles or secreted into the blood principally in the form of thyroxine, as is discussed below.

THE IODIDE PUMP (IODIDE TRAPPING)

The first stage in the formation of thyroid hormones, as shown in Figure 77–2, is the

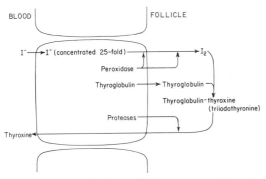

$$I_2 + HO - \bigcirc - CH_2 - CHNH_2 - COOH \xrightarrow{\text{iodinase}}$$
Tyrosine

$$HO - \bigcirc - CH_2 - CHNH_2 - COOH +$$
Monoiodotyrosine

$$HO - \bigcirc - CH_2 - CHNH_2 - COOH$$
Diiodotyrosine

Monoiodotyrosine + Diiodotyrosine $\longrightarrow$

$$HO - \bigcirc - O - \bigcirc - CH_2 - CHNH_2 - COOH$$
3,5,3' – Triiodothyronine

Diiodotyrosine + Diiodotyrosine $\longrightarrow$

$$HO - \bigcirc - O - \bigcirc - CH_2 - CHNH_2 - COOH$$
Thyroxine

Figure 77–2 Mechanisms of iodine transport, thyroxine formation, and thyroxine release into the blood. (Triiodothyronine formation and release parallels that of thyroxine.)

Figure 77–3 Chemistry of thyroxine and triiodothyronine formation.

transfer of iodides from the extracellular fluid into the thyroid glandular cells and thence into the follicle. The cell membranes have a specific ability to transport iodides actively to the interior of the follicle; this is called the *iodide pump,* or *iodide trapping.* In a normal gland, the iodide pump can concentrate the iodide to 25 to 50 times its concentration in the blood. However, when the thyroid gland becomes maximally active, the concentration can rise to as high as 350 times that in the blood.

CHEMISTRY OF THYROXINE AND TRIIODOTHYRONINE FORMATION

Oxidation of the Iodides. The first stage in the formation of the thyroid hormones is believed to be oxidative conversion of iodides either to *elemental iodine* or to *some other oxidized form* of iodine that is then capable of combining with the amino acid *tyrosine* to form the thyroid hormone. The reason for believing this is that iodination of tyrosine can be effected readily even in vitro by elemental iodine but not by iodides. Furthermore, large amounts of the enzyme *peroxidase* are present in the thyroid glandular cells. This enzyme is capable of oxidizing iodides, and, when it is absent from the cells, the rate of formation of thyroid hormones is greatly decreased.

Iodination of Tyrosine and Formation of the Thyroid Hormones. Figure 77–3 illustrates the successive stages of iodination of tyrosine and the final formation of the two most important thyroid hormones, thyroxine and triiodothyronine. Tyrosine is first iodized to *monoiodotyrosine* and then to *diiodotyrosine.* Two molecules of diiodotyrosine then conjugate, with the loss of the amino acid alanine, to form one molecule of *thyroxine.* Or one molecule of mono-

iodotyrosine combines with one molecule of diiodotyrosine to form *triiodothyronine.*

Storage of Thyroid Hormones in Thyroglobulin. In addition to synthesizing the thyroid hormones, the thyroid cells also secrete a large globulin called thyroglobulin (molecular weight 680,000) directly into the thyroid follicles and not into the circulating blood. Furthermore, it is the amino acid tyrosine in this molecule that is converted into monoiodotyrosine, diiodotyrosine, triiodothyronine, and thyroxine. This conversion of tyrosine into the thyroid hormones probably can occur both before and after the thyroglobulin has been secreted into the follicles, because oxidized iodine can act on tyrosine of thyroglobulin to form these hormones either before or after secretion. Thus, it is in the form of the complex molecule thyroglobulin that thyroid hormones are stored in the thyroid gland for a period of several weeks before being released into the circulating blood.

RELEASE OF THYROXINE FROM THE FOLLICLES

Thyroglobulin itself is not released into the circulating blood, but instead it is digested by *proteinases* that are also secreted by the thyroid cells. These split thyroxine and minute quantities of triiodothyronine from the thyroglobulin complex and allow these to diffuse through the glandular cells into the blood outside the follicles. The rate of activity of the proteinases (and, therefore, also, the rate of thyroxine secretion into the circulating body fluids) is controlled by thyrotropin from the adenohypophysis, which is discussed in detail later in the chapter. It is partly in this way that the overall rate of secretion of the thyroid gland is regulated.

TRANSPORT OF THYROXINE AND TRIIODOTHYRONINE TO THE TISSUES

Though approximately 25 per cent of the organic iodides in the thyroid gland are di-iodothyronine and approximately 6 per cent are triiodothyronine, almost none of these is released from the thyroid gland into the blood. Indeed, quantitatively, at least 95 per cent, and probably closer to 100 per cent, of the active thyroid hormones entering the body fluids are thyroxine, and the remainder is almost entirely triiodothyronine.

Binding of Thyroxine and Triiodothyronine with the Plasma Proteins. On entering the blood, both thyroxine and triiodothyronine immediately combine with several of the plasma proteins. They combine mainly with a globulin called *thyroxine-binding globulin,* which has physical characteristics midway between alpha$_1$ and alpha$_2$ globulins.

The affinity of thyroxine-binding globulin for thyroxine is approximately three times as great as its affinity for triiodothyronine. For this reason, whenever these two hormones are injected intravenously or secreted by the thyroid gland in equal quantities, the amount of *free* thyroxine that remains in the plasma is only one-third the amount of free triiodothyronine. This allows far more rapid entry of triiodothyronine into the cells than of thyroxine, which possibly accounts for much of the five-fold differential between the potencies of the two hormones.

About half the thyroxine in the blood is released to the tissue cells every eight days. Since thyroxine constitutes the major bulk of the thyroid hormones, this slow release of thyroxine is one of the causes of the very slow action of thyroxine, which is discussed later in the chapter.

FUNCTIONS OF THYROXINE IN THE TISSUES

GENERAL INCREASE IN METABOLIC RATE

The principal effect of thyroxine is to increase the metabolic activities of most tissues of the body (with a few notable exceptions, such as the brain, retina, spleen, testis, and lungs). The basal metabolic rate can increase to as much as 60 to 100 per cent above normal when large quantities of thyroxine are secreted. The rate of utilization of foods for energy is greatly accelerated. The rate of protein synthesis is at times increased, while at the same time the rate of protein catabolism is also increased. The growth rate of young persons is greatly accelerated. The mental processes are excited, and the activity of many other endocrine glands is often increased. Yet despite the fact that we know all these many changes in metabolism are under the influence of thyroxine, the basic mechanism (or mechanisms) by which thyroxine acts is almost completely unknown.

Effect of Thyroxine on Mitochondria. When thyroxine is given to an animal, the mitochondria in most cells of the body increase in size and also in number. Furthermore, the total membrane surface of the mitochondria increases almost directly in proportion to the increased metabolic rate of the whole animal. Therefore, it is an obvious deduction that the principal function of thyroxine might be simply to increase the number and activity of mitochondria, and these in turn increase the rate of formation of ATP to energize cellular function. Unfortunately, though, the increase in number and activity of mitochondria could as well be the *result* of increased activity of the cells as be the cause of the increase.

Possible Function of Thyroxine as an Antioxidant. Thyroxine in extremely small quantities can act as an antioxidant, preventing some of the peroxidation reactions that are known to occur in cells. Furthermore, since it is known that oxidations of this type frequently lead to lipid degradation, it has been postulated that the antioxidant effect of thyroxine prevents degradation of mitochondrial membranes and, therefore, allows mitochondria to persist much longer than normally in cells once they are formed. However, this theory is highly questionable.

Effect of Thyroxine on Cellular Enzyme Systems. Vast numbers of intracellular enzymes increase in quantity a week or so following administration of thyroxine. For instance, one enzyme, α-glycerophosphate dehydrogenase, can be increased to an activity six times its normal level. Since this enzyme is particularly important in the degradation of carbohydrates, its increase could help to explain the rapid utilization of carbohydrates under the influence of thyroxine. Also, the oxidative enzymes and the elements of the electron transport system, both of which are normally found in mitochondria, are greatly increased.

Since the enzymes that are increased are mainly those found also in mitochondria, it is possible that this effect is indirectly caused by the increase in mitochondria.

Other than for these still highly speculative clues, we still do not know the basic mechanism by which thyroxine performs its widespread stimulatory effects on almost all activities of the body.

Latency and Duration of Thyroid Hormone Activity. After injection of a relatively large quantity of thyroxine into a human being, essentially no effect on the metabolic rate can be discerned during the first 24 hours, which illustrates that there is a long *latent period* before thyroxine activity begins. Once activity does begin, it increases progressively and reaches a maximum in about 12 days, as shown in Figure 77–4. Thereafter, it decreases with a half-time of about 15 days. Some of the activity still persists as long as six weeks to two months later.

When triiodothyronine is injected into a person, the latent period may be as short as 6 to 12 hours. This short latent period is partly explained by considerably more rapid rate of absorption of triiodothyronine into the cells than of thyroxine, but it is probably also explained partially by differences in actions of thyroxine and triiodothyronine inside the cell.

The long time required for maximum effect of thyroxine to occur, about 12 days in the human being, is believed to be caused by the long time required for the cells to accumulate increased numbers of mitochondria and to synthesize the enzymes whose development is promoted by the thyroid hormones. Once these elements have been formed, they presumably continue to act for many weeks, and this would explain the long duration of thyroid hormone activity even after the thyroid hormone itself is gone.

EFFECTS OF THYROXINE ON METABOLISM OF SPECIFIC DIETARY SUBSTANCES

Effect on Carbohydrate Metabolism. In general, thyroxine has two specific effects on carbohydrate metabolism: first, it increases the rate of glucose absorption from the gastrointestinal tract, and, second, it increases the rate of glucose utilization by the cells. Both these effects probably result from the overall increase in activity of the enzyme systems caused by thyroxine. Indeed, the one enzyme that increases most abundantly in most thyroid-stimulated tissue is α-glycerophosphate dehydrogenase, an enzyme that plays a major role in carbohydrate metabolism.

Effect on Fat Metabolism. It is often stated that thyroid hormone accelerates the rate of fat me-

Figure 77–4 Prolonged effect on the basal metabolic rate of administering a large single dose of thyroid hormone.

tabolism more than it accelerates the rate of metabolism of any other foodstuff. Actually, in the normal person essentially all the carbohydrate from each meal is utilized anyway within a few hours, and thereafter most of the energy of the body is derived from fats. Administering thyroxine shortens the time that the carbohydrates last, and consequently fats must then be utilized for energy for a longer portion of the day than normally. Therefore, it is probable that thyroxine normally causes greater fat metabolism than carbohydrate metabolism, owing primarily to the increased portion of the day during which fat metabolism must occur. In fact, fat deposition can proceed even more rapidly than normally under the influence of thyroxine if excessive quantities of carbohydrates are ingested simultaneously. Yet it is also true that when carbohydrates are not in excess, fatty acid mobilization is stimulated by thyroxine.

EFFECT ON BLOOD AND LIVER FATS. In general, increased quantities of thyroid hormone decrease the quantity of circulating fats in the blood and also the quantity of fats in the liver. On the other hand, decreased thyroid secretion greatly increases the quantity of circulating fats, including the blood *cholesterol,* the *phospholipids,* and the *triglycerides,* and almost always causes excessive deposition of fat in the liver. The great increase in circulating blood lipids that occurs in hypothyroidism is often associated with arteriosclerosis, which was discussed in Chapter 68. Though the causes of these effects are not known, they might result partly from the fact that thyroxine greatly increases cholesterol degradation in the liver.

Effect on Protein Metabolism and on Growth. The rates of both protein anabolism and catabolism are increased by thyroxine, which is the expected effect because of the increased enzymatic activities in the cells. Thus, thyroid hormone is necessary for development of structural and other proteins of the body cells and therefore is necessary for growth in the young person.

On the other hand, thyroxine causes rapid oxidation of carbohydrates and fats, and, when these "protein sparers" are depleted, proteins must be utilized for energy. As a result, a negative nitrogen balance then ensues. Thyroxine also has a specific effect on the tissues to "mobilize" protein and thereby release amino acids into the extracellular fluids. In addition to making these amino acids available for energy purposes, this effect also increases the rate of gluconeogenesis.

Effect on Vitamin Metabolism. Because thyroxine increases the quantity of many of the different enzymes and because vitamins are essential parts of some of the enzymes or coenzymes, thyroxine causes increased need for vitamins. Therefore, a relative vitamin deficiency can occur when excess thyroxine is secreted, unless at the same time increased quantities of vitamins are available.

The vitamins most likely to become deficient include thiamine, vitamin B_{12}, to a lesser extent the other vitamin B compounds, and vitamin C. Also, at least a small amount of thyroxine is necessary for conversion of some of the carotenes into vitamin A by the liver. Consequently, lack of thyroid hormones sometimes causes the usual signs of vitamin A deficiency, including night blindness and xerophthalmia.

Effect on Bone and Calcium Metabolism — Thyrocalcitonin. Thyroxine increases osteoclastic activity in the bones more than it increases osteoblastic activity. As a result, the bones become porous, and greater than normal quantities of calcium and phosphate are emptied into the urine and excreted into the gastrointestinal tract. This same effect occurs when the rate of metabolism is increased as a result of fever, which indicates that the loss of calcium and phosphate from the bones following thyroxine administration could result partly from the increased rate of metabolism.

Recently, a new hormone called *thyrocalcitonin* (or simply, *calcitonin*) has been discovered; it is a polypeptide with a molecular weight of about 8700 secreted by the thyroid gland in response to elevated calcium concentration in the blood. It in turn causes the calcium concentration to diminish back toward normal, presumably functioning by increasing the rate of deposition of calcium in the bones. Calcitonin is also secreted by the parathyroid glands. It will be discussed further in Chapter 79 in relation to calcium and bone metabolism.

Figure 77–5 Relationship of thyroxine daily rate of secretion to the basal metabolic rate.

PHYSIOLOGIC EFFECTS OF THYROXINE ON DIFFERENT BODILY MECHANISMS

Effect on Basal Metabolic Rate. Because thyroxine increases the rate of metabolism in most cells of the body (with the exception of the brain, retina, spleen, and lungs), excessive quantities of thyroxine can occasionally increase the basal metabolic rate to as much as 100 per cent above normal. However, in most patients with severe hyperthyroidism the basal metabolic rate ranges between 40 and 60 per cent above normal. On the other hand, when no thyroid hormone is produced, the basal metabolic rate falls to approximately half normal; that is, the basal metabolic rate becomes −40 to −50, as discussed in Chapter 70. Figure 77–5 shows the approximate relationship between the daily supply of thyroxine and the basal metabolic rate. Extreme amounts of thyroxine are required to cause very high basal metabolic rates.

Effect on Body Weight. Greatly increased thyroxine production almost always decreases the body weight, and greatly decreased thyroxine production almost always increases the body weight; but these effects do not always occur, because thyroxine increases the appetite, and this may overbalance the change in metabolic rate.

Effect on Growth. Because anabolism of proteins cannot occur normally in the absence of thyroxine, the growth effect of growth hormone from the pituitary gland is not significant without the concurrent presence of thyroxine in the body fluids. There are two conditions in which this effect on growth is especially evident: First, in growing children who are hypothyroid, the rate of growth is greatly retarded. Second, in growing children who are hyperthyroid, excessive skeletal growth often occurs, causing the child to become considerably taller than otherwise.

Effect on the Cardiovascular System. Because increased metabolism induced by thyroxine increases the demand of the tissues for nutrient substances, the following effects occur in the cardiovascular system in hyperthyroidism and the opposite effects occur in hypothyroidism:

BLOOD FLOW AND CARDIAC OUTPUT. Increased metabolism in the tissues causes more rapid utilization of oxygen than normally and causes greater than normal quantities of metabolic end-products to be released from the tissues. These effects cause vasodilatation in most of the body tissues, thus increasing blood flow in almost all areas of the body. Especially does the rate of blood flow in the skin increase because of the increased necessity for heat elimination.

As a consequence of the increased blood flow to the constituent parts of the body, the cardiac output also increases, sometimes rising to 50 per cent or more above normal when excessive thyroxine is present.

HEART RATE. The heart rate increases considerably more under the influence of thyroxine than would be expected simply because of the increased cardiac output. Therefore, thyroxine probably has a direct effect on the excitability of the heart, which in turn increases the heart rate. This effect is of particular importance because the heart rate is one of the most sensitive indices that the clinician has for determining whether a patient has excessive or diminished thyroid hormone production.

BLOOD VOLUME. Thyroxine causes the blood volume to increase slightly. This effect probably results at least partly from the vasodilatation which allows increased quantities of blood to collect in the circulatory system.

ARTERIAL PRESSURE. The increased cardiac output resulting from thyroxine tends to increase the arterial pressure. On the other hand, dilatation of the peripheral blood vessels due to the local effects of thyroxine and to excessive body heat tends to decrease the pressure. Therefore, the mean arterial pressure often is unchanged or only slightly elevated. However, because of the increased rate of run-off of blood through the peripheral vessels, the pulse pressure is usually increased, with the systolic pressure elevated 10 to 20 mm. Hg and the diastolic pressure slightly reduced.

STRENGTH OF HEARTBEAT. The increased enzymatic activity caused by increased thyroid hormone production apparently increases the strength of the heart when only a slight excess of thyroxine is secreted. This is analogous to the increase in strength of heart beat that occurs in mild fevers and during exercise. However, when thyroid hormone is increased markedly, the heart muscle strength becomes depressed because of excessive protein catabolism. Indeed, some severely thyrotoxic patients die of cardiac decompensation secondary to myocardial failure and increased cardiac load imposed by the increased output.

Effect on Respiration. The increased rate of metabolism caused by thyroid hormone increases the utilization of oxygen and the formation of carbon dioxide; these effects activate all the mechanisms that increase the rate and depth of respiration.

Effect on the Gastrointestinal Tract. In addition to increased rate of absorption of foodstuffs, which has been discussed, thyroxine increases both the rate of secretion of the digestive juices and the motility of the gastrointestinal tract. Often, diarrhea results. Also, associated with this increased secretion and motility is increased appetite, so that the food intake usually increases. Lack of thyroxine causes constipation.

Effect on the Central Nervous System. In general, thyroxine increases the rapidity of cerebration, while, on the other hand, lack of thyroid hormone decreases this function. The hyperthyroid individual is likely to develop extreme nervousness and is likely to have many psychoneurotic tendencies, such as anxiety complexes, extreme worry, or paranoias.

In general, the reaction time for various functions requiring integration by the brain is greatly reduced by administration of thyroid hormone, but, on the other hand, the rate of conduction in nerves remains normal in the presence of excess thyroid hormone. Thus, thyroid hormone seems to increase synaptic activity but does not influence peripheral nerve activity.

Effect on the Function of the Muscles. Moderate increase in thyroxine usually makes the muscles react with considerable vigor, but when the quantity of thyroxine becomes extreme, the muscles become weakened because of excess protein catabolism. On the other hand, lack of thyroxine causes the muscles to become extremely sluggish; they contract readily but relax slowly after a contraction.

MUSCLE TREMOR. One of the most characteristic signs of hyperthyroidism is a fine muscle tremor. This is not the coarse tremor that occurs in Parkinson's disease or in shivering, for it occurs at the rapid frequency of 10 to 20 times per second. The tremor can be observed easily by placing a sheet of paper on the extended fingers and noting the degree of vibration of the paper. The cause of this tremor is not definitely known, but it is possibly due to increased activity in the areas of the cord that control muscle tone. The tremor is an excellent means for assessing the degree of thyroxine effect on the central nervous system.

Effect on Sleep. Because of the exhausting effect of thyroxine on the musculature and on the central nervous system, the hyperthyroid subject often has a feeling of constant tiredness; but, also, because of the excitable effects of thyroxine on the synapses, it is difficult for him to sleep. On the other hand, extreme somnolence is characteristic of hypothyroidism.

REGULATION OF THYROID SECRETION

To maintain a normal basal metabolic rate, precisely the right amount of thyroid hormone must be secreted all the time, and, to provide this, a specific feedback mechanism operates through the hypothalamus and adenohypophysis to control the rate of thyroid secretion in proportion to the metabolic needs of the body. This system can be explained as follows:

Effects of Thyrotropin on Thyroid Secretion— Function of Cyclic AMP. Thyrotropin is an adenohypophyseal hormone, a polypeptide with a molecular weight of about 10,000, that was discussed in Chapter 75; it increases the secretion of thyroxine by the thyroid gland. Thyrotropin is also known as *thyrotropic hormone,* or *thyroid-stimulating hormone* (TSH). Its specific effects on the thyroid gland are: (1) increased proteolysis of the thyroglobulin in the follicles with resultant release of thyroxine into the circulating blood and diminishment of the follicular substance itself; (2) increased activity of the iodide pump, which increases the rate of "iodide trapping" in the glandular cells, often increasing the ratio of thyroid intracellular to extracellular iodide concentration to as great as 350:1 during maximal stimulation; (3) increased size and increased secretory activity of the thyroid cells; and (4) increased number of thyroid cells, plus a change from cuboidal to columnar cells and much infolding of the thyroid epithelium into the follicles. In summary, thyrotropin *increases all the known activities of the thyroid glandular cells.*

The earliest effect following thyrotropin stimulation is proteolysis of the thyroglobulin, which causes release of thyroxine into the blood within 30 minutes. Therefore, this has frequently been said to be the primary effect of thyrotropin. However, entirely independently of this action, thyrotropin also increases iodide trapping and thyroxine formation from iodides and tyrosine. Therefore, it is likely that thyrotropin has some primary effect that causes all these generalized increases in thyroid function. Recent experiments indicate that this primary effect is to increase the quantity of cyclic AMP in the thyroid cell. This in turn has a stimulatory effect on many different cell functions such as increasing the trapping of iodine, increasing the secretion of proteases into the thyroid follicle, and even increasing the growth of the cellular elements. This method for control of thyroid cell activity is similar to the function of cyclic AMP in other target tissues of the body.

Hypothalamic Regulation of Thyrotropin Secretion by the Adenohypophysis. Electrical stimulation of the anterior hypothalamus, slightly above the median eminence and immediately posterior to the optic chiasm, increases the output of thyrotropin and thereby causes a corresponding increase in activity of the thyroid gland. Conversely, lesions in this same region reduce the normal output of thyrotropin and prevent the adenohypophysis from responding to cold, which ordinarily increases thyrotropin secretion.

Therefore, the hypothalamus has the ability to control at least partially the adenohypophyseal secretion of thyrotropin. This control is exerted by hypothalamic secretion of *thyrotropin-releasing factor* (TRF) into the hypophyseal portal blood. TRF has been obtained in an almost pure form by Guillemin, and it is probably an octapeptide. This factor in turn

acts on the adenohypophyseal glandular cells to increase the output of thyrotropin. Control of the adenohypophysis by the hypothalamus was discussed in detail in Chapter 75. When the portal system from the hypothalamus to the adenohypophysis is completely blocked, the output of thyrotropin is moderately decreased.

EFFECTS OF COLD AND OTHER NEUROGENIC STIMULI ON THYROTROPIN SECRETION. One of the best known stimuli for increasing the rate of thyrotropin secretion by the adenohypophysis is exposure of an animal to cold. Exposure for several weeks increases the output of thyroxine sometimes more than 100 per cent and can increase the basal metabolic rate of the animal 20 to 30 per cent. Indeed, even human beings moving to arctic regions develop basal metabolic rates 15 to 20 per cent above normal.

Various emotional reactions can also affect the output of thyrotropin and can, therefore, indirectly affect the secretion of thyroxine. Extreme states of excitement and anxiety — conditions that greatly stimulate the sympathetic nervous system — paradoxically cause acute decrease in secretion of thyrotropin. Yet, prolonged periods of anxiety, especially in young women soon after reaching adulthood, are frequently associated with the development of thyrotoxic goiter. On the basis of this, most clinicians believe that prolonged emotional states cause an effect exactly opposite to the acute effect: namely, increased, rather than decreased, secretion of thyrotropin.

Neither these emotional effects nor the effect of cold is observed when the hypophyseal stalk is cut, illustrating that both of these effects are mediated by way of the hypothalamus.

Inverse Effect of Thyroid Hormones on Adenohypophyseal Secretion of Thyrotropin — Feedback Regulation of the Metabolic Rate. Increased thyroxine in the body fluids decreases the secretion of thyrotropin by the adenohypophysis. Thus, when the rate of thyroxine secretion rises to about 1.75 times normal, the rate of thyrotropin secretion falls essentially to zero. Most of this depressant effect occurs even when the adenohypophysis has been completely separated from the hypothalamus, but the effect seems to be somewhat greater if the hypothalamus and hypothalamic-hypophyseal portal system are intact. Therefore, it is probable that increased thyroxine inhibits adenohypophyseal secretion of thyrotropin in two different ways: (1) by a direct effect on the adenohypophysis itself, and (2) by an indirect effect acting through the hypothalamus.

In most hormonal regulatory systems it is *not* the rate of hormone secretion itself that needs to be regulated at a constant level, but instead it is some desired effect in the body that needs to be regulated. Since thyroxine controls the overall metabolic activity of the body, we can probably assume that the factor that is controlled at a nearly constant rate by the thyrotropin-thyroxine control system is cellular metabolic rate. Thus, if the cellular metabolic rate should become too low, the thyrotropin-thyroxine system would become activated until enough thyroxine became available to increase the metabolic activities back to normal. Conversely, if these metabolic activities should become too great, the feedback system would become inactivated until the thyroxine level fell low enough to allow normal metabolic activities once again.

One of the ways by which increased metabolic rate could exert its feedback control over the thyrotropin-thyroxine system would be by causing increased heat formation in the cells of the temperature regulating center of the anterior hypothalamus; the increased heat in this center would decrease the output of thyrotropin-releasing factor, which would decrease the output of thyrotropin and thyroxine, thus returning the metabolic activities of the cells back to normal. These possible events are summarized in Figure 77–6.

ANTITHYROID SUBSTANCES

Drugs that suppress thyroid secretion are called antithyroid substances. The three best known of these are: thiocyanate, propylthiouracil, and high concentrations of inorganic iodides. The mechanism by which each of these blocks thyroid secretion is different from the others, and they can be explained as follows:

Decreased Iodide Trapping Caused by Thiocyanate Ions. Administration of thiocyanates (or perchlorates) decreases the rate at which iodide is pumped into the thyroid glandular cells, and, therefore, reduces the availability of iodides to the intracellular processes for the formation of thyroxine.

Lack of formation of thyroxine by the gland under the influence of thiocyanates causes the thyroid gland to enlarge — that is, to become a "goiter." The mechanism of this is the following: Lack of thyroxine decreases the feedback inhibition of the adenohypophysis, which in turn allows increased secretion of thyrotropin by the adenohypophysis. This then stimulates the glandular

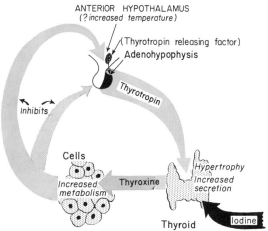

Figure 77–6 Regulation of thyroid secretion.

cells in the thyroid, making them secrete more and more thyroglobulin into the follicles, even though this does not contain significant quantities of conjugated thyroxine.

Depression of Thyroid Hormone Formation by Propylthiouracil. Propylthiouracil (or other thioureas, mercaptoimidazoles, and aminothiazoles) prevents formation of thyroid hormones from iodides and tyrosine. The precise mechanism by which this occurs is not known, though there is some reason to believe that propylthiouracil prevents the oxidation of iodides, the first step in iodination of tyrosine. Yet, for those reasons just discussed, large quantities of thyroglobulin, not containing thyroxine, begin to be secreted into the follicles. Therefore, under the influence of propylthiouracil, the thyroid gland also enlarges to form a goiter.

Decrease in Thyroid Activity Caused by Iodides. When iodides are present in the blood *in high concentration,* all activities of the thyroid gland seem to be decreased, but often they are decreased only for a few weeks. The rate of iodide trapping is reduced, the rate of thyroid hormone formation is decreased, the secretory activity of the thyroid cells is decreased, and the rate of thyroxine release from the thyroglobulin is decreased. Since these are almost exactly opposite to the effects of thyrotropin on the thyroid gland, it is believed that high concentrations of iodides in the blood inhibit thyrotropin directly, or it is even possible that iodides inhibit the hypothalamus to reduce the rate of thyrotropin production, which has been shown to occur in at least certain experimental preparations.

Because iodides in high concentrations decrease all phases of thyroid activity, they decrease the size of the thyroid gland in contradistinction to the increase in size caused by most of the other antithyroid agents. For this reason, iodides are frequently administered to patients for two or three weeks prior to surgical removal of the thyroid gland to reduce its size, and, therefore, to decrease the necessary amount of surgery.

INTERRELATIONSHIPS OF THE THYROID GLAND AND OTHER ENDOCRINE GLANDS

Increased thyroxine secretion increases the rates of secretion of most other endocrine glands, but it also increases the need of the tissues for hormones. For instance, increased thyroxine secretion increases the rate of glucose metabolism everywhere in the body and therefore causes a corresponding need for increased insulin secretion by the pancreas. Also, thyroxine increases many metabolic activities related to bone formation and, as a consequence, increases the need for parathyroid hormone. However, in addition to these general effects, thyroxine has relatively specific effects on the adrenal cortex and on the gonads.

Effect of Thyroxine on the Secretion of Adrenocortical Hormones. Administration of thyroxine increases the rate of glucocorticoid secretion by the adrenal cortex in the following manner: Thyroxine inhibits the production of thyrotropin by the adenohypophysis, but this causes the opposite effect on the secretion of corticotropin, increasing its secretion and thereby also increasing glucocorticoid secretion by the adrenal cortex.

Effect of Thyroxine on the Gonads. For normal sexual function to occur, thyroid secretion needs to be approximately normal: neither too great nor too little. In the male, lack of thyroxine is likely to cause complete loss of libido, while, on the other hand, great excesses of thyroxine frequently cause impotence. In the female, lack of thyroxine often causes *menorrhagia* and *polymenorrhea,* which mean, respectively, excessive and frequent bleeding. In some women the lack may cause irregular periods and, occasionally, even total *amenorrhea.* A hypothyroid female, like the male, is also likely to have greatly decreased libido. Conversely, in the hyperthyroid female, *oligomenorrhea,* which means greatly reduced bleeding, is usual, and occasionally, amenorrhea results.

Figure 77–7 Patient with exophthalmic hyperthyroidism. Note protrusion of the eyes and retraction of the superior eyelids. The basal metabolic rate was +40. (Courtesy of Dr. Leonard Posey.)

Therefore, the action of thyroxine on the gonads cannot be pinpointed to a specific function but probably results from a combination of direct metabolic effects of thyroxine on the gonads and of excitatory and inhibitory effects operating through the adenohypophysis.

DISEASES OF THE THYROID

HYPERTHYROIDISM

Most effects of hyperthyroidism are obvious from the preceding discussion of the various physiologic effects of thyroid hormone. However, some specific effects should be mentioned in connection especially with the development, diagnosis, and treatment of hyperthyroidism.

Many clinicians believe that essentially all patients with toxic, hyperplastic goiter develop the disease as a result of excessive production of thyrotropin by the adenohypophysis, though, unfortunately, it has not been possible to show increased thyrotropin levels in the blood, possibly because thyrotropin is too rapidly removed from the blood in the hyperthyroid state. Also, the cause of the excessive secretion of thyrotropin is usually unknown. Hyperthyroidism frequently follows severe physical or emotional disturbances; it occurs several times as frequently in cold climates as in hot climates, and it often occurs in successive generations of the same family.

Characteristics of Hyperthyroidism (Toxic Goiter, Thyrotoxicosis, Graves' Disease). In hyperthyroidism, the thyroid gland is usually increased to two to three times normal size. This increase in size is not an indication, however, of the actual increase in rate of thyroid hormone secretion, for there is tremendous infolding of the follicular cells into the follicles so that the number of cells is greatly increased without greatly increasing the size of the gland. Also, each cell increases its rate of secretion tremendously. Studies with radioactive iodine indicate that such hyperplastic glands secrete thyroxine at a rate as great as 5 to 15 times normal. These changes are all characteristic of increased thyrotropin stimulation, and this is the reason why thyrotropin is believed to be the cause of the hyperthyroidism.

The symptoms of toxic goiter – these are obvious from the preceding discussion of the physiology of the thyroid hormones – are intolerance to heat, increased sweating, mild to extreme weight loss (sometimes as much as 100 pounds), varying degrees of diarrhea, muscular weakness, nervousness or other psychic disorders, and tremor of the hands.

Exophthalmos and Exophthalmos-Producing Substance. Most, but not all, persons with hyperthyroidism develop some degree of protrusion of the eyeballs, as illustrated in Figure 77–7. This condition is called *exophthalmos*. A major degree of exophthalmos occurs in about one-third of the hyperthyroid patients, and the condition occasionally becomes severe enough that it causes blindness because the eyeball protrusion stretches the optic nerve. Much more often, the eyes are damaged because the eyelids do not close completely when the person blinks or when he is asleep. As a result, the dry epithelial surfaces of the eyes become irritated and often infected, resulting in ulceration of the cornea.

Although the cause of exophthalmos in hyperthyroidism is not completely understood, a strange phenomenon frequently occurs when the hyperthyroid patient is treated, whether this be done by surgical removal of the gland or by use of various drugs: The exophthalmos itself frequently becomes worse after thyroid hormone secretion has been returned to normal. A partial answer to this effect is that thyrotropin, in the absence of excess thyroid hormone, now is secreted in even greater quantities, and a substance called *exophthalmos-producing substance* is probably secreted in large quantities along with the enhanced secretion of thyrotropin. The present belief, then, is that most exophthalmos is caused by greatly enhanced secretion of exophthalmos-producing substance as a by-product of excessive thyrotropin secretion in hyperthyroidism.

The effect of exophthalmos-producing substance on the retro-orbital tissues is mainly to cause edematous swelling and deposition of large quantities of mucopolysaccharides in the extracellular spaces, followed by secondary fibrosis. When the hyperthyroidism is finally brought under control, the exophthalmos usually does not regress significantly, partly because the secretion of exophthalmos-producing substance may continue even after thyroid hormone secretion has been brought to normal, but partly also because the fibrosis that has occurred in the retro-orbital tissues will not resorb.

Diagnostic Tests for Hyperthyroidism. PROTEIN-BOUND IODINE. Essentially all the iodine bound with proteins of the blood is thyroxine iodine, and the normal person has between 4 and 7.5 μgm. of *plasma-bound iodine* per 100 ml. of plasma. In hyperthyroidism this level usually increases in proportion to the degree of hyperthyroidism—sometimes to 15 to 20 μgm. per 100 ml. of plasma. The plasma-bound iodine test is an effective means for making a diagnosis in approximately 85 to 90 per cent of patients with hyperthyroidism, but it occasionally is in error, owing to the presence of other iodine compounds in the plasma.

THE BASAL METABOLIC RATE. The basal metabolic rate is usually increased to approximately +40 to +60 in severe hyperthyroidism. A mild increase in basal metabolic rate does not necessarily mean hyperthyroidism. Indeed, an increased basal metabolic rate correctly indicates hyperthyroidism in only 60 per cent of the cases; other factors that were discussed in Chapter 70 can also cause changes in the basal metabolic rate even though thyroid hormone production is normal.

UPTAKE OF RADIOACTIVE IODINE. Another method for measuring the rate of activity of the thyroid gland is to determine its rate of radioactive iodine uptake. A small test dose of radioactive iodine is given intravenously, and the rate of uptake by the thyroid gland is measured by a calibrated radioactive detector placed over the neck. In the normal person, an average of 4 per cent of the injected radioactive iodine is assimilated by the thyroid gland per hour. In the hyperthyroid person, as much as 20 to 25 per cent is assimilated per hour.

Physiology of Treatment in Hyperthyroidism. The most direct treatment for hyperthyroidism is surgical removal of the thyroid gland. In general, however, it is desirable to prepare the patient for surgical removal of the gland prior to the operation. This is done by administering propylthiouracil, sometimes for as long as several months, until the basal metabolic rate of the patient has returned to normal. Then administration of high concontrations of iodides for two weeks immediately prior to operation causes the gland itself to recede in size. By using these preoperative procedures, the operative mortality is less than 1 in 1000 in the better hospitals, whereas prior to development of these procedures the operative mortality was as great as 1 in 25.

TREATMENT OF THE HYPERPLASTIC THYROID WITH RADIOACTIVE IODINE. As much as 80 to 90 per cent of an injected dose of iodide is absorbed by the hyperplastic, toxic thyroid gland within a day after injection. If this injected iodine is radioactive, it can destroy internally the secretory cells of the thyroid gland. Usually 5 millicuries of radioactive iodine is given to the patient, and then his condition is reassessed several weeks later. If he is still hyperthyroid, additional doses are repeated until he reaches a normal thyroid status. These quantities of radioiodine are about 1000 times as great as those used for diagnosis of hyperthyroidism as discussed above.

Thyroid Storm. Occasionally, patients with extremely severe thyrotoxicosis develop a state called *thyroid storm*, or *thyroid crisis* in which all the usual symptoms of thyrotoxicosis are excessively accentuated. The effects may be delirium, high fever, abnormal rhythm of the heart, extreme sweating, shock, vomiting, and dehydration. This condition is particularly likely to occur during the first day following surgical removal of a large hyperplastic thyroid gland, presumably because of excessive release of thyroxine into the circulatory system during the operative procedure. The extreme overactivity of the body's tissues can be so damaging that without treatment almost all persons entering this state die. However, by cooling the patient rapidly with ice or alcohol sponge baths and also administering large quantities of adrenocortical hormones, it is now possible to save approximately half the patients. The adrenocortical hormones seem to be especially important, for, in some way not understood, they allow rapid repair of the damaged metabolic systems of the body.

Thyroid Adenoma. A localized, adenomatous tumor occasionally develops in the thyroid gland and sometimes secretes large quantities of thyroid hormone, causing hyperthyroidism. Also, secretion by the tumor is usually independent of control by thyrotropic hormone from the adenohypophysis. Consequently, administration of iodine to block the action of thyrotropin is not effective in treatment of the condition, though other drugs, such as the thyroid inhibitors, may be partially effective. As long as the adenoma continues to secrete thyroxine, function of the remainder of the thyroid gland may be almost totally inhibited because thyroxine from the adenoma depresses the production of thyrotropin from the pituitary gland. After the adenoma is removed, several weeks to several months are required for return of the remainder of the gland to normal function.

HYPOTHYROIDISM

The effects of hypothyroidism in general are opposite to those of hyperthyroidism, but here again, a few physiological mechanisms peculiar to hypothyroidism alone are involved.

Endemic Colloid Goiter. The term "goiter" means a greatly enlarged thyroid gland. As pointed out in the discussion of iodine metabolism, 35 to 50 mg. of iodine is necessary each year for the formation of adequate quantities of thyroxine. In certain areas of the world, notably in the Swiss Alps and in the Great Lakes region of the United States, insufficient iodine is present in the soil for the foodstuffs to contain even this minute quantity of iodine. Therefore, in days prior to iodized table salt, many persons living in these areas developed extremely large thyroid glands called *endemic goiters*.

The mechanism for development of the large endemic goiters is the following: Lack of iodine prevents production of thyroxine by the thyroid gland, and, as a result, no thyroxine is available to inhibit production of thyrotropin by the adenohypophysis; this allows the adenohypophysis to secrete excessively large quantities of thyrotropin. The thyrotropin then causes the thyroid cells to secrete tremendous amounts of thyroglobulin (colloid) into the follicles, and the gland grows larger and larger. But unfortunately, due to lack of iodine, increased thyroxine production does not occur. The follicles become tremendous in size, and the whole thyroid gland may increase to as large as 300 to 500 grams or more.

Idiopathic Nontoxic Colloid Goiter. Enlarged thyroid glands almost identical to those of endemic colloid goiter frequently develop even though the affected persons receive sufficient quantities of iodine in their diets. These goitrous glands may secrete normal quantities of thyroxine but more frequently the secretion of thyroxine is depressed, as in endemic colloid goiter.

The exact cause of idiopathic colloid goiter is not known, though animals fed large quantities of certain foods are known to develop enlarged thyroid glands. These foods contain so-called goitrogenic substances, the actions of which are similar to those of propylthiouracil, which was discussed previously. Especially are goitrogenic substances found in some varieties of turnips and cabbages. However, the diet only rarely contains enough quantities of goitrogenic substances to cause a large, idiopathic colloid goiter.

Another possible cause of this type of goiter is altered enzyme systems of the thyroid cells. That is, the cells secrete the colloid substance, but because of altered intracellular enzyme systems, the cells fail to form the thyroid hormones or are unable to release the hormones from the thyroglobulin complex.

Regardless of the cause of the goiter, it is possible that depressed thyroxine production allows the adenohypophysis to secrete greater than normal quantities of thyrotropin and that it is actually the thyrotropin that causes the gland to enlarge.

Characteristics of Hypothyroidism. Whether hypothyroidism is due to endemic colloid goiter, idiopathic colloid goiter, destruction of the thyroid gland by irradiation, surgical removal of the thyroid gland, or destruction of the thyroid gland by various other diseases, the physiological effects are the same. These include extreme somnolence with sleeping 14 to 16 hours a day, extreme muscular sluggishness, slowed heart rate, decreased cardiac output, decreased blood volume, increased weight, constipation, mental sluggishness, failure of many trophic functions in the body evidenced by depressed growth of hair and scaliness of the skin, development of a frog-like husky voice, and, in severe cases, development of an edematous appearance throughout the body.

MYXEDEMA. The patient with almost total lack of thyroid function develops an edematous appearance, and the condition is known as "myxedema." Figure 77–8 shows such a patient with bagginess under the eyes and swelling of the face. Greatly increased quantities of mucopolysaccharides, mainly hyaluronic acid, collect in the interstitial spaces, and this causes the total quantity of interstitial fluid also to increase. The fluid is adsorbed to the mucopolysaccharides, thus greatly increasing the quantity of "ground substance" gel in the tissues. Because of the gel nature of the excess fluid, it is relatively immobile, and the edema is nonpitting in type.

ARTERIOSCLEROSIS IN HYPOTHYROIDISM. As pointed out in Chapter 68, lack of thyroid hormone increases the quantity of blood lipids, and an increase in blood lipids is usually associated with atherosclerosis and arteriosclerosis. Therefore, many hypothyroid patients, particularly those with myxedema, develop severe arteriosclerosis, which results in peripheral vascular disease, deafness, and often extreme coronary sclerosis and, consequently, early demise.

Figure 77–8 Patient with myxedema. (Courtesy of Dr. Herbert Langford.)

Diagnostic Tests in Hypothyroidism. The tests already described for diagnosis of hyperthyroidism give the opposite results in hypothyroidism. The basal metabolic rate in myxedema ranges between −40 and −50, the protein-bound iodine is less than 2 μgm./100 ml. of plasma instead of the normal 4 to 7.5, and the rate of radioactive iodine uptake by the thyroid gland (except in iodine deficiency hypothyroidism) ranges less than 1 per cent per hour rather than the normal of approximately 4 per cent per hour. Probably more important for diagnosis, however, than the various diagnostic tests are the characteristic symptoms of hypothyroidism as just discussed.

Treatment of Hypothyroidism. Figure 77–4 shows the effect of thyroxine on the basal metabolic rate, illustrating that thyroxine normally has a duration of action of more than one month. Consequently, it is easy to maintain a steady level of thyroid hormone activity in the body by daily oral ingestion of a single tablet of desiccated thyroid or thyroid extract. Furthermore, proper treatment of the hypothyroid patient results in such complete normality that formerly myxedematous patients properly treated have lived into their 90's after treatment for over 50 years.

When a person with myxedema is first treated, immediate diuresis ensues, presumably owing to removal of the myxedematous fluid from the interstitial spaces after the mucopolysaccharide in this fluid is metabolized. Obviously, treatment increases the activity in the circulatory system, and, if the person has already developed severe heart disease, the increased activity of the heart may lead to anginal pain. On the other hand, lack of treatment may lead to increased severity of the heart disease. In this instance the clinician is in a dilemma and must proceed with progressive treatment slowly if at all.

CRETINISM. Cretinism is the condition caused by extreme hypothyroidism during infancy and childhood, and it is characterized especially by failure of growth. Cretinism results from congenital lack of a thyroid gland (*congenital cretinism*), from failure of the thyroid gland to produce thyroxine because of genetic deficiency of the gland that prevents synthesis of thyroxine, or from iodine lack in the diet (*endemic cretinism*). The severity of endemic cretinism varies greatly in accordance with the amount of iodine in the diet, and whole populaces of an endemic area have been known to have cretinoid tendencies.

A newborn baby without a thyroid gland may have absolutely normal appearance and function because he has been supplied with thyroxine by the mother while *in utero,* but a few weeks after birth his movements become sluggish, and both his physical and mental growth are greatly retarded. Treatment of the cretin at any time usually causes normal return of physical growth, but, unless the cretin is treated within a few months after birth, his mental growth will be permanently retarded. This is probably due to the fact that physical development of the neuronal cells of the central nervous system is rapid during the first year of life so that any retardation at this point is extremely detrimental.

Skeletal growth in the cretin is characteristically more inhibited than is soft tissue growth, though both are inhibited to a certain extent. However, as a result of this disproportionate rate of growth, the soft tissues are likely to enlarge excessively, giving the cretin the appearance of a very obese and stocky, short child; this obesity is due to disproportionate growth of the soft tissue organs, such as the liver and the gastrointestinal tract. Indeed, occasionally the tongue becomes so large in relation to the skeletal growth that it obstructs swallowing and breathing, inducing a characteristic guttural breathing and sometimes even choking the baby.

REFERENCES

Andros, G., and Wollman, S. H.: Autoradiographic localization of radioiodide in the thyroid gland of the mouse. *Amer. J. Physiol., 213*:198, 1967.

Bastomsky, C. H., and McKenzie, J. M.: Cyclic AMP: a mediator of thyroid stimulation by thyrotropin. *Amer. J. Physiol., 213*: 753, 1967.

Challoner, D. R.: A direct effect of triiodothyronine on the oxygen consumption of rat fat cells. *Amer. J. Physiol., 16*:905, 1969.

Challoner, D. R.: Evidence for uncoupled respiration in thyrotoxic and epinephrine-stimulated myocardium. *Amer. J. Physiol., 214*:365, 1968.

Ciba Foundation Study Group No. 18: Brain-Thyroid Relationships. Boston, Little, Brown and Company, 1964.

Danowski, T. S.: Clinical Endocrinology. Vol. II: Thyroid. Baltimore, The Williams & Wilkins Co., 1962.

Greif, R. L.: Biosynthesis of thyroglobulin. *Proc. Int. Union of Physiol. Sciences, 6*:111, 1968.

Harrison, T. S.: Adrenal medullary and thyroid relationships. *Physiol. Rev., 44*:161, 1964.

Ismail, A. A.; El-Ridi, M. S.; Badran, I.; Khalifa, K.; Abdel-Hay, A. R.; and Talaat, M.: Extravascular circulation of thyroid hormones. *Amer. J. Physiol., 213*:1391, 1967.

McKenzie, J. M.: Humoral factors in the pathogenesis of Graves' disease. *Physiol. Rev., 48*:252, 1968.

Means, J. H.; DeGroot, L. J.; and Stanbury, J. B.: The Thyroid and Its Diseases. 3rd ed., New York, McGraw-Hill Book Company, 1963.

Piatnek-Leunissen, D., and Leunissen, R. L. A.: Heart mitochrondrial function in acute and in chronic hyperthyroidism in rats. *Circ. Res., 25*:171, 1969.

Piatnek-Leunissen, D., and Olson, R. E.: Cardiac failure in the dog as a consequence of exogenous hyperthyroidism. *Circ. Res., 20*:242, 1967.

Rawson, R. W.; Sonenberg, M.; and Money, W. L.: Diseases of the thyroid. *In* Duncan, G. G. (ed.): Diseases of Metabolism. 5th ed., Philadelphia, W. B. Saunders Company, 1964, p. 1159.

Rosenberg, I. S., and Bastomsky, C. H.: The thyroid. *Ann. Rev. Physiol., 27*:71, 1965.

Straw, J. A., and Fregly, M. J.: Evaluation of thyroid and adrenal-pituitary function during cold acclimation. *J. Appl. Physiol., 23*:825, 1967.

Szepesi, B., and Freedland, R. A.: Effect of thyroid hormones on metabolism. IV. Comparative aspects of enzyme responses. *Amer. J. Physiol., 216*:1054, 1969.

Van Middlesworth, L., and Jones, S. L.: Interference with de-iodination of some thyroxine analogues in the rat. *Endocrinology, 69*:1085, 1961.

Werner, S. C., and Nauman, J. A.: The thyroid. *Ann. Rev. Physiol., 30*:213, 1968.

Wolff, J.: Transport of iodide and other anions in the thyroid gland. *Physiol. Rev., 44*:45, 1964.

Young, J. W.: Effects of D- and L-thyroxine on enzymes in liver and adipose tissue of rats. *Amer. J. Physiol., 214*:378, 1968.

CHAPTER 78

Insulin, Glucagon, and Diabetes Mellitus

The pancreas, in addition to its digestive functions, secretes two hormones, *insulin* and *glucagon.* The purpose of this chapter is to discuss the functions of these hormones in regulating glucose, lipid, and protein metabolism, as well as to discuss the two diseases—*diabetes mellitus* and *hyperinsulinism*—caused, respectively, by hyposecretion of insulin and excess secretion of insulin.

Physiologic Anatomy of the Pancreas. The pancreas is composed of two major types of tissues, as shown in Figure 78-1: (1) the *acini,* which secrete digestive juices into the duodenum, and (2) the *islets of Langerhans,* which do not have any means for emptying their secretions externally but instead secrete insulin and glucagon directly into the blood. The digestive secretions of the pancreas were discussed in Chapter 64.

The islets of Langerhans of the human being contain two major types of cells, the *alpha* and *beta* cells, which are distinguished from one another by their morphology and staining characteristics. The beta cells produce insulin, and glucagon is secreted by the alpha cells.

Beta cells are often still present in the pancreas of a person who has severe diabetes, but these cells then usually have a hyalinized appearance and contain no secretory granules; also, they do not exhibit staining reactions for insulin in contrast to the normal beta cells, which do give staining reactions for insulin. Consequently, the hyalinized beta cells in diabetic patients are considered to be nonfunctional.

Chemistry of Insulin. Insulin is a small protein having a molecular weight of 5734. It is composed of two amino acid chains, illustrated in Figure 78-2, connected to each other by disulfide linkages. When the two amino acid chains are split apart, the functional activity of the insulin molecule is lost.

Before insulin can exert its function it must become "fixed" to the tissues, probably to the cell membranes, as explained later in the chapter. The mechanism by which insulin fixes to the tissues is to form bonds between the disulfide ring structure of the upper chain and sulfhydryl radicals of the tissues. Antidiuretic hormone fixes to the epi-thelial cells of the renal tubules by essentially the same type of linkage.

EFFECT OF INSULIN ON CARBOHYDRATE METABOLISM

The earliest studies of the effect of insulin on carbohydrate metabolism showed three basic effects of the hormone: (1) enhanced rate of glucose metabolism, (2) decreased blood glucose concentration, and (3) increased glycogen stores in the tissues. All of these effects probably result from the following basic mechanisms.

FACILITATION OF GLUCOSE TRANSPORT THROUGH THE CELL MEMBRANE

The single basic effect of insulin that has been demonstrated many times is its ability to increase the rate of glucose transport through the membranes of most cells in the body. This effect is illustrated in Figure 78-3. In the complete absence of insulin, the overall rate of glucose transport into the cells of the entire body becomes only about one-fourth the nor-

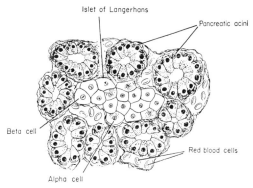

Figure 78-1 Physiologic anatomy of the pancreas.

915

NH₂ ⌐─S──S─⌐   NH₂   NH₂   NH₂
NH₂·Gly·Ileu·Val·Glu·Glu·Cy·Cy·Ala·Ser·Val·Cy·Ser·Leu·Tyr·Glu·Leu·Glu·Asp·Tyr·Cy·Asp
                    S   S
NH₂ NH₂   S   S
NH₂·Phe·Val·Asp·Glu·His·Leu·Cy·Gly·Ser·His·Leu·Val·Glu·Ala·Leu·Tyr·Leu·Val·Cy·Gly·Glu·Arg·Gly·Phe·Phe·Tyr·Thr·Pro·Lys·Ala

Figure 78–2   The insulin molecule.

mal value. On the other hand, when great excesses of insulin are secreted, the rate of glucose transport into the cells may be as great as five times normal. This means that, between the limits of no insulin at all and great excesses of insulin, the rate of glucose transport can be altered as much as 20-fold.

Some experiments that have been used to prove the accelerating effect of insulin on glucose transport through the cell membranes are the following:

1. Since glucose immediately becomes phosphorylated after entry into the cells and since almost no *free* glucose persists more than a few seconds after entering the cells, studies to prove an increase in intracellular glucose under the influence of insulin have been difficult. However, this has been accomplished in two different types of experiments: first, in experiments in which the concentration of extracellular glucose is so extreme that the glucose entering the cells cannot all be phosphorylated instantaneously and, second, in experiments in which the medium is at 4° C., a level at which the glucose-phosphorylating mechanism of the cells is almost blocked. Figure 78–4 shows increased "free" glucose inside the cells of a perfused heart under the

influence of insulin, illustrating a many-fold effect of the insulin.

2. Experiments have also been conducted on the transport of other monosaccharides similar to glucose but which are not metabolized in the cells. For instance, tissues in the body cannot metabolize *l*-arabinose; therefore, its intracellular concentration can be measured without difficulty. It has been easy to demonstrate that insulin causes marked increase in *l*-arabinose transport through the cellular membrane. The same is also true of a number of other nonmetabolizable monosaccharides, including especially xylose and galactose.

3. A third group of experiments has been performed to compare the effects of insulin on glucose metabolism in (a) suspensions of whole cells and (b) cellular homogenates. In these, insulin increases the metabolism of glucose by whole cells; but in the cellular homogenates, which have no intact cell membranes, the rate of glucose metabolism usually is not altered by the presence of insulin. This indicates that a cell membrane is necessary for full function of insulin.

Mechanism by which Insulin Accelerates Glucose Transport. As pointed out in the discussion of the cell membrane in Chapter 4,

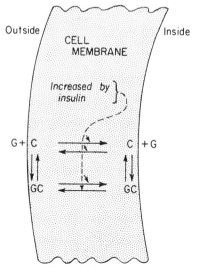

Figure 78–3   Effect of insulin in increasing glucose transport in either direction through the cell membrane.

Figure 78–4   Effect of insulin to enhance the concentration of glucose inside muscle cells. Note that in the absence of insulin (control) the intracellular glucose concentration remained near zero despite very high extracellular concentrations of glucose. (From Park, Morgan, Kaji, and Smith, in Eisenstein (ed.): The Biochemical Aspects of Hormone Action. Little, Brown and Co.)

glucose cannot pass into the cell through the cell pores but instead must enter by some transport mechanism through the membrane matrix. Figure 78–3 depicts the generally believed method by which glucose enters the cell, showing that glucose probably combines with a carrier substance in the cell membrane and then is transported to the inside of the membrane where it is released to the interior of the cell. The carrier then returns to the outer surface of the membrane to transport additional quantities of glucose. This process can occur *in either direction,* as shown by the reversible arrows in the diagram.

Glucose transport through the cell membrane does not occur against a concentration gradient. That is, once the glucose concentration inside the cell rises to as high as the glucose concentration on the outside, further glucose will not be transported to the interior. Therefore, the glucose transport process is one of *facilitated diffusion,* which means simply that glucose diffuses through the membrane by means of a facilitating carrier mechanism.

In the complete absence of insulin, the carrier mechanism for the transport of glucose to the interior of the cell operates very poorly. It is postulated that when insulin is not present, the transport of glucose is inhibited by an active process that requires continual energy from intracellular sources. The reason for believing this is that cellular hypoxia and other debilitating states of the cell that normally block active cellular processes increase glucose transport into the cell. Insulin, therefore, according to this theory, enhances glucose transport into cells by blocking this active inhibitory process, thus allowing full function of the facilitated diffusion mechanism for glucose transport.

Tissues in which Insulin Is Effective. Enhanced transport of glucose through the cell membrane is particularly effective in skeletal muscle and adipose tissue, and these two together make up approximately 65 per cent of the entire body by weight. In addition, insulin enhances glucose transport into the heart and at least into certain smooth muscle organs, such as the uterus.

On the other hand, insulin does not enhance glucose transport into the brain cells and does not enhance its transport through the intestinal mucosa or through the tubular epithelium of the kidney. However, these tissues together amount to less than 5 per cent of the total body mass. The effects of insulin in the remaining 30 per cent of the body are mainly unknown, but there is reason to believe that most of this tissue also responds at least partially to insulin.

In summary, insulin is extremely important for glucose transport into the cells of most tissues of the body. The most important exception to this is the brain; here, glucose transport is probably more dependent on *diffusion* through the blood-brain barrier (see Chap. 22) than through the cell membrane.

Other Transport Effects of Insulin that Affect Carbohydrate Metabolism. Insulin also enhances glucose transport through the mitochondrial membrane and probably also through at least certain portions of the membrane of the endoplasmic reticulum. These effects could play important roles in the utilization of glucose by the cell.

EFFECT OF INSULIN ON GLUCOSE UTILIZATION AND GLYCOGEN STORAGE IN EXTRAHEPATIC TISSUES

Glucose Utilization. Since insulin lack decreases glucose transport into most body cells to about one-fourth normal, without insulin most tissues of the body (with the major exception of the brain and, to a less extent, the heart) must depend on other metabolic substrates for energy.

Glycogen Storage. Another major effect of insulin is greatly enhanced storage of glycogen in the skeletal muscle cells throughout the body and moderate enhancement of glycogen storage in the skin and glandular tissues. Here again, this effect is undoubtedly caused by the availability of greatly increased intracellular glucose.

ACTION OF INSULIN ON CARBOHYDRATE METABOLISM IN THE LIVER

Knowledge of the action of insulin on carbohydrate metabolism in the liver is still somewhat confused because insulin does not cause an immediate increase in glucose transport into liver cells similar to that which occurs in skeletal muscle and most other tissues of the body. Yet insulin does have at least three major effects on liver metabolism of carbohydrates.

1. Insulin causes immediate release of glucose from the liver into the blood, with a corresponding decrease in the liver glycogen stores.

2. During the next few hours, this transport of glucose becomes reversed if excess glucose is infused into the animal along with the in-

sulin. Consequently, the stores of liver glycogen then begin to increase.

3. Insulin enhances the rate of glucose metabolism by the liver by as much as 300 to 400 per cent, but this effect occurs slowly, requiring 12 to 24 hours, unlike the effect that occurs in the extrahepatic tissues.

These differences between the effect of insulin on liver cells and on other cells of the body are yet unexplained, but they could result from the following three factors: (1) The liver cells are naturally highly permeable to glucose, so that insulin has little if any effect on the transport of glucose through the liver cell membrane. (2) Liver cells contain large quantities of glucose 6-phosphatase, an enzyme that continually dephosphorylates glucose that has previously been phosphorylated. Because of this, glucose is not trapped inside the liver cell by the phosphorylation mechanism as it is in muscle cells and most other cells of the body. (3) Liver cells contain a special enzyme, glucokinase, that is not present in large quantities in other cells. This enzyme causes phosphorylation of glucose, a function that is performed by hexokinase in other cells. However, the glucokinase of liver cells increases markedly over a period of 6 to 12 hours following insulin administration. Therefore, the ability of liver cells to phosphorylate glucose and thereby to utilize glucose in great quantities increases progressively for many hours following onset of insulin action.

On the basis of these characteristics of the liver cell, one could postulate that the early decrease in liver glycogen results from the following mechanism: The action of insulin on muscle and other peripheral tissues of the body causes rapid transport of glucose into the peripheral cells and consequently reduces

Figure 78–5 Effect on glycogen stores of simultaneous insulin and glucose infusion in liver and skeletal muscle, showing immediate glycogenolysis in the liver and immediate glycogenesis in skeletal muscles, followed by delayed glycogenesis in the liver as well.

blood glucose concentration. Therefore, glucose, dephosphorylated by glucose 6-phosphatase, diffuses rapidly out of the liver cells so that liver glycogen stores decrease.

The increase in liver glycogen stores that takes place a number of hours later is believed to result from the slow buildup of glucokinase in the liver cells, enhancing the phosphorylation of glucose and, therefore, promoting increased use of glucose, as well as increased storage of glycogen, by the liver cells.

Transfer of Glycogen from the Liver to Skeletal Muscles Following the Administration of Insulin. Figure 78–5 illustrates an application of the principles just discussed: Insulin administration causes early loss of glycogen from the liver and, yet, buildup of glycogen in skeletal muscle cells. Then, a number of hours later, the increase in glucokinase in the liver allows the liver stores of glycogen also to increase if sufficient glucose is available in the diet. This overall mechanism allows rapid transfer of glycogen from the liver to the muscles when the muscles are in need of glycogen.

REGULATION OF BLOOD GLUCOSE CONCENTRATION AND CONTROL OF INSULIN SECRETION

Glucose-Buffer Function of the Liver. In the preceding discussion it was pointed out that glucose can readily penetrate hepatic cells in either direction, passing from the extracellular fluid spaces into the hepatic cells to be stored as glycogen or the glycogen being split to glucose that then passes rapidly into the extracellular fluids. Thus, the liver acts as a large storage vault for glucose, and because of this it acts as a glucose-buffer system. When excess quantities of glucose enter the extracellular fluid, about two-thirds of this is stored almost immediately in the liver, and this prevents excessive increase in blood glucose concentration. Conversely, when the blood glucose concentration falls below normal, the stored glucose in the liver rapidly replenishes the blood glucose.

Insulin plays an important role in controlling the glucose-buffer function of the liver, for in the absence of insulin, glucokinase is greatly depressed in the liver cells, so that glucose is not readily stored in the liver in times of excess and, also, is not available for release in times of glucose need.

The liver glucose-buffer system reduces the variation in blood glucose concentration to approximately one-third what it is without the liver. In liver disease, the blood glucose concentration is poorly regulated, its concentration rising or decreasing much more than usual upon the slightest provocation.

GLUCONEOGENESIS FUNCTION OF THE LIVER. In addition to acting as a major storehouse for glucose, the liver also helps to regulate blood glucose concentration by increasing its rate of gluconeogenesis when glucose is needed in the extracellular fluid. Between meals, when glucose is not being absorbed from the gastrointestinal tract, the liver manufactures glucose from amino acids and from glycerol derived from fats. This is one of the principal means by which the blood glucose concentration is maintained in the normal range during the interdigestive period. As explained in Chapter 76, glucocorticoids probably play a major role in mobilizing amino acids and fats during this interdigestive period and thereby enhance the amount of glucose synthesized by gluconeogenesis in the hepatic cells. Glucagon possibly exerts a similar effect, as will be discussed later in the chapter.

Effect of Insulin on Blood Glucose Concentration. In the absence of insulin, little of the glucose absorbed from the gastrointestinal tract can be transported into the tissue cells. As a consequence, the blood glucose concentration rises very high—from a normal value of 90 mg./100 ml. sometimes to as high as 300 to 1200 mg./100 ml.

On the other hand, in the presence of great amounts of excess insulin, glucose is transported into the cells so rapidly that its concentration in the blood can fall to as low as 20 to 30 mg./100 ml. Therefore, the rate of insulin secretion by the pancreas must be regulated accurately so that the blood glucose concentration also will be regulated accurately at a constant and normal value.

Effect of Glucose on Insulin Secretion by the Pancreas—Feedback Mechanism for Blood Glucose Control. The glucose concentration of the plasma has a direct effect on the islets of Langerhans to control their rate of insulin secretion. Even the isolated pancreas perfused with a solution containing a high concentration of glucose secretes greatly increased quantities of insulin. On the contrary, it secretes decreased quantities of insulin when the blood glucose concentration falls. Therefore, the secretion of insulin by the pancreas affords an important feedback mechanism for continual regulation of blood glucose concentration.

Hypertrophy of the Pancreatic Islets in Response to Prolonged Elevation of Blood Glucose. Besides the immediate increase in insulin se-

cretion that results from increased blood glucose concentration, still an additional gradual increase in insulin secretion occurs if the elevated glucose concentration persists for several weeks. Over a period of one to three weeks, the islets of Langerhans hypertrophy, and secretion of insulin increases correspondingly. Therefore, if a person suddenly begins to eat a diet containing excessive amounts of carbohydrates, his normal pancreatic insulin secretion may not be capable of taking care of all the excess glucose, and his blood glucose concentration rises above normal. But after several weeks of pancreatic adaptation, an adequate amount of insulin is secreted, and the excess glucose is transported into the cells.

DIABETOGENIC EFFECT OF PROLONGED ELEVATION OF BLOOD GLUCOSE CONCENTRATION. Prolonged elevation of blood glucose concentration, caused either by excessive indulgence in a high carbohydrate diet or by administration of large amounts of one or more of the diabetogenic hormones mentioned in the preceding chapters—*growth hormone, glucocorticoids,* and rarely *luteotropic hormone* or *thyroxine*—can eventually "burn out" the beta cells of the islets of Langerhans, causing them to take on a hyalinized appearance with loss of their granules and with partial or complete cessation of insulin production. As a result, a typical case of diabetes mellitus ensues.

Regulation of Blood Glucose Concentration by the Sympathetic Nervous System and by Glucagon. Decreased blood glucose concentration excites the sympathetic nuclei of the hypothalamus, which then transmit impulses through the sympathetic nervous system to cause epinephrine release by the adrenal medullae and norepinephrine release both by the adrenals and by the sympathetic nerve endings. These hormones, especially epinephrine, exert a direct effect on the liver cells to increase the rate of glycogenolysis. They do this by activating the liver cell enzyme *adenyl cyclase,* which converts much of the ATP to cyclic adenylic acid. The cyclic adenylic acid in turn activates the normally inactive phosphorylase in the liver cells. This enzyme then causes glycogen to split into glucose phosphate, which is rapidly dephosphorylated by glucose 6-phosphatase and diffuses into the blood to elevate the blood glucose concentration back toward normal. Later in the chapter we shall see that a decrease in blood glucose concentration also causes glucagon secretion by the pancreas; the glucagon then has the same effect as that of epinephrine to cause glucose release from the liver. In addition, glucagon and glucocorticoids from the adrenal gland immediately enhance gluconeogenesis, which converts amino acids into glucose. These, then, are additional feedback mechanisms for regulation of blood glucose concentration.

Purpose of Blood Glucose Regulation. One might ask the question: Why is it important to maintain a constant blood glucose concentration, particularly since most tissues can shift to utilization of fats and proteins for energy in the absence of glucose? The answer is that glucose is the only nutrient that can be utilized by the brain, retina, and germinal epithelium in sufficient quantities to supply them with their required energy. Therefore, it is important to maintain a blood glucose concentration at a sufficiently high level to provide this necessary nutrition.

Approximately half of all the glucose formed by gluconeogenesis during the interdigestive period is used for metabolism in the brain. Indeed, it is important that the pancreas not secrete any insulin during this time, for otherwise the scant supplies of glucose that are available would all go into the muscles and other peripheral tissues and leave the brain without a nutritive source.

On the other hand, it is also important that the blood glucose concentration not rise too high for three reasons: First, glucose exerts a large amount of osmotic pressure in the extracellular fluid, and, if the glucose concentration rises to excessive values, this can cause considerable cellular dehydration. Second, an excessively high level of blood glucose concentration causes loss of glucose in the urine. And third, this causes osmotic diuresis in the kidneys, which can deplete the body of its fluids.

OTHER METABOLIC EFFECTS OF INSULIN

EFFECTS ON FAT METABOLISM

Since fat metabolism and carbohydrate metabolism share the major burden of supplying energy for the body, any factor that affects carbohydrate metabolism indirectly affects fat metabolism in the opposite direction. As long as adequate amounts of insulin and glucose are available, glucose is preferentially metabolized by the cells to supply essentially all the energy required. On the other hand, in the absence of adequate glucose or insulin, the major share of the energy required by the body is then supplied by fats.

Mechanism of Fat Mobilization in Insulin Lack. Figure 78–6 illustrates the effect of sudden insulin lack on the blood glucose and also on the fat mobilization products, the free fatty acids and acetoacetic acid, showing that all of these increase. The cause of this fat mobilization is two-fold. First, in the absence of insulin, glucose cannot enter the fat cell with ease. Several products of glucose metabolism are very important for causing, and also for maintaining, fat storage. These products are (1) α-glycerophosphate, which supplies the glycerol nucleus that combines with fatty acids to form the neutral fats in adipose tissue; (2) TPNH, which is required as a reducing agent in the process of fatty acid synthesis; and (3) DPNH, which is also required as a reducing substance in other stages of fatty acid synthesis. In the absence of these metabolic products of glucose, the process of storing triglycerides in fatty tissue is reversed, causing, instead, release of free fatty acids into the blood.

The second means by which insulin lack promotes fat mobilization is a direct one that is independent of the glucose effects of the hormone. Lack of insulin—in some way not yet understood, but perhaps by acting directly on the fat cell membrane—causes direct enhancement of lipolysis with resultant release of free fatty acids into the circulating body fluids.

Also, additional fat mobilization results from the action of cortisol on the fat cell; this hormone is usually secreted in large quantities in conditions of insulin lack.

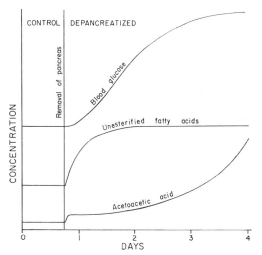

Figure 78–6 Effect of removing the pancreas on the concentrations of blood glucose, plasma free (unesterified) fatty acids, and acetoacetic acid.

Effect of Insulin Lack on Blood Lipid Concentration. In addition to the increase in free fatty acids in the circulating blood, essentially all the other lipid components of the plasma also become increased in the absence of insulin. This includes an increase in the overall quantity of lipoproteins and of their constituents, *triglycerides, cholesterol,* and *phospholipids.* At times the blood lipids increase as much as five-fold, giving a total concentration of plasma lipids of several per cent rather than the normal 0.6 per cent. These high lipid concentrations, especially the high concentration of cholesterol, have often been suggested as a possible factor in the development of extreme atherosclerosis in persons with serious diabetes.

Ketogenic Effect of Insulin Lack. Associated with increased fat mobilization from the tissues is usually also a ketogenic effect of insulin lack, which was discussed in Chapter 68 in relation to utilization of fats for energy. The reason for this effect is the following:

A major pathway by which fatty acids are used for energy is (a) conversion of fatty acids into acetoacetic acid in the liver, and then (b) transfer of the acetoacetic acid in the blood to the peripheral tissues, where (c) it enters the citric acid cycle and is used for energy. When large quantities of fats are being mobilized and utilized for energy, the concentration of acetoacetic acid in the extracellular fluids rises markedly, as illustrated in Figure 78–6, and some of this acid is converted into the other ketone bodies, β-hydroxybutyric acid and acetone, as explained in Chapter 68; thus the rapid mobilization and utilization of fats can cause *ketosis.*

However, the concentration of acetoacetic acid in the body fluids rises far higher than one would expect, unless there should exist some special mechanism that further enhances its concentration. It is possible that this high concentration results from the greatly increased adrenocortical secretion of glucocorticoids that occurs in the absence of insulin, because glucocorticoids further enhance the mobilization of fats from the fat storage areas.

As another explanation of the ketosis, various attempts have been made to show that acetoacetic acid is not utilized as fully as usual in the peripheral tissues in the absence of insulin, but thus far this has not been proved.

Obviously, the great increase in acetoacetic acid in the blood can cause acidosis, which is discussed in detail later in the chapter in relation to diabetes mellitus.

The Glucose–Fatty Acid Cycle. Since insulin promotes glucose transport into fat cells, which in turn causes fat deposition and reduced release of fatty acids into the body fluids, it is clear that glucose and insulin have a direct inhibitory effect on fatty acid utilization by the body.

Conversely, fatty acids have an inhibitory effect on the utilization of glucose, and this results primarily from the following two effects: (1) When excess fatty acids are available, they are degraded into acetyl Co-A, which is then degraded to carbon dioxide and hydrogen atoms in the citric acid cycle. In this process, the intermediate products of the citric acid cycle become greatly enhanced, and some of these products inhibit the conversion of fructose 6-phosphate into fructose 1,6-phosphate, one of the early steps essential to the degradation of glucose before it can be utilized for energy. (2) The excess acetyl Co-A derived from fatty acids also diminishes the conversion of pyruvate into acetyl Co-A.

This reciprocal inhibitory relationship between glucose and fatty acid utilization is called the *glucose–fatty acid cycle.* It plays a particularly important role in insulin lack because it compounds the problem of poor glucose utilization. That is, in addition to the direct effect of insulin lack in depressing glucose utilization, the resultant buildup of fatty acids also indirectly depresses glucose metabolism. This fatty acid feedback effect seems almost to double the glucose metabolic defect in the presence of insulin deficiency.

EFFECT OF INSULIN ON PROTEIN METABOLISM AND ON GROWTH

Effect of Insulin on Protein Anabolism—Transport of Amino Acids Through the Cell Membranes. Proteins are continually being formed from amino acids in all cells of the body, and they are also continually being split back into amino acids, which was discussed in Chapter 69 in relation to protein metabolism. The total quantity of protein stored in the tissues of the body is increased by insulin and greatly decreased by insulin lack. Indeed, a person with severe diabetes mellitus can become incapacitated almost as much from protein lack as from failing glucose metabolism.

Most of the effect of insulin on protein metabolism is probably secondary to the action of insulin on carbohydrate metabolism, for utilization of carbohydrates for energy acts as a protein sparer. When carbohydrates are not metabolized, both fats and proteins are mobilized in great abundance from the tissues and used for energy. Therefore, increased carbohydrate utilization under the influence of insulin would decrease catabolism of the tissue proteins while allowing the anabolic effect to proceed at full speed.

However, studies on isolated tissues in which carbohydrate metabolism has been strictly controlled have shown that insulin still has a moderate effect in promoting protein anabolism, particularly in the liver. Therefore, insulin does truly have a direct effect on protein synthesis. Part of this effect results from the fact that *insulin increases the transport of amino acids through the cell membranes* in a manner similar to that of its increasing glucose transport. This transport is carrier mediated, as is also true of glucose transport. Insulin enhances the transport several-fold, but the effect is generally far less intense than the effect on glucose transport.

Effect of Insulin on Growth—Potentiation of the Effects of Growth Hormone. Insulin is essential for growth of an animal, principally because of the protein-sparing effect of increased carbohydrate metabolism, but also because of the direct effect of insulin on protein anabolism.

In the absence of insulin, growth hormone has almost no effect in promoting growth of an animal. Figure 78–7 illustrates this, showing the cessation of growth after removing both the pancreas and the hypophysis of a growing rat. Administration of either growth hormone or insulin without the other causes only slight growth, but administration of both together causes the original rate of growth.

Though it is certain that growth hormone exerts most of its growth effect directly on the tissues of the body, a small part of the effect might be exerted through the pancreas, for experiments have shown that prolonged administration of growth hormone increases the rate of insulin production by the pancreas, and this insulin in turn can have its usual growth-promoting effect along with the direct action of growth hormone.

RELATION OF OTHER ENDOCRINE GLANDS TO INSULIN SECRETION

Hormones secreted by several other glands have either direct or indirect effects on the blood glucose concentration, which in turn affects the rate of insulin secretion by the pancreas. The following are some of the relationships between the functions of other endocrine glands and insulin secretion:

Relationship of the Anterior Pituitary Gland to Insulin Secretion. The anterior pituitary gland secretes at least four different hormones that have either direct or indirect effects on blood glucose concentration. *Growth hormone* greatly reduces the rate of glucose utilization by the tissues and therefore elevates the blood glucose concentration. At least one of the mechanisms by which growth hormone functions is to oppose glucose utilization by the cell, thus directly opposing the action of insulin. *Luteotropic hormone* seems to have essentially the same effect as growth hormone but to much less extent. *Corticotropin* increases the blood glucose concentration indirectly by first increasing adrenocortical secretion of glucocorticoids, these in turn promoting increased gluconeogenesis and decreasing glucose metabolism by the cell. Likewise, *thyrotropic hormone* increases the blood glucose concentration by increasing thyroid secretion of thyroxine, which in turn increases the rate of gluconeogenesis but to a much less extent than do the glucocorticoids.

The combined effects of a mixture of all the adenohypophyseal hormones, therefore, elevate the blood glucose concentration considerably. If sufficient quantities of these are injected into an animal over a long time, prolonged elevation of blood glucose concentration eventually overstimulates the beta cells of the islets of Langerhans so much that they burn out and cause frank diabetes mellitus. This is called the *diabetogenic effect* of the adenohypophyseal hormones. Growth hormone has an additional effect to stimulate the islets of Langerhans directly, therefore probably exerting more diabetogenic action than any of the other hormones.

On the contrary, a lack of these hormones reduces blood glucose concentration. Therefore, removal of the pituitary gland markedly ameliorates the diabetes in a depancreatized animal.

Relationship of the Adrenal Cortex to Insulin Secretion. Excess secretion of glucocorticoids by the adrenal cortex causes almost the same increase in blood glucose concentration as that seen following the administration of a mixture of adenohypophyseal hormones. Indeed, one of the major effects of administering the adenohypophyseal hormones results from the action of corticotropin to increase adrenocortical secretion of glucocorticoids, which increase blood glucose in two ways. First, they promote gluconeogenesis, probably mainly by mobilizing amino acids from the body proteins, thus making these available to the liver to be used in forming glucose. Second, they have an effect similar to that of growth hormone in opposing glucose utilization by the cells.

Removal of the adrenal cortex in a depancreatized dog causes almost the same amelioration of

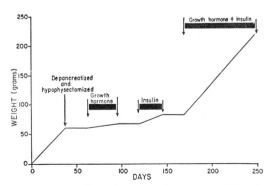

Figure 78–7 Effect of (a) growth hormone, (b) insulin, and (c) growth hormone plus insulin on growth in a depancreatized and hypophysectomized rat.

the animal's diabetes as removal of the adeno-hypophysis. However, when the adrenal cortex is removed rather than the adenohypophysis, the animal's electrolyte balance must be continually controlled by means of adequate mineralocorticoid therapy.

DIABETES MELLITUS

Experimental Diabetes Mellitus. In experimental animals diabetes mellitus has been induced by many different methods. The method universally effective is *depancreatectomy,* but in order to attain significant diabetes at least 90 to 95 per cent of the pancreas usually must be removed; otherwise, the islets of Langerhans in the remaining pancreatic tissue will be able to hypertrophy sufficiently to supply enough insulin for normal metabolic needs.

A second means for producing diabetes in animals is *administration of a poison known as alloxan.* This substance causes the beta cells of the islets of Langerhans to swell and finally to degenerate, but it is difficult to adjust the dosage so that most of the beta cells will be destroyed without killing the animal by other toxic effects.

A still less reliable method for inducing diabetes is *administration of anterior pituitary extract.* Especially when anterior pituitary extract is administered in conjunction with thyroxine, glucocorticoids, or even with some of the sex hormones can it result in extreme swelling and final degeneration of the beta cells of the islets of Langerhans, as explained earlier in the chapter. Though this is not a satisfactory means for producing diabetes in an experimental animal, it does indicate that nonpancreatic hormonal abnormalities probably cause the development of diabetes in some patients, such as that which occurs in giantism or in acromegaly, in which growth hormone is overly abundant.

Most factors that are known to destroy the beta cells are especially likely to destroy these cells when an animal is on a high carbohydrate diet, for such a diet also stimulates the beta cells. Therefore, it is easy to understand why diabetes is believed to result sometimes from excessive ingestion of sweets and other carbohydrates.

Hereditary Factors. Though various factors may cause the beta cells of the islets of Langerhans to "burn out," diabetes is, in general, a hereditary disease, for almost all persons who develop diabetes, especially those who develop it in early years, can trace the disease to one or more forebears. The disease is transmitted as a recessive genetic characteristic which is present in about 20 per cent of the population. This means that about 4 per cent of the population at some time during life develops at least some degree of diabetes mellitus.

Mild diabetes often occurs in older people and especially in older people who are overweight. As long as these people continue to eat excessively, the blood glucose level is elevated, but as soon as they reduce their diet their blood glucose level often reverts to normal. This diabetic effect of obesity is believed to be caused mainly by the depression of glucose metabolism in the presence of excess fatty acids in the blood.

PATHOLOGIC PHYSIOLOGY OF DIABETES

Most of the pathology of diabetes mellitus can be attributed to one of the following three major effects of insulin lack: (1) decreased utilization of glucose by the body cells, with a resultant increase in blood glucose concentration to as high as 300 to 1200 mg. per 100 ml.; (2) markedly increased mobilization of fats from the fat storage areas, causing abnormal fat metabolism and especially deposition of lipids in vascular walls to cause atherosclerosis; and (3) diminished deposition of protein in the tissues of the body, caused partly by failure of glucose to be used as a protein sparer and partly by loss of the direct effect of insulin to promote protein anabolism.

However, in addition, some special pathologic physiological problems occur in diabetes mellitus, and these are not so readily apparent. These are as follows:

Loss of Glucose in the Urine of the Diabetic Person. Whenever the quantity of glucose entering the kidney tubules in the glomerular filtrate rises above approximately 225 mg. per minute, a significant proportion of the glucose begins to spill into the urine; and when the quantity rises above about 325 mg. per minute, which is the tubular maximum for glucose, as explained in Chapter 35, all the excess above this is lost into the urine. If normal quantities of glomerular filtrate are formed per minute, 225 mg. of glucose will enter the tubules each minute when the blood glucose level rises to 180 mg. per cent. Consequently, it is frequently stated that the blood "threshold" for the appearance of glucose in the urine is approximately 180 mg. per cent. When the blood glucose level rises to 300 to 500 mg. per cent — common values in persons with diabetes — several hundred grams of glucose can be lost into the urine each day.

Dehydrating Effect of Elevated Blood Glucose Levels in Diabetes. Greatly elevated blood glucose levels, as high as 1500 to 2000 mg. per cent, can occur under certain conditions in diabetic persons. Yet the only significant effect of the elevated glucose is dehydration of the tissue cells, for glucose does not diffuse easily through the pores of the cell membrane, and the increased osmotic pressure in the extracellular fluids causes osmotic transfer of water out of the cells. Elevation of blood glucose from the normal level of approximately 90 mg. per cent to 400 mg. per cent increases the total osmotic pressure of the extracellular fluid 6 per cent. This alone causes slight dehydration of the tissues, and elevation to still higher values can cause considerable harm.

In addition to the direct dehydrating effect of excessive glucose, the loss of glucose in the urine causes *diuresis* because of the osmotic effect of glucose in the tubules; the diuresis in turn causes obligatory loss of electrolytes from the extracellular fluid. The overall effect is dehydration of the extracellular fluid, which then causes compensatory dehydration of the intracellular fluid for reasons discussed in Chapter 33. Thus, one of the important features of diabetes is a tendency for extracellular and intracellular dehydration to develop.

Acidosis in Diabetes. The shift from carbohydrate to fat metabolism in diabetes has already been discussed. When the body depends almost entirely on fats for metabolism, the level of acetoacetic acid and other keto acids in the body fluids may rise from 1 mEq./liter to as high as 30 mEq./liter. This, obviously, is likely to result in acidosis.

A second effect, which is usually even more important in causing acidosis than is the direct increase in keto acids, is a decrease in sodium concentration caused by the following effect: Keto acids have a low threshold for excretion by the kidneys; therefore, when the keto acid level rises in diabetes, as much as 100 to 200 grams of keto acids can be excreted in the urine each day. About half of the acid is excreted in the acidic form, but much of the other half is excreted combined with sodium derived from the extracellular fluid. As a result, the sodium concentration in the extracellular fluid usually decreases, and loss of this basic ion adds to the acidosis that is already caused by excessive keto acids in the extracellular fluid.

Obviously, all the usual reactions that occur in metabolic acidosis take place in diabetic acidosis. These include *rapid and deep breathing* called "Kussmaul respiration," which causes excessive expiration of carbon dioxide and *marked decrease in bicarbonate content of the extracellular fluids.*

Likewise, *large quantities of chloride ion are excreted by the kidneys* as an additional compensatory mechanism for correction of the acidosis. These extreme effects, however, occur only in the most severe degrees of untreated diabetes. The overall changes in the electrolytes of the blood as a result of severe diabetic acidosis are illustrated in Figure 78–8.

Relationship of Diabetic Symptoms to the Pathologic Physiology of Insulin Lack. *Polyuria* (excessive elimination of urine), *polydipsia* (excessive drinking of water), *polyphagia* (excessive eating), *loss of weight,* and *asthenia* (lack of energy) are the earliest symptoms of diabetes. As explained above, the polyuria is due to the osmotic diuretic effect of glucose in the kidney tubules. In turn, the polydipsia is due to dehydration resulting from polyuria. The failure of glucose utilization by the body causes loss of weight and a tendency toward polyphagia. The asthenia apparently also is caused mainly by lack of carbohydrate metabolism, but loss of body protein also contributes to muscle weakness.

DIAGNOSIS OF DIABETES

The usual methods for diagnosing diabetes are based on various chemical tests of the urine and the blood as follows:

Urinary Sugar. Simple office tests or more complicated quantitative laboratory tests may be used for determining the quantity of glucose lost in the urine. In general, the normal person loses undetectable amounts of glucose, whereas the diabetic loses glucose in small to large amounts, in proportion to the severity of his disease and his intake of carbohydrates. (However, a condition known as *renal glycosuria* sometimes occurs even in persons without diabetes mellitus. This condition results from a low tubular maximum for glucose, as explained in Chap. 38, so that even though the blood glucose level is perfectly normal, a large quantity of glucose may still be lost in the urine.)

The Fasting Blood Glucose Level. The fasting blood sugar level in the early morning, at least eight hours after any previous meal, is normally 80 to 90 mg. per cent, and 120 mg. per cent is generally considered to be the absolute upper limit of normal. A fasting blood sugar level above this value usually indicates diabetes mellitus.

The Glucose Tolerance Test. As illustrated by the middle curve in Figure 78–9, when a normal fasting person ingests 50 grams of glucose, his blood glucose level rises from approximately 90 mg. per cent to approximately 140 mg. per cent and falls back to below normal within three hours.

Though an occasional diabetic person has a normal fasting blood glucose concentration, it is usually above 120 mg. per cent, and his glucose tolerance test is almost always abnormal. On ingestion of 50 grams of glucose, these persons exhibit a progressive, slow rise in blood glucose level

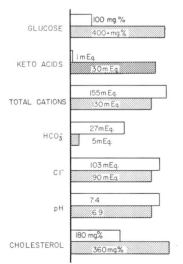

Figure 78–8 Changes in blood constituents in severe diabetes, showing normal values (light bars) and diabetic values (dark bars).

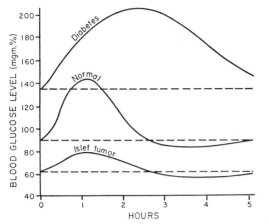

Figure 78–9 Glucose tolerance curve in the normal person, in a diabetic person, and in a person with an islet tumor (hyperinsulinism).

for two to three hours, as illustrated by the upper curve in Figure 78–9, and the glucose level falls back to the control value only after some five to six hours; yet it never falls below the control level. This slow fall of the curve and its failure to fall below the control level illustrates that the normal increase in insulin secretion following glucose ingestion does not occur in the diabetic, and a diagnosis of diabetes mellitus can usually be definitely established on the basis of such a curve.

Insulin Sensitivity.   To differentiate diabetes mellitus of pancreatic origin from high blood glucose levels resulting from excess secretion of adrenocortical or adenohypophyseal hormones, an *insulin sensitivity test* can be performed. When little insulin is produced by the pancreas, a test dose of insulin causes the blood glucose level to fall markedly, indicating greatly increased "insulin sensitivity." On the other hand, when the blood glucose level is high as a result of excessive adrenocortical or adenohypophyseal secretion, the glucose level responds very poorly to the test dose of insulin because the pancreas is already secreting large quantities of insulin.

Acetone Breath.   As pointed out in Chapter 68, small quantities of acetoacetic acid, which increase greatly in severe diabetes, can be converted to acetone, which is volatile and is vaporized into the expired air. Consequently, one frequently can make a diagnosis of diabetes mellitus simply by smelling acetone on the breath of a patient. Also, keto acids can be detected by chemical means in the urine, and their quantitation aids in determining the severity of the diabetes.

TREATMENT OF DIABETES

The theory of treatment in diabetes mellitus is to administer enough insulin so that the patient will have normal carbohydrate metabolism. If this is done, most consequences of diabetes can be prevented.

Figure 78–10 illustrates time-activity curves of different preparations of insulin following subcutaneous injection. Regular amorphous insulin and crystalline insulin have an activity lasting six to eight hours, whereas insulin that has been precipitated slowly with zinc to form large crystalline particles (lente insulin) or insulin that has been precipitated with various protein derivatives (globulin insulin and protamine zinc insulin) are relatively insoluble and absorbed slowly. For instance, protamine zinc insulin may continue to act for as long as 36 to 48 hours. Ordinarily, the diabetic patient is given a single dose of one of the long-acting insulins each day; this increases his overall carbohydrate metabolism throughout the day. Then additional quantities of regular insulin are given at those times of the day when his blood glucose level tends to rise too high, such as at meal times. Thus each patient is established on an individualized routine of treatment.

Diet of the Diabetic.   The insulin requirements of a diabetic are established with the patient on a standard diet containing normal, well-controlled amounts of carbohydrates, and any change in the quantity of carbohydrate intake changes the requirements for insulin. In the normal person, the pancreas has the ability to adjust the quantity of insulin produced to the intake of carbohydrate; but in the completely diabetic person, this control function is totally lost.

Control of the Diabetic Patient in Fever and Exercise.   Any abnormal state of metabolism in the diabetic patient alters the insulin requirement. Thus, fever, severe infection, and so forth frequently increase the requirement for insulin immensely, and failure to give the extra insulin can cause diabetic coma.

On the other hand, exercise frequently has exactly the opposite effect on insulin requirement, reducing the requirement considerably below that usually needed. The peculiar reason for this is that increased muscle activity increases the transport of glucose into the muscle cells even in the absence of insulin. Thus, *exercise actually has an insulin-like effect.* It is often remarked that a child with severe diabetes requires less insulin to control his

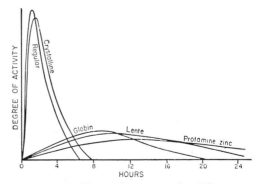

Figure 78–10   Time-action curves for different types of insulin.

diabetes if he lives a very active life than if he leads an overprotected life.

Relationship of Treatment to Arteriosclerosis. Diabetic patients develop atherosclerosis, arteriosclerosis, and severe coronary heart disease far more easily than do normal persons. Indeed, a person who has relatively poorly controlled diabetes throughout childhood is likely to die of heart disease in his 20's. Some clinicians have claimed that the treatment of the patient does not greatly affect the development of arteriosclerosis, but on the other hand there are indications that the method of treatment does make considerable difference.

In the early days of treating diabetes it was the tendency to reduce the carbohydrates in the diet so that the insulin requirements would be minimized. This procedure kept the blood sugar level down to normal values and prevented the loss of glucose in the urine, but it did not prevent the abnormalities of fat metabolism. Actually, it exacerbated these. Consequently, there is a tendency at present to allow the patient a normal carbohydrate diet and then to give simultaneously large quantities of insulin to metabolize the carbohydrates. This depresses the rate of fat metabolism and also depresses the high level of blood cholesterol that occurs in diabetes as a result of abnormal fat metabolism.

Because the complications of diabetes—such as atherosclerosis, greatly increased susceptibility to infection, diabetic retinopathy, cataracts, hypertension, and chronic renal disease—are more closely associated with the level of the blood lipids than with the level of blood glucose, it is the object of many clinics treating diabetes to administer sufficient glucose and insulin so that the quantity of blood lipids becomes normal.

Treatment with Drugs that Stimulate the Release of Insulin. Recently, several drugs that can be taken by mouth and that cause a hypoglycemic effect have been introduced for treatment of persons with mild diabetes. The most useful drugs that have thus far been tried *are sulfonylurea compounds,* the most important of which is *tolbutamide* (Orinase).

These drugs act by stimulating insulin production by the islets of Langerhans. Therefore, they are effective only in patients who still have active islets which for some reason are not secreting adequate quantities of insulin. Obviously, they are of no value in the treatment of severe diabetes, for the islets have already lost all potential ability to secrete insulin.

DIABETIC COMA

If diabetes is not controlled satisfactorily, severe dehydration and acidosis may result; and sometimes, even when the person is receiving treatment, sporadic changes in metabolic rates of the cells, such as might occur during bouts of fever, can also precipitate dehydration and acidosis.

If the pH of the body fluids falls below approximately 7.0 to 6.9, the diabetic person develops coma. Also, in addition to the acidosis, dehydration is believed to exacerbate the coma. Once the diabetic person reaches this stage, the outcome is usually fatal unless he receives immediate treatment.

Treatment of Diabetic Coma. The patient with diabetic coma is extremely refractory to insulin because acidic plasma has an *insulin antagonist,* an alpha globulin, that opposes the action of the insulin. Also, the very high free fatty acid and acetoacetic acid levels in the blood inhibit the usage of glucose, as was discussed earlier. Therefore, instead of the usual 60 to 80 units of insulin per day, which is the dosage usually necessary for control of severe diabetes, as much as 1500 to 2000 units of insulin must often be given the first day of treatment of coma. Administration of insulin alone may be sufficient to reverse the abnormal physiology and to effect a cure. However, it is usually advantageous to correct also the dehydration and acidosis immediately.

The dehydration is ordinarily corrected rapidly by administering large quantities of sodium chloride solution, and the acidosis is often corrected by administering sodium bicarbonate or sodium lactate solution; the bicarbonate is expired as carbon dioxide through the lungs, and the lactate is removed by the liver, thus providing sodium ions to neutralize the acidosis.

While correcting the dehydration of diabetic coma it is important to maintain the potassium ion of the extracellular fluids at a normal level. During the dehydration process, large quantities of potassium are removed from the tissue cells along with the removal of water. Consequently, on administering fluids and insulin to the comatose person, as these fluids and glucose are absorbed into the cells, large quantities of potassium are also reabsorbed into the cells to re-establish electrolyte balance. This may greatly depress the potassium level in the extracellular fluids and cause skeletal muscular paralysis; also, if this is carried to the extreme, excessive systolic contraction of the heart might occur. Patients have been known to die as a result of rapid loss of potassium from the extracellular fluids. Therefore, fluids administered for correction of diabetic dehydration are often fortified with a small quantity of extra potassium; yet when this is done, caution must be observed against overuse of potassium because a potassium level in the extracellular fluids above two times normal can cause cardiac standstill in diastole.

HYPERINSULINISM

Though much more rare than diabetes, increased insulin production, which is known as hyperinsulinism, does occasionally occur. This usually results from an adenoma of an islet of Langerhans. About 10 to 15 per cent of these adenomas are malignant, and occasionally metastases from the islets of Langerhans spread throughout the body, causing tremendous production of insulin by both the primary and the metastatic cancers. Indeed, in order to prevent hypoglycemia, in some of these patients more than 1000 grams of glucose have had to be administered each 24 hours.

Insulin Shock and Hypoglycemia. As already emphasized, the central nervous system derives essentially all its energy from glucose metabolism, and insulin is not necessary for this metabolism. However, if insulin causes the level of blood glucose to fall to low values, the metabolism of the central nervous system becomes depressed. Consequently, in patients with hyperinsulinism or in patients who administer too much insulin to themselves, the syndrome called "insulin shock" may occur as follows:

As the blood sugar level falls into the range of 50 to 70 per cent, the central nervous system usually becomes quite excitable, for this degree of hypoglycemia seems to facilitate neuronal activity. Sometimes various forms of hallucinations result, but more often the patient simply experiences extreme nervousness, and he may tremble all over. As the blood glucose level falls to 20 to 50 mg. per cent, clonic convulsions and loss of consciousness are likely to occur. As the glucose level falls still lower, the convulsions cease, and only a state of coma remains. Indeed, at times it is difficult to distinguish between diabetic coma as a result of insulin lack and coma due to hypoglycemia caused by excess insulin. However, the acetone breath and the rapid, deep breathing of diabetic coma are not present in hypoglycemic coma.

Obviously, proper treatment for a patient who has hypoglycemic shock or coma is immediate intravenous administration of large quantities of glucose. This usually brings the patient out of shock within a minute or more. Also, administration of epinephrine can cause glycogenolysis in the liver and thereby increase the blood glucose level extremely rapidly.

If treatment is not effected immediately, permanent damage to the neuronal cells of the central nervous system occurs; this happens especially in prolonged hyperinsulinism due to pancreatic tumors. Hypoglycemic shock induced by insulin administration is frequently used for treatment of psychogenic disorders. This type of shock, like electric shock therapy, frequently benefits especially the melancholic patient.

Depressed Glucose Tolerance Curve and Decreased Insulin Sensitivity in Hyperinsulinism. In addition to the usual signs of hypoglycemia that appear in hyperinsulinism, a definite diagnosis of the condition can usually be made by performing a glucose tolerance test and an insulin sensitivity test. The glucose tolerance curve is often (but not always) depressed, as shown by the lower curve of Figure 78-9, illustrating that the initial glucose level is low, and the increase after ingestion of glucose is slight. When the insulin sensitivity test is performed, insulin sensitivity is found to be greatly diminished because the injected dose of insulin is so slight in relation to the quantity of insulin already present in the body that almost no effect occurs.

FUNCTIONS OF GLUCAGON

Glucagon, like insulin, is a small protein; it has a molecular weight of 3482 and is composed of a chain of 29 amino acids. On injecting purified glucagon into an animal a profound hyperglycemic effect occurs, 1 microgram per kg. of glucagon elevating the blood glucose concentration approximately 20 mg./100 ml. of blood. For this reason, glucagon is frequently called *hyperglycemic* factor.

Glycogenolysis and Increased Blood Glucose Concentration Caused by Glucagon. The most dramatic effect of glucagon is its ability to cause glycogenolysis in the liver, which in turn increases the blood glucose concentration. This effect is similar to, but many times more powerful than, that of epinephrine in causing glycogenolysis. Like epinephrine, glucagon stimulates the formation of cyclic adenylic acid, which in turn activates the normally inactive phosphorylase in the liver cells. The active phosphorylase then, as explained in Chapter 67, causes rapid glycogenolysis followed by release of glucose into the blood.

However, glucagon does not cause glycogenolysis in the extrahepatic tissues. This, for reasons unknown, is in contrast to the action of epinephrine, which causes glycogenolysis both in the liver and in the extrahepatic tissues.

Infusion of glucagon for about four hours can cause such intensive liver glycogenolysis that all the liver stores of glycogen will be totally depleted.

PROLONGED HYPERGLYCEMIC EFFECT OF GLUCAGON. Even after all the glycogen in the liver has been exhausted under the influence of glucagon, continued infusion of this hormone causes continued hyperglycemia and glycosuria. This is caused by a yet unexplained effect of glucagon to increase the rate of gluconeogenesis. The increased gluconeogenesis could readily result secondarily from depletion of glycogen in the liver cells rather than as a direct effect of glucagon.

Regulatory Function of Glucagon To Prevent Hypoglycemia. When the blood glucose concentration falls to as low as 70 mg. per cent, the pancreas secretes large quantities of glucagon, which rapidly mobilizes glucose from the liver. Thus, glucagon protects against hypoglycemia. It has been claimed that patients dying of the rare disease called *idiopathic hypoglycemia* have no alpha cells in their islets of Langerhans and, therefore, secrete no glucagon.

It has also been suggested that glucagon is secreted during muscular activity and that it helps to cause glucose mobilization from the liver for use by the muscles; and in starvation, glucagon is abundantly secreted, perhaps in an attempt to maintain a normal blood glucose concentration.

REFERENCES

Bishop, J. S.; Steele, R.; Altszuler, N.; Dunn, A.; Bjerknes, C.; and deBodo, R. C.: Effects of insulin on liver glycogen synthesis and breakdown in the dog. *Amer. J. Physiol.,* 208:307, 1965.

Cahill, G. F., Jr.; Owen, O. E.; and Felig, P.: Insulin and fuel homeostasis. *Physiologist, 11*:97, 1968.

Diabetes mellitus and obesity. *Ann. N.Y. Acad. Sci., 148*:573, 1968.

Duncan, G. G.: Diabetes mellitus. *In* Duncan, G. G. (ed.): Diseases of Metabolism. 5th ed., Philadelphia, W. B. Saunders Company, 1964.

Foa, P. P., and Galansino, G.: Glucagon: Chemistry and Function in Health and Disease. Springfield, Illinois, Charles C Thomas, Publisher, 1962.

Frohman, L. A.: The endocrine function of the pancreas. *Ann. Rev. Physiol., 31*:353, 1969.

Grodsky, G. M., and Forsham, P. H.: Insulin and the pancreas. *Ann. Rev. Physiol., 28*:347, 1966.

Hales, C. N.: The glucose fatty acid cycle and diabetes mellitus. *In* Bittar, E. E., and Bittar, N. (eds.): The Biological Basis of Medicine. New York, Academic Press, Inc., 1968, Vol. I, p. 309.

Han, P. W.; Yu, Y. K.; and Chow, S. L.: Enlarged pancreatic islets of tube-fed hypophysectomized rats bearing hypothalamic lesions. *Amer. J. Physiol., 218*:769, 1970.

Henderson, M. J.; Morgan, H. E.; and Park, C. R.: Regulation of glucose uptake in muscle. IV. The effect of hypophysectomy on glucose transport, phosphorylation, and insulin sensitivity in the isolated, perfused heart. *J. Biol. Chem., 236*:273, 1961.

Issekutz, B., Jr., and Paul, P.: Intramuscular energy sources in exercising normal and pancreatectomized dogs. *Amer. J. Physiol., 215*:197, 1968.

Issekutz, B., Jr.; Paul, P.; and Miller, H. I.: Metabolism in normal and pancreatectomized dogs during steady-state exercise. *Amer. J. Physiol., 213*:857, 1967.

Kanazawa, Y.; Kuzuya, T.; and Ide, T.: Insulin output via the pancreatic vein and plasma insulin response to glucose in dogs. *Amer. J. Physiol., 215*:620, 1968.

LaRaia, P. J.; Craig, J.; and Reddy, W. J.: Glucagon: effect on adenosine 3'-5' monophosphate in the rat heart. *Amer. J. Physiol., 215*:968, 1968.

Levine, R.: Concerning the mechanisms of insulin action. *Diabetes, 10*:421, 1961.

Levine, R. A.: Hepatic glycogenolytic activity of cyclic 3',5'-AMP and its monobutyryl derivative. *Amer. J. Physiol., 213*:768, 1967.

Mallaisse, W. J.; Mallaisse-Lagae, F.; and Wright, P. H.: Effect of fasting upon insulin secretion in the rat. *Amer. J. Physiol., 213*:843, 1967.

Mayer, S. E.; Namm, D. H.; and Rice, L.: Effect of glucagon on 3',5'-AMP, phosphorylase activity and contractility of heart muscle of the rat. *Circ. Res., 26*:225, 1970.

Moir, T. W., and Nayler, W. G.: Coronary vascular effects of glucagon in the isolated dog heart. *Circ. Res., 26*:29, 1970.

Morgan, H. E.; Henderson, M. J.; Regan, D. M.; and Park, C. R.: Regulation of glucose uptake in muscle. I. The effects of insulin and anoxia on glucose transport and phosphorylation in the isolated perfused heart of normal rats. *J. Biol. Chem., 236*:253, 1961.

Park, C. R.; Morgan, H. E.; Henderson, M. J.; Regen, D. M.; Cadenase, E.; and Post, R. L.: The regulation of glucose uptake in muscle as studied in the perfused rat heart. *Recent Progr. Hormone Res., 17*:493, 1961.

Park, C. R.; Reinwein, D.; Henderson, M. J.; Cadenas, E.; and Morgan, H. E.: The action of insulin on the transport of glucose through the cell membrane. *Amer. J. Med., 26*:674, 1959.

Post, R. L.; Morgan, H. E.; and Park, C. R.: Regulation of glucose uptake in muscle. III. The interaction of membrane transport and phosphorylation in the control of glucose uptake. *J. Biol. Chem., 236*:269, 1961.

Pruett, E. D. R.: Glucose and insulin during prolonged work stress in men living on different diets. *J. Appl. Physiol., 28*:199, 1970.

Rieser, P.: Insulin, Membranes and Metabolism. Baltimore, The Williams & Wilkins Co., 1968.

Role of insulin in membrane transport (symposium). *Fed. Proc., 24*:1039, 1965.

Sutherland, E. W., and Robinson, G. A.: The role of cyclic AMP in the control of carbohydrate metabolism. *Diabetes, 18*:797, 1969.

Weber, G.: Hormonal control of gluconeogenesis. *In* Bittar, E. E., and Bittar, N. (eds.): The Biological Basis of Medicine. New York, Academic Press, Inc., 1968, Vol. 2, p. 263.

Woll, I. G.; Stirewalt, W. S.; and Moyer, A. N.: Effect of diabetes and insulin on nucleic acid metabolism of heart muscle. *Amer. J. Physiol., 214*:825, 1968.

CHAPTER 79

Parathyroid Hormone, Calcitonin, Calcium and Phosphate Metabolism, Vitamin D, Bone and Teeth

The physiology of parathyroid hormone and of the hormone calcitonin is closely related to calcium and phosphate metabolism, the function of vitamin D, and the formation of bone and teeth. Therefore, these are discussed together in the present chapter.

CALCIUM AND PHOSPHATE IN THE EXTRACELLULAR FLUID AND PLASMA

GASTROINTESTINAL ABSORPTION OF CALCIUM AND PHOSPHATE

By far the major source of calcium in the diet is milk or milk products, which are also major sources of phosphate, but phosphate is also present in many other dietary foods, including especially the meats.

Calcium is poorly absorbed from the gastrointestinal tract because of the relative insolubility of many of its compounds and also because bivalent cations are poorly absorbed through the gastrointestinal mucosa anyway. On the other hand, phosphate is absorbed exceedingly well most of the time except when excess calcium is in the diet; the calcium tends to form almost insoluble calcium phosphate compounds that fail to be absorbed but instead pass on through the bowels to be excreted in the feces. Therefore, the rate of phosphate loss in the feces is, in general, determined by the rate of calcium loss in the feces. In other words, the major problem in the absorption of calcium and phosphate is actually a problem of calcium absorption alone, for if this is absorbed, both are absorbed.

Promotion of Calcium Absorption by Vitamin D. The most important function of vitamin D is its ability to increase the rate of calcium absorption from the gastrointestinal tract. This effect results from a direct action of vitamin D on the epithelial cells of the mucosa in the duodenum and jejunum to increase their active transport of calcium through the intestinal membrane. Though the mechanism of this increased active transport is not known, it possibly results from the fact that vitamin D increases the quantity of a calcium-binding protein in the epithelial cells.

THE VITAMIN D COMPOUNDS. Several different sterol compounds belong to the vitamin D family, and all perform more or less the same functions. The two most common D vitamins are vitamins D_2 and D_3 (calciferol and activated 7-dehydrocholesterol, respectively). 7-Dehydrocholesterol is formed in the skin as a result of irradiation by ultraviolet rays from the sun. Consequently, appropriate exposure to the sun prevents vitamin D deficiency.

EFFECT OF VITAMIN D ON PHOSPHATE ABSORPTION. As just pointed out, phosphate is readily absorbed except when it combines with calcium to form insoluble calcium phosphate compounds. Consequently, when vitamin D increases the rate of calcium absorption it also increases the rate of phosphate absorption, though this is entirely a secondary effect.

Regulation of Calcium Absorption by the Calcium Ion Concentration in the Extracellular Fluids — Function of Parathyroid Hormone. Even when plenty of vitamin D is available, all the calcium in the gut still may not be absorbed, for the rate of calcium absorption is then determined almost entirely by the calcium ion concentration of the extracellular fluids. Even a fraction of a milliequivalent decrease in calcium ion concentration increases the rate of calcium absorption from the gut several-fold, whereas a similar slight increase in calcium ion concentration decreases the rate of absorption several-fold. Not all the factors that cause this extreme dependence of calcium absorption on extracellular fluid calcium ion concentration are known, but at least one of the factors is parathyroid hormone, which has a direct effect on the gastrointestinal mucosa to increase the rate of calcium absorption. Decreased calcium ion concentration in the extracellular fluid causes marked parathyroid hormone secretion. Thus, a feedback mechanism operates whereby decreased calcium ion concentration increases parathyroid hormone secretion, which in turn increases calcium absorption from the gut. This is part of the overall parathyroid control of calcium ion concentration in the extracellular fluid, as explained later in the chapter.

EXCRETION OF CALCIUM AND PHOSPHATE

Gastrointestinal Excretion of Calcium. About seven-eighths of the daily intake of calcium is excreted in the feces and the remaining one-eighth in the urine. The excretion of calcium in the feces is actually caused by a balance of calcium absorption and calcium secretion in the intestines. For instance, an average adult ingests about 750 mg. of calcium each day. Assuming that all this is absorbed and that 625 mg. of calcium is secreted in the intestinal juices, we find a net absorption of about 125 mg. per day and a loss in the feces of 625 mg.

Urinary Excretion of Calcium—Function of Parathyroid Hormone. Excretion of calcium in the urine conforms to much the same principles as the excretion of sodium. When the calcium ion concentration of the extracellular fluids is low, the rate at which calcium is excreted by the kidneys also becomes low, while even a minute increase in calcium ion concentration increases the calcium ion secretion markedly. One of the factors that helps to regulate this excretion of calcium ion is parathyroid hormone, for increased parathyroid hormone—which occurs when the extracellular calcium ion concentration falls—increases the rate of calcium reabsorption from the tubules, while decreased parathyroid hormone allows a greater proportion of the calcium to pass on into the urine. Thus, here again, the parathyroid feedback mechanism operates to control the concentration of calcium in the extracellular fluids.

The average amount of calcium excreted into the urine each day is approximately 125 mg., which is equal to the net absorption from the gut, except when bone is being deposited actively.

Excretion of Phosphate. Except for the portion of phosphate that is excreted in the feces in combination with calcium, almost all the dietary phosphate is absorbed into the blood from the gut and later excreted in the urine.

Phosphate is a *threshold substance;* that is, when its concentration in the plasma is below the critical value of approximately 1 millimol/liter, no phosphate at all is lost into the urine; but, above this critical concentration, the rate of phosphate loss is directly proportional to the additional increase. Thus, the kidney regulates the phosphate concentration in the extracellular fluid by altering the rate of phosphate secretion in accordance with the plasma phosphate concentration.

As discussed later in the chapter, phosphate excretion by the kidneys is also significantly affected by parathyroid hormone; increased parathyroid hormone causes increased excretion of phosphate and decreased parathyroid hormone causes retention of phosphate in the extracellular fluids.

THE CALCIUM IN THE PLASMA AND INTERSTITIAL FLUID

The concentration of calcium in the plasma is approximately 10 mg. per cent, normally varying between 9 and 11 mg. per cent. This is equivalent to approximately 5 mEq./liter. It is apparent from these narrow limits of normality that the calcium level in the plasma is regulated exactly—and mainly by parathyroid hormone, as discussed later in the chapter.

The calcium in the plasma is present in three different forms, as shown in Figure 79–1. (1) Approximately 50 per cent of the calcium is combined with the plasma proteins and consequently is nondiffusible through the capillary membrane. (2) Approximately 5 per cent of the calcium (0.2 mEq./liter) is diffusible through the capillary membrane but is combined with other substances of the plasma and interstitial fluids (citrate, for instance) in such a manner that it is not ionized. (3) The remaining 45 per cent of the calcium in the plasma is both diffusible through the capillary membrane and ionized. Thus, the plasma and interstitial fluids have a normal *calcium ion concentration of approximately 2.3 mEq./liter.* This ionic calcium is important for most functions of calcium in the body, including the effect of calcium on the heart, on the nervous system, and on bone formation.

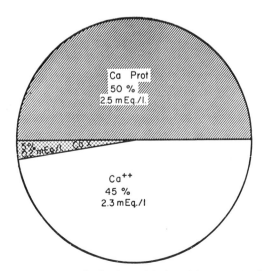

Figure 79–1 Distribution of ionic calcium (Ca^{++}), diffusible but un-ionized calcium ($Ca\,X$), and calcium proteinate ($Ca\,Prot$) in blood plasma.

THE INORGANIC PHOSPHATE IN THE EXTRACELLULAR FLUIDS

Inorganic phosphate exists in the plasma mainly in two forms: HPO_4^{--} and $H_2PO_4^-$. The concentration of HPO_4^{--} is approximately 2.1 mEq./liter, and the concentration of $H_2PO_4^-$ is approximately 0.26 mEq./liter. When the total quantity of phosphate in the extracellular fluid rises, so does the quantity of each of these two types of phosphate ions. Furthermore, when the pH of the extracellular fluid becomes more acid, there is a relative increase in $H_2PO_4^-$ and decrease in the HPO_4^{--} while the opposite occurs when the extracellular fluid becomes alkaline. These relationships were presented in the discussion of acid-base balance in Chapter 37.

Because it is difficult to determine chemically the exact quantities of HPO_4^{--} and $H_2PO_4^-$ in the blood, ordinarily the total quantity of phosphate is expressed in terms of milligrams of *phosphorus* per 100 ml. of blood. The average total quantity of inorganic phosphorus represented by both phosphate ions is about 4 mg./100 ml., varying between normal limits of 3.5 to 4 mg./100 ml. in adults and 4 to 5 mg./100 ml. in children.

EFFECTS OF ALTERED CALCIUM AND PHOSPHATE CONCENTRATIONS IN THE BODY FLUIDS

Changing the level of phosphate in the extracellular fluid from far below normal to as high as three to four times normal does not cause significant immediate effects on the body. There is a similar lack of effects for variations in most other anions, for even chloride ion can be substituted almost entirely by certain other anions without drastic effects.

Both prolonged hypocalcemia and hypophosphatemia greatly decrease bone mineralization, as explained later in the chapter. On the other hand, elevation or depletion of calcium ion in the extracellular fluid causes extreme immediate effects.

Tetany Resulting from Hypocalcemia. When the extracellular fluid concentration of calcium ion falls below normal, the nervous system becomes progressively more and more excitable because of increased neuronal membrane permeability. This increase in excitability occurs both in the central nervous system and in the peripheral nerves, though most symptoms are manifest peripherally. The nerve fibers become so excitable that they begin to discharge spontaneously, initiating nerve impulses that pass to the peripheral skeletal muscles where they elicit tetanic contraction. Consequently, hypocalcemia causes tetany.

Figure 79–2 illustrates tetany in the hand, which usually occurs before generalized tetany develops. This is called "carpopedal spasm."

Acute hypocalcemia in the human being ordinarily causes essentially no other serious effects besides tetany because tetany kills the patient before other effects can develop. Tetany ordinarily occurs when the blood concentration of calcium reaches approximately 7 mg. per cent, which is only 30 per cent below the normal calcium concentration.

When the calcium in the body fluids falls to a level not quite sufficient to cause tetany, "latent tetany" results; this can be diagnosed by weakly stimulating the nerves and noting the response. For instance, tapping on the 7th nerve where it passes over the angle of the jaw causes the facial muscles to twitch. Second, placing a tourniquet on the upper arm causes ischemia of the peripheral nerves and also increases the excitability of the nerves, thus causing the muscles of the lower arm and hand to go into spasm. Finally, if the person with latent tetany hyperventilates, the resulting alkalinization of his body fluids increases the irritability of the nerves, causing overt signs of tetany.

In experimental animals, in which the level of calcium can be reduced beyond the normal lethal stage, extreme hypocalcemia can cause marked dilatation of the heart, changes in cellular enzyme activities, increased cell membrane permeability in other cells in addition to nerve cells, and impaired blood clotting.

Hypercalcemia. When the level of calcium in the body fluids rises above normal, the nervous system is depressed, and reflex activities of the central nervous system become sluggish.

Figure 79–2 Hypocalcemic tetany in the hand, called "carpopedal spasm." (Courtesy Dr. Herbert Langford.)

The muscles, too, become sluggish and weak, probably because of calcium effects on the muscle cell membranes. Also, increased calcium ion concentration decreases the QT interval of the heart, and it causes constipation and lack of appetite, probably because of depressed contractility of the muscular walls of the gastrointestinal tract.

The depressive effects of increased calcium level begin to appear when the blood level of calcium rises above approximately 12 mg. per cent, and they can become marked as the calcium level rises above 15 mg. per cent. When the level of calcium rises above approximately 17 mg. per cent in the body fluids, calcium phosphate is likely to precipitate throughout the body; this condition is discussed shortly in connection with parathyroid poisoning.

BONE AND ITS RELATIONSHIPS WITH EXTRACELLULAR CALCIUM AND PHOSPHATES

Bone is composed of a tough *organic matrix* that is greatly strengthened by deposits of *calcium salts.* Average *compact bone* contains by weight approximately 25 per cent matrix and 75 per cent salts. However, *newly formed bone* may have a considerably higher percentage of matrix in relation to salts.

The Organic Matrix of Bone. The organic matrix of bone is approximately 97 per cent *collagen fibers,* and the remaining 3 per cent is a homogeneous medium called *ground substance.* The collagen fibers extend in all directions in the bone but to a great extent along the lines of tensional force. These fibers give bone its powerful tensile strength.

The ground substance is composed of extracellular fluid plus *mucoprotein, chondroitin sulfate,* and *hyaluronic acid.* The precise function of these is not known, though perhaps they help to provide a medium for deposition of calcium salts.

The Bone Salts. The crystalline salts deposited in the organic matrix of bone are composed principally of *calcium* and *phosphate,* and the formula for the major crystalline salts, known as *hydroxyapatites,* is the following:

$$Ca^{++}_{10-x}(H_3O^+)_{2x} \cdot (PO_4)_6(OH^-)_2$$

Each crystal, about 200 to 400 Å long and averaging 10 by 30 Å in cross-section, is shaped like a long, flat plate. The relative ratio of calcium to phosphorus can vary markedly under different nutritional conditions, the Ca/P ratio on a weight basis varying between 1.3 and 2.0.

Magnesium, sodium, potassium, and *carbonate* ions are also present among the bone salts, though x-ray diffraction studies fail to show definite crystals formed by these. Therefore, they are believed to be adsorbed to the surfaces of the hydroxyapatite crystals rather than organized into distinct crystals of their own. This ability of many different types of ions to adsorb to bone crystals extends to many ions normally foreign to bone, such as *strontium, uranium, plutonium, the other transuranic elements, lead, gold, other heavy metals,* and *at least 9 out of 14 of the major radioactive products released by explosion of the hydrogen bomb.* Deposition of radioactive substances in the bone can cause prolonged irradiation of the bone tissues, and, if a sufficient amount is deposited, an osteogenic sarcoma almost invariably eventually develops.

Tensile and Compressional Strength of Bone. The collagen fibers of bone have cross-striations approximately every 64 millimicrons along their length; and hydroxyapatite crystals, standing end on end like blocks in pillars, lie adjacent to each segment of the fiber bound tightly to it. This intimate bonding prevents "shear" in the bone; that is, it prevents the crystals and collagen fibers from slipping out of place, which is essential in providing strength to the bone.

The collagen fibers of bone, like those of tendons, have great tensile strength, while the calcium salts, which are similar in physical properties to marble, have great compressional strength. These combined properties, plus the degree of bondage between the collagen fibers and the crystals, provide a bony structure that has both extreme tensile and compressional strength. Thus, bones are constructed in exactly the same way that reinforced concrete is constructed. The steel of reinforced concrete provides the tensile strength, while the cement, sand, and rock provide the compressional strength. Indeed, the compressional strength of bone is greater than that of even the best reinforced concrete, and the tensile strength approaches that of steel in reinforced concrete.

PRECIPITATION AND ABSORPTION OF CALCIUM AND PHOSPHATE IN BONE— EQUILIBRIUM WITH THE EXTRACELLULAR FLUIDS

Supersaturated State of Calcium and Phosphate Ions in Extracellular Fluids with Respect

to Bone Salts. The concentrations of calcium and phosphate ions in extracellular fluid are considerably greater than those required to cause precipitation of hydroxyapatite. However, because of the large number of ions required to form a single molecule of hydroxyapatite, it is very difficult for all of these ions to come together simultaneously; therefore, the precipitation process is normally extremely slow; and, without the intermediation of some active process, it requires months to years to occur significantly. Furthermore, inhibitors are present in most tissues of the body, as well as in plasma, to prevent such precipitation; one such inhibitor is pyrophosphate. Therefore, hydroxyapatite crystals fail to precipitate in most normal tissues despite the state of supersaturation of the ions.

Mechanism for Causing Hydroxyapatite Precipitation in Bones. The initial stage in bone production is secretion by the osteoblasts of ground substance and of collagen. The collagen polymerizes rapidly to form collagen fibers, and the resultant tissue becomes cartilage. Shortly thereafter, hydroxyapatite crystals begin to precipitate in the cartilage, and most of the ground substance is resorbed. The crystals first appear within the substance of the collagen fibers, and then additional crystals form around the peripheries of these fibers until very little space remains for the ground substance. Thus, the final product is mainly collagen fibers and hydroxyapatite crystals, or bone.

It is believed that the collagen fibers of the cartilage are specially prepared in advance for causing precipitation of the hydroxyapatite crystals. Several theories have been offered as an explanation of this effect. One theory suggests that the osteoblasts secrete a substance into the cartilage to neutralize the inhibitor that normally prevents hydroxyapatite crystallization. Once this has occurred, a natural affinity of the collagen fibers for calcium salts supposedly causes precipitation. In support of this theory is the fact that purified collagen fibers prepared from other tissues of the body besides bone will also cause precipitation of hydroxyapatite in the presence of plasma.

A second theory suggests that phosphate radicals become bound at regular intervals along the collagen fiber and that these radicals then constitute nidi for growth of hydroxyapatite crystals. In support of this concept is the fact that osteoblasts secrete large quantities of alkaline phosphatase, which could split organic phosphates and thereby provide the required phosphate radicals. A third theory suggests that these phosphate radicals simply increase to such a high concentration that the tendency for precipitation of hydroxyapatite increases markedly, thereby overcoming the normal inhibitor factor against precipitation.

The formation of initial crystals within the collagen fibers is called *crystal seeding* or *nucleation.*

The growth of hydroxyapatite crystals in newly forming bones reaches 75 per cent completion in a few days, but it usually takes months for the bone to achieve full calcification.

PRECIPITATION OF CALCIUM IN NON-OSSEOUS TISSUES UNDER ABNORMAL CONDITIONS. Though calcium salts almost never precipitate in normal tissues besides bone, under abnormal conditions they do precipitate. For instance, they precipitate in arterial walls in the condition called arteriosclerosis and cause the arteries to become bone-like tubes. Likewise, calcium salts frequently deposit in degenerating tissues or in old blood clots. Presumably, in these instances, the inhibitor factors that normally prevent deposition of calcium salts disappear from the tissues, thereby allowing precipitation.

EXCHANGEABLE CALCIUM

If soluble calcium salts are injected intravenously, the calcium ion concentration can be made to increase immediately to very high levels. However, within minutes to an hour or more, the calcium ion concentration returns to normal. Likewise, if large quantities of calcium ions are removed from the circulating body fluids, the calcium ion concentration again returns to normal within minutes to hours. These effects result from the fact that the body contains a type of *exchangeable* calcium that is always in equilibrium with the calcium ions in the extracellular fluids. A small portion of this exchangeable calcium is that calcium found in all tissue cells, especially in highly permeable types of cells such as those of the liver and the gastrointestinal tract. However, most of the exchangeable calcium, as shown by studies using radioactively tagged calcium, is in the bone, and it normally amounts to about 0.4 to 1.0 per cent of the total bone calcium. This calcium is probably deposited in the bones by the process of adsorption or in the form of some readily mobilizable salt such as $CaHPO_4$. The radioactive calcium studies show that the exchangeable calcium occurs almost entirely at points in the bone that are currently undergoing active bone absorption.

The importance of exchangeable calcium to the body is that it provides a rapid buffering mechanism to keep the calcium ion concentration in the extracellular fluids from rising to excessive levels or falling to very low levels under transient conditions of excess or hypo-availability of calcium.

DEPOSITION AND ABSORPTION OF BONE

Deposition of Bone by the Osteoblasts. Bone is continually being deposited by *osteoblasts,* and it is continually being absorbed where *osteoclasts* are active. Osteoblasts are found on most surfaces of the bones and in many cavities. A small amount of osteoblastic activity occurs continually in all living bones so that at least some new bone is being formed constantly.

Absorption of Bone — Function of the Osteoclasts. Bone is also being continually absorbed in the presence of osteoclasts, which are located in many cavities throughout all bones. Osteoclasts can form from osteocytes or osteoblasts, or they can even form from fibroblasts in the bone marrow. Later in the chapter we will see that parathyroid hormone controls the bone absorptive activity of osteoclasts.

Histologically, bone absorption occurs immediately adjacent to the osteoclasts, as illustrated in Figure 79–3. Recent electron microscopic studies have shown the presence of a few minute bone crystals inside the osteoclasts, which indicates that the osteoclasts in some instances phagocytize and digest bony particles, and finally release calcium, phosphate, and the digestive end-products of the organic matrix into the extracellular fluids. However, this phagocytic process probably is not the major means by which bone is absorbed, for it is believed that most bone absorption results from osteoclastic secretion of enzymes or acids that digest or dissolute the organic matrix and simultaneously cause dissolution of the bone salts.

Equilibrium Between Bone Deposition and Absorption. Normally, except in growing bones, the rates of bone deposition and absorption are equal to each other so that the total mass of bone remains constant. However, osteoclasts actually eat holes in the bone in large areas, while in other areas new bone forms. Usually, osteoclasts exist in large masses, and once a mass of osteoclasts begins to develop, it usually eats away at the bone for about three weeks, eating out a tunnel that may be as great as 1 mm. in diameter. At the end of this time the osteoclasts are converted into osteoblasts, and new bone begins to develop. Bone deposition then continues for several months, the new bone being laid down in successive layers on the inner surfaces of the cavity until the tunnel is filled. Deposition of new bone ceases when the bone begins to encroach on the blood vessels supplying the area. The canal through which these vessels run, called the *haversian canal,* therefore, is all that remains of the original cavity. Each new area of bone deposited in this way is called an *osteon,* as shown in Figure 79–4.

VALUE OF CONTINUAL DEPOSITION AND ABSORPTION OF BONE. The continual deposition and absorption of bone has a number of physiologically important functions. First, bone ordinarily adjusts its strength in proportion to the degree of bone stress. Consequently, bones thicken when subjected to heavy loads. Second, even the shape of the bone can be rearranged for proper support of mechanical forces by deposition and absorption of bone in accordance with stress patterns. Third, since old bone becomes relatively weak and brittle, new organic matrix is needed as the old organic matrix degenerates. In this manner the normal toughness of bone is maintained. Indeed, the bones of children, in whom the rate of deposition and absorption is rapid, show little brittleness in comparison with the

Figure 79–3   Osteoblastic and osteoclastic activity in the same bone.

bones of old age, at which time the rates of deposition and absorption are slow.

Control of the Rate of Bone Deposition by Bone "Stress." Bone is deposited in proportion to the compressional load that the bone must carry. For instance, the bones of athletes become considerably heavier than those of nonathletes. Also, if a person has one leg in a cast but continues to walk on the opposite leg, the bone of the leg in the cast becomes thin and decalcified, while the opposite bone remains thick and normally calcified. Therefore, continual physical stress stimulates osteoblastic deposition of bone.

Bone stress also determines the shape of bones under certain circumstances. For instance, if a long bone of the leg breaks in its center and then heals at an angle, the compression stress on the inside of the angle causes increased deposition of bone, while increased absorption occurs on the outer side of the angle where the bone is not compressed. After many years of increased deposition on the inner side of the angulated bone and absorption on the outer side, the bone becomes almost straight. This is especially true in children because of the rapid remodeling of bone at younger ages.

The deposition of bone at points of compressional stress is believed to be caused by a *piezoelectric* effect, as follows: Compression of bone causes a negative potential at the compressed site and a positive potential elsewhere in the bone. It has been shown that minute quantities of current flowing in bone cause osteoblastic activity at the negative end of the current flow, and this would explain the increased bone deposition at compression sites. On the other hand, usual osteoclastic activity could account for absorption of bone at sites of tension.

Repair of a Fracture. A fracture of a bone in some way maximally activates all the periosteal and intraosseous osteoblasts involved in the break. Indeed, even fibroblasts in some surrounding tissues become osteoblasts. Therefore, within a short time a large bulge of fibrous tissue, osteoblastic tissue, and new organic bone matrix, followed shortly by the deposition of calcium salts, develops between the two broken ends of the bone. This is called a *callus*.

Many bone surgeons utilize the phenomenon of bone stress to accelerate the rate of fracture healing. This is done by use of special mechanical fixation apparatuses for holding the ends of the broken bone together so that the patient can use his bone immediately.

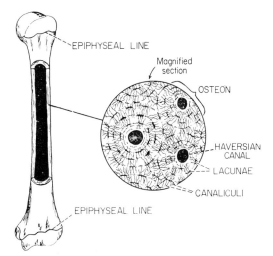

Figure 79–4 The structure of bone.

This obviously causes stress on the opposed ends of the broken bones, which accelerates osteoblastic activity at the break and often greatly shortens convalescence.

Blood Alkaline Phosphatase as an Indication of the Rate of Bone Deposition. The osteoblasts secrete large quantities of alkaline phosphatase when they are actively depositing bone matrix. This phosphatase is believed either to increase the local concentration of inorganic phosphate or to activate the collagen fibers in a way that they cause the deposition of calcium salts. Since some alkaline phosphatase diffuses into the blood, the blood level of alkaline phosphatase is usually a good indicator of the rate of bone formation.

The alkaline phosphatase level is below normal in only a few diseases; this includes especially hypoparathyroidism. On the other hand, it is greatly elevated (1) during growth of children, (2) following major bone fractures, and (3) in almost any bone disease that causes bone destruction and must be repaired by osteoblastic activity, such as rickets, osteomalacia, and osteitis fibrosa cystica.

PARATHYROID HORMONE

For many years it has been known that increased activity of the parathyroid gland causes rapid absorption of calcium salts from the bones with resultant hypercalcemia in the extracellular fluids; conversely, hypofunction of the parathyroid glands causes hypocalcemia, often with resultant tetany, as described earlier in the chapter. Also, parathyroid hormone is important in phosphate metabolism as well as in calcium metabolism.

Physiologic Anatomy of the Parathyroid Glands. Normally there are four parathyroid glands in the human being, and these are located immediately

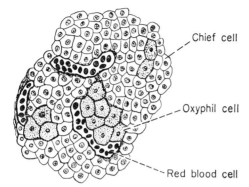

Figure 79–5 Histologic structure of a parathyroid gland.

behind the thyroid gland—one behind each of the upper and each of the lower poles of the thyroid. Each parathyroid gland is approximately 6 mm. long, 3 mm. wide, and 2 mm. thick, and has a macroscopic appearance of dark brown fat; therefore, the parathyroid glands are difficult to locate during thyroid operations. For this reason total or subtotal thyroidectomy frequently resulted in total removal of the parathyroid glands before the importance of these glands was generally recognized. Occasionally, parathyroid glands are located in the anterior mediastinum or, rarely, in the posterior mediastinum.

Removal of half the parathyroid glands usually causes little physiologic abnormality. However, removal of three out of four normal glands usually causes transient hypoparathyroidism. But even a small quantity of remaining parathyroid tissue is usually capable of hypertrophying satisfactorily to perform the function of all the glands.

The parathyroid gland of the adult human being, illustrated in Figure 79–5, contains mainly *chief cells* and *oxyphil cells,* but oxyphil cells are absent in many animals and in young human beings. The chief cells secrete most of the parathyroid hormone. The function of the oxyphil cells is not certain; they are probably aged chief cells that still secrete some hormones.

Chemistry of Parathyroid Hormone. Parathyroid hormone has been isolated in a pure form. It is a

small protein having a molecular weight of approximately 9000 and composed of 74 to 80 amino acids. Smaller compounds have also been isolated from the parathyroid gland that exhibit parathyroid hormone activity, but the activity is always less than that of the larger protein molecule. Therefore, the smaller compounds are almost certainly breakdown products of the normal parathyroid hormone.

EFFECT OF PARATHYROID HORMONE ON CALCIUM AND PHOSPHATE CONCENTRATIONS IN THE EXTRACELLULAR FLUID

Figure 79–6 illustrates the effect of injecting parathyroid hormone into a human being, showing marked elevation in calcium ion concentration of the extracellular fluids and depression of phosphate concentration. The maximum effect of parathyroid hormone on phosphate is reached within two to three hours, while the maximum effect on calcium ion concentration is reached in about eight hours and then lasts some 24 to 36 hours. The rise in calcium ion concentration is caused principally by a direct effect of parathyroid hormone to increase bone absorption. The decline in phosphate concentration, on the other hand, is caused principally by an effect of parathyroid hormone on the kidneys to increase the rate of phosphate excretion.

Bone Absorption Caused by Parathyroid Hormone. The absorptive effect of parathyroid hormone in bones is believed to result from the ability of this hormone to convert osteoblasts and osteocytes in bone into osteoclasts and, also, to increase their osteoclastic activity. In turn, the osteoclasts are believed to secrete either enzymes or acids that absorb the bone, as discussed earlier in the chapter.

Bone contains such great amounts of calcium in comparison with the total amount in all the extracellular fluids (about 1000 times as much) that even when parathyroid hormone causes a great rise in calcium concentration in the fluids, it is impossible to discern any immediate effect at all on the bones. Yet prolonged administration of parathyroid hormone finally results in evident absorption in all the bones with development of large cavities filled with very large, multinucleated osteoclasts.

Two of the experimental studies that have demonstrated the direct action of parathyroid hormone on bone are the following:

1. Implantation of small living sections of parathyroid gland adjacent to a bone causes large numbers of osteoclasts to develop in the immediate area, and this is followed by rapid local absorption of the bone.

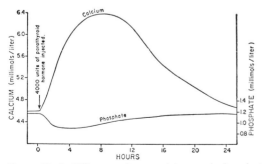

Figure 79–6 Effect on plasma calcium and phosphate concentrations of injecting 4000 units of parathyroid hormone into a human being.

2. Injection of parathyroid hormone into animals that have severe rickets causes reabsorption of the newly formed matrix even though no salts are present in the matrix. Therefore, parathyroid hormone acts on both the matrix and the salts at the same time, indicating once again that the primary effect of parathyroid hormone is to increase the activity of some absorptive factor related to the osteoclasts.

Effect of Parathyroid Hormone on Phosphate and Calcium Excretion by the Kidneys. Administration of parathyroid hormone causes immediate and rapid loss of phosphate in the urine. This effect is caused by diminished renal tubule reabsorption of phosphate ions. It was long believed that this was the most important function of parathyroid hormone. However, it is now known that this function is of secondary importance to the effect of parathyroid hormone on bone absorption.

Parathyroid hormone causes increased renal tubular reabsorption of calcium at the same time that it diminishes the rate of phosphate reabsorption. Moreover, it also increases the rate of reabsorption of magnesium ions and hydrogen ions, while it decreases the reabsorption of sodium, potassium, and amino acid ions in much the same way that it affects phosphate.

Were it not for the *long-term* effect of parathyroid hormone on the kidneys to increase calcium reabsorption, the continual loss of calcium into the urine would eventually deplete the bones of this mineral.

Effect of Parathyroid Hormone on Membrane Transport of Calcium and Phosphate. In an attempt to find some unifying concept that can explain the diverse functions of parathyroid hormone on bone, on the gut, on the kidney, and on certain other tissues, the effect of parathyroid hormone on membrane transport of calcium and phosphate has been studied in several different experimental preparations. The results demonstrate that parathyroid hormone causes transport of calcium outward through the mitochondrial membrane and transport of phosphates inward. Also, it causes outward transport of calcium from certain whole cells, such as from the cells making up the lens of the eye. Finally, parathyroid hormone opposes the secretion of calcium in milk and in saliva, presumably acting at the membranes of the glandular cells.

On the basis of these studies, it has even been suggested that osteoclastic activity in the bones is actually a membrane transport phenomenon. For instance, according to this hypothesis, transport of calcium away from the fluids in immediate contact with the bone could readily cause absorption of calcium salts.

Relationship of Vitamin D to Parathyroid Activity. Administration of large quantities of vitamin D to a person causes absorption of bone in much the same way that parathyroid hormone does; and in the absence of vitamin D, the effect of parathyroid hormone to cause bone absorption is greatly reduced. Also, enhanced transport of calcium through the intestinal mucosa and through the renal tubular epithelium by parathyroid hormone is dependent on an adequate supply of vitamin D.

Thus, in many ways vitamin D has functions similar to, or complementary to, those of parathyroid hormone. Yet the effect of vitamin D on calcium absorption from the gut is great in comparison with that of parathyroid hormone, and, conversely, the effect of parathyroid hormone on bone absorption is great in comparison with that of vitamin D. Therefore, the basic mechanisms of action are probably different despite the similarities of effects.

REGULATION OF EXTRACELLULAR CALCIUM ION CONCENTRATION BY THE PARATHYROID GLANDS

The calcium ion concentration in the extracellular fluids under normal conditions rarely varies more than 5 per cent above or below the normal concentration. Therefore, calcium ion regulation is one of the most highly developed homeostatic mechanisms of the entire body. This regulation is effected almost entirely by the parathyroid glands in the following way:

Control of Parathyroid Secretion by Calcium Ion Concentration. Even the slightest decrease in calcium ion concentration in the extracellular fluid causes the parathyroid glands to increase their rate of secretion and to hypertrophy. For instance, the parathyroid glands become greatly enlarged in *rickets*, in which the level of calcium is usually depressed only a few per cent; also they become greatly enlarged in pregnancy, even though the decrease in calcium ion concentration in the mother's extracellular fluid is hardly measurable; and, they are greatly enlarged during lactation because calcium is used for milk formation.

On the other hand, any condition that increases the calcium ion concentration causes decreased activity and reduced size of the parathyroid glands. Such conditions include (1) excess quantities of calcium in the diet, (2) increased vitamin D in the diet and (3) bone absorption caused by other factors besides parathyroid hormone (for example, bone absorption caused by disuse of the bones).

Feedback Control of Calcium Ion Concentration. It can now be understood how the parathyroid glands control the calcium ion concentration in the extracellular fluids, for reduced calcium concentration increases the rate of parathyroid secretion; and this in turn increases bone absorption, thereby raising the calcium ion concentration to normal. Conversely, increased calcium ion concentration inhibits the parathyroid glands, which automatically decreases the calcium ion concentration to normal.

Note particularly that the regulatory properties of parathyroid hormone do not derive only from its effect on bone, for in addition it *increases the absorption of calcium from the gut* and *from the renal tubules,* which also helps to raise the calcium ion concentration. These are particularly valuable effects when the bone salts have been so depleted that parathyroid hormone can no longer cause additional absorption of calcium from the bone.

Buffer Function of the Exchangeable Calcium in the Bones. The parathyroid hormone feedback control system is relatively slow to act, requiring hours and sometimes days for maximal effect when a new level of parathyroid secretion is required. Yet at times the amount of calcium absorbed into or lost from the body fluids is as much as 1 gram or more within several hours. For instance, in diarrhea, several grams of calcium can be secreted in the intestinal juices into the intestinal tract and lost into the feces each day. Conversely, after a person ingests large quantities of calcium, he may absorb 1 gram or more in an hour. This compares with a *total quantity of calcium in all the extracellular fluids of less than 1 gram.* Adding or subtracting 1 gram from such a small amount of calcium in the extracellular fluids would obviously cause serious hyper- or hypocalcemia. However, this does not occur for the following reasons:

The exchangeable calcium salts in the bone, which were discussed earlier in the chapter, are in reversible equilibrium with the calcium and phosphate in the extracellular fluids. That is, an increase in the product of extracellular calcium and phosphate ion concentrations causes immediate deposition of more bone salts in the bones. Conversely, a decrease in this product causes immediate absorption of some bone salts. This reaction is so rapid that *a single passage of blood containing a high concentration of calcium through a bone allows the bone*

to remove almost all the excess calcium. About 6 per cent of all the blood flows each minute through the bones, and this represents slightly over 1 per cent of all the extracellular fluid going through the bones each minute. Therefore, about half of any excess calcium that appears in the extracellular fluids is removed by this buffer function of the bones in approximately 70 minutes. Or, conversely, the half-time for making up an acute deficiency in extracellular calcium ion concentration is also about 70 minutes.

The total amount of exchangeable calcium phosphate in the bones is only 5 or 6 grams, yet before all this can be used up in buffering the calcium and phosphate concentrations in the blood, the parathyroid mechanism can come into play. Parathyroid absorption of the bone affects primarily the hydroxyapatite rather than the exchangeable calcium and therefore provides an almost unlimited amount of calcium for maintaining the calcium ion concentration of the body fluids. The ease with which calcium and phosphate ions, as well as other ions in the extracellular fluid, can exchange between the bone and extracellular fluids depends on the fact that the total surface area of all the bone crystals in all the bones is about 30 acres.

CALCITONIN

About ten years ago, a new hormone that has effects on blood calcium opposite to those of parathyroid hormone was discovered in the parathyroid gland of several lower animals. This hormone was named *calcitonin* because it reduces the blood calcium ion concentration. Soon after its initial discovery, it was found to be secreted in the human being not by the parathyroid glands but instead by the thyroid gland, for which reason it has also been called *thyrocalcitonin.* Still more recently it was discovered that calcitonin is secreted by the *ultimobranchial glands* of fish, amphibia, reptiles, and birds. Furthermore, its concentration in these glands is extremely great. In mammals, ultimobranchial glands do not exist as such but have become incorporated into either the parathyroid or thyroid glands. The so-called *parafollicular cells* in the interstitial tissue between the follicles of the human thyroid gland are probably remnants of the ultimobranchial glands of lower animals, and it is these cells that seem to secrete the calcitonin.

Physiologic Effects of Calcitonin. Calcitonin inhibits the absorption of bone. It presumably acts at the same site in bone as does parathyroid hormone, because it even inhibits the absorption of bone after administration of large quantities of parathyroid hormone. Unfortunately, the mechanism of this inhibition is unknown, though it occurs extremely rapidly. It causes almost total cessation of bone salt reabsorption within minutes; as a result, the calcium ion concentration in the body fluids begins to fall also within minutes. Thus, the action of calcitonin on calcium ion concentration occurs many times as rapidly as the effect of parathyroid hormone.

Role of Calcitonin in Feedback Homeostasis of Calcium Ion Concentration. Perfusion of the thyroid gland with plasma containing even slightly elevated calcium ion concentration causes an immediate, several-fold increase in rate of secretion of calcitonin. Conversely, decreased plasma calcium ion concentration can decrease calcitonin secretion to as little as one-fifth normal. Therefore, the calcitonin mechanism can act in a feedback system to control calcium ion concentration in a manner exactly opposite to that of the parathyroid hormone system.

However, there are two major differences between the calcitonin and the parathyroid feedback systems. First, the calcitonin mechanism operates extremely rapidly, reaching peak activity in less than an hour; this is in contrast to the many hours required for peak activity to be attained following parathyroid secretion. Therefore, the calcitonin mechanism could play a very important role in preventing acute excesses or deficiencies of calcium ions in the body fluids.

The second difference is that the calcitonin mechanism seems to act only as a short-term regulator and not to have significant function month in and month out as does the parathyroid hormone system. Therefore, over a prolonged period of time, it still seems to be the parathyroid system that sets the long-term level of calcium ions in the body fluids.

PHYSIOLOGY OF PARATHYROID AND BONE DISEASES

HYPOPARATHYROIDISM

When the parathyroid glands do not secrete sufficient parathyroid hormone, the osteoclasts of the bone become almost totally inactive. As a result, bone reabsorption is so depressed that the level of calcium in the body fluids decreases. Because calcium and phosphates are not being absorbed from the bone, the bone usually remains strong, and osteoblastic activity is concomitantly decreased, presumably because there is less bone strain in relation to the strength of the bone.

When the parathyroid glands are suddenly removed, the calcium level in the blood falls from the normal of 10 mg. per cent to 7 mg. per cent within two to three days. When this level is reached, the usual signs of tetany develop. Among the muscles of the body most sensitive to tetanic spasm are the laryngeal muscles. Spasm of these obstructs respiration, which is the usual cause of death in tetany unless appropriate treatment is applied.

If all four parathyroid glands are removed from an animal and the animal is prevented from dying of respiratory spasm by appropriate supportive measures, the total lack of parathyroid hormone secretion normally causes the blood level of calcium to fall as low as approximately 6 mg. per cent while at the same time the level of phosphorus increases from the normal of about 4 mg. per cent to approximately 12 mg. per cent.

Treatment of Hypoparathyroidism. PARATHYROID HORMONE (PARATHORMONE). Parathyroid hormone is occasionally used for treating hypoparathyroidism. However, because of the expense of this hormone, because its effect lasts only about 24 to 36 hours, and because the tendency of the body to develop immune bodies against it makes it progressively less and less active in the body, treatment of hypoparathyroidism with parathyroid hormone is rare in present-day therapy.

DIHYDROTACHYSTEROL AND VITAMIN D. In addition to its ability to cause increased absorption of calcium from the gastrointestinal tract, vitamin D also causes a weak effect similar to that of parathyroid hormone in promoting calcium and phosphate absorption from bones. Therefore, a person with hypoparathyroidism can be treated satisfactorily by administration of *large quantities* of vitamin D. One of the vitamin D compounds, dihydrotachysterol (A.T. 10), has a more marked ability to cause bone absorption than do most of the other vitamin D compounds. Administration of calcium plus vitamin D three or more times a week can almost completely control the calcium level in the extracellular fluid of a hypoparathyroid person.

HYPERPARATHYROIDISM

The cause of hyperparathyroidism ordinarily is a tumor of one of the parathyroid glands, and such tumors occur much more frequently in women than in men or children. Consequently, it is believed that pregnancy, lactation, and perhaps other causes of prolonged low calcium levels, all of which stimulate the parathyroid gland, may predispose to the development of such a tumor.

In hyperparathyroidism extreme osteoclastic activity occurs in the bones, and this elevates the calcium ion concentration in the extracellular fluid while usually (but not always) depressing slightly the concentration of phosphate ions.

Bone Disease in Hyperparathyroidism. Though in mild hyperparathyroidism new bone may be deposited rapidly enough to compensate for the osteoclastic reabsorption of bone, in severe hyperparathyroidism, the osteoclastic absorption soon far outstrips osteoblastic deposition, and the bone may be eaten away almost entirely. Indeed, the reason a hyperparathyroid person comes to the doctor is often a broken bone. X-ray of the bone shows extensive decalcification and occasionally large punched-out cystic areas of the bone that are filled with osteoclasts in the form of so-called giant-cell "tumors." Obviously, multiple fractures of the weakened bones result from only slight trauma, especially where cysts develop. The cystic bone disease of hyperparathyroidism is frequently called *osteitis fibrosa cystica,* or *von Recklinghausen's disease.*

As a result of increased osteoblastic activity that attempts to form new bone as rapidly as it is absorbed, the level of alkaline phosphatase in the body fluids rises markedly.

Effects of Hypercalcemia in Hyperparathyroidism. Hyperparathyroidism can at times cause the plasma calcium level to rise to as high as 15 to 20 mg. per cent. The effects of such elevated calcium levels, as detailed earlier in the chapter, are depression of the central and peripheral nervous systems, muscular weakness, constipation, abdominal pain, peptic ulcer, lack of appetite, and depressed relaxation of the heart during diastole.

PARATHYROID POISONING AND METASTATIC CALCIFICATION. When extreme quantities of parathyroid hormones are secreted, the level of calcium in the body fluids rises rapidly to very high values, and even the extracellular fluid phosphate concentration also often rises markedly, probably because the kidneys cannot excrete rapidly enough all the phosphate being absorbed from the bone. Therefore, the calcium and phosphate in the body fluids become greatly supersaturated even for the deposition of calcium phosphate ($CaHPO_4$) crystals. Therefore, these crystals begin to deposit in the alveoli of the lungs, in the tubules of the kidneys, in the thyroid gland, in the acid-producing area of the stomach mucosa, and in the walls of the arteries throughout the body. This extensive *metastatic* deposition of calcium phosphate can develop within a few days.

Ordinarily, the level of calcium in the blood must rise above 17 mg. per cent before there is danger of parathyroid poisoning, but once such elevation develops along with some concurrent elevation of phosphate, death frequently occurs in only a few days.

FORMATION OF KIDNEY STONES IN HYPERPARATHYROIDISM. Most patients with mild hyperparathyroidism show few signs of bone disease and few general abnormalities as a result of elevated calcium, but instead have an extreme tendency to form kidney stones. The reason for this is that almost all the excess calcium and phosphate mobilized from the bones in hyperparathyroidism is excreted by the kidneys, causing proportionate increase in the concentrations of these substances in the urine. As a result, crystals of calcium phosphate tend to precipitate in the kidney, forming calcium phosphate stones. Also, calcium oxalate stones develop as a result of the high level of calcium in the urine in association with normal levels of oxalate. Because the solubility of most renal stones is slight in alkaline media, the tendency for formation of renal calculi is considerably greater in alkaline urine than in acid urine. For this reason, acidotic diets and drugs are frequently used for treating renal calculi.

Secondary Hyperparathyroidism. Because a low level of calcium ions in the body fluids directly increases the secretion of parathyroid hormone, any factor that causes a low level of calcium initiates the condition known as secondary hyperparathyroidism. This may result from low calcium diet, pregnancy, lactation, rickets, or osteomalacia. The hyperplasia of the parathyroid glands is a corrective measure for maintaining the level of calcium in the extracellular fluids at a nearly normal value.

RICKETS

Rickets occurs mainly in children as a result of calcium or phosphate deficiency in the extracellular fluids. Ordinarily, rickets is due to lack of vitamin D rather than lack of calcium or phosphate in the diet. If the child is properly exposed to sunlight, the 7-dehydrocholesterol in his skin becomes activated by the ultraviolet rays and forms vitamin D_3, which prevents rickets by promoting calcium and phosphate absorption from the intestines as discussed earlier in the chapter.

Children who remain indoors through the winter in general do not receive adequate quantities of vitamin D without some supplementary therapy in the diet. Rickets tends to occur especially in the spring months because vitamin D formed during the preceding summer can be stored for several months in the liver, and, also, calcium and phosphorus absorption from the bones must take place for several months before clinical signs of rickets become apparent.

CALCIUM AND PHOSPHATE IN THE BLOOD OF PATIENTS WITH RICKETS. Ordinarily, the level of calcium in the blood in rickets is only slightly depressed, but the level of phosphate is greatly depressed. This is because the parathyroid glands prevent the calcium level from falling by promoting bone absorption every time the calcium level begins to fall. On the other hand, there is no good regulatory system for controlling a falling level of phosphate, and the increased parathyroid activity actually increases the excretion of phosphates in the urine.

EFFECT OF RICKETS ON THE BONE. During prolonged deficiency of calcium and phosphate in the body fluids, increased parathyroid hormone secretion protects the body against hypocalcemia by causing osteoclastic absorption of the bone; this in turn causes the bone to become progressively weaker and imposes marked physical stress on the bone, resulting in rapid osteoblastic activity. The osteoblasts lay down large quantities of organic bone matrix, which does not become calcified because even with the "crystal seeding" effect of the matrix the product of the calcium and phosphate ions is still insufficient to cause calcification. Consequently, the newly formed, uncalcified organic matrix gradually takes the place of other bone that is being reabsorbed.

Obviously, hyperplasia of the parathyroid glands is marked in rickets because of the decreased blood calcium level, and the alkaline phosphatase level in the blood is markedly increased as a result of the rapid osteoblastic activity.

TETANY IN RICKETS. In the early stages of rickets, tetany almost never occurs because the parathyroid glands continually stimulate osteoclastic absorption of bone and therefore maintain an almost normal level of calcium in the body fluids. However, when the bones become exhausted of calcium, the level of calcium may fall rapidly. As the blood level of calcium falls below 7 mg. per cent, the usual signs of tetany develop, and the child may die of tetanic respiratory spasm unless intravenous calcium is administered, which relieves the tetany immediately.

TREATMENT OF RICKETS. The treatment of rickets, obviously, depends on supplying adequate calcium and phosphate in the diet and also on administering vitamin D adequately. If vitamin D is not administered along with the calcium, little calcium is absorbed, and this calcium in turn carries large quantities of phosphate with it into the feces.

TETANY RESULTING FROM TREATMENT. Occasionally, tetany occurs when a child is treated for rickets, probably for the following reason: Administration of calcium and phosphate in the diet without sufficient vitamin D often results in better absorption of phosphate than of calcium. As a result, the phosphate level in the blood is likely to rise rapidly. Then, because the osteoblasts are already extremely active, the increase in phosphate level apparently causes rapid deposition of bone salts in the newly formed organic matrix, thus removing excessive quantities of calcium from the blood and causing the calcium level to fall so much that tetany supervenes. This difficulty can be prevented by administering sufficient quantities of vitamin D and larger quantities of calcium than phosphate.

Osteomalacia. Osteomalacia is rickets in adults and is frequently called "adult rickets."

Normal adults rarely have dietary lack of vitamin D or calcium because large quantities of calcium are not needed for bone growth as in children. However, lack of vitamin D and calcium occasionally occurs as a result of steatorrhea (failure to absorb fat), for vitamin D is fat soluble and calcium tends to form insoluble soaps with fat; consequently, in steatorrhea vitamin D and calcium tend to pass into the feces. Under these conditions an adult occasionally has such poor calcium and phosphate absorption that "adult rickets" can occur, though this rarely proceeds to the same stages as it does in childhood.

OSTEOMALACIA AND RICKETS CAUSED BY RENAL DISEASE. "Renal rickets" is a type of osteomalacia resulting from prolonged kidney damage. The kidney disease prevents normal excretion of acids and leads to acidosis, which is known to cause bone absorption and therefore is probably partly responsible for the bone disease.

Another type of renal disease that leads to rickets and osteomalacia is *congenital hypophosphatemia* resulting from congenitally reduced reabsorption of phosphates by the renal tubules. This type of rickets must be treated with phosphate compounds instead of calcium and vitamin D, and it is called *vitamin D resistant rickets.*

OSTEOPOROSIS

Osteoporosis is a different disease from osteomalacia and rickets, for it results from abnormal organic matrix formation rather than abnormal bone calcification. In general, in osteoporosis the osteoblastic activity in the bone is less than normal, and consequently the rate of bone deposition is depressed. The causes of osteoporosis are (1) lack of use of the bones; (2) malnutrition to the extent that sufficient protein matrix cannot be formed; (3) lack of vitamin C, which is necessary for the secretion of intercellular substances by all cells, including the osteoblasts; (4) postmenopausal lack of estrogen secretion, for estrogens have an osteoblastic stimulating activity; (5) old age, in which many of the protein anabolic functions are poor anyway so that bone matrix cannot be deposited satisfactorily; (6) Cushing's disease, because massive quantities of glucocorticoids cause decreased deposition of protein throughout the body and cause increased catabolism of protein; and (7) acromegaly, possibly because of lack of sex hormones, excess of adrenocortical hormones, and often lack of insulin because of the diabetogenic effect of adenohypophyseal hormones. Obviously, many diseases of protein metabolism can cause osteoporosis.

PHYSIOLOGY OF THE TEETH

The teeth cut, grind, and mix the food. To perform these functions the jaws have powerful muscles capable of providing an occlusive force between the front teeth of as much as 50 to 100 pounds and as much as 150 to 200 pounds for the jaw teeth. Also, the upper and lower teeth are provided with projections and facets which interdigitate so that each set of teeth fits with the other. This fitting is called *occlusion,* and it allows even small particles of food to be caught and ground between the tooth surfaces.

FUNCTION OF THE DIFFERENT PARTS OF THE TEETH

Figure 79–7 illustrates a sagittal section of a tooth, showing its major functional parts, the *enamel, dentine, cementum,* and *pulp.* The tooth can also be divided into the *crown,* which is the portion that protrudes out of the gum into the mouth, and the *root,* which is the portion that protrudes into the bony socket of the jaw. The collar between the crown and the root where the tooth is surrounded by the gum is called the *neck.*

Dentine. The main body of the tooth is composed of dentine, which has a strong, bony structure. Dentine is made up principally of hydroxyapatite crystals similar to those in bone, but much more dense. These are embedded in a strong meshwork of collagen fibers. In other words, the principal constituents of dentine are very much the same as those of bone. The major difference is its histologic organization, for dentine does not contain any osteoblasts, osteoclasts, or spaces for blood vessels or nerves. Instead, it is deposited and nourished by a layer of cells called *odontoblasts,* which line its inner surface along the wall of the pulp cavity.

The calcium salts in dentine make it extremely resistant to compressional forces, while the collagen fibers make it tough and resistant to tensional forces that might result when the teeth are struck by solid objects.

Enamel. The outer surface of the tooth is covered by a layer of enamel that is formed prior to eruption of the tooth by special epithelial cells called *ameloblasts.* Once the tooth has erupted, no more enamel is formed. Enamel is composed of small crystals of calcium phosphate salts (hydroxyapatite) with adsorbed carbonate, magnesium, sodium, potassium, and other ions embedded in a fine meshwork of very strong and almost completely insoluble protein fibers that are similar to (but not identical with) the keratin of hair. The smallness of the crystalline structure of the salts makes the enamel extremely hard, much harder than the dentine. Also, the protein fiber meshwork makes enamel very resistant to acids, enzymes, and other corrosive agents because this protein is one of the most insoluble and resistant proteins known.

Cementum. Cementum is a bony substance secreted by cells of the *periodontal membrane,* which lines the tooth socket. Many collagen fibers pass directly from the bone of the jaw, through the periodontal membrane, and then into the cementum. These collagen fibers and the cementum hold the tooth in place. When the teeth are exposed to excessive strain, the layer of cementum becomes thicker and stronger. Also, it increases in thickness and strength with age, causing the teeth to become progressively more firmly seated in the jaws as one reaches adulthood and older.

Pulp. The inside of each tooth is filled with pulp, which in turn is composed of connective tissue with an abundant supply of nerves, blood vessels, and lymphatics. The cells lining the surface of the pulp cavity are the odontoblasts, which, during the formative years of the tooth, lay down the dentine but at the same time encroach more and more on the pulp cavity, making it smaller. In later life the dentine stops growing and the pulp cavity remains essentially constant in size. However, the odontoblasts are still viable and send projections into small *dentinal tubules* that penetrate all the way through the dentine; these are of importance for providing nutrition.

DENTITION

Each human being and most other mammals develop two sets of teeth during a lifetime. The first teeth are called the *deciduous teeth,* or *milk teeth,* and they number 20 in the human being. These erupt between the seventh month and second year of life, and they last until the sixth to the thirteenth year. After each deciduous tooth is lost, a permanent tooth replaces it, and an additional 8 to 12 molars appear posteriorly in the jaw, making the total number of permanent teeth 28 to 32, depending on whether the four *wisdom teeth* finally appear, which does not occur in everyone.

Formation of the Teeth. Figure 79–8 illustrates the formation and eruption of teeth. Figure 79–8A shows invagination of the oral epithelium into the *dental lamina;* this is followed by the development of a tooth-producing organ. The epithelial cells above form ameloblasts, which secrete the enamel on the outside of the tooth. The epithelial cells below invaginate upward to form a pulp cavity and also to form the odontoblasts that secrete dentine. Thus, enamel is secreted on the outside of the tooth, and dentine is secreted on the inside, forming an early tooth, as illustrated in Figure 79–8B.

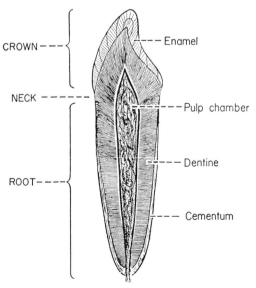

CROWN ---

NECK ------

ROOT ----

--- Enamel

---Pulp chamber

--- Dentine

--- Cementum

Figure 79–7   Functional parts of a tooth.

ERUPTION OF TEETH. During early childhood, the teeth begin to protrude upward from the jaw bone through the oral epithelium into the mouth. The cause of "eruption" is unknown, though several theories have been offered in an attempt to explain this phenomenon. One of these assumes that an increase of the material inside the pulp cavity of the tooth causes much of it to be extruded through the root canal and that this pushes the tooth upward. However, a more likely theory is that the bone surrounding the tooth progressively hypertrophies and in so doing shoves the tooth forward.

DEVELOPMENT OF THE PERMANENT TEETH. During embryonic life, a tooth-forming organ also develops in the dental lamina for each permanent tooth that will be needed after the deciduous teeth are gone. These tooth-producing organs slowly form the permanent teeth throughout the first 6 to 20 years of life. When each permanent tooth becomes fully formed, it, like the deciduous tooth, pushes upward through the bone of the jaw. In so doing it erodes the root of the deciduous tooth and eventually causes it to loosen and fall out. Soon thereafter, the permanent tooth erupts to take the place of the original one.

METABOLIC FACTORS IN DEVELOPMENT OF THE TEETH. The rate of development and the speed of eruption of teeth can be accelerated by both thyroid and growth hormones. Also, the deposition of salts in the early forming teeth is affected considerably by various factors of metabolism, such as the availability of calcium and phosphate in the diet, the amount of vitamin D present, and the rate of parathyroid hormone secretion. When all these factors are normal, the dentine and enamel will be correspondingly healthy, but, when they are deficient, the calcification of the teeth also may be defective so that the teeth will be abnormal throughout life.

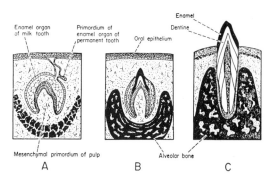

Figure 79–8 (A) Primordial tooth organ. (B) The developing tooth. (C) The erupting tooth. (Modified from Bloom and Fawcett: A Textbook of Histology, 8th Ed.)

MINERAL EXCHANGE IN TEETH

The salts of teeth, like those of bone, are composed basically of hydroxyapatite with adsorbed carbonates and various cations bound together in a hard crystalline substance. Also, new salts are constantly being deposited while old salts are being reabsorbed from the teeth, as also occurs in bone. However, experiments indicate that deposition and reabsorption occur mainly in the dentine and cementum, while very little occurs in the enamel. Much of that which does occur in the enamel occurs by exchange of minerals with the saliva instead of with the fluids of the pulp cavity. The rate of absorption and deposition of minerals in the cementum is approximately equal to that in the surrounding bone of the jaw, while the rate of deposition and absorption of minerals in the dentine is only one-third that of bone. The cementum has characteristics almost identical with those of usual bone, including the presence of osteoblasts and osteoclasts, while dentine does not

have these characteristics, as was explained above; this difference undoubtedly explains the different rates of mineral exchange.

The mechanism by which minerals are deposited and reabsorbed from the dentine is unknown. It is possible that the small processes of the odontoblasts protruding into the tubules of the dentine are capable of absorbing salts and then providing new salts to take the place of the old. It is also possible that the odontoblasts provide continuous replacement of the collagen fibers of the dentine. This would be comparable to the rejuvenation of the bone matrix by osteoblasts, which is necessary for adequate maintenance of bone strength.

In summary, rapid mineral exchange occurs in the dentine and cementum of teeth, though the mechanism of this exchange in dentine is unknown. On the other hand, enamel exhibits extremely slow mineral exchange so that it maintains much of its original mineral complement throughout life.

DENTAL ABNORMALITIES

The two most common dental abnormalities are *caries* and *malocclusion*. Caries means erosions of the teeth, while malocclusion means failure of the projections of the upper and lower teeth to interdigitate properly.

Caries. Two major but differing theories have been proposed to explain the cause of caries. One of these postulates that acids formed in crevices of the teeth by acid-producing bacteria cause erosion and absorption of the protein matrix of the enamel and dentine. This theory assumes that the acids are formed by splitting carbohydrates into lactic acid. For this reason it has been taught that eating a diet high in carbohydrate content and, particularly, eating large quantities of sweets between meals can lead to excessive development of caries.

The second theory proposes that proteolytic enzymes secreted by bacteria in the crevices of the teeth or in plaques that develop on the surfaces of unbrushed teeth digest the keratin matrix of the enamel. Then the calcium salts, unprotected by their protein fibers, are slowly dissolved by the saliva.

It is not known which of the above theories correctly describes the primary process for development of caries. Indeed, both might be operative simultaneously to cause carious teeth.

Some teeth are more resistant to caries than others. Studies in recent years indicate that teeth formed in children who drink water containing small amounts of fluorine develop enamel that is more resistant to caries than the enamel in children who drink water not containing fluorine. Fluorine does not make the enamel harder than usual, but instead it is said to inactivate proteolytic enzymes before they digest the protein matrix of the enamel. Regardless of the precise means by which fluorine protects the teeth, it is known that small amounts of fluorine deposited in enamel make teeth about twice as resistant to caries as are teeth without fluorine.

Malocclusion. Malocclusion is usually caused by a heredity abnormality that causes the teeth of one jaw to grow in an abnormal direction. In malocclusion, the teeth cannot perform their normal grinding or cutting action adequately; occasionally malocclusion results in abnormal displacement of the lower jaw in relation to the upper jaw, causing such undesirable effects as pressure on the anterior portion of the ear or pain in the mandibular joint.

The orthodontist can often correct malocclusion by applying prolonged gentle pressure against the teeth with appropriate braces. The gentle pressure causes absorption of alveolar jaw bone on the compressed side of the tooth and deposition of new bone on the tensional side of the tooth. In this way the tooth gradually moves to a new position as directed by the applied pressure.

REFERENCES

Arnaud, C. D., Jr.; Tenenhouse, A. M.; and Rasmussen, H.: Parathyroid hormone. *Ann. Rev. Physiol.,* 29:349, 1967.

Bell, N. H., and Stern, P. H.: Effects of changes in serum calcium on hypocalcemic response to thyrocalcitonin in the rat. *Amer. J. Physiol.,* 218:64, 1970.

Bronner, F.: Studies in calcium homeostasis. *Proc. Int. Union of Physiol. Sciences,* 6:271, 1968.

Bronner, F., and Aubert, J. P.: Bone metabolism and regulation of the blood calcium level in rats. *Amer. J. Physiol.,* 209:887, 1965.

Camerson, D. A.: The fine structure and function of bone cells. *In* Bittar, E. E., and Bittar, N. (eds.): The Biological Basis of Medicine. New York, Academic Press, Inc., 1969, Vol. 3, p. 391.

Copp, D. H.: Calcitonin – ultimobranchial hormone. *Proc. Int. Union of Physiol. Sciences,* 6:15, 1968.

Copp, D. H.: Endocrine regulation of calcium metabolism. *Ann. Rev. Physiol.,* 32:61, 1970.

Copp, D. H.; Cockcroft, D. W.; and Kueh, Y.: Ultimobrachial origin of calcitonin. Hypocalcemic effect of extracts from chicken glands. *Canad. J. Physiol. Pharmacol.,* 45:1095, 1967.

Danowski, T. S.: Clinical Endocrinology. Vol. III: Calcium, Phosphorus, Parathyroids and Bone. Baltimore, The Williams & Wilkins Co., 1962.

Fourman, P.: Calcium Metabolism and the Bone. 2nd ed., Philadelphia, F. A. Davis Co., 1968.

Gaillard, P.; Talmage, R. V.; and Budy, A. M.: The Parathyroid Glands. Chicago, University of Chicago Press, 1965.

Gitelman, H. J.; Kukolj, S.; and Welt, L. G.: Inhibition of parathyroid gland function of hypermagnesemia. *Amer. J. Physiol.,* 215:483, 1968.

Hac, L. R., and Freeman, S.: Effects of fluoride and vitamin D_2 on citrate in bone and serum. *Amer. J. Physiol.,* 216:179, 1969.

Hirsch, P. F., and Munson, P. L.: Thyrocalcitonin. *Physiol. Rev.,* 49:548, 1969.

Hurwitz, S.; Stacey, R. E.; and Bronner, F.: Role of vitamin D in plasma calcium regulation. *Amer. J. Physiol.,* 216:254, 1969.

Jackson, W. P. U.: Calcium Metabolism and Bone Disease. Baltimore, The Williams & Wilkins Co., 1967.

Jenkins, G. M.: The Physiology of the Mouth. 2nd ed., Oxford, Blackwell Scientific Publications, Ltd., 1960.

Krawitt, E. L., and Schedl, H. P.: In vivo calcium transport by rat small intestine. *Amer. J. Physiol.,* 214:232, 1968.

McLean, F. C. (ed.): Radioisotopes and Bone. Philadelphia, F. A. Davis Co., 1963.

Moorrees, C. F. A.: Orthodontics. *New Eng. J. Med., 279*:689, 1968.

Morii, H., and DeLuca, H. F.: Relationship between vitamin D deficiency, thyrocalcitonin, and parathyroid hormone. *Amer. J. Physiol.,* 213:358, 1967.

Posner, A. S.: Crystal chemistry of bone mineral. *Physiol. Rev.,* 49:760, 1969.

Rasmussen, H.: The parathyroid hormone. *Sci. Amer., 204*(4):56, 1961.

Richelle, L. J.: Calcification in biological systems: bone, mineral. *Proc. Int. Union of Physiol. Sciences,* 6:265, 1968.

Sammon, P. J.; Stacey, R. E.; and Bronner, F.: Role of parathyroid hormone in calcium homeostasis and metabolism. *Amer. J. Physiol.,* 218:479, 1970.

Schedl, H. P.; Osbaldiston, G. W.; and Mills, I. H.: Absorption, secretion, and precipitation of calcium in the small intestine of the dog. *Amer. J. Physiol.,* 214:814, 1968.

Shtacher, G., and Firschein, H. E.: Collagen and mineral kinetics in bone after fracture. *Amer. J. Physiol.,* 213:863, 1967.

Smith, F. G., Jr.; Tinglof, B.; Meuli, J.; and Borden, M.: Fetal response to parathyroid hormone in sheep. *J. Appl. Physiol.,* 27:276, 1969.

Snapper, I., and Bloom, M.: The parathyroid glands. *In* Duncan, G. G. (ed.): Diseases of Metabolism. 5th ed., Philadelphia, W. B. Saunders Company, 1964, p. 1280.

Urist, M. R.: Rarefying diseases of bone (with a comment on translocation of remodelled bone in osteoporosis). *In* Bittar, E. E., and Bittar, N. (eds.): The Biological Basis of Medicine. New York, Academic Press, Inc., 1969, Vol. 3, p. 425.

Vaes, G.: Biochemical aspects of bone resorption. *Proc. Int. Union of Physiol. Sciences,* 6:274, 1968.

CHAPTER 80

Reproductive Functions of the Male, and the Male Sex Hormones

The reproductive functions of the male can be divided into three major subdivisions: first, spermatogenesis, which means simply the formation of sperm; second, performance of the male sexual act; and third, regulation of male sexual functions by the various hormones. Associated with these reproductive functions are the effects of the male sex hormones on the accessory sexual organs, on cellular metabolism, on growth, and on other functions of the body.

Physiologic Anatomy of the Male Sexual Organs. Figure 80–1 illustrates the various portions of the male reproductive system. Note that the testis is

composed of a large number of *seminiferous tubules* where the sperm are formed. The sperm then empty into the *vasa recta* and from there into the *epididymis.* The epididymis leads into the *vas deferens,* which enlarges into the *ampulla of the vas deferens* immediately proximal to the prostate gland. A *seminal vesicle,* one located on each side of the prostate, empties into the prostatic end of the ampulla, and the contents from both the ampulla and the seminal vesicle pass into an *ejaculatory duct* leading through the body of the prostate gland to empty into the *internal urethra.* Finally, the *urethra* is the last connecting link from the testis to the exterior. The urethra is supplied with mucus derived from a large number of small *glands of Littré* located along its entire extent and also from large bilateral *bulbo-urethral glands* (Cowper's glands) located near the origin of the urethra.

SPERMATOGENESIS

Spermatogenesis occurs in all the seminiferous tubules during active sexual life, beginning at the age of approximately 13 as the result of stimulation by adenohypophyseal gonadotropic hormones and continuing throughout the remainder of life.

THE STEPS OF SPERMATOGENESIS

The seminiferous tubules, one of which is illustrated in Figure 80–2A, contain a large number of small to medium-sized cells called *spermatogonia* located in two to three layers along the outer border of the tubular epithelium. These continually proliferate and differentiate through definite stages of development to form sperm, as shown in Figure 80–2B.

The first stage in spermatogenesis is growth of some spermatogonia to form considerably enlarged cells called *primary spermatocytes.* Then each of the chromosomes in the primary spermatocyte is replicated, and the primary spermatocyte divides

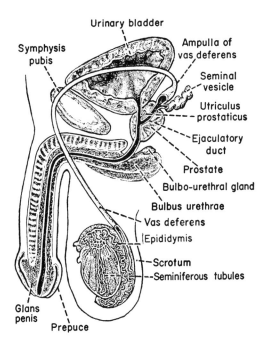

Figure 80–1 The male reproductive system. (Modified from Bloom and Fawcett: Textbook of Histology, 8th Ed.)

Urinary bladder

Symphysis pubis

Ampulla of vas deferens

Seminal vesicle

Utriculus prostaticus

Ejaculatory duct

Prostate

Bulbo-urethral gland

Bulbus urethrae

Vas deferens

Epididymis

Scrotum

Seminiferous tubules

Glans penis

Prepuce

A

B

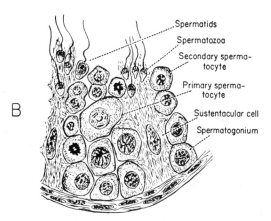

Figure 80–2 (A) Cross-section of a seminiferous tubule. (B) Spermatogenesis. (Modified from Arey: Developmental Anatomy, 7th Ed.)

head, neck, body, and *tail.* To form the head, the nuclear material is rearranged into a compact mass, and the cell membrane contracts around the nucleus. It is this nuclear material that fertilizes the ovum.

At the front of the sperm head is a small structure called the *acrosome,* which is believed to play a role in the entry of the sperm into the ovum.

The *centrioles* are aggregated in the neck of the sperm and the *mitochondria* are aggregated in the body.

Extending beyond the body is a long tail. This has the same structure as a cilium, which was described in detail in Chapter 2. The tail contains two paired fibrils down the center and nine double fibrils arranged around the border. It is covered by an extension of the cell membrane, and it contains large quantities of adenosine triphosphate, which undoubtedly energize the movement of the tail. Upon release of sperm from the male genital tract into the female tract, the tail begins to wave back and forth, providing snake-like propulsion that moves the sperm forward at a maximum velocity of about 1 foot per hour.

Function of the Sertoli Cells. The Sertoli cells of the geminal epithelium, known also as the *sustentacular cells,* are illustrated in Figure 80–2B. These cells are large, extending from the base of the seminiferous epithelium almost all the way to the interior of the tubule. The spermatids attach themselves to the Sertoli cells, and some specific relationship seems to exist between the

to form two *secondary spermatocytes,* each containing 46 chromosomes. Soon each of these cells divides by meiotic division (that is, without replication of the chromosomes, merely division of those already formed) to form two *spermatids,* each of which contains only 23 chromosomes, none of them paired. Thus, one of each of the original pairs of chromosomes is now in each spermatid. As noted in the following chapter, a similar reduction in chromosomes occurs in the ovum during its maturation. Then, when the spermatozoon combines with the ovum, the original complement of 46 chromosomes is again established.

The sex-determining pair of chromosomes in the spermatogonium is composed of one "X" chromosome, which is called the *female chromosome,* and one "Y" chromosome, the *male chromosome.* During meiotic division the sex-determining chromosomes divide among the spermatids so that half of the sperm become *male sperm* containing the "Y" chromosome and the other half *female sperm* containing the "X" chromosome. The sex of the offspring is determined by which of these two types of sperm fertilizes the ovum. This will be discussed further in Chapter 82.

When the spermatids are first formed, they still have the usual characteristics of epithelioid cells, but soon most of the cytoplasm disappears, and each spermatid begins to elongate into a spermatozoon, illustrated in Figure 80–3, composed of a

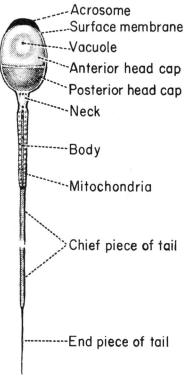

Figure 80–3 Structure of the human spermatozoon.

spermatids and the Sertoli cells that causes the spermatids to change into spermatozoa. For this reason it is believed that the Sertoli cells provide nutrient material, hormones, or enzymes that are necessary for causing appropriate changes in the spermatids.

THE SPERM

Maturation of Sperm in the Epididymis. Following formation in the seminiferous tubules, the sperm pass through the *vasa recta* into the *epididymis*. Sperm removed from the seminiferous tubules are completely nonmotile, and they cannot fertilize an ovum. However, after the sperm have been in the epididymis for some 18 hours to 10 days, they develop the power of motility and also become capable of fertilizing the ovum. It probably is not any special function of the epididymis that changes the sperm from the nonmotile state into the motile and fertile state but, instead, simply an aging process. In other words, the sperm when first released from the seminiferous tubules are still immature and must mature before they become fertile.

Storage of Sperm. A small quantity of sperm can be stored in the epididymis, but probably most sperm are stored in the vas deferens and possibly to some extent in the ampulla of the vas deferens. Though the sperm in these areas become motile if released to the exterior, they are relatively dormant as long as they are stored, probably for the following reason: The sperm, as a result of their own metabolism, secrete a considerable quantity of carbon dioxide into the surrounding fluid, and the resulting acidotic state of the fluid inhibits the activity of the sperm.

Sperm can be stored, maintaining their fertility, in the genital ducts for as long as 42 days, though it is doubtful that during normal sexual activity such prolonged storage ordinarily occurs. Indeed, with excessive sexual activity storage may not be longer than a few hours.

Physiology of the Mature Sperm. The usual motile and fertile sperm is capable of flagellated movement through fluid media at a rate of approximately 1 to 5 mm. per minute. Furthermore, sperm tend to travel in a straight line rather than with a circuitous movement. The activity of sperm is greatly enhanced in neutral and slightly alkaline media, but it is greatly depressed in mildly acid media, and strong acid media can cause rapid death of sperm. The activity of sperm increases greatly with increasing temperature, but so does the rate of metabolism, causing the life of the sperm to be considerably shortened. Though sperm can live for many weeks in the genital ducts, the life of sperm in normal ejaculated semen at normal body temperature is only 24 to 72 hours.

FUNCTION OF THE SEMINAL VESICLES

From early anatomic studies of the seminal vesicles it was erroneously believed that sperm were stored in these, whence came the name "seminal vesicles." However, these structures are only secretory glands instead of sperm storage areas.

The seminal vesicles are lined with a secretory epithelium that secretes a mucoid material containing an abundance of fructose and smaller amounts of ascorbic acid, inositol, ergothioneine, five of the amino acids, phosphorylcholine, and prostaglandin. During the process of ejaculation each seminal vesicle empties its contents into the ejaculatory duct at the same time that the vas deferens empties the sperm. This adds greatly to the bulk of the ejaculated semen, and the fructose and other substances in the seminal fluid probably are of considerable nutrient and protective value for the ejaculated sperm until one of them fertilizes the ovum. The prostaglandin perhaps promotes uterine contractions that help propel the sperm toward the fallopian tubes.

FUNCTION OF THE PROSTATE GLAND

The prostate gland secretes a thin, milky, alkaline fluid containing citric acid, calcium, acid phosphate, and fibrinolysin. During ejaculation, the capsule of the prostate gland contracts simultaneously with the contractions of the vas deferens and seminal vesicles so that the thin, milky fluid of the prostate gland adds to the bulk of the semen. The alkaline characteristic of the prostatic fluid may be quite important for successful fertilization of the ovum, because the fluid of the vas deferens is relatively acid owing to the presence of metabolic end-products of the sperm and, consequently, inhibits sperm fertility. Also, the vaginal secretions of the female are acidic (pH of 3.5 to 4.0). Sperm do not become optimally motile until the pH of the surrounding fluids rises to approximately 6 to 6.5. Consequently, it is probable that prostatic fluid neutralizes the acidity of these other fluids after ejaculation and greatly enhances the motility and fertility of the sperm.

SEMEN AND THE PHENOMENON OF CAPACITATION

Semen, which is ejaculated during the male sexual act, is composed of the fluids from the vas deferens, from the seminal vesicles, from the prostate gland, and from the mucous glands, especially the bulbo-urethral glands. The average pH of the combined semen is approximately 7.5, the alkaline prostatic fluid having neutralized the mild acidity of the other portions of the semen. The prostatic fluid gives the semen a milky appearance, while fluid from the seminal vesicles and from the mucous glands gives the semen a mucoid consistency. Within approximately one-half hour after ejaculation the mucoid consistency of semen disappears because of proteolytic enzymes in the fluid. In the early minutes after ejaculation, the sperm remain relatively immobile, possibly because of the viscosity of the mucus. However, after the mucus dissolves, the sperm simultaneously become highly motile.

For reasons not yet understood, sperm that are washed clear of the other components of semen become rapidly infertile but regain this fertility in a few minutes to an hour or more when remixed with the other components. This phenomenon is called *capacitation*.

Though sperm can live for many weeks in the male genital ducts, once they are ejaculated in the semen their maximal life span is only 24 to 72 hours at body temperature. At lowered temperatures, however, semen may be stored for several weeks; and when frozen at temperatures below $-100°$ C., sperm of some animals have been preserved for over a year.

MALE FERTILITY

The seminiferous tubular epithelium can be destroyed by a number of different diseases. For instance, bilateral orchiditis resulting from mumps usually causes sufficient degeneration of tubular epithelium that sterility results in a large percentage of males so afflicted. Another disease that frequently localizes in the testes and can cause severe tubular damage is typhus fever. Also, many male infants are born with degenerate tubular epithelium as a result of strictures in the genital ducts or as a result of unknown causes. Finally, a cause of sterility seems to be excessive temperature of the testes, as follows:

Effect of Temperature on Spermatogenesis. In addition to the direct effect of increased temperature in shortening the life of sperm as noted earlier, increasing the temperature of the testes can inhibit spermatogenesis.

It has often been stated that the reason the testicles are located in the dangling scrotum is to maintain the temperature of these glands below the temperature of the body. On cold days scrotal reflexes cause the musculature of the scrotum to contract, pulling the testicles close to the body, whereas on warm days the musculature of the scrotum becomes almost totally relaxed so that the testicles hang far from the body. Furthermore, the scrotum is well supplied with sweat glands which presumably aid in keeping the testicles cool. Thus the scrotum apparently is designed to act as a cooling mechanism for the testicles, without which spermatogenesis is said to be deficient.

CRYPTORCHIDISM. Cryptorchidism means failure of a testis to descend from the abdomen into the scrotum. During the development of the male fetus, the testes are derived from the genital ridges at the same points as the ovaries in the female. However, during the late stages of gestation, the testes descend through the inguinal canals into the scrotum. Occasionally this descent does not occur at all or occurs incompletely so that one or both testes remain in the abdomen, in the inguinal canal, or elsewhere along the route of descent.

A testicle that remains throughout life in the abdominal cavity is incapable of forming sperm. The tubular epithelium is completely degenerate, leaving only the interstitial structures of the testis. It is believed that even the few degrees higher temperature in the abdomen than in the scrotum is sufficient to cause degeneration of the tubular epithelium and consequently to cause sterility. For this reason operations to relocate the cryptorchid testes from the abdominal cavity into the scrotum prior to the beginning of adult sexual life are frequently performed on boys who have undescended testes. (As noted subsequently, testosterone secretion by the male fetus is the stimulus that causes the testes to move into the scrotum from the abdomen. Therefore, it may be that cryptorchidism is caused by abnormally formed testes that are unable to secrete enough testosterone, rather than cryptorchidism causing the degenerate testes.)

Effect of Sperm Characteristics on Fertility. Sperm from the fertile male are extremely uniform in size, and when such uniformity does not exist, even though the sperm may be motile and appear to be healthy, either total or "relative" sterility usually exists. Also, as illustrated in Figure 80–4, sperm frequently have grossly abnormal characteristics, such as two heads, two tails, or abnormally formed heads, and some sperm travel in circuitous routes rather than along straight lines. In general, when more than 25 per cent of the sperm are shown to be definitely abnormal even though the total number of sperm in each ejaculate is completely normal, the man nevertheless is likely to be sterile, for even the 75 per cent of the sperm that appear to be normal still may not be normal enough to fertilize the ovum.

Effect of Sperm Count on Fertility. The usual

quantity of semen ejaculated at each coitus averages approximately 3 ml., and in each milliliter of semen is an average of approximately 120,000,000 sperm, though even in "normal" persons this can vary from 35,000,000 to 200,000,000. This means an average total of 400,000,000 sperm are usually present in each ejaculate. When the number of sperm in each milliliter falls below approximately 35,000,000 sperm, the person is likely to be almost completely infertile. Thus, even though only a single sperm is necessary to fertilize the ovum, for reasons not yet completely understood, the ejaculate must contain a tremendous number of sperm for at least one to fertilize the ovum.

FUNCTION OF HYALURONIDASE SECRETED BY THE SPERM IN FERTILIZATION. Hyaluronidase is an enzyme that depolymerizes hyaluronic acid polymers, which are present in large quantity in the intercellular cementing substance, which means that hyaluronidase can cause cells cemented together to separate from each other. When the ovum is expelled from the follicle of the ovary into the abdominal cavity, it carries with it several layers of granulosa cells attached firmly to the surface of the ovum. Before a sperm can reach the ovum to fertilize it, these granulosa cells must be removed; hyaluronidase secreted into the semen by the sperm or hyaluronidase in the acrosomes of the sperm heads is postulated to cause the granulosa cells to break away from the ovum, allowing sperm to reach the surface of the ovum. It has been further postulated that when sperm are insufficient in number (below 35,000,000 per ml.), insufficient hyaluronidase is secreted for the granulosa cells to be removed, thus causing sterility.

It has been claimed, also, that semen contains a mucolytic enzyme, similar to hyaluronidase, that is capable of dissolving the mucus plug that frequently forms in the cervix of the uterus, and it is believed that lack of this mucolytic enzyme occasionally is responsible for male sterility.

THE MALE SEXUAL ACT

NEURONAL STIMULUS FOR PERFORMANCE OF THE MALE SEXUAL ACT

The most important source of impulses for initiating the male sexual act is the glans penis, for the glans contains a highly organized sensory end-organ system that transmits into the central nervous system a special modality of sensation that might be called *sexual sensation*. The massaging action of intercourse on the glans stimulates the sensory end-organs, and the sexual sensations in turn pass through the pudendal nerve, thence through the sacral plexus into the sacral portion of the spinal cord, and finally up the cord to undefined areas of the cerebrum. Impulses may also

Figure 80–4 Abnormal sperm.

enter the spinal cord from areas adjacent to the penis to aid in stimulating the sexual act. For instance, stimulation of the anal epithelium, the scrotum, and perineal structures in general may all send impulses into the cord which add to the sexual sensation. Sexual sensations can even originate in internal structures, such as irritated areas of the urethra, the bladder, the prostate, the seminal vesicles, the testes, and the vas deferens. Indeed, one of the causes of "sexual drive" is probably overfilling of the sexual organs with secretions. Infection and inflammation of these sexual organs sometimes cause almost continual sexual desire, and "aphrodisiac" drugs, such as cantharides, increase the sexual desire by irritating the bladder and urethral mucosa.

The Psychic Element of Male Sexual Stimulation. Appropriate psychic stimuli can greatly enhance the ability of a person to perform the sexual act. Simply thinking sexual thoughts or even dreaming that the act of intercourse is being performed can cause the male sexual act to occur and to culminate in ejaculation. Indeed, *nocturnal emissions* during dreams occur in many males during some stages of sexual life, especially during the teens.

Integration of the Male Sexual Act in the Spinal Cord. Though psychic factors usually play an important part in the male sexual act and can actually initiate it, the cerebrum is probably not absolutely necessary for its performance, for appropriate genital stimulation can cause ejaculation in some animals and in an occasional human being after their spinal cords have been cut above the lumbar region. Therefore, the male sexual act results from inherent reflex mechanisms integrated in the sacral and lumbar spinal cord, and these mechanisms can be initiated by either psychic stimulation or actual sexual stimulation.

STAGES OF THE MALE SEXUAL ACT

Erection.   Erection is the first effect of male sexual stimulation, and the degree of erection is proportional to the degree of stimulation, whether this be psychic or physical.

Erection is caused by parasympathetic impulses that pass from the sacral portion of the spinal cord through the nervi erigentes to the penis. These parasympathetic impulses dilate the arteries of the penis, thus allowing arterial blood to flow under high pressure into the *erectile tissue* of the penis, as illustrated in Figure 80–5. This erectile tissue is nothing more than large, cavernous venous sinusoids, which are normally relatively empty but which become dilated tremendously when arterial blood flows into them under pressure, since the venous outflow is partially occluded. Also, the erectile bodies are surrounded by strong fibrous coats; therefore, high pressure within the sinusoids causes ballooning of the erectile tissue to such an extent that the penis becomes hard and elongated.

Lubrication.   During sexual stimulation, parasympathetic impulses, in addition to promoting erection, cause the glands of Littré and the bulbo-urethral glands to secrete mucus. Thus mucus flows through the urethra during intercourse to aid in the lubrication of coitus. However, most of the lubrication of coitus is provided by the female sexual organs rather than by the male. Without satisfactory lubrication, the male sexual act is rarely successful because unlubricated intercourse causes pain impulses which inhibit rather than excite sexual sensations.

Ejaculation.   Ejaculation is the culmination of the male sexual act. When the sexual stimulus becomes extremely intense, the reflex centers of the spinal cord begin to emit rhythmic sympathetic impulses that leave the cord at L-1 and L-2 and pass to the genital organs through the hypogastric plexus to initiate emission, which is the forerunner of ejaculation.

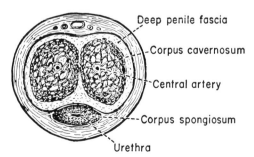

Figure 80–5   Erectile tissue of the penis.

Emission begins with peristaltic contractions in the ducts of the testis, the epididymus, and the vas deferens to cause expulsion of sperm into the internal urethra. Simultaneously, rhythmic contractions in the seminal vesicles and the muscular coat of the prostate gland expel seminal fluid and prostatic fluid along with the sperm. These mix with the mucus already secreted by the bulbo-urethral glands, and all of these different types of fluid combine to form the semen. The process to this point is emission.

Then, rhythmic nerve impulses are sent from the cord through S-1 and S-2 and thence through the pudendal nerves to skeletal muscles that encase the base of the erectile tissue, causing rhythmic increases in pressure in this tissue, which expresses the semen from the urethra to the exterior. This process is called *ejaculation* proper.

TESTOSTERONE AND OTHER MALE SEX HORMONES

SECRETION, METABOLISM, AND CHEMISTRY OF THE MALE SEX HORMONE

Secretion of Testosterone by the Interstitial Cells of the Testes.   Two different steroid male sex hormones have been isolated from the venous blood draining from the testes, *testosterone* and *Δ4-androstene-3,17,-dione.* However, the quantity of testosterone is so much greater than that of the second hormone that one can consider testosterone to be the single significant hormone responsible for the male hormonal effects caused by the testes.

Both hormones are formed by the *interstitial cells of Leydig,* which lie in the interstices between the seminiferous tubules, as illustrated in Figure 80–6. Interstitial cells in the testes are not numerous in a child, but they *are* numerous in a newborn male infant and also in the adult male anytime after puberty; at both these times the testes secrete large quantities of testosterone. Furthermore, when tumors develop from the interstitial cells of Leydig, great quantities of testosterone are secreted. Finally, when the germinal epithelium of the testes is destroyed by x-ray treatment or by excessive heat, the interstitial cells, which are less easily destroyed, continue to produce testosterone.

Secretion of "Androgens" Elsewhere in the Body.   The term "androgen" is used synonymously with the term male sex hormone, but it also includes male sex hormones produced elsewhere in the body besides the testes. For instance, the adrenal gland secretes at least five different androgens, though the total masculinizing activity of all these is normally so slight that they do not cause significant masculine characteristics even in women. But when an adrenal tumor of the androgen-producing cells occurs, the quantity of androgenic hormones may become great enough to cause all the usual male secondary sexual characteristics. These effects were described in connection with the adrenogenital syndrome in Chapter 76.

Rarely, embryonic rest cells in the ovary can develop into a tumor which produces androgens; one such tumor is the *arrhenoblastoma*. The normal ovary probably also produces minute quantities of androgens, but these are not significant.

Chemistry of the Androgens. All androgens are steroid compounds, as illustrated by the formula in Figure 80–7 for *testosterone*. Both in the testes and in the adrenals, the androgens can be synthesized either from cholesterol or directly from acetyl coenzyme A.

Metabolism of Testosterone. After secretion by the testes, testosterone circulates in the blood for not over 15 to 30 minutes before it either becomes fixed to the tissues or is degraded into inactive products that are subsequently excreted.

That portion of testosterone that becomes fixed to the tissues is converted within the cells to *dihydrotestosterone;* it is in this form that testosterone performs its intracellular functions.

DEGRADATION AND EXCRETION OF TESTOSTERONE. The testosterone that does not become fixed to the tissues is rapidly converted, mainly by the liver, into *androsterone* and *dehydroepiandrosterone* and simultaneously conjugated either as glucuronides or sulfates (glucuronides, particularly). These are excreted either into the gut in the bile or into the urine.

The 17-Ketosteroids in Relation to Testosterone Secretion. Both androsterone and dehydroepiandrosterone are 17-ketosteroids, which means simply that they have a keto oxygen in the 17 position of the chemical formulas. Furthermore, the metabolic end-products of the adrenal androgens are also excreted in the urine as slightly different types of 17-ketosteroids. On the other hand, essentially none of the other steroid hormones besides androgens is excreted as 17-ketosteroid. Therefore, the rate of excretion of 17-ketosteroids in the urine is an index of the rate of androgen production in the body and is often used clinically for diagnosis of androgen-producing tumors of the testes or adrenals.

Production of Estrogen by the Testes. In addition to testosterone, small amounts of estrogens are formed in the male (about one-fifth the amount in the nonpregnant female), and a reasonable quantity of these can be recovered from a man's urine. The functions of estrogens in the male are unknown.

The exact source of the estrogens in the male is

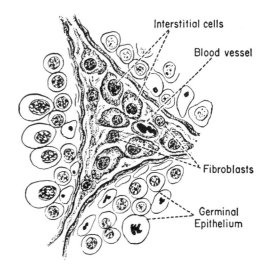

Figure 80–6 Interstitial cells located in the interstices between the seminiferous tubules. (Modified from Bloom and Fawcett: Textbook of Histology, 8th Ed.)

also still doubtful, but the following are known: (a) The quantity of estrogens decreases greatly when the germinal epithelium of the seminiferous tubules is destroyed. This indicates that the germinal cells synthesize estrogens in men. (b) Small amounts of estrogens are formed from testosterone during its degradation in other parts of the body. (c) The rate of estrogen secretion from the testes closely parallels the rate of testosterone secretion, for which reason it has also been suspected that the interstitial cells are the source (or a source) of the estrogens.

Thus, the problem of estrogen production in the male is unsettled except for the fact that small quantities of estrogens are produced in the testes or are formed from testosterone.

FUNCTIONS OF TESTOSTERONE

In general, testosterone is responsible for the distinguishing characteristics of the masculine body. The testes are stimulated by chorionic gonadotropin from the placenta to produce a small quantity of testosterone dur-

Testosterone

Figure 80–7 Testosterone.

ing fetal development, but essentially no testosterone is produced during childhood until approximately the age of 11 to 13. Then testosterone production increases rapidly at the onset of puberty and lasts throughout most of the remainder of life as illustrated in Figure 80–8, dwindling rapidly beyond the age of 40 to become perhaps one-fifth the peak value by the age of 80.

Functions of Testosterone During Fetal Development. Testosterone begins to be elaborated by the male at about the second month of embryonic life. Indeed, some embryologists believe that the major functional difference between the female and the male sex chromosomes is that the male chromosome causes the newly developing genital ridge to secrete testosterone while the female chromosome causes this ridge to secrete estrogens. Injection of large quantities of male sex hormone into gravid animals of certain species causes development of male sexual organs even though the fetus is female. Also, removal of the fetal testes in a male fetus causes development of female sexual organs. Therefore, probably testosterone secreted by the genital ridges and the subsequently developing testes is responsible for the development of the male sex characteristics, including the growth of a penis and a scrotum rather than the formation of a clitoris and a vagina. Also, it is almost certainly this hormone produced during fetal life that causes development of the prostate gland, the seminal vesicles, and the genital ducts.

Effect on the Descent of the Testes. The testes usually descend into the scrotum

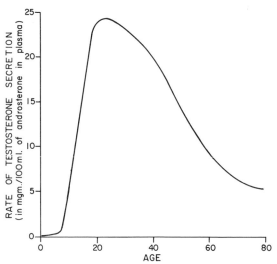

Figure 80–8 Rate of testosterone secretion at different ages, as judged from the concentrations of androsterone in the plasma.

during the last two months of gestation when the testes are secreting reasonable quantities of testosterone. If a male child is born with undescended testes, administration of testosterone causes the testes to descend in the usual manner if the inguinal canals are large enough to allow the testes to pass. Or, administration of gonadotropic hormones, which stimulate the interstitial cells of the testes to produce testosterone, also causes the testes to descend. Thus, the stimulus for descent of the testes is almost certainly testosterone, indicating again that testosterone is probably an important hormone for male sexual development during fetal life.

Effect of Testosterone on Development of Adult Primary and Secondary Sexual Characteristics. Testosterone secretion after puberty causes the penis, the scrotum, and the testes all to enlarge many-fold until about the age of 20. In addition, testosterone causes the "secondary sexual characteristics" of the male to develop at the same time, beginning at puberty and ending at maturity. These secondary sexual characteristics, in addition to the sexual organs themselves, distinguish the male from the female as follows:

Effect on the Distribution of Body Hair. Testosterone causes growth of hair (1) over the pubis, (2) upward along the linea alba sometimes to the umbilicus and above, (3) on the face, (4) usually on the chest and (5) less often on other regions of the body, such as the back. It also causes the hair on most other portions of the body to become more prolific.

Baldness. Testosterone decreases the growth of hair on the top of the head; a man who does not have functional testes does not become bald. However, many virile men never become bald, for baldness is a result of two factors: first, a *genetic background* for the development of baldness and, second, superimposed on this genetic background, *large quantities of androgenic hormones*. A woman who has the appropriate genetic background and who develops a long-sustained androgenic tumor becomes bald in the same manner as a man.

Effect on the Voice. Testosterone secreted by the testes or injected into the body causes hypertrophy of the laryngeal mucosa and enlargement of the larynx. These effects cause at first a relatively hoarse voice, but this gradually changes into the typical masculine bass voice.

Effect on the Skin. Testosterone increases the thickness of the skin over the entire body and increases the ruggedness of the subcutaneous tissues. Also, it causes increased quantities of melanin to be deposited in the skin, thereby deepening the hue of the skin.

Testosterone increases the rate of secretion by some or perhaps all the sebaceous glands. Especially important is the excessive secretion by the sebaceous glands of the face, for oversecretion of these can result in *acne*. Therefore, acne is one of the most common features of adolescence, when the male body is first becoming introduced to increased secretion of testosterone. After several years of testosterone secretion, the skin adapts itself to the testosterone in some way that allows it to overcome the acne.

EFFECT ON NITROGEN RETENTION AND MUSCULAR DEVELOPMENT. One of the most important male characteristics is the development of increasing musculature following puberty. This is associated with increased protein in other parts of the body as well. Many of the changes in the skin are probably also due to deposition of proteins in the skin, and the changes in the voice could even result from this protein anabolic function of testosterone.

Testosterone has often been considered to be a "youth hormone" because of its effect on the musculature, and it is occasionally used for treatment of persons who have poorly developed muscles.

EFFECT ON BONE GROWTH AND CALCIUM RETENTION. Following puberty or following prolonged injection of testosterone, the bones grow considerably in thickness and also deposit considerable calcium salts. Thus, testosterone increases the total quantity of bone matrix, and it also causes calcium retention. The increase in bone matrix is believed to result from the general protein anabolic function of testosterone, and the deposition of calcium salts to result from increased bone matrix available to be calcified.

Because of the ability of testosterone to increase the size and strength of bones, testosterone is often used in old age to treat osteoporosis.

When great quantities of testosterone (or any other androgen) are secreted, not only does the rate of bone growth increase, but the epiphyses of the long bones unite with the shafts of the bones at an early age in life. Therefore, despite the rapidity of growth, this early uniting of the epiphyses prevents the person from growing as tall as he would have grown had testosterone not been secreted at all. Even in normal men the final adult height is slightly less than that which would have been attained had the person been castrated prior to puberty.

EFFECT ON BASAL METABOLISM. Injection of large quantities of testosterone can increase the basal metabolic rate by as much as 15 per cent, and it is believed that even the usual quantity of testosterone secreted by the testes during active sexual life increases the rate of metabolism some 5 to 10 per cent above the value that it would be were the testes not active. This increased rate of metabolism is possibly an indirect result of the effect of testosterone on protein anabolism, the increased quantity of proteins — the enzymes especially — increasing the activities of all cells.

EFFECT ON THE RED BLOOD CELLS. When normal quantities of testosterone are injected into a castrated adult, the number of red blood cells per cubic millimeter of blood increases approximately 20 per cent. Also, the average man has 500,000 to 1,000,000 more red blood cells per cubic millimeter than the average woman. However, this difference may be due partly to the increased metabolic rate following testosterone administration rather than to a direct effect of testosterone on red blood cell production.

EFFECT ON ELECTROLYTE AND WATER BALANCE. As pointed out in Chapter 76, many different steroid hormones can increase the reabsorption of sodium in the distal tubules of the kidneys. Testosterone performs this function to a slight extent but only to a minor degree in comparison with the adrenal mineralocorticoids. Nevertheless, following puberty the blood and extracellular fluid volumes of the male in relation to his weight increase to a slight extent; this effect probably results at least partly from the sodium-retaining ability of testosterone.

BASIC MECHANISM OF ACTION OF TESTOSTERONE

Though it is not known exactly how testosterone causes all the effects just listed, it is believed that they result mainly from increased rate of protein formation in cells. This has been studied extensively in the prostate gland, one of the organs that is most affected by testosterone. In this gland, testosterone enters the cells within a few minutes after secretion and binds therein with cytoplasmic and nuclear proteins. Within 30 minutes the concentration of RNA begins to increase in the cells, and this is followed by progressive increase in cellular protein. After several days the quantity of DNA in the gland has also increased, and there has been a simultaneous increase in the number of prostatic cells.

Therefore, it is assumed that testosterone greatly stimulates production of proteins in general, though probably increasing more specifically those proteins related to development of secondary sexual characteristics. The obvious effect in the cell—a single type of action that would cause all these effects—would be to activate DNA molecules in the nucleus, but this is still speculation.

CONTROL OF MALE SEXUAL FUNCTIONS BY THE GONADOTROPIC HORMONES—FSH, LH, AND LTH

The anterior pituitary gland secretes three different gonadotropic hormones: (1) *follicle-stimulating hormone* (FSH); (2) *luteinizing hormone* (LH), also called *interstitial cell-stimulating hormone* (ICSH); and (3) *luteotropic hormone* (LTH), also called *lactogenic hormone,* or *prolactin.* FSH and LH play major roles in the control of male sexual functions. LTH is not known to play a significant role in male sexual function but does play significant roles in the normal monthly ovarian cycle of the female and in the secretion of milk.

Regulation of Testosterone Production by LH. Testosterone is produced by the interstitial cells of Leydig only when the testes are stimulated by LH from the pituitary gland, and the quantity of testosterone secreted varies approximately in proportion to the amount of LH available.

Injection of purified LH into a child causes fibroblasts in the interstitial areas of the testes to develop into interstitial cells of Leydig, though mature Leydig cells are not normally found in the child's testes until after the age of approximately 10. Also, simultaneous administration of a small amount of FSH along with LH greatly potentiates the effect of LH in promoting testosterone production.

EFFECT OF CHORIONIC GONADOTROPIN ON THE FETAL TESTES. During gestation the placenta secretes large quantities of chorionic gonadotropin, a hormone that has almost the same properties as LH. This hormone stimulates the formation of interstitial cells in the testes of the fetus and causes testosterone secretion. As pointed out earlier in the chapter, the secretion of testosterone during fetal life is important for promoting formation of male sexual organs.

Regulation of Spermatogenesis by Follicle-Stimulating Hormone (FSH). The conversion of primary spermatocytes into secondary spermatocytes in the seminiferous tubules is stimulated by FSH from the anterior pituitary gland; and in the absence of FSH, spermatogenesis will not proceed. However, FSH cannot by itself cause complete formation of spermatozoa. For spermatogenesis to proceed to completion, testosterone must be secreted simultaneously in small amounts by the interstitial cells. Thus, FSH seems to initiate the proliferative process of spermatogenesis, and testosterone apparently is necessary for final maturation of the spermatozoa. Because testosterone is secreted by the interstitial cells under the influence of LH, both FSH and LH must be secreted by the anterior pituitary gland if spermatogenesis is to occur.

Regulation of Gonadotropin Secretion by the Hypothalamus. The gonadotropins, like corticotropin and thyrotropin, are secreted by the anterior pituitary gland probably mainly in response to nervous activity in the hypothalamus. For instance, in the female rabbit, coitus with a male rabbit elicits nervous activity in the hypothalamus that in turn stimulates the adenohypophysis to secrete FSH and LH. These hormones then cause rapid ripening of follicles in the rabbit's ovaries, followed a few hours later by ovulation.

Many other types of nervous stimuli are also known to affect gonadotropin secretion. For instance, in sheep, goats, and deer, nervous stimuli in response to changes in weather and amount of light in the day increase the quantities of gonadotropins during one season of the year, the mating season, thus allowing birth of the young during an appropriate period for survival. Also, psychic stimuli can affect fertility of the male animal, as exemplified by the fact that transporting a bull under uncomfortable conditions can often cause almost complete temporary sterility. In the human being, it is known too that various psychogenic stimuli feeding into the hypothalamus can cause marked excitatory or inhibitory effects on gonadotropin secretion, in this way sometimes greatly altering the degree of fertility.

The hypothalamus controls gonadotropin secretion by way of the hypothalamic-hypophyseal portal system. Neurosecretory hormones called *gonadotropin-releasing factors* are carried in the hypophyseal portal blood to the adenohypophysis to control the secretion of LH and FSH but probably not LTH (see the discussion of this in the following chapter).

These gonadotropin-releasing factors are analogous to growth hormone-releasing factor, corticotropin-releasing factor, and thyrotropin-releasing factor, all of which have been noted in previous chapters to increase the secretion of respective adenohypophyseal hormones.

Yet the specific areas of the hypothalamus responsible for secretion of the two gonadotropin-releasing factors are not well established. Two particular areas seem to be most important: (1) the lowermost region of the hypothalamus in the vicinity of the *median eminence* and (2) an *area in the anterior hypothalamus located in the suprachiasmal region.* In addition, regions in the posterior hypothalamus have also been said to initiate secretion of gonadotropin-releasing factors.

Reciprocal Inhibition of the Hypothalamic-Anterior Pituitary Secretion of Gonadotropic Hormones by the Testicular Hormones. Injection of testosterone into either a male or a female animal inhibits the secretion of gonadotropins. This inhibition is dependent on normal function of the hypothalamus. Therefore, it is assumed that testosterone inhibits hypothalamic stimulation of the anterior pituitary gland to produce the gonadotropins. This inhibitory effect is much more marked on the production of LH than of FSH. Therefore, one can readily see that this inhibitory effect of testosterone provides a feedback control system for maintaining testosterone secretion at a constant level; that is, excess testosterone secretion inhibits LH secretion, which in turn reduces the testosterone secretion back to normal level. Conversely, the mechanism also operates in the opposite direction to protect against too little production of testosterone.

It is known, too, that spermatogenesis in the testes in some unexplained way inhibits the secretion of FSH. It is believed that some additional testicular hormone besides testosterone is responsible for this inhibition, and the hormone has frequently been called *inhibin,* or simply *"X" hormone.* It is also possible that this inhibition results from estrogen secreted by the active seminiferous tubules, which was discussed earlier in the chapter. Obviously, this feedback from the seminiferous tubules could control the rate of spermatogenesis. For instance, it might increase the rate of spermatogenesis when the male ejaculates often.

Puberty and Regulation of Its Onset. During the first 10 years of life, the male child secretes almost no gonadotropins and, consequently, almost no testosterone. Then, at the age of about 10, the anterior pituitary gland begins to secrete progressively increasing quantities of gonadotropins; and this is followed by a corresponding increase in testicular function. By approximately the age of 13, the male child reaches full adult sexual capability. This period of change is called *puberty.*

Initiation of the onset of puberty has long been a mystery. In earliest history of mankind, the belief was simply that the testicles "ripened" at this time, but with the discovery of the gonadotropins the belief shifted to ripening of the anterior pituitary gland. Now it is known from experiments in which both testicular and pituitary tissues have been transplanted from infant animals into adult animals that both the testes and the anterior pituitary of the infant are capable of performing adult functions if appropriately stimulated. Therefore, it is almost certain that *during childhood the hypothalamus, not the two glands, is at fault; that is, the hypothalamus simply does not secrete gonadotropin-releasing factors.* Also, it is now known that even the minutest amount of testosterone inhibits the production of gonadotropin-releasing factors by the childhood hypothalamus. This leads to the theory that the childhood hypothalamus is so extremely sensitive to inhibition that the minutest amount of testosterone secreted by the testicles inhibits the entire system. For reasons yet unknown, the hypothalamus loses this inhibitory sensitivity at the time of puberty, which allows the secretory mechanisms to develop full activity. Thus, puberty is now believed to result from an aging process of the hypothalamic sexual control centers.

The Male Adult Sexual Life and the Male Climacteric. Following puberty, gonadotropic hormones are produced by the male pituitary gland for the remainder of life, and at least some spermatogenesis usually continues until death. Most men, however, begin to exhibit slowly decreasing sexual functions in the late 40's or 50's, but it is not the same rapid decrease as that which occurs in women at the menopause. This is related to decrease in testosterone secretion as depicted in Figure 80-8. The decrease in male sexual function is called the male climacteric. Occasionally, the male climacteric is associated with symptoms of hot flashes, suffocation, and psychic disorders similar to the menopausal symptoms of the female. These symptoms can be abrogated by administration of testosterone, synthetic methyl testosterone, or even estrogens that are used for treatment of menopausal symptoms in the female.

The male climacteric is associated with greatly increased secretion of FSH by the anterior pituitary gland because testosterone is no longer available to inhibit its production.

ABNORMALITIES OF MALE SEXUAL FUNCTION

THE PROSTATE GLAND AND ITS ABNORMALITIES

The prostate gland remains relatively small throughout childhood and begins to grow at puberty. This gland reaches an almost stationary size by the age of about 20 and remains this size up to the age of approximately 40 to 50. At that time in some men it begins to degenerate along with the decreased production of testosterone by the testes. However, a benign prostatic fibroadenoma frequently develops in the prostate in older men and causes urinary obstruction. This effect is not caused by testosterone.

Cancer of the prostate gland is an extremely common cause of death, resulting in approximately 2 to 3 per cent of all male deaths.

Once cancer of the prostate gland does occur, the cancerous cells are usually stimulated to more rapid growth by testosterone and are inhibited by removal of the testes so that testosterone cannot be formed. Also, prostatic cancer can usually be inhibited by administration of estrogens. Some patients who have prostatic cancer that has already metastasized to almost all the bones of the body can be successfully treated for a few months to years by estrogen therapy, by removal of the testes, or by both; following this therapy the metastases degenerate and the bones heal. This treatment does not completely stop the cancer but does slow it down and greatly diminishes the severe bone pain. In general, the metastases return within a few months to a year or so.

HYPOGONADISM IN THE MALE

Hypogonadism is caused by any of several abnormalities: First, the person may be born without functional testes. Second, he may have undeveloped testes, owing to failure of the adenohypophysis to secrete gonadotropic hormones. Third, cryptorchidism (undescended testes) may occur, resulting in partial or total degeneration of the testes. Fourth, he may lose his testes, which is called *castration*.

When a boy loses his testes prior to puberty, a state of *eunuchism* ensues in which he continues to have neuter sexual characteristics throughout life. The height of the adult eunuch is slightly greater than that of the normal man, though the bones are quite thin, the muscles are considerably weaker than those of normal man, and, obviously, the sexual organs and secondary sexual characteristics are those of a child rather than those of an adult. The voice is child-like, there is no loss of hair on the head, and the normal masculine hair distribution on the face and elsewhere does not occur.

When a man is castrated following puberty, some male secondary sexual characteristics revert to those of a child, and others remain of masculine character. The sexual organs regress slightly in size but not to a child-like state, and the voice regresses from the bass quality only slightly. On the other hand, there is loss of masculine hair production, loss of the thick masculine bones, and loss of the musculature of the virile male.

In the castrated adult male, sexual desires are decreased but not totally lost, provided that sexual activities have been practiced previously. Indeed erection can still occur as before though with less ease, and ejaculation can still take place. Therefore, the male sexual act is not dependent on production of male sex hormone, though male sex hormone does enhance the psychic factors that promote successful culmination of the act.

Adiposogenital Syndrome (Fröhlich's Syndrome). Damage to certain areas of the hypothalamus greatly decreases the secretion of gonadotropic hormones by the adenohypophysis. If this occurs prior to puberty, it causes typical eunuchism. The damage often causes simultaneous overeating because of its effect on the feeding center of the hypothalamus. Consequently, the person develops severe obesity along with the eunuchism. This condition is illustrated in Figure 80–9 and is called *adiposogenital syndrome, Fröhlich's syndrome,* or *hypothalamic eunuchism.*

Figure 80–9 Adiposogenital syndrome in an adolescent male. Note the obesity and the childlike sexual organs. (Courtesy of Dr. Leonard Posey.)

TESTICULAR TUMORS AND HYPERGONADISM IN THE MALE

Interstitial cell tumors develop rarely in the testes, but when they do develop they sometimes produce as much as 100 times normal quantities of testosterone. When such tumors develop in young children, they cause rapid growth of the musculature and bones but also early uniting of the epiphyses so that the eventual adult height actually is less than that which would have been achieved otherwise. Obviously, such interstitial cell tumors cause excessive development of the sexual organs and of the secondary sexual characteristics. In the adult male, small interstitial cell tumors are difficult to diagnose because masculine features are already present. Diagnosis can be made, however, from urine tests that show greatly increased excretion of testosterone end-products such as the 17-ketosteroids.

Much more common than the interstitial cell tumors are tumors of the germinal epithelium. Because germinal cells are capable of differentiating into almost any type of cell, many of these tumors contain multiple tissues, such as placental tissue, hair, teeth, bone, skin, and so forth, all found together in the same tumorous mass called a *teratoma.* Often these tumors secrete no hormones, but if a significant quantity of placental tissue develops in the tumor, it may secrete large quantities of chorionic gonadotropin that has functions very similar to those of LH, which will be discussed in more detail in Chapter 82. Also, estrogenic hormones are frequently secreted by these tumors and cause the condition called *gynecomastia,* which means overgrowth of the breasts.

REFERENCES

Amelar, R. D.: Infertility in Men. Philadelphia, F. A. Davis Co., 1966.

Armstrong, D. T.: Reproduction. *Ann. Rev. Physiol.,* 32:439, 1970.

Bacci, G.: Sex Determination. New York, Pergamon Press, Inc., 1966.

Beech, F. A.: Cerebral and hormonal control of reflexive mechanisms involved in copulatory behavior. *Physiol. Rev.,* 47:289, 1967.

Bishop, D. W.: Sperm motility. *Physiol. Rev.,* 42:1, 1962.

Carey, H. M.: Modern Trends in Human Reproductive Physiology. London, Butterworth & Co., 1963.

Ciba Foundation Study Group No. 22: Gonadotrophins. Boston, Little, Brown and Company, 1965.

Donovan, B. T., and van der Werff ten Bosch, J. J.: Physiology of Puberty. Baltimore, The Williams & Wilkins Co., 1966.

Dukelow, W. R., and Chernoff, H. N.: Primate sperm survival and capacitation in a foreign uterine environment. *Amer. J. Physiol.,* 216:682, 1969.

Everett, J. W.: Central neural control of reproductive functions of the adenohypophysis. *Physiol. Rev.,* 44:373, 1964.

Everett, J. W.: Neuroendocrine aspects of mammalian reproduction. *Ann. Rev. Physiol.,* 31:383, 1969.

Federman, D. D.: Abnormal Sexual Development. Philadelphia, W. B. Saunders Company, 1967.

Kristen, B. E.: Effects of gonadotrophins on secretion of steroids by the testis and ovary. *Physiol. Rev.,* 44:609, 1964.

Lipsett, M. B.: Androgen secretion. *Proc. Int. Union of Physiol. Sciences,* 6:245, 1968.

Lloyd, C. W.: Human Reproduction and Sexual Behavior. Philadelphia, Lea & Febiger, 1964.

Mann, T.: The Biochemistry of Semen and of the Male Reproductive Tract. New York, John Wiley & Sons, Inc., 1964.

Pincus, G.: Reproduction. *Ann. Rev. Physiol.,* 24:57, 1962.

Segal, S. J.: Research in fertility regulation. *New Eng. J. Med.,* 279:364, 1968.

Villee, C. A., and Villee, D. B.: Reproduction. *Ann. Rev. Physiol.,* 25:263, 1963.

Zimmerman, E., and Critchlow, V.: Negative feedback and pituitary-adrenal function in female rats. *Amer. J. Physiol.,* 216:148, 1969.

Zogg, C. A.; Hays, R. L.; VanDemark, N. L.; and Johnson, A. D.: Effect of duration of experimental cryptorchidism on testis composition and metabolic activity. *Amer. J. Physiol.,* 215:985, 1968.

CHAPTER 81

Sexual Functions in the Female, and the Female Sex Hormones

The sexual and reproductive functions in the female can be divided into two major phases: first, preparation of the body for conception and gestation, and second, the period of gestation itself. The present chapter is concerned with the preparation of the body for gestation, and the following chapter presents the physiology of pregnancy.

PHYSIOLOGIC ANATOMY OF THE FEMALE SEXUAL ORGANS

Figure 81–1 illustrates the principal organs of the human female reproductive tract, including the *ovaries*, the *fallopian tubes*, the *uterus*, and the *vagina*. Reproduction begins with the development of ova in the ovaries. A single ovum is expelled from an ovarian follicle into the abdominal cavity in the middle of each monthly sexual cycle. This ovum then passes through one of the fallopian tubes into the uterus, and, if it has been fertilized by a sperm, it implants in the uterus where it develops into a fetus, a placenta, and fetal membranes.

The ovaries are composed of a *hilar connective tissue portion*, through which the blood vessels enter, and a *cortical portion*, which is concerned mainly with development of ova and secretion of ovarian hormones. The outer surface of the ovary is covered by a *germinal epithelium*, which embryologically is derived directly from the epithelium of the germinal ridges. During fetal life and to a lesser extent during childhood and early sexual life, *primordial ova* differentiate from the germinal epithelium and migrate into the substance of the ovarian cortex, carrying with them a layer of epithelioid granulosa cells. The ovum surrounded by a single layer of epithelioid granulosa cells is called a *primordial follicle*. At birth, approximately 500,000 primordial follicles are present in the two ovaries, and an undetermined number are formed after birth. However, during the entire sexual life of the female, only about 375 of these follicles develop enough to expel their ova, while the remainder degenerate (become *atretic*). At the end of sexual life, at the *menopause*, only a few primordial follicles remain in the ovaries.

THE MONTHLY OVARIAN CYCLE AND FUNCTION OF THE GONADO-TROPIC HORMONES

The normal sexual life of the female is characterized by monthly rhythmic changes in the rates of secretion of the sex hormones and corresponding changes in the sexual organs themselves. This rhythmic pattern is called the *female sexual cycle*. The duration of the cycle averages 28 days. It may be as short as 20 days or as long as 45 days even in completely normal women, though abnormal cycle length is often associated with decreased fertility.

The two significant results of the female sexual cycle are: First, only a *single* mature

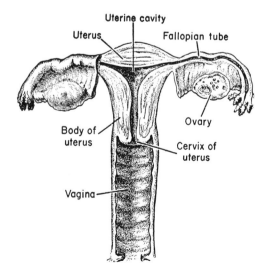

Figure 81–1 The female sexual organs.

Uterine cavity

Uterus

Fallopian tube

Ovary

Body of uterus

Cervix of uterus

Vagina

ovum is normally released from the ovaries each month so that only a single fetus can begin to grow at a time. Second, the uterine endometrium is prepared for implantation of the fertilized ovum at the required time of the month.

The Gonadotropic Hormones. The sexual cycle is completely dependent on gonadotropic hormones secreted by the anterior pituitary gland. Ovaries that are not stimulated by gonadotropic hormones remain completely inactive, which is essentially the case throughout childhood when almost no gonadotropic hormones are secreted. However, at the age of about eight, the pituitary begins secreting progressively more and more gonadotropic hormones, which culminates in the initiation of monthly sexual cycles between the ages of 11 and 15; this culmination is called *puberty.*

(Note that the ovaries function to a slight extent during fetal life because of stimulation by gonadotropic hormones from the placenta. But immediately after birth this stimulus is lost and the ovaries become almost totally dormant until puberty.)

The anterior pituitary secretes three different hormones that are known to be essential for full function of the ovaries: (1) *follicle-stimulating hormone* (FSH), (2) *luteinizing hormone* (LH), and (3) *luteotropic hormone* (LTH). FSH and LH are small glycoproteins having molecular weights of 30,000 and 26,000, respectively, and LTH is a single long polypeptide chain having a molecular weight of about 24,000. The only significant effects of FSH and LH are on the ovaries in the female and the testes in the male. However, LTH, in addition to its effects on the ovaries, helps to promote secretion of milk by the breasts and, therefore, is often called *lactogen, prolactin,* or *lactogenic hormone.* LH in the male is frequently called *interstitial cell-stimulating hormone* (ICSH) because it stimulates the interstitial cells.

Methods for assaying accurately the rates of secretion of the separate gonadotropic hormones are very difficult to perform, for which reason it has not been easy to determine exactly how much of each type of gonadotropic hormone is secreted during the different stages of the female sexual cycle. However, during each sexual month, there is a cyclic increase and decrease of FSH and LH and probably also of LTH. These cyclic variations are the cause of the cyclic ovarian changes, which are explained in the following sections.

FOLLICULAR GROWTH — FUNCTION OF FOLLICLE-STIMULATING HORMONE (FSH)

Figure 81–2 depicts the various stages of follicular growth in the ovaries, illustrating, first, the primordial follicle (primary follicle). Throughout childhood the primordial follicles do not grow, but at puberty, when FSH from the adenohypophysis begins to be secreted in large quantity, the entire ovaries and especially the follicles within them begin to grow. The first stage of follicular growth is enlargement of the ovum itself. This is followed by development of increasing numbers of granulosa cells and development of *layers of theca cells* around the outer surface of the granulosa cells. These epithelioid theca cells form the *theca interna,* and surrounding this develops a connective tissue capsule known as the *theca externa.*

The Vesicular Follicles — Function of Luteinizing Hormone (LH). The cells of the theca interna, and perhaps also the granulosa cells of the growing follicles, eventually begin secreting a *follicular fluid.* This causes an *antrum* to appear within the epithelioid mass of theca and granulosa cells. At this time the secretion of LH by the adenohypophysis begins to increase, and it then acts synergistically with the FSH to promote greatly accelerated further growth of the follicular tissues and increased secretion into the follicle. Therefore, the antrum becomes a frank vesicle, as illustrated in Figure 81–2. During growth of the *vesicular follicles,* the granulosa and theca cells continue to proliferate, and a large mass of these

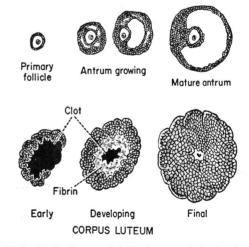

Primary follicle Antrum growing Mature antrum

Clot

Fibrin

Early Developing Final

CORPUS LUTEUM

Figure 81–2 Stages of follicular growth in the ovary, showing also formation of the corpus luteum. (Modified from Arey: Developmental Anatomy, 7th Ed.)

cells develops at one pole of the vesicular follicle; in this mass is located the ovum. This mass of cells secretes large quantities of estrogens, as explained later in the chapter.

During each monthly cycle, perhaps 20 or more primordial follicles develop antra and begin the vesicular phase of follicular growth, but ordinarily one of the follicles outgrows all the others and reaches a size of approximately 1 to 1.5 cm. This follicle ovulates, that is, expels its ovum, while the others involute.

OVULATION

Ovulation has never been observed visually in the human being, but it has been both observed and studied experimentally in rats and rabbits. Shortly before ovulation, the protruding outer wall of the follicle swells rapidly and a small area in the center of the swelling, called the *stigma*, protrudes like a nipple. The pressure in the follicle at this time is about 15 mm. Hg. Within another half hour or so, the stigma ruptures, and fluid begins to ooze from the follicle. About two minutes later, as the follicle becomes smaller because of loss of fluid, a more viscous fluid that has occupied the central portion of the follicle and that contains both the ovum and a mass of granulosa and theca cells is extruded into the abdomen.

The cause of ovulation seems to be necrosis or weakening of the cells of the stigma or of the cement substance between them. Several theories have been offered to explain this. One suggests that hyaluronidase is secreted into the follicle and causes decomposition of large amounts of mucopolysaccharides in the follicular fluid. This supposedly increases the osmotic pressure of the fluids and causes the follicle to swell until the stigma ruptures. However, measurements of pressure in the follicle do not show a preovulation rise in pressure. Another theory assumes that the granulosa and theca cells in the region of the stigma degenerate, which allows the rupture.

Necessity of Luteinizing Hormone (LH) for Ovulation; The Ovulatory Surge of LH. LH is necessary for final follicular growth and ovulation. Without this hormone, even though large quantities of FSH are available, the follicle will not progress to the stage of ovulation. LH acts synergistically with FSH to cause rapid swelling of the follicle shortly before ovulation. The swelling in itself undoubtedly contributes to the rupture, but LH also inhibits the growth of granulosa cells, and the cement substance between the granulosa cells disintegrates in the presence of LH. Therefore, it is likely that LH has a direct effect to cause the stigma to break open at the time of ovulation.

It is significant that an especially large amount of LH, called the *ovulatory surge*, is secreted by the pituitary during the day immediately preceding ovulation.

Atretic Follicles. In general only one ovum is expelled from the ovaries during each monthly sexual cycle. All the developing vesicular follicles that do not ovulate begin to become atretic either before or immediately after the single ovum is expelled. The ovum within each atretic follicle degenerates, and the follicular cells disappear completely, followed by ingrowth of connective tissue. This process of atresia is undoubtedly caused by a hormonal signal from the ovulating follicle, but the nature of this signal is unknown.

THE CORPUS LUTEUM

Within a few hours after expulsion of the ovum from the follicle, the granulosa and theca cells of the follicle undergo *luteinization*, and the mass of cells becomes a corpus luteum, which secretes mainly progesterone but also large amounts of estrogens. That is, these cells become greatly enlarged and develop lipid inclusions that give the cells a distinctive yellowish color, whence is derived the term *luteum*. A well-developed vascular supply grows into this mass of developing *lutein cells*.

In the normal female, the corpus luteum grows to approximately 1.5 cm., reaching this stage of development approximately seven or eight days following ovulation. After this, it begins to involute and loses its secretory function, as well as its lipid characteristics, approximately 12 days following ovulation, becoming then the so-called *corpus albicans;* during the ensuing few weeks this is replaced by connective tissue.

Luteinizing Function of Luteinizing Hormone (LH). The change of follicular cells into lutein cells is completely dependent on the secretion of LH by the adenohypophysis. In fact, this function gave LH its name "luteinizing."

Secretion by the Corpus Luteum: Function of Luteotropic Hormone (LTH) and LH. The corpus luteum is a highly secretory organ, secreting large amounts of both *progesterone* and *estrogens*. Unfortunately, we do not know the precise role of the individual gonadotropic hormones in causing secretion by the corpus luteum. However, total abrogation of gonadotropin secretion prevents further function of the corpus luteum. In some types of lower animals, injection of LH causes the corpus luteum to grow and to secrete progesterone and estrogen, while in other types of animals LTH performs this same function. In the rat, which is the animal most often used in endocrine studies, both hormones are required, LH to cause the initial luteinization and LTH to cause continued secretion. Therefore, it has been surmised that in the human being, too, LH is responsible for initial development of the corpus luteum while LTH is responsible for secretion by the corpus luteum. Until better means are available for assaying the secretion of LTH in the human being, this question will probably remain unanswered. Indeed, there is some question whether LTH plays any role at all in the ovarian cycle in the human being and also whether LTH and prolactin are precisely the same hormone.

SUMMARY

Approximately each 28 days, new follicles begin to grow in the ovary, one of which finally ovulates. The secretory cells of the follicle then develop into a corpus luteum that continues to secrete estrogens and progesterone under the influence of appropriate gonadotropic hormones. After another two weeks, the gonadotropin stimulus is lost, whereupon the ovarian hormones decrease greatly, and menstruation begins. A new sexual cycle then follows.

FEMALE SEX HORMONES

The two types of female sex hormones are the *estrogens* and *progesterone*. The estrogens mainly promote proliferation of specific cells in the body and are responsible for growth of sexual organs and most secondary sexual characteristics of the female. On the other hand, progesterone is concerned almost entirely with final preparation of the uterus for pregnancy and of the breasts for lactation.

CHEMISTRY OF THE SEX HORMONES

The Estrogens. In the normal, nonpregnant female, estrogens are secreted in significant quantities only by the ovaries, though a minute amount is secreted by the adrenal cortices. In pregnancy, tremendous quantities are also secreted by the placenta, indeed, up to 300 times the amount secreted by the ovaries during the normal sexual cycle.

At least six different natural estrogens have been isolated from the plasma of the human female, but only three are present in significant quantities, β-estradiol, estrone, and estriol, the formulas for which are illustrated in Figure 81–3. Both β-estradiol and estrone are present in large quantities in the venous blood from the ovaries, while estriol is an oxidative product derived from these first two, the conversion occurring mainly in the liver but also to some extent elsewhere in the body.

The estrogenic potency of β-estradiol is 12 times that of estrone and 80 times that of estriol. On the other hand, the quantity of estrone secreted by the ovaries averages four times the quantity of estradiol. Considering both the available quantities and their potencies, the total estrogenic effect of β-estradiol is about three times that of estrone. For this reason β-estradiol is considered to be the major estrogen secreted by the ovaries, though the estrogenic effects of estrone are certainly far from negligible.

SYNTHESIS OF THE ESTROGENS. Note from the formulas of the estrogens in Figure 81–3 that all these are steroids. They are synthesized in the ovaries mainly from acetyl coenzyme A, the acetate units of which can be conjugated to form the appropriate steroid nuclei. It is particularly interesting that testosterone, the male sex hormone, is probably synthesized first, and then converted into the estrogens.

SYNTHETIC ESTROGENS. Several different synthetic sterols have estrogenic activity, one of which, *ethynyl estradiol*, has proved to be particularly valuable for therapy because it, unlike the natural estrogens, is equally as potent when administered by mouth as by injection. Still another compound which is not a sterol, *stilbestrol*, exhibits even more estrogenic activity than β-estradiol, and, furthermore, is also active when administered by mouth. Therefore, this substance, too, has had wide clinical use. Also, it has been widely used in the animal industry to effect female characteristics in male animals.

FATE OF THE ESTROGENS; FUNCTION OF THE LIVER IN ESTROGEN DEGRADATION. Soon after they are secreted by the ovary, the estradiol and estrone that do not enter cells to perform physiologic functions are oxidized to estriol; this oxidation occurs principally in the liver but also to a slight extent elsewhere in the body. The liver also conjugates the estrogens to form glucuronides and sulfates, and about one-fifth of these conjugated products are excreted in the bile while larger quantities are excreted in the urine. The liver also combines estrogens loosely with a protein to form so-called *estroprotein*, and it is mainly in this form that the estrogens circulate in the extracellular fluids.

Figure 81-3 Chemical formulas of the principal female sex hormones.

Thus, the liver plays a key role in estrogen metabolism, and, because it converts the potent estrogens, estradiol and estrone, into an almost totally impotent estrogen, estriol, and because it excretes moderate quantities into the gut, diminished liver function actually *increases* the activity of estrogens in the body, sometimes causing *hyperestrinism*.

Progesterone. Almost all the progesterone secreted in the nonpregnant female is secreted by the corpus luteum during the latter half of each ovarian cycle. However, the adrenal glands form a minute quantity of progesterone or compounds that have progesteronic activity, and during pregnancy progesterone is formed in extreme quantities by the placenta, about 10 times the normal monthly amount, especially after the fourth month of gestation.

SYNTHESIS OF PROGESTERONE. Progesterone is a steroid having a molecular structure not far different from those of the other steroid hormones, the estrogens, testosterone, and the corticosteroids, as illustrated in Figure 81–3. Therefore, progesterone has some functions in common with all of these.

Progesterone is probably synthesized principally from acetyl coenzyme A. However, it can also be formed from cholesterol.

At least two other "progestin" hormones are secreted by the ovaries, but the quantities of these are so slight in comparison with that of progesterone that it is usually proper to consider progesterone as the single important progestin.

FATE OF PROGESTERONE. Progesterone is secreted in far greater quantities than the estrogens by the ovaries, but its potency per unit weight is much less than that of the estrogens. Within a few minutes after secretion, almost all the progesterone is degraded to other steroids that have no progesteronic effect. Here, as is also true with the estrogens, the liver is especially important for this metabolic degradation.

A major end-product of progesterone degradation is *pregnanediol*. Approximately 10 per cent of the original progesterone is excreted in the urine in this form. One can estimate the rate of progesterone formation in the body from the rate of this excretion, but because pregnanediol exerts no progesteronic effects it can be detected in the urine only by chemical means. Other fluids of the body, such as the blood, contain significant amounts of active progesterone, and the progesteronic activity can be assayed by the characteristic effect of progesterone in causing a hyperemic uterine endometrium, a function of progesterone that is discussed later in the chapter.

FUNCTIONS OF THE ESTROGENS — EFFECTS ON THE PRIMARY AND SECONDARY SEXUAL CHARACTERISTICS

The principal function of the estrogens is to cause cellular proliferation and growth of the tissues of the sexual organs and of other tissues related to reproduction.

Effect on the Sexual Organs. During childhood, estrogens are secreted only in small quantities, but following puberty the quantity of estrogens secreted under the influence of adenohypophyseal gonadotropic hormones increases some 20-fold or more. At this time the female sexual organs change from those of a child to those of an adult. The fallopian tubes, uterus, and vagina all increase in size. Also, the external genitalia develop, with deposition of fat in the mons pubis and labia majora and with enlargement of the labia minora.

In addition to increase in size of the vagina, estrogens change the vaginal epithelium from a cuboidal into a stratified type, which is considerably more resistant to trauma and infection than is the prepubertal epithelium. Infections in children, such as gonorrheal vaginitis, can actually be cured by administration of estrogens simply because of the resulting increased resistance of the vaginal epithelium to infection.

During the few years following puberty, the size of the uterus increases two- to threefold. More important, however, than increases in size are the changes that take place in the endometrium under the influence of estrogens, for estrogens cause marked proliferation of the endometrium and development of glands that will later be used to aid in nutrition of the implanting ovum. These effects are discussed later in the chapter in connection with the endometrial cycle.

Effect on the Fallopian Tubes. The estrogens have an effect on the mucosal lining of the fallopian tubes similar to that on the uterine endometrium: They cause the glandular tissues to proliferate; and, especially important, they cause the number of ciliated epithelial cells that line the fallopian tubes to increase. Also, the activity of the cilia is considerably enhanced, these always beating toward the uterus. This undoubtedly helps to propel the fertilized ovum toward the uterus.

Effect on the Breasts. The primordial breasts of both female and male are exactly alike, and under the influence of appropriate hormones, the masculine breast, at least during the first two decades of life, can develop sufficiently to produce milk in the same manner as the female breast.

Estrogens cause fat deposition in the breasts, development of the stromal tissues of the breasts, and growth of an extensive ductile system. The lobules and alveoli of the breast develop to a slight extent, but it is progesterone and prolactin that cause marked growth of these structures. In summary, the estrogens initiate growth of the breasts and the breasts' milk-producing apparatus, and they are also responsible for the characteristic external appearance of the mature female breast, but they do not complete the job of converting the breasts into milk-producing organs.

Effect on the Skeleton. Estrogens cause increased osteoblastic activity. Therefore, at puberty, when the female enters her sexual life, her growth rate becomes rapid for several years. However, estrogens have another effect on skeletal growth to cause early uniting of the epiphyses with the shafts of the long bones. This effect is much stronger in the female than is the similar effect of testosterone in the male. As a result, growth of the female usually ceases several years earlier than growth of the male. The female eunuch who is completely devoid of estrogen production usually grows several inches taller than the normal mature female because her epiphyses do not unite early under the influence of estrogens.

EFFECT ON THE PELVIS. Estrogens have a special effect to broaden the pelvis, changing the pelvic outlet from a narrow, funnel-like outlet into a broad, ovoid outlet. The exact means by which estrogens cause these changes in the pelvic bones is not known, but this effect presumably is a by-product of the general effects of estrogens on skeletal growth.

EFFECT ON CALCIUM AND PHOSPHATE RETENTION. As pointed out in the preceding chapter, testosterone causes greater than normal retention of calcium and phosphate. This is also true of the estrogens, for estrogens, like testosterone, promote growth of bones, entailing the deposition of increased amounts of bone matrix with subsequent retention of both calcium and phosphate.

Effect on Protein Deposition. Estrogens cause a slight increase in total body protein, which is evidenced by a slight positive nitrogen balance when estrogens are administered. This probably results from the growth-promoting effect of estrogen on the sexual organs and on a few other tissues of the body. The enhanced protein deposition caused by testosterone is much more general and many times as powerful as that caused by estrogens.

Effect on Metabolism and Fat Deposition. Estrogens increase the metabolic rate slightly but not so much as the male sex hormone testosterone. However, they cause deposition of increased quantities of fat in the subcutaneous tissues. As a result, the overall specific gravity of the female body, as judged by flotation in water, is considerably less than that of the male body, which contains more protein and less fat. In addition to deposition of fat in the breasts and subcutaneous tissues, estrogens cause especially marked deposition of fat in the buttocks and thighs, causing the broadening of the hips that is characteristic of the feminine figure.

Effect on Hair Distribution. Estrogens do not greatly affect hair distribution except in the pubic region. Here, they cause growth of the usual female pubic hair with a flat upper border rather than the triangular border characteristic of the male.

Effect on the Skin. Estrogens cause the skin to develop a texture that is soft and usually smooth but nevertheless thicker than that of the child or the female castrate. Also, estrogens cause the skin to become more vascular than normal; this effect is often associated with increased warmth of the skin, and it often results in greater bleeding of cut surfaces than is observed in men.

Effect on Electrolyte Balance. The similarity of estrogenic hormones to adrenocortical hormones has been pointed out; and estrogens, like adrenocortical hormones, cause sodium and water retention by the kidney tubules. However, this effect of estrogens is slight and rarely of significance except in pregnancy, as discussed in the following chapter.

Intracellular Functions of Estrogens. Thus far we have discussed the gross effects of estrogens on the body. The precise functions of estrogens in causing these effects are not known other than for the following clues: Estrogens circulate in the blood for only a few minutes before they are delivered to the target cells. On entry into these cells, the estrogens combine within 10 to 15 seconds with a protein in the cytoplasm and then are slowly released from this combination to cause intracellular effects. These effects are believed to occur mainly in the nucleus, for RNA begins to be produced in about 30 minutes; and over a period of many hours even DNA is produced, resulting eventually in division of the cell. The RNA diffuses to the cytoplasm where it causes greatly increased protein formation.

One of the principal differences between the protein anabolic effect of the estrogens and that of testosterone is that estrogen causes its effect almost exclusively in certain target organs, such as the uterus, the breasts, the skeleton, and certain fatty areas of the body; whereas testosterone has a more generalized effect throughout the body.

FUNCTIONS OF PROGESTERONE

Effect on the Uterus. By far the most important function of progesterone is *to promote secretory changes in the endometrium,* thus preparing the uterus for implantation of the fertilized ovum. This function is discussed specifically in connection with the endometrial cycle of the uterus.

In addition to this effect on the endometrium, progesterone decreases the frequency of uterine contractions, thereby helping to prevent expulsion of the implanted ovum, an effect discussed in the following chapter.

Effect on the Fallopian Tubes. Progesterone also promotes secretory changes in the mucosal lining of the fallopian tubes. These secretions are important for nutrition of the fertilized, dividing ovum as it traverses the fallopian tube prior to implantation.

Effect on the Breasts. Progesterone promotes development of the lobules and alveoli of the breasts, causing the alveolar cells to proliferate, to enlarge, and to become secretory in nature. However, progesterone does not cause the alveoli actually to secrete milk, for, as discussed in the following chapter, milk is secreted only after the prepared breast is further stimulated by prolactin (luteotropic hormone) from the pituitary.

Progesterone also causes the breasts to swell. Part of this swelling is due to the secretory development in the lobules and alveoli, but part also seems to result from increased fluid in the subcutaneous tissue itself.

Effect on Electrolyte Balance. Progesterone, like estrogens, testosterone, and adrenocortical hormones, enhances sodium, chloride, and water reabsorption from the distal tubules of the kidney. Indeed, it is even possible to use progesterone instead of deoxycorticosterone for treatment of animals in which the adrenal cortices have been removed completely.

Yet, strangely enough, when progesterone is injected in small quantities into a person, the rate of sodium and water retention by the kidneys is actually decreased. The cause of this is a competition between progesterone and aldosterone, which probably occurs as follows: It is believed that these two substances combine with the same reactive sites on the epithelial cells of the tubules. Therefore, when progesterone combines with these sites, aldosterone cannot combine. Yet progesterone exerts many hundred times less effect than does aldosterone. Therefore, despite the fact that progesterone causes sodium and water retention by the renal tubules, it blocks the far more potent effect of aldosterone, thus resulting in net loss of sodium and water from the body.

Protein Catabolic Effect. Progesterone exerts a mild catabolic effect on the body's protein similar to that of the glucocorticoids. Though in the normal sexual cycle this effect is probably not significant, it possibly is significant during pregnancy, when proteins must be mobilized for use by the fetus.

REGULATION OF THE FEMALE RHYTHM—INTERPLAY BETWEEN THE OVARIAN HORMONES AND THE PITUITARY GONADOTROPIC HORMONES

The lower part of Figure 81–4 illustrates the cyclic changes in estrogens and progesterone in the circulating blood of the normal female during the normal sexual cycle, while the upper part of the figure shows the approximate cyclic changes of gonadotropic hormones secreted by the anterior pituitary gland.

A repetitive, back-and-forth interplay is believed to occur between the ovarian hormones and the pituitary gonadotropins, causing a continual oscillatory increase and decrease of both the ovarian and gonadotropic hormones. The basis of this rhythmic oscillation can be explained as follows:

EFFECT OF THE OVARIAN HORMONES ON PITUITARY SECRETION OF GONADOTROPIC HORMONES

As noted earlier in this chapter, the gonadotropic hormones cause the ovaries to secrete

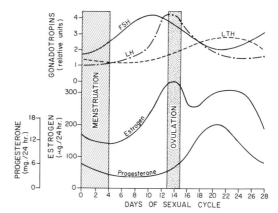

Figure 81–4 Approximate cyclic changes in ovarian and pituitary secretion of the different hormones involved in the female sexual cycle.

first estrogens and then progesterone. On the other hand, both estrogens and progesterone inversely affect the hypothalamus to decrease the rates of secretion of gonadotropic hormones as follows:

Negative Feedback of Estrogens and Progesterone on Pituitary Secretion of Follicle-Stimulating Hormone (FSH) and Luteinizing Hormone (LH). Both estrogens and progesterone, in large enough quantity, inhibit the production of FSH and LH. Much larger quantities of progesterone than estrogen must be administered to have this effect, and even then the effect of progesterone is poor unless estrogens have been secreted or injected prior to the progesterone. Thus, the feedback effect of progesterone supports that of estrogens rather than being an entirely separate effect.

The feedback effects of both estrogens and progesterone seem to operate entirely, or almost entirely, by actions of these hormones on the hypothalamus rather than by direct actions on the adenohypophysis, as is discussed below.

POSSIBLE POSITIVE FEEDBACK EFFECT OF ESTROGENS AND PROGESTERONE BEFORE OVULATION. For reasons not understood, the anterior pituitary gland secretes a greatly enhanced amount of LH on the day immediately before ovulation. This effect is believed to result from a positive feedback effect of estrogens and progesterone when these are present in low concentration. Thus, in the early part of the sexual month, the slight secretion of hormones, mainly estrogens, by the developing follicles causes markedly enhanced secretion of LH and slightly enhanced secretion of FSH. These in turn cause the follicles to grow even more rapidly until one of them ovulates. At this time, the corpus luteum begins to secrete large quantities of ovarian hormones that then supposedly surpass the critical concentration levels required to change from positive to negative feedback. Therefore, the ovarian hormones now inhibit pituitary gonadotropin secretion.

Reciprocal Relationship Between LH and FSH and the Third Gonadotropin, Luteotropic Hormone (LTH). When FSH and LH are secreted in large quantities, the pituitary secretes diminished quantities of LTH; conversely, when the pituitary secretes decreased quantities of FSH and LH, it simultaneously secretes greater quantities of LTH. Thus, there is a reciprocal relationship between the first two of these gonadotropins and the third, a relationship that probably plays a significant role in control of the monthly sexual cycle.

Function of the Hypothalamus in the Regulation of Gonadotropin Secretion. When the pituitary gland is transplanted from beneath the hypothalamus to some other area of the body, the secretion of FSH and LH decreases to less than one-fourth normal, while the secretion of LTH increases to as much as three times normal. Therefore, it is believed that the hypothalamus secretes an *FSH-releasing factor,* an *LH-releasing factor,* and an *LTH-inhibiting factor,* all of which are transported in the hypothalamic-hypophyseal portal system to the adenohypophysis to control the secretion of the gonadotropins. Small polypeptide substances have been isolated that stimulate both LH and FSH secretion, and an LTH-inhibiting factor has perhaps been isolated.

As is true in the male, destruction of the median eminence destroys essentially all hypothalamic control over the pituitary secretion of the gonadotropins. However, this perhaps results from the fact that the median eminence is the collecting point in the hypothalamus where the neurosecretory hormones are picked up by the hypothalamic-hypophyseal portal system. Another area for control of at least part of the gonadotropin secretion is located in the anterior hypothalamus in the region above the optic chiasm. Lesions in this region prevent the negative feedback effect of estrogens and progesterone on pituitary secretion of FSH and LH, which indicates that the neurons of this area are responsive to the effects of the gonadal hormones.

Remember that in lower animals, such as in the rabbit, coitus causes rapid release of gonadotropins from the adenohypophysis and results in ovulation about a day later. Furthermore, psychic disturbances in the human female can greatly alter the periodicity of her monthly sexual cycle as well as the character of the cycle.

FEEDBACK OSCILLATION OF THE HYPOTHALAMIC-PITUITARY-OVARIAN SYSTEM

Though the precise antagonistic relationships between ovarian hormones and the gonadotropic hormones are not known, it is clear from the preceding discussion that the gonadotropic hormones from the anterior pituitary gland are the basic cause of hormonal secretion by the ovaries; and, in turn, both the estrogens and progesterone of the ovaries, in large enough concentration, can decrease production of at least two of the gonadotropins, FSH and LH. Thus, conditions are appropriate for feedback oscillation to result, which can perhaps be explained as follows:

1. The adenohypophysis, when not affected by outside stimuli, secretes mainly FSH. Therefore, during the first part of the month, FSH begins to be secreted, and this causes the follicles of the ovaries to begin growing.

2. The adenohypophysis then begins to secrete LH, which, acting synergistically with FSH, causes the vesicular follicles of the ovaries to grow rapidly and to secrete progressively more and more estrogens. The estrogens then supposedly cause a *positive* feedback effect on the hypothalamic-pituitary system to enhance greatly the output of LH, which causes further rapid growth of the follicles, culminating in ovulation by one of them.

3. Immediately before and after ovulation, under the influence of LH, the follicular cells of the follicle take on lutein characteristics and become the corpus luteum. Then, the large amounts of estrogens and progesterone secreted by the corpus luteum exert their normal *negative* feedback effect on the hypothalamic-adenohypophysis complex to decrease the FSH and LH, but they simultaneously increase the output of LTH. Under the influence of LTH and the waning LH, the corpus luteum secretes large quantities of both progesterone and estrogens for 14 days, reaching peak outputs 7 to 8 days after ovulation.

4. From the 8th to the 12th day after ovulation, the corpus luteum involutes, probably because the output of LH has decreased greatly by this time. The secretion of estrogens and progesterone diminishes accordingly. A few days later, the diminished production of

estrogens and progesterone automatically allows the adenohypophysis to begin producing FSH once again in large quantities, followed shortly thereafter by increasing LH.

5. The increasing FSH and LH then stimulate development of new follicles, and a new cycle begins.

Oscillation in Anovulatory Cycles—The Sexual Cycle at Puberty. Often ovulation fails to occur, and the cycle is then said to be anovulatory. Indeed, in almost all girls immediately after puberty the first few cycles are anovulatory. Lack of ovulation also causes failure of development of a corpus luteum and consequently no enhanced secretion of progesterone during the latter part of the cycle. In such instances the rhythm is often considerably distorted, but an oscillatory cycle of sorts still occurs, usually prolonged and associated with abnormal rates of estrogen secretion as well as abnormal endometrial changes in the uterus. Yet hypothalamic-pituitary-ovarian oscillation can still occur even without enhanced progesterone secretion by the ovaries. Therefore, the female sexual cycle seems to result principally from reciprocal oscillation between FSH and LH secretion by the pituitary and estrogen secretion by the ovaries, whereas the other hormonal changes just discussed are mainly secondary effects.

PUBERTY

Puberty means the onset of adult sexual life, and, as pointed out earlier in the chapter, it is caused by a gradual increase in gonadotropic hormone secretion by the pituitary, beginning approximately the eighth year of life, as illustrated in Figure 81–5.

In the female, as in the male, the infantile pituitary gland and ovaries are capable of full function if appropriately stimulated. However, the hypothalamus is extremely sensitive to the inhibitory effects of estrogens and progesterone, which keep its stimulation of the pituitary almost completely suppressed throughout childhood. Then, at puberty, the hypothalamus matures; its excessive sensitivity to the negative feedback inhibition becomes greatly diminished, which allows enhanced production of gonadotropins and the onset of adult female sexual life.

Figure 81–6 illustrates (a) the increasing levels of estrogen secretion at puberty, (b) the cyclic variation during the monthly sexual cycles, (c) the further increase in estrogen secretion during the first few years of sexual life, (d) then progressive decrease in estrogen secretion toward the end of sexual life, and (e) finally almost no estrogen secretion beyond the menopause.

Figure 81–5 Total rates of secretion of gonadotropic hormones throughout the sexual lives of females and males, showing an especially abrupt increase in gonadotropic hormones at the menopause in the female.

THE MENOPAUSE

At an average age of approximately 45 to 50 years the sexual cycles usually become irregular, and ovulation fails to occur during many of these cycles. After a few months to a few years, the cycles cease altogether, as illustrated in Figure 81–6.

The cause of the menopause (cessation of the sexual cycles) apparently is "burning out" of the ovaries. In other words, throughout a woman's sexual life many of the primordial follicles grow into vesicular follicles with each sexual cycle, and eventually almost all the ova either degenerate or are ovulated. Therefore, at the age of about 45 only a few primordial follicles still remain to be stimulated by FSH and LH, and the production of estrogens by the ovary decreases as the number of primordial follicles approaches zero (which is illustrated in Fig. 81–6). When estrogen production falls below a critical value, the estrogens can no longer inhibit the production of FSH and LH sufficiently to cause oscillatory cycles.

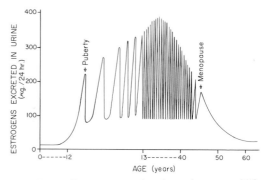

Figure 81–6 Estrogen secretion throughout sexual life.

Consequently, as illustrated in Figure 81–5, FSH and LH (mainly FSH) are produced thereafter in large and continuous quantities. Estrogens are produced in subcritical quantities for a short time after the menopause, but over a few years, as the final remaining primordial follicles become atretic, the production of estrogens by the ovaries falls almost to zero.

The Female Climacteric. The term "female climacteric" means the entire time, lasting from several months to several years, during which the sexual cycles become irregular and gradually stop. In this period the woman must readjust her life from one that has been physiologically stimulated by estrogen and progesterone production to one devoid of these feminizing hormones. The secretion of estrogens decreases rapidly, and essentially no progesterone is secreted after the last ovulatory cycle. The loss of the estrogens often causes marked physiologic changes in the function of the body, including (1) "hot flashes" characterized by extreme flushing of the skin, (2) psychic sensations of dyspnea, (3) irritability, (4) fatigue, (5) anxiety, and (6) occasionally various psychotic states. These symptoms are of sufficient magnitude in approximately 15 per cent of women to warrant treatment. If psychotherapy fails, daily administration of an estrogen in small quantities will reverse the symptoms, and by gradually decreasing the dose the menopausal woman is likely to avoid severe symptoms; unfortunately, such treatment prolongs the symptoms, however.

THE ENDOMETRIAL CYCLE AND MENSTRUATION

Associated with the cyclic production of estrogens and progesterone by the ovary is an endometrial cycle operating through the following stages: first, proliferation of the uterine endometrium; second, secretory changes in the endometrium; and third, desquamation of the endometrium, which is known as *menstruation.* The various phases of the endometrial cycle are illustrated in Figure 81–7.

Proliferative Phase (Estrogen Phase) of the Endometrial Cycle. At the beginning of each

menstrual cycle, most of the endometrium is desquamated by the process of menstruation. This includes complete loss of the epithelium and *stratum submucosum* and loss of most of the *stratum vasculare.* Therefore, after menstruation only a thin layer of endometrial stroma remains at the base of the original endometrium, and the only epithelial cells are those located in the remaining deep portions of the glands and crypts of the endometrium. *Under the influence of estrogens,* secreted in increasing quantities by the ovary during the first part of the ovarian cycle, the stromal cells and the epithelial cells proliferate rapidly. The endometrial surface is re-epithelialized within approximately four to five days after the beginning of menstruation. For the first two weeks of the sexual cycle—that is, until ovulation—the endometrium increases greatly in thickness, owing to increasing numbers of stromal cells and to progressive growth of the endometrial glands and blood vessels into the endometrium, all of which effects are promoted by the estrogens. At the time of ovulation the endometrium is approximately 2 to 3 mm. thick.

Secretory Phase (Progestational Phase) of the Endometrial Cycle. During the latter half of the sexual cycle, both estrogens and progesterone are secreted in large quantities by the corpus luteum. The estrogens cause additional cellular proliferation in the endometrium during this phase of the endometrial cycle; and progesterone, probably acting synergistically with the estrogens, causes considerable swelling of the endometrium. The glands increase in tortuosity, secretory substances develop in the glandular epithelial cells, and the glands secrete small quantities of endometrial fluid. Also, the cytoplasm of the stromal cells increases, lipid and glycogen deposits increase greatly in the stromal cells, and the blood supply to the endometrium increases in proportion to the developing secretory activity, the blood vessels becoming highly tortuous. The thickness of the endometrium approximately doubles during the secretory phase so that toward the end of the sexual cycle the endometrium has a thickness of 4 to 6 mm.

The whole purpose of all these endometrial changes is to produce a highly secretory endometrium containing large amounts of stored nutrients that can provide appropriate conditions for implantation of a fertilized ovum during the latter half of the monthly cycle. From the time fertilization first takes place until the ovum implants, the fallopian and uterine secretions, called "uterine milk," provide nutri-

Proliferative phase   Secretory phase   Menstrual phase
(11 days)   (12 days)   (5 days)

Figure 81–7 Phases of endometrial growth and menstruation during each monthly female sexual cycle.

tion for the early dividing ovum. Then, once the ovum implants in the endometrium, the trophoblastic cells on the surface of the morula begin to digest the endometrium and to absorb the substances digested, thus making still far greater quantities of nutrients available to the early embryo.

Menstruation. Approximately two days before the end of the sexual cycle the ovarian hormones, the estrogens and progesterone, decrease sharply to low levels of secretion, and menstruation follows.

Menstruation is caused by the sudden reduction in both estrogens and progesterone at the end of the monthly ovarian cycle. The first effect is decreased stimulation of the endometrial cells by these two hormones, followed rapidly by involution of the endometrium to about 65 per cent of its previous thickness. During the 24 hours preceding the onset of menstruation, the tortuous blood vessels leading to the mucosal layers of the endometrium become vasospastic, presumably because of some effect of the involution, such as release of a vasoconstrictor material, or perhaps because of a direct effect of estrogen withdrawal, since estrogens are an endometrial vasodilator. The vasospasm and loss of hormonal stimulation cause beginning necrosis in the endometrium, especially of blood vessels in the stratum vasculare. As a result, blood seeps into the vascular layer of the endometrium, the hemorrhagic areas growing over a period of approximately 24 to 36 hours. Gradually, the necrotic outer layers of the endometrium separate from the uterus at the site of the hemorrhages, until, at approximately 48 hours following the onset of menstruation, all the superficial layers of the endometrium have desquamated. The desquamated tissue and blood in the uterine vault initiate uterine contractions that expel the uterine contents.

During menstruation, approximately 35 ml. of blood and an additional 35 ml. of serous fluid are lost. This menstrual fluid is normally nonclotting, because a *fibrinolysin* is released along with the necrotic endometrial material. However, if excessive bleeding occurs from the uterine surface, the quantity of fibrinolysin may not be sufficient to prevent clotting. Consequently, the presence of clots during menstruation ordinarily is clinical evidence of uterine pathology.

Within approximately five days after menstruation starts, the loss of blood ceases, for by this time the endometrium has become completely re-epithelialized.

LEUKORRHEA DURING MENSTRUATION. During menstruation a tremendous number of leukocytes is released along with the necrotic material and blood. It is probable that some substance liberated by the endometrial necrosis, perhaps leukotaxine which was discussed in Chapter 9, causes this outflow of leukocytes. As a result of these many leukocytes and maybe still other factors, the uterus is resistant to infection during menstruation, even though the endometrial surfaces are denuded. Obviously, this is of extreme protective value.

INTERRELATIONSHIPS OF THE OVARIES WITH OTHER GLANDS

Relationship of the Ovaries to the Adrenal Glands. The adrenal glands even normally secrete small quantities of both estrogens and progesterone, though these quantities are too small to exert any major effect on the body. However, rare tumors of the adrenal gland secrete specifically increased quantities of female hormones and therefore cause feminizing characteristics.

Injection of estrogens causes the adrenal cortices to hypertrophy. This effect is mediated through the pituitary, for estrogens, while inhibiting production of follicle-stimulating hormones, increase the secretion of corticotropin.

Relationship of the Ovaries to the Thyroid Gland. Ovarian function does not have a marked effect on the thyroid gland, but, conversely, thyroid hormone exhibits considerable effect on the activity of the ovaries. For instance, when thyroid hormone production is deficient, the sexual cycles are likely to become anovulatory, often associated with excessive uterine bleeding, which is discussed later, or the cycles may even disappear entirely. This effect perhaps is due to decrease in the metabolic rate of the ovarian cells so that sufficient quantity of estrogens cannot be produced. Strangely enough, a marked increase in thyroid secretion is also often associated with ovulatory failure and amenorrhea. This paradoxical effect shows how delicate the monthly oscillation of the female sexual cycle is and that it can be stopped by either excesses or insufficiencies of the same extraneous substance.

Antagonistic Effects of Estrogens and Testosterone. Estrogens and testosterone exert opposite effects on the sexual organs and on secondary sexual tissues, such as the breasts or the prostate gland. These antagonistic effects are in part mediated through the pituitary because both estrogens and testosterone are capable of decreasing the production of follicle-stimulating hormone and luteinizing hormone, which in turn abrogates secretion of the normal gonadal secretions. In general, in order to antagonize the effect of estrogens on the breast, approximately 50 times as much testosterone as β-estradiol must be administered.

ABNORMALITIES OF SECRETION BY THE OVARIES

Hypogonadism. Less than normal secretion by the ovaries can result from poorly formed ovaries or lack of ovaries. When ovaries are absent from birth or when they never become functional, *female eunuchism* occurs. In this condition the usual secondary sexual characteristics do not appear, and the sexual organs remain infantile. Especially characteristic of this condition is excessive growth of the long bones because the epiphyses do not unite with the shafts of these bones at as early an age as in the normal adolescent woman. Consequently, the female eunuch is essentially as tall as, or perhaps even slightly taller than, her male counterpart of similar genetic background.

When the ovaries of a fully developed woman are removed, the sexual organs regress to some extent so that the uterus becomes almost infantile in size, the vagina becomes smaller, and the vaginal epithelium becomes thin and easily damaged. The breasts atrophy and become pendulous, and the pubic hair becomes considerably thinner. These are the same changes that occur in the woman after the menopause.

IRREGULARITY OF MENSES AND AMENORRHEA DUE TO HYPOGONADISM. As pointed out in the preceding discussion of the menopause, the quantity of estrogens produced by the ovaries must rise above a critical value if they are to be able to inhibit the production of follicle-stimulating hormone sufficiently to cause an oscillatory sexual cycle. Consequently, in hypogonadism or when the gonads are secreting small quantities of estrogens as a result of other factors, such as hypothyroidism, the ovarian cycle likely will not occur normally. Instead, several months may elapse between menstrual periods, or menstruation may cease altogether (amenorrhea). Characteristically, prolonged ovarian cycles are frequently associated with failure of ovulation, presumably due to insufficient secretion of luteinizing hormone, which is necessary for ovulation.

EXCESSIVE MENSTRUAL BLEEDING. Contrary to what might be expected, excessive menstrual bleeding most commonly occurs in hypogonadism rather than in hypergonadism. Such bleeding is especially likely to result at the end of anovulatory cycles, for these cycles are often prolonged, and estrogens continue to be produced by the ovaries for many weeks at a time, thereby building up a progressively thicker and thicker endometrium that does not slough away easily as occurs in normal menstruation; instead, it denudes slowly and bleeds severely. These persons can usually be treated satisfactorily by administering large quantities of progesterone for approximately 10 days, beginning two weeks after the onset of the previous menstruation and then suddenly withdrawing the progesterone. Normal menstruation usually ensues almost immediately after the progesterone is withdrawn, and a new cycle begins.

However, repetitive administration of progesterone with each cycle is often required.

Excessive bleeding also frequently results from pathologic conditions in the uterus, such as fibrous scars on the endometrial surface or tumors of the uterus that protrude into the endometrial cavity.

Hypersecretion by the Ovaries. Extreme hypersecretion of ovarian hormones by the ovaries is a rare clinical entity, for excessive secretion of estrogens and progesterone automatically decreases the production of gonadotropins by the pituitary, and this in turn limits the production of the ovarian hormones. Consequently, hypersecretion of feminizing hormones is recognized clinically only when a feminizing tumor develops.

A rare granulosa-theca cell tumor occasionally develops in an ovary, occurring more often after menopause than before. These tumors secrete large quantities of estrogens which exert the usual estrogenic effects, including hypertrophy of the uterine endometrium and irregular bleeding from this endometrium. In fact, bleeding is often the first indication that such a tumor exists.

Ovarian tumors that secrete progesterone are so rare that they are almost nonexistent. However, certain tumors of the adrenal gland have been known to secrete large quantities of progesterone. In general, the major effect is excessive retention of body fluids.

Endometriosis. Endometriosis is the development and growth of endometrium in the peritoneal cavity, this growth usually occurring in the pelvis closely associated with the sexual organs. There are two theories for explaining the origin of intraabdominal endometrial tissue that causes endometriosis. Some believe that this tissue results from endometrial "rests" that develop embryologically in the peritoneal cavity. Others believe that contraction of the uterus during menstruation occasionally expels viable endometrium backward through the fallopian tubes into the abdominal cavity and that this endometrial tissue then implants on the peritoneum. In support of this is the fact that much of the endometrium sloughed during menstruation will grow easily in tissue culture.

During each ovarian cycle the endometrium in the peritoneal cavity proliferates, secretes, and desquamates in the same manner that the intrauterine endometrium does. However, when desquamation occurs within the peritoneal cavity, the tissue and the hemorrhaging blood cannot be expelled to the exterior. Consequently, the quantity of endometrial tissue in the peritoneal cavity progressively increases with each subsequent menstrual cycle.

The presence of necrotic and hemorrhagic material in the abdominal cavity and also the swelling of the endometrial tissue during each ovarian cycle can cause considerable irritation of the peritoneum, sometimes producing severe abdominal pain. Also, fibrosis occurs in the areas of endometriosis, thereby promoting adhesions from one sexual organ to another and from the sexual organs to other intrapelvic and intra-abdominal structures. Endometriosis is one of the most prevalent causes of female infertility, owing especially to fibrotic immobilization of the sexual organs.

THE FEMALE SEXUAL ACT

Stimulation of the Female Sexual Act. As is true in the male sexual act, successful performance of the female sexual act depends on both psychic stimulation and local sexual stimulation.

The psychic factors that constitute "sex drive" in women are difficult to assess. The sex hormones, and the adrenocortical hormones as well, seem to exert a direct influence on the woman to create such a sex drive, but, on the other hand, the growing female child in modern society is often taught that sex is something to be hidden and that it is immoral. As a result of this training, much of the natural sex drive is inhibited, and whether the woman will have little or no sex drive ("frigidity") or will be more highly sexed probably depends partly on a balance between natural factors and previous training.

Local sexual stimulation in women occurs in more or less the same manner as in men, for massage, irritation, or other types of stimulating of the perineal region, sexual organs, and urinary tract create sexual sensations. The *clitoris* is especially sensitive for initiating sexual sensations. As in the male, the sexual sensations are mediated to the spinal cord through the pudendal nerve and sacral plexus. Once these sensations have entered the spinal cord, they are transmitted thence to the cerebrum. Also, local reflexes that are at least partly responsible for the female orgasm are integrated in the sacral and lumbar spinal cord.

Female Erection and Lubrication. Located around the introitus and extending into the clitoris is erectile tissue almost identical with the erectile tissue of the penis. This erectile tissue, like that of the penis, is controlled by the parasympathetic nerves that pass through the nervi erigentes from the sacral plexus to the external genitalia. In the early phases of sexual stimulation, the parasympathetics dilate the arteries to the erectile tissues, and this allows rapid inflow of blood into the erectile tissue so that the introitus tightens around the penis; this aids the male greatly in his attainment of sufficient sexual stimulation for ejaculation to occur.

Parasympathetic impulses also pass to the bilateral Bartholin's glands located beneath the labia minora to cause secretion of mucus immediately inside the introitus. This mucus, along with large quantities of mucus secreted by the vaginal mucosa itself, is responsible for most of the lubrication during sexual intercourse. The lubrication in turn is necessary for establishing during intercourse a satisfactory massaging sensation rather than an irritative sensation, which may be provoked by a dry vagina. A massaging sensation constitutes the optimal type of sensation for evoking the appropriate reflexes that culminate in both the male and female climaxes.

The Female Orgasm. When local sexual stimulation reaches maximum intensity, and especially when the local sensations are supported by appropriate conditioning impulses from the cerebrum, reflexes are initiated that cause the female orgasm, also called the *female climax*. The female orgasm is analogous to ejaculation in the male, and it probably is important for fertilization of the ovum. Indeed, the human female is known to be somewhat more fertile when inseminated by normal sexual intercourse rather than by artificial methods, thus indicating an important function of the female orgasm. Possible effects that could result in this are:

First, during the orgasm the perineal muscles of the female contract rhythmically, which presumably results from spinal reflexes similar to those that cause ejaculation in the male. It is possible, also, that these same reflexes increase uterine and fallopian tube motility during the orgasm, thus helping to transport the sperm toward the ovum, but the information on this subject is scanty.

Second, in many lower animals, copulation causes the posterior pituitary gland to secrete oxytocin; this effect is probably mediated through the amygdaloid nuclei and then through the hypothalamus to the pituitary. The oxytocin in turn causes increased contractility of the uterus, which also is believed to cause rapid transport of the sperm. Sperm have been shown to traverse the entire length of the fallopian tube in the cow in approximately five minutes, a rate at least 10 times as fast as that which the sperm themselves could achieve. Whether or not this occurs in the human female is unknown.

In addition to the effects of the orgasm on fertilization, the intense sexual sensations that develop during the orgasm also pass into the cerebrum and in some manner satisfy the female sex drive.

FEMALE FERTILITY

The Fertile Period of Each Sexual Cycle. The exact duration of time that the ovum remains viable and capable of being fertilized after it is expelled from the ovary is not known. However, this duration is probably not over 24 hours. Therefore, sperm must be available soon after ovulation if fertilization is to take place. On the other hand, a few sperm can remain viable in the female reproductive tract for up to 72 hours, though probably most of them for not more than 24 hours. Therefore, for fertilization to take place, intercourse usually must occur some time between one day prior to ovulation up to one day after ovulation. Thus, the period of female fertility during each sexual cycle is short.

One of the often practiced methods of contraception is to avoid intercourse near the time of ovulation. The difficulty with this method of contraception is the impossibility of predicting the exact time of ovulation. Yet the interval from ovulation until the next succeeding onset of menstruation is almost always between 13 and 15 days. Therefore, if the menstrual cycle is regular, the time of ovulation averages 14 days prior to the next onset of menstruation. In other words, if the periodicity of the menstrual cycle is 28 days, ovulation usually occurs within one day of the 14th day of the cycle. If, on the other hand, the periodicity of the cycle is 40 days, ovulation usually occurs within one day of the 26th day of the cycle. Finally, if the periodicity of the cycle is 21 days, ovulation usually occurs within one day of the 7th day of the cycle. Therefore, it is usually stated that avoidance of intercourse within 3 days on either side of the calculated day of ovulation prevents conception, and 5 days on either side of the calculated day of ovulation usually provides almost complete safety. Such a method of contraception can be used only when the periodicity of the menstrual cycle is regular, for otherwise it is impossible to determine the next onset of menstruation, and, therefore, it is impossible to predict the day of ovulation.

Hormonal Suppression of Fertility — "The Pill." It has long been known that administration of either estrogens or progesterone, if given in sufficient quantity, can inhibit ovulation. Though the exact mechanism of this effect is not clear, it is believed that in the presence of enough of either or both of these hormones, the hypothalamus fails to secrete the normal surge of LH-releasing factor that usually occurs about 13 days after onset of the monthly sexual cycle. From the discussion earlier in the chapter, it will be recalled that this surge is essential in causing the rapid growth of the follicle, which leads to subsequent ovulation.

The problem in devising methods for hormonal suppression of ovulation has been to develop appropriate combinations of estrogens and progestins that will suppress ovulation but that will not cause unwanted effects of these two hormones. For instance, too much of either of the hormones can cause abnormal menstrual bleeding patterns. However, use of a synthetic progestin in place of progesterone, especially the 19-norsteroids, along with small amounts of estrogens will usually prevent ovulation and yet, also, allow almost a normal pattern of menstruation. Therefore, almost all "pills" used for control of fertility consist of some combination of estrogen and one of the synthetic progestins.

Abnormal Conditions Causing Female Sterility. Approximately one out of every 6 marriages is infertile; in over half of these, the infertility is due to female sterility rather than to male sterility.

Occasionally, no abnormality whatsoever can be discovered in the female genital organs, in which case it must be assumed that the infertility is due either to abnormal physiologic function of the genital system or to abnormal genetic development of the ova themselves.

However, probably by far the most common cause of female sterility is failure to ovulate. This can result from either hyposecretion of gonadotropic hormones, in which case the intensity of the hormonal stimuli simply is not sufficient to cause ovulation, or it can result from abnormal ovaries which will not allow ovulation. For instance, thick capsules occasionally exist on the outside of the ovaries which prevent ovulation.

Because of the high incidence of anovulation in sterile women, special methods are often utilized to determine whether or not ovulation occurs. These are all based on the effects of progesterone on the body, for the normal increase in progesterone secretion does not occur during the latter half of anovulatory cycles. In the absence of progesteronic effects, the cycle can be assumed to be anovulatory. One of the tests is simply to analyze the urine for a surge in pregnanediol during the latter half of the sexual cycle, the lack of which indicates failure of ovulation. However, another common test is for the woman to chart her body temperature throughout the cycle. Secretion of progesterone during the latter half of the cycle raises the body temperature about one-half degree, the temperature rise coming abruptly at the time of ovulation. Such a temperature chart, showing the point of ovulation, is illustrated in Figure 81–8.

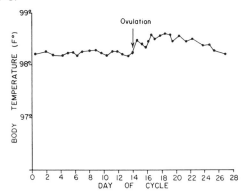

Figure 81–8 Elevation in body temperature shortly after ovulation.

Lack of ovulation caused by hyposecretion of the pituitary gonadotropic hormones can be treated by administration of *human chorionic gonadotropin,* a hormone that will be discussed in the following chapter and that is extracted from the human placenta. This hormone, though secreted by the placenta, has almost exactly the same effects as luteinizing hormone and, therefore, is a powerful stimulator of ovulation. However, excess use of this hormone can cause ovulation from many follicles simultaneously; and this results in multiple births, an effect that has caused as many as six children to be born to mothers treated for infertility with this hormone.

One of the most common causes of female sterility is *endometriosis,* for, as described earlier, endometriosis causes fibrosis throughout the pelvis; and this fibrosis frequently so enshrouds the ovaries that an ovum cannot be released into the abdominal cavity. Often, also, endometriosis occludes the fallopian tubes either at the fimbriated ends or elsewhere along their extent. Another common cause of female infertility is salpingitis, that is, inflammation of the fallopian tubes; this causes fibrosis in the tubes, thereby occluding them. In past years, such inflammation was extremely common as a result of gonococcal infection, but with modern therapy this is becoming a less prevalent cause of female infertility.

REFERENCES

Behrman, S. J., and Kistner, R. W.: Progress in Infertility. Boston, Little, Brown and Company, 1968.

Curet, L. B.; Mann, L.; Abrams, R.; Crenshaw, M. C.; and Barron, D. H.: Effect of estrogen on arterial blood levels of alpha-amino nitrogen. *J. Appl. Physiol., 28*:1, 1970.

Drill, V.: Oral contraceptives. New York. McGraw-Hill Book Company, 1966.

Dukelow, W. R.; Chernoff, H. N.; and Williams, W. L.: Fertilizable life of the rabbit ovum relative to sperm capacitation. *Amer. J. Physiol., 213*:1397, 1967.

Espey, L. L., and Lipner, H.: Enzyme-induced rupture of rabbit Graafian follicle. *Amer. J. Physiol., 208*:208, 1965.

Greenblatt, R. B.: Ovulation: Stimulation, Suppression, Detection. Philadelphia, J. B. Lippincott Co., 1966.

Henricks, D. M.; Oxenreider, S. L.; and Anderson, L. L.: Ovarian function after hypophysial stalk transection in the cow. *Amer. J. Physiol., 216*:1213, 1969.

Kleegman, S. J., and Kaufman, S. A.: Infertility in Women: Diagnosis and Treatment. Philadelphia, F. A. Davis Co., 1966.

Lawton, I. E., and Sawyer, C. H.: Role of amygdala in regulating LH secretion in the adult female rat. *Amer. J. Physiol., 218*: 622, 1970.

Masters, W. H., and Johnson, V. E.: Human Sexual Response. Boston, Little, Brown and Company, 1966.

McCann, S. M.: A hypothalamic luteinizing-hormone-releasing factor. *Amer. J. Physiol., 202*:601, 1962.

McCann, S. M.: Recent studies on the regulation of hypophysial luteinizing hormone secretion. *Amer. J. Med., 34*:379, 1963.

Pincus, G.: The Control of Fertility. New York, Academic Press, Inc., 1965.

Rogers, J.: Endocrine and Metabolic Aspects of Gynecology. Philadelphia, W. B. Saunders Company, 1963.

Savard, K.: Steroid synthesis in the ovary. *Proc. Int. Union of Physiol. Sciences, 6*:243, 1968.

Schmidt-Matthiesen, H.: The Normal Human Endometrium. McGraw-Hill Book Company, New York, 1968.

Shelesnyak, M. C.: Progestation: A biodynamist's view. *Proc. Int. Union of Physiol. Sciences, 6*:91, 1968.

Singhal, R. L., and Valadares, J. R. E.: Estrogenic regulation of uterine pyruvate kinase. *Amer. J. Physiol., 218*:321, 1970.

Vickery, B. H., and Bennett, J. P.: The cervix and its secretion in mammals. *Physiol. Rev., 48*:135, 1968.

Voogt, J. L.; Chen, C. L.; and Meites, J.: Serum and pituitary prolactin levels before, during, and after puberty in female rats. *Amer. J. Physiol., 218*:396, 1970.

CHAPTER 82

Pregnancy and Lactation

In the preceding two chapters the sexual functions of the male and the female were described to the point of fertilization of the ovum. If the ovum becomes fertilized, a completely new sequence of events called *gestation,* or *pregnancy,* takes place, and the fertilized ovum eventually develops into a full-term fetus. The purpose of the present chapter is to discuss the early stages of ovum development after fertilization and then to discuss the physiology of pregnancy. In the following chapter some special problems of fetal and early childhood physiology are discussed.

MATURATION OF THE OVUM AND OOGENESIS

In the female, the geminal epithelium lies on the outer surfaces of the ovaries. Throughout fetal development, small ingrowths of germinal epithelium invaginate into the ovarian stroma to form *primary oocytes* and *granulosa cells* that surround the oocytes. By the time of birth, approximately 500,000 such primary oocytes are present in the two ovaries. Some additional primary oocytes are probably formed after birth as well, and it is possible that a few are formed even throughout the sexual life of the female. Figure 82–1 illustrates this formation of primary oocytes from the germinal epithelium.

Figure 82–1 also shows the process by which the primary oocyte becomes a mature ovum, or *ootid.* Shortly before the ovum is released from the follicle, the so-called *first polar body* is expelled from the nucleus of the ovum, forming a *secondary oocyte.* Soon thereafter the secondary oocyte expels a *second polar body* almost coincidently with ovulation. In this process, each of the 23 pairs of chromosomes loses one of the partners so that 23 *unpaired* chromosomes remain in the ootid, the mature ovum that is now capable of being fertilized. This entire process is called *maturation of the ovum.* As explained in the preceding chapter, only one ovum normally matures and is ovulated into the abdominal cavity during each monthly sexual cycle of the female.

FERTILIZATION OF THE OVUM

After coitus, sperm are transported through the uterus and fallopian tubes to the region of the ovary within about five minutes. This is many times more rapid than the motility of the sperm themselves can account for, which indicates that propulsive movements of the uterus and fallopian tubes might be responsible for much of the sperm movement. It is known, for instance, that in some animals coitus causes the neurohypophysis to secrete oxytocin, which in turn enhances uterine contractions. Also, semen contains a prostaglandin that theoretically could add still further to the contractions.

Sperm can exist in the female genital tract for 24 to 72 hours but can remain healthy and highly fertile probably for only 24 hours. Furthermore, after ovulation a mature ovum also is fertilizable probably for 12 to 24 hours. Therefore, if coitus occurs when these two

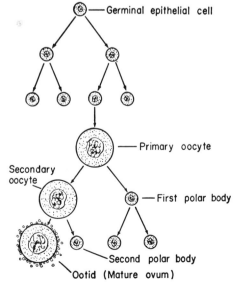

Figure 82–1 The stages of oogenesis. (Modified from Arey: Developmental Anatomy, 7th Ed.)

periods of fertility overlap, fertilization can take place.

Only one sperm is required for fertilization of the ovum, the process of which is illustrated in Figure 82–2. Furthermore, almost never does more than one sperm enter the ovum for the following reason: The zona pellucida of the ovum has a lattice-type structure, and once the ovum is punctured, some substance seems to diffuse out of the ovum into the lattice to prevent penetration by additional sperm. Indeed, microscopic studies show that many sperm do attempt to penetrate the zona pellucida but become inactivated before traveling only part way through.

Once a sperm enters the ovum, its head swells rapidly to form a *male pronucleus,* which is also illustrated in Figure 82–2. Later, the 23 chromosomes of the male pronucleus and the 23 of the *female pronucleus* align themselves to reform a complete complement of 46 chromosomes (23 pairs) in the fertilized ovum.

Sex Determination. The sex of the child is determined by the type of sperm that fertilizes the ovum — that is, whether it is a male sperm or a female sperm. It will be recalled from Chapter 80 that a male sperm carries a *Y sex chromosome* and 22 *autosomal chromosomes,* while a female sperm carries the same 22 autosomal chromosomes but an *X sex chromosome.* On the other hand, the ootid always has an X sex chromosome and never a Y chromosome. After recombination of the male and female pronuclei during fertilization, the fertilized ovum then contains 44 autosomal chromosomes and either 2 X chromosomes, which causes a female child to develop, or an X and a Y chromosome, which causes a male child to develop.

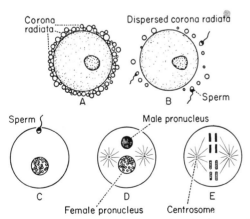

Figure 82–2 Fertilization of the ovum, showing (A) the mature ovum surrounded by the corona radiata, (B) dispersal of the corona radiata, (C) entry of the sperm, (D) formation of the male and female pronuclei, and (E) reorganization of a full complement of chromosomes and beginning division of the ovum. (Modified from Arey: Developmental Anatomy, 7th Ed.)

TRANSPORT AND IMPLANTATION OF THE DEVELOPING OVUM

Entry of the Ovum into the Fallopian Tube. When ovulation occurs, the ovum is expelled directly into the peritoneal cavity and must then enter one of the fallopian tubes. The fimbriated end of each fallopian tube falls naturally around the ovaries, and the inner surfaces of the fimbriated tentacles are lined with ciliated epithelium, the *cilia* of which continually beat toward the *abdominal ostium* of the fallopian tube. One can actually see a slow fluid current flowing toward the ostium. By this means the ovum enters one or the other fallopian tube.

It would seem likely that many ova might fail to enter the fallopian tubes. However, on the basis of conception studies it is probable that as many as 98 to 100 per cent succeed in this task. Indeed, cases are on record in which women with one ovary removed and the opposite fallopian tube removed have had as many as four children with relative ease of conception, thus illustrating that ova can even enter the opposite fallopian tube.

Transport of the Ovum Through the Fallopian Tube. Fertilization of the ovum normally takes place either before the ovum enters the fallopian tube or soon after entering. An additional three days is usually required for transport of the ovum through the tube into the cavity of the uterus. This transport is effected mainly by a feeble fluid current in the fallopian tube resulting from action of the ciliated epithelium, the cilia always beating toward the uterus. It is possible also that weak contractions of the fallopian tube aids in the passage of the ovum.

The fallopian tubes are lined with a rugged, cryptoid surface which actually impedes the passage of the ovum despite the fluid current. The impediment allows several stages of division to occur before the ovum enters the uterus. During this time, large quantities of secretions are formed by the glands lining the fallopian tube, and these are believed to be important for nutrition of the developing ovum. Indeed, fertilized ova continue to divide in vitro as long as they are bathed in a solution of homogenized fallopian tube mucosa, but they will not divide in almost all other media.

Implantation of the Ovum in the Uterus. After reaching the uterus, the developing ovum usually remains in the uterine cavity an additional four to five days before it implants in the endometrium, which means that implantation ordinarily occurs on the seventh to

Figure 82–3 Implantation of the early human embryo, showing trophoblastic digestion and invasion of the endometrium. (Courtesy of Dr. Arthur Hertig.)

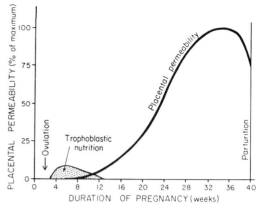

Figure 82–4 Nutrition of the fetus, illustrating that most of the early nutrition is due to trophoblastic digestion and absorption of nutrients from the endometrial decidua and that essentially all the later nutrition results from diffusion through the placental membrane.

eighth day following fertilization. During this time the ovum obtains its nutrition from the endometrial secretions, called the "uterine milk." Figure 82–3 shows a very early stage of implantation, illustrating that the developing ovum is in the *blastocyst stage.*

Implantation results from the action of trophoblastic cells that develop over the surface of the blastocyst. These cells secrete proteolytic enzymes that digest and liquefy the cells of the endometrium. Simultaneously, much of the fluid and nutrients thus released is actively absorbed into the blastocyst as a result of phagocytosis by the trophoblastic cells. Also at the same time, additional trophoblastic cells form cords of cells that extend to and attach to the edges of the digested endometrium. Thus, the blastocyst eats a hole in the endometrium and attaches to it at the same time.

Once implantation has taken place, the trophoblastic and sub-lying cells proliferate rapidly; and these, along with cells from the mother's endometrium, form the placenta and the various membranes of pregnancy.

EARLY INTRA-UTERINE NUTRITION OF THE EMBRYO

In the previous chapter it was pointed out that the progesterone secreted during the latter half of each sexual cycle has a special effect on the endometrium to convert the endometrial stromal cells into large swollen cells that contain extra quantities of glycogen, proteins, lipids, and even some necessary minerals for development of the ovum. If the ovum becomes fertilized, a sequence of hormonal changes occurs almost immediately that cause still more progesterone secretion, fur-

ther swelling of the endometrial cells, and further collection of nutrients in the cells. These cells are now called *decidual cells,* and the total mass of cells is called the *decidua.* It is into the decidua that the ovum implants.

As the trophoblastic cells invade the decidua, digesting and imbibing it, the stored nutrients in the decidua are used by the embryo for appropriate growth and development. During the first week after implantation, this is the only means by which the embryo can obtain any nutrients whatsoever, and the embryo continues to obtain a large measure of its total nutrition in this way for 8 to 12 weeks, though the placenta also begins to provide slight amounts of nutrition after approximately the sixteenth day beyond fertilization (a little over a week after implantation). Figure 82–4 illustrates this trophoblastic period of nutrition, which gradually gives way to placental nutrition.

FUNCTION OF THE PLACENTA

DEVELOPMENTAL AND PHYSIOLOGIC ANATOMY OF THE PLACENTA

While the trophoblastic cords from the blastocyst are attaching to the uterus, blood capillaries grow into the cords from the vascular system of the embryo, and, by the sixteenth day after fertilization, blood begins to flow. Simultaneously, blood sinuses supplied with blood from the mother develop between the surface of the uterine endometrium and the trophoblastic cords. The trophoblastic cells then gradually send out more and more projections, which become the *placental villi* into which capillaries from the fetus grow. Thus, the villi, carrying fetal blood, are surrounded by sinuses containing maternal blood.

The final structure of the placenta is illustrated in Figure 82–5. Note that the fetus' blood flows through the *umbilical arteries*, finally to the capillaries of the villi, and thence back through the *umbilical vein* into the fetus. On the other hand, the mother's blood flows from the *uterine arteries* into large *blood sinuses* surrounding the villi and then back into the *uterine veins* of the mother.

The lower part of Figure 82–5 illustrates the relationship between the fetal blood of the villus and the blood of the mother in the fully developed placenta. The capillaries of the villus are lined with an extremely thin endothelium and are surrounded by a layer of *mesenchymal tissue* which is covered on the outside of the villus by a thin layer of *syncytial trophoblastic* cells. During the first 16 weeks of pregnancy, still an additional layer of cells is present immediately beneath the syncytial trophoblastic layer. This layer is composed of distinct cuboidal cells called *cytotrophoblastic cells,* or *Langerhans' cells.*

The mature placenta has a total surface area of approximately 16 square meters, which is about one-fourth the total area of the pulmonary membrane. However, remember that even at full ma-

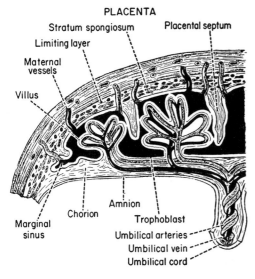

PLACENTA

Stratum spongiosum   Placental septum
Limiting layer
Maternal vessels
Villus

Amnion
Chorion
Marginal sinus   Trophoblast
Umbilical arteries
Umbilical vein
Umbilical cord

VILLUS

Fetal capillaries

Intervillous space

Chorionic epithelium

Figure 82–5 *Above:* Organization of the mature placenta. *Below:* Relationship of the fetal blood in the villus capillaries to the mother's blood in the intervillus spaces. (Modified from Gray and Goss: Anatomy of the Human Body. Lea & Febiger; and from Arey: Developmental Anatomy, 7th Ed.)

turity the placental membrane is still three cells thick, and the distance between the maternal blood and the fetal blood is many times the distance across the alveolar membranes of the lung. Nevertheless, many nutrients and other substances pass through the placental membrane by diffusion in very much the same manner as through the alveolar membranes and capillary membranes elsewhere in the body.

PERMEABILITY OF THE PLACENTAL MEMBRANE

Since the major function of the placenta is to allow diffusion of foodstuffs from the mother's blood into the fetus' blood and diffusion of excretory products from the fetus back into the mother, it is important to know the degree of permeability of the placental membrane. *Permeability is expressed as the total quantity of a given substance that crosses the entire placental membrane for a given concentration difference.*

In the early months of development, placental permeability is relatively slight, as illustrated in Figure 82–4, for two reasons: First, the total surface area of the placental membrane is still small at that time, and, second, the thickness of the membrane is great. However, as the placenta becomes older, the permeability increases progressively until the last month or so of pregnancy when it begins to decrease again. The increase in permeability is caused by both progressive enlargement of the surface area of the placental membrane and progressive thinning of the layers of the villi. On the other hand, the decrease shortly before birth results from deterioration of the placenta caused by its age and sometimes from destruction of whole segments due to infarction of isolated areas.

Occasionally, "breaks" occur in the placental membrane which allow fetal blood cells to pass into the mother, or more rarely, the mother's cells to pass into the fetus. Indeed, there are instances in which the fetus bleeds severely into the mother's circulation because of a ruptured placental membrane.

Diffusion of Oxygen Through the Placental Membrane. Almost exactly the same principles are applicable for the diffusion of oxygen through the placental membrane as through the pulmonary membrane; these principles were discussed in Chapter 40. The dissolved oxygen in the blood of the large placental sinuses simply passes through the villus membrane into the fetal blood because of a pressure gradient of oxygen from the mother's blood to the fetus' blood. The mean Po₂ in the mother's blood in the placental sinuses is approximately 50 mm. Hg toward the end of pregnancy, and the mean Po₂ in the blood leaving the villi and returning to the fetus is about 30 mm. Hg. The mean pressure gradient for diffusion of oxygen through the placental membrane is therefore about 20 mm. Hg.

One might wonder how it is possible for a fetus to obtain sufficient oxygen when the fetal blood leaving the placenta has a Po₂ of only 30 mm. Hg. However, there are three different reasons why even this low Po₂ is capable of allowing the fetal blood to transmit almost as much oxygen to the fetal tissues as is transmitted by the mother's blood to her tissues:

First, the hemoglobin of the fetus is primarily *fetal hemoglobin,* a type of hemoglobin synthesized in the fetus prior to birth. Figure 82–6 illustrates the comparative oxygen dissociation curves of maternal hemoglobin and fetal hemoglobin, showing that the curve for fetal hemoglobin is shifted considerably to the left of that for maternal hemoglobin. This means that at a given Po₂, the fetal hemoglobin can carry as much as 20 to 30 per cent more oxygen than can maternal hemoglobin.

Second, the hemoglobin concentration of the fetus is about 50 per cent greater than that of the mother; and this is an even more important factor than the first in enhancing the amount of oxygen transported to the fetal tissues.

Third, the Bohr effect, which was explained in relation to the exchange of carbon dioxide and oxygen in the lung in Chapter 41, provides another factor that enhances the transport of oxygen by the fetal blood. That is, hemoglobin can carry more oxygen at a high pH than it can at a low pH. The fetal blood entering the placenta carries large amounts of carbon dioxide, and, as explained later, this carbon dioxide diffuses from the fetal blood into the maternal blood. Loss of the carbon dioxide makes the fetal blood more alkaline while the increased carbon dioxide in the maternal blood makes this more acidic. These changes cause the combining capacity of fetal blood for oxygen to become increased while that of maternal blood becomes decreased. This forces still more oxygen from the maternal blood into the fetal blood. Thus, the Bohr shift operates both in the maternal blood and oppositely in the fetal blood, these two effects adding to make the Bohr effect twice as important here as it is for oxygen exchange in the lungs.

By these three means, the fetus is capable of receiving more than adequate oxygen through the placenta despite the fact that the fetal blood leaving the placenta has a Po₂ of only 30 mm. Hg.

The total diffusing capacity of the placenta for oxygen at term is about 1.2 ml. of oxygen per minute per mm. Hg oxygen gradient. This compares favorably with that of the lungs of the newborn baby.

Diffusion of Carbon Dioxide Through the Placental Membrane. Carbon dioxide is continually formed in the tissues of the fetus in the same way that it is formed in maternal tissues. And the only means for excreting the carbon dioxide is through the placenta. The Pco₂ builds up in the fetal blood until it is about 41 to 46 mm. Hg in contrast to about 40 to 45 mm. Hg in maternal blood. Thus, a low pressure gradient for carbon dioxide develops across the placental membrane, but this is sufficient to allow adequate diffusion of carbon dioxide from the fetal blood into the maternal blood, because the extreme solubility of carbon dioxide in the water of the placental membrane allows carbon dioxide to diffuse through this membrane rapidly, about 20 times as rapidly as oxygen.

Diffusion of Foodstuffs Through the Placental Membrane. Other metabolic substrates needed by the fetus diffuse into the fetal blood in the same manner as oxygen. For instance, the glucose level in the fetal blood ordinarily is approximately 20 to 30 per cent lower than the glucose level in the maternal blood, for glucose is being metabolized rapidly by the fetus. This in turn causes rapid diffusion of

Figure 82–6 Oxygen-hemoglobin dissociation curves for maternal (*M*) and fetal (*F*) bloods, illustrating the ability of the fetal blood to carry a much greater quantity of oxygen than can maternal blood for a given blood Po₂. (From Metcalfe, Moll, and Bartels: *Fed. Proc., 23:*775, 1964.)

additional glucose from the maternal blood into the fetal blood.

Because of the small molecular size of the amino acids and because of the high solubility of fatty acids in cell membranes, both these substances also diffuse from the maternal blood into the fetal blood. Also, such substances as calcium, phosphorus, and sodium chloride diffuse from the maternal blood into the fetal blood. As these substances are used by the fetal body, their concentrations in the fetal blood fall, and the increased concentration gradients then cause more of the same substrates to diffuse through the placental membrane.

Active Absorption by the Placental Membrane. As pointed out previously, early nutrition of the embryo depends on phagocytosis of fallopian tube and uterine secretions, and even on phagocytosis of endometrial tissues. The trophoblastic cells that line the outer surface of the villi can probably also actively absorb certain nutrients from the maternal blood in the placenta at least during the first half of pregnancy and perhaps even throughout the entire pregnancy. For instance, the measured amino acid content of fetal blood sometimes is greater than that of maternal blood, and calcium and inorganic phosphate occasionally occur in greater concentration in fetal blood than in maternal blood, while ascorbic acid is as much as three times as concentrated in the fetal blood. This indicates that the placental membrane does have the ability to absorb actively at least small amounts of certain substances.

Excretion Through the Placental Membrane. In the same manner that carbon dioxide diffuses from the fetal blood into the maternal blood, other excretory products formed within the fetus likewise diffuse into the maternal blood and then are excreted along with the excretory products of the mother. These include especially the nonprotein nitrogens, such as urea, uric acid, and creatinine. The level of urea in the fetal blood is only slightly greater than that in maternal blood because urea diffuses through the placental membrane with considerable ease. On the other hand, creatinine, which does not diffuse as easily, has a considerably higher concentration gradient, percentagewise, between the fetal blood and maternal blood. Therefore, insofar as is known, excretion from the fetus occurs entirely as a result of diffusion gradients across the placental membrane—that is, higher concentrations of the excretory products in the fetal blood than in the maternal blood.

STORAGE FUNCTION OF THE PLACENTA

During the first few months of pregnancy, the placenta grows tremendously in size while the fetus remains relatively diminutive. During this same time considerable quantities of metabolic substrates, including proteins, calcium, and iron, are stored in the placenta to be used in the latter months of pregnancy for growth by the fetus. Thus, in the early months of pregnancy, the placenta performs very much the same functions for the fetus that the liver performs for the adult human being, acting as a nutrient storehouse and helping to process some of the food substances that enter the fetus. For instance, the placenta is capable of storing glucose as glycogen and then secreting glucose into the embryonic blood stream in much the same manner that the liver can secrete glucose into the adult blood stream. By this process, the placenta actually controls the fetal blood glucose concentration. Later in the growth of the fetus, these metabolic functions of the placenta become less and less important, while the fetal liver becomes progressively more important.

HORMONAL FACTORS IN PREGNANCY

In pregnancy, the placenta secretes large quantities of *chorionic gonadotropin, estrogens, progesterone,* and *human placental lactogen,* the first three of which, and perhaps the fourth as well, are essential to the continuance of pregnancy. The functions of these hormones are discussed in the following sections.

CHORIONIC GONADOTROPIN AND ITS EFFECT TO CAUSE PERSISTENCE OF THE CORPUS LUTEUM AND TO PREVENT MENSTRUATION

Menstruation normally occurs approximately 14 days after ovulation, at which time most of the secretory endometrium of the uterus sloughs away from the uterine wall and is expelled to the exterior. If this should happen after an ovum has implanted, the pregnancy would terminate. However, this is prevented by the secretion of chorionic gonadotropin in the following manner:

Coincidently with the development of the trophoblastic cells from the early fertilized ovum, the hormone chorionic gonadotropin is secreted by the trophoblastic tissues into the fluids of the mother; and, as illustrated in Figure 82–7, the rate of secretion rises rapidly to reach a maximum approximately seven weeks after ovulation, then decreases to a relatively low value at 16 weeks after ovulation.

Function of Chorionic Gonadotropin. Chorionic gonadotropin is a glycoprotein having a molecular weight of 55,000 to 60,000 and having very much the same function as luteinizing hormone secreted by the pituitary. Most important, it prevents the normal involution of

Figure 82–7 Rates of secretion of estrogens, progesterone, and chorionic gonadotropin at different stages of pregnancy.

the corpus luteum at the end of the sexual month, and instead probably causes the corpus luteum to secrete even larger quantities of its usual hormones, progesterone and estrogens. These excess hormones cause the uterine endometrium to continue growing and to store large amounts of nutrients rather than to be passed in the menstruum. As a result, the *decidua-like cells* that develop in the endometrium during the normal female sexual cycle become actual *decidual cells* by the time the blastocyst begins to implant.

Under the influence of chorionic gonadotropin, the corpus luteum grows to about two times its initial size by a month or so after pregnancy begins, and its continued secretion of estrogens and progesterone maintains the decidual nature of the uterine endometrium, which is necessary to the early development of the placenta and other fetal tissues. If the corpus luteum is removed before approximately the eleventh week of pregnancy, spontaneous abortion usually occurs, though after this time the placenta itself secretes sufficient quantities of progesterone and estrogens to maintain pregnancy for the remainder of the gestation period.

EFFECT OF CHORIONIC GONADOTROPIN ON THE TESTES. Chorionic gonadotropin also exerts an *interstitial cell-stimulating effect* on the testes, thus resulting in the production of testosterone in male fetuses, and the small secretion of testosterone during gestation helps create the masculine sexual organs. Near the end of pregnancy, the testosterone secreted by the fetal testes causes the testicles to descend into the scrotum.

Clinically, chorionic gonadotropin administered to the cryptorchid male often causes the testicles to descend into the scrotum by causing testosterone to be secreted, which effects the actual testicular descent.

HUMAN PLACENTAL LACTOGEN

Recently, a new hormone called human placental lactogen has been discovered. This hormone begins to be secreted about the fifth week of pregnancy, and its rate of secretion increases progressively throughout pregnancy. In lower animals it has a luteotropic function, causing the corpora lutea to secrete large quantities of estrogens and progesterone. Though this function has not yet been proved in human beings, it is likely that this and many other functions previously ascribed to chorionic gonadotropin are perhaps performed by human placental lactogen. Thus, chorionic gonadotropin is analogous to luteinizing hormone, whereas human placental lactogen is analogous to luteotropic hormone.

Human placental lactogen also has a prolactin-like effect on the breasts and probably plays a major role in development of the mother's breasts prior to birth of the baby. It also has some similarities to growth hormone; for instance, inhibiting the action of insulin to cause glucose transport through cell membranes, and also promoting cellular growth. It is a small protein, having a molecular weight in the range of 20,000 to 30,000; and it is probably synthesized in the syncytial trophoblastic cells along with the other hormones secreted by the placenta.

SECRETION OF ESTROGENS BY THE PLACENTA

The placenta, like the corpus luteum, secretes both estrogens and progesterone. Both histochemical and physiologic studies indicate that these two hormones are secreted by the *syncytial trophoblastic cells,* along with the secretion of chorionic gonadotropin and human placental lactogen.

Figure 82–7 shows that the daily production of placental estrogens increases markedly toward the end of pregnancy, to as much as 300 times the daily production in the middle of a normal monthly sexual cycle.

In the discussions of estrogens in the preceding chapter it was pointed out that these hormones exert mainly a proliferative function on certain reproductive and associated organs. During pregnancy, the extreme quantities of estrogens cause (1) enlargement of the uterus, (2) enlargement of the breasts and growth of the breast glandular tissue, and (3) enlargement of the female external genitalia.

The estrogens also relax the various pelvic ligaments so that the sacroiliac joints become relatively limber and the symphysis pubis becomes elastic. These changes obviously make for easy passage of the fetus through the birth canal.

There is much reason to believe that estrogens also affect the development of the fetus during pregnancy, for example, by controlling the rate of cell reproduction in the early embryo.

SECRETION OF PROGESTERONE BY THE PLACENTA

Progesterone is also a hormone essential for pregnancy. In addition to being secreted in moderate quantities by the corpus luteum at the beginning of pregnancy, it is secreted in tremendous quantities by the placenta, sometimes as much as 1 gram per day, toward the end of pregnancy. Indeed, the rate of progesterone secretion increases by as much as 10-fold during the course of pregnancy, as illustrated in Figure 82–7.

The special effects of progesterone that are essential for normal progression of pregnancy are the following:

1. As pointed out earlier, progesterone causes decidual cells to develop in the uterine endometrium, and these then play an important role in the nutrition of the early embryo.

2. Progesterone has a special effect to decrease the contractility of the gravid uterus, thus preventing uterine contractions from causing spontaneous abortion.

3. Progesterone also contributes to the development of the ovum even prior to implantation, for it specifically increases the secretions of the fallopian tubes and uterus to provide appropriate nutritive matter for the developing *morula* and *blastocyst*. There are some reasons to believe, too, that progesterone even helps to control cell cleavage in the early developing embryo.

4. The progesterone secreted during pregnancy also helps to prepare the breasts for lactation, which is discussed in detail later in the chapter.

OTHER HORMONAL FACTORS IN PREGNANCY

Almost all the nonsexual endocrine glands of the mother react markedly to pregnancy. This results mainly from the increased metabolic load on the mother but also to some extent from inverse effects of placental hormones on the pituitary and other glands. Some of the most notable effects are the following:

Pituitary Secretion. The anterior pituitary gland enlarges at least 50 per cent during pregnancy and increases its production of corticotropin, thyrotropin, and probably also growth hormone. On the other hand, production of follicle-stimulating hormone and luteinizing hormone is greatly suppressed as a result of the inhibitory effects of estrogens and progesterone from the placenta.

Corticosteroid Secretion. The rate of adrenocortical secretion of the glucocorticoids is moderately increased throughout pregnancy. It is possible that the glucocorticoids help to mobilize amino acids from the mother's tissues so that these can be used for synthesis of tissues in the fetus.

Pregnant women also have increased secretion of aldosterone. This, along with the actions of the estrogens and progesterone, causes a tendency for even the normal pregnant woman to reabsorb excess sodium from the renal tubules and therefore to retain fluid.

Secretion by the Thyroid Gland. The thyroid gland ordinarily enlarges about 50 per cent during pregnancy and increases its production of thyroxine a corresponding amount. The increased thyroxine production is caused primarily by increased thyrotropic hormone from the adenohypophysis.

Secretion by the Parathyroid Glands. The parathyroid glands also often enlarge during pregnancy; this is especially true if the mother is on a calcium deficient diet. Enlargement of these glands causes calcium absorption from the mother's bones, thereby maintaining normal calcium ion concentration in the mother's extracellular fluids as the fetus removes calcium for ossifying its own bones.

Secretion of "Relaxin" by the Ovaries. An additional substance besides the estrogens and progesterone, a hormone called relaxin, can be isolated from the corpora lutea of the ovaries. This hormone, when injected, causes relaxation of the ligaments of the symphysis pubis in the estrous rat and guinea pig. However, this effect is very poor in the pregnant woman. Instead, this role is probably played by both the estrogens and progesterone, which also cause relaxation of the pelvic ligaments, though many days are required for these hormones to act.

It has also been claimed that relaxin has two other effects: (1) softening of the cervix of the pregnant woman at the time of delivery and (2) inhibition of uterine motility.

Relaxin is a polypeptide, having a molecular weight of about 9000.

In summary, relaxin is a substance that can be isolated from the ovaries, but its functional importance is almost totally unknown.

RESPONSE OF THE MOTHER TO PREGNANCY

Obviously, the presence of a growing fetus in the uterus adds an extra physiologic load on the mother, and much of the response of the mother to pregnancy is due to this increased load. The hormones secreted during pregnancy either by the placenta or by the endocrine glands can also cause many reactions in the mother. Among the reactions are increased size of the various sexual organs. For instance, the uterus increases from about 30 grams to about 700 grams, and the breasts approximately double in size. At the same time the vagina enlarges, and the introitus opens more widely. Also, the various hormones can cause marked changes in the appearance of the mother, sometimes resulting in the development of edema, acne, and masculine or acromegalic features.

CHANGES IN THE MATERNAL CIRCULATORY SYSTEM DURING PREGNANCY

Blood Flow Through the Placenta. About 750 ml. of blood flows through the maternal circulation of the placenta each minute during the latter phases of gestation. Obviously, the more rapidly this blood flows, the greater will be the concentration of oxygen and other metabolites in the fetal blood.

Cardiac Output of the Mother. The flow of blood through the placenta decreases the total peripheral resistance of the mother's circulatory system and, consequently, allows increased venous return of blood to the heart, which in turn tends to increase the cardiac output in the same manner that arteriovenous shunts increase the output. This factor, plus a general increase in metabolism, increases the cardiac output to 30 to 40 per cent above normal by the twenty-seventh week of pregnancy, but then, for reasons yet unexplained, the cardiac output falls near to normal during the last eight weeks of pregnancy, despite the high uterine blood flow.

Blood Volume of the Mother. The maternal blood volume shortly before term is approximately 30 per cent above normal. This increase occurs mainly during the latter half of pregnancy, as illustrated by the curve of Figure 82-8. The cause of the increased volume is probably mainly hormonal, for adrenocortical hormones, estrogens, and progesterone can all cause increased fluid retention by the kidneys, and all these hormones are secreted into the maternal body in very large quantities during the latter half of pregnancy. It is probable, also, that the shunting effect of blood through the placenta decreases the capillary pressures throughout the mother's body, which may be re-

sponsible for some increase in blood volume, for blood volume increases routinely when any type of arteriovenous shunt is prepared in an animal.

As a result of the increased blood volume in the mother, the hematocrit at first decreases because of dilution of the blood. However, toward the end of pregnancy the bone marrow becomes increasingly active, and the concentration of red blood cells returns almost to normal. Therefore, at the time of birth of the baby, the mother has approximately 1 to 2 liters of extra blood in her circulatory system. Only about one-fourth of this amount is normally lost during delivery of the baby, thereby allowing a considerable safety factor for the mother.

WEIGHT GAIN IN THE MOTHER

During the first months of pregnancy, the mother ordinarily loses a few pounds of weight, possibly as a result of nausea, but, during the entire pregnancy, the average weight gain is approximately 24 pounds, most of this gain occurring during the last two trimesters. Of this increase in weight, approximately 7 pounds is fetus, and approximately 4 pounds is amniotic fluid, placenta, and fetal membranes. This leaves 13 pounds increase in weight by the mother herself. The uterus increases approximately 2 pounds, and the breasts increase approximately 3 pounds, still leaving an average increase in weight of the mother herself of approximately 8 pounds. On the average, this extra 8 pounds of weight is accounted for by about 3 pounds gain in protein, 3 pounds of fluid, and 2 pounds of fat.

Often during pregnancy the mother has a greatly increased desire for food, partly as a result of fetal removal of food substrates from the mother's blood and partly because of hormonal factors. Without appropriate prenatal care some mothers eat tremendous quantities of food, and the weight gain, instead of averaging 24 pounds, may be as great as 75 pounds or more. In these instances, the weight gain is due to actual deposition of fat.

METABOLISM IN THE MOTHER DURING PREGNANCY

As a consequence of the increased secretion of many different hormones during pregnancy, including thyroxine, adrenocortical hormones, and the sex hormones, the basal metabolic rate of the mother increases about 15 per cent during the latter half of pregnancy. As a result, the mother frequently has sensations of becoming overheated. Also, owing to the extra load that the mother is carrying, greater amounts of energy than normally used must be expended for muscular activity.

Nutrition during Pregnancy. The supplemental food needed by the mother during pregnancy to

Figure 82-8 Effect of pregnancy on blood volume.

supply the needs of the fetus and fetal membranes includes especially extra dietary quantities of the various minerals, vitamins, and proteins. The growing fetus assumes priority in regard to many of the nutritional elements in the mother's body fluids, and many portions of the fetus continue to grow even though the mother does not eat a sufficient diet. For instance, lack of adequate nutrition in the mother hardly changes the rate of growth of the fetal nervous system, and the length of the fetus increases almost normally; on the other hand, lack of adequate nutrition can decrease the fetus' weight considerably, can decrease ossification of the bones, and can cause anemia, hypoprothrombinemia, and decreased size of many bodily organs of the fetus.

By far the greatest growth of the fetus occurs during the last trimester of pregnancy; the weight of the child almost doubles during the last two months of pregnancy. Ordinarily, the mother does not absorb sufficient protein, calcium, phosphorus, and iron from the gastrointestinal tract during the last month of pregnancy to supply the fetus. However, from the beginning of pregnancy the mother's body has been storing these substances to be used during the latter months of pregnancy. Some of this storage is in the placenta, but most of it is in the normal storage depots of the mother.

If appropriate nutritional elements are not present in the mother's diet, a number of maternal deficiencies can occur during pregnancy. Such deficiencies often occur for calcium, phosphates, iron, and the vitamins. For example, approximately 375 mg. of iron is needed by the fetus to form its blood and an additional 600 mg. is needed by the mother to form her own extra blood. The normal store of nonhemoglobin iron in the mother at the outset of pregnancy is often only 100 or so mg. and almost never over 700 mg. Therefore, without sufficient iron in the food the mother herself usually develops anemia. Therefore, in general, the obstetrician supplements the diet of the mother with the needed substances. It is especially important that the mother receive large quantities of vitamin D, for, even though the total quantity of calcium utilized by the fetus is small, calcium even normally is poorly absorbed by the gastrointestinal tract. Finally, shortly before birth of the baby vitamin K is often added to the diet so that the baby will have sufficient prothrombin to prevent postnatal hemorrhage.

RESPIRATION IN THE MOTHER DURING PREGNANCY

Because of the increased basal metabolic rate of the mother, and also because of the mother's increase in size, the total amount of oxygen utilized by the mother shortly before birth of the baby is usually approximately 20 per cent above normal. This causes the minute ventilation to increase about 50 per cent. Simultaneously, the growing uterus presses upward against the abdominal contents, and these in turn press upward against the diaphragm so that the total excursion of the diaphragm is decreased. Consequently, the respiratory rate must be increased to maintain adequate ventilation.

FUNCTION OF THE MATERNAL URINARY SYSTEM DURING PREGNANCY

The rate of urine formation by the pregnant mother is usually slightly increased because of increased load of excretory products. But, in addition, several special alterations of urinary function are as follows:

First, reabsorption of sodium, chloride, and water by the renal tubules tends to be increased greatly as a consequence of increased production of steroid hormones by the placenta and adrenal cortex.

Second, the glomerular filtration rate often increases as much as 50 per cent during pregnancy, which tends to increase the rate of water and electrolyte loss in the urine. This factor normally almost balances the first so that the mother ordinarily does not get into trouble from excess water and salt accumulation except when she develops *toxemia of pregnancy;* this disease is discussed later in the chapter.

Third, the ureters usually dilate during pregnancy because of two different factors: The enlarged uterus compresses the ureters as they pass over the pelvic rim, thereby increasing the intraureteral pressure. But, aside from this compression factor, the hormones secreted during pregnancy have a direct effect in relaxing the ureters. The two hormones that have been particularly implicated are progesterone and relaxin, though there is still much doubt about the ureteral effects of either of these. Ureteral distention and the accompanying renal pelvic distention that also results predisposes to infection in the urinary tract, frequently leading to serious renal debility, one of the common complications of pregnancy.

THE AMNIOTIC FLUID AND ITS FORMATION

Normally, the volume of amniotic fluid is between 500 ml. and 1 liter, but, it can be only a few milliliters or as much as several liters. Studies with isotopes of the rate of formation of amniotic fluid shows that on the average the water in amniotic fluid is completely replaced once every three hours, and the electrolytes sodium and potassium are replaced once every 15 hours. Yet, strangely enough, the sources of the fluid and the points of reabsorption are mainly unknown. A small portion of the fluid is derived from renal excretion by the fetus. Likewise, a certain amount of absorption occurs by way of the gastrointestinal tract and lungs of the fetus. However, even after death of the fetus, the rate of turnover of the amniotic fluid is still one-half as great as it is when the fetus is normal, which indicates that much of the fluid is formed and absorbed directly through the amniotic membranes. The total volume of amniotic fluid could be regulated by the amniotic membranes themselves, for as the volume increased the membranes would stretch and presumably would absorb increased quantities of fluid as a result, thus returning the volume to normal.

ABNORMAL RESPONSES OF THE MOTHER TO PREGNANCY

HYPEREMESIS GRAVIDARUM

In the earlier months of pregnancy, the mother frequently develops hyperemesis gravidarum, a condition characterized by nausea and vomiting and commonly known as "morning sickness." Occasionally, the vomiting becomes so severe that the mother becomes greatly dehydrated, and in rare instances the condition even causes death.

The cause of the nausea and vomiting is unknown, but it occurs to its greatest extent during the same time that chorionic gonadotropin is secreted in large quantities by the placenta. Because of this coincidence, many clinicians believe that chorionic gonadotropin is in some way responsible for the nausea and vomiting. It has been supposed that other tissues of the mother besides the corpus luteum are affected by chorionic gonadotropin, thereby predisposing to the development of hyperemesis gravidarum.

On the other hand, during the first few months of pregnancy rapid trophoblastic invasion of the endometrium also takes place, and, because the trophoblastic cells digest portions of the endometrium as they invade it, it is possible that degenerative products resulting from this invasion, instead of chorionic gonadotropin, are responsible for the nausea and vomiting. Indeed, degenerative processes in other parts of the body, such as following gamma ray irradiation and burns, can all cause similar nausea and vomiting.

TOXEMIA OF PREGNANCY

Approximately 7 per cent of all pregnant women experience rapid weight gain, edema, and often elevation of arterial pressure. This condition, known as toxemia of pregnancy, is characterized by inflammation and spasm of the arterioles in many parts of the body. It is also characterized by a decrease in both renal blood flow and glomerular filtration rate, which is exactly the opposite effect to that which occurs in the normal pregnant woman.

Various attempts have been made to prove that toxemia of pregnancy is caused by excessive secretion of placental or adrenal hormones, but proof of a hormonal basis for toxemia is yet completely lacking. Indeed, it is now believed that toxemia of pregnancy is not caused by abnormal hormonal balances. A more plausible theory is that toxemia of pregnancy results from some type of autoimmunity or allergy resulting from the presence of the fetus. Indeed, the acute symptoms disappear within a few days after birth of the baby.

Because of the diminished glomerular filtration rate in the kidneys, one of the major problems of toxemia is retention of salt and water. Therefore, it is a dictum among obstetricians that any pregnant woman who has a tendency toward toxemia must limit her salt intake. It usually is not necessary to limit the mother's water intake, for limitation of salt alone prevents excessive absorption of water by the kidney tubules for reasons discussed in Chapter 38.

Eclampsia. Eclampsia is a severe degree of toxemia that occurs in one out of several hundred pregnancies. Milder degrees of eclampsia are sometimes called *pre-eclampsia*.

Eclampsia is characterized by extreme vascular spasticity throughout the body, clonic convulsions followed by coma, greatly decreased kidney output, malfunction of the liver, hypertension, and a generalized toxic condition of the body. Usually, it occurs shortly before, or sometimes within a day or so, after parturition.

Even with the best treatment, some 5 per cent of eclamptic mothers still die. However, injection of vasodilator drugs plus dehydration of the patient can often reverse the vascular spasm and lower the blood pressure, bringing the mother out of the eclamptic state.

PARTURITION

INCREASED UTERINE IRRITABILITY NEAR TERM

Parturition means simply the process by which the baby is born. At the termination of pregnancy the uterus becomes progressively more excitable until finally it begins strong rhythmic contractions with such force that the baby is expelled. The exact cause of the increased activity of the uterus is not known, but at least two major categories of effects lead up to the culminating contractions responsible for parturition; these are, first, progressive hormonal changes that cause increased excitability of the uterine musculature, and, second, progressive mechanical changes.

Hormonal Factors That Cause Increased Uterine Contractility. RATIO OF ESTROGENS TO PROGESTERONE. Progesterone probably inhibits uterine contractility during pregnancy, thereby helping to prevent expulsion of the fetus. On the other hand, estrogens have a definite tendency to increase the degree of uterine contractility. Both these hormones are secreted in progressively greater quantities throughout pregnancy, but from the seventh month onward estrogen secretion increases more than progesterone secretion, and immediately before term relatively large quantities of nonconjugated estrogens ("free estrogens") appear in the extracellular fluids. Therefore, it has been postulated that the *estrogen to progesterone ratio* increases sufficiently toward the end of pregnancy to be at least partly responsible for the increased contractility of the uterus.

EFFECT OF OXYTOCIN ON THE UTERINE. Oxytocin is a hormone secreted by the neurohypophysis that specifically causes uterine contraction (see Chap. 75). There are four reasons for believing that oxytocin might be particularly important in increasing the contractility of the uterus near term. (1) The uterus increases its responsiveness to a given dose of oxytocin during the latter few months of pregnancy. (2) The rate of oxytocin secretion by the neurohypophysis, as judged from thus far rather incomplete studies, seems to be considerably increased at the time of labor. (3) Though hypophysectomized animals and human beings can still deliver their young at term, labor is prolonged. (4) Recent experiments in animals indicate that irritation or stretching of the uterine cervix, as occurs during labor, can cause a neurogenic reflex to the neurohypophysis to increase the rate of oxytocin secretion.

Mechanical Factors that Increase the Contractility of the Uterus. STRETCH OF THE UTERINE MUSCULATURE. Simply stretching smooth muscle organs usually increases their contractility. Furthermore, intermittent stretch, as occurs repetitively in the uterus because of movements of the fetus, can also elicit smooth muscle contraction.

Note especially that twins are born on the average *19 days* earlier than a single child, which emphasizes the importance of mechanical stretch in eliciting uterine contractions.

STRETCH OR IRRITATION OF THE CERVIX. There is much reason to believe that stretch or irritation of the uterine cervix is particularly important in eliciting uterine contractions. For instance, the obstetrician frequently induces labor by dilating the cervix or by rupturing the membranes so that the head of the baby stretches the cervix more forcefully than usual or irritates it in some other way.

The mechanism by which cervical irritation excites the body of the uterus is not known.

It has been supposed that stretch or irritation of neuronal cells in the cervix initiates reflexes to the body of the uterus, but the effect could also result simply from myogenic transmission from the cervix to the body of the uterus.

ONSET OF LABOR – A POSITIVE FEEDBACK THEORY FOR ITS INITIATION

During the last few months of pregnancy the uterus undergoes periodic episodes of weak and slow rhythmic contractions called *Braxton-Hicks contractions.* These become progressively stronger toward the end of pregnancy; and they eventually change rather suddenly, within hours, to become exceptionally strong contractions that start stretching the cervix and later force the baby through the birth canal, thereby causing parturition. This process is called *labor,* and the strong contractions that result in final parturition are called *labor contractions.*

Yet, strangely enough, we do not know what suddenly changes the slow and weak rhythmicity of the uterus into the strong labor contractions. Most obstetricians feel that one of the factors discussed above that increase uterine contractility suddenly becomes strong enough to promote these contractions. However, on the basis of experience during the past few years with other types of control systems, a theory has been proposed for explaining the onset of labor based on "positive feedback." This theory suggests that stretch of the cervix by the fetus' head finally becomes great enough to elicit a reflex increase in contractility of the uterine body. This pushes the baby forward, which stretches the cervix some more and initiates a new cycle. Thus, the process continues again and again until the baby is expelled. This theory is illustrated in Figure 82–9, and the data supporting it are the following:

First, labor contractions obey all the principles of positive feedback. That is, once the strength of uterine contraction becomes greater than a critical value, each contraction leads to subsequent contractions that become stronger and stronger until maximum effect is achieved. Referring to the discussion in Chapter 1 of positive feedback in control systems, we see that this is the precise nature of all positive feedback mechanisms that have a feedback gain of more than one.

Second, the next problem is to explain how positive feedback could set off the crescendo of strength of the uterine contractions. We have

1. Baby's head stretches cervix...

2. Cervical stretch excites fundic contraction...

3. Fundic contraction pushes baby down and stretches cervix some more...

4. Cycle repeats over and over again...

Figure 82-9 Theory for the onset of intensely strong contractions during labor.

two possible types of positive feedback in the uterus at term. First, the propensity for cervical irritation to cause contraction of the body of the uterus could cause positive feedback in the following way: An initial contraction of the uterine body could force the fetus' head or the fluid in the amniotic cavity against the cervix to stretch, tear, or irritate it. This stimulus of the cervix could in turn lead to feedback that would cause additional contraction of the uterine body. As a result, the next contraction would become stronger, and the process would proceed on and on with progressively stronger contractions until delivery.

A second positive feedback mechanism involves the secretion of oxytocin by the posterior pituitary gland. On the basis of studies in cows, this results from the following sequence of events: (a) uterine contraction, (b) cervical stretch or other type of cervical stimulation, (c) nerve signals from the cervix to the hypothalamic-pituitary axis, (d) increased secretion of oxytocin, and (e) increased contractility of the uterus, which leads to (f) a vicious cycle of increasing contraction.

Third, we now have to explain why the weak Braxton-Hicks contractions of midpregnancy do not lead to labor. Referring once more to the discussion of positive feedback in Chapter 1, we find that positive feedback causes a vicious cycle that goes to completion only when each feedback response is greater than the previous response. Thus, if each *increase* in strength of uterine contraction leads to a greater *increase* in strength of contraction the next time, the process would go to completion, but, if the second *increase* in contraction

is not greater than the first increase, the process would fade out. Therefore, the degree of contractility of the uterus must reach a certain stage of development before the positive feedback can become a vicious cycle. Up to that time, the natural contractions can continue without developing into frank labor contractions.

To summarize the theory, we can assume that multiple factors increase the contractility of the uterus toward the end of pregnancy. These are additive in their effects and they cause the Braxton-Hicks contractions to become progressively stronger. Eventually, one of these becomes strong enough that the contraction itself irritates the uterus, increases its contractility because of positive feedback, and results in a second contraction stronger than the first, and a third stronger than the second, and so forth. Once these contractions become strong enough to cause this type of feedback, with each succeeding contraction greater than the one preceding, the process proceeds to completion — all simply *because positive feedback becomes a vicious cycle when the degree of feedback is greater than one.*

One might immediately ask about the many instances of false labor in which the contractions become stronger and stronger and then fade away. Remember that for a vicious cycle to continue, *each* contraction with each new cycle must be stronger than the previous one. If, at any time after labor starts, some cycles fail to reexcite the uterus sufficiently, the positive feedback could go into a retrograde succession and the labor contractions would fade away.

ABDOMINAL CONTRACTION DURING LABOR

Once labor contractions become strong and painful, neurogenic reflexes, mainly from the birth canal to the spinal cord and thence back to the abdominal muscles, cause intense abdominal contraction. This additional contraction of the abdominal muscles and the reflexes causing it add greatly to the positive feedback that eventually causes expulsion of the fetus.

MECHANICS OF PARTURITION

The uterine contractions during labor begin at the top of the uterine fundus and spread downward over the body of the uterus. Also, the intensity of contraction is great in the top and body of the uterus but weak in the lower segment of the uterus adjacent to the cervix. Therefore, each uterine contraction tends to force the baby downward toward the cervix.

In the early part of labor, the contractions might occur only once every 30 minutes and last for about 1 minute. As labor progresses, the contractions finally appear as often as once every one to three minutes, and the intensity of contraction increases greatly with only a short period of relaxation between contractions.

The combined contractions of the uterine and abdominal musculature during delivery of the baby cause a downward force on the fetus of approximately 25 pounds during each strong contraction.

It is fortunate that the contractions of labor occur intermittently because strong contractions impede or sometimes even stop blood flow through the placenta and would cause death of the fetus were the contractions continuous. Indeed, in clinical use of various uterine stimulants, such as oxytocin, overuse of the drugs can cause uterine spasm rather than rhythmic contractions and can lead to death of the fetus.

In 19 out of 20 births the head is the first part of the baby to be expelled, and in most of the remaining instances the buttocks are presented first. The head acts as a wedge to open the structures of the birth canal as the fetus is forced downward from above.

The first major obstruction to expulsion of the fetus is the uterine cervix. Toward the end of pregnancy the cervix becomes soft, which allows it to stretch when labor pains cause the body of the uterus to contract. The so-called *first stage of labor* is the period of progressive cervical dilatation, lasting until the opening is as large as the head of the fetus. This stage usually lasts 8 to 24 hours in the first pregnancy but often only a few minutes after many pregnancies.

Once the cervix has dilated fully, the fetus' head moves rapidly into the birth canal, and, with additional force from above, continues to wedge its way through the canal until delivery is effected. This is called the *second stage of labor*, and it may last from as little as a minute after many pregnancies up to half an hour or more in the first pregnancy.

SEPARATION AND DELIVERY OF THE PLACENTA

During the succeeding 10 to 45 minutes after birth of the baby, the uterus contracts to a very small size, which causes a *shearing* effect between the walls of the uterus and the placenta, thus separating the placenta from its implantation site. Obviously, separation of the placenta opens the placental sinuses and causes bleeding. However, the amount of bleeding is limited to an average of 350 ml. by the following mechanism: The smooth muscle fibers of the uterine musculature are arranged in figures of 8 around the blood vessels as they pass through the uterine wall. Therefore, contraction of the uterus following delivery of the baby constricts the vessels that had previously supplied blood to the placenta.

LABOR PAINS

With each contraction of the uterus the mother experiences considerable pain. The pain in early labor is probably caused mainly by hypoxia of the uterine muscle resulting from compression of the blood vessels to the uterus. This pain is not felt when the *hypogastric nerves*, which carry the sensory fibers leading from the uterus, have been sectioned. However, during the second stage of labor, when the fetus is being expelled through the birth canal, much more severe pain is caused by cervical stretch, compression of neurons in the cervix, peritoneal stretch, and tearing of structures in the vaginal canal itself. This pain is conducted by somatic nerves instead of by the hypogastric nerves.

INVOLUTION OF THE UTERUS

During the first four to five weeks following parturition, the uterus involutes. Its weight becomes less than one-half its immediate postpartum weight within a week, and in four weeks the uterus may be as small as it had been prior to pregnancy — that is, if the mother lactates. This effect of lactation is discussed later. During early involution of the uterus the placental site on the endometrial surface autolyzes, causing a vaginal discharge known as "lochia," which is first bloody and then serous in nature, continuing in all for approximately a week and a half. After this time, the endometrial surface will have become re-epithelialized and ready for normal, nongravid sex life again.

LACTATION

DEVELOPMENT OF THE BREASTS

The breasts begin to develop at puberty; this development is stimulated by the estrogens of the monthly sexual cycles that stimulate growth of the stroma and ductile system plus deposition of fat to give mass to the breasts. However, much additional growth occurs during pregnancy, and the glandular tissue only then becomes completely developed for actual production of milk.

Growth of the Ductile System—Role of the Estrogens. All through pregnancy, the tremendous quantities of estrogens secreted by the placenta cause the ductile system of the breasts to grow and to branch. Simultaneously, the stroma of the breasts also increases in quantity, and large quantities of fat are laid down in the stroma.

Also, moderate quantities of growth hormone from the pituitary or *human placental lactogen* from the placenta are required for the estrogens to produce their effect on the breasts. These latter two hormones both cause protein deposition in the glandular cells, which is essential to their growth.

Development of the Lobule-Alveolar System— Role of Progesterone. The synergistic action of estrogens and growth hormone can cause only a primitive lobule-alveolar system to develop in the breasts at the same time that the ducts are growing, but the simultaneous action of progesterone causes growth of the lobules, budding of alveoli, and development of secretory characteristics in the cells of the alveoli. These changes, obviously, are analogous to the secretory effects of progesterone on the endometrium of the uterus.

Function of Prolactin and Other Hormones in Development of the Breasts. Prolactin is the hormone most probably concerned with causing milk secretion after birth of the baby, and it is probably the same hormone as luteotropic hormone, the importance of which for maintaining secretion by the corpus luteum was discussed in the preceding chapter. In lower animals prolactin has a powerful synergistic effect with estrogens and progesterone to stimulate development of the alveolar secretory system of the breast. Therefore, it is assumed that prolactin might play a similar role in the human being in aiding the final development of the breasts for lactation. It is possible that human placental lactogen from the placenta might also play a similar role during pregnancy. This hormone has physiologic properties almost the same as those of prolactin, though very little is yet known about it.

In addition to prolactin, at least three of the general hormones are necessary to provide appropriate background metabolism before the breasts will develop. These are thyroid hormone, the adrenal corticosteroids, and insulin. However, these hormones seem to have only a "permissive" effect rather than a specific effect in the development of the breasts.

INITIATION OF LACTATION — PROLACTIN

By the end of pregnancy, the mother's breasts are fully developed for nursing, but only a few milliliters of fluid is secreted each day until after the baby is born. This fluid is called *colostrum;* it contains essentially the same amounts of proteins and lactose as milk but almost no fat, and its maximum rate of production is about $1/100$ the subsequent rate of milk production.

The absence of lactation during pregnancy is believed to be caused by suppressive effects of progesterone and estrogens on the milk secretory process of the breasts and also on the secretion of prolactin by the pituitary.

However, immediately after the baby is born, the sudden loss of both estrogen and progesterone secretion by the placenta removes any inhibitory effects of these two hormones and allows marked production of prolactin by the pituitary, as illustrated in Figure 82–10. The prolactin stimulates synthesis of large quantities of fat, lactose, and casein by the mammary glandular cells, and within two to three days the breasts begin to secrete copious quantities of milk instead of colostrum. The sudden onset of milk secretion requires in addition to prolactin an adequate background secretion of both growth hormone and the adrenal corticosteroids. However, only prolactin has the specific effect of causing true lactation.

THE EJECTION OR "LET-DOWN" PROCESS IN MILK SECRETION — FUNCTION OF OXYTOCIN

Milk is secreted continuously into the alveoli of the breasts, but milk does not flow easily from the alveoli into the ductile system and therefore does not continually leak from the breast nipples. Instead, the milk must be "ejected" or "let-down" from the alveoli to the ducts before the baby can obtain it. This process is caused by a combined neurogenic and hormonal reflex involving the hormone *oxytocin* as follows:

When the baby suckles the breast, sensory impulses are transmitted through somatic nerves to the spinal cord and then to the hypothalamus, there causing *oxytocin* and, to a lesser extent, *vasopressin* secretion, as described in Chapter 75. These two hormones,

Figure 82-10 Changes in rates of secretion of estrogens, progesterone, and prolactin at parturition and during the succeeding weeks after parturition, showing especially the rapid increase in prolactin secretion immediately after parturition.

principally oxytocin, flow in the blood to the breasts where they cause the *myoepithelial cells* that surround the outer walls of the alveoli to contract, thereby expressing the milk from the alveoli into the ducts. Thus, within 30 seconds to a minute after a baby begins to suckle the breast, milk begins to flow. This process is called milk ejection, or milk let-down.

Suckling on one breast causes milk flow not only in that breast but also in the opposite breast.

Inhibition of Milk Ejection. A particular problem in nursing the baby comes from the fact that many psychogenic factors as well as generalized sympathetic stimulation throughout the body can inhibit oxytocin secretion and consequently depress milk ejection. For this reason, the mother must have an undisturbed puerperium if she is to be successful in nursing her baby.

CONTINUED SECRETION OF MILK FOR MANY MONTHS FOLLOWING PREGNANCY

If milk is not continually removed from the mother's breasts, the ability of the breasts to continue secreting milk is lost within one to two weeks. This is caused by cessation of prolactin secretion by the andenohypophysis. To state this another way, continued milking of the breasts causes continued secretion of prolactin and therefore continued lactation.

The stimulus that causes continued secretion of prolactin seems to be the suckling of the breasts, which, as is true in the milk ejection process, presumably results from impulses transmitted to the hypothalamus to suppress the release of prolactin-inhibiting factor. Decrease of this inhibitory substance allows a higher level of prolactin secretion by the anterior pituitary gland.

Milk production can continue for several years if the mother continues to be suckled, but the rate of secretion normally decreases rather markedly by seven to nine months.

EFFECT OF LACTATION ON THE UTERUS AND ON THE SEXUAL CYCLE

The uterus involutes after parturition far more rapidly in women who lactate than in women who do not lactate. This difference probably results from greatly diminished estrogen secretion by the ovaries during the period of lactation, for estrogens are known to enlarge the uterus and presumably also to prevent rapid uterine involution. The uterus of a lactating mother usually decreases to a size even smaller than that prior to pregnancy, whereas the uterus of a non-lactating mother is likely to remain considerably larger than the pregravid size and also to remain soft for many months.

Lactation usually prevents the sexual cycle for at least the first few months. Presumably this is caused by preoccupation of the pituitary with production of prolactin, which reduces the rate of secretion of the other gonadotropic hormones. However, after several months of lactating, the pituitary usually begins once again to produce sufficient quantities of follicle-stimulating hormone and luteinizing hormone to re-initiate the monthly sexual cycle. The rhythmic interplay between the ovarian and pituitary hormones during the sexual month does not necessitate marked reduction in prolactin secretion.

MILK AND THE METABOLIC DRAIN ON THE MOTHER CAUSED BY LACTATION

Table 82–1 gives the contents of human milk and cow's milk. The concentration of lactose in human milk is approximately 50 per cent greater than that in cow's milk, but on the other hand the concentration of protein in cow's milk is ordinarily two or more times as great as that in human milk. Finally, the ash, which contains the minerals, is only one-third as much in human milk as in cow's milk.

TABLE 82–1.   Percentage Composition of Milk

	Human Milk	Cow's Milk
Water	88.5	87
Fat	3.3	3.5
Sugar	6.8	4.8
Casein	0.9	2.7
Lactalbumin and other protein	0.4	0.7
Ash	0.2	0.7

At the height of lactation 1.5 liters of milk may be formed each day. With this degree of lactation great quantities of metabolic substrates are drained from the mother. For instance, approximately 50 grams of fat enter the milk each day, and approximately 100 grams of lactose, which must be derived from glucose, are lost from the mother each day. Also, some 2 to 3 grams of calcium phosphate may be lost each day, and, unless the mother is drinking large quantities of milk and has an adequate intake of vitamin D, the output of calcium and phosphate by the lactating mammae will be much greater than the intake of these substances. To supply the needed calcium and phosphate, the parathyroid glands enlarge greatly, and the bones become progressively decalcified. The problem of decalcification is usually not very great during pregnancy, but it can be a distinct problem during lactation.

REFERENCES

Assali, N. S. (ed.): Biology of Gestation. 2 volumes. New York, Academic Press, Inc., 1968.

Averill, R. L. W.: The hypothalamus and lactation. *Brit. Med. Bull.,* 22:261, 1966.

Battaglia, F. C.; Meschia, G.; and Makowski, E. L.: Comparison of in vitro and in vivo placental permeability measurements. *Amer. J. Physiol.,* 216:1590, 1969.

Ciba Foundation Study Group No. 23: Egg Implantation. Boston, Little, Brown and Company, 1966.

Ciba Foundation Study Group No. 26: The Effects of External Stimuli on Reproduction. Boston, Little, Brown and Company, 1967.

Glasser, S. R.; Wright, C.; and Heyssel, R. M.: Transfer of iron across the placenta and fetal membranes in the rat. *Amer. J. Physiol.,* 215:205, 1968.

Grosvenor, C. E.: Contraction of lactating rat mammary gland in response to direct mechanical stimulation. *Amer. J. Physiol.,* 208:214, 1965.

Hadek, R.: Mammalian Fertilization: An Atlas of Ultrastructure. Academic Press, Inc., New York, 1969.

Halasz, B.: Hypothalamic control of reproduction and lactation. *Proc. Int. Union of Physiol. Sciences,* 6:204, 1968.

Hytten, F. E., and Leitch, I.: The Physiology of Human Pregnancy. Philadelphia, F. A. Davis Co., 1964.

Kirschbaum, T. H.: Variability of magnitude of the Bohr effect in human fetal blood. *J. Appl. Physiol.,* 18:729, 1963.

Ladner, C.; Brinkman, C. R.; Weston, P.; and Assali, N. S.: Dynamics of uterine circulation in pregnant and nonpregnant sheep. *Amer. J. Physiol.,* 218:257, 1970.

Lloyd, C. W., and Weisz, J.: Some aspects of reproductive physiology. *Ann. Rev. Physiol.,* 28:267, 1966.

Locke, M.: Control Mechanisms in Developmental Processes. New York, Academic Press, Inc., 1968.

McLaren, A. (ed.): Advances in Reproductive Physiology. New York, Academic Press, Inc., 1968, Vol. 3.

Meites, J.: Inhibition by prolactin of pituitary prolactin secretion and mammary function in rats. *Proc. Int. Union of Physiol. Sciences,* 6:210, 1968.

Meites, J., and Nicoll, C. S.: Adenohypophysis: prolactin. *Ann. Rev. Physiol.,* 28:57, 1966.

Metcalfe, J.; Bartels, H.; and Moll, W.: Gas exchange in the pregnant uterus. *Physiol. Rev.,* 47:782, 1967.

Nalbandov, A. V., and Cook, B.: Reproduction. *Ann. Rev. Physiol.,* 30:245, 1968.

Parkes, A. S.: Marshall's Physiology of Reproduction. 3rd ed., 3 volumes. Boston, Little, Brown and Company, 1956–1966.

Peeters, G. J.: Recent studies on the milk-ejection reflex. *Proc. Int. Union of Physiol. Sciences,* 6:213, 1968.

Pinkerton, J. H.: Advances in Oxytocin Research. New York, Pergamon Press, Inc., 1965.

Portman, O. W.; Behrman, R. E.; and Soltys, P.: Transfer of free fatty acids across the primate placenta. *Amer. J. Physiol.,* 216:143, 1969.

Saunders, F. J.: Effects of sex steroids and related compounds on pregnancy and on development of the young. *Physiol. Rev.,* 48:601, 1968.

Sharman, A.: Reproductive Physiology in the Post-Partum Period. Baltimore, The Williams & Wilkins Co., 1966.

Short, R. V.: Reproduction. *Ann. Rev. Physiol.,* 29:373, 1967.

Stabenfeldt, G. H.; Osburn, B. I.; and Ewing, L. L.: Peripheral plasma progesterone levels in the cow during pregnancy and parturition. *Amer. J. Physiol.,* 218:571, 1970.

Symposium on Physiology of the Placenta: *Fed. Proc.,* 23:773, 1964.

CHAPTER 83

Special Features of Fetal
and Neonatal Physiology

A complete discussion of fetal development, function of the child immediately after birth, and growth and development through the early years of life lies in the province of formal courses in obstetrics and pediatrics. However, many aspects of these are strictly physiologic problems, some of which relate to the physiologic principles that we have discussed for the adult and some of which are peculiar to the infant himself. The present chapter delineates and discusses the most important of these special problems.

GROWTH AND FUNCTIONAL DEVELOPMENT OF THE FETUS

Early development of the placenta and of the fetal membranes occurs far more rapidly than development of the fetus itself. During the first two to three weeks the fetus remains almost microscopic in size, but thereafter, as illustrated in Figure 83–1, the dimensions of the fetus increase almost in proportion to age. At 12 weeks the length of the fetus is approximately 10 cm.; at 20 weeks, approximately 25 cm.; and at term (40 weeks), approximately 53 cm. (about 21 inches). Because the weight of the fetus is proportional to the cube of the length, the weight increases approximately in proportion to the cube of the age of the fetus. Note from Figure 83–1 that the weight of the

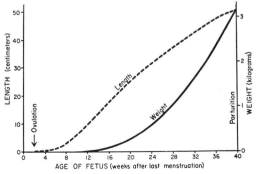

Figure 83–1   Growth of the fetus.

fetus remains almost nothing during the first few months and reaches 1 pound only at five and a half months of gestation. Then, during the last trimester of pregnancy, the fetus gains tremendously so that two months prior to birth the weight averages 3 pounds, one month prior to birth 4.5 pounds, and at birth 7 pounds, this birth weight varying from as low as 4.5 pounds to as high as 11 pounds in completely normal infants.

DEVELOPMENT OF THE ORGAN SYSTEMS

Within one month after fertilization of the ovum all the different organs of the fetus have already been "blocked out," and during the next two to three months the minute details of the different organs are established. Beyond the fourth month, the organs of the fetus are grossly the same as those of the newborn child, even including most of the smaller structures of the organs. However, cellular development of these structures is usually far from complete at this time and requires the full remaining five months of pregnancy for complete development. Even at birth certain structures, particularly the nervous system, the kidneys, and the liver, still lack full development, as is discussed in more detail later in the chapter.

The Circulatory System. The human heart begins beating during the fourth week following fertilization, contracting at the rate of about 65 beats per minute. This increases steadily as the fetus grows and reaches a rate of approximately 140 per minute immediately before birth.

FORMATION OF BLOOD CELLS. Nucleated red blood cells begin to be formed in the yoke sac and mesothelial layers of the placenta at about the third week of fetal development. This is followed a week later by the formation of non-nucleated red blood cells by the fetal mesenchyme and by the endothelium of the fetal blood vessels. Then at approximately six weeks, the liver begins to form blood cells, and in the third month the spleen and other lymphoid tissues of the body also begin forming blood cells. Finally, from approximately the third month on, the bone marrow forms more and more red and white blood cells. During the midportion of fetal life, the liver, spleen, and lymph nodes are the major sources of the fetus' blood cells, but, during the latter three months of fetal life, the bone marrow gradually takes over while these other structures lose their ability completely to form blood cells.

An especially interesting characteristic of fetal blood is that it contains an entirely different type of hemoglobin from that of adult blood, called *fetal hemoglobin*. Fetal hemoglobin combines with oxygen at a considerably lower Po_2 than does adult hemoglobin. This allows the fetal blood to carry as much as 30 per cent more oxygen in certain Po_2 ranges than can adult hemoglobin. This is of special importance in the fetus because its arterial Po_2 is always low, as was discussed in the previous chapter.

The Respiratory System. Obviously, respiration cannot occur during fetal life. However, respiratory movements do take place beginning at the end of the first trimester of pregnancy. Tactile stimuli or fetal asphyxia especially cause respiratory movements.

However, during the latter three to four months of pregnancy, the respiratory movements of the fetus are mainly inhibited, for reasons unknown. This could possibly result from (1) special chemical conditions in the body fluids of the fetus, (2) the presence of fluid in the fetal lungs, or (3) other possible unexplored stimuli.

The inhibition of respiration during the latter months of fetal life prevents filling of the lungs with debris from the meconium excreted by the gastrointestinal tract into the amniotic fluid. Yet the fluid samples from the lungs immediately after birth show that at least small amounts of amniotic fluid do move continually in and out of the pulmonary passageways.

FUNCTION OF THE NERVOUS SYSTEM

Most of the peripheral reflexes of the fetus are well formed by the third to fourth months of pregnancy. However, some of the more important higher functions of the central nervous system are still undeveloped even at birth. Indeed, myelinization of some major tracts of the central nervous system becomes complete only after approximately a year of postnatal life.

FUNCTION OF THE GASTROINTESTINAL TRACT

Even in midpregnancy the fetus ingests and absorbs large quantities of amniotic fluid, and during the latter two to three months, gastrointestinal function approaches that of the normal newborn infant. Small quantities of *meconium* are continually formed in the gastrointestinal tract and excreted from the bowels into the amniotic fluid. Excretion occurs particularly when the gastrointestinal tract becomes overly active as a consequence of fetal asphyxia. Meconium is composed partly of unabsorbed residue of amniotic fluid and partly of excretory products from the gastrointestinal mucosa and glands.

FUNCTION OF THE KIDNEYS

The fetal kidneys are capable of excreting urine during at least the latter half of pregnancy, and urination occurs normally *in utero*. However, the renal control systems for regulation of extracellular fluid electrolyte balances and acid-base balance are almost nonexistent until after midfetal life and do not reach full development until about a month after birth.

METABOLISM IN THE FETUS

The fetus utilizes mainly glucose for energy, and it has a high rate of storage of fat and protein, much if not most of the fat being synthesized from glucose rather than being absorbed from the mother's blood. Aside from these generalities there are some special problems of fetal metabolism in relation to calcium, phosphate, iron, and some vitamins as follows:

Metabolism of Calcium and Phosphate. Figure 83–2 illustrates the rate of calcium and phosphate accumulation in the fetus, showing that approximately 22.5 grams of calcium and 13.5 grams of phosphorus are accumulated in the average fetus during gestation. Approximately half of this accumulates during the last four weeks of gestation, which is also coincident with the period of rapid ossification of the fetal bones as well as with the period of rapid weight gain of the fetus.

During the earlier part of fetal life, the bones are relatively unossified, and have mainly a cartilaginous matrix. Indeed, x-ray pictures ordinarily will not show ossification until approximately the fourth month of pregnancy.

Note especially that the total amounts of calcium and phosphorus needed by the fetus during gestation represent only about one-fiftieth the quantities of these substances in the mother's bones. Therefore, this is a minimal drain from the mother. However, a great drain occurs after birth during lactation.

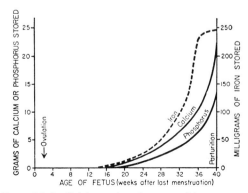

Figure 83–2 Calcium, phosphorus, and iron storage in the fetus at different stages of gestation.

Accumulation of Iron. Figure 83-2 shows that iron accumulates in the fetus somewhat more rapidly than calcium and phosphorus. Most of the iron is in the form of hemoglobin, which begins to be formed as early as the third week following fertilization of the ovum.

Small amounts of iron are concentrated in the progestational endometrium even prior to implantation of the ovum; this iron is ingested into the embryo by the trophoblastic cells for early formation of the red blood cells.

Approximately one-third of the iron in a fully developed fetus is normally stored in the liver. This iron can then be used for several months after birth by the newborn infant for formation of additional hemoglobin.

Utilization and Storage of Vitamins. The fetus needs vitamins equally as much as the adult and in some instances to a far greater extent. In general, the vitamins function the same in the fetus as in the adult, as discussed in Chapter 73. Special functions of several vitamins should be mentioned, however.

The B vitamins, especially vitamin B_{12} and folic acid, are necessary for formation of red blood cells and also for overall growth of the fetus.

Vitamin C is necessary for appropriate formation of intercellular substances, especially the bone matrix and fibers of connective tissue.

Vitamin D probably is not necessary for fetal growth, although the mother needs it for adequate absorption of calcium from her gastrointestinal tract. If the mother has plenty of this vitamin in her body fluids, large quantities will be stored by the fetal liver to be used by the newborn child for several months after birth.

Vitamin E, though its precise function is unknown, is necessary for normal development of the early ovum. In its absence in experimental animals, spontaneous abortion usually occurs at an early age.

Vitamin K is used by the fetal liver for formation of factor VII and prothrombin. When vitamin K is insufficient in the mother, factor VII and prothrombin become deficient in the child as well as in the mother. Since most vitamin K absorbed into the body is formed by bacterial action in the colon, the newborn child has no adequate source of vitamin K for the first week or so of life – that is, until he establishes a normal colonic bacterial flora. Therefore, prenatal storage of at least small amounts of vitamin K is helpful in preventing postnatal hemorrhage.

ADJUSTMENTS OF THE INFANT TO EXTRA-UTERINE LIFE

ONSET OF BREATHING

The most obvious effect of birth on the baby is loss of the placental connection with the mother, and therefore loss of this means for metabolic support. Especially important is loss of the placental oxygen supply and placental excretion of carbon dioxide. Therefore, by far the most important immediate adjustment of the infant to extra-uterine life is the onset of breathing.

Cause of Breathing at Birth. Following completely normal delivery from a mother who has not been depressed by anesthetics, the child ordinarily begins to breathe immediately and has a completely normal respiratory rhythm from the outset. The promptness with which the fetus begins to breath indicates that breathing is initiated by sudden exposure to the exterior world, probably resulting from sensory impulses originating in the suddenly cooled skin. However, if the infant does not breathe immediately, his body becomes progressively more hypoxic and hypercapnic, which provides additional stimulus to the respiratory center and usually causes breathing within a few seconds to a few minutes after birth.

Delayed and Abnormal Breathing at Birth – Danger of Hypoxia. If the mother has been depressed by an anesthetic during delivery, respiration by the child is likely to be delayed for several minutes, thus illustrating the importance of using as little obstetrical anesthesia as feasible. Also, many infants who have traumatic deliveries are slow to breathe or sometimes will not breathe at all. This can result from two possible effects: first, in a few infants, intracranial hemorrhage or brain contusion causes a typical concussion syndrome with a greatly depressed respiratory center. Second, and probably much more important, prolonged fetal hypoxia during delivery also causes serious depression of the respiratory center. Hypoxia frequently occurs during delivery because of (a) compression of the umbilical cord; (b) premature separation of the placenta; (c) excessive contraction of the uterus, which cuts off the blood flow to the placenta; or (d) excessive anesthesia of the mother, which depresses the oxygenation even of her blood.

DEGREE OF HYPOXIA THAT AN INFANT CAN STAND. In the adult, failure to breathe for only four minutes often causes death, but a newborn infant often survives as long as 15 minutes of failure to breathe after birth. Unfortunately, though, permanent mental impairment often ensues if breathing is delayed more than 7 to 10 minutes.

This ability of the neonatal infant to survive long periods of hypoxia probably results at least partly from special adaptation of its cellular respiratory enzymes to low Po₂. However, within a week or so after birth, the infant loses this resistance to hypoxia.

Expansion of the Lungs at Birth. At birth, the walls of the alveoli are held together by the surface tension of the viscid fluid that fills them. More than 25 mm. Hg of negative pressure is required to oppose the effects of this surface tension and therefore to open the alveoli for the first time. But once the alveoli are open, further respiration can be effected with relatively weak respiratory movements. Fortunately, the first inspirations of the newborn infant are extremely powerful, usually capable of creating as much as 50 mm. Hg negative pressure in the intrapleural space.

Figure 83–3 illustrates the tremendous forces required to open the lungs at the onset of breathing. To the left is shown the pressure-volume curve (compliance curve) for the first breath after birth. Observe, first, the lowermost curve, which shows that the lungs expand essentially not at all until the negative pressure has reached −40 cm. water (−30 mm. Hg). Then, as the negative pressure increases to −60 cm. water, only about 40 ml. air enters the lungs. Then, to deflate the lungs, considerable positive pressure is required, probably because of the viscous resistance offered by the fluid in the bronchioles.

Note that the second breath is much easier. However, breathing does not become completely normal until about 40 minutes after birth, as shown by the third compliance curve. The shape of this curve compares favorably with that for the normal adult, which was shown in Chapter 39.

RESPIRATORY DISTRESS SYNDROME. A small number of infants breathe satisfactorily immediately after birth but gradually develop severe respiratory distress during the ensuing few hours to several days and frequently succumb within the next day or so. The alveoli of these infants at death is filled with fluid containing large quantities of protein, almost as if pure plasma had leaked out of the capillaries into the alveoli. This condition is also called *hyaline membrane disease* because microscopic slides of the lung show the alveoli to be filled with a substance that looks like a hyaline membrane, but which is actually a proteinaceous fluid.

Unfortunately, the cause of the respiratory distress syndrome is not certain. However, the fluid from these lungs is deficient in *surfactant*, a substance secreted into the alveoli of normal lungs that decreases the surface tension of the alveolar fluid and therefore allows the alveoli to open easily. This substance was discussed in Chapter 39 in relation to the mechanics of respiration. It is notable that the disease occurs extremely frequently in two conditions: (1) in babies born of mothers who have untreated diabetes mellitus and (2) in premature babies.

CIRCULATORY READJUSTMENTS AT BIRTH

Equally as important as the onset of breathing at birth are the immediate circulatory readjustments that allow adequate blood flow through the lungs. Also, circulatory readjustments during the first week of life shunt more and more blood through the liver as well. To describe these readjustments we must first consider briefly the anatomic structure of the fetal circulation.

Specific Anatomic Structure of the Fetal Circulation. Because the lungs are mainly non-functional during fetal life and because the liver is only partially functional, it is not necessary for the fetal heart to pump much blood through either the lungs or the liver. On the other hand, the fetal heart must pump large quantities of blood through the placenta. Therefore, special anatomic arrangements cause the fetal circulatory system to operate considerably differently from that of the adult. First, as illustrated in Figure 83–4, blood returning from the placenta passes through the ductus venosus, mainly by-passing the liver. Then, most of the blood entering the right atrium from the inferior vena cava is directed in a straight pathway across the posterior aspect of the right atrium and thence through the *foramen ovale* directly into the left atrium. Thus, the well-oxygenated blood from the placenta enters the left side of the heart rather than the right side and is pumped mainly into the vessels of the head and forelimbs.

The blood entering the right atrium from the superior vena cava is directed downward through the tricuspid valve into the right ventricle. This blood is mainly deoxygenated blood from the head region of the fetus, and it is pumped by the right ventricle into the pulmonary artery, then mainly through the *ductus arteriosus* into the

Figure 83–3 Pressure-volume curves of the lungs (compliance curves) of a newborn baby immediately after birth, showing (a) the extreme forces required for breathing during the first two breaths of life and (b) development of a nearly normal compliance within 40 minutes after birth. (From Smith: *Sci. Amer., 209*(Oct.):32, 1963.)

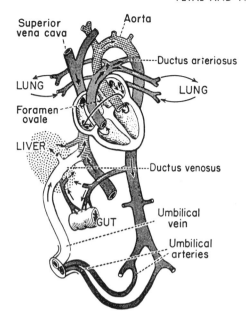

Figure 83-4 Organization of the fetal circulation. (Modified from Arey: Developmental Anatomy, 7th Ed.)

descending aorta and through the two umbilical arteries into the placenta. Thus, the deoxygenated blood becomes oxygenated.

Figure 83-5 illustrates the relative proportions of the total amount of blood pumped by the heart that passes through the different vascular circuits of the fetus. This figure shows that 55 per cent of all the blood goes through the placenta, leaving only 45 per cent to pass through all the peripheral tissues of the fetus. Furthermore, during fetal life, only 12 per cent of the blood flows through the

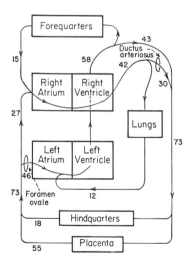

Figure 83-5 Diagram of the fetal circulatory system, showing relative distribution of blood flow to the different vascular areas. The numerals represent per cent of the total cardiac output flowing through the particular area. (Modified from Dawes, Mott, and Widdicombe: The fetal circulation in the lamb. *J. Physiol., 126:563, 1954.*)

lungs; yet immediately after birth almost 100 per cent passes through the lungs, which means a several-fold increase in blood flow at birth.

Changes in Fetal Circulation at Birth. The basic changes in the fetal circulation at birth were discussed in Chapter 30 in relation to congenital anomalies of the ductus arteriosus and foramen ovale that persist throughout life. Briefly, these changes are the following:

PRIMARY CHANGES IN PULMONARY AND SYS-TEMIC VASCULAR RESISTANCE AT BIRTH. The primary changes in the circulation at birth are: First, loss of the tremendous blood flow through the placenta *approximately doubles the systemic vascular resistance at birth*. This obviously *increases the aortic pressure* as well as the pressures in the left ventricle and left atrium.

Second, the *pulmonary vascular resistance greatly decreases* as a result of expansion of the lungs. In the unexpanded fetal lungs, the blood vessels are compressed because of the small volume of the lungs. Immediately upon expansion, these vessels are no longer compressed, and the resistance to blood flow decreases several-fold. Also, in fetal life the hypoxia and hypercapnia of the lungs cause marked tonic vasoconstriction of the lung blood vessels but vasodilation takes place when aeration of the lungs eliminates the hypoxia and hypercapnia. These changes reduce the resistance to blood flow through the lungs as much as 10-fold, which obviously *reduces the pulmonary arterial pressure*, the right ventricular pressure, and the right atrial pressure.

SECONDARY CHANGES IN THE VASCULAR SYSTEM AT BIRTH—CLOSURE OF THE FORAMEN OVALE AND CLOSURE OF THE DUCTUS ARTERI-OSUS. The *low right atrial pressure* and the *high left atrial pressure* that occur secondarily to the changes, respectively, in pulmonary and systemic resistance at birth cause a tendency for blood to flow backward from the left atrium into the right atrium rather than in the other direction, as occurred during fetal life. Consequently, the small valve that lies over the foramen ovale on the left side of the atrial septum closes over this opening, thereby preventing further flow. In two-thirds of all persons the valve becomes adherent over the foramen ovale within a few months to a few years and forms a permanent closure. But, even if permanent closure does not occur, the left atrial pressure throughout life remains 3 to 4 mm. Hg greater than the right atrial pressure, and the back pressure keeps the valve closed.

Similar effects occur in relation to the ductus arteriosus, for the increased systemic resistance *elevates the aortic pressure* while the decreased pulmonary resistance *reduces the pulmonary arterial pressure.* As a consequence, shortly after birth, blood begins to flow backward from the aorta into the pulmonary artery rather than in the other direction as in fetal life. This backward flow can sometimes become equally as great as all the blood pumped by the right ventricle, so that as much as two times the normal amount of blood flows through the lungs. However, within three to four days after birth, the muscular wall of the ductus arteriosus gradually constricts to stop all blood flow. This is called *functional closure* of the ductus arteriosus. Then, sometime during the second month of life the ductus arteriosus ordinarily becomes anatomically *occluded* by growth of fibrous tissue into its lumen.

The causes of either functional closure or anatomical closure of the ductus are not completely known. However, the most likely cause is increased oxygenation of the blood flowing through the ductus. In fetal life the P_{O_2} may be as low as 20 mm. Hg, but it increases to about 100 mm. Hg within a few minutes after birth. Furthermore, the degree of contraction of the ductus is highly related to the availability of oxygen.

In one out of several thousand infants, the ductus fails to close at all, resulting in a *patent ductus arteriosus,* the consequences of which were discussed in Chapter 30.

NUTRITION OF THE NEWBORN INFANT

The fetus obtains much of its energy from glucose and fatty acids obtained from the mother's blood. Immediately after birth, the amount of glucose stored in the infant's body is sufficient to supply the infant's needs for only a few hours. Unfortunately, the liver of the newborn infant is still far from functionally adequate at birth, which prevents significant gluconeogenesis. Therefore, the infant's blood glucose concentration frequently falls the first day to as low as 30 to 40 mg./100 ml. of plasma, and the infant must then utilize stored fats and proteins for metabolism until he can be provided with milk from the mother two to three days later.

Special problems are also frequently associated with adequate fluid supply for the newborn infant, because the infant's rate of body fluid turnover averages seven times that of an adult, and the mother's milk supply requires several days to develop. Ordinarily, the infant's weight decreases 5 to 10 per cent and sometimes as much as 20 per cent within the first two to three days of life. Most of this weight loss is loss of fluid rather than of body solids.

SPECIAL FUNCTIONAL PROBLEMS IN THE NEONATAL INFANT

The most important characteristic of the newborn infant is instability of his various hormonal and neurogenic control systems. This results partly from the immature development of the different organs of the body and partly from the fact that the control systems simply have not become adjusted to the completely new way of life.

The Respiratory System. The normal rate of respiration in the newborn is about 40 breaths per minute, and his tidal air with each breath averages 16 ml. This gives a total minute respiratory volume of 640 ml. per minute, which is about two times as great in relation to the body weight as that of an adult. *The functional residual capacity of the infant is only half that of the adult in relation to his body weight.* This causes rapid changes in blood gas concentration when the normal respiration becomes altered.

Blood Volume. The blood volume of a newborn infant immediately after birth averages about 300 ml., but, if the infant is left attached to the placenta for a few minutes after birth or if the umbilical cord is stripped to force blood out of its vessels into the baby, an additional 75 ml. of blood enters the infant to make a total of 375 ml. Then, during the ensuing few hours, fluid is lost into the tissue spaces from this blood, which increases the hematocrit but returns the blood volume once again to the normal value of about 300 ml. Some pediatricians feel that this extra blood volume in some instances causes mild pulmonary edema with consequent respiratory distress.

Cardiac Output. The cardiac output of the newborn infant averages 550 ml./minute, which, like respiration and body metabolism, is about two times as much in relation to body weight as in the adult. An occasional child is born with an especially low cardiac output caused by hemorrhage through the placental membrane into the mother's blood prior to birth.

Blood Characteristics. The red blood cell count in the newborn infant averages about 4 million per cubic millimeter. If blood is stripped from the cord into the infant, the red blood cell count rises an additional half to three-quarters million during the first few hours of life, giving a red blood cell count of about 4.75 million per cubic mm., as illustrated in Figure 83-6. Subsequent to this,

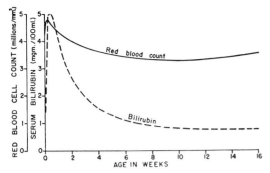

Figure 83–6 Changes in the red blood cell count and in the serum bilirubin concentration during the first 16 weeks of life, showing "physiologic anemia" at 6 to 12 weeks of life and "physiologic hyperbilirubinemia" during the first two weeks of life.

however, few new red blood cells are formed in the infant during the first few weeks of life, presumably because the hypoxic stimulus of fetal life is no longer present to stimulate red cell production. Thus, as shown in Figure 83–6, the average red blood cell count falls to 3.25 million per cubic mm. by about 8 to 10 weeks of age. From that time on, increasing activity by the fetus provides the appropriate stimulus for returning the red blood cell count to normal within another two to three months.

Immediately after birth, the white blood cell count of the infant is about 45,000 per cubic mm., which is about five times as great as that of the normal adult.

Neonatal Jaundice and Erythroblastosis Fetalis. Bilirubin formed in the fetus can cross the placenta and be excreted through the liver of the mother, but immediately after birth the only means for ridding the infant of bilirubin is through the infant's own liver, which, for the first week or so of life, still functions poorly and is incapable of conjugating significant quantities of bilirubin with glucuronic acid for excretion into the bile. Consequently, the plasma bilirubin concentration rises from a normal value of less than 1 mg./100 ml. to an average of 5 mg./100 ml. during the first three days of life and then gradually falls back to normal as the liver becomes functional. This condition, called *physiologic hyperbilirubinemia*, is illustrated in Figure 83–6, and it is associated with a mild *jaundice* (yellowness) of the infant's skin and especially of the sclerae of his eyes.

However, by far the most important cause of serious neonatal jaundice is *erythroblastosis fetalis*, which was discussed in detail in Chapter 8 in relation to Rh factor incompatibility between the infant and mother. Briefly, the baby inherits an Rh positive trait from its father while the mother is Rh negative. The mother then becomes immunized against the fetus' blood, and her antibodies in turn destroy the infant's red blood cells, releasing extreme quantities of bilirubin into the plasma. This condition occurs either mildly or seriously in 1 out of every 50 to 100 newborn infants.

Postnatal Bleeding Tendencies. Since most vitamin K in the body is synthesized by bacteria of the colon and since the newborn infant has not yet developed a bacterial flora in his colon, his vitamin K level falls rapidly the first few days after birth, causing factor VII and prothrombin production by the liver to be impaired for a week or so until a bacterial flora is established. During this time the infant is likely to develop a bleeding tendency. However, if the mother is given an adequate dose of vitamin K at least four hours prior to birth of the infant, sufficient vitamin K is usually stored in the infant's liver to see him through this possible period of hypovitaminosis K.

Fluid Balance, Acid-Base Balance, and Renal Function. The rate of fluid intake and fluid excretion in the infant is seven times as great in relation to weight as in the adult, which means that even a slight alteration of fluid balance can cause rapidly developing abnormalities. Second, the rate of metabolism in the infant is two times as great in relation to body mass as in the adult, which means that two times as much acid is normally formed, which leads to a tendency toward acidosis in the infant. Third, functional development of the kidneys is not complete until the end of approximately the first month of life. For instance, the kidneys of the newborn can concentrate urine to only one and a half times the osmolality of the plasma instead of the normal three to four times as in the adult.

Therefore, considering the immaturity of the kidney, together with the marked fluid turnover in the infant and rapid formation of acid, one can readily understand that among the most important problems of infancy are acidosis, dehydration, and rare instances of overhydration.

Changes in Total Body Water, Intracellular Fluid Volume, and Extracellular Fluid Volume During Fetal and Postnatal Life. Figure 83–7 illustrates volume changes in the different body fluid compartments throughout fetal and

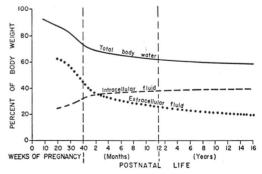

Figure 83–7 Changes in total body water, intracellular fluid volume, and extracellular fluid volume in the fetus and in the postnatal child. (Modified from Friis-Hansen: *Acta paediat., 46;* suppl. 110, 1957.)

postnatal life, showing especially that the early fetus is composed almost entirely of water, but the body accumulates progressively more solids throughout gestation and thereafter until adulthood. At birth the average percentage of water in the infant is 73 per cent; this compares with 58 per cent for the adult.

Another interesting aspect of the fetal body fluids is the high ratio of extracellular fluids to intracellular fluids in contrast to the opposite relationship in the adult. This change presumably results from progressive growth of the cells at the expense of the extracellular spaces. Since extracellular fluids contain mainly sodium chloride in contrast to a preponderance of potassium and magnesium phosphates in the intracellular fluids, the newborn infant has much higher total body sodium and chloride in relation to body weight and much lower total body potassium, magnesium, and phosphate than the adult.

Liver Function. Blood flow through the liver during fetal life is slight, but during the first few days after birth the ductus venosus gradually becomes occluded so that thereafter all portal blood flows through the liver. During the first few days of life, before all the portal flow goes through the liver, liver function may be quite deficient, as evidenced by the following effects:

1. The liver of the newborn conjugates bilirubin with glucuronic acid poorly and therefore excretes bilirubin only slightly during the first few days of life.

2. The liver of the newborn is deficient in forming plasma proteins, and the plasma protein concentration falls to 1 gram per cent less than that for older children. Occasionally, the protein concentration falls so low that the infant actually develops hypoproteinemic edema.

3. The gluconeogenesis function of the liver is particularly deficient. As a result, the blood glucose level of the unfed newborn infant falls to about 30 to 40 mg. per cent, and the infant must depend on its stored fats for energy until feeding can occur.

4. The liver of the newborn usually also forms too little of the factors needed for normal blood coagulation.

Digestion, Absorption, and Metabolism of Energy Foods. In general, the ability of the newborn infant to digest, absorb, and metabolize foods is not different from that of the older child, with the following three exceptions:

First, secretion of pancreatic amylase in the newborn infant is deficient so that the infant utilizes starches less adequately than do older children. However, the infant readily assimilates disaccharides and monosaccharides.

Second, absorption of fats from the gastrointestinal tract is somewhat less than in the older child. Consequently, milk with a high fat content, such as cow's milk, is frequently inadequately utilized.

Third, because the liver functions are imperfect

during at least the first week of life, the glucose concentration in the blood is unstable and also low.

The newborn is especially capable of synthesizing proteins and thereby storing nitrogen. Indeed, with a completely adequate diet, as much as 90 per cent of the ingested amino acids are utilized for formation of body proteins. This is a much higher percentage than in adults.

Metabolic Rate and Body Temperature. The normal metabolic rate of the newborn in relation to body weight is about two times that of the adult, which accounts also for the two times as great cardiac output and two times as great minute respiratory volume in the infant.

However, in relation to body surface area, the metabolic rate of the newborn infant is less than that of the adult. This is particularly important in body temperature regulation, for the body temperature of the newborn infant, particularly of premature infants, tends to fall severely. Figure 83–8 shows that the body temperature normally falls immediately during the first few hours after birth, but returns to normal in seven to eight hours. Still, the body temperature regulatory mechanisms remain poor during the early days of life, allowing marked deviations in temperature at first, which are also illustrated in Figure 83–8.

Nutritional Needs during the Early Weeks of Life. At birth, a newborn infant is usually in complete nutritional balance, provided its mother has had an adequate diet. Furthermore, function of the gastrointestinal system is usually more than adequate to digest and assimilate all the nutritional needs of the infant if these are provided in his diet. However, three specific problems do occur in the early nutrition of the infant as follows:

NEED FOR CALCIUM AND VITAMIN D. The newborn infant has only just begun rapid ossification of his bones at birth so that he needs a ready supply of calcium throughout infancy. This is ordinarily supplied adequately by his usual diet of milk. Yet absorption of calcium by the gastrointestinal tract is poor in the absence of vitamin D. Therefore, the vitamin D deficient infant can develop severe rickets in only a few weeks. This is particularly true in premature babies since

Figure 83–8 Fall in body temperature of the infant immediately after birth, and instability of body temperature during the first few days of life.

their gastrointestinal tracts absorb calcium even less effectively than those of normal infants.

NECESSITY FOR IRON IN THE DIET. If the mother has had adequate amounts of iron in her diet, the liver of the infant usually has stored enough iron to keep forming blood cells for four to six months after birth. But if the mother has had insufficient iron in her diet, anemia is likely to supervene in the infant after about three months of life. To prevent this possibility, early feeding of egg yolk, which contains reasonably large quantities of iron, or administration of iron in some other form is desirable by the second or third month of life.

VITAMIN C DEFICIENCY IN INFANTS. Ascorbic acid (vitamin C) is not stored in significant quantities in the fetal tissues; yet it is required for proper formation of cartilage, bone, and other intercellular structures of the infant. Furthermore, milk has poor supplies of ascorbic acid, especially cow's milk, which has only one-fourth as much as mother's milk. For this reason, orange juice or other sources of ascorbic acid are usually prescribed by the third week of life.

Immunity. Fortunately, the newborn inherits much immunity from its mother because many gamma globulins, which contain the antibodies, diffuse from the mother's blood through the placenta into the fetus. However, the newborn himself does not form gamma globulins to a significant extent. By the end of the first month, the baby's gamma globulins have decreased to less than one-half the original level, with corresponding decrease in immunity. Thereafter, the baby's own immunization processes begin to form antibodies, and the gamma globulin concentration returns essentially to normal by the age of 6 to 20 months.

Despite the decrease in gamma globulins soon after birth, the antibodies inherited from the mother still protect the infant for about six months against most major childhood infectious diseases, including diphtheria, measles, smallpox, and polio. Therefore, immunization against these diseases before six months is usually unnecessary. On the other hand, the inherited antibodies against whooping cough are normally insufficient to protect the newborn; therefore, for full safety he requires immunization against this disease within the first month or so of life.

ALLERGY. Fortunately, the newborn infant is rarely subject to allergy. Yet several months later when he first begins to form his own antibodies, extreme allergic states can develop, often resulting in serious eczema, gastrointestinal abnormalities, or even anaphylaxis. As the child grows still older and develops still higher degrees of immunity, these allergic manifestations usually disappear for the remainder of his life. The relationship of mild immunity to allergy was discussed in detail in Chapter 10.

Endocrine Problems. Ordinarily the endocrine system of the infant is highly developed at birth, and the infant rarely exhibits any immediate endocrine abnormalities. However, there are special instances in which endocrinology of infancy is important.

1. If a pregnant mother bearing a female child is treated with an androgenic hormone or if she develops an androgenic tumor during pregnancy, the child will be born with a high degree of masculinization of its sexual organs, thus resulting in a type of *hermaphroditism*.

2. The sex hormones secreted by the placenta and by the mother's glands during pregnancy occasionally cause the newborn's breasts to form milk during the first few days of life. Sometimes the breasts then become inflamed or even develop infectious mastitis.

3. An infant born of a diabetic mother will have considerable hypertrophy and hyperfunction of its islets of Langerhans. As a consequence, the infant's blood glucose concentration may fall to as low as 20 mg./100 ml. or even lower shortly after birth. Fortunately, the newborn infant, unlike the adult person, does not usually develop insulin shock or coma from this low level of blood glucose concentration.

Because of metabolic deficits in the diabetic mother, the fetus is often stunted in growth, and growth of the newborn infant and tissue maturation are often impaired. Also, there is a high rate of intra-uterine mortality, and of those fetuses that do come to term, there is still a high mortality rate. Two-thirds of the infants who die succumb to the respiratory distress syndrome, which was described earlier in the chapter.

4. Occasionally, a child is born with hypofunctional adrenal cortices, perhaps resulting from *agenesis* of the glands or *exhaustion atrophy*, which can occur when the adrenal glands have been overstimulated.

The fetal adrenal glands contain a type of cortical tissue, the so-called *fetal cortex*, which is different from the glandular tissue of the adult adrenal gland. Immediately after birth, the fetal cortex degenerates and is replaced by rapid growth of the adult type of adrenal cortex. The function of the fetal cortex is unknown.

5. If a pregnant woman is treated with thyroid hormone, the infant is likely to be born with a temporarily hyposecreting thyroid gland. On the other hand, if prior to pregnancy a woman had hyperthyroidism but had her thyroid gland removed, her pituitary usually still secretes great quantities of thyrotropin during gestation, and the child might be born with temporary hyperthyroidism.

SPECIAL PROBLEMS OF PREMATURITY

All the problems just noted for neonatal life are especially exacerbated in prematurity. These can be categorized under the following two headings: (1) immaturity of certain organ systems and (2) instability of the different homeostatic control systems. Because of these effects, a premature baby rarely lives if it is born more than two and a half to three months prior to term.

IMMATURE DEVELOPMENT

Almost all the organ systems of the body are immature in the premature infant, but some require particular attention if the life of the premature baby is to be saved.

Respiration. The respiratory system is especially likely to be underdeveloped in the premature infant. The vital capacity and the functional residual capacity of the lungs are especially small in relation to the size of the infant. As a consequence, respiratory distress is a common cause of death. Indeed, there is a very high incidence of the respiratory distress syndrome. Also, the low functional residual capacity in the premature infant is often associated with periodic breathing of the Cheyne-Stokes type.

Gastrointestinal Function. Another major problem of the premature infant is to ingest and absorb adequate food. If the infant is more than two months premature, the digestive and absorptive systems are amost always inadequate. The absorption of fat is so poor that the premature infant must have a low fat diet. Furthermore, the premature infant has unusual difficulty in absorbing calcium and therefore can develop severe rickets before one recognizes the difficulty. For this reason, special attention must be paid to adequate calcium and vitamin D intake.

Function of Other Organs. Immaturity of other organ systems that frequently causes serious difficulties in the premature infant includes: (a) immaturity of the liver, which results in poor intermediary metabolism and often also a bleeding tendency as a result of poor formation of coagulation factors; (b) immaturity of the kidneys, which are particularly deficient in their ability to rid the body of acids, thereby predisposing to acidosis as well as to many serious fluid balance abnormalities; (c) immaturity of the blood-forming mechanism of the bone marrow, which allows rapid development of anemia; and (d) depressed formation of gamma globulin by the reticuloendothelial system, which is often associated with serious infection.

INSTABILITY OF THE CONTROL SYSTEMS IN THE PREMATURE INFANT

Immaturity of the different organ systems in the premature infant creates a high degree of instability in the homeostatic systems of the body. For instance, the acid-base balance can vary tremendously, particularly when the food intake varies from time to time. Likewise, the blood protein concentration is usually somewhat low because of immature liver development, often leading to *hypoproteinemic edema*. And inability of the infant to regulate its calcium ion concentration frequently brings on hypocalcemic tetany. Also, the blood glucose concentration can vary between the extremely wide limits of 20 mg./100 ml. to over 100 mg./100 ml., depending principally on the regularity of feeding. It is no wonder, then, with these extreme variations in the internal environment of the premature infant, that mortality is high.

Instability of Body Temperature. One of the particular problems of the premature infant is inability to maintain normal body temperature. Its temperature tends to approach that of its surroundings. At normal room temperature the temperature may stabilize in the low 90's or even in the 80's. Statistical studies show that a body temperature maintained below 96° F. is associated with a particularly high incidence of death, which explains the common use of the incubator in the treatment of prematurity.

DANGER OF OXYGEN THERAPY IN THE PREMATURE INFANT

Because the premature infant frequently develops respiratory distress, oxygen therapy has often been used in treating prematurity. However, it has been discovered that use of high oxygen concentrations in treating premature infants, especially in early prematurity, causes vascular ingrowth into the vitreous humor of the eyes when the infant is withdrawn from the oxygen. This is followed later by fibrosis. This condition, known as *retrolental fibroplasia*, causes permanent blindness. For this reason, it is particularly important to avoid treatment of premature infants with high concentrations of respiratory oxygen. Physiologic studies indicate that the premature infant is probably safe in up to 40 per cent oxygen, but some child physiologists believe that complete safety can be achieved only by normal or even low oxygen concentration because this more nearly approaches intra-uterine conditions.

GROWTH AND DEVELOPMENT OF THE CHILD

The major physiologic problems of the child beyond the neonatal period are related to special metabolic needs for growth, which have been fully covered in the sections on metabolism and endocrinology.

Figure 83–9 illustrates the changes in heights of boys and girls from the time of birth until the age of 20 years. Note especially that these parallel each other almost exactly until the latter part of the first decade of life. At this time, the female estrogens cause rapid growth but early uniting of the epiphyses at about the fourteenth year of life,

Figure 83-9 Height of boys and girls from infancy to 20 years of age.

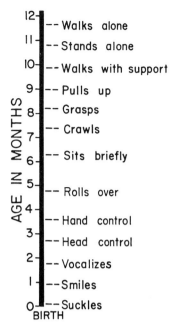

Figure 83-10 Behavioral development of the infant during the first year of life.

so that growth in height ceases. This contrasts with the effect of testosterone in the male, which causes growth at a slightly later age; however, the male undergoes a much more prolonged growth so that his final height is considerably greater than that of the female.

BEHAVIORAL GROWTH

Behavioral growth is principally a problem of maturity of the nervous system. Here, it is extremely difficult to dissociate maturity of the anatomical structures of the nervous system from maturity caused by training. Anatomical studies show that certain major tracts in the central nervous system are not completely myelinated until the end of the first year of life. For this reason we frequently state that the nervous system is not fully functional at birth. On the other hand, we know that most reflexes of even the fetus are fully developed by approximately the third to fourth month of intra-uterine life, and that unmyelinated nerve fibers can be equally as functional as myelinated nerve fibers. Therefore, much of the functional immaturity of the nervous system at birth might well be caused by lack of training rather than actual immaturity of the anatomical structures.

Nevertheless, the brain weight of the child increases rapidly during the first year and less rapidly during the second year, reaching almost adult proportions by the end of the second year. This is also associated with closure of the fontanels and sutures of the skull, which prevents much additional growth of the cranium beyond the first two years of life.

Figure 83-10 illustrates a normal progress chart for the infant during the first year of life. Comparison of a baby's actual development in relation to such a chart is frequently used for clinical assessment of a child's mental and behavioral growth.

REFERENCES

Abramson, H.: Resuscitation of the Newborn Infant. 2nd ed., St. Louis, The C. V. Mosby Co., 1966.

Adams, F. H.: Fetal and neonatal cardiovascular and pulmonary function. Ann. Rev. Physiol., 27:257, 1965.

Assali, N. S.; Morris, J. A.; and Beck, R.: Cardiovascular hemodynamics in the fetal lamb before and after lung expansion. Amer. J. Physiol., 208:122, 1965.

Barnes, A. C.: Intra-Uterine Development. Philadelphia, Lea & Febiger, 1968.

Cross, K. W.: Respiration and oxygen supplies in the newborn. In Handbook of Physiology. Baltimore, The Williams & Wilkins Co., 1965, Sec. III, Vol. II, p. 1329.

Cross, K. W., and Dawes, G. S.: The foetus and the newborn: recent research. Brit. Med. Bull., 22:1, 1966.

Dawes, G. S.: Control of the pulmonary circulation in the fetus and newborn. In Fishman, A. P., and Hecht, H. H. (eds.): The Pulmonary Circulation and Interstitial Space. Chicago, University of Chicago Press, 1969, p. 293.

Dawes, G. S.: Foetal and Neonatal Physiology. Chicago, Year Book Medical Publishers, Inc., 1968.

Engel, S.: The Prenatal Lung. New York, Pergamon Press, Inc., 1966.

Morris, I. G.: Gamma globulin absorption in the newborn. In Handbook of Physiology. Baltimore, The Williams & Wilkins Co., 1968, Sec. VI, Vol. III, p. 1491.

Morris, J. A.; Bekey, G. A.; Assali, N. S.; and Beck, R.: Dynamics of blood flow in the ductus arteriosus. Amer. J. Physiol., 208: 471, 1965.

Oski, F. A., and Naiman, J. L.: Hematologic Problems in the Newborn. Philadelphia, W. B. Saunders Company, 1966.

Saunders, F. J.: Effects of sex steroids and related compounds on pregnancy and on development of the young. Physiol. Rev., 48:601, 1968.

Smith, C. A.: The first breath. Sci. Amer., 209(4):27, 1963.

Stahlman, M.; Gray, J.; Young, W. C.; and Shepard, F. M.: Cardiovascular response of the neonatal lamb to hypoxia and hypercapnia. Amer. J. Physiol., 213:899, 1967.

Wynn, R. M.: Fetal Homeostasis. New York, New York Academy of Sciences, 1968.

INDEX